Year	Recipients	Nobel Prize	Research Topic
1972	G. M. Edelman R. R. Porter	Medicine or Physiology	Chemical structure of immunoglobulins
	C. Anfinsen	Chemistry	Relationship between primary and tertiary structure of proteins
1970	N. Borlaug	Peace Prize	Genetic improvement of Mexican wheat
1969	M. Delbruck A. D. Hershey S. E. Luria	Medicine or Physiology	Replication mechanisms and genetic structure of bacteriophages
1968	H. G. Khorana M. W. Nirenberg	Medicine or Physiology	Deciphering the genetic code
	R. W. Holley	Medicine or Physiology	Structure and nucleotide sequence of transfer RNA
1966	P. F. Rous	Medicine or Physiology	Viral induction of cancer in chickens
1965	F. Jacob A. M. L'woff J. L. Monod	Medicine or Physiology	Genetic regulation of enzyme synthesis in bacteria
1962	F. H. C. Crick J. D. Watson M. H. F. Wilkins	Medicine or Physiology	Double helical model of DNA
	J. C. Kendrew M. F. Perutz	Chemistry	Three-dimensional structure of globular proteins
1959	A. Kornberg S. Ochoa	Medicine or Physiology	Biological synthesis of DNA and RNA
1958	G. W. Beadle E. L. Tatum	Medicine or Physiology	Genetic control of biochemical processes
	J. Lederberg	Medicine or Physiology	Genetic recombination in bacteria
	F. Sanger	Chemistry	Primary structure of proteins
1954	L. Pauling	Chemistry	Alpha helical structure of proteins
1946	H. J. Muller	Medicine or Physiology	X-ray induction of mutations in *Drosophila*
1933	T. H. Morgan	Medicine or Physiology	Chromosomal theory of genetics
1930	K. Landsteiner	Medicine or Physiology	Discovery of human blood groups

Genetics: A Molecular Perspective

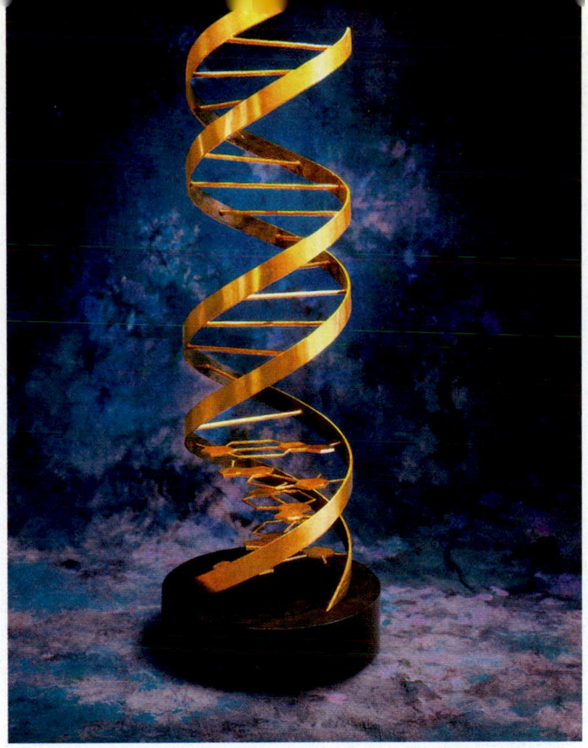

Genetics: A Molecular Perspective

William S. Klug
The College of New Jersey

Michael R. Cummings
University of Illinois at Chicago

With contributions by

Jon Herron, University of Washington
Charlotte Spencer, University of Alberta
Sarah M. Ward, Colorado State University

Prentice Hall

Pearson Education, Inc.
Upper Saddle River, New Jersey 07458

Library of Congress Cataloging-in-Publication-Data
Klug, William S.
Genetics: A molecular perspective / William S. Klug, Michael R. Cummings; with contributions by John Herron … [et al.]
p. cm.
Includes bibliographical references and index.
ISBN 0-13-008530-8
1. Molecular genetics. I. Cummings, Michael R. II. Title.

QH442 .K59 2003
572.8–dc21

2002074876

Editor in Chief, Life and Geosciences: Sheri L. Snavely
Executive Editor: Gary Carlson
Editorial Assistant: Lisa Tarabokjia
Production Editor and Composition: Aksen Associates
Production Assistant: Nancy Bauer
Photo Researcher: Yvonne Gerin
Photo Coordinator: Debbie Hewitson
Vice President of Production and Manufacturing: David W. Riccardi
Executive Managing Editor: Kathleen Schiaparelli
Assistant Managing Editor: Beth Sweeten
Assistant Managing Editor, Science Media: Nicole Bush
Media Editor: Andrew Stull
Executive Marketing Manager: Jennifer Welchans
Marketing Manager: Martha McDonald
Project Manager: Karen Horton
Manufacturing Manager: Trudy Pisciotti
Assistant Manufacturing Manager: Michael Bell
Director of Design: Carol Anson
Art Director: Kenny Beck
Interior Designer: Anne Flanagan
Art Editor: Adam Velthaus
Cover Designer: Imagineering Scientific and Technical Artworks
Managing Editor, Audio/Video Assets: Patty Burns
Director of Creative Services: Paul Belfanti
Illustrations: Imagineering Scientific and Technical Artworks
Copy Editor: Write With, Inc

Prentice Hall

Published by Pearson Education, Inc.
Upper Saddle River, NJ 07458

Printed in the United States of America

10 9 8 7 6 5 4 3 2

ISBN 0-13-008530-8

Pearson Education LTD., *London*
Pearson Education Australia PTY, Limited, *Sydney*
Pearson Education Singapore, Pte. Ltd.
Pearson Education North Asia Ltd, *Hong Kong*
Pearson Education Canada, Ltd, *Toronto*
Pearson Educación de Mexico, S.A. de C.V.
Pearson Education—Japan, *Tokyo*
Pearson Education Malaysia, Pte. Ltd

Dedication

We have long been derelict in not acknowledging the two women in our respective lives who have stood by us through thick and thin during the production of the numerous editions of our Genetics text series. Saving the best of the best for the dedication of our most recent text, our acknowledgement of the important role that they have played in our lives and our heartfelt appreciation go to them.

To Kathleen and Lee Ann

With all our love and thanks

For your support, patience, and understanding

Steve Klug
Mike Cummings

About the Authors

William S. Klug is currently Professor of Biology at The College of New Jersey (formerly Trenton State College) in Ewing, New Jersey. He served as Chairman of the Biology Department for 17 years, a position to which he was first elected in 1974. He received his B.A. degree in Biology from Wabash College in Crawfordsville, Indiana and his Ph.D. from Northwestern University in Evanston, Illinois. Prior to coming to Trenton State College, he returned to Wabash College as an Assistant Professor, where he first taught genetics as well as general biology and electron microscopy. His research interests have involved ultrastructural and molecular genetic studies of oogenesis in *Drosophila*. He has taught the genetics course as well as the senior capstone seminar course in human and molecular genetics to undergraduate Biology majors for each of the last 33 years. He was the recent recipient of the first annual teaching award given at The College of New Jersey as the faculty member who most challenges students to meet high standards.

Michael R. Cummings is currently Associate Professor in the Department of Biological Sciences and in the Department of Molecular Genetics at the University of Illinois at Chicago. He has also served on the faculty at Northwestern University and Florida State University. He received his B.A. from St. Mary's College in Winona, Minnesota, and his M.S. and Ph.D. from Northwestern University in Evanston, Illinois. He has also written textbooks in human genetics and general biology for non-majors. His research interests center on the molecular organization and physical mapping of human acrocentric chromosomes. At the undergraduate level, he teaches courses in Mendelian genetics, human genetics, and general biology for non-majors. He has received numerous teaching awards given by the university and by student organizations.

Brief Contents

Contents

3

DNA Replication and Recombination 53

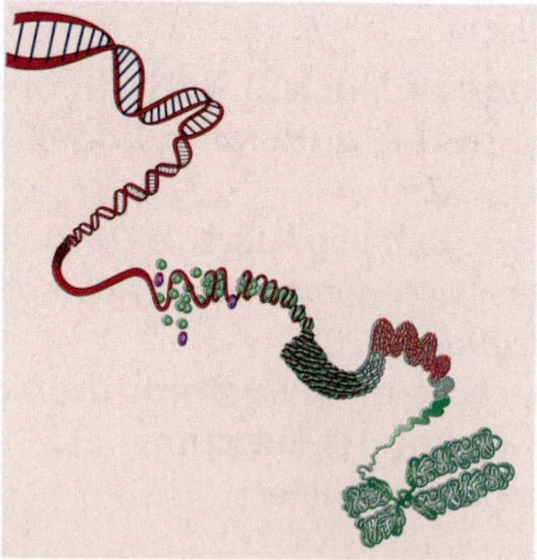

4

Chromosome Structure and DNA Sequence Organization 79

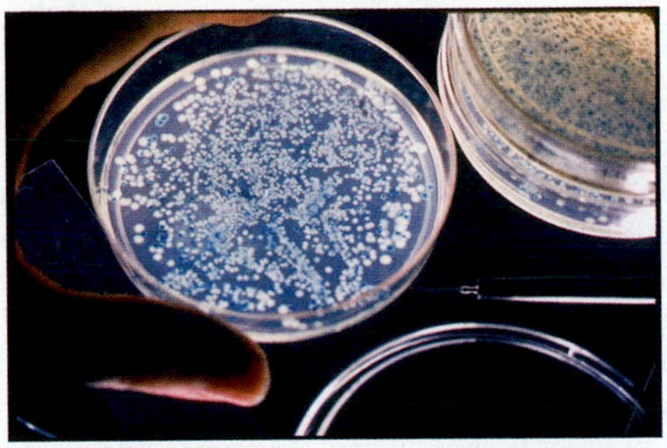

PART TWO
Expression of Genetic Information

5
The Genetic Code and Transcription 99

6
Translation And Proteins 127

PART THREE Transmission of Genetic Information

8

Mitosis and Meiosis 193

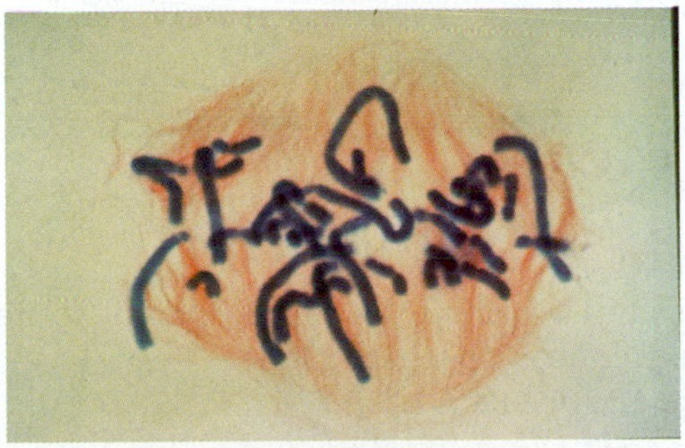

9

Mendelian Genetics 219

10

Extensions of Mendelian Genetics 247

11

Sex Determination and Sex Chromosomes 283

12

Linkage, Crossing Over, and Mapping in Eukaryotes 305

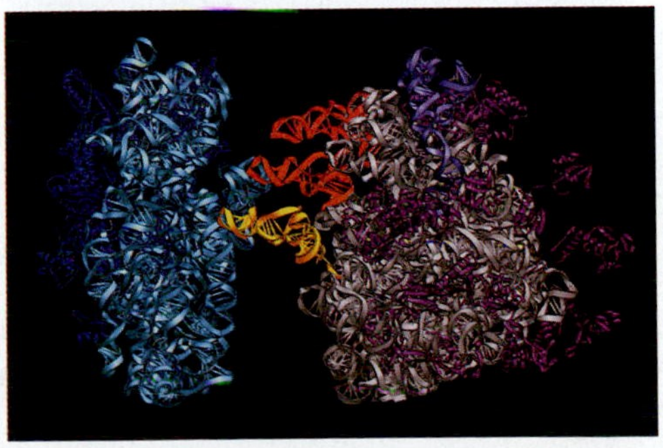

14

Extranuclear Inheritance 371

PART FOUR
DNA Biotechnology and Genomic Analysis

15

Genetics of Bacteria and Bacteriophages 389

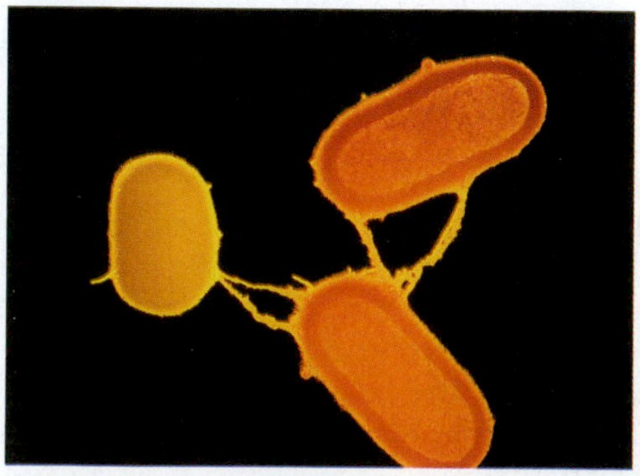

16

Recombinant DNA Technology 419

17

Genomics, Bioinformatics and Proteomics 449

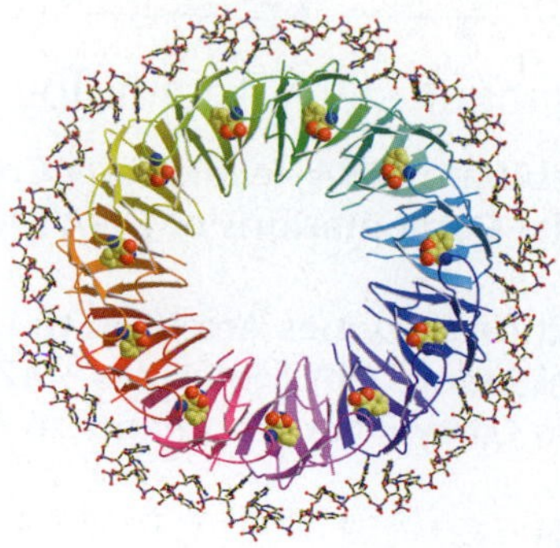

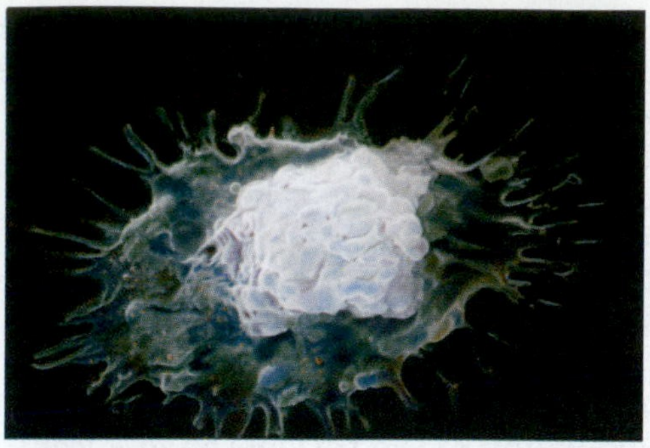

Preface

Genetics: A Molecular Perspective is the newest addition to the series of textbooks designed to provide support to students as they study one of the most fascinating scientific disciplines. This book is designed specifically for courses that begin with DNA as the focal point, and where a solid foundation in molecular genetics is provided before proceeding with coverage of the more traditonal areas of classical heredity.

Certainly no subject area has had a more sustained impact than Genetics on shaping our knowledge of the living condition. As a result of discoveries over the past 50 years, we now understand with reasonable clarity the underlying genetic mechanisms that explain how organisms develop into and then function as adults. We also better understand the basis of biological diversity and have greater insight into the evolutionary process.

As we edge into the new millennium, discoveries in this discipline continue to be numerous and profound. As geneticists and students of genetics, the thrill of being part of this era must be balanced by a strong sense of responsibility and careful attention to the many related issues that will undoubtedly arise. The formulation of proper laws and policies will depend on a comprehensive knowledge of genetics and measured responses to these issues. As a result, there has never been a higher premium or greater need for a useful and up-to-date genetics textbook.

In addition to the "DNA-first" approach taken, the first edition of *Genetics: A Molecular Perspective*, as with other texts in the series, has been designed to achieve six major goals. Specifically, we seek to:

- Emphasize concepts rather than excessive detail.
- Write clearly and directly to students in order to provide understandable explanations of complex, analytical topics.
- Establish a careful organization within and between chapters.
- Maintain constant emphasis on scientific analysis as the means to illlustrate how we know what we know.
- Propagate the rich history of genetics that so beautifully illustrates how information is acquired and extended within the discipline as it develops and grows.
- Create inviting, engaging, and pedagogically useful full-color figures enhanced by equally helpful photographs to support concept development.

These goals serve as the cornerstones of this new text. This pedagogic foundation allows the book to accommodate courses with varied approaches and lecture formats. While we assume that adopters of this text will initiate coverage by emphasizing DNA and molecular genetic expression, all chapters in the text are nevertheless written to be as independent of one another as possible. This allows instructors to utilize them in various sequences once the molecular foundation of genetics is firmly established. We believe that the varied approaches embodied in the goals listed above together provide students with optimal support for their study of genetics.

Writing a textbook that achieves these goals has been a labor of love for us. The creation of the initial edition of this text is a reflection not only of our passion for teaching genetics, but the constructive feedback and encouragement provided over the past two decades from adopters, reviewers, and our students.

Features of this Text

- Online Media Tutorials—Students are guided in their understanding of important concepts by working through what are simply the best animations, tutorial exercises, and self-assessment tools available.
- A Unique Organization—We provide a chapter sequence designed to immediately establish the importance of genetic function at the molecular level. Then, classical heredity is presented beginning with a consideration of mitosis and meiosis followed by extensive coverage of the findings of Gregor Mendel. Once solid coverage of transmission genetics is completed, we then return to the most recent findings embodied in the study of DNA biotechnology and genomic analysis. The book concludes with modern coverage of several advanced topics in eukaryotic genetics (gene regulation, development, and cancer) as well as a thorough consideration of genes and populations.
- Cutting Edge Topics—Beyond the up-to-date coverage of molecular genetics early in the text, the chapter *Genomics, Bioinformatics, and Proteomics* (Chapter 17) provides students with the most recent concepts and tools necessary to understand the information explosion occurring in these cutting-edge fields. The sister chapter, *Applications and Ethics of Genetic Technology* (Chapter 18) contains

important coverage of how the era of Genomics impacts on society. Unique to this text is Chapter 23—*Chromosome Genetics: Immunoglobulins, Isochores and Chromatid Dynamics*, where these advanced topics are presented. Finally, another addition to this series of texts is the chapter entitled *Conservation Genetics* (Chapter 27). This presentation represents the first coverage of this emerging discipline in any genetics textbook. The role of molecular biology in conservation is discussed.

- Modern Presentation of Topics—In addition to the the cutting edge topics discussed above, modern coverage is particularly evident in the discussions of recombinant DNA technology (Chapter 16), as well as in the coverage of the organization of repetitive DNA sequences in the human genome (Chapter 4), the TRAP protein that functions during regulation of the trptophan operon (Chapter 19), and the role of genetics in the origin of cancer (Chapter 22). Coverage of conservation genetics, and in particular, the genetic assessment of diversity in endangered species (Chapter 27), is also at the forefront of genetic studies.
- A New Art Program—The design of an entirely new art program will be evident to past users of the alternative edition, *Concepts of Genetics*. To all users, the pedagogic value of the well-designed and beautifully executed figures will become apparent, and constitutes one of the cornerstones of this new text.
- Excellent Photographs—A large number of photographs beautifully illustrate and enhance this textbook. Whenever possible, organisms that are subjects of genetic analysis are shown. Often, photographs are present as part of presentations in important figures, e.g., Taylor, Woods, and Hughes autoradiograms showing replication in *Vicia faba*, (Figure 3-5), the electron micrographs directly capturing transcription of rDNA in *Notophthalmus* and transcription and translation in *E. coli* (Figure 5-15), etc.
- Section Numbers—All major section titles are numbered making it easy for instructors to assign topics and for students to find topics within chapters
- Modern "Genetics, Technology, and Society" Essays—Short essays that link genetics and society accompany most chapters. The most modern topics pursued include consideration of stem cells, the genetic identification of anthrax strains, the attempts to create HIV vaccines, and emerging societal issues surroundingÒ human sex selection. These are in addition to essays that consider genetically modified foods, completion of the Human Genome Project, and endangered species--the Florida panther, among many others. These may be assigned even if they are not part of the formal lecture coverage.
- Extensive Problems to Solve—There are many entries in the "Problems and Discussion Questions" at the end of each chapter from which instructors may choose as student assignments are determined. These problems are written at various levels of difficulty, with those that are most challenging found in the "Extra Spicy Problems" sections. Students also can gain insights into different analytical approaches illustrated in each chapter by utilizing the "Insights and Solutions" section at the conclusion of each chapter (see below).

Emphasis on Concepts

Genetics: A Molecular Perspective emphasizes the conceptual framework of genetics. Our experience with this approach shows that students more easily comprehend and take with them to succeeding courses the most important ideas in genetics as well as an analytic view of biological problems. To aid students in identifying conceptual aspects of a major topic, each chapter begins with a section called "Key Concepts," which outlines the most important ideas about to be presented. Then, each chapter ends with a "Chapter Summary," which enumerates the five to ten key points that have been covered. These two features help to ensure that students focus on concepts and are not distracted by the many, albeit important, details of genetics. Specific examples and carefully designed figures support this approach throughout the book.

Problem Solving and Insights and Solutions

In order to optimize the opportunities for student growth in the important areas of problem solving and analytical thinking, each chapter concludes with an extensive collection of "Problems and Discussion Questions." These represent various levels of difficulty, with the most challenging problems located at the end of each section (the Extra Spicy entries). Brief answers to half the problems are in Appendix C. The Student Handbook is available when faculty wish to expose their students to detailed discussions and answers to all problems and questions.

As an aid to students in their development of analytical thinking skills, the Problems and Discussion Question section of each chapter is preceded by what has become an extremely popular and successful section called "Insights and Solutions." In this section we stress:

Problem solving
Quantitative analysis
Analytical thinking
Experimental rationale

Problems or questions are posed and detailed solutions or answers are provided. This feature primes students for moving on to the "Problems and Discussion Questions" section that concludes each chapter.

Acknowledgments

All comprehensive texts are dependent on the valuable input provided by many reviewers and colleagues. While we take full responsibility for any errors in this book, we gratefully acknowledge the help provided by those individuals who reviewed or otherwise contributed to the content and pedagogy of this and previous editions.

In particular, we thank Sarah Ward at Colorado State University for creating Chapter 27—Conservation Genetics and Jon Herron at the University of Washington for his input into Chapter 25—Population Genetics and Chapter 26—Genetics and Evolution. Charlotte Spencer at the Cross Cancer Institute in Alberta wrote or revised most of the Genetics, Technology, and Society essays. Others essays were previously contributed by Mark Shotwell at Slippery Rock University. We have also received extensive input from several colleagues who are deserving of special recognition for their writing efforts throughout the various editions in this series of texts. In particular, Elliott Goldstein at Arizona State University has provided valuable input in many areas of molecular genetics. We also express our thanks to Harry Nickla at Creighton University. In his role as author of the Student Handbook and the Instructor's Manual, he has reviewed and edited the problems at the end of each chapter. He also provided the brief answers to selected problems that appear in Appendix C. Additionally, Ruth Ballard at California State University, Sacramento has provided a number of new problems that appear in the Problem and Discussion Questions sections throughout the text. We appreciate the contributions of all of these geneticists and, particularly, the pleasant manner and dedication to this text displayed during our many interactions. We are forever grateful to them for their efforts.

At Prentice Hall, we express appreciation and high praise for the editorial guidance of Sheri Snavely and Gary Carlson, whose ideas and efforts have helped to shape and refine the features of this text. They have worked tirelessly to provide us with reviews from leading specialists who are also dedicated teachers, and to ensure that the pedagogy and design of the book are at the cutting edge of a rapidly changing discipline. We also appreciate the production efforts of Aksen Associates, whose quest for perfection is reflected throughout the text. In particular, Howard Aksen has provided an essential measure of sanity to the otherwise chaotic process of production. Without his work ethic and dedication, the text would never have come to fruition. Marketing is being handled with talent and enthusiasm by Jennifer Welchans and Marty McDonald. Karen Horton, biology project manager, worked to ensure that the book, supplements, and cutting-edge interactive media all work together seamlessly. Finally, the beauty and consistent presentation of the art work is the product of Imagineering, the Toronto-based company whose input is so much appreciated.

Finally, special thanks are due to those individuals who undertook the arduous task of proofing the manuscript. Mike Hoopman and Vince Heinrich examined the entire manuscript. At The College of New Jersey, a team of 15 students collectively agreed to undertake this task and together proofed most of the manuscript for errors. We thank Chris Regen, Sarah Jewell, Jagruti Patel, Melissa Greisemer, Shayna Davis, Susan Garfield, Keith Ritson, Subha Sundararajan, Arsallan Ahmad, Joe Balasius, Leanne Johnson, Rose Arone, Vijay Bhoj, Molly Sheridan, and Donna Morales for their dedication to this task. While it seems impossible to ever produce the "perfect" text, the efforts of all of these individuals have gone a very long way to meeting that goal.

Media reviewers for Genetics include:

Peggy Brickman, University of Georgia
David Kass, Eastern Michigan University
John Kemner, University of Washington
Arlene Larson, University of Colorado
John C. Osterman, University of Nebraska, Lincoln
Cheryl Ingram-Smith, Clemson University

Print reviewers for Genetics include:

Laurel F. Appel, Wesleyan University
Ruth Ballard, California State University, Sacramento
George Bates, Florida State University
Sidney L. Beck, DePaul University
Peta Bonham-Smith, University of Saskatchewan
Paul J. Bottino, University of Maryland
Philip Busey, University of Florida
Alan H. Christensen, George Mason University
Jim Clark, University of Kentucky
Garry Davis, University of Alaska, Anchorage
Johnny El-Rady, University of South Florida
Bert Ely, University of South Carolina
Lloyd M. Epstein, Florida State University
Dale Fast, Saint Xavier University
Donald Gailey, California State University, Hayward
George W. Gilchrist, Clarkson University
Sandra Gilchrist, University of South Florida
Thomas J. Glover, Hobart & William Smith Colleges
Elliott S. Goldstein, Arizona State University
Douglas Harrison, University of Kentucky
Don Hauber, Loyola University
Vincent Henrich, University of North Carolina, Greensboro
Philip L. Hertzler, Central Michigan University
Rebecca Jann, Queens College
Mitrick A. Johns, Northern Illinois University
Jocelyn Krebs, University of Alaska, Anchorage
Paul F. Lurquin, Washington State University
Clint Magill, Texas A & M University
Harry Nickla, Creighton University
Berl R. Oakley, Ohio State University
John C. Osterman, University of Nebraska–Lincoln
Dennis T. Ray, The University of Arizona
Joseph Reese, Pennsylvania State University
Thomas F. Savage, Oregon State University
Cathy Schaeff, American University
Rod Scott, Wheaton College
Thomas P. Snyder, Michigan Technical University
Paul Spruell, University of Montana
Christine Tachibana, University of Washington
Albrecht von Arnim, University of Tennessee
Laurence von Kalm, University of Central Florida
Tracy Whitford, East Stroudsburg University

As the above acknowledgments make clear, a text such as this is a collective enterprise. All of the above individuals deserve to share in any success this text enjoys. We want them to know that our gratitude is equaled only by the extreme dedication evident in their many efforts. Many, many thanks to you all.

MEDIA TUTORIAL Genetic variations

For the Student

All New Media Tutorials (CD-ROM)

The most sophisticated learning and tutorial package available for students of genetics, the new Student CD-ROM addresses students' most difficult concepts. Each concept or process begins with an overview that includes animations, proceeds to one or a series of interactive exercises, followed by self-quizzes. Each chapter contains a glossary, help function, search function, web links, plus additional - problem-solving questions. Students who experience difficulty are directed to specific sections of the text for review. Look for the icon on the outside margin of this page to identify media tutorials that appear on the CD-ROM.

The following topics are addressed in the interactive media tutorials:

Mitosis
Meiosis
Monohybrid Cross
Independent Assortment
Probability

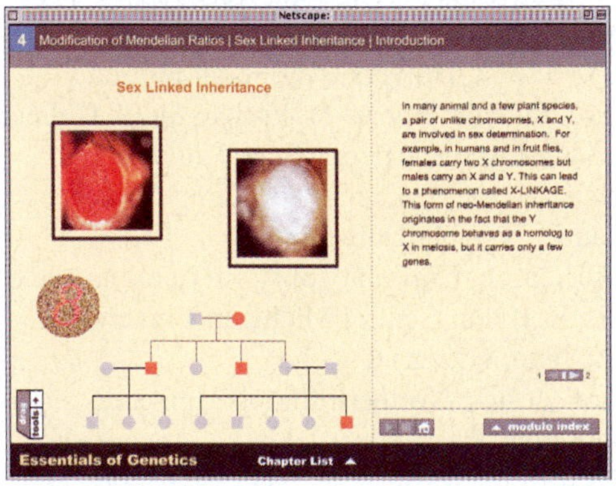

Analysis of Human Pedigrees
Extensions of Mendelian Inheritance
X-linked Inheritance
Linkage and Recombination
Linked Genes
Mapping a Three-Point Cross
More Three-Point Crosses
Virtual Crossover Laboratory
Chromosome Aberrations
Nondisjunction
DNA Structure
DNA Replication
DNA Recombination
Chromosome Structure
Transcription
Protein Translation
Mutation, Gene/Protein Colinearity
Mutations at the DNA Level
DNA Repair
Chemical Mutagenesis: Ames Test
Phage Genetics
Bacterial Genetics
Gene Expression: Prokaryotes
Restriction Enzymes and Vectors
Polymerase Chain Reaction
Restriction Mapping
Cloning Libraries
DNA Sequencing
RFLP
DNA Fingerprinting
Gene Expression: Eukaryotes
Population Genetics

Student Handbook and Solutions Manual

Harry Nickla, Creighton University
(0-13-100510-3)
This valuable handbook provides a detailed step-by-step solution or lengthy discussion for every problem in the text. The handbook also features additional study aids including extra study problems, chapter outlines, vocabulary exercises and an overview of how to study genetics.

New York Times Themes of the Times: Genetics and Molecular Biology

Compiled by Harry Nickla, Creighton University
(0-13-060462-3)
This exciting newspaper-format supplement brings together recent genetics and molecular biology articles from the pages of the New York Times. This free supplement, available through your local representative, encourages students to make the connections between genetic concepts and the latest research and breakthroughs that are making headlines in the field. This resource is updated regularly.

Science on the Internet: A Student's Guide

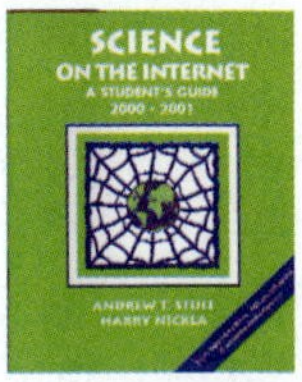

Andrew T. Stull
(I0-13-021308-X)
The perfect tool to help your students take advantage of the Concepts of Genetics Companion Website. This informative resource helps students locate and explore the myriad of science resources on the Web. It also provides an overview of the Web, general navigational

strategies, and brief student activities. It is available FREE when packaged with the text.

Companion Website—www.prenhall.com/klug

Navigating the Website is as easy as opening the textbook-all of the features are organized by text chapters. The Companion Website provides study aid tools, guided explorations, and updates on current topics relevant to genetics courses. Other features for the student include: Self-grading Chapter Problems; Chapter Search Terms; and Syllabus Manager.

In addition to providing resources for the student, the Companion Web site offers unique tools and support that make it easy to integrate Prentice Hall's on-line resources into the course.

For the Instructor

***Instructor's CD-ROM* (0-13-047350-2)**

For adopters of the *Genetics: A Molecular Perspective,* this CD-ROM contains

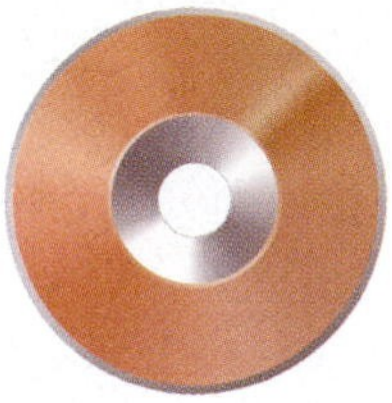

- A replicate version of the Student CD-ROM
- All animations from the Student CD-ROM in a separate file
- All text illustrations in PowerPoint format
- The complete Instructor's Manual

Instructors will be able to coordinate lecture presentations with the textbook knowing students will be studying using the same animations based upon the text. No more searching for the Instructor's Manual. It is on the CD-ROM.

Instructor's Resource Manual with Testbank

(0-13-009675-X)

This manual and test bank contains over 1000 questions and problems an instructor can use to prepare exams. The manual also provides optional course sequences, a guide to audiovisual supplements, and a section on searching the web. The testbank portion of the manual is also available in electronic format.

TestGen EQ Computerized Testing Software

(0-13-009677-6)

In addition to the printed volume, the test questions are also available as part of the TestGen EQ Testing Software, a text-specific testing program that is networkable for administering tests. It also allows instructors to view and edit questions, export the questions as tests, and print them out in a variety of formats.

Transparencies

(0-13-009678-4)

275 figures from the text are included in the transparency package: 200 four-color transparencies from the text plus 75 transparency masters. The font size of the labels has been increased for easy viewing from the back of the classroom.

WebCT Course for Genetics: A Molecular Perspective

(0-13-047367-7)

The Prentice Hall WebCT course content for *Genetics: A Molecular Perspective* helps you meet the challenge of creating robust, interactive, and educationally rich online courses. Our WebCT course material provides you with high quality, class-tested material pre-programmed and fully functional in the WebCT environment. Whether used as an online supplement to either a campus-based or distance learning course, our pre-assembled course content gives you a tremendous head start in developing your own online courses.

Blackboard Course for Genetics: A Molecular Perspective

(0-13-047363-4)

The *Genetics: A Molecular Perspective* Blackboard course contains web-based content and resources such as online study guides, assessment databanks, and lecture resource material. The abundant online content, combined with Blackboard's popular tools and easy-to-use interface, result in a robust web-based course that is easy to implement, manage, and use—taking your courses to new heights in student interaction and learning. The Blackboard course management solution enables you to quickly add an online component to your campus-based course to provide you with a sophisticated technology base for total customization, scalability, and integration into your distance learning course.

CourseCompass Course for Genetics: A Molecular Perspective

(0-13-047364-2)

The *Genetics: A Molecular Perspective* CourseCompass™ course is the perfect course management solution that combines quality content with state-of-the-art Blackboard technology! CourseCompass™ is a dynamic, interactive online course management tool powered by Blackboard but hosted by Pearson Education. This exciting product allows you to teach with market-leading *Concepts of Genetics* content in an easy-to-use customizable format.

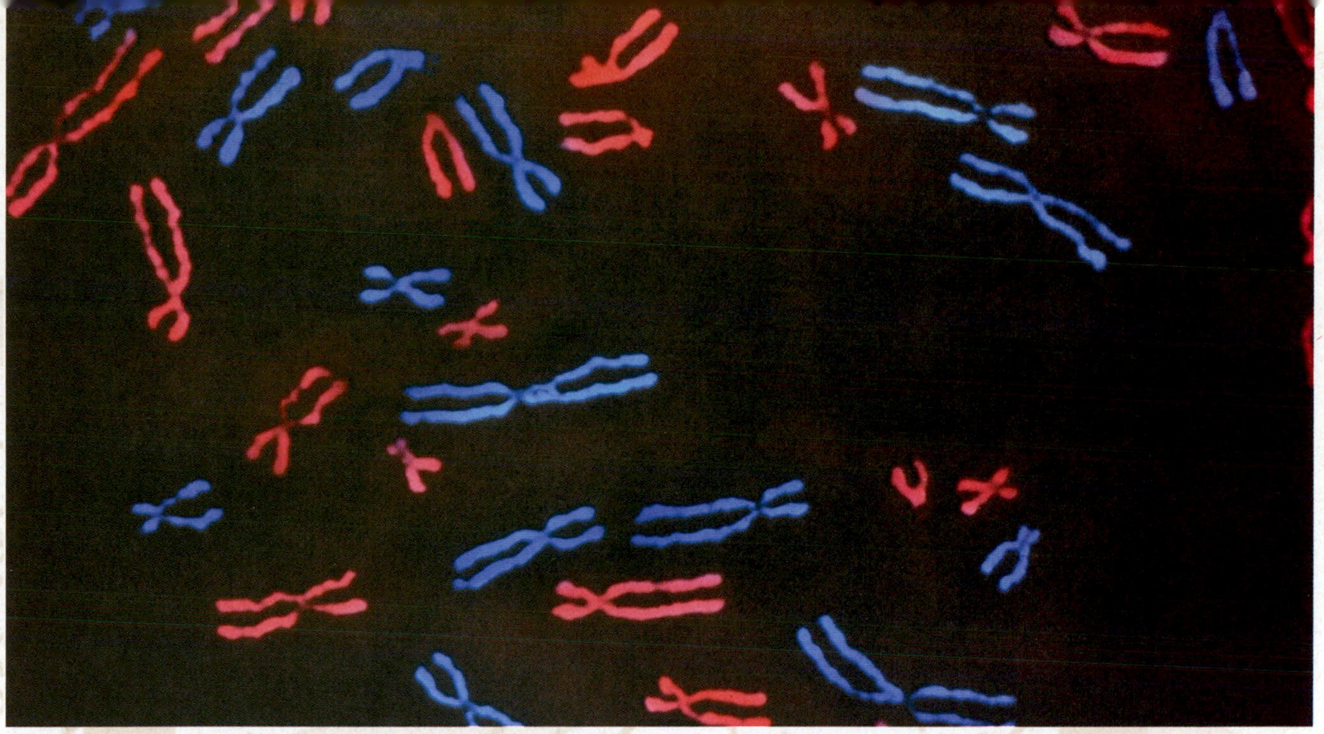

Human metaphase chromosomes, each composed of two sister chromatids joined at a common centromere.

1

Introduction to Genetics

KEY CONCEPTS

Genetics has always been of great interest to and had a profound effect on humankind. As knowledge in the discipline has grown, many different issues have arisen that have led to controversies at the interface between science and society. We begin this text with a short vignette that discusses briefly one recent instance where current advances in genetics directly affect all members of an entire country. This short story captures the flavor of how genetic technology may impact on society and provides a glimpse of how the future might be envisioned as further advances are made.

In early 1999, a controversy was growing among the 270,000 residents of the remote island nation of Iceland. As the new millennium approached, following months of fierce and sometimes heated debate, Iceland's parliament passed a law giving a local biotechnology company the rights to develop a large database containing detailed DNA profiles of every resident of the country. This genetic information will be correlated with the genealogical and medical records of each entry and marketed to researchers around the world over the next decade.

This is not a passage drawn from Aldous Huxley's *Brave New World*, but rather an example of the current interface between genetics and society as we enter the current century. There are many interesting reasons why such a potential invasion of genetic privacy is about to occur to the residents of this small outpost country. They represent a nearly unique case of genetic uniformity seldom seen or accessible to scientific investigation. Indeed, for a variety of reasons, most Icelanders share a close genetic resemblance, not only to one another, but to their Viking ancestors who settled this country over 1000 years ago. Because of this, geneticists believe that the Icelandic population is a tremendous asset in studying the link between genetics and disease. Due to the state-supported health-care system, medical records exist for all residents as far back as 1900. Genealogical information is available for every living resident, as well as for over 500,000 of the estimated 750,000 individuals who have ever lived in Iceland.

On the flip side are issues of privacy, consent, and commercialism—issues that will be central to most future dilemmas and controversies arising from applications of newly acquired genetic technology. The central questions currently being asked throughout the worldwide scientific community are what uses will be made of the vast amount of genetic information that we are acquiring and where society at large factors into the decision. For example, how will the knowledge of the complete nucleotide sequence of the human genome be used? More than at any other time in the history of science, addressing the ethical questions surrounding an emerging technology is clearly as important as the information that arises from this research.

As you launch your study of the discipline of genetics, try to remain sensitive to issues like those just described. While you proceed through this text, the knowledge gained will become a rich resource for achieving a thorough understanding of modern-day genetics. On the one hand, never has there been a more exciting time to be immersed in the study of this area of science. On the other hand, never has such a need for caution been so apparent to society. Along the way, enjoy your studies, but take your responsibility as a novice geneticist most seriously.

1.1 Genetics Has a Rich and Interesting History

Because genetic processes are fundamental to the comprehension of life itself, the discipline of genetics is thought by many to sit at the center of the field of biology. Genetic information directs cellular function, largely determines an organism's external appearance, and serves as the link between generations in every species. As such, knowledge of genetics is essential to the thorough understanding of other disciplines of biology, including molecular biology, cell biology, physiology, evolution, ecology, systematics, and behavior. Genetics therefore unifies biology and serves as its "core." Thus, it is not surprising that genetics has a long, rich history.

In the chapters that follow, we will discuss the nature of chromosomes, the way in which genetic information is transmitted from one generation to the next, and the way in which this information is stored, altered, expressed, and regulated. Many significant scientific findings, which serve as the foundation for our discussions, were obtained in the 19th century. As the 20th century dawned, still other highly relevant discoveries were made that began to clarify the understanding of the physical basis of living organisms and their relationship to one another. Several related ideas were gaining acceptance at that time and were particularly significant: (1) matter is composed of atoms, (2) cells are the fundamental units of living organisms, (3) nuclei somehow serve as the "life force" of cells, and (4) chromosomes housed within nuclei somehow play an important role in heredity. When these ideas were correlated with the newly rediscovered genetic findings of Gregor Mendel and integrated with Darwin's theory of natural selection and the origin of species, a more complete picture of life at the level of the individual and of the population emerged. The era of modern-day biology was initiated on this foundation.

But what of the many important ideas and hypotheses that served as forerunners of 19th-century thought? In the following short section, we consider some of these, several of which can be traced back well over 1000 years!

Prehistoric Times: Domesticated Animals and Cultivated Plants

While we don't know when people first recognized the existence of heredity, various types of archeological evidence (e.g., primitive art, preserved bones and skulls, and dried seeds) provide insights. Such evidence documents the successful domestication of animals and cultivation of plants thousands of years ago. These efforts represent artificial selection of genetic variants within populations. For example, between 8000 and 1000 B.C., horses, camels, oxen, and various breeds of dogs (derived from the wolf family) were domesticated to serve various roles. The cultivation of many plants, including maize, wheat, rice, and the date palm, is thought to have been initiated between 7000-5000 B.C. The remains of maize dating to this period have been recovered in caves in the Tehuacán Valley of Mexico. Assyrian art depicts artificial pollination of the date palm, thought to have originated in Babylonia (Figure 1–1). Such cultivation is believed to have been the source of date palms found in that region today, where over 400 varieties exist in just four oases in the Sahara Desert. They differ from one another in various traits, such as fruit taste.

Prehistoric evidence of cultivated plants and domesticated animals supports the hypotheses that our ancient ancestors learned that desirable and undesirable traits are passed to successive generations and that by influencing their breeding, many desirable varieties of animals and plants could be obtained. Human awareness of heredity was thus apparent during prehistoric times and successful attempts were made to manipulate the genetic material, even though it was unclear what it might be.

The Greek Influence: Hippocrates and Aristotle

While few, if any, significant ideas were put forward to explain heredity during prehistoric times, during the Golden Age of Greek culture, philosophers directed much attention to this subject, particularly as it relates to the origin of humans and to their reproduction. This is clearly evident in the writings of the Hippocratic school of medicine (500–400 B.C.) and subsequently of the philosopher and naturalist Aristotle (384–322 B.C.).

For example, the Hippocratic treatise *On the Seed* argues that male semen is formed in numerous parts of the body and is transported through blood vessels to the testicles. Active "humors" are the bearers of hereditary traits and are drawn from various parts of the body into the semen. These humors could be healthy or diseased, the latter condition accounting for the appearance of newborns exhibiting congenital disorders or deformities. Furthermore, it was believed that these humors could be altered in individuals and, in their new form, could be passed on to offspring. In this way, newborns could "inherit" traits that their parents had "acquired" because of their environment.

Aristotle (Figure 1–2), who was enormously interested in living organisms, and who had studied under Plato for some 20 years, was more expansive than Hippocrates in his analysis of human origins and heredity. Aristotle proposed that male semen was formed from blood, rather than from each organ, and that its generative power resided in a "vital heat" that it contained. This vital heat had the capacity to produce offspring of the same "form" (i.e., basic structure and capacities) as the parent. Aristotle believed that the vital heat cooked and shaped the menstrual blood produced by the female, which was the "physical substance" giving rise to an offspring. The embryo developed not because it already contained the parts in miniature (as some Hippocratics had

FIGURE 1–1 Relief carving depicting artificial pollination of date palms (800 B.C.). (*The Metropolitan Museum of Art, Gift of John D. Rockefeller, Jr., 1932. (32.143.3) Photograph © 1983)*

FIGURE 1–2 Aristotle describes the animals Alexander has sent him, as seen in a fresco in The Main Hall of The Assemblée Nationale in Paris, France. (*Painting by Eugene Delacroix. © Photo by Erich Lessing/Art Resource, NY.*)

thought), but because of the shaping power of the vital heat. These ideas constituted only one part of Aristotelian philosophy of order in the living world.

Although the ideas of Hippocrates and Aristotle sound primitive and naive today, we should recall that, prior to the 1800s, neither sperm nor eggs had yet been observed in mammals. Thus, the Greek philosophers' ideas were worthy ones in their time and for centuries to come. As we will see, their thinking was not so different from that of Charles Darwin in his formal proposal of the theory of pangenesis put forward during the 19th century.

1600–1850: The Dawn of Modern Biology

During the ensuing 1900 years (300 B.C.–1600 A.D.), the theoretical understanding of genetics was not extended by significant new ideas. However, in Roman times, plant grafting and animal breeding were common. By the Middle Ages, naturalists, well aware of the impact of heredity on organisms they studied, were faced with reconciling their findings with current religious beliefs. The theories of Hippocrates and Aristotle still prevailed and, when applied to humans, they no doubt conflicted with the prevailing religious doctrines.

Between 1600 and 1850, major strides were made that provided much greater insights into the biological basis of life, setting the scene for the revolutionary work and principles presented by Charles Darwin and Gregor Mendel. In the 1600s, the English anatomist William Harvey (1578–1657), better known for his experiments demonstrating that the blood is pumped by the heart through a circulatory system made up of the arteries and veins, also wrote a treatise on reproduction and development, patterned after Aristotle's work. In it, he is credited with the earliest statement of the **theory of epigenesis**—that an organism is derived from substances present in the egg, which differentiate into adult structures during embryonic development. Patterned after Aristotle's ideas, epigenesis holds that structures such as body organs are not initially present in the early embryo, but instead are formed *de novo* (anew). The theory of epigenesis conflicts directly with the 17th-century theory of preformation, which states that sex cells contain a complete miniature adult called the **homunculus** (Figure 1–3), perfect in every form. Preformation was popular well into the 18th century. However, work by the embryologist Casper Wolff (1733–1794) and others clearly disproved the preformation theory, thus favoring epigenesis. Wolff was convinced that several structures, such as the alimentary canal, were not initially present in the earliest embryos he studied, but instead were formed later during development.

At this same period, other significant chemical and biological findings affected future scientific thinking. In 1808, John Dalton expounded his **atomic theory**, which stated that all matter is composed of small, invisible units called atoms. Around 1830, Matthias Schleiden and Theodor Schwann, using improved microscopes, proposed the **cell theory**, stating that all organisms are composed of basic visible units

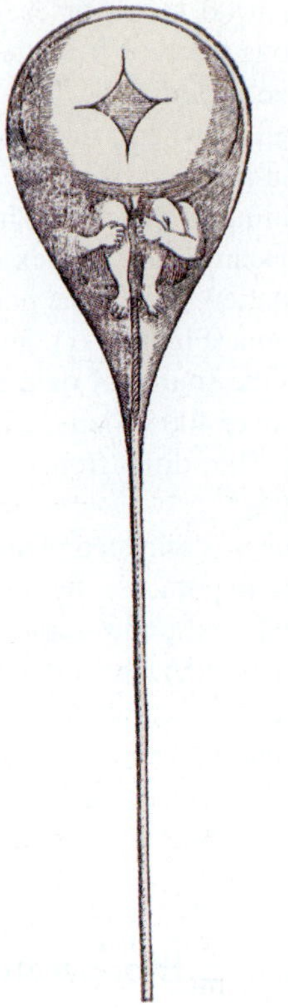

FIGURE 1–3 Depiction of the "homunculus," a sperm containing a miniature adult, perfect in proportion, and fully formed. (*Hartsoeker, N. Essay de dioptrique, Paris, 1694, p. 230*)

called cells, which are derived from similar preexisting structures. By this time, the idea of **spontaneous generation**, the creation of living organisms from nonliving components, had clearly been disproved, and living organisms were considered to be derived from preexisting organisms and to consist of cells made up of atoms.

Another notion prevalent in the 19th century, the **fixity of species**, was influential. According to this doctrine, animal and plant groups remain unchanged in form from the moment of their appearance on Earth. Embraced particularly by those who also adhered to a belief in special creation, this doctrine was popularized by several people, including the Swedish physician and plant taxonomist Carolus Linnaeus (1707–1778), who is better known for devising the binomial system of classification.

The influence of the theory of the fixity of species is illustrated by considering the work of the German plant breeder Joseph Gottlieb Kolreuter (1733–1806), who worked with tobacco. He crossbred two groups and derived a new hybrid form, which he then converted back to one of the parental types by repeated backcrosses. In other breeding experiments, using carnations, he clearly observed segregation of traits, which was to become one of Mendel's principles of genetics. These results seemed to contradict the idea of "fixed species" that do not change with time. Although Kolreuter was puzzled by these outcomes, he failed to recognize the real significance of his findings because of his belief in both special creation and the fixity of species.

Charles Darwin and Evolution

With the preceding information as background, we conclude our coverage of the historical context of genetics with a brief discussion of the work of Charles Darwin, who in 1859, published the book-length statement of his evolutionary theory, *On the Origin of Species*. Darwin's many geological, geographical, and biological observations convinced him that existing species arose by descent with modification from other ancestral species. Greatly influenced by his now-famous voyage on the H.M.S. *Beagle* (1831–1836), Darwin's thinking culminated in his formulation of the **theory of natural selection**, a theory that attempts to explain the causes of evolutionary change.

Formulated and proposed at the same time, but independently by Alfred Russel Wallace, natural selection is based on the observation that populations tend to consist of more offspring than the environment can support, leading to a struggle for survival among them. Those organisms with heritable traits that allow them to adapt to their environment are better able to survive and reproduce than those with less-adaptive traits. Over a long period of time, slight, but advantageous variations will accumulate. If a population bearing these inherited variations becomes reproductively isolated, a new species may result.

The primary gap in Darwin's theory was a lack of understanding of the genetic basis of variation and inheritance, a gap that left it open to reasonable criticism well into the 20th century. Aware of this weakness in his theory of evolution, in 1868, Darwin published a second book, *Variations in Animals and Plants under Domestication*, in which he attempted to provide a more definitive explanation of how heritable variation arises gradually over time. Two of his major ideas, pangenesis and the inheritance of acquired characteristics, have their roots in the theories involving "humors," as put forward by Hippocrates and Aristotle.

In his provisional hypothesis of **pangenesis**, Darwin coined the term *gemmules* (rather than humors) to describe the physical units representing each body part that were gathered by the blood into the semen. Darwin felt that these gemmules determined the nature or form of each body part. He further believed that gemmules could respond in an adaptive way to an individual's external environment. Once altered, such changes would be passed onto offspring, allowing for the inheritance of acquired characteristics. Jean-Baptiste Lamarck had previously formalized this idea in his 1809 treatise, *Philosophie Zoologique*. Lamarck's theory, which became known as the **doctrine of use and disuse**, proposed that organisms acquire or lose characteristics that then become heritable. Even though Darwin never understood the basis for inherited variation, his ideas concerning evolution may be the most influential theory ever put forward in the history of biology. He was able to distill his extensive observations and synthesize his ideas into a cohesive hypothesis describing the origin of diversity of organisms populating Earth.

As Darwin's work ensued, Gregor Johann Mendel (Figure 1–4) conducted his experiments between 1856 and 1863 and published his classic paper in 1866. In it, Mendel demonstrated a number of statistical patterns underlying inheritance and developed a theory involving hereditary factors in the germ cells to explain these patterns. His research was virtually ignored until it was partially duplicated and then cited by Carl Correns, Hugo de Vries, and Eric Von Tschermak around 1900, after which it was championed by William Bateson.

FIGURE 1–4 Gregor Johann Mendel, who in 1866 put forward the major postulates of transmission genetics as a result of experiments with the garden pea. *(Archiv/Photo Researchers, Inc.)*

By the early part of the 20th century, chromosomes were discovered and support for the epigenetic interpretation of development had grown considerably. It gradually became clear that heredity and development were dependent on "information" contained in chromosomes, which were contributed by gametes to each individual. The "gap" in Darwin's theory had narrowed considerably.

1.2 Genes and Chromosomes Are the Fundamental Units of Genetics

In ensuing chapters, we will consider in detail the experiments of Gregor Mendel and the subsequent research that led to a clear understanding of transmission genetics. Together, this body of work served as the basis of the **chromosomal theory of inheritance**, which states that inherited traits are controlled by genes that reside in chromosomes, which are faithfully transmitted through gametes to future generations. It is useful here to consider some fundamental issues underlying this theory. These issues provide a brief overview of the discipline of genetics and of genes and chromosomes. We shall set forth this overview by asking and answering a series of questions. You may wish to write or think through a response before reading our answer to each question. Throughout the text, they will be expanded on as more detailed information is presented.

What does "genetics" mean? **Genetics** is the branch of biology concerned with heredity and variation. This discipline involves the study of cells, individuals, their offspring, and the populations within which organisms live. Geneticists investigate all forms of inherited variation, as well as the molecular basis underlying such characteristics.

What is the center of heredity in a cell? In eukaryotic organisms, the **nucleus** contains the genetic material in the form of genes present on chromosomes. In prokaryotes, such as bacteria, the genetic material exists in an unenclosed, but recognizable area of the cell called the **nucleoid region** (Figure 1–5). In viruses, which are not true cells, the genetic material is ensheathed in the protein coat, together constituting the viral head or capsid.

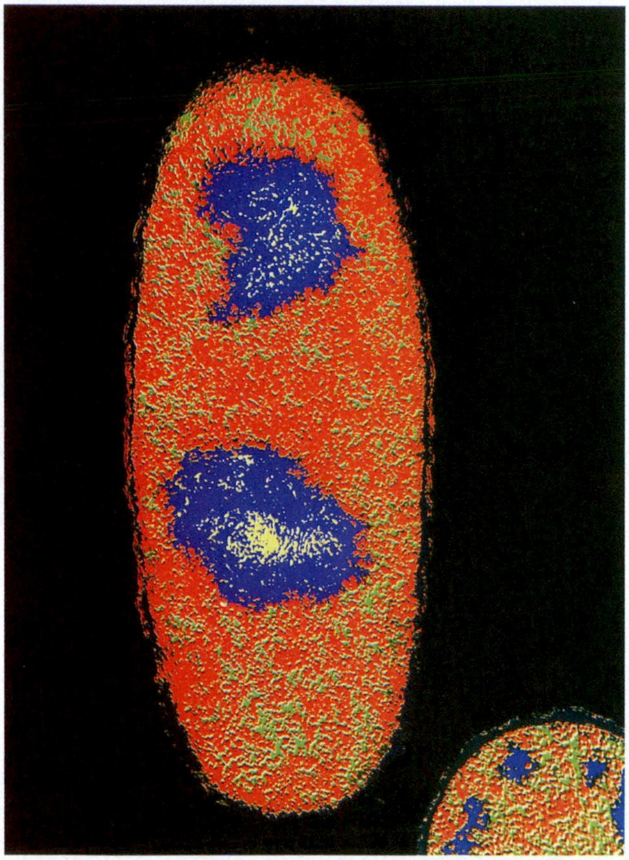

FIGURE 1–5 Enhanced electron micrograph of *Escherichia coli*, demonstrating the nucleoid regions (shown in blue). The bacterium has replicated its DNA and is about to begin cell division.

What is a gene? In simplest terms, the **gene** is the functional unit of heredity residing at a specific point along a chromosome. Conceptually, a gene is an informational storage unit capable of undergoing replication, expression, and mutation. As investigations have progressed, the gene has been found to be a very complex element. Biochemically, a gene is a length of DNA that specifies a product or action.

What is a chromosome? In viruses and bacteria, which have only a single **chromosome**, it is most simply thought of as a long, usually circular DNA molecule organized into genes. Most eukaryotes have many chromosomes (Figure 1–6) that are composed of linear DNA molecules intimately associated with proteins. In addition, eukaryotic chromosomes contain many nongenic regions. It is not yet clear what role, if any, is played by many of these regions. Our knowledge of the chromosome, like that of the gene, is continually expanding.

When and how can chromosomes be visualized? If the chromosomes are released from the viral head or the bacterial cell, they can be visualized under the electron microscope. In eukaryotes, chromosomes are most easily visualized under the light microscope when they are undergoing **mitosis** or **meiosis**. In these division processes, the material constituting chromosomes is tightly coiled and condensed, giving rise to the characteristic image of chromosomes. Following division, this material, called chromatin, uncoils during interphase, where it can be studied under the electron microscope.

How many chromosomes does an organism have? Although there are exceptions, members of most eukaryotic species have a specific number of chromosomes, called the **diploid number (2*n*)**, in each somatic cell. For example, humans have a diploid number of 46 (Figure 1–7). Upon close analysis, these chromosomes are found to occur in pairs, each member of which shares a nearly identical appearance when

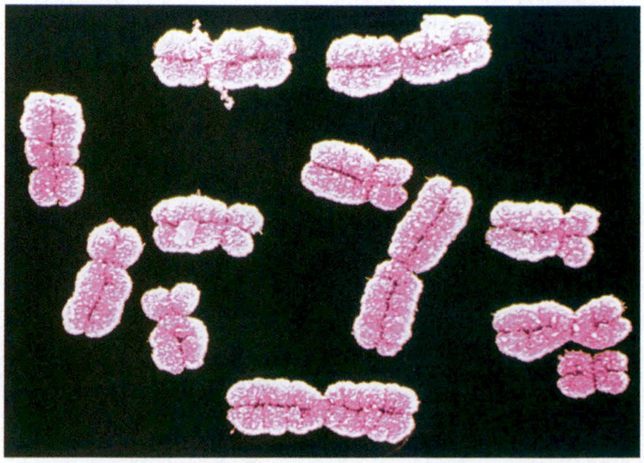

FIGURE 1–6 Human mitotic chromosomes visualized under the scanning electron microscope.

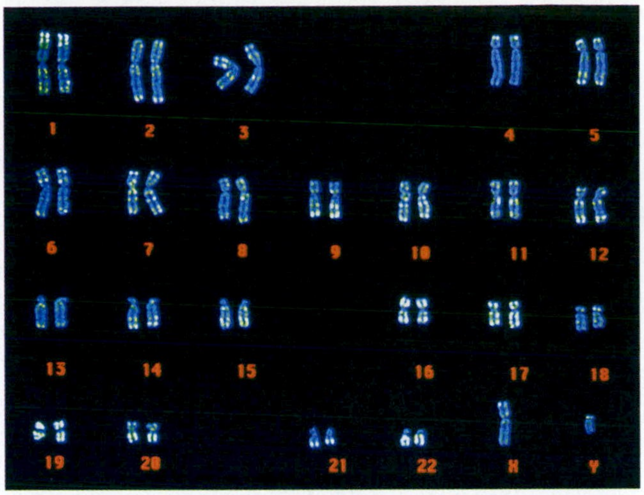

FIGURE 1–7 The human male karyotype. (*Sovereign/Phototake*)

visible during cell division. Called **homologous chromosomes**, the members of each pair are identical in their length and in the location of the **centromere**, the point of spindle-fiber attachment during division. They also contain the same sequence of gene sites, or **loci**, and pair with one another during gamete formation (the process of meiosis).

The number of different *types* of chromosomes in any diploid species is equal to half the diploid number and is called the **haploid number (*n*)**. Some organisms, such as yeast, are haploid during most of their life cycle and contain only one "set" of chromosomes. Other organisms, especially many plant species, are sometimes characterized by more than two sets of chromosomes and are said to be **polyploid**.

What is accomplished during the processes of mitosis and meiosis? Mitosis is the process by which the genetic material of eukaryotic cells is duplicated and distributed during cell division. Meiosis is the process whereby cell division produces gametes in animals and spores in most plants, which serve as the basis for transmission of the genetic information between generations. While mitosis occurs in somatic tissue and yields two progeny cells with an amount of genetic material identical to that of the progenitor cell, meiosis creates cells with precisely one-half of the genetic material. Each gamete receives one member of each homologous pair of chromosomes and is haploid. This reduction in chromosome number is essential if the offspring arising from two gametes are to maintain a constant number of chromosomes characteristic of their parents and other members of the species.

What are the sources of genetic variation? Classically, there are two sources of genetic variation, **chromosomal mutations** and **gene mutations**. The former, also called chromosomal aberrations, are the more substantial and include duplication, deletion, or rearrangement of chromosome segments. Gene mutations result from smaller changes in the stored chemical information in DNA, making up a part of the organism's **genotype**. Such a change may include substitution, duplication, or deletion of nucleotides, which compose this chemical information. Alternative forms of the gene, which result from mutation, are called **alleles**. Genetic variation frequently, but not always, results in a change in some characteristic of an organism, referred to as its **phenotype**.

1.3 Nucleic Acids and Proteins Serve as the Molecular Basis of Genetics

It will also be useful as we proceed through the first part of the text, which focuses on transmission genetics, to grasp the most basic tenets that underlie the molecular basis of genetic function. This will provide you with a foundation for a more comprehensive understanding of Mendel's work as well as that which followed it. As you will see, it is indeed remarkable that so much was learned without the knowledge of molecular genetics, which today, we take for granted.

The Trinity of Molecular Genetics

The way in which genes control inherited variation is best understood in terms of three molecules, sometimes referred to as the trinity of molecular genetics: **DNA**, **RNA**, and **protein**. The nucleic acid DNA (deoxyribonucleic acid) serves as the genetic material in all living organisms as well as in most viruses. DNA is organized into genes and stores genetic information. As part of the chromosomes, the information contained in genes can be transmitted faithfully by parents through gametes to their offspring. For the gene's DNA to subsequently influence an inherited trait, the stored genetic information in the DNA in most cases is first transferred to a closely related nucleic acid, RNA (ribonucleic acid). In eukaryotic organisms, RNA most often carries the genetic information out of the nucleus, where chromosomes reside, into the cytoplasm of the cell. There, the information in RNA is translated into proteins, which serve as the end products of most all genes. Ultimately, it is the diverse functions of proteins that determine the biochemical identity of cells and strongly influence the expression of inherited traits. The process of storage and expression of genetic information, upon which life on earth is based, is often summarized as

DNA makes RNA, which makes proteins

The process of transferring information from DNA to RNA is called **transcription**. The subsequent conversion of the genetic information contained in RNA into a protein is called **translation**.

The Structure of Nucleic Acids

Two depictions of the structure and components of DNA are shown in Figure 1–8. The molecule exists in cells as a long, coiled ladderlike structure described as a double helix. Each strand of the helix consists of a linear polymer made up of

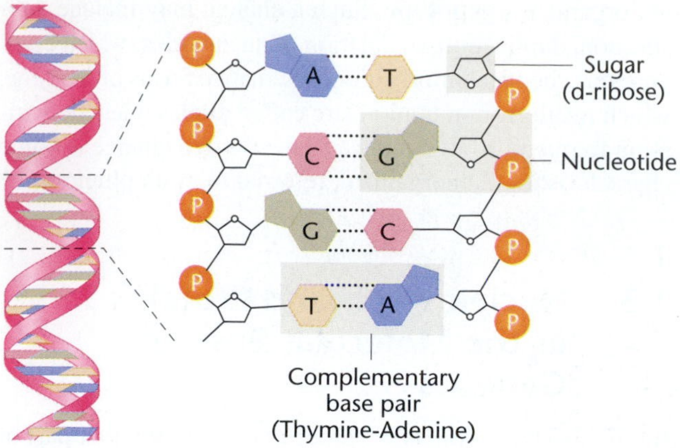

FIGURE 1–8 Summary of the structure of DNA, illustrating on the left, the nature of the double helix and on the right, the chemical components making up both strands.

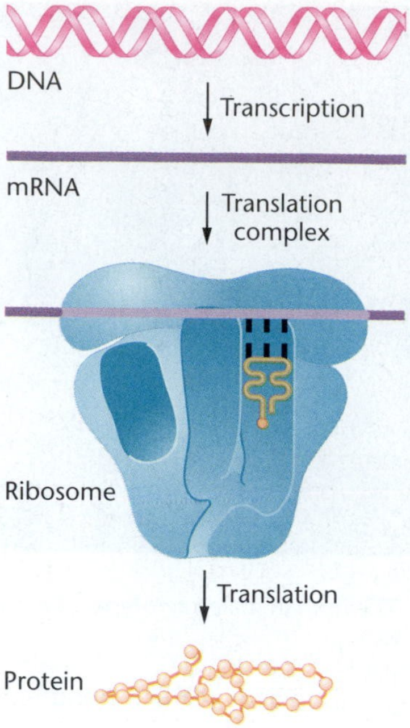

FIGURE 1–9 Depiction of genetic expression involving transcription of DNA into mRNA; and the translation of mRNA on a ribosome into a protein.

genetic building blocks called **nucleotides**, of which there are four types. Nucleotides vary, depending upon which of four nitrogenous bases is part of the molecule—A (adenine), G (guanine), T (thymine), or C (cytosine). These comprise the genetic alphabet, which in various combinations, ultimately specify the components of proteins. One of the great discoveries of the 20th century was made in 1953 by James Watson and Francis Crick, who established that the two strands of their proposed double helix are exact complements of one another, such that the rungs of the ladder always consist of either A═T or G≡C base pairs. As we shall see when we pursue molecular genetics in depth later in the text, this **complementarity** between adenine and thymine nitrogenous base pairs and between guanine and cytosine nitrogenous base pairs, attracted to one another by **hydrogen bonds**, is critical to genetic function. Complementarity serves as the basis for both the replication of DNA and for the transcription of DNA into RNA. During both processes, under the direction of the appropriate enzyme, DNA strands serve as templates for the synthesis of the complementary molecule.

RNA is chemically very similar to DNA. However, it demonstrates a small variation in its component sugar (ribose vs. deoxyribose), and it contains the nitrogenous base uracil in place of thymine. Additionally, in contrast to the double helix of DNA, RNA is generally single stranded. Importantly, it can form complementary structures with a strand of DNA. In such cases, uracil base pairs with adenine. As noted earlier, this complementarity is the basis for *transcription* of the chemical information in DNA into RNA, and the process is illustrated at the top of Figure 1–9

The Genetic Code and RNA Triplets

Once an RNA molecule complementary to one strand of a gene's DNA is transcribed, the RNA behaves as a messenger that directs the synthesis of proteins. This is accomplished during the association of this RNA molecule (called messenger RNA, or mRNA, for short) with a complex cellular structure called the **ribosome**. The process by which proteins are synthesized under the direction of mRNA, as mentioned earlier, is called *translation* (Figure 1–9). Ribosomes serve as nonspecific workbenches for protein synthesis.Proteins, as the end product of genes, are linear polymers made up of amino acids, of which there are 20 different types in living organisms. A major question is how information present in mRNA is encoded to direct the insertion of specific amino acids into protein chains as they are synthesized. The answer is now quite clear. The genetic code consists of a linear series of triplet nucleotides present in mRNA molecules. Each triplet reflects, through complementarity, the information stored in DNA and specifies the insertion of a specific amino acid as the mRNA is translated into the growing protein chain. A key discovery in how this is accomplished involved the identification of a series of adapter molecules called **transfer RNA (tRNA)**. Within the ribosome, these adapt the information encoded in the mRNA triplets to the specific amino acid during translation.

As the preceding discussion documents, DNA makes RNA, which most often makes protein. These processes occur with great specificity. Using an alphabet of only four letters (A, T, C, and G), a language exists that directs the synthesis of highly specific proteins that collectively serve as the basis for all biological function.

Proteins and Biological Function

As we have mentioned, proteins are the end products of genetic expression. They are the molecules responsible for imparting the properties that we attribute to the living process. The potential for achieving the diverse nature of biological function rests with the fact that the alphabet used to construct proteins consists of 20 letters (amino acids), which combine to create words that can be thousands of letters long

(Figure 1-10). If we consider a protein chain that is just 100 amino acids in length, and at each position there can be any one of 20 amino acids, then the number of different molecules, each with a unique sequence, is equal to

$$20^{100}$$

Since 20^{10} exceeds 5×10^{12}, or over 5 trillion, imagine how large 20^{100} is! Obviously, evolution has seized on a class of molecules that have the potential for enormous structural diversity as they serve as the mainstay in biological systems.

The main category of proteins is that which includes **enzymes**. These molecules serve as biological catalysts, essentially allowing biochemical reactions to proceed at rates that sustain life under the conditions that exist on earth. For

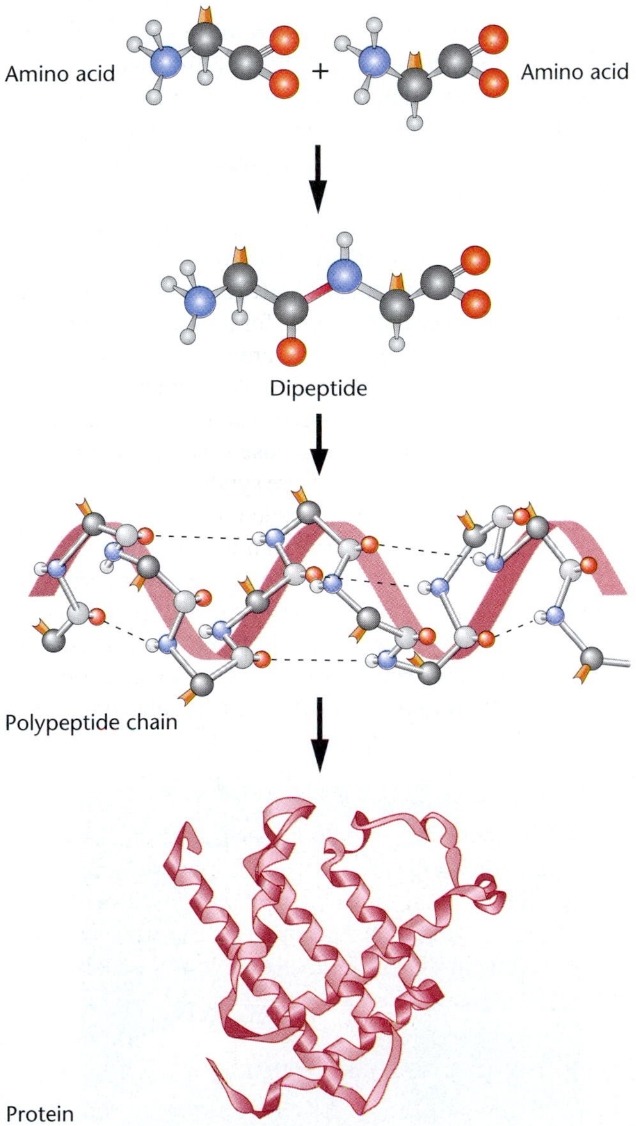

FIGURE 1–10 Depiction of three steps leading to the formation of a protein. Initially, two amino acids are joined to form a dipeptide. As amino acids are added, one by one, a longer polypeptide chain is formed which often coils into a right-handed alpha helical structure. This polypeptide then folds into a three-dimensional conformation specific to the protein's function.

example, by lowering the energy of activation in reactions, metabolism is able to proceed under the direction of enzymes at body temperature (37°C) in humans, where in the absence of enzymes, these chemical reactions would proceed at rates thousands of times more slowly. As a result, enzymes, each under the control of one or more specific genes, are capable of directing both the anabolism (the synthesis) and catabolism (the breakdown) of all organic molecules in the cell, including carbohydrates, lipids, nucleic acids, and proteins themselves.

There are countless proteins other than enzymes that are critical components of cells and organisms. These include such diverse examples as **hemoglobin**, the oxygen-binding pigment in red blood cells; **insulin**, the pancreatic hormone; **collagen**, the connective tissue molecule; **keratin**, the structural molecule in hair; **histones**, the proteins integral to chromosome structure in eukaryotes; **actin** and **myosin**, the contractile muscle proteins; and **immunoglobulins**, the antibody molecules of the immune system. Specific proteins are also critical components of all membranes and serve as molecules that regulate genetic expression. The potential for such diverse functions rests with the enormous variation of three-dimensional conformation that may be achieved by proteins. The final conformation of a protein is the direct result of the unique linear sequence of amino acids that constitute the molecule. To come full circle, this sequence is dictated by the stored information in the DNA of a gene that is transferred to RNA, which then directs the synthesis of a protein. DNA makes RNA that then makes protein.

1.4 Genetics Has Been Investigated Using Many Different Approaches

It will be useful in your study of genetics to know something about the various research approaches that have advanced our knowledge of the field. Investigations have involved viruses, bacteria, and a wide variety of plants and animals and have spanned all levels of biological organization, from molecules to populations. Although some overlap exists, most investigations have used one of four basic approaches.

The most classic investigative approach is the study of **transmission genetics**, in which the patterns of inheritance of traits are examined. Experiments are designed so that the transmission of traits from parents to offspring can be analyzed through several generations. Patterns of inheritance are sought that will provide insights into genetic principles. The first significant experimentation of this kind to have a major impact on the understanding of heredity was performed by Gregor Mendel in the middle of the 19th century. The information derived from his work serves today as the foundation of transmission genetics. In human studies, where designed matings are neither possible nor desirable, **pedigree analysis** is often useful. As illustrated in Figure 1-11, patterns of inheritance are traced through as many generations as possible, leading to inferences concerning the mode of inheritance of the trait under investigation.

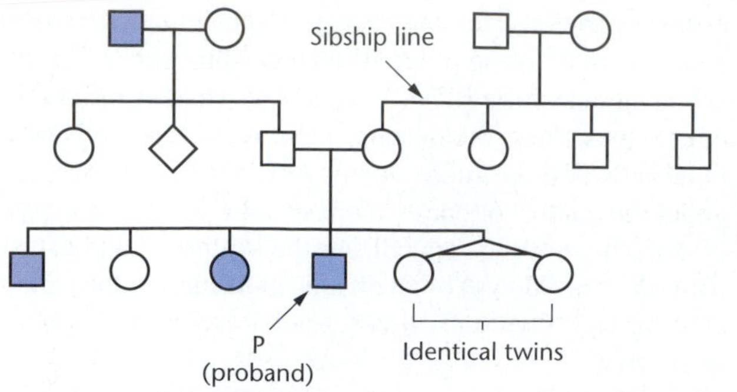

FIGURE 1–11 A representative human pedigree, tracing a genetic characteristic through three generations.

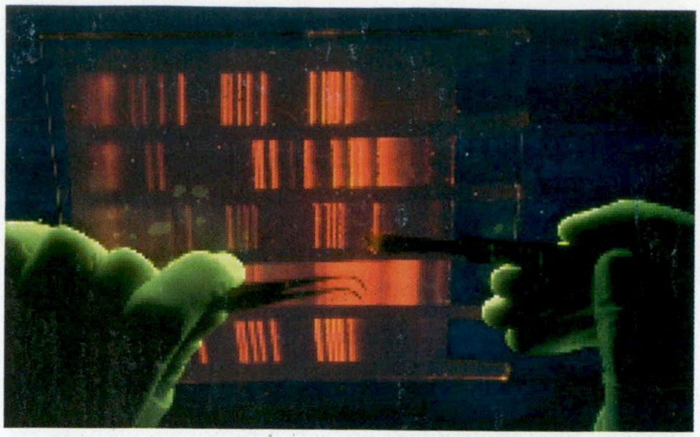

FIGURE 1–12 Visualization of DNA fragments under ultraviolet light. The bands were produced by using recombinant DNA technology. *(Dan McCoy/Rainbow)*

The second approach involves **cytogenetics**—the study of chromosomes. The earliest such studies used light microscopy. The initial discovery of chromosome behavior during mitosis and meiosis, late in the 19th century, was a critical event in the history of genetics, because of the important role these observations played in the rediscovery and acceptance of Mendelian principles. The light microscope continues to be useful in the investigation of chromosome structure and abnormalities and is instrumental in preparing **karyotypes**, which include all of the chromosomes characteristic of any species arranged in a standard sequence. (See the earlier Figure 1–7.) With the advent of electron microscopy, the repertoire of investigative approaches in genetics has grown. In high-resolution microscopy, genetic molecules and their behavior during gene expression can be visualized directly.

The third general approach involves **molecular genetic analysis**, which has had the greatest impact on the recent growth of genetic knowledge. Molecular studies, initiated in the early 1940s, have consistently expanded our knowledge of the role of genetics in life processes. Although experiments initially relied on bacteria and the viruses that infect them, extensive information is now available concerning the nature, expression, replication, and regulation of the genetic information in eukaryotes as well. The precise nucleotide sequence has been determined for many genes cloned in the laboratory. Recombinant DNA studies (Figure 1–12), in which genes from another organism are literally spliced into bacterial or viral DNA and then cloned, serve as the basis of a far-reaching research technology used in molecular genetic investigations. Building on this approach, the new fields of **genomics** and **bioinformatics** now exist whereby the entire genetic makeup of an organism may be cloned, sequenced, and the function of genes explored. Using this new technology, it is now possible to probe gene function in extreme detail. Such molecular and biochemical analysis has created the potential for gene therapy and has profound implications in medicine, agriculture, and bioethics.

One of the most striking achievements in the history of DNA biotechnology occurred in 1996 at the Roslin Institute in Scotland, when the world's most famous lamb, Dolly, was born (Figure 1–13). Representing the first animal ever to be cloned from an adult somatic cell, Dolly was the result of the research of Ian Wilmut, who fused the nucleus of an udder cell taken from a six-year-old sheep with an enucleated oocyte of another sheep. Following implantation into a surrogate mother, complete embryonic and fetal development was achieved under the direction of the genetic material of the udder cell. While the implications of this event are profound and raise numerous ethical concerns, the goal of Wilmut's research is to ultimately use cloned animals as models to study human disease and to produce therapeutic drugs beneficial to humans. Since this work, other organisms, including mice and a variety of livestock animals, have been successfully cloned.

The final investigative approach involves the study of **population genetics**. In these investigations, scientists attempt to define how and why certain genetic variation is maintained in populations, while other variation diminishes or is lost with time (Figure 1–14). Such information is critical to the understanding of evolutionary processes. Population genetics also allows us to predict gene frequencies in future generations.

Together, these varied approaches used in investigative genetics have transformed a subject that was only poorly understood in 1900 into one of the most advanced scientific

FIGURE 1–13 Dolly, a Finn Dorset sheep cloned from the genetic material of an adult mammary cell, shown next to her first-born lamb, Bonnie. *(Photo courtesy of Roslin Institute)*

FIGURE 1–14 Genetic variation exhibited in the skin present in populations of corn snakes. The wild type (normal) variety displays orange and black markings. (*Zig Leszczynski/Animals Animals/Earth Scenes*)

disciplines today. As a result, the impact of genetics on society has been, and will continue to be, immense. We shall discuss many examples of the applications of genetics in the following section and throughout the text.

1.5 Genetics Has a Profound Impact on Society

In addition to acquiring information for the sake of extending knowledge in any discipline of science—an experimental approach that is called **basic research**—scientists conduct investigations to solve problems facing society or simply to improve the well-being of members of our society—an approach that is called **applied research**. Together, both types of genetic research have combined to enhance the quality of our existence on this planet and to provide a more thorough understanding of life processes. As we shall see throughout this text, there is very little in our lives that genetics fails to touch.

Eugenics and Euphenics

There is always the danger that scientific findings will be used to formulate policies or actions that are unjust or even tragic. In this section, we review such a case that began near the end of the 19th century. At that time, Darwin's theory of natural selection provided a major influence on some people's thinking concerning the human condition. Our story recounts the initial attempt to apply genetic knowledge directly for the improvement of human existence. Championed in England by Sir Francis Galton, the general approach is called **eugenics**, a term Galton coined in 1883, that is derived from the Greek root meaning "well-born."

Galton, a cousin of Charles Darwin, believed that many human characteristics were inherited and could be subjected to artificial selection if human matings could be controlled. *Positive eugenics* encouraged parents displaying favorable characteristics, (e.g., superior intelligence, intellectual achievement, and artistic talent) to have large families. *Negative eugenics*, on the other hand, attempted to restrict the reproduction of parents displaying unfavorable characteristics (e.g., low intelligence, mental retardation, and criminal behavior).

In the United States, the eugenics movement was a significant social force and led to state and federal laws that required the sterilization of those considered "genetically inferior." Over half of the states passed such laws, commencing in 1907 with Indiana. Sterilization was mandated for "imbeciles, idiots, convicted rapists, and habitual criminals." Not without significant legal controversy, the issue rose to the U.S. Supreme Court, where in the 1927 case of Buck vs. Bell, Justice Oliver Wendel Holmes wrote in favor of upholding these laws:

> It is better for all the world, if instead of waiting to execute degenerate offspring for crime, or to let them starve for their imbecility, society can prevent those who are manifestly unfit from continuing their kind. ... Three generations of imbeciles are enough.*

By 1931, involuntary sterilization also applied to "sexual perverts, drug fiends, drunkards, and epileptics." Often, individuals were just deemed "feeble minded," a phrase applied to countless people displaying a plethora of characteristics or behavior deemed unacceptable. Sterilization programs continued in United States into the 1940s.

Immigration to the United States from certain areas of Europe and from Asia was also restricted to prevent the influx of what were regarded as genetically inferior people. In addition to the violation of individual human rights, such policies were seriously flawed by an inadequate understanding of the genetic basis of various characteristics. The formulation of eugenic policies was premised on the mistaken notions that "superior" and "inferior" traits are totally under genetic control and that genes deemed unfavorable could be removed from a population by selecting against (sterilizing) individuals expressing those traits. The potential impact of the environment and the genetic theory underlying population genetics were largely ignored as eugenic policies were developed.

In Nazi Germany in the 1930s, the concept of achieving a superior, racially pure group was an extension of the eugenics movement. Some historians feel that the Third Reich had modeled their early sterilization program after those that had become common in the United States. Initially applied to individuals considered socially and physically defective, the underlying rationale of negative eugenics was soon applied to entire ethnic groups, including Jews and Gypsies. Fueled by various forms of racial prejudice, Adolf Hitler and the Nazi regime took eugenics to its extreme by instituting policies aimed at the extinction of these "impure" human populations. The deplorable disregard for human life was preceded by incremental policies involving forced sterilization and mercy killings. This movement, based on scientifically invalid premises, culminated in the mass murder of the Holocaust.

*See *The sterilization of Carrie Buck* by J. D. Smith and K. R. Nelson, in the Selected Readings section.

Even before the Nazi party came to power in 1933, English and American geneticists began separating themselves from the eugenics movement. They were concerned about the validity of the premises underlying the movement and the evidence in support of these premises. Thus, many geneticists chose not to study human genetics for fear of being grouped with those who supported eugenics.

However, since the end of World War II, tremendous strides have been made in human genetics research. Today, a new term, **euphenics**, has replaced eugenics. Euphenics refers to medical or genetic intervention designed to reduce the impact of defective genotypes on individuals. The use of insulin by diabetics and the dietary control of newborn phenylketonurics are longstanding examples. Today, "genetic surgery" to replace defective genes looms clearly on the horizon. Furthermore, social policies now have a solid genetic foundation on which they may be based. Nevertheless, caution is still required to ensure that our expanded knowledge of human genetics does not obscure the role played by the environment in determining an individual's phenotype.

Genetic Advances in Agriculture and Medicine

As a result of research in genetics, major benefits have accrued to society in the fields of agriculture and medicine. Although the cultivation of plants and domestication of animals had begun long before, the rediscovery of Mendel's work in the early 20th century spurred scientists to apply genetic principles to these human endeavors. The use of selective breeding and hybridization techniques has had the most significant impact in agriculture.

Plants have been improved in four major ways: (1) enhanced potential for more vigorous growth and increased yields; (2) increased resistance to natural predators and pests, including insects and disease-causing microorganisms; (3) production of hybrids exhibiting a combination of superior traits derived from two different strains (often illustrating the genetic phenomenon of hybrid vigor), or even between two different species (Figure 1–15); and (4) selection of genetic variants with desirable qualities such as increased protein value, increased content of limiting amino acids, which are essential in the human diet, or smaller plant size, reducing vulnerability to adverse weather conditions.

Over the past five decades, these improvements have resulted in a tremendous increase in yield and nutrient value in such crops as barley, beans, corn, oats, rice, rye, and wheat. It is estimated that, in the United States, the use of improved genetic strains has led to a threefold increase in crop yield per acre. In Mexico, where corn is the staple crop, the plant's protein content and yield have increased significantly. A substantial effort has also been made to improve the growth of Mexican wheat. Led by Norman Borlaug, a team of researchers developed varieties of wheat that incorporated favorable genes from other strains found in various parts of the world, revolutionizing wheat production in Mexico and other underdeveloped countries. Because of this effort, which led to the well-publicized "Green Revolution," Borlaug received the Nobel Peace Prize in 1970. There is little question that this application of genetics, which continues even today, has contributed greatly to the well-being of our own species by improving the quality of nutrition worldwide.

FIGURE 1–15 *Triticale*, a hybrid grain derived from wheat and rye, produced as a result of applied genetic research breeding experiments. (*Grant Heilman/Grant Heilman Photography, Inc.*)

Applied research in genetics has also resulted in the development of superior breeds of livestock (Figure 1–16). Selective breeding has produced chickens that grow faster, yield more high-quality meat per chicken, and lay a greater number of larger eggs. In larger animals, including pigs and cows, the use of artificial insemination has been particularly important. Sperm samples derived from a single male with superior genetic traits can now be used to fertilize thousands

FIGURE 1–16 The effects of breeding and selection, as illustrated by the production of this Vietnamese potbellied pig. (*Renee Lynn/Photo Researchers, Inc.*)

Genetics, Technology, and Society

"The Frankenfood Debates: Genetically-Modified Foods"

Until recently, North Americans paid little attention to the nature or extent of genetic engineering of agricultural products. We have assumed that genetically engineered, or genetically modified (GM), foods are equivalent to nonengineered foods and that they pose no particular threats to human or environmental health. However, this attitude is gradually changing as we learn more about the controversies surrounding GM foods in Europe and as more countries ban the importation of GM seeds, crops, and foods.

Genetic engineering of plants involves recombinant DNA techniques. Scientists clone a gene of interest from a plant or animal, attach the gene to a suitable expression vehicle, and introduce the gene into plant cells that are growing in tissue culture. After the plant cells have successfully integrated the new gene, the cultured cells are grown into whole plants. In this way, every cell of the new GM plant contains the new gene. The GM plants are then extensively tested for growth and expression of the transgene.

GM crops burst on the agricultural scene in the mid-1990s. By 1999, half of U.S. soybeans, 40 percent of U.S. corn, and half of Canada's canola crop were grown from genetically engineered seeds. Other common GM crops include cotton, potatoes, and tomatoes. It is estimated that over 60 percent of all processed foods in North America contain ingredients derived from GM plants. A dozen other GM crops are in the pipeline for approval and marketing. Virtually all GM crops currently used have been developed and are marketed by large international agrochemical companies—companies that hold patents on the seeds and on the technology to generate the seeds.

At present, the two genetic modifications found in the majority of GM crops are *Bt* pest resistance and glyphosate herbicide resistance. The *Bt* gene is derived from a bacterium (*B. thuringiensis*) and encodes a toxin that kills certain insects. This toxin does not appear to affect other animals and plants. Plants that make their own *Bt* toxin do not require spraying for these insects. In theory, insecticide use will decrease, yields will increase, and farmer's profits will rise. Glyphosate (known commercially as Roundup®) is an herbicide that kills all plants, except for those with natural or engineered resistance. The use of glyphosate-resistant crops should allow farmers to kill weeds that compete with the GM crop, thereby increasing yields and profits.

Proponents of GM crops argue that genetic engineering of plants will reduce the use of pesticides and herbicides, improve the quality of foods, and increase yields sufficiently to feed the world's hungry population. Critics are less optimistic, even labeling GM foods as "Frankenfoods." The most frequent criticism of GM foods is a perceived threat to human health. Polls suggest that over half of North Americans worry about the safety of GM foods. In Europe, the figure is closer to 90 percent. The U.K. government recently banned all commercial GM crops until the year 2003. Many British and European supermarket chains are removing GM foods from their stores. The European Parliament is considering legislation to make agrochemical companies financially responsible for any ill effects from GM food consumption. Unfortunately, there has been little animal research, and no human research, on the long-term health effects of any GM food. So far, animal studies appear to show no ill effects of GM foods; however, each GM crop is different and few have been tested.

Environmental criticisms of GM crops are numerous. Many scientists argue that the continuous presence of *Bt* toxin in *Bt*-engineered crops will keep insect pests under constant pressure to evolve resistance. Once insects are resistant, the advantages of growing *Bt*-engineered crops will vanish, requiring farmers to resort to using other, more toxic, pesticides. Recent data suggest that *Bt*-expressing plants leak the *Bt* toxin through their roots into the soil, with unknown effects on soil ecology. And, *Bt*-expressing plants and their pollen may be toxic to beneficial insects, as suggested by recent studies of monarch butterflies and *Bt*-corn pollen. Another criticism is that glyphosate resistance may be transferred to wild relatives of the glyphosate-resistant plants through cross pollination, creating "superweeds" that cannot be killed with glyphosate. Similarly, engineered traits such as virus or fungus resistance may be transferred to weeds, with unpredictable effects. These scenarios are not impossible, as genetic exchange between cultivated and wild plants has been documented for decades. Interestingly, herbicide-resistant canola has already become a common weed in Canadian wheat fields. If wild and weedy plants acquire glyphosate resistance, it will be necessary to use higher levels of glyphosate, or other, more toxic, herbicides to control weeds. If these resistance scenarios are correct, the benefits of *Bt*-engineered and glyphosate-engineered crops will be short lived.

Critics argue that GM crops will do little to feed the world's hungry. They contend that the millions of people who are malnourished are too poor to afford patented seed—the petrochemicals required to produce the crop—and GM foods. They feel that GM technologies favor wealthier farmers, large monocultures, and exports. In a similar vein, critics of GM crops worry that an increasingly large proportion of agricultural biotechnology will be controlled by a small number of global corporations, leading to knowledge that is proprietary and patented. This may lead to a kind of "bioserfdom" of the world's farmers. Understandably, biotechnology companies want to protect their investments in genetic engineering by patenting the technology and preventing farmers from saving seeds for replanting. To ensure that farmers cannot replant GM seeds, some companies are developing "terminator technologies" that render GM seeds sterile. These restrictions, though, counter the argument that GM crops will feed the world's hungry. In addition, the concept that life can be patented and privatized is a profound moral issue for some people.

The major problem with the Frankenfood Debates is that there is very little solid information about either the benefits or perils of this new technology. GM foods have been developed and marketed so quickly that there has been little time for research on long-term medical or ecological effects of each genetically engineered crops. If this new genetic technology is to deliver benefits as promised, we need time to address the scientific, ethical, and political questions that surround it.

References

Chrispeels, M. J. 2000. Biotechnology and the poor. *Plant Physiol.* 124: 3–6.

Ferber, D. 1999. Risks and benefits: GM crops in the cross hairs. *Science* 286: 1662–66.

Serageldin, I. 1999. Biotechnology and food security in the 21st century. *Science* 285: 387–89.

Web site

"Living in a GM World" (a compilation of articles on GM crops and biotechnology). *New Scientist* Web site: http://www.newscientist.com/gm/gm.jsp

of females located in all parts of the world. More recently, at the extreme edge of biotechnological application, champion livestock, such as cows, are being cloned in order to ensure that the specific gene combinations associated with economically superior animals are preserved.

Equivalent strides have been made in medicine as a result of advances in genetics, particularly since 1950. Numerous disorders in humans have been discovered to result from either a single mutation or a specific chromosomal abnormality (Figure 1–17). For example, the genetic basis of disorders such as sickle-cell anemia, erythroblastosis fetalis, cystic fibrosis, hemophilia, muscular dystrophy, Tay–Sachs disease, Down syndrome, and many metabolic disorders is now well documented and often understood at the molecular level. The importance of acquiring knowledge of inherited disorders is underscored by the estimate that more than 10 million children or adults in the United States suffer from some form of genetic affliction and that every childbearing couple stands an approximately 3 percent risk of having a child with some form of genetic anomaly.

Additionally, it is now apparent that all forms of cancer have a genetic basis. Although cancer is not usually an inherited disorder, it is now very clear that *cancer is a genetic disorder at the somatic cell level.* That is, most cancers are derived from somatic cells that have undergone some type of genetic change; malignant tumors are then derived from the genetically altered cell. In some cases, a genetic predisposition to cancer also exists.

The recognition of the molecular basis of human genetic disorders and cancer has provided the impetus for the development of methods for detection and treatment. Based on advances in molecular genetics, particularly in the manipulation and analysis of DNA, prenatal detection of affected fetuses has become routine. Parents can also learn of their status as "carriers" of a large number of inherited disorders. Genetic counseling gives couples objective information on which they can base informed decisions about childbearing. In the case of cancer, recent genetic discoveries have already led to more effective early detection and more efficient treatment.

Applied research in genetics has also provided other medical benefits. Advances in immunogenetics have made possible compatible blood transfusions as well as organ transplants. In conjunction with immunosuppressive drugs, the number of successful transplant operations involving human organs, including the heart, liver, pancreas, and kidney, are increasing annually.

The most recent advances in human genetics have been dependent on the application of DNA biotechnology. First developed in the 1970s, recombinant DNA techniques paved the way for manipulating and cloning a variety of genes, including those that encode many medically important molecules, such as insulin, blood-clotting factors, growth hormone, and interferon. Human genes are isolated and spliced into vectors and transferred to host cells that serve as "production centers" for the synthesis of these proteins.

Recombinant DNA techniques have now been extended considerably. The DNA of any organism of interest can be routinely manipulated in the laboratory. The human genes responsible for inherited disorders, such as cystic fibrosis and Huntington disease, have been identified, isolated, cloned, and studied. It is hoped that such research will pave the way for gene therapy, whereby genetic disorders are treated by inserting normal copies of genes into the cells of afflicted individuals.

Perhaps the most far-reaching use of DNA biotechnology involves the Human Genome Project, in which the DNA sequence of the genome of several species, including our own, has been determined. The genomic sequence of several bacterial species, as well as that of yeast, the fruit fly, the mustard plant (*Arabidopsis*), the round worm (*Caenohabditis elegans*), and our own species is now known.

In later chapters, the applications of DNA biotechnology to agriculture and medicine are discussed in greater detail. Although other scientific disciplines are also expanding in knowledge, none has paralleled the growth of information that is occurring in genetics. As we pointed out at the outset of this chapter, while there never has been a more exciting time to be immersed in the study of genetics, the potential impact of this discipline on society has never been more profound. By the end of this course, we are confident you will agree that the present truly represents the "Age of Genetics."

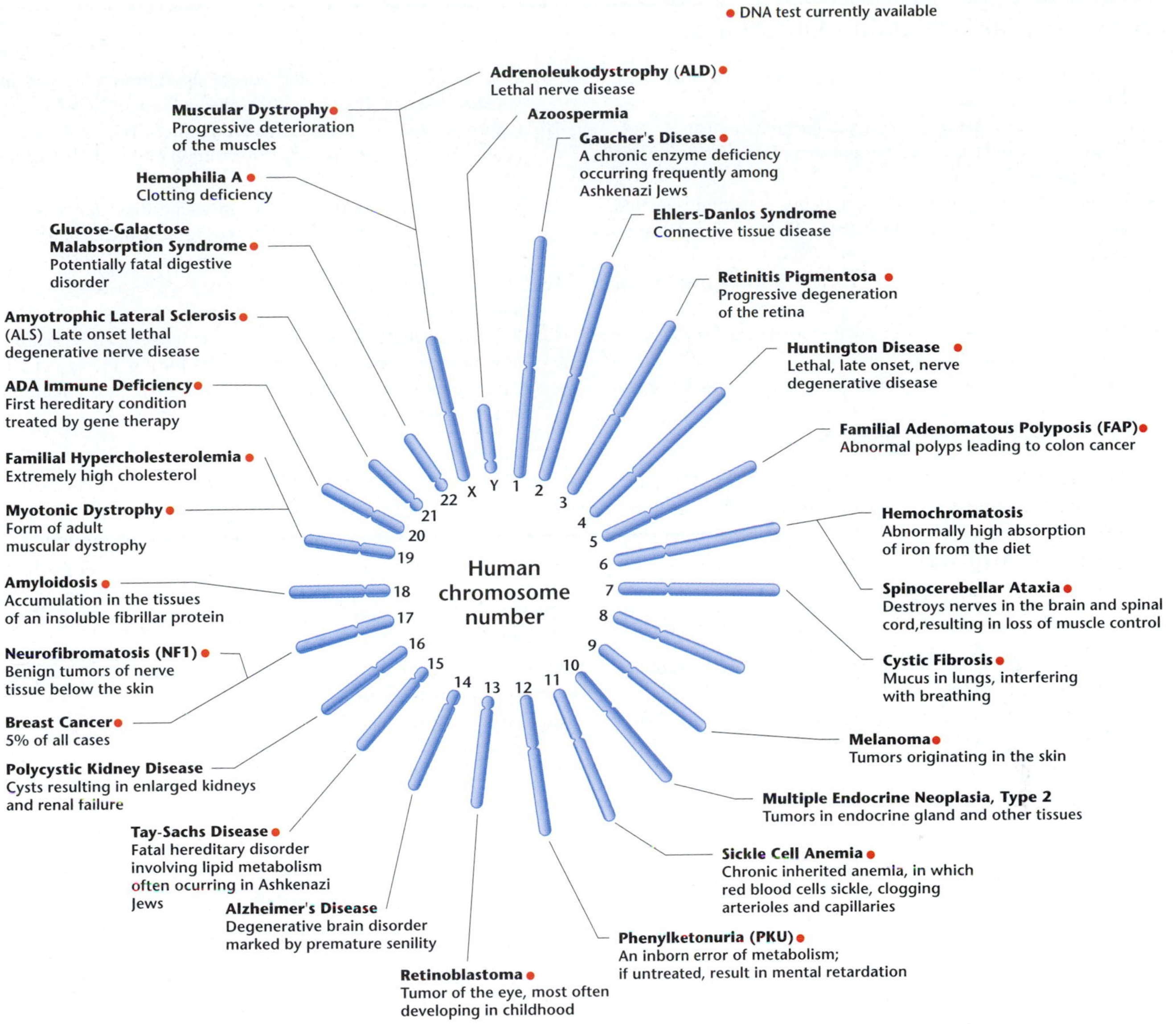

FIGURE 1–17 The chromosomes of a human being, showing the location of genes whose abnormal forms cause some of the better known hereditary diseases. Those inherited conditions that can be diagnosed using DNA analysis are indicated by a (●).

Chapter Summary

1. The history of genetics, which emerged as a fundamental discipline of biology early in the 20th century, dates back to prehistoric times.
2. Genes and chromosomes are the fundamental units in the chromosomal theory of inheritance, which explains the transmission of genetic information controlling phenotypic traits.
3. Molecular genetics, based on the general paradigm that DNA makes RNA which makes protein, serves as the underpinnings of the more classical work referred to as transmission genetics.
4. The four investigative approaches most often used in the study of genetics are (1) transmission genetic studies, (2) cytogenetic analyses, (3) molecular experimentation, and (4) inquiries into the genetic structure of populations.
5. Genetic research can be either basic or applied. Basic genetic research extends our knowledge of the discipline; the objective of applied genetics research is to solve specific problems affecting the quality of our lives and society in general.
6. Eugenics, the application of the knowledge of genetics for the improvement of human existence, has a long and controversial history. Euphenics, genetic intervention designed to ameliorate the impact of genotypes on individuals, represents the modern eugenic approach.
7. Genetic research has had a highly positive impact on many facets of agriculture and medicine.
8. DNA biotechnology is greatly expanding our research capability. It has also had a profound impact on the elucidation of inherited diseases, has made possible the mass production of medically important gene products, and will serve as the foundation on which gene therapy is developed.

Problems and Discussion Questions

1. Describe and contrast the ideas of Hippocrates and Aristotle relating to the genetic basis of life.
2. Define and contrast epigenesis and preformationism.
3. Which ideas and doctrines that preceded Darwin were central to his thinking?
4. Describe Darwin's and Wallace's theory of natural selection. What information was lacking from it; that is, what gap remained in it?
5. Contrast chromosomes and genes and describe their role in heredity.
6. Describe the "trinity" of molecular genetics and how it serves as the basis of modern genetics.
7. Describe the four major investigative approaches used in studying genetics.
8. Contrast basic and applied research.
9. Norman Borlaug received the Nobel Peace Prize for his work in genetics. Explain why his work resulted in form of Nobel award.
10. Contrast positive and negative eugenics. Which of these categories includes the approach called euphenics? Define this approach.
11. What were the basic flaws in the arguments that supported the eugenics movement as it developed in the United States between 1900 and 1945?
12. Describe several examples of how genetic research has been applied to agriculture and to medicine?
13. If you knew that a devastating late-onset inherited disease runs in your family and you could be tested for it at the age of 20, would you want to know? Would your answers be likely to change when you reach 40?

Selected Readings

Allen, G.E. 1996. Science misapplied: The eugenics age revisited. *Technol. Rev.* 99:23–31.

Anderson, W.F., and DIRCUMAKOS, E.G. 1981. Genetic engineering in mammalian cells. *Sci. Am.* (July) 245:106–21.

Borlaug, N.E. 1983. Contributions of conventional plant breeding to food production. *Science* 219:689–93.

Bowler, P.J. 1989. *The Mendelian revolution: The emergence of hereditarian concepts in modern science and society.* London: Athione.

Cocking, E.C., Davey, M.R., Pental, D., and Power, J.B. 1981. Aspects of plant genetic manipulation. *Nature* 293:265–70.

Day, P.R. 1977. Plant genetics: Increasing crop yield. *Science* 197:1334–39.

Dunn, L.C. 1965. *A short history of genetics.* New York: McGraw-Hill.

Gardner, E.J. 1972. *History of biology*, 3d ed. New York: Macmillan.

Garver, K.L., and Garver, B. 1991. Eugenics: Past, present, and future. *Am. J. Hum. Genet.* 49:1109–18.

Gasser, C.S., and Fraley, R.T. 1989. Genetically engineering plants for crop improvement. *Science* 244:1293–99.

———. 1992. Transgenic crops. *Sci. Am.* (June) 266:62–69.

Horgan, J. 1993. Eugenics revisited. *Sci. Am.* (June) 268:123–31.

Jones, S. 2000. *The origin of species—Updated.* New York: Random House.

Keller, E. F. 2000. *The century of the gene.* Cambridge, MA: Harvard University Press.

King, R.C., and Stansfield, W.D. 1997. *A dictionary of genetics*, 5th ed. New York: Oxford University Press.

Kolata, G. 1998. *Clone—The road to Dolly, and the path ahead.* New York: William Morrow and Co.

Olby, R.C. 1985. *Origins of Mendelism*, 2d ed. London: Constable.

Silver, L. 1998. *Remaking Eden: Cloning and beyond in a brave new world.* New York: Avon Books.

Smith, J.D. and Nelsen, K.R. 1989. *The sterilization of Carrie Buck.* Far Hills, NJ: New Horizon Press

Stubbe, H. 1972. *History of genetics: From prehistoric times to the rediscovery of Mendel*, trans. by T.R.W. Waters. Cambridge, MA: MIT Press.

Torrey, J.G. 1985. The development of plant biotechnology. *Am. Sci.* 73:354–63.

Vasil, I.K. 1990. The realities and challenges of plant biotechnology. *Bio/Technology* 8:296–301.

Weinberg, R.A. 1985. The molecules of life. *Sci. Am.* (Oct.) 253:48–57.

GENETICS MediaLab

This is the first in a series of special learning features, called *Genetics MediaLabs*. Within the MediaLab at the end of every chapter, you will find a list of media tools on your Student CD-ROM packaged with your book and at its Companion Web site. The Companion Web site for your textbook can be reached from the Web address **http://www.prenhall.com/klug.** To enter the site, select your book from the page, and then select the chapter to access the Web Problems listed in your textbook's MediaLabs. Each MediaLab includes several Web Problems, designed to help you learn, explore, and extend the knowledge you gain through reading the chapter.

CD Resources

The CD-ROM that is in the inside back cover of your textbook is the launch point for access to the MediaLab Web Problems that follow. In addition, you will discover many Animated Tutorials to help you explore the concepts in most of the chapters.

Web Problem 1

Estimated time for completion = 5 minutes

Should you believe everything that you read, see, or hear? At this point, you should be aware that fiction and misinformation are often disguised as truth; therefore, a healthy dose of skepticism is a good thing to have. Being wary of misinformation is only half the battle. Skills that enable you to critique what you read, see, and hear will prove valuable, especially in using the Internet. To learn more about techniques for critiquing information, visit Web Problem 1 in Chapter 1 of your Companion Web site, and select the Keyword **CRITIQUE**.

Web Problem 2

Estimated time for completion = 5 minutes

How do you cite and reference the valuable material that you do find on the Internet? Communicating is a large part of a scientist's career. An important discovery made, but never shared, is of little benefit to the world. After completing Web Problem 1, you should be able to identify truthful and valuable information, so now you need the knowledge to cite it or use it as a reference. To learn more about how to cite on-line information properly, visit Web Problem 2 in Chapter 1 of your Companion Web site, and select the Keyword **REFERENCE**.

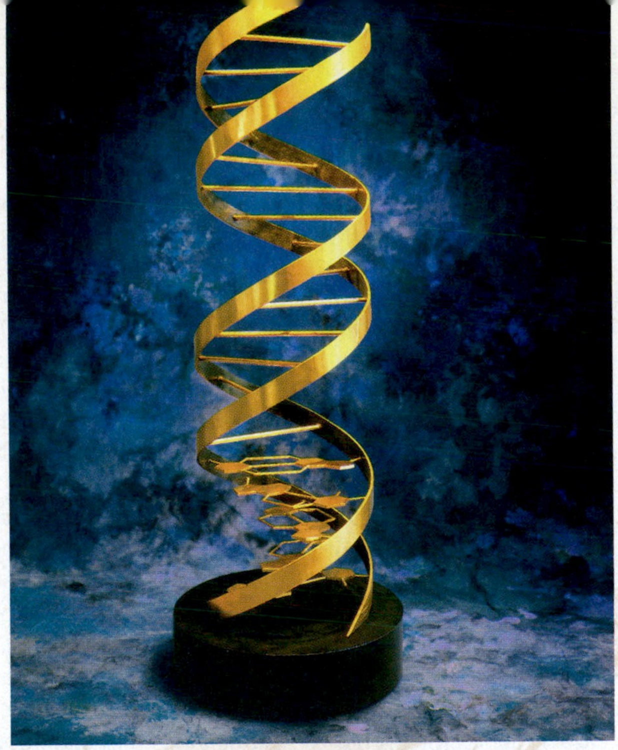
Bronze sculpture of double-helical DNA. *(Richard Megna/Fundamental Photographs)*

2
DNA Structure and Analysis

KEY CONCEPTS

2.1 The Genetic Material Must Exhibit Four Characteristics

2.2 Until 1944, Observations Favored Protein as the Genetic Material

2.3 Evidence Favoring DNA as the Genetic Material Was First Obtained during the Study of Bacteria and Bacteriophages
Transformation: Early Studies
Transformation: The Avery, MacLeod, and McCarty Experiment
The Hershey–Chase Experiment
Transfection experiments

2.4 Indirect and Direct Evidence Supports the Concept that DNA Is the Genetic Material in Eukaryotes
Indirect Evidence: Distribution of DNA
Indirect Evidence: Mutagenesis
Direct Evidence: Recombinant DNA Studies

2.5 RNA Serves as the Genetic Material in Some Viruses

2.6 Knowledge of Nucleic Acid Chemistry Is Essential to the Understanding of DNA Structure
Nucleotides: Building Blocks of Nucleic Acids
Nucleoside Diphosphates and Triphosphates
Polynucleotides

2.7 The Structure of DNA Holds the Key to Understanding Its Function
Base Composition Studies
X-Ray Diffraction Analysis
The Watson–Crick Model

2.8 Alternative Forms of DNA Exist

2.9 The Structure of RNA Is Chemically Similar to DNA, but Single Stranded

2.10 Many Analytical Techniques Have Been Useful during the Investigation of DNA and RNA
Absorption of Ultraviolet Light (UV)
Sedimentation Behavior
Denaturation and Renaturation of Nucleic Acids
Molecular Hybridization
Fluorescent *in situ* Hybridization (FISH)
Reassociation Kinetics and Repetitive DNA
Electrophoresis of Nucleic Acids

During the first third of the 20th century, the chromosomal theory of inheritance had been put forward and received substantial support. This theory correctly led geneticists to believe that chromosomes were the carriers of genetic information and the vehicles for transmission of traits between generations. Integral to this theory was the idea that chromosomes contained genes that influence such traits. The central question that remained completely unanswered in the 1930s involved the chemical basis of genetic processes. What molecule serves as the genetic information stored in genes that reside on chromosomes? The answer to this question would no doubt further the understanding of the phenomenon of life and lead to a much more precise explanation of genetic processes. Thus, it is not surprising that there was great interest in identifying this molecule.

The logical place to begin was to ask what chemical components are present in chromosomes that could make up genes and serve as the genetic material? Because chromosomes were known to have a nucleic acid and a protein component, both were candidates, and for various reasons, proteins were initially favored. However, in 1944, there emerged direct experimental evidence that the nucleic acid DNA, and not protein, serves as the informational basis for the process of heredity.

Once the importance of DNA in genetic processes was realized, work was intensified with the hope of discerning not only the structural basis of this molecule but also the relationship of its structure to its function. Between 1944 and 1953, many scientists sought information that might clarify how DNA serves as the genetic basis for living processes. The answer was believed to depend strongly on the nature of the chemical structure of the DNA molecule, given the complex but orderly functions ascribed to it.

These efforts were rewarded in 1953, when James Watson and Francis Crick set forth their hypothesis for the double-helical nature of DNA. The assumption that the molecule's functions would be clarified more easily once its general structure was determined, proved to be correct. This chapter initially reviews the evidence that DNA is the molecule that stores genetic information and thus, is the genetic material. We will then discuss the elucidation of its structure and a number of experimental techniques utilized in studying DNA.

2.1 The Genetic Material Must Exhibit Four Characteristics

For a molecule to serve the role of the genetic material, it must possess four major characteristics: **replication**, **storage of information**, **expression of information**, and **variation by mutation**. Replication of the genetic material is one facet of the cell cycle, a fundamental property of all living organisms. Once the genetic material of cells has been replicated, it must then be partitioned equally into daughter cells. During the formation of gametes, the genetic material is also replicated, but is partitioned so that each cell gets only one-half of the original amount of genetic material—the process of *meiosis* discussed in Chapter 9. Although the products of mitosis and meiosis are different, these processes are both part of the more general phenomenon of cellular reproduction.

The characteristic of storage can be viewed as being able to serve as a repository of genetic information, whether or not it is expressed. It is clear that, while most cells contain a complete complement of DNA, at any given point they express only a part of this genetic potential. For example, bacteria "turn on" many genes in response to specific environmental conditions and turn them off when such conditions change. In vertebrates, skin cells may display active melanin genes, but never activate their hemoglobin genes; digestive cells activate many genes specific to their function, but do not activate their melanin genes.

Inherent in the concept of storage is the need for the genetic material to be able to encode the nearly infinite variety of gene products found among the countless forms of life present on our planet. The chemical language of the genetic material must be capable of this potential task as it stores information and as that information is transmitted to progeny cells and organisms.

Expression of the stored genetic information is a complex process and is the basis for the concept of **information flow** within the cell. Figure 2–1 shows a simplified illustration of this concept. The initial event is the transcription of DNA, resulting in the synthesis of three types of RNA molecules: messenger RNA (mRNA), transfer RNA (tRNA), and ribosomal RNA (rRNA). Of these, mRNAs are translated into proteins. Each type of mRNA is the product of a specific gene and leads to the synthesis of a different protein. Translation occurs in conjunction with rRNA-containing ribosomes and involves tRNA, which acts as an adaptor to convert the chemical information in mRNA to the amino acids that make up proteins. Collectively, these processes

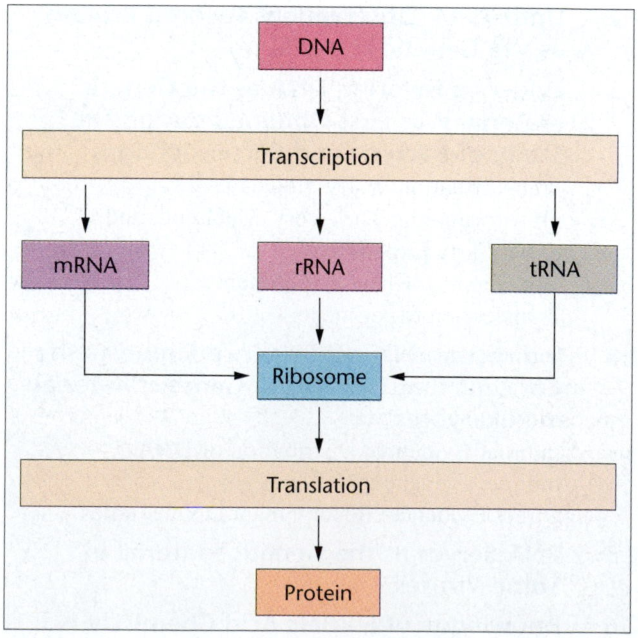

FIGURE 2–1 Simplified view of information flow involving DNA, RNA, and proteins within cells.

serve as the foundation for the **central dogma of molecular genetics**: *DNA makes RNA, which makes proteins.*

The genetic material must also serve as the basis of newly arising variability among organisms through the process of mutation. If a mutation—a change in the chemical composition of DNA—occurs, the alteration will be reflected during transcription and translation, often affecting the specified protein. If a mutation is present in gametes, it will be passed to future generations and, with time, may become distributed in the population. Genetic variation, which also includes rearrangements within and between chromosomes (see Chapter 8), provides the raw material for the process of evolution.

2.2 Until 1944, Observations Favored Protein as the Genetic Material

The idea that genetic material is physically transmitted from parent to offspring has been accepted for as long as the concept of inheritance has existed. Beginning in the late 19th century, research into the structure of biomolecules progressed considerably, setting the stage for describing the genetic material in chemical terms. Although proteins and nucleic acid were both considered major candidates for the role of the genetic material, many geneticists, until the 1940s, favored proteins. Three factors contributed to this belief.

First, proteins are abundant in cells. Although the protein content may vary considerably, these molecules account for over 50 percent of the dry weight of cells. Because cells contain such a large amount and variety of proteins, it is not surprising that early geneticists believed that some of this protein could function as the genetic material. The second factor was the accepted proposal for the chemical structure of nucleic acids during the early to mid-1900s. DNA was first studied in 1868 by a Swiss chemist, Friedrick Miescher. He was able to separate nuclei from the cytoplasm of cells and then isolate from these nuclei an acidic substance that he called **nuclein**. Miescher showed that nuclein contained large amounts of phosphorus and no sulfur, characteristics that differentiate it from proteins.

As analytical techniques were improved, nucleic acids, including DNA, were shown to be composed of four similar molecular building blocks called nucleotides. About 1910, Phoebus A. Levene proposed the **tetranucleotide hypothesis** to explain the chemical arrangement of these nucleotides in nucleic acids. He proposed that a very simple four-nucleotide unit, as shown in Figure 2–2, was repeated over and over in DNA. Levene based his proposal on studies of the composition of the four types of nucleotides. Although his actual data revealed proportions of the four nucleotides that varied considerably, he assumed a 1:1:1:1 ratio. The discrepancy was ascribed to inadequate analytical technique.

Because a single covalently bonded tetranucleotide structure was relatively simple, geneticists believed nucleic acids could not provide the large amount of chemical variation expected for the genetic material. Proteins, on the other hand, contain 20 different amino acids, thus affording the basis for substantial variation. As a result, attention was directed away

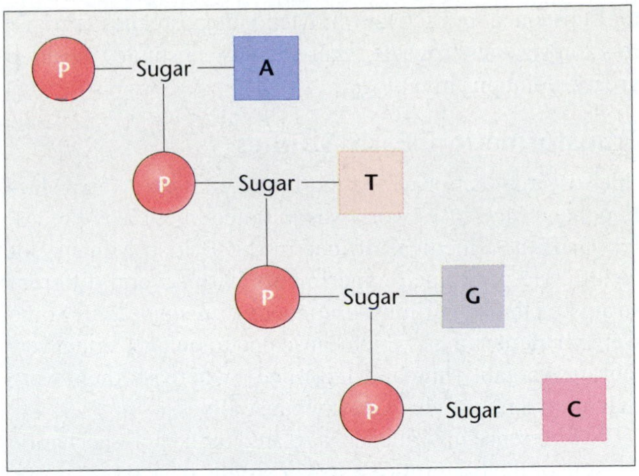

FIGURE 2–2 Diagrammatic depiction of Levene's proposed tetranucleotide containing one molecule each of the four nitrogenous bases: adenine (A), thymine (T), guanine (G), and cytosine (C). Other potential sequences of four nucleotides are possible.

from nucleic acids, strengthening the speculation that proteins served as the genetic material.

The third contributing factor simply concerned the areas of most active research in genetics. Before 1940, most geneticists were engaged in the study of transmission genetics and mutation. The excitement generated in these areas undoubtedly diluted the concern for finding the precise molecule that serves as the genetic material. Thus, proteins were the most promising candidate and were accepted rather passively.

Between 1910 and 1930, other proposals for the structure of nucleic acids were advanced, but they were generally overturned in favor of the tetranucleotide hypothesis. It was not until the 1940s that the work of Erwin Chargaff led to the realization that Levene's hypothesis was incorrect. Chargaff showed that, for most organisms, the 1:1:1:1 ratio was inaccurate, disproving Levene's hypothesis.

2.3 Evidence Favoring DNA as the Genetic Material Was First Obtained during the Study of Bacteria and Bacteriophages

The 1944 publication by Oswald Avery, Colin MacLeod, and Maclyn McCarty concerning the chemical nature of a "transforming principle" in bacteria was the initial event leading to the acceptance of DNA as the genetic material. Their work, along with the subsequent findings of other research teams, constituted direct experimental proof that DNA, and not protein, is the biomolecule responsible for heredity. It marked the beginning of the era of molecular genetics, a period of discovery in biology that has served as the foundation for current studies in biotechnology and has moved us closer to an understanding of the basis of life. The impact of the initial findings on future research and thinking paralleled that of the publication of Darwin's theory of evolution and the subsequent rediscovery of Mendel's postulates of transmission genetics. Together, these events constituted the three great revolutions in biology.

Transformation: Early Studies

The research that provided the foundation for Avery, MacLeod, and McCarty's work was initiated in 1927 by Frederick Griffith, a medical officer in the British Ministry of Health. He performed experiments with several different strains of the bacterium *Diplococcus pneumoniae*.* Some were **virulent** strains, which cause pneumonia in certain vertebrates (notably humans and mice), whereas others were **avirulent** strains, which do not cause illness.

The difference in virulence is related to the polysaccharide capsule of the bacterium. Virulent strains have this capsule, whereas avirulent strains do not. The nonencapsulated bacteria are readily engulfed and destroyed by phagocytic cells in the animal's circulatory system. Virulent bacteria, which possess the polysaccharide coat, are not easily engulfed. Hence, they are able to multiply and cause pneumonia.

The presence or absence of the capsule is the basis for another characteristic difference between virulent and avirulent strains. Encapsulated bacteria form smooth, shiny-surfaced colonies (S) when grown on an agar culture plate; nonencapsulated strains produce rough colonies (R). (see Figure 2–3.) This feature allows virulent and avirulent strains to be distinguished easily by standard microbiological culture techniques.

Each strain of *Diplococcus* may be one of dozens of different types called **serotypes**. The specificity of the serotype is due to the detailed chemical structure of the polysaccharide constituent of the thick, slimy capsule. Serotypes are identified by immunological techniques and are usually designated by Roman numerals. In the United States, types I and II are most common in causing pneumonia. Griffith used types II and III in the critical experiments that led to new concepts about the genetic material. Table 2–1 summarizes the characteristics of Griffith's two strains.

Griffith knew from the work of others that only living virulent cells would produce pneumonia in mice. If heat-killed virulent bacteria are injected into mice, no pneumonia results, just as living avirulent bacteria fail to produce the disease. Griffith's critical experiment (Figure 2–3) involved an injection into mice of living II*R* (avirulent) cells combined with heat-killed III*S* (virulent) cells. Since neither cell type caused death in mice when injected alone, Griffith expected that the double injection would not kill the mice. But, after five days, mice that received both types of cells were all dead. Analysis of their blood revealed a large number of living type III*S* (virulent) bacteria.

As far as could be determined regarding the capsular polysaccharide, these III*S* bacteria were identical to the III*S* strain from which the heat-killed cell preparation had been made. The control mice, injected only with living avirulent II*R* bacteria for this set of experiments, did not develop pneumonia and remained healthy. This finding ruled out the possibility that the avirulent II*R* cells had simply changed (or mutated) to virulent III*S* cells in the absence of the heat-killed III*S* fraction. Instead, some type of interaction was required between living II*R* and heat-killed III*S* cells.

Griffith concluded that the heat-killed III*S* bacteria were somehow responsible for converting live avirulent II*R* cells into virulent III*S* ones. Calling the phenomenon **transformation**, he suggested that the **transforming principle** might be some part of the polysaccharide capsule *or* some compound required for capsule synthesis, although the capsule alone did not cause pneumonia. To use Griffith's term, the transforming principle from the dead III*S* cells served as a "pabulum" for the II*R* cells.

Griffith's work led other physicians and bacteriologists to research the phenomenon of transformation. By 1931, Henry Dawson at the Rockefeller Institute had confirmed Griffith's observations and extended his work one step further. Dawson and his coworkers showed that transformation could occur in vitro (in a test tube). When heat-killed III*S* cells were incubated with living II*R* cells, living III*S* cells were recovered. Therefore, injection into mice was not necessary for transformation to occur. By 1933, Lionel J. Alloway had refined the in vitro system by using crude extracts of *S* cells and living *R* cells. The soluble filtrate from the heat-killed *S* cells was as effective in inducing transformation as were the intact cells. Alloway and others did not view transformation as a genetic event, but rather as a physiological modification of some sort. Nevertheless, the experimental evidence that a chemical substance was responsible for transformation was quite convincing.

Transformation: The Avery, MacLeod, and McCarty Experiment

The critical question, of course, was what molecule serves as the transforming principle? In 1944, after 10 years of work, Avery, MacLeod, and McCarty published their results in what is now regarded as a classic paper in the field of molecular genetics. They reported that they had obtained the transforming principle in a purified state and, that beyond reasonable doubt, the molecule responsible for transformation was DNA.

The details of their work, sometimes called the Avery, MacLeod, and McCarty experiment, are outlined in Figure 2–4. These researchers began their isolation procedure with large quantities (50–75 liters) of liquid cultures of type III*S* virulent cells. The cells were centrifuged, collected, and heat killed. Following homogenization and several extractions with the detergent deoxycholate (DOC), they obtained a soluble filtrate that, when tested, still contained the transforming principle. Protein was removed from the active filtrate by several chloroform extractions, and polysaccharides were enzymatically digested and removed. Finally, precipitation with ethanol yielded a fibrous mass that still retained the ability to induce transformation of type II*R* avirulent cells. From the original 75-liter sample, the procedure yielded 10 to 25 mg of the "active factor."

*This organism is now named *Streptococcus pneumoniae*.

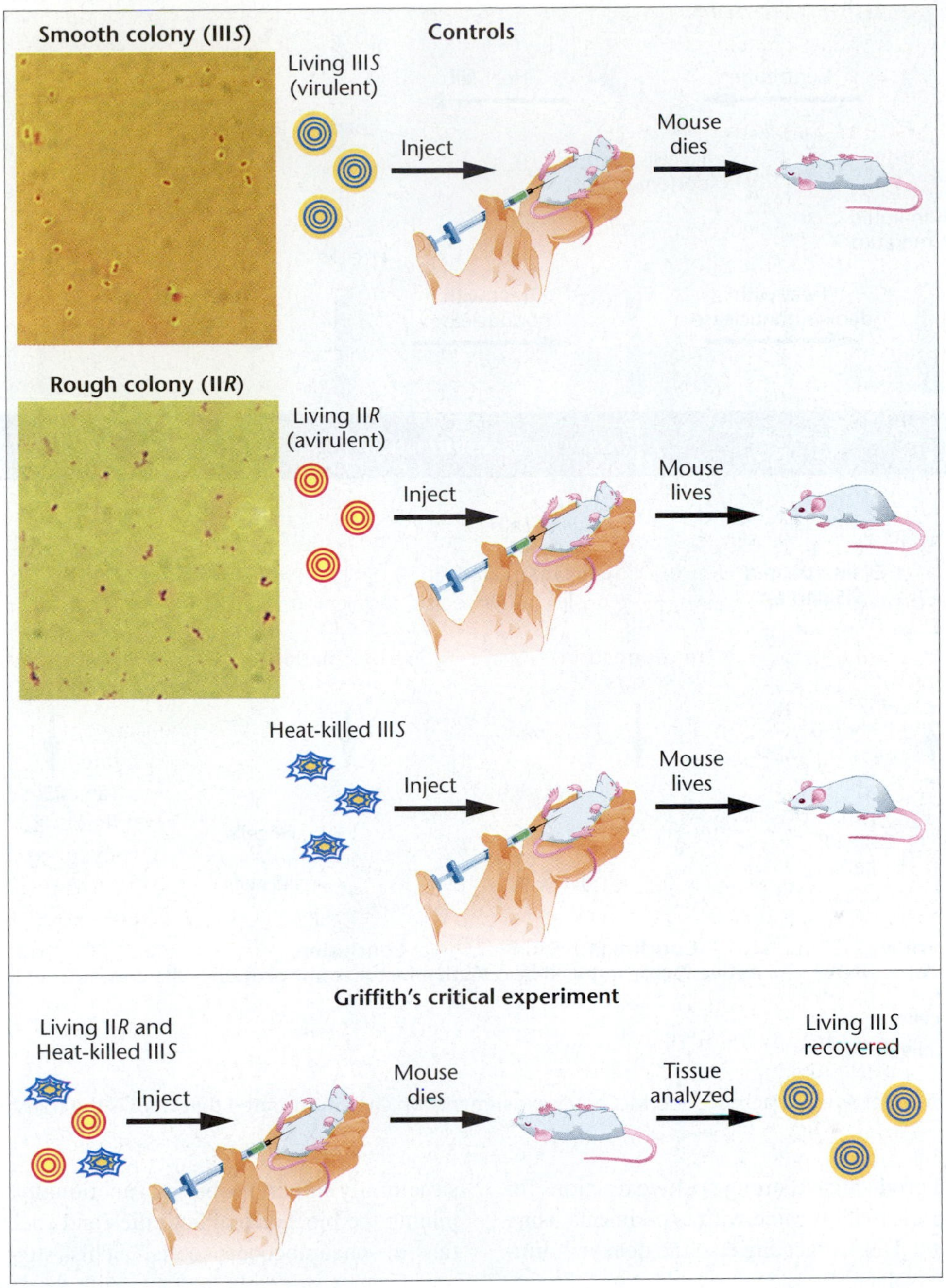

FIGURE 2–3 Griffith's transformation experiment. The photographs show bacterial colonies containing cells with capsules (type III*S*) and without capsules (type II*R*).

TABLE 2–1 Strains of *Diplococcus pneumoniae* Used by Frederick Griffith in His Original Transformation Experiments

Serotype	*Colony Morphology*	*Capsule*	*Virulence*
II*R*	Rough	Absent	Avirulent
III*S*	Smooth	Present	Virulent

Further testing clearly established that the transforming principle was DNA. The fibrous mass was first analyzed for its nitrogen–phosphorus ratio, which was shown to coincide with the ratio of "sodium desoxyribonucleate," the chemical name then used to describe DNA. To solidify their findings, Avery, MacLeod, and McCarty sought to eliminate, to the greatest extent possible, all probable contaminants from their final product. Thus, it was treated with the proteolytic enzymes trypsin and chymotrypsin and then with an RNA-digesting enzyme, called ribonuclease. Such treatments destroyed any remaining activity of proteins and RNA. Nevertheless, transforming activity still remained. Chemical

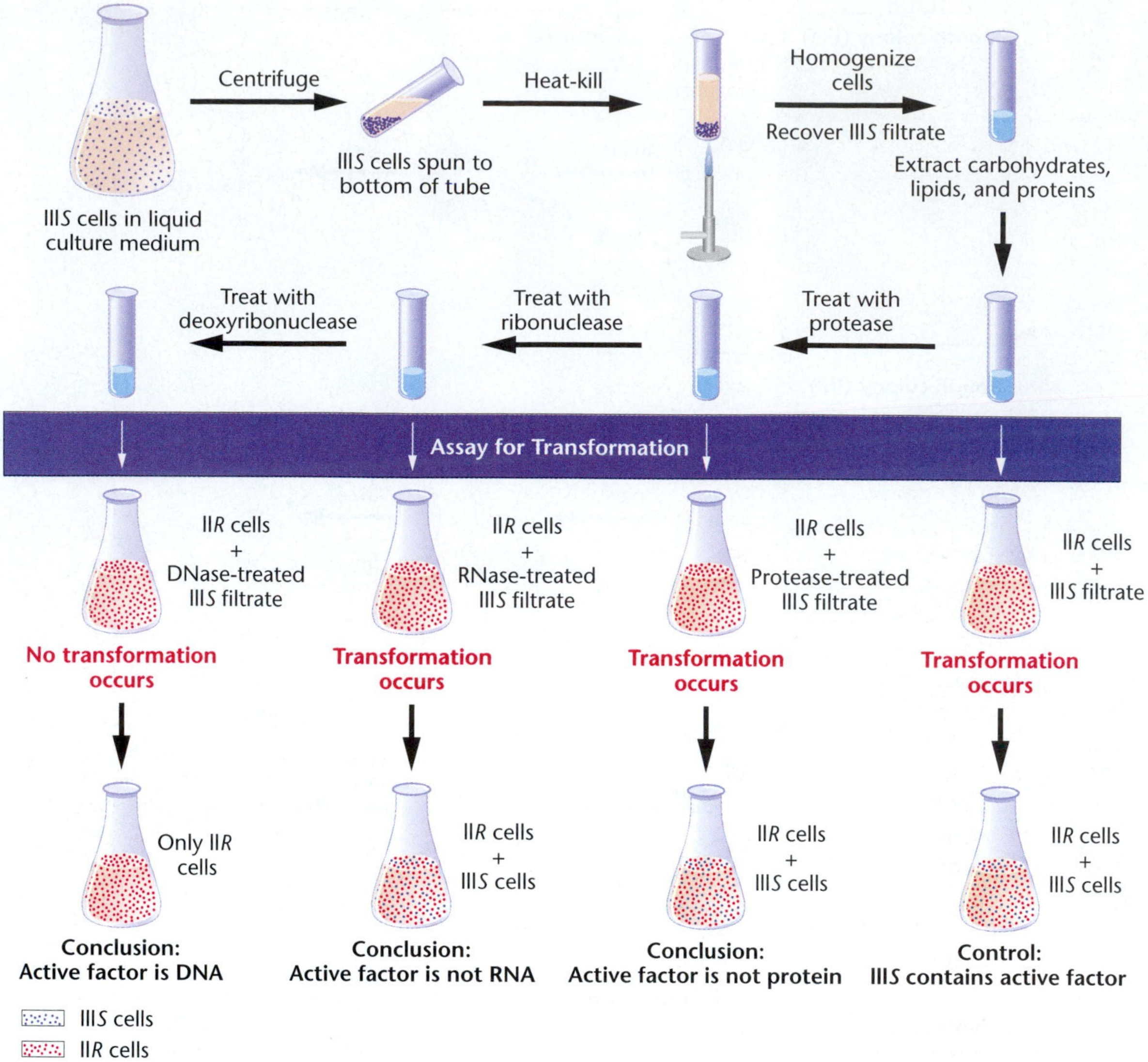

FIGURE 2–4 Summary of Avery, MacLeod, and McCarty's experiment, which demonstrated that DNA is the transforming principle.

testing of the final product gave strong positive reactions for DNA. The final confirmation came with experiments using crude samples of the DNA-digesting enzyme **deoxyribonuclease**, which was isolated from dog and rabbit sera. Digestion with this enzyme was shown to destroy transforming activity. There could be little doubt that the active transforming principle was DNA.

The great amount of work, the confirmation and reconfirmation of the conclusions drawn, and the unambiguous logic of the experimental design involved in the research of these three scientists are truly impressive. Their conclusion to the 1944 publication was, however, very simply stated: "The evidence presented supports the belief that a nucleic acid of the desoxyribose* type is the fundamental unit of the transforming principle of *Pneumococcus* Type III."

Avery and his colleagues recognized the genetic and biochemical implications of their work when they observed that "nucleic acids of this type must be regarded not merely as structurally important but as functionally active in determining the biochemical activities and specific characteristics of pneumococcal cells." This suggested that the transforming principle interacts with the II*R* cell and gives rise to a coordinated series of enzymatic reactions culminating in the synthesis of the type III*S* capsular polysaccharide. Avery, MacLeod, and McCarty emphasized that, once transformation occurs, the capsular polysaccharide is produced in successive generations. Transformation is therefore heritable, and the process affects the genetic material.

Immediately after the publication of the report, several investigators turned to, or intensified, their studies of transformation in order to clarify the role of DNA in genetic mechanisms. In particular, the work of Rollin Hotchkiss was instrumental in confirming that the critical factor in transformation was DNA and not protein. In 1949, in a separate study, Harriet Taylor isolated an **extremely rough *(ER)*** mutant strain from a rough (*R*) strain. This *ER* strain produced colonies that were more irregular than the *R* strain. The DNA from *R* accomplished the transformation of *ER* to *R*. Thus, the *R* strain, which served as the recipient in the

*Desoxyribose is now spelled deoxyribose.

Avery experiments, was shown to be able to serve as the DNA donor in transformation also.

Transformation has now been shown to occur in *Haemophilus influenzae, Bacillus subtilis, Shigella paradysenteriae*, and *Escherichia coli*, among many other microorganisms. Transformation of numerous genetic traits other than colony morphology has also been demonstrated, including ones involving resistance to antibiotics and the ability to metabolize various nutrients. These observations further strengthened the belief that transformation by DNA is primarily a genetic event, rather than simply a physiological change. This idea is pursued again in the "Insights and Solutions" section at the end of this chapter.

The Hershey–Chase Experiment

The second major piece of evidence supporting DNA as the genetic material was provided during the study of the bacterium *Escherichia coli* and one of its infecting viruses, **bacteriophage T2**. Often referred to simply as a **phage**, the virus consists of a protein coat surrounding a core of DNA. Electron micrographs have revealed the phage's external structure to be composed of a hexagonal head plus a tail. The general aspects of the life cycle of a T-even bacteriophage such as T2, as known in 1952, is shown in Figure 2–5. Briefly, the phage adsorbs to the bacterial cell and some component of the phage enters the bacterial cell. Following this infection step, the viral information "commandeers" the cellular machinery of the host and undergoes viral reproduction. In a reasonably short time, many new phages are constructed and the bacterial cell is lysed, releasing the progeny viruses.

In 1952, Alfred Hershey and Martha Chase published the results of experiments designed to clarify the events leading to phage reproduction. Several of the experiments clearly established the independent functions of phage protein and nucleic acid in the reproduction process associated with the bacterial cell. Hershey and Chase knew from existing data that

1. T2 phages consist of approximately 50 percent protein and 50 percent DNA.
2. Infection is initiated by adsorption of the phage by its tail fibers to the bacterial cell.
3. The production of new viruses occurs within the bacterial cell.

It appeared that some molecular component of the phage, DNA or protein (or both), enters the bacterial cell and directs viral reproduction. Which was it?

Hershey and Chase used the radioisotopes ^{32}P and ^{35}S to follow the molecular components of phages during infection (Figure 2–6). Because DNA contains phosphorus (P), but not sulfur, ^{32}P effectively labels DNA, and because proteins contain sulfur (S), but not phosphorus, ^{35}S labels protein. *This was a key point in the experiment.* If *E. coli* cells are first grown in the presence of ^{32}P or ^{35}S and then infected with T2 viruses, the progeny phages will have *either* a radioactively labeled DNA core *or* a radioactively labeled protein coat, respectively. These labeled phages can be isolated and used to infect unlabeled bacteria.

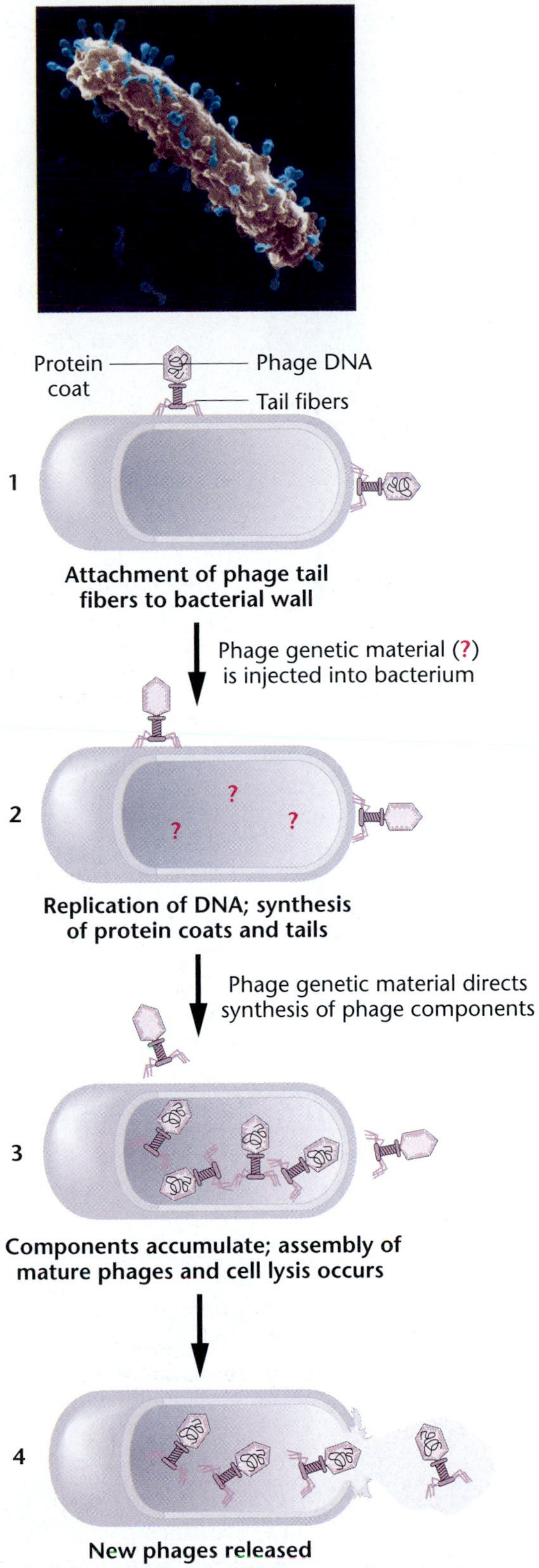
Protein coat
Phage DNA
Tail fibers
1
Attachment of phage tail fibers to bacterial wall
Phage genetic material (?) is injected into bacterium
2
?
?
?
Replication of DNA; synthesis of protein coats and tails
Phage genetic material directs synthesis of phage components
3
Components accumulate; assembly of mature phages and cell lysis occurs
4
New phages released

FIGURE 2–5 Life cycle of a T-even bacteriophage. The electron micrograph shows an *E. coli* cell during infection by numerous phages.

When labeled phages and unlabeled bacteria are mixed, an adsorption complex is formed as the phages attach their tail fibers to the bacterial wall. These complexes were isolated

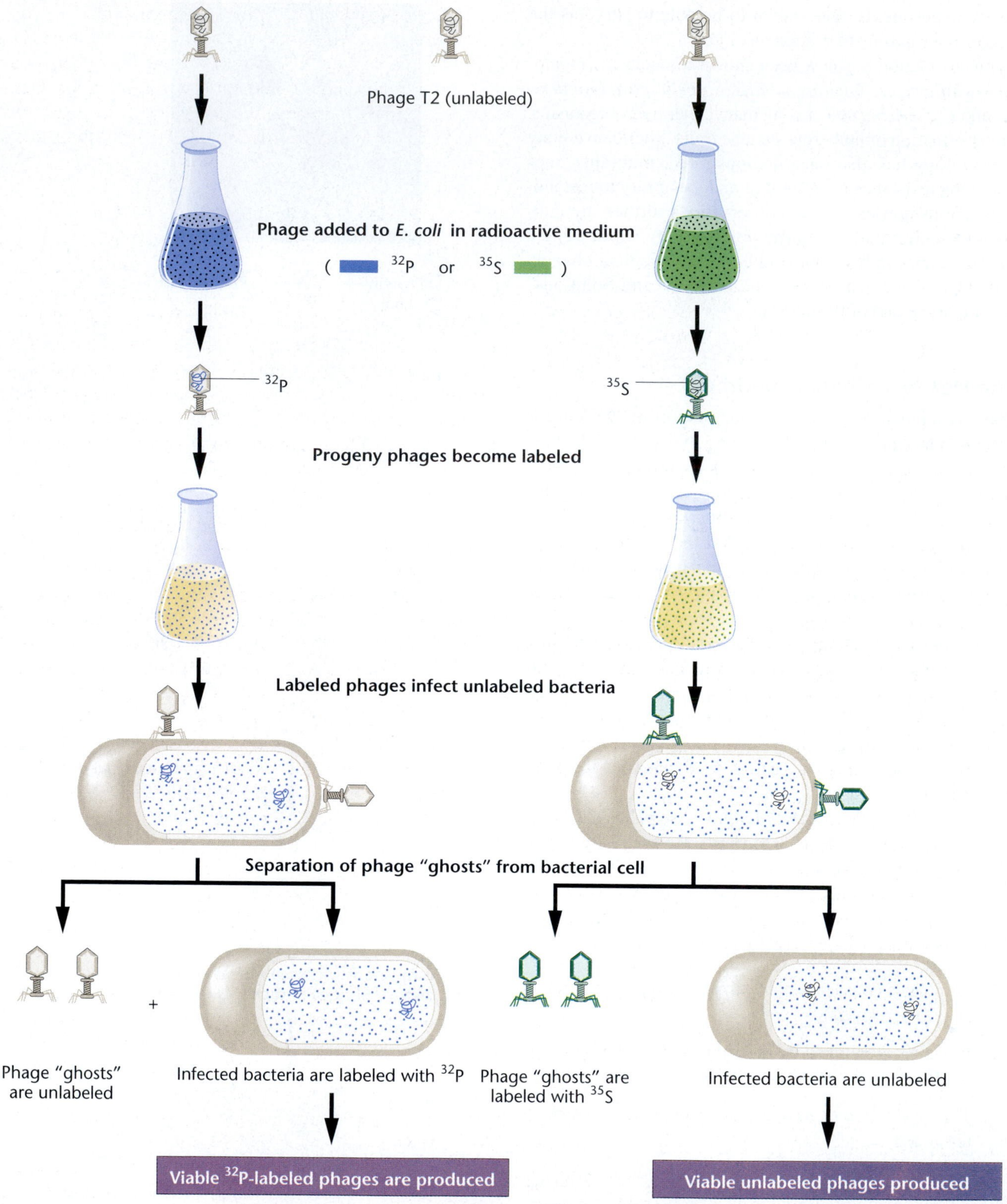

Phage T2 (unlabeled)
Phage added to E. coli in radioactive medium
(32P or 35S)
32P
35S
Progeny phages become labeled
Labeled phages infect unlabeled bacteria
Separation of phage "ghosts" from bacterial cell
+
Phage "ghosts" are unlabeled
Infected bacteria are labeled with 32P
Phage "ghosts" are labeled with 35S
Infected bacteria are unlabeled
Viable 32P-labeled phages are produced
Viable unlabeled phages produced

FIGURE 2–6 Summary of the Hershey–Chase experiment demonstrating that DNA, and not protein, is responsible for directing the reproduction of phage T2 during the infection of *E. coli*.

and subjected to a high shear force by placing them in a blender. This force stripped off the attached phages. When the mixture was centrifuged, the lighter phage particles separated from the heavier bacterial cells, allowing the isolation of both components (Figure 2–6). By tracing the radioisotopes, Hershey and Chase were able to demonstrate that most of the ^{32}P-labeled DNA had been transferred into the bacterial cell following adsorption; on the other hand, most of the ^{35}S-labeled protein remained outside the bacterial cell and was recovered in the phage "ghosts" (empty phage coats) after the blender treatment. Following this separation, the bacterial cells, which now contained viral DNA, were eventually lysed as new phages were produced. These progeny phages contained ^{32}P, but not ^{35}S.

Hershey and Chase interpreted these results to indicate that the protein of the phage coat remains outside the host cell and is not involved in directing the production of new phages. On the other hand, and most important, phage DNA enters the host cell and directs phage reproduction. Hershey and Chase had demonstrated the genetic material in phage T2 is DNA, not protein.

These experiments, along with those of Avery and his colleagues, provided convincing evidence to most geneticists that DNA was the molecule responsible for heredity. Since then, many significant findings have been based on this supposition. These many findings, constituting the field of molecular genetics, are discussed in detail in subsequent chapters.

Transfection Experiments

During the eight years following the publication of the Hershey-Chase experiment, additional research using bacterial viruses provided even more solid proof that DNA is the genetic material. In 1957, several reports demonstrated that if *E. coli* was treated with the enzyme lysozyme, the outer wall of the cell can be removed without destroying the bacterium. Enzymatically treated cells are naked, so to speak, and contain only the cell membrane as their outer boundary. Such structures are called **protoplasts** (or **spheroplasts**). John Spizizen and Dean Fraser independently reported that by using spheroplasts, they were able to initiate phage reproduction with disrupted T2 particles. That is, provided that the cell wall is absent, a virus does not have to be intact for infection to occur. Thus, the outer protein coat structure may be essential to the movement of DNA through the intact cell wall, but it is not essential for infection when spheroplasts are used.

Similar, but more refined, experiments were reported in 1960 by George Guthrie and Robert Sinsheimer. DNA was purified from bacteriophage ϕX174, a small phage that contains a single-stranded circular DNA molecule of some 5386 nucleotides. When added to *E. coli* protoplasts, the purified DNA resulted in the production of complete ϕX174 bacteriophages. This process of infection by only the viral nucleic acid, called **transfection**, proved conclusively that ϕX174 DNA alone contains all the necessary information for production of mature viruses. Thus, the evidence supporting the conclusion that DNA serves as the genetic material was further strengthened, even though all direct evidence thus far had been obtained from bacterial and viral studies.

2.4 Indirect and Direct Evidence Supports the Concept that DNA Is the Genetic Material in Eukaryotes

In 1950, eukaryotic organisms were not amenable to the types of experiments that in bacteria and viruses demonstrated that DNA is the genetic material. Nevertheless, it was generally assumed that the genetic material would be a universal substance and also serve this role in eukaryotes. Initially, support for this assumption relied on several circumstantial (indirect) observations that, taken together, inferred that DNA serves as the genetic material in eukaryotes. Subsequently, direct evidence established unequivocally the central role of DNA in genetic processes.

Indirect Evidence: Distribution of DNA

The genetic material should be found where it functions—in the nucleus as part of chromosomes. Both DNA and protein fit this criterion. However, protein is also abundant in the cytoplasm, whereas DNA is not. Both mitochondria and chloroplasts are known to perform genetic functions, and DNA is also present in these organelles. Thus, DNA is found only where primary genetic function occurs. Protein, on the other hand, is found everywhere in the cell. These observations are consistent with the interpretation favoring DNA over proteins as the genetic material.

Because it had earlier been established that chromosomes within the nucleus contain the genetic material, a correlation was expected to exist between the ploidy (n, $2n$, etc.) of a cell and the quantity of the molecule that functions as the genetic material. Meaningful comparisons can be made between the amount of DNA and protein in gametes (sperm and eggs) and somatic or body cells. The latter are recognized as being diploid ($2n$) and containing twice the number of chromosomes as gametes, which are haploid (n).

Table 2–2 compares the amount of DNA found in haploid sperm and diploid nucleated precursors of red blood cells from a variety of organisms. The amount of DNA and the number of sets of chromosomes is closely correlated. No consistent correlation can be observed between gametes and diploid cells for proteins. These data thus provide further circumstantial evidence favoring DNA over proteins as the genetic material of eukaryotes.

Indirect Evidence: Mutagenesis

Ultraviolet (UV) light is one of a number of agents capable of inducing mutations in the genetic material. Bacteria and other organisms can be irradiated with various wavelengths of ultraviolet light and the effectiveness of each wavelength measured by the number of mutations it induces. When the data are plotted, an **action spectrum** of UV light

TABLE 2–2 DNA Content of Haploid Versus Diploid Cells of Various Species (in picograms)

Organism	*n*	*2n*
Human	3.25	7.30
Chicken	1.26	2.49
Trout	2.67	5.79
Carp	1.65	3.49
Shad	0.91	1.97

Note: Sperm (n) and nucleated precursors to red blood cells ($2n$) were used to contrast ploidy levels.

as a mutagenic agent is obtained. This action spectrum can then be compared with the **absorption spectrum** of any molecule suspected to be the genetic material (Figure 2–7). *The molecule serving as the genetic material is expected to absorb at the wavelength(s) found to be mutagenic.*

UV light is most mutagenic at the wavelength λ of 260 nanometers (nm). Both DNA and RNA absorb UV light most strongly at 260 nm. On the other hand, protein absorbs most strongly at 280 nm, yet no significant mutagenic effects are observed at that wavelength. This indirect evidence supports the idea that a nucleic acid is the genetic material and tends to exclude protein.

Direct Evidence: Recombinant DNA Studies

Although the circumstantial evidence just described does not constitute direct proof that DNA is the genetic material in eukaryotes, these observations spurred researchers to forge ahead, basing their work on this hypothesis. Today, there is no doubt of the validity of this conclusion; DNA *is* the genetic material in all eukaryotes.

The strongest evidence is provided by molecular analysis utilizing **recombinant DNA technology**. In this procedure, segments of eukaryotic DNA corresponding to specific genes are isolated and literally spliced into bacterial DNA. Such a complex can be inserted into a bacterial cell and its genetic expression monitored. If a eukaryotic gene is introduced, the presence of the corresponding eukaryotic protein product demonstrates directly that this DNA is not only present, but functional in the bacterial cell. This has been shown to be the case in countless instances. For example, the human gene products insulin and interferon are produced by bacteria following the insertion of human genes that encode these proteins. As the bacterium divides, the eukaryotic DNA is replicated along with the host DNA and is distributed to the daughter cells, which also express the human genes and synthesize the corresponding proteins.

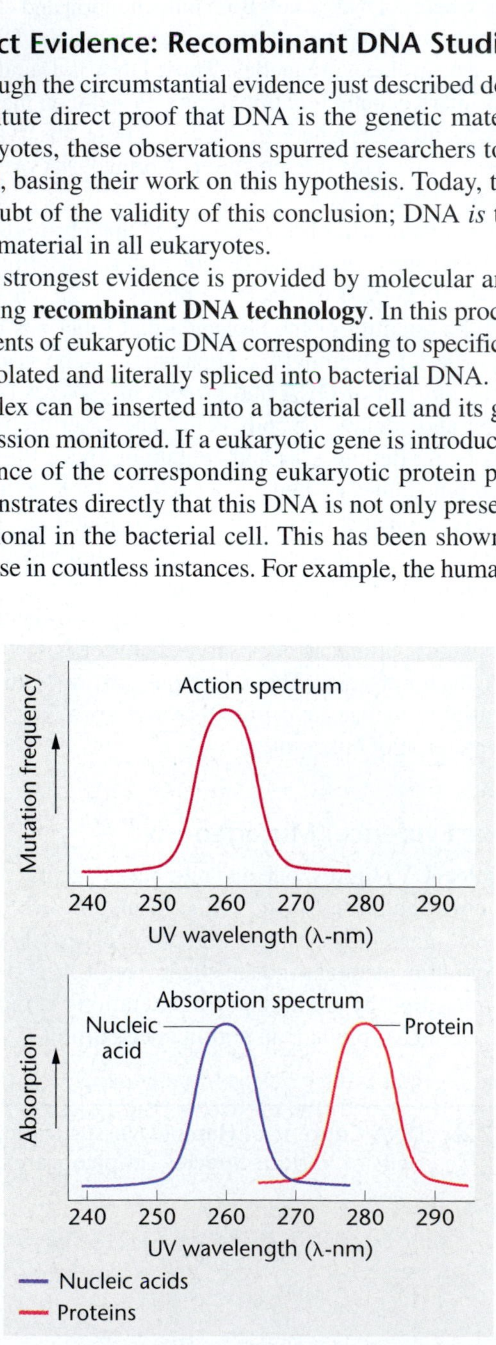

FIGURE 2–7 Comparison of the action spectrum (which determines the most effective mutagenic UV wavelength) and the absorption spectrum (which shows the range of wavelength where nucleic acids and proteins absorb UV light).

The availability of vast amounts of DNA coding for specific genes, attainable as a result of recombinant DNA research, has led to other direct evidence that DNA serves as the genetic material. Work in the laboratory of Beatrice Mintz has demonstrated that DNA encoding the human *β-globin* gene, when microinjected into a fertilized mouse egg, is later found to be present and expressed in adult mouse tissue and transmitted to and expressed in that mouse's progeny. These mice are examples of what are called **transgenic animals**.

More recent work has introduced *rat* DNA encoding a growth hormone into fertilized *mouse* eggs. About one-third of the resultant mice grew to twice their normal size, indicating that foreign DNA was present and functional in the experimental mice. Subsequent generations of mice inherited this genetic information and also grew to a large size.

We will return to the topic of recombinant DNA later in the text. (See Chapters 16–18.) The point to be made here is that in eukaryotes, DNA has been shown directly to meet the requirement of expression of genetic information. Later, we will see exactly how DNA is stored, replicated, expressed, and mutated.

2.5 RNA Serves as the Genetic Material in Some Viruses

Some viruses contain an RNA core rather than one composed of DNA. In these viruses, it would thus appear that RNA might serve as the genetic material—an exception to the general rule that DNA performs this function. In 1956, it was demonstrated that when purified RNA from **tobacco mosaic virus (TMV)** is spread on tobacco leaves, the characteristic lesions caused by viral infection subsequently appear on the leaves. It was concluded that RNA is the genetic material of this virus.

Soon afterwards, another type of experiment with TMV was reported by Heinz Fraenkel-Conrat and B. Singer, as illustrated in Figure 2–8. These scientists discovered that the RNA core and the protein coat from wild-type TMV and other viral strains could be isolated separately. In their work, RNA and coat proteins were separated and isolated from TMV and a second viral strain, **Holmes ribgrass (HR)**. Then, mixed viruses were reconstituted from the RNA of one strain and the protein of the other. When this "hybrid" virus was spread on tobacco leaves, the lesions that developed corresponded to the type of RNA in the reconstituted virus—that is, viruses with wild-type TMV RNA and HR protein coats produced TMV lesions and vice versa. Again, it was concluded that RNA serves as the genetic material in these viruses.

In 1965 and 1966, Norman R. Pace and Sol Spiegelman further demonstrated that RNA from the phage Q*β* could be isolated and replicated in vitro. Replication was dependent on

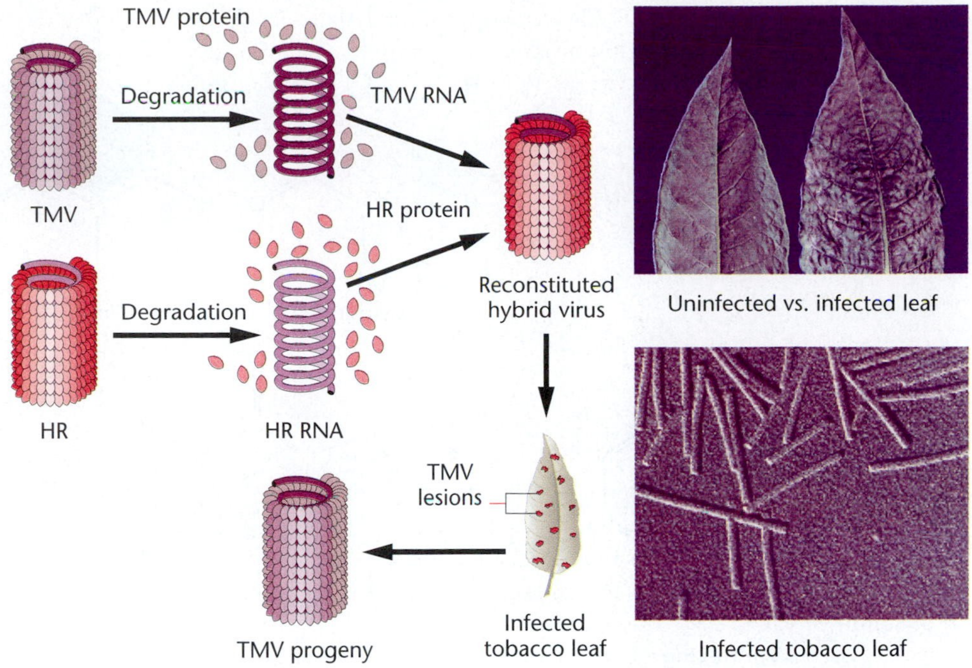

FIGURE 2–8 Reconstitution of hybrid tobacco mosaic viruses. In the hybrid, RNA is derived from the wild-type TMV virus, while the protein subunits are derived from the HR strain. Following infection, viruses are produced with protein subunits characteristic of the wild-type TMV strain and not those of the HR strain. The top photograph shows TMV lesions on a tobacco leaf compared with an uninfected leaf. At the bottom is an electron micrograph of mature viruses.

an enzyme, **RNA replicase**, which was isolated from host *E. coli* cells following normal infection. When the RNA replicated in vitro was added to *E. coli* protoplasts, infection and viral multiplication occurred. Thus, RNA synthesized in a test tube serves as the genetic material in these phages by directing the production of all the components necessary for viral replication.

Finally, one other group of RNA-containing viruses bears mentioning. These are the **retroviruses**, which replicate in an unusual way. Following infection of the host cell, their RNA serves as a template for the synthesis of the complementary DNA molecule. The process, **reverse transcription**, occurs under the direction of an RNA-dependent DNA polymerase enzyme called **reverse transcriptase**. This DNA intermediate can be incorporated into the genome of the host cell, and when the host DNA is transcribed, copies of the original retroviral RNA chromosomes are produced. Retroviruses include the human immunodeficiency virus (HIV), which causes AIDS, as well as the RNA tumor viruses.

2.6 Knowledge of Nucleic Acid Chemistry Is Essential to the Understanding of DNA Structure

Having established thus far the critical importance of DNA and RNA in genetic processes, we shall now provide a brief introduction to the chemical basis of these molecules. As we shall see, the structural components of DNA and RNA are very similar. This chemical similarity is important in the coordinated functions played by these molecules during gene expression. Like the other major groups of organic biomolecules (proteins, carbohydrates, and lipids), nucleic acid chemistry is based on a variety of similar building blocks that are polymerized into chains of varying lengths.

Nucleotides: Building Blocks of Nucleic Acids

DNA is a nucleic acid, and **nucleotides** are the building blocks of all nucleic acid molecules. Sometimes called mononucleotides, these structural units consist of three essential components: a **nitrogenous base**, a **pentose sugar** (a 5-carbon sugar), and a **phosphate group**. There are two kinds of nitrogenous bases: the nine-member double-ring **purines** and the six-member single-ring **pyrimidines**. Two types of purines and three types of pyrimidines are found commonly in nucleic acids. The two purines are **adenine** and **guanine**, abbreviated **A** and **G**. The three pyrimidines are **cytosine**, **thymine**, and **uracil**, abbreviated **C**, **T**, and **U**. The chemical structures of A, G, C, T, and U are shown in Figure 2–9(a). Both DNA and RNA contain A, C, and G; only DNA contains the base T, whereas only RNA contains the base U. Each nitrogen or carbon atom of the ring structures of purines and pyrimidines is designated by an unprimed number. Note that corresponding atoms in the two rings are numbered differently in most cases.

The pentose sugars found in nucleic acids give them their names. Ribonucleic acids (RNA) contain **ribose**, while deoxyribonucleic acids (DNA) contain **deoxyribose**. Figure 2–9(b) shows the ring structures for these two pentose sugars. Each carbon atom is distinguished by a number with a prime sign (e.g., C-1', C-2'). As you can see, compared with ribose, deoxyribose has a hydrogen atom at the C-2′ position rather than a hydroxyl group. The presence of a hydroxyl group at the C-2′ position thus distinguishes RNA from DNA.

If a molecule is composed of a purine or pyrimidine base and a ribose or deoxyribose sugar, the chemical unit is called a **nucleoside**. If a phosphate group is added to the nucleoside, the molecule is now called a **nucleotide**. Nucleosides and nucleotides are named according to the specific nitrogenous base (A, T, G, C, or U) that is part of the building block. The nomenclature and general structure of nucleosides and nucleotides are given in Figure 2–10.

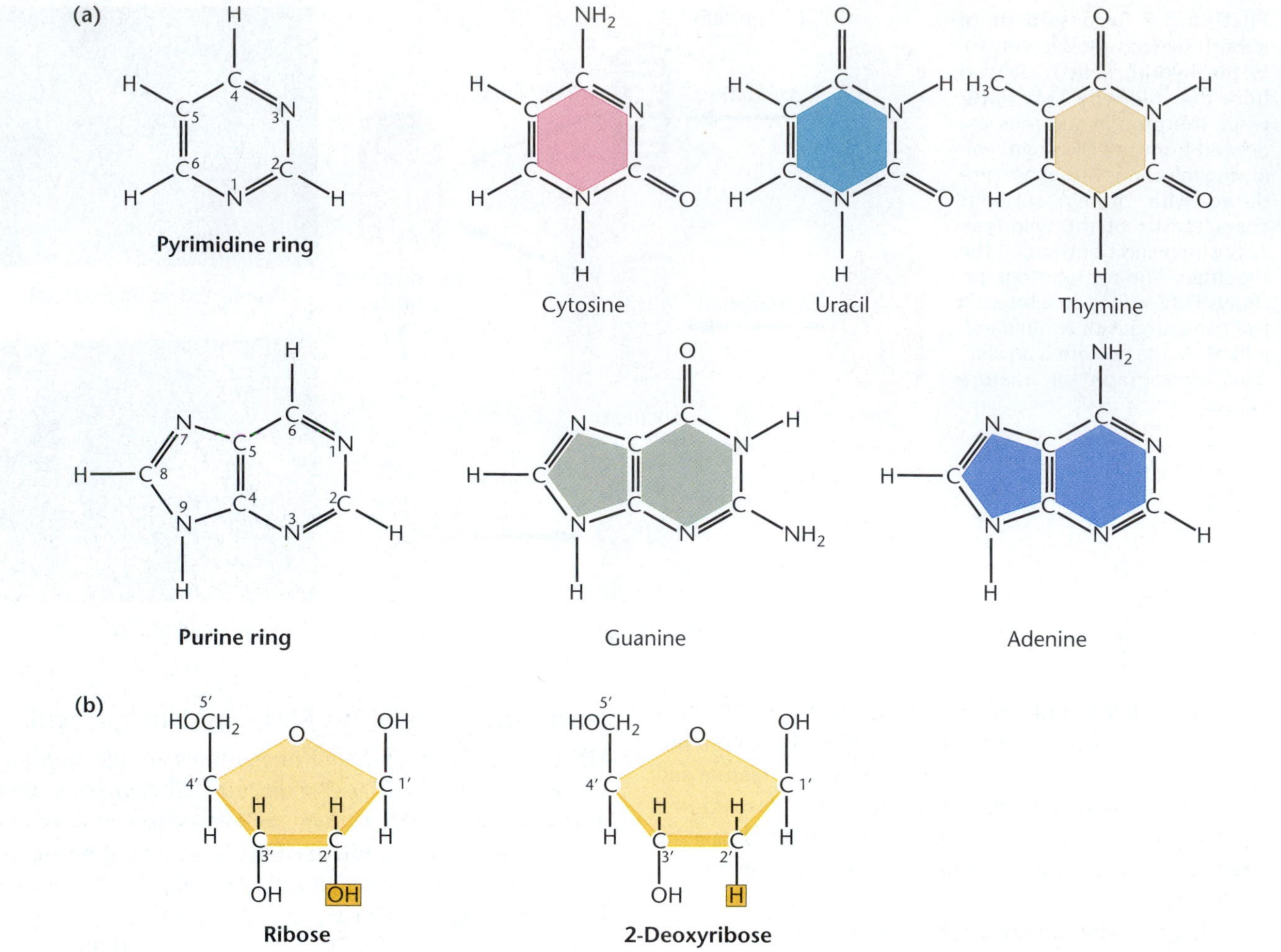

FIGURE 2–9 (a) Chemical structures of the pyrimidines and purines that serve as the nitrogenous bases in RNA and DNA. The nomenclature for numbering carbon and nitrogen atoms making up the two bases is shown within the structures shown on the left. (b) Chemical ring structures of ribose and 2-deoxyribose, which serve as the pentose sugars in RNA and DNA, respectively.

The bonding among the components of a nucleotide is highly specific. The C-1′ atom of the sugar is involved in the chemical linkage to the nitrogenous base. If the base is a purine, the N-9 atom is covalently bonded to the sugar. If the base is a pyrimidine, the bonding involves the N-1 atom. In a nucleotide, the phosphate group may be bonded to the C-2′, C-3′, or C-5′ atom of the sugar. The C-5' phosphate configuration is shown in Figure 2–10. It is by far the most prevalent one in biological systems and the one found in DNA and RNA.

Nucleoside Diphosphates and Triphosphates

Nucleotides are also described by the term **nucleoside monophosphate (NMP)**. The addition of one or two phosphate groups results in **nucleoside diphosphates (NDPs)** and **triphosphates (NTPs)**, as illustrated in Figure 2–11. The triphosphate form is significant because it serves as the precursor molecule during nucleic acid synthesis within the cell. (See Chapter 3.) Additionally, the triphosphates **adenosine triphosphate (ATP)** and **guanosine triphosphate (GTP)** are important in the cell's bioenergetics because of the large amount of energy involved in adding or removing the terminal phosphate group. The hydrolysis of ATP or GTP to ADP or GDP and inorganic phosphate P_i is accompanied by the release of a large amount of energy in the cell. When the chemical conversion of ATP or GTP is coupled to other reactions, the energy produced may be used to drive the reactions. As a result, both ATP and GTP are involved in many cellular activities, including numerous genetic events.

Polynucleotides

The linkage between two mononucleotides consists of a phosphate group linked to two sugars. A **phosphodiester bond** is formed, because phosphoric acid has been joined to two alcohols (the hydroxyl groups on the two sugars) by an ester linkage on both sides. Figure 2–12(a) shows the resultant phosphodiester bond in DNA. The same bond is found in RNA. Each structure has a **C-5′ end** and a **C-3′ end**. The joining of two nucleotides forms a **dinucleotide**; of three nucleotides, a **trinucleotide**; and so forth. Short chains consisting of approximately 20 nucleotides linked together are called **oligonucleotides**. Still longer chains are referred to as **polynucleotides**.

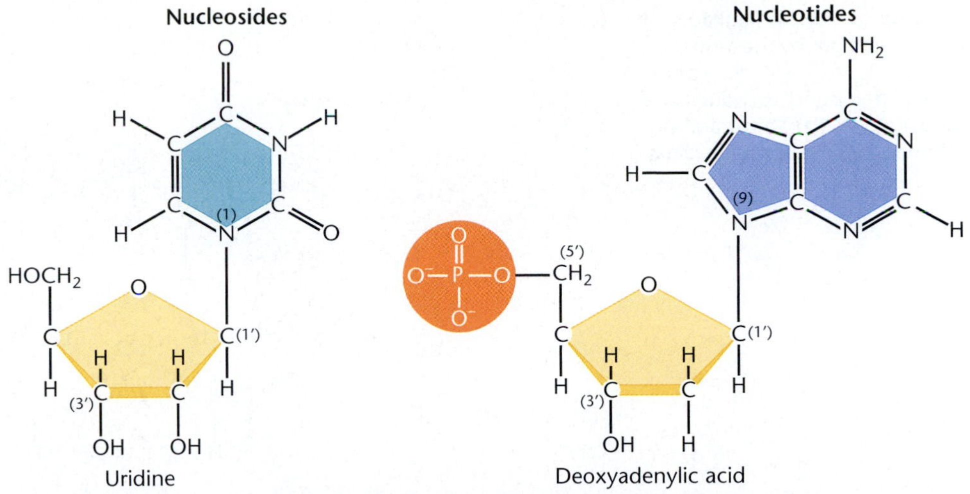

FIGURE 2–10 Structures and names of the nucleosides and nucleotides of RNA and DNA.

Ribonucleosides	Ribonucleotides
Adenosine Cytidine Guanosine Uridine	Adenylic acid Cytidylic acid Guanylic acid Uridylic acid
Deoxyribonucleosides	**Deoxyribonucleotides**
Deoxyadenosine Deoxycytidine Deoxyguanosine Deoxythymidine	Deoxyadenylic acid Deoxycytidylic acid Deoxyguanylic acid Deoxythymidylic acid

Because drawing the structures in Figure 2–12(a) is time consuming and complex, a schematic shorthand method has been devised [Figure 2–12(b)]. The nearly vertical lines represent the pentose sugar; the nitrogenous base is attached at the top, or the C-1′ position. The diagonal line, with the P in the middle of it, is attached to the C-3′ position of one sugar and the C-5′ position of the neighboring sugar; it represents the phosphodiester bond. Several modifications of this shorthand method are in use, and they can be understood in terms of these guidelines.

Although Levene's tetranucleotide hypothesis (described earlier in this chapter) was generally accepted before 1940, research in subsequent decades revealed it to be incorrect. It was shown that DNA does not necessarily contain equimolar quantities of the four bases. Additionally, the molecular weight of DNA molecules was determined to be in the range

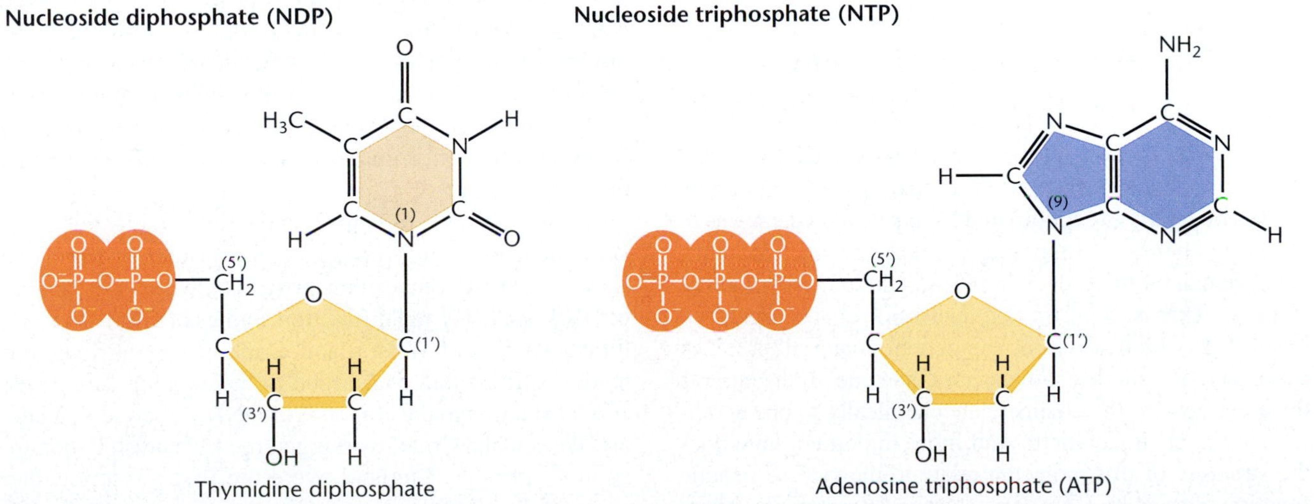

FIGURE 2–11 Basic structures of nucleoside diphosphates and triphosphates, as illustrated by thymidine diphosphate and adenosine triphosphate.

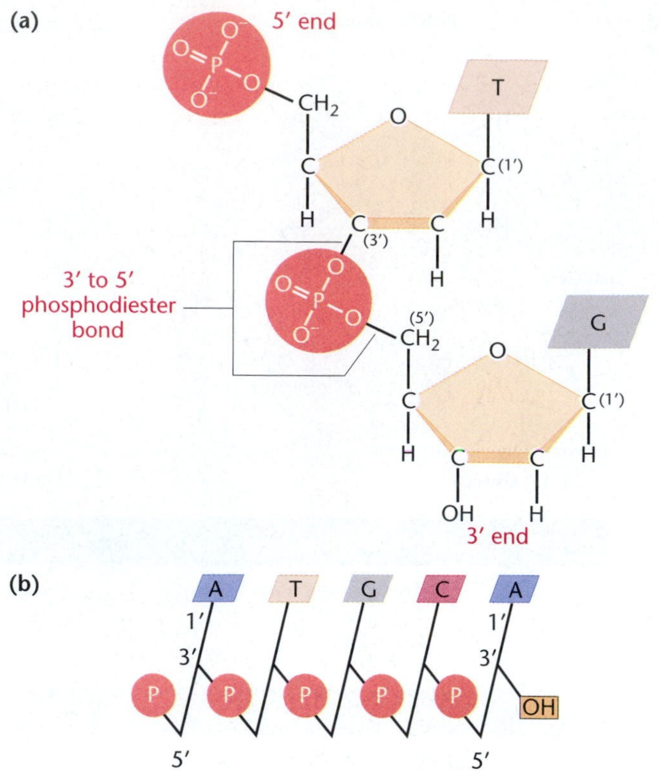

FIGURE 2–12 (a) Linkage of two nucleotides by the formation of a C-3′–C-5′ (3′–5′) phosphodiester bond, producing a dinucleotide. (b) Shorthand notation for a polynucleotide chain.

of 10^6 to 10^9 daltons, far in excess of that of a tetranucleotide. The current view of DNA is that it consists of exceedingly long polynucleotide chains.

Long polynucleotide chains would account for the observed molecular weight and provide the basis for the most important property of DNA—storage of vast quantities of genetic information. If each nucleotide position in this long chain is occupied by any one of four nucleotides, extraordinary variation is possible. For example, a polynucleotide that is only 1000 nucleotides in length can be arranged 4^{1000} different ways, each one different from all other possible sequences. This potential variation in molecular structure is essential if DNA is to serve the function of storing the vast amounts of chemical information necessary to direct cellular activities.

2.7 The Structure of DNA Holds the Key to Understanding Its Function

The previous sections in this chapter have established that DNA is the genetic material in all organisms (with certain viruses being the exception) and have provided details as to the basic chemical components making up nucleic acids. What remained to be deciphered was the precise structure of DNA. That is, how are polynucleotide chains organized into DNA, which serves as the genetic material? Is DNA composed of a single chain or more than one? If the latter is the case, how do the chains relate chemically to one another? Do the chains branch? And more important, how does the structure of this molecule relate to the various genetic functions served by DNA (i.e., storage, expression, replication, and mutation)?

From 1940 to 1953, many scientists were interested in solving the structure of DNA. Among others, Erwin Chargaff, Maurice Wilkins, Rosalind Franklin, Linus Pauling, Francis Crick, and James Watson sought information that might answer what many consider to be the most significant and intriguing question in the history of biology: "How does DNA serve as the genetic basis for the living process?" The answer was believed to depend strongly on the chemical structure and organization of the DNA molecule, given the complex but orderly functions ascribed to it.

In 1953, two young scientists, James Watson and Francis Crick, proposed that the structure of DNA is in the form of a double helix. Their proposal was published in a short paper in the journal *Nature*, which is reprinted in its entirety on page 37. In a sense, this publication constituted the finish line in a highly competitive scientific race to obtain what some consider to be the most significant finding in the history of biology. This "race," as recounted in Watson's book *The Double Helix* (see "Selected Readings"), demonstrates the human interaction, genius, frailty, and intensity involved in the scientific effort that eventually led to the elucidation of DNA structure.

The data available to Watson and Crick, crucial to the development of their proposal, came primarily from two sources: (1) base composition analysis of hydrolyzed samples of DNA and (2) X-ray diffraction studies of DNA. The analytical success of Watson and Crick can be attributed to model building that conformed to the existing data. If the correct solution to the structure of DNA is viewed as a puzzle, Watson and Crick, working at the Cavendish Laboratory in Cambridge, England, were the first to put together successfully all of the pieces. Given the far-reaching significance of this discovery, you may be interested to know that

Watson, who entered college (University of Chicago) at age 15, was a 24-year-old postdoctoral fellow in 1953. Crick, now considered one of the great theoretical biologists of our time, was still a graduate student in his mid-thirties.

Base Composition Studies

Between 1949 and 1953, Erwin Chargaff and his colleagues used chromatographic methods to separate the four bases in DNA samples from various organisms. Quantitative methods were then used to determine the amounts of the four bases from each source. Table 2–3(a) provides some of Chargaff's original data. Parts (b) and (c) of the table show more recently derived base-composition information from various organisms that reinforce Chargaff's findings. As we shall see, Chargaff's data were critical to the successful model of DNA put forward by Watson and Crick.

On the basis of these data, which you should examine, the following conclusions may be drawn:

1. The amount of adenine residues is proportional to the amount of thymine residues in the DNA of any species (columns 1, 2, and 5). Also, the amount of guanine residues is proportional to the amount of cytosine residues (columns 3, 4, and 6).
2. Based on the above proportionality, the sum of the purines (A + G) equals the sum of the pyrimidines (C + T), as shown in column 7.
3. The percentage of C + G does not necessarily equal the percentage of A + T. As we can see, the ratio between the two values varies greatly among species, as shown in column 8 and as is apparent in Table 2–3(c).

These conclusions indicate definite patterns of base composition of DNA molecules. The data provided the initial clue to "the puzzle." Additionally, they directly refute the tetranucleotide hypothesis, which stated that all four bases are present in equal amounts.

X-Ray Diffraction Analysis

When fibers of a DNA molecule are subjected to X-ray bombardment, these rays are scattered according to the molecule's atomic structure. The pattern of scatter (diffraction) can be captured as spots on photographic film and analyzed, particularly for the overall shape of and regularities within the molecule. This process, **X-ray diffraction** analysis, was successfully applied to the study of protein structure by Linus Pauling and other chemists. The technique had been attempted on DNA as early as 1938 by William Astbury. By 1947, he had detected a periodicity within the structure of the molecule of 3.4 angstroms (Å)*, which suggested to him that the bases were stacked like coins on top of one another.

* Today, measurement in nanometers (nm) is favored (1 nm = 10 Å)

TABLE 2–3 DNA Base Composition Data

*(a) Chargaff's data**

	Molar proportions[a]			
	1	*2*	*3*	*4*
Source	*A*	*T*	*G*	*C*
Ox thymus	26	25	21	16
Ox spleen	25	24	20	15
Yeast	24	25	14	13
Avian tubercle bacilli	12	11	28	26
Human sperm	29	31	18	18

(c) G + C content in several organisms

Organism	*%G + C*
Phage T2	36.0
Drosophila	45.0
Maize	49.1
Euglena	53.5
Neurospora	53.7

(b) Base compositions of DNAs from various sources

	Base composition				Base ratio		A + T/G + C ratio	
	1	*2*	*3*	*4*	*5*	*6*	*7*	*8*
Source	*A*	*T*	*G*	*C*	*A/T*	*G/C*	*(A + G)/(C + T)*	*(A + T)/(C + G)*
Human	30.9	29.4	19.9	19.8	1.05	1.00	1.04	1.52
Sea urchin	32.8	32.1	17.7	17.3	1.02	1.02	1.02	1.58
E. coli	24.7	23.6	26.0	25.7	1.04	1.01	1.03	0.93
Sarcina lutea	13.4	12.4	37.1	37.1	1.08	1.00	1.04	0.35
T7 bacteriophage	26.0	26.0	24.0	24.0	1.00	1.00	1.00	1.08

*Source: From Chargaff, 1950.

[a]Moles of nitrogenous constituent per mole of P (often, the recovery was less than 100 percent).

Between 1950 and 1953, Rosalind Franklin, working in the laboratory of Maurice Wilkins, obtained improved X-ray data from more purified samples of DNA (Figure 2–13). Her work confirmed the 3.4 Å periodicity seen by Astbury and suggested that the structure of DNA was some sort of helix. However, she did not propose a definitive model. Pauling had analyzed the work of Astbury and others and incorrectly proposed that DNA was a triple helix.

The Watson–Crick Model

Watson and Crick published their analysis of DNA structure in 1953 (their original paper is reproduced on p. 19). By building models under the constraints of the information just discussed, they proposed the double-helical form of DNA, as shown in Figure 2–14(a). This model has the following major features:

1. Two long polynucleotide chains are coiled around a central axis, forming a right-handed double helix.
2. The two chains are **antiparallel**; that is, their C-5′-to-C-3′ orientations run in opposite directions.
3. The bases of both chains are flat structures, lying perpendicular to the axis; they are "stacked" on one another, 3.4 Å (0.34 nm) apart and are located on the inside of the helix.
4. The nitrogenous bases of opposite chains are paired to one another as the result of the formation of hydrogen bonds (described shortly); in DNA, only A═T and G≡C pairs are allowed.
5. Each complete turn of the helix is 34 Å (3.4 nm) long; hence, 10 bases exist per turn in each chain.

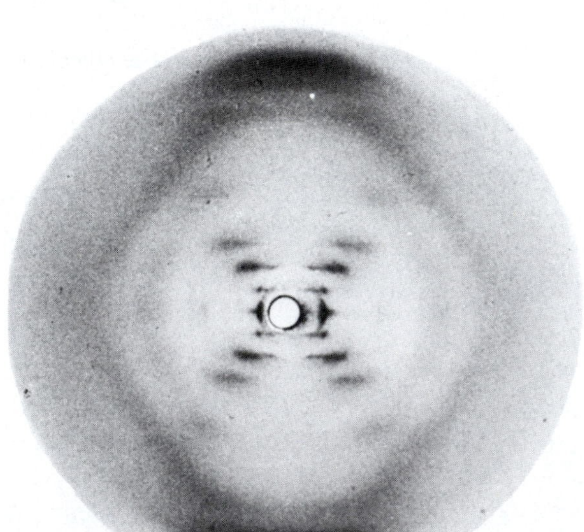

FIGURE 2–13 X-ray diffraction photograph of the B form of purified DNA fibers. The strong arcs on the periphery show closely spaced aspects of the molecule, providing an estimate of the periodicity of nitrogenous bases, which are 3.4 Å apart. The inner cross pattern of spots shows the grosser aspect of the molecule, indicating its helical nature.

6. In any segment of the molecule, alternating larger **major grooves** and smaller **minor grooves** are apparent along the axis.
7. The double helix measures 20 Å (2.0 nm) in diameter.

The nature of base pairing (Point 4) is the most genetically significant feature of the model. Before discussing it in detail, several other important features warrant emphasis. First, the antiparallel nature of the two chains is a key part of the double-helix model. While one chain runs in the 5′-to-3′ orientation (what seems right-side up to us), the other chain is in the 3′-to-5′ orientation (and thus appears upside down). This is illustrated in Figure 2–14(b). Given the constraints of the bond angles of the various nucleotide components, the double helix could not be constructed easily if both chains ran parallel to one another.

Second, the right-handed nature of the helix is best appreciated by comparing such a structure to its left-handed counterpart, which is a mirror image, as shown in Figure 2–15. The conformation in space of the right-handed helix is most consistent with the data that were available to Watson and Crick. As we shall see momentarily, an alternative form of DNA (Z-DNA) does exist as a left-handed helix.

The key to the model proposed by Watson and Crick is the specificity of base pairing. Chargaff's data had suggested that the amounts of A equaled T and that G equaled C. Watson and Crick realized that if A pairs with T and C pairs with G, thus accounting for these proportions, the members of each such base pair formed hydrogen bonds [Figure 2–14(b)], providing the chemical stability necessary to hold the two chains together. Arranged in this way, both **major** and **minor** grooves become apparent along the axis. Further, a purine (A or G) opposite a pyrimidine (T or C) on each "rung of the spiral staircase" of the proposed helix accounts for the 20 Å (2 nm) diameter suggested by X-ray diffraction studies.

The specific A═T and G≡C base pairing is the basis for the concept of **complementarity**, which describes the chemical affinity provided by the hydrogen bonds between the bases. As we will see, this concept is very important in the processes of DNA replication and gene expression.

Two questions are particularly worthy of discussion. First, why aren't other base pairs possible? Watson and Crick discounted the A═G and C═T pairs because these represent purine–purine and pyrimidine–pyrimidine pairings, respectively. These pairings would lead to alternating diameters of more than, and less than, 20 Å because of the respective sizes of the purine and pyrimidine rings; in addition, the three-dimensional configurations formed by such pairings do not produce the proper alignment leading to sufficient hydrogen-bond formations. It is for this reason that A═C and G═T pairings were also discounted, even though these pairs each consist of one purine and one pyrimidine.

The second question concerns hydrogen bonds. Just what is the nature of such a bond, and is it strong enough to stabilize the helix? A **hydrogen bond** is a very weak electrostatic attraction between a covalently bonded hydrogen atom and an atom with an unshared electron pair. The hydrogen atom assumes a partial positive charge, while the unshared

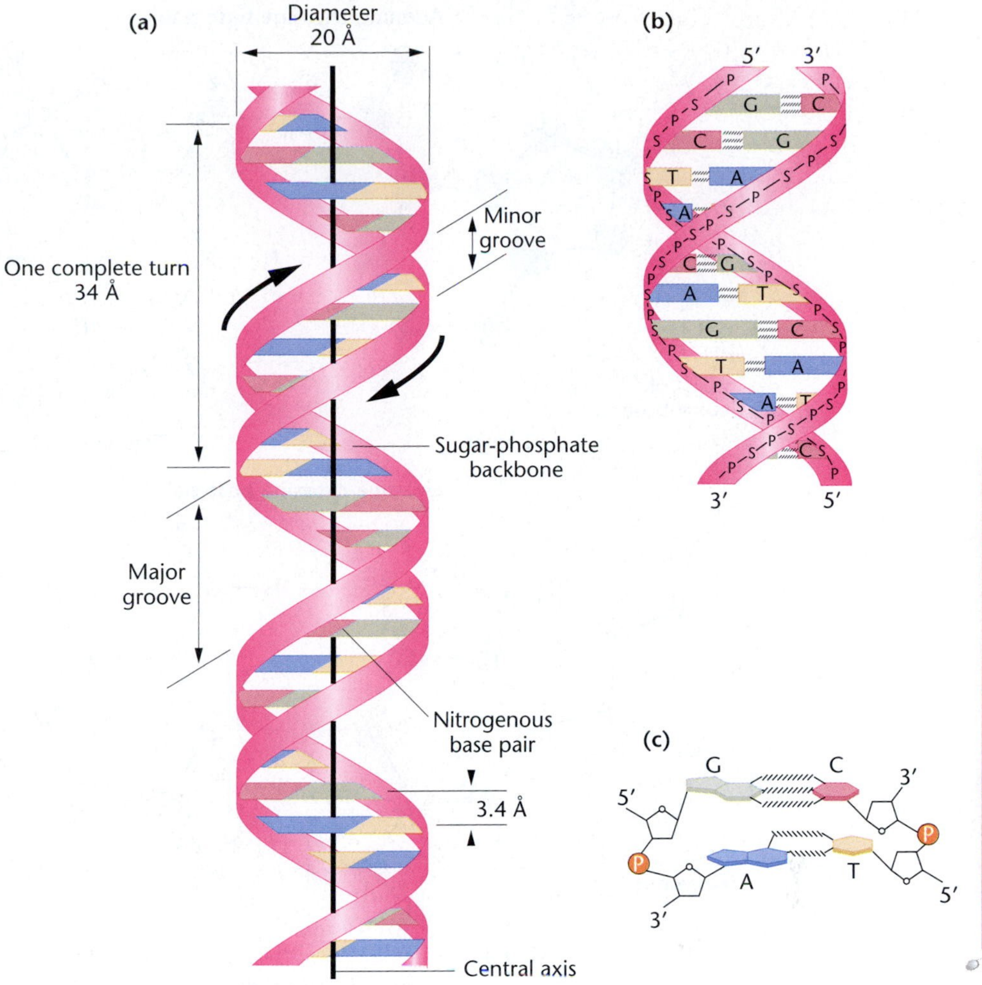

FIGURE 2–14 (a) The DNA double helix as proposed by Watson and Crick. The ribbonlike strands constitute the sugar-phosphate backbones, and the horizontal rungs constitute the nitrogenous base pairs, of which there are 10 per complete turn. The major and minor grooves are apparent. The solid vertical bar represents the central axis. (b) A detailed view depicting the bases, sugars, phophates, and hydrogen bonds of the helix. (c) A demonstration of the antiparallel nature of the helix and the horizontal stacking of the bases.

MEDIA TUTORIAL DNA Structure

electron pair—characteristic of covalently bonded oxygen and nitrogen atoms—assumes a partial negative charge. These opposite charges are responsible for the weak chemical attractions. As oriented in the double helix (Figure 2–16), adenine forms two hydrogen bonds with thymine, and guanine forms three hydrogen bonds with cytosine. Although two or three hydrogen bonds taken alone are energetically very weak, 2000–3000 bonds in tandem (which would be found in two long polynucleotide chains) are capable of providing great stability to the helix.

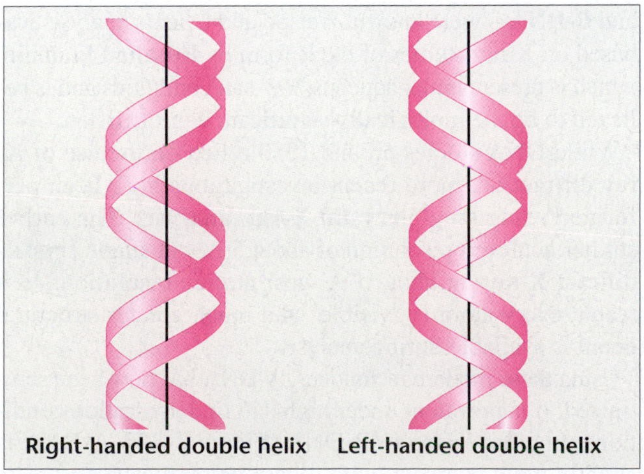

FIGURE 2–15 The right- and left-handed helical forms of DNA. Note that they are mirror images of one another.

Still, another stabilizing factor is the arrangement of sugars and bases along the axis. In the Watson–Crick model, the hydrophilic sugar-phosphate backbone is on the outside of the axis, where both components may interact with water. The relatively hydrophobic or "water-fearing" nitrogenous bases are "stacked" almost horizontally on the interior of the axis, shielded from water. These molecular arrangements provide significant chemical stabilization to the helix.

A more recent and accurate analysis of the form of DNA that served as the basis for the Watson–Crick model has revealed a minor structural difference. A precise measurement of the number of base pairs (bp) per turn has demonstrated a value of 10.4, rather than the 10.0 predicted by Watson and Crick. Where, in the classic model, each base pair is rotated around the helical axis 36°, relative to the adjacent base pair, the new finding requires a rotation of 34.6°. This results in slightly more than 10 base pairs per 360° turn.

The Watson–Crick model had an immediate effect on the emerging discipline of molecular biology. Even in their initial 1953 article, the authors observed, "It has not escaped our notice that the specific pairing we have postulated

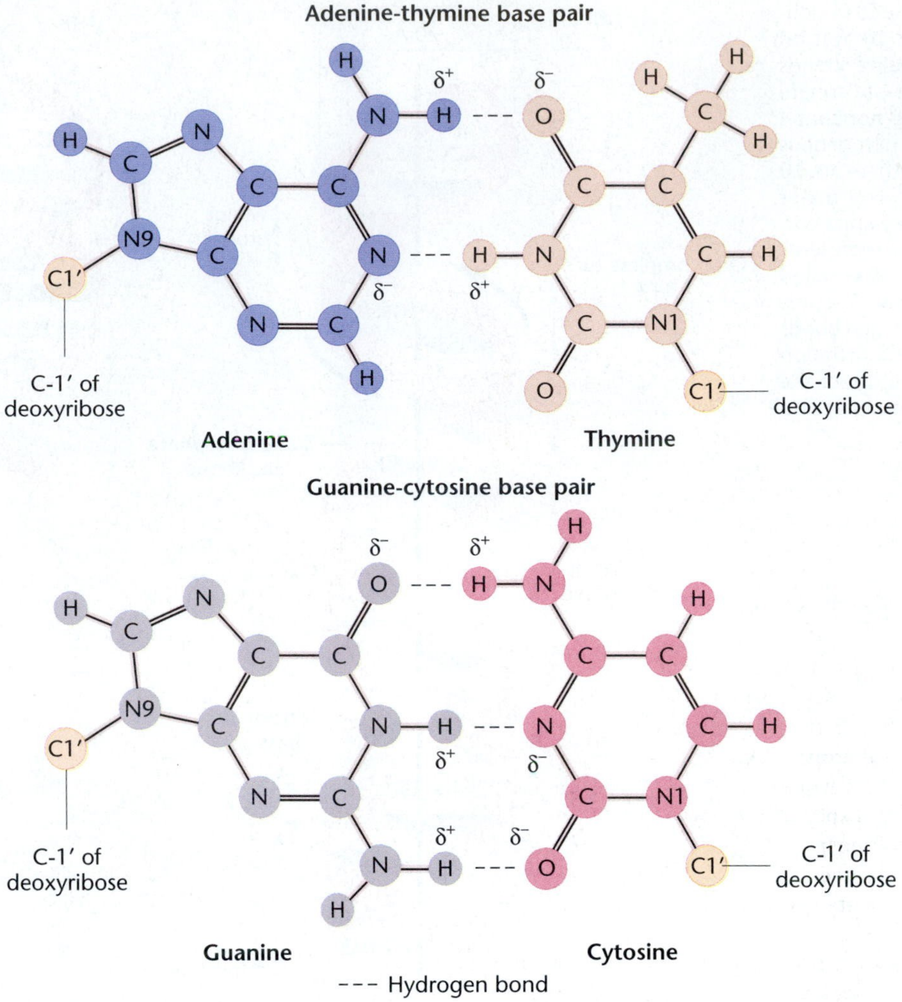

FIGURE 2–16 Ball-and-stick models of A=T and G≡C base pairs. The dashes (– – –) represent the hydrogen bonds that form between bases.

immediately suggests a possible copying mechanism for the genetic material." Two months later, in a second article in *Nature*, Watson and Crick pursued this idea, suggesting a specific mode of replication of DNA: the semiconservative model. The second article also alluded to two new concepts: (1) the storage of genetic information in the sequence of the bases and (2) the mutations or genetic changes that would result from an alteration of bases. These ideas have received vast amounts of experimental support since 1953 and are now universally accepted.

Watson and Crick's "synthesis" of ideas was highly significant with regard to subsequent studies of genetics and biology. The nature of the gene and its role in genetic mechanisms could now be viewed and studied in biochemical terms. Recognition of their work, along with that of Wilkins, led to their receipt of the Nobel Prize in Physiology and Medicine in 1962. Unfortunately, Rosalind Franklin had died in 1958 at the age of 37, making her contributions ineligible for consideration, since the award is not given posthumously. The Nobel Prize was to be one of many such awards bestowed for work in the field of molecular genetics.

2.8 Alternative Forms of DNA Exist

Under different conditions of isolation, several conformational forms of DNA have been recognized. At the time Watson and Crick performed their analysis, two forms—**A-DNA** and **B-DNA**—were known. Watson and Crick's analysis was based on X-ray studies of the B form by Rosalind Franklin, which is present under aqueous, low-salt conditions and is believed to be the biologically significant conformation.

While DNA studies around 1950 relied on the use of X-ray diffraction, more recent investigations have been performed using **single-crystal X-ray analysis**. The earlier studies achieved resolution of about 5 Å, but single crystals diffract X rays at about 1 Å, near atomic resolution. As a result, every atom is "visible" and much greater structural detail is available during analysis.

Using these modern techniques, A-DNA has now been scrutinized. It is prevalent under high-salt or dehydration conditions. In comparison to B-DNA (Figure 2–17), A-DNA is slightly more compact, with 9-base pairs in each complete turn of the helix, which is 23 Å in diameter. While it is also

Molecular Structure of Nucleic Acids: A Structure for Deoxyribose Nucleic Acid

We wish to suggest a structure for the salt of deoxyribose nucleic acid (D. N. A.). This structure has novel features which are of considerable biological interest. A structure for nucleic acid has already been proposed by Pauling and Corey.[1] They kindly made their manuscript available to us in advance of publication. Their model consists of three intertwined chains, with the phosphates near the fibre axis, and the bases on the outside. In our opinion, this structure is unsatisfactory for two reasons: (1) We believe that the material which gives the X-ray diagrams is the salt, not the free acid. Without the acidic hydrogen atoms it is not clear what forces would hold the structure together, especially as the negatively charged phosphates near the axis will repel each other. (2) Some of the van der Waals distances appear to be too small.

Another three-chain structure has also been suggested by Fraser (in the press). In his model the phosphates are on the outside and the bases on the inside, linked together by hydrogen bonds. This structure as described is rather ill-defined, and for this reason we shall not comment on it.

We wish to put forward a radically different structure for the salt of deoxyribose nucleic acid. This structure has two helical chains each coiled round the same axis. We have made the usual chemical assumptions, namely, that each chain consists of phosphate diester groups joining β-D-deoxyribofuranose residues with 3', 5' linkages. The two chains (but not their bases) are related by a dyad perpendicular to the fibre axis. Both chains follow right-handed helices, but owing to the dyad the sequences of the atoms in the two chains run in opposite directions. Each chain loosely resembles Furberg's[2] model No. 1; that is, the bases are on the inside of the helix and the phosphates on the outside. The configuration of the sugar and the atoms near it is close to Furberg's "standard configuration," the sugar being roughly perpendicular to the attached base. There is a residue on each chain every 3.4 Å in the *z*-direction. We have assumed an angle of 36° between adjacent residues in the same chain, so that the structure repeats after 10 residues on each chain, that is, after 34 Å the distance of a phosphorus atom from the fibre axis is 10 Å. As the phosphates are on the outside, cations have easy access to them.

The structure is an open one, and its water content is rather high. At lower water contents we would expect the bases to tilt so that the structure could become more compact.

The novel feature of the structure is the manner in which the two chains are held together by the purine and pyrimidine bases. The planes of the bases are perpendicular to the fibre axis. They are joined together in pairs, a single base from one chain being hydrogen-bonded to a single base from the other chain, so that the two lie side by side with identical *z*-coordinates. One of the pair must be a purine and the other a pyrimidine for bonding to occur. The hydrogen bonds are made as follows: purine position 1 to pyrimidine position 1; purine position 6 to pyrimidine position 6.

If it is assumed that the bases only occur in the structure in the most plausible tautomeric forms (that is, with the keto rather than the enol configurations) it is found that only specific pairs of bases can bond together. These pairs are: adenine (purine) with thymine (pyrimidine), and guanine (purine) with cytosine (pyrimidine).

In other words, if an adenine forms one member of a pair, on either chain, then on these assumptions the other member must be thymine; similarly for guanine and cytosine. The sequence of bases on a single chain does not appear to be restricted in any way. However, if only specific pairs of bases can be formed, it follows that if the sequence of bases on one chain is given, then the sequence on the other chain is automatically determined.

It has been found experimentally[3,4] that the ratio of the amounts of adenine to thymine, and the ratio of guanine to cytosine, are always very close to unity for deoxyribose nucleic acid.

It is probably impossible to build this structure with a ribose sugar in place of the deoxyribose, as the extra oxygen atom would make too close a van der Waals contact.

The previously published X-ray data[5,6] on deoxyribose nucleic acid are insufficient for a rigorous test of our structure. So far as we can tell, it is roughly compatible with the experimental data, but it must be regarded as unproved until it has been checked against more exact results. Some of these are given in the following communications. We were not aware of the details of the results presented there when we devised our structure, which rests mainly though not entirely on published experimental data and stereochemical arguments.

It has not escaped our notice that the specific pairing we have postulated immediately suggests a possible copying mechanism for the genetic material. Full details of the structure, including the conditions assumed in building it, together with a set of co-ordinates for the atoms, will be published elsewhere.

We are much indebted to Dr. Jerry Donohue for constant advice and criticism, especially on interatomic distances. We have also been stimulated by a knowledge of the general nature of the unpublished experimental results and ideas of Dr. M. H. F. Wilkins, Dr. R. E. Franklin and their co-workers at King's College, London. One of us (J. D. W.) has been aided by a fellowship from the National Foundation for Infantile Paralysis.

J.D. Watson
F.H.C. Crick
Medical Research Council Unit for the Study of the Molecular Structure of Biological Systems, Cavendish Laboratory, Cambridge, England.

[1]Pauling, L., and Corey, R. B., *Nature*, 171, 346 (1953); *Proc. U.S. Nat. Acad. Sci.*, 39, 84 (1953).
[2]Furberg, S., *Acta Chem. Scand.*, 6, 634 (1952).
[3]Chargaff, E., for references see Zamenhof, S., Brawerman, G., and Chargaff, E., *Biochim. et Biophys. Acta*, 9, 402 (1952).
[4]Wyatt, G. R., *J. Gen. Physiol*, 36, 201 (1952).
[5]Astbury, W. T., *Symp. Soc. Exp. Biol. 1, Nucleic Acid*, 66 (Camb. Univ. Press, 1947).
[6]Wilkins, M. H. F., and Randall, J. T., *Biochim. et Biophys. Acta*, 10, 192 (1953).

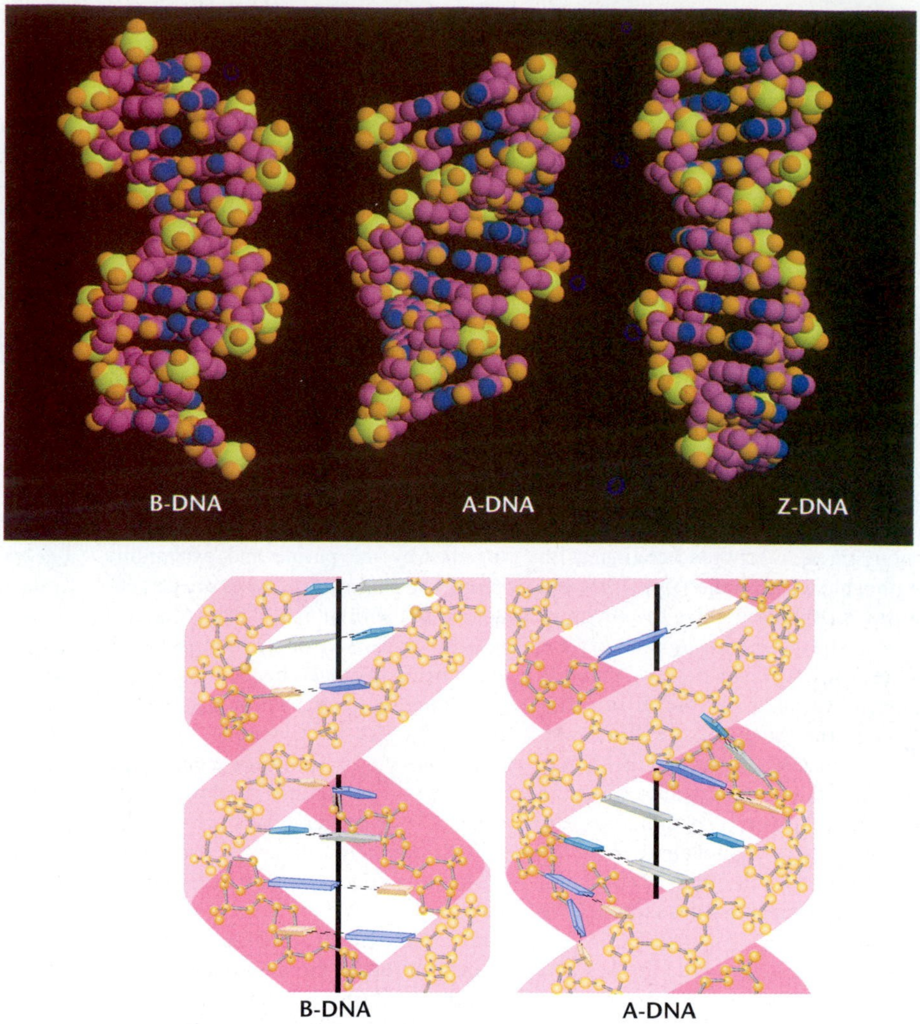

FIGURE 2–17 The top half of the figure shows computer-generated space-filling models of B-DNA (left), A-DNA (center), and Z-DNA (right). Below is an artist's depiction illustrating the orientation of the base pairs of B-DNA and A-DNA. (Note that in B-DNA the base pairs are perpendicular to the helix, while they are tilted and pulled away from the helix in A-DNA.)

a right-handed helix, the orientation of the bases is somewhat different. They are tilted and displaced laterally in relation to the axis of the helix. As a result of these differences, the appearance of the major and minor grooves is modified compared with those in B-DNA. It seems doubtful that A-DNA occurs under biological conditions.

Three other right-handed forms of DNA helices have been discovered when investigated under laboratory conditions. These have been designated C-, D-, and E-DNA. **C-DNA** is found under even greater dehydration conditions than those observed during the isolation of A- and B-DNA. It has only 9.3 base pairs per turn and is, thus, less compact. Its helical diameter is 19 Å. Like A-DNA, C-DNA does not have its base pairs lying flat; rather, they are tilted relative to the axis of the helix. Two other forms, **D-DNA** and **E-DNA**, occur in helices lacking guanine in their base composition. They have even fewer base pairs per turn: 8 and 7, respectively.

Still another form of DNA, called **Z-DNA**, was discovered by Andrew Wang, Alexander Rich, and their colleagues in 1979, when they examined a small synthetic DNA oligonucleotide containing only C-G base pairs. Z-DNA takes on the rather remarkable configuration of a left-handed double helix (Figure 2–17). Like A- and B-DNA, Z-DNA consists of two antiparallel chains held together by Watson–Crick base pairs. Beyond these characteristics, Z-DNA is quite different. The left-handed helix is 18 Å (1.8 nm) in diameter, contains 12 base pairs per turn, and assumes a zigzag conformation (hence its name). The major groove present in B-DNA is nearly eliminated in Z-DNA.

More recent research by Jean-Francois Allemand and colleagues has shown that if DNA is artificially stretched, still another form of the molecule is assumed, which they have called **P-DNA**. A model of P-DNA contrasted with the B form of DNA is quite interesting, since it is longer, more narrow, and the phosphate groups, found on the outside of B-DNA, are present inside the molecule. The nitrogenous bases, present inside the helix in B-DNA, are found closer to the external surface of the helix in P-DNA, and there are 2.62 bases per turn, in contrast to the 10.4 per turn in B-DNA.

The interest in alternative forms of DNA, such as Z and P, stems from the belief that DNA might have to assume a structure other that the B-form as it functions as the genetic material. During both replication and transcription (when its RNA complement is synthesized during gene expression), the strands of the helix must separate and become accessible to large enzymes, as well as a variety of other proteins involved in these processes. It is possible that changes in the shapes of the DNA facilitate these functions. Unique conformations might serve as points of molecular recognition for proteins. However, verification of the biological significance of alternative forms awaits the clear demonstration that they exist in vivo.

2.9 The Structure of RNA Is Chemically Similar to DNA, but Single Stranded

The second category of nucleic acids is ribonucleic acid, or RNA. The structure of these molecules is similar to DNA, with several important exceptions. Although RNA also has as its building blocks nucleotides linked into polynucleotide chains, the sugar ribose replaces deoxyribose and the nitrogenous base uracil replaces thymine. Another important difference is that most RNA is single stranded, although there are two important exceptions. First, RNA molecules sometimes fold back on themselves to form double-stranded regions of complementary base pairs. Second, some animal viruses that have RNA as their genetic material contain it in the form of double-stranded helices. Thus, there are several instances where RNA does not exist strictly as a linear single-stranded molecule.

Three major classes of cellular RNA molecules function during the expression of genetic information: **ribosomal RNA (rRNA)**, **messenger RNA (mRNA)**, and **transfer RNA (tRNA)**. These molecules all originate as complementary copies of one of the two strands of DNA segments during the process of transcription. That is, their nucleotide sequence is complementary to the deoxyribonucleotide sequence of DNA, which served as the template for their synthesis. Because uracil replaces thymine in RNA, uracil is complementary to adenine during transcription and during RNA base pairing.

Table 2–4 characterizes the major forms of RNA found in prokaryotic and eukaryotic cells. Different RNAs are distinguished according to their sedimentation behavior in a centrifugal field and their size, as measured by the number of nucleotides each contains. Sedimentation behavior depends on a molecule's density, mass, and shape, and its measure is called the **Svedberg coefficient** (*S*). While higher *S* values almost always designate molecules of greater molecular weight, the correlation is not direct; that is, a twofold increase in molecular weight does not lead to a twofold increase in *S*. This is because that, in addition to a molecule's mass, the size and the shape of the molecule also impact on its rate of sedimentation (*S*). As you can see, a wide variation exists in the size of the three classes of RNA.

Ribosomal RNA is generally the largest of these molecules (as is generally reflected in its *S* values) and usually constitutes about 80 percent of all RNA in an *E. coli* cell. Ribosomal RNAs are important structural components of **ribosomes**, which function as nonspecific workbenches during the synthesis of proteins during the process of translation. The various forms of rRNA found in prokaryotes and eukaryotes differ distinctly in size.

Messenger RNA molecules carry genetic information from the DNA of the gene to the ribosome, where translation occurs. They vary considerably in size, which is a reflection of the variation in the size of the protein encoded by the mRNA as well as the gene serving as the template for transcription of mRNA. While Table 2–4 shows that about 5% of RNA is mRNA in *E. coli*, this percentage varies from cell to cell and even at different times in the life of the same cell.

Transfer RNA, the smallest class of RNA molecules, carries amino acids to the ribosome during translation. Because more than one tRNA molecule interacts simultaneously with the ribosome, the molecule's smaller size facilitates these interactions.

We will discuss the functions of the three classes of RNA in much greater detail in Chapter 6. In addition, as we proceed through the text, we will encounter other unique RNAs that perform various roles. For example, **small nuclear RNA (snRNA)** participates in processing mRNAs (Chapter 6). **Telomerase RNA** is involved in DNA replication at the ends of chromosomes (Chapter 3), and **antisense RNA** is involved in gene regulation (Chapter 20). Our purpose in this section has been to contrast the structure of DNA, which stores

TABLE 2–4 RNA Characterization

RNA Class	*% Total RNA**	*Components (Svedberg Coefficient)*	*Eukaryotic (E) or Prokaryotic (P)*	*Number of Nucleotides*
Ribosomal (rRNA)	80	5*S*	P and E	120
		5.8*S*	E	160
		16*S*	P	1542
		18*S*	E	1874
		23*S*	P	2904
		28*S*	E	4718
Transfer (tRNA)	15	4*S*	P and E	75–90
Messenger (mRNA)	5	varies	P and E	100–10,000

*In *E. coli*

genetic information, with that of RNA, which most often functions in the expression of that information.

2.10 Many Analytical Techniques Have Been Useful during the Investigation of DNA and RNA

Since 1953, the role of DNA as the genetic material and the role of RNA in transcription and translation have been clarified through detailed analysis of nucleic acids. We will consider several methods of analysis of these molecules in this chapter. Some of these, as well as other research procedures, are presented in greater detail in Appendix A.

Absorption of Ultraviolet Light (UV)

Nucleic acids absorb ultraviolet (UV) light most strongly at wavelengths of 254 to 260 nm (see Figure 2–7) due to the interaction between UV light and the ring systems of the purines and pyrimidines. Thus, any molecule containing nitrogenous bases (i.e., nucleosides, nucleotides, and polynucleotides) can be analyzed by using UV light. This technique is especially important in the localization, isolation, and characterization of nucleic acids.

Ultraviolet analysis is used in conjunction with many standard procedures that separate molecules. As we shall see in the next section, the use of UV absorption is critical to the isolation of nucleic acids following their separation.

Sedimentation Behavior

Nucleic acid mixtures can be separated by subjecting them to one of several possible **gradient centrifugation** procedures (Figure 2–18). The mixture can be loaded on top of a solution prepared so that a concentration gradient has been formed from top to bottom. Then the entire mixture is centrifuged at high speeds in an ultracentrifuge. The mixture of molecules will migrate downward, with each component moving at a different rate. Centrifugation is stopped and the gradient eluted from the tube. Each fraction can then be measured spectrophotometrically for absorption at 260 nm. In this way, the previous position of a nucleic acid fraction along the gradient can be predicted and the fraction isolated and studied further.

The gradient centrifugations described rely on the sedimentation behavior of molecules in solution. Two major types of gradient centrifugation techniques are employed in the analysis of nucleic acids: sedimentation equilibrium and sedimentation velocity. Both require the use of high-speed centrifugation to create large centrifugal forces upon molecules in a gradient solution.

In **sedimentation equilibrium centrifugation** (sometimes called density gradient centrifugation), a density gradient is created that overlaps the densities of the individual components of a mixture of molecules. Usually, the gradient is made of a heavy metal salt, such as cesium chloride (CsCl). During centrifugation, the molecules migrate until they reach a point of neutral buoyant density. At this point, the centrifugal force on them is equal and opposite to the upward diffusion force, and no further migration occurs. If DNAs of different densities are used, they will separate as the molecules of each density reach equilibrium with the corresponding density of CsCl. The gradient may be fractionated and the components isolated (Figure 2–18). When properly executed, this technique provides high resolution in separating mixtures of molecules varying only slightly in density.

Sedimentation equilibrium centrifugation studies can also be used to generate data on the base composition of double-stranded DNA. The G≡C base pairs, compared with A=T pairs, are more compact and dense. As shown in Figure 2–19, the percentage of G≡C pairs in DNA is directly proportional to the molecule's buoyant density. By using this technique, we can make a useful molecular characterization of DNAs from different sources.

The second technique, **sedimentation velocity centrifugation**, employs an analytical centrifuge, which enables the migration of the molecules during centrifugation to be monitored with ultraviolet absorption optics. Thus, the "velocity of sedimentation" can be determined. This velocity has been standardized in units called Svedberg coefficients (*S*), as mentioned earlier.

In this technique, the molecules are loaded on top of the gradient, and the gravitational forces created by centrifugation drive them toward the bottom of the tube. Two forces work against this downward movement: (1) the viscosity of the solution creates a frictional resistance, and (2) part of the force of diffusion is directed upward. Under these conditions, the key variables are the mass and shape of the molecules being examined. In general, the greater the mass, the greater the sedimentation velocity. However, the molecule's shape affects the frictional resistance. Therefore, two molecules of equal mass, but different in shape, will sediment at different rates.

One use of the sedimentation velocity technique is the determination of **molecular weight (MW)**. If certain physical–chemical properties of a molecule under study are also known, the MW can be calculated based on the sedimentation velocity. The *S* values increase with molecular weight, but they are not directly proportional to it.

Denaturation and Renaturation of Nucleic Acids

When **denaturation** of double-stranded DNA occurs, the hydrogen bonds of the duplex structure break, the duplex unwinds, and the strands separate. However, no covalent bonds break. During strand separation, which can be induced by heat or chemical treatment, the viscosity of DNA decreases, and both the UV absorption and the buoyant density increase. Denaturation as a result of heating is sometimes referred to as melting. The increase in UV absorption of heated DNA in solution, called the **hyperchromic shift**, is easiest to measure. This effect is illustrated in Figure 2–20.

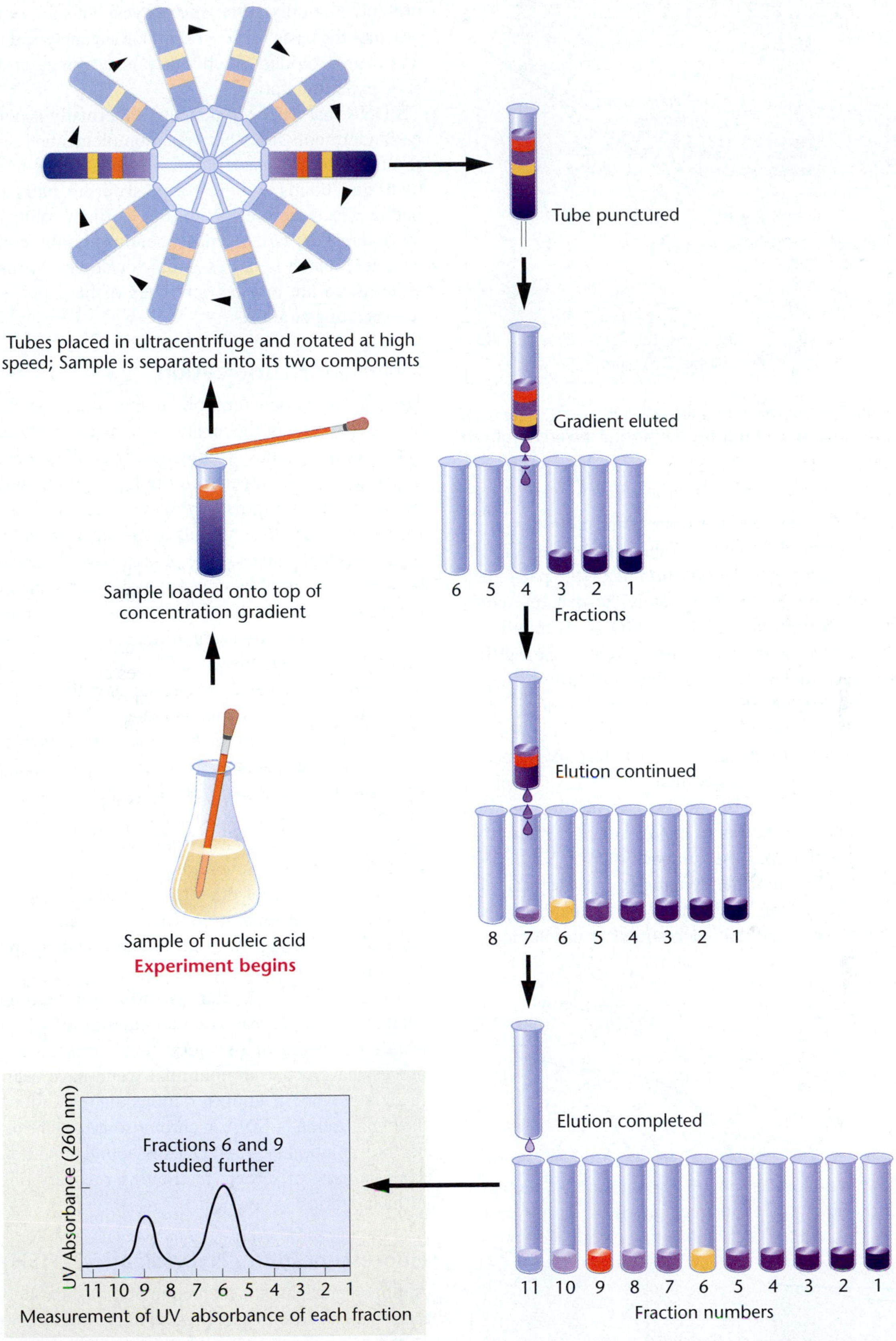

FIGURE 2–18 Separation of a mixture of two types of nucleic acid by gradient centrifugation. To fractionate the gradient, successive samples are eluted from the bottom of the tube. Each is measured for absorbance of ultraviolet light at 260 nm, producing a profile of the sample in graphic form.

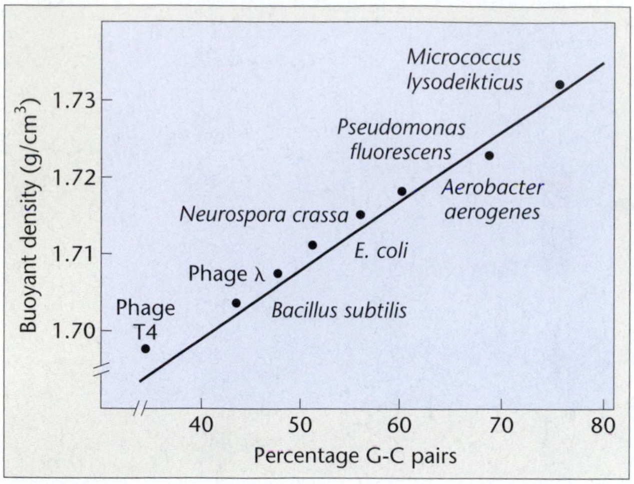

FIGURE 2–19 Percentage of guanine–cytosine G≡C base pairs in DNA, plotted against buoyant density for a variety of microorganisms.

Because G≡C base pairs have one more hydrogen bond than do A═T pairs, they are more stable to heat treatment. Thus, DNA with a greater proportion of G≡C pairs than A═T pairs requires higher temperatures to denature completely. When absorption at 260 nm (OD_{260}) is monitored and plotted against temperature during heating, a **melting profile** of DNA is obtained. The midpoint of this profile, or curve, is called the **melting temperature** (T_m) and represents the point at which 50 percent of the strands are unwound or denatured (Figure 2–20). When the curve plateaus at its maximum optical density, denaturation is complete, and only single strands exist. Analysis of melting profiles provides a characterization of DNA and an alternative method of estimating the base composition of DNA.

One might ask whether the denaturation process can be reversed; that is, can single strands of nucleic acids re-form a double helix, provided that each strand's complement is present? Not only is the answer yes, but such reassociation provides the basis for several important analytical techniques that have provided much valuable information during genetic experimentation.

If DNA that has been denatured thermally is cooled slowly, random collisions between complementary strands will result in their reassociation. At the proper temperature, hydrogen bonds will re-form, securing pairs of strands into duplex structures. With time during cooling, more and more duplexes will form. Depending on the conditions, a complete match is not essential for duplex formation, provided there are at least stretches of base pairing on two reassociating strands.

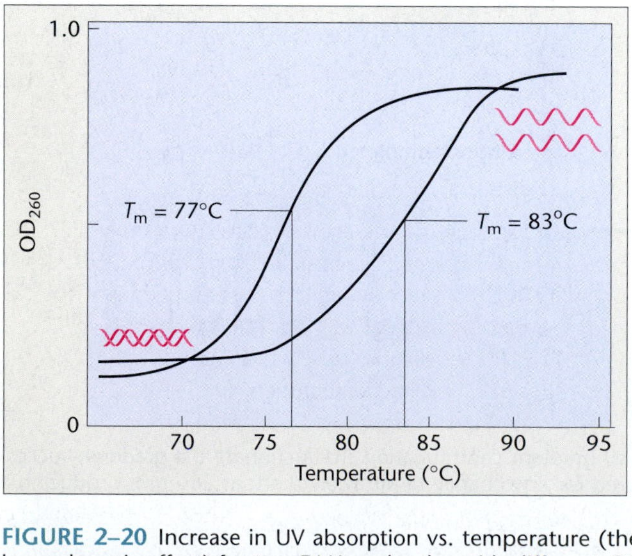

FIGURE 2–20 Increase in UV absorption vs. temperature (the hyperchromic effect) for two DNA molecules with different GC contents. The molecule with a melting point (T_m) of 83°C has a greater GC content than the molecule with a T_m of 77°C.

Molecular Hybridization

The property of denaturation–renaturation of nucleic acids is the basis for one of the most powerful and useful techniques in molecular genetics—**molecular hybridization**. This technique derives its name from the fact that renaturing single strands need not originate from the same nucleic acid source. For example, if DNA strands are isolated from two distinct organisms and some degree of base complementarity exists between them, double-stranded molecular hybrids will form during renaturation. Furthermore, when mixtures of DNA and RNA single strands are utilized, hybridization may also occur. A case in point is when RNA and the DNA from which it has been transcribed are present together (Figure 2–21). The RNA will find its single-stranded DNA complement and renature. In this example, the DNA strands are heated, causing strand separation, and then slowly cooled in the presence of single-stranded RNA. If the RNA has been transcribed on the DNA used in the experiment, and is therefore complementary to it, molecular hybridization will occur, creating a DNA:RNA duplex. Several methods are available for monitoring the amount of double-stranded molecules produced following strand separation. In early studies, radioisotopes were utilized to "tag" one of the strands and monitor its presence in hybrid duplexes that formed.

In the 1960s, molecular hybridization techniques contributed to our increased understanding of transcriptional events occurring at the gene level. Refinements of this process have occurred continually and have been the forerunners of work in studies of molecular evolution as well as the organization of DNA in chromosomes. Hybridization can occur in solution or when DNA is bound either to a gel or to a special type of screen, facilitating recovery of the newly formed hybrids.

Fluorescent *in situ* Hybridization (FISH)

A refinement using the technique of molecular hybridization has led to the use of DNA present in cytological preparations as the "target" for hybrid formation. When this approach is combined with the use of fluorescent probes to monitor hybridization, the technique is called **fluorescent *in situ* hybridization**, or simply the acronym **FISH**. In this procedure mitotic or interphase cells are fixed to slides and subjected to hybridization conditions. Single-stranded DNA or RNA is added, and hybridization is monitored. The nucleic

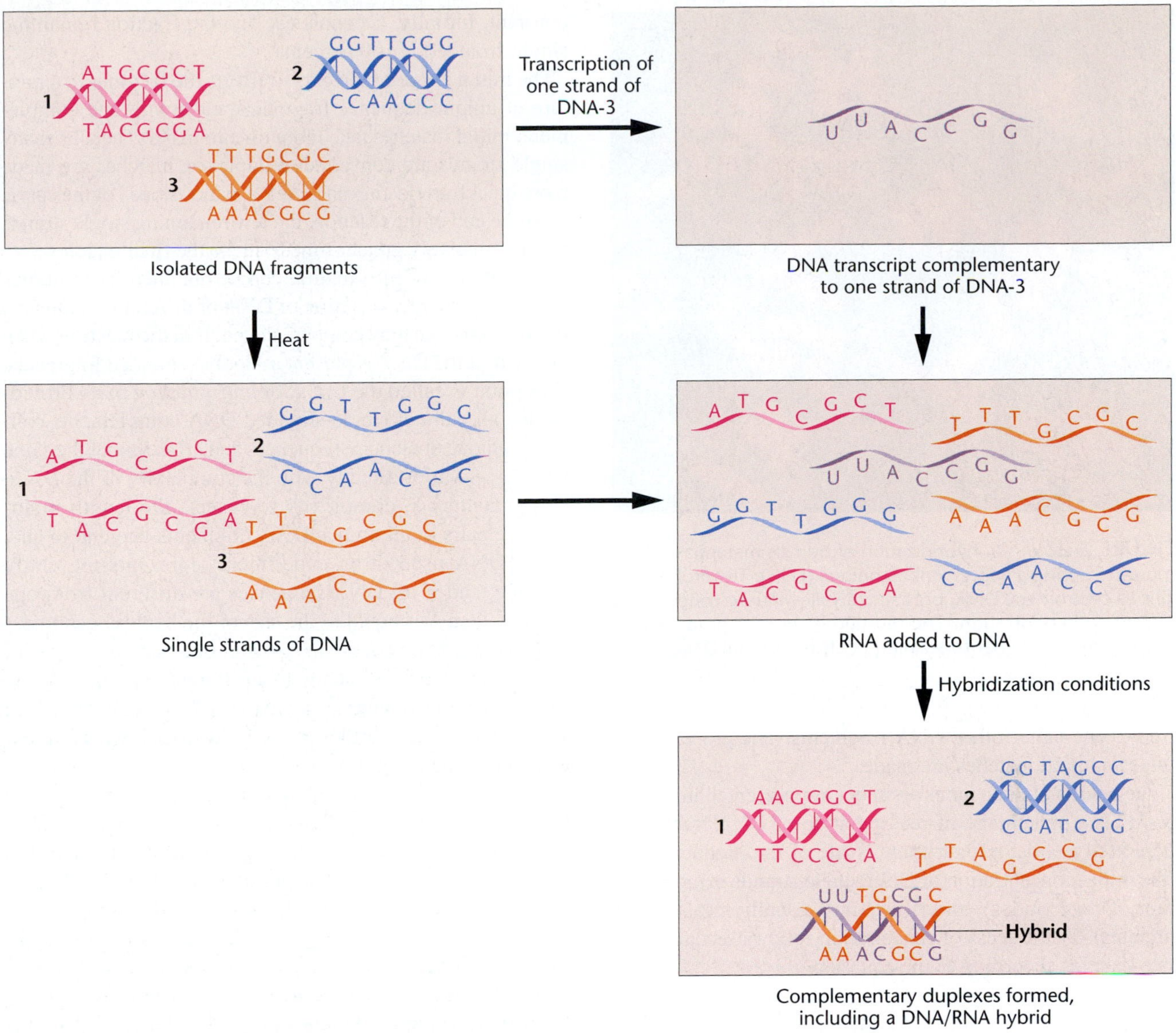

FIGURE 2–21 Diagrammatic representation of the process of molecular hybridization between DNA fragments and RNA that has been transcribed on one of the single-stranded fragments.

acid serves as a "probe," since it will hybridize only with the specific chromosomal areas for which it is complementary. Before the use of fluorescent probes was refined, radioactive probes were used in these *in situ* procedures to allow detection on the slide. In this approach, the technique of autoradiography was utilized. (See Appendix A.)

Fluorescent probes are prepared in a unique way. When DNA is used, it is first coupled to the small organic molecule biotin (creating biotinylated DNA). Once *in situ* hybridization is completed, another molecule (avidin or streptavidin) that has a high binding affinity for biotin is used. A fluorescent molecule such as fluorescein is linked to avidin (or streptavidin) and the complex is reacted with the cytological preparation. This procedure represents an extremely sensitive method for localizing the hybridized DNA.

Figure 2–22 illustrates the use of FISH in identifying the DNA specific to the centromeres of human chromosomes. The resolution of FISH is great enough to detect just a single gene within an entire set of chromosomes. (See Appendix A.) The use of this technique in identifying chromosomal locations housing specific genetic information has been a valuable addition to the repertoire of experimental geneticists.

Reassociation Kinetics and Repetitive DNA

In one extension of molecular hybridization procedures, the *rate of reassociation* of complementary single DNA strands is analyzed. This technique, called **reassociation kinetics**, was first refined and studied by Roy Britten and David Kohne.

The DNA used in such studies is first fragmented into small pieces as a result of shearing forces introduced during isolation. The resultant DNA fragments cluster around a uniform average size of several hundred base pairs. These fragments of DNA are then dissociated into single strands by heating. Next, the temperature is lowered and reassociation is monitored. During reassociation, pieces of single-stranded DNA collide randomly. If they are complementary, a stable double strand is formed; if not, they separate and are

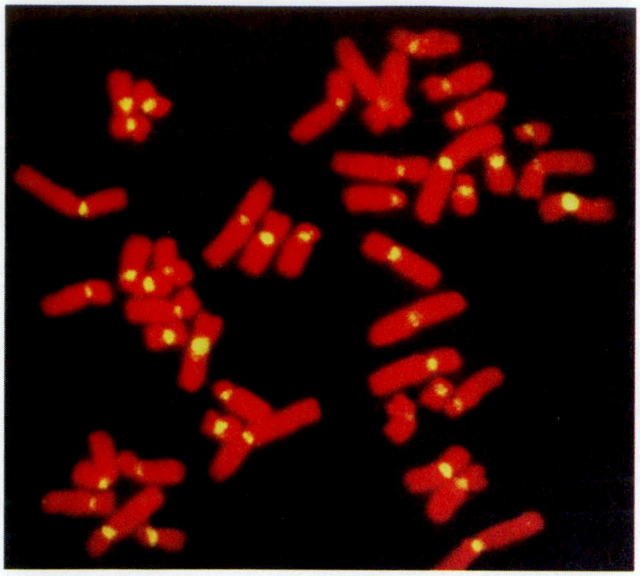

FIGURE 2–22 *In situ* hybridization of human metaphase chromosomes, using a fluorescent technique (FISH). The probe, specific to centromeric DNA, produces a yellow fluorescence signal indicating hybridization. The red fluorescence is produced by propidium iodide counterstaining of chromosomal DNA.

free to encounter other DNA fragments. The process continues until all matches are made.

The results of such an experiment are presented in Figure 2–23. The percentage of reassociation of DNA fragments is plotted against a logarithmic scale of the product of C_0 (the initial concentration of DNA single strands in moles per liter of nucleotides), and t (the time, usually measured in minutes). The process of renaturation follows second-order rate kinetics according to the equation

$$\frac{C}{C_0} = \frac{1}{1 + kC_0t}$$

where C is the single-stranded DNA concentration remaining after time t has elapsed and k is the second-order rate constant. Initially, C equals C_0, and the fraction remaining single stranded is 100 percent.

The initial shape of the curve reflects the fact that in a mixture of unique sequence fragments, each with one complement, initial matches take more time to make. Then, as many single strands are converted to duplexes, matches are made more quickly, reflecting an increase in the "slope" of the curve. Near the end of the reaction, the few remaining single strands require relatively greater time to make the final matches.

A great deal of information can be obtained from studies comparing the reassociation of DNA of different organisms. For example, we may compare the point in the reaction when one-half of the DNA is present as double-stranded fragments. This point is called the $C_0t_{1/2}$, or *half-reaction time*. Provided that all pairs of single-stranded DNA complements consist of unique nucleotide sequences and all are about the same size, $C_0t_{1/2}$ varies directly with the complexity of the DNA. Designated as X, complexity represents the length in nucleotide pairs of all unique DNA fragments laid end to end. If the DNA used in an experiment represents the entire genome, and if all DNA sequences are different from one another, then X is equal to the size of the haploid genome.

Figure 2–24 compares DNAs from two bacteriophages or one bacterial source, each with a different genome size. As can be seen, as genome size increases, the curves obtained are shifted farther and farther to the right, indicative of an extended reassociation time.

As shown in Figure 2–25, $C_0t_{1/2}$ is directly proportional to the size of the genome. Reassociation occurs at a reduced rate in larger genomes because it takes longer for initial matches if there are greater numbers of unique DNA fragments. This is so because collisions are random; the more sequences are present, the greater the number of mismatches before all correct matchings occur. The method has been useful in assessing genome size in viruses and bacteria.

When reassociation kinetics of DNA from eukaryotic organisms (whose genome sizes are much greater that phage or bacteria) were first studied, a surprising observation was made. Rather than exhibiting a reduced rate of reassociation, the data revealed that *some* DNA segments reassociate even

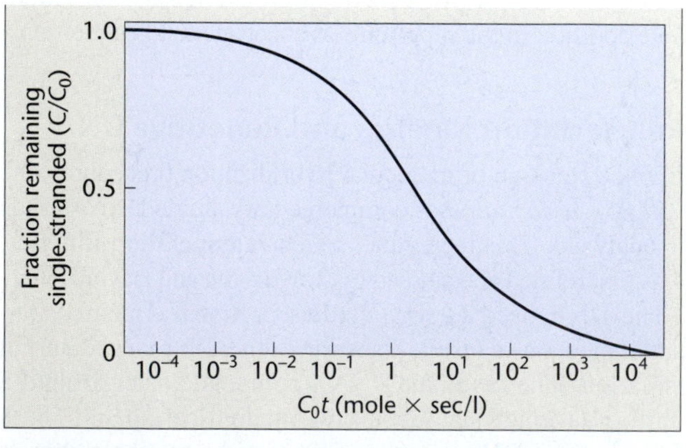

FIGURE 2–23 The ideal time course for reassociation of DNA (C/C_0) when, at time zero, all DNA consists of unique fragments of single-stranded complements. Note that the abscissa (C_0t) is plotted logarithmically.

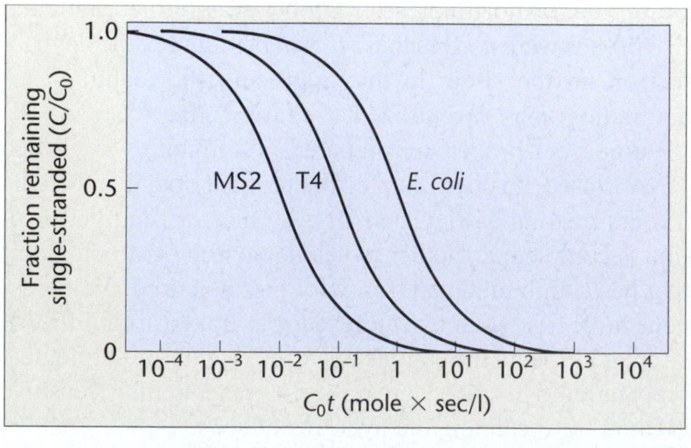

FIGURE 2–24 The reassociation rates (C/C_0) of DNA derived from phage MS2, phage T4, and *E. coli*. The genome of T4 is larger than MS2 and that of *E. coli* is larger than T4.

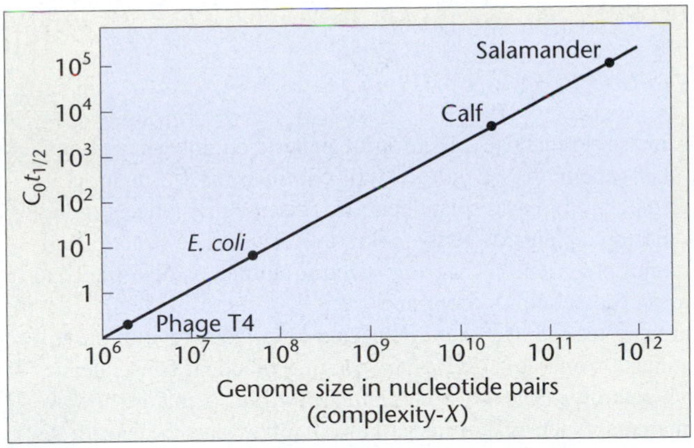

FIGURE 2–25 Comparison of $C_0t_{1/2}$ and genome size for phage T4, *E. coli*, calf, and salamander.

more rapidly than those derived from *E. coli*. The remaining DNA, as expected because of its greater complexity, took longer to reassociate. For example, Britten and Kohne examined DNA derived from calf thymus tissue (Figure 2–26). Based on these observations, they hypothesized that the rapidly reassociating fraction must represent **repetitive DNA sequences** present many times in the calf genome. This interpretation would explain why these DNA segments reassociate so rapidly. Multiple copies of the same sequence are much more likely to make matches, thus reassociating more quickly than single copies. On the other hand, they hypothesized that the remaining DNA segments consist of unique nucleotide sequences present only once in the genome; because there are more of these unique sequences, increasing the DNA complexity in calf thymus (compared with *E. coli*), their reassociation takes longer. The *E. coli* curve has been added to Figure 2–26 for the sake of comparison.

It is now clear that repetitive DNA sequences are prevalent in the genome of eukaryotes and is key to our understanding of how genetic information is organized in chromosomes. Careful study has shown that various levels of repetition exist. In some cases, short DNA sequences are repeated over a

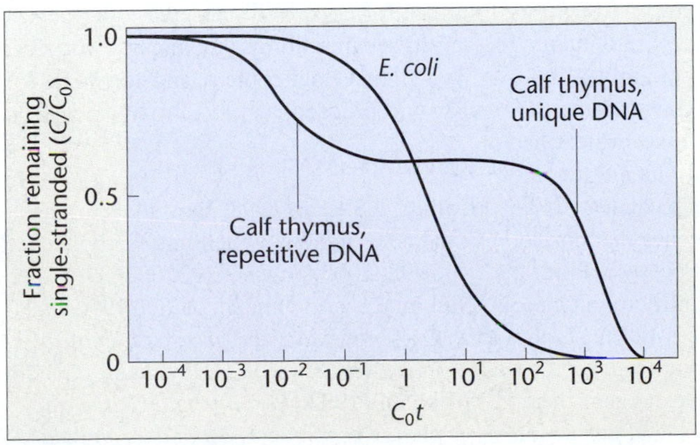

FIGURE 2–26 The C_0t curve of calf thymus DNA compared with *E. coli*. The repetitive fraction of calf DNA reassociates more quickly than that of *E. coli*, while the more complex unique calf DNA takes longer to reassociate than that of *E. coli*.

million times. In other cases, longer sequences are repeated only a few times, or intermediate levels of sequence redundancy are present. We will return to this topic in Chapter 4, where we will discuss the organization of DNA in genes and chromosomes. For now, we conclude the chapter by pointing out that the discovery of repetitive DNA was one of the first clues that much of the DNA in eukaryotes is not contained in genes that encode proteins. This concept will be developed and elaborated on as we expand our coverage of the molecular basis of heredity.

Electrophoresis of Nucleic Acids

We conclude the chapter by considering an essential technique involved in the analysis of nucleic acids, **electrophoresis**. This technique separates different-sized fragments of DNA and RNA chains and is invaluable in current research investigations in molecular genetics.

In general, electrophoresis separates, *or resolves*, molecules in a mixture by causing them to migrate under the influence of an electric field. A sample is placed on a porous substance (a piece of filter paper or a semisolid gel), which is placed in a solution that conducts electricity. If two molecules have approximately the same shape *and* mass, the one with the greatest net charge will migrate more rapidly toward the electrode of opposite polarity.

As electrophoretic technology developed from its initial application to protein separation, researchers discovered that using gels of varying pore sizes significantly improved the resolution of this research technique. Such advance is particularly useful for mixtures of molecules with a similar charge:mass ratio, but different sizes. For example, two polynucleotide chains of different *lengths* (e.g., 10 vs. 20 nucleotides) are both negatively charged based on the phosphate groups of the nucleotides. While they both move to the positively charged pole (the anode), the charge:mass ratio is the same for both chains, and separation based strictly on the electric field is minimal. However, using a porous medium such as **polyacrylamide gels** or **agarose gels**, which can be prepared with various pore sizes, allows us to separate these two molecules.

In such cases, *the smaller molecules migrate at a faster rate through the gel than the larger molecules* (Figure 2–27). The key to separation is based on the matrix (pores) of the gel, which restricts migration of larger molecules more than it restricts smaller molecules. The resolving power is so great that polynucleotides that vary by even one nucleotide in length are clearly separated. Once electrophoresis is complete, bands representing the variously sized molecules are identified either by autoradiography (if a component of the molecule is radioactive) or by the use of a fluorescent dye that binds to nucleic acids.

Electrophoretic separation of nucleic acids is at the heart of a variety of commonly used research techniques discussed later in the text (Chapters 16 and 18). Of particular note are the various "blotting" techniques (e.g., Southern blots and Northern blots), as well as DNA sequencing methods.

Genetics, Technology, and Society

Genetics and Society in the New Millennium

Since the beginning of human civilization, we have defined ourselves as the masters of the biological world. Civilization began when humans domesticated plants and animals and settled into societies—as recently as 12 millennia ago. Genetics, in the form of selective breeding, became the foundation of agricultural progress and contributed to the rise and fall of civilizations over thousands of years. Our ability to harness nature is reflected in religions and philosophies that place humans at the center of the universe, at the pinnacle of creation, above all other creatures.

Now, as we enter the 21st century, a revolution of biological thought has begun—a revolution that is changing both our mastery over living things, and our perception of ourselves.

The revolution began in 1953, with Watson and Crick's elucidation of the molecular structure of DNA. The structure of the DNA molecule immediately provided elegant solutions to age-old questions about the mechanisms of heredity, mutation, and evolution. Some of the greatest mysteries of life could suddenly be explained by the beauty and simplicity of a chemical that replicates and shuffles the code of life.

Over the next 30 years, DNA became the focus of laboratories around the world. Geneticists, biochemists, and molecular biologists quickly devised methods to purify, mutate, cut, and paste DNA in the test tube. DNA molecules from one organism were spliced into DNA molecules from another, and the chimeric molecules were introduced into bacteria or cells in culture. The nucleotide sequences of genes were defined and modified in vitro. Gene promoters, gene transcription units, and gene activities were assayed. The genetic traits of organisms such as bacteria, fungi, and fruit flies were modified by the removal or addition of genes from similar, or different, species. Genetic engineering had begun.

The DNA revolution has advanced at an explosive rate. In the 27 years since the first gene was cloned, scientists have discovered the genes that control hereditary diseases such as sickle-cell anemia, cystic fibrosis, and Tay–Sachs disease. Prenatal diagnosis for these and other genetic diseases is now available. Biotechnologists have genetically modified bacteria, plants, and animals to express proteins of agricultural and medical importance. Biotechnology now offers DNA forensic tests that have helped convict criminals, exonerate the innocent, and establish paternity. Scientists have even cloned mammals such as sheep and mice from adult somatic cells. DNA manipulation is now a powerful tool of medical research. Scientists are rapidly dissecting the molecular genetic mechanisms that control cell growth, aging, death, and cancer.

The effect that the DNA revolution has had on our views of ourselves and the world is reflected in everyday culture. Although scientists dismiss the idea that humans are simply the products of their genes, popular culture endows DNA with almost magical powers. In television situation comedies, magazines, and daily conversation, genes are said to explain personality, career choice, criminality, intelligence—even fashion preferences and political attitudes. Advertisements hijack the language of genetics in order to grant inanimate objects a "genealogy" or "genetic advantage." Popular culture speaks of DNA as an immortal force, with the ability to affect morality and fate. DNA is defined as the "essence of life" and the "immortal text," with the power to shape our future. Biological or genetic explanations for antisocial behavior appear to have more resonance for us than explanations involving social or economic factors.

But what of the future? Can we predict how DNA and genetics will help us shape the world in the new millennium? Although prophecy is certainly a risky business, some developments seem assured. The Human Genome Project has now decoded the entire human genome (ahead of schedule), which will lead to the identification of many genes that control normal and abnormal processes. In turn, this may enhance our ability to diagnose and predict genetic diseases. Scientists predict that the new millennium will bring us biotechnologies as complex as gene therapy and prenatal diagnosis and correction of genetic defects. Undoubtedly, the application of genetic engineering to agriculture will continue, as we manipulate plant genes for disease resistance, productivity, color, and flavor. The genetic engineering of farm animals is also likely to continue.

It is clear that the DNA revolution will have far-reaching practical consequences for humanity. It will also change how we think about ourselves and the world. As many more human genes are discovered, sequenced, and compared to those of other animals, it will become increasingly evident that we are closely related genetically to the rest of the animate world. The nucleotide sequence of our genome differs only about 1 percent from that of chimpanzees, and some of our genes are virtually identical to homologous genes in plants, animals, and bacteria. Will this knowledge alter our relationships with animals and with each other, as we realize the extent of our genetic kinship? As more genes are discovered that contribute to phenotypic traits as simple as eye color and as complicated as intelligence or sexual orientation, will this lead us to define ourselves more as genetic beings and less as creatures of free will or as the products of our environment?

As this millennium unfolds, we will inevitably be faced with the practical and philosophical consequences of the DNA revolution. Will society harness DNA for everyone's benefit, or will the new genetic knowledge be used as a vehicle for discrimination? At the same time that modern genetics grants us more dominion over life, will it paradoxically increase our feelings of powerlessness? Will our new genetic view of life increase our compassion for all life forms, or will it increase our perceived separation from the natural world? We will make our choices, and human history will proceed.

References:

Collins, F.S. et al. 1998. New goals for the U.S. human genome project: 1998–2003. *Science* 282: 682–89.

Nelkin, D., and Lindee, M.S. 1995. *The DNA mystique, the gene as a cultural icon*. New York: W.H. Freeman.

Wilkie, T. 1993. *Perilous knowledge, the human genome project and its implications*. London, UK: Faber and Faber.

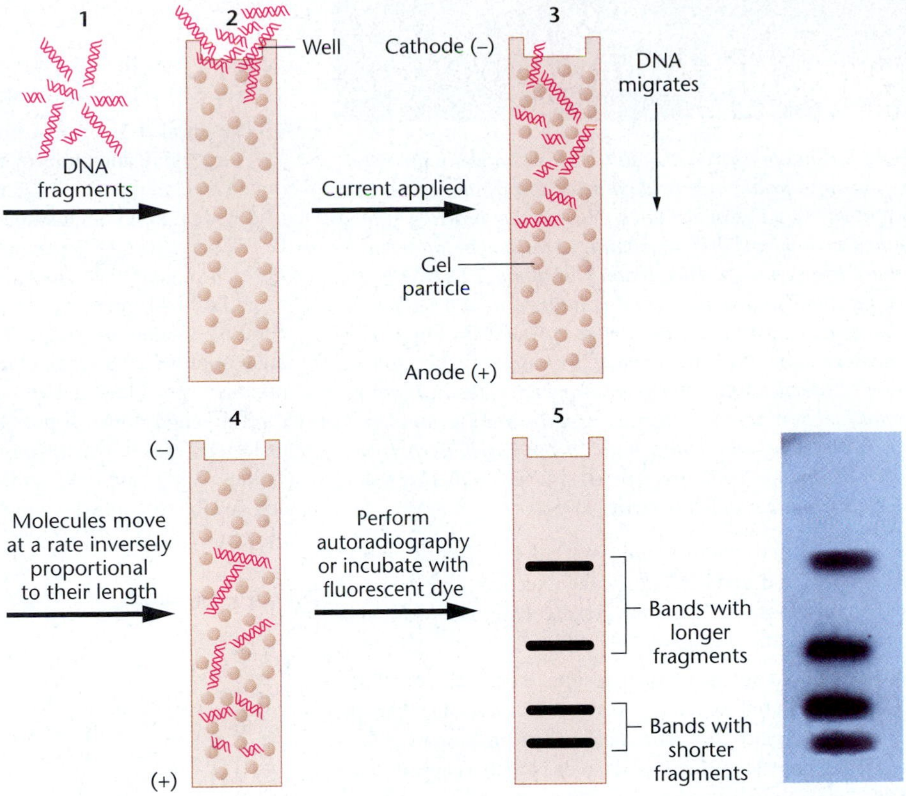

FIGURE 2–27 Electrophoretic separation of a mixture of DNA fragments that vary in length. The photograph shows an agarose gel with DNA bands corresponding to the diagram.

Chapter Summary

1. The existence of a genetic material capable of replication, storage, expression, and mutation is deducible from the observed patterns of inheritance in organisms.
2. Both proteins and nucleic acids were initially considered as the possible candidates for genetic material. Proteins are more diverse than nucleic acids (a requirement for the genetic material) and were favored owing to the advances being made in protein chemistry at the time. Additionally, Levene's tetranucleotide hypothesis had underestimated the magnitude of chemical diversity inherent in nucleic acids.
3. By 1952, transformation studies and experiments using bacteria infected with bacteriophages strongly suggested that DNA is the genetic material in bacteria and most viruses.
4. Initially, only circumstantial evidence supported the concept of DNA controlling inheritance in eukaryotes. These included DNA distribution in the cell, quantitative analysis of DNA, and UV-induced mutagenesis. More recent recombinant DNA techniques, as well as experiments with transgenic mice, have provided direct experimental evidence that the eukaryotic genetic material is DNA.
5. Numerous viruses provide an important exception to this general rule because many of them use RNA as their genetic material. In some viruses, RNA serves as the genetic material. These include bacteriophages as well as some plant and animal viruses.
6. Establishment of DNA as the genetic material paved the way for the expansion of molecular genetics research and has served as the cornerstone for further important studies for nearly half a century.
7. During the late 1940s and early 1950s, a considerable effort was made to integrate accumulated information on the chemical structure of nucleic acids into a model of the molecular structure of DNA. The X-ray crystallography data of Franklin and Wilkins suggested that DNA was some sort of helix. In 1953, Watson and Crick were able to assemble a model of the double-helical DNA structure based on these X-ray diffraction studies as well as on Chargaff's analysis of base composition of DNA.
8. The DNA molecule exhibits antiparallel orientation and adenine–thymine and guanine–cytosine base-pairing complementarity along the polynucleotide chains. This model of DNA presents an obvious straightforward mechanism for its replication. The structure assumed by the helix appears to be a function of the nucleotide sequence and its chemical environment. Several alternative forms of the DNA helical structure exist. Watson and Crick described a B configuration, one of several right-handed helices. Wang and Rich discovered the left-handed Z-DNA currently being investigated for its physiological and genetic significance.
9. The second category of nucleic acids important in genetic function is RNA, which is similar to DNA with the exceptions that it is usually single stranded, the sugar ribose replaces the deoxyribose, and the pyrimidine uracil replaces thymine. Classes of RNA—ribosomal, transfer, and messenger—facilitate the flow of information from DNA to RNA to proteins, which are the end products of most genes.
10. The structure of DNA lends itself to various forms of analyses, which have in turn led to studies of the functional aspects of the genetic machinery. Absorption of UV light, sedimentation properties, denaturation–reassociation, and electrophoresis procedures are among the important tools for the study of nucleic acids. Reassociation kinetics analysis enabled geneticists to postulate the existence of repetitive DNA in eukaryotes, where certain nucleotide sequences are present many times in the genome.

Insights and Solutions

With the initial appearance of the "Insights and Solutions section, it is appropriate to provide a brief description of its value to you in your studies of Genetics. This section provides sample solutions to problems that are similar to those that follow in the ensuing "Problems and Discussion Questions (PDQ)" section. The solutions will provide useful insights and approaches regarding genetic analysis as you engage in the PDQ section. Since the current chapter recounts some of the initial experiments that serve as the cornerstone of molecular genetics, the Insights and Solutions section here emphasizes experimental rationale and analytical thinking, an approach that is the basis of experimental genetics. In later chapters, we will also utilize this section to emphasize genetic problem solving.

1. (a) Based strictly on the analysis of the transformation data of Avery, MacLeod, and McCarty, what objection might be made to the conclusion that DNA is the genetic material? What other conclusion might be considered?

Solution: Based solely on their results, it may be concluded that DNA is essential for transformation. However, DNA might have been a substance that caused capsular formation by *directly* converting nonencapsulated cells to ones with a capsule. That is, DNA may simply have played a catalytic role in capsular synthesis, leading to cells displaying smooth type III colonies.

(b) What observations argue against this objection?

Solution: First, transformed cells pass the trait onto their progeny cells, thus supporting the conclusion that DNA is responsible for heredity, not for the direct production of polysaccharide coats. Second, subsequent transformation studies over a period of five years showed that other traits, such as antibiotic resistance, could be transformed. Therefore, the transforming factor has a broad general effect, not one specific to polysaccharide synthesis. This observation is more in keeping with the conclusion that DNA is the genetic material.

2. If RNA were the universal genetic material, how would it have affected the Avery experiment and the Hershey–Chase experiment?

Solution: In the Avery experiment, digestion of the soluble filtrate with RNase, rather than DNase would have eliminated transformation. Had this occurred, Avery and his colleagues would have concluded that RNA was the transforming factor. The Hershey and Chase results would not have changed, since ^{32}P would also label RNA, but not protein. Had they been using a bacteriophage with RNA as its nucleic acid, and had they known this, they would have concluded that RNA was responsible for directing the reproduction of their bacteriophage.

3. Sea urchin DNA, which is double stranded, was shown to contain 17.5 percent of its bases in the form of cytosine (C). What percentages of the other three bases are present in this DNA?

Solution: The amount of C = G, so guanine is also present as 17.5 percent. The remaining bases, A and T, are present in equal amounts, and together they represent the rest of the bases $(100 - 35)$. Therefore, $A = T = 65/2 = 32.5$ percent.

4. The quest to isolate an important disease-causing organism was successful, and molecular biologists were hard at work. The organism contained as its genetic material a remarkable nucleic acid with a base composition of A=21 percent, C=29 percent, G=29 percent, U=21 percent. When heated, it showed a major hyperchromic effect, and when kinetics were studied, the nucleic acid of this organism provided the C_0t curve shown next, in contrast to that of phage T4 and *E. coli*. T4 contains 10^5 nucleotide pairs and exhibits a $C_0t_{1/2}$ of 0.5. The unknown organism produced a $C_0t_{1/2}$ of 20. Analyze this information carefully, and draw *all* possible conclusions about the genetic material of this organism, based strictly on the preceding observations. What important, straightforward information is missing and needed to confirm your hypothesis about the nature of this molecule?

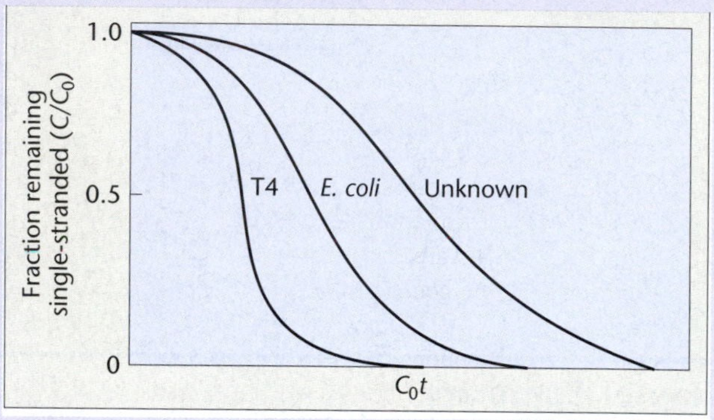

Solution: First of all, because of the presence of uracil (U), the molecule appears to be RNA. As A/U = G/C = 1, the molecule may be a double helix. The hyperchromic shift and reassociation kinetics support this hypothesis. The kinetic study demonstrates several things. First, the shape of the C_0t curve reveals that there is no repetitive sequence DNA. Further, the complexity (X), or total length of unique sequence DNA, is greater than that of either phage T4 or *E. coli*. X can, in fact, be calculated, since a direct proportionality between $C_0t_{1/2}$ and number of base pairs exists when there are only unique sequences present:

$$\frac{0.5}{10^5} = \frac{20}{X}$$

$$0.5X = 20(10^5)$$

$$X = 40(10^5)$$

$$= 4 \times 10^6 \text{ base pairs}$$

There *may* be a greater number of genes present compared to T4 or *E. coli*, but the excessive unique-sequence DNA may serve some other role, or simply play no genetic role. The missing information concerns the sugars. Our model predicts that ribose rather than d-ribose should be present. If not, the organism contains a very unusual molecule as its genetic material.

Problems and Discussion Questions

1. The functions ascribed to the genetic material are replication, expression, storage, and mutation. What does each of these terms mean?
2. Discuss the reasons why proteins were generally favored over DNA as the genetic material before 1940. What was the role of the tetranucleotide hypothesis in this controversy?
3. Contrast the various contributions made to an understanding of transformation by Griffith, by Avery and his colleagues, and by Taylor.
4. When Avery and his colleagues had obtained what was concluded to be purified DNA from the IIIS virulent cells, they treated the fraction with proteases, RNase, and DNase, followed by the assay for retention or loss of transforming ability. What were the purpose and results of these experiments? What conclusions were drawn?
5. Why were ^{32}P and ^{35}S chosen for use in the Hershey–Chase experiment? Discuss the rationale and conclusions of this experiment.
6. Does the design of the Hershey–Chase experiment distinguish between DNA and RNA as the molecule serving as the genetic material? Why or why not?
7. Would an experiment similar to that performed by Hershey and Chase work if the basic design were applied to the phenomenon of transformation? Explain why or why not.
8. What observations are consistent with the conclusion that DNA serves as the genetic material in eukaryotes? List and discuss them.
9. What are the exceptions to the general rule that DNA is the genetic material in all organisms? What evidence supports these exceptions?
10. Draw the chemical structure of the three components of a nucleotide, and then link the three together. What atoms are removed from the structures when the linkages are formed?
11. How are the carbon and nitrogen atoms of the sugars, purines, and pyrimidines numbered?
12. Adenine may also be named 6-amino purine. How would you name the other four nitrogenous bases, using this alternative system? (O is oxy, and CH_3 is methyl.)
13. Draw the chemical structure of a dinucleotide composed of A and G. Opposite this structure, draw the dinucleotide composed of T and C in an antiparallel (or upside-down) fashion. Form the possible hydrogen bonds.
14. Describe the various characteristics of the Watson–Crick double-helix model for DNA.
15. What evidence did Watson and Crick have at their disposal in 1953? What was their approach in arriving at the structure of DNA?
16. What might Watson and Crick have concluded, had Chargaff's data from a single source indicated the following?

	A	T	G	C
%	29	19	21	31

Why would this conclusion be contradictory to Wilkins and Franklin's data?

17. How do covalent bonds differ from hydrogen bonds? Define base complementarity.
18. List three main differences between DNA and RNA.
19. What are the three types of RNA molecules? How is each related to the concept of information flow?
20. What component of the nucleotide is responsible for the absorption of ultraviolet light? How is this technique important in the analysis of nucleic acids?
21. Distinguish between sedimentation velocity and sedimentation equilibrium centrifugation (density gradient centrifugation).
22. What is the basis for determining base composition, using density gradient centrifugation?
23. What is the physical state of DNA following denaturation?
24. Compare the following curves representing reassociation kinetics:

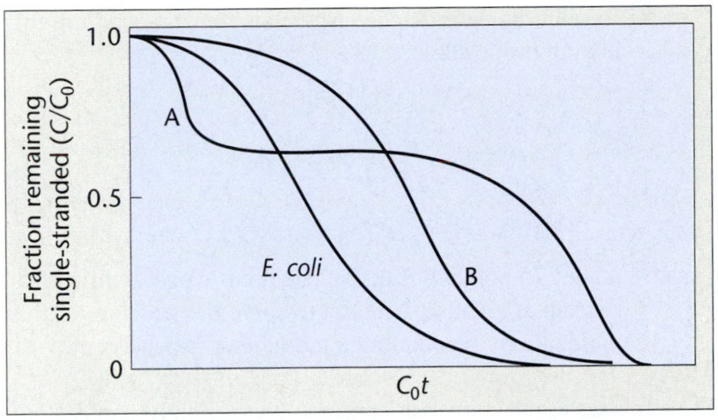

What can be said about the DNAs represented by each set of data compared with *E. coli?*

25. What is the hyperchromic effect? How is it measured? What does T_m imply?
26. Why is T_m related to base composition?
27. What is the chemical basis of molecular hybridization?
28. What did the Watson–Crick model suggest about the replication of DNA?
29. A genetics student was asked to draw the chemical structure of an adenine- and thymine-containing dinucleotide derived from DNA. His answer is shown here:

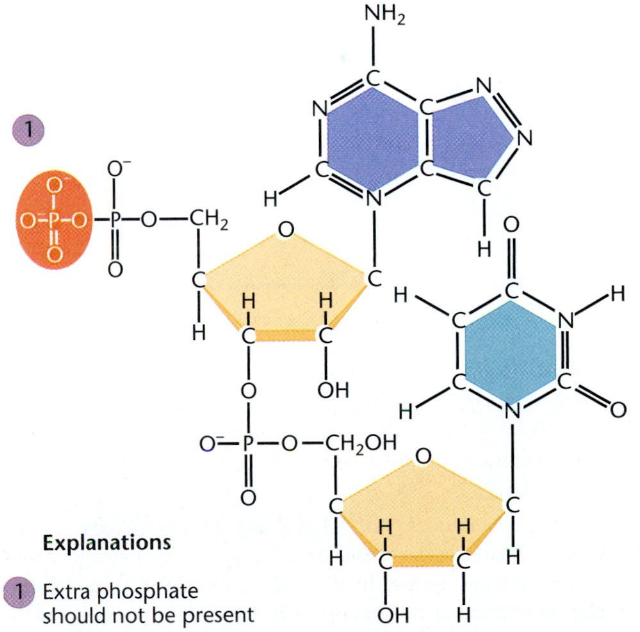

The student made more than six major errors. One of them is circled, numbered 1, and explained. Find five others. Circle them, number them 2–6, and briefly explain each by following the example given.

30. The DNA of the bacterial virus T4 produces a $C_0t_{1/2}$ of about 0.5 and contains 10^5 nucleotide pairs in its genome. How many nucleotide pairs are present in the genome of the virus MS2 and the bacterium *E. coli*, whose respective DNAs produce $C_0t_{1/2}$ values of 0.001 and 10.0?

31. A primitive eukaryote was discovered that displayed a unique nucleic acid as its genetic material. Analysis revealed the following observations:

(i) X-ray diffraction studies display a general pattern similar to DNA, but with somewhat different dimensions and more irregularity.

(ii) A major hyperchromic shift is evident upon heating and monitoring UV absorption at 260 nm.

(iii) Base composition analysis reveals four bases in the following proportions:

Adenine	=	8%
Guanine	=	37%
Xanthine	=	37%
Hypoxanthine	=	18%

(iv) About 75 percent of the sugars are d-ribose, while 25 percent are ribose. Attempt to solve the structure of this molecule by postulating a model that is consistent with the foregoing observations.

32. Considering the information in this chapter on B- and Z-DNA and right- and left-handed helices, carefully analyze the structures here, and draw conclusions about the helical nature of areas (a) and (b). Which is right handed and which is left handed?

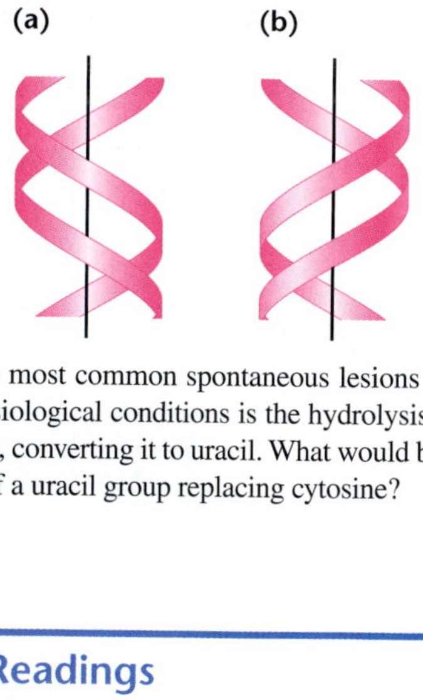

33. One of the most common spontaneous lesions that occurs in DNA under physiological conditions is the hydrolysis of the amino group of cytosine, converting it to uracil. What would be the effect on DNA structure of a uracil group replacing cytosine?

34. In some organisms, cytosine is methylated at carbon 5 of the pyrimidine ring after it is incorporated into DNA. If a 5-methyl cytosine is then hydrolyzed, as described in Problem 33, what base will be generated?

Extra-Spicy Problems

35. *Newsdate: March 1, 2015.* A unique creature has been discovered during exploration of outer space. Recently, its genetic material has been isolated and analyzed. This material is similar in some ways to DNA in its chemical makeup. It contains in abundance the 4-carbon sugar erythrose and a molar equivalent of phosphate groups. Additionally, it contains six nitrogenous bases: adenine (A), guanine (G), thymine (T), cytosine (C), hypoxanthine (H), and xanthine (X). These bases exist in the following relative proportions:

$$A = T = H \text{ and } C = G = X$$

X-ray diffraction studies have established a regularity in the molecule and a constant diameter of about 30 Å. Together, these data have suggested a model for the structure of this molecule.

(a) Propose a general model of this molecule. Describe it briefly.

(b) What base-pairing properties must exist for H and for X in the model?

(c) Given the constant diameter of 30 Å, do you think that *either* (i) both H and X are purines or both pyrimidines, *or* (ii) one is a purine and one is a pyrimidine?

36. You are provided with DNA samples from two newly discovered bacterial viruses. Based on the various analytical techniques discussed in this chapter, construct a research protocol that would be useful in characterizing and contrasting the DNA of both viruses. For each technique that you include in the protocol, indicate the type of information you hope to obtain.

37. During gel electrophoresis, DNA molecules can easily be separated according to size because all DNA molecules have the same charge to mass ratio and the same shape (long rod). Would you expect RNA molecules to behave in the same manner as DNA during gel electrophoresis? Why or why not?

Selected Readings

Adleman, L.M. 1998. Computing with DNA. *Sci. Am.* (Aug.)279:54–61.

Alloway, J.L. 1933. Further observations on the use of pneumococcus extracts in effecting transformation of type *in vitro. J. Exp. Med.* 57:265–78.

Avery, O.T., MacLeod, C.M., and McCarty, M. 1944. Studies on the chemical nature of the substance inducing transformation of pneumococcal types: Induction of transformation by a desoxyribonucleic acid fraction isolated from pneumococcus type III. *J. Exp. Med.* 79:137–58. (Reprinted in Taylor, J.H. 1965. *Selected papers in molecular genetics*. Orlando, FL: Academic Press.)

Britten, R.J., and Kohne, D.E. 1968. Repeated sequences in DNA. *Science* 161:529–40.

Cairns, J., Stent, G.S., and Watson, J.D. 1966. *Phage and the origins of molecular biology*. Cold Spring Harbor, NY: Cold Spring Harbor Laboratory Press.

Chargaff, E. 1950. Chemical specificity of nucleic acids and mechanism for their enzymatic degradation. *Experientia* 6:201–9.

Crick, F.H.C., Wang, J.C., and Bauer, W.R. 1979. Is DNA really a double helix? *J. Mol. Biol.* 129:449–61.

Davidson, J.N. 1976. *The biochemistry of the nucleic acids*, 8th ed. Orlando, FL: Academic Press.

Dawson, M.H. 1930. The transformation of pneumococcal types: I. The interconvertibility of type-specific *S. pneumococci. J. Exp. Med.* 51:123–47.

DeRobertis, E.M., and Gurdon, J.B. 1979. Gene transplantation and the analysis of development. *Sci. Am.* (Dec.) 241:74–82.

Dickerson, R.E. 1983. The DNA helix and how it is read. *Sci. Am.* (June) 249:94–111.

Dickerson, R.E., et al. 1982. The anatomy of A-, B-, and Z-DNA. *Science* 216:475–85.

Dubos, R.J. 1976. *The professor, the Institute and DNA: Oswald T. Avery, his life and scientific achievements.* New York: Rockefeller University Press.

Felsenfeld, G. 1985. DNA. *Sci. Am.* (Oct.) 253:58–78.

Fraenkel-Conrat, H., and Singer, B. 1957. Virus reconstruction: II. Combination of protein and nucleic acid from different strains. *Biochim. Biophys. Acta* 24:530–48.

(Reprinted in Taylor, J.H. 1965. *Selected papers in molecular genetics*, Orlando, FL: Academic Press.)

Franklin, R.E., and Gosling, R.G. 1953. Molecular configuration in sodium thymonucleate. *Nature* 171:740–41.

Geis, I. 1983. Visualizing the anatomy of A, B and Z-DNAs. *J. Biomol. Struc. Dynam.* 1: 581–91.

Griffith, F. 1928. The significance of pneumococcal types. *J. Hyg.* 27:113–59.

Guthrie, G.D., and Sinsheimer, R.L. 1960. Infection of protoplasts of *Escherichia coli* by subviral particles. *J. Mol. Biol.* 2:297–305.

Hershey, A.D., and Chase, M. 1952. Independent functions of viral protein and nucleic acid in growth of bacteriophage. *J. Gen. Phys.* 36:39–56. (Reprinted in Taylor, J.H. 1965. *Selected papers in molecular genetics.* Orlando, FL: Academic Press.)

Hotchkiss, R.D. 1951. Transfer of penicillin resistance in pneumococci by the desoxyribonucleate derived from resistant cultures. *Cold Spring Harbor Symp. Quant. Biol.* 16:457–61. (Reprinted in Adelberg, E.A. 1960. *Papers on bacterial genetics.* Boston: Little, Brown.)

Judson, H. 1979. *The eighth day of creation: Makers of the revolution in biology.* New York: Simon &ampersand; Schuster.

Levene, P.A., and Simms, H.S. 1926. Nucleic acid structure as determined by electrometric titration data. *J. Biol. Chem.* 70:327–41.

McCarty, M. 1980. Reminiscences of the early days of transformation. *Annu. Rev. Genet.* 14:1–16.

———. 1985. *The transforming principle: Discovering that genes are made of DNA.* New York: W. W. Norton.

McConkey, E.H. 1993. *Human genetics—The molecular revolution.* Boston: Jones and Bartlett.

Olby, R. 1974. *The path to the double helix.* Seattle: University of Washington Press.

Palmiter, R.D., and Brinster, R.L. 1985. Transgenic mice. *Cell* 41:343–45.

Pauling, L., and Corey, R.B. 1953. A proposed structure for the nucleic acids. *Proc. Natl. Acad. Sci. USA* 39:84–97.

Rich, A., Nordheim, A., and Wang, A.H.-J. 1984. The chemistry and biology of left-handed Z-DNA. *Annu. Rev. Biochem.* 53:791–846.

Schildkraut, C.L., Marmur, J., and Doty, P. 1962. Determination of the base composition of deoxyribonucleic acid from its buoyant density in CsCl. *J. Mol. Biol.* 4:430–43.

Spizizen, J. 1957. Infection of protoplasts by disrupted T2 viruses. *Proc. Natl. Acad. Sci. USA* 43:694–701.

Stent, G.S., ed. 1981. *The double helix: Text, commentary, review, and original papers.* New York: W. W. Norton.

Stewart, T.A., Wagner, E.F., and Mintz, B. 1982. Human b-globin gene sequences injected into mouse eggs, retained in adults, and transmitted to progeny. *Science* 217:1046–48.

Varmus, H. 1988. Retroviruses. *Science* 240:1427–35.

Watson, J.D. 1968. *The double helix.* New York: Atheneum.

Watson, J.D., and Crick, F.C. 1953a. Molecular structure of nucleic acids: A structure for deoxyribose nucleic acids. *Nature* 171:737–38.

———1953b. Genetic implications of the structure of deoxyribose nucleic acid. *Nature* 171:964.

Weinberg, R.A. 1985. The molecules of life. *Sci. Am.* (Oct.) 253:48–57.

Wilkins, M.H.F., Stokes, A.R., and Wilson, H.R. 1953. Molecular structure of desoxypentose nucleic acids. *Nature* 171:738–40.

Yung, J. 1996. New FISH probes—The end in sight. *Nature Genetics* 14:10–12.

Zimmerman, B. 1982. The three-dimensional structure of DNA. *Annu. Rev. Biochem.* 51:395–428.

GENETICS MediaLab

The resources that follow will help you achieve a better understanding of the concepts presented in this chapter. These resources can be found either on the CD packaged with this textbook or on the Companion Web site found at **http://www.prenhall.com/klug.**

CD Resources:

Module 2.1: DNA Structure

Web Problem 1:

Time for completion = 10 minutes

How is the DNA molecule suited for its function of encoding genetic information? The DNA molecule is simply a polymer of four kinds of nucleotides, and yet this molecule encodes information for diverse processes in a myriad of distinct organisms. In this exercise, you will examine the structure of DNA by viewing four interactive modules illustrating different aspects of DNA structure. You can manipulate each three-dimensional model with buttons that rotate, zoom in on and out from, highlight, and erase structural elements. Discuss how DNA is both uniform and variable. Include how the backbone, bases, major and minor grooves, and 5' and 3' ends of strands affect these qualities. To complete this exercise, visit Web Problem 1 in Chapter 2 of your Companion Web site, and select the keyword **DNA**.

Web Problem 2:

Time for completion = 10 minutes

In what ways do forms of DNA differ? Data from X-ray crystallography has contributed to an understanding of the three-dimensional properties of DNA. In this exercise, you will examine molecular models of nucleic acids for the A, B, and Z forms of DNA. Read the brief introduction to nucleic acids. Click on the hypertext links for A, B, and Z DNA to view the images. List the differences among these three major forms of DNA. Describe the effect of a base-pair mismatch. To complete this exercise, visit Web Problem 2 in Chapter 2 of your Companion Web site, and select the keywords **FORMS** and **DNA MODEL**. For a brief overview of the X-ray crystallography process, select the keyword **X-RAY**.

Web Problem 3:

Time for completion = 10 minutes

Many molecular biology methods are dependent on complementarity between strands of DNA or between DNA and RNA, but what is the relationship between sequences and denaturation? The melting temperature, T_m, of a sequence provides a measure of how easily a double-stranded sequence will denature. In this exercise, you will calculate the estimated T_m and other parameters for a given sequence. Enter a sequence in the query box, using the letters A, G, C, and T to represent the nucleotides. Click on the calculate button to perform your analysis. Repeating this process, design a set of input sequences to answer the following questions: What is the relation between T_m and the length of a sequence? Between T_m and GC content? What can you conclude about the optical density (OD) and molecular weight (MW) of different nucleotides? To complete this exercise, visit Web Problem 3 in Chapter 2LLL of your Companion Web site, and select the keyword **TM**.

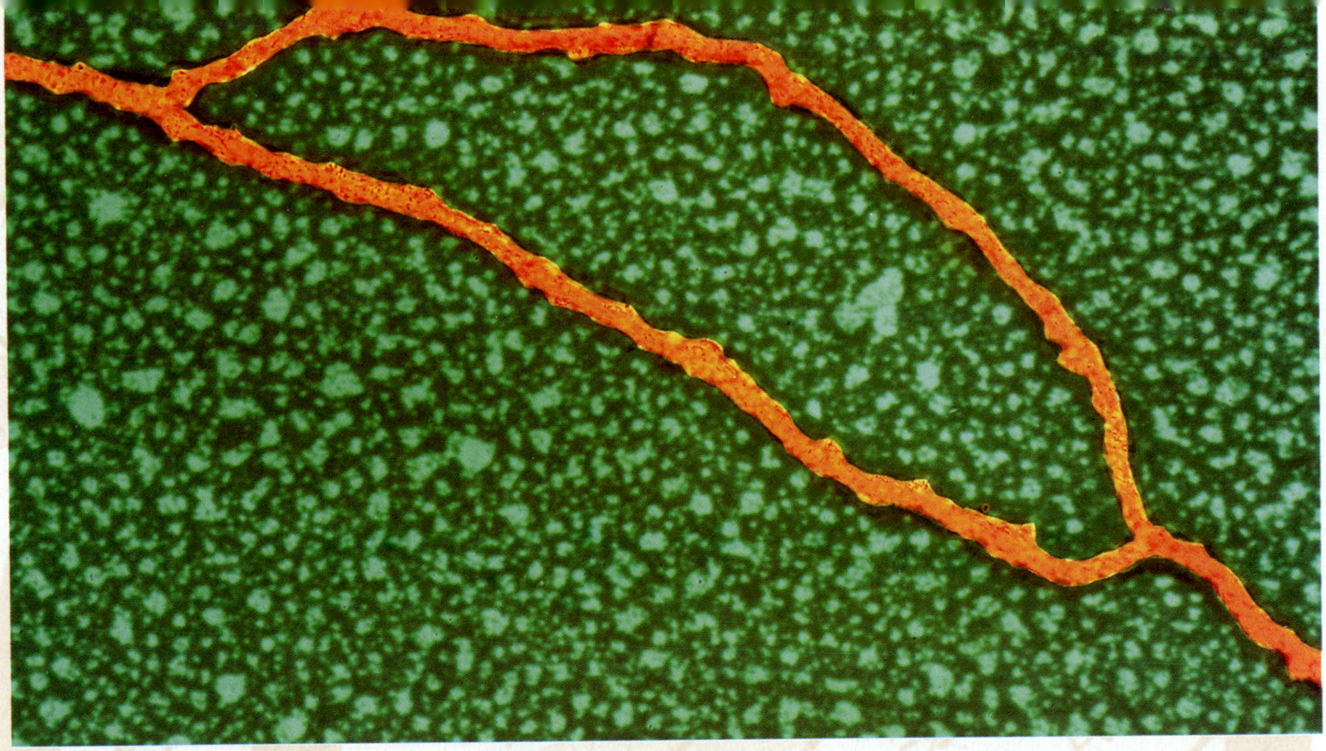

Transmission electron micrograph of human DNA from a HeLa cell, illustrating the replication bubble that characterizes DNA replication within a single replicon. (*Dr. Gopal Murti/Science Photo Library/Photo Researchers, Inc.*)

3

DNA Replication and Recombination

KEY CONCEPTS

Following Watson and Crick's proposal for the structure of DNA, scientists focused their attention on how this molecule is replicated. Replication is an essential function of the genetic material and must be executed precisely if genetic continuity between cells is to be maintained following cell division. It is an enormous, complex task. Consider for a moment that over 3×10^9 (3 billion) base pairs exist within the 23 chromosomes of the human genome. To duplicate faithfully the DNA of just one of these chromosomes requires a mechanism of extreme precision. Even an error rate of only 10^{-6} (one in a million) will still create 3000 errors (obviously an excessive number) during each replication cycle of the genome. While it is not error free, an extremely accurate system of DNA replication has evolved in all organisms.

As Watson and Crick noted in their 1953 paper, the model of the double helix provided their initial insight into how replication could occur. This mode, called *semiconservative replication*, is strongly supported from numerous studies of viruses, prokaryotes, and eukaryotes. Once the general mode of replication was clarified, research to determine the precise details of DNA synthesis intensified. What has since been discovered is that numerous enzymes and other proteins are needed to copy a DNA helix. Because of the complexity of the chemical events during synthesis, this subject remains an extremely active area of research.

In this chapter, we will discuss the general mode of replication, as well as the specific details of DNA synthesis. The research leading to such knowledge is another link in our understanding of life processes at the molecular level.

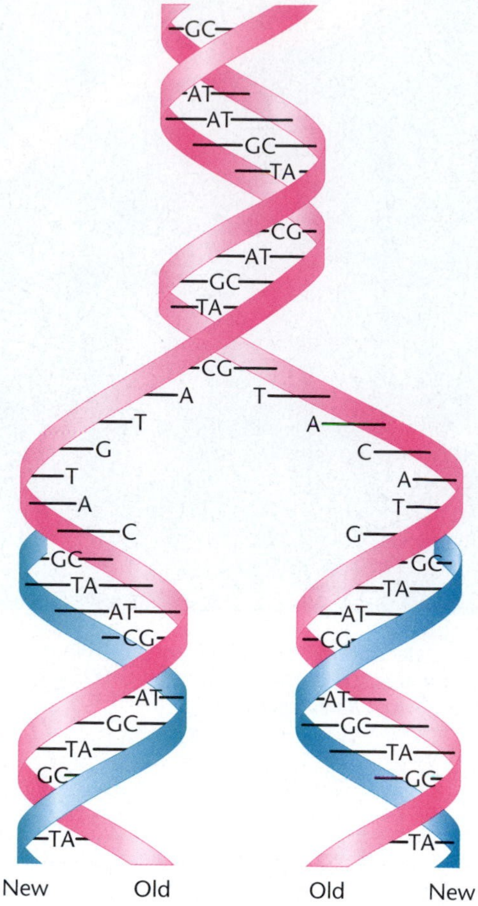

FIGURE 3–1 Generalized model of semiconservative replication of DNA. New synthesis is shown in teal.

3.1 DNA Is Reproduced by Semiconservative Replication

It was apparent to Watson and Crick that, because of the arrangement and nature of the nitrogenous bases, each strand of a DNA double helix could serve as a template for the synthesis of its complement (Figure 3–1). They proposed that, if the helix were unwound, each nucleotide along the two parent strands would have an affinity for its complementary nucleotide. As we learned in Chapter 2, the complementarity is due to the potential hydrogen bonds that can be formed. If thymidylic acid (T) were present, it would "attract" adenylic acid (A); if guanidylic acid (G) were present, it would "attract" cytidylic acid (C); likewise, A would attract T, and C would attract G. If these nucleotides were then covalently linked into polynucleotide chains along both templates, the result would be the production of two identical double strands of DNA. Each replicated DNA molecule would consist of one "old" and one "new" strand, hence the reason for the name **semiconservative replication**.

Two other possible modes of replication also rely on the parental strands as a template (Figure 3–2). In **conservative replication**, complementary polynucleotide chains are synthesized as described earlier. Following synthesis, however, the two newly created strands then come together and the parental strands reassociate. The original helix is thus "conserved."

In the second alternative mode, called **dispersive replication**, the parental strands are dispersed into two new double helices following replication. Hence, each strand consists of both old and new DNA. This mode would involve cleavage of the parental strands during replication. It is the most complex of the three possibilities and is therefore considered to be least likely to occur. It could not, however, be ruled out as an experimental model. Figure 3–2 shows the theoretical results of a single round of replication by each of the three different modes.

The Meselson–Stahl Experiment

In 1958, Matthew Meselson and Franklin Stahl published the results of an experiment providing strong evidence that semiconservative replication is the mode used by bacterial cells to produce new DNA molecules. They grew *E. coli* cells for many generations in a medium where $^{15}NH_4Cl$ (ammonium chloride) was the only nitrogen source. A "heavy" isotope of nitrogen, ^{15}N, contains one more neutron than the naturally occurring ^{14}N isotope. Unlike radioactive isotopes, ^{15}N is stable. After many generations, all nitrogen-containing

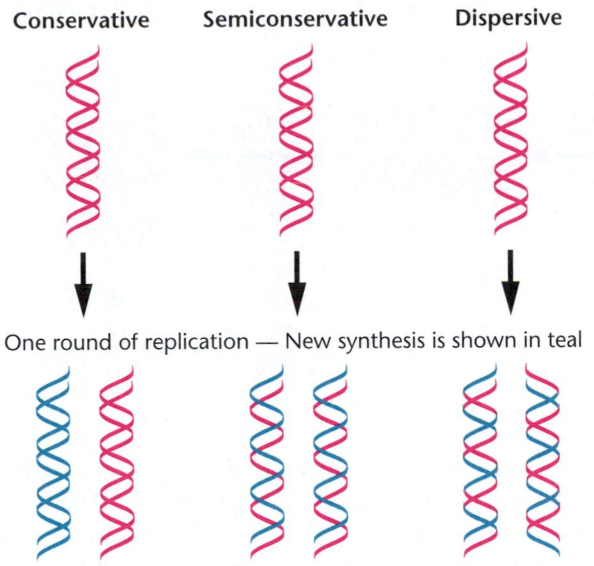

FIGURE 3–2 Results of one round of replication of DNA for each of the three possible modes by which replication could be accomplished.

molecules, including the nitrogenous bases of DNA, contained the heavier isotope in the *E. coli* cells.

Critical to the success of this experiment, DNA containing ^{15}N can be distinguished from ^{14}N-containing DNA. The experimental procedure involves the use of a technique referred to as **sedimentation equilibrium centrifugation**. Samples are "forced" by centrifugation through a density gradient of a heavy metal salt, such as cesium chloride. (See Chapter 9.) The more dense ^{15}N-DNA will reach equilibrium in the gradient at a point closer to the bottom (where the density is greater) than will ^{14}N-DNA.

In this experiment (Figure 3–3), uniformly labeled ^{15}N cells were transferred to a medium containing only $^{14}NH_4Cl$. Thus, all "new" synthesis of DNA during replication contained only the "lighter" isotope of nitrogen. The time of transfer to the new medium was taken as time zero ($t = 0$). The *E. coli* cells were allowed to replicate over several generations, with cell samples removed after each replication cycle. DNA was isolated from each sample and subjected to sedimentation equilibrium centrifugation.

After one generation, the isolated DNA was present in only a single band of intermediate density—the expected result for semiconservative replication in which each replicated molecule was composed of one new ^{14}N-strand and one old ^{15}N-strand (Figure 3–4). This result was not consistent with

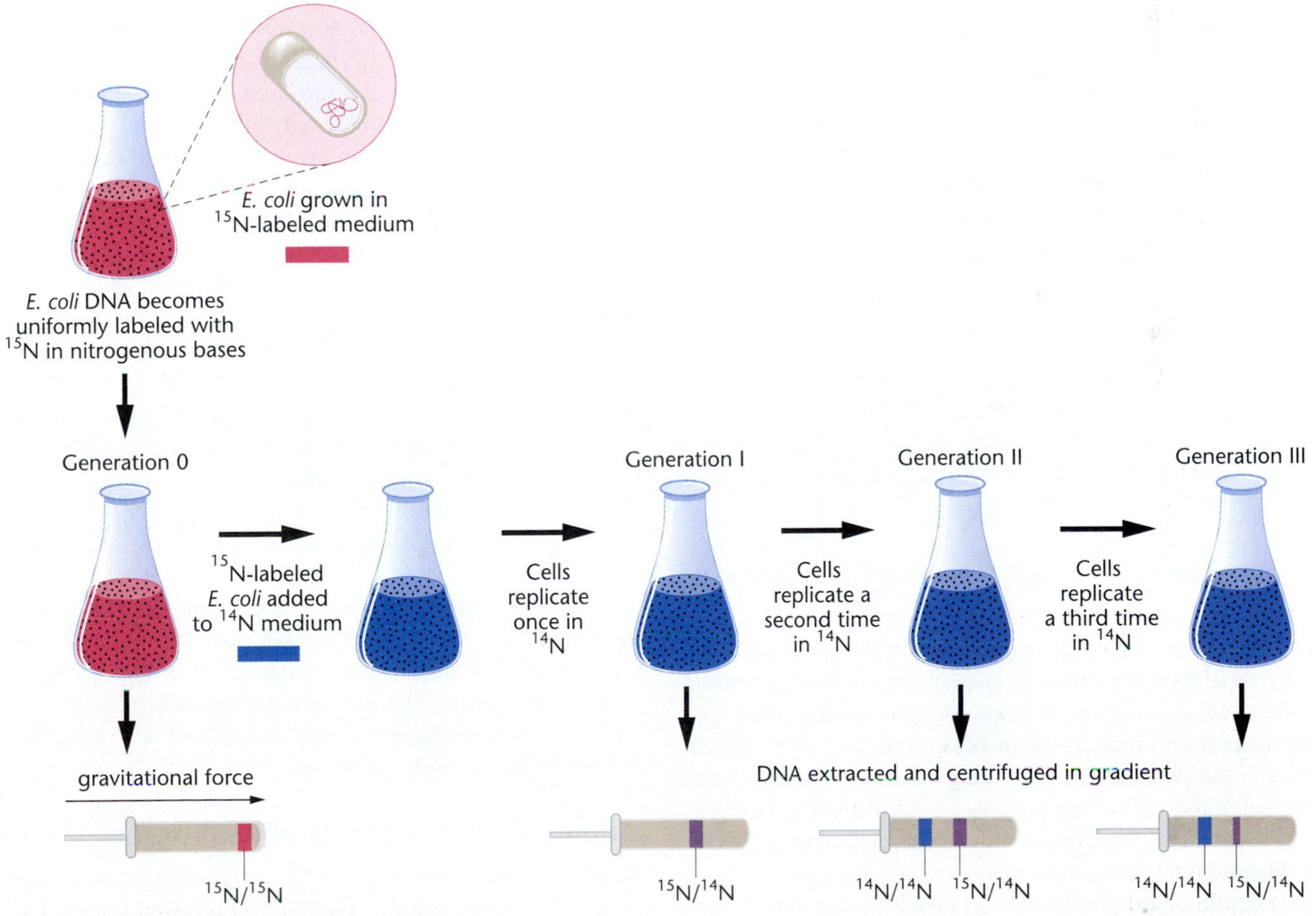

FIGURE 3–3 The Meselson–Stahl experiment.

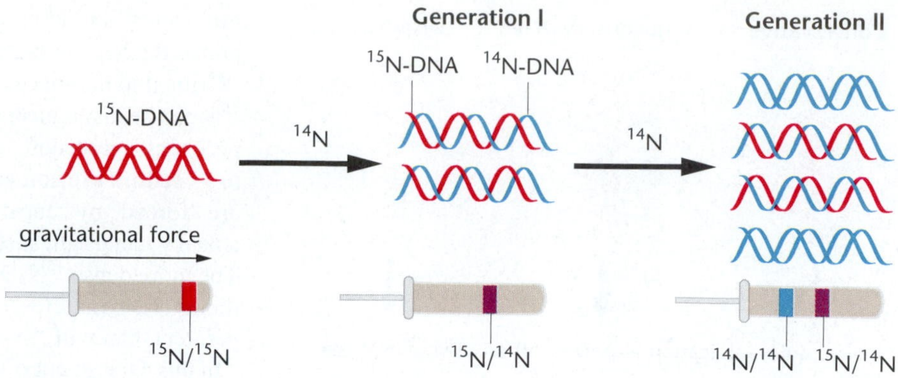

FIGURE 3–4 The expected results of two generations of semiconservative replication in the Meselson–Stahl experiment.

conservative replication, in which two distinct bands would have been predicted to occur.

After two cell divisions, DNA samples showed two density bands—one intermediate band and one lighter band corresponding to the ^{14}N position in the gradient. Similar results occurred after a third generation, except that the proportion of the ^{14}N band increased. This was again consistent with the interpretation that replication is semiconservative.

You may have realized that a molecule exhibiting intermediate density is also consistent with dispersive replication. However, Meselson and Stahl ruled out this mode of replication on the basis of two observations. First, after the first generation of replication in an ^{14}N-containing medium, they isolated the hybrid molecule and heat denatured it. Recall from Chapter 2 that heating will separate a duplex into single strands. When the densities of the single strands of the hybrid were determined, they exhibited *either* an ^{15}N profile *or* an ^{14}N profile, but *not* an intermediate density. This observation is consistent with the semiconservative mode, but inconsistent with the dispersive mode.

Furthermore, if replication were dispersive, *all* generations after $t = 0$ would demonstrate DNA of an intermediate density. In each generation after the first, the ratio of ^{15}N/^{14}N would decrease, and the hybrid band would become lighter and lighter, eventually approaching the ^{14}N band. This result was not observed. The Meselson–Stahl experiment provided conclusive support for semiconservative replication in bacteria and tended to rule out both the conservative and dispersive modes.

Semiconservative Replication in Eukaryotes

In 1957, the year before the work of Meselson and his colleagues was published, J. Herbert Taylor, Philip Woods, and Walter Hughes presented evidence that semiconservative replication also occurs in eukaryotic organisms. They experimented with root tips of the broad bean *Vicia faba*, which are an excellent source of dividing cells. These researchers were able to monitor the process of replication by labeling DNA with ^{3}H-thymidine, a radioactive precursor of DNA, and performing autoradiography.

Autoradiography, discussed in detail in Appendix A, is a common technique that, when applied cytologically, pinpoints the location of a radioisotope in a cell. In the procedure, a photographic emulsion is placed over a section of cellular material (root tips in this experiment), and the preparation is stored in the dark. The slide is then developed, much as photographic film is processed. Because the radioisotope emits energy, the emulsion turns black at the approximate point of emission following development. The end result is the presence of dark spots or "grains" on the surface of the section, identifying the location of newly synthesized DNA within the cell.

Taylor and his colleagues grew root tips for approximately one generation in the presence of the radioisotope and then placed them in unlabeled medium in which cell division continued. At the conclusion of each generation, they arrested the cultures at metaphase by adding colchicine (a chemical derived from the crocus plant that poisons the spindle fibers) and then examined the chromosomes by autoradiography. They found labeled thymidine only in association with chromatids that contained newly synthesized DNA. Figure 3–5 illustrates the replication of a single chromosome over two division cycles, including the distribution of grains.

These results are compatible with the semiconservative mode of replication. After the first replication cycle in the presence of the isotope, both sister chromatids show radioactivity, indicating that each chromatid contains one new radioactive DNA strand and one old unlabeled strand. After the second replication cycle, *which also takes place in unlabeled medium*, only one of the two sister chromatids of each chromosome should be radioactive, because half of the parent strands are unlabeled. With only the minor exceptions of sister chromatid exchanges (see Chapter 12), this result was observed.

Together, the Meselson–Stahl experiment and the experiment by Taylor, Woods, and Hughes soon led to the general acceptance of the semiconservative mode of replication. Later studies with other organisms reached the same conclusion and also strongly supported Watson and Crick's proposal for the double-helix model of DNA.

Origins, Forks, and Units of Replication

To enhance our understanding of semiconservative replication, let's briefly consider a number of relevant issues. The first concerns the **origin of replication**. Where along the chromosome is DNA replication initiated? Is there only a

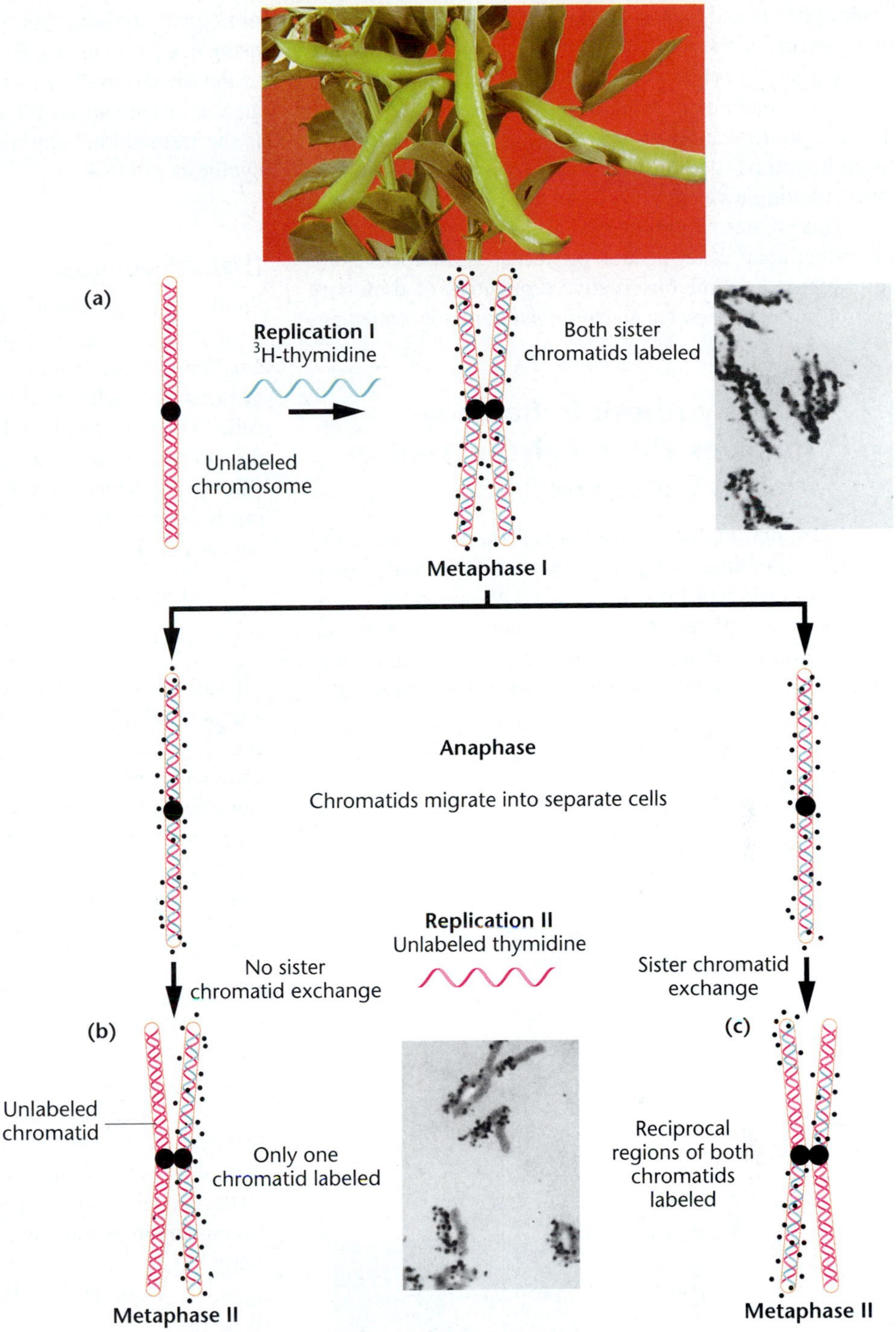

FIGURE 3–5 The Taylor–Woods–Hughes experiment, demonstrating the semiconservative mode of replication of DNA in root tips of *Vicia faba*. A portion of the plant is shown in the top photograph. (a) An unlabeled chromosome proceeds through the cell cycle in the presence of ^{3}H-thymidine. As it enters mitosis, both sister chromatids of the chromosome are labeled, as shown by autoradiography. After a second round of replication (b), this time in the absence of ^{3}H-thymidine only one chromatid of each chromosome is expected to be surrounded by grains. Except where a reciprocal exchange has occurred between sister chromatids (c), the expectation was upheld. The micrographs are of the actual autoradiograms obtained in the experiment.

single origin, or does DNA synthesis begin at more than one point? Is any given point of origin random, or is it located at a specific region along the chromosome? Second, once replication begins, does it proceed in a single direction or in both directions away from the origin? In other words, is replication **unidirectional** or **bidirectional**?

To address these issues, we need to introduce two terms. First, at each point along the chromosome where replication is occurring, the strands of the helix are unwound, creating what is called a **replication fork**. Such a fork will initially appear at the point of origin of synthesis and then move along the DNA duplex as replication proceeds. If replication is bidirectional, two such forks will be present, migrating in opposite directions away from the origin. Second, the length of DNA that is replicated following one initiation event at a single origin is a unit referred to as the **replicon**.

The evidence is clear regarding the origin and direction of replication. John Cairns tracked replication in *E. coli*, using both radioisotopes and autoradiography. He was able to demonstrate that there is only a single region in which replication is initiated. In *E. coli*, this specific region, called *oriC*, has been mapped along the chromosome. It consists of 245 base pairs, though only a small number are actually essential to the initiation of DNA synthesis. Since DNA synthesis in bacteriophages and bacteria originates at a single point, the entire chromosome constitutes one replicon. The presence of only a single origin is characteristic of bacteria, which have only one circular chromosome.

Results put forward by other researchers, again relying on autoradiography, demonstrated that replication is bidirectional, moving away from *oriC* in both directions (Figure 3–6). This creates two replication forks that migrate farther and farther apart as replication proceeds. These forks eventually merge as semiconservative replication of the entire chromosome is completed at a termination region, called *ter*.

3.2 DNA Synthesis in Bacteria Involves Three Polymerases, as well as Other Enzymes

The determination that replication is semiconservative and bidirectional indicates only the *pattern* of DNA duplication and the association of finished strands with one another once synthesis is completed. A more complex issue is how the actual *synthesis* of long complementary polynucleotide chains occurs on a DNA template. As in most molecular biological studies, this question was first approached by using microorganisms. Research began about the same time as the Meselson–Stahl work, and the topic is still an active area of investigation. What is most apparent in this research is the tremendous chemical complexity of the biological synthesis of DNA.

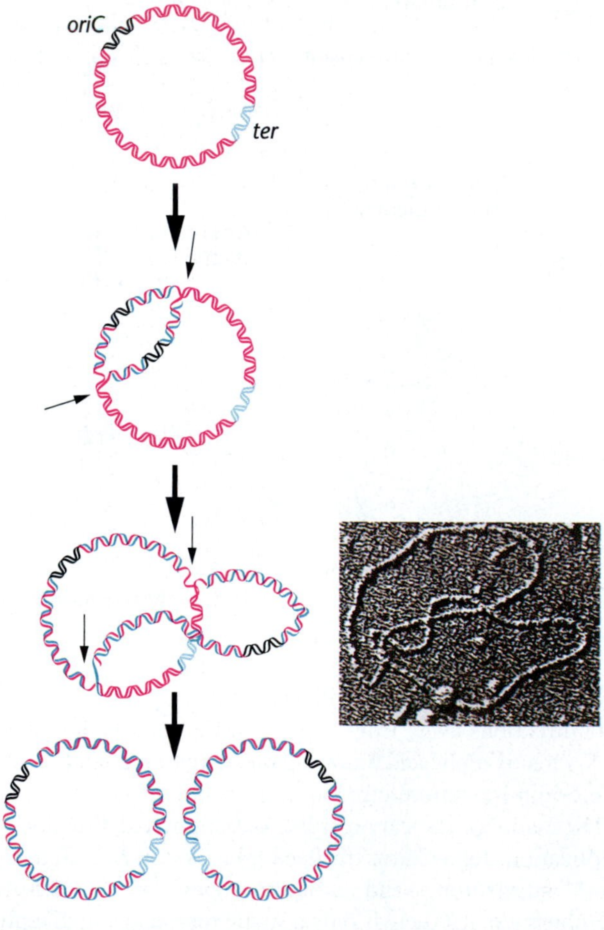

FIGURE 3–6 Bidirectional replication of the *E. coli* chromosome. The thin black arrows identify the advancing replication forks. The micrograph is of a bacterial chromosome in the process of replication, comparable to the figure next to it.

DNA Polymerase I

Studies of the enzymology of DNA replication were first reported by Arthur Kornberg and colleagues in 1957. They isolated an enzyme from *E. coli* that was able to direct DNA synthesis in a cell-free (in vitro) system. The enzyme is now called **DNA polymerase I**, as it was the first of several similar enzymes to be isolated.

Kornberg determined that there were two major requirements for in vitro DNA synthesis under the direction of DNA polymerase I:

1. all four deoxyribonucleoside triphosphates (dNTPs)*
2. template DNA

If any one of the four deoxyribonucleoside triphosphates was omitted from the reaction, no measurable synthesis occurred. If derivatives of these precursor molecules other than the nucleoside triphosphate were used (nucleotides or nucleoside diphosphates), synthesis also did not occur. If no template DNA was added, synthesis of DNA occurred, but was reduced greatly.

Most of the synthesis directed by Kornberg's enzyme appeared to be exactly the type required for semiconservative replication. The reaction is summarized in Figure 3–7, which depicts the addition of a single nucleotide. The enzyme has since been shown to consist of a single polypeptide containing 928 amino acids.

The way in which each nucleotide is added to the growing chain is a function of the specificity of DNA polymerase I. As shown in Figure 3–8, the precursor dNTP contains the three phosphate groups attached to the 5′-carbon of d-ribose. As the two terminal phosphates are cleaved during synthesis, the remaining phosphate attached to the 5′-carbon is covalently linked to the 3′-OH group of the d-ribose to which it is added. Thus, **chain elongation** occurs in the **5′-to-3′ direction** by the addition of one nucleotide at a time to the growing 3′ end. Each step provides a newly exposed 3′-OH group that can participate in the next addition of a nucleotide as DNA synthesis proceeds.

Having shown how DNA was synthesized, Kornberg sought to demonstrate the accuracy, or fidelity, with which the enzyme replicated the DNA template. Because the technology to determine the nucleotide sequences of the template and the product was not yet available in 1957, he had to rely initially on several indirect methods.

*dNTP designates the deoxyribose forms of the four nucleoside triphosphates; in a similar way, dNMP refers to the monophosphate forms.

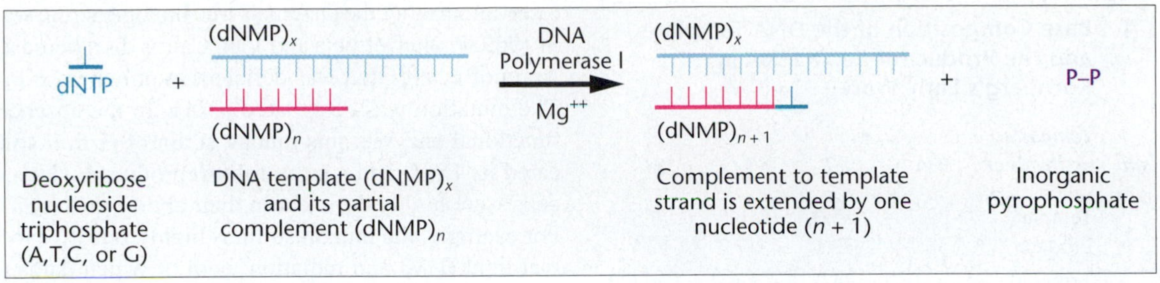

FIGURE 3–7 The chemical reaction catalyzed by DNA polymerase I. During each step, a single nucleotide is added to the growing complement of the DNA template, using a nucleoside triphosphate as the substrate. The release of inorganic pyrophosphate drives the reaction energetically.

One of Kornberg's approaches was to compare the nitrogenous base compositions of the DNA template with those of the recovered DNA product. Table 3–1 shows Kornberg's base composition analysis of three DNA templates. Within experimental error, the base composition of each product agreed with the template DNAs used. These data, along with other types of comparisons of template and product (see Problem 30 at the end of this chapter), suggested that the templates were replicated faithfully.

Synthesis of Biologically Active DNA

Despite Kornberg's extensive work, not all researchers were convinced that DNA polymerase I was the enzyme that replicates DNA within cells (in vivo). Their reservations were that *in vitro* synthesis was much slower than the *in vivo* rate, that the enzyme was much more effective replicating single-stranded DNA than double-stranded DNA, and that the enzyme appeared to be able to *degrade* DNA as well as to *synthesize* it—that is, the enzyme exhibits exonuclease activity.

Uncertain of the true cellular function of DNA polymerase I, Kornberg pursued another approach. He reasoned that if the enzyme could be used to synthesize **biologically active DNA** in vitro, then DNA polymerase I must be the major catalyzing force for DNA synthesis within the cell. The term *biological activity* means that the DNA synthesized is capable of supporting metabolic activities and directs the reproduction of the organism from which it was originally duplicated.

In 1967, Mehran Goulian, Kornberg, and Robert Sinsheimer experimented with the small bacteriophage ϕX174. This phage provides an ideal experimental system, because its genetic material is a very small (5386 nucleotides), circular single-stranded DNA molecule. Because the molecule is a closed circle, the experiment depended on a second enzyme, **DNA ligase** (also called the **polynucleotide joining enzyme**), to join the two ends of the linear molecule following replication.

In the normal course of ϕX174 infection, the circular single-stranded DNA, referred to as the (+) **strand**, enters the *E. coli* cell and serves as a template for the synthesis of the complementary (−) **strand**. The two strands (+ and −) remain together in a circular double helix, or duplex, called the **replicative form (RF)**. The RF serves as the template for its own replication, and subsequently, only (+) strands are produced. These strands are then packaged into viral coat proteins to form mature virus particles.

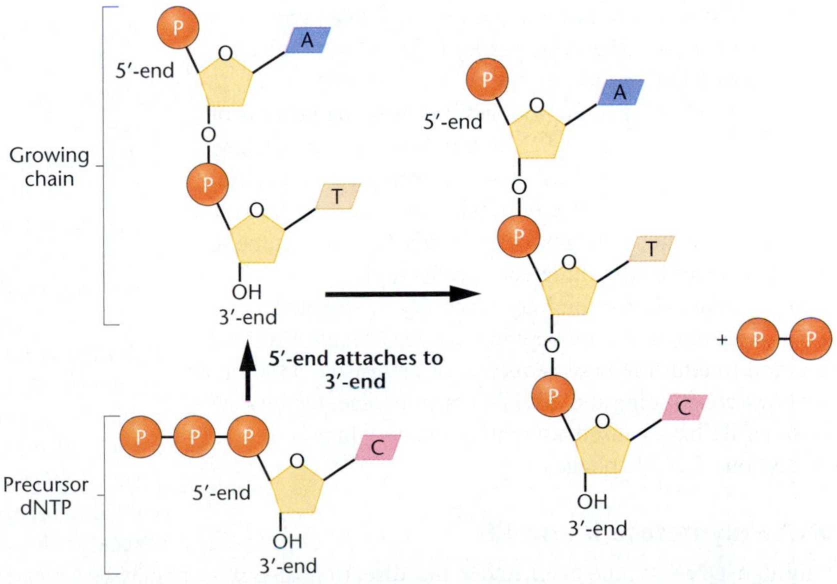

FIGURE 3–8 Demonstration of 5′-to-3′ synthesis of DNA.

TABLE 3–1 Base Composition of the DNA Template and the Product of Replication in Kornberg's Early Work

Organism	*Template or Product*	*%A*	*%T*	*%G*	*%C*
T2	Template	32.7	33.0	16.8	17.5
	Product	33.2	32.1	17.2	17.5
E. coli	Template	25.0	24.3	24.5	26.2
	Product	26.1	25.1	24.3	24.5
Calf	Template	28.9	26.7	22.8	21.6
	Product	28.7	27.7	21.8	21.8

Source: Kornberg (1960).

As shown in Figure 3–9, the experiment was carefully designed so that each newly synthesized strand could be distinguished and isolated from the template strand. Initially (step 1), the (+) strand was labeled with tritium (^{3}H)—a radioactive form of hydrogen. During synthesis, two "tags" were used to identify the new (−) strands. First, radioactive ^{32}P was present in the precursor nucleotides. Second, the base analogue 5-bromouracil (BU) was used in place of thymine. In the chemical structure of this analogue, a bromine atom is substituted for a carbon atom in the methyl group at the C-5 position of the pyrimidine ring, increasing the mass and therefore the density. As a result, newly synthesized (−) strands (step 2) were radioactive, and because their mass was greater, they could be isolated from the (+) strands by using sedimentation equilibrium centrifugation.

The duplex was "nicked" by the enzyme DNAase to open one strand and then heat denatured to separate the strands. The BU-containing strands, representing newly synthesized DNA, were then isolated by centrifugation. The process was repeated in the absence of any tags, using the BU-containing strands as templates. Strands synthesized after this point were detectable because they were "lighter" and nonradioactive. Eventually, newly synthesized infectious (+) strands were isolated. The protocol of the experiment dictated that these (+) strands must have been synthesized in vitro under the direction of Kornberg's enzyme.

The critical test of biological activity was the process of transfection (see Chapter 9), in which newly synthesized (+) strands were added to bacterial protoplasts (bacterial cells minus their cell wall). Following infection by the synthetic DNA, mature phages were produced—the synthetic DNA had successfully directed reproduction.

This demonstration of biological activity represented a precise assessment of faithful copying. If even a single error had occurred to alter the base sequence of any of the 5386 nucleotides constituting the ϕX174 chromosome, the change might easily have caused a mutation that would prohibit the production of viable phages.

DNA Polymerase II and III

Although DNA synthesized under the direction of polymerase I demonstrated biological activity, a more serious reservation about the enzyme's true biological role was raised in 1969. Paula DeLucia and John Cairns discovered a mutant strain of *E. coli* that was deficient in polymerase I activity. The mutation was designated *polA*1. In the absence of the functional enzyme, this mutant strain of *E. coli* still duplicated its DNA and successfully reproduced. However, the cells were highly deficient in their ability to "repair" DNA. For example, the mutant strain is highly sensitive to ultraviolet light (UV) and radiation, both of which damage DNA and are mutagenic. Nonmutant bacteria are able to repair a great deal of UV-induced damage.

These observations led to two conclusions:

1. At least one other enzyme that is responsible for replicating DNA in vivo is present in *E. coli* cells.
2. DNA polymerase I may serve a secondary function in vivo. This function is now believed by Kornberg and others to be critical to the *fidelity* of DNA synthesis, but the enzyme does not actually synthesize the entire complementary strand during replication.

To date, two other unique DNA polymerases have been isolated from cells lacking polymerase I activity and from normal cells that contain polymerase I. Table 3–2 shows that the two enzymes, called **DNA polymerase II** and **III**, share several characteristics with DNA polymerase I. While none of the three can *initiate* DNA synthesis on a template, all three can *elongate* an existing DNA strand, called a **primer**. As we shall see, RNA is also an adequate primer and is, in fact, used initially.

The DNA polymerase enzymes are all large complex proteins exhibiting a molecular weight in excess of 100,000 Da. All three possess 3′–5′ exonuclease activity, which means that they have the potential to polymerize in one direction and then pause, reverse their direction, and excise nucleotides just added. As we will discuss later in the chapter, this activity provides a capacity to proofread newly synthesized DNA and to remove incorrect nucleotides, which may then be replaced.

DNA polymerase I also demonstrates 5′–3′ exonuclease activity. This activity potentially allows the enzyme to excise nucleotides, starting at the end at which synthesis begins and proceeding in the same direction of synthesis. Thus, DNA polymerase I has the ability to remove the RNA primer. Two final observations probably explain why Kornberg isolated polymerase I and not polymerase III: Polymerase I is present in greater amounts than is polymerase III, and it is also much more stable.

What then are the roles of the three polymerases in vivo? Polymerase I is believed to be responsible for removing the primer, as well as for the synthesis that fills the gaps that naturally occur as primers are removed. Its exonuclease activity also allows for proofreading during this process, a form of DNA repair. Polymerase II appears to also be involved in repair of DNA that has been damaged by external forces, such as ultraviolet light. It is encoded by a gene that may be activated by disruption of DNA synthesis at the replication fork. Polymerase III is considered to be the enzyme

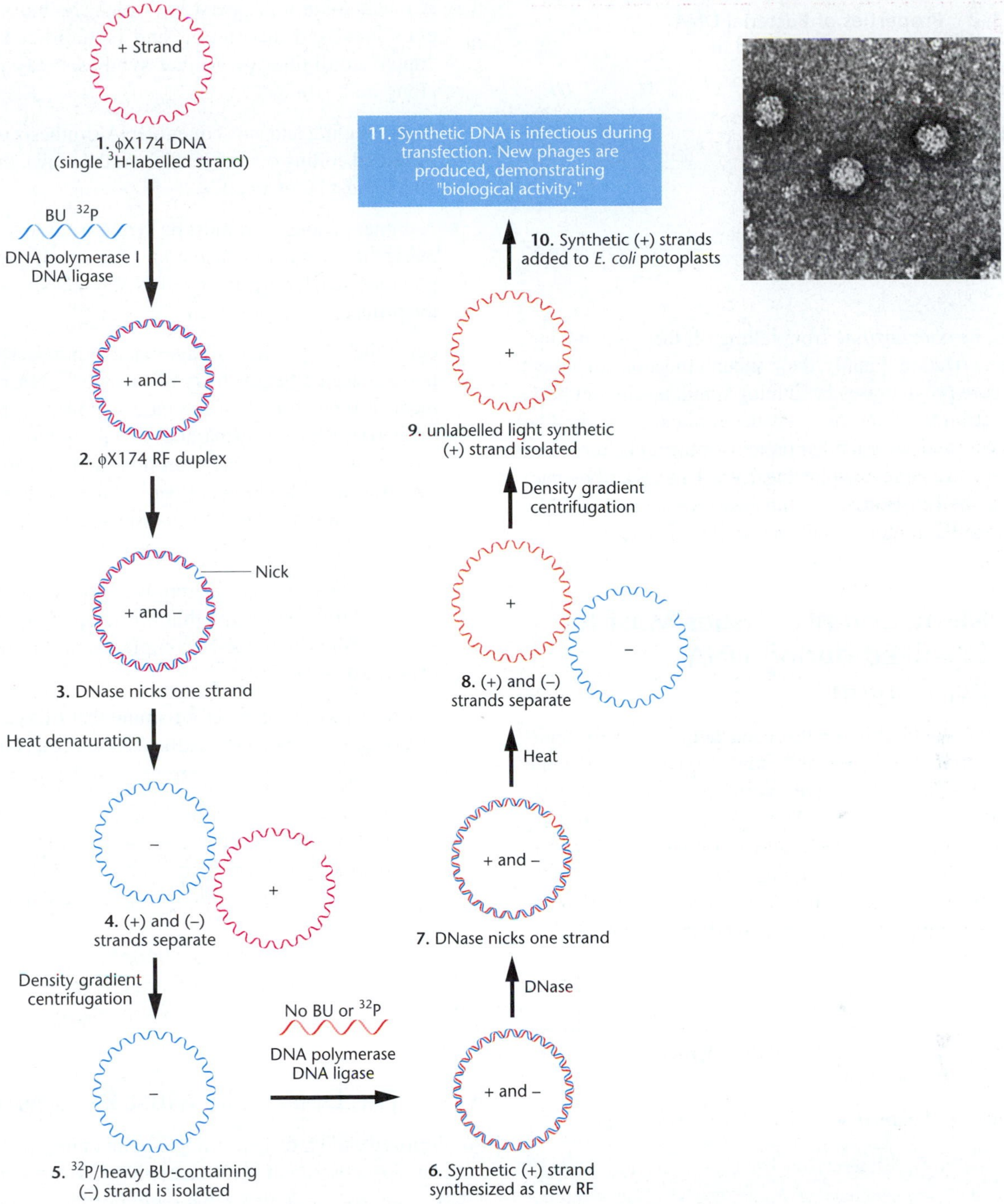

FIGURE 3–9 Schematic representation of Goulian, Kornberg, and Sinsheimer's experiment involving in vitro replication and isolation of synthetic (+) DNA strands of phage ϕX174. The DNA synthesized in vitro under the direction of DNA polymerase I successfully transfected *E. coli* protoplasts, thus demonstrating its biological activity. The electron micrograph shows ϕX174 particles.

responsible for the polymerization essential to replication. Its $3'$–$5'$ exonuclease activity provides its proofreading function, as just described.

We end this section by emphasizing the complexity of the DNA polymerase III molecule. Its active form, called a **holoenzyme**, consists of a dimer containing 10 different polypeptide subunits (Table 3–3) and has a molecular weight of 900,000 Da. The largest subunit, α, has a molecular weight of 140,000 Da and, along with subunits ϵ and θ constitutes the "core" enzyme responsible for the polymerization activity. The α subunit is responsible for nucleotide polymerization on the template strands, whereas the ϵ subunit of the core enzyme possesses the $3'$–$5'$ exonuclease activity.

A second group of five subunits (γ, δ, δ', χ, and ψ) forms what is called the γ complex, which is involved in "loading" the enzyme onto the template at the replication fork. This enzymatic function requires energy and is dependent on the hydrolysis of ATP. The β subunit serves as a "clamp" and

TABLE 3–2 Properties of Bacterial DNA Polymerases I, II, and III

Properties	*I*	*II*	*III*
Initiation of chain synthesis	−	−	−
5′–3′ polymerization	+	+	+
3′–5′ exonuclease activity	+	+	+
5′–3′ exonuclease activity	+	−	−
Molecules of polymerase/cell	400	?	15

prevents the core enzyme from falling off the template during polymerization. Finally, the τ subunit functions to dimerize two core polymerases facilitating simultaneous synthesis of both strands of the helix at the replication fork. The holoenzyme and several other proteins at the replication fork together form a huge complex (nearly as large as a ribosome) known as the **replisome**. We consider the function of DNA polymerase III in more detail later in this chapter.

3.3 Many Complex Issues Must Be Resolved during DNA Replication

We have thus far established that in bacteria and viruses replication is semiconservative and bidirectional along a single replicon. We also know that synthesis is in the 5′-to-3′ mode under the direction of DNA polymerase III, creating two replication forks. These move in opposite directions away from the origin of synthesis. As we can see in the points listed here, many issues must still be resolved in order to provide a comprehensive understanding of DNA replication:

TABLE 3–3 Subunits of the DNA Polymerase III Holoenzyme

Subunit	*Function*	*Groupings*
α	5′–3′ polymerization	"Core" enzyme: Elongates polynucleotide chain and proofreads
ϵ	3′–5′ exonuclease	
θ	core assembly	
γ, δ, δ', χ, ψ	Loads enzyme on template (Serves as clamp loader)	γ complex
β	Sliding clamp structure (processivity factor)	
τ	Dimerizes core complex	

1. A mechanism must exist by which the helix undergoes localized unwinding and is stabilized in this "open" configuration so that synthesis may proceed along both strands.
2. As unwinding and subsequent DNA synthesis proceed, increased coiling creates tension further down the helix, which must be reduced.
3. A primer of some sort must be synthesized so that polymerization can commence under the direction of DNA polymerase III. Surprisingly, RNA, not DNA, serves as the primer.
4. Once the RNA primers have been synthesized, DNA polymerase III begins to synthesize the DNA complement of both strands of the parent molecule. Because the two strands are antiparallel to one another, continuous synthesis in the direction that the replication fork moves is possible along only one of the two strands. On the other strand, synthesis is discontinuous in the opposite direction.
5. The RNA primers must be removed prior to completion of replication. The gaps that are temporarily created must be filled with DNA complementary to the template at each location.
6. The newly synthesized DNA strand that fills each temporary gap must be joined to the adjacent strand of DNA.
7. While DNA polymerases accurately insert complementary bases during replication, they are not perfect, and, occasionally, incorrect bases are added to the growing strand. A proofreading mechanism that also corrects errors is an integral process during DNA synthesis.

As we consider these points, examine Figures 3–10, 3–11, 3–12, and 3–13 to see how each issue is resolved. Figure 3–14 summarizes the model of DNA synthesis.

3.4 The DNA Helix Must Be Unwound

As discussed earlier, there is a single point of origin along the circular chromosome of most bacteria and viruses at which DNA synthesis is initiated. This region of the *E. coli* chromosome has been particularly well studied. Called ***oriC***, it consists of 245 base pairs characterized by repeating sequences of 9 and 13 bases (called **9mers** and **13mers**). As shown in Figure 3–10, one particular protein, called **DnaA** (because it is encoded by the gene called *dnaA*), is responsible for the initial step in unwinding the helix. A number of subunits of the DnaA protein bind to each of several 9mers. This step facilitates the subsequent binding of **DnaB** and **DnaC** proteins that further open and destabilize the helix. Proteins such as these, which require the energy normally supplied by the hydrolysis of ATP in order to break hydrogen bonds and denature the double helix, are called **helicases**. Other proteins, called **single-stranded binding proteins (SSBPs)**, stabilize this open conformation.

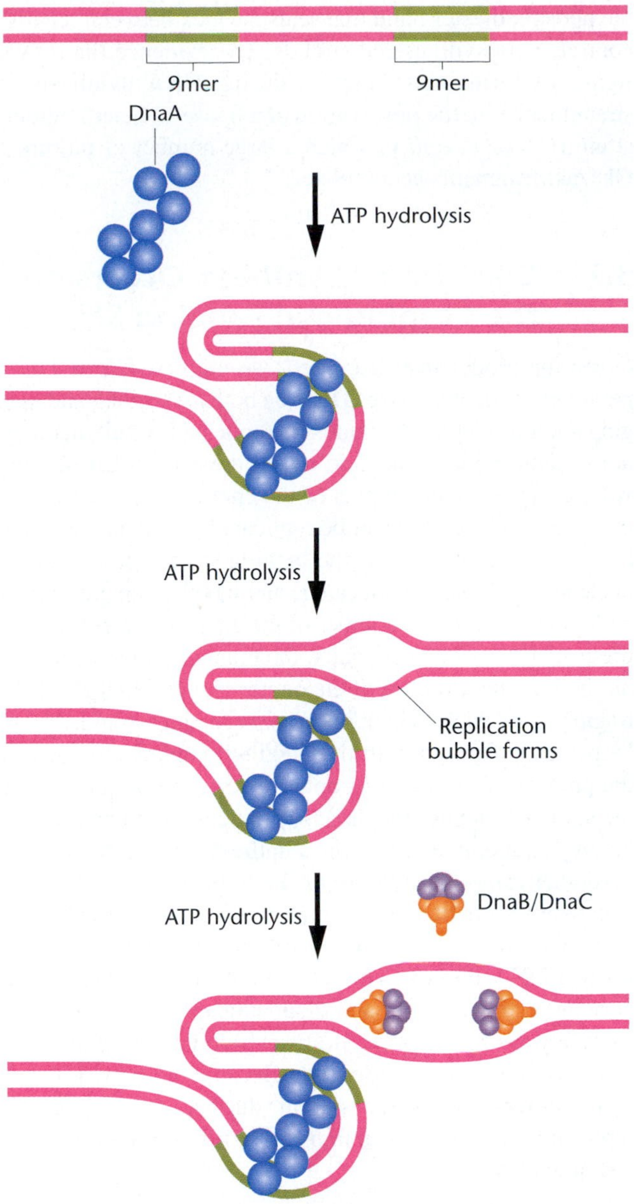

FIGURE 3–10 Helical unwinding of DNA during replication as accomplished by DnaA, DnaB, and DnaC proteins. Initial binding of many monomers of DnaA occurs at DNA sites containing repeating sequences of 9 nucleotides, called *9mers*. Not illustrated are 13mers, which are also involved.

As unwinding proceeds, a coiling tension is created ahead of the replication fork, often producing **supercoiling**. In circular molecules, supercoiling may take the form of added twists and turns of the DNA, much like the coiling you can create in a rubber band by stretching it out and then twisting one end. Such supercoiling can be relaxed by **DNA gyrase**, a member of a larger group of enzymes referred to as **DNA topoisomerases**. The gyrase makes either single- or double-stranded "cuts" and also catalyzes localized movements that have the effect of "undoing" the twists and knots created during supercoiling. The strands are then resealed. These various reactions are driven by the energy released during ATP hydrolysis.

Together, the DNA, the polymerase complex, and associated enzymes make up an array of molecules that participate in DNA synthesis and are part of what we have previously called the replisome.

3.5 Initiation of DNA Synthesis Requires an RNA Primer

Once a small portion of the helix is unwound, the initiation of synthesis may occur. As we have seen, DNA polymerase III requires a primer with a free 3′ end in order to elongate a polynucleotide chain. This prompted researchers to investigate how the first nucleotide could be added. Though no free 3′-hydroxyl group is initially present, it is now clear that, as previously pointed out, RNA is the primer that initiates DNA synthesis.

A short segment of RNA, (about 5 to 15 nucleotides long), complementary to DNA, is first synthesized on the DNA template. Synthesis of the RNA is directed by a form of RNA polymerase called **primase**, which does not require a free 3′ end to initiate synthesis. It is to this short segment of RNA that DNA polymerase III begins to add 5′-deoxyribonucleotides, initiating DNA synthesis. A conceptual diagram of initiation on a DNA template is shown in Figure 3–11. At a later point, the RNA primer must be clipped out and replaced with DNA. This occurs under the direction of DNA polymerase I. Recognized in viruses, bacteria, and several eukaryotic organisms, RNA priming is a universal phenomenon during the initiation of DNA synthesis.

3.6 Antiparallel Strands Require Continuous and Discontinuous DNA Synthesis

We must now reconsider the fact that the two strands of a double helix are **antiparallel** to each other—that is, one runs in the 5′–3′ direction, while the other has the opposite 3′–5′ polarity. Because DNA polymerase III synthesizes DNA in only the 5′–3′ direction, synthesis along an advancing replication fork occurs in one direction on one strand and in the opposite direction on the other.

As a result, as the strands unwind and the replication fork progresses down the helix (Figure 3–12), only one strand can serve as a template for **continuous DNA synthesis**. This strand is called the **leading DNA strand**. As the fork progresses, many points of initiation are necessary on the

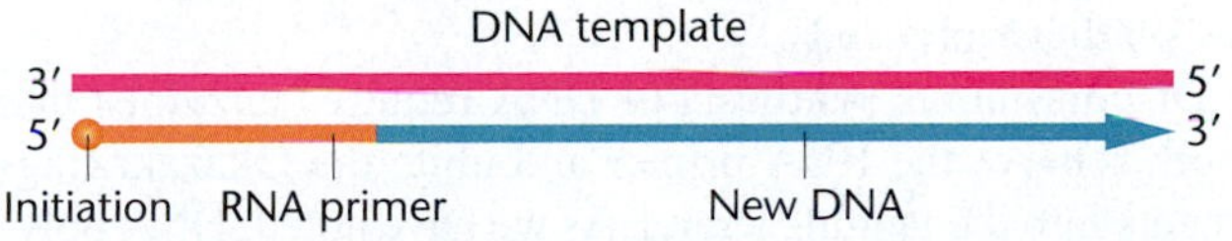

FIGURE 3–11 The initiation of DNA synthesis. A complementary RNA primer is first synthesized, to which DNA is added. All synthesis is in the 5′-to-3′ direction. Eventually, the RNA primer is replaced with DNA under the direction of DNA polymerase I.

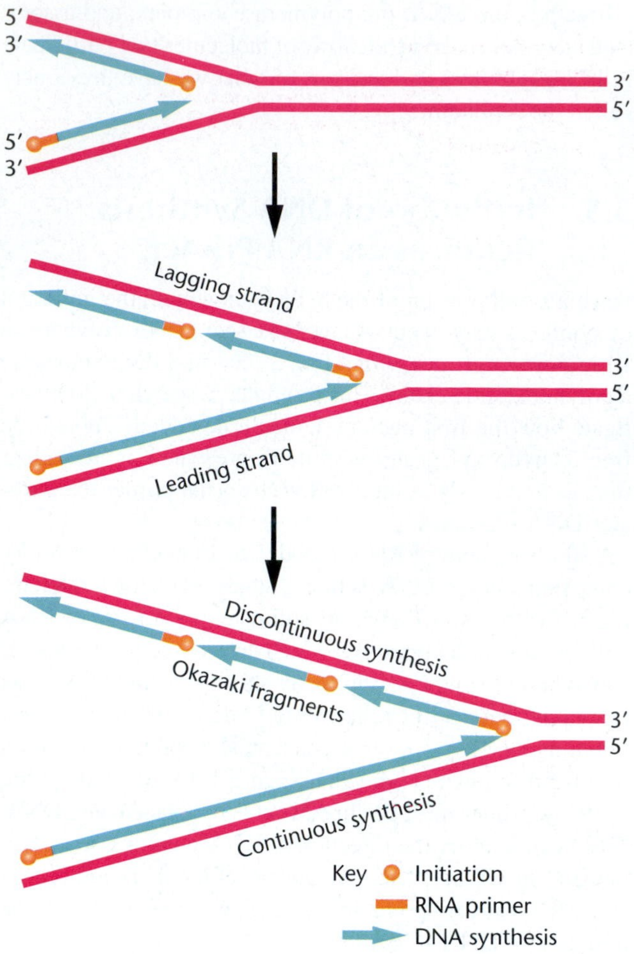

FIGURE 3–12 Opposite polarity of DNA synthesis along the two strands, necessary because the two strands of DNA run antiparallel to one another and DNA polymerase III synthesizes only in one direction (5′ to 3′). On the lagging strand, synthesis must be discontinuous, resulting in the production of Okazaki fragments. On the leading strand, synthesis is continuous. RNA primers are used to initiate synthesis on both strands.

opposite, or **lagging DNA strand**, resulting in **discontinuous DNA synthesis**.

Evidence supporting the occurrence of discontinuous DNA synthesis was first provided by Reiji Okazaki, Tuneko Okazaki, and their colleagues. They discovered that, when bacteriophage DNA is replicated in *E. coli*, some of the newly formed DNA that is hydrogen bonded to the template strand is present as small fragments containing 1000–2000 nucleotides. RNA primers are part of each such fragment. These pieces, now called **Okazaki fragments**, are converted into longer and longer DNA strands of higher molecular weight as synthesis proceeds.

Discontinuous synthesis of DNA requires enzymes that both remove the RNA primer and unite the Okazaki fragments into the lagging strand. As we have noted, DNA polymerase I removes the primer and replaces the missing nucleotides. Joining the fragments appears to be the work of DNA ligase, which is capable of catalyzing the formation of the phosphodiester bond that seals the nick between the discontinuously synthesized strands. The evidence that DNA ligase performs this function during DNA synthesis is strengthened by the observation of a ligase-deficient mutant strain (*lig*) of *E. coli* in which a large number of unjoined Okazaki fragments accumulate.

3.7 Concurrent Synthesis Occurs on the Leading and Lagging Strands

Given the model just discussed, we might ask how DNA polymerase III synthesizes DNA on both the leading and lagging strands. Can both strands be replicated simultaneously at the same replication fork, or are the events distinct, involving two separate copies of the enzyme? Evidence suggests that both strands can be replicated simultaneously. As Figure 3–13 illustrates, if the lagging strand forms a loop, nucleotide polymerization can occur on both template strands under the direction of a dimer of the enzyme. After the synthesis of 100 to 200 base pairs, the monomer of the enzyme on the lagging strand will encounter a completed Okazaki fragment, at which point it releases the lagging strand. A new loop is then formed with the lagging template strand, and the process is repeated. Looping inverts the orientation of the template, but not the direction of actual synthesis on the lagging strand, which is always in the 5′-to-3′ direction.

Another important feature of the holoenzyme that facilitates synthesis at the replication fork is a dimer of the β subunit that forms a clamplike structure around the newly formed DNA duplex. This β-subunit clamp prevents the **core enzyme** (the α, ϵ, and θ subunits that are responsible for catalysis of nucleotide addition) from falling off the template as polymerization proceeds. Because the entire holoenzyme moves along the parent duplex, advancing the replication fork, the β-subunit dimer is often referred to as a sliding clamp.

3.8 Proofreading and Error Correction Are an Integral Part of DNA Replication

The underpinning of DNA replication is the synthesis of a new strand that is precisely complementary to the template strand at each nucleotide position. Although the action of DNA polymerases is very accurate, synthesis is not perfect and a noncomplementary nucleotide is occasionally inserted erroneously. To compensate for such inaccuracies, the DNA polymerases all possess 3′–5′ exonuclease activity. This property imparts the potential for them to detect and excise a mismatched nucleotide (in the 3′–5′ direction). Once the mismatched nucleotide is removed, 5′–3′ synthesis can again proceed. This process, **exonuclease proofreading**, increases the fidelity of synthesis. In the case of the holoenzyme form of DNA polymerase III, the epsilon (ϵ)

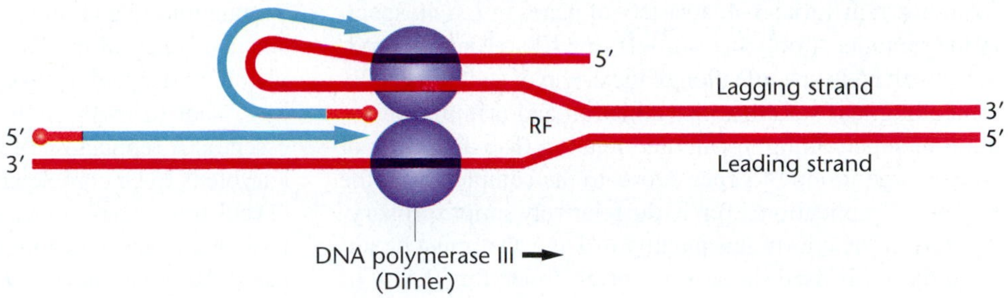

FIGURE 3–13 Illustration of how concurrent DNA synthesis may be achieved on both the leading and lagging strands at a single replication fork. The lagging template strand is "looped" in order to invert the physical direction of synthesis, but not the biochemical direction. The enzyme functions as a dimer, with each core enzyme achieving synthesis on one or the other strand.

subunit is directly involved in the proofreading step. In strains of *E. coli* which have a mutation that has rendered the ϵ subunit nonfunctional, the error rate (the mutation rate) during DNA synthesis is increased substantially.

Because the investigation of DNA synthesis is still an extremely active area of research, this model will no doubt be extended in the future. In the meantime, it provides a summary of DNA synthesis against which genetic phenomena can be interpreted.

3.9 A Coherent Model Summarizes DNA Replication

We can now combine the various aspects of DNA replication occurring at a single replication fork into a coherent model, as shown in Figure 3–14. At the advancing fork, a helicase is unwinding the double helix. Once unwound, single-stranded binding proteins associate with the strands, preventing the reformation of the helix. In advance of the replication fork, DNA gyrase functions to diminish the tension created as the helix supercoils. Each half of the dimeric polymerase is a core enzyme bound to one of the template strands by a β-subunit sliding clamp. Continuous synthesis occurs on the leading strand, while the lagging strand must "loop" around in order for simultaneous synthesis to occur on both strands. Not shown in the figure, but essential to replication on the lagging strand, is the action of DNA polymerase I and DNA ligase, which together replace the RNA primer with DNA and join the Okazaki fragments, respectively.

3.10 Replication Is Controlled by a Variety of Genes

Much of what we know about DNA replication in viruses and bacteria is based on genetic analysis of the process. For example, we have already discussed studies involving the *polA1* mutation, which revealed that DNA polymerase I is not the major enzyme responsible for replication. Many other mutations interrupt or seriously impair some aspect of replication, such as the ligase-deficient and the proofreading-deficient mutations mentioned previously. Genetic analysis frequently uses **conditional mutations**, which are expressed under one condition, but not under a different condition. For example, a **temperature-sensitive mutation** may not be expressed at a particular *permissive* temperature. When mutant cells are grown at a *restrictive* temperature, the mutation is expressed. The investigation of such temperature-sensitive mutants can provide insight into the product and the associated function of the normal, nonmutant gene.

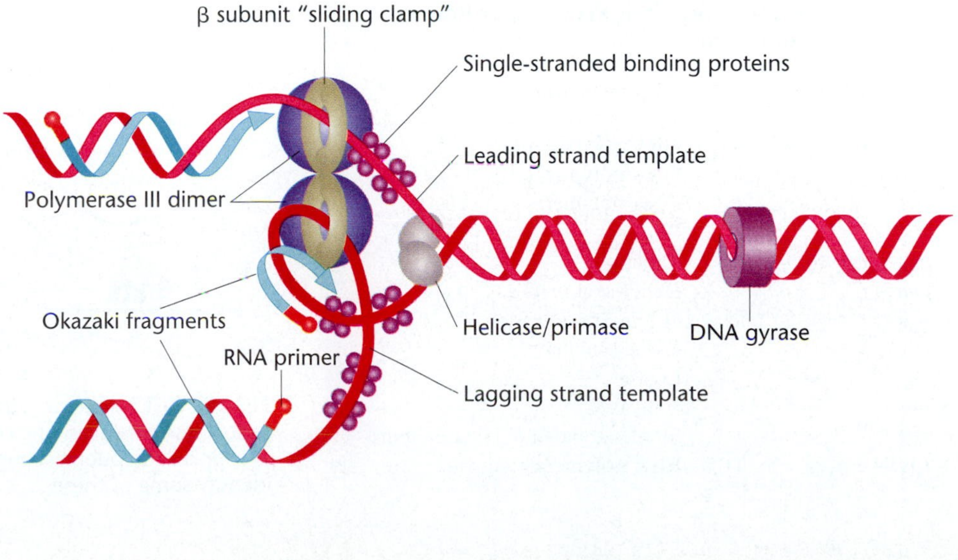

FIGURE 3–14 Summary of DNA synthesis at a single replication fork. Various enzymes and proteins essential to the process are shown.

As shown in Table 3–4, a variety of genes in *E. coli* specify the subunits of polymerases I, II, and III and encode products involved in specification of the origin of synthesis, helix unwinding and stabilization, initiation and priming, relaxation of supercoiling, repair, and ligation. The discovery of such a large group of genes attests to the complexity of the process of replication, even in the relatively simple prokaryote. Given the enormous quantity of DNA that must be unerringly replicated in a very brief time, this level of complexity is not unexpected. As we see next, the process is even more involved and therefore more difficult to investigate in eukaryotes.

3.11 Eukaryotic DNA Synthesis Is Similar to, but More Complex than, Synthesis in Prokaryotes

Research has shown that eukaryotic DNA is replicated in a manner similar to that of bacteria. In both systems, double-stranded DNA is unwound at a replication origin, two replication forks are formed, and bidirectional synthesis of DNA occurs on the leading and lagging strand templates under the direction of DNA polymerase. Eukaryotic polymerases have the same fundamental requirements for DNA synthesis as do bacterial systems: four deoxyribonucleoside triphosphates, a template, and a primer. However, because eukaryotic cells contain much more DNA per cell and because this DNA is complexed with proteins, eukaryotes face many problems not encountered by bacteria. As we might expect, these complications make the process of DNA synthesis much more complex in eukaryotes and more difficult to study. However, a great deal is now known about the process.

Multiple Replication Origins

The most obvious difference between eukaryotic and prokaryotic DNA replication is that eukaryotic chromosomes contain multiple replication origins, in contrast to the single site that is part of the *E. coli* chromosome. Multiple origins, visible under the electron microscope as "replication bubbles" that form as the helix opens up (Figure 3–15), are essential if replication of the entire genome of a typical eukaryote is to be completed in a reasonable time. Recall that (1) eukaryotes have much greater amounts of DNA than bacteria (e.g., yeast has 4 times as much DNA, and *Drosophila* has 100 times as much as *E. coli*), and (2) the rate of synthesis by eukaryotic DNA polymerase is much slower—only about 50 nucleotides per second, a rate 20 times less than the comparable bacterial enzyme. Under these conditions, replication from a single origin of a typical eukaryotic chromosome might take days to complete! However, replication of entire eukaryotic genomes is usually accomplished in a matter of hours.

Insights are now available concerning the molecular nature of the multiple origins and the initiation of DNA synthesis at these sites. Most information has originally been derived from the study of yeast (e.g., *Saccharomyces cerevisiae*), which has between 250–400 replicons per genome; subsequent studies have used mammalian cells, which have as many as 25,000 replicons. The origins in yeast have been isolated and are called **autonomously replicating sequences (ARSs)**. They consist of a unit containing a consensus sequence of 11 base pairs, flanked by other short sequences involved in efficient initiation. As we will learn in Chapter 8, DNA synthesis is restricted to the S phase of the eukaryotic cell cycle. Research has shown that the many origins are not all activated at once; instead, clusters of 20–80 adjacent replicons are activated sequentially throughout the S phase until all DNA is replicated.

How the polymerase finds the ARSs among so much DNA is an obvious recognition problem. The solution involves a mechanism that is initiated prior to the S phase. During the G1 phase of the cell cycle, all ARSs are initially bound by a group of specific proteins (six in yeast), forming what is called the **origin recognition complex (ORC)**. Mutations

TABLE 3–4 Some of the Various *E. coli* Mutant Genes and Their Products or Role in Replication

Mutant Gene	*Enzyme or Role*
polA	DNA polymerase I
polB	DNA polymerase II
dnaE, N, Q, X, Z	DNA polymerase III subunits
dnaG	Primase
dnaA, I, P	Initiation
dnaB, C	Helicase at *oriC*
oriC	Origin of replication
gyrA, B	Gyrase subunits
lig	Ligase
rep	Helicase
ssb	Single-stranded binding proteins
rpoB	RNA polymerase subunit

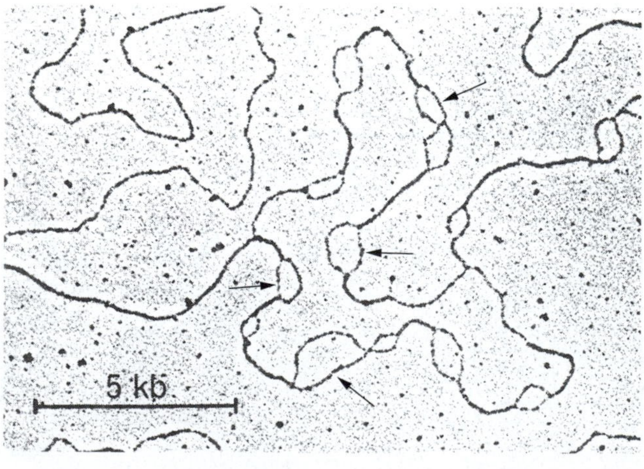

FIGURE 3–15 A demonstration of the multiple origins of replication along a eukaryotic chromosome. Each origin is apparent as a replication bubble along the axis of the chromosome. Arrows identify some of these.

in either the ARSs or in any of the genes encoding these proteins of the ORC abolish or reduce initiation of DNA synthesis. Since these recognition complexes are formed in G1, but synthesis is not initiated at these sites until S, there must be still other proteins involved in the actual initiation signal. The most important of these proteins are specific kinases, key enzymes involved in phosphorylation, that are integral parts of cell cycle control. When bound along with ORC, a prereplication complex is formed that is accessible to DNA polymerase. After these kinases are activated, they serve to complete the initiation complex, directing localized unwinding and triggering DNA synthesis. Activation also inhibits reformation of the pre-replication complexes once DNA synthesis has been completed at each replicon. This is an important mechanism, since it distinguishes segments of DNA that have completed replication from unreplicated DNA, thus maintaining orderly and efficient replication. Importantly, it ensures that replication only occurs once along each stretch of DNA during each cell cycle.

Eukaryotic DNA Polymerases

The most complex aspect of eukaryotic replication is the array of polymerases involved in directing DNA synthesis. As we will see, a total of six different forms of the enzyme has been isolated and studied. In order for the polymerases to have access to DNA, the topology of the helix must first be modified. As synthesis is triggered at each origin site, the double strands are opened up within an AT-rich region, allowing the entry of a helicase enzyme that proceeds to further unwind the double-stranded DNA. Before polymerases can begin synthesis, histone proteins complexed to the DNA (which form the characteristic nucleosomes of chromatin—see Chapter 11) also must be stripped away or otherwise modified. As DNA synthesis then proceeds, histones reassociate with the newly formed duplexes, reestablishing the characteristic nucleosome pattern (Figure 3–16). In eukaryotes, the synthesis of new histone proteins is tightly coupled to DNA synthesis during the S phase of the cell cycle.

The nomenclature and characteristics of the six polymerases are summarized in Table 3–5. Of these, three (Pol α, δ, and ϵ) are now considered to be essential to nuclear DNA replication in eukaryotic cells. Two others (Pol β and ζ) are thought to be involved in DNA repair. The sixth form (Pol γ) is involved in the synthesis of mitochondrial DNA. Presumably, its replication function is limited to that organelle even though it is encoded by a nuclear gene. As shown in Table 3–5, all but one of the six forms of the enzyme (the β form) consist of multiple subunits. Different subunits perform different functions during replication.

Pol α and δ may be the major forms of the enzyme involved in initiating nuclear DNA synthesis, so we concentrate our discussion on these. Two of the four subunits of the enzyme function as a primase in the synthesis of the RNA primers on both the leading and lagging template strands. Then, another subunit functions to elongate the RNA primer by adding complementary deoxyribonucleotides, constituting the initial phase of DNA synthesis. Pol α is said to possess low **processivity**, a term that essentially reflects the length of DNA that is synthesized by an enzyme before it dissociates from the template. Thus, after a short DNA sequence is added to the RNA primer, an event known as **polymerase switching** occurs, whereby Pol α dissociates from the template and is replaced by Pol δ. This form of the enzyme possesses high processivity as well as 3′-to-5′ exonuclease activity, which provides it with the potential to proofread. It is also capable of a hundredfold increase in the rate of synthesis in comparison to Pol α. Thus, under the direction of Pol δ, elongation and proofreading of the growing DNA strand occurs, as synthesis continues. Pol ϵ, the third essential form, possesses the same general characteristics as Pol δ, but it is believed to operate under different cellular conditions. In yeast, mutations that render Pol ϵ inactive are lethal, attesting to its essential function during replication.

The process described applies to both leading and lagging strand synthesis. On both strands, RNA primers must be replaced with DNA. On the lagging strand, the Okazaki fragments, which are about 10 times smaller (100–150 nucleotides) in eukaryotes than in prokaryotes, must be ligated.

To accommodate the increased number of replicons, eukaryotic cells contain many more DNA polymerase molecules than do bacteria. While *E. coli* has about 15 copies of DNA polymerase III per cell, there may be up to 50,000 copies of the a form of DNA polymerase in animal cells. As has been pointed out, the presence of greater numbers of smaller replicons in eukaryotes compared with bacteria compensates for the slower rate of DNA synthesis in eukaryotes. *E. coli* requires 20 to 40 minutes to replicate its chromosome, while *Drosophila*, with 40 times more DNA, is known during embryonic cell divisions to accomplish the same task in only 3 minutes.

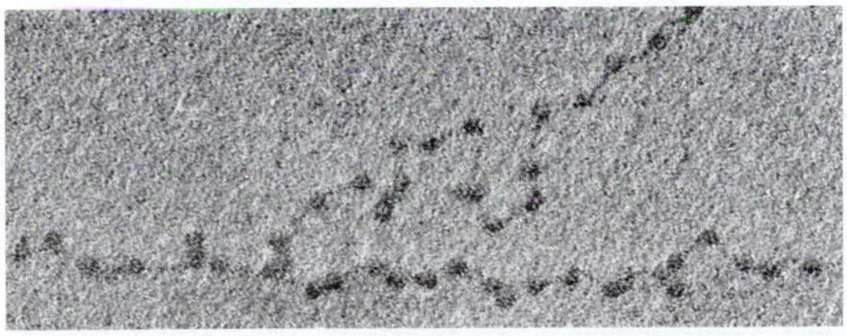

FIGURE 3–16 An electron micrograph of a eukaryotic-replicating fork demonstrating the presence of histone protein-containing nucleosomes on both branches.

3.12 The Ends of Linear Chromosomes Are Problematic during Replication

A final difference that exists between prokaryotic and eukaryotic DNA synthesis involves the nature of the chromosomes. Unlike the closed, circular DNA of bacteria and most bacteriophages, eukaryotic chromosomes are linear. During replication, they face a special problem at the "ends" of these linear molecules, called the **telomeres**. While synthesis

TABLE 3–5 Properties of Eukaryotic DNA Polymerases

	Polymerase α	*Polymerase* β	*Polymerase* δ	*Polymerase* ϵ	*Polymerase* γ	*Polymerase* ζ
Location	*Nucleus*	*Nucleus*	*Nucleus*	*Nucleus*	*Mitochondria*	*Nucleus*
Mass of subunits (kDa)	180, 70, 58, 48	39	125, 53, 50	250, 55	125, 35	173, 29
3′–5′ Exonuclease Activity	No	No	Yes	Yes	Yes	No
Essential to nuclear replication?	Yes	No	Yes	Yes	No	No

proceeds normally to the end of the leading strand, a difficulty arises on the lagging strand as the RNA primer is removed (Figure 3–17). Normally, the newly created gap would be filled by addition of a nucleotide to the existing 3′-OH group provided during discontinuous synthesis (to the right of gap (b) in Figure 3–17). However, since it is the end of the DNA molecule, there is no strand present to provide the 3′-OH group. As a result, each successive round of synthesis will theoretically shorten the end of the chromosome on the lagging strand by the length of the RNA primer. Because this is potentially such a significant problem, we can suppose that a molecular solution would have been developed early in evolution and subsequently be conserved and shared by most all eukaryotes, which is indeed the case.

A unique eukaryotic enzyme called **telomerase**, first discovered in the ciliated protozoan *Tetrahymena*, has helped us understand the solution. Before we look at how the enzyme actually works, let's look at what is accomplished. Telomeric DNA in eukaryotes is always found to consist of short, repeated nucleotide sequences. The telomeres of each end of each *Tetrahymena* chromosome contain many repeats of, and thus terminate in, the sequence 5′-TTGGGG-3′. As illustrated in Figure 3–18, telomerase is capable of adding several repeats of this sequence to the the 3′ end of the lagging strand (using 5′–3′ synthesis). These repeats are then thought to form a "hairpin loop," which may be stabilized by unorthodox hydrogen bonding between opposite guanine residues (G═G), creating a free 3′-OH group on the end

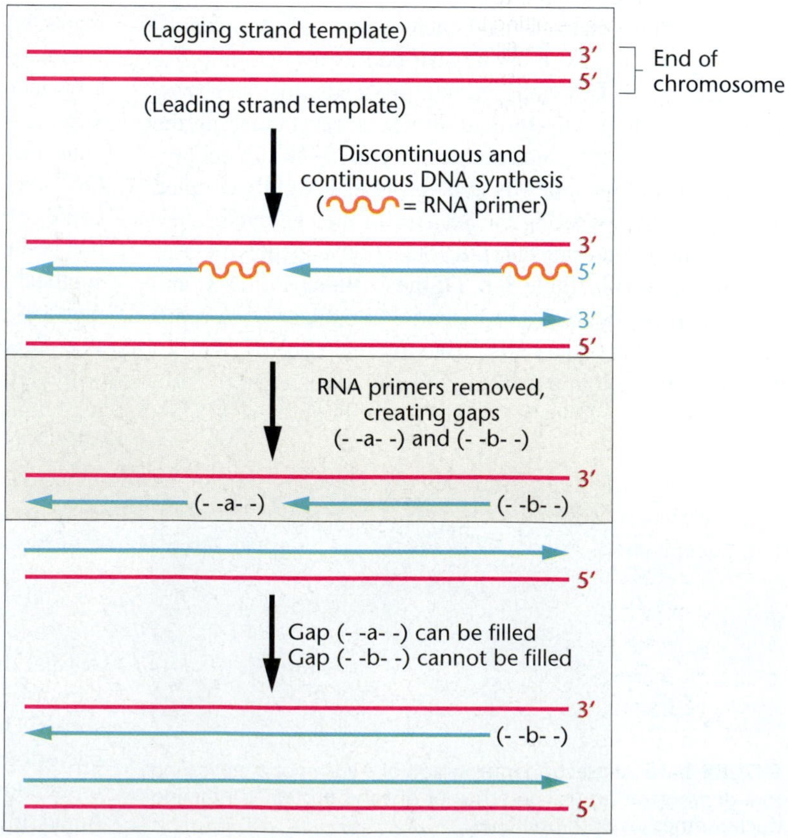

FIGURE 3–17 Diagram illustrating the difficulty encountered during the replication of the ends of linear chromosomes. A gap (--b--) is left following synthesis on the lagging strand.

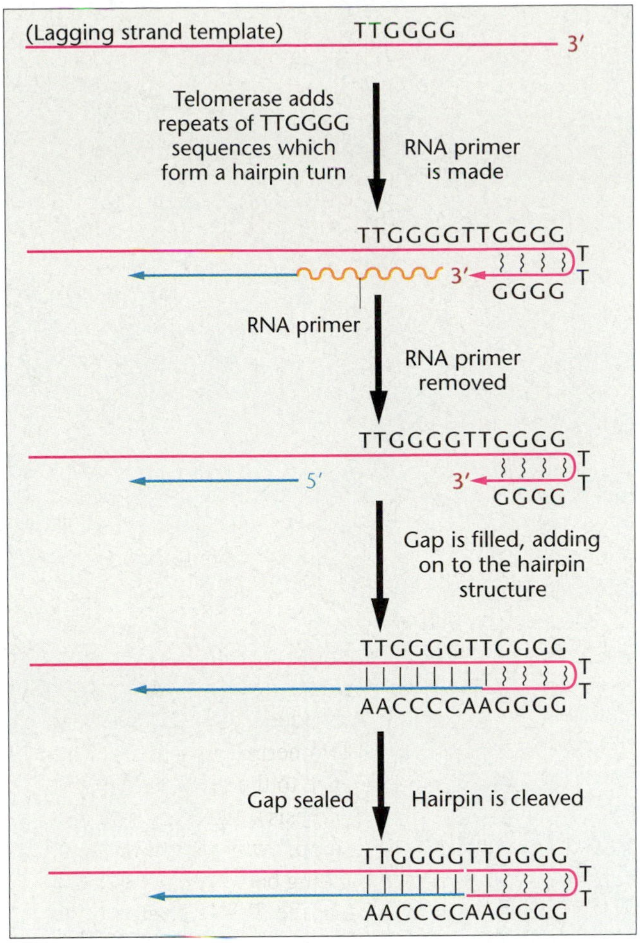
(Lagging strand template)
TTGGGG
3′
Telomerase adds repeats of TTGGGG sequences which form a hairpin turn
RNA primer is made
TTGGGGTTGGGG
T
T
3′
GGGG
RNA primer
RNA primer removed
TTGGGGTTGGGG
5′
3′
GGGG
Gap is filled, adding on to the hairpin structure
TTGGGGTTGGGG
AACCCCAAGGGG
Gap sealed
Hairpin is cleaved
TTGGGGTTGGGG
AACCCCAAGGGG

FIGURE 3–18 The predicted solution to the problem posed in Figure 3–17. The enzyme telomerase directs synthesis of the TTGGGG sequences, resulting in the formation of a hairpin structure. The gap can now be filled, and, following cleavage of the hairpin structure, the process averts the creation of a gap during replication of the ends of linear chromosomes.

of the hairpin that can serve as a substrate for DNA polymerase I. This makes it possible to fill the gap that would otherwise shorten the chromosome. The hairpin loop can then be cleaved off at its terminus, and the potential loss of DNA in each subsequent replication cycle is averted.

Detailed investigation by Elizabeth Blackburn and Carol Greider of how the *Tetrahymena* telomerase enzyme accomplishes the synthesis yielded an extraordinary finding. The enzyme is highly unusual in that it is a *ribonucleoprotein*, containing within its molecular structure a short piece of RNA that is essential to its catalytic activity. The RNA component serves as both a "guide" and a template for the synthesis of its DNA complement, a process referred to as *reverse transcription*. In *Tetrahymena*, the RNA contains several repeats of the sequence 5′-AACCCC-3′, the complement of the repeating 3′-TTGGGG-5′ telomeric DNA sequence that must be synthesized.

We can envision that part of the RNA sequence of the enzyme base pairs with the telomeric DNA, with the remainder of the RNA overlapping the end of the lagging strand. Reverse transcription of this overlapping sequence, which synthesizes DNA on an RNA template, will lead to the extension of the lagging strand, as depicted in Figure 10–18. It is believed that the enzyme is then translocated toward the end of the lagging strand and the sequence of events is repeated. In yeast, of which such synthesis has been investigated in detail, specific proteins are responsible for terminating synthesis, thus determining the number of repeating units added to the end of the lagging strand. Similar proteins no doubt function in the same way as in other eukaryotes.

Analogous enzyme functions have now been found in all eukaryotes studied. In humans, the telomeric DNA sequence that is repeated is 5′-TTAGGG-3′. Correspondingly, human telomerase contains RNA with the repeating sequence 3′-AAUCCC-5′.

As we shall see in Chapter 4, telomeric DNA sequences have been highly conserved throughout evolution, reflecting the critical function of telomeres. In the essay at the end of this chapter, we will see that telomere shortening has been linked to a molecular mechanism involved in the aging process of cells. In most eukaryotic somatic cells, telomerase is, in fact, not active, and thus, with each cell division, the telomeres of each chromosome shorten. After many divisions, the telomere is seriously eroded and the cell loses the capacity for further division. Malignant cells, on the other hand, maintain telomerase activity and are immortalized.

3.13 DNA Recombination, Like DNA Replication, Is Directed by Specific Enzymes

We conclude this chapter by turning to a topic to also be discussed in Chapter 12—genetic recombination. There, we point out that the process of crossing over between homologs depends on breakage and rejoining of the DNA strands. Now that we have discussed the chemistry and replication of DNA, we can consider how recombination occurs at the molecular level. In general, the information that follows pertains to genetic exchange between any two homologous double-stranded DNA molecules, whether they be viral or bacterial chromosomes or eukaryotic homologs during meiosis. Genetic exchange at equivalent positions along two chromosomes with substantial DNA sequence homology is referred to as **general** or **homologous recombination**.

Several models attempt to explain homologous recombination, and they all share certain common features. First, all are based on proposals put forth independently by Robin Holliday and Harold L. K. Whitehouse in 1964. They also depend on the complementarity between DNA strands for the precision of exchange. Finally, each model relies on a series of enzymatic processes in order to accomplish genetic recombination.

One such model is shown in Figure 3–19. It begins with two paired DNA duplexes or homologs (a), in each of which, an endonuclease introduces a single-stranded nick at an identical position (b). The ends of the strands produced by these cuts are then displaced and subsequently pair with their

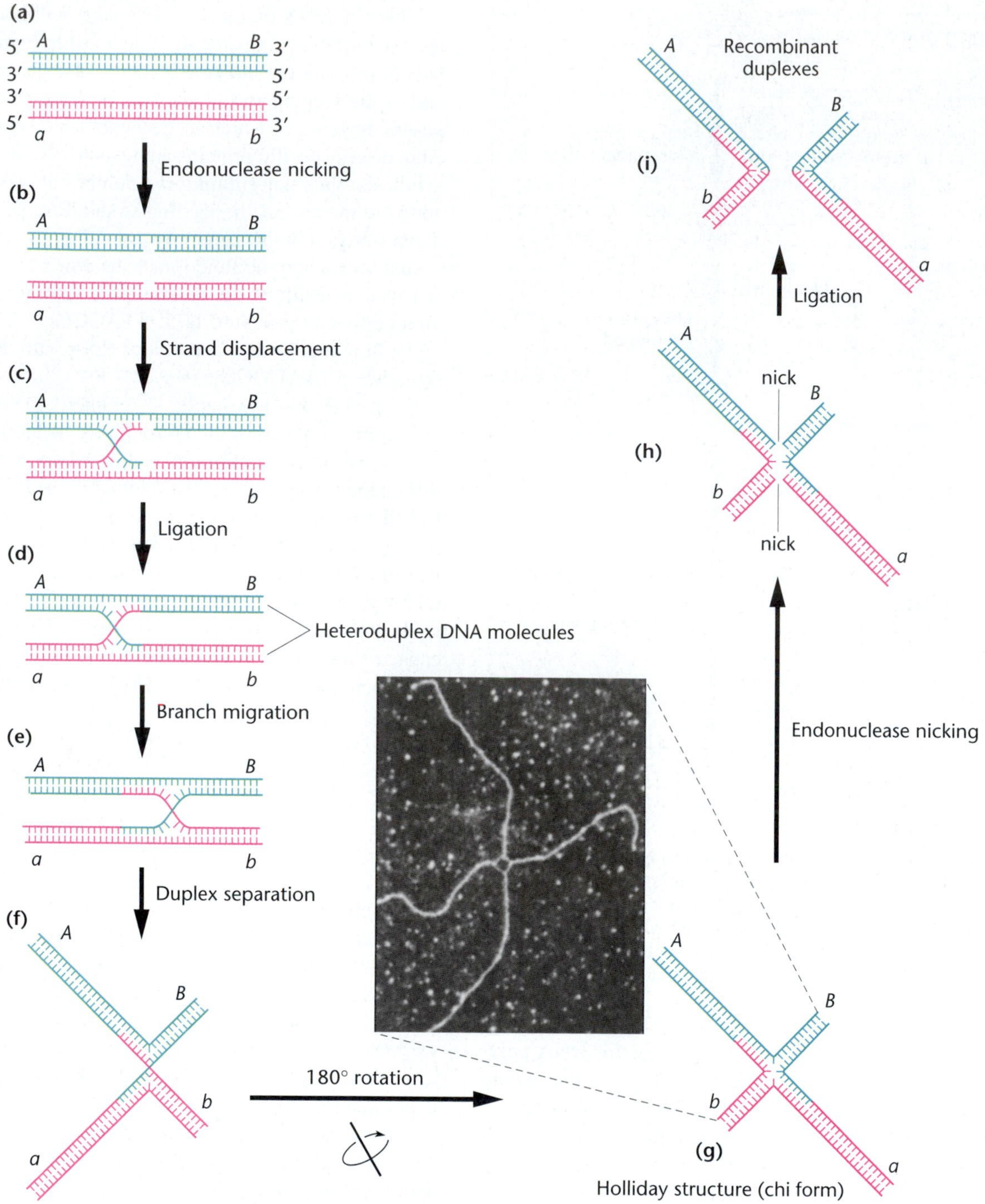

FIGURE 3–19 Model depicting how genetic recombination can occur as a result of the breakage and rejoining of heterologous DNA strands. Each stage is described in the text. The electron micrograph shows DNA in a χ-form structure similar to the diagram in (g); the DNA is an extended Holliday structure, derived from the *Col*E1 plasmid of *E. coli*.

MEDIA TUTORIAL DNA Recombination

complements on the opposite duplex (c). A ligase then seals the loose ends (d), creating hybrid duplexes called **heteroduplex DNA molecules**. The exchange creates a cross-bridged structure. The position of this cross bridge can then move down the chromosome by a process referred to as branch migration (e), which occurs as a result of a zipperlike action as hydrogen bonds are broken and then reformed between complementary bases of the displaced strands of each duplex. This migration yields an increased length of heteroduplex DNA on both homologs.

If the duplexes now separate (f), and the bottom portions rotate 180° (g), an intermediate planar structure called a χ form—the characteristic **Holliday structure**—is created. If the two strands on opposite homologs previously uninvolved in the exchange are now nicked by an endonuclease (h) and ligation occurs (i), recombinant duplexes are created. Note that the arrangement of alleles is altered as a result of recombination.

Evidence supporting this model includes the electron microscopic visualization of χ-form planar molecules from bacteria in which four duplex arms are joined at a single point of exchange [Figure 3–19(g)]. Further important evidence comes from the discovery of the **RecA protein** in *E. coli*. The molecule promotes the exchange of reciprocal single-stranded DNA molecules as occurs in Step (c) of the model. RecA also enhances the hydrogen-bond formation during strand displacement, thus initiating heteroduplex formation. Finally, many other enzymes essential to the nicking and ligation process have also been discovered and investigated. The products of the *recB*, *recC*, and *recD* genes are thought to be involved in the nicking and unwinding of DNA. Numerous mutations that prevent genetic recombination have been found in viruses and bacteria. These mutations represent genes whose products play an essential role in this process.

3.14 Gene Conversion Is a Consequence of DNA Recombination

A modification of the preceding model has helped us to better understand a unique genetic phenomenon known as **gene conversion**. Initially found in yeast by Carl Lindegren and in *Neurospora* by Mary Mitchell, gene conversion is characterized by a *nonreciprocal* genetic exchange between two closely linked genes. For example, if we were to cross two *Neurospora* strains, each bearing a separate mutation ($a+ \times b+$), a *reciprocal* recombination between the genes would yield spore pairs of the ++ *and* the *ab* genotypes. However, a nonreciprocal exhange yields one pair without the other. Working with pyridoxine mutants, Mitchell observed several asci-containing spore pairs with the ++ genotype, but not the reciprocal product (*ab*). Because the frequency of these events was higher than the predicted mutation rate and consequently could not be accounted for by mutation, they were called "gene conversions." They were so named because it appeared that one allele had somehow been "converted" to another in which genetic exchange had also occurred. Similar findings come from studies of other fungi as well.

Gene conversion is now considered to be a consequence of the process of DNA recombination. One possible explanation interprets conversion as a mismatch of base pairs during heteroduplex formation, as shown in Figure 3–20. Mismatched regions of hybrid strands can be repaired by the excision of one of the strands and synthesizing the complement by using the remaining strand as a template. Excision may occur in either one of the strands, yielding two possible "corrections." One repairs the mismatched base pair and "converts" it to restore the original sequence. The other also corrects the mismatch, but does so by copying the altered strand, creating a base-pair substitution. Conversion may have the effect of creating identical alleles on the two homologs that were different initially.

In our example in Figure 3–20, suppose that the G≡C pair on one of the two homologs was responsible for the mutant allele, while the A=T pair was part of the wild-type gene sequence on the other homolog. Conversion of the G≡C pair to A=T would have the effect of changing the mutant allele to wild type, just as Mitchell originally observed.

Gene-conversion events have helped to explain other puzzling genetic phenomena in fungi. For example, when mutant and wild-type alleles of a single gene are crossed, asci should yield equal numbers of mutant and wild-type spores. However, exceptional asci with 3:1 or 1:3 ratios are sometimes observed. These ratios can be understood in terms of gene conversion. The phenomenon has also been detected during mitotic events in fungi, as well as during the study of unique compound chromosomes in *Drosophila*.

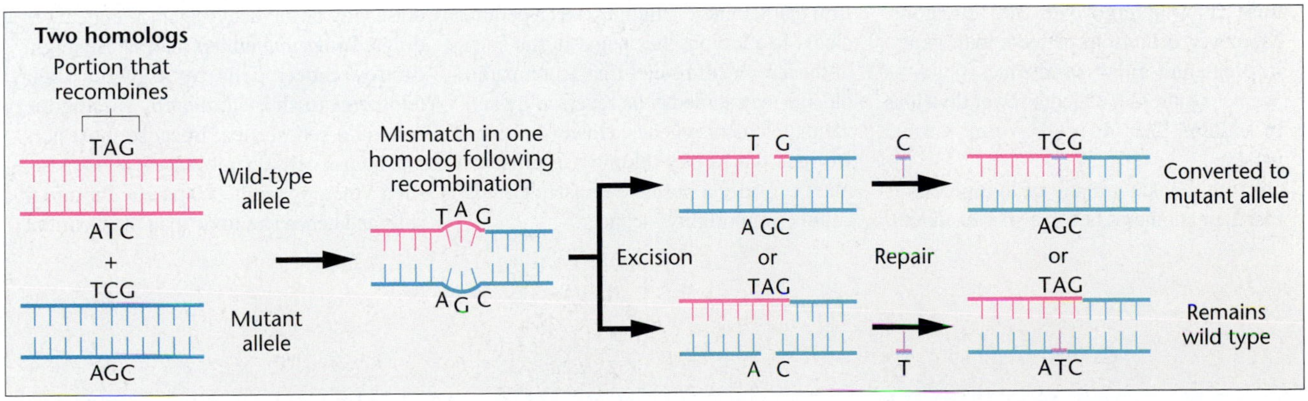

FIGURE 3–20 A proposed mechanism that accounts for the phenomenon of gene conversion. A base-pair mismatch occurs in one of the two homologs (bearing the mutant allele) during heteroduplex formation, which accompanies recombination in meiosis. During excision repair, one of the two mismatches is removed and the complement is synthesized. In one case, the mutant base pair is preserved. When it is subsequently included in a recombinant spore, the mutant genotype will be maintained. In the other case, the mutant base pair is converted to the wild-type sequence. When included in a recombinant spore, the wild-type genotype will be expressed, leading to a nonreciprocal exchange ratio.

Genetics, Technology, and Society

Telomerase: The Key to Immortality?

Humans, as do all multicellular organisms, grow old and die. As we age, our immune systems become less efficient, wound healing is impaired, and tissues and organs lose resilience. It has always been a mystery why we go through these age-related declines, and why each species has a characteristic finite lifespan. Why do we grow old? Can we reverse this march to mortality? Some recent discoveries suggest that the answers to these questions may lie at the ends of our chromosomes.

The study of human aging begins with a study of human cells growing in culture dishes. Like the organisms from which the cells are taken, cells in culture have a finite life span. This "replicative senescence" was noted over 30 years ago by Hayflick. He reported that normal human fibroblasts lose their ability to grow and divide after about 50 cell divisions. These senescent cells remain metabolically active, but can no longer proliferate. Eventually, they die. Although we don't know whether cellular senescence directly causes organismal aging, the evidence is suggestive. For example, cells from young people go through more divisions in culture than do cells from older people; human fetal cells divide 60–80 times before undergoing senescence, whereas cells from older adults divide only 10–20 times. In addition, cells from species with short life spans stop growing after fewer divisions than do cells from species with longer life spans; mouse cells divide 10–15 times in culture, but tortoise cells undergo over 100 divisions. Moreover, cells from patients with genetic premature aging syndromes (such as Werner syndrome) undergo fewer divisions in culture than do cells from normal patients.

Another characteristic of aging cells is that their telomeres become shorter. *Telomeres* are the tips of linear chromosomes and consist of several thousand repeats of a short DNA sequence (TTAGGG in humans). Telomeres help preserve the structural integrity of chromosomes by protecting their ends from degrading or from fusing to other chromosomes. Telomeres are created and maintained by *telomerase*—a remarkable RNA-containing enzyme that adds telomeric DNA sequences onto the ends of linear chromosomes. Telomerase also solves the "end-replication" problem—which asserts that linear DNA molecules become shorter at each replication, because DNA polymerase cannot synthesize new DNA at the 3′-ends of each parent strand. By adding numerous telomeric repeat sequences onto the 3′-ends of chromosomes, telomerase prevents the chromosomes from shrinking into oblivion. Unfortunately, normal somatic cells contain little, if any, telomerase. As a result, telomere length decreases by about 100 base pairs every time a normal cell divides. Telomere shortening may act as a clock that counts cell division and instructs the cell to stop dividing.

Could we gain perpetual youth and vitality by increasing our telomere lengths? A recent study suggests that it may be possible to reverse senescence by artificially increasing the amount of telomerase in our cells. When the investigators introduced cloned telomerase genes into normal human cells in culture, telomeres lengthened by thousands of base pairs and the cells continued to grow long past their senescence point. These observations confirm that telomere length acts as a cellular clock. In addition, they suggest that some of the atrophy of tissues that accompanies old age may someday be reversed by activating telomerase genes. However, before we rush out to buy telomerase pills, we must consider a possible consequence of cellular immortality—cancer.

Although normal cells undergo senescence after a specific number of cell divisions, cancer cells do not. It is thought that cancers arise after several genetic mutations accumulate in a cell. These mutations disrupt the normal checks and balances that control cell growth and division. It therefore seems logical that cancer cells would also stop the normal aging clock. If their telomeres became shorter after each cell division, tumor cells would eventually succumb to aging and cease growth. However, if they synthesized telomerase, they would arrest the ticking of the senescence clock and become immortal. In keeping with this idea, over 90% of human tumor cells contain telomerase activity and have stable telomeres. The correlation between uncontrolled tumor cell growth and the presence of telomerase activity is so good that telomerase assays are being developed as diagnostic markers for cancer. Although there is currently some debate about whether the presence of telomerase is a prerequisite for, or simply a consequence of, cell transformation, it is possible that acquiring telomerase activity may be an important step in the development of a cancer cell. In support of this hypothesis, recent studies show that the induction of telomerase activity, when combined with the inactivation of a tumor-suppressor gene ($p16^{INK4a}$), results in cell immortalization—an essential step toward tumor development. Therefore, any attempt to increase telomerase activity in normal cells carries the risk of enhancing the development of tumors.

An attractive possibility is that telomerase may be an ideal target for anticancer drugs. Drugs that inhibit telomerase might destroy cancer cells by allowing their telomeres to shorten, thereby forcing the cells into senescence. Because most normal human cells do not express telomerase, such a therapy might be specific for tumor cells and hence less toxic than most current anticancer drugs. Such antitelomerase anticancer drugs are currently under development by Geron Corporation. Although we don't yet know whether this approach will work in animals, it appears to work in cultured tumor cells. Tumor cells that are treated with an antitelomerase agent lose telomeric sequences and die after about 25 cell divisions.

Before antitelomerase drugs can be developed and used on humans, several questions must be answered. Is telomerase required by some normal human cells (such as lymphocytes and germ cells)? If so, antitelomerase drugs may be unacceptably toxic. Could some cancer cells compensate for the loss of telomerase by using other telomere-lengthening mechanisms (such as recombination)? If so, antitelomerase drugs may be doomed to failure. Even if we inhibit telomerase activity in tumor cells, could they undergo multiple divisions before reaching senescence and still damage the host?

Will telomerase allow us to both arrest cancers *and* reverse the descent into old age? Time will tell.

References

Bodnar, A.G., et al. 1998. Extension of life-span by introduction of telomerase into normal human cells. *Science* 279:349–52.

deLange, T. 1998. Telomeres and senescence: Ending the debate. *Science* 279:334–35.

Kiyono, T., et al. 1998. Both Rb/p16^{INK4a} inactivation and telomerase activity are required to immortalize human epithelial cells. *Nature* 396:84–88.

Chapter Summary

1. In theory, three modes of DNA replication are possible: semiconservative, conservative, and dispersive. Although all three rely on base complementarity, semiconservative replication is the most straightforward and was predicted by Watson and Crick.
2. In 1958, Meselson and Stahl resolved this question in favor of semiconservative replication in *E. coli*, showing that newly synthesized DNA consists of one old strand and one new strand. Taylor, Woods, and Hughes used root tips of the broad bean to demonstrate semiconservative replication in eukaryotes.
3. During the same period, Kornberg isolated the enzyme DNA polymerase I from *E. coli* and showed that it is capable of directing in vitro DNA synthesis, provided that a template and precursor nucleoside triphosphates are supplied.
4. The subsequent discovery of the *polA1* mutant strain of *E. coli*, capable of DNA replication despite its lack of polymerase I activity, cast doubt on the enzyme's in vivo replicative function. DNA polymerases II and III were then isolated. Polymerase III has been identified as the enzyme responsible for DNA replication in vivo.
5. During the process of DNA synthesis, the double helix unwinds, forming a replication fork at which synthesis begins. Proteins stabilize the unwound helix and assist in relaxing the coiling tension created ahead of the replication.
6. Synthesis is initiated at specific sites along each template strand by the enzyme primase, which results in a short segment of RNA that provides a suitable 3′ end upon which DNA polymerase III can begin polymerization.
7. Because of the antiparallel nature of the double helix, polymerase III synthesizes DNA continuously on the leading strand in a 5′–3′ direction. On the opposite strand, called the lagging strand, synthesis results in short Okazaki fragments that are later joined by DNA ligase.
8. DNA polymerase I removes and replaces the RNA primer with DNA, which is joined to the adjacent polynucleotide by DNA ligase.
9. The isolation of numerous phage and bacterial mutant genes affecting many of the molecules involved in the replication of DNA has helped to define the complex genetic control of the entire process.
10. DNA replication in eukaryotes is similar to, but more complex than, replication in prokaryotes. Multiple replication origins exist and multiple forms of DNA polymerase direct DNA synthesis.
11. Replication at the ends (telomeres) of linear molecules poses a special problem in eukaryotes that can be solved by a unique RNA-containing enzyme called telomerase.
12. Homologous recombination between genetic molecules relies on a series of enzymes that can cut, realign, and reseal DNA strands. The phenomenon of gene conversion may be best explained in terms of mismatch repair synthesis during these exchanges.

Insights and Solutions

1. Predict the theoretical results of conservative and dispersive replication of DNA under the conditions of the Meselson–Stahl experiment. Follow the results through two generations of replication after cells have been shifted to an ^{14}N-containing medium, using the following sedimentation pattern.

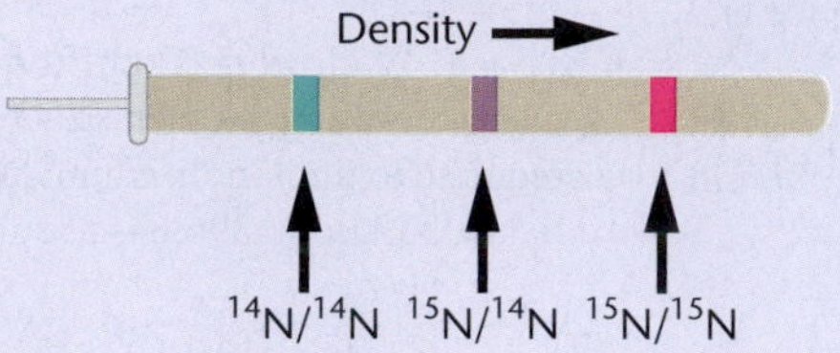

Solution:

Conservative replication

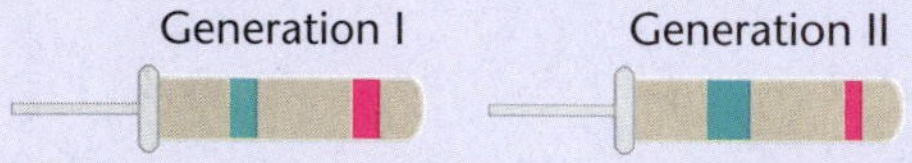

Dispersive replication

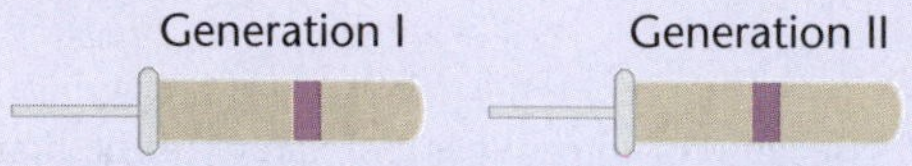

2. Mutations in the *dnaA* gene of *E. coli* are lethal and can only be studied following the isolation of conditional, temperature-sensitive mutations. Such mutant strains grow nicely and replicate their DNA at the permissive temperature of 18°C, but they do not grow or replicate their DNA at the restrictive temperature of 37°C. Two observations were useful in determining the function of the DnaA protein product. First, in vitro studies using DNA templates that have unwound do not require the DnaA protein. Second, if intact cells are grown at 18°C and are then shifted to 37°C, DNA synthesis continues at this temperature until one round of replication is completed and then stops. What do these observations suggest about the role of the *dnaA* gene product?

Solution: At 18°C (the permissive temperature), the mutation is not expressed and DNA synthesis begins. Following the shift to the restrictive temperature, the already-initiated DNA synthesis continues, but no new synthesis can begin. Because the DnaA protein is not required for synthesis of unwound DNA, these observations suggest that, in vivo, the DnaA protein plays an essential role in DNA synthesis by interacting with the intact helix and somehow facilitating the localized denaturing necessary for synthesis to proceed.

3. DNA is allowed to replicate in moderately radioactive ^{3}H-thymidine for several minutes and is then switched to a highly radioactive medium for several more minutes. Synthesis is stopped and the DNA is subjected to autoradiography and electron microscopy. Interpret as much as you can regarding DNA replication from the drawing of the micrograph presented here.

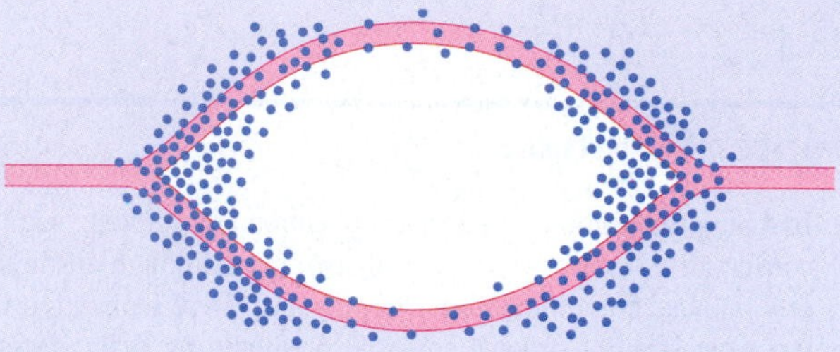

Solution: One interpretation is that there are two advancing replication forks proceeding in opposite directions and that replication is therefore bidirectional. The low density of grains represents synthesis that occurred during the first few minutes, beginning at the origin and proceeding outward in both directions. The higher density of grains represents the later minutes of synthesis when the level of radioactivity was increased.

Problems and Discussion Questions

1. Compare conservative, semiconservative, and dispersive modes of DNA replication.
2. Describe the role of ^{15}N in the Meselson–Stahl experiment.
3. In the Meselson–Stahl experiment, which of the three modes of replication could be ruled out after one round of replication? after two rounds?
4. Predict the results of the experiment by Taylor, Woods, and Hughes if replication were (a) conservative and (b) dispersive.
5. Reconsider Problem 35 in Chapter 2. In the model you proposed, could the molecule be replicated semiconservatively? Why? Would other modes of replication work?
6. What are the requirements for in vitro synthesis of DNA under the direction of DNA polymerase I?
7. In Kornberg's initial experiments, he actually grew *E. coli* in Anheuser-Busch beer vats. (He was working at Washington University in St. Louis.) Why do you think this was helpful to the experiment?
8. How did Kornberg test the fidelity of copying DNA by polymerase I?
9. Which of Kornberg's tests is the more stringent assay? Why?
10. Which characteristics of DNA polymerase I raised doubts that its in vivo function is the synthesis of DNA leading to complete replication?
11. Explain the theory of nearest neighbor frequency.
12. What is meant by "biologically active" DNA?
13. Why was the phage ϕX174 chosen for the experiment demonstrating biological activity?
14. Outline the experimental design of Kornberg's biological activity demonstration.
15. What was the significance of the *polA1* mutation?
16. Summarize and compare the properties of DNA polymerase I, II, and III.

17. List and describe the function of the 10 subunits constituting DNA polymerase III. Distinguish between the holoenzyme and the core enzyme.
18. Distinguish between (a) unidirectional and bidirectional synthesis and (b) continuous and discontinuous synthesis of DNA.
19. List the proteins that unwind DNA during in vivo DNA synthesis. How do they function?
20. Define and indicate the significance of (a) Okazaki fragments, (b) DNA ligase, and (c) primer RNA during DNA replication.
21. Outline the current model for DNA synthesis.
22. Why is DNA synthesis expected to be more complex in eukaryotes than in bacteria? How is DNA synthesis similar in the two types of organisms?
23. If the analysis of DNA from two different microorganisms demonstrated very similar base compositions are the DNA sequences of the two organisms also nearly identical?
24. Suppose that *E. coli* synthesizes DNA at a rate of 100,000 nucleotides per minute and takes 40 minutes to replicate its chromosome.
    (a) How many base pairs are present in the entire *E. coli* chromosomes?
    (b) What is the physical length of the chromosome in its helical configuration—that is, what is the circumference of the chromosome if it were opened into a circle?
25. Several temperature-sensitive mutant strains of *E. coli* display various characteristics. Predict what enzyme or function is being affected by each mutation.
    (a) Newly synthesized DNA contains many mismatched base pairs.
    (b) Okazaki fragments accumulate, and DNA synthesis is never completed.
    (c) No initiation occurs.
    (d) Synthesis is very slow.
    (e) Supercoiled strands are found to remain following replication, which is never completed.
26. Define gene conversion and describe how this phenomenon is related to genetic recombination.
27. Many of the gene products involved in DNA synthesis were initially defined by studying mutant *E. coli* strains that could not synthesize DNA.
    (a) The *dnaE* gene encodes the a subunit of DNA polymerase III. What effect is expected from a mutation in this gene? How could the mutant strain be maintained?
    (b) The *dnaQ* gene encodes the ϵ subunit of DNA polymerase. What effect is expected from a mutation in this gene?
28. In 1994, telomerase activity was discovered in human cancer cell lines. Although telomerase is not active in human somatic tissue, this discovery indicated that humans do contain the genes for telomerase proteins and telomerase RNA. Since inappropriate activation of telomerase can cause cancer, why do you think the genes coding for this enzyme have been maintained in the human genome throughout evolution? Are there any types of human body cells where telomerase activation would be advantageous or even necessary? Explain.

Extra-Spicy Problems

29. The genome of the fruit fly *Drosophila melanogaster* consists of approximately 1.6×10^8 base pairs. DNA synthesis occurs at a rate of 30 base pairs per second. In the early embryo, the entire genome is replicated in five minutes. How many *bidirectional origins of synthesis* are required to accomplish this feat?
30. To gauge the fidelity of DNA synthesis, Kornberg used the nearest neighbor-frequency test, which determines the frequency with which any two bases occur adjacent to each other along the polynucleotide chain. This test relies on the enzyme spleen phosphodiesterase. As we saw in Figure 3–8, during synthesis of DNA, 5′-nucleotides are inserted—that is, each nucleotide is added with the phosphate on the C-5′ of deoxyribose. As shown in the accompanying figure, however, the phosphodiesterase enzyme cleaves between the phosphate and the C-5′ atom, thereby producing 3′-nucleotides. The phosphates on only one of the four nucleotides (cytidylic acid, for example) are made radioactive with ^{32}P during DNA synthesis. Then, after enzymatic cleavage, the radioactive phosphate will be transferred to the base that is the "nearest neighbor" on the 5′ side of all cytidylic acid nucleotides.

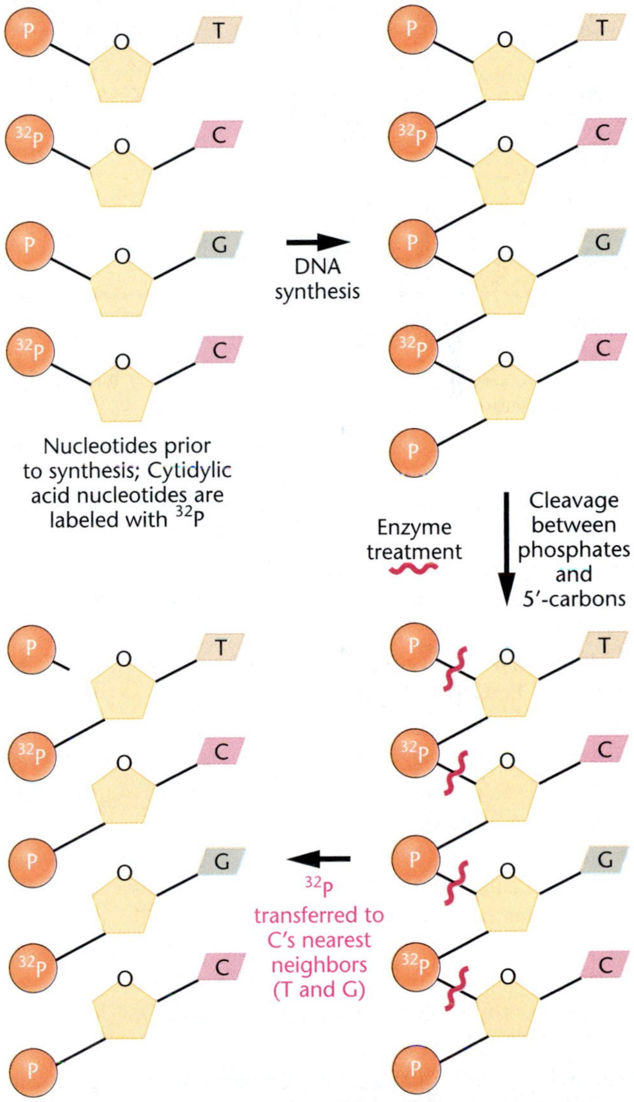

Following four separate experiments, in each of which only one of the four nucleotide types is made radioactive, the frequency of all 16 possible nearest neighbors can be calculated. When Kornberg applied the nearest neighbor-frequency test to the DNA template and the resultant product from a variety of experiments, he found general agreement between the nearest neighbor frequencies of the two.

Analysis of nearest neighbor data led Josse, Kaiser, and Kornberg (1961) to conclude that the two strands of the double helix are in opposite polarity to one another. Demonstrate their approach by determining the outcome of such an analysis if the strands of DNA shown here are (a) antiparallel versus (b) parallel:

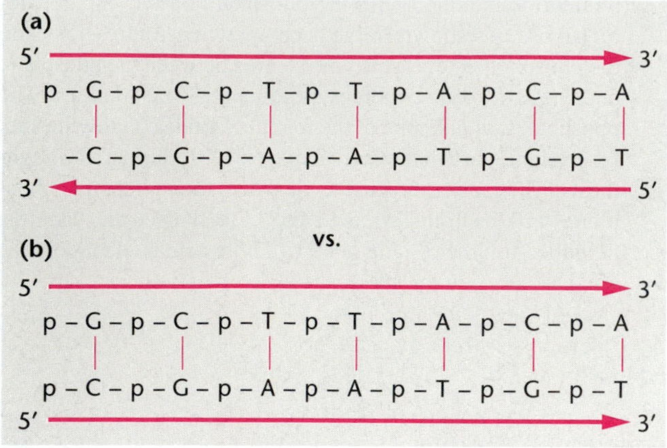

31. Assume a hypothetical organism in which DNA replication is conservative. Design an experiment similar to that of Taylor, Woods, and Hughes that will unequivocally establish this fact. Using the format established in Figure 10–5, draw sister chromatids and illustrate the expected results establishing this mode of replication.

32. DNA polymerases in all organisms only add 5′ nucleotides to the 3′ end of a growing DNA strand, never to the 5′ end. One possible reason for this is the fact that most DNA polymerases have a proof-reading function that would not be *energetically* possible if DNA synthesis occurred in the 3′ to 5′ direction.

(a) Draw out the reaction that DNA polymerase would have to catalyze if DNA synthesis occurred in the 3′ to 5′ direction.

(b) Consider the information present in your drawing, speculate as to why proofreading would be problematic, while it isn't if synthesis occurs in the 5′ to 3′ direction.

33. An alien organism was investigated. It displayed characteristics of eukaryotes. When DNA replication was examined, two unique features were apparent: 1) no Okazaki fragments were observed; and 2) there was a telomere problem, i.e., telomeres shortened, but only on one end of the chromosome. Put forward a model of DNA that is consistent with both of these observations.

Selected Readings

Bodnar, A.G., et al. 1998. Extension of life-span by introduction of telomerase into normal human cells. *Science* 279:349–52.

Blackburn, E.H. 1991. Structure and function of telomeres. *Nature* 350:569–72.

Bramhill, D., and Kornberg, A. 1988. A model for initiation at origins of DNA replication. *Cell* 54:915–18.

Camerini-Otero, R.D., and Hsieh, P. 1993. Parallel DNA triplexes, homologous recombination, and other homology-dependent DNA interactions. *Cell* 73:217–23.

Davidson, J.N. 1976. *The biochemistry of nucleic acids*, 8th ed. Orlando, FL: Academic Press.

DeLucia, P., and Cairns, J. 1969. Isolation of an *E. coli* strain with a mutation affecting DNA polymerase. *Nature* 224:1164–66.

Denhardt, D.T., and Faust, E.A. 1985. Eukaryotic DNA replication. *Bioessays* 2:148–53.

Diffley, J.F.X. 1996. Once and only once upon a time: Specifying and regulating origins of DNA replication in eukaryotic cells. *Genes and Development* 10:2819–30.

Dressler, D., and Potter, H. 1982. Molecular mechanisms in genetic recombination. *Annu. Rev. Biochem.* 51:727–61.

Gellert, M. 1981. DNA topoisomerases. *Annu. Rev. Biochem.* 50:879–910.

Gilbert, D.M. 2001. Making sense of eukaryotic DNA replication origins. *Science* 294:96–100.

Greider, C.W., and Blackburn, E.H. 1989. A telomeric sequence in the RNA of *Tetrahymena* telomerase required for telomere repeat synthesis. *Nature* 337:331–36.

Greider, C.W. 1996. Telomeres, telomerase, and cancer. *Sci. Am.* 274 (Feb.):92–97.

———. 1998. Telomerase activity, cell proliferation, and cancer. *Proc. Natl. Acad. Sci. USA* 95:90–92.

Herendeen, D.R., and Kelly, T.J. 1996. DNA polymerase III: Running rings around the fork. *Cell* 84:5–8.

Hindges, R., and Hübscher, U. 1997. DNA polymerase essential for DNA transactions *Biol. Chem.* 378:345–62.

Holliday, R. 1964. A mechanism for gene conversion in fungi. *Genet. Res.* 5:282–304.

Holmes, F.L. 2001. *Meselson, Stahl, and replication of DNA: A history of the "most beautiful experiment in biology"*. New Haven: Yale U. Press.

Huberman, J.C. 1987. Eukaryotic DNA replication: A complex picture partially clarified. *Cell* 48:7–8.

Josse, J., Kaiser, A.D., and Kornberg, A. 1961. Enzymatic synthesis of deoxyribonucleic acid: VIII. Frequencies of nearest-neighbor base sequences in deoxyribonucleic acid. *J. Biol. Chem.* 236:864–75.

Kornberg, A. 1960. Biological synthesis of DNA. *Science* 131:1503–8.

———. 1969. Active center of DNA polymerase. *Science* 163:1410–18.

———. 1974. *DNA synthesis.* New York: W.H. Freeman.

———. 1979. Aspects of DNA replication. *Cold Spring Harbor Symp. Quant. Biol.* 43:1–10.

Kornberg, A., and Baker, T. 1992. *DNA replication*, 2nd ed. New York: W.H. Freeman.

Lehman, I.R. 1974. DNA ligase: Structure, mechanism, and function. *Science* 186:790–97.

Lindegren, C.C. 1953. Gene conversion in *Saccharomyces*. *J. Genet.* 51:625–37.

Lodish, H., et al. 1999. *Molecular cell biology*, 4th ed. New York: Scientific American Books.

GENETICS MediaLab

The resources that follow will help you achieve a better understanding of the concepts presented in this chapter. These resources can be found either on the CD packaged with this textbook or on the Companion Web site found at **http://www.prenhall.com/klug**

CD Resources:

Module 3.1: DNA Replication

Module 3.2: DNA Recombination

Web Problem 1:

Time for completion = 15minutes

How are the roles of the molecules involved in DNA replication coordinated? Researchers have identified many of the components of the replication machinery and now understand the major steps in replication (local unwinding, leading and lagging strand synthesis using RNA primers, and proofreading). The coordination of these functions is still under investigation. At the linked Web site, you can manipulate replication-protein models derived from X-ray crystallography data. Directions on how to download the CHIME browser plug-in needed to view this module are found at the top of the Web page. Scroll to the Hall of DNA polymerization and click on the T7 DNA replication complex. Examine how replication proteins direct the addition of free nucleotides to the newly synthesized strand of DNA. Write a brief discussion about how physical attributes of these replication proteins contribute to their function. To complete this exercise, visit Web Problem 1 in Chapter 3 of your Companion Web site, and select the keyword **REPLICATION**.

Web Problem 2:

Time for completion = 10 minutes

How do scientists construct models of complex molecular processes such as recombination? Our understanding of this physical exchange of DNA strands is derived in part from experiments in which intermediate stages are isolated and identified by microscopy or biochemical analysis. Mutations in recombination proteins have been useful in teasing apart the roles of these molecules. The linked Web site contains electron micrographs of DNA bound by recombination proteins. Read the figure legend and click on the links to view the images. Describe one of the images you view. Include what stage of recombination you observe and any experimental conditions contributing to the "capture" of this intermediate. To complete this exercise, visit Web Problem 2 in Chapter 3 of your Companion Web site, and select the keyword **RECOMBINATION**.

Web Problem 3:

Time for completion = 10 minutes

What does a recombination event look like? A crucial step during recombination is the resolution of an intermediate form called a Holliday structure. Such a structure forms when the two homologous DNA molecules are joined in a cross-bridge. In this exercise, you will view a short animation of rotation of the Holliday junction. Describe what this rotation event accomplishes. The resolution of this structure requires the nicking of two strands and rejoining to form two separate DNA molecules. There are two possible pairs of DNA strands that can be nicked and rejoined; therefore, there are two distinct outcomes. Draw a diagram indicating the two pairs of DNA strands, and explain briefly what distinguishes one pair of nicked and rejoined DNA strands of the resulting DNA molecules from the other. To complete this exercise, visit Web Problem 3 in Chapter 3 of your Companion Web site, and select the keyword **HOLLIDAY**.

Meselson, M., and Stahl, F.W. 1958. The replication of DNA in *Escherichia coli*. *Proc. Natl. Acad. Sci. USA* 44:671–82.

Mitchell, M.B. 1955. Aberrant recombination of pyridoxine mutants of *Neurospora*. *Proc. Natl. Acad. Sci. USA* 41:215–20.

Ogawa, T., and Okazaki, T. 1980. Discontinuous DNA synthesis. *Annu. Rev. Biochem.* 49:421–57.

Okazaki, T., et al. 1979. Structure and metabolism of the RNA primer in the discontinuous replication of prokaryotic DNA. *Cold Spring Harbor Symp. Quant. Biol.* 43:203–22.

Radding, C.M. 1978. Genetic recombination: Strand transfer and mismatch repair. *Annu. Rev. Biochem.* 47:847–80.

Radman, M., and Wagner, R. 1988. The high fidelity of DNA duplication. *Sci. Am.* 259 (Aug.):40–46.

Schekman, R., Weiner, A., and Kornberg, A. 1974. Multienzyme systems of DNA replication. *Science* 186:987–93.

Stahl, F.W. 1979. *Genetic recombination: Thinking about it in phage and fungi*. New York: W.H. Freeman.

———. 1987. Genetic recombination. *Sci. Am.* (Feb.) 256:90–101.

Szostak, J.W., Orr-Weaver, T.L., and Rothstein, R.J. 1983. The double-strand-break repair model for recombination. *Cell* 33:25–35.

Taylor, J.H., Woods, P.S., and Hughes, W.C. 1957. The organization and duplication of chromosomes revealed by autoradiographic studies using tritium-labeled thymidine. *Proc. Natl. Acad. Sci. USA* 48:122–28.

Tömmes, P., and Hübscher, U. 1992. Eukaryotic DNA helicases: Essential enzymes for DNA transactions. *Chromosoma* 101:467–73.

Tomizawa, J., and Selzer, G. 1979. Initiation of DNA synthesis in *E. coli*. *Annu. Rev. Biochem.* 48:999–1034.

Wang, J.C. 1982. DNA topoisomerases. *Sci. Am.* 247 (July):94–108.

———. 1987. Recent studies of DNA topoisomerases. *Biochim. Biophys. Acta* 909:1–9.

Watson, J.D., et al. 1987. *Molecular biology of the gene, Vol. 1. General principles*, 4th ed. Menlo Park, CA: Benjamin/Cummings.

Whitehouse, H.L.K. 1982. *Genetic recombination: Understanding the mechanisms*. New York: Wiley.

Zubay, G.L., and Marmur, J., eds. 1973. *Papers in biochemical genetics*, 2d ed. New York: Holt, Rinehart & Winston.

Zyskind, J.W., and Smith, D.W. 1986. The bacterial origin of replication, *oriC*. *Cell* 46:489–90.

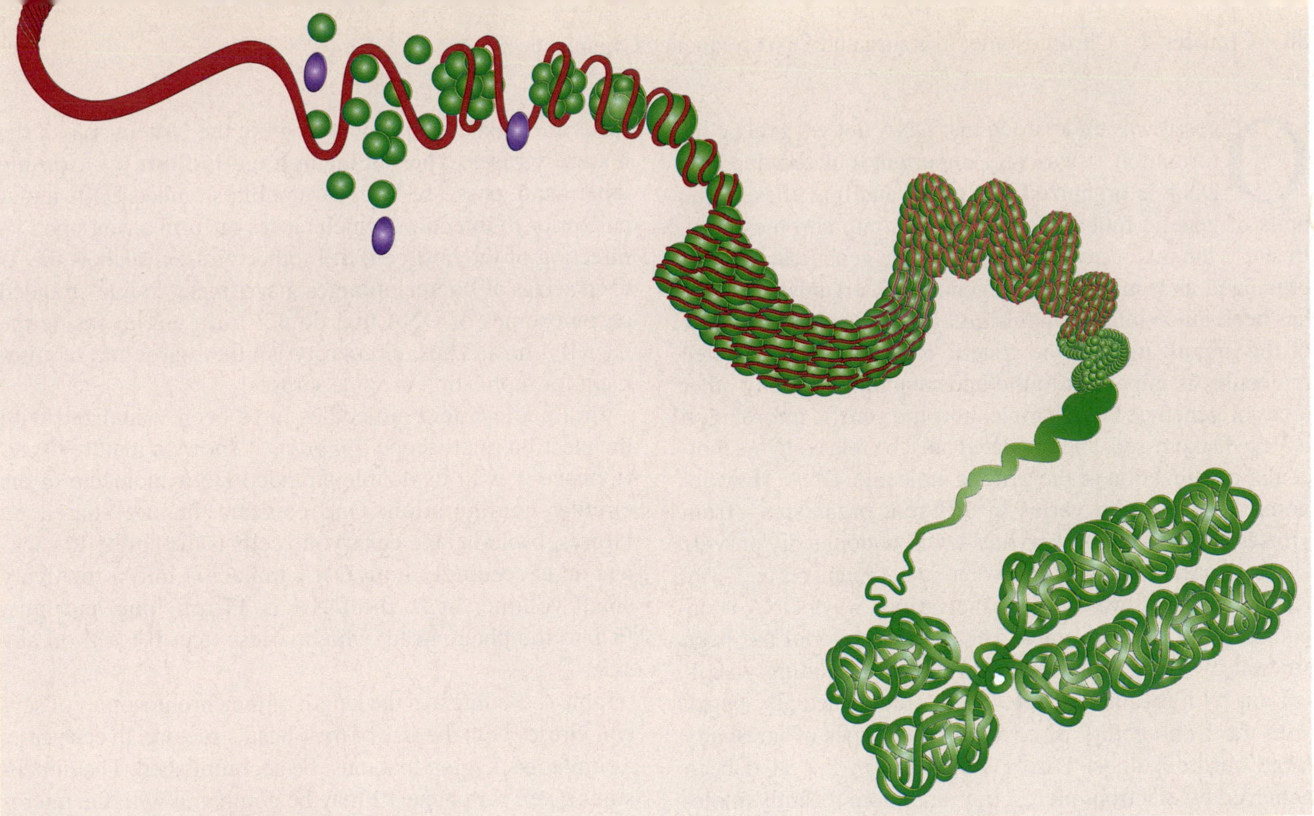

Levels of organization of the eukaryotic chromatin fiber as it is condensed into a metaphase chromosome.

4

Chromosome Structure and DNA Sequence Organization

KEY CONCEPTS

4.1 Viral and Bacterial Chromosomes Are Relatively Simple DNA Molecules

4.2 Supercoiling Is Common in the DNA of Viral and Bacterial Chromosomes

4.3 Specialized Chromosomes Reveal Variations in Structure
- Polytene Chromosomes
- Lampbrush Chromosomes

4.4 DNA Is Organized into Chromatin in Eukaryotes
- Chromatin Structure and Nucleosomes
- Chromatin Remodeling
- Heterochromatin

4.5 Chromosome Banding Differentiates Regions along the Mitotic Chromosome

4.6 Eukaryotic Genomes Demonstrate Complex Sequence Organization Characterized by Repetitive DNA
- Repetitive DNA and Satellite DNA
- Centromeric DNA Sequences
- Telomeric DNA Sequences
- Middle Repetitive Sequences: VNTRs and Dinucleotide Repeats
- Repetitive Transposed Sequences: SINES and LINES
- Middle Repetitive Multiple Copy Genes

4.7 The Vast Majority of a Eukaryotic Genome Does Not Encode Functional Genes

Once it was understood that DNA houses genetic information, it was very important to determine how DNA is organized into genes and how these basic units of genetic function are organized into chromosomes. In short, the major question was how the genetic material is organized as it makes up the genome of organisms. There has been much interest in this question, because knowledge of the organization of the genetic material and associated molecules is important to the understanding of many other areas of genetics. For example, how the genetic information is stored, expressed, and regulated must be related to the molecular organization of the genetic molecule, DNA. How genomic organization varies in different organisms—from viruses to bacteria to eukaryotes—will undoubtedly provide a better understanding of the evolution of organisms on Earth.

In this chapter, we focus on the various ways DNA is organized into chromosomes. The genetic material has been studied by using numerous approaches, including visualization by light microscopy. These studies actually began with the 19th-century discovery of two types of unusually large chromosomes. Direct visualization has also been achieved by electron microscopy, and more recently, molecular analysis has provided significant insights into chromosome organization.

We will first survey what is known about chromosomes in viruses and bacteria. Then, we will examine the large specialized structures called polytene and lampbrush chromosomes. In the second half of the chapter, we will discuss some of the basic questions of eukaryotic genomic organization. For example, how is DNA complexed with proteins to form chromatin, and how are the chromatin fibers characteristic of interphase condensed into chromosome structures visible during mitosis and meiosis?

We will conclude the chapter by examining some aspects of DNA sequence organization that are characteristic of eukaryotic genomes. Later, in Chapter 17, we will turn to a discussion of the very exciting studies that have elucidated how genes are organized in chromosomes.

4.1 Viral and Bacterial Chromosomes Are Relatively Simple DNA Molecules

Compared with eukaryotes, the chromosomes of viruses and bacteria are much less complicated. They usually consist of a single nucleic acid molecule, largely devoid of associated proteins and containing relatively little genetic information in comparison to the multiple chromosomes constituting the genome of higher forms. These characteristics have greatly simplified analysis, providing a fairly comprehensive view of the structure of viral and bacterial chromosomes.

The chromosomes of viruses consist of a nucleic acid molecule—either DNA or RNA—that can be either single or double stranded. They can exist as circular structures (covalently closed circles), or they can take the form of linear molecules. The single-stranded DNA of the **ϕX174 bacteriophage** and the double-stranded DNA of the **polyoma virus** are closed loops housed within the protein coat of the mature viruses. The **bacteriophage lambda** (λ), on the other hand, possesses a linear double-stranded DNA molecule prior to infection, which closes to form a ring upon its infection of the host cell. Still other viruses, such as the **T-even series of bacteriophages**, have linear double-stranded chromosomes of DNA that do not form circles inside the bacterial host. Thus, circularity is not an absolute requirement for replication in some viruses.

Viral nucleic acid molecules have been visualized with the electron microscope. Figure 4–1 shows a mature bacteriophage λ with its double-stranded DNA molecule in the circular configuration. One constant feature shared by viruses, bacteria, and eukaryotic cells is the ability to package an exceedingly long DNA molecule into a relatively small volume. In λ, the DNA is 17 μm long and must fit into the phage head, which is less than 0.1 μm on any side.

Table 4–1 compares the length of the chromosomes of several viruses with the size of their head structure. In each case, a similar packaging feat must be accomplished. The dimensions given for phage T2 may be compared with the micrograph of both the DNA and the viral particle shown in Figure 4–2. Seldom does the space available in the head of a virus exceed the chromosome volume by more than a factor of two. In many cases, almost all space is filled, indicating nearly perfect packing. Once packed within the head, the genetic material is functionally inert until it is released into a host cell.

Bacterial chromosomes are also relatively simple in form. They always consist of a double-stranded DNA molecule, compacted into a structure sometimes referred to as the **nucleoid**. *Escherichia coli*, the most extensively studied bacterium, has a large, circular chromosome measuring approximately 1200 μm (1.2 mm) in length. When the cell is gently lysed and the chromosome released, it can be visualized under the electron microscope (Figure 4–3).

DNA in bacterial chromosomes is found to be associated with several types of **DNA-binding proteins**. Two, called **HU** and **H**, are small, but abundant in the cell, and contain a high percentage of positively charged amino acids that can bond ionically to the negative charges of the phosphate groups in DNA. These proteins are structurally similar to molecules, called **histones**, that are associated with eukaryotic DNA. (We will discuss histone organization later in this chapter.) Unlike the tightly packed chromosome present in the head of a virus, the bacterial chromosome is not functionally inert. Despite its somewhat compacted condition in the bacterial cell, the chromosome can be readily replicated and transcribed.

4.2 Supercoiling Is Common in the DNA of Viral and Bacterial Chromosomes

One major insight into the way in which DNA is organized and packaged in viral and bacterial chromosomes has come

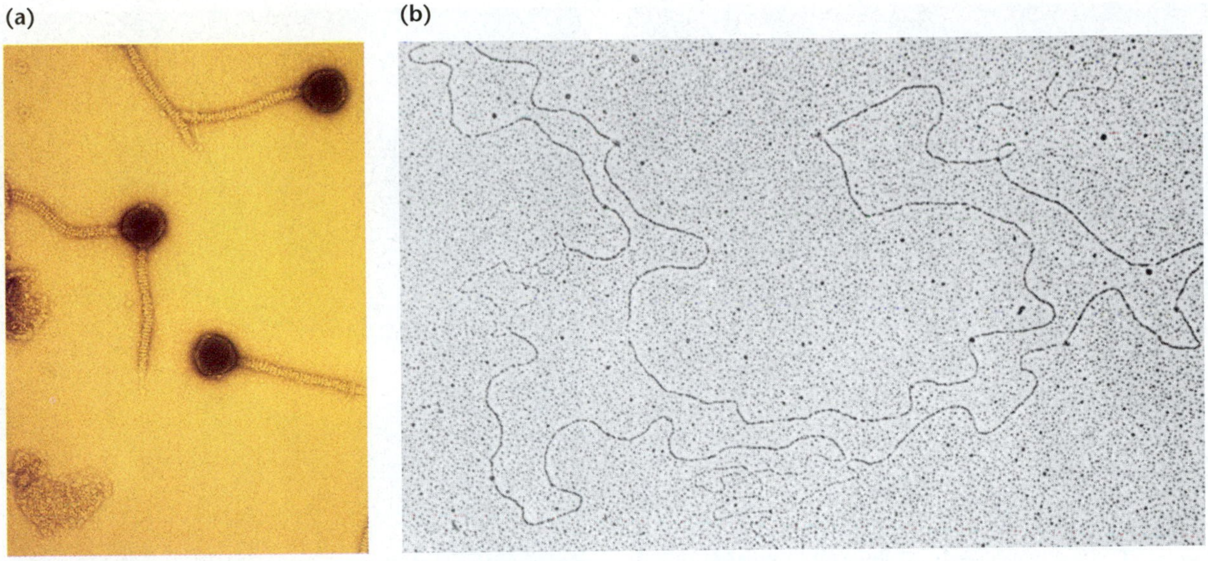

FIGURE 4–1 Electron micrographs of phage l (left) and the DNA isolated from it. The chromosome is 17 μm long. Note that the phages are magnified about five times more than the DNA (right).

from the discovery of **supercoiled DNA**, which is particularly characteristic of closed circular molecules. Supercoiled DNA was first proposed as a result of a study of double-stranded DNA molecules derived from the polyoma virus, which causes tumors in mice. In 1963, it was observed that when such DNA was subjected to high-speed centrifugation, it was resolved into three distinct components, each of different density and compactness. The one that was least compact, and thus least dense, demonstrated a decreased sedimentation velocity; the other two fractions each showed greater velocities owing to their greater compaction and density. All three were of identical molecular weight.

In 1965, Jerome Vinograd proposed an explanation for these observations. He postulated that the two fractions of greatest sedimentation velocity both consisted of polyoma DNA molecules that are circular, whereas the fraction of lower sedimentation contained polyoma DNA molecules that are linear. Closed circular molecules are more compact and they sediment more rapidly than do linear molecules of the same length and molecular weight.

Vinograd proposed further that the more dense of the two fractions of circular molecules consisted of covalently closed DNA helices that are slightly *underwound* in comparison to the less dense circular molecules. Energetic forces stabilizing the double helix resist this underwinding, causing it to **supercoil** in order to retain normal base pairing. Vinograd proposed that it is the supercoiled shape that causes tighter packing and thus the increase in sedimentation velocity.

The transitions just described are illustrated in Figure 4–4. Consider a double-stranded linear molecule that exists in the normal Watson–Crick right-handed helix [Figure 4–4(a)]. This helix contains 20 complete turns, which defines the **linking number** ($L = 20$) of this molecule. If the ends of the molecule are sealed, a closed circle is formed [Figure 4–4(b)], which is described as being *energetically relaxed*. Suppose, however, that the circle were cut open, underwound by several full turns, and resealed [Figure 4–4(c)]. Such a structure, where L is equal to 18, is *energetically "strained,"* and as a result, it will exist only temporarily in this form.

TABLE 4–1 The Genetic Material of Representative Viruses and Bacteria

		Nucleic Acid			
	Organism	*Type*	*SS or DS**	*Length (μm)*	*Overall Size of Viral Head or Bacteria (μm)*
Viruses	ϕX174	DNA	SS	2.0	0.025 × 0.025
	Tobacco mosaic virus	RNA	SS	3.3	0.30 × 0.02
	Lambda phage	DNA	DS	17.0	0.07 × 0.07
	T2 phage	DNA	DS	52.0	0.07 × 0.10
Bacteria	*Haemophilus influenzae*	DNA	DS	832.0	1.00 × 0.30
	Escherichia coli	DNA	DS	1200.0	2.00 × 0.50

*SS = single-stranded: DS = double-stranded

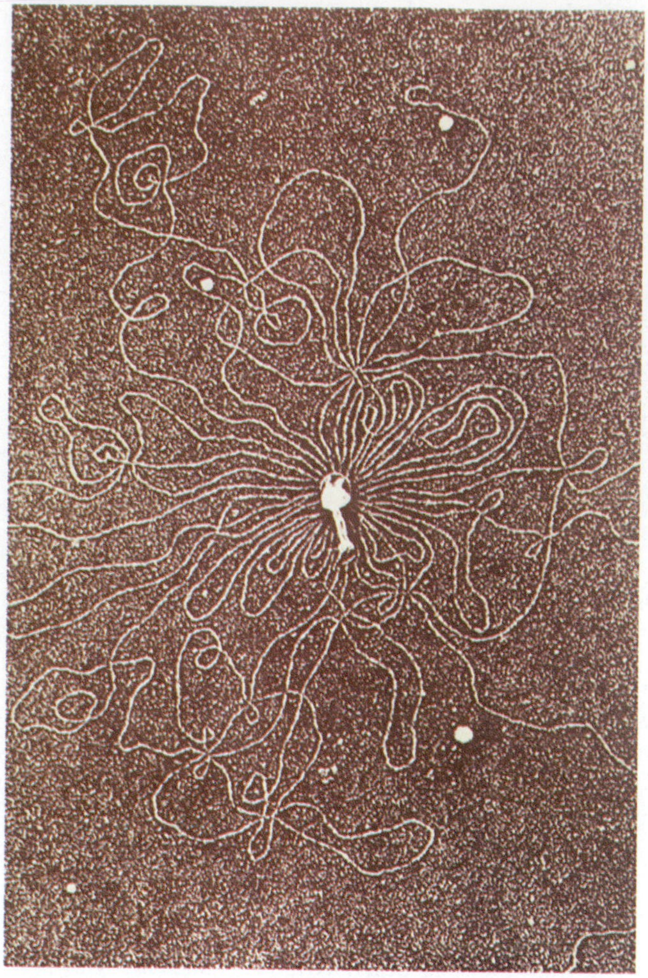

FIGURE 4–2 Electron micrograph of bacteriophage T2, which has had its DNA released by osmotic shock. The chromosome is 52 μm long.

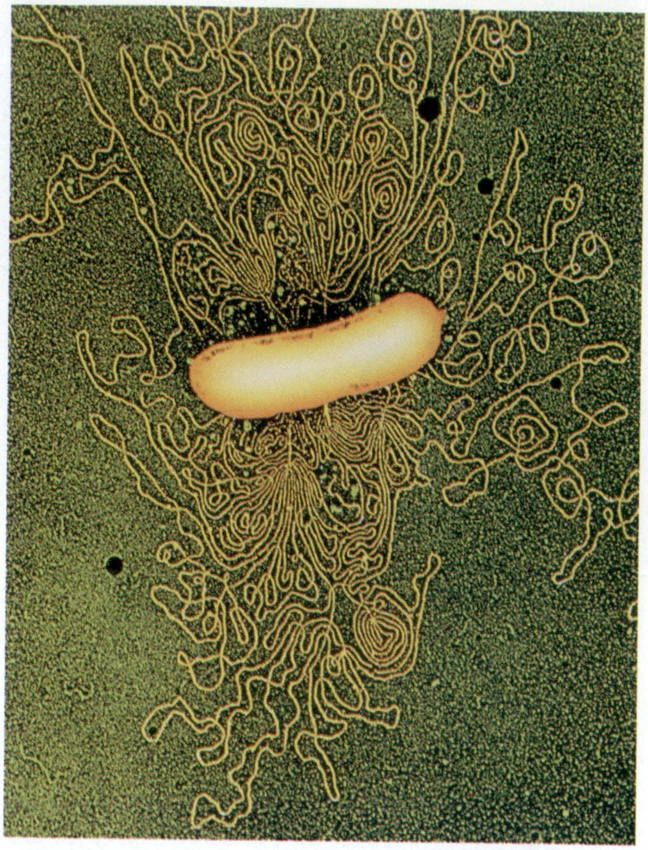

FIGURE 4–3 Electron micrograph of the bacterium *Escherichia coli*, which has had its DNA released by osmotic shock. The chromosome is 1200 μm long.

In order to assume a more energetically favorable conformation, the molecule can form supercoils in the opposite direction of the underwound helix. In our case [Figure 4–4(d)], two negative supercoils are introduced spontaneously, reestablishing the total number of original "turns" in the helix. The use of the term "negative" refers to the fact that, by definition, the supercoils are left handed. The end result is the formation of a more compact structure with enhanced physical stability.

In most closed circular DNA molecules in bacteria and their phages, DNA is slightly underwound [as in Figure 4–4(c)]. For example, the virus SV40 contains 5200 base pairs. If the DNA is relaxed, 10.4 base pairs occupy each complete turn of the helix, and the linking number can be calculated as

$$L = \frac{5200}{10.4} = 500.$$

When circular SV40 DNA is analyzed, it is underwound by 25 turns and L is equal to only 475. Predictably, 25 negative supercoils are observed. In *E. coli*, an even larger number of supercoils is observed, greatly facilitating chromosome condensation in the nucleoid region.

Two otherwise identical molecules that differ only in their linking number are said to be **topoisomers** of one another. But how can a molecule convert from one topoisomer to the other if there are no free ends, as is the case in closed circles of DNA? Biologically, this may be accomplished by any one of a group of enzymes that cut one or both of the strands and wind or unwind the helix before resealing the ends.

Appropriately, these enzymes are called **topoisomerases**. First discovered by Martin Gellert and James Wang, these catalytic molecules are known as either type I or type II, depending on whether they cleave one or both strands in the helix, respectively. In *E. coli*, topoisomerase I serves to reduce the number of negative supercoils in a closed-circular DNA molecule. Topoisomerase II introduces negative supercoils into DNA. This latter enzyme is thought to bind to DNA, twist it, cleave both strands, and then pass them through the "loop" that it has created. Once the phosphodiester bonds are reformed, the linking number is decreased and one or more supercoils form spontaneously.

Supercoiled DNA and topoisomerases are also found in eukaryotes. While the chromosomes in these organisms are not usually circular, supercoils can occur when areas of DNA are embedded in a lattice of proteins associated with the chromatin fibers. This association creates "anchored" ends that provide the stability for the maintenance of supercoils once they are introduced by topoisomerases. As occurs in

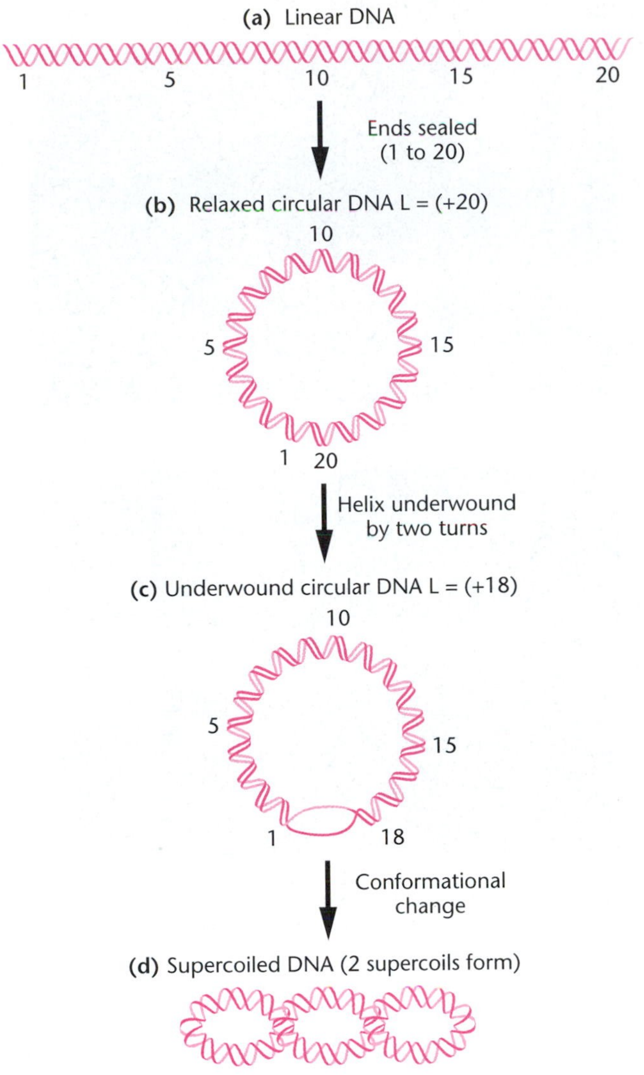

FIGURE 4–4 Depictions of the transformations leading to the supercoiling of circular DNA. *L* equals the linking number.

prokaryotes, DNA replication and transcription in eukaryotes creates supercoils downstream as the double helix unwinds and becomes accessible to the appropriate enzyme.

These enzymes may well play still other genetic roles involving eukaryotic DNA conformational changes. Interestingly, topoisomerases are involved in separating (decatenating) the DNA of sister chromatids following replication.

4.3 Specialized Chromosomes Reveal Variations in Structure

We move next to the consideration of chromosomes present in eukaryotes. Before discussing these at the molecular level, we first introduce two cases of highly specialized chromosomes. Both types, polytene chromosomes and lampbrush chromosomes, are so large that their organization was discerned using light microscopy long before we understood that DNA is the genetic material. The study of these chromosomes provided many of our initial insights into the arrangement and function of the genetic information.

Polytene Chromosomes

Giant **polytene chromosomes** are found in various tissues (salivary, midgut, rectal, and malpighian excretory tubules) in the larvae of some flies, as well as in several species of protozoans and plants. Such structures were first observed by E. G. Balbiani in 1881. The large amount of information obtained from studies of these genetic structures provided a model system for subsequent investigations of chromosomes. What is particularly intriguing about polytene chromosomes is that they can be seen in the nuclei of interphase cells.

Each polytene chromosome observed under the light microscope reveals a linear series of alternating bands and interbands (Figure 4–5). The banding pattern is distinctive for each chromosome in any given species. Individual bands are sometimes called **chromomeres**, a more generalized term describing lateral condensations of material along the axis of a chromosome. Each polytene chromosome is 200–600 μm long.

Extensive study using electron microscopy and radioactive tracers led to an explanation for the unusual appearance of these chromosomes. First, polytene chromosomes represent paired homologs. This is highly unusual, since they are found in somatic cells, where, in most organisms, chromosomal material is normally dispersed as chromatin and homologs are not paired. Second, their large size and distinctiveness result from the many DNA strands that compose them. The DNA of these paired homologs undergoes many rounds of replication, but without strand separation or cytoplasmic division. As replication proceeds, chromosomes are created having 1000–5000 DNA strands that remain in parallel register with one another. Apparently, it is the parallel register of so many DNA strands that gives rise to the distinctive band pattern along the axis of the chromosome.

The relationship between the structure of polytene chromosomes and the genes contained within them is intriguing. The presence of bands was initially interpreted as the visible manifestation of individual genes. The discovery that the strands present in bands undergo localized uncoiling during genetic activity further strengthened this view. Each such

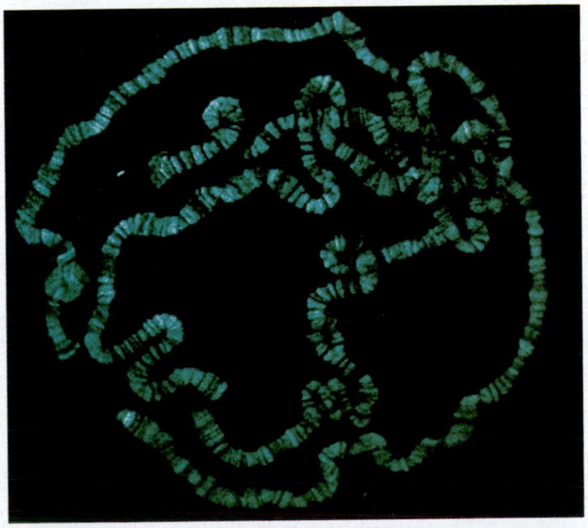

FIGURE 4–5 Polytene chromosomes derived from larval salivary gland cells of *Drosophila*.

uncoiling event results in what is called a **puff** because of its appearance (see Figure 4–6). That puffs are visible manifestations of gene activity (transcription that produces RNA) is evidenced by their high rate of incorporation of radioactively labeled RNA precursors, as assayed by autoradiography. Bands that are not extended into puffs incorporate fewer radioactive precursors or none at all.

The study of bands during development in insects, such as *Drosophila* and the midge fly *Chironomus*, reveals differential gene activity. A characteristic pattern of band formation that is equated with gene activation is observed as development proceeds. Despite attempts to resolve the issue of the number of genes contained within each band, it is still not clear how many there are. It is known that a band may contain up to 10^7 base pairs of DNA, certainly enough DNA to encode 50–100 average-size genes.

Lampbrush Chromosomes

Another type of specialized chromosome that has provided insights into chromosomal structure is the **lampbrush chromosome**, so named because its appearance is similar to the brushes used to clean kerosene lamp chimneys in the 19th century. Lampbrush chromosomes were first discovered in 1892 in the oocytes of sharks and are now known to be characteristic of most vertebrate oocytes, as well as the spermatocytes of some insects. Therefore, they are meiotic chromosomes. Most of the experimental work has been done with material taken from amphibian oocytes.

These unique chromosomes are easily isolated from oocytes in the diplotene stage of the first prophase of meiosis, where they are active in directing the metabolic activities of the developing cell. The homologs are seen as synapsed pairs held together by chiasmata. However, instead of condensing, as most meiotic chromosomes do, lampbrush chromosomes are often extended to lengths of 500–800 μm. Later, in meiosis, they revert to their normal length of 15–20 μm. Based on these observations, lampbrush chromosomes are interpreted as extended, uncoiled versions of the normal meiotic chromosomes.

The two views of lampbrush chromosomes in Figure 4–7 provide significant insights into their morphology. In Figure 4–7(a), the meiotic configuration under the light microscope is shown. The linear axis of each structure contains a large number of condensed areas, referred to generally as chromomeres, that are repeated along the axis. Emanating from each chromomere is a pair of lateral loops, which give the chromosome its distinctive appearance. In Figure 11–7(b), the scanning electron micrograph (SEM) is an enlargement that reveals adjacent loops present along one of the two axes of the chromosome. As with bands in polytene chromosomes,

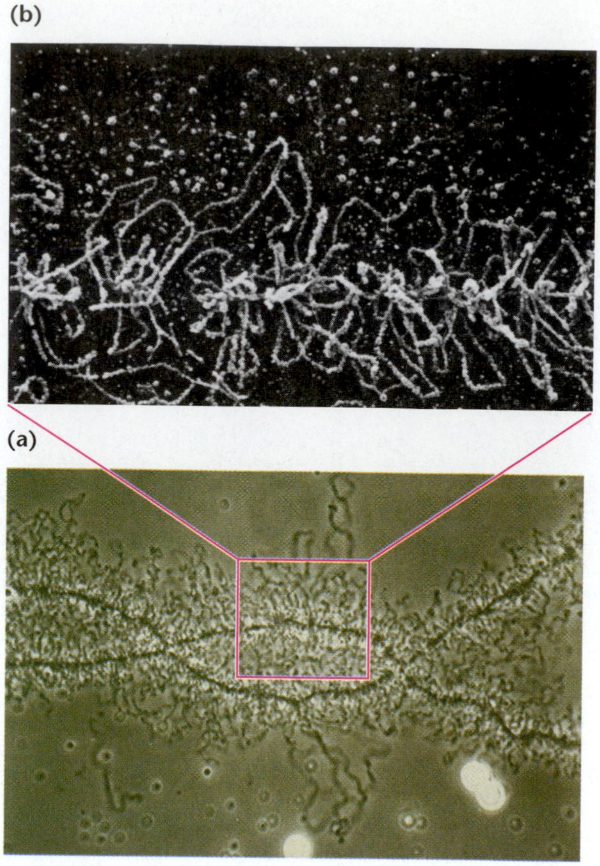

FIGURE 4–7 Lampbrush chromosomes derived from amphibian oocytes—(a) is a photomicrograph; (b) is a scanning electron micrograph.

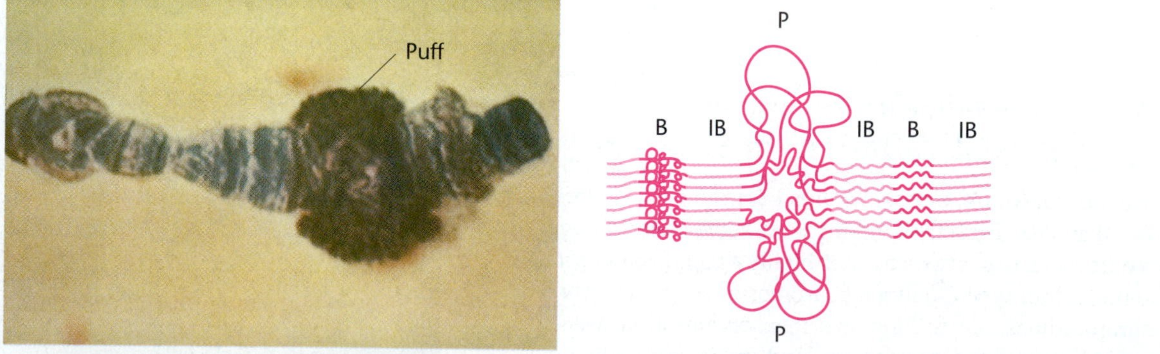

FIGURE 4–6 Photograph of a puff within a polytene chromosome. The diagram depicts the uncoiling of strands within a band (B) region to produce a puff (P) in polytene chromosomes. Interband regions (IB) are also labeled.

there is much more DNA present in each loop than is needed to encode a single gene. Such an SEM provides a clear view of the chromomeres and the chromosomal fibers emanating from them. Each chromosomal loop is thought to be composed of one DNA double helix, while the central axis is made up of two DNA helices. This hypothesis is consistent with the belief that each meiotic chromosome is composed of a pair of sister chromatids. Studies using radioactive RNA precursors reveal that the loops are active in the synthesis of RNA. The lampbrush loops, in a manner similar to puffs in polytene chromosomes, represent DNA that has been reeled out from the central chromomere axis during transcription.

4.4 DNA Is Organized into Chromatin in Eukaryotes

We now turn our attention to the way DNA is organized in eukaryotic chromosomes. Our focus will be on conventional eukaryotic cells, in which chromosomes are visible only during mitosis. Following chromosome separation and cell division, cells enter the interphase stage of the cell cycle, during which time the components of the chromosome uncoil and are present in the form referred to as **chromatin**. While in interphase, the chromatin is dispersed in the nucleus, and the DNA of each chromosome is replicated. As the cell cycle progresses, most cells reenter mitosis, whereupon chromatin coils into visible chromosomes once again. This condensation represents a contraction in length of some 10,000 times for each chromatin fiber.

The organization of DNA during the transitions just described is much more intricate and complex than in viruses or bacteria, which never exhibit a complex process similar to mitosis. This is due to the greater amount of DNA per chromosome, as well as the presence of a large number of proteins associated with eukaryotic DNA. For example, while DNA in the *E. coli* chromosome is 1200 μm long, the DNA in each human chromosome ranges from 19,000 to 73,000 μm in length. In a single human nucleus, all 46 chromosomes contain sufficient DNA to extend almost 2 meters. This genetic material, along with its associated proteins, is contained within a nucleus that usually measures about 5–10 μm in diameter.

Such intricacy parallels the structural and biochemical diversity of the many types of cells present in a multicellular eukaryotic organism. Different cells assume specific functions based on highly specific biochemical activity. While all cells carry a full genetic complement, different cells activate different sets of genes, so a highly ordered regulatory system governing the readout of the information must exist. Such a system must in some way be imposed on or related to the molecular structure of the genetic material.

Because of the limitation of light microscopy, early studies of the structure of eukaryotic genetic material concentrated on intact chromosomes, preferably large ones, such as the polytene and lampbrush chromosomes. Subsequently, new techniques for biochemical analysis, as well as the examination of relatively intact eukaryotic chromatin and mitotic chromosomes under the electron microscope, have greatly enhanced our understanding of chromosome structure.

Chromatin Structure and Nucleosomes

As we have seen, the genetic material of viruses and bacteria consists of strands of DNA or RNA nearly devoid of proteins. In eukaryotic chromatin, a substantial amount of protein is associated with the chromosomal DNA in all phases of the eukaryotic cell cycle. The associated proteins are divided into basic, positively charged **histones** and less positively charged **nonhistones**. Of the proteins associated with DNA, the histones clearly play the most essential structural role. Histones contain large amounts of the positively charged amino acids lysine and arginine, making it possible for them to bond electrostatically to the negatively charged phosphate groups of nucleotides. Recall that a similar interaction has been proposed for several bacterial proteins. There are five main types of histones (Table 4–2).

The general model for chromatin structure is based on the assumption that chromatin fibers, composed of DNA and protein, undergo extensive coiling and folding as they are condensed within the cell nucleus. X-ray-diffraction studies confirm that histones play an important role in chromatin structure. Chromatin produces regularly spaced diffraction rings, suggesting that repeating structural units occur along the chromatin axis. If the histone molecules are chemically removed from chromatin, the regularity of this diffraction pattern is disrupted.

A basic model for chromatin structure was worked out in the mid-1970s. Several observations were particularly relevant to the development of this model:

1. Digestion of chromatin by certain endonucleases, such as micrococcal nuclease, yields DNA fragments that are approximately 200 base pairs in length, or multiples thereof. This demonstrates that enzymatic digestion is not random, for if it were, we would expect a wide range of fragment sizes. Hence, chromatin consists of some type of repeating unit, each of which is protected from enzymatic cleavage, except for the point at which any two units are joined. It is the area between units that is attacked and cleaved by the nuclease.

2. Electron microscopic observations of chromatin have revealed that chromatin fibers are composed of linear

TABLE 4–2 Categories and Properties of Histone Proteins

Histone Type	*Lysine-Arginine Content*	*Molecular Weight (Da)*
H1	Lysine-rich	23,000
H2A	Slightly lysine-rich	14,000
H2B	Slightly lysine-rich	13,800
H3	Arginine-rich	15,300
H4	Arginine-rich	11,300

arrays of spherical particles (Figure 4–8). Discovered by Ada and Donald Olins, the particles occur regularly along the axis of a chromatin strand and resemble beads on a string. These particles, initially referred to as ν-bodies, are now called nucleosomes. These findings conform nicely to the earlier observation, which suggests the existence of repeating units.

3. Studies of precise interactions of histone molecules and DNA in the nucleosomes constituting chromatin show that histones H2A, H2B, H3, and H4 occur as two types of tetramers, $(H2A)_2 \cdot (H2B)_2$ and $(H3)_2 \cdot (H4)_2$. Roger Kornberg predicted that each repeating nucleosome unit consists of one of each tetramer in association with about 200 base pairs of DNA. Such a structure is consistent with previous observations and provides the basis for a model that explains the interaction of histones and DNA in chromatin.

4. When nuclease digestion time is extended, some of the 200 base pairs of DNA are removed from the nucleosome, creating what is called a **nucleosome core particle**, consisting of 146 base pairs. This number is consistent in all eukaryotes studied. The DNA lost in the prolonged digestion is responsible for linking nucleosomes together. This **linker DNA** is associated with the fifth histone, H1.

5. On the basis of this information, as well as on X-ray and neutron-scattering analyses of crystallized core particles by John T. Finch, Aaron Klug, and others, a detailed model of the nucleosome was put forward in 1984. In this model, the 146-bp DNA core is coiled around an octamer of histones in a left-handed superhelix, which completes about 1.7 turns per nucleosome.

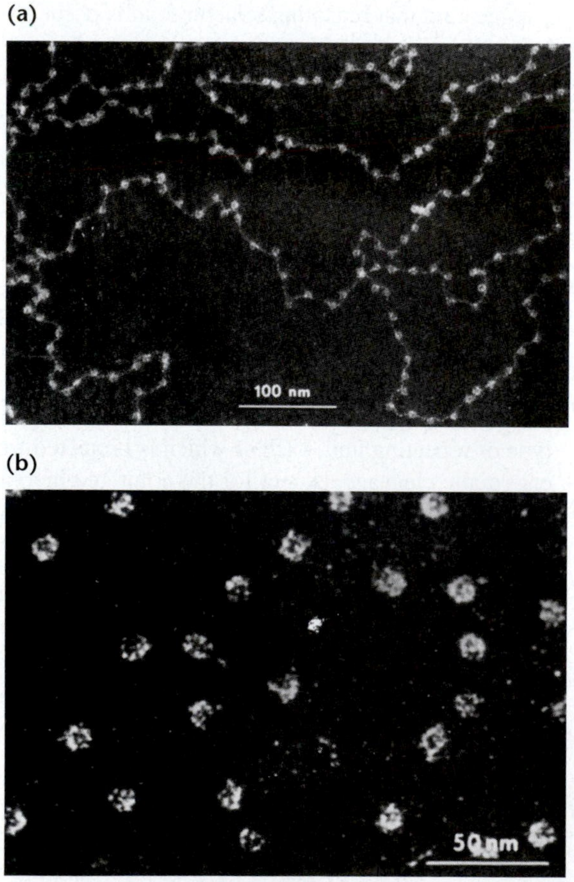

FIGURE 4–8 (a) Dark-field electron micrograph of nucleosomes present in chromatin derived from a chicken erythrocyte nucleus. (b) Dark-field electron micrograph of nucleosomes produced by micrococcal nuclease digestion.

The extensive investigation of nucleosomes now provides the basis for predicting how the chromatin fiber within the nucleus is formed and how it coils up into a mitotic chromosome. This model is illustrated in Figure 4–9. The 2-nm DNA molecule is initially coiled into a nucleosome that is about 11 nm in diameter [Figure 4–9(a)], consistent with the longer dimension of the ellipsoidal nucleosome. Significantly, the formation of the nucleosome represents the first level of packing, whereby the DNA helix is reduced to about one-third of its original length.

In the nucleus, the chromatin fiber seldom, if ever, exists in the extended form described in the previous paragraph. Instead, the 11-nm chromatin fiber is further packed into a thicker 30-nm structure, that was initially called a solenoid [Figure 4–9(b)]. This larger fiber consists of numerous nucleosomes coiled closely together, creating the second level of packing. The exact details of the structure are not completely clear, but 30-nm chromatin fibers are characteristically seen under the electron microscope.

In the transition to the mitotic chromosome, still another level of packing occurs. The 30-nm structure forms a series of looped domains that further condense into the chromatin fiber, which is 300 nm in diameter [Figure 4–9(c)]. The fibers are then coiled into the chromosome arms that constitute a chromatid, which is part of the metaphase chromosome [Figure 4–9(d)]. While we show the chromatid arms to be 700 nm in diameter, the value undoubtedly varies among different organisms. At a value of 700 nm, a pair of sister chromatids making up a chromosome measures about 1400 nm.

The importance of the organization of DNA into chromatin and chromatin into mitotic chromosomes can be illustrated by considering a human cell that stores its genetic material in a nucleus that is about 5–10 μm in diameter. The haploid genome contains 3.2×10^9 base pairs of DNA distributed among 23 chromosomes. The diploid cell contains twice that amount. At 0.34 nm per base pair, this amounts to an enormous length of DNA (as stated earlier, almost 2 meters)! One estimate is that about 25×10^6 nucleosomes per nucleus are complexed with the DNA present in a typical human nucleus.

In the overall transition from a fully extended DNA helix to the extremely condensed status of the mitotic chromosome, a packing ratio (the ratio of DNA length to the length of the structure containing it) of about 500 to 1 must be achieved. In fact, our model accounts for a ratio only of about 50 to 1. Obviously, the larger fiber can be further bent, coiled,

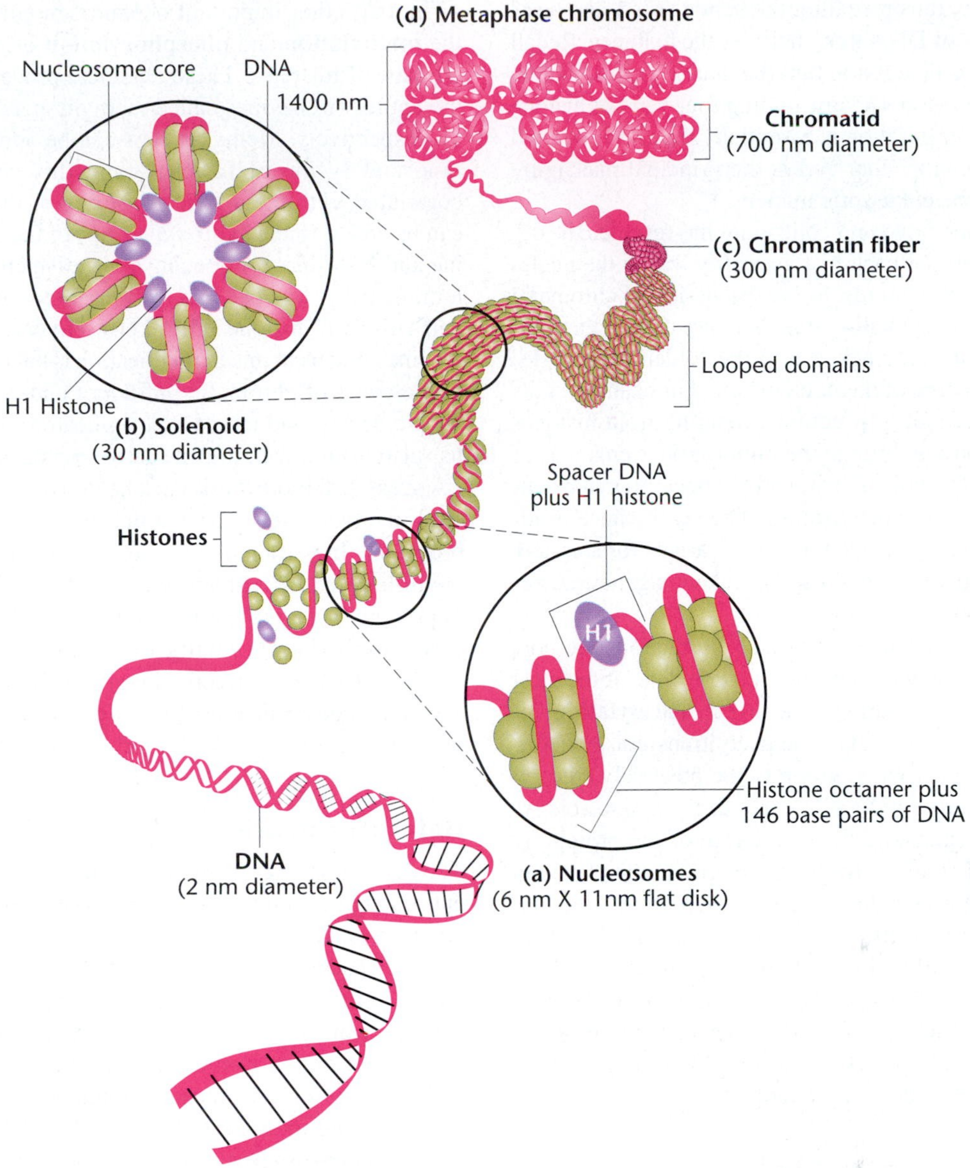

FIGURE 4–9 General model of the association of histones and DNA in the nucleosome, illustrating the way in which the chromatin fiber may be coiled into a more condensed structure, ultimately producing a mitotic chromosome.

MEDIA TUTORIAL, Chromosome Structure

and packed as even greater condensation occurs during the formation of a mitotic chromosome.

Chromatin Remodeling

As with many significant findings in genetics, the study of nucleosomes has answered some important questions, but at the same time, also led us to new ones. For example, in the preceding discussion, we have established that histone proteins play an important structural role in packaging DNA into the nucleosomes that make up chromatin. While solving this structural problem of how to organize a huge amount of DNA within the eukaryotic nucleus, a new problem is apparent. The chromatin fiber, when complexed with histones and folded into various levels of compaction, makes the DNA inaccessible to interaction with important nonhistone proteins. For example, the variety of proteins that function in enzymatic and regulatory roles during the processes of replication and gene expression must interact directly with DNA. To accommodate these protein–DNA interactions, chromatin must be induced to change its structure, a process now referred to as **chromatin remodeling**. In the case of replication and gene expression, chromatin must relax its compact structure, but be able to reverse the process during periods of inactivity.

Insights into how different states of chromatin structure may be achieved were forthcoming in 1997, when Timothy Richmond and members of his research team were able to significantly improve the level of resolution in X-ray-diffraction studies of nucleosome crystals (from 7Å in the 1984 studies to 2.8Å in the 1997 studies). One model based on their work is shown in Figure 4–10. At this resolution, most atoms are visible, thus revealing the subtle twists and turns of the superhelix of DNA that encircles the histones. Recall from our previous discussion that the double-helical ribbon represents 146 bp of DNA surrounding four pairs of histone proteins. This configuration is essentially repeated over and over in the chromatin fiber and is the principal packaging unit of DNA in the eukaryotic nucleus.

The work of Richmond and colleagues has revealed the details of the location of each histone entity within the nucleosome. Of particular interest to the discussion of chromatin remodeling is the observation that there are unstructured histone tails that are not packed into the folded histone domains within the core of the nucleosome. For example, tails devoid of any secondary structure extending from histones H3 and H2B protrude through the minor groove channels of the DNA helix. The tails of histone H4 appear to make connection with adjacent nucleosomes. The significance of histone tails is that they provide potential targets for a variety of chemical interactions that may be linked to genetic functions along the chromatin fiber.

Several of these potential chemical interactions are now recognized as important to genetic function. One of the most well-studied histone modifications involves **acetylation**, resulting from the action of histone acetyltransferase (HAT). This enzyme adds an acetyl group to the positively charged amino group present on the side chain of the amino acid lysine, effectively changing the net charge of the protein by neutralizing the positive charge. Lysine is in abundance in histones, and it has been known for some time that acetylation is linked to gene activation. It appears that high levels of acetylation open up the chromatin fiber, an effect that occurs in regions of active genes and decreases in inactive regions. A well-known example involves the inactive X chromosome forming a Barr body in mammals, in which histone H4 is known to be greatly underacetylated.

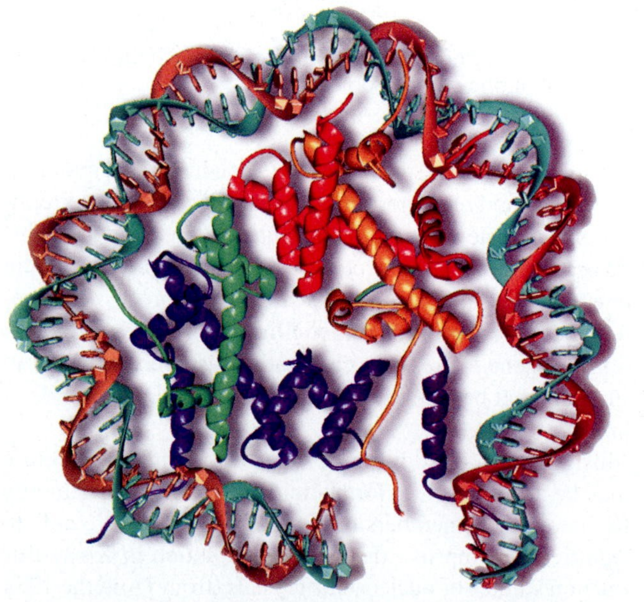

FIGURE 4–10 The nucleosome core particle derived from X-ray crystal analysis at 2.8 Å resolution. The double-helical DNA surrounds four pairs of histones.

The two other important chemical modifications include the **methylation** and **phosphorylation** of amino acids that are part of histones. These chemical processes result from the action of enzymes called methyltransferases and kinases, respectively. Methyl groups can be added to both arginine and lysine of histones, and this change has been correlated with the activation of genes. Phosphate groups can be added to the hydroxyl groups of the amino acids serine and histidine, introducing a negative charge on the protein. During the cell cycle, increased phosphorylation, particularly of histone H3, is known to occur at characteristic times. Such chemical modification is believed to be related to the cycle of chromatin unfolding and condensation that occurs during and after DNA replication. Phosphorylation has also been linked to gene activation during interphase.

A great deal more work must be done to elucidate the specific involvement of chromatin remodeling during genetic processes. In particular, the way in which the modifications are influenced by regulatory molecules within cells will be important. What is clear is that the dynamic forms in which chromatin exist are vitally important to the way that all genetic processes directly involving DNA are executed. These include replication, transcription, and its regulation—processes that are at the heart of genetic function.

Heterochromatin

The discussion in the previous section implies that, along its entire length, each eukaryotic chromosome consists of one continuous double-helical DNA molecule present in repeating nucleosomes, which might lead you to believe that the whole chromosome is structurally uniform. However, in the early part of the 20th century, it was observed that, while most of the chromosome is present as uncoiled chromatin, some parts of the chromosome remain condensed and stain deeply during interphase. In 1928, the terms **euchromatin** and **heterochromatin** were coined to describe the parts of chromosomes that are uncoiled and those that remain condensed, respectively.

Subsequent investigation revealed a number of characteristics that distinguish heterochromatin from euchromatin. Heterochromatic areas are genetically inactive because they either lack genes or contain genes that are repressed. Also, heterochromatin replicates later during the S phase of the cell cycle than does euchromatin. The discovery of heterochromatin provided the first clues that parts of eukaryotic chromosomes do not always encode proteins. Instead, some chromosome regions are thought to be involved in the maintenance of the chromosome's structural integrity and in other functions, such as chromosome movement during cell division.

Heterochromatin is characteristic of the genetic material of eukaryotes. Early cytological studies showed that areas of the centromeres are composed of heterochromatin. The ends of chromosomes, called telomeres, are also heterochromatic. In some cases, in fact, whole chromosomes are heterochromatic. Such is the case with the mammalian Y chromosome, which, for the most part, is genetically inert. And, as we will discuss in Chapter 11, the inactivated X chromosome in mammalian females is condensed into an inert heterochromatic Barr body. In some species, such as mealy bugs, all chromosomes of one entire haploid set are heterochromatic.

When certain heterochromatic areas from one chromosome are translocated to a new site on the same or another nonhomologous chromosome, genetically active areas sometimes become genetically inert if they now lie adjacent to the translocated heterochromatin. As we will see in Chapter 10, this influence on existing euchromatin is one example of what is more generally referred to as a **position effect**. That is, the position of a gene or groups of genes relative to all other genetic material may affect their expression.

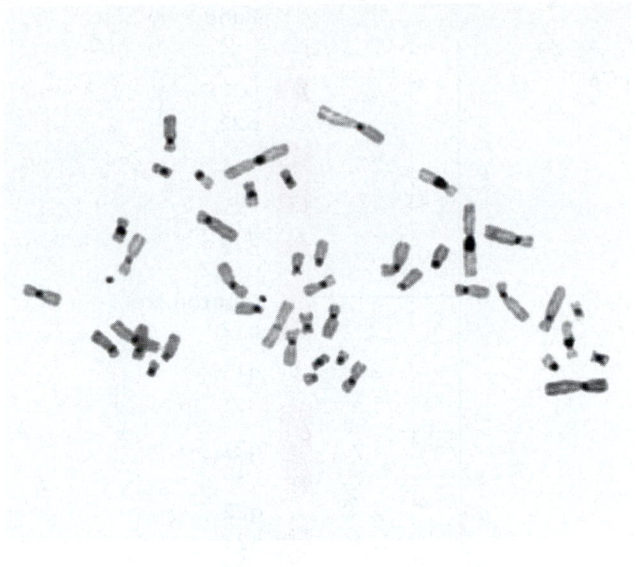

FIGURE 4–11 Karyotypes of a human mitotic chromosome preparation processed to demonstrate C-banding: Only the centromeres stain.

4.5 Chromosome Banding Differentiates Regions along the Mitotic Chromosome

Until about 1970, mitotic chromosomes viewed under the light microscope could be distinguished only by their relative sizes and the positions of their centromeres. Even in organisms with a low haploid number, two or more chromosomes are often indistinguishable from one another. However, new cytological procedures made possible differential staining along the longitudinal axis of mitotic chromosomes. Such methods are now referred to as "chromosome-banding" techniques, because the staining patterns resemble the bands of polytene chromosomes.

One of the first chromosome-banding techniques was devised by Mary Lou Pardue and Joe Gall. They found that if chromosome preparations from mice were heat denatured and then treated with Giemsa stain, a unique staining pattern emerged. Only the centromeric regions of mitotic chromosomes took up the stain! The staining pattern was thus referred to as **C-banding**. Relevant to our immediate discussion, this cytological technique identifies a specific area of the chromosome composed of heterochromatin. A micrograph of the human karyotype treated in this way is shown in Figure 4–11. Mouse chromosomes are all telocentric, thus localizing the stain at the end of each chromosome.

Other chromosome-banding techniques were developed about the same time. The most useful of these produces a staining pattern differentially along the length of each chromosome. This method, producing **G-bands** (Figure 4–12), involves the digestion of the mitotic chromosomes with the proteolytic enzyme trypsin, followed by Giemsa staining. The differential staining reactions reflect the heterogeneity and complexity of the chromosome along its length.

In 1971, a meeting was held in Paris to establish a uniform nomenclature for human chromosome-banding patterns based on G-banding. Figure 4–13 illustrates the application of this nomenclature to the X chromosome. On the left of the chromosome are the various organizational levels of banding of the p and q arms that can be identified, while the designations on the right side identify specific regions within each specific region of the chromosome.

Although the molecular mechanisms involved in producing the various banding patterns are not precisely defined, the significance of banding has been great in cytogenetic analysis, particularly in humans. The pattern of banding of each chromosome is unique, allowing a distinction to be made even between those chromosomes that are identical in size and centromere placement (e.g., human chromosomes 4 and 5 and chromosomes 21 and 22). So precise is the banding pattern of each chromosome that homologs can be distinguished from one another, and when a segment of one chromosome has been translocated to another chromosome, its origin can be determined with great precision.

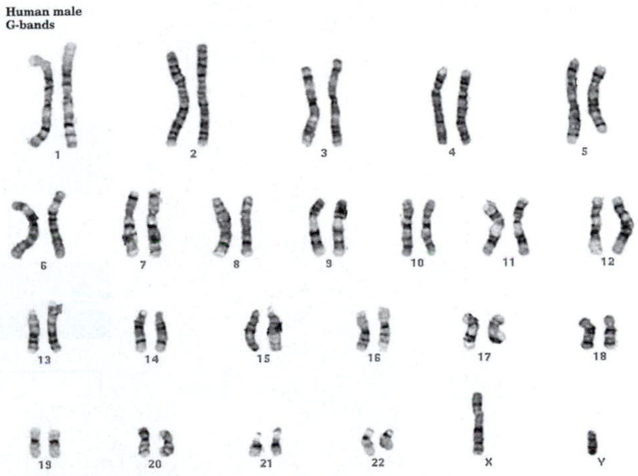

FIGURE 4–12 G-banded karyotype of a normal human male showing approximately 400 bands per haploid set of chromosomes. Chromosomes were derived from cells in metaphase

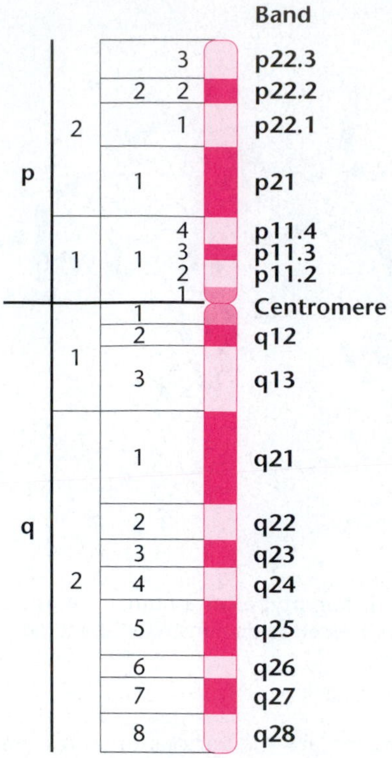

FIGURE 4–13 The regions of the human X chromosome distinguished by its banding pattern. The designations on the right identify specific bands.

4.6 Eukaryotic Genomes Demonstrate Complex Sequence Organization Characterized by Repetitive DNA

Thus far, we have examined how DNA is organized *into* chromosomes in bacteriophages, bacteria, and eukaryotes. We now begin an examination of what we know about the organization of DNA sequences *within* the chromosomes making up an organism's genome, placing our emphasis on eukaryotes. Having established the pattern of genome organization we will turn in a subsequent chapter (see Chapter 19) to a focus on how genes themselves are organized within chromosomes.

We learned in Chapter 2 that, in addition to single copies of unique DNA sequences that make up genes, a great deal of the DNA sequences within chromosomes is repetitive in nature and that various levels of repetition occur within the genome of organisms. Many studies have now provided insights into repetitive DNA, demonstrating various classes of these sequences and their organization within the genome. Figure 4–14 outlines the various categories of repetitive DNA, and you should make reference to it throughout the discussion in this section.

The figure also shows that some functional genes are present in more than one copy (referred to as multiple copy genes) and are therefore repetitive in nature. However, the majority of repetitive sequences is nongenic and, in fact, most repetitive sequences serve no known function. We will explore three main categories: (1) heterochromatin found associated with centromeres and making up telomeres; (2) tandem repeats of both short and longer DNA sequences; and (3) transposable sequences that are interspersed throughout the genome of eukaryotes.

Repetitive DNA and Satellite DNA

The nucleotide composition of the DNA (e.g., the percentage of G≡C versus A═T pairs) of a particular species is reflected in its density, which can be measured with sedimentation equilibrium centrifugation (introduced in Chapter 2). When eukaryotic DNA is analyzed in this way, the majority of it is present as a single main peak or band of fairly uniform density. However, one or more additional peaks represent DNA that differs slightly in density. One such component, called **satellite DNA**, represents a variable proportion of the total DNA, depending on the species. For example, a profile

FIGURE 4–14 An overview of the various categories of repetitive DNA.

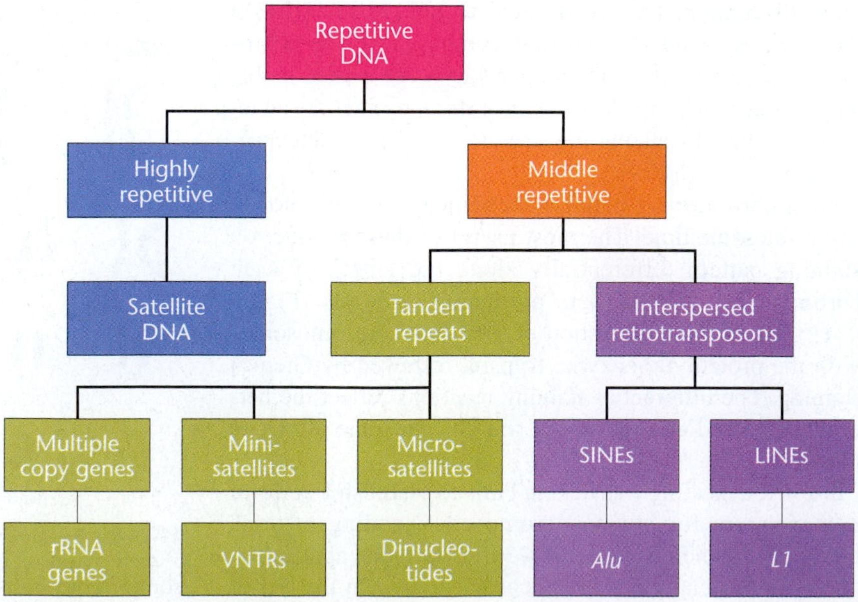

of main-band and satellite DNA from the mouse is shown in Figure 4–15. By contrast, prokaryotes contain only main-band DNA and are thus devoid of satellite DNA.

The significance of satellite DNA remained an enigma until the mid-1960s, when Roy Britten and David Kohne developed a technique for measuring the reassociation kinetics of DNA that had previously been dissociated into single strands (See Chapter 2). The researchers demonstrated that certain portions of DNA reannealed more rapidly than others and concluded that rapid reannealing was characteristic of multiple DNA fragments composed of identical or nearly identical nucleotide sequences—the basis for the descriptive term **repetitive DNA**. Recall that, in contrast, prokaryotic DNA is nearly devoid of anything other than unique, single-copy sequences.

When satellite DNA was subjected to analysis by reassociation kinetics, it fell into the category of *highly repetitive DNA* and is known to consist of short sequences repeated a large number of times. Further evidence suggested that these sequences are present as tandem repeats clustered in very specific chromosomal areas known to be heterochromatic—the regions flanking *centromeres*. This was discovered in 1969 when several researchers, including Mary Lou Pardue and Joe Gall, applied the technique of ***in situ* molecular hybridization** to the study of satellite DNA. The technique (see Appendix A) involves the molecular hybridization between an isolated fraction of radioactively-labeled DNA or RNA probes and the DNA contained in the chromosomes of a cytological preparation. Following the hybridization procedure, autoradiography is performed to locate the chromosome areas complementary to the fraction of DNA or RNA.

In their work, Pardue and Gall demonstrated that molecular probes made from mouse satellite DNA hybridize with DNA of centromeric regions of mouse mitotic chromosomes (Figure 4–16). Several conclusions can be drawn. Satellite DNA differs from main-band DNA in its molecular composition, as established by buoyant density studies. It is also composed of short repetitive sequences. Finally, most satellite DNA is found in the heterochromatic centromeric regions of chromosomes.

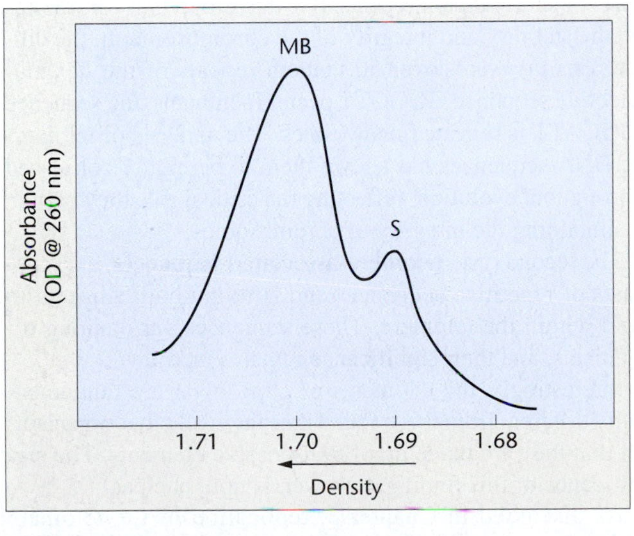

FIGURE 4–15 Separation of main-band (MB) and satellite (S) DNA from the mouse by using ultracentrifugation in a CsCl gradient.

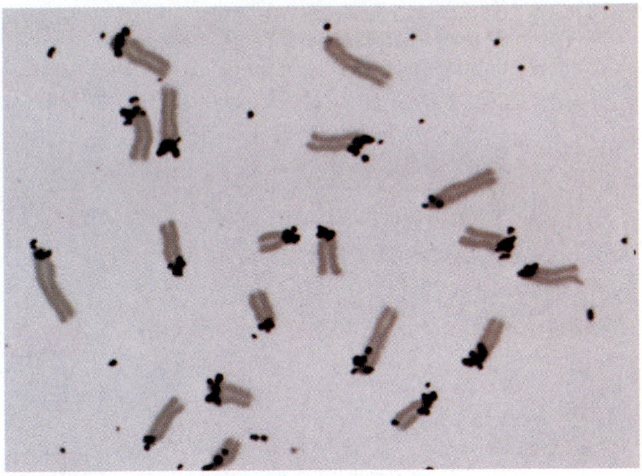

FIGURE 4–16 *In situ* hybridization between a radioactive probe representing mouse satellite DNA and mitotic chromosomes. The grains in the autoradiograph localize the chromosome regions (the centromeres) containing satellite DNA sequences.

Centromeric DNA Sequences

Centromeres, described cytologically in the late 19th century as the primary constrictions along eukaryotic chromosomes, play several crucial roles during mitosis and meiosis. First, they are responsible for the maintenance of sister chromatid cohesion prior to the anaphase stage. It is at the point of the centromere along the chromosome that sister chromatids remain paired together during the early stages of mitosis and meiosis. Second, centromeres are the site of the formation of the kinetochore, the proteinaceous platform that is organized around the centromere and attaches to the microtubules of spindle fibers. Hence, centromeres mediate chromosome migration during the anaphase stage. This process is essential to the separation of chromatids and thus the fidelity of chromosome distribution during cell division.

Most estimates of infidelity during mitosis are exceedingly low: 1×10^{-5} to 1×10^{-6}, or 1 error per 100,000 to 1,000,000 cell divisions. As a result, it has been generally assumed that the analysis of the DNA sequence of centromeric regions will provide insights into the rather remarkable features of this chromosomal region. This DNA region is designated the **CEN**, which has now been defined and investigated in a number of organisms.

The analysis of the CEN regions of yeast *Saccharomyces cerevisiae* chromosomes provided the basis for a model system first described by John Carbon and Louis Clarke. Because each centromere serves an identical function, it is not surprising that all CENs were found to be remarkably similar in their organization. The CEN region of all 16 yeast chromosomes in *Saccaromyces cerevisiae* consists of about 125 bp, which can be divided into three regions (Figure 4–17). The first and third regions (I and III) are relatively short and highly conserved on yeast chromosomes, consisting

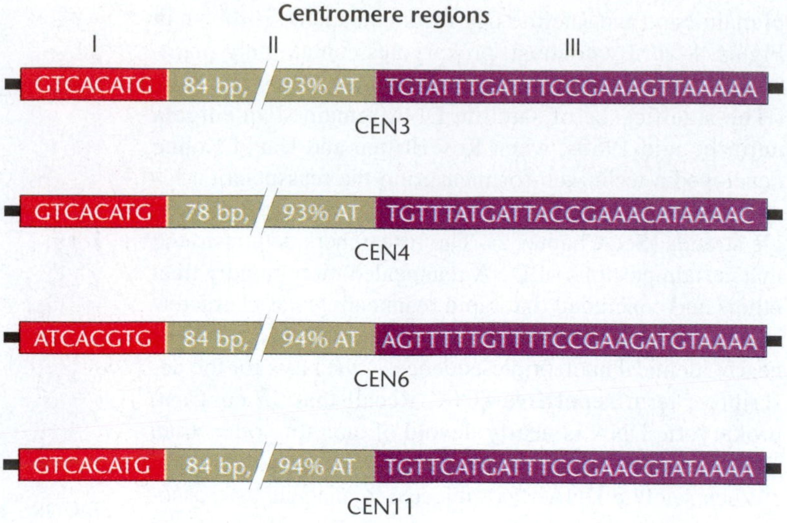

FIGURE 4–17 Nucleotide sequence information derived from DNA of the three major centromere regions of yeast chromosomes 3, 4, 6, and 11.

of only 8 and 26 bp, respectively. Region II, which is larger (80–85 bp) and extremely A=T rich (up to 95 percent), varies in sequence among different chromosomes.

Mutational analysis suggests that regions I and II are less critical to centromere function than is region III. Mutations in the former regions are often tolerated, but mutations in region III can disrupt centromere function. While the DNA of this region appears to be essential to the eventual binding to the spindle fiber, DNA sequences are not unique to specific chromosomes. They can be exchanged between chromosomes experimentally without altering centromere function.

Based on the studies in yeast, it was assumed that similar findings would be forthcoming in the multicellular eukaryotes. However, to the surprise of researchers, the DNA sequence in organisms such as mammals, including humans, varies considerably and has not been highly conserved. This and other observations have led to the conclusion that, in these organisms, the CEN sequences are not essential to centromere function, including the organization of kinetochores. Nevertheless, a large amount of information has been amassed about DNA within centromeres.

The amount of DNA associated with the centromeric region of the chromosome is much more extensive than in yeast. Recall from our discussion that highly repetitive "satellite" DNA is localized in the centromere regions of mice. Such sequences, absent from yeast, but characteristic of most multicellular organisms, vary considerably in size. For example, the 10 bp sequence AATAACATAG is tandemly repeated many times in the centromeres of all four chromosomes of *Drosophila*. In humans, one of the most recognized satellite DNA sequences is the **alphoid family**. Found mainly in the centromere regions, a motif of alphoid DNA of 171 bp is present in tandem head-to-tail repeating arrays, totaling up to 3 million base pairs. While such a motif is present in other closely related primates, neither the sequence nor the number of repeats of the 171 bp sequence is conserved. While the precise role of this highly repetitive DNA in centromere function remains unclear, it is known that the sequences are not transcribed.

Telomeric DNA Sequences

Another important structure is the **telomere**, a component of chromosomes that we have discussed earlier in this chapter (C-banding) and in Chapter 3 when we considered problems associated with replication at the telomere. Found at the ends of linear chromosomes, the function of telomeres is to provide stability to the chromosome by rendering chromosome ends generally inert in interactions with other chromosome ends. In contrast to broken chromosomes, whose ends may rejoin other such ends, telomere regions do not fuse with one another or with broken ends. It is thought that some aspect of the molecular structure of telomeres must be unique compared with most other chromosome regions.

As with centromeres, the analysis of telomeres was first approached by investigating the smaller chromosomes of simple eukaryotes, such as protozoans and yeast. The idea that all telomeres of all chromosomes in a given species might share a common nucleotide sequence has now been born out.

Two types of telomere sequences have been discovered. The first type, simply called **telomeric DNA sequences**, consists of short tandem repeats. It is this group that contributes to the stability and integrity of the chromosome. In the ciliate *Tetrahymena*, over 50 tandem repeats of the hexanucleotide sequence GGGGTT occur. In humans, the sequence GGGATT is repeated many times. The analysis of telomeric DNA sequences has shown them to be highly conserved throughout evolution, reflecting the critical role they play in maintaining the integrity of chromosomes.

The second type, **telomere-associated sequences**, also consists of repetitive sequences and is found both adjacent to and within the telomere. These sequences vary among organisms, and their significance remains unknown.

Interestingly, the telomeres of *Drosophila* are fundamentally different from those found thus far in all other organisms in that they are made up of transposable elements. The significance of this finding, however, is not yet clear.

As discussed in Chapter 3, replication of the telomere requires a unique RNA-containing enzyme **telomerase**. In its absence, the DNA at the ends of chromosomes becomes shorter during each replication. Because single-celled eukaryotes are immortalized cells, telomerase is critical for the survival of such species. In multicellular organisms, such as humans, telomerase is essential in germ-line cells, but is inactive in somatic cells. Chromosome shortening is considered part of the natural process of cell aging, serving as an internal clock. In human cancer cells, which have become immortalized, the transition to malignancy appears to require the activation of telomerase in order to overcome the normal senescence associated with chromosome shortening.

Middle Repetitive Sequences: VNTRs and Dinucleotide Repeats

A brief review of still another prominent category of repetitive DNA sheds more light on our understanding of the organization of the eukaryotic genome. In addition to highly repetitive DNA, which constitutes about 5% of the human genome (and 10 percent of the mouse genome), a second category, **middle** (or **moderately**) **repetitive DNA**, recognized by C_0t analysis, is fairly well characterized. Because we are learning a great deal about the human genome, we will use our own species to illustrate this category of DNA in genome organization.

Middle repetitive DNA most prominently consists of either tandemly repeated or interspersed sequences. No function has been ascribed to these components of the genome. An example includes those called **variable number tandem repeats (VNTRs)**. The repeating DNA sequence of VNTRs may be 15–100 bp long and is found within and between genes. Many such clusters are dispersed throughout the genome, and they are often referred to as **minisatellites**.

The number of tandem copies of each specific sequence at each location varies in individuals, creating localized regions of 1000–5000 bp (1–5 kb) in length. As we will see in Chapter 18, the variation in size (length) of these regions between individuals in humans was originally the basis for the forensic technique referred to as **DNA fingerprinting**.

Another group of tandemly repeated sequences consists of dinucleotides, also referred to as **microsatellites**. Like VNTRs, they are dispersed throughout the genome and vary among individuals in the number of repeats present at any site. For example, in humans, the most common microsatellite is the dinucleotide $(CA)_n$, where n equals the number of repeats. Most common, n is between 5 and 50. These clusters are also used forensically and, in addition, have served as useful molecular markers during genome analysis.

Repetitive Transposed Sequences: SINES and LINES

Still another category of repetitive DNA consists of sequences that are interspersed throughout the genome, rather than being tandemly repeated. They can be either short or long and many have the added distinction of being **transposable sequences**, which are mobile and can move to different locations within the genome. A large portion of eukaryotic genomes are composed of such sequences.

For example, **short interspersed elements**, called **SINES**, are less than 500 base pairs long and may be present 500,000 times or more in the human genome. The best characterized human SINE is a set of closely related sequences called the ***Alu* family** (the name is based on the presence of DNA sequences recognized by the restriction endonuclease *Alu*I). Members of this DNA family, also found in other mammals, are 200 to 300 base pairs long and are dispersed rather uniformly throughout the genome, both between and within genes. In humans, this family encompasses more than 5 percent of the entire genome.

Alu sequences are particularly interesting, although their function, if any, is yet undefined. Members of the *Alu* family are sometimes transcribed. The role of this RNA is not certain, but it is thought to be related to its mobility in the genome. In fact, *Alu* sequences are thought to have arisen from an RNA element whose DNA complement was dispersed throughout the genome as a result of the activity of reverse transcriptase (an enzyme that synthesizes DNA on an RNA template).

The group of **long interspersed elements (LINES)** represents still another category of repetitive transposable DNA sequences. In humans, the most prominent example is a family designated **L1**. Members of this sequence family are about 6400 base pairs long and are present up to 100,000 times, according to one estimate. Their 5′ end is highly variable, and their role within the genome has yet to be defined.

The basis for transposition of L1 elements is now clear. The L1 DNA sequence is first transcribed into an RNA molecule. The RNA then serves as the template for the synthesis of the DNA complement using the enzyme reverse transcriptase. This enzyme is encoded by a portion of the L1 sequence. The new L1 copy then integrates into the DNA of the chromosome at a new site. Because of the similarity of this transposition mechanism with that used by retroviruses, LINES are referred to as **retrotransposons**.

SINES and LINES represent a significant portion of human DNA. Both types of elements share the organizational feature of consisting of a mixture of about 70 percent unique and 30 percent repeating sequences within the DNA of each entity. Collectively, they constitute about 10 percent of the genome.

Middle Repetitive Multiple Copy Genes

In some cases, middle repetitive DNA includes functional genes present tandemly in multiple copies. For example, many copies exist of the genes encoding ribosomal RNA. *Drosophila* has 120 copies per haploid genome. Single genetic units encode a large precursor molecule that is processed into the 5.8*S*, 18*S*, and 28*S* rRNA components. In humans, multiple copies of this gene are clustered on the p arm of the acrocentric chromosomes 13, 14, 15, 21, and 22. Multiple copies of the genes encoding 5*S* rRNA are transcribed separately from multiple clusters found together on the terminal portion of the p arm of chromosome 1.

4.7 The Vast Majority of a Eukaryotic Genome Does Not Encode Functional Genes

Given the preceding information involving various forms of repetitive DNA in eukaryotes, it is of interest to pose an important question: *What proportion of the eukaryotic genome actually encodes functional genes?* Taken together, the various forms of highly repetitive and moderately repetitive DNA constitute up to 40 percent of the human genome. Such an observation is not uncommon in eukaryotes. In addition to repetitive DNA, there is a large amount of single-copy DNA sequences as defined by C_0t analysis that appear to be noncoding. A small portion of them are called **pseudogenes**, which represent evolutionary vestiges of duplicated copies of genes that have undergone sufficient mutations to render them nonfunctional. In some cases, multiple copies of such genes exist.

While the proportion of the genome consisting of repetitive DNA varies among organisms, one feature seems to be shared: *Only a very small part actually codes for proteins.* For example, the 20,000–30,000 genes encoding proteins in sea urchin occupy less than 10 percent of the genome. In *Drosophila*, only 5–10 percent of the genome is occupied by genes coding for proteins. In humans, it appears that the estimated 30,000 functional genes occupy less than 5 percent of the genome.

The study of the various forms of repetitive DNA has significantly enhanced our understanding of genome organization. In the next chapter, we will explore the organization of genes within chromosomes.

Chapter Summary

1. The organization of the molecular components that form chromosomes is essential to understanding the function of the genetic material. Largely devoid of associated proteins, bacteriophage and bacterial chromosomes contain DNA molecules in a form equivalent to the Watson–Crick model.
2. Polytene and lampbrush chromosomes are examples of specialized structures that have extended our knowledge of genetic organization and function.
3. The eukaryotic chromatin fiber is a nucleoprotein organized into repeating units called nucleosomes. Composed of 200 base pairs of DNA, an octamer of four types of histones plus one linker histone, the nucleosome is important in facilitating the conversion of the extensive chromatin fiber characteristic of interphase into the highly condensed chromosome seen in mitosis.
4. The structural heterogeneity of the chromosome axis has been established as a result of both biochemical and cytological investigation. Heterochromatin, prematurely condensed in interphase is, for the most part, genetically inert. The centromeric and telomeric regions, the Y chromosome, and the Barr body are examples.
5. DNA analysis has revealed unique nucleotide sequences in both the centromere and telomere regions of chromosomes, which no doubt impart the characteristic they share of being heterochromatic.
6. Eukaryotic genomes demonstrate complex sequence organization characterized by numerous categories of repetitive DNA.
7. Repetitive DNA consists of either tandem repeats clustered in various regions of the genome or single sequences interspersed uniformly throughout the genome. In the former group, the size of each cluster varies among individuals, providing one form of biochemical identity. The latter group of sequences may be short or long, such as *Alu* and L1, respectively, and are transposable elements.
8. The vast majority of a eukaryotic genome does not encode functional genes. In humans, for example, less than 5 percent of the genome is used to encode the 30,000 genes found in our genome.

Insights and Solutions

A previously undiscovered single-celled organism was found living at a great depth on the ocean floor. Its nucleus contained only a single linear chromosome containing 7×10^6 nucleotide pairs of DNA coalesced with three types of histonelike proteins.

1. A short micrococcal nuclease digestion yielded DNA fractions consisting of 700, 1400, and 2100 base pairs. Predict what these fractions represent. What conclusions can be drawn?

Solution: The chromatin fiber may consist of a variation of nucleosomes containing 700 base pairs of DNA. The 1400- and 2100-bp fractions represent two and three nucleosomes, respectively, linked together. Enzymatic digestion may have been incomplete, leading to the latter two fractions.

2. The analysis of individual nucleosomes revealed that each unit contained one copy of each protein, and that the short linker DNA contained no protein bound to it. If the entire chromosome consists of nucleosomes (discounting any linker DNA), how many are there, and how many total proteins are needed to form them?

Solution: Since the chromosome contains 7×10^6 base pairs of DNA, the number of nucleosomes, each containing 7×10^2 base pairs, is equal to

$$7 \times 10^6 / 7 \times 10^2 = 10^4 \text{ nucleosomes}$$

The chromosome thus contains 10^4 copies of each of the three proteins, for a total of 3×10^4 molecules.

3. Analysis then revealed the organism's DNA to be a double helix similar to the Watson–Crick model, but containing 20 base pairs per complete turn of the right-handed helix. The physical size of the nucleosome was exactly double the volume occupied by that found in all other known eukaryotes, by virtue of increasing the distance along the fiber axis by a factor of two. Compare the degree of compaction (the number of turns per nucleosome) of this organism's nucleosome with that found in other eukaryotes.

Solution: The unique organism compacts a length of DNA consisting of 35 complete turns of the helix (700 base pairs per nucleosome/20 base pairs per turn) into each nucleosome. The normal eukaryote compacts a length of DNA consisting of 20 complete turns of the helix (200 base pairs per nucleosome/10 base pairs per turn) into a nucleosome one-half the volume of that in the unique organism. The degree of compaction is therefore less in the unique organism.

4. No further coiling or compaction of this unique chromosome occurs in the unique organism. Compare this situation with that of a eukaryotic chromosome. Do you think an interphase human chromosome 7×10^6 base pairs in length would be a shorter or longer chromatin fiber?

Solution: The eukaryotic chromosome contains still another level of condensation in the form of "solenoids," which are dependent on the H1 histone molecule associated with linker DNA. Solenoids condense the eukaryotic fiber by still another factor of five. The length of the unique chromosome is compacted into 10^4 nucleosomes, each containing an axis length twice that of the eukaryotic fiber. The eukaryotic fiber consists of $7 \times 10^6/2 \times 10^2 = 3.5 \times 10^4$ nucleosomes, 3.5 more than the unique organism. However, they are compacted by the factor of five in each solenoid. Therefore, the chromosome of the unique organism is a longer chromatin fiber.

Problems and Discussion Questions

1. Contrast the size of the chromosome of bacteriophage λ and T2 with that of *E. coli*. How does this relate to the relative size and complexity of these phages and that bacterium?
2. Bacteriophages and bacteria almost always contain their DNA as circular (closed loops) chromosomes. Phage λ is an exception, maintaining its DNA in a linear chromosome within the viral particle. However, as soon as it is injected into a host cell, it circularizes before replication begins. Taking into account information in the previous chapter (see Chapter 10), what advantage exists in replicating circular DNA molecules compared to linear molecules?
3. How are giant polytene chromosomes formed? Describe their appearance.
4. What genetic process is occurring in a "puff" of a polytene chromosome? How do we know?
5. During what genetic process are lampbrush chromosomes present in vertebrates? What visual evidence supports this conclusion?
6. Why might it be predicted that the organization of eukaryotic genetic material would be more complex than that of viruses or bacteria?
7. Describe the sequence of research findings leading to the model of chromatin structure. What is the molecular composition and arrangement of the nucleosome? Describe the transitions that occur as nucleosomes are coiled and folded, ultimately forming a chromatid.
8. Provide a comprehensive definition of heterochromatin and list as many examples as you can.
9. Mammals contain a diploid genome consisting of at least 10^9 base pairs. If this amount of DNA is present as chromatin fibers where each group of 200 base pairs of DNA is combined with nine histones into a nucleosome, and each group of six nucleosomes is combined into a solenoid, achieving a final packing ratio of 50, determine
   (a) the total number of nucleosomes in all fibers,
   (b) the total number of solenoids in all fibers,
   (c) the total number of histone molecules combined with DNA in the diploid genome, and
   (d) the combined length of all fibers.
10. Assume that a viral DNA molecule is in the form of a 50-μm-long circular strand of a uniform 20 Å diameter. If this molecule is contained within a viral head that is a sphere with a diameter of 0.08 μm, will the DNA molecule fit into the viral head, assuming complete flexibility of the molecule? Justify your answer mathematically.
11. How many base pairs are in a molecule of phage T2 DNA, which is 52 μm long?

Extra-Spicy Problems

12. If a human nucleus is 10 μm in diameter and it must hold as much as 2 meters of DNA, which is complexed into nucleosomes that during full extension are 11 nm in diameter, what percentage of the volume of the nucleus is occupied by the genetic material?
13. Microsatellites are currently exploited as markers for paternity testing. A sample paternity test is shown on the next page, where eleven microsatellite markers were used to test samples from a mother, her child, and an alleged father.
    The name of the microsatellite locus is given in the left-hand column and the genotype of each individual is recorded as the number of repeats he or she carries at that locus. For example, at locus D9S302, the mother carries 30 repeats on one of her chromosomes and 31 on the other. In cases where an individual carries the same number of repeats on both chromosomes, only a single number is recorded. (Some of the numbers are followed by a decimal point, eg. 20.2 and include a partial repeat in addition to several complete repeats.)
    Assuming that these markers are inherited in a simple Mendelian fashion, can the alleged father be excluded as the source of the sperm that produced the child? Why or why not? Explain.

Microsatellite Locus Chromosome Location	*Mother*	*Child*	*Alleged Father*
D9S302	30	31	32
9q31-q33	31	32	33
D22S883	17	20.2	20.2
22pter-22qter	22	22	
D18S535	12	13	11
18q12.2-q12.3	14	14	13
D7SI 804	27	26	26
7pter-7qter	30	30	27
D3S2387	23	24	20.2
3p24.2.3pter	25.2	25.2	24
D4S2386	12	12	12
4pter-qter			16

Microsatellite Locus Chromosome Location	*Mother*	*Child*	*Alleged Father*
D5S1719	11	10.3	10
5pter-5qter	11.3	11	10.3
CSF1PO	11	11	10
5q33.3.q34		12	12
FESFPS	11	12	10
15q25-15qter	12	13	13
TH01	7	7	7
11p15.5			8
LIPOL	10	9	9

Selected Readings

Angelier, N., et al. 1984. Scanning electron microscopy of amphibian lampbrush chromosomes. *Chromosoma* 89:243–53.

Bauer, W.R., Crick, F.H.C., and White, J.H. 1980. Supercoiled DNA. *Sci. Am.* (July) 243:118–33.

Beerman, W., and Clever, U. 1964. Chromosome puffs. *Sci. Am.* (Apr.) 210:50–58.

Callan, H.G. 1986. *Lampbrush chromosomes*. New York: Springer–Verlag.

Carbon, J. 1984. Yeast centromeres: Structure and function. *Cell* 37:352–53.

Chen, T.R., and Ruddle, F.H. 1971. Karyotype analysis utilizing differential stained constitutive heterochromatin of human and murine chromosomes. *Chromosoma* 34:51–72.

Corneo, G., et al. 1968. Isolation and characterization of mouse and guinea pig satellite DNA. *Biochemistry* 7:4373–79.

DuPraw, E.J. 1970. *DNA and chromosomes*. New York: Holt, Rinehart & Winston.

Gall, J.G. 1963. Kinetics of deoxyribonuclease on chromosomes. *Nature* 198:36–38.

———1981. Chromosome structure and the *C*-value paradox. *J. Cell Biol.* 91:3s–14s.

Green, B.R., and Burton, H. 1970. *Acetabularia* chloroplast DNA: Electron microscopic visualization. *Science* 168:981–82.

Hewish, D.R., and Burgoyne, L. 1973. Chromatin substructure. The digestion of chromatin DNA at regularly spaced sites by a nuclear deoxyribonuclease. *Biochem. Biophys. Res. Comm.* 52:504–10.

Hill, R.J., and Rudkin, G.T. 1987. Polytene chromosomes: The status of the band–interband question. *BioEssays* 7:35–40.

Hsu, T.C. 1973. Longitudinal differentiation of chromosomes. *Annu. Rev. Genet.* 7:153–77.

Jeffreys, A.J., Wilson, V., and Thein, S.L. 1985. Hypervariable minisatellite regions in human DNA. *Nature* 314:66–73.

Korenberg, J.R. and Rykowski, M.C. 1988. Human genome organization: *Alu*, LINEs, and the molecular organization of metaphase chromosome bands. *Cell* 53:391–400.

Kornberg, R.D. 1975. Chromatin structure: A repeating unit of histones and DNA. *Science* 184:868–71.

Kornberg, R.D., and Klug, A. 1981. The nucleosome. *Sci. Am.* (Feb.) 244:52–64.

Lorch, Y., Zhang, N, and Kornberg, R.D. 1999. Histone octamer transfer by a chromatin-remodeling complex. *Cell.* 96:389–92.

Luger, K., et. al. 1997. Crystal structure of the nucleosome core particle at 2.8 resolution. *Nature* 389:251–56.

Moxon, E.R. and Wills, C. 1999. DNA microsatellites: Agents of evolution? *Sci. Am.* (Jan.) 280:94–99.

Moyzis, R.K. 1991. The human telomere. *Sci. Am.* (Aug) 265:48–55.

Olins, A.L., and Olins, D.E. 1974. Spheroid chromatin units (*n* bodies). *Science* 183:330–32.

———1978. Nucleosomes: The structural quantum in chromosomes. *Am. Sci.* 66:704–11.

Singer, M.F. 1982. SINES and LINES: Highly repeated short and long interspersed sequences in mammalian genomes. *Cell* 28:433–34.

Sullivan, B.A., Blower, M.D., and Karpen, G.H. 2001. Determining centromere identity: Cyclical stories and forking paths. *Nature Reviews: Genetics* 2:584–96.

Van Holde, K.E. 1989. *Chromatin*. New York: Springer–Verlag.

Verma, R.S., ed. 1988. *Heterochromatin: Molecular and structural aspects*. Cambridge, UK: Cambridge University Press.

Wolfe, A. 1998. *Chromatin: Structure and Function*. 3rd ed. San Diego: Academic Press.

Yunis, J.J., 1976. High resolution of human chromosomes. *Science* 191:1268–70.

Yunis, J.J., and Prakash, O. 1982. The origin of man: A chromosomal pictorial legacy. *Science* 215:1525–30.

GENETICS MediaLab

The resources that follow will help you achieve a better understanding of the concepts presented in this chapter. These resources can be found either on the CD packaged with this textbook or on the Companion Web site found at **http://www.prenhall.com/klug**

CD Resources:

Module 4.1: Chromosome Structure

Web Problem 1:
Time for completion = 10 minutes

How long is DNA? Stretched out end to end, the total length of DNA in the human genome would measure approximately 1.8 meters. Therefore, it is necessary for the cell to package DNA into a compact chromosome. The linked Web site illustrates aspects of chromosomal organization. After examining this Web site, list the levels of chromosomal organization that exist in a eukaryotic cell. Discuss briefly what additional kinds of chromosomal organization exist along the length of the chromosome. To complete this exercise, visit Web Problem 1 in Chapter 4 of your Companion Web site, and select the keyword **NUCLEUS.**

Web Problem 2:
Time for completion = 10 minutes

What happens when chromosomes get wound up? Replication and recombination alter the level of supercoiling that occurs within a DNA molecule. Proteins called topoisomerases regulate the level of supercoiling by cutting, winding, or unwinding the DNA helix and religating the DNA strands. In this exercise, you will examine molecular models describing how toposomerase I of E. coli modulates DNA supercoiling. Directions for downloading the CHIME browser plug-in needed to view this module are found at the top of the linked Web site. Scroll to the Hall of Recombination and click on the topoisomerase button. In a few words, describe the physical structure of topoisomerase I. Examine the section of DNA cleavage, and describe in a few sentences the model of how topoisomerase I allows the passage of DNA strands. To complete this exercise, visit Web Problem 2 in Chapter 4 of your Companion Web site, and select the keyword **TOPOISOMERASE.**

Web Problem 3:
Time for completion = 10 minutes

What is the role of DNA in organelles? Although research has often focused on nuclear DNA, mitochondrial and chloroplast DNA provide significant functions within the eukaryotic cell. The linked Web site describes the mitochondrial and chloroplast genomes and suggests evidence supporting the endosymbiont hypothesis for the origin of these genomes. Describing the general features, and briefly compare and contrast these two organellelike genomes with the nuclear genome. Discuss briefly whether you think that the endosymbiosis hypothesis seems reasonable from the data you have available at this Web site. To complete this exercise, visit Web Problem 3 in Chapter 4 of your Companion Web site, and select the keyword **ORGANELLE.**

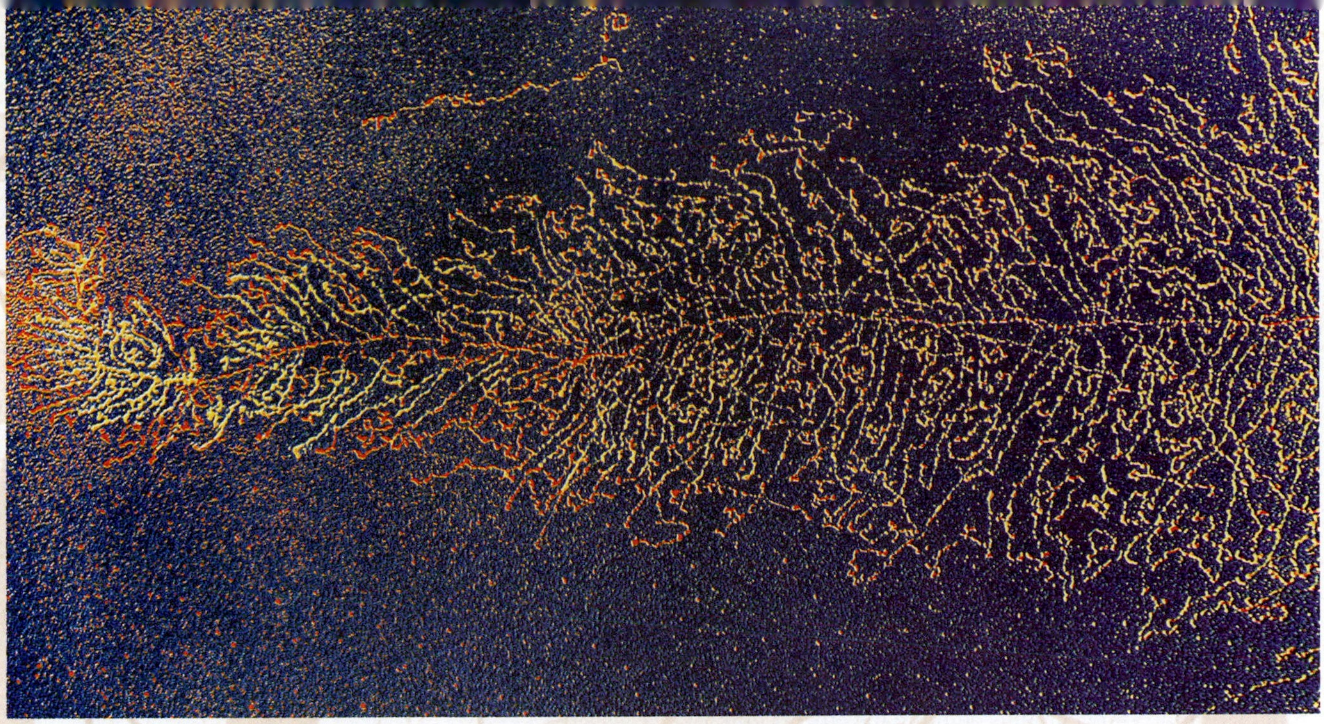

Electron micrograph visualizing the process of transcription. (*Professor Oscar L. Miller/Science Photo Library/Photo Researchers, Inc.*)

5

The Genetic Code and Transcription

KEY CONCEPTS

As we saw in Chapter 2, the structure of DNA consists of a linear sequence of deoxyribonucleotides. This sequence ultimately dictates the components constituting proteins, the end product of most genes. The central question has to do with how such information stored as a nucleic acid can be decoded into a protein. Figure 5–1 provides a simple overview of how this transfer of information occurs. The first step in gene expression involves the transfer of information present on one of the two strands of DNA (the template strand) into an RNA complement through the process of transcription. Once synthesized, this RNA acts as a "messenger" molecule, bearing the coded information—hence its name, messenger RNA (mRNA). Such RNAs then associate with ribosomes, in which decoding into proteins occurs.

In this chapter, we will focus on the initial phases of gene expression by addressing two major questions. First, how is genetic information encoded? Second, how does the transfer from DNA to RNA occur, thus defining the process of transcription? As we will see, ingenious analytical research established that the genetic code is written in units of three letters—ribonucleotides present in mRNA that reflect the stored information in genes. Each triplet code word directs the incorporation of a specific amino acid into a protein as it is synthesized. As we can predict based on our prior discussion of the replication of DNA, transcription is also a complex process dependent on a major polymerase enzyme and a cast of supporting proteins. We will explore what is known about transcription in bacteria and then contrast this prokaryotic model with the differences found in eukaryotes.

In Chapter 6, we will continue our discussion of gene expression by addressing how translation occurs and then discussing the structure and function of proteins. Together, the information in this and the next chapter provides a comprehensive picture of molecular genetics, which serves as the most basic foundation for the understanding of living organisms.

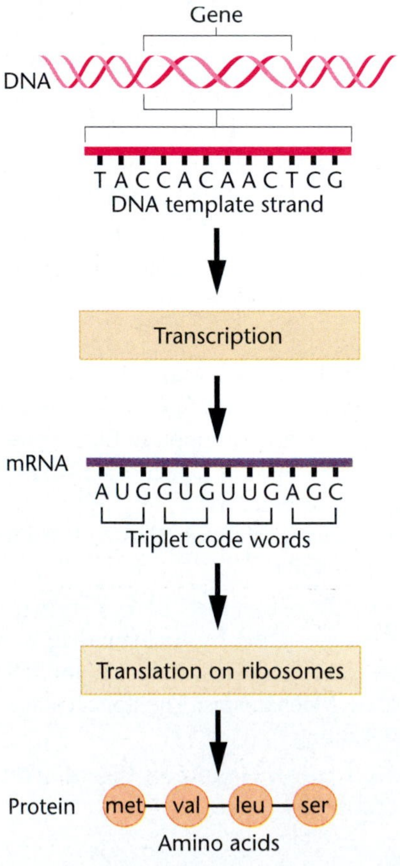

FIGURE 5–1 Flow of genetic information encoded in DNA to messenger RNA to protein.

5.1 The Genetic Code Exhibits a Number of Characteristics

Before we consider the various analytical approaches that led to our current understanding of the genetic code, let's summarize the general features that characterize it:

1. The genetic code is written in linear form, using the ribonucleotide bases that compose mRNA molecules as "letters." The ribonucleotide sequence is derived from the complementary nucleotide bases in DNA.
2. Each "word" within the mRNA contains three ribonucleotide letters. Each group of *three* ribonucleotides, called a **codon**, specifies *one* amino acid; thus, the code is a **triplet**.
3. The code is **unambiguous**, meaning that each triplet specifies only a single amino acid.
4. The code is **degenerate**, meaning that a given amino acid can be specified by more than one triplet codon. This is the case for 18 of the 20 amino acids.
5. The code contains "start" and "stop" signals, certain triplets that are necessary to **initiate** and to **terminate** translation.
6. No internal punctuation ("commas") is used in the code. Thus, the code is said to be **commaless**. Once translation of mRNA begins, the codons are read one after the other with no breaks between them.
7. The code is **nonoverlapping**. After translation commences, any single ribonucleotide at a specific location within the mRNA is part of only one triplet.
8. The code is nearly **universal**. With only minor exceptions, a single coding dictionary is used by almost all viruses, prokaryotes, archaea, and eukaryotes.

5.2 Early Studies Established the Basic Operational Patterns of the Code

In the late 1950s, before it became clear that mRNA serves as an intermediate in transferring genetic information from DNA to proteins, it was considered that DNA itself might directly participate during the synthesis of proteins. Because ribosomes had already been identified, the initial thinking was that the DNA of a gene might somehow become associated with ribosomes, in which the stored information was decoded during the synthesis of proteins. Most early models, faced with the question of how four nucleotides could encode 20 amino acids, embraced the idea that a DNA code would be overlapping. That is, each nucleotide would be part of more than one contiguous code word. Such a concept soon became untenable as accumulating evidence demonstrated inconsistencies resulting from constraints placed by an overlapping code model on the resultant amino acid sequences. We will address some of this evidence shortly. Additionally, there was growing evidence that RNA might be involved in the process. In 1961, François Jacob and Jacques Monod postulated the existence of **messenger RNA (mRNA)**. Once mRNA was discovered, it soon became clear that, even though genetic information is stored in DNA, the code which is translated into proteins resides in RNA. The central question then was how only four letters—the four nucleotides—could specify 20 words—the amino acids.

The Triplet Nature of the Code

In the early 1960s, Sidney Brenner argued on theoretical grounds that the code must be a triplet, since three-letter words represent the minimal use of four letters to specify 20 amino acids. A code of four nucleotides, taken two at a time, for example, would provide only 16 unique code words (4^2). a triplet code provides 64 words (4^3)—clearly more than the 20 needed—and it is much simpler than a four-letter code (4^4), which would specify 256 words.

The ingenious experimental work of Francis Crick, Leslie Barnett, Brenner, and R. J. Watts-Tobin represented the first solid evidence for the triplet nature of the code. These researchers induced insertion and deletion mutations in the *rII* locus of phage T4. (See Chapter 15.) Mutations in this locus cause rapid lysis and distinctive plaques. Such mutants will successfully infect strain B of *E. coli*, but cannot reproduce on a separate strain of *E. coli*, designated K12. Crick and his colleagues used the acridine dye proflavin to induce mutations. This mutagenic agent intercalates within the double helix of DNA, often causing the insertion or the deletion of one or more nucleotides during replication. As shown in Figure 5–2(a), an insertion of a single nucleotide causes the reading frame to shift, changing the specific sequence of all subsequent downstream triplets (to the right of the insertion in the figure). Thus, they are referred to as **frameshift mutations**. Upon translation, the amino acid sequence of the encoded protein will be altered. When these mutations are present at the *rII* locus, T4 will not reproduce on *E. coli* K12.

Crick and his colleagues reasoned that if phages with these induced mutations were treated again with proflavin, still other insertions or deletions would occur. A second change might result in a revertant phage, which would display wild-type behavior and successfully infect *E. coli* K12. For example, if the original mutant contained an insertion (+), a second event causing a deletion (−) close to the insertion would restore the original reading frame. In the same way, an event resulting in an insertion (+) might correct an original deletion (−).

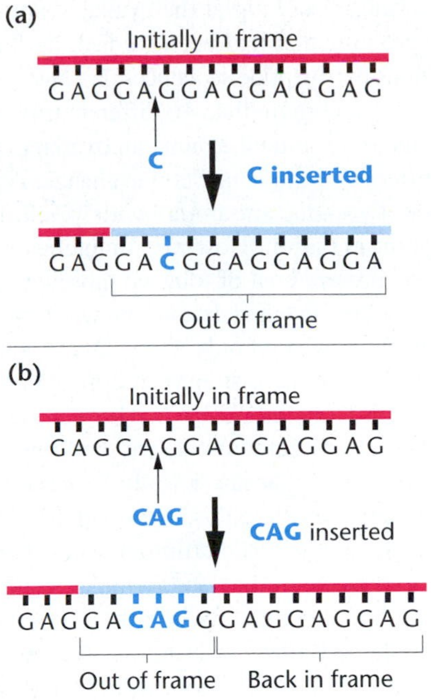

FIGURE 5–2 The effect of frameshift mutations on a DNA sequence repeating the triplet sequence GAG. (a) The insertion of a single nucleotide shifts all subsequent triplet reading frames. (b) The insertion of three nucleotides changes only two triplets, but the frame of reading is then reestablished to the original sequence.

In studying many mutations of this type, these researchers were able to compare various mutant combinations together within the same DNA sequence. They found that various combinations of one plus (+) and one minus (−) indeed caused reversion to wild-type behavior. Still other observations shed light on the number of nucleotides constituting the genetic code. When two pluses were together or when two minuses were together, the correct reading frame *was not* reestablished. This argued against a doublet (two letter) code. However, when three plusses [Figure 5–2(b)] or three minuses were present together, the original frame *was* reestablished. These observations strongly supported the triplet nature of the code.

The Nonoverlapping Nature of the Code

Work by Sydney Brenner and others soon established that the code could not be overlapping. For example, given the assumption of a triplet code, Brenner considered restrictions that might be placed on it if it were overlapping. He considered theoretical nucleotide sequences encoding a protein consisting of three amino acids. In the nucleotide sequence GTACA, for example, parts of the central triplet, TAC, are shared by the outer triplets, GTA and ACA. Brenner reasoned that if this were the case, then only certain amino acids should be found adjacent to the one encoded by the central triplet.

For any given central amino acid, only 16 combinations (2^4) of "three amino acid" sequences (a tripeptide) are theoretically possible. Brenner concluded that if the code were overlapping, tripeptide sequences within proteins should be somewhat limited. Looking at the available amino acid sequences of proteins that had been studied, he failed to find such restrictions in tripeptide sequences. For any central amino acid, he found many more than 16 different tripeptides.

A second major argument against an overlapping code involved the effect of a single nucleotide change. With an overlapping code, two adjacent amino acids would be affected by such a point mutation. However, mutations in the genes coding for the protein coat of tobacco mosaic virus (TMV), human hemoglobin, and the bacterial enzyme tryptophan synthetase invariably revealed only single amino acid changes.

The third argument against an overlapping code was presented by Francis Crick in 1957, when he predicted that DNA does not serve as a direct template for the formation of proteins. Crick reasoned that any affinity between nucleotides and an amino acid would require hydrogen bonding. Chemically, however, such specific affinities seemed unlikely. Instead, Crick proposed that there must be an **adaptor molecule** that could covalently bind to the amino acid, yet also be capable of hydrogen bonding to a nucleotide sequence. Because various adaptors would somehow have to overlap one another at nucleotide sites during translation, Crick reasoned that physical constraints would make the process overly complex and, perhaps, even inefficient during translation. As we will see later in this chapter, Crick's prediction was correct; transfer RNA (tRNA) serves as the adaptor in protein synthesis. And, the ribosome accommodates two tRNA molecules at a time during translation.

Crick's and Brenner's arguments, taken together, strongly suggested that, during translation, the genetic code is **nonoverlapping**. Without exception, this concept has been upheld.

The Commaless and Degenerate Nature of the Code

Between 1958 and 1960, information related to the genetic code continued to accumulate. In addition to his adaptor proposal, Crick hypothesized, on the basis of genetic evidence, that the code is **commaless**—that is, he believed no internal punctuation occurs along the reading frame. Crick also speculated that only 20 of the 64 possible triplets specify an amino acid and that the remaining 44 carry no coding assignment.

Was Crick wrong with respect to the 44 "blank" codes? That is, is the code **degenerate**, with more than one triplet coding for the same amino acid? Crick's frameshift studies suggested that, contrary to his earlier proposal, the code is degenerate. The reasoning leading to this conclusion is as follows: For the cases in which wild-type function is restored—that is, $(+)$ and $(-)$, $(++)$ and $(--)$, $(+ + +)$ and $(- - -)$—the original frame of reading is also restored. However, there may be numerous triplets between the various additions and deletions that would still be out of frame. If 44 of the 64 possible triplets were blank and did not specify an amino acid, one of these **nonsense triplets** would very likely occur in the length of nucleotides still out of frame. If a nonsense code were encountered during protein synthesis, it was reasoned that the process would stop or be terminated at that point. If so, the product of the *rII* locus would not be made, and restoration would not occur. Because the various mutant combinations were able to reproduce on *E. coli* K12, Crick and his colleagues concluded that, in all likelihood, most if not all of the remaining 44 codes were not blank. It follows that the genetic code is **degenerate**. As we shall see, this reasoning proved to be correct.

5.3 Studies by Nirenberg, Matthaei, and Others Led to Deciphering of the Code

In 1961, Marshall Nirenberg and J. Heinrich Matthaei characterized the first specific coding sequences, which served as a cornerstone for the complete analysis of the genetic code. Their success, as well as that of others who made important contributions in deciphering the code, was dependent on the use of two experimental tools, an **in vitro (cell-free) protein-synthesizing system** and an enzyme, **polynucleotide phosphorylase**, which allowed the production of synthetic mRNAs. These mRNAs served as templates for polypeptide synthesis in the cell-free system.

Synthesizing Polypeptides in a Cell-Free System

In the cell-free system, amino acids can be incorporated into polypeptide chains. This in vitro mixture, as might be expected, must contain the essential factors for protein synthesis in the cell: ribosomes, tRNAs, amino acids, and other molecules essential to translation. In order to follow (or trace) protein synthesis, one or more of the amino acids must be radioactive. Finally, an mRNA must be added, which serves as the template to be translated.

In 1961, mRNA had yet to be isolated. However, the use of the enzyme polynucleotide phosphorylase allowed the artificial synthesis of RNA templates, which could be added to the cell-free system. This enzyme, isolated from bacteria, catalyzes the reaction shown in Figure 5–3. Discovered in 1955 by Marianne Grunberg-Manago and Severo Ochoa, the enzyme functions metabolically in bacterial cells to degrade RNA. However, in vitro, with high concentrations of ribonucleoside diphosphates, the reaction can be "forced" in the opposite direction to synthesize RNA, as illustrated in the figure. In contrast to RNA polymerase, polynucleotide phosphorylase requires no DNA template. As a result, each addition of a ribonucleotide is random, based on the relative concentration of the four ribonucleoside diphosphates added to the reaction mixtures. The probability of the insertion of a specific ribonucleotide is proportional to the availability of that molecule, relative to other available ribonucleotides. *This point is absolutely critical to understanding the work of Nirenberg and others in the ensuing discussion.*

Taken together, the cell-free system for protein synthesis and the availability of synthetic mRNAs provided a means of deciphering the ribonucleotide composition of various triplets encoding specific amino acids.

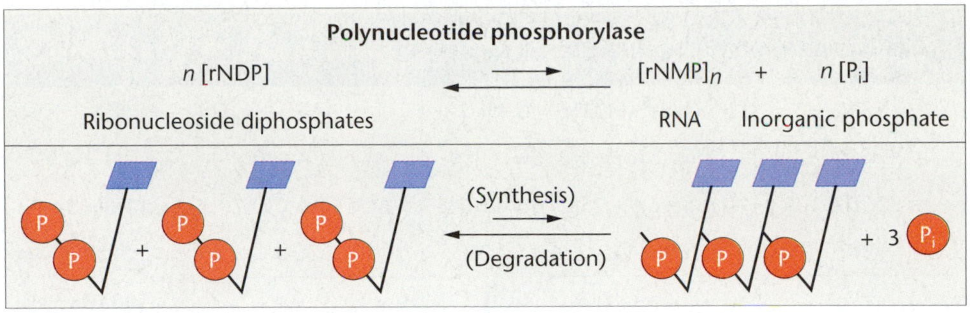

FIGURE 5–3 The reaction catalyzed by the enzyme polynucleotide phosphorylase. Note that the equilibrium of the reaction favors the degradation of RNA, but can be "forced" in the direction favoring synthesis.

Homopolymer Codes

In their initial experiments, Nirenberg and Matthaei synthesized **RNA homopolymers**, each consisting of only one type of ribonucleotide. Therefore, the mRNA added to the in vitro system was either UUUUUU..., AAAAAA..., CCCCCC..., or GGGGGG... . In testing each mRNA, they were able to determine which, if any, amino acids were incorporated into newly synthesized proteins. The researchers determined this by labeling one of the 20 amino acids added to the in vitro system and conducting a series of experiments, each with a different amino acid made radioactive.

For example, consider the experiments using ^{14}C-phenylalanine (Table 5–1). Nirenberg and Matthaei concluded that the message poly U (polyuridylic acid) directs the incorporation of only phenylalanine into the homopolymer polyphenylalanine. Assuming a triplet code, they had determined the first specific codon assignment: UUU codes for phenylalanine. In related experiments, they quickly found that AAA codes for lysine and CCC codes for proline. Poly G did not serve as an adequate template, probably because the molecule folds back on itself. Thus, the assignment for GGG had to await other approaches.

Note that the specific triplet codon assignments were possible only because homopolymers were used. The method yields only the composition of triplets, but since three Us, Cs, or As can have only one possible sequence (i.e., UUU, CCC, and AAA), the actual codon was identified.

TABLE 5–1 Incorporation of ^{14}C-Phenylalanine into Protein

Artificial mRNA	*Radioactivity (counts/min)*
None	44
Poly U	39,800
Poly A	50
Poly C	38

Source: After Nirenberg and Matthaei (1961).

Mixed Copolymers

With these techniques in hand, Nirenberg and Matthaei, and Ochoa and coworkers turned to the use of **RNA heteropolymers**. In this technique, two or more different ribonucleoside diphosphates are added in combination to form the artificial message. The researchers reasoned that if they knew the relative proportion of each type of ribonucleoside diphosphate, they could predict the frequency of any particular triplet codon occurring in the synthetic mRNA. If they then added the mRNA to the cell-free system and ascertained the percentage of any particular amino acid present in the new protein, they could analyze the results and predict the *composition* of triplets specifying specific amino acids.

This approach is illustrated in Figure 5–4. Suppose that A and C are added in a ratio of 1A:5C. The insertion of a ribonucleotide at any position along the RNA molecule during its synthesis is determined by the ratio of A:C. Therefore, there is a 1/6 possibility for an A and a 5/6 chance for a C to occupy each position. On this basis, we can calculate the frequency of any given triplet appearing in the message.

For AAA, the frequency is $(1/6)^3$, or about 0.4 percent. For AAC, ACA, and CAA, the frequencies are identical—that is, $(1/6)^2(5/6)$, or about 2.3 percent for each. Together, all three 2A:1C triplets account for 6.9 percent of the total three-letter sequences. In the same way, each of three 1A:2C triplets accounts for $(1/6)(5/6)^2$, or 11.6 percent (or a total of 34.8 percent). CCC is represented by $(5/6)^3$, or 57.9 percent of the triplets.

By examining the percentages of any given amino acid incorporated into the protein synthesized under the direction of this message, it is possible to propose probable base composition (Figure 5–4). Because proline appears 69 percent of the time, we could propose that proline is encoded by CCC (57.9 percent) and one triplet consisting of 2C:1A (11.6 percent). Histidine, at 14 percent, is probably coded by one 2C:1A (11.6 percent) and one 1C:2A (2.3 percent). Threonine, at 12 percent, is likely coded by only one 2C:1A. Asparagine and glutamine each appear to be coded by one of the 1C:2A triplets, and lysine appears to be coded by AAA.

Using as many as all four ribonucleotides to construct the mRNA, the researchers conducted many similar experiments.

Possible compositions	Probability of occurrence of any triplet	Possible triplets	Final %
3A	$(1/6)^3 = 1/216 = 0.4\%$	AAA	0.4
1C:2A	$(1/6)^2(5/6) = 5/216 = 2.3\%$	AAC ACA CAA	3 x 2.3 = 6.9
2C:1A	$(1/6)(5/6)^2 = 25/216 = 11.6\%$	ACC CAC CCA	3 x 11.6 = 34.8
3C	$(5/6)^3 = 125/216 = 57.9\%$	CCC	57.9
			100.0%

Chemical synthesis of message

RNA
C C C C C C C C A C C C C C C A A C C A C C C C C A C C C C C A C C C A A

Translation of message

Percentage of amino acids in protein		Probable base composition assignments
Lysine	<1%	AAA
Glutamine	2%	1C:2A
Asparagine	2%	1C:2A
Threonine	12%	2C:1A
Histidine	14%	2C:1A, 1C:2A
Proline	69%	CCC, 2C:1A

FIGURE 5–4 Results and interpretation of a mixed copolymer experiment in which a ratio of 1A:5C is used (1/6A:5/6C).

Although determination of the *composition* of triplet code words corresponding to all 20 amino acids represented a very significant breakthrough, the *specific sequences* of triplets were still unknown. Their determination awaited still other approaches.

The Triplet Binding Assay

It was not long before more advanced techniques were developed. In 1964, Nirenberg and Philip Leder developed the **triplet binding assay**, which led to specific assignments of triplets. The technique took advantage of the observation that ribosomes, when presented with an RNA sequence as short as three ribonucleotides, will bind to it and form a complex similar to what is found in vivo. The triplet acts like a codon in mRNA, attracting the complementary sequence within tRNA (Figure 5–5). Such a triplet sequence in tRNA that is complementary to a codon of mRNA is known as an **anticodon**.

Although it was not yet feasible to chemically synthesize long stretches of RNA, triplets of known sequence could be synthesized in the laboratory to serve as templates. All that was needed was a method to determine which tRNA–amino

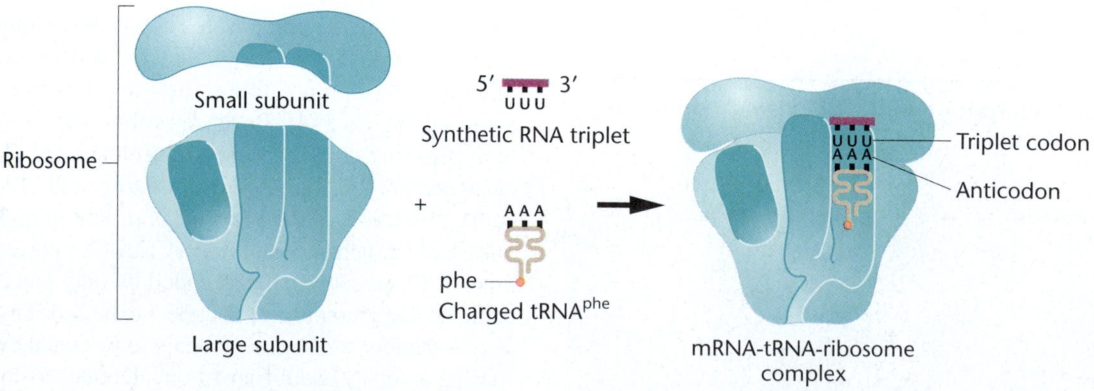

FIGURE 5–5 An example of the triplet-binding assay. The UUU triplet acts as a codon, attracting the complementary tRNA^{Phe} anticodon AAA.

acid was bound to the triplet RNA-ribosome complex. The test system devised was quite simple. The amino acid to be tested was made radioactive, and a charged tRNA was produced. Because code compositions were known, it was possible to narrow the decision as to which amino acids should be tested for each specific triplet.

The radioactively charged tRNA, the RNA triplet, and ribosomes are incubated together on a nitrocellulose filter, which will retain the larger ribosomes, but not the other smaller components, such as charged tRNA. If radioactivity is not retained on the filter, an incorrect amino acid has been tested. If radioactivity remains on the filter, it is retained because the charged tRNA has bound to the triplet associated with the ribosome. In such a case, a specific codon assignment can be made.

Work proceeded in several laboratories, and in many cases clear-cut, unambiguous results were obtained. Table 5–2, for example, shows 26 triplets assigned to nine amino acids. However, in some cases the degree of triplet binding was inefficient, and assignments were not possible. Eventually, about 50 of the 64 triplets were assigned. These specific assignments of triplets to amino acids led to two major conclusions. First, the genetic code is degenerate—that is, one amino acid may be specified by more than one triplet. Second, the code is also unambiguous—that is, a single triplet specifies only one amino acid. As we shall see later in this chapter, these conclusions have been upheld with only minor exceptions. The triplet binding technique was a major innovation in deciphering the genetic code.

Repeating Copolymers

Yet another innovative technique used to decipher the genetic code was developed in the early 1960s by Gobind Khorana, who was able to chemically synthesize long RNA molecules consisting of short sequences repeated many times. First, he created shorter sequences (e.g., di-, tri-, and tetranucleotides), which were then replicated many times and finally joined enzymatically to form the long polynucleotides. As illustrated in Figure 5–6, a dinucleotide made in this way is converted to a message with two repeating triplets. A trinucleotide is converted to a message with three potential triplets, depending on the point at which initiation occurs, and a tetranucleotide creates four repeating triplets.

When these synthetic mRNAs were added to a cell-free system, the predicted number of amino acids incorporated was upheld. Several examples are shown in Table 5–3. When such data were combined with those drawn from composition assignment and triplet binding, specific assignments were possible.

One example of specific assignments made from such data will illustrate the value of Khorana's approach. Consider the following three experiments in concert with one another: The repeating trinucleotide sequence UUCUUCUUC ... produces three possible triplets—UUC, UCU, and CUU, depending on the initiation point. When placed in a cell-free translation system, the polypeptides containing phenylalanine (phe), serine (ser), and leucine (leu) are produced. On the other hand, the repeating dinucleotide sequence UCUCUCUC ... produces the triplets UCU and CUC with the incorporation of leucine and serine into the polypeptide. The overlapping results indicate that the triplets UCU and CUC specify leucine and serine, but not which triplet specifies which amino acid. One can further conclude that *either* the CUU *or* the UUC triplet also encodes leucine *or* serine, while the other encodes phenylalanine.

To derive more specific information, let's examine the results of using the repeating tetranucleotide sequence UUAC, which produces the triplets UUA, UAC, ACU, and CUU. The CUU triplet is one of the two in which we are interested. Three amino acids are incorporated: leucine, threonine, and tyrosine. Because CUU must specify only serine or leucine, and because, of these two, only leucine appears, we may conclude that CUU specifies leucine.

Once the latter is established, we can logically determine all other assignments. Of the two triplet pairs remaining (UUC and UCU from the first experiment *and* UCU and CUC from the second experiment), whichever triplet is common to both must encode serine. This is UCU. By elimination, UUC is determined to encode phenylalanine and CUC is determined to encode leucine. While the logic must be carefully followed, four specific triplets encoding three different amino acids have been assigned from these experiments.

From such interpretations, Khorana reaffirmed triplets that were already deciphered and filled in gaps left from other approaches. For example, the use of two tetranucleotide sequences, GAUA and GUAA, suggested that at least two triplets were termination signals. He reached this conclusion because neither of these sequences directed the incorporation of any amino acids into a polypeptide. Since there are no triplets common to both messages, he predicted that each repeating sequence contained at least one triplet that terminates protein synthesis. Table 5–3 lists the possible triplets for the poly–(GAUA) sequence, of which UAG is a termination codon.

TABLE 5–2 Amino Acid Assignments to Specific Trinucleotides Derived from the Triplet Binding Assay

Trinucleotides	*Amino Acid*	*Trinucleotides*	*Amino Acid*	*Trinucleotides*	*Amino Acid*
UGU UGC	Cysteine	CUC CUA CUG	Leucine	CCU CCC	Proline
GAA GAG	Glutamic acid	AAA AAG	Lysine	CCG CCA	Proline
AUU AUG AUA	Isoleucine	AUG	Methionine	UCU UCC	Serine
UUA UUG CUU	Leucine	UUU UUC	Phenylalanine	UCA UCG	Serine

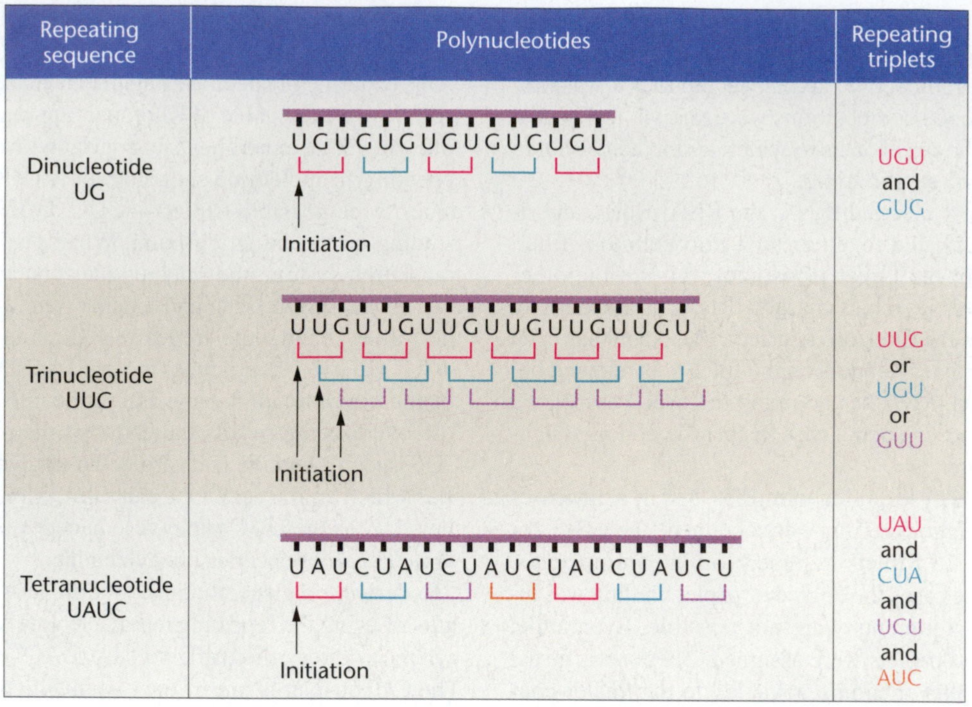

FIGURE 5–6 The conversion of di-, tri-, and tetranucleotides into repeating copolymers. The triplet codons produced in each case are shown.

5.4 The Coding Dictionary Reveals the Function of the 64 Triplets

The various techniques applied to decipher the genetic code have yielded a dictionary of 61 triplet codon–amino acid assignments. The remaining three triplets are termination signals, not specifying any amino acid. Figure 5–7 designates the assignments in a particularly illustrative form first suggested by Francis Crick.

TABLE 5–3 Amino Acids Incorporated Using Repeated Synthetic Copolymers of RNA

Repeating Copolyme	*Codons Produced*	*Amino Acids in Polypeptides*
UG	UGU	Cysteine
	GUG	Valine
AC	ACA	Threonine
	CAC	Histidine
UUC	UUC	Phenylalanine
	UCU	Serine
	CUU	Leucine
AUC	AUC	Isoleucine
	UCA	Serine
	CAU	Histidine
UAUC	UAU	Tyrosine
	CUA	Leucine
	UCU	Serine
	AUC	Isoleucine
GAUA	GAU	None
	AGA	None
	UAG	None
	AUA	None

Degeneracy and the Wobble Hypothesis

A general pattern of triplet codon assignments becomes apparent when we look at the genetic coding dictionary. Most evident is that the code is degenerate, as the early researchers predicted. That is, almost all amino acids are specified by two, three, or four different codons. Three amino acids (serine, arginine, and leucine) are each encoded by six different codons. Only tryptophan and methionine are encoded by single codons.

Also evident is the pattern of degeneracy. Most often in a set of codons specifying the same amino acid, the first two letters are the same, with only the third differing. Crick observed a pattern in the degeneracy at the third position, and in 1966, he postulated the **wobble hypothesis**.

Crick's hypothesis first predicted that the initial two ribonucleotides of triplet codes are more critical than is the third member in attracting the correct tRNA. He postulated that hydrogen bonding at the third position of the codon–anticodon interaction is *less* constrained and need not adhere as specifically to the established base-pairing rules. The wobble hypothesis thus proposes a more flexible set of base-pairing rules at the third position of the codon (Table 5– 4).

This relaxed base-pairing requirement, or "wobble," allows the anticodon of a single form of tRNA to pair with more than one triplet in mRNA. Consistent with the wobble hypothesis and the degeneracy of the code, U at the first position (the 5′ end) of the tRNA anticodon may pair with A or

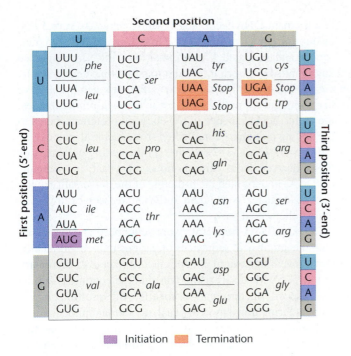

Second position

First position (5′-end)	U	C	A	G	Third position (3′-end)
U	UUU *phe*	UCU *ser*	UAU *tyr*	UGU *cys*	U
	UUC	UCC	UAC	UGC	C
	UUA *leu*	UCA	UAA *Stop*	UGA *Stop*	A
	UUG	UCG	UAG *Stop*	UGG *trp*	G
C	CUU *leu*	CCU *pro*	CAU *his*	CGU *arg*	U
	CUC	CCC	CAC	CGC	C
	CUA	CCA	CAA *gln*	CGA	A
	CUG	CCG	CAG	CGG	G
A	AUU *ile*	ACU *thr*	AAU *asn*	AGU *ser*	U
	AUC	ACC	AAC	AGC	C
	AUA	ACA	AAA *lys*	AGA *arg*	A
	AUG *met*	ACG	AAG	AGG	G
G	GUU *val*	GCU *ala*	GAU *asp*	GGU *gly*	U
	GUC	GCC	GAC	GGC	C
	GUA	GCA	GAA *glu*	GGA	A
	GUG	GCG	GAG	GGG	G

Initiation (AUG) Termination (UAA, UAG, UGA)

FIGURE 5–7 The coding dictionary. AUG encodes methionine, which initiates most polypeptide chains. All other amino acids except tryptophan, which is encoded only by UGG, are represented by two to six triplets. The triplets UAA, UAG, and UGA are termination signals and do not encode any amino acids.

G at the third position (the 3′ end) of the mRNA codon, and G may likewise pair with U or C. Inosine, one of the modified bases found in tRNA, may pair with C, U, or A. Applying these wobble rules, a minimum of about 30 different tRNA species is necessary to accommodate the 61 triplets specifying an amino acid. If nothing more, wobble can be considered a potential economy measure, provided that the fidelity of translation is not compromised. Current estimates are that 30 – 40 tRNA species are present in bacteria and up to 50 tRNA species in animal and plant cells.

TABLE 5-4 Anticocon–Codon Base-Pairing Rules

Base at 1st position (5′–end) of tRNA	*Base at 3rd position (3′–end) of mRNA*
A	U
C	G
G	C or U
U	A or G
I	A, U or C

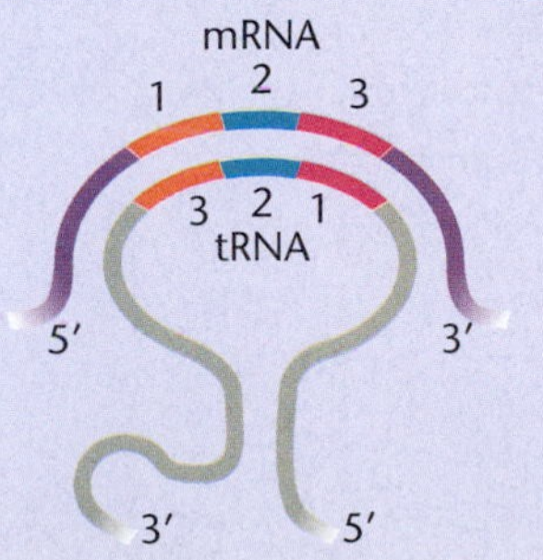

Initiation, Termination, and Suppression

Initiation of protein synthesis is a highly specific process. In bacteria (in contrast to the in vitro experiments discussed earlier), the initial amino acid inserted into all polypeptide chains is a modified form of methionine—***N*-formylmethionine (fmet)**. Only one codon, AUG, codes for methionine, and it is sometimes called the **initiator codon**. However, when AUG appears internally in mRNA rather than at an initiating position, unformylated methionine is inserted into the polypeptide chain. Rarely, another triplet, GUG, specifies methionine during initiation, though it is not clear why this happens, since GUG normally encodes valine.

In bacteria, either the formyl group is removed from the initial methionine upon the completion of synthesis of a protein, or the entire formylmethionine residue is removed. In eukaryotes, unformylated methionine is the initial amino acid during polypeptide synthesis. As in bacteria, it may be cleared from the polypeptide.

As mentioned in the preceding section, three other triplets (UAG, UAA, and UGA)* serve as **termination codons**, punctuation signals that do not code for any amino acid. They are not recognized by a tRNA molecule, and translation terminates when they are encountered. Mutations that produce any of the three triplets internally in a gene also result in termination. Consequently, only a partial polypeptide is synthesized, since it is prematurely released from the ribosome. When such a change occurs in the DNA, it is called a **nonsense mutation**.

Interestingly, a distinct mutation in a second gene may trigger suppression of premature termination resulting from a nonsense mutation (referred to as a **suppressor mutation**). These mutations cause the chain-termination signal to be read as a "sense" codon. The "correction" usually inserts an amino acid other than that found in the wild-type protein. But if the protein's structure is not altered drastically, it may function almost normally. Therefore, this second mutation has "suppressed" the mutant character resulting from the initial change to the termination codon.

Nonsense suppressor mutations occur in genes specifying tRNAs. If the mutation results in a change in the anticodon such that it becomes complementary to a termination code, there is the potential for insertion of an amino acid and suppression.

5.5 The Genetic Code Has Been Confirmed in Studies of Phage MS2

All aspects of the genetic code discussed so far yield a fairly complete picture. The code is triplet in nature, degenerate, unambiguous, and commaless, but it contains punctuation with respect to start and stop signals. These individual principles have been confirmed by the detailed analysis of the

*Historically, the terms *amber* (UAG), *ochre* (UAA), and *opal* (UGA) have been used to distinguish the three mutation possibilities.

RNA-containing bacteriophage MS2 by Walter Fiers and his coworkers.

MS2 is a bacteriophage that infects *E. coli*. Its nucleic acid (RNA) contains only about 3500 ribonucleotides, making up only three genes. These genes specify a coat protein, an RNA-directed replicase, and a maturation protein (the A protein). This simple system of a small genome and few gene products allowed Fiers and his colleagues to sequence the genes and their products. The amino acid sequence of the coat protein was completed in 1970, and the nucleotide sequence of the gene and a number of nucleotides on each end of it were reported in 1972.

When the chemical nature of the gene and protein are compared, a colinear relationship is evident. That is, based on the coding dictionary, the linear sequence of nucleotides (and thus the sequence of triplet codons) corresponds precisely with the linear sequence of amino acids in the protein. Furthermore, the codon for the first amino acid is AUG, the common initiator codon; and the codon for the last amino acid is followed by two consecutive termination codons, UAA and UAG.

By 1976, the other two genes and their protein products were sequenced, providing similar confirmation. The analysis clearly shows that the genetic code, as established in bacterial systems, is identical in this virus. Other evidence suggests that the code is also identical in eukaryotes.

5.6 The Genetic Code Is Nearly Universal

Between 1960 and 1978, it was generally assumed that the genetic code would be found to be universal, applying equally to viruses, bacteria, archaea, and eukaryotes. Certainly, the nature of mRNA and the translation machinery seemed to be very similar in these organisms. For example, cell-free systems derived from bacteria could translate eukaryotic mRNAs. Poly U was shown to stimulate translation of polyphenylalanine in cell-free systems when the components were derived from eukaryotes. Many recent studies involving recombinant DNA technology (see Chapter 16) reveal that eukaryotic genes can be inserted into bacterial cells and transcribed and translated. Within eukaryotes, mRNAs from mice and rabbits have been injected into amphibian eggs and efficiently translated. For the many eukaryotic genes that have been sequenced, notably those for hemoglobin molecules, the amino acid sequence of the encoded proteins adheres to the coding dictionary established from bacterial studies.

However, several 1979 reports on the coding properties of DNA derived from mitochondria of yeast and humans (mtDNA) undermined the principle of the universality of the genetic language. Since then, mtDNA has been examined in many other organisms.

Cloned mtDNA fragments were sequenced and compared with the amino acid sequences of various mitochondrial proteins, revealing several exceptions to the coding dictionary (Table 5–5). Most surprising was that the codon UGA, normally specifying termination, specifies the insertion of tryptophan during translation in yeast and human mitochondria. In human mitochondria, AUA, which normally specifies isoleucine, directs the internal insertion of methionine. In yeast mitochondria, threonine is inserted instead of leucine when CUA is encountered in mRNA.

In 1985, several other exceptions to the standard coding dictionary were discovered in the bacterium *Mycoplasma capricolum*, and in the nuclear genes of the protozoan ciliates *Paramecium*, *Tetrahymena*, and *Stylonychia*. For example, as shown in Table 5–5, one alteration converts the termination codons (UGA) to tryptophan. Several other code alterations convert a termination codon to glutamine (gln). These changes are significant because both a prokaryote and several eukaryotes are involved.

Note the apparent pattern in several of the altered codon assignments. The change in coding capacity involves only a shift in recognition of the third, or wobble, position. For example, AUA specifies isoleucine in the cytoplasm and methionine in the mitochondrion. In the cytoplasm, methionine is specified by AUG. In a similar way, UGA calls for termination in the cytoplasm, but calls for tryptophan in the mitochondrion. In the cytoplasm, tryptophan is specified by

TABLE 5–5 Exceptions to the Universal Code

Triplet	*Normal Code Word*	*Altered Code Word*	*Source*
UGA	termination	trp	Human and yeast mitochondria *Mycoplasma*
CUA	leu	thr	Yeast mitochondria
AUA	ile	met	Human mitochondria
AGA AGG	arg	termination	Human mitochondria
UAA	termination	gln	*Paramecium* *Tetrahymena* *Stylonychia*
UAG	termination	gln	*Paramecium*

UGG. It has been suggested that such changes in codon recognition may represent an evolutionary trend toward reducing the number of tRNAs needed in mitochondria; only 22 tRNA species are encoded in human mitochondria, for example. However, until more examples are revealed, the differences must be considered to be exceptions to the previously established general coding rules.

5.7 Different Initiation Points Create Overlapping Genes

Earlier we stated that the genetic code is nonoverlapping—each ribonucleotide in an mRNA is part of only one triplet. However, this characteristic of the code does not rule out the possibility that a single mRNA may have multiple initiation points for translation. If so, these points could theoretically create several different reading frames within the same mRNA, thus specifying more than one polypeptide. This concept of **overlapping genes** is illustrated in Figure 5–8(a).

That this might actually occur in some viruses was suspected when phage ϕX174 was carefully investigated. The circular DNA chromosome consists of 5386 nucleotides, which should encode a maximum of 1795 amino acids, sufficient for 5 or 6 proteins. However, this small virus in fact synthesizes 11 proteins consisting of more than 2300 amino acids. A comparison of the nucleotide sequence of the DNA and the amino acid sequences of the polypeptides synthesized has clarified the apparent paradox. At least four cases of multiple initiation have been discovered, creating overlapping genes [Figure 5–8(b)].

The sequences specifying the K and B polypeptides are initiated with separate reading frames within the sequence specifying the A polypeptide. The K sequence overlaps into the adjacent sequence specifying the C polypeptide. The E sequence is out of frame with, but initiated in that of the D polypeptide. Finally, the A′ sequence, while in frame, begins in the middle of the A sequence. They both terminate at the identical point. In all, seven different polypeptides are created from a DNA sequence that might otherwise have specified only three (A, C, and D).

A similar situation has been observed in other viruses, including phage G4 and the animal virus SV40. Like ϕX174, phage G4 contains a circular single-stranded DNA molecule. The use of overlapping reading frames optimizes the use of a limited amount of DNA present in these small viruses. However, such an approach to storing information has a distinct disadvantage in that a single mutation may affect more than one protein and thus increase the chances that the change will be deleterious or lethal. In the case we just discussed, a single mutation at the junction of genes *A* and *C* could affect three proteins (the A, C, and K proteins). It may be for this reason that overlapping genes are not common in other organisms.

5.8 Transcription Synthesizes RNA on a DNA Template

Even while the genetic code was being studied, it was quite clear that proteins were the end products of many genes. Hence, while some geneticists were attempting to elucidate the code, other research efforts were directed toward the nature of genetic expression. The central question was how DNA, a nucleic acid, is able to specify a protein composed of amino acids.

The complex, multistep process begins with the transfer of genetic information stored in DNA to RNA. The process by which RNA molecules are synthesized on a DNA template is called **transcription**. It results in an mRNA molecule complementary to the gene sequence of one of the two strands of the double helix. Each triplet codon in the mRNA is, in turn, complementary to the anticodon region of its corresponding tRNA as the amino acid is correctly inserted into the polypeptide chain during translation. The significance of the process of transcription is enormous, for it is the initial step in the process of information flow within the cell. The idea that RNA is involved as an intermediate molecule in the process of information flow between DNA and protein is suggested by the following observations:

1. DNA is, for the most part, associated with chromosomes in the nucleus of the eukaryotic cell. However, protein synthesis occurs in association with ribosomes located outside the nucleus in the cytoplasm. Therefore, DNA does not appear to participate directly in protein synthesis.

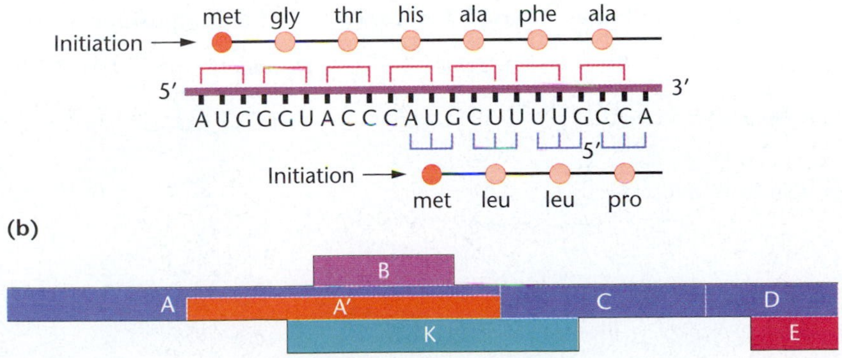

FIGURE 5–8 Illustration of the concept of overlapping genes. (a) An mRNA sequence initiated at two different AUG positions out of frame with one another will give rise to two distinct amino acid sequences. (b) The relative positions of the sequences encoding seven polypeptides of the phage ϕX174.

2. RNA is synthesized in the nucleus of eukaryotic cells, in which DNA is found, and is chemically similar to DNA.
3. Following its synthesis, most RNA migrates to the cytoplasm, in which protein synthesis (translation) occurs.
4. The amount of RNA is generally proportional to the amount of protein in a cell.

Like most new ideas in molecular genetics, the initial supporting experimental evidence for an RNA intermediate was based on studies of bacteria and their phages.

5.9 Studies with Bacteria and Phages Provided Evidence for the Existence of mRNA

In two papers published in 1956 and 1958, Elliot Volkin and his colleagues reported their analysis of RNA produced immediately after bacteriophage infection of *E. coli*. Using the isotope ^{32}P to follow newly synthesized RNA, they found that its base composition closely resembled that of the phage DNA, but was different from that of bacterial RNA (Table 5–6). This newly synthesized RNA was unstable, or short lived; however, its production was shown to precede the synthesis of new phage proteins. Thus, Volkin and his coworkers considered the possibility that synthesis of RNA is a preliminary step in the process of protein synthesis.

Although ribosomes were known to participate in protein synthesis, their role in this process was not clear. As we noted earlier, one possibility was that each ribosome is specific for the protein synthesized in association with it. That is, perhaps genetic information in DNA is transferred to the RNA of a ribosome during its synthesis so that different groups of ribosomes are restricted to the translation of particular proteins. The alternative hypothesis was that ribosomes are nonspecific "workbenches" for protein synthesis and that specific genetic information rests with a "messenger" RNA.

In an elegant experiment, using the *E. coli*–phage system, the results of which were reported in 1961, Sidney Brenner, François Jacob, and Matthew Meselson clarified this question. They labeled uninfected *E. coli* ribosomes with heavy isotopes and then allowed phage infection to occur in the presence of radioactive RNA precursors. By following these components during translation, the researchers demonstrated that the synthesis of phage proteins (under the direction of newly synthesized RNA) occurred on bacterial ribosomes that were present prior to infection. The ribosomes appeared to be nonspecific, strengthening the case that another type of RNA serves as an intermediary in the process of protein synthesis.

That same year, Sol Spiegelman and his colleagues reached the same conclusion when they isolated ^{32}P-labeled RNA following the infection of bacteria and used it in molecular hybridization studies. They tried hybridizing this RNA to the DNA of both phages and bacteria in separate experiments. The RNA hybridized only with the phage DNA, showing that it was complementary in base sequence to the viral genetic information.

The results of these experiments agree with the concept of a **messenger RNA (mRNA)** being made on a DNA template and then directing the synthesis of specific proteins in association with ribosomes. This concept was formally proposed by François Jacob and Jacques Monod in 1961 as part of a model for gene regulation in bacteria. Since then, mRNA has been isolated and thoroughly studied. There is no longer any question about its role in genetic processes.

5.10 RNA Polymerase Directs RNA Synthesis

To prove that RNA can be synthesized on a DNA template, it was necessary to demonstrate that there is an enzyme capable of directing this synthesis. By 1959, several investigators, including Samuel Weiss, had independently discovered such a molecule from rat liver. Called **RNA polymerase**, it has the same general substrate requirements as does DNA polymerase, the major exception being that the substrate nucleotides contain the ribose rather than the deoxyribose form of the sugar. Unlike DNA polymerase, no primer is required to initiate synthesis. The initial base remains as a nucleoside triphosphate (NTP). The overall reaction summarizing the synthesis of RNA on a DNA template can be expressed as:

$$n(\text{NTP}) \xrightarrow[\text{enzyme}]{\text{DNA}} (\text{NMP})_n + n(\text{PP}_i)$$

As the equation reveals, nucleoside triphosphates (NTPs) serve as substrates for the enzyme, which catalyzes the polymerization of nucleoside monophosphates (NMPs), or

TABLE 5-6 **Base Compositions (in mole percents) of RNA Produced Immediately Following Infection of *E. coli* by the Bacteriophages T2 and T7 in Contrast to the Composition of RNA of Uninfected *E. coli*.**

	Adenine	*Thymine*	*Uracil*	*Cystosine*	*Guanine*
Postinfection RNA in T2-infected cells	33	—	32	18	18
T2 DNA	32	32	—	17*	18
Postinfection RNA in T7-infected cells	27	—	28	24	22
T7 DNA	26	26	—	24	22
E. coli RNA	23	—	22	18	17

*5-hydroxymethyl cystosine.
Source: From Volkin and Astrachan (1956); and Volkin, Astrachan, and Countryman (1958).

nucleotides, into a polynucleotide chain $(NMP)_n$. Nucleotides are linked during synthesis by 5′-to-3′ phosphodiester bonds. (See Figure 3–8.) The energy created by cleaving the triphosphate precursor into the monophosphate form drives the reaction, and inorganic phosphates (PP_i) are produced.

A second equation summarizes the sequential addition of each ribonucleotide as the process of transcription progresses:

$$(NMP)_n + NTP \xrightarrow[\text{enzyme}]{\text{DNA}} (NMP)_{n+1} + PP_i$$

As this equation shows, each step of transcription involves the addition of one ribonucleotide (NMP) to the growing polyribonucleotide chain $(NMP)_{n+1}$, using a nucleoside triphosphate (NTP) as the precursor.

RNA polymerase from *E. coli* has been extensively characterized and shown to consist of subunits designated α, β, β', and σ. The active form of the enzyme, the **holoenzyme**, contains the subunits $\alpha_2\beta\beta'\sigma$ and has a molecular weight of almost 500,000 Da. Of these subunits, it is the β and β' polypeptides that provide the catalytic basis and active site for transcription. As we will see, the **σ subunit** [Figure 5–9(a)] plays a regulatory function involving the initiation of RNA transcription.

While there is but a single form of the enzyme in *E. coli*, there are several different σ factors, creating variations of the polymerase holoenzyme. On the other hand, eukaryotes display three distinct forms of RNA polymerase, each consisting of a greater number of polypeptide subunits than does the bacterial form of the enzyme. We shall return to a discussion of the eukaryotic form of the enzyme later in this chapter.

Promoters, Template Binding, and the Sigma Subunit

Transcription results in the synthesis of a single-stranded RNA molecule complementary to a region along only one of the two strands of the DNA double helix. For the purpose of discussion, we will call the DNA strand that is transcribed the template strand and call its complement the partner strand

The initial step is referred to as **template binding** [Figure 5–9(b)]. In bacteria, the site of this initial binding is established when the σ subunit of RNA polymerase recognizes specific DNA sequences called **promoters**. These regions are located in the 5′ region, upstream from the point of initial transcription of a gene. It is believed that the enzyme "explores" a length of DNA until it recognizes the promoter region and binds to about 60 nucleotide pairs of the helix, 40 of which are upstream from the point of initial transcription. Once this occurs, the helix is denatured or unwound locally, making the template strand of the DNA template accessible to the action of the enzyme.

The importance of promoter sequences cannot be overemphasized. They govern the efficiency of initiation of transcription. In bacteria, both strong promoters and weak promoters have been discovered, leading to a variation of initiation from once every 1 to 2 seconds to only once every 10 to 20 minutes. Mutations in promoter sequences may have the effect of severely reducing the initiation of gene expression.

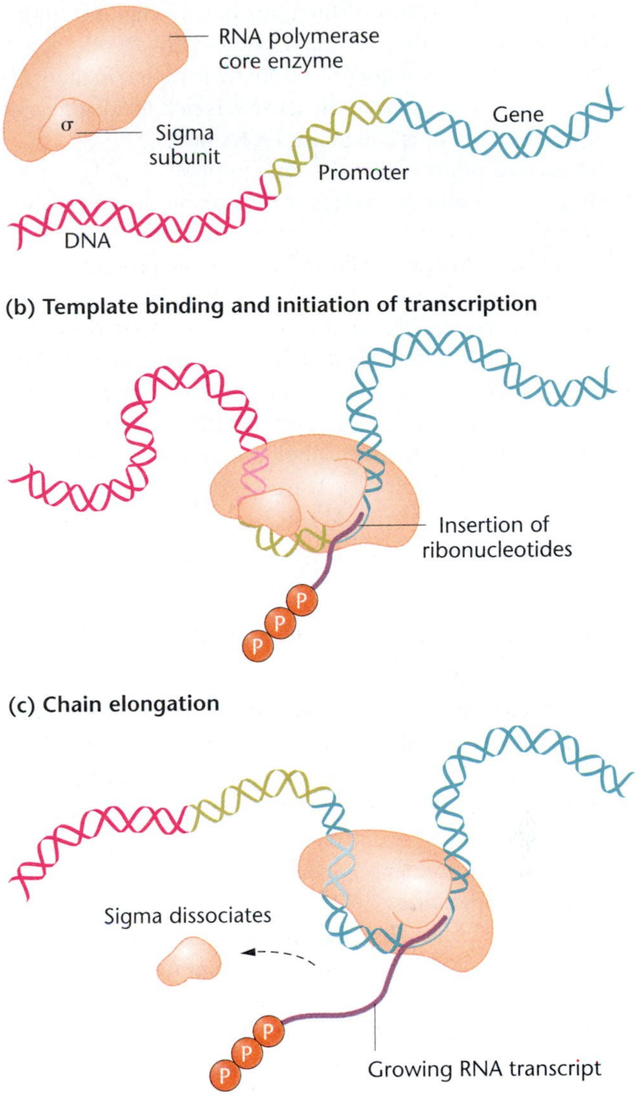

FIGURE 5–9 The early stages of transcription in prokaryotes, showing (a) the components of the process; (b) template binding at the −10 site involving the sigma subunit of RNA polymerase and subsequent initiation of RNA synthesis; and (c) chain elongation, after the sigma subunit has dissociated from the transcription complex and the enzyme moves along the DNA template.

MEDIA TUTORIAL Transcription

sion. Because the interaction of promoters with RNA polymerase governs transcription, the nature of the binding between them is at the heart of discussions concerning genetic regulation, the subject of Chapters 19 and 20. While we will pursue information involving promoter–enzyme interactions in greater detail in the aforementioned chapters, two points are appropriate here.

The first is the concept of **consensus sequences** of DNA. These are sequences that are similar (homologous) in different genes of the same organism or in one or more genes of related organisms. Their conservation throughout evolution attests to the critical nature of their role in biological processes. Two such sequences have been found in bacterial promoters. One, TATAAT, is located 10 nucleotides upstream from the site of initial transcription (the **−10 region**, or **Pribnow**

box). The other, TTGACA, is located 35 nucleotides upstream (the **−35 region**). Mutations in both regions diminish transcription, often severely. In most eukaryotic genes studied, a consensus sequence comparable to that in the −10 region has been recognized. Because it is rich in adenine and thymine residues, it is called the **TATA box**.

The second point is that the degree of RNA polymerase binding to different promoters varies greatly, which causes the variable gene expression mentioned earlier. Currently, this is attributed to sequence variation in the promoters.

A final general point to be made involves the sigma (σ) subunit in bacteria. The major form is designated as σ^{70}, based on its molecular weight of 70 kilodaltons (kDa). While the promoters of most bacterial genes recognize this form, there are several alternative forms of RNA polymerase in *E. coli* that have unique σ subunits associated with them (e.g., σ^{32}, σ^{54}, σ^{70}, and σ^{E}). These recognize different promoter sequences and provide specificity to the initiation of transcription.

Initiation, Elongation, and Termination of RNA Synthesis

Once it has recognized and bound to the promoter, RNA polymerase catalyzes **initiation**, the insertion of the first 5′-ribonucleoside triphosphate, which is complementary to the first nucleotide at the start site of the DNA template strand. As we noted, no primer is required. Subsequent ribonucleotide complements are inserted and linked together by phosphodiester bonds as RNA polymerization proceeds. This process, called **chain elongation** [Figure 5–9(c)], continues in the 5′-to-3′ direction, creating a temporary DNA/RNA duplex whose chains run antiparallel to one another.

After a few ribonucleotides have been added to the growing RNA chain, the σ subunit dissociates from the holoenzyme, and elongation proceeds under the direction of the core enzyme. In *E. coli*, this process proceeds at the rate of about 50 nucleotides/second at 37°C.

Eventually, the enzyme traverses the entire gene until it encounters a specific nucleotide sequence that acts as a termination signal. Such termination sequences, about 40 base pairs in length, are extremely important in prokaryotes because of the close proximity of the end of one gene and the upstream sequences of the adjacent gene. In some cases, the termination of synthesis is dependent upon the **termination factor, rho** (ρ). Rho is a large hexameric protein that physically interacts with the growing RNA transcript. At the point of termination, the transcribed RNA molecule is released from the DNA template, and the core polymerase enzyme dissociates. The synthesized RNA molecule is precisely complementary to a DNA sequence representing the template strand of a gene. Wherever an A, T, C, or G residue exists, a corresponding U, A, G, or C residue, respectively, is incorporated into the RNA molecule. Such RNA molecules ultimately provide the information leading to the synthesis of all proteins present in the cell.

It is important to note that bacteria groups of genes whose protein products are involved in the same metabolic pathway are often clustered together along the chromosome. In many such cases, the genes are contiguous and all but the last gene lack the encoded signals for termination of transcription. The result is that during transcription a large mRNA is produced, encoding more than one protein. Since genes in phage and bacteria were historically referred to as cistrons, the RNA is called a **polycistronic mRNA**. Since the products of genes transcribed in this fashion are usually all needed at the same time, this is an efficient way to transcribe and, subsequently, to translate the needed genetic information. In eukaryotes, **monocistronic** mRNAs are the rule.

5.11 Transcription in Eukaryotes Differs From Prokaryotic Transcription in Several Ways

Much of our knowledge of transcription has been derived from studies of prokaryotes. Most of the general aspects of the mechanics of these processes are similar in eukaryotes, but there are several notable differences:

1. Transcription in eukaryotes occurs within the nucleus under the direction of three separate forms of RNA polymerase. Unlike prokaryotes, in eukaryotes, the RNA transcript is not free to associate with ribosomes prior to the completion of transcription. For the mRNA to be translated, it must move out of the nucleus into the cytoplasm.

2. Initiation and regulation of transcription involve a more extensive interaction between upstream DNA sequences and protein factors involved in stimulating and initiating transcription. In addition to promoters, other control units, called enhancers, may be located in the 5′-regulatory region upstream from the initiation point, but they have also been found within the gene or even in the 3′ downstream region beyond the coding sequence.

3. Maturation of eukaryotic mRNA from the primary RNA transcript involves many complex stages referred to generally as "processing." An initial processing step involves the addition of a 5′ cap and a 3′ tail to most transcripts destined to become mRNAs. Other extensive modifications occur to the internal nucleotide sequence of eukaryotic RNA transcripts that eventually serve as mRNAs. The initial (or primary) transcripts are most often much larger than those that are eventually translated. Sometimes called **pre-mRNAs**, they are part of a group of molecules found only in the nucleus—a group referred to collectively as **heterogeneous nuclear RNA (hnRNA)**. Such RNA molecules are of variable but large size (up to 10^7 Da) and are complexed with proteins, forming **heterogeneous nuclear ribonucleoprotein particles (hnRNPs)**. Only about 25 percent of hnRNA molecules are converted to mRNA. In those that are converted, substantial amounts of the ribonucleotide sequence are excised and the remaining segments are spliced back together prior to translation. This

phenomenon has given rise to the concepts of **split genes** and **splicing** in eukaryotes.

In the remainder of this chapter, we will elaborate on these differences. We will also return to them and other topics directly related to regulation of eukaryotic transcription in Chapter 20.

Initiation of Transcription in Eukaryotes

The recognition of certain highly specific DNA regions by RNA polymerase is the basis of orderly genetic function in all cells. Eukaryotic RNA polymerase exists in three unique forms, each of which transcribes different types of genes, as indicated in Table 5–7. Each enzyme is larger and more complex than the single prokaryotic polymerase. For example, in yeast, the holoenzyme consists of two large subunits and 10 smaller subunits. In regard to the initial template binding step and promoter regions, most is known about polymerase II, which transcribes all mRNAs in eukaryotes.

At least three ***cis*-acting elements** of a eukaryotic gene aid the efficient initiation of transcription by polymerase II. In a future discussion of the *cis–trans* test in Chapter 10, we note that the use of the term *cis* is drawn from organic chemistry nomenclature, meaning "next to" or on the same side as, in contrast to being "across from," or *trans*, to other functional groups. In molecular genetics, then, *cis*-elements are adjacent parts of the same DNA molecule.

One such element found within the promoter region is called the **Goldberg–Hogness** or **TATA box**, located about 30 nucleotide pairs upstream (−30) from the start point of transcription. The consensus sequence is a heptanucleotide consisting solely of A and T residues (TATAAAA). The sequence and function are analogous to that found in the −10 promoter region of prokaryotic genes. Because such a region is common to most eukaryotic genes, the TATA box is thought to be nonspecific and simply to have the responsibility for fixing the site of initiation of transcription by facilitating denaturation of the helix. Such a conclusion is supported by the fact that A=T base pairs are less stable than G≡C pairs.

A second common *cis*-acting element is called the **CAAT box**, located upstream within the promoter of many genes at about 80 nucleotides from the start of transcription (−80). It contains the consensus sequence GGCCAATCT. Still other upstream regulatory regions are found, and most genes contain one or more of them. Along with the TATA and CAAT box, they influence the efficiency of the promoter.

TABLE 5–7 RNA Polymerases in Eukaryotes

Form	*Product*	*Location*
I	rRNA	Nucleolus
II	mRNA, snRNA	Nucleoplasm
III	5s rRNA	Nucleoplasm
	tRNA	Nucleoplasm

The location of each element is based on studies of deletions of particular regions of the promoter, each of which reduces the efficiency of transcription.

DNA regions called **enhancers** represent still another *cis*-acting element. Althought their locations can vary, enhancers are often found even farther upstream than the regions already mentioned, or even downstream or within the gene. Thus they have the potential to modulate transcription from a distance. Although they may not participate directly in template binding, they are essential to highly efficient initiation of transcription. We will return to a discussion of these elements in Chapter 21.

Complementing the *cis*-acting regulatory sequences are various ***trans*-acting factors** that facilitate template binding and, therefore, the initiation of transcription. These are proteins referred to as **transcription factors**. They are essential because RNA polymerase II cannot bind directly to eukaryotic promoter sites and initiate transcription without their presence. The transcription factors involved with human RNA polymerase II binding are well characterized and designated **TFIIA, TFIIB**, and so on. One of these, **TFIID**, binds directly to the TATA-box sequence and is sometimes called the **TATA-binding protein (TBP)**. TFIID consists of about 10 polypeptide subunits. Once initial binding to DNA occurs, at least seven other transcription factors bind sequentially to TFIID, forming an extensive preinitiation complex, which is then bound by RNA polymerase II.

Transcription factors with similar activity have been discovered in a variety of eukaryotes, including *Drosophila* and yeast. The nucleotide sequences in all organisms studied demonstrate a high degree of conservation. Transcription factors appear to supplant the role of the sigma factor in the prokaryotic enzyme. In Chapter 20, we will consider the role of transcription factors in eukaryotic gene regulation, as well as the various DNA-binding domains that characterize them.

Recent Discoveries Concerning RNA Polymerase Function

Interest in the process of transcription has most recently been focused on the details of the structure and function of RNA polymerase. The ability to crystallize large nucleic acid–protein structures and perform X-ray-diffraction analysis at a resolution below 5 Å has led to some remarkable observations. In particular, the work of Roger Kornberg and colleagues, using the enzyme isolated from yeast, has been particularly informative. It is useful to note here that achieving a resolution below 2.8 Å allows the visualization of each amino acid of every protein in the complex!

What Kornberg has discovered provides a highly detailed account of the most critical processes of transcription. RNA polymerase II in yeast contains two large subunits and ten smaller ones, forming a huge three-dimensional complex with a molecular weight of about 500 kDa. The promoter region of the DNA duplex that is to be transcribed enters a positively charged cleft formed between the two large subunits of the enzyme. The subunits form a structure resembling a pair of jaws that exist in two states. Prior to association with DNA, the cleft is open; once associated with DNA, the cleft partially closes, securing the duplex during the initiation of transcription. The critical region of the enzyme involved in the transition is about 50 kDa in size and is called the *clamp*.

Once secured by the clamp, a small duplex region of DNA separates at a position within the enzyme referred to as the *active center*, where complementary RNA synthesis is initiated on the DNA template strand. However, the entire complex remains unstable, and transcription most often terminates following the incorporation of only a few ribonucleotides. While it is not clear why, this so-called *abortive transcription* is repeated a number of times before a stable DNA:RNA hybrid that includes a transcript of 11 ribonucleotides is formed. Once this occurs, abortive transcription is overcome, a stable complex is achieved, and elongation of the RNA transcript proceeds in earnest. Transcription at this point is said to have achieved a level of highly processive RNA polymerization.

As transcription proceeds, the enzyme moves along the DNA, and at any given time, about 40 base pairs of DNA and 18 residues of the growing RNA chain are part of the enzyme complex. The RNA synthesized earliest runs through a groove in the enzyme and exits under a structure at the top and back designated as the *lid*. Another area, called the *pore*, has been identified at the bottom of the enzyme as the point through which the RNA precursors gain entry into the complex.

Eventually, as transcription proceeds, the portion of the DNA that signals termination is encountered, and the complex once again becomes unstable, much as it was during the earlier state of abortive transcription. The clamp opens and both DNA and RNA are released from the enzyme as transcription is terminated. This completes a cycle that constitutes the paradigm of transcription. An unstable complex is formed during the initiation of transcription, stability is established once elongation manages to create a duplex of sufficient size, elongation proceeds, and then instability again characterizes termination of transcription.

These findings extend our knowledge of transcription considerably. It is significant that of the 10 subunits that are part of Kornberg's yeast model, all are very similar to their counterparts in the human enzyme. Nine of the 10 have also been conserved in the other forms of eukaryotic RNA polymerase (I and III). Based on the preceding description, try to mentally visualize the process of transcription from the time DNA associates with the enzyme until the transcript is released from the large molecular complex. If you have developed a clear set of images, you no doubt have acquired a clear understanding of transcription in eukaryotes, which is more complex than in prokaryotes.

Heterogeneous Nuclear RNA and Its Processing: Caps and Tails

We now proceed with the assumption that an initial transcript has been produced. The genetic code is written in the ribonucleotide sequence of mRNA. This information originated, of course, in the template strand of DNA, in which complementary sequences of deoxyribonucleotides exist. In bacteria, the relationship between DNA and RNA appears to be quite direct. The DNA base sequence is transcribed into an mRNA sequence, which is then translated into an amino acid sequence according to the genetic code. In eukaryotes, by contrast, complex processing of mRNA occurs before it is transported to the cytoplasm to participate in translation.

By 1970, accumulating evidence showed that eukaryotic mRNA is transcribed initially as a precursor molecule much larger than that which is translated. This notion was based on the observation by James Darnell and his coworkers of heterogeneous nuclear RNA (hnRNA) in mammalian nuclei that contained nucleotide sequences common to the smaller mRNA molecules present in the cytoplasm. They proposed that the initial transcript of a gene results in a large RNA molecule that must first be processed in the nucleus before it appears in the cytoplasm as a mature mRNA molecule. The various processing steps, discussed in the sections that follow, are summarized in Figure 5–10.

The initial **posttranscriptional modification** of eukaryotic RNA transcripts destined to become mRNAs involves the 5′ end of these molecules, at which a **7-methylguanosine (7mG) cap** is added (Figure 5–10, Step 2). The cap, which is added even before the initial transcript is complete, appears to be important to the subsequent processing within the nucleus, perhaps by protecting the 5′ end of the molecule from nuclease attack. Subsequently, this cap may be involved in the transport of mature mRNAs across the nuclear membrane into the cytoplasm. The cap is fairly complex and is distinguished by a unique 5′–5′ bonding between the cap and the initial ribonucleotide of the RNA. Some eukaryotes also contain a methyl group (CH_3) on the 2′-carbon of the ribose sugars of the first two ribonucleotides of the RNA.

Further insights into the processing of RNA transcripts during the maturation of mRNA came from the discovery that both hnRNAs and mRNAs contain at their 3′ end a stretch of as many as 250 adenylic acid residues. Such **poly-A sequences** are added after the 5′–7mG cap has been added. First, the 3′ end of the initial transcript is cleaved enzymatically at a point some 10 to 35 ribonucleotides from a highly conserved AAUAAA sequence (Step 3). Then, polyadenylation occurs by the sequential addition of adenylic acid residues (Step 4). Poly A has now been found at the 3′ end of almost all mRNAs studied in a variety of eukaryotic organisms. The exceptions seem to be the RNAs that encode histone proteins.

While the AAUAAA sequence is not found on all eukaryotic transcripts, it appears to be essential to those that have it. If the sequence is changed as a result of a mutation, those transcripts that do have it cannot add the poly-A tail. In the absence of this tail, these RNA transcripts are rapidly degraded. Therefore, both the 5′ cap and the 3′ poly-A tail are critical if an RNA transcript is to be further processed and transported to the cytoplasm.

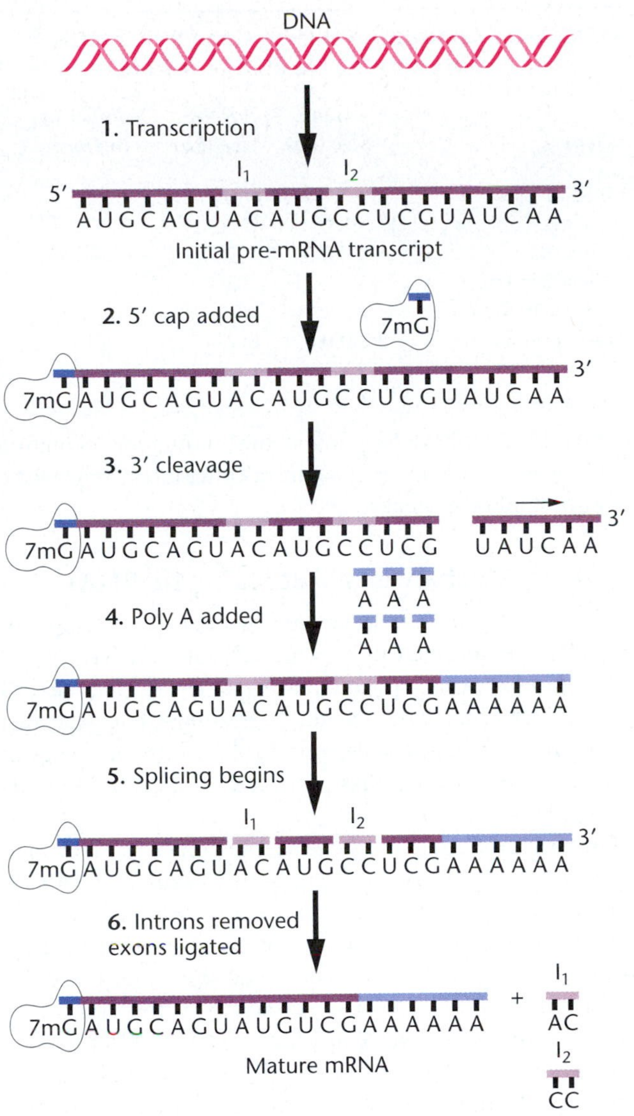

FIGURE 5–10 Posttranscriptional RNA processing in eukaryotes. Heterogeneous nuclear RNA (hnRNA) is converted to messenger (mRNA), which contains a 5′ cap and a 3′-poly-A tail, which then has introns spliced out.

5.12 The Coding Regions of Eukaryotic Genes Are Interrupted by Intervening Sequences

One of the most exciting discoveries in the history of molecular genetics occurred in 1977, when Susan Berget, Philip Sharp, and Richard Roberts presented direct evidence that the genes of animal viruses contain *internal* nucleotide sequences that are not expressed in the amino acid sequence of the proteins they encode. These internal DNA sequences are present in initial RNA transcripts, but they are removed before the mature mRNA is translated (Figure 5–10, Steps 5 and 6). Such nucleotide segments are called **intervening sequences**, and the genes that contain them are known as **split genes**. Those DNA sequences that are not represented in the final mRNA product are also called **introns** ("int" for intervening), and those retained and expressed are called **exons** ("ex" for expressed). Splicing involves the removal of the ribonucleotide sequences present in introns as a result of an excision process and the rejoining of exons.

Similar discoveries were soon made in a variety of eukaryotes. Two approaches have been most fruitful. The first involves the molecular hybridization of purified, functionally mature mRNAs with DNA containing the genes specifying that message. Hybridization between nucleic acids that are not perfectly complementary results in **heteroduplexes**, in which introns present in the DNA, but absent in the mRNA loop out and remain unpaired. Such structures can be visualized with the electron microscope, as shown in Figure 5–11. The heteroduplex in the figure contains seven loops (A–G), representing seven introns whose sequences are present in DNA, but not in the final mRNA.

The second approach provides more specific information (Figure 5–12). It involves a comparison of nucleotide sequences of DNA with those of mRNA and the correlation with amino acid sequences. Such an approach allows the precise identification of all intervening sequences.

Thus far, most eukaryotic genes have been shown to contain introns. One of the first so identified was the **beta-globin**

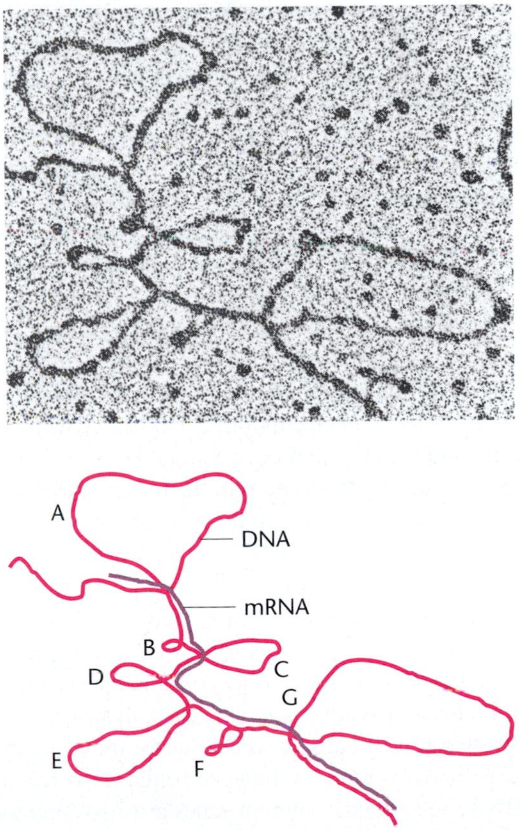

FIGURE 5–11 An electron micrograph and an interpretive drawing of the hybrid molecule (heteroduplex) formed between the template DNA strand of the chicken ovalbumin gene and the mature ovalbumin mRNA. Seven DNA introns, A–G, produce unpaired loops.

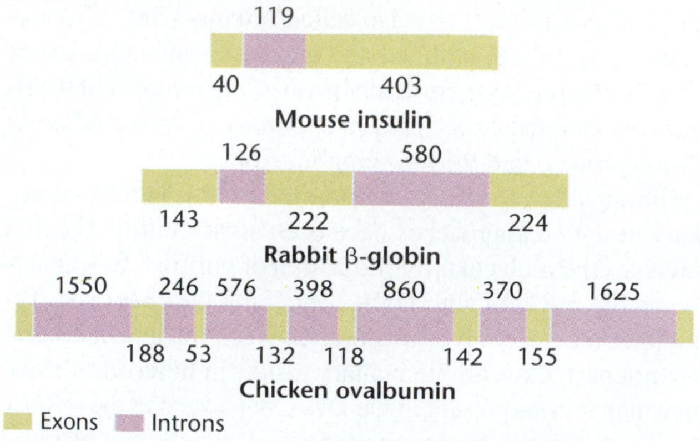

FIGURE 5–12 Intervening sequences in various eukaryotic genes. The numbers indicate the number of nucleotides present in various intron and exon regions.

TABLE 5–8 Contrasting Human Gene Size, mRNA Size, and the Number of Introns

Gene	*Gene Size (kb)*	*mRNA size (kb)*	*Number of Introns*
Insulin	1.7	0.4	2
Collagen [*pro*-α-2(1)]	38.0	5.0	50
Albumin	25.0	2.1	14
Phenylalanine hydroxylase	90.0	2.4	12
Dystrophin	2000.0	17.0	50

gene in mice and rabbits, studied independently by Philip Leder and Richard Flavell. The mouse gene contains an intron 550 nucleotides long, beginning immediately after the codon specifying the 104th amino acid. In the rabbit, there is an intron of 580 base pairs near the codon for the 110th amino acid. Additionally, a second intron of about 120 nucleotides exists earlier in both genes. Similar introns have been found in the beta-globin gene in all mammals examined.

The **ovalbumin gene** of chickens has been extensively characterized by Bert O'Malley in the United States and Pierre Chambon in France. As shown in Figure 5–12, the gene contains seven introns. In fact, as you can see, the majority of the gene's DNA sequence is "silent," being composed of introns. The initial RNA transcript is nearly three times the length of the mature mRNA. Compare the ovalbumin gene in Figures 5–11 and 5–5. Can you match the unpaired loops in Figure 5–11 with the sequence of introns specified in Figure 5–12?

Few eukaryotic genes seem to be without introns. An extreme example of the number of introns in a single gene is that found in the gene coding for one of the subunits of collagen, the major connective tissue protein in vertebrates. The *pro-α-2(1) collagen* gene contains 50 introns. The precision of cutting and splicing that occurs must be extraordinary if errors are not to be introduced into the mature mRNA. What is equally noteworthy is a comparison of the size of genes with the size of the final mRNA once introns are removed. As shown in Table 5–8, only about 15 percent of the collagen gene consists of exons that finally appear in mRNA. For other proteins, an even more extreme picture emerges. Only about 8 percent of the albumin gene remains to be translated, and in the largest human gene known, dystrophin (which is the protein product absent in Duchenne muscular dystrophy), less than 1 percent of the gene sequence is retained in the mRNA. Several other human genes are also contrasted in Table 5–8.

While the vast majority of eukaryotic genes examined thus far contain introns, there are several exceptions. Notably, those coding for histones and for interferon appear to contain no introns. It is not clear why or how the genes encoding these molecules have been maintained throughout evolution without acquiring the extraneous information characteristic of almost all other genes.

Splicing Mechanisms: Autocatalytic RNAs

The discovery of split genes led to intensive attempts to elucidate the mechanism by which introns of RNA are excised and exons are spliced back together. A great deal of progress has already been made. Interestingly, it appears that somewhat different mechanisms exist for different types of RNA, as well as for RNAs produced in mitochondria and chloroplasts.

Introns can be categorized into several groups based on their splicing mechanisms. In group I—represented by introns that are part of the primary transcript of rRNAs—no additional components are required for intron excision; the intron itself is the source of the enzymatic activity necessary for its own removal. This amazing discovery, which contradicted expectations, was made in 1982 by Thomas Cech and his colleagues during a study of the ciliate protozoan *Tetrahymena*. Reflecting their autocatalytic properties, RNAs that are capable of splicing themselves are sometimes called **ribozymes**.

The **self-excision process** is illustrated in Figure 5–13. Chemically, two nucleophilic reactions (called transesterification reactions) occur, the first involving an interaction between guanosine—which acts as a cofactor in the reaction—and the primary transcript [Figure 5–13(a)]. The 3′-OH group of guanosine is transferred to the nucleotide adjacent to the 5′ end of the intron. The second reaction involves the interaction of the newly acquired 3′-OH group on the left-hand exon and the phosphate on the 3′ end of the right intron [Figure 5–13(b)]. The intron is spliced out and the two exon regions are ligated, leading to the mature RNA [Figure 5–13(c)].

Self-excision of group I introns, as described, is now known to apply to pre-rRNAs from other protozoans. Self-excision also seems to govern the removal of introns present in the primary mRNA and tRNA transcripts produced in mitochondria and chloroplasts. These are referred to as group II introns. Like group I molecules, splicing involves two autocatalytic reactions leading to the excision of introns. However, guanosine is not involved as a cofactor.

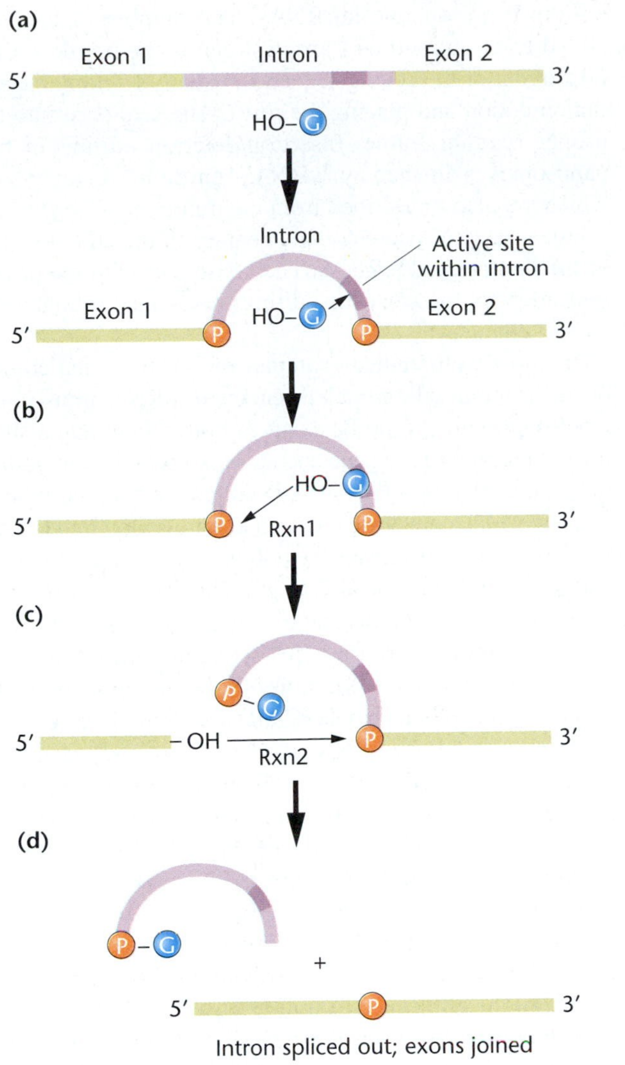

FIGURE 5–13 Splicing mechanism involved with group I introns removed from the primary transcript leading to rRNA. The process is one of self-excision involving two transesterification reactions.

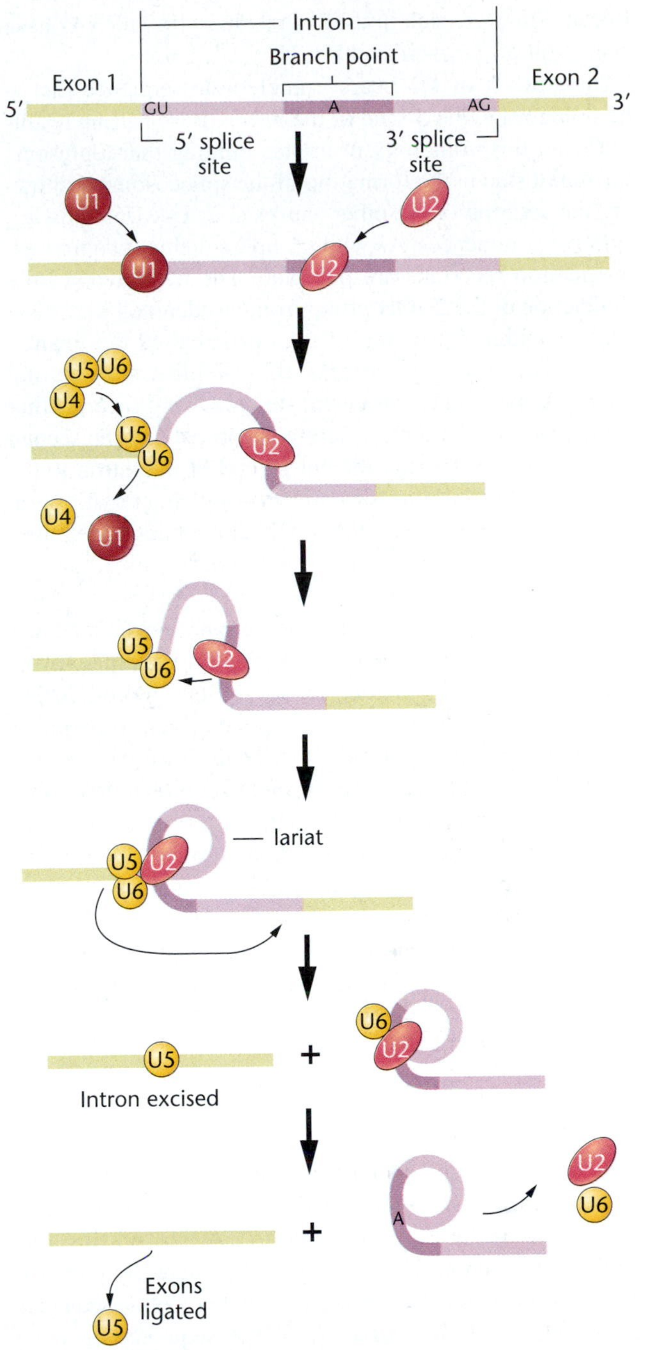

FIGURE 5–14 A model of the splicing mechanism involved with the removal of an intron from a pre-mRNA. Excision is dependent on various snRNAs (U1, U2, . . . , U6) that combine with proteins to form snurps, which function as part of a large structure referred to as the spliceosome. The lariat structure in the intermediate stage is characteristic of this mechanism.

Splicing Mechanisms: The Spliceosome

Introns are, of course, a major component of nuclear-derived pre-mRNA transcripts. Compared to the other RNAs we have been discussing, introns in nuclear-derived mRNA can be much larger—up to 20,000 nucleotides—and they are more plentiful. Their removal appears to require a much more complex mechanism, which has been more difficult to define.

Nevertheless, many clues have now emerged, and the model diagrammed in Figure 5–14 illustrates the removal of one intron. You should refer to this figure during this discussion. First, the nucleotide sequences near the perimeters of this type of intron are often similar. Many begin at the 5′ end with a GU dinucleotide sequence and terminate at the 3′ end with an AG dinucleotide sequence. These, as well as other consensus sequences shared by introns, attract specific molecules that form a molecular complex essential to splicing. Such a complex, called a **spliceosome**, has been identified in extracts of yeast, as well as in mammalian cells. It is very large, being 40*S* in yeast and 60*S* in mammals. Perhaps the most essential component of spliceosomes is the unique set of **small nuclear RNAs (snRNAs)**. These RNAs are usually 100 to 200 nucleotides or less and are often complexed with proteins to form **small nuclear ribonucleoproteins (snRNPs**, or **snurps)**. These are found only in the nucleus.

Because they are rich in uridine residues, the snRNAs have been arbitrarily designated U1, U2, ..., U6.

The snRNA of U1 bears a nucleotide sequence that is homologous to the 5′ end of the intron. Base pairing resulting from this homology promotes binding that represents the initial step in the formation of the spliceosome. Following the addition of the other snurps (U2, U4, U5, and U6), splicing commences. As with group I splicing, two transesterification reactions are involved. The first involves the interaction of the 2′-OH group from an adenine (A) residue present within the region of the intron called the **branch point**. The A residue attacks the 5′-splice site, cutting the RNA chain. In a subsequent step involving several other snurps, an intermediate structure is formed and the second reaction ensues, linking the cut 5′ end of the intron to the A. This results in the formation of a characteristic loop structure referred to as a lariat, which contains the excised intron. The exons are then ligated and the snurps are released.

The processing involved in splicing represents a potential regulatory step during gene expression. For example, several cases are known wherein introns present in pre-mRNAs *derived from the same gene* are spliced *in more than one way*, thereby yielding different collections of exons in the mature mRNA. This process, referred to as **alternative splicing**, yields a group of mRNAs that, upon translation, result in a series of related proteins called **isoforms**. A growing number of examples are now found in organisms ranging from viruses to *Drosophila* to humans. Alternative splicing of pre-mRNAs provides the basis for producing related proteins from a single gene. We shall return to this topic in our discussion of the regulation of gene expression in eukaryotes. (See Chapter 20.)

RNA Editing

In the late 1980s, still another quite unexpected form of posttranscriptional RNA processing was discovered in several organisms. In this form, referred to as **RNA editing**, the nucleotide sequence of a pre-mRNA is actually changed prior to translation. As a result, the ribonucleotide sequence of the mature RNA differs from the sequence encoded in the exons of the DNA from which the RNA was transcribed.

While there are numerous variations of RNA editing, the two main types studied initially include **substitution editing**, in which the identities of individual nucleotide bases are altered; and **insertion/deletion editing**, in which nucleotides are added to or subtracted from the total number of bases. Substitution editing is used in some nuclear-derived eukaryotic RNAs and is also very prevalent in mitochondrial and chloroplast RNAs transcribed in plants. *Physarum polycephalum*, a slime mold, uses both substitution and insertion/deletion editing for its mitochondrial mRNAs.

Trypanosoma, a parasite that causes African sleeping sickness, and its relatives use extensive insertion/deletion editing in mitochondrial RNAs. The number of uridines added to an individual transcript can make up more than 60 percent of the coding sequence, usually forming the initiation codon and placing the rest of the sequence into the proper reading frame. Insertion/deletion editing in trypanosomes is directed by "**gRNA**" (**guide RNA**) templates, which are also transcribed from the mitochondrial genome. These small RNAs are complementary to the edited region of the final, edited mRNAs. They base pair with the preedited mRNAs to direct the editing machinery to make the correct changes.

The most well-studied examples of substitutional editing occur in mammalian nuclear-encoded mRNA transcripts. Apolipoprotein B (apo B) exists in both a long and a short form, though a single gene encodes both proteins. In human intestinal cells, apo B mRNA is edited by a single C to U change, which converts a CAA glutamine codon into a UAA stop codon and terminates the polypeptide at approximately half its genomically encoded length. The editing is performed by a complex of proteins that bind to a "mooring sequence" on the mRNA transcript just downstream of the editing site. In a second system, the synthesis of subunits constituting the glutamate receptor channels (GluR) in mammalian brain tissue is also affected by RNA editing. In this case, adenosine (A) to inosine (I) editing occurs in pre-mRNAs prior to their translation, during which I is read as guanosine (G). The enzymes produced by a family of three *ADAR* (*a*denosine *d*eaminase *a*cting on *R*NA) genes are believed to be responsible for the editing of various sites within the glutamate channel subunits. The double-stranded RNAs required for editing by the *ADAR* enzymes are provided by intron–exon pairing of the GluR mRNA transcripts. The editing changes alter the physiological parameters (solute permeability and desensitization response time) of the receptors containing the subunits.

The importance of RNA editing resulting from the action of *ADAR*s is most apparent when one examines situations in which these enzymes have lost their functional capacity as a result of mutation. In several investigations, the loss of function was shown to have a lethal impact in mice. In one study, embryos heterozygous for a defective *ADAR1* gene die during embryonic development as a result of a defective hematopoietic system. In another study, mice with two defective copies of *ADAR2* progress through development normally, but are prone to epileptic seizures and die while still in the weaning stage. Their tissues contain the unedited version of one of the GluR products. The defect leading to death is believed to be in the brain. Heterozygotes for the mutation are normal.

Findings such as these in mammals have established that RNA editing provides still another important mechanism of posttranscriptional modification, and that this process is not restricted to small or asexually reproducing genomes, such as mitochondria. Several new examples of RNA editing have been found each year since its discovery, and this trend is likely to continue. Further, the process has important implications for the regulation of genetic expression.

5.13 Transcription Has Been Visualized by Electron Microscopy

We conclude this chapter by presenting a striking visual demonstration of the transcription process based on the electron microscope studies of Oscar Miller, Jr., Barbara Hamkalo, and Charles Thomas. Their combined work has captured the transcription process in both prokaryotes and eukaryotes. Figure 5–15 shows micrographs and interpretive drawings from two organisms, the bacterium *E. coli* and the newt *Notophthalmus viridescens*. In both cases, multiple strands of RNA are seen to emanate from different points along a central DNA template. Many RNA strands result because numerous transcription events are occurring simultaneously along each gene. Progressively longer RNA strands are found farther downstream from the point of initiation of transcription, whereas the shortest strands are closest to the point of initiation.

An interesting picture emerges from the study of *E. coli* [Figure 5–15(a)]. Because prokaryotes lack nuclei, cytoplasmic ribosomes are not separated physically from the chromosome. As a result, ribosomes are free to attach to *partially* transcribed mRNA molecules and initiate translation. The longer RNA strands demonstrate the greatest number of ribosomes. In the case of the newt [Figure 5–15(b)], the segment of DNA is derived from oocytes that are known to produce an enormous amount of ribosomal (rRNA), one of the major components of the ribosome. To accomplish this synthesis, the genes specific for ribosomal RNA (**rDNA**) are replicated many times in the oocyte, a process called **gene amplification**. The micrograph shows several of these genes, each undergoing simultaneous transcription events. For each rRNA gene, longer and longer strands of incomplete rRNA molecules are produced as the enzymes move along the DNA strand. Visualization of transcription confirms our expectations based on the biochemical analysis of this process.

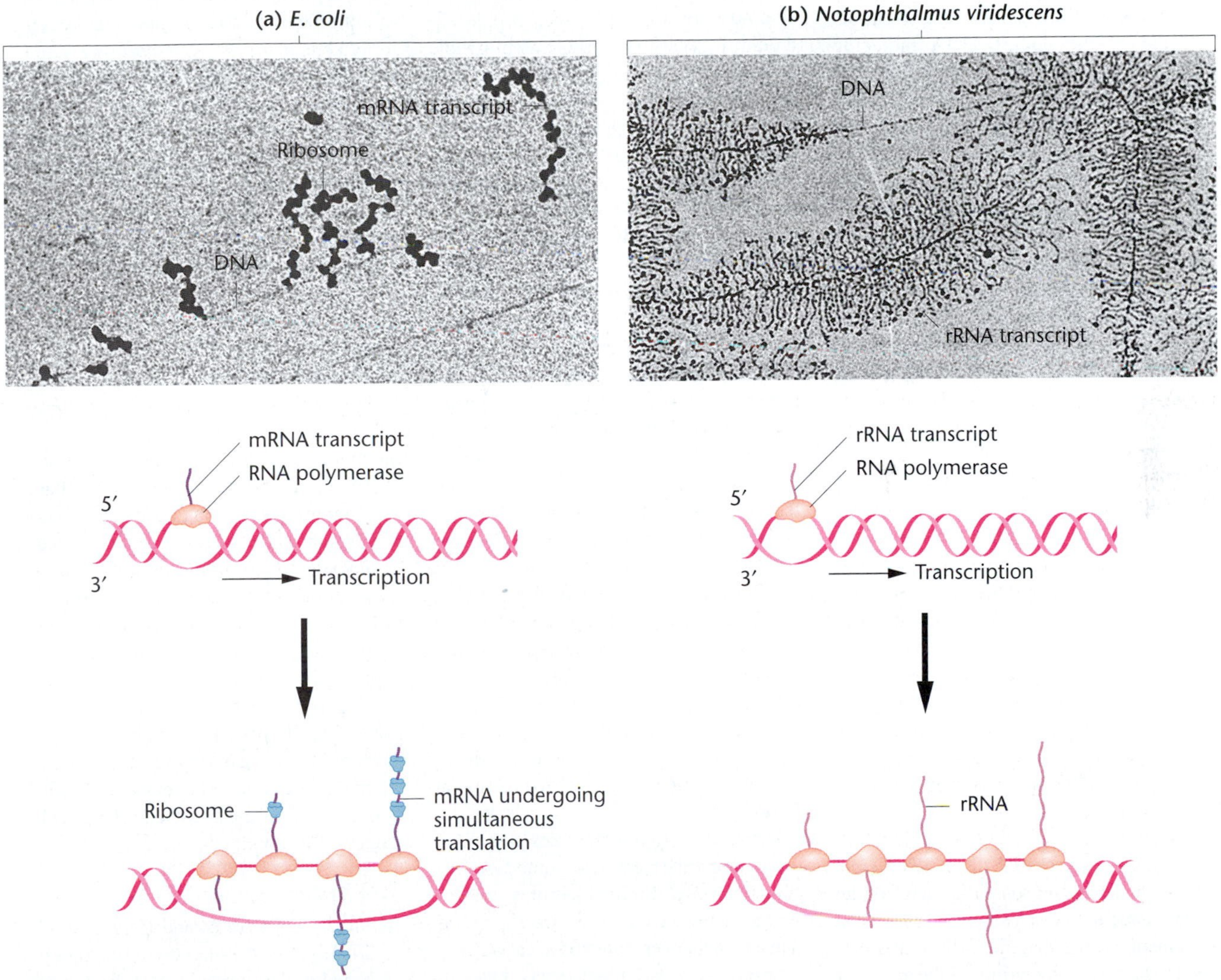

FIGURE 5–15 Electron micrographs and interpretive drawings of simultaneous transcription of genes in *E. coli* (a) and *Notophthalmus (Triturus) viridescens* (b).

Genetics, Technology, and Society

Antisense Oligonucleotides: Attacking the Messenger

Standard chemotherapies for diseases such as cancer and AIDS often have toxic side effects. These drugs affect both normal and diseased cells, with diseased cells being only slightly more susceptible than the patient's normal cells. Scientists have long wished for a magic bullet that could home-in and destroy infected or cancer cells, leaving normal cells alive and healthy. Over the last decade, one particularly promising candidate for magic-bullet status has emerged—the *antisense oligonucleotide*.

Antisense therapies have arisen from an understanding of the molecular biology of gene expression. Gene expression is a two-step process. First, a single-stranded messenger RNA (mRNA) is copied from one strand of the duplex DNA molecule. Second, the mRNA is transported to the cytoplasm and its genetic information is translated into the amino acid sequence of a polypeptide.

Normally, transcription occurs from only one strand of the DNA duplex. The resulting RNA is known as sense RNA. If transcription occurs from the opposite strand of DNA, the nucleotide sequence of the resulting RNA will be complementary to the sense RNA. This complementary RNA is known as antisense RNA. As with complementary strands of DNA molecules, complementary strands of RNA can form double-stranded molecules.

The formation of duplex structures between a sense RNA and an antisense oligonucleotide can affect the sense RNA in several ways. The binding of antisense molecules to sense RNA may physically block ribosome binding or elongation, thereby inhibiting translation. In addition, the binding of antisense molecules to sense RNA may trigger the degradation of the sense RNA, because double-stranded RNA molecules are attacked by intracellular ribonucleases. In both cases, gene expression is blocked.

What makes the antisense approach so exciting is its potential specificity. Scientists can design antisense oligonucleotides with a nucleotide sequence which is complementary to a specific mRNA, and then synthesize large quantities of the antisense oligonucleotide *in vitro*. Usually, single-stranded DNA oligonucleotides of approximately 20 nucleotides are synthesized for use as antisense drugs. It is theoretically possible for the antisense oligonucleotides to enter cells and to bind one specific target mRNA, and hence block the synthesis of one specific protein. If that protein is necessary for virus reproduction or for cancer cell growth (but is not necessary for normal cell functions), the antisense oligonucleotide should have only therapeutic effects.

In 1998, the first antisense oligonucleotide drug was approved for use in the U.S. The drug (Vitravene™, ISIS Pharmaceuticals and CIBA Vision) is a 21-nucleotide antisense molecule, complementary to a region of the cytomegalovirus (CMV) genome. It is used to treat CMV-induced retinitis in AIDS patients. CMV is a common virus that infects most people. Although it causes few problems in people with normal immune systems, it can cause serious symptoms in people with conditions (such as AIDS) that impair the immune system. Up to 40 percent of AIDS patients develop retinitis or blindness as a result of CMV infections of the eye. Clinical trials indicate that the introduction of antisense CMV oligonucleotides into the eyes of AIDS patients with CMV-induced retinitis significantly delays disease progression, with few side effects.

Antisense oligonucleotides may also act as anti-inflammatory drugs. Clinical trials are in progress to test antisense drugs in the treatment of asthma, rheumatoid arthritis, psoriasis and Crohn's disease. One interesting antisense oligonucleotide (now in clinical trials conducted by ISIS Pharmaceuticals and Boehringer Ingelheim) binds to the mRNA that encodes ICAM-1. ICAM-1 is a cell surface glycoprotein that activates the immune system and inflammatory cells. It is often overexpressed in tissues that suffer extreme inflammatory responses. It is hoped that antisense oligonucleotides will temper inflammatory responses by reducing the expression of ICAM-1.

Antisense oligonucleotides are also promising treatments for some types of cancer. These oligonucleotides are designed to reduce the synthesis of proteins that are either overexpressed in cancer cells or are present as mutant forms. One example of an antisense drug designed to attack cancer cells is known as Genasense (Genta, Inc.). The nucleotide sequence of Genasense is complementary to the first six codons of Bcl-2 mRNA. The Bcl-2 protein is an inhibitor of apoptosis (programmed cell death). Normal cells synthesize low levels of Bcl-2 protein. However, many types of cancer cells express high levels of this protein, and are resistant to cell death triggered by chemotherapy, radiation or immunotherapy. In theory, inhibiting the production of Bcl-2 in cancer cells should sensitize these cells to therapeutic agents. Recent clinical trials of Genasense have been encouraging. Almost half of patients with advanced malignant melanoma showed various degrees of anti-tumor effects after treatment with Genasense. In 2001, Genasense entered Phase III clinical trials for treatment of multiple myeloma, chronic lymphocytic leukemia, acute myelocytic leukemia and non-small cell lung cancer. Clinical trials will also assess this antisense oligonucleotide in the treatment of breast, colon, prostate and other solid tumors. Other antisense oligonucleotides in clinical trials are designed to attack mRNAs that encode oncogene products such as H-ras, RAF kinase and protein kinase C alpha.

Like most new technologies, antisense oligonucleotides are a source of scientific controversy. Although antisense drugs should act by binding one specific mRNA and preventing translation of one protein, it has been difficult to demonstrate this mechanism of action in clinical trials. Some oligonucleotides appear to stimulate the immune system in a non-specific manner. It is possible that some of their efficacy towards cancer, viruses or inflammation may operate by immune system stimulation. However, to those people who respond positively to antisense treatments, it is probably irrelevant whether the mechanism is specific or non-specific. In the next few years, if antisense therapeutics pass the scrutiny of scientific and clinical trials, we may have acquired a magic molecular bullet to use in the battle against a wide range of diseases.

References:

Tamm, I., DöRken, B. and Hartmann, G. 2001. Antisense therapy in oncology: New hope for an old idea? *The Lancet* 358:489–497.

Lebedeva, I. and Stein, C.A. 2001. Antisense oligonucleotides: Promise and reality. *Annu. Rev. Pharmacol. Toxicol.* 41:403–419.

Websites:

ISIS Pharmaceuticals
– **http://www.isip.com**

Genta Incorporated
–**http://www.genta.com**

Chapter Summary

1. The genetic code, stored in DNA, is copied to RNA, in which it is used to direct the synthesis of polypeptide chains. It is degenerate, unambiguous, nonoverlapping, and commaless.
2. The complete coding dictionary, determined using various experimental approaches, reveals that of the 64 possible codons—61 encode the 20 amino acids found in proteins, while three triplets terminate translation. One of these 61 is the initiation codon and specifies methionine.
3. The observed pattern of degeneracy often involves only the third letter of a triplet series and led Francis Crick to propose the "wobble hypothesis."
4. Confirmation for the coding dictionary, including codons for initiation and termination, was obtained by comparing the complete nucleotide sequence of phage MS2 with the amino acid sequence of the corresponding proteins. Other findings support the belief that, with only minor exceptions, the code is universal for all organisms.
5. In some bacteriophages, multiple initiation points may occur during the transcription of RNA, resulting in multiple reading frames and overlapping genes.
6. Transcription—the initial step in gene expression—describes the synthesis, under the direction of RNA polymerase, of a strand of RNA complementary to a DNA template.
7. The processes of transcription, like DNA replication, can be subdivided into the stages of initiation, elongation, and termination. Also like DNA replication, the process relies on base-pairing affinities between complementary nucleotides.
8. Initiation of transcription is dependent on an upstream (5′) DNA region, called the promoter, that represents the initial binding site for RNA polymerase. Promoters contain specific DNA sequences, such as the TATA box, that are essential to polymerase binding.
9. Transcription is more complex in eukaryotes than in prokaryotes. The primary transcript is a pre-mRNA that must be modified in various ways before it can be efficiently translated. Processing, which produces a mature mRNA, includes the addition of a 7-mG cap and a poly-A tail, and the removal, through splicing, of intervening sequences, or introns. RNA editing of pre-mRNA prior to its translation also occurs in some systems.

Insights and Solutions

1. Calculate how many triplet codons would be possible had evolution seized on six bases (three complementary base pairs) rather than four bases within the structure of DNA. Would six bases be sufficient using a two-letter code, assuming 20 amino acids and start-and-stop codons?

Solution: Six things taken three at a time will produce $(6)^3$, or 216, triplet codes. If the code was a doublet, there would be $(6)^2$, or 36, two-letter codes, more than enough to accommodate 20 amino acids and start–stop punctuation.

2. In a heteropolymer experiment using 1/2C:1/4A:1/4G, how many different triplets will occur in the synthetic RNA molecule? How often will the most frequent triplet occur?

Solution: There will be $(3)^3$, or 27, triplets produced. The most frequent will be CCC, present $(1/2)^3$, or 1/8, of the time.

3. In a regular copolymer experiment, in which UUAC is repeated over and over, how many different triplets will occur in the synthetic RNA, and how many amino acids will occur in the polypeptide when this RNA is translated? Be sure to consult Figure 5–7.

Solution: The synthetic RNA will repeat four triplets—UUA, UAC, ACU, and CUU—over and over.

Because both UUA and CUU encode leucine, while ACU and UAC encode threonine and tyrosine, respectively, the polypeptides synthesized under the directions of such an RNA contain three amino acids in the repeating sequence leu-leu-thr-tyr.

4. Actinomycin D inhibits DNA-dependent RNA synthesis. This antibiotic is added to a bacterial culture in which a specific protein is being monitored. Compared to a control culture, into which no antibiotic is added, translation of the protein declines over a period of 20 minutes, until no further protein is made. Explain these results.

Solution: The mRNA, which is the basis for the translation of the protein, has a lifetime of about 20 minutes. When actinomycin D is added, transcription is inhibited and no new mRNAs are made. Those already present support the translation of the protein for up to 20 minutes.

5. DNA and RNA base compositions were analyzed from a hypothetical bacterial species with the following results:

	(A + G)/(T + C)	(A + T)/(C + G)	(A + G)/(U + C)	(A + U)/(C + G)
DNA	1.0	1.2		
RNA			1.3	1.2

On the basis of these data, what can you conclude about the DNA and RNA of the organism? Are the data consistent with the Watson–Crick model of DNA? Is the RNA single stranded or double stranded, or can't we tell? If we assume that the entire length of DNA has been transcribed, do the data suggest that RNA has been derived from the transcription of one or both DNA strands, or can't we tell from these data?

Solution: This problem is a theoretical exercise designed to get you to look at the consequences of base complementarity as it affects the base composition of DNA and RNA. The base composition of DNA is consistent with the Watson–Crick double helix. In a double helix, we expect A + G to equal T + C (the number of purines should equal the number of pyrimidines). In this case, there is a preponderance of A=T base pairs (120 A=T pairs to every 100 G ≡ C pairs).

Given what we know about RNA, there is no reason to expect the RNA to be double stranded, but if it were double stranded, then we would expect that A = U and C ≡ G. If so, then (A + G)/(U + C) = 1. Since it doesn't equal unity, we can conclude that the RNA is not double stranded.

If all of the DNA is transcribed, from either one or both strands, the ratio of (A + U)/(C + G) in RNA should be 1.2, and, as predicted, it is. Note that this ratio will not change, regardless of whether only one or both of the strands are transcribed. This is the case, because for every A = T pair in DNA, for example, transcription of RNA will yield one A *and* one U if both strands are transcribed. If just one strand is transcribed, transcription will yield one A *or* one U. In either case, the (A + U)/(C + G) ratio in RNA will reflect the (A + T)/(C + G) ratio in the DNA from which it was transcribed. To prove this to yourself, draw out a DNA molecule with 12 A = T pairs and 10 C ≡ G pairs and transcribe *both* strands. Then, transcribe *either* strand. Count the bases in the RNAs produced in both cases and calculate the ratios. Thus, we cannot determine whether just one or both strands are transcribed from the (A + U)/(C + G) ratio.

However, if both strands are transcribed, then the ratio of (A + G)/(U + C) should equal 1.0, and it doesn't. It equals 1.3. To verify this conclusion, examine the theoretical data you drew out on paper. One explanation for the observed ratio of 1.3 is that only one of the two strands is transcribed. If this is the case, then the (A + G)/(U + C) will reflect the proportion of (A + T) pairs that are A and the proportion of the G ≡ C pairs that are *on the DNA strand that is transcribed*. Another explanation is that transcription occurs only on one strand at any given point (e.g., for one gene), but on the other strand at other points (for other genes).

Problems and Discussion Questions

1. Early proposals regarding the genetic code considered the possibility that DNA served directly as the template for polypeptide synthesis (see Gamow, 1954, in Selected Readings). In eukaryotes, what difficulties would such a system pose? What observations and theoretical considerations argue against such a proposal?
2. Crick, Barnett, Brenner, and Watts-Tobin, in their studies of frameshift mutations, found that either three pluses or three minuses restored the correct reading frame. If the code were a sextuplet (consisting of six nucleotides), would the reading frame be restored by either of the preceding combinations?
3. In a mixed copolymer experiment using polynucleotide phosphorylase, 3/4G:1/4C was added to form the synthetic message. The resulting amino acid composition of the ensuing protein was determined:

Glycine	36/64	(56 percent)
Alanine	12/64	(19 percent)
Arginine	12/64	(19 percent)
Proline	4/64	(6 percent)

From this information,

(a) indicate the percentage (or fraction) of the time each possible triplet will occur in the message.
(b) determine one consistent base-composition assignment for the amino acids present.
(c) considering the wobble hypothesis, predict as many specific triplet assignments as possible.

4. When repeating copolymers are used to form synthetic mRNAs, dinucleotides produce a single type of polypeptide that contains only two different amino acids. On the other hand, using a trinucleotide sequence produces three different polypeptides, each consisting of only a single amino acid. Why? What will be produced when a repeating tetranucleotide is used?
5. The mRNA formed from the repeating tetranucleotide UUAC incorporates only three amino acids, but the use of UAUC incorporates four amino acids. Why?
6. In studies using repeating copolymers, AC ... incorporates threonine and histidine, and CAACAA ... incorporates glutamine, asparagine, and threonine. What triplet code can definitely be assigned to threonine?
7. In a coding experiment using repeating copolymers (as shown in Table 12–3), the following data were obtained:

Copolymer	*Codons Produced*	*Amino Acids in Polypeptide*
AG	AGA, GAG	Arg, Glu
AAG	AGA, AAG, GAA	Lys, Arg, Glu

AGG is known to code for arginine. Taking into account the wobble hypothesis, assign each of the four remaining different triplet codes to its correct amino acid.

8. In the triplet-binding technique, radioactivity remains on the filter when the amino acid corresponding to the triplet is labeled. Explain the basis of this technique.

9. When the amino acid sequences of insulin isolated from different organisms were determined, some differences were noted. For example, alanine was substituted for threonine, serine was substituted for glycine, and valine was substituted for isoleucine at corresponding positions in the protein. List the single-base changes that could occur in triplets of the genetic code to produce these amino acid changes.

10. In studies of the amino acid sequence of wild-type and mutant forms of tryptophan synthetase in *E. coli*, the following changes have been observed:

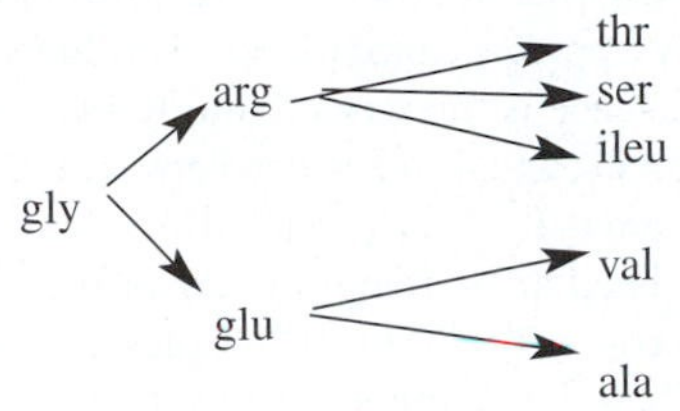

Determine a set of triplet codes in which only a single nucleotide change produces each amino acid change.

11. Why doesn't polynucleotide phosphorylase synthesize RNA in vivo?

12. Refer to Table 5–1. Can you hypothesize why a mixture of Poly U + Poly A would not stimulate incorporation of ^{14}C-phenylalanine into protein?

13. Predict the amino acid sequence produced during translation by the following short theoretical mRNA sequences (note that the second sequence was formed from the first by a deletion of only one nucleotide):

Sequence 1:	AUGCCGGAUUAUAGUUGA
Sequence 2:	AUGCCGGAUUAAGUUGA

What type of mutation gave rise to Sequence 2?

14. A short RNA molecule was isolated that demonstrated a hyperchromic shift indicating secondary structure. Its sequence was determined to be

AGGCGCCGACUCUACU

(a) Propose a two-dimensional model for this molecule.
(b) What DNA sequence would give rise to this RNA molecule through transcription?
(c) If the molecule were a tRNA fragment containing a CGA anticodon, what would the corresponding codon be?
(d) If the molecule were an internal part of a message, what amino acid sequence would result from it following translation? (Refer to the code chart in Figure 5–7.)

15. A glycine residue exists at position 210 of the tryptophan synthetase enzyme of wild-type *E. coli*. If the codon specifying glycine is GGA, how many single-base substitutions will result in an amino acid substitution at position 210? What are they? How many will result if the wild-type codon is GGU?

16. (a) Shown here is a theoretical viral mRNA sequence:

5'-AUGCAUACCUAUGAGACCCUUGGA-3'

Assuming that it could arise from overlapping genes, how many different polypeptide sequences can be produced? Using Figure 5–7, what are the sequences?

(b) A base substitution mutation that altered the sequence in (a) eliminated the synthesis of all but one polypeptide. The altered sequence is shown here:

5'-AUGCAUACCUAUGUGACCCUUGGA-3'

Using Figure 5–7, determine why.

17. Most proteins have more leucine than histidine residues, but more histidine than tryptophan residues. Correlate the number of codons for these three amino acids with this information.

18. Define the process of transcription. Where does this process fit into the central dogma of molecular genetics?

19. What was the initial evidence for the existence of mRNA?

20. Describe the structure of RNA polymerase in bacteria. What is the core enzyme? What is the role of the sigma subunit?

21. In a written paragraph, describe the abbreviated chemical reactions that summarize RNA polymerase-directed transcription.

22. Messenger RNA molecules are very difficult to isolate in prokaryotes because they are rather quickly degraded in the cell. Can you suggest a reason why this occurs? Eukaryotic mRNAs are more stable and exist longer in the cell than do prokaryotic mRNAs. Is this an advantage or disadvantage for a pancreatic cell making large quantities of insulin?

23. The following represent deoxyribonucleotide sequences derived from the template strand of DNA:

Sequence 1:	CTTTTTTGCCAT
Sequence 2:	ACATCAATAACT
Sequence 3:	TACAAGGGTTCT

(a) For each strand, determine the mRNA sequence that would be derived from transcription.
(b) Using Figure 5–7, determine the amino acid sequence that is encoded by these mRNAs.
(c) For Sequence 1, what is the sequence of the partner DNA strand?

Extra-Spicy Problems

24. In a mixed copolymer experiment, messengers were created with either 4/5C:1/5A or 4/5A:1/5C. These messages yielded proteins with the following amino acid compositions.

4/5C:1/5A		4/5A:1/5C	
Proline	63.0 percent	Proline	3.5 percent
Histidine 1	3.0 percent	Histidine	3.0 percent
Threonine	16.0 percent	Threonine	16.6 percent
Glutamine	3.0 percent	Glutamine	13.0 percent
Asparagine	3.0 percent	Asparagine	13.0 percent
Lysine	0.5 percent	Lysine	50.0 percent
	98.5 percent		99.1 percent

Using these data, predict the most specific coding composition for each amino acid.

25. Shown here are the amino acid sequences of the wild-type and three mutant forms of a short protein.

Use this information to answer the following questions:

Wild type:	mer-trp-tyr-arg-gly-ser-pro-thr
Mutant 1:	met-trp
Mutant 2:	met-trp-his-arg-gly-ser-pro-thr
Mutant 3:	met-cys-ile-val-val-val-gln-his

(a) Using Figure 5–7, predict the type of mutation that occurred leading to each altered protein.
(b) For each mutant protein, determine the specific ribonucleotide change that led to its synthesis.
(c) The wild-type RNA consists of nine triplets. What is the role of the ninth triplet?

(d) For the first eight wild-type triplets, which, if any, can you determine specifically from an analysis of the mutant proteins? In each case, explain why or why not.

(e) Another mutation (Mutant 4) is isolated. Its amino acid sequence is unchanged, but mutant cells produce abnormally low amounts of the wild-type proteins. As specifically as you can, predict where in the gene this mutation exists.

26. The genetic code is degenerate. Amino acids are encoded by either 1, 2, 3, 4, or 6 triplet codons. (See Figure 5–7.) An interesting question is whether the frequency of triplet codes is in any way correlated with the frequency that amino acids appear in proteins? That is, is the genetic code optimized for its intended use? Some approximations of the frequency of appearance of nine amino acids in proteins in *E. coli* are

	Percentage
Met	2
Cys	2
Gln	5
Pro	5
Arg	5
Ile	6
Glu	7
Ala	8
Leu	10

(a) Determine how many triplets encode each amino acid.

(b) Devise a way to graphically express the two sets of information (data).

(c) Analyze your data to determine what, if any, correlations can be drawn between the relative frequency of amino acids making up proteins with the number of triplets for each. Write a paragraph that states your specific and general conclusions.

(d) What would be the next steps in your analysis if you wanted to pursue this problem further?

27. As described in Chapter 4, Alu elements proliferate in the human genome by a process called retrotransposition, in which the Alu DNA sequence is transcribed into RNA, copied into double-stranded DNA, and then inserted back into the genome at a site distant from that of its "parent" *Alu*.

Clearly, this has been an extremely efficient process, since *Alu*s have proliferated to about 10^6 copies in the human genome! This efficiency is largely due to the fact that *Alu*s, like many small structural RNAs, carry their promoter sequences *within the transcribed region of the gene*, rather than 5′ to the transcription start site.

If *Alu*s carried promoters upstream to the transcription site, similar to those of protein-coding genes, what would happen once they were retrotransposed? Would a retrotransposed *Alu* be able to proliferate? Explain.

Selected Readings

Alberts, B., et al. 1998. *Essential cell biology*, 3d ed. New York: Garland Publishing.

Barralle, F.E. 1983. The functional significance of leader and trailer sequences in eukaryotic mRNAs. *Int. Rev. Cytol.* 81:71–106.

Barrell, B.G., Air, G., and Hutchinson, C. 1976. Overlapping genes in bacteriophage ϕX174. *Nature* 264:34–40.

Barrell, B.G., Banker, A.T., and Drouin, J. 1979. A different genetic code in human mitochondria. *Nature* 282:189–94.

Bass, B.L., ed. 2000. *RNA Editing*. Oxford: Oxford University Press

Beardsley, T. 1996. Vital data. *Sci. Am.* (Mar.) 274:100–105.

Birnstiel, M., Busslinger, M., and Strub, K. 1985. Transcription termination and 39 processing: The end is in site. *Cell* 41:349–59.

Bonitz, S.G., et al. 1980. Codon recognition rules in yeast mitochondria. *Proc. Natl. Acad. Sci. USA* 77:3167–70.

Breitbart, R.E., and Nadal-Ginard, B. 1987. Developmentally induced, muscle-specific *trans* factors control the differential splicing of alternative and constitutive troponin T-exons. *Cell* 49:793–803.

Brenner, S. 1989. *Molecular biology: A selection of papers*. Orlando, FL: Academic Press.

Brenner, S., Jacob, F., and Meselson, M. 1961. An unstable intermediate carrying information from genes to ribosomes for protein synthesis. *Nature* 190:575–80.

Brenner, S., Stretton, A.O.W., and Kaplan, D. 1965. Genetic code: The nonsense triplets for chain termination and their suppression. *Nature* 206:994–98.

Cattaneo, R. 1991. Different types of messenger RNA editing. *Annu. Rev. Genet.* 25:71–88.

Cech, T.R. 1986. RNA as an enzyme. *Sci. Am.* (Nov.) 255(5):64–75.

———. 1987. The chemistry of self-splicing RNA and RNA enzymes. *Science* 236:1532–39.

Cech, T.R., Zaug, A.J., and Grabowski, P.J. 1981. In vitro splicing of the ribosomal RNA precursor of *Tetrahymena*. Involvement of a guanosine nucleotide in the excision of the intervening sequence. *Cell* 27:487–96.

Chambon, P. 1975. Eucaryotic nuclear RNA polymerases. *Annu. Rev. Biochem.* 44:613–38.

———. 1981. Split genes. *Sci. Am.* (May) 244:60–71.

Cold Spring Harbor Laboratory. 1966. The genetic code. *Cold Spring Harbor Symp. Quant. Biol.* Vol. 31.

Cramer, P., et al. 2000. Architecture of RNA polymerase II and implications for the transcription mechanism. *Science* 288:640-49.

Crick, F.H.C. 1962. The genetic code. *Sci. Am.* (Oct.) 207:66–77.

———. 1966a. The genetic code: III. *Sci. Am.* (Oct.) 215:55–63.

———. 1966b. Codon–anticodon pairing: The wobble hypothesis. *J. Mol. Biol.* 19:548–55.

———. 1979. Split genes and RNA splicing. *Science* 204:264–71.

Crick, F.H.C., Barnett, L., Brenner, S., and Watts-Tobin, R.J. 1961. General nature of the genetic code for proteins. *Nature* 192:1227–32.

Darnell, J.E. 1983. The processing of RNA. *Sci. Am.* (Oct.) 249:90–100.

———. 1985. RNA. *Sci. Am.* (Oct.) 253:68–87.

Dickerson, R.E. 1983. The DNA helix and how it is read. *Sci. Am.* (Dec.) 249:94–111.

Dugaiczk, A., et al. 1978. The natural ovalbumin gene contains seven intervening sequences. *Nature* 274:328–33.

Fiers, W., et al. 1976. Complete nucleotide sequence of bacteriophage MS2 RNA: Primary and secondary structure of the replicase gene. *Nature* 260:500–507.

Gamow, G. 1954. Possible relation between DNA and protein structures. *Nature* 173:318.

Gnatt, A.L., et. al. 2001. Structural basis of transcription: An RNA polymerase II elongation complex at 3.3 Å resolution. Science 292:1876–82.

Guthrie, C., and Patterson, B. 1988. Spliceosomal snRNAs. *Annu. Rev. Genet.* 22:387–419.

Hall, B.D., and Spiegelman, S. 1961. Sequence complementarity of T2-DNA and T2-specific RNA. *Proc. Natl. Acad. Sci. USA* 47:137–46.

Hamkalo, B. 1985. Visualizing transcription in chromosomes. *Trends Genet.* 1:255–60.

Helman, J.D., and Chamberlin, M.J. 1988. Structure and function of bacterial sigma factors. *Annu. Rev. Biochem.* 57:839–72.

Hodges, P., and Scott, J. 1992. Apolipoprotein B mRNA editing: A new tier for the control of gene expression. *Trends Biochem. Sci.* 17:77–81.

Horton, H.R., et al. 2002. *Principles of biochemistry*, 3d ed. Upper Saddle River, NJ: Prentice Hall.

Humphrey, T., and Proudfoot, N.J. 1988. A beginning to the biochemistry of polyadenylation. *Trends Genet.* 4:243–45.

Judson, H.F. 1979. *The eighth day of creation*. New York: Simon & Schuster.

Jukes, T.H. 1963. The genetic code. *Am. Sci.* 51:227–45.

Kable, M.L., et al. 1996. RNA editing: A mechanism for gRNA-specified uridylate insertion into precursor mRNA. *Science* 273:1189–95.

Keegan, L.P., Gallo, A., and O'Connell, M.A. 2000. Survival is impossible without an editor. *Science* 290:1707-08.

Khorana, H.G. 1967. Polynucleotide synthesis and the genetic code. *Harvey Lectures* 62:79–105.

Lodish, H., et al. 2000. *Molecular cell biology*. New York: W. H. Freeman.

Maniatis, T., and Reed, R. 1987. The role of small nuclear ribonucleoprotein particles in pre-mRNA splicing. *Nature* 325:673–78.

Miller, O.L., and Beatty, B.R. 1969. Portrait of a gene. *J. Cell Physiol.* 74 (Suppl. 1): 225–32.

Miller, O.L., Hamkalo, B., and Thomas, C. 1970. Visualization of bacterial genes in action. *Science* 169:392–95.

Min Jou, W., Hageman, G., Ysebart, M., and Fiers, W. 1972. Nucleotide sequence of the gene coding for bacteriophage MS2 coat protein. *Nature* 237:82–88.

Nikolov, D.B. and Burley, S.K. 1997. RNA polymerase II transcription initiation: A structural view. *Proc. Natl. Acad. Sci. USA* 94:15–22.

Nilsen, T.W. 1994. RNA–RNA interactions in the spliceosome: Unraveling the ties that bind. *Cell* 78:1–4.

Nirenberg, M.W. 1963. The genetic code: II. *Sci. Am.* (March) 190:80–94.

Nirenberg, M.W., and Leder, P. 1964. RNA code words and protein synthesis. *Science* 145:1399–1407.

Nirenberg, M.W., and Matthaei, H. 1961. The dependence of cell-free protein synthesis in *E. coli* upon naturally occurring or synthetic polyribosomes. *Proc. Natl. Acad. Sci. USA* 47:1588–1602.

O'Malley, B., et al. 1979. A comparison of the sequence organization of the chicken ovalbumin and ovomucoid genes. In *Eucaryotic gene regulation*, ed. R. Axel, et al., pp. 281–99. Orlando, FL: Academic Press.

Padgett, R.A., Grabowski, P.J., Konarska, M.M., Seiler, S., and Sharp, P.A. 1986. Splicing of messenger RNA precursors. *Annu. Rev. Biochem.* 55:1119–50.

Reed, R., and Maniatis, T. 1985. Intron sequences involved in lariat formation during pre-mRNA splicing. *Cell* 41:95–105.

Ross, J. 1989. The turnover of mRNA. *Sci. Am.* (April) 260:48–55.

Sharp, P.A. 1987. Splicing of messenger RNA precursors. *Science* 235:766–71.

———1994. Nobel Lecture: Split genes and RNA splicing. *Cell* 77:805–15.

Sharp, P.A., and Eisenberg, D. 1987. The evolution of catalytic function. *Science* 238:729–30.

Shatkin, A.J. 1985. mRNA cap binding proteins: Essential factors for initiating translation. *Cell* 40:223–24.

Steitz, J.A. 1988. Snurps. *Sci. Am.* (June) 258(6):56–63.

Volkin, E., and Astrachan, L. 1956. Phosphorus incorporation in *E. coli* ribonucleic acids after infection with bacteriophage T2. *Virology* 2:149–61.

Volkin, E., Astrachan, L., and Countryman, J.L. 1958. Metabolism of RNA phosphorus in *E. coli* infected with bacteriophage T7. *Virology* 6:545–55.

Wang, Q., et.al. 2000. Requirement of the RNA editing deaminase *ADAR1* gene for embryonic erythropoiesis. *Science* 290:1765–68.

Watson, J.D. 1963. Involvement of RNA in the synthesis of proteins. *Science* 140:17–26.

Woychik, N.A. and Jampsey, M. 2002. The RNA polymerase II machinery: Structure illuminates function. *Cell* 108:453–64.

GENETICS MediaLab

The resources that follow will help you achieve a better understanding of the concepts presented in this chapter. These resources can be found either on the CD packaged with this textbook or on the Companion Web site found at **http://www.prenhall.com/klug**.

CD Resources:

Module 5.1: Transcription

Web Problem 1:

Time for completion = 5 minutes

How does the transcription machinery select where to begin transcription?

In eukaryotic cells, RNA polymerase II transcribes genes to produce mRNA. For transcription to begin, the RNA polymerase must be directed to the appropriate region of DNA at the 5' end of the gene by transcription activators. This is a complex process involving many proteins. Some of the proteins regulating the process are bound to DNA, while others are preassembled with the polymerase. This linked Web site presents the research of one laboratory in defining how the transcription process occurs. Describe one question the researchers are pursuing and what approach they are taking to answer that question. To complete this exercise, visit Web Problem 1 in Chapter 5 of your Companion Web site, and select the keyword **POLYMERASE**.

Web Problem 2:

Time for completion = 10 minutes

How universal is the universal genetic code? Not all organisms use this code at all times. The most prominent exception is the mitochondrion. In this exercise, you can view variations in the genetic code. In addition, codon use can vary: Sometimes, one codon is used more or less frequently than another codon that encodes the same amino acid. The linked Web site has a hypertext link to a searchable database for codon use in many organisms. Choose an example of a genetic code differing from the universal code, and list the distinctions. From the codon-use database, examine the codon use of one organism. Using the standard (universal) genetic code, discuss any difference in usage for Ser, Leu, and stop codons, including which codons are used most and least often. To complete this exercise, visit Web Problem 2 in Chapter 5 of your Companion Web site and select the keyword **CODON**.

Web Problem 3:

Time for completion = 10 minutes

How are introns removed from a transcript? The spliceosome consists of a complex of proteins and small nuclear RNA (together called snRNP, small nuclear ribonucleoprotein particles) that removes introns and splices exons together. The linked Web site describes the components of the spliceosome and their functions. Select the hypertext links to view the two splicing animations. The synthesis animation provides an overview of the process and the snRNPs involved, while the mechanism animation details the molecular events. Describe the steps involved in splicing a pre-mRNA transcript and what components interact with which portions of the transcript. To complete this exercise, visit Web Problem 3 in Chapter 5 of your Companion Web site, and select the keyword **SPLICEOSOME**.

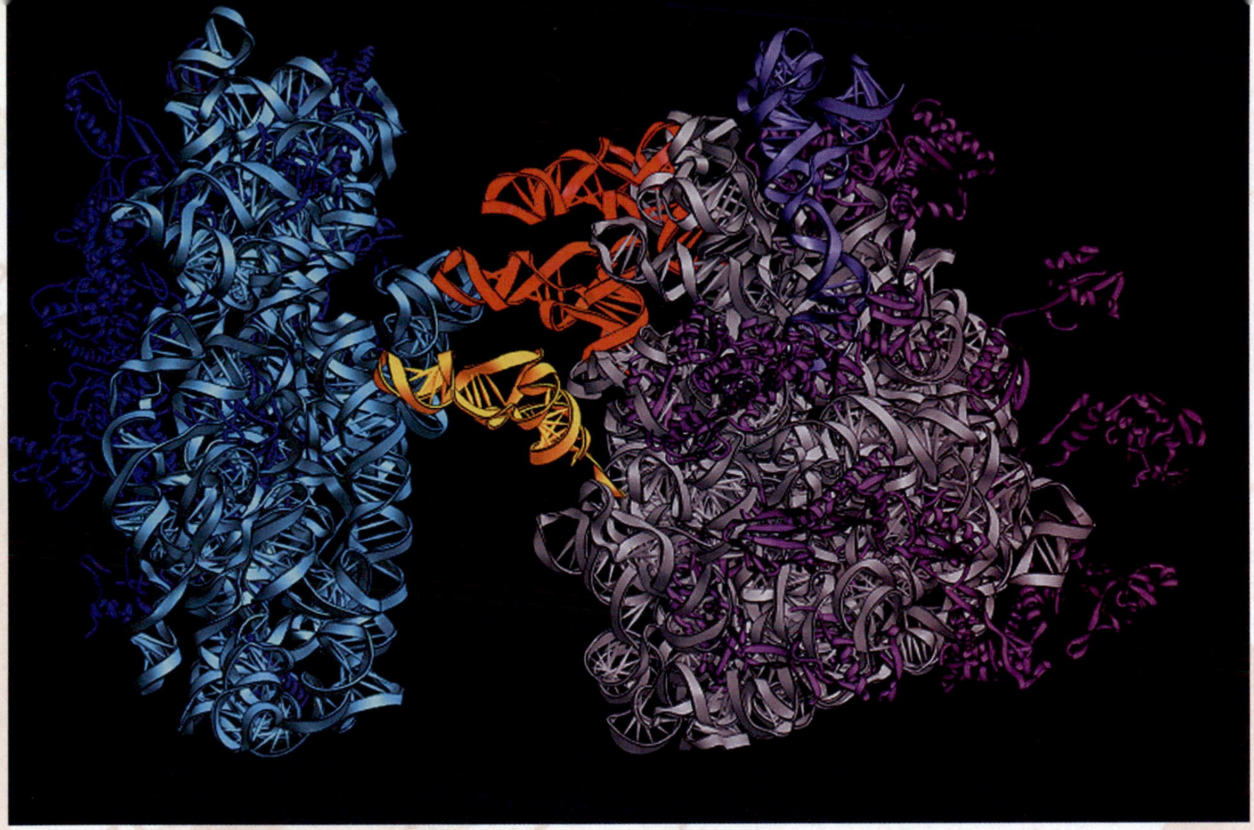

Crystal structure of a *Thermus thermophilus* 70S ribosome containing three bound transfer RNAs.

6

Translation and Proteins

KEY CONCEPTS

In Chapter 5, we established that there is a genetic code that stores information in the form of triplet nucleotides in DNA, and this information is initially expressed through the process of transcription into a messenger RNA that is complementary to one of the two strands of the DNA helix. However, the final product of gene expression, in almost all instances, is a polypeptide chain consisting of a linear series of amino acids whose sequence has been prescribed by the genetic code. In this chapter, we will examine how the information present in mRNA is processed in order to create polypeptides, which then fold into protein molecules. We will also review the evidence that confirmed that proteins are the end products of genes and discuss briefly the various levels of protein structure, diversity, and function. This information extends our understanding of gene expression and provides an important foundation for interpreting how the mutations that arise in DNA can result in the diverse phenotypic effects observed in organisms.

6.1 Translation of mRNA Depends on Ribosomes and Transfer RNAs

Translation of mRNA is the biological polymerization of amino acids into polypeptide chains. This process, alluded to in our earlier discussion of the genetic code, occurs only in association with ribosomes, which serve as nonspecific workbenches. The central question in translation is how triplet ribonucleotides of mRNA direct specific amino acids into their correct position in the polypeptide. That question was answered once **transfer RNA (tRNA)** was discovered. This class of molecules adapts specific triplet codons in mRNA to their correct amino acids. Such an adaptor role was postulated for tRNA by Francis Crick in 1957.

In association with a ribosome, mRNA presents a triplet codon that calls for a specific amino acid. A specific tRNA molecule contains within its nucleotide sequence three consecutive ribonucleotides complementary to the codon, called the **anticodon**, which can base-pair with the codon. Another region of this tRNA is covalently bonded to its corresponding amino acid.

Inside the ribosome, hydrogen bonding of tRNAs to mRNA holds the amino acids in proximity so that a peptide bond can be formed. The process occurs over and over as mRNA runs through the ribosome, and amino acids are polymerized into a polypeptide. Before we discuss the actual process of translation, we will first consider the structures of the ribosome and transfer RNA.

Ribosomal Structure

Because of its essential role in the expression of genetic information, the **ribosome** has been extensively analyzed. One bacterial cell contains about 10,000 of these structures, and a eukaryotic cell contains many times more. Electron microscopy has revealed that the bacterial ribosome is about 250 nm at its largest diameter and consists of two subunits, one large and one small. Both subunits consist of one or more molecules of rRNA and an array of **ribosomal proteins**. When the two subunits are associated with each other in a single ribosome, the structure is sometimes called a **monosome**.

The specific differences between prokaryotic and eukaryotic ribosomes are summarized in Figure 6–1. The subunit and rRNA components are most easily isolated and characterized on the basis of their sedimentation behavior in sucrose gradients (their rate of migration, abbreviated as *S*, as introduced in Chapter 2). In prokaryotes, the monosome is a 70*S* particle, and in eukaryotes it is approximately 80*S*. Sedimentation coefficients, which reflect the variable rate of migration of different-sized particles and molecules, are not additive. For example, the prokaryotic 70*S* monosome consists of a 50*S* and a 30*S* subunit, and the eukaryotic 80*S* monosome consists of a 60*S* and a 40*S* subunit.

The larger subunit in prokaryotes consists of a 23*S* RNA molecule, a 5*S* rRNA molecule, and 31 ribosomal proteins. In the eukaryotic equivalent, a 28*S* rRNA molecule is accompanied by a 5.8*S* and 5*S* rRNA molecule and about 50 proteins. The smaller prokaryotic subunits consist of a 16*S* rRNA component and 21 proteins. In the eukaryotic equivalent, an 18*S* rRNA component and about 33 proteins are found. The approximate molecular weights and number of nucleotides of these components are shown in Figure 6–1.

Regarding the components of the ribosome, it is now clear that the RNA molecules provide the basis for all important catalytic functions associated with translation. The many proteins, whose functions were long a mystery, are thought to promote the binding of the various molecules involved in translation, and in general, to fine-tune the process. This conclusion is based on the observation that some of the catalytic functions in ribosomes still occur in experiments involving "ribosomal protein-depleted" ribosomes.

Molecular hybridization studies have established the degree of redundancy of the genes coding for the rRNA components. The *E. coli* genome contains seven copies of a single sequence that encodes all three components—23*S*, 16*S*, and 5*S*. The initial transcript of these genes produces a 30*S* RNA molecule that is enzymatically cleaved into these smaller components. Coupling of the genetic information encoding these three rRNA components ensures that, following multiple transcription events, equal quantities of all three will be present as ribosomes are assembled.

In eukaryotes, many more copies of a sequence encoding the 28*S* and 18*S* components are present. In *Drosophila*, approximately 120 copies per haploid genome are each transcribed into a molecule of about 34*S*. This is processed into the 28*S*, 18*S*, and 5.8*S* rRNA species. These are homologous to the three rRNA components of *E. coli*. In *X. laevis*, over 500 copies per haploid genome are present. In mammalian cells, the initial transcript is 45*S*. The rRNA genes, called **rDNA**, are part of the moderately repetitive DNA fraction and are present in clusters at various chromosomal sites.

Each cluster in eukaryotes consists of **tandem repeats**, with each unit separated by a noncoding **spacer DNA** sequence. [See the micrograph in Figure 5.15(b).] In humans,

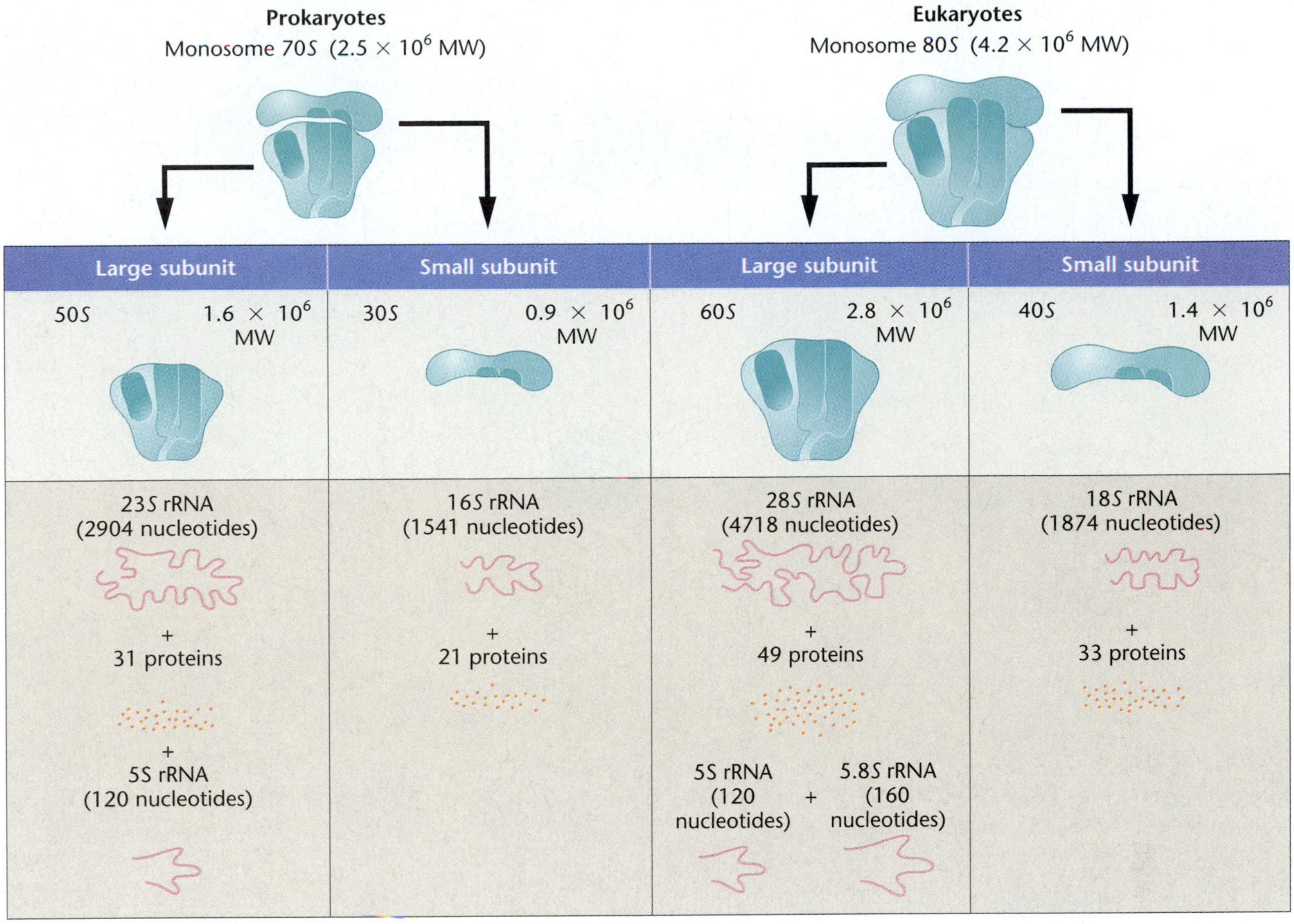

FIGURE 6–1 A comparison of the components in the prokaryotic and eukaryotic ribosome.

these gene clusters have been localized near the ends of chromosomes 13, 14, 15, 21, and 22. The unique 5*S* rRNA component of eukaryotes is not part of this larger transcript. Instead, genes coding for this ribosomal component are distinct and located separately. In humans, a gene cluster encoding 5*S* rRNA has been located on chromosome 1.

Despite the detailed knowledge available on the structure and genetic origin of the ribosomal components, a complete understanding of the function of these components has eluded geneticists. This is not surprising; the ribosome is perhaps the most intricate of all cellular structures. In bacteria, the monosome has a combined molecular weight of 2.5 million Da!

tRNA Structure

Because of their small size and stability in the cell, transfer RNAs (tRNAs) have been investigated extensively and are the best characterized RNA molecules. They are composed of only 75–90 nucleotides, displaying a nearly identical structure in bacteria and eukaryotes. In both types of organisms, tRNAs are transcribed as larger precursors, which are cleaved into mature 4*S* tRNA molecules. In *E. coli*, for example, $tRNA^{tyr}$ (the superscript identifies the specific tRNA and the cognate amino acid that binds to it) is composed of 77 nucleotides, yet its precursor contains 126 nucleotides.

In 1965, Robert Holley and his colleagues reported the complete sequence of $tRNA^{ala}$ isolated from yeast. Of great interest was the finding that a number of nucleotides are unique to tRNA. As shown in Figure 6–2, each contains a modification of one of the four nitrogenous bases expected in RNA (G, C, A, and U). For example, inosinic acid is present, which contains the purine hypoxanthine. Ribothymidylic acid and pseudouridine are examples among others. Variously referred to as *unusual, rare*, or *odd bases*, these modified structures are created *following* transcription, illustrating the more general concept of **posttranscriptional modification**. In this case, the unmodified base is inserted during transcription and, subsequently, enzymatic reactions catalyze the chemical modifications to the base.

Holley's sequence analysis led him to propose the two-dimensional **cloverleaf model of tRNA**. It had been known that tRNA demonstrates a secondary structure due to base pairing. Holley discovered that he could arrange the linear model in such a way that several stretches of base pairing would result. Such an arrangement created a series of paired stems and unpaired loops resembling the shape of a

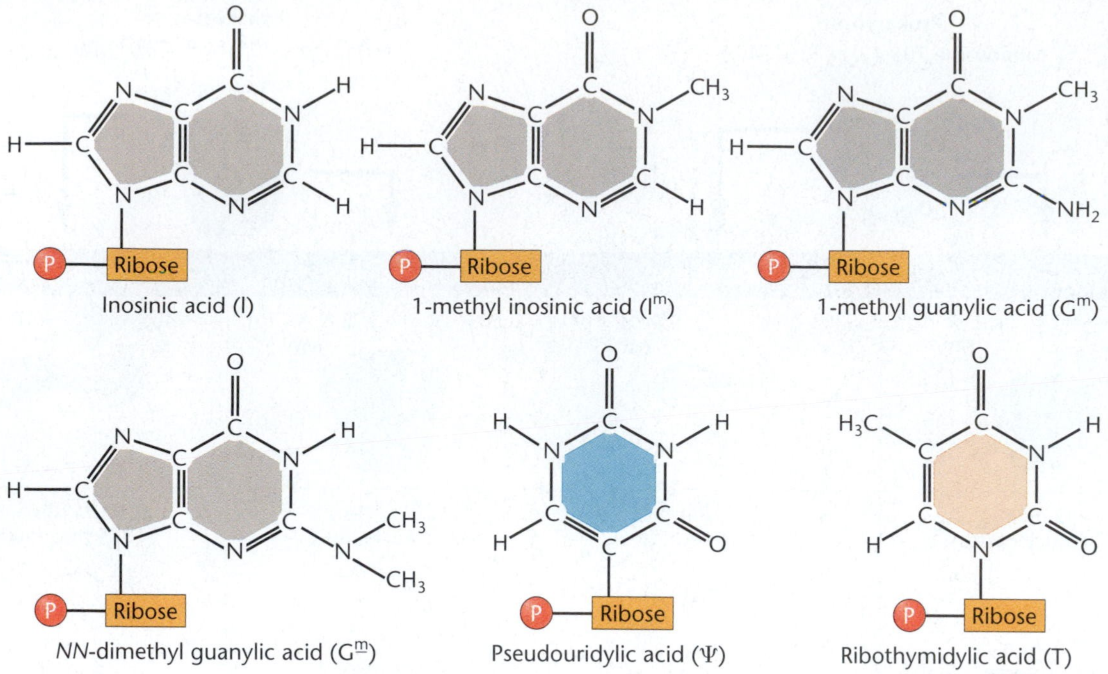

FIGURE 6–2 Unusual nitrogenous bases found in transfer RNA.

cloverleaf. Loops consistently contained modified bases that did not generally form base pairs. Holley's model is shown in Figure 6–3.

Because the triplets GCU, GCC, and GCA specify alanine, Holley looked for an anticodon sequence complementary to one of these codons in his tRNA^{ala} molecule. He found it in the form of CGI (the 3′-to-5′ direction), in one of the loops of the cloverleaf. The nitrogenous base I (inosinic acid) can form hydrogen bonds with U, C, or A, the third members of the triplets. Thus, the **anticodon loop** was established.

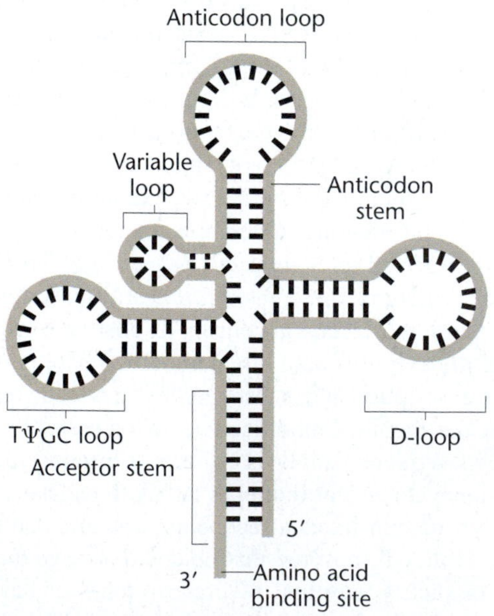

FIGURE 6–3 Holley's two-dimensional cloverleaf model of transfer RNA.

Studies of other tRNA species revealed many constant features. First, at the 3′ end, all tRNAs contain the sequence ...*pCpCpA*-3′. It is at this end of the molecule that the amino acid is covalently joined to the terminal adenosine residue. All tRNAs contain 5′-*Gp*... at the other end of the molecule.

In addition, the lengths of various stems and loops are very similar. Each tRNA examined also contains an anticodon complementary to the known amino acid codon for which it is specific, and all anticodon loops are present in the same position of the cloverleaf.

Because the cloverleaf model was predicted strictly on the basis of nucleotide sequence, there was great interest in the X-ray crystallographic examination of tRNA, which reveals a three-dimensional structure. By 1974, Alexander Rich and his colleagues in the United States, and J. Roberts, B. Clark, and Aaron Klug and their colleagues in England, had succeeded in crystallizing tRNA and performing X-ray crystallography at a resolution of 3 Å. At such resolution, the pattern formed by individual nucleotides is discernible.

As a result of these studies, a complete three-dimensional model of tRNA is now available (Figure 6–4). At one end of the molecule is the anticodon loop and stem, and at the other end is the 3′-acceptor region to which the amino acid is bound. It has been speculated that the shapes of the intervening loops may be recognized by the specific enzymes responsible for adding the amino acid to tRNA, a subject to which we now turn our attention.

Charging tRNA

Before translation can proceed, the tRNA molecules must be chemically linked to their respective amino acids. This activation process, called **charging**, occurs under the direction of enzymes called **aminoacyl tRNA synthetases**.

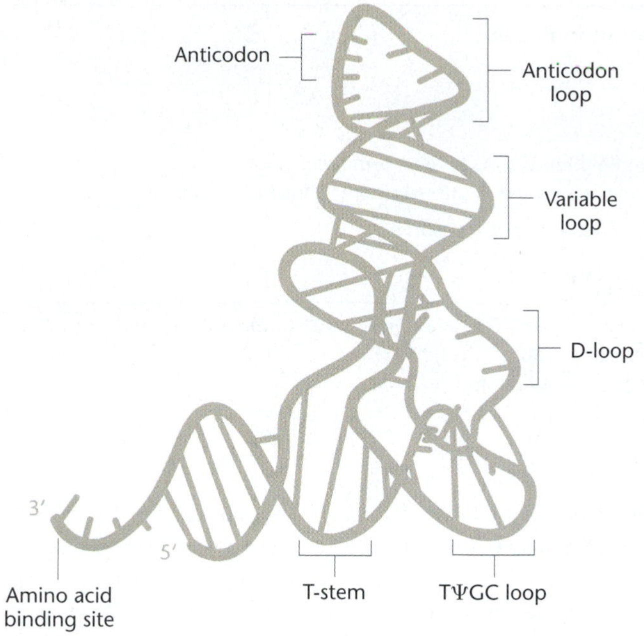

FIGURE 6–4 A three-dimensional model of transfer RNA.

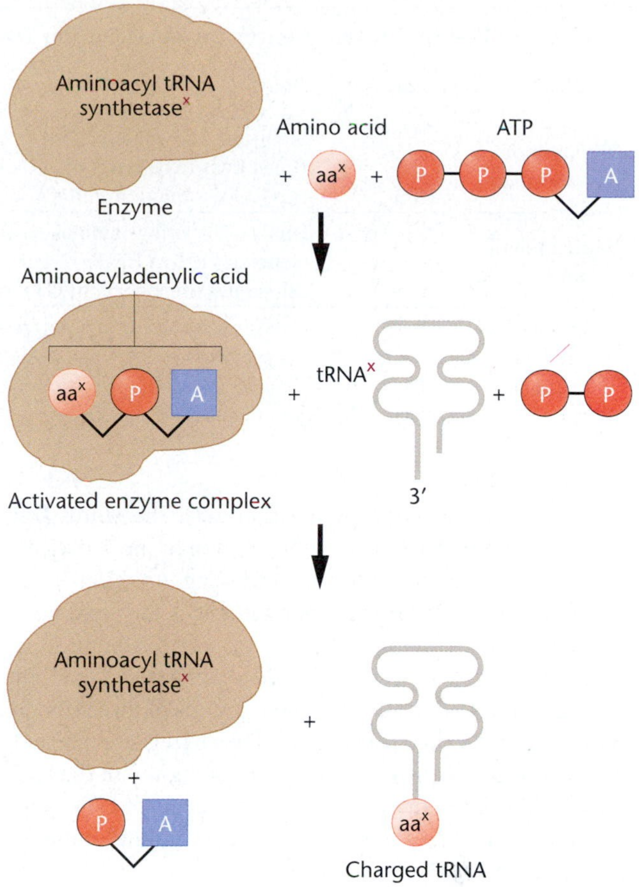

FIGURE 6–5 Steps involved in charging tRNA. The "x" denotes that for each amino acid only the corresponding specific tRNA and specific aminoacyl tRNA synthetase enzyme are involved in the charging process.

Because there are 20 different amino acids, there must be at least 20 different tRNA molecules and as many different enzymes. In theory, because there are 61 triplet codes, there could be the same number of specific tRNAs and enzymes. However, because of the ability of the third member of a triplet code to "wobble," it is now thought that there are at least 32 different tRNAs; it is also believed that there are only 20 synthetases, one for each amino acid, regardless of the greater number of corresponding tRNAs.

The charging process is outlined in Figure 6–5. In the initial step, the amino acid is converted to an activated form, reacting with ATP to create an **aminoacyladenylic acid**. A covalent linkage is formed between the 5′-phosphate group of ATP and the carboxyl end of the amino acid. This molecule remains associated with the synthetase enzyme, forming a complex that then reacts with a specific tRNA molecule. In the next step, the amino acid is transferred to the appropriate tRNA and bonded covalently to the adenine residue at the 3′ end. The charged tRNA may participate directly in protein synthesis. Aminoacyl tRNA synthetases are highly specific enzymes because they recognize only one amino acid and only a subset of corresponding tRNAs, called **isoaccepting tRNAs**. This is a crucial point if fidelity of translation is to be maintained.

6.2 Translation of mRNA Can Be Divided into Three Steps

In a way similar to transcription, the process of translation can be best described by breaking it into discrete phases. We will consider three such phases, each with its own set of illustrations (Figures 6–6, 6–7, and 6–8). But keep in mind that translation is a dynamic, continuous process. Correlate the discussion that follows with the step-by-step characterization in these figures. Many of the protein factors and their roles in translation are summarized in Table 6–1.

Initiation

Initiation of translation is depicted in Figure 6–6. Recall that the ribosome serves as a nonspecific workbench for the translation process. Most ribosomes, when they are not involved in translation, are dissociated into their large and small subunits. Initiation of translation in *E. coli* involves the small ribosome subunit, an mRNA molecule, a specific charged initiator tRNA, GTP, Mg^{2+}, and a number of proteinaceous initiation factors (IFs). These are initially part of the small subunit and are required to enhance the binding affinity of the various translational components. Unlike ribosomal proteins, IFs are released from the ribosome once initiation is completed. In prokaryotes, the initiation codon of mRNA—AUG—calls for the modified amino acid **formylmethionine (f-met)**.

The small ribosomal subunit binds to several initiation factors, and this complex in turn binds to mRNA (Step 1). In bacteria, such binding involves a sequence of up to six ribonucleotides (AGGAGG, not shown in Fig. 6-6), which *precedes* the initial AUG start codon of mRNA. This

TABLE 6–1 Various Protein Factors Involved During Translation in *E. coli*

Process	*Factor*	*Role*
Initiation of translation	IF1	Stabilizes 30*S* subunit
	IF2	Binds fmet-tRNA to 30*S*-mRNA complex; binds to GTP and stimulates hydrolysis
	IF3	Binds 30*S* subunit to mRNA; dissociates monosomes into subunits following termination
Elongation of polypeptide	EF-Tu	Binds GTP; brings aminoacyl-tRNA to the A site of ribosome
	EF-Ts	Generates active EF-Tu
	EF-G	Stimulates translocation; GTP-dependent
Termination of translation and release of polypeptide	RF1	Catalyzes release of the polypeptide chain from tRNA and dissociation of the translocation complex; specific for UAA and UAG termination codons
	RF2	Behaves like RF1; specific for UGA and UAA codons
	RF3	Stimulates RF1 and RF2

sequence (containing only purines and called the **Shine–Dalgarno sequence**) base-pairs with a region of the 16*S* rRNA of the small ribosomal subunit, facilitating initiation.

Another initiation protein then enhances the binding of charged formylmethionyl tRNA to the small subunit in response to the AUG triplet (Step 2). This step "sets" the reading frame so that all subsequent groups of three ribonucleotides are translated accurately. The aggregate represents the **initiation complex**, which then combines with the large ribosomal subunit. In this process, a molecule of GTP is hydrolyzed, providing the required energy, and the initiation factors are released (Step 3).

Elongation

The second phase of translation, elongation, is depicted in Figure 6–7. Once both subunits of the ribosome are assembled with the mRNA, binding sites for two charged tRNA molecules are formed. These are designated as the **P**, or **peptidyl**, and the **A**, or **aminoacyl**, **sites**. The charged initiator tRNA binds to the P site, provided that the AUG triplet of mRNA is in the corresponding position of the small subunit.

The increase of the growing polypeptide chain by one amino acid is called **elongation**. The sequence of the second triplet in mRNA dictates which charged tRNA molecule will become positioned at the A site (Step 1). Once it is present, **peptidyl transferase** catalyzes the formation of the peptide bond that links the two amino acids together (Step 2). The catalytic activity of peptidyl transferase is a function of rRNA of the large subunit, not one of the ribosomal proteins. At the same time, the covalent bond between the amino acid and the tRNA occupying the P site is hydrolyzed (broken). The product of this reaction is a dipeptide, which is attached to the 3′ end of tRNA still residing in the A site.

Before elongation can be repeated, the tRNA attached to the P site, which is now uncharged, must be released from the large subunit. The uncharged tRNA moves transiently through a third site on the ribosome called the **E site** (E stands for exit). The entire $\mathbf{mRNA{-}tRNA{-}aa_2{-}aa_1}$ complex then shifts in the direction of the P site by a distance of three nucleotides (Step 3). This event requires several protein elongation factors (EFs) as well as the energy derived from hydrolysis of GTP. The result is that the third triplet of mRNA is now in a position to accept another specific charged tRNA into the A site (Step 4). One simple way to distinguish the two sites in your mind is to remember that, *following the shift*, the P site contains a tRNA attached to a peptide chain (*P* for peptide), whereas the A site contains a tRNA with an amino acid attached (*A* for amino acid).

The sequence of elongation is repeated over and over (Steps 5 and 6). An additional amino acid is added to the growing polypeptide chain each time the mRNA advances through the ribosome. Once a polypeptide chain of reasonable size is assembled (about 30 amino acids), it begins to emerge from the base of the large subunit, as illustrated in Step 6. A tunnel exists within the large subunit, through which the elongating polypeptide emerges.

As we have seen, the role of the small subunit during elongation is one of "decoding" the triplets present in mRNA, while the role of the large subunit is peptide-bond synthesis. The efficiency of the process is remarkably high; the observed error rate is only about 10^{-4}. An incorrect amino acid will occur only once in every 20 polypeptides of an average length of 500 amino acids! In *E. coli*, elongation occurs at a rate of about 15 amino acids per second at 37°C.

Termination

Termination, the third phase of translation, is depicted in Figure 6–8. Termination of protein synthesis is signaled by one or more of three triplet codes in the A site: UAG, UAA, or UGA. These codons do not specify an amino acid, nor do they call for a tRNA in the A site. These codons are called **stop codons**, **termination codons**, or **nonsense codons**. The finished polypeptide is therefore still attached to the terminal tRNA at the P site, and the A site is empty. The termination codon signals the action of **GTP-dependent release factors**, which cleave the polypeptide chain from the terminal tRNA, releasing it from the translation complex (Step 1). Once this cleavage occurs, the tRNA is released from the ribosome, which then dissociates into its subunits (Step 2). If a termination codon should appear in the middle of an mRNA molecule as a result of mutation, the same process occurs, and the polypeptide chain is prematurely terminated.

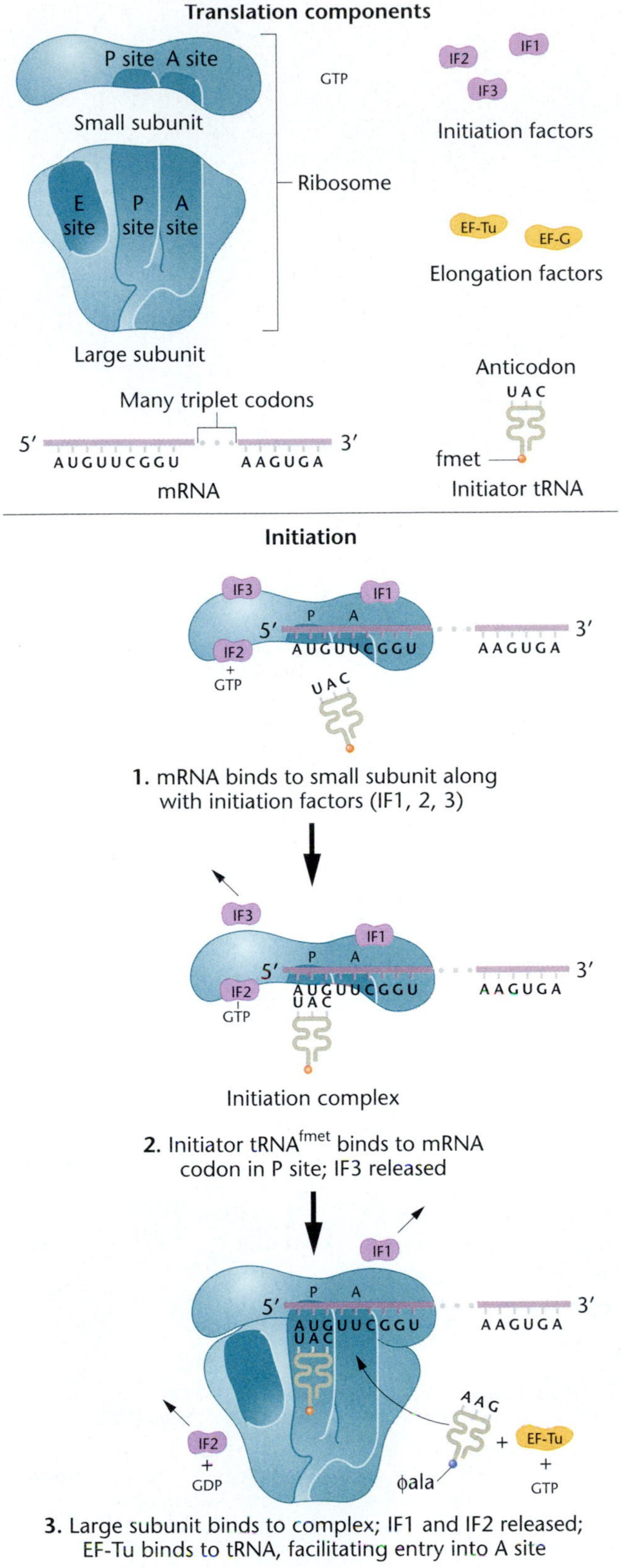

FIGURE 6–6 Initiation of translation. The components are depicted at the top of the figure.

Polyribosomes

As elongation proceeds and the initial portion of mRNA has passed through the ribosome, the message is free to associate with another small subunit to form a second initiation complex. This process can be repeated several times with a single mRNA and results in what are called **polyribosomes** or just **polysomes**.

Polyribosomes can be isolated and analyzed following a gentle lysis of cells. Figures 6–9(a) and (b) illustrate these complexes as seen under the electron microscope. In Figures 6–9(a), you can see mRNA (the thin line) between the individual ribosomes. The micrograph in Figure 6–9(b) is even more remarkable, for it shows the polypeptide chains emerging from the ribosomes during translation. The formation of polysome complexes represents an efficient use of the components available for protein synthesis during a unit of time. Using the analogy of a tape and a tape recorder, in polysome complexes, one tape (mRNA) would be played simultaneously by several recorders (the ribosomes). But at any given moment, each recording (the polypeptide being synthesized in each ribosome) would be at a different point of completion.

6.3 Crystallographic Analysis Has Revealed Many Details about the Functional Prokaryotic Ribosome

Our knowledge of the process of translation and the structure of the ribosome, as described in the previous sections, is based primarily on biochemical and genetic observations, in addition to the visualization of ribosomes under the electron microscope. Because of the tremendous size and complexity of the functional ribosome during active translation, obtaining crystals needed to perform X-ray diffraction studies has been extremely difficult. Nevertheless, great strides have been made in the past several years. First, the individual ribosomal subunits were crystallized and examined in several laboratories, most prominently that of V. Ramakrishnan. Then, in 2001, the crystal structure of the intact 70*S* ribosome, complete with associated mRNA and tRNAs, was examined by Harry Noller and his colleagues. In essence, the entire translational complex was visualized at the atomic level. Both Ramakrishnan and Noller derived the ribosomes from the bacterium *Thermus thermophilus*.

MEDIA TUTORIAL Protein Translation

Many noteworthy observations have been made from these investigations. A view of one of the models based on Noller's findings is shown (see page 127) as the opening photograph of this chapter. For example, the sizes and shapes of the subunits, measured at atomic dimensions, are in agreement with earlier estimates based on high-resolution electron micro-scopy. Further, the shape of the ribosome changes during different functional states, attesting to the dynamic nature of the process of translation. A great deal has also been learned about the prominence and location of the RNA components of the subunits. For example, about one third of

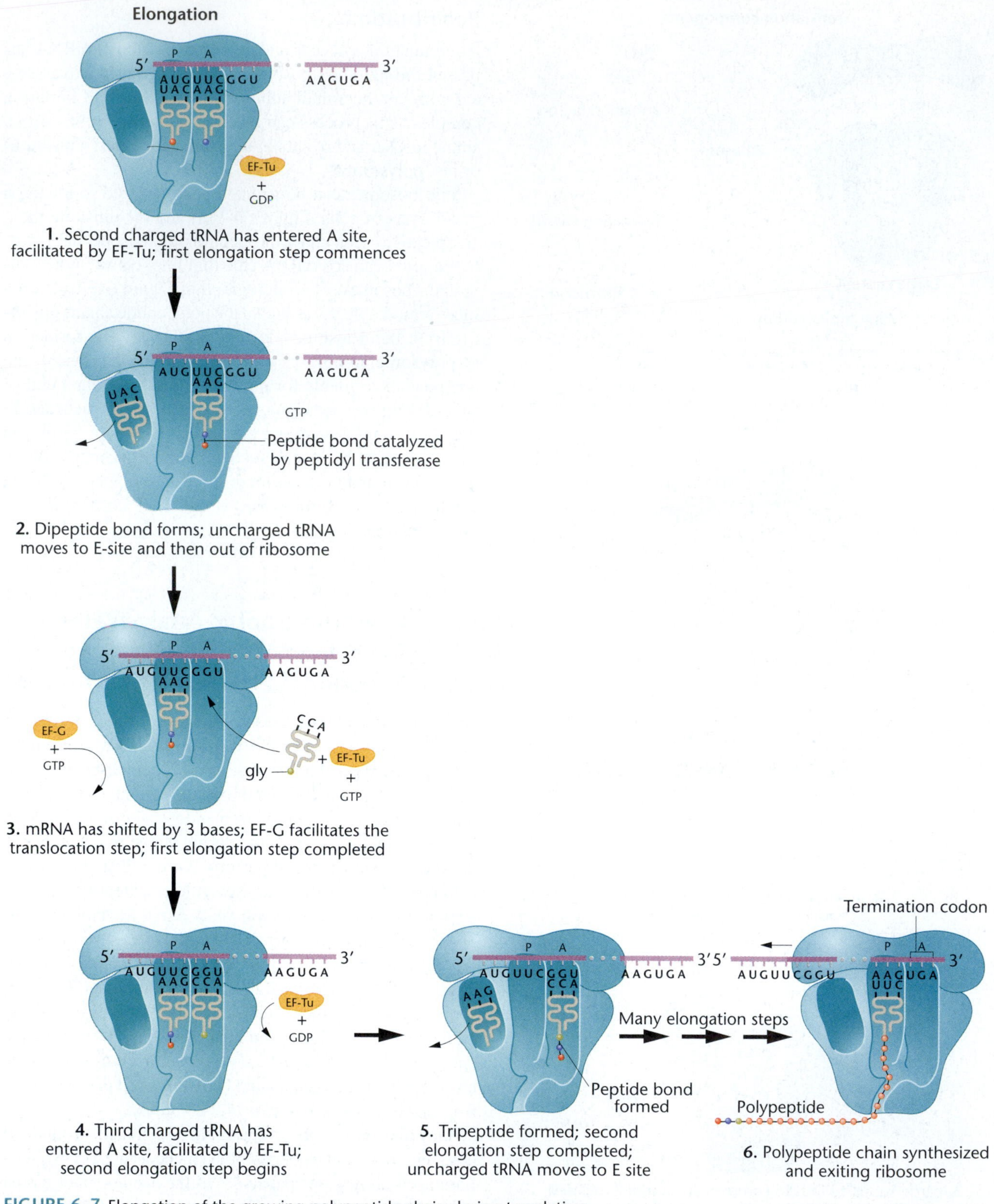

Elongation
5′
P
A
3′
AUGUUCGGU
UACAAG
AAGUGA
EF-Tu
+
GDP
1. Second charged tRNA has entered A site, facilitated by EF-Tu; first elongation step commences
UAC
GTP
Peptide bond catalyzed by peptidyl transferase
2. Dipeptide bond forms; uncharged tRNA moves to E-site and then out of ribosome
EF-G
+
GTP
CCA
gly
3. mRNA has shifted by 3 bases; EF-G facilitates the translocation step; first elongation step completed
AAGCCA
4. Third charged tRNA has entered A site, facilitated by EF-Tu; second elongation step begins
Peptide bond formed
5. Tripeptide formed; second elongation step completed; uncharged tRNA moves to E site
Many elongation steps
Termination codon
UUC
Polypeptide
6. Polypeptide chain synthesized and exiting ribosome

FIGURE 6–7 Elongation of the growing polypeptide chain during translation.

the 16*S* RNA is responsible for producing a flat projection within the smaller 30*S* subunit referred to as the platform, which modulates movement of the mRNA–tRNA complex during translocation.

More information supports the concept that RNA is the real "player" in the ribosome during translation. The interface between the two subunits, considered to be the location in the ribosome where polymerization of amino acids occurs, is composed almost exclusively of RNA. In contrast, the numerous ribosomal proteins are found mostly on the periphery of the ribosome. These observations confirm what has been predicted on genetic grounds—the catalytic steps that

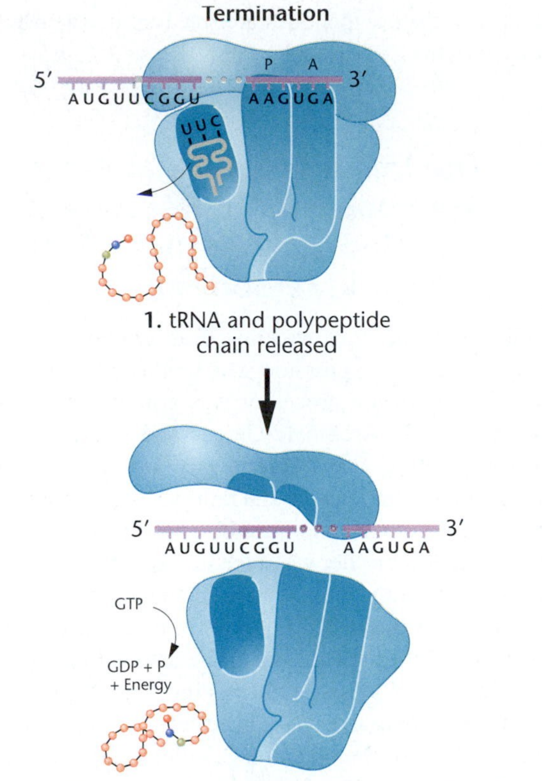

FIGURE 6–8 Termination of the process of translation.

join amino acids during translation occur under the direction of RNA, not proteins.

Another interesting finding involves the actual location of the three sites predicted to house tRNAs during translation. All three, the aminoacyl, the peptidyl, and the exit sites (designated the A, P, and E sites), have been identified, and in each case, the RNA of the ribosome makes direct contact with the various loops and domains of the tRNA molecule. This observation points to the importance of the different regions of tRNA and helps us understand why the specific three-dimensional conformation of all tRNA molecules has been preserved throughout evolution.

Still another observation is that the intervals between these A, P, and E sites are at least 20 Å, and perhaps as much as 50 Å, thus defining the atomic distance that the tRNA molecules must shift during each translocation event. This is considered a fairly large distance relative to the size of the tRNAs themselves. Analysis has led to the identification of molecular (RNA-protein) bridges that exist between the three sites that are involved in the translocation events. Other such bridges are present at other key locations and have been related to ribosome function. These observations provide us with a much better picture of the dynamic changes that must occur within the ribosome during translation.

A final observation takes us back almost 50 years, when Francis Crick proposed the **wobble hypothesis**, as introduced in Chapter 5. The Ramakrishnan group has identified the precise location along the 16*S* RNA of the 30*S* subunit involved in the decoding step between mRNA and tRNA. Two particular nucleotides of the 16*S* RNA actually flip out and probe the codon:anticodon region and are believed to check for accuracy of base pairing during this interaction. Related to the wobble hypothesis, the stringency of this step is high for the first two base pairs, but less stringent for the third (or wobble) base pair.

While these findings represent landmark studies, numerous questions still remain about ribosome structure and function. In particular, the role of the many ribosomal proteins is yet to be clarified. Nevertheless, the models that are emerging based on the work of Noller, Ramakrishnan, and their many colleagues provide us with a much better understanding of the mechanism of translation.

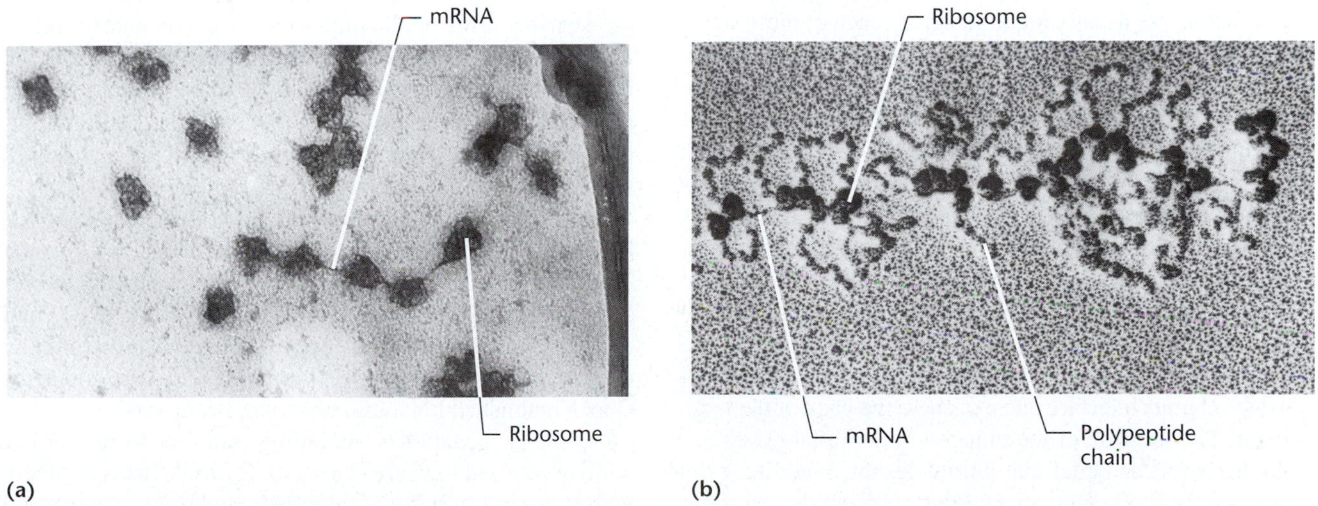

FIGURE 6–9 Polyribosomes visualized under the electron microscope. Those in (a) were derived from rabbit reticulocytes engaged in the translation of hemoglobin mRNA and in (b) from giant salivary gland cells of the midgefly, *Chironomus thummi*. In (b), the nascent polypeptide chain is apparent as it emerges from each ribosome. Its length increases as translation proceeds from left (5′) to right (3′) along the mRNA.

6.4 Translation Is More Complex in Eukaryotes

The general features of the model we just discussed were initially derived from investigations of the translation process in bacteria. As we saw, one of the main differences between translation in prokaryotes and eukaryotes is that, in the latter, translation occurs on ribosomes that are larger and whose rRNA and protein components are more complex than those of prokaryotes. (See Figure 6–1.)

Several other differences are also important. Eukaryotic mRNAs are much longer lived than are their prokaryotic counterparts. Most exist for hours rather than minutes prior to their degradation by nucleases in the cell, remaining available much longer to orchestrate protein synthesis.

Several aspects that involve the initiation of translation are different in eukaryotes. First, as we discussed in our consideration of mRNA maturation, the 5′ end is "capped" with a 7-methylguanosine residue. The presence of this "cap," absent in prokaryotes, is essential to efficient translation, as RNAs lacking the cap are translated poorly. In addition, most eukaryotic mRNAs contain a short recognition sequence that surrounds the initiating AUG codon—5′-ACCAUGG. Named after Marilyn Kozak, who discovered it, the Kozak sequence appears to function during initiation in the same way that the Shine–Dalgarno sequence functions in prokaryotic mRNA. Both greatly facilitate the initial binding of mRNA to the small subunit of the ribosome.

Another difference is that the amino acid formylmethionine is not required for the initiation of eukaryotic translation. However, as in prokaryotes, the AUG triplet, which encodes methionine, is essential to the formation of the translational complex, and a unique transfer RNA ($tRNA_i^{met}$) is used during initiation.

Protein factors similar to those in prokaryotes guide the initiation, elongation, and termination of translation in eukaryotes. Many of these eukaryotic factors are clearly homologous to their counterparts in prokaryotes. However, a greater number of factors are usually required during each of these steps, and some are more complex than in prokaryotes.

Finally, recall that, in eukaryotes, a large proportion of the ribosomes are found in association with the membranes that make up the endoplasmic reticulum (forming what is referred to as rough ER). Such membranes are absent from the cytoplasm of prokaryotic cells. This association in eukaryotes facilitates the transport of newly synthesized proteins from the ribosomes directly into the channels of the endoplasmic reticulum. Recent studies using cryoelectron microscopy have established how this occurs. There is a tunnel in the large subunit of ribosomes that begins near the point at which the two subunits interface and exits near the back of the large subunit. The location of the tunnel within the large subunit is the basis for the belief that it provides the conduit for the movement of the newly synthesized polypeptide chain out of the ribosome. In studies in yeast, newly synthesized polypeptides enter the ER through a membrane channel formed by a specific protein, Sec61. This channel is perfectly aligned with the exit point of the ribosomal tunnel. In prokaryotes, the polypeptides are released by the ribosome directly into the cytoplasm.

6.5 The Initial Insight that Proteins Are Important in Heredity Was Provided by the Study of Inborn Errors of Metabolism

Now, let's consider how we know that proteins are the end products of genetic expression. The first insight into the role of proteins in genetic processes was provided by observations made by Sir Archibald Garrod and William Bateson early in the 20th century. Garrod was born into an English family of medical scientists. His father was a physician with a strong interest in the chemical basis of rheumatoid arthritis, and his eldest brother was a leading zoologist in London. It is not surprising, then, that as a practicing physician, Garrod became interested in several human disorders that seemed to be inherited. Although he also studied albinism and cystinuria, we will describe his investigation of the disorder **alkaptonuria**. Individuals afflicted with this disorder cannot metabolize the alkapton 2,5-dihydroxyphenylacetic acid, also known as homogentisic acid. As a result, an important metabolic pathway (Figure 6–10) is blocked. Homogentisic acid accumulates in cells and tissues and is excreted in the urine. The molecule's oxidation products are black and easily detectable in the diapers of newborns. The products tend to accumulate in cartilaginous areas, causing a darkening of the ears and nose. In joints, this deposition leads to a benign arthritic condition. Alkaptonuria is a rare, but not serious disease that persists throughout an individual's life.

Garrod studied alkaptonuria by increasing dietary protein or adding to the diet the amino acids phenylalanine or tyrosine, both of which are chemically related to homogentisic acid. Under such conditions, homogentisic acid levels increase in the urine of alkaptonurics, but not in unaffected individuals. Garrod concluded that normal individuals can break down, or catabolize, this alkapton, but that afflicted individuals cannot. By studying the disorder's pattern of inheritance, Garrod further concluded that alkaptonuria was inherited as a simple recessive trait.

On the basis of these conclusions, Garrod hypothesized that hereditary information controls chemical reactions in the body, and that the inherited disorders he studied are the result of alternative modes of metabolism. While *genes* and *enzymes* were not familiar terms during Garrod's time, he used the corresponding concepts of *unit factors* and *ferments*. Garrod published his initial observations in 1902.

Only a few geneticists, including Bateson, were familiar with, or referred to, Garrod's work. Garrod's ideas fit nicely with Bateson's belief that inherited conditions were caused by the lack of some critical substance. In 1909, Bateson published *Mendel's Principles of Heredity*, in which he linked ferments with heredity. However, for almost 30 years, most geneticists failed to see the relationship between genes and

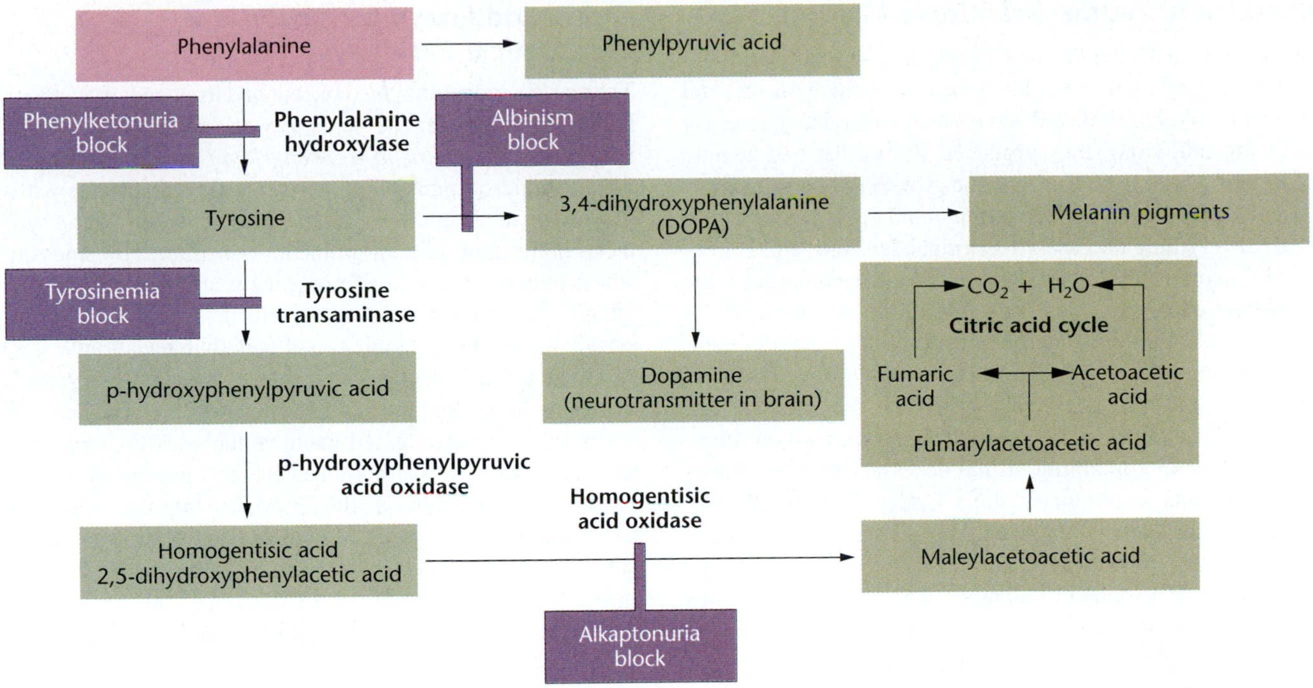

FIGURE 6–10 Metabolic pathway involving phenylalanine and tyrosine. Various metabolic blocks resulting from mutations lead to the disorders phenylketonuria, alkaptonuria, albinism, and tyrosinemia.

enzymes. Garrod and Bateson, like Mendel, were ahead of their time.

Phenylketonuria

The inherited human metabolic disorder, **phenylketonuria (PKU)**, results when another reaction in the pathway shown in Figure 6–10 is blocked. First described in 1934, this disorder can result in mental retardation and is transmitted as an autosomal recessive disease. Afflicted individuals are unable to convert the amino acid phenylalanine to the amino acid tyrosine. These molecules differ by only a single hydroxyl group (OH), present in tyrosine, but absent in phenylalanine. The reaction is catalyzed by the enzyme **phenylalanine hydroxylase**, which is inactive in affected individuals and active at a level of about 30 % in heterozygotes. The enzyme functions in the liver. While the normal blood level of phenylalanine is about 1 mg/100 ml, phenylketonurics show a level as high as 50 mg/100 ml.

As phenylalanine accumulates, it may be converted to phenylpyruvic acid and, subsequently, to other derivatives. These are less efficiently resorbed by the kidney and tend to spill into the urine more quickly than phenylalanine. Both phenylalanine and its derivatives enter the cerebrospinal fluid, resulting in elevated levels in the brain. The presence of these substances during early development is thought to cause mental retardation.

As a result of early detection based on PKU screening of newborns, retardation can be ameliorated. When the condition is detected in the analysis of an infant's blood, a strict dietary regimen is instituted. A low-phenylalanine diet can reduce by-products such as phenylpyruvic acid, and the abnormalities characterizing the disease can be diminished. The screening of newborns occurs routinely in all states of the United States. Phenylketonuria occurs in approximately 1 in 11,000 births.

Knowledge of inherited metabolic disorders, such as alkaptonuria and phenylketonuria, has caused a revolution in medical thinking and practice. Human disease, once thought to be attributed solely to the action of invading microorganisms, viruses, or parasites, clearly can have a genetic basis. We know now that literally thousands of medical conditions are caused by errors in metabolism that are the result of mutant genes. These human biochemical disorders include all classes of organic biomolecules.

6.6 Studies Of *Neurospora* Led to the One-Gene: One-Enzyme Hypothesis

In two separate investigations which began in 1933, George Beadle was to provide the first convincing experimental evidence that genes are directly responsible for the synthesis of enzymes. The first investigation, conducted in collaboration with Boris Ephrussi, involved *Drosophila* eye pigments. Together, Beadle and Ephrussi confirmed that mutant genes which altered the eye color of fruit flies could be linked to biochemical errors that, in all likelihood, involved the loss of enzyme function. Encouraged by these findings, Beadle then joined with Edward Tatum to investigate nutritional mutations in the pink bread mold *Neurospora crassa*. This investigation led to the **one-gene:one-enzyme hypothesis**.

Beadle and Tatum: *Neurospora* Mutants

In the early 1940s, Beadle and Tatum chose to work with *Neurospora* because much was known about its biochemistry, and mutations could be induced and isolated with relative ease. By inducing mutations, they produced strains that had genetic blocks of reactions essential to the growth of the organism.

Beadle and Tatum knew that this mold could manufacture nearly everything necessary for normal development. For example, using rudimentary carbon and nitrogen sources, the organism can synthesize nine water-soluble vitamins, 20 amino acids, numerous carotenoid pigments, and all essential purines and pyrimidines. Beadle and Tatum irradiated asexual conidia (spores) with X rays to increase the frequency of mutations and allowed them to be grown on "complete" medium containing all the necessary growth factors (e.g., vitamins, amino acids, etc.). Under such growth conditions, a mutant strain that would be unable to grow on minimal medium was able to grow by virtue of supplements present in the enriched complete medium. All the cultures were then transferred to minimal medium. If growth occurred on the minimal medium, the organisms were able to synthesize all the necessary growth factors themselves, and it was concluded that the culture did not contain a mutation. If no growth occurred, then it was concluded that the culture contained a nutritional mutation, and the only task remaining was to determine its type. Both cases are illustrated in Figure 6–11(a).

Many thousands of individual spores derived by this procedure were isolated and grown on complete medium. In subsequent tests on minimal medium, many cultures failed to grow, indicating that a nutritional mutation had been induced. To identify the mutant type, the mutant strains were tested on a series of different minimal media [Figure 6–11(b) and (c)], each containing groups of supplements, and subsequently on media containing single vitamins, amino acids, purines, or pyrimidines until one specific supplement that permitted growth was found. Beadle and Tatum reasoned that the supplement which restores growth is the molecule that the mutant strain could not synthesize.

The first mutant strain isolated required Vitamin B_6 (pyridoxine) in the medium, and the second one required vitamin B_1 (thiamine). Using the same procedure, Beadle and Tatum eventually isolated and studied hundreds of mutants deficient in the ability to synthesize other vitamins, amino acids, or other substances.

The findings derived from testing over 80,000 spores convinced Beadle and Tatum that genetics and biochemistry have much in common. It seemed likely that each nutritional mutation caused the loss of the enzymatic activity that facilitates an essential reaction in wild-type organisms. It also appeared that a mutation could be found for nearly any enzymatically controlled reaction. Beadle and Tatum had thus provided sound experimental evidence for the hypothesis that one gene specifies one enzyme, an idea alluded to over 30 years earlier by Garrod and Bateson. With modifications, this concept was to become a major principle of genetics.

Genes and Enzymes: Analysis of Biochemical Pathways

The one-gene:one-enzyme concept and its attendant methods have been used over the years to work out many details of metabolism in *Neurospora, Escherichia coli*, and a number of other microorganisms. One of the first metabolic pathways to be investigated in detail was that leading to the synthesis of the amino acid arginine in *Neurospora*. By studying seven mutant strains, each requiring arginine for growth (arg^-), Adrian Srb and Norman Horowitz were able to ascertain a partial biochemical pathway that leads to the synthesis of such a molecule. Their work illustrates how genetic analysis can be used to establish biochemical information.

Srb and Horowitz tested each mutant strain's ability to reestablish growth if either citrulline or ornithine, two compounds with close chemical similarity to arginine, was used as a supplement to minimal medium. If either was able to substitute for arginine, they reasoned that it must be involved in the biosynthetic pathway of arginine. The researchers found that both molecules could be substituted in one or more strains.

Of the seven mutant strains, four of them (*arg 4–7*) grew if supplied with either citrulline, ornithine, or arginine. Two of them (*arg 2* and *arg 3*) grew if supplied with citrulline or arginine. One strain (*arg 1*) would grow only if arginine were supplied; neither citrulline nor ornithine could substitute for it. From these experimental observations, the following pathway and metabolic blocks for each mutation were deduced:

$$\text{Precursor} \xrightarrow[\text{Enzyme A}]{arg\ 4-7} \text{Ornithine} \xrightarrow[\text{Enzyme B}]{arg\ 2-3} \text{Citrulline} \xrightarrow[\text{Enzyme C}]{arg\ 1} \text{Arginine}$$

The reasoning supporting these conclusions is based on the following logic: If mutants *arg 4* through *arg 7* can grow regardless of which of the three molecules is supplied as a supplement to minimal medium, the mutations preventing growth must cause a metabolic block that occurs *prior* to the involvement of ornithine, citrulline, or arginine in the pathway. When any one of these three molecules is added, its presence bypasses the block. As a result, it can be concluded that both citrulline and ornithine are involved in the biosynthesis of arginine. However, the sequence of their participation in the pathway cannot be determined on the basis of these data.

On the other hand, both the *arg 2* and the *arg 3* mutations grow if supplied citrulline, but not if they are supplied with only ornithine. Therefore, ornithine must occur in the pathway *prior* to the block. Its presence will not overcome the block. Citrulline, however, does overcome the block, so it must be involved beyond the point of blockage. Therefore, the conversion of ornithine to citrulline represents the correct sequence in the pathway.

Finally, we can conclude that *arg 1* represents a mutation preventing the conversion of citrulline to arginine. Neither ornithine nor citrulline can overcome the metabolic block because both participate earlier in the pathway.

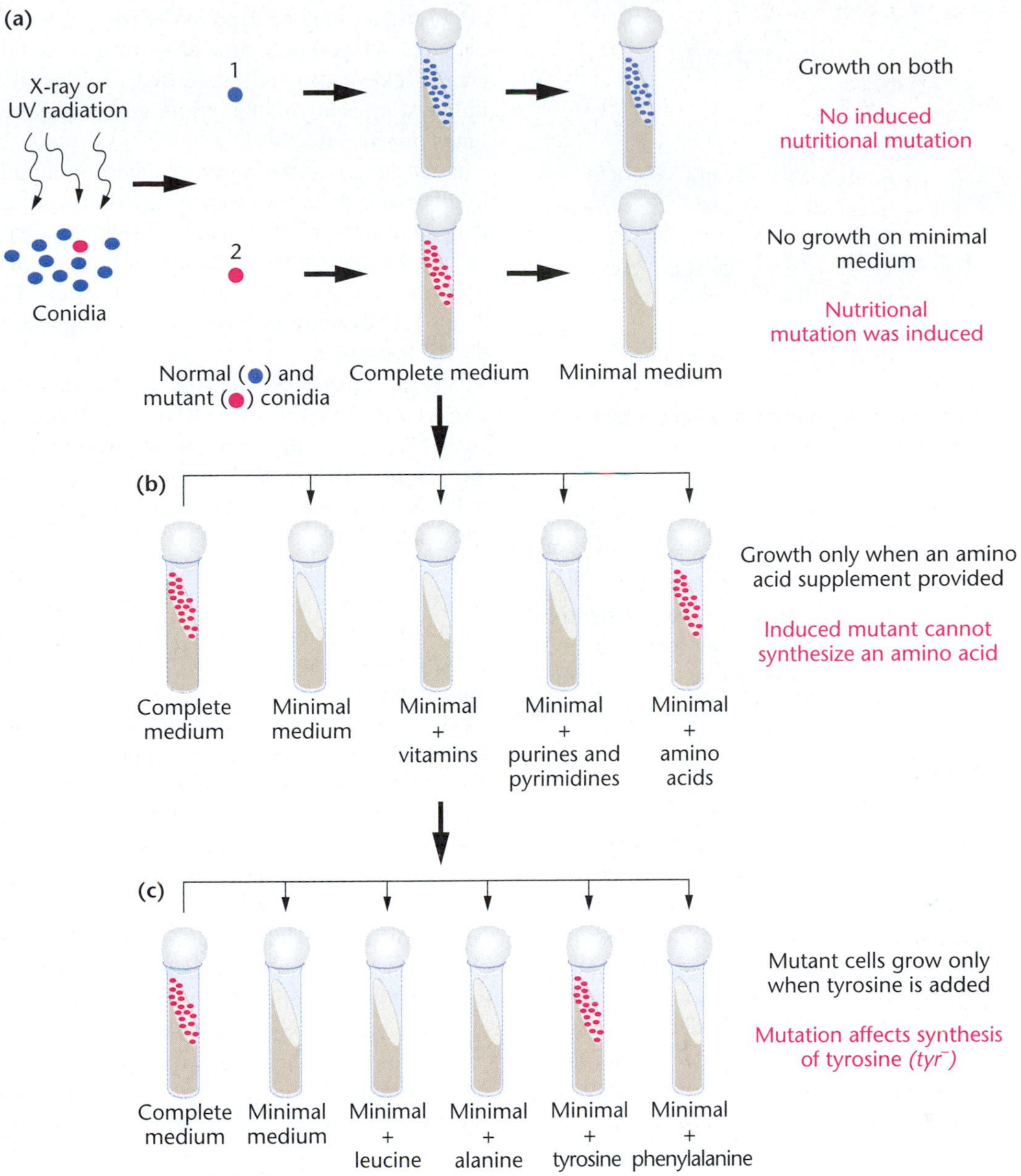

FIGURE 6–11 Induction, isolation, and characterization of a nutritional auxotrophic mutation in *Neurospora*. In (a), most conidia are not affected, but one conidium (shown in red) contains such a mutation. In (b) and (c), the precise nature of the mutation is determined to involve the biosynthesis of tyrosine.

Taken together, these reasons support the sequence of biosynthesis outlined here. Since Srb and Horowitz's work in 1944, the detailed pathway has been worked out and the enzymes controlling each step characterized. The abbreviated metabolic pathway is shown in Figure 6–12.

6.7 Studies of Human Hemoglobin Established that One Gene Encodes One Polypeptide

The one-gene:one-enzyme concept developed in the early 1940s was not immediately accepted by all geneticists. This is not surprising because it was not yet clear how mutant enzymes could cause variation in many phenotypic traits. For example, *Drosophila* mutants demonstrated altered eye size, wing shape, wing vein pattern, and so on. Plants exhibited mutant varieties of seed texture, height, and fruit size. How an inactive, mutant enzyme could result in such phenotypes was puzzling to many geneticists.

Two factors soon modified the one-gene:one-enzyme hypothesis. First, while nearly all enzymes are proteins, not all proteins are enzymes. As the study of biochemical genetics proceeded, it became clear that all proteins are specified by the information stored in genes, leading to the more accurate phraseology **one-gene:one-protein**. Second, proteins were often shown to have a subunit structure consisting of two or more polypeptide chains. This is the basis of the

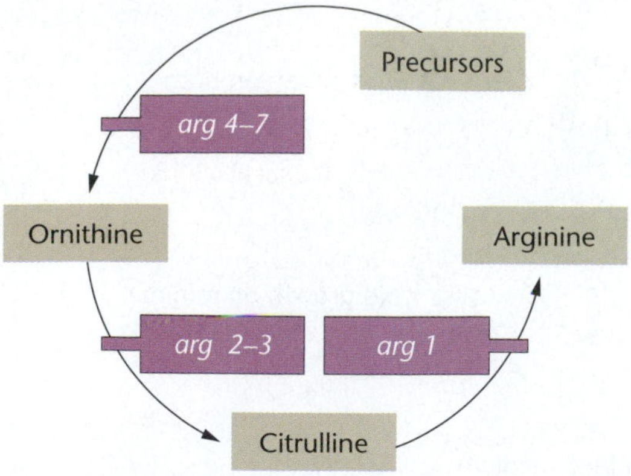

FIGURE 6–12 Abbreviated pathway resulting in the biosynthesis of arginine in *Neurospora*.

quarternary structure of proteins, which we will discuss later in the chapter. Because each distinct polypeptide chain is encoded by a separate gene, a more modern statement of Beadle and Tatum's basic principle is **one-gene:one-polypeptide chain**. These modifications of the original hypothesis became apparent during the analysis of hemoglobin structure in individuals afflicted with sickle-cell anemia.

Sickle-Cell Anemia

The first direct evidence that genes specify proteins other than enzymes came from the work on mutant hemoglobin molecules derived from humans afflicted with the disorder **sickle-cell anemia**. Affected individuals contain erythrocytes that, under low oxygen tension, become elongated and curved because of the polymerization of hemoglobin. The "sickle" shape of these erythrocytes is in contrast to the biconcave disc shape characteristic of those in normal individuals (Figure 6–13). Individuals with the disease suffer attacks when red blood cells aggregate in the venous side of capillary systems, where oxygen tension is very low. As a result, a variety of tissues may be deprived of oxygen and suffer severe damage. When this occurs, an individual is said to experience a sickle-cell crisis. If untreated, a crisis may be fatal. The kidneys, muscles, joints, brain, gastrointestinal tract, and lungs may be affected.

In addition to suffering crises, these individuals are anemic because their erythrocytes are destroyed more rapidly than are normal red blood cells. Compensatory physiological mechanisms include increased red-cell production by bone marrow and accentuated heart action. These mechanisms lead to abnormal bone size and shape as well as dilation of the heart.

In 1949, James Neel and E.A. Beet demonstrated that the disease is inherited as a Mendelian trait. Pedigree analysis revealed three genotypes and phenotypes controlled by a single pair of alleles, Hb^A and Hb^S. Normal and affected individuals result from the homozygous genotypes Hb^AHb^A and Hb^SHb^S, respectively. The red blood cells of heterozygotes, who exhibit the **sickle-cell trait**, but not the disease, undergo much less sickling because over half of their hemoglobin is normal. Although largely unaffected, such heterozygotes are "carriers" of the defective gene, which is transmitted, on average, to 50 percent of their offspring.

In the same year, Linus Pauling and his coworkers provided the first insight into the molecular basis of the disease. They showed that hemoglobins isolated from diseased and normal individuals differ in their rates of electrophoretic migration. In this technique (see Chapter 2 and Appendix A), charged molecules migrate in an electric field. If the net charge of two molecules is different, their rates of migration will be different. Hence, Pauling and his colleagues concluded that a chemical difference exists between normal and sickle-cell hemoglobin. The two molecules are now designated **HbA** and **HbS**, respectively.

Figure 6–14(a) illustrates the migration pattern of hemoglobin derived from individuals of all three possible genotypes when subjected to **starch gel electrophoresis**. The gel

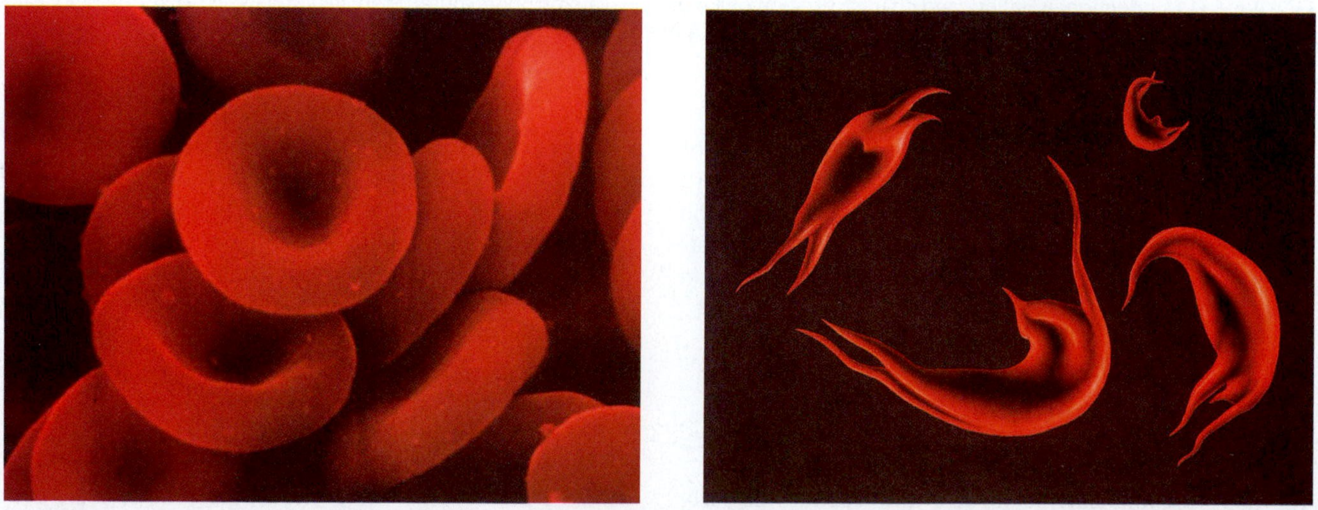

(a) (b)

FIGURE 6–13 A comparison of erythrocytes from normal individuals (a) and from individuals with sickle-cell anemia (b).

provides the supporting medium for the molecules during migration. In this experiment, samples are placed at a point of origin between the cathode $(-)$ and the anode $(+)$, and an electric field is applied. The migration pattern reveals that all molecules move toward the anode, indicating a net negative charge. However, HbA migrates farther than HbS, suggesting that its net negative charge is greater. The electrophoretic pattern of hemoglobin derived from carriers reveals the presence of both HbA and HbS, confirming their heterozygous genotype.

Pauling's findings suggested two possibilities. It was known that hemoglobin consists of four nonproteinaceous, iron-containing heme groups and a globin portion containing four polypeptide chains. The alternation in net charge in HbS could be due, theoretically, to a chemical change in either component.

Work carried out between 1954 and 1957 by Vernon Ingram resolved the question. He demonstrated that the chemical change occurs in the primary structure of the globin portion of the hemoglobin molecule. Using the **fingerprinting technique**, Ingram showed that HbS differs in amino acid composition compared with HbA. Human adult hemoglobin contains two identical alpha (α) chains of 141 amino acids and two identical beta (β) chains of 146 amino acids in its quaternary structure.

The fingerprinting technique involves the enzymatic digestion of the protein into peptide fragments. The mixture is then placed on absorbent paper and exposed to an electric field, where migration occurs according to net charge. The paper is then turned at a right angle and placed in a solvent, in which chromatographic action causes the migration of the peptides in the second direction. The end result is a two-dimensional separation of the peptide fragments into a distinctive pattern of spots or a "fingerprint." Ingram's work revealed that HbS and HbA differed by only a single peptide fragment [Figure 6–14(b)]. Further analysis then revealed a single amino acid change: Valine was substituted for glutamic acid at the sixth position of the β chain, accounting for the peptide difference [Figure 6–14(c)].

The significance of this discovery has been multifaceted. It clearly establishes that a single gene provides the genetic information for a single polypeptide chain. Studies of HbS also demonstrate that a mutation can affect the phenotype by directing a single amino acid substitution. Also, by providing the explanation for sickle-cell anemia, the concept of inherited **molecular disease** was firmly established. Finally, this

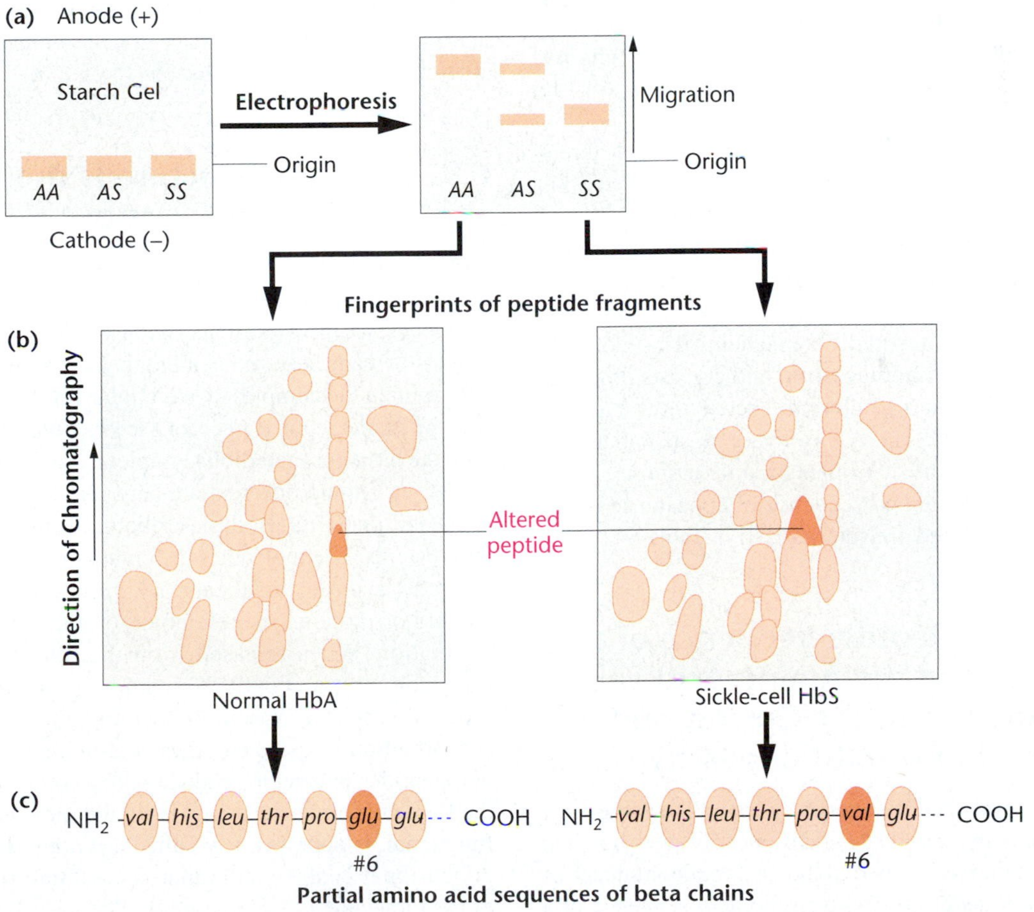

FIGURE 6–14 Investigation of hemoglobin derived from Hb^AHb^A and Hb^SHb^S individuals using electrophoresis, fingerprinting, and amino acid analysis. Hemoglobin from individuals with sickle-cell anemia (Hb^SHb^S) (a) migrates differently in an electrophoretic field, (b) shows an altered peptide in fingerprint analysis, and (c) shows an altered amino acid, valine, at the sixth position in the β chain. During electrophoresis, heterozygotes (Hb^AHb^S) reveal both forms of hemoglobin.

work led to a thorough study of human hemoglobins, which has provided valuable genetic insights.

In the United States, sickle-cell anemia is found almost exclusively in the African-American population. It affects about one in every 625 African-American infants. Currently, about 50,000 to 75,000 individuals are afflicted. In about 1 of every 145 African-American married couples, both partners are heterozygous carriers. In these cases, each of their children has a 25 percent chance of having the disease.

Human Hemoglobins

Having introduced human hemoglobins in an historical context, it may be useful to extend our discussion and provide an update of what is currently known about these molecules in our species. Molecular analysis reveals that a variety of hemoglobin molecules are produced in humans at different stages of the life cycle. All are tetramers consisting of numerous combinations of seven distinct polypeptide chains, each encoded by a separate gene. The expression of these various genes is developmentally regulated.

Almost all adult hemoglobin consists of **HbA**, which contains two $\boldsymbol{\alpha}$ and two $\boldsymbol{\beta}$ **chains**. Recall that the mutation in sickle-cell anemia involves the β chain, HbA represents about 98 percent of all hemoglobin found in an individual's erythrocytes after the age of six months. The remaining 2 percent consists of **HbA$_2$**, a minor adult component. This molecule contains two α chains and two **delta (δ) chains**. The δ chain is very similar to the β chain, consisting of 146 amino acids.

During embryonic and fetal development, quite a different set of hemoglobins is found. The earliest set to develop is called **Gower 1**, containing two **zeta (ζ) chains**, which are most similar to α chains, and two **epsilon (ϵ) chains**, which are most similar to β chains. By eight weeks gestation, the embryonic form is gradually replaced by still another hemoglobin molecule with still different chains. This molecule is called **HbF**, or **fetal hemoglobin**, and consists of two α chains and two **gamma (γ) chains**. There are two types of γ chains designated G_γ and A_γ. Both are most similar to β chains and differ from each other by only a single amino acid. The nomenclature and sequence of appearance of the five tetramers described are summarized in Table 6–2.

TABLE 6–2 Chain Compositions of Human Hemoglobins from Conception to Adulthood

Hemoglobin Type	*Chain Composition*
Embryonic-Gower 1	$\zeta_2\epsilon_2$
Fetal-HbF	$\alpha_2{}^{G}\gamma_2$
	$\alpha_2{}^{A}\gamma_2$
Adult-HbA	$\alpha_2\beta_2$
Minor adult-HbA$_2$	$\alpha_2\delta_2$

6.8 The Nucleotide Sequence of a Gene and the Amino Acid Sequence of the Corresponding Protein Exhibit Colinearity

Once it was established that genes specify the synthesis of polypeptide chains, the next logical question was how the genetic information contained in the nucleotide sequence of a gene can be transferred to the amino acid sequence of a polypeptide chain. It seemed most likely that a colinear relationship (the concept of **colinearity**) would exist between the two molecules. That is, the order of nucleotides in the DNA of a gene would correlate directly with the order of amino acids in the corresponding polypeptide.

The initial experimental evidence in support of this concept was derived from studies by Charles Yanofsky of the *trpA* gene that encodes the A subunit of the enzyme **tryptophan synthetase** in *E. coli*. Yanofsky isolated many independent mutants that had lost the activity of the enzyme, mapped them, and established their location with respect to one another within the gene. He then determined where the amino acid substitution had occurred in each mutant protein. When the two sets of data were compared, the colinear relationship was apparent. The location of each mutation in the *trpA* gene correlates with the position of the altered amino acid in the A polypeptide of tryptophan synthetase. This comparison is illustrated in Figure 6–15.

6.9 Protein Structure Is the Basis of Biological Diversity

Having established that the genetic information is stored in DNA and influences cellular activities through the proteins it encodes, we turn now to a brief discussion of protein structure. How can these molecules play such a critical role in determining the complexity of cellular activities? As we will see, the fundamental aspects of the structure of proteins provide the basis for incredible complexity and diversity. At the outset, we should differentiate between the terms **polypeptides** and **proteins**. Both describe molecules composed of amino acids. The molecules differ, however, in their state of assembly and functional capacity. *Polypeptides* are the precursors of proteins, as assembled on the ribosome during translation. When released from the ribosome following translation, a polypeptide folds up and assumes a higher order of structure. When this occurs, a three-dimensional conformation in space emerges, and in many cases, several polypeptides interact to produce such a conformation. When the final conformation is achieved, the molecule, now fully functional, is appropriately called a *protein*. It is the three-dimensional conformation that is essential to the function of the molecule.

The polypeptide chains of proteins, like nucleic acids, are linear nonbranched polymers. There are 20 amino acids that serve as the subunits (the building blocks) of proteins. Each

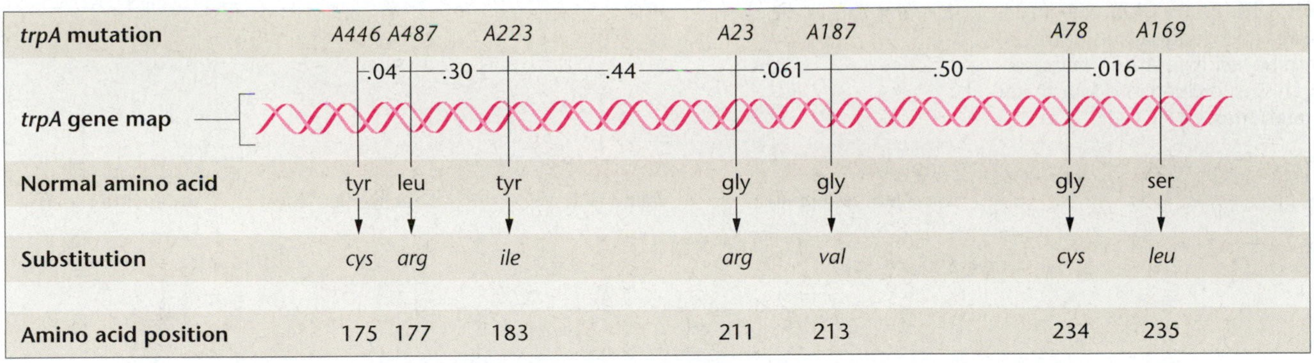

FIGURE 6–15 Demonstration of colinearity between the genetic map of various *trpA* mutations in *E. coli* and the affected amino acids in the protein product. The numbers shown between mutations represent linkage distances.

amino acid has a **carboxyl group**, an **amino group**, and a radical (R) **group** (a side chain) bound covalently to a **central carbon atom**. The R group gives each amino acid its chemical identity. Figure 6–16 illustrates the 20 R groups, which show a variety of configurations and can be divided into four main classes: (1) **nonpolar** (**hydrophobic**), (2) **polar** (**hydrophilic**), (3) **negatively charged**, and (4) **positively charged**. Because polypeptides are often long polymers, and because each position may be occupied by any one of 20 amino acids with unique chemical properties, enormous variation in chemical conformation and activity is possible. For example, if an average polypeptide is composed of 200 amino acids (a molecular weight of about 20,000 Da), 20^{200} different molecules, each with a unique sequence, can be created by using 20 different building blocks.

Around 1900, German chemist Emil Fischer determined the manner in which the amino acids are bonded together. He showed that the amino group of one amino acid can react with the carboxyl group of another amino acid during a dehydration reaction, releasing a molecule of H_2O. The resulting covalent bond is known as a peptide bond (Figure 6–17). Two amino acids linked together constitute a dipeptide, three a tripeptide, and so on. Once 10 or more amino acids are linked by peptide bonds, the chain may be referred to as a polypeptide. Generally, no matter how long a polypeptide is, it will contain an amino group at one end (the N-terminus) and a carboxyl group at the other end (the C-terminus).

Four levels of protein structure are recognized: primary, secondary, tertiary, and quaternary. The sequence of amino acids in the linear backbone of the polypeptide constitutes its **primary structure**. This sequence is specified by the sequence of deoxyribonucleotides in DNA through an mRNA intermediate. The primary structure of a polypeptide helps determine the specific characteristics of the higher orders of organization as a protein is formed.

The **secondary structure** refers to a regular or repeating configuration in space assumed by amino acids aligned closely to one another in the polypeptide chain. In 1951, Linus Pauling and Robert Corey predicted, on theoretical grounds, an **α-helix** as one type of secondary structure. The α-helix model [Figure 6–18(a)] has since been confirmed by X-ray crystallographic studies. It is rodlike and has the greatest possible theoretical stability. The helix is composed of a spiral chain of amino acids stabilized by hydrogen bonds.

The side chains (the R groups) of amino acids extend outward from the helix, and each amino acid residue occupies a vertical distance of 1.5 Å in the helix. There are 3.6 residues per turn. While left-handed helices are theoretically possible, all proteins demonstrating an α helix are right handed.

Also in 1951, Pauling and Corey proposed a second structure, **the β-pleated-sheet** configuration. In this model, a single polypeptide chain folds back on itself, or several chains run in either parallel or antiparallel fashion next to one another. Each such structure is stabilized by hydrogen bonds formed between atoms present on adjacent chains [Figure 6–18(b)]. A single zigzagging plane is formed in space with adjacent amino acids 3.5 Å apart.

As a general rule, most proteins demonstrate a mixture of α-helix and β-pleated-sheet structures. Globular proteins, most of which are round in shape and water soluble, usually contain a core of β-pleated-sheet structure, as well as many areas demonstrating helical structures. The more rigid structural proteins, many of which are water insoluble, rely on more extensive β-pleated-sheet regions for their rigidity. For example, **fibroin**, the protein made by the silk moth, depends extensively on this form of secondary structure.

While the secondary structure describes the arrangement of amino acids within certain areas of a polypeptide chain, **tertiary protein structure** defines the three-dimensional conformation of the entire chain in space. Each protein twists and turns and loops around itself in a very specific fashion, characteristic of the specific protein. Three aspects of this level of structure are most important in determining the conformation and in stabilizing the molecule:

1. Covalent disulfide bonds form between closely aligned cysteine residues to form the unique amino acid cystine.

2. Nearly all of the polar hydrophilic R groups are located on the surface, where they can interact with water.

FIGURE 6–16 Chemical structures and designations of the 20 amino acids found in living organisms, divided into four major categories.

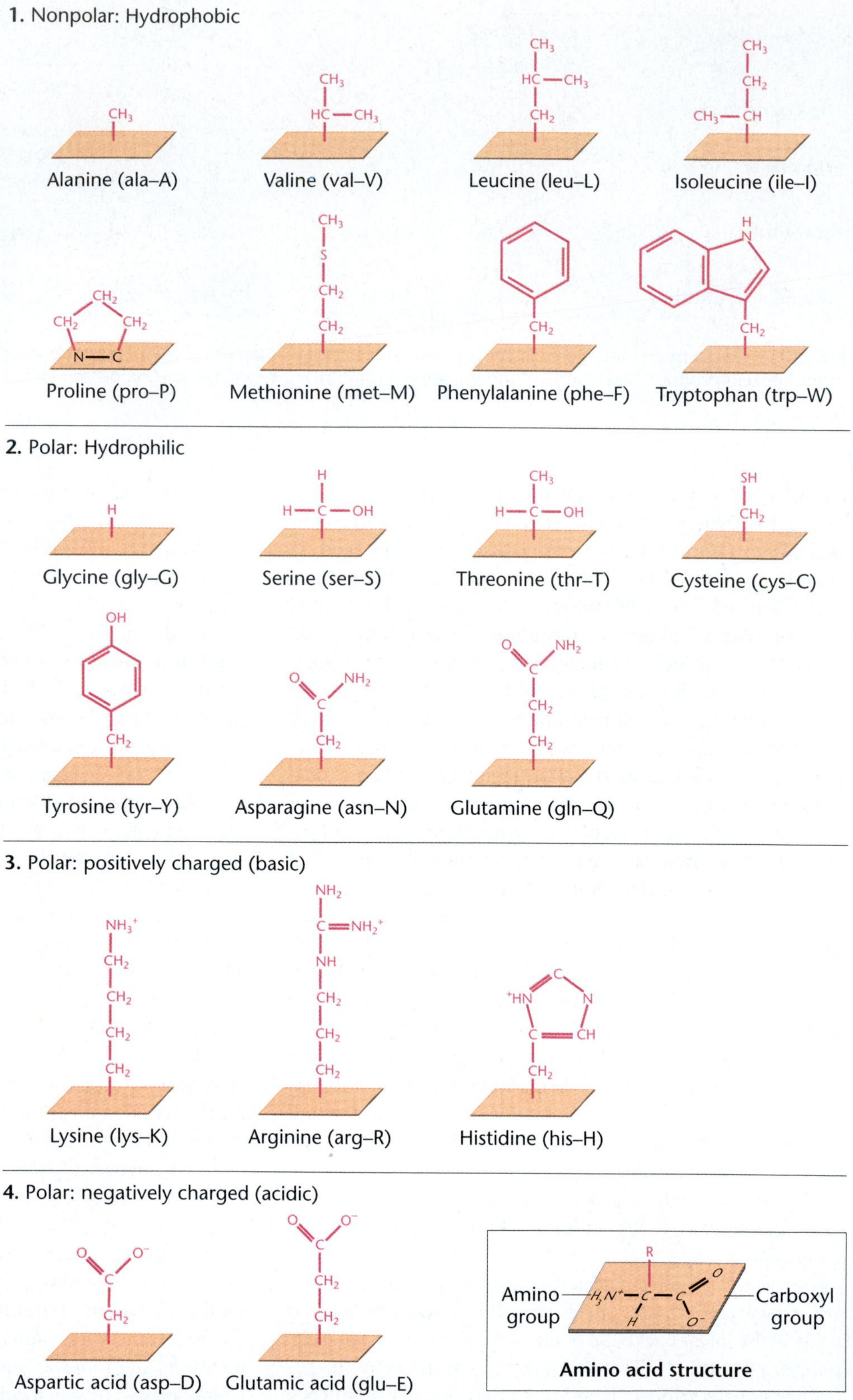

3. The nonpolar hydrophobic R groups are usually located on the inside of the molecule, where they interact with one another, avoiding interaction with water.

It is important to emphasize that the three-dimensional conformation achieved by any protein is a product of the primary structure of the polypeptide. Thus, the genetic code need only specify the sequence of amino acids in order to encode information that leads ultimately to the final assembly of proteins. The three stabilizing factors depend on the location of each amino acid relative to all others in the chain. As folding occurs, the most thermodynamically stable conformation possible results.

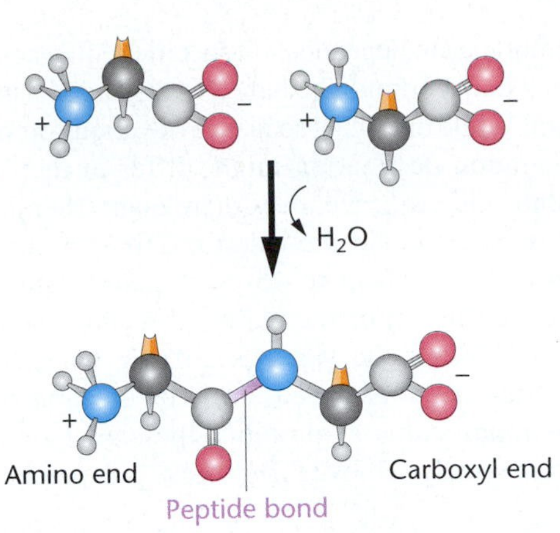

FIGURE 6–17 Peptide bond formation between two amino acids, resulting from a dehydration reaction.

A model of the three-dimensional tertiary structure of the respiratory pigment **myoglobin** is shown in Figure 6–19. This level of organization is extremely important because the specific function of any protein is directly related to its three-dimensional conformation.

The **quaternary level of organization** applies only to proteins composed of more than one polypeptide chain, indicating the conformation of the various chains in relation to one another. This type of protein is called oligomeric, and each chain is called a protomer, or, less formally, a subunit. The individual protomers have conformations that fit together with other subunits in a specific complementary fashion. Hemoglobin, an oligomeric protein consisting of four polypeptide chains, has been studied in great detail. Its quaternary structure is shown in Figure 6–20. Most enzymes, including DNA and RNA polymerase, demonstrate quaternary structure.

Posttranslational Modification

Before turning to a discussion of protein function, it is important to point out that polypeptide chains, like RNA transcripts, are often modified once they have been synthesized. This additional processing is broadly described as posttranslational modification. Although many of these alterations are detailed biochemical transformations and beyond the scope of our discussion, you should be aware that they occur and that they are critical to the functional capability of the final protein product. Several examples of posttranslational modification are presented here:

1. *The N-terminus and C-terminus amino acids are usually removed or modified.* For example, the initial N-terminal formylmethionine residue in bacterial polypeptides is usually removed enzymatically. Often, the amino group of the initial methionine residue is removed, and the amino group of the N-terminal residue is chemically modified (acetylated) in eukaryotic polypeptide chains.
2. *Individual amino acid residues are sometimes modified.* For example, phosphates may be added to the hydroxyl groups of certain amino acids, such as tyrosine. Modifications such as these create negatively charged residues that may ionically bond with other molecules. The process of phosphorylation is extremely important in regulating many cellular activities and is a result of the action of enzymes called **kinases**. In other proteins, methyl groups may be added enzymatically.

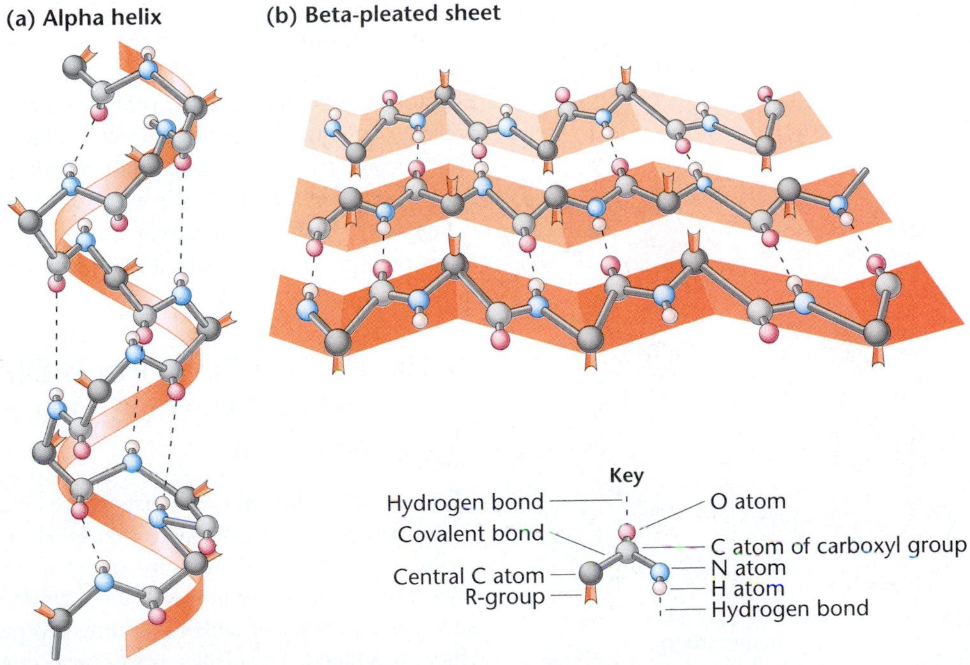

FIGURE 6–18 (a) The right-handed α-helix which represents one form of secondary structure of a polypeptide chain. (b) The β-pleated-sheet configuration, an alternative form of secondary structure of polypeptide chains. For the sake of clarity, some atoms are not shown.

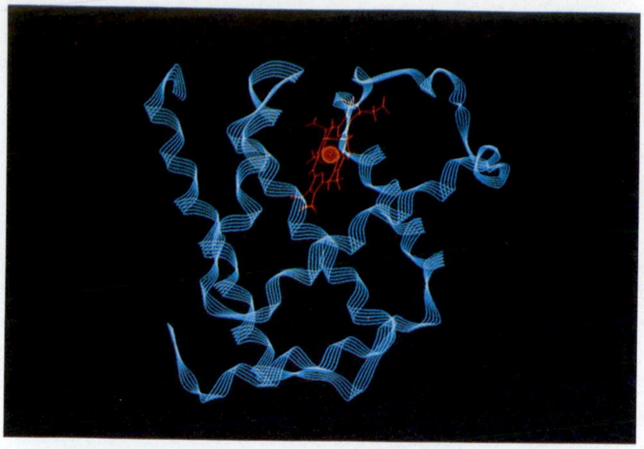

FIGURE 6–19 The tertiary level of protein structure in respiratory pigment myoglobin. The bound oxygen atom is shown in red.

3. *Carbohydrate side chains are sometimes attached.* These are added covalently, producing **glycoproteins**, an important category of molecules that includes many antigenic determinants, such as those specifying the antigens in the ABO blood-type system in humans.
4. *Polypeptide chains may be trimmed.* For example, insulin is first translated into a longer molecule that is enzymatically trimmed to its final 51-amino-acid form.
5. *Signal sequences are removed.* At the N-terminal end of some proteins, a sequence of up to 30 amino acids is found that plays an important role in directing the protein to the location in the cell in which it functions. This is called a **signal sequence**, and it determines the final destination of a protein in the cell. The process is called **protein targeting**. For example, proteins whose fate involves secretion or that are to become part of the plasma membrane are dependent on specific sequences for their initial transport into the lumen of the endoplasmic reticulum. While the signal sequence of various proteins with a common destination might differ in their primary amino acid sequence, they share many chemical properties. For example, those destined for secretion all contain a string of up to 15 hydrophobic amino acids preceded by a positively charged amino acid at the N-terminus of the signal sequence. Once the polypeptides are transported, but prior to achieving their functional status as proteins, the signal sequence is enzymatically removed from these polypeptides.
6. *Polypeptide chains are often complexed with metals.* The tertiary and quaternary levels of protein structure often include and are dependent on metal atoms. The function of the protein is thus dependent on the molecular complex that includes both polypeptide chains and metal atoms. Hemoglobin, containing four iron atoms along with four polypeptide chains, is a good example.

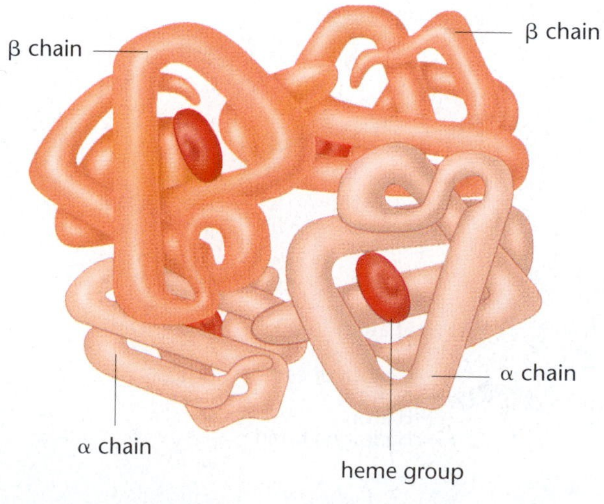

FIGURE 6–20 The quaternary level of protein structure as seen in hemoglobin. Four chains (two α and two β) interact with four heme groups to form the functional molecule.

These types of posttranslational modifications are obviously important in achieving the functional status specific to any given protein. Because the final three-dimensional structure of the molecule is intimately related to its specific function, how polypeptide chains ultimately fold into their final conformations is also an important topic. For many years, it was thought that protein folding was a spontaneous process whereby the molecule achieved maximum thermodynamic stability, based largely on the combined chemical properties inherent in the amino acid sequence of the polypeptide chain(s) composing the protein. However, numerous studies have shown that, for many proteins, folding is dependent upon members of a family of still other, ubiquitous proteins called **chaperones**. Chaperone proteins (sometimes called *molecular chaperones* or *chaperonins*) function to facilitate the folding of other proteins. While the mechanism by which chaperones function is not yet clear, like enzymes, they do not become part of the final product. Initially discovered in *Drosophila*, in which they are called heat-shock proteins, chaperones have been discovered in a variety of organisms, including bacteria, animals, and plants.

6.10 Protein Function Is Directly Related to the Structure of the Molecule

The essence of life on Earth rests at the level of diverse cellular function. One can argue that DNA and RNA simply serve as vehicles to store and express genetic information. However, proteins are at the heart of cellular function. And it is the capability of cells to assume diverse structures and functions that distinguishes most eukaryotes from less evolutionarily advanced organisms, such as bacteria. Therefore, an introductory understanding of protein function is critical to a complete view of genetic processes.

Proteins are the most abundant macromolecules found in cells. As the end products of genes, they play many diverse roles. For example, the respiratory pigments **hemoglobin** and myoglobin bind to oxygen. Hemoglobin transports the oxygen, which is essential for cellular metabolism. **Collagen** and **keratin** are structural proteins associated with the skin, connective tissue, and hair of organisms. **Actin** and **myosin** are contractile proteins, found in abundance in muscle tissue. Still other examples are the **immunoglobulins**, which function in the immune system of vertebrates; **transport proteins**, involved in movement of molecules across membranes; some of the **hormones** and their **receptors**, which regulate various types of chemical activity; and **histones**, which bind to DNA in eukaryotic organisms.

The largest group of proteins with a related function are the **enzymes**. Since we have referred to these molecules throughout this chapter, it may be useful to extend our discussion and include a more detailed description of their biological role. These molecules specialize in catalyzing chemical reactions within living cells. Enzymes increase the rate at which a chemical reaction reaches equilibrium, but they do not alter the end point of the chemical equilibrium. Their remarkable, highly specific catalytic properties largely determine the metabolic capacity of any cell type. The specific functions of many enzymes involved in the genetic and cellular processes of cells are described throughout the text.

Biological catalysis is a process whereby the **energy of activation** for a given reaction is lowered (Figure 6–21). The energy of activation is the increased kinetic energy state that molecules must usually reach before they react with one another. While this state can be attained as a result of elevated temperatures, enzymes allow biological reactions to occur at lower physiological temperatures. Thus, enzymes make life, as we know it, possible.

The catalytic properties and specificity of an enzyme are determined by the chemical configuration of the molecule's **active site**. This site is associated with a crevice, a cleft, or a pit on the surface of the enzyme, which binds the reactants, or substrates, facilitating their interaction. Enzymatically catalyzed reactions control metabolic activities in the cell. Each reaction is either **catabolic** or **anabolic**. Catabolism is the degradation of large molecules into smaller, simpler ones with the release of chemical energy. Anabolism is the synthetic phase of metabolism, yielding the various components that make up nucleic acids, proteins, lipids, and carbohydrates.

6.11 Proteins Are Made Up of One or More Functional Domains

We conclude this chapter by briefly discussing the finding that regions made up of specific amino acid sequences are associated with specific functions in protein molecules. Such sequences, usually between 50 and 300 amino acids, constitute what are called **protein domains** and are represented by modular portions of the protein that fold independently of the rest of the molecule into stable, unique conformations. Different domains impart different functional capabilities. Some proteins contain only a single domain, while others contain two or more.

The significance of domains rests at the tertiary structure level of proteins. Each such modular unit can be a mixture of secondary structures, including both α-helices and β-pleated-sheets. The unique conformation that is assumed in a single domain imparts a specific function to the protein. For example, a domain may serve as the catalytic basis of an enzyme, or it may impart the capability to bind to a specific ligand as part of a membrane or another molecule. Thus, in the study of proteins, you will hear of *catalytic domains*, *DNA-binding domains*, and so on. The result is that a protein must be looked at as being composed of a series of structural and functional modules. Obviously, the presence of multiple domains in a single protein increases the versatility of each molecule and adds to its functional complexity.

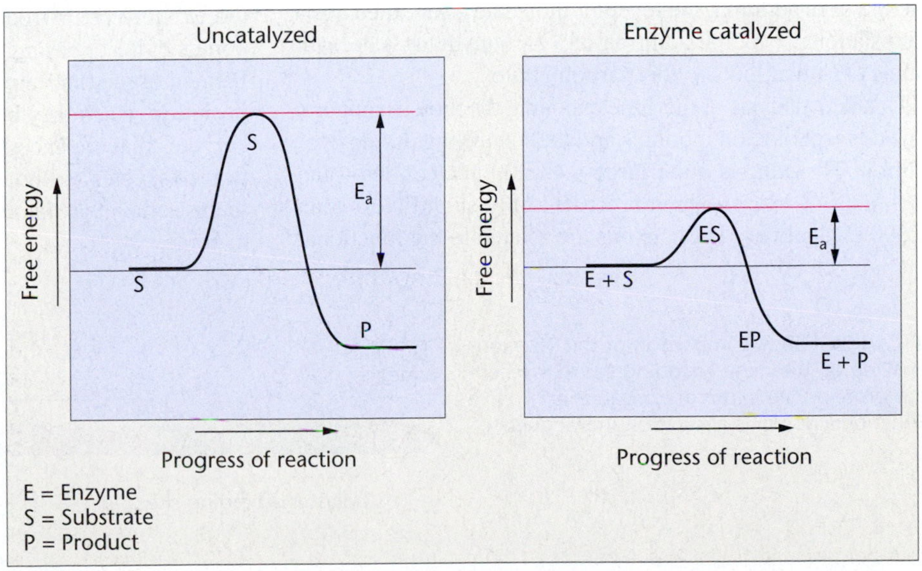

FIGURE 6–21 Energy requirements of an uncatalyzed versus an enzymatically catalyzed chemical reaction. The energy of activation (E_a) necessary to initiate the reaction is substantially lower as a result of catalysis.

Exon Shuffling and the Origin of Protein Domains

An interesting proposal to explain the genetic origin of protein domains was put forward by Walter Gilbert in 1977. Gilbert suggested that the functional regions of genes in higher organisms consist of collections of exons originally present in ancestral genes that are brought together through recombination during the course of evolution. Referring to the process as **exon shuffling**, Gilbert proposed that exons are modular in the sense that each might encode a single protein domain. Serving as the basis of the useful part of a protein, a single type of domain might function in a variety of proteins. In Gilbert's proposal, during evolution, many exons could be "mixed and matched" to form unique genes in eukaryotes.

Several observations lend support to this proposal. Most exons are fairly small, averaging about 150 nucleotides and encoding about 50 amino acids. Such a size is consistent with the production of many functional domains in proteins. Second, recombinational events that lead to exon shuffling would be expected to occur within areas of genes represented by introns. Because introns are free to accumulate mutations without harm to the organism, recombinational events would tend to further randomize their nucleotide sequences. Over extended evolutionary period, diverse sequences would tend to accumulate. This is, in fact, what is observed; introns range from 50 to 20,000 bases and exhibit fairly random base sequences.

Since 1977, a serious research effort has been directed toward the analysis of gene structure. In 1985, more direct evidence in favor of Gilbert's proposal of exon modules was presented. For example, the human gene encoding the membrane receptor for low-density lipoproteins (LDL) was isolated and sequenced. The LDL receptor protein is essential to the transport of plasma cholesterol into the cell. It mediates endocytosis and is expected to have numerous functional domains. These include the capability of this protein to bind specifically to the LDL substrates and to interact with other proteins at different levels of the membrane during transport across it. In addition, the receptor molecule is modified posttranslationally by the addition of a carbohydrate; a domain must exist that links to this carbohydrate.

Detailed analysis of the gene encoding the protein supports the concept of exon modules and their shuffling during evolution. The gene is quite large—45,000 nucleotides—and contains 18 exons, which represent only slightly less than 2600 nucleotides. These exons are related to the functional domains of the protein *and* appear to have been recruited from other genes during evolution.

Figure 6–22 illustrates these relationships. The first exon encodes a signal sequence that is removed from the protein before the LDL receptor becomes part of the membrane. The next five exons represent the domain specifying the binding site for cholesterol. This domain is made up of a 40-amino-acid sequence repeated seven times. The next domain consists of a sequence of 400 amino acids bearing a striking homology to the mouse peptide hormone epidermal growth factor (EGF). This region is encoded by eight exons and contains three repetitive sequences of 40 amino acids. A similar sequence is also found in three blood-clotting proteins. The 15th exon specifies the domain for the posttranslational addition of the carbohydrate, while the remaining 2 exons specify regions of the protein that are part of the membrane, anchoring the receptor to specific sites called coated pits on the cell surface.

These observations concerning the LDL exons are fairly compelling in support of the theory of exon shuffling during evolution. Certainly, there is no disagreement concerning the concept of protein domains being responsible for specific molecular interactions.

What remains controversial and evocative in the exon-shuffling theory is the question of when introns first appeared on the evolutionary scene. In 1978, W. Ford Doolittle proposed that these intervening sequences were part of the genome of the most primitive ancestors of modern-day eukaryotes. In support of the idea, Gilbert has argued that if intron similarities in a DNA sequence are found in identical positions within genes shared by totally unrelated eukaryotes (such as humans, chickens, and corn), they must have also been present in primitive ancestral genomes.

If Doolittle's proposal is correct, why are introns absent in most prokaryotes? Gilbert argues that they were present at one point during evolution, but as the genome of these primitive organisms evolved, they were lost. This occurred as a result of strong selection pressure to streamline their chromosomes to minimize the energy expenditure supporting replication and gene expression. Further, streamlining led to more error-free mRNA production. However, supporters of the opposing "intron-late" school, including Jeffrey Palmer, argue that introns first appeared much later during evolution, when they became a part of a single group of eukaryotes that are ancestral to modern-day members, but not to prokaryotes. Although it may be difficult to resolve, it seems certain that this controversy will persist for many years to come.

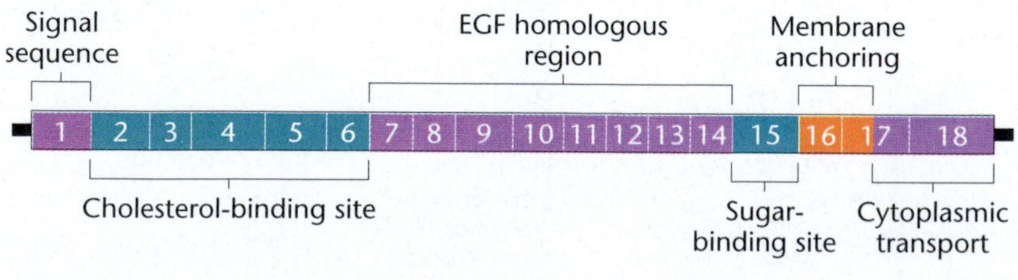

FIGURE 6–22 A comparison of the 18 exons making up the gene encoding the LDL receptor protein. The exons are organized into five functional domains and one signal sequence.

Genetics, Technology, and Society

Mad Cows and Heresies: The Prion Story

On March 20, 1996, the British government announced that a new brain disease had killed 10 young Britons, and that the victims might have caught the disease by eating infected beef. The disease, known as "mad cow disease" or bovine spongiform encephalopathy (BSE), slowly destroys brain cells and is always fatal. Recent studies confirm that BSE and the human disease, known as new variant Creutzfeldt–Jakob disease[1] (nvCJD), are so similar at the molecular and pathological levels that they are very likely the same disease. Mad cow disease triggered political turmoil in Europe, a worldwide ban on British beef, and the near collapse of the $8.9-billion British beef industry. The European union demanded the slaughter and incineration of 4.7 million British cattle, a campaign that cost the government over $12 billion to compensate farmers, import milk, and buy cattle for new herds. BSE and nvCJD have recently appeared in France, several other European countries, and South Africa. At least 90 Europeans have now died of nvCJD, and epidemiologists estimate that between 10,000 and 500,000 cases of nvCJD may appear in the next two decades.

BSE, nvCJD, and another condition, CJD, are all members of a group of neurological diseases known as spongiform encephalopathies, which affect animals (BSE) and humans (CJD and nvCJD). In this group of diseases, the infected brain tissue resembles a sponge (hence "spongiform") and is riddled with proteinaceous deposits. Victims of the disease lose motor function, become demented, and eventually die. CJD and nvCJD differ symptomatically. The incubation period for CJD is 20 to 30 years. By contrast, nvCJD has a shorter period of incubation, so its victims experience the onset of depression and dementia, as well as loss of coordination, at an earlier stage of the disease. nvCJD is linked to consuming BSE-contaminated cattle; no such link has been found for CJD. CJD strikes people over the age of 55, whereas, numerous nvCJD cases were seen in people under the age of 55, many even as young as the early teens. A number of CJD cases arise spontaneously and randomly, at a rate of 1 per million per year worldwide. A less common form of CJD is inherited as an autosomal dominant condition, but the transmitted forms (although not from cattle) are the strangest.

CJD can be transmitted through corneal or nervous tissue grafts or by injection of growth hormone derived from human pituitary glands. Kuru, a CJD-like disease of the Fore people of New Guinea, was transmitted from person to person through ritualistic cannibalism. Transmissible spongiform encephalopathies in animals include scrapie (sheep and goats), chronic wasting disease (deer and elk), and BSE. Like Kuru, BSE and scrapie are passed from animal to animal by ingestion of diseased animal remains, particularly neural tissue. The epidemic of BSE in Britain occurred because diseased cows and sheep were processed and fed to cattle as a protein supplement. In 1998, the British government banned the use of cows and sheep as feed for other cows and sheep, and the epidemic of BSE is subsiding. The European Union recently approved a temporary ban on the use of animal-containing feeds for all livestock, although the ban is nonbinding. U.S. and Canadian regulations still allow nonruminant animals to consume feed containing ruminant meat products, and ruminant animals to consume feed containing nonruminant animals as well as certain ruminant by-products, such as blood, gelatin, and fat. Regulations may change, as recent studies suggest that BSE may spread between cows, pigs, chickens, and other livestock, or from cows to calves. The U.S. Department of Agriculture has banned importation of cattle or meat products from countries that have BSE. At present, no cases of BSE have been reported in the United States. Due to concerns that nvCJD may be spread by blood transfusions, Canada and the United States banned blood donations from persons who spent more than 6 months in Britain between 1980 and 1986.

For many years, the spongiform encephalopathies defied the best efforts of scientists to analyze them. The diseases are difficult to study, because they require injection of infected brain material into the brains of experimental animals and the diseases take months or years to develop. In addition, the infectious agent is apparently not a virus or bacterium, and infected animals do not develop antibodies against these mysterious agents. There are no treatments for the disease, and the only way to make a firm diagnosis is to examine brain tissue after death. The infectious material is unaffected by radiation or nucleases that damage nucleic acids; however, it is destroyed by some reagents that hydrolyze or modify proteins. In the early 1980s, American scientist Stanley Prusiner purified the infectious agent and concluded that it consists of only protein. He proposed that scrapie is spread by an infectious protein particle that he called a prion. His hypothesis was dismissed by most scientists, as the idea of an infectious agent that contains no DNA or RNA as genetic material was heretical. However, Prusiner and others presented evidence supporting the prion hypothesis, and the notion that the disease can be transmitted by an infectious particle that contains no genetic material has gained acceptance.

If prions are composed only of protein, how do they cause disease? The answer may be as strange as the disease itself. The protein that makes up a prion (PrP) is a version of a normal protein that is synthesized in neurons and found in the brains of all adult animals. The difference between normal PrP and prion PrP lies in their protein secondary structures. Normal, noninfectious PrP folds into α-helices, whereas infectious prion PrP folds into β-sheets. When a normal PrP molecule contacts a prion PrP molecule, the normal protein is somehow unfolded and refolded into the abnormal PrP conformation. Once the normal PrP molecule has been transformed into an abnormal PrP molecule, it spreads its lethal conformation to neighboring normal PrP molecules, and the process takes off in a chain reaction. Normal PrP is a soluble protein that is easily destroyed by heat or enzymes that digest proteins. However, abnormal infectious PrP is insoluble in detergents, resists both heat and protease digestion, and is nearly indestructible. Hence, spongiform encephalopathies can be considered diseases of protein secondary structure.

Many urgent questions need to be addressed. How extensive is BSE contamination of the world's food supply? How many humans are infected with nvCJD, but

[1] Creutzfeldt–Jacob disease (CJD) was first identified by Hans Gerhand Creutzfeldt and Alphons Maria Jakob in the 1920s.

don't yet show symptoms? Can humans or animals act as asymptomatic carriers of prion diseases? Can prions exist in other parts of the body besides the brain and spinal cord, and if so, can nvCJD spread through blood transfusions, from mother to fetus, or by sterilized surgical instruments? Can we develop diagnostic tests and therapies for BSE and nvCJD? Are we near the end of the BSE story, or is it just beginning.

References

Almond, J., and Pattison, J. 1997. Human BSE. *Nature* 389:437–438.

Balter, M. 2000. Tracking the human fallout from "mad cow disease." *Science* 289:1452–1454.

Belay, E.D. 1999. Transmissible spongiform encephalopathies in humans. *Annu. Rev. Microbiol.* 53:283–314.

Prusiner, S.B. 1998. Prions. *Proc. Natl. Acad. Sci. U.S.A.* 95:15363–83.

Web sites

The BSE Inquiry Report (released October 26, 2000), *http://www.bse.org.uk*.

The Official Mad Cow Disease Home Page (a compilation of news reports), *http://www.mad-cow.org*.

Chapter Summary

1. Translation describes the synthesis in cells of polypeptide chains, under the direction of mRNA and in association with ribosomes. This process ultimately converts the information stored in the genetic code of the DNA making up a gene into the corresponding sequence of amino acids making up the polypeptide.
2. Translation is a complex energy-requiring process that also depends on charged tRNA molecules and numerous protein factors. Transfer RNA (tRNA) serves as the adaptor molecule between an mRNA triplet and the appropriate amino acid.
3. The processes of translation, like transcription, can be subdivided into the stages of initiation, elongation, and termination. The process relies on base-pairing affinities between complementary nucleotides and is more complex in eukaryotes than in prokaryotes.
4. The first insight that proteins are the end products of genes was provided by the study of inherited metabolic disorders in humans early in the 20th century by Garrod. Basic to his studies were inborn errors of metabolism leading to cystinuria, albinism, and alkaptonuria.
5. The investigation of nutritional requirements in *Neurospora* by Beadle and colleagues made it clear that mutations cause the loss of enzyme activity. Their work led to the concept of the one-gene:one-enzyme hypothesis.
6. The one-gene:one-enzyme hypothesis was later revised. Pauling and Ingram's investigations of hemoglobins from patients with sickle-cell anemia led to the discovery that one gene directs the synthesis of only one polypeptide chain.
7. Thorough investigations have revealed the existence of several major types of human hemoglobin molecules found in the embryo, fetus, and adult. Specific genes control each polypeptide chain constituting these various hemoglobin molecules.
8. The proposal suggesting that a gene's nucleotide sequence specifies in a colinear way the sequence of amino acids in a polypeptide chain was confirmed by experiments involving mutations in the tryptophan synthetase gene in *E. coli*.
9. Proteins, the end products of genes, demonstrate four levels of structural organization that together provide the chemical basis for their three-dimensional conformation, which is the basis of the molecule's function.
10. Of the myriad functions performed by proteins, the most influential role is assumed by enzymes. These highly specific, cellular catalysts play a central role in the production of all classes of molecules in living systems.
11. Proteins consist of one or more functional domains, which are shared by different molecules. The origin of these domains may be the result of exon shuffling during evolution.

Insights and Solutions

1. The growth responses that follow were obtained by using four mutant strains of *Neurospora* and the related compounds A, B, C, and D. None of the mutations grow on minimal medium. Draw all possible conclusions.

		Growth Product			
		C	*D*	*B*	*A*
Mutation	1	−	−	−	−
	2	−	+	+	+
	3	−	−	+	+
	4	−	−	+	−

Solution: First, nothing can be concluded about mutation 1, except that it is lacking some essential growth factor, perhaps even unrelated to the biochemical pathway represented by mutations 2–4; nor can anything be concluded about compound C. If it is involved in the pathway, it is a product synthesized prior to the synthesis of A, B, and D. We must now analyze these three compounds and the control of their synthesis by the enzymes encoded by genes 2, 3, and 4. Because product B allows growth in all three cases, it may be considered the "end product." It bypasses the block in all three instances. Using similar reasoning, product A precedes B in the pathway, since it allows a bypass in two of the three steps. Product D precedes B, yielding the more complete solution:

$$C\ (?) \rightarrow D \rightarrow A \rightarrow B$$

Now, determine which mutations control which steps. Since mutation 2 can be alleviated by products D, B, and A, it must control a step prior to all three products, perhaps the direct conversion to D, although we cannot be certain. Mutation 3 is alleviated by B and A, so its effect must precede them in the pathway. Thus, we will assign it as controlling the conversion of D to A. Likewise, we can assign mutation 4 to the conversion of A to B, leading to the more complete solution

$$C(?) \xrightarrow{2(?)} D \xrightarrow{3} A \xrightarrow{4} B$$

Problems and Discussion Questions

1. List and describe the role of all of the molecular constituents present in a functional polyribosome.
2. Contrast the roles of tRNA and mRNA during translation and list all enzymes that participate in the transcription and translation process.
3. Francis Crick proposed the "adaptor hypothesis" for the function of tRNA. Why did he choose that description?
4. During translation, what molecule bears the codon? The anticodon?
5. The α chain of eukaryotic hemoglobin is composed of 141 amino acids. What is the minimum number of nucleotides in an mRNA coding for this polypeptide chain? Assuming that each nucleotide is 0.34 nm long in the mRNA, how many triplet codes can, at one time, occupy space in a ribosome that is 20 nm in diameter?
6. Summarize the steps involved in charging tRNAs with their appropriate amino acids.
7. For it to carry out its role, each transfer RNA requires at least four specific recognition sites that must be inherent in its tertiary structure. What are they?
8. Discuss the potential difficulties involved in designing a diet to alleviate the symptoms of phenylketonuria.
9. Phenylketonurics cannot convert phenylalanine to tyrosine. Why don't these individuals exhibit a deficiency of tyrosine?
10. Phenylketonurics are often more lightly pigmented than are normal individuals. Can you suggest a reason why this is so?
11. Early detection and adherence to a strict dietary regime has relieved much of the mental retardation previously occurring in those afflicted with phenylketonuria (PKU). Affected individuals now often lead normal lives and have families. For various reasons, such individuals adhere less rigorously to their diet as they get older. Predict the effect of such a situation on newborns of mothers with PKU who neglect their diets.

12. A series of mutations in the bacterium *Salmonella typhimurium* results in the requirement of either tryptophan or some related molecule in order for growth to occur. From the data shown here, suggest a biosynthetic pathway for tryptophan:

	Growth Supplement				
Mutation	Minimal Medium	Anthra-nilic Acid	Indole Glycerol Phosphate	Indole	Tryptophan
trp-8	−	+	+	+	+
trp-2	−	−	+	+	+
trp-3	−	−	−	+	+
trp-1	−	−	−	−	+

13. The study of biochemical mutants in organisms such as *Neurospora* has demonstrated that some pathways are branched. The data shown here illustrate the branched nature of the pathway resulting in the synthesis of thiamine:

	Growth Supplement			
Mutation	*Minimal Medium*	*Pyrimidine*	*Thiazole*	*Thiamine*
thi-1	−	−	+	+
thi-2	−	+	−	+
thi-3	−	−	−	+

Why don't the data support a linear pathway? Can you postulate a pathway for the synthesis of thiamine in *Neurospora?*

14. Explain why the one-gene:one-enzyme concept is not considered totally accurate today.

15. Why is an alteration of electrophoretic mobility interpreted as a change in the primary structure of the protein under study?

16. Contrast the polypeptide-chain components of each of the hemoglobin molecules found in humans.

17. Using sickle-cell anemia as a basis, describe what is meant by a molecular or genetic disease. What are the similarities and dissimilarities between this type of a disorder and a disease caused by an invading microorganism?

18. Contrast the contributions of Pauling and Ingram to our understanding of the genetic basis for sickle-cell anemia.

19. Hemoglobins from two individuals are compared by electrophoresis and by fingerprinting. Electrophoresis reveals no difference in migration, but fingerprinting shows an amino acid difference. How is this possible?

20. Describe what colinearity means. Of what significance is the concept of colinearity in the study of genetics?

21. Certain mutations called *amber* in bacteria and viruses result in premature termination of polypeptide chains during translation. Many *amber* mutations have been detected at different points along the gene coding for a head protein in phage T4. How might this system be further investigated to demonstrate and support the concept of colinearity?

22. Define and compare the four levels of protein organization.

23. List as many different categories of protein functions as you can. Wherever possible, give an example of each category.

24. How does an enzyme function? Why are enzymes essential for living organisms on Earth?

25. Does Fiers' work with phage MS2, discussed in Chapter 5, constitute more direct evidence in support of colinearity than Yanofsky's work with the *trpA* locus in *E. coli* (discussed in this chapter)? Explain.

Hb Type	***Normal Amino Acid***	***Substituted Amino Acid***
HbJ Toronto	ala	asp (α-5)
HbJ Oxford	gly	asp (α-15)
Hb Mexico	gln	glu (α-54)
Hb Bethesda	tyr	his (β-145)
Hb Sydney	val	ala (β-67)
HbM Saskatoon	his	tyr (β-63)

26. Shown here are several amino acid substitutions in the α and β chains of human hemoglobin:
Using the code table (Figure 6–7), determine how many of them can occur as a result of a single nucleotide change.

Extra-Spicy Problems

27. In 1962, F. Chapeville and others reported an experiment in which they isolated radioactive ^{14}C-cysteinyl-tRNA^{cys} (charged tRNA^{cys} + cysteine). They then removed the sulfur group from the cysteine, creating alanyl-tRNA^{cys} (charged tRNA^{cys} + alanine). When alanyl-tRNA^{cys} was added to a synthetic mRNA calling for cysteine, but not alanine, a polypeptide chain was synthesized containing alanine. What can you conclude from this experiment? [*Reference*: Chapeville, et al., *Proc. Natl. Acad. Sci. USA* 48:1086–93 (1962).]

28. HbS results from the amino acid change of glutamic acid to valine at the number 6 position in the B chain of human hemoglobin. HbC is the result of a change at the same position in the B chain, but lysine replaces glutamic acid. Return to the genetic code table (Figure 12.7) and determine whether single nucleotide changes can account for these mutations. Then examine Figure 13-6 and examine the R groups in the amino acids glutamic acid, valine, and lysine. Describe the chemical differences between the three amino acids. Predict how the changes might alter the structure of the molecule and lead to altered hemoglobin function? HbS results in anemia and resistance to malaria, whereas in those with HbA, the parasite *Plasmodium falciparium* invades red blood cells and causes the disease. Predict whether those with HbC are likely to be anemic and whether they would be resistant to malaria.

Selected Readings

Anfinsen, C.B. 1973. Principles that govern the folding of protein chains. *Science* 181:223–30.

Bartholome, K. 1979. Genetics and biochemistry of phenylketonuria—Present state. *Hum. Genet.* 51:241–45.

Bateson, W. 1909. *Mendel's principles of heredity*. Cambridge: Cambridge University Press.

Beadle, G.W. 1945. Genetics and metabolism in *Neurospora*. *Physiol. Rev.* 25:643.

Beadle, G.W., and Tatum, E.L. 1941. Genetic control of biochemical reactions in *Neurospora*. *Proc. Natl. Acad. Sci. USA* 27:499–506.

Beckmann, R., et al. 1997. Alignment of conduits for the nascent polypeptide chain in the ribosome-Sec61 complex. *Science* 278:2123–26.

Beet, E.A. 1949. The genetics of the sickle-cell trait in a Bantu tribe. *Ann. Eugenics* 14:279–84.

Boyer, P.D., ed. 1974. *The enzymes*, Vol 10, 3d ed. Orlando, FL: Academic Press.

Branden, C., and Tooze, J. 1999. *Introduction to protein Structure*. NY: Garland Press.

Bray, D. 1995. Protein molecules as computational elements in living cells. *Nature* 376:307–12.

Brenner, S. 1955. Tryptophan biosynthesis in *Salmonella typhimurium*. *Proc. Natl. Acad. Sci. USA* 41:862–63.

Byers, P.H. 1989. Inherited disorders of collagen gene structure and expression. *Am. J. Med. Genet.* 34:72–80.

Chapeville, F., et al. 1962. On the role of soluble ribonucleic acid in coding for amino acids. *Proc. Natl. Acad. Sci. USA* 48:1086–93.

Cigan, A.M., Feng, L., and Donahue, T.F. 1988. $tRNA^{met}$ functions in directing the scanning ribosome to the start site of translation. *Science* 242:93–98.

Dahlberg, A.E. 1989. The functional role of ribosomal RNA in protein synthesis. *Cell* 57:525–29.

Dickerson, R.E. 1964. X-ray analysis and protein structure. In *The proteins*, ed. H. Neurath, Vol. 2, 2d ed. Orlando, FL: Academic Press.

Dickerson, R.E., and Geis, I. 1983. *Hemoglobin: Structure, function, evolution, and pathology*. Menlo Park, CA: Benjamin/Cummings.

Doolittle, R.F. 1985. Proteins. *Sci. Am.* (Oct.) 253:88–99.

Ezzell, C. 1994. Evolutions: Molecular chaperones and protein folding. *J. NIH Res.* 6:103.

Ephrussi, B. 1942. Chemistry of eye color hormones of *Drosophila*. *Quart. Rev. Biol.* 17:327–38.

Fisher, J., and Arnold, J. 1999. *Instant notes in chemistry for biologists*. NY: Springer Verlag.

Frank, J. 1998. How the ribosome works. *Amer. Scient.* 86:428–39.

Garrod, A.E. 1902. The incidence of alkaptonuria: A study in chemical individuality. *Lancet* 2:1616–20.

———. 1909. *Inborn errors of metabolism*. London: Oxford University Press. (Reprinted 1963, Oxford University Press, London.)

Garrod, S.C. 1989. Family influences on A.E. Garrod's thinking. *J. Inher. Metab. Dis.* 12:2–8.

Horton, H.R., et al. 1996. *Principles of biochemistry*, 2d ed. Upper Saddle River, NJ: Prentice Hall.

Ingram, V.M. 1957. Gene mutations in human hemoglobin: The chemical difference between normal and sickle-cell hemoglobin. *Nature* 180:326–28.

———. 1963. *The hemoglobins in genetics and evolution*. New York: Columbia University Press.

Koshland, D.E. 1973. Protein shape and control. *Sci. Am.* (Oct.) 229:52–64.

LaDu, B.N., Zannoni, V.G., Laster, L., and Seegmiller, J.E. 1958. The nature of the defect in tyrosine metabolism in alkaptonuria. *J. Biol. Chem.* 230:251.

Lake, J.A. 1981. The ribosome. *Sci. Am.* (Aug.) 245:84–97.

Lehninger, A.L., Nelson, D.L., and Cox, M.M. 1993. *Principles of biochemistry*, 2nd ed. New York: Worth Publishers.

Lodmell, J.S., and Dahlberg, A.E. 1997. A conformational switch in *Eschericia coli* 16S ribosomal RNA during decoding of messenger RNA. *Science* 277:1262–65.

Maniatis, T., et al. 1980. The molecular genetics of human hemoglobins. *Annu. Rev. Genet.* 14:145–78.

Moore, P.B. 1988. The ribosome returns. *Nature* 331:223–27.

Murayama, M. 1966. Molecular mechanism of red cell sickling. *Science* 153:145–49.

Neel, J.V. 1949. The inheritance of sickle-cell anemia. *Science* 110:64–66.

Nirenberg, M.W., and Leder, P. 1964. RNA codewords and protein synthesis. *Science* 145:1399–1407.

Noller, H.F. 1973. Assembly of bacterial ribosomes. *Science* 179:864–73.

———. 1984. Structure of ribosomal RNA. *Annu. Rev. Biochem.* 53:119–62.

Nomura, M. 1984. The control of ribosome synthesis. *Sci. Am.* (Jan.) 250:102–14.

Pauling, L., Itano, H.A., Singer, S.J., and Wells, I.C. 1949. Sickle-cell anemia: A molecular disease. *Science* 110:543–48.

Pfeffer, S.R., and Rothman, J.E. 1987. Biosynthetic protein transport and sorting by the endoplasmic reticulum and golgi. *Annu. Rev. Biochem.* 56:829–52.

Porce, B.T., and Garrett, R.A. 1999. Ribosomal mechanics, antibiotics, and GTP hydrolysis. *Cell* 97:423–26.

Prockop, D.J., and Kivirikko, K.I. 1995. Collagens: Molecular biology, diseases, and potentials for therapy. *Annu. Rev. Biochem.* 64:403–34.

Ramakrishnan, V. 2002. Ribosome structure and the mechanism of translation. *Cell* 108:557–72.

Rich, A., and Houkim, S. 1978. The three-dimensional structure of transfer RNA. *Sci. Am.* (Jan.) 238:52–62.

Rich, A., Warner, J.R., and Goodman, H.M. 1963. The structure and function of polyribosomes. *Cold Spring Harbor Symp. Quant. Biol.* 28:269–85.

Richards, F.M. 1991. The protein-folding problem. *Sci. Am.* (Jan.) 264:54–63.

Ross, J. 1989. The turnover of mRNA. *Sci. Am.* (Apr.) 260:48–55.

Rould, M.A., et al. 1989. Structure of *E. coli* glutaminyl-tRNA synthetase complexed with $tRNA^{gln}$ and ATP at 2.8 resolution. *Science* 246:1135–42.

Saks, M.E., Sampson, J.R., and Abelson, J.N. 1994. The transfer RNA identity problem: A search for rules. *Science* 263:191–97.

———. 1998. Evolution of a transfer RNA gene through a point mutation in the anticodon. *Science* 279:1665–70.

Scott-Moncrieff, R. 1936. A biochemical survey of some Mendelian factors for flower colour. *J. Genet.* 32:117–70.

Scriver, C.R., and Clow, C.L. 1980. Phenylketonuria and other phenylalanine hydroxylation mutants in man. *Annu. Rev. Genet.* 14:179–202.

Sharp, P.A., and , D. 1987. The evolution of catalytic function. *Science* 238:729–30.

Smith, J.D. 1972. Genetics of tRNA. *Annu. Rev. Genet.* 6:235–56.

Srb, A.M., and Horowitz, N.H. 1944. The ornithine cycle in *Neurospora* and its genetic control. *J. Biol. Chem.* 154:129–39.

Uchino, T., et al. 1995. Molecular basis of phenotypic variation in patients with argininemia. *Hum. Genet.* 96:255–60.

Wagner, R.P., and Mitchell, H.K. 1964. *Genetics and metabolism*, 2d ed. New York: John Wiley.

Warner, J., and Rich, A. 1964. The number of soluble RNA molecules on reticulocyte polyribosomes. *Proc. Natl. Acad. Sci. USA* 51:1134–41.

Wimberly, B.T., et al. 2000. Structure of the 30S ribosomal subunit. *Nature* 407:327–33.

Wittman, H.G. 1983. Architecture of prokaryotic ribosomes. *Annu. Rev. Biochem.* 52:35–65.

Yanofsky, C., Drapeau, G., Guest, J., and Carlton, B. 1967. The complete amino acid sequence of the tryptophan synthetase A protein and its colinear relationship with the genetic map of the A gene. *Proc. Natl. Acad. Sci. USA* 57:296–98.

Yusupov, M.M., et.al. 2001. Crystal structure of the ribosome at 5.5A resolution. *Science* 292:883–96.

Ziegler, I. 1961. Genetic aspects of ommochrome and pterin pigments. *Adv. Genet.* 10:349–403.

Zubay, G.L., and Marmur, J., eds. 1973. *Papers in biochemical genetics*, 2d ed. New York: Holt, Rinehart & Winston.

GENETICS MediaLab

The resources that follow will help you achieve a better understanding of the concepts presented in this chapter. These resources can be found either on the CD packaged with this textbook or on the Companion Web site found at **http://www.prenhall.com/klug**

CD Resources:

Module 6.1: Protein Translation

Module 6.2: Mutation, Gene/Protein Colinearity

Web Problem 1:

Time for completion = 5 minutes

What are the steps in the process of protein synthesis? As with DNA transcription, the process requires three steps. The translational machinery must initiate translation, elongate the polypeptide, and terminate the process. These stages of translation require the coordination of RNA and many protein molecules. The Web site provides an overview of this process and illustrates the dynamics involved by means of a short animation. After viewing the text and animation, describe the physical relationship between the tRNA, the mRNA, and the A and P sites of the ribosome. Discuss the role of the endoplasmic reticulum in translation. To complete this exercise, visit Web Problem 1 in Chapter 6 of your Companion Web site, and select the keyword **TRANSLATION**.

Web problem 2:

Time for completion = 15 minutes

How is protein synthesis regulated during early development? Many transcripts are produced during oogenesis for use at a later time in development. In this exercise, you will read about various mechanisms for controlling the use of transcripts in the oocytes of Xenopus, a genus of frog. Create a list of the different mechanisms controlling translation, indicate what portion of the transcript provides the regulatory information, and provide a specific example of each. To complete this exercise, visit Web Problem 2 in Chapter 6 of your Companion Web site, and select the keyword **TRANSLATIONAL REGULATION**.

Web problem 3:

Time for completion = 5 minutes

How is genetic information transferred in a cell? DNA replication, transcription, and translation are the events determining the flow of genetic information. While each process is distinct in many ways, there are similarities in the pattern of information processing. The examination of numerous cellular processes suggests that a general evolutionary strategy has been to repeat successful strategies in new contexts. In this exercise, you will view short animations of DNA transcription and protein translation. After watching these animations side by side, list as many common features of the processes as you can. To complete this exercise, visit Web Problem 3 in Chapter 6 of your Companion Web site, and select the keyword **INFORMATION**.

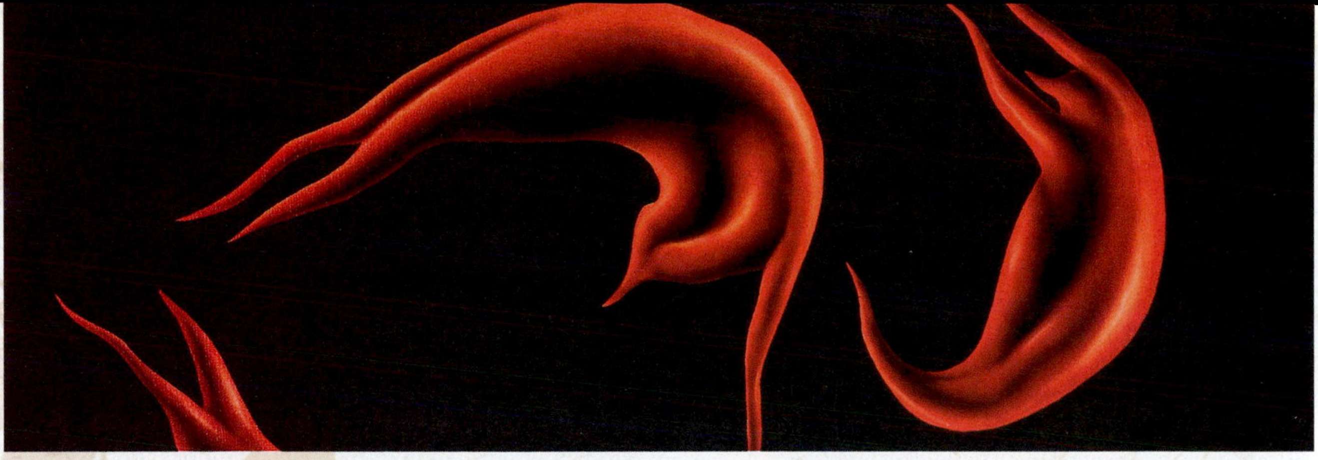

Mutant erythrocytes derived from an individual with sickle-cell anemia. (*Francis Leroy/ Biocosmos/Science Photo Library/Photo Researchers, Inc.*)

7

Gene Mutation, DNA Repair, and Transposable Elements

KEY CONCEPTS

In Chapter 2, we defined the four characteristics or functions ascribed to the genetic information: replication, storage, expression, and variation by mutation. In a sense, mutation is a failure to store genetic information faithfully. If a change occurs in the stored information, it may be reflected in the expression of that information and will be propagated by replication. Historically, the term mutation includes both chromosomal changes and changes within single genes. We will discuss the former alterations in Chapter 13, collectively referring to them as **chromosomal aberrations**. In this chapter, we are concerned with gene mutations. As you will see, a change may be as simple as the substitution, gain, or loss of a single nucleotide or as complex as the addition or deletion of many nucleotides within the normal sequence of DNA.

The term *mutation* was coined in 1901 by Hugo DeVries to describe the variation he observed in crosses involving the evening primrose, *Oenothera lamarckiana*. Most of the variation was due to multiple translocations, but two cases were subsequently shown to be caused by **gene mutations**—actual changes in the chemical composition of the DNA. As the studies of mutations progressed, it soon became clear that gene mutations serve as the source of most new alleles and thus are the origin of much of the genetic variability within populations. As new alleles arise, they constitute the raw material to be tested by the evolutionary process of natural selection, which will determine whether they are detrimental, neutral, or beneficial.

Mutations provide the basis for genetic studies. The resulting phenotypic variability provides the basis for geneticists to study the genes that control the modified traits. In genetic investigations, mutations serve as identifying "markers" of genes so that they may be followed during their transmission from parents to offspring. Without the phenotypic variability that mutations provide, genetic analysis would be impossible. For example, if all pea plants displayed a uniform phenotype, Mendel would have had no basis for his experimentation. Because of the importance of mutations, great attention has been given to their origin, induction, and classification.

Certain organisms lend themselves to the induction of mutations that can be easily detected and studied throughout reasonably short life cycles. Viruses, bacteria, fungi, fruit flies, certain plants, and mice fit these criteria. Thus, these organisms have been widely used to study mutation and mutagenesis, and through other studies, have also contributed to more general aspects of genetic knowledge.

Once we have completed our presentation of mutation, we will turn our attention to two related topics—DNA repair and transposable genetic elements. These topics are logical extensions of our consideration of gene mutation. Repair processes serve to counteract mutation. Transposable elements often disrupt the normal structure of the gene and therefore create mutations.

7.1 Mutations Are Usually, but Not Always, Spontaneous

Although it was known well before 1943 that pure cultures of bacteria could give rise to a small number of cells exhibiting heritable variation, particularly with respect to survival under different environmental conditions, the source of the variation was hotly debated. The majority of bacteriologists (now called microbiologists) believed that environmental factors induced changes in certain bacteria that led to their survival or adaptation to the new conditions. For example, strains of *Escherichia coli* are known to be sensitive to infection by the bacteriophage T1—that is, the virus reproduces at the expense of the bacterial cell, which is lysed or destroyed (See Figure 2–5). If a plate of *E. coli* is homogeneously sprayed with T1, almost all the cells are lysed. Rare *E. coli* cells, however, survive the infection and are not lysed. If these cells are isolated and established in pure culture, all of their descendants are *resistant* to T1 infection. The **adaptation hypothesis**, put forth to explain such observations, implied that the interaction of the phage and bacterium was essential to the acquisition of immunity. In other words, the phage had somehow "induced" resistance in the bacteria.

The occurrence of **spontaneous mutations** (sometimes referred to as random mutations) provided an alternative model to explain the origin of T1 resistance in *E. coli*. In 1943, Salvador Luria and Max Delbrück presented the first direct evidence that mutation in bacteria occurs spontaneously—that is, they arise independently of any external force. Thus, there is no way of predicting when, or where in the chromosome, mutation will occur. This experiment marked the initiation of modern bacterial genetic study.

The Luria–Delbrück Fluctuation Test

In a beautiful example of analytical and theoretical work, Luria and Delbrück carried out experiments with the *E. coli*–T1 systems to differentiate between the adaptation and spontaneous mutation hypotheses. They grew many small individual liquid cultures of phage-sensitive *E. coli* and then added numerous aliquots of each culture to petri dishes of agar medium containing T1 bacteriophages. To obtain precise quantitative data, they also determined the total number of bacteria added to each plate prior to incubation. Following incubation, each plate was scored for the number of phage-resistant bacterial colonies. This was easy to ascertain because only mutant cells were not lysed and thus survived to be counted.

The experimental rationale for distinguishing between the two hypotheses was as follows.

1. **Adaptation**. Every bacterium has a small, but constant, probability of acquiring resistance as a result of contact with the phages in the petri dish. Therefore, the number of resistant cells will depend only on the number of bacteria and phages added to each plate. The final results should be independent of all other experimental conditions. Therefore, the adaptation hypothesis

predicts that if a constant number of bacteria and phages is used for each culture and if the incubation time is constant, there should be little fluctuation in the number of resistant cells from plate to plate and from experiment to experiment.

2. **Spontaneous Mutation**. On the other hand, if resistance is acquired as a result of mutations that occur randomly, resistance will occur at a low rate during the incubation in liquid medium *prior to plating*—that is, before any contact with the phage occurs. When mutations occur *early* during incubation, the subsequent reproduction of the mutant bacteria will produce a relatively large number of resistant cells. When mutations occur *later* during incubation, far fewer resistant cells will be produced. The random mutation hypothesis therefore predicts that the number of resistant cells will fluctuate significantly from experiment to experiment, reflecting the varying times at which most spontaneous mutations occurred in liquid culture.

Table 7–1 shows a representative set of data from the Luria–Delbrück experiments. The middle column shows the number of mutants recovered from a series of aliquots derived from one large individual liquid culture. In a large culture, experimental differences are evened out because the culture is constantly mixed. As a result, these data serve as a control because the number in each aliquot should be nearly identical. As predicted, little fluctuation is observed. In contrast, the right-hand column shows considerable variation in the number of resistant mutants recovered from each of 10 independent liquid cultures. The amount of fluctuation in the data will support only one of the two alternative hypotheses. For this reason, the experiment has been designated the **fluctuation test**. Fluctuation is measured by the amount of variance, a statistical calculation.

TABLE 7–1 The Luria-Delbrück Experiment Demonstrating That Spontaneous Mutations are the Source of Phage-Resistant Bacteria

	Number of T1-Resistant Bacteria	
Sample No.	*Same Culture (Control)*	*Different Cultures*
1	14	6
2	15	5
3	13	10
4	21	8
5	15	24
6	14	13
7	26	165
8	16	15
9	20	6
10	13	10
Mean	16.7	26.2
Variance	15.0	2178.0

Source: After Luria and Delbrück, (1943).

A great fluctuation *was* observed among cultures in the Luria–Delbrück experiment, thus supporting the hypothesis that spontaneous mutations account for inherited variation in bacteria; further support came from other experimentation.

Adaptive Mutation in Bacteria

Although the concept of spontaneous mutation in viruses, bacteria, and even higher organisms has been accepted for some time, the possibility that organisms such as bacteria might also be capable of "selecting" a specific set of mutations occurring as a result of environmental pressures has long intrigued geneticists. Two independent investigations, published in 1988 by John Cairns and Barry Hall and their colleagues, have provided preliminary evidence that this might be the case. Though their findings are still controversial, this general topic is the subject of a current investigation of potentially great significance.

Cairns and his colleagues devised an experimental protocol that improves on the fluctuation test of Luria and Delbrück. The procedures were designed to detect *adaptive mutations* arising in response to factors in the environment in which bacteria are cultured. The results of this work suggest that some bacteria may "select" mutations that are adaptive to the environment.

Instead of using the characteristic of phage T1 resistance, in which the only surviving cells are those that have mutated, the Cairns work involved a strain of bacteria that contained a lactose mutation (lac^-). Such bacteria cannot use lactose as a carbon source. Cells first were grown in a rich liquid medium that provided an adequate carbon source other than lactose. Thus, the lac^- cells, as well as any spontaneous lac^+ mutants, were able to grow and reproduce quite well. Aliquots of cells were then plated on petri plates containing minimal medium to which lactose had been added. The lac^- cells, present in the vast majority, will survive on the plates, but because they lack a carbon source that they can metabolize, they cannot proliferate and form colonies. On the other hand, cells that have mutated to lac^+, whether in the original liquid culture or on the plate, will form colonies and will be detected. Those cells that have not mutated to lac^+ in liquid medium have the opportunity to do so *after* they are plated, and they may also be detected, since they will then form colonies. This is the major experimental difference between this approach and that of Luria and Delbrück some 45 years earlier.

Using a slightly more sophisticated mathematical analysis than Luria and Delbrück, Cairns attempted to identify the same sort of "fluctuation," or the lack of it, as an indication of either spontaneous or adaptive mutations, respectively. Cairns found both types of data! The distribution was calculated for both cases in which (1) spontaneous mutations occurred in the liquid cultures at random times (creating "fluctuation"), and (2) directed mutations arose as a response to the presence of lactose on the petri plates. Indeed, such a composite distribution was observed.

But how could it be proved that, in fact, the elevated frequency of mutations occurred in cells growing on the plates in response to lactose? To answer this question, Cairns and colleagues asked whether another mutation, unrelated to the metabolism of lactose, might have also occurred on the plates. They chose to assay for the production of Val^R mutations, which confer resistance to high concentrations of the amino acid valine. The assay was easily accomplished by overlaying the plates with agar containing valine and glucose because only Val^R mutants grew. Cairns and coworkers reasoned that if the lac^+ mutations did not arise specifically in response to the lactose, then Val^R mutations should arise with a similar distribution as earlier observed for the *lac* mutations. The result was that plates accumulating lac^- mutants were not simultaneously accumulating Val^R mutations. The researchers concluded that the higher frequency of lac^+ mutations resulted from the presence of lactose in the medium. This is, of course, an indirect inference.

Cairns's work suggests that the bacterial cell's genetic machinery can respond to its environment by producing adaptive mutations. This conclusion is contrary to the current thinking that mutations are spontaneous events, most of which occur as random errors during replication of DNA. As Cairns has pointed out, these findings, at first glance, seem to support the notion that life is mysterious and that certain observations cannot be explained in simple straightforward terms based on the laws of physics. Whatever the case may be, Cairns's observations are not isolated examples.

The work of Barry Hall and his colleagues demonstrated a similar adaptive response by *E. coli* to growth on the chemical salicin. In this case, a two-step genetic change was required. The first occurred in response to salicin even though that mutation offered no selective advantage without the second genetic alteration—the removal, or deletion, of a segment of nucleotides called an insertion sequence (IS). Further, the changes must occur sequentially. They do so at a much higher frequency than predicted, provided salicin is present in the growth medium. The observed frequency is several orders of magnitude higher.

What can possibly account for such observations? As Barry Hall has pointed out, our failure to explain such phenomena adequately may be attributable to our ignorance of fundamental mechanisms and rates of mutations *in nongrowing cells*, which are much more akin to a natural environment, as opposed to those found with cells grown in rich media under laboratory conditions. One suggestion is that under stressful nutritional conditions (starvation), bacteria may be capable of activating mechanisms that create a hypermutable state in genes that will enhance survival.

Though these ideas are still very much under debate, they highlight the fact that knowledge often may be derived from observations that initially cannot be explained. Such debates—and the often-surrounding controversies—are what constantly maintain intrigue and interest in science.

7.2 Mutations May Be Classified in Various Ways

Mutations are classified by various schemes. These organizing schemes are not mutually exclusive; they depend on which aspects of mutation are investigated or discussed. In this section, we describe three sets of distinctions that are used to classify mutations.

Spontaneous vs. Induced Mutations

Putting aside the case of "adaptive" mutations, all mutations are described as either **spontaneous** or **induced**. Although these two categories overlap to some degree, **spontaneous mutations** are those that happen naturally. No specific agents are associated with their occurrence, and they are generally assumed to be random changes in the nucleotide sequences of genes. Most such mutations are linked to normal chemical processes in the organism that alter the structure or the sequence of the nitrogenous bases that are part of the existing genes. Often, spontaneous mutations occur during the enzymatic process of DNA replication, an idea that we will discuss later in this chapter. Once an error is present in the genetic code, it may be reflected in the amino acid composition of the specified protein. If the changed amino acid is present in a part of the molecule critical to the structure or biochemical activity, a functional alteration can result.

In contrast to such spontaneous events, those that result from the influence of any artificial factor are considered to be **induced mutations**. It is generally agreed that any natural phenomenon that heightens chemical reactivity in cells will induce mutations. For example, radiation from cosmic and mineral sources and ultraviolet radiation from the sun are energy sources to which most organisms are exposed and, as such, may be factors that cause spontaneous mutations. The earliest demonstration of the artificial induction of mutation occurred in 1927, when Hermann J. Muller reported that X rays could cause mutations in *Drosophila*. In 1928, Lewis J. Stadler reported that X rays had the same effect on barley. In addition to various forms of radiation, a wide spectrum of chemical agents is also known to be mutagenic, as we will see later in the chapter.

Gametic vs. Somatic Mutations

When we consider the effects of mutation in eukaryotic organisms, it is important to distinguish whether the change occurs in somatic cells or in gametes. Mutations arising in somatic cells are not transmitted to future generations. Mutations occurring in somatic cells that create recessive autosomal alleles are rarely of any consequence to the organism. The expression of most such mutations is likely to be masked by the dominant allele. Somatic mutations will have a greater impact if they are dominant or if they are X-linked, since such mutations are most likely to be immediately expressed. Similarly, the impact will be more noticeable if such somatic mutations occur early in development, when undifferentiated cells will give rise to several differentiated tissues or organs. Mutations occurring in adult tissues are often masked by the thousands upon thousands of nonmutant cells performing the normal function.

Mutations in gametes or gamete-forming tissue are of greater significance because they are transmitted to offspring as part of the germ line. They have the potential of being expressed in all cells of an offspring. **Dominant autosomal mutations** will be expressed phenotypically in the first generation. **X-linked recessive mutations** arising in the gametes of a homogametic female may be expressed in hemizygous male offspring. This will occur provided that the male offspring receives the affected X chromosome. Because of heterozygosity, the occurrence of an **autosomal recessive mutation** in the gametes of either males or females (even one resulting in a lethal allele) may go unnoticed for many generations, until the resultant allele has become widespread in the population. The new allele will become evident only when a chance mating brings two copies of it together in the homozygous condition.

Other Categories of Mutation

In addition to their cause of occurrence or their point of origin, mutations may also be classified on the basis of their phenotypic effects. Note that a single mutation may well fall into more than one category. The most easily observed mutations are those affecting a **morphological trait**. Such variations are recognized on the basis of their deviation from the normal or wild-type phenotype. For example, all of Mendel's pea characters and many genetic variations encountered in the study of *Drosophila* fit this designation, since they cause obvious changes in the morphology of the organism.

A second broad category of mutations includes those that exhibit **nutritional** or **biochemical variations** in phenotype. In bacteria and fungi, a typical nutritional mutation is the inability to synthesize an amino acid or vitamin. In humans, sickle-cell anemia and hemophilia are examples of biochemical mutations. While such mutations in these organisms are not visible and do not always affect specific morphological characters, they can have a more general effect on the well-being and survival of the affected individual.

A third category consists of mutations that affect behavior patterns of an organism. For example, the mating behavior or circadian rhythms of animals can be altered. The primary effect of **behavior mutations** is often difficult to discern. For example, the mating behavior of a fruit fly may be impaired if it cannot beat its wings. However, the defect may be in (1) the flight muscles, (2) the nerves leading to them, or (3) the brain, where the nerve impulses that initiate wing movements originate.

Still another type of mutation may affect the regulation of genes. For example, as we will see in the *lac* operon discussed in Chapter 19, a regulatory gene can produce a product that controls the transcription of other genes. In other instances, a region of DNA either close to, or far away from, a gene may also modulate its activity. In either case, **regulatory mutations** can disrupt normal regulatory processes and permanently activate or inactivate a gene. Our knowledge of genetic regulation has been dependent on the study of mutations that disrupt this process.

Lethal and Conditional Mutations

Regardless of which of the aforementioned categories any given mutation may fall into, it is also possible that the mutation may interrupt a process that is essential to the survival of the organism. In this case, it is referred to as a **lethal mutation**. For example, a mutant bacterium that has lost the ability to synthesize an essential amino acid, when placed in a medium lacking that amino acid, will cease to grow and eventually will die. Various inherited human biochemical disorders are also good examples of lethal mutations. For example, Tay–Sachs disease and Huntington disease are caused by mutations that result in lethality, but at different points in the life cycle of humans.

Another interesting aspect of mutations is observed when their expression depends on the environment within which the organism finds itself. In such cases, we refer to them as **conditional mutations**. The mutation is present in the genome of an organism, but it is expressed and can be detected only under certain conditions. Among the best examples are **temperature-sensitive mutations**, found in a variety of organisms. At certain "permissive" temperatures, a mutant gene product functions normally, only to lose its functional capability at a "restrictive" temperature. When the organism is shifted from the permissive to the restrictive temperature, the impact of the mutation becomes apparent. In some cases, it might even be lethal. The study of conditional mutations has been extremely important in experimental genetics, particularly in understanding the function of genes essential to completion of the normal cell cycle.

7.3 The Spontaneous Mutation Rate Varies Greatly Among Organisms

Detection systems we have been discussing sometimes also allow geneticists to estimate mutation rates. However, we can only ascertain induced mutation when the induced rate clearly exceeds the spontaneous mutation rate for the organism under study. Determining the rate of spontaneous mutation provides the baseline for measuring the rate of experimentally induced mutation.

Examination of the spontaneous rate in a variety of organisms reveals many interesting points. (See Table 7–2.) First, the rate of spontaneous mutation is exceedingly low for all of the organisms studied. Second, the rate is seen to vary considerably in different organisms. Third, even within the same species, the spontaneous mutation rate varies from gene to gene.

TABLE 7–2 Rates of Spontaneous Mutations at Various Loci in Different Organisms

Organism	*Character*	*Gene*	*Rate*	*Units*
Bacteriophage T2	Lysis inhibition	$r \rightarrow r^+$	1×10^{-8}	Per gene replication
	Host range	$h^+ \rightarrow h$	3×10^{-9}	
	Lactose fermentation	$lac^- \rightarrow lac^+$	2×10^{-7}	
	Lactose fermentation	$lac^+ \rightarrow lac^-$	2×10^{-6}	
	Phage T1 resistance	$T1\text{-}s \rightarrow T1\text{-}r$	2×10^{-8}	
	Histidine requirement	$his^+ \rightarrow his^-$	2×10^{-6}	
	Histidine independence	$his^- \rightarrow his^+$	4×10^{-8}	
E. coli	Streptomycin dependence	$str\text{-}s \rightarrow str\text{-}d$	1×10^{-9}	Per cell division
	Streptomycin sensitivity	$str\text{-}d \rightarrow str\text{-}s$	1×10^{-8}	
	Radiation resistance	$rad\text{-}s \rightarrow rad\text{-}r$	1×10^{-5}	
	Leucine independence	$leu^- \rightarrow leu^+$	7×10^{-10}	
	Arginine independence	$arg^- \rightarrow arg^+$	4×10^{-9}	
	Tryptophan independence	$try^- \rightarrow try^+$	6×10^{-8}	
Salmonella typhimurium	Tryptophan independence	$try^- \rightarrow try^+$	5×10^{-8}	Per cell division
Diplococcus pneumoniae	Penicillin resistance	$pen^s \rightarrow pen^r$	1×10^{-7}	Per cell division
Chlamydomonas reinhardi	Streptomycin sensitivity	$str^r \rightarrow str^s$	1×10^{-6}	Per cell division
Neurospora crassa	Inositol requirement	$inos^- \rightarrow inos^+$	8×10^{-8}	Mutant frequency among asexual spores
	Adenine independence	$ade^- \rightarrow ade^+$	4×10^{-8}	
Zea mays	Shrunken seeds	$sh^+ \rightarrow sh^-$	1×10^{-6}	Per gamete per generation
	Purple	$pr^+ \rightarrow pr^-$	1×10^{-5}	
	Colorless	$c^+ \rightarrow c$	2×10^{-6}	
	Sugary	$su^+ \rightarrow su$	2×10^{-6}	
Drosophila melanogaster	Yellow body	$y^+ \rightarrow y$	1.2×10^{-6}	Per gamete per generation
	White eye	$w^+ \rightarrow w$	4×10^{-5}	
	Brown eye	$bw^+ \rightarrow bw$	3×10^{-5}	
	Ebony body	$e^+ \rightarrow e$	2×10^{-5}	
	Eyeless	$ey^+ \rightarrow ey$	6×10^{-5}	
Mus musculus	Piebald coat	$s^+ \rightarrow s$	3×10^{-5}	Per gamete per generation
	Dilute coat color	$d^+ \rightarrow d$	3×10^{-5}	
	Brown coat	$b^+ \rightarrow b$	8.5×10^{-4}	
	Pink eye	$p^+ \rightarrow p$	8.5×10^{-4}	
Homo sapiens	Hemophilia	$h^+ \rightarrow h$	2×10^{-5}	Per gamete per generation
	Huntington disease	$Hu^+ \rightarrow Hu$	5×10^{-6}	
	Retinoblastoma	$R^+ \rightarrow R$	2×10^{-5}	
	Epiloia	$Ep^+ \rightarrow Ep$	1×10^{-5}	
	Aniridia	$An^+ \rightarrow An$	5×10^{-6}	
	Achondroplasia	$A^+ \rightarrow A$	5×10^{-5}	

Viral and bacterial genes undergo spontaneous mutation on an average of about 1 in 100 million (10^{-8}) cell divisions. *Neurospora* exhibits a similar rate, but maize, *Drosophila*, and humans demonstrate a rate several orders of magnitude higher. The genes studied in these groups average between 1/1,000,000 and 1/100,000 (10^{-6} and 10^{-5}) mutations per gamete formed. Mouse genes are still another order of magnitude higher in their spontaneous mutation rate, 1/100,000 to 1/10,000 (10^{-5} and 10^{-4}). It is not clear why such a large variation occurs in the mutation rate. The variation might reflect the relative efficiency of the enzyme systems, whose function is to repair errors created during replication. We will discuss repair systems later in the chapter.

Deleterious Mutations in Humans

The vast majority of mutations are likely to occur in the large portions of the genome that do not encode genes or in intron regions found within genes. These are considered to be neutral mutations, since they do not usually affect gene products. While this information is of great interest to geneticists, perhaps a more significant piece of information regarding spontaneous mutations is the rate at which those mutations *that are deleterious* occur. Of greatest interest is the rate of occurrence of these mutations in our own species. For decades, the rate has been nearly impossible to determine. However, recent work, using molecular techniques, has allowed an accurate estimate to be made, and it is surprisingly high, at least 1.6 deleterious genetic changes per individual per generation.

The way in which this work was performed and its possible consequences to our species are clearly and insightfully discussed by James Crow in a short article that was published several years ago (January, 1999) in *Nature*. Dr. Crow is one of the foremost researchers and a true pioneer in mutation research. His article, "The Odds of Losing at Genetic Roulette" is reprinted here. We trust you will enjoy his perspectives on the general topic and be enlightened by his comments.

7.4 Mutations Occur in Many Forms and Arise in Different Ways

Even though we are aware that the gene is a reasonably complex genetic unit, particularly in eukaryotes, we will use a *simplified* definition of a gene in the description of the

The Odds of Losing at Genetic Roulette

The rate at which deleterious mutations occur in a genome is clearly a quantity of interest—not least when the genome is our own. It becomes particularly important if, as some have argued[1], the rate in some species is high, at several new mutations each generation. Yet the deleterious mutation rate has been notoriously difficult to measure, and no convincing estimates exist for any vertebrate. On page 344 of this issue[2], Eyre-Walker and Keightley give the first such estimates for ourselves, chimpanzees and gorillas.

Getting an estimate of the total mutation rate is relatively simple. Neutral mutations—those which neither enhance nor impair the organism carrying them—accumulate through the generations at a rate equal to the mutation rate[3]. The mutation rate can be determined by the rate of change of presumed neutral regions: areas of the genome, such as introns and pseudogenes, which are not translated into proteins. For mammals, these rates, extrapolated to the whole genome, lead to enormous numbers, on the order of 100 new mutations per individual[1]. Of course these can't all be deleterious, but no one knows what the proportion is. This proportion could be determined in principle by comparing the slower rate of evolutionary change of the genome as a whole with the faster rate expected under neutral assumptions; but this involves statistical uncertainties and extensive sequencing[4].

Eyre-Walker and Keightley[2] have made the analysis feasible by concentrating on protein-coding regions. They measured the amino-acid changes in 46 proteins in the human ancestral line after its divergence from the chimpanzee. Among 41,471 nucleotides, they found 143 nonsynonymous substitutions—mutations where swapping one DNA base for another changes an amino acid, and therefore the final protein made by that gene. If these had evolved at the neutral rate, 231 would be expected. The difference, 88 (38%), is an estimate of the number of deleterious mutations that have been eliminated by natural selection and have therefore made no contribution to contemporary populations.

Translating these numbers into mutation rates gave a total rate of 4.2 mutations per person per generation, and a deleterious rate of 1.6. The rates of chimpanzees and gorillas were very similar, the deleterious rates being 1.7 and 1.2, respectively. The authors took 60,000 as the gene number and 25 years as the generation length. The number 1.6 is probably an underestimate, for various reasons. For instance, mutations outside the coding region are not counted and some of these regions—such as those controlling gene expression—are expected to be subject to natural selection. The gene number may also be an underestimate. If there have been mutations that increase fitness, they would also cause the number of deleterious mutations to be underestimated. A less conservative, and probably more realistic, estimate doubles the value, giving 3 new deleterious mutations per person per generation.

What's the significance? Every deleterious mutation must eventually be eliminated from the population by premature death or reduced relative reproductive success, a 'genetic death'[5]. That implies three genetic deaths per person! Why aren't we extinct? If harmful mutations were eliminated independently, as in an asexual species, it has been estimated that this would lower population fitness to a fraction e^{-3}, or 5%, of the mutation-free value[6], leading to the inevitable extinction of species with limited reproductive capacity. A way out is for mutations to be eliminated in bunches. This happens if selection operates such that individuals with the most mutations are preferentially eliminated, for example if harmful mutations interact. But such a process can only work in sexual species, where mutations are shuffled each generation by genetic recombination[1,7]. The existence of a high deleterious mutation rate strengthens the argument that a major advantage of sex is that it is an efficient way to eliminate harmful mutations[1]. It also raises again the possibility of fitness decline or even extinction in rare species from too many harmful mutations[8].

Presumably, we humans have profited in the past by sexual reproduction's ability to reduce the effect of a high mutation rate on fitness. In the recent past the intensity of natural selection has been greatly reduced, especially where a high standard of living means that most infants reach reproductive age. From this it would seem that natural selection will weed out mutations more slowly than they accumulate. This effect may be accentuated by trends for males to start or continue reproducing later in life, because the sperm of older men contains more base-substitution mutations[8]. In a time of rapid environmental improvement, how this genetic decline will affect our health can only be guessed at.

Eyre-Walker and Keightley[2] noticed that the proportion of harmful mutations in the 46 genes in their study is greater in humans than in the equivalent genes of rodents. Their preferred explanation is that slightly deleterious mutations have become fixed in the population, by a process known as random genetic drift, during periods of human history when the breeding population size was low—especially during genetic 'bottlenecks.' This would increase whatever effect the accumulated mutations are having on current human welfare. Are some of our headaches, stomach upsets, weak eyesight and other ailments the result of mutation accumulation? Probably, but in our present state of knowledge, we can only speculate.

James F. Crow Department of Genetics University of Wisconsin Madison, WI 93706

[1]Kondrashov, A.S. *J. Theor. Biol.* **175**, 583–594 (1995).
[2]Eyre-Walker, A. & Keightley, P.D. *Nature* **397**, 344–347 (1999).
[3]Kimura, M. *The Neutral Theory of Molecular Evolution* (Cambridge University Press, 1983).
[4]Kondrashov, A.S. & Crow, J.F. *Hum. Mutat.* **2**, 229–234 (1993).
[5]Muller, H.J. *Am. J. Hum. Genet.* **2**, 111–176 (1950).
[6]Kimura, M. & Maruyama, T. *Genetics* **54**, 1337–1351 (1966).
[7]Crow, J.F. *Proc. Natl Acad. Sci. USA* **94**, 8380–8386 (1997).
[8]Lande, R. *Evolution* **48**, 1460–1469 (1994).

molecular basis of mutation that follows. In this context, it is easiest to consider a gene as a linear sequence of nucleotide pairs representing stored chemical information. Because the genetic code is a triplet, each sequence of three nucleotides specifies a single amino acid in the corresponding polypeptide. Any change that disrupts these sequences or the coded information provides sufficient basis for a mutation. The least complex change is the substitution of a single nucleotide. In Figure 7–1, such a change is compared with our own written language, using three-letter words to be consistent with the genetic code. A change of one letter can alter the meaning of the sentence "THE CAT SAW THE DOG," to "THE CAT SAW THE HOG" or "THE BAT SAW THE DOG," creating what is called *missense*. These are analogies to what are most appropriately referred to as **base substitutions** or **point mutations**. The mutation has changed the sense of the information into various forms of missense.

You will often see two other terms used to describe nucleotide substitutions. If a pyrimidine replaces a pyrimidine or a purine replaces a purine, a **transition** has occurred. If a purine and a pyrimidine are interchanged, a **transversion** has occurred. Another type of change is the insertion or deletion of one or more nucleotides at any point within the gene. As illustrated in Figure 7–1, the loss or addition of a single letter causes all of the subsequent three-letter words to be changed. These examples are called **frameshift mutations** because the frame of reading has been altered. A frameshift mutation will occur when any number of bases are added or deleted, except multiples of three, which would reestablish the initial frame of reading. We will return momentarily to a discussion of frameshift mutations as we introduce acridine dyes.

The analogy in Figure 7–1 demonstrates that insertions and deletions have the potential to change all the subsequent triplets in a gene. It is probable that, of the 64 possible triplets, one of the many altered triplets will be UAA, UAG, or UGA, the termination codons (see Chapter 5). When one of these triplets is encountered during translation, polypeptide synthesis is terminated. Obviously, the results of frameshift mutations can be very severe.

Tautomeric Shifts

In 1953, immediately after they had proposed the molecular structure of DNA, Watson and Crick published a paper in which they discussed the genetic implications of this structure. They recognized that the purines and pyrimidines found in DNA could exist in **tautomeric** forms—that is, a nitrogenous base can exist in alternate chemical forms called structural isomers, each differing by only a single proton

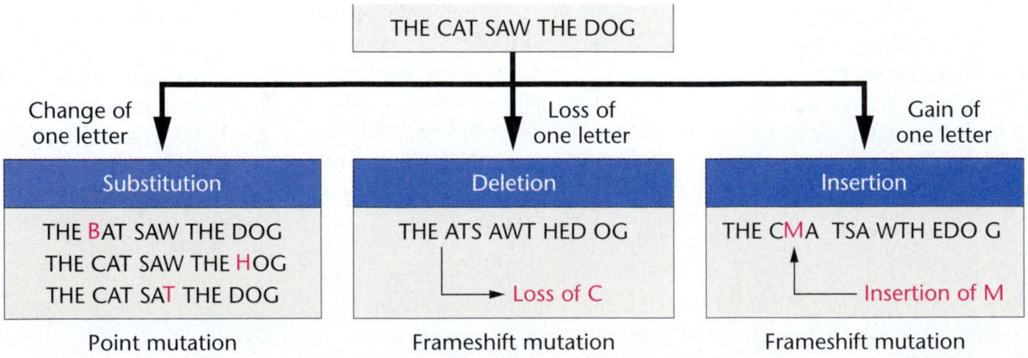

FIGURE 7–1 The impact of the substitution, deletion, and insertion of one letter in a sentence composed of three-letter words as analogies to point and frameshift mutations.

shift in the molecule. The biologically important tautomers involve the keto–enol forms of thymine and guanine, and the amino–imino forms of cytosine and adenine (Figure 7–2). Because such a shift changes the bonding structure of the molecule, Watson and Crick suggested that **tautomeric shifts** could result in base-pair changes or mutations.

The most stable tautomers of the nitrogenous bases result in the standard base pairings that serve as the basis of the double-helix model of DNA. The less frequently occurring, transient tautomers are capable of hydrogen bonding with noncomplementary bases. However, the pairing is always between a pyrimidine and a purine. Figure 7–3 compares the normal base-pairing relationships with the rare unorthodox pairings. Anomalous T═G and C═A pairs, among others, may be formed.

The effect leading to mutation occurs during DNA replication when a rare tautomer in the template strand matches with a noncomplementary base. In the next round of replication, the "mismatched" members of the base pair are separated, and each specifies its normal complementary base. The end result is a point mutation (Figure 7–4).

Base Analogs

Base analogs, which are mutagenic chemicals, are molecules that can substitute for purines or pyrimidines during nucleic acid biosynthesis. For example, **5-bromouracil (5-BU)*** a

*If 5-BU is chemically linked to d-ribose, the nucleoside analog bromodeoxyuridine (BUdR) is formed.

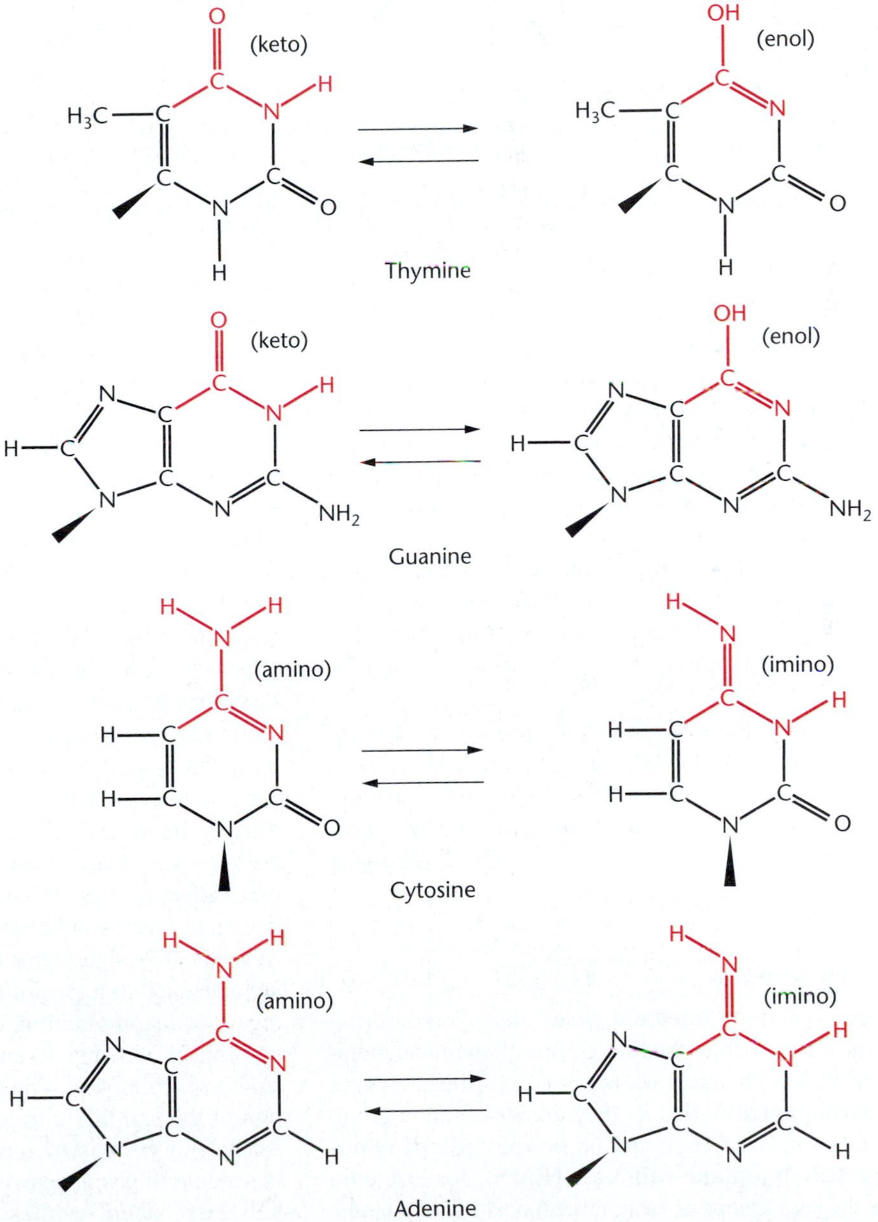

FIGURE 7–2 Rare tautomeric shifts that occur in the chemical structure of the four nitrogenous bases of DNA. The dense arrowheads indicate the point of bonding to the pentose sugar. Prior to bonding, an H atom exists at each of these points.

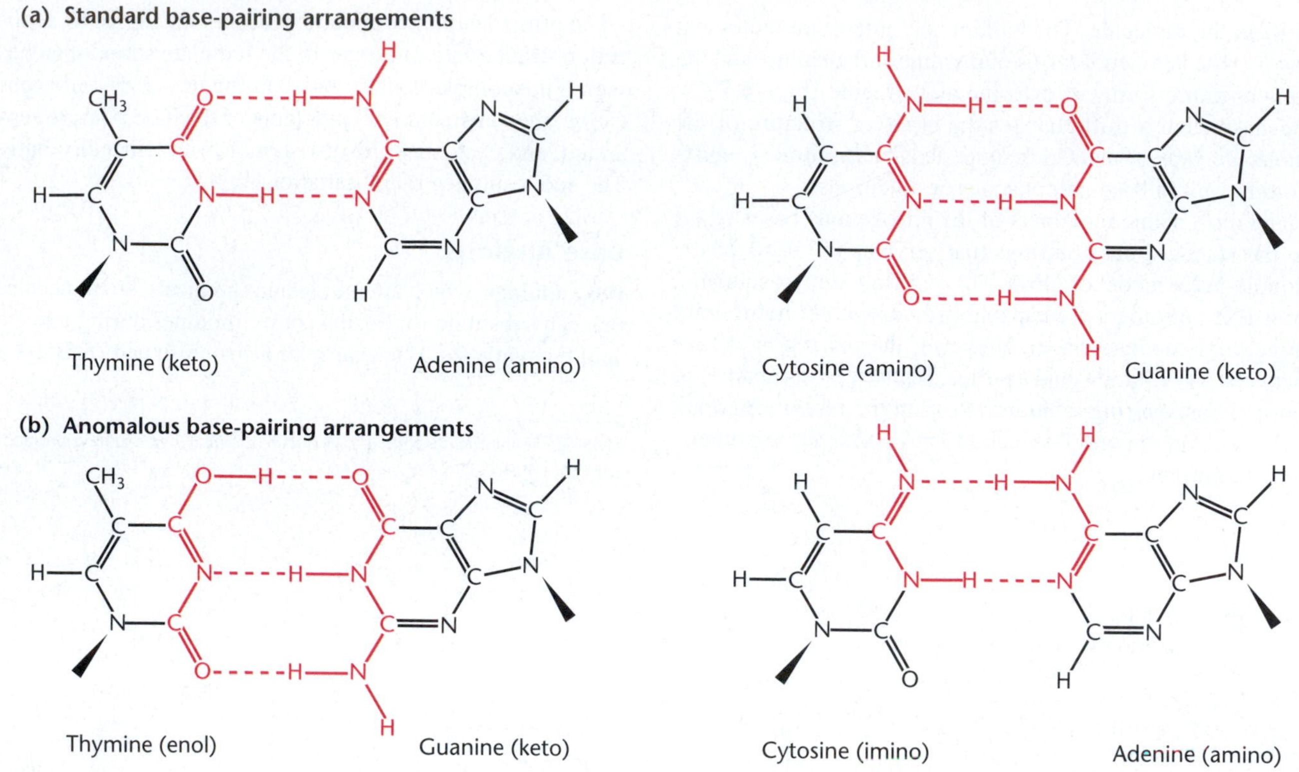

FIGURE 7–3 The standard base-pairing relationships compared with two anomalous arrangements occurring as a result of tautomeric shifts.

derivative of uracil, behaves as a thymine analog and is halogenated at the number 5 position of the pyrimidine ring. Figure 7–5 compares the structure of this analog with that of thymine. The presence of the bromine atom in place of the methyl group increases the probability that a tautomeric shift will occur. If 5-BU is incorporated into DNA in place of thymine and a tautomeric shift to the enol form occurs, 5-BU base pairs with guanine. After one round of replication, an A=T to G≡C transition results. (See Figure 14–8.) Furthermore, the presence of 5-BU within DNA increases the sensitivity of the molecule to ultraviolet (UV) light, which itself is mutagenic (as discussed in the next section).

There are other base analogs that are mutagenic. For example, **2-amino purine (2-AP)** can serve successfully as an analog of adenine. In addition to its base-pairing affinity with thymine, 2-AP can also base pair with cytosine, leading to possible transitions from A=T to G≡T following replication.

Alkylating Agents

The sulfur-containing **mustard gases**, discovered during World War I, were one of the first groups of chemical mutagens identified in chemical warfare studies. Mustard gases are **alkylating agents**—that is, they donate an alkyl group such as CH_3 or CH_3CH_2 to amino or keto groups in nucleotides. **Ethylmethane sulfonate (EMS)**, for example, alkylates the keto groups in the number 6 position of guanine and in the number 4 position of thymine (Figure 7–6). As with base analogs, base-pairing affinities are altered and transition mutations result. In the case of **6-ethyl guanine**, for example, the molecule acts like a base analog of adenine and pairs with thymine. Table 7–3 lists the chemical names and structures of several frequently used alkylating agents known to be mutagenic.

Other chemical mutagens called **acridine dyes** cause frameshift mutations. As illustrated in Figure 7–1 and introduced in an earlier discussion, these mutations result from the addition or removal of one or more base pairs (bp) in the polynucleotide sequence of the gene. Inductions of frameshift mutations have been studied in detail with a group of aromatic molecules known as **acridine dyes. Proflavin**, the most widely studied acridine mutagen, and **acridine orange** are examples, illustrated in Figure 7–7. Acridine dyes are roughly the same dimensions as a nitrogenous base pair and are known to intercalate or wedge between the purines and pyrimidines of intact DNA, inducing contortions in the DNA helix and causing deletions and insertions.

One model suggests that the resultant frameshift mutations are generated at gaps produced in DNA during replication, repair, or recombination. During these events, there is the possibility of slippage and improper base pairing of one strand with the other. The model suggests that the intercalation of the acridine into an improperly base-paired region can extend the existence of these slippage structures. If so, the probability increases that the mispaired configuration will exist when synthesis and rejoining occur, thereby

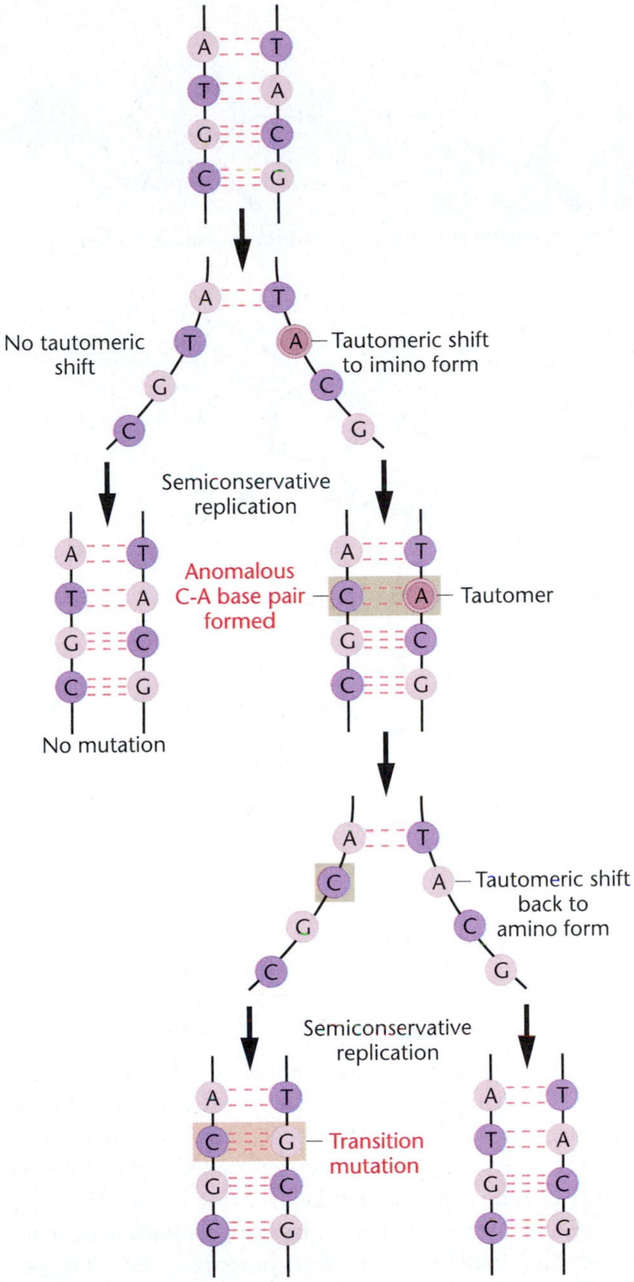

FIGURE 7–4 Formation of a T = A to a C≡G transition mutation as a result of a tautomeric shift in adenine.

resulting in an addition or deletion of one or more bases in one of the strands.

Apurinic Sites and Other Lesions

Still another type of mutation involves the spontaneous loss of one of the nitrogenous bases—most frequently, guanine or adenine—in an intact double-helix DNA molecule. These sites, created by the "breaking" of the glycosidic bond linking the 1′-C of d-ribose and the 9-N of the purine ring, are called **apurinic sites (AP sites)*** It has been estimated that, in the DNA of mammalian cells in culture, thousands of such spontaneous lesions are formed daily.

The absence of a nitrogenous base at an AP site will alter the genetic code if the involved strand is transcribed and translated. If replication occurs, the AP site is an inadequate template and may cause replication to stall. If a nucleotide is inserted, it is frequently incorrect, causing still another mutation. Fortunately, as we will see, cells contain repair systems that often counteract and correct this type of lesion.

Another type of lesion is known to be the source of some mutations. In the process of **deamination**, an amino group is converted to a keto group in cytosine and adenine (Figure 7–8). In these two cases, the mutagen is nitrous acid, known to be capable of inducing the deamination of bases in DNA. Cytosine is converted to uracil and adenine is changed to hypoxanthine. The major effect of these changes is to alter the base-pairing specificities of the two molecules during DNA replication. For example, cytosine normally pairs with guanine. Following its conversion to uracil, which pairs with adenine, the original G≡C pair is converted to an AU pair and, following an additional replication, to an A═T pair. When adenine is deaminated, an original A═T pair is converted to a G≡C pair because hypoxanthine pairs naturally with cytosine.

Ultraviolet Radiation and Thymine Dimers

All energy on the earth consists of a series of electromagnetic components of varying wavelength (Figure 7–9). Referred to as the **electromagnetic spectrum**, the energy associated with the various components varies inversely with wavelength. Everything longer than, and including, visible light is benign when it interacts with most organic molecules. However, everything of shorter wavelength than visible light, which is inherently more energetic, has a disruptive impact on organic molecules, such as those comprising living tissue. For example, in Chapter 2, we emphasized the fact that purines and pyrimidines absorb **ultraviolet (UV) radiation** most intensely at a wavelength of about 260 nm, a property that has been useful in the detection and analysis of nucleic acids. In 1934, as a result of studies involving *Drosophila* eggs, it was discovered that UV radiation is mutagenic, and by 1960, several studies concerning the in vitro effect on the components of nucleic acids had been completed. The major effect of UV radiation is on **pyrimidine dimers**, particularly between two thymine residues (Figure 7–10). While cytosine–cytosine and thymine–cytosine dimers may also be formed, they are less prevalent. The dimers distort the DNA conformation and inhibit normal replication. As a result, errors can be introduced in the base sequence of DNA during replication, and when extensive, these are responsible (at least in part) for the killing effects of UV radiation on microorganisms.

As we will see later, several mechanisms have evolved to correct UV-induced lesions. We will illustrate the essential

*Apyrimidinic sites, from which a pyrimidine has been lost, also occur and are also referred to as AP sites.

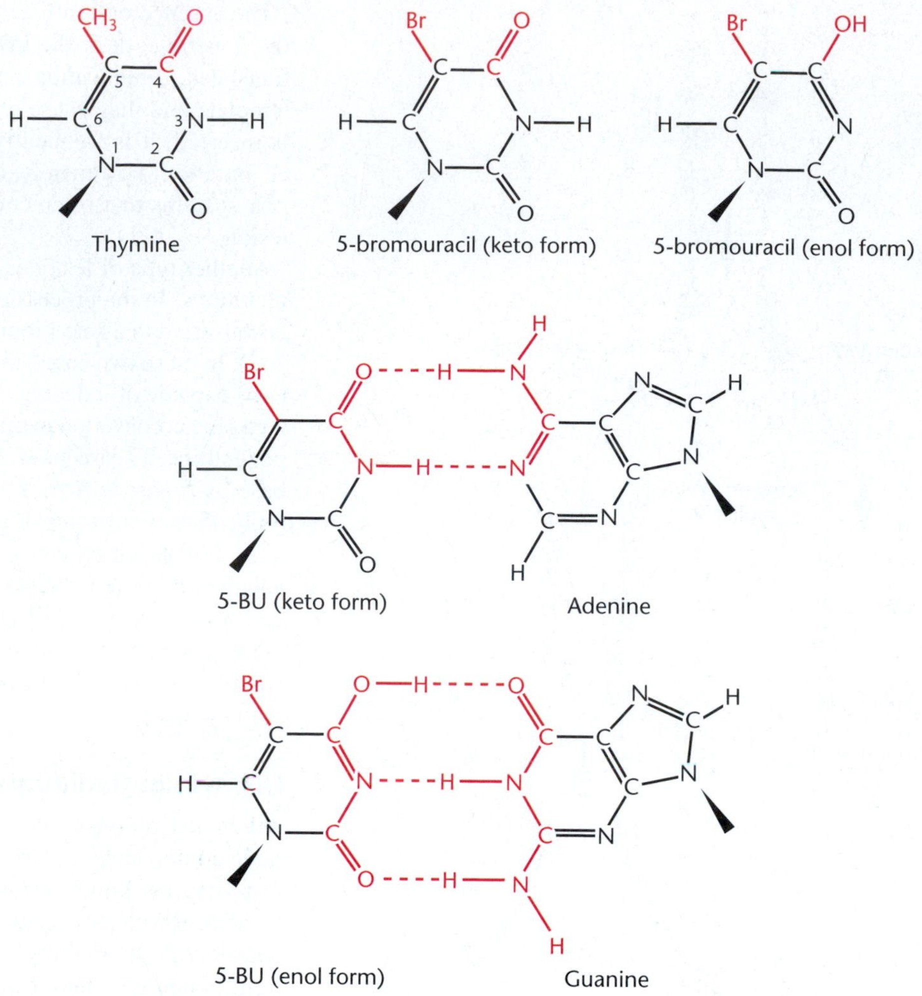

FIGURE 7–5 Similarity of 5-bromouracil (5-BU) structure to thymine structure. In the common keto form, 5-BU pairs normally with adenine, behaving as an analog. In the rare enol form, it pairs anomalously with guanine.

nature of such "repair" processes in humans by discussing the inherited disorder xeroderma pigmentosum, in which mutation has inactivated one UV-repair mechanism. Such individuals have been described as "children of the night," since they cannot risk exposure to the ultraviolet component of sunlight without the risk of epidermal malignancy.

7.5 High-Energy Radiation Penetrates Cells and Induces Mutations

Within the electromagnetic spectrum, energy varies inversely with wavelength. As shown in Figure 7–9, **X rays, gamma rays**, and **cosmic rays** have even shorter wavelengths than does UV radiation and are therefore more energetic. As a result, they are strong enough to penetrate deeply into tissues, causing ionization of the molecules encountered along the way. Hermann J. Muller and Lewis J. Stadler established in the 1920s that these sources of **ionizing radiation** are mutagenic. Since that time, the effects of ionizing radiation, particularly X rays, have been studied intensely.

As X rays penetrate cells, electrons are ejected from the atoms of molecules encountered by the radiation. Thus, stable molecules and atoms are transformed into free radicals and reactive ions. The trail of ions left along the path of a high-energy ray can initiate a variety of chemical reactions.

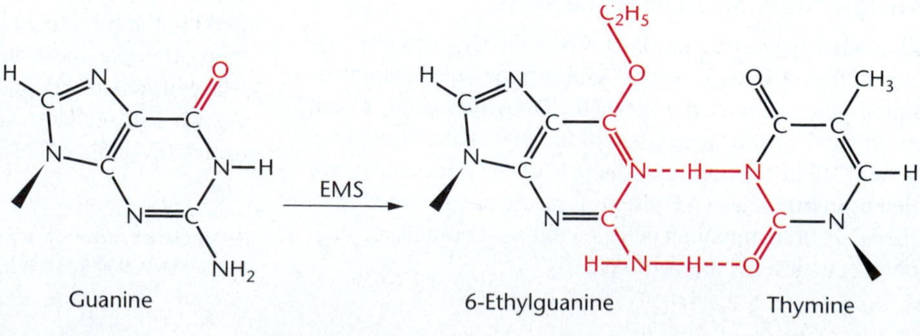

FIGURE 7–6 Conversion of guanine to 6-ethylguanine by the alkylating agent ethylmethane sulfonate (EMS). The 6-ethylguanine base pairs with thymine.

TABLE 7–3 Alkylating Agents

Common Name or Symbol	*Chemical Name*	*Chemical Structure*
Mustard gas (sulfur)	Di-(2-chloroethyl) sulfide	$\mathrm{Cl{-}CH_2{-}CH_2{-}S{-}CH_2{-}CH_2{-}Cl}$
EMS	Ethylmethane sulfonate	$\mathrm{CH_2{-}CH_2{-}O{-}\overset{\overset{\displaystyle O}{\Vert}}{\underset{\underset{\displaystyle O}{\vert}}{S}}{-}CH_3}$
EES	Ethylethane sulfonate	$\mathrm{CH_2{-}CH_2{-}O{-}\overset{\overset{\displaystyle O}{\Vert}}{\underset{\underset{\displaystyle O}{\vert}}{S}}{-}CH_3{-}CH_3}$

These reactions can directly or indirectly affect the genetic material, altering the purines and pyrimidines in DNA and resulting in point mutations. Ionizing radiation is also capable of breaking phosphodiester bonds, disrupting the integrity of chromosomes and producing a variety of aberrations.

Figure 7–11 shows a graph of the percentage of induced X-linked recessive lethal mutations versus the dose of X rays administered. A linear relationship is evident between X-ray dose and the induction of mutation; for each doubling of the dose, twice as many mutations are induced. Because the line intersects near the zero axis, this graph suggests that even very small doses of irradiation are mutagenic.

These observations can be interpreted in the form of the **target theory**, first proposed in 1924 by J. A. Crowther and F. Dessauer. The theory proposes that there are one or more sites, or targets, within cells and that a single event of irradiation at one site will bring about a damaging effect, or mutation. In effect, the target theory suggests that the X rays interact directly with the genetic material.

An observation concerning irradiation effects during the cell cycle is of particular interest. Cells that have entered mitosis are much more susceptible to radiation effects than cells in G1, S, or G2. This is because condensed chromosomes present a more susceptible target than dispersed chromatin. X rays can break chromosomes, resulting in terminal or intercalary deletions, translocations, and general chromosome fragmentation. This property is one of the reasons why radiation is used to treat human malignancy. Because tumorous cells are undergoing division more often than their nonmalignant counterparts, they are more susceptible to the immobilizing effect of radiation.

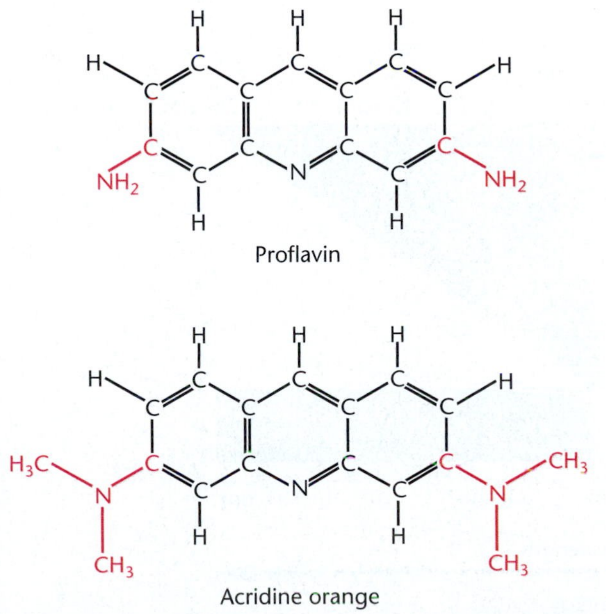

FIGURE 7–7 Chemical structures of proflavin and acridine orange, which intercalate with DNA and cause frameshift mutations.

7.6 Gene Sequencing Has Enhanced Understanding of Mutations in Humans

So far, we have been discussing the molecular basis of mutation largely in terms of nucleic acid chemistry. We know that various types of nucleotide conversions or frameshift mutations occur, primarily because our analyses of amino acid sequences of many proteins within the populations of a species show substantial diversity. This diversity, which has arisen during evolution, is a reflection of changes in the triplet codons following substitution, insertion, or deletion of one or more nucleotides in the DNA sequences constituting genes.

As our ability to analyze DNA more directly increases, we are better able to look specifically at the actual nucleotide sequence of genes and to gain greater insights into mutation. Several techniques capable of accurate, rapid sequencing of DNA (see Chapter 18 and Appendix A) have greatly extended our knowledge of molecular genetics. In this section, we examine the results of a number of studies that have investigated the actual gene sequence of various mutations that seriously affect humans.

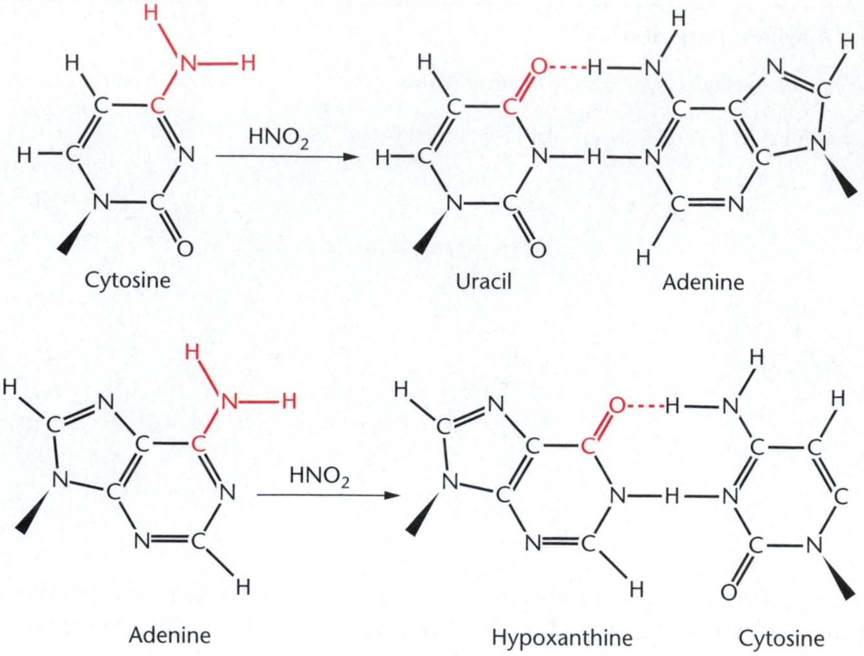

FIGURE 7–8 Deaminations caused by nitrous acid (HNO_2) leading to new base pairing arrangements and mutations. Cytosine may be converted to uracil, which base pairs with adenine. Adenine may be converted to hypoxanthine, which base pairs with cytosine.

ABO Blood Types

The ABO system is based on a series of antigenic determinants found on erythrocytes and other cells, particularly epithelial types. As we will discuss in Chapter 10, three alleles of a single gene exist, the product of which modifies the H substance. The modification involves glycosyltransferase activity, converting the H substance to either the A or B antigen, as a result of the product of the I^A or I^B allele, respectively; or failing to modify the H substance, as a result of the I^O allele. (See Figure 10–2.)

The responsible gene has been sequenced in 14 cases of varying ABO status. When the DNAs of the I^A and I^B alleles were compared, four consistent nucleotide substitutions were found. It is assumed that the resulting changes in the amino acid sequence of the glycosyltransferase gene product lead to the different modifications of the H substance.

The I^O allele situation is unique and interesting. Individuals homozygous for this allele have type O blood, lack glycosyltransferase activity, and fail to modify the H substance. Analysis of the DNA of this allele shows one consistent change that is unique compared with the sequence of the other alleles—the deletion of a single nucleotide early in the coding sequence, causing a frameshift mutation. A complete messenger RNA is transcribed, but at translation, the frame of reading shifts at the point of deletion and continues out of frame for about 100 nucleotides before a stop codon is

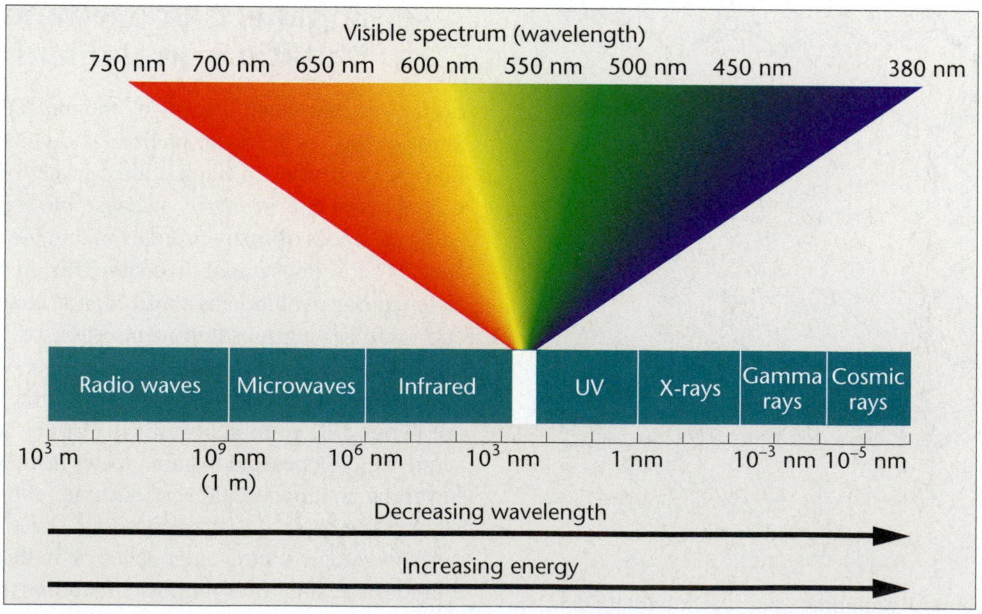

FIGURE 7–9 The components of the electromagnetic spectrum and their associated wavelengths.

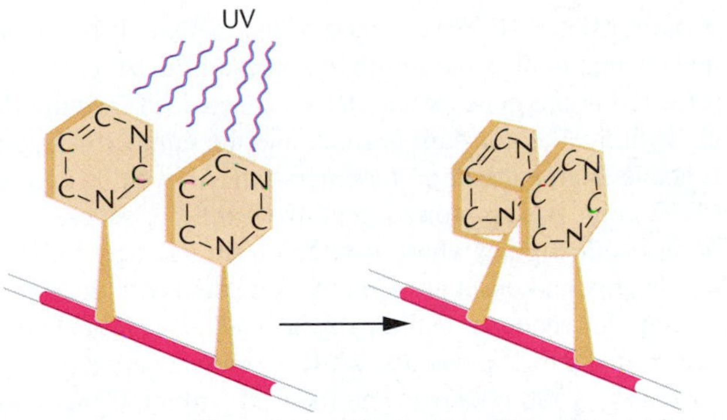

FIGURE 7–10 Induction of a thymine dimer by UV radiation, leading to distortion of the DNA. The atoms of the pyrimidine ring are shown to illustrate the formation of cross-links.

encountered. At this point, the polypeptide chain terminates prematurely, resulting in a nonfunctional product.

These findings provide a direct molecular explanation of the ABO allele system and the basis for the biosynthesis of the corresponding antigens. The molecular basis for the antigenic phenotypes is clearly the result of structural alterations, or mutations, of the nucleotide sequence of the gene encoding the glycosyltransferase enzyme.

Muscular Dystrophy

Muscular dystrophy is characterized by severe progressive muscular degeneration, or myopathy, resulting in the death of the affected individuals due to respiratory failure, usually in their early twenties. Because the condition is recessive and X-linked, and because affected males die before they can reproduce, females are rarely affected by the disorder. The incidence of 1/3500 live male births makes muscular dystrophy one of the most common life-shortening hereditary disorders known. Two related forms exist. **Duchenne muscular dystrophy (DMD)** is more common and more severe than the allelic form, **Becker muscular dystrophy (BMD)**.

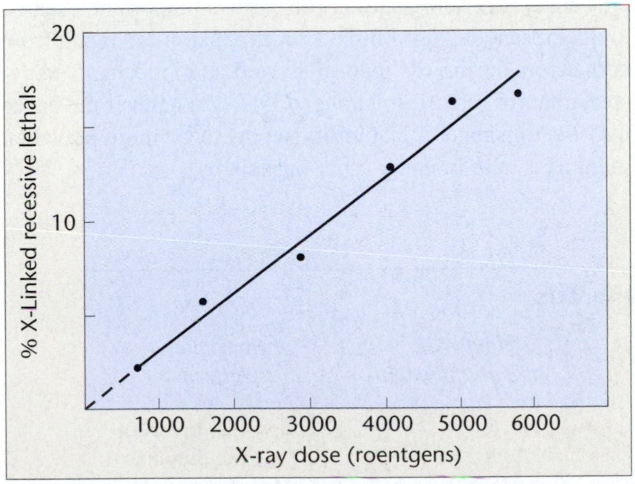

FIGURE 7–11 Plot of the percentage of X-linked recessive mutations induced by increasing doses of X rays. If extrapolated, the graph intersects the zero axis.

The gene, which is unusually large and consists of about 2.5 million bp, has been analyzed extensively. In normal (unaffected) individuals, transcription and subsequent processing of the initial transcript results in a messenger RNA containing only about 14,000 bases (14 kb). It is translated into the protein **dystrophin**, consisting of 3685 amino acids. In most cases of the less severe BMD, the dystrophin protein can be detected in the sarcolemma of smooth, striated, and cardiac myofibers. However, it is rarely detectable in the muscles of DMD patients. (See Figure 8–2.) This has led to the hypothesis that mutations causing BMD do not usually alter the reading frame of the gene, but that most DMD mutations change the reading frame early in the gene, resulting in the premature termination of dystrophin translation. We find that the hypothesis is consistent with the observed differences in severity of these two forms of the disorder.

In an extensive study performed in 1989, J. T. Den Dunnen and associates analyzed the DNA of 194 patients (160 DMD and 34 BMD). In most cases, the results were consistent with the "reading-frame" hypothesis. With few exceptions, DMD mutations changed the frame of reading, whereas BMD mutations usually did not.

Perhaps the most noteworthy of Den Dunnen's findings is the high percentage of deleterious mutations caused by the deletion or insertion of nucleotides within the gene. This observation reflects the fact that a mutation caused by a random single-nucleotide substitution within a gene is more likely to be tolerated without the devastating effect of muscular dystrophy than the addition or loss of numerous nucleotides that may alter the frame of reading. There are three reasons for this, as discussed in Chapter 5:

1. A nucleotide substitution may not change the encoded amino acid, since the code is degenerate.

2. If an amino acid substitution does result, the change may not be present at a location within the protein that is critical to its function.

3. Even if the altered amino acid is present at a critical region, it may still have little or no effect on the function of the protein. For example, an amino acid might be changed to another with nearly identical chemical properties or to one with very similar recognition properties, such as shape.

As a result, single-base substitutions sometimes have little or no effect on protein function, or they may simply reduce the efficiency, but not eliminate the functional capacity of the gene product. As more mutant genes are analyzed directly, our picture of mutation will become increasingly clear.

Trinucleotide Repeats in Fragile X Syndrome, Myotonic Dystrophy, and Huntington Disease

Beginning about 1990, the molecular analysis of the DNA representing the genes responsible for a number of inherited human disorders provided a remarkable set of observations. Mutant genes were characterized by an expansion of a simple **trinucleotide repeat sequence**, usually from fewer than 15 copies in normal individuals to a large number in affected individuals. For example, present in each of the three genes thought to be responsible for the X-linked fragile X syndrome and the autosomal disorders myotonic dystrophy and Huntington disease is a different trinucleotide DNA sequence repeated many times. While repeated sequences are also present in the nonmutant (normal) allele of each gene, the mutations are characterized by a significant variable increase in the number of times the trinucleotide is repeated. Table 7–4 summarizes the various cases that are discussed next, with particular emphasis on the size of the repeats in normal and diseased individuals.

In Chapter 10, we will discuss the onset of the expression of various phenotypes. In several cases, a correlation has been found between the number of repeats and the age of manifestation of mutant phenotypes. The greater the number of repeats, the earlier disease onset occurs. Further, in affected individuals, the number of repeats may increase in each subsequent generation. This general phenomenon, **genetic anticipation**, reflects a unique form of mutation related to an instability of the specific regions in each of the three different genes.

We begin with a short discussion of **fragile X syndrome**. The responsible gene, *FMR-1*, may have several hundred to several thousand copies of the trinucleotide sequence CGG. Individuals with up to 54 copies are normal and do not display the mental retardation associated with the syndrome. Individuals with 54–230 copies are considered "carriers." Although they are normal, their offspring may contain even more copies and express the syndrome.

Myotonic dystrophy, or DM (short for its original name, dystrophia myotonica), is the most common form of adult muscular dystrophy. Due to a dominant mutant gene located on the long arm of chromosome 19, the disorder is not as severe as as DMD, and it is highly variable both in symptoms and age of onset. Mild myotonia—atrophy and weakness—of the musculature of the face and extremities is most common. Cataracts, reduced cognitive ability, and cutaneous and intestinal tumors are also part of the syndrome.

The affected gene is thought to be *MDPK*, which encodes a serine–threonine protein kinase, MDPK. This protein is the product of 15 exons of the gene, the last of which encodes the 3′-untranslated RNA sequence of the mRNA. It is this sequence that houses the multiple copies of the trinucleotide, reflected in the gene as the DNA sequence CTG. Individuals with 5–37 copies are normal, and the number of copies is stable from generation to generation. Above this number (5–37 copies), symptoms range from mild to severe, with onset occurring anywhere between birth and age 60. Both the severity and onset are directly correlated with the size of the repeated sequence. Minimally affected patients are known to contain up to 150 repeats, while severely affected patients have up to 1500 copies of the the CTG triplet. Genetic anticipation is exhibited in the offspring.

More recently, the gene responsible for **Huntington disease**, has been found to demonstrate a similar mutational pattern. Behaving as an autosomal dominant and located on chromosome 4, the gene contains the trinucleotide CAG repeat 10 to 35 times in normal individuals. The sequence exists in significantly increased numbers (up to 120) in diseased individuals. Much earlier onset occurs when the number of copies present is closer to the upper range. Interestingly, in still another disorder, **spinobulbar muscular atrophy** (**Kennedy disease**), the involved gene (different from that in Huntington disease) also contains repeated copies of the CAG triplet. However, only 35 to 60 copies of the sequence cause individuals to be affected.

The role of such repeated sequences in normal and mutant genes remains a mystery. Their locations within the gene vary in each case. In Huntington and Kennedy diseases, the repeat lies within the coding portion of the gene. In the case of Huntington disease, this causes the mutant huntingtin protein to contain an excess of glutamine residues. Such is not the case in the other two disorders, however. In the gene responsible for fragile-X syndrome, the repeat is upstream (the 5' end) in an area that is most often involved in regulating gene expression. As pointed out earlier, in the case of myotonic dystrophy, the repeat is downstream (the 3'-end).

The mechanism by which the repeated sequence expands from generation to generation is also of great interest. How such an instability during DNA replication affects only specific areas of certain genes is currently an important research topic. Present day thinking is that expansion can result from both errors during replication as well as in mechanisms responsible for repairing damaged DNA. Whatever the cause may be, this general instability seems to be more prevalent in humans than in many other organisms.

TABLE 7–4 Summary of Trinucleotide-Repeat Disorders

	Trinucleotide Repeat	*Number in Normal Individuals*	*Inheritance Pattern*
Huntington disease	CAG	6–35	autosomal dominant
Myotonic dystrophy	CTG	5–37	autosomal dominant
Fragile-X syndrome	CGG	6–54	X-linked dominant
Spinobulbar muscular atrophy	CAG	10–35	X-linked dominant

7.7 The Ames Test Is Used to Assess the Mutagenicity of Compounds

There is great concern about the possible mutagenic properties of any chemical that enters the human body, whether through the skin, the digestive system, or the respiratory tract. For example, great attention has been given to residual materials of air and water pollution, food preservatives and additives, artificial sweeteners, herbicides, pesticides, and pharmaceutical products. As we have seen, mutagenicity may be tested in various organisms, including *Drosophila*, mice, and cultured mammalian cells. The most common test, which involves bacteria, was devised by Bruce Ames.

The **Ames test** (Figure 7–12), uses any of several strains of the bacterium *Salmonella typhimurium* that were selected for sensitivity and specificity for mutagenesis. One strain is used to detect base-pair substitutions, and the other three detect various frameshift mutations. Each mutant strain is unable to synthesize histidine, and therefore requires histidine for growth (his^-). The assay measures the frequency of reverse mutation, which yields wild-type (his^+) bacteria. Greater sensitivity to mutagens occurs because these strains bear other mutations that eliminate both the DNA excision-repair system (discussed later in this chapter) and the lipopolysaccharide barrier that coats and protects the surface of the bacteria.

It is very interesting to note that many substances entering the human body are relatively innocuous until activated metabolically, usually in the liver, to a more chemically reactive product. Thus, the Ames test includes a step in which the test compound is incubated in vitro in the presence of a mammalian liver extract. Or, test compounds may be injected into a mouse in which they are modified by liver enzymes and then recovered.

In the initial use of Ames testing in the 1970s, a large number of known carcinogens were examined, over 80 percent of which were shown to be strong mutagens! This is not surprising; the transformation of cells to the malignant state undoubtedly occurs as a result of some alteration of DNA. Although a positive response as a mutagen does not prove that a compound is carcinogenic, the Ames test is useful as a preliminary screening device. It is used extensively in conjunction with the industrial and pharmaceutical development of chemical compounds.

7.8 Organisms Can Counteract DNA Damage and Mutations by Activating Several Types of Repair Systems

In previous sections of this chapter, we established that replicating and nonreplicating DNA molecules are vulnerable to

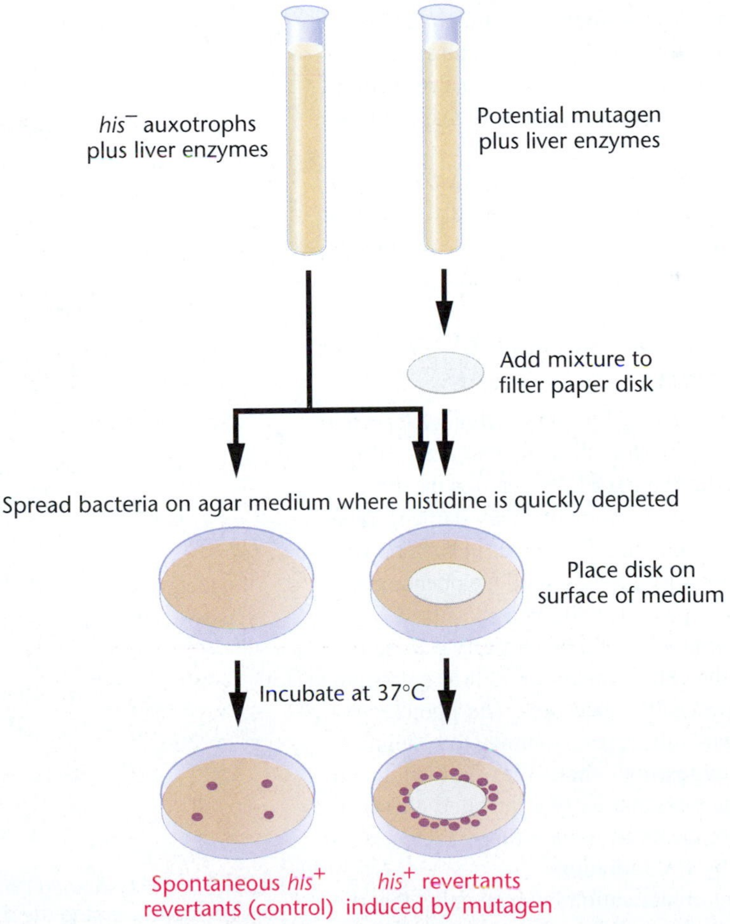
his⁻ auxotrophs
plus liver enzymes
Potential mutagen
plus liver enzymes
Add mixture to
filter paper disk
Spread bacteria on agar medium where histidine is quickly depleted
Place disk on
surface of medium
Incubate at 37°C
Spontaneous his⁺
revertants (control)
his⁺ revertants
induced by mutagen

FIGURE 7–12 The Ames test, which screens potential compounds for mutagenicity.

diverse chemical alterations and damage that can ultimately lead to gene mutations. Indeed, a perplexing variety of lesions can be induced in DNA. The sources of damage to DNA fall into three main categories. First are the exogenous, or so-called environmental, agents, several of which we have discussed previously. For example, UV and high-energy radiation, as well as genotoxic, or mutagenic, compounds, such as tobacco smoke, fall into this group. Second are endogenous by-products of normal cellular processes. These include reactive oxygen species (electrophilic oxidants) that are generated during normal aerobic respiration. For example, superoxides O_2^- hydroxyl radicals ($OH^·$), and hydrogen peroxide (H_2O_2) result from cellular metabolism and represent constant threats to the integrity of DNA. Reactive oxidants, also generated in the aftermath of exposure to high-energy radiation, are known to generate upwards of 100 types of chemical modifications of DNA. Third are the variety of spontaneous or induced chemical processes that directly affect the nitrogenous bases of DNA, either altering them or resulting in their removal from the DNA helix. For example, we have previously discussed the formation of apurinic sites and deamination as such processes.

Living systems have evolved a variety of elaborate repair systems that are able to counteract many of the forms of DNA damage resulting from these sources. As our knowledge of these systems continues to expand, the extensive variety and complex nature of the repair systems that are now apparent attest to the critical importance of DNA repair. As we will see, such repair systems are absolutely essential to the maintenance of the genetic integrity of organisms, and as such, to the survival of organisms on Earth. Of foremost concern in humans is the potential to counteract genetic damage that results in cancer.

We now embark on a review of the some of the systems of DNA repair. Since the field is expanding rapidly, our goal here will be to survey the major approaches that organisms have at their disposal to counteract genetic damage.

Photoreactivation Repair: Reversal of UV Damage in Prokaryotes

As illustrated in Figure 7–10, UV light is mutagenic as a result of the creation of pyrimidine dimers. The study of the mutagenicity of UV radiation paved the way for the discovery of many forms of natural repair of DNA damage. The first relevant discovery concerning UV repair in bacteria was made in 1949, when Albert Kelner observed the phenomenon of **photoreactivation repair**. He showed that the UV-induced damage to *E. coli* DNA could be partially reversed if, following irradiation, the cells were exposed briefly to light in the blue range of the visible spectrum. The photoreactivation repair process was subsequently shown to be temperature dependent, suggesting that the light-induced mechanisms involve an enzymatically controlled chemical reaction. Visible light appears to induce the repair process of the DNA damaged by UV radiation.

Further studies of photoreactivation have revealed that the process is dependent on the activity of a protein called the **photoreactivation enzyme (PRE)**. This molecule can be isolated from extracts of *E. coli* cells. The enzyme's mode of action is to cleave the bonds between thymine dimers, thus reversing the effect of UV radiation on DNA (Figure 7–13). Although the enzyme will associate with a dimer in the dark, it must absorb a photon of light to cleave the dimer. In spite of the potential for diminishing UV-induced mutations, photoreactivation repair is not absolutely essential in *E. coli*, since a mutation creating a null allele (a mutant allele producing no functional product) in the gene coding for the PRE is not lethal. In addition, the enzyme cannot be detected in humans and other eukaryotes, who must rely on other repair mechanisms to reverse the effects of UV radiation.

Base and Nucleotide Excision Repair

Investigations in the early 1960s suggested that, in addition to PRE, a *light-independent* repair system exists in prokaryotes such as *E. coli* that repairs damage to DNA caused by both exogenous and endogenous agents. The basic mechanism of this type of repair—referred to as **excision repair**—has been conserved throughout evolution, occurring in all prokaryotic and eukaryotic organisms. The process consists of the following three basic steps and can be described generally as a "cut-and-paste" system:

1. The distortion or error present on one of the two strands of the helix is recognized and then enzymatically clipped out by a nuclease. This "excision" usually includes a number of nucleotides adjacent to the error as well, leaving a gap on one strand in the helix.

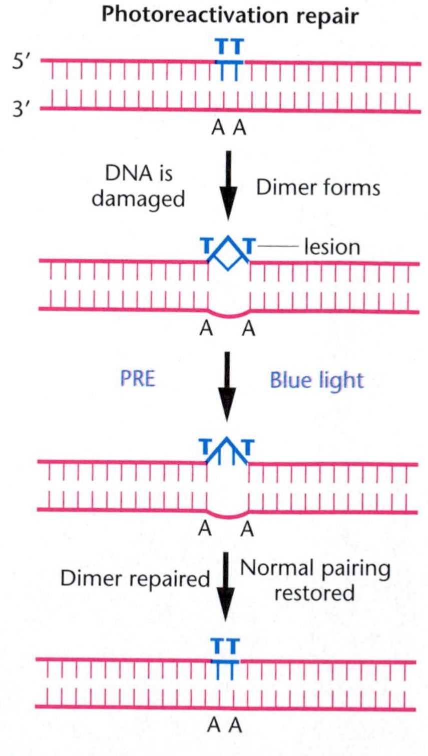

FIGURE 7–13 Damaged DNA repaired by photoreactivation repair. The bond creating the thymine dimer is cleaved by the photoreactivation enzyme (PRE), which must be activated by blue light.

2. A DNA polymerase fills this gap by inserting deoxyribonucleotides complementary to those on the intact strand, using it as a replicative template. The enzyme adds these bases to the free 3′-OH end of the clipped DNA. In *E. coli*, it is usually performed by DNA polymerase I.
3. The joining enzyme DNA ligase seals the final "nick" that remains at the 3′-OH end of the last base inserted, closing the gap.

DNA polymerase I, the enzyme discovered by Arthur Kornberg, was once assumed to be the universal DNA-replication enzyme. (See Chapter 3.) However, it was the discovery of the *polA1* mutation that demonstrated the role of this enzyme in repairing UV-induced lesions in *E. coli*. Cells carrying the *polA1* mutation lack functional polymerase I; yet, they replicate their DNA normally, and are unusually sensitive to UV radiation. Apparently, such cells are unable to fill the gap during excision repair of the lesions.

There are now known to be two types of excision repair: base excision repair and nucleotide excision repair. **Base excision repair (BER)** is involved in correcting minor chemical alterations of nitrogenous bases created by spontaneous hydrolysis or chemical agents. We describe such errors as minor, in that they do not tend to block DNA replication or transcription.

The first step of the BER pathway in *E. coli* involves the recognition of the chemically altered base by **DNA glycosylases**, which are specific to different types of DNA damage. For example, the enzyme uracil–DNA glycosylase recognizes the unusual presence of uracil when it is part of a nucleotide in DNA (Figure 7–14). The enzyme first cuts the glycosidic bond between the base and the sugar, creating an apyrimidinic (AP) site. Such a sugar with a missing base is then recognized by an enzyme called **AP endonuclease**. The endonuclease makes a cut in the sugar backbone at the AP site. This creates a distortion in the DNA helix that is recognized by the excision-repair system, which is then activated, leading ultimately to the correction of the error.

Although much has been learned about glycosylases in *E. coli*, much less is known about the DNA glycosylases in eukaryotes. In addition, unlike the nucleotide excision pathway we are going to discuss next, there is no human or animal disease model available that has a defective BER pathway, so it is difficult to assess the importance of BER in eukaryotes.

While base excision repair recognizes and replaces modified bases in DNA, the **nucleotide excision repair (NER)** pathway repairs lesions in DNA that distort the regular helix. The source of such damage is usually exogenous, such as UV light that induces pyrimidine dimers. Study of the repair of these dimers actually led to the discovery of NER, as we will see in the discussion that follows. However, all types of bulky lesions leading to distortions of the helix are subject to NER. Both replication and transcription may be impeded by these lesions.

The NER pathway, diagrammed in Figure 7–15, was first discovered in *E. coli* by Paul Howard-Flanders and coworkers, who were able to isolate several independent mutants that demonstrated sensitivity to UV radiation. One group of genes was designated *uvr* (ultraviolet repair) and included the *uvrA*, *uvrB*, and *uvrC* mutations. In the NER pathway, the *uvr* gene products are involved in recognizing and clipping out the lesions in the DNA. Usually, a very specific number of nucleotides are clipped out around both ends of the lesion. In *E. coli*, there are usually a total of 13 nucleotides removed, which include the lesion. The repair is then completed by DNA polymerase I and DNA ligase, in a manner similar to BER. The undamaged strand opposite the lesion is used as a template for the replication resulting in repair.

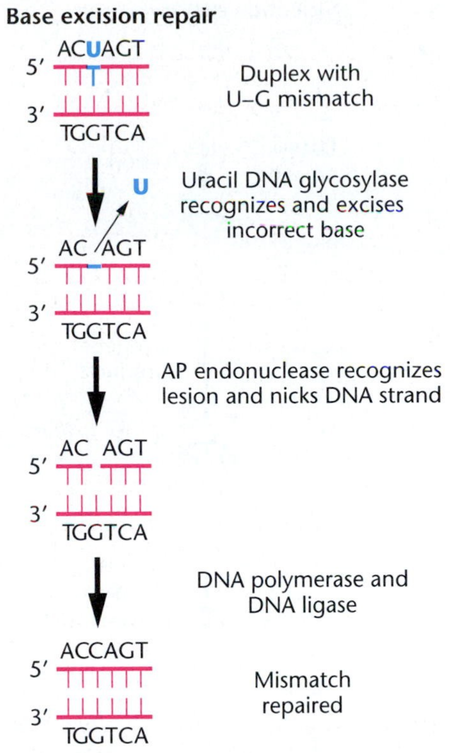
Base excision repair
5′ ACUAGT
3′ TGGTCA
Duplex with U–G mismatch
U
Uracil DNA glycosylase recognizes and excises incorrect base
5′ AC AGT
3′ TGGTCA
AP endonuclease recognizes lesion and nicks DNA strand
5′ AC AGT
3′ TGGTCA
DNA polymerase and DNA ligase
5′ ACCAGT
3′ TGGTCA
Mismatch repaired

FIGURE 7–14 Base excision repair (BER) accomplished by uracil DNA glycosylase, AP endonuclease, DNA polymerase, and DNA ligase. Uracil is recognized as a noncomplementary base, excised, and replaced with the complementary base (C).

Xeroderma Pigmentosum and Nucleotide Excision Repair in Humans

The mechanism of nucleotide excision repair (NER), as elucidated in eukaryotes, is much more complicated than prokaryotic NER because it involves many more proteins. Much of what is known about the system in humans has resulted from detailed studies of individuals with **xeroderma pigmentosum (XP)**, a rare genetic disorder that predisposes individuals to severe skin abnormalities. These individuals have lost their ability to undergo NER. As a result, individuals suffering from XP who are exposed to the UV radiation present in sunlight exhibit reactions that range from initial freckling and skin ulceration to the subsequent development of skin cancer. Figure 7–16 contrasts two XP individuals, one of whom has been detected early and protected from sunlight.

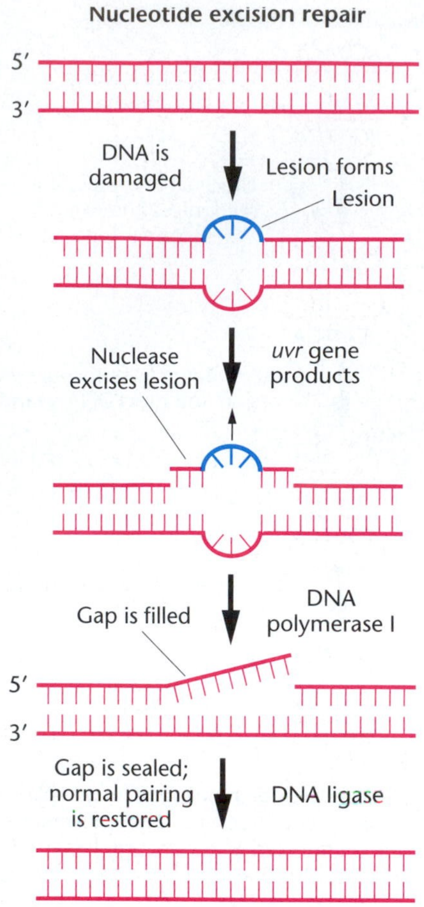

FIGURE 7–15 Nucleotide excision repair (NER) of a DNA lesion. In actuality, a total of 13 bases are excised in prokaryotes, and 29 bases are excised in eukaryotes during repair.

The condition is very severe and may be lethal, although early detection and protection from sunlight can arrest it. Because sunlight contains UV radiation, a causal relationship had been predicted between thymine dimer production and XP. Thus, the ability to repair UV-induced lesions was investigated in human fibroblast cultures derived from XP and normal individuals. (Fibroblasts are undifferentiated connective tissue cells.) The results suggested that the XP phenotype may be caused by more than one mutant gene.

In 1968, James Cleaver showed that cells from XP patients were deficient in **unscheduled DNA synthesis** (DNA synthesis other than that occurring during chromosome replication), which is elicited in normal cells by UV radiation. Because this type of synthesis is thought to represent the activity of the excision-repair system, the suggestion is that XP cells are deficient in excision repair.

The link between xeroderma pigmentosum and inadequate excision repair has been strengthened by the use of **somatic cell hybridization** studies, a technique that we will discuss in Chapter 12. Fibroblast cells from any two unrelated XP patients, when grown in tissue culture, may be induced to fuse together, forming what is called a heterokaryon, in which a single cell with two nuclei is formed that shares a common cytoplasm. After fusion, excision repair, as assayed by unscheduled DNA synthesis, may or may not be reestablished in the heterokaryon. When it is reestablished, the two variants are said to demonstrate **complementation**. Alone, neither cell type demonstrates excision repair, but together in a heterokaryon, the repair occurs. In genetic terms, this is strong evidence that in the two patients from whom the cells were derived, different affected genes led to the disease. Complementation occurs because the heterokaryon has at least one normal copy of each gene in the fused nucleus.

Based on many studies, patients have been divided into seven complementation groups, suggesting that at least seven different genes may be involved in excision repair. Any two in different complementation groups will complement one another. If two strains contain separate mutations, but in the same complementation group (gene), no complementation occurs.

These seven human genes, or their protein products, have now been identified. The genes are found in disparate regions of the genome, and a homologous gene for each has been identified in yeast. Approximately 20 percent of XP patients do not fall into any of the seven groups. They manifest similar symptoms, but their fibroblasts do not demonstrate defective excision repair. There is some evidence that they are less efficient in normal DNA replication.

As a result of the study of the defective genes in xeroderma pigmentosum, a great deal is now known about how NER counteracts DNA damage in normal cells. The first step

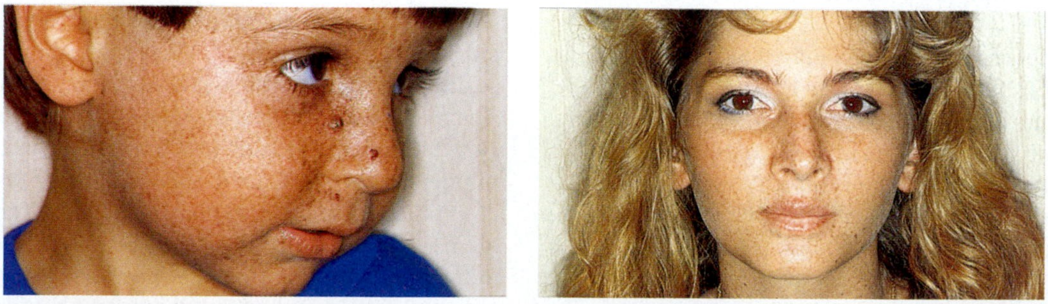

FIGURE 7–16 Two individuals with xeroderma pigmentosum. The 4-year-old boy on the left shows marked skin lesions induced by sunlight. Mottled redness (erythema) and irregular pigment changes that are a response to cellular injury are apparent. Two nodular cancers are present on his nose. The 18-year-old girl on the right has been carefully protected from sunlight since the diagnosis of xeroderma pigmentosum in infancy. Several cancers have been removed and she has worked as a successful model.

in humans is the recognition of the damaged DNA by a specific protein called XPA (*X*erodema *P*igmentosum gene *A*). The binding of XPA to the damaged DNA helix causes other proteins in the pathway to be recruited to the site. The transcription factor TFIIH is involved in creating the repair complex. This repair complex excises a 28-nucleotide-long fragment from the helix, including the lesion. Recall that NER in bacteria is also quite specific in the number of nucleotides excised, although less so than in eukaryotes.

Another defect in the human NER pathway has been linked to **Cockayne syndrome (CS)**, which is characterized by impairment of both physical and neurological development. While individuals are highly photosensitive, as in XP, no predisposition to tumor formation characterizes the disease. Genes separate from the ones involved in XP are affected in CS.

Proofreading and Mismatch Repair

One of the most common types of error in DNA is the one made during replication when an incorrect (noncomplementary) nucleotide is inserted by DNA polymerase. The enzyme in bacteria (DNA polymerase III) is known to make such an error approximately once every 100,000 insertions, leading to an error rate of 10^{-5}. Fortunately, the enzyme polices its own synthesis by **proofreading** each step, catching 99 percent of those errors. During polymerization, when an incorrect nucleotide is inserted, the enzyme complex has the potential to recognize the error and "reverse" its direction and behave as an exonuclease, cutting out the incorrect nucleotide and then replacing it. This improves the efficiency of replication one hundredfold, creating only $1/10^7$ mismatches immediately following DNA replication, for a final error rate of 10^{-7}.

To cope with those errors that remain after proofreading, still another mechanism, called **mismatch repair**, may be activated. Proposed over 20 years ago by Robin Holliday, the molecular basis of such a process is now well established. As in other DNA lesions, the alteration or mismatch must be detected, the incorrect nucleotide(s) must be removed, and the replacement with the correct nucleotide(s) must occur.

But a special problem exists with the correction of a mismatch in comparison to excision repair, in which a lesion signals its repair. In mismatch repair, the noncomplementary pair of nucleotides is recognized, but how does the repair system recognize which strand is correct (the template during replication) and which contains the mismatched base (the newly synthesized strand)?

How the repair system discriminates and recognizes the "new" nucleotide puzzled geneticists for decades. If the mismatch is recognized, but no discrimination occurred and excision was random, half of the time, the strand bearing the correct base would be clipped out. The concept of strand discrimination by a repair enzyme is thus a critical step.

At least in some bacteria, including *E. coli*, this process has been elucidated and is based on **DNA methylation**. These bacteria contain an enzyme, **adenine methylase**, that recognizes the DNA sequence

$$5' \ldots GATC \ldots 3'$$
$$3' \ldots CTAG \ldots 5'$$

as a substrate. Upon recognition, a methyl group is added to each of the adenine residues. This modification is stable throughout the cell cycle.

Following a further round of replication, the newly synthesized strands remain temporarily unmethylated. It is at this point that the repair enzyme recognizes the mismatch and preferentially binds to the unmethylated strand. A nick is made by an endonuclease protein, either 5′ or 3′, to the mismatch on the unmethylated strand. The nicked DNA strand is then unwound, degraded, and replaced until the mismatch is reached and excised.

A series of *E. coli* gene products, Mut H, L, S, and U, are involved in the overall process. Mut S recognizes the initial mismatch, and in fact, can also recognize small insertion and deletion loops created during DNA synthesis. Mutations in each gene result in bacterial strains deficient in mismatch repair. In bacteria, this repair process is remarkably efficient, improving the error rate one thousandfold (99.9 percent of the mismatches are corrected). While the preceding mechanism discussion is based on studies of *E. coli*, similar mechanisms involving homologous proteins are known to exist in yeast and in mammals.

Postreplication Repair and the SOS Repair System

Still other types of repair have been discovered, illustrating the diversity of such mechanisms that have evolved to overcome DNA damage. One system, called **postreplication repair**, was discovered in an excision-defective strain of *E. coli* and first proposed by Miroslav Radman. It is a system that responds *after* damaged DNA has escaped repair and failed to be completely replicated, thus its name. As illustrated in Figure 7–17, when DNA bearing a lesion of some sort (such as a pyrimidine dimer) is being replicated, DNA polymerase at first stalls at the lesion and then skips over it, leaving a gap along the newly synthesized strand. To counteract this, the RecA protein directs a recombinational exchange process whereby this gap is filled as a result of the insertion of a segment initially present on the undamaged complementary strand. That event creates a gap on the "donor" strand, which can then be filled by repair synthesis as replication proceeds. Because a recombinational event is involved, postreplication is also referred to as *homologous recombination repair*, a more general category.

Still another repair pathway in *E. coli* is called the **SOS repair system**, which also responds to damaged DNA, but in a different way. As a lesion is encountered during DNA synthesis that is sufficient to impede further synthesis, a newly discovered group of polymerases (e.g., DNA polymerase V in *E.coli*) have the capacity to respond to the "crisis" and allow replication to continue in spite of the lesion. Thus, they are referred to as **translesional polymerases**. While they overcome the cessation of synthesis, the down side of these enzymes is that they have less stringent base-pairing

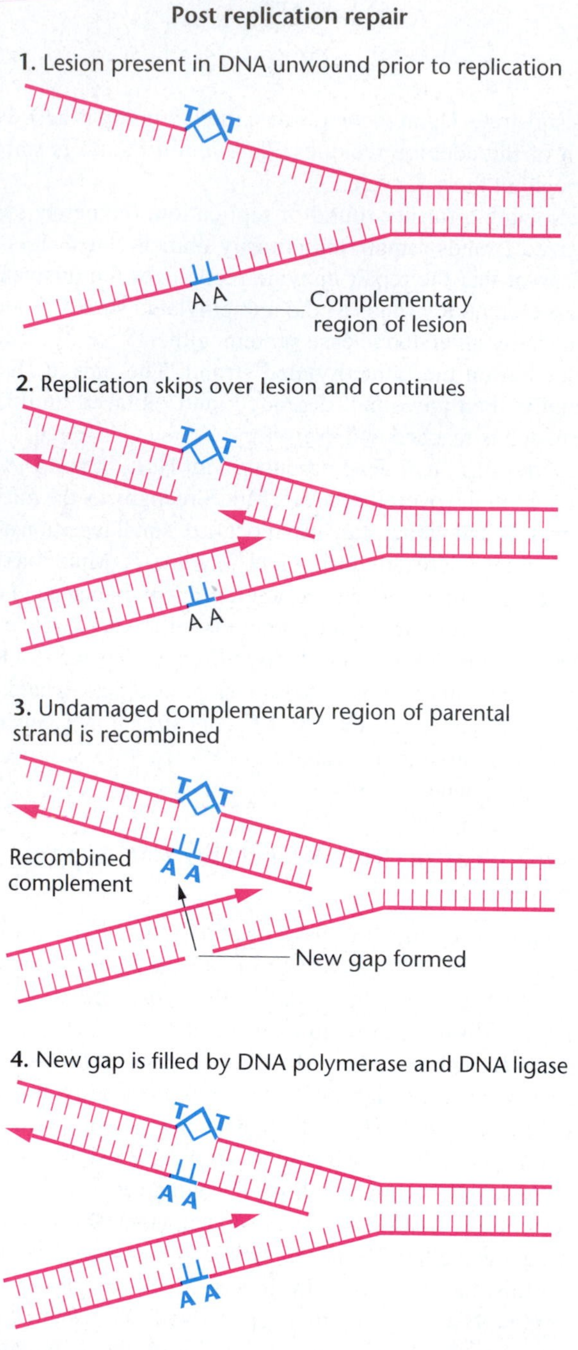
Post replication repair
1. Lesion present in DNA unwound prior to replication
T T
A A
Complementary region of lesion
2. Replication skips over lesion and continues
T T
A A
3. Undamaged complementary region of parental strand is recombined
T T
Recombined complement
A A
New gap formed
4. New gap is filled by DNA polymerase and DNA ligase
T T
A A
A A

FIGURE 7–17 Postreplication repair that occurs if DNA replication has skipped over a lesion, such as a thymine dimer. Through the process of recombination, the correct complementary sequence is recruited from the undamaged parental strand and inserted into the gap opposite the lesion. The new gap created is filled by DNA polymerase and DNA ligase.

requirements, making them, as they continue replication, highly prone to error. It is thought that all species contain this group of polymerases and that there may be a number of such enzymes, each designed to overcome particular lesions.

The activation of the SOS system of repair is believed to involve a number of genes, including many studied by Phil Hanawalt and Paul Howard-Flanders, both pioneers in the investigation of DNA repair mechanisms. The products of these genes are involved in the conversion of the lesion into an error-prone site, which enables the translesional DNA polymerase to replicate DNA past the lesion.

Double-Strand Break Repair in Mammals

Thus far, we have only discussed repair pathways that deal with damage *within* a particular strand of DNA. We conclude our discussion of DNA repair by considering what happens when both strands of the DNA helix are cleaved, for example, as a result of exposure to ionizing radiation, which causes what are called double-strand breaks. While such damage occurs in bacteria as well, we will focus our discussion on eukaryotic cells. Fortunately, a specialized form of DNA repair, the **DNA double-strand break repair (DSB repair)** pathway, is activated and is responsible for ultimately reannealing the two DNA segments. Recently, interest has grown in DNA DSB repair because defects in this pathway are associated with X-ray hypersensitivity and immune deficiency. Such defects may also underlie familial disposition to breast and ovarian cancer.

Similar to postreplication repair, one pathway involved in double-strand break repair is referred to as **homologous recombinational repair**, because damaged DNA is actually recombined and replaced with homologous undamaged DNA. This is necessary because, when both strands are broken, there is no undamaged parental strand available to use as the source of the complementary DNA sequence during repair. Therefore, the genetic information present in the homologous region of either sister homolog is "recruited" to replace the damaged double-stranded break. The undamaged homologous region is actually recombined into the damaged DNA molecule.The process usually occurs during the late S/G2 phase of the cell cycle, after DNA replication, thus allowing the recruitment of the undamaged copy of the "sister" chromosome to be utilized. In yeast, the system depends upon a complex of at least five proteins, called the RAD52 complex. As a result, the complex is able to restore the integrity of the broken double helix.

The second pathway, called **nonhomologous recombinational repair**, or **end joining**, achieves the same type of repair. However, as the name implies, the mechanism does not recruit a homologous region of DNA during repair. This system is activated in G1, prior to DNA replication Two different protein complexes are involved in nonhomologous repair.

7.9 Site-Directed Mutagenesis Allows Researchers to Investigate Specific Genes

This section introduces a useful experimental technique, **site-directed mutagenesis**, that allows researchers to introduce a designed mutation at a prescribed site within a gene of interest. The technique, developed by Michael Smith, relies on the availability of a cloned gene and uses a number of recombinant DNA technology manipulations discussed in Chapter 16. The underlying principles, however, are based on information we have presented in this chapter. Smith received the Nobel Prize in Chemistry in 1993 for his contribution.

The goal of the technique is to alter one or more specific nucleotides within a gene in order to change a specific triplet codon. Upon transcription and translation, the change causes the insertion of a "mutant" amino acid into the protein encoded by the original gene. Such designed mutations are particularly useful in studying the effects of specific mutations on both gene expression and protein function. Various approaches can be used to accomplish the goal; most are variations on the theme we are about to describe are illustrated in Figure 7–18.

The initial step (Step 1) is to determine the nucleotide sequence of the gene being studied. This is accomplished by DNA sequencing techniques, or, if the amino acid sequence of the protein is known, it can be predicted by extrapolating from the genetic code. The next step is to isolate one of the two complementary strands from the DNA of known sequence and decide which nucleotides are to be changed (Step 2). Then, a small piece of DNA, a synthetic oligonucleotide, is chemically synthesized, which is complementary to that region at all points, except in the triplet sequence or sequences that are to be altered (Step 3). Such a sequence includes the triplet encoding the amino acid that will change in the protein.

As illustrated in Step 4, this short piece of DNA is hybridized with the original parent strand, forming a partial duplex because of the complementarity along most of its length. If DNA polymerase and DNA ligase are then added to the hybrid complex, the short sequence is extended so that a duplex of the entire gene is formed. The two strands are perfectly complementary, except at the point of alteration.

If this DNA is replicated (Step 5), two types of duplexes are formed; one is like the original, unaltered gene and the other contains the newly designed sequence. Recombinant DNA technology makes it possible not only to complete these manipulations, but also to allow the altered gene to be expressed so that large amounts of the desired protein are available for study. Or, as we will see in the next section, the altered gene may be introduced into an organism and studied.

Still other approaches are utilized in conjunction with recombinant DNA techniques. These involve the use of the restriction enzymes and the polymerase chain reaction (PCR), discussed in Chapter 16.

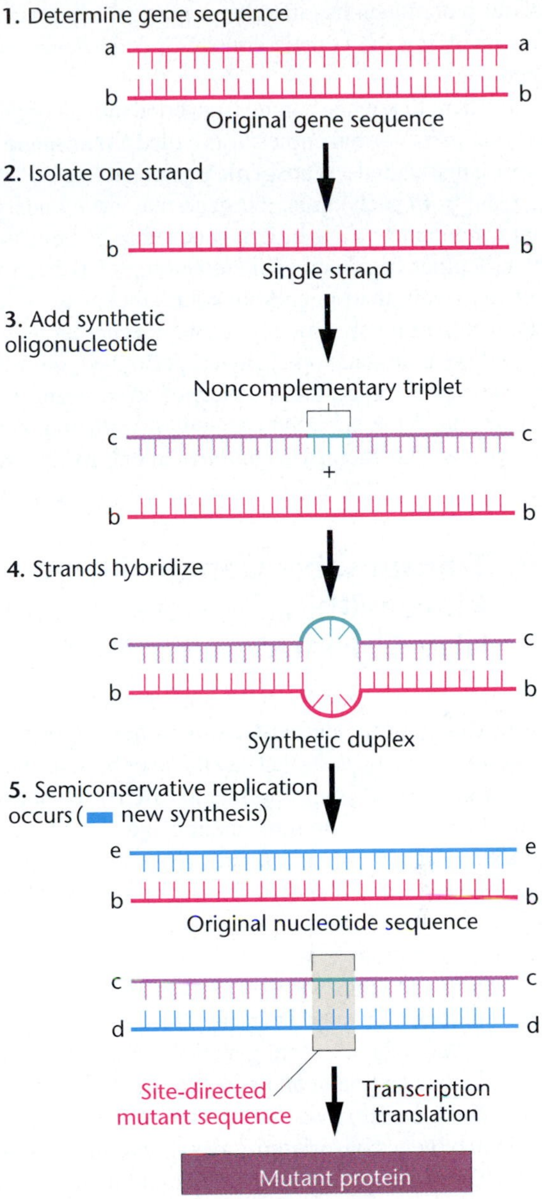

FIGURE 7–18 Site-directed mutagenesis. A single strand of DNA from a gene of interest is initially isolated. This is hybridized with a synthetic oligonucleotide containing a triplet altered so as to encode an amino acid of choice. Following semiconservative replication, a different complementary base pair is present in one of the new duplexes. Upon transcription and translation, a mutant protein, "designed" in the laboratory, will be produced.

Knockout Mutations and Transgenes

One of the most useful applications of site-directed mutagenesis is to alter a gene sufficiently so as to render it nonfunctional. Recall that such a "loss-of-function" condition is referred to as a null allele. Either deletions within the gene are created or a disruptive sequence is inserted. Techniques have been developed with which such a gene can be inserted into the germ line of an organism, allowing for the assessment of the gene's function. When the insertion into the genome involves the *replacement* of the comparable gene of the organism from which it originated, the process is described as a **gene knockout**. The organism now contains a **knockout mutation**, and the genetic alteration creates a knockout organism.

For example, a "knockout mouse" describes a member of a true-breeding strain that lacks the function of the gene that has been replaced with the null allele. Knockout mice now serve as models for studying some human genetic disorders. In such cases, the mouse gene comparable to the gene that causes the human disorder is isolated, subjected to site-directed mutagenesis, and used to replace the normal mouse gene. Cystic fibrosis and Duchenne muscular dystrophy have been investigated in this way. In mice, the application of gene knockout techniques has also been particularly fruitful in studies involving the genetic control of early development and behavior.

We note here that when a gene is inserted into an organism *in addition* to its normal copies, it is called a **transgene**, and the organism is called a **transgenic organism** (e.g., a "transgenic plant"). In such cases, the gene may have undergone site-directed mutagenesis, or it may be a foreign gene isolated from another organism. This technology has been used more extensively than gene knockout, since it is easier to accomplish because specific replacement is not required.

The study of transgenic organisms is performed routinely in plants, *Drosophila*, mice, and a variety of other organisms. Of particular note is the potential provided in agricultural studies, as well as gene therapy techniques in our own species.

7.10 Transposable Genetic Elements Move within the Genome and May Disrupt Genetic Function

We conclude this chapter by discussing the phenomenon of **transposable genetic elements**, sometimes referred to as **transposons**—genetic units that can move or be "transposed" within the genome. It is appropriate to discuss transposable elements here because the movement of genetic units from one place in the genome to another often disrupts genetic function and results in phenotypic variation. As such, the impact of the transposition of genetic units often fits into a broad definition of mutation.

Transposable elements were first studied in maize by Barbara McClintock almost 50 years ago. However, even in the 1950s and 1960s, the idea that genetic information was *not* fixed within the genome of an organism was slow to find acceptance. Such a notion was quite alien to the classical interpretation of genes on chromosomes. It was not until other transposable elements were discovered, and their molecular basis revealed, that the phenomenon was found to be nearly universal.

Insertion Sequences

Although the presence of transposable elements in maize had been predicted earlier, the first observation at the molecular level came in the early 1970s. A number of independent researchers, including Peter Starlinger and James Shapiro, visualized a unique class of mutations affecting different genes in various bacterial strains. For example, the expression of a cluster of related genes involving galactose metabolism in *E. coli* was repressed as a result of one such mutation; the phenotypic effect was heritable, but was found not to be caused by a base-pair change characteristic of conventional gene mutations. Instead, it was shown that a short specific DNA segment had been inserted into the bacterial chromosome at the beginning of the galactose gene cluster. When this segment was spontaneously excised from the bacterial chromosome, wild-type function was restored. Such a segment came to be called an **insertion sequence (IS)**.

It was subsequently revealed that several other distinct DNA segments could behave in a similar fashion, inserting into the chromosome and affecting gene function. These DNA segments are relatively short, not exceeding 2000 bp (2 kb). The first insertion sequence to be characterized in *E. coli*, IS1, is about 800 bp long; IS2, 3, 4, and 5 are about 1250–1400 bp in length.

The analysis of the DNA sequences of most IS units reveals a feature important to their mobility. The nucleotide sequences contain **inverted terminal repeats (ITRs)** of one another—that is, the final sequence at the 5' end of one strand is the same as the final sequence at the 5' end of the other strand, except that the sequences run in opposite directions (Figure 7–19). Although the figure shows the terminal-repeating unit to consist of only a few nucleotides, many more are actually involved. For example, in *E. coli*, the IS1 termini contain about 20 nucleotide pairs, IS2 and IS3 about 40 pairs, and IS4 about 18 pairs. It seems likely that these terminal sequences are an integral part of the mechanism of the insertion of IS units into DNA. That insertion of IS units is more likely to occur at certain DNA regions than at others suggests that IS termini can recognize certain target sequences in the DNA during the process of insertion.

Careful investigation has revealed IS units in the wild-type *E. coli* chromosome as well as in other autonomous segments of bacterial DNA, called plasmids. Thus, their presence does not always result in mutation. In the *E. coli* chromosome, five or more copies of IS1, IS2, and IS3 are present. The exact number of each varies, depending on the strain examined.

Bacterial Transposons

In addition to their potential mutational effects, IS units play an even more significant role in the formation and movement of the larger **transposon (Tn) elements**. Transposons in bacteria consist of IS units that contain genes whose functions are unrelated to the insertion process. Like IS units, Tn elements are mobile in both bacterial and viral chromosomes and in plasmids. The Tn elements provide a mechanism for the movement of genetic information from place to place both within and between organisms. Transposons were first

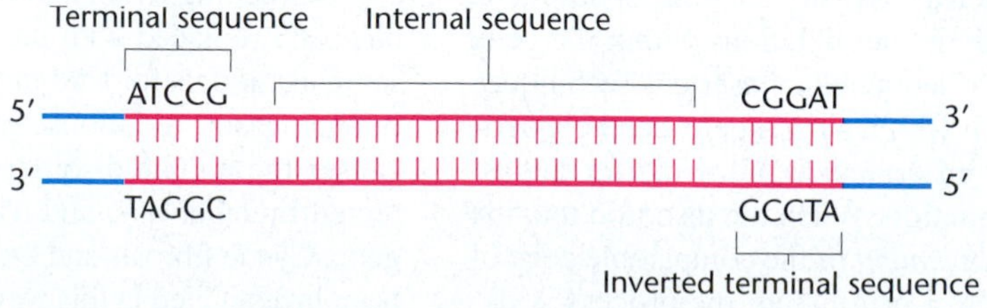

FIGURE 7–19 An insertion sequence (IS). The terminal sequences are perfect inverted repeats of one another.

discovered to move between DNA molecules as a result of observations of antibiotic-resistant bacteria. In the mid-1960s, Susumu Mitsuhashi first suggested that the genes responsible for resistance to several antibiotics were mobile and could move between bacterial plasmids and chromosomes.

Electron microscopic studies may be used to confirm the presence of the terminal-inverted repeat sequences within plasmids harboring transposons. When double-stranded DNA from such a plasmid is separated into single strands and each is allowed to reanneal separately, the inverted repeat units are complementary, forming a heteroduplex. As might be predicted, all areas other than the terminal-repeat units remain single stranded and form loops on either end of the double-stranded stem (Figure 7–20).

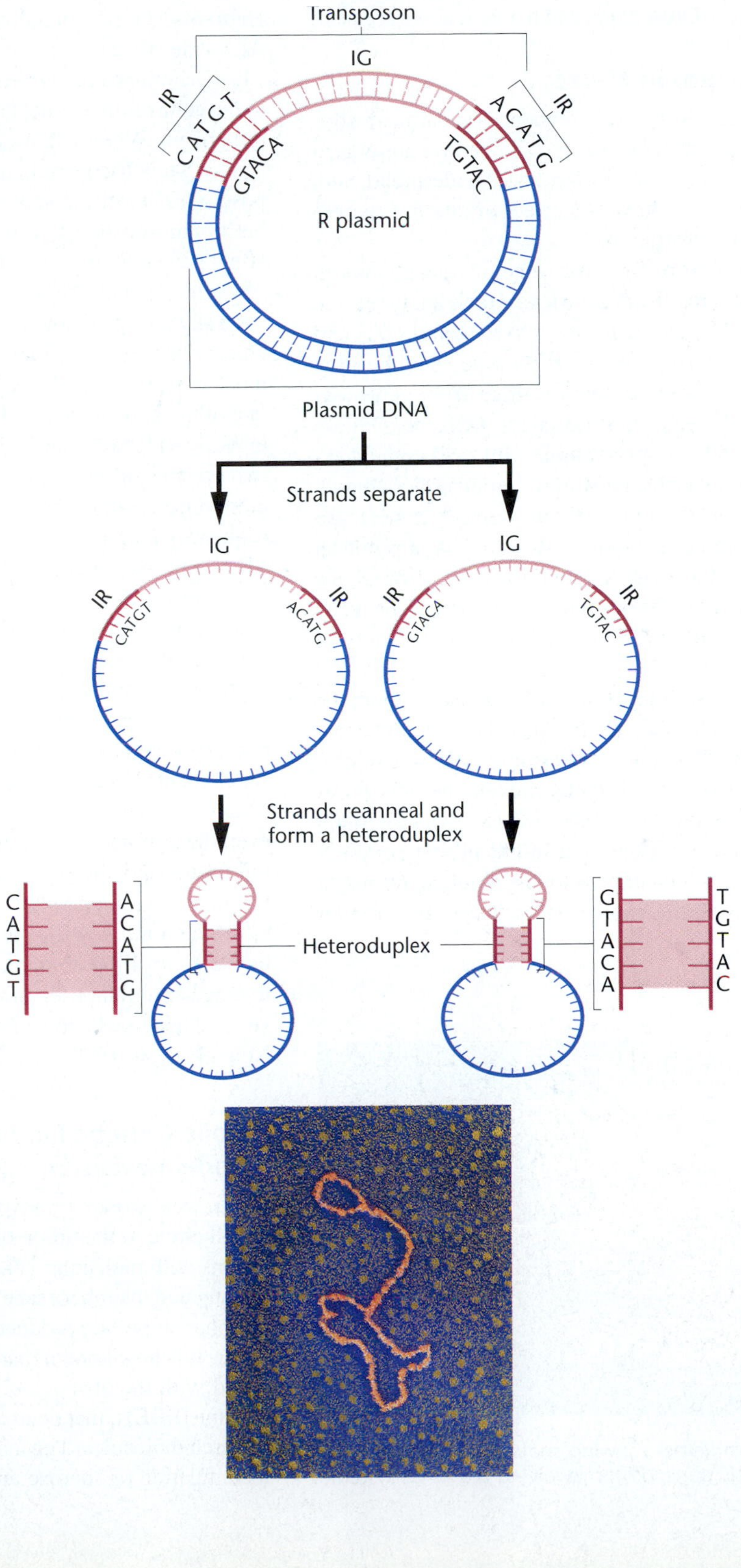

FIGURE 7–20 Heteroduplex formation as a result of inverted repeat sequences within a transposon inserted into a bacterial plasmid. The micrograph illustrates the final heteroduplex.

Transposons have become the focus of increased interest, particularly because they have been found in organisms other than bacteria. Bacteriophages that demonstrate the ability to insert their genetic material into the host chromosome behave in a similar fashion. The bacteriophage Mu, consisting of over 35,000 nucleotides, can insert its DNA at various places in the *E. coli* chromosome. Like IS units, if insertion occurs within a gene, mutant behavior results at that locus. Transposons have also been discovered in higher organisms, including yeast, maize, *Drosophila*, and humans, as we see next.

The Ac–Ds System in Maize

With our knowledge of insertion sequences and transposons in bacteria, it is no surprise that mobile genetic units also exist in eukaryotes. They are, in fact, more widespread, and, like bacteria, they often have the effect of altering the expression of genetic information.

About 20 years before the discovery of transposons in prokaryotic organisms, Barbara McClintock analyzed the genetic behavior of two mutations, *Dissociation* (*Ds*) and *Activator* (*Ac*), in corn plants (maize). She examined the phenotypes of the maize kernels resulting from genes expressed in either the endosperm or aleurone layers (Figure 7–21) and correlated her observations with a cytological examination of the maize chromosomes. Initially, McClintock determined that *Ds* is located on chromosome 9. If *Ac* is also present in the genome, *Ds* induces breakage at a point on the chromosome adjacent to its own location. If breakage occurs in somatic cells during their development, progeny cells often lose part of chromosome 9, causing a variety of phenotypic effects.

Subsequent analysis suggested to McClintock that both the *Ds* and *Ac* genes are sometimes transposed to different chromosomal locations. While *Ds* moves only if *Ac* is also present, *Ac* is capable of autonomous movement. Where *Ds* comes to reside determines its genetic effect—that is, it might cause chromosome breakage or it might inhibit gene expression. In cells in which expression is inhibited, *Ds* might move again, releasing this inhibition. In these cases, the *Ds* element is believed to insert into a gene and subsequently to depart from it, causing changes in gene expression.

FIGURE 7–21 Corn kernel showing spots of colored aleurone produced by genetic transposition involving the *Ac–Ds* system.

Figure 7–22 illustrates the sort of movements and effects of the *Ds* and *Ac* elements described. In McClintock's original observation, when the *Ds* element jumped out of chromosome 9, this excision event restored normal gene function. In such cells, pigment synthesis was restored. McClintock concluded that the *Ds* and *Ac* genes are mobile *controlling elements*. We now commonly refer to them as **transposable elements**, a term coined by another great maize geneticist, Alexander Brink.

It was not until many years later that anything comparable to the controlling elements in maize was recognized in other organisms. When bacterial insertion sequences and transposons were discovered, many parallels were evident. Transposons and insertion sequences were seen to move into and out of chromosomes, to insert at different positions, and to affect gene expression at the point of insertion.

Several *Ac* and *Ds* elements have now been isolated and carefully analyzed, and the relationship between the two elements has been clarified (Figure 7–23). The first *Ac* element sequence is 4563 bases long and strikingly similar to one of the known bacterial transposons. This sequence contains two 11-base-pair imperfect terminal-inverted repeats, two **open reading frames** (**ORFs**), and three noncoding regions. Open reading frames contain initiation and termination sequences and are considered to encode genetic products. The first *Ds* element studied (*Ds-a*) is nearly identical in structure to *Ac* except for a 194-bp segment that has been deleted from the largest open reading frame. There is some evidence that the gene encodes a **transposase enzyme**, essential to transposition of both the *Ac* and *Ds* elements. The deletion of part of the gene in the *Ds-a* element explains its dependence on the *Ac* element for transposition. Several other *Ds* elements have also been sequenced, and each reveals an even larger deletion in the same region. In each case, however, the terminal repeats are retained and seem to be essential for transposition, provided that a functional transposase enzyme is supplied by the gene in the *Ac* element.

Although the significance of Barbara McClintock's proposed controlling elements was not fully appreciated following her initial observations, molecular analysis has since verified her conclusions. For her work, she was awarded the Nobel Prize in 1983.

Mobile Genetic Elements and Wrinkled Peas: Mendel Revisited

More recent work on transposable elements in plants has led us full circle to the union of one of Gregor Mendel's observations with molecular genetics. Early in his work, Mendel investigated the inheritance of round and wrinkled peas. The two phenotypes are produced by alleles of a single gene, *rugosus*. It is now known that the wrinkled phenotype is associated with the absence of an enzyme, **starch-branching enzyme** (**SBEI**), that controls the formation of branch points in starch molecules. The lack of starch synthesis leads to the accumulation of sucrose and to a higher water content and

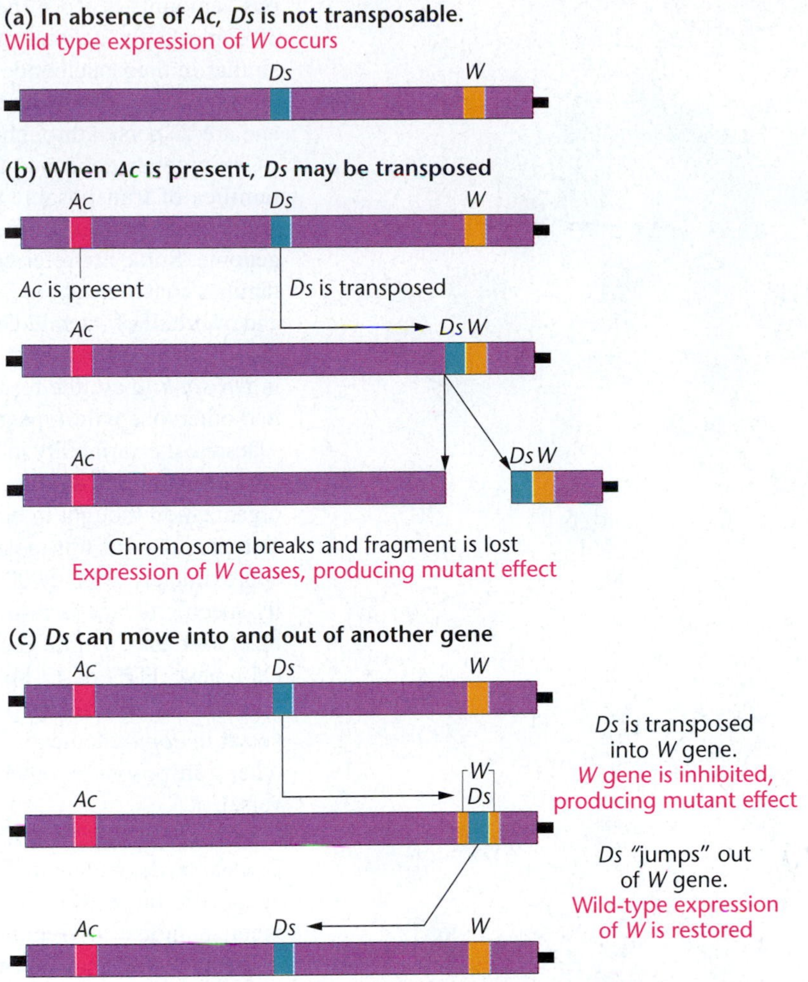
(a) In absence of Ac, Ds is not transposable.
Wild type expression of W occurs
Ds
W
(b) When Ac is present, Ds may be transposed
Ac
Ds
W
Ac is present
Ds is transposed
Ac
Ds W
Ac
Ds W
Chromosome breaks and fragment is lost
Expression of W ceases, producing mutant effect
(c) Ds can move into and out of another gene
Ac
Ds
W
Ds is transposed into W gene.
W gene is inhibited, producing mutant effect
Ac
W
Ds
Ds "jumps" out of W gene.
Wild-type expression of W is restored
Ac
Ds
W

FIGURE 7–22 Consequences of the influence of the *Activator* (*Ac*) element on the *Dissociation* (*Ds*) element. In (b), *Ds* is transposed to a region adjacent to a theoretical gene *W*. Subsequent chromosomal breakage is induced, the *W*-bearing segment is lost, and mutant gene expression occurs. In (c), *Ds* is transposed to a region within the *W* gene, causing immediate mutant expression. *Ds* may also "jump" out of the *W* gene, with the accompanying restoration of *W* gene activity and its wild-type expression.

osmotic pressure in the developing seeds. As the seeds mature, those that are wrinkled (genotype *rr*) lose more water than do the smooth seeds (*RR* or *Rr*), producing the wrinkled phenotype (Figure 7–24).

The structural gene for SBEI has been cloned and characterized in both wild-type and mutant genotypes. In the *rr* genotype, the SBEI protein is nonfunctional, presumably because the SBEI gene is interrupted by a 0.8-kb insertion, resulting in the production of an abnormal RNA transcript. The inserted DNA has 12-bp inverted repeats at each end that are highly homologous to the terminal sequences in the transposable element *Ac* from maize and to other *Ac*-like elements

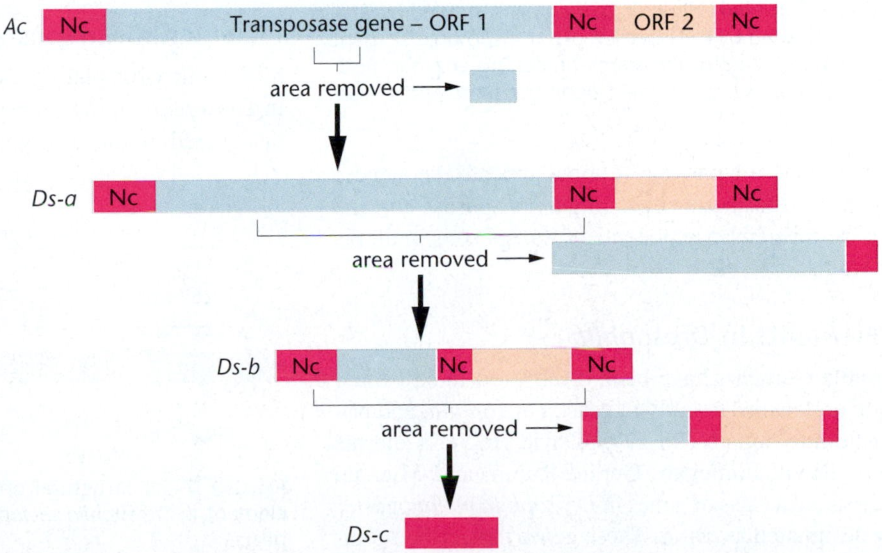
Ac
Nc
Transposase gene – ORF 1
Nc
ORF 2
Nc
area removed
Ds-a
Nc
Nc
Nc
area removed
Ds-b
Nc
Nc
Nc
area removed
Ds-c

FIGURE 7–23 A comparison of the structure of an *Ac* element with three *Ds* elements, all of which have been isolated and sequenced. The imperfect and inverted repeats are at the ends of the *Ac* element. The transposase gene is in an open reading frame (ORF 1). No function has yet been assigned to ORF 2. Noncoding regions are designated by Nc. As this scheme shows, *Ds-a* appears to be simply an *Ac* element containing a small deletion in the gene encoding the transposase enzyme.

FIGURE 7–24 The *wrinkled* trait in garden peas, studied by Gregor Mendel, is caused by the insertion of a transposable element into the structural gene for the starch-branching enzyme. Both *wrinkled* and *smooth* seeds can develop in the same pod.

from snapdragons and parsley. Terminal repeated sequences and the genetic information encoding a transposase enzyme appear to be universal components of transposons in all organisms studied.

Copia Elements in *Drosophila*

Transposable elements have been discovered in still other eukaryotic organisms, notably in yeast, *Drosophila*, and primates, including humans. In 1975, David Hogness and his colleagues David Finnegan, Gerald Rubin, and Michael Young identified a class of genes in *Drosophila melanogaster* that they designated as ***copia***. These genes transcribe "copious" amounts of RNA (hence their name). Present in up to 30 copies in the genome of cells, all types of *copia* genes are similar in their nucleotide sequence. Mapping studies show that they are transposable to different chromosomal locations and are dispersed throughout the genome.

Copia genes appear to be only one of approximately 30 families of transposable elements in *Drosophila*, each of which is present in from a few up to 20–50 copies in the genome. Some are referred to as *copia*-like. Together, these families constitute about 5 percent of the *Drosophila* genome and over half of the middle repetitive DNA of this organism. One estimate projects that 50 percent of all visible mutations in *Drosophila* are the result of the insertion of transposons into otherwise wild-type genes!

Despite the variability in DNA sequence between the members of different families, they share a common structural organization thought to be related to the insertion and excision processes of transposition. Each *copia* gene consists of approximately 5000–8000 bp of DNA, including a long family-specific **direct terminal repeat** (**DTR**) sequence of 276 bp at each end. Within each repeat is a short inverted terminal repeat (ITR) of 17 bp. These features are illustrated in Figure 7–25. The shorter ITR sequences are considered universal in *copia* elements. The DTR sequences are found in other transposons in other organisms, but they are not universal.

Insertion of *copia*, as with other transposable elements, appears to be dependent on ITR sequences and seems to occur at specific target sites in the genome. The *copia*-like elements demonstrate regulatory effects at the point of their insertion in the chromosome. Certain mutations, including those affecting eye color and segment formation, have been found to be due to insertions within genes. For example, the eye-color mutation *white–apricot* (w^a), an allele of the *white* (*w*) gene, contains a *copia* insertion element within the gene. Removing the transposable element sometimes restores the wild-type allele. In general, eukaryotic transposons are strikingly similar to one another and share many features with those in bacteria.

P Element Transposons in *Drosophila*

Still another interesting category of transposable elements in *Drosophila* is the family called **P elements**. These were discovered while studying the phenomenon of **hybrid**

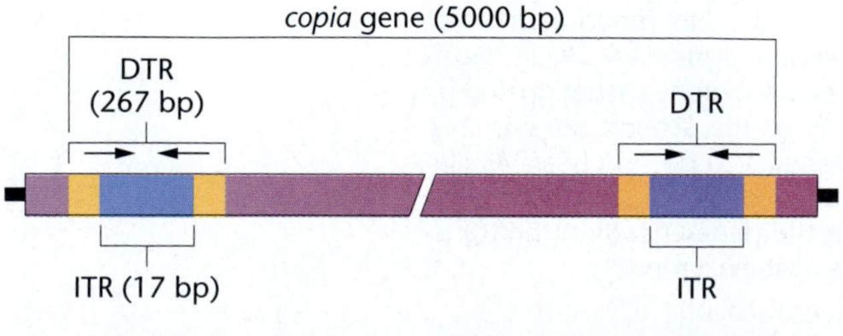

FIGURE 7–25 Structural organization of a *copia* transposable element in *Drosophila melanogaster*, showing the terminal repeats.

dysgenesis, a condition causing sterility, elevated mutation rate, and chromosome rearrangement in the offspring of crosses between certain strains of fruit flies. Hybrid dysgenesis is caused by high rates of P-element transposition in the germ line, a phenomenon wherein the mobile DNA elements insert themselves into or near genes, thereby causing mutations. P elements are 2.9 kb long, with 31-bp terminal inverted repeats. The elements encode at least two proteins, one of which is the transposase enzyme that is required for transposition. The transposase is expressed only in the germ line, accounting for the tissue specificity of P element transposition. Strains of flies that contain P elements inserted into their genomes are resistant to further transpositions due to the presence of a repressor protein, also encoded by the P elements.

P elements are useful as vectors to introduce transgenes into *Drosophila*. A common method of generating P-element-induced transformants is to inject *Drosophila* embryos with a mixture of P elements. One P element lacks the transposase gene and cannot move by itself. The other P element contains the transposase gene, but cannot insert into the genome. The two injected P elements will be taken up by the embryo's cells, transposase will be expressed in the germ line of the fly, and the P element will insert itself into the fly's DNA. The next generation will contain stable copies of P-element DNA, including the transgene, inserted at random locations.

Mutations can arise from several kinds of insertional events. If a P element inserts into the coding region of a gene, it can destroy the normal gene product. If it inserts into the promoter region of a gene, it can affect the level of expression of the gene. Insertions into introns can affect splicing or cause the premature termination of transcription.

Researchers screen the flies for mutant phenotypes of interest, isolate the fly's DNA, and clone the gene that was mutated. This technology has been used to identify genes involved in *Drosophila* development, behavior, and regulation of gene expression.

Another equally important P element technique is germ-line transformation. One of the most powerful ways to analyze a gene's activity and expression pattern is to clone the gene, manipulate its DNA sequences in a test tube, and reintroduce the mutated or modified gene into an organism. P elements provide a vehicle (or "vector") for inserting a cloned gene into a fly's genome. Embryos are injected with a mixture of P elements. One contains the transposase gene (but cannot insert). The other contains the marker gene such as *rosy* (an eye-color mutation) and the cloned gene of interest, but lacks a transposase gene. Flies that have been transformed are selected on the basis of their rosy eye color, and the behavior of the cloned and transposed gene is observed.

Various other elegant techniques, based on P element transposition, have been developed. These have allowed researchers to study the mechanics of DNA repair and DNA recombination. In addition, researchers are presently developing methods to target P element insertions to precise single-chromosomal sites, which should increase the precision of germ-line transformation in the analysis of gene activity.

Transposable Elements in Humans

The final class of transposable elements that we shall discuss is represented by the ***Alu* family** of **short interspersed elements (SINEs)**, which are characteristic of moderately repetitive DNA in mammals. We first introduced Alu and SINEs in Chapter 4. Human DNA contains some 500,000 copies of this 200- to 300-bp sequence in the genome. Originally detected through the use of reassociation kinetic analysis, these transposable elements received their name because they contain specific nucleotide sequences that are cleaved by a restriction endonuclease called *Alu*I.

The *Alu* family is considered to be a significant set of elements in the genome based on the fact that they have been found in the DNA of all the primates and rodents studied. Within the *Alu* elements in mammals, a specific 40-bp DNA sequence has been conserved. Finally, *Alu* sequences are represented in some nuclear transcripts. The *Alu* elements are considered transposable on the basis of several lines of evidence. The most important is the fact that their 200- to 300-bp sequence is flanked on either side by direct repeat sequences consisting of 7–20 bp, paralleling bacterial insertion sequences. These flanking regions are related to the insertion process during transposition. Secondly, clustered regions of *Alu* sequences vary in the DNA of both normal and diseased individuals and in different tissues of the same individual. Additionally, the sequences have been found extrachromosomally.

The potential mobility and mutagenic effects of these and other elements have far-reaching implications, as can be seen in a recent example of a transposon "caught in the act." The case involves a male child with hemophilia. One cause of hemophilia is a defect in blood-clotting factor VIII, the product of an X-linked gene. Haig Kazazian and his colleagues found a transposable element much longer than *Alu* sequences inserted at two points within the gene. Also present elsewhere in the genome, this sequence was shown to be a type of repetitive DNA referred to as a **long interspersed element (LINE)**.

Researchers were very interested in determining if one of the mother's X chromosomes also contained this specific LINE. If so, the unaffected mother would be heterozygous and pass the LINE-containing chromosome to her son . The startling finding was that the LINE sequence is *not* present on either of her X chromosomes, but *was* detected on chromosome 22 of both parents. This suggests that this mobile element may have "moved" from one chromosome to another in the gamete-forming cells of the mother, prior to being transmitted to the son.

Many questions remain concerning this and other transposable elements. What is their origin? Were they once some sort of retrovirus? Exactly how do they move, and what has been their role during evolution? These and other questions will intrigue researchers for many years to come.

Genetics, Technology, and Society

Chernobyl's Legacy

On April 26, 1986, Reactor 4 of the Chernobyl Nuclear Power Station exploded and ejected massive amounts of radioactive material into the surrounding countryside and throughout the northern hemisphere. The accident killed 31 emergency workers and caused acute radiation sickness in over 200 others. In the nine days following the initial explosion, as the reactor's temperature approached meltdown, radioactive fission products were discharged into the environment. High levels of radioactive iodine, xenon, strontium, and cesium were released into the atmosphere, along with over 5000 tons of vaporized boron carbide, dolomite, clay, sand, and lead. The plume of fallout traveled in a northwesterly direction through central Europe, reaching Finland and Sweden three days after the initial explosion. By May 5, the radioactive cloud reached the United Kingdom, and remnants of the cloud arrived in North America one week later. Millions of people in Europe and the former Soviet Union were exposed to measurable amounts of radioactivity. People living within 30 km of Chernobyl were exposed to high levels of radioactivity prior to their evacuation 36 hours following the accident. Equally high were the exposures of some of the 600,000 military and civilian workers who were sent to Chernobyl to decontaminate the area and to encase the shattered reactor in a sarcophagus.

Chernobyl was the world's largest accidental release of radioactive material. In addition, this disaster inflicted enormous economic and personal burdens on the people of the former Soviet Union. The question that remains is whether Chernobyl's radioactive pollution directly threatens the long-term health of millions of people.

More is known about the effects of ionizing radiation on human health than any other agent (except perhaps cigarette smoke). Radiation refers to the energy that is emitted from a source, and includes heat, light, radio waves, microwaves, X rays, and gamma rays from radioactive elements. Ionizing radiation (such as X rays and gamma rays) is a subset of radiation that possesses sufficient energy to eject electrons from atoms. Ionizing radiation can damage any cellular component, alter nucleotides, and induce strand breaks in DNA. These DNA lesions can lead to mutations or chromosomal translocations. It is this DNA-damaging capacity of ionizing radiation that makes it a mutagen.

High doses of ionizing radiation increase the risk of developing certain cancers. The survivors of the atomic bomb blasts of Hiroshima and Nagasaki showed an increased incidence of leukemia within 2 years of the bombing. Breast cancers increased 10 years after exposure, as did cancers of the lung, thyroid, colon, ovary, stomach, and nervous system. However, the incidence of other cancers, such as those of the gall bladder, pancreas, uterus, and bone, did not increase due to radiation exposure. As ionizing radiation induces DNA damage, the offspring of A-bomb survivors were expected to show increases in birth defects, yet these increases were not detected.

The problem in extrapolating from Hiroshima to Chernobyl is the difference in dose. In A-bomb survivors, the cancer rate increased among those exposed to at least 200 mSv (mSv = millisievert, a unit dose of absorbed radiation). It is estimated that people living in the most contaminated areas close to Chernobyl may have received about 50 mSv, and some cleanup workers were exposed to approximately 250 mSv. Outside the Chernobyl area, radiation doses were estimated at 0.4–0.9 mSv in Germany and Finland, 0.01 mSv in the United Kingdom, and 0.0006 mSv in the United States.

In order to put these numbers in context, the average dose for medical diagnostic procedures (such as chest and dental X rays) is 0.39 mSv per year. A person's radiation exposure from natural background sources (cosmic rays, rocks, and radon gas) is about 2–3 mSv per year. Smokers expose themselves to an average extra dose of about 2.8 mSv per year from the intake of naturally occurring radioactive materials in tobacco smoke. A whole-body X-ray dose of 3000 mSv or more can be fatal.

It appears that mutation rates among plants and animals exposed to Chernobyl's radioactive waste may be two- to tenfold higher than normal. Also, Chernobyl cleanup workers exhibit a 25-percent-higher mutation frequency at the *HPRT* locus. Mutation rates at minisatellites among children born in polluted areas are twofold higher than those from controls in the United Kingdom. But do these increased mutation rates translate into health effects?

In the 115,000 people evacuated from Chernobyl, it was estimated that 26 leukemias might appear above the 25–30 that would occur spontaneously. It was also estimated that up to 17,000 additional cancers might occur in Europe, above the 123 million that would occur normally. At present, however, there have been no detectable increases in leukemias or solid tumors in contaminated areas of the former Soviet Union, Finland, or Sweden, or in the 600,000 Chernobyl cleanup workers.

Despite the lack of detectable increases in leukemias and solid tumors, one type of cancer has increased in the Chernobyl area. The rate of childhood thyroid cancer has reached over 100 cases per million children per year, whereas the normal rates would be expected to be between 0.5 and 3 cases per million children per year. To date, approximately 1800 children have been diagnosed with thyroid cancer. The regions with the greatest contamination appear to have the highest rates of thyroid cancer. Although epidemiologists debate whether these increases are entirely due to radiation or to the increased reporting of cases, the scale of the increase and the fact that radioactive isotopes of iodine made up a significant portion of the Chernobyl fallout, make the link between thyroid cancer and Chernobyl fallout a plausible one.

The most profound effects of Chernobyl have been psychological. A recent study of Chernobyl cleanup workers found no increases in cancers, including leukemias or thyroid cancers. However, there was a 50-percent increase in suicides and a detectable increase in smoking- and alcohol-related disease. This mirrors other studies showing that 45 percent of people living within 300 km of Chernobyl believe that they have a radiation-induced illness. These people may be correct, as health effects such as depression, sleep disturbance, hypertension, and altered perception have been documented. People suffer from the stress of relocation, the loss of personal and real property, and the feeling of a lack of control over their radiation exposure or their future. Posttraumatic stress may be a greater threat to health than the actual radiation exposure from the accident. Without solid information about their real risks, people feel that they live in constant danger and are simply awaiting the results of a cancer "lottery." Even if cancer rates and genetic defects do not increase dramatically, the indirect health effects from the Chernobyl Nuclear Power Plant explosion have been, and continue to be, immense.

References

Anspaugh, L.R., Catlin, R.J. and Goldman, M. 1988. The global impact of the Chernobyl reactor accident. *Science* 242:1513–19.

Ginzberg, H.M. 1993. The psychological consequences of the Chernobyl accident—Findings from the International Atomic Energy Agency study. *Public Health Rep.* 108:184–92.

Jacob, P., et al. 2000. Thyroid cancer risk in Belarus after the Chernobyl accident: Comparison with external exposures. *Radiation and Environmental Biophysics* 39:25–31.

Rahu, M., et al. 1997. The Estonian study of Chernobyl cleanup workers: II. Incidence of cancer and mortality. *Radiation Res.* 147:653–57.

Williams, D. 1994. Chernobyl, eight years on. *Nature* 371:556.

Web site

UNSCEAR focuses on Chernobyl accident in General Assembly Report. June 6, 2000. Press release of the United Nations Information Service.
www.un.org/ha/chernobyl/unsceare.htm

Chapter Summary

1. The phenomenon of mutation not only provides the basis for most of the inherent variation present in living organisms, but it also serves as the working tool of the geneticist in studying and understanding the nature of genes and the mechanisms governing genetic processes.
2. Mutations are distinguished by the tissues affected. Somatic mutations are those that may affect the individual, but are not heritable. Mutations arising in gametes may produce new alleles that can be passed on to offspring and can enter the gene pool.
3. Another classification of mutations relies on their effect. Morphological mutations, for example, can be detected visibly. Other types include biochemical, lethal, conditional, and regulatory mutations; groups that are not mutually exclusive.
4. Spontaneous mutations may arise naturally as a result of rare chemical rearrangements of atoms, tautomeric shifts, or errors occurring during DNA replication. Spontaneous mutations are very rare.
5. The rate of mutagenesis can be increased experimentally by a variety of mutagenic agents. Agents such as nitrous acid, hydroxylamine, alkylating agents, and base analogs cause chemical changes in nucleotides that alter their base-pairing affinities. As a result, base-substitution mutations arise following DNA replication. The study of mutations caused by these agents has provided considerable insight into the molecular basis of mutations.
6. Frameshift mutations, induced specifically by acridine dyes, arise when the addition or deletion of one or more nucleotides (but not multiples of three) occurs.
7. Direct analysis of DNA from individuals with specific ABO blood types and muscular dystrophy has been informative. Complete loss of function, as in the case in blood type O and the Duchenne form of muscular dystrophy, occurs when deletions or duplications of nucleotides have shifted to reading frame during translation.
8. Another form of mutation, discovered in several human disorders, including fragile-X syndrome and myotonic dystrophy, involves unstable trinucleotide units that are repeated many times within certain genes. The number of repeats may increase in succeeding generations, and when the number of repeats exceeds a critical level, inherited disorders result. An increasing number of repeats often correlates with early onset and increased severity of the disease.
9. Ultraviolet light and high-energy radiation from gamma, cosmic, or X-ray sources are also potent mutagenic agents. UV light induces the formation of pyrimidine dimers in DNA, while high-energy radiation is able to penetrate deeper into tissues, causing the ionization of molecules in its path, including DNA.
10. Various forms of repair of DNA lesions, such as pyrimidine dimers, have been discovered, including photoreactivation, excision repair, mismatch repair, proofreading, and various forms of recombinational repair, including double-stranded break repair. The significance of repair mechanisms is apparent when, in humans, excision repair is lost due to mutation. The result is the severe human disorder xeroderma pigmentosum.
11. Site-directed mutagenesis is a technique that allows researchers to create specific alterations in the nucleotide sequence of the DNA of genes. The replacement of a normal gene with an altered "loss-of-function" allele creates knockout organisms.
12. The insertion sequences in bacteria and other transposable elements in eukaryotes are mobile genetic units characterizing the genomes of all organisms. They have a profound effect on genetic expression, thus serving as a distinct category of mutagenic agent.

Insights and Solutions

1. How could you isolate a mutant strain of bacteria that is resistant to penicillin, an antibiotic that inhibits cell-wall synthesis?

Solution: Grow a culture of bacterial cells in liquid medium and plate the cells on agar medium to which penicillin has been added. To enhance the chance of such a mutation arising, you might want to add a mutagen to the liquid culture. Only penicillin-resistant cells will reproduce and form colonies. Each colony will, in all likelihood, represent a cloned group of cells with the identical mutation. Isolate members of each colony.

2. The base analog 2-amino purine (2-AP) substitutes for adenine during DNA replication, but it may base pair with cytosine. The base analog 5-bromouracil (5-BU) substitutes for thymidine, but it may base pair with guanine. Follow the double-stranded trinucleotide sequence shown here through three rounds of replication, assuming that, in the first round, both analogs are present and become incorporated wherever possible. In the second and third round of replication, they are removed. What final sequences occur?

Solution:

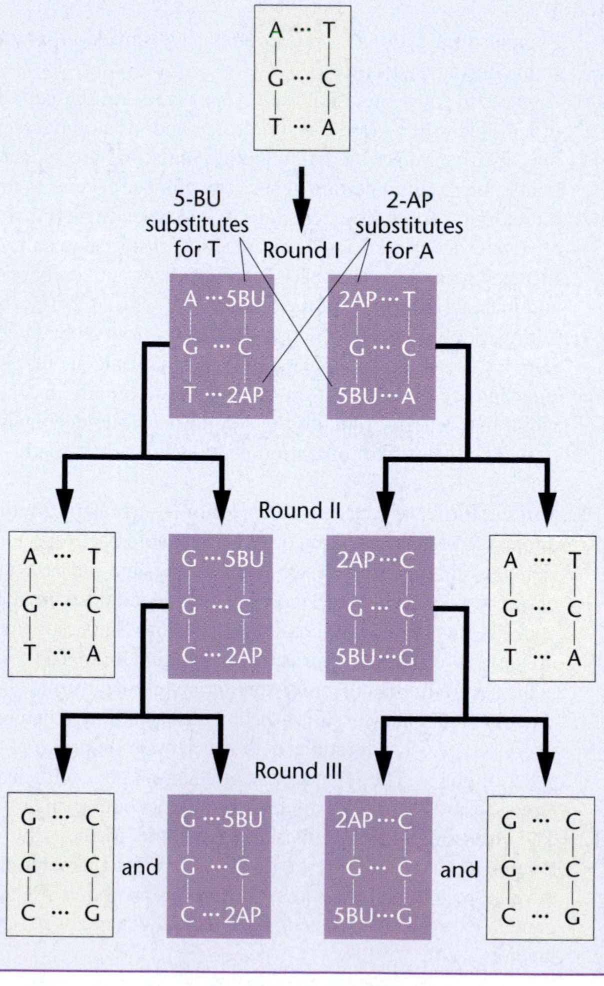

3. A rare dominant mutation expressed at birth was studied in humans. Records showed that six cases were discovered in 40,000 live births. Family histories revealed that in two cases, the mutation was already present in one of the parents. Calculate the spontaneous mutation rate for this mutation. What are some underlying assumptions that may affect our conclusions?

Solution: Only four cases represent a new mutation. Because each live birth represents two gametes, the sample size is from 80,000 meiotic events. The rate is equal to

$$\frac{4}{80{,}000} = \frac{1}{20{,}000} = 0.5 \times 10^{-4}$$

There are various reasons why this mutation rate may be inaccurate. We have assumed that the mutant gene is fully penetrant and is expressed in each individual bearing it. If it is not fully penetrant, our calculation may be an underestimate because one or more mutations may have gone undetected. We have also assumed that the screening was 100 percent accurate. One or more mutant individuals may have been "missed," again leading to an underestimate. Finally, we assume that the viability of the mutant and nonmutant individuals is equivalent and that they survive equally in utero. Therefore, our assumption is that the number of mutant individuals at birth is equal to the number at conception. If this were not true, our calculation would again be inaccurate.

4. Consider the following estimates:

(a) There are 5.5×10^9 humans living on this planet.

(b) Each individual has about 50,000 (0.5×10^5 genes).

(c) The average mutation rate at each locus is 10^{-5}.

How many spontaneous mutations are currently present in the human population? Assuming that these mutations are equally distributed among all genes, how many new alleles have arisen at each locus in the human population?

Solution: First, since each individual is diploid, there are two copies of each gene per person, each arising from a separate gamete. Therefore, the total number of spontaneous mutations is

$$\left(\frac{2 \times 0.5 \times 10^5 \text{genes}}{\text{individual}}\right)\left(\frac{5.5 \times 10^9 \text{ individuals}}{}\right)\left(\frac{10^{-5}\text{mutations}}{\text{gene}}\right)$$

$$= (10^5) \cdot (5.5 \times 10^9) \cdot (10^{-5})\text{mutations}$$

$$= 5.5 \times 10^9 \text{mutations}$$

On the average, in the entire human population there are

$$\frac{5.5 \times 10^9 \text{mutations}}{0.5 \times 10^5 \text{genes}} = \frac{11 \times 10^4 \text{mutations}}{\text{gene}}$$

Problems and Discussion Questions

1. What is the difference between a chromosomal aberration and a gene mutation?
2. Discuss the importance of mutations to the successful study of genetics.
3. Describe the technique for the detection of nutritional mutants in *Neurospora*.
4. Most mutations are thought to be deleterious. Why, then, is it reasonable to state that mutations are essential to the evolutionary process?
5. Why is a random mutation more likely to be deleterious than beneficial?
6. Most mutations in a diploid organism are recessive. Why?
7. What is meant by a conditional lethal mutation?
8. Contrast the concerns about mutation in somatic and gametic tissue.
9. Describe tautomerism and the way in which this chemical event may lead to mutation.
10. Contrast and compare the mutagenic effects of deaminating agents, alkylating agents, and base analogs.
11. Acridine dyes induce frameshift mutations. Is such a mutation likely to be more detrimental than a point mutation in which a single pyrimidine or purine has been substituted?
12. Why are X rays a more potent mutagen than is UV radiation?
13. Contrast the induction of mutations by UV radiation and X rays.
14. Contrast the various types of DNA repair mechanisms known to counteract the effects of UV radiation and other DNA damage. What is the role of visible light in photoreactivation?
15. Mammography is an accurate screening technique for the early detection of breast cancer in humans. Because this technique uses X rays diagnostically, it has been highly controversial. Can you explain why? What reasons justify the use of X rays for such a type of medical diagnosis?
16. Compose a short essay that relates the molecular basis of fragile-X syndrome, myotonic dystrophy, and Huntington disease to the severity of the disorders, as well as to the phenomenon called genetic anticipation.
17. Describe how the Ames assay screens for potential environmental mutagens. Why is it thought that a compound that tests positively in the Ames assay may also be carcinogenic?
18. Describe the general approach used in site-directed mutagenesis.
19. What genetic defect results in the disorder xeroderma pigmentosum (XP) in humans? How does this defect relate to the phenotype associated with the disorder?
20. Differentiate between point mutations that are transitions and those that are transversions. Using the DNA bases (A, T, C, and G), list the four types of transitions and the eight types of transversions.
21. In a bacterial culture in which all cells were unable to synthesize leucine (leu^-), a potent mutagen was added, and the cells were allowed to undergo one round of replication. At that point, samples were taken, a series of dilutions was made, and the cells were plated on either minimal medium or minimal medium to which leucine was added. The first culture condition (minimal medium) allowed the detection of mutations from leu^- to leu^+, while the second culture condition (minimum medium with leucine added) allowed the determination of the total cells, since all bacteria can grow. From the results are as follows:

Culture Condition	*Dilution*	*Colonies*
minimal medium	10^{-1}	18
minimal + leucine	10^{-7}	6

Determine the frequency of mutant cells. What is the rate of mutation at the locus involved with leucine biosynthesis?

22. Contrast the various transposable genetic elements in bacteria, maize, *Drosophila*, and humans. What properties do they share?
23. The initial discovery of IS units in bacteria involved their presence upstream (5′) to three genes controlling galactose metabolism. All three genes were affected despite not being contiguous with the IS unit. Offer an explanation as to why this might occur.
24. *Ty*, a transposable element in yeast, has been found to contain an open reading frame (ORF) that encodes the enzyme reverse transcriptase. This enzyme synthesizes DNA from an RNA template. Speculate the role of this gene product in the transposition of *Ty* within the yeast genome.

Extra-Spicy Problems

25. Presented here are hypothetical findings from studies of heterokaryons formed from seven human xeroderma pigmentosum cell strains:

	XP1	*XP2*	*XP3*	*XP4*	*XP5*	*XP6*	*XP7*
XP1	–						
XP2	–	–					
XP3	–	–	–				
XP4	+	+	+	–			
XP5	+	+	+	+	–		
XP6	+	+	+	+	–	–	
XP7	+	+	+	+	–	–	–

Note: + = complementing; – = noncomplementing.

These data represent the occurrence of unscheduled DNA synthesis in the fused heterokaryon when neither of the strains alone showed synthesis. What does unscheduled DNA synthesis represent? Which strains fall into the same complementation groups? How many different groups are revealed based on these limited data? How does one interpret the presence of these complementation groups?

26. Imagine yourself as one of the team of geneticists who launched the study of the genetic effects of high-energy radiation on the surviving Japanese population immediately following the atomic bomb attacks at Hiroshima and Nagasaki in 1945. Demonstrate your insights into both chromosomal and gene mutation by outlining a comprehensive short-term and long-term study that addresses this topic. Be sure to include strategies for considering the effects on both somatic and germ-line tissues.
27. Cystic fibrosis (CF) is a severe autosomal recessive disorder in humans that results from a chloride ion–channel defect in epithelial cells. Over 500 sequence alterations have been identified in the 24 exons of the responsible gene (*CFTR*, for cystic

fibrosis transmembrane regulator), including dozens of different missense mutations and frameshift mutations, as well as numerous splice-site defects. Although all affected CF individuals demonstrate chronic obstructive lung disease, there is variation in pancreatic enzyme insufficiency (PI). Speculate which types of observed mutations are likely to give rise to less severe symptoms of CF, including only minor PI.

Of the 300 sequence alterations, a number of them found within the exon regions of the *CFTR* gene do not give rise to cystic fibrosis. Taking into account your accumulated knowledge of the genetic code, gene expression, protein function, and mutation, explain to a freshman biology major how this might be.

28. Electrophilic oxidants are known to create the highly deleterious lesion in DNA named 7,8-dihydro-8-oxoguanine (oxoG). Whereas guanine base pairs with cytosine, oxoG base pairs with either cytosine or adenine,

(a) what are the sources of reactive oxidants within cells that cause this type of lesion?

(b) Drawing on your knowledge of nucleotide chemistry, draw the structure of oxoG, and, below it, draw guanine. Opposite guanine, draw cytosine, including the hydrogen bonds that allow these two molecules to base pair. Does the structure of oxoG, in contrast to guanine, provide any hint as to why it base pairs with adenine?

(c) Assume that an unrepaired oxoG lesion is present in the helix of DNA opposite cytosine. Following several rounds of replication, predict the type of mutation that will occur.

(d) Drawing on your knowledge of DNA repair mechanisms, consider which cellular approaches might work to counteract an oxoG lesion. Which of these is likely to be most effective.

(Reference:Bruner, S.D., et al. 2000. *Nature* 403: 859–62.)

29. Among Betazoids, the ability to read minds is under the control of a gene called "mindreader" (abbreviated *mr*.). Most Betazoids can read minds, but rare recessive mutations in the *mr* gene result in two alternative phenotypes: *delayed-receivers* and *insensitives*. Delayed-receivers have some mind-reading ability but perform the task much more slowly than normal Betazoids. Insensitives can not read minds at all.

Betazoid genes do not have introns, so the gene only contains coding DNA. It is 3,332 nucleotides in length and Betazoids use a *four letter* genetic code.

The table below shows some data from where the mutations carried by five unrelated *mr* mutations were analyzed.

Description of mutation	*Phenotype*
mr-1 Nonsense mutation in codon 829	delayed-receiver
mr-2 Missense mutaation in codon 52	delayed-receiver
mr-3 Deletion of nucleotides 83–150	delayed-receiver
mr-4 Missense mutation in codon 192	insensitive
mr-5 Deletion of nucleotides 83–93	insensitive

For each mutation, provide a plausible explanation for why it gives rise to its associated phenotype and not to the other phenotype, e.g., hypothesize why the nonsense mutation in codon 829 (*mr*-1) gives rise to the milder delayed-receiver phenotype rather than the more severe insensitive phenotype. Then repeat this type of analysis for the other mutations. (More than one explanation is possible, so be creative within *plausible* bounds!)

Selected Readings

Ames, B.N., McCann, J., and Yamasaki, E. 1975. Method for detecting carcinogens and mutagens with the *Salmonella*/mammalian microsome mutagenicity test. *Mut. Res.* 31:347–64.

Anderson, D.I, Slechta, E.S., and Roth, J. 1998. Evidence that gene amplification underlies adaptive mutability of the bacterial *lac* operon. *Science* 282:1133–35.

Auerbach, C. 1978. Forty years of mutation research: A pilgrim's progress. *Heredity* 40:177–87.

Auerbach, C., and Kilbey, B.J. 1971. Mutations in eukaryotes. *Annu. Rev. Genet.* 5:163–218.

Baltimore, D. 1981. Somatic mutation gains its place among the generators of diversity. *Cell* 26:295–96.

Bates, G., and Leharch, H. 1994. Trinucleotide repeat expansions and human genetic disease. *BioEssays* 16:277–84.

Beadle, G.W., and Tatum, E.L. 1945. *Neurospora* II. Methods of producing and detecting mutations concerned with nutritional requirements. *Am. J. Bot.* 32:678–86.

Becker, M.M., and Wang, Z. 1989. Origin of ultraviolet damage in DNA. *J. Mol. Biol.* 210:429–38.

Berg, D., and Howe, M., eds. 1989. *Mobile DNA*. Washington, DC: American Society of Microbiology.

Brenner, S., Barnett, L., Crick, F.H.C., and Orgel, L. 1961. The theory of mutagenesis. *J. Mol. Biol.* 3:121–24.

Cairns, J., Overbaugh, J., and Miller, S. 1988. The origin of mutants. *Nature* 335:142–45.

Capecchi, M.R. 1989. Altering the genome by homologous recombination. *Science* 244:1288–92.

Carter, P. 1986. Site-directed mutagenesis. *Biochem. J.* 237:1–7.

Cleaver, J.E. 1968. Defective repair of replication of DNA in xeroderma pigmentosum. *Nature* 218:652–56.

———. 1990. Do we know the cause of xeroderma pigmentosum? *Carcinogenesis* 11:875–82.

Cleaver, J.E., and Karentz, D. 1986. DNA repair in man: Regulation by a multiple gene family and its association with human disease. *Bioessays* 6:122–27.

Cohen, S.N., and Shapiro, J.A. 1980. Transposable genetic elements. *Sci. Am.* (Feb.) 242:40–49.

Comfort, N.C. 2001. *The tangled field: Barbara: McClintock's search for the patterns of genetic control*. Cambridge, MA: Harvard University Press.

Crow, J.F., and Denniston, C. 1985. Mutation in human populations. *Adv. Hum. Genet.* 14:59–216.

Culotta, E. 1994. A boost for "adaptive" mutation. *Science* 265:318–19.

Deering, R.A. 1962. Ultraviolet radiation and nucleic acids. *Sci. Am.* (Dec.) 207:135–44.

Den Dunnen, J.T., et al. 1989. Topography of the Duchenne muscular dystrophy (DMD) gene. *Am. J. Hum. Genet.* 45:835–47.

Devoret, R. 1979. Bacterial tests for potential carcinogens. *Sci. Am.* (Aug.) 241:40–49.

Doring, H.P., and Starlinger, P. 1984. Barbara McClintock's controlling elements: Now at the DNA level. *Cell* 39:253–59.

Drake, J.W. 1991. A constant rate of spontaneous mutation in DNA-based microbes. *Proc. Natl. Acad. Sci. USA* 88:7160–64.

Drake, J.W., et al. 1975. Environmental mutagenic hazards. *Science* 187:505–14.

Drake, J.W., Glickman, B.W., and Ripley, L.S. 1983. Updating the theory of mutation. *Am. Sci.* 71:621–30.

Eyre-Walker, A., and Keightley, P.D. 1999. High genomic deleterious mutation rates in hominids. *Nature* 397:344–47.

Fedoroff, N.V. 1984. Transposable genetic elements in maize. *Sci. Am.* (June) 250:85–98.

Finnegan, D.J. 1985. Transposable elements in eukaryotes. *Int. Rev. Cytol.* 93:281–326.

Foster, P. 1998. Adaptive mutation: Has the unicorn landed? *Genetics* 148:1453–59.

Friedberg, E.C., Walker, G.C., and Siede, W. 1995. *DNA repair and mutagenesis*. Washington, DC: ASM Press.

Hall, B.G. 1988. Adaptive evolution that requires multiple spontaneous mutations: I. Mutations involving an insertion sequence. *Genetics* 120:887–97.

———. 1990. Spontaneous point mutations that occur more often when advantageous than when neutral. *Genetics* 126:5–16.

Hanawalt, P.C., and Haynes, R.H. 1967. The repair of DNA. *Sci. Am.* (Feb.) 216:36–43.

Haseltine, W.A. 1983. Ultraviolet light repair and mutagenesis revisited. *Cell* 33:13–17.

Hendricikson, E.A. 1997. Cell-cycle regulation of mammalian DNA double-strand break repair. *Am. J. Hum. Genet.* 61:795–800.

Howard-Flanders, P. 1981. Inducible repair of DNA. *Sci. Am.* (Nov.) 245:72–80.

Hoeijmakers, J.H.J. 2001. Genome maintenance mechanisms for preventing cancer. *Nature* 411:366–73.

Jiricny, J. 1998. Eukaryotic mismatch repair: An update. *Mutation Research* 409:107–21.

Kelner, A. 1951. Revival by light. *Sci. Am.* (May) 184:22–25.

Knudson, A.G. 1979. Our load of mutations and its burden of disease. *Am. J. Hum. Genet.* 31:401–13.

Kraemer, F.H., et al. 1975. Genetic heterogeneity in xeroderma pigmentosum: Complementation groups and their relationship to DNA repair rates. *Proc. Natl. Acad. Sci. USA* 72:59–63.

Little, J.W., and Mount, D.W. 1982. The SOS regulatory system of *E. coli. Cell* 29:11–22.

Luria, S.E., and Delbrück, M. 1943. Mutations of bacteria from virus sensitivity to virus resistance. *Genetics* 28:491–511.

Massie, R. 1967. *Nicholas and Alexandra*. New York: Atheneum.

Massie, R., and Massie, S. 1975. *Journey*. New York: Knopf.

McCann, J., Choi, E., Yamasaki, E., and Ames, B. 1975. Detection of carcinogens as mutagens in the *Salmonella*/microsome test: Assay of 300 chemicals. *Proc. Natl. Sci. USA* 72:5135–39.

McClintock, B. 1956. Controlling elements and the gene. *Cold Spring Harbor Symp. Quant. Biol.* 21:197–216.

MacDonald, M.E., et al. 1993. A novel gene containing a trinucleotide repeat that is expanded and unstable in Huntington's disease chromosome. *Cell* 72:971–80.

McKusick, V.A. 1965. The royal hemophilia. *Sci. Am.* (Aug.) 213:88–95.

Miki, Y. 1998. Retrotransposal integration of mobile genetic elements in human disease. *J. Human Genet.* 43:77–84.

Muller, H.J. 1927. Artificial transmutation of the gene. *Science* 66:84–87.

——— 1955. Radiation and human mutation. *Sci. Am.* (Nov.) 193:58–68.

Newcombe, H.B. 1971. The genetic effects of ionizing radiation. *Adv. Genet.* 16:239–303.

Nickolofff, J.A. and Hoekstra, M.F., eds. 2001. *DNA damage and repair*, Vol. III: *Advances from phage to humans*. Totowa, NJ: Humana Press.

O'Hare, K. 1985. The mechanism and control of P element transposition in *Drosophila. Trends Genet.* 1:250–54.

Ossana, N., Peterson, K.R., and Mount, D.W. 1986. Genetics of DNA repair in bacteria. *Trends Genet.* 2:55–58.

Radman, M., and Wagner, R. 1988. The high fidelity of DNA duplication. *Sci. Am.* (Aug.) 259:40–46.

Sherratt, D.J., ed. 1995. *Mobile genetic elements*. New York: Oxford University Press.

Shortle, D., DiMario, D., and Nathans, D. 1981. Directed mutagenesis. *Annu. Rev. Genet.* 15:265–94.

Sigurbjorhsson, B. 1971. Induced mutations in plants. *Sci. Am.* (Jan.) 224:86–95.

Spradling, A.C., Stern, D.M., Kiss, I., Roote, J., Laverty, T., and Rubin, G.M. 1995. Gene disruptions using transposable P elements: An integral component of the *Drosophila* Genome Project. *Proc. Natl. Acad. Sci. USA* 92:10824–30.

Stadler, L.J. 1928. Mutations in barley induced by X-rays and radium. *Science* 66:84–87.

Sutherland, B.M. 1981. Photoreactivation. *Bioscience* 31:439–44.

Tomlin, N.V., and Aprelikova, O.N. 1989. Uracil DNA glycosylases and DNA uracil repair. *Int. Rev. Cytol.* 114:81–124.

Topal, M.D., and Fresco, J.R. 1976. Complementary base pairing and the origin of substitution mutations. *Nature* 263:285–89.

Vogel, F. 1992. Risk calculations for hereditary effects of ionizing radiation in humans. *Hum. Genet.* 89:127–46.

Wells, R.D. 1994. Molecular basis of genetic instability of triplet repeats. *J. Biol. Chem.* 271:2875–78.

Wills, C. 1970. Genetic load. *Sci. Am.* (March) 222:98–107.

Wolff, S. 1967. Radiation genetics. *Annu. Rev. Genet.* 1:221–44.

Yamamoto, F., et al. 1990. Molecular genetic basis of the histo-blood group ABO system. *Nature* 345:229–33.

GENETICS MediaLab

The resources that follow will help you achieve a better understanding of the concepts presented in this chapter. These resources can be found either on the CD packaged with this textbook or the Companion Web site found at **http://www.prenhall.com/klug**

CD Resources:

Module 14.1: Mutations at the DNA Level

Module 14.2: DNA Repair

Module 14.3: Chemical Mutagenesis: Ames Test

Web Problem 1:
Time for completion = 10 minutes

Do mutations arise in response to environmental conditions? The accepted answer for years was no. The work by Luria and Delbrück indicated that mutations occur spontaneously and that selective forces then act upon the variation that is present. Cairns and colleagues have recently reexamined this issue. They suggest that at times there may be an adaptive response to some conditions. To understand this controversy, you need to have a clear idea of the experimental conditions used to answer the question. The linked Web site discusses the basic design of such a fluctuation test. Describe the results you might expect in light of these two hypotheses. To complete this exercise, visit Web Problem 1 in Chapter 14 of your Companion Web site, and select the keyword **MUTATION**.

Web Problem 2:
Time for completion = 15 minutes

What is the relationship between a mutation and its phenotype? It is useful to have some sense of what types of mutations cause a particular phenotype and how the gene is affected. Researchers' increased ability to identify mutations by direct sequencing has led to an explosion of information. To sort the growing volume of information about human diseases in particular, researchers have established databases listing observed mutations. You will find a database of mutations in the androgen receptor gene at the linked Web site. Severe mutations in this gene disrupt its function and cause androgen insensitivity (AIS), leading to sex reversal in XY individuals. Examine the types of information available at this site, including the "selectable options" section. What conclusions can you draw about the effects of mutations in this gene? Discuss alternative ways to present such data or additional types of analyses you think would be useful. To complete this exercise, visit Web Problem 2 in Chapter 14 of your Companion Web site, and select the keyword **ANDROGEN RECEPTOR**.

Web Problem 3:
Time for completion = 10 minutes

How do transposons move from one place in the genome to another? To answer this question, researchers have devised genetic tests to distinguish among the different molecular steps required for transpositional movement. This field of research has provided insight into how the genome of an organism may change over time. There are two distinct methods of transposition illustrated at the linked Web site: replicative and nonreplicative. After examining the sections, compare and contrast these two modes of movement. To complete this exercise, visit Web Problem 3 in Chapter 14 of your Companion Web site, and select the keyword **TRANSPOSITION**.

Chromosomes in the prometaphase stage of mitosis, derived from a cell in the flower of *Haemanthus*. (*Dr. Andrew S. Bajer, University of Oregon*)

8
Mitosis and Meiosis

KEY CONCEPTS

8.1 **Cell Structure Is Closely Tied to Genetic Function**
- Cell Boundaries
- The Nucleus
- The Cytoplasm and Organelles

8.2 **Homologous Chromosomes Provide the Genetic Basis of Haploid and Diploid Cells**

8.3 **Mitosis Partitions Chromosomes into Dividing Cells**
- Interphase and the Cell Cycle
- Prophase
- Prometaphase and Metaphase
- Anaphase
- Telophase

8.4 **The Cell Cycle is Genetically Regulated**

8.5 **Meiosis Reduces the Chromosome Number from Diploid to Haploid in Germ Cells and Spores**
- An Overview of Meiosis
- The First Meiotic Division: Prophase I
- Metaphase, Anaphase, and Telophase I
- The Second Meiotic Division

8.6 **The Development of Gametes Varies during Spermatogenesis and Oogenesis**

8.7 **Meiosis Is Critical to the Successful Sexual Reproduction of All Diploid Organisms**

8.8 **Electron Microscopy Has Revealed the Cytological Nature of Mitotic and Meiotic Chromosomes**
- Chromatin vs. Chromosomes
- The Synaptonemal Complex

In earlier chapters, we established that in every living thing there exists a substance referred to as the genetic material. Except in certain viruses, this material is composed of the nucleic acid DNA, which is organized into units called *genes*, the products of which direct all metabolic activities of cells. DNA, with its array of genes, is organized into *chromosomes*, structures that serve as the vehicle for transmission of genetic information. The manner in which chromosomes are transmitted from one generation of cells to the next, and from organisms to their descendants, is exceedingly precise. In this chapter we consider exactly how genetic continuity is maintained between cells and organisms.

Two major processes of cell division are involved in eukaryotes: mitosis and meiosis. Although the mechanisms of the two processes are similar in many ways, the outcomes are quite different. **Mitosis** leads to the production of two cells, each with the same number of chromosomes as the parental cell and the identical genetic content. **Meiosis**, on the other hand, reduces the genetic content and the number of chromosomes by precisely half. This reduction is essential if sexual reproduction is to occur without doubling the amount of genetic material with each new generation. Strictly speaking, mitosis is that portion of the cell cycle during which the hereditary components are precisely and equally divided into daughter cells. Meiosis is part of a special type of cell division leading to the production of sex cells—gametes or spores—and is an essential step in the transmission of genetic information from an organism to its offspring. As we will see, meiosis plays an important role in scrambling allelic combinations and paternal and maternal homologs as genetically dissimilar gametes are formed.

Using microscopy, chromosomes are visible during a cell's life only during mitosis and meiosis. When cells are not undergoing division, the genetic material making up chromosomes unfolds and uncoils into a diffuse network within the nucleus, generally referred to as chromatin. We will briefly review the structure of cells, emphasizing those components that are of particular significance to genetic function. Then we devote the remainder of the chapter to the behavior of chromosomes during cell division.

8.1 Cell Structure Is Closely Tied to Genetic Function

Before describing mitosis and meiosis, a brief review of the structure of cells will be useful. As we shall see, many components, such as the nucleolus, ribosome, and centriole, are involved directly or indirectly in genetic processes. Other components, the mitochondria and chloroplasts, contain their own unique genetic information. It is also important to compare the structural differences between the prokaryotic bacterial cell and the eukaryotic cell. Variation in the structure and function of cells is dependent on specific genetic expression by each cell type.

Prior to 1940, our knowledge of cell structure was limited to what we could see with the light microscope. Around 1940, the transmission electron microscope was in its early stages of development, and by 1950, many details of cell ultrastructure had emerged. Under the electron microscope, cells were seen as highly organized, precise structures. A new world of whorled membranes, organelles, microtubules, granules, and filaments was revealed. These discoveries revolutionized thinking in the entire field of biology. We will be concerned with these aspects of cell structure that relate to genetic study. The typical animal cell shown in Figure 8–1 illustrates most of the structures we will discuss.

Cell Boundaries

The cell is surrounded by a **plasma membrane**, an outer covering that defines the cell boundary and delimits the cell from its immediate external environment. This membrane is not passive, but instead actively controls the movement of materials into and out of the cell. In addition to the plasma membrane, plant cells have an outer covering called the **cell wall**. One of the major components of this rigid structure is a polysaccharide called **cellulose**. Bacterial cells also have a cell wall, but its chemical composition is quite different from that of the plant cell wall, the major component being a complex macromolecule called a **peptidoglycan**. As its name suggests, the molecule consists of peptide and sugar units. Long polysaccharide chains are cross-linked with short peptides, which impart great strength and rigidity to the bacterial cell. Some bacterial cells have still another covering, a **capsule**. This mucus-like material protects these bacteria from phagocytic activity by the host during their pathogenic invasion of eukaryotic organisms. The presence of the capsule is under genetic control. In fact, as we will see in Chapter 2, its loss due to mutation in the pneumonia-causing bacterium *Diplococcus pneumoniae* provided the underlying basis for a critical experiment, proving that DNA is the genetic material.

Activities at cell boundaries are dynamic physiological processes. Both transport into and out of cells and communication between cells are critical to normal function. Because physiological processes are biochemical in nature, we would expect that many genes and their products are essential to these activities. This is indeed the case, and as such, mutations in these genes can alter or interrupt normal physiological functions, often with severe consequences. For example, the inherited disorder **Duchenne muscular dystrophy** is the result of complete loss of function of the gene product **dystrophin**, which is believed to function at the cell membrane of muscle cells. As shown in Figure 8–2, which uses immunofluorescent localization technology, dystrophin is completely absent from skeletal muscle of an afflicted individual compared with muscle from a control subject.

Many, if not most, animal cells have a covering over the plasma membrane called a **cell coat**. Consisting of glycoproteins (sometimes called the **glycocalyx**) and polysaccharides, the cell coat differs chemically from comparable structures in either plants or bacteria. One function served by the cell coat is to provide biochemical identity at the surface of cells. These forms of cellular identity at the cell surface are under genetic control. For example, the **AB** and **MN antigens** are found on the surface of red blood cells and

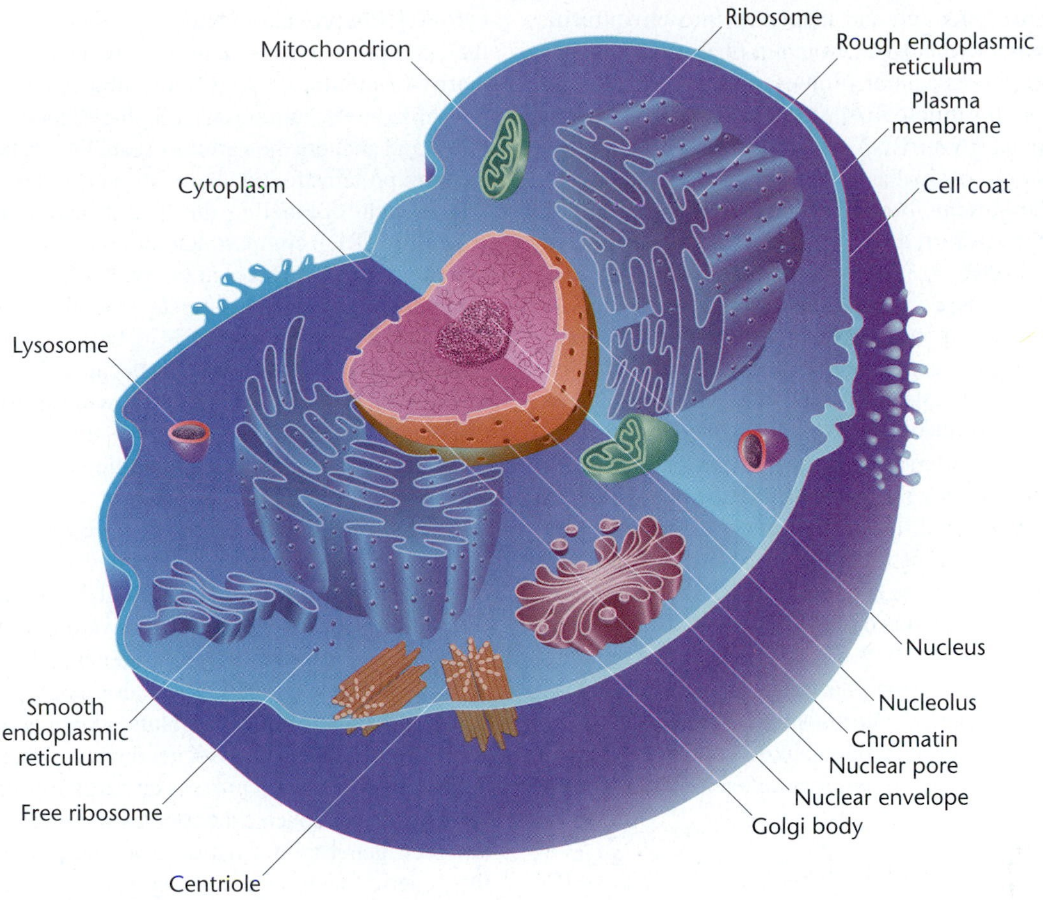

FIGURE 8–1 Drawing of a generalized animal cell. Emphasis has been placed on the cellular components discussed in the text.

elicit an immune response during blood transfusions. In other cells, the **histocompatibility antigens**, which elicit an immune response during tissue and organ transplants, are present as part of the cell coat. Also present as integral components of the cell surface are a variety of highly specific **receptor molecules**. These constitute recognition sites that receive and transfer chemical signals across the cell membrane into the cell. Such signals may initiate a variety of chemical activities and may ultimately signal specific genes to turn on. We may conclude that the cell surface links its host cell to its external environment and is tied in numerous ways to genetic processes.

The Nucleus

The nucleus houses the genetic material, **DNA**, which is complexed with an array of acidic and basic proteins into thin fibers. During nondivisional phases of the cell cycle,

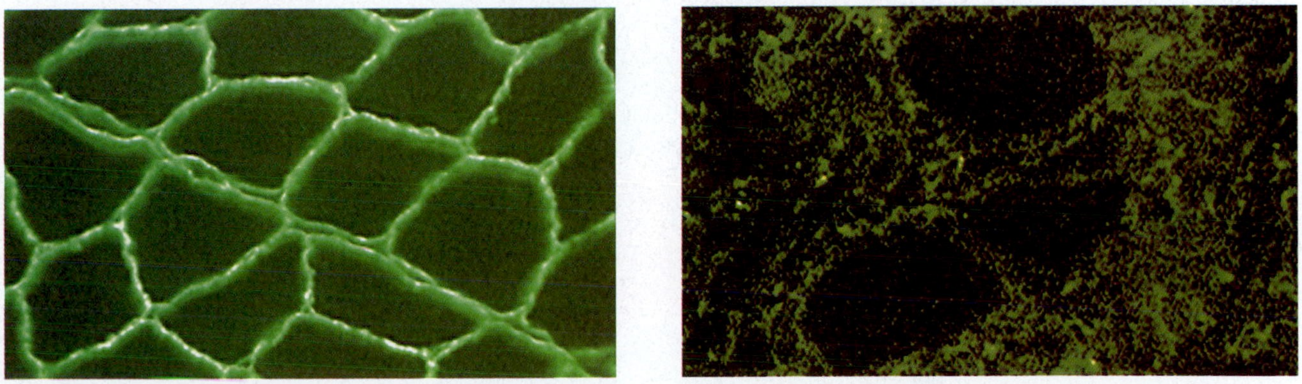

(a) (b)

FIGURE 8–2 (a) Localization of dystrophin in the sarcolemmal areas of normal skeletal muscle, using fluorescent immunoperoxidase staining. (b) Complete absence of immunoreactive dystrophin in the skeletal muscle of a patient with Duchenne muscular dystrophy.

these fibers are uncoiled and dispersed into **chromatin**. During mitosis and meiosis, chromatin fibers coil and condense into structures called chromosomes. Also present in the nucleus is the **nucleolus**, an amorphous component where ribosomal RNA (rRNA) is synthesized and where the initial stages of ribosomal assembly occur. One or more nucleoli may be present. The areas of DNA encoding rRNA are collectively referred to as the **nucleolus organizer region** or the **NOR**.

The lack of a nuclear envelope and membranous organelles is characteristic of prokaryotes. In bacteria such as *Escherichia coli*, the genetic material is present as a long, circular DNA molecule that is compacted into an area referred to as the **nucleoid region**. Part of the DNA may be attached to the cell membrane, but in general the nucleoid region constitutes a large area throughout the cell. Although the DNA is compacted, it does not undergo the extensive coiling characteristic of the stages of mitosis where, in eukaryotes, chromosomes become visible. Nor is the DNA in these organisms associated as extensively with proteins as is eukaryotic DNA. Figure 8–3, in which two bacteria are forming during cell division, illustrates the nucleoid regions that house the bacterial chromosome. Prokaryotic cells do not have a distinct nucleolus, but do contain genes that specify rRNA molecules, a major component of ribosomes.

The Cytoplasm and Organelles

The remainder of the eukaryotic cell enclosed by the plasma membrane, excluding the nucleus, is composed of **cytoplasm** and all associated cellular organelles. Cytoplasm consists of a nonparticulate, colloidal material referred to as the **cytosol**, which surrounds and encompasses numerous types of cellular **organelles**. Beyond these components, an extensive system of tubules and filaments comprising the **cytoskeleton** provides a lattice of support structures within the cytoplasm. This structural framework, consisting primarily of tubulin-derived microtubules and actin-derived microfilaments, maintains cell shape, facilitates cell mobility, and anchors the various organelles. Tubulin and actin are both proteins found abundantly in eukaryotic cells.

One such organelle, the membranous **endoplasmic reticulum (ER)**, compartmentalizes the cytoplasm, greatly increasing the surface area available for biochemical synthesis. The ER may appear smooth, in which case it serves as the site for synthesis of fatty acids and phospholipids, or it may appear rough, because it is studded with ribosomes. Ribosomes, which we will discuss in detail in Chapter 13, serve as sites for the translation of genetic information contained in messenger RNA (mRNA) into proteins.

Three other cytoplasmic structures are very important in the eukaryotic cell's activities: mitochondria, chloroplasts, and centrioles. **Mitochondria** are found in both animal and plant cells and are the sites of the oxidative phases of **cell respiration**. These chemical reactions generate large amounts of adenosine triphosphate (ATP), an energy-rich molecule. **Chloroplasts** are found in plants, algae, and some protozoans. This organelle is associated with **photosynthesis**, the major energy-trapping process on earth. Furthermore, these organelles can duplicate themselves and transcribe and translate their genetic information. It is interesting to note that the genetic machinery of mitochondria and chloroplasts closely resembles that of prokaryotic cells. In particular, mitochondria and chloroplasts contain DNA that lack associated proteins, making it distinct from that found in the nucleus. This and other observations have led to the proposal that mitochondria and chloroplasts were once primitive free-living organisms that established a symbiotic relationship with a primitive eukaryotic cell. The theory that

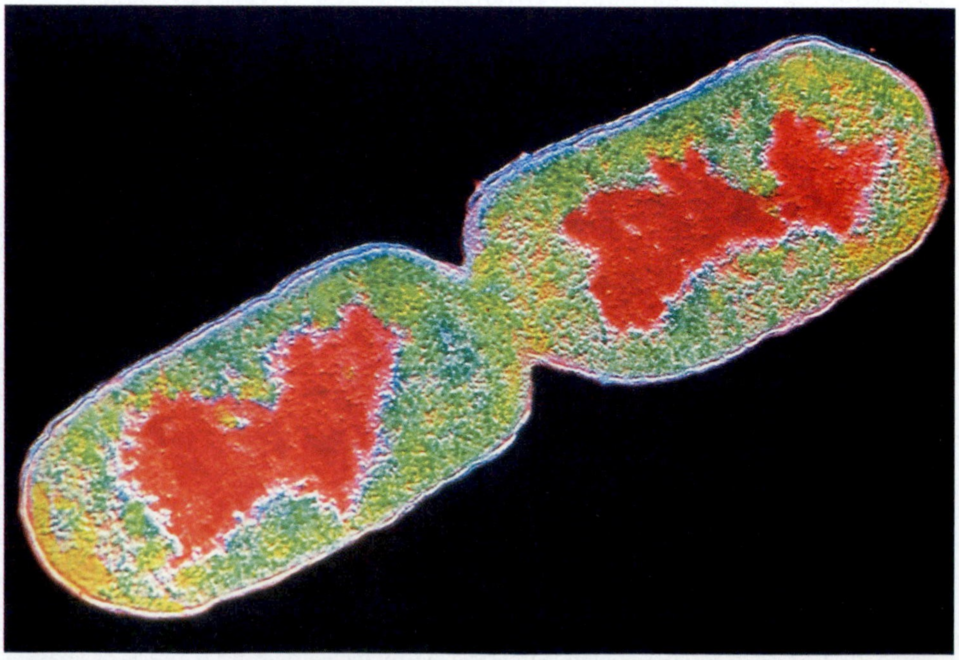

FIGURE 8–3 Color-enhanced electron micrograph of *E. coli* undergoing cell division. Particularly prominent are the two chromosomal areas (shown in red), called nucleoids, that have been partitioned into the daughter cells.

describes the evolutionary origin of these organelles, first put forward by Lynn Margolis, is called the **endosymbiotic hypothesis**.

Animal cells and some plant cells also contain a pair of complex structures called the **centrioles**. These cytoplasmic bodies, located in a specialized region called the **centrosome**, are associated with the organization of the spindle fibers that function in mitosis and meiosis. In some organisms, the centriole is derived from another structure, the **basal body**, which is associated with the formation of cilia and flagella. Over the years, many reports have suggested that centrioles and basal bodies contain DNA, which could be involved in their replication. Currently, this is thought not to be the case.

The organization of **spindle fibers** by the centrioles occurs during the early phases of mitosis and meiosis. Composed of arrays of microtubules, these fibers play an important role in the movement of chromosomes as they separate during cell division. The microtubules consist of polymers of alpha and beta polypeptide subunits of the protein tubulin. We will discuss the interaction of the chromosomes and spindle fibers later in the chapter.

8.2 Homologous Chromosomes Provide the Genetic Basis of Haploid and Diploid Cells

As we discuss the processes of mitosis and meiosis, understanding the concept of homologous chromosomes is essential. Such an understanding will also be critical to our future discussions of Mendelian genetics. Thus, before looking more closely at cell division,we will introduce some important terminology and then address this topic. Under the light microscope, chromosomes can be visualized during mitosis. When they are examined carefully, they are seen to take on distinctive lengths and shapes. Each contains a condensed or constricted region called the **centromere**, which establishes the general appearance of each chromosome. As Figure 8–4 illustrates, extending from either side of the centromere are the arms of the chromosome. Depending on the position of the centromere, different arm ratios are produced. Chromosomes are classified as **metacentric**, **submetacentric**, **acrocentric**, or **telocentric** on the basis of the centromere location. The shorter arm, by convention, is shown above the centromere and is called the **p arm** (p stands for "petite"). The longer arm is shown below the centromere and is called the **q arm** (q being the next letter in the alphabet).

When studying mitosis, several other observations are of particular relevance. First, all somatic cells derived from members of the same species contain an identical number of chromosomes. In most cases, this represents the **diploid number (2*n*)**. When the lengths and centromere placements of all such chromosomes are examined, a second general feature is apparent. Nearly all of the chromosomes exist in pairs with regard to these two criteria. The members of each pair are called **homologous chromosomes**. For each chromosome exhibiting a specific length and centromere placement, another exists with identical features. There are exceptions to

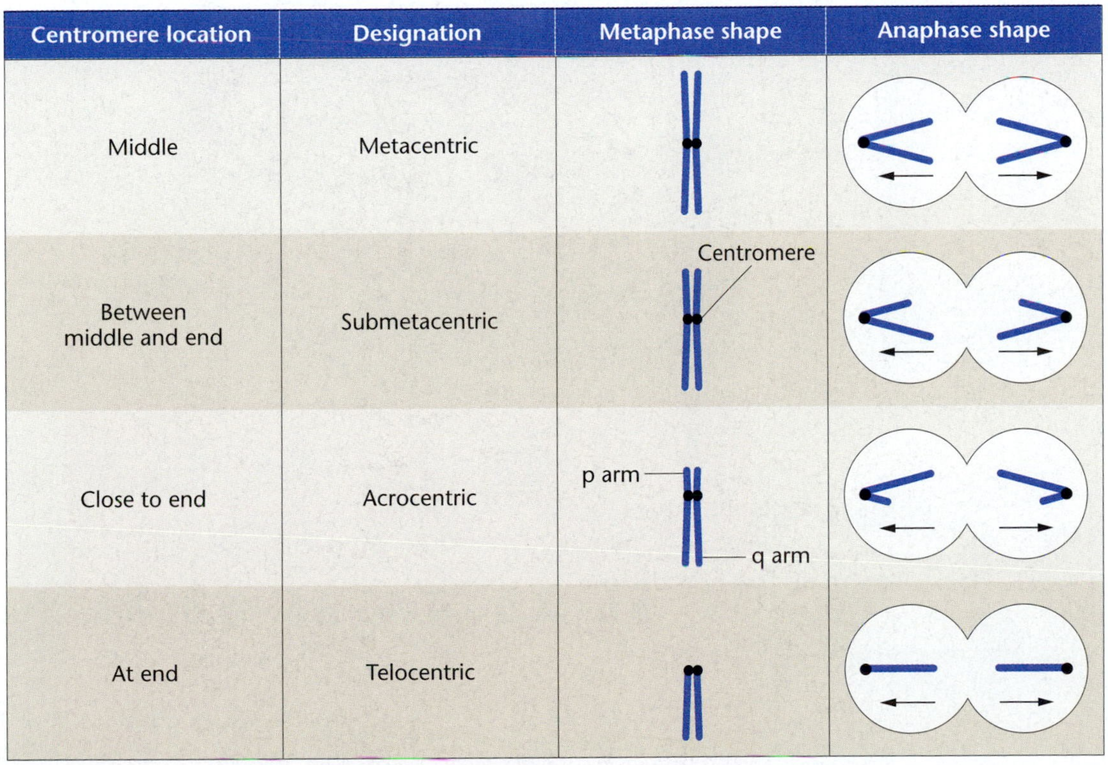

FIGURE 8–4 Centromere locations and designations of chromosomes based on centromere location. Note that the shape of the chromosome during anaphase of mitosis is determined by the position of the centromere.

the rule of chromosomes in pairs. Bacteria and viruses have but one chromosome, and organisms such as yeasts and molds and certain plants such as bryophytes (mosses) spend most of the life cycle in the haploid stage. That is, they contain only one member of each homologous pair of chromosomes during most of their lives.

Figure 8–5 illustrates the physical appearance of different pairs of homologous chromosomes. There, the human mitotic chromosomes have been photographed, cut out of the print, and matched up, creating a **karyotype**. As you can see, humans have a $2n$ number of 46 and, on close examination, exhibit a diversity of sizes and centromere placements. Note also that each one of the 46 chromosomes is clearly a double structure, consisting of two parallel sister chromatids connected by a common centromere. Had these chromosomes been allowed to continue through mitosis, each pair of the sister chromatids, which are replicas of one another, would have separated into two new cells as division continued.

The **haploid number** (***n***) of chromosomes is one-half of the diploid number. Collectively, the total set of genes and all accompanying DNA sequences contained in a haploid set of chromosomes constitutes the **genome** of the species. The examples listed in Table 8–1 demonstrates the wide range of n values found in plants and animals.

Homologous pairs of chromosomes have important genetic similarities. They contain identical gene sites along their lengths, each called a **locus** (pl. **loci**). Thus, they are identical in their genetic potential. In sexually reproducing organisms, one member of each pair is derived from the maternal parent (through the ovum) and one is derived from the paternal parent (through the sperm). Therefore, each diploid organism contains two copies of each gene as a consequence of **biparental inheritance**. As we will see in the following chapters on transmission genetics, the members of each pair of genes, while influencing the same characteristic or trait, need not be identical. In a population of members of the same species, many different alternative forms of the same gene, called **alleles**, can exist.

Having a firm grasp of the conceptual issues of haploid number, diploid number, and homologous chromosomes will aid you in studying meiosis. During the formation of gametes or spores, meiosis converts the diploid number of chromosomes to the haploid number. As a result, haploid gametes or spores contain precisely one member of each homologous pair of chromosomes—that is, one complete haploid set. Following fusion of two gametes in fertilization, the diploid number is reestablished, and the zygote contains two complete haploid sets of chromosomes. The constancy

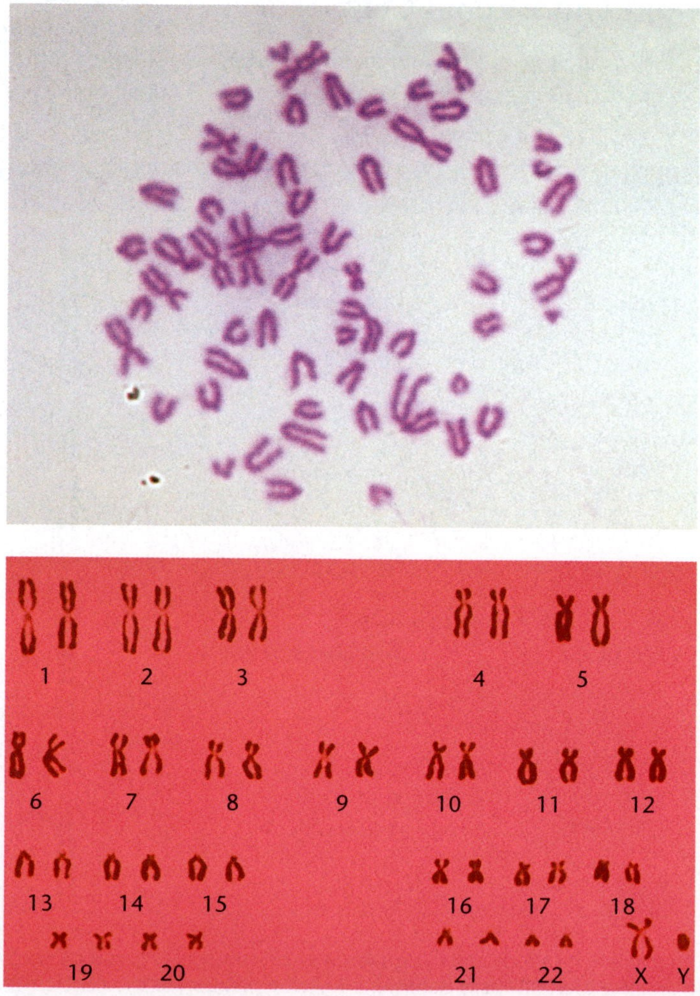

FIGURE 8–5 A metaphase preparation of chromosomes derived from a human male and the karyotype derived from it. All but the X and Y chromosomes (at the bottom right of the karyotype) are present in homologous pairs. Each chromosome is clearly a double structure, constituting a pair of sister chromatids joined at a common centromere.

TABLE 8–1 The Haploid Number of Chromosomes for Diverse Organisms

Common Name	*Scientific Name*	*Haploid Number*	*Common Name*	*Scientific Name*	*Haploid Number*
Black bread mold	*Aspergillus nidulans*	8	House mouse	*Mus musculus*	20
Broad bean	*Vicia faba*	6	Human	*Homo sapiens*	23
Cat	*Felis domesticus*	19	Jimson weed	*Datura stramonium*	12
Cattle	*Bos taurus*	30	Mosquito	*Culex pipiens*	3
Chicken	*Gallus domesticus*	39	Mustard plant	*Arabidopsis thaliana*	5
Chimpanzee	*Pan troglodytes*	24	Pink bread mold	*Neurospora crassa*	7
Corn	*Zea mays*	10	Potato	*Solanum tuberosum*	24
Cotton	*Gossypium hirsutum*	26	Rhesus monkey	*Macaca mulatta*	21
Dog	*Canis familiaris*	39	Roundworm	*Caenorhabditis elegans*	6
Evening primrose	*Oenothera biennis*	7	Silkworm	*Bombyx mori*	28
Frog	*Rana pipiens*	13	Slime mold	*Dictyostelium discoideum*	7
Fruit fly	*Drosophila melanogaster*	4	Snapdragon	*Antirrhinum majus*	8
Garden onion	*Allium cepa*	8	Tobacco	*Nicotiana tabacum*	24
Garden pea	*Pisum sativum*	7	Tomato	*Lycopersicon esculentum*	24
Grasshopper	*Melanoplus differentialis*	12	Water fly	*Nymphaea alba*	80
Green alga	*Chlamydomonas reinhardtii*	18	Wheat	*Triticum aestivum*	21
Horse	*Equus caballus*	32	Yeast	*Saccharomyces cerevisiae*	16
House fly	*Musca domestica*	6			

of genetic material is thus maintained from generation to generation.

There is one important exception to the concept of homologous pairs of chromosomes. In many species, the members of one pair, the **sex-determining chromosomes**, are not homologous. For example, in humans, females carry two homologous X chromosomes, whereas males contain a Y chromosome, in addition to one X chromosome (Figure 8–5). The X and Y chromosomes are not strictly homologous. The Y is considerably smaller and lacks most of the gene sites contained on the X. Nevertheless, in meiosis they behave as homologs so that gametes produced by males receive either the one X or one Y chromosome.

8.3 Mitosis Partitions Chromosomes into Dividing Cells

The process of **mitosis** is critical to all eukaryotic organisms. In many single-celled organisms that reproduce by cell division, such as protozoans, algae, and some fungi, mitosis provides the mechanism for their asexual reproduction. Multicellular diploid organisms begin life as single-celled fertilized eggs called **zygotes**. The mitotic activity of the zygote and the subsequent daughter cells is the foundation for development and growth of the organism. In adult organisms, mitotic activity is prominent in wound healing and other forms of cell replacement in certain tissues. For example, the epidermal skin cells of humans are continuously being sloughed off and replaced—as many as 100 billion (10^{11}) cells are lost daily according to one estimate! Mitosis is responsible for replacing them. Cell division also results in a continuous production of reticulocytes, cells that eventually shed their nuclei and are necessary for replenishing the supply of red blood cells in vertebrates. In abnormal situations, somatic cells may exhibit uncontrolled cell divisions, resulting in tumor formation.

The genetic material is partitioned into daughter cells during nuclear division or **karyokinesis**. This process is quite complex and requires great precision. The chromosomes first must be exactly replicated and then accurately partitioned. The end result is the production of two daughter nuclei, each with a chromosome composition identical to that of the parent cell.

Karyokinesis is followed by cytoplasmic division, or **cytokinesis**. The less complex division of the cytoplasm requires a mechanism that partitions the volume into two parts, then encloses both new cells in a distinct plasma membrane. Following partitioning of the cytoplasm, organelles either replicate themselves, arise from existing membrane structures, or are synthesized *de novo* (anew) in each cell. The subsequent proliferation of these structures is a reasonable and adequate mechanism for reconstituting the cytoplasm in daughter cells.

Following cell division, the initial size of each new daughter cell is approximately one-half the size of the parent cell. The nucleus of each new cell, however, is not appreciably smaller than the nucleus of the original cell. Quantitative measurements of DNA confirm that there is an equivalent amount of genetic material in both daughter nuclei as was initially present in the parent cell.

Interphase and the Cell Cycle

Many cells undergo a continuous alternation between division and nondivision. The events occurring from the completion of one division until the end of the next division constitute the **cell cycle** (Figure 8–6). We will consider the initial stage of the cycle, called **interphase**, as the interval between divisions. It was once thought that the biochemical activity during interphase was devoted solely to the cell's metabolism and growth. However, we now know that another biochemical step critical to the next mitotic division occurs during interphase: *the replication of the DNA of each chromosome.* Occurring before the cell enters mitosis, this period during which DNA is synthesized is called the **S phase**. The initiation and completion of synthesis can be detected by monitoring the incorporation of radioactive DNA precursors such as ^{3}H-thymidine. Their incorporation can be monitored using the technique of autoradiography (Appendix A).

Investigations of this nature show two periods during interphase when no DNA synthesis occurs, one before and one after S phase. These are designated **G1 (gap1)** and **G2 (gap2)**, respectively. Particularly during G1, intensive metabolic activity, cell growth, and cell differentiation occur. By the end of G2, the volume of the cell has roughly doubled, DNA has been replicated, and mitosis (M) is initiated. Following mitosis, continuously dividing cells repeat this cycle (G1, S, G2, M) over and over, as illustrated in Figure 8–6.

Much is known about the cell cycle based on *in vitro* (in glass) studies. When cultured outside of an organism, many cell types derived from mammals traverse the complete cell cycle in about 16 hours. The actual process of mitosis occupies only a small part of the cycle, often less than an hour. The lengths of the S and G2 phase of interphase are fairly consistent among different cell types. Most variation is seen in the length of time spent in the G1 phase. Figure 8–7 illustrates the relative length of these intervals in a typical human cell.

G1 is of great interest in the study of cell proliferation and its control. At a point late in G1, all cells follow one of two paths. They either withdraw from the cycle and enter a resting phase called **G0** (see Figure 8–6), or they become committed to initiating DNA synthesis and completing the cycle. Cells that enter G0 remain viable and metabolically active, but are nonproliferative. Cancer cells apparently avoid entering G0, or else they pass through it very quickly. Some cells enter G0 and never reenter the cell cycle. Others can be stimulated to return to G1, thereby the cycle.

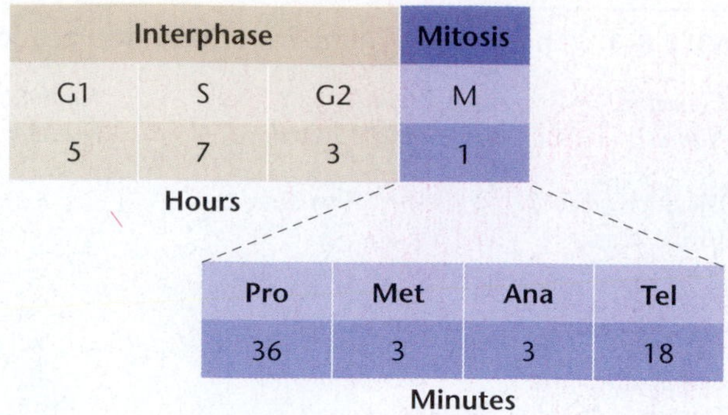

FIGURE 8–7 The time spent in each phase of one complete cell cycle of a typical human cell in culture. Actual times vary according to cell types and conditions.

Cytologically, interphase is characterized by the absence of visible chromosomes. Instead, the nucleus is filled with chromatin fibers that have formed as the chromosomes have uncoiled and dispersed following the previous mitosis. This is diagrammed in Figure 8–8(a) and in the comparable light micrograph in Figure 8–9(a).

Once the G1, S, and G2 phases of interphase are completed, mitosis is initiated. Mitosis is a dynamic period of vigorous and continual activity. For ease of discussion, the entire process has been subdivided into discrete phases, each characterized by specific events. These stages, in order of occurrence, are **prophase**, **prometaphase**, **metaphase**, **anaphase**, and **telophase**. Like interphase, these stages are also diagrammed in Figure 8–8. A photograph of each stage is shown in Figure 8–9.

Prophase

Often, over half of mitosis is spent in prophase, a stage characterized by several significant activities. One of the early events in prophase of all animal cells involves the migration of two pairs of centrioles to opposite ends of the cell. These structures are found just outside the nuclear envelope in an area of differentiated cytoplasm called the **centrosome**. It is thought that each pair of centrioles consists of one mature unit and a smaller, newly formed centriole.

The direction of migration of the centrioles is such that two poles are established at opposite ends of the cell. Following their migration, the centrioles are responsible for

FIGURE 8–6 The intervals comprising an arbitrary cell cycle. Following mitosis (M), cells enter the G1 stage of interphase, initiating a new cycle. Cells may become nondividing (G0) or continue through G1, where they become committed to begin DNA synthesis (S) and complete the cycle (G2 and Mitosis). Following mitosis, two daughter cells are produced and the cycle begins anew for both cells.

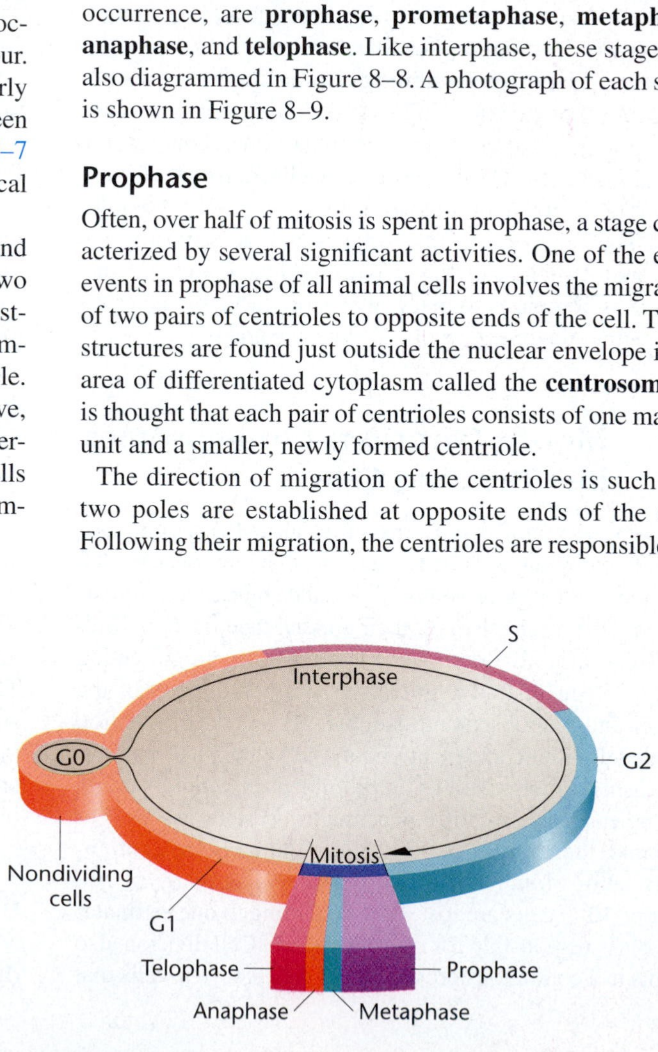

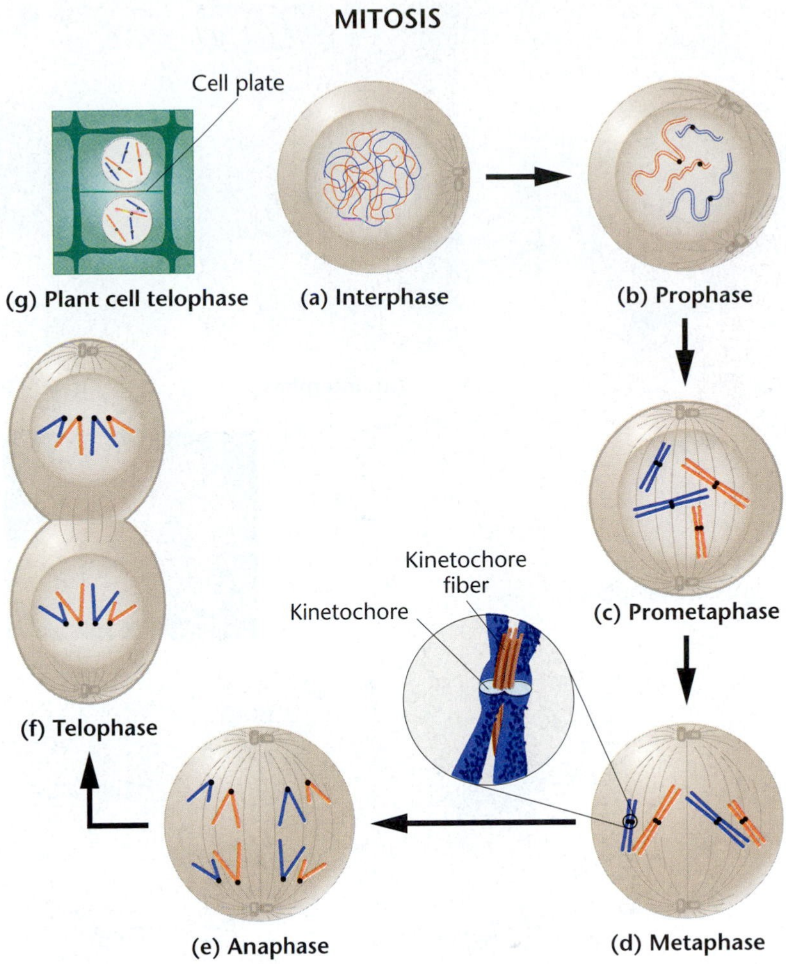

FIGURE 8–8 Mitosis in an animal cell with a diploid number of 4. The events occurring in each stage are described in the text. Of the two homologous pairs of chromosomes, one contains longer metacentric members and, the other, shorter submetacentric members. The maternal chromosome and the paternal chromosome of each pair are shown in different colors. In (g), the telophase stage in a plant cell illustrates the formation of the cell plate and lack of centrioles.

the organization of cytoplasmic microtubules into a series of **spindle fibers** that are formed and run between these poles, creating the axis along which chromosomal separation will occur. We will discuss the formation and role of these spindle fibers in greater detail next.

Interestingly, cells of most plants (there are a few exceptions), fungi, and certain algae seem to lack centrioles. Spindle fibers are nevertheless apparent during mitosis. Therefore, centrioles are not universally responsible for the organization of spindle fibers.

As the centrioles migrate, the nuclear envelope begins to break down and gradually disappears. In a similar fashion, the nucleolus disintegrates within the nucleus. While these events are taking place, the diffuse chromatin fibers begin to condense, continuing until distinct threadlike structures, or chromosomes, become visible. It becomes apparent near the end of prophase that each chromosome is actually a double structure, split longitudinally, except at a single point of constriction, called the **centromere**. (See Figure 8–5.) The two components of each chromosome are called **chromatids**. Because the DNA contained in each pair of chromatids represents the duplication of a single chromosome during the S phase of the previous interphase, these chromatids are genetically identical. Therefore they are called **sister chromatids**. In humans, with a diploid number of 46, a cytological preparation of late prophase will reveal 46 chromosomes randomly distributed in the area formerly occupied by the nucleus.

Prometaphase and Metaphase

The distinguishing event of the next stage of mitosis is the migration of each chromosome, led by its centromeric region, to the equatorial plane of the cell. In some descriptions the term **prometaphase** refers to the period of chromosome movement, and **metaphase** is applied strictly to the configuration of chromosomes following this movement, as depicted in Figures 8–8(c) and 8–9(c). The equatorial plane, also referred to as the metaphase plate, is the midline region of the cell, a plane that lies perpendicular to the axis established by the spindle fibers.

Migration is made possible by the binding of microtubules of the spindle fibers to a structure associated with the centromere of each chromosome called the **kinetochore** (Figure 8–10). The kinetechore, consisting of multilayered plates of proteins, forms on opposite sides of each centromere and becomes intimately associated with the two sister chromatids of each chromosome. Once attachment is complete, the sister chromatids are now ready to be pulled to opposite poles during the ensuing anaphase stage.

Spindle fibers consist of **microtubules**, which themselves consist of molecular subunits of the protein **tubulin**.

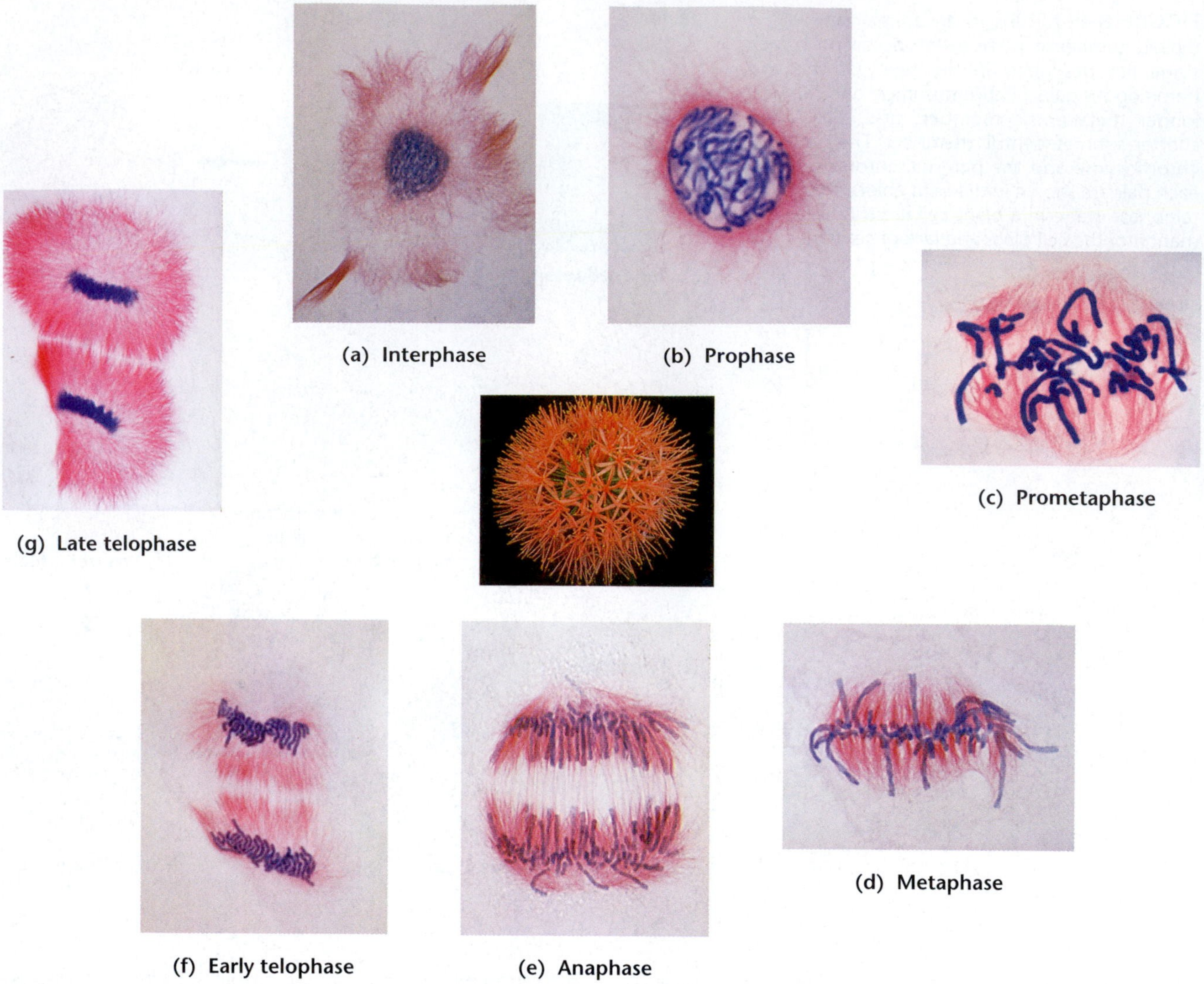

FIGURE 8–9 Light micrographs illustrating the stages of mitosis depicted in Figure 8–8. These stages are derived from the flower of *Haemanthus*, shown in the center.

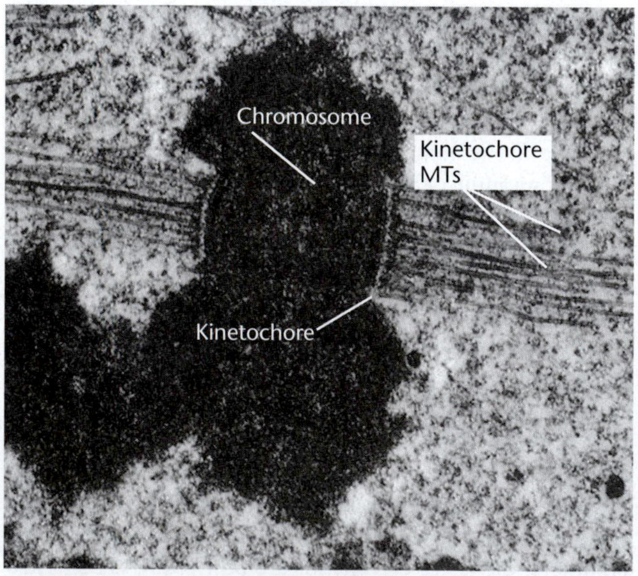

FIGURE 8–10 An electron micrograph showing the association between microtubules constituting the spindle fibers and the kinetochore (in association with a centromere) during mitosis.

Microtubules seem to originate and "grow" out of the two centrosome regions (containing the centrioles) at opposite poles of the cell. They are dynamic structures that lengthen and shorten as a result of the addition or loss of polarized tubulin subunits. There are two major categories of microtubules. Those most directly responsible for chromosome migration during anaphase make contact with, and adhere to, kinetochores as they grow from the centrosome region. They are referred to as **kinetochore microtubules** and have one end near the centrosome region (at one of the poles of the cell) and the other anchored to the kinetochore. It is interesting to note that the number of microtubules that bind to the kinetochore varies greatly between organisms. Yeast (*Saccharomyces*) have only a single microtubule bound to each platelike structure of the kinetochore. Mitotic cells of mammals, at the other extreme, reveal 30 to 40 microtubules bound to each portion of the kinetochore.

Those microtubules that do not adhere to kinetochores make contact with growing microtubules from the opposite pole of the cell, often interdigitating with one another. The polarized nature of the tubulin subunits provides the force that joins them together. They are referred to as **polar microtubules** (and sometimes, spindle microtubules) and provide the cytoplasmic framework of spindle fibers that establish and maintain the separation of the two poles during chromosome separation. Kinetochore microtubules are apparent in Figure 8–10.

At the completion of metaphase, each centromere is aligned at the plate with the chromosome arms extending outward in a random array. This configuration is shown in Figures 8–8(d) and 8–9(d).

Anaphase

Events critical to chromosome distribution during mitosis occur during the shortest stage of mitosis, **anaphase**. During this phase, sister chromatids of each chromosome *disjoin* (separate) from each other and migrate to opposite ends of the cell. For complete disjunction to occur, each centromeric region must be split in two, signaling the initiation of anaphase. Once anaphase occurs, each chromatid is referred to as a daughter **chromosome**.

As discussed above, movement of daughter chromosomes to the opposite poles of the cell is dependent upon the centromere–spindle fiber attachment. Investigations have revealed that chromosome migration results from the activity of a series of specific proteins, generally called motor proteins. These proteins use the energy generated by the hydrolysis of ATP, creating what are called **molecular motors** in the cell. These motors act at several positions within the dividing cell, but all are involved in the activity of microtubules and ultimately serve to propel the chromosomes to opposite ends of the cell. The centromeres of each chromosome *appear* to lead the way during migration, with the chromosome arms trailing behind. The location of the centromere determines the shape of the chromosome during separation, as you can see in Figure 8–4.

The steps occurring during anaphase are critical in providing each subsequent daughter cell with an identical set of chromosomes. In human cells there would now be 46 chromosomes at each pole, one from each original sister pair. Figures 8–8(e) and 8–9(e) show anaphase prior to its completion.

Telophase

Telophase is the final stage of mitosis and is depicted in Figures 8–8(f) and (g) and 8–9(f) and (g). At its beginning, there are two complete sets of chromosomes, one at each pole. The most significant event is **cytokinesis**, the division or partitioning of the cytoplasm. Cytokinesis is essential if two new cells are to be produced from one. The mechanism differs greatly in plant and animal cells. In plant cells, a **cell plate** is synthesized and laid down across the dividing cell in the region of the metaphase plate. Animal cells, however, undergo a constriction of the cytoplasm in much the same way a loop might be tightened around the middle of a balloon. The end result is the same: Two distinct cells are formed.

It is not surprising that the process of cytokinesis varies among cells of different organisms. Plant cells, which are more regularly shaped and structurally rigid, require a mechanism for the deposition of new cell wall material around the plasma membrane. The cell plate, laid down during telophase, becomes the **middle lamella**.. Subsequently, the primary and secondary layers of the cell wall are deposited between the cell membrane and middle lamella on both sides of the boundary between the two daughter cells. In animals, complete constriction of the cell membrane produces the **cell furrow** characteristic of newly divided cells.

Other events necessary for the transition from mitosis to interphase are initiated during late telophase. They represent a general reversal of those that occurred during prophase. In each new cell, the chromosomes begin to uncoil creating diffuse chromatin once again, while the nuclear envelope reforms around them. The nucleolus gradually re-forms and it becomes visible in the nucleus during early interphase. The spindle fibers also disappear. At the completion of telophase, the cell enters interphase.

8.4 The Cell Cycle is Genetically Regulated

The cell cycle, including mitosis, is fundamentally the same in all eukaryotic organisms. The similarity of the events leading to cell duplication in many diverse organisms suggests that the cell cycle is governed by a genetic program that has been conserved throughout evolution and is therefore genetically regulated. Elucidation of this genetic program therefore provides information basic to our understanding of the nature of living organisms. Furthermore, because disruption of this regulation may lead to uncontrolled cell division characterizing malignancy, interest in how genes regulate the cell cycle is keen.

A mammoth research effort over the past decade has paid high dividends, and we now have knowledge of many genes involved in the control of the cell cycle. This work was recognized by the receipt of the 2001 Nobel Prize awarded to Lee Hartwell, Paul Nurse, and Tim Hunt. As with other studies of genetic input into essential biological processes, investigation is focusing on the discovery of mutations that interrupt the cell cycle and the subsequent study of the effects of these mutations. As we shall return to this subject in Chapter 22, during our consideration of cancer, what follows is a brief overview.

Many mutations that exert their effect at various stages of the cell cycle are now known. First discovered in yeast, but now evident in all organisms, including humans, such mutations were originally designated as ***cell division cycle (cdc)* mutations**. The study of these mutations has established that during the cell cycle, at least three major **checkpoints** exist, where the cell is monitored or "checked" before it can proceed to the next stage of the cycle.

The products of many of these genes are enzymes called **kinases** that can add phosphates to other proteins. They serve as "master control" molecules that work in conjunction with proteins called **cyclins**. These kinases phosphorylate cyclins and influence their activity at the cell-cycle checkpoints. These activities thus regulate the cell cycle. When such a kinase works in conjunction with a cyclin, it is called a **Cdk protein**, for **Cyclin-dependent kinase protein**.

Figure 8–11 identifies the location of three "checkpoints" within the cell cycle. The first is the **G1/S checkpoint**, which monitors the size that the cell has achieved following the previous mitosis and whether the DNA has been damaged. If the cell has not achieved an adequate size or if the DNA has been damaged, further progress through the cycle is arrested until these conditions are "corrected," so to speak. If both conditions are initially "normal," then the checkpoint is traversed and the cell proceeds to the S phase of the cycle.

The second important checkpoint is the **G2/M checkpoint**, where physiological conditions in the cell are monitored prior to entering mitosis. If DNA replication or repair to any DNA damage has not been completed, the cell cycle is arrested until these processes are completed. The final checkpoint occurs during mitosis and is called the **M checkpoint**. Here, both the successful formation of the spindle fiber system and the attachment of spindle fibers to the kinetochores associated with the centromeres are monitored. If spindle fibers are not properly formed or attachment is inadequate, mitosis is arrested.

The importance of cell-cycle control and these checkpoints can be demonstrated by considering what happens when this regulatory system is impaired. If, for example, a cell that has incurred damage to its DNA is allowed to proceed through the cell cycle, the damage may lead to uncontrolled cell division, precisely the definition of a cancerous cell. As we have just observed, such a damaged cell would normally be arrested at either the G1/S or the G2/M checkpoint.

An interesting related finding involves the protein product of the ***p53* gene** in humans and its involvement during scrutiny at the G1/S checkpoint. This protein functions during the regulation of **apoptosis**, the genetic process whereby programmed cell death occurs. When the normal *p53* gene product is present, a proliferative cell that has incurred severe damage to its DNA will be targeted for programmed cell death at the G1/S checkpoint and, thus, effectively removed from the cell population. This surveillance is dependent on the product of the *p53* gene. However, if the gene has mutated, resulting in abnormal function of the *p53* gene product, the damaged cell may proceed through the checkpoint and continue to proliferate in an uncontrolled manner. In fact, a high percentage of human cancers has been found to contain mutations in the *p53* gene. These include a wide variety of cancers, including colon, breast, lung, and bladder malignancies. In the language of cancer genetics, *p53* is referred to as a *tumor-suppressor gene*. The role of this and other genes involved in the cell cycle has sparked great interest in cancer research, as discussed in Chapter 22.

8.5 Meiosis Reduces the Chromosome Number from Diploid to Haploid in Germ Cells and Spores

The process of meiosis, unlike mitosis, reduces the amount of genetic material by one-half. In diploids, mitosis produces daughter cells with a full diploid complement, whereas, meiosis produces gametes or spores with only one haploid set of chromosomes. During sexual reproduction, gametes then combine in fertilization to reconstitute the diploid complement found in parental cells. Figure 8–12 provides a comparison of the two processes, where two pairs of homologous chromosomes are followed.

Meiosis must be highly specific since, by definition, haploid gametes or spores must contain precisely one member of each homologous pair of chromosomes. Successfully completed, meiosis ensures genetic continuity from generation to generation. The process of sexual reproduction also ensures genetic variety among members of a species. As you study meiosis, you will see that this process results in gametes with many unique combinations of maternally and paternally derived chromosomes among the haploid complement. With such a tremendous genetic variation among the gametes, a large number of chromosome combinations are possible at fertilization. Furthermore, we will see that the meiotic event referred to as **crossing over** results in genetic exchange between members of each homologous pair of chromosomes. This creates intact chromosomes that

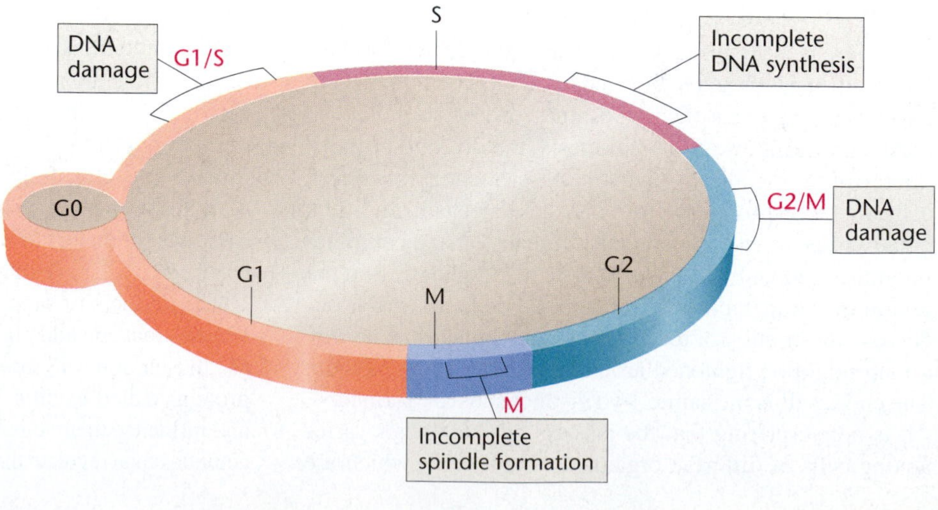

FIGURE 8–11 Illustration of the three major checkpoints within the cell cycle that regulate its progress.

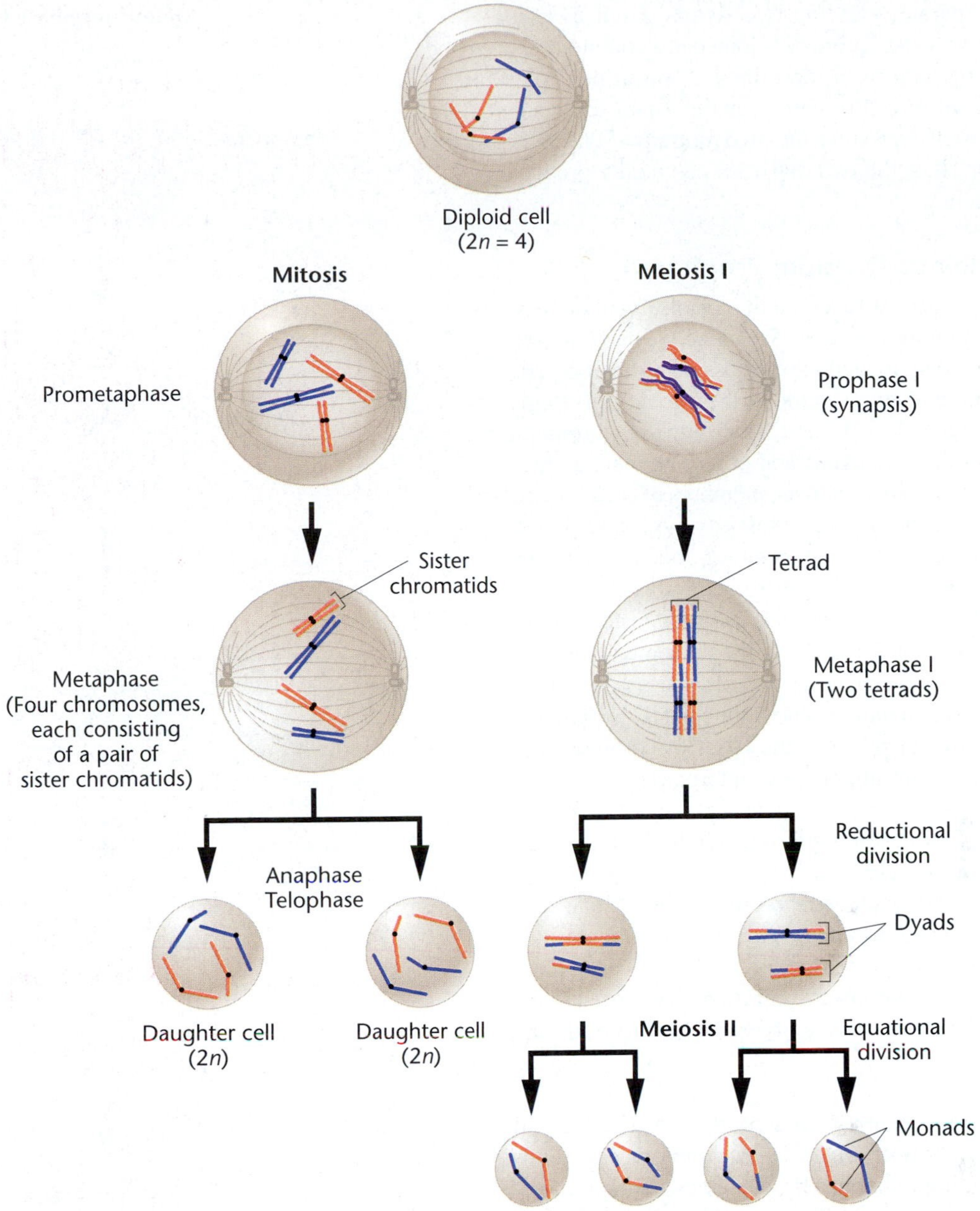

FIGURE 8–12 A comparison of the major events and outcomes of mitosis and meiosis. As in Figure 8–8, two pairs of homologous chromosomes are followed.

are mosaics of the maternal and paternal homologs from which they are derived, further enhancing the potential genetic variation in gametes and the offspring derived from them. Sexual reproduction, therefore, reshuffles the genetic material, producing offspring that often differ greatly from either parent. This process constitutes the major form of genetic recombination within species.

An Overview of Meiosis

In the preceding discussion, we established the goals of meiosis. Before we consider the phases of this process systematically, we will briefly examine how diploid cells give rise to haploid gametes or spores. You should refer to the meiotic portion of Figure 8–12 during the following discussion.

You have seen that in mitosis each paternally and maternally derived member of any given homologous pair of chromosomes behaves autonomously during division. By contrast, early in meiosis, homologous chromosomes pair up; that is, they **synapse**. Since each synapsed structure consists initially of two chromosomes, it is called a **bivalent**. Soon, however, as the bivalent condenses, it becomes apparent that both chromosomes making up the bivalent have, in fact, duplicated. This results in a unit, the **tetrad**, consisting of four components, two pairs of sister chromatids. Therefore, in order to achieve haploidy, two divisions are necessary. In meiosis I, described as a **reductional division** (because the number of centromeres, each representing one chromosome, is *reduced* by one-half following this division), components of each tetrad—representing the two homologs—separate, yielding two **dyads**. Each dyad is composed of two sister chromatids joined at a common centromere. During meiosis II, described as **equational** (because the number of centromeres remains *equal* following this division), each dyad splits into two **monads** of one chromosome each. Thus, the two divisions potentially produce four haploid cells.

The First Meiotic Division: Prophase I

We turn now to a detailed account of meiosis. As in mitosis, meiosis is a continuous process. We name the parts of each stage of division only to facilitate discussion. From a genetic standpoint, three events characterize the initial stage, **prophase I** (Figure 8–13). First, as in mitosis, chromatin present in interphase thickens and coils into visible chromosomes. Second, unlike mitosis, members of each homologous pair of chromosomes undergo synapsis. Third, crossing over, an exchange process, occurs between synapsed homologs. Because of the complexity of these genetic events, this stage of meiosis has been further subdivided into five substages: leptonema,* zygonema,* pachynema,* diplonema,* and diakinesis. As we discuss these stages, be aware that, even though it is not immediately apparent in the earliest phases of meiosis, the DNA of chromosomes has been replicated during the prior interphase.

Leptonema During the **leptotene stage**, the interphase chromatin material begins to condense, and the chromosomes, although still extended, become visible. Along each chromosome are **chromomeres**, localized condensations that resemble beads on a string. Recent evidence suggests that a process called **homology search**, which precedes and is essential to the initial pairing of homologs, begins during leptonema.

Media Tutorial Meiosis

Zygonema The chromosomes continue to shorten and thicken in the **zygotene stage**. During the process of homology search, homologous chromosomes undergo initial alignment with one another. This so-called rough pairing is complete by the end of zygonema. In yeast, homologs are separated by about 300 nm, and near the end of zygonema, structures referred to as lateral elements are visible between paired homologs. As meiosis proceeds, the overall length of the lateral elements increases and a more extensive ultrastructural component, the synaptonemal complex, begins to form between the homologs. We will discuss this meiotic component later in the chapter.

At the completion of zygonema, the paired homologs take the form of **bivalents**. As previously discussed, although both members of each bivalent have already replicated their DNA, it is not yet visually apparent at this stage that each member is a double structure. The number of bivalents in each species is equal to the haploid (n) number.

*These are the noun forms of these stages. The adjective forms (leptotene, zygotene, pachytene, and diplotene) are also used.

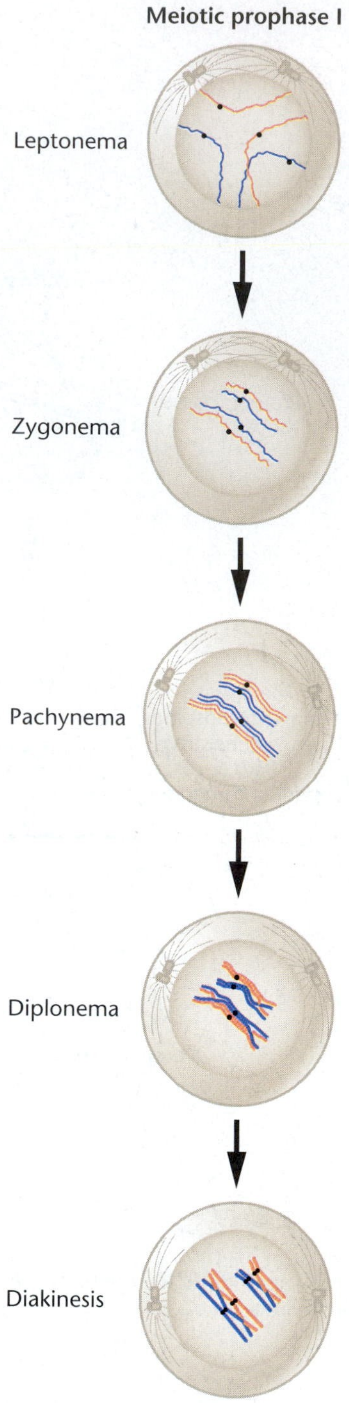

FIGURE 8–13 The substages of meiotic prophase I for the chromosomes depicted in Figure 8–8.

Pachynema In the transition from the zygotene to the **pachytene stage**, coiling and shortening of chromosomes continues, and further development of the synaptonemal complex occurs between the two members of each bivalent. This leads to a more intimate pairing, referred to as **synapsis**. Compared to the rough-pairing characteristic of zygonema in yeast, homologs are now separated by only 100 nm.

During pachynema, each homolog is first evident as a double structure, providing visual evidence of the earlier replication of the DNA of each chromosome. Thus, each bivalent contains four member chromatids. As in mitosis, replicates are called sister chromatids, while chromatids from maternal vs. paternal members of a homologous pair are called nonsister chromatids. The four-member structure is also referred to as a tetrad, and each tetrad contains two pairs of sister chromatids.

Diplonema During observation of the ensuing **diplotene stage**, it is even more apparent that each tetrad consists of two pairs of sister chromatids. Within each tetrad, each pair of sister chromatids begins to separate. However, one or more areas remain in contact where chromatids are intertwined. Each such area, called a **chiasma** (pl. **chiasmata**), is thought to represent a point where nonsister chromatids have undergone genetic exchange through the process of crossing over. Although the physical exchange between chromosome areas occurred during the previous pachytene stage, the result of crossing over is visible only when the duplicated chromosomes begin to separate. Crossing over is an important source of genetic variability. As indicated earlier, new combinations of genetic material are formed during this process.

Diakinesis The final stage of prophase I is **diakinesis**. The chromosomes pull farther apart, but nonsister chromatids remain loosely associated via the chiasmata. As separation proceeds, the chiasmata move toward the ends of the tetrad. The process, called **terminalization**, begins in late diplonema, and is completed during diakinesis. During this final period of prophase I, the nucleolus and nuclear envelope break down, and the two centromeres of each tetrad become attached to the recently formed spindle fibers. By the completion of prophase I, the centromeres of each tetrad structure are present on the equatorial plate of the cell.

Metaphase, Anaphase, and Telophase I

The remainder of the meiotic process is depicted in Figure 8–14. In the metaphase stage of the first division (**metaphase I**), the chromosomes have maximally shortened and thickened. The terminal chiasmata of each tetrad are visible and appear to be the only factor holding the nonsister chromatids together. Each tetrad interacts with spindle fibers, facilitating movement to the metaphase plate.

The alignment of each tetrad prior to this first anaphase stage is random. One-half of each tetrad will be pulled to one or the other pole at random, and the other half will move to the opposite pole. This random separation of dyads is the basis for the Mendelian postulate of **independent assortment**, which we discuss in Chapter 9. You may wish to return to the discussion when you study this principle.

During the stages of meiosis I, a single centromere holds each pair of sister chromatids together. It does *not* divide. At **anaphase I**, one-half of each tetrad (one pair of sister chromatids, called a **dyad**) is pulled toward each pole of the dividing cell. This separation process is the physical basis of what we refer to as **disjunction**, the separation of chromosomes from one another. Occasionally, errors in meiosis occur and separation is not achieved, as we will see later in the chapter. The term **nondisjunction** describes such an error. At the completion of the normal anaphase I, a series of dyads equal to the haploid number is present at each pole.

If crossing over had not occurred in the first meiotic prophase, each dyad at each pole would consist solely of either paternal or maternal chromatids. However, the exchanges produced by crossing over create mosaic chromatids of paternal and maternal origin.

In many organisms, **telophase I** reveals a nuclear membrane forming around the dyads. Next, the nucleus enters into a short interphase period. In other cases, the cells go directly from the first anaphase into the second meiotic division. If an interphase period occurs, the chromosomes do not replicate, since they already consist of two chromatids. In general, meiotic telophase is much shorter than the corresponding stage in mitosis.

The Second Meiotic Division

A second division, referred to as **meiosis II**, is essential if each gamete or spore is to receive only one chromatid from each original tetrad. The stages characterizing meiosis II are also diagrammed in Figure 8–14. During **prophase II**, each dyad is composed of one pair of sister chromatids attached by a common centromere. During **metaphase II**, the centromeres are positioned on the equatorial plate. When they divide, **anaphase II** is initiated, and the sister chromatids of each dyad are pulled to opposite poles. Because the number of dyads is equal to the haploid number, **telophase II** reveals one member of each pair of homologous chromosomes present at each pole. Each chromosome is referred to as a **monad**. Following cytokinesis in telophase II, four haploid gametes may result from a single meiotic event. At the conclusion of meiosis, not only has the haploid state been achieved, but if crossing over has occurred, each monad is a combination of maternal and paternal genetic information. As a result, the offspring produced by any gamete will receive a mixture of genetic information originally present in his or her grandparents. Meiosis thus creates a significant amount of genetic variation in each ensuing generation.

8.6 The Development of Gametes Varies during Spermatogenesis and Oogenesis

Although events that occur during the meiotic divisions are similar in all cells that participate in gametogenesis in most animal species, there are certain differences between the production of a male gamete (spermatogenesis) and a female gamete (oogenesis). Figure 8–15 summarizes these processes.

Spermatogenesis takes place in the testes, the male reproductive organs. The process begins with the expanded growth of an undifferentiated diploid germ cell called a **spermatogonium**. This cell enlarges to become a **primary spermatocyte**, which undergoes the first meiotic division.

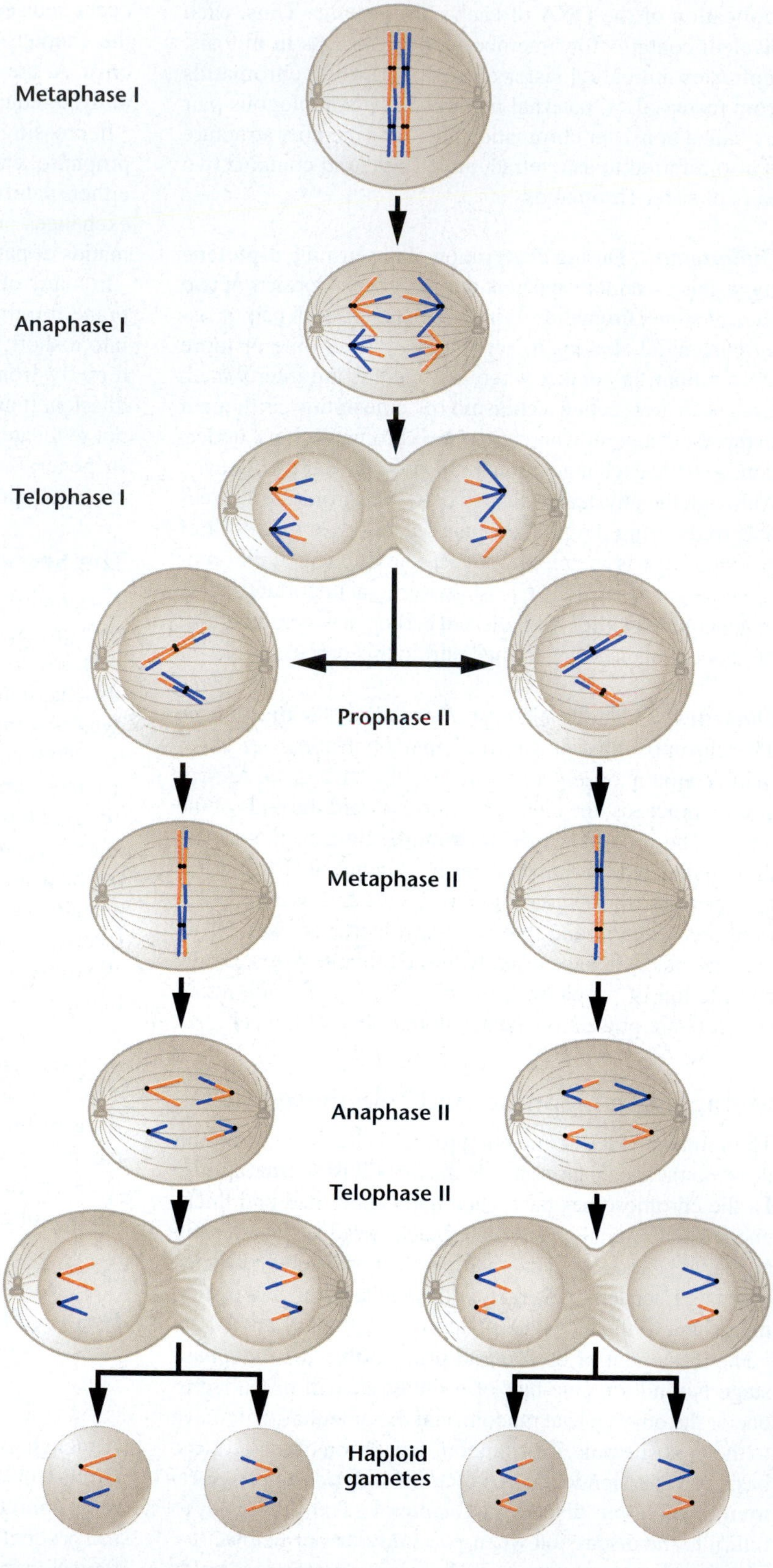

FIGURE 8–14 The major events occurring during meiosis in an animal with a diploid number of 4, beginning with metaphase I. Note that the combination of chromosomes contained in the cells produced following telophase II is dependent on the random alignment of each tetrad and dyad on the equatorial plate during metaphase I and metaphase II. Several other combinations, which are not shown, can also be formed. The events depicted here are described in the text.

The products of this division, called **secondary spermatocytes**, contain a haploid number of dyads. The secondary spermatocytes then undergo the second meiotic division, and each of these cells produces two haploid **spermatids**. Spermatids go through a series of developmental changes, **spermiogenesis**, and become highly specialized, motile **spermatozoa** or **sperm**. All sperm cells produced during spermatogenesis receive equal amounts of genetic material and cytoplasm. Spermatogenesis may be continuous or occur periodically in mature male animals, with its onset

FIGURE 8–15 A comparison of spermatogenesis and oogenesis in animal cells.

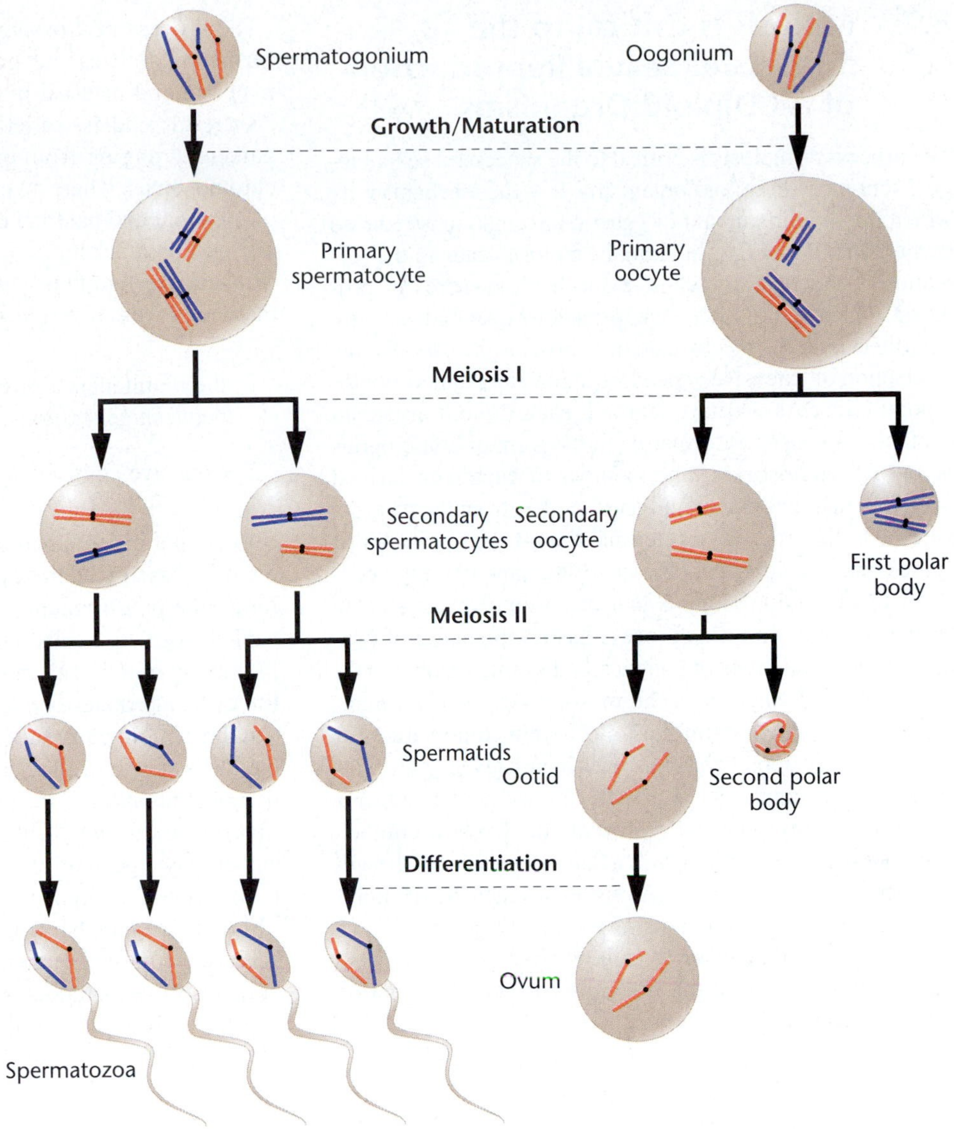

determined by the nature of the species' reproductive cycle. Animals that reproduce year-round produce sperm continuously, whereas those whose breeding period is confined to a particular season produce sperm only during that time.

In animal **oogenesis**, the formation of **ova** (singular: **ovum**), or eggs, occurs in the ovaries, the female reproductive organs. The daughter cells resulting from the two meiotic divisions receive equal amounts of genetic material, but they do *not* receive equal amounts of cytoplasm. Instead, during each division, almost all the cytoplasm of the **primary oocyte**, itself derived from the **oogonium**, is concentrated in one of the two daughter cells. The concentration of cytoplasm is necessary because a major function of the mature ovum is to nourish the developing embryo following fertilization.

During the first meiotic anaphase in oogenesis, the tetrads of the primary oocyte separate, and the dyads move toward opposite poles. During the first telophase, the dyads present at one pole are pinched off with very little surrounding cytoplasm to form the **first polar body**. The other daughter cell produced by this first meiotic division contains most of the cytoplasm and is called the **secondary oocyte**. The first polar body may or may not divide again to produce two small haploid cells. The mature ovum will be produced from the secondary oocyte during the second meiotic division. During this division, the cytoplasm of the secondary oocyte again divides unequally, producing an **ootid** and a **second polar body**. The ootid then differentiates into the mature ovum. Unlike the divisions of spermatogenesis, the two meiotic divisions of oogenesis may not be continuous. In some animal species, the two divisions may directly follow each other. In others, including humans, the first division of all oocytes begins in the embryonic ovary, but arrests in prophase I. Many years later, meiosis resumes in each oocyte just prior to its ovulation. The second division is completed only after fertilization.

8.7 Meiosis is Critical to the Successful Sexual Reproduction of All Diploid Organisms

The process of meiosis is critical to the successful sexual reproduction of all diploid organisms. It is the mechanism by which the diploid amount of genetic information is reduced to the haploid amount. In animals, meiosis leads to the formation of gametes, whereas in plants haploid spores are produced, which in turn lead to the formation of haploid gametes.

Furthermore, the mechanism of meiosis is the basis for the production of extensive genetic variation among members of a population. As we have learned, each diploid organism contains its genetic information in the form of homologous pairs of chromosomes, one member of each pair derived from the maternal parent and one member from the paternal parent. Following the reduction to haploidy, gametes or spores contain either the paternal or the maternal representative of every homologous pair of chromosomes. During sexual reproduction, this process has the potential of producing huge quantities of genetically dissimilar gametes. As the number of homologous chromosomes (the haploid number) increases, the possibilities of different combinations of maternal and paternal chromosomes in any given gamete increase. For example, an organism can produce 2^n number of combinations, where n represents the haploid number. For example, an organism with a haploid number of 10 will produce 2^{10} or 1024 combinations. Now calculate the number of different combinations of sperm or eggs in our own species: 2^{23}. When you arrive at the answer, you cannot help but be impressed with the potential for genetic variation resulting from meiosis.

The process of crossing over during meiotic prophase I further reshuffles the genetic information between the maternal and paternal members of each homologous pair. As a result, endless varieties of each homolog may occur in gametes, ranging from either intact maternal or paternal chromosomes, where no exchange occurred, to any mixture of maternal and paternal components, depending on where one or more exchanges occurred during crossing over.

In sum, the two most significant points about meiosis are that the process is responsible for

1. the maintenance of constancy of genetic information between generations, and
2. extensive genetic variation within populations.

It is important to touch briefly on the significant role that meiosis plays in the life cycles of fungi and plants. In many fungi, the predominant stage of the life cycle consists of haploid vegetative cells. They arise through meiosis and proliferate by mitotic cell division. In multicellular plants, the life cycle alternates between the diploid **sporophyte stage** and the haploid **gametophyte stage**. While one or the other predominates in different plant groups during this "alternation of generations," the processes of meiosis and fertilization constitute the "bridge" between the sporophyte and gametophyte generations (Figure 8–16). Therefore, meiosis is an essential component of the life cycle of plants.

Finally, we can ask what happens when meiosis fails to achieve the normal outcome. In rare cases during meiosis I or meiosis II, separation, or disjunction, of the chromatids of a tetrad or dyad fails to occur. Instead, both members move

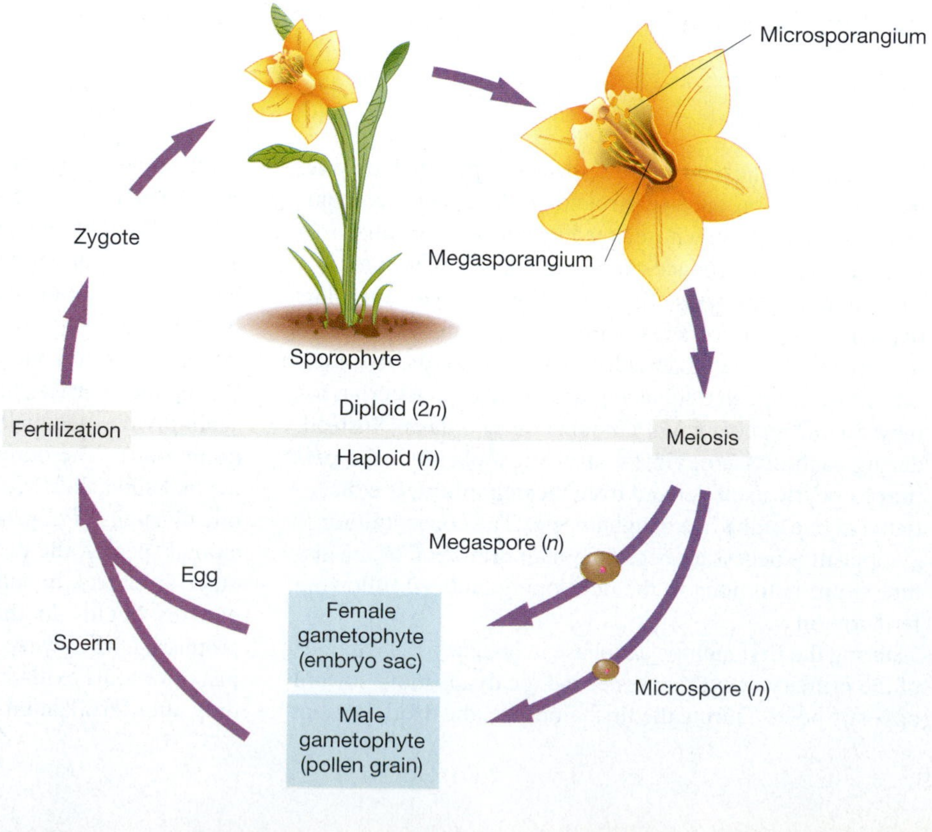

FIGURE 8–16 Alternation of generations between the diploid sporophyte ($2n$) and the haploid gametophyte (n) in a multicellular plant. The processes of meiosis and fertilization bridge the two phases of the life cycle. This is an angiosperm, where the sporophyte stage is the predominant phase.

to the same pole during anaphase. Such an event is called **nondisjunction**, because the two members fail to disjoin.

The results of nondisjunction during meiosis I and meiosis II for just one chromosome of a diploid genome are shown in Figure 8–17. As you can see, for the affected chromosome, abnormal gametes may be formed that contain either two members or none at all. Fertilization of these with a normal gamete produces a zygote with either three members (trisomy) or only one member (monosomy) of this chromosome. While these conditions are more frequently tolerated in plants, they usually have severe or lethal effects in animals. Trisomy and monosomy will be described in greater detail in Chapter 13.

8.8 Electron Microscopy Has Revealed the Cytological Nature of Mitotic and Meiotic Chromosomes

Thus far in this chapter, we have focused on mitotic and meiotic chromosomes, emphasizing their behavior during cell division and gamete formation. An interesting question is why chromosomes are invisible during interphase, but present during the various stages of mitosis and meiosis. Studies using electron microscopy clearly show why chromosomes are visible only during division stages.

Chromatin vs. Chromosomes

During interphase, only dispersed chromatin fibers are present in the nucleus [Figure 8–18(a)]. Once mitosis begins, however, the fibers coil and fold, condensing into typical mitotic chromosomes [Figure 8–18(b)]. If the fibers making up the mitotic chromosome are loosened, the areas of greatest spreading reveal individual fibers similar to those seen in interphase chromatin [Figure 8–18(c)]. Very few fiber ends seem to be present, and, in some cases, none can be seen. Instead, individual fibers always seem to loop back into the interior. Such fibers are obviously twisted and coiled around one another, forming the regular pattern of the mitotic chromosome.

Starting in late telophase of mitosis and continuing during G1 of interphase, chromosomes then unwind to form the long fibers characteristic of chromatin, which consist of DNA and associated proteins, particularly proteins called histones. It is in this physical arrangement that DNA can most efficiently function during transcription and replication.

Electron microscopic observations of mitotic chromosomes in varying states of coiling led Ernest DuPraw to postulate the **folded-fiber model**, illustrated in Figure 8–18(d). During metaphase, each chromosome consists of two sister chromatids joined at the centromeric region. Each arm of the chromatid appears to consist of a single fiber wound up much like a skein of yarn. The fiber is composed of tightly coiled double-stranded DNA and protein. An orderly coiling–twisting–condensing process appears to be involved

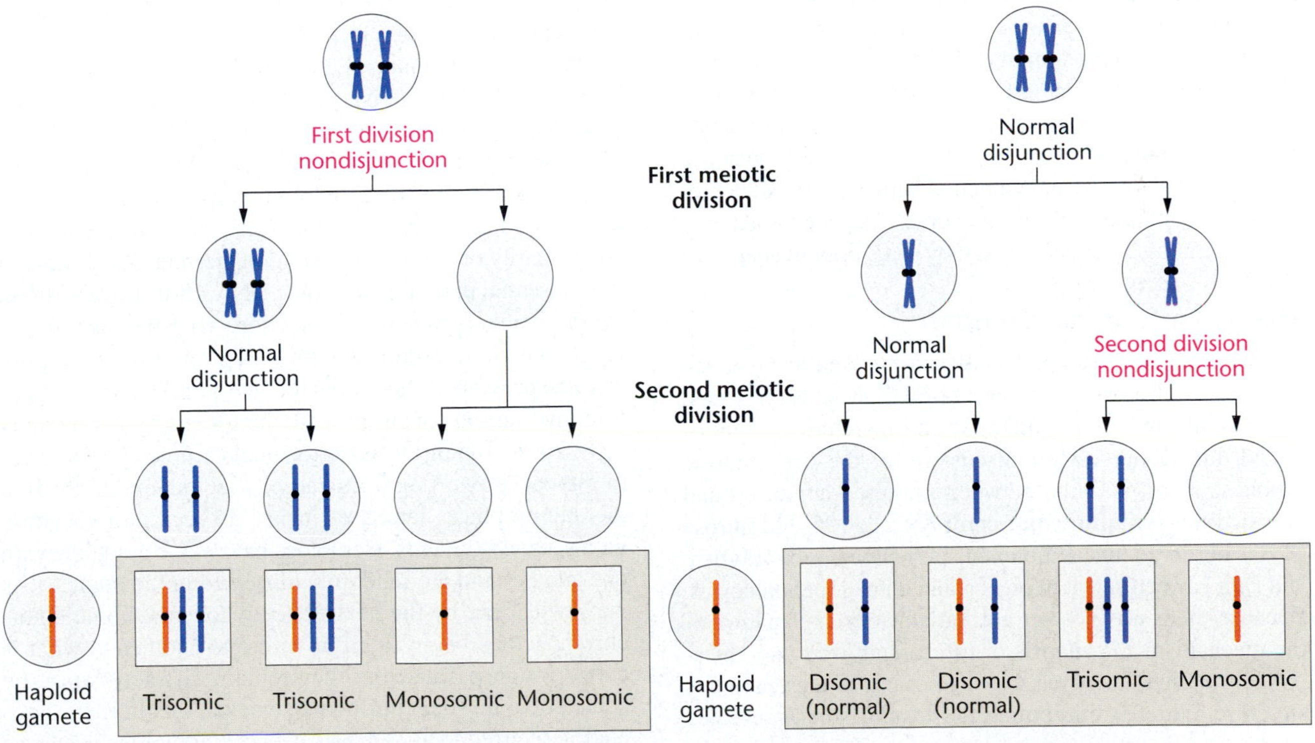

FIGURE 8–17 Diagram illustrating nondisjunction during the first and second meiotic divisions. In both cases, some gametes are formed either containing two members of a specific chromosome or lacking it altogether. Following fertilization by a normal haploid gamete, monosomic, disomic (normal), or trisomic zygotes result.

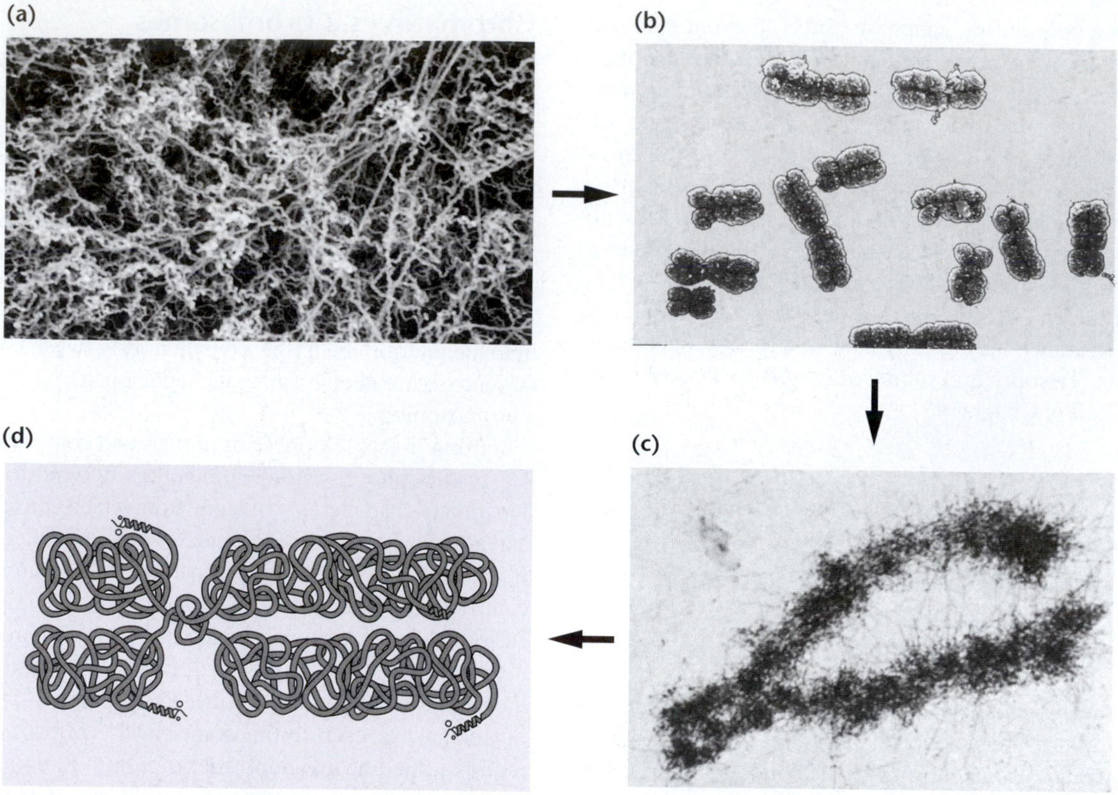

FIGURE 8–18 Comparison of (a) the chromatin fibers characteristic of the interphase nucleus with (b) and (c) metaphase chromosomes that are derived from chromatin during mitosis. Part (d) depicts the folded-fiber model, showing how chromatin is condensed into a metaphase chromosome. Parts (a) and (c) are transmission electron micrographs, while part (b) is a scanning electron micrograph.

in the transition of the interphase chromatin to the more condensed, mitotic chromosomes. It is estimated that during the transition from interphase to prophase, a 5000-fold contraction occurs in the length of DNA within the chromatin fiber! This process must indeed be extremely precise, given the highly ordered nature and consistent appearance of mitotic chromosomes in all eukaryotes. Note particularly in the micrographs the clear distinction between the sister chromatids constituting each chromosome. They are joined only by the common centromere that they share prior to anaphase.

The Synaptonemal Complex

The electron microscope has also been used to visualize another ultrastructural component of the chromosome found only in cells undergoing meiosis. This structure, first introduced during our earlier discussion of the first meiotic prophase stage, is found between synapsed homologs and is called the **synaptonemal complex**.* In 1956, Montrose Moses observed this complex in spermatocytes of crayfish, and Don Fawcett saw it in pigeon and human spermatocytes. Because there was not yet any satisfactory explanation of the mechanism of synapsis or of crossing over and chiasmata formation, many researchers became interested in this structure. With few exceptions, the ensuing studies revealed the synaptonemal complex to be present in most plant and animal cells visualized during meiosis.

As you can see in the electron micrograph in Figure 8–19(a) the synaptonemal complex is a tripartite structure. The central element is usually less dense and thinner (100–150 Å) than the two identical outer elements (500 Å). The outer structures, called lateral elements, are intimately associated with the synapsed homologs on either side. Selective staining has revealed that these lateral elements consist primarily of DNA and protein, suggesting that chromatin is an essential part of them. Some DNA fibrils traverse these lateral elements, making connections with the central element, which is composed primarily of protein. Figure 8–19(b) provides a diagrammatic interpretation of the electron micrograph consistent with the foregoing description.

The formation of the synaptonemal complex begins prior to the pachytene stage. As early as leptonema of the first meiotic prophase, lateral elements are seen in association with sister chromatids. Homologs have yet to associate with one another and are randomly dispersed in the nucleus. As we saw earlier, by the next stage, zygonema, homologous chromosomes begin to align with one another in what is called rough pairing, but they remain distinctly apart by some 300 nm. Then, during pachynema, the intimate association referred to as synapsis between homologs occurs as formation of the complex is completed. In some diploid organisms, this occurs in a zipperlike fashion, beginning at

*An alternative spelling of this term is synaptinemal complex

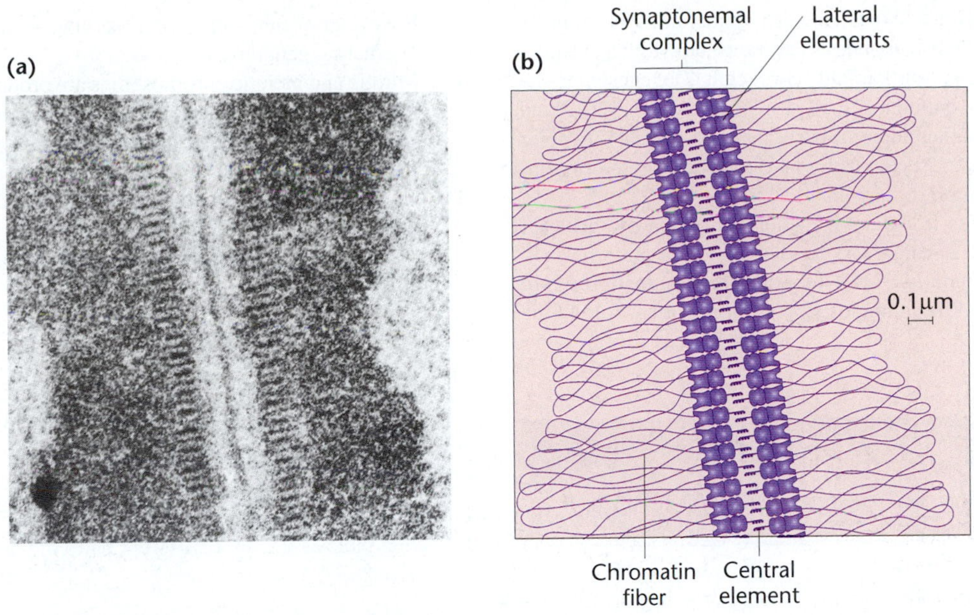

FIGURE 8–19 (a) Electron micrograph of a portion of a synaptonemal complex found between synapsed bivalents of *Neotiella rutilans.* (b) Schematic interpretation of the components making up the synaptonemal complex. The lateral elements, central element, and chromatin fiber are labeled.

the ends of the chromosomes, which may be attached to the nuclear envelope.

It is now agreed that the synaptonemal complex is the vehicle for the pairing of homologs and their subsequent segregation during meiosis. However, some degree of synapsis can occur in certain cases where no synaptonemal complexes are formed. Thus, it is possible that the function of this structure may go beyond its involvement in the formation of bivalents.

In certain instances where no synaptonemal complexes are formed during meiosis, synapsis is not complete and crossing over is reduced or eliminated. For example, in male *Drosophila melanogaster,* where synaptonemal complexes are not usually seen, meiotic crossing over rarely, if ever, occurs. This observation suggests that the synaptonemal complex may be important in order for chiasmata to form and crossing over to occur.

The study of *ZIP1*, a mutation in the yeast *Saccharomyces cerevisiae,* has provided further insights into chromosome pairing. Cells bearing this mutation can undergo the initial alignment stage (rough pairing) and full-length central and lateral element formation, but fail to achieve the intimate pairing characteristic of synapsis. It has been suggested that the gene product of the *ZIP1* locus is a protein component of the central element of the synaptonemal complex, since it is absent in mutant cells. This observation further suggests that a complete and intact synaptonemal complex is essential during the transition from the initial rough alignment stage to the intimate pairing characteristic of synapsis.

Chapter Summary

1. The structure of cells is elaborate and complex. Many components of cells are involved directly or indirectly with genetic processes.
2. In diploid organisms, chromosomes exist in homologous pairs. Each pair shares the same size, centromere placement, and gene loci. One member of each pair is derived from the maternal parent, and one is derived from the paternal parent.
3. Mitosis and meiosis are mechanisms by which cells distribute genetic information contained in their chromosomes to their descendants in a precise, orderly fashion.
4. Mitosis, or nuclear division, is part of the cell cycle and is the basis of cellular reproduction. Daughter cells are produced that are genetically identical to their progenitor cell.
5. Mitosis may be subdivided into discrete stages: prophase, prometaphase, metaphase, anaphase, and telophase. Condensation of chromatin into chromosome structures occurs during prophase. During prometaphase, chromosomes appear as double structures, each composed of a pair of sister chromatids. In metaphase, chromosomes line up on the equatorial plane of the cell. During anaphase, sister chromatids of each chromosome are pulled apart and directed toward opposite poles. Telophase completes daughter cell formation and is characterized by cytokinesis, the division of the cytoplasm.
6. The cell cycle is characteristic of all eukaryotes and is tightly regulated at three checkpoints: G1/S, G2/M, and M.
7. Meiosis, the underlying basis of sexual reproduction, results in the conversion of diploid cells to haploid gametes or spores. As a result of chromosome duplication and two subsequent divisions, each haploid cell receives one member of each homologous pair of chromosomes.

8. A major difference exists between animal meiosis in males and females. Spermatogenesis partitions the cytoplasm equally and produces four haploid sperm cells. Oogenesis, on the other hand, accumulates the cytoplasm in one egg cell and packages the other haploid sets of genetic material into polar bodies. The extra cytoplasm contributes to zygote development following fertilization.
9. Meiosis results in extensive genetic variation by virtue of the exchange during crossing over between maternal and paternal chromatids and their random segregation into gametes. In addition, meiosis plays an important role in the life cycles of fungi and plants, serving as the bridge between alternating generations.
10. Mitotic and meiotic chromosomes are produced as a result of the coiling and condensation of chromatin fibers characteristic of interphase. This transition is described by the folded-fiber model.
11. The synaptonemal complex is an ultrastructural component of the cell present during meiotic prophase I. It is important to the process of synapsis of homologs and may play a role in crossing over.

Insights and Solutions

1. In an organism with a diploid number of 6, how many individual chromosomal structures will align on the metaphase plate during (a) mitosis (b) meiosis I, and (c) meiosis II? Describe each configuration.

Solution:

(a) In mitosis, where homologous chromosomes do not synapse, there will be 6 double structures, each consisting of a pair of sister chromatids. The number of structures is equivalent to the diploid number.

(b) In meiosis I, the homologs have synapsed, reducing the number of structures to 3. Each is called a *tetrad* and consists of two pairs of sister chromatids.

(c) In meiosis II, the same number of structures exist (3), but in this case they are called *dyads*. Each consists of a pair of sister chromatids. When crossing over has occurred, each chromatid may contain part of one of its nonsister chromatids obtained during exchange in prophase I.

2. For the chromosomes illustrated in Figure 8.14, draw all possible alignment configurations that may occur during metaphase of meiosis I.

Solution: As shown in the following diagram, four configurations are possible when $n = 2$.

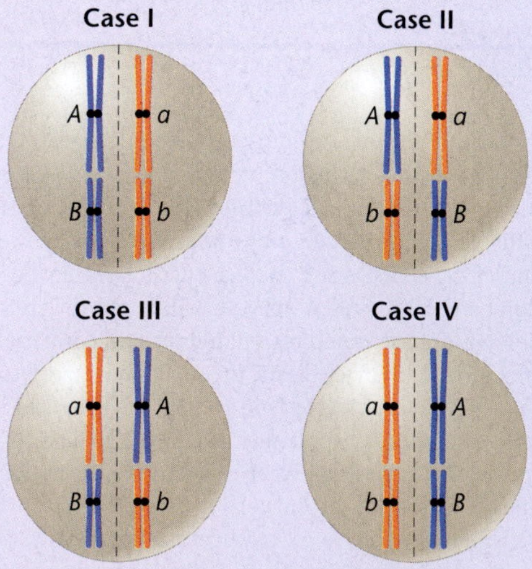

3. How many different chromosome configurations can occur following meiosis I if three different pairs of chromosomes are present (n = 3)?

Solution: If $n = 3$, then eight different configurations would be possible. The formula 2^n, where n equals the haploid number, will allow you to calculate the number of potential alignment patterns. As we will see in the next chapter, these patterns are produced as a result of the Mendelian postulate called *segregation*, and they serve as the physical basis of the Mendelian postulate of *independent assortment.*

4. Assuming that comparable chromosomes in different individuals are genetically dissimilar because of different alleles, how many unique zygotic combinations are possible following fertilization in an organism where $n = 3$?

Solution: Assuming that no crossing over occurs (and that maternal and paternal chromatids remain intact), then each organism can produce 2^n or 2^3 different chromosome combinations in their gametes. Following random fertilization, $(2^3)(2^3) = 2^6 = 64$ unique combinations are possible in the offspring.

5. Assume that there is one gene on both of the larger chromosomes and two alleles, *A* and *a*, as shown. Also assume a second gene with two alleles (*B*, *b*) is present on the smaller chromosomes. Calculate the probability of generating each gene combination (*AB, Ab, aB, ab*) following meiosis I.

Solution:

Case I	*AB* and *ab*
Case II	*Ab* and *aB*
Case III	*aB* and *Ab*
Case IV	*ab* and *AB*
Total: $AB = 2$	$(p = 1/4)$
$Ab = 2$	$(p = 1/4)$
$aB = 2$	$(p = 1/4)$
$ab = 2$	$(p = 1/4)$

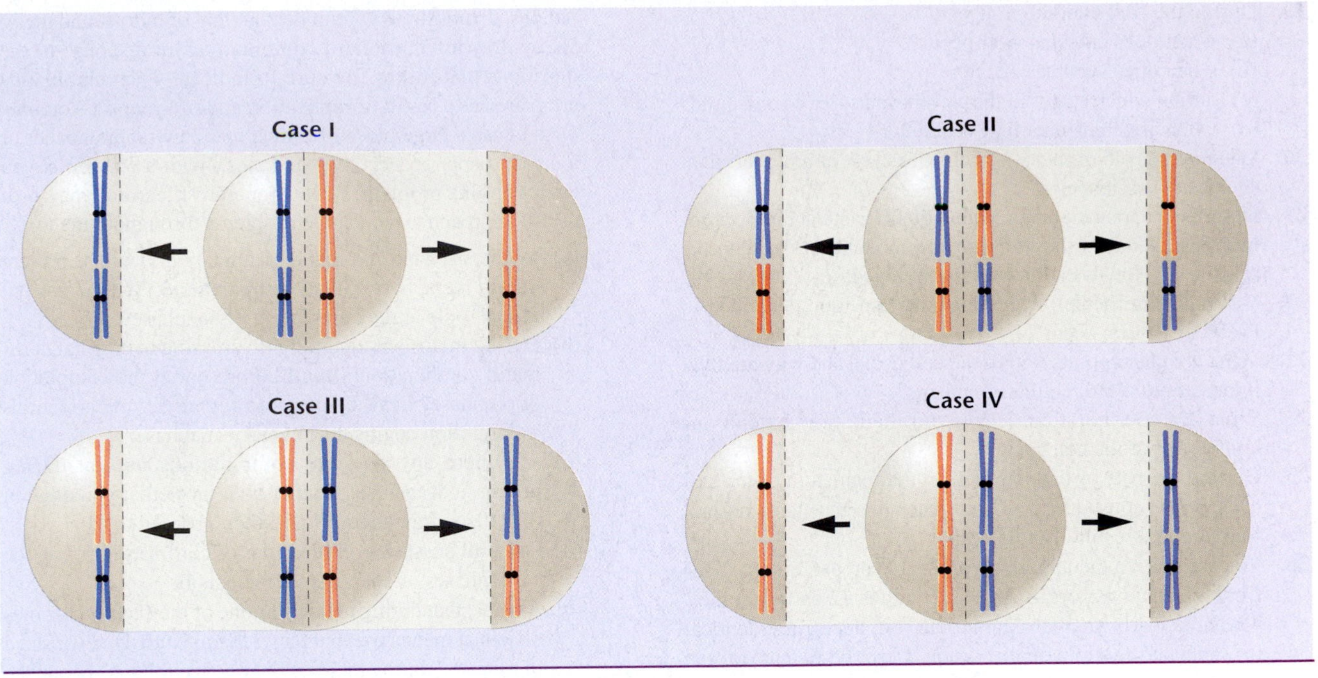

Problems and Discussion Questions

1. What role do the following cellular components play in the storage, expression, or transmission of genetic information: (a) chromatin, (b) nucleolus, (c) ribosome, (d) mitochondrion, (e) centriole, and (f) centromere?
2. Discuss the concepts of homologous chromosomes, diploidy, and haploidy. What characteristics are shared between two chromosomes considered to be homologous?
3. If two chromosomes of a species are the same length and have similar centromere placements, yet are not homologous, what is different about them?
4. Describe the events that characterize each stage of mitosis. Based on this description and using one pair of homologous chromosomes, draw each stage.
5. If an organism has a diploid number of 16, how many chromatids are visible at the end of mitotic prophase? How many chromosomes are moving to each pole during anaphase of mitosis?
6. How are chromosomes named on the basis of centromere placement?
7. Contrast telophase in plant and animal mitosis.
8. Outline and discuss the events, including regulatory checkpoints, of the cell cycle. What experimental technique was used to demonstrate the existence of the S phase?
9. Examine Figure 8.15 showing oogenesis in animal cells. Will the genetic composition of the second polar bodies (derived from meiosis II) always be identical to that of the ootid? Why or why not?
10. Contrast the end results of meiosis with those of mitosis.
11. Define the following terms and discuss their relevance to meiosis: (a) synapsis, (b) bivalents, (c) chiasmata, (d) crossing over, (e) chromomeres, (f) sister chromatids, (g) tetrads, (h) dyads, (i) monads, and (j) synaptonemal complex.
12. Contrast the genetic content and the origin of sister vs. nonsister chromatids during their earliest appearance in prophase I of meiosis. How might the genetic content of these change by the time tetrads have aligned at the equatorial plate during metaphase I?
13. Given the end results of the two types of division, why is it necessary for homologs to pair during meiosis and not desirable for them to pair during mitosis?
14. An organism has a diploid number of 16 in a primary oocyte.
    (a) How many tetrads are present in the first meiotic prophase?
    (b) How many dyads are present in the second meiotic prophase?
    (c) How many monads migrate to each pole during the second meiotic anaphase?
    (d) What is the probability that a gamete will contain only paternal chromosomes?
15. Contrast spermatogenesis and oogenesis. What is the significance of the formation of polar bodies?
16. Explain why meiosis leads to significant genetic variation, while mitosis does not.
17. During oogenesis in an animal species with a haploid number of 6, one dyad undergoes nondisjunction during meiosis II. Following the second meiotic division, the involved dyad ends up intact in the ovum. How many chromosomes are present in (a) the mature ovum and (b) the second polar body? (c) Following fertilization by a normal sperm, what chromosome condition is created?
18. What is the probability that, in an organism with a haploid number of 10, a sperm will be formed which contains all 10 chromosomes whose centromeres were derived from maternal homologs?

19. During the first meiotic prophase,
(a) when does crossing over occur?
(b) when does synapsis occur?
(c) during which stage are the chromosomes least condensed?
(d) when are chiasmata first visible?

20. What is the role of meiosis in the life cycle of a higher plant such as an angiosperm?

21. Describe the transition of a chromatin fiber into a mitotic chromosome. What is the name of the model that depicts this transition? How was this model developed?

22. When during the cell cycle does the transition described in Problem 21 first occur?

23. What are checkpoints within the cell cycle, and why are they important to multicellular organisms?

24. What are cdc mutations? How did their study extend our knowledge of the cell cycle?

25. Discuss the role of the *p53* gene in the regulation of the cell cycle. What consequences result when this gene loses normal function as a result of mutation?

26. You are given a metaphase chromosome preparation (a slide) from an unknown organism that contains 12 chromosomes. Two are clearly smaller than all the rest, appearing identical in length and centromere placement. Describe all that you can about these two chromosomes.

Extra-Spicy Problems

27. A diploid cell contains three pairs of chromosomes designated *A, B,* and *C*. Each pair contains a maternal and a paternal member (e.g., A^m and A^p, etc.). Using these designations, demonstrate your understanding of mitosis and meiosis by drawing chromatid combinations in response to the questions that follow. Be sure to indicate when chromatids are paired as a result of replication and/or synapsis. You may wish to use a large piece of brown manila wrapping paper or a cut-up paper grocery bag and work with a partner as you deal with this problem. Such cooperative learning may be a useful approach as you solve problems throughout the text.

(a) In mitosis, what chromatid combination(s) will be present during metaphase? What combination(s) will be present at each pole at the completion of anaphase?
(b) During meiosis I, assuming no crossing over, what chromatid combination(s) will be present at the completion of prophase? Draw all possible alignments of chromatids as migration begins during early anaphase.
(c) Are there any possible combinations present during prophase of meiosis II other than those that you drew in (b)? If so, draw them. If not, then proceed to (d).
(d) Draw all possible combinations of chromatids during the early phases of anaphase in meiosis II.
(e) Assume that during meiosis I none of the *C* chromosomes disjoin at metaphase, but they separate into dyads (instead of monads) during meiosis II. How would this change the alignments that you constructed during the anaphase stages in meiosis I and II? Draw them.
(f) Assume that each gamete resulting from (e) participated in fertilization with a normal haploid gamete. What combinations will result? What percentage of zygotes will be diploid, containing one paternal and one maternal member of each chromosome pair?

Selected Readings

Alberts, B. et al. 1994. *Molecular biology of the cell,* 3d ed. New York: Garland Publishing.

Baker, B.A. et al. 1976. The genetic control of meiosis. *Annu. Rev. Genet.* 10:53–134.

Baserga, R., and Kisieleski, W. 1963. Autobiographics of cells. *Sci. Am.* (Aug.) 209:103–10.

Brachet, J., and Mirsky, A. E. 1961. *The cell: Meiosis and mitosis,* Vol. 3. Orlando, FL: Academic Press.

Carpenter, A.T.C. 1994. Chiasma function. *Cell* 77:959–62.

DuPraw, E.J. 1970. *DNA and chromosomes.* New York: Holt, Rinehart & Winston.

Glover, D.M., Gonzalez, C., and Raff, J.W. 1993. The centrosome. *Sci. Am.* (June) 268:62–68.

Golomb, H.M., and Bahr, G.F. 1971. Scanning electron microscopic observations of surface structures of isolated human chromosomes. *Science* 171:1024–26.

Hall, J.L., Ramanis, Z., and Luck, D.J. 1989. Basal body/centriolar DNA: Molecular genetic studies in *Chlamydomonas. Cell* 59:121–32.

Hartwell, L.H. et al. 1974. Genetic control of the cell division cycle in yeast. *Science* 183:46–51.

Hartwell, L.H., and Kastan, M.B. 1994. Cell cycle control and cancer. *Science* 266:1821–28.

Hartwell, L.H., and Weinert, T.A. 1989. Checkpoint controls that ensure the order of cell cycle events. *Science* 246:629–34.

Hawley, R.S., and Arbel, T. 1993. Yeast genetics and the fall of the classical view of meiosis. *Cell* 72:301–303.

Kleckner, N. 1996. Meiosis: How could it work? *Proc. Natl. Acad. Sci.* 93:8167–74.

Koshland, D. 1994. Mitosis: Back to the basics. *Cell* 77:951–54.

Mazia, D. 1961. How cells divide. *Sci. Am.* (Jan.) 205:101–20.

——— 1974. The cell cycle. *Sci. Am.* (Jan.) 235:54–64.

McIntosh, J.R., and McDonald, K.L. 1989. The mitotic spindle. *Sci. Am.* (Oct.) 261:48–56.

McKim, K. et al. 1998. Meiotic synapsis in the absence of recombination. *Science* 279:876–78.

Moens, P.B. 1973. Mechanisms of chromosome synapsis at meiotic prophase. *Int. Rev. Cytol.* 35:117–34.

Murray, A.W., and Kirschner, M. 1991. What controls the cell cycle? *Sci. Am.* (March) 263:56–63.

——— 1993. *The cell cycle: An introduction.* New York: Oxford University Press.

Prescott, D.M. 1977. *Reproduction of eukaryotic cells.* Orlando, FL: Academic Press.

GENETICS MediaLab

The resources that follow will help you achieve a better understanding of the concepts presented in this chapter. These resources can be found either on the CD packaged with this textbook or on the Companion Web site found at **http://www.prenhall.com/klug**

CD Resources:

Module 8.1: Mitosis
Module 8.2: Meiosis

Web Problem 1:

Time for completion = 10 minutes

What are the phases of the cell cycle? What happens in each? How is G0 different from G1? How long does interphase last in a mammalian cell? How long is the M phase? Choose checkpoints in the upper left of the box showing the illustration of the cell cycle. What are the three checkpoints? Think about why a cell might have three such checkpoints. To complete this exercise, visit Web Problem 1 in Chapter 8 of your Companion Web site, and select the keyword **CELL CYCLE**.

Web Problem 2:

Time for completion = 10 minutes

What are some of the key differences between cellular events in mitosis and meiosis? Both processes begin in interphase, but the end products are quite different. Mitosis produces a pair of genetically identical cells, whereas meiosis produces four genetically different cells. At what phase in mitosis do sister chromatids separate? Do they separate at Anaphase I or II? At what phase does chiasma formation occur? Are the cells haploid after Meiosis I? Meiosis II? How many divisions occur in mitosis? How many in meiosis? To complete this exercise, visit Web Problem 2 in Chapter 8 of your Companion Web site, and select the keyword **COMPARISON**.

Web Problem 3:

Time for completion = 10 minutes

How does meiosis generate genetic variation among the gametes? Two processes occur during meiosis that generate this variation. After reading through the short tutorial on recombination during meiosis, identify the two sources of variation. Which of the processes generates variation among the four copies of a given chromosome? What would happen at the level of the genes if the homologues were not perfectly aligned during crossing-over, so that unequal parts of the chromosomes were exchanged? Which process accounts for Mendel's postulates of segregation and independent assortment (Chapter 9)? To complete this exercise, visit Web Problem 3 in Chapter 8 of your Companion Web site, and select the keyword **MEIOSIS**.

Prescott, D.M., and Flexer, A.S. 1986. *Cancer, the misguided cell,* 2d ed. Sunderland, MA: Sinauer.

Sawin, K.E. et al. 1992. Mitotic spindle organization by a plus end-directed microtubular motor. *Nature* 359:540–43.

Swanson, C.P., Merz, T., and Young, W.J. 1981. *Cytogenetics, the chromosome in division, inheritance, and evolution,* 2d ed. Englewood Cliffs, NJ: Prentice-Hall.

Wadsworth, P. 1993. Mitosis: Spindle assembly and chromosome movement. *Curr. Opin. Cell Biol.* 5:93–99.

Westergaard, M., and von Wettstein, D. 1972. The synaptinemal complex. *Annu. Rev. Genet.* 6:71–110.

Wheatley, D.N. 1982. *The centriole: A central enigma of cell biology.* New York: Elsevier/North-Holland Biomedical.

Yunis, J.J., and Chandler, M.E. 1979. Cytogenetics. In *Clinical diagnosis and management by laboratory methods,* ed. J.B. Henry, Vol. 1. Philadelphia: W.B. Saunders.

Mendel's garden, as seen in the 1980's. (*Courtesy of Allan Gotthelf*)

9
Mendelian Genetics

KEY CONCEPTS

Now that we have established a thorough grounding in molecular genetics and understand how genetic information is stored, may be altered by mutation, and is expressed, we turn to a consideration of more classical studies of inheritance. In this chapter, we will pursue the intial studies that shed light on how traits are transmitted from generation to generation, and in doing so, we will establish the basic principles of transmission genetics.

Although inheritance of biological traits has been recognized for thousands of years, the first significant insights into the mechanisms involved occurred about 130 years ago. In 1866, Gregor Johann Mendel published the results of a series of experiments that would lay the foundation for the formal discipline of genetics. Although Mendel's work went largely unnoticed until the turn of the century, following the rediscovery of his work, the concept of the gene as a distinct hereditary unit was established. Ways in which genes, as members of chromosomes, are transmitted to offspring and control traits were clairified. Research has continued unabated throughout the 20th century. Indeed, studies in genetics, most recently at the molecular level, have remained at the forefront of biological research since the early 1900s.

When Mendel began his studies of inheritance using *Pisum sativum*, the garden pea, there was no knowledge of chromosomes or of the role and mechanism of meiosis. Nevertheless, he was able to determine that discrete **units of inheritance** exist and to predict their behavior during the formation of gametes. Subsequent investigators, with access to cytological data, were able to relate their own observations of chromosome behavior during meiosis to Mendel's principles of inheritance. Once this correlation was made, Mendel's postulates were accepted as the basis for the study of what is known as **Mendelian** or **transmission genetics**. These principles describe how genes are transmitted from parents to offspring and were derived directly from Mendel's experimentation. Even today, they serve as the cornerstone of the study of inheritance. In this chapter, we focus on the development of Mendel's principles.

9.1 Mendel Used a Model Experimental Approach to Study Patterns of Inheritance

Johann Mendel was born in 1822 to a peasant family in the central European village of Heinzendorf. An excellent student in high school, he studied philosophy for several years afterward, and in 1843 was admitted to the Augustinian Monastery of St. Thomas in Brno, now part of the Czech Republic, taking the name of Gregor. In 1849, he was relieved of pastoral duties and received a teaching appointment that lasted several years. From 1851 to 1853, he attended the University of Vienna, where he studied physics and botany. In 1854, he returned to Brno, where for the next 16 years he taught physics and natural science. Mendel received support from the monastery for his studies and research throughout his life.

In 1856, Mendel performed his first set of hybridization experiments with the garden pea. The research phase of his career lasted until 1868, when he was elected abbot of the monastery. Although he retained his interest in genetics, his new responsibilities demanded most of his time. In 1884, Mendel died of a kidney disorder. The local newspaper paid him the following tribute: "His death deprives the poor of a benefactor, and mankind at large of a man of the noblest character, one who was a warm friend, a promoter of the natural sciences, and an exemplary priest."

Mendel first reported the results of some simple genetic crosses between certain strains of the garden pea in 1865. Although his was not the first attempt to provide experimental evidence pertaining to inheritance, Mendel's success, where others had failed can be attributed, at least in part, to his elegant model of experimental design and analysis.

Mendel showed remarkable insight into the methodology necessary for good experimental biology. First, he chose an organism that was easy to grow and to hybridize artificially. The pea plant is self-fertilizing in nature, but it is easy to cross-breed experimentally. The plant reproduces well and grows to maturity in a single season. Mendel then chose to follow seven visible features (unit characters), each represented by two contrasting forms or traits (Figure (9–1). For the character stem height, for example, he experimented with the traits *tall* and *dwarf*. He selected six other contrasting pairs of traits involving seed shape and color, pod shape and color, and pod and flower arrangement. From local seed merchants, Mendel obtained true-breeding strains, those in which each trait appeared unchanged generation after generation in self-fertilizing plants.

Several factors led to Mendel's success, in addition to his choice of a suitable organism. He restricted his examination to one or very few pairs of contrasting traits in each experiment. He also kept accurate quantitative records, a necessity in genetic experiments. From the analysis of his data, Mendel derived certain postulates that have become the principles of transmission genetics.

The results of Mendel's experiments went unappreciated until the turn of the century, well after his death. Once Mendel's publications were rediscovered by geneticists investigating the function and behavior of chromosomes, however, the implications of his postulates were immediately apparent. He had discovered the basis for the transmission of hereditary traits!

9.2 The Monohybrid Cross Reveals How One Trait Is Transmitted from Generation to Generation

Mendel's simplest crosses involved only one pair of contrasting traits. Each such breeding experiment is called a **monohybrid cross**. A monohybrid cross is made by mating individuals from two parent strains, each of which exhibits one of the two contrasting forms of the character under study. Initially, we examine the first generation of offspring of such

Character	Contrasting traits		F_1 results	F_2 results	F_2 ratio
Seeds	round/wrinkled		all round	5474 round 1850 wrinkled	2.96:1
	yellow/green		all yellow	6022 yellow 2001 green	3.01:1
Pods	full/constricted		all full	882 full 299 constricted	2.95:1
	green/yellow		all green	428 green 152 yellow	2.82:1
Flowers	violet/white		all violet	705 violet 224 white	3.15:1
Stem	axial/terminal		all axial	651 axial 207 terminal	3.14:1
	tall/dwarf		all tall	787 tall 277 dwarf	2.84:1

FIGURE 9–1 A summary of the seven pairs of contrasting traits and the results of Mendel's seven monohybrid crosses of the garden pea (*Pisum sativum*, shown in the photograph). In each case, pollen derived from plants exhibiting one trait was used to fertilize the ova of plants exhibiting the other trait. In the F_1 generation, one of the two traits, (dominant) was exhibited by all plants. The contrasting trait (recessive) then reappeared in approximately 1/4 of the F_2 plants.

MEDIA TUTORIAL Monohybrid Cross

a cross, and then we consider the offspring of selfing or self-fertilizing individuals from the first generation. The original parents are called the $\mathbf{P_1}$ or **parental generation**, their offspring are the $\mathbf{F_1}$ or **first filial generation**, and the individuals resulting from the selfing of the F_1 generation are the $\mathbf{F_2}$ or **second filial generation**. We can, of course, continue to follow subsequent generations if desired.

The cross between true-breeding pea plants with tall stems and dwarf stems is representative of Mendel's monohybrid crosses. *Tall* and *dwarf* represent contrasting forms or traits of the character of stem height. Unless tall or dwarf plants are crossed together or with another strain, they will undergo self-fertilization and breed true, producing their respective trait generation after generation. However, when Mendel crossed tall plants with dwarf plants, the resulting F_1 generation consisted of only tall plants. When members of the F_1 generation were selfed, Mendel observed that 787 of 1064 F_2 plants, were tall, while 277 of 1064 were dwarf. Note that in this cross (Figure 3–1), the dwarf trait disappeared in the F_1 generation, only to reappear in the F_2 generation. Mendel made similar crosses between pea plants, exhibiting each of the other pairs of contrasting traits. Results of these crosses are also shown in Figure 9–1. In every case, the outcome was similar to the tall/dwarf cross.

Genetic data are usually expressed and analyzed as ratios. In this particular example, many identical P_1 crosses were made and many F_1 plants—all tall—were produced. Of the 1064 F_2 offspring, 787 were tall and 277 were dwarf—a ratio of approximately 2.8:1.0, or about 3:1. Three-fourths appeared like the F_1 plants, while one-fourth exhibited the contrasting trait, which had disappeared in the F_1 generation.

It is important to point out one further aspect of the monohybrid crosses. In each cross, the F_1 and F_2 patterns of inheritance were similar regardless of which P_1 plant served as the source of pollen (sperm) and which served as the source of the ovum (egg). The crosses could be made either way—that is, pollen from the tall plant pollinating dwarf plants, or vice versa. These are called **reciprocal crosses**. Therefore, the results of Mendel's monohybrid crosses were not sex dependent.

To explain these results, Mendel proposed the existence of what he called particulate **unit factors** for each trait. He suggested that these factors serve as the basic units of heredity and are passed unchanged from generation to generation, determining various traits expressed by each individual plant. Using these general ideas, Mendel proceeded to hypothesize precisely how such factors could account for the results of the monohybrid crosses.

Mendel's First Three Postulates

Using the consistent pattern of results in the monohybrid crosses, Mendel derived the following three postulates, or principles, of inheritance:

1. UNIT FACTORS IN PAIRS

Genetic characters are controlled by unit factors that exist in pairs in individual organisms.

In the monohybrid cross involving tall and dwarf stems, a specific unit factor exists for each trait. Each diploid individual receives one factor from each parent. Because the factors occur in pairs, three combinations are possible: two factors for tallness, two factors for dwarfness, or one of each factor. Every individual possesses one of these three combinations, which determines stem height.

2. DOMINANCE/RECESSIVENESS

When two unlike unit factors responsible for a single character are present in a single individual, one unit factor is dominant to the other, which is said to be recessive.

In each monohybrid cross, the trait expressed in the F_1 generation results from the presence of the dominant unit factor. The trait that is not expressed in the F_1, but which reappears in the F_2, is under the genetic influence of the recessive unit factor. Note that this dominance–recessiveness relationship pertains only when unlike unit factors are present together in an individual. The terms **dominant** and **recessive** are also used to designate the traits. In the aforementioned case, the trait tall stem is said to be dominant to the recessive trait, dwarf stem.

3. SEGREGATION

During the formation of gametes, the paired unit factors separate, or segregate, randomly so that each gamete receives one or the other with equal likelihood.

These postulates provide a suitable explanation for the results of the monohybrid crosses. Let's use the tall/dwarf cross to illustrate. Mendel reasoned that P_1 tall plants contained identical paired unit factors, as did the P_1 dwarf plants. The gametes of tall plants all received one tall unit factor as a result of segregation. Likewise, the gametes of dwarf plants all received one dwarf unit factor. Following fertilization, all F_1 plants received one unit factor from each parent, a tall factor from one and a dwarf factor from the other, reestablishing the paired relationship. Because tall is dominant to dwarf, all F_1 plants were tall.

When F_1 plants form gametes, the postulate of segregation demands that each gamete randomly receive *either* the tall *or* dwarf unit factor. Following random fertilization events during F_1 selfing, four F_2 combinations will result in equal frequency:

(1) tall/tall

(2) tall/dwarf

(3) dwarf/tall

(4) dwarf/dwarf

Combinations (1) and (4) will clearly result in tall and dwarf plants, respectively. According to the postulate of dominance–recessiveness, combinations (2) and (3) will both yield tall plants. Therefore, the F_2 is predicted to consist of three-fourths tall and one-fourth dwarf plants, or a ratio of 3:1. This is approximately what Mendel observed in his cross between tall and dwarf plants. A similar pattern was observed in each of the other monohybrid crosses (Figure 9–1).

Modern Genetic Terminology

To illustrate the monohybrid cross and Mendel's first three postulates in a modern context, we must first introduce several new terms as well as a set of symbols for the unit factors. Traits such as tall or dwarf are physical expressions of the information contained in unit factors. We now call the physical expression of a trait the **phenotype** of the individual.

Mendel's unit factors represent units of inheritance called **genes** by modern geneticists. For any given character, such as plant height, the phenotype is determined by different combinations of alternative forms of a single gene called **alleles**. For example, tall and dwarf are alleles determining the height of the pea plant.

Geneticists use several different conventions involving gene symbols to represent genes. In Chapter 10, we will review some of these, but for now, we will adopt one that we can use consistently throughout this chapter. According to this convention, the first letter of the recessive trait is chosen to symbolize the character in question. The lowercase form of the letter designates the allele for the recessive trait, and the uppercase letter designates the allele for the dominant trait. Further, these gene symbols are italicized. Therefore, *d* stands for the dwarf allele and *D* represents the tall allele. When alleles are written in pairs to represent the two unit factors present in any individual (*DD*, *Dd*, or *dd*), these symbols are referred to as the **genotype**. This term reflects the genetic makeup of an individual, whether it is haploid or diploid. By following the principle of dominance and recessiveness, we can tell the phenotype of the individual from its genotype: *DD* and *Dd* are tall, and *dd* is dwarf. When identical alleles constitute the genotype (*DD* or *dd*), the individual is said to be **homozygous** or a **homozygote**; when alleles are different (*Dd*), we use the term **heterozygous** or **heterozygote**. Figure

9–2 illustrates the complete monohybrid cross, using modern terminology.

Mendel's Analytical Approach

What led Mendel to deduce unit factors in pairs? Because there were two contrasting traits for each character, it seemed logical that two distinct factors must exist. However, why does one of the two traits or phenotypes disappear in the F_1 generation? Observation of the F_2 generation helps to answer this question. The recessive trait and its unit factor do not actually disappear in the F_1; they are merely hidden or masked, only to reappear in one-fourth of the F_2 offspring. Therefore, Mendel concluded that one unit factor for tall and one for dwarf were transmitted to each F_1 individual; but because the tall factor or allele is dominant to the dwarf factor

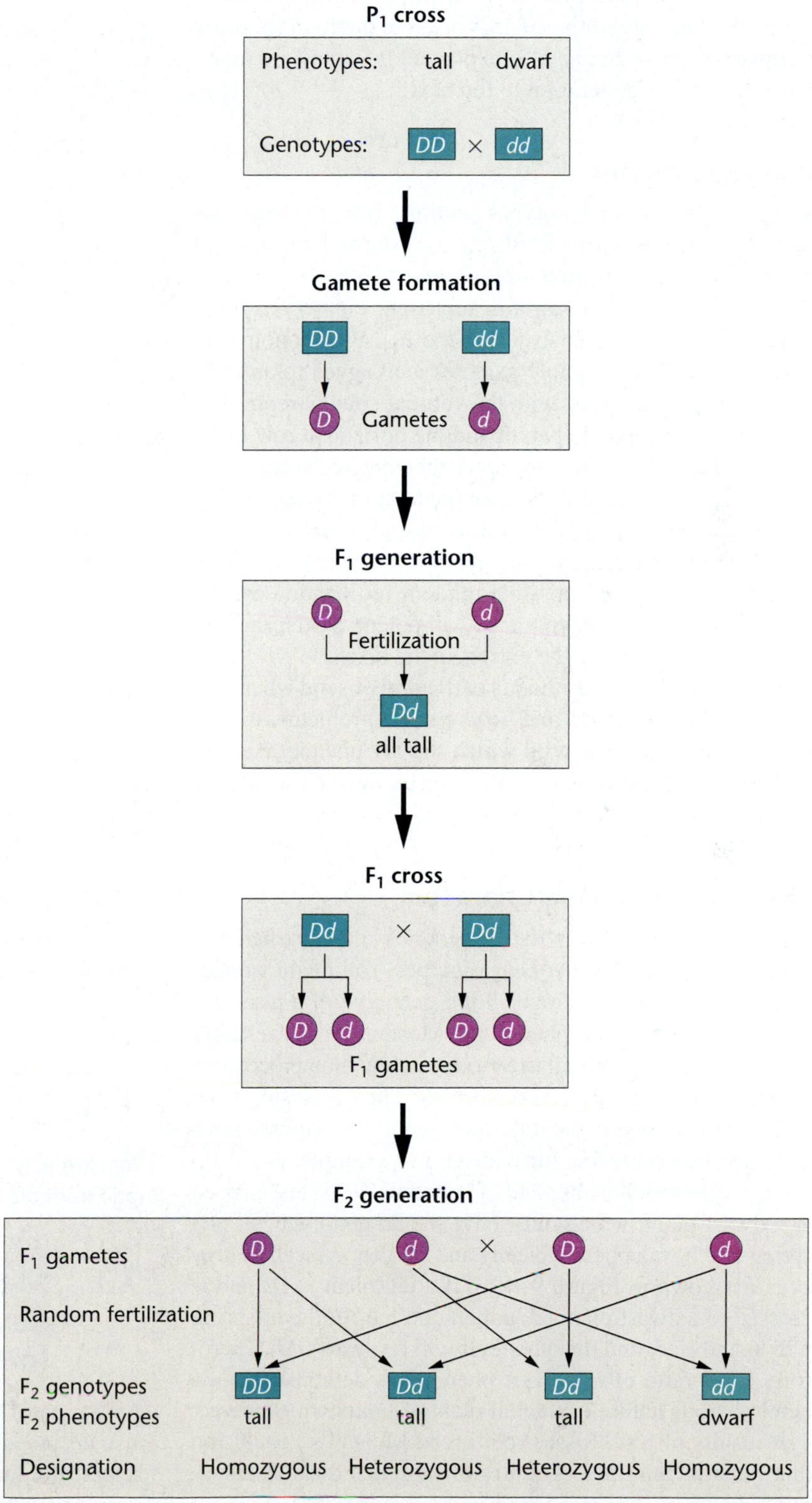

FIGURE 9–2 The monohybrid cross between tall and dwarf pea plants. The symbols *D* and *d* designate the tall and dwarf unit factors, respectively, in the genotypes of mature plants and gametes. All individuals are shown in rectangles, and gametes are shown in circles.

or allele, all F_1 plants are tall. Given this information, we can ask how Mendel explained the 3:1 F_2 ratio. As shown in Figure 9–2, Mendel deduced that the tall and dwarf alleles of the F_1 heterozygote segregate randomly into gametes. If fertilization is random, this ratio is predicted. If a large population of offspring is generated, the outcome of such a cross should reflect the 3:1 ratio.

Since Mendel operated without the hindsight that modern geneticists enjoy, his analytical reasoning must be considered a truly outstanding scientific achievement. On the basis of rather simple, but precisely executed, breeding experiments, he not only proposed that discrete **particulate units of heredity** exist, but he also explained how they are transmitted from one generation to the next!

Punnett Squares

The genotypes and phenotypes resulting from the recombination of gametes during fertilization can be easily visualized by constructing a **Punnett square**, named after Reginald C. Punnett, who first devised this approach. Figure 9–3 illustrates this method of analysis for the $F_1 \times F_1$ monohybrid cross. Each of the possible gametes is assigned to an individual column or a row, with the vertical column representing those of the female parent and the horizontal row those of the male parent. After we enter the gametes in rows and columns, we can predict the new generation by combining the male and female gametic information for each combination and entering the resulting genotypes in the boxes. This process represents all possible random fertilization events. The genotypes and phenotypes of all potential offspring are ascertained by reading the entries in the boxes.

The Punnett square method is particularly useful when first learning about genetics and how to solve problems. In Figure 9–3, note the ease with which the 3:1 phenotypic ratio and the 1:2:1 genotypic ratio may be derived in the F_2 generation.

The Test Cross: One Character

Tall plants produced in the F_2 generation are predicted to have either the *DD* or the *Dd* genotypes. You might wonder if there is a way to distinguish the genotype of a plant expressing the dominant phenotype. Mendel devised a rather simple method that is still used today in breeding procedures of plants and animals: the **test cross**. The organism of the dominant phenotype, but unknown genotype, is crossed to a **homozygous recessive individual**. For example, as shown in Figure 9–4(a), if a tall plant of genotype *DD* is test crossed to a dwarf plant, which must have the *dd* genotype, all offspring will be tall phenotypically and *Dd* genotypically. However, as shown in Figure 9–4(b), if a tall plant is *Dd* and is crossed to a dwarf plant (*dd*), then one-half of the offspring will be tall (*Dd*) and the other half will be dwarf (*dd*). Therefore, a 1:1 ratio of tall/dwarf phenotypes demonstrates the heterozygous nature of the tall plant of unknown genotype. The results of test crosses reinforced Mendel's conclusion that separate unit factors control the tall and dwarf traits.

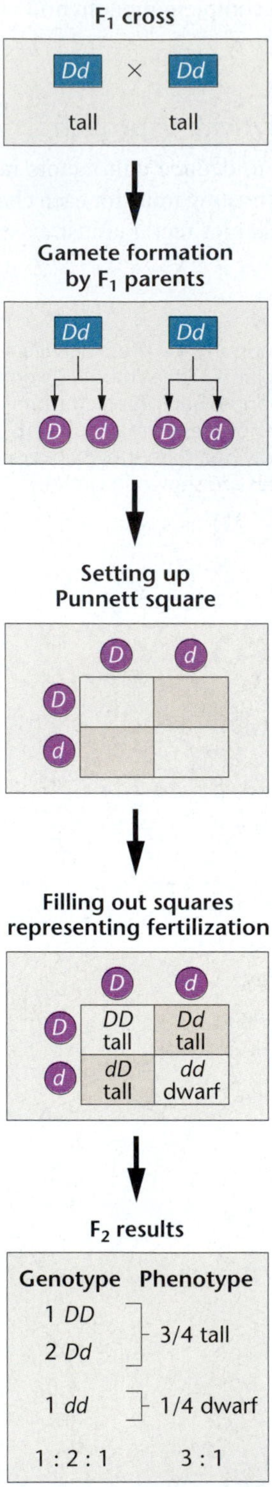

FIGURE 9–3 The use of a Punnett square in generating the F_2 ratio from the $F_1 \times F_1$ cross shown in Figure 9-2.

9.3 Mendel's Dihybrid Cross Revealed His Fourth Postulate: Independent Assortment

As a natural extension of the monohybrid cross, Mendel also designed experiments in which he examined two characters simultaneously. Such a cross, involving two pairs of

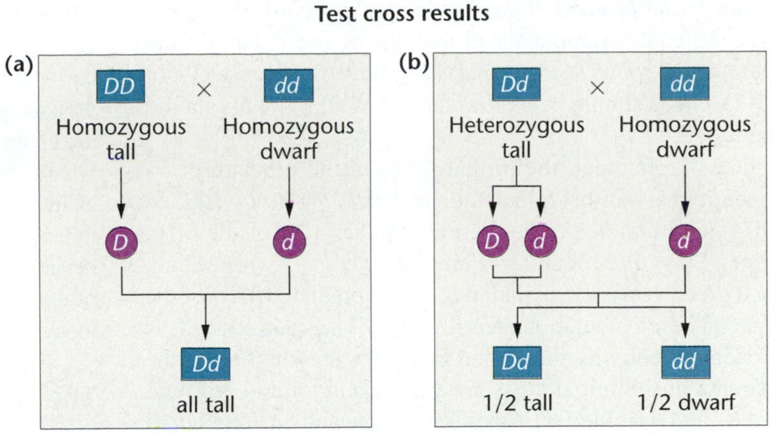

FIGURE 9–4 Test cross of a single character. In (a), the tall parent is homozygous. In (b), the tall parent is heterozygous. The genotype of each tall parent can be determined by examining the offspring when each is crossed to the homozygous recessive dwarf plant.

contrasting traits, is called a **dihybrid cross**, or a **two-factor cross**. For example, if pea plants having yellow seeds that are also round were bred with those having green seeds that are also wrinkled, the results shown in Figure 9–5 will occur. The F_1 offspring are all yellow and round. It is therefore apparent that yellow is dominant to green, and that round is dominant to wrinkled. In this dihybrid cross, the F_1 individuals are selfed, and approximately 9/16 of the F_2 plants express yellow and round, 3/16 express yellow and wrinkled, 3/16 express green and round, and 1/16 express green and wrinkled.

A variation of this cross is also shown in Figure 9–5. Instead of crossing one P_1 parent with both dominant traits (yellow, round) to one with both recessive traits (green, wrinkled), plants with yellow wrinkled seeds are crossed with those with green round seeds. Despite the change in the P_1 phenotypes, both the F_1 and F_2 results remain unchanged. It will become clear in the next section why this is so.

Independent Assortment

We can most easily understand the results of a dihybrid cross if we consider it theoretically as consisting of two monohybrid crosses conducted separately. Think of the two sets of traits as being inherited independently of each other; that is, the chance of any plant having yellow or green seeds is not at all influenced by the chance that this plant will also have round or wrinkled seeds. Thus, because yellow is dominant to green, all F_1 plants in the first theoretical cross would have yellow seeds. In the second theoretical cross, all F_1 plants would have round seeds, because round is dominant to wrinkled. When Mendel examined the F_1 plants of his dihybrid cross, all were yellow and round, as predicted.

The predicted F_2 results of the first cross are 3/4 yellow and 1/4 green. Similarly, the second cross should yield 3/4 round and 1/4 wrinkled. Figure 9–5 shows that in the dihybrid cross, 12/16 of all F_2 plants are yellow, while 4/16 are green, exhibiting the 3:1 (3/4:1/4) ratio. Similarly, 12/16 of all F_2 plants have round seeds, while 4/16 have wrinkled seeds, again revealing the 3:1 (3/4:1/4) ratio.

Because it is evident that the two pairs of contrasting traits are inherited independently, we can predict the frequencies of all possible F_2 phenotypes by applying the "product law" of probabilities: *When two independent events occur simultaneously, the combined probability of the two outcomes is*

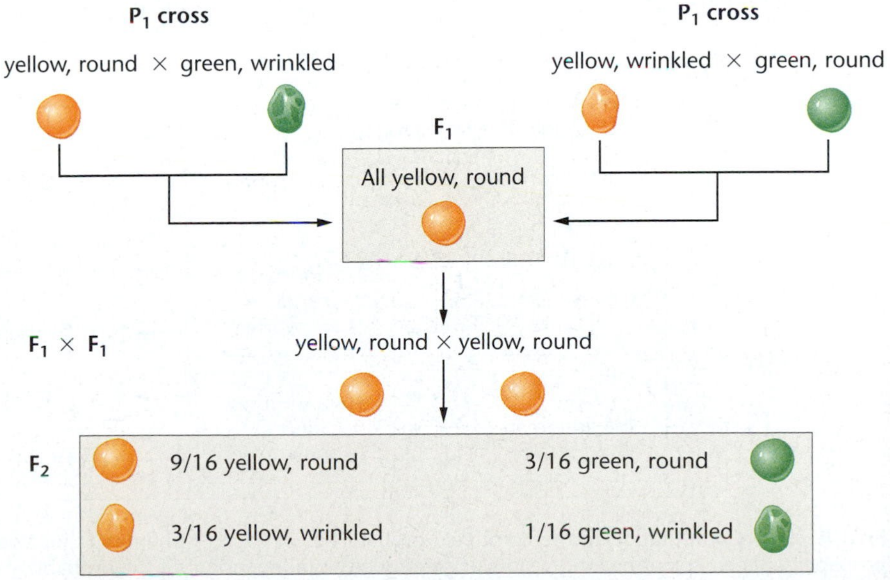

FIGURE 9–5 F_1 and F_2 results of Mendel's dihybrid crosses between yellow, round and green, wrinkled pea seeds and between yellow, wrinkled and green, round pea seeds.

MEDIA TUTORIAL Independent Assortment

equal to the product of their individual probabilities of occurrence. For example, the probability of an F_2 plant having yellow and round seeds is (3/4) (3/4) or 9/16, because 3/4 of all F_2 plants should be yellow and 3/4 of all F_2 plants should be round.

In a like manner, the probabilities of the other three F_2 phenotypes can be calculated: yellow (3/4) *and* wrinkled (1/4) are predicted to be present together 3/16 of the time; green (1/4) *and* round (3/4) are predicted 3/16 of the time; and green (1/4) *and* wrinkled (1/4) are predicted 1/16 of the time. These calculations are illustrated in Figure 9–6. It is now apparent why the F_1 and F_2 results are identical if the parents of the initial cross are yellow and round bred with green and wrinkled or if they are yellow and wrinkled bred with green and round. In both crosses, the F_1 genotype of all plants is identical. Each plant is heterozygous for both gene pairs. As a result, the F_2 generation is also identical in both crosses.

On the basis of similar results in numerous dihybrid crosses, Mendel proposed a fourth postulate:

4. INDEPENDENT ASSORTMENT

During gamete formation, segregating pairs of unit factors assort independently of each other.

This postulate stipulates that any pair of unit factors segregate independently of all other unit factors. Remember that, as a result of segregation, each gamete receives one member of every pair of unit factors. For one pair, whichever unit factor is received does not influence the outcome of segregation of any other pair. Thus, according to the postulate of **independent assortment**, all possible combinations of gametes will be formed in equal frequency.

The Punnett square in Figure 9–7 shows how independent assortment works in the formation of the F_2 generation. Examine the formation of gametes by the F_1 plants. Segregation prescribes that every gamete receives either a *G* or *g* allele and a *W* or *w* allele. Independent assortment stipulates that all four combinations (*GW, Gw, gW*, and *gw*) will be formed with equal probabilities.

In every $F_1 \times F_1$ fertilization event, each zygote has an equal probability of receiving one of the four combinations from each parent. If a large number of [many] offspring are produced, 9/16 are yellow and round, 3/16 are yellow and wrinkled, 3/16 are green and round, and 1/16 are green and wrinkled, yielding what is designated as **Mendel's 9:3:3:1 dihybrid ratio**. This is an ideal ratio based on probability events involving segregation, independent assortment, and random fertilization. Because of deviation due strictly to chance, particularly if small numbers of offspring are produced, actual results will seldom match the ideal ratio exactly.

The Test Cross: Two Characters

The test cross may also be applied to individuals that express two dominant traits, but whose genotypes are unknown. For example, the expression of the yellow round phenotype in the F_2 generation just described may result from the *GGWW*, *GGWw*, *GgWW*, and *GgWw* genotypes. If an F_2 yellow round plant is crossed with the homozygous recessive green, wrinkled plant (*ggww*), analysis of the offspring will indicate the exact genotype of that yellow round plant. Each of these genotypes will result in a different set of gametes and, in a testcross, a different set of phenotypes in the resulting offspring. Three cases are illustrated in Figure 9–8.

9.4 The Trihybrid Cross Demonstrates that Mendel's Principles Apply to Inheritance of Multiple Traits

Thus far, we have considered inheritance of up to two pairs of contrasting traits. Mendel demonstrated that the identical processes of segregation and independent assortment apply to three pairs of contrasting traits in what is called a **trihybrid cross**, also referred to as a **three-factor cross**.

Although a trihybrid cross is somewhat more complex than a dihybrid cross, its results are easily calculated if the principles of segregation and independent assortment are

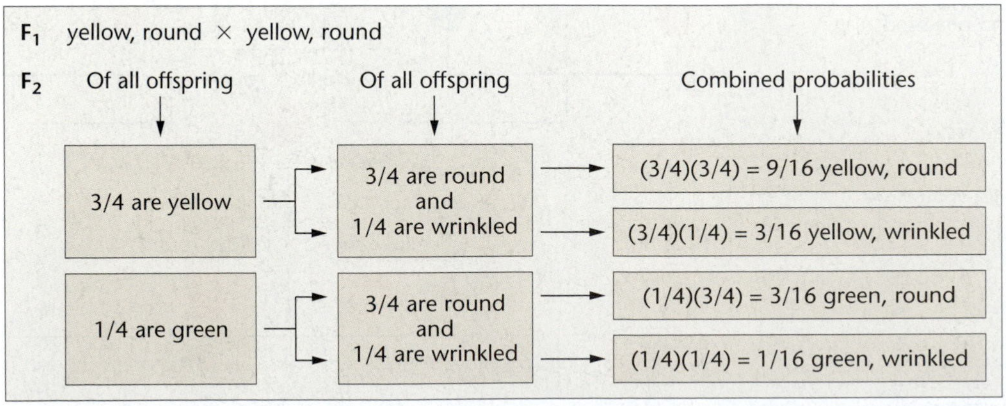

FIGURE 9–6 Computation of the combined probabilities of each F_2 phenotype for two independently inherited characters. The probability of each plant bearing yellow or green seeds is independent of the probability of it bearing round or wrinkled seeds.

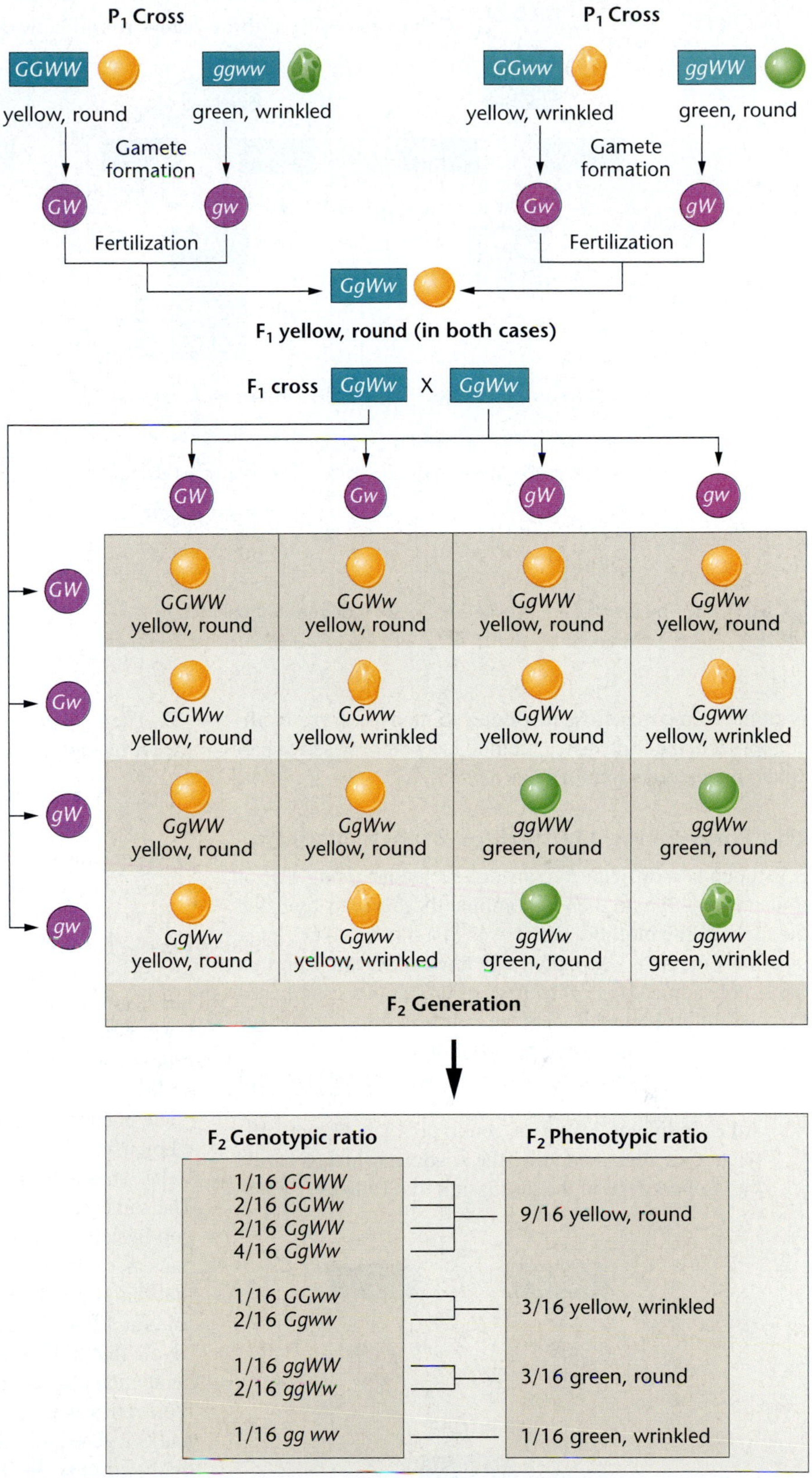

FIGURE 9–7 Analysis of the dihybrid crosses shown in Figure 9-5. The F_1 heterozygous plants are self-fertilized to produce an F_2 generation, which is computed using a Punnett square. Both the phenotypic and genotypic F_2 ratios are shown.

followed. For example, consider the cross shown in Figure 9–9 where the gene pairs representing theoretical contrasting traits are symbolized *A/a*, *B/b*, and *C/c*. In the cross between *AABBCC* and *aabbcc* individuals, all F_1 individuals are heterozygous for all three gene pairs. Their genotype, *AaBbCc*, results in the phenotypic expression of the dominant *A*, *B*, and *C* traits. When F_1 individuals are parents, each produces eight different gametes in equal frequencies. At this point, we could construct a Punnett square with 64 separate boxes and read out the phenotypes. Because such a method

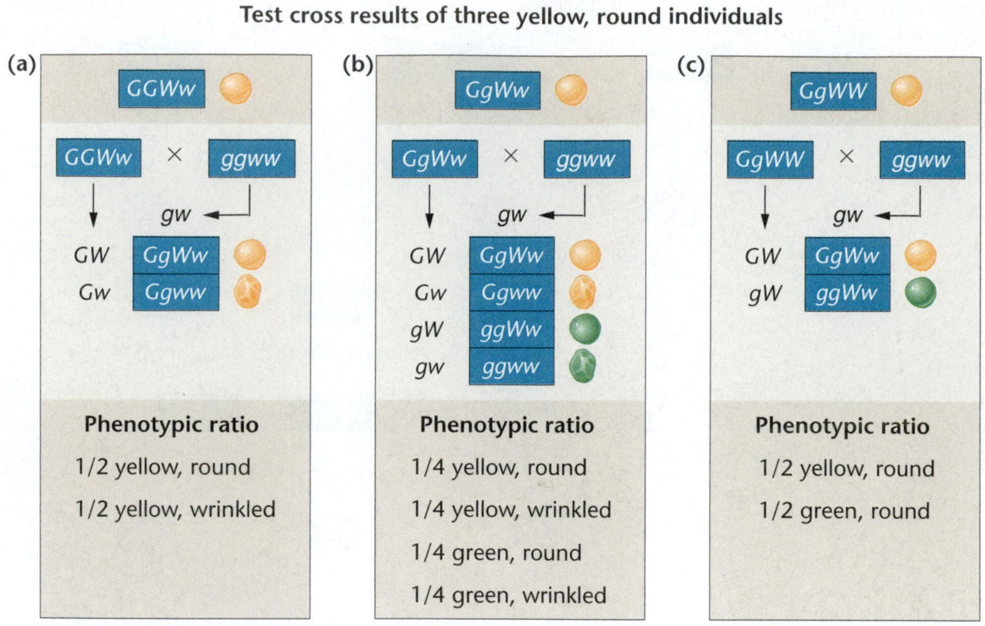

FIGURE 9–8 The test cross illustrated with two independent characters.

is cumbersome in a cross involving so many factors, another approach, the forked-line method has been devised to calculate the predicted ratio.

The Forked-Line Method, or Branch Diagram

It is much less difficult to consider each contrasting pair of traits separately and then to combine these results by using the **forked-line method**, which was first illustrated in Figure 9–6. This method, also called a **branch diagram**, relies on the simple application of the laws of probability established for the dihybrid cross. Each gene pair is assumed to behave independently during gamete formation.

When the monohybrid cross $AA \times aa$ is made, we know that

1. All F_1 individuals have the genotype *Aa* and express the phenotype represented by the *A* allele, which is called the *A* phenotype in the discussion that follows.

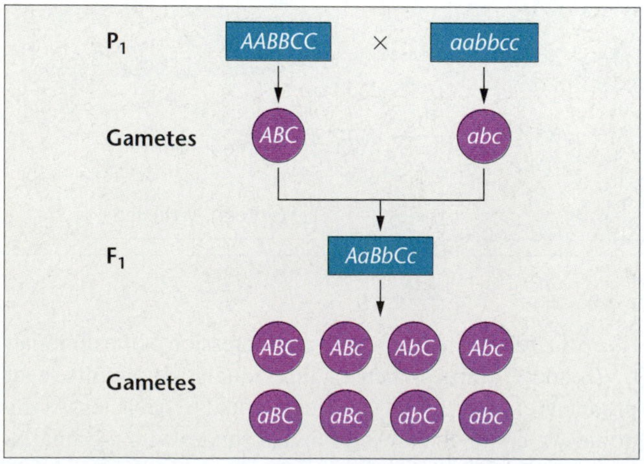

FIGURE 9–9 Formation of P_1 and F_1 gametes in a trihybrid cross.

2. The F_2 generation consists of individuals with either the *A* phenotype or the a phenotype in the ratio of 3:1.

The same generalizations apply to the $BB \times bb$ and $CC \times cc$ crosses. Thus, in the F_2 generation, 3/4 of all organisms will express phenotype *A*, 3/4 will express *B*, and 3/4 will express *C*. Similarly, 1/4 of all organisms will express phenotype *a*, 1/4 will express *b*, and 1/4 will express *c*. The proportions of organisms expressing each phenotypic combination can be predicted by assuming that fertilization, following the independent assortment of these three gene pairs during gamete formation, is a random process. We once again simply apply the product law of probabilities.

The phenotypic proportions of the F_2 generation calculated by using the forked-line method are illustrated in Figure 9–10. They fall into the trihybrid ratio of 27:9:9:9:3:3:3:1. The same method can be applied when solving crosses involving any number of gene pairs, *provided* that all gene pairs assort independently from each other. We shall see later that this is not always the case. However, it appeared to be true for all of Mendel's characters.

Note that in Figure 9–10, only phenotypic ratios of the F_2 generation have been derived. It is possible to generate genotypic ratios as well. To do so, we again consider the *A/a*, *B/b*, and *C/c* gene pairs separately. For example, for the *A/a* pair, the F_1 cross is $Aa \times Aa$. Phenotypically, an F_2 ratio of 3/4 *A*:1/4 *a* is produced. Genotypically, however, the F_2 ratio is different; 1/4 *AA*:1/2 *Aa*:1/4*aa* will result. Using Figure 9–10 as a model, we would enter these genotypic frequencies on the left side of the calculation. Each would be connected by three lines to 1/4 *BB*, 1/2 *Bb*, and 1/4 *bb*, respectively. From each of these nine designations, three more lines would extend to the 1/4 *CC*, 1/2 *Cc*, and 1/4 *cc* genotypes. On the right side of the completed diagram, 27 genotypes and their

Generation of F_2 trihybrid phenotypes

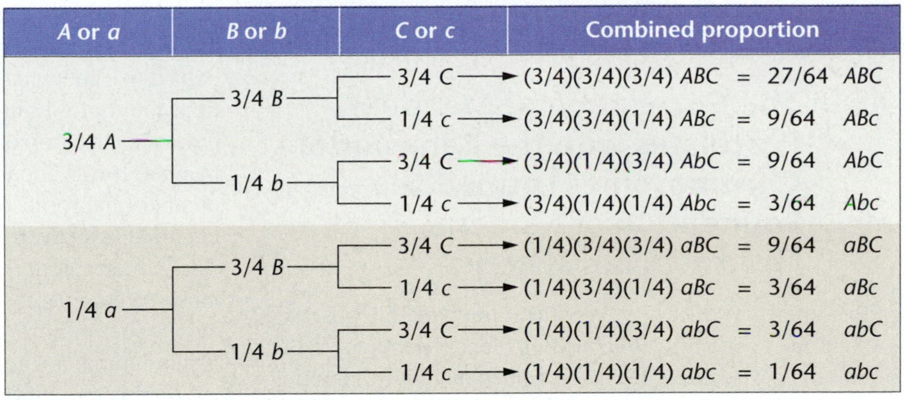

FIGURE 9–10 The generation of the F_2 trihybrid ratio using the forked-line, or branch diagram method, which is based on the expected probabability of occurrence of each phenotype.

frequencies of occurrence would appear. Problem 17 at the end of this chapter, asks you to use the forked-line or branch diagram method to determine the genotypic ratios generated in a trihybrid cross.

In crosses involving two or more gene pairs, the calculation of gametes and genotypic and phenotypic results is quite complex. Several simple mathematical rules will enable you to check the accuracy of various steps required in working genetic problems. First, you must determine the number of *heterozygous* gene pairs (n) involved in the cross. For example, where *AaBb* × *AaBb* represents the cross, $n = 2$; for *AaBbCc* × *AaBbCc*, $n = 3$; for *AaBBCcDd* × *AaBBCcDd*, $n = 3$ (because the *B* genes are not heterozygous). Once n is determined, 2^n is the number of different gametes that can be formed by each parent; 3^n is the number of different genotypes that result following fertilization; and 2^n is the number of different phenotypes that are produced from these genotypes. Table 9–1 summarizes these rules, which may be applied to crosses involving any number of genes, *provided that they assort independently from one another*.

9.5 Mendel's Work Was Rediscovered in the Early 20th Century

Mendel's work, initiated in 1856, was presented to the Brünn Society of Natural Science in 1865 and published the following year. However, his findings went largely unnoticed for about 35 years. Many reasons have been suggested to explain why the significance of his research was not immediately recognized.

First, Mendel's adherence to mathematical analysis of probability events was quite an unusual approach in that era and may have seemed foreign to his contemporaries. More important, his conclusions drawn from such analyses did not fit well with the existing hypotheses involving the source of variation among organisms. Students of evolutionary theory, stimulated by the proposals developed by Charles Darwin and Alfred Russel Wallace, believed in **continuous variation**, where offspring were a blend of their parents' phenotypes. By contrast, Mendel hypothesized that heredity was due to discrete or particulate units, resulting in **discontinuous variation**. For example, Mendel proposed that the F_2 offspring of a dihybrid cross are merely expressing traits produced by new combinations of previously existing unit factors. Thus, Mendel's hypotheses did not fit well with the evolutionists' preconceptions about causes of variation.

It is also likely that Mendel's contemporaries failed to realize that Mendel's postulates explained *how* variation was transmitted to offspring. Instead, they may have attempted to interpret his work in a way that addressed the issue of *why* certain phenotypes survive preferentially. It was this latter question that had been addressed in the theory of natural selection, but it was not addressed by Mendel. It may well be, therefore, that the collective vision of Mendel's scientific

TABLE 9–1 Simple Mathematical Rules Useful in Working Genetics Problems

Crosses between Organisms Heterozygous for Genes Exhibiting Independent Assortment

Number of Heterozygous Gene pairs	*Number of Different Types of Gametes Formed*	*Number of Different Genotypes Produced*	*Number of Different Phenotypes Produced**
n	2^n	3^n	2^n
1	2	3	2
2	4	9	4
3	8	27	8
4	16	81	16

*The fourth column assumes that dominance and recessiveness are operational for all green pairs.

colleagues was obscured by the impact of this extraordinary theory of organic evolution.

9.6 The Correlation of Mendel's Postulates with the Behavior of Chromosomes Formed the Foundation of Modern Transmission Genetics

In the latter part of the 19th century, a remarkable observation set the scene for the rebirth of Mendel's work: Walter Flemming's discovery of chromosomes in 1879. Flemming was able to describe the behavior of these threadlike structures in the nuclei of salamander cells during cell division. As a result of the findings of Flemming and many other cytologists, the presence of a nuclear component soon became an integral part of ideas surrounding inheritance. It was in this setting that scientists were able to reexamine Mendel's findings.

In the early 20th century, research led to renewed interest in Mendel's work. Hybridization experiments similar to Mendel's were independently performed by three botanists, Hugo DeVries, Karl Correns, and Erich Tschermak. DeVries's work, for example, had focused on unit characters, and he demonstrated the principle of segregation in his experiments with several plant species. He had apparently searched the existing literature and found that Mendel's work had anticipated his own conclusions. Correns and Tschermak had also reached conclusions similar to those of Mendel.

In 1902, two cytologists, Walter Sutton and Theodor Boveri, independently published papers linking their discoveries of the behavior of chromosomes during meiosis to the Mendelian principles of segregation and independent assortment. They pointed out that the separation of chromosomes during meiosis could serve as the cytological basis of these two postulates. Although they thought that Mendel's unit factors were probably chromosomes, rather than genes on chromosomes, their findings reestablished the importance of Mendel's work, which served as the foundation of ensuing genetic investigations.

Based on their studies, Sutton and Boveri are credited with initiating the **chromosomal theory of heredity**. As we will see in subsequent chapters, work by Thomas H. Morgan, Alfred H. Sturtevant, Calvin Bridges, and others using fruit flies established beyond a reasonable doubt that Sutton's and Boveri's hypothesis was correct.

Unit Factors, Genes, and Homologous Chromosomes

Because the correlation between Sutton's and Boveri's observations and Mendelian principles is the foundation for the modern interpretation of transmission genetics, we will examine it in some detail before moving on to other topics.

As we pointed out in Chapter 8, each species possesses a specific number of chromosomes in each somatic (body) cell nucleus. For diploid organisms, this number is called the **diploid number ($2n$)** and is characteristic of that species. During the formation of gametes, the number is precisely halved (n), and when two gametes combine during fertilization, the diploid number is reestablished. During meiosis, however, the chromosome number is not reduced in a random manner. It was apparent to early cytologists that the diploid number of chromosomes is composed of homologous pairs identifiable by their morphological appearance and behavior. The gametes contain one member of each pair. The chromosome complement of a gamete is thus quite specific, and the number of chromosomes in each gamete is equal to the haploid number.

With this basic information, we can see the correlation between the behavior of unit factors and chromosomes and genes. Figure 9–11 shows three of Mendel's postulates (in the left column) and the chromosomal explanation of each (in the right column). Unit factors are really genes located on homologous pairs of chromosomes [Figure 9–11(a)]. Members of each pair of homologs separate, or segregate, during gamete formation [Figure 9–11(b)]. Two different alignments are possible, both of which are shown.

To illustrate the principle of independent assortment, it is important to distinguish between members of any given homologous pair of chromosomes. One member of each pair is derived from the **maternal parent**, while the other comes from the **paternal parent**. We represent different parental origins by different colors. As shown in Figure 9–11(c), the two pairs of homologs undergo segregation independently of one another during gamete formation. Each gamete receives one chromosome from each pair. All possible combinations are formed. If we add the symbols used in Mendel's dihybrid cross (*G*, *g* and *W*, *w*) to the diagram, we see why equal numbers of the four types of gametes are formed. The independent behavior of Mendel's pairs of unit factors (*G* and *W* in this example) was due to the fact that they were on separate pairs of homologous chromosomes.

From observations of the phenotypic diversity of living organisms, we see that it is logical to assume that there are many more genes than chromosomes. Therefore, each homolog must carry genetic information for more than one trait. The currently accepted concept is that a chromosome is composed of a large number of linearly ordered, information-containing units called **genes**. Mendel's unit factors (which determine tall or dwarf stems, for example) actually constitute a pair of genes located on one pair of homologous chromosomes. The location on a given chromosome where any particular gene occurs is called its **locus** (pl. **loci**). The different forms taken by a given gene, called **alleles** (*G* or *g*), contain slightly different genetic information (green or yellow) that determines the same character (seed color). Alleles are alternative forms of the same gene. Although we have only discussed genes with two alternative alleles, most genes have *more* than two allelic forms. We discuss the concept of multiple alleles in Chapter 10.

We conclude this section by reviewing the criteria necessary to classify two chromosomes as a homologous pair:

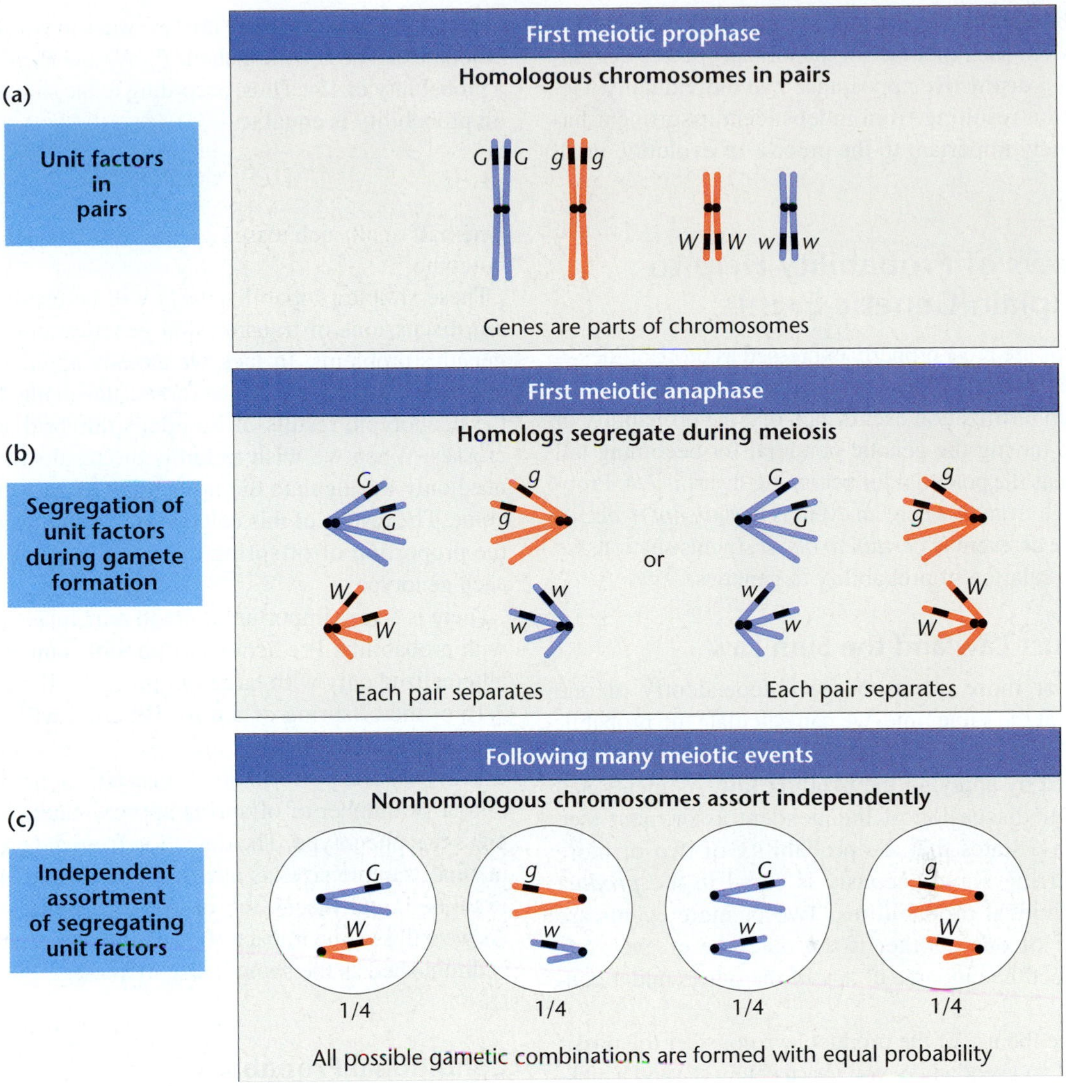

FIGURE 9–11 The correlation between the Mendelian postulates of (a) unit factors in pairs, (b) segregation, and (c) independent assortment, and the presence of genes located on homologous chromosomes and their behavior during meiosis.

1. During mitosis and meiosis, when chromosomes are visible as distinct structures, both members of a homologous pair are the same size and exhibit identical centromere locations.
2. During early stages of meiosis, homologous chromosomes pair together, or synapse.
3. Homologs contain the same linearly ordered gene loci.

9.7 Independent Assortment Leads to Extensive Genetic Variation

One of the major consequences of independent assortment is the production by an individual of genetically dissimilar gametes. Genetic variation results because the two members of any homologous pair of chromosomes are rarely, if ever, genetically identical. Therefore, because independent assortment leads to the production of all possible chromosome combinations, extensive genetic diversity results.

We have seen that for any individual, the number of possible gametes, each with different chromosome compositions, is 2^n, where n equals the haploid number. Thus, if a species has a haploid number of 4, then 2^4 or 16 different gamete combinations can be formed as a result of independent assortment. Although this number is not high, consider the human species, where $n = 23$. If we calculate 2^{23}, we find more than 8×10^6, or over 8 million, different types of gametes are possible. Because fertilization represents an event involving only one of approximately 8×10^6 possible gametes from each of two parents, each offspring represents only one of $(8 \times 10^6)^2$ or 64×10^{12} potential genetic combinations! This number of combinations of chromosomes is far greater than the number of humans who have ever lived on Earth! It is no wonder that, except for identical twins, each member of the human species demonstrates such a distinctive appearance and individuality. Genetic variation resulting from independent assortment has been extremely important to the process of evolution in all organisms.

9.8 Laws of Probability Help to Explain Genetic Events

Genetic ratios are most properly expressed as probabilities—for example, 3/4 tall:1/4 dwarf. These values predict the outcome of each fertilization event, such that the probability of each zygote having the genetic potential for becoming tall is 3/4, whereas the potential for becoming dwarf is 1/4. Probabilities range from 0, where an event *is certain not to occur*, to 1.0, where an event *is certain to occur*. In this section, we consider the relation of probability to genetics.

The Product Law and the Sum Law

When two or more events occur independently of one another, but at the same time, we can calculate the probability of possible outcomes when they occur together. This is accomplished by applying the **product law**. As mentioned in our earlier discussion of independent assortment (see p.xx), the law states that the probability of two or more events occurring simultaneously is equal to the *product* of their individual probabilities. Two or more events are independent of one another if the outcome of each one does not affect the outcome of any of the others under consideration.

To illustrate the use of the product law, consider the possible results of an event where you toss a penny (*P*) and a nickel (*N*) at the same time and examine all combinations of heads (*H*) and tails (*T*) that can occur. There are four possible outcomes:

$$(P_H{:}N_H) = (1/2)(1/2) = 1/4$$

$$(P_T{:}N_H) = (1/2)(1/2) = 1/4$$

$$(P_H{:}N_T) = (1/2)(1/2) = 1/4$$

$$(P_T{:}N_T) = (1/2)(1/2) = 1/4$$

The probability of obtaining a head or a tail in the toss of either coin is 1/2 and is unrelated to the outcome of the other coin. All four possible combinations are predicted to occur with equal probability.

If we were interested in calculating the probability of a generalized outcome that can be accomplished in more than one way, we would apply the **sum law** to the individual mutually exclusive outcomes. For example, we can ask the following: What is the probability of tossing our penny and nickel and obtaining one head and one tail? In such a case, we do not care whether it is the penny or the nickel that comes up heads, provided that the other coin has the alternative outcome. As we can see, there are two ways in which the desired outcome can be accomplished [$P_H{:}N_T$ and $P_T{:}N_H$], each with a probability of 1/4. Thus, according to the sum law, the overall probability is equal to

$$(1/4) + (1/4) = 1/2$$

One-half of all such tosses are predicted to yield the desired outcome.

These simple probability laws will be useful throughout our discussions of transmission genetics and as you solve genetics problems. In fact, we already applied the product law earlier when we used the forked-line method to calculate the phenotypic results of Mendel's dihybrid and trihybrid crosses. When we wish to know the results of a cross, we need only to calculate the probability of each possible outcome. The results of this calculation then allow us to predict the proportion of offspring expressing each phenotype or each genotype.

There is a very important point to remember when dealing with probability. Predictions of possible outcomes are usually realized only with large sample sizes. If we predict that 9/16 of the offspring of a dihybrid cross will express both dominant traits, it is very unlikely that, in a small sample, exactly 9 of every 16 will do so. Instead, our prediction is that of a large number of offspring approximately 9/16 will express this phenotype. The deviation from the predicted ratio in small sample sizes is attributed to chance, a subject we examine in our discussion of statistics in the next section. As we will see, the impact of deviation due strictly to chance is diminished as the sample size increases.

Conditional Probability

Sometimes we may wish to calculate the probability of an outcome that is dependent on a specific condition related to that outcome. For example, in the F_2 of Mendel's monohybrid cross involving tall and dwarf plants, we might wonder what the probability is that a tall plant is heterozygous (and not homozygous). The "condition" we have set is to consider only tall F_2 offspring since we know that all dwarf plants are homozygous. Of any F_2 tall plant, what is the probability of it being heterozygous?

Because the outcome and specific condition are not independent, we cannot apply the product law of probability. The likelihood of such an outcome is referred to as a **conditional probability**. In its simplest terms, we are asking what is the probability that one outcome will occur, given the specific condition upon which this outcome is dependent. Let us call this probability p_c.

To solve for p_c, we must consider both the probability of the outcome of interest and that of the specific condition which includes the outcome. These are (a) the probability of an F_2 plant being heterozygous as a result of receiving both a dominant and a recessive allele (p_a) and (b) the probability of the condition under which the event is being assessed, that is, being tall (p_b). Hence,

p_a = plant inheriting one dominant and one recessive allele (i.e., being a heterozygote)

$= 1/2$

p_b = probability of an F2 plant of a monohybird cross being tall

$= 3/4$

To calculate the conditional probability (p_c), we divide p_a by p_b:

$$p_c = \frac{p_a}{p_b}$$

$$= \frac{1/2}{3/4}$$

$$= (1/2)(4/3)$$

$$= 4/6$$

$$p_c = 2/3$$

The conditional probability of any tall plant being heterozygous is two-thirds (2/3). Thus, on the average, two-thirds of the F_2 tall plants will be heterozygous. We can confirm this calculation by reexamining Figure 9–3.

Conditional probability has many applications in genetics. During genetic counseling, for example, it is possible to calculate the probability (p_c) that an unaffected sibling of a brother or sister expressing a recessive disorder is a carrier of the disease-causing allele (i.e., a heterozygote). Assuming that both parents are unaffected (and are therefore carriers), the calculation of p_c is identical to the preceding example. The value of $p_c = 2/3$.

The Binomial Theorem

The final example of probability that we shall discuss involves cases where one of two alternative outcomes is possible during each of a number of trials. By applying the **binomial theorem**, we can rather quickly calculate the probability of any specific set of outcomes among a large number of potential events. For example, in families of any size, we can calculate the probability of any combination of male and female children, (e.g., in a family of four, we can calculate the probability of having two children of one sex and two children of the other sex).

The expression of the binomial theorem is

$$(a + b)^n = 1$$

where a and b are the respective probabilities of the two alternative outcomes and n equals the number of trials. For each value of n, the binomial must be expanded thusly:

n	*Binomial*	*Expanded Binomial*
1	$(a + b)^1$	$a + b$
2	$(a + b)^2$	$a^2 + 2ab + b^2$
3	$(a + b)^3$	$a^3 + 3a^2b + 3ab^2 + b^3$
4	$(a + b)^4$	$a^4 + 4a^3b + 6a^2b^2 + 4ab^3 + b^4$
5	$(a + b)^5$	$a^5 + 5a^4b + 10a^3b^2 + 10a^2b^3 + 5ab^4 + b^5$
	etc.	etc.

To expand any binomial, the various exponents (e.g., a^3b^2) are determined by using the pattern

$$(a + b)^n = a^n,\ a^{n-1}b,\ a^{n-2}b^2,\ a^{n-3}b^3, \ldots, b^n$$

The numerical coefficients preceding each expression can be most easily determined by using Pascal's triangle, Notice that all numbers other than the 1's are equal to the sum of the two numbers directly above them.

n	
	1
1	1 1
2	1 2 1
3	1 3 3 1
4	1 4 6 4 1
5	1 5 10 10 5 1
6	1 6 15 20 15 6 1
7	1 7 21 35 35 21 7 1
	etc.

Using these methods, we find that the initial expansion of $(a + b)^7$ is

$$a^7 + 7a^6b + 21a^5b^2 + 35a^4b^3 + \cdots + b^7$$

Applying the binomial theorem, we can return to our original question: *What is the probability that in a family of four children, two are male and two are female?*
First, assign initial probabilities to each outcome:

$$a = \text{male} = 1/2$$

$$b = \text{female} = 1/2$$

Then locate the appropriate term in the expanded binomial, where $n = 4$,

$$(a + b)^4 = a^4 + 4a^3b + 6a^2b^2 + 4ab^3 + b^4$$

In each term, the exponent of a represents the number of males, and the exponent of b represents the number of females. Therefore, the correct expression of p is

$$p = 6a^2b^2$$

$$= 6(1/2)^2(1/2)^2$$

$$= 6(1/2)^4$$

$$= 6(1/16)$$

$$= 6/16$$

$$p = 3/8$$

Thus, the probability of families of four children having two boys and two girls is 3/8. Of all families with four children, 3 out of 8 are predicted to have two boys and two girls.

Before examining one other example, we should note that a single formula can be applied in determining the numerical coefficient for any set of exponents,

$$n!/(s!t!)$$

where

n = the total number of events

s = the number of times outcome a occurs

t = the number of times outcome b occurs

Therefore, $n = s + t$.
The symbol "!"means "factorial." For example,

$$5! = (5)(4)(3)(2)(1) = 120.$$

Note that when using factorials, $0! = 1$.

Using the formula, let's determine the probability, in a family of seven, that five males and two females will occur. Thus, $n = 7$, $s = 5$, and $t = 2$. We first extend our equation to include five events of outcome a and two events of outcome b. The appropriate term is

$$p = \frac{n!}{s!t!}a^s b^t$$

$$= \frac{7!}{5!2!}(1/2)^5(1/2)^2$$

$$= \frac{(7)\cdot(6)\cdot(5)\cdot(4)\cdot(3)\cdot(2)\cdot(1)}{(5)\cdot(4)\cdot(3)\cdot(2)\cdot(1)\cdot(2)\cdot(1)}(1/2)^7$$

$$= \frac{(7)\cdot(6)}{(2)\cdot(1)}(1/2)^7$$

$$= \frac{42}{2}(1/2)^7$$

$$= 21(1/2)^7$$

$$= 21(1/128)$$

$$p = 21/128$$

Of families with seven children, on the average, 21/128 are predicted to have five males and two females.

Calculations using the binomial theorem have various applications in genetics, including the analysis of polygenic traits (see Chapter 24) and in population equilibrium studies (See Chapter 25).

9.9 Chi-Square Analysis Evaluates the Influence of Chance on Genetic Data

Mendel's 3:1 monohybrid and 9:3:3:1 dihybrid ratios are hypothetical predictions based on the following assumptions: (1) each allele is dominant or recessive, (2) segregation occurs normally, (3) independent assortment occurs, and (4) fertilization is random. The last three assumptions are influenced by chance events and therefore are subject to random fluctuation. This concept, called **chance deviation**, is most easily illustrated by tossing a single coin numerous times and recording the number of heads and tails observed. In each toss, there is a probability of 1/2 that a head will occur and a probability of 1/2 that a tail will occur. Therefore, the expected ratio of many tosses is 1:1. If a coin were tossed 1000 times, we would usually expect *about* 500 heads and 500 tails to be observed. Any reasonable fluctuation from this hypothetical ratio (e.g., 486 heads and 514 tails) would be attributed to chance.

As the total number of tosses is reduced, the impact of chance deviation increases. For example, if a coin were tossed only four times, you wouldn't be too surprised if all four tosses resulted in only heads or only tails. But, for 1000 tosses, 1000 heads or 1000 tails would be most unexpected. In fact, you might believe that such a result would be impossible. Actually, the probability of all heads or all tails in 1000 tosses can be predicted to occur with a probability of only $(1/2)^{1000}$ Because $(1/2)^{20}$ is equivalent to less than 1 in 2 million times, an event occurring with a probability as small as $(1/2)^{1000}$ would be virtually impossible.

Two major points are significant here:

1. The outcomes of segregation, independent assortment, and fertilization, like coin tossing, are subject to random fluctuations from their predicted occurrences as a result of chance deviation.

2. As the sample size increases, the average deviation from the expected results decreases. Therefore, a larger sample size diminishes the impact of chance deviation on the final outcome.

In genetics, being able to evaluate observed deviation is a crucial methodology. When we assume that data will fit a given ratio such as 1:1, 3:1, or 9:3:3:1, we establish what is called the **null hypothesis** $(\mathbf{H_0})$. It is so named because the hypothesis assumes that there is no *real difference* between the **measured values** (or ratio) and the **predicted values** (or ratio). The *apparent* difference can be attributed purely to chance. The null hypothesis is evaluated using statistical analysis. On this basis, the null hypothesis may either (1) be rejected or (2) fail to be rejected. If it is rejected, the observed deviation from the expected is *not* attributed to chance alone; the null hypothesis and the underlying assumptions leading to it must be reexamined. If the null hypothesis fails to be rejected, any observed deviations *are* attributed to chance.

One of the simplest statistical tests devised to assess the goodness of fit of the null hypothesis is **chi-square** (χ^2) **analysis**. This test takes into account the observed deviation in each component of an expected ratio as well as the sample size and reduces them to a single numerical value. The χ^2 value is then used to estimate how frequently the observed deviation, or even more deviation than was observed, can be expected to occur strictly as a result of chance. The formula used in chi-square analysis is

$$\chi^2 = \Sigma \frac{(o - e)^2}{e}$$

In this equation,

o = the observed value for a given category

e = the expected value for that category

and

Σ = the sum of the calculated values for each category in the ratio

Because $(o - e)$ is the deviation (d) in each case, the equation can be reduced to

$$\chi^2 = \frac{d^2}{e}$$

Table 9–2(a) shows the steps in the χ^2 calculation for the F_2 results of a hypothetical monohybrid cross. If you were analyzing these data, you would work from left to right, calculating and entering the appropriate numbers in each column. Regardless of whether the calculated deviation $(o-e)$ is initially positive or negative, it becomes positive after the number is squared. Table 9–2(b) illustrates the analysis of the F_2 results of a hypothetical dihybrid cross. Based on your study of the calculations involved in the monohybrid cross, check to make certain that you understand how each number was calculated in the dihybrid example.

The final step in chi-square analysis is to interpret the χ^2 value. To do so, you must initially determine the value of the **degrees of freedom (df)**, which is equal to $n - 1$, where n is the number of different categories into which each datum point may fall. For the 3:1 ratio, $n = 2$, so d/f = 1. For the 9:3:3:1 ratio, df = 3. Degrees of freedom must be taken into account because the greater the number of categories, the more deviation is expected as a result of chance.

Once you have determined the degrees of freedom, we can interpret the χ^2 value in terms of a corresponding **probability value** (p). Because this calculation is complex, we usually take the p value from a standard table or graph. Figure 9–12 shows the wide range of χ^2 and p values for numerous degrees of freedom in both a graph and a table. We will use the graph to explain how to determine the p value. The caption for Figure 9–12(b) explains how to use the table.

To determine p, execute the following steps:

1. Locate the χ^2 value on the abscissa (the horizontal or X axis).
2. Draw a vertical line from this point up to the line on the graph representing the appropriate df.
3. Extend a horizontal line from this point to the left until it intersects the ordinate (the vertical or Y axis).
4. Estimate, by interpolation, the corresponding p value.

TABLE 9–2 Chi-Square Analysis

(a) Monohybrid Cross

Expected Ratio	*Observed (o)*	*Expected (e)*	*Deviation (o − e)*	*Deviation² (d²)*	*d²/e*
3/4	740	3/4 (1000) = 750	740 − 750 = − 10	(− 10)² = 100	100/750 = 0.13
1/4	260	1/4 (1000) = 250	260 − 250 = + 10	(+ 10)² = 100	100/250 = 0.40
	Total = 1000				χ^2 = 0.53 p = 0.48

(b) Dihybrid Cross

Expected Ratio	*o*	*e*	*o − e*	*d²*	*d²/e*
9/16	587	567	+ 20	400	0.71
3/16	197	189	+ 8	64	0.34
3/16	168	189	− 21	441	2.33
1/16	56	63	− 7	49	0.78
	Total = 1008				χ^2 = 4.16 p = 0.26

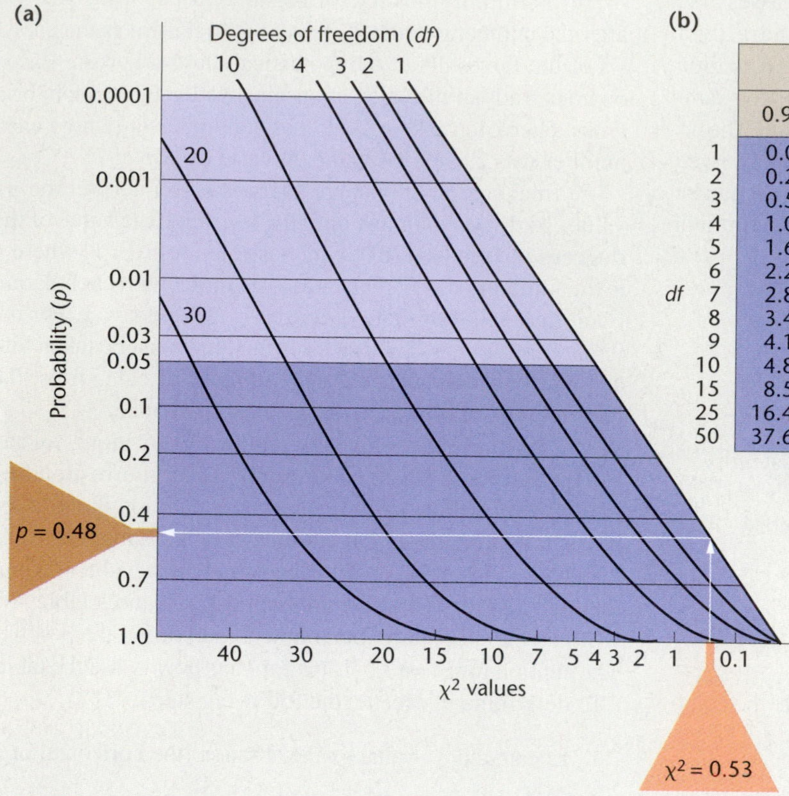

(b)

	Probability (p)					
df	0.90	0.50	0.20	0.05	0.01	0.001
1	0.02	0.46	1.64	3.84	6.64	10.83
2	0.21	1.39	3.22	5.99	9.21	13.82
3	0.58	2.37	4.64	7.82	11.35	16.27
4	1.06	3.36	5.99	9.49	13.28	18.47
5	1.61	4.35	7.29	11.07	15.09	20.52
6	2.20	5.35	8.56	12.59	16.81	22.46
7	2.83	6.35	9.80	14.07	18.48	24.32
8	3.49	7.34	11.03	15.51	20.09	26.13
9	4.17	8.34	12.24	16.92	21.67	27.88
10	4.87	9.34	13.44	18.31	23.21	29.59
15	8.55	14.34	19.31	25.00	30.58	37.30
25	16.47	24.34	30.68	37.65	44.31	52.62
50	37.69	49.34	58.16	67.51	76.15	86.60

χ^2 values

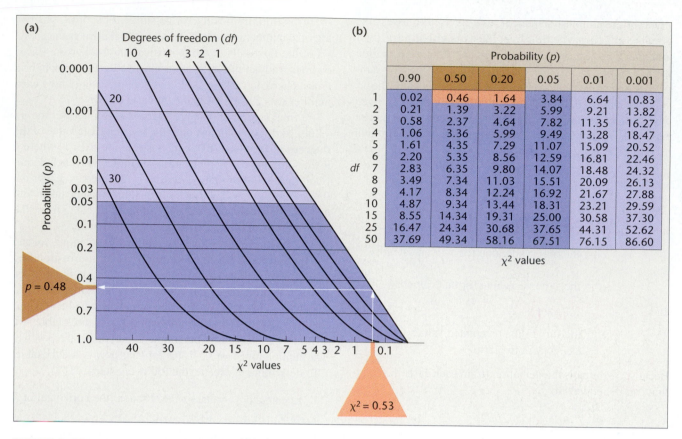

FIGURE 9–12 (a) Graph for converting χ^2 values to p values. (b) Table of χ^2 values for selected values of df and p. χ^2 values greater than those shown at $p = 0.05$ justify failing to reject the null hypothesis, while χ^2 values less than those at $p = 0.05$. All (the lighter blue areas) justify rejecting the null hypothesis. In our example, $\chi^2 = 0.53$ for 1 degree of freedom is converted to a p value between 0.20 and 0.50. The graph in (a) provides an estimated p value of 0.48 by interpolation. In this case, we fail to reject the null hypothesis.

Using our first example (the monohybrid cross) in Table 9–2, we estimate the p value of 0.48 in this manner [Figure 9–12(a)]. For the dihybrid cross, use this method to see if you can determine the p value. The X^2 value is 4.16 and df equals 3. The approximate p value is 0.26. Using the table, rather than the graph, confirms that both p values are between 0.20 and 0.50. Examine the table in Figure 9–12 (b) to confirm this.

Thus far, we have been concerned only with determining p. The most important aspect of X^2 analysis is understanding what the p value actually means. We will use the example of the dihybrid cross $p = 0.26$ to illustrate. In these discussions, it is simplest to think of the p value as a percentage (e.g., $0.26 = 26$ percent). In our example, the p value indicates that, if we repeat the same experiment many times, 26 percent of the trials would be expected to exhibit chance deviation as great as or greater than that seen in the initial trial. Conversely, 74 percent of the repeats would show less deviation as a result of chance than initially observed.

The preceding discussion of the p values reveal that a hypothesis (e.g., a 9:3:3:1 ratio) is never proved or disproved absolutely. Instead, a relative standard must be set that serves as the basis for either rejecting or failing to reject the null hypothesis. This standard is most often a p value of 0.05. When applied to chi-square analysis, a p value less than 0.05 means the probability is less than 5 percent that the observed deviation in the set of results could be obtained by chance alone. Such a p value indicates that the difference between the observed and predicted results is substantial and enables us *to reject the null hypothesis*.

On the other hand, p values of 0.05 or greater (0.05 to 1.0) indicate that the observed deviation will be obtained by chance alone 5 percent or more of the time. In such cases, this enables us *to fail to reject the null hypothesis*. Thus, the p value of 0.26, assessing the hypothesis that independent assortment accounts for the results, fails to be rejected. Therefore, the observed deviation can be reasonably attributed to chance.

A final note is relevant here concerning the case where the null hypothesis is rejected, that is, $p < 0.05$. Suppose, for example, the null hypothesis being tested was that the data represented a 9:3:3:1 ratio, indicative of independent assortment. If the null hypothesis is rejected, what are alternative interpretations of the data? Researchers first reassess the many assumptions that underlie the null hypothesis. In our case, we assumed that segregation operates faithfully for both gene pairs. We also assumed that fertilization is random and that the viability of all gametes is equal irrespective of genotype–that is, that all gametes are equally likely to participate in fertilization. Finally, following fertilization, we assumed that all preadult stages and adult offspring are equally viable, regardless of their genotype.

An example will clarify this: Suppose our null hypothesis is that a dihybrid cross between fruit flies will result in 3/16 mutant wingless flies (the proportion of mutant zygotes that may, in fact, occur at fertilization). However, these mutant embryos may not survive as well during their preadult development, or as young adults, compared with flies whose genotype gives rise to wings. As a result, when the data are gathered, there will be fewer than 3/16 wingless flies. Rejection of the null hypothesis is not cause for us to disregard the validity of the postulates of segregation and independent assortment, because other factors are operative.

The foregoing discussion serves to point out that statistical information must be assessed carefully on a case-by-case basis. When we reject a null hypothesis, we must examine all underlying assumptions. If there is no concern about their validity, then we must consider other alternative hypotheses to explain the results.

9.10 Pedigrees Reveal Patterns of Inheritance in Humans

In all the crosses we have discussed so far, one of the two traits for each character has been dominant to the other. Based on this observation, tw significant questions arise:

1. Does the expression of all genes occur in this fashion?
2. Is it possible to ascertain the mode of inheritance of genes in organisms where designed crosses and the production of large numbers of offspring are not practical?

The answer to the first question is no. As we will see in Chapter 10, many modes of genetic expression exist that modify the monohybrid and dihybrid ratios observed by Mendel.

The answer to the second question is yes. The pattern of inheritance of a specific phenotype can be studied even in humans.

Figure 9–13 shows a standard pedigree to illustrate the conventions to construct pedigrees. Circles represent females, and squares designate males. Parents are connected by a horizontal line, and vertical lines lead to their offspring. The offspring are called **sibs** and are connected by a horizontal **sibship line**. Sibs are placed from left to right according to birth order, and are labeled with Arabic numerals. Each generation is indicated by a Roman numeral. Thus, a designation such as I-2 represents the female of the first set of parents in the initial generation.

If the sex of an individual is unknown, a diamond is used. (See II-2.) If a pedigree traces only a single trait, as Figure 9–13 does, the circles, squares, and diamonds are shaded if the phenotype being considered is expressed. (See I-1, III-3, and III-4.) Those who fail to express a recessive trait, when known with certainty to be heterozygous, have only the left half of their square or circle shaded. (See II-3 and II-4.)

Twins are indicated by diagonal lines stemming from a vertical line connected to the sibship line. For **monozygotic** or **identical twins**, the diagonal lines are linked by a horizontal line. (See III-5,6.) **Dizygotic** or **fraternal twins**, lack this connecting line. (See III-8,9.) A number within one of the symbols (See II-10-13) represents numerous sibs of the same or unknown phenotypes. The individual whose phenotype

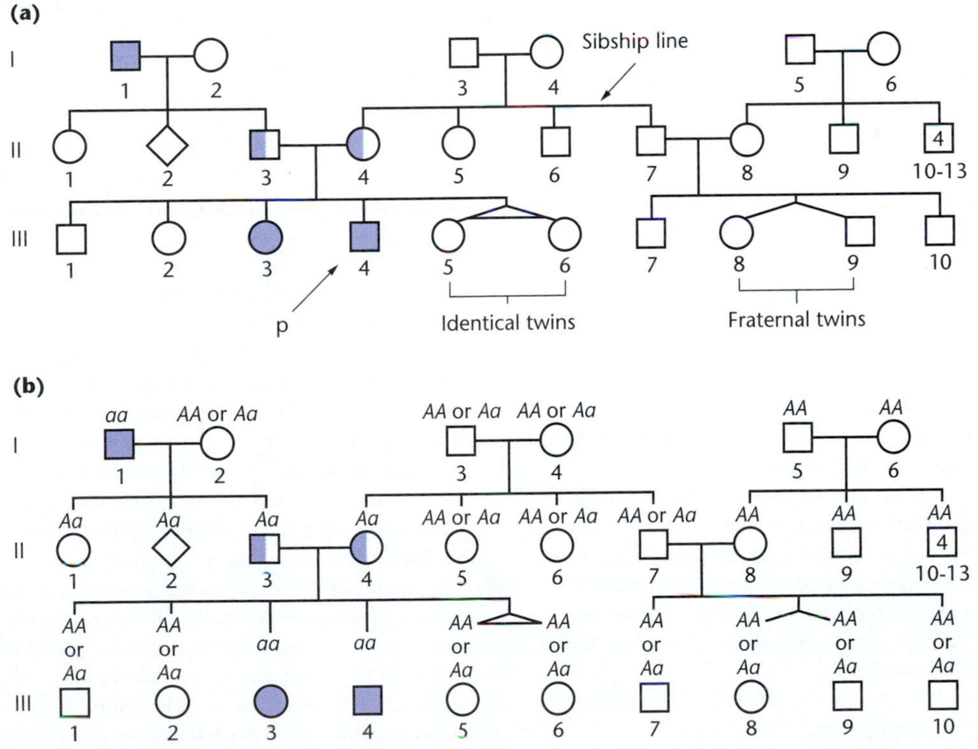

FIGURE 9–13 (a) A representative pedigree for a single character through three generations. (b) The most probable genotypes of each individual in the pedigree.

MEDIA TUTORIAL Analysis of Human Pedigrees

drew the attention of a physician or geneticist is called the **proband** and is indicated by an arrow connected to the designation **p** (See III-4). This term applies to both males and females. In this case, the proband happens to be a male.

The pedigree shown in Figure 9–13 traces the pattern of inheritance of the human trait **albinism**. As we analyze this pedigree next, we will see that albinism is inherited as a recessive trait.

The male parent of the first generation (I-1) is affected. Characteristic of a rare recessive trait with an affected parent, the trait "disappears" in the offspring. Assuming recessiveness, we might predict that the unaffected female parent (I-2) is a homozygous normal individual because none of the offspring show the disorder. Had she been heterozygous, one-half of the offspring would be expected to exhibit albinism, but none do. However, such a small sample (three offspring) prevents any certainty in the matter.

Further evidence supports the prediction of a recessive trait. If albinism were inherited as a dominant trait, individual II-3 would have to express the disorder in order to pass it to his offspring (III-3 and III-4). He does not. Inspection of the offspring constituting the third generation (row III) provides still further support for the hypothesis that albinism is a recessive trait. If it is, parents II-3 and II-4 are both heterozygous, and approximately one-fourth of their offspring should be affected. Two of the six offspring do show albinism. This deviation from the expected ratio is not unexpected in crosses with few offspring.

Based on the preceding analysis, we can predict the most probable genotypes of all the individuals in the pedigree. For both the first and second generations, in only a few cases can this can be done with some certainty. In the third generation, we cannot determine whether most phenotypically normal individuals are homozygous or heterozygous. Figure 9–13(b) shows the most probable genotypes for each individual.

Pedigree analysis of many traits has been an extremely valuable research technique in human genetic studies. However, the approach does not usually provide the certainty in drawing conclusions that is afforded by designed crosses yielding large numbers of offspring. Nevertheless, when many independent pedigrees of the same trait or disorder are analyzed, consistent conclusions can often be drawn. Table 9–3 lists numerous human traits and classifies them according to their recessive or dominant expression. As we will see in Chapter 10, the genes controlling some of these traits are located on the sex-determining chromosomes.

TABLE 9–3 Representative Recessive and Dominant Human Traits

Recessive Traits	*Dominant Traits*
Albinism	Achondroplasia
Alkaptonuria	Brachydactyly
Ataxia telangiectasia	Congenital stationary night
blindness	
Color blindness	Ehler–Danlos syndrome
Cystic fibrosis	Hypotrichosis
Duchenne muscular dystrophy	Huntington disease
Galactosemia	Hypercholesterolemia
Hemophilia	Marfan syndrome
Lesch–Nyhan syndrome	Neurofibromatosis
Phenylketonuria	Phenylthiocarbamide tasting (PTC)
Sickle-cell anemia	Porphyria (some forms)
Tay–Sachs disease	Widow's peak

Chapter Summary

1. Over a century ago, Gregor Mendel studied inheritance patterns in the garden pea, establishing the principles of transmission genetics.
2. Mendel's postulates help describe the basis for the inheritance of phenotypic expression. He showed that unit factors, later called alleles, exist in pairs and exhibit a dominant–recessive relationship in determining the expression of traits.
3. Mendel postulated that unit factors (now called alleles) must segregate during gamete formation, such that each gamete receives only one of the two factors with equal probability.
4. Mendel's postulate of independent assortment states that each pair of unit factors segregates independently of other such pairs. As a result, all possible combinations of gametes will be formed with equal probability.
5. Mendel designed the test cross to determine the exact genotype of peas expressing dominant traits.
6. The discovery of chromosomes in the late 1800s and subsequent studies of their behavior during meiosis led to the rebirth of Mendel's work, linking the behavior of his unit factors to that of chromosomes during meiosis.
7. The Punnett square and the forked-line methods are used to predict the probabilities of phenotypes (and genotypes) from crosses involving two or more gene pairs.
8. Genetic ratios are expressed as probabilities. Thus, deriving outcomes of genetic crosses relies on an understanding of the laws of probability, particularly the sum law, the product law, conditional probability, and the binomial theorem.
9. Statistical analysis is used to test the validity of experimental outcomes. In genetics, variations from the expected ratios due to chance deviations can be anticipated.
10. Chi-square analysis allows us to assess the null hypothesis, namely, that there is no real difference between the expected and observed values. As such, it tests the probability of whether observed variations can be attributed to chance deviation.
11. Pedigree analysis provides a method for analyzing the inheritance patterns in humans over several generations. Such analysis often provides the basis for determining the mode of inheritance of human traits and disorders.

Insights and Solutions

As a student, you will be asked to demonstrate your knowledge of transmission genetics by solving genetics problems. Success at this task represents not only comprehension of theory, but also its application to more practical genetic situations. Most students find problem solving in genetics to be challenging, but rewarding. This section is designed to provide basic insights into the reasoning essential to this process.

Genetics problems are in many ways similar to algebraic word problems. The approach taken should be identical: (1) analyze the problem carefully; (2) translate words into symbols, defining each one first; and (3) choose and apply a specific technique to solve the problem. The first two steps are the most critical. The third step is largely mechanical.

The simplest problems are those that state all necessary information about the P_1 generation and ask you to find the expected ratios of the F_1 and F_2 genotypes and/or phenotypes. Always follow these steps when you encounter this type of problem:

a. Determine insofar as possible the genotypes of the individuals in the P_1 generation.

b. Determine what gametes may be formed by the P_1 parents.

c. Recombine gametes by the Punnett square or the forked-line methods, or, if the situation is very simple, by inspection. From the genotypes of the F_1 generation, determine the phenotypes. Read the F_1 phenotypes directly.

d. Repeat the process to obtain information about the F_2 generation.

Performing the aforementioned steps requires an understanding of the basic theory of transmission genetics. For example, consider the following problem:

> A recessive mutant allele, *black*, causes a very dark body in *Drosophila* when homozygous. The normal wild-type color is described as gray. What F_1 phenotypic ratio is predicted when a black female is crossed to a gray male whose father was black?

To work this problem, you must understand dominance and recessiveness as well as the principle of segregation. Furthermore, you must use the information about the male parent's father. Here is one way to work this problem:

a. Because the female parent is black, she must be homozygous for the mutant allele (*bb*).

b. The male parent is gray; therefore, he must have at least one dominant allele (*B*). Since his father was black (*bb*) and he received one of the chromosomes bearing these alleles, the male parent must be heterozygous (*Bb*).

From here, the problem is simple:(See Figure below)

Apply the approach we just studied to the following problems:

1. In Mendel's work, using peas, he found that full pods are dominant to constricted pods, whereas round seeds are dominant to wrinkled seeds. One of his crosses was between full, round plants and constricted, wrinkled plants. From this cross, he obtained an F_1 that was all full and round. In the F_2, Mendel obtained his classic 9:3:3:1 ratio. Using this information, determine the expected F_1 and F_2 results of a cross between homozygous constricted, round and full, wrinkled plants.

Solution: First, define gene symbols for each pair of contrasting traits. Select the lowercase forms of the first letter of the recessive traits to designate those traits, and use the uppercase forms to designate the dominant traits. For example, use *C* and *c* to indicate full and constricted, and use *W* and *w* to indicate the round and wrinkled phenotypes, respectively.

Now, determine the genotypes of the P_1 generation, form gametes, reconstitute the F_1 generation, and determine the F_1 phenotype(s):

P_1	*ccWW* constisted, round	×	*CCww* full, wrinkled
	↓		↓
Gametes	*cW*		*Cw*
F_1		*CcWw* full, round	

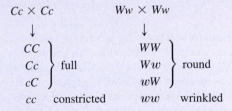

You can see immediately that the F_1 generation expresses both dominant phenotypes and is heterozygous for both gene pairs. We can expect that the F_2 generation will yield the classic Mendelian ratio of 9:3:3:1. Let's work it out anyway, just to confirm it, using the forked-line method. Because both gene pairs are heterozygous and can be expected to assort independently, we can predict the F_2 outcomes from each gene pair separately and then proceed with the forked-line method.

Every F_2 offspring is subject to the following probabilities:

Cc × *Cc*		*Ww* × *Ww*	
↓		↓	
CC, *Cc*, *cC*	full	*WW*, *Ww*, *wW*	round
cc	constricted	*ww*	wrinkled

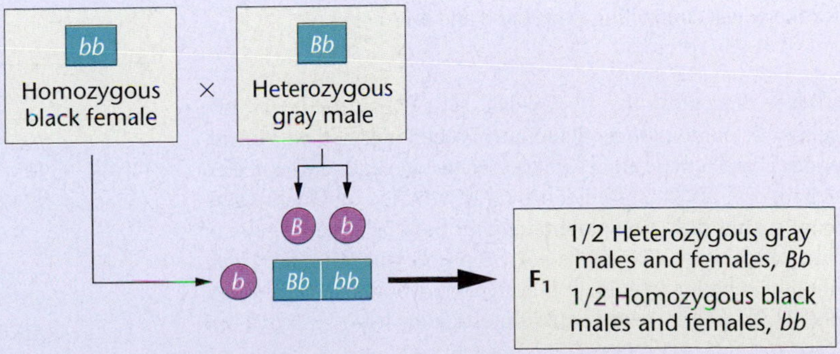

The forked-line method then allows us to confirm the 9:3:3:1 phenotypic ratio. Remember that this represents proportions of 9/16:3/16:3/16:1/16. Note that we are applying the product law as we compute the final probabilities:

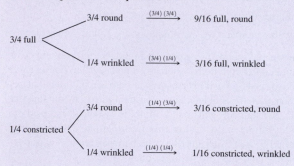

3/4 full
- 3/4 round $\xrightarrow{(3/4)\,(3/4)}$ 9/16 full, round
- 1/4 wrinkled $\xrightarrow{(3/4)\,(1/4)}$ 3/16 full, wrinkled

1/4 constricted
- 3/4 round $\xrightarrow{(1/4)\,(3/4)}$ 3/16 constricted, round
- 1/4 wrinkled $\xrightarrow{(1/4)\,(1/4)}$ 1/16 constricted, wrinkled

2. Determine the probability that a plant of genotype *CcWw* will be produced from parental plants of the genotypes *CcWw* and *Ccww*.

Solution: Because the two gene pairs independently assort during gamete formation, we need only calculate the individual probabilities of the two separate events (*Cc* and *Ww*) and apply the product law to calculate the final probability.

$Cc \times Cc$ 1/4 *CC*: 1/2 *Cc*: 1/4 *cc*

$Ww \times ww$ 1/2 *Ww* 1/2 *ww*

$$p = (1/2\ Cc)(1/2\ Ww) = 1/4\ CcWw$$

3. In another cross, involving parent plants of unknown genotype and phenotype, the following offspring were obtained.

3/8 full, round;

3/8 full, wrinkled;

1/8 constricted, round;

1/8 constricted, wrinkled.

Determine the genotypes and phenotypes of the parents.

Solution: This problem is more difficult and requires keener insights since you must work backwards. The best approach is to consider the outcomes of pod shape separately from those of seed texture.

Of all plants, 6/8 (3/4) are full and 2/8 (1/4) are constricted. Of the various genotypic combinations that can serve as parents, which will give rise to a ratio of 3/4:1/4? Because this ratio is identical to Mendel's monohybrid F_2 results, we can propose that both unknown parents share the same genetic characteristic as the monohybrid F_1 parents: They must both be heterozygous for the genes controlling pod shape and thus are

Cc

Before accepting this hypothesis, let's consider the possible genotypic combinations that control seed texture. If we consider this characteristic alone, we can see that the traits are expressed in a ratio of 4/8 (1/2) round: 4/8 (1/2) wrinkled. In order to generate such a ratio, the parents *cannot* both be heterozygous or their offspring would yield a 3/4:1/4 phenotypic ratio. They *cannot* both be homozygous or all offspring would express a single phenotype. Thus, we are left with testing the hypothesis that one parent is homozygous and one is heterozygous for the alleles controlling texture. The potential case of $WW \times Ww$ will not work, for it also would yield only a single phenotype. Hence, we are left with the potential case of $Ww \times ww$. Offspring in such a mating will yield 1/2 *Ww* (round): 1/2 *ww* (wrinkled), exactly the outcome we are seeking.

Now, let's combine our hypotheses and predict the outcome of crossing. In our solution, since we are only predicting phenotypes, we will use the dash symbol (−) to indicate that the second allele may be either dominant or recessive:

3/4 *C*−
- 1/2 *Ww* → 3/8 *C*− *Ww* full, round
- 1/2 *ww* → 3/8 *C*− *ww* full, wrinkled

1/4 cc
- 1/2 *Ww* → 1/8 *ccWw* constricted, round
- 1/2 *ww* → 1/8 *ccww* constricted, wrinkled

As you can see, this cross produces offspring in accordance with the information provided, thus solving the problem. Note that in the solution, we have used *genotypes* in the forked-line method, in contrast to the use of *phenotypes* in the solution of the first problem.

4. In the laboratory, a genetics student crossed flies with normal long wings to flies with mutant *dumpy* wings, which she believed was a recessive trait. In the F_1, all flies had long wings. In the F_2, the following results were obtained:

792 long-winged flies

and

208 dumpy-winged flies.

The student tested the hypothesis that the *dumpy* wing is inherited as a recessive trait by performing χ^2 analysis of the F_2 data.

(a) What ratio did the student hypothesize?

(b) Did the χ^2 analysis support the hypothesis?

(c) What do the data suggest about the *dumpy* mutation?

Solution:

(a) The student hypothesized that the F_2 data (792:208) fit Mendel's 3:1 monohybrid ratio for recessive genes.

(b) The initial step in χ^2 analysis is to calculate the expected results (*e*) by assuming a ratio of 3:1. Then calculate the deviations (*d*) between the expected values and observed values (the actual data):

Ratio	*o*	*e*	*d*	d^2	d^2/e
3/4	792	750	42	1764	2.35
1/4	208	250	−42	1764	7.06
Total =	1000				

$$\chi^2 = \sum \frac{d^2}{e}$$
$$= 2.35 + 7.06$$
$$= 9.41$$

Consulting Figure 9–12 allows us to determine the probability (p). This value will let us determine whether the deviations from the null hypothesis can be attributed to chance. There are two possible outcomes (n), so the degrees of freedom (df) $= n - 1$ or 1. The table in Figure 9–12 shows that $p = 0.01$ to 0.001. The graph gives an estimate of about 0.001. That p is less than 0.05 leads us to reject the null hypothesis. The data do not statistically fit a 3:1 ratio.

When we accept Mendel's 3:1 ratio as a valid expression of the monohybrid cross, we make numerous assumptions. One of these may explain why the null hypothesis was rejected. We must assume *that all genotypes are equally viable*. That is, at the time the data are collected, genotypes yielding long wings are equally likely to survive from fertilization through adulthood as the genotype yielding *dumpy* wings. Further study would reveal that *dumpy* flies are somewhat less viable than normal flies. As a result, we would expect *less* than $\frac{1}{4}$ of the total offspring to express dumpy. This observation is borne out in the data, although we have not proved it.

5. If two parents, both heterozygous carriers of the autosomal recessive gene causing cystic fibrosis, have five children, what is the probability that three will be normal?

Solution: First, the probability of having a normal child during each pregnancy is

$$p_a = \text{normal} = 3/4$$

while the probability of having an afflicted offspring is

$$p_b = \text{afflicted} = 1/4$$

Then apply the formula

$$\frac{n!}{s!t!}a^s b^t$$

where $n = 5$, $s = 3$, and $t = 2$

$$p = \frac{(5)\cdot(4)\cdot(3)\cdot(2)\cdot(1)}{(3)\cdot(2)\cdot(1)\cdot(2)\cdot(1)}(3/4)^3(1/4)^2$$

$$= \frac{(5)\cdot(4)}{(2)\cdot(1)}(3/4)^3(1/4)^2$$

$$= 10(27/64)\cdot(1/16)$$

$$= 10(27/1024)$$

$$= 270/1024$$

$$p = \sim 0.26$$

Problems and Discussion Questions

When working genetics problems in this and succeeding chapters, always assume that members of the P_1 generation are homozygous, unless the information given, or the data presented, indicates or requires otherwise.

1. In a cross between a black and a white guinea pig, all members of the F_1 generation are black. The F_2 generation is made up of approximately 3/4 black and 1/4 white guinea pigs.
   (a) Diagram this cross, showing the genotypes and phenotypes.
   (b) What will the offspring be like if two F_2 white guinea pigs are mated?
   (c) Two different matings were made between black members of the F_2 generation, with the following results:

Cross	*Offspring*
Cross 1	All black
Cross 2	3/4 black, 1/4 white

   Diagram each of the crosses.

2. Albinism in humans is inherited as a simple recessive trait. For the following families, determine the genotypes of the parents and offspring (when two alternative genotypes are possible, list both):
   (a) Two normal parents have five children, four normal and one albino.
   (b) A normal male and an albino female have six children, all normal.
   (c) A normal male and an albino female have six children, three normal and three albino.
   (d) Construct a pedigree of the families in (b) and (c). Assume that one of the normal children in (b) and one of the albino children in (c) are the parents of eight children.
3. Which of Mendel's postulates are illustrated by the pedigree in Problem 2? List and define these postulates.
4. Discuss the rationale relating Mendel's monohybrid results to his postulates.
5. What advantages were provided by Mendel's choice of the garden pea in his experiments?
6. Pigeons may exhibit a checkered or plain pattern. In a series of controlled matings, the following data were obtained:

	F_1 *Progeny*	
P_1 *Cross*	*Checkered*	*Plain*
(a) checkered × checkered	36	0
(b) checkered × plain	38	0
(c) plain × plain	0	35

   Then, F_1 offspring were selectively mated with the following results (the P_1 cross giving rise to each F_1 pigeon is indicated in parentheses):

	Progeny	
$F_1 \times F_1$ *Crosses*	*Checkered*	*Plain*
(d) checkered (a) × plain (c)	34	0
(e) checkered (b) × plain (c)	17	14
(f) checkered (b) × checkered (b)	28	9
(g) checkered (a) × checkered (b)	39	0

How are the checkered and plain patterns inherited? Select and define symbols for the genes involved, and determine the genotypes of the parents and offspring in each cross.

7. Mendel crossed peas having round seeds and yellow cotyledons (seed leaves) with peas having wrinkled seeds and green cotyledons. All the F_1 plants had round seeds with yellow cotyledons. Diagram this cross through the F_2 generation, using both the Punnett square and forked-line, or branch diagram, methods.
8. Based on the preceding cross, in the F_2 generation, what is the probability that an organism will have round seeds and green cotyledons *and* be true breeding?
9. Based on the same characters and traits as in Problem 7, determine the genotypes of the parental plants involved in the crosses shown here by analyzing the phenotypes of their offspring:

Parental Plants	*Offspring*
(a) round, yellow × round, yellow	3/4 round, yellow 1/4 wrinkled, yellow
(b) wrinkled, yellow × round, yellow	6/16 wrinkled, yellow 2/16 wrinkled, green 6/16 round, yellow 2/16 round, green
(c) round, yellow × round, yellow	9/16 round, yellow 3/16 round, green 3/16 wrinkled, yellow 1/16 wrinkled, green
(d) round, yellow × wrinkled, green	1/4 round, yellow 1/4 round, green 1/4 wrinkled, yellow 1/4 wrinkled, green

10. Are any of the crosses in Problem 9 test crosses? If so, which one(s)?
11. Which of Mendel's postulates can only be demonstrated in crosses involving at least two pairs of traits? State the postulate.
12. Correlate Mendel's four postulates with what is now known about homologous chromosomes, genes, alleles, and the process of meiosis.
13. What is the basis for homology among chromosomes?
14. Distinguish between homozygosity and heterozygosity.
15. In *Drosophila, gray* body color is dominant to *ebony* body color, while *long* wings are dominant to *vestigial* wings. Assuming that the P_1 individuals are homozygous, work the following crosses through the F_2 generation, and determine the genotypic and phenotypic ratios for each generation:

(a)	gray, long × ebony, vestigial
(b)	gray, vestigial × ebony, long
(c)	gray, long × gray, vestigial

16. How many different types of gametes can be formed by individuals of the following genotypes: (a) *AaBb*, (b) *AaBB*, (c) *AaBbCc*, (d) *AaBBcc*, (e) *AaBbcc*, and (f) *AaBbCcDdEe?* What are the gametes in each case?
17. Using the forked-line, or branch diagram, method, determine the genotypic and phenotypic ratios of the trihybrid crosses: (a) *AaBbCc* × *AaBBCC*, (b) *AaBBCc* × *aaBBCc*, and (c) *AaBbCc* × *AaBbCc*.
18. Mendel crossed peas with green seeds to those with yellow seeds. The F_1 generation produced only yellow seeds. In the F_2, the progeny consisted of 6022 plants with yellow seeds and 2001 plants with green seeds. Of the F_2 yellow-seeded plants, 519 were self-fertilized with the following results: 166 bred true for yellow and 353 produced a 3:1 ratio of yellow:green. Explain these results by diagramming the crosses.
19. In a study of black and white guinea pigs, 100 black animals were crossed individually to white animals and each cross was carried to an F_2 generation. In 94 of the cases, the F_1 individuals were all black and an F_2 ratio of 3 black:1 white was obtained. In the other 6 cases, half of the F_1 animals were black and the other half were white. Why? Predict the results of crossing the black and white F_1 guinea pigs from the 6 exceptional cases.
20. Mendel crossed peas with round green seeds to ones with wrinkled yellow seeds. All F_1 plants had seeds that were round and yellow. Predict the results of test crossing these F_1 plants.
21. Thalassemia is an inherited anemic disorder in humans. Individuals can be completely normal, they can exhibit a "minor" anemia, or they can exhibit a "major" anemia. Assuming that only a single gene pair and two alleles are involved in the inheritance of these conditions, which allele and which phenotype are recessive? Which are dominant?
22. The following are F_2 results of two of Mendel's monohybrid crosses:

(a)	Full pods	882
	Constricted pods	299
(b)	Violet flowers	705
	White flowers	224

State a null hypothesis to be tested using χ^2 analysis. Calculate the χ^2 value and determine the p value for both. Interpret the p values. Can the deviation in each case be attributed to chance or not? Which of the two crosses shows a greater amount of deviation?
23. In one of Mendel's dihybrid crosses, he observed 315 round yellow, 108 round green, 101 wrinkled yellow, and 32 wrinkled green F_2 plants. Analyze these data using the χ^2 test to see if
(a) They fit a 9:3:3:1 ratio.
(b) The round:wrinkled data fit a 3:1 ratio.
(c) The yellow:green data fit a 3:1 ratio.
24. In assessing data that fell into two phenotypic classes, a geneticist observed values of 250:150. She decided to perform a X^2 analysis by using two different null hypotheses: (a) The data fit a 3:1 ratio and (b) the data fit a 1:1 ratio. Calculate the X^2

values for each hypothesis. What can be concluded about each hypothesis?

25. The basis for rejecting any null hypothesis is arbitrary. The researcher can set more or less stringent standards by deciding to raise or lower the p value used to reject or fail to reject the hypothesis. In the case of the chi-square analysis of genetic crosses, would the use of a standard of $p = 0.10$ be more or less stringent in failing to reject the null hypothesis? Explain.

26. Consider the following pedigree.

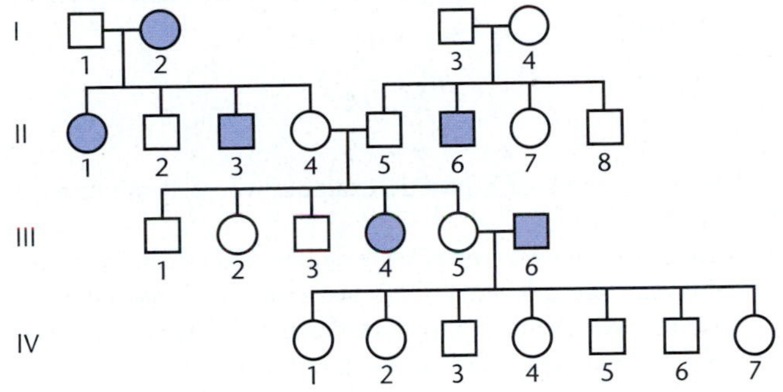

Predict the mode of inheritance and the most probable genotypes of each individual. Assume that the alleles A and a control the expression of the trait.

27. The following pedigree is for myopia (nearsightedness) in humans:

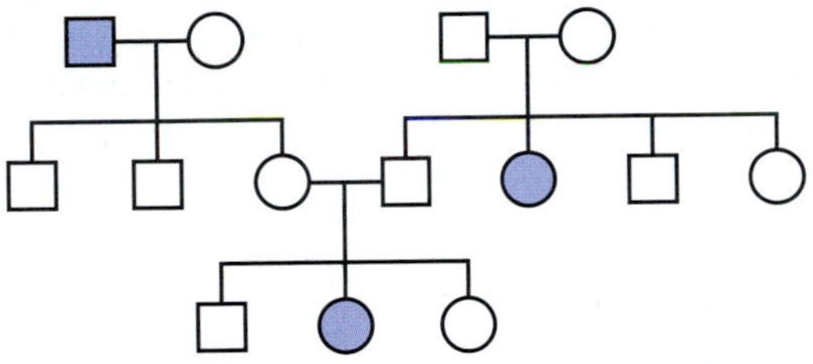

Predict whether the disorder is inherited as the result of a dominant or recessive trait. Determine the most probable genotype for each individual based on your prediction.

28. Consider three independently assorting gene pairs, A/a, B/b, and C/c. What is the probability of obtaining an offspring that is $AABbCc$ from parents that are $AaBbCC$ and $AABbCc$?

29. What is the probability of obtaining a triply recessive individual from the parents shown in Problem 28?

30. Of all offspring of the parents in Problem 28, what proportion will express all three dominant traits?

31. When a die (one of a pair of dice) is rolled, it has an equal probability of landing on any of its six sides.
(a) What is the probability of rolling a 3 with a single throw?
(b) When a die is rolled twice, what is the combined probability that the first throw will be a 3 and the second will be a 6?
(c) If two dice are rolled together, what is the combined probability that one will be a 3 and the other will be a 6?
(d) If one die is rolled and it comes up as an odd number, what is the probability that it is a 5?

32. Consider the F_2 offspring of Mendel's dihybrid cross. Determine the conditional probability that F_2 plants expressing both dominant traits are heterozygous at both loci.

33. Draw all possible conclusions concerning the mode of inheritance of the trait denoted in each of the following limited pedigrees: (each case is based on a different trait):

(a)

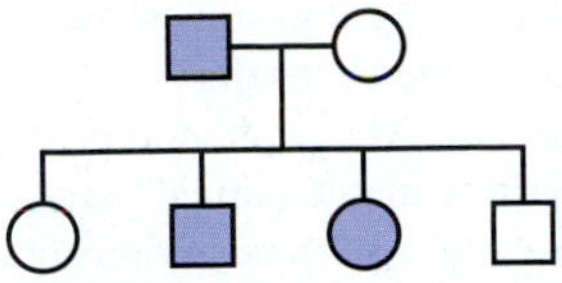

(b)

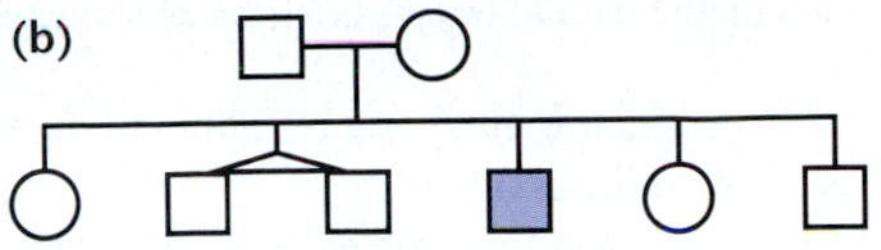

(c)

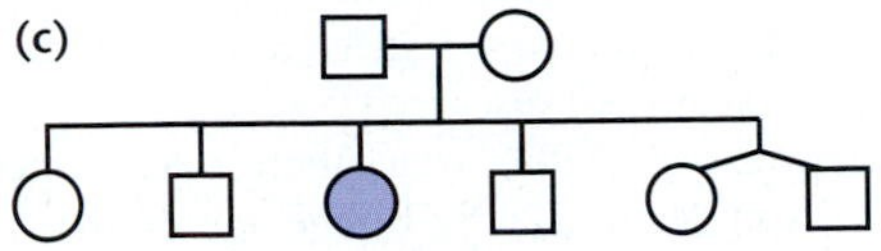

(d)

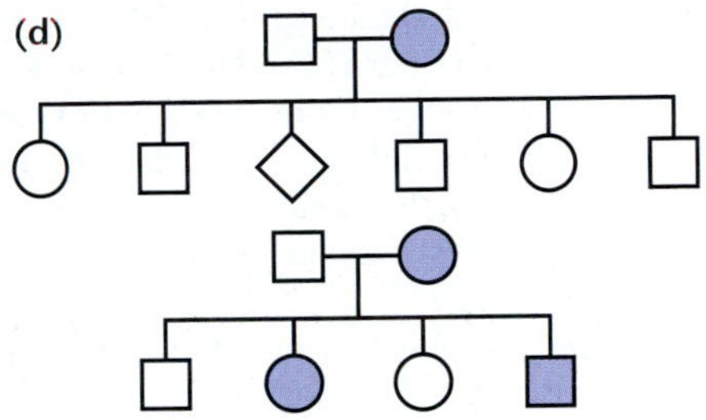

(e)

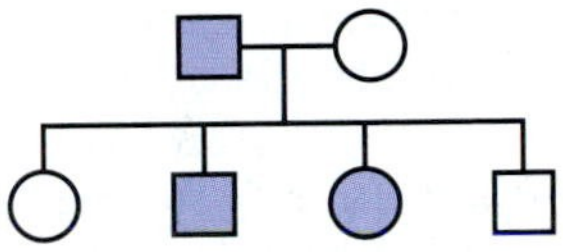

34. Cystic fibrosis is an autosomal recessive disorder. A male whose brother has the disease has children with a female whose sister has the disease. It is not known if either the male or the female is a carrier. If the male and female have one child, what is the probability that the child will have cystic fibrosis?

35. In a family of five children, what is the probability that
(a) All are males?
(b) Three are males and two are females?
(c) Two are males and three are females?
(d) All are the same sex?
Assume that the probability of a male child is equal to the probability of a female child ($p = 1/2$).

36. In a family of eight children, where both parents are heterozygous for albinism, what mathematical expression predicts the probability that six are normal and two are albinos?

Extra-Spicy Problems

37. Two true-breeding pea plants were crossed. One parent is round, terminal, violet, constricted, while the other expresses the respective contrasting phenotypes of wrinkled, axial, white, full. The four pairs of contrasting traits are controlled by four genes, each located on a separate chromosome. In the F_1, only round, axial, violet, and full were expressed. In the F_2, all possible combinations of these traits were expressed in ratios consistent with Mendelian inheritance.

(a) What conclusion about the inheritance of the traits can be drawn based on the F_1 results?
(b) In the F_2 results, which phenotype appeared most frequently? Write a mathematical expression that predicts the probability of occurrence of this phenotype.
(c) Which F_2 phenotype is expected to occur least frequently? Write a mathematical expression that predicts this probability.
(d) In the F_2 generation, how often is either of the P_1 phenotypes likely to occur?
(e) If the F_1 plants were test crossed, how many different phenotypes would be produced? How does this number compare with the number of different phenotypes in the F_2 generation test just discussed?

38. Tay–Sachs disease (TSD) is an inborn error of metabolism that results in death, usually by the age of two. You are a genetic counselor, and you interview a phenotypically normal couple, where the male had a female first cousin (on his father's side) who died from TSD, and where the female had a maternal uncle with TSD. There are no other known cases in either of the families, and none of the matings were or are between related individuals. Assume that this trait is very rare.
(a) Draw a pedigree of the families of this couple, showing the relevant individuals.
(b) Calculate the probability that both the male and female are carriers for TSD.
(c) What is the probability that neither of them are carriers?
(d) What is the probability that one of them is a carrier and the other is not? (*Hint:* The *p* values in (b), (c), and (d) should equal 1.)

39. The wild-type (normal) fruit fly, *Drosophila melanogaster*, has straight wings and long bristles. Mutant strains have been isolated that have either curled wings or short bristles. The genes representing these two mutant traits are located on separate autosomes. Carefully examine the data from the following five crosses shown below: (running across both columns at the bottom).
(a) For each mutation, determine whether it is dominant or recessive. In each case, identify which crosses support your answer; and (b) define gene symbols, and for each cross, determine the genotypes of the parents.

40. To what do you suppose the following mathematical expression applies?

$$\frac{n!}{s!t!u!}a^s b^t c^u$$

Can you think of a genetic example where it might have application?

41. To assess Mendel's law of segregation using tomatoes, a true-breeding tall variety (*SS*) is crossed with a true-breeding short variety (*ss*). The heterozygous tall plants (*Ss*) were crossed to produce two sets of F_2 data, as shown below.

Set I	*Set II*
30 tall 5 short	300 tall 50 short

(a) Using the χ^2 test, analyze the results for both data sets. Calculate χ^2 values and estimate the p values in both cases
(b) From the above analysis, what can you conclude about the importance of generating large data sets in experimental settings?

	Cross			Number of Progeny: *straight wings, long bristles*	*straight wings, short bristles*	*curled wings, long bristles*	*curled wings, short bristles*
1	straight, short	×	straight, short	30	90	10	30
2	straight, long	×	straight, long	120	0	40	0
3	curled, long	×	straight, short	40	40	40	40
4	straight, short	×	straight, short	40	120	0	0
5	curled, short	×	straight, short	20	60	20	60

Selected Readings

Carlson, E.A. 1987. *The gene: A critical history*, 2d ed. Philadelphia, PA: Saunders.

Cummings, M.R. 2000. *Human heredity: Principles and issues*, 5th ed. Belmont, CA: Wadsworth Publ. Co.

Dunn, L.C. 1965. *A short history of genetics*. New York: McGraw-Hill.

Miller, J.A. 1984. Mendel's peas: A matter of genius or of guile? *Sci. News* 125:108–109.

Olby, R.C. 1985. *Origins of Mendelism*, 2d ed. London: Constable.

Orel, V. 1996. *Gregor Mendel: The first geneticist*. Oxford: Oxford University Press.

Peters, J., ed. 1959. *Classic papers in genetics*. Englewood Cliffs, NJ: Prentice-Hall.

Snedecor, G.W., and Cochran, W.G. 1980. *Statistical methods*, 7th ed. Ames, IA: Iowa State University Press.

Sokal, R.R., and Rohlf, F.J. 1987. *Introduction to biostatistics*, 2d ed. New York: W.H. Freeman.

Soudek, D. 1984. Gregor Mendel and the people around him. *Am. J. Hum. Genet.* 36:495–98.

Stern, C., and Sherwood, E. 1966. *The origins of genetics: A Mendel source book*. San Francisco: W.H. Freeman.

Stubbe, H. 1972. *History of genetics: From prehistoric times to the rediscovery of Mendel's laws*. Cambridge, MA: MIT Press.

Sturtevant, A. H. 1965. *A history of genetics*. New York: Harper & Row.

Tschermak-Seysenegg, E. 1951. The rediscovery of Mendel's work. *J. Hered.* 42:163–72.

Voeller, B. R., ed. 1968. *The chromosome theory of inheritance: Classical papers in development and heredity*. New York: Appleton-Century-Crofts.

Welling, F. 1991. Historical study: Johann Gregor Mendel 1822–1884. *Am. J. Med. Genet.* 40:1–25.

GENETICS MediaLab

The resources that follow will help you achieve a better understanding of the concepts presented in this chapter. These resources can be found either on the CD packaged with this textbook or on the Companion Web site found at **http://www.prenhall.com/klug**

CD Resources:

Module 9.1: Monohybrid Cross
Module 9.2: Independent Assortment
Module 9.3: Probability
Module 9.4: Analysis of Human Pedigrees

Web Problem 1:

Time for completion = 10 minutes

In this exercise, you will practice making predictions about the genetic basis of flower color in peas, one of the traits studied by Mendel. Peas have either white or purple flowers. Mendel postulated that some factor present in the gametes affected a flower's color. He was confident that when the sperm and the ova united, two copies of the white gamete would make a white flower and two copies of the purple gamete would make a purple flower, but what if the zygote received one white and one purple? Clearly such flowers were either white or purple, not intermediate, so one factor must be dominant over the other one. Read about dominance in Chapter 9. How would you set up a test to see if the white factor was dominant over purple or vice versa? State your hypothesis regarding dominance, and outline the predictions you would make if a true-breeding white stock was crossed with a true-breeding purple stock. What genotype(s) would be present in each stock? What would be the genotype of their progeny? Interpret the results of the crosses. Which hypothesis was supported? To complete this exercise, visit Web Problem 1 in Chapter 9 of your Companion Web site, and select the keyword **MENDEL.**

Web Problem 2:

Time for completion = 10 minutes

In the last exercise, you determined dominance by crossing pure-breeding lines with white and purple flowers. Suppose you were asked to determine the genotype of a purple flowering plant that a fellow student found growing in a field of white flowering plants. How would you proceed? Assume that you have access to pure-breeding white and purple flowering peas. Read about Punnett squares and testcrosses in Chapter 9. When two or more outcomes are possible, scientists often specify multiple hypotheses and devise tests that will eliminate one or more of the hypotheses. State two hypotheses regarding the genetic makeup of your purple-flowered plant. Devise crosses using your purple plant or your true-breedin stocks (or both) to test your hypotheses. Use the monohybrid Punnett squares module to predict the outcomes of each cross under each of the two scenarios. How many crosses would you need to carry out to determine the genotype of the purple plant? Can you think of an alternative way to test your hypotheses? To complete this exercise, visit Web Problem 2 in Chapter 9 of your Companion Web site, and select the keyword **TESTCROSS.**

Web Problem 3:

Time for completion = 20 minutes

Now that you have practiced formulating hypotheses and tests for single traits, let's try a dihybrid cross. Wild-type *Drosophila melanogaster* have normal wings and red eyes. Suppose you have found mutants with vestigial wings and sepia eyes. Assume that these traits are not sex linked (Chapter 10) and are on separate chromosomes (Chapter 24). You will cross a female with vestigial wings and red eyes with a male that has sepia eyes and normal wings. Would it matter if one fly had both mutations? Write down your hypothesis regarding the dominance and recessiveness of these two traits, and derive a set of predictions using Punnett squares. Visit the Web site, using the link **DIHYBRID**, and observe the results of the F1. Does this provide enough information to test your hypothesis, or do you need to cross the F1 and observe the phenotypes of the F2? Compare the observed results with those you expected, and conduct a chi-square test to see if the data agree with your hypothesis. To complete this exercise, visit Web Problem 3 in Chapter 9 of your Companion Web site, and select the keyword **DIHYBRID**.

Mice expressing the black, yellow, and agouti coat phenotypes. (*Tom Cerniglio/Oak Ridge National Laboratory*)

10

Extensions of Mendelian Genetics

KEY CONCEPTS

In Chapter 9, we discussed the fundamental principles of transmission genetics. We saw that genes are present on homologous chromosomes and that these chromosomes segregate from each other and assort independently with other segregating chromosomes during gamete formation. These two postulates are the basic principles of gene transmission from parent to offspring. Once an offspring has received the total set of genes, however, it is the expression of genes that determines the organism's phenotype. When gene expression does not adhere to a simple dominant–recessive mode, or when more than one pair of genes influences the expression of a single character, the classic 3:1 and 9:3:3:1 F_2 ratios are usually modified. In this and the next several chapters, we consider more complex modes of inheritance. Although more complex modes of inheritance result, the fundamental principles set down by Mendel still hold true in these situations.

In this chapter, we restrict our initial discussion to the inheritance of traits that are controlled by only one set of genes. In diploid organisms, which have homologous pairs of chromosomes, two copies of each gene influence such traits. The copies need not be identical since alternative forms of genes, or **alleles**, occur within populations. How alleles act to influence a given phenotype will be our primary focus. We will then turn to **gene interaction**, a situation in which a single phenotype is affected by more than one gene. Numerous examples will be presented to illustrate a variety of heritable patterns observed in such situations.

Thus far, we have restricted our discussion to chromosomes other than the X and Y pair. By examining cases where genes are present on the X chromosome, illustrating **X-linkage**, we will see yet another modification of Mendelian ratios. Our discussion of modified ratios concludes with the consideration of sex-limited and sex-influenced inheritance, cases where the sex of the individual, but not necessarily the X chromosome, influences the phenotype. We conclude the chapter by entertaining the idea that the development of a given phenotype often varies depending on the overall environment in which a cell or an organism finds itself. This discussion points out that phenotypic expression depends on more that just the genotype of an organism.

10.1 Alleles Alter Phenotypes in Different Ways

Following the rediscovery of Mendel's work in the early 1900s, research focused on the many ways in which genes can influence an individual's phenotype. This course of investigation, stemming from Mendel's findings, is called neo-Mendelian genetics (*neo* from the Greek word meaning *since* or *new*).

Each type of inheritance described in this chapter was investigated when observations of genetic data did not precisely conform to the expected Mendelian ratios. Hypotheses that modified and extended the Mendelian principles were proposed and tested with specifically designed crosses. Explanations for these observations were in accord with the principle that a phenotype is under the influence of one or more genes located at specific loci on one or more pairs of homologous chromosomes.

To understand the various modes of inheritance, we must first examine the potential function of an allele. Alleles are alternative forms of the same gene. In some organisms where extensive populations have been studied, the allele that occurs most frequently in a population, the one that is arbitrarily designated as normal, is often referred to as the **wild-type allele**. This common allele is often dominant—the allele for tall plants in the garden pea, for example—and its product is therefore functional in the cell. Wild-type alleles are responsible, of course, for the corresponding wild-type phenotype and serve as standards for comparison against other mutations occurring at a particular locus.

A mutant allele contains modified genetic information and often specifies an altered gene product. For example, in human populations, there are many known alleles of the gene that encodes the β-chain of human hemoglobin. All such alleles store information necessary for the synthesis of the β-chain polypeptide, but each allele specifies a slightly different form of the same molecule. Once manufactured, the product of an allele may or may not have its function altered.

The process of mutation is the source of new alleles. A new allele may lead to a change in the phenotype. A new phenotype results from a change in functional activity of the cellular product specified by that gene. Usually, the alteration or mutation is expressed as a loss of the specific wild-type function. For example, if a gene is responsible for the synthesis of a specific enzyme, a mutation in the gene may change the conformation of this enzyme and reduce or eliminate its affinity for the substrate. Such a mutation will result in a loss of function. Conversely, another organism may have a different mutation in the same gene, representing a different allele. The resulting enzyme may demonstrate a reduced or an increased affinity for binding the substrate, or it may not have its affinity altered at all. Correspondingly, the functional capacity of the enzyme may be reduced, enhanced, or unchanged.

Although phenotypic traits may be affected by a single mutation, traits are often influenced by many gene products. In the case of enzymatic reactions, most are part of complex metabolic pathways. Therefore, phenotypic traits may be influenced by more than one gene and the allelic forms of each gene involved. In each of the many crosses discussed in the next few chapters, only one, or relatively few, gene pairs are involved. Keep in mind that in each cross, all genes that are not under consideration are assumed to have no effect on the inheritance patterns described.

10.2 Geneticists Use a Variety of Symbols for Alleles

In Chapter 9, we learned to symbolize alleles for very simple Mendelian traits. The initial letter of the name of a recessive trait, lowercased and italicized, denotes the recessive allele, and the same letter in uppercase refers to the dominant allele. Thus, for *tall* and *dwarf*, where *dwarf* is recessive, *D* and *d* represent the alleles responsible for these respective traits. Mendel used upper- and lowercase letters such as these to symbolize his unit factors.

Another useful system was developed in genetic studies of the fruit fly *Drosophila melanogaster* to discriminate between wild-type and mutant traits. This system uses the initial letter, or a combination of two or three letters, of the name of the mutant trait. If the trait is recessive, the lowercase form is used; if it is dominant, the uppercase form is used. The contrasting wild-type trait is denoted by the same letter, but with a superscript +.

For example, *ebony* is a recessive body color mutation in *Drosophila*. The normal wild-type body color is gray. According to the system just discussed, *ebony* is denoted by the symbol *e*, while gray is denoted by e^+. The responsible locus may be occupied by either the wild-type allele (e^+) or the mutant allele (*e*). A diploid fly may thus exhibit one of three possible genotypes:

e^+ / e^+:	gray homozygote (wild type);
e^+ / e:	gray heterozygote (wild type);
e / e:	ebony homozygote (mutant).

The slash between the letters indicates that the two allele designations represent the same locus on two homologous chromosomes. If we instead consider a dominant wing mutation such as *Wrinkled* (*Wr*) wing in *Drosophila*, the three possible designations would be Wr^+/Wr^+, Wr^+/Wr, and Wr/Wr. Both the latter two genotypes express the wrinkled-wing phenotype.

One advantage of this system is that further abbreviation can be used when convenient: The wild-type allele may simply be denoted by the + symbol. With *ebony* as an example, the designations of the three possible genotypes become

$+/+$:	gray homozygote (wild type);
$+/e$:	gray heterozygote (wild type);
e/e:	ebony homozygote (mutant).

This system works nicely for other organisms as well, provided there is a distinct wild-type phenotype for the character under consideration. As we will see in Chapter 13, it is particularly useful when two or three genes linked together on the same chromosome are considered simultaneously. If no dominance exists, we may simply use uppercase letters and superscripts to denote alternative alleles (e.g., R^1 and R^2, L^M and L^N, and I^A and I^B). Their use will become apparent in ensuing sections of this chapter.

Two other points are important. First, although we have adopted a standard convention for assigning genetic symbols, there are many diverse systems of genetic nomenclature used to identify genes in various organisms. Usually, the symbol selected reflects the function of the gene or, occasionally, a disorder caused by a mutant gene. For example, the *cdk* gene from yeast discussed in Chapter 8 refers to those encoding cyclin-dependent kinases. In bacteria, leu^- refers to a mutation that interrupts the biosynthesis of the amino acid leucine, where the wild-type gene is designated leu^+. The symbol *dnaA* represents a bacterial gene involved in DNA replication (and DnaA, without italics, designates the protein made by that gene). In humans, italicized capital letters are used to name genes: *BRCA1* represents one of the genes associated with susceptibility to *br*east *ca*ncer. Although these different systems may sometimes be confusing, they all represent different ways to symbolize genes.

10.3 In Incomplete, or Partial, Dominance, Neither Allele Is Dominant

A cross between parents with contrasting traits may generate offspring with an intermediate phenotype. For example, if plants such as four o'clocks or snapdragons with red-flowers are crossed with white-flowered plants, the offspring have pink flowers. Because some red pigment is produced in the F_1 intermediate-colored pink flowers, neither red nor white flower color is dominant. Such a situation is known as **incomplete** or **partial dominance**.

If the phenotype is under the control of a single gene and two alleles, where neither is dominant, the results of the F_1 (pink) × F_1 (pink) cross can be predicted. The resulting F_2 generation shown in Figure 10–1 confirms the hypothesis that only one pair of alleles determines these phenotypes. The genotypic ratio (1:2:1) of the F_2 generation is identical to that of Mendel's monohybrid cross. However, because neither allele is dominant, the phenotypic ratio is identical to the genotypic ratio. Note that since neither of the alleles is recessive, we have chosen not to use upper- and lowercase letters as symbols. Instead, we denote the red and white alleles as R^1 and R^2. We could have chosen W^1 and W^2 or still other designations such as C^W and C^R, where *C* indicates color and the *W* and *R* superscripts indicate white and red, respectively.

Clear-cut cases of incomplete dominance, which result in intermediate expression of the overt phenotype, are relatively rare. However, even when complete dominance seems apparent, careful examination of the gene product, rather than the phenotype, often reveals an intermediate level of gene expression. An example is the human biochemical disorder **Tay–Sachs disease**, in which homozygous recessive individuals are severely affected with a fatal lipid-storage disorder and neonates die during their first 1–3 years of life. In afflicted individuals, there is almost no activity of the responsible enzyme **hexosaminidase**, which is normally involved in lipid metabolism. Heterozygotes, with only a single copy of the mutant gene, are phenotypically normal, but with only about 50 percent of the enzyme activity found in homozygous normal individuals. Fortunately, this level of enzyme activity is adequate to achieve normal biochemical function. This situation is not uncommon in enzyme disorders.

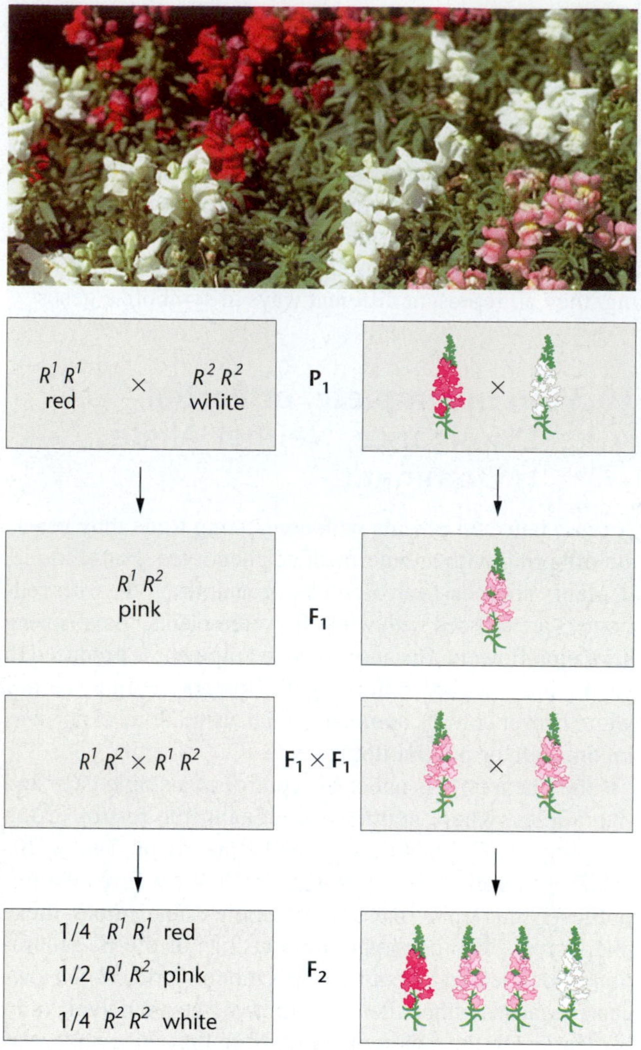

FIGURE 10–1 Incomplete dominance shown in the flower color of snapdragons.

MEDIA TUTORIAL Extensions of Mendelian Inheritance

10.4 In Codominance, the Influence of Both Alleles in a Heterozygote Is Clearly Evident

If two alleles of a single gene are responsible for producing two distinct and detectable gene products, a situation different from incomplete dominance or dominance/recessiveness arises. In such a case, the joint expression of both alleles in a heterozygote is called **codominance**. The **MN blood group** in humans illustrates this phenomenon. Karl Landsteiner and Philip Levin discovered a glycoprotein molecule found on the surface of red blood cells that acts as a native antigen, providing biochemical and immunological identity to individuals. In the human population, two forms of this glycoprotein exist, designated M and N. An individual may exhibit either one or both of them.

The MN system is under the control of an autosomal locus found on chromosome 4, with two alleles designated L^M and L^N. Because humans are diploid, three combinations are possible, each resulting in a distinct blood type:

Genotype	*Phenotype*
$L^M L^M$	M
$L^M L^N$	MN
$L^N L^N$	N

As predicted, a mating between two heterozygous MN parents may produce children of all three blood types:

$$L^M L^N \times L^M L^N$$
$$\downarrow$$
$$1/4\ L^M L^M$$
$$1/2\ L^M L^N$$
$$1/4\ L^N L^N$$

The preceding example shows that codominant inheritance is characterized by distinct expression of the gene products of both alleles. For codominance to be studied, both products must be phenotypically detectable. This characteristic distinguishes codominance from other modes of inheritance, such as incomplete dominance, where heterozygotes express an intermediate, or blended, phenotype.

10.5 Multiple Alleles of a Gene May Exist in a Population

Because the information stored in any gene is extensive, mutations can modify the gene in many ways. Each change has the potential for producing a different allele. Therefore, for any gene, the number of alleles within members of a population of individuals is not necessarily restricted to two. When three or more alleles of the same gene are found, **multiple alleles** are said to be present, creating a characteristic mode of inheritance. It is important to realize that *multiple alleles can be studied only in populations*. Any individual diploid organism has, at most, two homologous gene loci that may be occupied by different alleles of the same gene. However, among members of a species, many alternative forms of the same gene can exist.

The ABO Blood Groups

The simplest case of multiple alleles is that in which three alternative alleles of one gene exist. This situation is illustrated in the inheritance of the **ABO blood groups** in humans, discovered by Karl Landsteiner in the early 1900s. The ABO system, like the MN blood types, is characterized by the presence of antigens on the surface of red blood cells. However, the A and B antigens are distinct from the MN antigens and are under the control of a different gene, located on chromosome 9. As in the MN system, one combination of alleles in the ABO system exhibits a codominant mode of inheritance.

The ABO phenotype of any individual is ascertained by mixing a blood sample with antiserum containing type A or type B antibodies. If the antigen is present on the surface of the person's red blood cells, it will react with the corresponding antibody and cause clumping (or agglutination) of the red blood cells. When an individual is tested in this way, one of four phenotypes may be revealed. Each individual has either the A antigen (A phenotype), the B antigen (B phenotype), the A and B antigens (AB phenotype), or neither antigen (O phenotype).

In 1924, it was hypothesized that these phenotypes were inherited as the result of three alleles of a single gene. This hypothesis was based on studies of the blood types of many different families. Although different designations can be used, we will use the symbols I^A, I^B, and I^O to distinguish these three alleles. The *I* designation stands for **isoagglutinogen**, another term for antigen. If we assume that the I^A and I^B alleles are responsible for the production of their respective A and B antigens and that I^O is an allele that does not produce any detectable A or B antigens, we can list the various genotypic possibilities and assign the appropriate phenotype to each:

Genotype	*Antigen*	*Phenotype*
I^AI^A	A	A
I^AI^O	A	A
I^BI^B	B	B
I^BI^O	B	B
I^AI^B	A, B	AB
I^OI^O	Neither	O

Note that in these assignments, the I^A and I^B alleles behave dominantly to the I^O allele, but codominantly to each other.

We can test the hypothesis that three alleles control ABO blood groups by examining potential offspring from the various combinations of matings, as shown in Table 10–1. If we assume heterozygosity wherever possible, we can predict which phenotypes can occur. These theoretical predictions have been upheld in numerous studies examining the blood types of children of parents with all possible phenotypic combinations. The hypothesis that three alleles control ABO blood types in the human population is now universally accepted.

Our knowledge of human blood types has several practical applications. One of the most important is testing the compatibility of blood transfusions. Another application involves cases of disputed parentage, where newborns are inadvertently mixed up in the hospital, or when it is uncertain whether a specific male is the father of a child. An examination of the ABO blood groups as well as other inherited antigens of the possible parents and the child may help to resolve the situation. For example, of all the matings shown in Table 10–1, the only one that can result in offspring with all four phenotypes is that between two heterozygous individuals, one showing the A phenotype and the other showing the B phenotype. On genetic grounds alone, a male or female may be unequivocally ruled out as the parent of a certain child. This type of genetic evidence never proves parenthood, however.

The A and B Antigens

The biochemical basis of the ABO blood type system has now been carefully worked out. The A and B antigens are actually carbohydrate groups (sugars) that are bound to lipid molecules (fatty acids) protruding from the membrane of the red blood cell. The specificity of the A and B antigens is based on the terminal sugar of the carbohydrate group.

Almost all individuals possess what is called the **H substance**, to which one or two terminal sugars are added. As shown in Figure 10–2, the H substance itself consists of three sugar molecules, galactose (Gal), *N*-acetylglucosamine (AcGluNH), and fucose, chemically linked together. The I^A allele is responsible for an enzyme that can add the terminal sugar *N*-acetylgalactosamine, (AcGalNH) to the H substance. The I^B allele is responsible for a modified enzyme that cannot add *N*-acetylgalactosamine, but instead can add a

TABLE 10–1 Potential Phenotypes in the Offspring of Parents with All Possible ABO Blood Group Combinations, Assuming Heterozygosity Whenever Possible

Parents		*Potential Offspring*			
Phenotypes	*Genotypes*	*A*	*B*	*AB*	*O*
A × A	$I^AI^O \times I^AI^O$	3/4	—	—	1/4
B × B	$I^BI^O \times I^BI^O$	—	3/4	—	1/4
O × O	$I^OI^O \times I^OI^O$	—	—	—	all
A × B	$I^AI^O \times I^BI^O$	1/4	1/4	1/4	1/4
A × AB	$I^AI^O \times I^AI^B$	1/2	1/4	1/4	—
A × O	$I^AI^O \times I^OI^O$	1/2	—	—	1/2
B × AB	$I^BI^O \times I^AI^B$	1/4	1/2	1/4	—
B × O	$I^BI^O \times I^OI^O$	—	1/2	—	1/2
AB × O	$I^AI^B \times I^OI^O$	1/2	1/2	—	—
AB × AB	$I^AI^B \times I^AI^B$	1/4	1/2	1/2	—

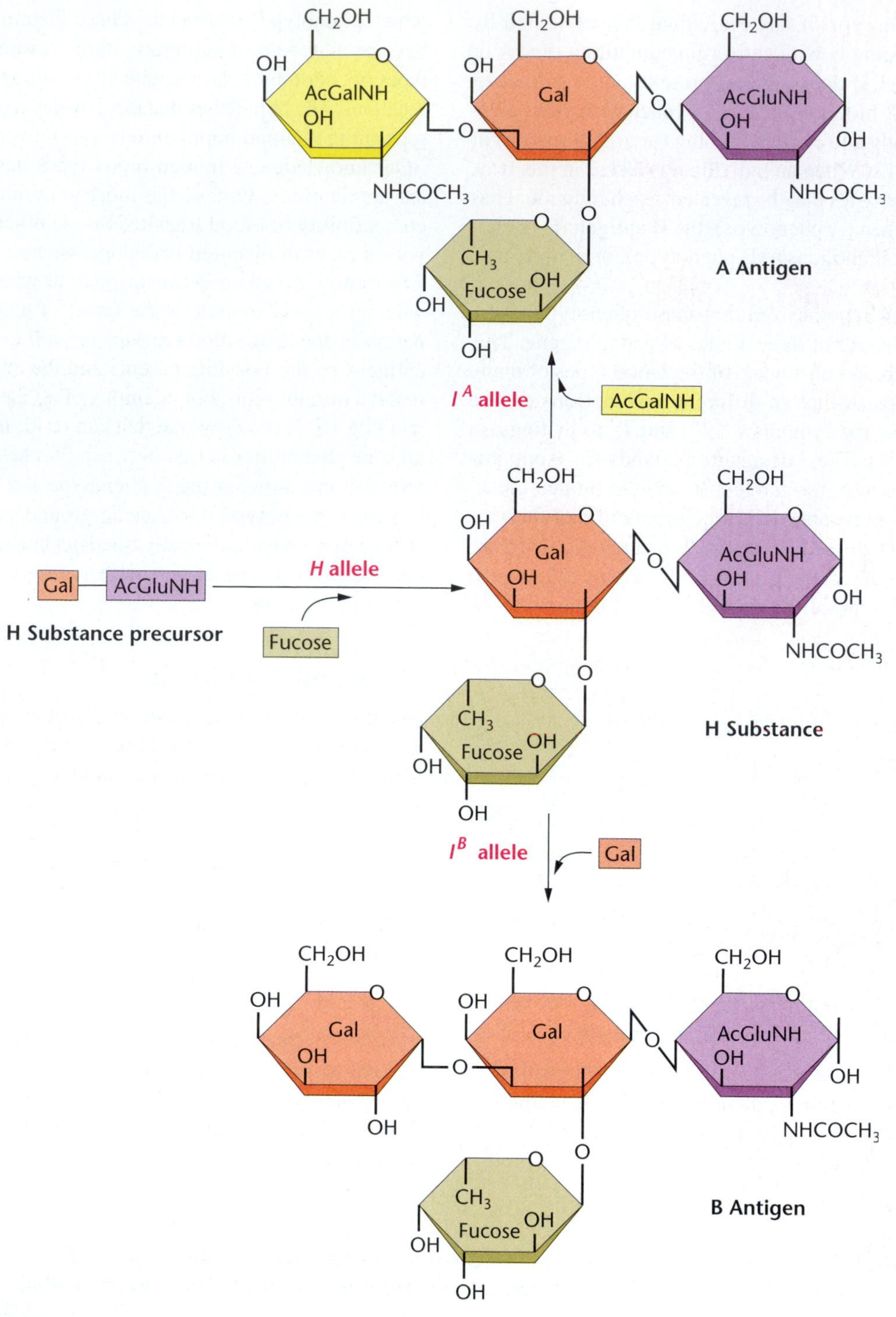

FIGURE 10–2 The biochemical basis of the ABO blood groups. The *H* allele, present in almost all humans, directs the conversion of a precursor molecule to the H substance by adding a molecule of fucose to it. The I^A and I^B alleles are then able to direct the addition of terminal sugar residues to the H substance. The I^O allele is unable to direct either of these terminal additions. Gal: galactose; AcGluNH: *N*-acetyl glucosamine; AcGalNH: *N*-acetylgalactosamine. Failure to produce the H substance results in the Bombay phenotype where individuals are type O, regardless of the presence of an I^A or I^B allele.

terminal galactose. Heterozygotes (I^AI^B) add either one or the other sugar at the many sites (substrates) available on the surface of the red blood cell, illustrating the biochemical basis of codominance in individuals of the AB blood type. Finally, persons of type O (I^OI^O) cannot add either terminal sugar, possessing only the H substance protruding from the surface of their red blood cells.

As we established in Chapter 7, the molecular genetic basis of the mutations leading to the I^A, I^B, and I^O alleles has been elucidated.

The Bombay Phenotype

In 1952, a very unusual situation provided information concerning the genetic basis of the H substance. A woman in

Bombay displayed a unique genetic history that was inconsistent with her blood type. In need of a transfusion, she was found to lack both the A and B antigens and was thus typed as O. However, as shown in the partial pedigree in Figure 10–3, one of her parents was type AB, and she was the obvious donor of an I^B allele to two of her offspring. Thus, she was genetically type B, but functionally type O!

The Bombay woman was subsequently shown to be homozygous for a rare recessive mutation, *h*, which prevented her from synthesizing the complete H substance. In this mutation, the terminal portion of the carbohydrate chain protruding from the red cell membrane was shown to lack fucose. In the absence of fucose, the enzymes specified by the I^A and I^B alleles apparently are unable to recognize the incomplete H substance as a proper substrate. Thus, neither the terminal galactose nor *N*-acetylgalactosamine can be added, even though the appropriate enzymes capable of doing so are present and functional. As a result, the ABO system genotype cannot be expressed in individuals of genotype *hh*, and they are functionally type O. To distinguish them from the rest of the population, they are said to demonstrate the Bombay phenotype. The frequency of the *h* allele is exceedingly low. Hence, the vast majority of the human population is of the *HH* or *Hh* genotype (almost all *HH*) and can synthesize the H substance.

The Secretor Locus

Expression of the ABO blood type system is affected by still a third gene, found at what is called the **secretor locus**. In about 80 percent of the human population, the A and B antigens are present in various body secretions as well as on the membrane of red blood cells. The ability to secrete these antigens in body fluids such as saliva, gastric juice, semen, and vaginal fluids is under the influence of the dominant allele, *Se*. The genotypes *Se*/*Se* and *Se*/*se* lead to "secretion." Nonsecretors (*se*/*se*) make the antigens as specified by the ABO loci, but do not secrete them. Secretion of these antigens is important in forensic science (the application of scientific knowledge to legal proceedings), when ABO typing is performed on tissue samples other than blood.

The Rh Antigens

Another set of antigens thought by some geneticists to illustrate multiple allelism includes those designated Rh. Discovered by Landsteiner, Levine, and others about 1940, the **Rh antigens** have received a great deal of attention because of their direct involvement in the disorder **erythroblastosis fetalis**. Also referred to as **hemolytic disease of the newborn (HDN)**, erythroblastosis fetalis is a form of anemia. It occurs in an Rh-positive fetus whose mother is Rh-negative and whose father is Rh-positive, contributing that allele to the fetus. Such a genetic combination results in a potential immunological incompatibility between the mother and fetus. If fetal blood passes through the ruptured placenta at birth and enters the maternal circulation, the mother's immune system recognizes the Rh antigen as foreign and builds antibodies against it. During a second pregnancy, the antibody concentration becomes high enough that when maternal antibodies pass across the placenta, they enter the fetus's circulation and begin to destroy the fetus's red blood cells. This causes the hemolytic anemia.

About 10 percent of all human pregnancies demonstrate Rh incompatibility. However, for numerous reasons, less than 0.5 percent actually result in anemia. Currently, modern medical practice provides incompatible mothers with anti-Rh antibodies immediately after giving birth to all Rh-positive babies. This destroys any Rh-positive cells that have entered the mother's circulation so that she does not produce her own anti-Rh antibodies. Before this treatment was developed, many fetuses failed to survive to term. For those that did, complete blood transfusions were often necessary, but even then many newborns failed to survive.

Initial genetic investigations led to the belief that in the human population only two alleles controlled the presence or absence of the antigen. It was thought that the Rh^+ allele determined the presence of the antigen and behaved as a dominant gene. The Rh^- allele resulted in the absence of the antigen.

With the development of more refined tests determining the presence of the antigen, it became apparent that the genetic control of Rh antigens was much more complex than originally thought. For example, two closely related antigens are sometimes detected in individuals who originally test negatively for the standard Rh antigen. Alexander Wiener subsequently proposed that a series of multiple alleles at a single Rh locus accounted for these variants. Other investigators, including Ronald A. Fisher and Robert R. Race, proposed an alternative type of inheritance involving three closely linked genes, each with two alleles, involved in the inheritance of Rh factors. The term linkage is used to describe genes located on the same chromosome. We will discuss this concept in detail in Chapter 12.

Table 10–2 presents the two systems and the nomenclature each uses to designate the alleles. In the Wiener system, the presence of at least one of the four dominant alleles is sufficient to yield the Rh-positive blood type. In the Fisher-Race system, the genetic locus bearing the set of *D* and *d* alleles is most critical. The presence of at least one dominant *D* allele results in the Rh-positive phenotype, whereas the *dd* genotype ensures the Rh-negative blood type. While the *C*, *c*, *E*, and *e* alleles specify related antigens, they are less immunologically significant. Because of the com-plexity of

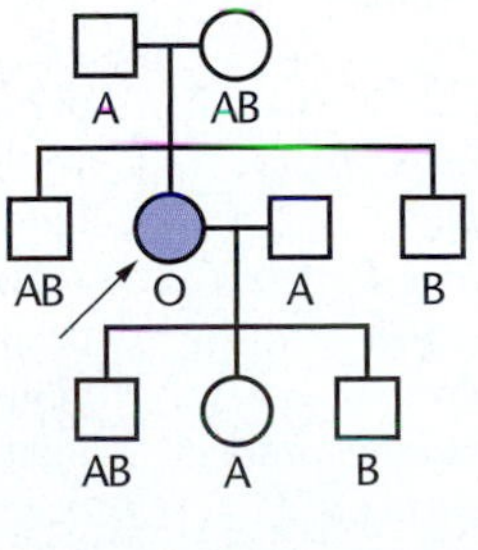

FIGURE 10–3 A partial pedigree of a woman displaying the Bombay phenotype. Functionally, her ABO blood group behaves as type O. Genetically, she is type B.

TABLE 10–2 A Comparison of the Alleles in the Wiener System With Those of the Fisher–Race System Used to Explain the Genetic Basis of the Rh Blood Groups

Wiener Nomenclature	*Fisher–Race Nomenclature*	*Phenotype*
R^1	CDE	
R^2	CDe	
R^0	cDE	Rh^+
R^z	cDe	
r	CdE	
r'	Cde	
r''	cdE	Rh^-
r^y	cde	

the antigenic patterns, even modern molecular analysis has yet to provide strong favor of one system over the other.

The *white* Locus in *Drosophila*

Many other phenotypes in both plants and animals are influenced by multiple allelic inheritance at a single locus. In *Drosophila*, for example, where the induction of mutations has been used extensively as a method of investigation, many alleles are present at practically every locus. The recessive eye mutation, *white*, discovered by Thomas H. Morgan and Calvin Bridges in 1912, represents only one of over 100 alleles that can occupy this locus. In this allelic series, eye colors range from complete absence of pigment in the *white* allele, to deep ruby in the *white-satsuma* allele, to orange in the *white-apricot* allele, to a buff color in the *white-buff* allele. These alleles are designated w, w^{sat}, w^a, and w^{bf}, respectively (Table 10–3). In each case, the total amount of pigment in these mutant eyes is reduced to less than 20 percent of that found in the brick-red wild-type eye.

10.6 Lethal Alleles May Be Recessive or Dominant

Many gene products are essential to an organism's survival. Mutations resulting in a gene product that is nonfunctional can sometimes be tolerated in the heterozygous state; that is, one wild-type allele may produce enough of the essential product for the organism to survive. However, such a mutation behaves as a **recessive lethal allele**, and homozygous recessive individuals will not survive. The time of death will depend on when the product is essential. In mammals, for example, this might occur during development, early childhood, or even during adulthood.

In some cases, the allele responsible for a lethal effect when it is homozygous may also result in a distinctive mutant phenotype when it is present heterozygously. Such an allele is behaving as a recessive lethal, but is dominant with respect to the phenotype. For example, a mutation that causes a yellow coat in mice was discovered in the early part of the 20th century. The yellow coat varies from the normal agouti (wild-type) coat phenotype, as shown in Figure 10–4. Crosses between the various combinations of the two strains yield unusual results:

	Crosses		
A	agouti × agouti	⟶	all agouti
B	yellow × yellow	⟶	2/3 yellow: 1/3 agouti
C	agouti × yellow	⟶	1/2 yellow: 1/2 agouti

These results are explained on the basis of a single pair of alleles. With regard to coat color, the mutant yellow allele A^Y is dominant to the wild-type agouti allele (A), so heterozygous mice will have yellow coats. However, the yellow allele also behaves as a homozygous recessive lethal. Mice of the genotype A^YA^Y die before birth, so no homozygous yellow mice are ever recovered. The genetic basis for these three crosses is provided in Figure 10–4.

Many genes are known to behave similarly in other organisms. In *Drosophila*, *Curly* wing (*Cy*), *Plum* eye (*Pm*), *Dichaete* wing (*D*), *Stubble* bristle (*Sb*), and *Lyra* wing (*Ly*) behave as homozygous lethals, but are dominant with respect to the expression of the mutant phenotype when heterozygous.

TABLE 10–3 Some of the Alleles Present at the *white* Locus of *Drosophila*

Allele	*Name*	*Eye Color*
w	*white*	pure white
w^a	*white-apricot*	yellowish orange
w^{bf}	*white-buff*	light buff
w^{bl}	*white-blood*	yellowish ruby
w^{cf}	*white-coffee*	deep ruby
w^e	*white-eosin*	yellowish pink
w^{mo}	*white-mottled orange*	light mottled orange
w^{sat}	*white-satsuma*	deep ruby
w^{sp}	*white spotted*	fine grain, yellow mottling
w^t	*white-tinged*	light pink

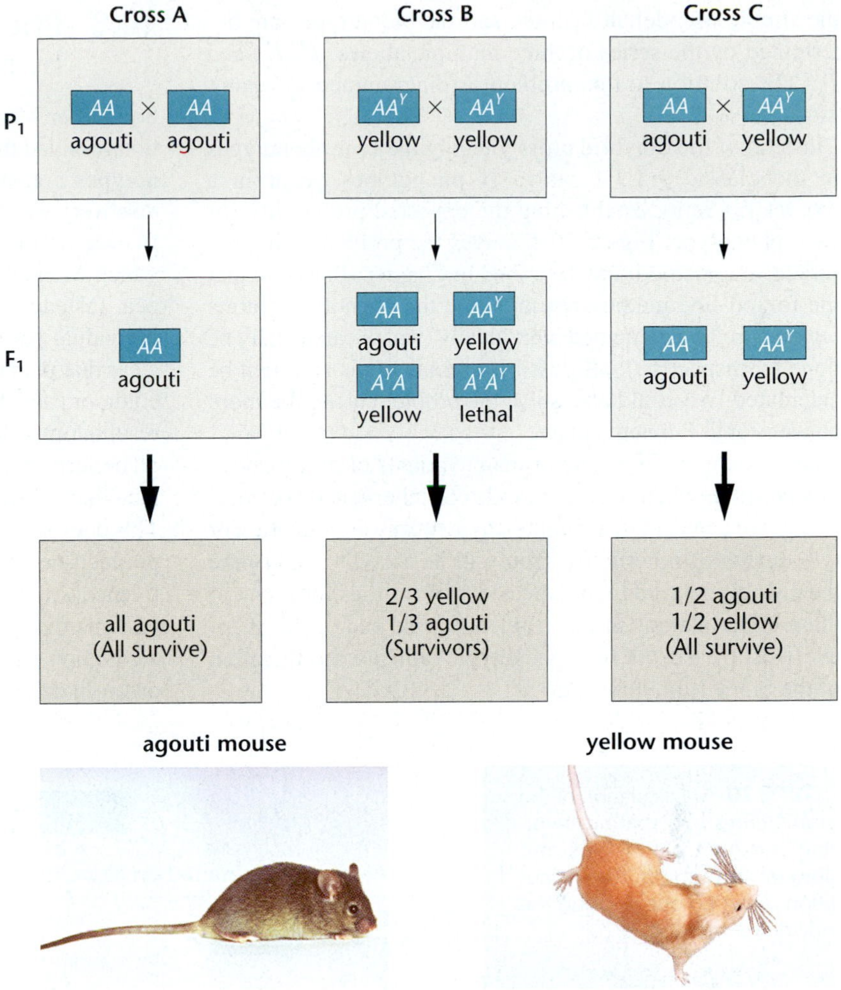

FIGURE 10–4 Inheritance patterns in three crosses involving the normal wild-type agouti allele (A) and the mutant yellow allele (A^Y) in the mouse. Note that the mutant allele behaves dominantly to the normal allele in controlling coat color, but it also behaves as a homozygous lethal allele. The genotype A^YA^Y does not survive.

In instances where one copy of the wild-type gene is not sufficient for normal development, even the heterozygote will not survive. In this case, the mutation is behaving as a **dominant lethal allele**, because its presence somehow overrides the expression of the wild-type product, or the amount of wild-type product is simply insufficient to support its essential function. For example, in humans, a disorder called **Huntington disease** (previously referred to as Huntington's chorea) is due to a dominant allele, H. The onset of the disease in heterozygotes (Hh) is delayed, usually well into adulthood, typically at about age 40. Affected individuals then undergo gradual nervous and motor degeneration and eventually succumb to the disorder a number of years later. This lethal disorder is particularly tragic, because an affected individual may have produced a family before discovering the condition. Each child has a 50 percent probability of inheriting the lethal allele and transmitting it to his or her offspring. The American folk singer and composer Woody Guthrie, father of Arlo Guthrie, died from this disease at age 39.

Dominant lethal alleles are very rare. For them to exist in a population, the affected individual must reproduce before the affected individual dies, as can occur in Huntington disease. If all affected individuals die before reaching the reproductive age, the mutant allele will not be passed to future generations and will disappear from the population unless it arises again as a result of a new mutation.

10.7 Combinations of Two Gene Pairs Involving Two Modes of Inheritance Modify the 9:3:3:1 Ratio

Each example discussed so far modifies Mendel's 3:1 F_2 monohybrid ratio. Therefore, combining any two of these modes of inheritance in a dihybrid cross will also modify the classic 9:3:3:1 ratio. Having established the foundation of the modes of inheritance of incomplete dominance, codominance, multiple alleles, and lethal alleles, we can now deal with the situation of two modes of inheritance occurring simultaneously. Mendel's principle of independent assortment applies to these situations, provided that the genes controlling each character are not linked on the same chromosome.

Consider, for example, a mating between two humans who are both heterozygous for the autosomal recessive allele that causes albinism and who are both of blood type AB. What is the probability of any particular phenotypic combination occurring in each of their children? Albinism is inherited in the simple Mendelian fashion, and the blood types are determined by the series of three multiple alleles, I^A, I^B, and I^O. The solution to this problem is diagrammed in Figure 10–5.

Instead of the dihybrid cross yielding the four phenotypes in the classic 9:3:3:1 ratio, six phenotypes occur in a 3:6:3:1:2:1 ratio, establishing the expected probability for each phenotype. Figure 10–5 solves the problem using the forked-line method first described in Chapter 10. Recall that the forked-line method requires that the phenotypic ratios for each trait be computed individually (which can usually be done by inspection); all possible combinations can then be calculated. We could also solve the problem using the more conventional Punnett square.

This example is just one of many variants of modified ratios possible when different modes of inheritance are combined. We can deal in a similar way with any combination of two modes of inheritance. You will be asked to determine the phenotypes and their expected probabilities for many of these combinations in the problems at the end of the chapter. In each case, the final phenotypic ratio is a modification of the 9:3:3:1 dihybrid ratio.

10.8 Phenotypes Are Often Affected by More Than One Gene

Soon after Mendel's work was rediscovered, experimentation revealed that many traits characterized by discrete phenotypes are often affected by more than one gene. This discovery was significant because it revealed that genetic influence on the phenotype is much more complex than that which Mendel encountered in his crosses with the garden pea. Instead of single genes affecting the development of individual parts of a plant or animal body, it soon became clear that phenotypic characteristics, such as eye color, hair color, or fruit shape, are influenced by many genes and their resultant products.

The term **gene interaction** is often used to describe the idea that several genes influence a particular characteristic. This does not mean, however, that two or more genes or their products necessarily interact directly with one another to influence a particular phenotype. Rather, the cellular function of numerous gene products contributes to the development of a common phenotype. For example, the development of an organ like the eye of an insect is exceedingly complex and

FIGURE 10–5 Calculation of the probabilities in a mating involving the ABO blood type and albinism in humans. The calculation is carried out by using the forked-line method.

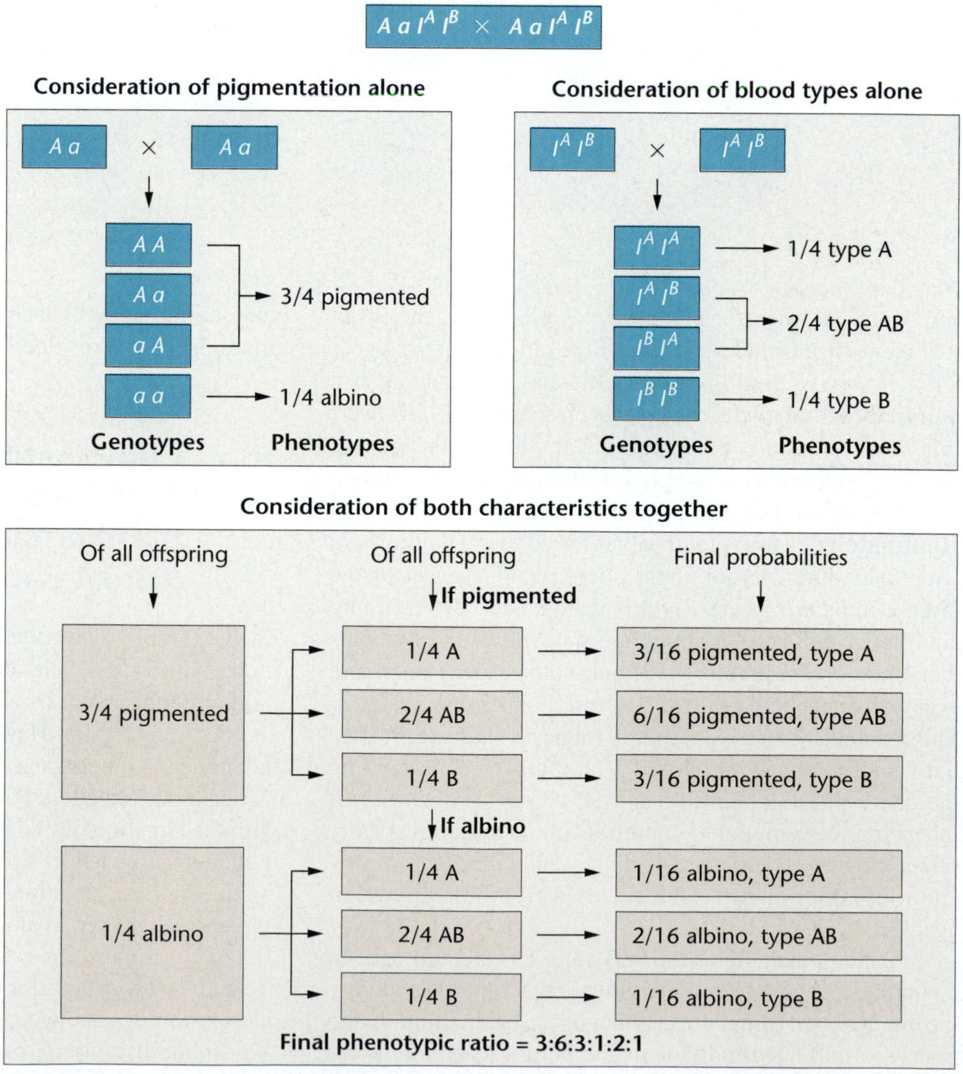

leads to a structure with multiple phenotypic manifestations—most simply described as a specific size, shape, texture, and color. The development of the eye can best be envisioned as a complex cascade of developmental events. This process illustrates the developmental concept of **epigenesis** (first introduced in Chapter 1), whereby each ensuing step of development increases the complexity of the sensory organ and is under the control and influence of many genes. To clarify, we will present several examples that illustrate gene interaction at the biochemical level.

Epistasis

Some of the best examples of gene interaction are those that show the phenomenon of epistasis. Derived from the Greek word for "stoppage," epistasis occurs when the effect of one gene or gene pair masks or modifies the effect of another gene or gene pair. The genes involved influence the same general phenotypic characteristic, sometimes in an antagonistic manner, as when masking occurs. In other cases, however, the genes involved exert their influence on one another in a complementary, or cooperative, fashion.

For example, the homozygous presence of a recessive allele may prevent or override the expression of other alleles at a second locus (or several other loci). In this case, the alleles at the initial locus are said to be **epistatic** to those at the second locus. The alleles at the second locus, which are masked, are described as being **hypostatic** to those at the first locus. In another example, a single dominant allele at the first locus may influence the expression of the alleles at a second gene locus. In a third case, two gene pairs may complement one another such that at least one dominant allele at each locus is required to express a particular phenotype.

The Bombay phenotype discussed earlier is an example of the homozygous recessive condition at one locus masking the expression of a second locus. There, we established that the homozygous condition (*hh*) masks the expression of the I^A and I^B alleles. Only individuals containing at least one *H* allele (*HH* or *Hh*) can form the A or B antigens. As a result, individuals whose genotypes include the I^A or I^B allele and who are also *hh* express the type O phenotype, regardless of their potential to make either antigen. An example of the outcome of matings between individuals heterozygous at both loci is illustrated in Figure 10–6. If many individuals of the genotype I^AI^BHh have children, the phenotypic ratio of 3 A: 6 AB: 3 B: 4 O is expected in their offspring.

It is important to note two things when examining this cross and the predicted phenotypic ratio:

1. A key distinction exists in this cross compared to the modified dihybrid cross shown in Figure 10–5: *Only one characteristic—blood type—is being followed.* In the modified dihybrid cross in Figure 10–5, blood type *and* skin pigmentation are followed as separate phenotypic characteristics.

2. Even though only a single character was followed, the phenotypic ratio was expressed in sixteenths. If we knew nothing about the H substance and the genes controlling it, we could still be confident that a second gene pair, other than that controlling the A and B antigens, was involved in the phenotypic expression. *When studying a single character, a ratio that is expressed in 16 parts (e.g., 3:6:3:4) suggests that two gene pairs are "interacting" during the expression of the phenotype under consideration.*

The study of gene interaction has revealed a number of inheritance patterns that modify the Mendelian dihybrid F_2 ratio (9:3:3:1) in other ways. In several of the subsequent examples, epistasis has the effect of combining one or more of the four phenotypic categories in various ways. The generation of these four groups is reviewed in Figure 10–7, along with several modified ratios.

As we discuss these and other examples (Figure 10–8), we will make several assumptions and adopt certain conventions:

1. In each case, distinct phenotypic classes are produced, each clearly discernible from all others. Such traits illustrate discontinuous variation, where phenotypic categories are discrete and qualitatively different from one another.

2. The genes considered in each cross are not linked and therefore assort independently of one another during gamete formation. So that you may easily compare the results of different crosses, we designate alleles as *A*, *a* and *B*, *b* in each case.

3. When we assume that complete dominance exists between the alleles of any gene pair, such that *AA* and *Aa* or *BB* and *Bb* are equivalent in their genetic effects, we use the designations *A–* or *B–* for both combinations, where the dash (–) indicates that either allele may be present, without consequence to the phenotype.

4. All P_1 crosses involve homozygous individuals (e.g., $AABB \times aabb$, $AAbb \times aaBB$, or $aaBB \times AAbb$). Therefore, each F_1 generation consists of only heterozygotes of genotype *AaBb*.

5. In each example, the F_2 generation produced from these heterozygous parents is our main focus of analysis. When two genes are involved (Figure 10–7), the F_2 genotypes fall into four categories: 9/16 $A{-}B{-}$, 3/16 *A–bb*, 3/16 $aaB{-}$, and 1/16 *aabb*. Because of dominance, all genotypes in each category are equivalent in their effect on the phenotype.

The first case to be discussed is the inheritance of coat color in mice (Case 1 of Figure 10–8). As we saw in our discussion of lethal alleles, normal wild-type coat color is agouti, a grayish pattern formed by alternating bands of pigment on each hair. (See Figure 10–4.) Agouti is dominant to black (nonagouti) hair, which results from the homozygous expression of a recessive mutation, *a*. Thus, *A–* results in agouti, whereas *aa* yields black coat color. When homozygous, a recessive mutation, *b*, at a separate locus, eliminates pigmentation altogether, yielding albino mice (*bb*), regardless of the genotype at the *a* locus. The presence of at least one *B* allele allows pigmentation to occur in much the

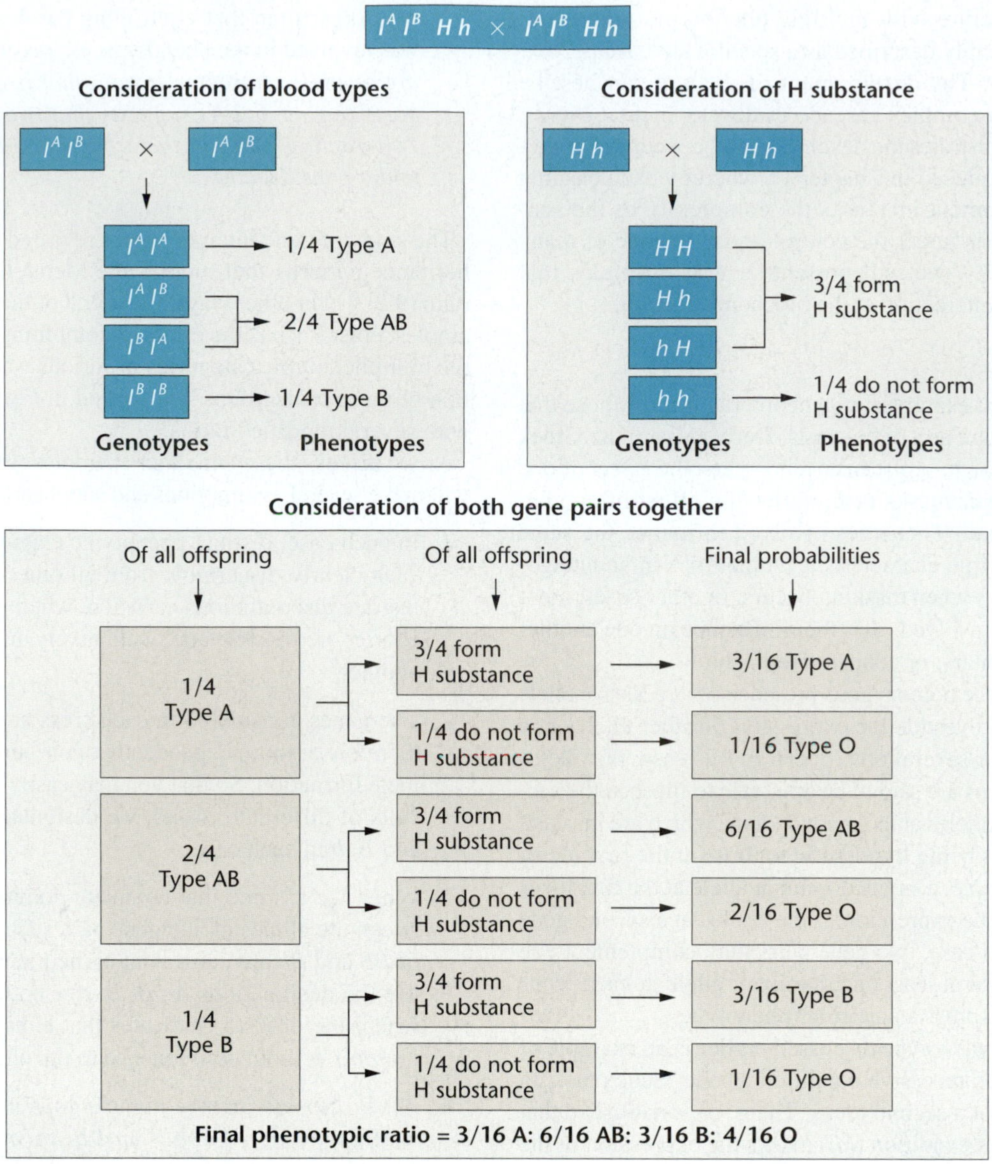

FIGURE 10–6 The outcome of a mating between individuals heterozygous at two genes that determine each individuals ABO blood type. Final phenotypes are calculated by considering both genes separately and then combining the results using the forked-line method.

same way that the *H* allele in humans allows the expression of the ABO blood types. In a cross between agouti (*AABB*) and albino (*aabb*), members of the F_1 are all *AaBb* and have agouti coat color. In the F_2 progeny of a cross between two F_1 double heterozygotes, the following genotypes and phenotypes are observed:

$$F_1\text{: } AaBb \times AaBb$$

$$\downarrow$$

F_2Ratio	*Genotype*	*Phenotype*	*Final Phenotypic Ratio*
9/16	*A– B–*	agouti	9/16 agouti
3/16	*A– bb–*	albino	4/16 albino
3/16	*aa B–*	black	3/16 black
1/16	aa bb	albino	

We can envision gene interaction yielding the observed 9:3:4 F_2 ratio as a two-step process:

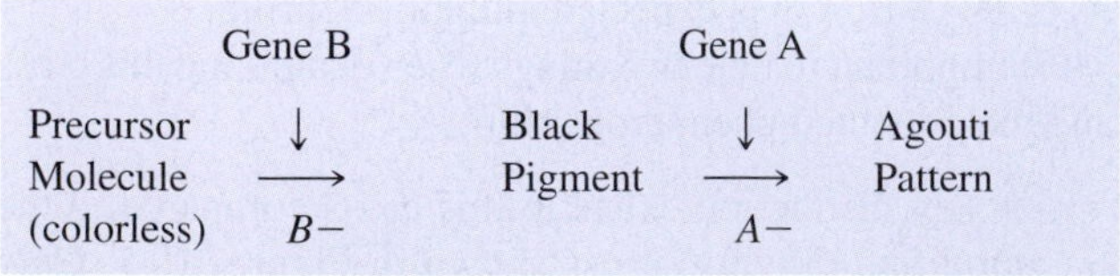

In the presence of a *B* allele, black pigment can be made from a colorless substance. In the presence of an *A* allele, the black pigment is deposited during the development of hair in a pattern that produces the agouti phenotype. If the *aa* genotype occurs, all of the hair remains black. If the *bb* genotype occurs, no black pigment is produced, regardless of the presence of the *A* or *a* alleles, and the mouse is albino.

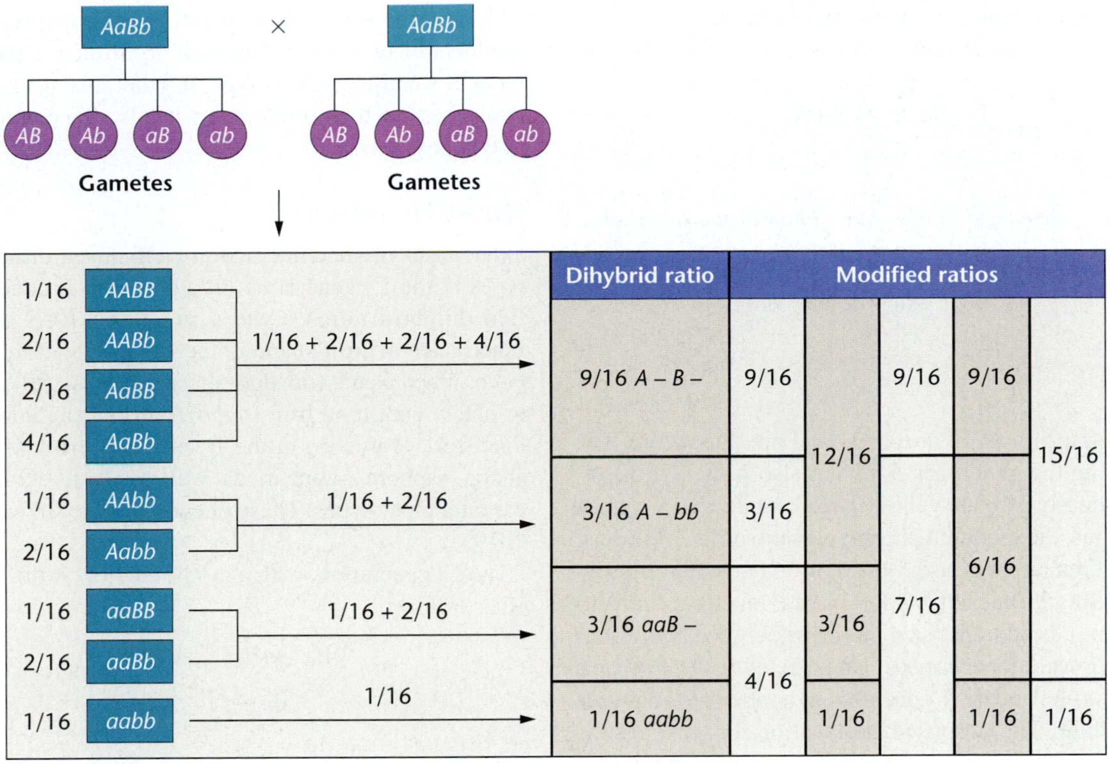

FIGURE 10–7 Generation of the various modified dihybrid ratios from the nine unique genotypes produced in a cross between individuals heterozygous at two genes.

Therefore, the *bb* genotype masks or suppresses the expression of the *A* allele, demonstrating epistasis.

A second type of epistasis occurs when a dominant allele at one genetic locus masks the expression of the alleles of a second locus. For instance, Case 2 of Figure 10–8 deals with the inheritance of fruit color in summer squash. Here, the dominant allele *A* results in white fruit color regardless of the genotype at a second locus, *B*. In the absence of a dominant *A* allele (the *aa* genotype), *BB* or *Bb* results in yellow color, while *bb* results in green color. Therefore, if two

Case	Organism	Character	F$_2$ Phenotypes 9/16	3/16	3/16	1/16	Modified ratio
1	Mouse	Coat color	agouti	albino	black	albino	9:3:4
2	Squash	Color	white		yellow	green	12:3:1
3	Pea	Flower color	purple	white			9:7
4	Squash	Fruit shape	disc	sphere		long	9:6:1
5	Chicken	Color	white		colored	white	13:3
6	Mouse	Color	white-spotted	white	colored	white-spotted	10:3:3
7	Shepherd's purse	Seed capsule	triangular			ovoid	15:1
8	Flour beetle	Color	red | sooty | red | sooty	black	jet	black	6:3:3:4

FIGURE 10–8 The basis of modified dihybrid F$_2$ phenotypic ratios, resulting from crosses between doubly heterozygous F$_1$ individuals. The four groupings of the F$_2$ genotypes shown in Figure 10–7 and across the top of this figure are combined in various ways to produce these ratios.

white-colored double heterozygotes (*AaBb*) are crossed, an interesting phenotypic ratio occurs because of this type of epistasis:

$$F_1\text{: } AaBb \times AaBb$$

$$\downarrow$$

F_2 Ratio	Genotype	Phenotype	Final Phenotypic Ratio
9/16	*A− B−*	white	12/16 white
3/16	*A− bb−*	white	
3/16	*aa B−*	yellow	3/16 yellow
1/16	aa bb	green	1/16 green

Of the offspring, 9/16 are *A–B–* and are thus white. The 3/16 bearing the genotypes *A–bb* are also white. Of the remaining squash, 3/16 are yellow (*aaB–*), while 1/16 are green (*aabb*). Thus, the modified phenotypic ratio of 12:3:1 occurs.

Our third example (Case 3 of Figure 10–8), first discovered by William Bateson and Reginald Punnett (of Punnett-square fame), is demonstrated in a cross between two strains of white-flowered sweet peas. Unexpectedly, the F_1 plants were all purple, and the F_2 occurred in a ratio of 9/16 purple to 7/16 white. The proposed explanation for these results suggests that the presence of at least one dominant allele of each of two gene pairs is essential in order for flowers to be purple. All other genotype combinations yield white flowers because the homozygous condition of either recessive allele masks the expression of the dominant allele at the other locus.

The cross is shown as follows:

$$P_1\text{: } AAbb \times aaBB$$

white white

$$\downarrow$$

F_1: All *AaBb* (purple)

$$\downarrow$$

F_2 Ratio	Genotype	Phenotype	Final Phenotypic Ratio
9/16	*A− B−*	purple	9/16 purple
3/16	*A− bb*	white	
3/16	*aa B−*	white	7/16 white
1/16	*aa bb*	white	

We can now envision how two gene pairs might yield such results:

	Gene A		*Gene B*	
Precursor Substance (colorless)	↓ ⟶ *A−*	Intermediate Product (colorless)	↓ ⟶ *B−*	Final Product (purple)

At least one dominant allele from each pair of genes is necessary to ensure both biochemical conversions to the final product, yielding purple flowers. In the preceding cross, this will occur in 9/16 of the F_2 offspring. All other plants (7/16) have flowers that remain white.

These three examples illustrate in a simple way how the products of two genes "interact" to influence the development of a common phenotype. In other instances, more than two genes and their products are involved in controlling phenotypic expression.

Novel Phenotypes

Other cases of gene interaction yield novel, or new, phenotypes in the F_2 generation, in addition to producing modified dihybrid ratios. Case 4 in Figure 10–8 depicts the inheritance of fruit shape in the summer squash *Cucurbita pepo*. When plants with disc-shaped fruit (*AABB*) are crossed to plants with long fruit (*aabb*), the F_1 generation all have disc fruit. However, in the F_2 progeny, fruit with a novel shape—sphere—appear, as well as fruit exhibiting the parental phenotypes. These phenotypes are shown in Figure 10–9.

The F_2 generation, with a modified 9:6:1 ratio, is generated as follows:

$$F_1\text{: } AaBb \times AaBb$$

disc ↓ disc

F_2 Ratio	Genotype	Phenotype	Final Phenotypic Ratio
9/16	*A− B−*	disc	9/16 disc
3/16	*A− bb*	sphere	
3/16	*aa B−*	sphere	6/16 sphere
1/16	aa bb	long	1/16 long

In this example of gene interaction, both gene pairs influence fruit shape equivalently. A dominant allele at either locus ensures a sphere-shaped fruit. In the absence of dominant alleles, the fruit is long. However, if both dominant alleles (*A* and *B*) are present, the fruit is flattened into a disc shape.

Another interesting example of an unexpected phenotype arising in the F_2 generation is the inheritance of eye color in *Drosophila melanogaster*. The wild-type eye color is brick

FIGURE 10–9 Summer squash exhibiting various fruit-shape phenotypes, where disc (white), long (orange gooseneck), and sphere (bottom left) are apparent.

red. When two autosomal recessive mutants, *brown* and *scarlet*, are crossed, the F_1 generation consists of flies with wild-type eye color. In the F_2 generation, wild, scarlet, brown, and white-eyed flies are found in a 9:3:3:1 ratio. While this ratio is numerically the same as Mendel's dihybrid ratio, the *Drosophila* cross involves only one character: eye color. This is an important distinction to make as modified dihybrid ratios resulting from "gene interaction" are studied.

The *Drosophila* cross is an excellent example of gene interaction because the biochemical basis of eye color in this organism has been determined (Figure 10–10). *Drosophila*, as a typical arthropod, has compound eyes made up of hundreds of individual visual units called ommatidia. The wild-type eye color is due to the deposition and mixing of two separate pigments in each ommatidium—the bright-red pigment **drosopterin** and the brown pigment **xanthommatin**. Each pigment is produced by a separate biosynthetic pathway. Each step of each pathway is catalyzed by a separate enzyme and is thus under the control of a separate gene. As shown in Figure 10–10, the *brown* mutation, when homozygous, interrupts the pathway leading to the synthesis of the bright-red pigment. Because only xanthommatin pigments are present, the eye is brown. The *scarlet* mutation, affecting a gene located on a separate autosome, interrupts the pathway leading to the synthesis of the brown xanthommatins and renders the eye color bright red in homozygous mutant flies. Each mutation apparently causes the production of a nonfunctional enzyme. Flies that are double mutants and thus homozygous for both *brown* and *scarlet* lack both functional enzymes and can make neither of the pigments; they represent the novel white-eyed flies appearing in 1/16 of the F_2 generation.

Other Modified Dihybrid Ratios

The remaining cases (5–8) in Figure 10–8 illustrate additional modifications of the dihybrid ratio and provide still other examples of gene interactions. As you will note, ratios of 13:3, 10:3:3; 15:1, and 6:3:3:4 are illustrated. These cases, like the four preceding them, have two things in common. First, in arriving at a suitable explanation of the inheritance pattern of each one, we have not violated the principles of segregation and independent assortment. Therefore, the added complexity of inheritance in these examples does not detract from the validity of Mendel's conclusions. Second, the F_2 phenotypic ratio in each example has been expressed in sixteenths. When similar observations are made in crosses where the inheritance pattern is unknown, they suggest to geneticists that two gene pairs are controlling the observed phenotypes. You should make the same inference in your analysis of genetics problems. Other insights into solving genetics problems are provided in the "Insights and Solutions" section at the conclusion of this chapter.

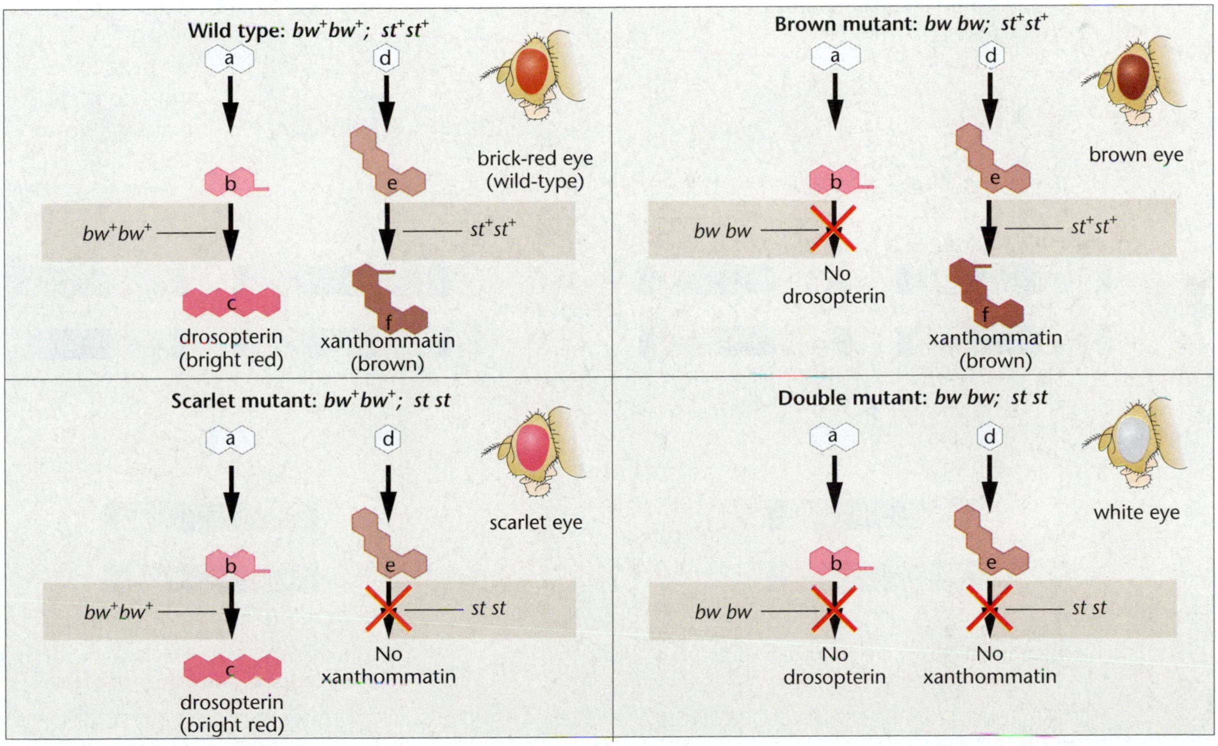

FIGURE 10–10 A theoretical explanation of the biochemical basis of the four eye-color phenotypes produced in a cross between *Drosophila* with brown eyes and scarlet eyes. In the presence of at least one wild-type bw^+ allele, an enzyme is produced that converts substance b to c, and the pigment drosopterin is synthesized. In the presence of at least one wild-type st^+ allele, substance e is converted to f, and the pigment xanthommatin is synthesized. The homozygous presence of the recessive *bw* and *st* mutant alleles blocks the synthesis of these respective pigment molecules. Either one, both, or neither of these pathways can be blocked, depending on the genotype.

10.9 Complementation Analysis Can Determine Whether or Not Two Similar Mutations Are Alleles

An interesting situation arises when two mutations, both of which produce a similar phenotype, are isolated independently. Suppose that two investigators, one in a genetics laboratory in the United States and the other in a genetics laboratory in Canada, independently isolated and established a true-breeding strain of wingless *Drosophila* and demonstrated that each was due to a recessive mutation. We might assume that both strains contained mutations in the same gene. However, since we know that many genes are involved in the formation of wings, mutations in any one of them might inhibit wing formation during development. The experimental approach called **complementation analysis** allows us to determine whether two such mutations are in the same gene—that is, whether they are alleles or whether they represent mutations in separate genes.

Our analysis seeks to answer this simple question: *Are two mutations that yield similar phenotypes present in the same gene or in two different genes?* To find the answer, we cross the two mutant strains together and analyze the F_1 generation. There are two alternative outcomes and interpretations of such a cross, as illustrated in Figure 10–11 and discussed next. The mutation isolated in the United States is designated m^{usa}, while the mutation isolated in Canada is designated m^{can}.

Case 1. *All offspring develop normal wings.*

Interpretation: The two recessive mutations are in separate genes and are not alleles of one another. Following the cross, all F_1 flies are heterozygous for both genes. **Complementation** is said to occur. Because each mutation is in a separate gene and each F_1 fly is heterozygous at both loci, the normal products of both genes are produced (by the one normal copy of each gene). As a result, wings develop.

Case 2. *All offspring fail to develop wings.*

Interpretation: The two mutations affect the same gene and are alleles of one another. *Complementation does not occur*. Because the two mutations affect the same gene, the F_1 flies are homozygous for the two mutant alleles (one is the m^{usa} allele and the m^{can} allele). No normal product of the gene is produced. In the absence of this essential product, wings do not form.

Complementation analysis, as originally devised by the *Drosophila* geneticist Edward B. Lewis, is often called the ***cis-trans*** **test**. Borrowed from nomenclature used in organic

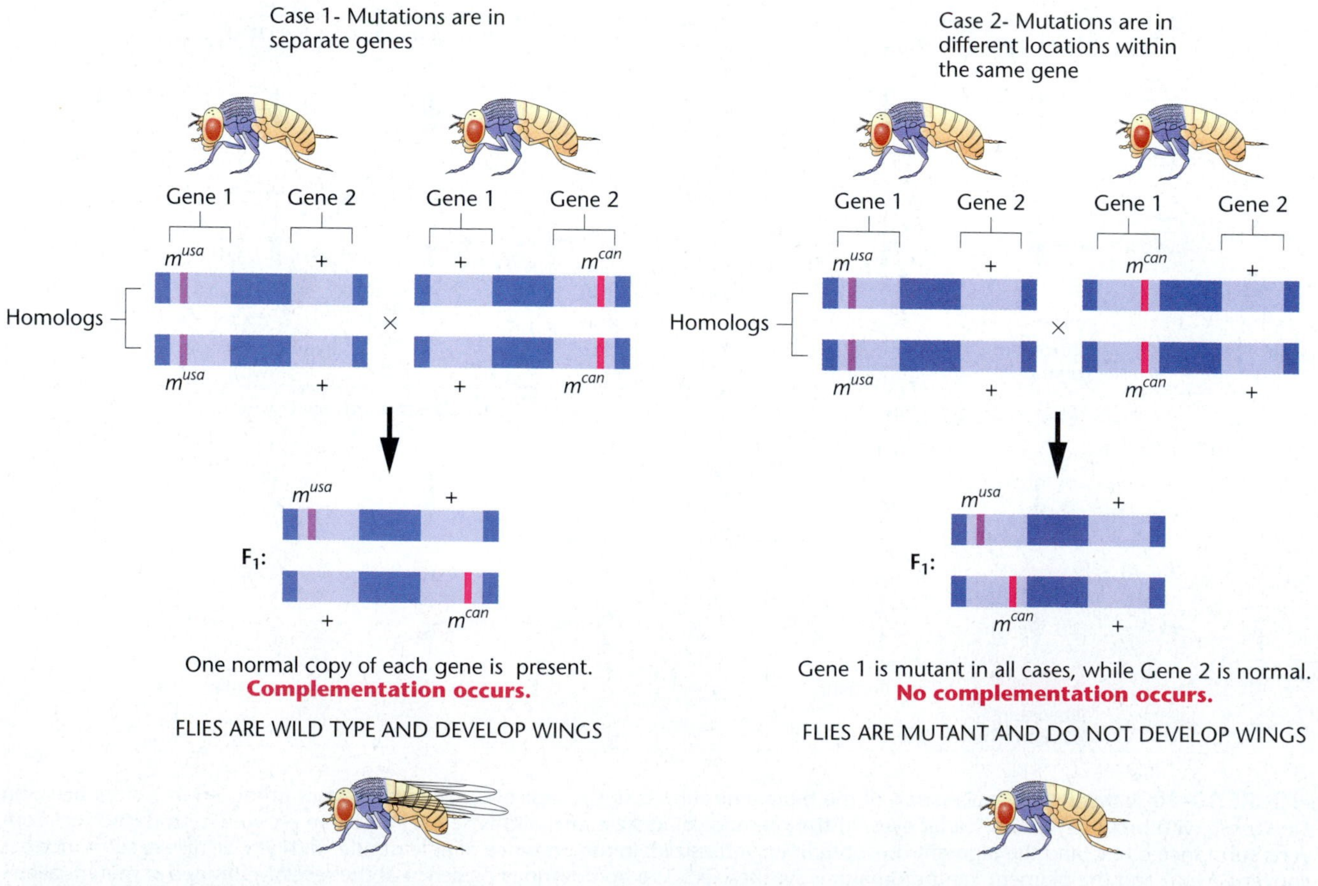

FIGURE 10–11 Complementation analysis of alternative outcomes of two wingless mutations in *Drosophila* (m^{usa} and m^{can}). In case 1, the mutations are not alleles of the same gene, while in case 2, the mutations are alleles of the same gene.

chemistry, *cis* (along side of one another) refers to the case where the two mutations are on the same homolog. Likewise, *trans* (opposite one another) refers to the case where the mutations are on separate homologs. As shown in Figure 10–11, the *trans* configuration is critical in determining whether the two mutations are alleles of the same gene or not. If we determine that the two mutations are indeed alleles, their presence as heterozygotes in the *cis* configuration serves as an important control in complementation analyses. In this configuration, flies will develop wings.

Complementation analysis may be used to screen any number of individual mutations that result in the same phenotype. Such an analysis may reveal that only a single gene is involved or that two or more genes are involved. All mutations determined to be present in any single gene are said to fall into the same **complementation group**, and they will complement mutations in all other groups.

If large numbers of mutations affecting the same trait are available and studied by using complementation analysis, it is possible to predict the total number of genes involved in the determination of that trait. Recall that we established this in Chapter 7, when we discussed the inherited human disorder xeroderma pigmentosum. There, complementation analysis reveals that mutations in any of seven complementation groups (genes) lead to the disorder.

10.10 X-Linkage Describes Genes on the X Chromosome

In many animal and some plant species, one of the sexes contains a pair of unlike chromosomes that are involved in sex determination. In many cases, these are designated as the X and Y. For example, in both *Drosophila* and humans, males contain an X and a Y chromosome, whereas females contain two X chromosomes. The Y chromosome must contain a region of pairing homology with the X chromosome if the two are to synapse and segregate during meiosis. For example, on the human X and Y, two such areas have been identified, one on each end of both chromosomes, referred to as **pseudoautosomal regions (PARs)**. However, the remainderof the Y chromosome in humans as well as other species is considered to be largely blank genetically, lacking almost all genes present on the X chromosome. As a result, genes present on the X chromosome exhibit unique patterns of inheritance in comparison to autosomal genes. The term **X-linkage** is used to describe such situations.

We consider the role of the X and Y chromosomes in sex determination in Chapter 11. There, we will also learn that some organisms lack a chromosome equivalent to the Y and that it is not always the case that the male has unlike sex-determining chromosomes. We will also see that modern molecular techniques have allowed the identification of a few human Y-linked genes that have counterparts on the X chromosome. Below, we will focus on genes present on the X, but absent from the Y chromosome, causing a modification of Mendelian ratios, the central theme of this chapter.

X-Linkage in *Drosophila*

One of the first cases of X-linkage was documented in 1910 by Thomas H. Morgan during his studies of the *white* eye mutation in *Drosophila*. (Figure 10–12). We will use this case to illustrate X-linkage. The normal wild-type red eye color is dominant to white eye color.

Morgan's work established that the inheritance pattern of the white-eye trait was clearly related to the sex of the parent carrying the mutant allele. Unlike the outcome of the typical Mendelian monohybrid cross where F_1 and F_2 data were very similar regardless of which P_1 parent exhibited the recessive mutant trait, reciprocal crosses between white-eyed and red-eyed flies did not yield identical results. Morgan's analysis led to the conclusion that the *white* locus is present on the X rather than on one of the autosomes. As such, both the gene and the trait are said to be X-linked.

Results of reciprocal crosses between white-eyed and red-eyed flies are shown in Figure 10–12. The obvious

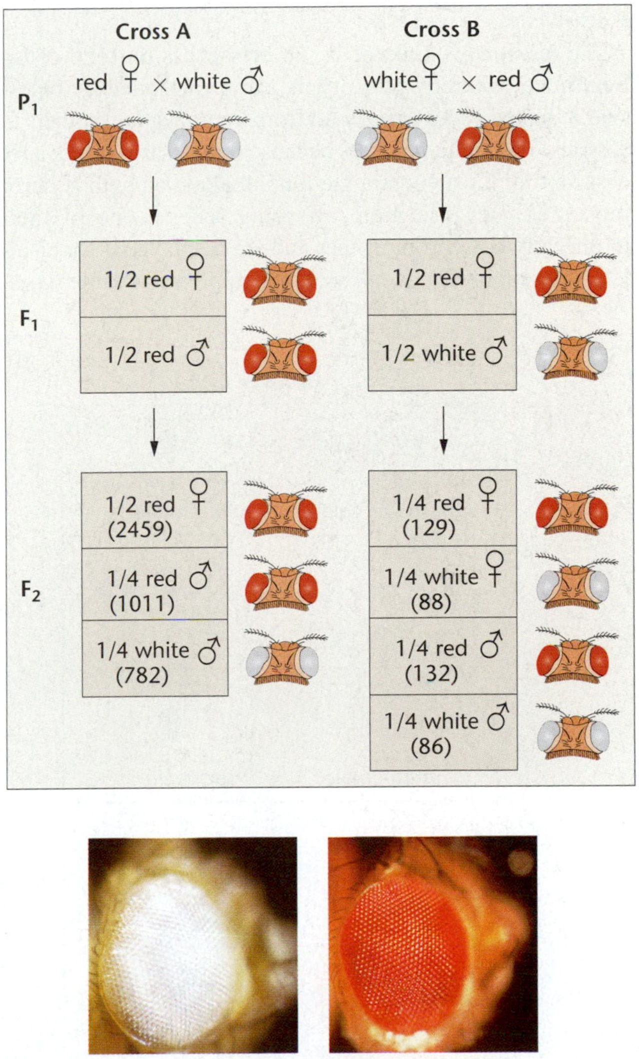

FIGURE 10–12 The F_1 and F_2 results of T. H. Morgan's reciprocal crosses involving the X-linked *white* mutation in *Drosophila melanogaster*. The actual F_2 data are shown in parentheses. The photographs show white eyes and the brick-red wild-type eye color.

differences in phenotypic ratios in both the F_1 and F_2 generations are dependent on whether or not the P_1 white-eyed parent was male or female.

Morgan was able to correlate these observations with the difference found in the sex chromosome composition between male and female *Drosophila*. He hypothesized that in males with white eyes, the recessive allele for white eye is found on the X chromosome, but its corresponding locus is absent from the Y chromosome. Females thus have two available gene loci, one on each X chromosome, while males have only one available locus on their single X chromosome.

Morgan's interpretation of X-linked inheritance, shown in Figure 10–13, provides a suitable theoretical explanation for his results. Since the Y chromosome lacks homology with most genes on the X chromosome, whatever alleles are present on the X chromosome of the males will be directly expressed in the phenotype. Because males cannot be either homozygous or heterozygous for X-linked genes, this condition is referred to as **hemizygous**. In such cases, no alternative alleles are present, and the concept of dominance and recessiveness is irrelevant.

One result of X-linkage is the **crisscross pattern of inheritance**, whereby phenotypic traits controlled by recessive X-linked genes are passed from homozygous mothers to all sons. This pattern occurs because females exhibiting a recessive trait must contain the mutant allele on both X chromosomes. Because male offspring receive one of their mother's two X chromosomes and are hemizygous for all alleles present on that X, all sons will express the same recessive X-linked traits as their mother. This pattern of inheritance is apparent in pedigree in Figure 10–14.

In addition to documenting the phenomenon of X-linkage, Morgan's work has taken on great historical significance. By 1910, the correlation between Mendel's work and the behavior of chromosomes during meiosis had provided the basis for the **chromosome theory of inheritance**, as postulated by Sutton and Boveri. (See Chapter 9.) Morgan's work, and subsequently that of his student Calvin Bridges, providing direct evidence that genes are transmitted on specific chromosomes, is considered the first solid experimental evidence in support of this theory. In the ensuing two decades, these findings inspired further research, the outcomes of which provided indisputable evidence in support of this theory.

X-Linkage in Humans

In humans, many genes and the traits controlled by them are recognized as being linked to the X chromosome. These X-linked traits can be easily identified in pedigrees, characterized by a crisscross pattern of inheritance. A pedigree for one form of human color blindness is shown in Figure 10–14. The mother in generation I passes the trait onto all her sons, but to none of her daughters. If the offspring in generation II have children by normal individuals, the color-blind sons will produce all normal male and female offspring (III-1, 2, and 3); the normal-vision daughters will produce normal-vision female offspring (III-4, 6, and 7), as well as color-blind (III-8) and normal-vision (III-1 and 5) male offspring.

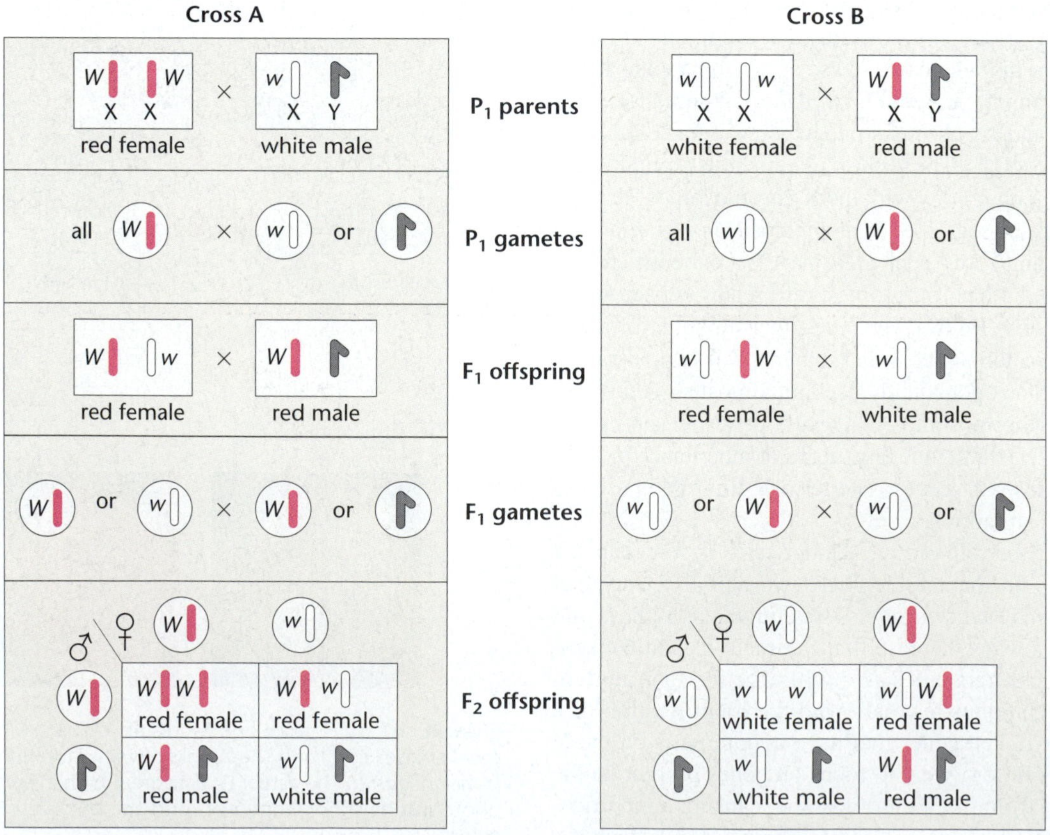

MEDIA TUTORIAL X-linked Inheritance

FIGURE 10–13 The chromosomal explanation of the results of the X-linked crosses shown in Figure 10–12.

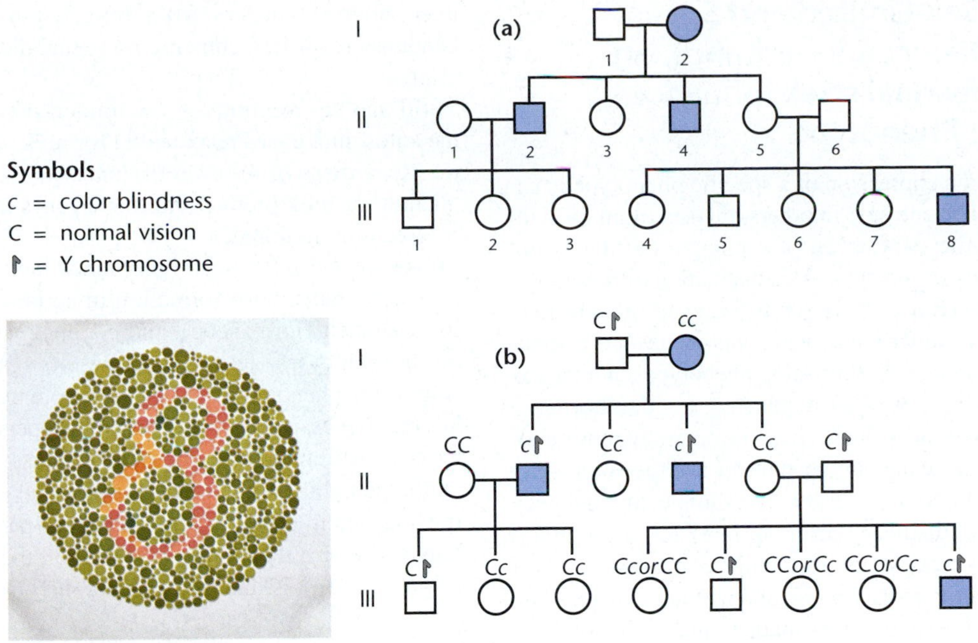

FIGURE 10–14 (a) A human pedigree of the X-linked color-blindness trait. (b) The most probable genotypes of each individual in the pedigree. The photograph is of an Ishihara color-blindness chart. Red-green color-blind individuals see a 3, rather than an 8 visualized by those with normal color vision.

Many X-linked human genes have now been identified, as shown in Table 10–4. For example, the genes controlling two forms of hemophilia and two forms of muscular dystrophy are located on the X chromosome. In addition, numerous genes whose expression yields enzymes are X-linked. Glucose-6-phosphate dehydrogenase and hypoxanthine-guanine-phosphoribosyl transferase are two examples. In the latter case, the severe Lesch–Nyhan syndrome results from the mutant form of the X-linked gene product.

Because of the way in which X-linked genes are transmitted, unusual circumstances may be associated with recessive X-linked disorders in comparison to recessive autosomal disorders. For example, if an X-linked disorder debilitates or is lethal to the affected individual prior to reproductive maturation, the disorder occurs exclusively in males. This is the case because the only sources of the lethal allele in the population are heterozygous females who are "carriers" and do not express the disorder. They pass the allele to one-half of their sons, who develop the disorder because they are hemizygous but who rarely, if ever, reproduce. Heterozygous females also pass the allele to one-half of their daughters, who become carriers but do not develop the disorder. An example of such an X-linked disorder is the Duchenne form of muscular dystrophy. The disease has an onset prior to age 6 and is often lethal prior to age 20. It normally occurs only in males.

TABLE 10–4 Human X-Linked Traits

Condition	*Characteristics*
Color blindness, deutan type	Insensitivity to green light.
Color blindness, protan type	Insensitivity to red light.
Fabry's disease	Deficiency of galactosidase A; heart and kidney defects, early death
G-6-PD	Deficiency of glucose-6-phosphate dehydrogenase, severe anemic reaction following intake of primaquines in drugs and certain foods, including fava beans.
Hemophilia A	Classical form of clotting deficiency; deficiency of clotting factor VIII.
Hemophilia B	Christmas disease; deficiency of clotting factor IX.
Hunter syndrome	Mucopolysaccharide storage disease resulting from iduronate sulfatase enzyme deficiency; short stature, clawlike fingers, coarse facial features, slow mental deterioration, and deafness
Ichthyosis	Deficiency of steroid sulfatase enzyme; scaly dry skin, particularly on extremities.
Lesch–Nyhan syndrome	Deficiency of hypoxanthine-guanine phosphoribosyl transferase enzyme (HGPRT) leading to motor and mental retardation, self-mutilation, and early death.
Muscular dystrophy (Duchenne type)	Progressive, life-shortening disorder characterized by muscle degeneration and weakness; sometimes associated with mental retardation; deficiency of the protein dystrophin.

10.11 In Sex-Limited and Sex-Influenced Inheritance, an Individual's Sex Influences the Phenotype

In some cases, the expression of a specific phenotype is absolutely limited to one sex; in others, the sex of an individual influences the expression of a phenotype that is not limited to one sex or the other. This distinction differentiates **sex-limited inheritance** from **sex-influenced inheritance**.

In both types of inheritance, autosomal genes are responsible for the existence of contrasting phenotypes, but the expression of these genes is dependent on the hormone constitution of the individual. Thus, the heterozygous genotype may exhibit one phenotype in males and the contrasting one in females. In domestic fowl, for example, tail and neck plumage is often distinctly different in males and females (Figure 10–15), demonstrating sex-limited inheritance. Cock feathering is longer, more curved, and pointed, whereas hen feathering is shorter and more rounded. Inheritance of these feather phenotypes is controlled by a single pair of autosomal alleles whose expression is modified by the individual's sex hormones.

As shown here, hen feathering is due to a dominant allele, *H*:

Genotype	*Phenotype*	
	♀	♂
HH	Hen feathered	Hen feathered
Hh	Hen feathered	Hen feathered
hh	Hen feathered	Cock feathered

However, regardless of the homozygous presence of the recessive *h* allele, all females remain hen-feathered. Only in males does the *hh* genotype result in cock feathering.

In certain breeds of fowl, the hen-feathering or cock-feathering allele has become fixed in the population. In the Leghorn breed, all individuals are of the *hh* genotype; as a result,

FIGURE 10–15 Hen feathering (left) and cock feathering (right) in domestic fowl. The feathers in the hen are shorter and less curved.

male plumage differs from female plumage. Seabright bantams are all *HH*, showing no sexual distinction in feathering.

Still another example of sex-limited inheritance involves the autosomal genes responsible for milk yield in dairy cattle. Regardless of the overall genotype that influences the quantity of milk production, those genes are obviously expressed only in females.

Cases of sex-influenced inheritance include pattern baldness in humans, horn formation in certain breeds of sheep (e.g., Dorsett Horn sheep), and certain coat patterns in cattle. In such cases, autosomal genes are responsible for the contrasting phenotypes displayed by both males and females, but the expression of these genes is dependent on the hormone constitution of the individual. Thus, the heterozygous genotype exhibits one phenotype in one sex and the contrasting one in the other. For example, pattern baldness in humans, where the hair is very thin on the top of the head (Figure 10–16), is inherited in the following way:

Genotype	*Phenotype*	
	♀	♂
BB	Bald	Bald
Bb	Not bald	Bald
bb	Not bald	Not bald

Even though females can display pattern baldness, this phenotype is much more prevalent in males. When females do inherit the *BB* genotype, the phenotype is much less pronounced than in males and is expressed later in life.

10.12 Phenotypic Expression Is Not Always a Direct Reflection of the Genotype

We conclude this chapter with the consideration of phenotypic expression. In Chapter 9, we assumed that the genotype of an organism is always directly expressed in its phenotype. For example, peas homozygous for the recessive *d* allele (*dd*) will always be dwarf. We have discussed gene expression as though the genes operate in a closed "black-box" system in which the presence or absence of functional products directly determines the collective phenotype of an individual. The situation is actually much more complex. Most gene products function within the internal milieu of the cell, and cells interact with one another in various ways. Further, the organism must survive under diverse environmental influences. Thus, gene expression and the resultant phenotype are often modified through the interaction between an individual's particular genotype and the internal and external environment.

The degree of environmental influence can vary from inconsequential to subtle to very strong. Subtle interactions are the most difficult to detect and document, and they have led to "nature vs. nurture" questions in which scientists

FIGURE 10–16 Pattern baldness, a sex-influenced autosomal trait in humans

debate the relative importance of genes versus environment—usually inconclusively. In this final section of this chapter, we will deal with some of the variables known to modify gene expression.

Penetrance and Expressivity

Some mutant genotypes are always expressed as a distinct phenotype, whereas others produce a proportion of individuals whose phenotypes cannot be distinguished from normal (wild type). The degree of expression of a particular trait may be studied quantitatively by determining the penetrance and expressivity of the genotype under investigation. The percentage of individuals that show at least some degree of expression of a mutant genotype defines the **penetrance** of the mutation. For example, the phenotypic expression of many mutant alleles in *Drosophila* is indistinguishable from wild type. If 15 percent of mutant flies show the wild-type appearance, the mutant gene is said to have a penetrance of 85 percent.

By contrast, **expressivity** reflects the *range of expression* of the mutant genotype. Flies homozygous for the recessive mutant gene *eyeless* yield phenotypes that range from the presence of normal eyes to a partial reduction in size to the complete absence of one or both eyes (Figure 10–17). Although the average reduction of eye size is one-fourth to one-half, expressivity ranges from complete loss of both eyes to completely normal eyes.

Examples such as the expression of the *eyeless* gene have provided the basis for experiments to determine the causes of phenotypic variation. If the laboratory environment is held constant and extensive variation is still observed, other genes may be influencing or modifying the *eyeless* phenotype. On the other hand, if the genetic background is not the cause of the phenotypic variation, environmental factors such as temperature, humidity, and nutrition may be involved. In the case of the *eyeless* phenotype, experiments have shown that both genetic background and environmental factors influence its expression.

Genetic Background: Suppression and Position Effects

Though it is difficult to assess the specific effect of the **genetic background** and the expression of a gene responsible for determining a potential phenotype, two effects of genetic background have been well characterized.

First, the expression of other genes throughout the genome may have an effect on the phenotype produced by the gene in question. The phenomenon of **genetic suppression** is an example. Mutant genes such as *suppressor of sable* (*su-s*), *suppressor of forked* (*su-f*), and *suppressor of Hairy-wing* (*su-Hw*) in *Drosophila* completely or partially restore the normal phenotype in an organism that is homozygous (or hemizygous) for the *sable*, *forked*, and *Hairy-wing* mutations, respectively. For example, flies hemizygous for both *forked* (a bristle mutation) and *su-f* have normal bristles. In each case, the suppressor gene causes the complete reversal of the expected phenotypic expression of the original mutation. Suppressor genes are excellent examples of the genetic background modifying primary gene effects.

Second, the physical location of a gene in relation to other genetic material may influence its expression. Such a situation is called a **position effect**. For example, if a region of a chromosome is relocated or rearranged (called a translocation or inversion event), normal expression of genes

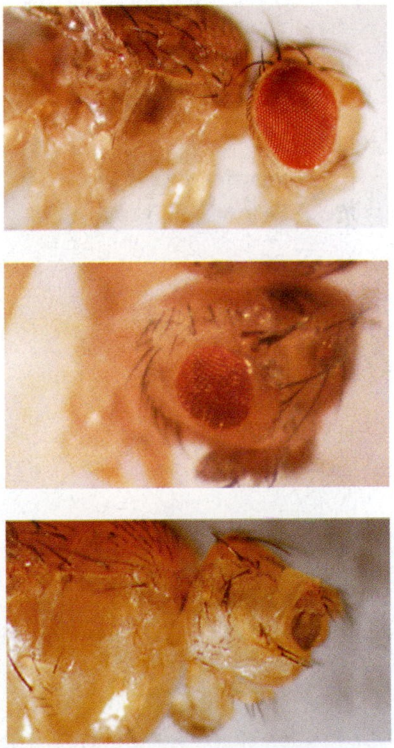

FIGURE 10–17 Variable penetrance as illustrated by the expression of the *eyeless* mutation in *Drosophila*. Gradations in phenotype range from wild type to partial reduction to eyeless.

in that chromosomal region may be modified. This is particularly true if the gene is relocated to or near certain areas of the chromosome that are prematurely condensed and genetically inert, referred to as **heterochromatin**.

An example of a position effect involves female *Drosophila* heterozygous for the X-linked recessive eye-color mutant *white* (w). The w^+/w genotype normally results in a wild-type brick-red eye color. However, if the region of the X chromosome containing the wild-type w^+ allele is translocated so that it is close to a heterochromatic region, expression of the w^+ allele is modified. Instead of having a red color, the eyes are variegated, or mottled with red and white patches (Figure 10–18). Therefore, following translocation, the dominant effect of the normal w^+ allele is reduced. A similar position effect is produced if a heterochromatic region is relocated next to the *white* locus on the X chromosome. Apparently, heterochromatic regions inhibit the expression of adjacent genes. Loci in many other organisms also exhibit position effects, providing proof that alteration of the normal arrangement of genetic information can modify its expression.

Temperature Effects

Because chemical activity depends on the kinetic energy of the reacting substances, which in turn depends on the surrounding temperature, we can expect temperature to influence phenotypes. One example is the evening primrose, which produces red flowers when grown at 23°C and white flowers when grown at 18°C. An even more striking example is seen in Siamese cats and Himalayan rabbits, which exhibit dark fur in certain body regions where the body temperature is slightly cooler, particularly the nose, ears, and paws (Figure 10–19). In these cases, it appears that the wild-type enzyme responsible for pigment production is functional at the lower temperatures present in the extremities, but it loses its catalytic function at the slightly higher temperatures found throughout the rest of the body.

Mutations that are affected by temperature are said to be **conditional** and are called **temperature sensitive**. Examples are known in viruses and a variety of organisms, including bacteria, fungi, and *Drosophila*. In extreme cases, an organism carrying a mutant allele may express a mutant phenotype when grown at one temperature, but express the wild-type phenotype when reared at another temperature. This type of temperature effect is useful in studying mutations that interrupt essential processes during development and are thus normally detrimental to the organism. For example, if bacterial viruses carrying certain temperature-sensitive mutations infect bacteria cultured at 42°C, known as the *restrictive condition*, infection progresses up to the point where the essential gene product is required (e.g., for viral assembly) and then arrests. If cultured under *permissive conditions* of 25°C, the gene product is functional, infection proceeds normally, and new viruses are produced. The use of temperature-sensitive mutations, which may be induced and isolated, has added immensely to the study of viral genetics.

(a)

(b)

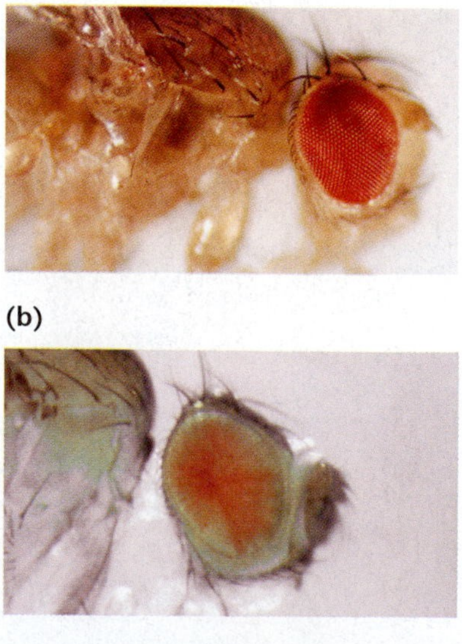

FIGURE 10–18 Position effect, as illustrated in the eye phenotype in two female *Drosophila* heterozygous for the gene *white*. (a) Normal dominant phenotype showing brick-red eye color. (b) Variegated color of an eye caused by rearrangement of the *white* gene to another location in the genome.

Nutritional Effects

Another example where the phenotype is not a direct reflection of the organism's genotype involves **nutritional mutations**. In microorganisms, mutations that prevent synthesis of nutrient molecules are quite common, for example, when an enzyme essential to a biosynthetic pathway becomes inactive. A microorganism bearing such a mutation is called an **auxotroph**. If the end product of a biochemical pathway can no longer be synthesized, and if that molecule is essential to normal growth and development, the mutation prevents growth and may be lethal. For example, if the bread mold *Neurospora* can no longer synthesize the amino acid leucine, proteins cannot be synthesized. If leucine is present in the growth medium, the detrimental effect is overcome. Nutritional mutants have been crucial to genetic studies in bacteria and also served as the basis for George Beadle and Edward Tatum's proposal, in the early 1940s, that one gene functions to produce one enzyme. (See Chapter 6).

A slightly different set of circumstances exists in humans. The presence or absence of certain dietary substances, which normal individuals may consume without harm, can adversely affect individuals with abnormal genetic constitutions. Often, a mutation may prevent an individual from metabolizing some substance commonly found in normal diets. For example, as we saw in Chapter 6, those afflicted with the genetic disorder **phenylketonuria** cannot metabolize the amino acid phenylalanine. Those with **galactosemia** cannot metabolize galactose. However, if the dietary intake of the molecule is drastically reduced or eliminated, the associated phenotype may be ameliorated.

The fairly common case of **lactose intolerance**, in which individuals are intolerant of the milk sugar lactose, illustrates

FIGURE 10–19 (a) A Himalayan rabbit. (b) A Siamese cat. Both show dark fur color on the snout, ears, and paws. These patches are due to the effect of a temperature-sensitive allele responsible for pigment production at the lower temperatures of the extremities, but, which is inactive at slightly higher temperatures.

the general principles involved. Lactose is a disaccharide consisting of a molecule of glucose linked to a molecule of galactose and makes up 7 percent of human milk and 4 percent of cow's milk. To metabolize lactose, humans require the enzyme **lactase**, which cleaves the disaccharide. Adequate amounts of lactase are produced during the first few years after birth. However, in many people, the level of this enzyme soon drops drastically, and as adults, they become intolerant of milk. The major phenotypic effect involves severe intestinal diarrhea, flatulence, and abdominal cramps. This condition (while not limited to these groups), is particularly prevalent in Eskimos, Africans, Asians, and Americans with these heritages. In some of these cultures, milk is converted to cheese, butter, and yogurt, where the amount of lactose is reduced significantly and the adverse effects can be reduced. In the United States, milk low in lactose is commercially available, and to aid in the digestion of other lactose-containing foods, lactase is now a commercial product that can be ingested.

Onset of Genetic Expression

Not all genetic traits are expressed at the same time during an organism's life span. In most cases, the age at which a gene is expressed corresponds to the normal sequence of growth and development. In humans, the prenatal, infant, preadult, and adult phases require different genetic information. As a result, many severe inherited disorders are often not manifested until after birth. For example, **Tay–Sachs disease**, inherited as an autosomal recessive, is a lethal lipid-metabolism disease involving an abnormal enzyme, hexosaminidase A. Newborns appear phenotypically normal for the first few months. Then, developmental retardation, paralysis, and blindness ensue, and most affected children die by the age of three.

The **Lesch–Nyhan syndrome**, inherited as an X-linked recessive disease, is characterized by abnormal nucleic acid metabolism (biochemical salvage of nitrogenous purine bases), leading to the accumulation of uric acid in blood and tissues, mental retardation, palsy, and self-mutilation of the lips and fingers. The disorder is due to a mutation in the gene encoding hypoxanthine-guanine phosphoribosyl transferase (HPRT). Newborns are normal for six to eight months prior to the onset of the first symptoms.

Still another example involves **Duchenne muscular dystrophy (DMD)**, an X-linked recessive disorder associated with progressive muscular wasting. It is not usually diagnosed until the age of 3 to 5 years. Even with modern medical intervention, the disease is often fatal in the early twenties.

Perhaps the most variable of all inherited human disorders regarding age of onset is **Huntington disease**. Inherited as an autosomal dominant, Huntington disease affects the frontal lobes of the cerebral cortex, where progressive cell death occurs over a period of more than a decade. Brain deterioration is accompanied by spastic uncontrolled movements, intellectual and emotional deterioration, and ultimately death. While onset has been reported at all ages, it most frequently occurs between ages 30 and 50, with a mean onset of 38 years.

Conditions such as these support the concept that the critical expression of normal genes varies throughout the life cycle of organisms, including humans. Gene products may play more essential roles at certain times. Further, it is likely that the internal physiological environment of an organism changes with age.

Genetic Anticipation

Interest in studying the genetic onset of phenotypic expression has intensified with the discovery of heritable disorders that exhibit a progressively earlier age of onset and an increased severity of the disorder in each successive generation. This general phenomenon is referred to as **genetic anticipation**.

Myotonic dystrophy (DM), the most common type of adult muscular dystrophy, clearly illustrates genetic anticipation. Individuals afflicted with this autosomal dominant disorder exhibit extreme variation in the severity of symptoms. Mildly affected individuals develop cataracts as adults, but have little or no muscular weakness. Severely affected individuals demonstrate more extensive myopathy, myotonia (muscle hyperexcitability) and may be mentally retarded. In its most extreme form, the disease is fatal just after birth. A great deal of excitement was generated in 1989, when C. J. Howeler and colleagues confirmed the correlation of increased severity with earlier onset. They studied 61 parent–child pairs that expressed the disorder and, in 60 of the cases, age of onset was earlier in the child than in his or her parent.

In 1992, an explanation was put forward to explain both the molecular cause of the mutation responsible for DM as well as the basis of genetic anticipation. As we will see in Chapter 14, a short (3 nucleotide) DNA sequence of the DM gene is repeated a variable number of times and is unstable. Normal individuals have from 5-35 copies of this sequence, minimally affected individuals reveal about 150 copies, and severely affected individuals demonstrate as many as 1500 repeats. The most remarkable observation was that, in successive generations of DM individuals, the size of the repeated segment increases. Although it is not yet clear exactly how the expansion in size affects onset and phenotypic expression, the increased number of copies is now accepted as the basis of genetic anticipation in this disorder. Several other inherited human disorders, including the fragile-X syndrome, Kennedy disease, and Huntington disease also reveal an association between the size of specific regions of the responsible gene and disease severity. You may recall that we previously discussed the molecular explanation in detail in Chapter 7.

Genomic (Parental) Imprinting

Our final example of modification of the laws of Mendelian inheritance involves the variation of phenotypic expression depending strictly on the parental origin of the chromosome carrying a particular gene, a phenomenon called **genomic** (or **parental**) **imprinting**. In some species, certain chromosomal regions and the genes contained within them somehow retain a memory, or an "imprint," of their parental origin that influences whether specific genes either are expressed or remain genetically silent—that is, are not expressed.

The imprinting step is thought to occur before or during gamete formation, leading to differentially marked genes (or chromosome regions) in sperm-forming versus egg-forming tissues. The process is clearly different from mutation because the imprint can be reversed in succeeding generations as genes pass from mother to son to granddaughter, and so on.

An example of imprinting involves the inactivation of one of the X chromosomes in mammalian females. In mice, prior to development of the embryo, imprinting occurs in tissues such that the X chromosome of paternal origin is genetically inactivated in all cells, while the genes on the maternal X chromosome remain genetically active. As embryonic development is subsequently initiated, the imprint is "released" and random inactivation of either the paternal *or* the maternal X chromosome occurs.

In 1991, more specific information established that three specific mouse genes undergo imprinting. One is the gene encoding insulinlike growth factor II (*Igf2*). A mouse that carries two nonmutant alleles of this gene is normal in size, whereas a mouse that carries two mutant alleles lacks a growth factor and is a dwarf. The size of a heterozygous mouse (one allele normal and one mutant; see Figure 10–20) depends on the parental origin of the normal allele. The mouse is normal in size if the normal allele came from the father, but dwarf if the normal allele came from the mother. From this, we can deduce that the normal *Igf2* gene is imprinted to function poorly during the course of egg production in females, but functions normally when it has passed through sperm-producing tissue in males.

Imprinting continues to depend on whether the gene passes through sperm-producing or egg-forming tissue leading to the next generation. For example, a heterozygous normal-sized male will donate to half his offspring a "normal-functioning" wild-type allele that will counteract a mutant allele received from the mother. In humans, two distinct genetic disorders are thought to be caused by differential imprinting of the same region of chromosome 15 (15q1). In both cases, the disorders *appear* to be due to an identical deletion of this region in one member of the chromosome 15 pair. The first disorder, **Prader–Willi syndrome (PWS)**, results when only an undeleted maternal chromosome remains. If only an undeleted paternal chromosome remains, an entirely different disorder, **Angelman syndrome (AS)**, results.

The two conditions are clearly different phenotypically. PWS entails mental retardation as well as a severe eating disorder marked by an uncontrollable appetite, obesity, and diabetes. AS involves distinct behavioral manifestations as well as mental retardation. We can conclude that the involved region of chromosome 15 is imprinted differently in male versus female gametes and that both a maternal and paternal region are required for normal development.

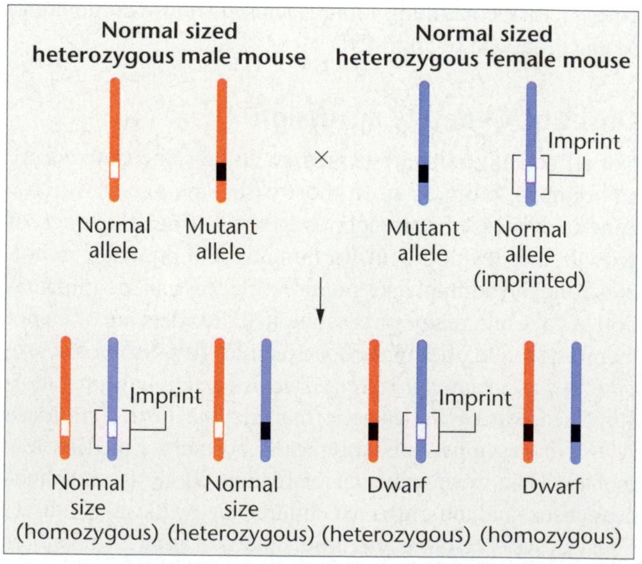

FIGURE 10–20 The effect of imprinting on the mouse *Igf2* gene, which produces dwarf mice in the homozygous condition. Heterozygous offspring that receive the normal allele from their father are normal in size. Heterozygotes that receive the normal allele from their mother, which has been imprinted, are dwarf.

Genetics, Technology, and Society

Improving the Genetic Fate of Purebred Dogs

For dog lovers, nothing is quite so heartbreaking as watching a dog slowly go blind, standing by helplessly as he struggles to adapt to a life of perpetual darkness. That's what happens in progressive retinal atrophy (PRA), an inherited disorder first described in Gordon setters in 1911. Since then, PRA has been found in many other breeds of dogs, including Irish setters, border collies, Norwegian elkhounds, toy poodles, miniature schnauzers, cocker spaniels, and Siberian huskies.

The products of many genes are required for the development and maintenance of a healthy retina, and a defect in any one of them has the potential to cause retinal dysfunction. Decades of research have led to the identification of four such genes (*rcd1, rcd2, erd,* and *prcd*), and more are likely to be discovered. Different genes are mutated in different breeds (e.g., *rcd1* in Irish setters and *prcd* in Labrador retrievers). In all dogs examined so far, PRA shows a recessive pattern of inheritance.

Whichever mutation is responsible, PRA is more common in certain purebred dogs than in mixed breeds. The development of distinct breeds of dogs has involved the intensive selection for desirable attributes, whether a particular size, shape, color, or behavior. Many desired characteristics are determined by recessive alleles. The fastest way to increase the homozygosity of these alleles and establish the characteristics in the population is to mate close relatives, which are likely to carry the same alleles. In the practice known as line breeding, for example, dogs may be mated to a cousin or a grandparent.

Unfortunately, the generations of inbreeding that have established favorable characteristics in purebreeds have also increased the homozygosity of certain harmful recessive alleles such as in PRA, resulting in other inherited diseases. Many breeds are plagued with inherited hip dysplasia, which is particularly prevalent in German shepherds. Deafness and kidney disorders are common genetic maladies in dalmatians. Over 300 such diseases have been characterized in purebred dogs, and most breeds of dogs are affected by one or more of them. By some estimates, fully 25 percent of purebred dogs are affected with one genetic ailment or another. While mutations are the cause of these genetic diseases, there is general agreement that breeding practices, if not executed properly, increase the frequency of disease.

Inbreeding is the likely explanation for the elevated frequency of PRA in certain breeds. Fortunately, advances in canine genetics are beginning to provide new tools for the breeding of healthy dogs. Since 1995, a genetic test has been available to identify mutations in the *rcd1* gene, which is responsible for the form of PRA that occurs in Irish setters. This test is now being used to identify heterozygous carriers of *rcd1* mutations, dogs that show no symptoms of PRA, but, if mated with other carriers, pass the trait on to about 25 percent of their offspring. Eliminating PRA carriers from breeding programs could, in theory, eradicate this condition from Irish setters in just a few generations.

It took many years to identify *rcd1* and other genes responsible for PRA. In the future, the isolation of genes underlying canine inherited disease should be faster, thanks to the Dog Genome Project, a collaborative effort involving scientists at the University of California, the University of Oregon, the Fred Hutchinson Cancer Research Center in Seattle, and other research centers. The project's goal is the creation of a complete genetic map of the 39 chromosomes in the dog. In late 1997, preliminary stages were completed, paving the way for the development of a more comprehensive map. The identification of genes that cause inherited disease will allow genetic tests for the diagnosis of disease before symptoms develop, as well as for ensuring that breeding animals are free of harmful recessive alleles.

The Dog Genome Project may have benefits for humans beyond the reduction of disease in their canine companions. This is because about 85 percent of the genes in the dog genome have equivalents in humans, so the identification of a disease-causing dog gene may be a shortcut to the isolation of the corresponding gene in humans. For example, some forms of PRA in dogs appear to be equivalent to retinitis pigmentosa (RP) in humans, which afflicts about 1.5 million people worldwide. Despite much research, RP remains poorly understood, and current treatments only slow its progress. Understanding the genetic basis of PRA could lead to breakthroughs in the diagnosis and treatment of RP, potentially saving the sight of thousands of people every year. By contributing to the cure of human diseases, dogs may prove to be "man's best friend" in an entirely new way.

References

Berson, E.L. 1996. Retinitis pigmentosum: Unfolding its mystery. *Proc. Natl. Acad. Sci. USA* 93:4526–28.

Ray, K., Baldwin, V., Acland, G., and Aquirre, G. 1995. Molecular diagnostic tests for ascertainment of genotype at the rod cone dysplasia (*rcd1* locus) in Irish setters. *Curr. Eye Res.* 14:243.

Smith, C.A. 1994. New hope for overcoming canine inherited disease. *Am. J. Vet. Med. Assoc.* 204:41–46.

Many questions remain unanswered in the area of genomic imprinting. It is not known how many genes are subject to imprinting, nor its developmental role. While it appears that regions of chromosomes, rather than specific genes are imprinted, the molecular mechanism of imprinting is still a matter for conjecture. It has been hypothesized that DNA methylation may be involved. In vertebrates, methyl groups can be added to the carbon atom at position 5 in cytosine (see Chapter 2) as a result of the activity of the enzyme DNA methyl transferase. Methyl groups are added when the dinucleotide CpG or groups of CpG units (called CpG islands) are present along a DNA chain. DNA methylation is a reasonable mechanism for establishing a molecular imprint, since there is some evidence that a high level of methylation can inhibit gene activity and that active genes (or their regulatory sequences) are often undermethylated. Additionally, variation in methylation has been noted in the mouse genes that undergo imprinting. Whatever the cause of this phenomenon, it is a fascinating topic and one that will receive significant attention in future research studies.

Chapter Summary

1. Since Mendel's work was rediscovered, the study of transmission genetics has expanded to include many alternative modes of inheritance. In many cases, phenotypes may be influenced by two or more genes in a variety of ways.
2. Incomplete, or partial, dominance is exhibited when intermediate phenotypic expression of a trait occurs in an organism that is heterozygous for two alleles.
3. Codominance is exhibited when distinctive effects of two alleles occur simultaneously in a heterozygous organism.
4. The concept of multiple allelism applies to populations, since a diploid organism may host only two alleles at any given locus. Within a population, however, many alternative alleles of the same gene occur.
5. Lethal mutations usually result in the impairment, inactivation, or the lack of synthesis of gene products essential during development. Such mutations may be recessive or dominant. The manifestation of some lethal genes, such as the gene causing Huntington disease, are not expressed until adulthood.
6. Phenotypes are often affected by more than one gene, leading to a variety of modifications of the Mendelian dihybrid and trihybrid ratios. Such cases are often classified under the general heading of gene interaction.
7. Epistasis may occur when two or more genes influence a single characteristic. Usually, the expression of one of the genes masks the expression of the other gene or genes.
8. Complementation analysis determines whether independently isolated mutations producing similar phenotypes are alleles of one another or whether they represent separate genes.
9. Genes located on the X chromosome result in a characteristic mode of inheritance referred to as X-linkage. X-linked phenotypic ratios result because hemizygous individuals (those with an X and a Y chromosome) express all alleles present on their X chromosome.
10. Sex-limited and sex-influenced inheritance occur when the sex of the organism affects the phenotype controlled by a gene located on an autosome.
11. Phenotypic expression is not always the direct reflection of the genotype. Penetrance measures the percentage of organisms in a given population that exhibits evidence of the corresponding mutant phenotype. Expressivity, on the other hand, measures the range of phenotypic expression of a given genotype.
12. Phenotypic expression can be modified by genetic background, temperature, and nutrition. Position effects illustrate the existence of genetic background that affects phenotypic expression.
13. Genetic anticipation refers to the phenomenon where the onset of phenotypic expression occurs earlier and becomes more severe in each ensuing generation.
14. Genomic imprinting is a process whereby a region of either the paternal or maternal chromosome is modified (marked or imprinted), thereby affecting phenotypic expression. Expression therefore depends upon which parent contributes a mutant allele.

Insights and Solutions

Genetic problems take on added complexity if they involve two independent characters and multiple alleles, incomplete dominance, or epistasis. The most difficult types of problems are those that pioneering geneticists faced during laboratory or field studies. They had to determine the mode of inheritance by working backward from the observations of offspring to parents of unknown genotype.

1. Consider the problem of comb-shape inheritance in chickens, where walnut, rose, pea, and single are observed as distinct phenotypes. *How is comb shape inherited, and what are the genotypes of the* P_1 *generation of each cross?* Use the following data to answer these questions:

Cross 1:	single × single ⟶ all single
Cross 2:	walnut × walnut ⟶ all walnut
Cross 3:	rose × pea ⟶ all walnut
Cross 4:	$F_1 \times F_1$ of Cross 3

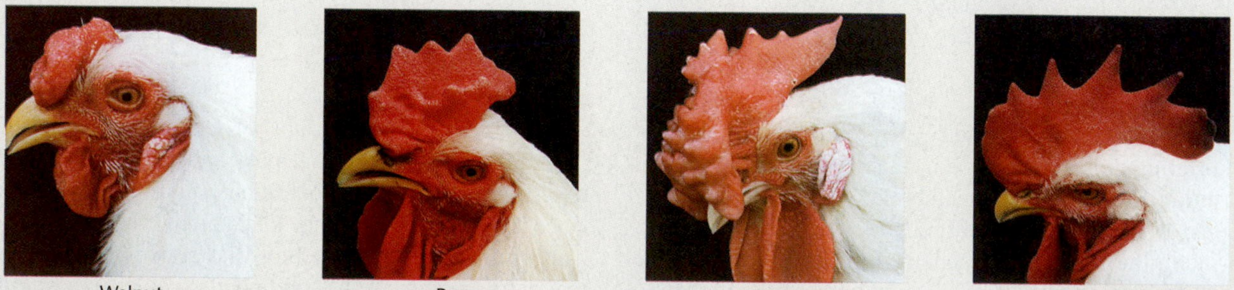

Walnut Pea Rose Single

(Single and pea photos: J. James Bitgood, University of Wisconsin Animal Sciences Dept., Madison, WI. Walnut and rose photos: Courtesy of Dr. Ralph Somes.)

walnut × walnut ⟶ 93 walnut
28 rose
32 pea
10 single

Solution: At first glance, this problem appears quite difficult. However, as with other seemingly difficult problems, applying a systematic approach and breaking the analysis into steps usually saves the day. The approach involves two steps. First, analyze the data carefully for any useful information. Once you identify something that is clearly helpful, follow an empirical approach; that is, formulate an hypothesis and, in a sense, test it against the given data. Look for a pattern of inheritance that is consistent with all cases.

For example, this problem gives two immediately useful facts. First, in Cross 1, P_1 singles breed true. Second, while P_1 walnut breeds true (Cross 2), a walnut phenotype is also produced in crosses between rose and pea (Cross 3). When these F_1 walnuts are mated (Cross 4), all four comb shapes are produced in a ratio that approximates 9:3:3:1. This observation should immediately suggest a cross involving two gene pairs, because the resulting data closely resemble the ratio of Mendel's dihybrid crosses. Since only one character is involved and discontinuous phenotypes occur (comb shape), perhaps some form of epistasis is occurring. This may serve as a working hypothesis, and you must now propose how the two gene pairs "interact" to produce each phenotype.

If you call the allele pairs *A*, *a* and *B*, *b*, you might predict that, since walnut represents 9/16 in Cross 4, *A–B–* will produce walnut. You might also hypothesize that in the case of Cross 2, the genotypes were *AABB* × *AABB* where walnut was seen to breed true. (Recall that *A–* and *B–* mean *AA* or *Aa* and *BB* or *Bb*, respectively.) Because single is the phenotype representing $\frac{1}{16}$ of the offspring of Cross 4, we could predict that the phenotype is the result of the *aabb* genotype, which is consistent with Cross 1.

Now we have only to determine the genotypes for rose and pea. A logical prediction would be that at least one dominant *A* or *B* allele, combined with the double recessive condition of the other allele pair, can account for these phenotypes; that is,

A–bb rose

and

aa B– pea

If, in Cross 3, *AAbb* (rose) were crossed with *aaBB* (pea), all offspring would be *AaBb* (walnut). This is consistent with the data, and we must now look at Cross 4. We predict these walnut genotypes to be *AaBb* (as before), and from the cross

AaBb (walnut) × *AaBb* (walnut)

we expect

9/16 *A–B–* (walnut)
3/16 *A–bb* (rose)
3/16 *aa B–* (pea)

and

1/16 *aa bb* (single)

Our prediction is consistent with the data. The initial hypothesis of the epistatic interaction of two gene pairs proves consistent throughout, and the problem is solved.

This example demonstrates the need for a basic theoretical knowledge of transmission genetics. Then, you must search for the appropriate clues, so that you can proceed in a stepwise fashion toward a solution. Mastering problem solving requires practice, but will give you a great deal of satisfaction. Apply this general approach to the following problems.

2. In radishes, flower color may be red, purple, or white. The edible portion of the radish may be long or oval. When only flower color is studied, no dominance is evident, and red × white yields all purple. If these F_1 purples are interbred, the F_2 generation consists of 1/4 red:1/2 purple:1/4 white. Regarding radish shape, long is dominant to oval in a normal Mendelian fashion.

(a) Determine the F_1 and F_2 phenotypes from a cross between a true-breeding red, long radish and one that is white oval. Be sure to define all gene symbols initially.

Solution: This is a modified dihybrid cross where the gene pair controlling color exhibits incomplete dominance. Shape is controlled conventionally. First, establish gene symbols:

RR = red *Rr* = purple *rr* = white
O– = long *oo* = oval

P_1: *RROO* × *rroo*
(red long) (white oval)

F_1: all *RrOo* (purple long)

$F_1 \times F_1$: *RrOo* × *RrOo*

F_2:	1/4 *RR*	3/4 *O–*	3/16 *RR O–*	red long
		1/4 *oo*	1/16 *RR oo*	red oval
	2/4 *Rr*	3/4 *O–*	6/16 *Rr O–*	purple long
		1/4 *oo*	2/16 *Rr oo*	purple oval
	1/4 *rr*	3/4 *O–*	3/16 *rr O–*	white long
		1/4 *oo*	1/16 *rr oo*	white oval

Note that to generate the F_2 results, we have used the forked-line method. First, we consider the outcome of crossing F_1 parents for the color genes (*Rr* × *Rr*). Then the outcome of shape is considered (*Oo* × *Oo*).

(b) A red oval plant was crossed with a plant of unknown genotype and phenotype, yielding the data shown here:

Offspring: 103 red long:101 red oval
98 purple long: 100 purple oval

Determine the genotype and phenotype of the unknown plant.

Solution: The two characters appear to be inherited independently, so consider them separately. The data indicate a 1/4:1/4:1/4:1/4 proportion. First, consider color:

P_1: red × ??? (unknown)
F_1: 204 red (1/2)
198 purple (1/2)

Because the red parent must be *RR*, the unknown must have a genotype of *Rr* to produce these results. It is thus purple. Now, consider shape:

P_1: oval × ??? (unknown)
F_1: 201 long (1/2)
201 oval (1/2)

Since the oval plant must be *oo*, the unknown plant must have a genotype of *Oo* to produce these results. It is thus long. The unknown plant is here

RrOo purple long

3. In humans, red–green color blindness is inherited as an X-linked recessive trait. A woman with normal vision whose father is color blind marries a male who has normal vision. Predict the color vision of their male and female offspring.

Solution: The female is heterozygous since she inherited an X chromosome with the mutant allele from her father. Her husband is normal. Therefore, the parental genotypes are

Cc × C ↾ (↾ is the Y chromosome)

All female offspring are normal (*CC* or *Cc*). One-half of the male children will be color blind (*c* ↾), and the other half will have normal vision (*C* ↾).

Problems and Discussion Questions

1. In shorthorn cattle, coat color may be red, white, or roan. Roan is an intermediate phenotype expressed as a mixture of red and white hairs. The following data were obtained from various crosses:

red	×	red	⟶	all red
white	×	white	⟶	all white
red	×	white	⟶	all roan
roan	×	roan	⟶	1/4 red: 1/2 roan: 1/4 white

How is coat color inherited? What are the genotypes of parents and offspring for each cross?

2. Contrast incomplete dominance and codominance.

3. In foxes, two alleles of a single gene, *P* and *p*, may result in lethality (*PP*), platinum coat (*Pp*), or silver coat (*pp*). What ratio is obtained when platinum foxes are interbred? Is the *P* allele behaving dominantly or recessively in causing lethality? In causing platinum coat color?

4. In mice, a short-tailed mutant was discovered. When it was crossed to a normal long-tailed mouse, 4 offspring were short tailed and 3 were long tailed. Two short-tailed mice from the F_1 generation were selected and crossed. They produced 6 short-tailed and 3 long-tailed mice. These genetic experiments were repeated three times with approximately the same results. What genetic ratios are illustrated? Hypothesize the mode of inheritance and diagram the crosses.

5. List all possible genotypes for the A, B, AB, and O phenotypes. Is the mode of inheritance of the ABO blood types representative of dominance? Of recessiveness? Of codominance?

6. With regard to the ABO blood types in humans, determine the genotype of the male parent and female parent shown here:

Male Parent: Blood type B; mother type O;

Female Parent: Blood type A; father type B.

Predict the blood types of the offspring that this couple may have and the expected proportion of each.

7. In a disputed parentage case, the child is blood type O, while the mother is blood type A. What blood type would exclude a male from being the father? Would the other blood types prove that a particular male was the father?

8. The A and B antigens in humans may be found in water-soluble form in secretions, including saliva, of some individuals (*Se/Se* and *Se/se*) but not in others (*se/se*). The population thus contains "secretors" and "nonsecretors."

(a) Determine the proportion of various phenotypes (blood type and ability to secrete) in matings between individuals that are blood type AB and type O, both of whom are *Se/se*.

(b) How will the results of such matings change if both parents are heterozygous for the gene controlling the synthesis of the H substance (*Hh*)?

9. Describe what is meant by the phrase "gene interaction" as it pertains to the influence of genes on the phenotype.

10. In rabbits, a series of multiple alleles controls coat color in the following way: *C* is dominant to all other alleles and causes full color. The chinchilla phenotype is due to the c^{ch} allele, which is dominant to all alleles other than *C*. The c^h allele, dominant only to c^a (albino), results in the Himalayan coat color. Thus, the order of dominance is $C > c^{ch} > c^h > c^a$. For each of the following three cases, the phenotypes of the P_1 generations of two crosses are shown, as well as the phenotype of one member of the F_1 generation.

	P_1 Phenotypes		*F_1 Phenotypes*
(a)	Himalayan × Himalayan	→	albino
			× ⟶ ??
	full color × albino	⟶	chinchilla
(b)	albino × chinchilla	⟶	albino
			× ⟶ ??
	full color × albino	⟶	fullcolor
(c)	chinchilla × albino	⟶	Himalayan
			× ⟶ ??
	full color × albino	⟶	Himalayan

Chinchilla rabbit

Himalayan rabbit

For each case, determine the genotypes of the P_1 generation and the F_1 offspring, and predict the results of making each cross between F_1 individuals as previously shown.

11. In the guinea pig, one locus involved in the control of coat color may be occupied by any of four alleles: C (black), c^k (sepia), c^d (cream), or c^a (albino). Like coat color in rabbits (Problem 10), an order of dominance exists: $C > c^k > c^d > c^a$. In the following crosses, write the parental genotypes and predict the phenotypic ratios that would result:

(a) sepia × cream where both guinea pigs had an albino parent;

(b) sepia × cream where the sepia guinea pig had an albino parent and the cream guinea pig had two sepia parents;

(c) sepia × cream, where the sepia guinea pig had two full color parents and the cream guinea pig had two sepia parents;

(d) sepia × cream, where the sepia guinea pig had a full color parent and an albino parent and the cream guinea pig had two full color parents.

12. Three gene pairs located on separate autosomes determine flower color and shape as well as plant height. The first pair exhibits incomplete dominance, where the color can be red, pink (the heterozygote), or white. The second pair leads to personate (dominant) or peloric (recessive) flower shape, while the third gene pair produces either the dominant tall trait or the recessive dwarf trait. Homozygous plants that are red, personate, and tall are crossed to those that are white, peloric, and dwarf. Determine the F_1 genotype(s) and phenotype(s). If the F_1 plants are interbred, what proportion of the offspring will exhibit the same phenotype as the F_1 plants?

personate **peloric**

13. As in Problem 12, color may be red, white, or pink, and flower shape may be personate or peloric. For the following crosses, determine the P_1 and F_1 genotypes:

(a) red peloric × white personate	F1: all pink personate;
(b) red personate × white peloric	F1: all pink personate;
(c) pink personate × red peloric	F1: 1/4 red personate 1/4 red peloric 1/4 pink peloric 1/4 pink personate
(d) pink personate× white peloric	F1: 1/4 white personate 1/4 white peloric 1/4pink personate 1/4 pink peloric

What phenotypic ratios would result from crossing the F_1 of (a) to the F_1 of (b)?

14. Horses can be cremello (a light cream color), chestnut (a brownish color), or palomino (a golden color with white in the horse's tail and mane). Of these phenotypes, only palominos never breed true.

cremello × palomino	1/2 cremello 1/2 palomino
chestnut × palomino	1/2 chestnut 1/2 palomino
palomino × palomino	1/4 chestnut 1/2 palomino 1/4 cremello

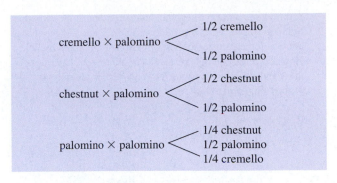

(a) From the results shown in the previous column, determine the mode of inheritance by assigning gene symbols and indicating which genotypes yield which phenotypes:

b) Predict the F_1 and F_2 results of many initial matings between cremello and chestnut horses.

15. With reference to the eye-color phenotypes produced by the recessive, autosomal, unlinked *brown* and *scarlet* loci in *Drosophila* (see Figure 10–10), predict the F_1 and F_2 results of the following P_1 crosses (recall that when both the *brown* and *scarlet* alleles are homozygous, no pigment is produced, and the eyes are white):

(a) wild type × white

(b) wild type × scarlet

(c) brown × white

16. Pigment in the mouse is only produced when the *C* allele is present. Individuals of the *cc* genotype have no color. If color is present, it may be determined by the *A*, *a* alleles. *AA* or *Aa* results in agouti color, while *aa* results in black coats.

(a) What F_1 and F_2 genotypic and phenotypic ratios are obtained from a cross between *AACC* and *aacc* mice?

(b) In three crosses between agouti females whose genotypes were unknown and males of the *aacc* genotype, the following phenotypic ratios were obtained:

(1) 8 agouti 8 colorless	(2) 9 agouti 10 black	(3) 4 agouti 5 black 10 colorless

What are the genotypes of these female parents?

17. In some plants a red pigment, cyanidin, is synthesized from a colorless precursor. The addition of a hydroxyl group (OH^-) to the cyanidin molecule causes it to become purple. In a cross between two randomly selected purple plants, the following results were obtained:

94 purple
31 red
43 colorless

How many genes are involved in the determination of these flower colors? Which genotypic combinations produce which phenotypes? Diagram the purple × purple cross.

18. In rats, the following genotypes of two independently assorting autosomal genes determine coat color:

A–B–	(gray)
A–bb	(yellow)
aaB–	(black)
aabb	(cream)

A third gene pair on a separate autosome determines whether or not any color will be produced. The *CC* and *Cc* genotypes allow color according to the expression of the *A* and *B* alleles. However, the *cc* genotype results in albino rats regardless of the *A* and *B* alleles present. Determine the F_1 phenotypic ratio of the following crosses:

(a) *AAbbCC*	×	*aaBBcc*
(b) *AaBBCC*	×	*AABbcc*
(c) *AaBbCc*	×	*AaBbcc*
(d) *AaBBCc*	×	*AaBBCc*
(e) *AABbCc*	×	*AABbcc*

19. Given the inheritance pattern of coat color in rats, as described in Problem 18, predict the genotype and phenotype of the parents who produced the following F_1 offspring:

(a) 9/16 gray: 3/16 yellow: 3/16 black: 1/16 cream
(b) 9/16 gray: 3/16 yellow: 4/16 albino
(c) 27/64 gray: 16/64 albino: 9/64 yellow: 9/64 black: 3/64 cream
(d) 3/8 black: 3/8 cream: 2/8 albino
(e) 3/8 black: 4/8 albino: 1/8 cream

20. In a species of the cat family, eye color can be gray, blue, green, or brown, and each trait is true breeding. In separate crosses involving homozygous parents, the following data were obtained:

Cross	P_1	F_1	F_2
A	green × gray	all green	3/4 green: 1/4 gray
B	green × brown	all green	3/4 green: 1/4 brown
C	gray × brown	all green	9/16 green: 3/16 brown 3/16 gray: 1/16 blue

(a) Analyze the data: How many genes are involved? Define gene symbols and indicate which genotypes yield each phenotype.
(b) In a cross between a gray-eyed cat and one of unknown genotype and phenotype, the F_2 generation was not observed. However, the F_2 resulted in the same F_2 ratio as in cross C. Determine the genotypes and phenotypes of the unknown P_1 and F_1 cats.

21. In a plant, a tall variety was crossed with a dwarf variety. All F_1 plants were tall. When $F_1 \times F_1$ plants were interbred, 9/16 of the F_2 were tall and 7/16 were dwarf.
(a) Explain the inheritance of height by indicating the number of gene pairs involved and by designating which genotypes yield tall and which yield dwarf. (Use dashes where appropriate.)
(b) Of the F_2 plants, what proportion of them will be true breeding if self-fertilized? List these genotypes.

22. In a unique species of plants, flowers may be yellow, blue, red, or mauve. All colors may be true breeding. If plants with blue flowers are crossed to red-flowered plants, all F_1 plants have yellow flowers. When carried to an F_2 generation, the following ratio was observed:

9/16 yellow: 3/16 blue: 3/16 red: 1/16 mauve

In still another cross using true-breeding parents, yellow-flowered plants are crossed with mauve-flowered plants. Again, all F_1 plants had yellow flowers and the F_2 showed a 9:3:3:1 ratio, as just shown.
(a) Describe the inheritance of flower color by defining gene symbols and designating which genotypes give rise to each of the four phenotypes.
(b) Determine the F_1 and F_2 results of a cross between true-breeding red and true-breeding mauve-flowered plants.

23. Shown in the following table are five human matings (1–5), including both maternal and paternal phenotypes for ABO, MN, and Rh blood-group antigen status:

Parental Phenotypes		*Offspring*
(1) A, M, Rh^-	× A, N, Rh^-	(a) A, N, Rh^-
(2) B, M, Rh^-	× B, M, Rh^+	(b) O, N, Rh^+
(3) O, N, Rh^+	× B, N, Rh^+	(c) O, MN, Rh^-
(4) AB, M, Rh^+	× O. N, Rh^+	(d) B, M, Rh^+
(5) AB, MN, Rh^-	× AB, MN, Rh^-	(e) B, MN, Rh^+

Each mating resulted in one of the five offspring shown to the right (a–e). Match each offspring with one correct set of parents, using each parental set only once. Is there more than one set of correct answers?

24. A husband and wife have normal vision, although both of their fathers are red–green color-blind, which is inherited as an X-linked recessive condition. What is the probability that their first child will be (a) a normal son? (b) a normal daughter? (c) a color-blind son? (d) a color-blind daughter?

25. In humans, the ABO blood type is under the control of autosomal multiple alleles. Color blindness is a recessive X-linked trait. If two parents who are both type *A* and have normal vision produce a son who is color blind and is type O, what is the probability that their next child will be a female who has normal vision and is type O?

26. In *Drosophila*, an X-linked recessive mutation, scalloped (*sd*), causes irregular wing margins. Diagram the F_1 and F_2 results if (a) a scalloped female is crossed with a normal male; (b) a scalloped male is crossed with a normal female. Compare these results with those that would be obtained if scalloped were not X-linked.

27. Another recessive mutation in *Drosophila*, ebony (*e*), is on an autosome (chromosome 3) and causes darkening of the body compared with wild-type flies. What phenotypic F_1 and F_2 male and female ratios will result if a scalloped-winged female with normal body color is crossed with a normal-winged ebony male? Work this problem by both the Punnett square method and the forked-line method.

28. In *Drosophila*, the X-linked recessive mutation *vermilion* (*v*) causes bright red eyes, which is in contrast to brick-red eyes of wild type. A separate autosomal recessive mutation, *suppressor of vermilion* (*su-v*), causes flies homozygous or hemizygous for *v* to have wild-type eyes. In the absence of *vermilion* alleles, *su-v* has no effect on eye color. Determine the F_1 and F_2 phenotypic ratios from a cross between a female with wild-type alleles at the *vermilion* locus, but who is homozygous for *su-v*, with a *vermilion* male who has wild-type alleles at the *su-v* locus.

29. While *vermilion* is X-linked and brightens the eye color, *brown* is an autosomal recessive mutation that darkens the eye. Flies carrying both mutations, lose all pigmentation, and are white eyed. Predict the F_1 and F_2 results of the following crosses:
(a) *vermilion* females × *brown* males
(b) *brown* × *vermilion* males
(c) *white* females × wild-type males

30. In a cross in *Drosophila* involving the X-linked recessive eye mutation *white* and the autosomally linked recessive eye mutation *sepia* (resulting in a dark eye), predict the F_1 and F_2 results of crossing true-breeding parents of the following phenotypes:

(a) white females × sepia males;

(b) sepia females × white males.

Note that white is epistatic to the expression of sepia.

31. Consider the following three pedigrees all involving a single human trait:

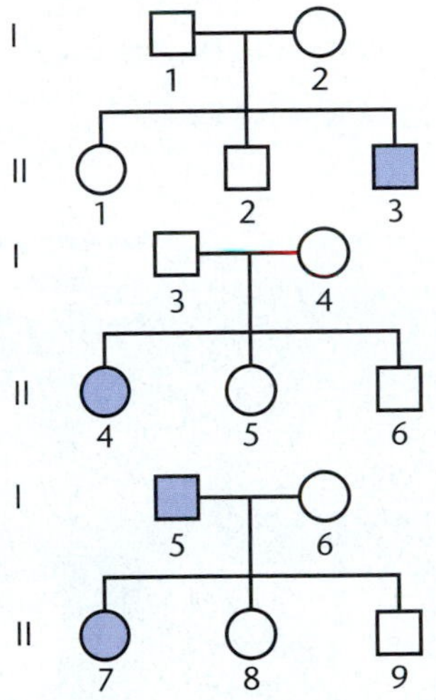

(a) Which conditions, if any, can be excluded?

Conditions: dominant and X-linked
dominant and autosomal
recessive and X-linked
recessive and autosomal

(b) For any condition that you excluded, indicate the single individual in generation II (e.g., II-1, II-2) that was most instrumental in your decision to exclude that condition. If none were excluded, answer "none apply."

(c) Given your conclusions in part (a), indicate the genotype of the individuals listed here.

II-1; II-6; II-9.

If more than one possibility applies, list all possibilities. Use the symbols *A* and *a* for the genotypes.

32. In spotted cattle, the colored regions may be mahogany or red. If a red female and a mahogany male, both derived from separate true-breeding lines, are mated and the cross carried to an F_2 generation, the following results are obtained:

F_1:	1/2 mahogany males
	1/2 red females
F_2:	3/8 mahogany males
	1/8 red males
	1/8 mahogany males
	3/8 red remales

When the reciprocal of the initial cross is performed (mahogany female and red male), identical results are obtained. Explain these results by postulating how the color is genetically determined. Diagram the crosses.

33. Predict the F_1 and F_2 results of crossing a male fowl that is cock feathered with a true-breeding hen-feathered female fowl. Recall that these traits are sex limited.

34. Two mothers give birth to sons at the same time at a busy urban hospital. The son of couple 1 is afflicted with hemophilia, a disease caused by an X-linked recessive allele. Neither parent has the disease. Couple 2 have a normal son, despite the fact that the father has hemophilia. Several years later, couple 1 sues the hospital, claiming that these two newborns were swapped in the nursery following their birth. You are called, as a genetic counselor, to testify. What information can you provide the jury concerning the allegation?

35. Discuss the topic of phenotypic expression and the many factors that impinge on it.

36. Contrast penetrance and expressivity as the terms relate to phenotypic expression.

37. Contrast the phenomena of genetic anticipation and genomic imprinting.

Extra-Spicy Problems

38. Labrador retrievers may be black, brown, or golden in color. While each color may breed true, many different outcomes occur if many litters are examined from a variety of matings, where the parents are not necessarily true breeding. Shown below are just some of the many possibilities.

(a) black	×	brown	⟶	all black
(b) black	×	brown	⟶	1/2 black 1/2 brown
(c) black	×	brown	⟶	3/4 black 1/4 golden
(d) black	×	golden	⟶	all black
(e) black	×	golden	⟶	4/8 golden 3/8 black 1/8 brown
(f) black	×	brown	⟶	2/4 golden 1/4 black 1/4 brown
(g) brown	×	brown	⟶	3/4 brown 1/4 golden
(h) black	×	black	⟶	9/16 black 4/16 golden 3/16 brown

Propose a mode of inheritance that is consistent with these data, and indicate the corresponding genotypes of the parents in each mating. Indicate as well the genotypes of dogs that breed true for each color.

39. In watermelons (*Citrullus lanatus*), the flesh may exhibit numerous colors, including canary, red, salmon, and orange. Three crosses are shown below involving four true-breeding strains:

Yellow Baby	⟶	canary flesh
Tendersweet	⟶	orange flesh
Golden Honey	⟶	salmon flesh
Sweet Princess	⟶	red flesh

(a) Consider initially the first two crosses:

Cross I: Yellow Baby × Tendersweet
(canary) (orange)

F_1: all canary

F_2: 9/16 canary
4/16 orange
3/16 red

Cross II: Yellow Baby × Golden Honey
(canary) (salmon)

F_1: all canary

F_2: 9/16 canary
4/16 salmon
3/16 red

Analyze the results and put forward a proposal that is consistent with them that explains the inheritance of the four flesh colors.

(b) Now consider the third cross:

Cross III: Yellow Baby × Sweet Princess
(canary) (red)

F_1: all canary

F_2: 9/16 canary
7/16 red

How can you extend your proposal in Part a) to accommodate the results in Cross III?

(c) Based on your proposals in Part (a) and (b), what are the complete genotypes of the four true-breeding strains?

(d) Propose two further crosses and the predicted results that test and confirm your proposals in Part a) and *b).*
[*Reference*: Henderson, W.R., Scott, G.H., and Whener, T.C. 1998. *J. Heredity* 89:50-53.]

40. A true-breeding purple-leafed plant isolated from one side of the rain forest in Puerto Rico (El Yunque) was crossed to a true-breeding white variety found on the other side of the rain forest. The F_1 offspring were all purple. A large number of $F_1 \times F_1$ crosses produced in the following results:

purple: 4219; white: 5781 (Total = 10,000)

Propose an explanation for the inheritance of leaf color. As a geneticist, how might you go about testing your hypothesis? Describe the genetic experiments that you would conduct.

41. In Dexter and Kerry cattle, animals may be polled (hornless) or horned. The Dexter animals have short legs, whereas the Kerry animals have long legs. When many offspring were obtained from matings between polled Kerrys and horned Dexters, one-half were found to be polled Dexters and one-half polled Kerrys. When these two types of F_1 cattle were tied to one another, the following F_2 data were obtained:

3/8 polled Dexters
3/8 polled Kerrys
1/8 polled Dexters
1/8 polled Kerrys

Kerry cow

Dexter cow

A geneticist was puzzled by these data and interviewed farmers who had bred these cattle for decades. She learned that Kerrys were true breeding. Dexters, on the other hand, were not true breeding and never produced as many offspring as Kerrys. Provide a genetic explanation for these observations.

42. An alien geneticist escaping from a planet where genetic research had recently been prohibited brought with him to Earth two pure-breeding lines of pet frogs. One line croaked by uttering *rib-it rib-it* and had purple eyes. The other line croaked more softly by muttering *knee-deep knee-deep* and had green eyes. With a new-found freedom of inquiry, he mated the two types of frogs. In the F_1, all frogs had blue eyes and uttered *rib-it rib-it*. He proceeded to make many $F_1 \times F_1$ crosses, and when he fully analyzed the F_2 data, he realized they could be reduced to the following ratio:

27/64 blue-eyed, rib-it utterer

12/64 green-eyed, rib-it utterer

9/64 blue-eyed, knee-deep mutterer

9/64 purple-eyed, rib-it utterer

4/64 green-eyed, knee-deep mutterer

3/64 purple-eyed, knee-deep mutterer

(a) How many total gene pairs are involved in the inheritance of both traits? Support your answer.
(b) Of these, how many are controlling eye color? How can you tell? How many are controlling croaking?
(c) Assign gene symbols for all phenotypes and indicate the genotypes of the P_1 and F_1 frogs.
(d) Indicate the genotypes of the six F_2 phenotypes.
(e) After years of experiments, the geneticist isolated pure-breeding strains of all six F_2 phenotypes. Indicate the F_1 and F_2 phenotypic ratios of the following cross, using these pure-breeding strains:

blue-eyed, knee-deep mutterer × purple-eyed, rib-it utterer

(f) One set of crosses with his true-breeding lines initially caused the geneticist some confusion. When he crossed true-breeding purple-eyed, knee-deep mutterers with true-breeding green-eyed, knee-deep mutterers, he often got different results. In some matings, all offspring were blue-eyed, knee-deep mutterers, but in other matings all offspring were purple-eyed, knee-deep mutterers. In still a third mating, $\frac{1}{2}$ blue-eyed, knee-deep mutterers and $\frac{1}{2}$ purple-eyed, knee-deep mutterers were observed. Explain why the results differed.
(g) In another experiment, the geneticist crossed two purple-eyed, rib-it utterers together with the results shown here:

9/16 purple-eyed, rib-it utterer

3/16 purple-eyed, knee-deep mutterer

3/16 green-eyed, rib-it utterer

1/16 green-eyed, knee-deep mutterer

What were the genotypes of the two parents?

43. The preceding pedigree is characteristic of an inherited condition known as male precocious puberty, where affected males show signs of puberty by age 4:

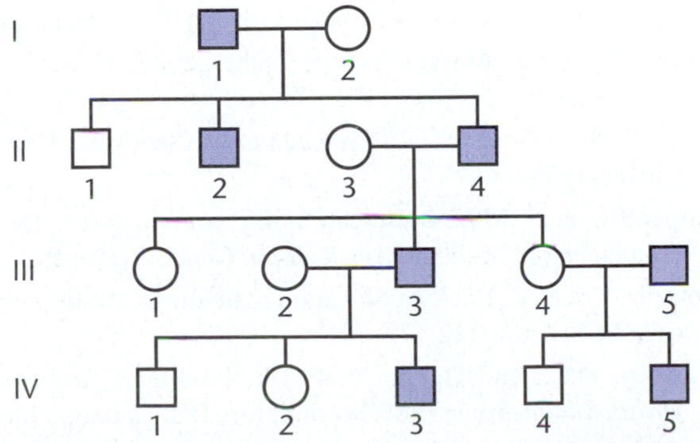

Propose a genetic explanation of this phenotype [Reference: A. Shenker (1993)].

44. In birds, the male contains two like sex chromosomes and is designated ZZ. The female contains one Z as well as a nearly blank chromosome, designated W (females are ZW). In parakeets, two genes control feather color. The presence of the dominant *Y* allele at the first gene results in the production of a yellow pigment. The dominant *B* allele at the second gene controls melanin production. When both genes are active, a green pigment results. If only the *Y* gene is active, the feathers are yellow. If only the *B* gene is active, a blue color is exhibited. If neither gene is active, the birds are albinos. Therefore, with our conventional designations, phenotypes are produced as follows:

$Y{-}B{-}$	green
$Y{-}bb$	yellow
$yyB{-}$	blue
$yybb$	albino

(a) A series of crosses established that one of the genes is autosomal and one is Z-linked. Based on the results of the following cross shown here, where both parents are true-breeding, determine which gene is Z-linked:

P_1: green male × albino female

F_1: 1/2 green males, 1/2 green females

F_2: 6/16 green males

2/16 yellow males

3/16 green females

1/16 yellow females

3/16 blue females

1/16 albino females

Support your answer by establishing the genotypes of the P_1 parents and working the cross through the F_2 generation.

(b) In a cross where the parental genotypes and whether the parents were true breeding were unknown, the offspring from repeated matings were recorded, as shown here:

13 green males

3 yellow males

11 blue females

5 albino females

Based on the results, determine the phenotypes and genotypes of the parents.

45. An animal whose pelts are extremely valuable was studied. Pelts were found to be either brown (B-) or white (bb), with white being rare, but much more valuable than brown. However, animals with white pelts breed very poorly, if at all. A breeder buys a male and female brown animal and mates them, ultimately recovering 15 brown and 5 white offspring. He sells three of the valuable white animals and consults you as a geneticist for advice on how to maximize his profits in marketing these animals.

a) When he randomly mates many F_1 brown animals, some matings produce no white offspring. Explain to him why.

b) He now has available P_1, F_1 and F_2 animals, as well as the records of all the $F_1 \times F_2$ matings. Provide him with the most efficient strategy for maximizing the production of white animals in future offspring.

46. Three independently assorting genes are known to control the biochemical pathway here that provides the basis for flower color in a hypothetical plant:

$$\text{colorless} \xrightarrow{A-} \text{yellow} \xrightarrow{B-} \text{green} \xrightarrow{C-} \text{speckled}$$

Homozygous recessive mutations, which interrupt each step, are known. Determine the phenotypic results in the F_1 and F_2 generations resulting from the P_1 crosses given, involving true-breeding plants:

(a) speckled	(*AABBCC*)	×	yellow	(*AAbbCC*)
(b) yellow	(*AAbbCC*)	×	green	(*AABBcc*)
(c) colorless	(*aaBBCC*)	×	green	(*AABBcc*)

47. Deep in a previously unexplored South American rain forest, a species of plants was discovered with true-breeding varieties whose flowers were either pink, rose, orange, or purple. A very astute plant geneticist made a single cross, carried to the F_2 generation, as shown here:

P_1:	purple × pink
F_1:	all purple
F_2:	27/64 purple
	16/64 pink
	12/64 rose
	9/64 orange

Based solely on these data, he was able to propose both a mode of inheritance for flower pigmentation and a biochemical pathway for the synthesis of these pigments.

Carefully study the data. Create a hypothesis of your own to explain the mode of inheritance. Then propose a biochemical pathway consistent with your hypothesis. How could you test the hypothesis by making other crosses?

Selected Readings

Bartolomei, M.S., and Tilghman, S.M. 1997. Genomic imprinting in mammals. *Annu. Rev. Genet.* 31:493–525.

Brink, R.A., ed. 1967. *Heritage from Mendel.* Madison: University of Wisconsin Press.

Bultman, S.J., Michaud, E.J., and Woychik, R.P. 1992. Molecular characterization of the mouse *agouti* locus. *Cell* 71:1195–1204.

Carlson, E.A. 1987. *The gene: A critical history*, 2nd ed. Philadelphia: Saunders.

Cattanach, B.M., and Jones, J. 1994. Genetic imprinting in the mouse: Implications for gene regulation. *J. Inherit. Metab. Dis.* 17:403–20.

Chapman, A.B. 1985. *General and quantitative genetics.* Amsterdam: Elsevier.

Clarke, C.A. 1968. The prevention of "Rhesus" babies. *Sci. Am.* (Nov.) 219:46–52.

Corwin, H.O., and Jenkins, J.B. 1976. *Conceptual foundations of genetics: Selected readings.* Boston: Houghton-Mifflin.

Crow, J.F. 1983. *Genetics notes*, 8th ed. New York: Macmillan.

Drayna, D., and White, R. 1985. The genetic linkage map of the human X chromosome. *Science* 230:753–58.

Dunn, L.C. 1966. *A short history of genetics.* New York: McGraw-Hill.

East, E.M. 1910. A Mendelian interpretation of variation that is apparently continuous. *Am. Naturalist* 44:65–82.

———. 1916. Studies on size inheritance in *Nicotiana. Genetics* 1:164–76.

Falconer, D.S. 1981. Introduction to quantitative genetics, 2d ed. New York: Longman.

Feil, R., and Kelsey, G. 1997. Genomic imprinting: A chromatin connection. *Am. J. Hum. Genet.* 61:1213–19.

Foster, H.L. et al., eds. 1981. The mouse in biomedical research, Vol. 1. *History, genetics, and wild mice.* Orlando, FL: Academic Press.

Foster, M. 1965. Mammalian pigment genetics. *Adv. Genet.* 13:311–39.

Grant, V. 1975. *Genetics of flowering plants.* New York: Columbia University Press.

Harper, P.S. et al. 1992. Anticipation in myotonic dystrophy: New light on an old problem. *Am. J. Hum. Genet.* 51:10–16.

Howeler, C.J. et al. 1989. Anticipation in myotonic dystrophy: Fact or fiction? *Brain* 112:779–97.

Lindsley, D.C., and Grell, E.H. 1967. *Genetic variations of Drosophila melanogaster.* Washington, DC: Carnegie Institute of Washington.

GENETICS MediaLab

The resources that follow will help you achieve a better understanding of the concepts presented in this chapter. These resources can be found either on the CD packaged with this textbook or on the Companion Web site found at **http://www.prenhall.com/klug**

CD Resources:

Module 10.1: Extensions of Mendelian Inheritance
Module 10.2: X-linked Inheritance

Web Problem 1:

Time for completion = 15 minutes

Is dominance always complete? You will study flower color in the four-o'clock plant, *Mirabilis*. As in peas, flower color is determined by a single locus with two alleles. In this case, however, dominance is incomplete. Read about incomplete dominance in Chapter 10 of your text. Once you complete the F1 cross in the tutorial, go on to the next module to study the F2's. Suppose you begin by crossing true-breeding stocks with red and white flowers. What genotypic ratio would you predict in the F1? What is the expected phenotypic ratio in the F1? How do these results differ from the crosses of the peas you did in the last chapter? Suppose you cross the F1's; what ratios of phenotypes and genotypes would you expect in the F2 generation? Do the ratios of genotypes differ from a cross of heterozygotes in which one allele is dominant? To complete this exercise, visit Web Problem 1 in Chapter 10 of your Companion Web site, and select the keyword **INCOMPLETE DOMINANCE**.

Web Problem 2:

Time for completion = 10 minutes

Epistasis is the modification of the phenotypic expression of one locus due to the genotype at another locus. Epistatic effects are common among the genes affecting mammalian fur color. You will examine a table of loci that influence the coloration of horses and search for cases of epistatic interaction. In horses, the main gene affecting color is the *E* locus. The black allele (*E*) is dominant over the red allele (*e*). One example of epistasis is the *A* locus; if a horse is black, then the presence of the dominant *A* allele will restrict the black coloring to the "points" (that is, legs, nose, ears, and tail) of the horse, whereas a recessive homozygote will be black all over. Suppose you encounter an all-black horse. Use the table to determine what the genotype might be at the C and D loci. What might the genotype of an all-white horse be at the *E* locus? Is the W locus epistatic? Why might you not want to breed two all-white horses? To complete this exercise, visit Web Problem 2 in Chapter 10 of your Companion Web site, and select the keyword **EPISTASIS**.

Web Problem 3:

Time for completion = 10 minutes

The allele for red-green color blindness is on the X chromosome in humans. Visit Web Problem 3 in Chapter 10 of your Companion Web site, and select the keyword **SEX LINKAGE**. There you will find a series of problems that let you explore and discover the genotypes of the members of a family with red-green color-blindness.

Mahedevan, M. et al. 1992. Myotonic dystrophy mutation: An unstable CTG repeat in the 39 untranslated region of the gene. *Science* 255:1253–58.

McKusick, V.A. 1962. On the X chromosome of man. *Quart. Rev. Biol.* 37:69–175.

Morgan, T.H. 1910. Sex-limited inheritance in *Drosophila. Science* 32:120–22.

Nolte, D.J. 1959. The eye-pigmentary system of *Drosophila. Heredity* 13:233–41.

Pawelek, J.M., and Körner, A.M. 1982. The biosynthesis of mammalian melanin. *Am. Sci.* 70:136–45.

Peters, J.A., ed. 1959. *Classic papers in genetics*. Englewood Cliffs, NJ: Prentice-Hall.

Phillips, P.C. 1998. The language of gene interaction. *Genetics* 149:1167-71.

Race, R.R., and Sanger, R. 1975. *Blood groups in man*, 6th ed. Oxford: Blackwell Scientific Publishers.

Sapienza, C. 1990. Parental imprinting of genes. *Sci. Am.* (Oct.) 363:52–60.

Shenker, A. 1993. A constitutively activating mutation of the luteinizing hormone receptor in familial male precocious puberty. *Nature* 365:652–54.

Siracusa, L.D. 1994. The *agouti* gene: Turned on to yellow. *Cell* 10:423–28.

Smith, C.A. 1995. New hope for overcoming canine inherited disease. *J. Am. Veter. Med. Assoc.* 204:41–46.

Stern, C. 1973. *Principles of human genetics*, 3d ed. New York: W.H. Freeman.

Tearle, R.G. et al. 1989. Cloning and characterization of the scarlet gene of *Drosophila melanogaster. Genetics* 122:595–606.

Voeller, B.R., ed. 1968. *The chromosome theory of inheritance—Classic papers in development and heredity*. New York: Appleton-Century-Crofts.

Vogel, F., and Motulsky, A.G. 1997. *Human genetics: Problems and approaches*, 3d ed. New York: Springer-Verlag.

Watkins, M.W. 1966. Blood group substances. *Science* 152:172–81.

Wiener, A.S., ed. 1970. *Advances in blood groupings*, Vol. 3. New York: Grune and Stratton.

Yoshida, A. 1982. Biochemical genetics of the human blood group ABO system. *Am. J. Hum. Genet.* 34:1–14.

Ziegler, I. 1961. Genetic aspects of ommochrome and pterin pigments. *Adv. Genet.* 10:349–403.

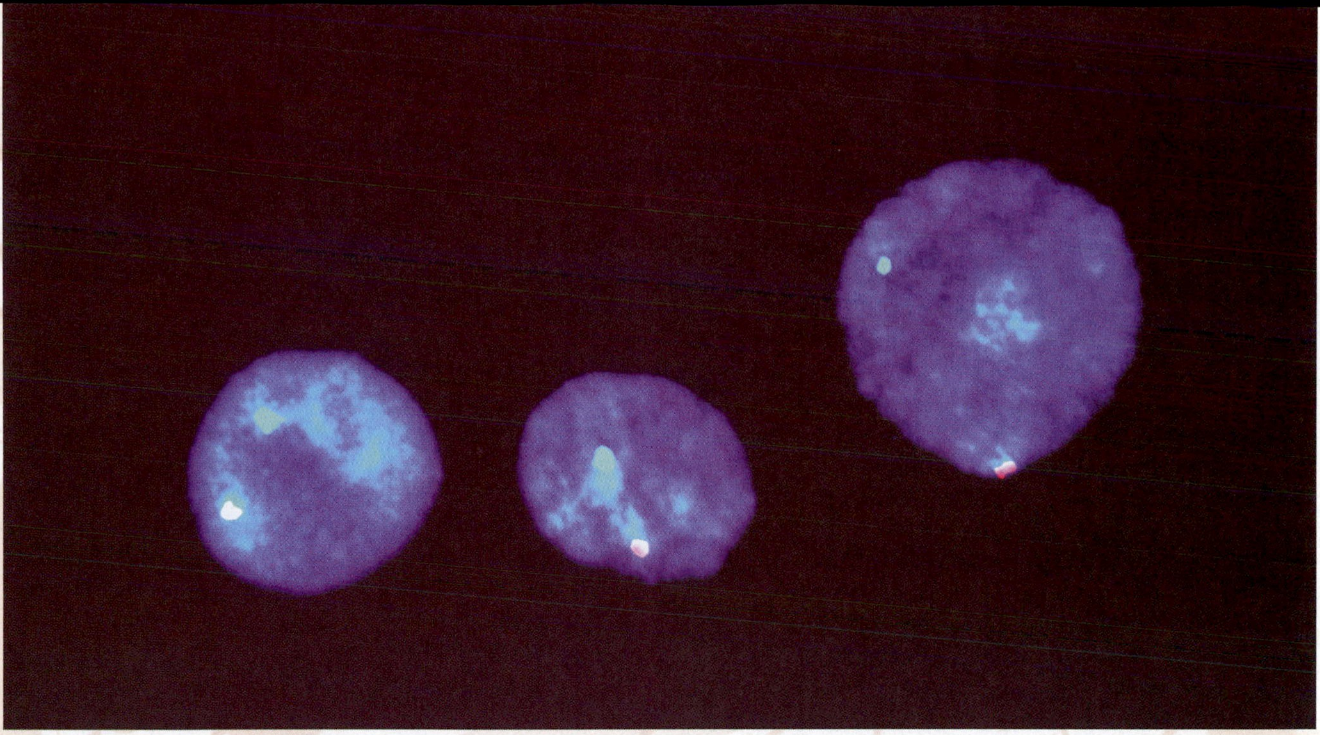

Demonstration of the X and Y chromosomes (the green and the pink dots respectively) in mammalian fetal cells using fluorescent *in situ* hybridization (FISH).

11

Sex Determination and Sex Chromosomes

KEY CONCEPTS

11.1 Sexual Differentiation Plays a Varying Role in the Life Cycle of Organisms
Chlamydomonas
Zea mays
Caenorhabditis elegans

11.2 X and Y Chromosomes Were First Linked to Sex Determination Early in the 20th Century

11.3 The Y Chromosome Determines Maleness in Humans
Klinefelter and Turner Syndromes
47,XXX Syndrome
47,XYY Condition
Sexual Differentiation in Humans
The Y Chromosome and Male Development

11.4 The Ratio of Males to Females in Humans Is Not 1.0

11.5 Dosage Compensation Prevents Excessive Expression of X-Linked Genes in Humans and Other Mammals
Barr Bodies
The Lyon Hypothesis
The Mechanism of Inactivation

11.6 The Ratio of X Chromosomes to Sets of Autosomes Determines Sex in *Drosophila*
Dosage Compensation in *Drosophila*
Drosophila Mosaics

11.7 Temperature Variation Controls Sex Determination in Reptiles

In the biological world, a wide range of reproductive modes and life cycles are recognized. Organisms exist that display no evidence of sexual reproduction. Other species alternate between short periods of sexual reproduction and prolonged periods of asexual reproduction. In most diploid eukaryotes, however, sexual reproduction is the only natural mechanism resulting in new members of an existing species. Orderly transmission of genetic units from parents to offspring, and thus any phenotypic variability, relies on the processes of segregation and independent assortment that occur during meiosis. Meiosis produces haploid gametes so that, following fertilization, the resulting offspring maintain the diploid number of chromosomes characteristic of their species. Hence, meiosis ensures genetic constancy within members of the same species.

These events, which are involved in the perpetuation of all sexually reproducing organisms, depend ultimately on an efficient union of gametes during fertilization. In turn, successful mating between organisms, the basis for fertilization, depends on some form of sexual differentiation in organisms. Even though it is not overtly evident, this differentiation occurs as low on the evolutionary scale as bacteria and single-celled eukaryotic algae. In evolutionarily higher forms of life, the differentiation of the sexes is more evident as phenotypic dimorphism in the males and females of each species. The shield and spear ♂, the ancient symbols of iron and Mars, and the mirror ♀ the symbol of copper and Venus, represent the maleness and femaleness acquired by individuals.

Dissimilar, or **heteromorphic chromosomes**, such as the X–Y pair, often characterize one sex or the other, resulting in their label as **sex chromosomes**. Nevertheless, it is genes, rather than chromosomes, that ultimately serve as the underlying basis of sex determination. As we will see, some of these genes are present on sex chromosomes, but others are autosomal. Extensive investigation has revealed a wide variation in sex chromosome systems, even in closely related organisms, suggesting that mechanisms controlling sex determination have undergone rapid evolution in many instances.

In this chapter, we will first review several representative modes of sexual differentiation by examining the life cycle of three organisms often studied in genetics: the green alga *Chlamydomonas;* the maize plant, *Zea mays;* and the nematode (roundworm), *Caenorhabditis elegans* (most often referred to as *C. elegans*). These will serve to contrast the different roles that sexual differentiation plays in the lives of diverse organisms. Then, we will delve more deeply into what is known about the genetic basis for the determination of sexual differences, with a particular emphasis on two other organisms: our own species, representative of mammals; and *Drosophila,* where pioneering sex-determining studies were performed.

11.1 Sexual Differentiation Plays a Varying Role in the Life Cycle of Organisms

In multicellular organisms, it is important to distinguish between **primary sexual differentiation**, which involves only the gonads where gametes are produced, and **secondary sexual differentiation**, which involves the overall appearance of the organism, including clear differences in such organs as mammary glands and external genitalia. In plants and animals, the terms **unisexual**, **dioecious**, and **gonochoric** are equivalent; they all refer to an individual containing only male *or* only female reproductive organs. Conversely, the terms **bisexual**, **monoecious**, and **hermaphroditic** refer to individuals containing both male *and* female reproductive organs, a common occurrence in both the plant and animal kingdoms. These organisms can produce fertile gametes of both sexes. The term **intersex** is usually reserved for individuals of intermediate sexual differentiation, who are most often sterile.

Chlamydomonas

The life cycle of the green alga *Chlamydomonas,* shown in Figure 11–1, is representative of organisms that exhibit only infrequent periods of sexual reproduction. Such organisms spend most of their life cycle in the haploid phase, asexually producing daughter cells by mitotic divisions. However, under unfavorable nutrient conditions, such as nitrogen depletion, certain daughter cells function as gametes. Following fertilization, a diploid zygote, which can withstand the unfavorable environment, is formed. When conditions become more suitable, meiosis ensues and haploid vegetative cells are again produced. In such species, there is little visible difference between the haploid vegetative cells that reproduce asexually and the haploid gametes that are involved in sexual reproduction. The two gametes that fuse together during mating are not usually morphologically distinguishable. Such gametes are called **isogametes**, and species producing them are said to be **isogamous**.

In 1954, Ruth Sager and Sam Granik demonstrated that gametes in *Chlamydomonas* could be subdivided into two mating types. Working with clones derived from single haploid cells, they showed that cells from a given clone would mate with cells from some, but not all other clones. When they tested mating abilities of large numbers of clones, all could be placed into one of two mating categories, either mt^+ or mt^- cells. "Plus" cells would only mate with "minus" cells, and vice versa, as represented in Figure 11–2. Following fertilization and meiosis, the four haploid cells (**zoospores**) produced were found to consist of two plus types and two minus types.

Further experimentation established that there is a chemical difference between plus and minus cells. When extracts were prepared from cloned *Chlamydomonas* cells (or their flagella) and then added to cells of the opposite mating type, clumping or agglutination occurred. No such agglutination occurred if the extract were added to cells of the mating type from which it was derived. These observations suggest that despite the morphological similarities between isogametes, a chemical differentiation has occurred between them. Therefore, in this alga, a primitive means of sex differentiation exists even though there is no morphological indication that such differentiation has occurred.

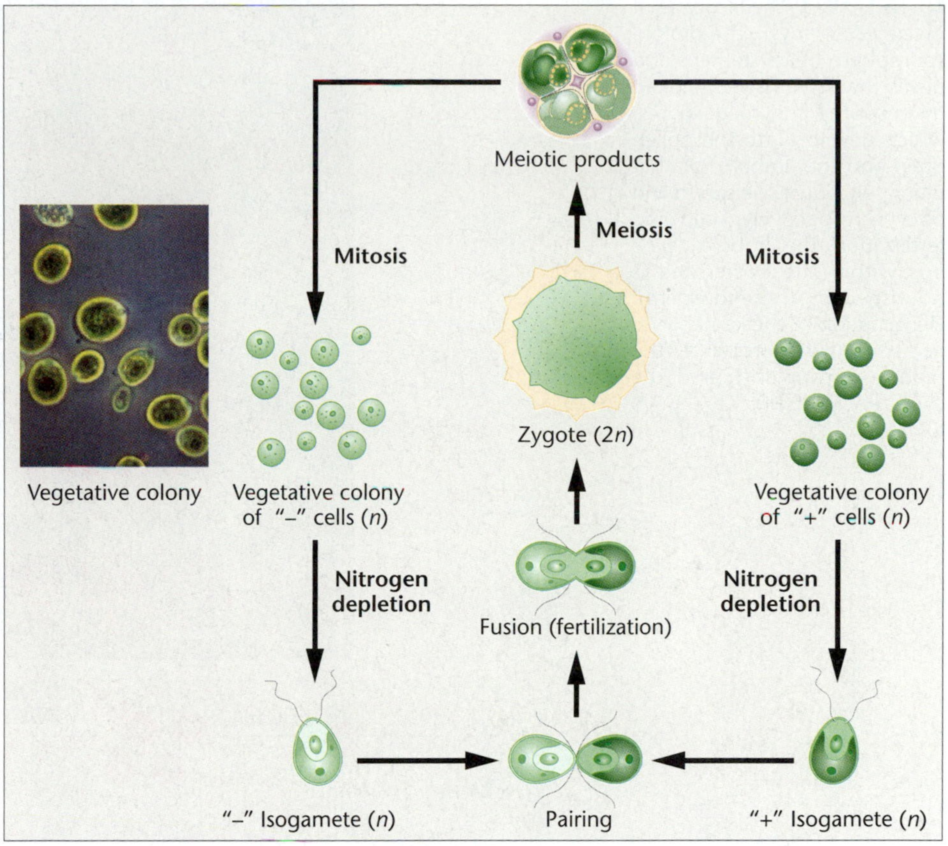

FIGURE 11–1 The life cycle of *Chlamydomonas.* Unfavorable conditions stimulate the formation of isogametes of opposite mating type that may fuse in fertilization. The resulting zygote undergoes meiosis, producing two haploid cells of each mating type. The photograph shows vegetative cells of this green alga.

Zea mays

As we first discussed in Chapter 2 (see Figure 2–16), life cycles of many plants alternate between the haploid gametophyte stage and the diploid sporophyte stage. The processes of meiosis and fertilization link the two phases during the life cycle. The relative amount of time spent in the two phases varies between the major plant groups. In some nonseed plants, such as mosses, the haploid gametophyte phase and the morphological structures representing this stage predominate. The reverse is true in seed plants.

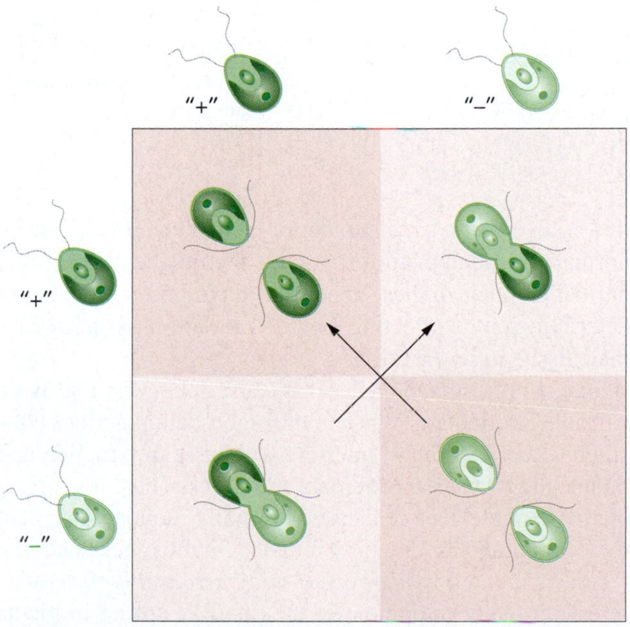

FIGURE 11–2 Illustration of mating types during fertilization in *Chlamydomonas*. Only when plus (+) and minus (−) cells are together will mating occur.

Maize (*Zea mays*), familiar to you as corn, exemplifies a monoecious seed plant, where the sporophyte phase and the morphological structures representing this stage predominate during the life cycle. Both male and female structures are present on the adult plant. Thus, sex determination must occur differently in different tissues of the same organism, as illustrated in the life cycle of this familiar plant (Figure 11–3). The **stamens**, or **tassels**, produce diploid microspore mother cells, each of which undergoes meiosis and gives rise to four haploid microspores. Each haploid microspore in turn develops into a mature male microgametophyte—the pollen grain—which contains two sperm nuclei with identical genotypes.

Comparable female diploid cells, known as megaspore mother cells, exist in the **pistil** of the sporophyte. Following meiosis, only one of the four haploid megaspores survives. It usually divides mitotically three times, producing a total of eight genetically identical haploid nuclei enclosed in the embryo sac. Two of these nuclei unite near the center of the embryo sac, becoming the endosperm nuclei. At the micropyle end of the sac where the sperm enters, three nuclei remain: the oocyte nucleus and two synergids. The other three antipodal nuclei are clustered at the opposite end of the embryo sac.

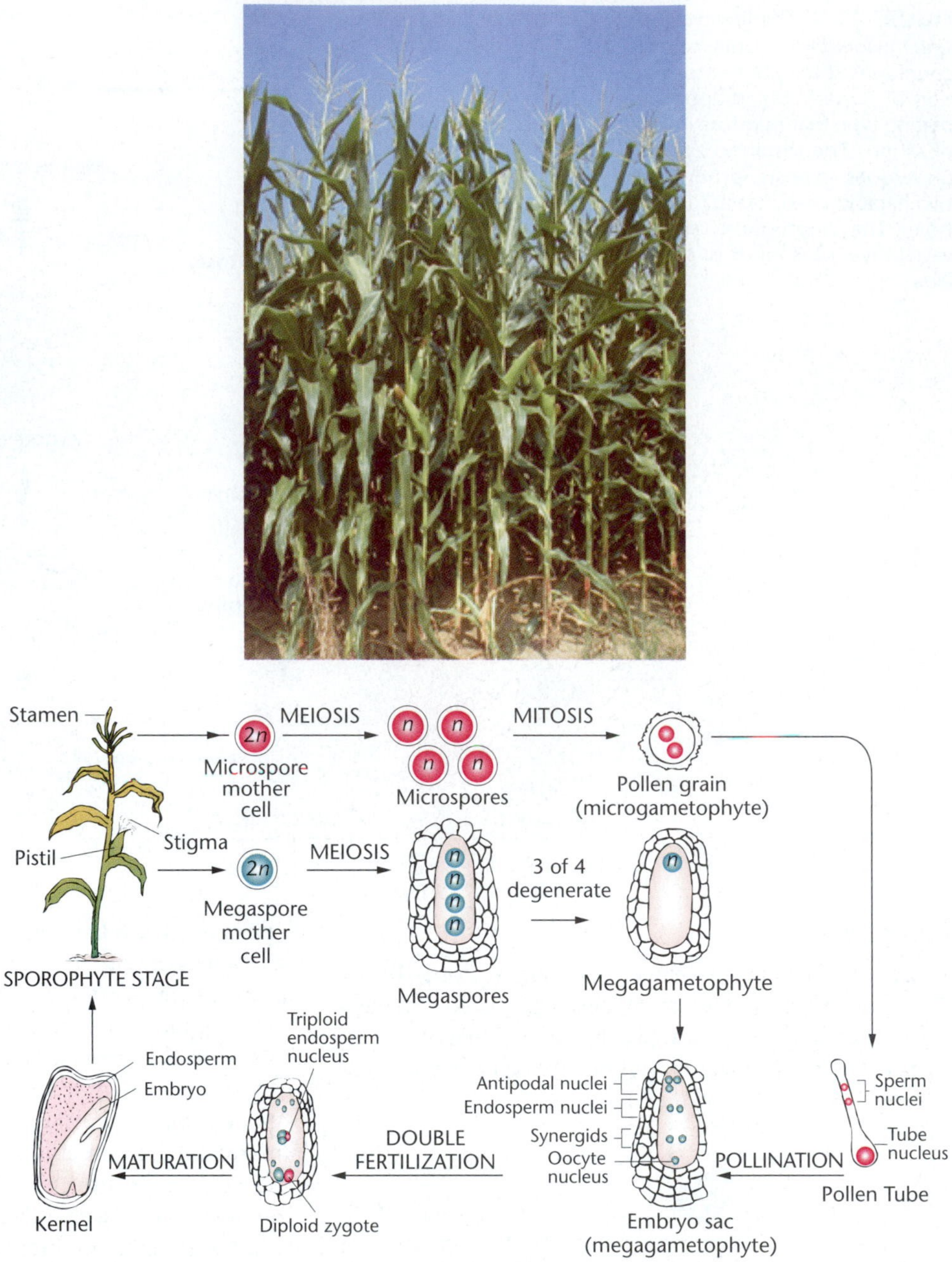

FIGURE 11–3 The life cycle of maize (*Zea mays*). The diploid sporophyte bears stamens and pistils that give rise to haploid microspores and megaspores, which develop into the pollen grain and the embryo sac that ultimately house the sperm and oocyte, respectively. Following fertilization, the embryo develops within the kernel and is nourished by the endosperm. Germination of the kernel gives rise to a new sporophyte (the mature corn plant), and the cycle repeats itself.

Pollination occurs when pollen grains make contact with the silks (or stigma) of the pistil and develop extensive pollen tubes that grow toward the embryo sac. When contact is made at the micropyle, the two sperm nuclei enter the embryo sac. One sperm nucleus unites with the haploid oocyte nucleus and the other sperm nucleus unites with two endosperm nuclei. This process, known as double fertilization, results in the diploid zygote nucleus and the triploid endosperm nucleus, respectively. Each ear of corn may contain as many as 1000 of these structures, each of which develops into a single kernel. Each kernel, if allowed to germinate, gives rise to a new plant, the sporophyte.

The mechanism of sex determination and differentiation in a monoecious plant like *Zea mays*, where the tissues that form both male and female gametes are of the same genetic constitution, was difficult to comprehend at first. However, the discovery of a large number of mutant genes that disrupt normal tassel and pistil formation supports the concept that normal products of these genes play an important role in sex determination by affecting the differentiation of male or female tissue in several ways.

For example, mutant genes that cause sex reversal provide valuable information. When homozygous, all mutations classified as *tassel seed* (*ts*) interfere with tassel production and induce the formation of female structures. Thus, it is possible for a single gene to cause a normally monoecious plant to become functionally only female. On the other hand, the recessive mutations *silkless* (*sk*) and *barren stalk* (*ba*) interfere with the development of the pistil, resulting in plants with only functional male reproductive organs.

Data gathered from studies of these and other mutants suggest that the products of many wild-type alleles of these genes interact in controlling sex determination. During development, certain cells are "determined" to become male or female structures. Following sexual differentiation into either male or female structures, male or female gametes are produced.

Caenorhabditis elegans

The nematode worm *Caenorhabditis elegans* [*C. elegans*, for short—Figure 11–4(a)] has become a popular organism in genetic studies, particularly during the investigation of the genetic control of development. Its usefulness is based on the fact that the hermaphroditic adult consists of exactly 959 cells, and the precise lineage of each cell can be traced back to specific embryonic origins. Among many interesting mutant phenotypes that have been studied, behavioral modifications have also been a favorite topic of inquiry.

There are two sexual phenotypes in these worms: males, which have only testes, and hermaphrodites that contain both testes and ovaries. During larval development of hermaphrodites, testes form that produce sperm, which is then stored. Ovaries are also produced, but oogenesis does not occur until the adult stage is reached several days later. The developed eggs are fertilized by the stored sperm during the process of self-fertilization.

The outcome of this process is quite interesting [Figure 11–4(b)]. The vast majority of organisms that result, like the parental worm, are hermaphrodites; less than 1 percent of the offspring are males. As adults, they can mate with hermaphrodites, producing about half male and half hermaphrodite offspring.

(a)

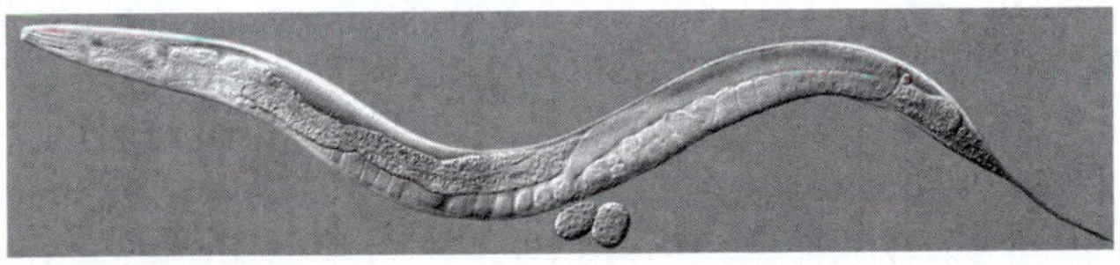

(b)

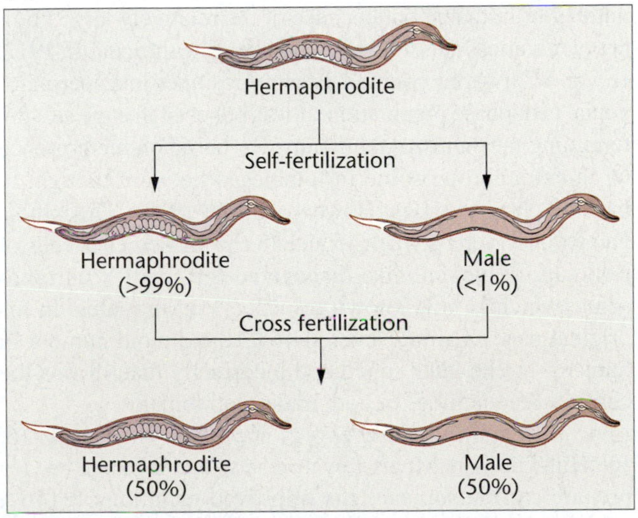

FIGURE 11–4 (a) Photomicrograph of an hermaphroditic nematode, *C. elegans*; (b) The outcomes of self-fertilization in an hermaphrodite and a mating of a hermaphrodite and a male worm.

The genetic signal that determines maleness in contrast to hermaphroditic development is provided by genes located on both the X chromosome and autosomes. *C. elegans* lacks a Y chromosome altogether. Hermaphrodites have two X chromosomes, while males have only one X chromosome. It is believed that the ratio of X chromosomes to the number of sets of autosomes ultimately determines the sex of these worms. A ratio of 1.0 (two X chromosomes and two copies of each autosome) results in hermaphrodites and a ratio of 0.5 results in males. The absence of a heteromorphic Y chromosome is not uncommon in organisms.

11.2 X and Y Chromosomes Were First Linked to Sex Determination Early in the 20th Century

How sex is determined has long intrigued geneticists. In 1891, H. Henking identified a nuclear structure in the sperm of certain insects, which he labeled the X-body. Several years later, Clarance McClung showed that some grasshopper sperm contain an unusual genetic structure, which he called a heterochromosome, but the rest lack this structure. He mistakenly associated its presence with the production of male progeny. In 1906, Edmund B. Wilson clarified the findings of Henking and McClung when he demonstrated that female somatic cells in the insect *Protenor* contain 14 chromosomes, including 2 X chromosomes. During oogenesis, an even reduction occurs, producing gametes with 7 chromosomes, including one X. Male somatic cells, on the other hand, contain only 13 chromosomes, including a single X chromosome. During spermatogenesis, gametes are produced containing either 6 chromosomes, without an X, or 7 chromosomes, one of which is an X. Fertilization by X-bearing sperm results in female offspring, and fertilization by X-deficient sperm result in male offspring [Figure 11–5(a)].

The presence or absence of the X chromosome in male gametes provides an efficient mechanism for sex determination in this species and also produces a 1:1 sex ratio in the resulting offspring. The mechanism, now called the **XX/XO** or ***Protenor* mode of sex determination**, depends on the random distribution of the X chromosome into one-half of the male gametes during segregation.

Wilson also experimented with the hemipteran insect *Lygaeus turicus,* in which both sexes have 14 chromosomes. Twelve of these are autosomes. In addition, the females have 2 X chromosomes, while the males have only a single X and a smaller heterochromosome labeled the **Y chromosome**. Females in this species produce only gametes of the (6A + X) constitution, but males produce two types of gametes in equal proportions: (6A + X) and (6A + Y). Therefore, following random fertilization, equal numbers of male and female progeny will be produced with distinct chromosome complements. This mode of sex determination is called the ***Lygaeus*** or **XX/XY** type [Figure 11–5(b)].

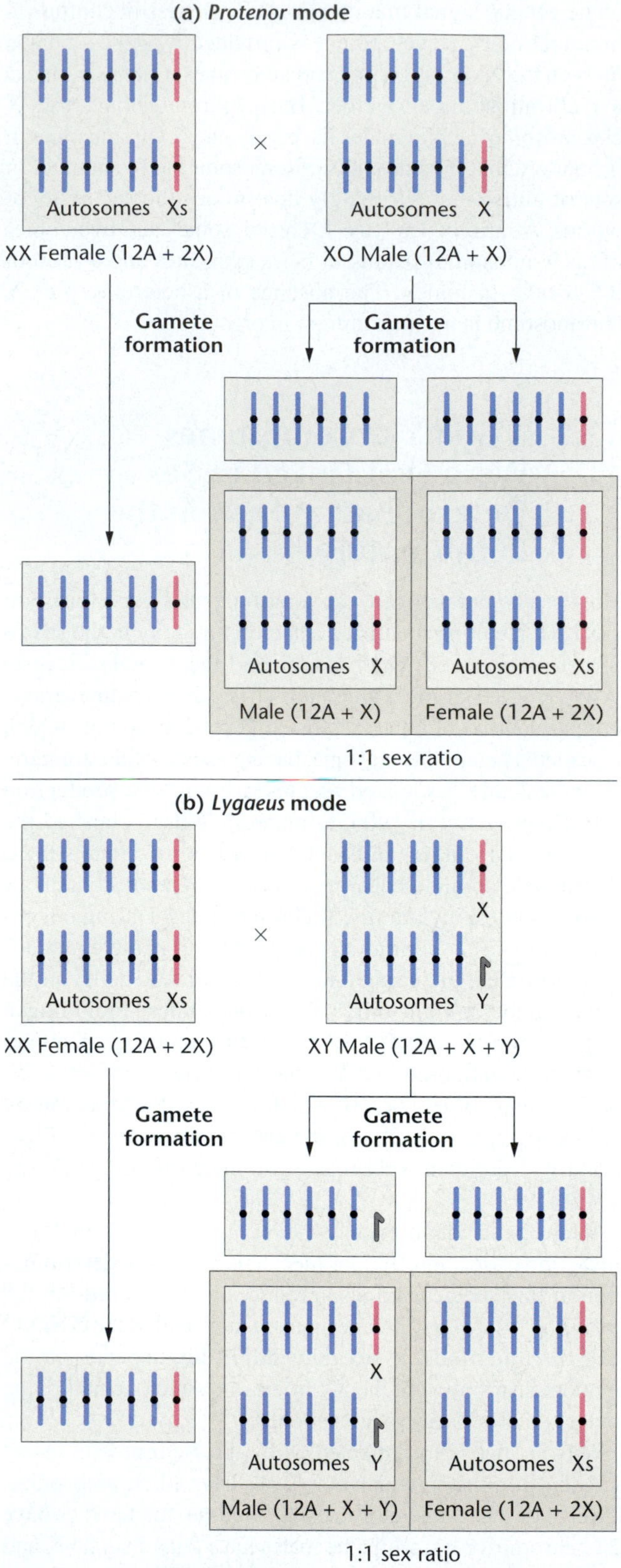

FIGURE 11–5 (a) The *Protenor* mode of sex determination where the heterogametic sex (the male in this example) is XO and produces gametes with or without the X chromosome; (b) The *Lygaeus* mode of sex determination, where the heterogametic sex (again, the male in this example) is XY and produces gametes with either an X or a Y chromosome. In both cases, the chromosome composition of the offspring determines its sex.

In *Protenor* and *Lygaeus* insects, males produce unlike gametes. As a result, they are described as the **heterogametic sex**, and in effect, their gametes ultimately determine the sex of the progeny in those species. In such cases, the female, who has like sex chromosomes, is the **homogametic sex**, producing uniform gametes with regard to chromosome numbers and types.

The male is not always the heterogametic sex. In other organisms, the female produces unlike gametes, exhibiting either the *Protenor* (XX/XO) or *Lygaeus* (XX/XY) mode of sex determination. Examples include moths and butterflies, most birds, some fish, reptiles, amphibians, and at least one species of plants (*Fragaria orientalis*). To immediately distinguish situations in which the female is the heterogametic sex, some workers use the notation **ZZ/ZW**, where ZW is the heterogamous female, instead of the XX/XY notation.

The situation with fowl (chickens) illustrates the difficulty in establishing which sex is heterogametic and whether the *Protenor* or *Lygaeus* mode is operable. While genetic evidence supported the hypothesis that the female is the heterogametic sex, the cytological identification of the sex chromosome was not accomplished until 1961, because of the large number of chromosomes (78) characteristic of chickens. When the sex chromosomes were finally identified, the female was shown to contain an unlike chromosome pair, including a heteromorphic chromosome (the W chromosome). Thus, in fowl, the female is indeed heterogametic and is characterized by the *Lygaeus* type of sex determination.

11.3 The Y Chromosome Determines Maleness in Humans

The first attempt to understand sex determination in our own species occurred almost 100 years ago and involved the examination of chromosomes present in dividing cells. Efforts were made to accurately determine the diploid chromosome number of humans, but because of the relatively large number of chromosomes, this proved to be quite difficult. In 1912, H. von Winiwarter counted 47 chromosomes in a spermatogonial metaphase preparation. It was believed that the sex-determining mechanism in humans was based on the presence of an extra chromosome in females, who were thought to have 48 chromosomes. However, in the 1920s, Theophilus Painter observed between 45 and 48 chromosomes in cells of testicular tissue and also discovered the small Y chromosome, which is now known to occur only in males. In his original paper, Painter favored 46 as the diploid number in humans, but he later concluded incorrectly that 48 was the chromosome number in both males and females.

For 30 years, this number was accepted. Then, in 1956, Joe Hin Tjio and Albert Levan discovered a better way to prepare chromosomes. This improved technique led to a strikingly clear demonstration of metaphase stages showing that 46 was indeed the human diploid number. Later that same year, C. E. Ford and John L. Hamerton, also working with testicular tissue, confirmed this finding. The familiar karyotype of humans (Figure 11–6) is based on Tjio and Levan's technique.

Within the normal 23 pairs of human chromosomes, one pair was shown to vary in configuration in males and females. These two chromosomes were designated the X and Y sex chromosomes. The human female has two X chromosomes, and the human male has one X and one Y chromosome.

We might believe that this observation is sufficient to conclude that the Y chromosome determines maleness. However, several other interpretations are possible. The Y could play no role in sex determination; the presence of two X chromosomes could cause femaleness; or maleness could result from the lack of a second X chromosome. The evidence that clarified which explanation was correct awaited the study of variations in the human sex chromosome composition. As such investigations revealed, the Y chromosome does indeed determine maleness in humans.

Klinefelter and Turner Syndromes

About 1940, scientists identified two human abnormalities characterized by aberrant sexual development, **Klinefelter syndrome** and **Turner syndrome**.* Individuals with Klinefelter syndrome have genitalia and internal ducts that are usually male, but their testes are rudimentary and fail to produce sperm. They are generally tall and have long arms and legs and large hands and feet.

Although some masculine development does occur, feminine sexual development is not entirely suppressed. Slight enlargement of the breasts (gynecomastia) is common, and the hips are often rounded. This ambiguous sexual development, referred to as intersexuality, may lead to abnormal social development. Intelligence is often below the normal range.

* Although the possessive form of the names of most syndromes (eponyms) is sometimes used (e.g., Klinefelter's), the current preference is to use the nonpossessive form, which we have adopted for all human syndromes.

In Turner syndrome, the affected individual has female external genitalia and internal ducts, but the ovaries are rudimentary. Other characteristic abnormalities include short stature (usually under 5 feet), skin flaps on the back of the neck, and underdeveloped breasts. A broad, shieldlike chest is sometimes noted. Intelligence is usually normal.

In 1959, the karyotypes of individuals with these syndromes were determined to be abnormal with respect to the sex chromosomes. Individuals with Klinefelter syndrome have more than one X chromosome. Most often they have an XXY complement in addition to 44 autosomes [Figure 11–7(a)]. People with this karyotype are designated **47,XXY**. Individuals with Turner syndrome are most often monosomic and have only 45 chromosomes, including just a single X chromosome. They are designated **45,X** [Figure 11–7(b)]. Note the convention used in designating the above chromosome compositions. The number indicates the total number of chromosomes present, and the information after the comma designates the relevant deviation from the normal diploid content. Both conditions result from nondisjunction, the failure of the X chromosomes to segregate properly during meiosis. (See Figures 9–17.)

These Klinefelter and Turner karyotypes and their corresponding sexual phenotypes allow us to conclude that the Y chromosome determines maleness in humans. In its absence, the sex of the individual is female, even if only a single X chromosome is present. The presence of the Y chromosome in the individual with Klinefelter syndrome is sufficient to determine maleness, even though male development is not complete. Similarly, in the absence of a Y chromosome, as in the case of individuals with Turner syndrome, no masculinization occurs.

Klinefelter syndrome occurs in about 2 of every 1000 male births. The karyotypes **48,XXXY**, **48,XXYY**, **49,XXXXY**, and **49,XXXYY** are similar phenotypically to 47,XXY, but manifestations are often more severe in individuals with a greater number of X chromosomes.

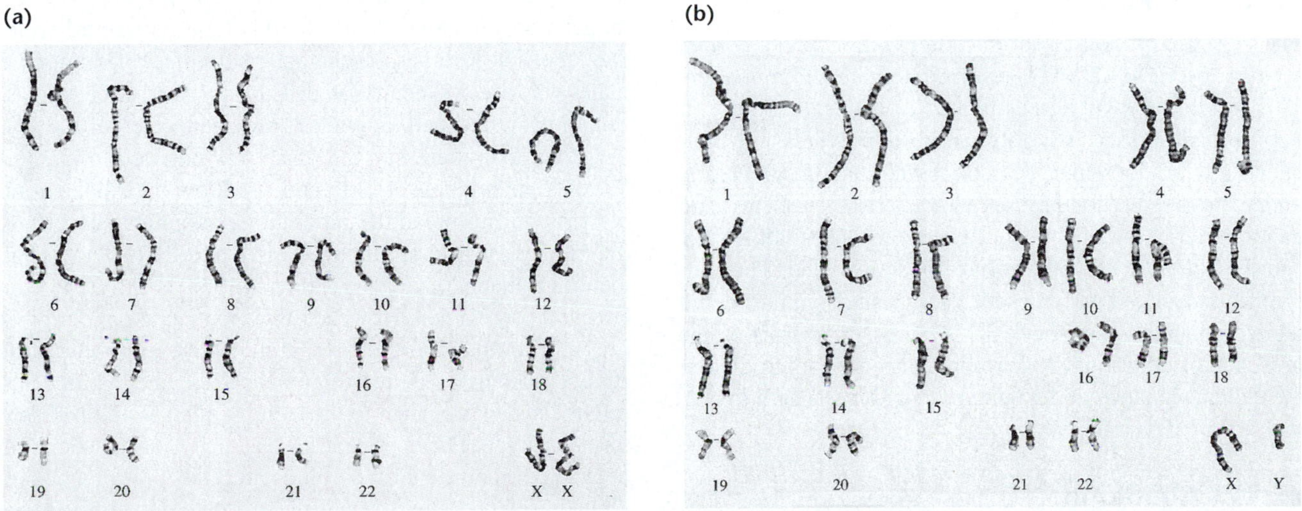

FIGURE 11–6 The traditional human karyotypes derived from a normal female and a normal male. Each contains 22 pairs of autosomes and two sex chromosomes. The female (a) contains two X chromosomes, while the male (b) contains one X and one Y chromosome.

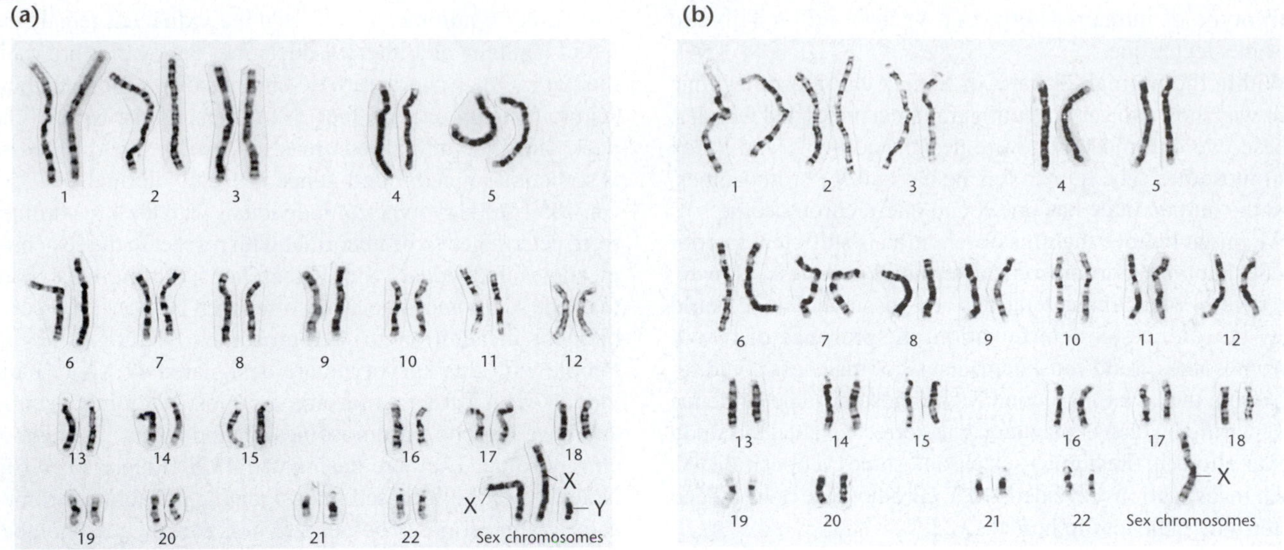

FIGURE 11–7 The karyotypes and phenotypic depictions of individuals with (a) Klinefelter syndrome (47,XXY) and (b) Turner syndrome (45,X).

Turner syndrome can also result from karyotypes other than 45,X, including individuals called **mosaics** whose somatic cells display two different genetic cell lines, each exhibiting a different karyotype. Such cell lines result from a mitotic error during early development, the most common chromosome combinations being **45,X/46,XY** and **45,X/46,XX**. Thus, an embryo that began life with a normal karyotype can give rise to an individual whose cells show a mixture of karyotypes and who exhibits this syndrome.

Turner syndrome is observed in about 1 in 2000 female births, a frequency much lower than that for Klinefelter syndrome. One explanation for this difference is the observation that a substantial majority of 45,X fetuses die *in utero* and are aborted spontaneously. Thus, a similar frequency of the two syndromes may occur at conception.

47,XXX Syndrome

The presence of three X chromosomes along with a normal set of autosomes (**47,XXX**) results in female differentiation. This syndrome, which is estimated to occur in about 1 of 1200 female births, is highly variable in expression. Frequently, 47,XXX women are perfectly normal. In other cases, underdeveloped secondary sex characteristics, sterility, and mental retardation may occur. In rare instances, **48,XXXX** and **49,XXXXX** karyotypes have been reported. The syndromes associated with these karyotypes are similar to, but more pronounced than, the 47,XXX. Thus, in many cases, the presence of additional X chromosomes appears to disrupt the delicate balance of genetic information essential to normal female development.

47,XYY Condition

Another human condition involving the sex chromosomes, **47,XYY**, has also been intensively investigated. Studies of this condition, where the only deviation from diploidy is the presence of an additional Y chromosome in an otherwise normal male karyotype, have led to an interesting controversy.

In 1965, Patricia Jacobs discovered 9 of 315 males in a Scottish maximum security prison to have the 47,XYY karyotype. These males were significantly above average in height and had been incarcerated as a result of antisocial (nonviolent) criminal acts. Of the 9 males studied, 7 were of subnormal intelligence, and all suffered personality disorders. Several other studies produced similar findings. The possible correlation between this chromosome composition and criminal behavior piqued considerable interest and extensive investigations of the phenotype and frequency of the 47,XYY condition in both criminal and noncriminal populations ensued. Above-average height (usually over 6 feet) and subnormal intelligence have been generally substantiated, and the frequency of males displaying this karyotype is indeed higher in penal and mental institutions compared with unincarcerated males. (See Table 11–1.) A particularly relevant question involves the characteristics displayed by XYY males who are not incarcerated. The only nearly constant association is that such individuals are over 6 feet tall!

A study that addressed this issue was initiated to identify 47,XYY individuals at birth and to follow their behavioral patterns during preadult and adult development. By 1994, the two investigators, Stanley Walzer and Park Gerald, had identified about 20 XYY newborns in 15,000 births at Boston Hospital for Women. However, they soon came under great pressure to abandon their research. Those opposed to the study argued that the investigation could not be justified and might cause great harm to those individuals who displayed this karyotype. The opponents argued that (1) no association between the additional Y chromosome and abnormal behavior had been previously established in the population at large and (2) "labeling" these individuals in the study might create a self-fulfilling prophecy. That is, as a result of participation in the study, parents, relatives, and friends might treat

TABLE 11–1 Frequency of XYY Individuals in Various Settings

Setting	*Restriction*	*Number Studied*	*Number XYY*	*Frequency XYY*
Control population	Newborns	28,366	29	0.10%
Mental-penal	No height restriction	4,239	82	1.93
Penal	No height restriction	5,805	26	0.44
Mental	No height restriction	2,562	8	0.31
Mental-penal	Height restriction	1,048	48	4.61
Penal	Height restriction	1,683	31	1.84
Mental	Height restriction	649	9	1.38

Source: Compiled from data presented in Hook, 1973, Tables 1–8. Copyright 1973 by the American Association for the Advancement of Science.

individuals identified as 47,XYY differently, ultimately producing the expected antisocial behavior. Despite the support of a government funding agency and the faculty at Harvard Medical School, Walzer and Gerald abandoned the investigation in 1995.

Since Walzer and Gerald's work, it has become apparent that many XYY males are present in the population who do not exhibit antisocial behavior and who lead normal lives. Therefore, we must conclude that there is no consistent correlation between the extra Y chromosome and the predisposition of males to behavioral problems.

Sexual Differentiation in Humans

Once researchers had established that, in humans, it is the Y chromosome that houses genetic information necessary for maleness, they made efforts to pinpoint a specific gene or genes capable of providing the "signal" responsible for sex determination. Before we delve into this topic, it is useful to consider how sexual differentiation occurs in order to better comprehend how humans develop into sexually dimorphic males and females. During early development, every human embryo undergoes a period when it is potentially hermaphroditic. By the fifth week of gestation, gonadal primordia arise as a pair of ridges associated with each embryonic kidney. Primordial germ cells migrate to these ridges, where an outer cortex and inner medulla form. The **cortex** is capable of developing into an ovary, while the inner **medulla** may develop into a testis. In addition, two sets of undifferentiated male (Wolffian) and female (Mullerian) ducts exist in each embryo.

If the cells of the genital ridge have the XY constitution, development of the medullary region into a testis is initiated around the seventh week. However, in the absence of the Y chromosome, no male development occurs, and the cortex of the genital ridge subsequently forms ovarian tissue. Parallel development of the appropriate male or female duct system then occurs, and the other duct system degenerates. A substantial amount of evidence indicates that in males, once testes differentiation is initiated, the embryonic testicular tissue secretes two hormones that are essential for continued male sexual differentiation.

In the absence of male development, as the 12th week of fetal development approaches, the oogonia within the ovaries begin meiosis and primary oocytes can be detected. By the 25th week of gestation, all oocytes become arrested in meiosis and remain dormant until puberty is reached some 10 to 15 years later. In males, on the other hand, primary spermatocytes are not produced until puberty is reached.

The Y Chromosome and Male Development

In Chapter 10, we alluded to the fact that the human Y chromosome, unlike the X, has long been thought to be nearly blank genetically. It is now known that this is not true, even though the Y chromosome contains many fewer genes than does the X. Current analysis has revealed numerous genes and regions with potential genetic function, some with and some without homologous counterparts on the X. For example, present on both ends of the Y chromosome are the so-called **pseudoautosomal regions (PARs)** that share homology with regions on the X chromosome and which synapse and recombine with it during meiosis. The presence of such a pairing region is critical to the segregation of the X and Y during male gametogenesis. The remainder of the chromosome, about 95 percent of it, is referred to as the **nonrecombining region of the Y (NRY)**. As we will see next, some portions of the NRY are also homologous to genes on the X chromosome, and some are not.

The human Y chromosome is diagrammed in Figure 11–8. The NRY is divided about equally between euchromatic regions that contain functional genes and heterochromatic regions that lack genes. Within the euchromatin, adjacent to the PAR of the short arm of the Y chromosome, is a critical gene that controls male sexual development. It is called the ***sex-determining region Y (SRY)***. Since in humans, the absence of a Y chromosome inevitably leads to female development, as expected, this gene is absent from the X chromosome. *SRY* encodes a gene product that somehow triggers the undifferentiated gonadal tissue of the embryo to form testes. This product, called the **testis-determining factor (TDF)**, is also present in the adult male testis. *SRY* (usually a closely related version) is present in all mammals thus far examined, indicative of its essential function throughout this diverse group of animals.

Evidence proving that *SRY* is the responsible gene for male sex determination has relied on the molecular geneticist's ability to identify the presence or absence of DNA sequences

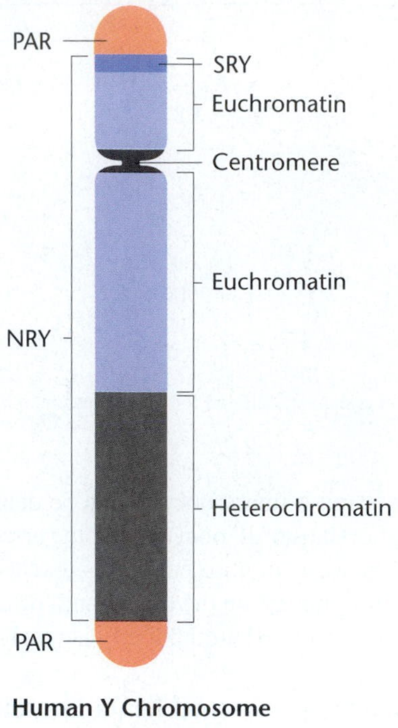

FIGURE 11–8 The various regions of the human Y chromosome.

in rare individuals whose expected sex chromosome composition does not correspond to their sexual phenotype. For example, there are human males who demonstrate two X and no Y chromosomes. Often, they have attached to one of their X's the region of the Y chromosome containing *SRY*. There are also females who have one X and one Y chromosome. Their Y is almost always missing the *SRY* gene. These observations argue strongly in favor of the role of *SRY* in providing the primary signal for male development.

Further support of this conclusion involves an experiment using **transgenic mice**. Such animals are produced from fertilized eggs injected with foreign DNA that is subsequently incorporated into the genetic composition of the developing embryo. In normal mice, a chromosome region designated *Sry* has been identified that is comparable to *SRY* in humans. When DNA containing only mouse *Sry* is injected into normal XX mouse eggs, most of the offspring develop into males!

How specifically the product of this gene triggers the embryonic gonadal tissue to develop into testes rather than ovaries is a question still under extensive investigation. A number of other autosomal genes are believed to be part of a cascade of genetic expression initiated by *SRY*. One such autosomal gene, *SOX9* in humans, is the subject of Problem 20 at the end of this chapter. Others include *WT1*, found on human chromosome 11, originally identified as an oncogene associated with Wilms tumor, which affects the kidney and gonads. Another, *SF1*, is involved in the regulation of enzymes affecting steroid metabolism. In mice, this gene is initially active in both the male and female bisexual genital ridge, persisting until the point in development when testis formation is apparent. At that time, its expression persists in males, but is extinguished in females. The link between these various genes and sex determination brings us closer to the complete understanding of how males and females arise in humans.

We conclude this section by relating some recent findings by David Page and his colleagues that have provided a more complete picture of the human Y chromosome (Figure 11–8). Using molecular probes, they have now discovered about a dozen genes in the PAR regions, most on the short arm, and almost two dozen genes in the euchromatin of the NRY. These latter genes can be sorted into two groups. The first category consists of those that have a homolog on the X chromosome and are expressed in a wide range of tissues in both sexes. Genes in this group seem to encode information of general cellular function, thus serving so-called "housekeeping" functions. The genes in the second group lack a functional homolog on the X chromosome, often are present in multiple copies, and are expressed only in the testis. They appear to encode proteins specific to testis development and function. Thus, the products of many of these genes are directly related to fertility in males. It is currently believed that a great deal of male sterility in our population can be linked to mutations in these genes.

Together, the preceding information clearly refutes the so-called "wasteland" theory that depicts the human Y chromosome as genetically blank. Interestingly, as we will see in a subsequent section of this chapter, the group of Y-linked housekeeping genes, with homologs on the X, must be accounted for in any mechanism of dosage compensation that equalizes the expression of genes present on the sex chromosomes.

11.4 The Ratio of Males to Females in Humans Is Not 1.0

The presence of heteromorphic sex chromosomes in one sex of a species but not the other provides a potential mechanism for producing equal proportions of male and female offspring. This potential is premised on the segregation of the X and Y (or Z and W) chromosomes during meiosis, such that one-half of the gametes of the heterogametic sex receive one of the chromosomes and one-half receive the other one. As we just learned in Section 11.3, in humans, small pseudoautosomal regions of pairing homology do exist at both ends of the X and the Y chromosomes. Provided that both types of gametes are equally successful in fertilization and that the two sexes are equally viable during development, a one to one ratio of male and female offspring results.

Given the potential for the production of equal numbers of both sexes, the actual proportion of male to female offspring has been investigated and is referred to as the **sex ratio**. We can assess it in two ways. The **primary sex ratio** reflects the proportion of males to females conceived in a population. The **secondary sex ratio** reflects the proportion of each sex that is born. The secondary sex ratio is much easier to determine, but has the disadvantage of not accounting for any disproportionate embryonic or fetal mortality.

When the secondary sex ratio in the human population was determined in 1969 by using worldwide census data, it was found not to equal 1.0. For example, in the Caucasian population in the United States, the secondary ratio was a little less than 1.06, indicating that about 106 males are born for each 100 females. In 1995, this ratio dropped to slightly less than 1.05. In the African-American population in the United States, the ratio was 1.025. In other countries the excess of male births is even greater than reflected in these values. For example, in Korea, the secondary sex ratio was 1.15.

Despite these ratios, it is possible that the *primary sex ratio* is 1.0, and that it is altered between conception and birth. For the secondary ratio to exceed 1.0, then, prenatal female mortality would have to be greater than prenatal male mortality. However, this hypothesis has been examined and shown to be false. In fact, just the opposite occurs. In a Carnegie Institute study, reported in 1948, the sex of approximately 6000 embryos and fetuses recovered from miscarriages and abortions was determined, and fetal mortality was actually higher in males. On the basis of the data derived from that study, the primary sex ratio in U.S. Caucasians was estimated to be 1.079. More recent data have estimated that this figure is much higher—between 1.20 and 1.60, suggesting that many more males than females are conceived in the human population.

It is not clear why such a radical departure from the expected primary sex ratio of 1.0 occurs. To come up with a suitable explanation, we must examine the assumptions upon which the theoretical ratio is based:

1. Because of segregation, males produce equal numbers of X- and Y-bearing sperm.
2. Each type of sperm has equivalent viability and motility in the female reproductive tract.
3. The egg surface is equally receptive to both X- and Y-bearing sperm.

While no direct experimental evidence contradicts any of these assumptions, the human Y chromosome is smaller than the X chromosome and therefore of less mass. Thus, it has been speculated that Y-bearing sperm are more motile than X-bearing sperm. If this is true, then the probability of a fertilization event leading to a male zygote is increased, providing one possible explanation for the observed primary ratio.

11.5 Dosage Compensation Prevents Excessive Expression of X-Linked Genes in Humans and Other Mammals

The presence of two X chromosomes in normal human females and only one X in normal human males is unique compared with the equal numbers of autosomes present in the cells of both sexes. On theoretical grounds alone, it is possible to speculate that this disparity should create a "genetic dosage" problem between males and females for all X-linked genes. Recall that in Chapter 10 we discussed the topic of X-linkage, the inheritance of traits under the control of genes located on one of the sex chromosomes. There, we saw that during meiosis, sex chromosomes, like autosomes, are subject to the laws of segregation and independent assortment during their distribution into gametes. Since females have two copies of the X chromosome and males only one, there is the potential for females to produce twice as much of each gene product for all X-linked genes. The additional X chromosomes in both males and females exhibiting the various syndromes discussed earlier in this chapter should compound this dosage problem even more. In this section, we will describe certain research findings regarding X-linked gene expression that demonstrate a genetic mechanism allowing for **dosage compensation**.

Barr Bodies

Murray L. Barr and Ewart G. Bertram's experiments with female cats, as well as Keith Moore and Barr's subsequent study with humans, demonstrate a genetic mechanism in mammals that compensates for X chromosome dosage disparities. Barr and Bertram observed a darkly staining body in interphase nerve cells of female cats that was absent in similar cells of males. In humans, this body can be easily demonstrated in female cells derived from the buccal mucosa or in fibroblasts, but not in similar male cells (Figure 11–9). This highly condensed structure, about 1μm in diameter, lies against the nuclear envelope of interphase cells. It stains positively in the Feulgen reaction for DNA.

Current experimental evidence demonstrates that this body, called a **sex chromatin body** or simply a **Barr body**, is an inactivated X chromosome. Susumo Ohno was the first to suggest that the Barr body arises from one of the two X chromosomes. This hypothesis is attractive because it provides a mechanism for dosage compensation. If one of the two X chromosomes is inactive in the cells of females, the dosage of genetic information that can be expressed in males and females is equivalent. Convincing, but indirect evidence for this hypothesis comes from the study of the sex chromosome syndromes described earlier in this chapter. Regardless of how many X chromosomes exist, all but one of them appear to be inactivated and can be seen as Barr bodies. For example, no Barr body is seen in Turner 45,X females; one is seen in Klinefelter 47,XXY males; two in 47,XXX females; three in 48,XXXX females; and so on (Figure 11–10). Therefore, the number of Barr bodies follows an $N - 1$ rule where N is the total number of X chromosomes present.

Although this mechanism of inactivation of all but one X chromosome increases our understanding of dosage compensation, it further complicates our perception of other matters. Because one of the two X chromosomes is inactivated in normal human females, why then is the Turner 45,X individual not entirely normal? Why aren't females with the triplo-X and tetra-X karyotypes (47,XXX and 48,XXXX) completely unaffected by the additional X chromosome? Further, in Klinefelter syndrome (47,XXY), X chromosome inactivation effectively renders such individuals 46,XY. Why aren't these males unaffected by the extra X chromosome in their nuclei?

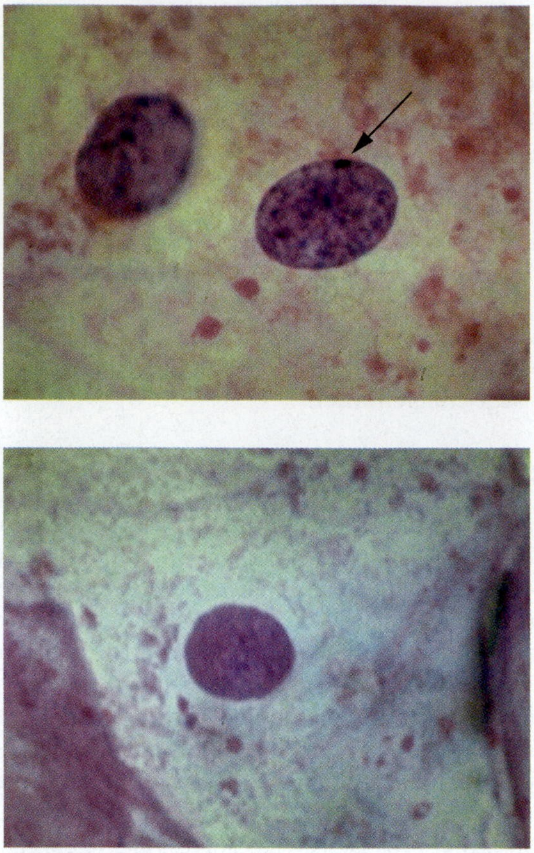

FIGURE 11–9 Photomicrographs comparing cheek epithelial cell nuclei from a male that fails to reveal Barr bodies (bottom) with a female that demonstrates Barr bodies (indicated by an arrow in the top image). This structure, also called a sex chromatin body, represents an inactivated X chromosome.

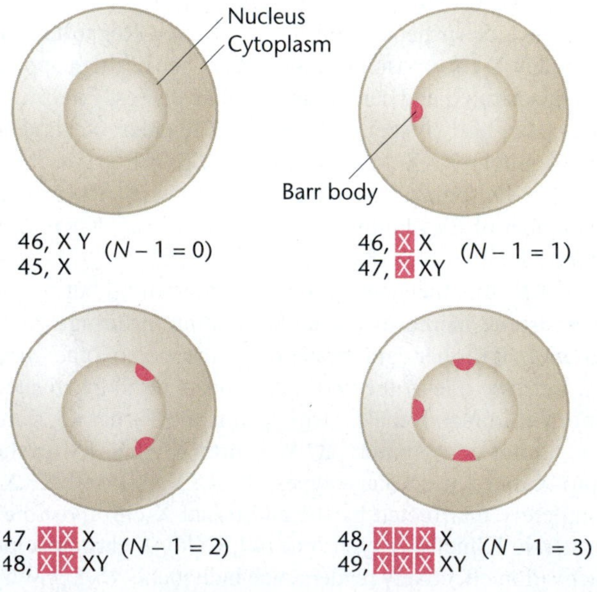

FIGURE 11–10 Barr body occurrence in various human karyotypes, where all X chromosomes except one ($N - 1$) are inactivated.

One possible explanation is that chromosome inactivation does not normally occur in the very early stages of development of those cells destined to form gonadal tissues. Another possible explanation is that not all of each X chromosome forming a Barr body is inactivated. If either hypothesis is correct, over expression of certain X-linked genes might occur at critical times during development despite apparent inactivation of additional X chromosomes.

The Lyon Hypothesis

In mammalian females, one X chromosome is of maternal origin, and the other is of paternal origin. Which one is inactivated? Is the inactivation random? Is the same chromosome inactive in all somatic cells? In 1961, Mary Lyon and Liane Russell independently proposed a hypothesis that answers these questions. They postulated that the inactivation of X chromosomes occurs randomly in somatic cells at a point early in embryonic development. Further, once inactivation has occurred, all progeny cells have the same X chromosome inactivated.

This explanation, which has come to be called the **Lyon hypothesis**, was initially based on observations of female mice heterozygous for X-linked coat color genes. The pigmentation of these heterozygous females was mottled, with large patches expressing the color allele on one X and other patches expressing the allele on the other X. Indeed, such a phenotypic pattern would result if different X chromosomes were inactive in adjacent patches of cells. Similar mosaic patterns occur in the black and yellow-orange patches of female tortoiseshell and calico cats (Figure 11–11). Such X-linked coat color patterns do not occur in male cats because all their cells contain the single maternal X chromosome and are therefore hemizygous for only one X-linked coat color allele.

The most direct evidence in support of the Lyon hypothesis comes from studies of gene expression in clones of human fibroblast cells. Individual cells may be isolated following biopsy and cultured in vitro. If each culture is derived from a single cell, it is referred to as a clone. The synthesis of the enzyme **glucose-6-phosphate dehydrogenase (G6PD)** is controlled by an X-linked gene. Numerous mutant alleles of this gene have been detected, and their gene products can be differentiated from the wild-type enzyme by their migration pattern in an electrophoretic field.

Fibroblasts have been taken from females heterozygous for different allelic forms of G6PD and studied. The Lyon hypothesis predicts that if inactivation of an X chromosome occurs randomly early in development and is permanent in all progeny cells, such a female should show two types of clones, each showing only one electrophoretic form of G6PD, in approximately equal proportions.

In 1963, Ronald Davidson and colleagues performed an experiment involving 14 clones from a single heterozygous female. Seven showed only one form of the enzyme, and 7 showed only the other form. What was most important was that none of the 14 showed both forms of the enzyme. Studies of G6PD mutants thus provide strong support for the random permanent inactivation of either the maternal or paternal X chromosome.

FIGURE 11–11 (a) A calico cat, where the random distribution of orange and black patches illustrates the Lyon hypothesis. The white patches are due to another gene; (b) A tortoiseshell cat, which lacks the white patches characterizing calicos.

The Lyon hypothesis is generally accepted as valid; in fact, the inactivation of an X chromosome into a Barr body is sometimes referred to as **lyonization**. One extension of the hypothesis is that mammalian females are mosaics for all heterozygous X-linked alleles—some areas of the body express only the maternally derived alleles, and others express only the paternally derived alleles. Two especially interesting examples involve **red–green color blindness** and **anhidrotic ectodermal dysplasia**, both X-linked recessive disorders. In the former case, hemizygous males are fully color blind in all retinal cells. However, heterozygous females display mosaic retinas with patches of defective color perception and surrounding areas with normal color perception. Males hemizygous for anhidrotic ectodermal dysplasia show absence of teeth, sparse hair growth, and lack of sweat glands. The skin of females heterozygous for this disorder reveals random patterns of tissue with and without sweat glands (Figure 11–12). In both examples, random inactivation of one or the other X chromosome early in the development of heterozygous females has led to these occurrences.

The Mechanism of Inactivation

The least understood aspect of the Lyon hypothesis is the mechanism of chromosome inactivation in mammals. How are almost all genes of an entire chromosome inactivated? Recent investigations are beginning to clarify this issue. A single region of the mammalian X chromosome, called the **X-inactivation center** (*Xic*),* is the major control unit. Genetic expression of this region, located on the proximal end of the p arm, occurs only on the X chromosome that is inactivated. The constant association of expression of *Xic* and X chromosome inactivation supports the conclusion that this region is an important genetic component in the inactivation process.

The *Xic* is about 1 Mb (10^6 base pairs) in length and is known to contain several putative regulatory units and four genes. One of these, **X-inactive specific transcript** (***Xist***)*, is now believed to represent the critical locus within the *Xic*. Several interesting observations have been made regarding the RNA that is transcribed from it, with much of the underlying work having been done by using the mouse *Xist* gene. First, the RNA product is quite large and lacks what is called an extended **open reading frame** (**ORF**). An ORF

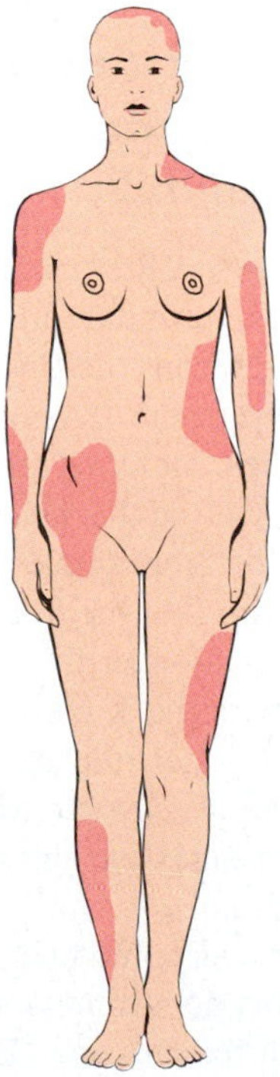

FIGURE 11–12 Depiction of the absence of sweat glands (shaded regions) in a female heterozygous for the X-linked condition anhidrotic ectodermal dysplasia. The locations vary from female to female, based on the random pattern of X chromosome inactivation during early development, resulting in unique mosaic distributions of sweat glands in heterozygotes.

* On the human X chromosome, the inactivation center and the key gene are designated as XIC and *XIST*, respectively.

includes the information necessary for translation of the RNA product into a protein. Thus, the RNA is not translated, but instead serves a structural role in the nucleus, presumably in the mechanism of chromosome inactivation. This finding has led to the belief that the RNA products of *XIST* and *Xist* spread over and coat the X chromosome bearing the gene that produced it, creating some sort of molecular "cage" that entraps it, leading to its inactivation. Inactivation is therefore said to be *cis*-acting.

Second, transcription of *Xist* occurs initially at low levels on all X chromosomes. As the inactivation process begins, however, transcription continues and is enhanced only on the X chromosome(s) that becomes inactivated.

In 1996, a research group led by Graeme Penny provided convincing evidence that transcription of *Xist* is the critical event in chromosome inactivation. These researchers were able to introduce a targeted deletion (7 kb) into this gene that destroyed its activity. As a result, the chromosome bearing the mutation lost its ability to become inactivated. Several interesting questions remain unanswered. First, in cells with more than two chromosomes, what sort of "counting" mechanism exists that designates all but one X chromosome to be inactivated? Second, what "blocks" the Xic of the active chromosome, preventing transcription of *Xist?* Third, how is inactivation of the same X chromosome or chromosomes maintained in progeny cells, as the Lyon hypothesis calls for? The inactivation signal must be stable as cells proceed through mitosis. Whatever the answers to these questions, we have taken an exciting step toward understanding how dosage compensation is accomplished in mammals.

11.6 The Ratio of X Chromosomes to Sets of Autosomes Determines Sex in *Drosophila*

Because males and females in *Drosophila melanogaster* (and other *Drosophila* species) have the same general sex chromosome composition as humans (males are XY and females are XX), we might assume that the Y chromosome also causes maleness in these flies. However, the elegant work of Calvin Bridges in 1916 showed this not to be true. He studied flies with quite varied chromosome compositions, leading him to the conclusion that the Y chromosome is not involved in sex determination in this organism. Instead, Bridges proposed that both the X chromosomes and autosomes together play a critical role in sex determination. Recall that in the nematode, *C. elegans,* which lacks a Y chromosome, the sex chromosomes and autosomes are also critical to sex determination.

Bridges' work can be divided into two phases: (1) A study of offspring resulting from nondisjunction of the X chromosomes during meiosis in females and (2) subsequent work with progeny of females containing three copies of each chromosome, called triploid ($3n$) females. As we have seen previously in this chapter and earlier (see Figure 2–17), nondisjunction is the failure of paired chromosomes to segregate or separate during the anaphase stage of the first or second meiotic divisions. The result is the production of two types of abnormal gametes, one of which contains an extra chromosome ($n + 1$) and the other of which lacks a chromosome ($n - 1$). Fertilization of such gametes with a haploid gamete produces ($2n + 1$) or ($2n - 1$) zygotes. As in humans, if nondisjunction involves the X chromosome, in addition to the normal complement of autosomes, both an XXY and an X"0" sex chromosome composition may result. (The "0" signifies that there is neither a second X or Y chromosome present.) Contrary to what was later discovered in humans, Bridges found that the XXY flies were normal females, and the X"0" flies were sterile males. The presence of the Y chromosome in the XXY flies did not cause maleness, and its absence in the X"0" flies did not produce femaleness. From these data, he concluded that the Y chromosome in *Drosophila* lacks male-determining factors, but since the X"0" males were sterile, it does contain genetic information essential to male fertility.

Bridges was able to clarify the mode of sex determination in *Drosophila* by studying the progeny of triploid females ($3n$), which have three copies each of the haploid complement of chromosomes. *Drosophila* has a haploid number of 4, thereby displaying three pairs of autosomes in addition to its pair of sex chromosomes. Triploid females apparently originate from rare diploid eggs fertilized by normal haploid sperm. Triploid females have heavy-set bodies, coarse bristles, and coarse eyes, and they may be fertile. Because of the odd number of each chromosome (3), during meiosis, a wide range of chromosome complements is distributed into gametes that give rise to offspring with a variety of abnormal chromosome constitutions. A correlation among the sexual morphology, chromosome composition, and Bridges' interpretation is shown in Figure 11–13.

Bridges realized that the critical factor in determining sex is the **ratio of X chromosomes to the number of haploid sets of autosomes (A)** present. Normal (2X:2A) and triploid (3X:3A) females each have a ratio equal to 1.0, and both are fertile. As the ratio exceeds unity (3X:2A, or 1.5, for example), what was originally called a superfemale is produced. Because this type of female is rather weak, is infertile, and has lowered viability, it is now more appropriately called a **metafemale**.

Normal (XY:2A) and sterile (X0:2A) males each have a ratio of 1:2, or 0.5. When the ratio decreases to 1:3, or 0.33, as in the case of an XY:3A male, infertile **metamales** result. Other flies recovered by Bridges in these studies contained an X:A ratio intermediate between 0.5 and 1.0. These flies were generally larger, and they exhibited a variety of morphological abnormalities and rudimentary bisexual gonads and genitalia. They were invariably sterile and expressed both male and female morphology, thus being designated as **intersexes**.

Bridges' results indicate that in *Drosophila,* factors that cause a fly to develop into a male are not localized on the sex chromosomes, but are instead found on the autosomes. Some female-determining factors, however, are localized on the X chromosomes. Thus, with respect to primary sex determination, male gametes containing one of each autosome plus a

Normal diploid male

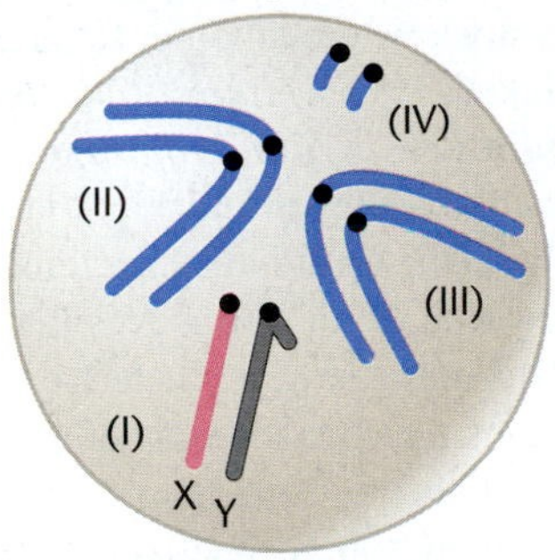

2 sets of autosomes
+
X Y

Chromosome composition	Chromosome formulation	Ratio of X chromosomes to autosome sets	Sexual morphology
	3X/2A	1.5	Metafemale
	3X/3A	1.0	Female
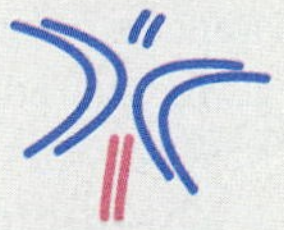	2X/2A	1.0	Female
	3X/4A	0.75	Intersex
	2X/3A	0.67	Intersex
	X/2A	0.50	Male
	XY/2A	0.50	Male
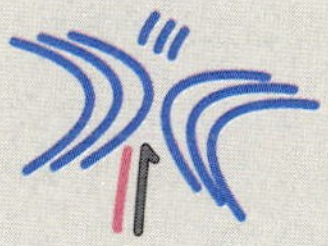	XY/3A	0.33	Metamale

FIGURE 11–13 Chromosome compositions, the ratios of X chromosomes to sets of autosomes, and the resultant sexual morphology in *Drosophila melanogaster.* The normal diploid male chromosome composition is shown as a reference on the left (XY/2A).

Y chromosome result in male offspring, not because of the presence of the Y, but because of the lack of a second X chromosome. This mode of sex determination is explained by the **genic balance theory**. Bridges proposed that a threshold for maleness is reached when the X:A ratio is 1:2 (X:2A), but that the presence of an additional X (XX:2A) alters the balance and results in female differentiation.

Numerous mutant genes have been identified that are involved in sex determination in *Drosophila*. The recessive autosomal gene *transformer* (*tra*), discovered over 50 years ago by Alfred H. Sturtevant, clearly demonstrated that a single autosomal gene could have a profound impact on sex determination. Females homozygous for *tra* are transformed into sterile males, but homozygous males are unaffected.

More recently, another gene, *Sex-lethal* (*Sxl*), has been shown to play a critical role, serving as a "master switch" in sex determination. Activation of the X-linked *Sxl* gene, which relies on a ratio of X chromosomes to sets of autosomes that equals 1.0, is essential to female development. In the absence of activation, resulting, for example, from an X:A ratio of 0.5, male development occurs. It is interesting to note that mutations that inactivate the *Sxl* gene, as originally studied in 1960 by Hermann J. Muller, kill female embryos, but have no effect on male embryos, consistent with the role of the gene, as just described.

While it is not yet exactly clear how this ratio influences the *Sxl* locus, we do have some insights into the question. The *Sxl* locus is part of a hierarchy of gene expression and exerts control over still other genes, including *tra* (discussed in the previous paragraph) and *dsx* (*doublesex*), as well as others. Only in females is the wild-type allele of *tra* activated by the product of *Sxl* which, in turn, influences the expression of *dsx*. Depending on how the initial RNA transcript of *dsx* is processed (spliced), the resultant dsx protein activates either male- or female-specific genes required for sexual differentiation. Each step in this regulatory cascade requires a form of processing called **RNA splicing**, in which portions of the RNA are removed and the remaining fragments "spliced" back together prior to translation into a protein. In the case of the *Sxl* gene, its transcript may be spliced in several different ways, a phenomenon called **alternative splicing**. Two different RNA transcripts are produced in females and males, respectively. In potential females, the transcript is active and initiates a cascade of regulatory gene expression, ultimately leading to female differentiation. In potential males, the transcript is inactive, leading to a different pattern of gene activity, whereby male differentiation occurs. We will return to this topic in Chapter 20, when alternative splicing is again addressed as one of the mechanisms involved in the regulation of genetic expression in eukaryotes.

Dosage Compensation in *Drosophila*

Since *Drosophila* females contain two copies of X-linked genes whereas males contain only one copy, a dosage problem exists, as it does in mammals such as humans and mice. However, the mechanism of dosage compensation in *Drosophila* differs considerably from that in mammals, since X chromosome inactivation is not observed. Instead, male X-linked genes are transcribed at twice the level of the comparable genes in females. Interestingly, if groups of X-linked genes are moved (translocated) to autosomes, dosage compensation also affects them even when they are no longer part of the X chromosome.

As in mammals, considerable gains have been made recently in understanding the process of dosage compensation in *Drosophila.* At least four autosomal genes are known to be involved, under the same master-switch gene, *Sxl*, that induces female differentiation during sex determination. Mutations in any of these genes severely reduce the increased expression of X-linked genes in males, causing lethality.

Evidence supporting a mechanism of increased genetic activity in males is now available. The well-accepted model proposes that one of the autosomal genes, *mle* (*maleless*), encodes a protein that binds to numerous sites along the X chromosome, causing enhancement of genetic expression. The products of the other three autosomal genes also participate in and are required for *mle* binding.

This model predicts that the master-switch *Sxl* gene plays an important role during dosage compensation. In XY flies, *Sxl* is inactive; therefore, the autosomal genes are activated, causing enhanced X chromosome activity. On the other hand, *Sxl* is active in XX females and functions to inactivate one or more of the male-specific autosomal genes, perhaps *mle.* By dampening the activity of these autosomal genes, it ensures that they will not serve to double gene expression of X-linked genes in females, which would further compound the dosage problem.

Tom Cline has proposed that, before the aforementioned dosage compensation mechanism is activated, *Sxl* acts as a sensor for the expression of several other X-linked genes. In a sense, *Sxl* counts X chromosomes. When it registers the dose of their expression to be high, for example, as the result of two X chromosomes, the *Sxl* gene product is modified, and it dampens the expression of the autosomal genes. Although this model may yet be modified or refined, it is useful in guiding future research.

Clearly, an entirely different mechanism of dosage compensation exists in *Drosophila* (and probably many related organisms) than that in mammals. The development of elaborate mechanisms to equalize the expression of X-linked genes demonstrates the critical nature of gene expression. A delicate balance of gene products is necessary to maintain normal development of both males and females.

Drosophila Mosaics

Our knowledge of sex determination and X-linkage in *Drosophila* (Chapter 10) helps us to understand the unusual appearance of a unique fruit fly, shown in Figure 11–14. This fly was recovered from a stock where all other females were heterozygous for the X-linked genes *white* eye (w) and *miniature* wing (m). It is a **bilateral gynandromorph**, which means that one-half of its body (the left half) has developed as a male and the other half (the right half) as a female.

We can account for the presence of both sexes in a single fly in the following way. If a female zygote (heterozygous for *white* eye and *miniature* wing) were to lose one of its X chromosomes during the first mitotic division, the two cells would be of the XX and X"0" constitution, respectively. Thus, one cell would be female and the other would be male. Each of these cells is responsible for producing all progeny cells that make up either the right half or the left half of the body during embryogenesis.

In the case of the bilateral gynandromorph, the original cell of X0 constitution apparently produced only identical progeny cells and gave rise to the left half of the fly, which, because of its chromosomal constitution, was male. Since the male half demonstrated the *white, miniature* phenotype, the X chromosome bearing the w^+, m^+ alleles was lost, while the w, m-bearing homolog was retained. All cells on the right side of the body were derived from the original XX cell, leading to female development. These cells, which remained heterozygous for

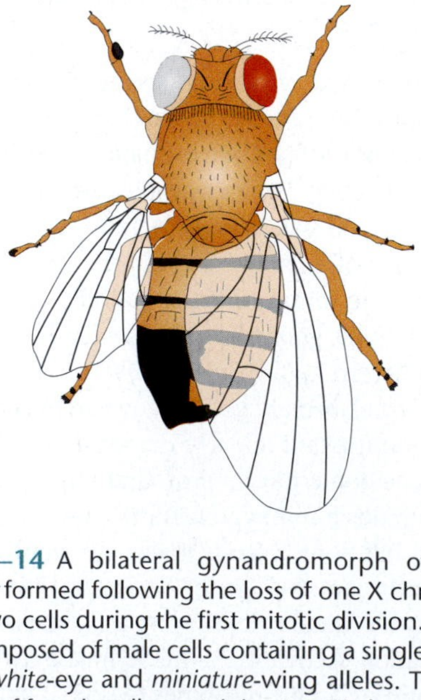

FIGURE 11–14 A bilateral gynandromorph of *Drosophila melanogaster* formed following the loss of one X chromosome in one of the two cells during the first mitotic division. The left side of the fly, composed of male cells containing a single X, expresses the mutant *white*-eye and *miniature*-wing alleles. The right side is composed of female cells containing two X chromosomes heterozygous for the two recessive alleles.

Genetics, Technology, and Society

A Question of Gender: Sex Selection in Humans

The desire to choose a baby's gender is as pervasive as human nature itself. Throughout time, people have resorted to varied and sometimes bizarre methods to achieve the preferred gender of their offspring. In medieval Europe, prospective parents would place a hammer under the bed to help them conceive a boy, or a pair of scissors to conceive a girl. Equally effective were practices based on the ancient belief that semen from the right testicle created male offspring and that from the left testicle created females. Men in ancient Greece would lie on their right side during intercourse in order to conceive a boy. Up until the 18th century, European men would tie off (or remove) their left testicle to increase the chances of getting a male heir.

In some cultures, efforts to control the sex of offspring has a darker side — female infanticide. In ancient Greece, the murder of female infants was so common that the male:female ratio in some areas approached 4:1. Some societies, even to present times, continue to practice female infanticide. In some parts of rural India, hundreds of families are reported to admitting to this practice, even as late as the 1990s. In 1997, the World Health Organization reported population data showing that about 50 million women were "missing" in China, likely due to institutionalized neglect of female children. The practice of female infanticide arises from poverty and age-old traditions. In these cultures, sons work and provide income and security, whereas daughters not only contribute no income but require large dowries when they marry. Under these conditions, it is easy to see why females are held in low esteem.

In recent times, sex-specific abortion has replaced much of the traditional female infanticide. Amniocentesis and ultrasound techniques have become lucrative businesses that provide prenatal sex determination. Studies in India estimate that hundreds of thousands of fetuses are aborted each year because they are female. As a result of sex-selective abortion, the female:male ratio in India was 927:1000 in 1991. In some Northern states, the ratio is as low as 600:1000. Although sex determination and selective abortion of female fetuses was outlawed in India and China in the mid-1990s, the practice is thought to continue.

In Western industrial countries, advances in genetics and reproductive technology offer parents ways to select their children's gender prior to implantation. Following in vitro fertilization, embryos can be biopsied and assessed for gender. Only sex-selected embryos are then implanted. The simplest and least invasive method for sex selection is preconception gender selection (PGS), which involves separating X- and Y-chromosome bearing spermatozoa. The only effective PGS method so far devised involves sorting the sperm based on their DNA content. Due to the different size of the X- and Y-chromosomes, X-bearing sperm contain 2.8-3.0% more DNA than Y-bearing sperm. Sperm samples are treated with a fluorescent DNA stain, then passed single-file through a laser beam in a Fluorescence-Activated Cell Sorter (FACS) machine. The machine separates the sperm into two fractions based on the intensity of their DNA-fluorescence. Using this method, human sperm can be separated into X- and Y-chromosome fractions, with enrichments of about 85% and 75%, respectively. The sorted sperm are then used for standard intrauterine insemination. The Genetics and IVF Institute (Fairfax, Virginia) are presently using this PGS technique in an FDA-approved clinical trial. As of January, 2002, 419 human pregnancies have resulted from the method. The company reports an approximately 80-90% success rate in producing the desired gender.

The emerging PGS methods raise a number of legal and ethical issues. Some feel that prospective parents have the legal right to use sex-selection techniques as part of their fundamental procreative liberty. Others believe that this liberty does not extend to custom-designing a child to the parents' specifications. Proponents state that the benefits far outweigh any dangers to offspring or society. The medical uses of PGS are a clear case for benefit. People at risk for transmitting X-linked diseases such as hemophilia or Duchenne muscular dystrophy can now enhance their chance of conceiving a female child who will not express the disease. As there are over 500 known X-linked diseases and they are expressed in about 1 in 1000 live births, PGS could greatly reduce suffering for many families.

The greatest number of people undertake PGS for non-medical reasons, to "balance" their families. It is possible that the ability to intentionally select the desired sex of an offspring may reduce overpopulation and economic burdens for families who would repeatedly reproduce to get the desired gender. In some cases, PGS may reduce the number of abortions of female fetuses. It is also possible that PGS may increase the happiness of both parents and children, as the child would be more "wanted."

On the other hand, some argue that PGS serves neither the individual nor the common good. It is argued that PGS is inherently sexist, based on the concept of superiority of one sex over another, and leads to an increase in linking a child's worth to gender. Some fear that large-scale PGS will reinforce sex discrimination and lead to sex ratio imbalances. Others feel that sexism and discrimination are not caused by sex ratios and would be better addressed through education and economic equality measures for men and women. The experience so far in Western countries suggest that sex ratio imbalances would not result from PGS. Over half of couples in the US who use PGS request female offspring. However, the consequences of widespread PGS in some Asian countries may be more problematic. Both India and China already have sex ratio imbalances, which contribute to some socially undesirable side-effects, such as the presence of millions of adult men who are unable to marry.

Some critics of PGS argue that this technology may contribute to social and economic inequality, if it will be available only to those who can afford it. Other critics fear that our approval of PGS will open the door to accepting other genetic manipulations of children for socially-acceptable characteristics such as skin color. It is difficult to predict the full effects that PGS will bring to the world. But the gender-selection genie is now out of the bottle, and will be unwilling to step back in.

References

Sills, E.S., Kirman, I., Thatcher, S.S.III, and Palermo, G.D. 1998. Sex-selection of human spermatozoa: evolution of current techniques and applications. *Arch. Gynecol. Obstet.* 261:109-115.

Robertson, J.A. 2001. Preconception Gender Selection. *Am. J. Bioethics* 1:2-9.

Web Sites

Microsort technique, Genetics & IVF Institute, Fairfax, Virginia.
http://www.microsort.net

Female Infanticide, Gendercide Watch.
http://www.gendercide.org/case_infanticide.html

both mutant genes, expressed the wild-type eye–wing phenotypes.

Depending on the orientation of the spindle during the first mitotic division, gynandromorphs can be produced where the "line" demarcating male versus female development occurs along or across any axis of the fly's body.

11.7 Temperature Variation Controls Sex Determination in Reptiles

We conclude this chapter by discussing several cases involving reptiles where the environment, specifically temperature, has a profound influence on sex determination. In contrast to **chromosomal** or **genotypic sex determination** (**CSD** or **GSD**), which describes all examples thus far presented, the cases that we will discuss shortly are categorized as **temperature-dependent sex determination** (**TSD**). As we shall see, the investigations leading to this information may well come closer to revealing the true nature of the primary basis of sex determination than any finding previously discussed.

In many species of reptiles, sex is predetermined at conception by sex chromosome composition, as is the case in many organisms already considered in this chapter. For example, in many snakes, including vipers, a ZZ/ZW mode is in effect, where the female is the heterogamous sex. However, in boas and pythons, it is impossible to distinguish one sex chromosome from the other in either sex. In lizards, both the XX/XY and ZZ/ZW systems are found, depending on the species. In other reptilian species, including all crocodiles, most turtles, and some lizards, sex determination is achieved according to the incubation temperature of eggs during a critical period of embryonic development. Since 1980, it has become clear that TSD is quite widespread among reptiles.

Interestingly, there are three distinct patterns of TSD, as illustrated in Figure 11–15. In the first two, low temperatures yield 100 percent males, and high temperatures yield 100 percent females (Case I), or just the opposite occurs (Case II). In the third pattern (Case III), low *and* high temperatures yield 100 percent females, while intermediate temperatures yield various proportions of males. The third pattern is seen in various species of crocodiles, turtles, and lizards, although other members of these groups are known to exhibit the other patterns. Two observations are noteworthy. First, under certain temperatures in all three patterns, both male and female offspring result. Secondly, the pivotal temperature (P) is fairly narrow, usually less than 5°C, and sometimes only 1°C.

The central question raised by these observations is what metabolic or physiological parameters that lead to the differentiation of one sex or the other are being affected by temperature? The answer is thought to involve steroids (mainly estrogens) and the enzymes involved in their synthesis. Studies have clearly demonstrated that the effects of temperature on estrogens, androgens, and inhibitors of the enzymes controlling their synthesis are involved in the sexual differentiation of ovaries and testes. One enzyme in particular,

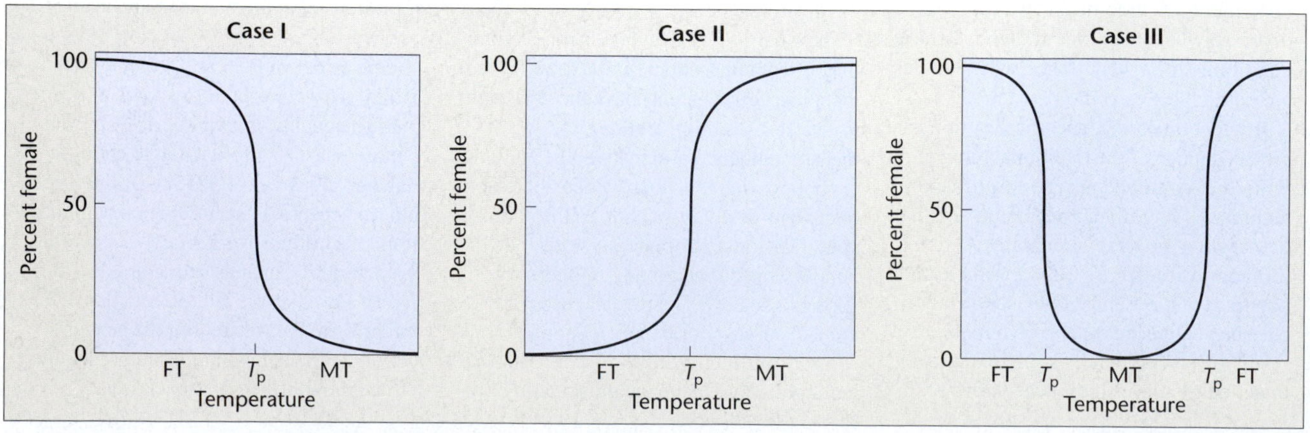

FIGURE 11–15 Three different patterns of temperature-dependent sex determination (TSD) in reptiles, as described in the text. The relative pivotal temperature T_P is crucial to sex determination during a critical point during embryonic development (FT = female-determining temperature; MT = male-determining temperature).

aromatase, converts androgens (male hormones such as testosterone) to estrogens (female hormones such as estradiol). The activity of this enzyme is correlated with the pathway that is followed during gonadal differentiation activity and is high in developing ovaries and low in developing testes. Researchers in this field, including Claude Pieau and colleagues, have proposed that a thermosensitive factor mediates the transcription of the reptilian aromatase gene which leads to temperature-dependent sex determination. Several other genes are likely to be involved in this mediation.

The involvement of sex steroids in gonadal differentiation has also been documented in birds, fishes, and amphibians. Thus, sex-determining mechanisms involving estrogens seem to be characteristic of nonmammalian vertebrates. The regulation of such a system, while temperature dependent in many reptiles, appears to be controlled by sex chromosomes (XX/XY or ZZ/ZW) in many of these other organisms. A final intriguing thought on this matter is that the product of *SRY*, a key component in mammalian sex determination, has been shown to bind in vitro to a regulatory portion of the aromatase gene, whereby it could act as a repressor of ovarian development.

Chapter Summary

1. In sexually reproducing organisms, meiosis, which both creates genetic variability and ensures genetic constancy, depends on fertilization. Fertilization ultimately relies on some form of sexual differentiation, which is achieved by a variety of sex-determining mechanisms.
2. The genetic basis of sexual differentiation is usually related to different chromosome compositions in the two sexes. The heterogametic sex either lacks one chromosome or contains a unique heteromorphic chromosome, usually referred to as the Y chromosome.
3. In humans, the study of individuals with altered sex chromosome compositions has established that the Y chromosome is responsible for male differentiation. The absence of the Y leads to female differentiation. Similar studies in *Drosophila* have excluded the Y in such a role, instead demonstrating that a balance between the number of X chromosomes and sets of autosomes is the critical factor.
4. The primary sex ratio in humans substantially favors males at conception. During embryonic and fetal development, male mortality is higher than that of females. The secondary sex ratio at birth still favors males by a small margin.
5. Dosage compensation mechanisms limit the expression of X-linked genes in females, who have two X chromosomes, as compared with males who have only one X. In mammals, compensation is achieved by the inactivation of either the maternal or paternal X early in development. This process results in the formation of Barr bodies in female somatic cells. In *Drosophila*, compensation is achieved by stimulation of genes located on the single X chromosome to double their genetic expression.
6. The Lyon hypothesis states that, early in development, inactivation is random between the maternal and paternal X. All subsequent progeny cells inactivate the same X as their progenitor cell. Mammalian females thus develop as genetic mosaics with respect to their expression of heterozygous X-linked alleles.
7. In many reptiles, the incubation temperature at a critical time during embryogenesis is often responsible for sex determination. Temperature influences the activity of enzymes involved in the metabolism of steroids related to sexual differentiation.

Insights and Solutions

1. In *Drosophila,* the X chromosomes may become attached to one another ($\widehat{XX}$) such that they always segregate together. Some flies contain both an attached X chromosome and a Y chromosome.

(a) What sex would such a fly be? Explain why this is so.

Solution: The fly would be a female. The ratio of X chromosomes to sets of autosomes would be 1.0, leading to normal female development. The Y chromosome has no influence on sex determination in *Drosophila.*

(b) Given this answer, predict the sex of the offspring that would occur in a cross between this fly and a normal one of the opposite sex.

Solution: All flies would have two sets of autosomes, but each would have one of the following sex chromosome compositions:

(1) $\widehat{XX}X$ → a metafemale with 3 X's (called a trisomic)
(2) $\widehat{XX}Y$ → a female like her mother
(3) XY → a normal male
(4) YY → no development occurs

(c) If the offspring just mentioned were allowed to interbreed, what would be the outcome?

Solution: A stock would be created that would generate the attached X females generation after generation.

2. The *Xg* cell-surface antigen is coded for by a gene that is located on the X chromosome. There is no equivalent gene on the Y chromosome. Two codominant alleles of this gene have been identified: *Xg1* and *Xg2*. A woman of genotype *Xg2/Xg2* bears children with a man of genotype *Xg1*/Y, and they produce a son with Klinefelter syndrome of genotype *Xg1/Xg2*/Y. Using proper genetic terminology, briefly explain how this individual was generated. In which parent and in which meiotic division did the mistake occur?

Solution: Because the son with Klinefelter syndrome is *Xg1/Xg2*/Y, he must have received both the *Xg1* allele and the Y chromosome from his father. Therefore, nondisjunction must have occurred during meiosis I in the father.

Problems and Discussion Questions

1. Define the terms heteromorphic and heterogamy as they are related to sex determination.
2. Contrast the *Protenor* and *Lygaeus* modes of sex determination.
3. Contrast the evidence explaining the different modes of sex determination in *Drosophila* and humans.
4. Devise a method of nondisjunction in human female gametes that would give rise to Klinefelter and Turner syndrome offspring following fertilization by a normal male gamete.
5. An insect species is discovered in which the heterogametic sex is unknown. An X-linked recessive mutation for *reduced wing* (*rw*) is discovered. Contrast the F_1 and F_2 generations from a cross between a female with reduced wings and a male with normal-sized wings when (a) the female is the heterogametic se; (b) the male is the heterogametic sex; (c) Is it possible to distinguish between the *Protenor* and *Lygaeus* mode of sex determination based on the outcome of these crosses?
6. When cows have twin calves of unlike sex (fraternal twins), the female twin is usually sterile and has masculinized reproductive organs. This calf is referred to as a freemartin. In cows, twins may share a common placenta and thus fetal circulation. Predict why a freemartin develops.
7. An attached X female fly, as described in the Insights and Solutions section of this chapter, expresses the recessive X-linked *white* eye phenotype. It is crossed to a male fly that expresses the X-linked recessive miniature wing phenotype. Determine the outcome of this cross regarding the sex, eye color, and wing size of the offspring.
8. It has been suggested that any male-determining genes contained on the Y chromosome in humans cannot be located in the limited region that synapses with the X chromosome during meiosis. What might be the outcome if such genes were located in this region?
9. What is a Barr body, and where is it found in a cell?
10. Indicate the expected number of Barr bodies in interphase cells of the following individuals: Klinefelter syndrome; Turner syndrome; and karyotypes 47,XYY, 47,XXX, and 48,XXXX.
11. Define the Lyon hypothesis.
12. Relate the potential effect of the Lyon hypothesis on the retina of a human female heterozygous for the X-linked red-green color-blindness trait.
13. Cat breeders are aware that kittens expressing the X-linked calico coat pattern are almost invariably females. Why?
14. What does the apparent need for dosage compensation mechanisms suggest about the expression of genetic information in normal diploid individuals?
15. The marine echiurid worm *Bonellia viridis* is an extreme example of the environment's influence on sex determination. Undifferentiated larvae either remain free swimming and differentiate into females, or they settle on the proboscis of an adult female and become males. If larvae that have been on a female proboscis for a short period are removed and placed in seawater, they develop as intersexes. If larvae are forced to develop in an aquarium where pieces of proboscises have been placed, they develop into males. Contrast this mode of sexual differentiation with that of mammals. Suggest further experimentation to elucidate the mechanism of sex determination in *Bonellia*.
16. Discuss the possible reasons why the primary sex ratio in humans is as high as 1.40 to 1.60.
17. In mice, the *Sry* gene (described in this chapter) is located on the Y chromosome very close to one of the pseudoautosomal regions that pairs with the X chromosome during male meiosis. Given this information, propose a model to explain the generation of unusual males who have two X chromosomes (with an *Sry*-containing piece of the Y attached to one X chromosome).

Extra Spicy Problems

18. The genes encoding the red and green color-detecting proteins of the human eye are located next to one another on the X chromosome and probably arose during evolution from a common ancestral pigment gene. The two proteins demonstrate 76 percent homology in their amino acid sequences. A normal woman with one copy of each gene on each of her two X chromosomes has a red color blind son who was shown to contain one copy of the green gene and no copies of the red gene. Devise an explanation at the chromosomal level (during meiosis) that explains these observations.
19. The X-linked dominant mutation in the mouse, *Testicular feminization* (*Tfm*), eliminates the normal response to the testicular hormone testosterone during sexual differentiation. An XY animal bearing the *Tfm* allele on the X chromosome develops testes, but no further male differentiation occurs. The external genitalia of such an animal are female. From this information, what might you conclude about the role of the *Tfm* gene product and the X and Y chromosomes in sex determination and sexual differentiation in mammals? Can you devise an experiment, assuming you can "genetically engineer" the chromosomes of mice, to test and confirm your explanation?
20. Campomelic dysplasia (CMD1) is a congenital human syndrome, featuring malformation of bone and cartilage. It is caused by an autosomal dominant mutation of a gene located on chromosome 17. Consider the following observations in sequence, and in each case, draw whatever appropriate conclusions are warranted:

    (a) Of those with the syndrome who are karyotypically 46,XY, approximately 75 percent are sex reversed, exhibiting a wide range of female characteristics.

    (b) The nonmutant form of the gene, called *SOX9,* is expressed in the developing gonad of the XY male, but not the XX female.

    (c) The *SOX9* gene shares 71 percent amino acid coding sequence homology with the Y-linked *SRY* gene.

    (d) CMD1 patients who exhibit a 46,XX karyotype develop as females, with no gonadal abnormalities.

    (e) The product of the *SRY* gene is believed to be a transcription factor, which by definition (see Chapter 21) activates the expression of other genes. Can you hypothesize how *SRY* and *SOX9* might be related?

21. In the wasp, *Bracon hebetor,* a form of parthenogenesis (where unfertilized eggs initiate development) resulting in haploid organisms is not uncommon. All haploids are males. When offspring arise from fertilization, females almost invariably result. P. W. Whiting has shown that an X-linked gene with nine multiple alleles (X_a, X_b,etc.) controls sex determination. Any homozygous or hemizygous condition results in males and any heterozygous condition results in females. If an X_a/X_b female mates with an X_a male and lays 50 percent fertilized and 50 percent unfertilized eggs, what proportion of male and female offspring will result?

22. Shown next are two graphs that plot the percentage of males occurring against the atmospheric temperature during the early development of fertilized eggs in (a) snapping turtles and (b) most lizards. Interpret these data as they relate to the effect of temperature on sex determination.

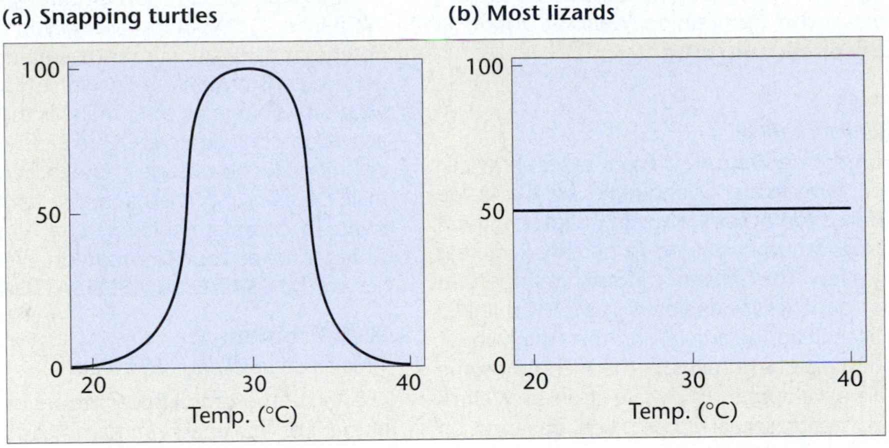

23. CC (Carbon Copy), the cloned kitten, was created from an ovarian cell taken from her genetic donor, Rainbow. The diploid nucleus from the cell was extracted and then injected into an enucleated egg. The resulting "zygote" was then allowed to divide in a petri dish and the cloned embryo was implanted in the uterus of a "surrogate" mother cat, who gave birth to CC. Since Rainbow is a calico cat, scientists expected that any clone produced from her would also be calico. They also expected that a clone would not have the same exact coat color pattern as Rainbow due to random X-inactivation during the clone's development.

When CC was born, the scientists were gratified to see that their prediction concerning the calico nature of CC's coat color pattern were verified. However, they were somewhat surprised that CC completely lacked any orange patches; all her color patches were black. Based on your understanding of X-inactivation and the cloning process (as described above), provide an explanation for CC's unusual coat.

Selected Readings

Amory, J.K., et al. 2000. Klinefelter's syndrome. *Lancet* 356:333–35.

Avner, P., and Heard, E. 2001. X-chromosome inactivation: Counting, choice, and initiation. *Nature Reviews* 2:59–67.

Barr, M.L. 1966. The significance of sex chromatin. *Int. Rev. Cytol.* 19:35–39.

Burgoyne, P.S. 1998. The mammalian Y chromosome: A new perspective. *Bioessays* 20:363–66.

Carrel, L., and Willard, H.F. 1998. Counting on Xist. *Nature Genetics* 19:211–12.

Court-Brown, W.M. 1968. Males with an XYY sex chromosome complement. *J. Med. Genet.* 5:341–59.

Davidson, R., Nitowski, H., and Childs, B. 1963. Demonstration of two populations of cells in human females heterozygous for glucose-6-phosphate dehydrogenase variants. *Proc. Natl. Acad. Sci. USA* 50:481–85.

Dellaporta, S.L., and Calderon-Urrea, A. 1994. The sex determination process in maize. *Science* 266:1501–05.

Erickson, J.D. 1976. The secondary sex ratio of the United States, 1969–71: Association with race, parental ages, birth order, paternal education and legitimacy. *Ann. Hum. Genet.* (London) 40:205–12.

Freije, D., et al. 1992. Identification of a second pseudoautosomal region near the Xq and Yq telomeres. *Science* 258:1784–87.

Gorman, M., Kuroda, M., and Baker, B.S. 1993. Regulation of sex-specific binding of maleness dosage compensation protein to the male X chromosome in *Drosophila*. Cell 72:39–49.

Haseltine, F.P., and Ohno, S. 1981. Mechanisms of gonadal differentiation. *Science* 211:1272–78.

Hodgkin, J. 1990. Sex determination compared in *Drosophila* and *Caenorhabditis*. *Nature* 344:721–28.

Hook, E.B. 1973. Behavioral implications of the humans XYY genotype. *Science* 179:139–50.

Irish, E.E. 1996. Regulation of sex determination in maize. *BioEssays* 18:363–69.

Jacobs, P.A., et al. 1974. A cytogenetic survey of 11,680 newborn infants. *Ann. Hum. Genet.* 37:359–76.

Jegalian, K, and Lahn, B.T. 2001. Why the Y is so weird. *Sci. Am.* (Feb.) 284:56-61.

Koopman, P., et al. 1991. Male development of chromosomally female mice transgenic for *Sry*. *Nature* 351:117–21.

Lahn, B.T., and Page, D.C. 1997. Functional coherence of the human Y chromosome. *Science* 278:675–80.

Lahn, B.T., Pearson, N.M., and Jegalian, K. 2001. The human Y chromosome, in the light of evolution. *Nature Reviews—Genetics* 2:207–16.

Lucchesi, J. 1983. The relationship between gene dosage, gene expression, and sex in *Drosophila*. *Dev. Genet.* 3:275–82.

Lyon, M.F. 1961. Gene action in the X-chromosome of the mouse (*Mus musculus L.*). *Nature* 190:372–73.

———. 1972. X-chromosome inactivation and developmental patterns in mammals. *Biol. Rev.* 47:1–35.

———. 1988. X-chromosome inactivation and the location and expression of X-linked genes. *Am. J. Hum. Genet.* 42:8–16.

GENETICS MediaLab

The resources that follow will help you achieve a better understanding of the concepts presented in this chapter. These resources can be found either on the CD packaged with this textbook or on the Companion Web site found at **http://www.prenhall.com/klug**

Web Problem 1:

Time for completion = 15 minutes

Why is the Y chromosome so small? In many cases of genetic sex determination, females are XX and males are XY. As you know from our discussion of sex linkage in Chapter 4, the X chromosome carries a number of loci in humans, whereas the Y carries only a few. The difference in size has important consequences for genetics, as explored in the Web linked article by Ken Miller. Read the article, and then think about the following questions: In humans, is the Y chromosome necessary for male development? Why are errors on X chromosome repaired through sexual reproduction, but errors on the Y are not? How does this difference relate to the size of the chromosomes? Bill Rice set up an experiment in which specific autosomes were not allowed to recombine. Did his results support the idea that mistakes accumulating on the Y might be responsible for its reduction in size? To complete this exercise, visit Web Problem 1 in Chapter 11 of your Companion Web site, and select the keyword **Y CHROMOSOME**.

Web Problem 2:

Time for completion = 10 minutes

Human females have twice as many copies of genes on the X chromosome as do males; however, it is unlikely that they need twice as many of the proteins encoded by those genes. *Dosage compensation* refers to the mechanism by which transcription of the genetic material on the X chromosome is altered in one or the other sex. Read the article about dosage compensation in female mammals and the related material in Chapter 11. What did Mary Lyon's hypothesis regarding dosage compensation in female mammals predict about the genetic constitution of male versus female cells? Is X inactivation random in all mammals? Is there evidence that X inactivation is not complete? Are Barr bodies ever found in the cells of males? Is dosage compensation implemented in female *Drosophila*, as it is in female mammals? Explain your answer. To complete this exercise, visit Web Problem 2 in Chapter 11 of your Companion Web site, and select the keyword **DOSAGE COMPENSATION**.

Web Problem 3:

Time for completion = 10 minutes

The Y chromosome appears to be necessary for the development of "maleness" in mammals, but is it also sufficient? Recent research suggests that some autosomal genes are necessary for complete male development. Read the linked Web-site article on sex determination in mice, and use the information there and in Chapter 7 to answer the following questions: Dr. Eicher and her colleagues observed an excess of female mice in their colony. What was the genotype of the excess females? Does the *Sry* gene on the Y chromosome appear to regulate the expression of any autosomal genes? What are some hypotheses explaining the function of those genes? Why would information about sex determination in mice be important for an understanding of human sex reversal disorders? To complete this exercise, visit Web Problem 3 in Chapter 11 of your Companion Web site, and select the keyword **GENDER GENES**.

———. 1998. X-chromosome inactivation spreads itself: Effects in autosomes. *Am. J. Hum. Genet.* 63:17–19.

Marin, I., and Baker, S.S. 1998. The evolutionary dynamics of sex determination. *Science* 281:1990–94.

Marshall-Graves, J.A. 1998. Interaction between *SRY* and *SOX* genes in mammalian sex determination. *BioEssays* 20:264–69.

McMillen, M.M. 1979. Differential mortality by sex in fetal and neonatal deaths. *Science* 204:89–91.

Page, D.C., et al. 1987. The sex-determining region of the human Y chromosome encodes a finger protein. *Cell* 51:1091–1104.

Penny, G.D., et al. 1996. Requirement for Xist in X chromosome inactivation. *Nature* 379:131–37.

Pieau, C. 1996. Temperature variation and sex determination in reptiles. *BioEssays* 18:19–26.

Reddy, K.S., and Sulcova, V. 1998. Pathogenetics of 45,X/46,XY gonadal mosaicism. *Cytogenet. Cell Genet.* 82:52–57.

Schafer, A.J. 1996. Sex determination in humans. *BioEssays* 18:955–63.

Vainio. S., et al. 1999. Female development in mammals is regulated by Wnt-4 signalling. *Nature* 397:405–09.

Westergaard, M. 1958. The mechanism of sex determination in dioecious flowering plants. *Adv. Genet.* 9:217–81.

Whiting, P.W. 1939. Multiple alleles in sex determination in *Habrobracon. J. Morphology* 66:323–55.

Witkin, H.A., et al. 1996. Criminality in XYY and XXY men. *Science* 193:547–55.

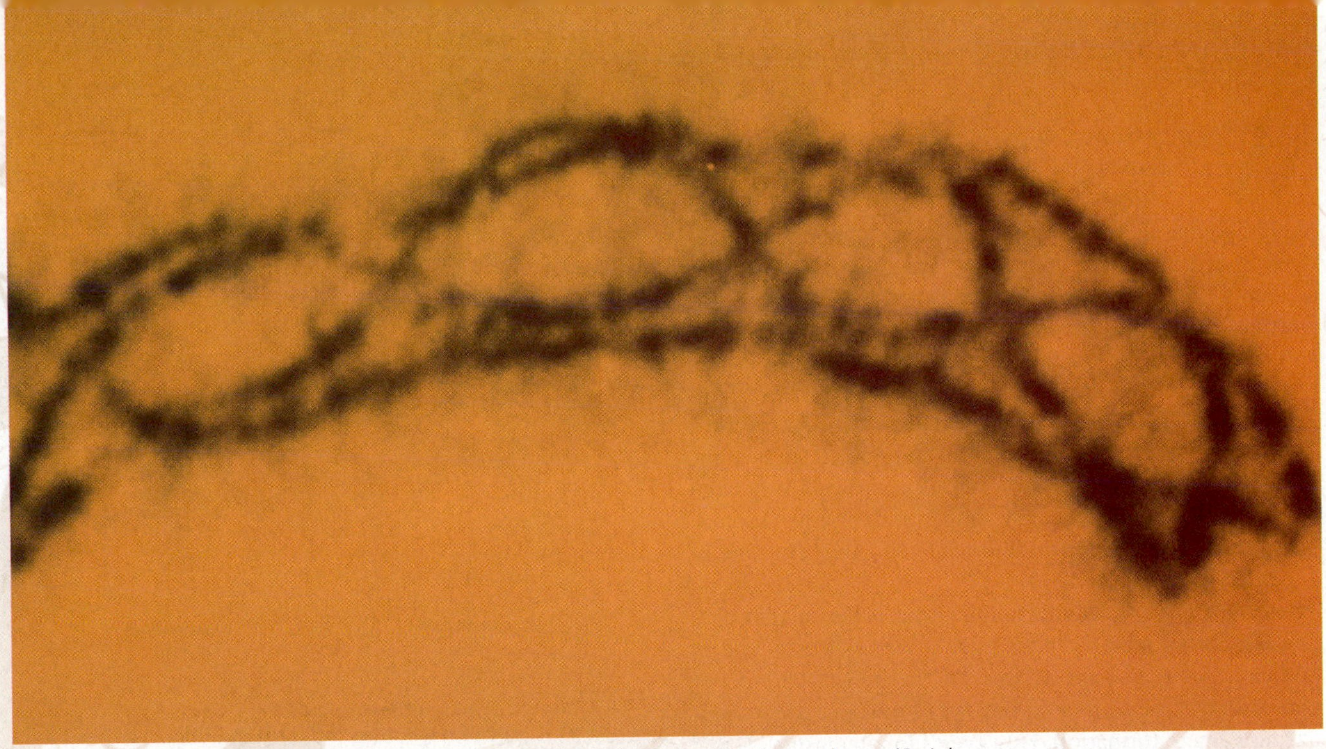

Chiasmata present between synapsed homologs during the first meiotic prophase. (*B. John /Cabisco/Visuals Unilimited*)

12

Linkage, Crossing Over, and Mapping in Eukaryotes

KEY CONCEPTS

Walter Sutton, along with Theodor Boveri, was instrumental in uniting the fields of cytology and genetics. As early as 1903, Sutton pointed out the likelihood that there must be many more "unit factors" than chromosomes in most organisms. Soon thereafter, genetic studies with several organisms revealed that certain genes were not transmitted according to the law of independent assortment, rather these genes seemed to segregate as if they were somehow joined or linked together. Further investigations showed that such genes were part of the same chromosome, and they were indeed transmitted as a single unit.

We now know that most chromosomes consist of very large numbers of genes. Those that are part of the same chromosome are said to be *linked* and to demonstrate **linkage** in genetic crosses.

Because the chromosome, not the gene, is the unit of transmission during meiosis, linked genes are not free to undergo independent assortment. Instead, the alleles at all loci of one chromosome should, in theory, be transmitted as a unit during gamete formation. However, in many instances this does not occur. As we saw in Chapter 8, during the first meiotic prophase, when homologs are paired (or synapsed), a reciprocal exchange of chromosome segments may take place. This event, called **crossing over**, results in the reshuffling, or **recombination**, of the alleles between homologs.

Crossing over is currently viewed as an actual physical breaking and rejoining process that occurs during meiosis. You can see an example in the micrograph that opens this chapter. The exchange of chromosome segments provides an enormous potential for genetic variation in the gametes formed by any individual. This type of variation, in combination with that resulting from independent assortment, ensures that all offspring will contain a diverse mixture of maternal and paternal alleles.

The degree of crossing over between any two loci on a single chromosome is proportional to the distance between them, known as the **interlocus distance**. Thus, depending on which loci are being considered, the percentage of recombinant gametes varies. This correlation serves as the basis for the construction of **chromosome maps**, which indicate the relative locations of genes on the chromosomes.

In this chapter, we will discuss linkage, crossing over, and chromosome mapping. We will also consider a variety of other topics involving the exchange of genetic information, concluding the chapter with the rather intriguing question of why Mendel, who studied seven genes, did not encounter linkage. Or did he?

MEDIA TUTORIAL Linkage and Recombination

12.1 Genes Linked on the Same Chromosome Segregate Together

To provide a simplified overview of the major theme of this chapter, Figure 12–1 illustrates and contrasts the meiotic consequences of (a) independent assortment, (b) linkage

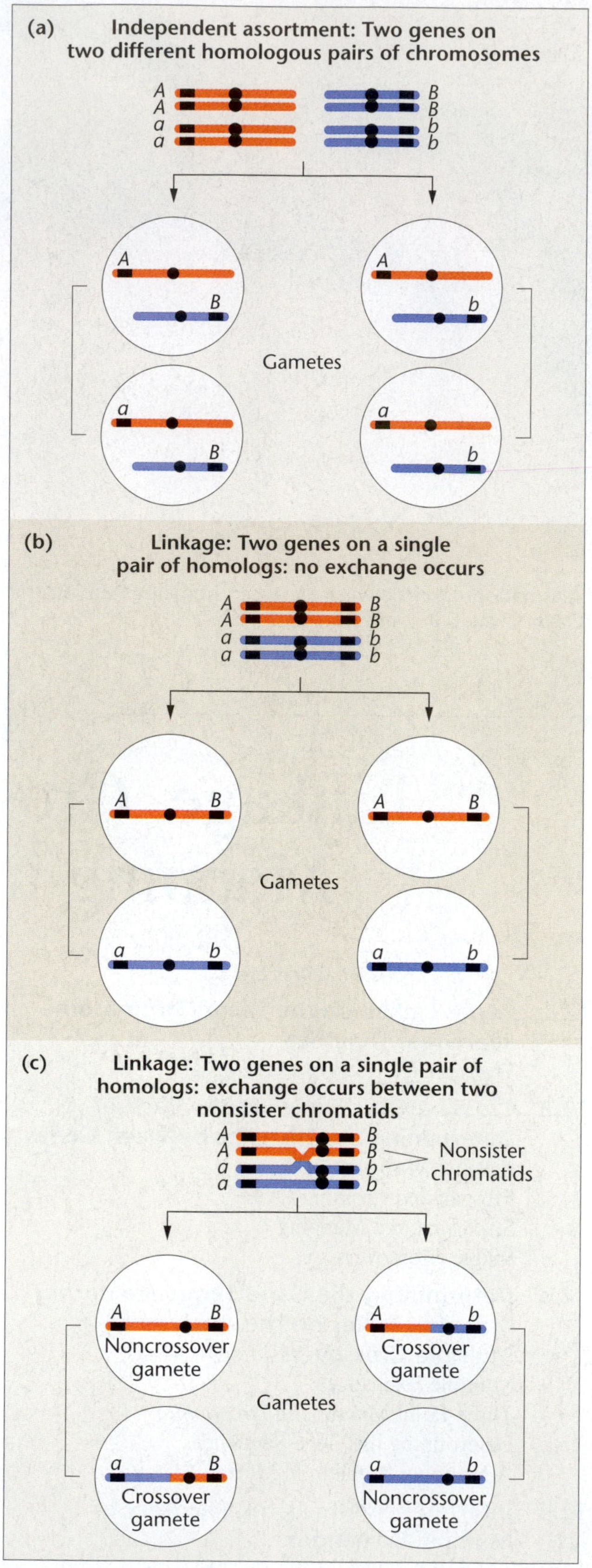

FIGURE 12–1 Results of gamete formation where two heterozygous genes are (a) on two different pairs of chromosomes; (b) on the same pair of homologs, but where no exchange occurs between them; and (c) on the same pair of homologs, where an exchange occurs between two nonsister chromatids.

without crossing over, and (c) linkage *with* crossing over. Figure 12.1(a) illustrates the results of independent assortment of two pairs of chromosomes, each containing one heterozygous gene pair. No linkage is exhibited. When a large number of meiotic events are observed, four genetically different gametes are formed in equal proportions. Each contains a different combination of alleles of the two genes.

We can compare these results with those that occur if the same genes are linked on the same chromosome. If no crossing over occurs between the two genes [Figure 12–1(b)], only two genetically different gametes are formed. Each gamete receives the alleles present on one homolog or the other, which has been transmitted intact as the result of segregation. This case illustrates **complete linkage**, which produces only **parental** or **noncrossover gametes**. The two parental gametes are formed in equal proportions. Though complete linkage between two genes seldom occurs, it is useful to consider the theoretical consequences of this concept when learning about crossing over.

Figure 12–1(c) illustrates the results of crossing over between two linked genes. As you can see, this crossover involves only two nonsister chromatids of the four chromatids present in the tetrad. This exchange generates two new allele combinations, called **recombinant** or **crossover gametes**. The two chromatids not involved in the exchange result in noncrossover gametes, like those in Figure 12–1(b).

The frequency with which crossing over occurs between any two linked genes is generally proportional to the distance separating the respective loci along the chromosome. In theory, two randomly selected genes can be so close to each other that crossover events are too infrequent to be easily detected. As shown in Figure 12–1(b), this circumstance, complete linkage, produces only parental gametes. On the other hand, if a small, but distinct distance separates two genes, few recombinant and many parental gametes will be formed. As the distance between the two genes increases, the proportion of recombinant gametes increases and that of the parental gametes decreases.

As we will discuss again later in this chapter, when the loci of two linked genes are far apart, the number of recombinant gametes approaches, but does not exceed, 50 percent. If 50 percent recombinants occurred, a 1:1:1:1 ratio of the four types (two parental and two recombinant gametes) would result. In such a case, transmission of two linked genes would be indistinguishable from that of two unlinked, independently assorting genes. That is, the proportion of the four possible genotypes would be identical, as shown in Figure 12–1(a) and (c).

The Linkage Ratio

If complete linkage exists between two genes because of their close proximity and organisms heterozygous at both loci are mated, an F_2 phenotypic ratio results that is unique, which we shall designate the **linkage ratio**. To illustrate this ratio, we will consider a cross involving the closely linked recessive mutant genes *brown* (*bw*) eye and *heavy* (*hv*) wing vein in *Drosophila melanogaster* (Figure 12–2). The normal wild-type alleles bw^+ and hv^+ are both dominant and result in red eyes and thin wing veins, respectively.

In this cross, flies with mutant brown eyes and normal thin veins are mated to flies with normal red eyes and mutant heavy veins. To be more concise, we will discuss the flies simply in terms of their mutant phenotypes and say that brown-eyed flies are crossed with heavy-veined flies. Using the genetic symbols established in Chapter 10, we can represent linked genes by placing their allele designations above and below a single or double horizontal line. Those placed above the line are located at loci on one homolog and those below the line are at the homologous loci on the other homolog. Thus, we represent the P_1 generation as follows:

$$P_1: \quad \frac{bw\ hv^+}{bw\ hv^+} \times \frac{bw^+\ hv}{bw^+\ hv}$$

brown, thin   red, heavy

Because the genes are located on an autosome, and not the X chromosome, no designation between males and females is necessary.

In the F_1 generation, each fly receives one chromosome of each pair from each parent; all flies are heterozygous for both gene pairs and exhibit the dominant traits of red eyes and thin veins:

$$F_1: \quad \frac{bw\ hv^+}{bw^+ hv}$$

red, thin

As shown in Figure 12–2, because of complete linkage, when the F_1 generation is interbred, each F_1 individual forms only parental gametes. Following fertilization, the F_2 generation will be produced in a 1:2:1 phenotypic and genotypic ratio. One-fourth of this generation will show brown eyes and thin veins; one-half will show both wild-type traits, namely, red eyes and thin veins; and one-fourth will show red eyes and heavy veins. In more concise terms, the ratio is 1 brown: 2 wild:1 heavy. Such a ratio is characteristic of complete linkage, which is observed only when two genes are very close together and the number of progeny is relatively small.

Figure 12–2 also demonstrates the results of a test cross with the F_1 flies. Such a cross produces a 1:1 ratio of brown, thin and red, heavy flies. Had the genes controlling these traits been incompletely linked or located on separate autosomes, the test cross would have produced four phenotypes, rather than two.

When we consider large numbers of mutant genes in any given species, genes located on the same chromosome will show evidence of linkage to one another. As a result, we can establish a **linkage group** for each chromosome. In theory, the number of linkage groups should correspond to the haploid number of chromosomes. In organisms with large numbers of mutant genes available for genetic study, this correlation has been confirmed.

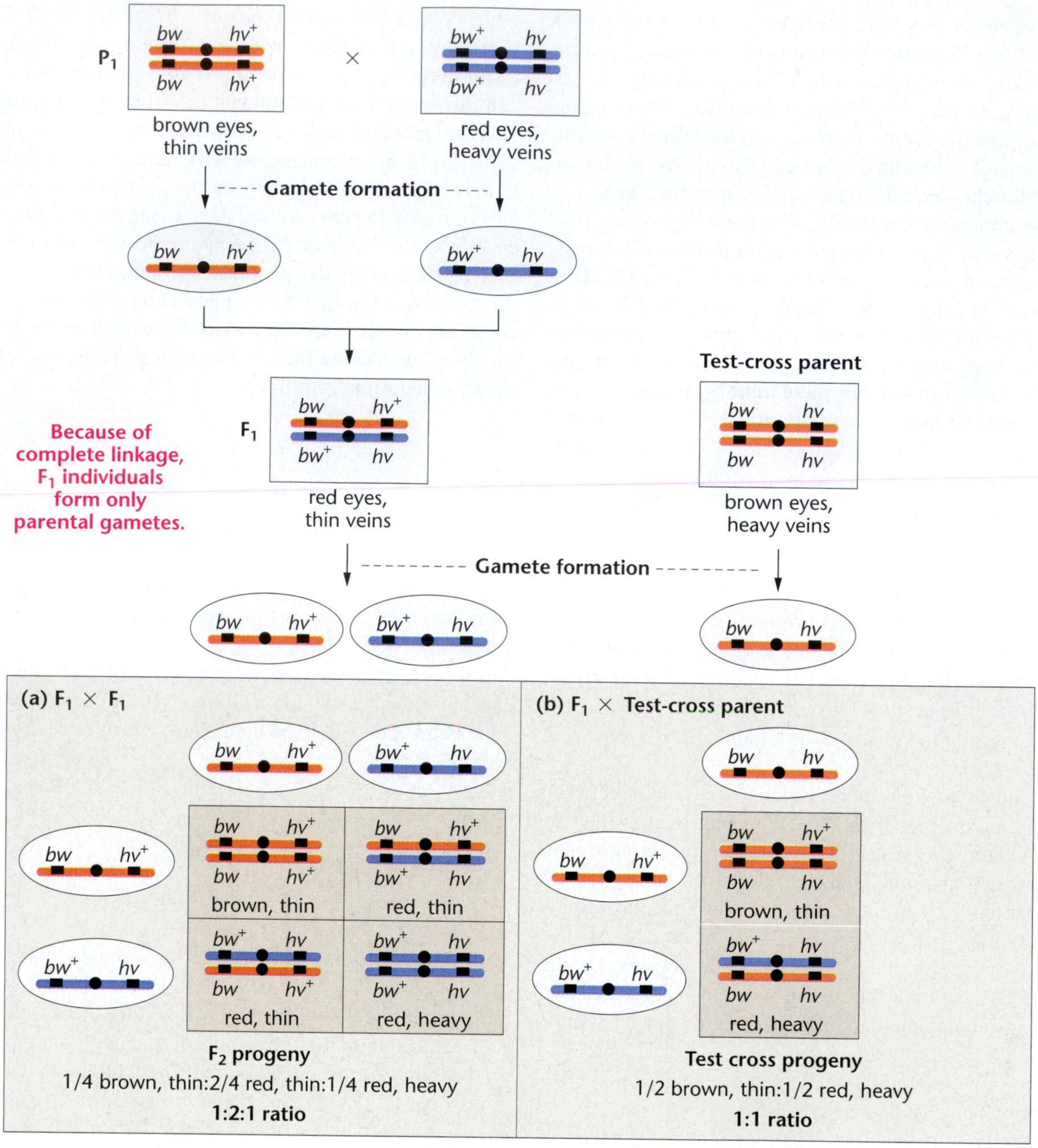

FIGURE 12–2 Results of a cross involving two genes located on the same chromosome where complete linkage is demonstrated. (a) The F_2 results of the cross. (b) The results of a test cross involving the F_1 progeny.

12.2 Crossing Over Serves as the Basis of Determining the Distance between Genes during Chromosome Mapping

It is highly improbable that two randomly selected genes that are linked on the same chromosome will be so close to one another along the chromosome that they demonstrate complete linkage. Instead, crosses involving two such genes will almost always produce a percentage of offspring resulting from recombinant gametes. The percentage is variable and depends upon the distance between the two genes along the chromosome. This phenomenon was first explained in 1911, by two *Drosophila* geneticists, Thomas H. Morgan and his undergraduate student, Alfred H. Sturtevant.

Morgan and Crossing Over

As you may recall from our discussion in Chapter 4, Morgan, was the first to discover the phenomenon of X-linkage. In his studies, he investigated numerous *Drosophila* mutations located on the X chromosome. When he analyzed crosses involving only one trait, he was able to deduce the mode of X-linked inheritance. However, when he made crosses that simultaneously involved two X-linked genes, his results were at first puzzling. For example, as shown in cross A of Figure 12–3, he crossed mutant *yellow*-bodied (y) and *white*-eyed (w) females with wild-type males (gray body and red eyes). The F_1 females were wild type, whereas the F_1 males expressed both mutant traits. In the F_2 98.7 percent of the total offspring showed the parental phenotypes—yellow-bodied, white-eyed flies and wild-type flies (gray bodied, red eyed).

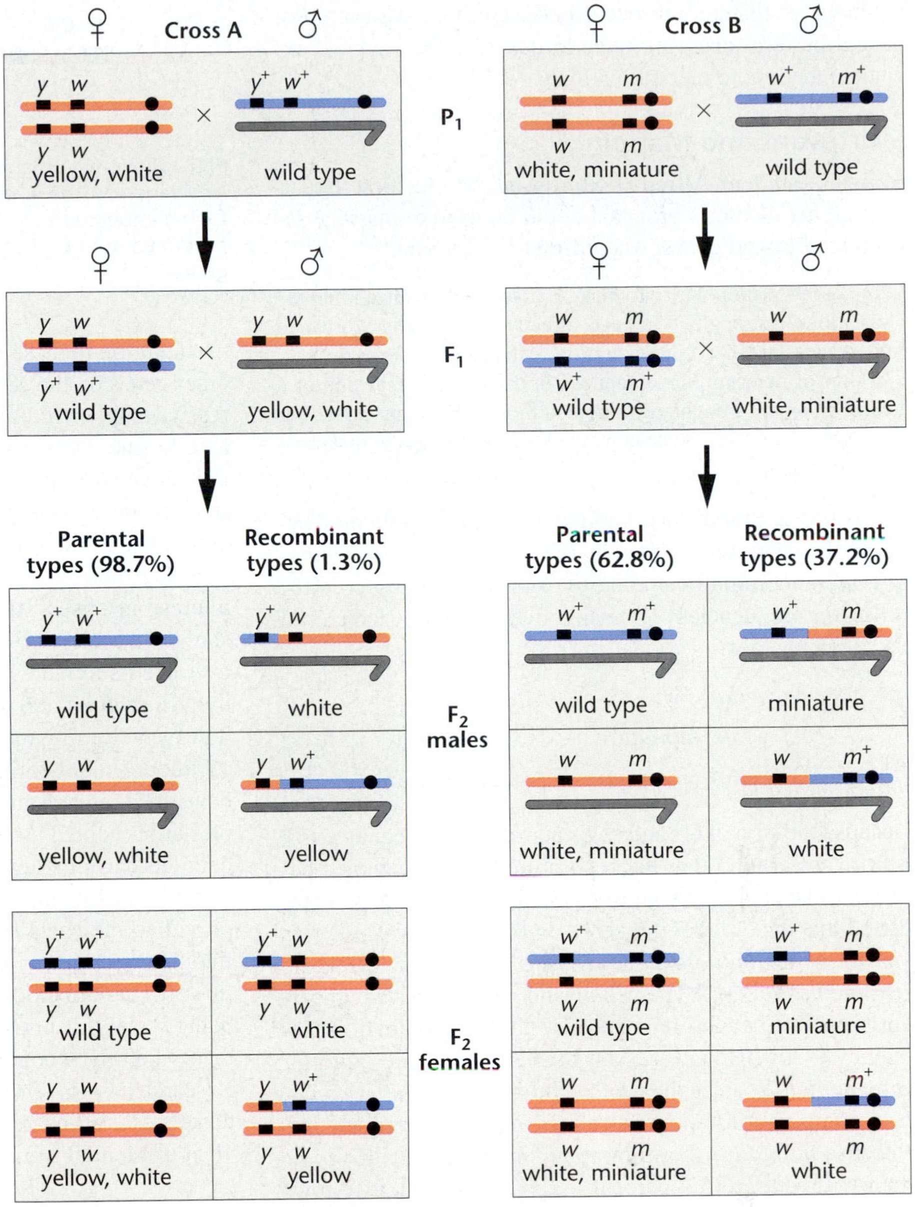

FIGURE 12–3 The F_1 and F_2 results of crosses involving the *yellow*-body, *white*-eye mutations and the *white*-eye, *miniature*-wing mutations. In cross A, 1.3 percent of the F_2 flies (males and females) demonstrate recombinant phenotypes, which express either *white* or *yellow*. In cross B, 37.2 percent of the F_2 flies (males and females) demonstrate recombinant phenotypes, which are either miniature or white mutants.

The remaining 1.3 percent of the flies were either yellow bodied with red eyes or gray bodied with white eyes. It was as if the genes had somehow separated from each other during gamete formation in the F_1 flies.

When Morgan made crosses involving other X-linked genes, the results were even more puzzling (cross B of Figure 12–3). The same basic pattern was observed, but the proportion of F_2 phenotypes differed. For example, when he crossed *white*-eye, *miniature*-wing mutants with wild-type flies, only 62.8 percent of all the F_2 flies showed the parental phenotypes, while 37.2 percent of the offspring appeared as if the mutant genes had been separated during gamete formation.

As a result of these observations, Morgan was faced with two questions: (1) What was the source of gene separation? and (2) Why did the frequency of the apparent separation vary depending on the genes being studied? The answer Morgan proposed to the first question was based on his knowledge of earlier cytological observations made by F. A. Janssens and others. Janssens observed that synapsed homologous chromosomes in meiosis wrapped around each other, creating **chiasmata** (sing., **chiasma**) where points of overlap are evident. Morgan proposed that these chiasmata could represent points of genetic exchange.

In the crosses shown in Figure 12–3, Morgan postulated that if an exchange occurred between the mutant genes on the two X chromosomes of the F_1 females, it would lead to the percentages in the observed results. He suggested that such exchanges led to 1.3 percent recombinant gametes in the *yellow–white* cross and 37.2 percent in the *white–miniature* cross. On the basis of this and other experimentation, Morgan concluded that if linked genes exist in a linear order along the chromosome, then a variable amount of exchange occurs between any two genes.

As an answer to the second question, Morgan proposed that two genes located relatively close to each other along a chromosome are less likely to have a chiasma form between them than if the two genes are farther apart on the chromosome. Therefore, the closer two genes are, the less likely it is that a genetic exchange will occur between them. Morgan proposed the term **crossing over** to describe the physical exchange leading to recombination.

Sturtevant and Mapping

Morgan's student, Alfred H. Sturtevant, was the first to realize that his mentor's proposal could be used to map the sequence of linked genes. According to Sturtevant,

> "In a conversation with Morgan ... I suddenly realized that the variations in strength of linkage, already attributed by Morgan to differences in the spatial separation of the genes, offered the possibility of determining sequences in the linear dimension of a chromosome. I went home and spent most of the night (to the neglect of my undergraduate homework) in producing the first chromosomal map. ..."

For example, Sturtevant compiled data on recombination between the genes represented by the *yellow, white,* and *miniature* mutants initially studied by Morgan and observed the following frequencies of crossing over between each pair of these three genes:

(1) *yellow, white*	0.5 percent
(2) *white, miniature*	34.5 percent
(3) *yellow, miniature*	35.4 percent

Because the sum of (1) and (2) is approximately equal to (3), Sturtevant argued that the recombination frequencies between linked genes are additive. On this assumption, he predicted that the order of the genes on the X chromosome was *yellow–white–miniature.* In arriving at this conclusion, he reasoned as follows: The *yellow* and *white* genes are apparently close to each other because the recombination frequency is low. However, both of these genes are quite far from *miniature,* since the *white, miniature* and *yellow, miniature* combinations show large recombination frequencies. Because *miniature* shows more recombination with *yellow* than with *white* (35.4 percent vs. 34.5 percent), it follows that *white* is between the other two genes, not outside of them.

Sturtevant knew from Morgan's work that the frequency of exchange could be taken as an estimate of the relative distance between two genes or loci along the chromosome. He constructed a map of the three genes on the X chromosome, with one map unit equated with 1 percent recombination between two genes.* In the preceding example, the distance between *yellow* and *white* would be 0.5 map unit(mu), and between *yellow* and *miniature* 35.4 mu. It follows that the distance between *white* and *miniature* should be (35.4 - 0.5), or 34.9 mu. This estimate is close to the actual frequency of recombination between *white* and *miniature* (34.5 mu). The simple map for these three genes is shown in Figure 12–4. It is important to keep in mind that map units such as those shown represent *relative* measurements of distance.

*In honor of Morgan's work, map units are often referred to as centimorgans (cM).

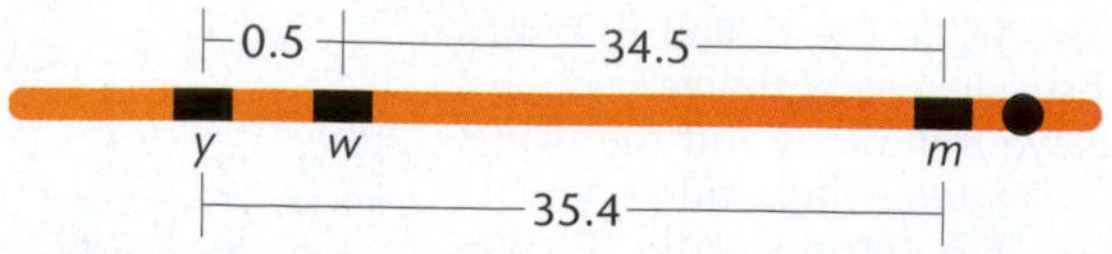

FIGURE 12–4 A map of the *yellow* (*y*), *white* (*w*), and *miniature* (*m*) genes on the X chromosome of *Drosophila melanogaster.* Each number represents the percentage of recombinant offspring produced in one of three crosses, each involving two different genes.

In addition to these three genes, Sturtevant considered two other genes on the X chromosome and produced a more extensive map, including all five genes. Soon thereafter, he and a colleague, Calvin Bridges, began a search for autosomal linkage in *Drosophila.* By 1923, they had clearly shown that linkage and crossing over were not restricted to X-linked genes, but could also be demonstrated with autosomes.

During their work, Sturtevant and Bridges made another interesting observation. In *Drosophila,* crossing over was shown to occur only in females. The fact that no crossing over occurs in males made the analysis of genetic mapping in *Drosophila* much less complex. In most other organisms, however, crossing over does occur in both sexes.

Although many refinements in chromosome mapping have developed since Sturtevant's initial work, his basic principles are accepted as correct and have been used to produce detailed chromosome maps of organisms for which large numbers of linked mutant genes are known. In addition to providing the basis for chromosome mapping, Sturtevant's findings were historically significant to the field of genetics. In 1910, the **chromosomal theory of inheritance** was still being widely disputed. Even Morgan was skeptical of this theory before he conducted the bulk of his experimentation. Research has now firmly established that chromosomes contain genes in a linear order and that these genes are the equivalent of Mendel's unit factors.

Single Crossovers

Why should the relative distance between two loci influence the amount of recombination and crossing over observed between them? The basis for this variation is explained in the following analysis:

During meiosis, a limited number of crossover events occur in each tetrad. These recombinant events occur randomly along the length of the tetrad. Therefore, the closer that two loci reside along the axis of the chromosome, the less likely it is that any **single crossover** event will occur in between them. The same reasoning suggests that the farther apart two linked loci are, the more likely it is that a random crossover event will occur in between them.

In Figure 12–5(a), a single crossover occurs between two nonsister chromatids, but in an area of the chromosome that is not in between the two loci. Therefore, the crossover goes undetected because no recombinant gametes are produced. In Figure 12–5(b), where two loci are quite far apart, the crossover occurs in an area of the chromosome between them. This separates them, yielding recombinant gametes.

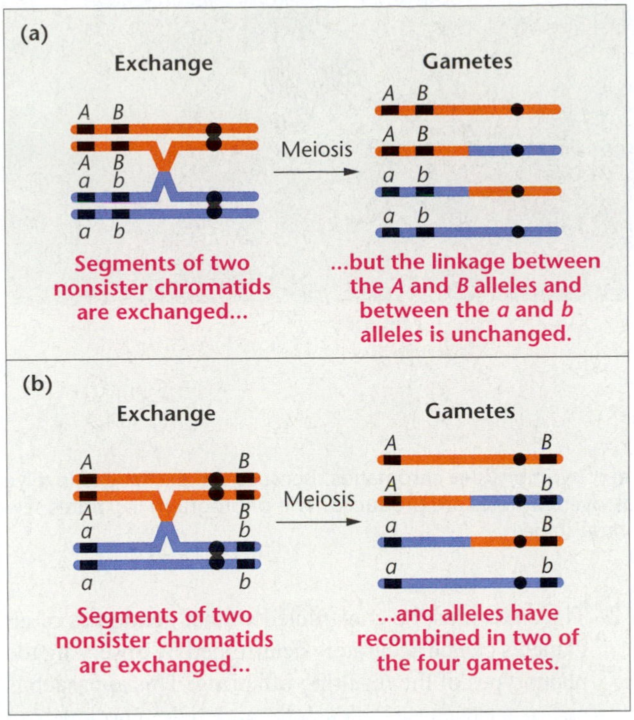

FIGURE 12–5 Two examples of a single crossover between two nonsister chromatids and the gametes subsequently produced. In (a) the exchange does not alter the linkage arrangement between the alleles of the two genes, only parental gametes are formed, and the exchange goes undetected. In (b) the exchange separates the alleles, resulting in recombinant gametes, which are detectable.

When a single crossover occurs between two nonsister chromatids, the other two chromatids of the tetrad are not involved in the exchange, and they enter the gamete unchanged. Following the completion of meiosis, they represent noncrossover gametes. In this case, even if a single crossover occurs 100 percent of the time between two linked genes, recombination will subsequently be observed in only 50 percent of the potential gametes formed. This concept is diagrammed in Figure 12–6. Theoretically, if we consider only single exchanges and observe 20 percent recombinant gametes, crossing over actually occurs between these two loci in 40 percent of the tetrads. The general rule is that under these conditions, the percentage of tetrads involved in an exchange between two genes is twice as great as the percentage of recombinant gametes produced. Therefore, the theoretical limit of recombination due to crossing over is 50 percent.

When two linked genes are more than 50 map units apart, a crossover can theoretically be expected to occur between them in 100 percent of the tetrads. If this prediction were achieved, each tetrad would yield equal proportions of the four gametes shown in Figure 12–6, just as if the genes were on different chromosomes and assorting independently. For a variety of reasons, this theoretical limit is seldom achieved.

12.3 Determining the Gene Sequence during Mapping Relies on the Analysis of Multiple Crossovers

The study of single crossovers between two linked genes provides the basis of determining the *distance* between them. However, when many linked genes are studied, their *sequence* along the chromosome is more difficult to determine. Fortunately, the discovery that multiple exchanges occur between the chromatids of a tetrad has facilitated the process of producing more extensive chromosome maps. As we shall see next, when three or more linked genes are investigated simultaneously, it is possible to determine first the sequence of and then the distances between genes.

Multiple Exchanges

It is possible that in a single tetrad, two, three, or more exchanges will occur between nonsister chromatids as a result of several crossing over events. Double exchanges of genetic material result from **double crossovers (DCOs)**, as shown in Figure 12–7. To study a double exchange, three gene pairs must be investigated, each heterozygous for two alleles. Before we can determine the frequency of recombination among all three loci, we must review some simple probability calculations.

As we have seen, the probability of a single exchange occurring in between the *A* and *B* or the *B* and *C* genes is directly related to the physical distance separating the loci. The closer *A* is to *B* and *B* is to *C*, the less likely it is that a single exchange will occur in between either of the two sets of loci. In the case of a double crossover, two separate and independent events or exchanges must occur simultaneously. The

FIGURE 12–6 The consequences of a single exchange between two nonsister chromatids occurring in the tetrad stage. Two noncrossover (parental) and two crossover (recombinant) gametes are produced.

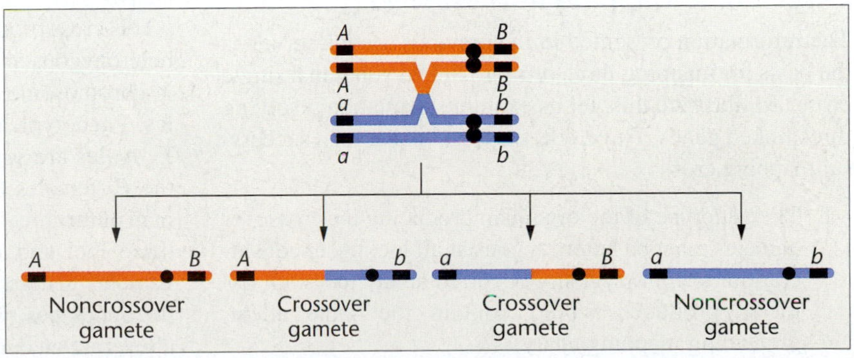

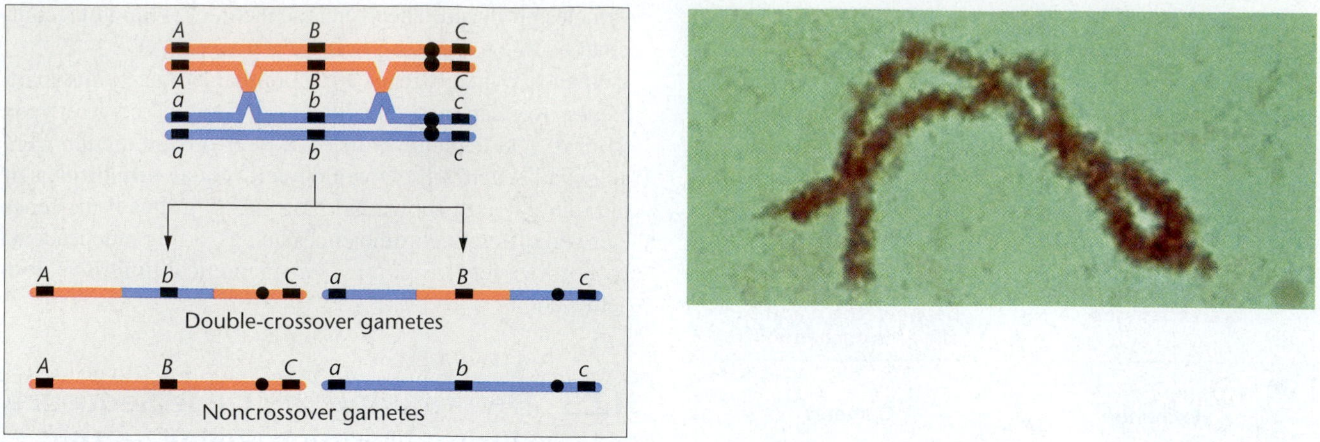

FIGURE 12–7 Consequences of a double exchange occurring between two nonsister chromatids. Because the exchanges involve only two chromatids, two noncrossover gametes and two double-crossover gametes are produced. The photograph illustrates several chiasmata found in a tetrad isolated during the first meiotic prophase stage.

mathematical probability of two independent events occurring simultaneously is equal to the product of the individual probabilities. This is the product law we introduced in Chapter 9.

Suppose that crossover gametes result from single exchanges between *A* and *B* 20 percent of the time $(p = 0.20)$ and between *B* and *C* 30 percent of the time $(p = 0.30)$. The probability of recovering a double-crossover gamete arising from two exchanges, between *A* and *B* and between *B* and *C*, is predicted to be $(0.20)(0.30) = 0.06$, or 6 percent. This calculation makes apparent that the expected frequency of double-crossover gametes is always much lower than that of either single-crossover class of gametes alone.

If three genes are relatively close together along one chromosome, the expected frequency of double-crossover gametes is extremely low. For example, assume the *A–B* distance in Figure 12–7 to be 3 mu and the *B–C* distance to be 2 mu. The expected double-crossover frequency would be $(0.03)(0.02) = 0.0006$, or 0.06 percent. This translates to only 6 events in 10,000. In such a mapping experiment, where closely linked genes are involved, very large numbers of offspring are required to detect double-crossover events. In this example, it would be unlikely that a double crossover would be observed even if 1000 offspring are examined. If we extend these probability considerations, it is evident that if four or five genes were being mapped, even fewer triple and quadruple crossovers could be expected to occur.

Three-Point Mapping in *Drosophila*

The information presented in the previous section serves as the basis for mapping three or more linked genes in a single cross. To illustrate this, let us examine a situation involving three linked genes. Three criteria must be met for a successful mapping cross:

1. The genotype of the organism producing the crossover gametes must be heterozygous at all loci under consideration. If homozygosity occurred at any locus, all gametes produced would contain the same allele, precluding mapping analysis.
2. The cross must be constructed so that genotypes of all gametes can be accurately determined by observing the phenotypes of the resulting offspring. This approach is necessary because the gametes and their genotypes can never be observed directly. To overcome this problem, each phenotypic class must reflect the genotype of the gametes of the parents producing it.
3. A sufficient number of offspring must be produced in the mapping experiment to recover a representative sample of all crossover classes.

These criteria are met in the three-point mapping cross from *Drosophila melanogaster* shown in Figure 12–8. The cross involves three X-linked recessive mutant genes—*yellow* body color, *white* eye color, and *echinus* eye shape. To diagram the cross, we must assume some theoretical sequence, even though we do not yet know if it is correct. In Figure 12–8, we initially assume the sequence of the three genes to be *y–w–ec*. If this is incorrect, our analysis shall demonstrate it and reveal the correct sequence.

In the P_1 generation, males hemizygous for all three wild-type alleles are crossed to females that are homozygous for all three recessive mutant alleles. Therefore, the P_1 males are wild type with respect to body color, eye color, and eye shape, and they are said to have a wild-type phenotype. The females, on the other hand, exhibit the three mutant traits: *yellow* body color, *white* eyes, and *echinus* eye shape.

This cross produces an F_1 generation consisting of females heterozygous at all three loci and males that, because of the Y chromosome, are hemizygous for the three mutant alleles. Phenotypically, all F_1 females are wild type, while all F_1 males are *yellow, white,* and *echinus*. The genotype of the F_1 females fulfills the first criterion for constructing a map of the three linked genes; that is, it is heterozygous at the three loci and may serve as the source of recombinant gametes generated by crossing over. Note that, because of the genotypes of the P_1 parents, all three of the F_1 mutant alleles are on one homolog and all three wild-type alleles are

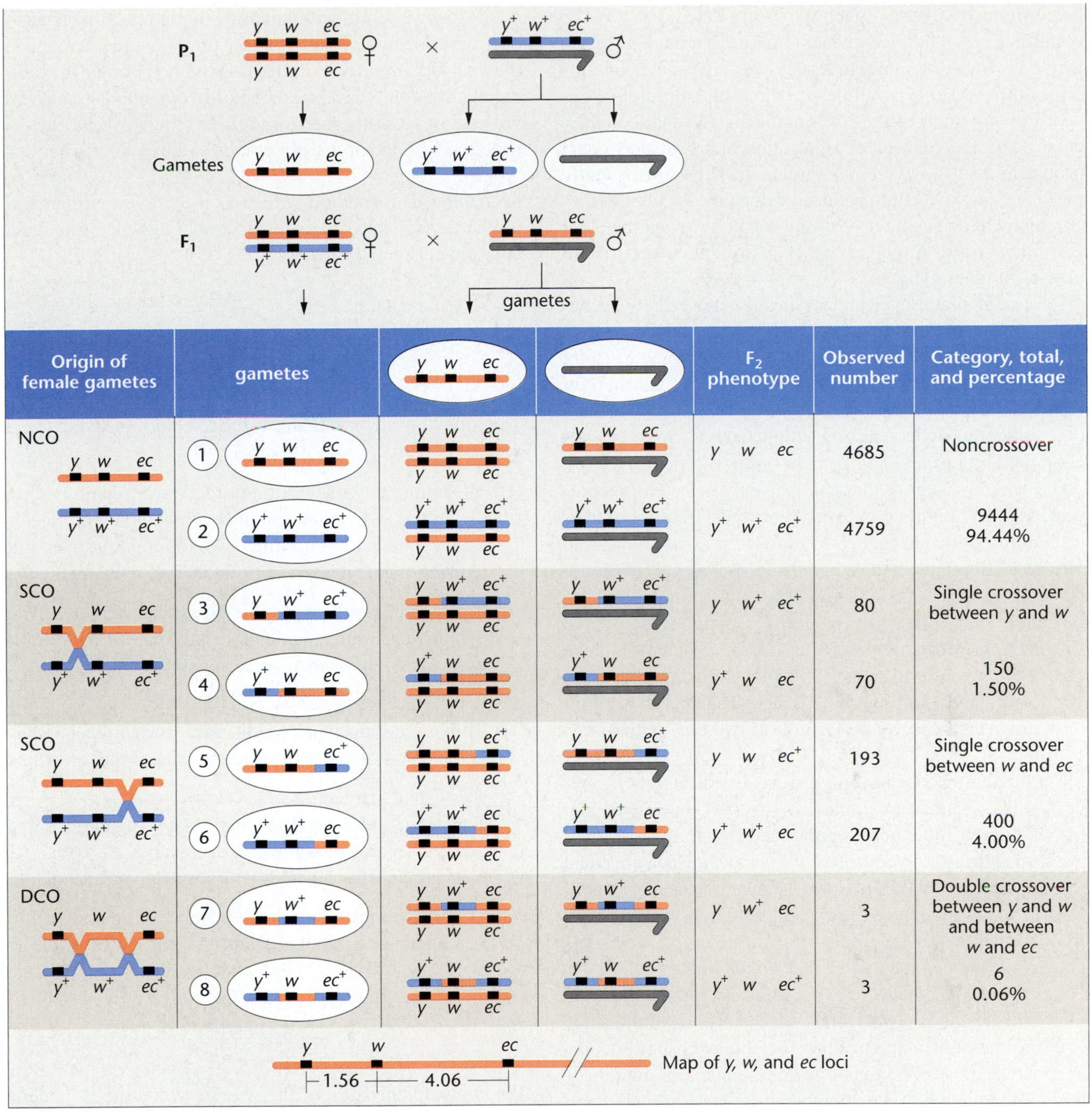

FIGURE 12–8 A three-point mapping cross involving the *yellow* (y or y^+), *white* (w or w^+), and *echinus* (ec or ec^+) genes in *Drosophila melanogaster*. NCO, SCO, and DCO refer to noncrossover, single-crossover, and double-crossover groups, respectively. Because of the complexity of this and several of the ensuing figures, centromeres have not been included on the chromosomes, and only two nonsister chromatids are initially shown in the left-hand column.

on the other homolog. *Other arrangements are possible.* For example, the heterozygous F_1 female might have the *y* and *ec* mutant alleles on one homolog and the *w* allele on the other. This would occur if, in the P_1 cross, one parent was *yellow* and *echinus* and the other parent was *white.*

In our cross, the second criterion is met by virtue of the gametes formed by the F_1 males. Every gamete will contain either an X chromosome bearing the three mutant alleles or a Y chromosome, which does not contain any of the three loci being considered. Whichever type participates in fertilization, the genotype of the gamete produced by the F_1 female will be expressed phenotypically in the F_2 female and male offspring. As a result, all noncrossover and crossover gametes produced by the F_1 female parent can be determined by observing the F_2 phenotypes.

With these two criteria met, we can construct a chromosome map from the crosses illustrated in Figure 12–8, but first, we must determine which F_2 phenotypes correspond to the various noncrossovers and crossover categories. Two of these can be determined immediately.

The **noncrossover** F_2 phenotypes are determined by the combination of alleles present in the parental gametes formed by the F_1 female. In this case, each gamete contains either three wild-type alleles *or* three mutant alleles, depending on which of the X chromosomes is unaffected by crossing over. As a result of segregation, approximately equal proportions of the two types of gametes, and subsequently the F_2 phenotypes, are produced. Since the F_2 phenotypes complement one another (i.e., one is wild type and the other is mutant for all three genes), they are called **reciprocal classes** of phenotypes.

The two noncrossover phenotypes are most easily recognized because *they occur in the greatest proportion of offspring.* Figure 12–8 shows that gametes (1) and (2) are present in the greatest numbers. Therefore, flies that are *yellow, white,* and *echinus* and those that are normal or wild type for all three characters constitute the non-crossover category and represent 94.44 percent of the F_2 offspring.

The second category that can be easily detected is represented by the double-crossover phenotypes. Because of their probability of occurrence, *they must be present in the least numbers.* Remember that this group represents two independent, but simultaneous single-crossover events. Two reciprocal phenotypes can be identified: gamete (7), which shows the mutant traits *yellow* and *echinus,* but normal eye color; and gamete (8), which shows the mutant trait *white,* but normal body color and eye shape. Together these double-crossover phenotypes constitute only 0.06 percent of the F_2 offspring.

The remaining four phenotypic classes fall into two categories that result from single crossovers. Gametes (3) and (4), reciprocal phenotypes produced by single-crossover events occurring between the *yellow* and *white* loci, are equal to 1.50 percent of the F_2 offspring. Gametes (5) and (6), constituting 4.00 percent of the F_2 offspring, represent the reciprocal phenotypes resulting from single-crossover events occurring between the *white* and *echinus* loci.

We can now calculate the map distances between the three loci. The distance between *y* and *w*, or between *w* and *ec*, is equal to the percentage of all detectable exchanges occurring between them. For any two genes under consideration, this will include all related single crossovers as well as all double crossovers. The latter are included because they represent two simultaneous single crossovers. For the *y* and *w* genes, this includes gametes (3), (4), (7), and (8), totaling 1.50 percent + 0.06 percent, or 1.56 mu. Similarly, the distance between *w* and *ec* is equal to the percentage of offspring resulting from an exchange between these two loci, including gametes (5), (6), (7), and (8), totaling 4.00 percent + 0.06 percent, or 4.06 mu. The map of these three loci on the X chromosome, based on these data, is shown at the bottom of Figure 12–8.

Determining the Gene Sequence

In the preceding example, we assumed the order, or sequence, of the three genes along the chromosome to be *y–w–ec*. Our analysis established that the sequence is consistent with the data. However, in most mapping experiments, the gene sequence is not known, and this constitutes another variable in the analysis. In our example, had the gene order been unknown, we could have used one of two methods (which we will study next) to determine it. In your own work, you should select one of these methods and use it consistently.

Method I This method is based on the fact that there are only three possible orders, each containing one of the three genes in between the other two:

(I)	*w–y–ec*	(*y* is in the middle)
(II)	*y–ec–w*	(*ec* is in the middle)
(III)	*y–w–ec*	(*w* is in the middle)

The following steps will allow you to determine which gene order is correct:

1. Assuming any of the three orders, first determine the *arrangement of alleles* along each homolog of the heterozygous parent giving rise to noncrossover and crossover gametes (the F_1 female in our example).
2. Determine whether a double-crossover event occurring within that arrangement will produce the *observed double-crossover phenotypes.* Remember that these phenotypes occur least frequently and can be identified easily.
3. If this order does not produce the correct phenotypes, try each of the other two orders. One of the three must work!

The next steps are shown in Figure 12–9, using the cross just discussed. The three possible orders are labeled I, II, and III, as shown previously. Either *y, ec,* or *w* must be in the middle:

1. If we assume order I with *y* between *w* and *ec*, the arrangement of alleles along the homologs of the F_1 heterozygote is

$$\frac{w \quad y \quad ec}{w^+ \quad y^+ \quad ec^+}$$

We know this because of the way in which the P_1 generation was crossed. The P_1 female contributed an X chromosome bearing the *w, y,* and *ec* alleles, while the P_1 male contributed an X chromosome bearing the w^+, y^+, and ec^+ alleles.

2. A double crossover within the previous arrangement would yield the following gametes:

$$\underline{w \quad y^+ \quad ec} \quad \text{and} \quad \underline{w^+ \quad y \quad ec^+}$$

If *y* is in the middle, following fertilization, the F_2 double-crossover phenotypes will correspond to the foregoing gametic genotypes, yielding offspring that express the white, echinus phenotype and offspring that express the yellow phenotype. Determination of the actual double crossovers, however, reveals them to be yellow, echinus flies and white flies. Therefore, our assumed order is incorrect.

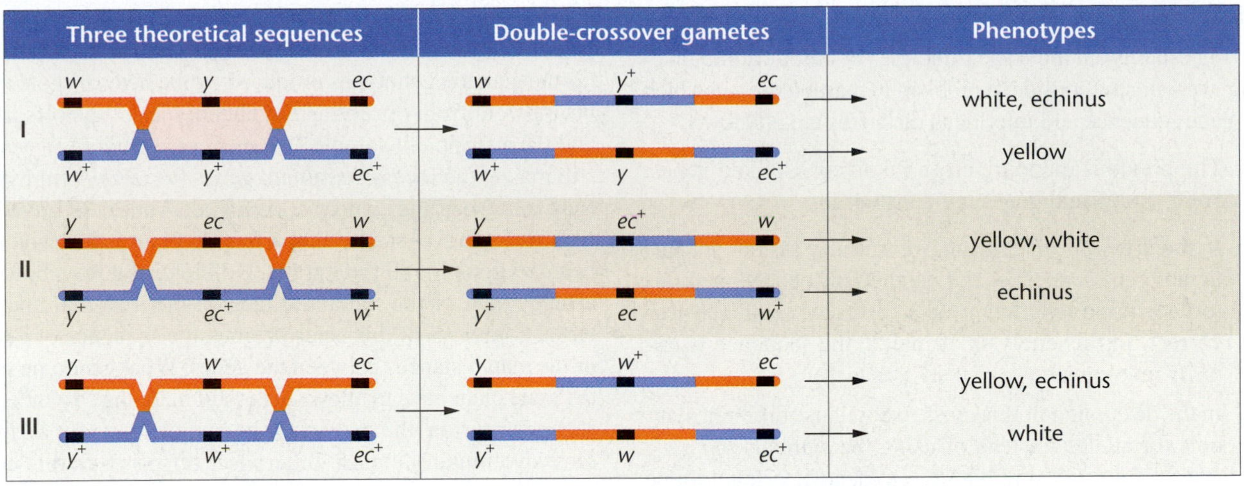

FIGURE 12–9 The three possible sequences of the *white, yellow,* and *echinus* genes, the results of a double crossover in each case, and the resulting phenotypes produced in a test cross. For simplicity, the two noncrossover chromatids of each tetrad are omitted.

3. If we consider the other orders, one with ec/ec^+ alleles in the middle (II) or one with the w/w^+ alleles in the middle (III),

$$\text{(II)}\ \frac{y \quad ec \quad w}{y^+ \quad ec^+ \quad w^+} \quad \text{or} \quad \text{(III)}\ \frac{y \quad w \quad ec}{y^+ \quad w^+ \quad ec^+}$$

we see that arrangement II again provides *predicted* double-crossover phenotypes that do not correspond to the *actual* double-crossover phenotypes. The predicted phenotypes are yellow, white flies and echinus flies in the F_2 generation. Therefore, this order is also incorrect. However, arrangement III *will* produce the *observed* phenotypes—*yellow, echinus* flies and *white* flies. Therefore, this order, where the *w* gene is in the middle, is correct.

To summarize method I, first, determine the arrangement of alleles on the homologs of the heterozygote yielding the crossover gametes. This is done by locating the reciprocal noncrossover phenotypes. Then test each of three possible orders to determine which one yields the observed (actual) double-crossover phenotypes. Whichever of the three does so represents the correct order. This method is summarized in Figure 12–9.

Method II Method II also begins by assuming the arrangement of alleles along each homolog of the heterozygous parent. In addition, it requires one further assumption:

Following a double-crossover event, the allele in the middle position will fall between the outside, or flanking, alleles that were present on the opposite parental homolog.

To illustrate, assume order (I), *w–y–ec,* in the following arrangement:

$$\frac{w \quad y \quad ec}{w^+ \quad y^+ \quad ec^+}$$

Following a double-crossover event, the *y* and y^+ alleles would be switched to this arrangement:

$$\frac{w \quad y^+ \quad ec}{w^+ \quad y \quad ec^+}$$

After segregation, two gametes would be formed:

$$\underline{w \quad y^+ \quad ec} \quad \text{and} \quad \underline{w^+ \quad y \quad ec^+}$$

Because the genotype of the gamete will be expressed directly in the phenotype following fertilization, the double-crossover phenotypes will be:

white, echinus flies, and yellow flies

Note that the *yellow* allele, assumed to be in the middle, is now associated with the two outside markers of the other homolog, w^+ and ec^+. However, these predicted phenotypes do not coincide with the observed double-crossover phenotypes. The *yellow* gene, therefore, is not in the middle.

This same reasoning can be applied to the assumption that the *echinus* gene or the *white* gene is in the middle. In the former case, we will reach a negative conclusion. If we assume that the *white* gene is in the middle, the *predicted* and *actual* double crossovers coincide. Therefore, we conclude that the *white* gene is located between the *yellow* and *echinus* genes.

To summarize method II, determine the arrangement of alleles on the homologs of the heterozygote yielding crossover gametes. Then determine the actual double-crossover phenotypes. Simply select the single allele that has been switched so that it is now no longer associated with its original neighboring alleles.

In our example *y, ec,* and *w* are together in the F_1 heterozygote, as are y^+, ec^+, and w^+. In the F_2 double-crossover classes, it is *w* and w^+ that have been switched. The *w* allele is now associated with y^+ and ec^+, while the w^+ allele is now associated with the *y* and *ec* alleles. Therefore, the *white* gene is in the middle, and the *yellow* and *echinus* genes are the flanking markers.

A Mapping Problem in Maize

Having established the basic principles of chromosome mapping, we will now consider a problem in maize (corn), in which the gene sequence and interlocus distances are unknown:

1. The previous mapping cross involved X-linked genes. Here, autosomal genes are considered.
2. In the previous discussion, we initially assumed a particular gene sequence, and we then introduced methods to determine whether it or a different sequence was correct. In this analysis of maize, the sequence is *initially* unknown.
3. In the discussion of this cross, we will use different symbols for alleles. Instead of using the symbols bm^+, v^+, and pr^+, we will simply use + to denote each wild-type allele. As first described in Chapter 10, this annotation of symbols is easier to manipulate, but it requires a strong understanding of mapping procedures.

When we consider three autosomally linked genes in maize, the experimental cross must still meet the same three criteria established for the X-linked genes in *Drosophila:* (1) one parent must be heterozygous for all traits under consideration, (2) the gametic genotypes produced by the heterozygote must be apparent from observing the phenotypes of the offspring, and (3) a sufficient sample size must be available.

In maize, the recessive mutant genes *bm* (*brown* midrib), *v* (*virescent* seedling), and *pr* (*purple* aleurone) are linked on chromosome 5. Assume that a female plant is known to be heterozygous for all three traits. Nothing is known about the arrangement of the mutant alleles on the maternal and paternal homologs of this heterozygote, the sequence of genes, or the map distances between the genes. What genotype must the male plant have to allow successful mapping? To meet the second criterion, the male must be homozygous for all three recessive mutant alleles. Otherwise, offspring of this cross showing a given phenotype might represent more than one genotype, making accurate mapping impossible. Note that this is equivalent to performing a test cross.

Figure 12–10 diagrams this cross. As shown, we do not know either the arrangement of alleles or the sequence of loci in the heterozygous female. Some of the possibilities

(a) Some possible allele arrangements and gene sequences in a heterozygous female

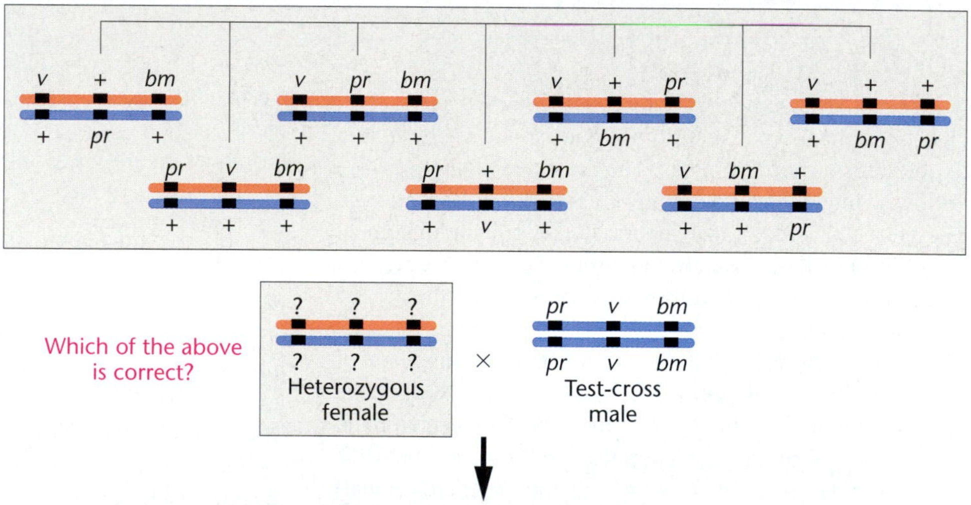

(b) Actual results of mapping cross *

Phenotypes of offspring	Number	Total and percentage	Exchange classification
+ *v* *bm*	230	467 42.1%	Noncrossover (NCO)
pr + +	237		
+ + *bm*	82	161 14.5%	Single crossover (SCO)
pr *v* +	79		
+ *v* +	200	395 35.6%	Single crossover (SCO)
pr + *bm*	195		
pr *v* *bm*	44	86 7.8%	Double crossover (DCO)
+ + +	42		

*The sequence *pr* – *v* – *bm* may or may not be correct

FIGURE 12–10 (a) Some possible allele arrangements and gene sequences in a heterozygous female. The data from a three-point mapping cross, depicted in (b), where the female is test crossed, provide the basis for determining which combination of arrangement and sequence is correct. [See Figure 12–11 (d).]

are shown, but we have yet to determine which is correct. In the testcross male parent and the offspring, *we do not know the sequence*, and so we must designate it randomly. Note that we have chosen to place *v* in the middle. This may or may not be correct.

The offspring have been arranged in groups of two for each pair of reciprocal phenotypic classes. The two members of each reciprocal class are derived from either no crossing over (NCO), one of two possible single-crossover events (SCO), or a double-crossover event (DCO).

To solve this problem, it will help to refer to Figure 12–10 and Figure 12–11, as you consider the following questions:

1. What is the correct heterozygous arrangement of alleles in the female parent?
   Determine the two noncrossover classes, those that occur with the highest frequency. In this case, they are $\underline{+ \quad v \quad bm}$ and $\underline{pr \quad + \quad +}$. Therefore, the alleles on the homologs of the female parent must be arranged as shown in Figure 12–11(a). These homologs segregate into gametes, unaffected by any recombination event. Any other arrangement of alleles could not yield the observed noncrossover classes. (Remember that $\underline{+ \quad v \quad bm}$ is equivalent to $\underline{pr^+ v\ bm}$ and that $\underline{pr \quad + \quad +}$ is equivalent to $\underline{pr\ v^+ \ bm^+}$

2. What is the correct sequence of genes?
   To answer this question, we will first use the approach described in Method I. We know that the arrangement of alleles is

$$\frac{+ \quad v \quad bm}{pr \quad + \quad +}$$

But is the assumed sequence correct? That is, will a double-crossover event yield the observed double-crossover phenotypes following fertilization? Simple observation shows that it will not [Figure 12–11(b)]. Try the other two orders [Figure 12–11(c) and (d)], *keeping the same arrangement*:

$$\frac{+ \quad bm \quad v}{pr \quad + \quad +} \quad \text{and} \quad \frac{v \quad + \quad bm}{+ \quad pr \quad +}$$

Only the case on the right yields the observed double-crossover classes [Figure 12–11(d)]. Therefore, the *pr* gene is in the middle.

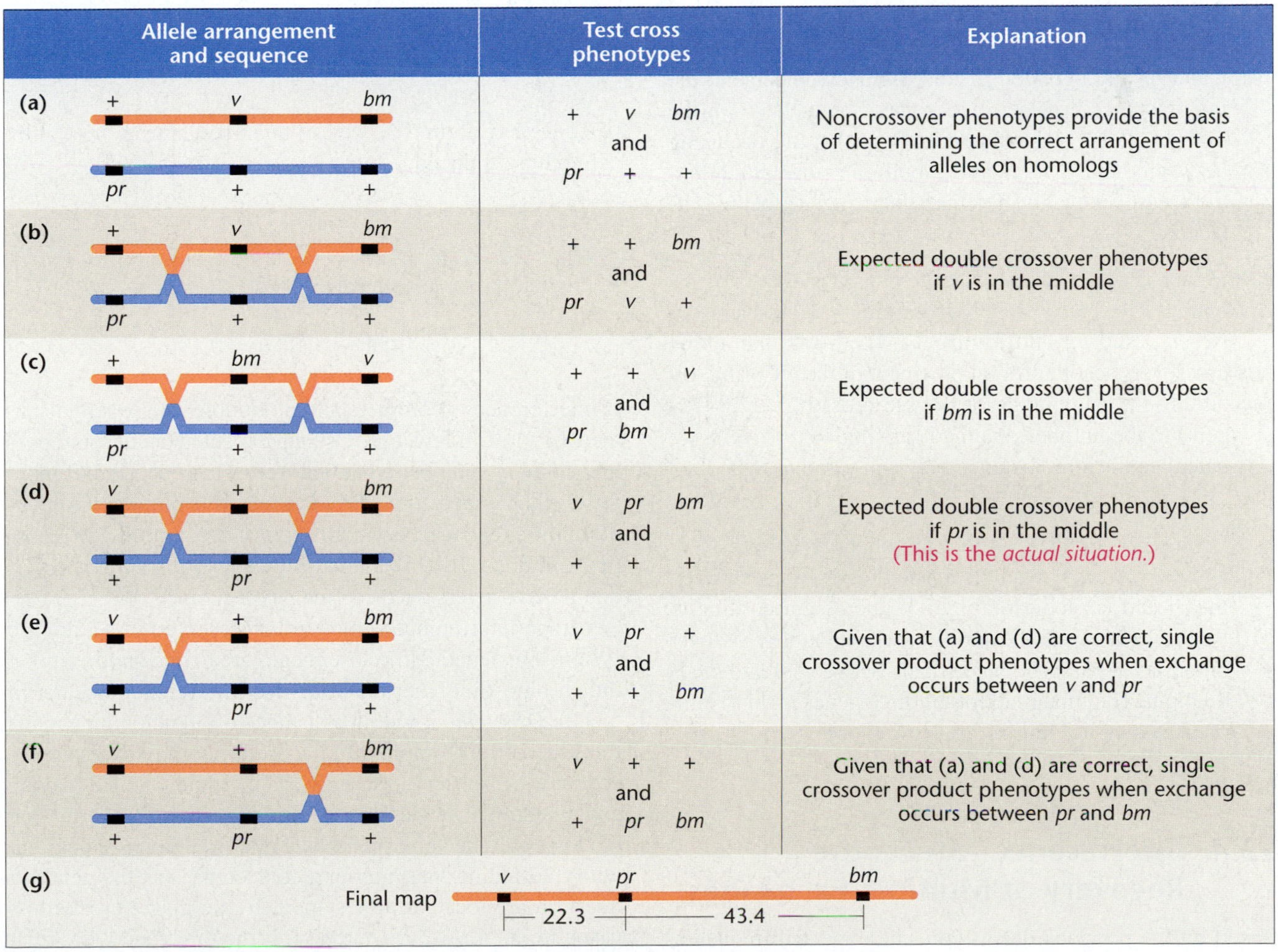

Allele arrangement and sequence
Test cross phenotypes
Explanation
(a) + v bm / pr + +
+ v bm and pr + +
Noncrossover phenotypes provide the basis of determining the correct arrangement of alleles on homologs
(b) + v bm / pr + +
+ + bm and pr v +
Expected double crossover phenotypes if v is in the middle
(c) + bm v / pr + +
+ + v and pr bm +
Expected double crossover phenotypes if bm is in the middle
(d) v + bm / + pr +
v pr bm and + + +
Expected double crossover phenotypes if pr is in the middle
(This is the actual situation.)
(e) v + bm / + pr +
v pr + and + + bm
Given that (a) and (d) are correct, single crossover product phenotypes when exchange occurs between v and pr
(f) v + bm / + pr +
v + + and + pr bm
Given that (a) and (d) are correct, single crossover product phenotypes when exchange occurs between pr and bm
(g) Final map
v pr bm
22.3
43.4

FIGURE 12–11 Producing a map of the three genes in the cross in Figure 12–10, where neither the arrangement of alleles nor the sequence of genes in the heterozygous female parent is known.

MEDIA TUTORIAL More Three-Point Crosses

The same conclusion is reached if we used Method II to analyze the problem. In this case, no assumption of gene sequence is necessary. The arrangement of alleles in the heterozygous parent is

$$\frac{+ \quad v \quad bm}{pr \quad + \quad +}$$

The double-crossover gametes are also known:

$$\underline{pr\ v\ bm} \quad \text{and} \quad \underline{+\ +\ +}$$

We can see that it is the *pr* allele that has shifted, so as to be associated with *v* and *bm* following a double crossover. The latter two alleles (*v* and *bm*) were present together on one homolog, and they stayed together. Therefore, *pr* is the odd gene, so to speak, and it is in the middle. Thus, we arrive at the same arrangement and sequence as we did with Method I:

$$\frac{v \quad + \quad bm}{+ \quad pr \quad +}$$

3. What is the distance between each pair of genes?

Having established the correct sequence of loci as *v–pr–bm,* we can now determine the distance between *v* and *pr* and between *pr* and *bm.* Remember that the map distance between two genes is calculated on the basis of all detectable recombinational events occurring between them. This includes both the single- and double-crossover events involving the two genes being considered.

Figure 12–11(e) shows that the phenotypes $\underline{v\ \ pr\ \ +}$ and $\underline{+\ \ +\ \ bm}$ result from single crossovers between *v* and *pr,* and Figure 12–10 shows that single crossovers account for 14.5 percent of the offspring. By adding the percentage of double crossovers (7.8 percent) to the number obtained for single crossovers, we calculate the total distance between *v* and *pr* to be 22.3 mu.

MEDIA TUTORIAL Virtual Crossover Laboratory

Figure 12–11(f) shows that the phenotypes $\underline{v\ \ +\ \ +}$ and $\underline{+\ \ pr\ \ bm}$ result from single crossovers between the *pr* and *bm* loci, totaling 35.6 percent according to Figure 12.10. With the addition of the double-crossover classes (7.8 percent) the distance between *pr* and *bm* is 43.4 mu. The final map for all three genes in this example is shown in Figure 12–11(g).

12.4 Interference Affects the Recovery of Multiple Exchanges

Based on the previous discussion, the expected frequency of multiple exchanges, such as double crossovers, can be predicted once the distance between genes is established. For example, in the maize cross of the previous section, the distance between *v* and *pr* is 22.3 mu, and the distance between *pr* and *bm* is 43.4 mu. If the two single crossovers that make up a double crossover occur independently of one another, we can calculate the expected frequency of double crossovers (DCO_{exp}):

$$DCO_{exp} = (0.223) \times (0.434) = 0.097 = 9.7 \text{ percent}$$

Most often in mapping experiments, the observed DCO frequency is less than the expected number of DCOs. In the maize cross, for example, only 7.8 percent DCOs are observed when 9.7 percent are expected. The phenomenon called **interference** is used to explain this reduction, which is when a crossover event in one region of the chromosome inhibits a second event in nearby regions.

To quantify the disparities that result from interference, we can calculate the **coefficient of coincidence (*C*)**:

$$C = \frac{\text{Observed DCO}}{\text{Expected DCO}}$$

In the maize cross, we have

$$C = \frac{0.078}{0.097} = 0.804$$

Once we have found *C*, we can quantify interference (*I*) by using this simple equation:

$$I = 1 - C$$

In the maize cross, we have

$$I = 1.000 - 0.804 = 0.196$$

If interference is complete and no double crossovers occur, then $I = 1.0$. If fewer DCOs than expected occur, *I* is a positive number and **positive interference** has occurred. If more DCOs than expected occur, *I* is a negative number and **negative interference** has occurred. In this example, *I* is a positive number (0.196), indicating that 19.6 percent fewer double crossovers occurred than expected.

Positive interference is most often observed in eukaryotic systems. In general, the closer genes are to one another along the chromosome, the more positive interference occurs. In fact, in *Drosophila,* within a distance of 10 map units, interference is often complete and no multiple crossovers are recovered. Our observation suggests that interference may be explained by physical constraints that prevent the formation of closely aligned chiasmata. The interpretation is consistent with the finding that interference decreases as the genes in question are located farther apart. In the maize cross illustrated in Figures 12–10 and 12–11, the three genes are relatively far apart, and 80 percent of the expected double crossovers are observed.

12.5 As the Distance between Two Genes Increases, Mapping Experiments Become More Inaccurate

In theory, the frequency of crossing over between any two genes in a mapping experiment is expected to be directly proportional to the actual distance between two genes. However, in most cases, the experimentally derived mapping distance between two genes is an underestimate, and the farther apart the two genes are, the greater the inaccuracy. The discrepancy is due primarily to multiple exchanges that are predicted to occur between the two genes, but which are not recovered during experimental mapping. As we will explain next, the inaccuracy is the result of probability events that can be described using the **Poisson distribution**.

First, let us examine a mapping experiment that involves two exchanges between two genes that are far apart on a chromosome. As shown in Figure 12–12, there are three possible ways that two exchanges (equivalent to a double-crossover event) can occur between nonsister chromatids within a tetrad. A **two-strand double exchange** yields no recombinant chromatids, a **three-strand double exchange** yields 50 percent recombinant chromatids, and a **four-strand double exchange** yields 100 percent recombinant chromatids. In the aggregate, therefore, these uncommon multiple events "even out" and two genes that are far apart on the chromosome theoretically yield the maximum of 50 percent recombination, essential for *accurate* gene mapping.

In such a mapping experiment, double exchanges that occur between two genes (and, for that matter, all multiple exchanges) are relatively infrequent in comparison to the total number of single crossovers. As such, the *actual* occurrence of the infrequent events is subject to probability considerations based on the Poisson distribution. In our case, this distribution allows us to predict mathematically the frequency of samples that will *actually* undergo double exchanges. It is the failure of such exchanges to occur that leads to the underestimate of mapping distance.

The Poisson distribution is a mathematical function that assigns probabilities of observing various numbers of a specific event in a sample. To illustrate the effect of Poisson distribution, consider the analogy of an Easter egg hunt where 1000 children randomly search a large area for 1000 randomly hidden eggs. In one hour, all eggs are recovered. If all children are equally adept in the search, we can safely predict that many children will have one egg, but also that many will have either no eggs or more than one egg. The Poisson distribution allows us to predict mathematically the frequency (probability) of each outcome, that is, the frequency of children within the sample that recovered 0, 1, 2, 3, 4, ... eggs. Poisson distribution applies when the average number of events is small (a child finds an egg), while the total number of times the event that can occur within the sample is relatively large (1000 eggs can be found).

The Poisson terms used to calculate predicted distributions of events are

Distribution of Events	*Probability*
0	e^{-m}
1	me^{-m}
2	$(m^2/2)(e^{-m})$
3	$(m^3/6)(e^{-m})$
etc.	

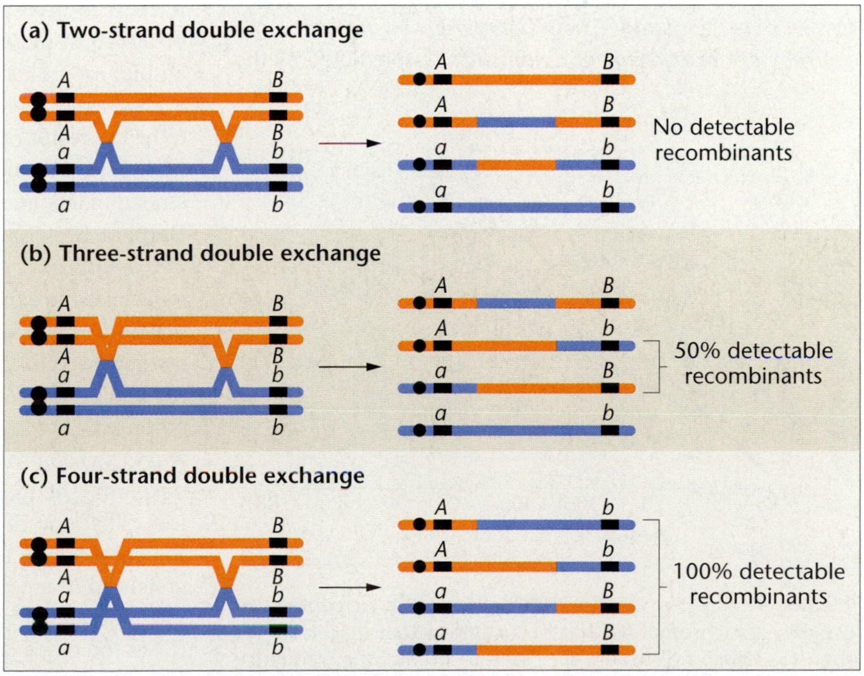

FIGURE 12–12 Three types of double exchanges that may occur between two genes. Two of them, (b) and (c), involve more than two chromatids. In each case, the detectable recombinant chromatids are bracketed.

where the mean number of independently occurring events is m and e represents the base of natural logarithms (e = about 2.7). For the Easter egg analogy, a calculation will reveal that over 300 children will fail to find an egg. Had we attempted to estimate the total number of youngsters in the hunt by tabulating the number who found at least one egg, we would have seriously underestimated the number of participants in the hunt.

In relation to chromosome mapping, we are interested in the cases where double exchanges may potentially occur between two genes, but because of the predictions based on Poisson distribution, no such exchanges actually occur within the data sample. Such an analysis creates what is called a **mapping function** that relates recombination (crossover) frequency (RF) to map distance.

We can now apply the Poisson distribution, assuming no interference occurs. (See the previous section.) Any class where m is one or more (one or more random crossovers) will yield, on average, 50 percent recombinant chromatids. Thus, we are interested in the zero term, which effectively reduces the number of recombinant chromatids. The proportion of meioses with one or more crossovers is equal to one minus the fraction of zero crossovers $(1 - e^{-m})$, whereby 50 percent recombinant chromatids will occur. Therefore,

$$\text{percent observed recombination (RF)} = 0.5(1 - e^{-m}) \times 100$$

Solving this equation, we find that the curve (mapping function) shown in Case (a) of Figure 12–13 is generated. This is compared with the hypothetical Case (b) of Figure 12–13, where recombination is directly proportional to mapping distance—that is, where interference is complete and no multiple exchanges occur.

Careful examination of the graph reveals two important observations. When the actual map distance is low (i.e., 0–7 mu), the two lines coincide. *When two genes are close together, the accuracy of a mapping experiment is very high! However, as the distance between two genes increases, the accuracy of the experiment diminishes.* As predicted by the Poisson distribution, the absence of multiple exchanges has a very significant impact. For example, when 25 percent recombinant chromatids are detected, actual map distance is almost 35 percent! When just over 30 percent recombinants are detected, the true distance, discounting any interference, is 50 mu! Such inaccuracy has been well documented in a number of studies involving various organisms, including maize, *Drosophila,* and *Neurospora.*

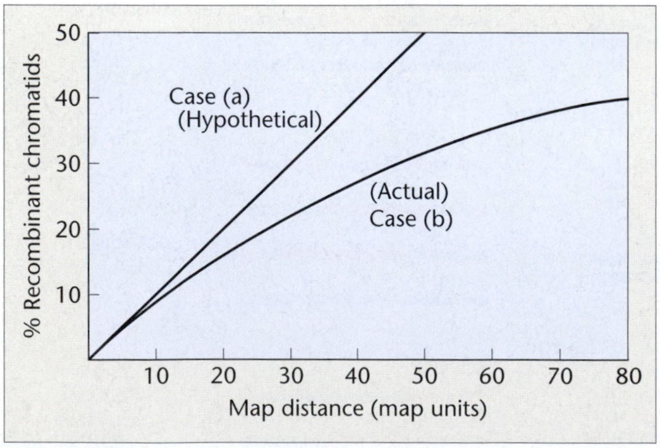

FIGURE 12–13 The relationship between the percentage of recombinant chromatids that occur and actual map distance when (a) Poisson distribution is used to predict the frequency of recombination in relation to map distance; and (b) there is a direct relationship between recombination and map distance.

12.6 *Drosophila* Genes Have Been Extensively Mapped

In organisms such as *Drosophila,* maize, and the mouse, where large numbers of mutants have been discovered and experimental crosses are easy to perform, extensive maps of each chromosome have been made. Figure 12–14 presents partial maps of the four chromosomes of *Drosophila.* Virtually every morphological feature of the fruit fly has been subjected to mutations. Each locus affected by mutation is first localized to one of the four chromosomes, or linkage groups, and then mapped in relation to other linked genes of that group. As you can see, the genetic map of the X chromosome is somewhat less extensive than that of autosome 2 or 3. In comparison to these three, autosome 4 is miniscule. Cytological evidence has shown that the relative lengths of the genetic maps correlate roughly with the relative physical lengths of these chromosomes.

12.7 Crossing Over Involves a Physical Exchange between Chromatids

Once genetic mapping was understood, it was of great interest to investigate the relationship between chiasmata observed in meiotic prophase I and crossing over. For example, are chiasmata visible manifestations of crossover events? If so, then crossing over in higher organisms appears to be the result of an actual physical exchange between homologous chromosomes. That this is the case was demonstrated independently in the 1930s by Harriet Creighton and Barbara McClintock in *Zea mays* and by Curt Stern in *Drosophila.*

Because the experiments are similar, we will consider only one of them: the work with maize. Creighton and McClintock studied two linked genes on chromosome 9. At one locus, the alleles *colorless* (c) and *colored* (C) control endosperm coloration. At the other locus, the alleles *starchy* (Wx) and *waxy* (wx) control the carbohydrate characteristics of the endosperm. The maize plant studied was heterozygous at both loci. That one of the homologs contains two unique cytological markers is the key to the experiment. The markers consisted of a densely stained knob at one end of the chromosome and a translocated piece of another chromosome (8) at the other end. The arrangements of alleles and cytological markers could be detected cytologically and are shown in Figure 12–15.

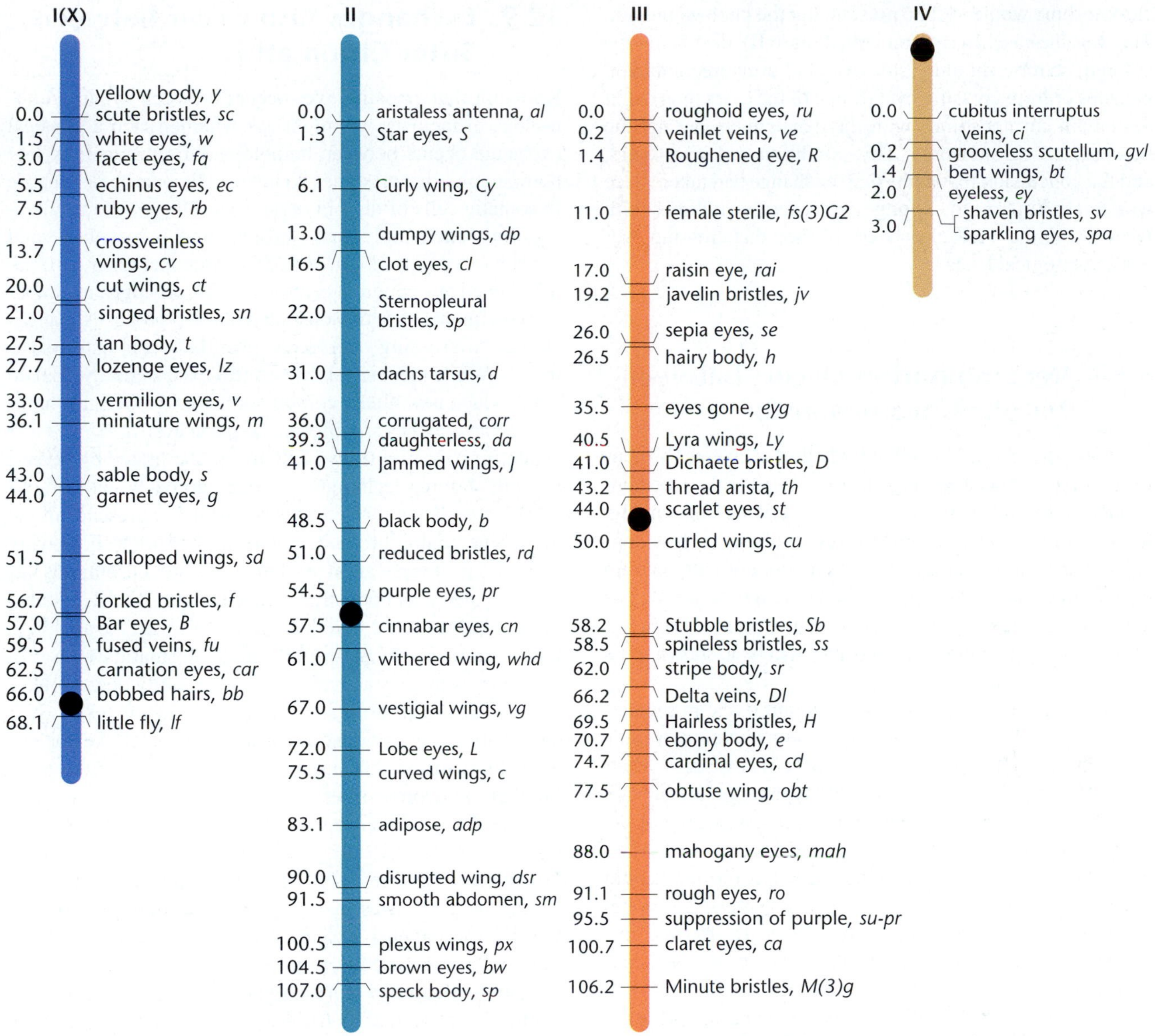

FIGURE 12–14 A partial genetic map of the four chromosomes of *Drosophila melanogaster*. The circle on each chromosome represents the position of the centromere.

Creighton and McClintock crossed this plant to one homozygous for the color allele (*c*) and heterozygous for the endosperm alleles. They obtained a variety of different phenotypes in the offspring, but they were most interested in one that occurred as a result of crossing over involving the chromosome with the unique cytological markers. They examined the chromosomes of this plant, with the colorless, waxy phenotype (case I in Figure 12–15) for the presence of the cytological markers. If genetic crossing over is accompanied by a physical exchange between homologs, the translocated

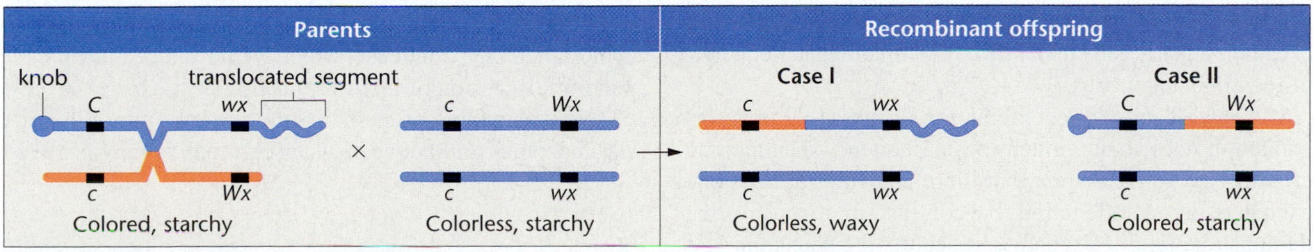

FIGURE 12–15 The phenotypes and chromosome compositions of parents and recombinant offspring in Creighton and McClintock's experiment in maize. The knob and translocated segment served as cytological markers which established that crossing over involves an actual exchange of chromosome arms.

chromosome would still be present, but the knob would not. This was the case! In a second plant (case II), the phenotype colored, starchy should result from either nonrecombinant gametes or from crossing over. Some of the cases then ought to contain chromosomes with the dense knob, but not the translocated chromosome. This condition was also found, and the conclusion that a physical exchange had taken place was again supported. Along with Stern's findings with *Drosophila,* the work clearly established that crossing over has a cytological basis.

12.8 Recombination Occurs between Mitotic Chromosomes

In 1936, Curt Stern considered whether exchanges similar to crossing over occur during mitosis. He was able to demonstrate that this indeed is the case in *Drosophila.* This finding, the first to demonstrate **mitotic recombination**, was considered unusual because homologs do not normally pair up during mitosis in most organisms. However, such synapsis appears to be the rule in *Drosophila.* Since Stern's discovery, genetic exchange during mitosis has also been shown to be a general event in certain fungi.

Stern observed small patches of mutant tissue in females heterozygous for the X-linked recessive mutations *yellow* body and *singed* bristles. Under normal circumstances, a heterozygous female is completely wild type (gray-bodied with straight, long bristles). He explained the appearance of the mutant patches by postulating that, during mitosis in certain cells during development, homologous exchanges could occur between the loci for *yellow* (*y*) and *singed* (*sn*) or between *singed* and the centromere. Figure 12–16 diagrams the nature of the proposed exchanges, in contrast to the case where no exchange occurs. The mutant patches that will occur are also depicted. When no exchange occurs, all tissue is wild type (gray body and gray, straight bristles). After the two types of exchanges, tissues are produced with either a *yellow* patch *or* with an adjacent *yellow* and *singed* patch (called a **twin spot**).

In 1958, Guido Pontecorvo and others described a similar phenomenon in the fungus *Aspergillus.* Although the vegetative stage is normally haploid, some cells fuse. The resultant diploid cells then divide mitotically. As in *Drosophila,* crossing over rarely occurs between linked genes during mitosis in this diploid stage, resulting in recombinant cells. Pontecorvo referred to these events that produce genetic variability as the **parasexual cycle**. On the basis of such exchanges, genes can be mapped by estimating the frequency of recombinant classes.

As a rule, if mitotic recombination occurs at all in an organism, it does so at a much lower frequency than meiotic crossing over. We assume that there is always at least one exchange per meiotic tetrad. By contrast, in organisms that demonstrate mitotic exchange, it occurs in 1 percent or less of mitotic divisions.

12.9 Exchanges Also Occur between Sister Chromatids

Knowing that crossing over occurs between synapsed homologs in meiosis, we might ask whether such a physical exchange occurs between homologs during mitosis. While homologous chromosomes do not usually pair up or synapse in somatic cells of diploid organisms (*Drosophila* is an exception), each individual chromosome in prophase and metaphase of mitosis consists of two identical sister chromatids, joined at a common centromere. Surprisingly, a number of experiments have demonstrated that reciprocal exchanges similar to crossing over occur even between sister chromatids. While these **sister chromatid exchanges (SCEs)** do not produce new allelic combinations, evidence is accumulating that attaches significance to these events.

Identification and study of SCEs are facilitated by several modern staining techniques. In one approach, cells are allowed to replicate for two generations in the presence of the thymidine analog bromodeoxyuridine (BUdR)* Following two rounds of replication, each pair of sister chromatids has one member with both strands labeled with BUdR. Using a differential stain, we find that chromatids with the analog in both strands stain less brightly than chromatids with it in only one strand. As a result, SCEs are readily detectable, if they occur. In Figure 12–17, numerous instances of SCE events are clearly evident. Because of the patchlike appearance, these sister chromatids are sometimes referred to as **harlequin chromosomes**.

While the significance of SCEs is still uncertain, several observations have led to great interest in this phenomenon. We know, for example, that agents that induce chromosome damage (e.g., viruses, X rays, ultraviolet light, and certain chemical mutagens) also increase the frequency of SCEs. Further, an elevated frequency of SCEs is characteristic of **Bloom syndrome**, a human disorder caused by a mutation in the chromosome 15 *BLM* gene. This rare, recessively inherited disease is characterized by prenatal and postnatal retardation of growth, a great sensitivity of the facial skin to the sun, immune deficiency, a predisposition to malignant and benign tumors, and abnormal behavior patterns. The chromosomes from cultured leukocytes, bone marrow cells, and fibroblasts derived from homozygotes are very fragile and unstable when compared with those derived from homozygous and heterozygous normal individuals. Increased breaks and rearrangements between nonhomologous chromosomes are observed in addition to excessive amounts of sister chromatid exchanges. Recent work by James German and colleagues suggest that the *BLM* gene encodes an enzyme called **DNA helicase**, which we introduced in Chapter 3 in our discussion of DNA replication.

The mechanisms of exchange between nonhomologous chromosomes and between sister chromatids may prove to

* The abbreviation BrdU is also used to denote bromodeoxyuridine.

FIGURE 12–16 The production of mutant tissue in a female *Drosophila* heterozygous for the recessive *yellow* (*y*) and *singed* (*sn*) alleles as a result of mitotic recombination.

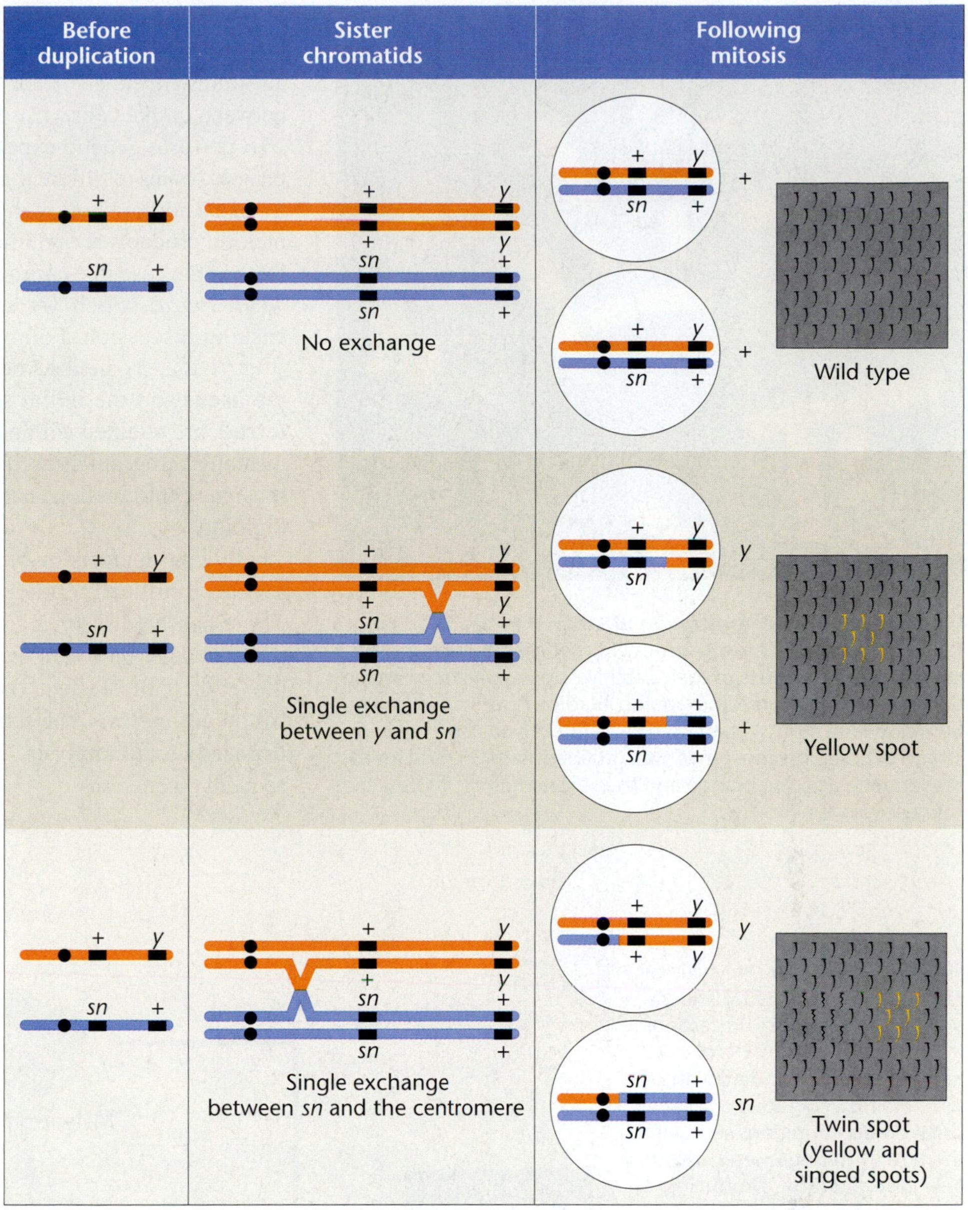

share common features because the frequency of both events increases substantially in individuals with genetic disorders. These findings suggest that further study of sister chromatid exchange may contribute to an increased understanding of recombination mechanisms and the relative stability of normal and genetically abnormal chromosomes. We encountered another demonstration of SCEs in Chapter 3 when we considered replication of DNA (See Figure 3–5).

12.10 Linkage Analysis and Mapping Can Be Performed in Haploid Organisms

We turn now to still another topic that is an extension of our study of transmission genetics: linkage analysis and chromosome mapping in haploid eukaryotes. As we shall see, even though analysis of the location of genes relative to one another throughout the genome of haploid organisms may *seem* a bit more complex than in diploid organisms, the basic underlying principles are the same. In fact, many basic principles of inheritance were established during the study of haploid fungi.

While many single-celled eukaryotes are haploid during the vegetative stages of their life cycle, they also form reproductive cells that fuse during fertilization, forming a diploid zygote. The zygote then undergoes meiosis and reestablishes haploidy. The haploid meiotic products are the progenitors of the subsequent members of the vegetative phase of the life cycle. Figure 12–18 illustrates this type of cycle in the green alga *Chlamydomonas*. Even though the haploid cells that fuse during fertilization *look* identical (and

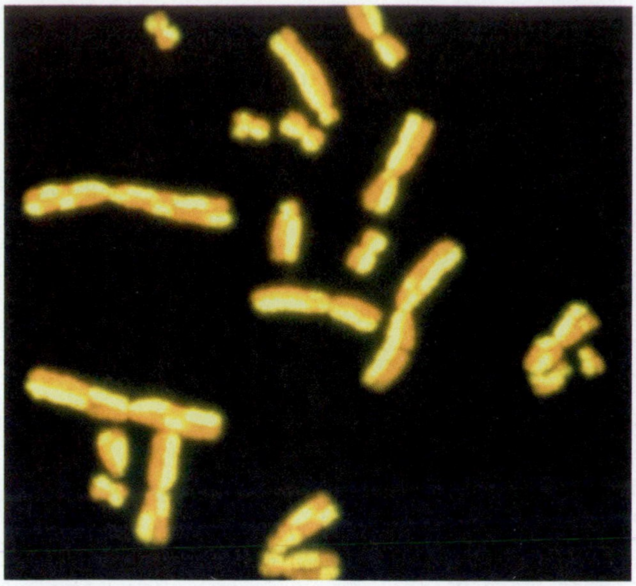

FIGURE 12–17 Demonstration of sister chromatid exchanges (SCEs) in mitotic chromosomes. Sometimes called harlequin chromosomes because of their patchlike appearance, chromatids containing the thymidine analog BUdR in both DNA strands fluoresce *less* brightly than do those with the analog in only one strand. These chromosomes were stained with 33258-Hoechst reagent and acridine orange and then viewed under fluorescence microscopy.

are thus called **isogametes**), a chemical identity that distinguishes two distinct types exists on their surface. As a result, all strains are either "+" or "–", and fertilization occurs only between unlike cells.

To perform genetic experiments with haploid organisms, genetic strains of different genotypes are isolated and crossed to one another. Following fertilization and meiosis, the meiotic products are retained together and can be analyzed. Such is the case in *Chlamydomonas* as well as in the fungus *Neurospora,* which we shall use as an example in the ensuing discussion. Following fertilization in *Neurospora* (Figure 12–19), meiosis occurs in a saclike structure called the **ascus** and the initial set of haploid products, called a **tetrad**, are retained within it. This term has a quite different meaning here from its use to describe the four-stranded chromosome configuration characteristic of meiotic prophase I in diploids.

Following meiosis in *Neurospora,* each cell in the ascus divides mitotically, producing eight haploid **ascospores**. These can be dissected and examined morphologically or tested to determine their genotypes and phenotypes. Because the eight cells reflect the *sequence* of their formation following meiosis, the tetrad is "ordered" and we can do **ordered tetrad analysis**. This process is critical to our subsequent discussion.

FIGURE 12–18 The life cycle of *Chlamydomonas.* The diploid zygote (in the center) undergoes meiosis, producing "+" or "–" haploid cells that undergo mitosis, yielding vegetative colonies. Unfavorable conditions stimulate them to form isogametes, which fuse in fertilization, producing a zygote and repeat the cycle. Two of these stages are illustrated photographically.

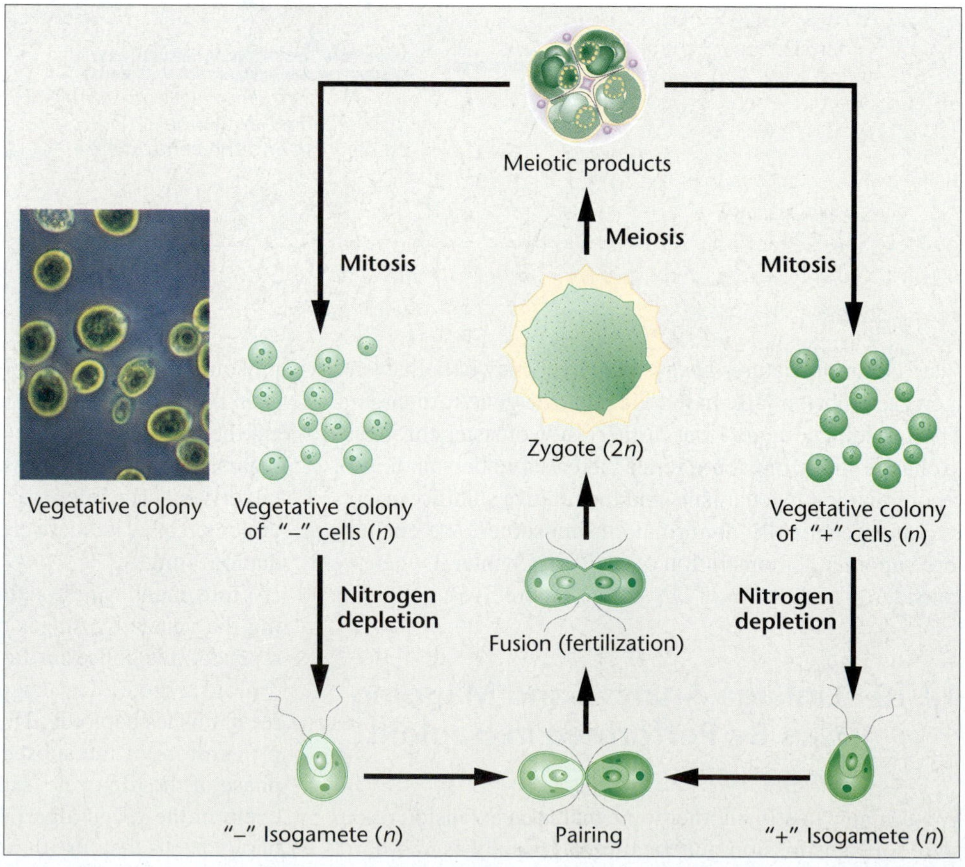

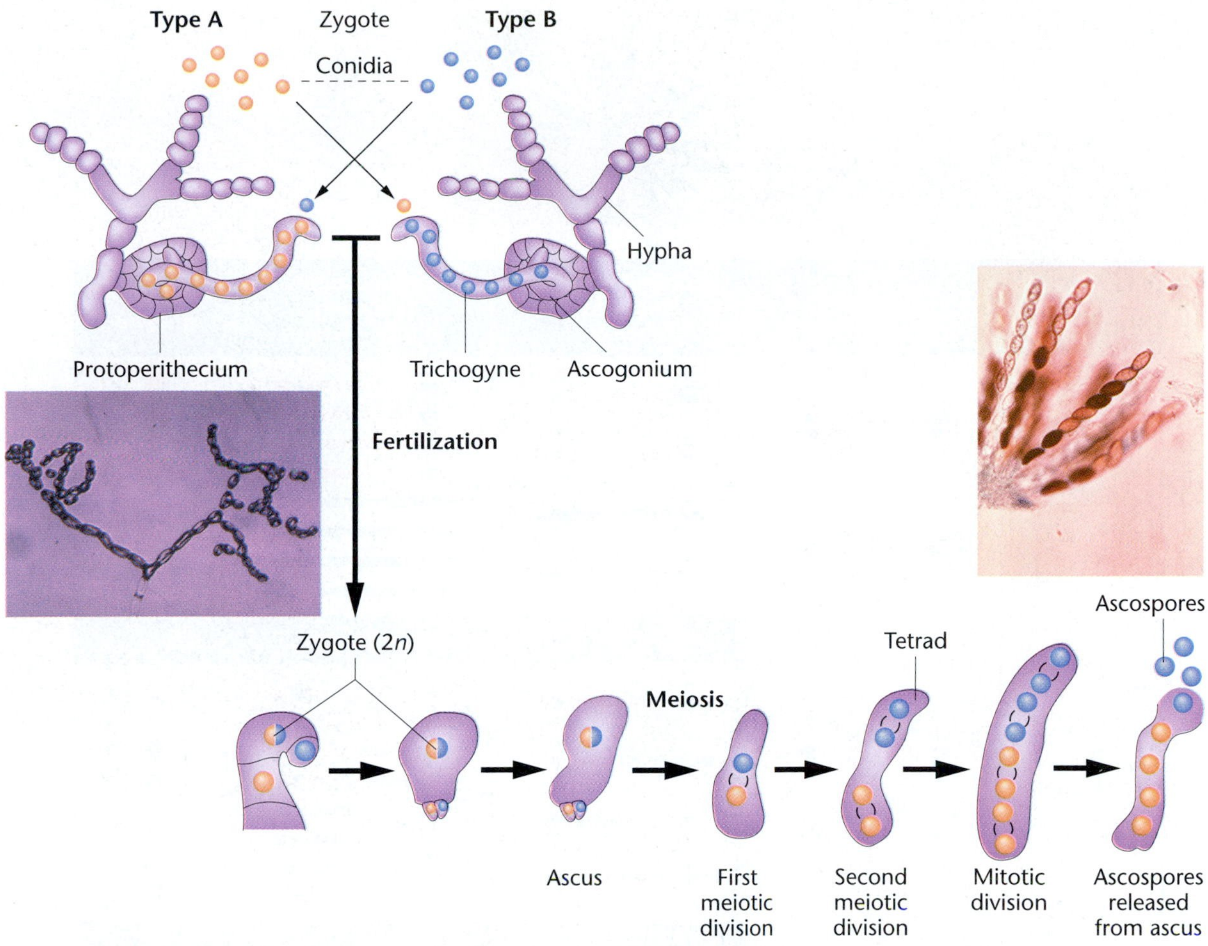

FIGURE 12–19 Sexual reproduction during the life cycle of *Neurospora* is initiated following fusion of conidia of opposite mating types. After fertilization, each diploid zygote becomes enclosed in an ascus where meiosis occurs, leading to four haploid cells, two of each mating type. A mitotic division then occurs and the eight haploid ascospores are later released. Upon germination, the cycle may be repeated. The photographs show the vegetative stage of the organism and several asci that may form in a single structure, even though we have illustrated the events occurring in only one ascus.

Gene-to-Centromere Mapping

When a single gene (*a*) is analyzed in *Neurospora,* as diagrammed in Figure 12–20, the data can be used to calculate the map distance between that gene and the centromere. This process is sometimes referred to as **mapping the centromere**. It can be accomplished by experimentally determining the frequency of recombination using tetrad data. Note that once the four meiotic products of the tetrad are formed, a mitotic division occurs, resulting in eight ordered products (ascospores).

If no crossover event occurs between the gene under study and the centromere, the pattern of ascospores (contained within an ascus, pl. asci) appears as shown in Figure 12–20(a) (*aaaa*++++). Note that the pattern (++++*aaaa*) can also be formed. However, it is indistinguishable from (*aaaa*++++). These patterns represent **first-division segregation**, because the two alleles are separated during the first meiotic division. However, crossover events will alter the pattern, as shown in Figure 12–20(b) (*aa*++*aa*++) and 12–20(c) (++*aaaa*++). Two other recombinant patterns also occur, depending on the chromatid orientation during the second meiotic division: (++*aa*++*aa*) and (*aa*++++*aa*). All four patterns, resulting from a crossover event between the *a* gene and the centromere, reflect **second-division segregation,** because the two alleles are not separated until the second meiotic division. Since the mitotic division simply replicates the patterns (from 4 to 8 ascospores), ordered tetrad data are usually condensed to reflect the genotypes reflected in ascospore pairs that may be distinguished from one another. Six unique combinations are possible:

	First-Division Segregation			
(1)	*a*	*a*	+	+
(2)	+	+	*a*	*a*
	Second-Division Segregation			
(3)	*a*	+	*a*	+
(4)	+	*a*	+	*a*
(5)	+	*a*	*a*	+
(6)	*a*	+	+	*a*

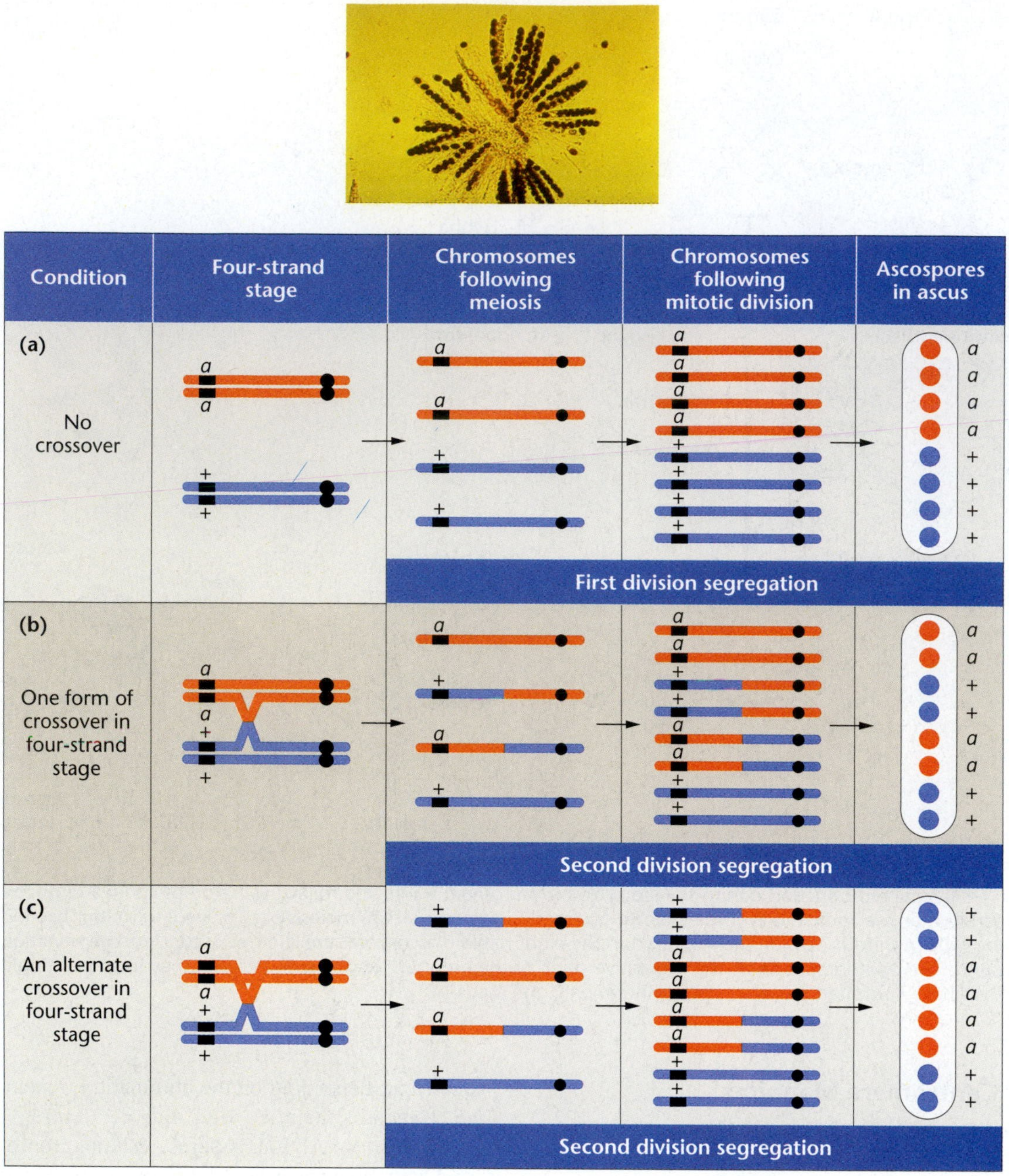

FIGURE 12–20 Three ways in which different ascospore patterns can be generated in *Neurospora*. Analysis of these patterns can serve as the basis of gene-to-centromere mapping. The photograph shows a variety of ascopore arrangements within *Neurospora* asci.

To calculate the distance between the gene and the centromere, data must be tabulated from a large number of asci resulting from a controlled cross. We then use these data to calculate the distance (d):

$$d = \frac{1/2(\textit{second division segregant asci})}{\textit{total asci scored}} \times 100$$

The distance (d) reflects the percentage of recombination and is only one-half the number of second-division segregant asci. This is because crossing over in each occurs in only two of the four chromatids during meiosis.

To illustrate, assume that a represents albino and + represents wild type in *Neurospora*. In crosses between the two genetic types, suppose the following data are observed:

65 first-division segregants

70 second-division segregants

The distance between a and the centromere is thus

$$d = \frac{(1/2)(70)}{135} = 0.259 \times 100 = 25.9$$

or about 26 mu.

As the distance increases up to 50 units, in theory all asci should result from second-division segregation. However, numerous factors prevent this. As in diploid organisms, accuracy is greatest when the gene and the centromere are relatively close together. As we discuss next, we can also analyze haploid organisms in order to distinguish between linkage and independent assortment of two genes. Mapping distances between gene loci are calculated once linkage is established. As a result, detailed maps of organisms such as *Neurospora* and *Chlamydomonas* are now available.

Ordered versus Unordered Tetrad Analysis

In our previous discussion, we assumed that the genotype of each ascospore and its position in the tetrad can be determined. To perform such an **ordered tetrad analysis**, individual asci must be dissected and each ascospore must be tracked as it germinates. This is a tedious process, but it is essential for two types of analysis:

1. To distinguish between first-division segregation and second-division segregation of alleles in meiosis.
2. To determine whether recombinational events are reciprocal or not. In the first case, such information is essential to "map the centromere," as we have just discussed. Thus, ordered tetrad analysis must be performed in order to map the distance between a gene and the centromere.

In the second case, ordered tetrad analysis has revealed that recombinational events are not always reciprocal, particularly when closely linked genes are studied in *Ascomycetes*. This observation has led to the investigation of the phenomenon called **gene conversion**. Recall that we previously discussed this phenomenon in detail in Chapter 3.

It is much less tedious to isolate individual asci, allow them to mature, and then determine the genotypes of each ascospore, but not in any particular order. This approach is referred to as **unordered tetrad analysis**. As we shall see in the next section, such an analysis can be used to detemine whether or not two genes are linked on the same chromosome, and if so, to determine the map distance between them.

Linkage and Mapping

Analysis of genetic data derived from haploid organisms can be used to distinguish between linkage and independent assortment of two genes; it further allows mapping distances to be calculated between gene loci once linkage is established. We shall next consider tetrad analysis in the alga *Chlamydomonas*. With the exception that the four meiotic products are not ordered and *do not* undergo a mitotic division following the completion of meiosis, the general principles discussed for *Neurospora* also apply to *Chlamydomonas*.

To compare independent assortment and linkage, we will consider two theoretical mutant alleles, *a* and *b*, representing two distinct loci in *Chlamydomonas*. Suppose that 100 tetrads derived from the cross *ab* × ++ yield the tetrad data shown in Table 12–1. As you can see, all tetrads produce one of three patterns. For example, all tetrads in category I produce two ++ cells and two *ab* cells and are designated as **parental ditypes (P)**. Category II tetrads produce two *a*+ cells and two +*b* cells and are called **nonparental ditypes (NP)**. Category III tetrads produce one cell of each of the four possible genotypes and are thus termed **tetratypes (T)**.

TABLE 12–1 Tetrad Analysis in *Chlamydomonas*

Category	*I*	*II*	*III*
Tetrad type	Parental (P)	Nonparental (NP)	Tetratypes (T)
Genotypes present	+ + + + *a b* *a b*	*a* + *a* + + *b* + *b*	+ + *a* + + *b* *a b*
Number of tetrads	43	43	14

These data support the hypothesis that the genes represented by the *a* and *b* alleles are located on separate chromosomes. To understand why, you should refer to Figure 12–21. In parts (a) and (b) of that figure, the origin of parental (P) and nonparental (NP) ditypes is demonstrated for two unlinked genes. According to the Mendelian principle of independent assortment of unlinked genes, approximately equal proportions of these tetrad types are predicted. Thus, when the parental ditypes are equal to the nonparental ditypes, then the two genes are not linked. The data in Table 12–1 confirm this prediction. Because independent assortment has occurred, it can be concluded that the two genes are located on separate chromosomes.

The origin of category III, the tetratypes, is diagrammed in Figure 12–21(c,d). The genotypes of tetrads in this category can be generated in two possible ways. Both involve a crossover event between one of the genes and the centromere. In Figure 12–21(c), the exchange involves one of the two chromosomes and occurs between gene *a* and the centromere; in Figure 12–21(d), the other chromosome is involved, and the exchange occurs between gene *b* and the centromere.

Production of tetratype tetrads does not alter the final ratio of the four genotypes present in all meiotic products. If the genotypes from 100 tetrads (which yield 400 cells) are computed, 100 of each genotype are found. This 1:1:1:1 ratio is predicted according to independent assortment.

Now consider the case where the genes *a* and *b* are linked (Figure 12–22). The same categories of tetrads will be produced. However, parental and nonparental ditypes will not necessarily occur in equal proportions; nor will the four genotypic combinations be found in equal numbers if the genotypes of all meiotic products are computed. For example, the following data might be encountered:

Category I	*Category II*	*Category III*
P 64	NP 6	T 30

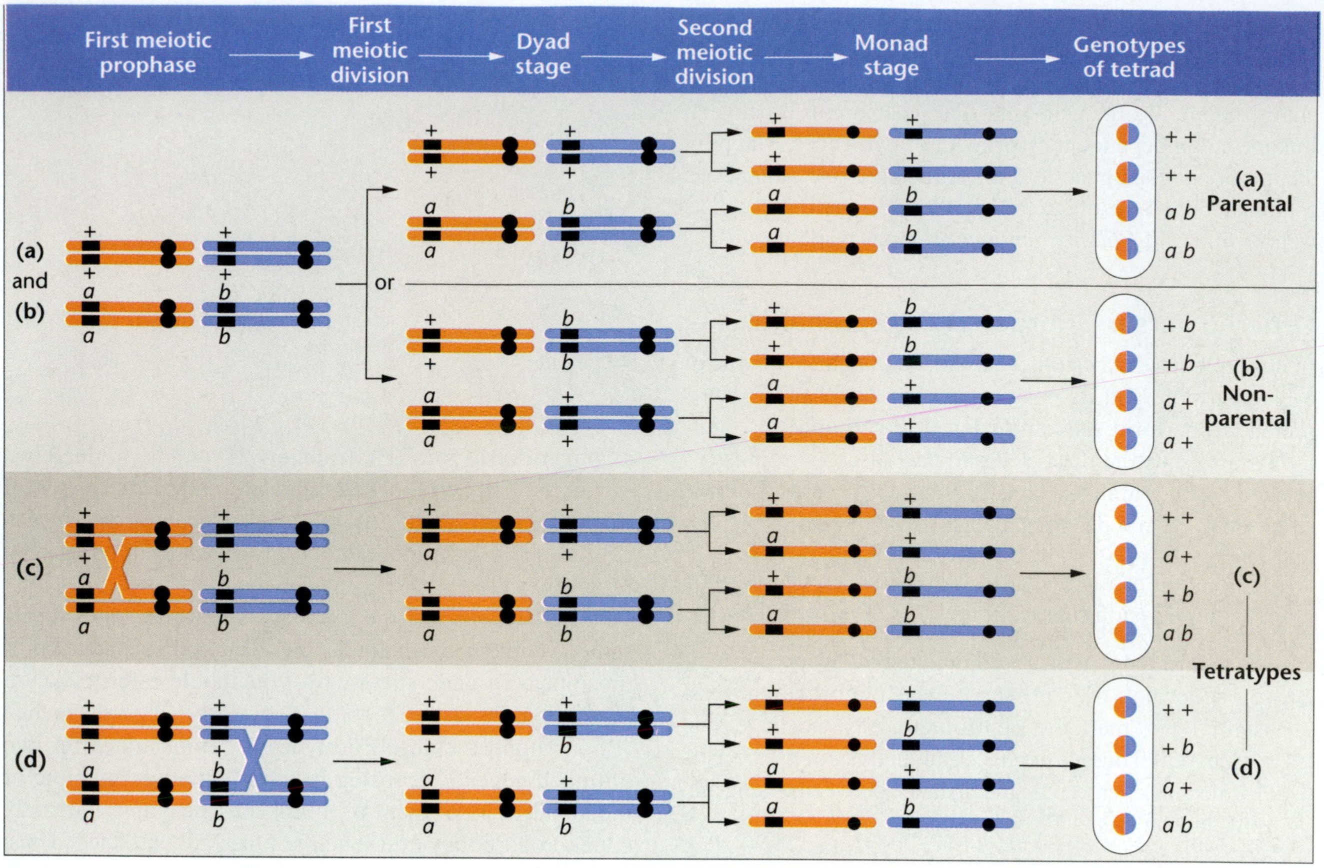

FIGURE 12–21 The origin of various genotypes found in tetrads in *Chlamydomonas* when two genes located on separate chromosomes are considered.

FIGURE 12–22 The various types of exchanges leading to the genotypes found in tetrads in *Chlamydomonas* when two genes located on the same chromosome are considered.

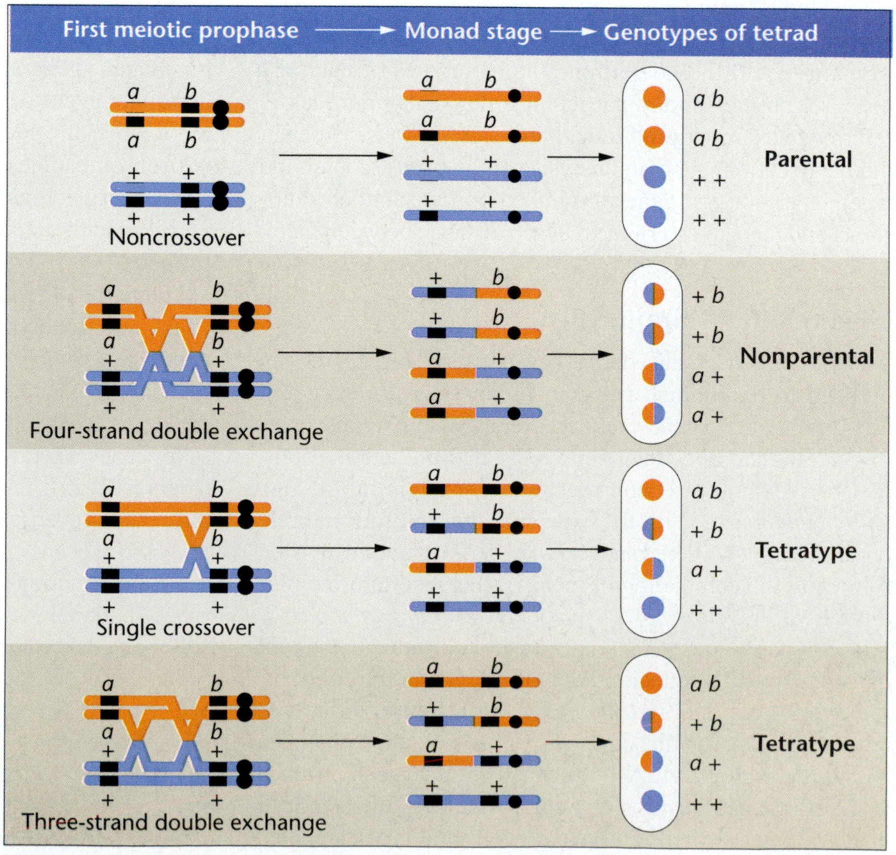

Since the parental and nonparental categories are not produced in equal proportions, we can conclude that independent assortment is not in operation and that the two genes are linked. We can then proceed to determine the map distance between them.

In the analysis of these data, we are concerned with the determination of which tetrad types represent genetic exchanges between the two genes. The parental ditype tetrads (P) arise only when no crossing over occurs between the two genes. The nonparental ditype tetrads (NP) arise only when a double exchange involving all four chromatids occurs between two genes. The tetratype tetrads (T) arise when either a single crossover occurs or when an alternative type of double exchange occurs between the two genes. The various types of exchanges described here are diagrammed in Figure 12–22.

When the proportion of the three tetrad types has been determined, it is possible to calculate the map distance between the two linked genes. The following formula computes the exchange frequency, which is proportional to the map distance between the two genes:

$$\text{exchange frequency }(\%) = \frac{\text{NP} + 1/2(\text{T})}{\text{total number of tetrads}} \times 100$$

In this formula, NP represents the nonparental tetrads; all meiotic products represent an exchange. The tetratype tetrads are represented by T; assuming only single exchanges, one-half of the meiotic products represents exchanges. The sum of the scored tetrads that fall into these categories is then divided by the total number of tetrads examined $(\text{P} + \text{NP} + \text{T})$. If this calculated number is multiplied by 100, it is converted to a percentage, which is directly equivalent to the map distance between the genes.

In our example, the calculation reveals that genes *a* and *b* are separated by 21 mu:

$$\frac{6 + 1/2(30)}{100} = \frac{6 + 15}{100} = \frac{21}{100} = 0.21 \times 100 = 21\%$$

Although we have considered linkage analysis and mapping of only two genes at a time, such studies often involve three or more genes. In these cases, both gene sequence and map distances can be determined.

12.11 Lod Score Analysis and Somatic Cell Hybridization Were Historically Important in Creating Human Chromosome Maps

For obvious reasons, our own species is not a good source of data for the types of extensive linkage analysis performed with experimental organisms. Thus, in humans, the earliest linkage studies had to rely on pedigree analysis. Attempts were made to establish whether a trait was X-linked or autosomal. As we established in Chapter 10, traits determined by genes located on the X chromosome result in characteristic pedigrees, and thus, such genes were easier to identify. For autosomal traits, geneticists tried to distinguish clearly whether pairs of traits demonstrated linkage or independent assortment. When extensive pedigrees are available, it is possible to ascertain that the genes under consideration are closely linked (i.e., rarely separated by crossing over) from the fact that the two traits segregate together. For example, this approach established linkage between the genes encoding the **Rh antigens** and the gene responsible for the phenotype referred to as **elliptocytosis** (where the shape of erythrocytes is oval).

The difficulty arises, however, when two genes of interest are separated on a chromosome such that recombinant gametes are formed, obscuring linkage in a pedigree. In such cases, an approach relying on probability calculations, called the **lod score method**, helps to demonstrate linkage. First devised by J. B. S. Haldane and C. A. Smith in 1947 and refined by Newton Morton in 1955, the lod score (standing for *l*og of the *od*ds favoring linkage) assesses the probability that a particular pedigree involving two traits reflects linkage. First, the probability is calculated that family data (pedigrees) concerning two or more traits conform to the transmission of traits without linkage. Then the probability is calculated that the identical family data following these same traits result from linkage with a specified recombination frequency. The ratio of these probability values expresses the "odds" for, and against, linkage. The lod score method represented an important advance in assigning human genes to specific chromosomes and in constructing preliminary human chromosome maps. Its accuracy is limited by the extent of the family data, however, and the initial results were discouraging because of these limitations and because of the relatively high haploid number of human chromosomes (23). By 1960, short of the assignment of some genes to the X chromosome, very little autosomal linkage or mapping information had become available.

In the 1960s, a newly developed technique, **somatic cell hybridization**, proved to be an immense aid in assigning human genes to their respective chromosomes. This technique, first discovered by Georges Barsky, relies on the fact that two cells in culture can be induced to fuse into a single hybrid cell. Barsky used two mouse cell lines, but it soon became evident that cells from different organisms will also fuse together. When fusion occurs, an initial cell type called a **heterokaryon** is produced. The hybrid cell contains two nuclei in a common cytoplasm. Using the proper techniques, it is possible to fuse human and mouse cells, for example, and isolate the hybrids from the parental cells.

As the heterokaryons are cultured *in vitro,* two interesting changes occur. Eventually, the nuclei fuse together, creating what is termed a **synkaryon**. Then, as culturing is continued for many generations, chromosomes from one of the two parental species are gradually lost. In the case of the human–mouse hybrid, human chromosomes are lost randomly until, eventually, the synkaryon has a full complement of mouse chromosomes and only a few human chromosomes. As we will see, it is the preferential loss of human chromosomes rather than mouse chromosomes that makes possible the assignment of human genes to the chromosomes upon which they reside.

The experimental rationale is straightforward. For example, if a specific human gene product is synthesized in a synkaryon containing three human chromosomes, then the gene responsible for that product must reside on one of the three human chromosomes remaining in the hybrid cell. Or, if the human gene product is absent, the responsible gene cannot be present on any of the remaining three human chromosomes. Ideally, a panel of 23 hybrid cell lines, each with only one unique human chromosome, would allow the immediate assignment to a particular chromosome of any human gene for which the product could be characterized.

In practice, a panel of cell lines, each with several remaining human chromosomes, is most often used. The correlation of the presence or absence of each chromosome with the presence or absence of each gene product is called **synteny testing**. Consider, for example, the hypothetical data provided in Figure 12–23, where four gene products (*A, B, C,* and *D*) are tested in relationship to eight human chromosomes. Let us carefully analyze the gene that produces product *A*:

1. Product *A* is not produced by cell line 23, but chromosomes 1, 2, 3, and 4 are present in cell line 23. Therefore, we can rule out the presence of gene *A* on those four chromosomes and conclude that it must be on chromosome 5, 6, 7, or 8.

2. Product *A* is produced by cell line 34, which contains chromosomes 5 and 6, but not 7 and 8. Therefore, gene *A* is on chromosome 5 or 6, but cannot be on 7 or 8 because they are absent, even though product A is produced.

3. Product *A* is also produced by cell line 41, which contains chromosome 5, but not chromosome 6. Therefore, gene *A* is on chromosome 5, according to this analysis.

Using a similar approach, we can assign gene *B* to chromosome 3. You should perform the same analysis to demonstrate for yourself that this is correct.

Gene *C* presents a unique situation. The data indicate that it is not present on any of the first seven chromosomes (1–7) because it is not produced by any of the three cell lines that collectively contain those chromosomes. While it might be on chromosome 8, no direct evidence supports this conclusion. Other panels are needed. We leave gene *D* for you to analyze. Upon what chromosome does it reside?

By using the approach just described, literally hundreds of human genes were assigned to specific chromosomes. Figure 12–24 illustrates some gene locations on two human chromosomes: X and 1. The gene assignments shown were either derived or confirmed with the use of somatic-cell hybridization techniques. To map genes for which the products have yet to be discovered, researchers have had to rely on other approaches. For example, by combining recombinant DNA technology with pedigree analysis, it has been possible to assign the genes responsible for Huntington disease, cystic fibrosis, and neurofibromatosis to their respective chromosomes, 4, 7, and 17. We will discuss this approach in Chapter 16.

We conclude this discussion by addressing the next step in human gene mapping: assigning genes to different regions of a given chromosome. Sometimes in hybrid cell lines, fragments of a particular chromosome become transferred to another chromosome, resulting in a translocation. It is possible using chromosome banding techniques to identify the exact origin of the translocation and correlate the presence of a chromosomal segment in hybrid cells with specific gene expression. In that way, gene maps of human chromosomes may be compiled. The partial maps shown in Figure 12–24 illustrate this point. Although human chromosome maps are not as specific as the genetic map of Drosophila, these initial studies provided a great deal of information concerning the chromosome locations of a multitude of human genes. As we will see in Part IV (Chapters 15, 16, 17 and 18), modern technology involving recombinant DNA and the Human Genome Project have greatly extended our knowledge of gene locations within the human genome.

12.12 Did Mendel Encounter Linkage?

We conclude this chapter by examining a modern-day interpretation of the experiments that form the cornerstone of transmission genetics: Mendel's crosses with garden peas. Some observers believe that Mendel had extremely good fortune in his classic experiments with the garden pea. He did not encounter any *apparent* linkage relationships between the seven mutant characters in any of his crosses. Had Mendel obtained highly variable data characteristic of linkage

Hybrid cell lines	Human chromosomes present								Gene products expressed			
	1	2	3	4	5	6	7	8	*A*	*B*	*C*	*D*
23	●	●	●	●					−	+	−	+
34	●	●			●	●			+	−	−	+
41	●		●		●		●		+	+	−	+

FIGURE 12–23 A hypothetical grid of data used in synteny testing to assign genes to their appropriate human chromosomes. Three somatic hybrid cell lines, designated 23, 34, and 41, have each been scored for the presence, or absence, of human chromosomes 1–8, as well as for their ability to produce the hypothetical human gene products *A, B, C,* and *D*.

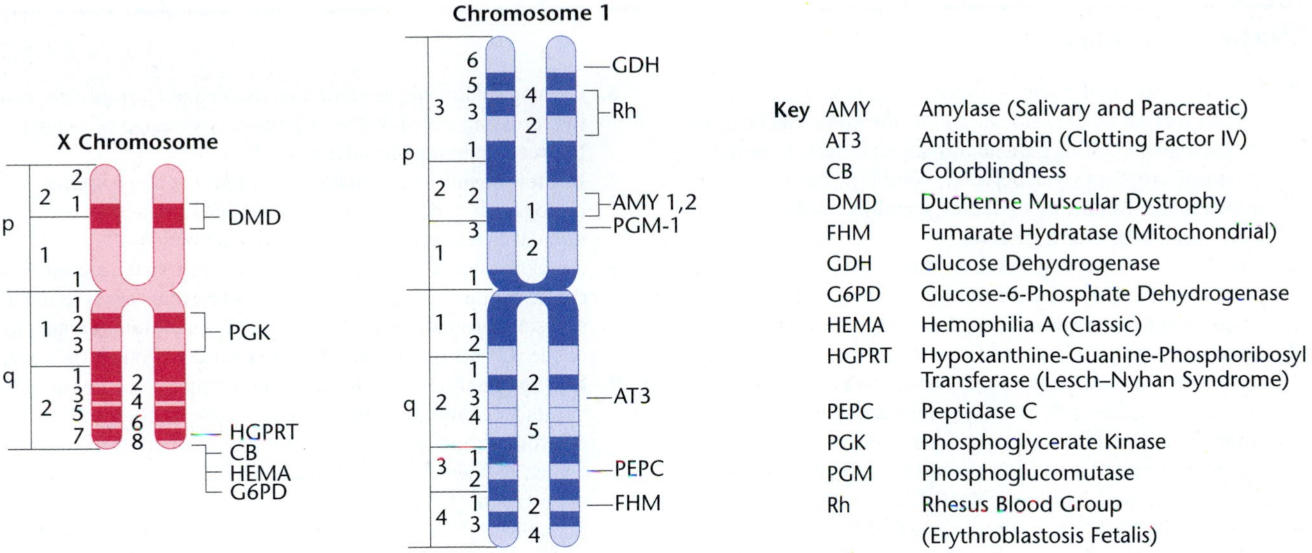

FIGURE 12–24 Representative regional gene assignments for human chromosome 1 and the X chromosome. Many assignments were initially derived by using somatic cell hybridization techniques.

and crossing over, these unorthodox observations might have hindered his successful analysis and interpretation.

The article by Stig Blixt, reprinted in its entirety in the box below, presents a modern-day interpretation of Mendel's experiments, demonstrating the inadequacy of this hypothesis. As we shall see, some of Mendel's genes were indeed linked. We shall leave it to Stig Blixt to enlighten you as to why Mendel did not detect linkage.

Why Didn't Gregor Mendel Find Linkage?

It is quite often said that Mendel was very fortunate not to run into the complication of linkage during his experiments. He used seven genes, and the pea has only seven chromosomes. Some have said that had he taken just one more, he would have had problems. This, however, is a gross oversimplification. The actual situation, most probably, is that Mendel worked with three genes in chromosome 4, two genes in chromosome 1, and one gene in each of chromosomes 5 and 7. (See Table 1). It seems at first glance that, out of the 21 dihybrid combinations Mendel theoretically could have studied, no less than four (that is, *a–i, v–fa, v–le, fa–le*) ought to have resulted in linkages. As found, however, in hundreds of crosses and shown by the genetic map of the pea, a and i in chromosome 1 are so distantly located on the chromosome that no linkage is normally detected. The same is true for v and le on the one hand, and fa on the other, in chromosome 4. This leaves v–le, which ought to have shown linkage.

Mendel, however, seems not to have published this particular combination and thus, presumably, never made the appropriate cross to obtain both genes segregating simultaneously. It is therefore not so astonishing that Mendel did not run into the complication of linkage, although he did not avoid it by choosing one gene from each chromosome.

STIG BLIXT

Weibullsholm Plant Breeding Institute, Landskrona, Sweden, and Centro de Energia Nuclear na Agricultura, Piracicaba, SP, Brazil.

TABLE 1 **Relationship between Modern Genetic Terminology and Character Pairs Used by Mendel**

Character Pair Used by Mendel	*Alleles in Modern Terminology*	*Located in Chromosome*
Seed color, yellow–green	*I–i*	1
Seed coat and flowers, colored–white	*A–a*	1
Mature pods, smooth expanded–wrinkled indented	*V–v*	4
Inflorescences, from leaf axis–umbellate in top of plant	*Fa–fa*	4
Plant height, 0.5–1 m	*Le–le*	4
Unripe pods, green yellow	*Gp–gp*	5
Mature seeds, smooth wrinkled	*R–r*	7

Chapter Summary

1. Genes located on the same chromosome are said to be linked. Alleles located on the same homolog, therefore, can be transmitted together during gamete formation. However, the mechanism of crossing over between homologs during meiosis results in the reshuffling of alleles, thereby contributing to genetic variability within gametes.
2. Early in this century, geneticists realized that crossing over provides an experimental basis for mapping the location of linked genes relative to one another along the chromosome.
3. Interference is a phenomenon describing the extent to which a crossover in one region of a chromosome influences the occurrence of a crossover in an adjacent region of the chromosome. The coefficient of coincidence (C) is a quantitative estimate of interference calculated by dividing the observed double crossovers by the expected double crosssovers.
4. Due to statistical considerations described by the Poisson distribution, as the actual distance between two genes increases, experimentally determined mapping distances become more and more inaccurate (underestimated).
5. Extensive genetic maps have been created in organisms such as corn, mice, and *Drosophila*.
6. Cytological investigations of both maize and *Drosophila* reveal that crossing over involves a physical exchange of segments between nonsister chromatids.
7. In a few organisms, including *Drosophila* and *Aspergillus*, homologs pair during mitosis and crossing over occurs between them at a frequency far less than during meiosis.
8. An exchange of genetic material between sister chromatids can occur during mitosis as well. These events are referred to as sister chromatid exchanges (SCEs). An elevated frequency of such events is seen in the human disorder Bloom syndrome.
9. Linkage analysis and chromosome mapping are possible in haploid eukaryotes. Our discussion has included gene-to-centromere and gene-to-gene mapping as well as the consideration of how to distinguish between linkage and independent assortment.
10. Somatic cell hybridization techniques have made possible linkage and mapping analysis of human genes.
11. Evidence now suggests that several of the genes studied by Mendel are, in fact, linked. However, in such cases, the genes are sufficiently far apart to prevent the detection of linkage.

Insights and Solutions

1. In a series of two-point map crosses involving three genes linked on chromosome 3 in *Drosophila,* the following distances were calculated:

cd–sr 13 mu

cd–ro 16 mu

(a) Determine the sequence and construct a map of these three genes.

Solution: It is impossible to do so. There are two possibilities based on these limited data,

Case 1: *cd* (13) *sr* (3) *ro*

or

Case 2: *ro* (16) *cd* (13) *sr*

(b) What mapping data will resolve this?

Solution: The map distance determined by crossing over between *ro* and *sr.* If case 1 is correct, it should be 3 mu, and if case 2 is correct, it should be 29 mu. In fact, this distance is 29 mu, demonstrating that case 2 is correct.

(c) Can we tell which of the sequences shown here is correct?

ro 16 cd 13 sr

or

sr 13 *cd* 16 *ro*

Solution: No; based on the mapping data, they are equivalent.

2. In *Drosophila, Lyra (Ly)* and *Stubble (Sb)* are dominant mutations located at loci 40 and 58, respectively, on chromosome 3. A recessive mutation with bright-red eyes was discovered and shown also to be on chromosome 3. A map was obtained by crossing a female who was heterozygous for all three mutations to a male homozygous for the bright-red mutation (which we refer to here as *br*). The following data are obtained:

Phenotype				*Number*
(1)	*Ly*	*Sb*	*br*	404
(2)	+	+	+	422
(3)	*Ly*	+	+	18
(4)	+	*Sb*	*br*	16
(5)	*Ly*	+	*br*	75
(6)	+	*Sb*	+	59
(7)	*Ly*	*Sb*	+	4
(8)	+	+	*br*	2
				Total = 1000

Determine the location of the *br* mutation on chromosome 3. By referring to Figure 6–14, predict what mutation has been discovered. How could you be sure?

Solution: First determine the *arrangement* of the alleles on the homologs of the heterozygous crossover parent (the female in this case) by locating the most frequent reciprocal phenotypes, which arise from the noncrossover gametes. These are phenotypes (1) and (2). Each one represents the arrangement of alleles on one of the homologs. Therefore, the arrangement is

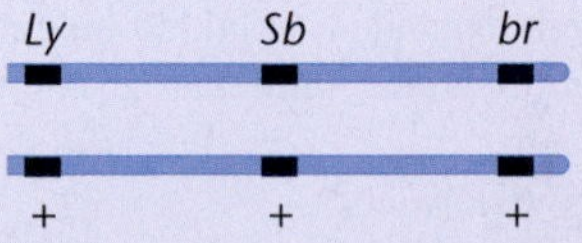

Second, determine the correct *sequence* of the three loci along the chromosome by determining which sequence will yield the

observed double-crossover phenotypes and which are the least frequent reciprocal phenotypes (7 and 8).

If the sequence is correct as written, then a double crossover depicted here,

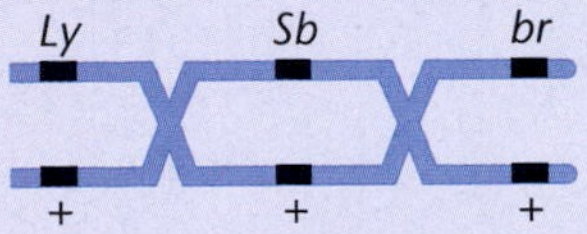

will yield *Ly* + *br* and + *Sb* + as phenotypes. Inspection shows that these categories (5 and 6) are actually single crossovers, not double crossovers. Therefore, the sequence, as written, is incorrect. There are only two other possible sequences. The *br* gene is either to the left of *Ly*, or it is between *Ly* and *Sb*:

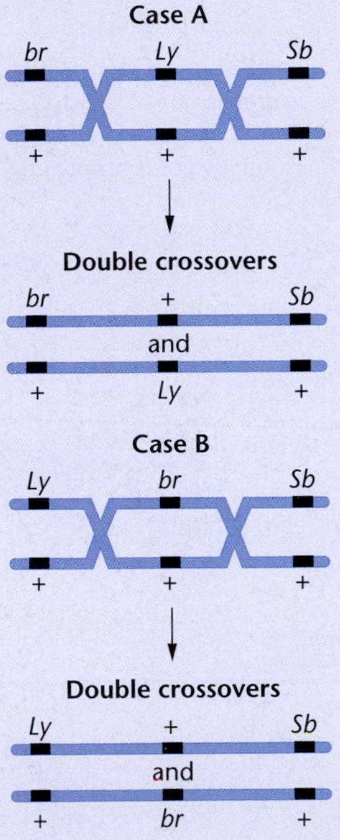

Comparison with the actual data shows that case B is correct. The double-crossover gametes (7) and (8) yield flies that express *Ly* and *Sb*, but not *br*, *or* express *br*, but not *Ly* and *Sb*. Therefore, the correct arrangement and sequence are as follows:

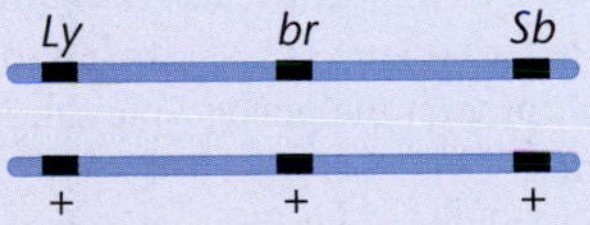

Once the correct arrangement and sequence are established, it is possible to determine the location of *br* relative to *Ly* and *Sb*. A single crossover between *Ly* and *br*, as shown here

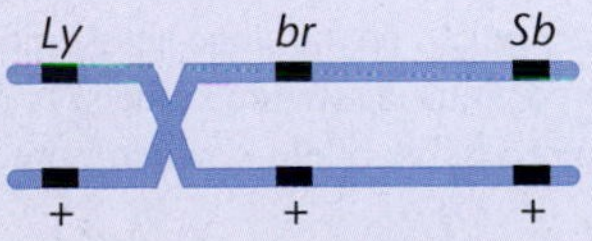

yields flies that are *Ly* + + and + *br Sb* (categories 3 and 4). Therefore, the distance between the *Ly* and *br* loci is equal to

$$\frac{18 + 16 + 4 + 2}{1000} = \frac{40}{1000} = 0.04 = 4 \text{ mu}$$

Remember that, because we need to know the frequency of all crossovers between *Ly* and *br*, we must add in the double crossovers, since they represent two single crossovers occurring simultaneously. Similarly, the distance between the *br* and *Sb* loci is derived mainly from single crossovers between them:

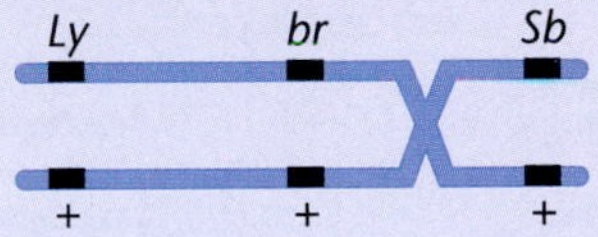

This event yields *L y br* + and + + *Sb* phenotypes (categories 5 and 6). Therefore, the distance equals

$$75 + 59 + 4 + 2/1000 = 140/1000 = 0.14 = 14 \text{ mu}$$

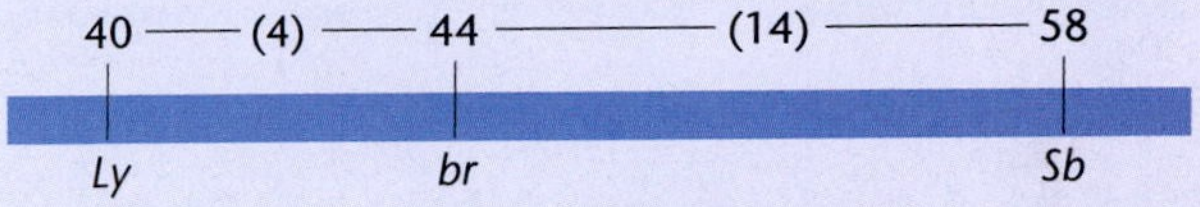

The final map shows that *br* is located at locus 44, since *Lyra* and *Stubble* are known:

Inspection of Figure 12–14 reveals that the mutation *scarlet*, which has bright-red eyes, is known to exist at locus 44, so it is reasonable to hypothesize that the bright-red eye mutation is an allele of *scarlet*. To test this hypothesis, we could cross females of our bright-red mutant with known *scarlet* males. If the two mutations are alleles, no complementation will occur, and all progeny will reveal a bright-red mutant eye phenotype. If complementation occurs, all progeny will show normal brick-red (wild-type) eyes, since the bright red mutation and *scarlet* are at different loci. (They are probably very close together.) In such a case, all progeny will be heterozygous at both the bright eye and the *scarlet* loci and will not express either mutation because they are both recessive. This cross represents what is called an **allelism test**.

3. In rabbits, *black* (*B*) is dominant to *brown* (*b*), while *full color* (*C*) is dominant to *chinchilla* (c^{ch}). The genes controlling these traits are linked. Rabbits that are heterozygous for both traits and express *black, full color* were crossed to rabbits that express *brown, chinchilla,* with the following results:

31 brown, chinchilla
34 black, full color
16 brown, full color
19 black, chinchilla

Determine the arrangement of alleles in the heterozygous parents and the map distance between the two genes.

Solution: This is a two-point map problem, where the two reciprocal noncrossover phenotypes are recognized as those present in the highest numbers (*brown, chinchilla* and *black, full*). The less frequent reciprocal phenotypes (*brown, full* and *black, chinchilla*) arise from a single crossover. The arrangement of alleles is derived from the noncrossover phenotypes because they enter gametes intact. The cross is shown as follows:

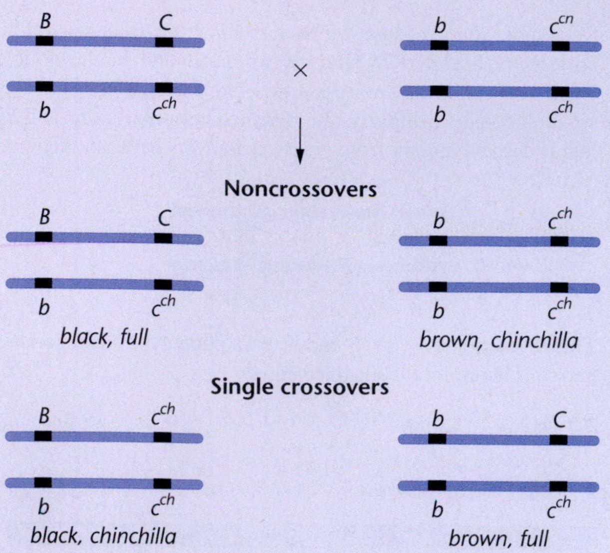

The single crossovers give rise to 35/100 offspring (35 percent). Therefore, the distance between the two genes is 35 mu.

4. In a cross in *Neurospora* where one parent expresses the mutant allele *a* and the other expresses a wild-type phenotype (+), the following data were obtained in the analysis of ascospores:

Asci Types

Sequence of ascospores in ascus	*1*	*2*	*3*	*4*	*5*	*6*
	+	*a*	*a*	+	*a*	+
	+	*a*	*a*	+	*a*	+
	+	*a*	+	*a*	+	*a*
	+	*a*	+	*a*	+	*a*
	a	+	*a*	+	+	*a*
	a	+	*a*	+	+	*a*
	a	+	+	*a*	*a*	+
	a	+	+	*a*	*a*	+
	39	33	5	4	9	10

Total = 100

Calculate the gene-to-centromere distance.

Solution: Ascus types 1 and 2 represent first-division segregants (fds) where no crossing over occurred between the *a* locus and the centromere. All others (3–6) represent second-division segregation (sds). By applying the formula

$$\text{distance} = \frac{1/2 \text{ sds}}{\text{total asci}}$$

we obtain the following result:

$$d = 1/2(5 + 4 + 9 + 10)/100$$
$$= 1/2\,(28)/100$$
$$= 0.14$$
$$= 14 \text{ mu}$$

Problems and Discussion Questions

1. What is the significance of genetic recombination to the process of evolution?
2. Describe the cytological observation that suggests that crossing over occurs during the first meiotic prophase.
3. Why does more crossing over occur between two distantly linked genes than between two genes that are very close together on the same chromosome?
4. Why is a 50 percent recovery of single-crossover products the upper limit, even when crossing over *always* occurs between two linked genes?
5. Why are double-crossover events expected in lower frequency than single-crossover events?
6. What is the proposed basis for positive interference?
7. What two essential criteria must be met in order to execute a successful mapping cross?
8. The genes *dumpy* (*dp*), *clot* (*cl*), and *apterous* (*ap*) are linked on chromosome 2 of *Drosophila*. In a series of two-point mapping crosses, the following genetic distances were determined:

dp – *ap*	42
dp – *cl*	3
ap – *cl*	39

What is the sequence of the three genes?

9. Consider two hypothetical recessive autosomal genes *a* and *b*. Where a heterozygote is test crossed to a double-homozygous mutant, predict the phenotypic ratios under the following conditions:
   (a) *a* and *b* are located on separate autosomes.
   (b) *a* and *b* are linked on the same autosome, but are so far apart that a crossover always occurs between them.
   (c) *a* and *b* are linked on the same autosome, but are so close together that a crossover almost never occurs.
   (d) *a* and *b* are linked on the same autosome about 10 mu apart.
10. In corn, colored aleurone (in the kernels) is due to the dominant allele *R*. The recessive allele *r*, when homozygous, produces

colorless aleurone. The plant color (not the kernel color) is controlled by another gene with two alleles, *Y* and *y*. The dominant *Y* allele results in green color, whereas the homozygous presence of the recessive *y* allele causes the plant to appear yellow. In a testcross between a plant of unknown genotype and phenotype and a plant that is homozygous recessive for both traits, the following progeny were obtained:

Colored green	88
Colored yellow	12
Colorless green	8
Colorless yellow	92

Explain how these results were obtained by determining the exact genotype and phenotype of the unknown plant, including the precise association of the two genes on the homologs (i.e., the arrangement).

11. In the cross shown here, involving two linked genes, *ebony* (*e*) and *claret* (*ca*), in *Drosophila,* where crossing over does not occur in males, offspring were produced in a 2 + :1 *ca*:1 *e* phenotypic ratio:

♀		♂
$\frac{e\ ca^+}{e^+\ ca}$	×	$\frac{e\ ca^+}{e^+\ ca}$

These genes are 30 units apart on chromosome 3. What contribution did crossing over in the female make to these phenotypes?

12. With two pairs of genes involved (*P*/*p* and *Z*/*z*), a test cross (*ppzz*) with an organism of unknown genotype indicated that the gametes produced were in the following proportions:

PZ, 42.4 percent; *Pz*, 6.9 percent; *pZ*, 7.1 percent; and *pz*, 43.6 percent

Draw all possible conclusions from these data.

13. In a series of two-point map crosses involving five genes located on chromosome II in *Drosophila,* the following recombinant (single-crossover) frequencies were observed:

pr–adp	29
pr–vg	13
pr–c	21
pr–b	6
adp–b	35
adp–c	8
adp–vg	16
vg–b	19
vg–c	8
c–b	27

(a) If *adp* gene is present near the end of chromosome II (locus 83), construct a map of these genes.

(b) In another set of experiments, a sixth gene, *d,* was tested against *b* and *pr:*

d–b	17 percent
d–pr	23 percent

Predict the results of two-point maps between *d* and *c*, *d* and *vg*, and *d* and *adp*.

14. Two different female *Drosophila* were isolated, each heterozygous for the autosomally linked genes *b* (*black body*), *d* (*dachs tarsus*), and *c* (*curved wings*). These genes are in the order *d–b–c,* with *b* being closer to *d* than to *c*. Shown here is the genotypic arrangement for each female along with the various gametes formed by both:

Female A		***Female B***	
$\frac{d\ b\ +}{+\ +\ c}$		$\frac{d\ +\ +}{+\ b\ c}$	
↓	***Gamete formation***	↓	
(1) *d b c*	(5) *d* + +	(1) *d b* +	(5) *d b c*
(2) + + +	(6) + *b c*	(2) + + *c*	(6) + + +
(3) + + *c*	(7) *d* + *c*	(3) *d* + *c*	(7) *d* + +
(4) *d b* +	(8) + *b* +	(4) + *b* +	(8) + *b c*

Identify which categories are noncrossovers (NCO), single crossovers (SCO), and double crossovers (DCO) in each case. Then, indicate the relative frequency in which each ill be produced.

15. In *Drosophila,* a cross was made between females expressing the three X-linked recessive traits, *scute* (*sc*) *bristles, sable body* (*s*), and *vermilion eyes* (*v*), and wild-type males. In the F_1, all females were wild type, while all males expressed all three mutant traits. The cross was carried to the F_2 generation and 1000 offspring were counted, with the results shown here:

Phenotype			***Offspring***
sc	*s*	*v*	314
+	+	+	280
+	*s*	*v*	150
sc	+	+	156
sc	+	*v*	46
+	*s*	+	30
sc	*s*	+	10
+	+	*v*	14

No determination of sex was made in the F_2 data.

(a) Using proper nomenclature, determine the genotypes of the P_1 and F_1 parents.

(b) Determine the sequence of the three genes and the map distance between them.

(c) Are there more or fewer double crossovers than expected? Calculate the coefficient of coincidence. Does this represent positive or negative interference?

16. Another cross in *Drosophila* involved the recessive, X-linked genes *yellow* (*y*), *white* (*w*), and *cut* (*ct*). A yellow-bodied, white-eyed female with normal wings was crossed to a male whose eyes and body were normal, but whose wings were cut. The F_1 females were wild type for all three traits, while the F_1 males expressed the yellow-body, white-eye traits. The cross was carried to an F_2 progeny, and only male offspring were tallied. On the basis of the data shown here, a genetic map was constructed:

Phenotype			Male Offspring
y	+	*ct*	9
+	*w*	+	6
y	*w*	*ct*	90
+	+	+	95
+	+	*ct*	424
y	*w*	+	376
y	+	+	0
+	*w*	*ct*	0

(a) Diagram the genotypes of the F_1 parents.
(b) Assuming that *white* is at locus 1.5 on the X chromosome, construct a map.
(c) Were any double-crossover offspring expected?
(d) Could the F_2 female offspring be used to construct the map? Why or why not?

17. In *Drosophila, Dichaete* (*D*) is a mutation on chromosome 3 with a dominant effect on wing shape. It is lethal when homozygous. The genes *ebony* (*e*) and *pink* (*p*) are recessive mutations on chromosome 3 affecting the body and eye color, respectively. Flies from a *Dichaete* stock were crossed to homozygous *ebony, pink* flies, and the F_1 progeny, with a *Dichaete* phenotype, were backcrossed to the *ebony, pink* homozygotes. The following were the results of this backcross:

Phenotype	Number
Dichaete	401
ebony, pink	389
Dichaete, ebony	84
pink	96
Dichaete, pink	2
ebony	3
Dichaete, ebony, pink	12
wild type	13

(a) Diagram this cross, showing the genotypes of the parents and offspring of both crosses.
(b) What is the sequence and interlocus distance between these three genes?

18. *Drosophila* females homozygous for the third chromosomal genes *pink* and *ebony* (the same genes from Problem 17) were crossed with males homozygous for the second chromosomal gene *dumpy.* Because these genes are recessive, all offspring were wild type (normal). F_1 females were test-crossed to triply recessive males. If we assume that the two linked genes, *pink* and *ebony,* are 20 mu apart, predict the results of this cross. If the reciprocal cross were made (F_1 males—where no crossing over occurs—with triply recessive females), how would the results vary, if at all?

19. In *Drosophila,* two mutations, *Stubble* (*Sb*) and *curled* (*cu*), are linked on chromosome 3. *Stubble* is a dominant gene that is lethal in a homozygous state, and *curled* is a recessive gene. If a female of the genotype

$$\frac{Sb \quad cu}{+ \quad +}$$

is to be mated to detect recombinants among her offspring, what male genotype would you choose as a mate?

20. In *Drosophila,* a heterozygous female for the X-linked recessive traits *a, b,* and *c* was crossed to a male that phenotypically expressed *a, b,* and *c.* The offspring occurred in the following phenotypic ratios:

+	*b*	*c*	460
a	+	+	450
a	*b*	*c*	32
+	+	+	38
a	+	*c*	11
+	*b*	+	9

No other phenotypes were observed.
(a) What is the genotypic arrangement of the alleles of these genes on the X chromosome of the female?
(b) Determine the correct sequence and construct a map of these genes on the X chromosome.
(c) What progeny phenotypes are missing? Why?

21. Why did Stern observe more "twin spots" than *singed* spots in his study of somatic crossing over? If he had been studying *tan* body color (locus 27.5) and *forked* bristles (locus 56.7) on the X chromosome of heterozygous females, what relative frequencies of tan spots, forked spots, and "twin spots" would you predict might occur?

22. Are mitotic recombinations and sister chromatid exchanges effective in producing genetic variability in an individual? In the offspring of individuals?

23. What possible conclusions can be drawn from the observations that no synaptonemal complexes are observed in male *Drosophila* and female *Bombyx* and that no crossing over occurs in these organisms?

24. An organism of the genotype *AaBbCc* was test crossed to a triply recessive organism (*aabbcc*). The genotypes of the progeny were as follows:

20	*AaBbCc*	20	*AaBbcc*
20	*aabbCc*	20	*aabbcc*
5	*AabbCc*	5	*Aabbcc*
5	*aaBbCc*	5	*aaBbcc*

(a) Assuming simple dominance and recessiveness in each gene pair, if these three genes were all assorting independently, how many genotypic and phenotypic classes would result in the offspring, and in what proportion?
(b) Answer the same question assuming the three genes are so tightly linked on a single chromosome that no crossover gametes were recovered in the sample of offspring.
(c) What can you conclude from the *actual* data about the location of the three genes in relation to one another?

25. Based on our discussion of the potential inaccuracy of mapping (see Figure 12–13), would you revise your answer to Problem 24? If so, how?

26. In a plant, fruit was either red or yellow and was either oval or long, where *red* and *oval* are the dominant traits. Two plants, both heterozygous for these traits, were testcrossed, with the following results:

Phenotype	Progeny Plant A	Progeny Plant B
red, long	46	4
yellow, oval	44	6
red, oval	5	43
yellow, long	5	47
	100	100

Determine the location of the genes relative to one another and the genotypes of the two parental plants.

27. In a plant heterozygous for two gene pairs (*Ab/aB*), where the two loci are linked and 25 mu apart, two such individuals were crossed together. Assuming that crossing over occurs during the formation of both male and female gametes and that the *A* and *B* alleles are dominant, determine the phenotypic ratio of the offspring.

28. In a cross in *Neurospora* involving two alleles *B* and *b*, the following tetrad patterns were observed:

Tetrad Pattern	Number
BBbb	36
bbBB	44
BbBb	4
bBbB	6
BbbB	3
bBBb	7

Calculate the distance between the gene and the centromere.

29. In *Neurospora,* the cross a + × + *b* yielded only two types of ordered tetrads in approximately equal numbers:

	Spore Pair 1–2		3–4		5–6		7–8	
Tetrad Type 1	a	+	a	+	+	*b*	+	*b*
Tetrad Type 2	+	+	+	+	a	*b*	a	*b*

What can be concluded?

30. Here are two sets of data derived from crosses in *Chlamydomonas,* involving three genes represented by the mutant alleles *a, b,* and *c*:

Genes	Cross	P	NP	T
1	*a* and *b*	36	36	28
2	*b* and *c*	79	3	18
3	*a* and *c*	?	?	?

Determine as much as you can concerning genetic arrangement of these three genes relative to one another. Assuming that *a* and *c* are linked and are 38 mu and that 100 tetrads are produced, describe the expected results of Cross 3.

31. In *Chlamydomonas,* a cross *ab* x ++ yielded the following unordered tetrad data where *a* and *b* are linked:

	(1)	(2)	(3)	(4)	(5)	(6)
	+ +	+ +	*a* +	*a b*	*a b*	*a b*
	+ +	*a b*	*a* +	*a* +	+ +	+ *b*
	a b	+ +	+ *b*	+ *b*	+ *b*	*a* +
	a b	*a b*	+ *b*	+ +	*a* +	+ +
Number	38	5	6	17	2	3

(a) Identify the categories representing parental ditypes (P), nonparental ditypes (NP), and tetratypes (T).
(b) Explain the origin of category (2).
(c) Determine the map distance between *a* and *b*.

32. The following results are ordered tetrad pairs from a cross between strain *cd* and strain++:

Tetrad Class 1	2	3	4	5	6	7
c +	*c* +	*c d*	+ *d*	*c* +	*c d*	*c* +
c +	*c d*	*c d*	*c* +	+ +	+ +	+ *d*
+ *d*	+ +	+ +	*c* +	*c d*	*c d*	*c d*
+ *d*	+ *d*	+ +	+ *d*	+ *d*	+ +	+ +
1	17	41	1	5	3	1

They are summarized by tetrad classes.
(a) Name the ascus type of each class from 1 to 7 (P, NP, or T).
(b) The data support the conclusion that the *c* and *d* loci are linked. State the evidence in support of this conclusion.
(c) Calculate the gene–centromere distance for each locus.
(d) Calculate the distance between the two linked loci.
(e) Draw a linkage map, including the centromere, and explain the discrepancy between the distances determined by the two different methods in parts (c) and (d).
(f) Describe the arrangement of crossovers needed to produce the ascus class 6.

33. In a cross in *Chlamydomonas, AB* × *ab*, 211 unordered asci were recovered:

10	*AB,*	*Ab,*	*aB,*	*ab*
102	*Ab,*	*aB,*	*Ab,*	*aB*
99	*AB,*	*AB,*	*ab,*	*ab*

(a) Correlate each of the three tetrad types in the problem with their appropriate tetrad designations (names).
(b) Are genes *A* and *B* linked?
(c) If they are linked, determine the map distance between the two genes. If they are unlinked, provide the maximum information you can about why you drew this conclusion.

34. A number of human-mouse somatic-cell hybrid clones were examined for the expression of specific human genes and the presence of human chromosomes. The results are summarized in this table:

	Hybrid cell clone A	B	C	D	E	F
Genes expressed						
ENO1 (enolase-1)	−	+	−	+	+	−
MDH1 (malate dehydrogenase-1)	+	+	−	+	−	+
PEPS (peptidase S)	+	−	+	−	−	−
PGM1 (phosphoglucomutase-1)	−	+	−	+	+	−
Chromosomes (present or absent)						
1	−	+	−	+	+	−
2	+	+	−	+	−	+
3	+	+	−	−	+	−
4	+	−	+	−	−	−
5	−	+	+	+	+	+

Assign each gene to the chromosome upon which it is located

Extra-Spicy Problems

35. A female of genotype

$$\frac{a\quad b\quad c}{+\quad +\quad +}$$

produces 100 meiotic tetrads. Of these, 68 show no crossover events. Of the remaining 32, 20 show a crossover between *a* and *b*, 10 show a crossover between *b* and *c*, and 2 show a double crossover between *a* and *b* and between *b* and *c*. Of the 400 gametes produced, how many of each of the 8 different genotypes will be produced? Assuming the order *a–b–c* and the allele arrangement previously shown, what is the map distance between these loci?

36. In laboratory, a genetics student was assigned an unknown mutation in *Drosophila* that had a whitish eye. He crossed females from his true-breeding mutant stock to wild-type (brick-red-eyed) males, recovering all wild-type F_1 flies. In the F_2 generation, the following offspring were recovered in the following proportions:

wild type	5/8
bright red	1/8
brown eye	1/8
white eye	1/8

The student was stumped until the instructor suggested that perhaps the whitish eye in the original stock was the result of homozygosity for a mutation causing brown eyes *and* a mutation causing bright-red eyes, illustrating gene interaction (See Chapter 10). After much thought, the student was able to analyze the data, explain the results, and learn several things about the location of the two genes relative to one another. One key to his understanding was that crossing over occurs in *Drosophila* females, but not in males. Based on his analysis, what did the student learn about the two genes?

37. *Drosophila melanogaster* has one pair of sex chromosomes (XX or XY) and three autosomes, referred to as chromosomes 2, 3, and 4. A genetics student discovered a male fly with very short legs. Using this male, the student was able to establish a pure breeding stock of this mutant and found that it was recessive. She then incorporated the mutant into a stock containing the recessive gene *black* (body color located on chromosome 2) and the recessive gene *pink* (eye color located on chromosome 3). A female from the homozygous *black, pink, short* stock was then mated to a wild-type male. The F_1 males of this cross were all wild type and were then backcrossed to the homozygous *b p sh* females. The F_2 results appeared as shown in the table that follows. No other phenotypes were observed.

	Wild	Pink*	Black, Short	Black, Pink, Short
Females	63	58	55	69
Males	59	65	51	60

**Pink* indicates that the other two traits are wild type, and so on.

(a) Based on these results, the student was able to assign *short* to a linkage group (a chromosome). Which one was it? Include a step-by-step reasoning.

(b) The student repeated the experiment, making the reciprocal cross, F_1 females backcrossed to homozygous *b p sh* males. She observed that 85 percent of the offspring fell into the given classes, but that 15 percent of the offspring were equally divided among *b* + *p*, *b* + +, + *sh p*, and + *sh* + phenotypic males and females. How can these results be explained and what information can be derived from the data?

38. In *Drosophila*, a female fly is heterozygous for three mutations, *Bar* eyes (*B*), *miniature* wings (*m*), and *ebony* body (*e*). Note that *Bar* is a dominant mutation. The fly is crossed to a male with normal eyes, miniature wings, and ebony body. The results of the cross are shown below:

111	wild, miniature,wild	101	Bar, wild,ebony
29	wild, wild, wild	31	Bar, miniature, ebony
117	Bar, wild, wild	35	wild, wild, ebony
26	Bar, miniature, wild	115	wild, miniature, ebony

a. Interpret the results of this cross. If you conclude that linkage is involved between any of the genes, determine the map distance(s) between them.

Selected Readings

Allen, G.E. 1978. *Thomas Hunt Morgan: The man and his science.* Princeton, NJ: Princeton University Press.

Blixt, S. 1975. Why didn't Gregor Mendel find linkage? *Nature* 256:206.

Catcheside, D.G. 1977. *The genetics of recombination.* Baltimore, MD: University Park Press.

Chaganti, R., Schonberg, S., and German, J. 1974. A manyfold increase in sister chromatid exchange in Bloom's syndrome lymphocytes. *Proc. Natl. Acad. Sci. USA* 71:4508–12.

Creighton, H.S., and McClintock, B. 1931. A correlation of cytological and genetical crossing over in *Zea mays. Proc. Natl. Acad. Sci. USA* 17:492–97.

Douglas, L., and Novitski, E. 1977. What chance did Mendel's experiments give him of noticing linkage? *Heredity* 38:253–57.

Ellis, N.A. et al. 1995. The Bloom's syndrome gene product is homologous to RecQ helicases. *Cell* 83:655–66.

Ephrussi, B., and Weiss, M.C. 1969. Hybrid somatic cells. *Sci. Am.* (April) 220:26–35.

Garcia-Bellido, A. 1972. Some parameters of mitotic recombination in *Drosophila melanogaster. Molec. Genet.* 115:54–72.

Hotta, Y., Tabata, S., and Stern, H. 1984. Replication and nicking of zygotene DNA sequences: Control by a meiosis-specific protein. *Chromosoma* 90:243–53.

King, R.C. 1970. The meiotic behavior of the *Drosophila* oocyte. *Int. Rev. Cytol.* 28:125–68.

Latt, S.A. 1981. Sister chromatid exchange formation. *Annu. Rev. Genet.* 15:11–56.

Lindsley, D.L., and Grell, E.H. 1972. *Genetic variations of Drosophila melanogaster.* Washington, DC: Carnegie Institute of Washington.

Moens, P.B. 1977. The onset of meiosis. In *Cell biology—A comprehensive treatise, Vol. 1. Genetic mechanisms of cells,* ed. L. Goldstein and D.M. Prescott, pp. 93–109. Orlando, FL: Academic Press.

Morgan, T.H. 1911. An attempt to analyze the constitution of the chromosomes on the basis of sex-linked inheritance in *Drosophila. J. Exp. Zool.* 11:365–414.

Morton, N.E. 1955. Sequential test for the detection of linkage. *Am. J. Hum. Genet.* 7:277–318.

——— 1995. *LODs*—past and present. *Genetics* 140:7–12.

Neuffer, M.G., Jones, L., and Zober, M. 1968. *The mutants of maize.* Madison, WI: Crop Science Society of America.

Perkins, D. 1962. Crossing-over and interference in a multiply marked chromosome arm of *Neurospora. Genetics* 47:1253–74.

Ruddle, F.H., and Kucherlapati, R.S. 1974. Hybrid cells and human genes. *Sci. Am.* (July) 231:36–49.

Stahl, F.W. 1979. *Genetic recombination.* New York: W.H. Freeman.

Stern, C. 1936. Somatic crossing over and segregation in *Drosophila melanogaster. Genetics* 21:625–31.

Stern, H., and Hotta, Y. 1973. Biochemical controls in meiosis. *Annu. Rev. Genet.* 7:37–66.

——— 1974. DNA metabolism during pachytene in relation to crossing over. *Genetics* 78:227–35.

Sturtevant, A.H. 1913. The linear arrangement of six sex-linked factors in *Drosophila,* as shown by their mode of association. *J. Exp. Zool.* 14:43–59.

——— 1965. *A history of genetics.* New York: Harper & Row.

Taylor, J.H., ed. 1965. *Selected papers on molecular genetics.* Orlando, FL: Academic Press.

Voeller, B.R., ed. 1968. *The chromosome theory of inheritance: Classical papers in development and heredity.* New York: Appleton-Century-Crofts.

von Wettstein, D., Rasmussen, S.W., and Holm, P.B. 1984. The synaptonemal complex in genetic segregation. *Annu. Rev. Genet.* 18:331–414.

Wolff, S., ed. 1982. *Sister chromatid exchange.* New York: Wiley–Interscience.

GENETICS MediaLab

The resources that follow will help you achieve a better understanding of the concepts presented in this chapter. These resources can be found either on the CD packaged with this textbook or the Companion Web site found at **http://www.prenhall.com/klug**

CD Resources:

Module 12.1: Linkage and Recombination
Module 12.2: Linked Genes
Module 12.3: Mapping a Three-Point Cross
Module 12.4: More Three-Point Crosses
Module 12.5: Virtual Crossover Laboratory

Web Problem 1:

Time for completion = 10 minutes

What are the consequences of crossing over, which occurs during prophase of meiosis I when the homologous chromosomes are paired? Read this article from NIH about the physical events associated with crossing over. How do we define linkage between genes? How does crossing over affect variation among gametes? What other process causes variation among gametes? After examining the figure under "Crossing over and recombination in meiosis," note that equal portions of the chromosomes have been exchanged. What might happen if unequal portions were exchanged? To complete this exercise, visit Web Problem 1 in Chapter 12 of your Companion Web site, and select the keyword **CROSSING OVER.**

Web Problem 2:

Time for completion = 10 minutes

How can we determine the order and distance between genes? The analysis of recombination between linked marker loci forms the basis for constructing chromosome maps. In the first part of this exercise, you will review material on linkage. (See Chapter 12 for more information.) The second part gives an excellent overview of how one analyzes a suite of mutations by using the techniques of classical genetics. How do you distinguish between parental and recombinant chromosomes? What is the largest distance, in centimorgans, that you can measure between two linked loci, using nothing but mutant and wild-type stocks for those two loci? Why is the recombination frequency between unlinked loci not 100%? How do double crossovers affect your estimate of distance between loci? To complete this exercise, visit Web Problem 2 in Chapter 12 of your Companion Web site, and select the keyword **LINKAGE**.

Web Problem 3: from Web site

Time for completion = 20 minutes

Arthur Henry Sturtevant devised the principles of linkage analysis while he was an undergraduate at Columbia University. After reading the biographical sketch of Sturtevant, you should be able to answer the following questions: How did Sturtevant propose to relate the frequency of crossing over to the distance separating two genes on a chromosome? Why are double crossover events so important? How do inversions (sections of the chromosome that have been excised and inverted, Chapter 3) alter the frequency of crossing over? Does crossing over always result in an even exchange of genetic material between the two chromosomes? What did Sturtevant's studies of the Bar mutation in Drosophila contribute to our knowledge of the genetics of human hemoglobin? To complete this exercise, visit Web Problem 3 in Chapter 12 of your Companion Web site, and select the keyword **STURTEVANT**.

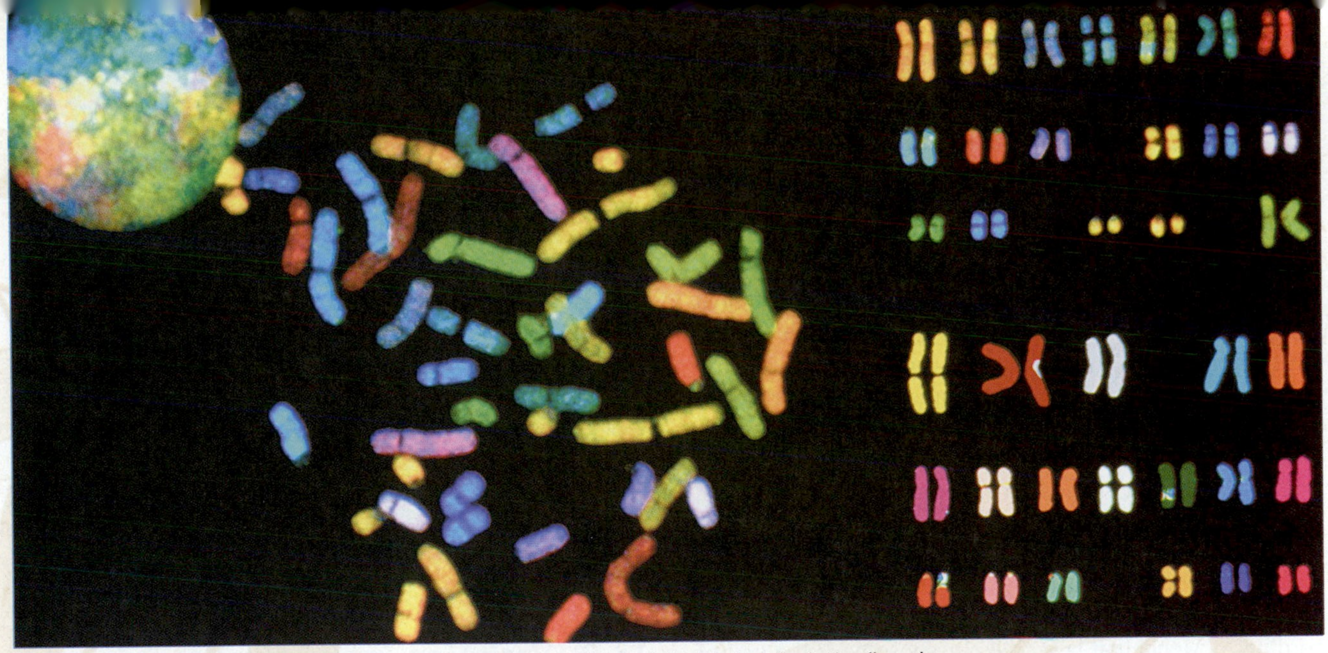

Spectral karyotyping of human chromosomes utilizing differentially labeled "painting" probes. *(Evelin Schrock, Stan du Manoir, and Tom Reid. National Institutes of Health)*

13

Chromosome Mutations: Variation in Chromosome Number and Arrangement

KEY CONCEPTS

Thus far, we have emphasized how mutations and the resulting alleles affect an organism's phenotype and how traits are passed from parents to offspring according to Mendelian principles. In this chapter, we shall look at phenotypic variation that occurs as a result of changes that are more substantial than alterations of individual genes—modifications at the level of the chromosome.

Although most members of diploid species normally contain precisely two haploid chromosome sets, there are many known cases of variations from this pattern. Modifications include a change in the total number of chromosomes, the deletion or duplication of genes or segments of a chromosome, and rearrangements of the genetic material either within or among chromosomes. Taken together, such changes are called **chromosome mutations** or **chromosome aberrations**, to distinguish them from gene mutations. Because, according to Mendelian laws, the chromosome is the unit of genetic transmission, chromosome aberrations are passed on to offspring in a predictable manner, resulting in many unique genetic outcomes.

Since the genetic component of an organism is delicate-ly balanced, even minor alterations of either the content or location of genetic information within the genome may result in some form of phenotypic variation. More substantial changes may be lethal, particularly in animals. Throughout the chapter, we shall consider the many types of chromosomal aberrations, the phenotypic consequences for the organism that harbors an aberration, and the impact of the aberration on offspring of the affected individual. We will also discuss the role of chromosome aberrations in the evolutionary process.

13.1 Specific Terminology Describes Variations in Chromosome Number

Variation in chromosome number ranges from the addition or loss of one or more chromosomes to the addition of one or more haploid sets of chromosomes. Before we embark on our discussion, it is useful to clarify the terminology that describes such changes. In the general condition known as **aneuploidy**, an organism gains or loses one or more chromosomes, but not a complete set. The loss of a single chromosome from an otherwise diploid genome is called *monosomy*. The gain of one chromosome results in *trisomy*. Such changes are contrasted with the condition of **euploidy**, where complete haploid sets of chromosomes are present. If more than two sets are present, the term **polyploidy** applies. Organisms with three sets are specifically *triploid*; those with four sets are *tetraploid*, and so on. Table 13–1 provides an organizational framework for you to follow as we discuss each of these categories of aneuploid and euploid variation and the subsets within them.

TABLE 13–1 Terminology for Variation in Chromosome Numbers

Term	*Explanation*
Aneuploidy	$2n \pm x$ chromosomes
Monosomy	$2n - 1$
Trisomy	$2n + 1$
Tetrasomy, pentasomy, etc.	$2n + 2$, $2n + 3$, etc.
Euploidy	Multiples of n
Diploidy	$2n$
Polyploidy	$3n$, $4n$, $5n$, ...
Triploidy	$3n$
Tetraploidy, pentaploidy, etc.	$4n$, $5n$, etc.
Autopolyploidy	Multiples of the same genome
Allopolyploidy (Amphidiploidy)	Multiples of different genomes

13.2 Variation in the Number of Chromosomes Results from Nondisjunction

It is useful, as we consider cases that include the gain or loss of chromosomes, to examine how such aberrations originate. For instance, how do the syndromes arise where the number of sex-determining chromosomes in humans is altered as first described in Chapter 11? As you may recall, in spite of a mechanism in somatic cells that inactivates all X chromosomes in excess of one, the gain (47,XXY), or the loss (45,X) of a sex-determining chromosome from an otherwise diploid genome alters the normal phenotype, resulting in **Klinefelter syndrome** and **Turner syndrome**, respectively. (See Figure 11–7.) Human females are also known who have extra X chromosomes (e.g., 47,XXX, 48,XXXX), and males may have an extra Y chromosome (47,XYY).

Such chromosomal variation originates as a random error during the production of gametes. As first introduced in Chapter 8, **nondisjunction** is the failure of chromosomes or chromatids to disjoin and move to opposite poles during division. When this occurs in meiosis, the normal distribution of chromosomes into gametes is disrupted. The results of nondisjunction during meiosis I and meiosis II for a single chromosome of a diploid organism are shown in Figure 13–1. As you can see, for the affected chromosome, abnormal gametes can form that contain either two members or none at all. Fertilizing these with a normal haploid gamete produces a zygote with either three members (trisomy) or only one member (monosomy) of this chromosome. As we shall see, nondisjuction leads to a variety of autosomal aneuploid conditions in humans and other organisms.

13.3 Monosomy, the Loss of a Single Chromosome, May Have Severe Phenotypic Effects

We turn now to a consideration of variations in the number of autosomes and the genetic consequence of such changes. The most common examples of aneuploidy, where an organism has a chromosome number other than an exact

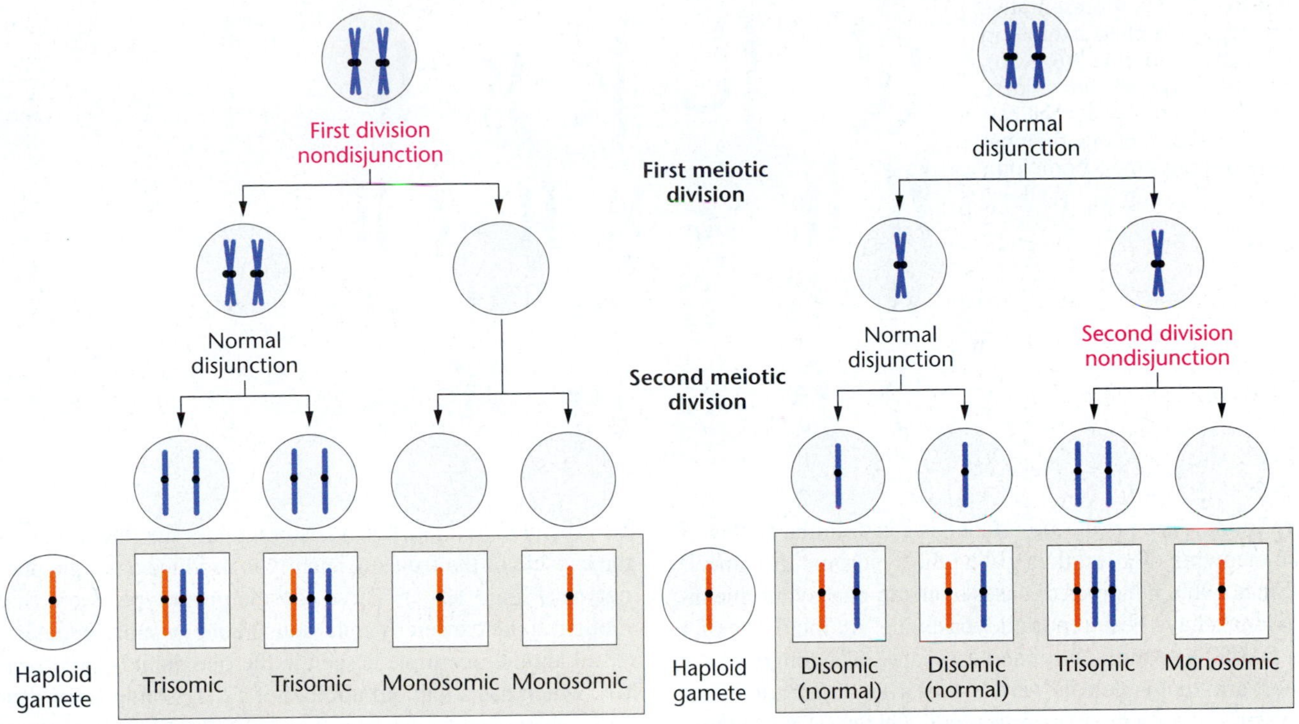

FIGURE 13–1 Nondisjunction during the first and second meiotic divisions. In both cases, some of the gametes formed either contain two members of a specific chromosome or lack that chromosome. Following fertilization by a gamete with a normal haploid content, monosomic, disomic (normal), or trisomic zygotes are produced.

multiple of the haploid set, are cases in which a single chromosome is either added to or lost from a normal diploid set. The loss of one chromosome produces a $2n - 1$ complement and is called **monosomy**.

Although monosomy for the X chromosome occurs in humans, as we have seen in 45,X Turner syndrome, monosomy for any of the autosomes is not usually tolerated in humans or other animals. In *Drosophila*, flies monosomic for the very small chromosome 4—a condition referred to as **Haplo-IV**—survive, but they develop more slowly, exhibit a reduced body size, and have impaired viability. Chromosome 4 contains no more than 5 percent of the genome of *Drosophila*. Monosomy for the larger chromosomes 2 and 3 is apparently lethal because such flies have never been recovered.

The failure of monosomic individuals to survive in many animal species is at first quite puzzling, since at least a single copy of every gene is present in the remaining homolog. One possible explanation involves the unmasking of recessive lethals that are tolerated in heterozygotes carrying the corresponding wild-type alleles. If an organism heterozygous for just one recessive lethal allele loses the homologous chromosome bearing the normal allele (which prevents lethality), the unpaired chromosome condition will lead to the death of the organism.

Another possible explanation is that the expression of genetic information during early development is carefully regulated such that a delicate equilibrium of gene products is required to ensure normal development. While this is thought to be true during animal development, such a requirement does not appear to be so stringent in the plant kingdom, where aneuploidy is tolerated. Monosomy for autosomal chromosomes has been observed in maize, tobacco, the evening primrose *Oenothera*, and the Jimson weed *Datura*, among other plants. Nevertheless, such monosomic plants are almost always less viable than their diploid derivatives. Since both pollen grains and ovules must undergo extensive development after undergoing meiosis, but before participating in fertilization, they are particularly sensitive to the lack of one chromosome and are seldom viable.

Partial Monosomy in Humans: The Cri-du-Chat Syndrome

In humans, autosomal monosomy has not been reported beyond birth. Individuals with such chromosome complements are undoubtedly conceived, but none apparently survive embryonic and fetal development. There are, however, examples of survivors with **partial monosomy**, where only part of one chromosome is lost. These cases are also referred to as **segmental deletions**. One such case was first reported by Jerome LeJeune in 1963, when he described the clinical symptoms of the **cri-du-chat syndrome** ("cry of the cat"). This syndrome is associated with the loss of a small part of the short arm of chromosome 5 (Figure 13–2). Thus, the genetic constitution may be designated as **46, –5p**, meaning that such an individual has all 46 chromosomes, but that some or all of the p arm (the petite or short arm) of one member of the chromosome 5 pair is missing.

Infants with this syndrome may exhibit anatomic malformations, including gastrointestinal and cardiac complications, and they are often mentally retarded. Abnormal development of the glottis and larynx is characteristic of this

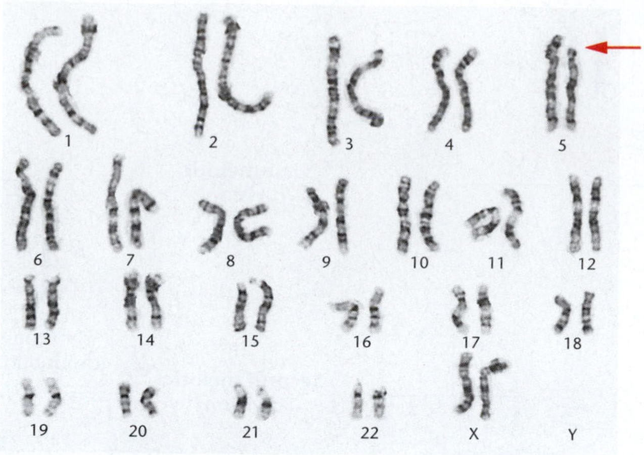

FIGURE 13–2 A representative karyotype of a child exhibiting cri-du-chat syndrome (46, −5p). In the karyotype, the arrow identifies the absence of a small part of the short arm of one member of the chromosome 5 homologs.

syndrome. As a result, the infant has a cry similar to that of the meowing of a cat, thus giving the syndrome its name.

Since 1963, hundreds of cases of cri-du-chat syndrome individuals have been reported worldwide. An incidence of 1 in 50,000 live births has been estimated. The length of the short arm that is deleted varies somewhat; longer deletions appear to have a greater impact on the physical, psychomotor, and mental skill levels of those children who survive. Although the effects of the syndrome are severe, many individuals achieve a level of social development in the trainable range. Those who receive home care and early special schooling are ambulatory, develop self-care skills, and learn to communicate verbally.

13.4 Trisomy Involves the Addition of a Chromosome to a Diploid Genome

In general, the effects of trisomy $(2n + 1)$ parallel those of monosomy. However, the addition of an extra chromosome produces somewhat more viable individuals in both animal and plant species than does the loss of a chromosome. In animals, this is often true, provided that the chromosome involved is relatively small.

As in monosomy, the sex chromosome variation of the trisomic type has a less dramatic effect on the phenotype than does autosomal variation. Recall from our previous discussion that *Drosophila* females with three X chromosomes and a normal complement of two sets of autosomes (3X:2A) survive and reproduce, but are less viable than normal 2X:2A females. In humans, the addition of an extra X or Y chromosome to an otherwise normal male or female chromosome constitution (47,XXY, 47,XYY, and 47,XXX) leads to viable individuals exhibiting various syndromes. However, the addition of a large autosome to the diploid complement in both *Drosophila* and humans has severe effects and is usually lethal during development.

In plants, trisomic individuals are usually viable, but their phenotype may be altered. A classic example involves the Jimson weed *Datura*, a plant long known for its narcotic effect, whose diploid number is 24. Twelve different primary trisomic conditions are possible, and examples of each one have been recovered. Each trisomy alters the phenotype of the capsule of the fruit sufficiently to produce a unique phenotype (Figure 13–3). These capsule phenotypes were first thought to be caused by mutations in one or more genes.

Still another example is seen in the rice plant (*Oryza sativa*), which has a haploid number of 12. Trisomic strains for each chromosome have been isolated and studied. The plants of 11 of them can be distinguished from one another and from wild type. Trisomics for the longer chromosomes are the most distinctive and grow more slowly than the rest. This is in keeping with the belief that larger chromosomes cause greater genetic imbalance than smaller ones. In addition to growth rate, leaf structure, foliage, stems, grain morphology, and plant height vary between the various trisomies.

In plants as well as animals, trisomy may be detected during cytological observations of meiotic divisions. Since three copies of one of the chromosomes are present, pairing configurations are usually irregular. At any particular region along the chromosome length, only two of the three homologs may synapse, though different regions of the trio may be paired. When three copies of a chromosome are synapsed, the configuration is called a **trivalent**, which may be arranged on the spindle so that, during anaphase, one member moves to one pole and two go to the opposite pole (Figure 13–4). In some cases, one bivalent and one univalent (an unpaired chromosome) may be present instead of a trivalent prior to the first meiotic division. Meiosis thus produces gametes with a chromosome composition of $(n + 1)$, which can perpetuate the trisomic condition.

Down Syndrome

The only human autosomal trisomy in which a significant number of individuals survive longer than a year past birth was discovered in 1866 by John Langdon Down. The condition is now known to result from trisomy of chromosome 21, one of the G group* (Figure 13–5), and is called **Down syndrome** or simply **trisomy 21** (designated **47, +21**). This

* On the basis of size and centromere placement, human autosomal chromosomes are divided into seven groups: A (1–3), B (4–5), C (6–12), D (13–15), E (16–18), F (19–20), and G (21–22).

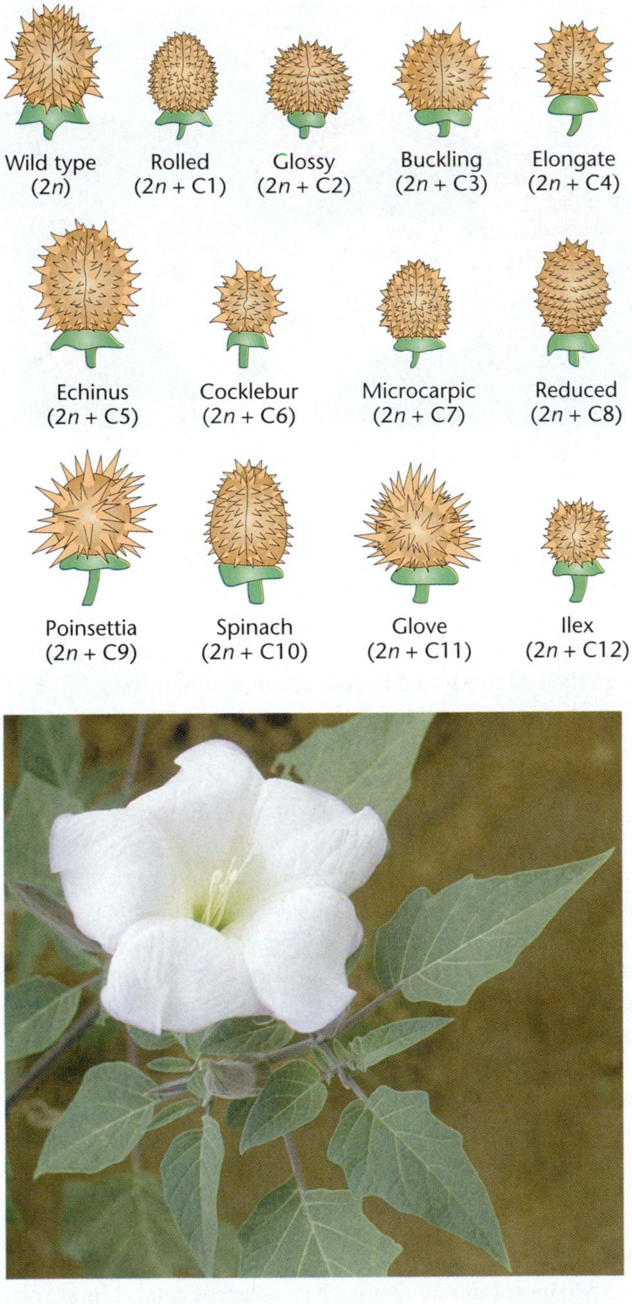

FIGURE 13–3 Drawings of capsule phenotypes of the fruits of the Jimson weed *Datura stramonium*. In comparison with wild type, each phenotype is the result of trisomy of 1 of the 12 chromosomes characteristic of the haploid genome. The photograph illustrates the plant itself.

trisomy is found in approximately one infant in every 800 live births.

Typical of other conditions referred to as a syndrome, there are many phenotypic characteristics that may be present, but any single affected individual usually expresses only a subset of these. In the case of Down syndrome, there are 12–14 such characteristics, but each individual, on average, expresses 6–8 of them. Nevertheless, the outward appearance of these individuals is very similar, and they bear a striking resemblance to one another. This is, for the most part, due to a prominent epicanthic fold in the corner of each eye and their typically flat face and round head. They are also characteristically short. They may also have protruding, furrowed tongues, which cause the mouth to remain partially open; and short, broad hands with fingers showing characteristic palm and fingerprint patterns. Physical, psychomotor, and mental development is retarded, and poor muscle tone is characteristic. Their life expectancy is shortened, although individuals are known to survive into their 50s.

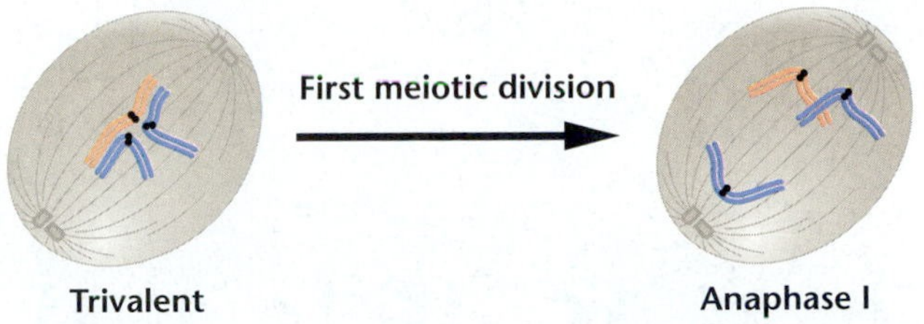

FIGURE 13–4 Diagrammatic representation of one possible pairing arrangement during meiosis I of three copies of a single chromosome, forming a trivalent configuration. During anaphase I, two chromosomes move toward one pole and one chromosome toward the other pole.

Children afflicted with Down Syndrome are prone to respiratory disease and heart malformations, and they show an incidence of leukemia approximately 15 times higher than that of the normal population. However, careful medical scrutiny and treatment throughout their lives has extended their survival significantly. A striking observation is that death of older Down syndrome adults is frequently due to Alzheimer's disease. Because of the reduced lifespan, the onset of this disease occurs at a much earlier age than it does in the normal population.

The most characteristic origin of this trisomic condition is through nondisjunction of chromosome 21 during meiosis. Failure of paired homologs to disjoin during either anaphase I or II may lead to gametes with the $n + 1$ chromosome composition. About 75 percent of these errors leading to Down syndrome are attributed to nondisjunction during meiosis I. Following fertilization with a normal gamete, the trisomic condition is created.

Chromosome analysis has shown that, while the additional chromosome may be derived from either the mother or father, the ovum is the source in about 95 percent of 47, +21 trisomy cases. Before the development of techniques involving polymorphic markers that clearly distinguish paternal from maternal homologs, this conclusion was supported by the more indirect evidence derived from studies of the age of mothers giving birth to infants afflicted with Down syndrome. Figure 13–6 shows the relationship between the incidence of Down syndrome births and maternal age, illustrating the dramatic increase as the age of the mother increases. While the frequency is about 1 in 1000 at maternal age 30, a tenfold increase to a frequency of 1 in 100 is noted at age 40. The frequency increases still further to about 1 in 50 at age 45. A very alarming statistic is that when childbearing women over 45 are included for all births to women over age 40, there is an average probability of 0.35 (35 percent) of a Down syndrome birth. Even so in spite of this high probability, because the overwhelming proportion of pregnancies involve women under 35, more than half of

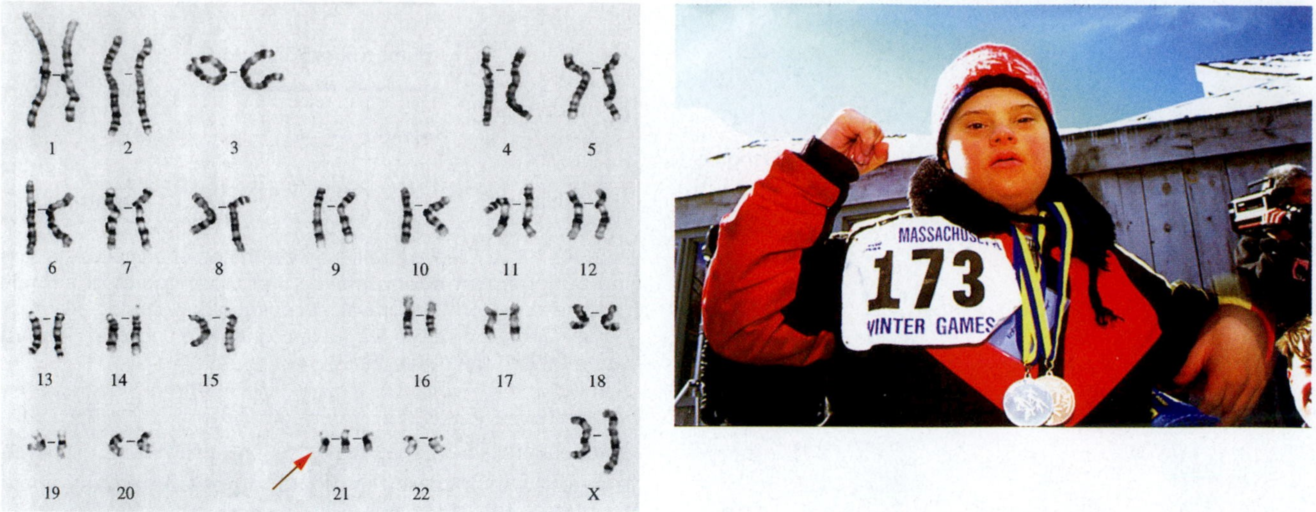

FIGURE 13–5 The karyotype and a photograph of a child with Down syndrome. In the karyotype, three members of the G-group chromosome 21 are present, creating the 47, +21 condition.

Down syndrome births occur to women in this younger cohort.

While the nondisjunctional event that produces Down syndrome seems more likely to occur during oogenesis in women between the ages of 35 and 45, we do not know with certainty why this is so. However, one observation may be relevant. In human females, meiosis in all eggs is initiated during fetal development. Synapsis of homologs has occurred and recombination has been initiated. Then oocyte development is arrested in meiosis I. Thus, all primary oocytes have been formed by birth. Then, once ovulation begins at puberty, meiosis is reinitiated in one egg during each ovulatory cycle and continues into meiosis II. The process is once again arrested after ovulation and is not completed unless fertilization occurs.

The end result of this progression is that each succeeding ovum has been arrested in meiosis I for about a month longer than the one preceding it. As a result, women 30 or 40 years old produce ova that are significantly older and arrested longer than those they ovulated 10 or 20 years previously. However, no direct evidence proves that ovum age is the cause of the increased incidence of nondisjunction leading to Down syndrome.

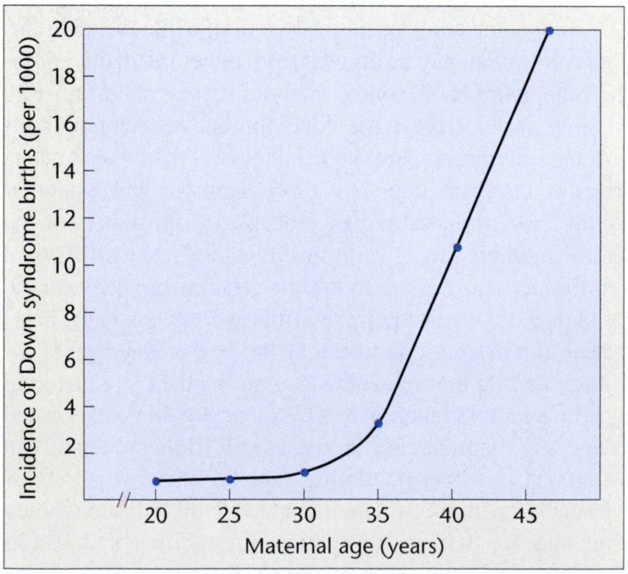

FIGURE 13–6 Incidence of Down syndrome births contrasted with maternal age.

These statistics obviously pose a serious problem for the woman who becomes pregnant late in her reproductive years. Genetic counseling early in such pregnancies is highly recommended. Such counseling informs prospective parents about the probability that their child will be affected and educates them about Down syndrome. Although some individuals with Down syndrome must be institutionalized, others benefit greatly from special education programs and may be cared for at home. Further, these children are noted for their affectionate, loving natures.

A genetic counselor may recommend a prenatal diagnostic technique, where fetal cells are isolated and cultured. **Amniocentesis** and **chorionic villus sampling (CVS)** are the two most familiar approaches, whereby fetal cells are obtained from the amniotic fluid or the chorion of the placenta, respectively. In a more current approach, fetal cells are derived directly from the maternal circulation. Once perfected, this approach will be preferable, because it is noninvasive, posing no risk to the fetus. Once fetal cells are obtained, the karyotype can then be determined by cytogenetic analysis. If the fetus is diagnosed as having Down syndrome, a therapeutic abortion is one option currently available to parents. Obviously, this is a difficult decision involving a number of religious and ethical issues.

Since Down syndrome is caused by a random error—nondisjunction of chromosome 21 during maternal or paternal meiosis—the occurrence of the disorder is not expected to be inherited. Nevertheless, Down syndrome occasionally runs in families. These instances, referred to as **familial Down syndrome**, involve a translocation of chromosome 21, another type of chromosomal aberration, which we will discuss later in the chapter.

Patau Syndrome

In 1960, Klaus Patau and his associates observed an infant with severe developmental malformations and a karyotype of 47 chromosomes (Figure 13–7). The additional chromosome was medium sized, one of the acrocentric D group. It is now designated as chromosome 13. The trisomy 13 condition has since been described in many newborns and is called **Patau syndrome** (**47, +13**). Affected infants are not mentally alert, are thought to be deaf, and characteristically have a harelip, cleft palate, and demonstrate polydactyly. Autopsies have revealed congenital malformation of most organ systems, a condition indicative of abnormal developmental events occurring as early as five to six weeks of gestation. The average survival of these infants is about three months.

The average maternal and paternal ages of parents of Patau infants are higher than the ages of parents of normal children, but they are not as high as the average maternal age in cases of Down syndrome. Both male and female parents average about 32 years of age when the affected child is born. Because the condition is so rare, occurring as infrequently as 1 in 19,000 live births, it is not known whether the origin of the extra chromosome is more often maternal, paternal, or whether it arises equally from either parent.

Edwards Syndrome

In 1960, John H. Edwards and his colleagues reported on an infant trisomic for a chromosome in the E group, now known to be chromosome 18 (Figure 13–8). This aberration has been named **Edwards syndrome** (**47, +18**). The phenotype of this child, like that of individuals with Down and Patau syndromes, illustrates that the presence of an extra autosome produces congenital malformations and reduced life expectancy. These infants are smaller than the average newborn. Their skulls are elongated in an anterior-posterior direction, and their ears are set low and malformed. A webbed neck, congenital dislocation of the hips, and a receding chin are often characteristic of such individuals. Although the frequency of trisomy 18 is somewhat greater than that of trisomy 13, the average survival time is about the same, less than four months. Death is usually caused by pneumonia or heart failure.

Again, the average maternal age is high—34.7 years by one calculation. In contrast to Patau syndrome, the preponderance of Edwards syndrome infants are females. In one set of observations based on 143 cases, 80 percent were female. Overall, about 1 in 8000 live births exhibits this malady.

Viability in Human Aneuploidy

The reduced viability of individuals with recognized monosomic and trisomic conditions suggests that many other aneuploid conditions may arise, but the affected fetuses do not survive to term. This observation has been confirmed by karyotypic analysis of spontaneously aborted fetuses. In an extensive review of this subject in 1971 by David H. Carr, it

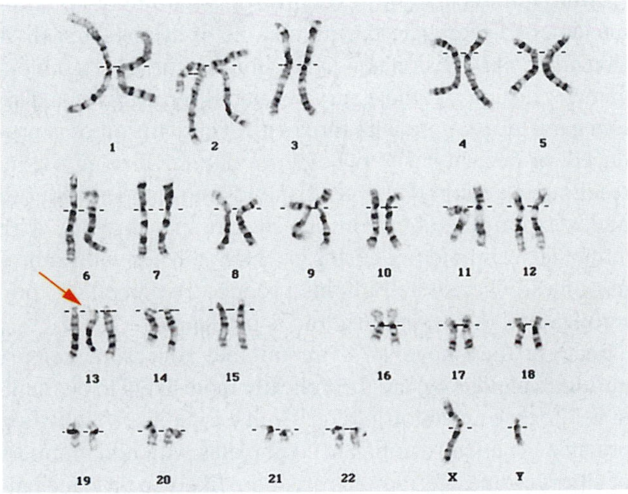

Mental retardation	Microcephaly
Growth failure	Cleft lip and palate
Low set deformed ears	Polydactyly
Deafness	Deformed finger nails
Atrial septal defect	Kidney cysts
Ventricular septal defect	Double ureter
Abnormal polymorphonuclear granulocytes	Umbilical hernia
	Developmental uterine abnormalities
	Cryptorchidism

FIGURE 13–7 The karyotype and phenotypic depiction of an infant with Patau syndrome, where three members of the D group chromosome 13 are present, creating the 47, +13 condition.

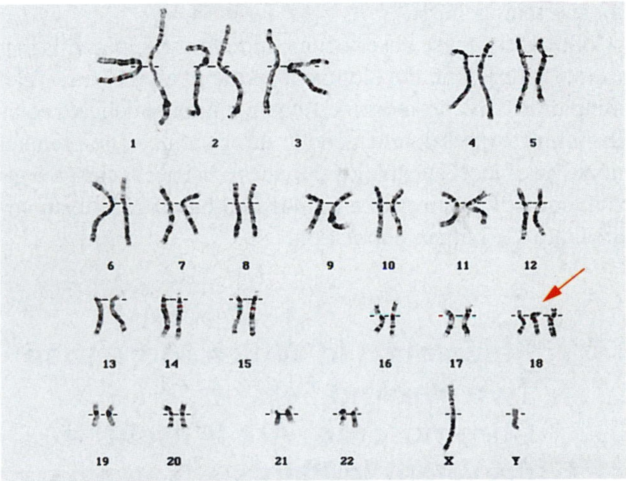

Growth failure	Abnormal kidneys
Mental retardation	Persistent ductus arteriosus
Open skull sutures at birth	Deformity of hips
High arched eyebrows	Prominent external genitalia
Low set deformed ears	Muscular hypertonus
Short sternum	Prominent heel
Ventricular septal defect	Dorsal flexion of big toes
Flexion deformities of fingers	

FIGURE 13–8 The karyotype and phenotypic characteristics of Edwards syndrome. Three members of the E group chromosome 18 are present, creating the 47, +18 condition.

was shown that a significant percentage of abortuses are trisomic for one or another of the autosomal chromosomes. Trisomies for every human chromosome were recovered. Monosomies, however, were seldom found in the Carr study, even though nondisjunction should produce $n - 1$ gametes with a frequency equal to $n + 1$ gametes. This finding leads us to believe that gametes lacking a single chromosome are either so functionally impaired that they never participate in fertilization or that the monosomic embryo dies so early in its development that recovery occurs infrequently. Various forms of polyploidy and other miscellaneous chromosomal anomalies were also found in Carr's study.

Studies that included Carr's work revealed some striking statistics. It was estimated that about 15–20 percent of all human conceptions are terminated by spontaneous abortion and that about 30 percent of all resultant abortuses demonstrate some form of chromosomal anomaly. A calculation using these figures $(0.20 \times 0.30 = 0.06)$ predicts that up to 6 percent of all pregnancies originate with an abnormal number of chromosomes. Most all of these are terminated prior to birth. More recently derived figures are even more dramatic. It is now estimated that as many as 10–30 percent of all fertilized eggs in humans contain some error in chromosome number.

The largest percentage of chromosomal abnormalities are aneuploids. Surprisingly, an aneuploid with one of the highest incidence rates among abortuses is the 45,X condition, which produces an infant with Turner syndrome if the fetus survives to term. About 70–80 percent of aborted and liveborn 45,X conditions include the maternal X chromosome. Thus, the meiotic error leading to this syndrome occurs during spermatogenesis.

Collectively, these observations support the hypothesis that normal embryonic development requires a precise diploid complement of chromosomes that maintains a delicate equilibrium of expression of genetic information. The prenatal mortality of most aneuploids provides a barrier against the introduction of a variety of chromosome-based genetic anomalies into the human population.

13.5 Polyploidy, in Which More than Two Haploid Sets of Chromosomes Are Present, Is Prevalent in Plants

The term *polyploidy* describes instances where more than two multiples of the haploid chromosome set are found. The naming of polyploids is based on the number of sets of chromosomes found: a **triploid** has $3n$ chromosomes; a **tetraploid** has $4n$; a **pentaploid**, $5n$; and so forth. Several general statements may be made about polyploidy. This condition is relatively infrequent in many animal species, but is well known in lizards, amphibians, and fish. It is much more common in plant species. Odd numbers of chromosome sets are not usually maintained reliably from generation to generation because a polyploid organism with an uneven number of homologs usually does not produce genetically balanced gametes. For this reason, triploids, pentaploids, and so on are not usually found in species that depend solely upon sexual reproduction for propagation.

Polyploidy can originate in two ways: (1) The addition of one or more extra sets of chromosomes, identical to the normal haploid complement of the same species, results in **autopolyploidy**; and (2) the combination of chromosome sets from different species may occur as a consequence of interspecific matings resulting in **allopolyploidy** (from the Greek word *allo*, meaning other or different). The distinction between auto- and allopolyploidy is based on the genetic origin of the extra chromosome sets, as illustrated in Figure 13–9.

In our discussion of polyploidy, we will use certain symbols to clarify the origin of additional chromosome sets. For example, if A represents the haploid set of chromosomes of any organism, then

$$A = a_1 + a_2 + a_3 + a_4 + \cdots + a_n$$

where a_1, a_2, and so on represent individual chromosomes, and where n is the haploid number. Using this nomenclature, a normal diploid organism would be represented simply as AA.

Autopolyploidy

In autopolyploidy, each additional set of chromosomes is identical to the parent species. Therefore, triploids are represented as AAA, tetraploids are $AAAA$, and so forth.

Autotriploids arise in several ways. A failure of all chromosomes to segregate during meiotic divisions (first-division or second-division nondisjunction) can produce a diploid gamete. If such a gamete survives and is fertilized by a haploid gamete, a zygote with three sets of chromosomes is produced, or occasionally, two sperm may fertilize an ovum, resulting in a triploid zygote. Triploids can also be produced under experimental conditions by crossing diploids with tetraploids. Diploid organisms produce gametes with n chromosomes, whereas tetraploids produce $2n$ gametes. Upon fertilization, the desired triploid is produced.

Because they have an even number of chromosomes, **autotetraploids** ($4n$) are theoretically more likely to be found in nature than are autotriploids. Unlike triploids, which often produce genetically unbalanced gametes with odd numbers of chromosomes, tetraploids are more likely to produce balanced gametes when involved in sexual reproduction.

How polyploidy arises naturally is of great interest. In theory, if chromosomes have replicated, but the parent cell never divides and reenters interphase, the chromosome number will be doubled. That this very likely occurs is supported by the observation that tetraploid cells can be produced experimentally from diploid cells by applying cold or heat shock to meiotic cells, or by applying colchicine to somatic cells undergoing mitosis. **Colchicine**, an alkaloid derived from the autumn crocus, interferes with spindle formation, and thus, replicated chromosomes that cannot be separated at anaphase do not migrate to the poles. When colchicine is removed, the cell can reenter interphase. When the paired sister chromatids separate and uncoil, the nucleus will contain

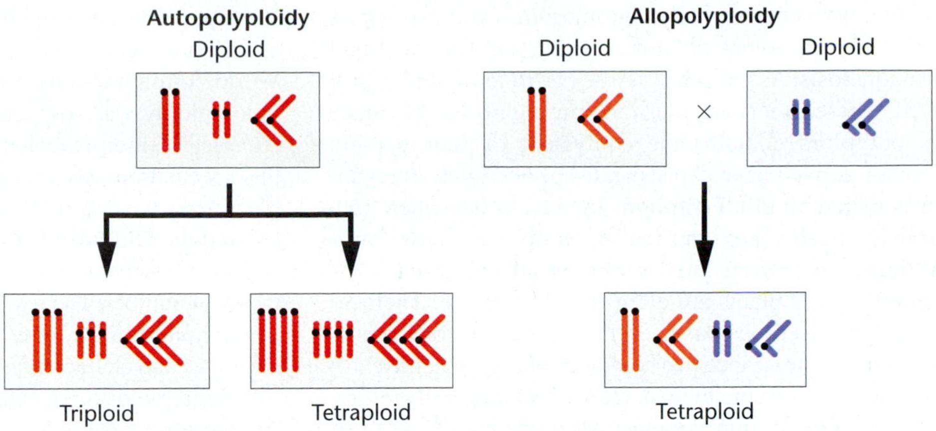

FIGURE 13–9 Contrasting chromosome origins of an autopolyploid versus an allopolyploid karyotype.

twice the diploid number of chromosomes and is therefore $4n$. This process is illustrated in Figure 13–10.

In general, autopolyploids are larger than their diploid relatives. Such an increase seems to be due to a larger cell size, rather than a greater number of cells. Although autopolyploids do not contain new or unique information compared with the diploid relative, the flower and fruit of plants are often increased in size, making such varieties of greater horticultural or commercial value. Economically important triploid plants include several potato species of the genus *Solanum*, Winesap apples, commercial bananas, seedless watermelons, and the cultivated tiger lily *Lilium tigrinum*. These plants are propagated asexually. Diploid bananas contain hard seeds, but the commercial, triploid, "seedless" variety has edible seeds. Tetraploid alfalfa, coffee, peanuts, and McIntosh apples are also of economic value, because they are either larger or grow more vigorously than do their diploid or triploid counterparts. The commercial strawberry is an octoploid.

We have long been curious how cells with increased ploidy values, where no new genes are present, express different phenotypes than their diploid counterparts. Our current ability to examine gene expression using modern biotechnology has provided some interesting insights. For example, Gerald Fink and his colleagues have been able to create strains of the yeast *Saccaromyces cerevisiae* with one, two, three, or four copies of the genome. Thus, each strain contains identical genes (they are said to be isogenic) but different ploidy values. They then proceeded to examine expression levels of all genes during the entire cell cycle of the organism. Using the rather stringent standards of a ten-fold increase or decrease of gene expression, Fink and coworkers proceeded to identify ten cases where, as ploidy increased, gene expression was increased at least ten-fold and seven cases where it was reduced by a similar level.

One of these genes provides insights into how polyploid cells are larger than their haploid or diploid counterparts. In polyploid yeast, two G1 cyclins, Cln1 and Pc11 are repressed as ploidy increases, while the size of the yeast cells increases. This is explained based on the observation that G1 cyclins facilitate the cell's movement through G1, which is delayed when expression of these genes is repressed. The cell stays in G1 longer and, on average, grows to a larger size before it moves beyond the G1 stage of the cell cycle. Yeast cells also show different morphology as ploidy increases. Several of the other genes, repressed as ploidy increases, have been linked to cytoskeletal dynamics, which accounts for the morphological changes.

Allopolyploidy

Polyploidy can also result from hybridization of two closely related species. If a haploid ovum from a species with chromosome sets *AA* is fertilized by a haploid sperm from a species with sets *BB*, the resulting hybrid is *AB*, where $A = a_1, a_2, a_3, \ldots, a_n$ and $B = b_1, b_2, b_3, \ldots, b_n$. The hybrid plant may be sterile because of its inability to produce viable gametes. Most often, this occurs when some, or all, of the *a* and *b* chromosomes are not homologous and therefore cannot synapse in meiosis. As a result, unbalanced genetic conditions result. If, however, the new *AB* genetic combination undergoes a natural or induced chromosomal

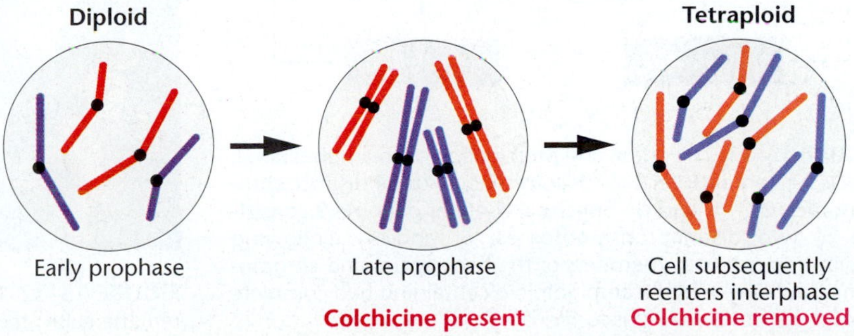

FIGURE 13–10 The potential involvement of colchicine in doubling the chromosome number, as occurs during the production of an autotetraploid. Two pairs of homologous chromosomes are followed. While each chromosome has replicated its DNA earlier during interphase, the chromosomes do not appear as double structures until late prophase. When anaphase fails to occur normally, the chromosome number doubles if the cell reenters interphase.

doubling, two copies of all *a* chromosomes and two copies of all *b* chromosomes are now present, and they will pair during meiosis. As a result, a fertile *AABB* tetraploid is produced. These events are illustrated in Figure 13–11. Since this polyploid contains the equivalent of four haploid genomes derived from two separate species, such an organism is called an **allotetraploid**. In such cases, when both original species are known, an equivalent term, **amphidiploid**, is preferred to describe the allotetraploid.

Amphidiploid plants are often found in nature. Their reproductive success is based on their potential for forming balanced gametes. Since two homologs of each specific chromosome are present, meiosis can occur normally (Figure 13–11), and fertilization can successfully propagate the plant sexually. This discussion assumes the simplest situation, where none of the chromosomes in set *A* are homologous to those in set *B*. In amphidiploids formed from closely related species, some homology between *a* and *b* chromosomes will no doubt exist. In this case, meiotic pairing is more complex. During synapsis, multivalents will be formed, resulting in the production of unbalanced gametes. In such cases, aneuploid varieties of amphidiploids may arise. Allopolyploids are rare in most animals because mating behavior is most often species specific, and thus, the initial step in hybridization is unlikely to occur.

A classic example of amphidiploidy in plants is the cultivated species of American cotton, *Gossypium* (Figure 13–12).

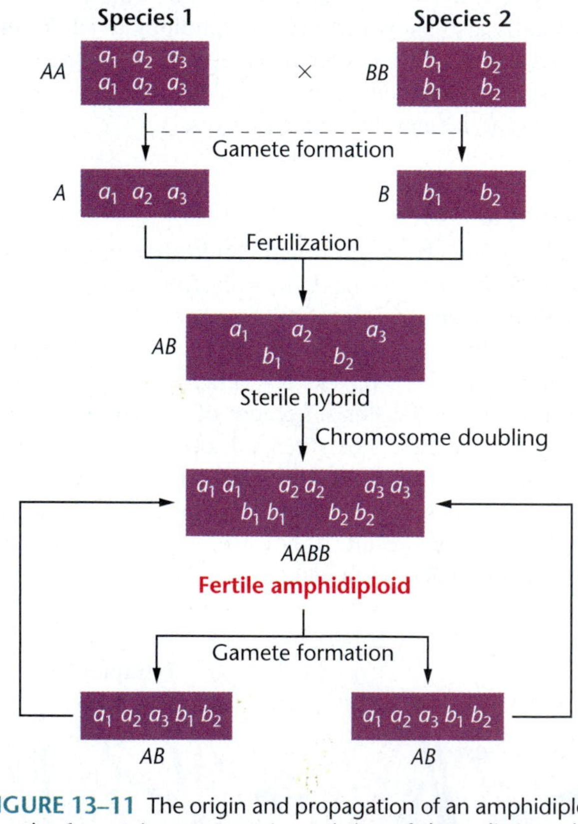

FIGURE 13–11 The origin and propagation of an amphidiploid. Species 1 contains genome A consisting of three distinct chromosomes, a_1, a_2, and a_3. Species 2 contains genome B consisting of two distinct chromosomes, b_1 and b_2. Following fertilization between members of the two species and chromosome doubling, a fertile amphidiploid containing two complete diploid genomes (*AABB*) is formed.

This species has 26 pairs of chromosomes: Thirteen are large and 13 are much smaller. When it was discovered that Old World cotton had only 13 pairs of large chromosomes, allopolyploidy was suspected. After an examination of wild American cotton revealed 13 pairs of small chromosomes, this speculation was strengthened. J. O. Beasley was able to reconstruct the origin of cultivated cotton experimentally. He crossed the Old World strain with the wild American strain and then treated the hybrid with colchicine to double the chromosome number. The result of these treatments was a fertile amphidiploid variety of cotton. It contained 26 pairs of chromosomes and characteristics similar to the cultivated variety.

Amphidiploids often exhibit traits of both parental species. An interesting example, but one with no practical economic importance, is that of the hybrid formed between the radish *Raphanus sativus* and the cabbage *Brassica oleracea*. Both species have a haploid number of 9. The initial hybrid consists of 9 *Raphanus* and 9 *Brassica* chromosomes (9R + 9B). While hybrids are almost always sterile, some fertile amphidiploids (18R + 18B) have been produced. Unfortunately, the root of this plant is more like the cabbage and its shoot more like the radish. Had the converse occurred, the hybrid might have been of economic importance.

A much more successful commercial hybridization has been performed using the grasses wheat and rye. Wheat (genus *Triticum*) has a basic haploid genome of 7 chromosomes. In addition to normal diploids ($2n = 14$), cultivated allopolyploids exist, including tetraploid ($4n = 28$) and hexaploid ($6n = 42$) species. Rye (genus *Secale*) also has a genome consisting of 7 chromosomes. The only cultivated species is the diploid plant ($2n = 14$).

Using the technique outlined in Figure 13–11, geneticists have produced various hybrids. When tetraploid wheat is crossed with diploid rye, and the F_1 treated with colchicine, a hexaploid variety ($6n = 42$) is derived. The hybrid, designated *Triticale* (see Figure 1–15), represents a new genus. Fertile hybrid varieties derived from various wheat and rye species can be crossed together or backcrossed. These crosses have created many variations of the genus *Triticale*. For

FIGURE 13–12 The pods of the amphidiploid form of *Gossypium*, the cultivated cotton plant.

example, a useful octaploid (an allooctaploid) variety with 56 chromosomes has been produced from a diploid rye and a hexaploid wheat plant.

The hybrid plants demonstrate characteristics of both wheat and rye. For example, certain hybrids combine the high protein content of wheat with the high content of the amino acid lysine in rye. The lysine content is low in wheat and thus is a limiting nutritional factor. Wheat is considered a high-yielding grain, whereas rye is noted for its versatility of growth in unfavorable environments. *Triticale* species, combining both traits, have the potential of significantly increasing grain production. Programs designed to improve crops through hybridization have long been underway in several underdeveloped countries of the world, as discussed in Chapter 1.

Recall that in a previous chapter, we discussed the use of **somatic cell hybridization** to map human genes (See Chapter 12). This technique has also been applied to the production of amphidiploid plants (Figure 13–13). Cells from the developing leaves of plants can be treated to remove their cell wall, resulting in **protoplasts**. These altered cells can be maintained in culture and stimulated to fuse with other protoplasts, producing somatic cell hybrids. If cells from different plant species are fused in this way, hybrid amphidiploid cells can be produced. Since protoplasts can be induced to divide and differentiate into stems that develop leaves, the potential for producing allopolyploids is available in the research laboratory. In some cases, entire plants can be derived from cultured protoplasts. If only stems and leaves are produced, these can be grafted onto the stem of another plant. If flowers are formed, fertilization may yield mature seeds, which, upon germination, yield an allopolyploid plant.

There are now many examples of allopolyploids that have been created commercially by using the aforementioned approach. Although most of them are not true amphidiploids

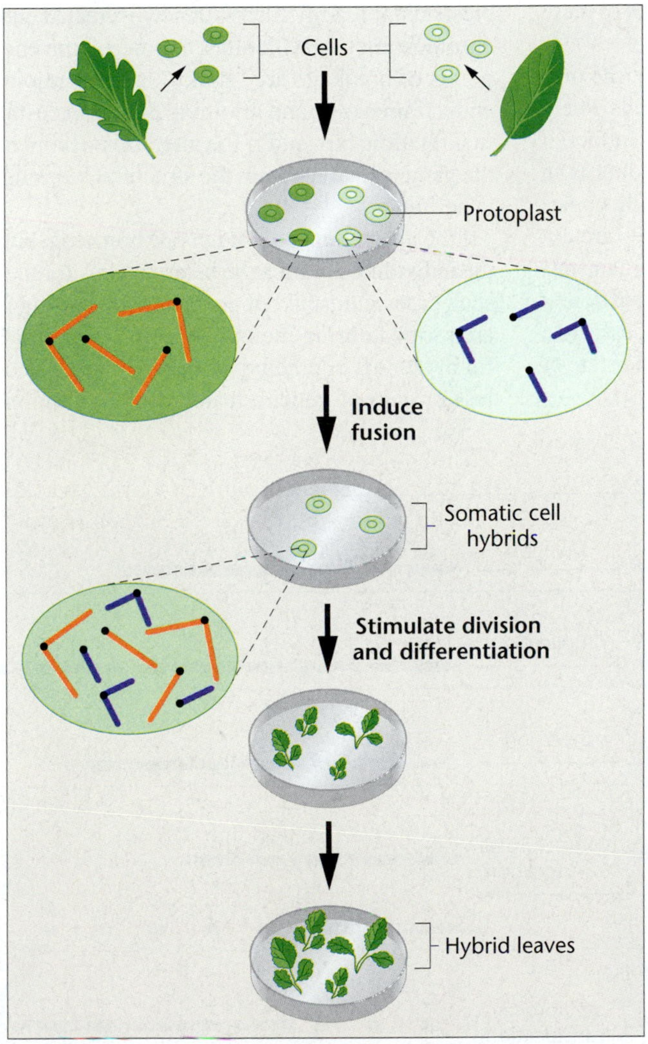

FIGURE 13–13 Application of the somatic cell hybridization technique in the production of an amphidiploid. Cells from the leaves of two species of plants are removed and cultured. The cell walls are digested away and the resultant protoplasts are induced to undergo cell fusion. The hybrid cell is selected and stimulated to divide and differentiate, as illustrated in the photograph. An amphidiploid has a complete set of chromosomes from each parental cell type and displays phenotypic characteristics of each. Two pairs of chromosomes from each species are depicted.

(one or more chromosomes are missing), precise allotetraploids are sometimes produced.

Endopolyploidy

Endopolyploidy is the condition in which only certain cells in an otherwise diploid organism are polyploid. In such cells, replication and separation of chromosomes occur without nuclear division. Endopolyploidy is the result of a process called endomitosis.

Numerous examples of naturally occurring endopolyploidy have been observed. For example, vertebrate liver cell nuclei, including human ones, often contain $4n$, $8n$, or $16n$ chromosome sets. The stem and parenchymal tissue of apical regions of flowering plants are also often endopolyploid. Cells lining the gut of mosquito larvae attain a $16n$ ploidy, but during the pupal stages, such cells undergo very quick reduction divisions, giving rise to smaller diploid cells. In the water strider *Gerris*, wide variations in chromosome numbers are found in different tissues, with as many as 1024 to 2048 copies of each chromosome in the salivary gland cells. Since the diploid number in this organism is 22, the nuclei of these cells may contain over 40,000 chromosomes.

Although the role of endopolyploidy is not clear, the proliferation of chromosome copies often occurs in cells where high levels of certain gene products are required. In fact, it is well established that certain genes whose product is in high demand in *every* cell exist naturally in multiple copies in the genome. Ribosomal and transfer RNA genes are examples of multiple-copy genes. In certain cells of organisms, where even this condition may not allow for a sufficient amount of a particular gene product, it may be necessary to replicate the entire genome, allowing an even greater rate of expression of that gene.

13.6 Variation Occurs in the Structure and Arrangement of Chromosomes

The second general class of chromosome aberrations includes structural changes that delete, add, or rearrange substantial portions of one or more chromosomes. Included in this broad category are deletions and duplications of genes or part of a chromosome and rearrangements of genetic material in which a chromosome segment is inverted, exchanged with a segment of a nonhomologous chromosome, or merely transferred to another chromosome. Exchanges and transfers are called translocations, in which the location of a gene is altered within the genome. These types of chromosome alterations are illustrated in Figure 13–14.

In most instances, these structural changes are due to one or more breaks along the axis of a chromosome, followed by either the loss or rearrangement of genetic material. Chromosomes can break spontaneously, but the rate of breakage may increase in cells exposed to chemicals or radiation. Although the actual ends of chromosomes, known as telomeres, do not readily fuse with newly created ends of "broken" chromosomes or with other telomeres, the ends produced at points of breakage are "sticky" and can rejoin other broken ends. If breakage and rejoining does not reestablish the original relationship, and if the alteration occurs in germ plasm, the gametes will contain the structural rearrangement, which is heritable.

If the aberration is found in one homolog, but not the other, the individual is said to be heterozygous for the aberration. In such cases, unusual but characteristic pairing configurations are formed during meiotic synapsis. These patterns are useful in identifying the type of change that has occurred. If no loss or gain of genetic material occurs, individuals bearing

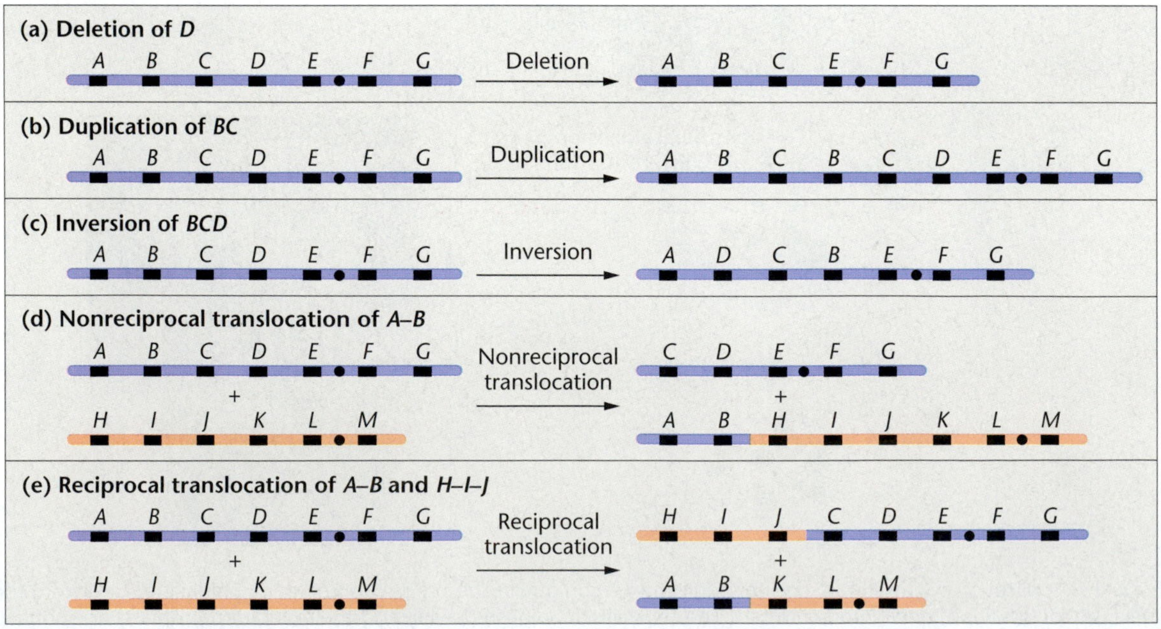

FIGURE 13–14 Overview of the five different types of rearrangement of chromosome segments.

the aberration "heterozygously" are likely to be unaffected phenotypically. However, the unusual pairing arrangements often lead to gametes that are duplicated or deficient for some chromosomal regions. When this occurs, the offspring of "carriers" of certain aberrations often have an increased probability of demonstrating phenotypic manifestations.

13.7 A Deletion Is a Missing Region of a Chromosome

When a chromosome breaks in one or more places, and a portion of it is lost, the missing piece is referred to as a **deletion** (or a **deficiency**). The deletion can occur either near one end or from the interior of the chromosome. These are called **terminal** or **intercalary deletions**, respectively [Figure 13–15(a) and (b)]. The portion of the chromosome retaining the centromere region will usually be maintained when the cell divides, whereas the segment without the centromere will eventually be lost in progeny cells following mitosis or meiosis. For synapsis to occur between a chromosome with a large intercalary deficiency and a normal complete homolog, the unpaired region of the normal homolog must "buckle out" into a **deletion** or **compensation loop** [Figure 13–15(c)].

As seen earlier in our discussion of the cri-du-chat syndrome, where only a small part of the short arm of chromosome 5 is lost, a deletion of a portion of a chromosome need not be very great before the effects become severe. If even more genetic information is lost as a result of a deletion, the aberration is often lethal. Such chromosome mutations never become available for study. Note that when a deletion involves a noticeable piece of a chromosome, as in cri-du-chat syndrome, we may also refer to the resulting condition as partial monosomy.

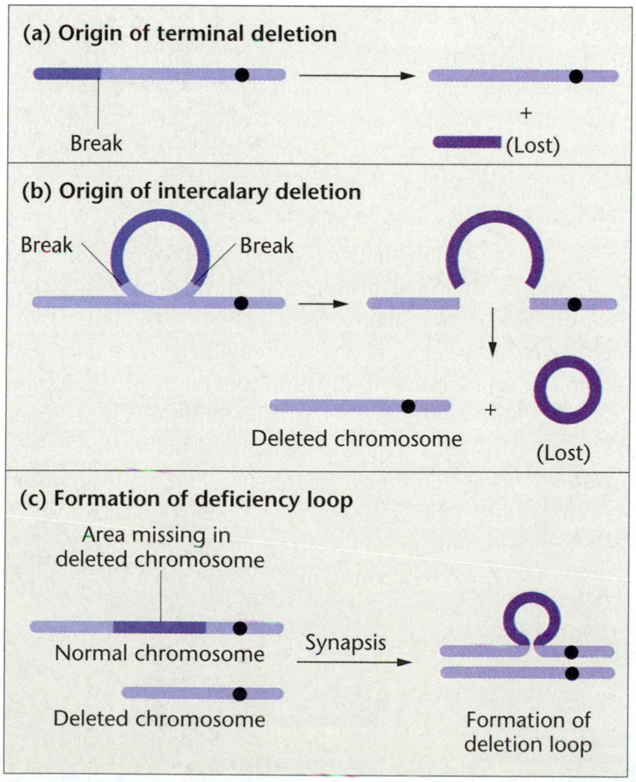

FIGURE 13–15 Origins of (a) a terminal and (b) intercalary deletion. In part (c), pairing occurs between a normal chromosome and one with an intercalary deletion by looping out the undeleted portion to form a deficiency (or a compensation) loop.

A final consequence of deletions can be noted in organisms heterozygous for a deficiency. Consider the mutant *Notch* phenotype in *Drosophila*. In these flies, the wings are notched on the posterior and lateral margins. Data from breeding studies indicate that the phenotype is controlled by an X-linked dominant mutation because heterozygous females have notched wings and transmit this allele to one-half of their female progeny. The mutation also appears to behave as a homozygous and hemizygous lethal because such females and males are never recovered. It has also been noted that if notched-winged females are also heterozygous for the closely linked recessive mutations *white*-eye, *facet*-eye, or *split*-bristle, they express these mutant phenotypes as well as *Notch*. Because these mutations are recessive, heterozygotes should express the normal, wild-type phenotypes. These genotypes and phenotypes are summarized in Table 13–2.

These observations have been explained through a cytological examination of the particularly large chromosomes that characterize certain larval cells of many insects. Called polytene chromosomes (see Chapter 4), they demonstrate a banding pattern specific to each region of each chromosome. In such cells of heterozygous *Notch* females, a deficiency loop was found along the X chromosome from the band designated 3C2 through band 3C11, as shown in Figure 13–16. These bands had previously been shown to include the loci for the *white, facet*, and *split* genes, among others. This region's deficiency in one of the two homologous X chromosomes has two distinct effects. First, it results in the *Notch* phenotype. Second, by deleting the loci for genes whose mutant alleles are present on the other X chromosome, the deficiency creates a partially hemizygous condition whereby the recessive *white, facet*, or *split* alleles are expressed

TABLE 13–2 Notch Genotypes and Phenotypes

Genotype	*Phenotype*
$\dfrac{N^+}{N}$	*Notch* female
$\dfrac{N}{N}$	Lethal
$\dfrac{N}{\rightharpoondown}$	Lethal
$\dfrac{N^+w}{N(w^+)^*}$	*Notch, white* female
$\dfrac{N^+fa\,spl}{N(fa^+spl^+)^*}$	*Notch, facet, split* female

*(deleted)

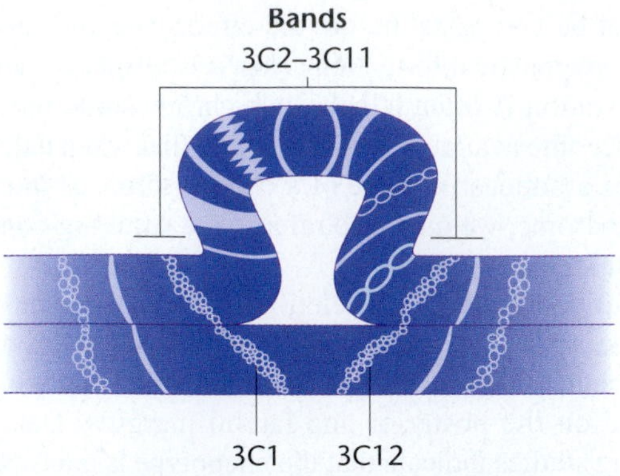

FIGURE 13–16 Deficiency loop formed in salivary chromosomes of *Drosophila melanogaster* where the fly is heterozygous for a deletion. The deletion encompasses bands 3C2 through 3C11, corresponding to the region associated with the Notch phenotype.

whenever present. This type of phenotypic expression of recessive genes in association with a deletion is an example of the phenomenon called **pseudodominance**.

Many independently arising *Notch* phenotypes have been investigated. The common deficient band for all *Notch* phenotypes has now been designated as 3C7. In every case that *white* was also expressed pseudodominantly, the band 3C2 was also missing. The bands that cytologically distinguish the *Notch* locus from the *white* locus have been confirmed in this manner.

The *Notch* Gene Product

Aside from the above discussion concerning the impact when the *Notch* gene is deleted, it is of interest to note that the wild type form of the gene has also been nvestigated. The gene present in *Drosophila* is widespread evolutionarily, with homologous counterparts being found in most multicellular organisms. It encodes a large protein that serves as a cell surface receptor that is important during morphogenesis and cell fate determination.

For example, in *Drosophila*, the protein plays a role in regulating the fate of cells whose potential is to become either epidermal or neural structures. The *Drosophila* receptor contains an amino acid domain that is conserved and repeated in the *lin-12* transmembrane receptor protein of *C. elegans* and the peptide hormone called the epidermal growth factor (EPG) in mammals. *Notch* receptors are activated by a specific type of transmembrane ligand that signals between adjacent cells. Signaling causes proteolytic cleavage of the intracellular domain of the receptor and its subsequent movement to the nucleus. There, it interacts with transcription factors, and presumably, this alters cell specific gene expression, influencing morphogenesis in cells.

13.8 A Duplication Is a Repeated Segment of the Genetic Material

When any part of the genetic material—a single locus or a large piece of a chromosome—is present more than once in the genome, it is called a **duplication**. As in deletions, pairing in heterozygotes may produce a compensation loop. Duplications can arise as the result of unequal crossing over between synapsed chromosomes during meiosis (Figure 13–17) or through a replication error prior to meiosis. In the former case, both a duplication and a deficiency are produced.

We consider next three interesting aspects of duplications. First, they may result in gene redundancy. Second, as with deletions, duplications can produce phenotypic variation. Third, according to one convincing theory, duplications have also been an important source of genetic variability during evolution.

Gene Redundancy and Amplification: Ribosomal RNA Genes

Although many gene products are not needed in every cell of an organism, other gene products are known to be essential components of all cells. For example, ribosomal RNA must be present in abundance to support protein synthesis. The more metabolically active a cell is, the higher the demand for this molecule, which becomes a part of each ribosome. We might hypothesize that a single copy of a gene encoding rRNA is inadequate in many cells. Studies using the technique of molecular hybridization, which allows the determination of the percentage of the genome coding for specific RNA sequences, show that our hypothesis is correct. Indeed, it is the norm for organisms to have multiple copies of genes coding for rRNA. Collectively such DNA is called **rDNA**, and the general phenomenon is called gene redundancy. For example, in the common intestinal bacterium *Escherichia coli* (*E. coli*) about 0.4 percent of the haploid genome consists of rDNA. This is equivalent to 5–10 copies of the gene. In *Drosophila melanogaster*, 0.3 percent of the haploid genome, equivalent to 130 copies, consists of rDNA. Although the presence of multiple copies of the same gene is not restricted to those coding for rRNA, we focus on them in this section.

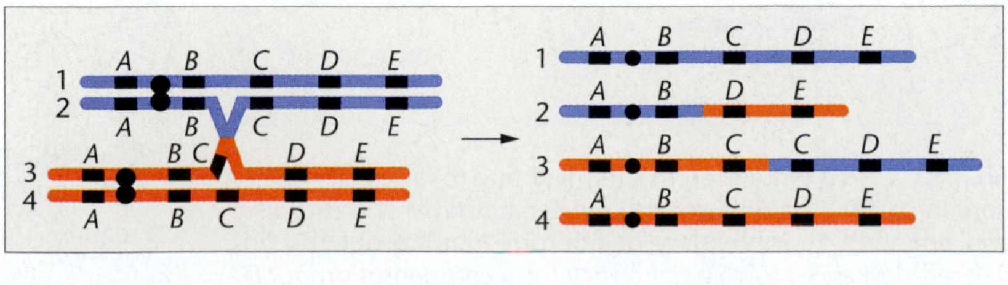

FIGURE 13–17 The origin of duplicated and deficient regions of chromosomes as a result of unequal crossing over. The tetrad at the left is mispaired during synapsis. A single crossover between chromatids 2 and 3 results in deficient and duplicated chromosomal regions. (See chromosomes 2 and 3, respectively, on the right.) The two chromosomes uninvolved in the crossover event remain normal in their gene sequence and content.

Studies of *Drosophila* have documented the critical importance of the extensive amounts of rRNA and ribosomes that is made possible by multiple copies of these genes. In *Drosophila*, the X-linked mutation *bobbed* has, in fact, been shown to be due to a deletion of a variable number of the 130 genes coding for rRNA. When the number of these genes is reduced and inadequate rRNA is produced, the result is mutant flies that have low viability, are underdeveloped, and have bristles reduced in size. Both general development and bristle formation, which occur very rapidly during normal pupal development, apparently depend on a great number of ribosomes to support protein synthesis. Many *bobbed* alleles have been studied, and each has been shown to involve a deletion, often of a unique size. The reduction of both viability and bristle length correlates well with the relative number of rRNA genes deleted. We may conclude that the normal rDNA redundancy observed in wild-type *Drosophila* is near the minimum required for adequate ribosome production during normal development.

In some cells, particularly oocytes, even the normal redundancy of rDNA may be insufficient to provide adequate amounts of rRNA and ribosomes. Oocytes store abundant nutrients in the ooplasm for use by the embryo during early development. In fact, more ribosomes are included in the oocytes than in any other cell type. By considering how the amphibian *Xenopus laevis* (the South African clawed frog) acquires this abundance of ribosomes, we will see a second way in which the amount of rRNA is increased. This phenomenon is referred to as **gene amplification**.

The genes that code for rRNA are located in an area of the chromosome known as the **nucleolar organizer region (NOR)**. The NOR is intimately associated with the nucleolus, which is a processing center for ribosome production. Molecular hybridization analysis has shown that each NOR in *Xenopus* contains the equivalent of 400 redundant gene copies coding for rRNA. Even this number of genes is apparently inadequate to synthesize the vast amount of ribosomes that must accumulate in the amphibian oocyte to support development following fertilization. To further amplify the number of rRNA genes, the rDNA is selectively replicated, and each new set of genes is released from its template. Because each new copy is equivalent to an NOR, multiple small nucleoli are formed around each NOR in the oocyte. As many as 1500 of these "micronucleoli" have been observed in a single oocyte. If we multiply the number of micronucleoli (1500) by the number of gene copies in each NOR (400), we see that amplification in *Xenopus* oocytes can result in over half a million gene copies. If each copy is transcribed only 20 times during the maturation of a single oocyte, in theory, sufficient copies of rRNA will be produced to result in well over one million ribosomes.

The *Bar* Eye Mutation in *Drosophila*

Duplications can cause phenotypic variations that might at first appear to be caused by a simple gene mutation. The *Bar* eye phenotype (Figure 13–18) in *Drosophila* is a classic example. Instead of the normal oval eye shape, *Bar*-eyed flies have narrow, slitlike eyes. This phenotype appears to be inherited as a dominant X-linked mutation. However, because both heterozygous females and hemizygous males exhibit the trait, but homozygous females show a more pronounced phenotype than either of these two cases, the inheritance is more accurately illustrates the phenomenon of **semidominance**.

In the early 1920s, Alfred H. Sturtevant and Thomas H. Morgan discovered and investigated this "mutation." As illustrated in Figure 13–18(a), normal wild-type females (B^+/B^+) have about 800 facets in each eye. Heterozygous females (B^+/B) have about 350 facets, while homozygous females (*B/B*) average only about 70 facets. Females are occasionally recovered with even fewer facets and are designated as *double Bar* (B^D/B^+).

Some 14 years later, Calvin Bridges and Herman J. Muller compared the polytene X chromosome-banding pattern of the *Bar* fly with that of the wild-type fly. Recall from our discussion of deletions that such chromosomes contain specific banding patterns which characterize specific chromosomal regions (see Figure 13–16). The Bridges–Muller studies revealed that one copy of region 16A of the X chromosome was present in wild-type flies and that this region was duplicated in *Bar* flies and triplicated in *double-Bar* flies. These observations provided evidence that the *Bar* phenotype is not the result of a simple chemical change in the gene, but is instead a duplication. The *double-Bar* condition originates as a result of unequal crossing over, which produces the triplicated 16A region [Figure 13–18(b)].

Figure 13–18 also illustrates what is referred to as a **position effect**. You may recall from our introduction of this term in Chapter 10 that a position effect refers to altered gene expression resulting from new "positioning" of a gene within the genome. In this case, when the eye facet phenotypes of *B/B* and B^D/B^+ flies are compared, an average of 68 and 45 facets are found, respectively. In both cases, there are two extra 16A regions. However, when the repeated segments are distributed on the same homolog instead of being positioned on two homologs, the phenotype is more pronounced. Thus, the same amount of genetic information produces an altered *phenotype*, depending on the *position* of the genes.

The Role of Gene Duplication in Evolution

During the study of evolution, it is intriguing to speculate on the possible mechanisms of genetic variation. The origin of unique gene products present in phylogenetically advanced organisms, but absent in less advanced, ancestral forms is a topic of particular interest. In other words, how do "new" genes arise?

In 1970, Susumo Ohno published the provocative monograph *Evolution by Gene Duplication* in which he elaborated on the importance of gene duplication to the origin of new genes during evolution. While he was not the first to suggest that duplications might provide a "reservoir" from which new genes might arise, Ohno provided a detailed account of this idea. Ohno's thesis is based on the supposition that the gene products of essential genes, present as only a single copy in the genome, are indispensable to the survival of members of any species during evolution. Therefore, unique

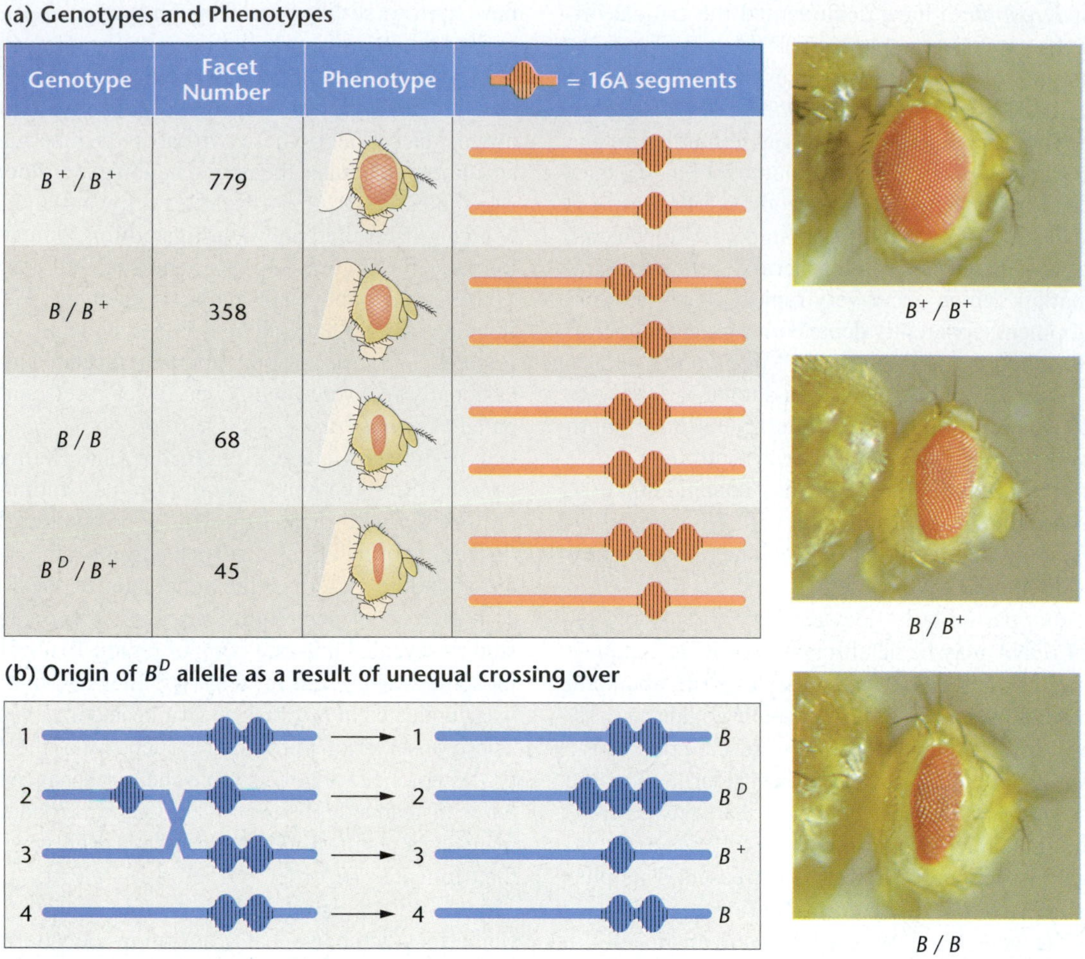

FIGURE 13–18 (a) The duplication genotypes and resultant *Bar* eye phenotypes in *Drosophila*. Photographs show two *Bar* eye phenotypes and the wild type (B^+/B^+). (b) The origin of the B^D (double-Bar) allele.

genes are not free to accumulate mutations that alter their primary function and potentially give rise to new genes.

However, if an essential gene were to become duplicated in the germ line, major mutational changes in this extra copy would be tolerated in future generations because the original gene provides the genetic information for its essential function. The duplicated copy would be free to acquire many mutational changes over extended periods of time. Over short intervals, the new genetic information may be of no practical advantage. However, over long evolutionary periods, the duplicated gene may change sufficiently so that its product assumes a divergent role in the cell. The new function may impart an "adaptive" advantage to organisms, enhancing their fitness. Ohno has outlined a mechanism through which sustained genetic variability may have originated.

Ohno's thesis is supported by the discovery of genes that have a substantial amount of their DNA sequence in common, but whose gene products are distinct. For example, the vertebrate digestive enzymes, trypsin and chymotrypsin, fit this description, as do the respiratory proteins, myoglobin and hemoglobin. The DNA sequence homology is great enough in these cases to conclude that members of each gene pair arose from a common ancestral gene through duplication. During evolution, the related genes diverged sufficiently so that their products became unique.

Other support includes the presence of **gene families**, regional groups of genes whose products perform the same general function. Again, members of a family show DNA sequence homology sufficient to conclude that they share a common origin. The various types of globin chains that are part of hemoglobin is one example. As we saw in Chapter 6, various globin chains function at different times during development, but they are all a part of hemoglobin, functioning to transport oxygen. Other examples include the immunologically important T-cell receptors and antigens encoded by the major histocompatibility complex (MHC).

Many recent findings derived from our ability to sequence entire genomes have continued to support the idea that gene duplication has been a common feature of evolutionary progression. For example, Jurg Spring has compared a large number of genes in *Drosophila* and their counterparts in humans. In 50 genes studied, the fruit fly has one copy, while there are multiple copies present in the human genome. There are many other investigations that have provided similar findings.

An interesting debate is now occurring concerning a second aspect of Ohno's thesis—that major evolutionary jumps, such as the transition from invertebrates to vertebrates, may have involved the duplication of entire genomes. Ohno suggested that this might have occurred several times during the course of evolution. While it seems clear that genes, and even segments of chromosomes, have been duplicated, there is not yet any compelling evidence that has convinced evolutionary biologists that this was clearly the case. However, from the standpoint of gene expression, duplicating all genes proportionately seems more likely to be tolerated than duplicating just a portion of them. Whatever may be the case, it is interesting to examine evidence based on new technology that tests an hypothesis that was put forward over 40 years ago, long before that technology was available.

13.9 Inversions Rearrange the Linear Gene Sequence

The **inversion**, another class of structural variation, is a type of chromosomal aberration in which a segment of a chromosome is turned around 180° within a chromosome.

An inversion does not involve a loss of genetic information, but simply rearranges the linear gene sequence. An inversion requires two breaks along the length of the chromosome and subsequent reinsertion of the inverted segment. Figure 13–19 illustrates how an inversion might arise. By forming a chromosomal loop prior to breakage, the newly created "sticky ends" are brought close together and rejoined. The inverted segment may be short or quite long and may or may not include the centromere. If the centromere *is* included in the inverted segment, (as it is in Figure 13–19), the term **pericentric** is used to describe the inversion.

If the centromere is *not* part of the rearranged chromosome segment, the inversion is said to be **paracentric**. Although the gene sequence has been reversed in the paracentric inversion, the ratio of arm lengths extending from the centromere is unchanged (Figure 13–20). In contrast, some pericentric inversions create chromosomes with arms of different lengths from those of the noninverted chromosome, thereby changing the arm ratio. The change in arm lengths may sometimes be detected during the metaphase stage of mitotic or meiotic divisions.

Although inversions may seem to have a minimal impact on individuals bearing them, their consequences are of great interest to geneticists. Organisms that are heterozygous for inversions may produce aberrant gametes that have a major impact on their offspring. Inversions may also result in position effects and might play an important role in the evolutionary process.

Consequences of Inversions during Gamete Formation

If only one member of a homologous pair of chromosomes has an inverted segment, normal linear synapsis during meiosis is not possible. Organisms with one inverted chromosome and one noninverted homolog are called **inversion heterozygotes**. Pairing between two such chromosomes in meiosis may be accomplished only if they form an **inversion loop** (Figure 13–21). In other cases, if no loop is formed, the homologs are seen to synapse everywhere, but along the length of the inversion, where they appear separated.

If crossing over does not occur within the inverted segment of the inversion heterozygote, the homologs will segregate and result in two normal and two inverted chromatids that are distributed into gametes, and the inversion will be passed on to one-half of the offspring. However, if crossing over occurs within the inversion loop, abnormal chromatids are produced. For example, the effect of a single exchange event within a paracentric inversion is diagrammed in Figure 13–22(a). As in any meiotic tetrad, a single crossover between nonsister chromatids produces two parental chromatids and two recombinant chromatids. As shown, one recombinant chromatid is **dicentric** (two centromeres), and one recombinant chromatid is **acentric** (lacking a centromere). Both contain duplications and deletions of chromosome segments, as well. During anaphase, an acentric chromatid moves randomly to one pole or the other, or it may be lost, whereas a dicentric chromatid is pulled in two directions. This polarized movement produces a **dicentric bridge** that is cytologically recognizable. A dicentric chromatid will usually break at some point so that part of the chromatid goes into one gamete and part into another gamete during the reduction divisions. Therefore, gametes containing either broken chromatid are deficient in genetic material.

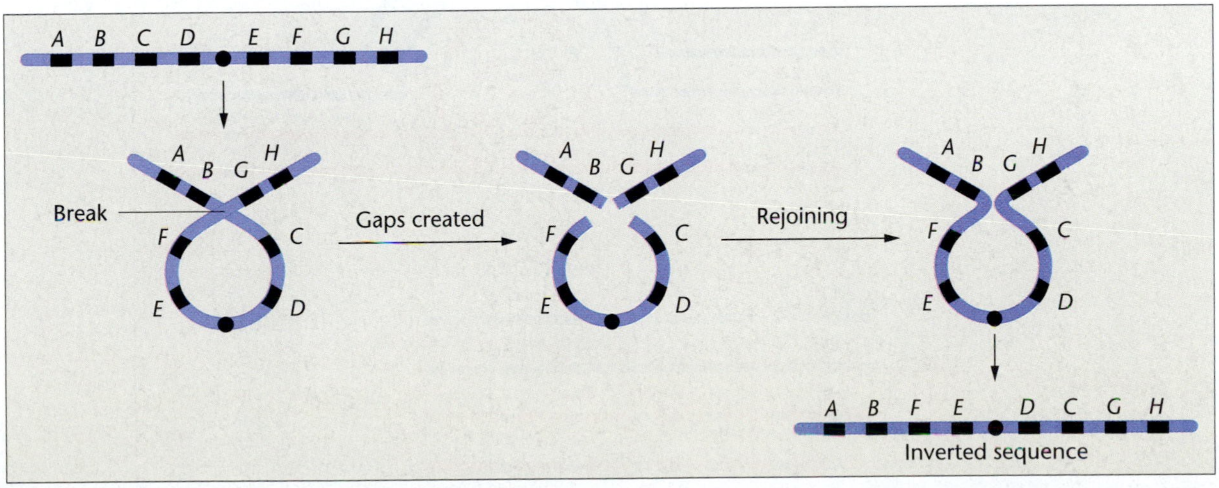

FIGURE 13–19 One possible origin of a pericentric inversion.

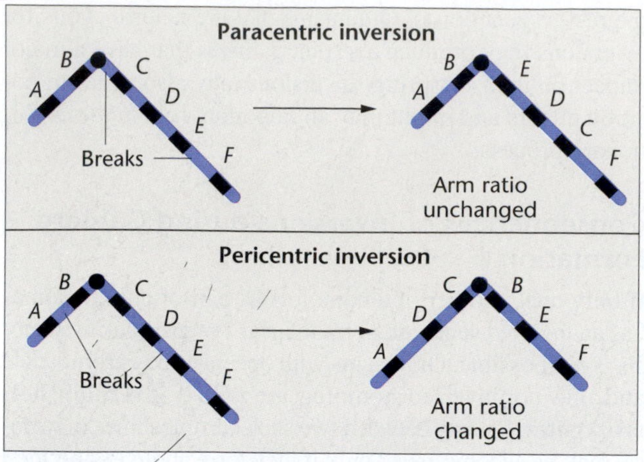

FIGURE 13–20 A comparison of the arm ratios of a submetacentric chromosome before and after the occurrence of a paracentric and pericentric inversion. Only the pericentric inversion results in an alteration of the original ratio.

A similar chromosomal imbalance is produced as a result of a crossover event between a chromatid bearing a pericentric inversion and its noninverted homolog, as illustrated in Figure 13–22(b). The recombinant chromatids that are directly involved in the exchange have duplications and deletions. However, no acentric or dicentric chromatids are produced.

In plants, gametes receiving such aberrant chromatids usually fail to develop normally, leading to aborted pollen or ovules. Thus, lethality occurs prior to fertilization, and inviable seeds result. In animals, the gametes have developed prior to the meiotic error, so fertilization is more likely to occur in spite of the chromosome error. However, the end result is the production of inviable embryos following fertilization. In both cases, viability is reduced.

Because viable offspring do not result in either plants or animals, it appears as if the inversion suppresses crossing over, since offspring bearing crossover gametes are not recovered. *Actually*, in inversion heterozygotes, the inversion has the effect of *suppressing the recovery of crossover products* when chromosome exchange occurs within the inverted region. If crossing over always occurred within a paracentric or pericentric inversion, 50 percent of the gametes would be ineffective. The viability of the resulting zygotes is therefore greatly diminished. Furthermore, up to one-half of the viable gametes have the inverted chromosome, and the inversion will be perpetuated within the species. Moreover, the cycle will be repeated continuously during meiosis in future generations.

Position Effects of Inversions

Another consequence of inversions involves the new positioning of genes relative to other genes and particularly

FIGURE 13–21 Illustration of how synapsis occurs in a paracentric inversion heterozygote by virtue of the formation of an inversion loop.

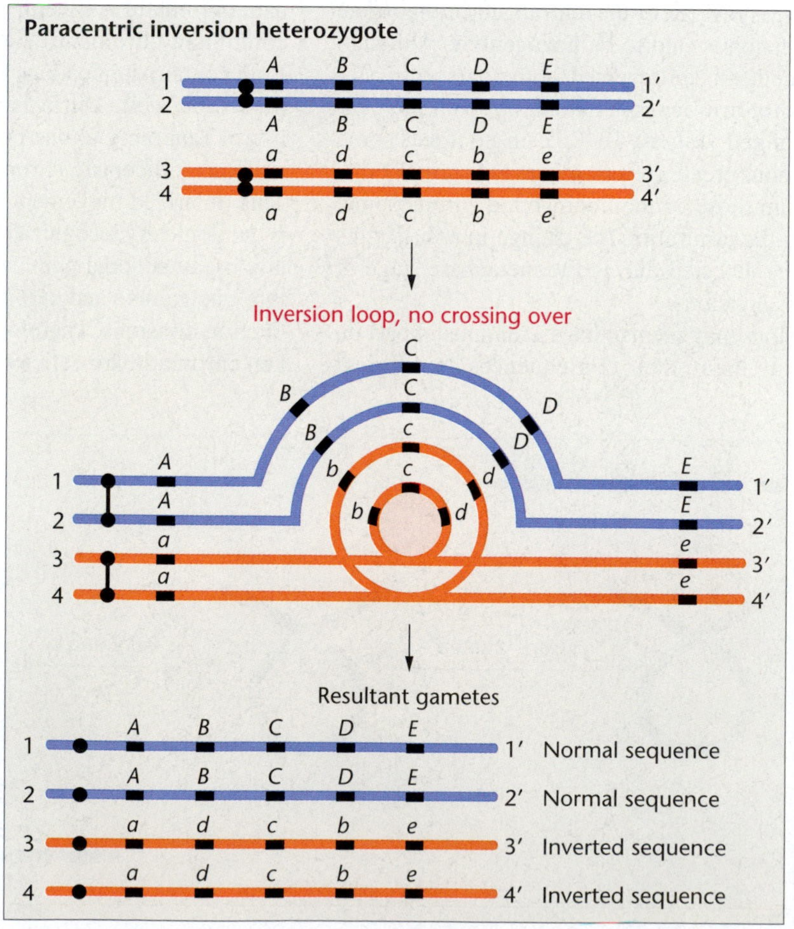

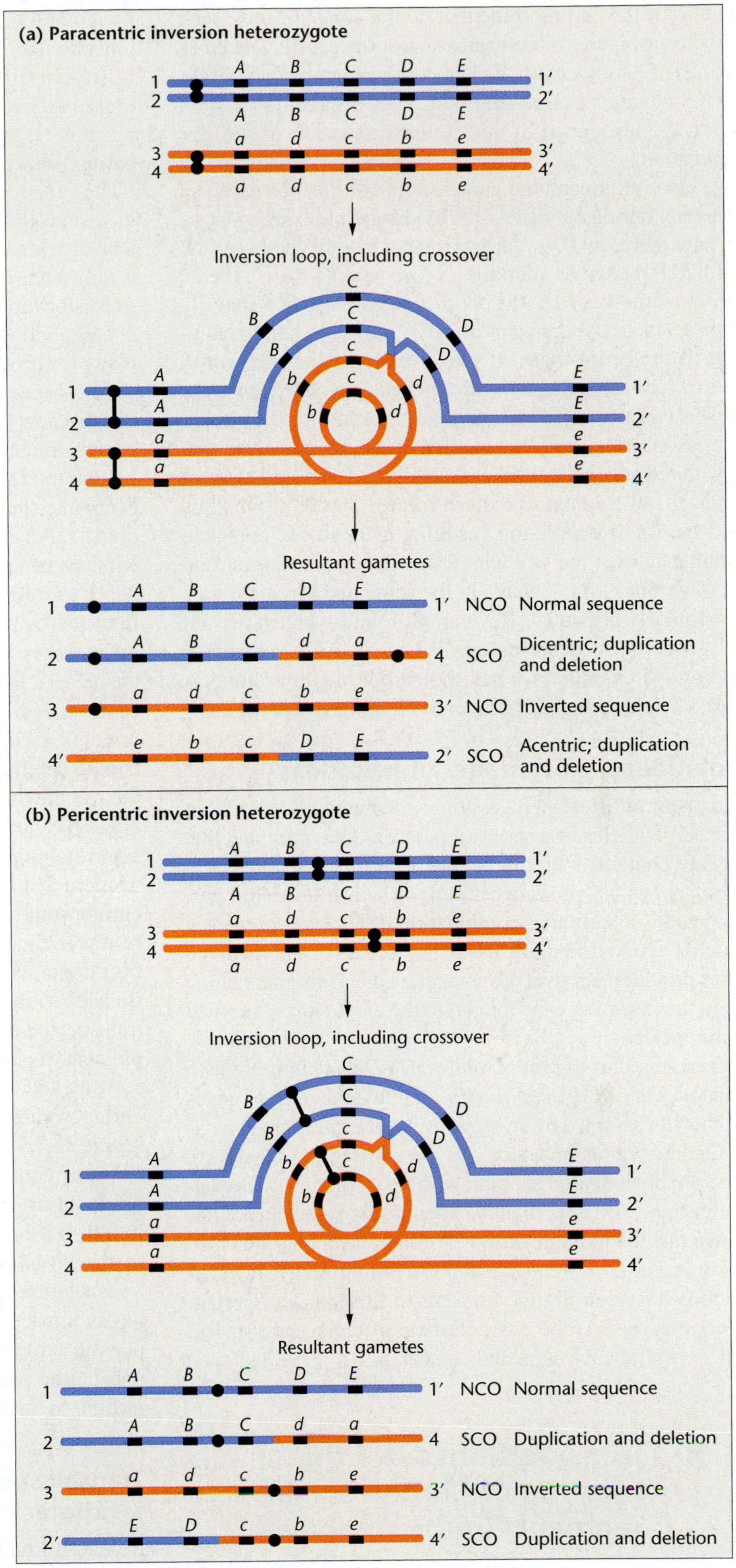

FIGURE 13–22 The effects of a single crossover within an inversion loop in cases involving (a) a paracentric inversion and (b) a pericentric inversion. In (a), two altered chromosomes are produced, one that is acentric and one that is dicentric. Both chromosomes also contain duplicated and deficient regions. In (b), two altered chromosomes are produced, both with duplicated and deficient regions. Ultimately, the number of viable gametes resulting from these events is reduced by about 50 percent since the abnormal outcomes are likely to be lethal.

to areas of the chromosome that do not contain genes, such as the centromere. If the expression of the gene is altered as a result of its relocation, a change in phenotype may result. Such a change is an example of what is called a position effect, as introduced in our earlier discussion of the *Bar* duplication.

We also introduced the general topic of position effect in Chapter 4 during our discussion of phenotypic expression. As we saw there, in *Drosophila* females heterozygous for the X-linked recessive mutation *white* eye (w^+/w), the X chromosome bearing the wild-type allele (w^+) may be inverted such that the *white* locus is moved to a point adjacent to the centromere. If the inversion is not present, a heterozygous female has wild-type red eyes, since the *white* allele is recessive. Females with the X chromosome inversion just described have eyes that are mottled or variegated, that is, with red and white patches (see Figure 10–18). Placement of the w^+ allele next to the centromere apparently inhibits wild-type gene expression, resulting in the loss of complete dominance over the *w* allele. Other genes, also located on the X chromosome, behave in the same manner when they are similarly relocated. Reversion to wild-type expression has sometimes been noted. When this has occurred, cytological examination has shown that the inversion has reestablished the normal sequence along the chromosome.

Evolutionary Advantages of Inversions

One major effect of an inversion is the maintenance of a set of specific alleles at a series of adjacent loci, provided that they are contained within the inversion. Because the recovery of crossover products is suppressed in inversion heterozygotes, a particular combination of alleles is preserved intact in the viable gametes. If the alleles of the involved genes provide a survival advantage to organisms maintaining them, the inversion is beneficial to the evolutionary survival of the species.

For example, if the set of alleles *ABcDef* is more adaptive than the sets *AbCdeF* or *abcdEF*, the favorable set of genes will not be disrupted by crossing over if it is maintained within a heterozygous inversion. As we will see in Chapter 26, there are documented examples where in versions are adaptive in this way. Specifically, Theodosius Dobzhansky has shown that the maintenance of different inversions on chromosome 3 of *Drosophila pseudoobscura* through many generations has been highly adaptive to this species. Certain inversions seem to be characteristic of enhanced survival under specific environmental conditions.

13.10 Translocations Alter the Location of Chromosomal Segments in the Genome

Translocation, as the name implies, is the movement of a chromosomal segment to a new location in the genome. Reciprocal translocation, for example, involves the exchange of segments between two nonhomologous chromosomes. The least complex way for this event to occur is for two nonhomologous chromosome arms to come close to each other so that an exchange is facilitated. Figure 13–23(a) illustrates a simple reciprocal translocation in which only two breaks are required. If the exchange includes internal chromosome segments, four breaks are required, two on each chromosome.

The genetic consequences of reciprocal translocations are, in several instances, similar to those of inversions. For example, genetic information is not lost or gained. Rather, there is only a rearrangement of genetic material. The presence of a translocation does not, therefore, directly alter the viability of individuals bearing it. Like an inversion, a translocation may also produce a position effect, because it may realign certain genes in relation to other genes. This exchange may create a new genetic linkage relationship that can be detected experimentally.

Homologs heterozygous for a reciprocal translocation undergo unorthodox synapsis during meiosis. As shown in Figure 13–23(b), pairing results in a crosslike configuration. As with inversions, genetically unbalanced gametes are also produced as a result of this unusual alignment during meiosis. In the case of translocations, however, aberrant gametes are not necessarily the result of crossing over. To see how unbalanced gametes are produced, focus on the homologous centromeres in Figure 13–23(b) and (c). According to the principle of independent assortment, the chromosome containing centromere 1 migrates randomly toward one pole of the spindle during the first meiotic anaphase; it travels with *either* the chromosome having centromere 3 *or* the chromosome having centromere 4. The chromosome with centromere 2 moves to the other pole, along with *either* the chromosome containing centromere 3 *or* centromere 4. This results in four potential meiotic products. The 1,4 combination contains chromosomes that are not involved in the translocation. The 2,3 combination, however, contains translocated chromosomes. These contain a complete complement of genetic information and are balanced. The other two potential products, the 1,3 and 2,4 combinations, contain chromosomes displaying duplicated and deleted (deficient) segments.

When incorporated into gametes, the resultant meiotic products are genetically unbalanced. If they participate in fertilization, lethality often results. As few as 50 percent of the progeny of parents heterozygous for a reciprocal translocation survive. This condition, called **semisterility**, has an impact on the reproductive fitness of organisms, thus playing a role in evolution. Furthermore, in humans, such an unbalanced condition results in partial monosomy or trisomy, leading to a variety of birth defects.

Translocations in Humans: Familial Down Syndrome

Research conducted since 1959 has revealed numerous translocations in members of the human population. One common type, called a **Robertsonian translocation**, or **centric fusion**, involves breaks at the extreme ends of the short arms of two nonhomologous acrocentric chromosomes (13, 14, 15, 21, and 22). The small acentric fragments are

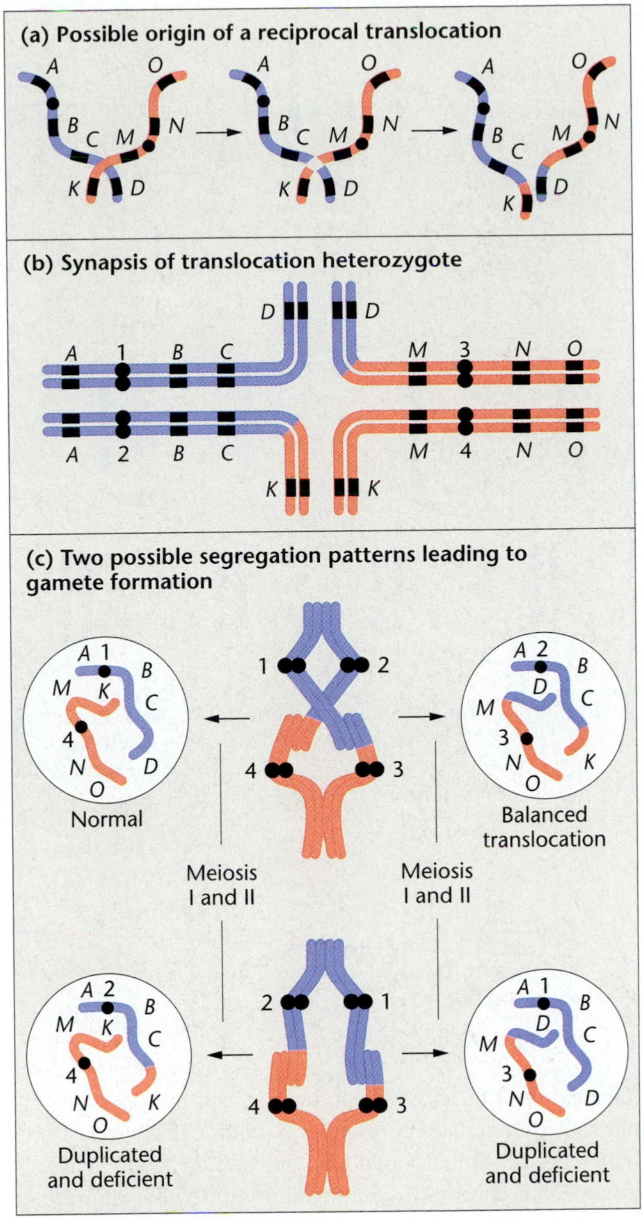

FIGURE 13–23 (a) The possible origin of a reciprocal translocation; (b) the synaptic configuration formed during meiosis in an individual that is heterozygous for the translocation. (c) Two possible segregation patterns, one of which leads to a normal and a balanced gamete (called alternate segregation) and one that leads to gametes containing duplications and deficiencies (called adjacent segregation).

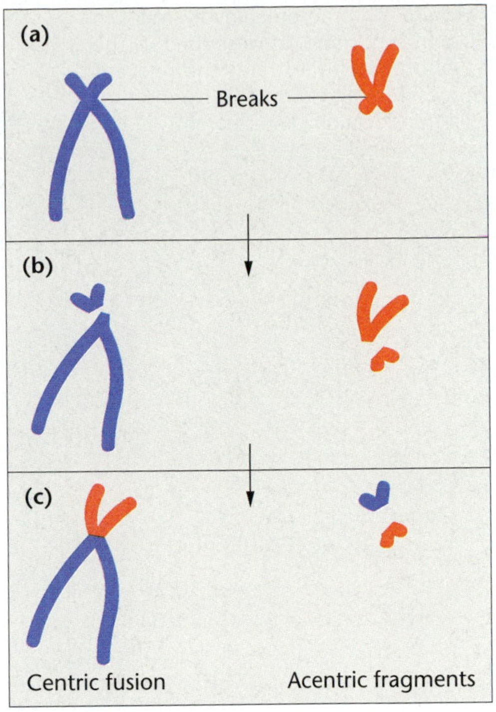

FIGURE 13–24 The possible origin of a Robertsonian translocation. Two independent breaks occur within the centromeric region on two nonhomologous chromosomes. Centric fusion of the long arms of the two acrocentric chromosomes creates the unique chromosome. Two acentric fragments remain.

lost, and the larger chromosomal segments fuse at their centromeric region, producing a new, larger submetacentric or metacentric chromosome (Figure 13–24). These are the most common types of chromosome rearrangement in humans.

One such translocation accounts for cases in which Down syndrome is inherited or familial. Earlier in this chapter, we pointed out that most instances of Down syndrome are due to trisomy 21, resulting from nondisjunction during meiosis in one parent. Trisomy accounts for over 95 percent of all cases of Down syndrome. In such instances, the chance of the same parents producing a second afflicted child is extremely low. However, in the remaining families stricken with a Down child, the syndrome occurs in a much higher frequency over several generations.

Cytogenetic studies of the parents and their offspring from these unusual cases explain the cause of familial Down syndrome. Analysis reveals that one of the parents contains a **14/21 D/G translocation** (Figure 13–25). That is, one parent has the majority of the G-group chromosome 21 translocated to one end of the D-group chromosome 14. This individual is phenotypically normal even though he or she has only 45 chromosomes.

During meiosis, one-fourth of the individual's gametes will have two copies of chromosome 21: a normal chromosome and most of a second copy translocated to chromosome 14. When such a gamete is fertilized by a standard haploid gamete, the resulting zygote has 46 chromosomes, but three copies of chromosome 21. These individuals exhibit Down syndrome. Other potential surviving offspring contain either the standard diploid genome (without a translocation) or the balanced translocation like the parent. Both cases result in phenotypically normal individuals. Knowledge of translocations has allowed geneticists to resolve the seeming paradox of an inherited trisomic phenotype in an individual with an apparent diploid number of chromosomes.

It is interesting to note that the "carrier," who has 45 chromosomes and exhibits a normal phenotype, does not contain the *complete* diploid amount of genetic material. A small region is lost from both chromosomes 14 and 21 during the translocation event. This occurs because the ends of both chromosomes have broken off prior to their fusion. These

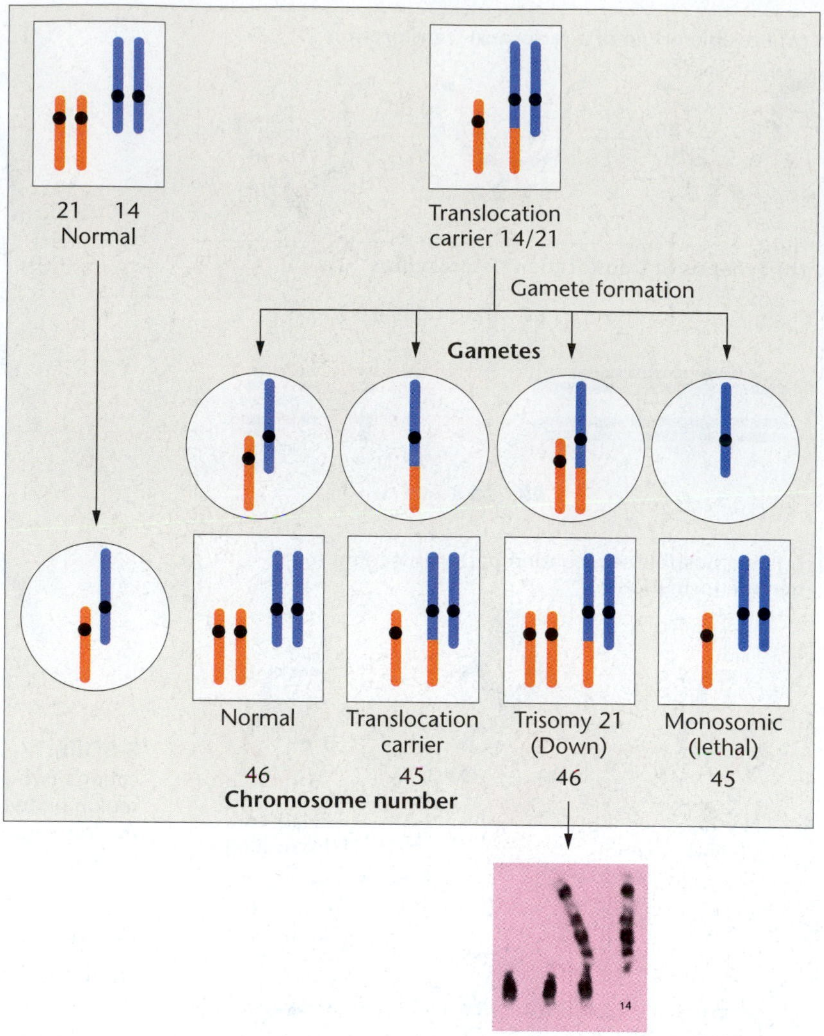

FIGURE 13–25 Chromosomal involvement in familial Down syndrome, as described in the text. The photograph illustrates the relevant chromosomes from a trisomy 21 offspring produced by a translocation carrier parent.

specific regions are known to be two of many chromosomal locations housing multiple copies of the genes encoding rRNA, the major component of ribosomes. Despite the loss of up to 20 percent of these genes, the carrier is unaffected.

13.11 Fragile Sites in Humans Are Susceptible to Chromosome Breakage

We conclude this chapter with a brief discussion of the results of an interesting discovery made around 1970, during observations of metaphase chromosomes prepared following human cell culture. In cells derived from certain individuals, a specific area along one of the chromosomes failed to stain, giving the appearance of a gap. In other individuals whose chromosomes displayed such morphology, the gaps appeared at other places within the chromosome set. Such areas eventually became known as **fragile sites**, since they appeared to be susceptible to chromosome breakage when cultured in the absence of certain chemicals such as folic acid, which is normally present in the culture medium. Fragile sites were at first considered curiosities, until a strong association was subsequently shown to exist between one of the sites and a form of mental retardation.

The cause of the fragility at these sites (and others) is unknown. Because they represent points along the chromosome that are susceptible to breakage, these sites may indicate regions where chromatin is not tightly coiled or compacted. Note that even though almost all studies of fragile sites have been carried out in vitro, using mitotically dividing cells, clear associations have been established between several of these sites and the corresponding altered phenotypes, including mental retardation and cancer.

Fragile X Syndrome (Martin–Bell Syndrome)

Most fragile sites do not appear to be associated with any clinical syndrome. However, individuals bearing a *folate-sensitive site* on the X chromosome (Figure 13–26) exhibit **fragile X syndrome** (or **Martin–Bell syndrome**), the most common form of inherited mental retardation. This syndrome affects about 1 in 4000 males and 1 in 8000 females. Because it is a dominant trait, females carrying only one fragile X chromosome can be mentally retarded. Fortunately, the trait is not fully penetrant, and many individuals with the genetic defect are not affected. In addition to mental deficiency, affected individuals often exhibit attention deficit disorder or autistic-like behavior. They also have characteristic long, narrow faces with protruding chins, enlarged ears, and males have increased testicular size.

FIGURE 13–26 A normal human X chromosome (left) contrasted with a fragile X chromosome (right). The "gap" region (near the bottom of the chromosome) is associated with the fragile X syndrome.

A gene that spans the fragile site is responsible for this syndrome. This gene, known as *FMR1*, is one of a growing number of genes that have been discovered in which a sequence of three nucleotides is repeated many times, expanding the size of the gene. This phenomenon, called **trinucleotide repeats**, has been recognized in other human disorders, including Huntington disease. In *FMR1*, the trinucleotide sequence CGG is repeated in a transcribed, but an untranslated area called the 5′ UTR—The "untranslated region" that is upstream from the coding sequence of the gene. The number of repeats varies immensely within the human population, and a high number correlates directly with expression of fragile X syndrome. Normal individuals have between 6 and 54 repeats, whereas those with 55–230 repeats are considered to be "carriers" of the disorder. A level above 230 repeats leads to expression of the syndrome.

It is thought that when the number of repeats reaches this level, the CGG regions of the gene become chemically modified so that the bases within and around the repeat are methylated, causing inactivation of the gene. The normal product of the gene (FMRP - fragile X mental retardation protein) is an RNA-binding protein that is known to be expressed in the brain. Evidence is now accumulating that directly links the absence of the protein with the cognitive defects associated with the syndrome.

The protein is prominent in cells of the developing brain and normally shuttles between the nucleus and cytoplasm, transporting mRNAs to ribosomal complexes. It is believed that its absence in dendritic cells prevents the translation of other critical proteins during the development of the brain, presumably those produced from mRNAs that are transported by FMRP. Ultimately, the result is impairment of synaptic activity related to learning and memory. Molecular analysis has shown that FMRP is very selective, binding specifically to mRNAs that have a high guanine content. These RNAs form unique secondary structures called **G-quartets**, for which the protein has a strong binding affinity. That possession of such quartets is important to recognition by FMRP is evident by examining cases where the formation of these stacked quartets is inhibited (using Li^+). In such cases, FMRP binding to these RNAs is abolished.

The *Drosophila* homolog (*dfrx*) of the human *FRM1* gene has now been identified and used as a model for experimental investigation. The normal protein product of the fruit fly displays chemical properties that parallel its human counterpart. Mutations have been generated and studied that either overexpress the gene or eliminate its expression altogether. The latter case is designated a null mutation, where the gene has been "knocked out" using a molecular procedure now common in DNA biotechnology. The results of these studies show that the normal gene functions during the formation of synapses and that abnormal synaptic function accompanies the loss of *dfrx* expression. Knockout mice lacking this gene also display a similar phenotype. Ultimately, then, a closely related version of this gene is conserved in humans, *Drosophila*, and mice. In all three species, it has been shown to be essential to synaptic maturation and the subsequent establishment of proper cell communication in the brain. In humans, the neurological features of the syndrome are linked to the absence of its genetic expression.

From a genetic standpoint, one of the most interesting aspects of fragile X syndrome is the instability of the CGG repeats. An individual with 6–54 repeats transmits a gene containing the same number to his or her offspring. However, those with 55–230 repeats, while not at risk to develop the syndrome, may transmit to their offspring a gene with an increased number of repeats. The number of repeats continues to increase in future generations, illustrating the phenomenon known as **genetic anticipation**, first introduced in Chapter 4. Once the threshold of 200 is exceeded, expression of the malady becomes more severe in each successive generation as the number of trinucleotide repeats increases.

While the mechanism that leads to the trinucleotide expansion has not yet been established, several factors that influence the instability are known. Most significant is the observation that expansion from the carrier status (55–200 repeats) to the syndrome status (over 200 repeats) occurs during the transmission of the gene by the maternal parent, but not by the paternal parent. Furthermore, several reports suggest that male offspring are more likely to receive the increased repeat size leading to the syndrome than are female offspring. Obviously, we have more to learn about the genetic basis of instability and expansion of DNA sequences.

Fragile Sites and Cancer

A second link between a fragile site and a human disorder was reported in 1996 by Carlo Croce, Kay Huebner, and their colleagues, who demonstrated an association between an autosomal fragile site and cancer. They showed that the gene *FHIT* (standing for *f*ragile *hi*stidine *t*riad), located within a well-defined fragile site on chromosome 3, is often altered or missing in cells taken from tumors of individuals with lung cancer. A variety of mutations were found in cells derived from the tumors where the DNA had apparently been broken and incorrectly fused back together, resulting in deletions within the gene. In most cases, these mutations caused the *FHIT* gene to become inactivated.

This gene is part of the fragile region of the autosome designated *FRA3B*, which has also been linked to other cancers, including the esophagus, breast, cervix, kidney, colon, and stomach. The nature of the genetic alterations found in cancer cells suggests that the *FHIT* gene, because it is within a fragile region, may be highly susceptible to induced breaks in DNA. If these breaks are incorrectly repaired, cancer-specific chromosome alterations may occur. Thus, this region of the chromosome appears to be particularly sensitive to carcinogen-induced damage, creating a susceptibility to cancer.

Still in question, however, is whether the alteration in the fragile site, leading to inactivation of what appears to be a tumor-suppressor gene, causes cancer, *or* whether the behavior of malignant cells somehow induces breaks at fragile sites, subsequently inactivating this gene. Another important question under investigation is whether molecular polymorphism at the fragile site exists within the human population, causing some individuals to be more susceptible to the effects of carcinogens than others. Whatever the answers, these are significant and exciting studies.

Chapter Summary

1. Investigations into the uniqueness of each organism's chromosomal constitution have further enhanced our understanding of genetic variation. Alterations of the precise diploid content of chromosomes are referred to as chromosomal aberrations or chromosomal mutations.
2. Deviations from the expected chromosomal number, or mutations in the structure of the chromosome, are inherited in predictable Mendelian fashion. They often result in inviable organisms or substantial changes in the phenotype.
3. Aneuploidy is the gain or loss of one or more chromosomes from the diploid content, resulting in conditions of monosomy, trisomy, tetrasomy, etc. Studies of monosomic and trisomic disorders are increasing our understanding of the delicate genetic balance that enables normal development.
4. When complete sets of chromosomes are added to the diploid genome, polyploidy occurs. These sets can have identical or diverse genetic origin, creating either autopolyploidy or allopolyploidy, respectively.
5. Large segments of the chromosome can be modified by deletions or duplications. Deletions can produce serious conditions such as the cri-du-chat syndrome in humans, whereas duplications can be particularly important as a source of redundant or new genes.
6. Inversions and translocations, while altering the gene order along chromosomes, initially cause little or no loss of genetic information or deleterious effects. However, heterozygous combinations may cause genetically abnormal gametes following meiosis, with death a common result.
7. Fragile sites in human mitotic chromosomes have sparked research interest because one such site on the X chromosome is associated with the most common form of inherited mental retardation. Another fragile site, located on chromosome 3, has been linked to lung cancer.

Insights and Solutions

1. In a cross, involving three linked genes, *a*, *b*, and *c* in maize, a heterozygote (*abc*/+++) was test crossed (to *abc*/*abc*). Even though the three genes were separated along the chromosome, hence predicting that crossover gametes and the resultant phenotype should be observed, only two phenotypes were recovered: *abc* and +++. Additionally, the cross produced significantly fewer viable seeds (and thus plants) than expected. Can you propose why no other phenotypes were recovered and why viability was reduced?

Solution: One of the two chromosomes contains an inversion that overlaps all three genes, effectively precluding the recovery of any "crossover" offspring. If this is a paracentric inversion and the genes are clearly separated (assuring that a significant number of crossovers will occur between them), then numerous acentric and dicentric chromosomes will be formed, resulting in the observed reduction in viability.

2. A male *Drosophila* from a wild-type stock was discovered to have only seven chromosomes, whereas the normal $2n$ number is eight. Close examination revealed that one member of chromosome IV (the smallest chromosome) was attached to (translocated to) the distal end of chromosome II and was missing its centromere, consequently, accounting for the reduction in chromosome number.

(a) Diagram all members of chromosomes II and IV during synapsis in Meiosis I.

Solution:

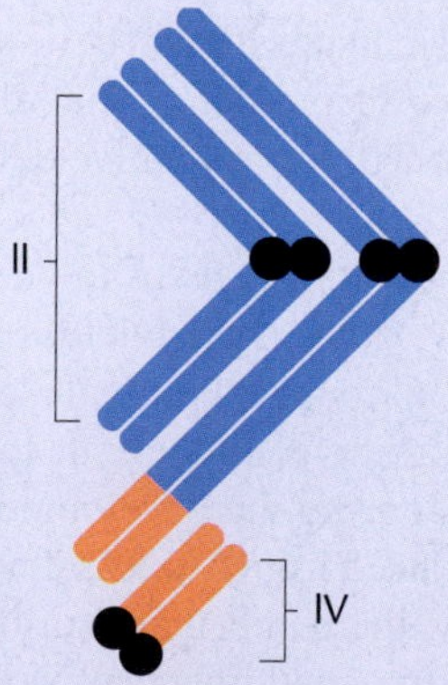

(b) If this male mates with a female with a normal chromosome composition who is homozygous for the recessive chromosome IV mutation *eyeless* (*ey*), what chromosome compositions will occur in the offspring regarding chromosomes II and IV?

Solution:

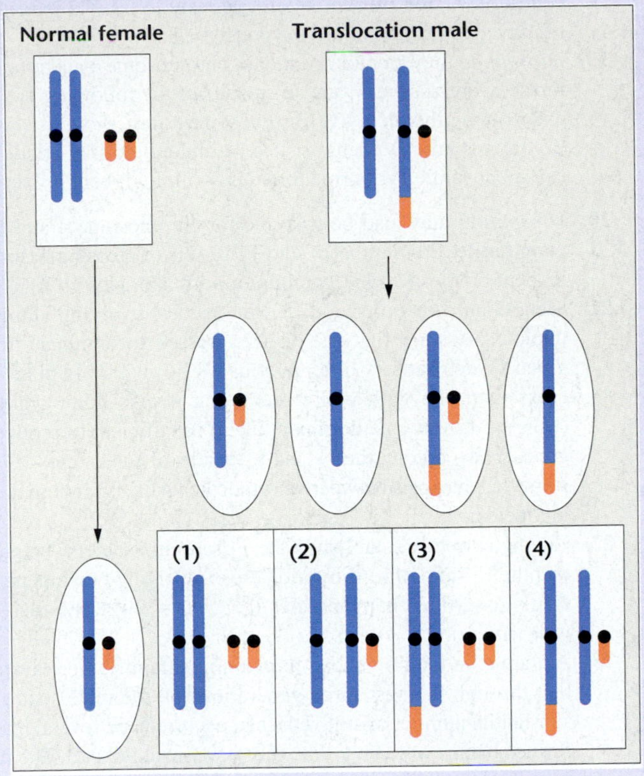

(c) What phenotypic ratio will result regarding the presence of eyes, assuming all abnormal chromosome compositions survive?

Solution:

(a) normal (heterozygous)

(b) eyeless (monosomic, contains chromosome IV from mother)

(c) normal (heterozygous)

(d) normal (heterozygous)

The final ratio is 3/4 normal: 1/4 eyeless.

3. If a Haplo-IV female *Drosophila* (containing only one chromosome 4, but an otherwise normal set of chromosomes) that has white eyes (an X-linked trait) and normal bristles is crossed with a male with a diploid set of chromosomes and normal red eye color, but who is homozygous for the recessive chromosome 4 bristle mutation, *shaven* (*sv*), what F_1 phenotypic ratio might be expected?

Solution: Let us first consider only the eye color phenotypes. This is a straightforward X-linked cross. Offspring will appear as 1/2 red females: 1/2 white males as shown here:

P_1: ww (white female) × w^+/Y (red male)

F_1: 1/2 ww^+ (red female) : 1/2 w/Y (white male)

The bristle phenotypes will be governed by the fact that the normal-bristled P_1 female produces gametes, one-half of which contain a chromosome 4(sv^+) and one-half that have no chromosome 4 ("−" designates no chromosome). Following fertilization by sperm from the *shaven* male, one-half of the offspring will receive two members of chromosome 4 and be heterozygous for *sv*, expressing normal bristles. The other half will have only one copy of chromosome 4. Because its origin is from the male parent, where the chromosome bears the *sv* allele, these flies will express *shaven* because there is no wild-type allele present to mask this recessive allele:

P_1: $sv^+/-$ (normal female) × sv/sv (shaven male)

F1 1/2sv^+/sv males and females = 1/2 wild type

1/2 $sv/-$ males and females = 1/2 shaven

Using the forked-line method, we can consider both eye color and bristle phenotypes together:

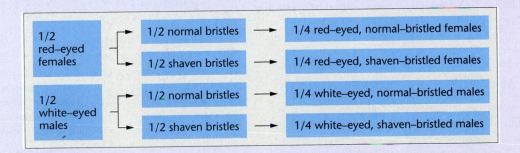

1/2 red–eyed females → 1/2 normal bristles → 1/4 red–eyed, normal–bristled females
1/2 red–eyed females → 1/2 shaven bristles → 1/4 red–eyed, shaven–bristled females
1/2 white–eyed males → 1/2 normal bristles → 1/4 white–eyed, normal–bristled males
1/2 white–eyed males → 1/2 shaven bristles → 1/4 white–eyed, shaven–bristled males

Problems and Discussion Questions

1. For a species with a diploid number of 18, indicate how many chromosomes will be present in the somatic nuclei of individuals who are haploid, triploid, tetraploid, trisomic, and monosomic.
2. Define and distinguish between the following pairs of terms:

   aneuploidy/euploidy
   monosomy/trisomy
   Patau syndrome/Edwards syndrome
   autopolyploidy/allopolyploidy
   autotetraploid/amphidiploid
   paracentric inversion/pericentric inversion

3. Contrast the relative survival times of individuals with Down syndrome, Patau syndrome, and Edwards syndrome. Speculate as to why such differences exist.
4. What evidence suggests that Down syndrome is more often the result of nondisjunction during oogenesis rather than during spermatogenesis?
5. What evidence exists that humans with aneuploid karyotypes occur at conception, but are usually inviable?
6. Contrast the fertility of an allotetraploid with an autotriploid and an autotetraploid.
7. When two plants belonging to the same genus, but different species were crossed together, the F_1 hybrid was more viable and had more ornate flowers. Unfortunately, this hybrid was sterile and could only be propagated by vegetative cuttings. Explain the sterility of the hybrid. How might a horticulturist attempt to reverse its sterility?
8. Describe the origin of cultivated American cotton.
9. Predict how the synaptic configurations of homologous pairs of chromosomes might appear when one member is normal and the other member has sustained a deletion or duplication.
10. Inversions are said to "suppress crossing over." Is this terminology technically correct? If not, restate the description accurately.
11. Contrast the genetic composition of gametes derived from tetrads of inversion heterozygotes where crossing over occurs within a paracentric and pericentric inversion.
12. Contrast the *Notch* locus with the *Bar* locus in *Drosophila*. What phenotypic ratios would be produced in a cross between *Notch* females and *Bar* males?
13. Discuss Ohno's hypothesis on the role of gene duplication in the process of evolution.
14. What roles have inversions and translocations played in the evolutionary process?
15. A human female with Turner syndrome also expresses the X-linked trait hemophilia, as did her father. Which of her parents underwent nondisjunction during meiosis, giving rise to the gamete responsible for the syndrome?
16. The primrose, *Primula kewensis*, has 36 chromosomes that are similar in appearance to the chromosomes in two related species, *Primula floribunda* ($2n = 18$) and *Primula verticillata* ($2n = 18$). How could *P. kewensis* arise from these species? How would you describe *P. kewensis* in genetic terms?
17. Varieties of chrysanthemums are known that contain 18, 36, 54, 72, and 90 chromosomes; all are multiples of a basic set of 9 chromosomes. How would you describe these varieties genetically? What feature is shared by the karyotypes of each variety? A variety with 27 chromosomes was discovered, but it was sterile. Why?
18. What is the effect of a rare double crossover within a pericentric inversion present heterozygously? Within a paracentric inversion present heterozygously?
19. *Drosophila* may be monosomic for chromosome 4 and remain fertile. Contrast the F_1 and F_2 results of the following crosses involving the recessive chromosome 4 trait, *bent bristles*:
   (a) monosomic IV, bent bristles × diploid, normal bristles
   (b) monosomic IV, normal bristles × diploid, bent bristles
20. *Drosophila* may also be trisomic for chromosome 4 and remain fertile. Predict the F_1 and F_2 results of crossing trisomic, bent bristles ($b/b/b$) × diploid, normal brisles (b^+/b^+)
21. Mendelian ratios are modified in crosses involving autotetraploids. Assume that one plant expresses the dominant trait green (seeds) and is homozygous (*WWWW*). This plant is crossed to one with white seeds that is also homozygous (*wwww*). If only one dominant allele is sufficient to produce green seeds, predict the F_1 and F_2 results of such a cross. Assume that synapsis between chromosome pairs is random during meiosis.
22. Having correctly established the F_2 ratio in Problem 21, predict the F_2 ratio of a "dihybrid" cross involving two independently assorting characteristics (e.g., P_1 = *WWWWAAAA* × *wwwwaaaa*).
23. A couple, in looking ahead to planning a family, were aware that through the past three generations on the male's side a substantial number of stillbirths had occurred and several malformed babies were born who died early in childhood. The female had studied genetics and urged her husband to visit a genetic counseling clinic, where a complete karyotype-banding analysis was subsequently performed. Although it was found that he had a normal complement of 46 chromosomes, banding analysis revealed that one member of the chromosome 1 pair (in Group A) contained an inversion covering 70 percent of its length. The homolog of chromosome 1 and all other chromosomes showed the normal banding sequence.
   (a) How would you explain the high incidence of past stillbirths?
   (b) What would you predict about the probability of abnormality/normality of their future children?
   (c) Would you advise the female that she would have to "wait out" each pregnancy to term so as to determine whether the fetus was normal? If not, what would you suggest she do?

Extra-Spicy Problems

24. In a cross in *Drosophila*, a female heterozygous for the autosomally linked genes *a*, *b*, *c*, *d*, and *e* (*abcde*/+++++) was testcrossed to a male homozygous for all recessive alleles. Even though the distance between each of the these loci was at least 3 map units, only four phenotypes were recovered, yielding the following data:

Phenotype	*No. of Flies*
+ + + + +	440
a b c d e	460
+ + + + *e*	48
a b c d +	52
	Total = 1000

Why are many expected crossover phenotypes missing? Can any of these loci be mapped from the data given here? If so, determine map distances.

25. A woman, seeking genetic counseling, was found to be heterozygous for a chromosomal rearrangement between the second and third chromosomes. Her chromosomes, compared with those in a normal karyotype, are diagrammed here:

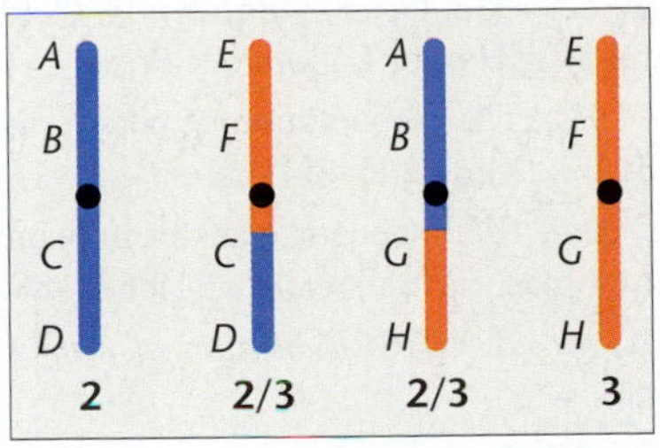

(a) What kind of chromosomal aberration is shown?
(b) Using a drawing, demonstrate how these chromosomes would pair during meiosis. Be sure to label the different segments of the chromosomes.
(c) This woman is phenotypically normal. Does that surprise you? Why or why not? Under what circumstances might you expect a phenotypic effect of such a rearrangement?
(d) This woman has had two miscarriages. She has come to you, an established genetic counselor, for advice. She raises the following questions:

Is there a genetic explanation of her frequent miscarriages? Should she abandon her attempts to have a child of her own? If not, what is the chance that she could have a normal child? Provide an informed response to each of her concerns.

26. A son with Klinefelter syndrome is born to a mother who is phenotypically normal and a father who has the X-linked skin condition called anhidrotic ectodermal dysplasia. The mother's skin is completely normal with no signs of the skin abnormality. In contrast, her son has patches of normal skin and patches of abnormal skin.
(a) Which parent contributed the abnormal gamete?
(b) Using the appropriate genetic terminology, describe the meiotic mistake that occurred. Be sure to indicate which division the mistake occurred in.
(c) Using the appropriate genetic terminology, explain the son's skin phenotype.

27. 200 human oocytes that had failed to be fertilized during in vitro fertilization procedures were subsequently examined in order to investigate the origin of nondisjunction (Angel, R. 1997. *Am. J. Hum. Genet.* 61:23–32.). These oocytes had completed meiosis I and were arrested in metaphase II (MII). The majority (67 percent) had a normal MII-metaphase complement, showing 23 chromosomes, each consisting of two sister chromatids joined at a common centromere. The remaining oocytes all had abnormal chromosome compositions. Surprisingly, when trisomy is considered, none of the abnormal oocytes had 24 chromosomes.
(a) Interpret these results in regard to the origin of trisomy, as it relates to nondisjunction and when it occurs. Why are the results surprising?
A large number of the abnormal oocytes contained 22 1/2 chromosomes, that is, 22 chromosomes plus a single chromatid representing the 1/2 chromosome.
(b) What chomosome compositions will result in the zygote if such oocytes proceed through meiosis and are fertilized by normal sperm?
(c) How could the complement of 22 1/2 chromosomes arise? Answer this question with a drawing that includes several pairs of MII chromosomes.
(d) Do these findings support or dispute the generally accepted theory regarding nondisjunction and trisomy, as outlined in Figure 8–1?

28. In a human genetic study, a family was investigated with 5 phenotypically normal children. Two were "homozygous" for a Robertsonian translocation between chromosome 19 and 20 (they contained two identical copies of the fused chromosome). These children have only 44 chromosomes, but a complete genetic complement. Three of the children were "heterozygous" for the translocation, and contained 45 chromosomes, with one translocated chromosome, plus a normal copy of both chromosome 19 and 20. Two other pregnancies resulted in stillbirths. It was discovered that the parents were first cousins. Based on the above information, determine the chromosome compositions of the parents. What led to the stillbirths? Why was the discovery that the parents were first cousins a key piece of information in understanding the genetics of this family.

Selected Readings

Antonarakis, S.E. 1998. Ten years of genomics, chromosome 21, and Down syndrome. *Genomics* 51:1–16.

Ashley-Koch, A.E., et al. 1997. Examination of factors associated with instability of the *FMR1* CGG repeat. *Am. J. Hum. Genet.* 63:776–85.

Beasley, J.O. 1942. Meiotic chromosome behavior in species, species hybrids, haploids, and induced polyploids of *Gossypium. Genetics.* 27:25–54.

Blakeslee, A.F. 1934. New jimson weeds from old chromosomes. *J. Hered.* 25:80–108.

Borgaonker, D.S. 1989. *Chromosome variation in man: A catalogue of chromosomal variants and anomalies*, 5th ed. New York: Alan R. Liss.

Boue, A. 1985. Cytogenetics of pregnancy wastage. *Adv. Hum. Genet.* 14:1–58.

Burgio, G.R., et al., eds. 1981. *Trisomy 21.* New York: Springer-Verlag.

Carr, D.H. 1971. Genetic basis of abortion. *Annu. Rev. Genet.* 5:65–80.

Croce, C.M., et al. 1996. The *FHIT* genKe at 3p14.2 is abnormal in lung cancer. *Cell.* 85:17–26.

Cummings, M.R. 2000. *Human heredity: Principles and issues.* 5th ed. Pacific Grove, CA: Brooks/Cole.

DeArce, M.A., and Kearns, A. 1984. The fragile X syndrome: The patients and their chromosomes. *J. Med. Genet.* 21:84–91.

Feldman, M., and Sears, E.R. 1981. The wild gene resources of wheat. *Sci. Am.* (Jan.) 244:102–12.

Galitski, T., et al. 1999. Ploidy regulation of gene expression. *Science* 285:251–54

Gersh, M., et al. 1995. Evidence for a distinct region causing a cat-like cry in patients with 5p deletions. *Am. J. Hum. Genet.* 56:1404–10.

Gupta, P.K., and Priyadarshan, P.M. 1982. *Triticale*: Present status and future prospects. *Adv. Genet.* 21:256–346.

Hassold, T.J., et al. 1980. Effect of maternal age on autosomal trisomies. *Ann. Hum. Genet.* (London). 44:29–36.

Hassold, T.J., and Hunt, P. 2001. To err (meiotically) is human:The genesis of human aneuploidy. *Nature Reviews Genetics* 2: 280–91.

Hassold, T.J., and Jacobs, P.A. 1984. Trisomy in man. *Annu. Rev. Genet.* 18:69–98.

Hecht, F. 1988. Enigmatic fragile sites on human chromosomes. *Trends Genet.* 4:121–22.

Henikoff, S. 1994. A reconsideration of the mechanism of position effect. *Genetics.* 138:1–5.

Hulse, J.H., and Spurgeon, D. 1974. Triticale. *Sci. Am.* (Aug.) 231:72–81.

Jacobs, P.A., et al. 1974. A cytogenic survey of 11,680 newborn infants. *Ann. Hum. Genet.* 37:359–76.

Kaiser, P. 1984. Pericentric inversions: Problems and significance for clinical genetics. *Hum. Genet.* 68:1–47.

Kaytor, M.D. and Orr, H. 2001. RNA targets of the fragile X protein. *Cell* 107:555–57.

Khush, G.S. 1973. *Cytogenetics of aneuploids.* Orlando, FL: Academic Press.

Khush, G.S., et al. 1984. Primary trisomics of rice: Origin, morphology, cytology and use in linkage mapping. *Genetics.* 107:141–63.

Lewis, E.B. 1950. The phenomenon of position effect. *Adv. Genet.* 3:73–115.

Lewis, W.H., ed. 1980. *Polyploidy: Biological relevance.* New York: Plenum Press.

Lupski, J.R., Roth, J.R., and Weinstock, G.M. 1996. Chromosomal duplications in bacteria, fruit flies, and humans. *Am. J. Hum. Genet.* 58: 21–26.

Lynch, M., and Conery, J.S. 2000. The evolutionary fate and consequences of duplicated genes. *Science* 290:1151–54.

Madan, K. 1995. Paracentric inversions: a review. *Hum. Genet.* 96:503–515.

Mantell, S.H., Mathews, J.A., and McKee, R.A. 1985. *Principles of plant biotechnology: An introduction to genetic engineering in plants.* Oxford: Blackwell.

Obe, G., and Basler, A. 1987. *Cytogenetics: Basic and applied aspects.* New York: Springer-Verlag.

Ohno, S. 1970. *Evolution by gene duplication.* New York: Springer-Verlag.

Oostra, B.A., and Verkerk, A.J. 1992. The fragile X syndrome: Isolation of the *FMR-1* gene and characterization of the fragile X mutation. *Chromosoma.* 101:381–87.

Page, S.L., and Shaffer, L.G. 1997. Nonhomologous Robertsonian translocations form predominantly during female meiosis. *Nature Genetics* 15:231–32.

Patterson, D. 1987. The causes of Down syndrome. *Sci. Am.* (Aug.) 257:52–61.

Schimke, R.T., ed. 1982. *Gene amplification.* Cold Spring Harbor, NY: Cold Spring Harbor Laboratory Press.

Shepard, J.F. 1982. The regeneration of potato plants from protoplasts. *Sci. Am.* (May) 246:154–166.

Shepard, J.F., et al. 1983. Genetic transfer in plants through interspecific protoplast fusion. *Science* 219:683–88.

Simmonds, N.W., ed. 1976. *Evolution of crop plants.* London: Longman.

Smith, G.F., ed. 1984. *Molecular structure of the number 21 chromosome and Down syndrome.* New York: New York Academy of Sciences.

Stebbins, G.L. 1966. Chromosome variation and evolution. *Science* 152:1463–69.

Strickberger, M.W. 2000. *Evolution.* 3d ed. Boston: Jones and Bartlett.

Sutherland, G. 1984. The fragile X chromosome. *Int. Rev. Cytol.* 81:107–43.

———. 1985. The enigma of the fragile X chromosome. *Trends Genet.* 1:108–11.

Swanson, C.P., Merz, T., and Young, W.J. 1981. *Cytogenetics: The chromosome in division, inheritance, and evolution.* 2d ed. Englewood Cliffs, NJ: Prentice-Hall.

Taylor, A.I. 1968. Autosomal trisomy syndromes: A detailed study of 27 cases of Edwards syndrome and 27 cases of Patau syndrome. *J. Med. Genet.* 5:227–52.

Therman, E. and Susman, B. 1995. *Human chromosomes.* 3d ed. New York: Springer-Verlag.

Tjio, J.H., and Levan, A. 1956. The chromosome number of man. *Hereditas.* 42:1–6.

Turpin, R., and LeJeune, J. 1969. *Human afflictions and chromosomal aberrations.* Oxford: Pergamon Press.

Wharton, K.A., et al. 1985. *opa*: A novel family of transcribed repeats shared by the *Notch* locus and other developmentally regulated loci in *D. melanogaster. Cell* 40:55–62.

Wilkins, L.E., Brown, J.A., and Wolf, B. 1980. Psychomotor development in 65 home-reared children with cri-du-chat syndrome. *J. Pediatr.* 97:401–5.

Yunis, J.J., ed. 1977. *New chromosomal syndromes.* Orlando, FL: Academic Press.

GENETICS MediaLab

The resources that follow will help you achieve a better understanding of the concepts presented in this chapter. These resources can be found either on the CD packaged with this textbook or the Companion Web site found at **http://www.prenhall.com/klug**

CD Resources:

Module 13.1: Chromosome Aberrations

Module 13.2: Nondisjunction

Web Problem 1:

Time for completion = 15 minutes

How do changes in chromosome number (aneuploidy, polyploidy) result in new species or varieties of plants and agricultural cultivars? You will learn about the genetics of polyploids and aneuploids, as well as some of the morphological and physiological changes that occur because of extra copies of the chromosomes. Read through the linked Website article on "Chromosome number, translocations, and their consequences." Additional information can be found in Chapter 3. Characterize the difference between autopolyploids and allopolyploids. What are the potential meiotic problems with triploid autopolyploids? Why do you think so many agricultural cultivars are polyploids? Why are trisomics more common among aneuploids than are monosomics? Timothy grass, *Phleum pratense*, at first appeared to be an allopolyploid (from crossing *Phleum nodosum* and *Phleum alpinum*), but later was shown to be an autopolyploid. How was this possible? To complete this exercise, visit Web Problem 1 in Chapter 13 of your Companion Web site, and select the keyword **POLYPLOIDY**.

Web Problem 2:

Time for completion = 10 minutes

How do chromosomal mutations affect human biology? Down syndrome is a relatively common birth defect, occurring at a rate of about 1 in 800 births, that results from extra chromosome 21 material. Read the linked Web-site article referenced on the genetics and risks of Down syndrome. The most common form is a simple trisomic 21, with an entire extra copy of the chromosome. What are the other genetic mechanisms associated with Down syndrome? How many chromosomes are present in a phenotypically normal carrier of a Robertsonian translocation of chromosomes 14 and 21? Robertsonian translocations are inherited from a carrier parent in only about 25% of translocation Down syndrome patients. How did the translocation arise in these cases? What would be the genotypes of the gametes produced by a translocation carrier? Are there testing methods with a lower risk factor than amniocentesis that could detect a fetus with Down syndrome? To complete this exercise, visit Web Problem 2 in Chapter 13 of your Companion Web site, and select the keyword **DOWN SYNDROME**.

Web Problem 3:

Time for completion = 10 minutes

How are some genetic diseases caused by repeated trinucleotide DNA sequences that increase in copy number from generation to generation? Fragile X syndrome and myotonic distrophy (introduced in Chapter 14) are two examples of such diseases. Read the three linked Web-site pages about the genetics of myotonic dystrophy. Additional information can be found in Chapter 13. What is the genetic basis of myotonic dystrophy? The disease is associated with CTG tandem repeats within the myotonin protein kinase gene. What is the difference between normal and afflicted individuals? The number of repeats varies from generation to generation. What mechanism could be responsible for this kind of variation? What is the relationship between the number of repeats and the onset and severity of the disease? What particular problems might a genetic counselor want to consider when discussing whether or not a person with a very mild form of the disease should have children? To complete this exercise, visit Web Problem 3 in Chapter 13 of your Companion Web site, and select the keyword **MYOTONIC DYSTROPHY**.

The variegated leaves of the shrub, *Acanthopanax.*

14

Extranuclear Inheritance

KEY CONCEPTS

Throughout the history of genetics, occasional reports have challenged the basic tenet of Mendelian transmission genetics—that the phenotype is transmitted by nuclear genes located on chromosomes of both parents. Observations have revealed inheritance patterns that reflect neither Mendelian nor neo-Mendelian principles, and some indicate an apparent extranuclear or extrachromosomal influence on the phenotype. Before the role of DNA in genetics was firmly established, such observations were commonly regarded with skepticism. However, with the increasing knowledge of molecular genetics and the discovery of DNA in mitochondria and chloroplasts, **extranuclear inheritance** is now recognized as an important aspect of genetics.

There are several varieties of extranuclear inheritance. One major type, previously referred to as *maternal inheritance*, is more accurately described as *organelle heredity*. In such cases, DNA contained in mitochondria or chloroplasts determines the phenotype of the offspring. Examples are recognized on the basis of the uniparental transmission of these organelles from the female parent through the egg to progeny. A second type involves infectious heredity, resulting from the symbiotic or parasitic association of a microorganism. In such cases, an inherited phenotype is affected by the presence of the microorganism in the cytoplasm of the host cells. The third variety involves the maternal effect on the phenotype, whereby nuclear gene products are stored in the egg and then transmitted through the ooplasm to offspring.

14.1 Organelle Heredity Involves DNA in Chloroplasts and Mitochondria

In this section, we will examine examples of inheritance patterns related to chloroplast and mitochondrial function. Before DNA was discovered in these organelles, however, the exact mechanism of transmission of the traits to be discussed next was not clear, except that their inheritance appeared to be linked to something in the cytoplasm rather than linked to genes in the nucleus. Transmission was most often from the maternal parent through the ooplasm, causing the results of reciprocal crosses to vary. Such a pattern is now more appropriately referred to as **organelle heredity**.

Analysis of the inheritance patterns resulting from mutant alleles in chloroplasts and mitochondria has been difficult for two reasons. First, the function of these organelles is dependent upon gene products from both nuclear and organelle DNA. Second, many organelles are contributed to each progeny. In such cases, if only one or a few of these contain a mutant gene, the corresponding mutant phenotype may not be revealed. Analysis is thus much more complex than for Mendelian characters.

First, we will discuss several of the classical examples that ultimately called attention to organelle heredity. Then, in the subsequent section, we will discuss information concerning DNA and the resultant genetic function in each organelle.

Chloroplasts: Variegation in Four O'Clock Plants

In 1908, Carl Correns (one of the rediscoverers of Mendel's work) provided the earliest example of inheritance linked to chloroplast transmission. Correns discovered a variant of the four o'clock plant, *Mirabilis jalapa*, that had branches with either white, green, or variegated leaves. As shown in Figure 14–1, inheritance in all possible combinations of crosses is strictly determined by the phenotype of the ovule source. For example, if the seeds (representing the progeny) were derived from ovules on branches with green leaves, all progeny plants bore only green leaves, regardless of the phenotype of the source of pollen. Correns concluded that inheritance was transmitted through the cytoplasm of the maternal parent because the pollen, which contributes little or no cytoplasm

FIGURE 14–1 Offspring from crosses between flowers from various branches of variegated four o'clock plants. The photograph shows the four o'clock plant with a green branch.

	Location of Ovule		
Source of Pollen	*White branch*	*Green branch*	*Variegated branch*
White branch	White	Green	White, green, or variegated
Green branch	White	Green	White, green, or variegated
Variegated branch	White	Green	White, green, or variegated

to the zygote, had no apparent influence on the progeny phenotypes. Since leaf coloration is related to the chloroplast, then genetic information contained either in that organelle or somehow present in the cytoplasm and influencing the chloroplast must be responsible for this inheritance pattern.

Chloroplast Mutations in *Chlamydomonas*

The unicellular green alga *Chlamydomonas reinhardi* has provided an excellent system for the investigation of plastid inheritance. This haploid eukaryotic organism (Figure 14–2) has a single large chloroplast containing over 50 copies of a circular double-stranded DNA molecule. Matings that reestablish diploidy are immediately followed by meiosis, and the various stages of the life cycle are easily studied in culture in the laboratory. The first cytoplasmic mutation, streptomycin resistance (str^R), was reported in 1954 by Ruth Sager. Although *Chlamydomonas*' two mating types—mt^+ and mt^-—appear to make equal cytoplasmic contributions to the zygote, Sager determined that the *str* phenotype is transmitted only through the mt^+ parent (Figure 14–2). Reciprocal crosses between sensitive and resistant strains yield different results depending on the genotype of the mt^+ parent, which is expressed in all offspring. As shown, one half of the offspring are mt^+ and one half of them are mt^-, indicating that mating type is controlled by a nuclear gene that segregates in a Mendelian fashion.

Since Sager's discovery, a number of other *Chlamydomonas* mutations (including resistance to, or dependence on, a variety of bacterial antibiotics) that show a similar uniparental inheritance pattern have been discovered. These mutations

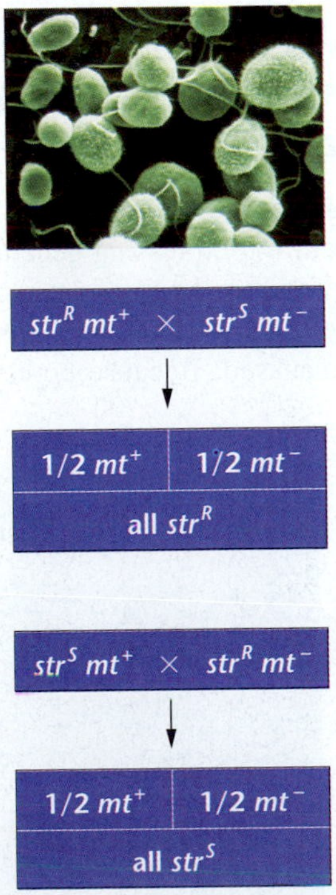

FIGURE 14–2 The results of reciprocal crosses between streptomycin-resistant (str^R) and streptomycin-sensitive (str^S) strains in the green alga *Chlamydomonas (*Shown in the photograph).

have all been linked to the transmission of the chloroplast, and their study has extended our knowledge of chloroplast inheritance.

Following fertilization, which involves the fusion of two cells of opposite mating type, the single chloroplasts of the two mating types fuse. After the resulting zygote has undergone meiosis and haploid cells are produced, it is apparent that the genetic information of the chloroplasts of progeny cells is derived only from the mt^+ parent. The genetic information present within the mt^- chloroplast has degenerated.

It is interesting to note that the inheritance of phenotypes affected by mitochondria is also uniparental in *Chlamydomonas*. However, studies of the transmission of several cases of antibiotic resistance have shown that it is the mt^- parent that transmits the genetic information to progeny cells. This is just the opposite of what occurs with chloroplast-derived phenotypes, such as str^R. The significance of inheriting one organelle from one parent and the other organelle from the other parent is not yet established.

Mitochondrial Mutations: The Case of *poky* in *Neurospora*

Mutations affecting mitochondrial function have been discovered and studied, revealing that mitochondria also contain a distinctive genetic system. As with chloroplasts, mitochondrial mutations are transmitted through the cytoplasm. In the current discussion, we will emphasize the link between mitochondrial mutations and the resultant extranuclear inheritance patterns.

In 1952, Mary B. Mitchell and Hershel K. Mitchell studied the bread mold *Neurospora crassa* (Figure 14–3). They discovered a slow-growing mutant strain and named it *poky*. (It is also designated *mi-1* for *maternal inheritance*.) Slow growth is associated with impaired mitochondrial function, specifically in relation to certain cytochromes essential for electron transport. Results of genetic crosses between wild-type and *poky* strains suggest that the trait is maternally inherited. If the female parent is *poky* and the male parent is wild type, all progeny colonies are *poky*. The reciprocal cross produces normal wild-type colonies.

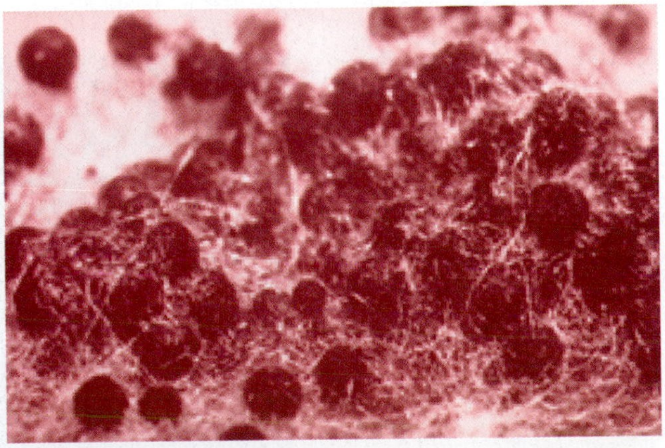

FIGURE 14–3 Micrograph illustrating the growth of the pink bread mold *Neurospora crassa*.

Studies with *poky* mutants show that occasionally hyphae from separate mycelia fuse with one another, giving rise to structures containing two or more nuclei in a common cytoplasm. If the hyphae contain nuclei of different genotypes, the structure is called a **heterokaryon**. The cytoplasm will contain mitochondria derived from both initial mycelia. A heterokaryon may give rise to haploid spores, or conidia, that produce new mycelia.

Heterokaryons produced by the fusion of *poky* and wild-type hyphae initially show normal rates of growth and respiration. However, mycelia produced through conidia formation become progressively more abnormal until they show the *poky* phenotype. This occurs despite the presumed presence of both wild-type and *poky* mitochondria in the cytoplasm of the hyphae.

To explain the initial growth and respiration pattern, we assume that the wild-type mitochondria support the respiratory needs of the hyphae. The subsequent expression of the *poky* phenotype suggests that the presence of the *poky* mitochondria may somehow prevent or depress the function of these wild-type mitochondria. Perhaps the *poky* mitochondria replicate more rapidly and "wash out" or dilute wild-type mitochondria numerically. Another possibility is that *poky* mitochondria produce a substance that inactivates the wild-type organelle or interferes with the replication of its DNA (mtDNA). This type of interaction makes *poky* an example of a broader category referred to as suppressive mutations, which may also include many other mitochondrial mutations of *Neurospora* and yeast.

Petites in *Saccharomyces*

Another extensive study of mitochondrial mutations has been performed with the yeast *Saccharomyces cerevisiae*. The first such mutation, described by Boris Ephrussi and his coworkers in 1956, was named *petite* because of the small size of the yeast colonies (Figure 14–4). Many independent petite mutations have since been discovered and studied, and all have a common characteristic—a deficiency in cellular respiration involving abnormal electron transport. Because this organism is a facultative anaerobe and can grow by fermenting glucose through glycolysis, it may survive the loss of mitochondrial function by generating energy anaerobically.

The complex genetics of petite mutations is diagrammed in Figure 14–5. A small proportion of these mutants are the result of nuclear mutations. They exhibit Mendelian inheritance and are thus called ***segregational petites***. The remainder demonstrates cytoplasmic transmission, producing one of two effects in matings. ***Neutral petites***, when crossed to wild type, produce meiotic products (called ascospores) that give rise only to wild-type, or normal, colonies. The same pattern continues if progeny of such cross are backcrossed to neutral petites. This is because the majority of neutrals lack mtDNA completely or have lost a substantial portion of it, so the normal mitochondria capable of reproduction must come from the wild-type cell.

A third mutational type, the ***suppressive petite***, behaves like *poky* in *Neurospora*. Crosses between mutant and wild type give rise to mutant diploid zygotes, which, after meiosis, immediately yield all mutant cells. Under these conditions, the petite mutation behaves "dominantly" and seems to suppress the function of the wild-type mitochondria. *Suppressive petites* also have deletions of mtDNA, but they are not nearly as extensive as deletions in the *neutral petites*.

Two major hypotheses that center around the organellar DNA have been advanced to explain suppressiveness. One explanation suggests that the mutant (or deleted) DNA in the mitochondria (mtDNA) replicates more rapidly, resulting in the mutant mitochondria "taking over" or dominating the phenotype by numbers alone. The second explanation suggests that recombination occurs between the mutant and wild-type mtDNA, introducing errors into or disrupting the normal mtDNA. It is not yet clear which one, if either, of these explanations is correct.

14.2 Knowledge of Mitochondrial and Chloroplast DNA Helps Explain Organelle Heredity

That both mitochondria and chloroplasts contain their own DNA and a system for expressing genetic information was first suggested by the discovery of the mutations and the resultant inheritance patterns in plants, yeast, and other fungi, as previously discussed. Because both mitochondria and

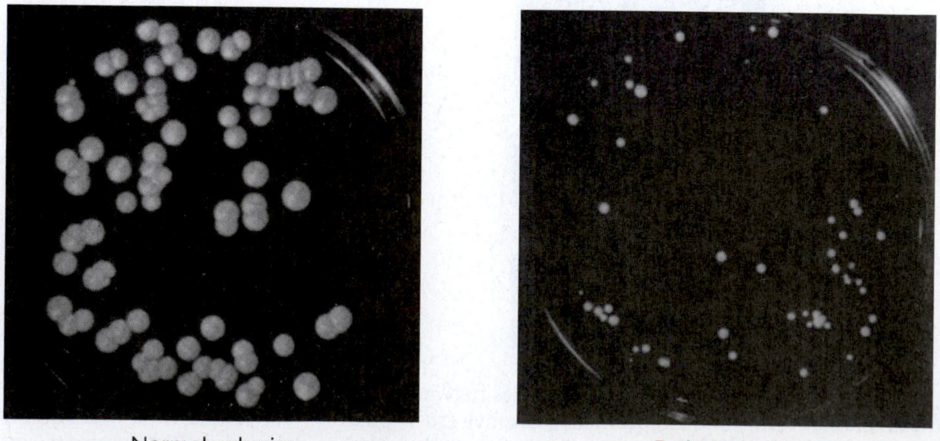

Normal colonies   Petite colonies

FIGURE 14–4 A comparison of normal vs. petite colonies in the yeast *Saccharomyces cerevisiae*.

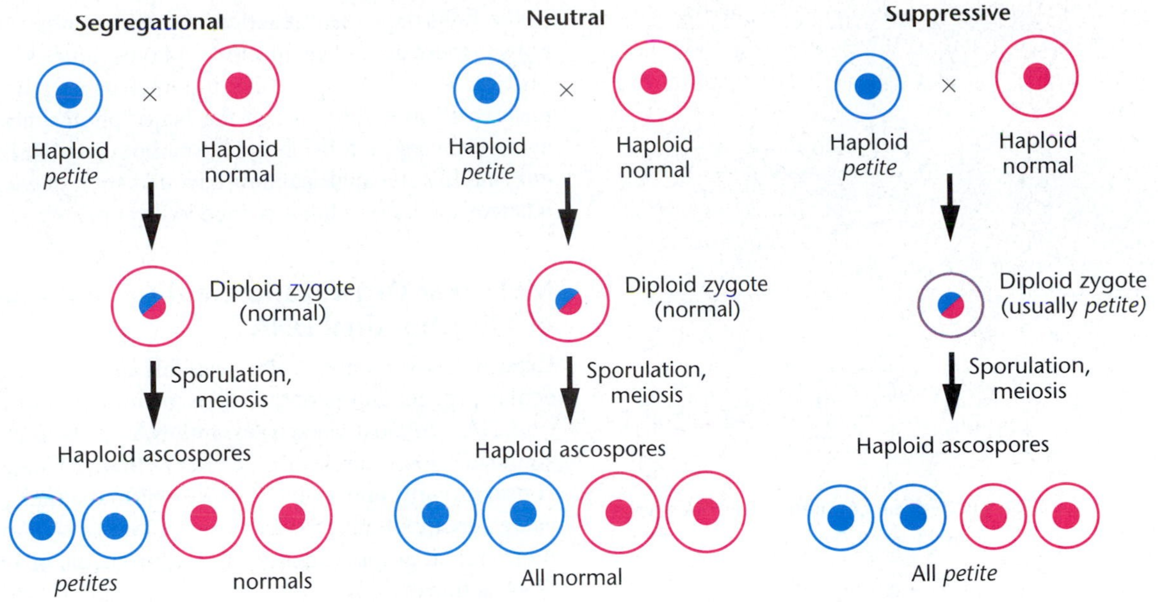

FIGURE 14–5 The outcome of crosses involving the three types of *petite* mutations affecting mitochondrial function in the yeast *Saccharomyces cerevisiae.*

chloroplasts are inherited through the maternal cytoplasm in most organisms, and since each of the earlier examples can be linked theoretically to the altered function of either chloroplasts or mitochondria, geneticists set out to look for more direct evidence of DNA in these organelles.

Electron microscopists not only documented the presence of DNA in both organelles, they also saw DNA in a form quite unlike that seen in the nucleus of the eukaryotic cells that house these organelles. This DNA looks remarkably similar to that seen in viruses and bacteria! This similarity, along with many other observations, led to the idea that mitochondria and chloroplasts arose independently about 2 billion years ago from free-living, protobacteria that possessed the abililties now attributed to these organelles—aerobic respiration and photosynthesis, respectively. This idea, championed by Lynn Margulis and others and called the **endosymbiotic hypothesis**, proposes that the ancient bacteria were engulfed by larger primitive eukaryotic cells, which originally lacked the ability to respire aerobically or to capture energy from sunlight. A beneficial, symbiotic relationship subsequently developed, whereby the bacteria eventually lost their ability to function autonomously, while the eukaryotic host cells gained the ability to undergo either oxidative respiration or photosynthesis, as the case may be. Although some questions remain unanswered, evidence continues to accumulate in support of this theory, and its basic tenets are now widely accepted.

A brief examination of the development of modern-day mitochondria will help us better understand endosymbiotic theory. During the subsequent course of evolution following the invasion event, distinct branches of diverse eukaryotic organisms arose. As evolution progressed, the companion bacteria also underwent their own independent changes. The primary alteration was the transfer of many of the genes from the invading bacterium to the nucleus of the host. The products of these genes, while still functioning in the organelle, are nevertheless now encoded and transcribed in the nucleus and translated in the cytoplasm prior to their transport into the organelle. What DNA remains today in the typical mitochondrial genome is miniscule compared with the free-living bacteria from which it was derived. The most gene-rich organelles now have fewer than 10 percent of the genes present in the smallest bacterium known.

Similar changes have characterized the evolution of chloroplasts. In the subsequent sections, we will explore in some detail what is known about modern-day chloroplasts and mitochondria.

Molecular Organization and Gene Products of Chloroplast DNA

As further support of the endosymbiont theory, the details regarding the autonomous genetic system of chloroplasts have now been worked out. This organelle, responsible for photosynthesis, contains DNA as a source of genetic information and a complete protein-synthesizing apparatus. The molecular components of the chloroplast translation apparatus are jointly derived from both nuclear and organelle genetic information.

Chloroplast DNA (cpDNA), shown in Figure 14–6, is fairly uniform in size among different organisms, ranging between 100–225 kb in length. It shares many similarities to DNA found in prokaryotic cells. It is circular, double stranded, replicated semiconservatively, and free of the associated proteins characteristic of eukaryotic DNA. Compared with nuclear DNA from the same organism, it invariably shows a different buoyant density and base composition.

The size of cpDNA is much larger than that of mtDNA. This can be accounted for, to some extent, by an increased number of genes. However, the biggest difference appears to be due to the presence in cpDNA of many long noncoding nucleotide sequences both between and within genes, the

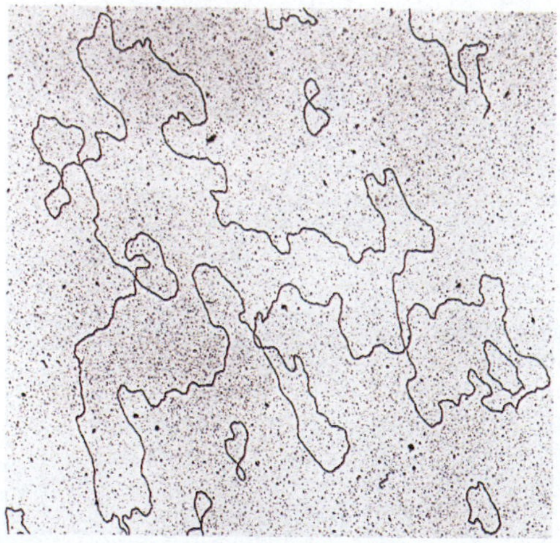

FIGURE 14–6 Electron micrograph of chloroplast DNA derived from lettuce.

latter being introns. Duplications of many DNA sequences are also present. Since such noncoding sequences vary in different plants, these observations are indicative of the independent evolution occurring in chloroplasts following their initial invasion of a primitive eukaryoticlike cell.

In the green alga *Chlamydomonas*, there are about 75 copies of the chloroplast DNA molecule per organelle. Each copy consists of a length of DNA that contains 195,000 bp (195 kb). In higher plants, such as the sweet pea, multiple copies of the DNA molecule also are present in each organelle, but the molecule (134 kb) is considerably smaller than that in *Chlamydomonas*. Interestingly, genetic recombination between the multiple copies of DNA within chloroplasts has been documented in some organisms.

Numerous gene products encoded by chloroplast DNA function during translation within the organelle. For example, in a variety of higher plants (beans, lettuce, spinach, maize, and oats), two sets of the genes coding for the ribosomal RNAs—5*S*, 16*S*, and 23*S* rRNA—are present. Additionally, chloroplast DNA encodes numerous tRNAs, as well as many ribosomal proteins specific to the chloroplast ribosomes. For example, in the liverwort, whose DNA was the first to be sequenced, there are genes encoding 30 tRNAs, RNA polymerase, multiple rRNAs, and numerous ribosomal proteins. The variation of gene products encoded in cpDNA in different plants again attests to the independent evolution that occurred within chloroplasts.

Chloroplast ribosomes have a sedimentation coefficient slightly less than 70*S*, similar, but not identical, to bacterial ribosomes. Even though some chloroplast ribosomal proteins are encoded by chloroplast DNA and some by nuclear DNA, most, if not all, such proteins are distinct from their counterparts present in cytoplasmic ribosomes. Both observations provide direct support for the endosymbiont hypothesis.

Still other chloroplast genes specific to the photosynthetic function have been identified. For example, in the liverwort, there are 92 chloroplast genes encoding proteins that are part of the thylakoid membrane, the cellular component integral to the light-dependent reactions of photosynthesis. Mutations in these genes may inactivate photosynthesis. A typical distribution of genes between the nucleus and the chloroplast is illustrated by one of the major photosynthetic enzymes, ribulose-1-5-bisphosphate carboxylase (RuBP). This enzyme has its small subunit encoded by a nuclear gene, whereas the large subunit is encoded by cpDNA.

Molecular Organization and Gene Products of Mitochondrial DNA

Extensive information is also available concerning the molecular aspects and gene products of mitochondrial DNA (mtDNA). In most eukaryotes, mtDNA exists as a double-stranded, closed circle (Figure 14–7) that replicates semiconservatively and is free of the chromosomal proteins characteristic of eukaryotic chromosomal DNA. An exception is found in some ciliates, such as protozoa, in which the DNA is linear.

In size, mtDNA is much smaller than cpDNA and varies greatly among organisms, as demonstrated in Table 14–1. In a variety of animals, including humans, mtDNA consists of about 16,000–18,000 bp (16–18 kb). However, yeast (*Saccharomyces*) mtDNA consists of 75 kb. Plants typically exceed this amount—367 kb is present in mitochondria in the mustard plant, *Arabidopsis*. Vertebrates have 5–10 such DNA molecules per organelle, while plants have 20–40 copies per organelle.

We can say several other things about mtDNA. With only rare exceptions, introns are absent from mitochondrial genes and gene repetitions are seldom present. Nor is there usually much in the way of intergenic spacer DNA. This description is applicable particularly to species whose mtDNA is fairly small in size, such as humans. However, in *Saccharomyces*, with a much larger DNA molecule, much of the excess DNA is accounted for by introns and intergenic spacer DNA.

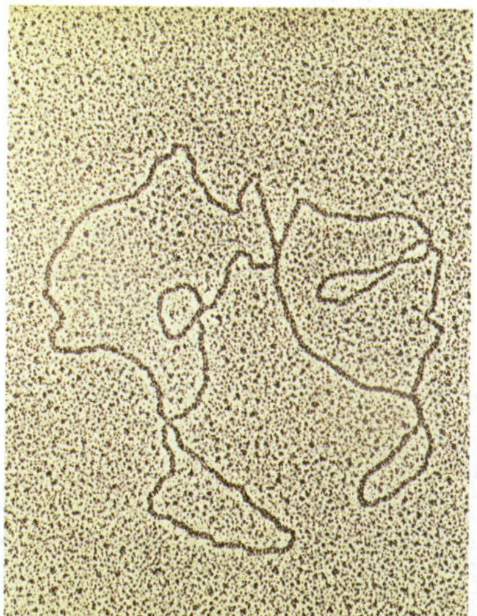

FIGURE 14–7 Electron micrograph of mitochondrial DNA derived from *Xenopus laevis*.

TABLE 14–1 The Size of mtDNA in Different Organisms

Organisms	*Size (kb)*
Human	16.6
Mouse	16.2
Xenopus (frog)	18.4
Drosophila (fruit fly)	18.4
Saccharomyces (yeast)	75.0
Pisum sativum (pea)	110.0
Arabidopsis (mustard plant)	367.0

As noted in Chapter 5, expression of mitochondrial genes uses several modifications of the otherwise standard genetic code. Replication is dependent on enzymes encoded by nuclear DNA. In humans, mtDNA encodes two ribosomal RNAs (rRNAs), 22 transfer RNAs (tRNAs), as well as 13 polypeptides essential to the oxidative respiratory functions of the organelle. In most cases, these polypeptides are part of multichain proteins, where the other subunits of each protein are often encoded in the nucleus, synthesized in the cytoplasm, and transported into the organelle. Thus, the protein-synthesizing apparatus and the molecular components for cellular respiration are jointly derived from nuclear and mitochondrial genes.

As might be predicted by the endosymbiont hypothesis, ribosomes found in the organelle differ from those present in the neighboring cytoplasm. Table 14–2 shows that mitochondrial ribosomes of different species vary considerably in their sedimentation coefficients, ranging from 55*S*–80*S*, while cytoplasmic ribosomes are uniformly 80S.

Nuclear-coded gene products essential to biological activity in mitochondria are quite numerous. They include, for example, DNA and RNA polymerases, initiation and elongation factors essential for translation, ribosomal proteins, aminoacyl tRNA synthetases, and several tRNA species. These imported components are distinct from their cytoplasmic counterparts, even though both sets are coded by nuclear genes. For example, the synthetase enzymes essential for charging mitochondrial tRNA molecules (a process essential to translation) show a distinct affinity for the mitochondrial tRNA species as compared with the cytoplasmic tRNAs.

TABLE 14–2 Sedimentation Coefficients of Mitochondrial Ribosomes

Kingdom	*Examples*	*Sedimentation Coefficient (S)*
Animals	Vertebrates	55–60
	Insects	60–71
Protists	*Euglena*	71
	Tetrahymena	80
Fungi	*Neurospora*	73–80
	Saccharomyces	72–80
Plants	Maize	77

Similar affinity has been shown for the initiation and elongation factors. Furthermore, while bacterial and nuclear RNA polymerases are known to be composed of numerous subunits, the mitochondrial variety consists of only one polypeptide chain. This polymerase is generally sensitive to antibiotics that inhibit bacterial RNA synthesis, but not to eukaryotic inhibitors. The various contributions of nuclear and mitochondrial gene products are contrasted in Figure 14–8.

14.3 Mutations in Mitochondrial DNA Cause Human Disorders

The DNA found in human mitochondria has been completely sequenced and contains 16,569 base pairs. As mentioned earlier, the mitochondrial gene products include the following:

13 proteins, required for aerobic cellular respiration;

22 transfer RNAs (tRNAs), required for translation;

two ribosomal RNAs (rRNAs), required for translation.

Because a cell's energy supply is largely dependent on aerobic cellular respiration, disruption of any mitochondrial gene by mutation may potentially have a severe impact on that organism. We have seen this in our previous discussion of *petite* mutants in yeast, which would be a lethal mutation were it not for this organism's ability to respire anaerobically. In fact, mtDNA is particularly vulnerable to mutations. There are two possible reasons for this. First, the ability to repair mtDNA damage may not be equivalent to that of nuclear DNA. Second, the concentration of highly mutagenic free radicals generated by cell respiration that accumulate in a such a confined space very likely raises the mutation rate in mtDNA.

Fortunately, a zygote receives a large number of organelles through the egg, so if only one organelle or a few of them contains a mutation, its impact is greatly diluted by the many mitochondria that lack the mutation and function normally. During early development, cell division disperses the initial population of mitochondria present in the zygote, and, in the newly formed cells, these organelles reproduce autonomously. Therefore, if a deleterious mutation arises or is present in the initial population of organelles, adults will have cells with a variable mixture of both normal and abnormal organelles. This condition is called **heteroplasmy**.

In order for a human disorder to be attributable to genetically altered mitochondria, several criteria must be met:

1. Inheritance must exhibit a maternal rather than a Mendelian pattern.
2. The disorder must reflect a deficiency in the bioenergetic function of the organelle.
3. There must be a specific genetic mutation in one or more of the mitochondrial genes.

Thus far, several disorders in humans are known to demonstrate these characteristics. For example, **myoclonic epilepsy**

FIGURE 14–8 Gene products that are essential to mitochondrial function. Those shown entering the organelle are derived from the cytoplasm and encoded by nuclear genes.

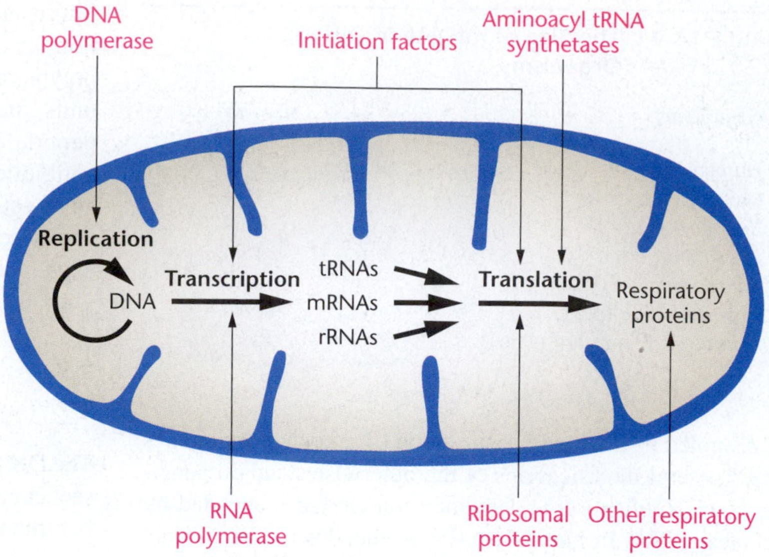

and ragged red fiber disease (MERRF) demonstrates a pattern of inheritance consistent with maternal inheritance. Only offspring of affected mothers inherit the disorder; the offspring of affected fathers are all normal. Individuals with this rare disorder express ataxia (lack of muscular coordination), deafness, dementia, and epileptic seizures. The disease is so named because of the presence of "ragged-red" skeletal muscle fibers that exhibit blotchy red patches resulting from the proliferation of aberrant mitochondria. (Figure 14–9). Brain function, which also has a high energy demand, is also affected in this disorder, leading to the neurological symptoms.

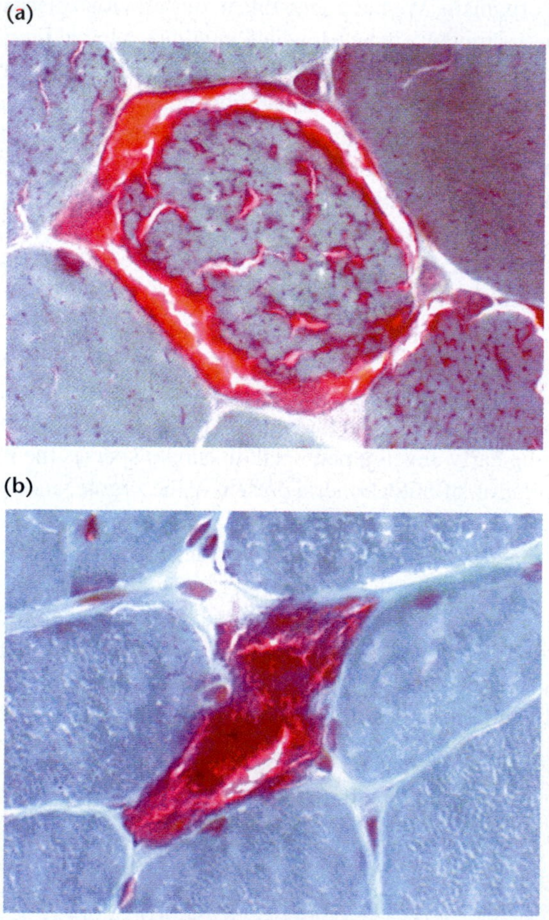

FIGURE 14–9 Ragged red fibers in skeletal muscle cells from patients with a mitochondrial disease. (a) The muscle fiber has mild proliferation (see red rim and speckled cytoplasm). (b) Marked proliferation where mitochondria have replaced most cellular structures.

Analysis of mtDNA from patients with MERFF has revealed a mutation in one of the 22 mitochondrial genes encoding a transfer RNA. Specifically, the gene encoding $tRNA^{Lys}$ contains an A-to-G transition within its sequence. This genetic alteration apparently interferes with the capacity for translation within the organelle, which in turn leads to the various manifestations of the disorder.

The cells of affected individuals exhibit heteroplasmy, containing a mixture of normal and abnormal mitochondria. Different patients contain different proportions of the two, and even different tissues from the same patient exhibit various levels of abnormal mitochondria. Were it not for heteroplasmy, the mutation would very likely be lethal, testifying to the essential nature of mitochondrial function and its reliance on the genes encoded by mtDNA.

A second disorder, **Leber's hereditary optic neuropathy (LHON)**, also exhibits maternal inheritance as well as mtDNA lesions. The disorder is characterized by sudden bilateral blindness. The average age of vision loss is 27, but onset is quite variable. Four mutations have been identified, all of which disrupt normal oxidative phosphorylation. Over 50 percent of cases are due to a mutation at a specific position in the mitochondrial gene encoding a subunit of NADH dehydrogenase. This mutation is transmitted maternally to all offspring. It is interesting to note that, in many instances of LHON, there is no family history; a significant number of cases appear to result from "new" mutations.

Individuals severely affected by a third disorder, **Kearns–Sayre syndrome (KSS)**, lose their vision, experience hearing loss, and display heart conditions. The genetic basis of KSS involves deletions at various positions within mtDNA. Many KSS patients are symptom-free as children, but display progressive symptoms as adults. The proportion of mtDNAs that reveal deletion mutations increases as the severity of symptoms increases.

The study of hereditary mitochondrial-based disorders provides insights into the critical importance of this organelle during normal development, as well as the relationship between mitochondrial function and neuromuscular and neurological disorders. Such study has also suggested a hypothesis for aging based on the progressive accumulation of mtDNA mutations and the accompanying loss of mitochondrial function.

14.4 Infectious Heredity Is Based on a Symbiotic Relationship between Host Organism and Invader

Examples abound of cytoplasmically transmitted phenotypes in eukaryotes that are due to an invading microorganism or particle. The foreign invader coexists in a symbiotic relationship, is usually passed through the maternal ooplasm to progeny cells or organisms, and confers a specific phenotype. We shall consider several examples of this phenomenon.

Kappa in *Paramecium*

First described by Tracy Sonneborn, certain strains of *Paramecium aurelia* are called **Killers** because they release a cytoplasmic substance called **paramecin** that is toxic and sometimes lethal to sensitive strains. This substance is produced by kappa particles that replicate in the Killer cytoplasm; one cell may contain 100 to 200 such particles. The kappa particles contain DNA and protein, and depend on a dominant nuclear gene, *K*, for their maintenance. To understand how the *K* gene and kappa particles are transmitted, we must look at the way *Parameciam* reproduce.

Paramecia are diploid protozoans that can undergo sexual exchange of genetic information through the process of conjugation, shown in Figure 14–10. Each *Paramecium* contains one macronucleus and two diploid micronuclei. Early in conjugation, both micronuclei in each mating pair undergo meiosis, resulting in eight haploid micronuclei. Seven of these degenerate, and the remaining one undergoes a single mitotic division. Each cell then donates one of the two haploid micronuclei to the other, re-creating the diploid condition in both cells. As a result, exconjugates are typically identical genotypes. In some instances, conjugation may be accompanied by cytoplasmic exchange.

Autogamy is a similar process, involving only a single cell. Following meiosis of both micronuclei, seven products degenerate and one survives. This nucleus divides, and the resulting nuclei fuse to re-create the diploid condition. If the original cell was heterozygous, autogamy results in homozygosity, because the newly formed diploid nucleus is derived solely from a single haploid meiotic product. In a population of cells originally heterozygous, half of the new cells express one allele and half express the other allele.

Figure 14–11 illustrates the results of crosses between *KK* and *kk* cells, with and without cytoplasmic exchange. Sometimes, no cytoplasmic exchange occurs, even though the resultant cells may be *Kk* (or *KK*, following autogamy). When none occurs, no kappa particles are transmitted and cells remain sensitive. When exchange occurs, the cells become Killers provided the kappa particles are supported by at least one dominant *K* allele.

Kappa particles are bacterialike and may contain temperate bacteriophages. One theory holds that these viruses of kappa may enter vegetative state and proceed to reproduce; during this multiplication, they produce the toxic products that are released and kill sensitive strains.

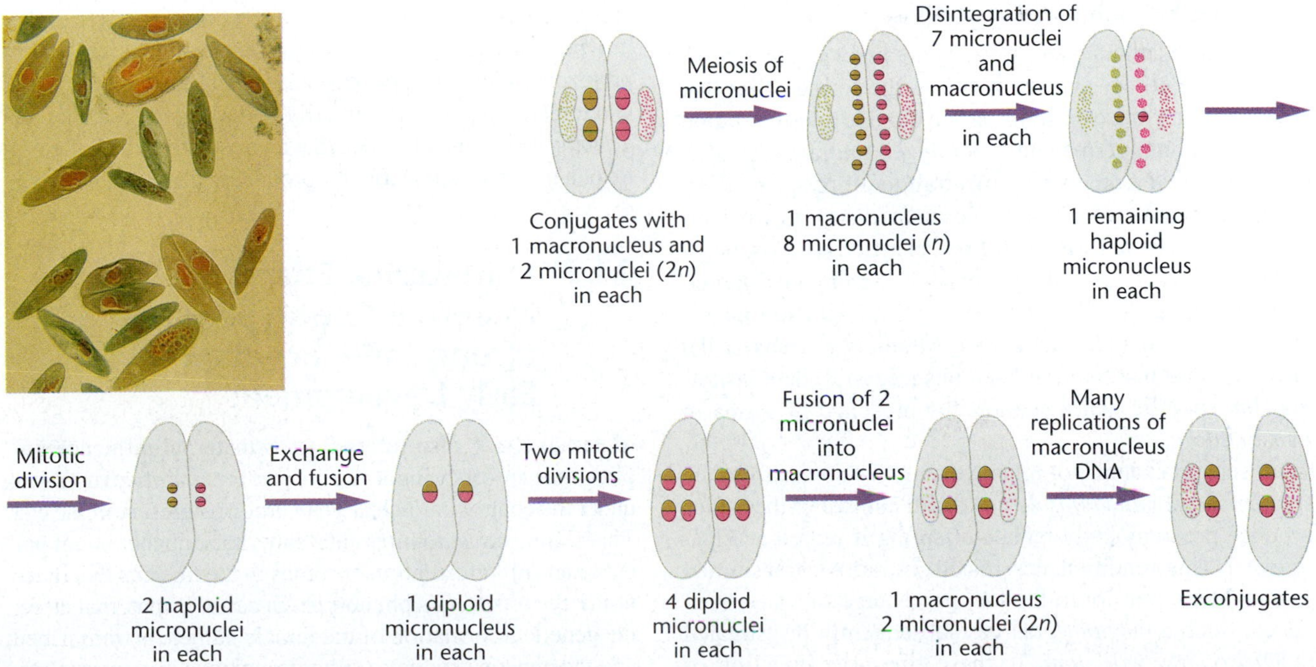

FIGURE 14–10 Genetic events occurring during conjugation in *Paramecium*. The photographic inset shows several pairs of organisms undergoing conjugation.

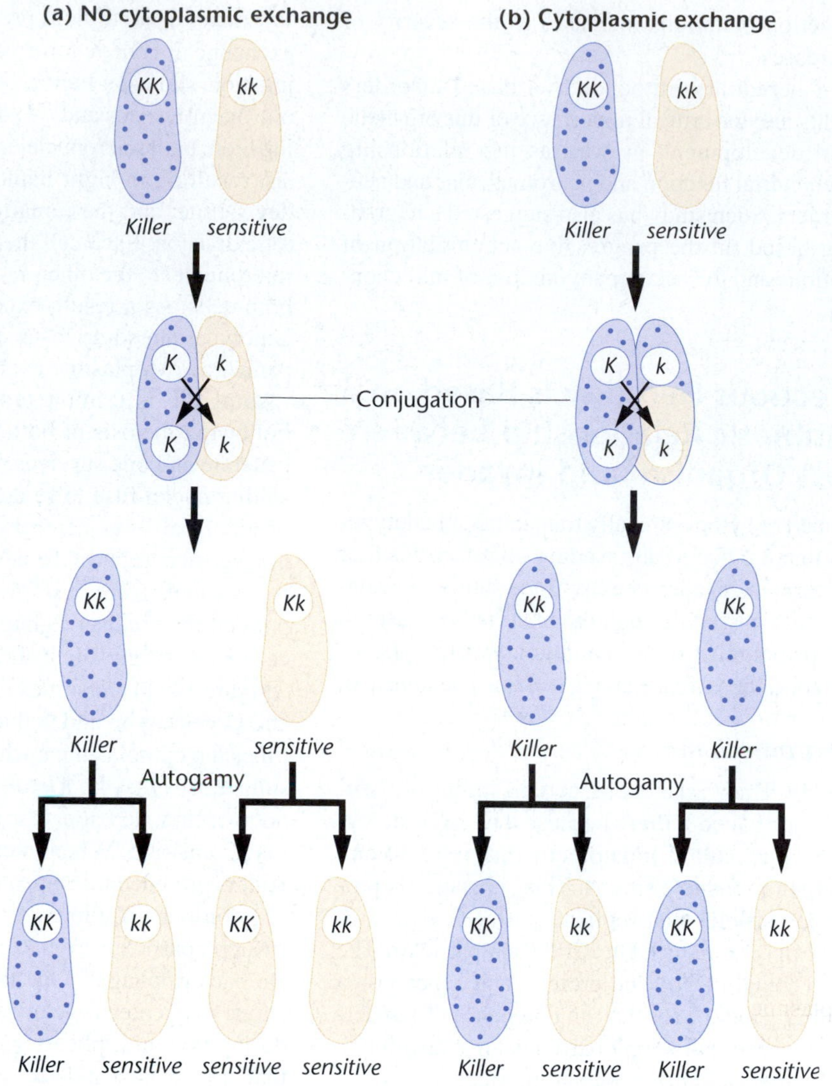

FIGURE 14–11 Results of crosses between Killer (*KK*) and sensitive (*kk*) strains of *Paramecium*, with and without cytoplasmic exchange during conjugation. The kappa particles (dots) are maintained only when a *K* allele is present.

Infective Particles in *Drosophila*

Two examples of infectious heredity are known in *Drosophila*, **CO_2 sensitivity** and **sex ratio**. In the former, flies that would normally recover from carbon dioxide anesthetization instead become permanently paralyzed and are killed by CO_2. Sensitive mothers pass this trait to all offspring. Furthermore, extracts of sensitive flies induce the trait when injected into resistant flies. Phillip L'Heritier has postulated that sensitivity is due to the presence of a virus called **sigma**. The particle has been visualized under the electron microscope and is smaller than kappa. Attempts to transfer the virus to other insects have been unsuccessful, demonstrating that specific genes support the presence of sigma in *Drosophila*.

Our second example of infective particles comes from the study of *Drosophila bifasciata*. A small number of these flies produce predominantly female offspring if reared at 21°C or lower. This condition, designated sex ratio, is transmitted to daughters, but not to the low percentage of males produced. Such a phenomenon was subsequently investigated in *Drosophila willistoni*. In these flies, the injection of ooplasm from sex-ratio females into the abdomens of normal females induced the condition, suggesting that an extra-chromosomal element is responsible for the sex-ratio phenotype. The agent has now been isolated and shown to be a protozoan. While the protozoan has been found in both males and females, it is lethal primarily to developing male larvae. There is now some evidence that a virus harbored by the protozoan may be responsible for producing a male-lethal toxin.

14.5 In Maternal Effect, the Maternal Genotype Has a Strong Influence during Early Development

Maternal effect, also referred to as maternal influence, implies that an offspring's phenotype for a particular trait is under the control of nuclear gene products present in the egg. This is in contrast to biparental inheritance, where both parents transmit information on genes in the nucleus that determines the offspring's phenotype. In cases of maternal effect, the genetic information of the female gamete is transcribed, and the genetic products (either proteins or yet untranslated mRNAs) are present in the egg cytoplasm. Following fertilization, these products influence patterns or traits established during early development. Three examples will illustrate the influence of the maternal genome on particular traits.

Ephestia Pigmentation

A very straightforward illustration of a maternal effect is seen in the Mediterranean meal moth *Ephestia kuehniella*. The wild-type larva of this moth has a pigmented skin and brown eyes as a result of the dominant gene *A*. The pigment is derived from a precursor molecule, kynurenine, which is in turn a derivative of the amino acid tryptophan. A mutation, *a*, interrupts the synthesis of kynurenine and, when homozygous, may result in red eyes and little pigmentation in larvae. However, as illustrated in Figure 14–12, results of the cross *Aa* × *aa*, depend on which parent carries the dominant gene. When the male is the heterozygous parent, a 1:1 brown–red-eyed ratio is observed in larvae, as predicted by Mendelian segregation. When the female is heterozygous for the *A* gene, however, all larvae are pigmented and have brown eyes, in spite of half of them being *aa*. As these larvae develop into adults, one-half of them gradually develop red eyes, reestablishing the 1:1 ratio.

One explanation for these results is that the *Aa* oocytes synthesize kynurenine or an enzyme necessary for its synthesis and accumulate it in the ooplasm prior to the completion of meiosis. Even in *aa* progeny whose mothers were *Aa*, this pigment is distributed in the cytoplasm of the cells of the developing larvae, thus, they develop pigmentation and brown eyes. Eventually, the pigment is diluted among many cells and depleted, resulting in the conversion to red eyes as adults. The *Ephestia* example demonstrates the maternal effect in which a cytoplasmically stored nuclear gene product influences the larval phenotype and, at least temporarily, overrides the genotype of the progeny.

Limnaea Coiling

Shell coiling in the snail *Limnaea peregra* is an excellent example of maternal effect on a permanent rather than a transitory phenotype. Some strains of this snail have left-handed, or sinistrally, coiled shells (*dd*), while others have right-handed, or dextrally, coiled shells (*DD* or *Dd*). These snails are hermaphroditic and may undergo either cross- or self-fertilization, providing a variety of types of matings.

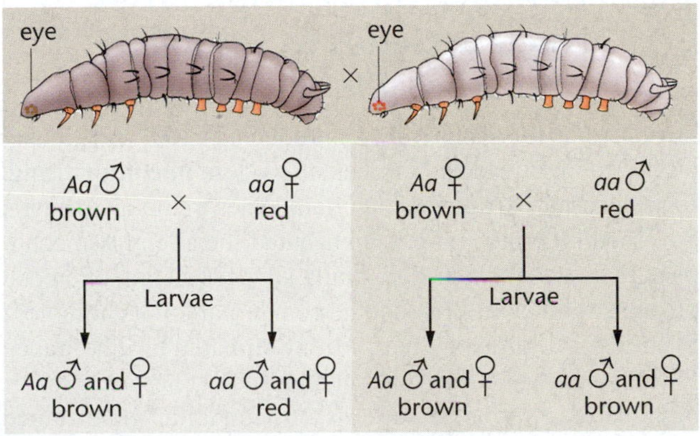

FIGURE 14–12 Maternal influence in the inheritance of eye pigment in the meal moth *Ephestia kuehniella*. Multiple light receptor structures (eyes) are present on each side of the anterior portion of larvae.

Figure 14–13 illustrates the results of reciprocal crosses between true-breeding snails. As you can see, the crosses yield different outcomes, even though both are between sinistral and dextral organisms and produce all heterozygous offspring. Examination of the progeny reveals that their phenotypes depend on the genotypes of the female parents. If we adopt that conclusion as a working hypothesis, we can test it by examining the offspring in subsequent generations of self-fertilization events. In each case, the hypothesis is upheld. Maternal parents that are *DD* or *Dd* produce only dextrally coiled progeny. Maternal parents that are *dd* produce only sinistrally coiled progeny. The coiling pattern of the progeny snails is determined by the genotype of the parent producing the egg, regardless of the phenotype of that parent.

Investigation of the developmental events in *Limnaea* reveals that the orientation of the spindle in the first cleavage division after fertilization determines the direction of coiling. Spindle orientation appears to be controlled by maternal genes acting on the developing eggs in the ovary. The orientation of the spindle, in turn, influences cell divisions following fertilization and establishes the permanent adult coiling pattern. The dextral allele (*D*) produces an active gene product that causes right-handed coiling. If ooplasm from dextral eggs is injected into uncleaved sinistral eggs, they cleave in a dextral pattern. However, in the converse experiment, sinistral ooplasm has no effect when injected into dextral eggs. Apparently the sinistral allele is the result of a classic recessive mutation that encodes an inactive gene product.

We can conclude, therefore, that females that are either *DD* or *Dd* produce oocytes that synthesize the *D* gene product, which is stored in the ooplasm. Even if the oocyte contains only the *d* allele following meiosis and is fertilized by a *d*-bearing sperm, the resulting *dd* snail will be dextrally coiled (right handed).

Embryonic Development in *Drosophila*

A more recently documented example of maternal effect involves various genes that control embryonic development in *Drosophila melanogaster*. The genetic control of embryonic development in *Drosophila*, discussed in greater detail in Chapter 21, is a fascinating story. The protein products of the maternal-effect genes function to activate other genes, which may in turn activate still other genes. This cascade of gene activity leads to a normal embryo whose subsequent development yields a normal adult fly. The extensive work by Edward B. Lewis, Christiane Nüsslein-Volhard, and Eric Wieschaus (who shared the 1995 Nobel Prize for Physiology or Medicine for their findings) has clarified how these and other genes function. Genes that illustrate maternal effect have products that are synthesized by the developing egg and stored in the oocyte prior to fertilization. Following fertilization, these products specify molecular gradients that determine spatial organization as development proceeds.

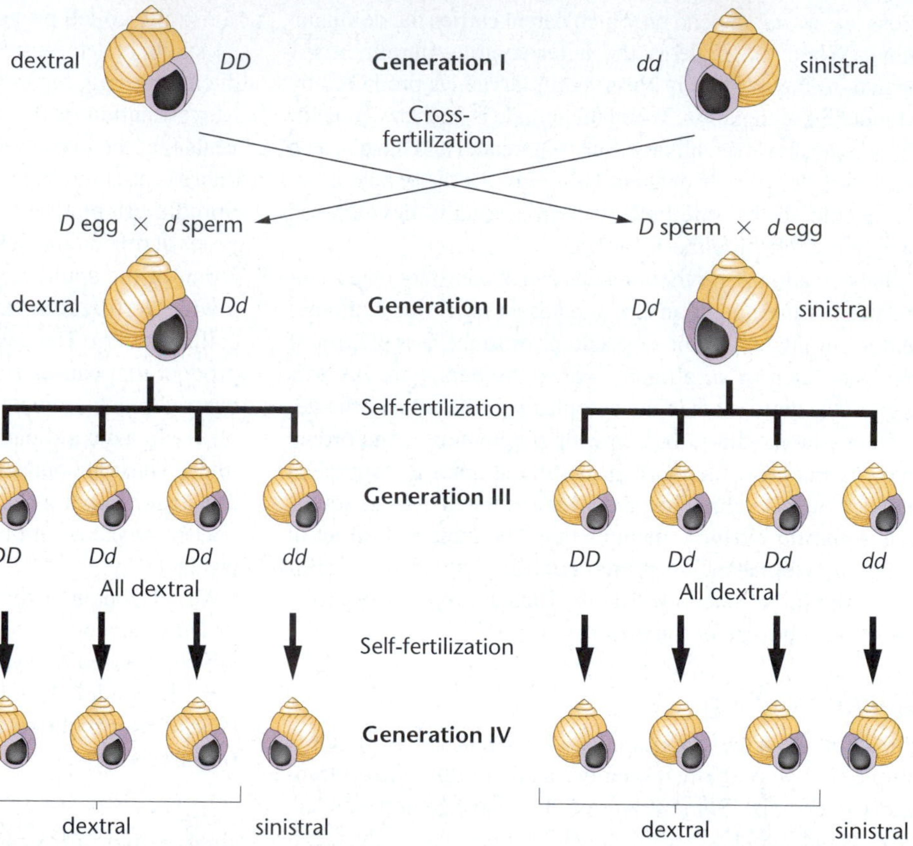

FIGURE 14–13 Inheritance of coiling in the snail *Limnaea peregra*. Coiling is either dextral (right handed) or sinistral (left handed). A maternal effect is evident in generations II and III, where the genotype of the maternal parent, rather than the offspring's own genotype, controls the phenotype of the offspring. The photograph illustrates a mixture of right- vs. left-handed coiled snails.

For example, the gene *bicoid (bcd)* plays an important role in specifying the development of the anterior portion of the fly. Embryos derived from mothers who are homozygous for this mutation (bcd^-/bcd^-) fail to develop anterior areas that normally give rise to the head and thorax of the adult fly. Embryos whose mothers contain at least one wild-type allele (bcd^+) develop normally, even if the genotype of the embryo is homozygous for the mutation. Consistent with the concept of *maternal effect*, the *genotype of the female parent*, not the *genotype of the embryo*, determines the phenotype of the offspring.

When we return to our discussion of this general topic in Chapter 21, we will see examples of other genes illustrating maternal effect, as well as many "zygotic" genes whose expression occurs during early development and that behave genetically in the conventional Mendelian fashion.

14.6 Genomic Imprinting Causes Nuclear Genes to Exhibit Uniparental Inheritance

We conclude this chapter by reintroducing a topic previously discussed in some detail in Chapter 10, **genomic imprinting**. In such cases, expression of certain genes in the offspring is governed by which parent contributed the allele. A mechanism is in place that preferentially silences a copy of a gene on one chromosome, but not the counterpart on its homolog. Such a situation leads to a form of uniparental inheritance for the phenotypic trait governed by such genes and creates obvious exceptions to patterns of Mendelian inheritance characterized by segregation. The condition is reversible, whereby the "silent" allele becomes reactivated or the active allele becomes silenced when passed through the germline to an offspring of the sex opposite its parent.

Genetics, Technology, and Society

Mitochondrial DNA and the Mystery of the Romanovs

By most accounts, Nicholas II, the last Tsar of Russia, was a substandard monarch. He was accused of bungling during the Russo-Japanese War of 1904–1905, and his regime was plagued by corruption and incompetence. Even so, he probably didn't deserve the fate that befell him and his family one summer night in 1918. As we shall see, a full understanding of that event has relied on, of all things, mitochondrial DNA.

After being forced to abdicate in 1917, ending 300 years of Romanov rule, Tsar Nicholas and the imperial family were banished to Ekaterinburg in western Siberia. There, it was believed, they would be out of reach of the fiercely anti-imperialist Bolsheviks, who were then fighting to gain control of the country. But the Bolsheviks eventually caught up with the Romanovs. On a July night in 1918, Tsar Nicholas, Tsarina Alexandra (grand-daughter of Queen Victoria), their five children (Olga, 22; Tatiana, 21; Marie, 19; Anastasia, 17; and Alexis, 13), their family doctor, and three of their servants were awakened and brought to a downstairs room of the house where they were held prisoner. There they were made to form a double row against the wall, presumably so that a photograph could be taken. Instead, 11 men with revolvers burst into the room and opened fire. After exhausting their ammunition, they proceeded to bayonet the bodies and smash their faces in with rifle butts. The corpses were then hauled away and flung down a mineshaft, only to be pulled out two days later and dumped into a shallow grave, doused with sulfuric acid, and covered over. There they rested for more than 60 years.

An air of mystery soon developed around the demise of the Romanovs. Did all of the children of Nicholas and Alexandra die with their parents that bloody night, or did the youngest daughter, Anastasia, get away? Over the years, the possibility that Anastasia miraculously escaped execution has inspired countless books, a Hollywood movie, a ballet, a Broadway play, and, most recently, an animated movie. In all of these retellings, Anastasia re-emerges to claim her birthright as the only surviving member of the Romanovs. Adding to the puzzle was one Anna Anderson. Two years after being dragged from a Berlin canal after a suicide attempt in 1920, she began to claim to be the Grand Duchess Anastasia. Despite a history of mental instability and a curious inability to speak Russian, she managed to convince many people.

The unraveling of the mystery began in 1979, when a Siberian geologist and a Moscow filmmaker discovered four skulls they believed to belong to the Tsar's family. It wasn't until the summer of 1991, after the establishment of glasnost, that exhumation began. Altogether, almost 1000 bone fragments were recovered, which were re-assembled into nine skeletons, five females and four males. Based on measurements of the bones and computer-assisted superimposition of the skulls onto photographs, the remains were tentatively identified as belonging to the murdered Romanovs. But there were still two missing bodies, one of the daughters (believed to be Anastasia) and the boy Alexis.

The next step in authenticating the remains involved studies of DNA that were conducted by Pavel Ivanov, the leading Russian forensic DNA analyst, in collaboration with Peter Gill of the British Forensic Science Service. Their goals were to establish family relationships among the remains, and then to determine, by comparisons with living relatives, whether the family group was in fact the Romanovs. They froze bone fragments from the nine skeletons in liquid nitrogen, ground them into a fine powder, and extracted small amounts of DNA, which comprised both nuclear DNA and mitochondrial DNA (mtDNA). Genomic DNA typing of each skeleton confirmed the familial relationships and showed that, indeed, one of the princesses and Alexis were missing. The final proof that the bones belonged to the Romanovs awaited the analysis of mtDNA, however.

Mitochondrial DNA is ideal for forensic studies for several reasons. Since all the mitochondria in a human cell are descended from the mitochondria present in the egg, mtDNA is transmitted strictly from mother to offspring, never from father to offspring. Therefore, mtDNA sequences can be used to trace maternal lineages without the complicating effects of meiotic crossing over, which recombines maternal and paternal nuclear genes every generation. In addition, mtDNA is small (16,600 base pairs) and present in 500 to 1000 of copies per cell, so it is much easier to recover intact than nuclear DNA.

Ivanov's group amplified two highly variable regions from the mtDNA isolated from all nine bone samples and determined the nucleotide sequences of these regions. By comparing these sequences with living relatives of the Romanovs, they hoped to establish the identity of the Ekaterinburg remains once and for all. The sequences from Tsarina Alexandra were an exact match with those from Prince Philip of England, who is her grandnephew, verifying her identity. Authentication was more complicated for the Tsar, however.

The Tsar's sequences were compared with those from the only two living relatives who could be persuaded to participate in the study, Countess Xenia Cheremeteff-Sfiri (his great-grandniece) and James George Alexander Bannerman Carnegie, third Duke of Fife (a more distant relative, descended from a line of women stretching back to the Tsar's grandmother). These comparisons produced a surprise. At position 16169 of the mtDNA, the Tsar seemed to have either one or another base, a C or a T. The Countess and the Duke, in contrast, both had only T. The conclusion was that Tsar Nicholas had two different populations of mitochondria in his cells, each with a different base at position 16169 of its DNA. This condition, called heteroplasmy, is now believed to occur in 10 to 20 percent of humans.

This ambiguity between the presumed Tsar and his two living maternal relatives cast doubt on the identification of the remains. Fortunately, the Russian government granted a request to analyze the remains of the Tsar's younger brother, Grand Duke Georgij Romanov, who died in 1899 of tuberculosis. The Grand Duke's mtDNA was found to have the same heteroplasmic variant at position 16169, either a C or a T. It was concluded that the Ekaterinburg bones were the doomed imperial family. With years of controversy finally resolved, the now-authenticated remains of Tsar Nicholas II, Tsarina Alexandra, and three of their daughters were buried in the St. Peter and Paul Cathedral in St. Petersburg on July 17, 1998, 80 years to the day after they were murdered.

This DNA analysis did not solve the mystery of the fate of Anastasia, however. Did she die with her parents, sisters, and brother in 1918? Or is it possible that Anna Anderson was telling the truth, that she was the escaped duchess? In a separate study, Anna Anderson's nuclear and mtDNA was recovered from intestinal tissue preserved from an operation performed five years before her death in 1984. Analysis of this DNA proved that she was not Anastasia, but rather a Polish peasant named Franziska Schanzkowska.

If Anastasia's remains were not among those of her parents and sisters and if Anna Anderson was an imposter, what did happen to Anastasia? Most evidence suggests that Anastasia and her brother Alexis were not found with the others in the mass grave because their bodies were burned over the grave site two days after the killings and the ashes scattered, never to be found again. Not exactly a Hollywood ending.

References

Gibbons, A. 1998. Calibrating the mitochondrial clock. *Science* 279:28–29.

Ivanov, P. et al. 1996. Mitochondrial DNA sequence heteroplasmy in the Grand Duke of Russia Georgij Romanov establishes the authenticity of the remains of Tsar Nicholas II. *Nat. Genet.* 12:417–20.

Masse, R. (1996). *The Romanovs: The final chapter*. New York: Ballantine Books.

Stoneking, M. et al. 1995. Establishing the identity of Anna Anderson Manahan. *Nat. Genet.* 9:9–10.

An example of imprinting involves the X chromosomes in mammalian females. As we have discussed in Chapter 11, a mechanism of **dosage compensation** exists, whereby the random inactivation of either the paternal or maternal X chromosome occurs during embryonic development. In mice, however, prior to the development of the embryo proper, imprinting occurs in tissues such that the X chromosome of paternal origin is genetically inactivated in all cells, while the genes on the maternal X chromosome remain genetically active. As embryonic development is subsequently initiated, the imprint is "released" and either the paternal or the maternal X chromosome is subsequently inactivated randomly later during development.

A more specific example involves the mouse gene that encodes insulinlike growth factor II (*Igf2*). A mouse that carries two nonmutant alleles of this gene is normal in size, whereas a mouse that carries two mutant alleles lacks a growth factor and is a dwarf. Imprinting is apparent in the size of heterozygous mice (that have one normal allele and one mutant allele). Their size depends on the parental origin of the normal allele. The mouse is average in size if the normal allele came from the father, but dwarf if the normal allele came from the mother. From this it can be deduced that the normal *Igf2* gene is imprinted to function poorly during the course of egg production in females, but functions normally when it has passed through sperm-producing tissue in males. Imprinting continues to depend on whether the gene passes through sperm-producing or egg-forming tissue leading to the next generation.

In humans, two distinct genetic disorders are thought to be caused by the imprinting of a region of chromosome 15. The disorders appear to both be due to the same deletion in a region of one member of the chromosome 15 pair. The first disorder, **Prader–Willi syndrome (PWS)**, results when only an undeleted maternal chromosome remains. If only an undeleted paternal chromosome remains, a different disorder, **Angelman syndrome (AS)**, results.

The molecular mechanism of imprinting is thought to involve DNA methylation. In vertebrates, methyl groups can be added to the carbon atom at position 5 in cytosine, one of the nitrogenous bases (See Chapter 2). A mechanism involving methylation of DNA is a reasonable candidate for explaining molecular imprinting, since there is some evidence that a high level of methylation can inhibit gene activity, and that active genes (or their regulatory sequences) are often undermethylated. Additionally, variation in methylation has been noted in the mouse genes that undergo imprinting. Whatever the cause of this phenomenon, it is a fascinating topic that illustrates extrachromosomal inheritance.

Chapter Summary

1. Patterns of inheritance sometimes vary from that expected from the biparental transmission of nuclear genes. In such instances, phenotypes most often appear to result from genetic information transmitted through the egg.
2. Organelle heredity is based on the genotypes of chloroplast and mitochondrial DNA as these organelles are transmitted to offspring. Chloroplast mutations affect the photosynthetic capabilities of plants, whereas mitochondrial mutations affect cells highly dependent on ATP generated through cellular respiration. The resulting mutants display phenotypes related to the loss of function of these organelles.
3. Both chloroplasts and mitochondria originated in eukaryotic cells some 2 billion years ago as invading protobacteria. As evolution proceeded, many of the genes of the bacteria were transferred to the nucleus of the cell and a symbiotic relationship developed where these organelles enhanced the energetic capacity of the cell. Evidence in support of this endosymbiont hypothesis is extensive and centers around many observations involving the DNA and genetic machinery present in modern-day chloroplasts and mitochondria.
4. Another form of extranuclear inheritance is attributable to the transmission of infectious microorganisms. Kappa particles and CO_2-sensitivity and sex-ratio determinants are examples.
5. Maternal-effect patterns result when nuclear gene products controlled by the maternal genotype of the egg influence early development. *Ephestia* pigmentation, coiling in snails, and gene expression during early development in *Drosophila* are examples.
6. Genomic imprinting, in which one copy of a gene is preferentially silenced, is still another illustration of extranuclear inheritance.

Insights and Solutions

1. Analyze the following hypothetical pedigree and determine the most consistent interpretation of how the trait is inherited and any inconsistencies:

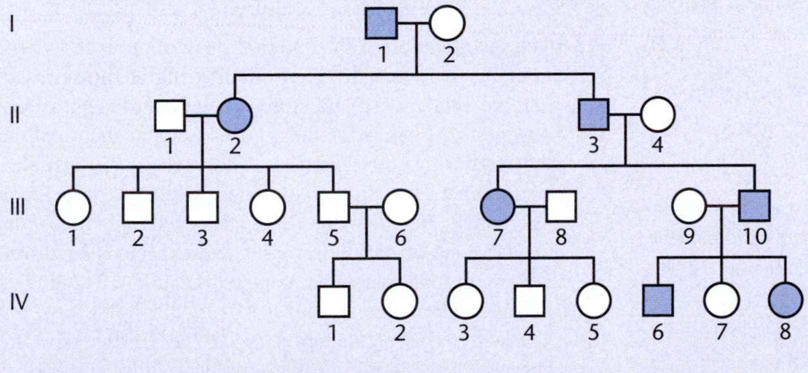

Solution: The trait is passed from all affected male parents to all but one offspring, but *never* passed maternally. Individual IV-7 (a female) is the only exception.

2. Can the explanation in Problem 1 be attributed to a gene on the Y chromosome? Defend your answer.

Solution: No, because male parents pass the trait to their daughters as well as to their sons.

3. Is the case in Problem 1 an example of a paternal effect or of paternal inheritance?

Solution: It has all the earmarks of paternal inheritance because males pass the trait to almost all of their offspring. To assess whether the trait is due to a paternal effect (resulting from a nuclear gene in the male gamete), analysis of further matings would be needed.

Problems and Discussion Questions

1. What genetic criteria distinguish a case of extrachromosomal inheritance from a case of Mendelian autosomal inheritance? From a case of X-linked inheritance?
2. Streptomycin resistance in *Chlamydomonas* may result from a mutation in either a chloroplast gene or a nuclear gene. What phenotypic results would occur in a cross between a member of an mt^+ strain resistant in both genes and a member of an mt^- strain sensitive to the antibiotic? What results would occur in the reciprocal cross?
3. A plant may have green, white, or green-and-white (variegated) leaves on its branches owing to a mutation in the chloroplast that prevents color from developing. Predict the results of the following crosses:

	Ovule Source		*Pollen Source*
(a)	Green branch	×	White branch
(b)	White branch	×	Green branch
(c)	Variegated branch	×	Green branch
(d)	Green branch	×	Variegated branch

4. In aerobically cultured yeast, a *petite* mutant is isolated. To determine the type of mutation causing this phenotype, the *petite* and wild-type strains are crossed. Shown here are three potential outcomes of such a cross. For each set of results, what conclusion about the type of *petite* mutation is justified?
   (a) all wild type.
   (b) some *petite*: some wild type.
   (c) all *petite*.
5. In diploid yeast strains, sporulation and subsequent meiosis can produce haploid ascospores, which may fuse to reestablish diploid cells. When ascospores from a *segregational petite* strain fuse with those of a normal wild-type strain, the diploid zygotes are all normal. Following meiosis, ascospores are *petite* and normal. Is the *segregational petite* phenotype inherited as a dominant or a recessive trait?
6. Predict the results of a cross between ascospores from a *segregational petite* strain and a *neutral petite* strain. Indicate the phenotype of the zygote and the ascospores it may subsequently produce.
7. Described here are the results of three crosses between strains of *Paramecium*:

(a) Killer × sensitive	⟶	1/2 Killer: 1/2 sensitive
(b) Killer × sensitive	⟶	all Killer
(c) Killer × sensitive	⟶	3/4 Killer: 1/4 sensitive

Determine the genotypes of the parental strains.

8. *Chlamydomonas*, a eukaryotic green alga, is sensitive to the antibiotic erythromycin, which inhibits protein synthesis in prokaryotes.
   (a) Explain why.
   (b) There are two mating types in this alga, mt^+ and mt^-. If an mt^+ cell sensitive to the antibiotic is crossed with an mt^- cell that is resistant, all progeny cells are sensitive. The reciprocal cross (mt^+ resistant and mt^- sensitive) yields all resistant progeny cells. Assuming that the mutation for resistance is in the chloroplast DNA, what can you conclude from the results from these crosses?
9. In *Limnaea*, what results would you expect in a cross between a *Dd* dextrally coiled and a *Dd* sinistrally coiled snail, assuming cross-fertilization occurs as shown in Figure 14–13? What results would occur if the *Dd* dextral produced only eggs and the *Dd* sinistral produced only sperm?
10. In a cross of *Limnaea*, the snail contributing the eggs was dextral, but of unknown genotype. Both the genotype and the phenotype of the other snail are unknown. All F_1 offspring exhibited dextral coiling. Ten of the F_1 snails were allowed to undergo self-fertilization. One-half produced only dextrally coiled offspring, whereas the other half produced only sinistrally coiled offspring. What were the genotypes of the original parents?
11. In *Drosophila subobscura*, the presence of a recessive gene called *grandchildless* (*gs*) causes the offspring of homozygous females, but not those of homozygous males, to be sterile. Can you offer an explanation as to why females and not males are affected by the mutant gene?
12. A male mouse from a true-breeding strain of hyperactive animals is crossed with a female mouse from a true-breeding strain of lethargic animals. (These are both hypothetical strains.) All the progeny are lethargic. In the F_2 generation, all offspring are lethargic. What is the best genetic explanation for these observations? Propose a cross to test your explanation.

Extra-Spicy Problems

13. The specification of the anterior–posterior axis in *Drosophila* embryos is initially controlled by various gene products that are synthesized and stored in the mature egg following oogenesis. Mutations in these genes result in abnormalities of the axis during embryogenesis. These mutations illustrate *maternal effect*. How do such mutations vary from those involved in organelle heredity that illustrate *cytoplasmic inheritance?* Devise a set of parallel crosses and expected outcomes involving mutant genes that contrast maternal effect and cytoplasmic inheritance.
14. The maternal-effect mutation, *bicoid (bcd)*, is recessive. In the absence of the bicoid protein product, embryogenesis is not completed. Consider a cross between a female heterozygous for the bicoid (bcd^+/bcd^-) and a male homozygous for the mutation (bcd^-/bcd^-).
    (a) How is it possible for a male homozygous for the mutation to exist?
    (b) Predict the outcome (normal vs. failed embryogenesis) in the F_1 and F_2 generations of the cross described.
15. Shown here is a pedigree for a hypothetical human disorder:

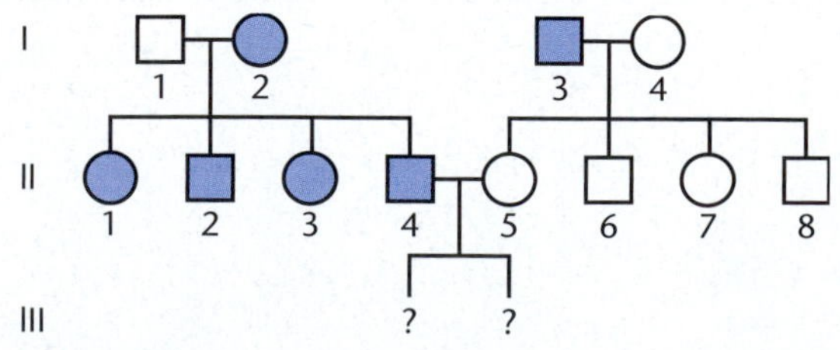

Analyze the pedigrees and propose a genetic explanation for the nature of the disorder. Consistent with your explanation, predict the outcome of a mating between individuals II-4 and II-5.

16. In the late 1950s, Yuichiro Hiraizumi, a postdoctoral fellow at the University of Wisconsin, crossed a laboratory strain of *Drosophila melanogaster* that was homozygous for the recessive second chromosomal mutations *cinnabar* (*cn*) and *brown* (*bw*) with a wild-type strain collected in Madison, Wisconsin. The lab strain had white eyes, while the wild strain had red eyes and was subsequently designated *SD*. The resulting progeny, being heterozygous for *cn* and *bw*, had red eyes. When Hiraizumi backcrossed F_1 females with *cn bw/cn bw* males, 50 percent of the offspring had white eyes, as expected. When the reciprocal backcross was made (F_1 males with *cn bw/cn bw* females), less than 2 percent of the flies had white eyes.

(a) Propose an explanation that is consistent with these results.
(b) Design a genetic experiment to test your hypothesis.
(c) *SD* stands for "segregation distortion." What is the significance of this description? [*Reference: Genetics* 44:232–50 (1959).]

Selected Readings

Adams, K. L., et al. 2000. Repeated, recent and diverse transfers of a mitochondrial gene to the nucleus in flowering plants. *Nature* 408:354–57.

Bogorad, L. 1981. Chloroplasts. *J. Cell Biol.* 91:256s–70s.

Cattanach, B.M., and Jones, J. 1994. Genetic imprinting in the mouse: Implications for gene regulation. *J. Inherit. Metab. Dis.* 17:403–20.

Choman, A. 1998. The myoclonic epilepsy and ragged-red fiber mutation provides new insights into human mitochondrial function and genetics. *Am. J. Hum. Genet.* 62:745–51.

Cohen, S. 1973. Mitochondria and chloroplasts revisited. *Am. Sci.* 61:437–45.

Feil, R., and Kelsey, G. 1997. Genomic imprinting: A chromatin connection. *Am. J. Hum. Genet.* 61:1213–19.

Freeman, G., and Lundelius, J.W. 1982. The developmental genetics of dextrality and sinistrality in the gastropod *Lymnaea peregra. Wilhelm Roux Arch.* 191:69–83.

Gillham, N.W. 1978. *Organelle heredity*. New York: Raven Press.

Goodenough, U., and Levine, R.P. 1970. The genetic activity of mitochondria and chloroplasts. *Sci. Am.* (Nov.) 223:22–29.

Green, B.R., and Burton, H. 1970. Acetabularia chloroplast DNA: Electron microscopic visualization. *Science* 168:981–82.

Grivell, L.A. 1983. Mitochondrial DNA. *Sci. Am.* (Mar.) 248:78–89.

Lander, E.S., et al. 1990. Mitochondrial diseases: Gene mapping and gene therapy. *Cell* 61:925–26.

Larson, N.G., and Clayton, D.A. 1995. Molecular genetic aspects of human mitochondrial disorders. *Ann. Rev. Genet.* 29:151–78.

Levine, R.P., and Goodenough, U. 1970. The genetics of photosynthesis and of the chloroplast in *Chlamydomonas reinhardi. Annu. Rev. Genet.* 4:397–408.

Manfredi, G., et al. 1997. The fate of human sperm-derived mtDNA in somatic cells. *Am. J. Hum. Genet.* 61:953–60.

Margulis, L. 1970. *Origin of eukaryotic cells*. New Haven, CT: Yale University Press.

Mitchell, M.B., and Mitchell, H.K. 1952. A case of maternal inheritance in *Neurospora crassa. Proc. Natl. Acad. Sci. USA* 38:442–49.

Nüsslein-Volhard, C. 1996. Gradients that organize embryo development. *Sci. Am.* (Aug.) 275:54–61.

Preer, J.R. 1971. Extrachromosomal inheritance: Hereditary symbionts, mitochondria, chloroplasts. *Annu. Rev. Genet.* 5:361–406.

Rosing, H.S., et al. 1985. Maternally inherited mitochondrial myopathy and myoclonic epilepsy. *Ann. Neurol.* 17:228–37.

Sager, R. 1965. Genes outside the chromosomes. *Sci. Am.* (Jan.) 212:70–79.

——— 1985. Chloroplast genetics. *BioEssays* 3:180–84.

Schwartz, R.M., and Dayhoff, M.O. 1978. Origins of prokaryotes, eukaryotes, mitochondria and chloroplasts. *Science* 199:395–403.

Slonimski, P. 1982. *Mitochondrial genes*. Cold Spring Harbor, NY: Cold Spring Harbor Laboratory Press.

Sonneborn, T.M. 1959. Kappa and related particles in *Paramecium. Adv. Virus Res.* 6:229–56.

Strathern, J.N., et al., eds. 1982. *The molecular biology of the yeast Saccharomyces: Life cycle and inheritance*. Cold Spring Harbor, NY: Cold Spring Harbor Laboratory Press.

Sturtevant, A.H. 1923. Inheritance of the direction of coiling in *Limnaea. Science* 58:269–70.

Tzagoloff, A. 1982. *Mitochondria*. New York: Plenum Press.

Wallace, D.C. 1992. Mitochondrial genetics: A paradigm for aging and degenerative diseases. *Science* 256:628–32.

——— 1997. Mitochondrial DNA in aging and disease. *Sci. Am.* Aug. 277:40–59.

——— 1999. Mitochondrial diseases in man and mouse. *Science* 283: 1482–88.

Wallace, D.C., et al. 1988. Familial mitochondrial encephalomyopathy (MERRF): Genetic, pathophysiological and biochemical characterization of a mitochondrial DNA disease. *Cell* 55:601–10.

GENETICS MediaLab

The resources that follow will help you achieve a better understanding of the concepts presented in this chapter. These resources can be found either on the CD packaged with this textbook or on the Companion Web site found at **http://www.prenhall.com/klug**

Web Problem 1:

Time for completion = 10 minutes

Can infectious elements alter Mendelian ratios? Certain microbes alter the genetics of sexual reproduction. One of the most interesting of these is Wolbachia, a rickettsialike organism that has been found in the gametes of many insect species. Read the Science News article about Wolbachia and answer the following questions: What is cytoplasmic incompatibility? What is the genetic mechanism by which Wolbachia produces all-male broods in jewel wasps? Would you expect to find genetic variation within a brood of parthenogenetically produced offspring? How are these parasites transferred across generations? To complete this exercise, visit Web Problem 1 in Chapter 14 of your Companion Web site, and select the keyword **WOLBACHIA**.

Web Problem 2:

Time for completion = 10 minutes

Where did the organelles in eukaryotic cells come from? Dr. Lynn Margulis's endosymbiotic theory of organelle inheritance explains why organelles, such as chloroplasts and mitochondria, are maternally inherited and have their own DNA. Read this article about the origin of these organelles, and think about the genetic consequences of this extraordinary event. Hypothesize about the genetic changes implied by Dr. Jeon's discovery of symbiosis between amoeba and intracellular bacteria. Where are the genes for mitochondria? Why are mitochondria maternally inherited? To complete this exercise, visit Web Problem 2 in Chapter 14 of your Companion Web site, and select the keyword **MITOCHONDRIA**.

Web Problem 3:

Time for completion = 10 minutes

Are maternal and paternal chromosomes always equally expressed? Genetic imprinting describes the silencing of either the maternal or paternal copy of a gene, as described in Chapter 8 of your text. Davor Solter, M.D., of the Max Plank Institute, has studied the developmental consequences of imprinting in mice. After you read about his research, answer the following questions: Given that genes can be maternally or paternally imprinted, is the imprinted gene expressed in the zygote? Have imprinted genes been implicated in any human diseases? Does imprinting alter Mendelian inheritance? Why is the March of Dimes so interested in Dr. Solter's research? To complete this exercise, visit Web Problem 3 in Chapter 14 of your Companion Web site, and select the keyword **IMPRINTING**.

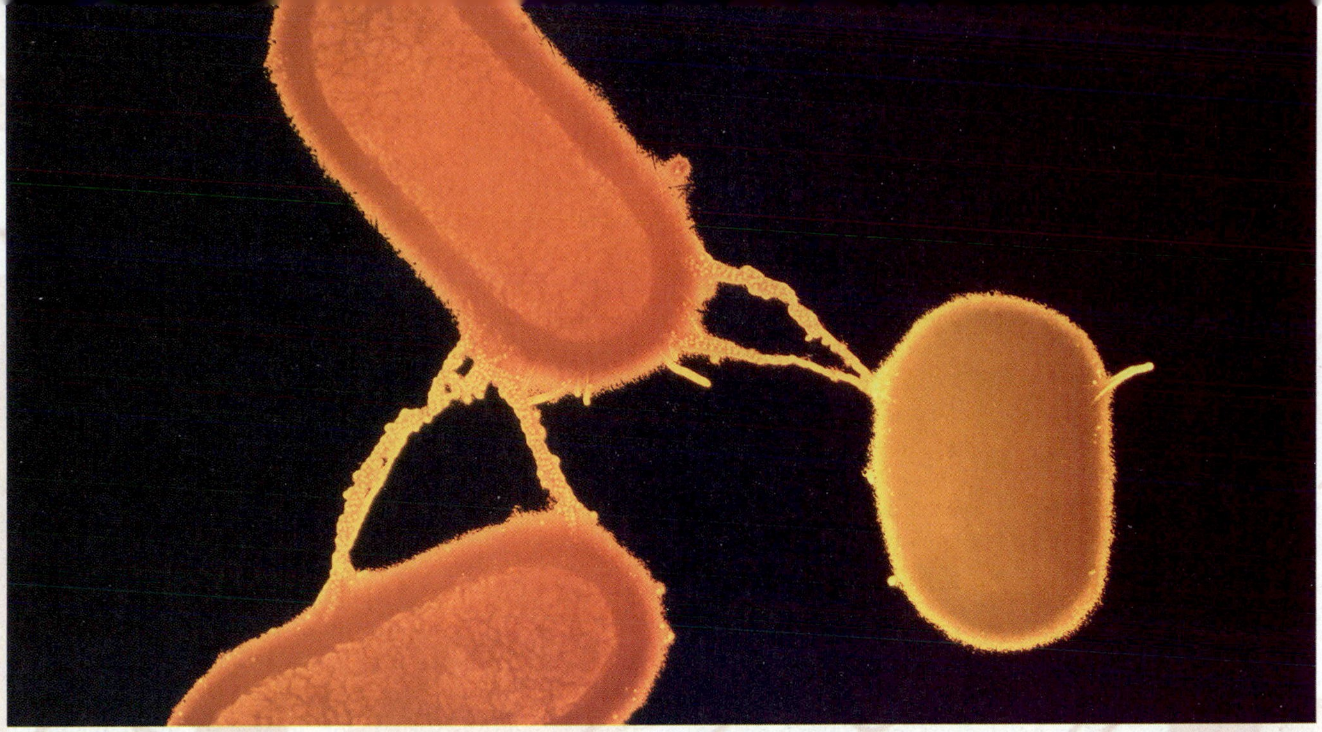

Transmission electron micrograph of conjugating *E. coli*. (*Dr. L. Caro/Science Photo Library/Photo Researchers, Inc.*)

15

Genetics of Bacteria and Bacteriophages

KEY CONCEPTS

In this chapter, we shift our attention to a discussion of various genetic phenomena in **bacteria** (prokaryotes) and **bacteriophages**, viruses that have bacteria as their host. The study of bacteria and bacteriophages has been essential to the accumulation of knowledge in many areas of genetic study. For example, much of what we have previously discussed about molecular genetics initially was derived from experimental work with these organisms. Furthermore, as we shall see in Chapter 16, our knowledge of bacteria and their resident plasmids has served as the basis for their widespread use in DNA cloning and other recombinant DNA studies.

Bacteria and their viruses have been especially useful research organisms in genetics for a number of reasons. First, they have extremely short reproductive cycles. Literally hundreds of generations, giving rise to billions of genetically identical bacteria or phages, can be produced in short periods of time. Furthermore, they can be studied in pure cultures. That is, a single species or mutant strain of bacteria or one type of virus can be isolated and investigated independently of other similar organisms.

In this chapter, we focus on genetic recombination in both bacteria and bacteriophages. Complex processes have evolved in these microorganisms and viruses that facilitate genetic recombination within populations. As we shall see, these processes are the basis for the chromosome mapping analysis that forms the cornerstone of molecular genetic investigations of bacteria and the viruses that invade them.

15.1 Bacteria Mutate Spontaneously and Grow at an Exponential Rate

Genetic studies using bacteria depend upon our ability to isolate mutations in these organisms. It has long been known that pure cultures of bacteria give rise to cells that exhibit heritable variation, particularly with respect to growth under unique environmental conditions. Prior to 1943, the source of this variation was hotly debated. The majority of bacteriologists believed that environmental factors induced changes in certain bacteria that led to their survival or adaptation to the new conditions. For example, strains of *E. coli* are known to be sensitive to infection by the bacteriophage T1. Infection by the bacteriophage T1 leads to the reproduction of the virus at the expense of the bacterial cell, which is lysed or destroyed. If a plate of *E. coli* is uniformly sprayed with T1, almost all cells are lysed. Rare *E. coli* cells, however, survive infection and are not lysed. If these cells are isolated and established in pure culture, all their descendants are resistant to T1 infection. The adaptation hypothesis, put forth to explain this type of observation, implies that the interaction of the phage and bacterium is essential to the acquisition of immunity. In other words, the phage "induces" resistance in the bacteria. On the other hand, the occurrence of **spontaneous mutations**, which occur in the presence or the absence of phage T1, suggested an alternative model to explain the origin of resistance in *E. coli*. In 1943, Salvador Luria and Max Delbruck presented the first convincing evidence that bacteria, like eukaryotic organisms, are capable of spontaneous mutation. This experiment, referred to as the **fluctuation test**, marks the initiation of modern bacterial genetic study. Spontaneous mutation is now considered the primary source of genetic variation in bacteria. We explored this discovery in some detail in Chapter 7.

Mutant cells that arise spontaneously in otherwise pure cultures can be isolated and established independently from the parent strain by using established selection techniques. As a result, mutations for almost any desired characteristic can now be induced and isolated. Because bacteria and viruses usually contain only a single chromosome and are therefore haploid, all mutations are expressed directly in the descendants of mutant cells, adding to the ease with which these microorganisms can be studied.

Bacteria are grown in either a liquid culture medium or in a petri dish on a semisolid agar surface. If the nutrient components of the growth medium are very simple and consist only of an organic carbon source (such as a glucose or a lactose) and a variety of inorganic ions, including Na^+, K^+, Mg^{++}, Ca^{++}, and NH_4^+ present as inorganic salts, it is called **minimal medium**. To grow on such a medium, a bacterium must be able to synthesize all essential organic compounds (e.g., amino acids, purines, pyrimidines, sugars, vitamins, and fatty acids). A bacterium that can accomplish this remarkable biosynthetic feat—one that we ourselves cannot duplicate—is termed a **prototroph**. It is said to be wild type for all growth requirements and can grow on minimal medium. On the other hand, if a bacterium loses, through mutation, the ability to synthesize one or more organic components, it is said to be an **auxotroph**. For example, a bacterium that loses the ability to make histidine is designated as a his^- auxotroph, in contrast to its prototrophic his^+ counterpart. For the his^- bacterium to grow, this amino acid must be added as a supplement to the minimal medium. Note that medium that has been extensively supplemented is referred to as complete medium.

To study mutant bacteria in a quantitative fashion, an inoculum (e.g., 0.1 ml, 1.0 ml) of bacteria is placed in liquid culture medium. The bacteria exhibit a characteristic growth pattern, as illustrated in Figure 15–1. Initially, during the **lag phase**, growth is slow. Then, a period of rapid growth, called the **log phase**, ensues. During this phase, cells divide continually with a fixed time interval between cell divisions, resulting in exponential growth. When the bacteria reach a cell density of about 10^9 cells/ml, nutrients and oxygen become limiting and cells enter the **stationary phase**. Because the doubling time during the log phase can be as short as 20 minutes, an initial inoculum of a few thousand cells added to the culture easily achieves maximum cell density during an overnight incubation.

Cells grown in liquid medium can be quantified by transferring them to semisolid medium in a petri dish. Following incubation and many divisions, each cell gives rise to a colony visible on the surface of the medium. From the number of colonies that subsequently grow, it is possible to estimate the number of bacteria present in the original culture. If the number of colonies is too great to count, then a

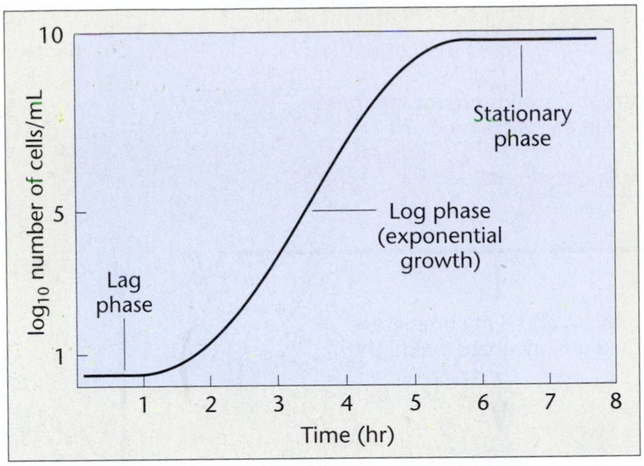

FIGURE 15–1 Typical bacterial population growth curve illustrating the initial lag phase, the subsequent log phase where exponential growth occurs, and the stationary phase that occurs when nutrients are exhausted.

series of successive dilutions (a technique called serial dilution) of the original liquid culture is made and plated, until the colony number is reduced to the point where it can be counted (Figure 15–2). For example, assume that the three petri dishes in Figure 15–2 represent dilutions of 10^{-3}, 10^{-4}, and 10^{-5}, respectively (left to right). We need only select the dish in which the number of colonies can be counted accurately. Because each colony presumably arose from a single bacterium, the number of colonies times the dilution factor represents the number of bacteria in the initial milliliter. In this case, the dish farthest to the right contains 15 colonies. Since it represents a dilution of 10^{-5}, we can estimate the initial number of bacteria to be 15×10^5 per milliliter. Calculations such as these are useful in a number of studies.

15.2 Conjugation Is One Means of Genetic Recombination in Bacteria

Development of techniques that allowed the identification and study of bacterial mutations led to detailed investigations of the arrangement of genes on the bacterial chromosome. Such studies began in 1946, when Joshua Lederberg and Edward Tatum showed that bacteria undergo **conjugation**, a parasexual process in which the genetic information from one bacterium is transferred to, and recombined with, that of another bacterium. (See the photograph that opens this chapter.) Like meiotic crossing over in eukaryotes, genetic recombination in bacteria provided the basis for the development of methodology for chromosome mapping. Note that the term **genetic recombination**, as applied to bacteria and bacteriophages, leads to the *replacement* of one or more genes present in one strain with those from a genetically distinct strain. While this is somewhat different from our use of genetic recombination in eukaryotes—where, the term describes crossing over that results in *reciprocal exchange events*—the overall effect is the same: Genetic information is transferred from one chromosome to another, resulting in an altered genotype. Two other phenomena that result in the transfer of genetic information from one bacterium to another, **transformation** and **transduction**, have also served as a basis for determining the arrangement of genes on the bacterial chromosome. We shall discuss these processes in later sections of this chapter.

Lederberg and Tatum's initial experiments were performed with two multiple auxotroph strains (nutritional mutants) of *E. coli* K12. As shown in Figure 15–3, strain A required methionine (met) and biotin (bio) in order to grow, whereas strain B required threonine (thr), leucine (leu), and thiamine (thi). Neither strain would grow on minimal medium. The two strains were first grown separately in supplemented media, and then cells from both were mixed and grown together for several more generations. They were then plated on minimal medium. Any bacterial cells that grew on minimal medium were prototrophs (wild-type bacteria that did not need nutritional supplements). It is highly improbable that any of the cells that contained two or three mutant genes would undergo spontaneous mutation simultaneously at two or three independent locations, leading to wild-type cells. Therefore, the researchers assumed that any prototrophs recovered must have arisen as a result of some form of genetic exchange and recombination between the two mutant strains.

In this experiment, prototrophs were recovered at a rate of $1/10^7$ (10^{-7}) cells plated. The controls for this experiment involved separate plating of cells from strains A and B on minimal medium. No prototrophs were recovered. On the basis of these observations, Lederberg and Tatum proposed

FIGURE 15–2 Results of the serial dilution technique and subsequent culture of bacteria. Each of the dilutions varies by a factor of 10. Each colony was derived from a single bacterial cell.

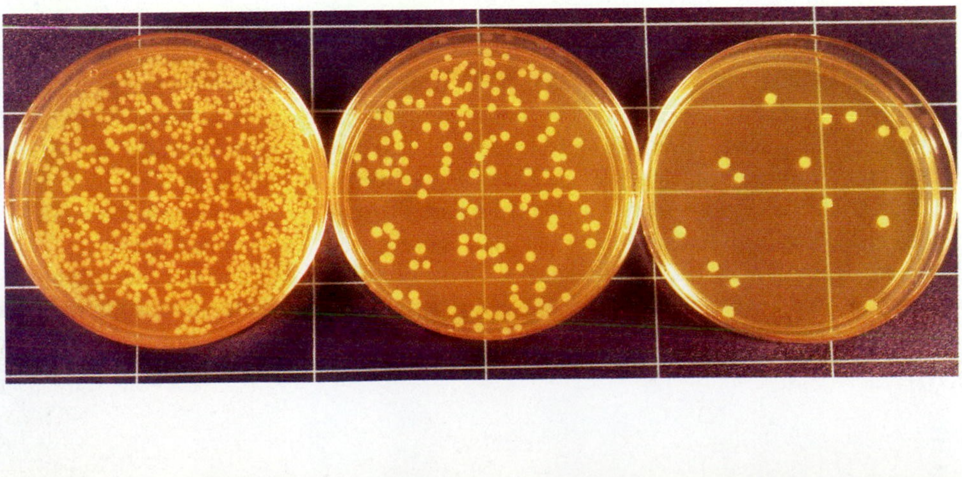

FIGURE 15–3 Genetic recombination involving two auxotrophic strains producing prototrophs. Neither auxotrophic strain will grow on minimal medium, but prototrophs do, suggesting that genetic recombination has occurred.

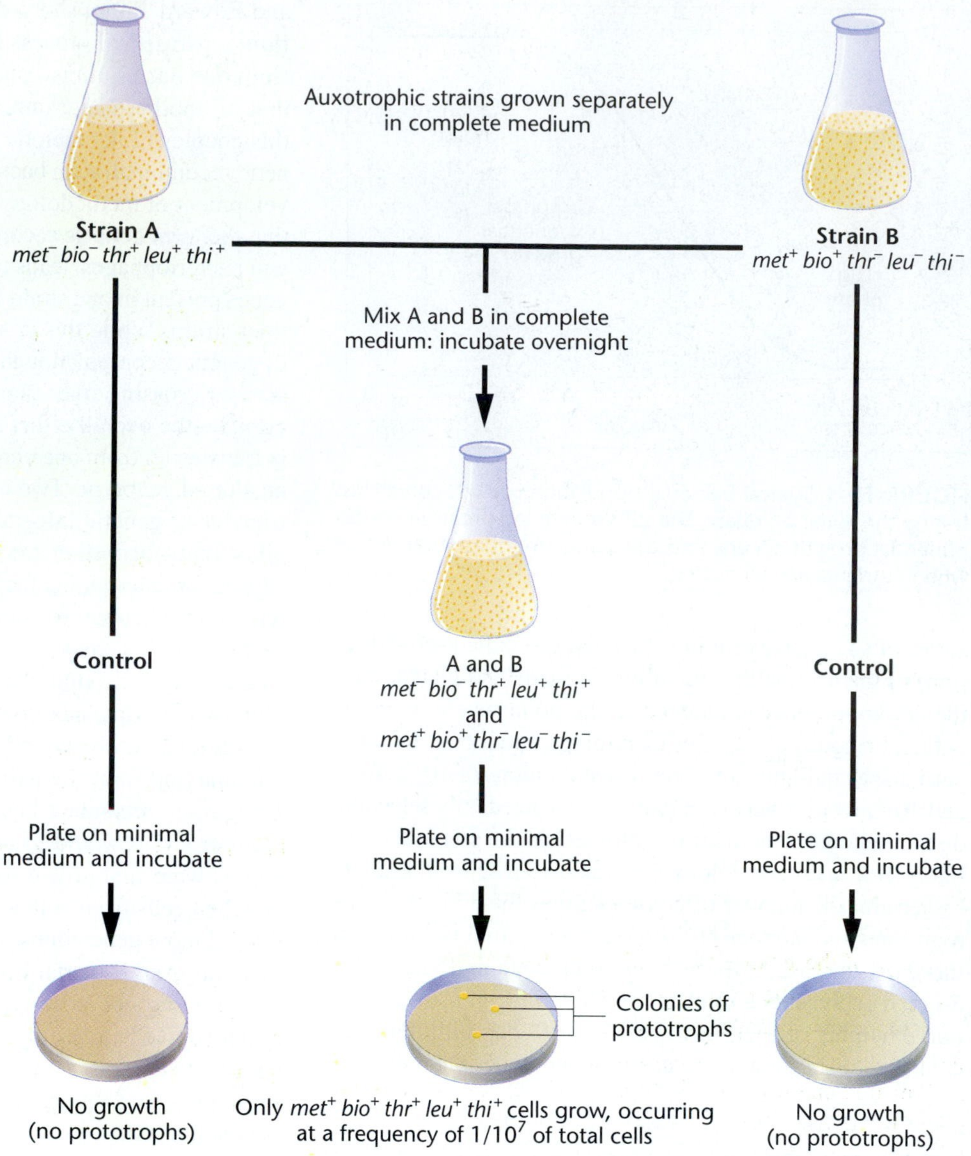

MEDIA TUTORIAL Bacterial Genetics

that, while the events were indeed quite rare, genetic recombination had occurred.

F^+ and F^- Bacteria

Lederberg and Tatum's findings were soon followed by numerous experiments designed to elucidate the genetic basis of conjugation. It quickly became evident that different strains of bacteria were involved in a unidirectional transfer of genetic material. When cells serve as donors of parts of their chromosomes, they are designated as **F^+ cells** (F for "fertility"). Recipient bacteria receive the donor chromosome material (now known to be DNA), and recombine it with part of their own chromosome. They are designated as **F^- cells.**

It was established subsequently that cell contact is essential to chromosome transfer. Support for this concept was provided by Bernard Davis, who designed a U-tube in which to grow F^+ and F^- cells (Figure 15–4). At the base of the tube is a sintered glass filter with a pore size that allows passage of the liquid medium, but that is too small to allow the passage of bacteria. The F^+ cells are placed on one side of the filter and F^- cells on the other side. The medium is moved back and forth across the filter so that the bacterial cells essentially share a common medium during incubation. Davis plated samples form both sides of the tube on minimal medium, but no prototrophs were found. He logically concluded that *physical contact is essential to genetic recombination.* We now know that this physical interaction is the initial stage of the process of conjugation and is mediated through a conjugation tube called the **F** (or **sex pilus**, pl. pili). Bacteria often have many pili, which are microscopic tubelike extensions of the cell. Different types of pili perform different cellular functions, but all pili are involved in some way with adhesion. After contact has been initiated between mating pairs through the F pili (Figure 15–5), transfer of the chromosome begins.

Later evidence established that F^+ cells contained a **fertility factor** (called the **F factor**) that confers the ability to donate part of their chromosome during conjugation. Experiments by Joshua and Esther Lederberg and by William

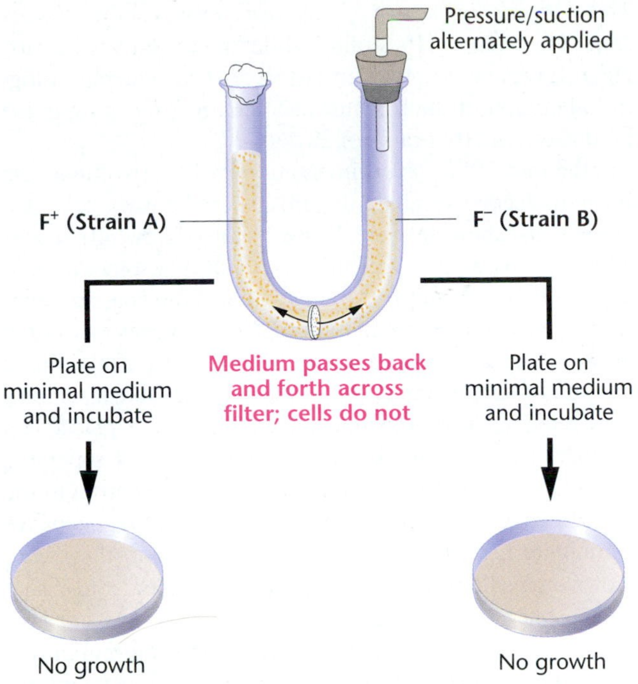

FIGURE 15–4 When strain A and strain B auxotrophs are grown in a common medium, but separated by a filter, no genetic recombination occurs and no prototrophs are produced. The apparatus shown is a Davis U-tube.

Hayes and Luca Cavalli-Sforza showed that certain conditions could eliminate the F factor in otherwise fertile cells. However, if these "infertile" cells were then grown with fertile donor cells, the F factor was regained.

The conclusion that the F factor is a mobile element was further supported by the observation that, following conjugation and genetic recombination, recipient cells always become F^+. Thus, in addition to the *rare* cases of transfer of genes from the bacterial chromosome (genetic recombination), the F factor itself is passed to *all* recipient cells. On this basis, the initial crosses of Lederberg and Tatum (Figure 15–3) may be designated as follows:

Strain A		**Strain B**
F^+	×	F^-
(DONOR)		(RECIPIENT)

Isolation of the F factor confirmed these conclusions. Like the bacterial chromosome, though distinct from it, the F factor has been shown to consist of a circular, double-stranded DNA molecule, equivalent to about 2 percent of the bacterial chromosome (about 100,000 nucleotide pairs). Contained in the F factor, among others, are 19 genes, the products of which are involved in the transfer of genetic information (*tra* genes). These include those essential to the formation of the sex pilus.

As we soon shall see, the F factor is in reality an autonomous genetic unit referred to as a **plasmid**. However, in our historical coverage of its discovery, we will continue in this chapter to refer to it as a "factor."

It is believed that the transfer of the F factor during conjugation involves separation of the two strands of its double helix and the movement of one of the two strands into the recipient cell. The other strand remains in the donor cell. Both strands, one moving across the conjugation tube and one remaining in the donor cell, are replicated. The result is that both the donor *and* the recipient cells are F^+. This process is diagrammed in Figure 15–6.

In sum, *E. coli* cells may or may not contain the F factor. When it is present, the cell is able to form a sex pilus and potentially serve as a donor of genetic information. During conjugation, a copy of the F factor is almost always transferred from the F^+ cell to the F^- recipient, converting it to the F^+ state. The question remained as to exactly why such a low proportion of cells involved in these matings (10^{-7}) also resulted in genetic recombination. The answer awaited further experimentation. Subsequent discoveries not only clarified how genetic recombination occurs, but also defined a mechanism by which the *E. coli* chromosome could be mapped. We shall first address chromosome mapping.

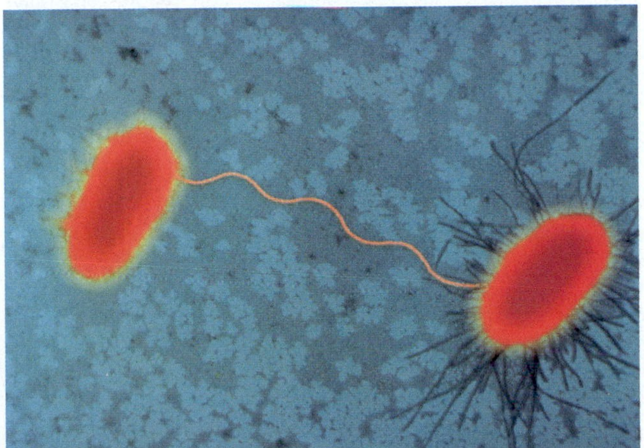

FIGURE 15–5 An electron micrograph of conjugation between an F^+ *E. coli* cell and an F^- cell. The sex pilus linking them is clearly visible.

Hfr Bacteria and Chromosome Mapping

In 1950, Cavalli-Sforza treated an F^+ strain of *E. coli* K12 with nitrogen mustard, a potent chemical known to induce mutations. From these treated cells, he recovered a strain of donor bacteria that underwent recombination at a rate of $1/10^4$ (or 10^{-4}), 1000 times more frequently than the original F^+ strains. In 1953, William Hayes isolated another strain that demonstrated an elevated frequency. Both strains were designated **Hfr**, for **high-frequency recombination**. Because Hfr cells behave as chromosome donors, they are a special class of F^+ cells.

Another important difference was noted between Hfr strains and the original F^+ strains. If the donor is from an Hfr strain, recipient cells, while sometimes displaying genetic recombination, never become Hfr—that is, they remain F^-. In comparison, then,

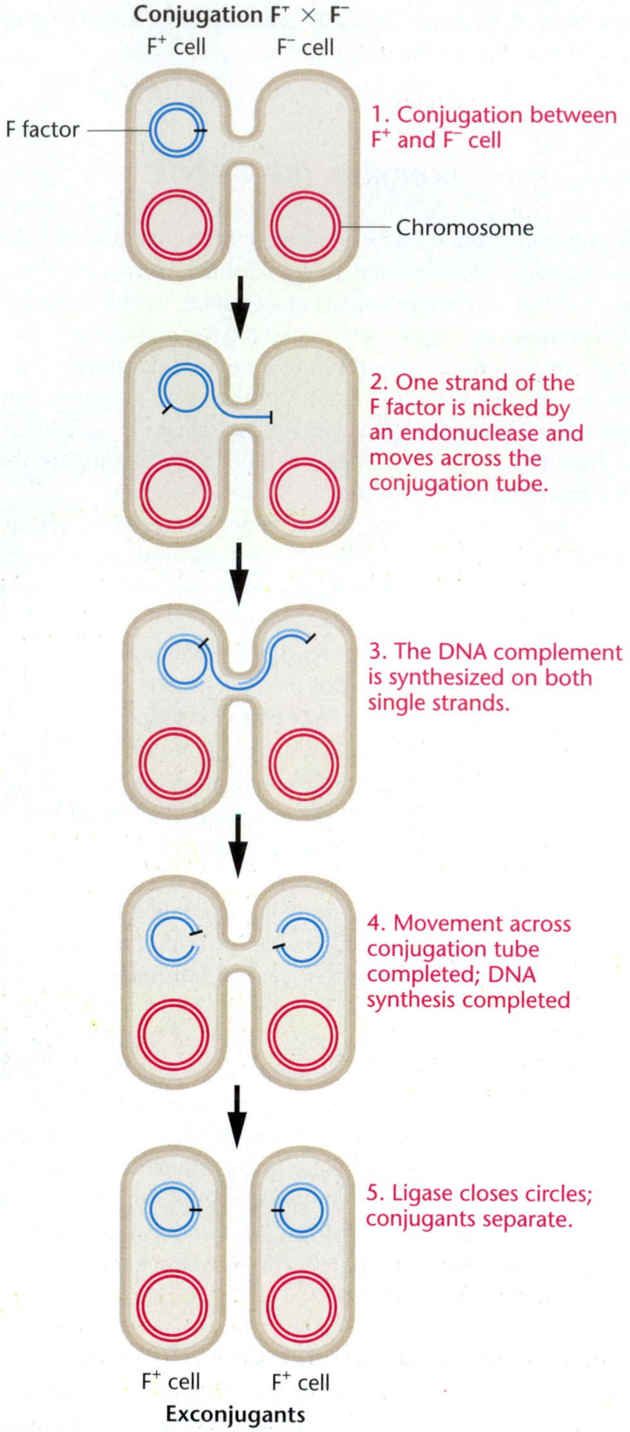

FIGURE 15–6 An $F^+ \times F^-$ mating, demonstrating how the recipient F^- cell is converted to F^+. During conjugation, the DNA of the F factor is replicated with one new copy entering the recipient cell, converting it to F^+. To indicate the clockwise rotation during replication, a black bar is shown on the F factor.

$F^+ \times F^- \longrightarrow$ Recipient becomes F^+

$Hfr \times F^- \longrightarrow$ Recipient remains F^-

Perhaps the most significant characteristic of Hfr strains is the *nature of recombination*. In any given strain, certain genes are more frequently recombined than others, and some not at all. This *nonrandom pattern of gene transfer* was shown to vary from Hfr strain to Hfr strain. While these results were puzzling, Hayes interpreted them to mean that some physiological alteration of the F factor had occurred, resulting in the production of Hfr strains of *E. coli*.

In the mid-1950s, experimentation by Ellie Wollman and François Jacob explained the differences between cells that are Hfr and those that are F^+ and showed how Hfr strains allow genetic mapping of the *E. coli* chromosome. In their experiments, Hfr and F^- strains with suitable marker genes were mixed and recombination of specific genes assayed at different times. To accomplish this, a culture containing a mixture of an Hfr and an F^- strain was incubated, samples were removed at various intervals, and each was placed in a blender. The shear forces created in the blender separated conjugating bacteria so that the transfer of the chromosome was effectively terminated. To assay the cells for genetic recombination following the blender treatment, they were grown on medium *containing* the antibiotic in order to ensure the recovery of only recipient cells.

This process, called the **interrupted mating technique**, demonstrated that specific genes of a given Hfr strain were transferred and recombined sooner than others. Figure 15–7 illustrates this point. During the first 8 minutes after the two strains are mixed, no genetic recombination can be detected. At about 10 minutes, recombination of the *aziR* gene can be detected, but no transfer of the ton^S, lac^+, or gal^+ genes is noted. By 15 minutes, 70 percent of the recombinants are *aziR*; 30 percent are now also ton^S; but none is lac^+ or gal^+. Within 20 minutes, the lac^+ gene is found among the

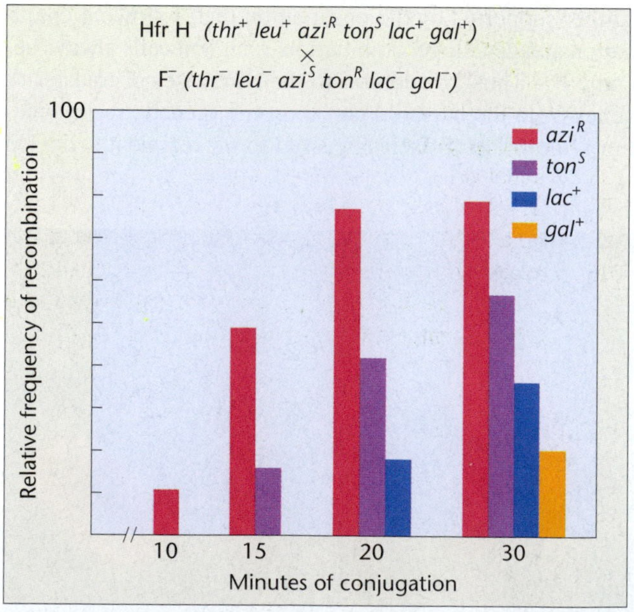

FIGURE 15–7 The progressive transfer during conjugation of various genes from a specific Hfr strain of *E. coli* to an F^- strain. In this strain certain genes (*azi* and *ton*) are transferred sooner than others and recombine more frequently. Others (*lac* and *gal*) take longer to transfer and recombine with a lower frequency. Still others (*thr* and *leu*) are always transferred and are used in the initial screen for recombinants.

recombinants; and within 30 minutes, gal^+ is also being transferred. Wollman and Jacob had demonstrated an *oriented transfer of genes* that correlated with the length of time conjugation proceeded.

It appeared that the chromosome of the Hfr bacterium was transferred linearly and that the gene order and distance between genes, as measured in minutes, could be predicted from such experiments (Figure 15–8). This information served as the basis for the first genetic map of the *E. coli* chromosome. "Minutes" in bacterial mapping are equivalent to "map units" in eukaryotes.

Wollman and Jacob repeated the same type of experiment with other Hfr strains, obtaining similar results, but with one important difference. Although genes were always transferred linearly with time (as in their original experiment), which genes entered first and which followed later seemed to vary from Hfr strain to Hfr strain [Figure 15–9(a)]. When they reexamined the rate of entry of genes, and thus the different genetic maps for each strain, a definite pattern emerged. The major difference between each strain was simply the point of the origin and the direction in which entry proceeded from that point [Figure 15–9(b)].

To explain these results, Wollman and Jacob postulated that the *E. coli* chromosome is circular (a closed circle, with no free ends). If the point of origin (*O*) varied from strain to strain, a different sequence of genes would be transferred in each case. But what determines *O*? They proposed that, in various Hfr strains, the F factor integrates into the chromosome at different points and its position determines the *O* site. One such case of integration is shown in Figure 15–10 (Step 1). During conjugation between this Hfr and an F^- cell, the position of the F factor determines the initial point of transfer (Steps 2 and 3). Those genes adjacent to *O* are transferred first. *The F factor becomes the last part to be transferred* (Step 4). However, conjugation rarely, if ever, lasts long enough to allow the entire chromosome to pass across the conjugation tube (Step 5). This proposal explains why recipient cells, when mated with Hfr cells, remain F^-.

Figure 15–10 also depicts the way in which the two strands making up a DNA molecule unwind during transfer, allowing for the entry of one of the strands of DNA into the recipient (Step 3). Following replication, the entering DNA now has the potential to recombine with its homologous region of the host chromosome. The DNA strand that remains in the donor also undergoes replication.

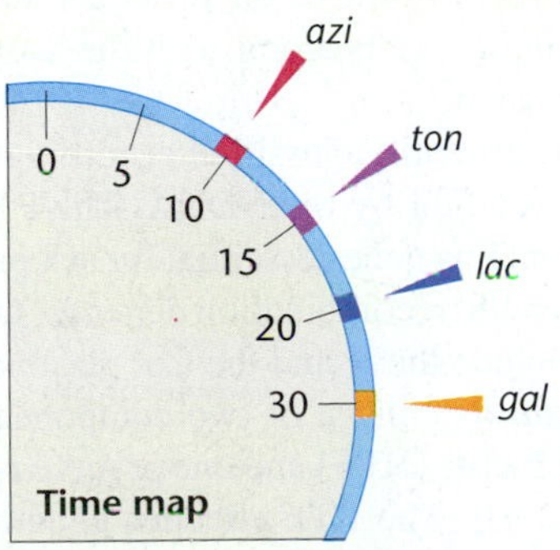

FIGURE 15–8 A time map of the genes studied in the experiment depicted in Figure 15–7.

The use of interrupted mating technique with different Hfr strains allowed researchers to map the entire *E. coli* chromosome. Mapped in time units, strain K12 (or *E. coli* K12) is 100 minutes long. Over 900 genes have now been placed on the map. In most instances, only a single copy of each gene exists.

Recombination in $F^+ \times F^-$ Matings: A Reexamination

The preceding model helped geneticists better understand how genetic recombination occurs during the $F^+ \times F^-$ matings. Recall that recombination occurs much less frequently in them than in Hfr $\times$ F^- matings and that random gene transfer is involved. The current belief is that, when F^+ and F^- cells are mixed, conjugation occurs readily and that each F^- cell involved in conjugation with an F^+ cell receives a copy of the F factor, *but that no genetic recombination occurs*. However, at an extremely low frequency in a population of F^+ cells, the F factor integrates spontaneously from the cytoplasm to a random point in the bacterial chromosome, converting the F^+ cell to the Hfr state, as we saw in Figure 15–10. Therefore, in $F^+ \times F^-$ crosses, the extremely low frequency of genetic recombination (10^{-7}) is attributed to the rare, newly formed Hfr cells, which then undergo conjugation with F^- cells. Because the point of integration of the F factor is random, the genes that are transferred by any newly formed Hfr donor will also appear to be random within the larger F^+/F^- population. The recipient bacterium will appear as a recombinant cell, but will, in fact, remain F^-. If it subsequently undergoes conjugation with an F^+ cell, it will be converted to F^+.

The F′ State and Merozygotes

In 1959, during experiments with Hfr strains of *E. coli*, Edward Adelberg discovered that the F factor could lose its integrated status, causing the cell to revert to the F^+ state (Figure 15–11, Step 1). When this occurs, the F factor frequently carries several adjacent bacterial genes along with it (Step 2). Adelberg labeled this condition **F′** to distinguish it from F^+ and Hfr. F′, like Hfr, is thus another special case of F^+. This conversion is described as one from Hfr to F′.

The presence of bacterial genes within a cytoplasmic F factor creates an interesting situation. An F′ bacterium behaves like an F^+ cell, initiating conjugation with F^- cells (Figure 15–11, Step 3). When this occurs, the F factor, containing chromosomal genes, is transferred to the F^- cell (Step 4). As a result, whatever chromosomal genes are part of the F factor are now present in duplicate in the recipient cell (Step 5), because the recipient still has a complete chromosome. This creates a partially diploid cell called a **merozygote**. Pure cultures of F′ merozygotes can be established. They have been extremely useful in the study of bacterial genetics, particularly in genetic regulation.

(a)

Hfr strain	Order of transfer (Earliest)							(Latest)
H	*thr*	*leu*	*azi*	*ton*	*pro*	*lac*	*gal*	*thi*
1	*leu*	*thr*	*thi*	*gal*	*lac*	*pro*	*ton*	*azi*
2	*pro*	*ton*	*azi*	*leu*	*thr*	*thi*	*gal*	*lac*
7	*ton*	*azi*	*leu*	*thr*	*thi*	*gal*	*lac*	*pro*

(b)

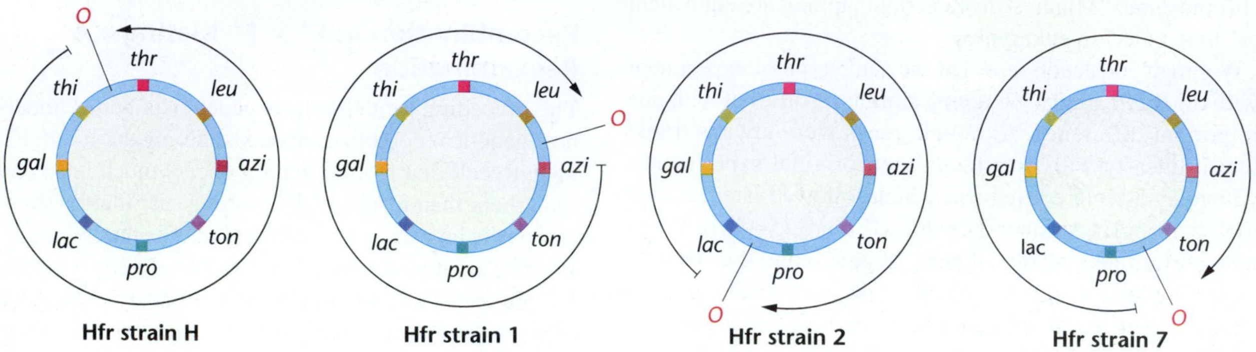

FIGURE 15–9 (a) The order of gene transfer in four Hfr strains, suggesting that the *E. coli* chromosome is circular. (b) The point where transfer originates (*O*) is identified in each strain. Note that transfer can proceed in either direction, depending on the strain. The origin is determined by the point of integration of the F factor into the chromosome, and the direction of transfer is determined by the orientation of the F factor as it integrates.

15.3 Rec Proteins Are Essential to Bacterial Recombination

Once researchers established that a unidirectional transfer of DNA occurs between bacteria, they were interested in determining how the actual recombination event occurs in the recipient cell. Just how does the donor DNA replace the comparable region in the recipient chromosome? As with many systems, the biochemical mechanism by which recombination occurs was deciphered through genetic studies. Major insights were gained as a result of the isolation of a group of mutations representing *rec* genes.

The first relevant observation in this case involved a series of mutant genes labeled *recA*, *recB*, *recC*, and *recD*. The first mutant gene, *recA*, was found to diminish genetic recombination in bacteria a thousandfold, nearly eliminating it altogether. The other *rec* mutations reduced recombination by about 100 times. Clearly, the normal wild-type products of these genes play some essential role in the process of recombination.

By looking for a functional gene product present in normal cells, but missing in mutant cells, researchers subsequently isolated several gene products and showed that they played a role in genetic recombination. The first is called the **RecA protein*** The second is a more complex protein called the **RecBCD protein**, an enzyme consisting of polypeptide subunits encoded by three other *rec* genes. This genetic research has extended our knowledge of the process of recombination considerably and underscores the value of isolating mutations, establishing their phenotypes, and determining the biological role of the normal, wild-type gene.

*Note that the names of bacterial genes begin with lowercase letters and are italicized. The names of the corresponding gene products (proteins) begin with an uppercase letter and are not italicized. For example, the *recA* gene encodes the RecA protein.

15.4 F Factors Are Plasmids

In the preceding sections, we have examined the extrachromosomal heredity unit called the F factor. When it exists autonomously in the bacterial cytoplasm, the F factor is composed of a double-stranded closed circle of DNA [Figure 15–12(a)]. These characteristics place the F factor in the more general category of genetic structures called **plasmids**. These structures contain one or more genes—often, quite a few. Their replication depends on the same enzymes that replicate the chromosome of the host cell, and they are distributed to daughter cells along with the host chromosome during cell division.

Plasmids are generally classified according to the genetic information specified by their DNA. The F factor confers fertility and contains genes essential for sex pilus formation, upon which genetic recombination depends. Other examples of plasmids include the R and the Col plasmids.

Most **R plasmids** consist of two components: the **resistance transfer factor (RTF)** and one or more **r-determinants** [Figure 15–12(b)]. The RTF encodes genetic information essential to transfer of the plasmid between bacteria, and the r-determinants are genes conferring resistance to antibiotics.

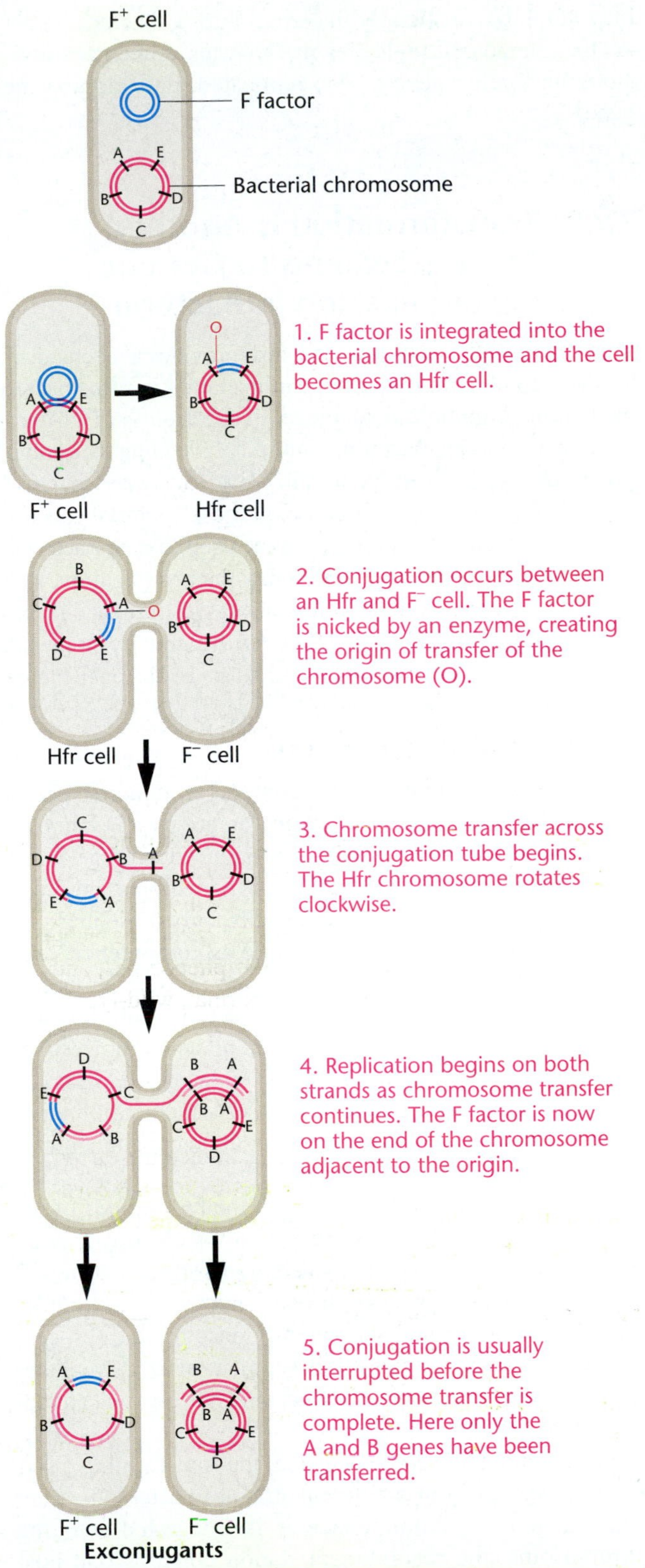

FIGURE 15–10 Conversion of F^+ to an Hfr state occurs by integration of the F factor into the bacterial chromosome. The point of integration determines the origin (*O*) of transfer. During a subsequent conjugation (Steps 2.–4.), an enzyme nicks the F factor, now integrated into the host chromosome, initiating the transfer of the chromosome at that point. Conjugation is usually interrupted prior to complete transfer. Above, only the *A* and *B* genes are transferred to the F^- cell, which may recombine with the host chromosome.

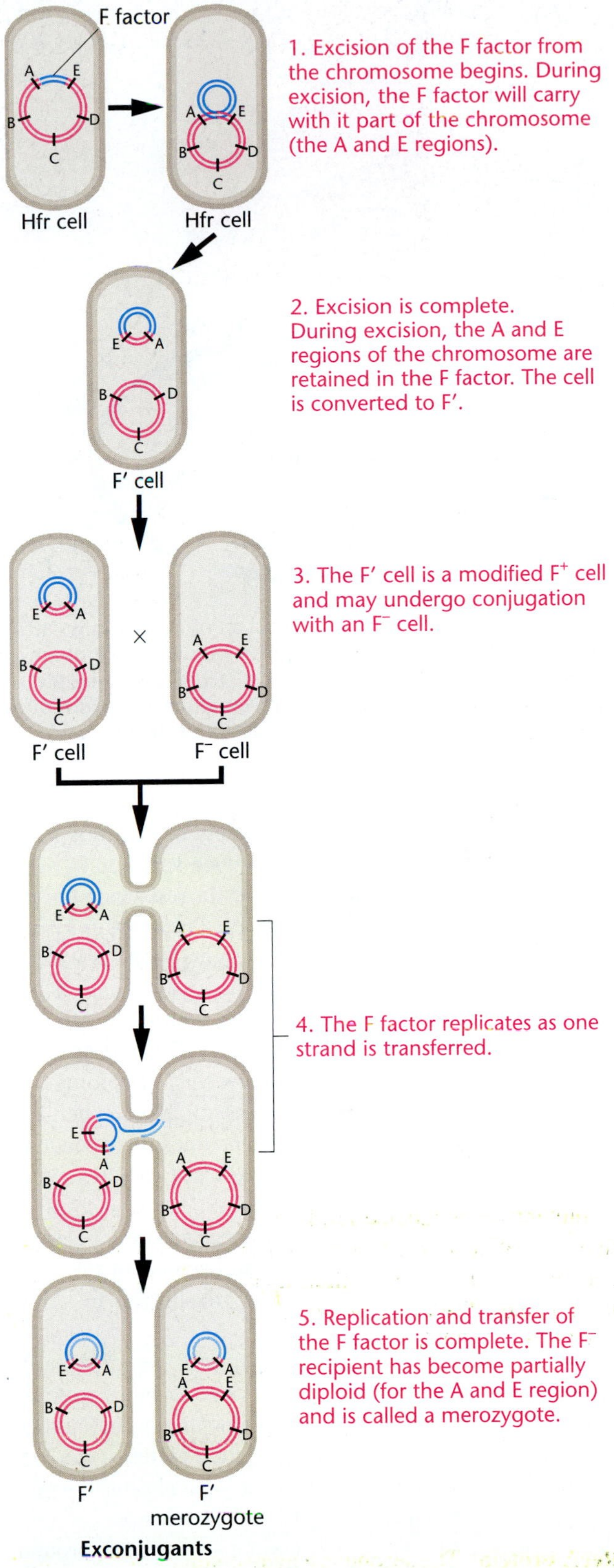

FIGURE 15–11 Conversion of an Hfr bacterium to F' and its subsequent mating with an F^- cell. The conversion occurs when the F factor loses its integrated status. During excision from the chromosome, the F factor may carry with it one or more chromosomal genes (*A* and *E*). Following conjugation with an F^- cell, the recipient cell becomes partially diploid and is called a merozygote. It also behaves as an F^+ donor cell.

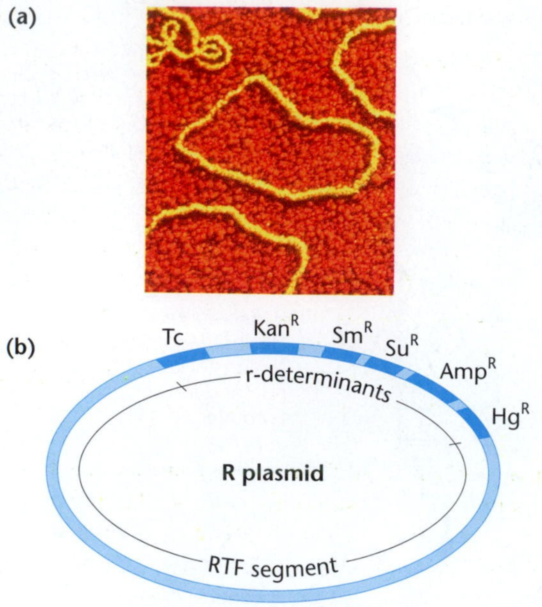

FIGURE 15–12 (a) Electron micrograph of a plasmid isolated from *E. coli*; (b) Diagrammatic representation of an R plasmid containing resistance transfer factors (RTFs) and multiple r-determinants (Tc, tetracycline; Kan, kanamycin; Sm, streptomycin; Su, sulfonamide; Amp, ampicillin; and Hg, mercury).

While RTFs are quite similar in a variety of plasmids from different bacterial species, there is a wide variation in r-determinants, each of which is specific for resistance to one class of antibiotic. Sometimes, a bacterial cell contains r-determinant plasmids, but no RTF is present. Such a cell is resistant, but cannot transfer the genetic information for resistance to recipient cells. The most commonly studied plasmids, however, contain the RTF as well as one or more r-determinants. Resistance to tetracycline, streptomycin, ampicillin, sulfonamide, kanamycin, and chloramphenicol are most frequently encountered. Sometimes these occur in a single plasmid, conferring multiple resistance to several antibiotics [Figure 15–12(b)]. Bacteria bearing such plasmids are of great medical significance, not only because of their multiple resistance, but also because of the ease with which the plasmids may be transferred to other bacteria.

The **Col plasmid**, ColE1, derived from *E. coli*, is clearly distinct from R plasmids. It encodes one or more proteins that are highly toxic to bacterial strains that do not harbor the same plasmid. These proteins, called **colicins**, can kill neighboring bacteria. Bacteria that carry the plasmid are said to be colicinogenic. Present in 10 to 20 copies per cell, col plasmid also contains a gene encoding an immunity protein that protects the host cell from the toxin. Unlike an R plasmid, the Col plasmid is not usually transmissible to other cells.

Interest in plasmids has increased dramatically because of their role in the genetic technology known as recombinant DNA research, which we discuss in Chapters 16. Specific genes from any source can be inserted into a plasmid, which may then be artificially inserted into a bacterial cell. As the altered cell replicates its DNA and undergoes division, the foreign gene is also replicated, thus cloning the genes.

15.5 Transformation Is Another Process Leading to Genetic Recombination in Bacteria

Transformation is a process that also provides a mechanism for the recombination of genetic information in some bacteria. In transformation, small pieces of extracellular DNA are taken up by a living bacterium, ultimately leading to a stable genetic change in the recipient cell. We are interested in transformation in this chapter because, in those bacterial species in which it occurs, the process can be used to map bacterial genes, although in a more limited way than conjugation. Recall that the process of transformation was instrumental in experiments that established DNA as the genetic material (Chapter 2).

The Transformation Process

Transformation (Figure 15–13) consists of numerous steps that can be divided into two main categories: (1) entry of DNA into a recipient cell, and (2) recombination of the donor DNA with its homologous region in the recipient chromosome. In a population of bacterial cells, only those in a particular physiological state, referred to as **competence**, take up DNA. Entry is thought to occur at a limited number of receptor sites on the surface of the bacterial cell. Passage across the cell wall and membrane is an active process requiring energy and specific transport molecules. This model is supported by the fact that substances which inhibit energy production or protein synthesis in the recipient cell also inhibit the transformation process.

During the process of entry, one of the two strands of the invading DNA molecule is digested by nucleases, leaving only a single strand to participate in transformation (Steps 2 and 3). The surviving DNA strand aligns with its complementary region of the bacterial chromosome. In a process involving several enzymes, this segment of DNA replaces its counterpart in the chromosome, which is excised and degraded (Step 4).

For recombination to be detected, the transforming DNA must be derived from a different strain of bacteria, bearing some genetic variation. Once it is integrated into the chromosome, the recombinant region contains one host strand (present originally) and one mutant strand. Because these strands are from different sources, this helical region is referred to as a **heteroduplex**. Following one round of replication, one chromosome is restored to its original configuration, identical to that of the recipient cell, and the other contains the mutant gene. Following cell division, one nonmutant (untransformed) cell and one mutant (transformed) cell are produced (Step 5).

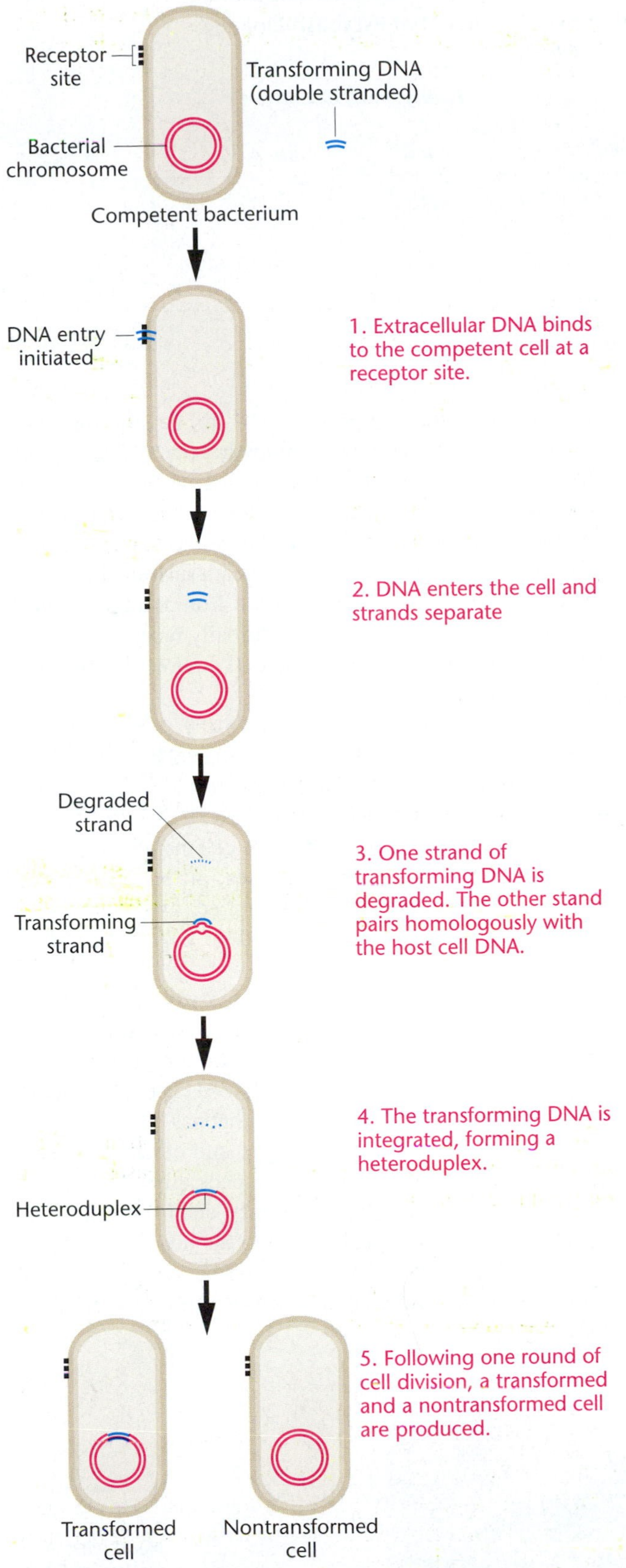
Receptor site
Transforming DNA (double stranded)
Bacterial chromosome
Competent bacterium
DNA entry initiated
1. Extracellular DNA binds to the competent cell at a receptor site.
2. DNA enters the cell and strands separate
Degraded strand
Transforming strand
3. One strand of transforming DNA is degraded. The other stand pairs homologously with the host cell DNA.
4. The transforming DNA is integrated, forming a heteroduplex.
Heteroduplex
5. Following one round of cell division, a transformed and a nontransformed cell are produced.
Transformed cell
Nontransformed cell

FIGURE 15–13 Proposed steps for transformation of a bacterial cell by exogenous DNA. Only one of the two strands of the entering DNA is involved in the transformation event, which is completed following cell division.

Transformation and Mapping

For DNA to be effective in transformation, it must include between 10,000 to 20,000 nucleotide pairs, about 1/200 of the *E. coli* chromosome. This size is sufficient to encode several genes. Genes that are adjacent or very close to one another on the bacterial chromosome can be carried on a single segment of DNA of this size. Because of this fact, a single transfer event can result in the **cotransformation** of several genes simultaneously. Genes that are close enough to each other to be cotransformed are said to be *linked*. In contrast to the use of the term *linkage* in eukaryotes, which indicates all genes on a single chromosome, note that here linkage refers to the proximity of genes, i.e., genes next to, or close to one another.

If two genes are not linked, simultaneous transformation can occur only as a result of two independent events involving two distinct segments of DNA. As in double crossing over in eukaryotes, the probability of two independent events occurring simultaneously is equal to the product of the individual probabilities. Thus, the frequency of two unlinked genes being transformed simultaneously is much lower than if they are linked.

Linked bacterial genes were first demonstrated in 1954 during studies of *Pneumococcus* by Rollin Hotchkiss and Julius Marmur. They were examining transformation at the *streptomycin* and *mannitol* loci. Recipient cells were str^s and mtl^-, meaning that they were sensitive to streptomycin and could not ferment mannitol. Cells that were $str^r mtl^+$, which are resistant to streptomycin and able to ferment mannitol, were used to derive the transforming DNA. If the *str* and *mtl* genes are not linked, simultaneous transformation for both genes will occur with such a minimal probability as to be nearly undetectable. However, as shown in Table 15–1, a low, but detectable, rate (0.17 percent) of cotransformation did occur.

A second experiment confirmed that the genes were indeed linked. Instead of $str^r\ mtl^+$ DNA, a mixture of $str^r mtl^-$ and $str^s mtl^+$ DNA served as the donor DNA. Thus, both the str^r and mtl^+ genes were available to recipient cells, but on separate segments of DNA. If the observed rate of double transformants in the previous experiment had been due to two independent events and the genes were not linked, then a similar rate would also be observed under these conditions. Instead, the number of double transformants was 25 times fewer, confirming linkage. Careful examination of the data in Table 15–1 confirms the linkage of these loci.

In addition to establishing linkage relationships, relative mapping distances between linked genes can also be determined from the recombination data provided by transformation experiments. Though the analysis is more complex, such data are interpreted in a manner analogous to chromosome mapping in eukaryotes. In the "Problems and Discussion Questions" section at the end of the chapter, we consider data that provides mapping information.

Subsequent studies of transformation have shown that, in addition to *Diplococcus pneumonia*, a variety of bacteria

TABLE 15–1 Results of Several Transformation Experiments That Establish Linkage between the *str* and *mtl* Loci in *Pneumococcus*

Donor DNA	Recipient Cel Genotype	Transformed Genotypes (%)		
		$str^r mtl^-$	$str^s mtl^+$	$str^r mtl^+$
$str^r mtl^+$	$str^s mtl^-$	4.3	0.40	0.17
$str^r mtl^-$ and $str^s mtl^+$		2.8	0.85	0.0066

Source: Data from Hotchkiss and Marmur, 1954, p. 55.

readily undergo this process of recombination (e.g., *Hemophilus influenzae, Bacillus subtilis, Shigella paradysenteriae*, and *E. coli*). Other investigations have established that culture conditions can be adjusted to "induce" artificial transformation, an important component of recombinant DNA technology.

15.6 Bacteriophages Are Bacterial Viruses

Bacteriophages, or **phages** as they are commonly known, are viruses that have bacteria as their hosts. During their reproduction, phages can be involved in still another mode of bacterial genetic recombination called transduction. To understand this process, we first must consider the genetics of bacteriophages, which themselves also undergo recombination. In this section, we will first examine the structure and life cycle of one type of bacteriophage. We will then discuss how these phages are studied during their infection of bacteria. Finally, we will contrast two possible modes of behavior once initial phage infection occurs. This information will serve as background for our subsequent discussion of transduction and bacteriophage recombination. Furthermore, a great deal of genetic research has been done using bacteriophages as a model system, making them a worthy subject of discussion.

Phage T4: Structure and Life Cycle

Bacteriophage T4 is one of a group of related bacterial viruses referred to as T-even phages. It exhibits an intricate structure, as illustrated in Figure 15–14. Its genetic material, DNA, is contained within an icosahedral (a polyhedron with 20 faces) protein coat, together making up the head of the virus. The DNA is sufficient in quantity to encode more than 150 average-sized genes. The head is connected to a tail that contains a collar and a contractile sheath surrounding a central core. Tail fibers, which protrude from the tail, contain binding sites in their tips that specifically recognize unique areas of the outer surface of the cell wall of the bacterial host, *E. coli*.

The life cycle of phage T4 (Figure 15–15) is initiated when the virus binds by adsorption to the bacterial host cell. Then, an ATP-driven contraction of the tail sheath causes the central core to penetrate the cell wall. The DNA in the head is extruded, and it then moves across the cell membrane into the bacterial cytoplasm. Within minutes, all bacterial DNA, RNA, and protein synthesis is inhibited, and synthesis of viral molecules begins. At the same time, degradation of the host DNA is initiated.

A period of intensive viral gene activity characterizes infection. Initially, phage DNA replication occurs, leading to a pool of viral DNA molecules. Then, the components of the head, tail, and tail fibers are synthesized. The assembly of mature viruses is a complex process that has been well studied by William Wood, Robert Edgar, and others. Three sequential pathways occur: (1) DNA packaging as the viral heads are assembled, (2) tail assembly, and (3) tail-fiber assembly. Once DNA is packaged into the head, it combines with the tail components, to which tail fibers are added. Total construction is a combination of self-assembly and enzyme-directed processes.

MEDIA TUTORIAL Phage Genetics

FIGURE 15–14 The structure of bacteriophage T4, including an icosahedral head filled with DNA, a tail consisting of a collar, tube, sheath, base plate, and tail fibers. During assembly, the tail components are added to the head and then tail fibers are added.

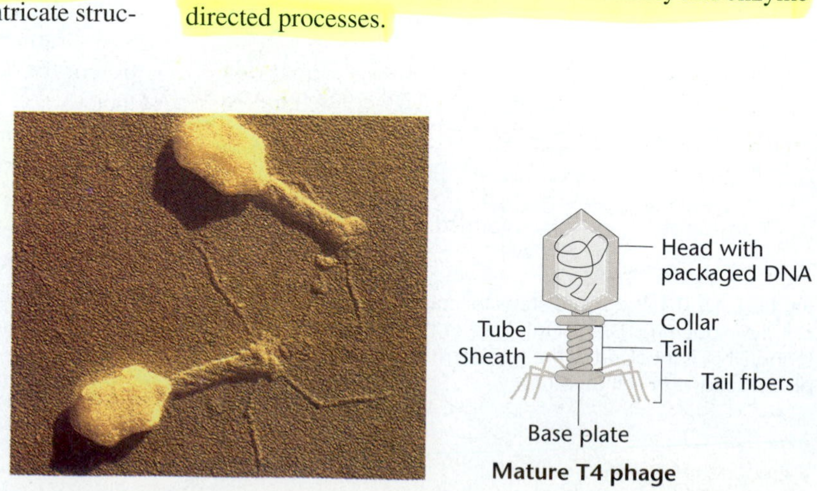

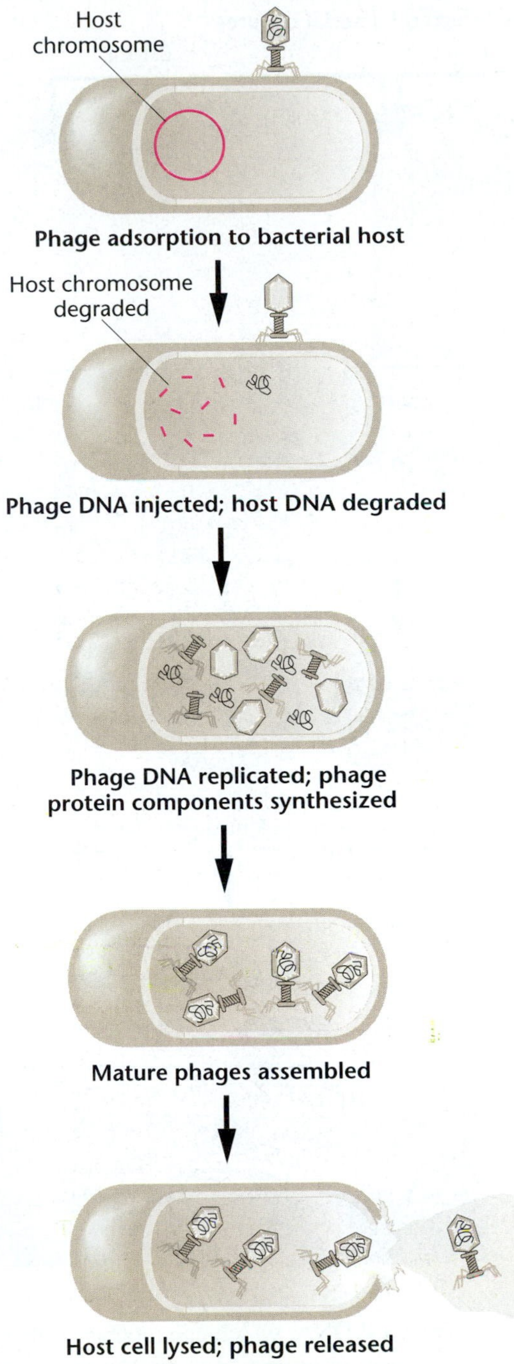
Host chromosome
Phage adsorption to bacterial host
Host chromosome degraded
Phage DNA injected; host DNA degraded
Phage DNA replicated; phage protein components synthesized
Mature phages assembled
Host cell lysed; phage released

FIGURE 15–15 Life cycle of bacteriophage T4.

When approximately 200 viruses are constructed, the bacterial cell is ruptured by the action of lysozyme (a phage gene product), and the mature phages are released from the host cell. The 200 new phages will infect other available bacterial cells, and the process will be repeated over and over again.

The Plaque Assay

The experimental study of bacteriophages and other viruses has played a critical role in our understanding of molecular genetics. During infection of bacteria, enormous quantities of bacteriophages may be obtained for investigation. Often, over 10^{10} viruses per milliliter of culture medium are produced. Many genetic studies have relied on our ability to quantitate the number of phages produced following infection under specific culture conditions using a technique called the plaque assay.

This assay is illustrated in Figure 15–16, in which actual plaque morphology is also shown. A serial dilution of the original virally infected bacterial culture is first performed. Then, a 0.1-ml aliquot from one or more dilutions is added to a small volume of melted nutrient agar (about 3 ml) into which a few drops of a healthy bacterial culture have been mixed. The solution is then poured evenly over a base of solid nutrient agar in a Petri dish and allowed to solidify prior to incubation. A clear area, called a **plaque**, is found where bacteria are absent. Plaques occur wherever a single virus has initially infected one bacterium in the lawn that has grown up during incubation. The plaque represents multiple clones of the single infecting T4 bacteriophage, created as reproduction cycles are repeated. If the dilution factor is too low, many phage are present and the plaques are plentiful. In such cases, they will fuse, lysing the entire lawn. This has occurred in the 10^{-3} dilution in Figure 15–16. On the other hand, if the dilution factor is increased, fewer phages are present in a given aliquot and plaques can be counted. From such data, the density of viruses in the initial culture can be estimated. The calculation is similar to that used for determining bacterial density by counting colonies following serial dilution of an initial culture:

$$(\text{plaque number/ml}) \times (\text{dilution factor})$$

Using the results shown in Figure 15–16, it is observed that there are 23 phage plaques derived from the 0.1-ml aliquot of the 10^{-5} dilution. Therefore, we can estimate that there are 230 phages per milliliter at this dilution. This initial viral density in the undiluted sample, in which 23 plaques are observed from 0.1 ml of the 10^{-5} dilution, is calculated as

$$(230/\text{ml}) \times (10^5) = 230 \times 10^5/\text{ml} = 2.3 \times 10^7/\text{ml}$$

Because this figure is derived from the 10^{-5} dilution, we can estimate that there will be only 0.23 phage per 0.1 ml in the 10^{-7} dilution. As a result, when 0.1 ml from this tube is assayed, it is predicted that no phage particles will be present. This possibility is borne out in Figure 15–16, in which an intact lawn of bacteria is depicted. The dilution factor is simply too great.

The use of the plaque assay has been invaluable in mutational and recombinational studies of bacteriophages. We will apply this technique more directly later in this chapter, when we discuss Seymour Benzer's elegant genetic analysis of a single gene in phage T4.

Lysogeny

The relationship between virus and bacterium does not always result in viral reproduction and lysis. As early as the 1920s, it was known that some bacteriophages could enter a bacterial cell and establish a symbiotic relationship with it. The precise molecular basis of this symbiosis is now well understood. Upon entry, the viral DNA, instead of

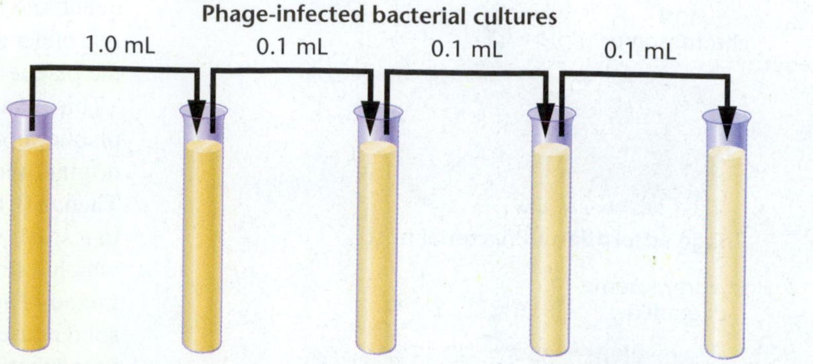

Total volume	10 mL	10 mL	10 mL	10 mL	10mL
Dilution	0	10^{-1}	10^{-3}	10^{-5}	10^{-7}
Dilution factor	0	10	10^3	10^5	10^7

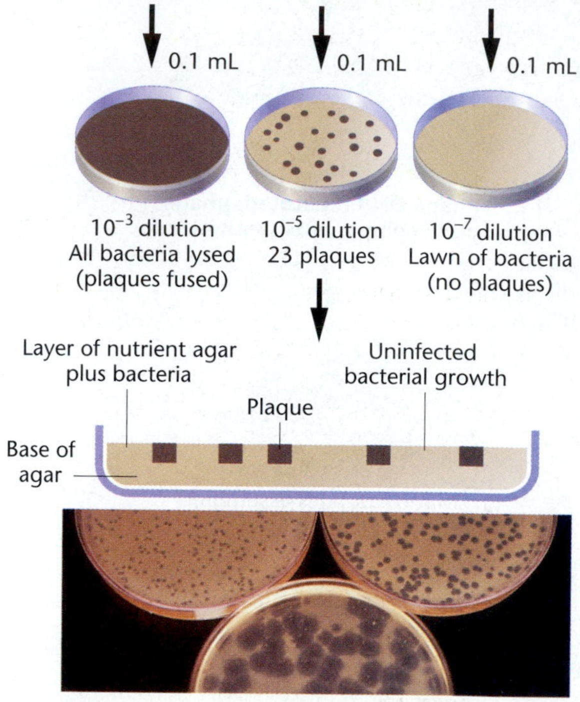

FIGURE 15–16 The plaque assay for bacteriophage analysis. Serial dilutions of a bacterial culture infected with bacteriophages are first made. Then, three of the dilutions (10^{-3}, 10^{-5}, and 10^{-7}) are analyzed using the plaque assay technique. In each case, 0.1 ml of the diluted culture is used. Each plaque represents the initial infection of one bacterial cell by one bacteriophage. In the 10^{-3} dilution, so many phages are present that all bacteria are lysed. In the 10^{-5} dilution, 23 plaques are produced. In the 10^{-7} dilution, the dilution factor is so great that no phages are present in the 0.1 ml sample, and thus no plaques form. From the 0.1 ml sample of the 10^{-5} dilution, the original bacteriophage density is calculated to be $23 \times 10 \times 10^5$ phages/ml (23×10^6 or 2.3×10^7). The photograph illustrates phage T2 plaques on lawns of *E. coli*.

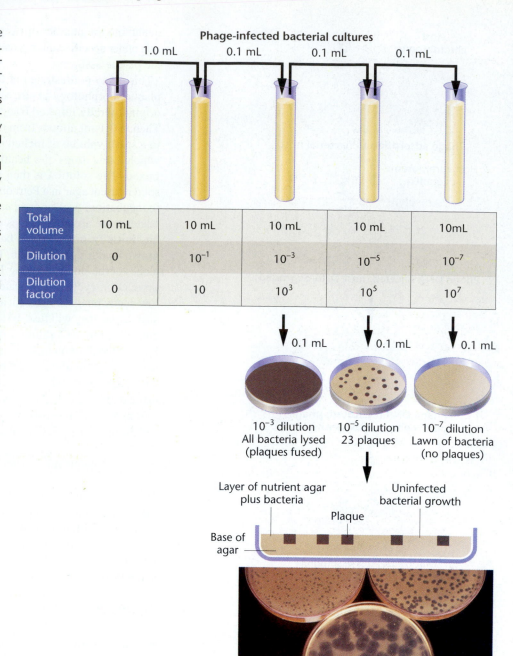

replicating in the bacterial cytoplasm, is integrated into the bacterial chromosome—a step that characterizes the developmental stage referred to as **lysogeny**. Subsequently, each time the bacterial chromosome is replicated, the viral DNA is also replicated and passed to daughter bacterial cells following division. No new viruses are produced and no lysis of the bacterial cell occurs. However, in response to certain stimuli, such as chemical or ultraviolet-light treatment, the viral DNA may lose its integrated status and initiate replication, phage reproduction, and lysis of the bacterium.

Several terms are used to describe this relationship. The viral DNA integrated into the bacterial chromosome is called a **prophage**. Viruses that can either lyse the cell or behave as a prophage are called **temperate**. Those that can only lyse the cell are referred to as **virulent**. A bacterium harboring a prophage has been **lysogenized** and is said to be **lysogenic;** that is, it is capable of being lysed as a result of induced viral reproduction. The viral DNA, which can replicate either in the bacterial cytoplasm or as part of the bacterial chromosome, is sometimes classified as an **episome**.

15.7 Transduction Is Virus-Mediated Bacterial DNA Transfer

In 1952, Norton Zinder and Joshua Lederberg were investigating possible recombination in the bacterium *Salmonella typhimurium*. Although they recovered prototrophs from mixed cultures of two different auxotrophic strains, subsequent investigations revealed that recombination was occurring in a manner different from that attributable to the presence of an F factor, as in *E. coli*. What they discovered was a process of bacterial recombination mediated by bacteriophages and now called **transduction**.

The Lederberg–Zinder Experiment

Lederberg and Zinder mixed the *Salmonella* auxotrophic strains LA-22 and LA-2 together, and, when the mixture was plated on minimal medium, they recovered prototrophic cells. The LA-22 was unable to synthesize the amino acids phenylalanine and tryptophan ($phe^- \ trp^-$), and LA-2 could not synthesize the amino acids methionine and histidine ($met^- \ his^-$). Prototrophs ($phe^+ \ trp^+ \ met^+ \ his^+$) were recovered at a rate of about $1/10^5$ (or 10^{-5}) cells.

Although these observations at first suggested that the recombination involved was the type observed earlier in conjugative strains of *E. coli*, experiments using the Davis U-tube soon showed otherwise (Figure 15–17). The two auxotrophic strains were separated by a sintered glass filter, thus preventing cell contact but allowing growth to occur in a common medium. Surprisingly, when samples were removed from both sides of the filter and plated independently on minimal medium, prototrophs were recovered only from the side of the tube containing LA-22 bacteria. Recall that, if conjugation were responsible, the conditions in the Davis U-tube would be expected to prevent recombination altogether (See Figure 15–4).

Since LA-2 cells appeared to be the source of the new genetic information (phe^+ and trp^+), how that information crossed the filter from the LA-2 cells to the LA-22 cells, allowing recombination to occur, was a mystery. The unknown source was designated simply as a **filterable agent (FA)**.

Three subsequent observations were useful in identifying the FA:

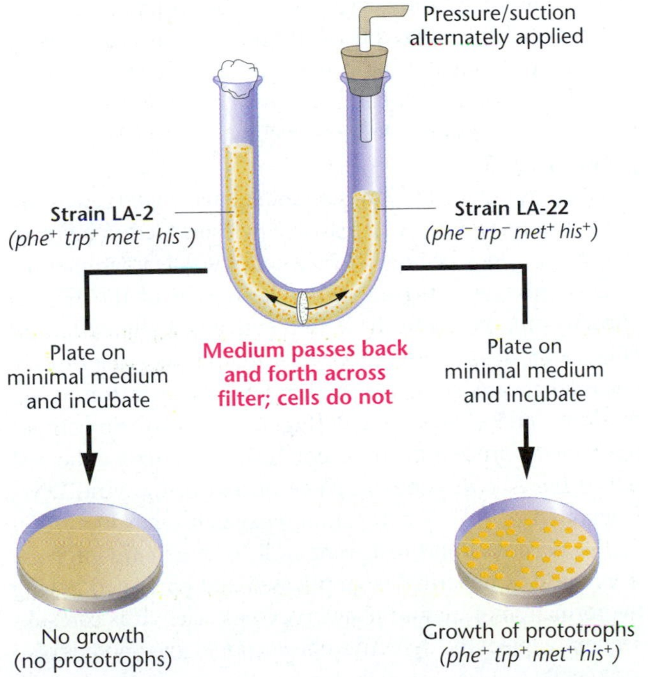

FIGURE 15–17 The Lederberg–Zinder experiment using *Salmonella*. After placing two auxotrophic strains on opposite sides of a Davis U-tube, Lederberg and Zinder recovered prototrophs from the side containing the LA-22 strain, but not from the side containing the LA-2 strain. These initial observations led to the discovery of the phenomenon called transduction.

1. The FA was produced by the LA-2 cells only when they were grown in association with LA-22 cells. If LA-2 cells were grown independently and that culture medium was then added to LA-22 cells, recombination did not occur. Therefore, LA-22 cells play some role in the production of FA by LA-2 cells and do so only when the two share common growth medium.

2. The presence of DNase, which enzymatically digests DNA, did not render the FA ineffective. Therefore, the FA is not naked DNA, ruling out transformation.

3. The FA could not pass across the filter of the Davis U-tube when the pore size was reduced below the size of bacteriophages.

Aided by these observations and aware of temperate phages that could lysogenize *Salmonella*, researchers proposed that the genetic recombination event was mediated by bacteriophage P22, present initially as a prophage in the chromosome of the LA-22 *Salmonella* cells. It was hypothesized that, rarely, P22 prophages might enter the vegetative or lytic phase, reproduce, and be released by the LA-22 cells. Such phages, being much smaller than a bacterium, then cross the filter of the U-tube and subsequently infect and lyse some of the LA-2 cells. In the process of lysis of LA-2, the P22 phages occasionally packaged in their heads a region of the LA-2 chromosome. If this region contained the phe^+ and trp^+ genes, and the phages subsequently passed back across the filter and infected LA-22 cells, these newly lysogenized cells would behave as prototrophs. This process of transduction, whereby bacterial recombination is mediated by bacteriophage P22, is diagrammed in Figure 15–18.

The Nature of Transduction

Further studies revealed the existence of transducing phages in other species of bacteria. For example, *E. coli* can be transduced by phages P1 and λ. *Bacillus subtilis* and *Pseudomonas aeruginosa* can be transduced by the phages SPO1 and F116, respectively. The details of several different modes of transduction have also been established. Even though the initial discovery of transduction involved a temperate phage and a lysogenized bacterium, the same process can occur during the normal lytic cycle. Sometimes, a small piece of bacterial DNA is packaged *along with* the viral chromosome, so that the transducing phage contains both viral and bacterial DNA. In such cases, only a few bacterial genes are present in the transducing phage. However, when *only* bacterial DNA is packaged, regions as large as 1 percent of the bacterial chromosome may become enclosed in the viral head. In either case, the ability to infect is unrelated to the type of DNA in the phage head, making transduction possible.

When bacterial rather than viral DNA is injected into the bacterium, it either remains in the cytoplasm or recombines with the homologous region of the bacterial chromosome. If the bacterial DNA remains in the cytoplasm, it does not replicate, but is transmitted to one of the progeny cells following each division. When this happens, only a single cell, partially diploid for the transduced genes, is produced—a

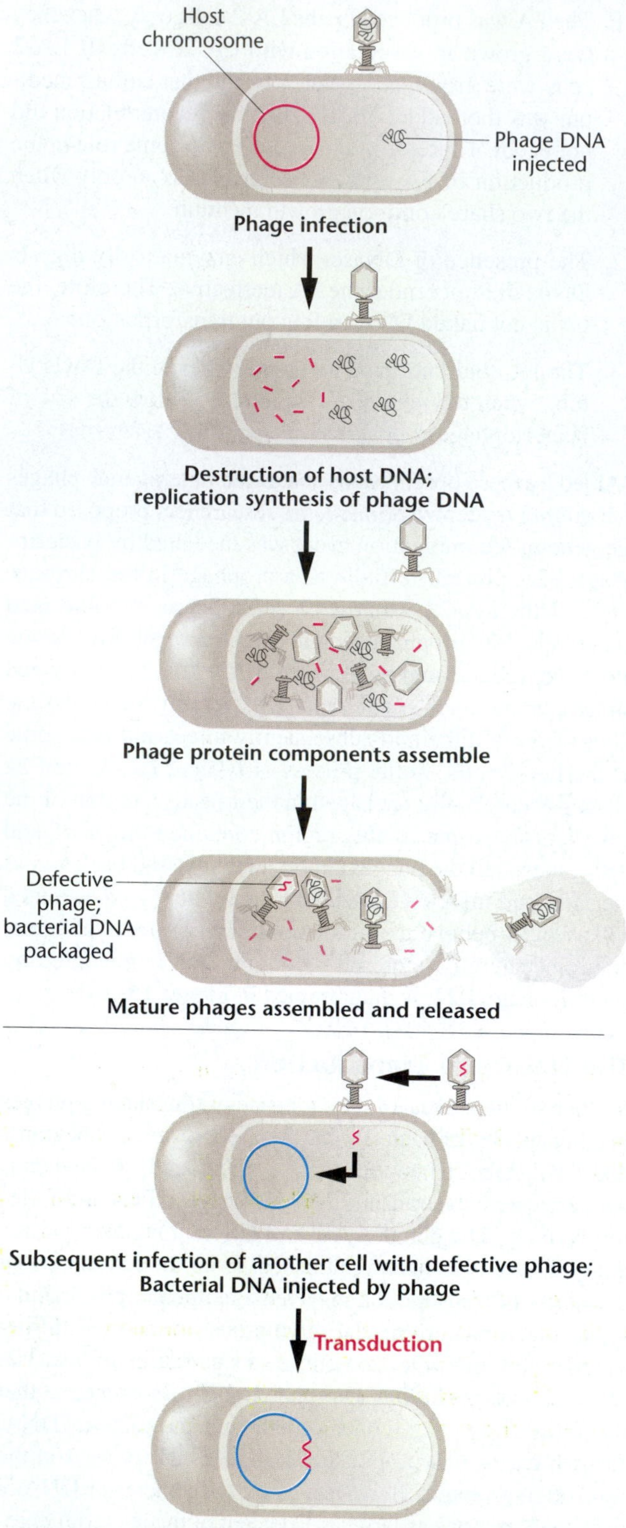
Host chromosome
Phage DNA injected
Phage infection
Destruction of host DNA; replication synthesis of phage DNA
Phage protein components assemble
Defective phage; bacterial DNA packaged
Mature phages assembled and released
Subsequent infection of another cell with defective phage; Bacterial DNA injected by phage
Transduction
Integration of bacterial DNA into recipient chromosome

FIGURE 15–18 Generalized transduction.

phenomenon called **abortive transduction**. If the bacterial DNA recombines with its homologous region of the bacterial chromosome, the transduced genes are replicated as part of the chromosome and passed to all daughter cells. This process is called **complete transduction**.

Both abortive and complete transduction are subclasses of the broader category of **generalized transduction**, which is characterized by the random nature of DNA fragments and genes transduced. Each fragment of the bacterial chromosome has a finite, but small, chance of being packaged in the phage head. Most cases of generalized transduction are of the abortive type; some data suggest that complete transduction occurs 10–20 times less frequently.

Transduction and Mapping

Like transformation, generalized transduction was used in linkage and mapping studies of the bacterial chromosome. The fragment of bacterial DNA involved in a transduction event is large enough to include numerous genes. As a result, two genes that are closely aligned (i.e., are linked) on the bacterial chromosome may be simultaneously transduced, a process called **cotransduction**. Two genes that are not close enough to one another along the chromosome to be included on a single DNA fragment require two independent events in order to be transduced into a single cell. Since this occurs with a much lower probability than cotransduction, linkage can be determined.

By concentrating on two or three linked genes, transduction studies can also determine the precise order of these genes. The closer linked genes are to each other, the greater the frequency of cotransduction. Mapping studies involving three closely aligned genes can be executed. The analysis of such an experiment is predicated on the same rationale underlying other mapping techniques.

Specialized Transduction

In some instances, only certain genes are recombined, a situation called **specialized transduction**. This is in contrast to generalized transduction previously described, where all genes have an equal probability of being recombined. One of the best examples involves transduction of *E. coli* by the temperate phage λ.

In this case, transduction is restricted to the *gal* (galactose) or *bio* (biotin) genes. The reason why transduction involves only these genes became clear when it was learned that the λ DNA always integrates into the region of the *E. coli* chromosome between these two genes at a site called *att* [Figure 15–19(a)]. Phage λ DNA that is integrated in the bacterial chromosome can subsequently detach from it, reproduce, and lyse the host cell [Figure 15–19(b)]. Sometimes the excision process occurs incorrectly and carries either the *gal* or *bio E. coli* genes in place of part of the viral DNA [Figure 15–19(b)]. The resulting phage chromosome is defective because it has lost some of its own genetic information, but it is nevertheless replicated and packaged during the formation of mature phage particles. The virus can subsequently inject the defective chromosome into another bacterial cell.

In this case of specialized transduction, the defective phage chromosome is integrated into the bacterial chromosome and is replicated along with it. Such bacterial cells contain the defective phage DNA, making them diploid for the *gal* or *bio* genes. The presence of this transducing gal^+ or bio^+ DNA causes these auxotrophs to revert to a gal^+ or bio^+ phenotype.

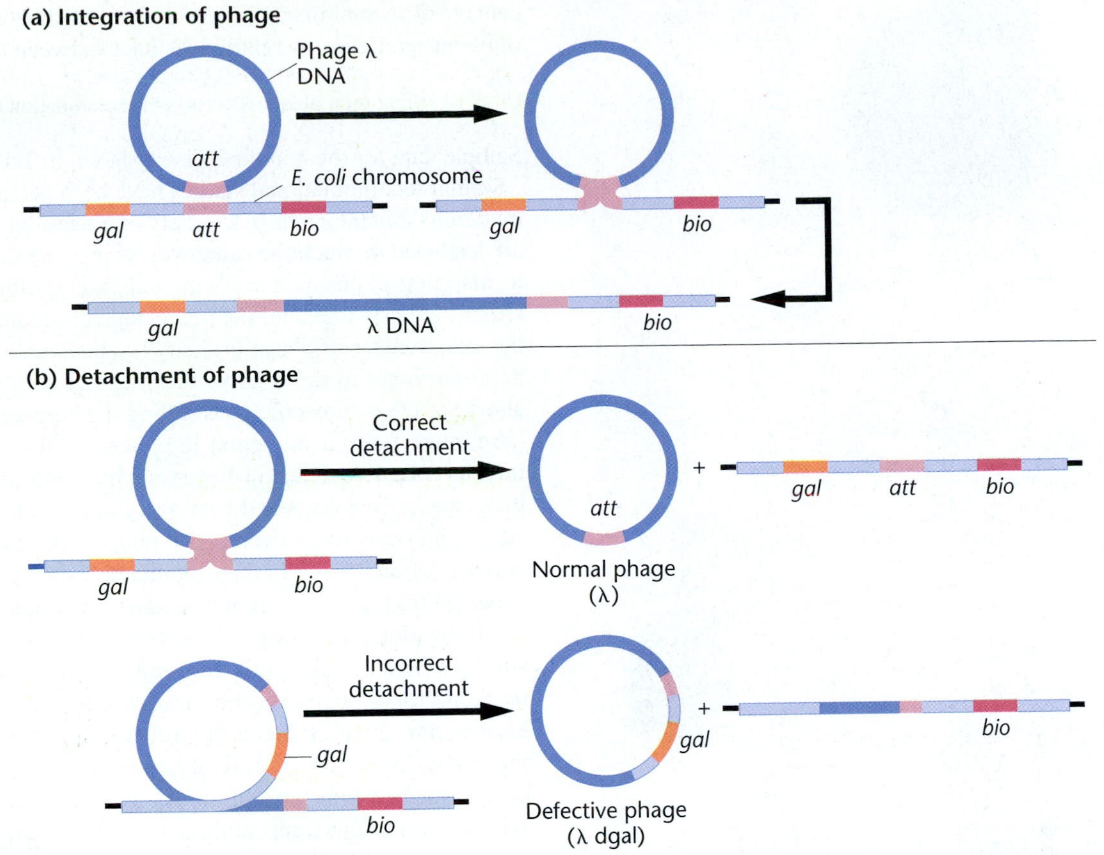

FIGURE 15–19 The production of defective phage $(\lambda)dgal$, which can result in specialized transduction following another round of infection of *E. coli*. If detachment occurs correctly, no transduction will result.

15.8 Bacteriophages Undergo Intergenic Recombination

Around 1947, several research teams demonstrated that genetic recombination can be detected in bacteriophages. This led to the discovery that gene mapping can be performed in these viruses. Such studies relied on finding numerous phage mutations that could be visualized or assayed. As in bacteria and eukaryotes, these mutations allow genes to be identified and followed in mapping experiments. Thus, before considering recombination and mapping in these bacterial viruses, we will briefly introduce several of the mutations that were studied.

Phage mutations often affect the morphology of the plaques formed following lysis of bacterial cells. For example, in 1946, Alfred Hershey observed unusual T2 plaques on plates of *E. coli* strain B. Where the normal T2 plaques are small and have a clear center surrounded by a diffuse (nearly invisible) halo, the unusual plaques were larger and possessed a more distinctive outer perimeter (Figure 15–20). When the viruses were isolated from these plaques and replated on *E. coli* B cells, the resulting plaque appearance was identical. Thus, the plaque phenotype was an inherited trait resulting from the reproduction of mutant phages. Hershey named the mutant *rapid lysis* (*r*) because the plaques were larger, apparently resulting from a more rapid or more efficient life cycle of the phage. It is now known that, in wild-type phages, reproduction is inhibited once a particular-sized plaque has been formed. The *r* mutant T2 phages are able to overcome this inhibition, producing larger plaques.

Luria also discovered another bacteriophage mutation *host range* (*h*). This mutation extends the range of bacterial hosts that the phage can infect. Although wild-type T2 phages can infect *E. coli* B, they normally cannot attach or be adsorbed to the surface of *E. coli* B-2. The *h* mutation, however, provides the basis for adsorption and subsequent infection of *E. coli* B-2. When grown on a mixture of *E. coli* B and B-2, the center of the *h* plaque appears much darker than the h^+ plaque (Figure 15–20).

Table 15–2 lists other types of mutations that have been isolated and studied in the T-even series of bacteriophages (e.g., T2, T4, T6). These mutations are important to the study of genetic phenomena in bacteriophages.

Intergenic Recombination and Mapping in Bacterial Viruses

Genetic recombination in bacteriophages was discovered during **mixed infection experiments**, in which two distinct mutant strains were allowed to *simultaneously* infect the same bacterial culture. These studies were designed so that the number of viral particles sufficiently exceeded the number

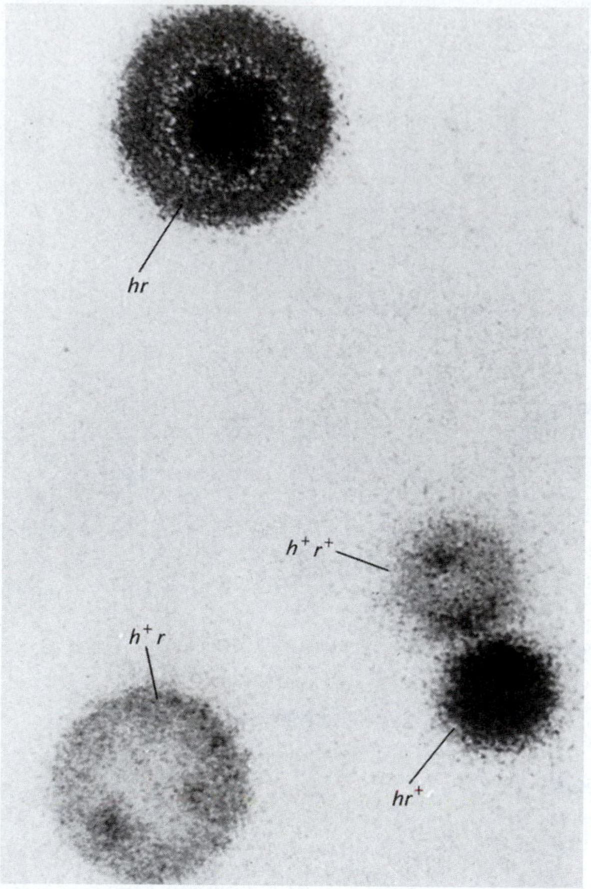

FIGURE 15–20 Plaque morphology phenotypes observed following simultaneous infection of *E. coli* by two strains of phage T2, h^+r and hr^+. In addition to the parental genotypes, recombinant plaques *hr* and h^+r^+ were also recovered.

of bacterial cells so as to ensure simultaneous infection of most cells by both viral strains. Because two loci are involved, recombination is referred to as intergenic. For example, in one study using the T2/*E. coli* system, the parental viruses were of either the h^+r (wild-type host range, rapid lysis) or the hr^+ (extended host range, normal lysis) genotype. If no recombination occurred, these two parental genotypes would be the only expected phage progeny. However, the recombinants h^+r^+ and *hr* were detected in addition to the parental genotypes. (See Figure 15–20.) As with eukaryotes, the percentage of recombinant plaques divided by the total number of plaques reflects the relative distance between the genes:

$$(h^+r^+) + (hr) \,/\, \text{total plaques} \times 100 = \text{recombinational frequency}$$

Sample data for the *h* and *r* loci are shown in Table 15–3.

Similar recombinational studies have been performed with numerous mutant genes in a variety of bacteriophages. Data are analyzed in much the same way as they are in eukaryotic mapping experiments. Two- and three-point mapping crosses are possible, and the percentage of recombinants in the total number of phage progeny is calculated. This value is proportional to the relative distance between two genes along the DNA molecule constituting the chromosome.

An interesting phenomenon in phage crosses is **negative interference**. Recall that in eukaryotic mapping crosses, positive interference leads to the recovery of fewer than expected double-crossover events. In phage crosses, just the opposite often occurs. In three-point analysis, a greater than expected frequency of double exchanges is often observed.

Investigations into phage recombination support a model similar to that of eukaryotic crossing over—a breakage and reunion process between the viral chromosomes. A fairly clear picture of the dynamics of viral recombination is emerging. Following the early phase of infection, the chromosomes of the phages begin replication. As this stage progresses, a pool of chromosomes accumulates in the bacterial cytoplasm. If double infection by phages of two genotypes has occurred, then the pool of chromosomes initially consists of the two parental types. Genetic exchange between these two types will occur before, during, and after replication, producing recombinant chromosomes.

In the case of the h^+r–hr^+ example discussed here, recombinant h^+r^+ and *hr* chromosomes are produced. Each of these chromosomes may undergo replication, with new replicates undergoing exchange with each other and with parental chromosomes.

Furthermore, recombination is not restricted to exchange between two chromosomes—three or more may be involved simultaneously. As phage development progresses, chromosomes are randomly removed from the pool and packed into the phage head, forming mature phage particles. Thus, a variety of parental and recombinant genotypes are represented in progeny phages.

As we will see in the next section, powerful selection systems have made it possible to detect *intragenic* recombination

TABLE 15–2 Some Mutant Types of T-Even Phages

Name	*Description*
minute	Small plaques
turbid	Turbid plaques on *E. coli* B
star	Irregular plaques
uv-sensitive	Alters UV sensitivity
acriflavin-resistant	Forms plaques on acriflavin agar
osmotic shock	Withstands rapid dilution into distilled water
lysozyme	Does not produce lysozyme
amber	Grows in *E. coli* K12, but not B
temperature-sensitive	Grows at 25°C, but not at 42°C

TABLE 15–3 Results of a Cross Involving the *h* and *r* Genes in Phage T2 ($hr^+ \times h^+r$)

Genotype	*Plaques*	*Designation*
$h\ r^+$	42	Parental progeny
$h^+\ r$	34	76%
$h^+\ r^+$	12	Recombinants
$h\ r$	12	24%

Source: Data derived from Hershey and Rotman, 1949.

in viruses, where exchanges occur at points within a single gene. Such studies have led to the fine-structure analysis of the gene, as discussed in an ensuing section.

15.9 Intragenic Recombination Occurs in Phage T4

We conclude this chapter with an account of an ingenious example of genetic analysis. In the early 1950s, Seymour Benzer undertook a detailed examination of a single locus, *rII*, in phage T4. Benzer successfully designed experiments to recover the extremely rare genetic recombinants arising as a result of **intragenic exchange**. Such recombination is equivalent to eukaryotic crossing over, but in this case, *within a gene rather than between two genes*. Benzer demonstrated that such recombination occurs between the DNA of individual bacteriophages during simultaneous infection of the host bacterium *E. coli*.

The end result of Benzer's work was the production of a detailed map of the *rII* locus. Because of the extremely detailed information provided from his analysis, Benzer's work was described as the "fine-structure analysis" of the gene. Because these experiments occurred decades before DNA-sequencing techniques were developed, the insights concerning the internal structure of the gene were particularly noteworthy.

The *rII* Locus of Phage T4

The primary requirement in genetic analysis is the isolation of a large number of mutations in the gene being investigated. Mutants at the *rII* locus produce distinctive plaques when plated on *E. coli* strain B, allowing their easy identification. Figure 15–20 illustrates mutant *r* plaques compared to their wild type r^+ counterparts in the related T2 phage. Benzer's approach was to isolate many independent *rII* mutants—he eventually obtained about 20,000—and to perform recombinational studies so as to produce a genetic map of this locus.

The key to Benzer's analysis was that *rII* mutant phages, though capable of infecting and lysing *E. coli* B, could not successfully lyse a second related strain, *E. coli* K12(λ).* Wild-type phages, by contrast, could lyse both the B and the K12 strains. Benzer reasoned that these conditions provided the potential for a highly sensitive screening system. If phages from any two different mutant strains were allowed to simultaneously infect *E. coli* B, exchanges between the two mutant sites within the locus would produce rare wild-type recombinants (Figure 15–21). If the phage population, which contained over 99.9 percent *rII* phages and less than 0.1 percent wild-type phages, were then allowed to infect strain K12, the wild-type recombinants would successfully reproduce and produce wild-type plaques. *This is the critical step in recovering and quantifying rare recombinants.*

*The inclusion of (λ) in the designation of K12 indicates that this bacterial strain is lysogenized by phage λ. This, in fact, is the reason that *rII* mutants cannot lyse such bacteria. Note that, as we refer to this strain in future discussions, we will abbreviate it simply as *E. coli* K12.

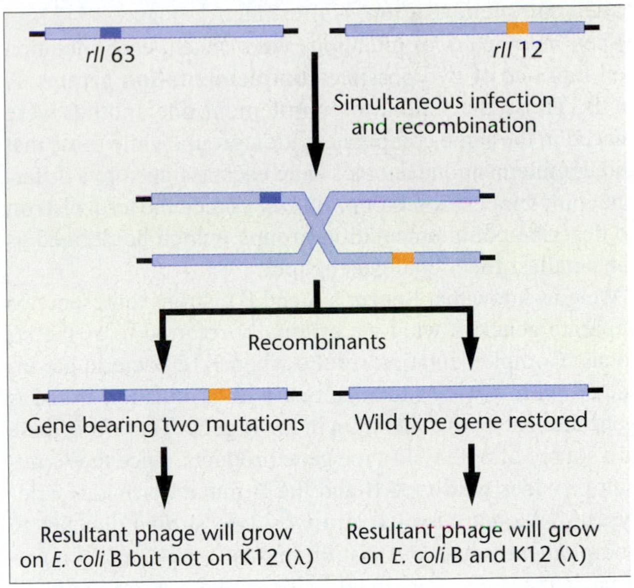

FIGURE 15–21 Illustration of intragenic recombination between two mutations in the *rII* locus of phage T4. The result is the production of a wild-type phage that will grow on both *E. coli* B and K12 and a phage that has incorporated both mutations into the *rII* locus. It will grow on *E. coli* B, but not on *E. coli* K12.

By using serial dilution techniques, Benzer was able to determine the total number of mutant *rII* phages produced on *E. coli* B and the total number of recombinant wild-type phages that would lyse *E. coli* K12. These data provided the basis for calculating the frequency of recombination, a value proportional to the distance within the gene between the two mutations being studied. As we will see, this experimental design was extraordinarily sensitive. Remarkably, it was possible for Benzer to detect as few as one recombinant wild-type phage among 100 million mutant phages. When information from many such experiments is combined, a detailed map of the locus is possible.

Before we discuss this mapping, we need to describe an important discovery Benzer made during the early development of his screen, for this discovery led to the development of a technique used widely in genetics labs today, the **complementation assay**. Recall that we first introduced the theory behind complementation analysis in Chapter 10. You may wish to review this discussion before proceeding.

Complementation by *rII* Mutations

Before Benzer was able to initiate intragenic recombination studies, he had to resolve a problem encountered during the early stages of his experimentation. While doing a control for his experiment in which K12 bacteria were simultaneously infected with pairs of different *rII* mutant strains, Benzer sometimes found that the K12 bacteria were lysed. This was initially quite puzzling, since only the wild-type *rII* was supposed to be capable of lysing K12 bacteria. How could two mutant strains of *rII*, each of which was thought to contain a defect in the same gene, show a wild-type function?

Benzer reasoned that, during simultaneous infection, each mutant strain provided something that the other lacked, thus restoring wild-type function. This phenomenon, which he called **complementation**, is illustrated in Figure 15–22(a). When many pairs of mutations were tested, each mutation fell into one of two possible **complementation groups**, A or B. Those that failed to complement one another were placed in the same complementation group, while those that did complement one another were each assigned to a different complementation group. Benzer coined the term **cistron** to describe complementation groups, which he defined as the smallest functional genetic unit.

We now know that Benzer's A and B cistrons represent two separate genes in what we originally referred to as the *rII* locus. Complementation results when K12 bacteria are infected with two *rII* mutants, one with a mutation in the A gene and one with a mutation in the B gene. Therefore, there is a source of both wild-type gene products, since the A mutant provides wild-type B and the B mutant provides wild-type A. We can also explain why two strains that fail to complement, say two A cistron mutants, are actually mutations in the same gene. In this case, if two A cistron mutants are combined, there will be a source of the wild-type B product, but no source of the wild-type A product [Figure 15–22(b)].

Once Benzer was able to place all *rII* mutations in either the A or the B cistron, he was set to return to his intragenic recombination studies, testing mutations in the A cistron against each other and testing mutations in the B cistron against each other.

Recombinational Analysis

Of the approximately 20,000 *rII* mutations, roughly half fell into each cistron. Benzer set about mapping the mutations within each one. For example, focusing initially on the A cistron, if two *rIIA* mutants were first allowed to infect *E. coli* B in a liquid culture, and if a recombination event occurred between the mutational sites in the A cistron, then wild-type progeny viruses would be produced at low frequency. If samples of the progeny viruses from such an experiment were then plated on *E. coli* K12, only the wild-type recombinants would lyse the bacteria and produce plaques. The total number of nonrecombinant progeny viruses was also determined by plating samples on *E. coli* B.

This experimental protocol is illustrated in Figure 15–23. The percentage of recombinants can be determined by counting the plaques at the appropriate dilution in each case. As in

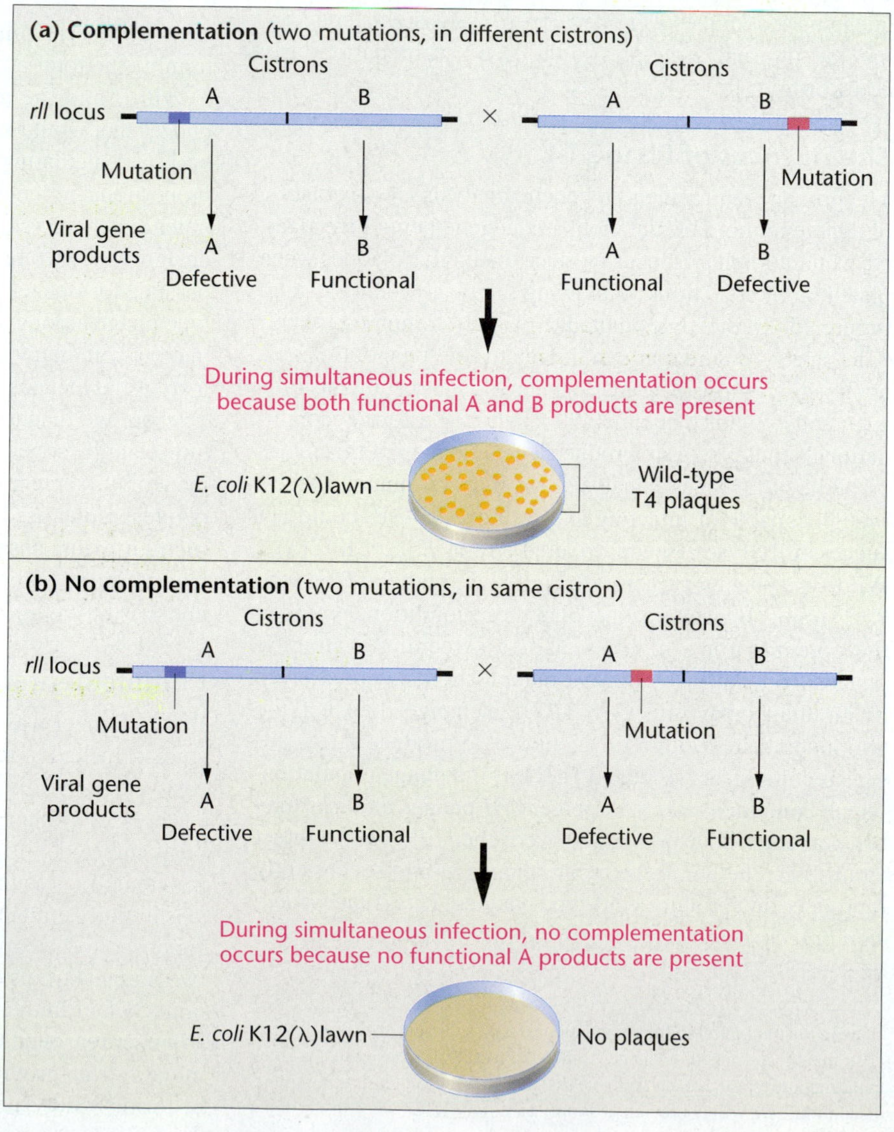

FIGURE 15–22 Comparison of two *rII* mutations that either (a) complement one another, or (b) do not complement one another. Complementation occurs when each mutation is in a separate cistron. Failure to complement occurs when the two mutations are in the same cistron.

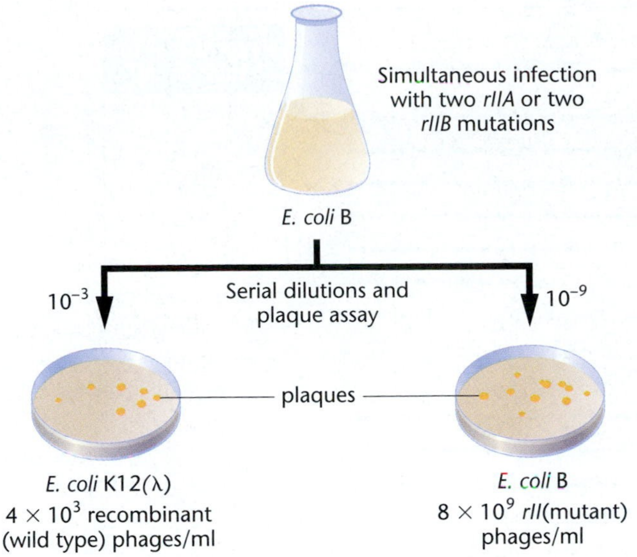

FIGURE 15–23 The experimental protocol for recombinational studies between pairs of *rIIA* or *rIIB* mutations.

eukaryotic mapping experiments, the frequency of recombination is an estimate of the distance between the two mutations within the cistron. For example, if the number of recombinants is equal to 4×10^3 ml and the total number of progeny is 8×10^9 ml, then the frequency of recombination between the two mutants is

$$2\left(\frac{4 \times 10^3}{8 \times 10^9}\right) = 2(0.5 \times 10^{-6})$$
$$= 10^{-6}$$
$$= 0.000001$$

Multiplying by 2 is necessary because each recombinant event yields two reciprocal products, only one of which—the wild type—is detected.

Deletion Testing of the rII Locus

Although the system for assessing recombination frequencies described earlier allowed for mapping mutations within each cistron, testing 1000 mutants two at a time in all combinations would have required millions of experiments. Fortunately, Benzer was able to overcome this obstacle when he devised an analytical approach referred to as **deletion testing**. He discovered that some of the *rII* mutations were, in reality, deletions of small parts of each cistron. That is, the genetic changes giving rise to the *rII* properties were not a characteristic of point mutations. Most important, when a deletion mutation was tested during simultaneous infection with a point mutation located in the deleted part of the same cistron, it *never* yielded wild-type recombinants. The basis for the failure to do so is illustrated in Figure 15–24. Thus, a method was available that could roughly, but quickly, localize any mutation, provided it was contained within a region covered by a deletion.

The use of deletion testing could thus serve as the basis for the initial localization of each mutation. For example, as shown in Figure 15–25, seven overlapping deletions spanning various regions of the A and B cistrons were used for the initial screening of the point mutations. Depending on whether the viral chromosome bearing a point mutation does or does not undergo recombination with the chromosome bearing a deletion, each point mutation can be assigned to a specific area of the cistron. Further deletions within each of the seven areas can be used to localize, or map, each *rII* point mutation more specifically. Remember that, in each case, a point mutation is localized in the area of a deletion when it fails to give rise to any wild-type recombinants.

The *rII* Gene Map

After several years of work, Benzer produced a genetic map of the two cistrons composing the *rII* locus of phage T4 (Figure 15–26). Of the 20,000 mutations analyzed, 307 distinct sites within this locus were mapped in relation to one another. Areas containing many mutations, designated as **hot spots**, were apparently more susceptible to mutation than were areas in which only one or a few mutations were found. Additionally, Benzer discovered areas within the cistrons in which no mutations were localized. He estimated that as many as 200 recombinational units had not been localized by his studies.

The significance of Benzer's work is his application of genetic analysis to what had previously been considered an abstract unit—the gene. Benzer had demonstrated in 1955 that a gene is not an indivisible particle, but instead consists of mutational and recombinational units that are arranged in a specific order—today, we know these are nucleotides composing DNA. His analysis, performed prior to the detailed molecular studies of the gene in the 1960s, is considered one of the classic examples of genetic experimentation.

FIGURE 15–24 Demonstration that recombination between a phage chromosome with a deletion in the A cistron and another phage with a point mutation overlapped by that deletion cannot yield a chromosome with wild-type A and B cistrons.

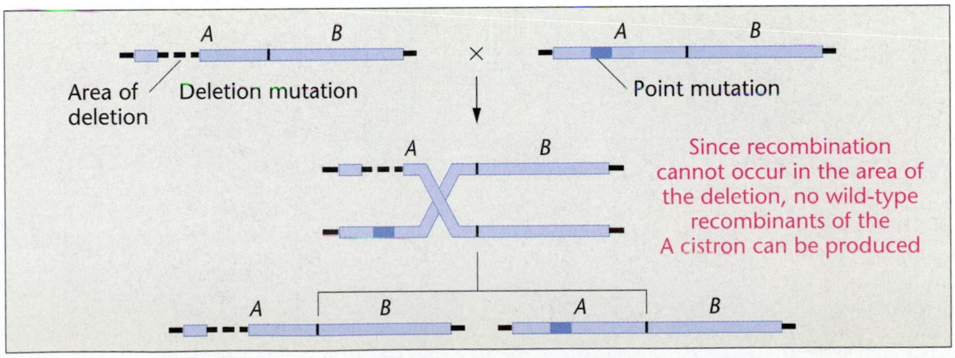

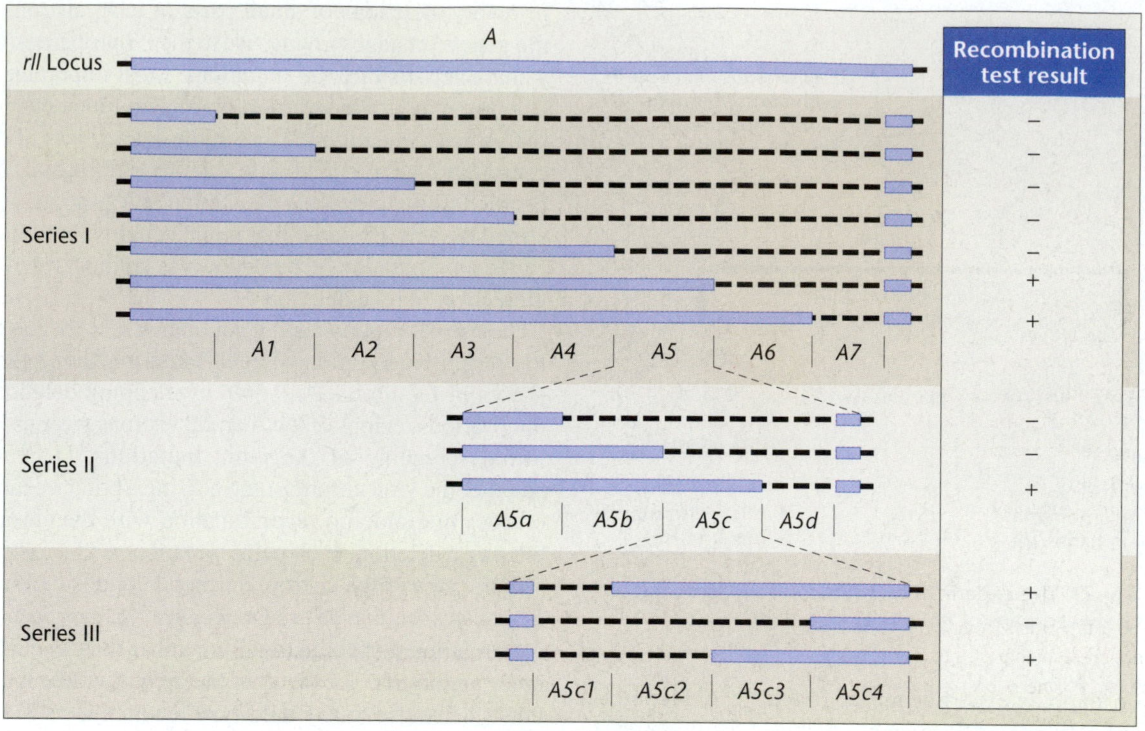

FIGURE 15–25 Three series of overlapping deletions in the A cistron of the *rII* locus used to localize the position of an unknown *r* mutation. For example, if a mutant strain tested against each deletion (dashed areas) in Series 1 for the production of recombinant wild-type progeny shows the results at the right (+ or −), the mutation must be in segment A5. In Series II, the mutation is further narrowed to segment A5c, and in Series III to segment A5c3.

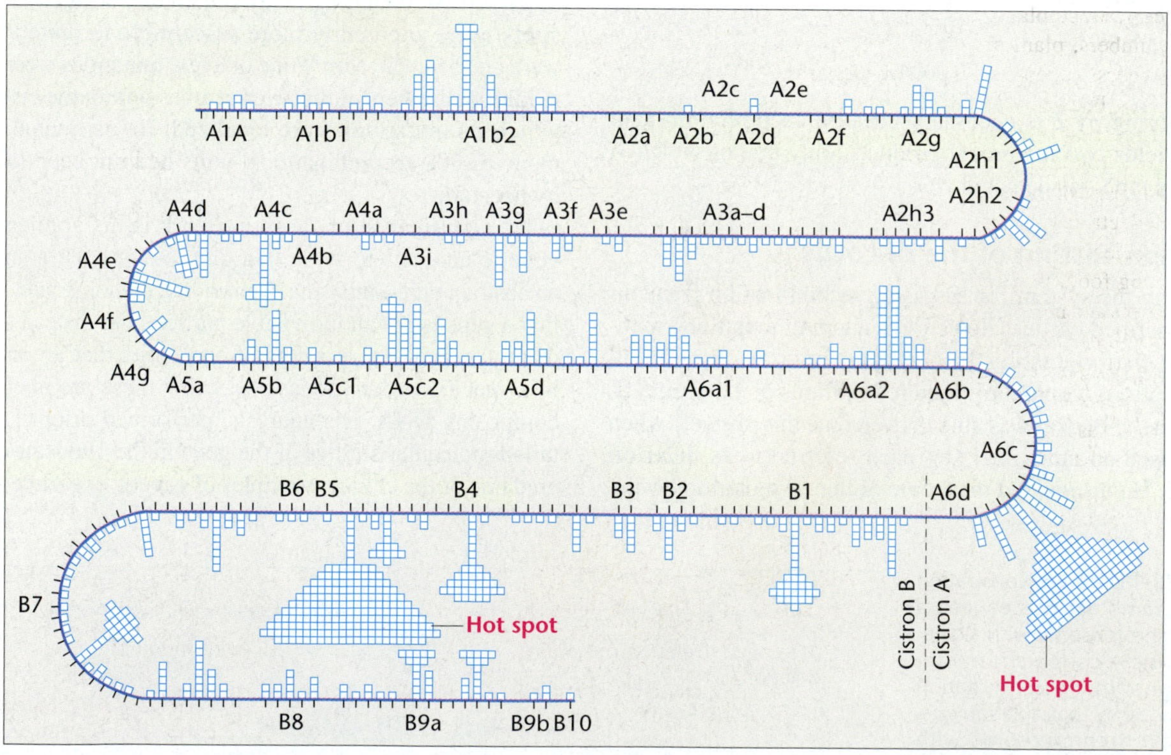

FIGURE 15–26 A partial map of mutations in the A and B cistrons of the *rII* locus of phage T4. Each square represents an independently isolated mutation. Note the two areas in which the largest number of mutations are present, referred to as "hot spots" (A6cd and B5).

Genetics, Technology, and Society

Eradicating Cholera: Edible Vaccines

Almost lost amid the furor over the cloned sheep Dolly was the original purpose of the undertaking: to genetically engineer an animal that would produce a valuable pharmaceutical product, eventually leading to a herd of genetically identical drug-producing animals. The goal, in other words, was to be able to turn sheep, cattle, and other animals into living drug factories.

Meanwhile, almost unnoticed by the general public, the genetic engineering of plants to serve a simlar role is much closer to reality. When foreign genes are inserted into the plant genome, transgenic plants are created that produce the foreign gene product. One of the most intriguing and potentially life-saving objectives is the genetic engineering of plants to produce vaccines against human diseases. Immunization would then involve eating foods altered to contain a protein that acts as an antigen and stimulates the production of antibodies to protect against bacterial or viral infection.

Plants have many advantages as vaccine-producers. Since plants can be grown in large numbers, plant-produced vaccines should be less expensive than conventional vaccines. Further, extensive purification and refrigerated transport and storage of vaccines will not be required. This is important in parts of the world where the supply of electricity is unreliable. Finally, since people would simply eat the vaccine-containing food, there would be no need for syringes and needles, or medical staff to give injections.

Leading the effort to develop edible vaccines in plants is Charles J. Arntzen, formerly at Texas A&M University and now president of the Boyce Thompson Institute for Plant Research at Cornell University. Arntzen's research team is focusing on intestinal diseases, especially cholera. Cholera may at first seem an odd target, since it is a disease that has not been a major public health problem in this country for over a century. But cholera remains a leading cause of death of infants and children throughout the Third World, where basic sanitation is lacking and water supplies are often contaminated. For example, in July of 1994, 70,000 cases of cholera were reported among the Rwandans crowded into refugee camps in Goma, Zaire, leading to 12,000 fatalities. A cholera epidemic struck East Africa in the fall of 1997 killing nearly 3000. And after an absence of over 100 years, cholera reappeared in Latin America in 1991, spreading from Peru to Mexico and claiming more than 10,000 lives.

The causative agent of cholera is *Vibrio cholerae*, a curved, rod-shaped bacterium found mostly in rivers and oceans. Most strains of *V. cholerae* are harmless; only one strain, called O1, is pathogenic. Infection occurs when a person drinks water or eats food contaminated with this strain. Once in the digestive system, the bacteria colonize the small intestine and begin producing proteins called enterotoxins. The cholera enterotoxin binds to the surface of the mucosal cells lining the intestine, triggering the massive secretion of water and dissolved salts from these cells. This results in violent diarrhea, which, if untreated, is followed by severe dehydration, muscle cramps, lethargy, and often death.

The cholera enterotoxin consists of two polypeptides, called the A and B subunits, which individually have no effect. For the toxin to be active, one A subunit must be linked to five B subunits. Since it is the B subunit of this complex that binds to intestinal cells, Arntzen's group decided to use this polypeptide as their antigen, reasoning that antibodies against it would potentially prevent toxin binding and render the bacteria harmless.

Their test of this idea involved the B subunit of an *E. coli* enterotoxin, which is similar in structure and immunological properties to the cholera protein. The DNA coding sequence of the B-subunit gene was obtained, to which they attached a promoter that would prompt transcription in all tissues of the plant. They then introduced this hybrid gene into potato plants by means of *Agrobacterium*-mediated transformation. They chose the potato not only because methods for transformation and regeneration of this plant are fairly routine, but also so they could assay the effectiveness of the antigen in the edible part of the plant, the tuber. Analysis showed that the engineered plants expressed their new gene and produced the enterotoxin.

After feeding mice a few grams of these genetically engineered tubers that had produced the B subunit, they found that the mice began to produce specific antibodies against it and to secrete them into the small intestine. And, most critically, mice later fed purified enterotoxin were protected from its effects. They did not develop the symptoms of cholera. Clinical trials are now planned to test the efficacy of the potato-produced vaccine in humans.

In the meantime, the Arntzen group is also working towards producing edible vaccines in bananas, which have several advantages over potatoes. For one thing, bananas can be grown almost anywhere throughout the tropical or sub-tropical developing countries of the world, exactly where they are needed the most. And unlike potatoes, bananas are usually eaten raw, avoiding the potential inactivation of the antigenic proteins by cooking. Finally, bananas are well liked by infants and children, making this approach to immunization a more feasible one.

Procedures are now being perfected for the transformation of banana cells with genes encoding the cholera enterotoxin and the regeneration of transgenic plants. It will be some time before the engineered bananas are ready to test, however, since it takes three years to grow a banana crop. If all goes as planned, it may someday be possible to immunize all Third World children against cholera and other intestinal diseases, saving untold thousands of young lives.

References

Arntzen, C. J. 1997. Edible vaccines. *Public Health Rep.* 112: 190–197.

Haq, T. A., Mason, H. S., Clements, J. D., and Arntzen, C. J. 1995. Oral immunization with a recombinant bacterial antigen produced in transgenic plants. *Science* 268: 714–716.

Sanchez, J. L., and Taylor, D. N. 1997. Cholera. *Lancet* 349: 1825–1830.

Chapter Summary

1. Inherited phenotypic variation in bacteria results from spontaneous mutation.
2. Genetic recombination in bacteria may result from three different modes: conjugation, transformation, or transduction.
3. Conjugation is initiated by a bacterium housing a plasmid called the F factor. If the F factor is in the cytoplasm of a donor cell (F^+), the recipient F^- cell receives a copy of the F factor, converting it to the F^+ status.
4. If the F factor is integrated into the donor cell chromosome (Hfr), conjugation is initiated with the recipient cell, and the donor chromosome moves unidirectionally into the recipient, leading to recombination. Time mapping of the bacterial chromosome is based on the orientation and position of the F factor in the donor chromosome.
5. The products of a group designated as the *rec* genes are directly involved in the process of recombination between the invading DNA and the recipient bacterial chromosome.
6. Plasmids, such as the F factor, are autonomously replicating DNA molecules found in the bacterial cytoplasm. Plasmids contain unique genes, such as those conferring antibiotic resistance, as well as those necessary for their transfer during conjugation.
7. In the phenomenon of transformation, which does not require cell contact, exogenous DNA enters the host chromosome of a recipient bacterial cell. Linkage mapping of closely aligned genes is performed using this process.
8. Bacteriophages (viruses that infect bacteria) demonstrate a defined life cycle, during which they reproduce within the host cell. They can be studied using the plaque assay.
9. Bacteriophages may be lytic—that is, they infect, reproduce, and lyse the host cell; or they may lysogenize the host cell, where they infect it, integrate their DNA into the host chromosome, and do not reproduce.
10. Transduction is virus-mediated bacterial DNA recombination. When a lysogenized bacterium subsequently reenters the lytic cycle, the new bacteriophages serve as vehicles for the transfer of host (bacterial) DNA. In the process of generalized transduction, a random part of the bacterial chromosome is transferred. In specialized transduction, only specific genes adjacent to the point of insertion of the prophage are transferred.
11. Transduction is also used for bacterial linkage and mapping studies.
12. Various mutant phenotypes, including plaque morphology and host range, have been studied in bacteriophages. These have served as the basis for investigating genetic exchange and mapping between these viruses.
13. Genetic analysis of the *rII* locus in bacteriophage T4 allowed Seymour Benzer to study intragenic recombination. By isolating *rII* mutants, performing complementation analysis, recombinational studies, and deletion mapping, Benzer was able to locate and map over 300 distinct sites within the two cistrons of the *rII* locus.

Insights and Solutions

1. Time mapping was performed in a cross involving the genes *his, leu, mal,* and *xyl*. The recipient cells were auxotrophic for all four genes. After 25 minutes, mating was interrupted with the following results in recipient cells:

(a) 90 percent were xyl^+.

(b) 80 percent were mal^+.

(c) 20 percent were his^+.

(d) none were leu^+.

What are the positions of these genes relative to the origin (*O*) of the F factor and to one another?

Solution: Because the *xyl* gene was transferred most frequently, it is closest to *O* (very close). The *mal* gene is next and reasonably close to *xyl*, followed by the *his* gene. The *leu* gene is well beyond these three, since no recombinants are recovered that include it. The diagram here illustrates these relative locations along a piece of the circular chromosome:

2. Three strains of bacteria, each bearing a separate mutation, a^-, b^-, or c^-, were used as the sources of donor DNA in a transformation experiment. Recipient cells were wild type for those genes, but expressed the mutant d^-.

(a) Based on the data, and assuming that the location of the *d* gene precedes the *a, b,* and *c* genes, propose a linkage map for the four genes.

DNA Donor	Recipient	Transformants	Frequency of Transformants
a^-d^+	a^+d^-	a^+d^+	0.21
b^-d^+	b^+d^-	b^+d^+	0.18
c^-d^+	c^+d^-	c^+d^+	0.63

Solution: These data reflect the relative distances between each of the *a, b, c* genes and the *d* gene. The *a* and *b* genes are about the same distance away from the *d* gene and are thus tightly linked to one another. The *c* gene is more distant. Assuming that the *d* gene precedes the others, the map looks like this:

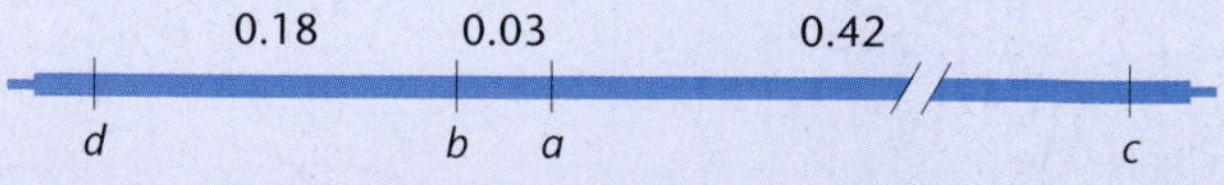

(b) If the donor DNA were wild type and the recipient cells were either a^-b^-, a^-c^-, or b^-c^-, in which case would wild-type transformants be expected most frequently?

Solution: Because the *a* and *b* genes are closely linked, they are most likely to be cotransformed in a single event. Thus, recipient cells of a^-b^- are most likely to be converted to wild type.

3. In four Hfr strains of bacteria, all derived from an original F^+ culture grown over several months, a group of hypothetical genes was studied and shown to be transferred in the following orders:

Hfr Strain	*Order of Transfer*					
1	*E*	*R*	*I*	*U*	*M*	*B*
2	*U*	*M*	*B*	*A*	*C*	*T*
3	*C*	*T*	*E*	*R*	*I*	*U*
4	*R*	*E*	*T*	*C*	*A*	*B*

Assuming that *B* is the first gene along the chromosome, determine the sequence of all genes shown. One strain creates an apparent dilemma. Which one is it? Explain why the dilemma is only apparent and not real.

Solution: The solution is obtained by overlapping the genes in each strain in the sequence in which they were transferred:

	2	*U*	*M*	*B*	*A*	*C*	*T*						
Strain:	3					*C*	*T*	*E*	*R*	*I*	*U*		
	1							*E*	*R*	*I*	*U*	*M*	*B*

Starting with *B*, the sequence of genes is *BACTERIUM*.

Strain 4 creates an apparent dilemma, which is resolved by realizing that the F factor is integrated in the opposite orientation; thus, the genes enter in the opposite sequence, starting with gene *R*:

$$\overrightarrow{MUIRETCAB}$$

4. In Benzer's fine-structure analysis of the *rII* locus in phage T4, he was able to perform complementation testing of any pair of mutations once it was clear that the locus contained two cistrons. Complementation was assayed by simultaneously infecting *E. coli* K12 with two phage strains, each with an independent mutation, neither of which could alone lyse K12. From the data that follow, determine which mutations are in which cistron, assuming that mutation 1 (M-1) is in the A cistron and mutation 2 (M-2) is in the B cistron. Are there any cases where the mutation cannot be properly assigned?

Test Pair	*Results*[*]
1, 2	+
1, 3	−
1, 4	−
1, 5	+
2, 3	−
2, 4	+
2, 5	−

[*] + or − indicate complementation or the failure of complementation, respectively

Solution: M-1 and M-5 complement one another and, therefore, are not in the same cistron. Thus, M-5 must be in the B cistron. M-2 and M-4 complement one another. Using the same reasoning, M-4 is not with M-2 and, therefore, is in the A cistron. M-3 fails to complement either M-1 or M-2, and so it would seem to be in both cistrons. One explanation is that the physical basis of M-3 somehow overlaps both the A and the B cistrons. It might be a double mutation with one in each cistron. It might also be a deletion that overlaps both cistrons, making it impossible for it to complement either M-1 or M-2.

5. Another mutation, M-6, was tested with the results shown here:

Test Pair	*Results*
1, 6	+
2, 6	−
3, 6	−
4, 6	+
5, 6	−

Draw all possible conclusions about M-6.

Solution: These results are consistent with assigning M-6 to the B cistron.

6. Recombination testing was then performed for M-2, M-5, and M-6 so as to map the B cistron. Recombination analysis using both *E. coli* B and K12 showed that recombination occurred between M-2 and M-5 and between M-5 and M-6, but not between M-2 and M-6. Why not?

Solution: Either M-2 and M-6 represent identical mutations or one of them may be a deletion that overlaps the other, but not M-5. Furthermore, the data cannot rule out the possibility that both are deletions.

7. In recombination studies, what is the significance of the value determined by calculating growth on the K12 versus the B strains of *E. coli* following simultaneous infection in *E. coli*? Which value is always greater?

Solution: By performing plaque analysis on *E. coli* B, where wild-type and mutant phages are both lytic, the total number of phages per milliliter can be determined. Because almost all cells are *rII* mutants of one type or another, this value is much larger. To avoid total lysis of the plate, extensive dilutions are necessary. On K12, *rII* mutations will not grow, but wild-type phages will. Since wild-type phages are the rare recombinants, there are relatively few of them and extensive dilution is not required.

Problems and Discussion Questions

1. Distinguish among the three modes of recombination in bacteria.
2. With respect to F^+ and F^- bacterial matings, answer the following questions:
   (a) How was it established that physical contact was necessary?
   (b) How was it established that chromosome transfer was unidirectional?
   (c) What is the physical–chemical basis of a bacterium being F^+?
3. List all major differences between (a) the $F^+ \times F^-$ and the Hfr $\times F^-$ bacterial crosses; and (b) the F^+, F^-, Hfr, and F′ bacteria.
4. Describe the basis for chromosome mapping in the Hfr $\times F^-$ crosses.
5. Why are the recombinants produced from an Hfr $\times F^-$ cross never F^+?
6. Describe the origin of F′ bacteria and merozygotes.
7. Describe what is known about the mechanisms of the transformation process.
8. In a transformation experiment involving a recipient bacterial strain of genotype a^-b^-, the following results were obtained:

Transforming DNA	*Transformants (%)*		
	a^+b^-	a^-b^+	a^+b^+
a^+b^-	3.1	1.2	0.04
a^+b^- and a^-b^+	2.4	1.4	0.03

What can you conclude about the location of the *a* and *b* genes relative to each other?

9. In a transformation experiment, donor DNA was obtained from a prototroph bacterial strain $(a^+b^+c^+)$, and the recipient was a triple auxotroph $(a^-b^-c^-)$. The following transformant classes were recovered:

$a^+\ b^-\ c^-$	180
$a^-\ b^+\ c^-$	150
$a^+\ b^+\ c^-$	210
$a^-\ b^-\ c^+$	179
$a^+\ b^-\ c^+$	2
$a^-\ b^-\ c^+$	1
$a^-\ b^+\ c^+$	3

What general conclusions can you draw about the linkage relationships among the three genes?

10. Explain the observations that led Zinder and Lederberg to conclude that the prototrophs recovered in their transduction experiments were not the result of F^- mediated conjugation.
11. Define plaque, lysogeny, and prophage.
12. Differentiate between generalized and restricted transduction. Which can be used in mapping and why?
13. Two theoretical genetic strains of a virus ($a^-\ b^-\ c^-$ and $a^+\ b^+\ c^+$) are used to simultaneously infect a culture of host bacteria. Of 10,000 plaques scored, the following genotypes were observed:

$a^+\ b^+\ c^+$	4100	$a^-\ b^+\ c^-$	160
$a^-\ b^-\ c^-$	3990	$a^+\ b^-\ c^+$	140
$a^+\ b^-\ c^-$	740	$a^-\ b^-\ c^+$	90
$a^-\ b^+\ c^+$	670	$a^+\ b^+\ c^-$	110

Determine the genetic map of these three genes on the viral chromosome. Decide whether interference was positive or negative.

14. Describe the conditions under which genetic recombination may occur in bacteriophages.
15. If a single bacteriophage infects one *E. coli* cell present on a lawn of bacteria and, upon lysis, yields 200 viable viruses, how many phages will exist in a single plaque if only three more lytic cycles occur?
16. A phage-infected bacterial culture was subjected to a series of dilutions, and a plaque assay was performed in each case, with the results shown here. What conclusion can be drawn in the case of each dilution?

	Dilution Factor	*Assay Results*
(a)	10^4	All bacteria lysed
(b)	10^5	14 plaques
(c)	10^6	0 plaques

17. In complementation studies of the *rII* locus of phage T4, three groups of three different mutations were tested. For each group, only two combinations were tested. On the basis of each set of data (shown here), predict the results of the third experiment

Group A	*Group B*	*Group C*
$d \times e$—lysis	$g \times b$—no lysis	$j \times k$—lysis
$d \times f$—no lysis	$g \times i$—no lysis	$j \times l$—lysis
$e \times f$—?	$b \times i$—?	$k \times l$—?

18. In an analysis of other *rII* mutants, complementation testing yielded the following results:

Mutants	*Results (lysis)*
1, 2	+
1, 3	+
1, 4	−
1, 5	−

(a) Predict the results of testing 2 and 3, 2 and 4, and 3 and 4 together.
(b) If further testing yielded the following results, what would you conclude about mutant 5?

Mutants	*Results*
2, 5	−
3, 5	−
4, 5	−

(c) Following mixed infection of mutants 2 and 3 on *E. coli* B, progeny viruses were plated in a series of dilutions on both *E. coli* B and K12 with the following results:

Strain Plated	*Dilution*	*Plaques*
E. coli B	10^{-5}	2
E. coli K12	10^{-1}	5

What is the recombination frequency between the two mutants?

(d) Another mutation, 6, was then tested in relation to mutations 1 through 5. In initial testing, mutant 6 complemented mutants 2 and 3. In recombination testing with 1, 4, and 5, mutant 6 yielded recombinants with 1 and 5, but not with 4. What can you conclude about mutation 6?

19. When the interrupted mating technique was used with five different strains of Hfr bacteria, the following order of gene entry and recombination was observed:

Hfr Strain	*Order*				
1	*T*	*C*	*H*	*R*	*O*
2	*H*	*R*	*O*	*M*	*B*
3	*M*	*O*	*R*	*H*	*C*
4	*M*	*B*	*A*	*K*	*T*
5	*C*	*T*	*K*	*A*	*B*

On the basis of these data, draw a map of the bacterial chromosome. Do the data support the concept of circularity?

Extra-Spicy Problems

20. During the analysis of seven *rII* mutations in phage T4, mutants 1, 2, and 6 were in cistron A, while mutants 3, 4, and 5 were in cistron B. Of these, mutant 4 was a deletion overlapping mutant 5. The remainder were point mutations. Nothing was known about mutant 7.

(a) Predict the results of complementation (+ or −): 1 and 2; 1 and 3; 2 and 4; and 4 and 5.

(b) In recombination studies between 1 and 2, the following results were obtained:

Strain	*Dilution*	*Plaques*	*Phenotypes*
E. coli B	10^{-7}	4	r
E. coli K12	10^{-2}	8	+

Calculate the recombination frequency.

(c) When mutant 6 was tested for recombination with mutant 1, the data were the same for strain B, as shown in (b), but not for K12. The researcher lost the K12 data, but remembered that recombination was 10 times more frequent than when mutants 1 and 2 were tested. What were the lost values (dilution and colony numbers)?

(d) Mutant 7 failed to complement any of the other mutants (1–6). Define the nature of mutant 7.

21. In *Bacillus subtilis*, linkage analysis of two mutant genes affecting the synthesis of the two amino acids, tryptophan (trp_2^-) and tyrosine (trp_1^-) was performed using transformation. Examine the following data and draw all possible conclusions regarding linkage:

Donor DNA	*Recipient Cell*	*Transformants*	*No.*
A. $trp_2^+\ tyr_1^+$	$trp_2^-\ tyr_1^-$	$trp^+\ tyr^-$	196
		$trp^-\ tyr^+$	328
		$trp^+\ tyr^+$	367
B. $trp_2^+\ tyr_1^-$ and $trp_2^-\ tyr_1^+$	$trp_2^-\ tyr_1^-$	$trp^+\ tyr^-$	190
		$trp^-\ tyr^+$	256
		$trp^+\ tyr^+$	2

What is the role of Part B of the experiment? [*Reference*: E. Nester, M. Schafer, and J. Lederberg (1963).]

22. An Hfr strain is used to map three genes in an interrupted mating experiment. The cross is $Hfr/a^+b^+c^+rif \times F^-/a^-b^-c^-rif^r$ (no map order is implied in the listing of the alleles; *rif* = the antibiotic rifampicin). The a^+ gene is required for the biosynthesis of nutrient A, the b^+ gene for nutrient B, and c^+ for nutrient C. The minus alleles are auxotrophs for these nutrients. The cross is initiated at time = 0, and, at various times, the mating mixture is plated on three types of medium. Each plate contains minimal medium (MM) *plus* rifampicin *plus* specific supplements that are indicated in the following table (the results for each time point are shown as number of colonies growing on each plate):

Supplements Added to MM	*Time of Interruption*			
	5 min	*10 min*	*15 min*	*20 min*
Nutrients A and B	0	0	4	21
Nutrients B and C	0	5	23	40
Nutrients A and C	4	25	60	82

(a) What is the purpose of rifampicin in the experiment?

(b) Based on these data, determine the approximate location on the chromosome of the *a*, *b*, and *c* genes relative to one another and to the F factor.

(c) Can the location of the *rif* gene be determined in this experiment? If not, design an experiment to determine the location of *rif* relative to the F factor and to gene *B*.

Selected Readings

Adelberg, E.A. 1960. *Papers on bacterial genetics*. Boston: Little, Brown.

Benzer, S. 1955. Fine structure of a genetic region in bacteriophage. *Proc. Natl. Acad. Sci. USA* 42:344–54.

———. 1961. On the topography of the genetic fine structure. *Proc. Natl. Acad. Sci. USA* 47:403–15.

———. 1962. The fine structure of the gene. *Sci. Am.* (Jan.) 206:70–816.

Birge, E.A. 1988. *Bacterial and bacteriophage genetics—An introduction*. New York: Springer–Verlag.

Brock, T. 1990. *The emergence of bacterial genetics*. Cold Spring Harbor, NY: Cold Spring Harbor Laboratory Press.

Broda, P. 1979. *Plasmids*. New York: W.H. Freeman.

Bukhari, A.I., Shapiro, J.A., and Adhya, S.L., eds. 1977. *DNA insertion elements, plasmids, and episomes*. Cold Spring Harbor, NY: Cold Spring Harbor Laboratory.

Cairns, J., Stent, G.S., and Watson, J.D., eds. 1966. *Phage and the origins of molecular biology*. Cold Spring Harbor, NY: Cold Spring Harbor Laboratory Press.

Campbell, A.M. 1976. How viruses insert their DNA into the DNA of the host cell. *Sci. Am.* (Dec.) 235:102–13.

Fox, M.S. 1966. On the mechanism of integration of transforming deoxyribonucleate. *J. Gen. Physiol.* 49:183–96.

Hayes, W. 1953. The mechanisms of genetic recombination in *Escherichia coli. Cold Spring Harbor Symp. Quant. Biol.* 18:75–93.

———. 1968. *The genetics of bacteria and their viruses*, 2nd ed. New York: Wiley.

Hershey, A.D. 1946. Spontaneous mutations in a bacterial virus. *Cold Spring Harbor Symp. Quant. Biol.* 11:67–76.

Hershey, A.D., and Chase, M. 1951. Genetic recombination and heterozygosis in bacteriophage. *Cold Spring Harbor Symp. Quant. Biol.* 16:471–79.

Hershey, A.D., and Rotman, R. 1949. Genetic recombination between host range and plaque-type mutants of bacteriophage in single cells. *Genetics* 34:44–71.

Hotchkiss, R.D., and Marmur, J. 1954. Double marker transformations as evidence of linked factors in deoxyribonucleate transforming agents. *Proc. Natl. Acad. Sci. USA* 40:55–60.

Jacob, F., and Wollman, E.L. 1961a. *Sexuality and the genetics of bacteria*. Orlando, FL: Academic Press.

———. 1961b. Viruses and genes. *Sci. Am.* (June) 204:92–106.

Lederberg, J. 1986. Forty years of genetic recombination in bacteria: A fortieth anniversary reminiscence. *Nature* 324:627–28.

Luria, S.E., and Delbruck, M. 1943. Mutations of bacteria from virus sensitivity to virus resistance. *Genetics* 28:491–511.

Lwoff, A. 1953. Lysogeny. *Bacteriol. Rev.* 17:269–3316.

Meselson, M., and Weigle, J.J. 1961. Chromosome breakage accompanying genetic recombination in bacteriophage. *Proc. Natl. Acad. Sci. USA* 47:857–68.

Miller, J.H. 1992. *A short course in bacterial genetics*. Cold Spring Harbor, NY: Cold Spring Harbor Press.

Morse, M.L., Lederberg, E.M., and Lederberg, J. 1956. Transduction in *Escherichia coli* K12. *Genetics* 41:141–56.

Nester, E., Schafer, M., and Lederberg, J. 1963. Gene linkage in DNA transfer: A cluster of genes in *Bacillus subtilis. Genetics* 48:529–51.

Novick, R.P. 1980. Plasmids. *Sci. Am.* (Dec.) 243:102–216.

Ozeki, H., and Ikeda, H. 1968. Transduction mechanisms. *Annu. Rev. Genet.* 2:245–78.

Smith, H.O., Danner, D.B., and Deich, R.A. 1981. Genetic transformation. *Annu. Rev. Biochem.* 50:41–68.

Smith-Keary, P.F. 1989. *Molecular genetics of Escherichia coli*. New York: Guilford Press.

Stahl, F.W. 1979. *Genetic recombination: Thinking about it in phage and fungi*. New York: W.H. Freeman.

———. 1987. Genetic recombination. *Sci. Am.* (Nov.) 256:91–101.

Stent, G.S. 1963. *Molecular biology of bacterial viruses*. New York: W.H. Freeman.

———. 1966. *Papers on bacterial viruses*, 2nd ed. Boston: Little, Brown.

Tessman, I. 1965. Genetic ultrafine structure in the T4 *rII* region. *Genetics* 51:63–75.

Visconti, N., and Delbruck, M. 1953. The mechanism of genetic recombination in phage. *Genetics* 38:5–33.

Wollman, E.L., Jacob, F., and Hayes, W. 1956. Conjugation and genetic recombination in *Escherichia coli* K-12. *Cold Spring Harbor Symp. Quant. Biol.* 21:141–62.

Zinder, N.D. 1958. Transduction in bacteria. *Sci. Am.* (Nov.) 199:38–416.

———. 1992. Forty years ago: The discovery of bacterial transduction. *Genetics* 132:291–94.

Zinder, N.D., and Lederberg, J. 1952. Genetic exchange in *Salmonella. J. Bacteriol.* 64:679–99.

GENETICS MediaLab

The following resources will help you achieve a better understanding of the concept presented in this chapter. These resources can be found either on the CD packaged with this textbook or the Companion Web site found at **http://www.prenhall.com/klug**

CD Resources:

Module 15.1: Phage Genetics

Module 15.2: Bacterial Genetics

Web Problem 1:

Time for completion = 10 minutes

Do bacterial genomes recombine? This exercise uses an animation of bacterial conjugation to explore some of the ways that genetic material is transferred from one cell to another. What are the differences between a F+ cell, an Hfr cell, and an F' cell? Which strand of DNA in the rolling circle model of gene transfer and replication is used as the template for replicating the donor genome? Which strand is replicated in the recipient's genome? Once the recipient cell has replicated the donor's DNA, how is the genetic information incorporated into the genome? To complete this exercise, visit Web Problem 1 in Chapter 15 of your Companion Web site and select the keyword **CONJUGATION**.

Web Problem 2:

Time for completion = 10 minutes

How do viruses participate in bacterial recombination? Transduction refers to the transfer of genetic material between two bacterial cells using a bacteriophage as a vector. Follow the tutorial linked on the text web page and use material in Chapter 11 to answer the following questions. What is the key difference between the genes transferred in generalized transduction and those transferred through specialized transduction? Must the viral DNA incorporate into the host's genome (as a prophage) in order to recombine with the bacterial chromosome? What is the role of the rec proteins in transduction? Do they play any role in either conjugation or transformation? Would specialized transduction or generalized transduction be more useful in linkage and mapping studies? Why? To complete this exercise, visit Web Problem 2 in Chapter 15 of your Companion Web site and select the keyword **TRANSDUCTION**.

Web Problem 3:

Time for completion = 10 minutes

Can viruses themselves recombine? If two phage infect a single cell, it is possible that, within that cell, the phage DNA may recombine to create new bacteriophage genotypes. Seymour Benzer conducted an amazing set of experiments on a single locus in the bacteriophage T4. Using the system of RII mutants, Benzer constructed the first map of a gene. Why did Benzer use two strains of *E. coli* (the B and the K12 strains)? Why, when Benzer mixed the mutant and wild-type bacterial strains together and plated them on K12, did he sometimes get no plaques and other times only a few plaques? Were Benzer's recombinant phage detected in cells that were in the lysogenic or the lytic phase of the phage life cycle? To complete this exercise, visit Web Problem 3 in Chapter 15 of your Companion Web site and select the keyword **VIRAL RECOMBINATION**.

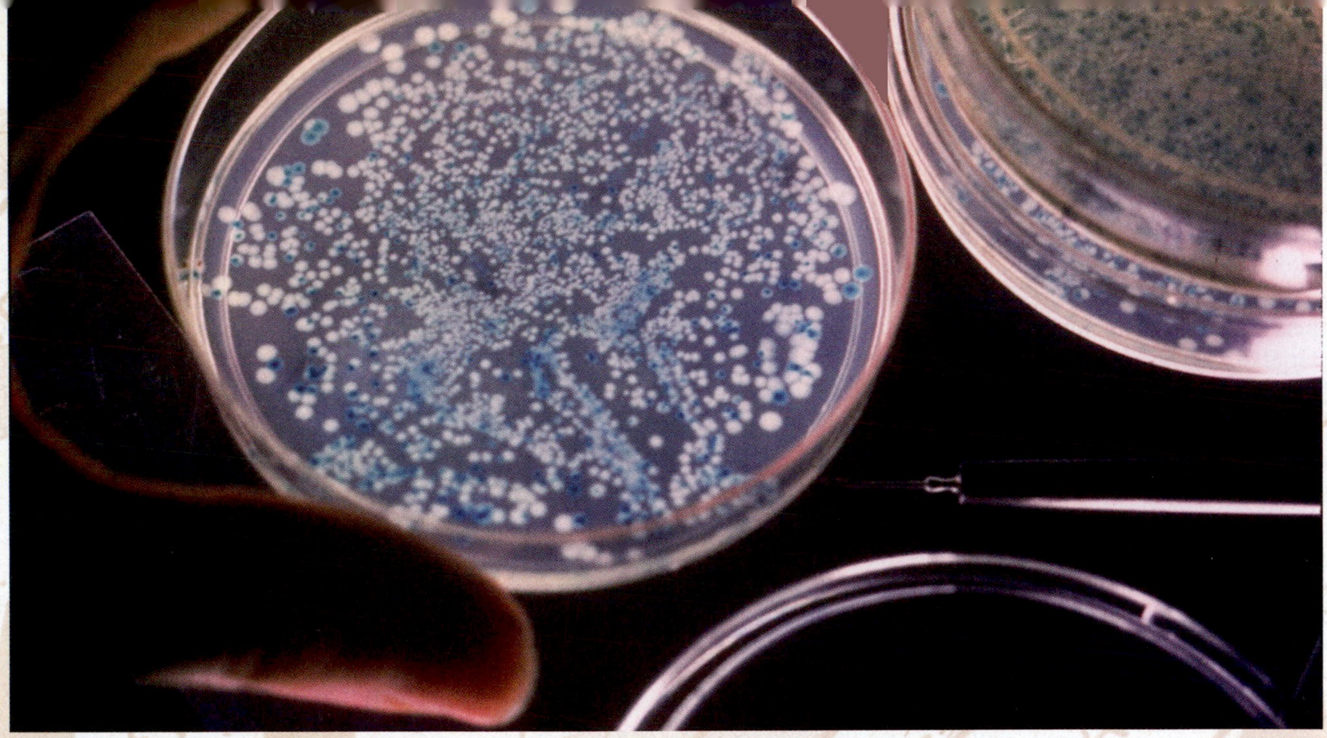

A Petri plate showing the growth of host cells after uptake of recombinant plasmids. (*Michael Gabridge/Visuals Unlimited*)

16

Recombinant DNA Technology

The techniques used to create, replicate, and analyze recombinant DNA molecules were developed in the mid- and late 1970s as a way for researchers to isolate and study specific DNA sequences. The first commercial application of recombinant DNA technology was approved in 1982, when the Food and Drug Administration permitted the marketing of human insulin produced by bacterial cells. This was a milestone in the development of the biotechnology industry. Today, recombinant DNA technology is used to produce medicines, vaccines, industrial chemicals, genetically modified organisms, and food products. The biotechnology industry is now a multibillion-dollar component of the national and world economy.

Recombinant DNA technology makes it possible to identify and isolate a single gene from the tens of thousands present in the human genome and to produce large quantities of this gene in the form of cloned DNA molecules. In addition, the gene can be transferred into cells that will synthesize its encoded gene product, which can then be recovered and purified for use in research, medicine, or industry.

Clones are identical organisms, cells, or molecules descended from a single ancestor. Cloning a gene involves producing many identical copies that can be used for numerous purposes, including research into the structure and organization of a gene and the commercial production of proteins such as insulin. In this chapter, we review the basic methods of recombinant DNA technology used to isolate, replicate, and analyze genes. In Chapter 18, we discuss some applications of this technology to research, medicine, the legal system, agriculture, and industry.

16.1 Recombinant DNA Technology Creates Artificial Combinations of DNA Molecules

The term **recombinant DNA** refers to a new combination of DNA molecules that are not found together naturally. Although processes such as crossing over technically produce recombinant DNA, the term is generally reserved for DNA produced by joining molecules derived from different biological sources.

MEDIA TUTORIAL Restriction Enzymes and Vectors

Recombinant DNA technology uses methods derived from nucleic acid biochemistry, coupled with genetic techniques originally developed for the study of bacteria and viruses. As described here, recombinant DNA technology is a powerful tool for isolating potentially unlimited quantities of a gene. Although several methods are available, the basic procedure involves these steps:

1. DNA is purified from cells or tissues.
2. Enzymes called **restriction endonucleases** or **restriction enzymes** generate specific DNA fragments. These enzymes recognize and cut DNA molecules at specific nucleotide sequences.
3. The fragments produced by treatment with restriction enzymes are joined to other DNA molecules that serve as **vectors**, or carrier molecules. A vector joined to a DNA fragment is a **recombinant DNA molecule**.
4. The recombinant DNA molecule, consisting of a vector carrying an inserted DNA fragment, is transferred to a host cell. Within the host cell, the recombinant molecule replicates, producing dozens of identical copies known as clones.
5. As host cells replicate, the recombinant DNA molecules within them are passed on to all their progeny, creating a population of host cells, each of which carries copies of the cloned DNA sequence.
6. The cloned DNA can be recovered from host cells, purified, and analyzed, or;
7. The cloned DNA in host cells is transcribed, its mRNA translated, and the gene product isolated and used for research, or sold commercially.

16.2 Restriction Enzymes Are Used to Construct Recombinant DNA Molecules

The cornerstone of recombinant DNA technology is a class of enzymes called restriction endonucleases. These enzymes, isolated from bacteria, are so named because they restrict or prevent viral infection by degrading the invading viral DNA. Restriction enzymes recognize a specific nucleotide sequence and cut both strands of the DNA within that sequence. The 1978 Nobel Prize was awarded to Werner Arber, Hamilton Smith, and Daniel Nathans for their work on restriction enzymes. To date, over 200 restriction enzymes have been identified. Their usefulness in cloning derives from their ability to reproducibly cut DNA into fragments.

One of the first restriction enzymes identified was isolated from *E. coli* and is designated *Eco*RI (pronounced echo-r-one). Its recognition sequence and point of cleavage are shown in Figure 16–1. Note that the recognition sequence for

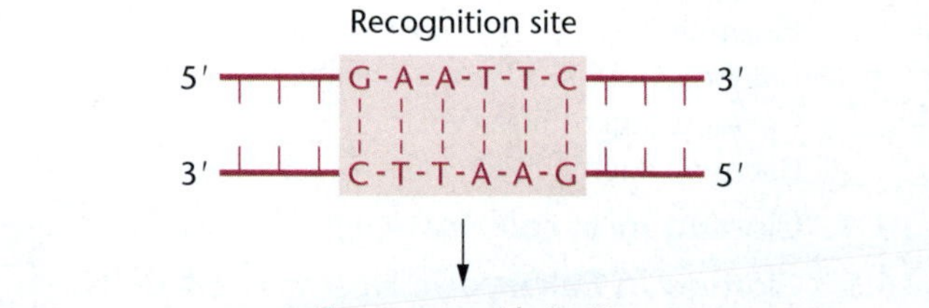

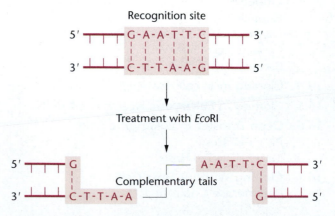

FIGURE 16–1 The restriction enzyme *Eco*RI recognizes and binds to the palindromic nucleotide sequence GAATTC. Cleavage of the DNA at this site produces complementary single-stranded tails. These single-stranded tails anneal with single-stranded tails from other DNA fragments to form recombinant DNA molecules.

*Eco*RI reads the same in the 5′-to-3′ direction on one strand as it does in the 5′-to-3′ direction on the complementary strand. Sequences that read the same on both strands of DNA are known as **palindromes**.

The DNA fragments produced by *Eco*RI digestion have overhanging single-stranded tails (called "sticky ends") that reanneal with complementary single-stranded tails on other DNA fragments. If they are mixed under the proper conditions, DNA fragments from two sources form recombinant molecules by hydrogen bonding of their sticky ends. The enzyme **DNA ligase** covalently links these fragments to form recombinant DNA molecules (Figure 16–2). Some common restriction enzymes and their recognition sequences are shown in Figure 16–3.

16.3 Vectors Carry DNA Fragments to Be Cloned

Fragments of DNA produced by restriction enzyme digestion cannot directly enter bacterial cells for cloning; when a DNA fragment is joined to a vector, however, it can gain entry to a host cell, where it can be replicated or cloned into many copies. Vectors are, in essence, carrier DNA molecules. To serve as a vector, a DNA molecule must have several properties:

1. It must be able to replicate itself independently and the DNA segment it carries.
2. It should contain several different restriction enzyme cleavage sites that are present only once in the vector. One site is cleaved with a restriction enzyme and used to insert a DNA segment cut with the same enzyme.
3. It should carry a selectable marker (usually in the form of antibiotic resistance genes or genes for enzymes missing in the host cell) to distinguish host cells that carry vectors from host cells that do not contain vectors.
4. It should be easy to recover from the host cell.

A number of vectors are currently in use, including those derived from **plasmids** and **bacteriophages**.

Plasmids

The first vectors were genetically modified plasmids, which are versatile and still widely used for cloning, but are unsuitable for cloning large fragments of DNA. Plasmids are naturally occurring, extrachromosomal, double-stranded DNA molecules that replicate autonomously within bacterial cells (Figure 16–4). The genetics of plasmids and their host bacterial cells has been discussed in Chapter 15 and Chapter 19. In this section, we emphasize the role of plasmids as vectors. Although only a single plasmid may enter a host cell, many plasmids can increase their copy number in the host cell, so that several hundred copies are present. When used as vectors, such plasmids allow more copies of cloned DNA to be produced. To use them as vectors, they are modified so that they contain a limited number of convenient restriction sites and marker genes that detect the presence of plasmids in host cells.

Many genetically engineered plasmid vectors are now available, and certain features make it easier to identify host

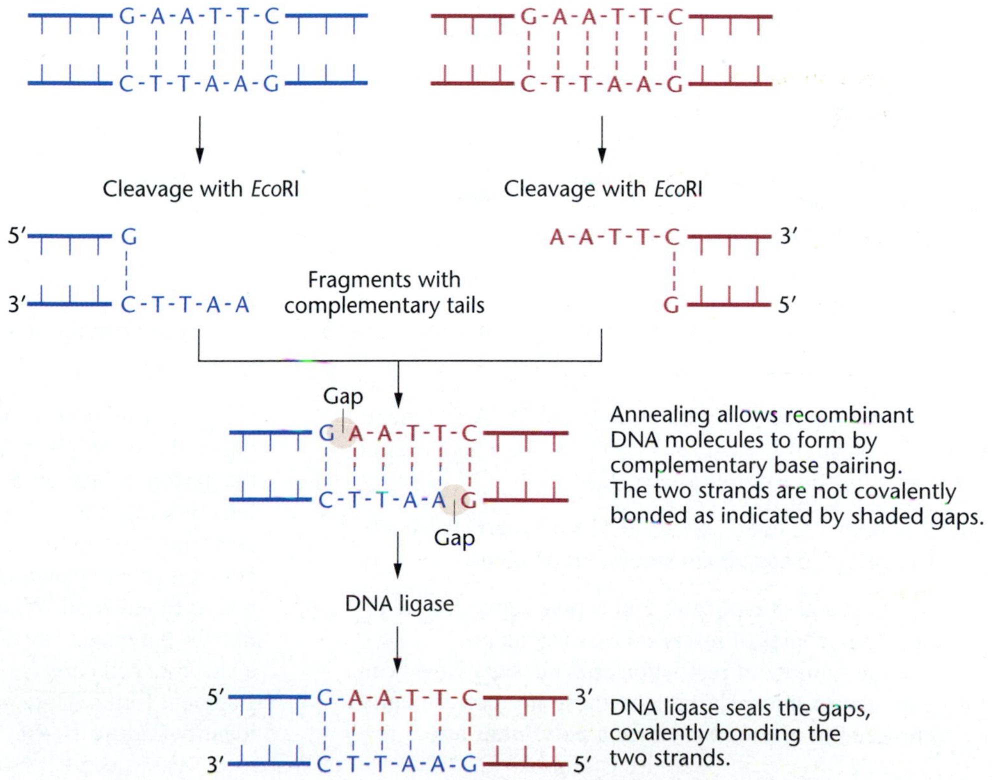

FIGURE 16–2 DNA from different sources is cleaved with *Eco*RI and mixed to allow annealing to form recombinant molecules. The enzyme DNA ligase then chemically bonds these annealed fragments into an intact recombinant DNA molecule.

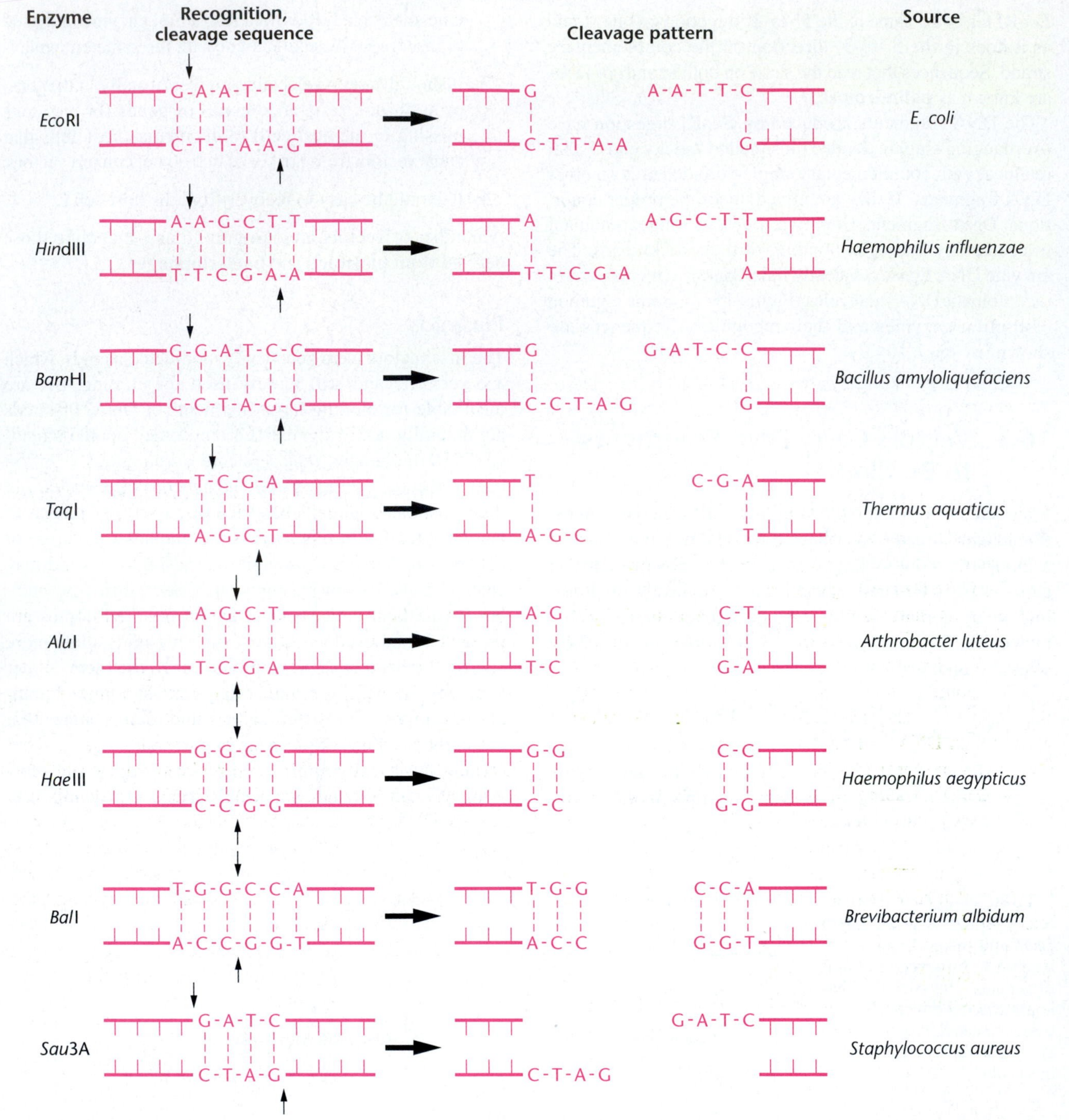

FIGURE 16–3 Some common restriction enzymes, with their recognition sequences, cutting sites, cleavage patterns, and sources.

cells carrying a plasmid with an inserted DNA fragment. One such plasmid is pUC18 (Figure 16–5). This plasmid has several useful features as a vector:

1. It is small (2686 bp), which allows it to carry relatively large DNA inserts up to several kb in length.
2. In a host cell, it replicates 500 copies per cell, producing many copies of inserted DNA fragments.
3. A large number of restriction enzyme sites have been engineered into pUC18, and these are conveniently clustered in one region, called a **polylinker site**.
4. It has a selection system that allows recombinant plasmids to be identified. It carries a fragment of the bacterial *lacZ* gene, and the polylinker is inserted into this fragment. Through the *lacZ* gene's action, bacterial host cells carrying pUC18 produce blue colonies when grown on medium containing a compound called Xgal. When a DNA fragment is inserted into the polylinker site, the *lacZ* gene is inactivated, and a bacterial cell carrying pUC18 with an inserted DNA fragment forms white colonies, making them easy to identify (Figure 16–6).

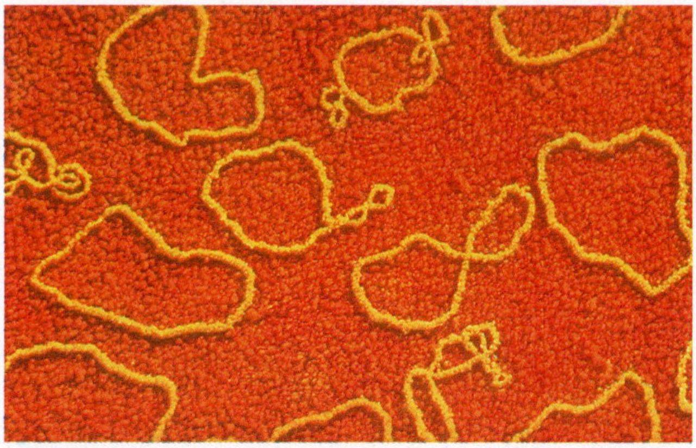

FIGURE 16–4 A color-enhanced electron micrograph of circular plasmid molecules isolated from the bacterium *E. coli*. Geneticallly engineered plasmids are used as vectors for cloning DNA. *(K. G. Murti/Visuals Unlimited)*

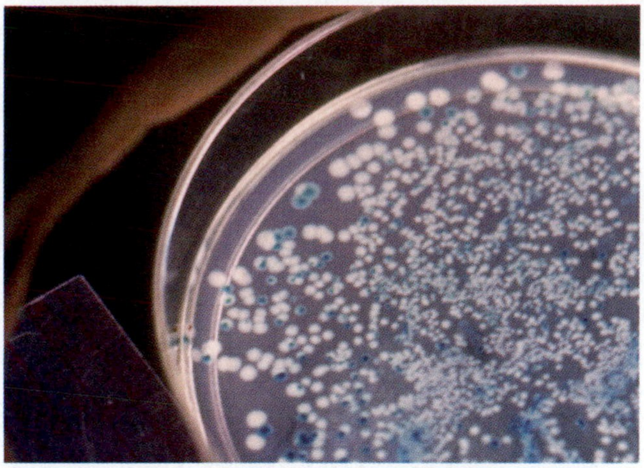

FIGURE 16–6 A Petri plate showing the growth of bacterial cells after uptake of recombinant plasmids. The medium on the plate contains a compound called Xgal. DNA inserts into the pUC18 vector disrupt the gene responsible for the formation of blue colonies. Plasmids in the blue colonies do not carry any cloned DNA inserts, whereas the white colonies contain plasmids with DNA inserts. *(Michael Gabridge/Visuals Unlimited)*

Lambda and M13 Bacteriophage Vectors

Plasmid vectors generally carry 5 to 10 kb of inserted DNA, but for many experiments, larger pieces of DNA are necessary. For these purposes, genetically modified strains of the bacterial virus lambda (λ), also called **lambda phage**, are used as vectors (Figure 16–7). Lambda genes have all been identified and mapped, and the nucleotide sequence of the entire genome is known. The central third of the lambda chromosome is dispensable, and this region can be replaced with foreign DNA without affecting the phage's ability to infect cells and form plaques (Figure 16–8).

To clone DNA using a lambda-replacement vector, the phage DNA is cut with a restriction enzyme (e.g., *Eco*RI), resulting in a left arm, a right arm, and the dispensable central region. The arms are isolated and combined with DNA fragments that have been generated by cutting genomic DNA with *Eco*RI. Ligation with DNA ligase follows. Lambda vectors containing inserted DNA are then introduced into bacterial host cells, where they reproduce to form many particles of infective phage, each of which carries a DNA insert. As they reproduce, the phages form plaques from which the cloned DNA can be recovered.

Other bacteriophages are used as vectors for specialized functions. One of these is the single-stranded phage known as M13 (Figure 16–9). When M13 infects a bacterial cell, its single-stranded DNA replicates to produce a double-stranded molecule, known as the replicative form (RF). RF molecules can be regarded as being similar to plasmids, and foreign DNA can be inserted into the restriction enzyme sites in the phage DNA. When reinserted into bacterial cells, the RF molecules replicate to produce single strands, which include one strand of the inserted DNA. The single-stranded DNA produced by M13 is extruded from the cell and can be recovered and used directly for DNA sequencing or as a template for mutational alteration of the cloned sequence.

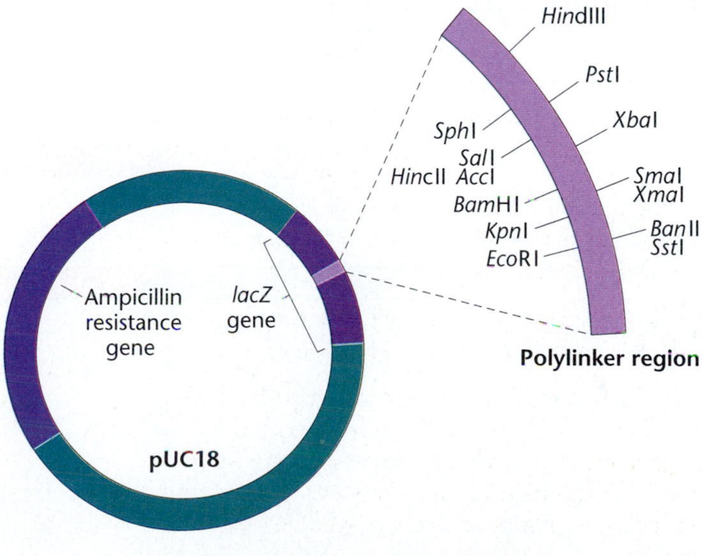

FIGURE 16–5 The plasmid pUC18 offers several advantages as a vector for cloning. Because of its small size, it accepts relatively large DNA fragments for cloning; it replicates to a high copy number, and has a large number of restriction sites in the polylinker, located within a *lacZ* gene. Bacteria carrying pUC18 produce blue colonies when grown on media containing Xgal. DNA inserted into the polylinker site disrupts the *lacZ* gene; this results in white colonies and allows direct identification of colonies carrying cloned DNA inserts.

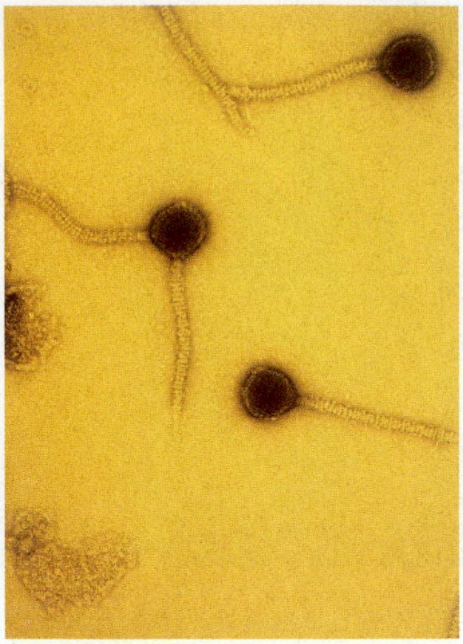

FIGURE 16–7 A colorized electron micrograph of a cluster of bacteriophage lambda, widely used as a vector in recombinant DNA work. *(M. Wurtz/Biozentrum, University of Bassel/Science Photo Library/Photo Researchers, Inc.)*

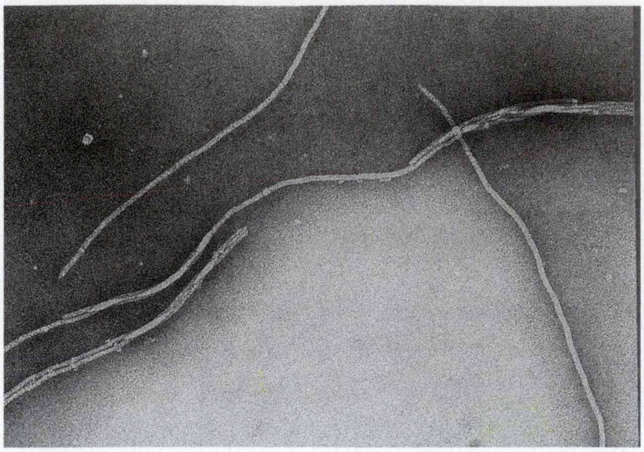

FIGURE 16–9 M13 bacteriophages are long thin filaments containing a single strand of DNA. They infect *E. coli* host cells by attaching to the cellular extensions of the bacterium called *F pili*.

Phage vectors can carry inserts of about 10–15 kb, almost twice as long as the DNA inserts in plasmid vectors. This is an important advantage when cloning entire genomes. In addition, some phage vectors only accept inserts of a minimum size. This means that in a cloning experiment where large inserts must be isolated, vectors do not end up carrying relatively useless inserts of a few dozen or a few hundred nucleotides in length.

Cosmids and Shuttle Vectors

Cosmids are hybrid vectors constructed by using parts of the lambda chromosome and plasmid DNA. Cosmids

FIGURE 16–8 Phage lambda as a vector. DNA is extracted from a preparation of lambda phage, and the central gene cluster is removed by treatment with a restriction enzyme. The DNA to be cloned is cut with the same enzyme and ligated into the arms of the lambda chromosome. The recombinant chromosome is then packaged into phage proteins to form a recombinant virus. This virus infects bacterial cells and replicates its chromosome, including the DNA insert.

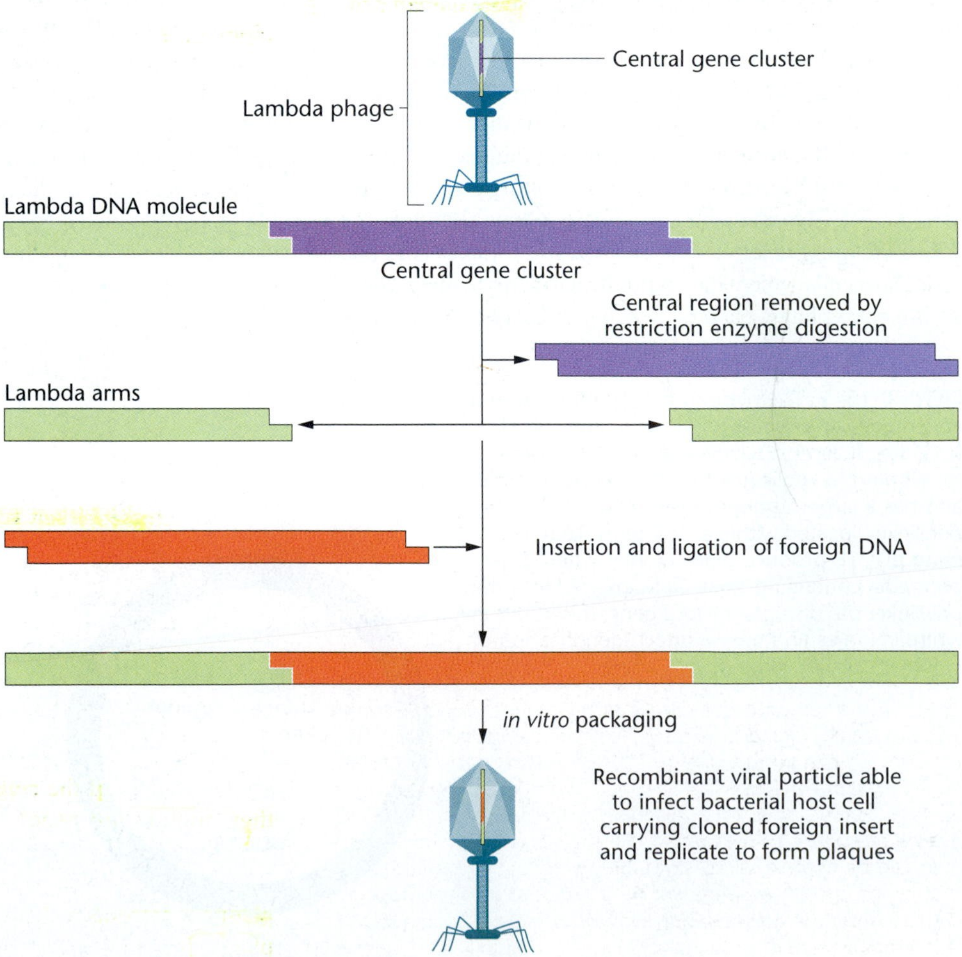

contain the *cos* sequence of phage lambda, necessary for packing phage DNA into phage protein coats and plasmid sequences for replication. They also contain an antibiotic resistance gene to identify host cells carrying cosmids (Figure 16–10). Cosmids carrying inserted DNA fragments are packaged into lambda protein heads, forming infective phage particles. After the cosmid infects a bacterial host cell, it replicates as a plasmid. Because almost the entire lambda genome has been deleted, cosmids can carry DNA inserts that are much larger than those carried by lambda vectors. Cosmids can carry almost 50 kb of inserted DNA, phage vectors can accommodate DNA inserts of about 10 through 15 kb, and plasmids are usually limited to inserts 5 through 10 kb in length.

Other hybrid vectors that have been constructed with origins of replication derived from different sources (e.g., plasmids and animal viruses such as SV40) can replicate in more than one type of host cell. These shuttle vectors usually contain genetic markers that are selectable in both host systems and can be used to shuttle DNA inserts back and forth between *E. coli* and another host cell, such as yeast. Often such vectors are employed in studying gene expression. Other vectors, those used in specific applications, will be described in the next sections.

Bacterial Artificial Chromosomes

The mapping and analysis of large complex eukaryotic genomes require cloning vectors that can accommodate very large DNA fragments. In addition, since some human genes range from 1000 kb to over 2000 kb, vectors with large cloning capacities are useful in studying the organization of these genes. Recently, a number of vectors that use bacterial host cells have been developed.

One of these vectors is based on the fertility plasmid (F factor) of bacteria and is called a **bacterial artificial chromosome (BAC)**. Recall from Chapter 15 that F factors are independently replicating plasmids that are involved in the transfer of genetic information during bacterial conjugation. Because F factors can carry fragments of the bacterial chromosome up to 1 Mb in length, they have been engineered to act as vectors for eukaryotic DNA and can carry inserts of about 300 kb (Figure 16–11). BAC vectors carry the F factor genes for replication and copy number, and incorporate an antibiotic resistance marker and restriction enzyme sites for inserting foreign DNA to be cloned. In addition, the cloning site is flanked by promoter sites that can be used to generate RNA molecules for the expression of the cloned gene, for use as probes in chromosome walking, and for sequencing the cloned insert.

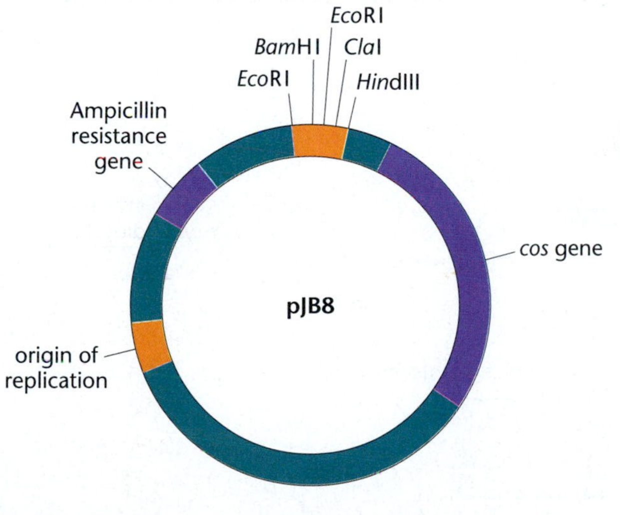

FIGURE 16–10 The cosmid pJB8 contains a bacterial origin of replication (*ori*), a single cos site (*cos*), an ampicillin resistance gene (*amp*, for selection of colonies that have taken up the cosmid), and a region containing four restriction sites for cloning (*Bam*HI, *Eco*RI, *Cla*I, and *Hind*III). Because the vector is small (5.4 kb long), it can accept foreign DNA segments between 33 and 46 kb in length. The *cos* site allows cosmids carrying large inserts to be packaged into lambda viral coat proteins as though they were viral chromosomes. The viral coats carrying the cosmid can be used to infect a suitable bacterial host, and the vector, carrying a DNA insert, will be transferred into the host cell. Once inside, the *ori* sequence allows the cosmid to replicate as a bacterial plasmid.

16.4 Cloning in *E. coli* Host Cells

As discussed earlier, scientists use biotechnology to construct *and* replicate recombinant DNA molecules, to produce many copies, or clones. This is accomplished by transferring recombinant molecules into host cells where replication takes place. The first cloning method, cell-based cloning, has been widely used to copy DNA fragments and used to study gene organization and function.

A variety of prokaryotic cells serve as hosts for recombinant vector replication. One of the most commonly used hosts is a laboratory strain of the bacterium *E. coli* known as K12. *E. coli* strains such as K12 are genetically well characterized, and they serve as hosts for a wide range of vectors. Several steps are required to create recombinant DNA molecules (Figure 16–12) in an *E. coli* host cell:

1. The DNA to be cloned is isolated and treated with a restriction enzyme to create fragments that end in a specific sequence.
2. The fragments are ligated to plasmid molecules that have been cut with the same restriction enzyme, creating a recombinant vector.
3. The recombinant vector is transferred into an *E. coli* host cell, where it replicates to form dozens of copies.
4. The bacteria are grown on a nutrient plate where they form colonies and are screened to identify those that have taken up the recombinant plasmids.

Because the cells in each colony are derived from a single ancestral cell, all the cells in the colony and the plasmids they contain are genetically identical clones. Similarly, phages containing foreign DNA are used to infect *E. coli* host cells, and, when plated on solid medium, each resulting plaque represents a cloned descendant of a single ancestral phage.

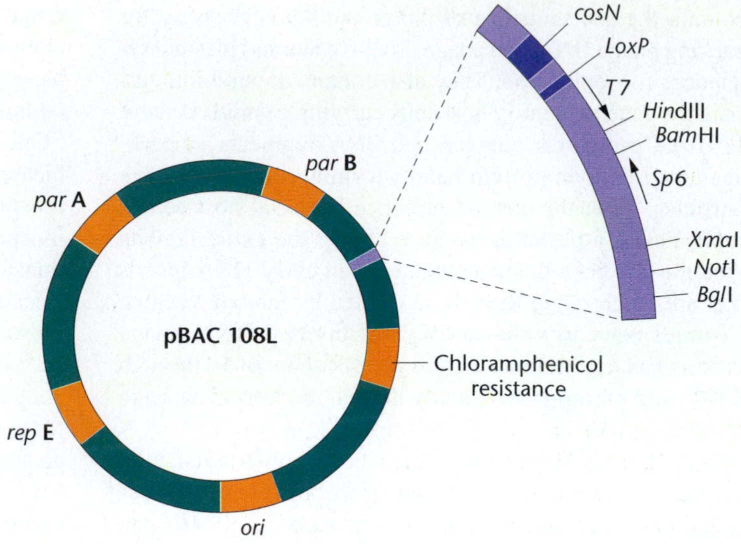

FIGURE 16–11 A bacterial artificial chromosome (BAC). The polylinker carries a number of unique restriction sites for the insertion of foreign DNA. The arrows labeled T7 and Sp6 are promoter regions that allow expression of genes cloned between these regions.

16.5 Cloning in Eukaryotic Hosts: Yeast Cells

The yeast *Saccharomyces cerevisiae* is extensively used as a host cell for the growth and expression of cloned eukaryotic genes for five reasons. (1) Although yeast is a eukaryotic organism, it can be grown and manipulated in much the same way as bacterial cells. (2) The genetics of yeast has been intensively studied, providing a large catalog of mutations and a highly developed genetic map. (3) The entire

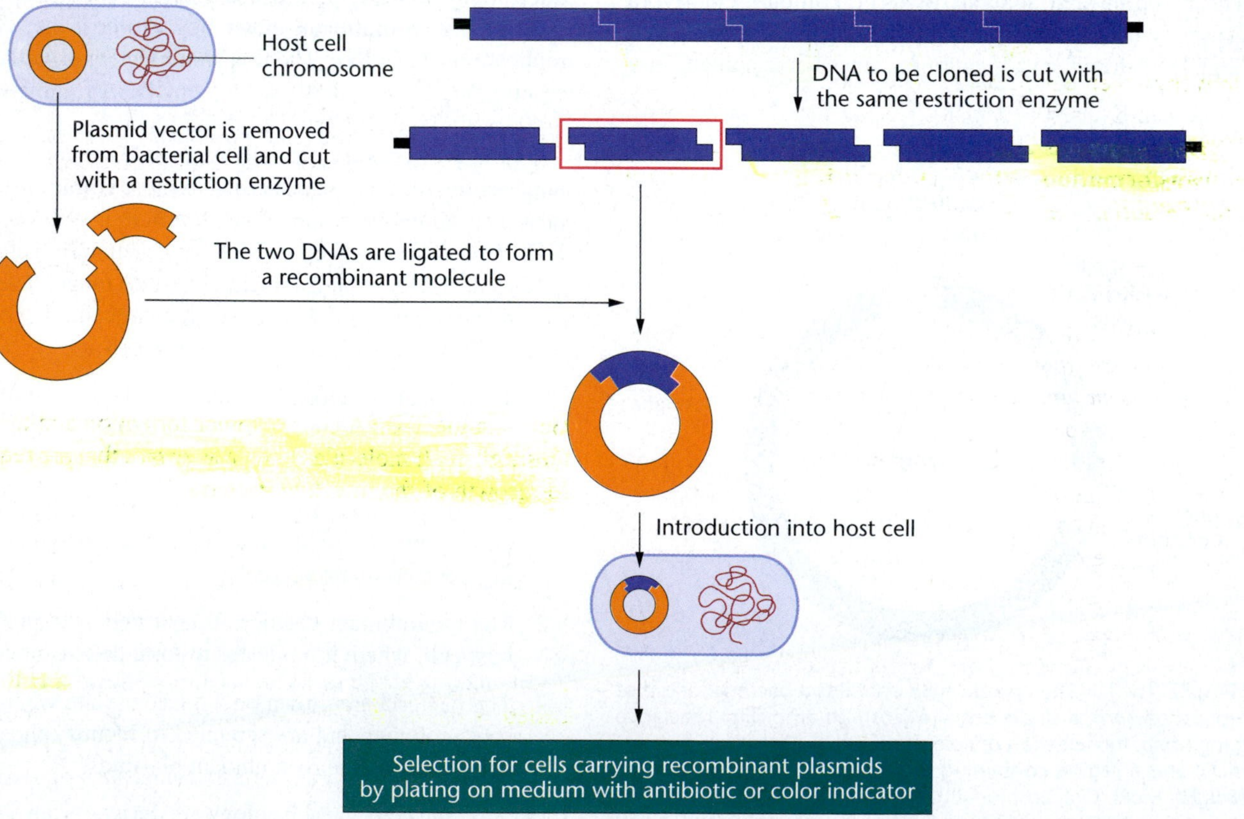

FIGURE 16–12 Cloning with a plasmid vector. Plasmid vectors are isolated and cut with a restriction enzyme. The DNA to be cloned is cut with the same restriction enzyme, producing a collection of fragments. These fragments are spliced into the vector and transferred to a bacterial host for replication. Bacterial cells carrying plasmids with DNA inserts are identified by growth on selective medium and isolated. The cloned DNA is then recovered from the bacterial host for further analysis.

yeast genome has been sequenced, and most genes in the organism have been identified. (4) To study the function of some eukaryotic proteins, it is necessary to use a host cell that can posttranslationally modify the product of the cloned gene to convert it into a functional form. Bacterial host cells cannot carry out these modifications. (5) Because yeast has been used for centuries in the baking and brewing industries, it is considered a safe organism for producing proteins for vaccines and therapeutic agents (Table 16–1).

Several types of yeast cloning vectors have been developed, one of which is the **yeast artificial chromosome (YAC)** (Figure 16–13). Like natural chromosomes, a YAC contains telomeres at each end, an origin of replication (which initiates DNA synthesis), and a centromere. These components are joined to selectable markers (*TRP1* and *URA3*) and a cluster of restriction sites for DNA inserts. Yeast chromosomes range from 230 kb to over 1900 kb, making it possible to clone DNA inserts ranging from 100 kb to 1000 kb in YACs. The ability to clone large pieces of DNA into these vectors makes them an important tool in genome projects, including the Human Genome Project. (Discussed in Chapters 17 and 18.)

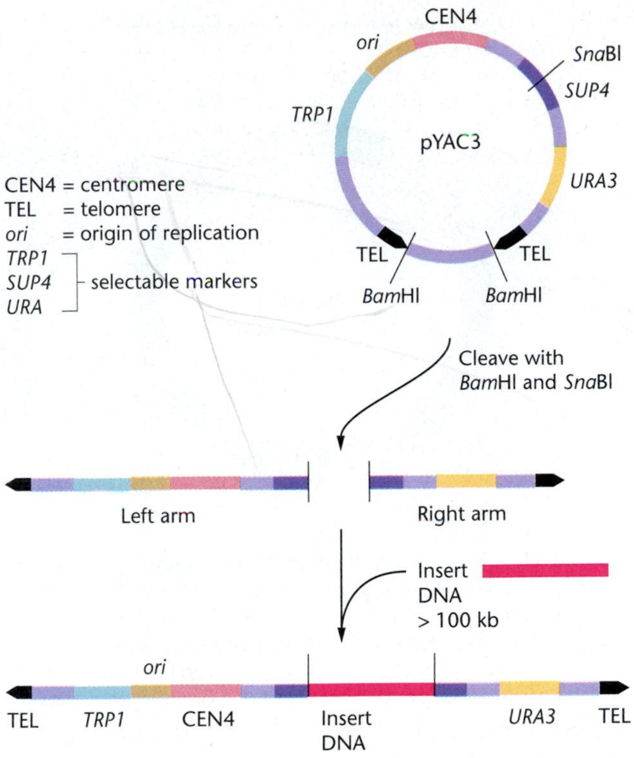

FIGURE 16–13 Cloning into a yeast artificial chromosome. The synthetic chromosome contains telomere sequences (TEL), a centromere (CEN4) derived from yeast chromosome 4, and an origin of replication (*ori*). These elements give the cloning vector the properties of a chromosome. *TRP1*, *SUP4*, and *URA3* are yeast genes that are selectable markers for the left and right arms of the chromosome, respectively. Within the *SUP4* gene is a restriction site for the enzyme *Sna*B1. Two *Bam*H1 sites flank a spacer segment. Cleavage with *Sna*B1 and *Bam*H1 breaks the artificial chromosome into two arms. The DNA to be cloned is treated with the same enzyme, producing a collection of fragments. The arms and fragments are ligated together, and the artificial chromosome inserted into yeast host cells. Because yeast chromosomes are large, the artificial chromosome accepts inserts in the million base pair range (Mb = megabase).

16.6 Gene Transfer in Eukaryotic Cells

In addition to yeast, both plant and animal cells can take up DNA from their environment. Moreover, vectors, including YACs, can be used to transfer DNA into eukaryotic cells. When the vector is a plasmid, DNA uptake is referred to as **transformation**. When the vector is a virus, the term **transfection** is used to describe uptake.

Plant Cell Hosts

Gene transfer into higher plants has been achieved using bacterial plasmids as vectors. The infectious soil bacterium *Agrobacterium tumifaciens* produces tumors (called plant galls) in many species of plants. Tumor formation is associated with the presence of a tumor-inducing (Ti) plasmid in the bacteria (Figure 16–14). When these bacteria infect plant cells, a segment of the Ti plasmid, known as *T-DNA*, is transferred into the chromosomal DNA of the host plant cell. Genes in the T-DNA control tumor formation and the synthesis of small molecules known as *opines* that are required for growth of the infecting bacteria. Foreign genes can be inserted into the T-DNA segment of Ti, and the insert-carrying Ti can be transferred into a plant cell by the infecting bacteria. The foreign DNA is inserted into the plant genome when the T-DNA is integrated into a host cell chromosome. Such genetically altered single cells can be induced to grow in tissue culture to form a cell mass called a *callus*. By manipulating the culture medium, the callus can be induced to form roots and shoots, and eventually to develop into mature plants carrying a foreign gene.

Plants (or animals) carrying a foreign gene are called **transgenic** organisms. In Chapter 18, we will see how gene transfer in plants has been used to alter agriculturally important crop plants genetically.

TABLE 16–1 Recombinant Proteins Synthesized in Yeast Cells

Hepatitis B virus surface protein
Malaria parasite protein
Insulin
Epidermal growth factor
Platelet-derived growth factor
α_1-antitrypsin
Clotting factor XIIIA

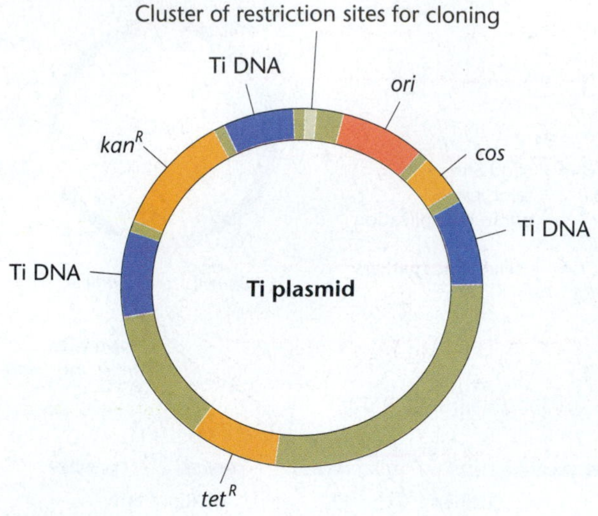

FIGURE 16–14 A Ti plasmid designed for cloning in plants. Segments of Ti DNA, including those necessary for opine synthesis and integration, are combined with bacterial segments that incorporate cloning sites and antibiotic resistance genes (*kan*^{R} and *tet*^{R}). The vector also contains an origin of replication *(ori)* and a lambda *cos* site that permits recovery of cloned inserts from the host plant cell.

Mammalian Cell Hosts

DNA can be transferred to mammalian cells by several methods, including endocytosis or encapsulation of DNA into artificial membranes (liposomes), followed by fusion with cell membranes. Also, DNA can be transferred using YACs and vectors based on retroviruses. The DNA introduced into a mammalian cell by any of these methods is usually integrated into the host genome. DNA transfer methods have been used to transfer genes into fertilized eggs, producing transgenic animals. These methods have also been used to study molecular aspects of gene expression and to replicate cloned genes using mammalian cells as hosts.

YACs are used to increase the efficiency of transferring genes into the germ line of mice. Gene transfer is dependent on the introduction of DNA sequences into the nucleus of an appropriate mouse cell, such as a fertilized egg or an embryonic stem cell, followed by the integration of the DNA into a chromosomal site.

YACs are transferred to mice in several ways. The first involves the microinjection of purified YAC DNA into the nucleus of a mouse oocyte (Figure 16–15). The zygotes are then transferred to the uteri of foster mothers for normal development. Other techniques use mouse embryonic stem (ES) cells as hosts. One of these methods fuses a yeast cell carrying a YAC with a stem cell, transferring the YAC and all, or a portion of, the yeast genome to the stem cell. The transgenic ES cells are then transferred into mouse blastocyst-stage embryos, where they participate in the formation of adult tissues, including those that form germ cells. The ability to transfer large DNA segments into mice has applications in many areas of research. Some of these are described in Chapter 18.

Vectors for the delivery of functional genes into mammalian cells have been developed using avian and mouse retroviruses. The genome of such retroviruses is a single-stranded RNA that is transcribed by reverse transcriptase into a double-stranded DNA molecule. This DNA is integrated into the host genome and passed on to daughter cells as part of the host chromosome. The retroviral genome can be engineered to remove some viral genes, creating vectors that accept foreign DNA, including human structural genes. These vectors are being used in the treatment of genetic disorders by gene therapy, a topic that is also discussed in Chapter 18.

FIGURE 16–15 Cloned DNA can be transferred in mammals by direct injection into the oocytes.

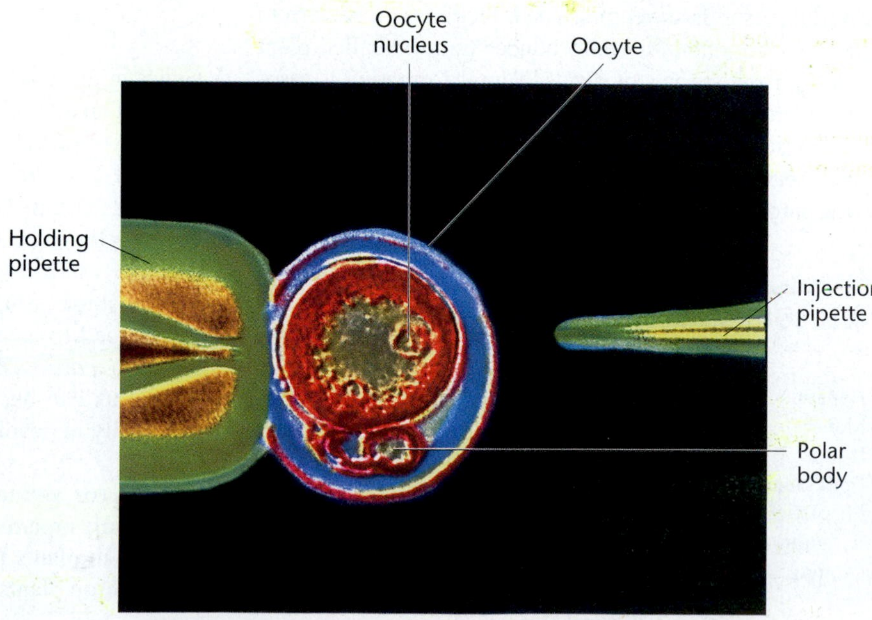

16.7 Cloning Without Host Cells: The Polymerase Chain Reaction (PCR)

Developed in the early 1970s, recombinant DNA techniques revolutionized the way geneticists and molecular biologists conduct research and gave birth to the booming biotechnology industry. However, they are often labor-intensive and time-consuming. In 1986, another technique, called the **polymerase chain reaction (PCR)**, was developed. This advance again revolutionized recombinant DNA methodology and further accelerated the pace of biological research. The significance of this method is underscored by the fact that the 1993 Nobel Prize in Chemistry was awarded to Kary Mullis for developing the PCR technique.

PCR is a rapid method of DNA cloning that has extended the power of recombinant DNA research and does away with the use of host cells as vehicles for cloning DNA fragments. Although host cell cloning is still widely used, PCR is now the method of choice in many applications, including molecular biology, human genetics, evolution, development, conservation, and forensics (the gathering of evidence in criminal cases). Some of these applications are discussed in Chapter 18.

PCR: Rationale and Techniques

PCR generates many copies of a specific DNA sequence through a series of reactions in a test tube (*in vitro*), and amplifies target DNA sequences initially present in infinitesimally small quantities in a population of other DNA molecules. To do this, some information about the nucleotide sequence of the target DNA is required. The information is used to synthesize two oligonucleotide primers that are added to a denatured (single-stranded) DNA sample. In the reaction, the primers hybridize to nucleotides that flank the sequence to be amplified. A heat-stable form of DNA polymerase (called *Taq* polymerase) synthesizes a second strand of the target DNA (Figure 16–16).

The PCR reaction involves three basic steps, and the amount of amplified DNA produced is limited theoretically only by the number of times these steps are repeated:

1. The DNA to be amplified is denatured into single strands. This DNA does not have to be purified and can come from any number of sources, including genomic DNA, forensic samples such as dried blood or semen, dried samples stored as part of medical records, single hairs, mummified remains, and fossils. Heating (at 90–95°C) denatures the double-stranded DNA, which dissociates into single strands (usually about 5 minutes).

2. The temperature of the reaction is lowered to somewhere between 50°C and 70°C, and at this temperature, the primers bind to the denatured DNA. These primers are synthetic oligonucleotides (15–30 nucleotides long) that bind specifically to sequences

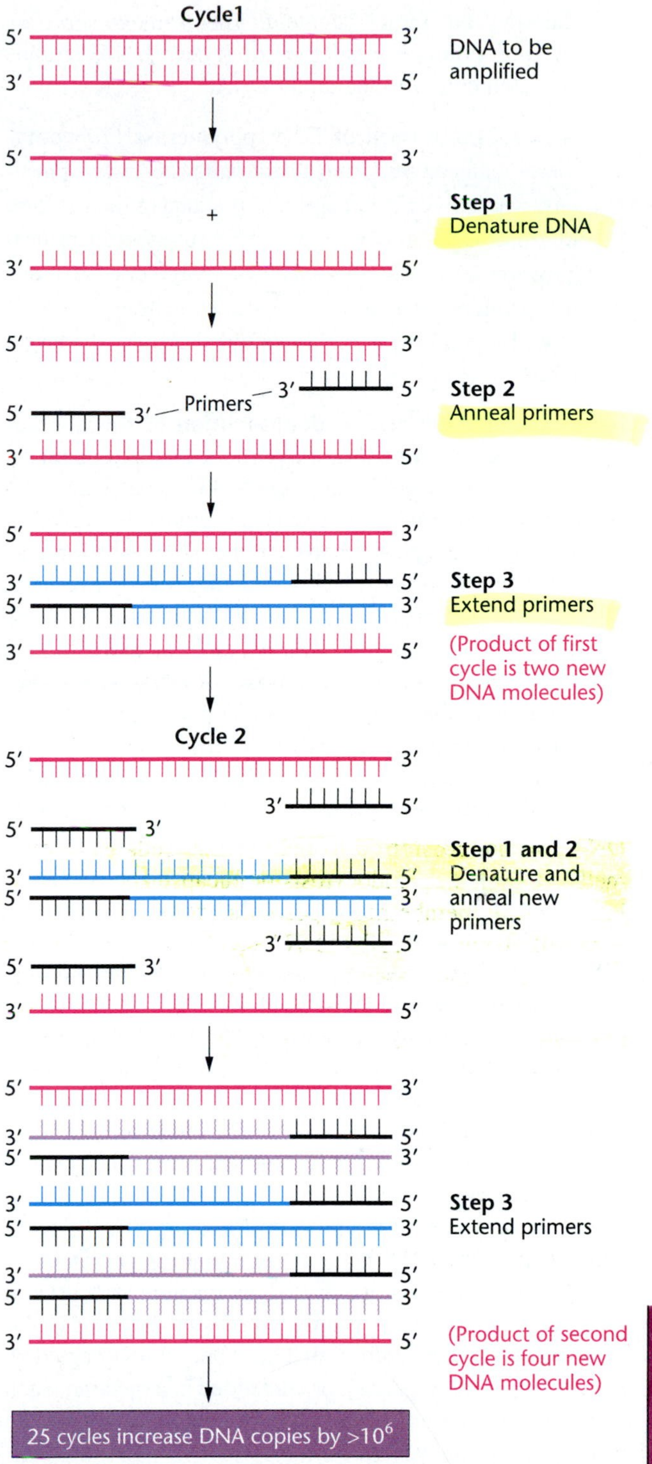

FIGURE 16–16 PCR amplification. In the polymerase chain reaction, the target DNA is denatured into single strands; each strand is then annealed to a short, complementary primer. The primers are synthetic oligonucleotides that are complementary to sequences flanking the region to be amplified. DNA polymerase and nucleotides extend the primers in the 3' direction, using the single-stranded DNA as a template. The result is a double-stranded DNA molecule with the primers incorporated into the newly synthesized strand. In a second PCR cycle, the products of the first cycle are denatured into single strands, primers are annealed, and DNA polymerase then synthesizes new strands. Repeated cycles can amplify the original DNA sequence by more than a millionfold.

MEDIA TUTORIAL Polymerase Chain Reaction

flanking the target segment. The primers serve as starting points for synthesizing new DNA strands complementary to the target DNA.

3. A heat-stable form of DNA polymerase (*Taq* polymerase, an enzyme from a bacterium, *Thermus aquaticus*, which lives in hot springs) is added to the reaction mixture. DNA synthesis is carried out at temperatures between 70°C and 75°C. The *Taq* polymerase extends the primers by adding nucleotides in the $5' \rightarrow 3'$ direction, making a double-stranded copy of the target DNA.

Each set of three steps—**denaturation** of the double-stranded product, **annealing** of primers, and **extension** by polymerase—is referred to as a **cycle**. PCR is a chain reaction because the number of new DNA strands is doubled in each cycle, and the new strands and the old strands serve as templates in the next cycle. Each cycle, which takes 4–5 minutes, can be repeated, and 25–30 cycles (which can be completed in less than 3 hours) result in more than a 1,000,000-fold increase in the amount of DNA. The limitations on the number of molecules that can be produced by amplification is largely a function of the amount of primer present or the number of nucleotides in the mixture. Machines called *thermocyclers* have automated the process. They can be programmed to carry out a predetermined number of cycles, yielding large amounts of the target DNA, which can be used in cloning, sequencing, clinical diagnosis, and genetic screening.

PCR-based DNA cloning has several advantages over host cell-based cloning. The PCR reaction is fast and can be carried out in a few hours, rather than the weeks required for host cell-based cloning. In addition, the design of PCR primers uses computer software, and the commercial synthesis of the oligonucleotides is also fast and economical. If desired, the products of PCR can be cloned into plasmid vectors for further use.

PCR is very sensitive and amplifies target DNA from vanishingly small DNA samples, including the DNA in a single cell. This feature of PCR is invaluable in several areas, including genetic testing, forensics, and molecular paleontology. DNA samples that have been partly degraded, mixed with other materials, or embedded in a medium (such as amber) can be used where conventional cloning would be difficult or impossible.

Limitations of PCR

Although PCR is a valuable technique, it does have limitations. Some information about the nucleotide sequence of the target DNA must be known, and even minor contamination of the sample with DNA from other sources can cause difficulties. For example, cells shed from the skin of a laboratory worker while performing PCR can contaminate samples gathered from a crime scene or taken from fossils. Such contamination makes it difficult to obtain accurate results. PCR reactions must always be run with carefully designed and appropriate controls.

Other Applications of PCR

PCR is now one of the most widely used techniques in genetics and molecular biology. PCR and its variations have many other applications. It quickly identifies restriction-site variants and variations in tandemly repeated DNA sequences that can serve as genetic markers for gene mapping studies. Using gene-specific primers, PCR screens for mutations in genetic disorders, allowing the location and nature of the mutation to be determined quickly. As we discuss in Chapter 18, primers can be designed to distinguish target sequences that differ by only a single nucleotide, making it possible to develop allele-specific probes for genetic testing.

Random primers allow the indiscriminate amplification of DNA, which is particularly advantageous when studying samples from single cells, from fossils, or from crime scenes where a single hair or even a saliva-licked postage stamp is the source of the DNA. Researchers also explore uncharacterized DNA regions adjacent to known regions and even sequence DNA with PCR. PCR has been used to enforce the worldwide ban on the sale of certain whale products and to settle arguments about the pedigree background of purebred dogs. In short, PCR is one of the most versatile techniques in modern genetics.

16.8 Libraries Are Collections of Cloned Sequences

Because each cloned DNA segment is relatively small and may represent only a single gene or a portion of a gene, many separate clones are needed to cover even a small fraction of an organism's genome. A set of DNA clones derived from a single individual is called a cloned *library*. These libraries can represent an entire genome, a single chromosome, or a set of genes that are expressed in a single cell type.

Genomic Libraries

Ideally, a **genomic library** contains at least one copy of all the sequences in an individual's genome. Genomic libraries are constructed using host cell cloning methods, since PCR-cloned DNA fragments are relatively small. In the process, DNA is extracted from cells or tissues, then cut with restriction enzymes and the fragments ligated into vectors. Since some vectors (such as plasmids) carry only a few thousand base pairs (kilobases) of inserted DNA, selecting the vector so that it contains a genome in the smallest number of clones is an important consideration in preparing a genomic library.

The number of clones required to carry all sequences in a genome depends on the average size of the cloned inserts and the size of the genome to be cloned. The number of required clones in a library can be calculated as

$$N = \frac{\ln(1 - P)}{\ln(1 - f)}$$

where P is the probability of recovering a given sequence and f is the fraction of the genome in each clone.

Suppose we wish to prepare a library of the human genome by using a lambda phage vector. The human genome contains approximately 3.0×10^6 kb of DNA, and the average size of a cloned insert that can be carried by the vector is 17 kb. Therefore, about 8.1×10^5 phages are required to establish a library containing all genomic sequences. If we select a plasmid vector with a capacity of 5 kb, several million clones will be needed. Thus, in this example, a phage vector is the best choice for constructing a library of the human genome.

Chromosome-Specific Libraries

A library made from a subgenomic fraction such as a single chromosome can be of great value in the selection of specific genes and the study of chromosome organization. For example, DNA from a small segment of the X chromosome of *Drosophila* corresponding to a region of about 50 polytene bands was isolated by microdissection. The DNA in this chromosomal fragment was extracted, cut with a restriction endonuclease, and cloned into a lambda vector. This region of the chromosome contains the genes *white*, *zeste*, and *Notch*, as well as the original site of a transposable element that can translocate a chromosomal segment to more than 100 other loci. Although technically difficult, this procedure produced a library that contains only the genes of interest and their adjacent sequences.

Libraries derived from individual human chromosomes have been prepared by using a technique known as flow cytometry. In this procedure, chromosomes from mitotic cells are stained with two fluorescent dyes, one that binds to AT pairs, the other to GC pairs. The stained chromosomes flow past a laser beam that stimulates them to fluoresce, and a photometer sorts and fractionates the chromosomes by differences in dye binding and light scattering (Figure 16–17).

By using this technique, cloned libraries for each of the human chromosomes have been prepared. Having cloned libraries for each human chromosome has greatly facilitated the mapping and analysis of individual chromosomes as part of the Human Genome Project.

A modification of electrophoresis known as pulsed-field gel electrophoresis has been used to isolate yeast chromosomes for the construction of chromosome-specific libraries (Figure 16–18). The isolation and construction of a cloned library of yeast chromosome III (315 kb) was the starting point for the formation of a consortium of 35 laboratories to determine the nucleotide sequence of this chromosome.

One of the unexpected results of this project was the finding that about half of all the genes on this chromosome were previously unknown. It was difficult for many geneticists to accept the fact that the conventional methods of systematic mutagenesis and mapping of genes used for decades was so inefficient. However, this finding was confirmed and extended in 1994 with the publication of the nucleotide sequence of yeast chromosome XI (664 kb) and the sequencing of the rest of the yeast genome in 1996. These results indicate that single-chromosome libraries can be used to gain access to genetic loci where conventional methods, such as mutagenesis and genetic analysis, have proven unsuccessful and where other probes, such as mRNA or gene products, are unavailable or unknown. In addition, chromosome-specific libraries provide a means for studying the molecular organization and even the nucleotide sequence in a defined region of the genome.

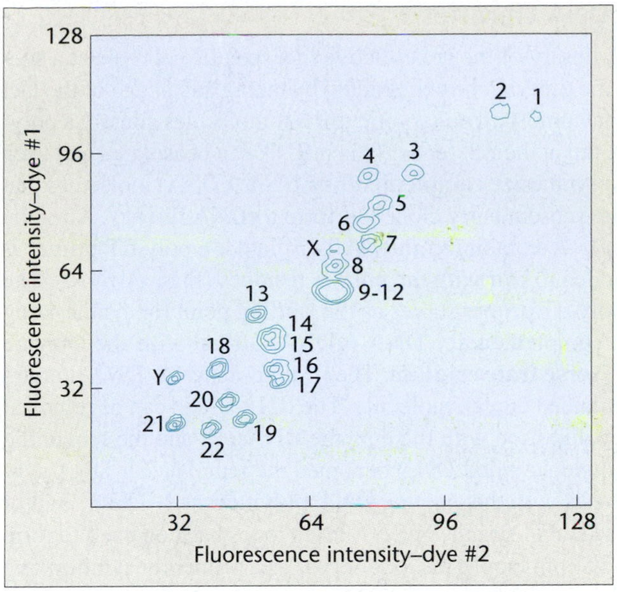

FIGURE 16–17 Each human chromosome has a unique profile, resulting from the absorption of two fluorescent dyes. Based on this differential fluorescence, chromosomes can be separated from each other by flow sorting.

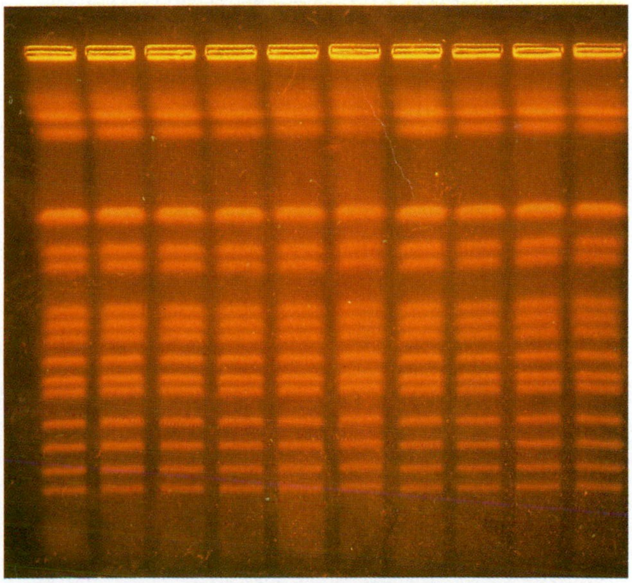

FIGURE 16–18 Intact yeast chromosomes separated using a method of electrophoresis employing contour-clamped homogeneous electric field (CHEF). In each lane, 15 of the 16 yeast chromosomes are visible, separated by size, with the largest chromosomes at the top.

cDNA Libraries

A library of the genes active in a specific cell type at a specific time can be constructed by taking advantage of the fact that almost all eukaryotic mRNA molecules contain a poly-A tail at their 3′-ends. This mRNA can be isolated and used to synthesize **complementary DNA (cDNA)** molecules that are subsequently cloned to form a **cDNA library**. After the poly-A-containing mRNA is isolated, a poly-dT primer is added to pair with the poly-A residues (Figure 16–19). The poly-dT primer serves as the starting point for synthesizing a complementary DNA (cDNA) strand with the enzyme **reverse transcriptase**. The result is an RNA–DNA double-stranded duplex molecule. The RNA strand can be removed by digestion with the enzyme RNAseH, and the remaining single-stranded DNA becomes the template for DNA synthesis with the enzyme **DNA polymerase I**. The 3′-end of the single-stranded cDNA often loops back on itself to form a hairpin loop. This loop serves as a primer for synthesizing the second strand by DNA polymerase I. The product is a double-stranded DNA molecule with the strands joined at one end. The loop can be opened using the enzyme S_1 nuclease, producing a double-stranded DNA molecule (called **complementary DNA** or **cDNA**), that can be cloned into a plasmid or phage vector.

PCR methods also generate cDNA from the 3′- or the 5′-end of mRNA molecules. This strategy, called RACE (*r*apid *a*mplification of *c*DNA *e*nds), depends on knowing a short nucleotide sequence contained in the coding region of the mRNA to be amplified. RACE identifies mRNAs that may be present in only 1–2 copies per cell, and the ends of cDNA molecules generated by RACE are used to search for cloned genes in genomic libraries.

A cDNA library is different from a genomic library in that it represents a subset of all the genes in a genome. cDNA libraries use mRNA as a starting point and thus represent only the expressed sequences in a given cell type, tissue, or stage of development. The decision whether to construct a genomic or a cDNA library depends on the question at hand. If you are interested in a particular gene, it may be easier to prepare a cDNA library from a tissue that expresses that gene. For example, red blood cells make large amounts of hemoglobin, and most of the mRNA in these cells encodes globin molecules. A cDNA library prepared from red blood cell mRNA is a direct approach to isolating a globin gene. If the regulatory sequences adjacent to the globin gene are of interest, then you need to construct a genomic library, since these regulatory sequences are not present in globin mRNA and would not be represented in the cDNA library.

16.9 Recovering Cloned Sequences from a Library

A genomic library can contain up to several hundred thousand clones. To select a gene of interest from this library, we need to identify and isolate only the clone or clones that contain that gene. We also must be able to determine whether a clone contains all or only part of that gene. Several methods allow us to sort through a library to recover the clones of interest; the choice often depends on the circumstances and available information about the gene being sought.

Using Probes to Identify Specific Cloned Sequences

Many procedures employ probes to select specific clones from a library. A **probe** is any piece of DNA or RNA that has been labeled and is complementary to some part of a cloned

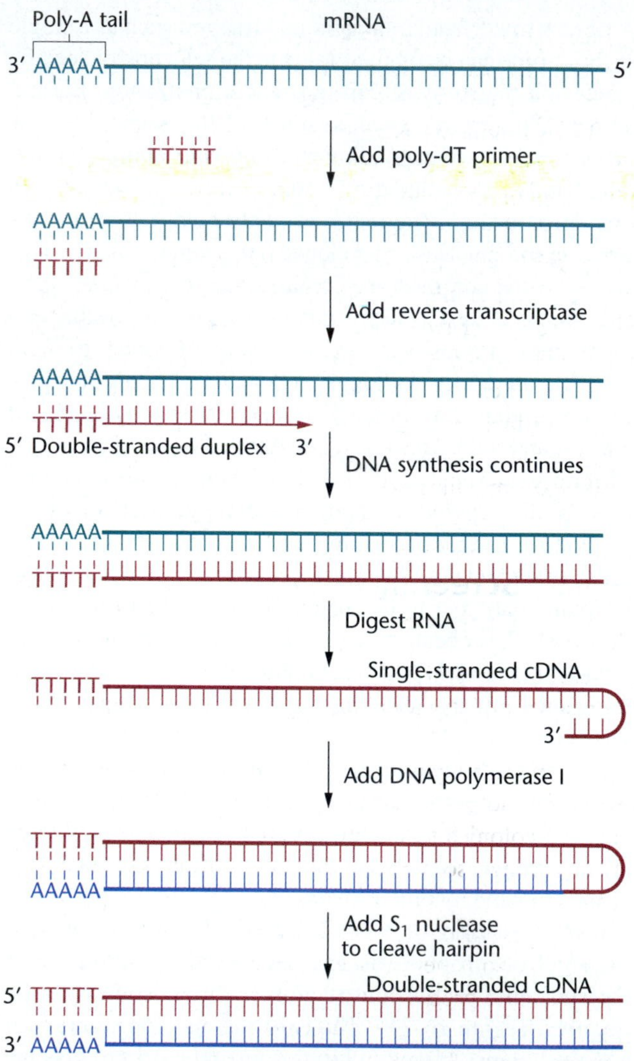

FIGURE 16–19 Producing cDNA from mRNA. Because many eukaryotic mRNAs have a polyadenylated tail (A) of variable length at their 3′-end, a short poly-dT oligonucleotide annealed to this tail serves as a primer for the enzyme reverse transcriptase. Reverse transcriptase uses the mRNA as a template to synthesize a complementary DNA strand (cDNA) and forms an mRNA/cDNA double-stranded duplex. The mRNA is digested with alkali treatment, or by the enzyme RNAse H. The 3′-end of the cDNA often folds back to form a hairpin loop. The loop serves as a primer for DNA polymerase, which is used to synthesize the second DNA strand. The S_1 nuclease opens the hairpin loop; the result is a double-stranded cDNA molecule that can be cloned into a suitable vector, or used as a probe for library screening.

sequence present in the library. The probe identifies complementary nucleic acid sequences present in one or more clones. Probes can be prepared as single- or double-stranded molecules, but are used in single-stranded form in hybridization reactions. (Hybridization reactions were described in Chapter 2.) Often, probes are radioactive polynucleotides, but some probes depend on chemical or color reactions to indicate the location of a specific clone.

Probes are derived from a variety of sources; even related genes isolated from other species can be used if enough of the DNA sequence has been conserved. For example, extrachromosomal copies of the ribosomal genes of the clawed frog *Xenopus laevis* can be isolated by centrifugation and cloned into plasmid vectors. Because ribosomal gene sequences have been highly conserved during evolution, clones carrying the human ribosomal genes can be recovered from a human genomic library by using cloned *Xenopus* ribosomal DNA fragments as probes.

If the gene to be selected from a genomic library is expressed in certain cell types, a cDNA probe can be used. This technique is particularly helpful when purified or enriched mRNA for a gene product can be obtained. For example, β-globin mRNA is present in high concentrations in certain stages of red blood cell development. The mRNA purified from these cells can be copied by reverse transcriptase into a cDNA molecule, or amplified by PCR RACE. In fact, a cloned cDNA probe was originally used to recover the structural gene for human β-globin from a cloned genomic library.

16.10 Screening a Library for Specific Clones

To screen a plasmid library, clones from the library are grown on nutrient plates, forming hundreds or thousands of colonies (Figure 16–20). A replica of the colonies on the plate is made by gently pressing a nylon or nitrocellulose filter onto the plate's surface; this transfers the pattern of bacterial colonies from the plate to the filter. The filter is passed through solutions to lyse the bacteria, denature the double-stranded DNA into single strands, and bind these strands to the filter.

The DNA from the colonies is screened by incubation with a nucleic acid probe. If a radioactive DNA probe is used, it is denatured to form single strands and then added to a solution containing the filter for incubation. If the DNA sequence of any of the cloned DNA on the filter is complementary to the probe, a double-stranded DNA–DNA hybrid molecule will form between the probe and the cloned DNA. After incubation, unbound or excess probe is washed away, and the filter is assayed to detect hybrid molecules that may have formed. If a radioactive probe has been used, the filter is overlaid with a piece of X-ray film. Radioactive decay in the probe molecules hybridized to DNA on the filter will expose the film and produce dark spots on it; they represent colonies on the plate containing the cloned gene of interest (Figure 16–20). The corresponding colony is identified and recovered from the original nutrient plate, and the cloned DNA it contains can be used in further experiments. In nonradioactive probes, a chemical reaction emits photons of light (chemiluminescence) to expose the photographic film and reveal the location of colonies carrying the gene of interest.

To screen a phage library, a slightly different method, called **plaque hybridization**, is used. A solution of phages carrying DNA inserts is spread over a lawn of bacteria growing on a plate. The phages infect the bacterial cells on the plate and form plaques as they replicate. Each plaque, which appears as a clear spot on the plate, represents the progeny of a single phage and is a clone. The plaques are transferred to a nylon or nitrocellulose membrane by pressing the membrane onto the plate's surface. The phage DNA is denatured into single strands and screened with a labeled probe. Phage plaques are much smaller than plasmid colonies, and many plaques can be screened on a single filter, making this method more efficient for screening large genomic libraries.

Identifying Nearby Genes: Chromosome Walking

In some cases, when the approximate location of a gene is known, it is possible to clone the gene by first cloning nearby sequences. Often these nearby sequences are identified by linkage analysis and serve as starting points for **chromosome walking**, in which adjacent clones are isolated from a library. In a chromosome walk, the end piece of a cloned DNA fragment is subcloned and used as a probe to recover another set of overlapping clones from the genomic library (Figure 16–21). These clones are analyzed to determine their degree of overlap. A subfragment from one end of the overlapping cloned DNA is used to recover another set of overlapping clones, and the analysis is repeated. In this way, it is possible to "walk" along the chromosome, clone by clone.

The gene in question can be identified by nucleotide sequencing of the recovered clones and searching for an **open reading frame (ORF)**. An ORF is a stretch of nucleotides that begins with a start codon followed by amino acid–encoding codons, and ends with one or more stop codons. Although laborious and time consuming, chromosome walking has identified genes for several human genetic disorders, including those for cystic fibrosis and muscular dystrophy.

Chromosome walking has several limitations in complex eukaryotic genomes, including the human genome. If a probe contains a repetitive sequence, such as an *Alu* sequence (see Chapter 4 for a discussion of these and other repetitive sequences), it hybridizes to other clones in the genomic library containing that sequence. Most of these clones will not be adjacent to the clone from which the probe was derived, and the walk terminates. In some cases, a technique called **chromosome jumping** can be used to skip over the region containing the repetitive sequences and continue the walk.

MEDIA TUTORIAL Cloning Libraries

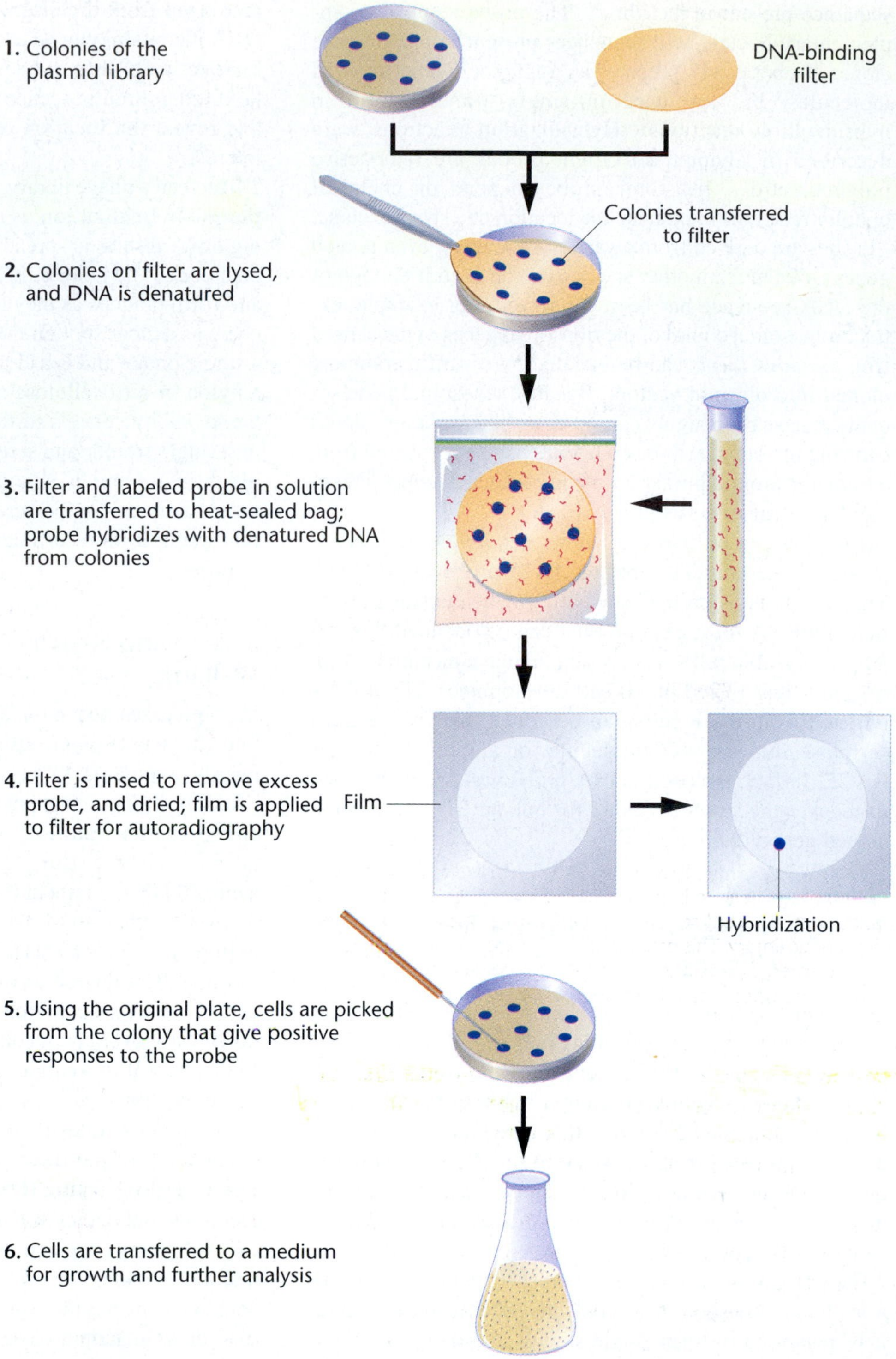

1. Colonies of the plasmid library
DNA-binding filter
Colonies transferred to filter
2. Colonies on filter are lysed, and DNA is denatured
3. Filter and labeled probe in solution are transferred to heat-sealed bag; probe hybridizes with denatured DNA from colonies
4. Filter is rinsed to remove excess probe, and dried; film is applied to filter for autoradiography
Film
Hybridization
5. Using the original plate, cells are picked from the colony that give positive responses to the probe
6. Cells are transferred to a medium for growth and further analysis

FIGURE 16–20 Screening a plasmid library to recover a cloned gene. The library, present in bacteria on Petri plates, is overlaid with a DNA binding filter, and colonies are transferred to the filter. Colonies on the filter are lysed, and the DNA denatured to single strands. The filter is placed in a hybridization bag along with buffer and a labeled single-stranded DNA probe. During incubation, the probe forms a double-stranded hybrid with complementary sequences on the filter. The filter is removed from the bag and washed to remove excess probe. Hybrids are detected by placing a piece of X-ray film over the filter and exposing it for a short time. The film is developed, and hybridization events are visualized as spots on the film. Colonies containing the insert that hybridized to the probe are identified from the orientation of the spots. Cells are picked from this colony for growth and further analysis.

16.11 Characterizing Cloned Sequences

The recovery and identification of genes and other specific DNA sequences by cloning or by PCR is a powerful tool for analyzing genomic structure and function. In fact, much of the Human Genome Project is based on such techniques. In the following sections, we consider some of these methods, which are answering questions about the organization and function of cloned sequences.

Restriction Mapping

One of the first steps in characterizing a DNA clone is the construction of a **restriction map**. A restriction map is a compilation of the number, order, and distance between

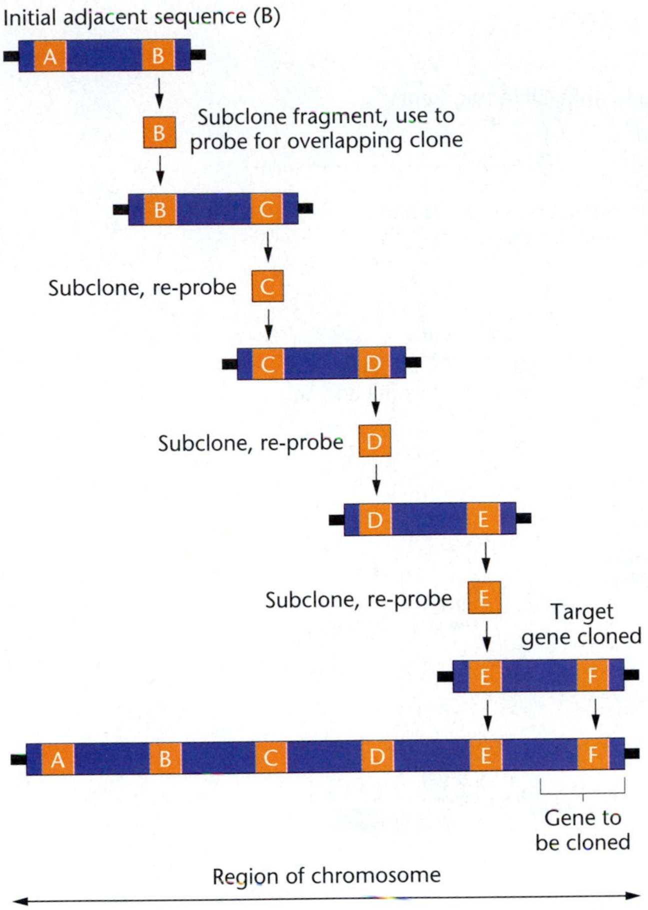

FIGURE 16–21 In chromosome walking, the approximate location of the gene to be cloned is known. A subcloned fragment of a linked adjacent sequence recovers overlapping clones from a genomic library. This process of subcloning and probing the genomic library is repeated to recover overlapping clones until the gene in question has been reached.

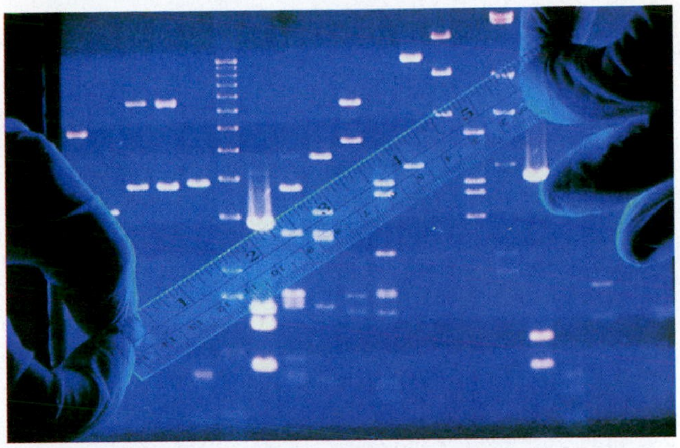

FIGURE 16–22 An agarose gel containing separated DNA fragments stained with a dye (ethidium bromide) and visualized under ultraviolet light. Smaller fragments migrate faster and farther than do larger fragments, resulting in the distribution shown.

restriction enzyme-cutting sites along a cloned segment of DNA. Map units are expressed in **base pairs (bp)** or, for longer distances, **kilobase pairs (kb)**. Restriction maps provide information about the length of a cloned insert and the location of restriction enzyme sites within the DNA. These data can be used to subclone fragments of a gene or compare its internal organization with that of other cloned sequences.

Fragments generated by cutting DNA with restriction enzymes can be separated by gel electrophoresis, a method that separates fragments by size, with the smallest pieces moving farthest. (See Appendix A for a detailed description of electrophoresis.) The fragments appear as a series of bands that can be visualized by staining the DNA with ethidium bromide and viewing under ultraviolet illumination (Figure 16–22).

Figure 16–23 shows the construction of a restriction map from a cloned DNA segment. For this map, let's begin with a cloned DNA segment 7.0 kb in length. Three samples of the cloned DNA are digested with restriction enzymes—one with *Hind*III, one with *Sal*I, and one with both *Hind*III and *Sal*I. The generated fragments are separated by electrophoresis, stained with ethidium bromide, and photographed. They are then measured by comparing them with a set of standards run in adjacent lanes. The map is constructed by analyzing the fragments:

1. When the DNA is cut with *Hind*III, two fragments (0.8 and 6.2 kb) are produced; this confirms that the cloned insert is 7.0 kb in length and that there is only one cutting site for this enzyme (located 0.8 kb from one end).

2. When the DNA is cut with *Sal*I, two fragments (1.2 and 5.8 kb) result, indicating that this enzyme has one cutting site, located 1.2 kb from one end of the insert.

Taken together, the results show that each enzyme has one restriction site, but the relationship between the two sites is unknown. From the information available, two different maps are possible. In one map (model 1), the *Hind*III site is located 0.8 kb from one end and the *Sal*I site 1.2 kb from the same end. In the alternative map (model 2), the *Hind*III site is located 0.8 kb from one end and the *Sal*I site 1.2 kb from the other end.

The correct model is selected by considering the results from the sample digested with both *Hind*III and *Sal*I. Model 1 predicts that digestion with both enzymes will generate three fragments: 0.4, 0.8, and 5.8 kb in length; model 2 predicts that these fragments will be 0.8, 5.0, and 1.2 kb in length. The actual fragment pattern, observed after cutting with both enzymes, indicates that model 1 is correct (Figure 16–23).

Restriction maps are an important way to characterize a cloned DNA segment and can be constructed in the absence of information about the coding capacity or function of the mapped DNA. In conjunction with other techniques, restriction mapping can define the boundaries of a gene, dissect the molecular organization of a gene and its flanking regions, and locate mutational sites within genes.

Restriction maps can also refine genetic maps. To a large extent, the accuracy of genetic maps depends on the frequency of recombination between genetic markers and on

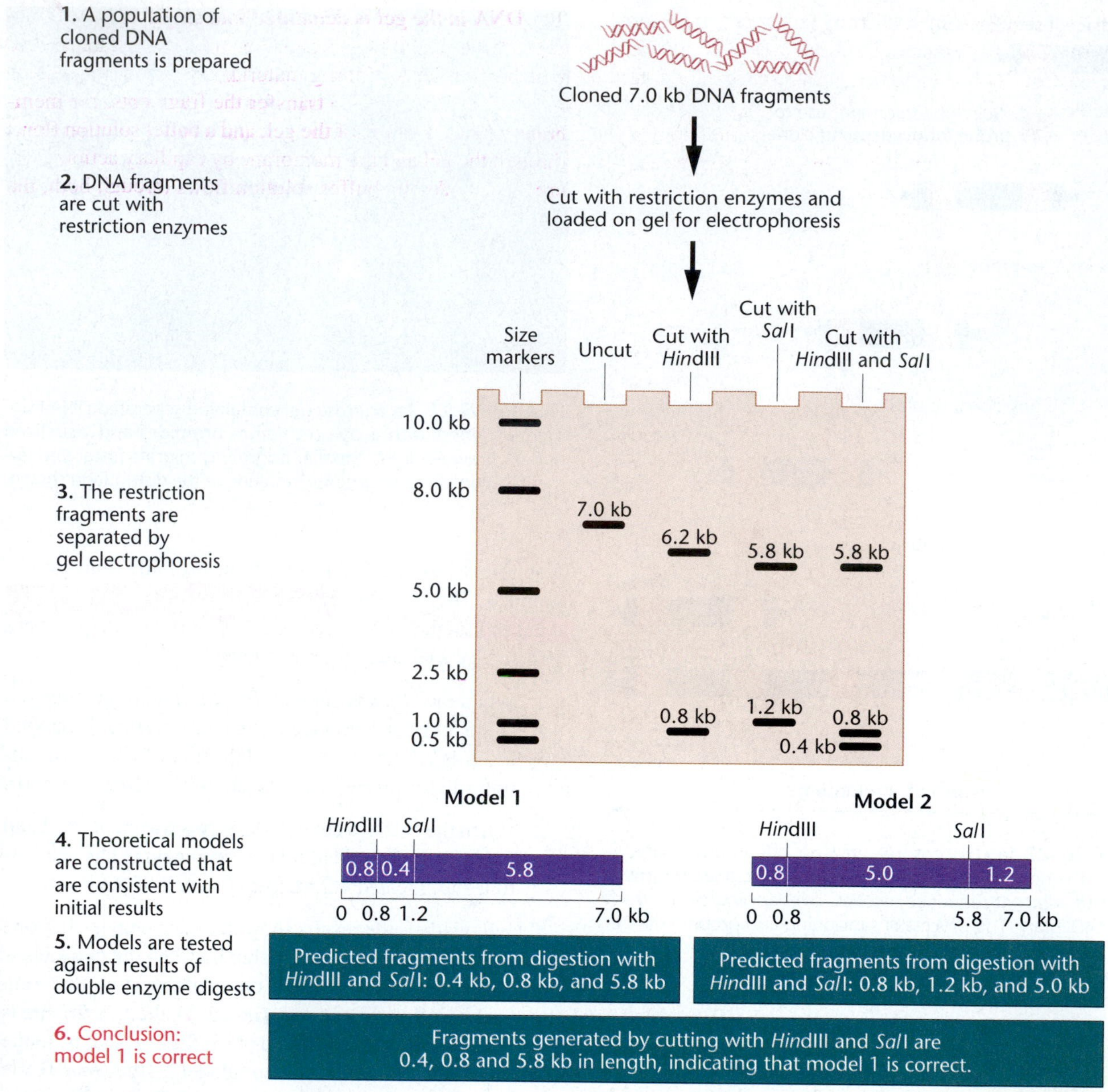
1. A population of cloned DNA fragments is prepared
Cloned 7.0 kb DNA fragments
2. DNA fragments are cut with restriction enzymes
Cut with restriction enzymes and loaded on gel for electrophoresis
Size markers
Uncut
Cut with HindIII
Cut with SalI
Cut with HindIII and SalI
10.0 kb
8.0 kb
5.0 kb
2.5 kb
1.0 kb
0.5 kb
7.0 kb
6.2 kb
5.8 kb
5.8 kb
0.8 kb
1.2 kb
0.8 kb
0.4 kb
3. The restriction fragments are separated by gel electrophoresis
Model 1
HindIII SalI
0.8 0.4 5.8
0 0.8 1.2 7.0 kb
Model 2
HindIII SalI
0.8 5.0 1.2
0 0.8 5.8 7.0 kb
4. Theoretical models are constructed that are consistent with initial results
5. Models are tested against results of double enzyme digests
Predicted fragments from digestion with HindIII and SalI: 0.4 kb, 0.8 kb, and 5.8 kb
Predicted fragments from digestion with HindIII and SalI: 0.8 kb, 1.2 kb, and 5.0 kb
6. Conclusion: model 1 is correct
Fragments generated by cutting with HindIII and SalI are 0.4, 0.8 and 5.8 kb in length, indicating that model 1 is correct.

MEDIA TUTORIAL Restriction Mapping

FIGURE 16–23 Constructing a restriction map. Samples of the 7.0-kb DNA fragments are digested with restriction enzymes: One sample is digested with *Hind*III, one with *Sal*I, and one with both *Hind*III and *Sal*I. The resulting fragments are separated by gel electrophoresis. The separated fragments are measured by comparing them with molecular-weight standards in an adjacent lane. Cutting the DNA with *Hind*III generates two fragments: 0.8 kb and 6.2 kb. Cutting with *Sal*I produces two fragments: 1.2 kb and 5.8 kb. Models are constructed to predict the fragment sizes generated by cutting with *Hind*III and with *Sal*I. Model 1 predicts that 0.4-, 0.8, and 5.8-kb fragments will result from cutting with both enzymes. Model 2 predicts that 0.8-, 1.2-, and 5.0-kb fragments will result. Comparing the predicted fragments with those observed on the gel indicates that model 1 is the correct restriction map.

the number of markers used to construct the map. In humans, for example, the genome size is large (3.2×10^9 bp) and the number of known genes is small, resulting in long distances between markers. Restriction enzyme-cutting sites are inherited and can be used as genetic markers, reducing the distance between markers, increasing their accuracy, and providing reference points for correlating genetic and physical maps.

Restriction sites play an important role in mapping genes to specific human chromosomes and to defined regions of individual chromosomes. In addition, if a restriction site maps close to a mutant gene, it can be used as a marker in a diagnostic test. These sites, known as **restriction fragment length polymorphisms**, or **RFLPs**, are described in Chapter 18. RFLPs have proven especially useful where the mutant genes underlying a human genetic disorder is poorly characterized at the molecular level.

Nucleic Acid Blotting

Many of the techniques described in this chapter rely on hybridization between complementary nucleic acid (DNA or RNA) molecules. We now describe one of the most widely used methods for detecting such hybrids. This technique, called **Southern blotting** (after Edward Southern, who devised it), separates DNA fragments by gel electrophoresis, transfers them to filters, and screens the filters with labeled probes.

To make a Southern blot, cloned DNA is cut into fragments with one or more restriction enzymes, and the fragments are separated by gel electrophoresis (Figure 16–24). The DNA in the gel is denatured into single-stranded fragments by treatment with an alkaline solution, and transferred to a sheet of DNA-binding material, usually nitrocellulose or a nylon derivative. To transfer the fragments, the membrane is placed on top of the gel, and a buffer solution flows through the gel and the membrane by capillary action (Figure 16–24). As the buffer solution flows through both, the

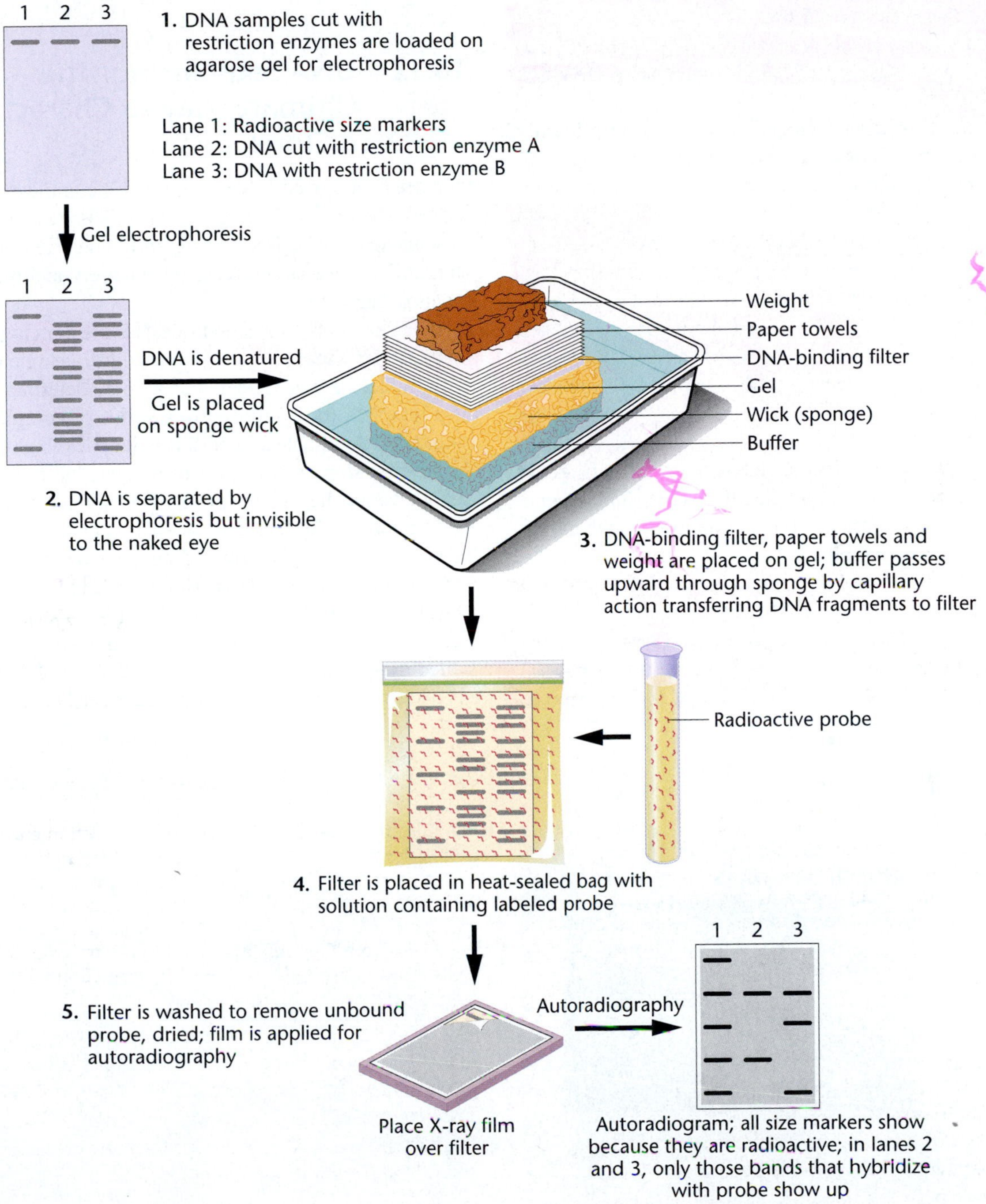
1 2 3
1. DNA samples cut with restriction enzymes are loaded on agarose gel for electrophoresis
Lane 1: Radioactive size markers
Lane 2: DNA cut with restriction enzyme A
Lane 3: DNA with restriction enzyme B
Gel electrophoresis
1 2 3
DNA is denatured
Gel is placed on sponge wick
2. DNA is separated by electrophoresis but invisible to the naked eye
Weight
Paper towels
DNA-binding filter
Gel
Wick (sponge)
Buffer
3. DNA-binding filter, paper towels and weight are placed on gel; buffer passes upward through sponge by capillary action transferring DNA fragments to filter
Radioactive probe
4. Filter is placed in heat-sealed bag with solution containing labeled probe
5. Filter is washed to remove unbound probe, dried; film is applied for autoradiography
Autoradiography
1 2 3
Place X-ray film over filter
Autoradiogram; all size markers show because they are radioactive; in lanes 2 and 3, only those bands that hybridize with probe show up

FIGURE 16–24 The Southern blotting technique. Samples of the DNA to be probed are cut with restriction enzymes and the fragments separated by gel electrophoresis. The pattern of fragments is visualized and photographed under ultraviolet illumination by staining the gel with ethidium bromide. The gel is then placed on a sponge wick in contact with a buffer solution and covered with a DNA-binding filter. Layers of paper towels or blotting paper are placed on top of the filter and held in place with a weight. Capillary action draws the buffer through the gel, transferring the pattern of DNA fragments from the gel to the filter. The DNA fragments on the filter are then denatured into single strands and hybridized with a labeled DNA probe. The filter is washed to remove excess probe and overlaid with a piece of X-ray film for autoradiography. The hybridized fragments show up as bands on the X-ray film.

DNA fragments move out of the gel and become immobilized on the membrane.

The DNA fragments on the membrane are hybridized with a labeled single-stranded DNA probe. Only DNA fragments complementary to the probe's nucleotide sequence will form double-stranded hybrids. Excess probe is washed away, and the hybridized fragments are visualized on a piece of film (Figure 16–25). In addition to characterizing cloned DNAs, Southern blots map restriction sites within and near a gene and identify DNA fragments carrying a single gene from a mixture of fragments and from related genes in different species. Southern blots also detect rearrangements and duplications in genes associated with human genetic disorders and cancers.

To determine whether a cloned gene is transcriptionally active in a given cell or tissue type, a related blotting technique probes for the presence of mRNA that is complementary to a cloned gene. This technique first extracts mRNA from one or several cell or tissue types and fractionates the RNA by gel electrophoresis. The resulting pattern of RNA bands is transferred to a sheet of membrane, as in the Southern blot. The membrane is then hybridized to a single-stranded DNA probe, derived from the cloned gene. If RNA complementary to the DNA probe is present on the filter, it is detected as a band on the film. Because the original procedure (DNA bound to a filter) is known as a Southern blot, this procedure (RNA bound to a filter) is called a **northern blot**. (Following this somewhat perverse logic, another procedure involving proteins bound to a filter is known as a **western blot**.)

Northern blots provide information about the expression of specific genes and are used to study patterns of gene expression in embryonic and adult tissues. Northern blots also detect alternatively spliced mRNAs and multiple types of transcripts derived from a single gene and are used to derive other information about transcribed mRNAs. If marker RNAs of known size are run in an adjacent lane, the size of a gene's mRNA can be measured. In addition, the amount of transcribed RNA present in the cell or tissue being studied is related to the density of the RNA band on the film. Measuring the density of the band gives the relative transcriptional activity. Thus, northern blots characterize and quantify the transcriptional activity of genes in different cells, tissues, and organisms.

16.12 DNA Sequencing: The Ultimate Way to Characterize a Clone

In a sense, a cloned DNA molecule is completely characterized when its nucleotide sequence is known. The ability to sequence cloned DNA has greatly enhanced our understanding of gene structure, gene function, and the mechanisms of regulation.

The most common method of **DNA sequencing** is based on **chain termination**. In this procedure, a single stranded-DNA molecule whose sequence is to be determined serves as a template for synthesizing a series of complementary strands that terminate at specific nucleotides (Figure 16–26). In the first step, a short primer is annealed to the single-stranded template, and the primer-bound DNA is distributed into four tubes. In the second step, this primer is elongated in the $5' \rightarrow 3'$ direction by adding the enzyme DNA polymerase and the four deoxyribonucleotide triphosphates (dATP, dCTP, dGTP, and dTTP). In addition, each tube contains a small amount of one of the four base-specific analogs, called **dideoxynucleotides** (e.g., ddATP). As DNA synthesis takes place, the DNA polymerase occasionally inserts a

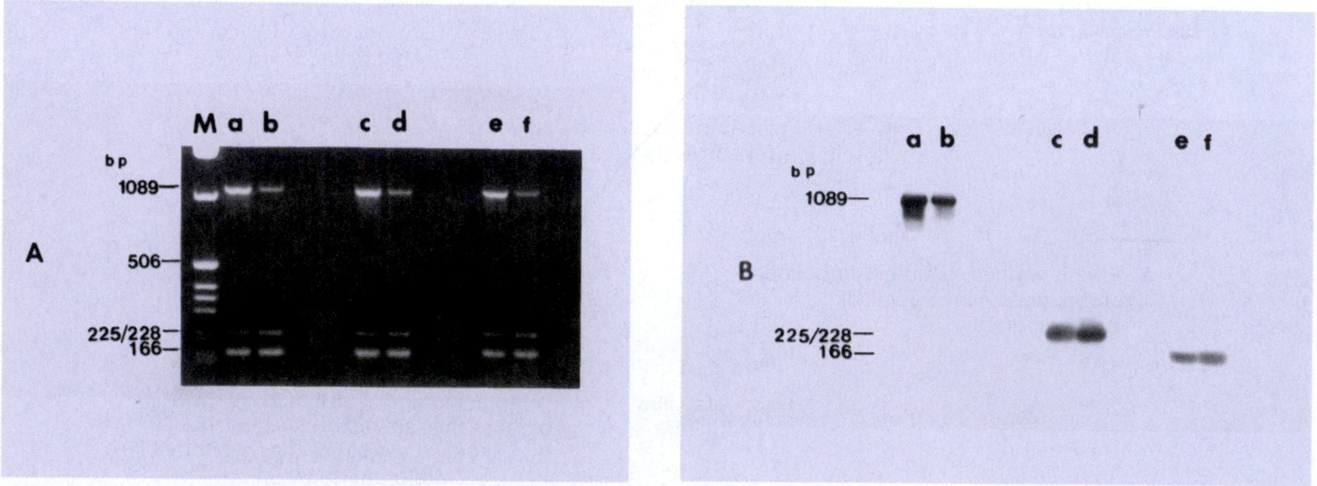

(a) (b)

FIGURE 16–25 (a) Agarose gel stained with ethidium bromide to show DNA fragments. M contains size markers. Lanes a, c, and e contain DNA from one bacterial strain of *E coli,* and lanes b, d, and f contain DNA from another strain. (b) A Southern blot prepared from the gel in (a). Only those bands containing DNA sequences complementary to the probe show hybridization.

dideoxynucleotide (instead of a deoxynucleotide) into a growing DNA strand. Since the analog lacks a 3′-hydroxyl group, it cannot form 3′ bonds to add another nucleotide to the strand, and DNA synthesis terminates. For example, if ddATP is present, termination takes place at sites opposit to thymidine in the template strand (Figure 16–26). As the reaction proceeds, the tubes accumulate a series of DNA molecules that differ in length by one nucleotide at their 3′-ends. The fragments from each reaction tube are separated in four adjacent lanes (one for each tube) by gel electrophoresis.

Electrophoresis separates DNA fragments in each lane that differ in size by a single nucleotide. The result is a series of bands forming a ladderlike pattern (Figure 16–27). The nucleotide sequence of the DNA can be read directly from bottom to top, corresponding to the 5′→3′ sequence of the DNA strand complementary to the template (Figure 16–27).

For genome sequencing, large-scale projects use automated machines that sequence several hundred thousand nucleotides per day. In this procedure, the four dideoxy-nucleotide analogs are labeled with a fluorescent dye (Figure 16–28), so that chains terminating in A are labeled with one color, those ending in C with another color, and so forth. The reaction is carried out in a single tube and loaded into one lane on the gel. The fluorescent detector in the sequencing machine reads the color of each band and determines whether it represents an A, a T, a C, or a G. The data are stored and analyzed, using the appropriate software, or printed for reading (Figure 16–29).

DNA sequencing projects have now been completed on the genomes of dozens of organisms. Sequencing provides information about the number, nature, and organization of genes in a genome and elucidates the mutational events that alter both genes and gene products, confirming that genes and proteins are colinear molecules.

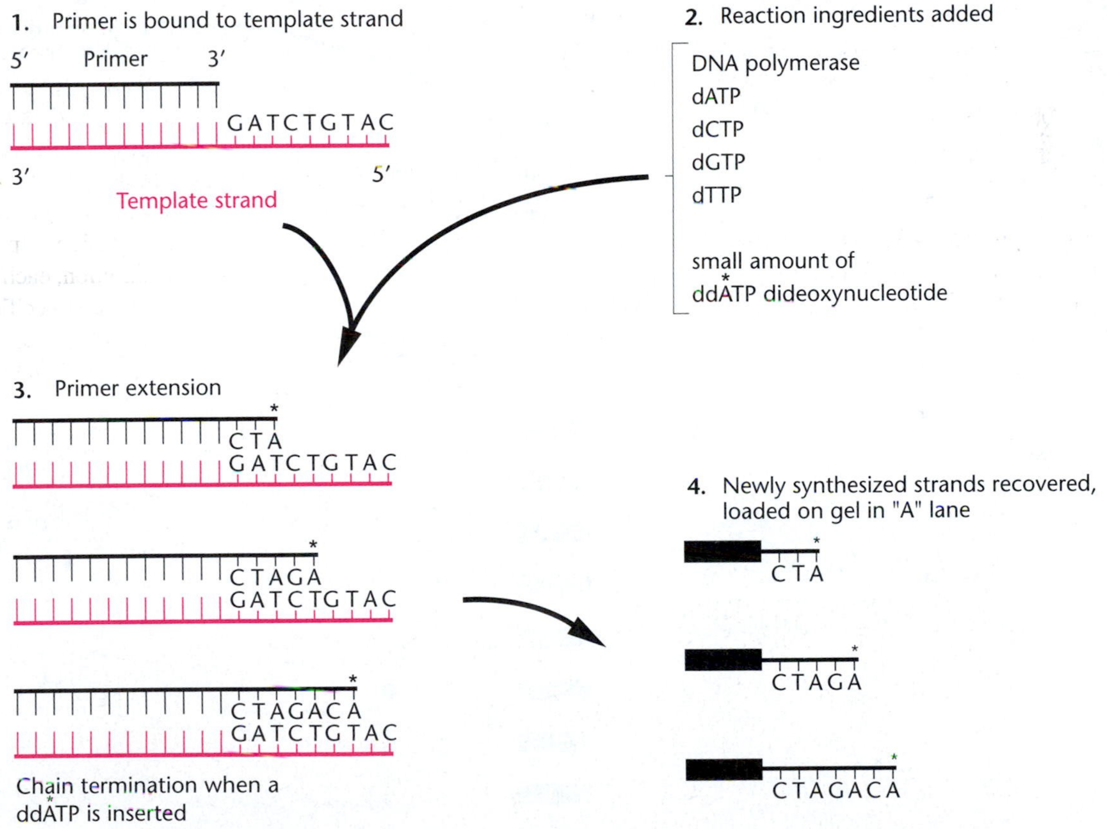

FIGURE 16–26 DNA sequencing using the chain termination method. (1) A primer is annealed to a sequence adjacent to the DNA being sequenced (usually at the insertion site of a cloning vector). (2) A reaction mixture is added to the primer–template combination. This includes DNA polymerase, the four dNTPs (one of which is radioactively labeled) and a small amount of one dideoxynucleotide. Four tubes are used, each containing a different dideoxynucleotide (ddATP, ddCTP, etc.). (3) During primer extension, the polymerase occasionally inserts a ddNTP instead of a dNTP, terminating the synthesis of the chain, because the ddNTP does not have the 3′-OH group needed to attach the next nucleotide. In the figure, ddATP and the A inserted from this didexoynucleotide are indicated with an asterisk. Over the course of the reaction, all possible termination sites will have a ddNTP inserted. (4) The newly synthesized chains are removed from the template, and the mixture is placed on a gel. Chains terminating in A are loaded in the A lane, those ending in C are loaded in the C lane, and so forth.

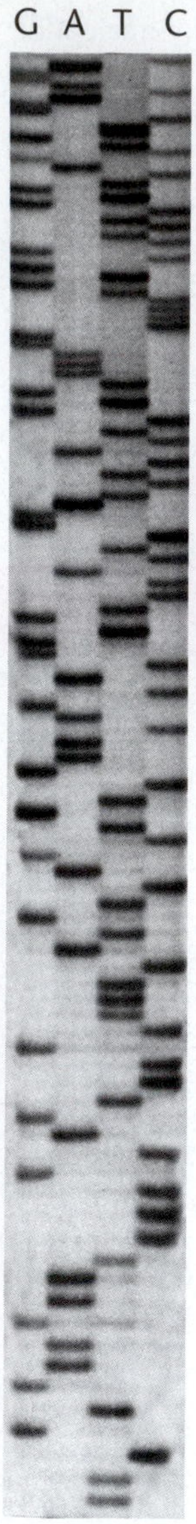

FIGURE 16–27 DNA sequencing gel showing the separation of fragments in the four sequencing reactions (one per lane). To obtain the base sequence of the DNA fragment, the gel is read from the bottom, beginning with the lowest band in any lane, then the next lowest, then the next, and so on. For example, the sequence of the DNA on this gel begins with TTCGTGAAGAA. *(Dr. Suzanne McCutcheon)*

MEDIA TUTORIAL DNA Sequencing

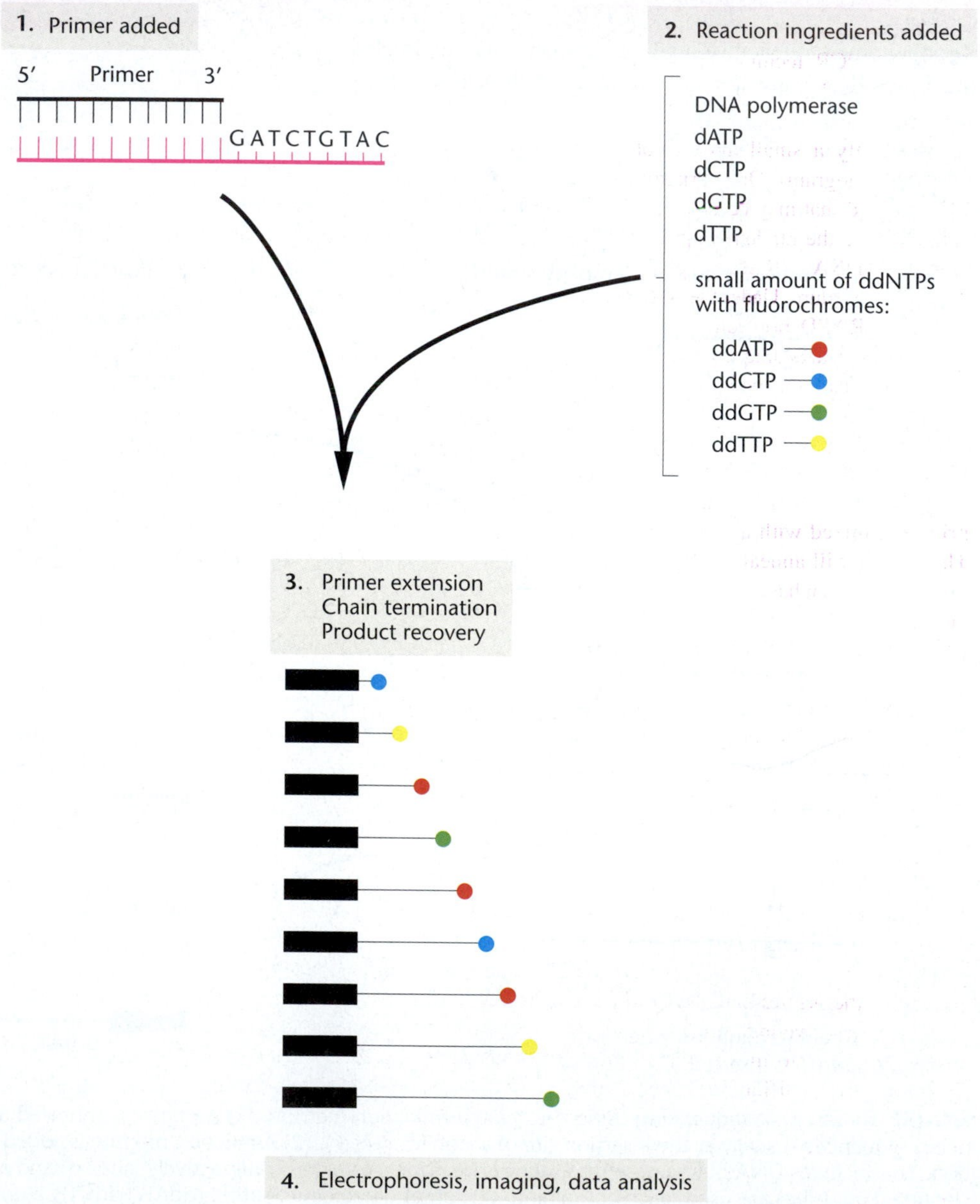

FIGURE 16–28 In DNA sequencing using dideoxynucleotides labeled with fluorescent dyes, all four ddNTPs are added to the same tube, and during primer extension, all combinations of chains are produced. The products of the reaction are added to a single lane on a gel, and the bands are read by a detector and imaging system. This process is now automated, and robotic machines, such as those used in the Human Genome Project, sequence several hundred thousand nucleotides in a 24-hour period and then store and analyze the data automatically.

Genetics, Technology, and Society

DNA Fingerprints in Forensics: The Case of the Telltale Palo Verde

The use of DNA analysis in forensics is making it harder and harder to get away with many crimes. It's now a simple matter to isolate DNA from tissue left at a crime scene, a splattering of blood, some skin left under a victim's fingernails, or even the cells clinging to the base of a hair shaft. A variety of techniques can be used to determine the likelihood that the sample came from a suspect in the case. In recent years, the PCR technique has been increasingly used in forensics, both because it is fast (taking only hours) and because it requires only a small amount of DNA (about a nanogram). One variation of the PCR method that may be especially valuable is called the random amplified polymorphic DNA (RAPD, pronounced "rapid") procedure. Under the right conditions, the RAPD procedure generates a DNA profile that is unique to each individual and that can thus be used for personal identification.

The RAPD procedure involves the amplification of a set of DNA fragments of unknown sequence. First, an amplification primer is mixed with a sample of DNA. The primer will anneal to all sites in the DNA sample that have a matching base sequence. For primers 10 nucleotides in length, binding will occur at a few thousand sites scattered randomly throughout the genome. When primers happen to bind to DNA sites that are not too far apart (within about 2000 nucleotides), DNA amplification can occur between these sites, eventually producing millions of copies of the DNA sequence lying between the binding sites. When the products of such a reaction are separated by electrophoresis on an agarose gel, each amplified DNA segment will appear as a distinct band. The locations of the primer binding sites in the genome of any two individuals in a population are likely to differ, due to minor DNA sequence differences (e.g., base changes, insertions, and deletions). Since these affect the initial binding to the primers, the number and locations of the DNA bands on the gel will vary from one individual to another. The resulting DNA profile is referred to as a DNA fingerprint. It represents a characteristic "snapshot" of the genome of an individual, allowing it to be distinguished from those of other individuals in the population.

When a DNA profile from tissue found at a crime scene matches that of a suspect, it does not prove that the tissue belongs to the suspect; instead, it excludes all those who have a different pattern. The strategy, therefore, is to generate five or more DNA profiles from the same sample, using different amplification primers. The more profiles that match between the sample and the suspect, the more unlikely it is that the sample at the crime scene came from someone other than the suspect. The likelihood that a complete match of all the banding patterns occurs simply by chance can be calculated, giving, for example, a 1-in-100,000 to a 1-in-1,000,000 probability of a "random match."

One of the most interesting uses of DNA fingerprinting in a criminal case did not involve the suspect's own DNA, but rather the DNA of plants growing at the crime scene. On the night of May 2, 1992, a Phoenix woman was strangled and her body dumped near an abandoned factory in the surrounding desert. Investigators discovered a pager near the body, making its owner a prime suspect. When questioned, this man admitted being with the woman the day of the murder, but claimed he had not been near the factory and suggested that the woman must have stolen his pager from his pickup truck. A search of the pickup provided the crucial clue, placing the suspect at the factory: In its bed were two seed pods from a palo verde tree.

The palo verde tree (*Cercidium floridum*), native to the desert of the Southwest, is a member of the bean family. In an adaptation to the hot, dry climate, it puts out very small leaves. To compensate for this, its stems become green and carry out photosynthesis. This gives the plant its name; palo verde is Spanish for "green stem." The tree may not resemble a bean plant in some respects, but, like a typical bean plant, it produces seeds in pods that drop to the ground when mature.

The homicide investigators assigned to the case wondered whether it could be proved that the seed pods found in the bed of the suspect's truck had fallen from a palo verde tree near the place where the body was found. If so, it would be a key piece of evidence placing the suspect at the scene. But how could this be demonstrated? The investigators turned to Dr. Timothy Helentjaris, then at the University of Arizona in nearby Tucson. Helentjaris proceeded to determine whether the DNA profile of the seed pods from the pickup truck matched that of any of the palo verde trees near the scene. He understood that for a match between a seed pod and a tree to have significance, he would first have to show that palo verdes differ genetically from one tree to the next. Otherwise, a pod could never be unambiguously assigned to one particular tree. Fortunately, he found considerable variation in DNA profiles among the palo verde trees.

Helentjaris was then given the two seed pods from the suspect's truck along with pods collected from 12 palo verde trees from the vicinity of the factory. The investigators knew which of the 12 trees was the key one at the crime scene, but they did not tell Helentjaris. He extracted DNA from seeds from each pod and performed RAPD reactions using an amplification primer that he had earlier found to reproducibly yield profiles of 10–15 distinct bands. The result of this controlled experiment was unmistakable: The DNA profile from one of the pods found in the truck exactly matched that of only 1 of the 12 trees, the one nearest to where the body was found. In an important additional test, Helentjaris showed that this DNA profile differed from that of pods collected from 18 other trees located at random sites around Phoenix. Collectively, Helentjaris's analysis led to the estimate of the likelihood of a random match at a little less than 1 in 1,000,000.

Helentjaris's analysis was admitted as evidence in the trial, placing the suspect at the crime scene. At the completion of the five-week trial, the suspect was found guilty of first-degree murder. His conviction was upheld upon appeal, and he is currently serving a life sentence without parole.

References

Ayala, F.J., and Black, B. 1993. Science and the courts. *Am. Sci.* 81:230–39.

Krawczak, M., and Schmidtke, J. 1994. *DNA fingerprinting*. Oxford: BIOS Scientific.

Marx, L. 1988. DNA fingerprinting takes the witness stand. *Science* 240: 1616–18.

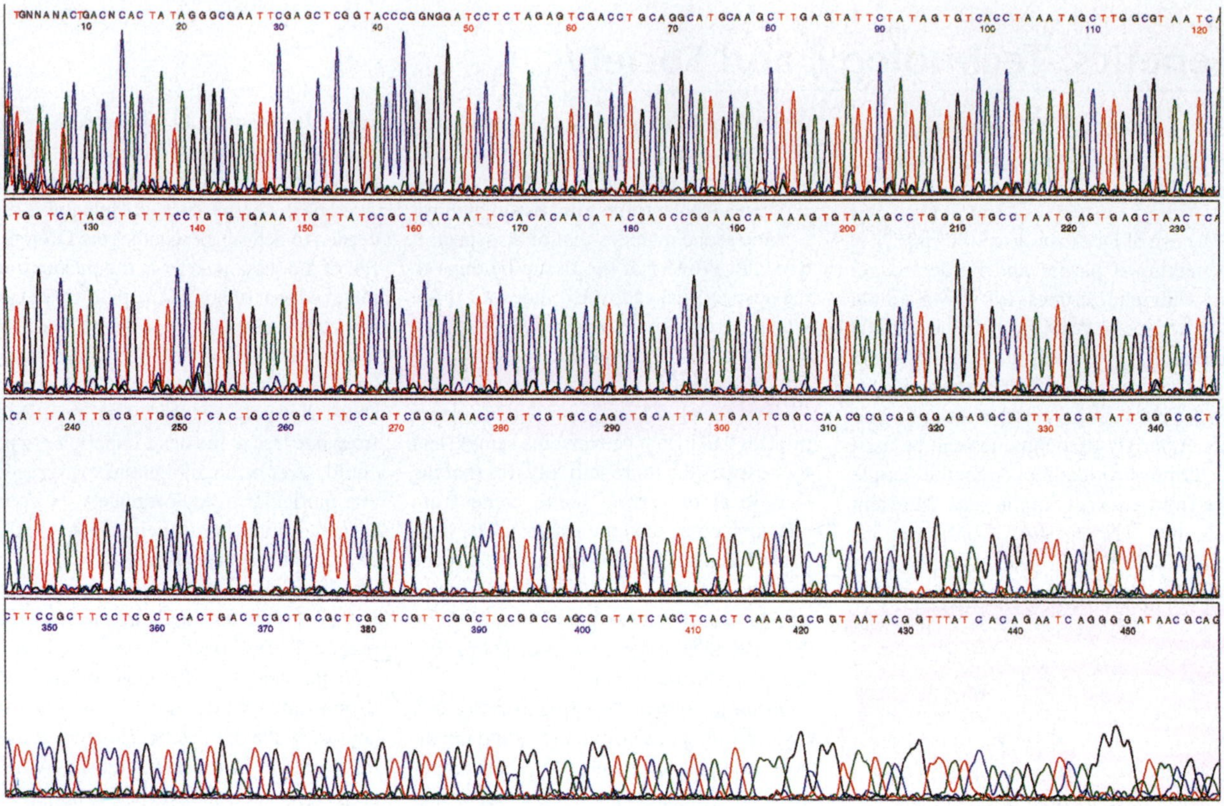

FIGURE 16–29 Automated DNA sequencing using fluorescent dyes, one for each base. The separated bases are read in order along the axis from left to right.

Chapter Summary

1. The cornerstone of recombinant DNA technology is a class of enzymes called restriction endonucleases, which cut DNA at specific recognition sites. The fragments produced are joined with DNA vectors to form recombinant DNA molecules.
2. Vectors replicate autonomously in host cells and facilitate the manipulation of the newly created recombinant DNA molecules. Vectors are constructed from many sources, including bacterial plasmids and viruses.
3. Recombinant DNA molecules are transferred into a host, and cloned copies are produced during host cell replication. A variety of host cells may be used for replication, including bacteria, yeast, and mammalian cells. Cloned copies of foreign DNA sequences are recovered, purified, and analyzed.
4. The polymerase chain reaction (PCR) is a method for amplifying a specific DNA sequence that is present in a collection of DNA sequences, such as genomic DNA. The PCR method allows DNA to be cloned without host cells and is a rapid, sensitive method with wide-ranging applications.
5. Once cloned, DNA sequences are analyzed through a variety of methods, including restriction mapping and DNA sequencing. Other methods such as Southern blotting hybridization identify genes and flanking regulatory regions within the cloned sequences.

Insights and Solutions

The recognition site for the restriction enzyme *Sau*3A is GATC (Figure 16–3). The recognition site for the enzyme *Bam*HI is GGATCC, where the four internal bases are identical to the *Sau*3A site. This means that the single-stranded ends produced by the two enzymes are identical. Suppose you have a cloning vector that contains a *Bam*HI site and foreign DNA that you have cut with *Sau*3A.

1. Can this DNA be ligated into the *Bam*HI site of the vector? Why?

Solution: DNA cut with *Sau*3A can be ligated into the vector's *Bam*HI site because the single-stranded ends generated by the two enzymes are identical.

2. Can the DNA segment cloned into this site be cut from the vector with *Sau*3A? With *Bam*HI? What potential problems do you see with the use of *Bam*HI?

Solution: The DNA can be cut from the vector with *Sau*3A because the recognition sequence for this enzyme (GATC) is maintained on each side of the insert. Recovering the cloned insert with *Bam*HI is more problematic. In the ligated vector, the conserved sequences are GGATC (left) and GATCC (right). Only about 25 percent of the time will the correct base follow the conserved sequence (to produce GGATCC on the left) and only about 25 percent of the time will the correct base precede the conserved sequence (and produce GGATCC on the right). Thus, only about 6 percent of the time (0.25×0.25) will *Bam*HI be able to cut the insert from the vector.

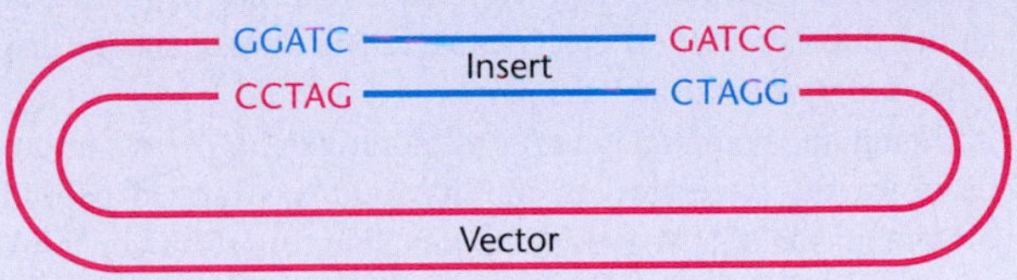

Problems and Discussion Questions

1. In recombinant DNA studies, what is the role of each of the following: restriction endonucleases, terminal transferases, vectors, calcium chloride, and host cells?
2. Why is poly dT an effective primer for reverse transcriptase?
3. An ampicillin-resistant, tetracycline-resistant plasmid is cleaved with *Eco*RI, which cuts within the ampicillin gene. The cut plasmid is ligated with *Eco*RI-digested *Drosophila* DNA to prepare a genomic library. The mixture is used to transform *E. coli* K12.
   (a) Which antibiotic should be added to the medium to select cells that have incorporated a plasmid?
   (b) What antibiotic-resistance pattern should be selected to obtain plasmids containing *Drosophila* inserts?
   (c) How can you explain the presence of colonies that are resistant to both antibiotics?
4. Clones from Problem 3 are found to have an average length of 5 kb. Given that the *Drosophila* genome is 1.5×10^5 kb long, how many clones would be necessary to give a 99 percent probability that this library contains all genomic sequences?
5. Restriction enzymes recognize palindromic sequences in intact DNA molecules and cleave the double-stranded helix at these sites. Inasmuch as the bases are internal in a DNA double helix, how is this recognition accomplished?
6. In a control experiment, a plasmid containing a *Hin*dIII site within a kanamycin-resistant gene is cut with *Hin*dIII, religated, and used to transform *E. coli* K12 cells. Kanamycin-resistant colonies are selected, and plasmid DNA from these colonies is subjected to electrophoresis. Most of the colonies contain plasmids that produce single bands that migrate at the same rate as the original intact plasmid. A few colonies, however, produce two bands, one of original size and one that migrates much higher in the gel. Diagram the origin of this slow band during the religation process.
7. When making cDNA, the single-stranded DNA produced by reverse transcriptase can be made double stranded by treatment with DNA polymerase I. However, no primer is required with the DNA polymerase. Why is this?
8. What facts should you consider in deciding which vector to use in constructing a genomic library of eukaryotic DNA?
9. Using DNA sequencing on a cloned DNA segment, you recover the following nucleotide sequence:

   CAGTATCCTAGGCAT

   Does this segment contain a palindromic recognition site for a restriction enzyme? What is the double-stranded sequence of the palindrome? What enzyme would cut at this site? (Consult Figure 16–3 for a list of restriction enzyme recognition sites.)
10. Figure 16–3 lists restriction enzymes that recognize sequences of four bases and six bases. How frequently should four- and six-base recognition sequences occur in a genome? If the recognition sequence consists of eight bases, how frequently would such sequences be encountered? Under what circumstances would you select such an enzyme for use?
11. You are given a cDNA library of human genes prepared in a bacterial plasmid vector. You are also given the cloned yeast gene that encodes EF-1a, a protein that is highly conserved in protein sequence among eukaryotes. Outline how you would use these resources to identify the human cDNA clone encoding EF-1a.

12. You have recovered a cloned DNA segment of interest and determined that the insert is 1300 bp in length. To characterize this cloned segment, you have isolated the insert and decide to construct a restriction map. Using enzyme I and enzyme II, followed by gel electrophoresis, you determine the number and size of the fragments produced by enzymes I and II alone and in combination as shown here:

Enzymes	*Restriction Fragment Sizes*
I	350 bp, 950 bp
II	200 bp, 1100 bp
I and II	150 bp, 200 bp, 950 bp

Construct a restriction map from these data, showing the positions of the restriction sites relative to one another and the distance between them in base pairs.

13. Although the potential benefits of cloning in higher plants are obvious, the development of this field has lagged behind cloning in bacteria, yeast, and mammalian cells. Can you think of any reason for this?

14. cDNA can be cloned into vectors to create a cDNA library. In analyzing cDNA clones, it is often difficult to find clones that are full length, that is, that extend to the 5′-end of the mRNA. Why is this so?

15. List the steps involved in screening a genomic library. What needs to be known before starting such a procedure? What are the potential problems with such a procedure? How can they be overcome or minimized?

16. Although the capture and trading of great apes has been banned in 112 countries since 1973, it is estimated that about 1000 chimpanzees are removed annually from Africa and smuggled into Europe, the United States, and Japan. Private owners (such as zoos or circuses) often disguise this illegal trade by simulating births in captivity. Until recently, genetic identity tests to uncover these illegal activities have not been used because of the lack of availability of highly polymorphic markers and the difficulties of obtaining chimp blood samples. Recently, a study was reported in which DNA samples were extracted from freshly plucked chimp hair roots and used as templates for the polymerase chain reaction. The primers used in these studies flank highly polymorphic sites in human DNA, resulting from variable numbers of tandem nucleotide repeats. Several suspect chimp offspring and their supposed parents were tested to determine if the offspring were `legitimate" or were the "product of the illegal trading" and not the offspring of the putative parents. A sample of the data is shown here.

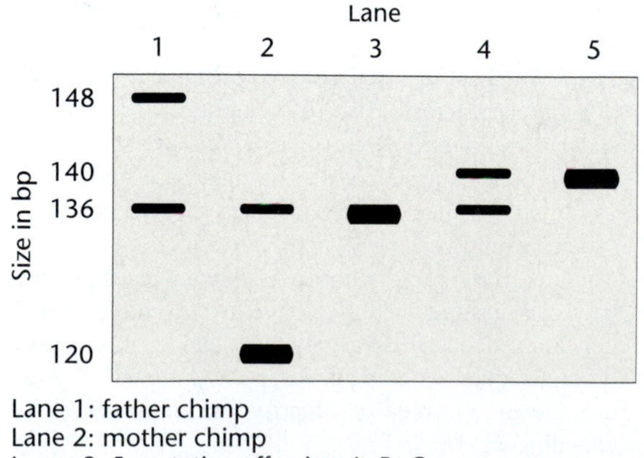

Lane 1: father chimp
Lane 2: mother chimp
Lanes 3–5: putative offspring A, B, C

Examine the data carefully and choose the best conclusion.

(a) None of the offspring are legitimate.
(b) Offspring B and C are not the products of these parents and were probably purchased on the illegal market. The data are consistent with offspring A being legitimate.
(c) Offspring A and B are products of the parents shown, but C is not and was therefore probably purchased on the illegal market.
(d) There is not enough data to draw any conclusions. Additional polymorphic sites should be examined.
(e) No conclusion can be drawn because ``human" primers were used.

17. Briefly describe the problem that a stretch of repeated sequences would cause in a chromosome walk and name the procedure that is used to overcome this problem.

18. You have obtained a clone of a human gene A that is linked (within 200 kb) to a gene that causes breast cancer. Choose and correctly order six of the items from the following list to outline how you could use the technique of chromosome walking to obtain clones of all genes within 200 kb of gene A:

(1) Partially digest DNA with restriction enzyme *Bam*HI to obtain overlapping fragments of about 20 kb.
(2) Completely digest DNA with restriction enzyme *Eco*RI.
(3) Isolate human genomic DNA.
(4) Isolate *E. coli* genomic DNA.
(5) Probe northern blot with A.
(6) Repeat steps 1–5 until the desired number of clones is obtained.
(7) Insert fragment mixture into lambda to make a phage bank.
(8) Screen bank (or library) with gene probe A and isolate a hybridizing clone.
(9) Subclone a small fragment from the end of this clone, use it to rescreen the library, and isolate a hybridizing clone.

Extra-Spicy Problems

19. The accompanying partial restriction map shows a recombinant plasmid, pBIO220, formed by cloning a piece of *Drosophila* DNA (striped box), including the gene *rosy* into the vector pBR322, which also contains the penicillin-resistance gene, *pen*. The vector part of the plasmid contains only the two E sites shown and no A or B sites. The gel shows several restriction digests of pBIO220.

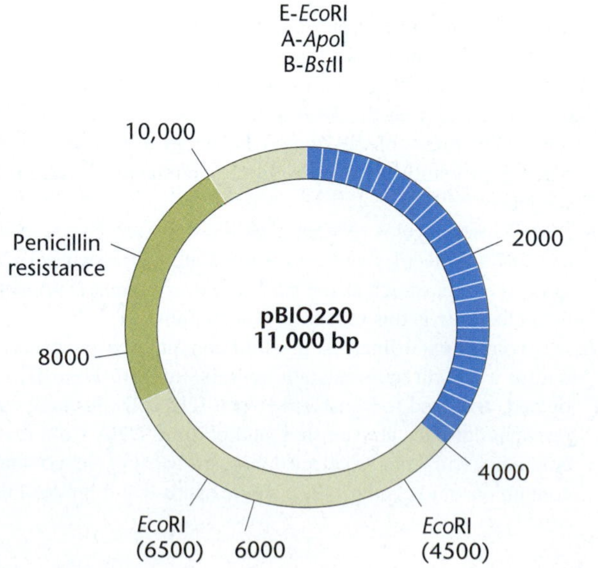

The enzymes used are shown here.

(a) Use the stained gel pattern to deduce where restriction sites are located in the cloned fragment.

(b) A PCR-amplified copy of the entire 2000-bp *rosy* gene was used to probe a Southern blot of the same gel. Use the Southern-blot results to deduce the locations of *rosy* in the cloned fragment. Redraw the map showing the location of *rosy*.

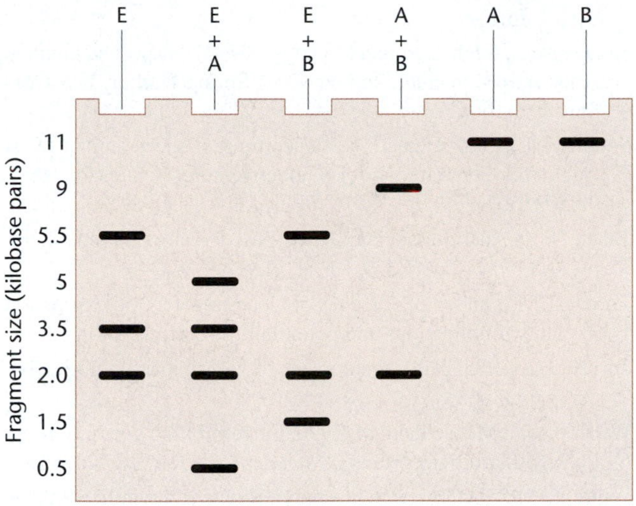

Stained gel

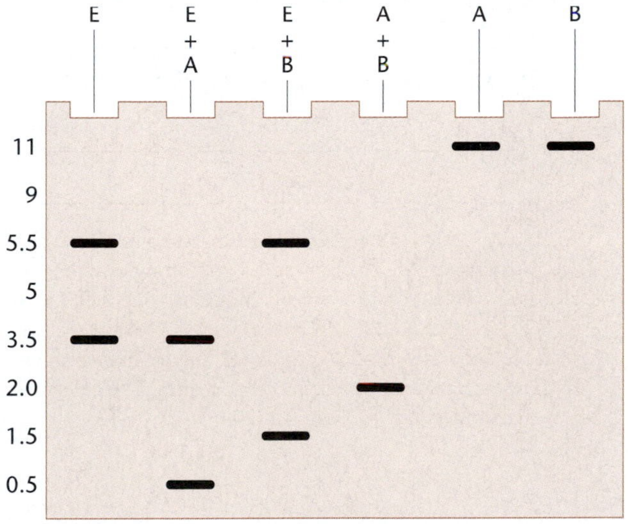

Southern blot

20. One of the restriction maps shown is consistent with the pattern of bands shown in the accompanying figure in the gel after digestion with several restriction endonucleases. The enzymes used are shown in this figure: (Top of next column)

E-*Eco*RI
N-*Nco*I
A-*Aat*II

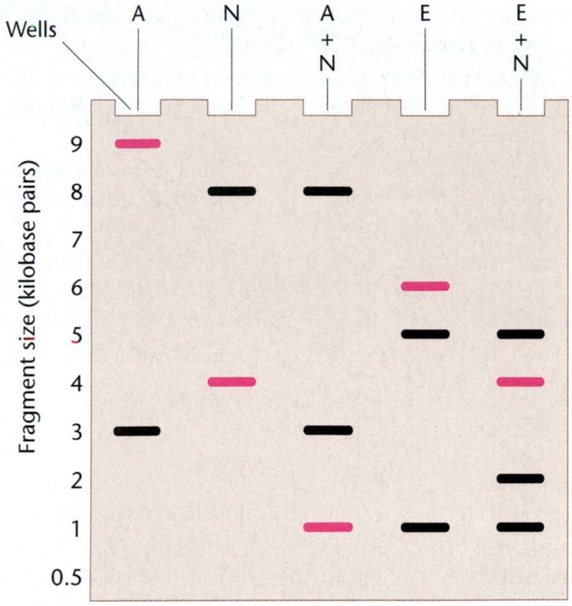

(a) From your analysis of the pattern of bands on the gel, select the correct map and explain your reasoning.

(b) In a Southern blot prepared from this gel, the highlighted bands (purple) hybridized with the gene *pep*. Where is the *pep* gene located?

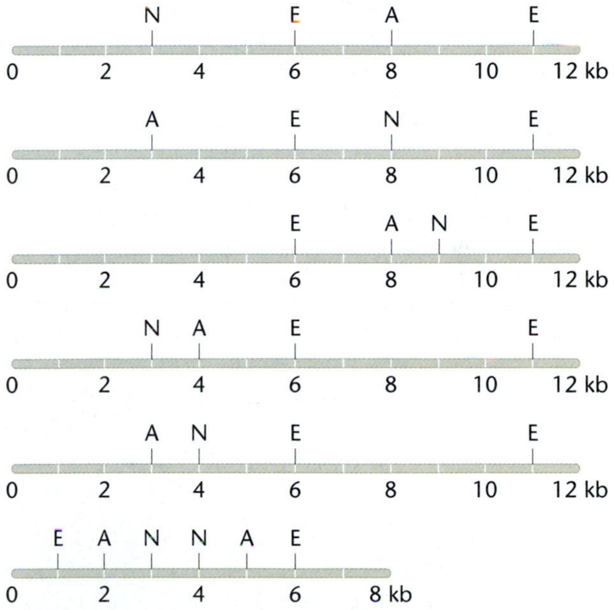

Selected Readings

Alton, E.W., and Geddes, D. M. 1995. Gene therapy for cystic fibrosis: A clinical perspective. *Gene Ther.* 2:88–95.

Antonarakis, S. 1989. Diagnosis of genetic disorders at the DNA level. *N. Engl. J. Med.* 320:153–63.

Barker, D., et al. 1987. Gene for von Recklinghausen neurofibromatosis is in the pericentromeric region of chromosome 17. *Science* 236:1001–162.

Burger, S.L., and Kimmel, A.R. 1987. Guide to molecular cloning techniques. Methods in enzymology, vol. 152. San Diego, CA: Academic Press.

Carter, P.J. and Samulski, P.J. 2000. Adeno-associated viral vectors as gene delivery vehicles. *Int. J. Mol. Med.* 6:17–27.

Colosimo, A., Goncz, K.K., Holmes, A.R., et al. 2000. Transfer and expression of foreign genes in mammalian cells. *BioTechniques* 29:314–331.

Dube, I.D., and Cournoyer, D. 1995. Gene therapy: Here to stay. *Can. Med. Assoc. J.* 152:1605–13.

Eisensmith, R.C., and Woo, S. 1995. Molecular genetics of phenylketonuria: From molecular anthropology to gene therapy. *Adv. Genet.* 32:199–271.

Guyer, M.S., and Collins, F.S. 1993. The Human Genome Project and the future of medicine. *Am. J. Dis. Child.* 147:1145–52.

Knorr, D., and Sinskey, A.J. 1985. Biotechnology in food production and processing. *Science* 229:1224–29.

McKusick, V.A. 1988. The new genetics and clinical medicine. *Hosp. Pract.* 23:177–91.

Mullis, K.B. 1990. The unusual origin of the polymerase chain reaction. *Sci. Am.* (April) 262:56–65.

Old, R.W., and Primrose, S.B. 1994. Principles of genetic manipulation: An introduction to genetic engineering. 5th ed. Palo Alto, CA: Blackwell Scientific.

Oste, C. 1988. Polymerase chain reaction. *BioTechniques* 6:162–67.

Reiss, J., and Cooper, D.N. 1990. Application of the polymerase chain reaction to the diagnosis of human genetic disease. *Hum. Genet.* 85:1–8.

Sambrook, J., Fitch, E.F., and Maniatis, T. 1989. Molecular cloning: A laboratory manual, 2nd ed. Cold Spring Harbor, NY: Cold Spring Harbor Press.

Southern, E. 1975. Detection of specific sequences among DNA fragments separated by gel electrophoresis. *J. Mol. Biol.* 98:503–507.

Tal, J. 2000. Adeno-associated virus-based vectors in gene therapy. *J. Biomed. Sci.* 7:279–91.

Thomas, T.L., and Hall, T.C. 1985. Gene transfer and expression in plants: Implications and potential. *BioEssays* 3:149–53.

Torrey, J.G. 1985. The development of plant biotechnology. *Am. Sci.* 73:354–63.

Welsh, J., and McClelland, M. 1990. Fingerprinting genomes using PCR with arbitrary primers. *Nucleic Acids Res.* 18:7213–18.

White, R. 1985. DNA sequence polymorphisms revitalize linkage approaches in human genetics. *Trends Genet.* 1:177–80.

Williams, J.G.K., et al. 1990. DNA polymorphisms amplified by arbitrary primers are useful as genetic markers. *Nucleic Acids Res.* 18:6531–35.

GENETICS MediaLab

The resources that follow will help you achieve a better understanding of the concepts presented in this chapter. These resources can be found either on the CD packaged with this textbook or on the Companion Web site found at **http://www.prenhall.com/klug**

CD Resources:

Module 16.1: Restriction Enzymes and Vectors

Module 16.2: Polymerase Chain Reaction

Module 16.3: Restriction Mapping

Module 16.4: Cloning Libraries

Module 16.5: DNA Sequencing

Web Problem 1:

Time for completion = 20 minutes

How can the physical features of a molecule be used to distinguish it from a group of similar molecules? Many of the techniques employed in molecular biology use characteristics such as charge, size, shape, and complementarity to separate molecules. Of these features, perhaps the most useful is complementarity between hybridizing molecules. The hybridization of target and probe molecules is a central theme in many molecular techniques. In this exercise, you will review some of these methods. Describe the similarities and differences among Southern, Northern, and Western blots. For each, list what aspects of the probe or target molecules are essential to performing the technique. To complete this exercise, visit Web Problem 1 in Chapter 16 of your Companion Website, and select the keyword **HYBRIDIZATION**. The tutorial takes about 20 minutes to read through. It is followed by three problems that may require some thought and will take time, but that should give you the feel for how techniques like this are used to discover details about mutant genes and their protein products.

Web Problem 2:

Time for completion = 5 minutes

The polymerase chain reaction (PCR) has radically changed the field of molecular genetics. The ability to amplify DNA quickly without a cellular host has dramatically increased the pace of molecular genetics. In this exercise, you will view a short animation about PCR, after which you will list the different colored molecules and describe what each one represents. Indicate how many of each type of colored molecule are present after the three rounds of PCR shown and how many you predict would be present after five rounds of PCR. To complete this exercise, visit Web Problem 2 in Chapter 16 of your Companion Web site, and select the keyword **PCR**.

Web Problem 3:

Time for completion = 10 minutes

What is a recombinant DNA clone? DNA clones are copies of a small fragment of DNA, amplified in a host cell. In contrast, Dolly the cloned sheep, was a case of cloning an entire organism. In this recombinant DNA problem set, you can test your knowledge of recombinant DNA techniques and use that knowledge to interpret pedigree information on a family with Huntington disease. Feedback will indicate whether your answers are correct, and tutorial information is provided to clarify your understanding. To complete this exercise, visit Web problem 3 in Chapter 16 of your Companion Web site, and select the keyword **RECOMBINANT DNA**.

Arabidopsis thaliana serves as a model system for molecular and developmental genetic investigations. (*J. Mylne & J. Botella. Department of Botany, University of Queensland, Australia*)

17

Genomics, Bioinformatics, and Proteomics

KEY CONCEPTS

From the early years of the 20th century, geneticists worked to identify and map genes in organisms of interest. A wide range of organisms continues to be studied; however, emphasis has gradually been placed on a few organisms, such as *Drosophila*, maize, mice, bacteria, and yeast. For mapping studies, geneticists developed a set of methods by using two main approaches. The first approach looked for and identified spontaneous mutations or mutants created by using chemical or physical means. Once a set of mutations was available, the second approach, the generation of genetic maps by linkage analysis was started. In some organisms, such as *Drosophila*, physical maps of genes' locations on chromosomes were also created. These approaches, developed 70–90 years ago, were efficient and widely used in genetics. The drawback of such approaches is that at least one mutation for each gene in the genome is required, and obtaining mutations is a difficult, labor-intensive endeavor. In addition, mutations often have a lethal phenotype, making it difficult or impossible to map the mutated gene.

Beginning in the mid-1980s, geneticists moved away from the classical approaches of mutagenesis and mapping and began using recombinant DNA technology for genetic analysis. In this approach, a collection of clones called a genomic library is established. The clones are pieced together into overlapping sets and assembled into genetic and physical maps that encompass the entire genome. In the final step, the clones are sequenced, with all genes in the genome identified by their nucleotide sequence. Collectively, these methods are called **genomics**. A number of genomic methods were discussed in Chapter 16, and more will be described here. Once obtained, genomic sequence data must be stored and analyzed. This is a formidable task: The human genome sequence is large enough to fill more than 2000 standard diskettes. **Bioinformatics** is an emerging field concerned with the development and application of computer hardware and software to the acquisition, storage, analysis, and visualization of biological information. Databases for the storage and analysis of genome information are now essential tools for geneticists. **Proteomics** is the study of gene products encoded by a genome, including a list of which genes are expressed, their time of expression, the type and extent of any posttranslational modification of the gene product, the function of the encoded protein, and its location in various cellular compartments.

In many cases, gathering and analyzing genome information is a large-scale, labor-intensive endeavor, often requiring the coordinated effort of many laboratories. For this reason, geneticists have formed genome projects. One of the largest and best known of these efforts is the Human Genome Project (HGP). The **Human Genome Project** is a coordinated, international effort whose ultimate goal is to determine the sequence of the 3.2 billion base pairs in the haploid human genome. In the United States, a project to map and sequence the human genome was proposed in 1986, and in 1988, the National Institutes of Health and the U.S. Department of Energy created a joint committee to develop a plan for the project. The HGP got under way in 1990. Other countries, notably France, Britain, and Japan, began similar projects, all of which are now coordinated by an international organization, the Human Genome Organization (HUGO). Under the umbrella of the HGP, there are genome projects for a number of organisms, including bacteria (*E. coli*), yeast (*S. cerevisiciae*), nematode (*C. elegans*), fruit fly (*D. melanogaster*), and the mouse (*M. musculus*). The timeline for these projects is shown in Figure 17–1.

Because sequencing our own genome has widespread implications, the HGP is also studying the ethical, legal, and social implications of genomic research. Issues under study include uses of genetic testing, privacy, and the fair use of genetic information in insurance, employment, and health care. The Ethical, Legal, and Social Implications (ELSI) Program of the Human Genome Project addresses these issues and involves scientists, health professionals, policy makers, and the public in formulating policy recommendations and legislation. This program and its goals are discussed in Chapter 18.

In this chapter, we outline the technology of genome sequence analysis and review the significant findings from a variety of genome projects using prokaryotic and eukaryotic organisms. This review includes insights into the organization of genes in chromosomes, genome evolution, findings from comparative genomics, and the minimum number of genes necessary for life, as well as the evolution of gene families. We also discuss the emerging field of proteomics and examine how geneticists are studying patterns of gene expression as well as the applications of this knowledge in various fields.

17.1 Genomics: Sequencing Identifies and Maps All Genes in a Genome

Geneticists use two different methods for sequencing genomes (Figure 17–2). The **clone-by-clone method** was the first to be developed. It begins with the construction of genomic libraries of restriction fragments covering all the genomic DNA (genomic clones) of an organism. By using genetic markers, overlapping clones are assembled into genetic and physical maps encompassing the entire genome. The nucleotide sequence is determined on a clone-by-clone basis until the entire genome is sequenced. The clone-by-clone method was chosen for the publicly funded Human Genome Project sponsored by the National Institutes of Health and the Department of Energy.

In the second method, called **shotgun cloning**, genomic libraries are prepared and randomly selected clones are sequenced until all clones in the library are analyzed. Assembler software organizes the nucleotide sequence information into a genome sequence. This method, developed by Craig Venter and his colleagues at The Institute for Genome Research (TIGR) was used to sequence the genome of *Haemophilus influenzae* in 1995, the first organism to have its genome completely sequenced. After refining the method and using it to sequence the genomes of other prokaryotes, the shotgun method was used to sequence eukaryotic

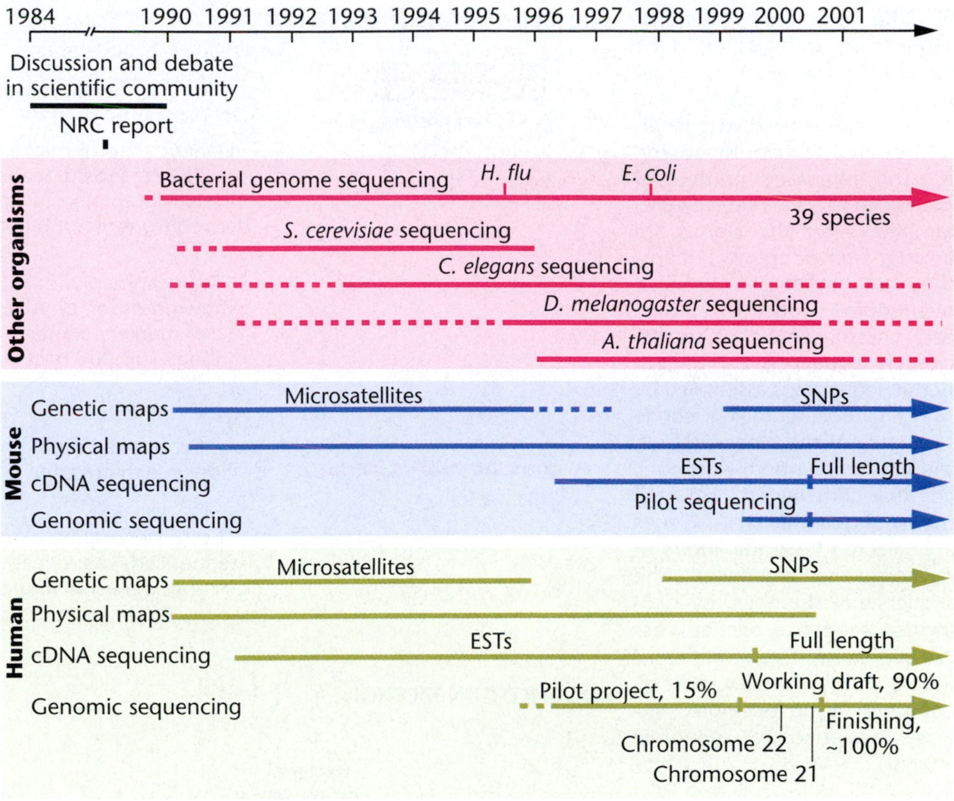

FIGURE 17–1 A timeline of genome projects funded by the National Institutes of Health (NIH), including the Human Genome Project (International Human Genome Sequencing Consortium. 2001. Initial sequencing and analysis of the human genome. *Nature* 409: 860–921, Figure 1, p. 862.)

genomes, including *Drosophila* and humans. Using the shotgun method, Venter and his colleagues started a privately funded human genome project. The project began in September 1999, and sequencing was finished in June 2000.

17.2 Bioinformatics Provides Tools for Analyzing Genomic Information

Traditionally, geneticists use laboratory notebooks to record and analyze the results of their experiments. Technology has now generated methods for high-throughput experiments, sequencing entire genomes, or measuring expression of genes on a whole-genome scale. These technologies produce data on a scale that was unimaginable only a few years ago. The ability to generate large amounts of experimental data has fundamentally changed the way results from genetic experiments are stored and analyzed. Instead of lab notebooks, geneticists and other biologists now use large-scale public and private databases to store, analyze, and visualize their experimental results. As mentioned earlier, bioinformatics unites computer sciences with biology. One of the basic challenges in bioinformatics is managing the flood of information from genome projects. In the next sections, we will discuss the analysis of data from genome projects, including compiling genome sequences, identifying genes, and assigning functions to identified genes. Access to genomic databases as well as some databases explaining more about bioinformatics are provided at the book's Web site.

Compiling the Sequence Ensures Accuracy

To ensure that the nucleotide sequence of a genome is complete and error free, the genome is sequenced more than once. For example, using the shotgun method on the genome of the bacterium *Pseudomonas aeruginosa*, researchers sequenced the 6.3 million nucleotides seven times to ensure that the sequence was accurate and free from errors. Even with this level of redundancy, the assembler software recognized 1604 regions that required further clarification. These regions were reanalyzed and resequenced to improve accuracy. Finally, the accuracy of the shotgun method's sequence was compared with the sequence of two widely separated genomic regions obtained by conventional cloning. The sequence of the 81,843 nucleotides cloned and sequenced by conventional methods (clone by clone) was in perfect agreement with the sequence obtained by the shotgun method. This level of care is not unusual. Similar precautions are used in all genome projects.

The HGP sequenced the 3.2 billion base pairs of the human genome a total of 12 times. The privately based shotgun-cloning project based at the biotechnology company, Celera, used a strategy of sequencing from both ends of DNA fragments and covered the genome 35.6 times. Although a draft of the human genome sequence is finished, several other tasks are yet to be completed. These include obtaining the remaining sequence and correcting errors (proofreading the genome), filling sequence gaps (which amounted to about 150 Mb in mid-2001), and then sequencing the 7–15 percent of the genome that contains heterochromatin.

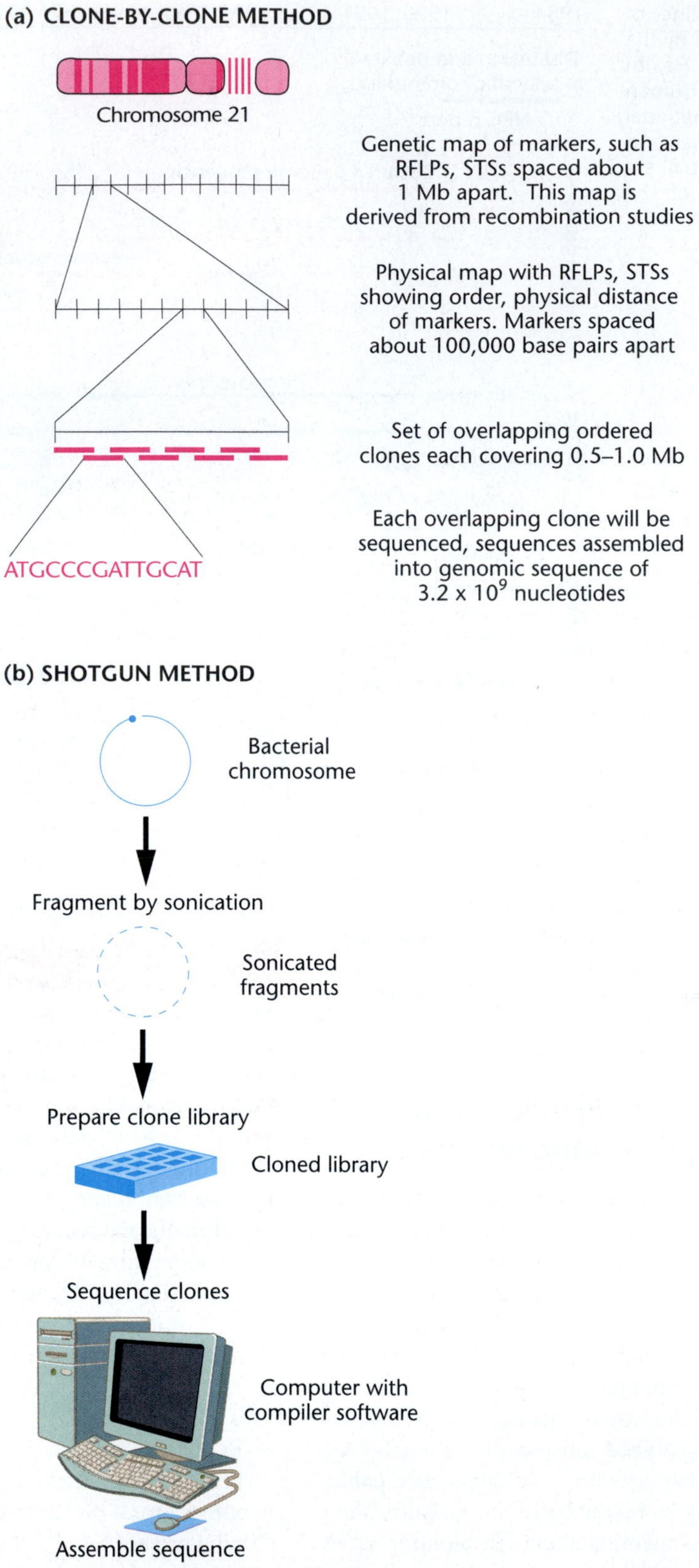

(a) CLONE-BY-CLONE METHOD
Chromosome 21
Genetic map of markers, such as RFLPs, STSs spaced about 1 Mb apart. This map is derived from recombination studies
Physical map with RFLPs, STSs showing order, physical distance of markers. Markers spaced about 100,000 base pairs apart
Set of overlapping ordered clones each covering 0.5–1.0 Mb
ATGCCCGATTGCAT
Each overlapping clone will be sequenced, sequences assembled into genomic sequence of 3.2 x 10⁹ nucleotides
(b) SHOTGUN METHOD
Bacterial chromosome
Fragment by sonication
Sonicated fragments
Prepare clone library
Cloned library
Sequence clones
Computer with compiler software
Assemble sequence

FIGURE 17–2 The two genome sequencing strategies. (a) In the clone-by-clone method, a genomic library is prepared, and clones are organized into genetic and physical maps by observing the inheritance pattern of genetic markers in heterozygous families. After the clones are arranged into physical maps, they are broken into smaller, overlapping clones that cover each chromosome. Each smaller clone is sequenced, and the genomic sequence is assembled by stringing together the nucleotide sequence of the clones. (b). In the shotgun method, a genomic library is constructed from fragments of genomic DNA. Clones are selected from the library at random, and sequenced. The sequence is assembled by looking for sequence overlaps between clones from different libraries. This is usually done by computer, using assembler software designed for genomic analysis.

Heterochromatic regions of the genome were excluded by design, as they contain long stretches of repetitive DNA sequences, and were initially thought to contain no genes. However, in sequencing the genome of *Drosophila*, researchers discovered that heterochromatic regions do contain a small number of genes (about 50 in *Drosophila*). As a result of this discovery, heterochromatic regions of the human genome must be sequenced to ensure that all genes are identified. Once the human genome or any other genome is sequenced, compiled, and proofread, the next stage—annotation—begins.

Annotating the Sequence Identifies Genes

After a genome sequence has been obtained, organized, and checked for accuracy, the next task is to find all the genes that encode proteins. This is the first step in annotation, a process that identifies genes, their regulatory sequences, and their function(s). Annotation also identifies nonprotein coding genes (including ribosomal RNA, transfer RNA, and small nuclear RNAs), finds and characterizes mobile genetic elements and repetitive sequence families that may be present in the genome.

Locating protein-coding genes is done by inspecting the sequence, using computer software or by eye. Genes have several identifying features. Protein-coding genes are composed of **open reading frames (ORFs)**, a series of nucleotides that specify an amino acid sequence. ORFs begin with an initiation sequence (usually ATG) and end with a termination sequence (TAA, TAG, or TGA). Scanning a DNA sequence for ORFs beginning with an ATG and followed by a termination codon is one strategy for finding genes. ORF scanning, usually by computer, is an effective method for annotating bacterial genomes. (Figure 17–3). (The genetic code is discussed in Chapter 5.)

As shown for the *lacY* gene, identification of longer genes by ORF scanning presents few problems. However, one of the problems in annotation is discriminating between short genes and short, random ORFs that are not genes. If all the annotated ORFs in a genome database are not really genes, the number of genes in a genome may be over-estimated. In addition, before attempting to assign function to newly discovered ORFs, it is important to determine whether they are in fact, genes.

By analyzing the length distribution of ORFs reported in a genome using several different methods, including comparison to databases containing the sequence of known proteins (SWISS-PROT and others), it appears that many sequenced genomes contain annotated ORFs that are not genes (Table 17.1). For example, the *E. coli* genome database reports 4289 ORFs for this organism. Matching the ORFs to known proteins reduces that number by 12 percent to 3771. Using stop codon frequency to identify genuine ORFs reduces the gene number to 3463. If phylogenetic considerations are used in classifying the ORFs, the gene number is reduced to 3327, an overall 22 percent reduction in gene number. This example illustrates that annotation is not a simple procedure, and that the process and its outcome depend on the underlying assumptions employed in the analysis.

Genes in eukaryotic genomes (including the human genome), have several features that make annotation even more difficult. First, many genes in eukaryotes have a pattern of exons (coding regions) interrupted by introns (non-coding regions). As a result, these genes are not organized as continuous ORFs. Scanning software often interprets each exon as a separate gene, because stop codons are often found in introns. Second, as we shall see in a subsequent section, genes in humans and other eukaryotes are often widely spaced, increasing the chances of finding false ORFs (over 70 percent of the human genome is composed of intergenic spacer DNA).

Newer versions of ORF scanning software for eukaryotic genomes make scanning more efficient. These programs search for features including codon bias, intron/exon junctions, upstream regulatory sequences, the 3' AATAAA poly A signal, and in some genomes, CpG islands. CpG islands are DNA regions containing many copies of this nucleotide doublet. CpG islands are found in gene-rich regions of the mammalian genome, often adjacent to genes. Codon bias is the selective use of one or two codons to encode amino acids that can be encoded by a number of different codons. For example, the codons GCA, GCT, GCC, and GCG can encode alanine. Yet in the human genome, GCC is used 41 percent of the time, and GCG only 11 percent of the time. Codon bias is present in exons, but should not be present in introns or intergenic spacers.

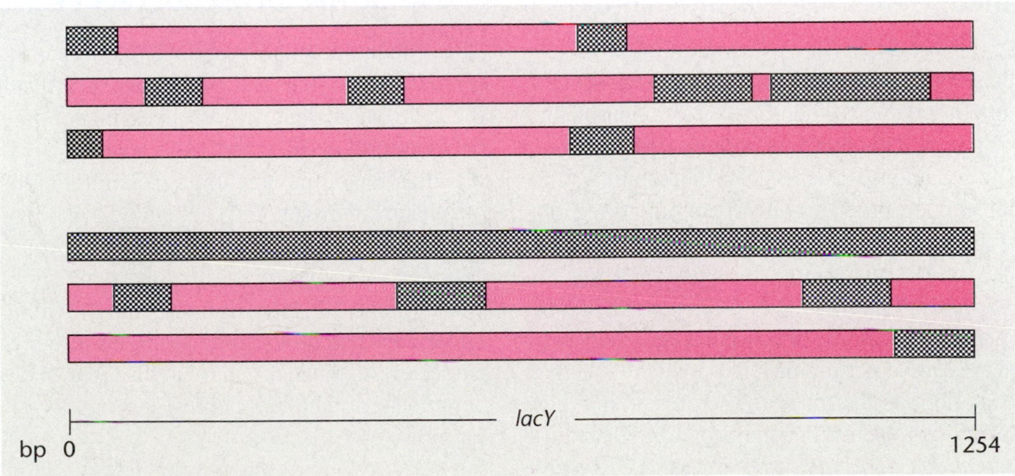

FIGURE 17–3 ORF analysis of the *lacY* gene of *E. coli*. ORFs for each of the six possible reading frames, three on each DNA strand, are shown as dark, stippled regions. Only ORFs longer than 50 codons are marked. Because most genes are much longer than 50 codons, ORF length is one indication of an actual gene. In this case, only one ORF covers the entire sequence, and shows the location and orientation of the *lacY* gene. (*From a public database at* http://www.ncbi.nlm.gov/gorf/)

TABLE 17–1 Gene Estimates in Microbial Genomes

	A+T (%)	*Number of Annotated Genes*	*Number of Genes Estimated from*		*Number of COGs*
			SWISS-PROT	*Stop triplet*	
Archaea					
A.pernix	44	2694	1376	1423	1169
A. fulgidus	51	2407	1818	1927	1849
M. Jannaschii	69	1715	1350	1573	1320
Bacteria					
D. radiodurans	33	2937	2323	1904	2176
P. aeruginosa	33	5565	4753	3508	4191
M. tuberculosis H37Rv	34	3918	3410	2537	2668
N. meningitidis A	48	2121	1539	1447	1455
V. cholerae	48	3828	2991	2931	2745
E. coli K–12	49	4289	3771	3463	3327
B. subtilis	56	4100	3263	3330	2803
A. aeolicus	57	1522	1337	1412	1317
C. pneumoniae CWL029	59	1052	903	909	647
M. pneumoniae M129	60	677	610	617	423
H. pylori 26695	61	1566	1303	1384	1081
H. influenzae Rd	62	1709	1479	1526	1504
M. genitalium	68	480	461	474	376
B. burgdorferi	71	850	756	772	694
R. prowazekii	71	834	759	795	674
Eukaryota					
S. cerevisiae	62	6269	5560	5728	2175

*The table contains the number of annotated proteins, the A+T content, the number of proteins estimated from matches to SWISS-PROT and from stop triplet frequency, and the number of clusters of orthologous genes (COGs) for the organism. The list is ordered by kingdom and A+T contents.

To facilitate the annotation of the human genome, several organizations have started Ensembl, a database to coordinate the efforts of individual scientists, genome projects and companies in identifying the 30,000 to 40,000 human genes. The database software integrates sequence information with chromosome gene maps, and protein sequence information, and provides links to other databases.

Once Identified, Genes Are Classified into Functional Groups

After a genomic sequence is annotated, the predicted ORFs are examined one at a time to identify the function of the encoded gene product. This analysis is done using several methods: searching databases such as GenBank to find similar genes isolated from other organisms; comparing predicted ORFs with those from known, well-characterized bacterial genes; and lastly, looking for functional motifs, regions of DNA that encode protein domains such as ion channels, DNA binding regions, or secretion or export signals. In addition, researchers examine ORFs to identify genes with specialized functions that may not be present in other organisms. Table 17–2 presents the results of an ORF analysis in the *Pseudomonas aeruginosa* genome; the ORFs are classified on a scale according to confidence level. As in other bacterial genomes sequenced to date, about half the ORFs have no known or proposed function (confidence level 4). The remaining ORFs are assigned to functional classes 1-3 (Table 17–3). The remaining task in the analysis of the *P. aeruginosa* genome, as in many other genomes, is to assign functions to all the genes and to understand how the unknown genes and the known genes interact in the biology of the organism.

17.3 Prokaryotic Genomes Have Some Unexpected Features

The sequencing of several dozen genomes is now complete for both prokaryotes and eukaryotes. The results indicate that there are significant differences in genome organization between prokaryotes and eukaryotes. We will consider the organization of these genomes separately. Because prokaryotes have small genomes amenable to shotgun cloning and sequencing, more than 50 genome projects have been completed for two types of prokaryotes, eubacteria and archaea (Table 17–4), with more than 200 additional projects underway. The prokaryotic genomes already sequenced include several that cause human diseases, such as cholera, tuberculosis, and genital herpes.

Genomes of Eubacteria Have a Wide Range of Sizes

Based on genome project results, we can make a number of generalizations concerning the size and organization of

TABLE 17–2 Analysis of the *Pseudomonas aeruginosa* Genome

General Features			
Genome size (bp)	6,264,403		
[G+C content	66.6%		
Coding regions	89.4%		
Coding Sequences			
Confidence level	***ORFs***	***(%)***	***Definition***
1	372	(6.7)	*P. aeruginosa* genes with demonstrated function
2	1,059	(19.0)	Strong homologs of genes with demonstrated function from other organisms
3	1,590	(28.5)	Genes with proposed function based on motif searches or limited homology
4	769	(13.8)	Homologs of reported genes of unknown function
4	1,780	(32.0)	No homology to any reported sequences
Total	5,570	(100)	

Source: Stover, et al., 2000. Complete genome sequence of *Pseudomonas aeruginosa* PA01, an opportunistic pathogen. *Nature* 406:959–964. Table is derived from Table 1, p. 961.

TABLE 17–3 Functional Classes of Predicted Genes in *Pseudomonas*

	ORFs	
Functional Class	***Number***	***%***
Adaptation, protection (for example, cold shock proteins)	60	1.1
Amino acid biosynthesis and metabolism	150	2.7
Antibiotic resistance and susceptibility	19	0.3
Biosynthesis of cofactors, prosthetic groups and carriers	119	2.1
Carbon compound catabolism	130	2.3
Cell division	26	0.5
Cell wall	83	1.5
Central intermediary metabolism	64	1.1
Chaperones & heat shock proteins	52	0.9
Chemotaxis	43	0.8
DNA replication, recombination, modification and repair	81	1.5
Energy metabolism	166	3.0
Fatty acid and phospholipid metabolism	56	1.0
Membrane proteins	7	0.1
Motility & attachment	65	1.2
Nucleotide biosynthesis and metabolism	60	1.1
Protein secretion/export apparatus	83	1.5
Putative enzymes	409	7.3
Related to phage, transposon or plasmid	38	0.7
Secreted factors (toxins, enzymes, alginate)	58	1.0
Transcription, RNA processing and degradation	45	0.8
Transcriptional regulators	403	7.2
Translation, post-translational modification, degradation	149	2.7
Transport of small molecules	555	10.0
Two-component regulatory systems	118	2.1
Hypothetical	1,774	31.8
Unknown (conserved hypothetical)	757	13.6
Total	5,570	100

Source: Stover, et al., 2000. Complete genome sequence of *Pseudomonas aeruginosa* PA01, an opportunistic pathogen. *Nature* 406:959–964.

TABLE 17–4 Genome Size and Gene Number in Selected Prokaryotes

	Genome Size (Mb)	*Number of Genes*
Archaea		
Archaeoglobus fulgidis	2.17	2,493
Methanococcus jannaschii	1.66	1,813
Thermoplasma acidophilum	1.56	1,509
Eubacteria		
Escherichia coli	4.64	4,397
Bacillus subtilis	4.21	4,212
Haemophilus influenzae	1.83	1,791
Aquifex aeolicus	1.55	1,552
Rickettsia prowazekii	1.11	834
Mycoplasma pneumoniae	0.82	710
Mycoplasma genitalium	0.58	503

bacterial genomes (eubacteria, or true bacteria). Traditionally, the bacterial genome has been thought of as relatively small (less than 5 Mb), and contained in a single circular DNA molecule. The flood of genomic information now available has challenged this viewpoint. Although most prokaryotic genomes are small, genome sizes vary widely. In fact, there is some overlap between larger bacterial genomes (30 Mb in *Bacillus megaterium*) and smaller eukaryotic genomes (12.1 Mb in yeast).

Most prokaryotic genomes sequenced to date are circular DNA molecules, but an increasing number of genomes composed of linear DNA molecules are being identified, including *Borrelia burgdorferi*, the organism that causes Lyme disease. Perhaps more importantly, new findings about plasmids are redefining our view of the bacterial genome as a single, circular DNA molecule. Most plasmids carry nonessential genes, can be transferred from one cell to another, and the same plasmid is often present in bacteria from different species, all of which suggests that plasmid genes should not be included as part of a bacterial genome. However, *B. burgdorferi* carries approximately 17 plasmids, which contain at least 430 genes. Some of these genes appear to be essential to the bacterium, including genes for purine biosynthesis and membrane proteins. Recently, sequencing the genome of *Vibrio cholerae*, the organism responsible for cholera, revealed the presence of two circular chromosomes (Figure 17–4). Chromosome 1 (2.96 Mb) contains 2770 ORFs, and chromosome 2 (1.07 Mb) contains 1115 ORFs, some of which encode essential genes, such as ribosomal proteins. Based on its nucleotide sequence, chromosome 2 is apparently derived from a plasmid captured by an ancestral species. In addition, two chromosomes are present in other species of *Vibrio*. The existence of prokaryotic genomes containing multiple chromosomes, where at least one chromosome is derived from a plasmid raises several important questions. For example, when should a plasmid be considered a chromosome, and what regulatory mechanisms control gene expression and metabolism in a multichromosome

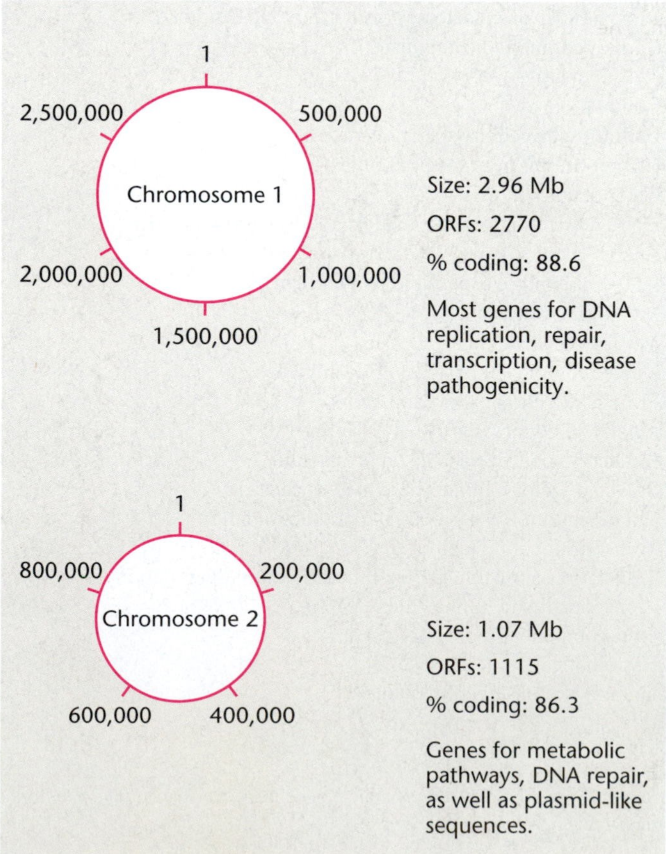

FIGURE 17–4 The *Vibrio cholerae* genome is contained in two chromosomes. The larger chromosome (chromosome 1) contains most of the genes for essential cellular functions and infectivity. Most of the genes on chromosome 2 (52 percent of 1115) are of unknown function. The bias in gene content and the presence of plasmidlike sequences on chromosome 2 suggest that this chromosome was a megaplasmid captured by an ancestral *Vibrio* species. (Redrawn from Heidelberg, JF, et al., 2000. DNA sequence of both chromosomes of the cholera pathogen, *Vibrio cholerae. Nature* 406:477–483, *Figure 2, p. 478.*)

prokaryotic genome? Answers to these questions and others will redefine some ideas about prokaryotic genomes, the nature of plasmids, and may provide clues about the evolution of multichromosome eukaryotic genomes.

Genomes of Eubacteria Have Many Unexpected Features

We can make a number of generalizations about the organization of protein-coding genes in bacterial chromosomes. First, gene density is very high, averaging about one gene per kilobase pair of DNA. *E. coli* has a large genome (4.6 Mb), containing 4288 protein-coding genes, a density close to one gene per kilobase pair (Figure 17–5). *M. genitalium* has a small genome (0.6 Mb), with 503 genes. Gene density is also close to one gene per kilobase pair. This close packing of genes in prokaryotic genomes results in a very high proportion of the DNA (approximately 85–90 percent) serving as coding DNA. In *M. genitalium*, there is only about 110–125 bp of DNA between genes. Typically, less than 1 percent of bacterial DNA is noncoding DNA, usually in the form of transposable elements, that can move from one place to another in the genome (discussed in Chapter 7). Introns are extremely rare in bacterial genomes.

A second generalization we can make is that bacterial genomes are characterized by the presence of operons. (Operons are discussed in Chapter 15.) In *E. coli*, 27 percent of the predicted transcription units are contained in operons (almost 600 operons).

In other bacterial genomes, the organization of genes into transcriptional units is challenging our ideas about operons. In *Aquifex aeolicus*, most genes are in polycistronic transcription units we would usually call operons. However, our working definition that operons contain genes that are part of a single biochemical pathway does not hold true in this organism. In *A. aeolicus* (Figure 17–6), one operon contains six genes: two for DNA recombination, one for lipid synthesis, one for nucleic acid synthesis, one for protein synthesis, and one that encodes a protein for cell motility. Other operons in this species also contain genes with several different functions. This finding, combined with similar results from other genome projects, provides new insights into the nature of operons, their role in biochemical processes in bacterial cells, and may redefine our concept of the operon.

Figure 17–7 shows a small portion of the *E. coli* gene map. The lines above the genes show how those genes are organized into transcription units with promoters that locate the start of transcription for each unit. Some of these units are organized as operons. Note that there are three cases of overlapping genes, where one gene is nested inside another on the same DNA strand (shown in green). Overlapping genes are common in viruses (where space is at a premium), but are uncommon in bacteria.

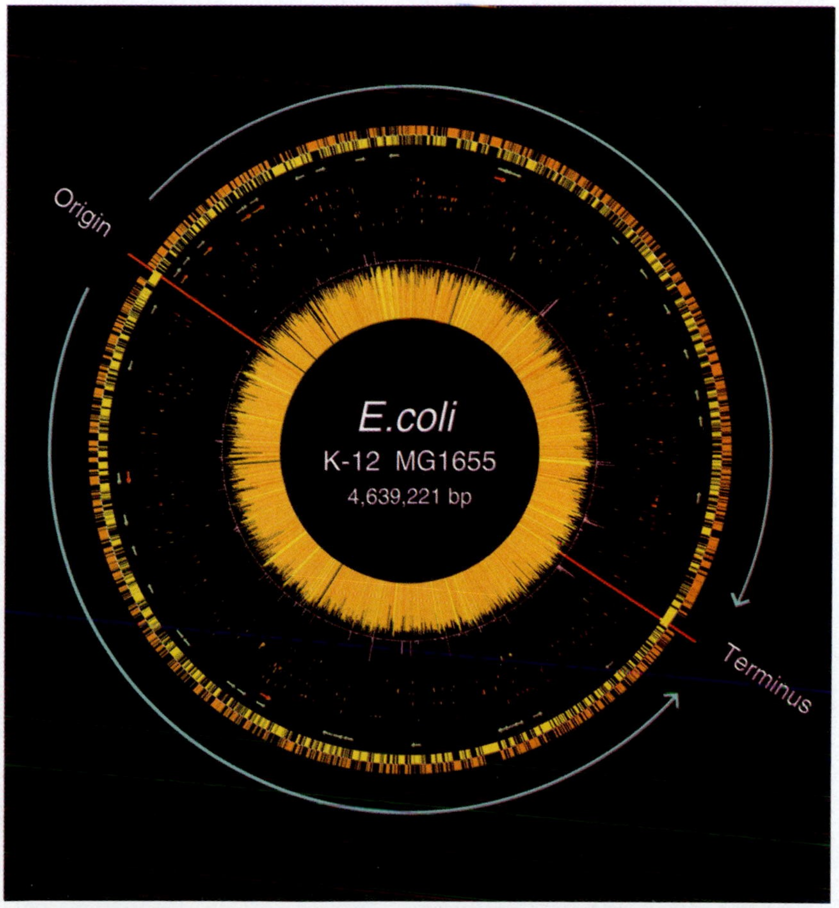

FIGURE 17–5 The *E. coli* genome. The origin and terminus of replication is shown. The outer circle of bars represents genes transcribed in a clockwise direction, and the inner circle represents genes transcribed in a counterclockwise direction. (*Provided by Dr. Fred Blattner, Laboratory of Genetics, University of Wisconsin, Madison WI.*)

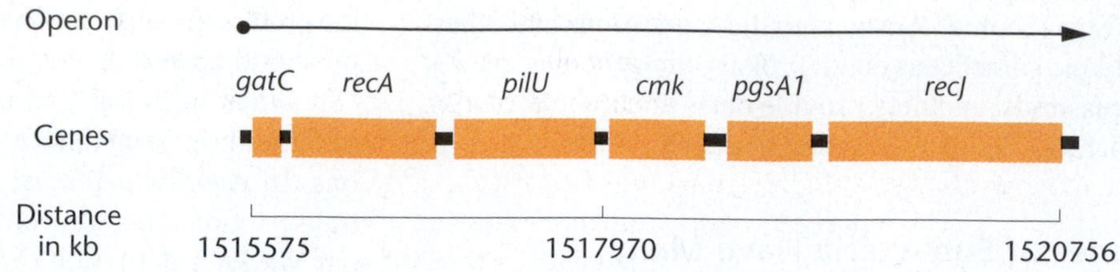

FIGURE 17–6 An operon from the *A. aeolicus* genome. This operon contains genes for protein synthesis (*gatC*), for DNA recombination (*recA* and *recJ*), for a motility protein (*pilU*), for nucleotide biosynthesis (*cmk*), and for lipid biosynthesis (*pgsA1*). This organization challenges the conventional idea that genes in an operon encode products that control a common biochemical pathways. (*From a public database at* http://www.ncbi.nlm.nih.gov/Entrez; *and from Deckert, G. et al., 1998. The complete genome of the hyperthermophilic bacterium* Aquifex aeolicus. Nature 392:353–358, *Figure 1, p. 358.*)

Genomes of Archaea Share Features with Eukaryotes

Archaea (formerly known as archaebacteria) are one of the three major divisions of living organisms. The other two are the Eubacteria (true bacteria, without a membrane-bound nucleus, such as *Haemophilus* and *Mycoplasma*) and Eukarya (containing a true membrane-bound nucleus). Archaea, like eubacteria, are prokaryotes; that is, they have no nucleus. They have only recently been recognized as a separate kingdom. Carl Woese proposed the original classification of Archaea as a separate kingdom in 1977, based on nucleotide sequence of the DNA encoding ribosomal RNA (rDNA). This classification was later confirmed by protein sequences and metabolic pathway analysis.

Typically, Archaea are extremophiles; they live in extreme environments with very high temperature, high salt, high pressure, or extreme pH. Even though Archaea are structurally similar to eubacteria, it was recognized early on that in many aspects of metabolism, they more closely resemble eukaryotes. With the completion of several genome projects from the Archaea, we can now examine this relationship more closely. The sequence of the achaeon, *Methanococcus jannaschii*, was completed in 1996. It has a circular double-stranded DNA genome of 1.7 Mb, with 1738 protein-coding genes. This organism was originally isolated from a high-temperature deep-sea vent 2600 meters below sea level. It has an optimum growth temperature of 85°C and can survive at 94°C (water boils at 100°C.) The *M. jannaschii* genome contains three chromosomes: a large, circular chromosome of 166 Mb and two small, circular chromosomes of 58.4 and 16.5 kb. Most genes in this organism (58 percent) do not match any other known genes. The majority of genes involved in energy production, cell division, and general metabolism closely resemble those of eubacteria. Gene organization also resembles eubacteria; they are densely packed, have operons, and do not have introns.

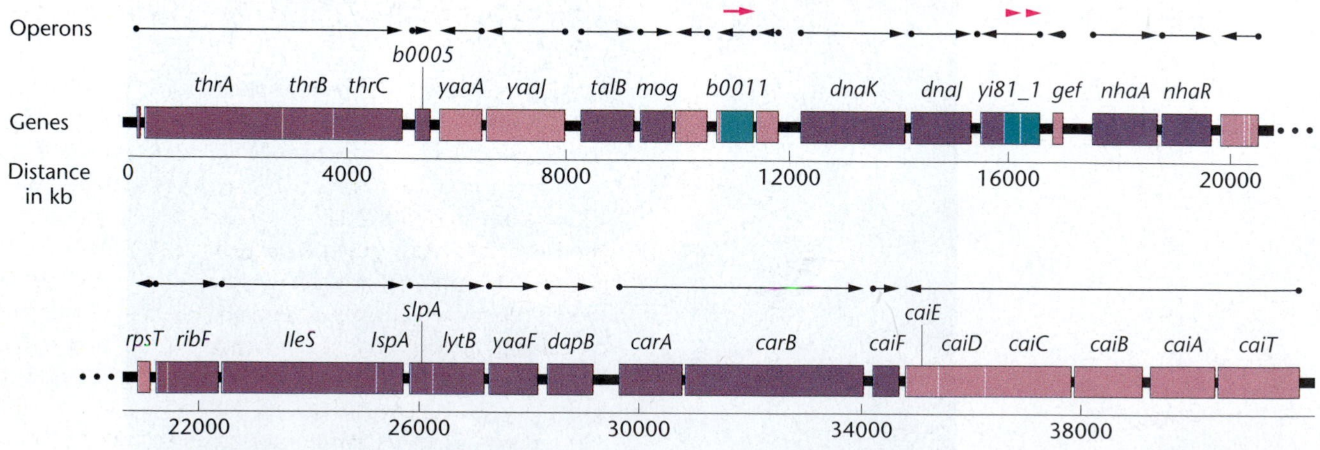

FIGURE 17–7 A portion of the *E. coli* chromosome showing genes and operons. A dot indicates the promoter for each gene or operon. Arrows and color indicate the direction of transcription: dark purple genes are transcribed left to right, light purple are transcribed right to left. Overlapping genes are shown in green. (*From a public database at* http://www.ncbi.nlm.nih.gov/Entrez)

On the other hand, archaea genes also have significant similarities to eukaryotic genes. The genes involved in RNA synthesis, protein synthesis, and DNA synthesis more closely resemble those in eukaryotes. Most surprising is the presence of histone chromosomal proteins and evidence that chromosomal DNA is organized into chromatin. Although they have no introns in their protein-coding genes, archaea do have introns in their tRNA genes, as do eukaryotes.

17.4 Eukaryotic Genomes Have a Mosaic of Organizational Patterns

Eukaryotic nuclear genomes are usually divided into a series of linear DNA molecules, each contained in an individual chromosome. In addition, eukaryotes have a second genome, the mitochondrial genome, present as a circular DNA molecule. Plants and other photosynthetic organisms also have a third genome, a circular DNA molecule carried in the chloroplast. Our focus in this chapter is the nuclear genome. Although the basic features of the eukaryotic genome are similar in different species, genome size is highly variable in eukaryotes (Table 17–5). Genome sizes range from about 10 Mb in fungi to over 100,000 Mb in some flowering plants (a ten-thousandfold range), while the number of chromosomes per genome ranges from two into the hundreds (about a hundredfold range).

Eukaryotic Genomes Share Some General Features

Compared with prokaryotes, eukaryotes have a relatively low gene density as shown in Figure 17–8. In general, more complex eukaryotes have less compact genomes, with lower levels of gene density. If we compare regions of different eukaryotic genomes, several features become apparent:

- **Gene density**. A 50-kb region of the yeast genome [Figure 17–8(b)] contains over 20 genes, while a 50-kb segment of the human genome [Figure 17–8(c)] contains only 6 genes.

TABLE 17–5 Genome Size and Gene Number in Selected Eukaryotes

Organism	*Genome Size (Mb)*	*Number of Genes*
S. cerevisiae (yeast)	12	6,548
P. falciparum (malaria)	30	~ 6,500
C. elegans (nematode)	97	> 20,000
A. thalania (mustard plant)	120	~ 20,000
D. melanogaster (fruit fly)	170	~ 16,000
O. sativa (rice)	415	~ 20,000
Z. mays (maize)	2,500	~ 20,000
H. sapiens (human)	3,300	~ 35,000
H. vulgare (barley)	5,300	~ 20,000

- **Introns**. None of the yeast genes shown in Figure 17–8 contain introns, whereas each human gene does contain introns. In fact, the entire yeast genome has only 239 introns, while some single genes in the human genome contain over 100 introns.

- **Repetitive sequences**. The presence of introns and repetitive sequences are two major reasons for the wide range of genome sizes in eukaryotes. In some plants, such as maize, repetitive sequences are the dominant feature of the genome. (Repetitive sequences are discussed in Chapter 4.) The maize genome has about 5000 Mb of DNA, over 80 percent of which is composed of repetitive DNA, resulting in very low gene density in the genome of this species [Figure 17–8(d)].

The *C. elegans* Genome Has a Surprising Organization

Although eukaryotic genomes share many features, specific genomic features vary widely. The genome of the nematode *C. elegans* contains 97 Mb of DNA organized into 6 chromosomes, with about 20,000 genes, 19,099 of which are protein-coding genes. The gene density is much lower than in yeast—three times more genes and eight times more DNA, with an average density of about 1 gene per 5 kb. About half the *C. elegans* genome is composed of intergenic spacer DNA, and a significant fraction of the genome is highly repetitive simple-sequence DNA (sequences such as ATAT repeated tens of thousands of times).

The organization of *C. elegans* genes differs dramatically from most eukaryotes. Approximately 25 percent of the genes in *C. elegans* genes are organized into polycistronic transcription units, or operons, like those in bacteria (Figure 17–9). In addition, compared with yeast, introns are far more prevalent in *C. elegans* genes; the average gene has five introns. In fact, 26 percent of the *C. elegans* genome is introns; in addition, many genes are found within introns of other genes (Figure 17–10).

Genomes of Higher Plants Have Clusters of Genes

To geneticists, the small flowering plant *Arabidopsis thaliana* is the *Drosophila* of the plant world. Its very small genome, 120 Mb distributed in five chromosomes, is comparable in gene number and size to those of *C. elegans* and *Drosophila*. It contains an estimated 20,000 genes with a gene density of 1 gene per 5 kb, which also resembles *C. elegans* and *Drosophila*. In fact, at least half its genes are closely related to genes found in bacteria and humans. Because of its compact genome, *Arabidopsis* is a model organism for studying other plants with larger genomes, such as the economically important grasses, rice, maize, and barley.

The sequencing of the *Arabidopsis* genome has been completed. Analysis of the nucleotide sequence of chromosomes 2 and 4 reveals a dynamic genome. Both chromosomes contain many tandem gene duplications (239 duplications

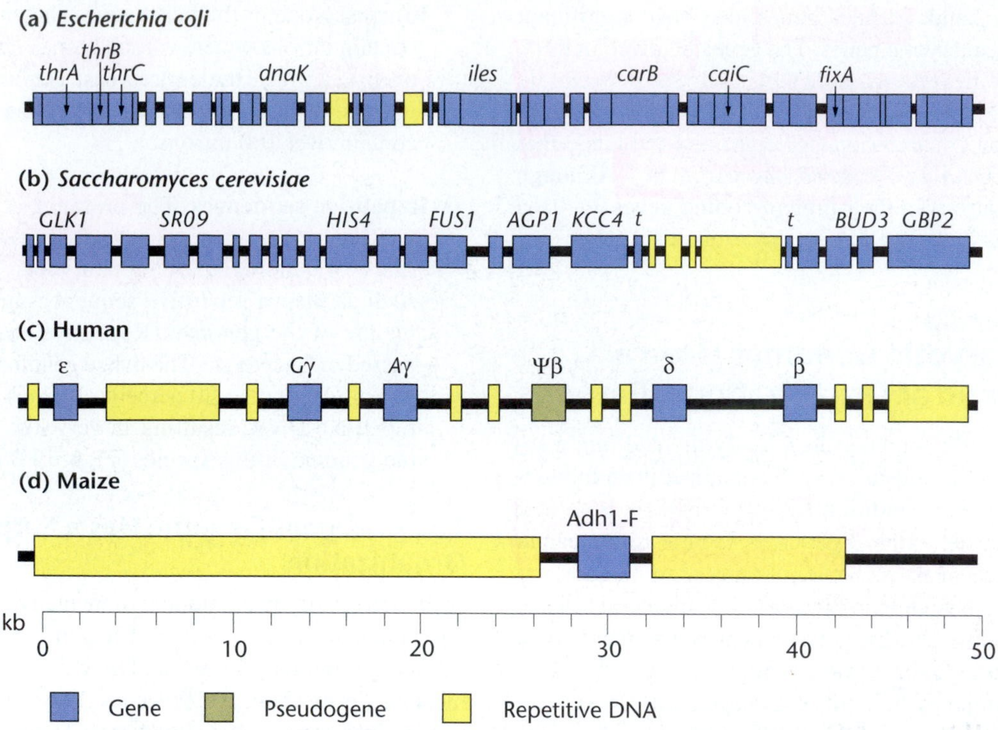

FIGURE 17–8 Gene density in four organisms.(a) In *E. coli*, gene density is high, and there are very few repetitive sequences. In eukaryotes (b-d), gene density is lower, and portions of the genome are occupied by repetitive DNA sequences. (b) A-50 kb region from chromosome III of yeast contains over 20 genes and little repetitive DNA. (c) A 50-kb region from human chromosome 11 contains 6 genes and stretches of repetitive DNA. (d) 50 kb of the maize genome surrounding the *Adh* locus. This gene is surrounded by long stretches of repetitive DNA. [*Parts (a) and (b) are from a public database at* http://www.ncbi.nlm.nih.gov/Entrez. Part (c) *is* redrawn from *Margot. J.B., et al., 1989. Complete nucleotide sequence of the rabbit β-like globin gene cluster. Analysis of intergenic sequences and comparison with the human β-like gene cluster.* J. Mol. Biol. 205: 15–40, *Figure 8. p. 37. Part (d) is redrawn from SanMiguel, P.. et al.. 1996. Nested retrotransposons in the intergenic region of the maize genome.* Science 274:756–68 *Figure 1 and Figure 3. pp. 756, 766.*]

on chromosome 2, involving 539 genes) as well as larger duplications involving four blocks of DNA sequences, spanning 2.5 Mb. There is some interchromosomal gene duplication as well, with genes on chromosome 4 also present on chromosome 5. Genomic and gene duplications have played a large role in evolution that will be discussed in a later section.

Other plants such as maize and barley have genomes more than an order of magnitude larger than *Arabidopsis*, but have about the same number of genes (Table 17–5). These large-genome plants have genes clustered in stretches of DNA that are separated by long stretches of intergenic DNA (Figure 17–11). Collectively, the gene clusters are known as the gene space; they occupy only 12–24 percent of the genome. In maize, the intergenic DNA is composed mainly of transposons. (See Chapter 7.) Because plants closely related to *Arabidopsis* have large genomes, the small genome of *Arabidopsis* is thought to result from a genomic contraction,

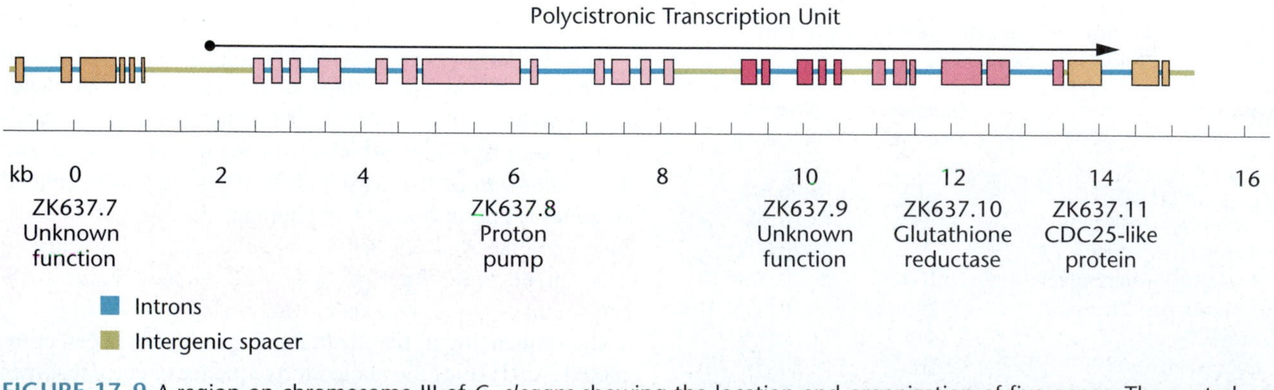

FIGURE 17–9 A region on chromosome III of *C. elegans* showing the location and organization of five genes. The central genes, *ZK637.8*, *ZK637.9*, and *ZK637.10* are part of an operon and are transcribed as a set to form a polycistronic mRNA. Operons are common in bacterial genomes, but rare in most eukaryotic genomes. In *C. elegans*, however, about 25 percent of all genes are part of operons.

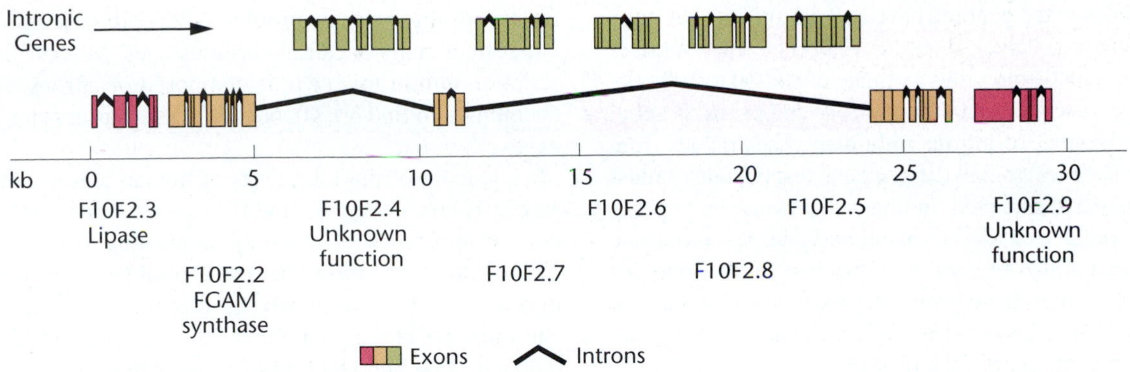

FIGURE 17–10 A 30-kb section of the *C. elegans* genome showing the location and organization of eight genes. One gene, *F10F2.3*, which encodes a lipase, spans just over 25 kb and contains two introns. Five other genes (*FIOF2.4. F10F2. 7, F10F2.6. F10F2.8*, and *FIOF2.5*) are located in the introns of the lipase gene.

in which almost all of the gene-empty regions disappeared, along with most of the transposon sequences. In addition, there may have been a reduction in the size and number of introns, accounting for the compact genome size in this plant.

Organization of the Human Genome: The Human Genome Project

In June 2000, the public and private genome projects announced jointly that they had completed a draft of the human genome sequence, and in February 2001, they each published an analysis, covering about 96 percent of the euchromatic region of the genome. Much work remains to be done in completing the sequence, filling in gaps, and in the analysis of the vast amount of data gathered in this project. The human genome is about 25 times larger than the genome of any organism sequenced to date, and is the first vertebrate genome to be completed. In this section, we will present a summary of the major features of the genome analysis completed to date.

Although the human genome contains over 3 billion nucleotides, the DNA sequences that encode proteins make up only about 5 percent of the genome. At least 50 percent of the genome is derived from transposable elements, such as LINE and Alu sequences. Genes are distributed over 24 chromosomes with clusters of gene-rich regions separated by gene-poor regions, often referred to as gene deserts. Gene-poor regions correlate with the G bands seen in karyotypes. Chromosome 19 has the highest gene density, and chromosomes 13 and the Y chromosome have the lowest gene density. The chromosomal organization of genes in the human genome is discussed in the next section.

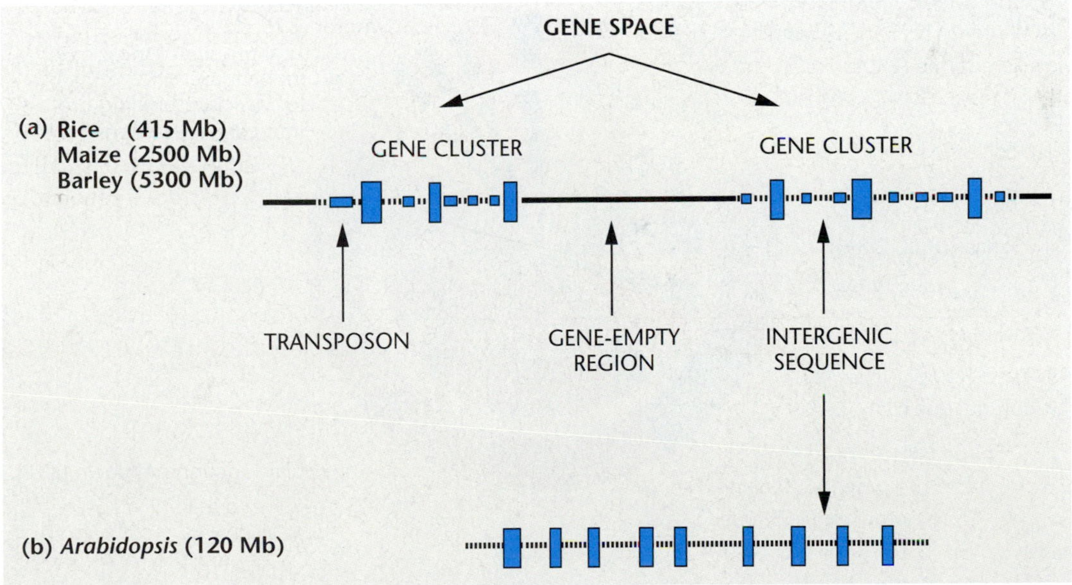

FIGURE 17–11 Genome organization in (a) several large-genome plants and (b) the compact genome of *Arabidopsis*. In (a), genes (vertical boxes) are located in clusters, separated by long, gene-empty spaces of repetitive DNA sequences. Within the gene clusters, the intergenic spaces contain many transposons (small horizontal boxes). In the *Arabidopsis* genome (b), gene-empty regions have been lost, and transposable elements have been lost or reduced. The result is a much smaller genome with genes at a much higher density throughout the genome. (*Barakat. et al. 1998*. Proc. Nat. Acad. Sci. USA 95:10044–10049. *Figure 4. p. 10048*.)

Not all genes in the genome have been identified, but it appears that humans have 30,000 to 40,000 genes, a number much lower than the previous estimate of 80,000 to 100,000. The average gene size, including introns and exons, is about 27 kb. The number of introns in human genes ranges from zero (histone genes) to 234 (titin, a gene that encodes a muscle protein). An unexpected finding is that hundreds of genes have been transferred directly from bacteria into vertebrate genomes, including our own, by a mechanism that remains unknown. Functions have been assigned to about 60 percent of the identified genes (Figure 17–12), and annotation of the rest is an on-going part of the project.

The Chromosomal Organization of Human Genes Shows Diversity

Although sequencing is not complete, the human genome appears to be a typical eukaryotic genome. About 3000 Mb in length, it is organized into 24 chromosomes, ranging from about 55 Mb to 250 Mb in size. It was originally estimated that the human genome might contain between 80,000 and 100,000 genes. More recently, results (discussed next) from chromosome sequencing indicate that there may only be between 35,000 and 40,000 genes in the genome. Human genes tend to be larger and to contain more and larger introns than the genes and introns in invertebrate genomes such as *Drosophila*. The largest human gene known so far is the gene-encoding dystrophin. This gene, associated in mutant form with muscular dystrophy, is 2.5 Mb in length and is larger than many bacterial chromosomes. Most of the transcription unit in this gene is composed of introns. It is not uncommon to find 30, 40, or even 50 introns in some human genes.

Sequencing of the long arms of human chromosomes 21 and 22 is now complete, and the genetic landscape of these two chromosomes reveals some interesting features (Figure 17–13). One of the most interesting findings is the difference in gene density on these two chromosomes. The long arm of chromosome 21 is about 33.65 Mb, and contains 225 genes (about 1 gene per 150 kb of DNA), while the long arm of chromosome 22 is about 34.65 Mb, and has 541 genes (about 1 gene per 64 kb of DNA).

Closer examination of gene distribution reveals that genes on these chromosomes are not evenly spaced. The proximal half of the long arm of chromosome 21 (corresponding to the large G band) is gene poor and averages 1 gene per 304 kb of DNA (Figure 17–14). The distal (telomeric) half of the long arm has a much higher gene density, with a gene every 95 kb of DNA. In addition, two regions of chromosome 21 can be regarded as almost empty of genes. One region spanning 7 Mb contains only one gene, and another three regions of 1 Mb each, contain no genes. Together, these gene-poor regions add up to 10 Mb, or about one-third the length of the long arm. Chromosome 22 has a 2.5-Mb region near the telomere and two smaller regions (about 1 Mb each) located elsewhere that lack genes.

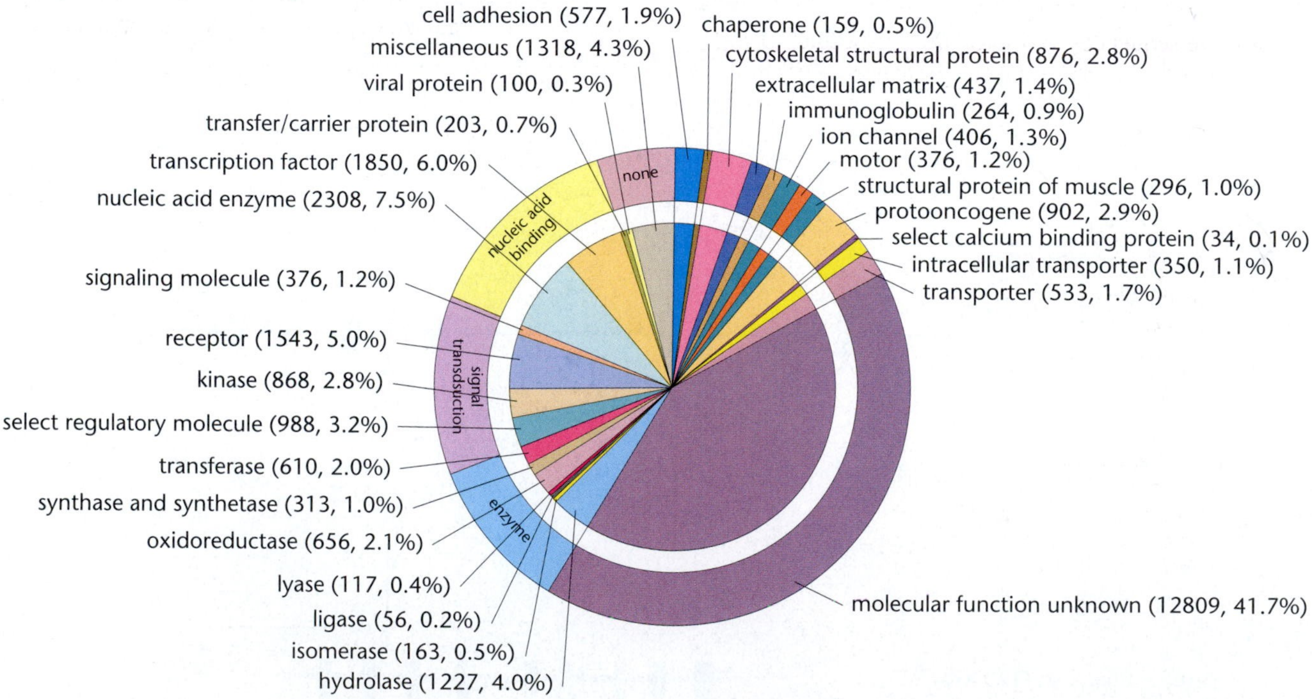

FIGURE 17–12 A preliminary list of assigned functions for 26,588 genes in the human genome. These are based on similarity to proteins of known function. Among the most common genes are those involved in nucleic acid metabolism (7.5 percent of all genes identified), receptors (5 percent), protein kinases (2.8 percent) and cytoskeletal structural proteins (2.8 percent) A total of 12,809 predicted proteins (41 percent) have unknown functions, reflecting the work needed to fully decipher our genome. (*Venter, et al. 2001. The sequence of the human genome.* Science 291:1304–1351. *Figure 15, p. 1335.*)

FIGURE 17–13 Chromosomes 21 and 22 are the smallest chromosomes in the human genome. The regions already sequenced are shown in red adjacent to each chromosome. Some of the disease genes identified on each chromosome are shown below the chromosome.

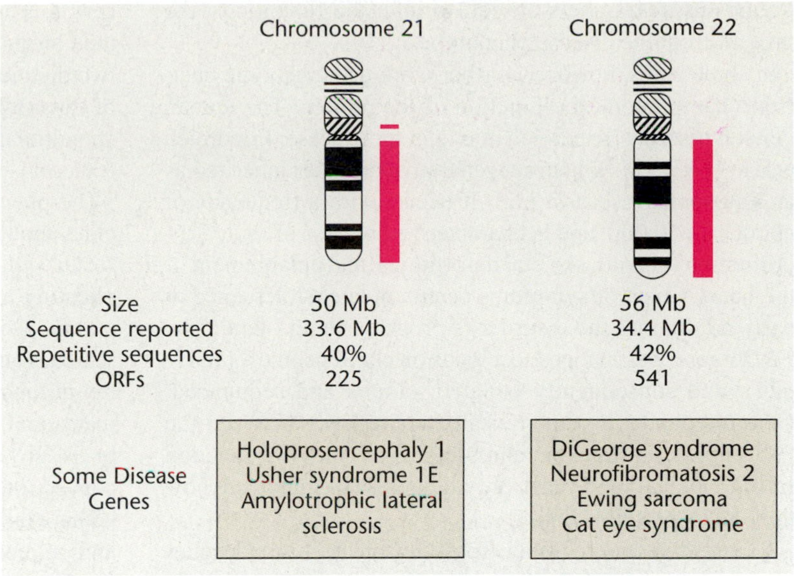
Chromosome 21
Chromosome 22
Size
Sequence reported
Repetitive sequences
ORFs
50 Mb
33.6 Mb
40%
225
56 Mb
34.4 Mb
42%
541
Some Disease Genes
Holoprosencephaly 1
Usher syndrome 1E
Amylotrophic lateral sclerosis
DiGeorge syndrome
Neurofibromatosis 2
Ewing sarcoma
Cat eye syndrome

Chromosomes 21 and 22 each contain duplicated regions. On chromosome 21, a 220-kb region is duplicated near both ends of the long arm, and another 10-kb region is duplicated in the region around the centromere. On chromosome 22, a 60-kb segment is duplicated. These duplicated regions are separated only by a 12-Mb segment. A study of chromosome breakpoints associated with translocations and deletions on chromosome 21 indicates that breakpoints cluster within, and near, duplicated regions, suggesting that these regions may mediate events involved in chromosomal rearrangement. Future analysis of the duplicated regions and their associated repetitive sequences may provide a molecular explanation for the events in chromosome breakage.

17.5 Genomics Provides Insight into Human Genetic Disorders

Once genomic sequence data has been compiled and annotated, it can be used to identify and characterize genes that control cellular functions. For humans, much interest centers on genes related to human genetic disorders. Unfortunately, even when a disease gene has been mapped, cloned and sequenced, several important questions may remain unanswered: what does the protein do, and how does mutation alter or abolish its function? If searches of genome databases show no similarity between the human protein encoded at a disease locus and known proteins from other

FIGURE 17–14 Gene density in the long arms of human chromosomes 21 and 22 (genes per Mb of DNA). In chromosome 21, the region in the dark staining band is relatively gene poor, and the region near the distal tip is gene rich. On chromosome 22, there is a small region near the chromosome tip that is gene poor and two other regions, closer to the centromere that are relatively gene poor. (*Redrawn from Saccone, S. et al., 2001. Genes, isochores and bands in human chromosomes 21 and 22. Chromosome Res. 9:533–539, Figure 1, p. 536.*)

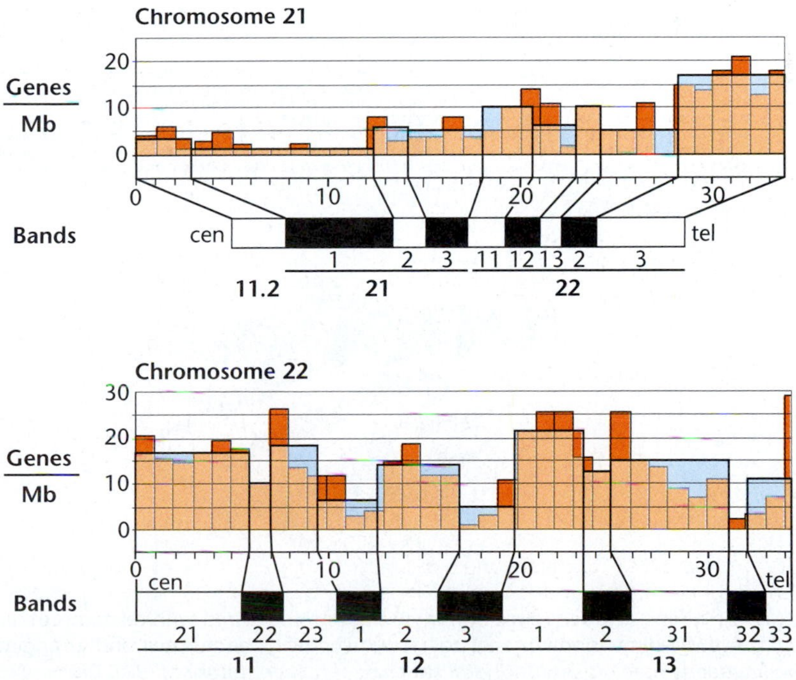

organisms, researchers are left to infer the function of the protein from the disease phenotype.

In some cases, however, further work using genome databases can pinpoint the function of the protein. The human genetic disorder Friedreich ataxia is such a case. Friedreich ataxia (FRDA) is a neurodegenerative disorder inherited as an autosomal recessive trait. It occurs with a frequency of about 1 in 50,000, and is characterized by loss of motor coordination (ataxia), skeletal deformities, and enlargement of the heart. Onset of symptoms occurs in late adolescence or early adulthood; the disorder is progressive and fatal. The *FRDA* gene was mapped to a locus on chromosome 9 (9q13-q21), and subsequently isolated, cloned and sequenced. Genome database scans revealed a similarity between the FRDA protein, frataxin, and proteins of unknown function from *C. elegans* and yeast, providing little insight about how the FRDA protein works.

To carry the search further, Toby Gibson and his colleagues explored genome databases using sequence comparison software with a high sensitivity for certain protein domains. They found a partial match between FRDA and the CyaY protein from *E. coli* and two other bacterial species. Reciprocal searches, beginning with the CyaY amino acid sequence also showed a degree of matches with the FRDA protein.

A phylogenetic analysis of CyaY revealed that the gene is present in eight bacterial species, all belonging to a subdivision of purple bacteria. Because purple bacteria are the closest living relatives to the prokaryotic ancestor to mitochondria (see Chapter 26 for a detailed discussion of mitochondrial origins), Gibson hypothesized that the FRDA gene evolved from a CyaY gene present in a prokaryote mitochondrial ancestor. Alignment of frataxin with CyaY shows alignment at the C-terminus of frataxin, with most of the conservation in two distinct domains. The 55 amino acids at the N-terminus of frataxin not present in CyaY, are compatible with a mitochondrial targeting peptide, strengthening the hypothesis that frataxin is a nuclear-encoded mitochondrial protein.

The phenotype of affected individuals did provide a few clues about FRDA function. After exercise, the muscles of affected individuals show decreased production of ATP, implicating a mitochondrial defect as the cause of Friedreich ataxia. Using the clues provided by the genome database searches and the phenotype, other research groups confirmed the mitochondrial location of frataxin, solved its three-dimensional structure (Figure 17.15), and showed that frataxin is an important regulator of mitochondrial energy conversion.

This example illustrates the power of comparative genomics and sequence analysis to explain the molecular basis of genetic disorders. It also shows that database analysis can be as important as laboratory work in explaining biological phenomena.

17.6 Genomics Helps Reconstruct Events in Genome Evolution

The fossil record indicates that about 3.5 billion years ago, cells similar to bacteria were present on our planet. We can only assume that the genomes of these organisms, or organisms that appeared shortly thereafter, were composed of a double-stranded DNA molecule. The first eukaryotic

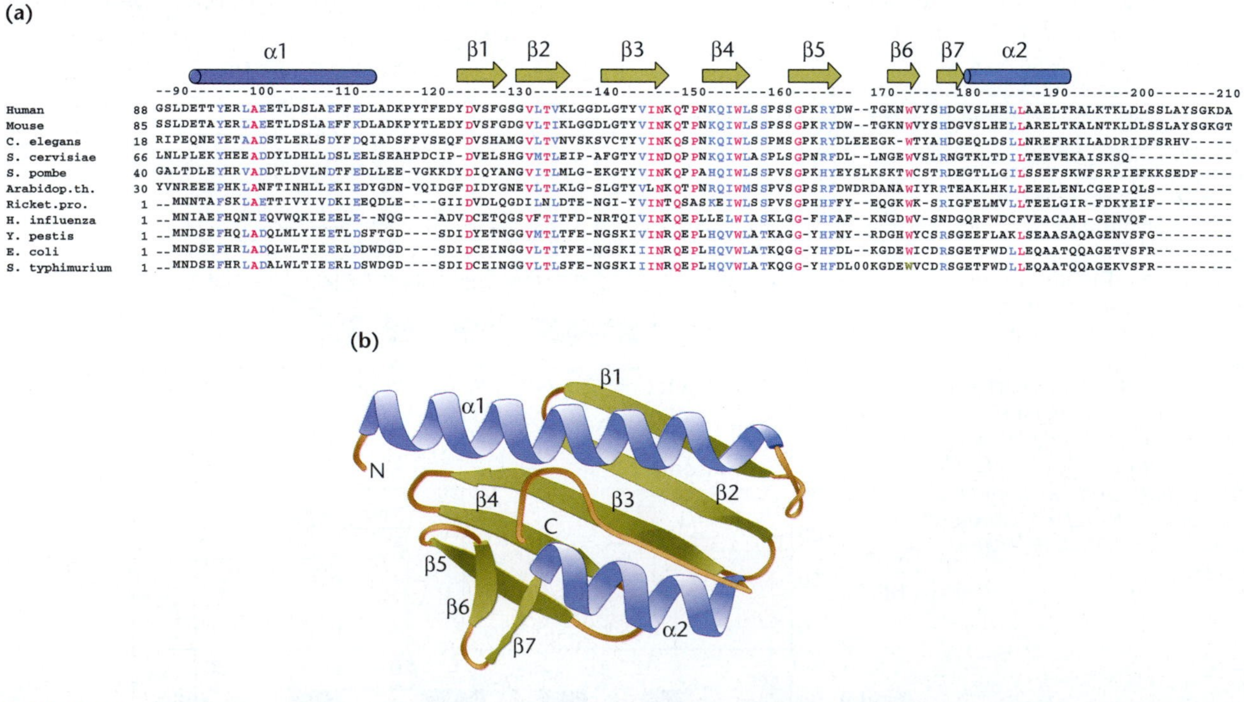

FIGURE 17–15 (a) The amino acid sequence of human frataxin compared with related proteins from eukaryotes and prokaryotes, including *E. coli.* (b) A three-dimensional model of frataxin showing the number and arrangement of the helical and pleated domains. From: *Dhe-Paganon, S., et al., 2000. Crystal structure of human frataxin. J. Biol. Chem. 275: 30753–56; Figures 1–2, p. 30755, with permission.*

fossils (resembling single-celled algae) date from about 1.4 billion years ago. As both prokaryotes and eukaryotes underwent changes in size, shape, and complexity, their genomes also changed dynamically. These changes were driven by genetic mechanisms that include mutation, recombination, transposition, gene transfer, as well as gene deletion and duplication. By examining and comparing genomes of organisms that exist today, we gain insights into how these mechanisms have shaped genomes.

What Is the Minimum Genome for Living Cells?

What is the minimum number of genes necessary to support life? We cannot fully answer this question yet, because we don't know the functions of each gene in the genomes sequenced to date. However, using genes with known functions, we can speculate on the minimum number of genes required to maintain life. To do this, we can compare sequence information from two of the smallest bacterial genomes, *Mycoplasma genitalium* and *M. pneumoniae*. These two closely-related organisms are among the simplest self-replicating prokaryotes known. *M. genitalium* has a genome of 0.6 Mb, while that of *M. pneumoniae* is 0.8 Mb. They are members of a group of bacteria that lack a cell wall and invade, often causing disease in a wide range of hosts, including insects, plants, and humans (genital and respiratory infections).

The *M. genitalium* genome has 480 protein-coding genes. The genome of its close relative, *M. pneumoniae* includes those 480 genes, and an additional 197 genes, for a total of 677 protein-coding genes. In contrast, *E. coli* has a 4.6-Mb genome with 4288 ORFs, and *H. influenzae* has a 1.8-Mb genome with 1727 ORFs. Table 17–6 summarizes the functions of some genes in these bacteria. Obviously, cells must contain genes that encode products required for DNA replication and repair, transcription and translation, transport proteins; general cellular processes, including cell division and secretion; and for the myriad biochemical pathways involving metabolism.

It is interesting to compare the genes required for a given function among species. For example, *E. coli* has 131 genes for amino acid metabolism, *H. influenzae* has 68, and *M. genitalium* has only 1. Despite having nearly four times as many genes, the physiological capabilities of *Haemophilus* are very similar to *Mycoplasma*. The greatest difference between *Haemophilus* and *Mycoplasma* involves a marked increase in biosynthetic capability. *Haemophilus*, which is a more complex bacterium, has 68 genes involved in amino acid biosynthesis, whereas *Mycoplasma* has only one such gene. Without this capability, *Mycoplasma* must rely on numerous metabolic products from its host.

These comparisons combined with experimental results enable us to speculate on the minimal number of genes necessary for organisms to exist as independent, self-reproducing organisms. Using transposon mutagenesis to disrupt genes in *M. genitalium* and *M. pneumoniae* and testing for viability, researchers have estimated that living organisms require a minimum of 250–350 genes.

Genome Reduction during Evolution in Disease-Causing Bacteria

Another way of gaining insight into which genes are most essential for life comes from comparing the genomes of two other closely related species of *Mycoplasma*: *M. tuberculosis*, which causes tuberculosis, and *M. leprae*, which causes leprosy. *M. tuberculosis* has 3959 protein-coding genes in a 4.41 Mb genome; *M leprae* has a 3.26 Mb genome with 1604 protein-coding genes. Over evolutionary time, *M. leprae* has lost over 2,000 genes, or more than 50 percent of its ancestral gene set. Deletions and decay, caused by mutation and insertion of transposable elements have inactivated many genes (Figure 17–16). As a result, metabolic functions important for cell growth have been eliminated, and the bacterium is very slow growing. (It divides every 14 days, while *E. coli* divides every 20 minutes.) The deletion of an operon for NADH-dependent energy transfer has greatly impaired energy metabolism and may explain the organism's slow growth. In

TABLE 17–6 Functional Classes of Genes in Three Bacterial Species

Functional Class	*E. coli*	*H. influenzae*	*M. genitalium*
Protein-coding genes	4,288	1,727	470
DNA replication, repair	115	87	32
Transcription	55	27	12
Translation	182	141	101
Regulatory proteins	178	64	7
Amino acid biosynthesis	131	68	1
Nucleic acid biosynthesis	58	53	19
Lipid metabolism	48	25	6
Energy metabolism	243	112	31
Uptake, transport proteins	427	123	34

Source: Blattner, F., et al., 1997. The complete genome sequence of *Escherichia coli* K–12. *Science 277*: 1453–1462, Table 4, p. 1458.
Source: Fraser, C. M., et al., 1995. The minimal gene complement of *Mycoplasma genitalium*. Science 270: 397–403. Table 2, p. 400.

dnaA	*dnaB*	*dnaE1*	*dnaE2*	*dnaG*	*dnaN*	*dnaQ*	*dnaZX*
recA	*recB*	*recC*	*recD*	*recF*	*recG*	*recN*	*recR*
recX	*dinF*	*dinG*	*dinP*	*dinX*	*helY*	*helZ*	*polA*
uvrA	*uvrB*	*uvrC*	*uvrD*	*uvrD2*	*gyrA*	*gyrB*	*priA*
ruvA	*ruvB*	*ruvC*	*xthA*	*alkA*	*fpg*	*mfd*	*ung*
mutT1	*mutT2*	*mutT3*	*mutY*	*ligA*	*ligB*	*ligC*	*nei*
radA	*ogt*	*topA*	*nth*	*ssb*	*tagA*	*mrr*	*ihr*

FIGURE 17–16 Genes involved in DNA replication, repair and recombination in *M. tuberculosis* and *M. leprae*. Genes deleted or mutated in *M. leprae* are shown as white boxes. As shown, many of the genes needed for DNA repair and recombination have been lost from the genome of *M. leprae*. Maintenance of the genome has been compromised in this organism, and may lead to its extinction. (*From: Young, D., and Robertson, B. 2001. Genomics: Leprosy-a degenerative disease of the genome.* Curr. Biol. 11: R381-R383, *Figure 2, p. 382.*)

addition, mutations in genes for recombination and DNA repair are unable to prevent further damage to the *M. leprae* genome. This combination of mutations may have doomed *M. leprae* to extinction, as epidemiologists have estimated that the organism is at the edge of being able to sustain itself by infecting new individuals.

Duplications Are Important in the Origin and Evolution of the Eukaryotic Genome

Eukaryotes are traditionally distinguished from prokaryotes by several distinctive features, including a membrane-bound nucleus, cytoplasmic membrane systems such as the endoplasmic reticulum, and a cytoskeleton. As genome projects provide more information, comparisons between prokaryotic and eukaryotic genomes are providing clues to the origins of the eukaryotic genome and how eukaryotes arose from prokaryotes. Several lines of evidence, including amino acid analysis of proteins, gene sequences, and metabolic pathways, indicate that the eukaryotic genome is actually a mosaic that has received major contributions from both the Archaea and the Eubacteria. For example, the eukaryotic nuclear genome has few, if any, operons, and its genes contain introns, both of which are features of the Archaea.

The eukaryotic mitochondrial genome strongly resembles that of alpha proteobacteria. To explain these observations, it has been proposed that eukaryotes arose as a consequence of a symbiotic association, or fusion, between an anaerobic archaebacterial host and an alpha proteobacterium (like Rickettsia), which evolved into the mitochondrion. The monophyletic nature of all eukaryotes suggests that this event occurred successfully only once in the history of the earth. This topic is discussed in detail in Chapter 26.

Genome Duplications Occurred Throughout Eukaryote Evolution

The events just outlined may account for the origin of the eukaryotic genome, but not for the size and complexity that distinguish eukaryote genomes from those of prokaryotes. As discussed earlier in this chapter, gene duplication played an important role in the evolution of eukaryotic genomes. Most attention has focused on duplications of single genes, but Susumo Ohno has proposed that whole genome duplications are an important evolutionary mechanism. Analysis of nucleotide sequence data from genome projects supports the idea that much of the difference in gene number that separates prokaryotes from eukaryotes resulted from genome expansions. We now recognize that a major expansion in eukaryote genomic size resulted from a genome duplication event that accompanied the appearance of vertebrates in the fossil record. However, genome duplications have occurred at other times throughout eukaryotic evolution.

A recent analysis of the yeast genome shows traces of an ancient expansion by genomic duplication. Kenneth Wolfe and Denis Shields searched through the yeast genome and compared the sequence of each yeast gene with every other yeast gene. They uncovered 55 duplicated regions containing 376 genes, which covers 50 percent of the genome. Of the 55 regions, 50 are in the same relative position on different chromosomes (Figure 17–17). For example, chromosomes XI and XIII contain a duplicated block of genes (block 43) in which gene order and orientation with respect to the centromere has been conserved. Other chromosomes contain internal duplications, such as block 53, located on either side of the centromere on chromosome XII. Phylogenetic analysis (see Chapter 26) indicates that this genome duplication event took place about 100 million years ago.

Analysis of the human genome also shows evidence of an ancient large-scale duplication, followed by rearrangements and gene loss in some duplicated regions. These duplications range from small segments covering only a few genes up to large stretches that cover almost an entire chromosome. The larger duplications date to the origin of the vertebrates, about 500 million years ago. Altogether, there are 1077 blocks of duplicated regions in the human genome, containing just over 10,000 genes (about one-third of the genome). One such duplication between chromosomes 18 and 20 is shown in Figure 17–18.

Gene Duplications Increase Genetic Diversity

The almost uninterrupted flow of nucleotide sequence data from genome projects is providing evidence that multigene families are present in many, if not all, genomes. In addition to genomewide duplications, small blocks of genes and single genes can be duplicated by several mechanisms, including

- **unequal crossing over**—a recombination event between members of a homologous pair of chromosomes in which a DNA segment is duplicated in one of the recombination products.

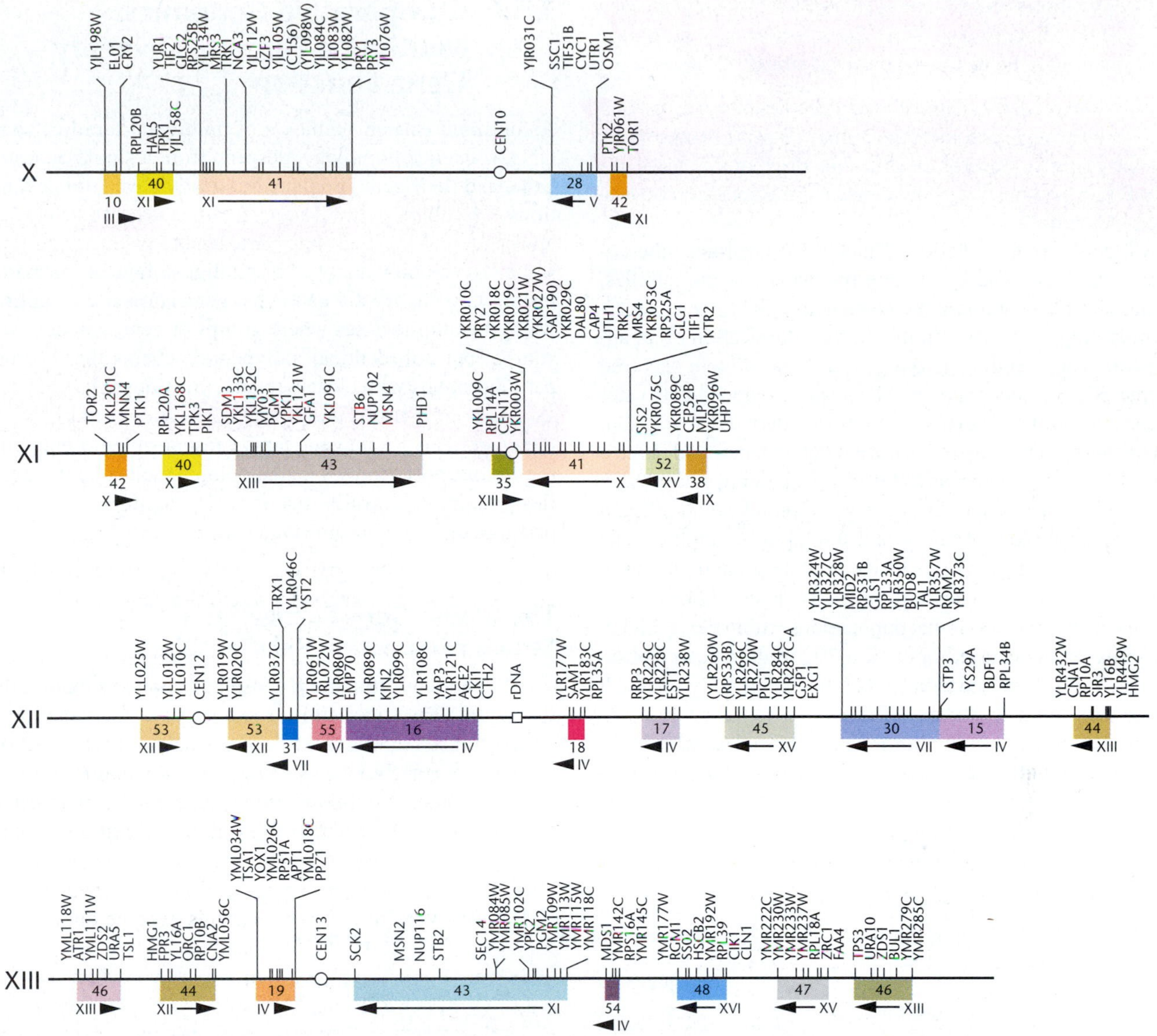

FIGURE 17–17 Location of duplicated regions on yeast chromosomes X, XI, XII, and XIII. The duplicated blocks are numbered, and roman numerals below the blocks give the location of the duplicate block. Arrows show the relative orientation of the blocks. Genome wide orientation relative to the centromere is conserved, except for five blocks. (*Wolfe, and Shields. 1997. Molecular evidence for an ancient duplication of the entire yeast genome* Nature 387:708–713. *Figure 2. p. 710.*)

FIGURE 17–18 A segmental duplication of genes on human chromosomes l8 and 20. This duplication involves a total of 64 genes. For clarity, only 12 duplicated pairs are shown here. In the genome, there are 1077 duplicated blocks of genes, containing a total of 10,310 genes. (*Venter, et al. 2001. The sequence of the human genome.* Science 291:1304–1351, *Figure 13, p. 1332.*)

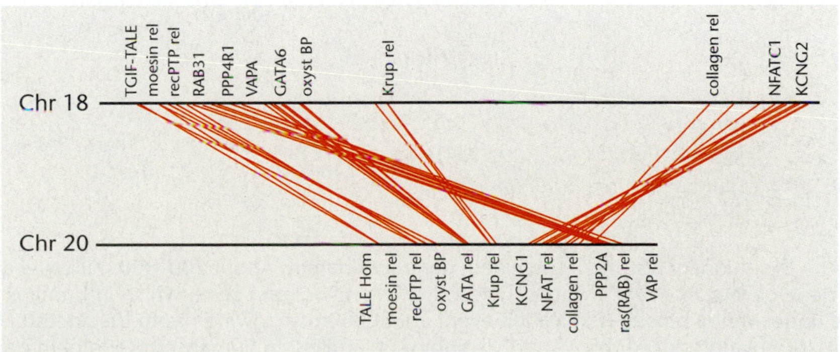

- **replication errors**—during replication of a template molecule, slippage can cause the insertion of a short segment into the newly synthesized strand.

Once generated, members of multigene families may remain linked on a single chromosome or may disperse to other parts of the genome. Several mechanisms drive this process, including inversions, translocations, and transposition by mobile elements.

Molecular phylogenetics (Chapter 26) has traced the ancestry and relationships among members of gene families. One of the best studied examples is the globin gene superfamily (Figure 17–19). In this family, duplication of an ancestral gene encoding an oxygen transport protein occurred some 800 million years ago. This split produced two sister genes, one of which evolved into the modern day myoglobin gene. Myoglobin is an oxygen-carrying protein found in muscle. The other gene became the ancestral globin gene. About 500 million years ago, the ancestral globin gene duplicated to form prototypes of the α- and β-globin subfamilies. The alpha- and beta-globin genes encode the proteins found in hemoglobin, the oxygen-carrying molecule in red blood cells. (See Chapter 6.) Additional duplications within the α- and β -globin genes occurred in the last 200 million years Subsequent events dispersed members of this superfamily, and each is now on separate chromosomes.

Similar patterns of evolution are observed in other gene families, including the trypsin–chymotrypsin family of proteases, the homeotic selector genes of animals, and the rhodopsin family of visual pigments.

17.7 Comparative Genomics: Multigene Families Diversify Gene Function

As outlined earlier, members of multigene families share DNA-sequence homology, descend from a single ancestral gene, and their gene products frequently have similar functions. Members of multigene families are often, but not always, found together in a single location along a chromosome. To see how analysis of multigene families provides insight into eukaryotic genome organization and evolution, we first examine cases where groups of genes encode very similar, but not identical, polypeptide chains that become part of proteins with closely related functions. Multiple proteins that arise from single-gene duplications are known as **paralogs**. The globin gene family responsible for encoding the various polypeptides in hemoglobin molecules exemplifies a multigene family that arose by duplication and dispersal to different chromosomal sites.

The Globin Gene Family Has a Structure/Function Relationship

The human **α-globin** and **β-globin** genes are subfamilies of the globin gene superfamily and are two of the most intensively studied regions of the human genome. There is a cluster of three α-globin genes on the short arm of chromosome 16, and another with five β-globin genes on the short arm of chromosome 11. Members of both subfamilies share

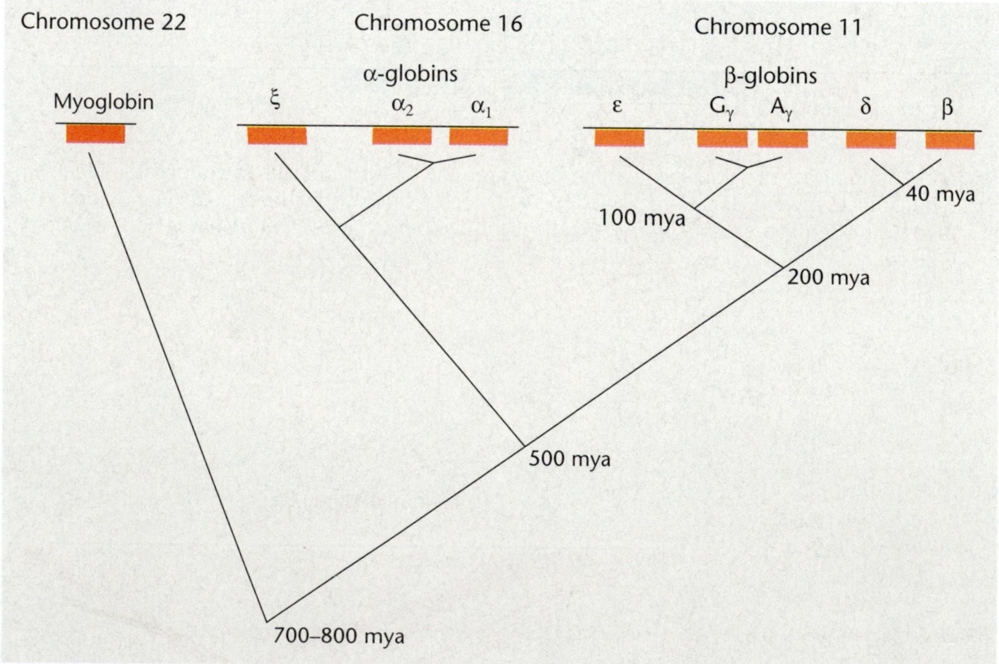

FIGURE 17–19 Evolutionary history of the globin gene superfamily. About 700–800 million years ago (mya), a duplication event in an ancestral gene gave rise to two lineages. One led to the myoglobin gene, which in humans is located on chromosome 22. The other lineage underwent a second duplication event about 500 mya, giving rise to the ancestors of the α and β subfamilies. Duplications about 200 mya produced the α- and β-globin subfamilies. In humans, the α-globin genes are located on chromosome 16 and the β-globin genes are on chromosome 11.

nucleotide sequence similarity, but members of the same subfamily have the greatest amount of sequence similarity.

Hemoglobin is a tetramer, containing two α- and two β-polypeptides. Each polypeptide incorporates a heme group that reversibly binds oxygen. Within each subfamily, genes are coordinately turned on and off during embryonic, fetal, and adult stages of development. For both the alpha and beta subfamilies, this expression occurs in the same order in which the genes are arranged on the chromosome.

The alpha subfamily [Figure 17–20(a)], spans more than 30 kb and contains three genes: the ζ (zeta) gene, expressed only in the early embryonic stage, and two copies of the α gene, expressed during the fetal (α_1) and adult stages (α_2). In addition, two nonfunctional **pseudogenes** ($\psi\zeta$ and $\psi\alpha_1$) are present in the cluster. Pseudogenes are designated by the prefix ψ (psi), followed by the symbol of the gene they most resemble. Thus, the designation $\psi\alpha_1$ indicates a pseudogene of the adult α_1 gene. Pseudogenes are nonfunctional versions of genes that resemble other gene sequences, but contain significant nucleotide substitutions, deletions, and duplications that prevent their expression.

The organization of the alpha subfamily members and the location of their introns and exons reveal several interesting features. First, as is common in eukaryotes, the DNA encoding the three functional α genes occupies only a small portion of the region containing the subfamily. Most of the DNA in this region is intergenic spacer. Second, each functional gene in this subfamily contains two introns at precisely the same positions. Third, the nucleotide sequences within corresponding exons are nearly identical in the ζ and α genes. Both genes encode polypeptide chains of 141 amino acids. However, their intron sequences are highly divergent, even though they are about the same size. Significantly, much of the nucleotide sequence of each gene is contained in these noncoding introns.

The human β-globin gene cluster is longer than the α-globin cluster and contains five genes spaced over 60 kb of DNA [Figure 17–20(b)]. As with the alpha subfamily, the order of genes on the chromosome parallels their order of expression during development. Of the five genes, three are expressed prior to birth. The ε (epsilon) gene is expressed only during embryogenesis, while the two nearly identical γ (gamma) genes (G_γ and A_γ) are expressed only during fetal development. The polypeptide products of the two γ genes differ only by a single amino acid. The two remaining genes, δ (delta) and β (beta), are expressed following birth. Finally, a single pseudogene $\psi\beta_1$ is present within the subfamily. All five functional genes encode proteins with 146 amino acids and have two similarly sized introns at exactly the same positions. The second intron in the β genes is significantly larger than its counterpart in the functional α genes. These similarities reflect the evolutionary history of each subfamily and the events such as gene duplication, nucleotide substitution, and chromosome translocations that produced the present-day globin superfamily.

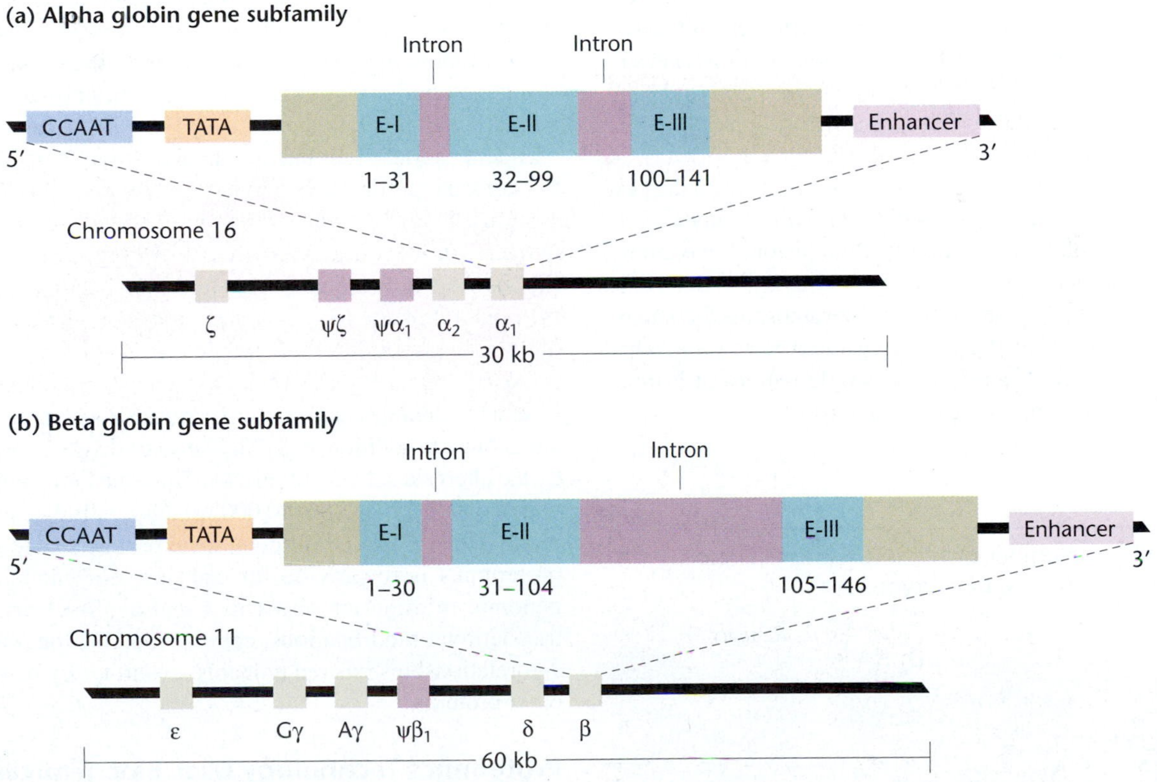

FIGURE 17–20 Organization of the alpha-globin gene subfamily (a) on chromosome 16 and the β-globin gene subfamily on chromosome 11 (b). Also shown is the internal organization of the α_1 gene and the β gene. Each gene contains three exons (E-I, E-II, E-III) and two introns. The numbers below the exons indicate the amino acids in the gene product encoded by each exon.

The Histone Gene Family Is Organized into Clusters

Organization of the histone genes is a variation on the theme established in the globin and immunoglobulin gene families. Here a cluster of five related, but nonidentical genes, separated from each other by highly divergent intergenic spacer regions, is tandemly repeated many times. Recall from Chapter 4 that histones are positively charged (basic) proteins which interact with the negatively charged phosphate groups of DNA to form nucleosomes. In rapidly dividing cells, histone synthesis must keep pace with DNA replication, and the necessary quantity of histones must appear each time, as one cell cycle runs into the next. Multiple copies of the histone cluster allow the coordinate synthesis of DNA and histones during S phase of the cell cycle.

Many interesting correlations support this idea. Yeast cells, with much less DNA than other eukaryotes, have only two clusters of four histone genes. (They lack histone H1.) In contrast, clusters of five histone genes are tandemly repeated from 10 to 800 times in many complex organisms (Table 17–7). Cell division is very rapid during development in sea urchins, and these organisms have one of the largest sets of histone gene clusters.

Beyond its tandem organization, the histone gene family shows several other differences from other families. First, almost all histone genes lack introns. Second, individual genes within clusters of a given species often orient in opposite directions with respect to transcription. As shown in Figure 17–21, the arrangement of genes and their polarity of transcription, shown by the direction of the arrow, vary in the sea urchin and *Drosophila*, two organisms where this cluster has been well studied. Studies also show polarity differences in other organisms. Despite polarity differences, the amino acid sequences of the various histones (reflecting the DNA encoding each one) are remarkably similar in highly divergent organisms. The homology of histone genes is one of the best examples of sequence conservation through evolution.

In mice and humans, members of the histone gene family are clustered, but do not form tandem arrays. A region at the distal tip of the long arm of human chromosome 7 contains all copies of the histone gene family, interspersed with other nonhistone genes. Therefore, no single pattern of histone gene organization applies to all organisms.

TABLE 17–7 Number of Histone Gene Clusters in Selected Eukaryotes

Organism	*Repeats*
Yeast	2
Chickens	10
Mammals	20
Xenopus	40
Drosophila	100
Sea urchin	300–600
Newt	600–800

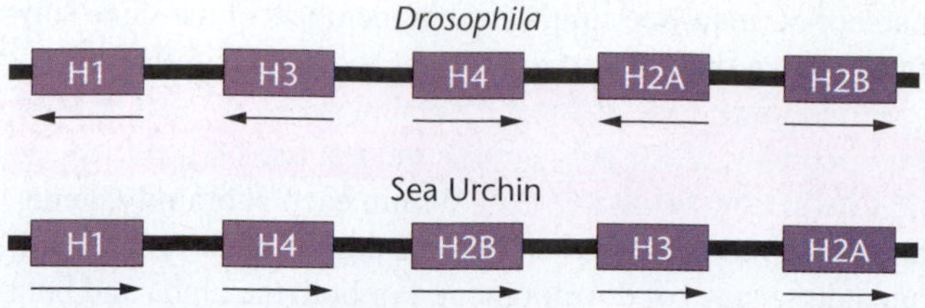

FIGURE 17–21 The histone gene cluster in *Drosophila* and in the sea urchin. Arrows indicate the direction of transcription.

17.8 Proteomics Identifies and Analyzes the Proteins Expressed in a Cell

Proteome is a relatively new term, coined to define the complete set of proteins expressed and modified during a cell's entire lifetime. In a narrower sense, it also describes the set of proteins expressed in a cell at a given time. **Proteomics**, the study of the proteome, uses technologies ranging from genetic analysis to mass spectrometry. As genome projects provide information about the number and kinds of genes present in prokaryotic and eukaryotic genomes, attention is turning to the question of protein function. In all genomes sequenced to date, many newly discovered genes have no known function, and others have only presumed functions assigned by analogy with known genes. For example, in two intensively studied organisms, *E. coli* and *S. cerevisiae*, more than half the genes encoded by their genomes have no known function. About 41 percent of the predicted proteins identified in the Human Genome Project are of unknown function. Some of these may be derived from retroviruses and not be real genes, but many newly identified genes remain to have a function assigned to them.

Understanding gene function involves more than identifying the gene products. Once made, many gene products are modified by the cleavage of end groups (such as signal sequences, propeptides, or initiator methionine residues) by the addition of chemical groups (methyl-, acetyl-, phosphoryl), or by linkage to sugars and lipids. Proteins are internally and externally cross-linked and even processed by removing internal amino acid sequences (called **inteins**). Over a hundred mechanisms of posttranslational modification are known, in addition to the high level of diversity produced by the alternate splicing of mRNA. Thus, the human genome, which may have 35,000–40,000 protein-coding genes, may produce over 350,000 different gene products. The goal of proteomics is to provide for each protein encoded in a genome, information about its function, structure, posttranslational modifications, cellular localization, variants, and relationships (shared domains, evolutionary history) to other proteins.

Proteomics Technology Uses Biochemical and Physical Methods

Genome projects have a highly developed array of techniques, including vectors, PCR, automated DNA sequencing,

Genetics, Technology, and Society

Footprints of a Killer

For millennia, anthrax has afflicted humans. It is likely that anthrax was one of the deadly Egyptian plagues at the time of Moses, and cases of anthrax have been documented since Roman times. Throughout most of history, human anthrax cases were sporadic and resulted from contact with animal products contaminated with spores of *Bacillus anthracis*—a bacterium that preferentially infects grazing herbivores. Spores can enter the body through cuts in the skin, by inhalation or by ingestion. Although skin anthrax is usually curable, inhalation anthrax and ingestion anthrax are fatal. But it wasn't until the dawning of the age of biological weaponry that anthrax took on its role as a man-made agent of terror.

The hostile use of infectious biological agents such as anthrax reaches back into antiquity. The Romans deliberately contaminated food and water supplies with animal carcasses during the Carthaginian Wars in the 5th century BC. In the middle ages, bodies of plague-infected people and animals were hurled by catapults into cities under siege and were used to contaminate the enemy's drinking wells. The use of infectious biological projectiles continued into the 20th century, including during the South African Boer wars and the Russian Revolution.

Anthrax has been a particularly popular vehicle for psychological and strategic terror during human conflicts. During World War I, Germany infected livestock with *Bacillus anthracis*, with the aim of exporting infected meat to the Allies. During World War II, Japan killed up to 10,000 prisoners by infecting them with numerous biological and chemical agents including anthrax. The Japanese biological warfare program also involved spraying anthrax and other biological agents over 11 Chinese cities. From 1940 to 1942, approximately 700 Chinese people (and 1,700 Japanese soldiers) died in these attacks. The Allies also developed extensive biological weapons programs during this time. Britain manufactured 5 million anthrax-containing cattle cakes, created a 500-lb. anthrax bomb, and conducted anthrax explosives testing on Gruinard Island, off the coast of Scotland. This small island remained contaminated with anthrax spores and uninhabitable for over 4 decades. In 1986, Gruinard was decontaminated with 280 tons of formaldehyde and 2,000 tons of seawater. The United States conducted major biological weapons programs beginning in the early 1940s, and manufactured large quantities of anthrax by 1946, including 5,000 bombs filled with *Bacillus anthracis*. Between 1949 and 1968, US cities were secretly used as experimental sites to test aerosolization and dispersal of biological agents. Bacteria of similar size to *Bacillus anthracis* (such as *S. marcescens* and *B. globi*) were sprayed over San Francisco and into the ventilator shafts of the New York City subway system. These tests showed that small amounts of bacteria could be rapidly and easily dispersed over urban areas.

In 1969, President Nixon terminated the US offensive biological weapons program. Stocks of biological weapons were destroyed between 1971 and 1973. However, small quantities of anthrax and other pathogens were retained at Fort Detrick, in order to develop therapies to biological weapons attacks. In addition, the US Army Medical Research Institute of Infectious Diseases (USAMRIID) was established to conduct research into biological weapons defenses. USAMRIID programs were conducted in collaboration with numerous universities and research institutes.

Although the US has halted its offensive biological weapons program, over 15 countries currently have, or are developing, biological weapons. These include Russia, Israel, Egypt, China, Iran, Iraq, Libya, Syria, India, and North Korea. For example, in the 1980s, the Soviet Union was capable of manufacturing thousands of tons of weaponized anthrax annually, in 20- to 50-ton reactors. In 1979, anthrax spores were accidentally released from a Soviet biological weapons facility in Sverdlovsk. The anthrax cloud dispersed for 3 miles from the site of release, killing at least 70 people. In the 1990s, it was found that Iraq had manufactured 8,000 liters of anthrax spores - enough to kill every human on earth. Less is known about non-government sources of anthrax weapons. The Japanese terrorist group, Aum Shinrikyo, conducted research on anthrax and is thought to have made unsuccessful attempts to use anthrax as a weapon. In addition, the Egyptian Islamic Jihad apparently obtained anthrax from an East Asian country for a small fee. In the US, hundreds of research labs store anthrax in order to study its biology and to develop treatments and vaccines. There are so many biological weapons programs in the world, that thousands of scientists know how to weaponize anthrax spores.

In addition to its widespread use as a biological weapon, and the widespread knowledge of how to turn it into a lethal weapon, *Bacillus anthracis* itself provides a formidable foe. Anthrax spores are easy to transport, conceal and disperse. Anthrax is relatively easy to produce in large quantities, and the spores are extremely stable. One gram of weaponized anthrax spores contains 100 million lethal doses. In human hands, anthrax is a stealthy, efficient killer.

Against this backdrop of worldwide bioterror emerged the post-September 11 anthrax attacks, and the scientific efforts to track the source of the attacks. Between October 2001 and January 2002, a total of 23 anthrax cases were reported to the US Centers for Disease Control. Eleven of these were inhalation anthrax cases and 12 were skin anthrax cases. There were five fatalities, and approximately 32,000 people were given prophylactic antibiotic therapy. Anthrax spores were discovered in post offices, government buildings and media centers in Florida, Washington DC, New Jersey and New York. The anthrax appeared to come from a handful of letters that passed through high-speed mail sorting equipment, contaminating postal workers as well as people in the offices to which the letters were addressed. But who sent the letters, and where did they get the highly milled weapons-grade bacterial spores?

Few physical clues emerged from the anthrax-laced envelopes. Although the handwriting and threats were similar in all letters, investigators found no fingerprints, hair or fibers on the letters. As the envelopes were sealed with tape, there was no saliva with which to perform DNA typing tests to identify the sender.

The anthrax bacterium itself may yield the clearest evidence with which to apprehend the killers. Using a PCR-based technique to identify restriction fragment length polymorphisms in bacterial DNA, scientists discovered that the same strain - the Ames strain - was present in all of the letters. This strain was first isolated in Iowa and has been maintained by the US Army and many US and overseas research labs since 1980. But identifying the specific lab from which the anthrax was obtained may be a more challenging task. Most organisms undergo random genetic variation over generations, and these variations can distinguish isolates grown in one location from isolates grown in another. However, *Bacillus anthracis* is one of the most genetically homogeneous organisms known. Scientists have been unable to distinguish the Ames strain used in the attacks from other Ames strains, using genetic fingerprinting of 15 different DNA fragment markers. At present, scientists are sequencing the genomes of the Ames strain used in the attacks and comparing the sequence to that of other Ames strains maintained in research labs and government facilities. If the source of the anthrax strain used in the post-September 11 attacks can be identified by DNA sequence comparison, it could lead investigators directly down the path to the perpetrators of this latest incident of anthrax bioterrorism.

References

Enserink, M. 2001. Taking anthrax's genetic fingerprints. *Science* 294:1810-1812.

Enserink, M. 2002. TIGR begins assault on the anthrax genome. *Science* 295:1442-1443.

Lesho, E., Dorsey, D. and Bunner, D. 1998. Feces, dead horses, and fleas: evolution of the hostile use of biological agents. *Western Journal of Medicine* 168:512.

and bioinformatics software. Proteomics requires a different range of techniques, many of which are still in various stages of development. The basic techniques in proteomics involve separating and identifying proteins isolated from cells. One way to do it is by two-dimensional gel electrophoresis (Figure 17–22). In this technique, proteins extracted from a cell are placed on a polyacrylamide gel and an electric charge is applied across the gel, separating the proteins according to their molecular weight. When completed, the gel is rotated 90 degrees, and during another round of electrophoresis, the proteins are separated in a second dimension according to their molecular mass. When the gels are stained, proteins are revealed as spots; typical gels show 200–10,000 spots (Figure 17–22). To identify individual proteins, spots are cut from the gel, and digested with enzymes such as trypsin, to produce a characteristic set of fragments. The fragments are analyzed by mass spectrometry, which is called peptide mass fingerprinting (Figure 17–23). To identify the proteins, the mass of the peptides is compared with the masses predicted from data in genetic or protein databases. To increase the speed and accuracy of proteome analysis, researchers are working to adapt DNA chip technology. (See Chapter 18 for a description of DNA chips.) In the next sections, we discuss two applications of proteomics, an analysis of a bacterial proteome, and a description of the architecture of a subcellular structure, the nuclear pore.

The Bacterial Proteome Changes with Alterations in the Environment

As outlined earlier, *M genitalium*, with a reduced genome of 480 genes, represents one of the simplest, independently living organisms known. Several groups are analyzing the proteome of *M. genitalium* to help define the minimum set of biochemical and metabolic reactions needed for a living system. Valerie Wasinger and her colleagues used proteomics

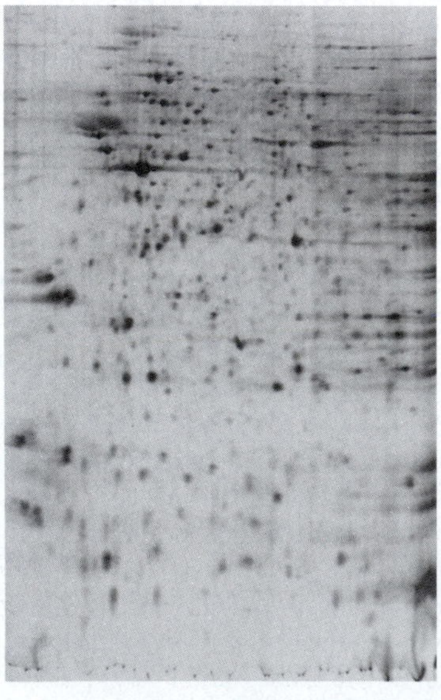

FIGURE 17–22 A two-dimensional protein gel, showing the separated proteins as spots. Several thousand proteins can be displayed on such gels. (*Laboratory of Biochemical Genetics/National Institute of Mental Health, NIH.*)

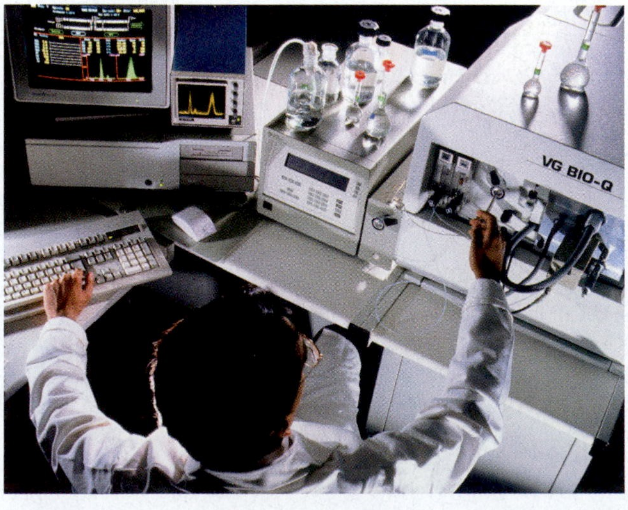

FIGURE 17–23 Proteins are digested and analyzed by mass spectrometry to identify the gene products present in a cell at a given time. (*Geoff Tompkinson/Science Photo Library/Photo Researchers, Inc.*)

to provide a snapshot of what genes are expressed during rapid growth and in the stationary phase following exponential growth.

Using two-dimensional electrophoresis, they identified 427 protein spots in exponentially growing cells. Of these, 201 were analyzed and identified by peptide digestion, mass spectrometry, and comparison to known proteins. The analysis uncovered 158 known proteins (33 percent of the proteome) and 17 unknown proteins. The remaining spots included fragments derived from larger proteins, different forms of the same protein (isoforms), and posttranslationally modified products. The identified proteins included enzymes involved in energy metabolism, DNA replication, transcription, translation, and transport of materials across the cell membrane.

During the transition from exponential growth to stationary phase there was a 42 percent reduction in the number of proteins synthesized. In addition, some new proteins appeared, and other proteins underwent dramatic changes in abundance. These changes are apparently a consequence of nutrient depletion, increased acidity of the growth medium, and other adaptations to environmental changes.

Wasinger's analysis helps establish the minimum number of expressed genes required for independent existence and the changes in gene expression that accompany the transition to the stationary phase. This study also shows that it is unlikely that there are conditions under which a cell will express its entire proteome. In *M. genitalium*, only 33 percent of the proteome was expressed during maximal growth. By virtue of technical limitations, these represent the most abundantly expressed proteins. The remaining 67 percent of the proteome are proteins likely to be expressed under different environmental conditions—those present in very low abundance (too low to be detected on gels)—or proteins that are not solubilized and recovered by the extraction methods used in Wasinger's study. In spite of these limitations, proteomic analysis provides a wide range of information that cannot be obtained by genome sequencing.

Architecture of the Nuclear Pore Complex: Matching Genes, Structure, and Function

The nuclear pore complex (NPC) is composed of proteins embedded in the nuclear membrane (Figure17–24). They connect the cytoplasmic and nuclear compartments of the cell and allow exchange of materials between the nucleus and the cytoplasm. Since this traffic includes mRNA destined for the cytoplasm, nuclear pores are important control points for the regulation of gene expression. The structural and functional complexity of such structures has been a barrier to understanding their molecular organization. As a result, how the NPC regulates the flow of ions and macromolecules is not well understood. Genomics, combined with proteomics, have provided researchers with the tools to investigate the three-dimensional architecture of the nuclear pore and to identify its proteins.

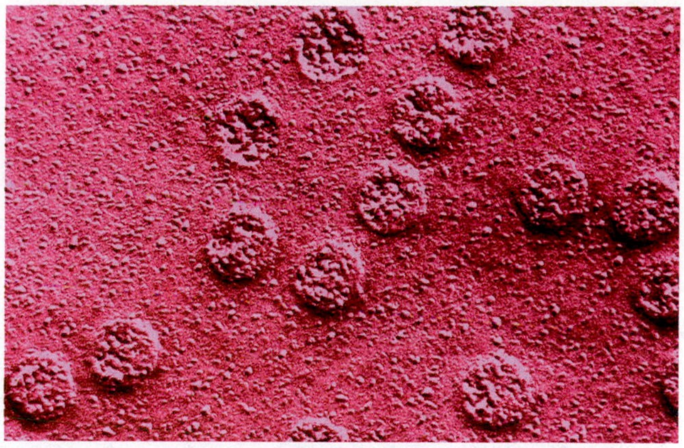

FIGURE 17–24 A transmission electron micrograph of the cytoplasmic side of the nuclear pore complex. (*Richard Kessel/Visuals Unlimited, Inc.*).

On the basis of the size of the yeast's NPC, it was estimated to contain up to 200 different proteins. Using the yeast genome database as a resource, researchers were able to identify potential NPC proteins (called nucleoporins) by sequence similarity to known NPC proteins from other organisms. In another approach, Michael Rout and his colleagues at the Rockefeller Institute used proteomics to define the molecular architecture of this subcellular structure and to provide insight into its functions. They first developed a method to isolate and purify NPCs and then used proteomics to identify the nucleoporins. Using mass spectrometry on peptide digests of proteins purified from nuclear pore complexes, they discovered that the yeast's NPC contains only about 30 different proteins, with each protein present in multiple copies. The nucleoporins are organized into 16 subunits, 8 on the pore's nuclear side and 8 on its cytoplasmic side [Figure 17–25(a)]. In addition, there are filament-base structures on both sides of the pore, organized into a basket on the nuclear side. Using a combination of mutant analysis, microscopy, and proteomics, the location of NPC proteins in the three-dimensional architecture of the pore has been mapped [Figure 17–25(b)]. Most of the nucleoporins are symmetrically distributed on the cytoplasmic and nuclear side of the pore. Five proteins are found only on one side or the other, and seven are distributed more on one side than the other.

To understand how nucleoporins function in nucleocytoplasmic transport or the regulation of transport, it will be necessary to study protein-protein interactions within the pore, as well as interactions of pore proteins with transport proteins, signal molecules, and other cellular components.

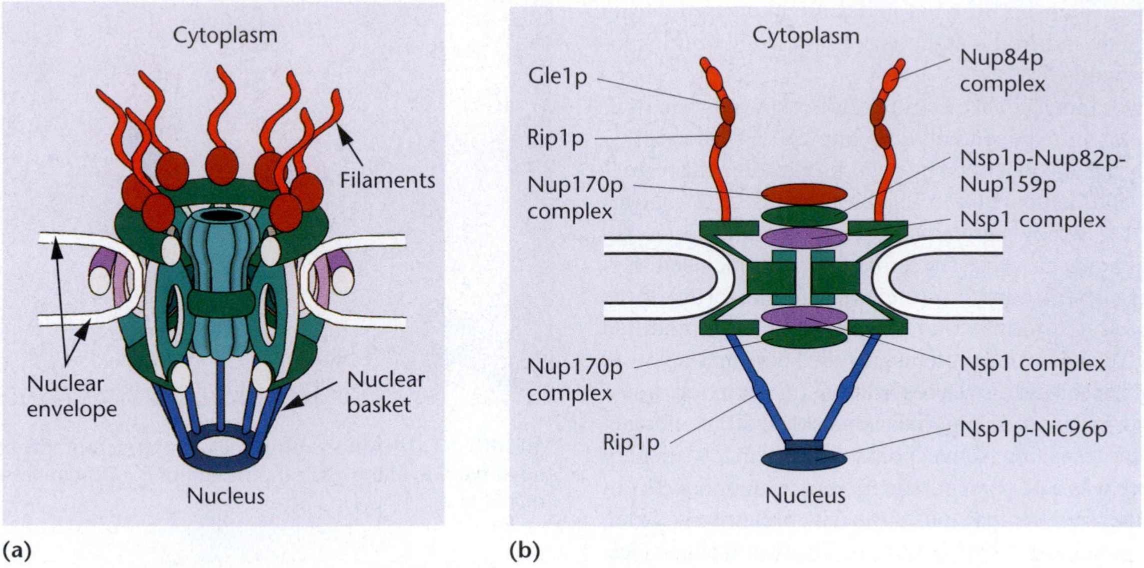

FIGURE 17–25 The yeast nuclear pore complex reconstructed from proteomic and genomic analysis. (a) The pore is made up of subunits, each of which contains about 30 different proteins. Each pore has 16 subunits arranged in an octagonal structure extending across the nuclear envelope. Additional nucleoporins that bind transport-factor proteins form filaments on the cytoplasmic side and on the nuclear basket on the nuclear side of the pore. (*Blobel, G., and Wozniak, R.W. 2000. Proteomics for the pore.* Nature 403:835–836, *Figure 1, p. 835.*) (b) The location of nucleoporins in the three-dimensional architecture of the nuclear pore. (*From Stoffler, D., et al., 1999 The nuclear pore complex: from molecular architecture, to functional dynamics.* Curr. Opin. Cell Biol. 11:391–401, *Figure 3b, p. 394.*)

Chapter Summary

1. Once obtained, genome sequences are analyzed in several steps to ensure that the sequence is accurate, to identify all encoded genes, and to classify known genes into functional categories.
2. Bacterial chromosomes are usually small, circular, double-stranded DNA. Gene density is very high, averaging one gene per kilobase pair of DNA. Typically, as much as 90 percent of the chromosome is gene coding. Many genes are organized into polycistronic transcription units that do not contain introns. Archaea are a major prokaryotic kingdom. Their chromosome and gene organization strongly resemble eukaryotic genes and genomes.
3. Eukaryotic genomes are organized into two or more chromosomes, each containing a linear double-stranded DNA molecule. The gene density is much lower than that in bacteria. Genes typically are not organized into operons, rather each is a separate transcription unit. However, the nematode *C. elegans* has many genes organized into operons. Eukaryotic genes are often interrupted with introns.
4. Complex multicellular eukaryotes differ from the less complex yeast in a number of ways. They have more genes and much more DNA. This results in gene densities falling to 1 gene per 5 kb and even 1 gene per 10-20 kb or more. A higher proportion of genes have introns, the number of introns per gene increases, and the size of introns increases as complexity increases from yeast to *C. elegans* to humans. Some plants, such as *Arabidopsis*, have a gene structure and organization that is indistinguishable from animals. Other plants, such as maize, have a different organization, with vast blocks of transposable elements separating islands of genes.
5. In the human genome, large differences exist in gene density on different chromosomes, with gene-rich regions alternating with gene-poor regions. Duplicated segments are a common feature of the chromosomes sequenced to date.
6. Many eukaryotic genes have undergone duplication followed by sequence divergence, leading to multigene families. The globin, immunoglobulin, and histone gene clusters are prime examples of this phenomenon.
7. Proteomics is used to study the expression of genes in bacterial cells under different growth conditions and is providing insight into the gene sets cells use during growth. These techniques have also been used to study the architecture of subcellular elements, such as the nuclear pore complex.

Insights and Solutions

1. Gene duplication has been an important process in the evolution of eukaryotes as a source of "new" genes. Duplication can arise by replication errors prior to meiosis, causing the insertion of a short, duplicated DNA sequence into the newly replicated DNA strand. Duplications can also arise by unequal crossing over between members of a homologous pair of chromosomes or between sister chromatids. New copies of essential genes produced by duplication can acquire mutations over time and gradually acquire related or new functions that provide a selective advantage.

Unequal crossing over is a form of recombination between nonallelic sequences on mispaired chromatids. What are the consequences of unequal crossing over for the participating chromatids? What factors can influence the chances that an unequal crossing over will occur?

Solution: Unequal crossing over is a reciprocal event. Such an exchange results in the deletion of sequences on one of the participating chromatids, and the insertion of a duplicated region on the other chromatid. This is true whether the unequal crossing over occurs between non-sister or sister chromatids. Thus, unequal crossing over transfers a gene to a different chromatid, no new gene is created; it is moved to a new location.

Sequence homology at the site of the unequal crossover event is often responsible for the mispairing of chromatids. Regions of the genome containing repeated DNA sequences (such as Alu) can be hot spots for mispairing and recombination.

Problems and Discussion Questions

1. Compare and contrast the chemical nature, size, and form assumed by the genetic material of bacteria and yeast.
2. Why might we predict that the organization of eukaryotic genetic material is more complex than that of viruses or bacteria?
3. Compare the gene organization of bacterial genes to that of eukaryotic genes.
4. The β-globin gene family consists of 60 kb of DNA, yet only 5 percent of the DNA encodes β-globin gene products. Account for as much of the remaining 95 percent of the DNA as you can.
5. How do assumptions about ORF size change the estimated number of genes in a genome?
6. *Arabidopsis* and rice each have about 20,000 genes, but the rice genome is about three times laarger than that of *Arabidopsis*? Why?
7. The minimum genome for free-living cells has been estimated at 250 genes. Could the minimum genome for a parasitic cell be lower? Is there a lower size limit for such a genome?

Selected Readings

Baltimore, D. 2001. Our genome unveiled. *Nature* 409: 814–816.

Blackstock, W.P., et al. 1999. Proteomics: quantitative and physical mapping of cellular proteins. *Trends Biotechnol.* 17:121–127.

Blattner, F.R., et al. 1997. The complete genome sequence of *Escherichia coli* K-12. *Science* 277:1453–74.

Brown, P.O., and Botstein, D. 1999. Exploring the world of the genome with DNA microarrays. *Nature Genetics* (Suppl.) 21:33–37.

Bult, C.J., et al. 1996. Complete genome sequence of the methanogenic Archaeon, *Methanococcus jannaschii. Science* 273: 1058–72.

Coller, H., et al. 2000. Expression analysis with oligonucleotide arrays reveals that MYC regulates genes involved in growth, cell cycle, signaling, and adhesion. *Proc. Nat. Acad. Sci.* 97:3260–65.

Fraser, C.M., et al. 1995. The minimal gene complement of *Mycoplasma genitalium. Science* 270:397–403.

Freeman, W.M., et al. 2000. Fundamentals of DNA hybridization arrays for gene expression analysis. *BioTechniques* 29:1042–1055.

Gall, J.G. 1981. Chromosome structure and the C-value paradox. *J. Cell Biol.* 91:3s-14s.

Gardner, M.J., et al. 1998. Chromosome 2 sequence of the human malaria parasite *Plasmodium falciparum. Science* 282.1126–32.

Gellert, M. 1996. A new view of V-D-J recombination. *Genes to Cells* 1:269–75.

Goffeau, A., et al. 1996. Life with 6000 genes. *Science* 274:546–67.

Golub, T., et al. 2000. Molecular classification of cancer: class discovery and class prediction by gene expression monitoring. *Science* 286:531–537.

Hamedeh, H., and Mshari, C. 2000. Gene chips and functional genomics. *Am. Scientist* 88: 508–515.

Heintz, N. 1991. The regulation of histone gene expression during the cell cycle. *Biochim. Biophys. Acta* 1088.327–39.

International Human Genome Sequencing Consortium. 2001. Initial sequencing and analysis of the human genome. *Nature* 409.860–921.

Iyer, V.R., et al. 1999. The transcriptional program in the response of human fibroblasts to serum. *Science* 283:83–87.

Meinke, D.W., et al. 1998. *Arabidopsis thaliana*: A model plant for genome analysis. *Science* 282.662–82.

GENETICS MediaLab

The resources that follow will help you achieve a better understanding of the concepts presented in this chapter. These resources can be found either on the CD packaged with this textbook or on the Companion Web site found at **http://www.prenhall.com/klug**

Web Problem 1:

Time for completion = 10 minutes

What is my protein? On a regular basis, researchers use the program BLAST to align a peptide sequence of interest against a database of other sequences. A good match to a protein of known function may provide clues to the identity of the protein in question. In this exercise, you will use the Web-based tool at the National Center for Biotechnology Information to perform a trial analysis. In the search file, enter P02144. This code references the sequence for human myoglobin. Myoglobin is a member of the large globin gene family and is well conserved among mammals. Click the 'BLAST' button, and then click the 'FORMAT' button to view the results on your browser. What types of proteins appear in the report? How good are the alignments? What organisms are these proteins from? Now, repeat the analysis from the beginning, using the protein NP_078945. This protein is predicted on the basis of a gene on human chromosome 5, and no known function is ascribed to it. How informative are the results? What would you do next if you were studying NP_078945? To complete this exercise, visit Web problem 1 in Chapter 17 of your companion Web site, and select the keyword **BLAST**.

Web Problem 2

Time for completion = 15 minutes

How can the human genome be usefully represented? The 3 billion base pairs of the human genome can be easily loaded onto a home computer, but extracting the information contained in the sequence presents a challenge. In this exercise, you will use one of several current "genome browsers" to search for a human gene. At the linked Web site (keyword **MAP VIEWER**), enter a gene name in the "Search" field, and press Return. Choose a gene that interests you, or enter "myoglobin" from the previous exercise. Progressively zoom in the chosen gene; a Help document provided at the site will aid you in navigating. Once you obtain a view of the sequence, reproduce a diagram of the intron/exon structure of the gene. Which chromosome is it on? How does its structure compare with that of neighboring genes? With that of other genes you select? To complete this exercise, visit Web problem 2 in Chapter 17 of your companion Web site, and select the keyword **MAP VIEWER**.

Web Problem 3

Time for completion = 15 minutes

How are entire genomes organized? Many bacterial genomes have been sequenced, offering an opportunity for this kind of analysis. *Pseudomonas aeruginosa* is a significant bacterial pathogen whose genome is now available. This organism is notable in that it codes for an unusually high number of small-molecule transporters. At the linked Web site (keyword **PSEUDOMONAS**), browse the color-coded function classes, paying particular attention to these transporters. Do they tend to cluster or are they dispersed? Are there other classes of genes with this pattern? With different patterns? How might the overall organization of genes contribute to the evolution of their function? To complete this exercise, visit Web problem 2 in Chapter 17 of your companion Web site, and select the keyword **PSEUDOMONAS**.

Mewes, H.W., et al. 1997. Overview of the yeast genome. *Nature* 387:5–105.

Myers, E.W., et al. 2000. A whole-genome assembly of *Drosophila. Science* 287: 2196–2204

Orkin, S.H. 1995. Regulation of globin gene expression in erythroid cells. *Eur. J. Biochem.* 231: 271–81.

San Miguel, P, et al 1998. The paleontology of intergene retrotransposons of maize. *Nature Genet.* 20: 43–47.

Stover, C.K., et al., 2000. Complete genome sequence of *Pseudomonas aeruginosa* PAO1, an opportunistic pathogen. *Nature* 406: 959–964.

The *C. elegans* Sequencing Consortium 1998. Genome sequence of the nematode *C. elegans*: A platform for investigating biology. *Science* 282: 2012–18.

Venter, J.C., et al. 1998. Shotgun sequencing of the human genome. *Science* 280: 1540–42.

———. 2001. The sequence of the human genome. *Science* 291:1304–1351.

Williams, S.C., et al. 1996. Sequence and evolution of the human germline V lambda repertoire. *J. Mol. Bio.* 264:200–210.

Wolfe, K. H., and Shields, D. C., 1997. Molecular evidence for an ancient duplication of the entire yeast genome. *Nature* 387: 708–713.

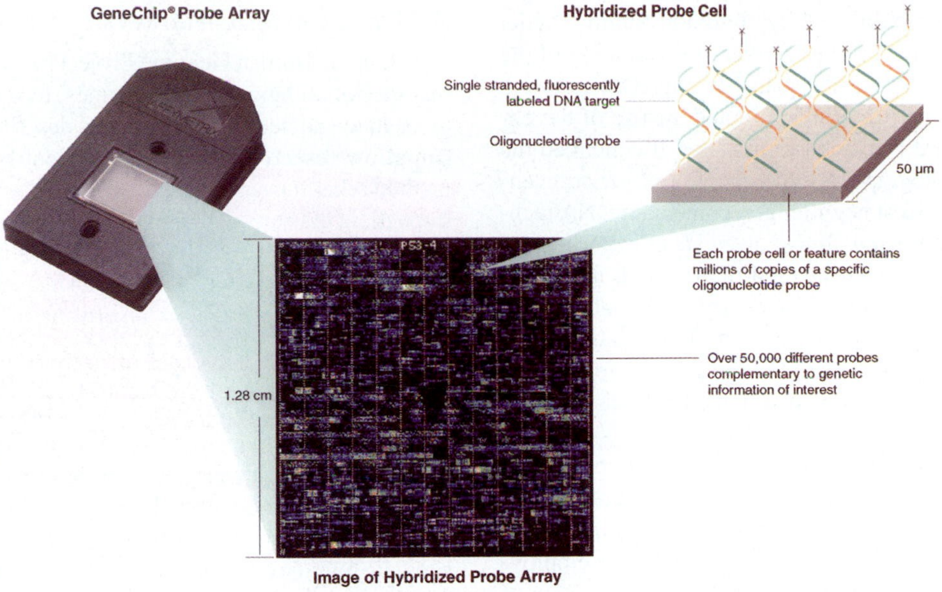

Affymetrix Gene Chip® probe arrays are used to detect DNA mutations and monitor gene expression. (*Affymetrix, Inc.*)

18

Applications and Ethics of Genetic Technology

KEY CONCEPTS

18.1 Mapping Human Genes at the Molecular Level
- RFLPs as Genetic Markers
- Linkage Analysis Using RFLPs
- Positional Cloning: The Gene for Neurofibromatosis
- Fluorescent *In Situ* Hybridization (FISH) Gene Mapping

18.2 Genetic Disorders: Diagnosis and Screening
- Prenatal Genotyping for Mutations in the β-globin Gene
- Prenatal Diagnosis of Sickle-Cell Anemia
- Single-Nucleotide Polymorphisms and Genetic Screening
- DNA Microarrays and Genetic Screening
- Genetic Testing and Ethical Dilemmas

18.3 Treating Disorders with Gene Therapy
- Gene Therapy for Severe Combined Immunodeficiency (SCID)
- Problems and Failures in Gene Therapy
- The Future of Gene Therapy: New Vectors and Target-Cell Strategies
- Ethical Issues and Gene Therapy

18.4 DNA Fingerprints
- Minisatellites, VNTRs, and STRs
- Forensic Applications of DNA Fingerprints

18.5 Genome Projects Use Recombinant DNA Technology
- The Human Genome Project: An Overview
- The Ethical, Legal, and Social Implications (ELSI) Program
- After the Genome Projects
- Studying Gene Expression with Microarrays

18.6 Biotechnology Is an Outgrowth of Recombinant DNA Technology
- Insulin Production by Bacteria
- Transgenic Animal Hosts and Pharmaceutical Products
- Transgenic Crop Plants and Herbicide Resistance
- Transgenic Plants and Edible Vaccines

In 1971, a paper published by Hamilton Smith, Daniel Nathans, and Walter Arber marked the beginning of the recombinant DNA era. The paper described the isolation of an enzyme from a bacterial strain and the use of the enzyme to cleave viral DNA at specific sites. It contained the first published photograph of DNA cut with a restriction enzyme. From this modest beginning, recombinant DNA technology has revolutionized almost all fields of experimental biology and has spread far beyond the research lab to become the foundation of a major industry. In the intervening years, this technology has become part of everyday life; recombinant DNA techniques are used to produce medicines, industrial chemicals, and even milk and other foods found in the supermarket.

In Chapter 16, we described the methods used to create and analyze recombinant DNA molecules. This chapter describes how these tools are being used in research and commercial applications and the ethical problems posed by the employment of this technology. We will consider a cross-section of applications that illustrate the power of recombinant DNA technology in mapping and identifying human genes, diagnosing and treating disease, and generating new plants and animals. We begin by explaining how recombinant DNA has changed basic methods of genetic mapping and generated projects that are sequencing the genomes of many organisms. Next, we examine the impact of recombinant DNA on human genetics and medicine, from the prenatal analysis of genotypes to the treatment of genetic disorders. Finally, we will discuss some of the products of the multi-billion dollar biotechnology industry.

18.1 Mapping Human Genes at the Molecular Level

When the first human genetic disorders were mapped, the phenotype was usually coupled to the presence of a mutant gene product that could be identified in affected individuals. By using the mutant protein as a marker, pedigree analysis was used to establish the pattern of inheritance and, in a few cases, to establish the chromosomal locus of the gene. However, in the majority of human genetic disorders, the function of the normal gene product is unknown, and without a marker, conventional methods of mapping cannot be used. Although the genetic and molecular basis for a handful of diseases was known before the recombinant DNA revolution, real progress in the mapping of these genes has come only in the last two decades. Now, with our increasingly detailed knowledge of the genome, we can map a gene without information about its product.

The first step in this process identifies a molecular marker linked to the phenotype of the disease. With this linkage information in hand, researchers either look in the chromosomal region where the marker resides for potential gene sequences or, with an increasing number of genes localized to chromosomal sites, check the identified chromosomal area for genes that have already been mapped.

RFLPs as Genetic Markers

Now that the Human Genome Project has sequenced the vast majority of all human chromosomes, looking for coding regions in the nucleotide sequence, as described in Chapter 17, can allow researchers to directly identify genes. Another method relies on variations in nucleotide sequence that occur throughout the human genome (mostly in noncoding regions) with a frequency of about 1 in 200 nucleotides. These nucleotide changes occur at specific sites and are created by substitutions, deletions, or insertions of one or more nucleotide pairs. Such variations can create or destroy restriction enzyme cutting sites. If a restriction enzyme site created by nucleotide variation is present on one chromosome, but absent on its homolog, the two chromosomes can be distinguished from one another by their pattern of restriction fragments on a DNA blot (Figure 18–1). The region of chromosome *A* shown in the figure contains three *Bam*HI sites; its homolog, chromosome *B*, contains two such sites. When chromosome *A* is cut with *Bam*HI, 3-kb and 7-kb fragments are generated, whereas only a single 10-kb fragment is generated when chromosome *B* is cut. By using a probe from this chromosomal region, these fragments can be visualized on a Southern blot (Figure 18–1).

Variations in DNA fragment length generated by cutting with a restriction enzyme are called **restriction fragment-length polymorphisms (RFLPs)**. RFLPs are quite common; thousands have been identified in the human genome, and most have been assigned to specific regions of individual chromosomes. These variations are inherited as codominant alleles. They can be mapped to specific regions on individual chromosomes and can be used as markers to follow the inheritance of genetic disorders from generation to generation in an affected family.

Using RFLPs to map the chromosomal locus of a genetic disorder involves identifying an RFLP marker that cosegregates with the genetic disorder in a multigenerational family (three generations or more). Selecting which RFLP to use in the family being studied is a matter of trial and error. Many different RFLPs are tested in order to find one for which most members of the family are heterozygous. This process allows each member of the chromosome pair being tested to be identified as it passes from generation to generation and provides a way to establish linkage between a specific chromosome (identified by the RFLP marker) and the phenotype of the disease.

Linkage Analysis Using RFLPs

To map a genetic disorder, the inheritance of a given RFLP and the disorder are traced through a family (Figure 18–2). If loci for the RFLP and the disorder are near each other on the same chromosome, they will show linkage. In such studies, most RFLPs will not show linkage with the phenotype of the disease, because either the RFLP and the disease loci are on different chromosomes or there is frequent recombination between the RFLP and the disease loci. However, these data are not useless, because they help to identify chromosomes that do *not* carry the disease locus.

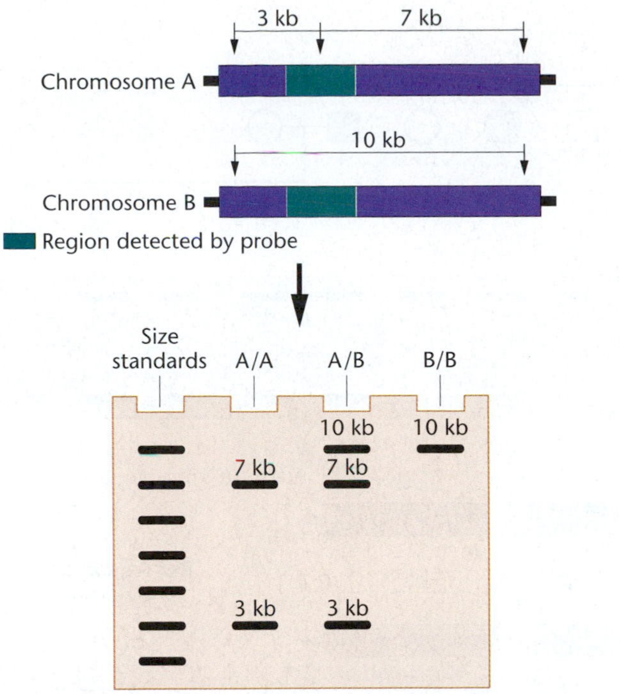

Genotypes	Fragment sizes
Homozygous for chromosome A (A/A)	3 kb, 7 kb
Heterozygous (A/B)	3 kb, 7 kb, 10 kb
Homozygous for chromosome B (B/B)	10 kb

FIGURE 18–1 Restriction fragment-length polymorphisms (RFLPs). The alleles on chromosome *A* and chromosome *B* represent DNA segments from homologous chromosomes. The region that hybridizes to a probe is shown in green. Arrows indicate the location of restriction enzyme cutting sites that define the alleles. On chromosome *A*, three cutting sites generate fragments of 7 kb and 3 kb. On chromosome *B*, only two cutting sites are present, generating a single fragment that is 10 kb in length. The absence of the cutting site in chromosome *B* could be the result of a single base mutation within the enzyme recognition or cutting site. Because these differences in restriction cutting sites are inherited in a codominant fashion, there are three possible genotypes: *AA, AB*, and *BB*. The allele combination carried by any individual can be detected by restriction digestion of genomic DNA (obtained, for example, from a blood sample or skin fibroblasts), followed by gel electrophoresis, transfer to a DNA binding filter, and hybridization to the appropriate probe. The fragment patterns for the three possible genotypes are shown as they would appear on a Southern blot.

To analyze the results of such experiments, researchers use probability to determine whether an RFLP marker and a genetic disorder are linked. The probabilities are ultimately expressed as the **logarithm of the odds**, and the method of analysis is referred to as the **lod score method**. (See Chapter 12.) A lod score of three means that it is 1000 times more likely that the gene and the RFLP marker are linked than not linked; a lod score of four means that the odds are 10,000 times greater in favor of linkage. Lod scores of three to four are usually taken as evidence that the gene and the marker are linked.

In mapping human chromosomes, the unit of linkage is the **centimorgan (cM)**, named after the geneticist T. H. Morgan. One centimorgan is equal to a recombination frequency of 1 percent between two loci. The genetic distance between two genes, expressed in centimorgans, is not directly correlated with the physical distance expressed in nucleotides, but, in human chromosomes, a distance of 1 cM corresponds roughly to 1–3 million nucleotides of DNA. Many genes have been mapped in humans by using techniques such as RFLP analysis, and, by compiling studies from many families, the location of many markers can be determined and genetic maps for human chromosomes can be constructed (Figure 18–3).

Positional Cloning: The Gene for Neurofibromatosis

The search for the chromosomal locus for the gene causing **type 1 neurofibromatosis (*NF1*)** is an excellent example of the use of RFLP analysis in gene mapping. NF1 is inherited as an autosomal dominant condition, with an incidence of about 1 in 3000. NF1 is associated with a range of nervous-system defects, including benign tumors and an increased incidence of learning disorders. The spontaneous mutation rate at this locus is very high, and up to 50 percent of all cases may represent new mutations. To map this gene by conventional methods would require the identification of large families with a long, multigenerational history of NF1. To be useful, each family should also carry one or more genetic disorders already mapped to an autosome. In practice, it would be almost impossible to assemble such family groups, and the gene was mapped using recombinant DNA techniques instead.

The *NF1* gene was mapped by RFLP analysis in several steps. First, research laboratories around the world compared the inheritance of NF1 and dozens of RFLP markers in affected families over multiple generations. Each RFLP marker represents a specific human chromosome or chromosome region. The resulting **exclusion map** indicated which RFLPs are not linked to the disease and, therefore, which chromosomes do *not* carry the NF1 locus. This work also produced evidence of linkage and pointed to chromosomes 5, 10, and 17 as sites where the *NF1* gene might reside. The next stage focused on these chromosomes and produced conclusive evidence that the disorder is closely linked to an RFLP that maps to a site near the centromere of chromosome 17 (Figure 18–4). Then, more than 30 RFLP markers from this region of chromosome 17 were used to analyze 13,000 individuals from NF1 families, and the gene was mapped to region 17q11.2. Finally, by using a collection of genomic clones that spanned a small region of 17q11.2, the locus for NF1 was identified in 1990 by chromosome walking (see Chapter 16) and DNA sequencing. The gene's identity was confirmed when mutant versions of the gene were found in individuals with NF1.

Once the gene was identified, the amino acid sequence of the gene product was reconstructed from the DNA sequence and found to consist of 2485 amino acids. Databases with

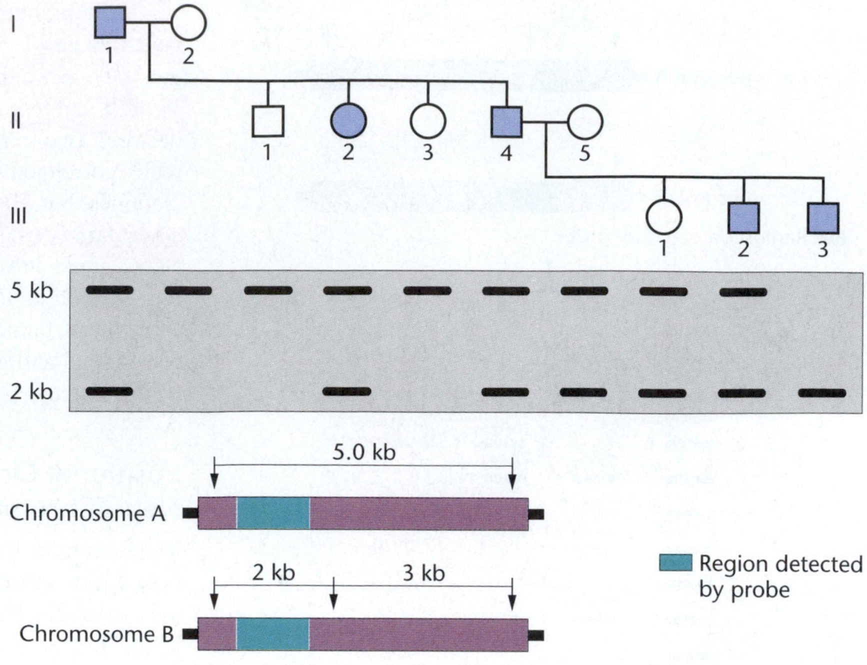

FIGURE 18–2 Establishing linkage between a dominant trait and an RFLP allele. The pedigree shows a family with members affected by a dominant trait (filled symbols). Family members also carry two alleles of an RFLP locus assigned to a specific chromosome. A 5.0-kb allele (allele *A*) is present on one homolog. The *B* allele on the other homolog consists of two fragments (2.0 kb and 3.0 kb). The probe used in the Southern blot detects the 5-kb *A* allele and the 2.0-kb portion of the *B* allele. The RFLP pattern for each family member is shown below the appropriate pedigree symbol. Individual III-1, who is unaffected, probably received an *A* allele from her father and a *B* allele from her mother. Individual III-2 is affected and probably received an *A* allele from his mother and a *B* allele from his father. The youngest son (III-3), who is affected, received a *B* allele from each parent. Taken together, the pedigree and the Southern blot suggest that the mutant allele for the dominant trait and the RFLP *B* allele are on the same homolog and are therefore linked. Assigning a mutant allele to a chromosome by RFLP analysis is the first step in mapping a gene.

protein-sequence information show that the NF1 gene product is similar to proteins that play a role in signal transduction. Further analysis confirmed that the NF1 protein, **neurofibromin**, is involved in the transduction of intracellular signals and the down-regulation of a gene that controls cell growth. Mutation in the NF1 gene leads to a loss of control of cell growth, causing the production of the small tumors that are characteristic of this disorder.

The mapping, cloning, and sequencing of the NF1 gene and the identification of the gene product took a little over 3 years. Researchers began this project with no direct knowledge of the nature of the gene product or the mutational events that produce the NF1 phenotype. The mapping and cloning process of the gene for NF1 described here is an example of **positional cloning**. This recombinant DNA-based method is a departure from previous methods of mapping, which worked from an identified gene product to the gene locus. In positional cloning, the gene can be mapped, isolated, and cloned with no knowledge of the gene product. Through use of this strategy, an ever-increasing number of human genes are being mapped and isolated.

Fluorescent *in Situ* Hybridization (FISH) Gene Mapping

Genes can be mapped directly to metaphase chromosomes in a process called *in situ* hybridization (see Appendix A). Chromosomal preparations on microscope slides are denatured to convert the double-stranded DNA in the chromosomes into single strands. A labeled probe, containing all or part of a cloned gene, is hybridized to the chromosomal preparation. The probe can be labeled with a radioactive isotope, antibodies, or fluorescent dyes. In contrast to the use of gels in Southern blotting, the resulting hybrids are examined and detected under a microscope.

When fluorescent dyes are used to label the probes, a fluorescence microscope is used to examine the chromosome preparation and to determine the site of hybridization. This technique, known as FISH (fluorescent in situ hybridization; see Chapter 4), can be used to directly assign a cloned gene to a chromosomal locus [Figure 18–5(a)]. The technique has a resolution of about 1Mb and is now being used routinely to map human genes.

An even higher level of resolution of around 100kb can be obtained by using fiber FISH, a variation of the FISH technique. In fiber FISH, chromosome fibers (see Chapter 4 for a discussion of chromosome organization) are stretched along the glass slide during chromosome preparation. Fluorescent probes are hybridized to the fibers [Figure 18–5(b)]. This method is used to directly establish the order and distance between two or more genes on a chromosome.

18.2 Genetic Disorders: Diagnosis and Screening

In many cases, genetic diseases can be diagnosed prenatally. The most widely used methods for prenatal diagnosis of genetic disorders are **amniocentesis** and **chorionic villus sampling (CVS)**. In amniocentesis, a needle is used to withdraw amniotic fluid (Figure 18–6), and the fluid and the cells it contains are analyzed for chromosomal or single-gene disorders. In CVS, a catheter is inserted into the uterus, and a small tissue sample of the fetal chorion is retrieved.

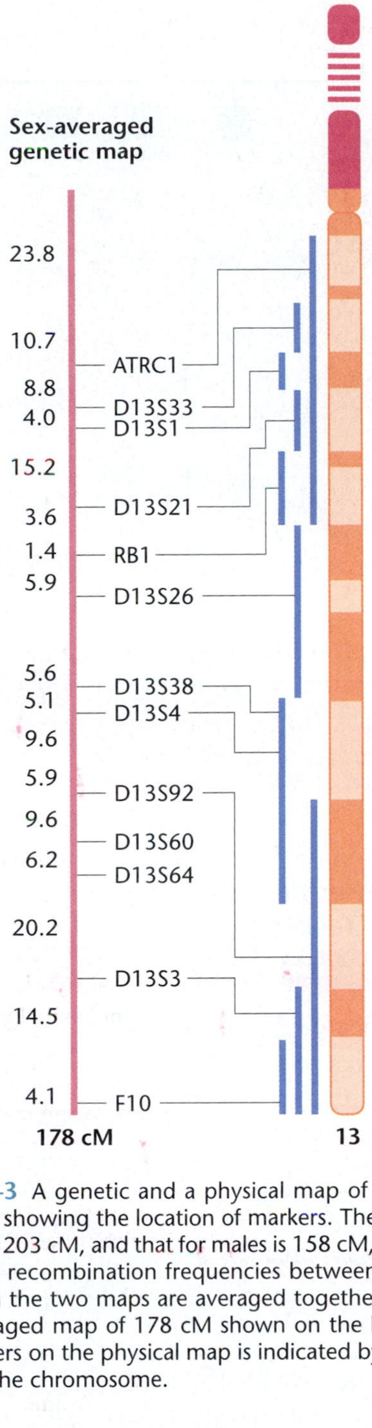

FIGURE 18–3 A genetic and a physical map of human chromosome 13, showing the location of markers. The genetic map for females is 203 cM, and that for males is 158 cM, reflecting the difference in recombination frequencies between females and males. When the two maps are averaged together, the result is the sex-averaged map of 178 cM shown on the left. The location of markers on the physical map is indicated by the brackets adjacent to the chromosome.

Cytogenetic, biochemical, and recombinant DNA-based testing is then performed on the tissue.

Coupled with these sample-recovery methods, recombinant DNA technology has proven to be a highly sensitive and accurate tool for the prenatal detection of genetic disorders. Cloned DNA sequences have expanded the range of prenatal testing, because the fetal genotype is examined directly, rather than relying on the few tests available for normal or mutant gene products. This capability is particularly important because, frequently, the gene product cannot be detected before birth, even when tests are available. For example, defects in the adult β-globin protein cannot be detected prenatally, because the β-globin gene is not expressed until a few days after birth.

Prenatal Genotyping for Mutations in the β-Globin Gene

Mutations in the β-globin locus lead to a wide range of genetic disorders. One of these disorders, β-thalassemia, is an autosomal recessive disorder associated with decreased or absent β-globin production. At the molecular level, this condition can be caused by a variety of deletions or point mutations. One such deletion covers about 600 nucleotides and includes the third exon of the gene. Restriction enzyme digestion and Southern blot hybridization (see Chapter 16) to a cloned β-globin probe can be used to diagnose the genetic status of a fetus and other members of an affected family (Figure 18–7).

Deletions, however, are relatively rare mutational events. Mutations are more commonly associated with changes in one or a small number of nucleotides (point mutations). However, if single-nucleotide changes take place in an exon, they can have devastating clinical and phenotypic effects. Therefore, more sensitive techniques (discussed later) have been developed to screen for these small-scale changes in sequence.

Prenatal Diagnosis of Sickle-Cell Anemia

Sickle-cell anemia is an autosomal recessive condition common in people with family origins in areas of West Africa, the Mediterranean basin, and parts of the Middle East and India. Sickle-cell anemia is caused by a single amino acid substitution in the β-globin gene. This change is caused by a single-nucleotide substitution that, coincidentally, eliminates a cutting site for the restriction enzymes *Mst*II and *Cvn*I. As a result, the mutation alters the pattern of restriction fragments seen on Southern blots. These differences in RFLP patterns are used to diagnose sickle-cell anemia prenatally and to determine the genotypes of parents and other family members who may be heterozygous carriers of this condition.

For prenatal diagnosis, fetal cells are obtained by amniocentesis or by CVS. DNA is extracted from these cells and digested with a restriction enzyme, such as *Mst*II. This enzyme cuts three times in the region of the normal β-globin gene, producing two small DNA fragments. In the mutant allele, the middle *Mst*II site has been destroyed by the mutation, and one large restriction fragment is produced by digestion with *Mst*II (Figure 18–8). The restriction-digested DNA fragments are separated by gel electrophoresis, transferred to a nylon membrane, and visualized by Southern blot hybridization.

In Figure 18–8, the parents (I-1 and I-2) are both heterozygous carriers of the mutation. Digestion of DNA from each parent produces a large band (the mutant allele) and two smaller bands (the normal allele). The parents' first child (II-1) is homozygous normal, because she has only the two

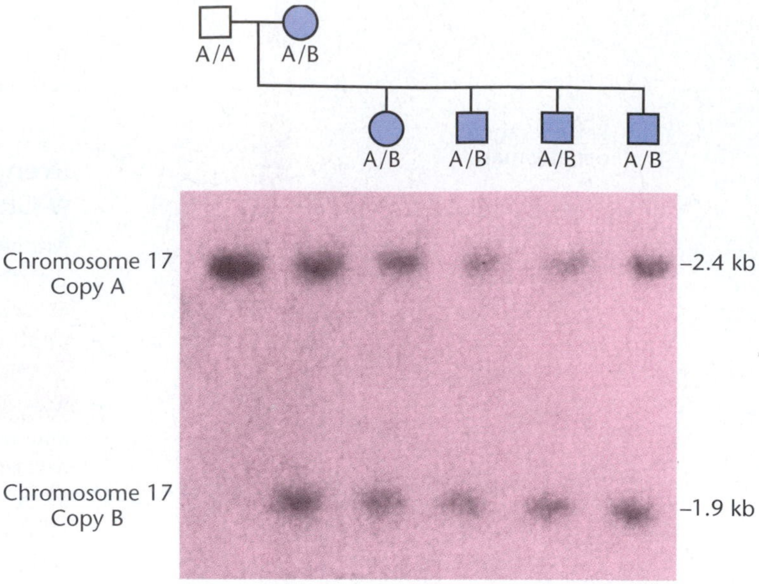

FIGURE 18–4 Segregation of a 2.4-kb RFLP allele with type 1 neurofibromatosis (NF1) in each of four affected offspring and their mother. This RFLP is detected by probe pA10-41, which is a DNA segment from a region near the centromere of human chromosome 17. On the basis of this result and results from tests using other probes, the locus for NF1 was assigned to chromosome 17.

smaller bands. The second child (II-2) has sickle-cell anemia; he has only one large band and is homozygous for the mutant allele. The fetus (II-3) has a large band and two small bands and is, therefore, heterozygous for sickle-cell anemia. He or she will be unaffected, but will be a carrier.

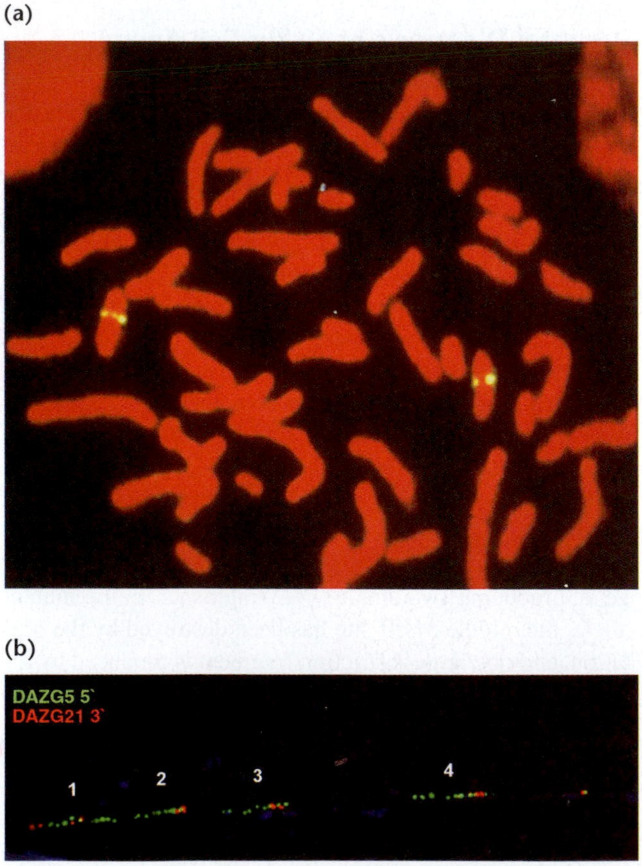

FIGURE 18–5 (a) Localization of a gene by fluorescent *in situ* hybridization (FISH) to metaphase chromosomes. The sites of hybridization appear as yellow dots on the sister chromatids of the chromosome pair. (b) The location of four clusters of two DNA sequences (DAZG5 and DAZG2) along a chromosome fiber.

Only about 5 to 10 percent of all point mutations can be detected by restriction enzyme analysis. However, if a mutant gene has been well characterized and the mutated region has been sequenced, synthetic oligonucleotides can be used as probes to detect mutant alleles, as described in the next section.

Single-Nucleotide Polymorphisms and Genetic Screening

Alleles that differ by as little as a single nucleotide can be identified by synthetic probes known as **allele-specific oligonucleotides (ASO)**. In contrast to restriction enzyme analysis, which is limited to cases for which a mutation changes a restriction site, ASOs detect single-nucleotide changes of all types, including those that do not affect restriction enzyme cutting sites. As a result, this method offers increased resolution and wider application. Under proper conditions, an ASO will hybridize only with its complementary sequence and not with other sequences, which might vary by as little as a single nucleotide. A method using ASOs and PCR analysis is now available to screen for many disorders, including sickle-cell anemia. In this procedure, DNA from white blood cells is extracted and denatured into single strands. This DNA is used to amplify a region of the β-globin by PCR. A small amount of the amplified DNA is spotted onto filters, and each filter is hybridized to an ASO (Figure 18–9). After visualization, the genotype can be read directly from the filters. With an ASO used for the normal sequence [Figure 18–9(a)], the homozygous normal (AA) genotype produces a dark spot (two copies of the normal allele), and the heterozygous genotype (AS) produces a light spot (one copy of the normal allele). The homozygous recessive sickle-cell genotype will not bind the probe, and no spot will be visualized. With a probe used for the mutant allele [Figure 18–9(b)], the pattern is reversed. This rapid, inexpensive, and highly accurate technique is becoming the method of choice for the diagnosis of a wide range of genetic disorders caused by point mutations.

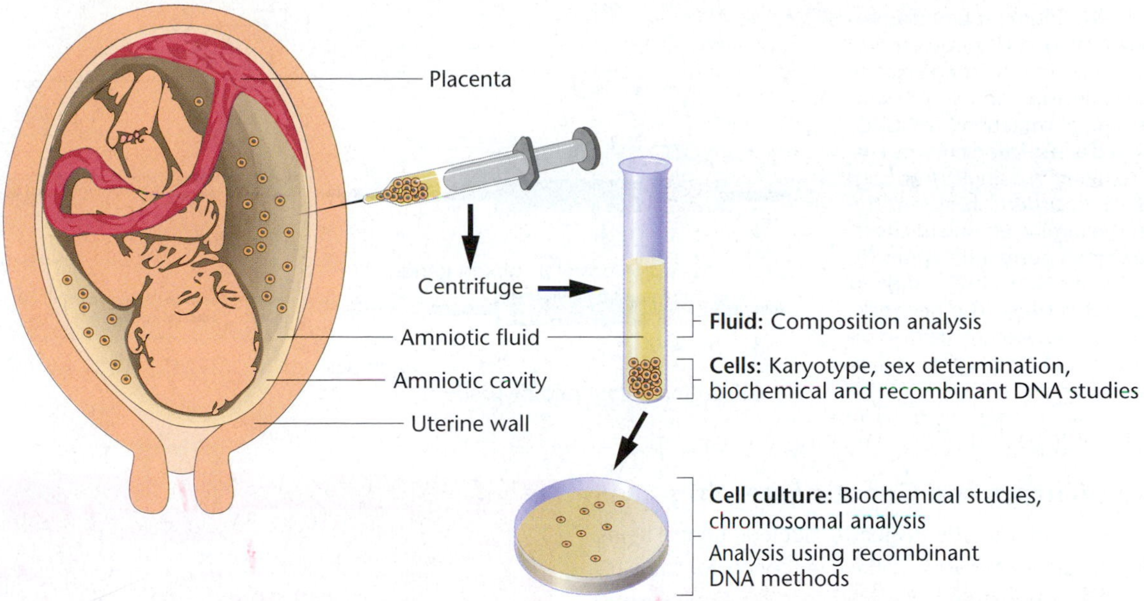

FIGURE 18–6 The technique of amniocentesis. The position of the fetus is first determined by ultrasound, and then a needle is inserted through the abdominal and uterine wall to recover fluid and fetal cells for cytogenetic or biochemical analysis.

In cases where the nucleotide sequence of the normal allele is known and where the molecular nature of the mutant allele has been identified, ASOs are synthesized directly from normal and mutant copies of the gene in order to screen for heterozygous carriers of the genetic disorder. For example, in people with cystic fibrosis, a deletion (called *Δ508*) is found in 70 percent of all mutant copies of the gene. Cystic fibrosis (CF) is an autosomal recessive disorder associated with a defect in a protein called the **cystic fibrosis transmembrane conductance regulator** (**CFTR**), which regulates chloride-ion transport across the plasma membrane. To detect heterozygous carriers of the *Δ508* mutation, allele-specific oligonucleotides are made by PCR from cloned samples of the normal allele and the mutant allele. DNA extracted from white blood cells of the individuals to be tested is spotted on a nylon filter and hybridized to each ASO (Figure 18–10). In affected individuals, only the ASO made from the mutant allele hybridizes; in heterozygotes, both ASOs hybridize; and in normal homozygotes, only the ASO from the normal allele hybridizes.

Because CF affects approximately 1 in 2000 individuals of northern European descent, screening for CF can be used in these populations to detect and counsel people about their genetic status with respect to CF. However, not all of the known mutations for this gene (over 500 mutations have been identified) can be screened, so a negative result does not eliminate someone as a heterozygous carrier, and it is likely that more CF mutations remain to be identified. Consequently, CF screening is not widepread, but will no doubt become commonplace once tests can cover 98–99 percent of all possible CF mutations.

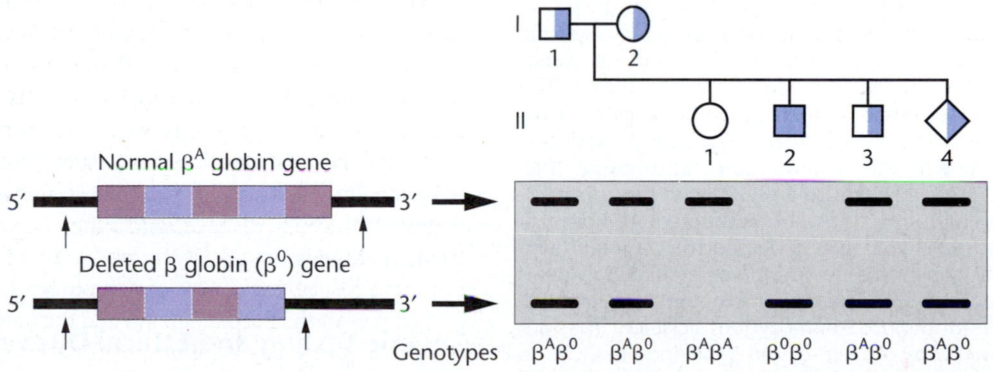

FIGURE 18–7 Diagnosis of β-thalassemia caused by a partial deletion of the β-globin gene. The family pedigree is shown above each individual's genotype on a Southern blot. The normal β-globin allele (β^A) contains three exons and two introns. The deleted β-globin allele (β^0) has the third exon deleted. Arrows indicate the cutting sites for restriction enzymes used in this analysis. The normal gene produces a larger fragment (shown as the top row of fragments on the Southern blot); the smaller fragments produced by the deleted gene are represented at the bottom of the gel. The genotypes of each individual in the pedigree can be determined from the pattern of bands on the blot and is given below the blot.

FIGURE 18–8 Southern blot diagnosis of sickle-cell anemia. Arrows represent the location of restriction enzyme cutting sites. In the mutant (β^S) globin allele, a point mutation (GAGGTG) has destroyed a restriction enzyme cutting site, resulting in a single large fragment on a Southern blot. In the pedigree, the family has one unaffected homozygous normal daughter (II-1), an affected son (II-2), and an unaffected fetus (II-3). The genotype of each family member can be read directly from the blot and is given below the blot.

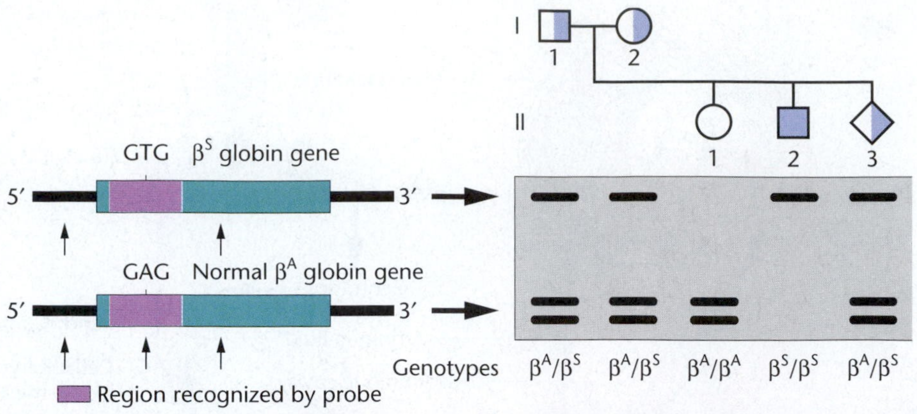

DNA Microarrays and Genetic Screening

DNA probes similar to allele-specific nucleotides are being coupled with the technology of the semiconductor industry to produce **DNA microarrays** (also called *DNA chips*) (Figure 18–11). The microarrays are made of glass and are divided into fields (small squares); each field can be as small as half the width of a human hair. A field contains copies of a specific synthetic DNA probe about 20 nucleotides in length attached to the glass (Figure 18–12). Along a row of fields, the sequence of the probe differs by one nucleotide from field to field. Thus, a set of four fields (one for each nucleotide) is needed in order to test the nucleotide content of a given position in a DNA molecule. The current generation of chips can hold between 280,000 and 560,000 fields, but chips with several million fields are now being developed.

For genetic testing, DNA is extracted from cells and cut with one or more restriction enzymes. The resulting fragments are tagged with a fluorescent dye, melted into single strands, and pumped into the microarray. Fragments with a nucleotide sequence that exactly matches the probe sequence will bind, and those with a sequence that doesn't match are washed off. The microarray is then scanned by a laser; fields where hybridization occurs fluoresce (Figure 18–13). Software linked to the microarray analyzes the pattern of hybridization, and the data can be presented in several forms.

DNA microarrays are already used to scan for mutations in the *p*53 gene, which is mutated in 60 percent of all cancers, and to screen for mutations in the *BRCA*1 gene, which predispose women to breast cancer. In addition to testing for mutations in single genes, DNA chips can be made to screen thousands of different genes simultaneously. Chips have been programmed with up to 30,000 genes in order to study gene expression during embryonic development in the mouse, and chips that carry all the genes in the human genome (now estimated to be 35,000–45,000) are in the planning stages. This technology will make it possible to analyze someone's DNA for dozens or hundreds of diseases, including those that predispose the person to heart attacks, diabetes, Alzheimer disease, and other genetically defined disease subtypes.

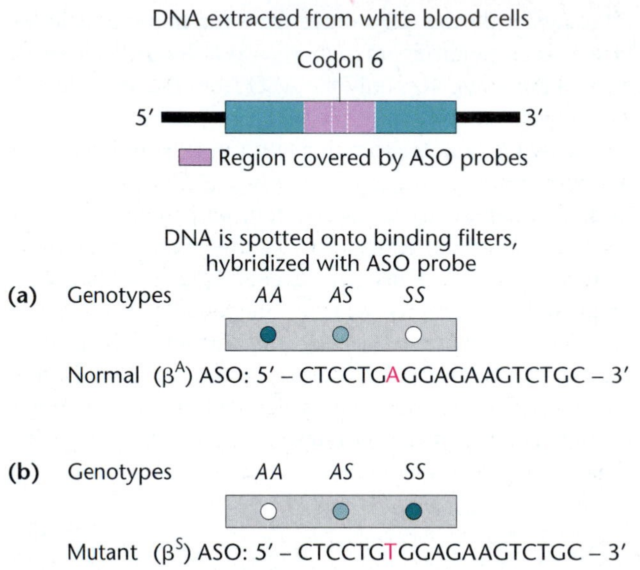

FIGURE 18–9 Genotype determinations, using allele-specific oligonucleotides (ASOs). In this technique, the β-globin gene is amplified by PCR, using DNA extracted from blood cells. The amplified DNA is denatured and spotted onto strips of DNA-binding filters. Each strip is hybridized to a specific ASO and visualized on X-ray film after hybridization and exposure. If all three genotypes are hybridized to an ASO from the normal β-globin allele, the pattern in (a) will be observed: AA-homozygous individuals have two copies of the normal β-globin gene and will show heavy hybridization; AS-heterozygous individuals carry one normal β-globin allele and one mutant allele and will show weaker hybridization; SS-homozygous sickle-cell individuals carry no normal copy of the β-globin gene and will show no hybridization to the ASO probe for the normal β-globin allele. (b) The same genotypes hybridized to the probe for the sickle-cell β-globin allele will show the reverse pattern: no hybridization by the AA genotype, weak hybridization by the heterozygote (AS), and strong hybridization by the homozygous sickle-cell genotype (SS).

Genetic Testing and Ethical Dilemmas

In an earlier section, we described examples of two types of genetic testing: prenatal diagnosis and screening for heterozygous carriers of recessive disorders. By using current technology, genetic testing can also be used to predict someone's risk of disease, thereby identifying people who

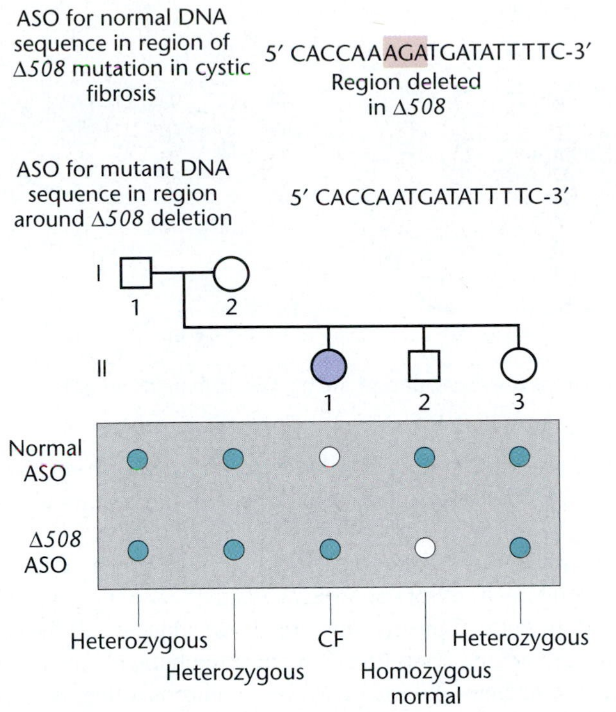

FIGURE 18–10 Screening for cystic fibrosis (CF) by allele-specific oligonucleotides (ASOs). ASOs for the region spanning the most common mutation in CF, a three-nucleotide deletion *Δ508*, are prepared from a normal CF allele and a *Δ508* CF allele. In screening, the CF alleles carried by an individual are amplified by PCR, using DNA extracted from blood samples, and spotted on a DNA-binding membrane. The membrane is hybridized to a mixture of the two ASOs. The genotype of each family member can be read directly from the filter. DNA from I-1 and I-2 hybridizes to both ASOs, indicating that these individuals carry a normal allele and a mutant allele and are, therefore, heterozygous. The DNA from II-1 hybridizes only to the *Δ508* ASO, indicating that this individual is homozygous for the mutation and has cystic fibrosis. The DNA from II-2 hybridizes only to the normal ASO, indicating that this individual carries two normal alleles. II-3 has two hybridization spots and is heterozygous.

FIGURE 18–11 An Affymetrix Gene Chip®. The glass square in the center covers the fields containing single-stranded probe DNAs.

are presently healthy, but are at high risk of contracting a genetic disease in the future. In the next few years, DNA microarrays may be used to test for 50–100 diseases at a time, including many that may not develop for years or decades. These advances will affect our health, reproductive patterns, and medical care in fundamental ways. The use of this technology also raises legal, social, and ethical issues that will not be easy to resolve. For example, what should people know before deciding to have a genetic test? How can we protect the information revealed by a genetic test? How can we define and prevent genetic discrimination? We know that heterozygotes for sickle-cell anemia are more resistant to malaria than people who are homozygous for the normal allele. This type of situation may be true of other genetic disorders as well. How do we protect beneficial aspects of mutations as we strive to eliminate their destructive aspects? Some mutations have horrific consequences; others gave rise to our very existence as humans. We know a great deal, and yet we know very little. Thoughtful and wide-ranging public debate is important as we explore the use of biotechnology.

Many of the potential risks and benefits of genetic testing are still unknown. We can test for many genetic diseases for which there are no effective treatments to cure or mitigate the clinical consequences. Should we test people for these disorders? With present technology, the fact that a genetic test gives a negative result does not rule out the future development of the disease, nor does a positive result always mean that one will get the disease. How can we effectively communicate the results of testing and the risks to those being tested?

Public policy and laws on genetic testing are being formulated more slowly than the technology and use of genetic testing. The **Ethical, Legal, and Social Implications (ELSI)** Program of the Human Genome Project (described later in this chapter) has set up task forces to identify issues related to genetic testing. This program is charged with developing recommendations for policy makers and lawmakers that will retain the benefits of genetic testing, but reduce or eliminate the potential harm. In addition, other groups made up of scientists, health-care professionals, lawmakers, ethicists, and consumers are debating these issues and formulating policy options.

18.3 Treating Disorders with Gene Therapy

Gene *products* such as insulin have been used for decades in therapeutic treatment of genetic disorders. Methods for transferring specific *genes* into mammalian cells, originally developed as a research tool, are now being used to treat genetic disorders, a process known as **gene therapy**. In theory, gene therapy transfers a normal allele into a somatic cell that carries one or more mutant alleles. Expression of the normal allele results in a functional gene product whose action produces a normal phenotype. Delivery of these structural genes and their regulatory sequences is accomplished by using a vector or gene transfer system.

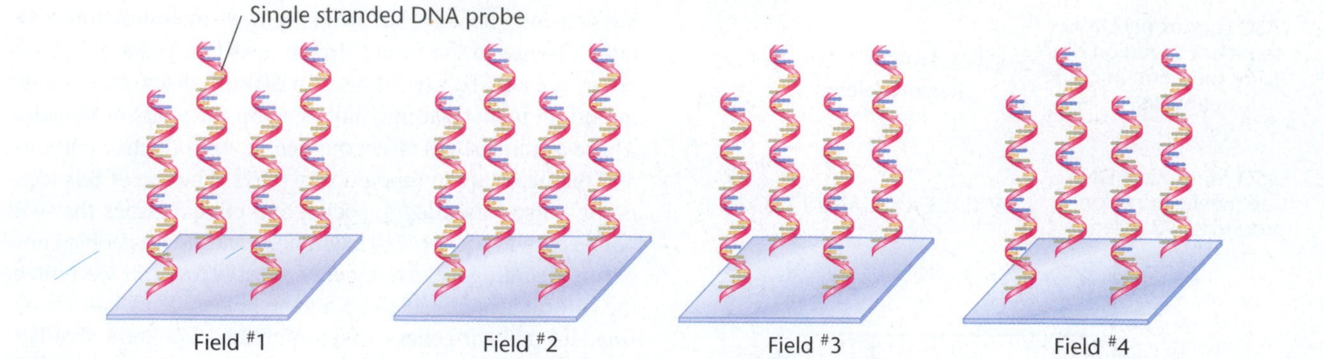

FIGURE 18–12 A single-stranded DNA molecule with bases extending from the sugar/phosphate backbone in fields on a DNA chip.

Several methods are used to transfer genes into human cells. These methods include using viral vectors, chemically assisting transfer of genes across cell membranes, and fusing cells with artificial vesicles containing cloned DNA sequences. In the first generation of gene therapy trials, the most common method of gene transfer used genetically modified retroviruses as vectors. The first retroviral vectors were based on a mouse virus called *Moloney murine leukemia virus* (Figure 18–14). To create this vector, a cluster of three genes was removed from the virus, allowing a cloned human gene to be inserted. After being packaged into a viral protein coat, the recombinant vector can infect cells, but cannot replicate itself, because of the missing viral genes. Once inside the cell, the recombinant virus with the inserted human gene moves to the nucleus and integrates into a chromosome, where it becomes part of the cell's genome. In initial attempts at gene therapy, several heritable disorders, including **severe combined immunodeficiency (SCID)**, **familial hypercholesterolemia**, and **cystic fibrosis**, were treated. In the upcoming sections, we will describe the first attempt to use gene therapy to treat a young girl with SCID and then give an overview of the trials that are currently underway in several animal-model systems, using a new generation of viral vectors. Finally, we will discuss several of the ethical issues raised by the use of gene therapy.

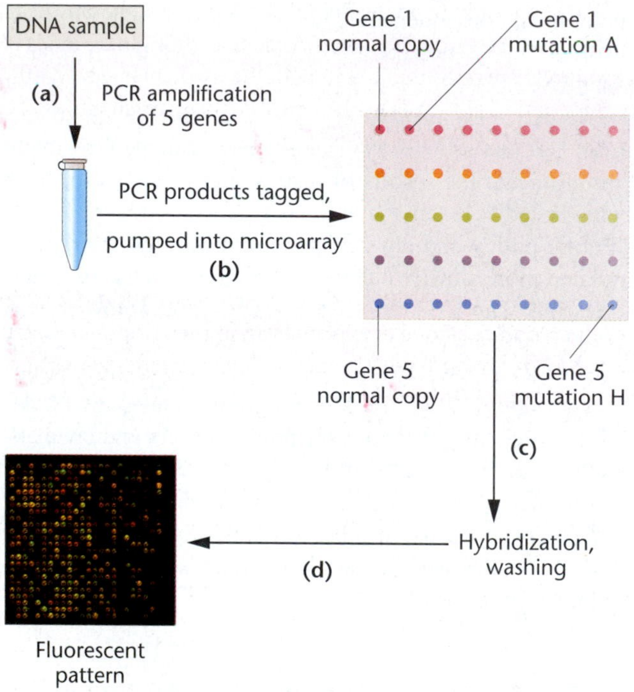

FIGURE 18–13 Gene screening, using a DNA microarray. (a) DNA extracted from a blood sample is amplified by PCR. In this example, primers for five genes are used, but in practice, many more are used. (b) The microarray contains single-stranded probes for the normal allele of each of the genes (Column I) and eight mutant alleles for each of the five genes (1 row = 1 gene). (c) The single-stranded PCR products are tagged with fluorescent probes and pumped into the microarray. (d) The resulting hybridization is revealed by the pattern and color of the spots on the microarray. The microarrays are scanned by a laser and analyzed by software, and the results can be presented in several formats. In practice, microarrays containing several hundred thousand probes are used. Each probe is attached to the glass substrate and occupies a different field on the microarray.

Gene Therapy for Severe Combined Immunodeficiency (SCID)

The first gene therapy trials began in 1990 with the treatment of a young girl, Ashanti DeSilva, who has a genetic disorder called severe combined immunodeficiency (SCID). Affected individuals have no functional immune system and usually die from what would otherwise be minor infections (Figure 18–15). An autosomal form of SCID is caused by a mutation in the gene encoding the enzyme **adenosine deaminase (ADA)**. Treatment started with isolation of white blood cells called *T cells* from the patient (Figure 18–16). These cells, which are part of the immune system, are mixed with a genetically modified retrovirus carrying a normal copy of the *ADA* gene (Figure 18–16). The virus infects the T cells and inserts a normal copy of the *ADA* gene into the cells' genome. The genetically modified T cells are grown in the laboratory in order to ensure that the transferred gene is expressed, and the patient is treated by injecting a billion or so of the altered T cells into the bloodstream.

After treatment, Ashanti has maintained normal ADA protein levels in 25–30 percent of her T cells, and she now leads a normal life. Unfortunately, a second child treated a short time later had the normal gene in only 0.1 to 1 percent of her

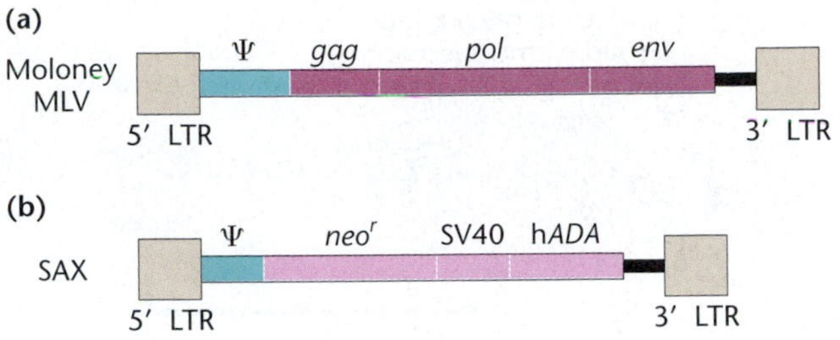

FIGURE 18–14 Retroviral vectors constructed from the Moloney murine leukemia virus (Moloney MLV). The native MLV genome contains a ψ sequence required for encapsulation and genes that encode viral coat proteins (*gag*), an RNA-dependent DNA polymerase (*pol*), and surface glycoproteins (*env*). At each end, the genome is flanked by long terminal repeat (LTR) sequences that control transcription and integration into the host genome. The SAX vector retains the LTR and ψ sequences, and it includes a bacterial neomycin resistance (neo^r) gene that can be used as a selective marker. As shown, the vector carries a cloned human adenosine deaminase (h*ADA*) gene, which is fused to an SV40 early region promoter–enhancer. The SAX construct is typical of retroviral vectors that are used in human gene therapy.

white blood cells after treatment, a level not high enough to be effective. In later trials, attempts were made to transfer the *ADA* gene into bone marrow cells that form T cells, but were mostly unsuccessful.

Problems and Failures in Gene Therapy

Over a 10-year period from 1990–1999, more than 4000 people underwent gene transfer. Unfortunately, these trials were largely failures and led to a loss of confidence in gene therapy. Hopes for gene therapy plummeted even further in September 1999 when a teenager, Jesse Gelsinger, died during gene therapy. His death was triggered by a massive immune response to the vector, a modified adenovirus. (Adenoviruses cause colds and respiratory infections.)

FIGURE 18–15 David, a boy born with severe combined immunodeficiency (SCID), survived by being isolated in a germ-free environment. He died at the age of 12 following a bone marrow transplant undertaken to provide him with a functional immune system.

Many of the problems with gene therapy have been traced to inefficient vectors. These first-generation vectors, such as Moloney virus and adenovirus, have several drawbacks. First, integration of the retroviral genome (including the cloned human gene) into the host cell's genome occurs only if the host cells are replicating their DNA. Under *in vivo* conditions, there is little DNA synthesis in many highly differentiated cell types that are suitable target tissues. Second, most of these viruses eventually elicit an immune response in the host, as happened in Jesse's case. Third, insertion of viral genomes into the host chromosome can inactivate or mutate an indispensable gene. Fourth, retroviruses have a low cloning capacity and cannot carry inserted sequences much larger than 8 kb; many human genes, even without introns, exceed this size. Finally, there is the possibility of producing an infectious virus if a recombination event takes place between the vector and retroviral genomes already present in the host cell. For these reasons, other viral vectors and strategies for targeting cells are being developed.

The Future of Gene Therapy: New Vectors and Target-Cell Strategies

The disappointing and deadly results from gene therapy trials that used first-generation vectors led to a widespread crisis of confidence in gene therapy in both the medical and scientific communities. However, with some promising results in several animal model systems, the tide may finally have turned. This renewed hope is due primarily to the use of new vectors that both circumvent several of the problems encountered with earlier vectors and have several ingenious design features that allow the levels of gene product to be regulated.

In early 1999, scientists reported the successful production of the blood hormone erythropoeitin in rhesus monkeys and mice whose muscle cells had been injected with a viral vector containing the erythropoeitin gene. These animals produced the hormone only when they received the antibiotic rapamycin. The results are particularly encouraging because the levels of the hormone are quite high, the presence of the foreign DNA has not triggered an immune response, and the gene can be repeatedly activated.

The success of these trials can be attributed in large part to a new viral vector used in the study. Adeno-associated virus (see Table 18–1 for a list of several second-generation vectors) is related to adenovirus, but has several key advantages: It can incorporate into nonreplicating DNA, and it is a very small virus, so it does not typically elicit an immune response. Furthermore, the vector incorporated a rapamycin-responsive promoter that is used to switch on the erythropoietin gene. The gene is expressed only when cells are exposed to rapamycin. This therapy is being considered for patients who have low blood counts, such as dialysis patients, who currently receive regular injections of erythropoietin itself.

In 2000, a group of French researchers modified a first-generation retroviral vector and used it successfully to treat an X-linked form of SCID, and other researchers, using new, second-generation vectors, have had initial success treating

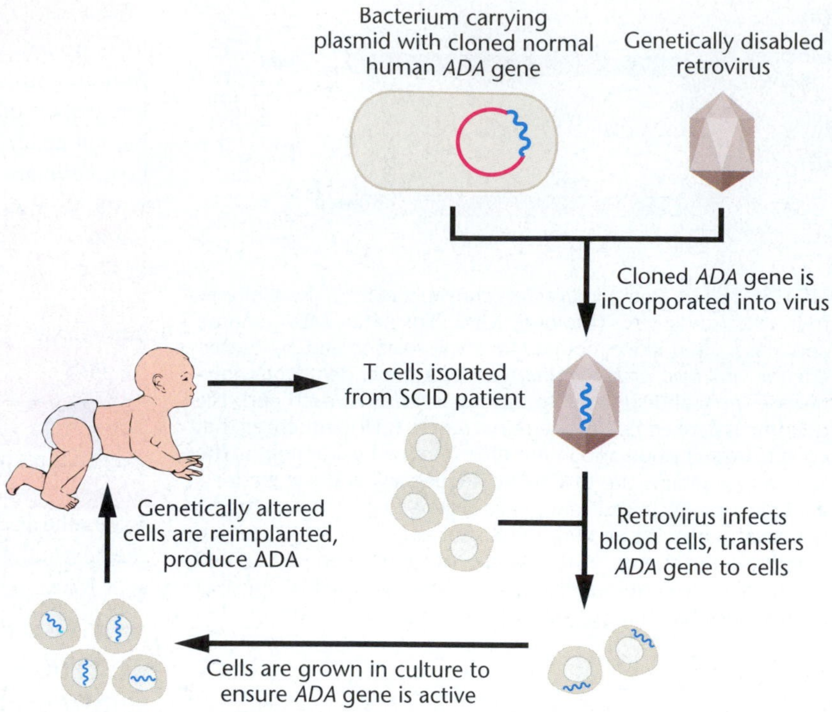

FIGURE 18–16 Gene therapy for treatment of severe combined immunodeficiency (SCID), a fatal disorder of the immune system caused by lack of the enzyme adenosine deaminase (ADA). The cloned human *ADA* gene is transferred into a viral vector, which is used to infect white blood cells removed from the patient. The transferred *ADA* gene is incorporated into a chromosome and becomes active. After growth to enhance their numbers, the cells are reimplanted into the patient, where they produce ADA, allowing the development of an immune response.

hemophilia. In addition to treating heritable genetic disorders and patients on dialysis, gene therapy is now in clinical trials for treating several forms of cancer, infectious diseases, and cardiovascular diseases (Table 18–2). In the last few years, dozens of biotechnology companies have been founded specifically to develop products for gene therapy, and it is possible that, sometime in the 21st century, gene therapy will be a commonplace method of treatment for a large number of genetic disorders.

Ethical Issues and Gene Therapy

Gene therapy obviously raises many ethical concerns, and these are still sources of intense debate. As a result, all gene therapy trials currently underway or in the planning stages are restricted to the use of somatic cells as targets for gene transfer. The ethical guidelines for gene therapy in its present form have been revised and strengthened in the wake of Jesse Gelsinger's death. This experimental treatment is initiated only after careful review at several levels, and the trials are monitored to protect the interests of the patient. This form of gene therapy is called **somatic gene therapy**; only one individual is affected, and the therapy is done with the permission and informed consent of the patient.

Two other forms of gene therapy have not been approved, primarily because of the unresolved ethical issues surrounding them. The first is called **germ line therapy**, where germ cells (the cells that give rise to the gametes, i.e., the sperm and eggs) or mature gametes are used as targets for gene transfer. In this approach, the transferred gene will be incorporated into all of the individual's cells produced from the genetically altered gamete, including his or her own germ cells. This means that individuals in future generations will also be affected, without their consent. Is this procedure ethical? Do we have the right to make this decision for future generations? Thus far, the concerns have outweighed the potential benefits, and such research is prohibited.

The second form of gene therapy, which raises an even greater ethical dilemma, is termed **enhancement gene therapy**, whereby human potential might be enhanced for some desired trait. Such use of gene therapy is extremely controversial and strongly opposed by many. Should genetic technology be used to enhance human potential? For example, if such genes were identified and cloned, should it be

TABLE 18–1 New Vectors for Gene Therapy

Vector	*Cell Targets*	*Cloning Capacity*	*Advantages*	*Disadvantages*
Adenovirus	Lung, respiratory tract	7.5 kb	Efficient transfection	Strong immune response
Adeno-associated virus	Fibroblasts, T cells, others	4.5 kb	Transfects many cell types	Small insert size
Retroviruses	Proliferating cells	8 kb	Prolonged expression	Low transfection efficiency
Lentiviruses	Stem cells, proliferating cells	8 kb	Efficient transfection	Related to HIV

TABLE 18–2 Ongoing Gene Therapy Trials

Disease/Disorder	*Number of Trials*
Cancer	216
Single-gene disorders	49
Infectious diseases	24
Cardiovascular diseases	8

permissible to use gene therapy to increase height, to enhance athletic ability, or to extend intellectual potential? Presently, the consensus is that enhancement therapy, like germ line therapy, is an unacceptable use for gene therapy. However, the issues are still being debated and are unresolved. The outcome of these debates may affect not only the fate of individuals, but that of our species as well.

18.4 DNA Fingerprints

As discussed earlier, the presence or absence of restriction sites in the human genome can be used as genetic markers. Another type of nucleotide sequence variation, discovered in the mid-1980s, depends on variation in the length of repetitive DNA sequence clusters. These polymorphisms in human DNA serve as the basis for a technique called **DNA fingerprinting**. DNA fingerprinting is now used for a wide variety of applications, from identifying individuals for paternity to forensics and conservation (see Chapter 27).

Minisatellites (VNTRs) and Microsatellites (STRs)

One of the most useful forms of RFLPs arises from variations in the number of tandemly repeated DNA sequences present between two restriction enzyme sites. These sequences are **minisatellites**, or clusters of 10–100 nucleotides. For example, the sequence

5'-GACTGCCTGCTAAGATGACTGCCTGCTAAGATC
GACTGCCTGCTAAGATGACTGCCTGCTAAGATCT
GACTGCCTGCTAAGATGACTGCCTGCTAAGATAA
GACTGCCTGCTAAGATGACTGCCTGCTAAGATGA
TGAC TGCCTGCTAAGAT-3'

is composed of nine tandem repeats of the 16 base-pair sequence GACTGCCTGCTAAGAT. Clusters of such sequences are widely dispersed in the human genome and the number of repeats at each locus ranges from 2 to more than 100. These loci, known as **variable-number tandem repeats (VNTRs)** were introduced in Chapter 4 as examples of middle repetitive DNA. The number of repeats at a given locus is variable, and each variation constitutes a VNTR allele. Many loci have dozens of alleles each; as a result, heterozygosity is common.

A pattern of bands is produced when DNA is cut with restriction enzymes and visualized by Southern blotting using VNTR sequences as probes. This pattern is known as a DNA fingerprint (Figure 18–17). These patterns are the equivalent of actual fingerprints because the pattern of bands is always the same for a given individual, no matter what tissue is used as the source of the DNA, but the pattern varies from individual to individual, as do real fingerprints. In fact, there is so much variation in the band pattern from individual to individual that, theoretically, each person's pattern is essentially unique. This technique was developed during the 1980s and was first used to solve the murders of two school girls in a high-profile case in Great Britain.

An important limitation of DNA fingerprint analysis is that it requires a relatively large sample of DNA (10,000 cells or 50ng), more than is usually found at a typical crime scene, and the DNA must be relatively intact (non-degraded). Thus, DNA fingerprinting has been more useful in the area of paternity testing than in forensics. In paternity testing, blood draws from the child, mother, and alleged father provide an unlimited source of fresh, intact cells for DNA extraction and analysis.

Forensic scientists have recently developed a different set of similar markers that are analyzed by the polymerase chain reaction, rather than RFLP analysis, allowing trace evidence samples to be typed. These markers, **short tandem repeats (STRs)** are very similar to VNTRs but the repeated motif is shorter, between 2 and 9 base-pairs. Thirteen tetrameric (four base-pair repeat) STRs have been developed into a marker panel that is currently being used by the FBI to create a database of DNA profiles of convicted felons (Figure 18–18). It is also used now used routinely in forensic casework to generate DNA profiles from trace samples (e.g. single hairs, saliva from a cigarette butt or a toothbrush, etc.) and from samples that are old and/or degraded (e.g. a skull in a field, Egyptian mummies that are over 2400 years old). As a result, STRs have replaced VNTRs in most forensic laboratories. In addition, because STR typing is less expensive, less labor-intensive, and much quicker to perform than RFLP analysis, it is fast replacing VNTR typing in most paternity laboratories as well.

The results of STR analysis are analyzed and interpreted using statistics, probability, and population genetics. The population frequency of each STR allele in the standard set has been measured in many groups across the U.S. Using this information, the probability of having any combination of these alleles can be calculated. For example, if an allele of locus 1 is carried by 1 in 333 individuals, and an allele of locus 2 is carried by 1 in 83 individuals, the probability that someone will carry both alleles is equal to the product of their individual frequencies, or 1 in 28,000. This overall probability may not be especially convincing, but if an allele of a third locus (carried by 1 in 100 individuals) and an allele of a fourth locus (carried by 1 in 25 individuals) are included in the calculations, the combined frequency becomes 1 in 70 million. That is, the chance that anyone has this combination of alleles is about 1 in 70 million, or about 4 individuals in the U.S. population. When the genotype of

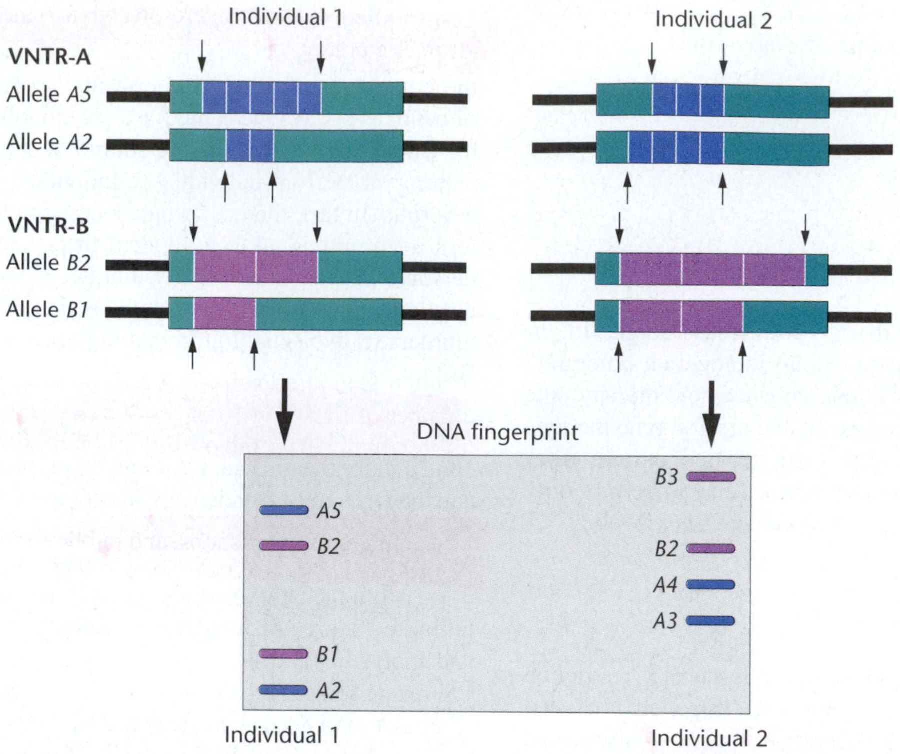

FIGURE 18–17 VNTR loci and DNA fingerprints. VNTR alleles at two loci (*A* and *B*) are shown for each individual. Arrows mark restriction enzyme cutting sites flanking the VNTRs. Restriction enzyme digestion produces a series of fragments that can be detected as bands on a Southern blot (bottom). Because of differences in the number of repeats at each locus differ, the overall pattern of bands is distinct for each individual, even though one band is shared (the B2 allele band). Such a pattern is known as a *DNA fingerprint*.

MEDIA TUTORIAL DNA Fingerprinting

the alleles of all thirteen CODIS STR loci (26 alleles in all) are generated to produce a complete STR profile, a typical chance that anyone has this same combination is 1 in 100 trillion. Since the planet's population is only about 6 billion, it is easy to see why STR analysis is often referred to as **human DNA identification testing**.

Forensic Applications of DNA Fingerprints

In a criminal case, if the suspect's DNA fingerprints do not match that of the evidence, they can be excluded as the criminal (exclusions occur in about 30% of all cases). When a match is made between DNA obtained at a crime scene and the DNA of a suspect, there are two possible interpretations: The DNA fingerprint came from the suspect, or it is from someone else with the same pattern of bands. The DNA fingerprint evidence does not prove the suspect is guilty, but is one piece of evidence that must be considered along with other facts in the case. However, with the advent of STR typing, more forensic scientists, particularly in the United States, are willing to testify that the profile could only have come from one person and, in their expert opinion, provides a direct link between a crime scene sample and a suspect that has the same STR profile as the sample.

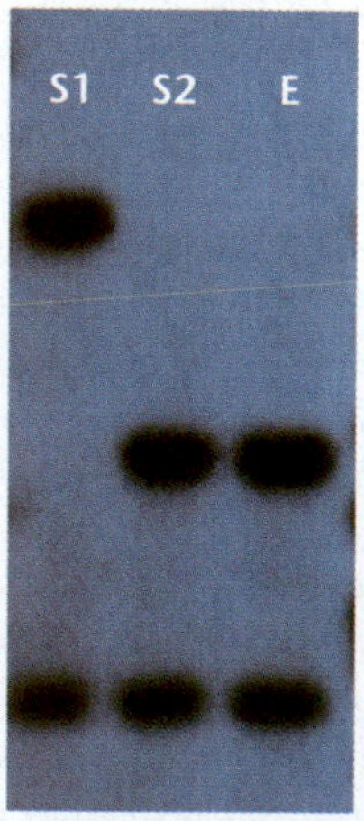

FIGURE 18–18 DNA fingerprinting in a forensic case. The DNA profile of suspect 2 (S2) matches that of the blood sample obtained as evidence *E*.

18.5 Genome Projects Use Recombinant DNA Technology

Over the past 90 years, geneticists have developed efficient methods to generate and map mutations as a way of identifying and locating genes. One drawback to this approach is that genes can be identified and mapped only if mutants are isolated. To identify all genes in a genome by this method, at least one mutation for each gene is required, and because some mutant alleles are lethal, as well as for other technical reasons, this goal is probably impossible.

Geneticists now use recombinant DNA technology to identify all the genes in a genome directly. Instead of screening for mutants and constructing genetic maps, a genomic library of clones is established, and the nucleotide sequence of all the clones is determined. The result is a genomic map, with all genes in the genome identified by nucleotide sequence and location. (See Chapter 17 for a discussion of genomic mapping.) Analysis of the derived amino acid sequence and the use of gene-specific mutants help establish the function of the encoded proteins. In addition, sequencing provides information about nongene components of a genome—regions that are beyond the reach of mutation and mapping strategies.

Genomes of hundreds of organisms have been sequenced, and hundreds more are partially completed. The genomes of several model organisms have been sequenced by the publicly funded Human Genome Project, while other genomes have been sequenced by organizations such as The Institute for Genome Research (TIGR) and companies such as Celera. Nonhuman organisms included in the Human Genome Project were selected for several reasons, including the genome's size; the availability of detailed genetic maps; and the value of constructing comparative maps to study gene number, location, and function across species ranging from bacteria to hum-ans.

The Human Genome Project: An Overview

The **Human Genome Project** (HGP) is an international, coordinated effort to determine the sequence of the 3.3 billion nucleotide pairs in the haploid human genome. In the United States, the project began in 1990 as a joint program of the National Institutes of Health and the U.S. Department of Energy. Other countries, notably France, Britain, and Japan, began similar projects, which are now coordinated by an international organization, the Human Genome Organization (HUGO).

The task of sequencing the human genome began with a series of well-planned steps. (See Chapter 17 for a detailed discussion.) Work progressed more or less simultaneously on all phases of the project in laboratories around the world (Figure 18–19). In addition to publicly funded genome projects, a private effort to sequence the human genome was launched in 1998 as a joint venture between two companies, PE BioSystems and Celera. In February 2001, each genome project published an analysis of the approximately 96 percent of the sequence available at that time. Hundreds of small gaps in the sequence remain to be closed. Details of the findings are discussed in Chapter 17.

The HGP has also finished sequencing genomes of four of the five model organisms: *E. coli* (bacteria), *S. cerevisiae* (yeast), *C. elegans* (a metazoan roundworm), and the fruit fly (*D. melanogaster*). Both the HGP and the private joint venture are sequencing the genome of the mouse, which should be completed by 2003.

Knowledge gained by sequencing the human genome will greatly advance our understanding of human genetics and will have a great impact on biomedical research and health care. However, applications of the knowledge gained from the project raise ethical, social, and legal issues that should be identified, debated, and resolved, often in the form of laws or public policy.

The Ethical, Legal, and Social Implications (ELSI) Program

At the time that the Human Genome Project (HGP) was first being discussed, scientists and the public raised concerns about how genome information would be used and how the interests of both individuals and society can be protected. To address these concerns, the Ethical, Legal, and Social Implications (ELSI) Program was established as part of the Human Genome Project. The ELSI program considers a number of issues, including the impact of genetic information on individuals, the privacy and confidentiality of genetic information, implications for medical practice, genetic counseling, and reproductive decision making. Through research grants, workshops, and public forums, ELSI is formulating policy options to address these issues.

Currently, ELSI is focusing on four areas: (1) privacy and fairness in the use and interpretation of genetic information, (2) ways of transferring genetic knowledge from the research laboratory to clinical practice, (3) ways to ensure that participants in genetic research know and understand the potential risks and benefits of their participation and give informed consent, and (4) public and professional education. Hopefully, as the HGP moves from generating information about the genetic basis of disease to improving treatments, promoting prevention, and developing cures, ethical issues will be identified and an international consensus will be developed on appropriate policies and laws.

After the Genome Projects

As outlined here and in Chapter 17, several genome projects have already reached their goals. Although analysis of the human genome is still in its early stages, important insights into human genetic disorders are already emerging. For example, a new mechanism of mutation, the expansion of trinucleotide repeats, has been discovered. (See Chapter 7.) Expansion of trinucleotide repeats is responsible for several neurodegenerative diseases, including Huntington disease, myotonic dystrophy, and spinal/bulbar muscular atrophy. Another unexpected finding is that duplicating small regions of chromosomes can increase gene dosage; this condition leads to disorders such as Charcot-Marie-Tooth syndrome. More intriguing is the finding that mutations in different parts of a single gene can give rise to different genetic disorders. For example, multiple endocrine neoplasia, thyroid carcinoma, and Hirschprung disease arise from different mutations in a single gene.

The shift in research focus that occurred after genome projects were completed in model organisms such as yeast and *Drosophila* gives us a preview of what to expect once a genome has been sequenced. In these organisms, about half of the genes in the genome are newly discovered. Attention is shifting to two areas: identifying the function of the newly discovered genes and analyzing patterns of gene expression. For example, what genes are active in a cell at a given time?

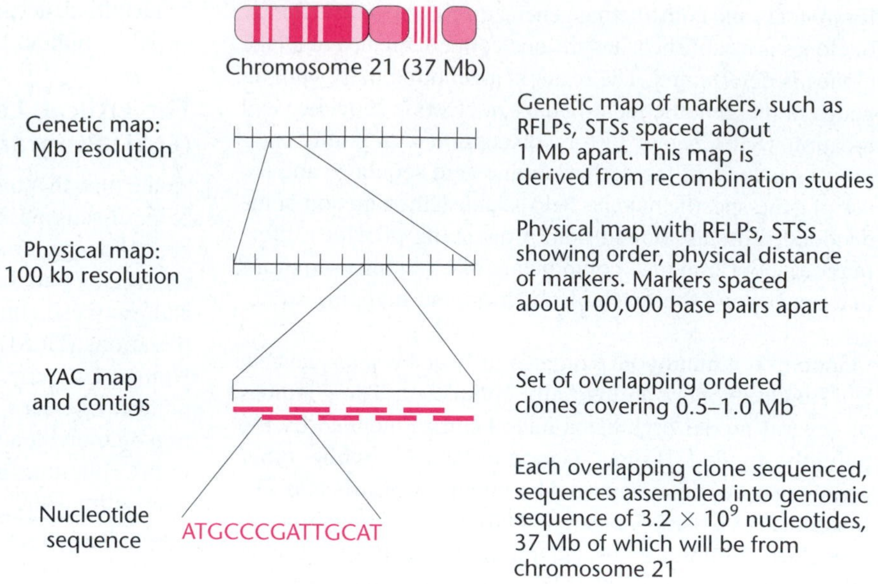
Chromosome 21 (37 Mb)
Genetic map: 1 Mb resolution
Genetic map of markers, such as RFLPs, STSs spaced about 1 Mb apart. This map is derived from recombination studies
Physical map: 100 kb resolution
Physical map with RFLPs, STSs showing order, physical distance of markers. Markers spaced about 100,000 base pairs apart
YAC map and contigs
Set of overlapping ordered clones covering 0.5–1.0 Mb
Nucleotide sequence
ATGCCCGATTGCAT
Each overlapping clone sequenced, sequences assembled into genomic sequence of 3.2 × 10⁹ nucleotides, 37 Mb of which will be from chromosome 21

FIGURE 18–19 An overview of the original strategy used in the Human Genome Project. The first goal, achieved in 1995, was to have a genetic map of each chromosome, with markers spaced at distances of about 1 Mb (1 million base pairs of DNA). This work was accomplished by finding markers such as RFLPs and STSs and assigning them to chromosomes. Once assigned to chromosomes, the markers' inheritance was observed in heterozygous families in order to establish the order and distance between them (a genetic map). In the second stage, the goal was to prepare a physical map of each chromosome (our example uses chromosome 21, the smallest chromosome), containing the location of markers spaced about 100,000 base pairs apart. This goal has now been achieved. The third stage involved the construction of a set of overlapping clones in yeast artificial chromosomes (YACs) or other vectors that cover the length of the chromosome. The last stage was the sequencing of the entire genome. Sequencing by means of this method and another method has produced a first draft of the genome sequence.

When are they active in developmental or physiological processes? What protein functions do they encode? As outlined in Chapter 17, researchers are now analyzing the **transcriptome**, the set of mRNA molecules produced by a cell at any given time, and the **proteome**, the expressed set of proteins present in a cell at a given time. As these fields develop and are applied to the human genome, we can expect to see new generations of therapeutic drugs, derived from functional analysis of the expressed genome and the proteome.

Studying Gene Expression with Microarrays

Although the emphasis in this chapter is on applications of biotechnology in medicine, pharmaceuticals, and agriculture, biotechnology has also revolutionized areas of basic research. Differential gene expression is what makes a muscle cell different from a liver cell, a cell in one stage of embryonic development different from a cell in another stage, or a cancer cell different from a normal cell. In addition, a given cell type responds to signals from the environment by changing patterns of gene expression. To determine which genes are expressed in a cell at any given time, or in response to a given signal, researchers typically isolate all the cytoplasmic mRNA. Using this mRNA, researchers then analyze for the presence or absence of transcripts one gene at a time. Analysis of each gene's pattern of expression is a different experiment, requiring several days to complete.

DNA microarrays, developed by Patrick Brown and his colleagues at Stanford University have revolutionized studies of gene expression and many other areas of research. Instead of measuring patterns of gene expression one gene at a time, researchers can analyze hundreds or thousands of different genes in a single day.

Figure 18–20 shows an experiment, to analyze patterns of gene expression in normal cells and cancer cells derived from the same cell type. In this experiment, mRNA is isolated from the normal cell and the cancer cell. The mRNA population represents all the genes expressed in the normal cell and the cancer cell. By using reverse transcriptase, the mRNA is converted to cDNA and labeled with a fluorescent dye. The normal cells have a green fluorescent tag; the cDNA from the cancer cells have a red tag. The labeled cDNAs are mixed and pumped through a DNA microarray. The cDNAs can bind to complementary single-stranded probe DNAs on the chip, but not to other DNAs. A laser scans the chip, and the resulting pattern of hybridization is presented as a series of colored dots, with each dot corresponding to a field on the microarray.

If the DNA in a field binds only the cDNA from the cancer cell, the dot will be red; if only the cDNA from the normal cell binds, the dot will be green. If DNA on the microarray binds cDNAs from both cell types equally, the dot will be yellow (Figure 18–21). Other colors represent differential amounts of the two cDNAs and indicate different levels of gene expression in the normal cell and in the cancerous cell.

DNA microarrays and the research tools developed for genome projects are fundamentally changing the way basic research is done, allowing data to be collected on a scale that was unimaginable only a few years ago. This approach, called high throughput or systems biology, links the methods of genomics, proteomics, and microarrays and will speed the

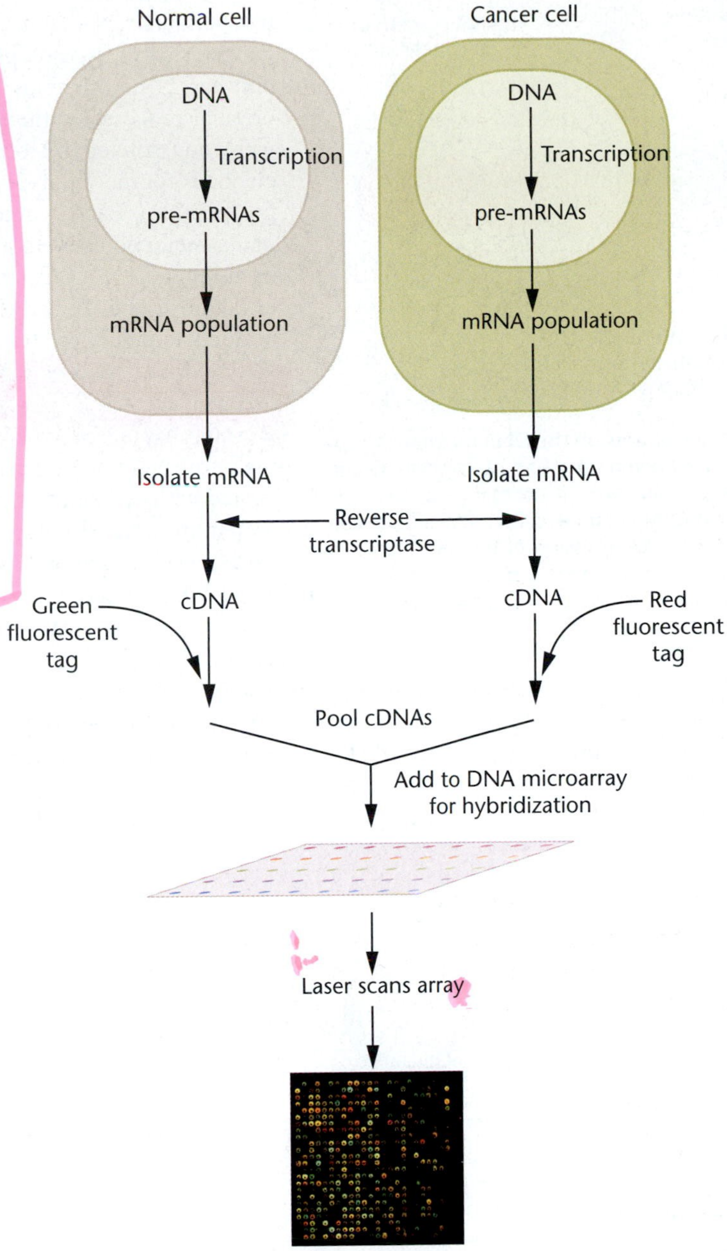

FIGURE 18–20 Analysis of gene expression in normal and cancer cells using a DNA microarray. Messenger RNA is isolated from each cell type and used to make cDNA which is labeled with a fluorescent dye (red for cDNA from cancer cells, and green for cDNA from normal cells). The cDNAs are pooled and pumped through a DNA microarray containing probes for several thousand genes. The cDNAs complementary to the probes will bind, while all others will be washed off the chip. The pattern of hybridization is visible as a series of colored dots, one for each field on the microarray. A laser is used to scan the array and software programs are used to determine the pattern of expression for normal and cancer cells.

rate at which we can study and understand fundamental processes in genetics and biology.

18.6 Biotechnology Is an Outgrowth of Recombinant DNA Technology

Although recombinant DNA techniques were originally developed to facilitate basic research on gene organization and regulation of expression, gene transfer methods are also the foundation of the biotechnology industry. Recombinant DNA technology is used to manufacture a wide range of products, including hormones, clotting factors, herbicide-resistant plants, enzymes for food production, and vaccines. In the last decade, the biotechnology industry has grown into a multibillion-dollar segment of the economy. In this section, we will review some of the current uses of biotechnology and the issues raised by this application of recombinant DNA technology.

Insulin Production by Bacteria

The first human gene product manufactured by using recombinant DNA and licensed for therapeutic use was human insulin, which became available in 1982. Insulin is a protein hormone that regulates sugar metabolism. Individuals who cannot produce insulin have diabetes, a disease that, in its more severe form, affects more than 2 million individuals in the United States.

Clusters of cells embedded in the pancreas synthesize a precursor peptide known as *preproinsulin*. As this polypeptide is secreted from the cell, amino acids are cleaved from the end and the middle of the chain. This process produces

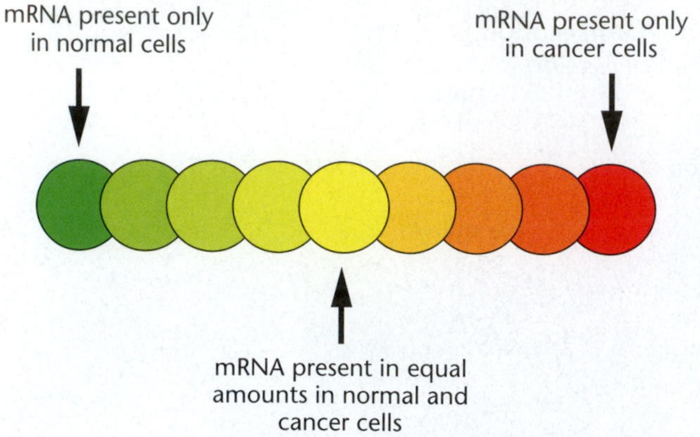

FIGURE 18–21 The color of dots on the DNA microarray represent levels of gene expression in normal and cancer cells. Green dots represent genes expressed only in normal cells, red dots represent genes expressed only in cancer cells. Intermediate colors represent different levels of expression of the same gene in normal and cancer cells. From: *Hamadeh, H. and Afshari, C.A. 2000. Gene chips and functional genomics. Am. Sci. 88: 508–515, Fig. 6, P. 513.*

the mature insulin molecule, which consists of two polypeptide chains (*A* and *B* chains), joined by disulfide bonds.

Although human insulin is now produced synthetically by another process, a look at the original method is instructive, as it shows both the promises and the difficulties of applying recombinant DNA technology. The functional insulin protein consists of two polypeptide chains, *A* and *B*. The *A* subunit has 21 amino acids, and the *B* subunit has 30. In the original process, synthetic genes for the *A* and *B* subunits were constructed by oligonucleotide synthesis (63 nucleotides for the *A* polypeptide and 90 nucleotides for the *B* polypeptide). Each synthetic oligonucleotide was inserted into a vector adjacent to a gene encoding the bacterial form of the enzyme, β-galactosidase. When transferred to a bacterial host, the β-*gal* gene and the adjacent synthetic oligonucleotide were transcribed and translated as a unit. The product, a **fusion polypeptide**, consisted of the amino acid sequence for β-galactosidase attached to the amino acid sequence for one of the insulin subunits (Figure 18–22). The fusion proteins were purified from bacterial extracts and treated with cyanogen bromide in order to cleave the fusion protein from the β-galactosidase.

Each insulin subunit was produced separately by this process. When mixed, the two subunits spontaneously unite, forming an intact, active insulin molecule. The purified insulin is then packaged for use by diabetics who must take insulin injections.

A number of genetically engineered proteins for therapeutic use have been produced by similar methods or are in clinical trials (Table 18–3). In most cases, cloning a human gene into a plasmid and inserting the recombinant vector into

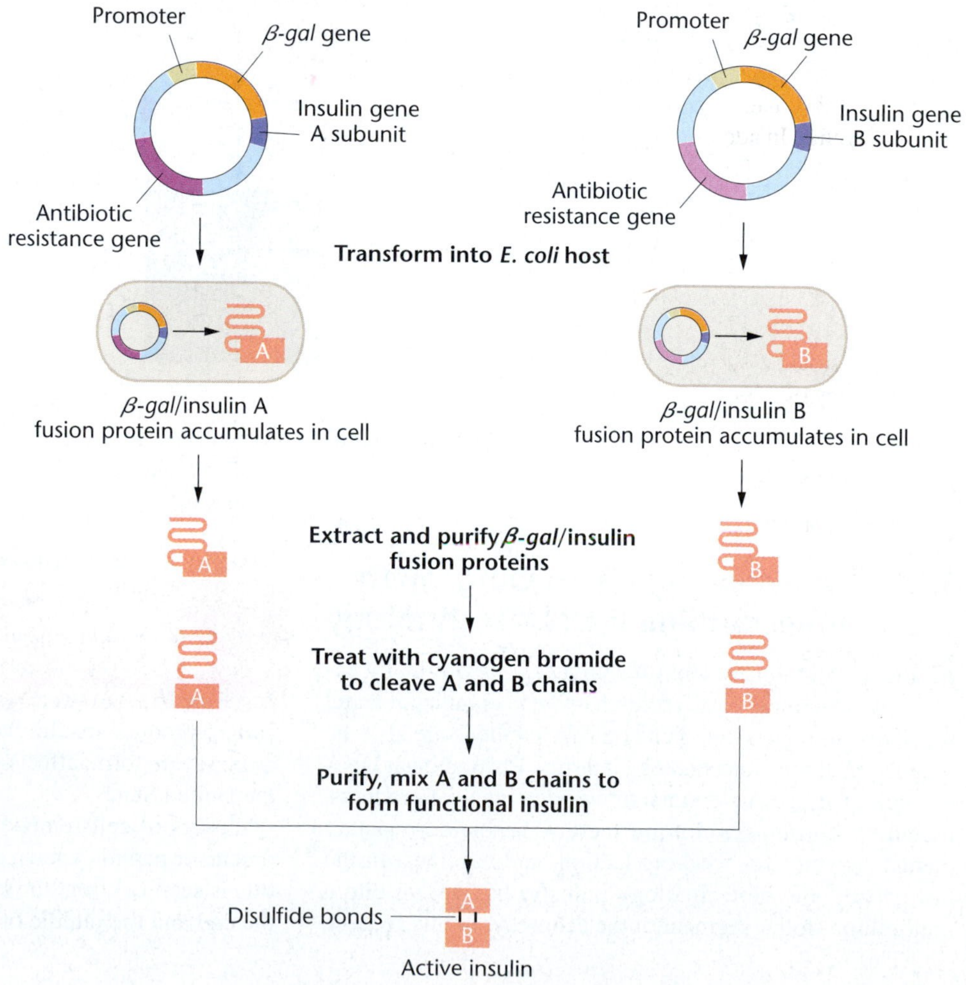

FIGURE 18–22 Method used to synthesize recombinant human insulin. Synthetic oligonucleotides encoding the insulin *A* and *B* chains were inserted at the tail end of a cloned *E. coli* β-*galactosidase* (β-*gal*) gene. The recombinant plasmids were transferred to *E. coli* hosts, where the β-gal/insulin fusion protein was synthesized and accumulated in the host cells. Fusion proteins were extracted from the host cells and purified. Insulin chains were released from the β-galactosidase by treatment with cyanogen bromide. The insulin subunits were purified and mixed to produce a functional insulin molecule.

TABLE 18–3 Genetically Engineered Pharmaceutical Products Now Available or in Clinical Testing

Gene Product	*Condition Treated*
Atrial natriuretic factor	Heart failure, hypertension
Epidermal growth factor	Burns, skin transplants
Erythropoietin	Anemia
Factor VIII	Hemophilia
Gamma interferon	Cancer
Granulocyte colony-stimulating factor	Cancer
Hepatitis B vaccine	Hepatitis
Human growth hormone	Dwarfism
Insulin	Diabetes
Interleukin-2	Cancer
Superoxide dismutase	Transplants
Tissue plasminogen activator	Heart attacks

a bacterial host produces the proteins. After ensuring that the transferred gene is expressed, large quantities of the transformed bacteria are produced, and the human protein is recovered and purified.

Transgenic Animal Hosts and Pharmaceutical Products

Bacterial hosts were used to produce the first generation of recombinant proteins, even though there are some disadvantages in using prokaryotic hosts to synthesize eukaryotic proteins. For example, bacterial cells are unable to process and modify many eukaryotic proteins. Thus, they cannot add sugars and phosphate groups that are often needed for full biological activity. In addition, eukaryotic proteins produced in prokaryotic cells often don't fold into the proper three-dimensional configuration and, as a result, are inactive. To overcome these difficulties and to increase yields, second-generation methods use eukaryotic hosts. Rather than being produced by host cells grown in tissue culture, human proteins such as alpha-1-antitrypsin are being produced in the milk of livestock, as described in the following discussion.

A deficiency of the enzyme alpha-1-antitrypsin is associated with the heritable form of emphysema, a progressive and fatal respiratory disorder common among people of European ancestry. To produce alpha-1-antitrypsin by genetic engineering, the human gene was cloned into a vector at a site adjacent to a sheep promoter sequence that regulates the expression of milk-associated proteins. Genes placed next to this promoter are expressed only in mammary tissue. This fusion gene was microinjected into sheep zygotes fertilized *in vitro* (Figure 18–23), which, in turn, were implanted into surrogate mothers. The resulting **transgenic** sheep developed normally and, after mating, produced milk that contained high concentrations of functional human alpha-1-antitrypsin. This human protein is present in concentrations up to 35 g/L of milk. It is easy to envision that a small herd of lactating sheep could easily provide an adequate supply of this protein and that herds of other transgenic animals, acting as biofactories, might become part of the pharmaceutical industry. In fact, the cloning of Dolly the sheep (see "Genetics, Technology and Society" in this chapter) was done in order to facilitate the establishment of a flock of sheep that will produce high levels of human proteins.

Transgenic Crop Plants and Herbicide Resistance

Damage from weed infestation destroys about 10 percent of crops worldwide. To combat this problem, more than 100 different herbicides are used, at an annual cost of more than 10 billion dollars. Some of these herbicides also kill crop plants, while others remain in the environment or are present in runoff and contaminate water supplies. Creating herbicide-resistant plants, using recombinant DNA technology, is one way to improve crop yields. In this procedure, vectors transfer traits for herbicide resistance to crop plants. For example, the herbicide glyphosate is effective at very low concentrations, is not toxic to humans, and is rapidly degraded by soil microorganisms. Glyphosate works by inhibiting the action of a chloroplast enzyme called *EPSP synthase*. This

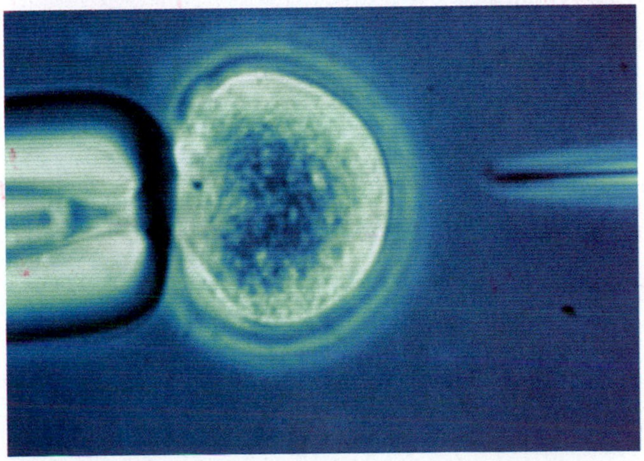

FIGURE 18–23 Injection of cloned DNA. A micropipette is used to transfer cloned genes into the nucleus of a mammalian zygote. The injected zygote will then be transferred to the uterus of a surrogate mother for development.

enzyme is important in amino acid biosynthesis in both bacteria and plants. Without the ability to synthesize vital amino acids, plants wither and die.

The development of glyphosate-resistant crop plants began with the isolation and cloning of an EPSP synthase gene from a glyphosate-resistant strain of *E. coli*. This gene was cloned into a vector adjacent to plant promoter sequences and upstream from plant transcription-termination sequences. The recombinant vector was transferred into the bacterium *Agrobacterium tumifaciens* (Figure 18–24). Plasmid-carrying bacteria were, in turn, used to infect cells in discs cut from plant leaves. Calluses formed from these discs were then selected for their ability to grow on glyphosate. Transgenic plants generated from glyphosate-resistant calluses were grown and sprayed with glyphosate at concentrations four times higher than that needed to kill wild-type plants. The transgenic plants overproducing the EPSP synthase grew and developed, while the control plants withered and died. Glyphosate-resistant corn and soybeans developed in this way are now on the market.

While use of these genetically engineered crops should reduce herbicide costs to the farmer, the crops do present an important ethical dilemma: Is it appropriate to generate crops that encourage the use of herbicides? Another ethically debatable application of genetic engineering to agriculture is the introduction of genes into crop plants that transfer resistance to viruses, bacteria, fungi, and insects.

Other work in this field has been directed at improving drought resistance or the nutritional value of crops such as soybeans and corn. Many of these projects are in the developmental stage, and these transgenic products will reach the marketplace over the next few years.

Transgenic Plants and Edible Vaccines

One of the most beneficial applications of recombinant DNA technology may be in the production of vaccines. Vaccines stimulate the immune system to produce antibodies against a disease-causing organism and thereby confer immunity against the disease. Two types of vaccines are commonly used—**inactivated vaccines**, prepared from killed samples of the infectious virus or bacteria, and **attenuated vaccines**, which are live viruses or bacteria that can no longer reproduce and cause disease when present in the body.

Recombinant DNA technology is currently being used to produce a new type of vaccine called a **subunit vaccine**. These vaccines consist of one or more surface proteins of the virus or bacterium. The protein acts as an antigen to stimulate the immune system to make antibodies against the virus or bacterium. One of the first licensed subunit vaccines was for a surface protein of hepatitis B, a virus that causes liver damage and cancer. (See Chapter 22 for a discussion of hepatitis B.) The gene for the hepatitis B surface protein was cloned into a yeast-expression vector and produced by using yeast as a host. The protein is extracted and purified from the host cells and then packaged for use as a vaccine.

Vaccination programs in developing countries are faced with problems of transporting and storing vaccines under

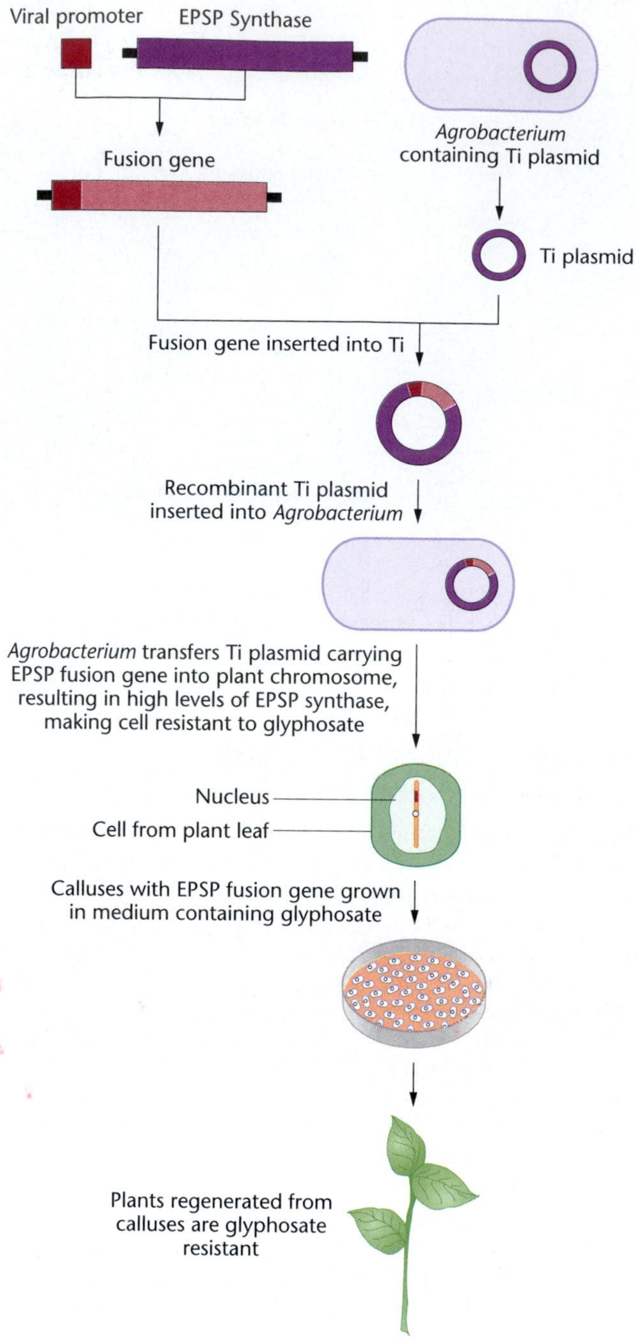

FIGURE 18–24 Gene transfer of glyphosate resistance. The *EPSP* gene is fused to a promoter from cauliflower mosaic virus. This chimeric, or fusion gene, is then transferred to a Ti plasmid vector, and the recombinant vector is inserted into an *Agrobacterium* host. *Agrobacterium* infection of cultured plant cells transfers the *EPSP* fusion gene into a plant-cell chromosome. Cells that acquire the gene are able to synthesize large quantities of EPSP synthase, making them resistant to the herbicide glyphosate. Resistant cells are selected by growth in herbicide-containing medium. Plants regenerated from these cells are herbicide resistant.

refrigeration and of maintaining sterile conditions for injection. To overcome these problems, recombinant DNA methods are being used with the goal of developing cheap, edible vaccines synthesized in edible food plants. Such vaccines would be inexpensive to produce, would not require

Genetics, Technology, and Society

Beyond Dolly: The Cloning of Humans

The birth of Dolly, a healthy white-faced lamb, took the world by surprise. Until February 1997, the idea that an animal could be cloned from the cells of another adult animal was science fiction—something from *Brave New World* or *The Boys from Brazil*. For many people, the notion that animals could be cloned conjured up scenes of multiple replicated Adolf Hitlers or creations gone astray, as in Shelley's *Frankenstein*. For decades, the public had been comforted by scientists who asserted that it would be impossible to clone a new mammal from cells of an adult. It was thought that cells from adult animals could not be reprogrammed in such a way as to form a new, entire organism. But then Dolly appeared.

Dolly, the first mammal cloned from adult cells, was brought into being by a group of embryologists led by Ian Wilmut and Keith Campbell at the Roslin Institute in Scotland. Their reason for developing methods to clone farm animals was to provide identical transgenic animals that secrete pharmaceutical products, such as blood clotting factors or insulin, into their milk. In this way, animals could be used as bioreactors to synthesize proteins that are difficult or expensive to synthesize in bacteria or in the test tube.

The cloning method that Wilmut and Campbell used to create Dolly—a procedure called *nuclear transfer*—was first suggested by embryologist Hans Spemann in 1938. The idea is simply to replace the nucleus of a newly fertilized (or unfertilized) egg with the nucleus from an adult cell, thereby creating a "reconstructed" zygote containing the nucleus from one animal and the egg cytoplasm from another. In theory, the genetic information in the donor nucleus should direct all further embryonic development, and the new organism should be a genetic replica of the adult that donated the nucleus. Although this procedure sounds simple in theory, it proved to be extremely difficult in practice. Even when the technical procedures such as collecting eggs, removing nuclei, and transplanting donor nuclei were perfected, the cloning procedure would work only if the donor nucleus came from an embryo. Because adult nuclei come from specialized structures such as the skin, liver, and kidney, they are "differentiated" and express only a small subset of all the genes in the nuclear genome. As all the genes in the genome must be actively expressed in an embryo, adult nuclei proved to be inadequate for the job.

Wilmut and Campbell accomplished what was perceived to be the impossible by genetically reprogramming the adult nuclei before transferring them into recipient eggs. They removed donor cells from the udder of an adult ewe and starved the cells so that they became quiescent and entered the G0 phase of the cell cycle. For unknown reasons, this starvation procedure allowed the silent genes within the udder cell nucleus to be turned on after the nucleus was transplanted into the cytoplasm of a sheep's egg. In addition, the researchers passed an electric current through the egg to facilitate nuclear transfer and to stimulate the new "zygote" to begin dividing. To create Dolly, over 200 udder-cell nuclei were transferred. Of these nuclei, only 29 developed into embryos, and 13 were implanted into surrogate mother ewes. One pregnancy resulted, which culminated in the birth of Dolly.

Although Dolly was the first mammal to be cloned from adult cells, the approach has since been used to create cloned mice from adult mouse cells. In addition, transgenic sheep and cattle have been cloned by similar procedures, except that fetal-cell nuclei were used as donors, rather than adult nuclei. One of these transgenic sheep ("Polly") bears the human gene for blood-clotting factor IX and secretes factor IX into her milk, paving the way for the production of pharmaceuticals from cloned farm animals.

Besides the benefits for drug production, cloning promises other advantages. For example, cloning might allow scientists to preserve and replicate endangered species, cure genetic diseases in farm animals, and provide animal models of human diseases for which there are presently no research models.

Despite the obvious advantages for agriculture and medicine, the cloning of mammals has led to outcry and concern. The idea that humans could be cloned has been termed immoral, repugnant, and ethically wrong. Within days of the announcement about Dolly, bills were introduced into the U.S. Congress to prohibit research into human cloning, and worldwide bans were called for. Frightening scenarios were proposed: rich and powerful people cloning themselves for reasons of vanity; people with serious illnesses cloning replicas to act as organ donors; and legions of human clones suffering loss of autonomy, individuality, and kinship ties. But is it really possible to clone humans? And if so, should human cloning ever be done?

The answer to the first question is simple: The same technology used to create Dolly could be used to clone a human. Most of the necessary technical procedures are being used now for *in vitro* fertilization. And it seems likely that adult human nuclei could be reprogrammed similarly to the adult sheep nuclei that created Dolly.

The answer to the second question is not as simple. Both scientific and ethical issues cloud our judgment about human cloning. Present technology cannot guarantee the health of any cloned individual, human or animal. It is possible that clones could exhibit a higher risk of genetic disease or cancer, because of the accumulation of mutations in the donated adult nucleus. Clones might also show accelerated aging, due to shortened telomeres that are present in the donor's chromosomes. If the donor's nucleus is not completely reprogrammed prior to cloning, the clone could undergo abnormal development as well. Until the safety issues are resolved, it may be prudent to suspend any attempts to clone humans.

The ethical issues are even more problematic. It is possible to foresee positive aspects to human cloning. Infertile couples or couples who suffer from genetic disease on one side of the family could choose to clone one of the partners in order to raise a child who is biologically related. Cloning a person's cells *in vitro* could provide a source of cells or tissues to treat a number of serious diseases. On the other hand, it is equally possible to foresee negative aspects. Would cloning seriously alter what it means to be a unique human being? What would be the fate of clones created for organ transplantation? Could the technique be misused by the unscrupulous for social or political goals? It is important for society to consider these issues now in order to ensure responsible and beneficial outcomes of this new technology.

References

Cibelli, J.B., et al. 1998. Cloned transgenic calves produced from nonquiescent fetal fibroblasts. *Science* 280:1256–58.

Pennisi, E. 1998. After Dolly, a pharming frenzy. *Science* 279: 646–48.

Robertson, J.A. 1998. Human cloning and the challenge of regulation. *N. Eng. J. Med.* 339:118–25.

Solter, D. 1998. Dolly is a clone—and no longer alone. *Nature* 394: 315–16.

Wilmut, I. 1998. Cloning, for medicine. *Sci. Am.* (Dec.) 58–63.

Wilmut, I. et al. 1997. Viable offspring derived from fetal and adult mammalian cells. *Nature* 385:810–13.

refrigeration, and would not have to be given under sterile conditions by trained medical personnel.

As a model system, the antigenic subunit of hepatitis B vaccine has been transferred to tobacco plants and expressed in the leaves (Figure 18–25). For use as a source of vaccine, the gene would be inserted into food plants such as grains or vegetables. In clinical trials with a vaccine against diarrhea, genetically engineered potatoes that carry recombinant bacterial antigens were used successfully to vaccinate human volunteers who ate small quantities (50–100 g) of the potatoes. This proof of principle establishes that edible vaccines may be feasible. Similar tests that use genetically engineered bananas are currently underway. If such tests are successful, genetically engineered edible plants will soon be used to vaccinate infants, children, and adults against many infectious diseases.

Plants and animals were domesticated some 8000 to 10,000 years ago, and we have been modifying these organisms by selective breeding ever since, producing the diversity of domesticated plants and animals present today. Recombinant DNA technology has changed the rate at which new plants and animals can be developed and, by enabling the transfer genes between species, has altered the types of changes that can be made. Unlike selective-breeding programs, however, biotechnology has generated concerns about the release of genetically modified organisms into the environment and about the safety of eating such products. If biotechnology is to achieve a new, green revolution, these concerns need to be addressed through prudent research and education.

FIGURE 18–25 Tobacco leaves expressing the hepatitis B antigen. Transgenic tobacco plants carrying an antigenic subunit of hepatitis B virus were generated. Leaves from transgenic plants were treated with antibodies against hepatitis B antigen, showing that the plants produced the antigen. The central leaf in the photograph is a leaf from a normal tobacco plant and is unstained.

Chapter Summary

1. Recombinant DNA technology offers a new approach to genetic analysis. Instead of relying on the isolation and mapping of mutant genes, large cloned segments of the genome are manipulated to make genetic and physical maps that use molecular markers rather than phenotypes visible at the level of the organism.
2. Genome projects are underway or completed for an ever-increasing number of organisms, including humans. The goals of these genome projects are to identify and map all genes in the genome and to determine the complete nucleotide sequence of the genome.
3. Cloned DNA is finding a wide range of applications, including gene mapping and the identification and isolation of genes responsible for genetic disorders. The method known as *positional cloning*, which is based on recombinant DNA technology, allows a gene to be mapped and identified with no knowledge of the nature or function of the gene product.
4. Recombinant DNA techniques are being used in the prenatal diagnosis of human genetic disorders. This method allows the direct examination of the genotype, whereas previous methods relied on gene expression and the identification of the gene product. Recombinant DNA methods can also identify carriers of genetic disorders, and they are the basis for proposals to screen the population for a number of genetic disorders, including sickle-cell anemia and cystic fibrosis.
5. The availability of cloned human genes has led to their use in gene therapy. In somatic gene therapy, a cloned normal copy of a gene is transferred into a vector; the vector then transfers

the gene to a target tissue that takes up and expresses the cloned copy of the gene, thereby altering the mutant phenotype. The development of new and more effective vector systems probably means that gene therapy will become a standard method for treatment of genetic disorders in the near future.

6. Recombinant DNA techniques that detect allelic variants of variable tandem nucleotide repeats (DNA fingerprints) have found applications in forensics, paternity testing, and a wide range of other fields, including archeology, conservation biology, and public health.

7. The biotechnology industry uses recombinant DNA methods to produce human gene products in a variety of hosts, ranging from bacteria to farm animals. In addition, gene transfer techniques are improving crop plants by transferring herbicide resistance. In the near future, it may be possible to use food plants to vaccinate people against infectious disease, making resistance to infectious agents almost universal.

Insights and Solutions

1. DNA fingerprints have been used to test forensic specimens, to identify criminal suspects, and to settle immigration cases and paternity disputes. Probes for DNA fingerprinting can be derived from a single locus or multiple loci. Two multiple-loci probes have been widely employed in both criminal and civil cases and derive from minisatellite loci on chromosome 1 (1cen-q24) and chromosome 7 (7q31.3). These probes, which are widely used because they produce a highly individual fingerprint, have determined paternity in thousands of cases over the last few years.

(a) Shown in the accompanying figure are the results of DNA fingerprinting of a mother (*M*), putative father (*F*), and child (*C*), using the aforementioned probes. As shown, the child has 6 maternal bands, 11 paternal bands, and 5 bands shared between the mother and the alleged father. Based on this fingerprint, can you conclude that the man tested is the father of the child?

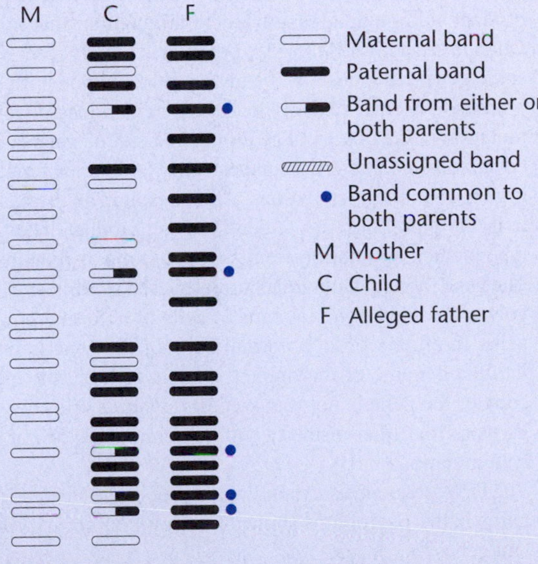

Solution: All bands present in the child can be assigned as coming from either the mother or the father. In other words, all the bands in the child's DNA fingerprint that are not maternal are present in the father. Because the father and the child share 11 bands and the child has no unassigned bands, paternity can be assigned with confidence. In fact, the chance that the man tested is not the father is on the order of 10^{-13}.

(b) The DNA fingerprints of the mother, the alleged father, and the child for a second case are shown in the accompanying figure.

In this case, the child has 8 maternal bands, 15 paternal bands, 6 bands that are common to both the mother and the alleged father, and 1 band that is not present in either the mother or the alleged father. What are the possible explanations for the presence of the lattermost band? Based on your analysis of the band pattern, which explanation is most likely?

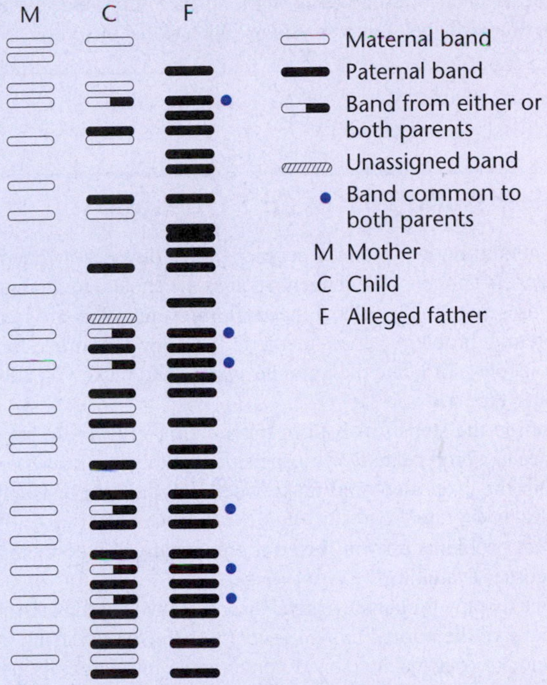

Solution: In this case, one band in the child cannot be assigned to either parent. Two possible explanations for this finding are that the child is mutant for one band or that the man tested is not the father. In estimating the probability of paternity, the mean number of resolved bands (n) is determined, and the mean probability (x) that a band in individual A matches that in a second, unrelated individual B is calculated. In this case, because the child and the father share 15 bands in common, the probability that the man tested is not the father is very low (probably 10^{-7} or lower). As a result, the most likely explanation is that the child is a mutant for a single band. In fact, in 1419 cases of genuine paternity resolved by the minisatellite probes on chromosomes 1 and 7, single-mutant bands in the children were recorded in 399 cases, accounting for 28 percent of all cases.

2. Infection by HIV-1 (human immunodeficiency virus) is responsible for the destruction of cells in the immune system and results in the symptoms of AIDS (acquired immunodeficiency syndrome). HIV infects and kills cells of the immune system that carry a cell-surface receptor known as CD4. An HIV surface protein known as gp120 binds to the CD4 receptor and allows the virus entry into the cell. The gene encoding the CD4 protein has been cloned. How can this clone be used along with recombinant DNA techniques to combat HIV infection?

Solution: Several methods that use the CD4 gene are being explored to combat HIV infection. First, because infection depends on an interaction between the viral gp120 protein and the CD4 protein, the cloned CD4 gene has been modified to produce a soluble form of the protein (sCD4). The idea is that HIV can be prevented from infecting cells if the gp120 protein of the virus is bound up with the soluble form of the CD4 protein and thus is unable to bind to CD4 proteins on the surface of immune-system cells. Studies in cell-culture systems indicate that the presence of sCD4 effectively prevents HIV infection of tissue-culture cells. However, studies in HIV-positive humans have been somewhat disappointing, mainly because the strains of HIV used in the laboratory are different than those found in infected individuals. HIV-infected cells carry the viral gp120 protein on their surface. To kill such cells, the CD4 gene has been fused with those encoding bacterial toxins. The resulting fusion protein contains the CD4 regions that bind to gp120 and the toxin regions that kill the infected cell. In tissue-culture experiments, cells infected with HIV are killed by the fusion protein, whereas uninfected cells survive. It is hoped that the targeted delivery of drugs and toxins can be used in therapeutic applications to treat HIV infection.

Problems and Discussion Questions

1. In attempting to vaccinate people against diseases by having them eat antigens, the antigen (such as the cholera toxin) must be presented to the cells of the small intestine. What are some potential problems of this method? Why don't absorbed food molecules stimulate the immune system and make you allergic to the food you eat?
2. Outline the steps involved in transferring glyphosate resistance to a crop plant. Do you envision that this trait can escape from the crop plant and make weeds glyphosate resistant? Why or why not?
3. What problems do you forsee if enhancement gene therapy becomes available?
4. Gene therapy for human genetic disorders involves transferring a copy of the normal human gene into a vector and using the vector to transfer the cloned human gene into target tissues. Presumably, the gene enters the target tissue and becomes active, and the gene product relieves the symptoms. Although this method is now being used to treat several disorders, there are some unresolved problems with it.
   (a) Why are disorders such as muscular dystrophy difficult to treat by gene therapy?
   (b) What are the potential problems in using retroviruses as vectors?
   (c) In the long run, should gene therapy involve germ tissue instead of somatic tissue? What are some of the potential ethical problems associated with the former approach?
5. In producing physical maps of markers and cloned sequences, what advantage does *Drosophila* offer that other organisms, including humans, do not?
6. Outline the steps involved in identifying a gene by positional cloning. What steps may cause difficulty in this process? Once a region on a chromosome has been identified as containing a given gene, what kind of mutations would speed the process of identifying the locus?
7. The phenotype of many behavioral traits, such as manic depression and schizophrenia, may be controlled by several genes, each at a different locus. Can positional cloning be used to map and isolate such genes? What if a trait is controlled by six genes, each contributing equally in an additive way to the phenotype? Can positional cloning be used in this case? Why or why not?
8. Suppose that you develop a screening method for cystic fibrosis that allows you to identify the predominant mutation *Δ508* and the next six most prevalent mutations. What do you need to consider before using this method in screening the population at large for this disorder?
9. Dominant mutations can be categorized according to whether they increase or decrease the overall activity of a gene or gene product. Although a loss-of-function mutation (that is, a mutation that inactivates the gene product) is usually recessive, for some genes, one dose of the gene product is not sufficient to produce a normal phenotype. In this case, a loss-of-function mutation in the gene will be dominant, and the gene is said to be *haploinsufficient*. A second category of dominant mutations is gain-of-function mutations, which result in an increased activity or expression of the gene or gene product. The phenotype of such a mutation results from too much gene product. The gene therapy technique currently used in clinical trials involves the "addition" to somatic cells of a normal copy of a gene. In other words, a normal copy of the gene is inserted into the genome of the mutant somatic cell, but the mutated copy of the gene is not removed or replaced. Will this strategy work for either of the two aforementioned types of dominant mutation?
10. The DNA sequence surrounding the site of the sickle-cell mutation in the β-globin is shown as follows for normal and mutant genes:

5′ G A C T C C T G A G G A G A A G T 3′

3′ C T G A G G A C T C C T C T T C A 5′

Normal DNA

5′ G A C T C C T G T G G A G A A G T - 3′

3′ C T G A G G A C A C C T C T T C A - 5′

Sickle-cell DNA

Each type of DNA is denatured into single strands and applied to a filter. The paper containing the two spots is hybridized to an ASO of the following sequence:

5'-GACTCCTGAGGAGAAGT-3'

Which spot, if either, will hybridize to this probe?

11. One form of hemophilia, an X-linked disorder of blood clotting, is caused by mutation in clotting factor VIII. Many single-nucleotide mutations of this gene have been described, making the detection of mutant genes by Southern blots inefficient. There is, however, an RFLP for the enzyme *Hin*dIII contained in an intron of the factor VIII gene that can often be used in screening, as shown in the following figure:

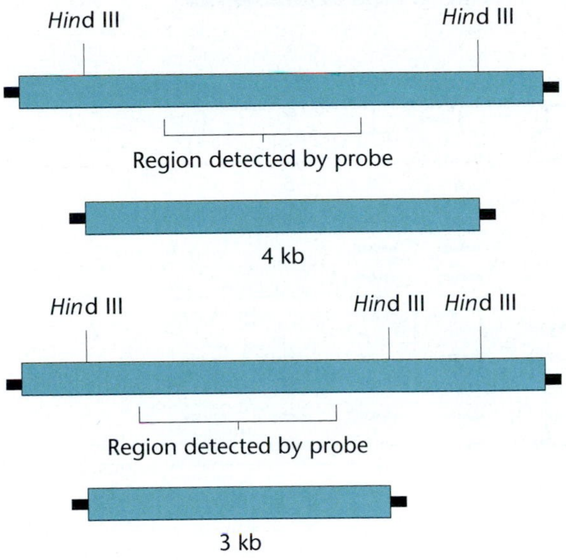

A female whose brother has hemophilia is at 50-percent risk of being a carrier of this disorder. To test her status, DNA is obtained from her white blood cells and those of family members, cut with *Hin*dIII, and the fragments are probed and visualized by Southern blotting. The results are shown as follows:

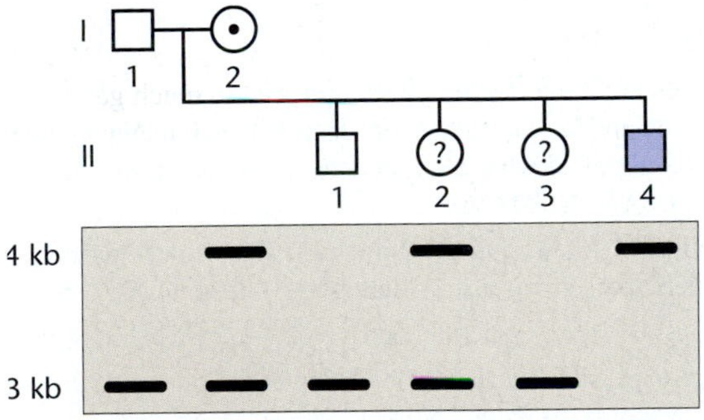

Determine whether any of the females in generation II are carriers for hemophilia.

12. The human insulin gene contains a number of introns. Despite the fact that bacterial cells will not excise introns from mRNA, how can a gene like this be cloned into a bacterial cell and produce insulin?

13. In mice transfected with the rabbit β-globin gene, the rabbit gene is active in a number of tissues, including the spleen, brain, and kidney. In addition, some mice suffer from thalassemia, caused by an imbalance in the coordinate production of α- and β-globins. What problems associated with gene therapy are illustrated by these findings?

Extra-Spicy Problems

14. You are asked to help in prenatal genetic testing of a couple, each of whom is found to be a carrier for a deletion in the β-globin gene that produces β-thalassemia when homozygous. The couple already has one child who is unaffected and is not a carrier. The woman is pregnant, and they wish to know the status of the fetus. You receive DNA samples obtained from the fetus by amniocentesis and from the rest of the family by extraction from white blood cells. Using a probe the deletion, you obtain the following blot:

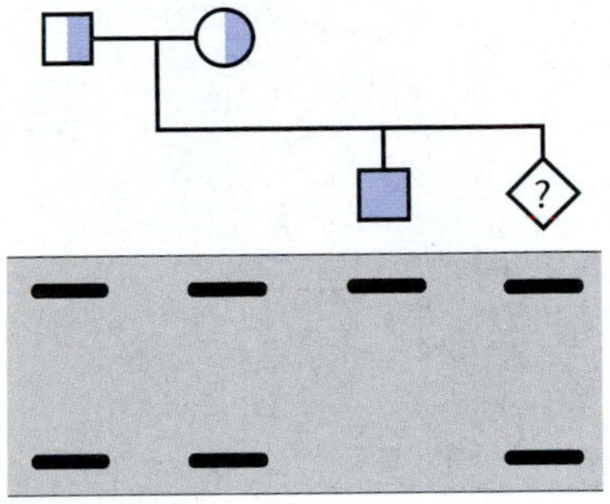

Is th fetus affected? What is its genotype for the β-globin gene?

15. The following is a pedigree that shows the inheritance of a rare disease state:

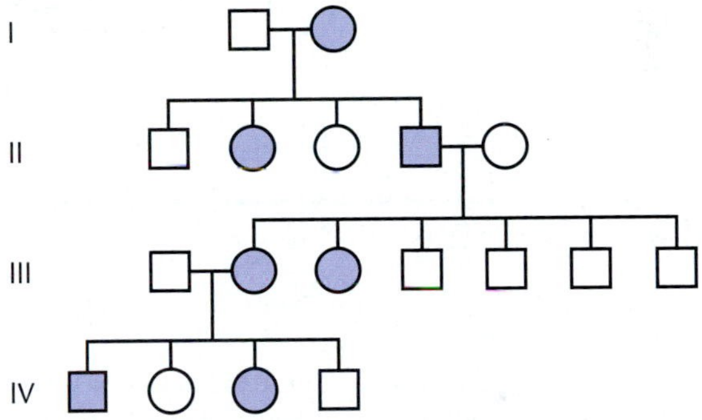

(a) Which mode or modes of inheritance are excluded by or consistent with this pedigree? DNA samples from generations II and III (in the pedigree) are obtained and subjected to RFLP linkage analysis. One RFLP examined is found on chromosome 10 (identified by probe *A*), and the other is found on chromosome 21 (identified by probe *B*). The results of the analysis are shown below.

In answering the remaining questions, assume that additional data were gathered on this family and that they were consistent with the data shown and statistically significant.

(b) On which chromosome is the disease gene located?

(c) Individual III-1 is married to a normal man whose RFLP genotype is A1A2 and B1B1. What kind of prenatal diagnostic test can be done to determine whether the child they are expecting will be normal? Be sure to describe what you can conclude regarding the result. Indicate the accuracy of such a test.

(d) Individual III-4 marries a woman of genotype B1B1. They have a child who is B1B1. The father assumes that the child is illegitimate, but the mother, who has taken a genetics course, argues that the child could be the result of an event that occurred in the father's germ line, and she cites two possibilities. What are they?

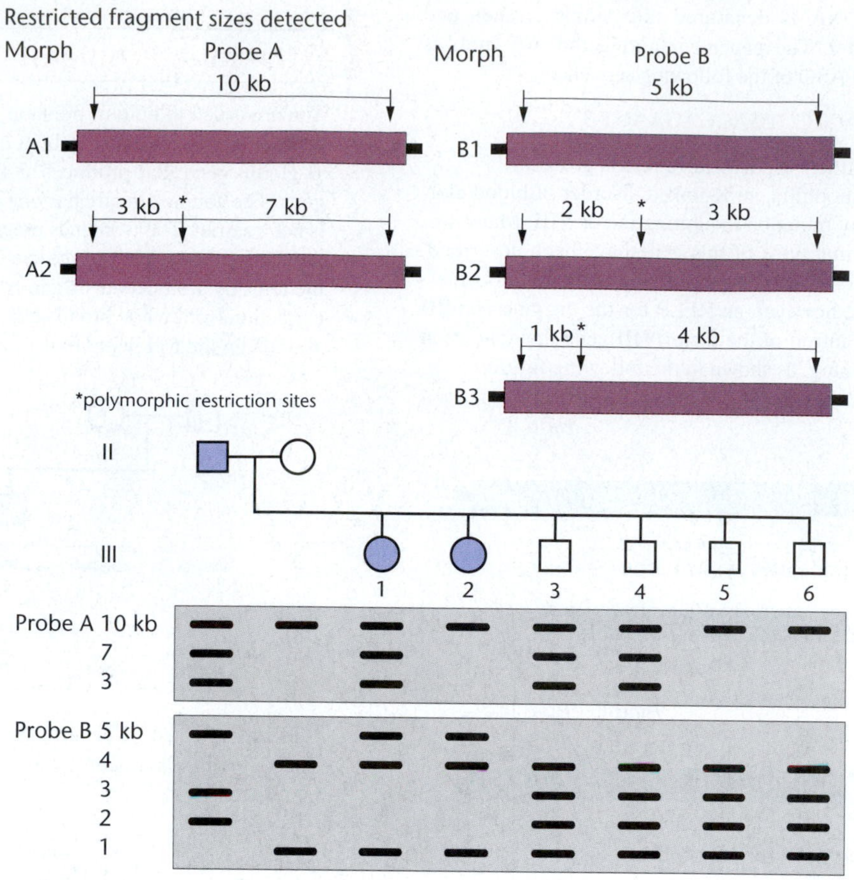

Selected Readings

Anderson, W.F. 2000. The best of times, the worst of times. *Science* 288: 627–28.

Barker, D., et al. 1987. Gene for von Recklinghausen neurofibromatosis is in the pericentromeric region of chromosome 17. *Science* 236:1001–102.

Beaudet, A.L. 1999. Making genomic medicine a reality. *Am. J. Hum. Genet.* 64:1–13.

Bleck, O., McGrath, J.A., and South, A.P. 2001. Searching for candidate genes in the new millenium. *Clin. Exp. Dermatol.* 26: 279–83.

Cavanna-Calvo, M., et al., 2000. Gene therapy of severe combined immunodeficiency (SCID)-XI disease. *Science* 288: 669–672.

Cawthon, R.M., et al. 1990. A major segment of the neurofibromatosis type 1 gene: cDNA sequence, genomic structure, and point mutations. *Cell* 62:193–201.

Chang, J.C., and Kan, Y.W. 1981. Antenatal diagnosis of sickle-cell anemia by direct analysis of the sickle mutation. *Lancet* 2:1127–29.

Crystal, R.G. 1995. Transfer of genes to humans: Early lessons and obstacles to success. *Science* 270:404–10.

Daniell, H., Streatfield, S.J., and Wycoff, K. 2001. Medical molecular farming: Production of antibodies, biopharmaceuticals and edible vaccines in plants. *Trends Plant Sci.* 6: 219–26.

Danna, K., and Nathans, D. 1971. Specific cleavage of simian virus 40 DNA by restriction endonuclease of *Hemophilus influenzae*. *Proc. Natl. Acad. Sci. USA* 68:2913–17.

Engler, O.B., et al., 2001. Peptide vaccines against hepatitis B virus: From animal model to human studies. *Mol. Immunol.* 38: 457–65.

Estruch, J.J. 1997. Transgenic plants: An emerging approach to pest control. *Nature Biotech.* 15:137–41.

Fleischmann, R.D., et al. 1995. Whole genome random sequencing and assembly of *Haemophilus influenzae*. *Science* 269:496–512.

Guyer, M., and Collins, F.S. 1995. How is the Human Genome Project doing, and what have we learned so far? *Proc. Natl. Acad. Sci. USA* 92:10841–48.

Karpati, G., Pari, G., and Molnar, M.J. 1999. Molecular therapy for genetic muscle diseases—status 1999. *Clin. Genet.* 55:1–8.

Kmiec, E.B. 1999. Gene therapy. *Amer. Scient.* 87:240–47.

Lewontin, R., and Hartl, D. 1991. Population genetics in forensic DNA typing. *Science* 254:1745–50.

Lissens, W., and Sermon, K. 1997. Preimplantation genetic diagnosis: Current status and new developments. *Human Reproduction* 12:1756–61.

Mason, H., Lam, D., and Arntzen, C. 1992. Expression of hepatitis B surface antigen in transgenic plants. *Proc. Natl. Acad. Sci. USA* 89:11745–49.

Moldoveanu, Z., et al. 1993. Oral immunization with influenza virus in biodegradable microspheres. *J. Infect. Dis.* 167:84–90.

Phillips, M.I. 2001. Gene therapy for hypertension: The preclinical data. *Hypertension* 38: 543–48.

GENETICS MediaLab

The resources that follow will help you achieve a better understanding of the concepts presented in this chapter. These resources can be found either on the CD packaged with this textbook or on the Companion Web site found at **http://www.prenhall.com/klug**

CD Resources:

Module 18.1: RFLP

Module 18.2: DNA Fingerprinting

Web Problem 1:
Time for completion = 15 minutes

How should we treat the information gained from genotyping individuals? Who should have access to your sequence? As testing becomes more routine, medical and insurance records will contain greater amounts of genetic information. Laws regarding the privacy of this information are now taking shape, offering an opportunity for public input. The Human Genome Project has gathered information and opinions on these issues at the linked Web site (keyword **LEGISLATION**). Read some of the legislation introduced in the 106th Congress. Pay particular attention to the specifics of who is granted access. Do you agree or disagree? Can you foresee an unintended abuse by emerging technologies that is not covered? Would a prohibition have an unacceptable cost in terms of public benefit? Are existing laws sufficient to offer protection? To complete this exercise, visit Web Problem 1 in Chapter 18 of your Companion Web site, and select the keyword **LEGISLATION**.

Web Problem 2:
Time for completion = 10 minutes

How can the average person deal with the rapidly expanding amount of genetic information available to the public and to health care providers? This wealth of genetic data spurred the development of the field of genetic counseling. The goal was to train individuals in both genetics and counseling so that they would be able to convey genetic information to individuals in an accurate and objective manner. This Web site describes a case study of a family with a child diagnosed with neurofibromatosis. Briefly discuss some of the issues that a genetic counselor might encounter in counseling such a family. To complete this exercise, visit Web Problem 2 in Chapter 18 of your Companion Web site, and select the keyword **GENETIC COUNSELING**.

Web Problem 3:
Time for completion = 20 minutes

How is DNA testing used in criminal or paternity cases? Nucleotide sequence provides a unique identifier of each person's cells. Similarly, because a child inherits half of each parent's genetic information, this sequence can be used to determine the likelihood of two individuals being related. Test your understanding of DNA testing methods with a series of interactive questions at the Web site indicated. Through feedback, learn whether your answers are correct. Tutorial information is provided to clarify your understanding. Discuss whether forensics information is used to prove that someone is guilty or to prove that he or she is innocent. To complete this exercise, visit Web Problem 3 in Chapter 18 of your Companion Web site, and select the keyword **FORENSICS**.

Reiss, J., and Cooper, D.N. 1990. Application of the polymerase chain reaction to the diagnosis of human genetic disease. *Hum. Genet.* 85:1–8.

Riley, J.H., et al., 2000. The use of single nucleotide polymorphisms in the isolation of common disease genes. *Pharmacogenomics* 1: 39–47.

Southern, E.M. 1975. Detection of specific sequences among DNA fragments separated by gel electrophoresis. *J. Mol. Biol.* 98:503–17.

Velander, W.H., Lubon, H., and Drohan, W.N. 1997. Transgenic livestock as drug factories. *Sci. Amer.* (Jan.) 276:70–75.

Villa-Komaroff, L., et al. 1978. A bacterial clone synthesizing proinsulin. *Proc. Natl. Acad. Sci. USA* 75:3727–31.

Weier, H.U. 2001. DNA fiber mapping techniques for the assembly of high-resolution physical maps. *J. Histochem. Cytochem.* 49: 939–48.

Williams, J.G.K., et al. 1990. DNA polymorphisms amplified by arbitrary primers are useful as genetic markers. *Nucl. Acids Res.* 18:6531–35.

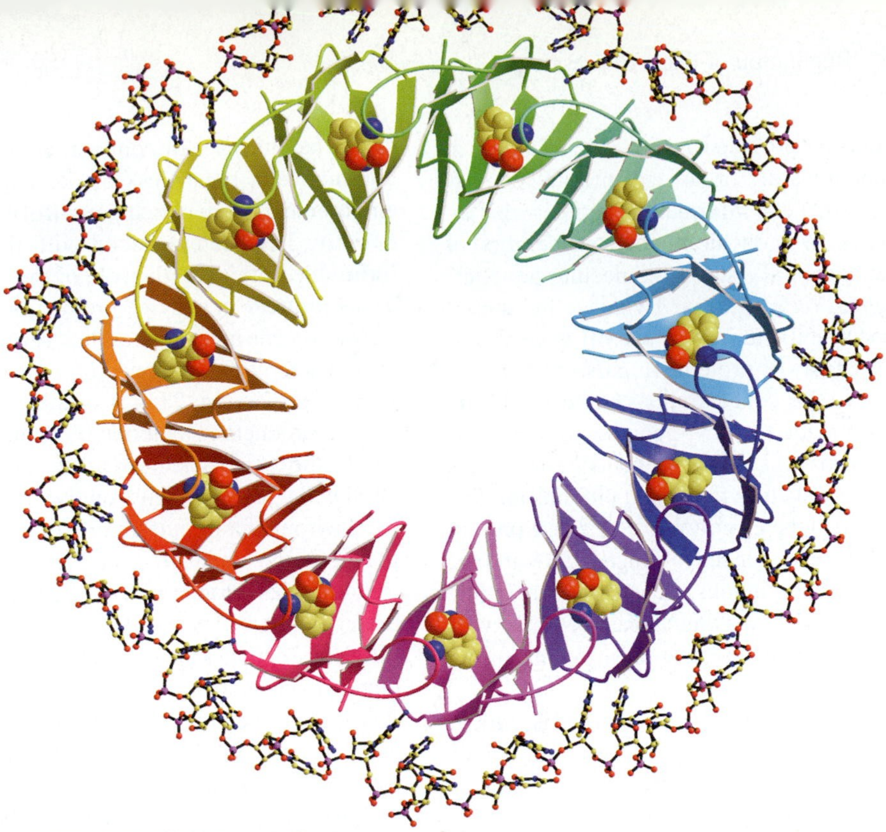

Model of the *trp* RNA-binding attenuation protein (TRAP).

19

Regulation of Gene Expression in Prokaryotes

We have thus far provided a thorough grounding in how geneomes are organized into genes, how genes store genetic information, and how this information is expressed. We now turn to the consideration of one of the most fundamental issues in molecular genetics: *How is genetic expression regulated?* Evidence in support of the idea that genes can be turned on and off is very convincing. Detailed analysis of proteins in *Escherichia coli*, for example, has shown that concentrations of the 4000 or so polypeptide chains encoded by the genome vary widely. Some proteins may be present in as few as 5 to 10 molecules per cell, whereas others, such as ribosomal proteins and the many proteins involved in the glycolytic pathway, are present in as many as 100,000 copies per cell. Though in prokaryotes most gene products exist continuously at a basal level (a few copies), the level can be increased dramatically. Clearly, fundamental regulatory mechanisms must exist to control the expression of the genetic information.

In this chapter, we will explore some of what is known about the regulation of genetic expression in bacteria. As we have seen in a number of previous chapters, these organisms served as excellent research organisms during many seminal investigations in molecular genetics. Relevant to our current topic, bacteria also served as an excellent model system for studies involving the induction of genetic transcription in response to changes in environmental conditions. Our focus will be on regulation at the level of the gene. Keep in mind that posttranscriptional regulation also occurs in bacteria. However, we will defer discussion of this level of regulation to a subsequent chapter when we consider eukaryotic regulation.

19.1 Prokaryotes Exhibit Efficient Genetic Mechanisms to Respond to Environmental Conditions

Regulation of gene expression has been extensively studied in prokaryotes, particularly in *E. coli*. We have learned that highly efficient genetic mechanisms have evolved to turn genes on and off, depending on the cell's metabolic need for the respective gene products. Not only do bacteria respond to changes in their environment, but they also regulate gene activity involved in a variety of normal cellular responses (including the replication, recombination, and repair of their DNA) and in cell division and development.

The idea that microorganisms regulate the synthesis of gene products is not a new one. As early as 1900, it was shown that when lactose (a galactose–glucose-containing disaccharide) is present in the growth medium of yeast, the organisms produce enzymes specific to lactose metabolism. When lactose is absent, the enzymes are not manufactured. Soon thereafter, investigators were able to generalize that bacteria "adapt" to their environment, producing certain enzymes only when specific chemical substrates are present. Such enzymes were thus referred to as **adaptive** (sometimes also called facultative). In contrast, enzymes that are produced continuously, regardless of the chemical makeup of the environment, were called **constitutive**. Since then, the term *adaptive* has been replaced with the more accurate term **inducible**, reflecting the role of the substrate, which serves as the **inducer** in enzyme production.

More recent investigation has revealed a contrasting system, whereby the presence of a specific molecule inhibits genetic expression. This is usually true for molecules that are end products of anabolic biosynthetic pathways. For example, the amino acid tryptophan can be synthesized by bacterial cells. If a sufficient supply of tryptophan is present in the environment or culture medium, it is energetically inefficient for the organism to synthesize the enzymes necessary for tryptophan production. A mechanism has evolved whereby tryptophan plays a role in repressing transcription of mRNA essential to the production of the appropriate biosynthetic enzymes. In contrast to the inducible system controlling lactose metabolism, the system governing tryptophan expression is said to be **repressible**, and the substrate of the enzyme is the **repressor**.

As we will soon see, regulation, whether it is inducible or repressible, may be under either **negative** or **positive control**. Under negative control, genetic expression occurs *unless it is shut off by some form of a regulator molecule*. In contrast, under positive control, transcription occurs *only if a regulator molecule directly stimulates RNA production*. In theory, either type of control can govern inducible or repressible systems. Our discussion in the ensuing sections of this chapter will help clarify these contrasting systems of regulation. For the enzymes involved in lactose and tryptophan, negative control is operative.

19.2 Lactose Metabolism in *E. coli* Is Regulated by an Inducible System

Beginning in 1946 with the studies of Jacques Monod and continuing through the next decade with significant contributions by Joshua Lederberg, François Jacob, and Andre L'woff, genetic and biochemical evidence involving lactose metabolism was amassed. Insights were provided into the way in which the gene activity is repressed when lactose is absent, but induced when it is available. In the presence of lactose, the concentration of the enzymes responsible for its metabolism increases rapidly from a few molecules to thousands per cell. The enzymes responsible for lactose metabolism are thus *inducible*, and lactose serves as the *inducer*.

In prokaryotes, genes that code for enzymes with related functions (e.g., genes involved with lactose metabolism) tend to be organized in clusters, and they are often under the coordinated genetic control of a single regulatory unit. The location of this unit is almost always upstream to the gene cluster it controls and is known as a ***cis*-acting element**. Interactions at the site of such an element involve binding molecules that control transcription of the gene cluster. Such molecules are called ***trans*-acting elements**. Actions at the regulatory site determine whether the genes are expressed and thus whether the corresponding enzymes or other protein products are present. Binding a *trans*-acting element at a *cis*-acting site can regulate the gene cluster either negatively (by turning the genes off) or positively (by turning genes in the cluster on). In this section, we discuss how such bacterial gene clusters are coordinately regulated.

The discovery of a regulatory gene and a regulatory site that are part of the gene cluster was paramount to the understanding of how gene expression is controlled in the system. Neither of these regulatory elements encodes enzymes necessary for lactose metabolism—that is, the function of the three genes in the cluster. As illustrated in Figure 19–1, the three structural genes and the adjacent regulatory site constitute the **lactose**, or ***lac*****, operon**. Together, the entire gene cluster functions in an integrated fashion to provide a rapid response to the presence or absence of lactose.

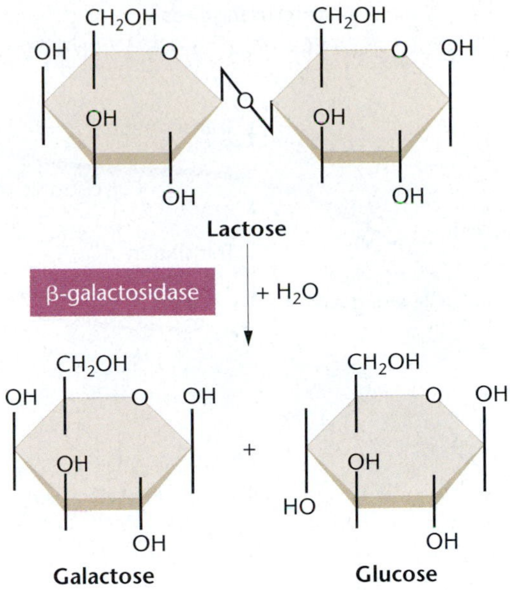

FIGURE 19–2 The catabolic conversion of the disaccharide lactose into its monosaccharide units, galactose, and glucose.

Structural Genes

Genes coding for the primary structure of the enzymes are called **structural genes**. There are three structural genes in the *lac* operon. The *lacZ* gene encodes β-galactosidase, an enzyme that converts the disaccharide lactose to the monosaccharides glucose and galactose (Figure 19–2). This conversion is essential if lactose is to serve as the primary energy source in glycolysis. The second gene, *lacY*, specifies the primary structure of **permease**, an enzyme that facilitates the entry of lactose into the bacterial cell. The third gene, *lacA*, codes for the enzyme **transacetylase**. While its physiological role is still not completely clear, there is some thought that it may be involved in the removal of toxic by-products of lactose digestion from the cell.

To study the genes coding for these three enzymes, researchers isolated numerous mutations in order to eliminate the function of one or the other enzyme. Such lac^- mutants were first isolated and studied by Joshua Lederberg. Mutant cells that fail to produce active β-galactosidase ($lacZ^-$) or permease ($lacY^-$) are unable to use lactose as an energy source. Mutations also were found in the transacetylase gene. Mapping studies by Lederberg established that all three genes are closely linked or contiguous to one another in the order *Z–Y–A* (See Figure 19–1).

Another observation is relevant to what became known about the structural genes. Knowledge of their close linkage led to the discovery that all three genes are transcribed as a single unit, resulting in a polycistronic mRNA (Figure 19–3). This results in the coordinate regulation of all three genes, since a single message serves as the basis for translation of all three gene products.

The Discovery of Regulatory Mutations

How does lactose activate structural genes and induce the synthesis of the related enzymes? A partial answer comes from the discovery and study of **gratuitous inducers**, chemical analogues of lactose such as the sulfur analogue **isopropylthiogalactoside (IPTG)**, shown in Figure 19–4. Gratuitous inducers behave like natural inducers, but they do not serve as substrates for the enzymes that are subsequently synthesized. Their discovery provides strong evidence that the primary induction event does *not* depend on the interaction between the inducer and the enzyme.

What, then, is the role of lactose in induction? The answer to this question required the study of another class of mutations called **constitutive mutants**. In this type of mutation, the enzymes are produced regardless of the presence or absence of lactose. Maps of the first type of constitutive mutation, $lacI^-$, showed that it is located at a site on the DNA close to, but distinct from, the structural genes. The *lacI* gene

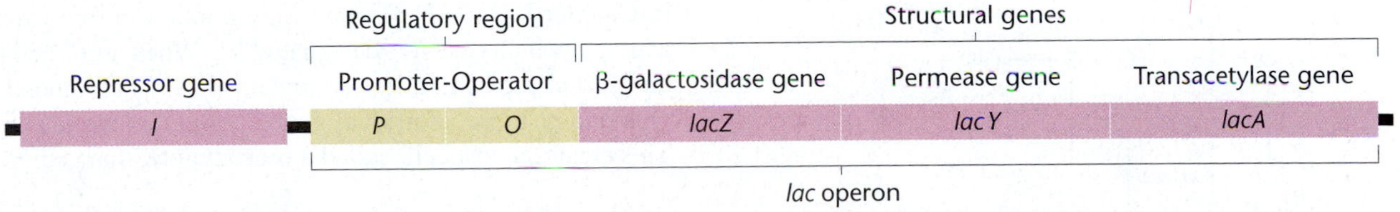

FIGURE 19–1 A simplified overview of the genes and regulatory units involved in the control of lactose metabolism. (This region of DNA is not drawn to scale.) A more detailed model will be developed later in this chapter. (See Figure 19–10.)

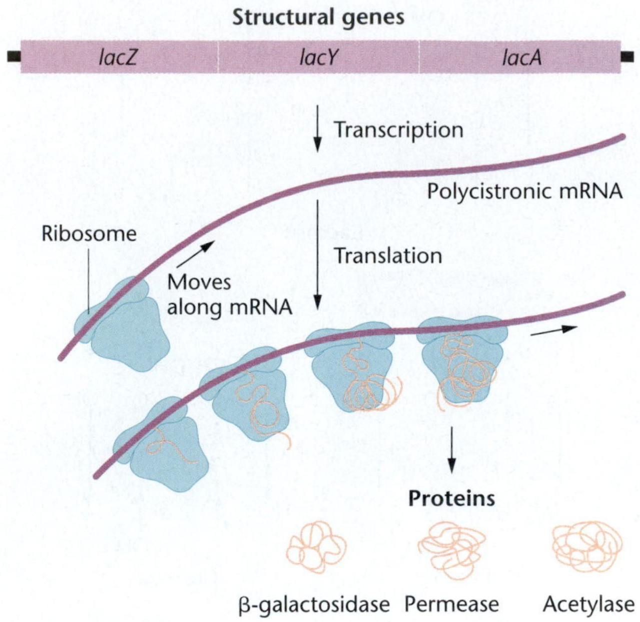

FIGURE 19–3 The structural genes of the *lac* operon are transcribed into a single polycistronic mRNA, which is translated simultaneously by several ribosomes into the three enzymes encoded by the operon.

is appropriately called a **repressor gene**. We will soon see why this name is appropriate. A second set of constitutive mutations producing identical effects was found in a region immediately adjacent to the structural genes. This class of mutations, designated $lacO^C$, identifies the **operator region** of the operon. Because inducibility has been eliminated in both types of constitutive mutation (the enzymes are continually produced), clearly, regulation has been disrupted by genetic changes.

The Operon Model: Negative Control

Around 1960, Jacob and Monod proposed a scheme involving negative control called the **operon model**, whereby a group of genes is regulated and expressed together as a unit. As we saw in Figure 19–1, the *lac* operon they proposed consists of the *Z, Y*, and *A* structural genes, as well as the adjacent sequences of DNA referred to as the operator region. They argued that the *lacI* gene regulates the transcription of the structural genes by producing a **repressor molecule**, and that the repressor is **allosteric**, meaning that the molecule

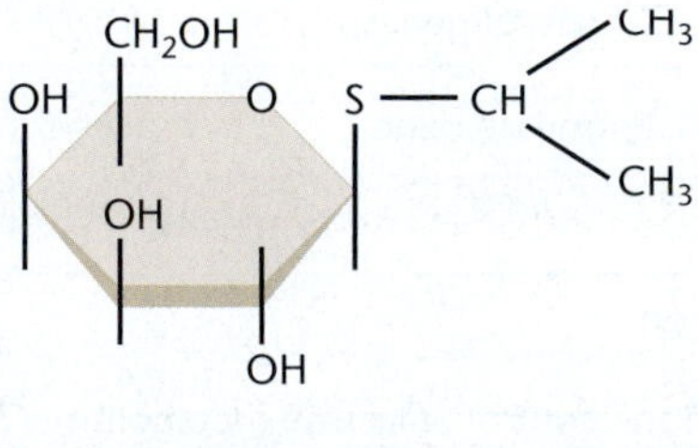

FIGURE 19–4 The gratuitous inducer isopropylthiogalactoside (IPTG).

reversibly interacts with another molecule, causing both a conformational change in three-dimensional shape and a change in chemical activity. Figure 19–5 illustrates the components of the *lac* operon as well as the action of the *lac* repressor in the presence and absence of lactose. You should refer to it as you read the next section.

Jacob and Monod suggested that the repressor normally interacts with the DNA sequence of the operator region. When it does so, it inhibits the action of RNA polymerase, effectively repressing the transcription of the structural genes [Figure 19–5(b)]. However, when lactose is present, this sugar binds to the repressor and causes an allosteric conformational change. This change alters the binding site of the repressor, rendering it incapable of interacting with operator DNA [Figure 19–5(c)]. In the absence of the repressor–operator interaction, RNA polymerase transcribes the structural genes, and the enzymes necessary for lactose metabolism are produced. Because transcription occurs only when the repressor *fails* to bind to the operator region, *negative control* is exerted.

The operon model uses these potential molecular interactions to explain the efficient regulation of the structural genes. In the absence of lactose, the enzymes encoded by the genes are not needed and are repressed. When lactose is present, it indirectly induces the activation of the genes by binding with the repressor.* If all lactose is metabolized, none is available to bind to the repressor, which is again free to bind to operator DNA and repress transcription.

Both the I^- and O^C constitutive mutations interfere with these molecular interactions, allowing continuous transcription of the structural genes. In the case of the I^- mutant, seen in Figure 19–6(a), the repressor protein is altered and cannot bind to the operator region, so the structural genes are always turned on. In the case of the O^C mutant [Figure 19–6(b)], the nucleotide sequence of the operator DNA is altered and will not bind with a normal repressor molecule. The result is the same: Structural genes are always transcribed.

Genetic Proof of the Operon Model

The operon model is a good one because it leads to three major predictions that can be tested to determine its validity. The major predictions to be tested are that (1) the *I* gene produces a diffusible cellular product; (2) the *O* region is involved in regulation, but does not produce a product; and (3) the *O* region must be adjacent to the structural genes in order to regulate transcription.

The construction of partially diploid bacteria allows us to assess these assumptions, particularly those that predict *trans*-acting regulatory elements. For example, as introduced in Chapter 15, the F plasmid may contain chromosomal genes, in which case it is designated F^+. When an F′ cell acquires such a plasmid, it now contains its own chromosome plus one or more additional genes present in the plasmid. This creates a host cell, called a **merozygote**, that is diploid

*Technically, allolactose, an isomer of lactose, is the inducer. Allolactose is produced during the initial step in the metabolism of lactose by β-galactosidase.

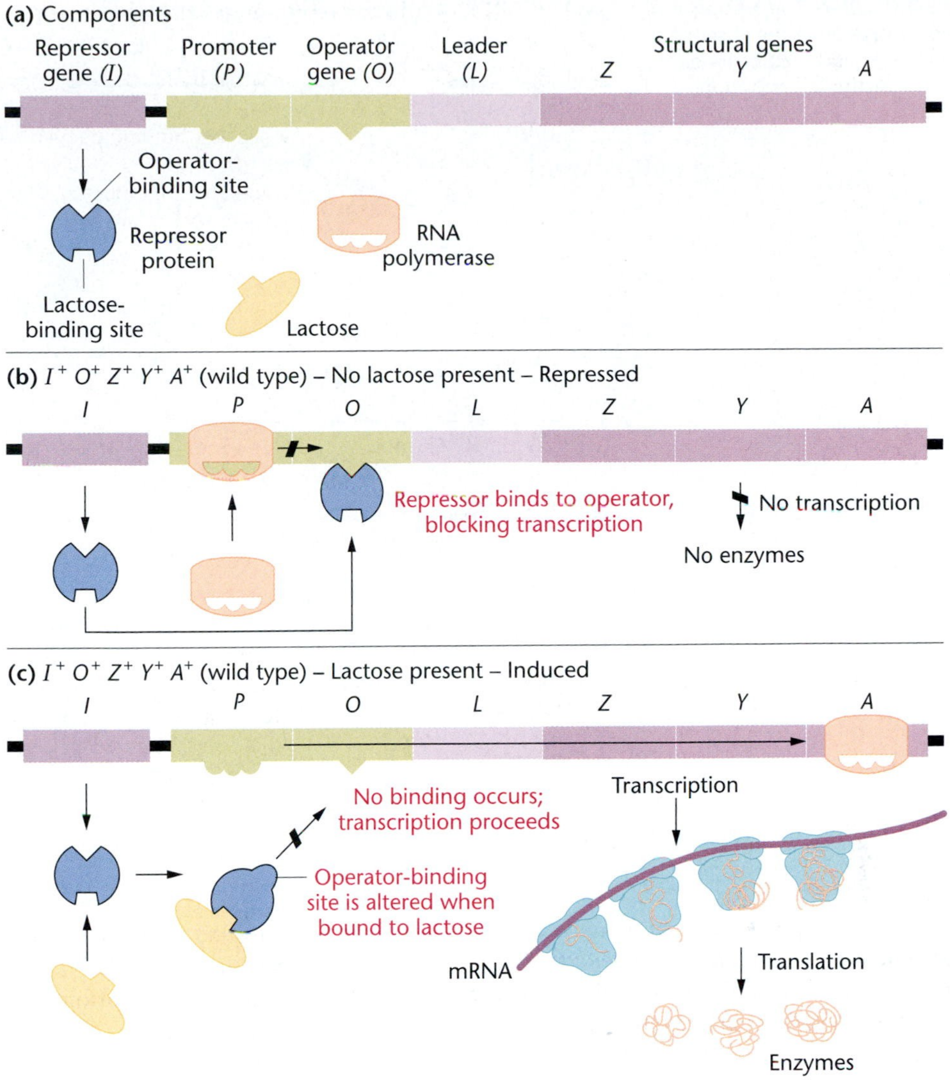

FIGURE 19–5 The components of the wild-type *lac* operon and the response in the absence and the presence of lactose, as described in the text.

for those genes. The use of such a plasmid makes it possible, for example, to introduce an I^+ gene into a host cell whose genotype is I^-, or to introduce an O^+ region into a host cell of genotype O^C. The Jacob–Monod operon model predicts how regulation should be affected in such cells. Adding an I^+ gene to an I^- cell should restore inducibility, because a normal repressor, which is a *trans*-acting factor, would again be produced. Adding an O^+ region to an O^C cell should have no effect on constitutive enzyme production, since regulation depends on an O^+ region immediately adjacent to the structural genes—that is, O^+ is a *cis*-acting regulator.

Results of these experiments are shown in Table 19–1, where *Z* represents the structural genes. The inserted genes are listed after the designation F′. In both cases described here, the Jacob–Monod model is upheld (part B of Table 19–1). Part C shows the reverse experiments, where either an I^- gene or an O^C region is added to cells of normal inducible genotypes. As the model predicts, inducibility is maintained in these partial diploids.

Another prediction of the operon model is that certain mutations in the *I* gene should have the opposite effect of I^-. That is, instead of being constitutive because the repressor can't bind the operator, mutant repressor molecules should be produced that cannot interact with the inducer, lactose. As a result, the repressor would always bind to the operator sequence, and the structural genes would be permanently repressed (Figure 19–7). If this were the case, the presence of an additional I^+ gene would have little or no effect on repression.

In fact, such a mutation, I^S was discovered wherein the operon is "superrepressed," as shown in part D of Table 19–1. An additional I^+ gene does not effectively relieve repression of gene activity. These observations again support the operon model for gene regulation.

Isolation of the Repressor

Although the operon theory of Jacob and Monod succeeded in explaining many aspects of genetic regulation in prokaryotes, the nature of the repressor molecule was not known when their landmark paper was published in 1961. While they had assumed that the allosteric repressor was a protein, RNA was also a candidate because activity of the molecule required the ability to bind to DNA. Despite many attempts to isolate and characterize the hypothetical repressor

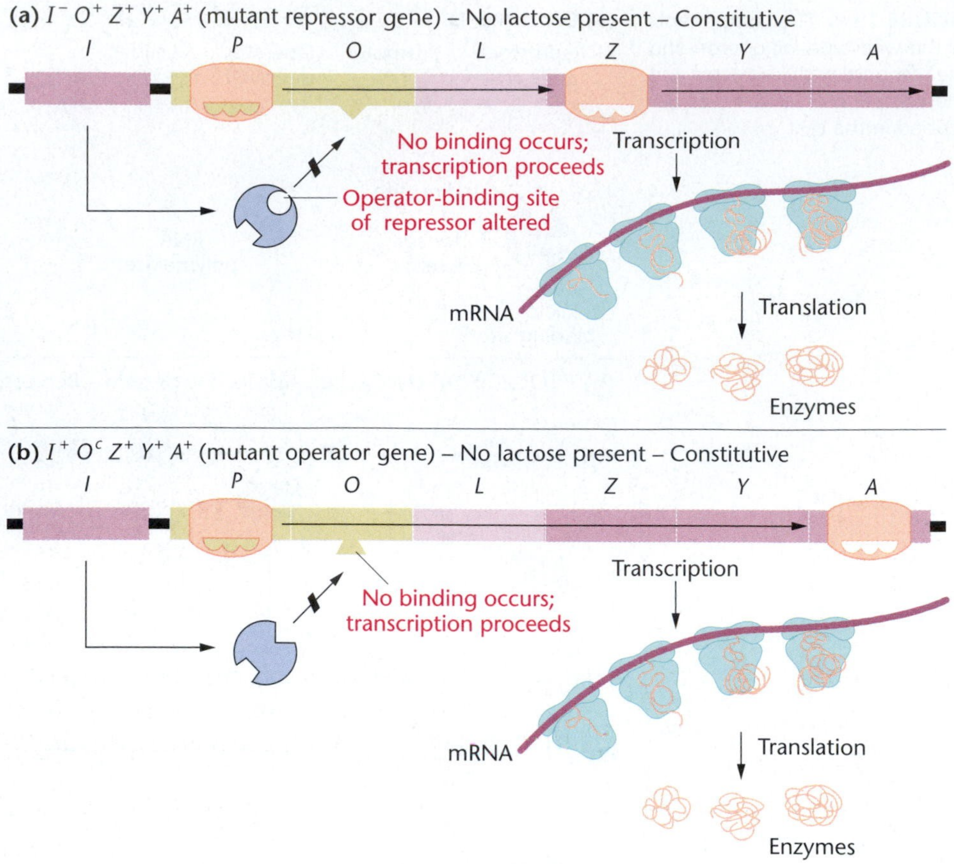

FIGURE 19–6 The response of the *lac* operon in the absence of lactose when a cell bears either the I^- or the O^C mutation.

molecule, no direct chemical evidence was immediately forthcoming. A single *E. coli* cell contains no more than 10 or so copies of the *lac* repressor; direct chemical identification of 10 molecules in a population of millions of proteins and RNAs in a single cell presented a tremendous challenge.

TABLE 19–1 A Comparison of Gene Activity (+ or −) in the Presence or Absence of Lactose for Various *E. coli* Genotypes

Genotype	*Presence of β-Galactosidase Activity* — *Lactose Present*	*Lactose Absent*
$I^+O^+Z^+$	+	−
A. $I^+O^+Z^-$	−	−
$I^-O^+Z^+$	+	+
$I^+O^cZ^+$	+	+
B. $I^-O^+Z^+/F'I^+$	+	−
$I^+O^cZ^+/F'O^+$	+	+
C. $I^+O^+Z^+/F'I^-$	+	−
$I^+O^+Z^+/F'O^c$	+	−
D. $I^sO^+Z^+$	−	−
$I^sO^+Z^+/F'I^+$	−	−

Note: In parts B to D, most genotypes are partially diploid, containing an F factor plus attached genes (F′).

In 1966, Walter Gilbert and Benno Müller-Hill reported the isolation of the *lac* repressor in partially purified form. To achieve the isolation, they used a *regulator quantity* (I^q) mutant strain that contains about 10 times as much repressor as do wild-type *E. coli* cells. Also instrumental in their success was the use of the gratuitous inducer, IPTG, which binds to the repressor, and the technique of **equilibrium dialysis**. In this technique, extracts of I^q cells were placed in a dialysis bag and allowed to attain equilibrium with an external solution of radioactive IPTG, which is small enough to diffuse freely in and out of the bag. At equilibrium, the concentration of IPTG was higher inside the bag than in the external solution, indicating that an IPTG-binding material was present in the cell extract and that this material was too large to diffuse across the wall of the bag.

Ultimately, the IPTG-binding material was purified and shown to have various characteristics of a protein. In contrast, extracts of I^- constitutive cells having no *lac* repressor activity did not exhibit IPTG-binding activity, strongly suggesting that the isolated protein was the repressor molecule.

To confirm this thinking, Gilbert and Müller-Hill grew *E. coli* cells in a medium containing radioactive sulfur and then isolated the IPTG-binding protein, which was labeled in its sulfur-containing amino acids. This protein was mixed with DNA from a strain of phage lambda (λ), which carries the $lacO^+$ gene. The DNA sediments at 40*S*, while the IPTG-binding protein sediments at 7*S*. The DNA and protein were mixed and sedimented in a gradient, using ultracentrifugation. The radioactive protein sediments at the same rate as

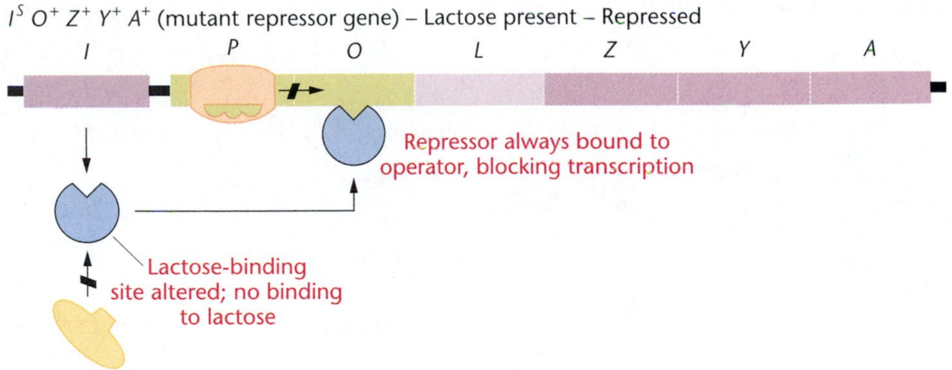

FIGURE 19–7 The response of the *lac* operon in the presence of lactose in a cell bearing the I^S mutation.

does DNA, indicating that the protein binds to the DNA. Further experiments showed that the IPTG-binding, or repressor, protein binds only to DNA containing the *lac* region and does not bind to *lac* DNA containing an operator-constitutive (O^C) mutation.

19.3 The Catabolite-Activating Protein (CAP) Exerts Positive Control over the *lac* Operon

As is apparent from the preceding discussion of the *lac* operon, the role of β-galactosidase is to cleave lactose into its components glucose and galactose. Then, for galactose to be used by the cell, it is converted to glucose. What if the cell found itself in an environment that contained an ample amount of lactose *and* glucose? It would not be energetically efficient for a cell to be "induced" by lactose to make β-galactosidase, since what it really needs, glucose, is already present. As we shall see next, still another molecular component, called the **catabolite-activating protein (CAP)**, is involved in effectively repressing the expression of the *lac* operon when glucose is present. This inhibition, called **catabolite repression**, reflects the greater simplicity with which glucose may be metabolized in comparison to lactose. The cell "prefers" glucose, and if it is available, the *lac* operon is not activated, even when lactose is present.

To understand CAP and its role in regulation, let's backtrack for a moment. When the *lac* repressor is bound to the inducer, the *lac* operon is activated and RNA polymerase transcribes the structural genes. As we have learned in Chapter 5, transcription is initiated as a result of the binding that occurs between RNA polymerase and the nucleotide sequence of the **promoter region**, found upstream $(5')$ from the initial coding sequences. Within the *lac* operon, the promoter is found between the *I* gene and the operator region (*O*). (See Figure 19–1.) Careful examination has revealed that polymerase binding is never very efficient unless CAP is also present to facilitate the process.

The mechanism is summarized in Figure 19–8. In the absence of glucose and under inducible conditions, CAP exerts **positive control** by binding to the CAP site, facilitating RNA polymerase binding at the promoter, and thus transcription. Therefore, for maximal transcription, the repressor must be bound by lactose (so as not to repress operon expression), *and* CAP must be bound to the CAP-binding site.

This leads to the central question about CAP. What role does glucose play in inhibiting CAP binding when it is present? The answer involves still another molecule, **cyclic adenosine monophosphate (cAMP)**, upon which CAP binding is dependent. In order to bind to the promoter, CAP must be bound to cAMP. The level of cAMP is itself dependent on an enzyme, **adenyl cyclase**, which catalyzes the conversion of ATP to cAMP.* (See Figure 19–9.)

The role of glucose in catabolite repression is now clear. It inhibits the activity of adenyl cyclase, causing a decline in the level of cAMP in the cell. Under this condition, CAP cannot form the CAP–cAMP complex essential to the positive control of transcription of the *lac* operon.

Like the *lac* repressor, CAP and cAMP–CAP have been examined by using X-ray crystallography. CAP is a dimer that inserts into adjacent regions of a specific nucleotide sequence of the DNA making up the promoter. The cAMP–CAP complex, when bound to DNA, bends it, causing it to assume a new conformation.

Binding studies in solution further clarify the mechanism of gene activation. Alone, neither cAMP–CAP nor RNA polymerase has a strong affinity to bind to *lac* promoter DNA. Nor does either molecule have a strong affinity to bind to the other. However, when both are together in the presence of the *lac* promoter DNA, a tightly bound complex is formed, an example of what, in biochemical terms, is called **cooperative binding**. In the case of cAMP–CAP and the *lac* operon, the phenomenon illustrates the high degree of specificity that is involved in the genetic regulation of just one small group of genes.

Regulation of the *lac* operon by catabolite repression results in efficient energy use, because the presence of glucose will override the need for the metabolism of lactose, should it also be available to the cell. Catabolite repression involving CAP has also been observed for other inducible operons, including those controlling the metabolism of galactose and arabinose.

*Because of its involvement with CAMP, CAP is also called cyclic AMP receptor protein (CRP), and the gene encoding the protein is named *crp*. Since the protein was first named CAP, we will adhere to the initial nomenclature.

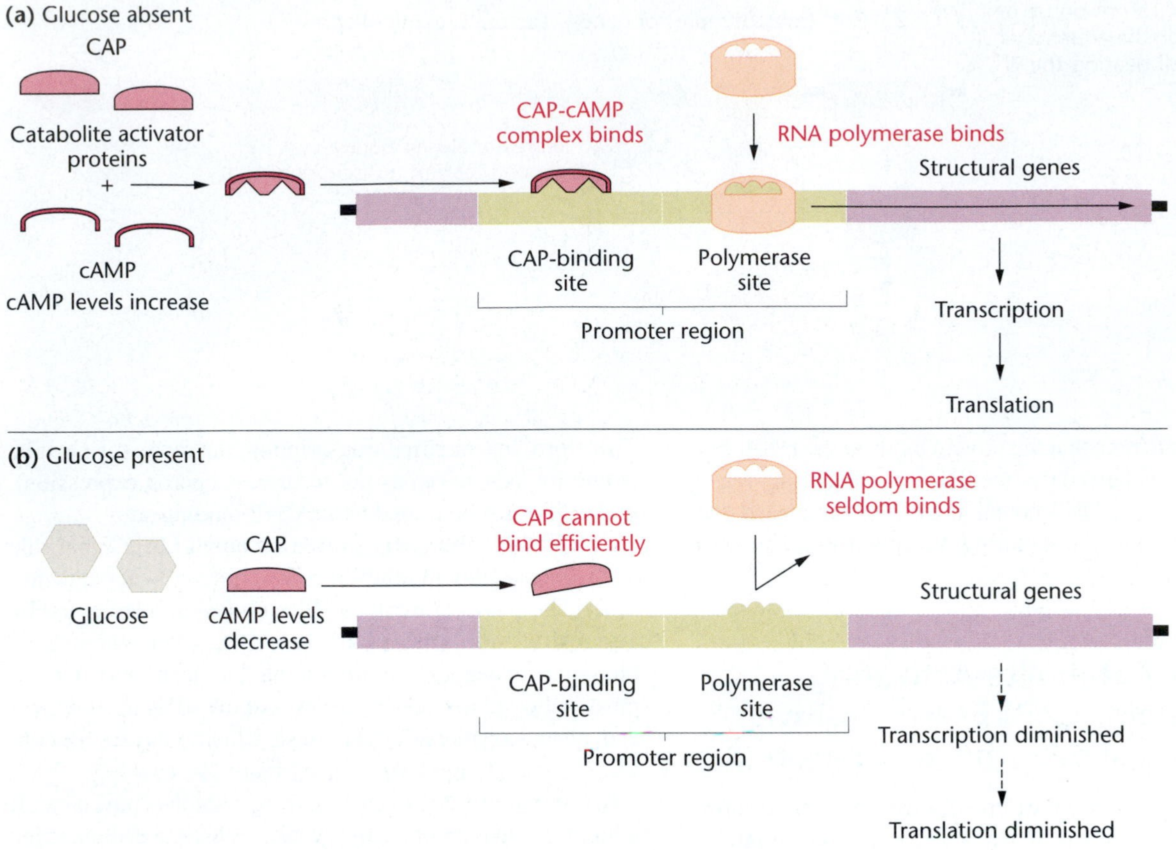

FIGURE 19–8 Catabolite repression. (a) In the absence of glucose, cAMP levels increase, resulting in the formation of a CAP–cAMP complex, which binds to the CAP site of the promoter, stimulating transcription. (b) In the presence of glucose, cAMP levels decrease, CAP–cAMP complexes are not formed, and transcription is not stimulated.

19.4 Crystal Structure Analysis of Repressor Complexes Has Confirmed the Operon Model

We now have available thorough knowledge of the biochemical nature of the regulatory region of the *lac* operon, identifying the precise locations of its various components relative to one another (Figure 19–10). Just a few years ago, in 1996, Mitchell Lewis, Ponzy Lu, and their colleagues succeeded in determining the crystal structure of the *lac* repressor, as well as the structure of the repressor bound to the inducer and to operator DNA. As a result, previous information that was based on genetic and biochemical data has now been complemented with the missing structural

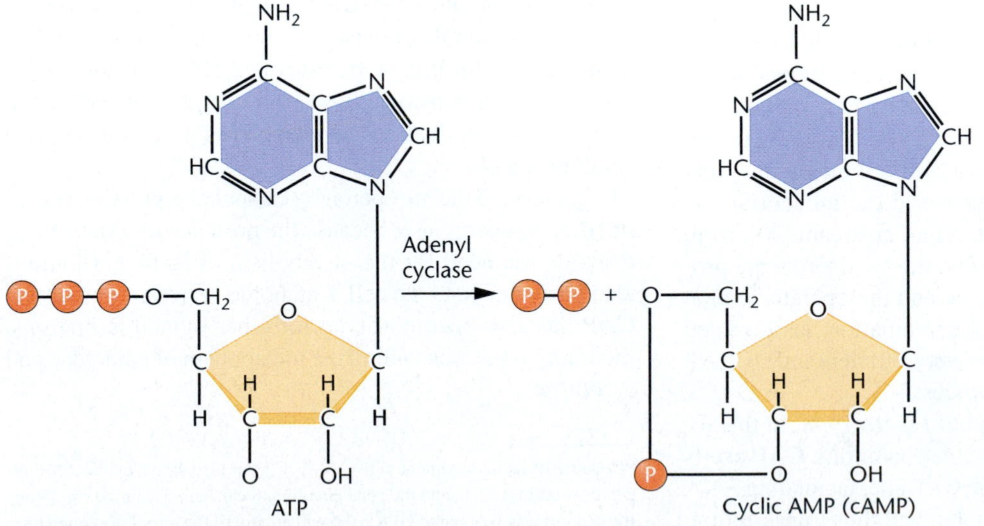

FIGURE 19–9 The formation of cAMP from ATP, catalyzed by adenyl cyclase.

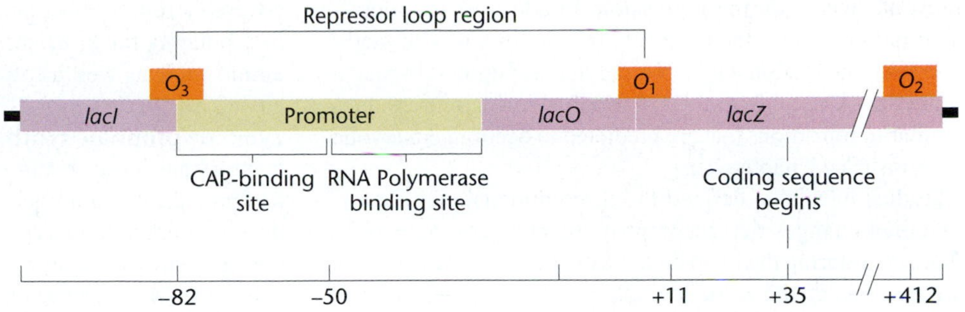

FIGURE 19–10 A detailed depiction of various regulatory regions involved in the control of genetic expression of the *lac* operon, as described in the text. The numbers on the bottom scale represent nucleotide sites upstream and downstream from the initiation of transcription.

interpretation. Together, a nearly complete picture of the regulation of the operon has emerged.

The repressor, as the gene product of the *I* gene, is a monomer consisting of 360 amino acids. Within the monomer, the region of inducer binding has been identified [Figure 19–11(a)]. The functional repressor contains four such subunits, creating a homotetramer. The tetramer can be cleaved with a protease under controlled conditions and yields five fragments. Four are derived from the N-terminal ends of the tetramer, and they bind to operator DNA. The fifth fragment is the remaining core of the tetramer, derived from the COOH-terminus ends; it binds to lactose and gratuitous inducers such as IPTG. Analysis has revealed that, at any single time, each tetramer can bind to two symmetrical operator DNA helices [Figure 19–11(b)].

The operator DNA that was previously defined by mutational studies ($lacO^C$) and confirmed by DNA-sequencing analysis is located just downstream from the start of the *lacZ* gene, but upstream from the beginning of the actual coding sequence. The crystallographic studies show that the actual region of repressor binding of this primary operator O_1 consists of 21 base pairs. Two other auxiliary operator regions have been identified, as shown in Figure 19–10. One, O_2, is 401 base pairs downstream from the primary operator within the *lacZ* gene. The other, O_3 is 93 base pairs upstream from O_1 just above the CAP site. In vivo, all three operators must be bound for maximum repression.

Binding by the repressor at two operator sites distorts the conformation of DNA, causing it to bend away from the repressor. When a model is created involving dual binding of operators O_1 and O_3 [Figure 19–11(c)], the 93 base pairs of DNA that intervene must jut out, forming what is called a **repression loop**. This model positions the promoter region that binds RNA polymerase on the inside of the loop, which

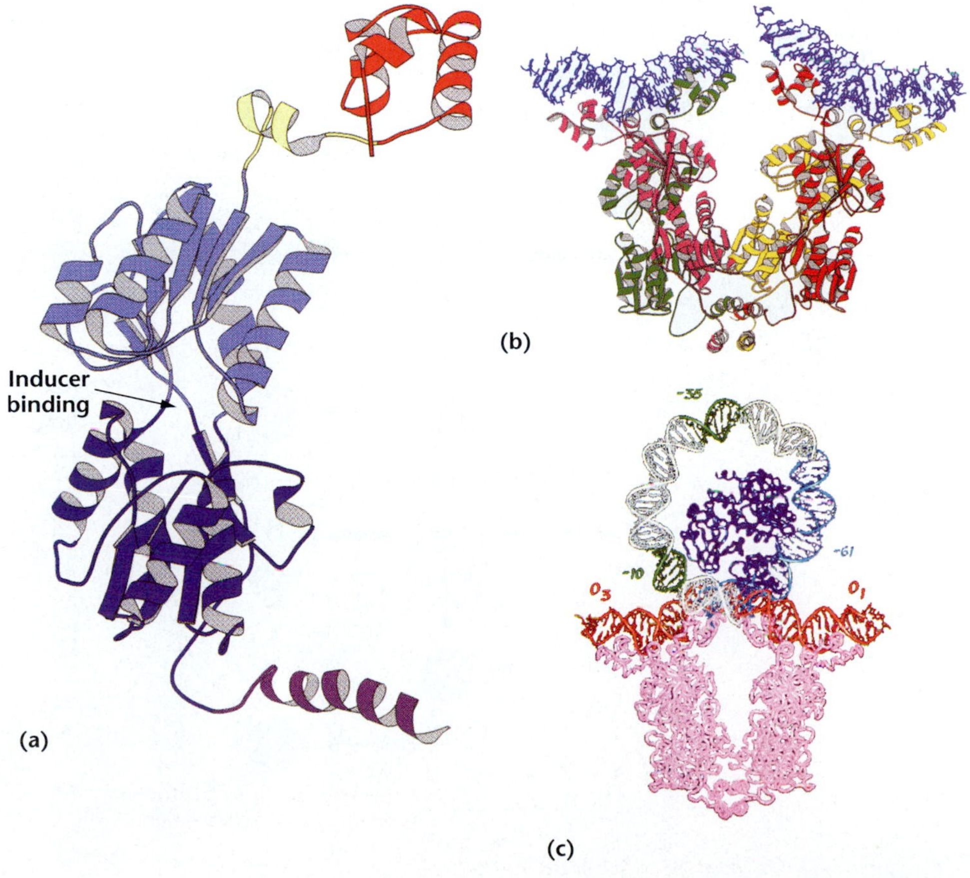

FIGURE 19–11 Models of the *lac* repressor and its binding to operator sites with DNA, as generated from crystal structure analysis: (a) The repressor monomer. The arrow points to the inducer-binding site. The DNA-binding region is shown in red. (b) The repressor dimer bound to two 21 base-pair segments of operator DNA (shown in blue). (c) The repressor and CAP (shown in dark blue) bound to the *lac* DNA. Binding to operator regions O_1 and O_3 creates a 93 base-pair repression loop of promoter DNA.

prevents access during repression. In addition, the repression loop positions the CAP-binding site in a way to facilitate CAP interaction with RNA polymerase upon subsequent induction. The finding of DNA looping during repression is similar to transitions that are predicted to occur in eukaryotic systems (See Chapter 20).

Studies have also defined the three-dimensional conformational changes that accompany the allosteric transitions that occur during the interactions with the inducer molecules. Taken together, the crystallographic studies bring us to a new level of understanding of the regulatory process occurring within the *lac* operon, confirming the findings and predictions of Jacob and Monod in their model set forth over 40 years ago and based strictly on genetic grounds.

19.5 The tryptophan (trp) Operon in *E. coli* Is a Repressible Gene System

Although the process of induction had been known for some time, it was not until 1953 that Monod and colleagues discovered a repressible operon. Wild-type *E. coli* are capable of producing the enzymes necessary for the biosynthesis of amino acids as well as other essential macromolecules. Focusing his studies on the amino acid tryptophan and the enzyme **tryptophan synthetase**, Monod discovered that if tryptophan is present in sufficient quantity in the growth medium, the enzymes necessary for its synthesis are not produced. Energetically, repression of the genes involved in the production of these enzymes is highly economical for the cell when ample tryptophan is present.

Further investigation showed that a series of enzymes encoded by five contiguous genes on the *E. coli* chromosome is involved in tryptophan synthesis. These genes are part of an operon, and in the presence of tryptophan, all are coordinately repressed, and none of the enzymes are produced. Because of the great similarity between this repression and the induction of enzymes for lactose metabolism, Jacob and Monod proposed a model of gene regulation analogous to the *lac* system (Figure 19–12).

To account for repression, they suggested the presence of a *normally inactive repressor* that alone cannot interact with

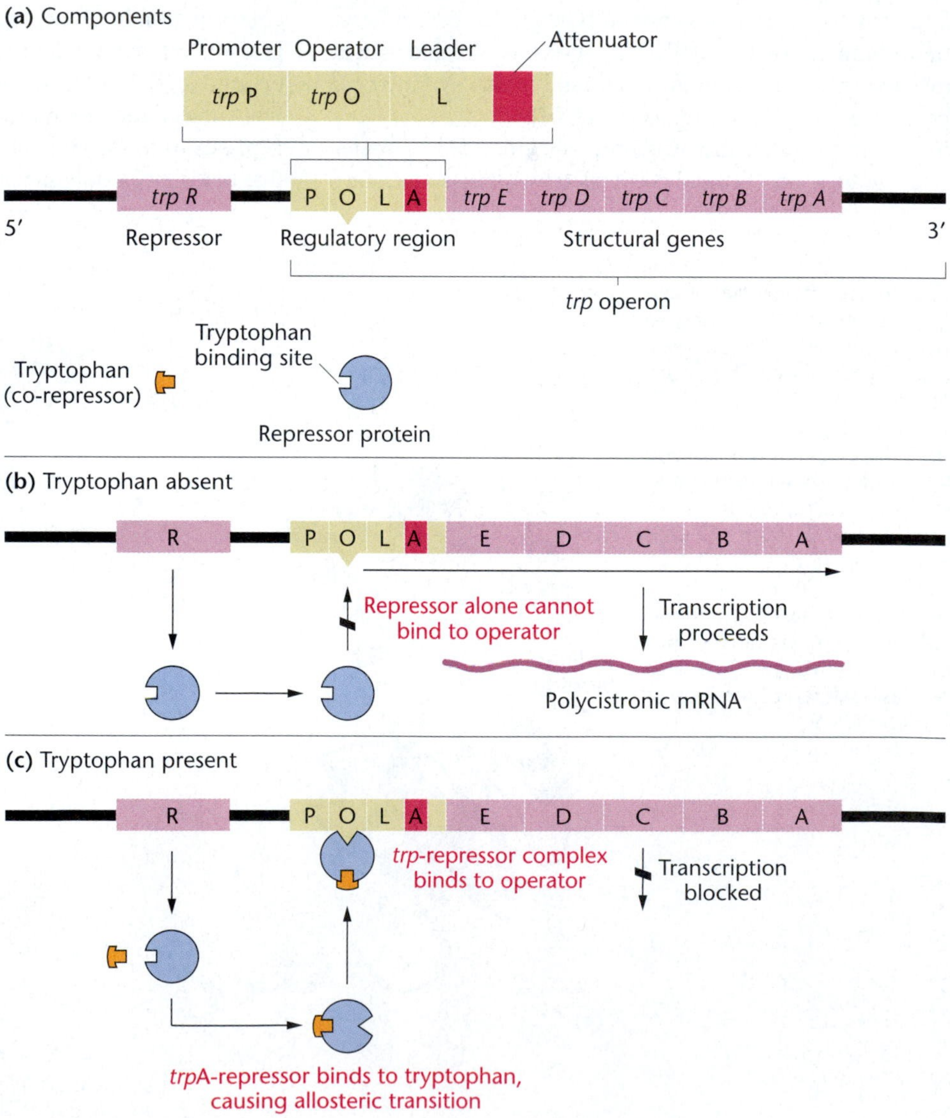

FIGURE 19–12 (a) The components involved in the regulation of the tryptophan operon. (b) Regulatory conditions are depicted that involve either activation or (c) repression of the structural genes. In the absence of tryptophan, an inactive repressor is made that cannot bind to the operator (*O*), thus allowing transcription to proceed. In the presence of tryptophan, it binds to the repressor, causing an allosteric transition to occur. This complex binds to the operator region, leading to repression of the operon.

the operator region of the operon. However, the repressor is an allosteric molecule that can bind to tryptophan. When this amino acid is present, the resultant complex of repressor and tryptophan attains a new conformation that binds to the operator, repressing transcription. Thus, when tryptophan, the end product of this anabolic pathway, is present, the system is repressed and enzymes are not made. Since the regulatory complex inhibits transcription of the operon, this repressible system is under negative control. And, as tryptophan participates in repression, it is referred to as a **corepressor** in this regulatory scheme.

Evidence for the *trp* Operon

Support for the concept of a repressible operon was soon forthcoming, based primarily on the isolation of two distinct categories of constitutive mutations. The first class, $trpR^-$, maps at a considerable distance from the structural genes. This locus represents the gene coding for the repressor. Presumably, the mutation either inhibits the interaction of the repressor with tryptophan or inhibits repressor formation entirely. Whichever the case, no repression ever occurs in cells with the $trpR^-$ mutation. As expected, if the $trpR^+$ gene encodes a functional repressor molecule, the presence of a copy of this gene will restore repressibility.

The second constitutive mutant is analogous to that of the operator of the lactose operon, because it maps immediately adjacent to the structural genes. Furthermore, the addition of a wild-type operator gene into mutant cells (as a *trans*-acting element) does not restore enzyme repression. This is predictable if the mutant operator can no longer interact with the repressor–tryptophan complex.

The entire *trp* operon has now been well defined, as shown in Figure 19–12. Five contiguous structural genes (*trp E, D, C, B,* and *A*) are transcribed as a polycistronic message that directs translation of the enzymes that catalyze the biosynthesis of tryptophan. As in the *lac* operon, a promoter region (*trpP*) represents the binding site for RNA polymerase, and an operator region (*trpO*) binds the repressor. In the absence of binding, transcription is initiated within the overlapping *trpP–trpO* region and proceeds along a **leader sequence** 162 nucleotides prior to the first structural gene (*trpE*). Within that leader sequence, still another regulatory site has been demonstrated, called an attenuator, the subject of the next section of this chapter. As we shall see, this regulatory unit is an integral part of the control mechanism of the operon.

19.6 Attenuation Is a Critical Process during the Regulation of the *trp* Operon in *E. coli*

Charles Yanofsky, his coworker Kevin Bertrand, and their colleagues observed that, even when tryptophan is present and the *trp* operon is repressed, initiation of transcription can still occur, leading to the intial portion of the mRNA (the 5′ leader sequence). Hence, while the activated repressor binds to the operator region, it does not strongly inhibit the *initial expression* of the operon, suggesting that there must be a subsequent mechanism by which tryptophan somehow inhibits transcription and thus enzyme synthesis. Yanofsky discovered that, following initiation of transcription, *in the presence of high concentrations of tryptophan*, mRNA synthesis is usually terminated at a point about 140 nucleotides along the transcript. This process is called **attenuation**,* indicative of the effect of diminishing genetic expression of the operon. However, when tryptophan is absent, or present in very low concentrations, transcription is initiated and *not* subsequently terminated, instead continuing beyond the leader sequence along the DNA encoding the structural genes, starting with the trpE gene. As a result, a polycistronic mRNA is produced and the enzymes essential to the biosynthesis of tryptophan are subsequently translated. The *trp* operon and the genetic components involved in the attenuation process are diagrammed in Figure 19–13(a).

Identification of the site involved in attenuation was made possible by the isolation of various deletion mutations in the region 115 to 140 nucleotides into the leader sequence. Such mutations abolish attenuation. This site is referred to as the **attenuator**. An explanation of how attenuation occurs and how it is overcome, put forward by Yanofsky and colleagues, is summarized in Figure 19–13(b). The initial DNA sequence that is transcribed gives rise to a mRNA molecule that has the potential to fold into two mutually exclusive stem-loop structures referred to as "hairpins." In the presence of excess tryptophan, the hairpin that is formed behaves as a **terminator** structure and transcription is almost always abolished prematurely. On the other hand, if tryptophan is scarce, the alternative hairpin that is formed behaves as an **antiterminator** structure. Transcription is allowed to proceed past the involved DNA sequence, and the entire mRNA is subsequently produced. These hairpins are illustrated in Figure 19–13(b) and (c).

The question, of course, is how the absence (or a low concentration) of tryptophan allows attenuation to be somehow bypassed. A key point in Yanofsky's model is that the leader transcript must be translated in order for the antiterminator hairpin to form. He discovered that the leader transcript includes two triplets (UGG) that encode tryptophan preceded upstream by an initial AUG sequence that prompts the initiation of translation by ribosomes. When adequate tryptophan is present, charged $tRNA^{trp}$ is also present. As a result, translation proceeds past these triplets, and the *terminator hairpin* is formed, as illustrated in Figure 19–13(b). If cells are starved of tryptophan, charged $tRNA^{trp}$ is unavailable. The ribosome then "stalls" during the translation of triplets as charged $tRNA^{trp}$ is called for, but is unavailable because of the lack of tryptophan. This event induces the formation of the *antiterminator hairpin* within the transcript, as shown in Figure 19–13(c). As a result, attenuation is overcome, and transcription proceeds, leading to expression of the entire set of structural genes.

*Attenuation is derived from the verb attenuate, meaning "to reduce in strength, weaken, or impair."

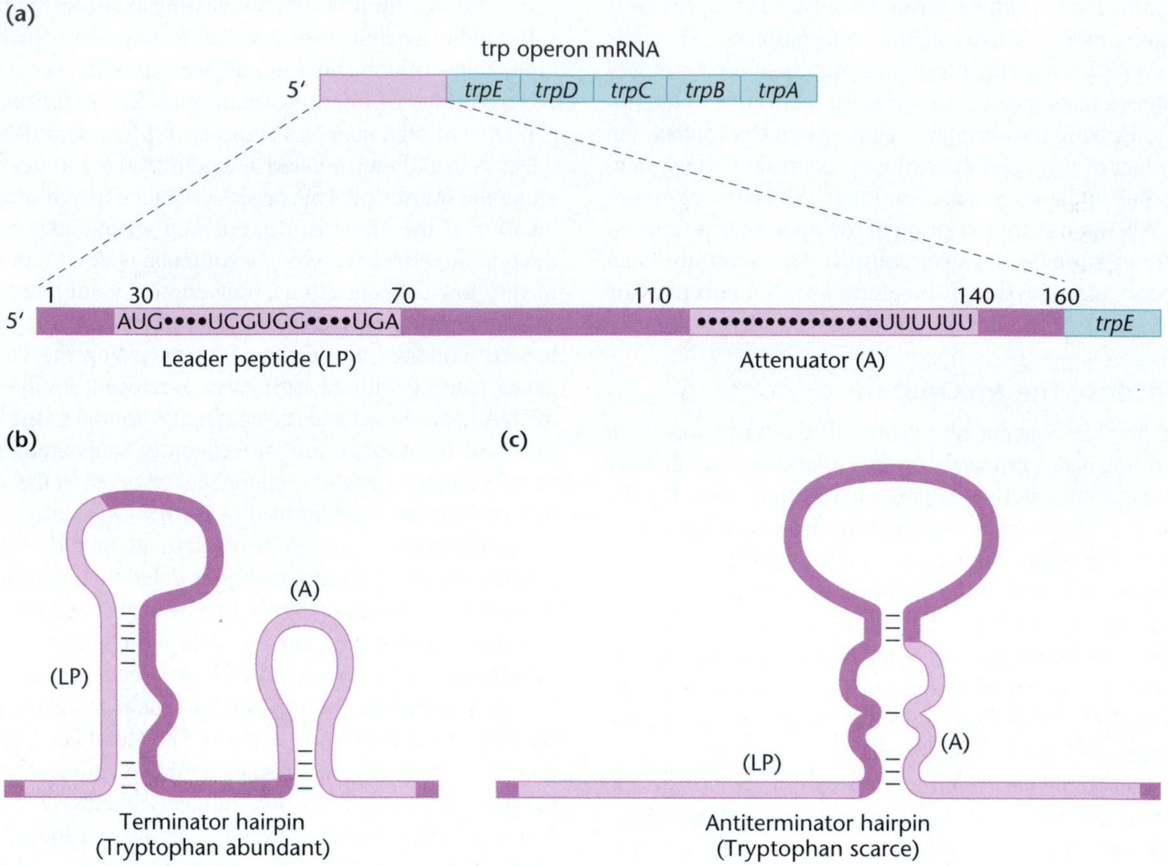

FIGURE 19–13 Diagram of the involvement of the leader sequence of the mRNA transcript of the *trp* operon of *E. coli* during attenuation. (a) The 5′ leader sequence of the *trp* operon is expanded to identify the portions encoding the leader peptide (LP) and the attenuator region (AT). Critical nucleotide sequences, including the UGG triplets that encode tryptophan, are also identified; (b) the terminator hairpin, which forms when the ribosome proceeds during translation of the *trp* codons; and (c) the antiterminator hairpin, which forms when the ribosome stalls at the *trp* codons because tryptophan is scarce.

Details of the secondary structures of the transcript have now been worked out and described. They satisfy conditions leading either to continued transcription or termination (attenuation). Additionally, other mutations in the leader sequence have been isolated that are predicted to alter the secondary structure of the transcript and its impact on attenuation. In each case, the predicted result has been upheld. Yanofsky's model, while complex, has significantly extended our knowledge of genetic regulation. Furthermore, the experimental evidence supporting it represents an integrated approach involving genetic and biochemical analysis, which is becoming commonplace in molecular genetic research.

The phenomenon of attenuation appears to be a mechanism common to other bacterial operons in *E. coli* that regulate the enzymes essential to the biosynthesis of amino acids. In addition to tryptophan, operons involved in threonine, histidine, leucine, and phenylalanine display attenuators in their leader sequences. As with the *trp* operon, each is known to contain multiple codons calling for the amino acid being regulated, which prompt "stalling" of translation if the appropriate amino acid is missing. For example, the leader sequence in the histidine operon codes for seven histidine residues in a row. The threonine operon leader sequence calls for eight threonine residues. As with tryptophan, when the amino acid is present, stalling does not occur, a terminator hairpin structure is formed, and attenuation occurs.

19.7 *TRAP* and *AT* Proteins Govern Attenuation in *B. subtilis*

It is useful to point out that not all organisms, even those of the same type, solve problems such as gene regulation in exactly the same way. Thus, it is not unusual for us to discover new strategies arising during evolution. Often, the new approach is a variation on the same theme. Such is the case regarding the regulation of the *trp* operon in bacteria.

As we saw previously, *E. coli*, a Gram negative bacterium, utilizes charged $tRNA^{trp}$ and a terminator hairpin in the leader sequence of the transcript as the underlying basis for attenuating transcription of its *trp* operon. The Gram positive bacterim, *Bacillus subtilis*, also utilizes attenuation and hairpins to regulate its *trp* operon. In fact, *B. subtilis* relies on attenuation as the sole mechanism for regulation, lacking a mechanism that represses transcription entirely in this operon, as is present in *E. coli*.

However, the molecular signals that cause attenuation in *B. subtilis* do not invoke the process of translation and "stalling," as does *E. coli*, to induce the hairpin that terminates transcription. Instead, a specific protein, isolated in the 1990s by Charles Yanofsky and coworkers, either binds or does not bind to the attenuator leader sequence, thereby inducing the alternative terminator or antiterminator configurations, respectively.

The way attenuation is accomplished in *B. subtilis* is unique. The protein, *trp* **RNA-binding attenuation protein (TRAP)**, binds to tryptophan if it is present in the cell. TRAP consists of 11 subunits, forming a symmetrical quaternary protein structure. Each subunit can bind one molecule of tryptophan, incorporating it into a deep pocket within the protein [Figure 19–14(a)]. When fully saturated with tryptophan, this protein can bind to the 5′ leader sequence of the RNA transcript, which contains 11 triplet repeats of either GAG or UAG, each separated by several spacer nucleotides. Each triplet is linked to one of the subunits, which contains a binding pocket for the triplet. The binding [Figure 19–14(b)] forms an RNA belt around TRAP, preventing the antiterminator hairpin from forming. This results in the formation of the terminator configuration, thus leading to premature termination of transcription and, consequently, attenuation of expression of the operon.

The preceding attenuation strategy involves an interesting mechanism of regulation. There are two observations which suggested to Yanofsky and his colleagues that it may be even more complex than it might appear. First, regulation of the *trp* operon in *B. subtilis* is extremely sensitive to a wide range of tryptophan concentrations. Such a fine tuning suggests that something more than a simple on–off mechanism

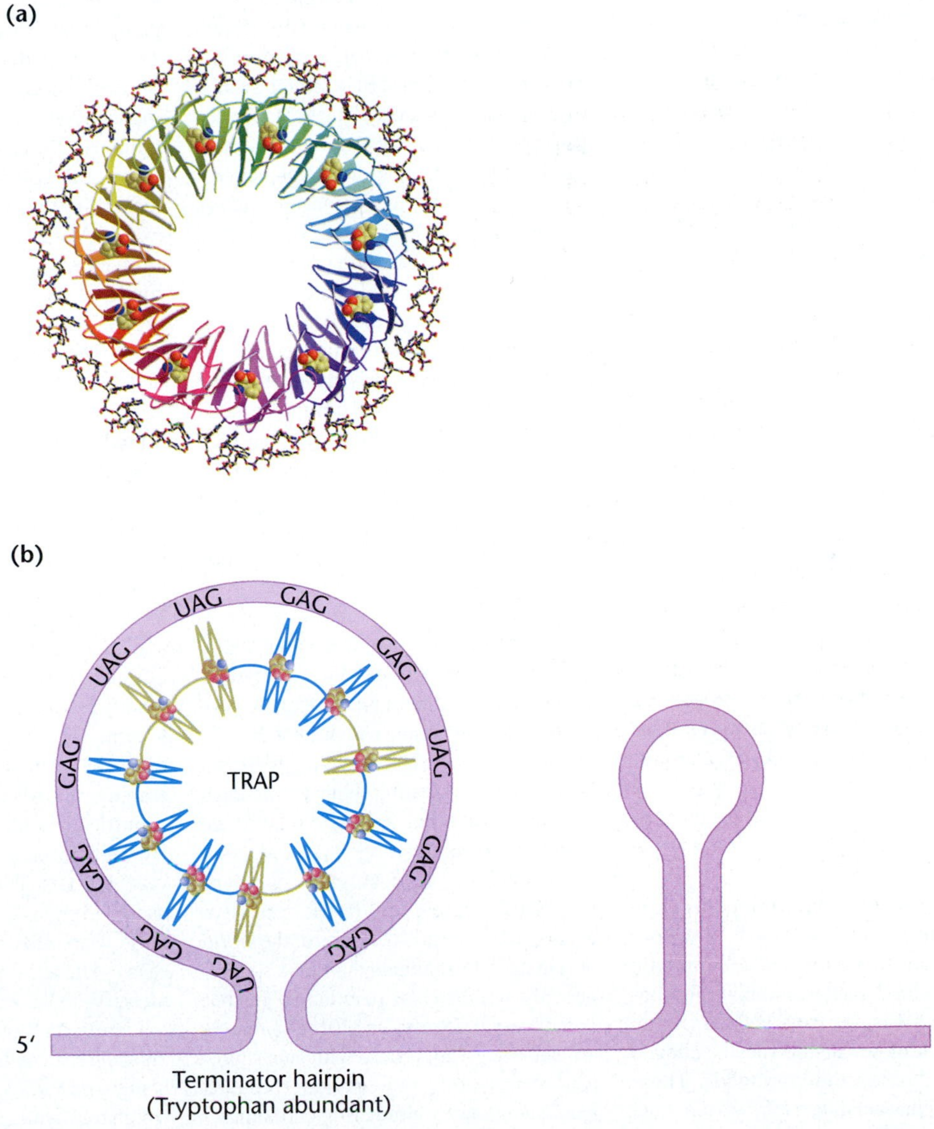

FIGURE 19–14 (a) Model of the *trp* RNA-binding attenuation protein (TRAP). The symmetrical molecule consists of 11 subunits, each capable of binding to one molecule of tryptophan (shown as spheres) (b) The interaction of a tryptophan-bound TRAP molecule with the leader sequence of the *trp* operon of *B. subtilis*. This interaction induces the formation of the terminator hairpin, attenuating expression of the *trp* operon.

Genetics, Technology, and Society

Why Is There No Effective Aids Vaccine?

AIDS is the deadliest plague of the last century, surpassing the great influenza epidemic of 1918. It causes unimaginable human suffering, and threatens to destabilize vast areas of the world--economically, socially, and politically. Since the epidemic began in the 1970's, it has progressed into every country and continent in the world. Over 6 million people are infected each year. At present, 36 million people are infected with HIV worldwide and over 100 million will be infected by the end of the decade. Approximately 90% of AIDS cases occur in developing countries, particularly in sub-Saharan Africa, and Southeast Asia. One tenth of these are children and 50% are 15-24 year olds. According to a recent United Nations report, one-third of the work force in Zimbabwe will die from AIDS by the year 2005. Since the beginning of the epidemic, 13 million children have been orphaned following the death of their HIV-infected parents. It is estimated that there will be over 40 million AIDS orphans by the end of the decade. Since the beginning of the epidemic, over 21 million people have died. Despite preventive measures, AIDS continues to increase in North America. Approximately 45,000 people are infected each year. At present, about 1 million people in North America are living with AIDS.

Ironically, we know how to prevent the spread of HIV infection: modification of sexual practices, needle exchanges, screening of the blood supply, treatment of infected pregnant women. But these measures have only slowed, not stopped, HIV spread. The powerful drug combinations now used to treat AIDS are extremely costly, preventing their use in parts of the world that need treatments the most. In addition, the virus is developing drug-resistance.

The need for a vaccine to stop this global pandemic is obvious. If medical science can eradicate smallpox, and control polio and tuberculosis, then why can we not control AIDS? It is not from lack of trying. Since 1987, more than 60 vaccine clinical trials were conducted worldwide. These trials show that candidate HIV vaccines are safe and result in production of antibodies that recognize the protein used as the vaccine. But no vaccine to date has been able to effectively neutralize a real virus infection. There is currently one HIV vaccine in Phase III clinical trials, with results to be presented by the end of 2002. Another Phase III trial is scheduled to begin in 2002. About 20 candidate vaccines are in preclinical development.

Historically, the best viral vaccines have been prepared from live-attenuated viruses or whole viruses that have been killed. However, in the case of HIV, there are obvious safety concerns; hence, no human clinical trials have been performed with either live-attenuated or whole-killed viruses.

To date, most candidate HIV vaccines use the HIV envelope protein–called gp120 or gp160–as the immunogen. This protein is present in the external coating of the virus and is the viral surface protein which binds to the human cells that HIV infects. The gp120/160 vaccines take many forms: small synthetic peptides, recombinant proteins synthesized within organisms such as vaccinia virus or bacteria, or even naked DNA encoding gp160 protein. So far, tests of these vaccines have yielded disappointing results. Although the gp120/160 vaccines elicit antibody responses to the protein in the vaccine, they are not effective against real viruses isolated from the wild.

Why is it so difficult to develop an HIV vaccine? The reasons are a combination of biological and social factors. Perhaps the greatest impediment to developing an effective AIDS vaccine is the genetic plasticity of the virus. The reverse transcriptase of HIV-1 is extremely error-prone, introducing one or more mutations into the HIV genome during each round of replication. In addition, the highest rate of mutation occurs in the *env* gene, which encodes the gp120/160 protein. As a result, gp120/160 is continually mutating, both within each infected individual and throughout the world. A vaccine that triggers production of antibodies against one type of gp120/160 protein will be less effective against subtypes in other parts of the world or against HIV mutants that arise spontaneously within each person. To further complicate the story, gp120/160 proteins are heavily glycosylated, with sugar molecules protecting the envelope proteins from detection by circulating antibodies.

HIV's stealthy life cycle also interferes with vaccine effectiveness. Although HIV is transmitted as free virus in body fluids, it rapidly becomes sequestered within host cells. After infection, HIV's genetic material integrates into the host cell's genome, remaining quiescent for long periods of time. During this latent period, the virus is undetectable by the host's immune system. The final punch in HIV's biological assault is its choice of host cell. By infecting and destroying the host's CD4+ helper T cells, the virus disables the immune system and ensures its ultimate escape from immune clearance.

HIV vaccine development has been handicapped by the lack of a relevant animal model system in which to test candidate vaccines. The only non-human animal that can be infected with HIV is the chimpanzee; however, serious economic and ethical considerations affect the use of these higher primates as experimental animals. Scientists also test HIV vaccines in monkeys using an HIV-like virus called simian immunodeficiency virus (SIV) or a recombinant virus of SIV and HIV. However, SIV may be too genetically removed from HIV, and may not be a relevant model for HIV vaccine action. Many scientists argue that the only true experimental animal for testing HIV vaccines is the human, and that extensive clinical trials of multiple vaccine types should be conducted as quickly as possible.

But this is not as simple as it sounds. Phase III clinical trials (efficacy trials) must be conducted on thousands of volunteers who are not presently infected with HIV, but are at risk of infection. These volunteers must be followed over many years, and such trials require tens of millions of dollars to conduct. Volunteers may ultimately test positive for HIV as a result of vaccine injection. This has raised fears that they may suffer discrimination in obtaining employment or insurance. In industrialized countries, people with the highest risk of HIV infection are the hardest to recruit to vaccine trials. They include injection drug users, high-risk male homosexuals and those with other sexually transmitted diseases. These populations are difficult to identify and do not trust government or academia. In some developing countries, the incidence of HIV infection is extremely high, providing a larger population from which to recruit for vaccine trials. However, the lack of trained investigators and medical infrastructure make trials difficult to conduct in these countries. In addition, most candidate vaccines have been developed by Western researchers and are based on strains of HIV found predominantly in industrialized countries. Understandably, few developing countries want to participate in trials when there is no certainty that they will benefit from, or have access to, any vaccine that emerges from these trials. The scientific uncertainties and difficulties in recruiting for clinical trials has led to a vicious circle. Pharmaceutical companies are reluctant to invest millions of dollars in research and clinical trials when it is uncertain whether effective vaccines can be devised against HIV. However, without extensive clinical trials of many types of candidate vaccines, it is unlikely that we will ever know whether an AIDS vaccine will be effective. The type of global, high risk research and clinical testing required to develop effective AIDS vaccines will involve extensive private and public sector cooperation and high levels of investment from industrialized nations. Given the horrific human and social consequences of the global AIDS pandemic, there appears to be little choice.

References

Vastag, B. 2001. HIV vaccine efforts inch forward. J.A.M.A. 286:1826–1828.

Johnston, M.I. and Flores, J. 2001. Progress in HIV vaccine development. Current Opinion in Pharmacology 1:504–510.

Nabel, G.J. 2001. Challenges and opportunities for development of an AIDS vaccine. Nature 410:1002–1007.

Web Site

HIV vaccine development status report, NIH
http://www.niaid.nih.gov/aidsvacine/whsummarystatus.htm

attributed to TRAP may be at work. Second, a mutation in the gene encoding tryptophanyl-tRNA sythetase leads to the overexpression of the *trp* operon, even in the presence of excess tryptophan. This enzyme is responsible for charging $tRNA^{trp}$, and, as uncharged $tRNA^{trp}$ molecules accumulate, regulation is interrupted. The interruption of regulation would suggest that uncharged $tRNA^{trp}$ plays some essential role in attenuation, leading Yanofsky and his coworker Angela Valbruzzi to ask how $tRNA^{trp}$ might be involved with TRAP. What they found extends our knowledge of the system of regulation.

They discovered that there exists still another protein, **anti-TRAP (AT)**, which provides a metabolic signal that $tRNA^{trp}$ is uncharged, thereby indicating that tryptophan is very scarce in the cell. Yanofsky and Valbruzzi theorized that uncharged $tRNA^{trp}$ induces a separate operon to express the AT gene. The AT protein then associates with TRAP, specifically when it is in the tryptophan-activated state, inhibiting binding to its target leader RNA sequence.

Such a finding, not only adds still another dimension to our understanding of this system, but it also explains the original observation of the mutation in the $tRNA^{trp}$ synthetase gene that leads to the overexpression of the *trp* operon. The mutation prevents charging of $tRNA^{trp}$, even in the presence of tryptophan. As uncharged $tRNA^{trp}$ accumulates, it induces AT, which binds to TRAP. This prevents TRAP from binding to the leader RNA sequence, thus overinducing the *trp* operon.

The preceding description provides insights into the complexity of regulatory mechanisms in bacteria. That such intricate strategies have resulted during the evolutionary process attests to the critical importance of carefully regulating gene expression in bacteria.

19.8 The *ara* Operon Is Controlled by a Regulator Protein that Exerts Both Positive and Negative Control

We conclude this chapter with a brief discussion of the **arabinose (*ara*) operon** as studied in *E. coli.* This inducible operon is unique because the same regulatory protein is capable of exerting both positive or negative control, and as a result, either induces or represses gene expression. The various panels in Figure 19–15 accompany the following description of the conditions in which the operon is defined and shown to be either active or inactive:

1. The metabolism of the sugar arabinose is under the direction of the enzymatic products of three structural genes, *ara B, A,* and *D*. Their transcription is controlled by the regulatory protein AraC, encoded by the *ara C* gene, that interacts with two regulatory regions, *araI* and $araO_2$ *[Figure 19–15(a)]. These sites can be bound individually or coordinately by the AraC protein.

2. The *I* region bears that designation, since, when only it is bound by AraC, the system is induced [Figure 19–15(b)]. For this to occur, both arabinose and cAMP must be present. Thus, like induction of the *lac* operon, a CAP-binding site is present in the promoter region that modulates catabolite repression in the presence of glucose.

*An additional operator region (O1) is also present, but is not involved in the regulation of the ara structural gene.

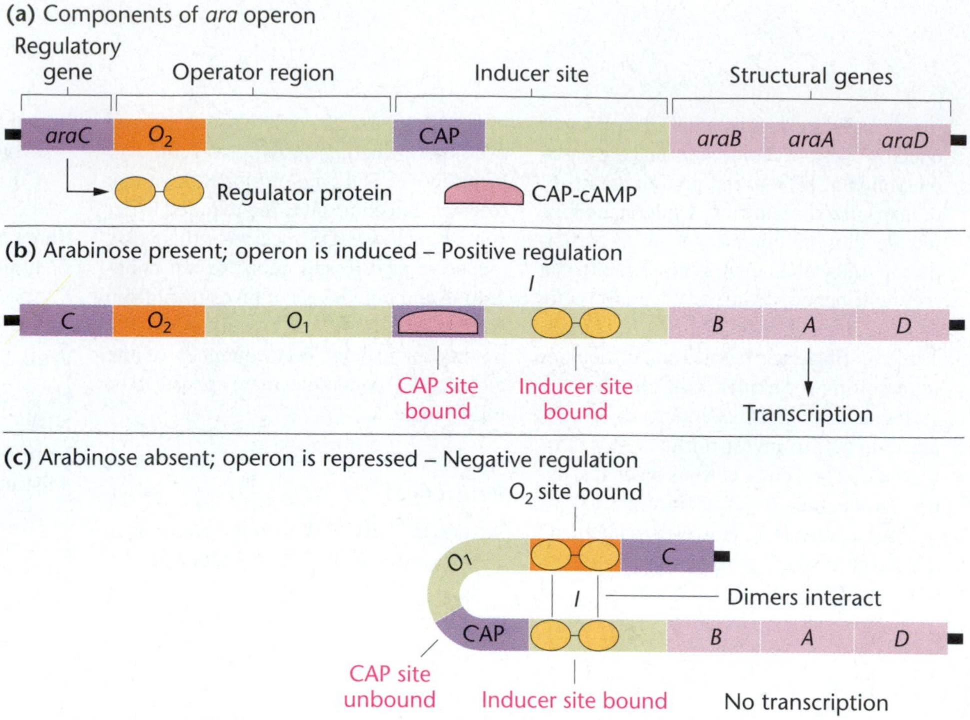

FIGURE 19–15 Genetic regulation of the *ara* operon. The regulatory protein of the *araC* gene acts as either an inducer (in the presence of arabinose) or a repressor (in the absence of arabinose).

3. In the absence of both arabinose and cAMP, the AraC protein *also* binds to the O_2 site (so named because it was the second operator region to be discovered in this operon). When both I and O_2 are bound by AraC, a conformational change occurs in the DNA, whereby a tight loop is formed [Figure 19–15c)] and the structural genes are repressed.

4. The O_2 region is found about 200 nucleotides upstream from the I region. Binding at either regulatory region involves a dimer of AraC. When both regions are bound, the dimers interact leading to the loop that causes repression. Presumably, this loop inhibits the access of RNA polymerase to the promoter region. The region of DNA located between I and O_2, which loops out, is critical to the formation of the repression complex. Genetic alteration through insertion or deletion of even a few nucleotides is sufficient to interfere with loop formation and, therefore, repression.

These findings involving the *ara* operon serve to illustrate the degree of complexity that exists in the regulation of a group of related genes. As we alluded to at the beginning of this chapter, the development of regulatory mechanisms has provided evolutionary advantages to bacterial systems that allow them to adjust to a variety of natural environments. Without question, these systems are keenly equipped genetically not only to survive under varying physiological conditions, but to do so with great biochemical efficiency.

Chapter Summary

1. A system of genetic regulation must exist if the complete genome is not to be continuously active in transcription throughout the life of every cell of all species. Highly refined mechanisms have evolved that regulate transcription, optimizing genetic efficiency.
2. Genetic analysis of bacteria has been pursued successfully since the 1940s. The ease of obtaining large quantities of pure cultures of mutant strains of bacteria as experimental material has made bacteria an organism of choice in numerous types of genetic studies.
3. The discovery and study of the *lac* operon in *E. coli* pioneered the study of gene regulation in bacteria. Genes involved in the metabolism of lactose are coordinately regulated by a negative control system, whereby gene expression is induced by lactose. In its absence, the system is shut down.
4. The catabolite-activating protein (CAP) is essential to the binding of RNA polymerase to the promoter and subsequent transcription in the *lac* and other operons. Therefore, CAP exerts positive control over gene expression. For CAP to stimulate the *lac* operon, it must be bound to cyclic AMP.
5. When glucose is present, cyclic AMP levels are low, CAP binding does not occur, and the structural genes are not expressed. This phenomenon is called catabolite repression. Such a mechanism is efficient energetically, since glucose is preferred, even in the presence of lactose.
6. The *lac* repressor has been isolated and studied. Crystal structure analysis has demonstrated how it interacts with the DNA of the operon as well as with inducers. These studies have demonstrated conformational changes in DNA leading to the formation of a repression loop that inhibits binding between RNA polymerase and the promoter region of the operon.
7. The biosynthesis of tryptophan in *E. coli* involves a number of enzymes whose presence or absence is controlled by another distinct operon. In contrast to the inducible operon controlling lactose metabolism, the *trp* operon is

repressible. In the presence of tryptophan, an active repressor is made that shuts off the expression of the structural genes. Like the *lac* operon, the *trp* operon functions under negative control.

8. An additional regulatory step, referred to as attenuation, has been studied in the *trp* operon. The mechanism varies between *E. coli* and *B. subtilis*, but both depend on alternative hairpin structures that form under different condition in the leader sequence of the mRNA.
9. The *ara* operon is unique in that the regulator protein exerts both positive and negative control over expression of the genes specifying the enzymes that metabolize the sugar arabinose.

Insights and Solutions

1. A theoretical operon (*theo*) in *E. coli* contains several structural genes encoding enzymes that are involved sequentially in the biosynthesis of an amino acid. Unlike the *lac* operon, in which the repressor gene is separate from the operon, the gene encoding the regulator molecule is contained within the *theo* operon. When the end product (the amino acid) is present, it combines with the regulator molecule, and this complex binds to the operator, repressing the operon. In the absence of the amino acid, the regulatory molecule fails to bind to the operator, and transcription proceeds.

Characterize this operon; then consider the following mutations, as well as the situation in which the wild-type gene is present along with the mutant gene in partially diploid cells (F′):

(a) Mutation in the operator region.

(b) Mutation in the promoter region.

(c) Mutation in the regulator gene.

In each case, will the operon be active or in active in transcription, assuming that the mutation affects the regulation of the *theo* operon? Compare each response with the equivalent situation of the *lac* operon:

Solution: The *theo* operon is repressible and under negative control. When there is no amino acid present in the medium (or the environment), the product of the regulatory gene cannot bind to the operator region, and transcription proceeds under the direction of RNA polymerase. The enzymes necessary for the synthesis of the amino acid are produced, as is the regulator molecule. If the amino acid *is* present, or after sufficient synthesis occurs, the amino acid binds to the regulator, forming a complex that interacts with the operator region, causing repression of transcription of the genes within the operon.

The *theo* operon is similar to the tryptophan system, except that the regulator gene is within the operon, rather than being located separate from it. Therefore, in the *theo* operon, the regulator gene is itself regulated by the presence or absence of the amino acid.

(a) As in the *lac* operon, a mutation in the *theo* operator gene inhibits binding with the repressor complex, and transcription occurs constitutively. The presence of an F′ plasmid bearing the wild-type allele would have no effect, since it is not adjacent to the structural genes.

(b) A mutation in the *theo* promoter region would no doubt inhibit binding to RNA polymerase and therefore inhibit transcription. This would also happen in the *lac* operon. A wild-type allele present in an F′ plasmid would have no effect.

(c) A mutation in the *theo* regulator gene, as in the *lac* system, may inhibit either its binding to the repressor or its binding to the operator gene. In both cases, transcription will be constitutive, because the *theo* system is repressible. Both cases result in the failure of the regulator to bind to the operator, allowing transcription to proceed. In the *lac* system, failure to bind the corepressor lactose would permanently repress the system. The addition of a wild-type allele would restore repressibility, provided that this gene was transcribed constitutively.

Problems and Discussion Questions

1. Contrast the need for the enzymes involved in lactose and tryptophan metabolism in bacteria when lactose and tryptophan, respectively, are (a) present and (b) absent.
2. Contrast positive vs. negative control of gene expression.
3. Contrast the role of the repressor in an inducible system and in a repressible system.
4. Even though the *lac Z, Y*, and *A* structural genes are transcribed as a single polycistronic mRNA, each gene contains the appropriate initiation and termination signals essential for translation. Predict what will happen when a cell growing in the presence of lactose contains a deletion of one nucleotide (a) early in the *Z* gene and (b) early in the *A* gene.
5. For the *lac* genotypes shown in the accompanying table, predict whether the structural genes (*Z*) are constitutive, permanently repressed, or inducible in the presence of lactose.
6. For the genotypes and condition (lactose present or absent) shown in the accompanying table, predict whether functional enzymes, nonfunctional enzymes, or no enzymes are made.

Genotype	*Constitutive*	*Repressed*	*Inducible*
$I^+O^+Z^+$			X
$I^-O^+Z^+$			
$I^+O^cZ^+$			
$I^-O^+Z^+/F'I^+$			
$I^+O^cZ^+/F'O^+$			
$I^sO^+Z^+$			
$I^sO^+Z^+/F'I^+$			

Genotype	Condition	Functional Enzyme Made	Nonfunctional Enzyme Made	No Enzyme Made
$I^+O^+Z^+$	No lactose			X
$I^+O^cZ^+$	Lactose			
$I^-O^+Z^-$	No lactose			
$I^-O^+Z^-$	Lactose			
$I^-O^+Z^+/F'I^+$	No lactose			
$I^+O^cZ^+/F'O^+$	Lactose			
$I^+O^+Z^-/F'I^+O^+Z^+$	Lactose			
$I^-O^+Z^-/F'I^+O^+Z^+$	No lactose			
$I^sO^+Z^+/F'O^+$	No lactose			
$I^+O^cZ^+/F'O^+Z^+$	Lactose			

7. Describe how the *lac* repressor was isolated. What properties demonstrate it to be a protein? Describe the evidence that it indeed serves as a repressor within the operon scheme.
8. Predict the level of genetic activity of the *lac* operon as well as the status of the *lac* repressor and the CAP protein under the cellular conditions listed in the accompanying table.

	Lactose	Glucose
(a)	−	−
(b)	+	−
(c)	−	+
(d)	+	+

9. Predict the effect on the inducibility of the *lac* operon of a mutation that disrupts the function of
   (a) the *crp* gene, which encodes the CAP protein.
   (b) the CAP-binding site within the promoter.

Extra-Spicy Problems

10. In a theoretical operon, genes *A, B, C,* and *D* represent the repressor gene, the promoter sequence, the operator gene, and the structural gene, *but not necessarily in that order*. This operon is concerned with the metabolism of a theoretic molecule (tm). From the data provided in the accompanying table, first decide whether the operon is inducible or repressible. Then, assign *A, B, C,* and *D* to the four parts of the operon. Explain your rationale. (AE = active enzyme; IE = inactive enzyme; NE = no enzyme).

Genotype	tm Present	tm Absent
$A^+B^+C^+D^+$	AE	NE
$A^-B^+C^+D^+$	AE	AE
$A^+B^-C^+D^+$	NE	NE
$A^+B^+C^-D^+$	IE	NE
$A^+B^+C^+D^-$	AE	AE
$A^-B^+C^+D^+/F'A^+B^+C^+D^+$	AE	AE
$A^+B^-C^+D^+/F'A^+B^+C^+D^+$	AE	NE
$A^+B^+C^-D^+/F'A^+B^+C^+D^+$	AE + IE	NE
$A^+B^+C^+D^-/F'A^+B^+C^+D^+$	AE	NE

11. A bacterial operon is responsible for the production of the biosynthetic enzymes needed to make the theoretical amino acid tisophane (tis). The operon is regulated by a separate gene, *R*, deletion of which causes the loss of enzyme synthesis. In the wild-type condition, when tis is present, no enzymes are made. In the absence of tis, the enzymes are made. Mutations in the operator gene (O^-) result in repression regardless of the presence of tis.
    Is the operon under positive or negative control? Propose a model for (a) repression of the genes in the presence of tis in wild-type cells and (b) the O^- mutations.
12. A marine bacterium is isolated and shown to contain an inducible operon whose genetic products metabolize oil when it is encountered in the environment. Investigation demonstrates that the operon is under positive control and that there is a *reg* gene whose product interacts with an operator region (*o*) to regulate the structural genes designated *sg*.
    In an attempt to understand how the operon functions, a constitutive mutant strain and several partial diploid strains were isolated and tested with the results shown here:

Host Chromosome	F′ Factor	Phenotype
wild type	none	inducible
wild type	*reg* gene from mutant strain	inducible
wild type	operon from mutant strain	constitutive
mutant strain	*reg* gene from wild type	constitutive

    Draw all possible conclusions about the mutation as well as the nature of regulation of the operon. Is the constitutive mutation in the *trans*-acting *reg* element or in the *cis*-acting *o* cooperator element?
13. The SOS repair genes in *E. coli* (see Chapter 7) are negatively regulated by the *lexA* gene product, called the LexA repressor. When a cell sustains extensive damage to its DNA, the LexA repressor is inactivated by the *recA* gene product (RecA), and transcription of the SOS genes is increased dramatically.
    One of the SOS genes is the *uvrA* gene. You are studying the function of the *uvrA* gene product in DNA repair. You isolate a mutant strain that shows constitutive expression of the UvrA protein. You name this mutant strain $uvrA^C$. The following simple diagram shows the *lexA* and *uvrA* operons:

(a) Describe two different mutations that would result in a *uvrA* constitutive phenotype. Indicate the actual genotypes involved.

(b) Outline a series of genetic experiments, using partial diploid strains that would allow you to determine which of the two possible mutations you have isolated.

(c) A fellow student considers this problem and argues that there is a more straightforward, nongenetic experiment that could differentiate between the two types of mutations. While the experiment requires no fancy genetics, you must be able to easily assay the products of the other SOS genes. Propose such an experiment.

14. In Figure 19–13, numerous critical regions of the leader sequence of mRNA that play important roles during the process of attenuation in the *trp* operon are depicted. Shown here, beginning at about position 30 downstream from the 5′-end, is the sequence of ribonucleotides that make up the leader sequence:

Within this molecule are the sequences that serve as the basis of the formation of the secondary structures. These structures are critical in the formation of the alternative hairpins, as well as the successive triplets that encode tryptophan, wherein "stalling' during translation occurs. Obtain a large piece of paper (such as manila wrapping paper) and, along with several other students from your genetics class, work through the base sequence, attempting to identify the *trp* codons and which parts of the molecule represent the base-pairing regions that form the terminator and antiterminator hairpins shown in Figure 19–13.

15. Consider the regulation of the *trp* operon in *B. subtilis*. For each of the following cases, indicate whether the structural genes in the operon are being expressed or not expressed:

(a) TRAP present, $tRNA^{trp}$ abundant, tryptophan abundant.

(b) TRAP present, $tRNA^{trp}$ abundant, tryptophan scarce.

(c) TRAP present, $tRNA^{trp}$ scarce, tryptophan abundant.

(d) TRAP absent, $tRNA^{trp}$ abundant, tryptophan abundant.

In each case, also indicate whether AT is present. Describe the role of TRAP and AT in attenuation.

AUGAAAGCAAUUUUCGUACUGAAAGGUUGGUGGCGCACUUCCUGAAACGGGCAGUGUAUUCACCAUGCGUAAAGCAAUCAGAUACCCAGCCCGCCUAAUGAGCGGGCUUUUUUUU

Selected Readings

Antson, A.A., et. al. 1999. Structure of the trp RNA-binding attenuation protein, TRAP, bound to RNA. *Nature* 401:235–42.

Beckwith, J.R., and Zipser, D., eds. 1970. *The lactose operon.* Cold Spring Harbor, NY: Cold Spring Harbor Laboratory Press.

Bertrand, K., et al. 1975. New features of the regulation of the tryptophan operon. *Science* 189:22–26.

Cohen, J.S., and Hogan, M.E. 1994. New genetic medicines. *Sci. Am.* (Dec.) 271:76–82.

Englesberg, E., and Wilcox, G. 1974. Regulation: Positive control. *Annu. Rev. Genet.* 8:219–42.

Gilbert, W., and MüLLER-HILL, b. 1966. Isolation of the *lac* repressor. *Proc. Natl. Acad. Sci. USA* 56:1891–98.

——— 1967. The *lac* operator is DNA. *Proc. Natl. Acad. Sci. USA* 58:2415–21.

Jacob, F., and Monod, J. 1961. Genetic regulatory mechanisms in the synthesis of proteins. *J. Mol. Biol.* 3:318–56.

Keller, E.B., and Calvo, J.M. 1979. Alternative secondary structures of leader RNAs and the regulation of the *trp, phe, his, thr,* and *leu* operons. *Proc. Natl. Acad. Sci. USA* 76:6186–90.

Lee, D. and Schleif, R. 1989. In vivo DNA loops in *araCBAD:* Size limits and helical repeats. *Proc. Natl. Acad. Sci. USA* 86:476–80.

Lewin, B. 1974. Interaction of regulator proteins with recognition sequences of DNA. *Cell* 2:1–7.

Lewis, M., et al. 1996. Crystal structure of the lactose operon repressor and its complexes with DNA and inducer. *Science* 271:1247–54.

Miller, J.H., and Reznikoff, W.S. 1978. *The operon.* Cold Spring Harbor, NY: Cold Spring Harbor Laboratory Press.

Olson, K.E., et al. 1996. Genetically engineered resistance to dengue-2 virus transmission in mosquitoes. *Science* 272:884–86.

Platt, T. 1981. Termination of transcription and its regulation in the tryptophan operon of *E. coli. Cell* 24:10–23.

Ptashne, M. and Gann, A. 2002. *Genes and signals.* Cold Spring Harbor, NY: Cold Spring Harbor Laboratory Press.

Stroynowski, I., and Yanofsky, C. 1982. Transcript secondary structures regulate transcription termination at the attenuator of *S. marcescens* tryptophan operon. *Nature* 298:34–38.

Studier, F.W. 1972. Bacteriophage T7. *Science* 176:367–76.

Umbarger, H.E. 1978. Amino acid biosynthesis and its regulation. *Annu. Rev. Biochem.* 47:533–606.

Valbuzzi, V. and Yanofsky, C. 2001. Inhibition of the *B. subtilis* regualtory protein TRAP by the TRAP-inhibitory protein, AT. *Science* 293:2057–61.

Wagner, R.W. 1994. Gene inhibition using antisense oligodeoxynucleotides. *Nature* 372:333–35.

Yanofsky, C. 1981. Attenuation in the control of expression of bacterial operons. *Nature* 289:751–58.

Yanofsky, C., and Kolter, R. 1982. Attenuation in amino acid biosynthetic operons. *Annu. Rev. Genet.* 16:113–134.

GENETICS MediaLab

The resources that follow will help you achieve a better understanding of the concepts presented in this chapter. These resources can be found either on the CD packaged with this textbook or on the Companion Web site found at **http://www.prenhall.com/klug**

CD Resources:

Module 19.1: Gene Expression: Prokaryotes

Web Problem 1:
Time for completion = 15 minutes

How are decisions made at the cellular level? To choose among the alternatives, the cell must be able to sense particular conditions and respond differentially. A classic example of such a decision is observed when a bacteriophage infects a bacterial cell. The bacteriophage has two choices: lysis or lysogeny. Ultimately, the option chosen reflects differences in gene regulation that are established immediately after infection. Read about the life cycle of bacteriophages depicted on the linked Web site mentioned subsequently (keyword **PHAGE**). Phage lambda is described in more detail at **LAMBDA**. View animations of the regulatory cascade, which results in **LYSOGENY** or **LYSIS**. Discuss the key features determining whether a phage enters either of these pathways. To complete this exercise, visit Web problem 1 in Chapter 19 of your Companion Web site, and select the keyword **PHAGE**.

Web problem 2:
Time for completion = 20 minutes

How does a cell respond to changes in environmental conditions? One of the best-described examples of cellular response is the lactose operon, examined by Jacob, Monod, and Wollman. Test your understanding of lactose operon regulation with a series of interactive questions at the Web site indicated. Through feedback, learn whether your answers are correct, with tutorial information provided to clarify your understanding. Construct a chart listing various genotypes, conditions (with and without lactose), and the resulting effect. To complete this exercise, visit Web Problem 2 in Chapter 19 of your Companion Web site, and select the keyword **OPERON**.

Web problem 3:
Time for completion = 5 minutes

What mechanisms are used to regulate bacterial genes? Different operons illustrate positive or negative regulation through various mechanisms. Regulatory control reflects the basal state of the cell and how it responds to changing conditions. This Web site describes current research involving the pyrR operon of *Bacillus subtilis*. Compare and contrast the proposed model of regulation at this operon with those discussed in your text. State how you might test the proposed model. To complete this exercise, visit Web Problem 3 in Chapter 19 of your Companion Web site, and select the keyword **MECHANISM**.

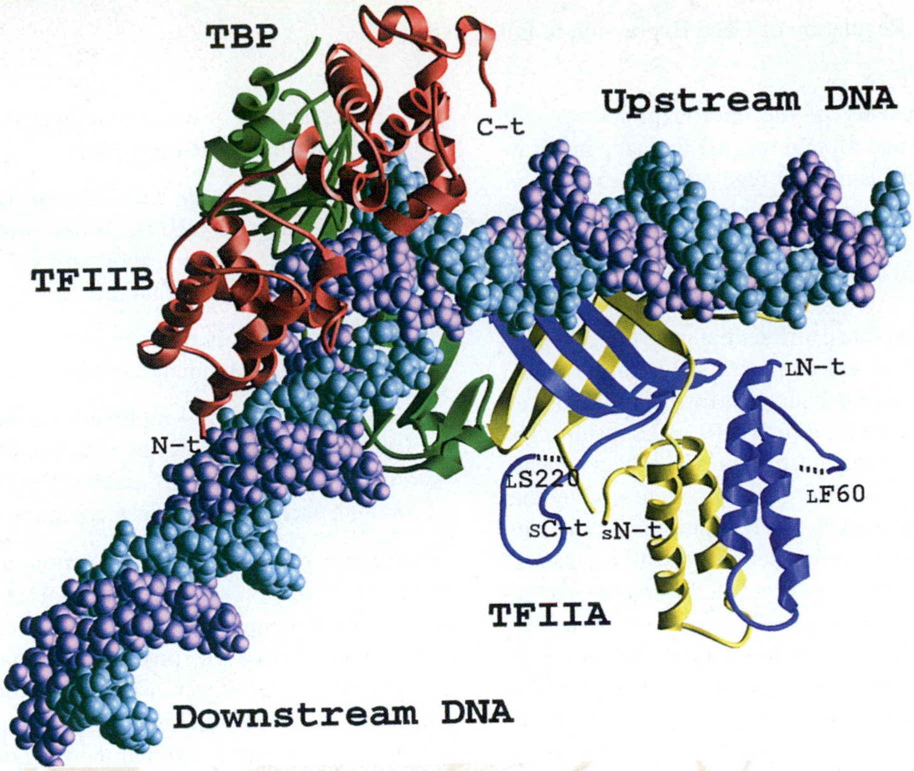

Molecular complex formed between TATA box-binding protein (TBP), transcription factors TFIIA and TFIIB, and TATA box DNA during the initiation of transcription. (*Yale University and the Howard Hughes Medical Institute. Dr. Paul Sigler*)

20

Regulation of Gene Expression in Eukaryotes

KEY CONCEPTS

The 1960 discovery of the operon in *Escherichia coli* by Jacob and Monod was the first step in unraveling the mechanisms that regulate gene expression. While some of the principles of regulation found in the *lac* operon are also present in eukaryotes, it is clear that eukaryotes have evolved a more complex system of gene regulation.

In multicellular eukaryotes, differential gene expression is at the heart of embryonic development and maintenance of the adult state. Cells of the pancreas, for example, do not make retinal pigment, nor do retinal cells make insulin. The question is, How does an organism express a subset of genes in one cell type and a different subset of genes in another cell type? At the cellular level, this regulation is *not* accomplished by eliminating unused genetic information; instead, mechanisms activate specific portions of the genome and repress the expression of other genes. The activation and repression of selected loci represent a delicate balancing act for an organism; expression of a gene at the wrong time, in the wrong cell type, or in abnormal amounts can lead to a deleterious phenotype—cancer or cell death—even when the gene itself is normal.

In this chapter, we will review the general features of eukaryotic gene regulation, outline the components needed to control transcription, and discuss how these components interact. We will also consider the role of posttranscriptional mechanisms in regulating gene expression in eukaryotes. When appropriate, we will compare the more complex approaches exhibited by eukaryotes to the less complex counterparts in prokaryotes.

20.1 Eukaryotic Gene Regulation: Different from Regulation in Prokaryotes

There are several reasons why gene regulation is more complex in eukaryotes than in prokaryotes:

1. Eukaryotic cells contain a much greater amount of genetic information than do prokaryotic cells, and this DNA is complexed with histones and other proteins to form chromatin. As we shall see, the structure of chromatin—whether it is open (decondensed) and available to be transcribed or closed (condensed) and not available—is the major on/off switch for gene regulation. (In prokaryotes, remember that the operator is the on/off switch.)

2. Genetic information in eukaryotes is carried on many chromosomes (rather than just on one), and these chromosomes are enclosed within a double-membrane-bound nucleus.

3. Since the genetic information in eukaryotes is segregated from the cytoplasm, transcription is spatially and temporally separated from translation—transcription occurs in the nucleus and translation occurs later in the cytoplasm. Because of this, attenuation control, a regulatory mechanism in prokaryotes, is not possible.

4. The transcripts of eukaryotic genes are processed before transport to the cytoplasm.

5. Eukaryote mRNA has a much longer half-life ($t_{1/2}$) than does prokaryotic mRNA. When prokaryotes want to stop making a protein, they turn off transcription and the mRNA decays within minutes.

6. Because mRNA is much more stable, eukaryotes have a series of translational controls.

7. Most eukaryotes are multicellular with differentiated cell types. These different cell types typically use different sets of genes to make different proteins, even though each cell contains a complete set of genes.

Hence, it is evident that the regulation of eukaryotic gene expression can potentially occur at many levels (Figure 20–1), which include (1) transcriptional control, (2) posttranscription control (i.e., processing of the pre-mRNA), (3) transport to the cytoplasm, (4) stability of the mRNA, (5) translational control (i.e., selecting which mRNAs are translated), and (6) posttranslational modification of the protein product. As most eukaryotic genes are regulated in part at the transcriptional level, in the next sections, we will emphasize transcriptional control, though we will discuss other levels of control as well.

20.2 Functional Architecture of the Interphase Nucleus

During interphase of the cell cycle, chromosomes are unwound and cannot be seen as intact structures by microscopy. The development of chromosome-painting techniques has revealed that the interphase nucleus is not a bag of tangled chromosome arms, but is, instead, a highly organized structure. Further, it is becoming clear that nuclear organization of chromosomes and dynamic changes in chromatin structure are key elements in controlling gene expression in eukaryotic cells. Before we examine events at the molecular level that initiate gene expression, we will first examine the structural mechanisms that determine whether or not a gene will be transcribed.

In the interphase nucleus, chromosomes occupy discrete compartments called **territories** (Figure 20–2). The channels between the chromosomes are called **interchromosomal domains**. Chromosome structure is continuously rearranged so that transcriptionally active genes are cycled to the edge of chromosome territories at the border of the interchromosomal domain channels (Figure 20–2).

In addition to being moved to the edge of a territory, transcriptionally active genes are differentially located with respect to other chromosome components. For example, the CD4 locus is active in T cells of the immune system, but inactive in B cells. In Figure 20–3, the transcriptionally inactive CD4 locus in B cells is associated with gamma satellite, a component of heterochromatin. (Heterochromatin can silence transcription of adjacent loci.) However, when CD4 is transcriptionally active (in T cells), it is not associated with gamma satellite.

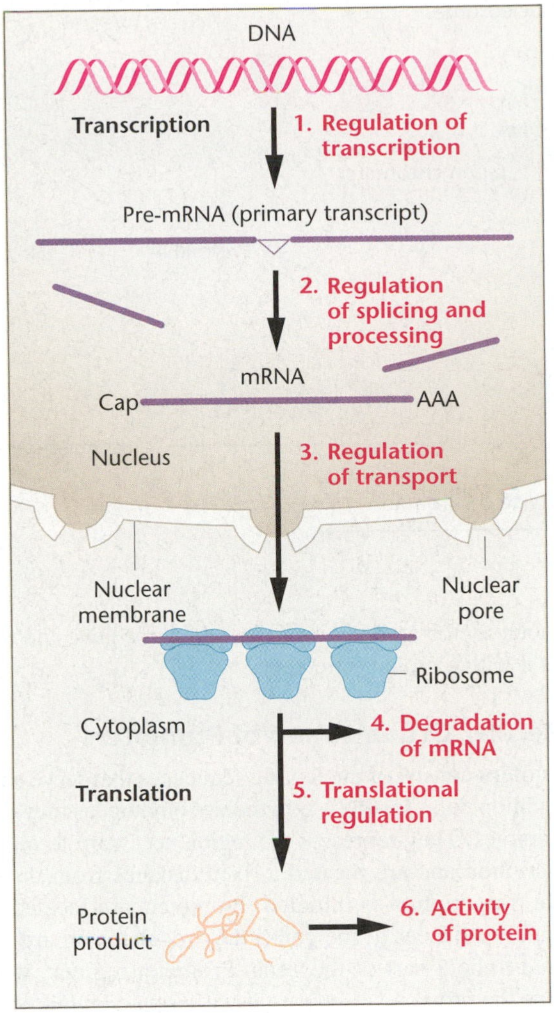

FIGURE 20–1 Various levels of regulation that are possible during the expression of the genetic material.

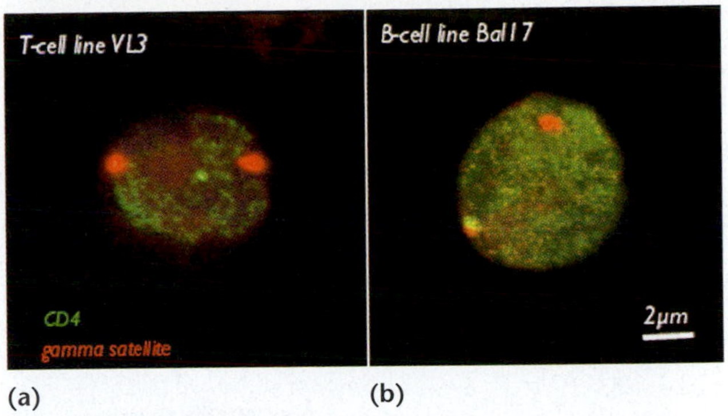

FIGURE 20–3 Transcriptional activity is associated with differential location within the nucleus. (a) In the T-cell line VL3, the CD4 gene is active (bright green spot) and is not associated with heterochromatic gamma satellite sequences (red spots). (b) in the B-cell line Bal17, the CD4 gene is transcriptionally inactive. and shows as a yellow spot (green combined with red) associated with gamma satellite sequences (red spots).

Because the transcriptionally active genes are preferentially localized at channel edges, pre-mRNA transcripts are likely to be formed on territory borders, to accumulate in the interchromosomal channels in which they are processed, and made ready for transport to the cytoplasm. But after a gene has been moved to the edge of a chromosome territory, and before the molecular events of transcription begin, chromatin remodeling is required.

20.3 Chromatin Remodeling Is an Epigenetic Process

MEDIA TUTORIAL Gene Expression: Eukaryotes

The DNA in eukaryotic chromosomes is combined with histones and nonhistone proteins to form chromatin. Some chromosomal regions are highly condensed in the interphase nucleus and form transcriptionally inert heterochromatin. Other regions (euchromatic regions) have an open configuration, and genes in these regions can be transcribed. Levels of chromatin organization are a reflection of the higher order structure of nucleosomes, the basic repeating unit of chromatin. (Review nucleosomes and chromosome structure in Chapter 4.) Changes in chromatin organization are essential for many processes, including transcription, replication, DNA repair, and recombination. Many of these changes are mediated by chemical modifications of the N-terminal tails of the core histones in a nucleosome. Histone tails are exposed on the surface of the nucleosome and can be modified by acetylation, phosphorylation, and methylation to allow access to the underlying DNA. Acetylation modifies certain lysine residues in the histone tails by addition of an acetyl group. Acetylated histones are found in regions of open chromatin where transcription occurs; this modification allows transcriptional enzymes access to the DNA template (Figure 20–4). Acetylation is mediated by histone

FIGURE 20–2 (a) In the nucleus, each chromosome occupies a discrete territory, and is separated from other chromosomes by an interchromosomal domain, where mRNA transcription and processing is thought to occur. (b) FISH probes hybridized to human chromsome 7. At left is hybridization to metaphase chromosomes. In the nucleus (right) the probes reveal the location of the territories occupied by chromosome 7.

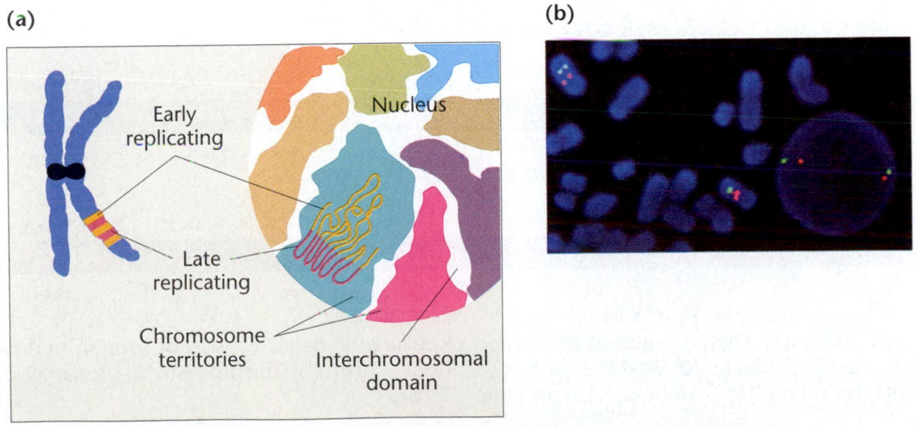

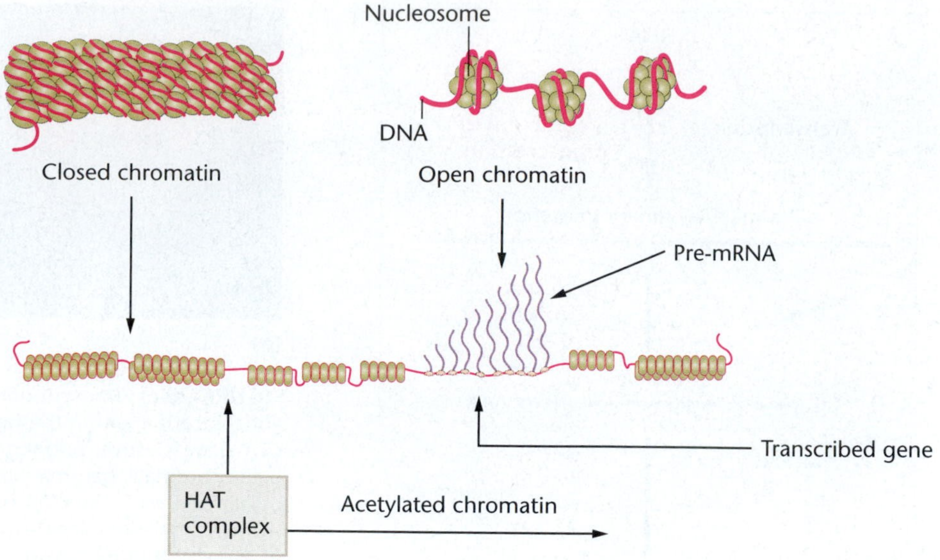

FIGURE 20–4 Histone acetyltransferases (HATs) are enzymes that add acetate groups to the tails of histones in nucleosomes of closed chromatin (top left). Action of a HAT complex converts closed chromatin to open chromatin (top right), allowing transcription of mRNA to take place.

acetyltransferases (HATs), and deacetylation by histone deacetylases (HDACs). Both enzymes are important in the regulation of gene expression.

Gene expression in eukaryotes requires two steps: (1) the recruitment and activation of chromatin remodeling enzymes (called chromatin remodeling machines) that alter nucleosome structure and make promoter sites accessible to the transcription machinery, (2) the recruitment of coactivators that help assemble the factors necessary for transcription, including general transcription factors, RNA polymerase II, and so on. Each of these mechanisms will be described in detail in later sections of the chapter. In the next sections, we will describe the components required for transcription and the events leading to the production of mRNA molecules from a transcribed gene.

20.4 Assembly of the Basal Transcription Complex Occurs at the Promoter

The structure of eukaryotic genes was first introduced in Chapters 5 and 6. Recall that these genes have two types of regulatory sequences that control their transcription, promoters, and enhancers. Figure 20–5 shows the structure of the promoter region at the 5′ end of a eukaryotic gene and compares it with a prokaryotic gene.

Molecular Organization of Promoters

Promoters consist of nucleotide sequences that serve as the recognition point for RNA polymerase binding, as they do in prokaryotes. They represent the region necessary to *initiate* transcription and are located a fixed distance from the site where transcription is initiated. Promoters are located immediately adjacent to the genes they regulate and are considered to be a part of the gene. Promoter regions, which include the promoter itself, are usually several hundred nucleotides in length.

Eukaryotic promoters require the binding of a number of protein factors to initiate transcription. Promoter regions for genes that are recognized by RNA polymerase II, which transcribes primarily mRNA, consist of short modular DNA sequences usually located within 100 bp upstream (in the 5′ direction) of the gene. The promoter region of most genes contains several elements [Figure 20–5(a)]. The promoter itself, to which the RNA polymerase will bind, is a sequence called the **TATA box**. Located about 25 to 30 bases upstream from the initial point of transcription (designated as −25 to −30), this box consists of an 8-bp consensus sequence (a sequence conserved in most genes studied) composed only of

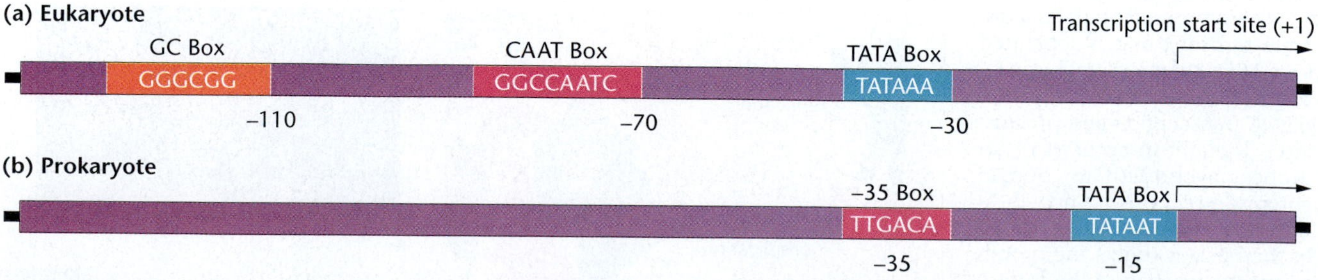

FIGURE 20–5 (a) The promoter at the 5′-end of eukaryotic genes consists of several modular elements, including the TATA box (−30), the CAAT box (−70), and the GC box (at about −110). (b) The promoter in prokaryotic genes consists of several modular elements, including the TATA box (−15) and the −35 box.

AT base pairs, often flanked on either side by GC-rich regions. Mutations in the TATA box reduce transcription, and deletions may alter the initiation point of transcription (Figure 20–6).

Many promoter regions also contain other components; one of the more common, called the **CAAT box**, has the consensus sequence CAAT or CCAAT. The CAAT box frequently appears in the region 70 to 80 bp from the start site, and it can function in a 5′-to-3′ or a 3′-to-5′ orientation. Mutational analysis suggests that the CAAT box and the protein it binds are critical to the promoter's ability to facilitate transcription. Mutations on either side of this element have no effect on transcription, whereas mutations within the CAAT sequence dramatically lower the rate of transcription (Figure 20–6). Another element often seen in some promoter regions, called the **GC box**, has the consensus sequence GGGCGG and is often found at about position −110. This module is found in either orientation and often occurs in multiple copies. The CAAT and GC elements bind transcription factors and function more like enhancers, which we will cover in the next section.

Compare the structure of the eukaryotic promoter region to the structure of the prokaryotic promoter region [Figure 20–5(b)]. The eukaryotic promoter has a single element, the TATA box, which is set further back from the start of transcription than is the prokaryotic promoter. The prokaryotic promoter has two elements, the TATA box at −10 and a −35 box.

Sequences that make up the upstream regulatory elements in eukaryotic genes vary in location and organization (Figure 20–7). Note that, in different genes, there are significant differences in the number and orientation of promoter elements and the distance between them. In addition, eukaryotic genes use three RNA polymerases for transcription (prokaryotic genes are all transcribed by a single RNA polymerase). These three are classified as type I (ribosomal RNAs), type II (mRNAs and snRNAs), and type III (tRNAs, 5*S* rRNA, and several other small cellular RNAs). The promoter for each type of polymerase has a different nucleotide sequence and binds different transcription factors.

RNA Polymerases and Promoters

There are two essential points to remember when comparing the ways in which eukaryotic and prokaryotic polymerases bind to the promoter. First of all, the prokaryote chromosome is essentially naked DNA. Its RNA polymerase has no difficulty finding promoter sequences. Eukaryotic chromosomes are chromatin with the vast majority of promoters hidden within the nucleosomes and not available to be bound by RNA polymerase. The chromatin structure must be altered and opened up. (We will discuss chromatin structure in the next section.) Second, prokaryotic RNA polymerase can read the DNA sequence with its sigma factor, as well as find and bind to promoters directly. Eukaryotic RNA polymerases cannot. A set of transcription factors must be preassembled on the promoter to serve as a platform for RNA polymerase to bind.

We now have some insight into the transcriptional apparatus of type II genes, those transcribed by RNA polymerase II, mainly mRNA coding genes. A series of **transcription factors**, proteins that are not part of the RNA polymerase molecule itself, but are needed to control the initiation of transcription, are assembled at the promoter in a specific order to provide a platform for the RNA polymerase to recognize and bind (Figure 20–8). To initiate formation of the apparatus, the TFIID complex binds to the TATA box through its TBP subunit (Figure 20–9). TFIID is the TBP plus a set of additional factors called TAFs (TATA Associated Factors). About 20 base pairs of DNA are involved in binding the TBP and the other subunits in the TFIID transcription factor. This has been described as the *commitment* stage. TFIID responds to contact with activator proteins by conformational changes that expedite the binding of additional factors, such as TFIIA and TFIIB. RNA polymerase II and some additional factors such as TFIIF then bind, followed by TFIIE, TFIIH, and

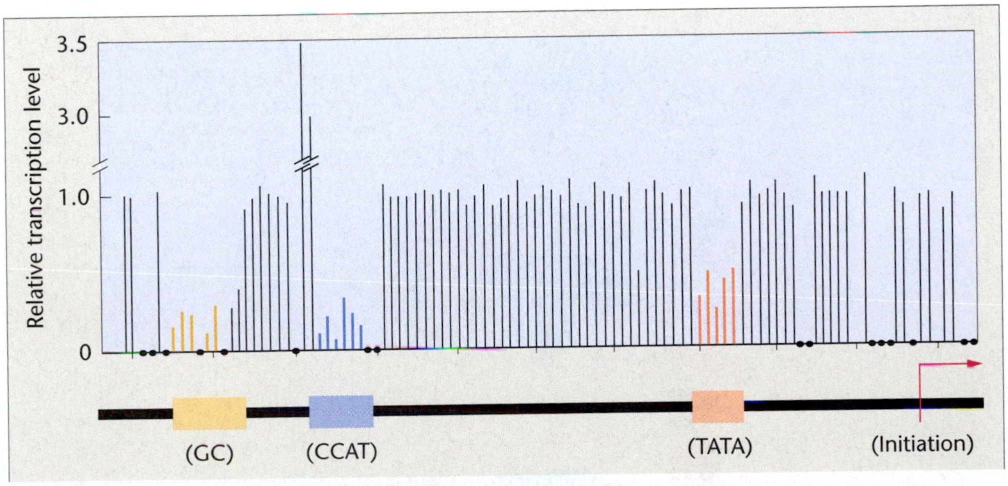

FIGURE 20–6 Summary of the effects of point mutations in the promoter region on transcription of the β-globin gene. Each line represents the level of transcription produced by a single nucleotide mutation (relative to wild type) in a separate experiment. Dots represent nucleotides in which no mutation was obtained. Note that mutations within the specific elements of the promoter have the greatest effect on the level of transcription.

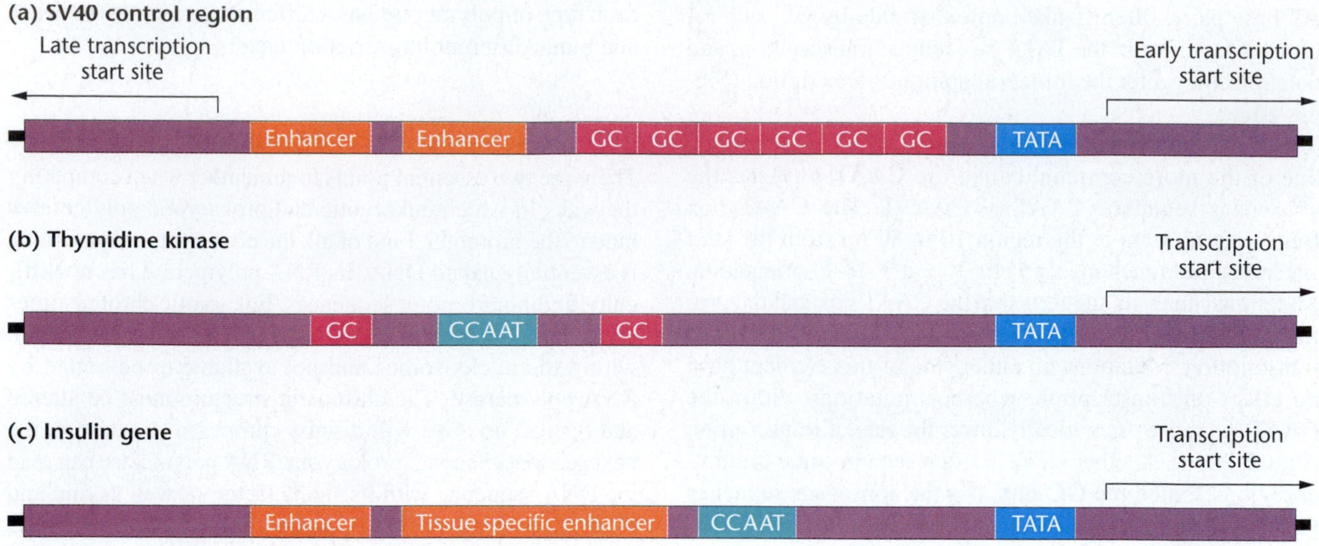

FIGURE 20–7 Upstream organization of several eukaryotic genes, illustrating the variable nature, number, and arrangement of controlling elements.

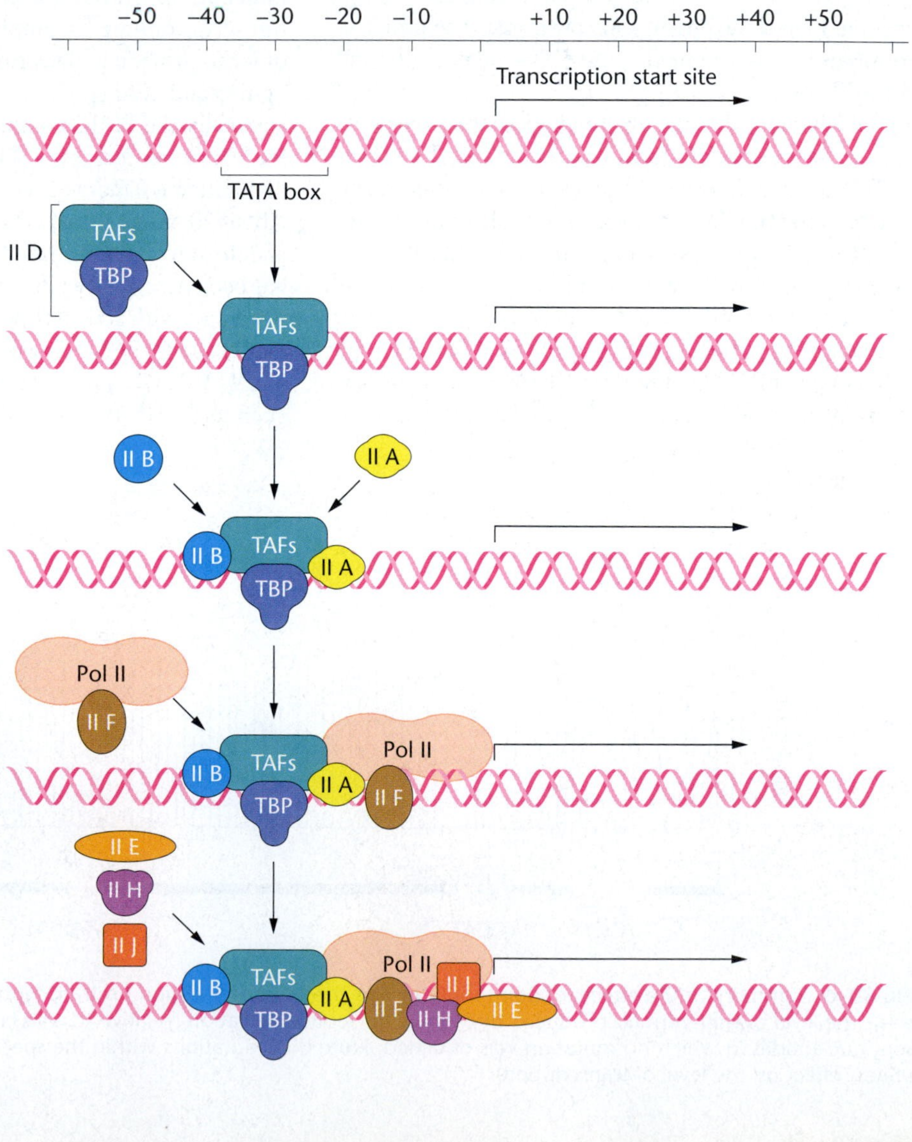

FIGURE 20–8 The assembly of transcription factors required for the initiation of transcription by RNA polymerase II.

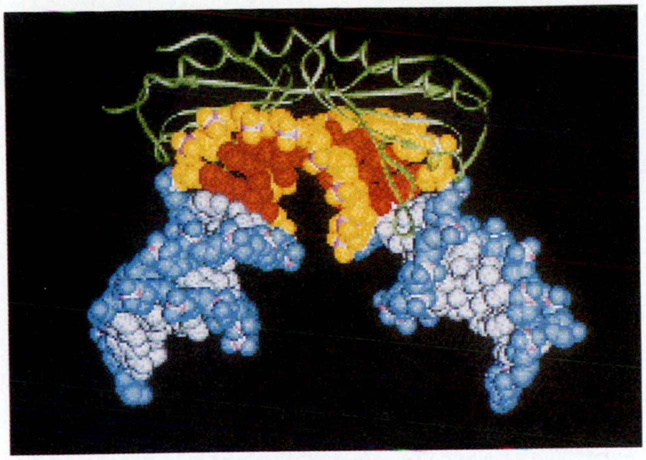

FIGURE 20–9 Molecular complex formed between the TATA-box binding protein (green) and the 8-bp TATA box of DNA (red), itself part of DNA (yellow).

TFIIJ. The final stage is *promoter clearance* during which the RNA polymerase leaves the TATA box, and the transcription of the DNA downstream ensues at a minimal *basal* level. The *induced* state, in which transcription is stimulated above the basal level, is not yet well defined, but involves other areas of the promoter region, enhancers, and numerous transcription factors, which control the assembly of the transcription complex and the rate of promoter clearance.

20.5 Enhancers Control Chromatin Structure and the Rate of Transcription

Transcription of most, if not all, eukaryotic genes is regulated not only by the promoter region, but also by additional DNA sequences called **enhancers**. Enhancers can be on either side of a gene, at some distance from the gene, or even within the gene. They are called *cis* regulators because they are found adjacent to the structural genes they regulate, as opposed to *trans* regulators (e.g., binding proteins), which can regulate a gene on any chromosome. Enhancers typically interact with multiple regulatory proteins and transcription factors, and they can increase the efficiency of transcription initiation or activate the promoter. Within enhancer sites, binding sites are often found for positive as well as negative gene regulators. Thus, there is some degree of analogy between enhancers and operator regions in prokaryotes. However, enhancers appear to be much more complex in both structure and function. Enhancers greatly stimulate the transcriptional activity of a promoter and can be distinguished from promoters by several characteristics:

1. The position of the enhancer need not be fixed; it can be upstream, downstream, or within the gene it regulates.
2. Its orientation can be inverted without significant effect on its action.
3. If an enhancer is experimentally moved to another location in the genome, or if an unrelated gene is placed near an enhancer, the transcription of the adjacent gene is enhanced.

An example of an enhancer located *within* the gene it regulates is found in the immunoglobulin heavy-chain gene, in which an enhancer is located in an intron between two coding regions. This enhancer is also active only in cells in which the immunoglobulin genes are being expressed, indicating that tissue-specific gene expression can be modulated through enhancers. Internal enhancers have been discovered in other eukaryotic genes, including the immunoglobulin light-chain gene.

Downstream enhancers are found in the human β-globin gene and the chicken thymidine kinase gene. In chickens, an enhancer located between the β-globin and the ε-globin genes apparently works in one direction to control the ε-globin gene during embryonic development and in the opposite direction to regulate the β-globin gene during adult life. In yeast, regulatory sequences similar to enhancers, called **upstream activator sequences (UAS)**, can function upstream at variable distances and in either orientation. They differ from enhancers in that they cannot function downstream of the transcription start point.

Like the promoter region, enhancers contain modules composed of short DNA sequences. The enhancer of the SV40 virus has a complex structure consisting of two adjacent 72-bp sequences located some 200 bp upstream from a transcriptional start point (Figure 20–10). Each of the two 72-bp regions contains five sequences that contribute to the

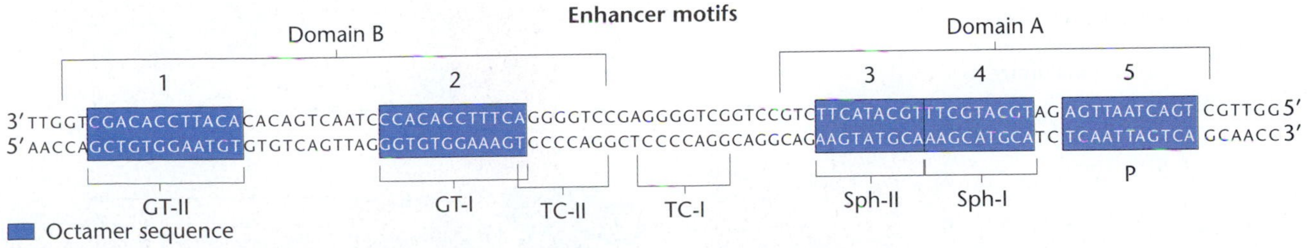

FIGURE 20–10 DNA sequence for the SV40 enhancer. Screened sequences are those required for maximum enhancer effect. The brackets below the sequence are the various sequence motifs within this region. The two domains of the enhancer (A and B) are indicated.

maximum rates of transcription. If one or the other of these regions is deleted, there is no effect on transcription; but if both are deleted, *in vivo* transcription is greatly reduced.

Enhancers differ from promoters in a number of significant ways. While promoter sequences are essential for basal-level transcription, enhancers are necessary for the full level of transcription. In addition, enhancers are responsible for time- and tissue-specific gene expression. Since enhancers are able to stimulate levels of transcription at a distance, an intriguing question is, how they are able to do this? As we will see, they perform at least two functions. First, transcription factors bind to enhancers and alter the configuration of chromatin, and, second, by bending or looping the DNA, they bring distant enhancers and their promoters into direct contact in order to form complexes with transcription factors and polymerases (Figure 20–11). In the new configuration, transcription is stimulated to a higher level, increasing the overall rate of RNA synthesis.

As in prokaryotes, transcriptional control in eukaryotes involves the interaction between DNA sequences adjacent to genes and DNA-binding proteins. The regulatory sequences adjacent to a wide range of genes have been identified, mapped, and sequenced. The development of DNA-protection and gel-retardation assays have allowed the detection of DNA-binding proteins in cell extracts, and new affinity chromatography techniques have been used to isolate transcriptional factors present in very low concentrations. The result has been a virtual explosion of information about eukaryotic transcription factors. This section will review some of the new information, including the upstream organization of eukaryotic genes, the characteristics of transcription factors, their interaction with DNA sequences, other regulatory factors, and the initiation of transcription by RNA polymerase.

We have already discussed one class of transcription factors, those that work at the promoter to form a platform for the RNA polymerase to which to bind. We will now discuss a second class, those that bind to enhancer sites. These may act as *positive factors* to increase transcription or as *negative factors* to decrease transcription. These proteins control where and when genes are expressed and the rate of expression. The positive class of transcription factors can also be divided into two categories, true activators and antirepressors. True activators function by contacting elements of the transcription complex at the promoter and activating transcription. Antirepressors function by altering the chromatin structure to allow other factors to bind.

True Activator Transcription Factors

True activators are modular proteins with at least two functional domains (clusters of amino acids that carry out a specific function). One binds to DNA sequences present in the enhancer (**DNA-binding domain**), and another activates transcription through protein–protein interaction (***trans*-activating domain**). The *trans*-activating domain binds to RNA polymerase or to other transcription factors at the promoter. The existence of separate domains in eukaryotic transcription factors was first demonstrated by the genetic analysis of mutants in yeast. The domains of eukaryotic transcription factors take on several forms. The DNA-binding domains have distinctive three-dimensional structural patterns or **motifs**. There are a number of major types of these structural motifs, including helix–turn–helix (HTH), zinc finger, and basic leucine zipper (bZIP). This classification is not exhaustive, and other new groups will undoubtedly be established as new factors are characterized.

The first DNA-binding domain to be discovered was the helix–turn–helix (HTH) motif. In prokaryotes, HTH motifs have been identified in the *cro* repressor (in the DNA-binding domain), *lac* repressor, *trp* repressor, and other proteins. Studies indicate that the HTH motif is present in many prokaryotic DNA-binding proteins. This motif is characterized by its geometric conformation, rather than a distinctive amino acid sequence (Figure 20–12). Two adjacent α helices separated by a "turn" of several amino acids enables the protein to bind to DNA (hence the name of the motif). In the *cro* repressor, one of the helices binds within a major groove containing six bases, while the other helix stabilizes the interaction by binding to the DNA backbone. Unlike several of the other DNA-binding motifs, the HTH pattern cannot fold or function alone, but is always part of a larger DNA-binding domain. Amino acid residues outside the HTH motif are important in regulating DNA recognition and binding.

The potential for forming HTH geometry has been recognized in distinct regions of a large number of eukaryotic

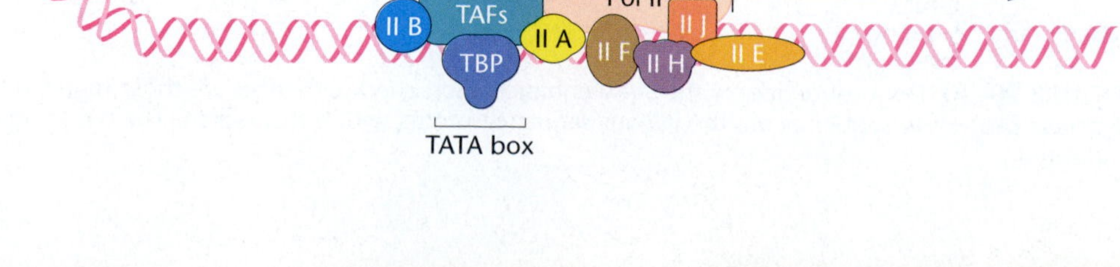

FIGURE 20–11 DNA looping allows factors that bind to enhancers at a distance from the promoter to interact with regulatory proteins in the transcription complex and to maximize transcription.

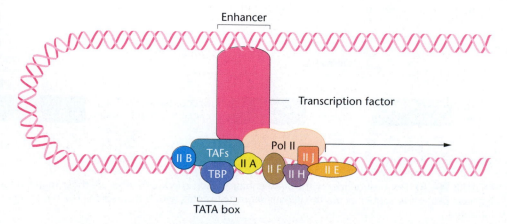

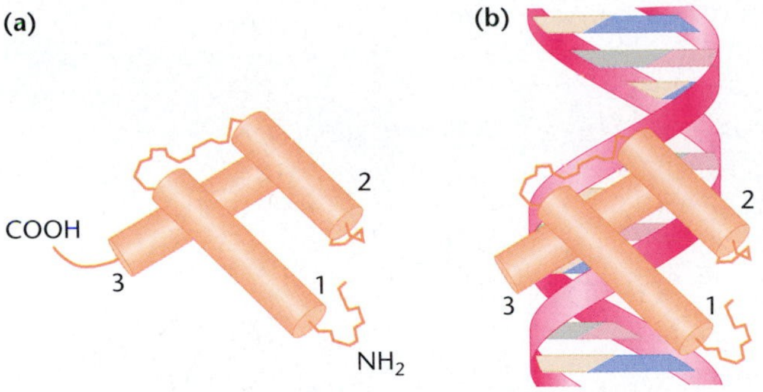

FIGURE 20–12 A *helix–turn–helix* or *homeodomain* in which (a) three planes of the α-helix of the protein are established, and (b) these domains bind in the grooves of the DNA molecule.

genes known to regulate developmental processes. Present almost universally in eukaryotic organisms is the **homeobox**, a stretch of 180 bp specifying a 60-amino-acid **homeodomain** sequence that can form an HTH structure. Of the 60 amino acids, many are basic (arginine and lysine), and a conserved sequence is found among these genes. We shall discuss homeobox-containing genes in Chapter 21 because of their significance to animal developmental processes.

Zinc fingers, one of the major structural families of eukaryotic transcription factors, are involved in many aspects of gene regulation. The zinc-finger motif was originally discovered in the *Xenopus* transcription factor TFIIIA. This structural motif has now been identified in proto-oncogenes (Chapter 22), in genes that regulate development in *Drosophila* (the *Krüppel* gene, see Chapter 21), in proteins whose synthesis is induced by growth factors and differentiation signals, and in transcription factors. There are several types of zinc-finger proteins, each with a distinctive structural pattern.

One zinc-finger protein contains clusters of two cysteine and two histidine residues at repeating intervals (Figure 20–13). The consensus amino acid repeat is $CysN_{2-4}CysN_{12-14}HisN_3His$. The interspersed cysteine and histidine residues covalently bind zinc atoms, folding the amino acids into loops known as zinc fingers. Each finger consists of approximately 23 amino acids (with a loop of 12 to 14 amino acids between the Cys and His residues) and a linker between loops consisting of 7 or 8 amino acids. The amino acids in the loop interact with, and bind to, specific DNA sequences. Studies have shown that zinc fingers bind in the major groove of the DNA helix and wrap at least partway around the DNA. Within the major groove, the zinc finger contacts, and may form, hydrogen bonds with a set of bases, especially in G-rich strands. The number of fingers in a zinc-finger transcription factor ranges from 2 to 13, as does the length of the DNA-binding sequence.

A third type of domain, a DNA-binding domain, is represented by the **basic leucine zipper (bZIP)**. This DNA-binding domain is adjacent to the **leucine zipper**, a domain that allows protein–protein dimerization. The leucine zipper, first described as a stretch of 35 amino acids in a nuclear rat liver protein, contains four leucine residues spaced 7 amino acids apart and flanked by basic amino acids. This region forms a helix with leucine residues protruding at every other turn. When two such molecules dimerize, the leucine residues "zip" together (Figure 20–14). The dimer contains two basic α-helical regions adjacent to the zipper that bind to phosphate residues and specific bases in DNA, making the dimer look like a pair of scissors.

Transcription factors also contain domains that activate transcription. These regions, which are distinct from the DNA-binding domains, can occupy from 30 to 100 amino acids. These domains interact with other transcription factors (such as those that bind to the promoter) or directly with RNA polymerase. In addition, many transcription factors also contain domains that bind coactivators, such as hormones or small metabolites that regulate their activity.

Overall, the picture of transcriptional regulation in eukaryotes is complex, but several important generalizations can be drawn. The regulation of transcription by transcription factors is largely positive, although transcription repressors are now being recognized as important components of gene regulation. The structural organization of chromatin and alterations in chromatin structure to allow the binding of transcription factors are considered to be the primary levels of regulation.

In addition, different transcription factors may compete for binding to a given DNA sequence or to two overlapping sequences, or the same site may bind different factors in two different tissues. Transcription factor concentration and the

FIGURE 20–13 (a) A zinc finger in which cysteine and histidine residues bind to a Zn^{++} atom. (b) This loops the amino acid chain out into a fingerlike configuration.

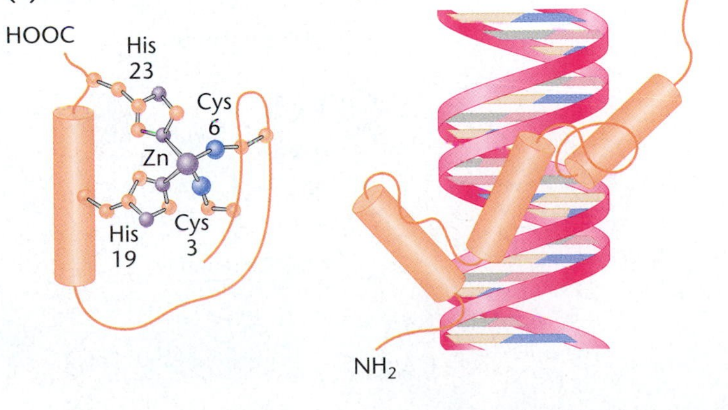

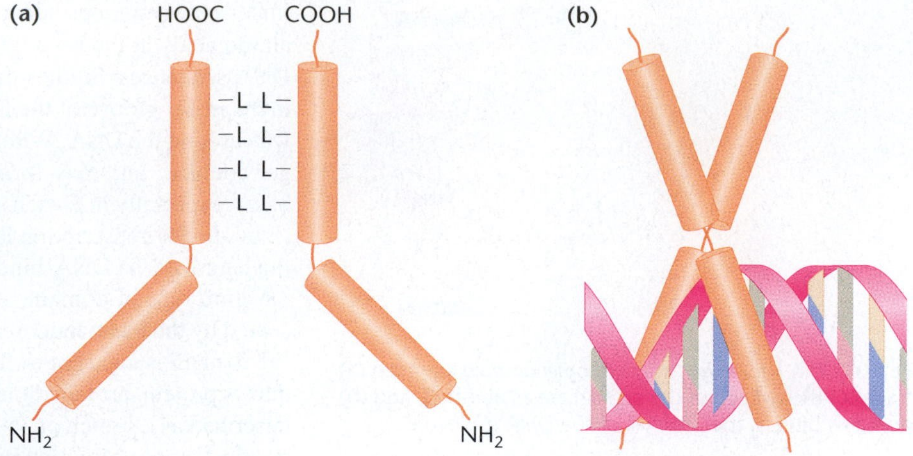

FIGURE 20–14 (a) A leucine zipper is the result of dimers from leucine residue at every other turn of the α-helix in facing stretches of two polypeptide chains. (b) When the α-helical regions form a leucine zipper, the regions beyond the zipper form a Y-shaped region that grips the DNA in a scissorlike configuration.

strength with which each factor binds to the DNA will dictate which factor binds. Finally, because of the variability in the location and modular nature of enhancers, and because of the variety of transcription factors, an initiation complex at the promoter can interact with multiple enhancers, with individual proteins contacting the basal-transcription complex at the promoter in a number of ways, all involving protein–protein interaction.

Antirepressor Transcription Factors

The second class of transcription factors, **antirepressors**, work in a totally different fashion; they *remodel chromatin*. Chromatin is not merely DNA wound around nucleosomes; it is a dynamic structure. Histones make up only half of the chromosomal proteins; nonhistone proteins are also important components of chromatin. Chromatin structure changes as the cell goes from interphase into mitosis and then back to interphase. Euchromatin has a different structure from heterochromatin. Most important for our discussion, chromatin remodeling must occur when genes are turned on or turned off (Figure 20–15). Investigators have shown that genes that are actively transcribed or *can* be transcribed are sensitive or hypersensitive to *in vitro* digestion by DNase I, while inactive or repressed genes are relatively resistant to DNase I digestion. Open chromatin is more easily digested and more DNase sensitive than is closed chromatin, which is not easily digested and is DNase insensitive. As described in an earlier

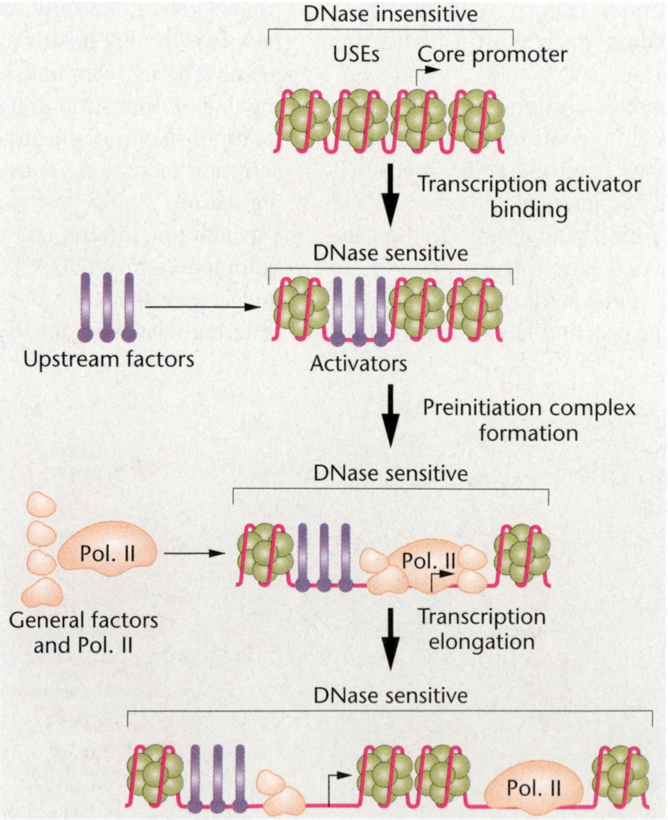

FIGURE 20–15 Nucleosomes can inhibit multiple steps required for gene transcription. The binding of enhancer transcription factors requires accessing nucleosomal DNA and may result in displacement of the nucleosome. Similarly the formation of the basal transcription complex at the TATA box and transcription start site is also suppressed by nucleosomes. Finally, RNA polymerase II is slowed by the presence of nucleosomes.

section of this chapter, chromatin structure is therefore the on/off switch for gene activity.

Chromatin is altered by two different processes. The first is catalyzed by an ATP hydrolysis-dependent remodeling complex, of which **SWI/SNF** is an example. This is a large 11-subunit complex first described in yeast and subsequently found widely distributed in eukaryotes. Remodeling complexes are targeted to specific gene regions by transcription factors, with different transcription factors associating with different remodeling complexes. Some transcription factors can bind to closed chromatin and initiate its opening. They may alter nucleosome structure by several different mechanisms. As seen in Figure 20–16(a), this may entail altering the contacts between the DNA and the nucleosome histone proteins. Alternatively, the path of the DNA around the nucleosome core particle may be altered, as seen in Figure 20–16(b). Another possibility is that the structure of the nucleosome core particle itself may be altered, as seen in Figure 20–16(c).

The second mechanism of chromatin alteration is histone modification catalyzed by one of many histone acetyltransferase enzymes (**HAT**). When an acetate group is added to specific basic amino acids on the histone tails, the attraction between the basic histone protein and acidic DNA is lessened (Figure 20–17). As in the remodeling complexes, HATs are targeted to genes by specific transcription factors. The remodeling of a chromatin region containing a gene entails cooperative interaction between a remodeling complex and a HAT. Of course, what can be opened can also be closed. In this case, **deacetylases** (**HD** complexes) can remove acetate groups from histone tails (Figure 20–17, bottom). Typically, from the focal point of an enhancer, remodeling spreads toward the promoter of the gene and then into the transcription unit. **Insulator elements**, short DNA sequences that bind specific proteins, can act as barriers to prevent the spread of remodeling into neighboring genes.

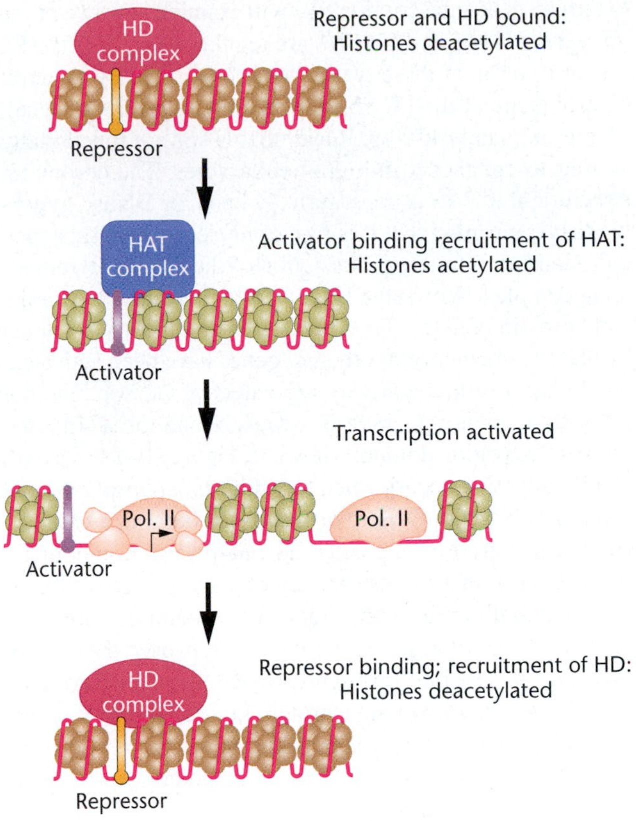

FIGURE 20–17 Proposed model of the action of HAT and HD complexes. Transcription factors recruit the complex to the gene, which either adds or removes acetyl groups, aiding in either opening or closing the chromatin structure.

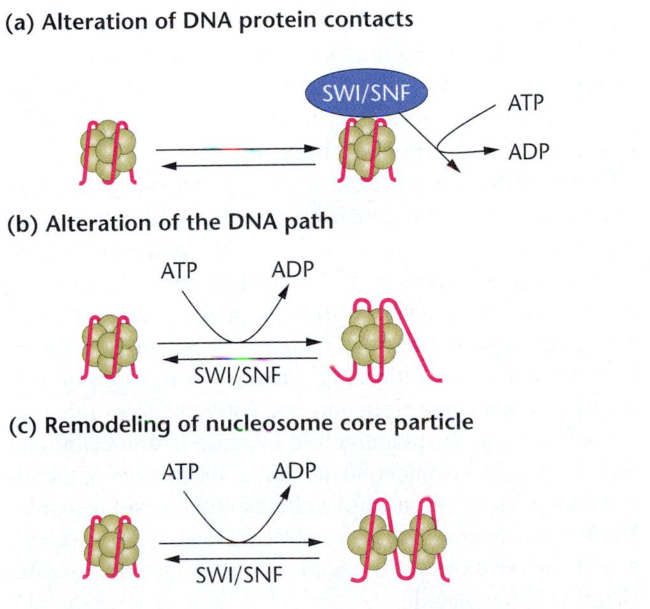

FIGURE 20–16 Three mechanisms that might be used to alter nucleosome structure by the ATP hydrolysis-dependent remodeling complex SWI/SNF. (a) DNA–histone contacts may be loosened. (b) The path of the DNA around an unaltered nucleosome core particle may be altered. (c) The conformation of the nucleosome core particle may be altered.

20.6 Yeast *gal* Genes: Positive Induction and Catabolite Repression

One of the first model systems used to study eukaryotic genetic regulation was the genes in yeast that encode the enzymes for galactose breakdown, or catabolism. These *gal* genes serve a function similar to the *lac* operon and *ara* operon in *E. coli*. Expression of the *gal* genes is **inducible**—that is, they are regulated by the presence or absence of their substrate, galactose. In the absence of galactose, the genes are not transcribed. If galactose is added to the growth medium, transcription begins immediately, and the mRNA concentration of the transcripts increases a thousandfold. However, transcription is activated only if the concentration of glucose is low. So, like the *lac* and *ara* operons, the *gal* genes are under a second level of control, **catabolite repression** (see Chapter 15). A mutation in the regulator of the *gal* genes, *GAL4*, prevents activation, indicating that transcription is under **positive control**—that is, the regulator must be present to turn on gene transcription. We will examine how two of the *gal* genes, *GAL1* and *GAL10*, are regulated (Figure 20–18).

Transcription of these two genes is controlled by a central control region, called **UAS$_G$** (upstream activating sequence), of approximately 170 bp. Recall that UASs are functionally similar to enhancers in higher eukaryotes. The chromatin structure of a UAS is constitutively open, or **DNase hypersensitive**, meaning that it is free of nucleosomes. This open chromatin structure requires action of the SWI/SNF remodeling complex. Within the UAS are four binding sites for the Gal4 protein (Gal4p). These sites are permanently occupied by Gal4p, whether or not the *gal* genes have been activated. Gal4p is, in turn, negatively regulated by Gal80p, another *gal* gene regulator. Gal80p is always bound to Gal4p, covering its activation domain, shown in Figure 20–18 as a dark patch. Induction occurs when the inducer, a phosphorylated galactose, binds to Gal80p or Gal4p (or both) and causes a structural alteration, exposing the Gal4p activation domain. As in *E. coli*, these genes are under a second level of control—catabolite repression—and thus cannot be turned on in the presence of glucose. In this case, however, the catabolite does not operate through cAMP, but rather through a complex system involving a protein kinase. Through activation, additional factors are recruited to further remodel the chromatin, encompassing the two promoters exposing the TATA boxes to the TBP.

Gal4p is a protein consisting of 881 amino acids. It includes a DNA-binding domain that recognizes and binds to sequences in the UAS$_G$ and a *trans*-activating domain that activates transcription [Figure 20–19(a)]. These functional domains were identified by cloning and expressing truncated *GAL4* genes and by assaying the gene products for their ability to bind DNA and to activate transcription. Because the protein contains two functional domains, it is possible to delete one of these regions while the other remains intact and functional.

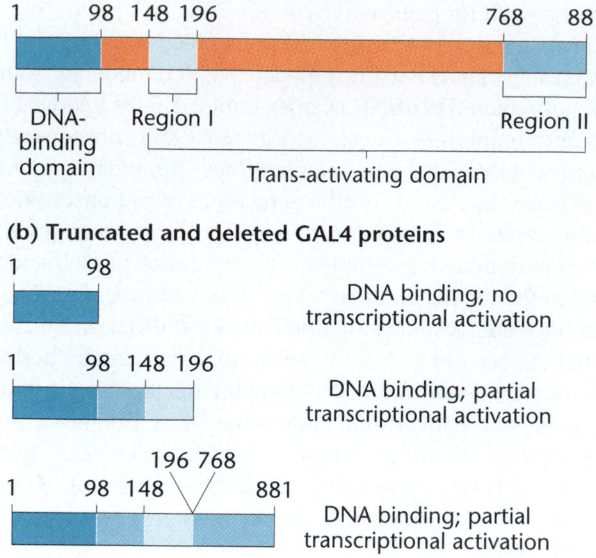

FIGURE 20–19 The Gal4p protein has three domains that participate in the activation of the *GAL1* and *GAL10* loci. Amino acids 1–98 or 1–147 bind to DNA-recognition sites in UAS$_G$. Amino acids 148–196 and 768–881 are required for transcriptional activation. Gal80p binding activity resides in the amino acid region 851–881.

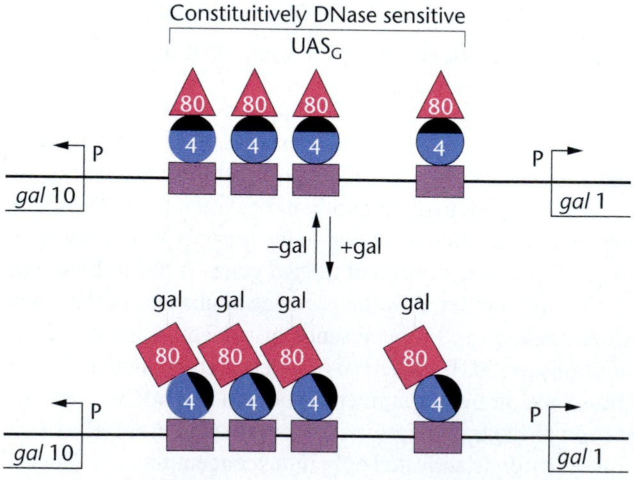

FIGURE 20–18 Model of *GAL1* and *GAL10* UAS structures showing the binding sites for the Gal4p positive regulator and their interaction with the Gal80p negative regulator. Induction is shown as the structure of the Gal80p becomes altered, exposing the activation domain of Gal4p.

A region at the N-terminus of the protein was found to be the DNA-binding domain, responsible for recognition of and binding to the nucleotide sequences in UAS$_G$. Proteins with deletions from amino acid 98 to the C-terminus (amino acid 881) retain the ability to bind to the UAS$_G$ region. Similar experiments also identified two regions of the protein involved in transcriptional activation: region I (amino acids 148–196) and region II (amino acids 768–881). Constructs that contain the DNA-binding domain (amino acids 1–96) and region I (amino acids 148–196), or the DNA-binding domain and region II (amino acids 768–881), have reduced transcriptional activity [Figure 21–19(b)].

To explore the nature of the activating regions, researchers started with a protein containing the DNA-binding region and the activating region I (*GAL4* 1–238), and they isolated point mutations in region I that result in increased activation. Most single mutations that increase activation result from amino acid substitutions that increase the negative charge of region I and cause a threefold increase in transcription. A multiple mutant with four more negative charges than the starting protein causes a ninefold increase in transcriptional activation and has almost 80 percent of the activity of the intact Gal4p. However, negative charge alone is not responsible for the activating function; some mutants with decreased activity do not have fewer negative charges than the parental *GAL4* 1–238 molecule.

These results strongly suggest that activation directs contact between the activating domain of the transcription factor and other proteins. How does this protein–protein interaction activate transcription? There are several possibilities, including chromatin remodeling, stabilization of binding between the promoter DNA and RNA polymerase, increasing the rate of unwinding the double-stranded DNA within the transcribed region, accelerating release of RNA polymerase from the promoter, or attracting and stabilizing other factors that bind to the promoter or to the RNA polymerase.

An attractive candidate for an activator target is TFIID—a complex of proteins that contains the TATA binding protein (TBP)—and about 10 other TAFs (TBP-associated factors). The idea that TFIID is a target for activation is supported by the finding that some *GAL4* constructs stimulate transcription in mammalian cell extracts and change the conformation of DNA-bound TFIID. In addition, other studies show that although TBP alone can bind to the promoter, TAFs must be present for the complex to respond to activators.

20.7 DNA Methylation and Regulation of Gene Expression

Alteration of chromatin conformation is one way eukaryotic gene expression can be regulated or modulated. In this section, we will consider another mechanism that plays a role in gene regulation (the chemical modification of DNA by adding or removing methyl groups from the DNA bases).

The DNA of most eukaryotic organisms is modified after replication by the enzyme-mediated addition of methyl groups to bases and sugars. Base methylation most often involves addition of methyl groups to cytosine. In the genome of any given eukaryotic species, approximately 5 percent of the cytosine residues are methylated. However, the extent of methylation can be tissue specific and can vary from less than 2 percent to over 7 percent. The ability of base methylation to alter gene expression is known from studies on the *lac* operon in *E. coli*. The methylation of DNA in the operator region, even at a single cytosine residue, can cause a marked change in the affinity of the repressor for the operator. Methylation of cytosine occurs at the 5′ position, causing the methyl group to protrude into the major groove of the DNA helix where it can alter the binding of proteins to the DNA. Methylation occurs most often in the cytosine of CG doublets in DNA, usually in both strands:

$$5'\text{—}{}^{m}\text{CpG—}3'$$
$$3'\text{—GpC}^{m}\text{—}5'$$

The methylated state of DNA can be determined by restriction enzyme analysis. The enzyme *Hpa*II cleaves at the recognition sequence CCGG; however, if the second cytosine is methylated, the enzyme will not cut the DNA. The enzyme *Msp*I cuts at the same CCGG site, regardless of whether the second cytosine is methylated. If a segment of DNA is unmethylated, both enzymes produce the same restriction pattern of bands. As shown in Figure 20–20, if one site is methylated, digestion with *Hpa*II produces an altered pattern of fragments. Using this method in eukaryotes to analyze the methylation status of a given gene in different tissues shows that if a gene is expressed, it is usually not methylated or has only a low level of methylation.

Evidence for the role of methylation in eukaryotic gene expression is somewhat indirect and is based on a number of observations. First, an inverse relationship exists between the degree of methylation and the degree of expression. That is, low amounts of methylation are associated with high levels of gene expression, and high levels of methylation are

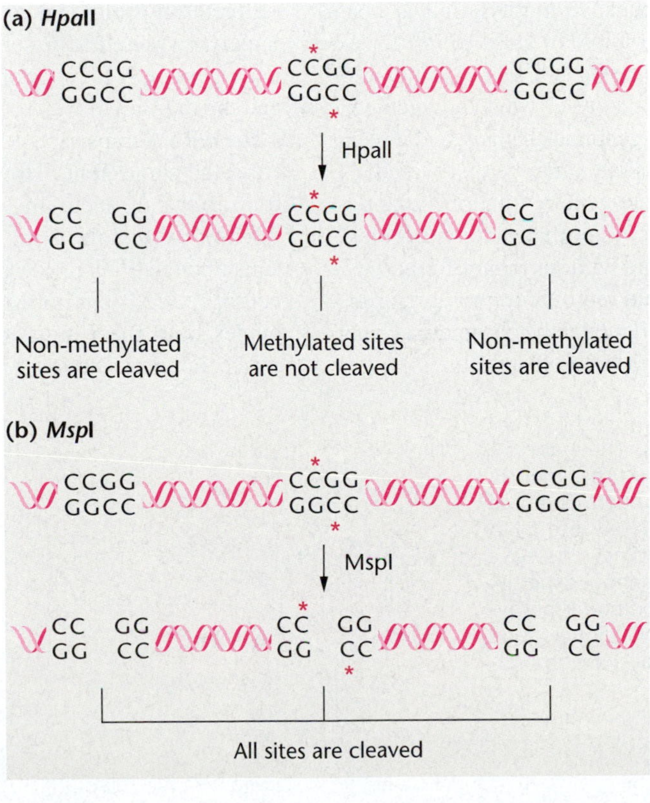

FIGURE 20–20 The restriction enzymes *Hpa*II and *Msp*I recognize and cut at CCGG sequences. (a) If the second cytosine is methylated (indicated by an asterisk), *Hpa*II will not cut. (b) The enzyme *Msp*I cuts at all CCGG sites, whether or not the second cytosine is methylated. Thus, the state of methylation of a given gene in a given tissue can be determined by cutting DNA extracted from that tissue with *Hpa*II and *Msp*I.

associated with low levels of gene expression. In mammalian females, the inactivated X chromosome, which is almost totally inactive, has a higher level of methylation than does the active X chromosome. Within the inactive X, regions that escape inactivation have much lower levels of methylation than those seen in adjacent inactive regions.

Second, methylation patterns are tissue specific and, once established, are heritable for all cells of that tissue. Perhaps the strongest evidence for the role of methylation in gene expression comes from studies using base analogs. The nucleotide 5′-azacytidine is incorporated into DNA in place of cytidine and cannot be methylated (Figure 20–21), causing the undermethylation of the sites where it is incorporated. The incorporation of 5′-azacytidine causes changes in the pattern of gene expression and can stimulate expression of alleles on inactivated X chromosomes.

Recently, 5′-azacytidine has been used in clinical trials for the treatment of sickle-cell anemia. In this autosomal recessive condition, a mutant hemoglobin protein with a defect in β-globin causes changes in shape of red blood cells that, in turn, generate a cascade of clinical symptoms. During embryogenesis, the ε- and γ^g-globin genes are expressed, but normally become transcriptionally inactive after birth, when β-globin synthesis begins. Treatment of affected individuals with 5′-azacytidine causes a reduction in the amount of methylation in the ε-globin and γ^g-globin genes and initiates reexpression of these embryonic and fetal genes. The ε- and γ-proteins replace β globin in hemoglobin molecules, bringing about a reduction in the amount of sickling in the red blood cells.

Though the available evidence indicates that the absence of methyl groups in DNA is related to increases in gene expression, methylation cannot be regarded as a general mechanism for gene regulation because methylation is not a general phenomenon in eukaryotes. In *Drosophila*, for example, there is no methylation of DNA. Thus, methylation may represent only one of a number of ways in which gene expression can be regulated by genomic changes, and it seems likely that similar mechanisms remain to be discovered.

How might methylation affect gene regulation? One possibility comes from the observation that there are proteins that bind to 5′-methyl cytosine without regard to the DNA sequence. These proteins could recruit corepressors or histone (or both) deacetylases to remodel the chromatin, changing it from an open structure to a closed structure.

20.8 Posttranscriptional Regulation of Gene Expression

As we have seen, regulation of genetic expression can occur at many points along the pathway from DNA to protein. Although transcriptional control is perhaps the most obvious and widely used mode of regulation in eukaryotes, **posttranscriptional** modes of regulation also occur in many organisms. Eukaryotic nuclear RNA transcripts are modified prior to translation, noncoding introns are removed, the remaining exons are precisely spliced together, and the mRNA is modified by the addition of a cap at the 5′ end and a poly-A tail at the 3′ end. The message is then complexed with proteins and exported to the cytoplasm. Each of these processing steps offers several possibilities for regulation. We will examine two that are especially important in eukaryotes, alternative splicing of a single mRNA transcript to give multiple mRNAs and regulation of the stability of mRNA itself.

Alternative Splicing Pathways for mRNA

Alternative splicing can generate different forms of a protein, so that expression of one gene can give rise to a family of related proteins. Figure 20–22 illustrates an example in which the polypeptide products derived from a single type of pre-mRNA are distinct from one another. The initial bovine pre-mRNA transcript is processed into one or two **preprotachykinin mRNAs (PPT mRNAs)**. This precursor mRNA molecule potentially includes the genetic information specifying two neuropeptides, called **P** and **K**. These two peptides, are members of the family of sensory neurotransmitters referred to as **tachykinins** and have different physiological roles. While the P neuropeptide is largely restricted to tissues of the nervous system, the K neuropeptide is found more predominantly in the intestine and the thyroid.

The mRNA transcripts for both neuropeptides are derived from the same gene. However, processing of the initial transcript can occur in two ways. In one case (Figure 20–22), exclusion of the K exon during processing results in the α-PPT mRNA, which, upon translation, yields neuropeptide P, but not K. Conversely, processing that includes both the P and K exons yields β-PPT mRNA, which, upon translation, results in the synthesis of both the

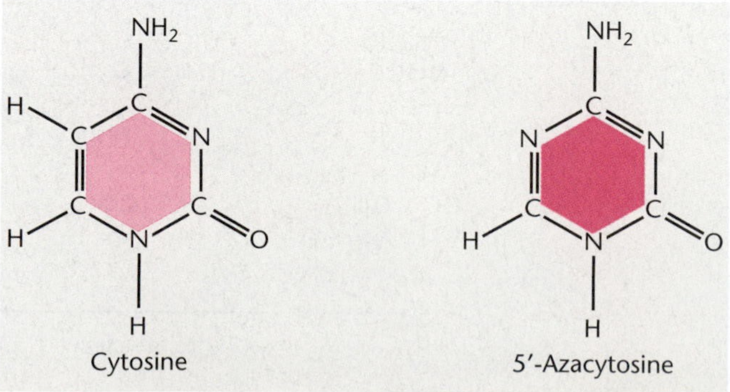

FIGURE 20–21 The base 5′-azacytosine, which has a nitrogen at the 5′ position. The base can be incorporated into DNA in place of cytidine. The base 5′-azacytosine cannot be methylated,causing the undermethylated of the CpG dinucleotide wherever it has been incorporated.

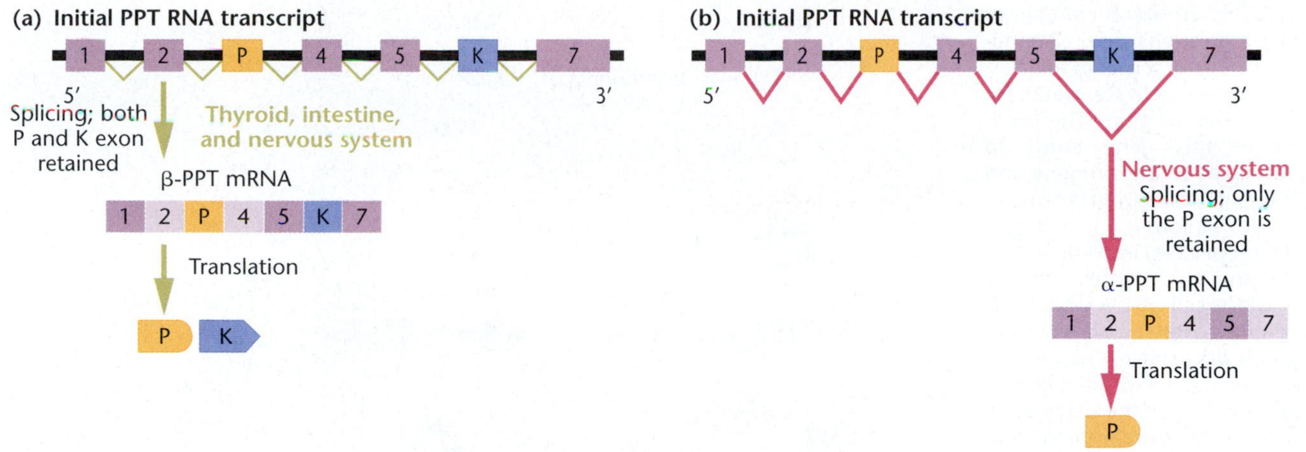

FIGURE 20–22 Alternative splicing of the initial RNA transcript of the preprotachykinin gene (PPT). Exons are either numbered or designated by the letters P or K. When the K exon is excluded, α-PPT mRNA is produced, and only the P neuropeptide is synthesized. The inclusion of P and K exons leads to β-PPT mRNA, which, upon translation, yields both the P and K tachykinin neuropeptides.

P and K neuropeptides. The analysis of the relative levels of the two types of RNA has demonstrated striking differences between tissues. In nervous-system tissues, α-PPT mRNA predominates by as much as a threefold factor, while β-PPT mRNA is the predominant type in the thyroid and the intestine.

Given the existence of alternate splicing, how many different polypeptides can be derived from the same pre-mRNA? Work on α-tropomyosin, an accessory protein that regulates muscle contraction, has provided a partial answer to this question. In rats, the α-tropomyosin gene contains a total of 14 exons, 6 of which make up three pairs that are alternatively spliced. Only one member of each pair ends up in the finished mRNA, never both. Alternative splicing of this pre-mRNA results in 10 different forms of α-tropomyosin, many of which are expressed in a tissue-specific manner. Another gene, for the muscle form of troponin T (a protein that regulates the calcium needed for contraction) produces 64 known forms of the protein from a single pre-mRNA by alternative splicing.

Alternative Splicing and *Drosophila* Sex Determination

Alternative splicing not only has the potential to produce many different forms of a protein from a single gene, but can also affect the outcome of developmental processes. Perhaps the most striking example involves sex determination in *Drosophila*, in which splicing involving a single exon eventually determines whether the embryo will develop as a male or a female.

As outlined in Chapter 11, sex in *Drosophila* is determined by the ratio of X chromosomes to sets of autosomes (X:A). When the ratio is 0.5 (1X:2A), males are produced, even when no Y chromosome is present; when the ratio is 1.0 (2X:2A), females are produced. Intermediate ratios (2X:3A) produce intersexes. The chromosomal ratios are interpreted by a small number of genes that initiate a cascade of developmental events resulting in the production of male or female somatic cells and the corresponding male or female phenotypes. The major genes in this pathway are *sex lethal* (*Sxl*), *transformer* (*tra*), and *doublesex* (*dsx*).

Figure 20–23 illustrates this process. In females, the chromosomal ratio activates transcription of the *Sxl* gene and results in the production of Sxl protein (the mechanism is not completely understood). In males, the chromosomal ratio does not result in production of Sxl protein. When the Sxl protein is present, it binds to and controls the splicing of the pre-mRNA for the next gene in the pathway, *transformer* (*tra*).

The pre-mRNA of *tra* includes a termination codon in exon 2. The female-specific Sxl protein binds to the *tra* pre-mRNA and directs the splicing machinery to remove exon 2 from the mature mRNA. In males with no Sxl protein, splicing proceeds along the default pathway, and the stop codon is incorporated into the mature mRNA. In males, the translation of *tra* mRNA is prematurely terminated by the stop codon, resulting in an inactive gene product, whereas in females, translation produces a functional gene product. The result is a female-specific tra protein that leads to the female phenotype.

The *dsx* gene is a critical control point in the development of sexual phenotype, for it produces a functional mRNA and protein in both females and males. However, the pre-mRNA is processed in a sex-specific manner to yield different transcripts. In females, the tra protein binds to the *dsx* pre-mRNA and directs its processing in an alternative, female-specific manner. In males, default splicing of the *dsx* pre-mRNA results in a male-specific mRNA and protein product. The female dsx protein causes female sexual differentiation in somatic tissues by repressing male sexual development through coordinated action with the *ix* gene product. The male dsx protein brings about male sexual development by suppressing the female pathway. In the absence of *dsx* function, flies develop into phenotypic intersexes.

In sum, the *Sxl* gene acts as a switch that selects the pathway of sexual development by controlling the processing of the *dsx* transcript in a female-specific fashion. The Sxl protein is produced only in embryos with an X:A ratio

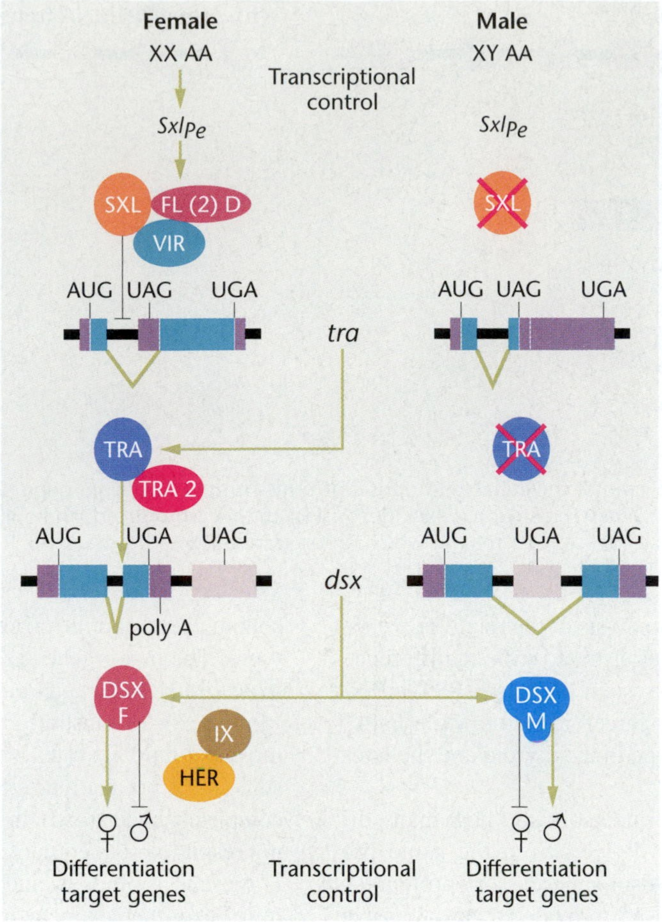

FIGURE 20–23 Hierarchy of gene regulation for sex determination in *Drosophila*. In females, the X:A ratio activates transcription of the *Sxl* gene. The product of this gene binds to pre-mRNA of the *tra* gene and directs its splicing in a female-specific fashion. The female-specific tra protein, in combination with the tra-2 protein, directs female-specific splicing of *dsx* pre-mRNA, resulting in a female-specific dsx protein. This protein, in combination with the ix protein, suppresses the pathway of male sexual development and activates the female pathway. In males, the X:A ratio does not activate *Sxl*. The result is a male-specific processing of *tra* pre-mRNA, resulting in no functional tra protein, which in turn leads to male-specific processing of *dsx* transcripts, resulting in a male-specific dsx protein that activates the pathway for male sexual development.

of 1. Failure to control the splicing of the *dsx* transcript in a female mode results in the default splicing of the transcript in a male mode, leading to the production of a male phenotype.

Controlling mRNA Stability

After mRNA precursors are processed and transported, they enter the population of cytoplasmic mRNA molecules, from which messages are recruited for translation. All mRNA molecules have a characteristic life span (called a half-life, $t_{1/2}$); they are degraded some time after they are synthesized, usually in the cytoplasm. The lifetimes of different mRNA molecules vary widely. Some are degraded within minutes after synthesis, whereas others last hours, or even months and years (in the case of mRNAs stored in oocytes). The mRNA stability, and thus its *turnover rate*, is intrinsic to the sequence of the mRNA.

An mRNA molecule is more than an open reading frame that codes for a protein. It also includes 5′ and 3′ untranslated sequences that contain regulatory information. This information can include **stability sequences** as well as an **address sequence** to locate the mRNA in a particular compartment of the cell. In addition, there are **instability elements**, which can be general (for all cells) or cell-type specific. A typical instability sequence, AUUUA, is found in the mRNA of many oncogenes, such as *fos*.

Although the regulation of gene expression at the level of translation may seem inefficient, a growing body of evidence indicates that altering the stability of mRNA engaged in translation may be a significant point for controlling eukaryotic gene expression. Why would such an inefficient mechanism exist? Clearly, factors that influence the half-life of particular mRNAs have an important effect on the number of protein molecules that can be produced from a single mRNA molecule. Longer lived mRNAs can produce more protein than those that are short lived. Altering the half-life of an mRNA molecule already engaged in translation is one way a cell can respond to rapidly changing conditions inside or outside the cell.

Another way to control mRNA stability is through **translation level** control—that is, the use of the message controls its stability. One of the best studied examples of translational regulation is the synthesis of α- and β-tubulins, the subunit components of eukaryotic microtubules. Treatment of a cell with the drug colchicine leads to a rapid disassembly of microtubules and an increase in the concentration of the α and β subunits. Under these conditions, the synthesis of α- and β-tubulins drops dramatically. However, when cells are treated with vinblastine, a drug that also causes microtubule disassembly, the synthesis of tubulins is increased. Though the two drugs both cause microtubule disassembly, vinblastine precipitates the subunits, lowering the concentration of free α and β subunits. At low concentrations, the synthesis of tubulins is stimulated, whereas at higher concentrations, synthesis is inhibited. This type of translational regulation is known as **autoregulation**.

Work by Don Cleveland and his colleagues has described the conditions under which the stability of tubulin mRNA is regulated. Fusion of tubulin gene fragments to a cloned thymidine kinase gene shows that the first 13 nucleotides at the 5′ end of the mRNA following the transcription start site (encoding the amino terminal region of the protein) cause the hybrid tubulin–thymidine kinase mRNA to be regulated as efficiently as intact tubulin mRNA. These first 13 nucleotides encode the amino acids Met-Arg-Glu-Ile (abbreviated as MREI). Deletions, translocations, or point mutations in this 13-nucleotide segment abolish regulation. The examination of the cytoplasmic distribution of mRNAs demonstrates that only tubulin mRNAs in polysomes are depleted by drug treatment. Copies of tubulin mRNAs not bound to polysomes were protected from degradation. Also, translation must proceed to at least codon 41 of the mRNA in order for regulation to occur.

A model, shown in Figure 20–24, has been proposed to explain these observations. In this model, regulation occurs after the process of translation has begun. The first four amino acids (Met-Arg-Glu-Ile, MREI) of the tubulin gene product (in this case, β-tubulin) constitute a recognition element to which the regulatory factors bind. The concentration of α- and β-tubulin subunits in the cytoplasm may serve this regulatory function. The protein–protein interaction invokes the action of an RNase that may be a ribosomal component or a nonspecific cytoplasmic RNase. Action by the RNase degrades the tubulin mRNA in the act of translation, shutting down tubulin biosynthesis. As an alternative, the binding of the regulatory element may cause the ribosome to stall in its translocation along the mRNA, leaving it exposed to the action of RNase. Translation must proceed to at least codon 41 because the first 30 to 40 amino acids translated occupy a tunnel within the large ribosomal subunit, and only the translation of this number of amino acids will make the MREI recognition sequence accessible for binding.

Translation-coupled mRNA turnover has been proposed as a regulatory mechanism for other genes, including histones, some transcription factors, lymphokines, and cytokines. This suggests the existence of a regulatory mechanism that acts on mRNAs in the act of translation, probably by protein–protein interaction between a regulatory element and the nascent polypeptide chain, resulting in the activation of RNases.

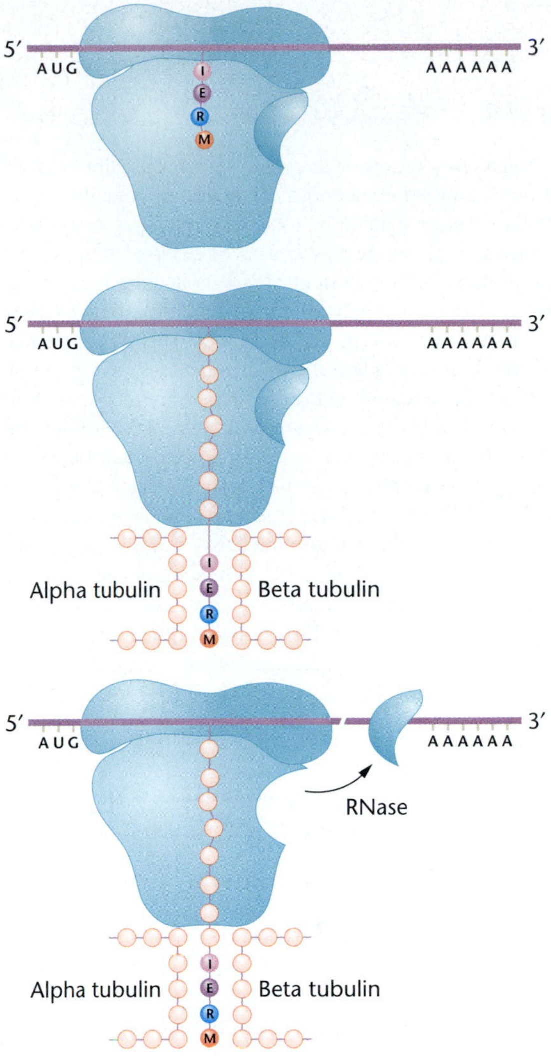

FIGURE 20–24 Model of posttranslational regulation of tubulin synthesis.

Chapter Summary

1. Mechanisms controlling gene regulation in eukaryotes are governed by the properties of expanded genome size, the spatial and temporal separation between transcription and translation, and, in most organisms, multicellularity.
2. The roles played by chromatin structure, transcription factors, and signal transduction, as well as the expression of specific gene sets, are studied as models for gene regulation in higher organisms.
3. Transcription in eukaryotes is controlled by the interaction between promoters and enhancers. Other regulatory squences, such as the CCAAT box, are elements of promoter regions found near the promoter itself. Enhancer elements, which appear to control the degree of transcription, can be located before, after, or within the gene expressed.
4. Transcription factors are proteins that bind to DNA-recognition sequences within the promoters and enhancers and activate transcription through protein–protein interaction.
5. A gene may have a different methylation pattern in different tissues, which is often correlated with different regulatory patterns.
6. Several types of posttranscriptional control of gene expression are possible in eukaryotes. One such mechanism is alternative processing of a single class of pre-mRNAs to generate different mRNA species. Another mechanism controls mRNA stability, either through intrinsic sequence elements or through a feedback system in response to the cytoplasmic concentration of the protein.

Insights and Solutions

1. Regulatory sites for eukaryotic genes are usually located within a few hundred nucleotides of the transcriptional starting site, but they can be located up to several kilobases away. DNA sequence-specific binding assays have been used to detect and isolate protein factors present at low concentrations in nuclear extracts. In these experiments, DNA-binding sequences are bound to material on a column, and nuclear extracts are passed over the column. The idea is that if proteins that specifically bind to the sequence represented on the column are present, they will bind to the DNA, and they can be recovered from the column after all other nonbinding material has been washed away. Once a DNA-binding protein has been identified and isolated, the problem is to devise a general method for screening cloned libraries for the genes encoding these and other DNA-binding factors. Determining the amino acid sequence of the protein and constructing synthetic oligonucleotide probes is time consuming and useful only for one factor at a time. Knowing the strong affinity for binding between the protein and its DNA-recognition sequence, how would you screen for genes encoding binding factors?

Solution: Several general strategies have been developed, and one of the most promising has been devised by Steve McKnight's laboratory at the Fred Hutchinson Cancer Center. The cDNA isolated from cells expressing the binding factor is cloned into the lambda vector, gt11. Plaques of this library, containing proteins derived from expression of cDNA inserts, are adsorbed onto nitrocellulose filters and probed with double-stranded radioactive DNA corresponding to the binding site. If a fusion protein corresponding to the binding factor is present, it will bind to the DNA probe. After the unbound probe is washed off, the filter is subjected to autoradiography and the plaques corresponding to the DNA-binding signals can be identified. An added advantage of this strategy is filter recycling by washing the bound DNA from the filters for reuse. Such an ingenious procedure is similar to the colony-hybridization and plaque-hybridization procedures described for library screening in Chapter 16, and it provides a general method for isolating genes encoding DNA-binding factors.

Problems and Discussion Questions

1. Why is gene regulation assumed to be more complex in a multicellular eukaryote than in a prokaryote? Why is the study of this phenomenon in eukaryotes more difficult?
2. List and define the levels of gene regulation discussed in this chapter.
3. Distinguish between the regulatory elements referred to as promoters and enhancers.
4. Is the binding of a transcription factor to its DNA recognition sequence necessary and sufficient for an initiation of transcription at a regulated gene? What else plays a role in this process?
5. Contrast transcription factors designated as true activators with those designated as antirepressors.
6. Contrast and compare the regulation of the *gal* genes in yeast with the *lac* genes in *E. coli*.
7. Write an essay comparing the control of gene regulation in eukaryotes and prokaryotes at the level of initiation of transcription. How do the regulatory mechanisms work? What are the similarities and differences in these two types of organisms regarding the specific components of the regulatory mechanisms? Address how the differences or similarities relate to the biological context of the control of gene expression.

Extra-Spicy Problems

8. In the autoregulation of tubulin synthesis, two models for the mechanism were proposed: (1) that the tubulin subunits bind to the mRNA and (2) that the subunits interact with the nascent tubulin polypeptides. To distinguish between these two models, Cleveland and colleagues introduced mutations into the 13-base regulatory element of the β-tubulin gene. Some of the mutations resulted in amino acid substitution, while others did not. In addition, they shifted the reading frame of the intact 13-base sequence. Presented here are the results of a mutagenesis study of the mRNA:

	met	arg	glu	lys	*Autoregulation*
Wild type	AUG	AGG	GAA	ATC	+
Second codon mutations	____	UGG	____		–
	____	GGG	____		–
	____	CGG	____		+
	____	AGA	____		+
	____	AGC	____		–
Third codon mutations		____	GAC	____	+
		____	AAC	____	–
		____	UAU	____	–
		____	UAC	____	–

Which of the two models is supported by the results? What experiments would you do to confirm this?

9. You are interested in studying transcription factors and have developed an *in vitro* transcription system by using a defined segment of DNA that is transcribed under the control of a eukaryotic promoter. The transcription of this DNA occurs when you add purified RNA polymerase II, TFIID (the TATA binding factor), and TFIIB and TFIIE (which bind to RNA polymerase). You perform a series of experiments that compare the efficiency of transcription in the "defined system" with the efficiency of transcription in a crude nuclear extract. You test the two systems with your template DNA and with various deletion templates that you have generated. The following are the results of your study:

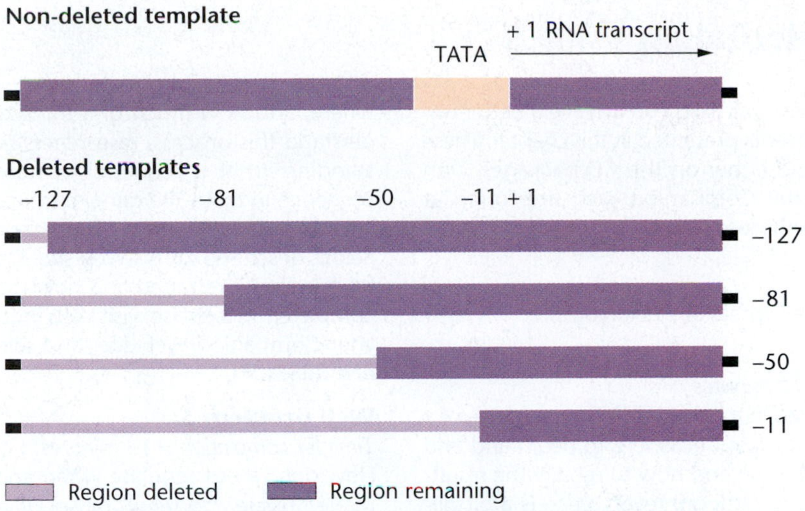

DNA added	Nuclear extract	Purified system
undeleted	++++	+
−127 deletion	++++	+
−81 deletion	++++	+
−50 deletion	+	+
−11 deletion	o	o

\+ Low efficiency transcription
++++ High efficiency transcription
o No transcription

(a) Why is there no transcription from the −11 deletion template?

(b) How do the results for the nuclear extract and the defined system differ from the *undeleted* template? How would you interpret these results?

(c) For the various deleted templates, compare the results with the nuclear extract and the purified system. How would you interpret the results with the deleted templates? Be very specific about what you can conclude from these data.

Selected Readings

Ayoubi, T.A., and Van de Ven, W.J. 1996. Regulation of transcription by alternate promoters. *FASEB J.* 10:453–60.

Cleveland, D. 1988. Autoregulated instability of tubulin mRNAs: A novel eukaryotic regulatory mechanism. *Trends Biochem. Sci.* 13:339–43.

Cline, T.W., and Meyer, B.J. 1996. Vive la difference: Males vs. females in flies vs worms. *Ann. Rev. Genet.* 30:637–702.

Drapkin, R., Merino, A., and Reinberg, D. 1993. Regulation of RNA polymerase II transcription. *Curr. Opin. Cell Biol.* 5:469–76.

Dillon, N., and Sabbattini, P. 2000. Functional gene expression domains: Defining the functional unit of eukaryotic gene regulation. *Bioessays* 22:657–665.

Dynan, W.S. 1988. Modularity in promoters and enhancers. *Cell* 58:1–4.

Francastel, C., Schübeler, D., Martin, D., and Groudine, M. 2000. Nuclear compartmentalization and gene activity. *Nature Rev. Genet.* 1:137–143.

Fry, C.J., and Peterson, C.L. 2001. Chromatin remodeling enzymes: Who's on first? *Curr. Biol.* 11:R185–R197.

Gregory, P.D., Wagner, K., and Hürz, W. 2001. Histone acetylation and chromatin remodeling. *Exp. Cell Res.* 265:195–202.

Goodrich, J.A., Cutler, G., and Tjian, R. 1996. Contacts in context: Promoter specificity and macromolecular interaction in transcription. *Cell* 84:825–30.

Jacobsen, A., and Peltz, S. 1996. Interrelationships of the pathway of mRNA decay and translation in eukaryotic cells. *Ann. Rev. Biochem.* 65:693–739.

Jacobson, R.H., and Tijian, R. 1996. Transcription factor IIA: A structure with multiple functions. *Science* 272:830–36.

Jenuwein, T., and Allis, C.D. 2001. Translating the histone code. *Science* 293:1074–1080.

Kakidani, H., and Kass, S.U., Pruss, D., and Wolffe, A.P. 1997. How does DNA methylation repress transcription? *Trends in Genet.* 13:444–49.

Lamond, A.I. and Earnshaw, W.C. 1998. Structure and function in the nucleus. *Science* 280:547–553.

Lee, T.I., and Young, R.A. 2000. Transcription of eukaryotic protein-coding genes. *Ann. Rev. Genet.* 34:77–137.

GENETICS MediaLab

The resources that follow will help you achieve a better understanding of the concepts presented in this chapter. These resources can be found either on the CD packaged with this textbook or on the Companion Web site found at **http://www.prenhall.com/klug**

CD Resources:

Module 20.1: Gene Expression: Eukaryotes

Web problem 1:
Time for completion = 10 minutes
How is a gene organized in a eukaryote? The sequence of a gene includes information about where to begin and end transcription and translation and how to modify the resulting transcript or protein. Although every gene is a bit different, some general features are shared among many eukaryotic genes. In this exercise, you will review the basic features of the eukaryotic gene. Draw a representative eukaryotic gene, pointing out the location of any common features. Indicate on which strand of this double-stranded region those elements are located. To complete this exercise, visit Web Problem 1 in Chapter 20 of your Companion Web site, and select the keyword **TRANSCRIPTIONAL REGULATION**.

Web problem 2:
Time for completion = 10 minutes
How does a cell know which genes to transcribe? The promoter region of a gene is important for designating when, where, and how much of a transcript to produce. To understand this process, researchers have focused on understanding what is required for function of the promoter sequence in *cis* to the gene, and how *trans* factors interact with the promoter region. The work of one researcher in this field is described at this Web site. Discuss what experimental questions are raised and how they are being tested. To complete this exercise, visit Web Problem 2 in Chapter 20 of your Companion Web site, and select the keyword **GENE EXPRESSION**.

Web problem 3:
Time for completion = 15 minutes
How does a cell regulate when and where a transcript is made? Tissue and temporal specificity of transcription has been an area of intense work in the last two decades. In particular, much attention has been given to the role of transcriptional regulation in establishing and maintaining the developmental identity of a cell. In this exercise, you will view a database of genes under tissue-specific transcriptional regulation. Select at least one example from two different types of transcription factors, and for each one, create a précis that includes the function of the gene product, when and where the gene is expressed, what transcriptional factors regulate the gene, and where in the promoter these factors bind. To complete this exercise, visit Web Problem 3 in Chapter 20 of your Companion Web site, and select the keyword **TISSUE SPECIFICITY**.

Lohr, D. 1997. Nucleosome transactions on the promoters of the yeast *GAL* and *PHO* genes. *J. Biol. Chem.* 272:26795–98.

MacDougall, C., Harbison, D., and Bownes, M. 1995. The developmental consequences of alternate splicing in sex determination and differentiation in *Drosophila. Dev. Biol.* 172:353–76.

Peterson, C.L., and Logie, C. 2000. Recruitment of chromatin-remodeling machines. *J. Cell. Biochem.* 78:179–185.

Ptashne, M. 1988. GAL4 activates gene expression in mammalian cells. *Cell* 52:161–67.

Ptashne, M., and Gann, A.A.F. 1990. Activators and targets. *Nature* 346:329–31.

Qumsiyeh, M.B. 1999. Structure and function of the nucleus: Anatomy and physiology of chromatin. *Cell. Mol. Life Sci.* 55:11129–1140.

Sudarsanam, P., and Winston, F. 2000. The SWI/SNF family of nucleosome-remodeling complexes and transcriptional control. *Trends in Genet.* 16: 345–351.

Turner, B.M. 2000. Histone acetylation and an epigenetic code. *Bioessays* 22: 836–845.

Workman, J.L., and Kingston, R.E. 1998. Alteration of nucleosome structure as a mechanism of transcriptional control. *Ann. Rev. Biochem.* 67:545–79.

Wu, J., and Grunstein, M. 2000. 25 years after the nucleosome model: Chromatin modifications. *Trends Biochem Sci.* 619–623.

Yen, T.J., Gay, D.A., Pachter, J.S., and Cleveland, D.W. 1988. Autoregulated changes in stability of polyribosome-bound beta-tubulin mRNAs are specified by the first 13 translated nucleotides. *Mol. Cell Biol.* 8:1224–35.

This unusual four-winged *Drosophila* has developed an extra set of wings as a result of a homeotic mutation. (*Courtesy of E. B. Lewis, California Institute of Technology Archives*)

21
Developmental Genetics

KEY CONCEPTS

In multicellular plants and animals, a fertilized egg, without further stimulus, begins a cycle of developmental events that ultimately give rise to an adult member of the species from which the egg and sperm were derived. Thousands, millions, or even billions of cells are organized into a cohesive and coordinated unit that we perceive as a living organism (Figure 21–1). The series of events through which organisms attain their final adult forms is studied by developmental biologists. This area of study is perhaps the most intriguing in biology. Not only does the comprehension of developmental processes require the knowledge of many different biological disciplines—such as molecular, cellular, and organismal biology—but also an understanding of how biological processes change over time and ultimately transform the organism.

In the past hundred years, investigations in embryology, genetics, biochemistry, molecular biology, cell physiology, biophysics, and evolution have all contributed to the study of development. The findings have largely pointed out the tremendous complexity of developmental processes. Unfortunately, a description of *what* happens does not answer the "why" and "how" of development. Over the last two decades, however, genetic analysis and molecular biology have identified genes that regulate developmental processes, and we are beginning to understand how the action and interaction of these genes control basic developmental processes in a wide range of organisms.

In this chapter, the primary emphasis will be on the role of gene action in regulating development. This area of study, called developmental genetics, has contributed tremendously to our understanding of developmental processes, because genetic information is both required for the molecular and cellular functions mediating developmental events and contributes to the determination of the final phenotype of the newly formed organism.

(a)

(b)

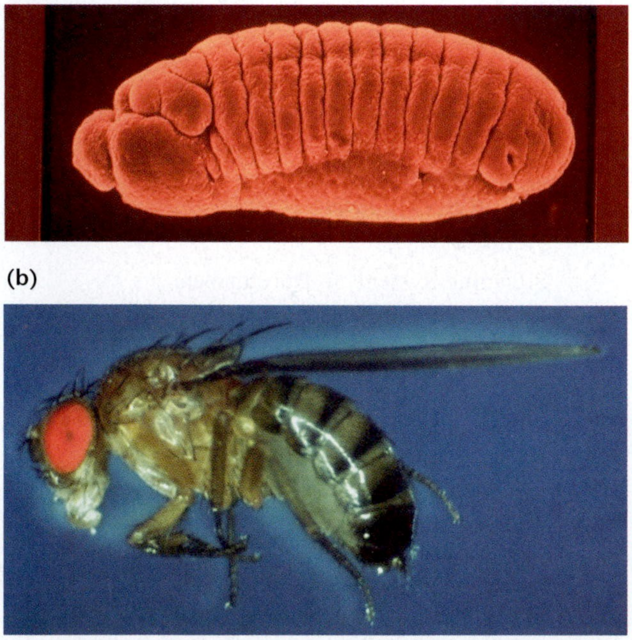

FIGURE 21–1 (a) A *Drosophila* embryo and (b) the adult fly that develops from it.

21.1 Basic Concepts in Developmental Genetics

In higher eukaryotes, development begins with the formation of the **zygote**, or the cell that is generated by the fusion of the sperm and oocyte. The oocyte is a single cell with a heterogeneous cytoplasm that is nonuniform in distribution. Following fertilization and early cell divisions, the nuclei of progeny cells find themselves in different environments as the maternal cytoplasm is distributed into the new cells. Evidence suggests that in different cells, the cytoplasm exerts different influences on the genetic material, causing differential transcription at specific points during development. Hence, **cytoplasmic localization**, or the inheritance of particular cytoplasmic components by individual embryonic cells, plays a major regulatory role during development. Early gene products synthesized by the zygote's genome further alter the cytoplasm of each cell, producing a still different cellular environment that, in turn, leads to the activation of other genes, and so on. As the number of cells in the embryo increases, they influence one another by **cell–cell interaction**. The environmental factors acting on the genetic material, therefore, now include the cytoplasm of individual cells, as well as signals from other cells. Although early in embryogenesis most cells show no evidence of structural or functional specialization, the combination of their localized cytoplasmic components and their position in the developing embryo determine the ultimate structure and functions they will assume. It is as if their fate has been programmed prior to the actual events leading to specialization.

As cells respond to their continually changing external and internal environments, they embark on a pathway composed of several developmental stages. The most important stage in this pathway is **determination**, when, in response to the many external and internal cues, a specific developmental fate for a cell becomes fixed. It is now clear that determination precedes the actual events of **differentiation**, the process by which the cell achieves its final form and function.

In the next sections, we will examine how patterns of gene expression relate to the processes of determination, differentiation, and cell–cell interaction during development. We will begin by examining the role of differential gene expression in the progressive restriction of developmental options available to cells. This topic will be introduced with binary switch genes as an example illustrating that differential gene expression leads to different cell types and will be expanded upon as we discuss the two common mechanisms for regulating gene expression, cytoplasmic localization, and cell–cell communication.

Several model systems, including *Drosophila*, *C. elegans*, and *Arabidopsis thalania* will be used to illustrate the mechanisms of cytoplasmic localization and cell-cell communications, as well as the contribution that genetic analysis has made to our understanding of development in multicellular organisms. Finally, the chapter will conclude with a discussion of the most dramatic restriction of cell fate during development, namely, programmed cell death.

21.2 The Role of Binary Switch Genes in Development

As part of the progressive restriction of transcriptional capacity that accompanies development, certain genes act as switches, decreasing the number of alternative developmental pathways. Each decision point is usually binary—that is, there are two alternatives possible—and the action of a switch gene programs the cell to follow only one of these pathways. Thus, such genes are referred to as **master regulatory genes** or **binary switch genes**. These genes are defined by their ability to initiate the complete development of an organ, or a tissue type. We will briefly describe two examples: (1) *myo*D-related genes, which control muscle cell determination and differentiation in both invertebrates and vertebrates, and (2) *eyeless/Small eye*, which is a master regulator of eye formation, again in both invertebrates and vertebrates.

Myoblast-Determining Genes

One of the best characterized classes of switch genes includes those that control skeletal muscle cell formation. These are the four-member set of **myoblast-determining** or ***myo*D** genes. The nucleotide and amino acid sequences of this gene family are highly conserved in vertebrates (Table 21–1). The vertebrate *myoD* genes are less similar to *myo*D-related genes in invertebrates, suggesting a more distant evolutionary relationship. The *myo*D genes are major regulatory genes that control events leading to skeletal muscle cell differentiation in embryonic muscle cells, the *myoblasts*.

Four members of the *myoD* family have been identified: *myoD, myogenin, myf5*, and *MRF4* (Table 21–1). Each of these genes encodes a DNA-binding protein with a helix–loop–helix (HLH) motif. (See Chapter 20 for a review of eukaryotic transcription factors.) The *myo*D proteins contain an 80-amino-acid basic region (Figure 21–2) that mediates gene-specific DNA binding and transcriptional activation. These proteins work by forming heterodimers with other proteins (the E-box proteins). The heterodimers

TABLE 21–1 The *myo*D Gene Family

Xenopus	*Rodent*	*Human*
XMyoD	*myoD*	*myf3*
XMyogenin	*myogenin*	*myf4*
XMyf5	*myf5*	*myf5*
XMRF4	*MRF4*	*myf6*

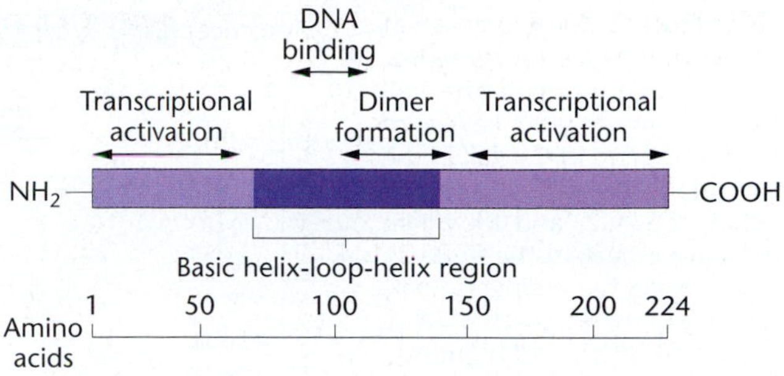

FIGURE 21–2 Structure of myogenin, a 224-amino-acid protein belonging to the *myo*D family. This protein has a characteristic basic region and helix–loop–helix domain near the center. An adjacent region controls dimer formation necessary for activity. Regions near either end control transcriptional activation.

recognize the sequence CANNTG (where N is any nucleotide) that are found in the enhancers and promoters of most skeletal muscle genes.

Myocyte formation can be experimentally induced in cells that normally do not form muscle, such as fibroblasts (connective tissue cells of the skin), adipocytes (fat cells), neurons (nerve cells), and melanocytes (pigment cells) by activating genes of the *myoD* family. These experiments provide strong evidence for the central role of the *myoD* genes in the formation of skeletal muscle cells (myocytes).

During myocyte formation, these genes are sequentially expressed: *myf5* and *myoD* are expressed in early stages. In late stages of muscle formation, *myogenin* expression is followed by *MRF4*. Differences in the timing of expression for the *myo*D-related proteins suggest that they each have separate and distinct functions. According to this model,

1. *myoD* plays a role in withdrawing the cell from the cell cycle.
2. *myf5* is the switch gene involved in *determination.*
3. *myogenin* controls the process of differentiation.
4. *MRF4* is thought to be involved in myotube maturation and to maintain the differentiated adult phenotype.

Studies on cultured myoblasts indicate that MyoD proteins are under the control of cell growth factors. Myoblasts grown in the presence of high concentrations of growth factors undergo mitosis, and action of the gene products, MyoD and Myf5, is repressed. Growth factors in the nutrient medium induce a number of genes (*c-jun, c-mos, c-myc*) that repress expression of *myoD*. When growth factors are removed from the nutrient medium, the myoblasts exit the cell cycle, and enter G0. When this occurs, *myoD* and *myf5* initiate a program of determination, leading to myogenin expression, which triggers expression of a cascade of genes encoding contractile proteins, causing the myoblast to differentiate as a muscle cell.

Synchronization of the cell cycle in cultured myoblasts indicates that levels of MyoD and Myf5 are synchronized with phases of the cell cycle (Figure 21–3). MyoD is at its highest

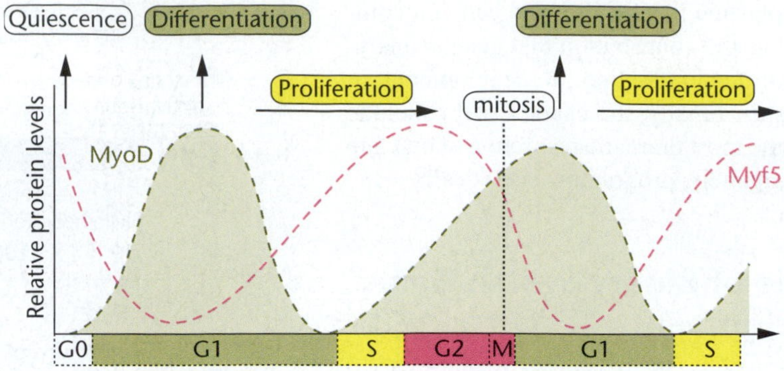

FIGURE 21–3 The expression of MyoD and Myf5 are synchronized with phases of the cell cycle in myoblasts. Levels of MyoD are high in G1, fall in the transition to S phase, and rise again in S, G2, and M, again reaching a peak in the next G1 phase. Myf5 has an inverse pattern of expression. Levels of Myf5 are lowest in G1, and highest in the S/G2 transition.

level in early G1, and its lowest level is at the G1/S transition, while Myf5 has the opposite pattern. The trigger for withdrawal from the cell cycle and entry into the pathway of differentiation is in G1 when MyoD levels are high and Myf5 levels are low. This selection, is, in turn, under the control of cell cycle genes, including cyclin-dependent kinases. (Review the cell cycle in Chapter 8.)

The Control of Eye Formation

A second binary switch gene, *eyeless*, triggers the differentiation of a specific cell type and an entire organ, the compound eye, in *Drosophila* and in most other species with eyes. As a result of its actions, the *eyeless* gene has been dubbed the master regulator of eye development. The eyeless protein acts very early in development and is expressed in all embryonic cells that will give rise to the eye. In loss of function mutants, cells normally destined to become part of the eye degenerate late in development and undergo programmed cell death (discussed at the end of this chapter). Like *myo*D, *eyeless* codes for a transcription factor, although in this case it is a member of another gene family, called the paired-box family.

Expression of *eyeless* in other tissues, such as legs, wings, and antennae can result in the formation of eyes in other locations [Figure 21–4(a)]. This indicates that *eyeless* is able to direct the program of gene expression for eye formation in cells that do not normally form eyes and is able to override the program of determination and differentiation present in these cells.

The *eyeless* gene is part of a network of genes [Figure 21–4(b)]. This network is regarded as the master regulator of eye formation. Some of the genes in the network are required for the expression of other genes, while others reinforce the transcription of network members.

The discovery that *eyeless* directs the formation of eyes in vertebrates has forced reevaluation of the long-held belief that the compound eye of insects and the single-lens eye of vertebrates evolved separately. Such an assumption was based on the observation that the compound insect eye and the vertebrate camera eye have different embryonic origins, develop by different pathways, and are structurally very different. In 1995, Walter Gehring and his colleagues generated transgenic flies that carried copies of the mouse homologue of *eyeless*, placed under the control of a strong promoter. The results were striking and led to two inescapable conclusions. First, this single switch gene was capable of triggering the formation of extra eyes on the wings, antennae, and legs of flies [Figure 21–4(a)], and second, because the mouse gene works in flies, the invertebrate and vertebrate eye *are* in fact homologous at the molecular level. Therefore, these eyes are related evolutionarily. The downstream targets of these transcription factors are also conserved, indicating that steps in the genetic regulation of eye development are shared between species that diverged over half a billion years ago (from a common ancestor without eyes).

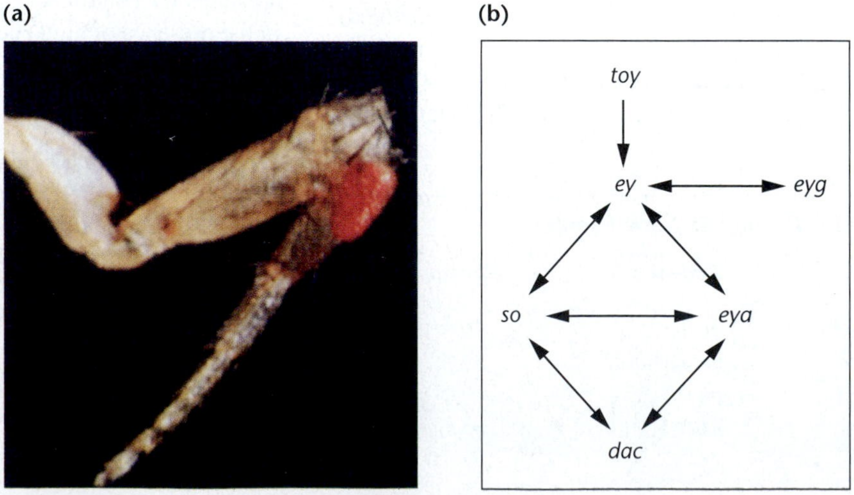

FIGURE 21–4 (a)Expression of the *eyeless* gene in transgenic flies. Complete eyes are formed at ectopic locations, including the leg. (b) The network of genes that regulate eye formation in *Drosophila*.

21.3 Genetics of Embryonic Development in *Drosophila*: The Role of Cytoplasmic Localization

Why a certain cell turns on or off specific genes during development is a central question in developmental biology. There is no simple answer to this question. However, information derived from the study of many different organisms gives us a starting point. One example, embryonic development in *Drosophila*, reflects the direction and scope of our knowledge in this area and highlights the key role played by molecular components placed in the oocyte cytoplasm during oogenesis. We will begin with an overview of *Drosophila* development and then examine how analysis of mutations has provided insight into the number and action of genes that regulate the processes of determination and differentiation.

Overview of *Drosophila* Development

Beginning with the fertilized egg, *Drosophila* passes through a preadult period of development with five distinct phases: the embryo, three larval stages, and the pupal stage. The adult fly emerges from the pupal case about 10 days after fertilization (Figure 21–5). Externally, the *Drosophila* egg has a number of structures that delineate the anterior, posterior, dorsal, and ventral regions (Figure 21–6). The anterior end of the egg contains the micropyle, a specialized conical structure for the entrance of sperm into the egg, while the posterior end is rounded and marked by a series of aeropyles (openings that allow gas exchange during development). The dorsal side of the egg is flattened and contains the chorionic appendages, while the ventral side is curved. Internally, the egg cytoplasm is organized into a series of maternally derived molecular gradients. These gradients play a key role in establishing the developmental fates of nuclei that migrate into specific regions of the embryo.

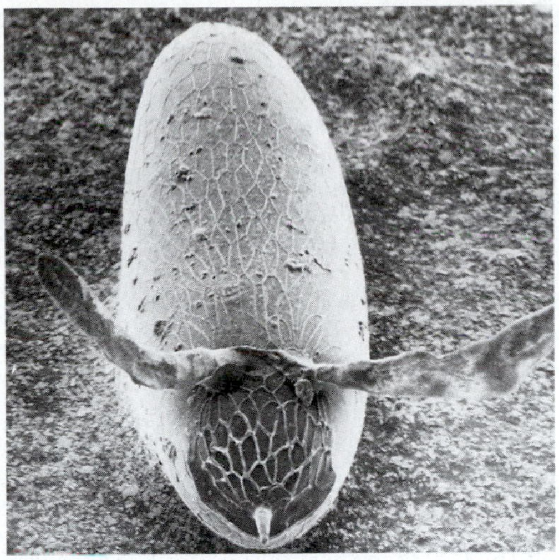

FIGURE 21–6 Scanning electron micrograph of a newly oviposited *Drosophila* egg. The micropyle is the small, white projection at the anterior tip (at the bottom). The dorsal surface (with the two chorionic appendages) is facing up.

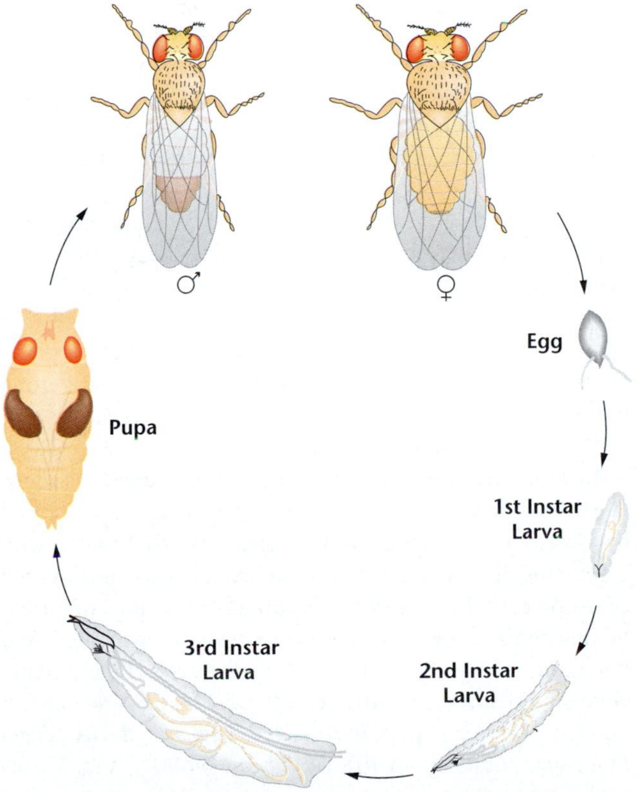

FIGURE 21–5 *Drosophila* life cycle.

Immediately after fertilization, the zygote nucleus undergoes a series of divisions. After nine rounds of division, most of the approximately 512 nuclei migrate to the egg's outer surface, or cortex, where further divisions take place [Figure 21–7(c) and (d)]. This is the **syncytial blastoderm** stage of embryonic development. (A syncytium is any cell with more than one nucleus.) The nuclei are now surrounded by cytoplasm containing regionally localized maternally derived transcripts and proteins. Under the influence of these cytoplasmic components, a program of gene expression is initiated in the nuclei, leading to various developmental outcomes.

The formation of germ cells at the posterior pole of the embryo illustrates the regulatory role of these localized cytoplasmic components [Figure 21–7(d) and (e)]. Experiments have shown that any nucleus that is placed in the posterior cytoplasm (the pole plasm) will form germ cells. Hence, the pole plasm contains maternal components that direct these posterior nuclei to give rise to germ cells.

Transcriptional programs triggered in migrating nuclei also result in the establishment of a body plan organized along anterior–posterior and dorsal–ventral axes of symmetry. Following formation of the syncytial blastoderm, cell membranes form around the nuclei, creating the **cellular blastoderm** [Figure 21–7(e)]. Later, the embryo becomes organized into a series of segments, called parasegments, defined by the expression patterns of various embryonic genes (Figure 21–8). Parasegments later give rise to the adult segments, though they are shifted back by one-half segment. Cells within segments first become determined to form either an anterior or a posterior portion (called a **compartment**) of

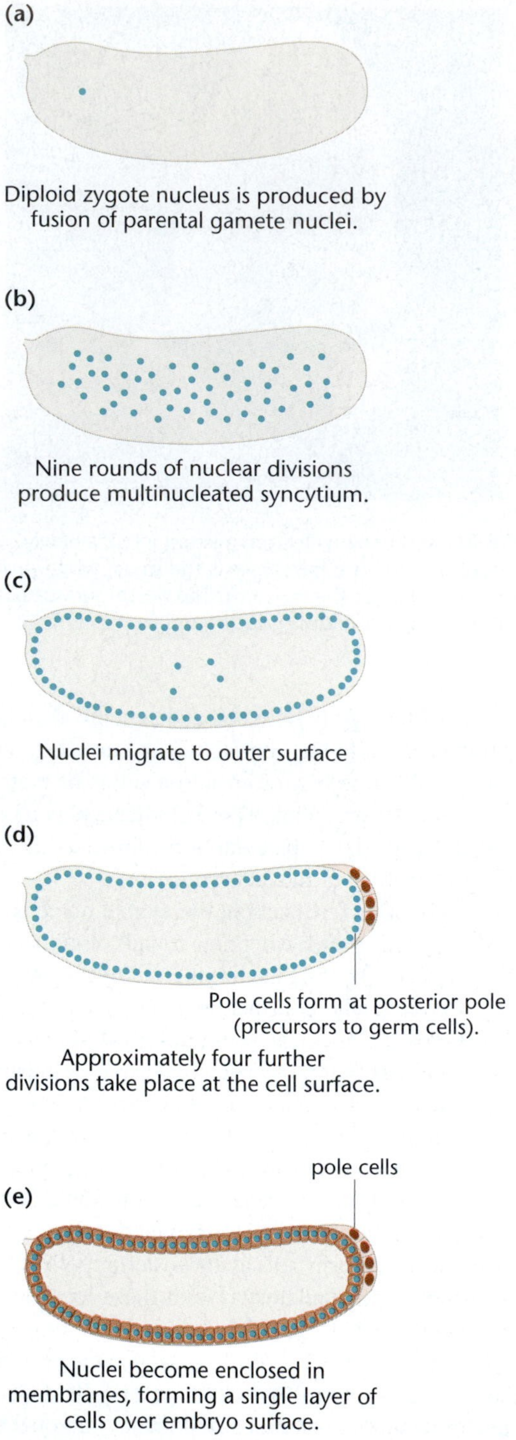
(a)
Diploid zygote nucleus is produced by fusion of parental gamete nuclei.
(b)
Nine rounds of nuclear divisions produce multinucleated syncytium.
(c)
Nuclei migrate to outer surface
(d)
Pole cells form at posterior pole (precursors to germ cells).
Approximately four further divisions take place at the cell surface.
pole cells
(e)
Nuclei become enclosed in membranes, forming a single layer of cells over embryo surface.

FIGURE 21–7 Early stages of embryonic development in *Drosophila*. (a) Fertilized egg with zygotic nucleus, about 30 minutes after fertilization. (b) Nuclear divisions occur about every 10 minutes, producing a multinucleate cell, the syncytial blastoderm. (c) After approximately nine divisions (512 nuclei), the nuclei migrate to the outer surface or cortex of the egg. (d) At the surface, four additional rounds of nuclear division occur. A small cluster of cells, the pole cells, form at the posterior pole about 2.5 hours after fertilization. These cells will form the germ cells of the adult. (e) About 3 hours after fertilization, the nuclei become enclosed in membranes, forming a single layer of cells over the embryo surface, creating the cellular blastoderm.

the segment. In subsequent stages, further restrictions of developmental options occur so that cells in each compartment are eventually programmed to form a single structure in the adult body.

The adult body plan of *Drosophila* derives its overall organization from the larval body plan and is composed of head, thoracic, and abdominal segments. In addition, many of the adult body structures are formed from small groups of cells, or **imaginal discs**, which developed during the larval stages. There are 12 bilaterally paired discs—eye-antennal discs, leg discs, wing discs, and so forth—and one genital disc. Figure 21–9 shows which imaginal discs will give rise to which adult structures. During metamorphosis, most of the larval body parts histolyze, or break down, and the imaginal discs undergo mitosis and differentiate to form the head, thorax, and abdomen of the adult fly.

Genetic Analysis of Embryogenesis

Although our knowledge of the anatomy of development in *Drosophila* has been useful, genetic analysis has provided the most significant information about the events in embryogenesis. Genes controlling embryonic development are either **maternal-effect genes** or **zygotic genes**. Maternal-effect genes deposit products (mRNA and proteins) in the developing egg during oogenesis. Although these products may be distributed evenly throughout the egg cytoplasm, they are often distributed in a gradient or concentrated in specific regions of the egg. The phenotype of flies with mutations in maternal-effect genes is expected to exhibit female sterility, since none of the embryos of females homozygous for a recessive mutation would receive wild-type gene products from their mother, and therefore, they would not develop normally. In *Drosophila*, these maternal-effect genes encode transcription factors, receptors, and proteins that regulate translation. During embryogenesis, these gene products activate or repress expression of zygotic genes in a time/space matrix.

Zygotic genes are those expressed in the developing embryo, and are therefore transcribed after fertilization. The phenotype of mutations in this class show embryonic lethality. In a cross between two flies heterozygous for a recessive zygotic mutation, one-fourth of the embryos (the homozygotes) fail to develop. In *Drosophila*, many of the zygotic genes are transcribed in patterns that depend on the distribution of maternal-effect proteins.

Much of what is known about the zygotic genes that regulate development came from a mutagenic analysis performed by Christiane Nüsslein-Volhard and Eric Wieschaus, who systematically screened for mutations that affect embryonic development. They examined thousands of dead F_2 offspring of mutagenized flies, looking for recessive embryonic lethal mutations with defects in external structures. The parents were thus identified as carriers of these mutations, which fall into three classes: *gap, pair-rule*, and *segment polarity* genes. In a paper published in 1980, Nüsslein-Volhard and Wieschaus proposed a model in which embryonic development is initiated by gradients of maternal-effect gene products. The

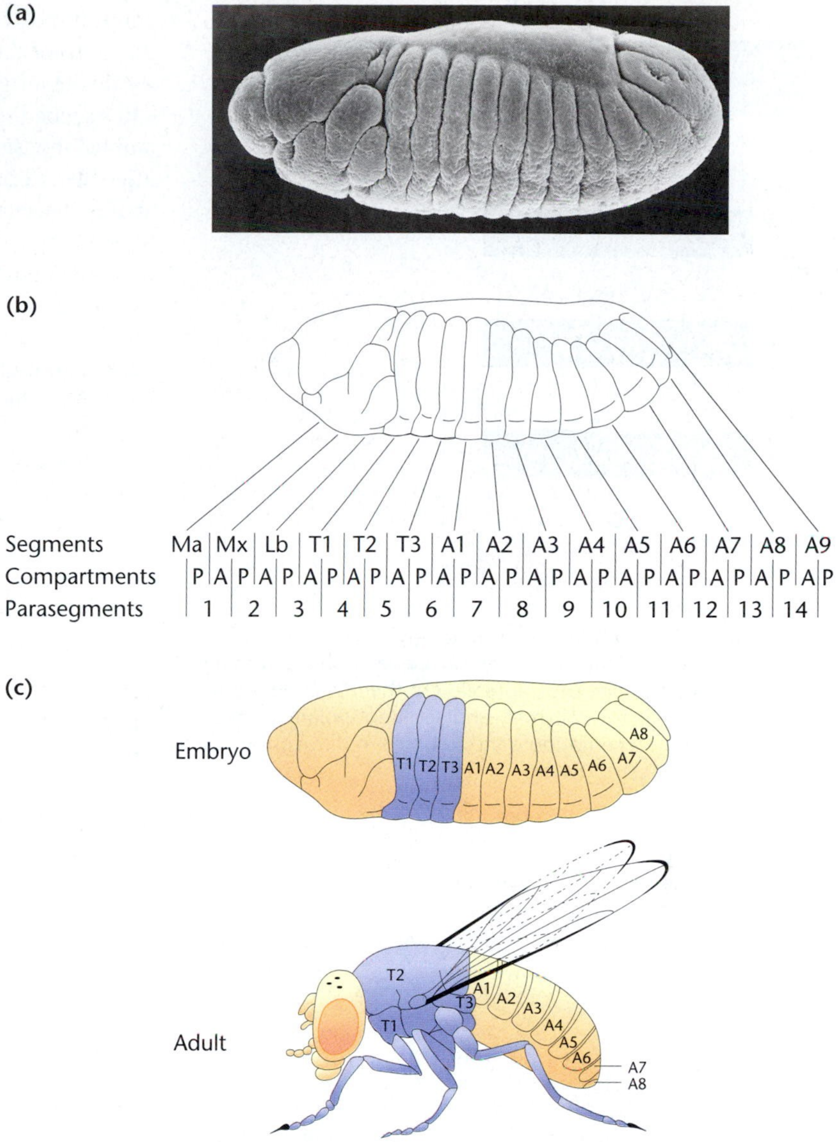

FIGURE 21–8 Segmentation in *Drosophila*. (a) Scanning electron micrograph of a *Drosophila* embryo at about 10 hours after fertilization. At this stage, the segmentation pattern of the body is clearly established. (b) The segments, compartments, and parasegments of the *Drosophila* embryo. Ma, Mx, and Lb represent the segments that will form head structures. T1-T3 are thoracic segments, and A1-A9 are abdominal segments. Each segment is divided into anterior (A) and posterior (P) compartments. Parasegments represent an early pattern specification that is later refined into the segmental plan of the body. Note that all the parasegments are shifted forward by one compartment to form the segments. (c) The segmented embryo and the adult structures that will form each segment.

positional information laid down by these molecular gradients along the anterior-posterior axis of the embryo is interpreted by two sets of zygotic genes: (1) those genes identified in their mutant screen (gap, pair-rule, and segment polarity genes), collectively called the **segmentation genes** (which divide the embryo into a series of stripes or segments and define the number, size, and polarity of each segment), and (2) **selector genes**, (which specify the identity or fate of each segment) (Figure 21–10).

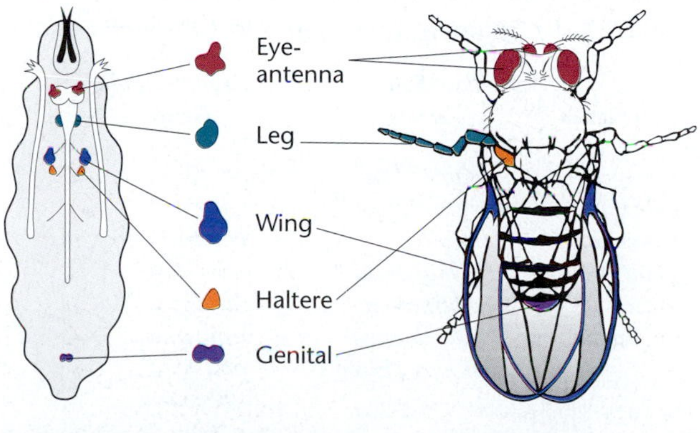

FIGURE 21–9 Imaginal discs of *Drosophila* larva and the adult structures derived from them.

Figure 21–11 shows the model developed by Nüsslein-Volhard and Wieschaus. The process begins when maternal-effect gene products (placed in the egg during oogenesis) are activated immediately after fertilization and form the anterior-posterior axis [Figure 21–11(a)]. Their activity restricts cells to forming either anterior or posterior structures. Gene products along this gradient activate transcription of the gap genes that divide the embryo into a limited number of broad regions [Figure 21–11(b)]. Gap proteins are transcription factors that activate pair-rule genes, whose products divide the embryo into smaller regions about two segments wide [Figure 21–11(c)]. The combined action of all gap genes defines segment borders. The pair-rule genes in turn activate the segment polarity genes, which then divide the segments into anterior and posterior compartments [Figure 21–11(d)]. The collective action of the maternal genes that form the

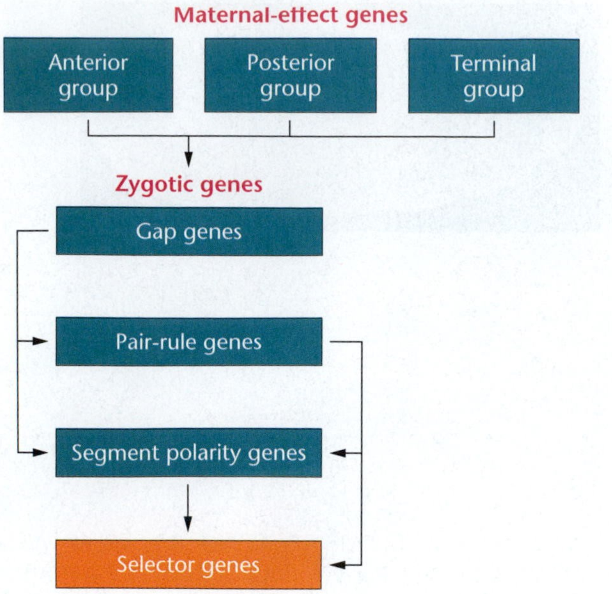

FIGURE 21–10 The hierarchy of genes involved in establishing the segmented body plan in *Drosophila*. Gene products from the maternal genes regulate the expression of the first three groups of zygotic genes (gap, pair-rule, and segment polarity; collectively called the segmentation genes), which in turn control expression of the selector genes.

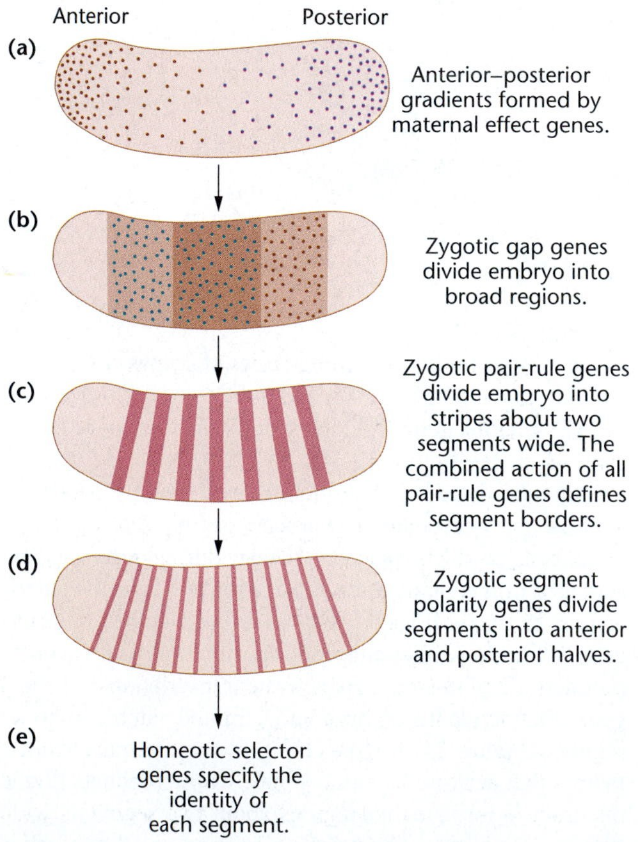

FIGURE 21–11 (a) Progressive restriction of cell fate during development in *Drosophila*. Gradients of maternal proteins are established along the anterior-posterior axis of the embryo. (b), (c), and (d) Three groups of segmentation genes progressively define the body segments. (e) Individual segments are given identity by the selector genes.

anterior–posterior axis and the segmentation genes define the fields of action for the homeotic selector genes that specify the identity of each segment [Figure 21–11(e)].

In a second screening, Wieschaus and Trudi Schüpbach examined thousands of flies for maternal-effect mutations that affected external embryonic structures. They estimated that about 40 maternal-effect genes and 50–60 zygotic genes regulate embryonic development in *Drosophila*. This means that only about 100 genes control normal embryogenesis, a surprisingly low number given the complexity of the structures formed from the fertilized egg. For their work on the genetic control of development in *Drosophila*, Nüsslein-Volhard, Wieschaus, and E. B. Lewis—the geneticist who initially identified and studied many of these genes in the 1970s—were awarded the 1995 Nobel Prize for Physiology or Medicine.

21.4 Zygotic Genes and Segment Formation

Zygotic genes are activated or repressed in a positional gradient by the maternal-effect gene products. Three subsets of these genes (gap, pair-rule, and segment polarity)—the segmentation genes—divide the embryo into a series of segments along the anterior-posterior axis. These segmentation genes are normally transcribed in the developing embryo, and mutations in these genes are seen as embryonic lethal phenotypes.

Over 20 segmentation loci have been identified (Table 21–2). They are classified on the basis of their mutant phenotypes: (1) gap genes delete a group of adjacent segments, (2) pair-rule genes affect every other segment, and (3) segment polarity genes cause defects in homologous portions of each segment.

Gap Genes

Transcription of **gap genes** is activated or inactivated by gene products previously expressed along the anterior-posterior axis and by other genes of the maternal gradient system. When gap genes are mutated, embryos have large gaps in their segmentation pattern. *hunchback* mutants lose head and

TABLE 21–2 Segmentation Genes in *Drosophila*

Gap Genes	*Pair-Rule Genes*	*Segment Polarity Genes*
Krüppel	*hairy*	*engrailed*
knirps	*even-skipped*	*wingless*
hunchback	*runt*	*cubitis interruptus^D*
giant	*fushi-tarazu*	*hedgehog*
tailless	*odd-paired*	*fused*
huckebein	*odd-skipped*	*armadillo*
	sloppy-paired	*patched*
		gooseberry
		paired
		naked
		disheveled

thorax structures, *Krüppel* mutants lose thoracic and abdominal structures, and mutations in *knirps* result in the loss of most abdominal structures. Gap-gene transcription divides the embryo into a series of broad regions (roughly, the head, thorax, and abdomen). Within these regions, different combinations of gene activity eventually specify both the type of segment that forms and the proper order of segments in the body of the larva, pupa, and adult. To date, all gap genes that have been cloned encode transcription factors with zinc finger DNA-binding motifs. The expression of the gap genes correlates roughly with their mutant phenotypes: *hunchback* at the anterior, *Krüppel* in the middle, and *knirps* at the posterior (Figure 21–12). Gap genes control the transcription of pair-rule genes.

Pair-Rule Genes

Pair-rule genes divide the broad regions established by gap genes into sections about one segment wide. Mutations in pair-rule genes eliminate segment-size sections at every other segment. The pair-rule genes are expressed in narrow bands or stripes of nuclei that extend around the circumference of the embryo. The expression of this gene set first establishes the boundaries of segments and then establishes the developmental fate of the cells within each segment by controlling the segment polarity genes. At least eight pair-rule genes act to divide the embryo into a series of stripes. However, the boundaries of these stripes overlap, meaning that cells in the stripes express different combinations of pair-rule genes in an overlapping fashion (Figure 21–13).

Many pair-rule genes encode transcription factors containing helix–turn–helix homeodomains. The transcription of the pair-rule genes is mediated by the action of gap-gene products, but the pattern is resolved into highly delineated stripes by the interaction among the gene products of the pair-rule genes themselves (Figure 21–14).

Segment Polarity Genes

Expression of **segment polarity genes** is controlled by transcription factors encoded by pair-rule genes. Within each segment, segment polarity genes become active in a single band of cells that extends around the embryo's circumference (Figure 21–15). This divides the embryo into 14 segments and the products of the segment polarity genes control the cellular identity within each segment. Some segment polarity genes, including *engrailed*, encode transcription factors. Rather than activating transcription, however, the engrailed protein competitively inhibits activation by other homeodomain proteins, resulting in the establishment of a segment border.

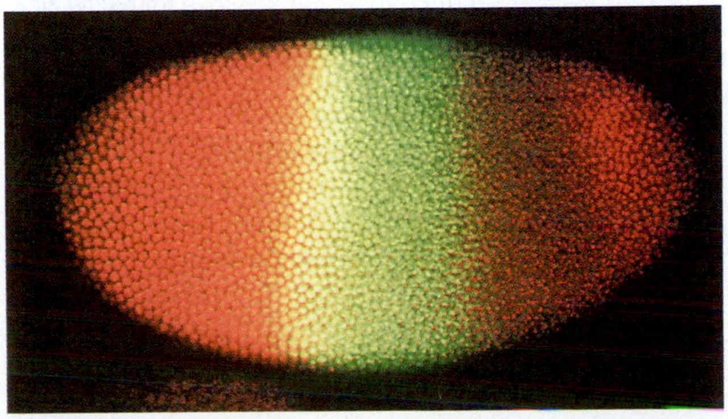

FIGURE 21–12 Expression of several gap genes in a *Drosophila* embryo. The hunchback protein is shown in orange and Krüppel is indicated in green. The yellow stripe is created when cells contain both hunchback and Krüppel proteins.

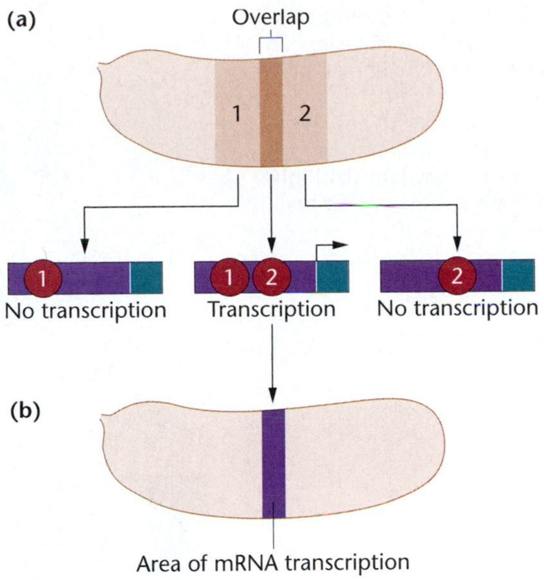

FIGURE 21–13 New patterns of gene expression can be generated by overlapping regions containing two different gene products. (a) Transcription factors 1 and 2 are present in an overlapping region of expression. If both transcription factors must bind to the promoter of a target gene to trigger expression, the gene will be active only in cells containing both factors (most likely in the zone of overlap). (b) The expression of the target gene in the restricted region of the embryo.

Selector Genes

As segment boundaries are established by action of the segmentation genes, **selector genes** are activated. These genes determine the adult structures to be formed by each body segment, including the antennae, mouth parts, legs, wings, thorax, and abdomen. Mutants of these genes are known as **homeotic mutants** (from the Greek word for "same"), because the identity of one segment is transformed so that it is the same as that of a neighboring segment. For example, the wild-type allele of the *Antennapedia (Antp)* gene specifies structures in the second thoracic segment (which carries a leg). Dominant gain-of-function *Antp* mutations lead to the expression of the gene in the head, transforming the antennae of mutant flies into legs (Figure 21–16).

Drosophila homeotic selector genes are found in two clusters on chromosome 3 (Table 21–3). The ***Antennapedia (Antp-C)* complex** contains five genes required for specifying structures in the head and first two thoracic segments (Figure 21–17). The ***bithorax (BX-C)* complex** contains three genes that specify structures in the posterior portion of the second thoracic segment, all of the third thoracic segment, and abdominal segments (Figure 21–17).

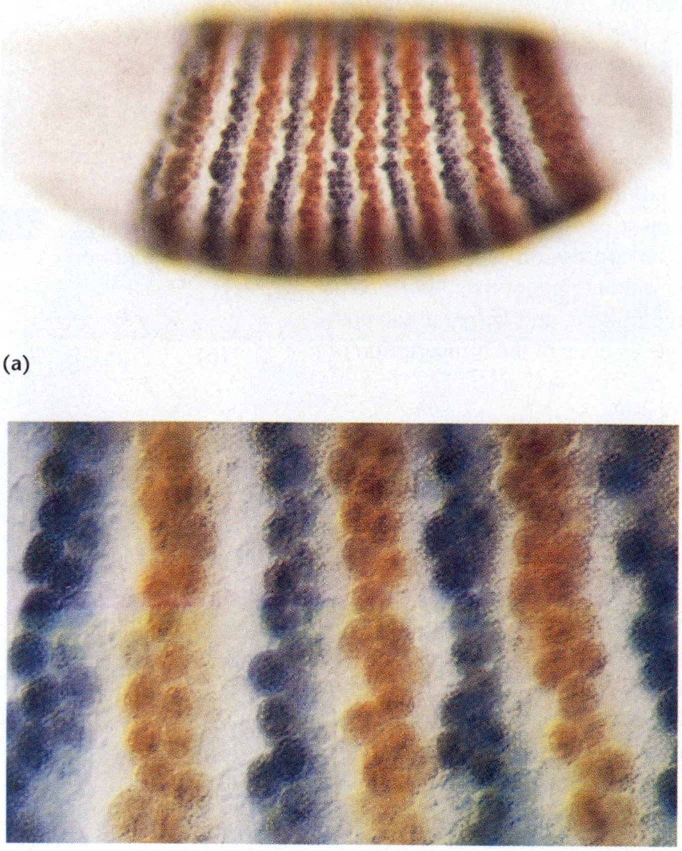

(a)

(b)

FIGURE 21–14 Stripe pattern of pair-rule gene expression in *Drosophila* embryo. This embryo is stained to show patterns of expression of the genes *even-skipped* and *fushi-tarazu*; (a) low-power view, and (b) high-power view of the same embryo.

Gap genes and pair-rule genes activate the selector genes. Although selector gene expression is confined to certain domains in the embryo, the timing and patterns of expression are rather complex and involve interactions between segmentation and selector genes, as well as interactions among the various selector genes.

A number of genes that control selector gene expression have been identified, including *extra sex combs (esc), Polycomb (Pc), supersex combs* (sxc)*, and *trithorax (trx)*. In the mutant *extra sex combs (esc)*, some of the head and all of the thoracic and abdominal segments develop as posterior segments (Figure 21–18), indicating that this gene normally controls the expression of *BX-C* genes in all body segments. The mutation does not affect either the number or the polarity of the segments. It does affect their developmental fate, indicating that the esc^{+} gene product stored in the egg by the maternal genome may be required to interpret the information gradient correctly in the egg cortex.

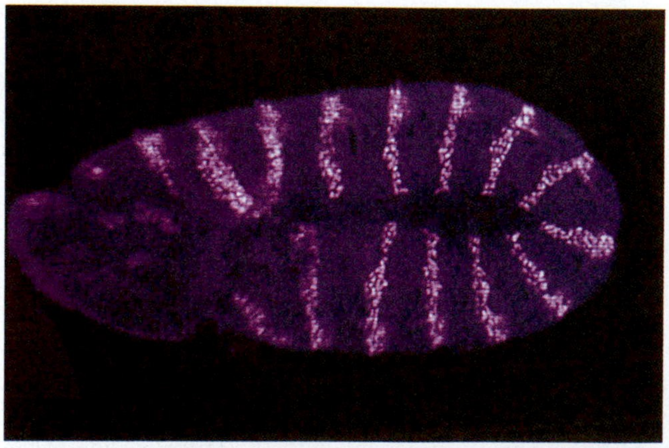

FIGURE 21–15 The 14 stripes of expression of the segment polarity gene *engrailed* in a *Drosophila* embryo.

Several genes that regulate selector gene expression, including the *Polycomb* genes, operate by binding to many sites along chromosomes, rather than just to the promoters and enhancers of individual genes. This may explain an intriguing observation about the expression pattern of selector genes. When embryos are stained to reveal mRNAs for each selector gene, the anterior borders for each gene's expression parallels its position on the chromosome. Genes at the 3′ end of the two gene clusters are expressed at the anterior end of the embryo, while in genes located toward the 5′ end of the clusters, the patterns of expression move toward the embryo's posterior end (Figure 21–19). In addition, expression patterns overlap in the more posterior regions, such that several genes are expressed in the same segments. These overlapping patterns of expression suggest two things: (1) Regulation of selector gene transcription is coordinated during development and, (2) the formation of each segment requires the activity of a particular combination of homeotic genes.

*The sex combs are a set of bristles found only on the legs of males.

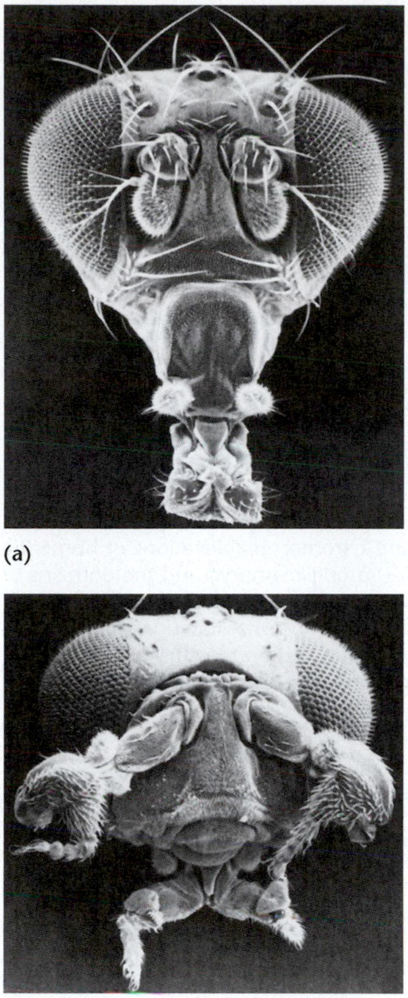

(a)

(b)

FIGURE 21–16 *Antennapedia (Antp)* mutation in *Drosophila.* (a) Head from wild-type *Drosophila*, showing the antenna and other head parts. (b) Head from an *Antp* mutant, showing the replacement of normal antenna structures with a leg. This is caused by activation of the *Antp* gene in the head region.

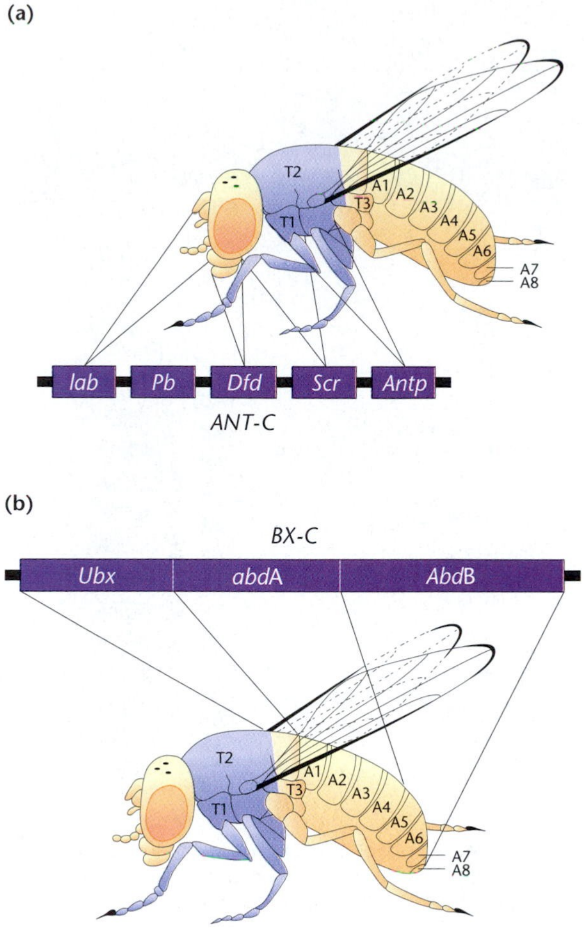

FIGURE 21–17 Genes of the *Antennapedia* complex and the adult structures they specify. The *labial (lab)* and *Deformed (Dfd)* genes control the formation of head segments. The *Sex comb reduced (Scr)* and *Antennapedia (Ant)* genes specify the identity of the first two thoracic segments. The remaining gene in the complex, *Proboscipedia (Pb)*, may not act during embryogenesis, but may be required to maintain the differentiated state in adults. In mutants, the labial palps are transformed into legs.

E. B. Lewis proposed that genes in the *bithorax* complex and other genes involved in segmentation arose from a common ancestral gene by tandem duplication and subsequent divergence of structure and function. In fact, each homeotic selector gene listed in Table 21–3 encodes a transcription factor that includes a DNA-binding domain encoded by a 180-bp sequence known as a **homeobox**. The homeobox encodes a 60-amino-acid sequence, the **homeodomain** (see Chapter 20). Similar sequences have been identified in the genomes of other eukaryotes with segmented-body plans, including *Xenopus*, chicken, mice, and humans. In mammals, the complexes of selector genes are called **Hox gene clusters** (Figure 21–20). Homeodomains from all organisms examined to date are very similar in amino acid sequence and encode a protein associated with the transcriptional regulation of a specific gene set. This suggests that the segmented body plan may have evolved only once.

TABLE 21–3 Homeotic Selector Genes of *Drosophila*

Antennapedia Complex	*Bithorax* Complex
labial	*Ultrabithorax*
Antennapedia	*Abdominal a*
Sex comb reduced	*Abdominal B*
Deformed	
proboscipedia	

To summarize, genes that control development in *Drosophila* act in a temporally and spatially ordered cascade, beginning with the genes that establish the anterior-posterior and dorsal-ventral axes of the egg and early embryo. Gradients of maternal mRNAs and proteins along the anterior-posterior axis activate the gap genes, which subdivide the embryo into broad bands. Gap genes in turn activate the pair-rule genes, which divide the embryo into segments. The final group of segmentation genes, the segment polarity genes, divides each segment into anterior and posterior regions arranged linearly along the anterior-posterior axis. These segments are then given identity by the selector genes. Such a

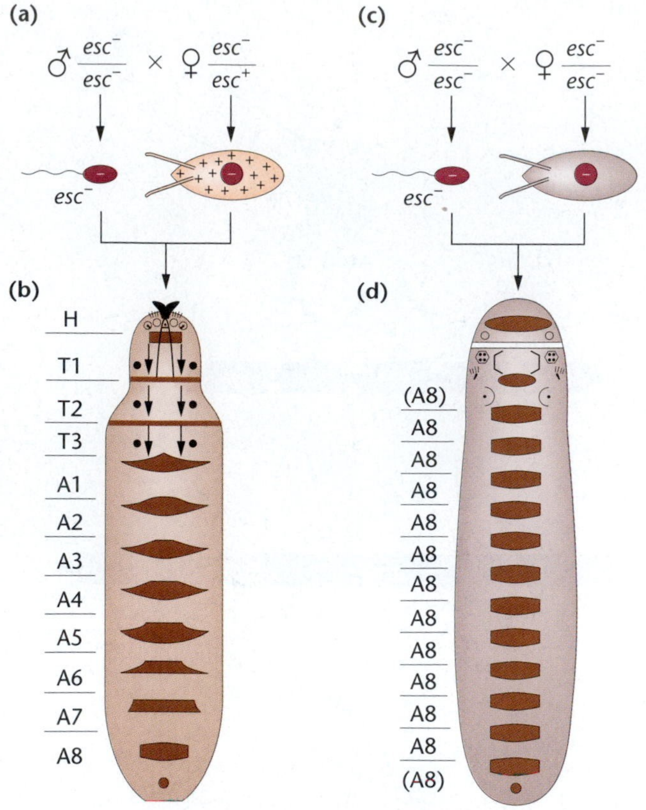

FIGURE 21–18 Action of the *extra sex combs* mutation in *Drosophila*. (a) Heterozygous females form wild-type *esc* gene product and store it in the oocyte. (b) When the esc^- egg formed at meiosis by the heterozygous female is fertilized by esc^- sperm, it produces wild-type larva with a normal segmentation pattern (maternal rescue). Borderlines between the head and thorax and between the thorax and abdomen are marked with arrows. (c) Homozygous esc^- females produce defective eggs, which, when fertilized by esc^- sperm, (d) produce a larva in which most of the segments of the head, thorax, and abdomen are transformed into the eighth abdominal segment. Maternal rescue demonstrates that the *esc* gene product is produced by the maternal genome and is stored in the oocyte for use in the embryo. (H, head; T1-T3, thoracic segments; A1-A8, abdominal segments).

progressive restriction of developmental potential in cells of the *Drosophila* embryo (all of which occurs during the first one-third of embryogenesis), involves a cascade of gene action triggered by localized cytoplasmic components stored by the maternal genome.

Scientists are now turning their attention to axis formation and pattern formation in vertebrate embryos. Christiane Nüsslein-Volhard and Wolfgang Driever carried out large-scale mutagenesis and screening for mutants in the zebrafish, a vertebrate model system. They isolated nearly 400 zygotic mutations that affect embryogenesis. Work is now underway to establish the pathways of regulation that mediate the development of zebrafish. These studies will allow scientists to compare the developmental programs in invertebrates and vertebrates and determine when, during evolution, pathways branched off and created the diverse array of species now present.

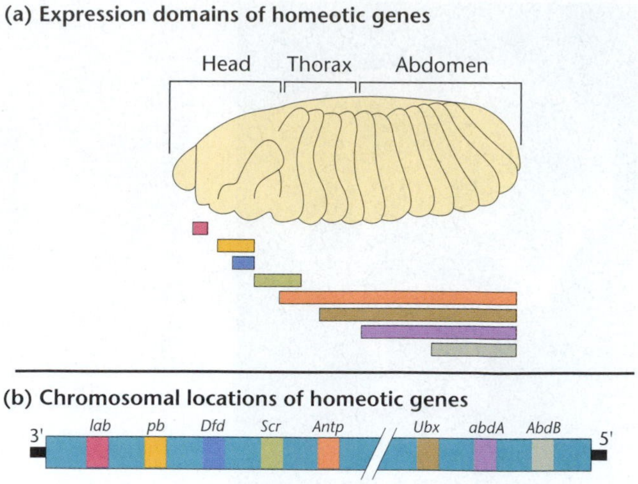

FIGURE 21–19 The colinear relationship between the spatial pattern of expression and chromosomal locations of homeotic genes in *Drosophila*. (a) *Drosophila* embryo and the domains of homeotic gene expression in the embryonic epidermis and central nervous system. (b) Chromosomal location of homeotic genes. Note that the order of genes on the chromosome correlates with the sequential anterior borders of their expression domains.

21.5 Flower Development in *Arabidopsis* Is Also Regulated by the Activation of Homeotic Selector Genes

Flower development in *Arabidopsis thalania* (Figure 21–21), a small plant in the mustard family, has been used to study pattern formation in plants. A cluster of undifferentiated cells, called the *floral meristem*, gives rise to flowers (Figure 21–22). Each flower consists of four organs: sepals, petals, stamens, and carpels that develop from concentric rings of cells within the meristem (Figure 21–23). Each organ develops from a different whorl. Three classes of floral homeotic genes control the development of these organs: Class A genes specify sepals, class A and class B genes specify petals, and class B and class C genes control stamen formation. Class C genes alone specify carpels [Figure 21–24(a)]. The genes in each class are listed in Table 21–4. Class A genes are active in the two outermost whorls (sepals and petals), class B are expressed in the second and third whorls (petals and stamens) and class C genes act in the third and fourth whorls (stamens and carpels). The organ formed depends on the expression pattern of the three gene classes. If only class A genes are expressed, sepals form. If class A *and* class B genes are expressed, petals form. Activity of class B *and* class C genes leads to stamen formation. If only class C genes are expressed, carpels form.

As in *Drosophila*, mutations in homeotic genes cause organs to form in abnormal locations. For example, in *AG* mutants (lacking activity of a C class gene), the order of organs is sepal, petal, petal, sepal, instead of the normal order, sepal, petal, stamen, and carpel [Figure 21–24(b)]. If both class A and class B gene activity are eliminated by mutation, then

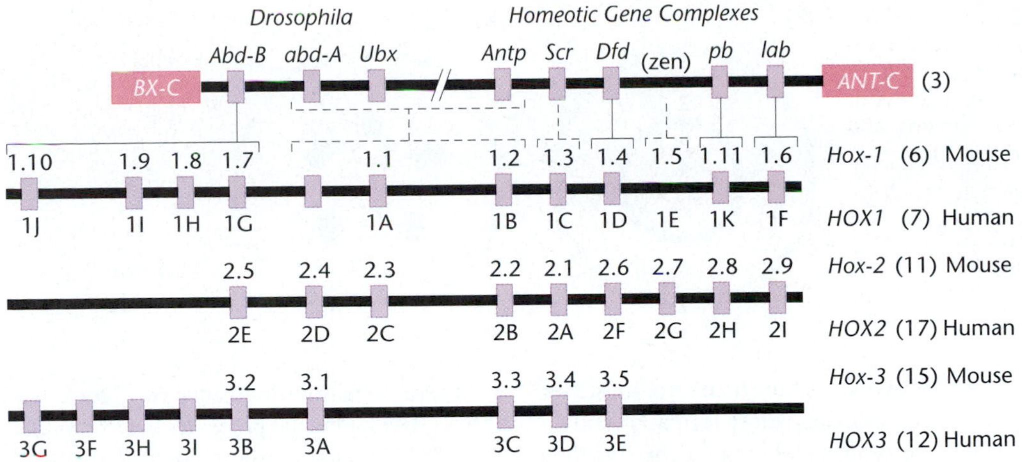

FIGURE 21–20 The *Hox* genes of mammals (human and mouse) and their organizational alignment with the *BX-C* and *ANT-C* complexes of *Drosophila*.

all whorls will have only class C genes active (which will spread to the first and second whorls, because class A is absent), and only carpels will form.

The evolutionary conservation of developmental processes and the homology between distantly related species is emphasized by the fact that the ABC gene classes of *Arabidopsis* are transcription factors, expressed in overlapping patterns, as are the homeotic genes in *Drosophila*. In *Arabidopsis*, however, the floral homeotic genes are members of a different family of transcription factors, called the **MADS-box proteins**. Each member of this family contains a common sequence of 58 amino acids. The action of the floral homeotic genes is controlled by a gene called *CURLY LEAF* that shares significant homology with members of the *Drosophila Polycomb* gene family. Thus, both plants and animals use the same mechanism (the activation of unique gene sets in each neighboring region of a particular structure) to determine its developmental pathway.

FIGURE 21–21 The flowering plant, *Arabidopsis thalania*, used as a model organism in plant genetics. *(Elliot M. Meyerowitz/California Institute of Technology, Division of Biology)*

21.6 Cell-Cell Interactions in *C. elegans* Development

During development in multicellular organisms, cells alter the transcriptional patterns and developmental fate of neighboring cells. These interactions can be short range or long range, and involve the generation and reception of signal molecules. **Cell–cell signaling** is an important process in the development of most eukaryotic organisms, including *Drosophila*, as well as vertebrates such as Xenopus, mice, and humans.

Signaling Systems in Development

In early development, metazoans use five signaling systems; after organogenesis begins, another five signal systems are added to those already in use. These systems act both independently and in coordinated networks to send and receive developmental signals that elicit specific transcriptional

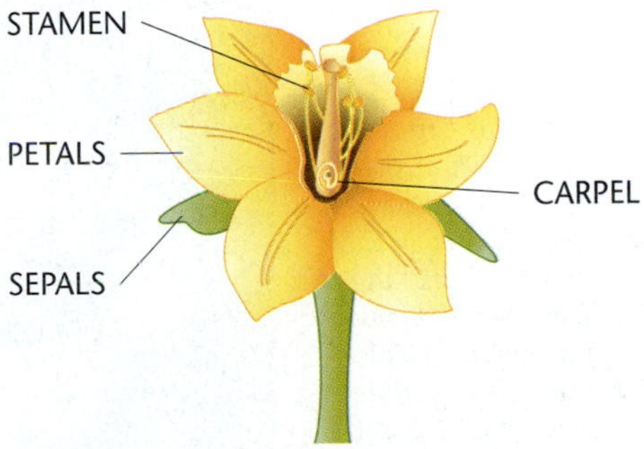

FIGURE 21–22 Parts of a flower. The floral organs are arranged concentrically. The sepals form the outermost ring, followed by petals and stamens, with carpels on the inside.

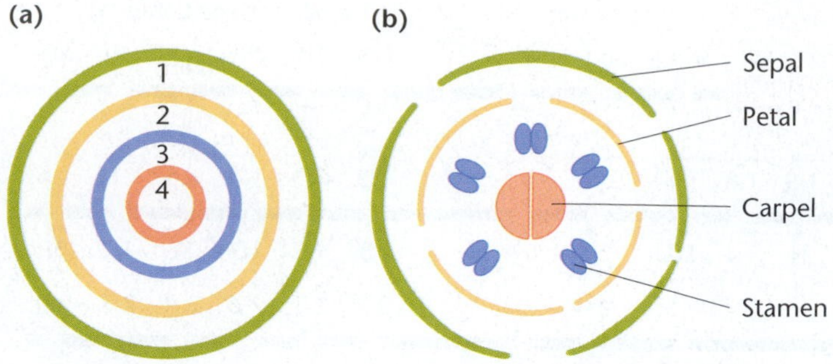

FIGURE 21–23 Cell arrangement in the floral meristem. (a) The four concentric rings or whorls, labeled 1–4, give rise to the sepals, petals, stamens, and carpels, respectively (b).

responses. The signal networks establish anterior–posterior polarity and body axes, coordinate pattern formation and direct the differentiation of tissues and organs. The signaling pathways used in early development and some of the developmental processes they control are listed in Table 21–5. After an introduction to the components and interactions of one of these systems—the Notch signaling pathway—we will briefly examine its role in the development of the vulva in the nematode, *Caenorhabditis elegans*.

The Notch Signaling Pathway.

The genes in the Notch pathway are named after the *Drosophila* mutants, in which the parts of the pathway were discovered. Notch is a short-range signaling system that works through direct cell–cell interactions to control the developmental fate of the interacting cells. The *Notch* gene (and the equivalent gene in other organisms) encodes a transmembrane protein that is a signal receptor (Figure 21–25). The signal is another transmembrane protein encoded by the *Delta* gene (and its equivalents). Because both the signal and receptor are membrane bound, the Notch signal system works only between adjacent cells. When the Delta protein binds to the Notch receptor, it triggers cleavage of the cytoplasmic tail of the Notch protein. This released fragment binds to the Su(H) (suppressor of Hairy) protein, and the protein complex moves from the cytoplasm into the nucleus. In the nucleus, the complex binds to transcriptional cofactors and activates transcription for a specific set of genes that control a developmental pathway.

Several variations of this pathway control a number of different developmental processes in *Drosophila*, thus establishing the dorsal–ventral boundary in the wing disk and directing the fate of cells in the nervous system, muscle, and gut. One of the main roles of the Notch signal system is to specify the fate of equivalent cells in a population. In its simplest form, such an interaction involves two neighboring cells that are developmentally equivalent. We will explore this role of the Notch signaling system in development of the vulva in *C. elegans*, after a brief introduction to nematode embryogenesis.

Overview of *C. elegans* Development

The nematode ***Caenorhabditis elegans*** is widely used to study the genetic control of development. This organism has several advantages for such studies: (1) The genetics of the organism are well known, (2) the genome sequence is available, and (3) adults are formed from a small number of cells that follow a developmental program which is unchanged from individual to individual. Adult nematodes are about 1 mm long and mature from a fertilized egg in about 2 days (Figure 21–26). The life cycle consists of an embryonic stage (about 16 hours), four larval stages (L1 through L4), and the adult stage. Adults are of two sexes: XX self-fertilizing hermaphrodites that can make both eggs and sperm and XO males. Self-crossing mutagen-treated hermaphrodites quickly results in homozygous stocks of mutant strains, and hundreds of mutants have been generated, catalogued, and mapped.

FIGURE 21–24 (a) Wild-type flowers of *Arabidopsis* have four petals, six stamens, a pistil fused from the two carpels, and sepals (not visible in this photo.) (b) Homeotic *agamous* mutants have petals and sepals at places where stamens and carpels should form. *(Left photo: Elliot M. Meyerowitz/ California Institute of Technology, Division of Biology; right photo: J. Bowman/Elliot M. Meyerowitz/ California Institute of Technology, Division of Biology)*

TABLE 21–4 Homeotic Selector Genes in *Arabidopsis*

Class A	*APETALA1 (AP1)**
	APETALA2 (AP2)
Class B	*APETALA3 (AP3)*
	PISTILLATA (PI)
Class C	*AGAMOUS (AG)*

*By convention, wild-type genes in *Arabidopsis* use capital letters

The adult hermaphrodite consists of 959 somatic cells (and about 2000 germ cells). The exact cell lineage from fertilized egg to adult has been mapped (Figure 21–27) and is invariant from individual to individual. With knowledge of the lineage of each cell, we can easily follow the events that result either from mutational alterations in cell fate or the killing of cells by laser microbeams or ultraviolet irradiation.

In *C. elegans* hermaphrodites, the fate of cells in the development of the reproductive system is determined by a cell–cell interaction, and insight is provided into how gene expression and cell-cell interactions are linked in the specification of developmental pathways.

Genetic Analysis of Vulva Formation

Adult hermaphrodites lay eggs through the **vulva**, an opening located about midbody (see Figure 21–26). The vulva is formed in stages during larval development and the process involves several rounds of cell-cell interactions.

During development in *C. elegans*, two neighboring cells, Z1.ppp and Z4.aaa, interact with each other so that one becomes the gonadal anchor cell and the other becomes a precursor to the ventral uterus. The determination of which is which comes during the second larval stage (L2) and is controlled by the *lin-12* gene, which encodes the Notch cell surface receptor protein. In recessive *lin-12(0)* mutants (a loss-of-function mutant), both cells become anchor cells. The dominant mutant *lin-12(d)* (a gain-of-function mutation) causes both to become uterine precursors. Thus, it appears that the expression of the *lin-12* gene causes the selection of the uterine pathway, since in the absence of the LIN-12 (Notch) receptor, both cells become anchor cells. However, as shown in Figure 21–28, the situation is more complex than it first appears. Initially, the neighboring cells are developmentally equivalent. Each cell displays low levels of both components: the signal protein for uterine development encoded by the *lag-2* (Delta)gene, and low levels of the LIN-12 (Notch) receptor. This situation is unstable, and both cells cannot remain as senders and receivers of the signal. At a critical time in L2, by chance, the cell producing more LAG-2 (Delta) signal causes its neighbor to increase transcription of the LIN-12 receptor. The cell with more LIN-12 receptor becomes the ventral uterine precursor cell, and the other cell, producing more LAG-2 signal, becomes the anchor cell. The critical factor in this first round of cell-cell interaction and

TABLE 21.5 Signaling Systems Used in Early Embryonic Development

Wnt Pathway
Dorsalization of body
Female reproductive development
Dorso–ventral differences
TGF-β Pathway
Mesoderm induction
Left–right asymmetry
Bone development
Hedgehog Pathway
Notochord induction
Somitogenesis
Gut/visceral mesoderm
Receptor Tyrosine Kinase Pathway
Mesoderm maintenance
Notch/Delta Pathway
Blood cell development
Neurogenesis
Retina development

Taken from Gerhart, J. 1999. 1998 Warkany lecture:Signaling pathways in development. *Teratology* 60:226–39.

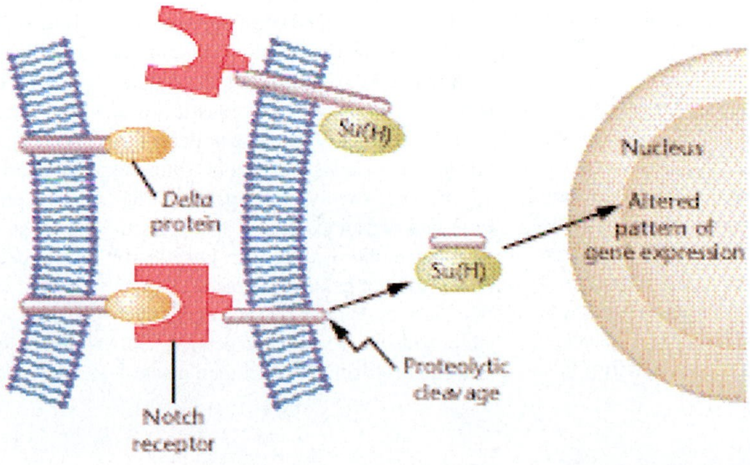

FIGURE 21–25 Components of the Notch signaling system. The cell carrying the Delta transmembrane protein is the sending cell; the cell carrying the transmembrane Notch protein receives the signal. Binding of Delta to Notch triggers a proteolytic-mediated activation of transcription. The fragment cleaved from the cytoplasmic side of the Notch protein combines with the Su(H) protein, moves to the nucleus where it activates a program of gene transcription.

Genetics, Technology, and Society

Stem Cell Wars

Stem cell research is at the center of a battle fought by scientists, politicians, advocacy groups, religious leaders and ethicists. Proponents fight for the right to carry out stem cell research, claiming that it is revolutionary, if not miraculous, — with the potential to cure diabetes, Parkinson's disease and spinal cord injuries, and improve the quality of life for millions. Critics lobby for an end to stem cell research, warning that it will propel us down the slippery slope toward disregard for human life. Although stem cell research is the focus of presidential proclamations, highly publicized media campaigns and legislative bans, few of us understand it sufficiently to evaluate its pros and cons. What is stem cell research and why does it spark such intense controversy?

Stems cells are primitive cells that replicate indefinitely and have the unique capacity to differentiate into cells with specialized functions, such as those found in heart, brain, liver and muscle. Stem cells are the origin of all the cells that make up the approximately 200 distinct types of tissues in our bodies. In contrast to stem cells, mature, fully-differentiated cells do not replicate or undergo transformations into different cell types. Some types of stem cells are defined as pluripotent, meaning that they have the ability to differentiate into any mature cell type in the body. Other types of stem cells are not pluripotent, and are able to differentiate into only one of several mature cell types.

In the last few years, several research teams have isolated and cultured human pluripotent stem cells. These cells remain undifferentiated and grow indefinitely in culture dishes. When treated with growth factors or hormones, these pluripotent stem cells differentiate into cells that have characteristics of neurons, bone, kidney, liver, heart or pancreatic cells.

The fact that pluripotent stem cells grow prolifically in culture and differentiate into more specialized cells has created great excitement. Some foresee a day when stem cells may be a cornucopia from which to harvest unlimited numbers of specialized cells to replace cells in damaged, diseased tissues. Hence, stem cells could be used to treat Parkinson's disease, type 1 diabetes, chronic heart disease, kidney and liver failure, Alzheimer, Duchenne muscular dystrophy, and spinal cord injuries. Some predict that stem cells will be genetically modified to eliminate transplant rejection, or to deliver specific gene products, thereby correcting genetic defects or treating cancers. The excitement about stem cell therapies has been fueled by reports of dramatically successful experiments in animals. For example, mice with spinal cord injuries regained their mobility, bowel and bladder control after they were injected with human stem cells. Both proponents and critics of stem cell research agree that stem cell therapies could be revolutionary. Why, then, should stem cell research be so contentious?

The answer to that question lies in the source of pluripotent stem cells. To date, all pluripotent stem cell lines have been derived from 5-day embryonic blastocysts. Blastocysts at this stage consist of about 200 cells, most of which will develop into placental and supporting tissues for the early embryo. The inner cell mass of the blastocyst consists of about 30-40 pluripotent stem cells that develop into all tissues of the embryo. *In vitro* fertilization clinics grow fertilized eggs to the 5-day blastocyst stage prior to implanting them into the uterine wall. Embryonic stem (ES) cell lines are created by dissecting out the inner cell mass of 5-day blastocysts and growing the undifferentiated cells in culture dishes. All human ES cell lines have been derived from unused 5-day blastocysts that were discarded by *in vitro* fertilization clinics.

The fact that early embryos are destroyed in the process of establishing human ES cell lines disturbs people who believe that preimplantation embryos are persons with rights; however, it does not disturb people who believe that these embryos are too primitive to have an inherent moral status. Both sides in the debate put forth lengthy arguments revolving around the fundamental question of what constitutes a human being. For example, a widely held philosophical view states that personhood requires a nervous system capable of cognition and consciousness. If true, preimplantation embryos (although genetically human) are not persons. Some argue that early blastocysts must be protected due to their potential to develop into persons. Others argue that, if this is true, then discarding extra blastocysts from *in vitro* fertilization clinics is wrong. In addition, the genetic potential of each cell in the body to form a cloned human may have equally symbolic meaning. Some critics predict that ES cell research will lead to therapeutic cloning, a technique involving transplanting a nucleus from a patient's somatic cell into an enucleated egg. The resulting ES cells are harvested from the blastocyst, the ES cells are grown and differentiated in culture and then transplanted back to the patient. They worry that therapeutic cloning will lead directly to human reproductive cloning, which is considered by many to be ethically wrong.

Critics of ES cell research argue that we may be able to benefit from stem cell therapies without resorting to the use of ES cells. This argument is based on recent reports about the plasticity of adult stem cells. Adult stem cells are undifferentiated cells that are present in differentiated tissues such as blood and brain. They divide within the differentiated tissue, and differentiate into mature cells that make up the tissue in which they are found. Adult stem cells have been found in bone marrow, blood, retina, brain, skeletal muscle, liver, skin, and pancreas. The best-known adult stem cells are hematopoietic stem cells (HSCs) which are found in bone marrow, peripheral blood and umbilical cords. HSCs differentiate into mature blood cell types such as red blood cells, lymphocytes and macrophages. HSCs have been used clinically for many years, as transplant material to reconstitute the immune systems of patients undergoing treatment for cancer and autoimmune diseases. Interestingly, recent studies suggest that adult stem cells from bone marrow, brain, fat and skin may have the capacity to differentiate into other cell types such liver, muscle, bone and neurons respectively.

Although these reports are intriguing, it is still too early to know whether adult stem cells will hold the same pluripotent promise as ES cells. Adult stem cells are rare, difficult to identify and isolate, and grow poorly, if at all, in culture. However, if these obstacles can be overcome, adult stem cells may provide an ethical alternative to ES cells, and calm the raging debate. On the other hand, it could simply create new philosophical dilemmas. If adult stem cells are found to exhibit the same pluripotency as ES cells, they could have the same potential to create a human embryo — dragging critics and proponents of stem cell research back into the same moral quagmire. At the present time, it is impossible to predict whether either adult or embryonic stem cells will be as miraculous as predicted by scientists and the popular press. But if stem cell research progresses at its current rapid pace, we won't have long to wait.

References

Robertson, J.A. 2001. Human embryonic stem cell research: ethical and legal issues. Nature Reviews Genetics 2:74-78.

Freed, C.R. 2002. Will embryonic stem cells be a useful source of dopamine neurons for transplant into patients with Parkinson's disease? Proc. Natl. Acad. Sci. USA 99:1755-1757.

Web Sites

National Institutes of Health. 2001. "Stem Cells: Scientific Progress and Future Research Directions". **http://www.nih.gov/news/stemcell/scireport.htm**

National Institutes of Health. 2000. "Stem Cells: A Primer". **http://www.nih.gov/news/stemcell/primer.htm**

determination is the balance between the LAG-2 (Delta) gene product and the LIN-12 (Notch) gene product.

A second round of cell-cell interactions in L3 involves the anchor cell (located in the gonad) and six precursor cells (located in the skin—hypodermis) adjacent to the gonad. The precursor cells are named P3.p, P4.p, P5.p, P6.p, P7.p, and P8.p and collectively are called Pn.p cells. The fate of each of the Pn.p cells is specified by its position relative to the anchor cell. The determination pathway is illustrated in Figure 21–29 and described in the subsequent paragraphs.

Sometime in L3, the *lin-3* gene is activated in the anchor cell. The secreted gene product, LIN-3, is structurally related to the vertebrate epidermal growth factor (EGF). All six Pn.p cells express a cell surface receptor encoded by the *let-23* gene, which is homologous to the vertebrate EGF receptor. The binding of the LIN-3 protein to the LET-23 receptor triggers an intracellular cascade of events that causes the Pn.p cells to become determined to form either vulval precursor cells or secondary vulvar cells. Expression of *let-23* is required to establish the primary and secondary fates of Pn.p cells. Recessive loss-of-function mutations in *let-23* cause all the Pn.p cells to develop as hypodermis (a tertiary fate). In *let-23* mutants, the Pn.p cells act as though they have not received a signal from the anchor cell, and no vulva is formed.

The signal transduction cascade from the anchor cells to the Pn.p cells also involves the *let-60* gene. Recessive mutations in *let-60* cause Pn.p cells to develop as though they have not received a signal from the anchor cell. Dominant

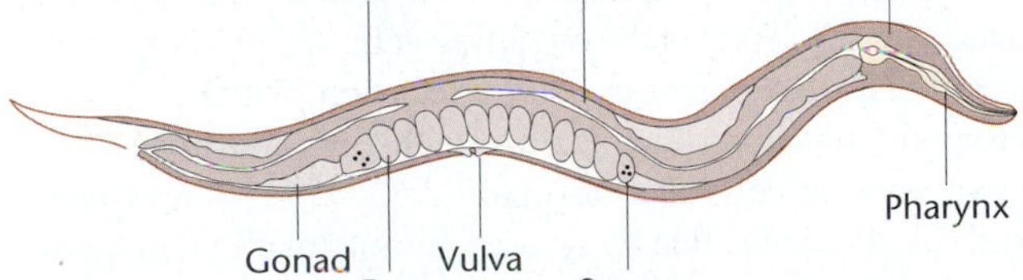

FIGURE 21–26 An adult *Caenorhabditis elegans*. This nematode, about 1 mm in length, consists of 959 cells and has been used to study many aspects of the genetic control of development.

let-60 alleles have the opposite phenotype: They cause all Pn.p cells to form vulval cells, resulting in the formation of multiple vulvas. The *let-60* gene has been cloned, and it is the *C. elegans* homolog of the *ras* gene, which in mammals acts downstream of receptor proteins in signal transduction. The *let-60* dominant gain-of-function mutant that causes multiple vulva formation has a Gly–Glu mutation at amino acid 13, the same mutation that converts *c-ras* to an oncogene. (see Chapter 22).

The data on the signal transduction cascade initiated by the binding of LIN-3 to LET-23 suggest that the Pnp cell (P6.p) closest to the anchor cell receives the strongest signal. In P6.p, expression of the *Vulvaless (Vul)* gene (the gene is named for its mutant phenotype) is activated. The cell then divides three times to produce vulva cells. The two neighboring Pnp cells (P5.p and P7.p) receive a lower amount of signal, which specifies a secondary fate—asymmetric division to form more vulva cells. To reinforce this determination, the primary vulval precursor activates *lin-12* in the two neighboring cells. This lateral inhibition signal prevents the neighboring secondary cells from adopting the division pattern of the primary cell. In other words, cells in which both *Vul* and *lin-12* are active cannot become primary vulva cells. The three remaining Pn.p cells receive no signal from the anchor cell. In these cells (P3.p, P4.p, and P8.p) the *Multivulva (Muv)* gene is activated. *Muv* represses *Vul*, and the three cells develop as hypodermal (skin) cells.

Thus, three levels of cell-cell interactions are required to specify the developmental pathway leading to vulva formation in *C. elegans*. First, two neighboring cells interact during L2 to establish the identity of the anchor cell. Second, in L3, the anchor cell interacts with three Pn.p cells to establish the primary vulvar precursor cell and two secondary cells. Third, the primary vulvar cell interacts with the secondary cells to suppress their selection of the primary vulvar pathway. Each of these interactions is accompanied by the production of molecular signals and by the reception and processing of these signals by neighboring cells.

This example of cell-cell interactions acting in a spatial and temporal cascade to specify the developmental fates of individual cells is a developmental theme repeated over and

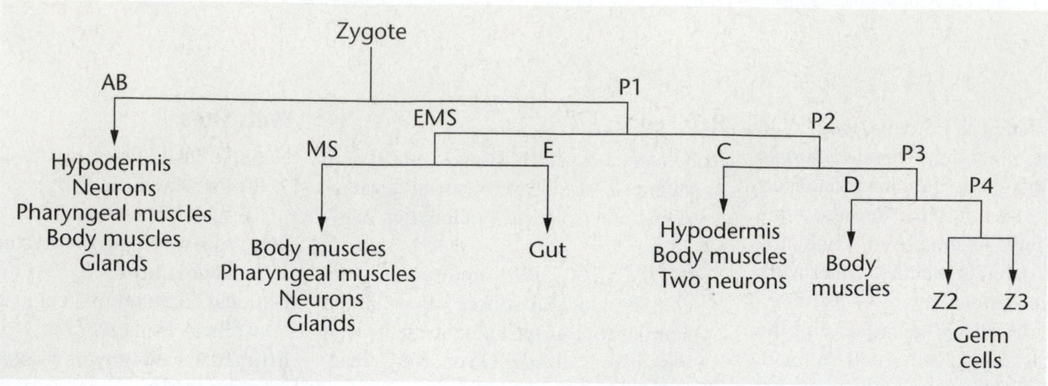

FIGURE 21–27 A truncated cell lineage chart for *C. elegans*, showing early divisions and the tissues and organs. Each vertical line represents a cell division, and horizontal lines connect the two cells produced. For example, the first cell division creates two new cells from the zygote, AB and P1. The cells in this chart refer to those present in the first-stage larva L1.During subsequent larval stages, further cell divisions will produce the 959 somatic cells of the adult hermaphrodite worm.

over in developing organisms from prokaryotes to higher vertebrates, including mice and humans.

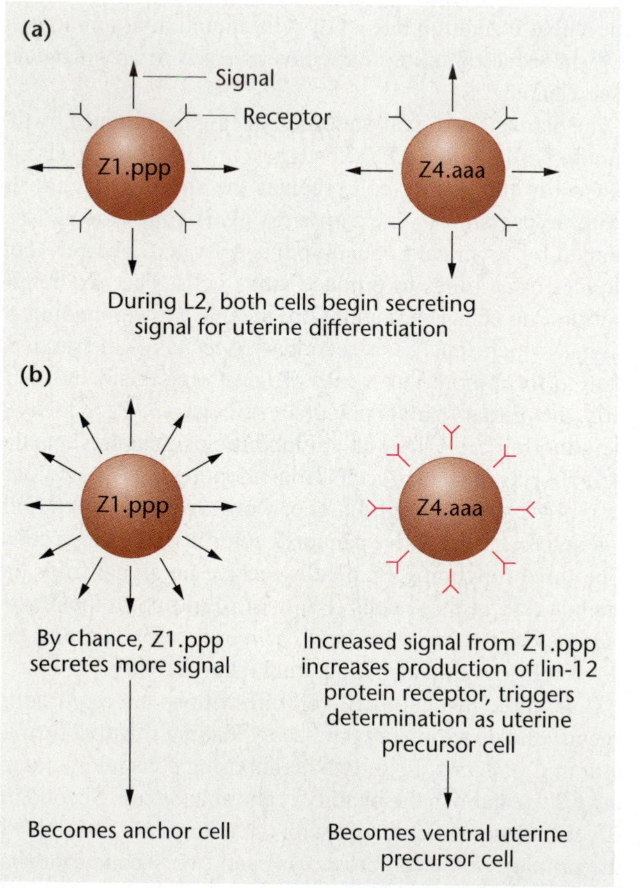

FIGURE 21–28 Cell-cell interaction in anchor cell determination. (a) During L2, two neighboring cells begin the secretion of chemical signals for the induction of uterine differentiation. (b) By chance, cell Z1.ppp secretes more of these signals, causing cell Z4.aaa to increase production of the receptor for signals. The action of increased signals causes Z4.aaa to become the ventral uterine precursor cell and allows Z1.ppp to become the anchor cell.

21.7 Programmed Cell Death Is Required for Normal Development

While there is now ample evidence that most cells retain a complete genome and can therefore theoretically be reprogrammed, there is at least one cellular fate that is irreversible: cell death. **Programmed cell death**, or **apoptosis**, is a normal process whereby cells undergo a characteristic series of events, such as chromosome condensation and blebbing of the plasma membrane, that leads to death. During development these programmed cell suicides contribute to the shaping and molding of tissues and organs. For example, the formation of digits in the vertebrate limb requires the death of the cells between the digits. In *C. elegans*, development relies on programmed cell death. The number of cells that die during the worm's development is always the same: 131 of 1090 in hermaphrodites and 147 of 1178 in males. In addition, the point in development at which a given cell dies and the identity of the cells that die are always the same.

Genetic analysis of mutants indicates that although programmed cell death occurs in cells of different developmental origins, all the cells die by using the same genetic pathway. In *C. elegans*, 15 genes are involved in cell death. These genes form four groups: (1) decision makers, (2) execution of the decision, (3) engulfment of dying cells, and (4) degradation of cell debris in the engulfing cells. Expression of *ced-3* and *ced-4* are necessary for execution of the cell death program; mutations that inactivate either of these genes result in survival of cells that normally die. Expression of *ced-3* and *ced-4* is controlled by *ced-9*. Gain-of-function mutations that cause constitutive expression or overexpression of *ced-9* prevent cell death. Conversely, loss-of-function mutants that inactivate *ced-9* cause embryonic lethality. Based on these observations, it has been concluded that *ced-9* works

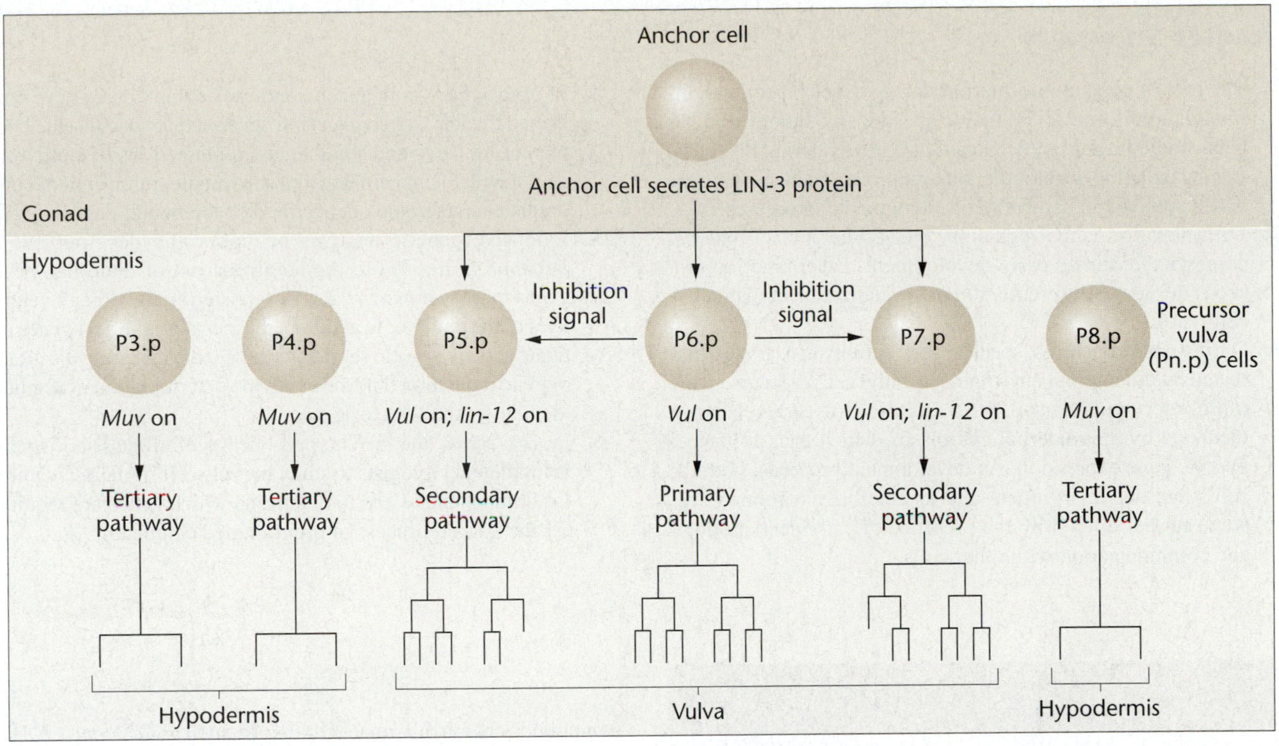

FIGURE 21–29 Cell lineage determination in *C. elegans* vulva formation. A signal from the anchor cell in the form of LIN-3 protein is received by three precursor vulval cells (Pn.p cells). The cells closest to the anchor cell become primary vulval precursor cells, and adjacent cells become secondary precursor cells. Primary cells secrete a signal that activates the *lin-12* gene in secondary cells, preventing them from becoming primary cells. Flanking Pn.p cells, which receive no signal from the anchor cell activity of the *Muv* gene, cause these cells to become hypodermis cells, instead of vulval cells.

by preventing expression of *ced-3* and *ced-4* in cells that survive (Figure 21–30). This means that *ced-9* is a *binary switch gene* for programmed cell death. Cells that express *ced-9* survive and those that do not, die.

The *ced-9* gene of *C. elegans* has a human homolog, *bcl-2*, a protooncogene that plays a role in programmed cell death in mammals. Overexpression of *bcl-2* prevents cell death in cells that would normally die. In humans, overexpression of *bcl-2* is found in follicular lymphoma, a form of cancer. Transfer of a cloned *bcl-2* gene into *ced-9* mutant *C. elegans* embryos prevents programmed cell death, indicating that nematodes and mammals share a common pathway for programmed cell death. Thus, the homology in molecules and mechanisms among species across the phylogenetic tree, illustrated throughout this chapter, extends to the molecules and mechanisms required for cells to die.

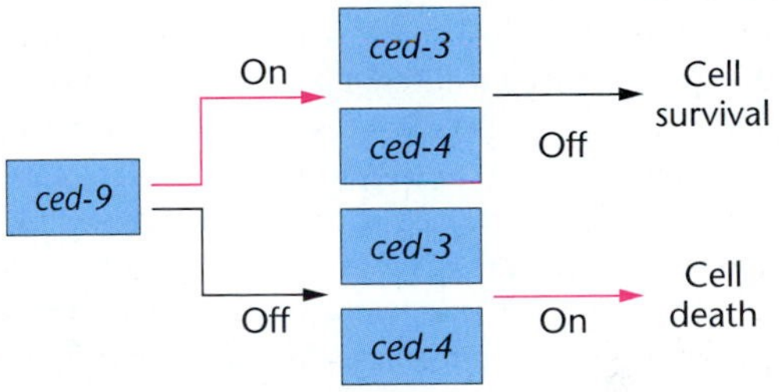

FIGURE 21–30 In the genetic pathway controlling cell death, the gene *ced-9* acts as a binary switch. If *ced-9* is active, it represses the expression of *ced-3* and *ced-4*, and the cell lives. If *ced-9* is inactive, *ced-3* and *ced-4* are expressed, and the cell dies.

Chapter Summary

1. The role of genetic information during development and differentiation is one of the major questions in biology and has been studied extensively. Geneticists are exploring this question by isolating developmental mutations and identifying the genes involved in controlling developmental processes.
2. Determination is the regulatory event whereby cell fate becomes fixed during early development. Determination precedes the actual differentiation or specialization of distinctive cell types.
3. During embryogenesis, specific gene activity appears to be affected by the internal environment of the cell, or localized cytoplasmic components. The regulation of early events is mediated by the maternal cytoplasm, which then influences zygotic gene expression. As development proceeds, both the cell's internal environment and its external environment become further altered by the presence of early gene products and communication with other cells.
4. In *Drosophila*, both genetic and molecular studies have confirmed that the egg contains information that specifies the body plan of the larva and adult and that interactions of embryonic nuclei with the maternal cytoplasm initiate transcriptional programs characteristic of specific developmental pathways.
5. Extensive genetic analysis of embryonic development in *Drosophila* has led to the identification of maternal-effect genes that lay down the anterior–posterior and dorsal–ventral axes of the embryo. In addition, these maternal-effect genes activate sets of zygotic segmentation genes, initiating a cascade of gene regulation that ends with the determination of segment identity by the selector genes.
6. In *C. elegans*, the stereotyped lineage of all cells allows developmental biologists to study the cell-cell signaling required for organogenesis and to determine which genes are required for the normal process of programmed cell death.

Insights and Solutions

1. In the slime mold *Dictyostelium*, experimental evidence suggests that cyclic AMP (cAMP) plays a central role in the developmental program leading to spore formation. The genes encoding the cAMP cell-surface receptor have been cloned, and the amino acid sequence of the protein components is known. To form reproductive structures, free-living individual cells aggregate together and then differentiate into one of two cell types, prespore cells or prestalk cells. Aggregating cells secrete waves or oscillations of cAMP to foster the aggregation of cells, and then continuously secrete cAMP to activate genes in the aggregated cells at later stages of development. It has been proposed that cAMP controls cell-cell interaction and gene expression. It is important to test this hypothesis by using several experimental techniques. What different approaches can you devise to test this hypothesis, and what specific experimental systems would you employ to test them?

Solution: Two of the most powerful forms of analysis in biology involve the use of biochemical analogues (or inhibitors) to block gene transcription or the action of gene products in a predictable way and the use of mutations to alter the gene and its products. These two approaches can be used to study the role of cAMP in the developmental program of *Dictyostelium*. First, compounds chemically related to cAMP, such as GTP and GDP, can be used to test whether they have any effect on the processes controlled by cAMP. In fact, both GTP and GDP lower the affinity of cell-surface receptors for cAMP, effectively blocking the action of cAMP. To inhibit the synthesis of the cAMP receptor, it is possible to construct a vector that contains a DNA sequence that transcribes an antisense RNA (a molecule that has a base sequence complementary to the mRNA). Antisense RNA forms a double-stranded structure with the mRNA, preventing it from being transcribed. If normal cells are transformed with a vector that expresses antisense RNA, no cAMP receptors will be produced. It is possible to predict that such cells will fail to respond to a gradient of cAMP and, consequently, will not migrate to an aggregation center. In fact, that is what happens. Such cells remain dispersed and nonmigratory in the presence of cAMP. Similarly, it is possible to determine whether this response to cAMP is necessary to trigger changes in the transcriptional program by assaying for the expression of developmentally regulated genes in cells expressing this antisense RNA.

Mutational analysis can be used to dissect components of the cAMP receptor system. One approach is to use transformation with wild-type genes to restore mutant function. Similarly, because the genes for the receptor proteins have been cloned, it is possible to construct mutants with known alterations in the component proteins and transform them into cells to assess their effects.

Problems and Discussion Questions

1. Carefully distinguish between the terms *differentiation* and *determination*. Which phenomenon occurs initially during development?
2. The *Drosophila* mutant *spineless aristapedia* (ss^a) results in the formation of a miniature tarsal structure (normally part of the leg) on the end of the antenna. This is a *homeotic* mutation. From your knowledge of imaginal discs, what insight is provided by ss^a concerning the role of genes during determination?
3. In the sea urchin, early development up to gastrulation may occur even in the presence of actinomycin D, which inhibits RNA synthesis. However, if actinomycin D is present early on but removed at the end of blastula formation, gastrulation does not proceed. In fact, if actinomycin D is present only between the 6th and 11th hours of development, gastrulation (normally occurring at the 15th hour) is arrested. What conclusions can be drawn concerning the role of gene transcription between hours 6 and 15?
4. How can you determine whether a particular gene is being transcribed in different cell types?
5. Observing that a particular gene is being transcribed during development, how can you tell whether the expression of this gene is under transcriptional or translational control?
6. In *Drosophila*, both *ftz* and *engrailed* encode homeobox transcription factors and are capable of eliciting the expression of other genes. Both genes work at about the same time during development and in the same region to specify cell fate in body segments. The question is: Does *ftz* regulate the expression of *engrailed*, or does *engrailed* regulate *ftz*? Or are they both regulated by another gene? To answer these questions, mutant analysis is performed. In ftz^- embryos *(ftz/ftz)*, engrailed protein is absent; in $engrailed^-$ embryos *(eng/eng)*, *ftz* expression is normal. What does this tell you about the regulation of these two genes? Does the *engrailed* gene regulate *ftz*? Does the *ftz* gene regulate *engrailed?*
7. Define what is meant by a homeotic mutant. If it were possible to introduce one of the homeotic genes in *Drosophila* into an *Arabidopsis* embryo that was homozygous for a homeotic flowering gene, would you expect any of the *Drosophila* genes to be able to rescue the *Arabidopsis* mutation? Why or why not?
8. Nuclei from almost any source may be injected into *Xenopus* oocytes. Studies have shown that these nuclei remain active in transcription and translation. How can such an experimental system be useful in developmental genetic studies?
9. The concept of epigenesis indicates that an organism develops by forming cells that acquire new structures and functions, which become greater in number and complexity as development proceeds. This theory is in contrast to the preformationist doctrine that miniature adult entities are contained in the egg that must merely unfold and grow to give rise to a mature organism. What sorts of isolated evidence presented in this chapter might have led to the preformation doctrine? Why is the epigenetic theory held as correct today?

Extra-Spicy Problems

10. In studying gene action during development, it is desirable to be able to position genes in a hierarchy or pathway of action, to establish which genes are primary and in what order genes act. There are several ways of doing this. One way is to make double mutants and study the outcome. The gene *fushi-tarazu (ftz)* is expressed in early embryos at the seven-stripe stage. All genes involved in forming the anterior-posterior pattern affect the expression of this gene, as do the gap genes. However, expression of segment-polarity genes is affected by *ftz*. What is the location of *ftz* in this hierarchy?
11. (a) The identification and characterization of genes that control sex determination has been another focus of investigators working with *C. elegans*. As with *Drosophila*, sex in this organism is determined by the ratio of X chromosomes to sets of autosomes. A diploid wild-type male has one X chromosome and a diploid wild-type hermaphrodite has two X chromosomes. Many different mutations have been identified that affect sex determination. Loss-of-function mutations in a gene called *her-1* cause an XO animal to develop into a hermaphrodite and have no effect on XX development. (That is, XX animals are normal hermaphrodites.) In contrast, loss-of-function mutations in a gene called *tra-1* cause an XX animal to develop into a male. From this information, deduce the roles of these genes in wild-type sex determination.

    (b) Based on the phenotypes of single- and double-mutant strains, a model for sex determination in *C. elegans* has been generated. This model proposes that the *her-1* gene controls sex determination by establishing the level of activity of the *tra-1* gene, which, in turn, controls the expression of genes involved in generating the various sexually dimorphic tissues. Given this information, does the *her-1* gene product have a negative or a positive effect on the activity of the *tra-1* gene? What would be the phenotype of a *tra-1, her-1* double mutant?

Selected Readings

Adams, J.M., and Cory, S. 2001. Life-or-death decisions by the Bcl-2 protein family. *Trends in Biochem.* 26:61–66.

Baker, N.E. 2001. Master regulatory genes; telling them what to do. *BioEssays* 23:763–766.

Briggs, R., and King, T. 1952. Transplantation of living nuclei from blastula cells into enucleated frog eggs. *Proc. Natl. Acad. Sci. USA* 38:455–63.

Davidson, E. 1976. *Gene activity in early development*. New York: Academic Press.

Driever, W., Thoma, G., and Nüsslein-Volhard, C. 1989. Determination of spatial domains of zygotic gene expression in the *Drosophila* embryo by the affinity of binding sites for the bicoid morphogen. *Nature* 340:363–67.

Duke, R.C., Ojcius, D.M., and Young, J.D. 1996. Cell suicide in health and disease. *Sci. Am.* (Dec.) 275:80–87.

Gehring, W. 1968. The stability of the differentiated state in cultures of imaginal disks in *Drosophila*. In *The stability of the differentiated state*, ed. H. Ursprung. New York: Springer-Verlag.

Gilbert, S. 1997. *Developmental biology*, 5th ed. Sunderland, MA: Sinauer Associates.

Goodrich, J., et al. 1997. A Polycomb-group gene regulates homeotic gene expression in *Arabidopsis. Nature* 386:44–51.

Gurdon, J.B. 1968. Transplanted nuclei and cell differentiation. *Sci. Am.* (Dec.) 219:24–35.

———. 1974. *The control of gene expression in animal development*. Cambridge, MA: Harvard U. Press.

Gurdon, J.B., Laskey, R.A., and Reeves, O.R. 1975. The developmental capacity of nuclei transplanted from keratinized skin cells of adult frogs. *J. Embryol. Exp. Morphol.* 34:93–112.

Halder, G., Callaerts, P., and Gehring, W.J. 1995. Induction of ectopic eyes by targeted expression of the *eyeless* gene in *Drosophila. Science* 267:1788–92.

Hempel, F.D., Welch, D.R., and Feldman, L.J. 2000. Floral induction and determination; where is flowering controlled? *Trends Plant Sci.* 5:17–21.

Hengartner, M.O. 1995. Life and death decisions: *ced-9* and programmed cell death in *Caenorhabditis elegans. Science* 270:931.

———. 1999. Programmed cell death in the nematode *C. elegans. Recent Prog. Horm. Res.* 54:213–22.

Kaufmann, T., Seeger, M., and Olson, G. 1990. Molecular organization of the *Antennapedia* gene cluster of *Drosophila melanogaster. Adv. Genet.* 27:309–62.

King, T.J. 1966. Nuclear transplantation in amphibia. In *Methods in cell physiology*, ed. D. Prescott, Vol. 2. Orlando, FL: Academic Press.

Kitzmann, M., and Fernandez, A. 2001. Crosstalk between cell cycle regulators and the myogenic factor MyoD in skeletal myoblasts. *Cell. Mol. Life Sci.* 58:571–579.

Koornneef, M., et al. 1998. Genetic control of flowering time in *Arabidopsis. Ann. Rev. Plant Mol. Biol.* 49:345–370.

Kumar, J.P., and Moses, K. 2001. EGF receptor and Notch signaling act upstream of eyeless/PAX6 to control eye specification. *Cell* 104:687–697.

Levine, M., Rubin, G., and Tjian, R. 1984. Human DNA sequences homologous to a protein coding region conserved between homeotic genes of *Drosophila. Cell* 38:667–73.

Lewis, E.B. 1976. A gene complex controlling segmentation in *Drosophila. Nature* 276:565–70.

———. 1994. Homeosis: The first 100 years. *Trends Genet.* 10:341–43.

Lockshin, R.A., Zakeri, A., and Tilly, J. 1998. *When cells die.* New York: John Wiley.

Ma., H. 1998. To be, or not to be, a flower - control of floral meristem identity. *Trends Genet.* 14: 26–32.

Meyerowitz, E. 1997. Plants and the logic of development. *Genetics* 145:5–9.

Mumm. J.S., and Kopan, R. 2000. Notch Signaling: From the outside in. *Develop. Biol.* 228:151–165.

Nüsslein-Volhard, C. 1994. Of flies and fishes. *Science* 266:572–74.

Nüsslein-Volhard, C., and Weischaus, E. 1980. Mutations affecting segment number and polarity in *Drosophila. Nature* 287:795–801.

Sathe, S.S., and Harte, P.J. 1995. *Drosophila* extra sex combs protein contains WD motifs essential for its function as a repressor of homeotic genes. *Mech. Dev.* 52:77–87.

Small, S., and Levine, M. 1991. The initiation of pair-rule stripes in the *Drosophila* blastoderm. *Curr. Opin. Genet. Dev.* 1:255–60.

Struhl, G. 1981. A gene product required for correct initiation of segmental determination in *Drosophila. Nature* 293:36–41.

Tabata, T. 2001. Genetics of morphogen gradients. *Nature Reviews Genet.* 2:620–630.

Tautz, D. 1996. Selector genes, polymorphisms, and evolution. *Science* 271:160–61.

Wang, M., and Sternberg, P.W. 2001. Pattern formation during *C. elegans* vulval induction. *Curr. Top. Dev. Biol.* 51:189–220.

Wei, Q., and Patterson, B.M. 2001. Regulation of MyoD function in the dividing myoblast. *FEBS Letters* 490:171–178.

Weigel, D., and Meyerowitz, E. 1994. The ABCs of floral homeotic genes. *Cell* 78:203–9.

Wieschaus, E. 1996. Embryonic transcription and the control of developmental pathways. *Genetics* 142:5–10.

Wilmut, I., et al. 1997. Viable offspring from fetal and adult mammalian cells. *Nature* 385:810–13.

GENETICS MediaLab

The resources that follow will help you achieve a better understanding of the concepts presented in this chapter. These resources can be found either on the CD packaged with this textbook or on the Companion Web found at **http://www.prenhall.com/klug**

Web Problem 1:

Time for completion = 10 minutes

Does cellular differentiation alter the genetic material within the nucleus such that totipotency is lost? Dolly, the sheep cloned in a Scottish genetics laboratory, was the first of several successful mammalian cloning experiments which suggest that the answer is no. After reading the article "Virtual Embryo" about mammalian cloning, you should be able to answer the following questions: Is the differentiation of a particular cell most likely a result of changes in the nucleus or changes in the cytoplasm? Are cells that are cloned by injecting a nucleus into an enucleated egg truly clones of the donor cells? What, if any, genetic components of the cell might be different between the clone and the donor? What are some important implications of cloning for human medical treatment? To complete this exercise, visit Web Problem 1 in Chapter 21 of your Companion Web site, and select the keyword **CLONING**.

Web Problem 2:

Time for completion = 10 minutes

What combination of technology and ingenuity allows certain scientists to make critical breakthroughs in understanding how organisms function? The German scientists Christiane Nüsslein-Volhard and Eric Wieschaus won the 1995 Nobel prize in medicine for discovering the role of genetics in the development of the *Drosophila* embryo. While reading and thinking about Dr. Nüsslein-Volhard's training and achievements, pay particular attention to the critical factors that led her to the discovery. Did the discovery of these genes depend on the development of sophisticated molecular techniques or on the ingenuity of the investigators? How did Nüsslein-Volhard decide to study *Drosophila* embyrogenesis? Were the events leading her to the field more a result of careful planning, or were they serendipitous? Were Nüsslein-Volhard's major collaborators in the first part of this research her professors or her fellow students? The article presents two views of the particular challenges facing women in science. What do you think of Dr. Nüsslein-Volhard's approach? To complete this exercise, visit Web Problem 2 in Chapter 21 of your Companion Web site, and select the keyword **NÜSSLEIN-VOLHARD**.

Web Problem 3:

Time for completion = 15 minutes

How can the expression of a single homeotic gene cause the differentiation of different appendages in different body segments? The homeotic genes discovered in *Drosophila* have homologues in many other taxa, including plants and mammals, and appear to regulate the differentiation of various body segments. For example, the *Ultrabithorax* (*Ubx*) locus regulates the expression of thoracic appendages in *Drosophila* and other insects. In this exercise, you will read an article on *Ubx,* use information in Chapter 21 in order to answer the following questions: What is the function of the proteins encoded by *Ubx*? Flies (order Diptera) such as *Drosophila* have one pair of wings, whereas other insect orders, such as butterflies and dragonflies, have two pairs. Is the suppression of wing development on the third thoracic segment in Diptera associated with the suppression of expression of *Ubx*? How does the expression of this gene differ between flies and butterflies? A single RNA transcript of *Ubx* appears to serve different roles in different tissues. How is the transcript modified in different tissues? To complete this exercise, visit Web Problem 3 in Chapter 21 of your Companion Web site, and select the keyword **HOMEOTIC**.

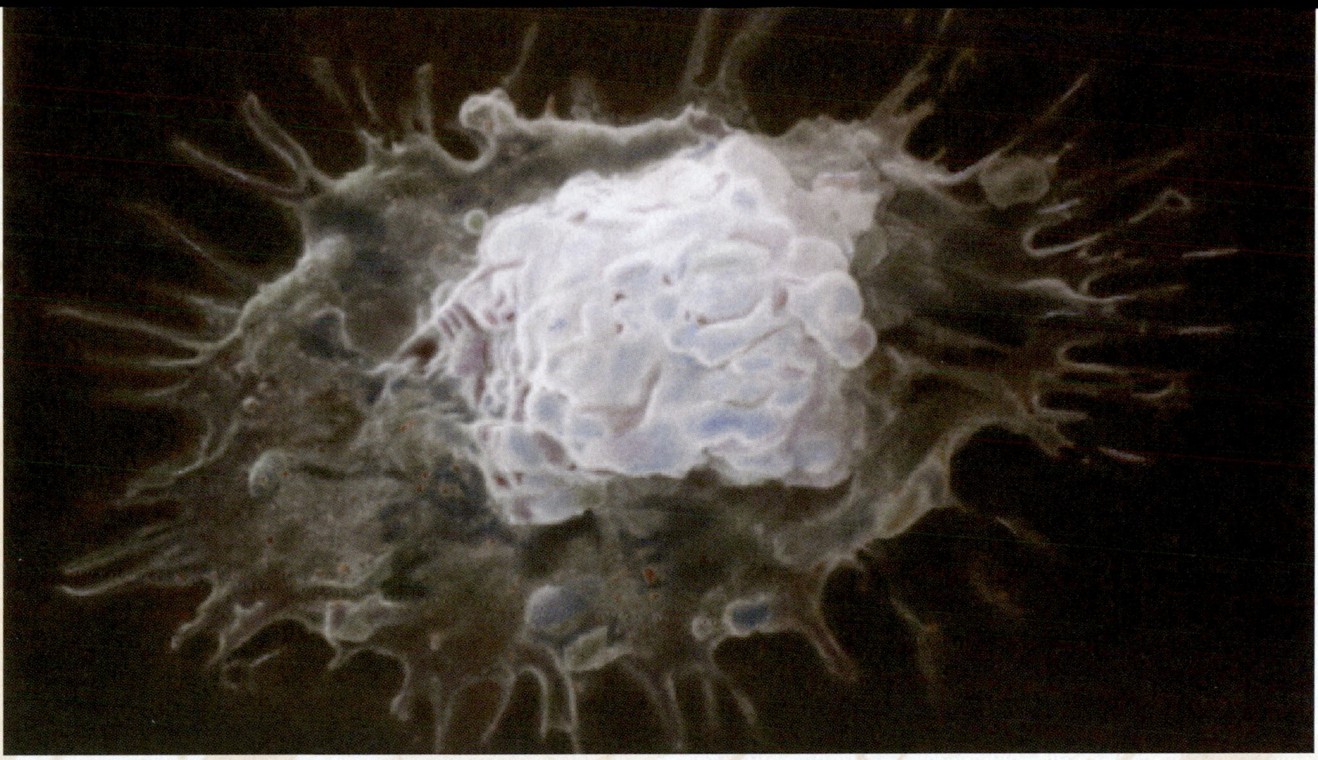

A human breast cancer cell. (*AMC/Albany Medical College/Custom Medical Stock Photo, Inc.*)

22
Genetics and Cancer

Although often viewed as a single disease, cancer is actually a complex group of diseases affecting a wide range of cells and tissues. A genetic link to cancer was first proposed early in the 20th century, and this idea has served as one of the foundations of cancer research. Mutations that alter gene expression are now regarded as a common feature of all cancers. In most cancers, mutations arise in somatic cells and are not passed on to future generations through the germ cells. However, in about 1 percent of all cancer cases, germ-line mutations are transmitted to offspring and are responsible for susceptibility to cancer. Although the frequency of these cases is low, these mutations have provided insights into the origins of cancer. For cancer to occur, the presence of inherited mutation is insufficient and must be accompanied by an additional somatic mutation at the homologous locus, creating a homozygous mutant condition. *Cancer is now considered a genetic disorder at the cellular level.*

Mutations associated with cancer can involve small-scale changes, such as a single nucleotide substitution, or large-scale events, including chromosome rearrangement, chromosome gain or loss, or even the integration of viral genomes into chromosomal sites. Large-scale genomic alterations are a common feature of cancer; the majority of human tumors are characterized by visible chromosomal changes. Some of these chromosomal changes, particularly in leukemia, are so characteristic that they can be used to diagnose the disorder and make accurate predictions about the severity and course of the disease.

That cancer runs in families has been known for over 200 years. Most often, no clear-cut pattern of inheritance can be determined for these familial forms. This is primarily because affected individuals in these families inherit only one mutant allele, which predisposes them to cancer. In such cases, the likelihood that an individual will ultimately develop cancer depends on several factors: which gene is mutated, mutations in other genes, and environmental factors. These variables may influence the age of onset and the severity of the disease. From studies of familial cancers, we can identify a class of genes called **cancer susceptibility genes** that increase the risk of cancer. Mutant alleles of these cancer susceptibility genes have an important role in sporadic cancers as well as familial forms of cancer.

Because of the background rate of spontaneous mutation, there will always be a baseline rate of cancer. Over and above this baseline rate, environmental agents that promote mutation can also contribute to cancer rates. Almost all known environmental **carcinogens** (cancer-causing agents), such as ionizing radiation, chemicals, and viruses, act by generating mutations. Given that mutations play such a central role in cancer, in this chapter, we explore questions about how mutations convert normal cells into malignant tumors, which mutant genes are most likely to result in cancer, and how many mutations are required to cause cancer.

To answer these questions, let's consider the properties of cancer cells that distinguish them from normal cells and ask what genes control these properties. Cancer cells have two properties in common: (1) uncontrolled growth and (2) the ability to **metastasize** or spread, from their original site to other locations in the body. Cell division is the result of cells traversing the cell cycle; in cancer cells, control over the cell cycle is lost, and cells proliferate rapidly. Investigations into the genetic regulation of the cell cycle are providing many insights into the mechanisms of cancer.

Gene products localized on the cell surface control metastasis of cancer cells. The genetics of metastasis is related to an understanding of how cells interact with the extracellular matrix and with neighboring cells through cell surface molecules. Though this field is less well developed than the study of the cell cycle, it is beginning to provide some insights into the secondary events in tumor progression.

In this chapter, we consider the interaction between genes and cancer, with emphasis on the cell cycle and genetic disorders associated with cancer. We will also examine the relationship between mutation and cancer, the relationship between chromosomal changes and cancer, and the role of environmental agents in the genesis of cancer.

22.1 Cell Cycle Regulation Is Closely Related to the Genetics of Cancer

The cell cycle represents the sequence of events occurring between mitotic divisions in a eukaryotic cell. Because this cycle is closely related to the genetics of cancer, we first discuss what we currently know about events in the cell cycle and the genes that regulate progression through the cycle.

The Cell Cycle

As outlined in Chapter 8, the cell cycle progresses from chromosomal DNA replication (S phase) to the segregation of chromosomes into two nuclei during mitosis (M phase). Interspersed between these stages are two gaps, called G1 and G2. Together, G1, S, and G2 make up interphase (Figure 22–1). G1 begins after mitosis; the synthesis of many cytoplasmic elements, including ribosomes, enzymes, and membrane-derived organelles, occurs at this time. In S, DNA replication produces a copy of each chromosome. Then the second period of growth and synthesis—the G2 phase—occurs as a prelude to mitosis.

While the cycles of some cells, such as those in human skin is continuous, other cell types, including many nerve cells, withdraw from G1 and permanently enter a nondividing state known as G0. Still, other cell types such as white blood cells can be recruited from G0 and reenter the cycle. Taken together, these observations suggest that the cell cycle is tightly regulated and is dependent on a cell's life history and its differentiated state. We next look at what we currently know about the genetic regulation of the cell cycle.

Cell Cycle Checkpoints and Control of the Cycle

Much of the basic work on the cell cycle has been conducted mainly by two groups—geneticists working with yeasts, especially *Saccharomyces cerevisiae* and *Schizosaccharomyces pombe*; and developmental biologists studying

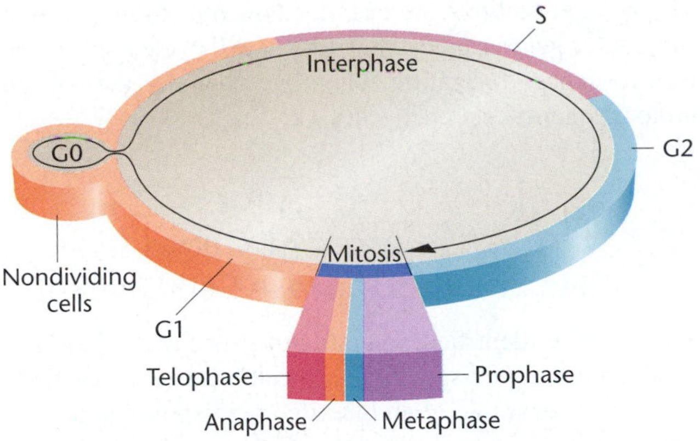

FIGURE 22–1 The cell cycle is controlled at several checkpoints, including one at the G2/M transition, and another in late G1 phase before entry into S phase. These checkpoints involve the interaction between transitory proteins, called cyclins, and kinases that add phosphate groups to proteins. Phosphorylation of target proteins triggers a cascade of events allowing progress through the cell cycle.

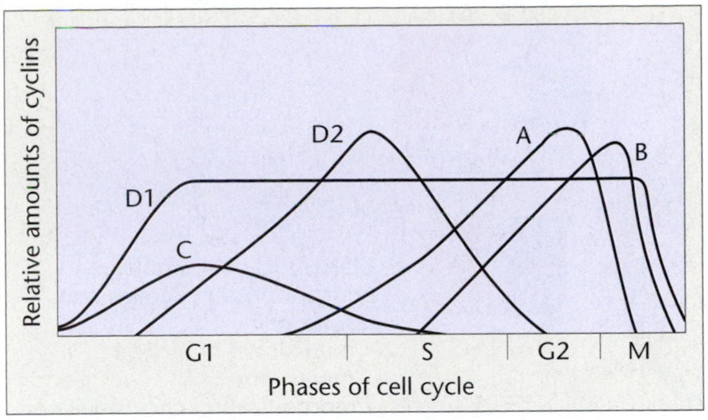

FIGURE 22–2 Relative expression times and amounts of cyclins during the cell cycle. D1 accumulates early in G1 and is expressed at a constant level through most of the cycle. Cyclin C accumulates in G1, reaches a peak, and declines by mid-S phase. Cyclin D2 begins accumulating in the last half of G1, reaches a peak just after the beginning of S, and then declines by early G2. Cyclin A appears in late G1, accumulates through S, reaches a peak in G2, and is degraded rapidly as M phase begins. Cyclin B appears in mid-S phase, peaks at the G2/M transition, and is rapidly degraded.

newly fertilized eggs of organisms such as frogs, sea urchins, and newts. Both groups have succeeded in identifying and characterizing genes involved in the cell cycle, and their work is now converging and overlapping with important areas of cancer biology, particularly studies on growth factors and the genes that suppress or promote tumor formation. In recognition of this work, the 2001 Nobel Prize for Physiology or Medicine was awarded to Lee Hartwell, Tim Hunt, and Paul Nurse for their work on the cell cycle.

Although details of all the molecular events or even the exact number and sequence of steps are not yet known, it appears that all eukaryotic cells employ a common series of biochemical pathways to regulate events in the cell cycle. This information can be used to understand and predict events in normal human cells and in mutated cells that have become cancerous.

The cell cycle is regulated at several points: one, a point near the G1/S transition and another near the G2/M transition. (See Figure 22–1.) At both points, a decision is made to proceed or halt progression through the cycle. This decision is controlled by physical interaction between proteins belonging to two different classes. One consists of enzymes called **protein kinases** that selectively phosphorylate target proteins. Although many different kinases are present in the cell, only a few, called **cyclin-dependent kinases (CDKs)** are involved in the regulation of the cell cycle. The second class is proteins called **cyclins**, which are involved in controlling the progression through the cell cycle. First identified in the embryos of developing invertebrates, cyclins are synthesized and degraded in a synchronous pattern related to stages of the cell cycle (Figure 22–2). Altogether, almost a dozen different cyclins have been identified, and a growing number of CDKs are being described, indicating that multiple checkpoints in the cell cycle exist, or that these molecules have multiple functions.

Physical interaction between kinases and cyclins produces a regulatory molecule that controls the cell's movement through the cycle. Different CDKs control the initiation of the S and M phases. At the G1/S control point, the CDK4 kinase binds to a cyclin D molecule and activates the transcription of a set of genes required for S phase. The onset of mitosis (M phase) in most cells is controlled by CDK1, originally isolated from maturing amphibian eggs and genetically identified in yeast as the product of the *cdc2* gene. This protein has a highly conserved sequence in all eukaryotes examined; its function is necessary for entry into M phase.

Several events mark the entry from G2 into mitosis (M), including the condensation of chromatin into chromosomes, breakdown of the nuclear membrane, and reorganization of the cytoskeleton. Major events in this transition are regulated by the formation of active CDK1/cyclin B complexes. When bound to cyclin B, CDK1 catalyzes phosphorylation of a cytoplasmic protein, caldesmon, which brings about nuclear membrane breakdown and rearrangement of the cytoskeleton. CDK1 also phosphorylates histone H1, which may play a role in chromatin condensation (Figure 22–3). Although experimental results suggest that cyclin A is involved in the progression from G2 to M, its functions are not clearly understood.

Once mitosis begins, there is an internal checkpoint called the **M checkpoint**. Proteins that control this checkpoint regulate entry into anaphase. When active, these checkpoint proteins interfere with assembly of the spindle fibers and their attachment to the kinetochore. A number of kinases and other proteins have been implicated in this spindle-assembly checkpoint.

Mutations that disrupt any step in cell cycle regulation are candidates for cancer-causing genes. Evidence is accumulating that the G1 checkpoint is aberrant in many forms of cancer, and mutant G1 cyclins and kinases are thought to be the best candidates for cancer-promoting genes. Recently, a form of cyclin D called D1 has been shown to be identical to an overexpressed protein found in certain forms of leukemia.

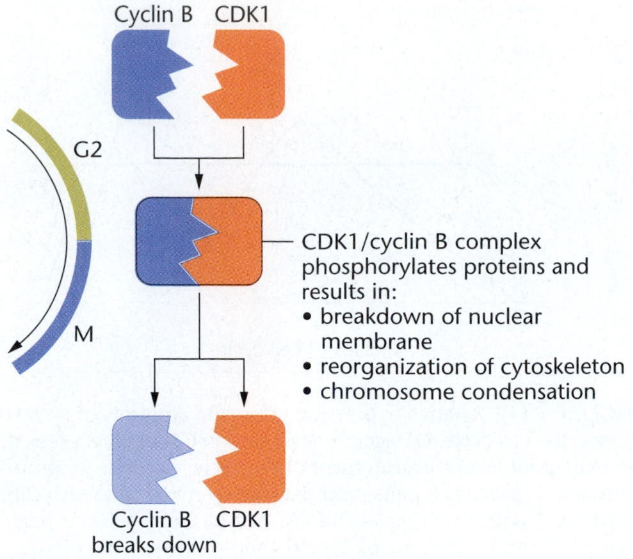

FIGURE 22–3 Transition from G2 to M is controlled by CDK1 and cyclin B. These molecules interact to form a complex that adds phosphate groups to cellular components that break down the nuclear membrane (lamins A, B, and C), reorganize the cytoskeleton (caldesmon), and initiate chromosome condensation (histone H1). Cyclin B may specify cellular localization of target molecules. Other cyclins (especially cyclin A) are thought to be involved at this stage, but these functions are not yet known.

22.2 Mutations That Confer a Predisposition to Cancer

It is clear that the two main properties of cancer, uncontrolled cell division and metastasis, result from mutations. As mentioned earlier, these mutations can involve large-scale genomic instability, chromosome loss, chromosome rearrangement, or the insertion of foreign (often viral) DNA sequences into loci. Smaller scale mutations, including nucleotide substitutions or deletions, or more subtle modifications that alter only the amount of a gene product or the time over which it is active may also be involved.

In general, control of cell division is regulated in two ways: (1) by genes that normally function to suppress cell division and (2) by genes that normally function to promote cell division. The first group of regulatory genes is called **tumor-suppressor genes**. When expressed, these genes halt passage through the cell cycle and prevent mitosis. For cell division to occur, these genes or their gene products (or both) must be inactive or absent. If tumor suppressor genes become permanently inactivated or deleted through mutation, control over cell division is lost, and the mutant cells begin to proliferate in an uncontrolled fashion.

Genes that normally promote cell division are called **protooncogenes**. These genes can be "on" or "off," and when they are "on," they promote cell division. To stop mitosis, these genes or their gene products (or both) must be inactivated. If these genes become permanently switched on, then uncontrolled cell division occurs, leading to tumor formation. Mutant forms of protooncogenes are called **oncogenes**.

In the next sections, we examine how mutations in tumor suppressor genes can lead to a loss of cell cycle control and lead to cancer formation. Then, we consider the role of protooncogenes and oncogenes.

22.3 Tumor-Suppressor Genes Normally Suppress Cell Division

Studies have identified a number of genes that when mutated, confer a predisposition to specific cancers (see Table 22–1). One example of an inherited predisposition to cancer is retinoblastoma, a an inherited cancer of the retinal cells of the eye.

Retinoblastoma

Retinoblastoma (RB) occurs with a frequency that ranges from 1 in 14,000 to 1 in 20,000, and most often appears between the ages of 1 and 3 years. Two forms of retinoblastoma are known. In the familial form (about 40 percent of all cases), individuals who carry one mutant *RB* allele are far more susceptible to developing retinoblastoma than those with two normal wild-type *RB* alleles. Predisposition to this form of cancer is inherited as an autosomal dominant trait although the mutation itself is actually recessive, as we will explain later. About 90 percent of those who inherit the mutant *RB* allele will develop retinal tumors, usually in both eyes. In addition, these individuals are predisposed to other forms of cancer, including osteosarcoma, a bone cancer, even if they do not develop retinoblastoma.

The second form of retinoblastoma (the remaining 60 percent) is not familial, and tumors develop spontaneously. This *sporadic form* is characterized by the appearance of tumors only in one eye, and onset occurs at a much later age than in the familial form.

By studying both types of retinoblastoma, Alfred Knudson and his colleagues developed a model that requires the presence of two mutated copies of the *RB* gene in the same retinal cell for tumor formation; in other words, tumor

TABLE 22–1 **Inherited Predispositions to Cancer**

Tumor Predisposition Gene	*Chromosome*
Early-onset familial breast cancer	17q
Familial adenomatous polyposis (FAP)	5q
Familial melanoma	9p
Gorlin syndrome	9q
Hereditary nonpolyposis colon cancer (HNPCC)	2p
Li–Fraumeni syndrome	17p
Multiple endocrine neoplasia, type 1	11q
Multiple endocrine neoplasia, type 2	22q
Neurofibromatosis, type 1	17q
Neurofibromatosis, type 2	22q
Retinoblastoma (RB)	13q
Von Hippel–Lindau syndrome	3p
Wilms tumor	11p

development is a recessive trait. In the familial form, one mutant *RB* allele is inherited and is carried by all cells of the body, including cells of the retina [(Figure 22–4(a)]. If the second *RB* allele becomes mutated in a retinal cell, retinoblastoma will develop. Therefore, individuals carrying an inherited mutation of the *RB* gene are predisposed to develop retinoblastoma, as only one additional mutational event is required to cause tumor formation. This does not happen in all cases. About 10 percent of those inheriting a mutant *RB* allele do not develop cancer. Presumably, the wild-type allele of the *RB* gene does not mutate in any retinal cells.

In the *nonfamilial sporadic cases* [Figure 22–4(b)], independent mutations in both normal *RB* alleles must occur in the same retinal cell for a tumor to develop. As might be expected, these events are far less frequent and occur at a much later age. As predicted by Knudson's model, such sporadic forms of retinoblastoma are more likely to occur in a single eye.

Similar studies on predisposition to other cancers have led to the conclusion that the number of mutations necessary for the development of cancer ranges from 2 to perhaps as many as 20 (Table 22–2).

The Retinoblastoma (RB) Gene

The *retinoblastoma* (*RB*) gene, located on chromosome 13, encodes a protein designated as **pRB**. This protein is found in the nuclei of all cell and tissue types examined so far and also is found in both resting (G0) cells and actively dividing cells. Moreover, pRB is present at all stages of the cell cycle and is part of a regulatory pathway that is important in cell cycle progression. It acts as a molecular switch that controls the passage of cells from G1 into S phase. Because the normal gene is expressed ubiquitously in the body, cells can progress through the G1/S transition only when pRB is inactivated by phosphorylation. pRB is regulated by the action of CDKs. When CDK4 binds to cyclin D, the kinase phosphorylates pRB. pRB is phosphorylated in the S and the G2/M stages of the cell cycle, but is not phosphorylated in G0 and G1. When pRB is unphosphorylated (its active state), it binds to members of the E2F family of transcription factors. Recall from Chapter 20 that transcription factors are proteins that bind to the promoter region of genes and regulate transcription. E2F transcription factors control the expression of some 20–30 genes required to move the cell from G1 to S.

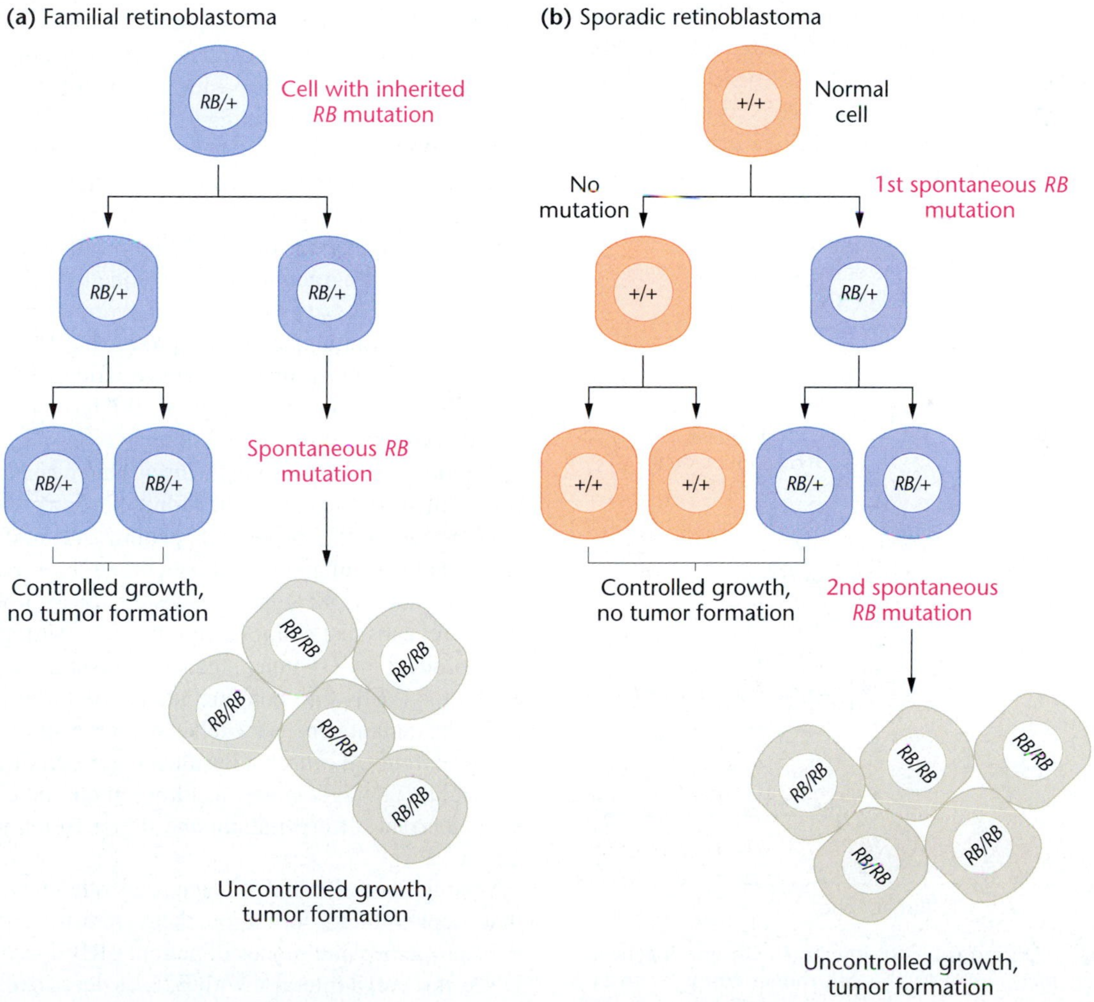

FIGURE 22–4 (a) In familial retinoblastoma, one mutation is inherited and present in all cells. A second mutation at the retinoblastoma locus in any retinal cell will result in uncontrolled cell growth and tumor formation. (b) In spontaneous retinoblastoma, two mutations in the retinoblastoma gene in a single cell are acquired sequentially, causing uncontrolled cell growth and division, and resulting in tumor formation.

TABLE 22–2 Number of Mutations Associated with Some Cancers

Cancer	*Chromosome Sites*	*Minimum Number of Mutations Required*
Retinoblastoma	13q	2
Wilms tumor	11p	2
Colon cancer	5p, 12p, 17p, 18q	4–5
Small-cell lung cancer	3p, 11p, 13q, 17p	10–15

In its unphosphorylated form, pRB binds to E2F, blocking transcription, and keeping the cell in G1 (Figure 22–5).

The phosphorylation of pRB by CDK4 takes place in late G1. When phosphate groups are added, pRB releases E2F, allowing transcription of genes that move the cell from G1 into S.

Direct evidence for the role of pRB in regulating the cell cycle comes from several lines of experimental work. Osteosarcoma (bone cancer) cells are homozygous for mutant *RB* alleles and do not produce pRB. When osteosarcoma cells are injected into a cancer-prone strain of mice, tumors are formed. If a normal *RB* gene is transferred to the cancer cells, pRB is produced and no tumors are formed when these genetically modified cells are injected into mice.

In a separate two-step experiment, a normal *RB* gene was transferred into osteosarcoma cells. These cells produced pRB and stopped cell division. When D or E cyclins were added to these pRB-blocked cells, cell division resumed. The addition of cyclins resulted in phosphorylation of pRB, presumably by activating CDK4 or another kinase. The results demonstrate that active pRB stops cell division, that pRB is a target of a G1 CDK/cyclin complex, and that inactivating pRB allows passage through the cell cycle and subsequent division, confirming the role of pRB in G1 control.

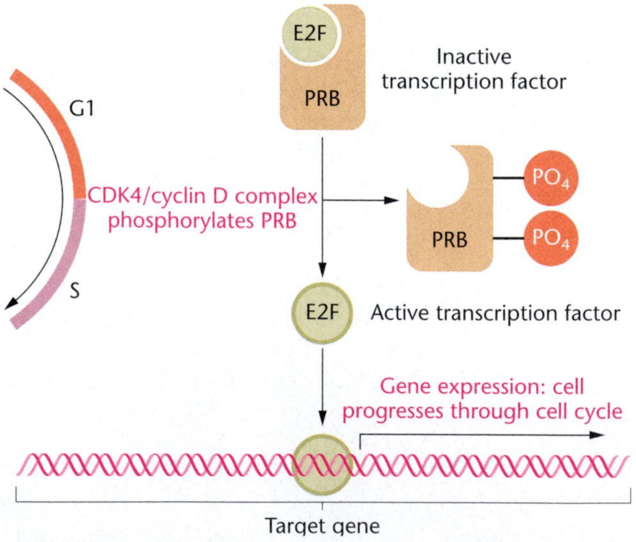

FIGURE 22–5 During G1, pRB interacts with and inactivates transcription factor E2F. As the cell moves from G1 to S, a CDK/cyclin complex forms and adds phosphate groups to pRB. As pRB becomes hyperphosphorylated, E2F is released and becomes transcriptionally active, allowing the cell to pass through S phase. Phosphorylation of pRB is transitory; as cyclin is degraded, phosphorylation declines.

In normal retinal cells, pRB is active and prevents passage into S by interacting with the transcription factor E2F. In retinoblastoma cells, both copies of the RB gene are mutated, and no functional pRB is present. As a result, E2F activates the genes required for passage through the G1/S checkpoint into the S phase. This important checkpoint is continually overridden in these cells, resulting in uncontrolled growth and tumor formation.

The Wilms Tumor Gene

Wilms tumor (WT) is a kidney cancer found primarily in children. It has a frequency of about 1 in 10,000 births, and, like retinoblastoma, it is found in two forms, a noninherited sporadic form and a familial form conferring a predisposition to WT. The familial predisposition is inherited as an autosomal dominant trait. According to Knudson's model, familial cases inherit one mutant allele through the germ line and develop a mutation in the remaining normal allele in a somatic cell. Sporadic cases, on the other hand, require two independent mutations of the *WT* gene within the same cell, and tumors more often involve only one kidney.

The *WT* gene is on the short arm of chromosome 11 and encodes a transcription factor. This protein has four contiguous zinc finger domains, motifs characteristic of DNA-binding proteins that regulate transcription. *WT* has a restricted pattern of expression; it is normally active only in cells of the fetal kidney and only during the brief time when the nephron (the basic filtration unit of the kidney) is being formed. The structure of the WT protein, its pattern and time of expression, and the resultant mutant phenotype have led to the hypothesis that the *WT* gene encodes a nuclear protein which turns off genes that maintain cell division. Alternatively, the gene product may switch on genes that begin differentiation of the fetal cells into components of the kidney. In tumor cells, the mutant gene is switched on at the appropriate time in the appropriate cells, but the altered gene product is unable to regulate its target genes, resulting in continued cell division, aberrant differentiation, and tumor formation (Figure 22–6).

Although both *RB* and *WT* genes encode nuclear proteins that suppress tumor formation, there are some differences in their properties and modes of action. pRB does not bind to DNA; instead, it interacts with E2F, as described in the previous section. WT has all the characteristics of a DNA-binding transcription factor. pRB is expressed in all dividing cells, whereas the WT protein is found only in a restricted period of kidney development. pRB is a general regulator of cell

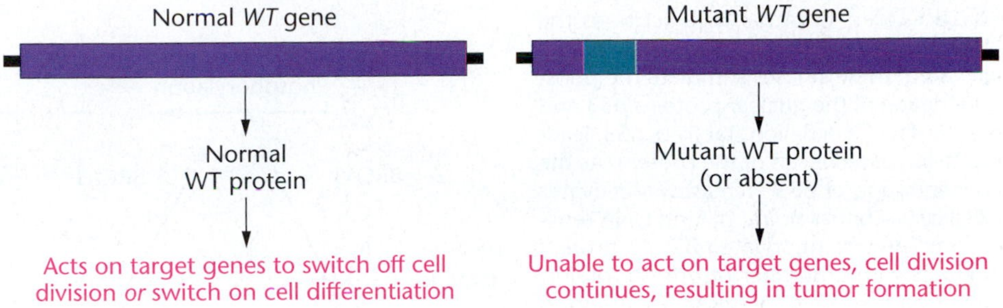

FIGURE 22–6 The proposed role of the WT protein in regulating cell division. In normal cells (at left), the WT protein is produced and acts on a set of target genes to switch off cell division or switch on cell differentiation. Any mutation that causes the loss or inactivation of the WT protein (at right) will result in a failure to regulate the target genes, allowing cell division and tumor formation to occur.

division; the WT protein is a cell- or tissue-specific regulator of gene activity during fetal or neonatal development.

p53: Guardian of the Genome

Another tumor suppressor gene, ***p53***, has been implicated in human cancer. *p53* encodes a nuclear transcription factor. *p53* mutations are found in a wide range of cancers, including breast, lung, bladder, and colon cancers. It is estimated that 50–60 percent of all cancers are associated with mutations in the *p53* gene, suggesting that it controls one or more key events in the proliferation of all cells and is not involved in a cell- or tissue-specific form of regulation. Inherited mutations of *p53* are associated with the Li-Fraumeni syndrome, an autosomal dominant condition with a predisposition to develop cancers in several tissues at a high frequency.

Normally the p53 protein is present in cells at a low level, in a rapidly degraded, inactive form. Several types of signals can shut down p53 degradation, causing a rapid increase in p53 protein levels. Chemical damage to DNA, double-stranded breaks in DNA induced by ionizing radiation, or the presence of DNA repair intermediates generated by exposure to ultraviolet light all cause a shutdown of p53 turnover. Activation of p53 protein can lead to several responses, including (1) DNA repair; (2) cell cycle arrest; and (3) **apoptosis**, a genetically programmed pathway of cell death. These tasks are accomplished by target genes activated by *p53* in its role as a transcription factor. The expression of more than 20 target genes are controlled by the action of *p53*.

Cell cycle arrest by *p53* can occur at several points in the cycle, including G1. To arrest the cycle at this phase, *p53* activates transcription of a CDK inhibitor called p21. The p21 protein targets a number of CDK/cyclin complexes, including the CDK4/cyclin D complex (which was discussed earlier).

Inhibiting the CDK4/cyclin D complex keeps pRB in its active configuration, repressing the transcription of genes needed to move the cell from G1 into S phase. Cells without functional p53 protein cannot arrest in G1 following irradiation and move immediately from G1 into S. These cells do not repair the DNA damage caused by the radiation and, as a result, have a high rate of mutation. Thus, the *p53* gene is often referred to as the "guardian of the genome."

As just mentioned, more than 50 percent of all human cancers carry a mutation in the *p53* gene. The active p53 protein is a tetramer, so mutation of one *p53* allele usually abolishes all *p53* activity, because almost all tetramers will contain at least one defective subunit. As a result, *p53* mutations act as dominant negative mutations, and individuals heterozygous for a *p53* mutation will develop cancer with a frequency of 90–95 percent. The central role of the *p53* gene in cell cycle control emphasizes the relationships between cancer and the cell cycle and between genes that regulate cell growth and cancer.

Breast Cancer Genes

Mutations in *BRCA1*, a gene on the long arm of chromosome 17, are associated with a predisposition to breast cancer; this predisposition is inherited as an autosomal dominant trait. About 85 percent of women carrying one mutant *BRCA1* allele will develop a mutation in the second allele and get breast cancer. These women also have an increased risk of ovarian cancer.

A second breast cancer gene, *BRCA2*, on the long arm of chromosome 13, is also inherited as an autosomal dominant predisposition to breast cancer, but is not associated with an increased risk of ovarian cancer. Together, these genes account for a large majority of breast cancer associated with a genetic predisposition (about 10 percent of all cases of breast cancer). The mutant forms of these genes play no role in sporadic cases of breast cancer, which make up 90 percent of all cases.

Both *BRCA1* and *BRCA2* encode large proteins that are confined to the nucleus and expressed in many tissues. Expression is at its highest during S phase of the cell cycle. Circumstantial evidence, including their common pattern of expression, similar phenotype of the mutant alleles (breast cancer), cellular localization, and experimental evidence from studies in mice, suggests that both genes have similar functions and are involved in DNA repair. Direct evidence for the role of the BRCA1 protein in DNA repair also exists. In cells exposed to ionizing radiation (which produces double-stranded DNA breaks), BRCA1 is phosphorylated (a sign of molecular activation). Current efforts center on identifying the kinase proteins that carry out this phosphorylation, in an attempt to place the BRCA1 and the BRCA2 proteins in a pathway associated with DNA repair.

Recent evidence suggests that the BRCA1 protein can be phosphorylated by two different kinases and that either kinase is activated by double-stranded DNA breaks (Figure 22–7). In this model, DNA damage activates the ATM kinase or the Chk2 kinase (or both). These kinases phosphorylate the BRCA1 and p53 proteins. The activated p53 protein arrests replication during S phase to allow DNA repair. The activated

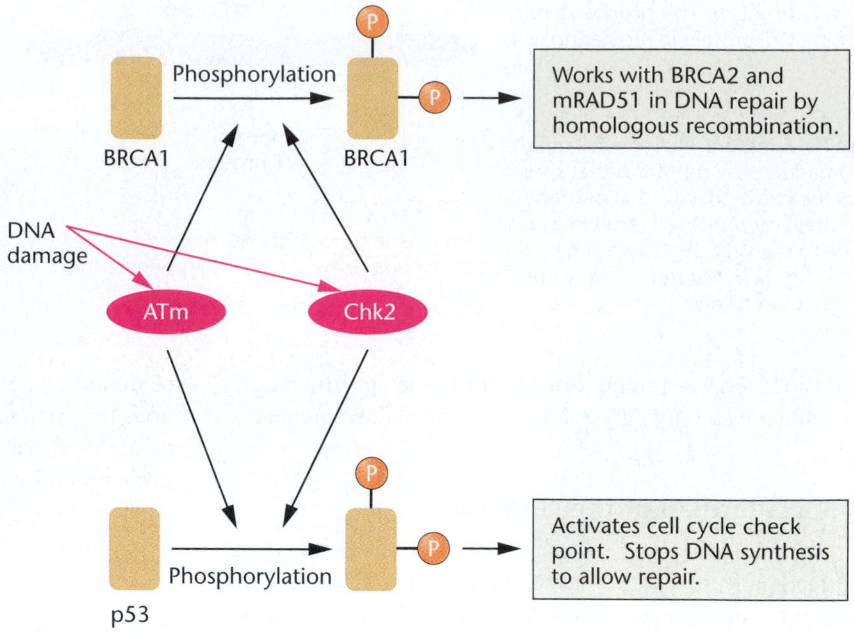

FIGURE 22–7 DNA damage activates the kinase ATM, and/or the kinase Chk2. Once activated, these kinases stimulate the phosphorylation of the nuclear proteins p53 and BRCA1. Phosphorylation stabilizes p53, leading to increased levels of the protein. As the concentration of p53 increases, it activates a cell cycle control point, halting DNA replication. The phosphorylated BRCA1 protein interacts with a number of other proteins, including BRCA2 and mRAD51 to bring about repair of double-stranded DNA breaks by homologous recombination.

BRCA1 protein participates in DNA repair with the BRCA2 protein, the mRAD51 protein, and other nuclear proteins involved in DNA repair.

The role of BRCA1 and BRCA2 proteins in DNA repair may also explain why they are not associated with sporadic cases of breast cancer. In sporadic cases, at least three mutations would have to accumulate in a breast cell for it to become cancerous: Both copies of the *BRCA1* or *BRCA2* genes would have to be mutant, and at least one other mutation in a cell-cycle regulation gene would have to occur. On the other hand, those who inherit a mutant copy of either the *BRCA1* or *BRCA2* gene already have one mutation and need only two more to trigger breast cancer.

22.4 Oncogenes Induce Uncontrolled Cell Division

We now discuss the second category of genes involved in cell cycle regulation: those that normally promote cell division, called protooncogenes. When expressed, these genes promote or maintain cell division. To stop division, these genes or their gene products (or both) must be inactivated. As mutant alleles called oncogenes, they induce or maintain the uncontrolled cell division associated with cancer. The protein products of protooncogenes are found throughout the cell, including the plasma membrane, cytoplasm, and nucleus (Table 22–3). In spite of their wide-ranging locations within the cell, suggesting varying functions, all protooncogene proteins characterized to date alter gene expression either directly or indirectly.

Unlike tumor suppressor genes, where mutations in both alleles of a gene are necessary to promote the development of cancer, only one of the two copies of a protooncogene needs to mutate to induce malignancy, resulting in a dominant phenotype.

Rous Sarcoma Virus and Oncogenes

In 1910, Francis Peyton Rous first inferred the existence of specific genes associated with the transformation of normal cells into cancerous cells. Rous studied a tumor in chickens known as a **sarcoma**. He injected cell-free extracts from these tumors into healthy chickens and induced the formation of sarcomas. He postulated the existence of an agent that transmitted the disease, which decades later was shown by other investigators to be a virus, now known as the **Rous sarcoma virus (RSV)**. Rous received the Nobel Prize in 1966

TABLE 22–3 Cellular Location of Protooncogene and Oncogene Proteins

Gene	*Location of protooncogene Protein*	*Location of oncogene Protein*
src	Membranes	Membranes
ras	Membranes	Membranes
myc	Nucleus	Nucleus
fps	Cytoplasm	Cytoplasm and membranes
abl	Nucleus	Cytoplasm
erbB	Plasma membrane	Plasma membrane and Golgi

for his pioneering work in establishing the relationship between viruses and cancer.

RSV infects cells and reproduces within them. Once inside, the single-stranded RNA genome of RSV is transcribed by the enzyme **reverse transcriptase**, converting the RNA genome into a single-stranded DNA molecule. The single-stranded DNA is used as a template to synthesize a complementary strand, creating a double-stranded DNA molecule that integrates into the host genome, forming a **provirus**. Later, the DNA is transcribed into RNA, which is translated into viral proteins. Packaging RNA molecules into these proteins forms new virus particles. Because the replication cycle of viruses like RSV "reverses" the flow of genetic information, they are called **retroviruses**.

In RSV, tumors result from a single gene, *src*, present in the viral genome. This gene is an oncogene. Retroviruses that carry oncogenes are known as **acute transforming viruses**. Retroviruses that do not carry oncogenes but can induce transcription of cellular genes that cause tumor formation are known as **nonacute viruses**.

Origin of Oncogenes

Oncogenes (*onc*) carried by acute transforming viruses, like RSV, are acquired from the host genome during infection, when a portion of the viral genome is exchanged for a cellular protooncogene (Figure 22–8). Oncogenes in the retroviral genome are called *v-onc*; the normal, cellular version of the gene is *c-onc* (also called a protooncogene). Retroviruses with a *v-onc* gene infect and transform a specific type of host cell into a tumor cell. For RSV, *v-src*, the oncogene from the chicken genome, transforms chicken cells into sarcomas. The cellular version of the same gene, found in the chicken genome, is *c-src*. More than 20 oncogenes have been identified by their presence in retroviral genomes, with over 60 oncogenes identified to date. Some of these are listed in Table 22–4. Although oncogenes were first identified in retroviruses, not all oncogenes are mutant versions of cellular genes carried by retroviruses. Some oncogenes arise by mutagenic events that occur spontaneously in cells. We must, therefore, consider a broad range of mechanisms in oncogenetic events.

Formation of Oncogenes

At least three mechanisms can explain the conversion of protooncogenes into oncogenes: **point mutations**, **translocations**, and **overexpression**. (See Table 22–5.) Some of these events are mediated by viruses; others by intracellular events that occur in the absence of retroviruses.

Mutations in Ras proteins demonstrate how a point mutation converts a protooncogene into an oncogene. The ***ras* gene family** encodes signal transduction proteins with a major role in the regulation of cell growth and division. Ras proteins are molecular switches that transmit signals from the extracellular environment to the cytoplasm. More than 30 percent of all human cancers carry a mutant *ras* oncogene. Ras proteins are embedded in the plasma membrane, and cycle between an inactive (switch "off") state and an active (switch "on") state. Active Ras proteins interact with cytoplasmic proteins, starting a cascade of events that transmits a signal from the cytoplasm to the nucleus. In the nucleus, the signal activates the transcription of genes that initiate cell division.

Comparing the amino acid sequences of Ras proteins from a number of different human carcinomas (tumors of epithelial tissue) shows that *ras* mutations generate single amino acid substitutions at either position 12 or 61 (Figure 22–9) in the 189 amino acid Ras protein. Each of these changes can be created by a single nucleotide substitution in the *ras* gene. Mutant Ras proteins cannot cycle from the active to the inactive form and are locked into the "on" position, continually signaling for cell division (Figure 22–10).

A well-characterized oncogene activation by translocation involves *c-abl*, an oncogene associated with chronic myelogenous leukemia (CML). In this case, described in detail in a later section, the translocation results in altered gene activity that causes tumor formation.

At least three mechanisms of protooncogene activation are linked to overexpression of a gene product. First, the protooncogene may acquire a new promoter, causing transcription levels to increase or a silent locus to become activated. This happens in avian leukosis, where strong viral promoters integrate preferentially adjacent to a protooncogene, causing an increase in mRNA and protein levels. Overexpression can result from acquisition of new upstream regulatory sequences, including enhancers. The third mechanism involves amplification of the protooncogene. In human tumors, members of the *myc* oncogene family are frequently amplified. The *c-myc* protooncogene is found amplified up to several hundred copies in some human tumors.

A final note involves the link between oncogenes and apoptosis. As described earlier, apoptosis is associated

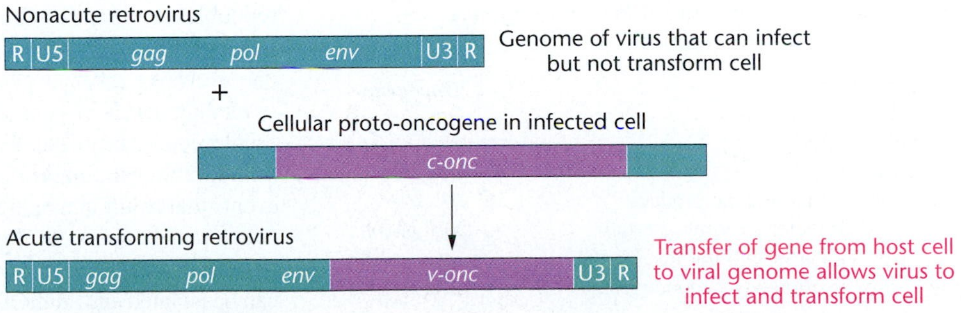

FIGURE 22–8 A transforming retrovirus has acquired a copy of a gene from the host genome, converting it from a *c-onc* or protooncogene into an oncogene (*v-onc*) that confers on the virus the ability to transform a specific type of host cell into a cancerous cell.

TABLE 22–4 Origin and Function of Representative Oncogenes

Oncogene	*Origin*	*Species*	*Cellular Function*
src	Rous sarcoma virus	Chicken	Tyrosine kinase signal protein
fos	FBJ osteosarcoma virus	Mouse	Transcription factor
sis	Simian sarcoma virus	Monkey	Platelet-derived growth factor
abl	Abelson murine leukemia virus	Mouse	Tyrosine kinase
myc	Avian myelocytomatosis virus	Chicken	Transcription factor
erbB	Avian erythroblastosis virus	Chicken	EGF receptor
N-ras	Neuroblastoma, leukemia	Human	GTP-binding protein

with a *p53*-induced pathway that responds to DNA damage. Recent evidence shows that oncogenes and their corresponding oncoproteins can also induce apoptosis by sensitizing the cell to apoptotic triggers (such as DNA damage). It is currently believed that the processes mediating growth and apoptosis are coupled. In a current model, the activation of cell proliferation (e.g., protooncogene conversion to an oncogene) primes the cellular apoptotic program. As a result, many cells containing oncogenes do not proliferate, but instead undergo apoptosis. This model helps us to understand why cancer does not develop every time a protooncogene changes into an oncogene.

22.5 Metastasis Is Genetically Controlled

Cancer cells cannot regulate their growth and division. As a result, they develop into malignant tumors that may potentially invade neighboring tissue. Sometimes, cancer cells detach from the primary tumor and settle elsewhere in the body, where they grow and divide, producing secondary tumors. This process, called **metastasis**, is often the cause of death in cancer patients.

A metastatic cancer cell spreads from the primary tumor by entering the blood or lymphatic circulatory system. These cells are carried in the circulation until they lodge in a capillary bed. Normally, more than 99 percent of the cells die. The surviving cells invade tissue adjacent to the capillary bed and begin dividing to form a secondary tumor. To reach a new site, the tumor cells pass through the layer of epithelial cells lining the interior wall of the capillary (or lymph vessel) and penetrate the surrounding extracellular matrices.

TABLE 22–5 Conversion of Protooncogenes to Oncogenes

Mechanism	*Oncogene*
Point mutation	*ras*
Translocation	*abl*
Overexpression of gene product	
New promoter by viral insertion	*mos, myb*
New enhancer by viral insertion	*myc*
Amplification of protooncogene	*myc*

The extracellular matrix is a meshwork of proteins and carbohydrate molecules separating tissues; it is a scaffold for tissue growth and inhibits the migration of cells. To establish a secondary tumor, metastatic cells secrete enzymes that digest this meshwork, creating holes through which they can migrate.

The ability to invade extracellular matrix is also a property of some normal cell types. Implantation of the embryo in the uterine wall during pregnancy, for example, requires cell migration across the extracellular matrix. White blood cells reach the site of infection by penetrating capillary walls. The mechanism of invasion is probably the same in these cells as in cancer cells. The difference is that in normal cells, the invasive ability is controlled and tightly regulated, while in tumor cells, regulation has been lost.

Metastasis is regulated by genes that encode or regulate matrix-cutting enzymes. Protein-cutting enzymes called **metalloproteinases** are necessary for cell invasion. Tumor cell lines with high metastatic capacity produce more metalloproteinases than do tumor cell lines with low invasive properties. Activated metalloproteinases are inhibited by the presence of a tissue inhibitor of metalloproteinase (TIMP). Normal cells produce TIMP and thereby suppress inappropriate activity of metalloproteinase. In tumor cells, TIMP inhibits metalloproteinase only when the number of TIMP molecules is greater than the number of metalloproteinase molecules. On the basis of its action and regulation, TIMP can be considered as a protein that suppresses metastasis.

22.6 Colon Cancer Results from a Series of Mutations

Cancer is clearly a multistep process resulting from a number of specific mutational events. Studies of tumors such as retinoblastoma have established that in some cases only a few steps are required to transform a normal cell into a malignant one. However, most cancers develop in several steps, with intermediate levels of genetic and cellular transformation.

Colorectal cancer has been used to obtain detailed information about the nature and order of genetic and cellular events that result in a malignant growth. The malignant form of colorectal tumors develops from preexisting benign tumors. Several discrete precancerous stages occur, and these stages can be isolated and studied. Furthermore, both hereditary and

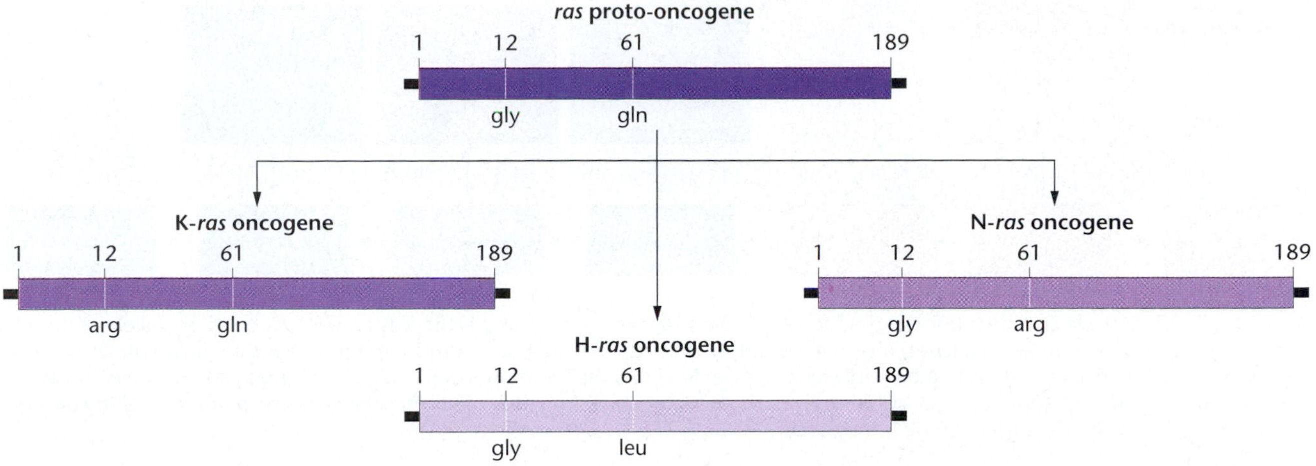

FIGURE 22–9 The *ras* protooncogene encodes a protein of 189 amino acids. In the normal protein, glycine is encoded at position 12, and glutamine at position 61. Analysis of *ras* oncogene proteins from several tumors shows a single amino acid substitution at one of these positions, which can convert a protooncogene into a tumor-promoting oncogene.

sporadic forms of colorectal cancer exist. As a result, colorectal cancer is a useful model for studying the interaction of genetic and environmental factors in tumor formation.

Two forms of genetic predisposition to colon cancer are known: (1) an autosomal dominant trait, known as **familial adenomatous polyposis (FAP)** and (2) a genetically complex trait, known as **hereditary nonpolyposis colorectal cancer (HNPCC)**. FAP is present in about 1 percent of all cases, and HNPCC is responsible for 2–4 percent of all colon cancers. The remaining 95 percent of all cases are sporadic.

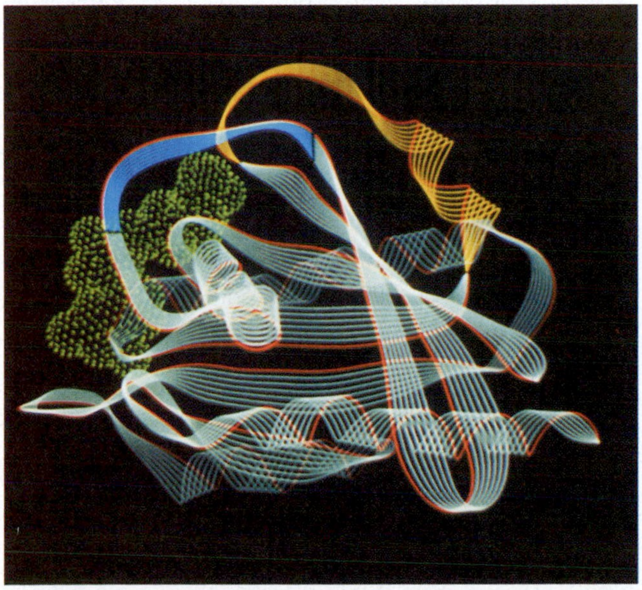

FIGURE 22–10 A three-dimensional computer-generated image of ras proteins in two different conformations. Normal ras proteins act as molecular switches controlling cell growth and differentiation. The switch is "on" (conformation shown in blue) when GTP binds to the protein, and "off" (conformation shown in yellow) when the GTP is hydrolyzed to GDP. Switching the protein between states alters the conformation of the protein in the two regions (blue and yellow). Oncogenic mutations of *ras* are stuck in the "on" state, continuously signaling for cell growth.

Researchers have developed genetic models of both forms of colon cancer, which provide insight into the events that cause normal cells to become cancerous.

FAP and Colon Cancer

In the FAP-associated model, spontaneous cases of colon cancer begin with a mutation in the *APC* gene, which maps to the long arm of chromosome 5. This mutation occurs in a normal epithelial cell in the lining of the colon. In spontaneous cases, an *APC* mutation causes epithelial cells to partially escape cell cycle control. These cells divide to form a small cluster called a **polyp** (Figure 22–11). Those who inherit a mutant *APC* allele have the mutation in all cells of their body. As a result, *APC* heterozygotes form hundreds or thousands of polyps. In either case, polyps are clones, all of which carry an *APC* mutation. The loss of the corresponding allele on the homologous copy of chromosome 5 is not necessary for polyp formation. The first step and the relative order of subsequent mutations are shown in Figure 22–12.

Subsequent mutations in polyp cells cause intermediate stages of tumor formation. Mutations in the *ras* oncogene on chromosome 12 (combined with a mutant *APC* allele) cause the polyp to grow larger and to develop fingerlike

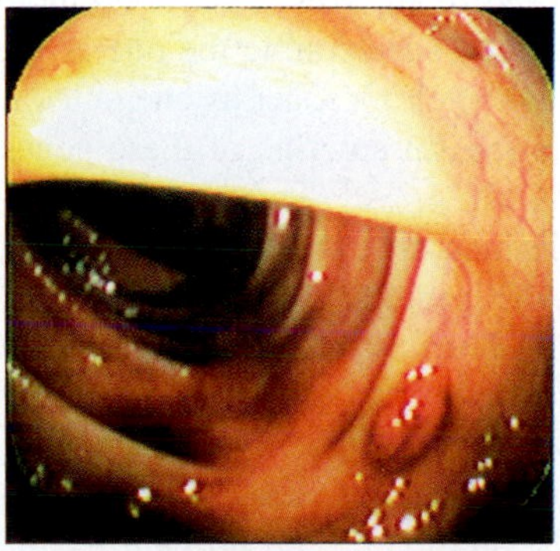

FIGURE 22–11 A polyp in the colon. (*Albert Paglialunga/Phototake*)

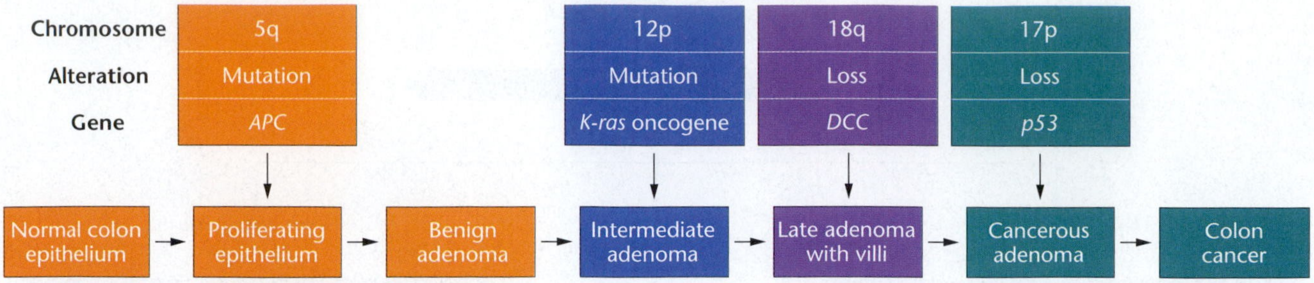

FIGURE 22–12 A model for the multistep production of colon cancer. The first step is the loss or inactivation of one allele of the *APC* gene on chromosome 5. In familial cases, one mutation of the *APC* gene is inherited. Subsequent mutations involving genes on chromosomes 12, 17, and 18 in cells of the benign adenomas can lead to a malignant transformation resulting in colon cancer. Although the mutations in chromosomes 12, 17, and 18 usually occur at a later stage than those involving chromosome 5, the sum of the changes is more important than the order in which they occur.

(villous) outgrowths. This is the intermediate adenoma stage. To progress further, a polyp cell must acquire mutations in the *DCC* (deleted in colon cancer) gene on chromosome 18. Mutations in both *DCC* alleles results in late-stage adenomas. Finally, mutation of the *p53* gene causes the transition to a cancerous cell. As described earlier, mutations in *p53* are pivotal to the development of a number of cancers, including lung, brain, and breast cancers as well as colon cancer. Metastasis occurs after the formation of colon cancer, and it involves an unknown number of mutational steps.

HNPCC and Colon Cancer

HNPCC-associated colon cancer develops after a small number of polyps are formed, rather than the hundreds or thousands associated with FAP. Genes associated with HNPCC have been mapped to loci at 2p16 and 3p21. Mutations at these loci generate a cascade of mutations in short, tandemly repeated microsatellite sequences located throughout the genome.

The gene on chromosome 2 is a DNA repair gene called *MSH2*. Inactivation of this gene causes the rapid accumulation of mutations and the development of colorectal cancer. A second DNA repair gene, *MLH1*, associated with HNPCC has been mapped to chromosome 3. At least two other DNA repair genes related to *MLH1* and *MSH2* have been identified and remain to be mapped. These genes are called mismatch repair genes (MMR) and generate a condition called microsatellite instability. (Microsatellites are discussed in Chapter 4.) Mutations in any MMR genes promote genomewide genetic instability, accelerating the rate at which mutations accumulate, with colon cancer as one outcome. It may turn out that mutations in any one of these genes is enough to cause HNPCC.

Mutations leading to HNPCC are similar to FAP-associated colon cancer (Figure 22–12), but have several important differences. FAP begins with a mutation in *APC*, and the formation of hundreds or thousands of polyps. Each polyp progresses slowly toward cancer by accumulating mutations in other genes (*ras*, *DCC*, etc.). However, because there are so many polyps, there is a high probability that at least one polyp will accumulate all the mutations necessary to cause colon cancer. HNPCC begins with mutation in DNA repair genes in an epithelial cell of the colon, generating many genomewide mutations, one of which may include *APC*. Mutation of *APC* is a relatively rare event, so only a few polyps will form. However, even though there are few polyps, the continuing high mutation rate ensures that at least one polyp will acquire all the mutations necessary to become cancerous. FAP-associated colon cancer is therefore a disease of rapid tumor initiation, with slow progression to cancer, while HNPCC is a disease with slow initiation of tumors, but rapid progression to cancer. Both forms of colon cancer are mediated by mutations in the *APC* gene.

Gatekeeper and Caretaker Genes

The two pathways to colon cancer in FAP and HNPCC offer an insight into the nature of genes that control cancer predisposition. These differences indicate that mutations in two types of genes cause predisposition to cancer: **gatekeeper genes** and **caretaker genes**. The FAP gene is a gatekeeper; it normally inhibits cell growth. In general, tumor suppressor genes are gatekeepers. In different cell types, only one or a few genes serve as gatekeepers, and if both copies of the gatekeeper mutate, a specific cancer, such as retinoblastoma or colon cancer, develops. Individuals predisposed to a specific form of cancer inherit one mutant copy of such a gatekeeper and need only one additional mutation to initiate tumor formation. In spontaneous cases, both copies of the gatekeeper must mutate for cancer to develop.

Caretaker genes maintain the integrity of the genome, including DNA repair genes such as *MSH2* and *MLH1*. These genes repair DNA damage caused by environmental agents such as ultraviolet light, or mismatches that occur during DNA replication. Mutation of a caretaker gene does not promote tumor formation directly, but leads to genetic instability that increases the mutation rate of all genes, including gatekeeper genes. If a gatekeeper gene mutates and begins tumor formation, the process is accelerated by the high rate of mutation in the cell.

22.7 Chromosome Translocations Are Often Associated with Cancer

Alterations in chromosome structure or number (or both) are associated with many forms of cancer. In most cases, the relationship between changes in chromosome number or structure, and the development of cancer is not clear. For example, individuals with Down syndrome carry an extra copy of chromosome 21. This quantitative alteration in genetic content is also associated with a 15–20 fold increased risk of leukemia compared with the risk to the general population. In a limited number of cases in which specific alterations in chromosome structure are associated with cancer, more direct information indicates how structural rearrangements are associated with the development or maintenance (or both) of the cancerous state.

Changes in chromosome structure or number are often associated with various forms of cancer. In selected cases, the relationship between the chromosome aberration and the development or maintenance (or both) of the cancerous state is clear. Such a connection is most clearly seen in leukemias, where the presence of a specific chromosome translocation is well defined (Table 22–6). One of the best studied examples is the translocation between chromosomes 9 and 22, which is associated with **chronic myelogenous leukemia (CML)** (Figure 22–13). Originally, this translocation was mistakenly described as an abnormal chromosome 21 and called the **Philadelphia chromosome** (because it was discovered in that city). Later, Janet Rowley showed that the Philadelphia chromosome actually results from the exchange of genetic material between chromosomes 9 and 22. Such a translocation is never seen in a normal cell and is found only in the white blood cells involved in CML. These observations indicate that the translocation is a primary and causal event in the generation of CML and that this form of cancer may originate from a single cell bearing this translocated chromosome.

A proto-oncogene, *c-abl* maps to the breakpoint on chromosome 9, and the *bcr* gene is at the breakpoint on chromosome 22. The c-Abl protein is a kinase involved in cellular growth control and the normal Bcr protein activates a phosphorylation reaction. In the translocation, part of the *bcr* gene is fused to a site upstream of the second exon in the *c-abl* gene. The hybrid *bcr/abl* gene is transcriptionally active, and produces a functional 210-kDa Bcr-Abl fusion protein. Experiments demonstrate that the bcr/abl gene is an oncogene and produces CML-like disease in mice. The ability of the Bcr-Abl fusion protein to transform normal cells depends on the kinase activity of the Abl portion of the molecule.

TABLE 22–6 Specific Chromosome Aberration and Cancer

Cancer	*Chromosome Alteration**
Chronic myelogenous leukemia	t(9;22)
Acute promyelocytic leukemia	t(15;17)
Acute lymphocytic leukemia	t(4;11)
Acute myelogenous leukemia	t(8;21)
Prostate cancer	del(10q)
Synovial sarcoma	t(X;18)
Testicular cancer	inv(12p)
Retinoblastoma	del(13q)
Wilms tumor	del(11p)

*t = translocation, del = deletion, inv = inversion.

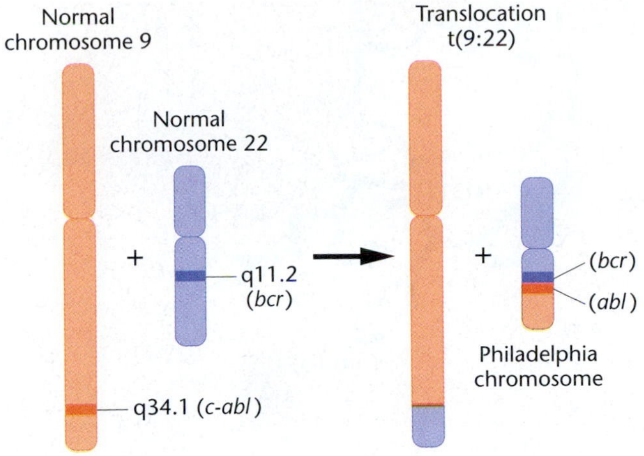

FIGURE 22–13 The Philadelphia chromosome. A reciprocal translocation involving the long arms of chromosomes 9 and 22 results in the production of a characteristic chromosome, the Philadelphia chromosome, which is associated with cases of chronic myelogenous leukemia (CML). The t(9;22) translocation results in the fusion of the *c-abl* oncogene on chromosome 9 with the *bcr* gene on chromosome 22. The fusion protein is a powerful hybrid molecule that allows cells to escape control of the cell cycle, resulting in leukemia.

Cancer Chemotherapy and the Genetics of Drug Resistance in CML

The presence of the Bcr-Abl fusion protein only in cancer cells presents a unique opportunity to design therapeutic drugs that will affect only the cancer cells. Most anticancer drugs have been discovered by isolating and screening thousands of chemicals to check their ability to kill cancer cells, or inhibit their growth.

The active form of the Bcr-Abl fusion protein phosphorylates Crkl, a protein linked to signaling pathways, leading to cell division. This step requires ATP, which binds to a pocket in the Abl portion of the protein. Using this knowledge, scientists engineered a protein kinase inhibitor with a high affinity for the ATP binding site in the Ccr-Abl fusion protein (Figure 22–14). This drug, STI-571, has been highly effective, and in clinical trials, induced remission in more than 96 percent of early-stage patients. The success of STI-571 as a drug designed to target specific proteins in cancer cells is being used as a model for the development of other cancer drugs.

However, STI-571 shares one feature with conventional anticancer drugs: development of drug resistance. Patients in advanced stages of CML respond well to treatment with STI-571, but 80 percent relapse within 12 months and become resistant to drugs. Relapse is accompanied by high levels of Bcr-Abl kinase activity, leading investigators to focus on changes in the *brc-abl* gene in drug-resistant patients.

In one group of 11 relapsed patients, the *brc-abl* gene was amplified in three cases. In these cases, the explanation was

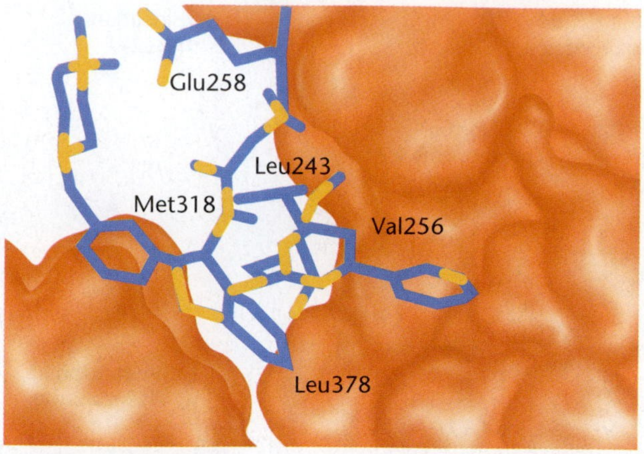

FIGURE 22–14 The binding site in the hybrid kinase produced by the *Bcr-Abl* oncogene occupied by the anti-cancer drug STI-571, which inhibits kinase activity. From: *Buchdinger, E., et al., 2001. Bcr-Abl inhibition as a modality of CML therapeutics. Biochim. Biophys. Acta 1551: M11–M18, Figure 5A. p. M16.*

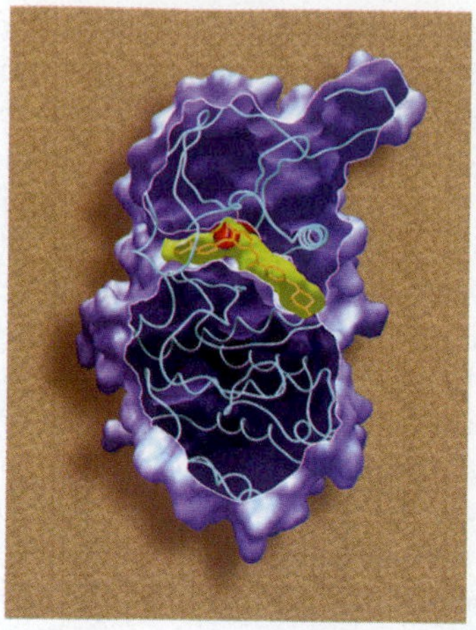

FIGURE 22–15 In some cases, drug resistance to STI-571 is caused by the replacement of threonine with isoleucine (red) in the ATP-binding pocket, which inhibits binding of STI-571 (yellow). Courtesy of Lore Leighton. From: *Marx, J. 2001. Why some leukemia cells resist STI-571. Science 292: 2231–32.*

straightforward: Extra copies of the gene resulted in higher levels of the fusion protein, overpowering the ability of the drug to effectively inhibit kinase activity. In 6 other patients, drug resistance was caused by a mutation in the *bcr-abl* gene. Interestingly, all 6 patients carried the same mutation: replacement of threonine with isoleucine in the ATP binding pocket. In all six cases, the mutation was caused by a single nucleotide (C→T) change at nucleotide 944. Threonine at amino acid position 315 is critical in binding the STI-571 molecule (Figure 22–15). Replacement of threonine with isoleucine prevents STI-571 binding, but allows ATP binding, making the protein active in Crkl phosphorylation even in the presence of the drug. It is not clear whether the mutation arose after the start of drug treatment or was present in a small fraction of cells before treatment began.

Advances in proteomics is expected to accelerate the development of drugs designed to target specific proteins, but the example here points out two challenges for cancer chemotherapy: genomic instability of cancer cells and the role of natural selection in the response of cancer cells to anticancer drugs.

22.8 Environmental Factors and Cancer

The relationship between environmental agents and the genesis of cancer is often elusive, and in the early stages of an investigation, this relationship is based on indirect evidence. Such studies often begin with an epidemiological survey comparing the cancer death rates among different geographic populations. When differences are found in death rates for a type of cancer, an age group, or a cluster of related occupations such as chemical workers, further work is necessary to identify one or more environmental factors that correlate with these cancer deaths. Such correlations are not conclusive, but they identify factors that, upon additional investigation, may directly relate to the development of cancer. Finally, extensive laboratory investigations may establish the mechanism by which an environmental agent generates cancer. In other cases, such as for certain viruses, the relationship between cancer and the environment is more straightforward, as we see next.

Hepatitis B and Cancer

Epidemiological surveys show that individuals who develop a form of liver cancer known as **hepatocellular carcinoma (HCC)** are infected with the **hepatitis B virus (HBV)** Figure 22–16). In fact, for those carrying HBV, the risk of cancer increases by a factor of 100. Aside from the risk of cancer, HBV infection is a public health risk affecting some 300 million people worldwide; it is most prevalent in Asia

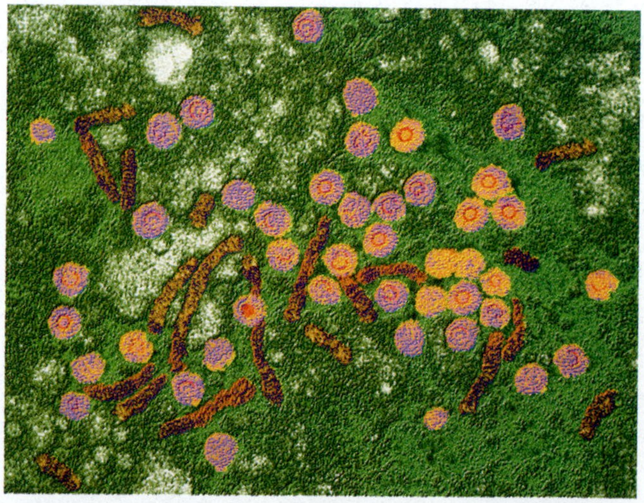

FIGURE 22–16 A false-color transmission electron micrograph of a hepatitis B virus. Infection with this virus often results in cancer of the liver. Worldwide, viral infections of all kinds are thought to be responsible for about 15 percent of the total number of cancers.

and tropical regions of Africa. The infection produces a wide range of responses, from a chronic, self-limiting infection with few symptoms, to active hepatitis and cirrhosis, to fatal liver disorders and hepatocellular carcinoma. In the last few years, efforts have centered on understanding the mechanism by which HBV replicates and its role in causing liver cancer.

The genome of HBV is a mostly double-stranded DNA molecule of 3200 nucleotides. After cellular infection, the DNA moves to the nucleus, where it inserts into a chromosome. The viral DNA is transcribed into an RNA molecule and packaged into a viral capsid. The capsid containing the RNA pregenome moves to the cytoplasm, where it is reverse-transcribed into a DNA strand, which in turn is made double-stranded. The copied DNA genome is repackaged into a new capsid for release from the cell, or it returns to the nucleus for another round of replication.

The key to the role of HBV in carcinogenesis apparently resides in its entry into the nucleus. While in the nucleus, the HBV genome can insert itself at many different sites into human chromosomes (Figure 22–17). Insertion often results in cytogenetic alterations of the host genome that include translocations, deletions, or amplifications of adjacent regions. As outlined earlier, most forms of cancer are associated with chromosomal rearrangements of these types, and such rearrangements may trigger the development of a cancerous transformation. The HBV genome has been shown to integrate into the *cyclin A* gene (*CCNA2*), and abnormal expression of this gene can disrupt the cell cycle and normal control of proliferation. Similarly, HBV integration results in activation of protooncogene loci, and altered expression of such loci is clearly implicated in the development of malignant cell growth. Thus, a strong link exists between the HBV virus as an environmental agent and the development of cancer in infected individuals.

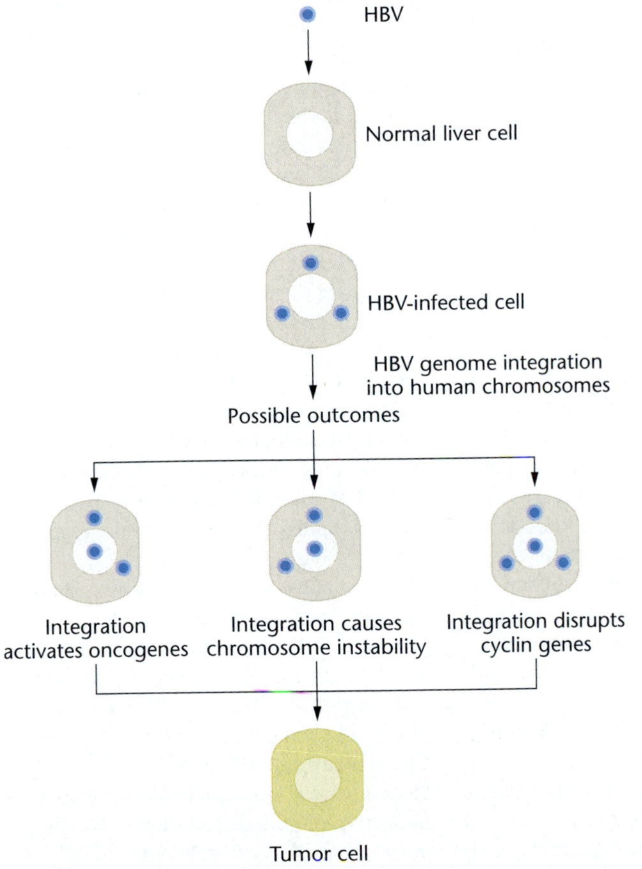

FIGURE 22–17 The possible outcomes following cellular infection with hepatitis B virus (HBV). Following the integration of the viral chromosome into the human genome, oncogenes can be activated, resulting in tumor formation. Integration of the HBV genome can also cause chromosome instability, including the production of translocations and deletions that trigger a malignant cyclin A gene, triggering abnormal regulation of the cell cycle and cellular proliferation, resulting in cancer formation.

Environmental Agents

Many surveys of cancer incidence (Table 22–7) point to the role of the physical environment and personal behavior as factors contributing to the development of cancer. The precise role of the environment in the genesis of cancer can be difficult to ascertain. Obviously, the induction of cancer involves an interaction between the genotype and environmental agents. The differential expression of a specific genotype in various environments is often neglected in addressing the role of environment in cancer, but researchers estimate that at least 50 percent of all cancers are environmentally induced. Environmental agents responsible for cancer include background levels of radiation, occupational exposure to physical and chemical agents, exposure to sunlight, and personal behavior such as diet and the use of tobacco. To show how environmental factors are identified, let's consider a recent study on the risk of colon cancer.

Epidemiological surveys reveal that cancer of the colon occurs with a much higher frequency in North America and Western Europe than in Asia, Africa, or other parts

TABLE 22–7 Epidemiology of Cancer in Various Countries

	Death Rate per 100,000	
Country	***Male***	***Female***
Australia	212	125
Canada	214	136
Dominican Republic	54	48
Egypt	39	18
England	248	156
Greece	188	103
Israel	174	145
Japan	190	109
Nicaragua	22	35
Portugal	180	108
Singapore	249	130
United States	216	137
Venezuela	135	128

Genetics, Technology, and Society

The Double-Edged Sword of Genetic Testing: The Case of Breast Cancer

The world is increasingly amazed—and perhaps bewildered—by the remarkable achievements in genetics and molecular biology over the last two decades. Both scientists and the media speculate that we will soon be able to test a person's gene sequences, and accurately predict that person's risk of developing diseases such as heart disease and cancers. Of course, the prospect of using molecular biology to prevent and cure a whole range of diseases that have genetic components is exciting. In our enthusiasm for the new genetic technologies, however, we often forget that these achievements have significant limitations and profound ethical concerns. The story of breast cancer genetic testing illustrates how we must temper our high expectations with respect for uncertainty.

Breast cancer is the most common type of cancer among women and the second leading cause of cancer deaths (after lung cancer). Each year, more than 180,000 new cases are diagnosed in the United States. Breast cancer is not limited to women; about 1400 men are diagnosed with breast cancer each year. The likelihood of developing breast cancer is low before age 35, but the risk increases after that. A woman's lifetime risk of developing breast cancer is about 10 percent.

Approximately 5–10 percent of breast cancers are familial, defined by the appearance of several cases of breast or ovarian cancer amongst near blood relatives and by an early onset of the disease. The clustering of breast and ovarian cancers in certain families suggests that genetic mutations are involved in the development of these cancers. In 1994, two genes were identified that appear to be linked to familial breast cancers. Germline mutations in these genes (*BRCA1* and *BRCA2*) are associated with the majority of familial breast cancers; however, mutations in several other genes may also contribute. The molecular functions of *BRCA1* and *BRCA2* are still uncertain, although they may involve repair of damaged DNA or transcription regulation of hormone-responsive genes. Mutations in these genes act as autosomal dominants with variable penetrance. Women who bear mutations in *BRCA1* have a 50–80 percent lifetime risk of developing breast cancer and a 20–40 percent risk of developing ovarian cancer. Men with germline mutations in *BRCA2* have a 6 percent lifetime breast cancer risk—a 100-fold increase over the general male population.

Recently, genetic testing for mutations in the *BRCA1* and *BRCA2* genes has become available to families at high risk of familial breast cancer. These tests are extremely accurate, detecting any of the hundreds of mutations that can occur within the coding region of these genes. But the tests have limitations. They do not detect mutations in regulatory regions outside of the coding region, mutations that could result in aberrant expression of these genes. Also, little is known about how any particular mutation manifests itself in terms of cancer risk, as penetrance can be affected by environmental factors and interactions with other susceptibility genes.

Some familial breast cancer patients and their families opt to take the *BRCA1* and *BRCA2* genetic tests. These patients feel that test results will help them to prevent recurrence of the disease, will guide them in childbearing decisions, and allow them to inform family members at risk. But none of these options has clear-cut positive benefits.

Women whose test results are negative are relieved that "they are not subject to familial breast cancer." However, their risk of developing breast cancer is still 10 percent (the population risk), and they should continue to monitor for the disease. Also, negative *BRCA1* and *BRCA2* tests do not eliminate the possibility that the patient bears a familial breast cancer mutation in another, unknown gene or that *BRCA1* or *BRCA2* mutations do exist, but occur outside of the gene coding regions, inaccessible to current genetic tests.

Those whose tests are positive face difficult choices. Treatment options are poor, consisting of close monitoring, prophylactic mastectomy and oophorectomy (removal of breasts and ovaries), and taking birth control pills or tamoxifen. Prophylactic surgery can reduce the risks but cannot eliminate them, as cancers still occur in tissues that remain after surgery. Drugs such as tamoxifen can also reduce the risk but have serious side effects. Genetic tests do not only affect the patient, they also affect the patient's entire family. People often experience fear, anxiety, and guilt on learning that they are carriers of a genetic disease. Studies suggest that even people who refuse genetic test results suffer from increased anxiety. Confidentiality is a major concern. Patients fear that their genetic information may be leaked to insurance companies or employers, jeopardizing their prospects for jobs and affordable health insurance.

Genetic testing is so new that the health system is lagging behind the science. Because genetic test results have both psychological side effects and medical ambiguities, genetic counseling is imperative for patients and their families. However, there are insufficient numbers of qualified genetic counselors with experience in genetic testing, and the issues in the most qualified of hands are still complex and difficult. Physicians often have limited knowledge of human clinical genetics and feel inadequate to advise their patients, and insurance companies have yet to develop policies concerning genetic tests and genetic information. It is incumbent on governments to develop regulations and standards for the safety and efficacy of genetic tests and for the uses of genetic information, to limit discrimination and ensure confidentiality. Given the unclear interpretation of *BRCA1/2* genetic tests, the reasonably ineffective treatment options, and the potential psychological and societal side effects, it is not surprising that only about 60 percent of familial breast cancer patients and their families decide to take the genetic tests.

The unanswered questions about *BRCA1/2* genetic testing are many and important. What risks are associated with which mutations? What types of cancers result from which mutations? Should all people be allowed access to *BRCA1/2* tests? Can we develop effective treatments for these diseases? How can we ensure that the high costs of genetic tests and counseling do not limit this new technology to only a portion of the population? Our struggle with these and similar questions is just beginning as we develop genetic tests for more and more diseases over the next few decades.

References

Kahn, P. 1996. Coming to grips with genes and risk. *Science* 274:496–98.

Martin, A-M., and Weber, B. L. 2000. Genetic and hormonal risk factors in breast cancer. *Journal of the National Cancer Institute* 92:1126–1135.

Miki, Y., et al. 1994. A strong candidate for the breast and ovarian cancer - susceptibility gene *BRCA1*. *Science* 266:66–71.

Wopster, R., et al. 1995. Identification of the breast cancer susceptibility gene BRCA2. *Nature* 378:789–92.

Website

Health Technology Advisory Committee, Minnesota Department of Health. 1998. Genetic testing for susceptibility to breast cancer.

http://www.health.state.mn.us/htac/gt.htm

of the world. When people migrate from low-risk areas to high-risk areas, their risk of colon cancer increases to match that of the native residents. This finding suggests that environmental factors, including diet, may play a role in the incidence of colon cancer. In a long-range health study, the dietary habits of more than 88,000 registered nurses across the United States have been monitored since 1980. In that population, 150 cases of colon cancer were observed through 1986. Detailed analysis indicated that nurses who consumed daily meals of pork, beef, or lamb had a 2.5-fold higher risk of colon cancer than those who consumed such meals less than once a month. A more detailed analysis of diets strongly suggests that animal fat in the diet is the environmental risk factor for colon cancer. At the same time, a negative correlation exists between eating skinless chicken meat and the incidence of colon cancer. Laboratory studies have been undertaken to identify the mechanisms by which fat brings about a cancerous transformation of the intestinal epithelium. However, even in the absence of information about the mechanism of action, it would seem prudent to reduce one's intake of animal fat in order to reduce the risk of colon cancer.

Research on the role of external factors as a cause of cancer indicate that neither the environment in general nor pollution in particular is responsible for a large fraction of cancer cases. Rather, diet, tobacco, and other agents, including medical and dental X-rays, as well as drugs, play significant roles in cancer development. It is estimated that these agents cause 50 percent of all cancer cases; thus, personal choices (e.g., the composition of food in the diet, tobacco use, and suntanning) figure prominently in the cancer equation. Education and judicious changes in lifestyles could prevent a high percentage of all human cancers.

Chapter Summary

1. Cancer is a genetic disorder at the cellular level that can result from the mutation of a given subset of genes or from alterations in the timing and amount of gene expression. Although some forms of cancer show familial patterns of inheritance, few show clear evidence for Mendelian inheritance.
2. Cancer results from the uncontrolled proliferation of cells and from the ability of such cells to metastasize or migrate to other sites and form secondary tumors. Mutant forms of genes involved in the regulation of the cell cycle are obvious candidates for cancer-causing genes. The cell cycle is apparently regulated at several sites, with two types of gene products primarily involved, cyclins and kinases. The actions of these genes and the genes they control are important in regulating cell division, and links between these genes and the process of tumor formation are being discovered.
3. Although Mendelian patterns of cancer development are not clear cut, there are genes that predispose to cancer. The study of genes such as that for retinoblastoma has provided insight into the number and sequence of mutations that result in the development of tumors.
4. Tumor-suppressor genes normally act to suppress cell division. When these genes are mutated or have altered expression, control over cell division is lost. Molecular analysis of two such genes, *retinoblastoma* (*RB*) and *Wilms tumor* (*WT*), indicates that these genes work to control gene expression in different ways, either by affecting the activity of transcription factors or by themselves acting as transcription factors. Mutations in both alleles of a tumor-suppressor gene are necessary to promote the development of cancer.
5. Oncogenes are genes that normally function to maintain cell division, and these genes must be mutated or inactivated to halt cell division. If these genes escape control and become permanently switched on, cell division occurs in an uncontrolled fashion. In contrast to tumor-suppressor genes, a mutation in only one copy of an oncogene can promote the development of cancer.
6. In most cases of cancer, a series of mutations is necessary to cause the malignant state. Colon cancer is a useful model illustrating the multistep nature of cancer.
7. The cells of most tumors have visible chromosomal alterations, and the study of these aberrations has provided some insight into the steps involved in the development of cancer. This relationship between cancer and chromosome alterations has been best studied in leukemias, where the formation of hybrid genes or substitution of regulatory sequences is associated with the transformation of normal cells into malignant tumors.
8. Although cancer is the result of genetic alterations, the environment appears to be an important factor in cancer induction. Environmental agents include occupational exposure to physical or chemical substances, viruses, diet, and other personal choices such as the use of tobacco and suntanning.

Insights and Solutions

1. In disorders such as retinoblastoma, a mutation in one allele of the *retinoblastoma* (*RB*) gene can be inherited from the germ line, causing an autosomal dominant predisposition to the development of eye tumors. To develop tumors, a somatic mutation in the second copy of the *RB* gene is necessary, indicating that the mutation itself acts as a recessive trait. In sporadic cases, two independent mutational events, involving both *RB* alleles, are necessary for tumor formation. Given that the first mutation can be inherited, what are the ways in which a second mutational event can occur?

Solution: In considering how this second mutation can arise, several levels of mutational events need to be considered, including changes in nucleotide sequence and events involving whole chromosomes or chromosome parts. Retinoblastoma results when both copies of the *RB* locus have been lost or inactivated. With this in mind, perhaps the best way to proceed is to prepare a list of the phenomena that can result in a mutational loss or inactivation of a gene.

One way for the second *RB* mutation to occur is by a nucleotide alteration that converts the remaining normal *RB* allele to a mutant form of the gene. This alteration may occur through a nucleotide substitution or by a frameshift mutation caused by the insertion or deletion of one or more nucleotides. A second mechanism of mutation can involve the loss of the chromosome carrying the normal allele. This event would take place during mitosis, resulting in chromosome 13 monosomy, leaving the mutant copy of the gene as the only *RB* allele. This mechanism does not necessarily involve loss of the entire chromosome; deletion of the long arm (*RB* is on 13q) or an interstitial deletion involving the *RB* locus and some surrounding material would have the same result. As an alternative, a chromosome aberration involving loss of the normal copy of the *RB* gene might be followed by a duplication of the chromosome carrying the mutant allele. Two copies of chromosome 13 would be restored to the cell, but no normal allele of the *RB* gene would be present. Finally, a recombination event followed by chromosome segregation could produce a homozygous combination of mutant *RB* alleles.

More can be discovered about the mechanisms involved in RB by analyzing the cells from tumors by using a combination of cytogenetic and molecular techniques (such as RFLP analysis and hybridizations to look for deletions) to see which mechanisms are actually found in tumors and to what extent they play a role in generating the second mutation. Although such analysis is still in the preliminary stages, all the mechanisms proposed have been found in tumors, indicating that a variety of spontaneous events can bring about the second mutation that triggers retinoblastoma.

Problems and Discussion Questions

1. As a genetic counselor, you are asked to assess the risk for a couple who plans to have children, but where there is a family history of retinoblastoma. In this case, both the husband and wife are phenotypically normal, but the husband has a sister with familial retinoblastoma in both eyes. What is the probability that this couple will have a child with retinoblastoma? Are there any tests that you could recommend to help in this assessment?
2. Review the stages of the cell cycle. What events occur in each of these stages? Which stage is most variable in length?
3. Where are the major regulatory points in the cell cycle?
4. Progression through the cell cycle depends on the interaction of two types of regulatory proteins, kinases and cyclins. List the functions of each, and describe how they interact with each other to cause cells to move through the cell cycle.
5. What is the difference between saying that cancer is inherited and saying that predisposition to cancer is inherited?
6. Define tumor-suppressor genes. Why are most tumor suppressor genes expected to be recessive?
7. Review the differences among transcriptional activity, tissue distribution, and function of the tumor-suppressor genes associated with retinoblastoma and Wilms tumor. What properties do they have in common? Which are most different?
8. Distinguish between oncogenes and protooncogenes. In what ways can protooncogenes be converted to oncogenes?
9. How do translocations such as the Philadelphia chromosome lead to oncogenesis?
10. Given that 50 percent of all cancers are environmentally induced, and most of these are caused by actions such as smoking, suntanning, and diet choices, how much of the money spent on cancer research do you think should be devoted to research and education on the prevention of cancer, rather than on finding a cure?
11. The compound benzo[*a*]pyrene is found in cigarette smoke. This compound chemically modifies guanine bases in DNA. Such abnormal bases are typically removed by an enzyme, which hydrolyzes the base, leaving an apurinic site. If such a site is left unrepaired, an adenine is preferentially inserted *across from* the apurinic site. In a study of lung cancer patients, tumor cells from 15 out of 25 patients had a G to T transversion in the *p53* gene, which has a known role in cancer formation. Imagine that you are asked to testify as an expert witness in the following court case: A widow of a man who died of lung cancer is suing R. J. Reynolds for selling tobacco products that killed her husband (who was a life-long smoker). What do you tell the jury? (Science only; no personal expositions on lawyers or the legal system.) [*Reference: Nature* 350:377–78 (1991).]

Extra-Spicy Problems

12. About 5 to 10 percent of breast cancer cases are caused by an inherited susceptibility. An inherited mutation in a gene called *BRCA1* is thought to account for approximately 80 percent of

families with a high incidence of both early-onset breast and ovarian cancer. Table 22–8 summarizes some of the data that has been collected on *BRCA1* mutations in such families. Table 22–9 shows neutral polymorphisms found in control families (not showing an increased frequency of breast and ovarian cancer). Study these data and then answer the following questions:

(a) Note the coding effect of the mutation found in kindred group 2082 in Table 22–8. This results from a single base-pair substitution. Starting with a drawing of the normal double-stranded DNA sequence for this codon (with the 5′ and 3′ ends labeled), show the sequence of events that generated this mutation, assuming that it resulted from an uncorrected mismatch event during DNA replication.

(b) Examine the types of mutations that are listed in Table 22–8. Is the *BRCA1* gene likely to be a tumor-suppressor gene or an oncogene?

(c) Although the mutations in Table 22.8 are clearly deleterious, causing breast cancer in women at very young ages, each of the kindred groups examined had at least one woman who carried the mutation but lived until age 80 without developing cancer. Name at least two different mechanisms (or variables) that could underlie variation in the expression of a mutant phenotype. Then, in the context of the current model for cancer formation, propose an explanation for the incomplete penetrance of this mutation. How do these variables relate to this explanation? [Reference: *Science* 266:66–71 (1994).]

13. (a) Examine Table 22–9. What is meant by a neutral polymorphism? What is the significance of this table in the context of examining a family or population for *BRCA1* mutations that predispose an individual to cancer?

(b) Examine Table 22–9. Is the polymorphism PM2 likely to result in a neutral missense mutation or a silent mutation?

(c) Answer question (b) for the polymorphism PM3. [Reference: Science 266:66–71 (1994)].

TABLE 22–8 Predisposing Mutations in *BRCA1*

		Mutation		
Kindred	***Codon***	***Nucleotide Change***	***Coding Effect***	***Frequency in Control Chromosomes***
1901	24	−11bp	Frameshift or splice	0/180
2082	1313		Gln → Stop	0/170
1910	1756	Extra C	Frameshift	0/162
2099	1775	T → G	Met → Arg	0/120
2035	NA*	?	Loss of transcript	NA*

*NA indicates not applicable, as the regulatory mutatuon is inferred and the position has not been identified.

TABLE 22–9 Neutral Polymorphisms in *BRCA1*

			*Frequency in Control Chromosomes**			
Name	Codon Location	Base in Codon†	A	C	G	T
PM1	317	2	152	0	10	0
PM6	878	2	0	55	0	100
PM7	1190	2	109	0	53	0
PM2	1443	3	0	115	0	58
PM3	1619	1	116	0	52	0

*The number of chromosomes with a particular base at the indicated polymorphic site (A, C, G, or T) is shown.

†That is, position 1, 2, or 3 of the codon.

Selected Readings

Aaltonen, L.A. 1993. Clues to the pathogenesis of familial colorectal cancer. *Science* 260:812–16.

Ames, B., Magraw, R., and Gold, L. 1990. Ranking possible cancer hazards. *Science* 236:71–80.

Ames, B., Profet, M., and Gold, L. 1990. Dietary pesticides (99.99% all natural). *Proc. Natl. Acad. Sci. USA* 87:7777–81.

Anwar, S., et al. 2000. Hereditary non-polyposis colorectal cancer: an updated review. *Eur. J. Surg. Oncol.* 26:635–45.

Beijersbergen, R.L., and Bernards, R. 1996. Cell cycle regulation by the retinoblastoma family of growth inhibitory proteins. *Biochim. Biophys. Acta* 1287:103–20.

Benedict, W., et al. 1990. Role of the retinoblastoma gene in the initiation and progression of a human cancer. *J. Clin. Invest.* 85:988–93.

Bieche, I., and Lidereau, R. 1995. Genetic alterations in breast cancer. *Genes Chromosomes Cancer* 14:227–51.

Birrer, M., and Minna, J. 1989. Genetic changes in the pathogenesis of lung cancer. *Annu. Rev. Med.* 40:305–17.

Brown, M.A. 1997. Tumor suppressor genes and human cancer. *Adv. Genet.* 36:45–135.

Cavenee, W.K., and White, R.L. 1995. The genetic basis of cancer. *Sci. Am.* (Mar.) 272:72–79.

Chen, J., et al. 1998. Stable interaction between the products of the *BRCA1* and *BRCA2* tumor suppressor genes in mitotic and meiotic cells. *Molec. Cell* 2:317–28.

Compagni, A., and Christofori, G. 2000. Recent advances in research on multistage tumorigenesis. *Brit. J. Cancer* 83:1–5.

Cornelis, J.F., et al. 1998. Metastasis. *Am. Scient.* 86:130–41.

Croce, C., and Klein, G. 1985. Chromosome translocations and human cancer. *Sci. Am.* (Mar.) 252:54–60.

Damm, K. 1993. ErbA: Tumor suppressor turned oncogene? *FASEB J.* 7:904–09.

Easton, D., et al. 1993. Genetic linkage analysis in familial breast and ovarian cancer: Results from 214 families. *Am. J. Hum. Genet.* 52:678–701.

Elledge, S.J. 1996. Cell cycle checkpoints: Preventing an identity crisis. *Science* 274:1664–72.

Evan, G., and Littlewood, T. 1998. A matter of life and cell death. *Science* 281:1317–22.

Fearon, E.R. 1997. Human cancer syndromes: Clues to the origin and nature of cancer. *Science* 278:1043–50.

Feunteun, J., and Lenoir, G.M. 1996. *BRCA1*, a gene involved in inherited predisposition to breast cancer. Biochim. Biophys. *Acta* 1242:177–180.

Giaccone, G. 1996. Oncogenes and antioncogenes in lung tumorigenesis. *Chest 109* (Suppl. 5):130S–35S.

Hatkeyama, M., et al. 1994. The cancer cell and the cell cycle clock. *Cold Spring Harbor Symp. Quant. Biol.* 59:1–10.

Huan, B., and Siddique, A. 1993. Regulation of hepatitis B virus gene expression. *J. Hepatol.* 17(Suppl. 3): 520–S23.

Kinzler, K. W., and Vogelstein, B. 1996. Lessons from hereditary colorectal cancer. *Cell* 87:156–70.

——————. 1997. Gatekeepers and caretakers. *Nature* 386: 761–63.

Lengauer, C., Kinzler, K.W., and Vogelstein, B. 1997. Genetic instability in colorectal cancer. *Nature* 386: 623–27.

Nurse, P. 1997. Checkpoint pathways come of age. *Cell* 91:865–67.

Olsson, H., and Borg, A. 1996. Genetic predisposition to breast cancer. *Acta Oncol.* 35:1–8.

Peltomaki, P., et al. 1993. Genetic mapping of a locus predisposing to human colorectal cancer. *Science* 260:810–12.

Pines, L. 1996. Cyclin from sea urchin to HeLas: Making the human cell cycle. *Biochem. Soc. Trans.* 24:15–33.

Raff, M. 1998. Cell suicide for beginners. *Nature* 396:119–22.

Sherr, C. J. 1996. Cancer cell cycles. *Science* 274:1672–77.

Shibata, D., and Aaltonen, L.A. 2001. Genetic predisposition and somatic diversification in tumor development and progression. *Adv. Cancer Res.* 80:83–114.

Shito, K., et al. 2000. Pathogenesis of non-familial colorectal carcinomas with high microsatellite instability. *J. Clin. Pathol.* 53:841–45.

Smith, M.L., and Fornace, A.J., Jr. 1996. The two faces of tumor suppressor *p53*. *Am. J. Pathol.* 148:1019–22.

Solomon, E., et al. 1987. Chromosome 5 allele loss in human colorectal carcinomas. *Nature* 328:616–29.

Soussi, T. 2000. The p53 tumor suppressor gene: from molecular biology to clinical investigation. *Ann. N.Y. Acad. Sci.* 910:121–37.

Stahl, A., et al. 1994. The genetics of retinoblastoma. *Ann. Genet.* 37:172–178.

Timar, J., et al. 1995. Interaction of tumor cells with elastin and the metastatic phenotype. *Ciba Found. Symp.* 192:321–35.

von Lindern, M., et al. 1992. Translocation t(6;9) in acute non-lymphocytic leukaemia results in the formation of a DEK-CAN fusion gene. *Baillieres Clin. Haematol.* 5:857–79.

Weinberg, R.A. 1995. The molecular basis of oncogenes and tumor suppressor genes. *Ann. NY Acad. Sci.* 758:331–338.

——————. 1995. The retinoblastoma protein and cell cycle control. *Cell* 81:323–30.

——————. 1996. How cancer arises. *Sci. Am.* (Sept.) 275:62–70.

——————. 1996. E2F and cell proliferation: A world turned upside down. *Cell* 85:457–59.

Zang, H., Tombline, G., and Weber, B. L. 1998. *BRCA1*, *BRCA2*, and DNA damage response: Collision or collusion? *Cell* 92:433–36.

GENETICS MediaLab

The resources that follow will help you achieve a better understanding of the concepts presented in this chapter. These resources can be found either on the CD packaged with this textbook or on the Companion Web site found at **http://www.prenhall.com/klug**

Web Problem 1:

Time for completion = 15 minutes

What is the problem encountered by cells as they replicate their DNA during the S phase of the cell cycle? Does the problem involve the 3′ end or the 5′ end? What are the two parts of the telomerase molecule? Why is telomerase important? To answer these questions and view an animation of telomerase activity, visit Web Problem 1 in Chapter 22 of your Companion Web site, and select the keyword **TELOMERASE**.

Web Problem 2:

Time for completion = 10 minutes

What role does heredity play in susceptibility to cancer? The article on the characteristics of inherited forms of breast cancer, along with information in Chapter 23, will help you answer the following questions: 1. Given that individuals normally inherit a predisposition to cancer, not the cancer itself, how is the predisposition transmitted? 2. If individuals with *BRCA1* and *BRCA2* mutants have an increased risk of breast cancer; is the risk associated with a specific protein product of these genes or the lack of a functional product? 3. Are individuals with *BRCA1* and *BRCA2* mutations at higher-than-average risk for cancers other than breast cancer? To complete this exercise, visit Web Problem 2 in Chapter 22 of your Companion Web site, and select the keyword **BREAST CANCER**.

Web Problem 3:

Time for completion = 15 minutes

Where are the genes associated with cancer in the human genome? Dr. Philip McClean has produced a map of the human genome with the names and locations of oncogenes and tumor suppressor genes hot linked to the Online Mendelian Inheritance in Man database. Are the genes associated with human cancers clustered together or distributed throughout the genome? Pick a chromosome with four or more cancer genes, and examine the Online Mendelian Inheritance description of those linked genes. Do closely linked genes seem to cause similar or different cancers? Are most of the mutations you examined oncogenes or tumor suppressor genes? Do any of the oncogenes you found have an apparent viral origin? How did the virus's genes get into the human DNA? To complete this exercise, visit Web Problem 3 in Chapter 22 of your Companion Web site, and select the keyword **CANCER MAP**.

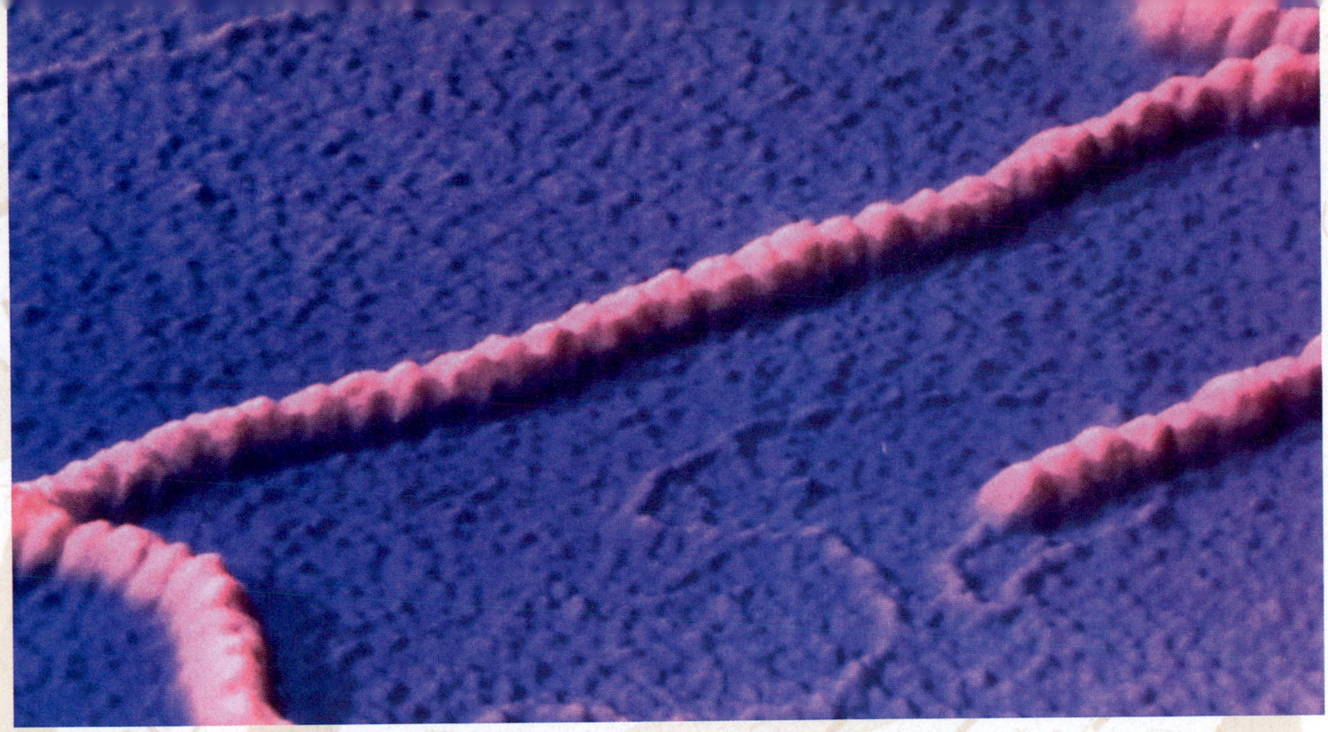

Chromatin fibers viewed using a scanning transmission electron microscope (STEM). (*Science VU/BMRL/Visuals Unlimited*)

23

Chromosome Genetics: Immunoglobulins, Isochores, and Chromatid Dynamics

KEY CONCEPTS

23.1 Chromosomes and Chromatin: A Brief History

23.2 Genetic Diversity in the Immune System
The Immunoglobulin Gene Family
The TCR Gene Family

23.3 Diversity in the Immune System Is Produced by Recombination
Light Chain Production in B Cells
Heavy Chain Production in B Cells
T Cell Receptors

23.4 Recombination in the Immune System
V(D)J Recombination and the RAG Proteins
Chromatin Remodeling Regulates V(D)J Recombination

23.5 The Chromosomal Landscape: Organization of Chromosome Regions and Genes

23.6 Vertebrate Genomes Are Mosaics of DNA Segments Called Isochores
Isochores and Chromosome Bands
Isochores, Chromosome Bands, and Gene Distribution

23.7 Chromosome Dynamics in Mitosis
Sister Chromatid Cohesion Begins in S Phase
Sister Chromatid Separation at Anaphase
Regulation of Separin Activity in Mitosis
Chromatid Separation in Meiosis

Our understanding of the relationship between DNA, genes, and chromosomes has developed by means of intermittent bursts of insight, interspersed with periods of relatively slow progress. Historically, the mechanics of mitosis and meiosis were described in the last part of the 19th century. Early in the 20th century, by proposing the chromosome theory of inheritance, Walter Sutton and Theodor Boveri first linked the behavior of chromosomes in meiosis to the principles of segregation and independent assortment discovered by Gregor Mendel. The structure of chromatin as a component of chromosomes came into focus in the 1970s with the discovery of nucleosomes. In spite of this and other advances, much of the basic level of chromosome organization between the nucleosome and the chromosome remain unknown. Development of recombinant DNA technology in the 1970s made it possible to isolate, clone, and characterize individual genes and their regulatory sequences. Using this technology, the Human Genome Project has provided us with a draft of the nucleotide sequence of the human genome that is being used to create a complete map of our genome.

Until recently, little progress has been made in integrating our knowledge of nucleotide sequence, gene organization, and gene regulation with chromosomal structure or in gaining insights into the molecular mechanisms involved in chromosome behavior in cell division. This situation is now changing, and we are now in the early stages of understanding the dynamic relationship between structure and function, from the nucleotide sequence of a gene up to the behavior of chromosomes in mitosis and meiosis. This chapter will explore these relationships at three levels. First, we will describe the immunoglobulin and T cell receptor gene families, and outline the molecular events associated with recombination and deletion in these genes. Next, we will examine the evidence for isochores as a unit of genome organization, and lastly, consider the molecular events that accompany the chromosome dynamics of cell division.

23.1 Chromosomes and Chromatin: A Brief History

It is now clear that nucleosomes are far more than a structural component of chromosomes; they are dynamic regulators of gene expression, and nucleosome structure and chromatin remodeling are important factors in determining which genes are expressed and when they are expressed. We will explore gene expression in genes of the immune system and the role of chromatin structure and chromatin remodeling in regulating expression of this gene set. At a higher level of organization, information from the Human Genome Project has been used to assemble the nucleotide sequence of several chromosomes and to integrate the data with structural landmarks such as chromosome bands, centromeres, and telomeres. Such integration permits construction of a bottom-up view of the chromosome, from the nucleotide sequence level to the gene level, and it provides some insight into how many genes are on specific chromosomes, what those genes are, where they are located, and their relationship to the chromosome landmarks mentioned. In this chapter, we will consider the isochore organization of the human genome and how such an organization is related to the discovery of gene-rich and gene-poor regions of human chromosomes, as well as how these regions can be read from the Giemsa banding patterns seen at the cytogenetic level. Finally, we will investigate and summarize what is known about the molecular events that occur during chromosome replication and division in mitosis and meiosis. Although the picture is still incomplete, recent advances in the area provide a glimpse of the molecular mechanisms that regulate genome duplication and the distribution of genome copies to daughter cells.

23.2 Genetic Diversity in the Immune System

The immune system is an effective barrier against invasion of potentially harmful foreign agents, which are recognized as nonself and destroyed by the immune system. Genomic alterations are a hallmark of normal events in the maturation and function of the immune system. These events include rounds of recombination-associated deletion of DNA sequences in specific cells.

One component of the immune system produces antibodies against foreign substances. **Antibodies** are proteins produced and secreted by the **B cells** of the immune system. Agents that trigger antibody production are called **antigens**. Antigens can be free molecules, or they can be part of the surface of a cell, a microorganism, or a virus. Among vertebrates, each individual can produce millions of different antibodies, with each antibody responding to a different antigen. Coincidentally, there are millions of different antigen recognition sites on the surfaces of the immune system's **T cells**. Antibodies made by B cells and the receptors of T cells are encoded by a small family of genes. From a set of less than 300 genes, recombination events and associated deletions, as well as mismatching of rejoined DNA segments in immature cells of the immune system create a vast potential for an immune response.

The Immunoglobulin Gene Family

In humans, antibodies are produced by B cells, a type of lymphocyte (white blood cell). Five classes of antibodies or **immunoglobulins (Ig)** are recognized (Table 23–1). The IgG class represents about 80 percent of the antibodies found in the blood and is a well-studied class of antibodies. A typical antibody molecule (Figure 23–1) consists of two different polypeptide chains, each present in two copies and held together by cross-linking disulfide bonds. Each larger or **heavy chain (H)** contains approximately 440 amino acids. The sequence of the first 110 amino acids at the N-terminus differs among heavy chains produced by different B cells and is known as the **variable region (VH)**. The remaining amino acids are the same in all heavy chains and make up the **constant region (CH)** of the heavy chain. In humans, genes on the long arm of chromosome 14 encode the H chains.

TABLE 23–1 Categories and Components of Immunoglobulins

Ig Class	*Light Chain*	*Heavy Chain*	*Tetramers*	
IgM	κ or λ	μ	$\kappa_2\mu_2$,	$\lambda_2\mu_2$
IgD	κ or λ	δ	$\kappa_2\delta_2$,	$\lambda_2\delta_2$
IgG	κ or λ	γ	$\kappa_2\gamma_2$,	$\lambda_2\gamma_2$
IgE	κ or λ	ϵ	$\kappa_2\epsilon_2$,	$\lambda_2\epsilon_2$
IgA	κ or λ	α	$\kappa_2\alpha_2$,	$\lambda_2\alpha_2$

Each **light chain (L)** consists of 220 amino acids, with the first 110 amino acids making up the variable region (VL). The 110 amino acids at the C-terminal end of the molecule are the same in all light chains and form the constant region (CL) of the light chain. Two types of L chains are present in humans: **kappa (κ) chains**, encoded by genes on chromosome 2, and **lambda (λ) chains**, encoded by genes on chromosome 22. Together, the variable regions of the heavy and light chains form the **antibody combining site**. Each combining site has a unique structural configuration that allows it to bind to a specific antigen, like a key in a lock.

The TCR Gene Family

T cells play a vital role in the immune response. T cell subtypes act as the on and off switches for the immune response. The ability of T cells to participate in an immune response depends on the presence of **T cell receptors (TCRs)** that interact with cell-associated antigens. The receptors are membrane-bound proteins (Figure 23–2) made up of two chains, alpha (α) and beta (β) or gamma (γ) and delta (δ). Each TCR chain is made up of an N-terminal variable region that protrudes from the cell surface, and a C-terminal constant region embedded in the plasma membrane. Together, the variable regions of the α and β (or γ and δ)

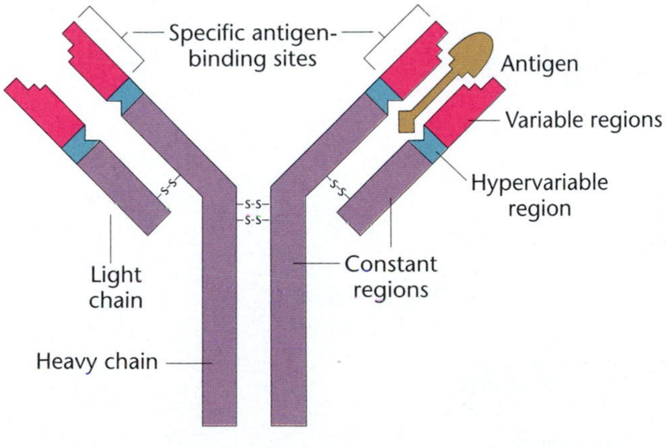

FIGURE 23–1 A typical antibody (immunoglobulin) molecule. The molecule is Y-shaped and contains four polypeptide chains. The longer arms are heavy chains, and the shorter arms are light chains. Each chain contains a constant region and a variable region. The variable and hypervariable regions form an antibody combining site that interacts with a specific antigen, similar to a lock-and-key mechanism. The chains are joined together by disulfide bonds.

chains form the **antigen combining site**, that recognizes and binds to antigens on the surface of cells, microorganisms or viruses. Once bound to an antigen, the T cell activates the immune response and the production of B cells making an antibody against the antigen.

One important characteristic of the immune system is its ability to generate new cells that contain different antigen receptors and different antibodies. Since there are billions of such combinations, it is impossible for separate genes to encode each combination. There is simply not enough DNA in the human genome to encode the hundreds of millions or billions of receptors and antibodies we produce. One of the long-standing questions in immunogenetics is how this vast molecular variability in antibodies and antigen receptors is encoded in the genome.

23.3 Diversity in the Immune System Is Produced by Recombination

Antibody diversity results from a number of special features of the genome, including a unique form of recombination. To understand how this process of recombination results in the production of millions of different antibodies from a small number of antibody genes, it is necessary to understand the organization of immunoglobulin genes and the events associated with the production of L chains and H chains.

Light Chain Production in B Cells

In humans, the kappa L-chain genes (Figure 23–3) consist of several elements: the **L–V (leader–variable region**, the **J (joining) region**, and a **C (constant) region**. There are 70–100 DNA segments in the L–V region, each with a different nucleotide sequence adjacent to a promoter. A significant number of the L–V segments are pseudogenes. The J region contains six different DNA segments, and the C region contains only one DNA segment. A regulatory region but no promoter, is located between the last J segment and the C segment, ensuring that segments cannot be transcribed individually.

During B cell maturation, one of the 70–100 L–V segments (in our example, it is L_2V_2) and its promoter are *randomly* joined by a recombination event to one of the six J regions (in our example, J_3), and the C region. This recombination event forms a functional L-chain gene [Figure 23–3 (line 2)]. All DNA between the randomly selected L–V segment and the J_3 segment is excised and destroyed. As we will see in a section later in the chapter, this form of recombination is different from the recombination associated with meiosis and involves different enzymes.

The joining event between L–V segments and J segments is imprecise and can occur over a span of about 6 bp. In addition, during this recombination event, the ends of the DNA are subject to a **break-nibble-add mechanism**, in which a few nucleotides are randomly removed and a few nucleotides are randomly added. The newly created gene shown in Figure 23–3 contains three exons, L_2, V_2–J_3, and C, and two introns, one between L_2 and V_2–J_3, and one

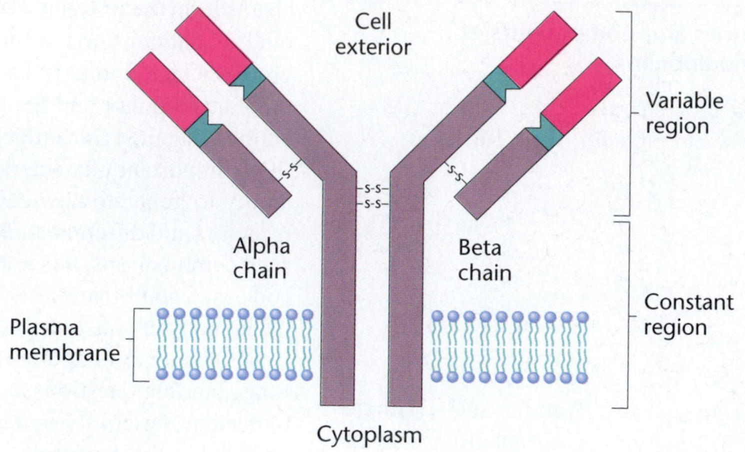

FIGURE 23–2 T cells produce surface receptors that recognize antigens. This T cell receptor is composed of two chains: an alpha (α) chain and a beta (β) chain. Each chain has a constant region and a variable region, and each is encoded by a different gene, containing L–V, D, J, and C regions.

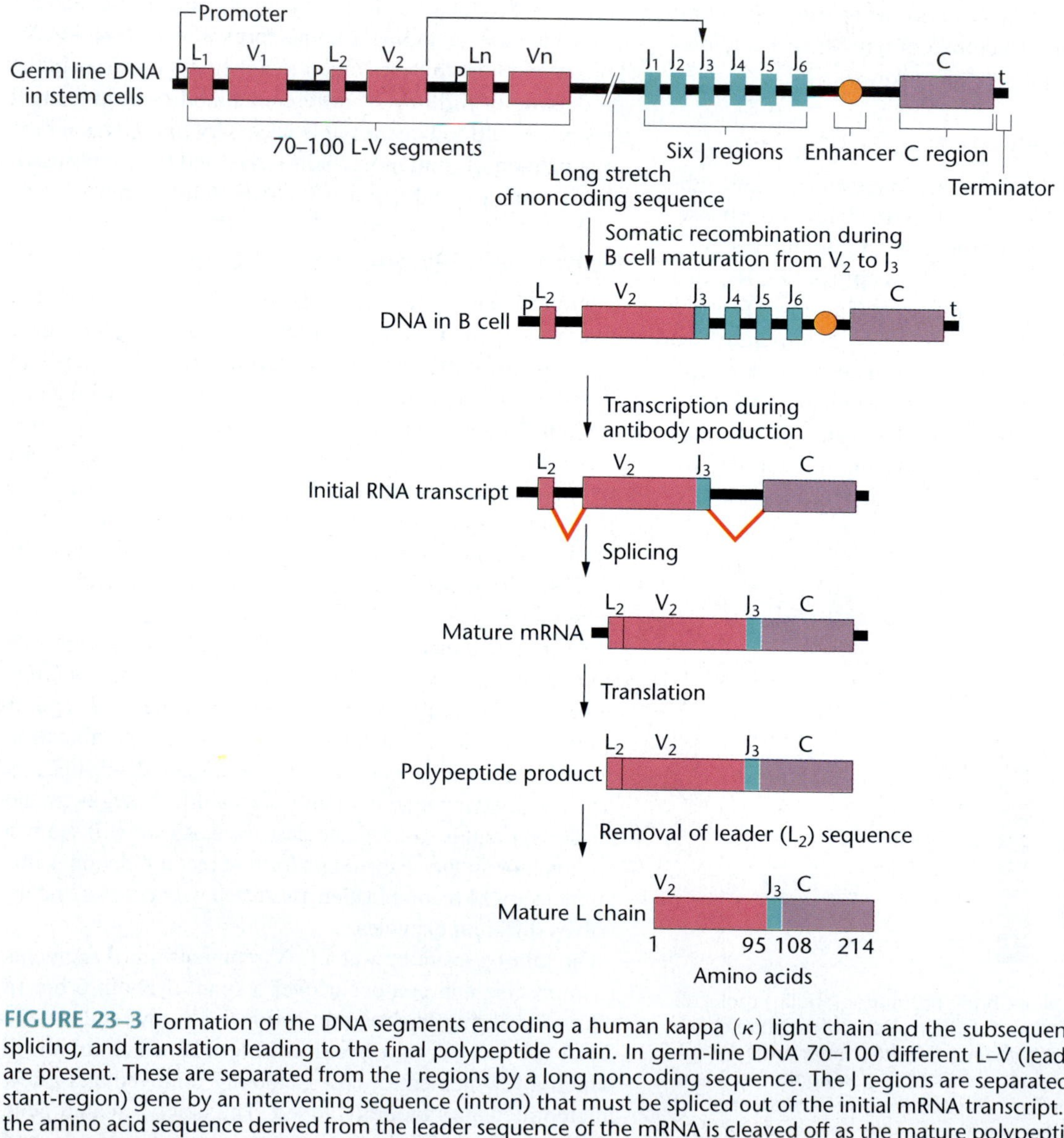

FIGURE 23–3 Formation of the DNA segments encoding a human kappa (κ) light chain and the subsequent transcription, mRNA splicing, and translation leading to the final polypeptide chain. In germ-line DNA 70–100 different L–V (leader–variable) segments are present. These are separated from the J regions by a long noncoding sequence. The J regions are separated from a single C (constant-region) gene by an intervening sequence (intron) that must be spliced out of the initial mRNA transcript. Following translation, the amino acid sequence derived from the leader sequence of the mRNA is cleaved off as the mature polypeptide chain passes across the cell membrane.

between J_3 and C. Segments J_4–J_6 become part of the second intron. This new L-chain gene is transcribed, its mRNA processed and translated to form kappa polypeptides that become part of an antibody molecule. The new gene, created by recombination, is stable and is passed on to all progeny of the B cell.

Antibody diversity in the kappa L chains is the result of combining any of the 70–100 L–V segments with any of the six J segments and generates about 600 different kappa genes. Since each joining reaction during recombination can occur over several base pairs, it increases the possible number of kappa genes to several thousand. Additional diversity is generated by the break-nibble-add mechanism and the combination of κ L chains with the diverse H chains, produced in B cells.

Heavy Chain Production in B Cells

The heavy chain genes extend over a large region of DNA in the human genome, and each gene includes four types of segments: L–V (leader–variable), D (diversity), J (joining), and C (constant). There are approximately 300 different V segments, 10–30 different D segments, and 6 different J segments (Figure 23–4). In addition, there are 9 different C segments, for the five classes of immunoglobulins. (See Table 23–1.)

During B cell maturation, recombination randomly joins one L–V segment with a D segment and then with a J segment. In Figure 23–4, the L–V–D–J sequence, assembled by recombination, lies adjacent to the C segment of the μ H-chain gene. This joining resembles that described earlier for the L chain, and includes a break-nibble-add step.

The random recombination of H-chain components forms a large number of H-chain genes. If we assume there are 25 D segments, then combining any of these D segments with any of the 300 V segments and 6 J segments, we find that about 45,000 H-chain genes are formed. Add in the imprecise joining and break-nibble-add for J–D and V to D–J recombination events, and we find that the total number of different H-chain genes climbs markedly. By multiplying the combination of all H-chain genes with all L-chain genes, hundreds of millions of combinations are possible, all derived from a few hundred coding sequences.

In sum, antibody diversity is generated by a combination of special features of the immunoglobulin genes: (1) multiple sets of variable DNA segments, (2) multiple sets of diversity and joining segments, (3) multiple splice locations with break-nibble-add joining, and (4) multiple combinations of L chains and H chains. As B cells mature, DNA recombination events rearrange these genes so that each mature B cell comes to encode, synthesize, and secrete only one specific type of antibody. In other words, each B cell can make only one type of L chain and one type of H chain. When an antigen is present, it stimulates a T cell, which in turn activates a B cell that encodes an antibody against that antigen to divide and differentiate. As a result, plasma cells are produced, all of which synthesize one type of antibody that interacts with the antigen.

T Cell Receptors

The generation of diversity in T-cell receptors is accomplished through somatic recombination events that occur during T-cell maturation, in a manner similar to the generation of antibody diversity in B-cell maturation. The V region of the β-receptor gene (Figure 23–5) is composed of about 25 different V genes, one D gene, and approximately six J genes, associated with each of the two C genes. The organization of the variable regions of the α, δ, and γ regions of T-cell receptor genes are similar to those of the β gene.

The diversity of T cell antigen receptors derives from the recombination of a V segment with a D and J segment and the association of this VDJ segment with one of the C segments. As with the immunoglobulin genes, there is some imprecision in the cutting and joining of segments, contributing to receptor diversity.

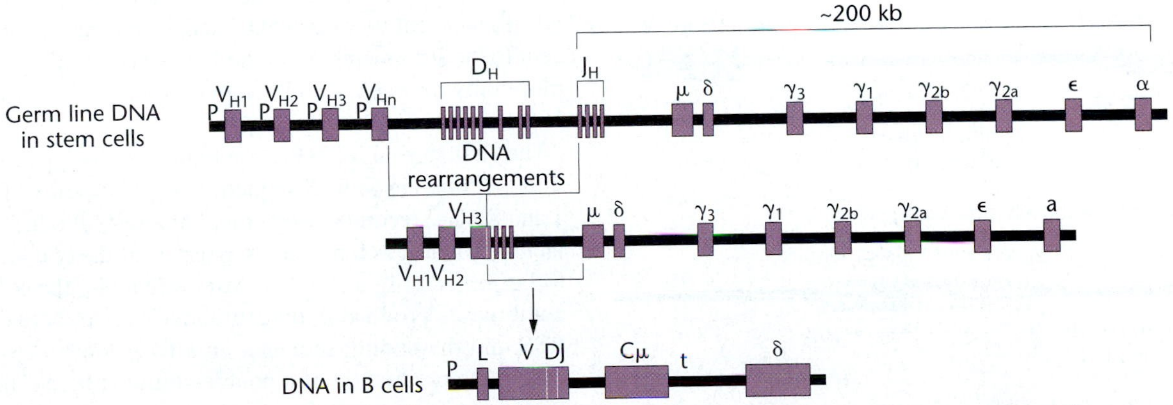

FIGURE 23–4 The variable region of the H chain is assembled by joining three different DNA segments together: L–V, D, and J. In embryonic DNA, these segments occur in clusters separated by long intervals of other DNA sequences adjacent to the mu (μ) gene C region, which will produce the μ heavy chain. In a maturing B cell, a random combination of one L–V region with one of the 20 D regions, and one of the four J regions produces the μ heavy chain gene. This gene is transcribed from its promoter (P) to the terminator (t) and translated in the B cell. In other B cells, different combinations of H-chain segments are joined together, producing a large number of different μ heavy chains.

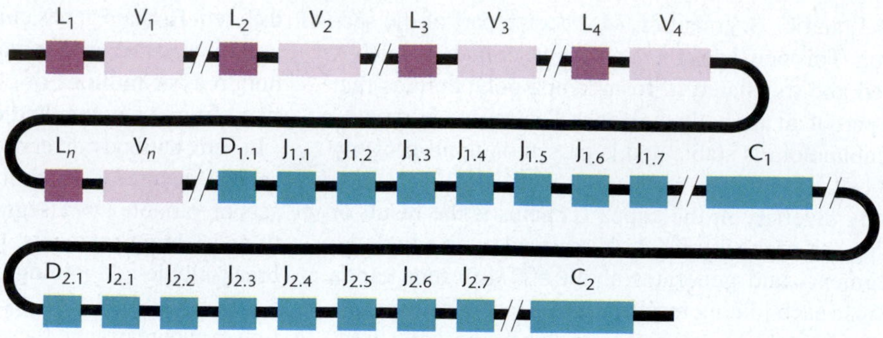

FIGURE 23–5 Organization of the T cell beta (β) receptor gene. There are about 25 different L–V gene segments, one D gene, and seven J genes associated with each of the two C (constant-region) genes. During T cell differentiation, recombination combines one L–V segment, one D segment, and one J segment adjacent to each C gene.

23.4 Recombination in the Immune System

Clear evidence for recombination as the mechanism for generating antibody diversity came from the work of Susumu Tonegawa and colleagues in the mid-1970s. Using recombinant DNA techniques, they demonstrated that DNA segments coding for the variable regions of L chains are distant from J segments in embryonic cells, but adjacent to one another in antibody-producing cells. In their experiments, Tonegawa and his colleagues isolated, cloned, and characterized a DNA fragment encoding a λV segment (VL) derived from embryonic cells. The organization of such a segment was compared with an equivalent region isolated from an antibody-producing cell. In the embryo, the region encoding the first 95 amino acids of the VL region was separated by 4.5 kb from the region encoding the remaining 13 amino acids (the J region). However, in the DNA isolated from the antibody-producing cell, the V and J segments were joined to form a single transcription unit encoding a specific VL gene (Figure 23–6). This work inspired a large-scale effort to study the recombination events associated with the maturation of the immune system.

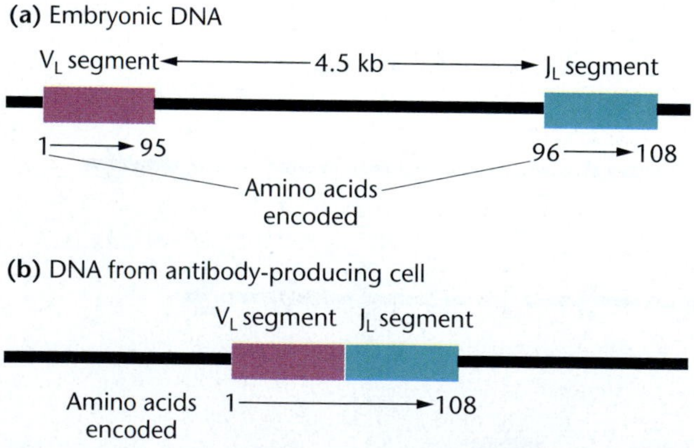

FIGURE 23–6 (a) In DNA, isolated and cloned from embryonic germ line DNA, the V region of an L gene encoding the first 95 amino acids of the L chain are located 4.5 kb away from the DNA encoding the 13 amino acids of the J region. (b) In DNA isolated from an antibody-producing cell, the V and J regions are contiguous, forming a single transcription unit. This evidence provided support for the hypothesis that recombination is involved in forming antibody genes in maturing B cells.

V(D)J Recombination and the RAG Proteins

At the molecular level, the process of recombination in the formation of antibody genes and antigen receptor genes is complex and involves a number of participating proteins, some of which remain to be identified. Proteins encoded by the recombination-activating genes **RAG-1** and **RAG-2** play important roles in these events. These adjacent loci on chromosome 11 in the human genome encode recombinases, are cotranscribed, contain no introns, and are both required for recombination activity. The genes are normally expressed only in lymphocytes and only in stages of maturation when V(D)J recombination is taking place. Recombination is cell specific: Only antibody genes (Ig genes) are rearranged in maturing B cells, but not in T cells, and TCR genes are recombined in T cells, but not B cells.

During V(D)J recombination, gene segments located at distances of up to several hundred kilobases are brought into alignment, forming a synaptic complex. This process depends on the presence of **recombination signal sequences (RSSs)** located adjacent to all coding regions of immunoglobulin and TCR genes. These sequences consist of a conserved heptamer (CACAGTG) flanking the coding segment and a nonamer (ACAAAAACC), separated from the heptamer by a spacer region. The nonconserved spacer can be 12 nucleotides or 23 nucleotides in length. Recombination and joining between coding segments require one spacer of each type. For example, in the κ L-chain gene, 12 bp spacers are adjacent to V segments, and 23 bp spacers are adjacent to the J segments. With the 12/23 rule, joining can take place only between V and J segments and not between two V segments or two J segments.

The first step in V(D)J recombination involves recognition and binding to RSS sequences by a complex of RAG-1 and RAG-2 recombinases, containing several other protein factors. Binding is followed by synapsis of the two segments to be joined (Figure 23–7). After synapsis, the RAG recombinases produce double-stranded breaks between the RSS and the coding sequence in a two step process. The first is the production of a double-stranded break; the second is the formation of a hairpin structure on the coding ends (Figure 23–7).

The RAG proteins and other proteins usually involved in repair of radiation damage open the hairpins on the ends of the coding segments. Opening the hairpins can occur at the tip, or slightly off center. If the hairpin is opened off center, a short 3′ or 5′ overhang is produced. The coding ends to be

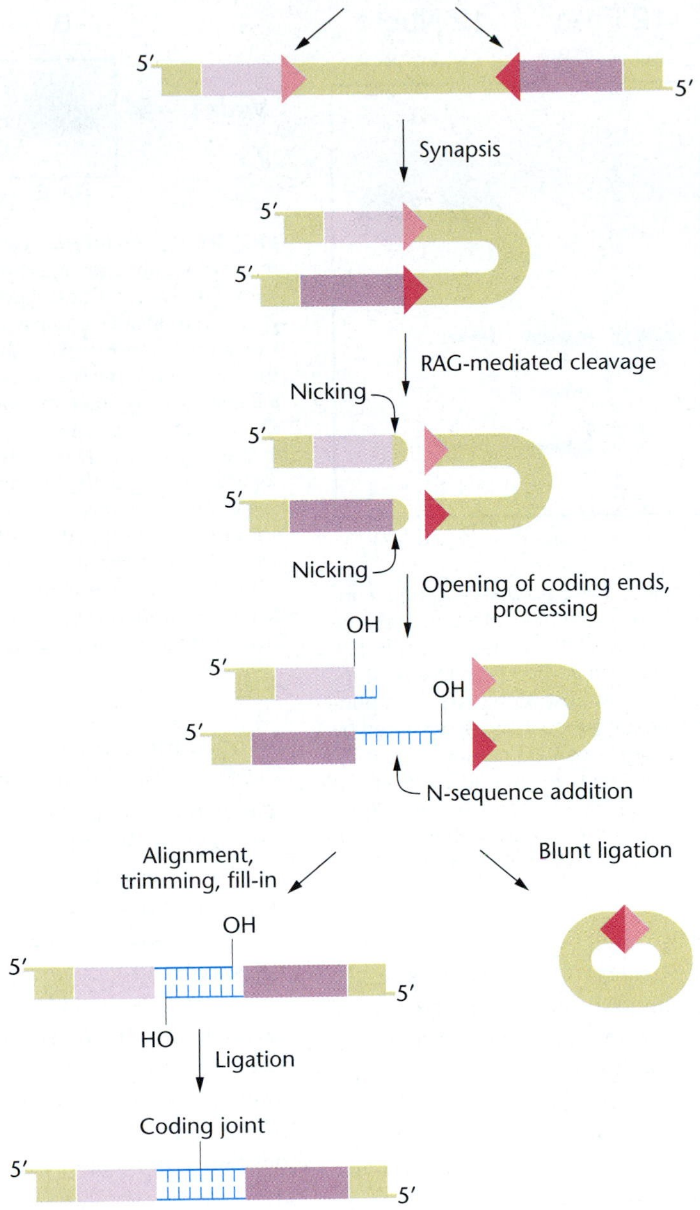

FIGURE 23–7 Steps in V(D)J recombination. In the first step, a matching set of RSSs is recognized by the RAG complex, and synapsed to form a loop. In the second step, double stranded breaks in the DNA are produced by the RAG complex on one side of the RSSs, forming hairpin loops at the coding ends of the DNA. The coding ends are opened by nicking at or near the hairpin tip. This event is often followed by addition of nucleotides, filling in of single stranded regions, alignment of the cut ends, and ligation to form a mature antibody gene. The fragment carrying the RSS signals is ligated with the RSS sequences reversed, which prevents rebinding of the RAG complex.

joined might not be complementary, and the details of events preceding ligation are not yet known in detail. It is known that the two coding ends anneal by short, 1–3 bp homologies, followed by further processing, trimming and filling in prior to ligation (Figure 23–8). These steps, along with filling overhangs, create additional diversity in the ligated segments of the Ig or TCR gene. The fragment marked with the RSS signals is ligated to form a circle, which is later broken down.

Production of a functional Ig gene requires that several recombination events occur in H-chain and L-chain genes in a single B cell. These events take place in a preferred order. In H-chain genes, D to J recombination occurs before V to DJ recombination. Similarly, several recombination events are required to produce the two genes encoding the polypeptides for T cell receptors. This remarkable form of recombination is limited to seven loci (α, β, δ, and γ in T cells; IgH, κ, and λ in B cells) in two cell types (B and T) only during maturation of the immune system.

Chromatin Remodeling Regulates V(D)J Recombination

The question of how V(D)J recombination is regulated by developmental time, cell type, and loci remains unresolved. In some cases, recombination occurs at only one allele (allelic exclusion), adding to the question of how the RAG recombinases target the correct RSSs in the right cell at the right loci at the right time. Fred Alt and his colleagues originally proposed that accessibility of the loci and their RSSs is the key to regulating V(D)J recombination. According to this idea, the TCR and Ig loci are differentially packaged into chromatin in T and B cells. Physical changes in such packaging make the loci more or less accessible to the RAG proteins at specific times. Recent work using *in vitro* systems has provided some clues about the relationship between chromatin structure and RAG-mediated recombination, and confirmed the role of chromatin remodeling in regulating V(D)J recombination.

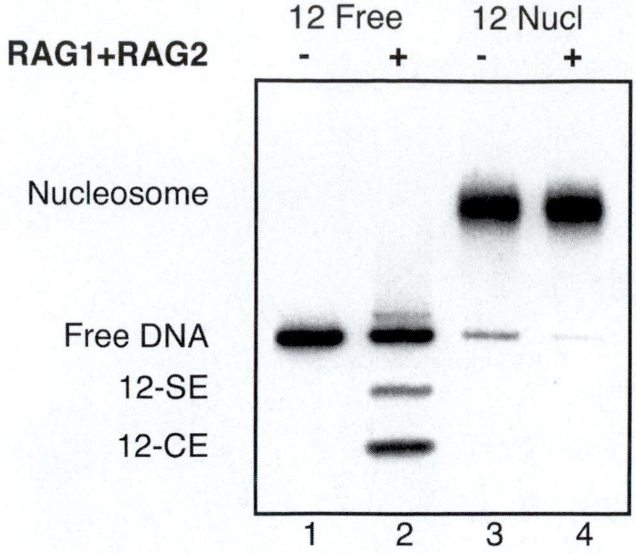

FIGURE 23–8 In this experiment, a 196 bp DNA fragment containing a functional RSS signal was assembled into a nucleosome covering the RAG binding site (lanes 3, 4). Nucleosome-free 196 bp fragments (lanes 1, 2) served as a control. The DNA was incubated in the presence (RAG+) or absence (RAG−) of purified RAG-1 and RAG-2. The reaction products were separated by gel electrophoresis. In the presence of RAG complex (lane 2), nucleosome-free DNA is cleaved, releasing the RSS spacer (12 SE and 12 CE). When nucleosome DNA is incubated with RAG-1 and RAG-2, the RSS sequence is not cut, indicating that nucleosomes block recognition and cutting by RAG complexes.

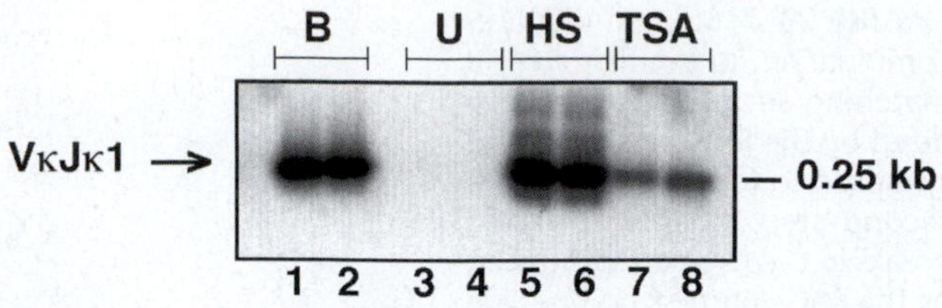

FIGURE 23–9 Histone acetylation in pre-B cells stimulates immunoglobulin κ recombination. The pre-B cell line was treated with the histone deacetylase inhibitor, trichostatin A (TSA) (lanes 7, 8). As a positive control, recombination was induced by heat shock (HS) (lanes 5, 6) in the same cell line. Another sample of the same pre-B cell line was untreated (U) (lanes 3, 4). DNA from a B cell line carrying a rearrangement of the κ gene (B) (lanes 1, 2) was also used as a control. After incubation, rearrangements in the κ gene (VκJκ1) were assayed by PCR, and the products separated by gel electrophoresis. Keeping the DNA acetylated (by inhibiting removal of acetyl groups, lanes 7 and 8) activated VκJκ1 recombination. Other experiments showed that this activation was not due to an increase in expression of the RAG genes, nor increased levels of RAG proteins. The results are consistent with increased acetylation at the VκJκ1, providing increased access to RAG proteins already present in the cells.

As outlined in Chapter 20, nucleosomes limit access of *trans*-acting regulatory molecules to DNA and as a result, transcription is inhibited. In addition, distortion of DNA wound around the histone core can prevent binding of transcription factors, and the higher order structure that results from the formation of chromatin fibers can further hide DNA sequences from proteins such as transcription factors and other regulatory molecules. In general, then, nucleosomes and higher levels of chromatin organization inhibit events such as transcription and recombination.

The first question is whether nucleosomes actually inhibit V(D)J recombination. A number of *in vitro* systems have been developed and used to answer this question. In one of these systems, a 196-bp restriction fragment containing a functional 12-spacer RSS fragment was assembled into a nucleosome that covered the RAG cleavage site (Figure 23–8, lanes 3,4). Nucleosome-free 196 bp fragments served as a control (Figure 23–8, lanes 1, 2). Purified RAG-1 and RAG-2 were added to test whether the initiation of a double-stranded break (DSB), the first step in V(D)J recombination, would occur. As shown in Figure 23–8, in the presence of RAG-1 and RAG-2, DSBs are produced only in free DNA (lane 2). The presence of a nucleosome blocks the formation of DSBs.

The second question is whether or not acetylation of nucleosome histones by histone acetyltransferases (HATs) can stimulate V(D)J recombination. Under *in vivo* conditions, inhibiting histone deacetylases (HDACs) activates V(D)J recombination in pre-B cells (Figure 23–9). Other researchers have confirmed that histone acetylation is tightly coupled to V(D)J recombination in transgenic mice, and that chromatin remodeling is important in providing access to RSSs sites.

In contrast, *in vitro* experiments from several laboratories have showed no change in RSS cleavage when histones in individual nucleosomes are acetylated or even hyperacetylated. The differences between these *in vitro* experiments and the successful *in vivo* experiments just outlined may indicate that acetylation of nucleosome histones is only one step in stimulating V(D)J recombination and, that other remodeling activities are present *in vivo*, but not in the *in vitro* systems. The larger questions of how V(D)J recombination is stimulated in a cell-specific manner at a certain time in development remain to be answered.

23.5 The Chromosomal Landscape: Organization of Chromosome Regions and Genes

Metaphase chromosomes have several distinguishing landmarks, including telomeres, centromeres and blocks of heterochromatin that can be seen with a light microscope. In addition, the chromosomes have a characteristic series of dark and light bands when trypsin and Giemsa are used for staining. (See Chapter 4.) The presence of these features indicate that chromosomes are not uniform in structure, but instead, have a unique landscape that extends from the nucleotide sequence of DNA up to the level of the intact chromosome. In the previous section, we explored the relationship between chromatin remodeling and the expression of genes that encode components of the immune system. In this section, we will take a more global view of chromosomes and examine the association between chromosome organization and the distribution of genes.

Nucleotide sequence data gathered as part of the Human Genome Project has been combined with earlier work on DNA composition and chromosome banding studies, revealing that the human genome is organized into subchromosomal domains with significant transcriptional and structural differences. We will examine the organizational differences, including GC/AT content, CpG islands, gene-rich and gene-poor regions, the noncoding repetitive sequences, and matrix attachment sites.

23.6 Vertebrate Genomes Are Mosaics of DNA Segments Called Isochores

Before DNA sequencing, investigators studied the nucleotide composition of DNA by using equilibrium density gradient centrifugation. (See Chapters 2 and 4 to review this form of centrifugation.) In this technique, the DNA forms a peak at a location in the gradient that is directly proportional to its GC content (Figure 23–10). Smaller peaks are called satellites. Using compounds that bind differentially to GC-rich or AT-rich sequences, we find that the main band DNA (the nonsatellite, nonribosomal DNA) of humans and other vertebrates can be separated into several different components, called **isochores** (Figure 23–10). In other words, the human genome is a mosaic of long (> 300 kb) DNA segments (isochores) belonging to a number of families, with different GC content (Figure 23–10). About 63% of the genome consists of GC-poor, light (L1 and L2) isochores, and a number of GC-rich, heavy (H1, H2, and H3) isochores that make up 24%, 7.5%, and 4.7%, respectively, of the genome. Not all eukaryotic genomes have the same isochore organization. The isochore pattern of the human genome and several other organisms is shown in Figure 23–11.

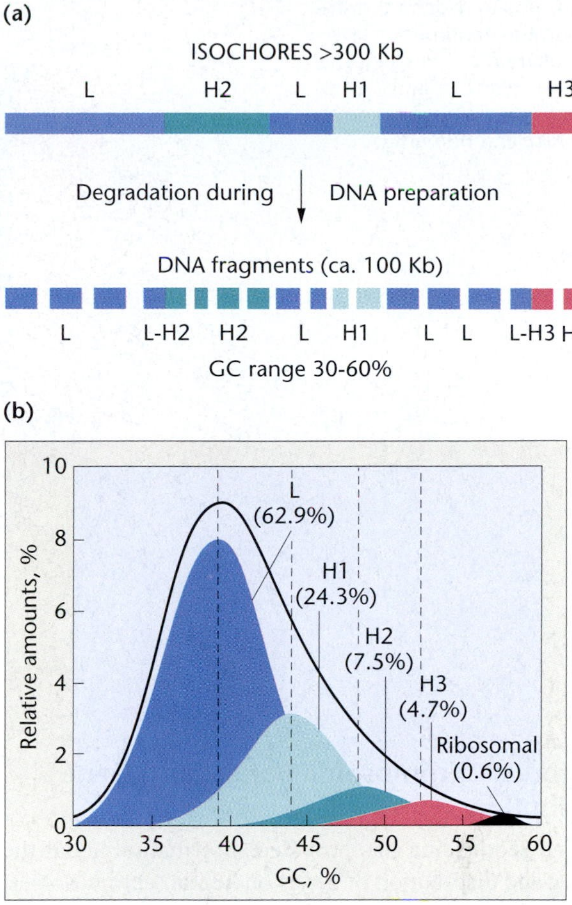

FIGURE 23–10 Isochore organization of the human genome. (a) The genome is a mosaic of large (> 300 kb) segments of DNA (isochores) that differ in CG content. During DNA extraction and purification, the DNA is broken into smaller fragments (~100 kb) (b) When separated by CsCl density gradient centrifugation, the DNA fragments can be separated by their density (directly proportional to GC content) into a number of families: L (L1 + L2), H1, H2, and H3. The blue line represents the density profile of human DNA. The broken vertical lines indicates the modal GC content of each family, and the percentages indicate the relative amounts of each isochore family in the human genome.

Isochores and Chromosome Bands

The relationship between isochores isolated by centrifugation and the organization of chromosomes became apparent when investigators began to assemble chromosome maps that correlated base composition with chromosome bands. Human metaphase chromosomes have three different types of bands: C bands (centromeric regions), G bands (Giemsa bands), and R bands (reverse bands). A subset of R bands, called T bands, are mainly located at telomeric regions. (For further information, review chromosome banding in Chapter 4.) Compositional maps of chromosomes combine the GC composition of cloned genes with a physical map of the chromosome, revealing the genomic distribution of genes and isochores. Figure 23–12 is a compositional map, showing the distribution of genes and isochores on the long arm of human chromosome 21.

Compositional maps show that G bands (the dark bands in most chromosome diagrams) are composed of mainly of isochores from the L1 and L2 family (GC-poor DNA); R′ bands (R bands that are not T bands) are composed of almost equal amounts of L1/L2 isochores and GC-rich H1 isochores. T bands are composed of the GC-rich H3 family, with some contribution by some GC-rich H1 and H2 families. In other words, the human genome is actually a mosaic of long DNA segments with vastly different GC content, and the staining pattern seen in Giemsa-banded chromosomes is a reflection of this molecular variation.

The larger question, however, is what, if anything, do isochores and mosaic organization of GC sequences on chromosomes have to do with the distribution of genes in the genome? Are genes more or less distributed equally along the length of a chromosome, or are they clustered in regions that are interspersed with regions containing relatively few genes? If so, do these regions correspond to isochores that are GC-rich or GC-poor? Compositional maps provide clues to this relationship, but definitive answers require nucleotide sequence information spanning the length of a chromosome.

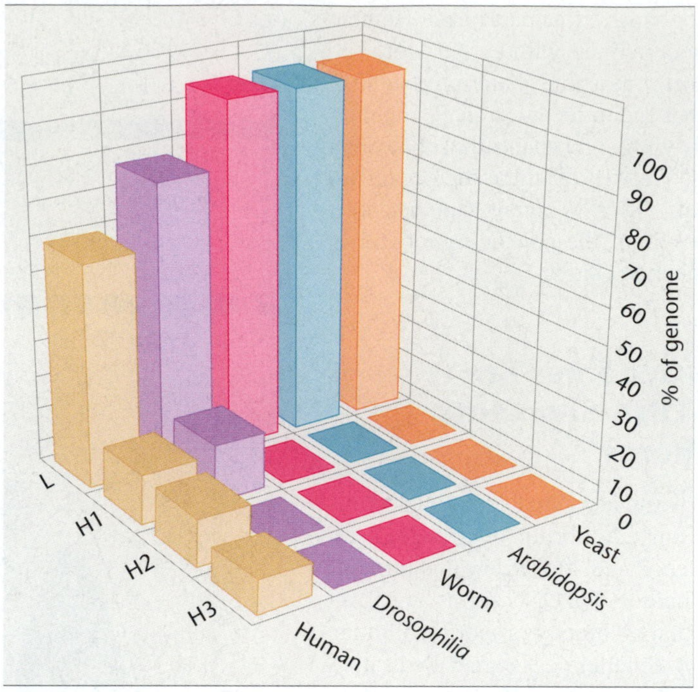

FIGURE 23–11 Isochore distribution in the genomes of several eukaryotic organisms, including yeast, plants (*Arabidopsis*), *C. elegans* (worm), *Drosophila*, and humans.

Isochores, Chromosome Bands, and Gene Distribution

Genome sequencing has provided information about the number and distribution of genes on human chromosomes, and this information can be correlated with our knowledge of isochore distribution and their relationship to chromosome bands. Before exploring how genes are distributed on chromosomes, let's review some features of eukaryotic genes. First, many genes in advanced eukaryotes are associated with CpG islands. These islands, about 1 kb in length, are characterized by high GC levels, contain regulatory sequences, and are located 5′ to the genes they regulate. Second, if we examine how often G or C is used in the third codon position, genes can be grouped into compositional patterns. This GC_3 value is also called a genome phenotype, since it differs among different classes of vertebrates and is one of the important clues in understanding genome evolution.

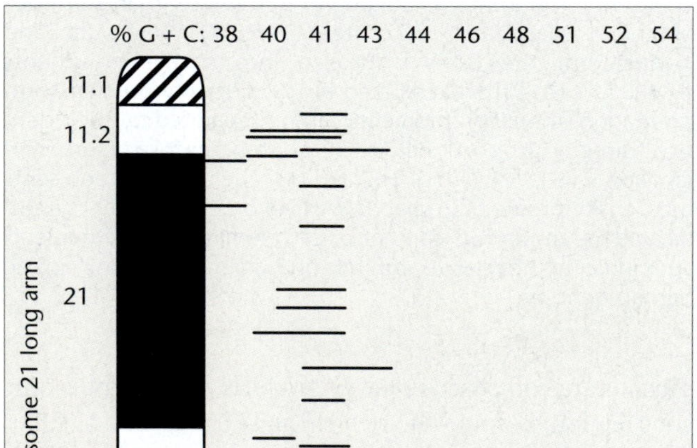

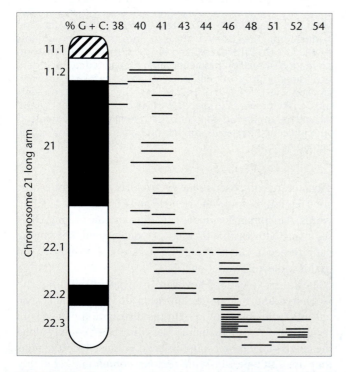

FIGURE 23–12 A compositional map of human chromosome 21, correlating the distribution of genes (horizontal lines), isochore families (% GC), and chromosomal bands on the long arm. Note that the highest GC isochores (H3) and highest number of genes are concentrated in the R band (22.3) at the distal tip of the long arm. Areas containing isochores of lower GC content have fewer genes.

Combining what is known about isochore composition and gene distribution, a profile of gene concentration can be drawn for the human genome [Figure 23–13(a)]. As shown in Figure 23–13, there are very few genes in low GC isochores (such as L1), and much higher levels in GC-rich H3 isochores (about twentyfold higher). That is, the distribution of genes in the human genome is remarkably nonuniform, with a small fraction of the genome containing most of the genes. The sharp change in slope in the gene concentration curve defines two "gene spaces" in the human genome. The gene-poor portion (one gene per 50–150 kb) of the genome is formed by L and H1 isochore families. These isochores, below about 60% GC_3 and 45% CG content, are located in G bands and in R bands lacking H3 isochores, and make up the majority of the genome (about 88%). The gene-rich region of the genome (one gene per 5–15 kb) is formed by isochores H2 and H3 (Figure 23–13b). These isochores are located in two small sets of R bands, the $H3^+$ bands and the $H3^*$ bands. Altogether, about 54% of the genes are located in the gene-rich fraction of the genome (12% of the genome), while the remaining 46% of the genes are distributed in the remaining 88% of the genome.

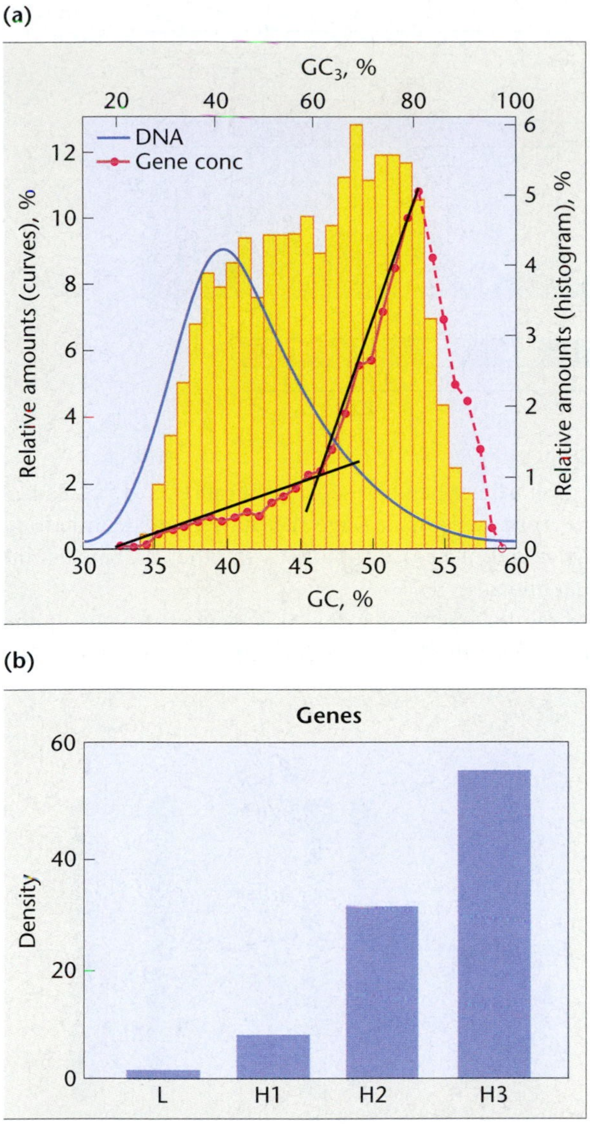

FIGURE 23–13 (a) A gene concentration profile of the human genome. Gene distribution is nonuniform in the genome, with a break in the slope at about 60% GC_3 and 47% GC. This break defines two gene spaces in the human genome. About 46% of all genes reside in low GC isochores, making up what is called the empty space (88% of the genome). The genome core (12% of the genome) contains 54% of all the genes. The broken line represents a decrease in gene concentration caused by ribosomal DNA. The blue line represents the relative amounts of DNA at each position. The yellow bars represent a histogram of gene distribution. (b) Relative gene density in each isochore family.

The relationships among GC content (isochores), chromosome bands, and genes are shown for the long arm of human chromosome 21 in Figure 23–14. The first large Giemsa band, 21.1, is composed of L1 isochores, has a low GC content and very few genes. At the telomeric end of the long arm, the $H3^+$ R band is composed of high GC H3 isochores and, at its peak, has more than 20 genes/Mb.

Understanding the organization of genes, isochores, and chromosome bands is important in unraveling the association between intranuclear chromosomal domains, the interchromosomal spaces, and the regulation of gene transcription. (See Chapter 20 for a discussion of nuclear organization.) If most genes are confined to small regions of chromosomes, placing these regions at the border of the interchromosomal spaces in the interphase nucleus allows transcription of a large number of genes to take place with a minimum of chromatin rearrangement. As we are discovering, the interphase nucleus is not a bag of chromosomes, but has a highly organized internal structure; and chromosomes are not DNA/protein complexes with genes uniformly distributed along their length, but are complex structures with gene-rich and gene-poor regions. Learning how these organizational paradigms are related to gene expression and nuclear function is one of the challenges facing geneticists and cell biologists in the coming years.

23.7 Chromosome Dynamics in Mitosis

During the cell cycle, a cell replicates its chromosome set and during anaphase of mitosis, distributes a complete chromosome set to each daughter cell. This process occurs over and over in a precise fashion during development, maintenance and repair of tissues, and wound healing. To prevent errors, the process of sister chromatid separation must be tightly regulated. Mistakes in chromosome distribution can result in cell death, or the development of cancer.

The observation that sister chromatids separate during anaphase of mitosis raises questions about what holds sister chromatids together in the first place, when does this cohesion begin, what molecular processes are involved, and what events trigger the separation of the chromatids? In recent years, work on mitosis in yeast has provided a framework for answering these and other questions about the molecular events that control the chromosome dynamics of mitosis. This work has been extended to vertebrate organisms, including humans, and provides an outline for a general mechanism for chromatid separation in eukaryotes.

Sister Chromatid Cohesion Begins in S Phase

The isolation and analysis of mutants in yeast that fail to maintain sister chromatid cohesion have allowed identification of the molecular components of cohesion and separation. These mutant studies identified a protein complex, called **cohesin**, that holds sister chromatids together until their separation in anaphase. Cohesin contains four polypeptide subunits: Smc1, Smc3, Scc1, and Scc3 (Table 23–2). In vertebrates, there are two subtypes of Scc3, called SA1 and SA3. Meiosis-specific forms of cohesin subunits also exist in yeast and other eukaryotes. Mutations in any of the four cohesin polypeptides cause sister chromatid separation prior to anaphase.

In yeast, cohesin complexes bind to DNA as early as the G1 stage of the cell cycle. Binding takes place at closely-spaced regions along the chromatin. Linking to sister chromatids is facilitated by the action of another protein, called Eco1/Ctf7. In mammalian cells, cohesin can be loaded onto chromatin at all stages of the cell cycle. In these cells, cohesin is associated with chromatin throughout interphase, dissociates from chromosomes during prophase, and is reloaded onto chromatin during telophase.

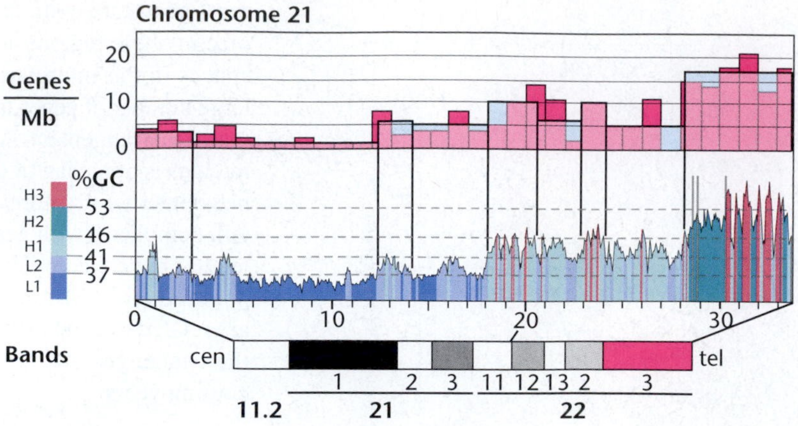

FIGURE 23–14 Correlation among chromosome bands, isochores, and gene distribution in human chromosome 21. Bottom to top: Ideogram of the long arm showing four classes of G bands (gray to black), based on staining intensity and two classes of R bands (H3+, red, H3 −, white). GC: GC profiles for human isochore families, 37%, 41%, 46%, and 53% are the upper limits for the L1, L2, H1, and H2 families, respectively. Genes/Mb: gene density calculated as number of genes per Mb of DNA.

Evidence suggests that cohesin links sister chromatids together during S phase, as soon as the replication fork has passed. In *eco1/ctf7* mutants, cohesion between sister chromatids does not form during S phase, even though cohesin is loaded onto chromosomes at normal levels. In temperature-sensitive *eco1/ctf7* mutants, if S phase occurs at the permissive temperature, sister chromatid binding takes place. Shifting to the restrictive temperature after S phase does not reverse linking of sister chromatids by cohesin.

After S phase in yeast, cohesin can be detected between sister chromatids and at centromeric regions, at specific loci spaced about 5–10 kb along the chromosome [Figure 23–15(a)]. In human cells, sister chromatids are not as tightly bound to each other during G2 as they are in yeast cells. In parallel with this observation, cohesin is more concentrated in centromeric regions of human chromosomes and is less abundant along chromosome arms [Figure 23–15(b)]. Thus, although there are variations in the amount of cohesin bound to sister chromatids, the cohesin complex clearly has two roles: It forms bridges between sister chromatids during S phase and maintains cohesion between sister chromatids through metaphase of mitosis.

Sister Chromatid Separation at Anaphase

In yeast, the Scc1 and Scc3 subunits of cohesin disappear suddenly during the transition from metaphase to anaphase. This observation focused attention on the role of these subunits in the mechanism of sister chromatid separation. When chromosomes are aligned on the metaphase plate, Scc1 is cleaved at two sites by a protease called **separin**. Once cleaved, Scc1 and Scc3 dissociate from the chromatin, and the chromosomes begin to separate (Figure 23–15). In addition to yeasts, separin cleavage sites have been identified and characterized in Scc1 orthologues in other organisms, including humans.

To show that Scc1 cleavage is sufficient to trigger the anaphase separation of newly formed chromosomes, Frank

TABLE 23–2 Subunits of Cohesin

Saccharomyces cerevisiae	*Schizosaccharomyces pombe*	*Homo sapiens*
Scc1/Rec8*	Rad21/Rec8*	Scc1
Scc3	Psc3	SA1, SA2
Smc1	Psm1	Smc1
Smc3	Psm3	Smc3

* meiosis-specific subunit

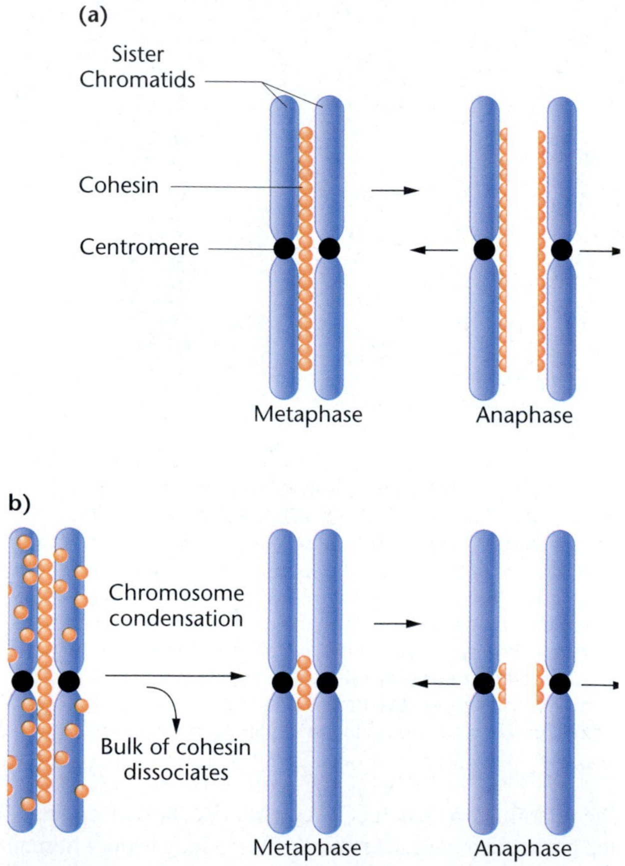

FIGURE 23–15 Sister chromatid cohesion in mitosis. (a) In yeast, cohesin is present at the centromere and along the arms of the sister chromatids. At anaphase, the cohesin is cleaved, releasing the chromosomes and allowing movement toward opposite spindle poles. (b) In humans, cohesin is lost in two steps. Between prophase and metaphase, cohesin dissociates from the arms of the sister chromatids, but remains concentrated in the centromeric region. At the metaphase to anaphase transition, this cohesin is cleaved, freeing the chromosomes to migrate to opposite spindle poles.

Uhlmann, Kim Nasmyth and colleagues genetically engineered yeast cells to replace one of the Scc1 cleavage sites with one recognized by the tobacco-etch virus (TEV) protease. This modified Scc1 was called Scc1-TEV268. Induction of the TEV protease at metaphase in cells with Scc1-TEV268 causes cleavage and loss of the modified Scc1 from the chromosomes and triggers sister chromatid separation and movement of chromosomes to opposite poles of the cells. These experiments and others demonstrate that cleavage of the Scc1 subunit of cohesin is the event that triggers anaphase in yeast.

In vertebrates, including humans, most of the cohesin along the sister chromatids dissociates during prophase, leaving the highest levels of cohesin concentrated in centromeric regions. In spite of this difference, purified separin cleaves Scc1 *in vitro*, and, in human tissue culture cells, Scc1 is cleaved and disappears from human chromosomes at anaphase, suggesting that the separin/Scc1 system may control anaphase in many, if not all eukaryotes.

Regulation of Separin Activity in Mitosis

To initiate anaphase, separin must work after chromosomes have assembled on the metaphase plate and after each chromosome has made contact with spindle fibers that extend from both poles of the cell. The timing of cohesin cleavage is a critical event that must be closely regulated. Because cohesin is cleaved by the protease separin, regulation of separin activity is a key step in mitosis. To ensure that anaphase separation of chromosomes occurs at the proper time, separin activity must be precisely regulated. One regulator of separin is a protein called **securin**. The name securin derives from its function as a molecule that binds to or `secures' separin, inhibiting its activity until the events of metaphase are completed.

In yeast, the degradation of securin occurs at the metaphase to anaphase transition. Securin is targeted for destruction by action of the anaphase promoting complex (APC), a cluster of eight proteins. APC activity is controlled by the spindle assembly checkpoint, which inhibits APC activity until all chromosomes have formed attachments to spindle fibers from both spindle poles late in metaphase. This ensures that separin becomes active only when the metaphase chromosomes are ready for separation. Once activated, APC starts a cascade of events. APC degrades securin, which in turn activates separin, which then cleaves the Scc1 subunit of cohesin, causing sister chromatid separation and the segregation of the newly formed chromosomes to opposite poles of the cell (Figure 23–16). The exact pathway of cohesin breakdown in animal cells has not been established, but is probably similar to that in yeast.

Chromatid Separation in Meiosis

In mitosis, one round of DNA synthesis in the S phase is followed by one round of cell division. In meiosis, one round of DNA synthesis is followed by two rounds of cell division. Meiosis II and mitosis are similar in that cleavage of cohesin separates sister chromatids and allows migration of newly formed chromosomes to opposite poles of the cell. Meiosis

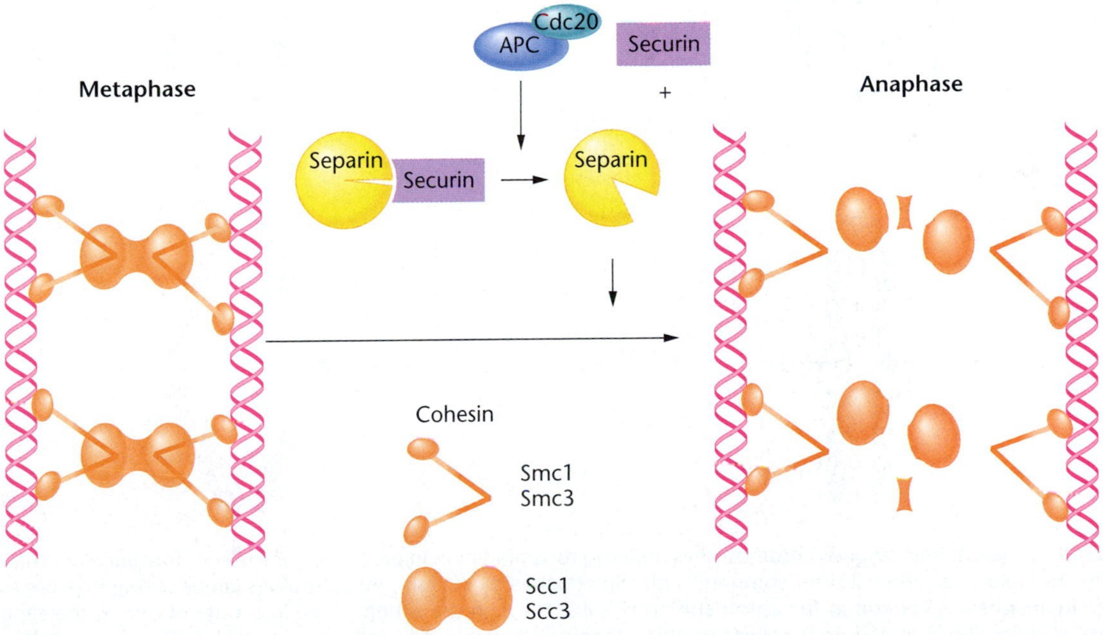

FIGURE 23–16 Control of separin activity during metaphase of mitosis. In yeast, sister chromatids are held together by cohesin complexes, with the Smc1 and Smc2 subunits bound to the chromatin and the Scc1 and Scc3 subunits forming the bridge between the sister chromatids. A protein, separin, cleaves the Scc1 subunit of cohesin, releasing the Scc1 and Scc3 subunits, and allowing separation of the sister chromatids at the beginning of anaphase. The activity of separin is controlled by another protein, securin. Securin binds to separin, inactivating it. This prevents premature separation of sister chromatids. At the metaphase to anaphase transition, the APC complex is activated, and targets the destruction of securin. Freed from securin's inhibition, separin then cleaves the Scc1 subunit of cohesin, releasing the sister chromatids.

I on the other hand, is completely different than mitosis or meiosis II. In meiosis I, pairs of homologous chromosomes align on the metaphase plate. Each chromosome attaches to spindle fibers from only one pole. In contrast to meiosis II or mitosis, it is homologous chromosomes that are separated from each other at anaphase of meiosis I. At the molecular level, many of the events in meiosis use the same mechanisms as mitosis, but with some important differences that are specific to meiosis. In meiotic yeast cells, the Scc1 component of cohesin is replaced by a meiosis-specific subunit called Rec8. Separin breaks down the meiotic form of cohesin by cleaving Rec8. The Rec8 form of cohesin appears in the centromere and sister chromatid arms during the S phase prior to meiosis. Rec8 cohesin disappears from the chromatid arms during anaphase I and remains at the centromere until anaphase II. This two step process may seem puzzling at first, but is required for the successful separation of homologous chromosomes in the first meiotic division.

If you think about it, cleavage of cohesin is obviously required for mitosis and meiosis II, but since the two sister chromatids will move together toward the same spindle pole, why does Rec8 cohesin disappear from the chromatid arms in meiosis I? The reason for this is that homologous chromosomes undergo recombination in meiosis I, forming chiasmata. If sister chromatid arms do not separate at anaphase of meiosis I, the homologues cannot separate from each other (Figure 23–17). However, to ensure that sister chromatids remain together until anaphase II, cohesion at centromere regions must be maintained. In other words, in meiosis, sister chromatids separate in a two step process. In anaphase I, Rec8 cohesin is cleaved along the chromatid arms, but remains intact at the centromere. At anaphase II, Rec8 cohesin at centromeres is digested by separin, allowing the chromosomes to move to opposite poles. The sequence of events in meiosis I trigger separin activity, and how securin bound to centromeric cohesin is protected from digestion by separin are not known in detail, but remain as areas of active research.

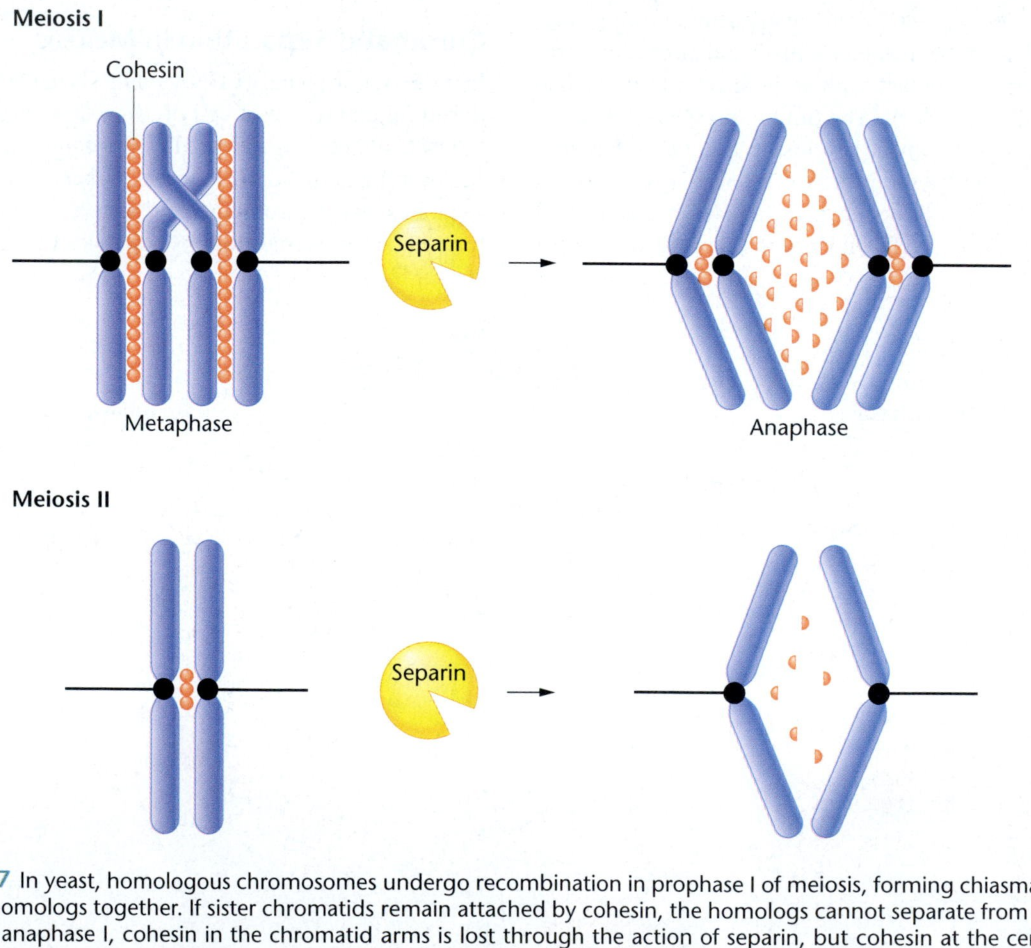

FIGURE 23–17 In yeast, homologous chromosomes undergo recombination in prophase I of meiosis, forming chiasmata that physically link the homologs together. If sister chromatids remain attached by cohesin, the homologs cannot separate from each other in anaphase I. In anaphase I, cohesin in the chromatid arms is lost through the action of separin, but cohesin at the centromeres remains intact, allowing the homologs to segregate to opposite spindle poles. In meiosis II, separin cleaves the cohesin at the centromere, converting the sister chromatids into chromosome, which migrate to opposite spindle poles. As in mitosis, the action of separin is regulated by securin.

Chapter Summary

1. Genomic alterations are a hallmark of normal events in the maturation and function of the immune system. These events include rounds of recombination-associated deletion of DNA sequences in specific cells. From a set of less than 300 genes, recombination events and associated deletions and mismatching of rejoined DNA segments in immature cells of the immune system create a vast potential for an immune response.
2. Antibody diversity results from a number of special features of the genome, including a unique form of recombination. At the molecular level, the process of recombination in the formation of antibody genes and antigen receptor genes involves RAG-1 and RAG-2, recombinases normally expressed only in B and T cells and only in stages of maturation when V(D)J recombination is taking place. Recombination is cell specific; only antibody genes (Ig genes) are rearranged in maturing B cells but not in T cells, and TCR genes are recombined in T cells but not B cells.
3. TCR and Ig loci are differentially packaged into chromatin in T and B cells. Physical changes in this packaging make the loci more or less accessible to the RAG proteins at specific times. Recent work has confirmed the role of chromatin remodeling in regulating V(D)J recombination.
4. The human genome is a mosaic of long (> 300 kb) DNA segments (isochores) belonging to a number of families, with different GC content. About 63% of the genome consists of GC-poor, light (L1 and L2) isochores, and a number of GC-rich, heavy (H1, H2, and H3) isochores that make up 24%, 7.5%, and 4.7%, respectively, of the genome.
5. In the human genome genes are more widely spaced in low GC isochores (such as L1), and much higher levels in GC-rich H3 isochores (about twentyfold higher). In other words, the distribution of genes in the human genome is remarkably nonuniform, with a small fraction of the genome containing most of the genes.
6. After S phase in yeast, molecular links can be detected between sister chromatids and at centromeric regions, at specific loci spaced about 5–10 kb along the chromosome. In human cells, sister chromatids are not as tightly bound to each other during G2 as they are in yeast cells, and cohesin is more concentrated in centromeric regions of human chromosomes than along chromosome arms.
7. A cascade of closely regulated events at anaphase cleaves the cohesin, separating the chromosomes, thus allowing them to move to opposite poles of the spindle. A meiosis-specific form of cohesin uses the same molecular machinery to ensure the distribution of chromosomes and chromatids in the two meiotic divisions.

Insights and Solutions

1. An antibody molecule contains two identical H chains and two identical L chains. This results in antibody specificity, with the antibody binding to a specific antigen. Recall that there are five classes of H-chain genes and two classes of L-chain genes. Because of the high degree of variability in the genes that encode the H and L chains, it is possible (even likely) that an antibody-producing cell contains two different alleles of the H-chain gene and two different alleles of the L-chain gene. Yet, the antibodies produced by the plasma cell contain only a single type of H chain and a single type of L chain. How can you account for this, based on the number of classes of H-chain and L-chain genes and the possibility of heterozygosity?

Solution: Although an antibody-producing cell contains different classes of H-chain and L-chain genes, and although it is entirely possible, and even likely, that a given antibody-producing cell will contain different alleles for the H-chain gene and for the L-chain gene, a phenomenon known as allelic exclusion allows the expression of only one H-chain allele and one L-chain allele in any given plasma cell at a given time. That is not to say that the same antibody is produced over the life span of the antibody-producing cell, however. At first, many plasma cells produce antibodies of the IgM class, and, at later times, they may produce antibodies of a different class (e.g., IgG). Even in switching between different H-chain genes, allelic exclusion is maintained, with only one allele of an H-chain gene (or L-chain gene) being expressed at a given time.

Problems and Discussion Questions

1. What is the difference between B and T cells?
2. What is a function of the cysteine residues in antibody structure?
3. What do the symbols V_L, C_H, IgG, J, and D represent in immunoglobulin structure?
4. If germ-line DNA contains 10 V, 30 D, 50 J, and 3 C segments, how many unique DNA sequences can be formed by recombination?
5. If there are 5 V, 10 D, and 20 J regions available to form a heavy-chain gene, and 10 V and 100 J regions available to form a L-chain gene, how many unique antibodies can be formed?
6. How does the inherent imprecision of the RAG 1and 2 recombination mechanism contribute to the diversity of the immune response?
7. What does the data in Figure 23–11 tell you about the DNA sequence composition of the yeast genome versus the human genome?
8. How has sequencing of the human genome helped solidify the theory that the human genome is a mosaic?
9. What effect would mutations in the subunit of the APC have on chromosome behavior?

10. How does meiosis I differ from mitosis?

Extra Spicy Problems

11. You are studying immunodeficiency syndromes and have mapped the mutation causing the disease to a gene encoding the histone H3. Sequencing the mutant gene reveals that it contains a mutation in the region encoding the N-terminal tail of the protein. How could mutation in the N-terminal region affect the function of the immune system?

12. In yeast, genes can be overexpressed at any time during the cell cycle by fusing the coding region to a highly inducible promoter such as *GAL1*. What do you think would happen to chromosome behavior if the gene encoding separin were overexpressed during G1? During mitosis?

Selected Readings

Amon, A. 2001. Together until separin do us part. *Nature Cell Biol.* 3:E12–E14.

Bernardi, G. 2000. Isochores and the evolutionary genomics of vertebrates. *Gene* 241:3–17.

Bernardi, G. 2001. Misunderstandings about isochores. *Gene* 276:3–13.

Cherry, S., and Baltimore, D. 1999. Chromatin remodeling directly activates V(D)J recombination. *Proc. Nat. Acad. Sci.* 96:10788–10793.

Dej. K., and Orr-Weaver, T. 2000. Separation anxiety at the centromere. *Trends Cell Biol.* 10:392–399.

Cohen-Fix, O. 2000. Sister chromatid separation: Falling apart at the seams. *Curr. Biol.* 10:R816–R819.

Cohen-Fix, O. 2001. The making and breaking of sister chromatid cohesion. *Cell* 106:137–140.

Gardiner, K., Aissani, B., and Bernardi, G. 1990. A compositional map of human chromosome 21. *EMBO J.* 9:1853–1858.

Gardiner, K., Horisberger, M., Kraus, J., Tantravahi, U., Korenberg, J., Rao, V., Reddy, S., and Patterson, D. 1990. Analysis of human chromosome 21: Correlation of physical and cytogenetic maps; gene and CpG island distribution. *EMBO J.* 9:25–34.

Grawunder, U., and Harfst, E. 2001. How to make ends meet in V(D)J recombination. *Curr. Opin. Immunol.* 13:186–94.

Holmquist, G. 1992. Chromosome bands, their chromatin flavors, and their functional features. *Am. J. Hum. Genet.* 51:17–37.

Kim, D.R., Dai, Y., Mundy, C.L., Yang, W., and Oettinger, M. A. 1999. Mutation of acidic residues in RAG1 define the active site of the V(D)J recombinase. *Genes and Develop.* 13:3070–80.

Landree, M.A., Wibbenmeyer, J.A., and Roth, D.B. 1999. Mutational analysis of RAG1 and RAG2 identifies three catalytic amino acids in RAG1 critical for both cleavage steps of V(D)J recombination. *Genes Develop.* 13:3059–69.

Losada, A., and Hirano, T. 2001. Intermolecular DNA interactions stimulated by the cohesin complex *in vitro*: Implications for sister chromatid cohesion. *Curr. Biol.* 11:268–272.

McBlane, F., and Boyes, J. 2000. Stimulation of V(D)J recombination by histone acetylation. *Curr. Biol.* 10:483–486.

McMurray, M., and Krangel, M. 2000. A role for histone acetylation in the developmental regulation of V(D)J recombination. *Science* 287:495–98.

Nasmyth, K. 2001. Disseminating the genome: joining, resolving, and separating sister chromatids during mitosis and meiosis. *Annu. Rev. Genet.* 35:673–745.

Nasmyth, K., Peters, J-M., and Uhlmann, F. 2000. Splitting the chromosome: Cutting the ties that bind sister chromatids. *Science* 288:1379–1384.

Niimura, Y., and Gojobori, T. 2002. *In silico* chromosome staining: Reconstruction of Giemsa bands from the whole genome sequence. *Proc. Nat. Acad. Sci.* 99:797–802.

Notarangelo, L.D., Villa, A., and Schwarz, K. 1999. RAG and RAG defects. *Curr. Opin. Immunol.* 11:435–442.

Oettinger, M.A. 1999. V(D)J recombination: On the cutting edge. *Curr. Opin. Cell Biol.* 11:325–329.

Oliver, J.L., Bernaola-Galavan,P., Carpena, P., and Roman-Roldan, R. 2001. Isochore chromosome maps of eukaryotic genomes. *Gene* 276:47–56.

Roth, D.B., and Roth, S.Y. 2000. Unequal access: Regulating V(D)J recombination through chromatin remodeling. *Cell* 103:699–702.

Saccone, S., De Sario, A., Wiegant, J., Raap, A.K., Della Valle, G., and Bernardi, G. 1993. Correlations between isochores and chromosomal bands in the human genome. *Proc. Nat. Acad. Sci.* 90:11929–11933.

Saccone, S., Pavlicek, A., Federico, C., Paces, J., and Bernardi, G. 2001. Genes, isochores and bands in human chromosomes 21 and 22. *Chromosome Res.* 9:533–539.

Tanaka, T., Cosma, M.P., Wirth, LK., and Nasmyth, K. 1999. Identification of cohesin association sites at centromeres and along chromosome arms. *Cell* 98:847–858.

Uhlmann, F. 2001. Chromosome condensation: Packaging the genome. *Curr. Biol.* 11:R384–R387.

Uhlmann, F., Lottspeich, F., and Nasmyth, K. 1999. Sister chromatid separation at anaphase onset is promoted by cleavage of the cohesin subunit Scc1. *Nature* 400:37–42.

Van Heemst, D., and Heytig, C. 2000. Sister chromatid cohesion and recombination in meiosis. *Chromosoma.* 109:10–26.

Wanatabe, Y., and Nurse, P. 1999. Cohesin Rec8 is required for reductional chromosome segregation at meiosis. *Nature* 400:461–464.

Yamamoto, A., and Hiraoka, Y. 2001. How do meiotic chromosomes meet their homologous partners? Lessons from fission yeast. *BioEssays* 23:526–533.

Yu, J., Tong, S., Shen, Y., and Kao, F-T. 1997. Gene identification and DNA sequence analysis in the GC-poor 20 megabase region of human chromosome 21. *Proc. Nat. Acad. Sci.* 94:6862–6867.

A field of pumpkins, where size is under polygenic control.

24
Quantitative Genetics

KEY CONCEPTS

In Chapter 10, numerous examples were discussed that illustrated gene interaction. In each case, the resultant phenotypic variation was classified into distinct traits. Pea plants were tall or dwarf; squash shape was spherical, disc shaped, or elongated; and fruit fly eye color was red or white. These phenotypes are examples of discontinuous variation, in which discrete phenotypic categories exist. Many other traits in a population demonstrate considerably more variation and are not as easily categorized into distinct classes. Such phenotypes are thus said to demonstrate continuous variation.

Traits exhibiting continuous variation are most often controlled by two or more genes that provide an additive component to the phenotype that can be quantified. In this chapter, we will examine cases that illustrate such patterns of inheritance and outline some statistical techniques used to study such traits. They illustrate what is described as **quantitative**, or **polygenic**, **inheritance**. In addition, we will consider how geneticists assess the relative importance of genetic versus environmental factors as they contribute to phenotypic variation, and we will discuss an approach used to map these "quantitative" genes within the genome.

24.1 Continuous Variation Characterizes the Inheritance of Quantitative Traits

During the 19th century, geneticists often found themselves studying traits that exhibited a continuous gradation of phenotypes. For example, Sir Francis Galton investigated the diameter of sweet peas. When plants with large peas were crossed to those with small peas, the F_1 plants contained peas that were all of an intermediate diameter. When the F_2 generation was examined, peas were of many sizes, as large or as small as the original parents, and many sizes in between.

This example illustrates a pattern of inheritance encountered by other investigators, including at least one cross made by Mendel. The F_1 generations were intermediate blends of the parental phenotypes, and the F_2 generation exhibited more or less continuous variation of phenotypic expression. In each case, the traits under investigation behaved in a quantitative fashion expressed as size, height, weight, color, and so on.

Not surprisingly, these traits were difficult to study, and their mode of inheritance was not clarified until early in the 20th century. Until then, they posed apparent exceptions to the patterns observed by Mendel in most of his crosses and clouded the initial understanding of transmission genetics. Because these results were exceptions to the patterns observed by Mendel in most of his crosses, they failed to support his hypotheses and no doubt delayed the acceptance of his work. Nevertheless, the genetic explanation of continuous variation serves as the foundation for our current understanding of the field of genetics now called quantitative, or polygenic, inheritance.

24.2 Quantitative Traits Can Be Explained in Mendelian Terms

One of the first cases of continuous phenotypic variation was encountered by Josef Gottlieb Kölreuter when he crossed tall and dwarf varieties of the tobacco plant *Nicotiana longiflora.* The plants of the F_1 generation were all intermediate in height. When the F_2 generation was examined, individuals showed continuous variation in height, ranging from tall to dwarf, like the original parents, including many heights in between. A critical observation involved the distribution of phenotypes in the second generation: The majority of the F_2 plants were intermediate like the F_1, but only a few were as tall or dwarf as the P_1 parents. These distributions are depicted in histograms in Figure 24–1. Note that the F_2 data demonstrate a normal distribution, as evidenced by the bell-shaped curve in the histogram.

At the beginning of the 20th century, geneticists noted that many characters in different species had similar patterns of inheritance, such as height and stature in humans, seed size in the broad bean, grain color in wheat, and kernel number and ear length in corn. In each case, offspring in the succeeding generation seemed to be a blend of their parents' characteristics.

The issue of whether continuous variation could be accounted for in Mendelian terms caused considerable controversy in the early 1900s. William Bateson and G. Udny Yule, who adhered to the Mendelian explanation of inheritance, suggested that a large number of factors or genes could account for the observed patterns. This proposal, called the **multiple-factor** or **multiple-gene hypothesis**, implied that many factors or genes contribute to the phenotype in a *cumulative* or *quantitative* way. However, other geneticists argued that Mendel's unit factors could not account for the blending of parental phenotypes characteristic of these patterns of inheritance and were thus skeptical of this hypothesis.

By 1920, the conclusions of several critical sets of experiments largely resolved the controversy and demonstrated that Mendelian factors could account for continuous variation. In one experiment, Edward M. East performed crosses between two strains of tobacco plant. The fused inner petals of the flower, or corollas, of strain A were decidedly shorter than the corollas of strain B. With only minor variation, each strain was true breeding. Thus, the differences between them were clearly under genetic control.

When plants from the two strains were crossed, examination of the F_1, F_2, and selected F_3 data revealed a very distinct pattern (Figure 24–2). The F_1 generation displayed corollas that were intermediate in length compared with the P_1 varieties and showed only minor variation among individuals. While corolla lengths of the P_1 plants were about 40 mm and 94 mm, the F_1 generation showed much less variation, containing a vast majority of plants with corollas that were about 64 mm. In the F_2 generation, lengths varied much more, ranging from 52 mm to 82 mm. However, the majority of individuals were similar to their F_1 parents, and as the deviation from this average increased, fewer and fewer plants were

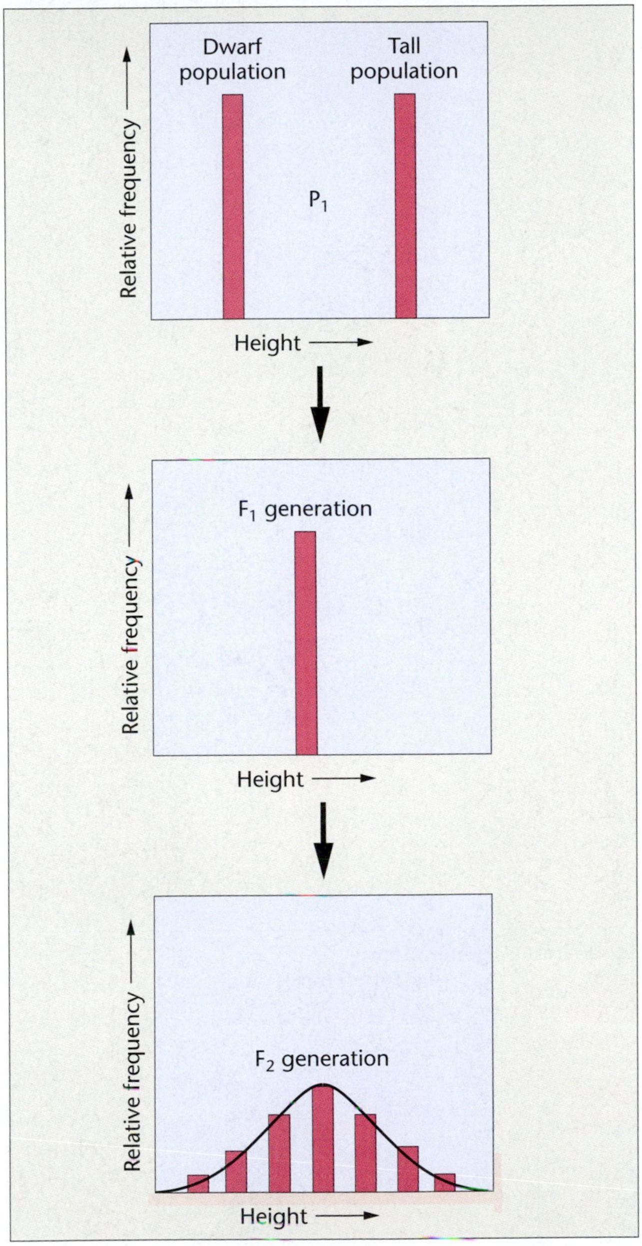

FIGURE 24–1 Histograms showing the relative frequency of individuals expressing various height phenotypes derived from Kölreuter's cross between dwarf and tall tobacco plants carried to the F_2 generation. The photograph shows a tobacco plant.

observed. When the F_2 data are plotted comparing frequency vs. length, a normal distribution (a bell-shaped curve) results.

East further experimented with this population by selecting F_2 plants of various corolla lengths and allowing them to produce separate F_3 generations. Several are illustrated in Figure 24–2. In each case, a bell-shaped distribution was observed, with most individuals similar in height to the selected F_2 parents, but with considerable variation around this value.

East's experiments demonstrated that although the variation in corolla length, at first glance, seemed continuous, experimental crosses nevertheless resulted in the segregation of distinct phenotypic classes, as observed in the three independent F_3 categories. As we will see next, this key finding was the basis for East's explanation of how the multiple-factor hypothesis could account for traits that deviate considerably in their expression.

Additive Alleles: The Basis of Continuous Variation

The multiple-factor hypothesis, suggested by the observations of East and others, embodies the following major points:

1. Characters that exhibit continuous variation can usually be quantified by measuring, weighing, counting, and so on.
2. Two or more pairs of genes, located throughout the genome, account for the hereditary influence on the phenotype in an *additive way.* Because many genes may be involved, inheritance of this type is often called *polygenic.*
3. Each gene locus may be occupied by either an **additive allele**, which contributes a constant amount to the phenotype, or by a **nonadditive allele**, which does not contribute quantitatively to the phenotype.
4. The total effect on the phenotype of each additive allele, while small, is approximately equivalent to all other additive alleles at other gene sites.
5. Together, the genes controlling a single character produce substantial phenotypic variation.
6. Analysis of polygenic traits requires the study of large numbers of progeny from a population of organisms.

These points center around the concept that additive alleles at numerous loci control quantitative traits. To illustrate this, let's examine Herman Nilsson-Ehle's experiments involving grain color in wheat performed early in the 20th century. In one set of experiments, wheat with red grain was crossed to wheat with white grain (Figure 24–3). The F_1 generation demonstrated an intermediate color. In the F_2, approximately 15/16 of the plants showed some degree of red grain, while 1/16 of the plants showed white grain. Because the ratio occurred in sixteenths, we can hypothesize that two gene pairs control the phenotype and, if so, that they segregate independently from one another in a Mendelian fashion.

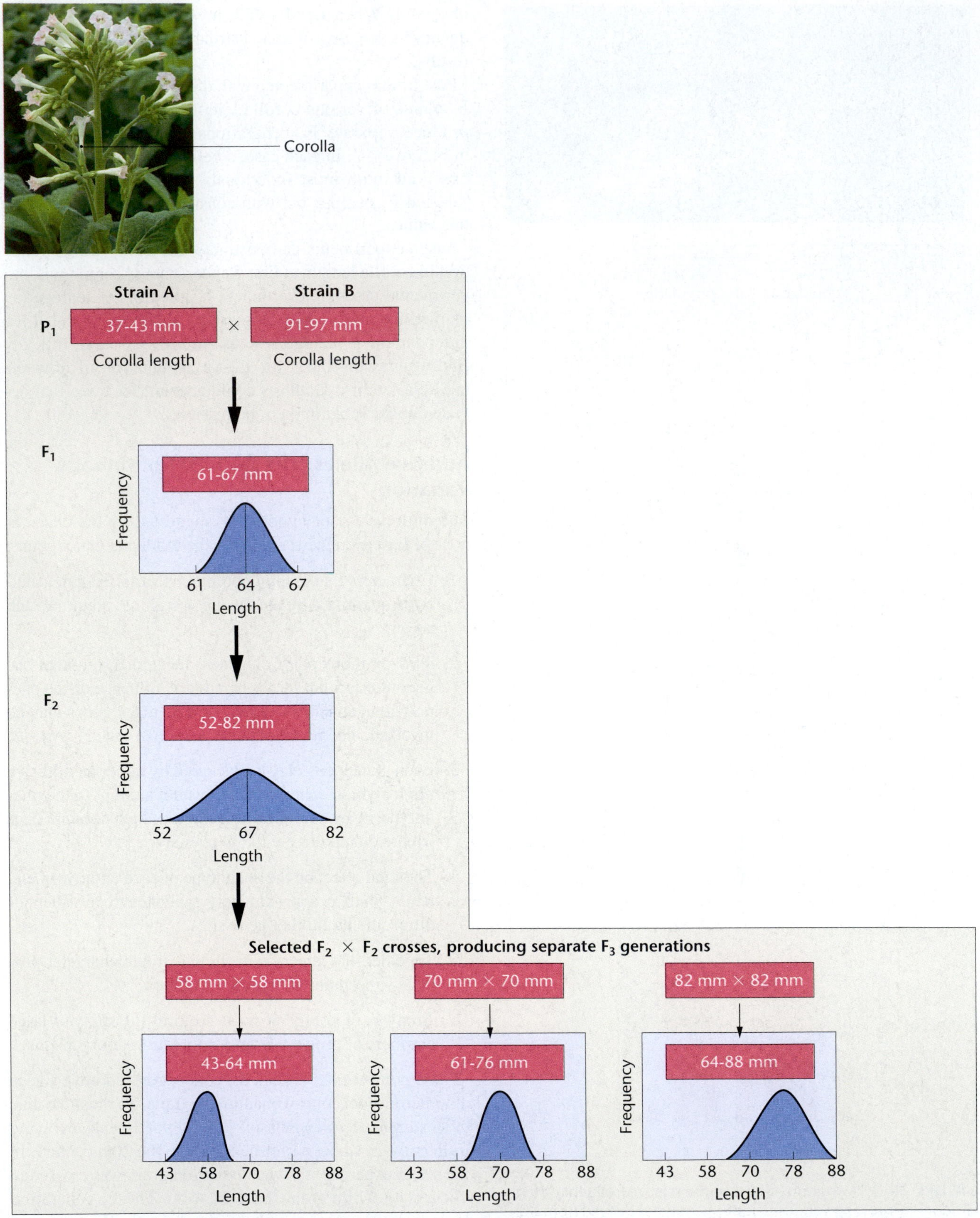

FIGURE 24–2 The F_1, F_2, and selected F_3 results of East's cross between two strains of *Nicotiana* with different corolla lengths. Corollas of strain A plants vary from 37 to 43 mm, while corollas of strain B plants vary from 91 to 97 mm. The photograph shows the flower and corolla of a tobacco plant.

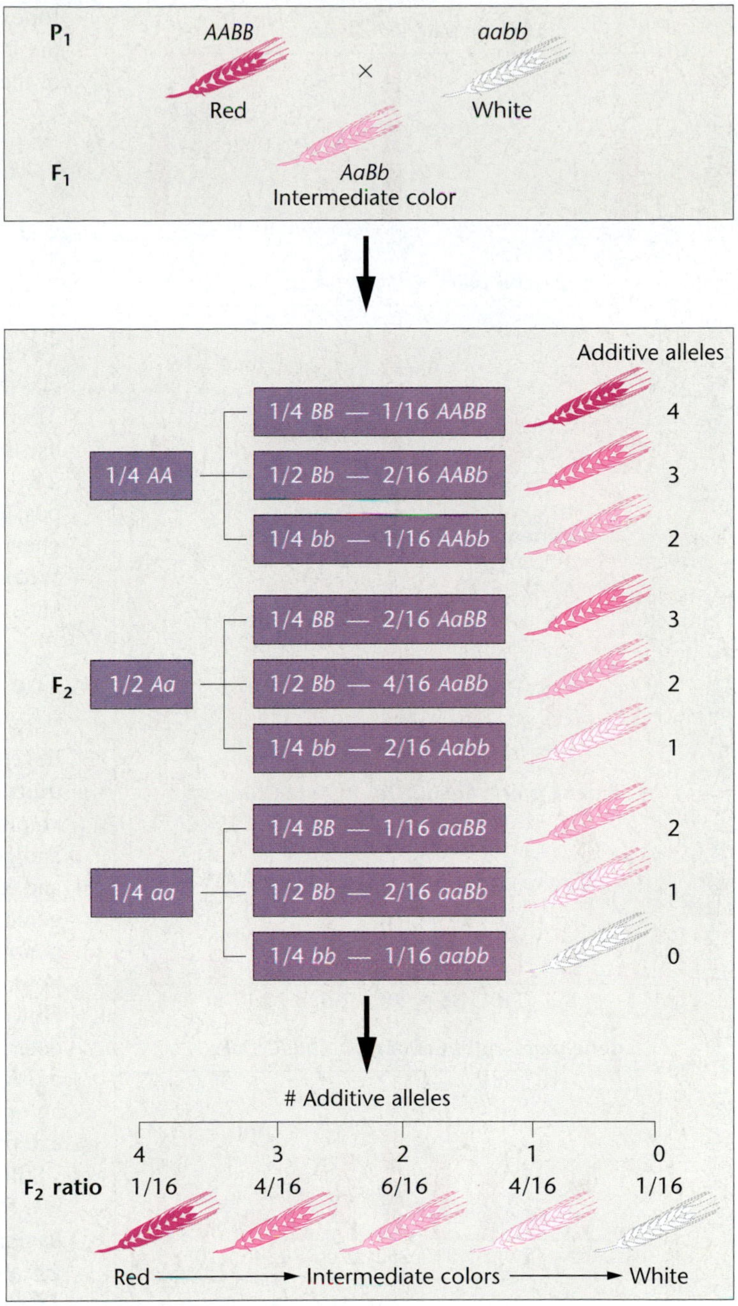

FIGURE 24–3 How the multiple-factor hypothesis accounts for the 1:4:6:4:1 phenotypic ratio of grain color when all alleles designated by an uppercase letters are additive and contribute an equal amount of pigment to the phenotype.

Careful examination of the F_2 revealed that grain with color could be classified into four different shades of red. If two gene pairs were operating, each with one potential additive allele and one potential nonadditive allele, we can envision how the multiple-factor hypothesis could account for this variation. In the P_1, both parents were homozygous; the red parent contains only additive alleles (uppercase letters in Figure 24–3), while the white parent contains only nonadditive alleles (lowercase letters). The F_1, being heterozygous, contains only two additive alleles and expresses an intermediate phenotype. In the F_2, each offspring has either 4, 3, 2, 1, or 0 additive alleles (Figure 24–3). Wheat with no additive alleles (1/16) is white like one of the P_1 parents, while wheat with 4 additive alleles is red like the other P_1 parent. Plants with 3, 2, or 1 additive alleles constitute the other three categories of red color observed in the F_2, with most (6/16) having 2 additive alleles like the F_1 plants.

Multiple-factor inheritance, where additive alleles influence the phenotype in a quantitative manner, results in continuous variation. Therefore, continuous variation can be explained in a Mendelian fashion. As we saw in Nilsson-Ehle's initial cross, if two gene pairs were involved, only five F_2 phenotypic categories, in a 1:4:6:4:1 ratio, would be expected. There is no reason, however, why three, four, or more gene pairs cannot function in a similar fashion in controlling various phenotypes. As more gene pairs become involved, greater and greater numbers of classes would be expected to appear in more complex ratios. The number of phenotypes and the expected F_2 ratios of crosses involving up to five gene pairs are illustrated in Figure 24–4.

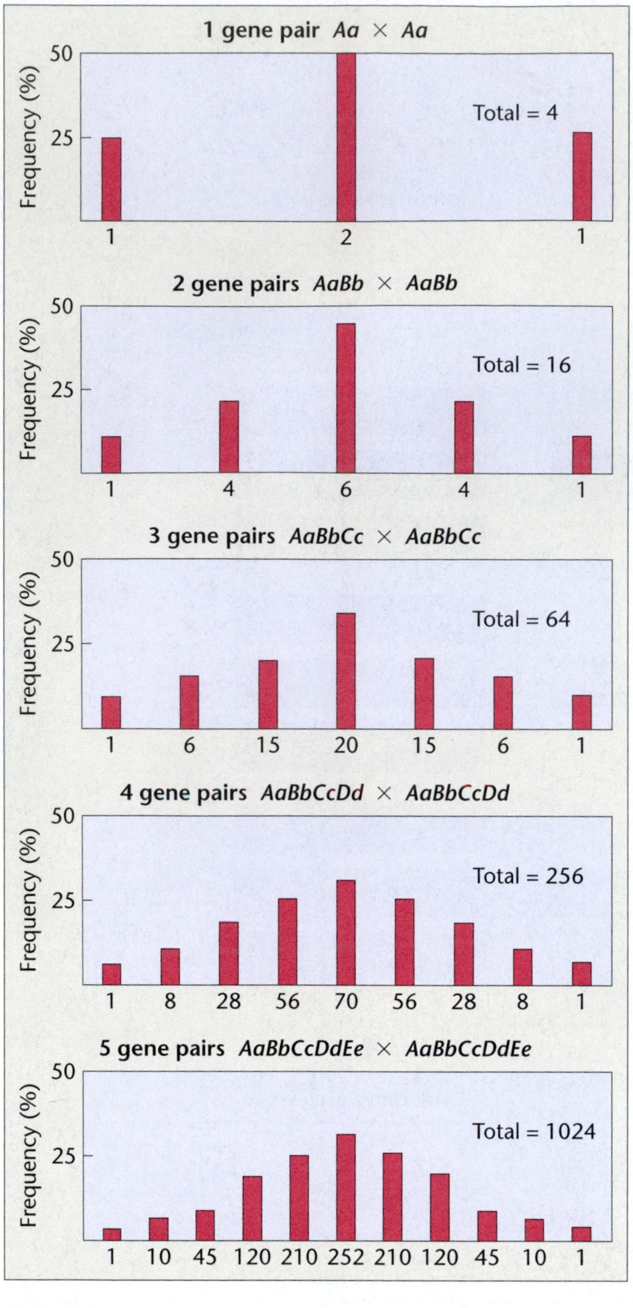

FIGURE 24–4 The results of crossing two heterozygotes when polygenic inheritance is in operation with 1–5 gene pairs. Each histogram bar indicates a distinct F_2 phenotypic class from one extreme (left end) to the other extreme (right end). Each phenotype results from a different number of additive alleles.

Calculating the Number of Polygenes

When additive effects control polygenic traits, it is of interest to determine the number of genes that are involved. If the ratio (proportion) of F_2 individuals resembling *either* of the two most extreme phenotypes (the parental phenotypes) can be determined, then the number of gene pairs involved (n) may be calculated by using the following simple formula:

$$1/4^n = \text{ratio of } F_2 \text{ individuals expressing either extreme phenotype}$$

In our previous example, the P_1 phenotypes shown in Figure 24–3 represent these two extremes. In the F_2 generation, for example, 1/16 of the progeny are either red *or* white like the P_1 classes; this ratio can be substituted on the right side of the equation to solve for n:

$$\frac{1}{4^n} = \frac{1}{16}$$

$$\frac{1}{4^2} = \frac{1}{16}$$

$$n = 2$$

Table 24–1 lists the ratio and the number of F_2 phenotypic classes produced in crosses involving up to five gene pairs.

For low numbers of gene pairs, it is sometimes easier to use the $(2n + 1)$ rule. If n equals the number of gene pairs, $2n + 1$ will determine the total number of categories of possible phenotypes. When $n = 2, 2n + 1 = 5$. Each phenotypic category can have 4, 3, 2, 1, or 0 additive alleles. When $n = 3, 2n + 1 = 7$. Each phenotypic category could have 6, 5, 4, 3, 2, 1, or 0 additive alleles, and so on.

The Significance of Polygenic Control

Polygenic control is a significant concept because it is believed to be the mode of inheritance for a vast number of traits involved in animal breeding and agriculture. For example, height, weight, and physical stature in animals, size and grain yield in crops, beef and milk production in cattle, and egg production in chickens are thought to be under polygenic control. In most cases, it is important to note that the genotype, which is fixed at fertilization, establishes the potential range in which a particular phenotype may fall. However, as we saw in Chapter 10, environmental factors determine how much of the potential will be realized. In the crosses described thus far in this chapter, we have assumed an optimal environment, which minimizes variation from external sources.

Still other examples can be drawn from human genetic studies. Skin pigmentation, intelligence, various forms of behavior, obesity, and even the predisposition to certain diseases are all thought to be under the control of numerous genes. The latter two conditions, obesity and predisposition to disease (e.g., coronary heart disease) are good examples of **complex traits**. Unlike the cases of where continuous variation is observed, no clear Mendelian pattern of inheritance is observable. However, both conditions run in families such

TABLE 24–1 Determination of the Number of Gene Pairs (n) Involved in Polygenic Crosses

n	*Individuals Expressing either Extreme Phenotype*	*Distinct F_2 Phenotypic Classes*
1	1/4	3
2	1/16	5
3	1/64	7
4	1/256	9
5	1/1024	11

that sons and daughters of affected parents are much more likely to also be affected than are children of unaffected parents. In both cases, many genes are now known to be involved, but the environment substantially influences the ultimate appearance of these traits. In such examples, when the combination of genetic and environmental factors are both substantial influences, **multifactorial** analysis is sometimes used to describe the genetic basis of the phenotype.

24.3 The Study of Polygenic Traits Relies on Statistical Analysis

Analysis of any given polygenic trait involves quantitative measurements, usually from many offspring generated from many crosses. The outcome can be expressed as a frequency diagram that often demonstrates a normal (bell-shaped) distribution (Figure 24–5). While it is hoped that each series of crosses is representative of the population at large, variation in samples due strictly to chance may influence the data gathered. To assess the validity of the experimental data, geneticists use statistical techniques, which were first devised by Galton early in the 20th century. Galton's efforts to assess the inheritance of traits exhibiting continuous variation served as the initial basis of the field of study called **biometry**, a quantitative, statistics-based approach to the study of biology.

Statistical analysis serves three purposes:

1. Data can be mathematically reduced to provide a descriptive summary of the sample.
2. Data from a small, but random sample, can be used to infer information about groups larger than those from which the original data were obtained (statistical inference).
3. Two or more sets of experimental data can be compared to determine whether they represent significantly different populations of measurements.

Several statistical methods are useful in the analysis of traits that exhibit a normal distribution, including the mean, variance, standard deviation, and standard error of the mean.

The Mean

The distributions of the two sets of phenotypic measurements graphed in Figure 24–6 cluster around a central value. This clustering is called a **central tendency**, one measurement of which is the **mean** ($\overline{X}$).

The mean is simply the arithmetic average of a set of measurements or data and is calculated as

$$\overline{X} = \frac{\Sigma X_i}{n}$$

where $\overline{X}$ is the mean, ΣX_i represents the sum of all individual values in the sample, and n is the number of individual values.

Although the mean provides a descriptive summary of the sample, it is of itself of limited value. As illustrated in Figure 24–6, a symmetrical distribution of values in the sample may, in one case, be clustered near the mean. Or, a set of values may have the same mean, but be distributed widely around it. These contrasting conditions represent different types of variation within each sample called the **frequency distribution**. Whether due to chance or to one or more experimental variables, such variation creates the need for methods to describe sample measurements statistically.

Variance

As shown in Figure 24–6, the range and distribution of values on either side of the mean determines the shape of the distribution curve. The degree to which values within this distribution diverge from the mean is called the sample **variance** (s^2) and is used as an estimate of the variation present in an infinitely large population. The variance for a sample is calculated as

$$s^2 = \frac{\Sigma(X_i - \overline{X})^2}{n - 1}$$

where the sum (Σ) of the squared differences between each measured value (X_i) and the mean ($\overline{X}$) is divided by one less than the total sample size ($n - 1$). To avoid the numerous subtraction functions necessary in calculating for a large sample, we can convert the equation to its algebraic equivalent:

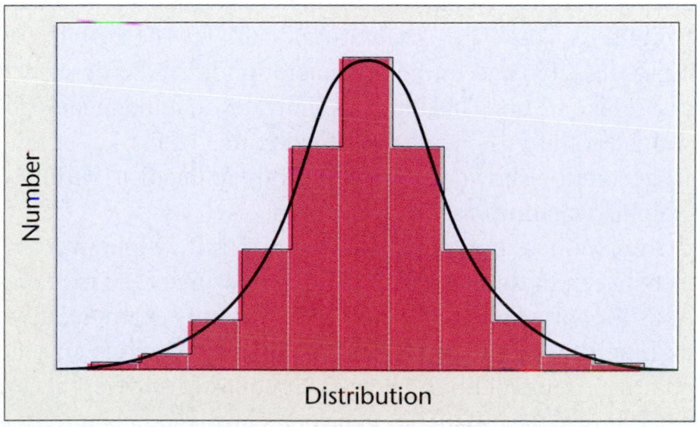

FIGURE 24–5 A normal frequency distribution characterized by a bell-shaped curve.

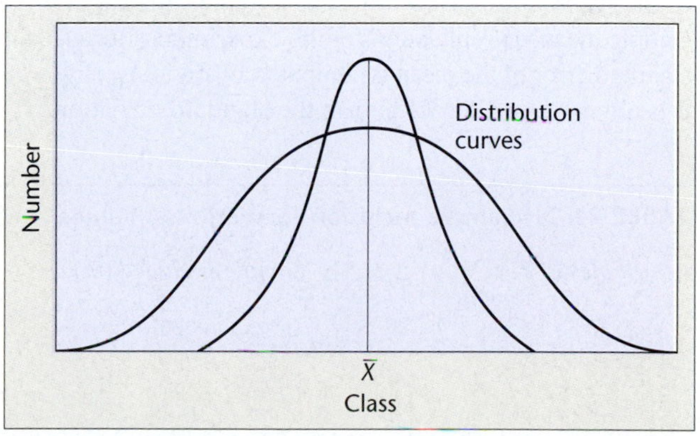

FIGURE 24–6 Two normal frequency distributions with the same mean, but different amounts of variation.

$$s^2 = \frac{\Sigma X_i^2 - n\overline{X}^2}{n - 1}$$

The variance is a valuable measure of sample variability. Two distributions may have identical means ($\overline{X}$), yet vary considerably in their frequency distribution around the mean. The variance represents the average squared deviation of the measurements from the mean. Estimation of variance is particularly valuable in determining the degree of genetic control of traits when the immediate environment also influences the phenotype.

Standard Deviation

Because the variance is a squared value, its unit of measurement is also squared (e.g., cm^2, mg^2, etc.). To express variation around the mean in the original units of measurement, it is necessary to calculate the square root of the variance, a term called the **standard deviation** (s):

$$s = \sqrt{s^2}$$

Table 24–2 shows the percentage of the individual values within a normal distribution that is included with different multiples of the standard deviation. The mean, plus or minus one standard deviation ($\overline{X} \pm 1s$), includes 68 percent of all values in the sample. Over 95 percent of all values are found within two standard deviations ($\overline{X} \pm 2s$) As such, the standard deviation provides an important descriptive summary of a set of data. Furthermore, s can be interpreted as a probability. The mean, plus or minus one standard deviation ($\overline{X} \pm 1s$) indicates that there is a 68 percent probability that a measured value picked at random will fall within that range.

Standard Error of the Mean

To estimate how much the means of other similar samples drawn from the same population might vary, we can calculate the **standard error of the mean** ($S_{\overline{X}}$), namely,

$$S_{\overline{X}} = \frac{s}{\sqrt{n}}$$

where s is the standard deviation and is the square root of the sample size. The standard error of the mean is a measure of the accuracy of the sample mean—that is, the variation of sample means in replications of the experiment. Because the standard error of the mean is computed by dividing s by $\sqrt{n}$, it is always a smaller value than the standard deviation.

TABLE 24–2 Sample Inclusion for Various s Values

Multiples of s	*Sample Included (%)*
$\overline{X} \pm 1s$	68.3%
$\overline{X} \pm 1.96s$	95.0
$\overline{X} \pm 2s$	95.5
$\overline{X} \pm 3s$	99.7

Analysis of a Quantitative Character

To illustrate how biometric methods are used to analyze quantitative characters statistically, we will consider a simplified example involving fruit weight in tomatoes. Let us assume that fruit weight is a quantitative character and that one highly inbred strain (one that is highly homozygous) produces tomatoes averaging 18 oz in weight, and another highly inbred strain produces fruit averaging 6 oz in weight. These two varieties are crossed and produce an F_1 generation with weights ranging from 10 to 14 oz. The F_2 population contains individuals that produce fruit ranging from 6 to 18 oz. The results characterizing both generations are shown in Table 24–3.

The mean value for the fruit weight in the F_1 generation can be calculated as

$$\overline{X} = \frac{\Sigma X_i}{n} = \frac{626}{52} = 12.04$$

Similarly, the mean value for fruit weight in the F_2 generation is calculated as

$$\overline{X} = \frac{\Sigma X_i}{n} = \frac{872}{72} = 12.11$$

Average fruit weight is 12.04 oz in the F_1 generation and 12.11 oz in the F_2 generation. Although these mean values are similar, it is apparent from the frequency distributions (Table 24–3) that there is more variation present in the F_2 generation. Fruit weight ranges from 6 to 18 oz in the F_2 generation, but only from 10 to 14 oz in the F_1 generation.

To assist in quantifying the amount of variation present in each generation, we can calculate the variance (Table 24–4). As we have observed, the sample variance can be calculated as the sum of the squared differences between each value and the mean, divided by one less than the total number of observations. However, in the case where a number of observations (f) have been grouped into representative classes (x), the variance is calculated according to the formula

$$s^2 = \frac{n\Sigma f(x^2) - (\Sigma fx)^2}{n(n - 1)}$$

As shown in Table 24–4, the variance is 1.29 for the F_1 generation and 4.27 for the F_2 generation. When converted to the standard deviation ($s = \sqrt{s^2}$) the values become 1.13 and 2.06, respectively. Therefore, the distribution of tomato weight in the F_1 generation can be described as 12.04 ± 1.13, and in the F_2 generation it can be described as 12.11 ± 2.06. This analysis indicates that the mean fruit weight of the F_1 is nearly identical to that of the F_2, but the F_2 generation shows greater variability in the distribution of weights than does the F_1 generation.

Observations about the inheritance of fruit weight in crosses between these two strains of tomatoes meet the expectations for polygenic traits. For the sake of this example, if we assume that each parental strain is homozygous for the additive or nonadditive alleles that control fruit weight, we can estimate the number of gene pairs involved in controlling this trait in these two strains of tomatoes. Since 1/72 of the

TABLE 24–3 Distribution of F_1 and F_2 Progeny

		Weight (oz.)												
		6	7	8	9	10	11	12	13	14	15	16	17	18
Number of Individuals	F_1:					4	14	16	12	6				
	F_2:	1	1	2	0	9	13	17	14	7	4	3	0	1

F_2 offspring have a phenotype that overlaps one of the parental strains (72 total F_2 offspring; one weighs 6 oz, one weighs 18 oz; see Table 24–3), the use of the formula $1/4^n = 1/72$ indicates that n is between 3 and 4, indicative of the number of genes that control fruit weight in these tomato strains. If the experiment were repeated many times, with similar results, our confidence in this conclusion would be bolstered.

24.4 Heritability Is a Measure of the Genetic Contribution to Phenotypic Variability

Having just introduced several ways in which quantitative or continuous variation can be measured and characterized in populations, we now consider how we assess the extent to which genetic factors contribute to such phenotypic variation. Often, much of the variation can be attributed to genetic factors, with the total environment having less impact. In other cases, the environment may have a greater impact on phenotypic variation within a population. The following discussion considers how geneticists attempt to define the relative impact of heredity and environment on phenotypic variation.

Broad-Sense Heritability

Provided that a trait can be measured quantitatively, experiments on many plants and animals can test the origin of variation. One approach is to use inbred strains containing individuals of a relatively homogeneous (highly homozygous) genetic background. Experiments are then designed to test the effects of the range of prevailing environmental conditions on phenotypic variability. Variation observed *between* different inbred strains reared in a constant environment is due predominantly to genetic factors. Variation observed

TABLE 24–4 Calculation of Variance

F_1				F_2			
x	f	$f(x)$	$f(x)^2$	x	f	$f(x)$	$f(x)^2$
6				6	1	6	36
7				7	1	7	49
8				8	2	16	128
9				9	0	0	0
10	4	40	400	10	9	90	900
11	14	154	1,694	11	13	143	1,573
12	16	192	2,304	12	17	204	2,448
13	12	156	2,028	13	14	182	2,366
14	6	84	1,176	14	7	98	1,372
15				15	4	60	900
16				16	3	48	768
17				17	0	0	0
18				18	1	18	324
	$n = 52$	$\Sigma fx = 626$	$\Sigma fx^2 = 7{,}602$		$n = 72$	$\Sigma fx = 872$	$\Sigma fx^2 = 10{,}864$

For F_1:

$$s^2 = \frac{52 \times 7602 - (626)^2}{52(52 - 1)}$$

$$= \frac{395{,}304 - 391{,}876}{2652}$$

$$= 1.29$$

For F_2:

$$s^2 = \frac{72 \times 10{,}864 - (872)^2}{72(72 - 1)}$$

$$= \frac{782{,}208 - 760{,}384}{5112}$$

$$= 4.27$$

among members of the same inbred strain reared under different environmental conditions is due to nongenetic factors, which are generally categorized as "environmental."

The relative importance of genetic and environmental factors can be formally assessed by examining the **heritability index** (H^2), which we calculate using an analysis of variance among individuals of a known genetic relationship. (In the ensuing discussion, we will assign the term V to designate variance.) This is an important approach when investigating organisms with long generation times. Also called **broad-sense heritability**, H^2 measures the degree to which **phenotypic variance** (V_P) is due to variation in genetic factors for a single population under the limits of environmental variation during the study. It *cannot be overemphasized* that calculation of H^2 *does not* determine the proportion of the total phenotype attributable to genetic factors. It *does* estimate the proportion of observed variation in the phenotype attributable to genetic factors in comparison to environmental factors.

Phenotypic variance is due to the sum of three components: **environmental variance** (V_E), **genetic variance** (V_G), and variance resulting from the **interaction of genetics and environment** (V_{GE}). Therefore, phenotypic variance (V_P) is theoretically expressed as

$$V_P = V_E + V_G + V_{GE}$$

Because V_{GE} is usually negligible, it is often omitted. Thus, this simpler equation is generally used:

$$V_P = V_E + V_G$$

Broad-sense heritability expresses that proportion of variance due to the genetic component:

$$H^2 = \frac{V_G}{V_P}$$

An H^2 value that approaches 1.0 indicates that the environmental conditions have had little impact on phenotypic variance in the population studied. An H^2 value close to 0.0 indicates that variation in the environment has been almost solely responsible for the observed phenotypic variation within the population studied. Rarely do phenotypic characteristics demonstrate H^2 values that are close to either extreme, indicative of the joint role that both genetics and environment play in expression of most phenotypic characteristics.

It is not possible to obtain an absolute H^2 value for any given character. If measured in a different population under a greater or lesser degree of environmental variability, H^2 might well be different for that character. For that reason, broad-sense heritability estimates are most useful in highly inbred strains or genetic clones such as artificially selected plant populations that are asexually propagated by cuttings. Furthermore, broad-sense heritability estimates are not very accurate in estimating the selection potential of quantitative traits, since H^2 calculations take into account all forms of genetic variation not specifically additive genetic effects. Therefore, another type of calculation, narrow-sense heritability, has been devised that is of more practical use.

Narrow-Sense Heritability

Information regarding heritability is most useful in animal and plant breeding as a measure of potential response to selection. In this case, a different estimate of heritability must be used, based on a subcomponent of V_G referred to as **additive variance** V_A:

$$V_G = V_A + V_D + V_I.$$

Here V_A represents additive variance that results from the average effect of additive components of genes. **Dominance variance**, V_D, is the deviation from the additive components that results when phenotypic expression in heterozygotes is not precisely intermediate between the two homozygotes. **Interactive variance**, V_I, is the deviation from the additive components that occurs when two or more loci behave epistatically. Interactive variance is not associated with the average effect of V_A and is often negligible. Thus, it is often excluded from calculations.

When V_G is partitioned into V_A and V_D, a new assessment of heritability, h^2, or **narrow-sense heritability** can be calculated. Thus, h^2 values are useful in assessing selection potential in randomly breeding animal and plant populations:

$$h^2 = \frac{V_A}{V_P}$$

Because $V_P = V_E + V_G$ and $V_G = V_A + V_D$, we obtain

$$h^2 = \frac{V_A}{V_E + V_A + V_D}$$

As we shall subsequently see, h_2 values are useful in predicting the phenotypes of offspring during selection procedures. Based on our knowledge of the parental phenotypes, the closer a value is to 1.0, the greater is our ability to make an accurate prediction.

Artificial Selection

The process of selecting a specific group of organisms from an initially heterogeneous population for future breeding purposes is referred to as **artificial selection**. A relatively high h^2 value is a prediction of the impact that selection will have in altering a population. As you can imagine based on our preceding discussion, measuring the components necessary to calculate h^2 is a complex task. A more simplified approach involves measurement of the central tendencies (the means) of a trait from (1) a parental population exhibiting a bell-shaped distribution (M), (2) a "selected" segment of the parental population that expresses the most desirable quantitative phenotypes (M1), and (3) the offspring (M2) resulting from interbreeding the selected M_1 group. When this is accomplished, the following relationship of the three means and h^2 exists:

$$M2 = M + h^2(M1 - M)$$

Solving this equation for h^2 gives us

$$h^2 = \frac{M2 - M}{M1 - M}$$

Once these relationships are established, the equation can be further simplified by defining M2 − M as the **response** (R) and M1 − M as the **selection differential** (S), where h^2 reflects the ratio of the response observed to the total response possible. Thus,

$$h^2 = \frac{R}{S}$$

To illustrate the assessment of selection, suppose that we measure the diameter of corn kernels in a population where the mean diameter (M) was much larger than desirable (20 mm), and from that population, we select a group with the smallest diameters, for which the mean (M_1) equals 10 mm. Plants that yielded this selected population are interbred, and the progeny kernels yield a mean (M_2) of 13 mm. Then we can calculate h^2 in order to estimate the potential for artificial selection on kernel size:

$$h^2 = \frac{13 - 20}{10 - 20}$$

$$= \frac{-7}{-10}$$

$$= 0.70$$

On the basis of this calculation, we may conclude that the selection potential for kernel size is relatively high.

The longest running artificial selection experiment known is still being conducted at the State Agricultural Laboratory in Illinois. Since 1896, corn has been selected for both high and low oil content. After 76 generations, selection continues to result in increased oil content (Figure 24–7). Predictably, as selection has progressed successfully, a greater number of plants possess a higher percentage of additive alleles involved in oil production. Thus, heritability (h^2) of increased oil content in succeeding generations has declined (see parenthetical values at generations 9, 25, 52, and 76 in Figure 24–7) as artificial selection comes closer and closer to optimizing the genetic potential for oil production. Theoretically, the process will continue until all individuals in the population contain a uniform genotype for the additive alleles responsible for oil content. At that point, heritability will be reduced to zero, and the response to artificial selection will cease. Examination of the selection for low oil content shows that heritability is approaching this point.

Similar calculations involving artificial selection for other traits in a variety of organisms have established whether artificial selection will be an effective approach in obtaining populations exhibiting desirable phenotypes. Table 24–5 lists estimates of narrow-sense heritability for a variety of traits in various organisms. These proportions are expressed as percentage values. As you can see, heritability varies considerably among traits.

In general, heritability is low for traits that are essential to an organism's survival, primarily because the genetic component must be largely optimized during evolution. Egg production, litter size, and conception rate are examples

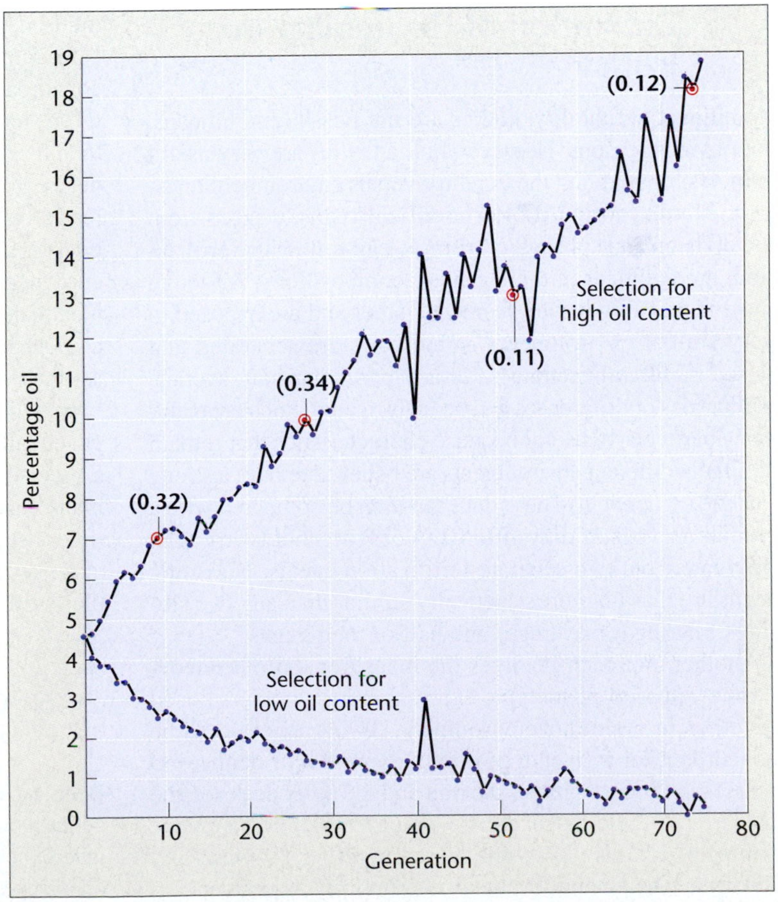

FIGURE 24–7 Response of corn selected for high and low oil content over 76 generations. The numbers in parentheses at generations 9, 25, 52, and 76 for the "high oil" line indicate the calculation of heritability at these points in the continuing experiment.

TABLE 24–5 Estimates of Heritability for Traits in Different Organisms

Trait	*Heritability* (h^2)
Mice	
Tail length	60%
Body weight	37
Litter size	15
Chickens	
Body weight	50
Egg production	20
Egg hatchability	15
Cattle	
Birth weight	51
Milk yield	44
Conception rate	3

where such physiological limitations on selection have already been established. Traits that are less critical to survival, such as body weight, tail length, and wing length, show higher heritability values. Narrow-sense heritability estimates are more valuable when they have been based on data collected in many populations and environments and where a clear trend is established. Based on such estimates, selection techniques have led to vast improvements in the quality of animal and plant products.

24.5 Twin Studies Allow an Estimation of Heritability in Humans

Traditional heritability studies are not possible in humans, for obvious reasons. However, human twins are very useful subjects for studying the heredity versus environment question. **Monozygotic (MZ)** or **identical twins**, derived from the division and splitting of a single egg following fertilization, are identical in their genetic compositions. Although most identical twins are reared together and are exposed to very similar environments, some pairs are separated and raised in different settings. For any particular trait, average similarities or differences can be investigated. Such an analysis is particularly useful because characteristics that remain similar in different environments are believed to have a strong genetic component. These data can then be compared with a similar analysis of **dizygotic (DZ)** or **fraternal twins**, who originate from two separate fertilization events. Dizygotic twins are thus no more genetically similar than any two siblings, sharing (on average) one-half of their genes.

Another approach involves the measurement of **concordance** values of phenotypic expression in twin pairs raised together in a similar environment. Twins are said to be concordant for a given trait if both express it or neither expresses it. If one expresses the trait and the other does not, the pair is said to be **discordant**. Comparison of concordance values of MZ vs. DZ twins reared together (Table 24–6) illustrates the potential value for heritability assessment.

TABLE 24–6 A Comparison of Concordance of Various Traits Between Monozygotic (MZ) and Dizygotic (DZ) Twins

	Concordance	
Trait	*MZ*	*DZ*
Blood Types	100%	66%
Eye Color	99	28
Mental retardation	97	37
Measles	95	87
Hair color	89	22
Handedness	79	77
Idiopathic epilepsy	72	15
Schizophrenia	69	10
Diabetes	65	18
Identical allergy	59	5
Cleft lip	42	5
Club foot	32	3
Mammary cancer	6	3

These data must be examined very carefully before any conclusions are drawn. If the concordance value approaches 90 to 100 percent in monozygotic twins, we might be inclined to interpret that value as indicating a large genetic contribution to the expression of the trait. In some cases—for example, blood types and eye color—we know that this is indeed true. In the case of measles, however, a high concordance value merely indicates that the trait is almost always induced by a factor in the environment—in this case, a virus.

It is more meaningful to compare the *difference* between the concordance values of monozygotic and dizygotic twins. If these values are significantly higher for monozygotic twins than for dizygotic twins, we suspect that there is a strong genetic component involved in the determination of the trait. We reach this conclusion because monozygotic twins, with identical genotypes, would be expected to show a greater concordance than genetically related, but not genetically identical, dizygotic twins. In the case of measles, where concordance is high in both types of twins, the environment is assumed to contribute significantly.

Even though a particular trait may demonstrate considerable genetically based variation, it is often difficult to formulate a precise mode of inheritance based on available data. In many cases, the trait is considered to be controlled by multiple-factor inheritance. However, when the environment is also exerting a partial influence, such a conclusion is particularly difficult to prove.

24.6 Quantitative Trait Loci Can Be Mapped

Because quantitative traits are influenced by numerous genes, geneticists would like to identify them and determine where these genes are located in the genome. Are they linked on a single chromosome or scattered throughout the genome among many chromosomes? As we shall see in subsequent chapters, locating, or "mapping" genes within the genome is an important step toward establishing the genetic identity of organisms. The initial approach is to localize genes controlling quantitative traits on a particular chromosome or chromosomes. In the context of this discussion, these genes are called **quantitative trait loci (QTLs)**. As we shall see, some rather ingenious methods have been devised to map these genes.

As an example of QTL analysis, we consider the phenotypic trait in *Drosophila* of resistance to the insecticide DDT, which has been shown to be under polygenic control. To find the loci responsible, strains selected for resistance to DDT and strains selected for sensitivity to DDT were crossed to flies carrying dominant genes that serve as markers on each of *Drosophila*'s four chromosomes. Following a variety of crosses, offspring were produced that contained many different combinations of marker chromosomes and chromosomes from either resistant or sensitive strains. As shown in Figure 24–8, flies with various chromosome combinations were then tested for resistance (their survival) when exposed to DDT. Results indicate that each chromosome in *Drosophila* contains genes that contribute to resistance. In other words, the loci bearing the genes that control DDT resistance are scattered throughout the genome.

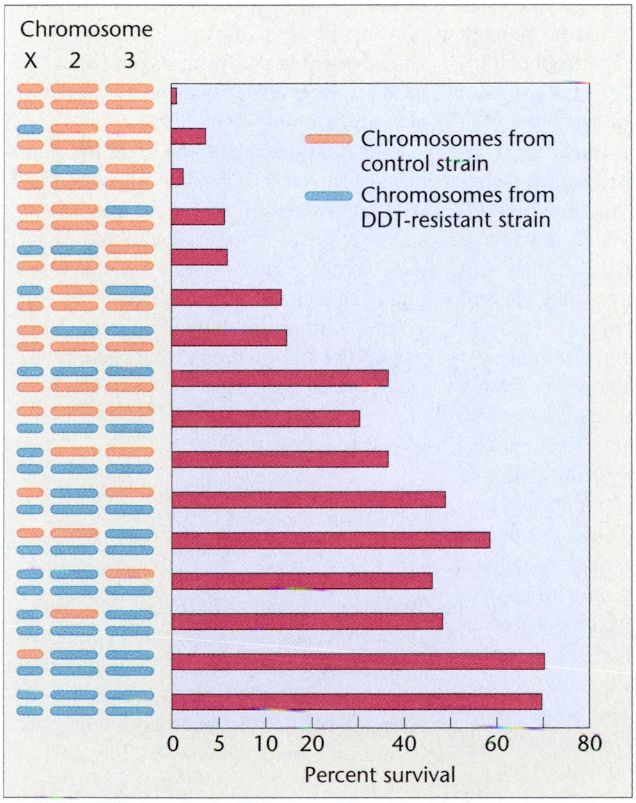

FIGURE 24–8 Survival rates of *Drosophila* carrying combinations of chromosomes from DDT-resistant and DDT-sensitive (control) strains when exposed to DDT. The results indicate that DDT resistance is polygenic, with genes on each of the chromosomes making a major contribution. Chromosome 4 carries only a few *Drosophila* genes and was omitted from this analysis.

We can now map the positions of these genes more specifically along each chromosome in *Drosophila* because of the presence of molecular markers on each chromosome. Using an approach called **interval mapping**, initially worked out by Eric Lander and David Botstein in 1989, the locations of QTLs are determined relative to the known positions of these markers. Known as **restriction fragment length polymorphisms (RFLPs)** these markers represent specific nuclease cleavage sites. (See Chapter 18 for a detailed description of RFLPs.) This approach enumerates and maps the loci responsible for quantitative traits.

RFLP markers are now available for many organisms of agricultural importance, making possible systematic mapping of QTLs. For example, hundreds of RFLP markers have been located in the tomato. They are spaced along all 12 chromosomes of this organism. Analysis is performed by crossing plants with extreme, but opposite, phenotypes and following the crosses through several generations. When both a marker and a phenotypic trait of interest are expressed together, they are said to *cosegregate*. Consistent cosegregation establishes the presence of a QTL at or near the RFLP marker along the chromosome. Whenever both an RFLP marker and a QTL responsible for the trait under investigation are closely linked along a single chromosome, they are more likely to demonstrate an association throughout a pedigree than if they are not closely linked. When numerous QTLs are located, a genetic map is created for the involved genes. While this approach sounds rather straightforward, it is important to note that the technique relies heavily on the application of statistical methods to validate final assignments.

RFLP analysis has resulted in extensive mapping of QTLs in the tomato, *Lycopersicon esculentum*. Many loci have been identified that are responsible for fruit weight, soluble solid content, and acidity. These loci are distributed on all 12 chromosomes representing the haploid genome of this plant. Several chromosomes contain loci for all three traits. Note that the RFLP method initially allows the identification of small chromosomal regions, not individual genes, although each locus may well house only a single gene.

Fruit weight in the tomato has been the focus of a highly successful research effort conducted by Steven Tanksley and his colleagues. Currently, 28 QTLs responsible for phenotypic variation in fruit weight have been identified. One of these, *fw2.2*, has now been isolated, cloned, and transferred between plants, with most interesting results. While the cultivated tomato can weigh up to 1000 grams, the progenitor of the modern tomato is thought to have weighed only a few grams. Two distinct alleles of *fw2.2* are now recognized as a result of RFLP mapping studies. One allele is present in all wild small-fruited varieties of tomatoes investigated. The other allele is present in all domesticated large-fruited varieties. When the cloned allele from small-fruited varieties is transferred to a plant that normally produces large tomatoes, a remarkable result is achieved. The transformed plant produces fruits that are reduced in weight by about 30 percent (Figure 24–9). In the varieties employed in this study, the reduction averaged 17 grams, a significant phenotypic change caused by the action of a single gene.

Genetics, Technology, and Society

The Green Revolution Revisited

Of the greater than 6 billion people now living on Earth, about 750 million don't have enough to eat. And despite efforts to limit population growth, an additional 1 million people are expected to go hungry each year for the next several decades.

Will we be able to solve this problem? The past gives us some reasons to be optimistic. In the 1950s and 1960s, in the face of looming population increases, plant scientists around the world set about to increase the production of crop plants, including the three most important grains, wheat, rice, and maize. These efforts became known as the Green Revolution. The approach was three pronged: (1) to increase the use of fertilizers, pesticides, and irrigation water, (2) to bring more land under cultivation, and (3) to develop improved varieties of crop plants by intensive plant breeding. While highly successful in recent years, the rate of increase in grain yields has slowed. If food production is to keep pace with the projected increase in the world's population, plant breeders will have to depend more and more on the genetic improvement of crop plants to provide higher yields. But is this possible? Are we approaching the theoretical limits of yield in important crop plants? Recent work with rice suggests that the answer to this question is a resounding no.

Rice ranks third in worldwide production, just behind wheat and maize. About 2 billion people, fully one-third of Earth's population, depend on rice for their basic nourishment. The majority of the world's rice crop is grown and consumed in Asia, but it is also a dietary staple in Africa and Central America. The Green Revolution for rice began in 1960, with the establishment of the International Rice Research Institute (IRRI), headquartered at Los Baños, Philippines. The goal was to breed rice with improved disease resistance and higher yield. Breeders were almost too successful: The first high-yield varieties were so top heavy with grain that they tended to fall over. (Plant breeders call this "lodging.") To reduce lodging, IRRI breeders crossed a high-yield line with a dwarf native variety to create semidwarf lines, which were introduced to farmers in 1966. Due in large part to the adoption of the semidwarf lines, the world production of rice doubled in the next 25 years.

Rice breeders cannot afford to rest on their laurels, however, as the yield of modern rice varieties has not improved much in recent years. Predictions suggest that a 70 percent increase in the annual rice harvest may be necessary to keep pace with anticipated population growth during the next 30 years. Breeders are now looking to wild rice varieties for further crop improvement. Leading the way are Susan McCouch, Steven Tanksley, and their coworkers at Cornell University. To test the hypothesis that wild-rice species carry genes that will improve the yield of cultivated varieties, they crossed cultivated rice (*Oryza sativa*) with a low-yield wild ancestor species (*Oryza rufipogon*) and then successively backcrossed the interspecific hybrid to cultivated rice for three generations. In theory, this would create lines whose genomes were about 95 percent from *O. sativa* and 5 percent from *O. rufipogon*. When testing these backcrossed lines for grain yield, they found that several of them outproduced cultivated rice by as much as 30 percent. These results demonstrated strikingly that even though wild-rice relatives have low yields and appear to be inferior to cultivated rice, they still carry genes that will increase the yield of elite rice varieties. It will now be up to breeders to exploit the wild-rice relatives.

But introducing favorable genes from a wild relative into a cultivated variety by conventional breeding is a long and involved process, often requiring a decade or more of crossing, selection, backcrossing, and more selection. Future improvements in cultivated species must be quicker if crop yields are to keep up with population growth. Fortunately, modern gene-mapping techniques are now leading to the identification of QTLs that control complex traits such as yield and disease resistance. This makes possible a more direct approach to crop improvement, which has been termed the *advanced backcross QTL method*.

First, a cultivated variety is crossed with a wild relative, just as *O. sativa* was interbred with *O. rufipogon*. The hybrid is then backcrossed to the cultivated variety to generate lines that contain only a small fraction of the "wild" genome. The backcross lines with the best qualities (e.g., highest yield and most disease resistance) are selected, and the "wild QTL" responsible for the superior performance are identified, using a detailed molecular linkage map. Once the beneficial QTLs are discovered by this method, they may be introduced into other cultivated varieties. In order for this strategy to succeed, it is essential that wild crop relatives be preserved as a storehouse of potentially useful genes. Efforts were begun in the 1970s to protect the existing crop relatives of many plants in their natural habitats and to preserve them in seed banks. As the work of McCouch and Tanksley and others has shown, it is not possible to predict which wild varieties may be needed decades or even centuries from now to contribute their beneficial alleles to cultivated varieties. To prevent the loss of superior genes, the widest possible spectrum of wild species must be preserved, even those that have no obvious favorable characteristics.

Almost 60 years ago, the great Russian plant geneticist N. I. Vavilov suggested that wild relatives of crop plants could be the source of genes to improve agriculture. In the coming century, Vavilov's vision may finally be realized, as genes from long-neglected wild crop relatives, identified by new molecular methods, spark a revitalized Green Revolution.

References

Mann, C. 1997. Reseeding the Green Revolution. *Science* 277:1038–43.

Ronald, P.C. 1997. Making rice disease-resistant. *Sci. Am.* (Nov.) 277:98–105.

Tanksley, S.D., and McCouch, S.R. 1997. Seed banks and molecular maps: Unlocking genetic potential from the wild. *Science* 277:1063–66.

Xiao, J. et al. 1996. Genes from wild rice improve yield. *Nature* 384:223–24.

FIGURE 24–9 Phenotypic effect of the *fw2.2* transgene in the tomato. When the allele causing small fruit is transferred to a plant that normally produces large fruit(+), the fruit is reduced in size. A control fruit(−), which is normally larger, is shown for comparison.

Thus, for the first time, the genetic basis of quantitative variation is available for meaningful investigation. Tanksley's research group has now established that the *fw2.2* locus encodes a gene, *ORFX*, that is involved in the control of the number of carpels. Carpels are ovule-bearing units within the ovary of the flower that, following fertilization, develop into a fruit. The number of carpel units influences fruit size. That the allele for small fruit is partially dominant to the large fruit allele suggests that the genetic alteration between the two alleles is involved in the regulation of floral development, ultimately determining carpel number. This is in contrast to a genetic change that alters the sequence and structure of a protein that somehow imparts weight to the fruit. Of added interest to these findings is the observation that the *ORFX* gene is structurally similar to a human oncogene in the *ras* family (see Chapter 22) implicated in malignancy.

Mapping QTLs and defining the function of genes present in these regions in agriculturally important plants will greatly enhance programs designed to improve their yield. Knowledge gained from the study of so-called "quantitative genes" in plants will no doubt pave the way for investigations of similar genes in animals, including our own species. Since such loci are thought to be critical components of what we have previously referred to as complex traits, our knowledge of how genes actually control phenotypes will be greatly extended.

Chapter Summary

1. Continuous variation—variation that is not easily categorized into distinct phenotypic classes—is exhibited in crosses involving traits under polygenic control. Such traits are quantitative in nature and are inherited as a result of the cumulative impact of additive alleles.
2. Polygenic characteristics can be analyzed using statistical methods, which include the mean, the variance, the standard deviation, and the standard error of the mean. Such statistical analysis can be descriptive, can be used to make inferences about a population, or can be used to compare and contrast sets of data.
3. For many phenotypic characteristics, it is difficult to ascertain when variation is due to genetic or to environmental factors. Heritability, the estimate of the relative importance of genetic versus nongenetic factors in determining genetic variation in populations, can be calculated for many characters and is especially useful in selective breeding of commercially valuable plants and animals.
4. Studies involving twins are aimed at resolving the question of heredity versus environment in human traits. The degree of concordance of a trait may be compared in monozygotic (identical) and dizygotic (fraternal) twins raised together or apart.
5. Loci containing genes that control a quantitative trait are called quantitative trait loci (QTLs). Using either genetic or molecular markers, the location and distribution within the genome of QTLs can be ascertained.

Insights and Solutions

1. In a plant, height varies from 6 to 36 cm. When 6-cm and 36-cm plants were crossed, all F_1 plants were 21 cm. In the F_2 generation, a continuous range of heights was observed. Most were around 21 cm, and 3 of 200 were as short as the 6-cm P_1 parent.

(a) What mode of inheritance is illustrated, and how many gene pairs are involved?

Solution: Polygenic inheritance is illustrated where a continuous trait is involved and where alleles contribute additively to the phenotype. The 3/200 ratio of F_2 plants is the key to determining the number of gene pairs. This reduces to a ratio of 1/66.7, very close to 1/64. Using the formula $1/4^n = 1/64$ (where 1/64 is equal to the proportion of F_2 phenotypes as extreme as either P_1 parent), $n = 3$. Therefore, three gene pairs are involved.

(b) How much does each additive allele contribute to height?

Solution: The variation between the two extreme phenotypes is

$$36 - 6 = 30 \text{ cm}.$$

Because there are six potential additive alleles (*AABBCC*), each contributes

$$30/6 = 5 \text{ cm}$$

to the base height of 6 cm, which results when no additive alleles (*aabbcc*) are part of the genotype.

(c) List all genotypes that give rise to plants that are 31 cm.

Solution: All genotypes that include 5 additive alleles will be 31 cm (5 alleles $\times$ 5 cm + 6 cm base height = 31 cm). Therefore, *AABBCc, AABbCC,* and *AaBBCC* are the genotypes that will result in plants that are 31 cm.

(d) In a cross separate from the earlier mentioned F_1 a plant of unknown phenotype and genotype was testcrossed with the following results:

1/4 11 cm

2/4 16 cm

1/4 21 cm

An astute genetics student realized that the unknown plant could be only one phenotype, but could be any of three genotypes. What were they?

Solution: When testcrossed (with *aabbcc*), the unknown plant must be able to contribute either one, two, or three additive alleles in its gametes in order to yield the three phenotypes in the offspring. Since no 6-cm offspring are observed, the unknown plant never contributes all nonadditive alleles (*abc*). Only plants that are homozygous at one locus and heterozygous at the other two loci will meet these criteria. Therefore, the unknown parent can be any of three genotypes, all of which have a phenotype of 26 cm:

AABbCc

AaBbCC

AaBBCc

For example, in the first genotype (*AABbCc*),

AABbCc $\times$ *aabbcc*

yields

1/4 *AaBbCc* 21 cm

1/4 *AaBbcc* 16 cm

1/4 *AabbCc* 16 cm

1/4 *Aabbcc* 11 cm

which is the ratio of phenotypes observed.

2. The results shown in the following table were recorded for ear length in corn:

For each of the parental strains and the F_1, calculate the mean values for ear length.

Solution: The mean values can be calculated as follows:

$$\overline{X} = \frac{\Sigma X_i}{n} \qquad P_A: \overline{X} = \frac{\Sigma X_i}{n} = \frac{378}{57} = 6.63$$

$$P_B: \overline{X} = \frac{\Sigma X_i}{n} = \frac{1697}{101} = 16.80$$

$$F_1: \overline{X} = \frac{\Sigma X_i}{n} = \frac{836}{69} = 12.11$$

3. For the corn plant described in Problem 2, compare the mean of the F_1 with that of each parental strain. What does this tell you about the type of gene action involved?

Solution: The F_1 mean (12.11) is almost midway between the parental means of 6.63 and 16.80. This indicates that the genes in question may be additive in effect.

4. The mean and variance of corolla length in two highly inbred strains of *Nicotiana* and their progeny are shown here:

One parent (P_1) has a short corolla and the other parent (P_2) has a long corolla. Calculate the heritability (H^2) of corolla length in this plant. One parent (P_1) has a short corolla and the other parent (P_2) has a long corolla.

Strain	Mean (mm)	Variance (mm)
P_1 short	40.47	3.12
P_2 long	93.75	3.87
F_1 ($P_1 \times P_2$)	63.90	4.74
F_2 ($F_1 \times F_1$)	68.72	47.70

Solution: The formula for estimating heritability is $H^2 = V_G/V_P$, where V_G and V_P are the genetic and phenotypic components of variation, respectively. The main issue in this problem is obtaining some estimate of two components of phenotypic variation: genetic and environmental factors. V_P is the combination of genetic and environmental variance. Because the two parental strains are true breeding, they are assumed to be homozygous, and the variance of 3.12 and 3.87 is considered to be the result of environmental influences. The average of these two values is 3.50. The F_1 is also genetically homogeneous and gives us an additional estimate of the impact of environmental factors. By averaging this value along with that of the parents,

$$\frac{4.74 + 3.50}{2} = 4.12$$

we obtain a relatively good idea of environmental impact on the phenotype. The phenotypic variance in the F_2 is the sum of the genetic (V_G) and environmental (V_E) components. We have estimated the environmental input as 4.12, so 47.70 minus 4.12 gives us an estimate of V_G, which is 43.58. Heritability then becomes 43.58/47.70 or 0.91. This value, when viewed in percentage form, indicates that about 91 percent of the variation in corolla length is due to genetic influences.

Length of Ear in cm

	5	*6*	*7*	*8*	*9*	*10*	*11*	*12*	*13*	*14*	*15*	*16*	*17*	*18*	*19*	*20*	*21*
Parent A	4	21	24	8													
Parent B									3	11	12	15	26	15	10	7	2
F_1					1	12	12	14	17	9	4						

Problems and Discussion Questions

1. Distinguish between discontinuous and continuous variation. Which type relates to inheritance of a quantitative nature?
2. Define the following: (a) polygenes, (b) additive alleles, (c) multiple-factor hypothesis, (d) monozygotic and dizygotic twins, (e) concordance and discordance, and (f) heritability.
3. A dark-red strain and a white strain of wheat are crossed and produce an intermediate, medium-red F_1. When the F_1 plants are interbred, an F_2 generation is produced in a ratio of 1 dark red: 4 medium-dark-red: 6 medium red: 4 light red: 1 white. Further crosses reveal that the dark-red and white F_2 plants are true breeding.
   (a) Based on the ratio of offspring in the F_2, how many genes are involved in the production of color?
   (b) How many additive alleles are needed to produce each possible phenotype?
   (c) Assign symbols to these alleles and list possible genotypes that give rise to the medium-red and the light-red phenotypes.
   (d) Predict the outcome of the F_1 and F_2 generations in a cross between a true-breeding medium-red plant and a white plant.
4. Height in humans depends on the additive action of genes. Assume that this trait is controlled by the four loci R, S, T, and U and that environmental effects are negligible. Instead of additive versus nonadditive alleles, assume that additive and partially additive alleles exist. Additive alleles contribute two units, and partially additive alleles contribute one unit to height.
   (a) Can two individuals of moderate height produce offspring that are much taller or shorter than either parent? If so, how?
   (b) If an individual with the minimum height specified by these genes marries an individual of intermediate or moderate height, will any of their children be taller than the tall parent? Why or why not?
5. An inbred strain of plants has a mean height of 24 cm. A second strain of the same species from a different geographical region also has a mean height of 24 cm. When plants from the two strains are crossed together, the F_1 plants are the same height as the parent plants. However, the F_2 generation shows a wide range of heights; the majority are like the P_1 and F_1 plants, but approximately 4 of 1000 are only 12 cm high, and about 4 of 1000 are 36 cm high.
   (a) What mode of inheritance is occurring here?
   (b) How many gene pairs are involved?
   (c) How much does each gene contribute to plant height?
   (d) Indicate one possible set of genotypes for the original P_1 parents and the F_1 plants that could account for these results.
   (e) Indicate three possible genotypes that could account for F_2 plants that are 18 cm high and three that account for F_2 plants that are 33 cm high.
6. Erma and Harvey were a compatible barnyard pair, but a curious sight. Harvey's tail was only 6 cm, while Erma's was 30 cm. Their F_1 piglet offspring all grew tails that were 18 cm. When inbred, an F_2 generation resulted in many piglets (Erma and Harvey's grandpigs), whose tails ranged in 4-cm intervals from 6 to 30 cm (6, 10, 14, 18, 22, 26, and 30). Most had 18-cm tails, while 1/64 had 6-cm tails and 1/64 had 30-cm tails.
   (a) Explain how tail length is inherited by describing the mode of inheritance, indicating how many gene pairs are at work and designating the genotypes of Harvey, Erma, and their 18 cm offspring.
   (b) If one of the 18 cm F_1 pigs is mated with the 6 cm F_2 pigs, what phenotypic ratio would be predicted if many offspring resulted? Diagram the cross.
7. In the following table, average differences of height and weight between monozygotic twins (reared together and apart), dizygotic twins, and nontwin siblings are compared:
   Draw as many conclusions as you can concerning the effects of genetics and the environment in influencing these human traits.

Trait	*MZ Reared Together*	*MZ Reared Apart*	*DZ Reared Together*	*Sibs Reared Together*
Height (cm)	1.7	1.8	4.4	4.5
Weight (kg)	1.9	4.5	4.5	4.7

8. Discuss how monozygotic and dizygotic twins reared together and apart are useful in assessing the genetic component responsible for phenotypic variation in humans.
9. List as many human traits as you can that are likely to be under the control of a polygenic mode of inheritance.
10. Corn plants from a test plot are measured, and the distribution of heights at 10-cm intervals is recorded in the following table: Calculate (a) the mean height, (b) the variance, (c) the standard deviation, and (d) the standard error of the mean. Plot a rough graph of plant height against frequency. Do the values represent a normal distribution? Based on your calculations, how would you assess the variation within this population?

Height (cm)	*Plants (no.)*
100	20
110	60
120	90
130	130
140	180
150	120
160	70
170	50
180	40

11. Contrast broad-sense heritability (H^2) and narrow-sense heritability (h^2). To what type of population is each calculation applicable? Which is useful in artificial selection procedures? Why?
12. The mean and variance of plant height of two highly inbred strains (P_1 and P_2) and their progeny (F_1 and F_2) are shown here:
   Calculate the broad-sense heritability (H^2) of plant height in this species.

Strain	*Mean (cm)*	*Variance*
P_1	34.2	4.2
P_2	55.3	3.8
F_1	44.2	5.6
F_2	46.3	10.3

13. A hypothetical study investigated the vitamin A content and the cholesterol content of eggs from a large population of chickens. The variances (V) were calculated, as shown here:

	Trait	
Variance	**Vitamin A**	**Cholesterol**
V_P	123.5	862.0
V_E	96.2	484.6
V_A	12.0	192.1
V_D	15.3	185.3

(a) Calculate the narrow heritability (h^2) for both traits.
(b) Which trait, if either, is likely to respond to selection?

14. In an assessment of learning in *Drosophila,* flies may be trained to avoid certain olfactory cues. In one population, a mean of 8.5 trials was required. A subgroup of this parental population that was trained most quickly (mean = 6.0) was interbred and their progeny examined. These flies demonstrated a mean training value of 7.5. Calculate and interpret narrow-sense heritability for olfactory learning in *Drosophila.*

15. In a population of tomato plants, mean fruit weight is 60 g and h^2 is 0.3. Predict the results of artificial selection (the mean weight of the progeny) if tomato plants whose fruit averaged 80 g were selected from the original population and interbred.

Extra-Spicy Problems

16. A mutant strain of *Drosophila* was isolated and shown to be resistant to an experimental insecticide, whereas normal (wild-type) flies were sensitive to the chemical. Following a cross between resistant flies and sensitive flies, isolated populations were derived that had various combinations of chromosomes from the two strains. Each was tested for resistance, as shown in the figure in the lower left column. Analyze the data and draw any appropriate conclusion about which chromosome(s) contain a gene responsible for inheritance of resistance to the insecticide.

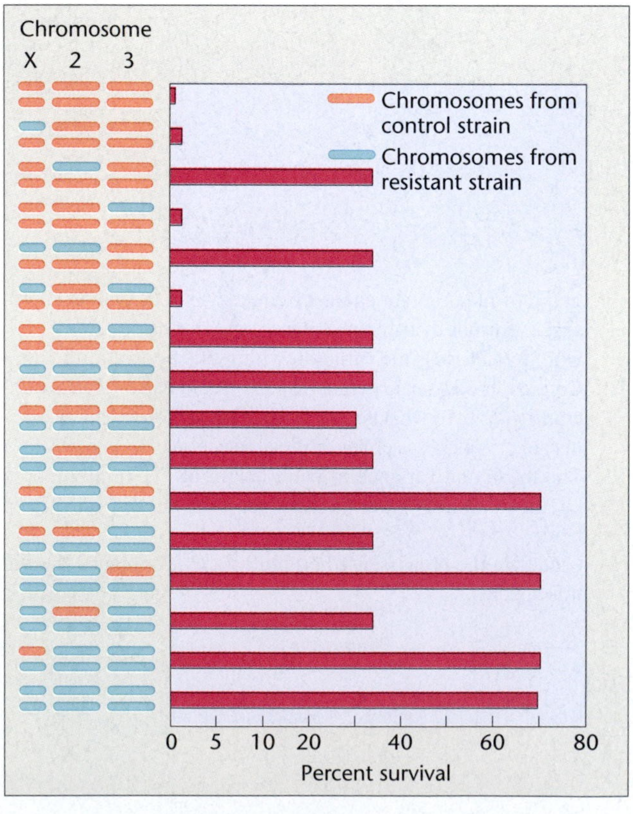

17. In 1988, Horst Wilkens investigated blind cavefish, comparing them with members of a sibling species with normal vision that are found in a lake. (We will call them cavefish and lakefish.) Wilkens found that cavefish eyes are about seven times smaller than lakefish eyes. F_1 hybrids have eyes of intermediate size. These data as well as the $F_1 \times F_1$ cross and those from backcrosses ($F_1 \times$ cavefish and $F_1 \times$ lakefish) are depicted here:

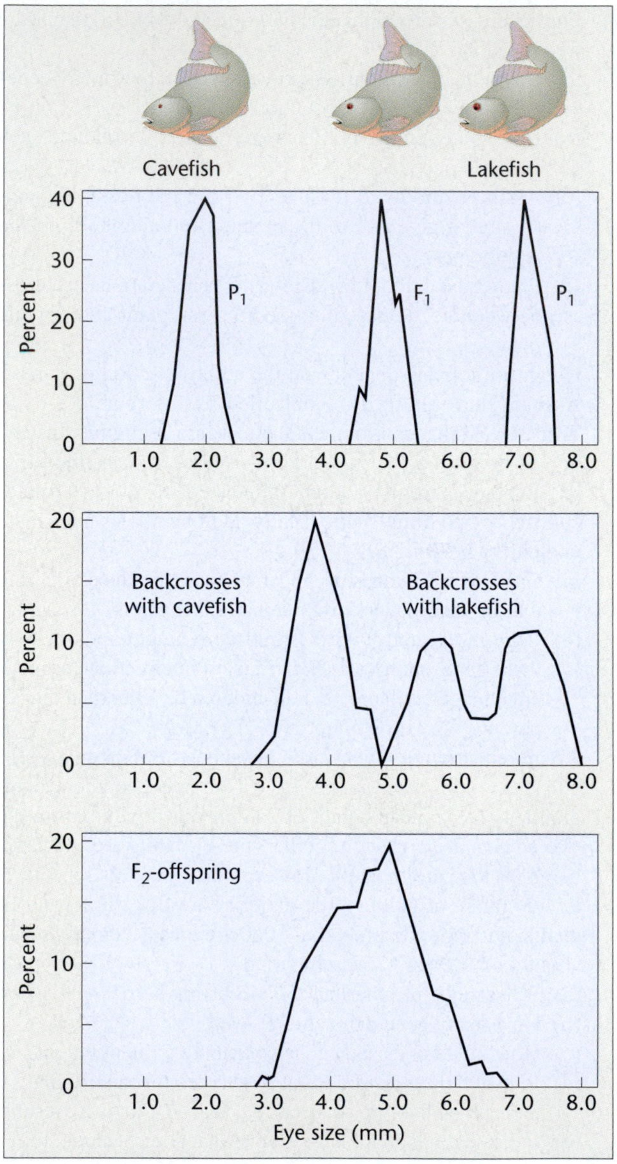

Examine Wilkens' results and respond to the following questions:

(a) Based strictly on the F_1 and F_2 results of Wilkens' initial crosses, what possible explanation concerning the inheritance of eye size seems most feasible?
(b) Based on the results of the F_1 backcross with cavefish, is your explanation supported? Explain.
(c) Based on the results of the F_1 backcross with lakefish, is your explanation supported? Explain.

(d) Wilkens examined about 1000 F_2 progeny and estimated that 6–7 genes are involved in determining eye size. Is the sample size adequate to justify this conclusion? (You may want to refer to Chapter 9.) Propose an experimental protocol to test the hypothesis.

(e) A comparison of the embryonic eye in cavefish and lakefish revealed that both reach approximately 4 mm in diameter. However, lakefish continue to grow, while cavefish eye size is greatly reduced. Speculate on the role of the genes involved in this problem.
[*Reference:* Wilkens, H. (1988). *Ecol. Biol.* 23:271–367.]

18. A 3" plant was crossed with a 15" plant and all F_1 plants were 9". In the F_2, plants exhibited a "normal distribution" with heights of 3, 4, 5, 6, 7, 8, 9, 10, 11, 12, 13, 14, and 15".

(a) What ratio will constitute the "normal distribution" in the F_2?

(b) What outcome will occur if the F_1 plants are test crossed with plants that are homozygous for all non-additive alleles?

Selected Readings

Brink, R., ed. 1967. *Heritage from Mendel.* Madison: University of Wisconsin Press.

Browman, K. W. 2001. Review of statistical methods of QTL mapping in experimental crosses. *Lab Animal* 30:44-52.

Crow, J.F. 1993. Francis Galton: Count and measure, measure and count. *Genetics* 135:1.

Dudley, J.W. 1977. 76 generations of selection for oil and protein percentage in maize. In *Proc. Intern. Conf. on Quant. Genet.* E. Pollack, O. Kempthorne, and T. Bailey eds. pp. 459–73. Ames: Iowa State University Press.

Falconer, D.S., and Mackay, F. C. 1996. *Introduction to quantitative genetics,* 4th ed. Essex, England: Longman.

Farber, S. 1980. *Identical twins reared apart.* New York: Basic Books.

Feldman, M.W., and Lewontin, R.C. 1975. The heritability hangup. *Science* 190:1163–66.

Fowler, C., and Mooney, P. 1990. *Shattering: Food, politics, and the loss of genetic diversity.* Tucson: University of Arizona Press.

Frary, A., et al. 2000. *fw2.2*: A quantitative trait locus key to the evolution of tomato fruit size. *Science* 289:85–88.

Haley, C. 1991. Use of DNA fingerprints for the detection of major genes for quantitative traits in domestic species. *Anim. Genet.* 22:259–77.

———1996. Livestock QTLs: Bringing home the bacon. *Trends Genet.* 11:488–90.

Lander, E., and Botstein, D. 1989. Mapping Mendelian factors underlying quantitative traits using RFLP linkage maps. *Genetics* 121:185–99.

Lander, E., and Schork, N. 1994. Genetic dissection of complex traits. *Science* 265:2037–48.

Lewontin, R.C. 1974. The analysis of variance and the analysis of causes. *Am. J. Hum. Genet.* 26:400–11.

Lynch, M., and Walsh, B. 1998. *Genetics and analysis of quantitative traits.* Sunderland, MA: Sinauer Associates.

Macy, T.F.C. 2001. Quantitative trait loci in *Drosophila. Nature Reviews Genetics* 2:11–19.

Newman, H.H., Freeman, F.N., and Holzinger, K.T. 1937. *Twins: A study of heredity and environment.* Chicago: University of Chicago Press.

Paterson, A., Deverna, J., Lanini, B., and Tanksley, S. 1990. Fine mapping of quantitative traits loci using selected overlapping recombinant chromosomes in an interspecific cross of tomato. *Genetics* 124:735–42.

Plomin, R., McClearn, G., Gora-Maslak, G., and Neiderhiser, J. 1991. Use of recombinant inbred strains to detect quantitative trait loci associated with behavior. *Behav. Genet.* 21:99–116.

Stuber, C.W. 1996. Mapping and manipulating quantitative traits in maize. *Trends Genet.* 11:477–87.

Zar, J.H. 1999. *Biostatistical analysis*, 4th ed. Upper Saddle River, NJ: Prentice Hall.

GENETICS MediaLab

The resources that follow will help you achieve a better understanding of the concepts presented in this chapter. These resources can be found either on the CD packaged with this textbook or on the Companion Web site for this book at **http://www.prenhall.com/klug**

Web Problem 1:

Time for completion = 10 minutes

How do we study polygenic traits? The genes underlying quantitative traits are assumed to behave like Mendelian factors, following the same rules of segregation and assortment as the genes we studied in Chapters 3 and 4. Quantitative traits also are assumed to be influenced by the environment. Read Dr. Phillip McClean's explanation of genetic and environmental effects on quantitative traits. What does "additive genetic effects" mean? Examine the data presented on the yields of winter wheat. Does one genotype produce a higher yield in most years than the other two? Can you tell whether any one genotype is less variable from year to year? Which genotype would offer the best performance over the broadest range of conditions? To complete this exercise, visit Web Problem 1 in Chapter 5 of your Companion Web site, and select the keyword **WHEAT**.

Web Problem 2:

Time for completion = 10 minutes

What does heritability measure? Genetic and environmental variances can be statistically estimated through a carefully planned breeding design. The proportion of the total phenotypic variance that can be associated with genetic effects is termed heritability. Read the article about estimating heritability in swine. The expected breeding value (EBV) is used to predict future genetic potential and is computed as the difference between the mean of a given trait in one individual and the mean of that trait in the individual's contemporary group (animals of similar breed, age, and sex, raised under similar environmental conditions), multiplied by the heritability of the trait. Thus, a boar with a lower EBV for back fat is leaner than most of his contemporaries raised under identical conditions. Consider the broad-sense heritability described in your text in Chapter 5. Would a breed of swine with a high genetic variance have a higher or lower heritability than one with a low genetic variance, assuming that the environmental variances are equal? If the heritability of weight gain was high, would sibling piglets be expected to be more or less similar than piglets from a population in which the heritability was low? Would such a comparison, where the populations are drawn from different environments, be valid? (Hint: Think about the components of variance.) To complete this exercise, visit Web Problem 2 in Chapter 5 of your Companion Web site, and select the keyword **HERITABILITY**.

Web Problem 3:

Time for completion = 10 minutes

How can we study quantitative genetic traits in humans? With experimental organisms, we can set up families and breeding experiments to estimate the genetic influences on quantitative traits, but it would be unethical to perform experiments of this sort on people. Many researchers studying physical and mental disease, personality traits, and estimates of intelligence use data on monozygotic (MZ) and dizygotic (DZ) twins to infer what portion of such traits might have a genetic basis. Read the summary of research on cigarette smoking involving twins. Why does the similarity in concordance of smoking in MZ and DZ twins imply that there is a greater genetic rather than environmental influence? How do the adoption studies, studies of twins reared apart, or comparisons of biological and adoptive parents with twins further support a strong genetic influence in becoming a cigarette smoker? To complete this exercise, visit Web Problem 3 in Chapter 5 of your Companion Web site, and select the keyword **TWINS**.

These lady-bird beetles from the Chiricahua Mountains in Arizona show considerable phenotypic variation. (*Edward S. Ross/California Academy of Sciences*)

25 Population Genetics

KEY CONCEPTS

Alfred Russel Wallace and Charles Darwin first identified natural selection as the mechanism of adaptive evolution in the mid-19th century, based on a series of observations on populations of organisms: (1) Phenotypic variations exist among individuals within populations; (2) these differences are passed from parents to offspring; (3) more offspring are born than will survive and reproduce; and (4) some variants are more successful at surviving and reproducing than others. In populations where all four factors operate, the relative abundance of the population's different phenotypes changes across generations. In other words, the population evolves. Although Wallace and Darwin described how organisms evolve by natural selection, there was no accurate model of the mechanisms responsible for variation and inheritance. Gregor Mendel published his work on the inheritance of traits in 1866, but it received little notice at the time. When Mendel's work was rediscovered in 1900, biologists immediately recognized its importance and began a 30-year effort to reconcile Mendel's concept of genes and alleles with the theory of evolution. In a key insight, biologists realized that changes in the relative abundance of different phenotypic traits in a population are tied to changes in the relative abundance of the alleles that influence the traits. The discipline within evolutionary biology that studies changes in allele frequencies is known as **population genetics**.

Early in the 20th century a number of workers, including G. Udny Yule, William Castle, Godfrey Hardy, and Wilhelm Weinberg, formulated the basic principles of population genetics. For many years, theorists focused on developing mathematical models that would describe the genetic structure of populations. Prominent among the theoreticians who developed these models were Sewall Wright, Ronald Fisher, and J. B. S. Haldane. Following their work, experimentalists and field workers tested the models by using biochemical and molecular techniques that measure variation directly at the protein and DNA levels. These experiments examined allele frequencies and the forces that alter the frequencies, such as selection, mutation, migration, and random genetic drift. In this chapter, we consider some general aspects of population genetics and also discuss other areas of genetics that relate to evolution.

25.1 Populations and Gene Pools

Members of a species often range over a wide geographic area. A **population** is a group of individuals from the same species that lives in the same geographic area and that actually or potentially interbreeds. If we consider a single genetic locus in this population, we may find that individuals within the population have different genotypes. To study population genetics, we compute frequencies at which various alleles and genotypes occur and how these frequencies change from one generation to the next.

When we consider generational changes in alleles and genotypes, we look at gene pools. A **gene pool** consists of all gametes made by all the breeding members of a population in a single generation. These gametes will combine to form the zygotes that become the next generation. The eggs and sperm that constitute the gene pool are haploid and thus contain only a single allele for each genetic locus. When we consider a single locus, we may find that different gametes carry different alleles. The proportion of gametes in a gene pool that carry a particular allele represents the frequency with which this allele occurs in the population.

Populations are dynamic; they expand and contract through changes in birth and death rates, migration, or contact with other populations. The dynamic nature of populations has important consequences and can, over time, lead to changes in the population's gene pool.

25.2 Calculating Allele Frequencies

Most population genetic researchers first measure the frequencies at which alleles occur at a particular locus. To do this, the genotypes of a large number of individuals in the population must be determined. In some cases, researchers can infer genotypes directly from phenotypes. In other cases, proteins or DNA sequences are analyzed to determine genotypes. To understand how allele frequencies are calculated, we consider an example that involves HIV infection rates.

Our example comes from research into the genetic factors that influence an individual's susceptibility to infection by HIV-1, the virus responsible for the bulk of AIDS cases worldwide. Of particular interest to AIDS researchers is the small number of individuals who make high-risk choices (such as unprotected sex with HIV-positive partners), yet remain uninfected. In 1996, Rong Liu and colleagues discovered that two exposed, but uninfected, individuals were homozygous for a mutant allele of a gene called *CC-CKR5*.

The *CC-CKR-5* gene, located on chromosome 3, encodes a protein called the C–C chemokine receptor-5, often abbreviated CCR5. Chemokines are signaling molecules associated with the immune system. When white blood cells with CCR5 receptor proteins bind to chemokines, the cells respond by moving into inflamed tissues to fight infection.

The CCR5 protein is also a receptor for strains of HIV-1. To gain entry into cells, a protein called Env (short for envelope protein) on the surface of HIV-1 binds to the CD4 protein on the surface of the host cell. Binding to CD4 causes Env to change shape and form a second binding site. This second site binds to CCR5, which, in turn, initiates the fusion of the viral protein coat with the host cell membrane. The merging of viral envelope with cell membrane transports the HIV viral core into the host cell's cytoplasm.

The mutant allele of the *CCR5* gene contains a 32-bp deletion in one of its coding regions. As a result, the protein encoded by the mutant allele is shortened and made nonfunctional. The protein never makes it to the cell membrane, so HIV-1 cannot enter these cells. The gene's normal allele is called *CCR51* (or *1*), and its allele with the 32-bp deletion is called *CCR5-Δ32* (or *Δ32*). The two uninfected individuals described by Liu both had the genotype *Δ32/Δ32* As a result, they had no CCR5 receptors on the surface of their cells and were resistant to infection by strains of HIV-1 that require CCR5 as a coreceptor.

At least three *Δ32/Δ32* individuals who are infected with HIV-1 have been found. Curiously, researchers have not discovered any adverse effects associated with this genotype. Heterozygotes with genotype *1/Δ32* are susceptible to HIV-1 infection, but evidence suggests that they progress more slowly to AIDS. Table 25–1 summarizes the genotypes possible at the *CCR5* locus and the phenotypes associated with each.

The discovery of the *CCR5-Δ32* allele, and the fact that it provides some protection against AIDS, generates two important questions: Which human populations harbor the *Δ32* allele, and how common is it? To address these questions, several teams of researchers surveyed a large number of people from a variety of populations. Genotypes were determined by direct analysis of DNA (Figure 25–1). We'll look at one sample population of 100 French individuals from a survey in Brittany in which 79 individuals have genotype *1/1*, 20 have genotype *1/Δ32* and 1 has genotype *Δ32/Δ32*. This sample has 158 *1* alleles (carried by the *1/1* individuals) plus 20 *1* alleles carried by *1/Δ32* individuals, for a total of 178. The frequency of the *CCR51* allele in the sample population is thus 178/200 = 0.89 = 89 percent. The *CCR5-Δ32* allele count shows 20 carried by *1/Δ32* individuals, plus 2 carried by the *Δ32/Δ32* individual, for a total of 22. The frequency of the *CCR5-Δ32* allele is thus 22/200 = 0.11 = 11 percent. Notice that 0.89 + 0.11 = 1.00, which confirms that we have accounted for the entire gene pool. Table 25–2 shows two methods for computing the frequencies of the *1* and *Δ32* alleles in the Brittany sample.

Figure 25–2 shows the frequency of the *CCR5-Δ32* allele in the 18 European populations surveyed. The studies show that populations in Northern Europe around the Baltic Sea have the highest frequencies of the *Δ32* allele. There is a sharp gradient in allele frequency from north to south, and populations in Sardinia and Greece have very low frequencies. In populations without European ancestry, the *Δ32* allele is essentially absent. The highly patterned global distribution of the *Δ32* allele presents an evolutionary puzzle that we'll return to later in the chapter.

TABLE 25–1 ***CCR5* Genotypes and Phenotypes**

Genotype	*Phenotype*
1/1	Susceptible to sexually transmitted strains of HIV-1
1/Δ 32	Susceptible, but may progress to AIDS slowly
Δ32/Δ32	Resistant to most sexually transmitted strains of HIV-1

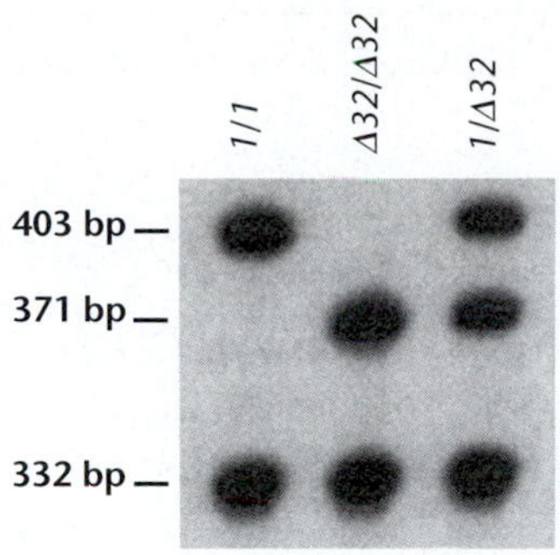

FIGURE 25–1 Allelic variation in the *CCR5* gene. Michel Samson and colleagues used PCR to amplify a part of the *CCR5* gene containing the site of the 32-bp deletion, cut the resulting DNA fragments with a restriction enzyme, and ran the fragments on an electrophoresis gel. Each lane reveals the genotype of a single individual. The *1* allele produces a 332-bp fragment and a 403-bp fragment; the *Δ32* allele produces a 332-bp fragment and a 371-bp fragment. Heterozygotes produce three bands. (From Samson, et al. 1996.) (Michel Samson, Frederick Libert, et al., Resistance to HIV-1 infection in Caucasian individuals bearing alleles of the *CCR-5* chemokine receptor gene. Reprinted with permission from *Nature* 382 (Aug. 22, 1996: p. 725): Fig. 3. © 1996 Macmillan Magazines, Ltd.)

25.3 The Hardy-Weinberg Law

The large variation in the frequency of the *CCR5-Δ32* allele among populations raises a number of questions. For example, can we expect the allele to increase in populations in which it is currently rare? Population genetics explores such questions by using a mathematical model developed independently by the British mathematician Godfrey H. Hardy and the German physician Wilhelm Weinberg. This model, called the **Hardy–Weinberg law**, shows what happens to alleles and genotypes in an "ideal" population (free of many of the complications that affect real populations) by using a set of simple assumptions:

1. Individuals of all genotypes have equal rates of survival and equal reproductive success—that is, there is no selection.
2. No new alleles are created or converted from one allele into another by mutation.
3. Individuals do not migrate into or out of the population.
4. The population is infinitely large, which in practical terms means that the population is large enough that sampling errors and other random effects are negligible.
5. Individuals in the population mate randomly.

The Hardy-Weinberg law demonstrates that an "ideal" population has these properties:

1. The frequency of alleles does not change from generation to generation.
2. After one generation, of random mating, offspring genotype frequencies can be predicted from the parent allele frequencies, and would be expected to remain constant from that point.

TABLE 25–2 Methods of Determining Allele Frequencies from Data on Genotypes

A. Counting Alleles

Genotype	*1/1*	*1/Δ32*	*Δ32/Δ32*	*Total*
Number of individuals	79	20	1	100
Number of *1* alleles	158	20	0	178
Number of Δ32 alleles	0	20	2	22
Total number of alleles	158	40	2	200

Frequency of *CCR51* in sample: 178/200 = 0.89 = 89%
Frequency of CCRS-Δ32 in sample: 22/200 = 0.11= 11%

B. From Genotype Frequencies

Genotype	*1/1*	*1/Δ32*	*Δ32/Δ32*	*Total*
Number of individuals	79	20	1	100
Genotype frequency	79/100 = 0.79	20/100 = 0.20	1/100 = 0.01	1.00

Frequency of *CCR51* in sample: 0.79 + (0.5) 0.20 = 0.89
Frequency of *CCR5-Δ32* in sample: (0.5) 0.20 + 0.01 = 0.11

What makes the Hardy-Weinberg law useful is its assumptions. By specifying the assumptions under which the population cannot evolve, the Hardy-Weinberg law identifies the real-world forces that cause allele frequencies to change. In other words, by holding certain conditions constant, the Hardy-Weinberg law isolates the forces of evolution and allows them to be quantified. The Hardy-Weinberg law also reveals "neutral genes," those not being operated on by the forces of evolution.

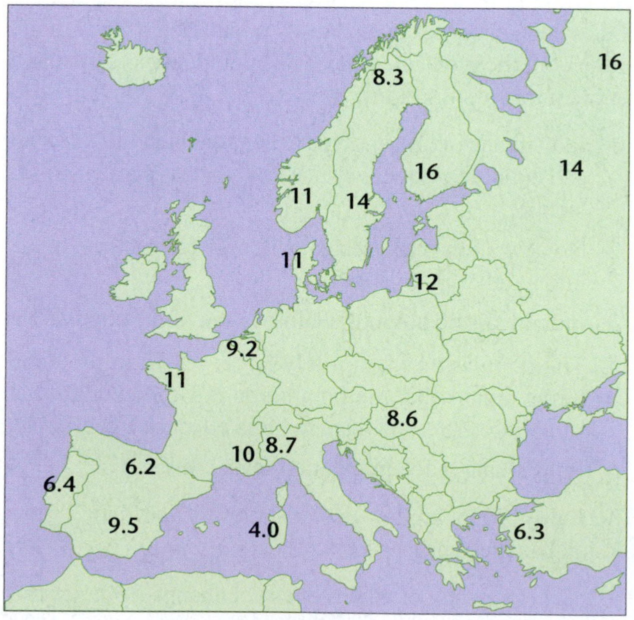

FIGURE 25–2 The frequency (percentage) of the *CCR5-Δ32* allele in 18 European populations. (From F. Leibert, et al. 1998. The *CCR5* mutation conferring protection against HIV-1 in Caucasian populations has a single and recent origin in northeastern Europe. *Hum. Molec. Genet.* 7:399–406 by permission of Oxford University Press.)

Demonstration of the Hardy–Weinberg Law

To demonstrate how the Hardy–Weinberg law works, we begin with a specific case and then consider the general case. In both examples, we focus on a single locus with two alleles, *A* and *a*.

Imagine a population in which the frequency of allele *A*, in both eggs and sperm, is 0.7, and the frequency of allele *a* is 0.3. Note that 0.7 + 0.3 = 1, indicating that all the alleles for gene *A* present in the gene pool are accounted for. We assume, per Hardy–Weinberg requirements, that individuals mate randomly, which we visualize as follows: We place all the gametes in the gene pool in a barrel and stir. We then randomly draw eggs and sperm from the barrel and pair them to make zygotes. What genotype frequencies does this give us? For any one zygote, the probability or chance that the egg will contain *A* is 0.7, and the probability that the sperm will contain *A* is 0.7. The probability that both egg *and* sperm will contain *A* is 0.7 × 0.7 = 0.49. In other words, we predict that genotype *AA* will occur 49 percent of the time. The probability that a zygote will be formed from an egg carrying *A* and a sperm carrying *a* is 0.7 × 0.3 = 0.21, and the probability that a zygote will be formed from an egg carrying *a* and a sperm carrying *A* is 0.7 × 0.3 = 0.21. Thus, the frequency of genotype *Aa* is 0.21 + 0.21 = 0.42 = 42 percent. Finally, the probability that a zygote will be formed from an egg carrying *a* and a sperm carrying *a* is 0.3 × 0.3 = 0.09 meaning that the predicted frequency of genotype *aa* is 9 percent. As a check on our calculations, note that 0.49 + 0.42 + 0.09 = 1.0, which confirms that we have accounted for all of the zygotes. These calculations are summarized in Figure 25–3.

We started with the frequency of a particular allele in a specific gene pool and calculated the probability that certain genotypes would be produced from this pool. When the zygotes develop into adults and reproduce, what will be the frequency distribution of alleles in the new gene pool? Recall that under the Hardy–Weinberg law, we assume that all genotypes have equal rates of survival and reproduction. This means that in the next generation, all genotypes contribute equally to the new gene pool. The *AA* individuals constitute 49 percent of the population, and we can predict that the gametes they produce will constitute 49 percent of the gene pool. These gametes all carry allele *A*. Likewise, *Aa* individuals constitute 42 percent of the population, so we predict that their gametes will constitute 42 percent of the new gene pool. Half (0.5) of these gametes will carry allele *A*. Thus, the frequency of allele *A* in the gene pool is $0.49 + (0.5)\,0.42 = 0.7$. The other half of the gametes produced by *Aa* individuals will carry allele *a*. The *aa* individuals constitute 9 percent of the population, so their gametes will constitute 9 percent of the new gene pool. These gametes all carry allele *a*. Thus, we can predict that the allele *a* in the new gene pool is $(0.5)\,0.42 + 0.09 = 0.3$. As a check on our calculation, note that $0.7 + 0.3 = 1.0$ accounting for all of the gametes in the gene pool of the new generation.

We have arrived back where we began, with a gene pool in which the frequency of allele *A* is 0.7 and the frequency of allele *a* is 0.3. These calculations demonstrate the Hardy–Weinberg law: Allele frequencies in our population do not change from one generation to the next, and, after just one generation of random mating, the genotype frequencies can be predicted from the allele frequencies. In other words, this population does not evolve with respect to the locus we have examined.

For the general case of the Hardy–Weinberg law, we use variables instead of numerical values for the allele frequencies. Imagine a gene pool in which the frequency of allele *A* is *p*, and the frequency of allele *a* is *q*, such that $p + q = 1$. If we randomly draw an egg and sperm from the gene pool and pair them to make a zygote, the probability that both egg and sperm will carry allele *A* is $p \times p$ Thus, the frequency of genotype *AA* among the zygotes is p^2. The probabilities that the egg carries *A* and the sperm carries *a* is $p \times q$, and that the egg carries *a* and the sperm carries *A* is $q \times p$. Thus, the frequency of genotype *Aa* among the zygotes is $2pq$. Finally, the probability that the egg and sperm will both carry *a* is $q \times q$, making the frequency of genotype *aa* among the zygotes q^2. Therefore, the distribution of genotypes among the zygotes is

$$p^2 + 2pq + q^2 = 1$$

Figure 25–4 shows these calculations.

For our general case, we ask what the allele frequencies in the new gene pool will be when these zygotes develop into adults and reproduce. All gametes from *AA* individuals carry allele *A*, as do half of the gametes from *Aa* individuals. Hence, we predict that the frequency of allele *A* in the new gene pool will be

$$p^2 + \frac{1}{2}2pq = p^2 + pq$$

Recall that $p + q = 1$. We can therefore substitute $(1 - p)$ for *q* in the previous equation, which gives

$$p^2 + p(1 - p) = p^2 + p - p^2 = p$$

Likewise, half the gametes from *Aa* individuals carry allele *a*, as do all gametes from *aa* individuals. Thus, we predict that the frequency of allele *a* in the new gene pool will be

$$\left(\frac{1}{2}\right)2pq + q^2 = pq + q^2$$

We substitute $(1 - q)$ for *p* in this equation and get

$$(1 - q)q + q^2 = q - q^2 + q^2 = q$$

Eggs \ Sperm	fr(*A*) = 0.7	fr(*a*) = 0.3
fr(*A*) = 0.7	fr(*AA*) = 0.7 × 0.7 = 0.49	fr(*Aa*) = 0.7 × 0.3 = 0.21
fr(*a*) = 0.3	fr(*aA*) = 0.3 × 0.7 = 0.21	fr(*aa*) = 0.3 × 0.3 = 0.09

FIGURE 25–3 Calculating genotype frequencies from allele frequencies. Gametes represent withdrawals from the gene pool to form the genotypes of the next generation. In this population, the frequency of the *A* allele is 0.7, and the frequency of the *a* allele is 0.3. The frequencies of the genotypes in the next generation are calculated as 0.49 for *AA*, 0.42 for *Aa*, and 0.09 for *aa*. Under the Hardy–Weinberg law, the frequencies of *A* and *a* remain constant from generation to generation.

Eggs \ Sperm	fr(*A*) = *p*	fr(*a*) = *q*
fr(*A*) = *p*	fr(*AA*) = p^2	fr(*Aa*) = *pq*
fr(*a*) = *q*	fr(*aA*) = *qp*	fr(*aa*) = q^2

FIGURE 25–4 The general case of allele and genotype frequencies under Hardy–Weinberg assumptions. The frequency of allele *A* is *p* and the frequency of allele *a* is *q*. After mating, the three genotypes *AA*, *Aa*, and *aa* have the frequencies p^2, $2pq$, and q^2 respectively.

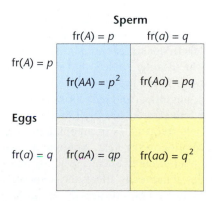

MEDIA TUTORIAL Population Genetics

Once again, we arrive back where we began, with a gene pool in which the predicted proportion of allele A is p and that of allele a is q. The Hardy–Weinberg law holds true for any frequencies of A and a, as long as the frequencies add to 1 and the five assumptions are invoked. A population in which the allele frequencies remain constant from generation to generation and in which the genotype frequencies can be predicted from the allele frequencies is said to be in a state of **Hardy–Weinberg equilibrium** for that locus.

Consequences of the Law

The Hardy–Weinberg law has several important consequences. First, it shows that dominant traits do not necessarily increase from one generation to the next. Second, it demonstrates that **genetic variability** can be maintained in a population since, once established in an ideal population, allele frequencies remain unchanged. Third, if we invoke Hardy–Weinberg assumptions, then knowing the frequency of just one genotype enables us to calculate the frequencies of all other genotypes. This relationship is particularly useful in human genetics because we can now calculate the frequency of heterozygous carriers for recessive genetic disorders even when all we know is the frequency of affected individuals.(See section 25.5).

We began this discussion by asking whether we can expect the *CCR5-Δ32* allele to increase in populations in which it is currently rare. From what we now know about the Hardy–Weinberg law, we can say for the general case that if (1) individuals of all genotypes have equal rates of survival and reproduction, (2) there is no mutation, (3) no differential migration into or out of the population, (4) the population is extremely large, (5) and the individuals in the population choose their mates randomly with respect to the locus of interest, then the frequency of the *Δ32* allele will not change.

The general case demonstrates the most important role of the Hardy–Weinberg law: It is the foundation upon which population genetics is built. By showing that within populations there are loci that do not evolve, we can use the Hardy–Weinberg law to identify forces that do cause populations to evolve by examining those loci that do show changes in allele frequency over time. As we shall see in the last half of the chapter, when the assumptions of the Hardy–Weinberg law are broken—because of natural selection, mutation, migration, and reductions in population size leading to random sampling errors (also known as genetic drift)—the allele frequencies in a population may change from one generation to the next. Nonrandom mating does not, by itself, alter *allele* frequencies, but by altering *genotype* frequencies, it indirectly affects the course of evolution. The Hardy–Weinberg law tells geneticists where to look to find the causes of evolution in populations. We return to the *CCR5-Δ32* allele later in the chapter to see how a population genetics perspective has generated fruitful hypotheses for further research.

Testing for Equilibrium

One way we establish whether one or more of the Hardy–Weinberg assumptions do not hold in a given population is by determining whether the population's genotypes are in equilibrium. To do this, we first determine the frequencies of the genotypes, either directly from the phenotypes (if heterozygotes are recognizable) or by analyzing proteins or DNA sequences. We then calculate the allele frequencies from the genotype frequencies, as demonstrated earlier. Finally, we use the allele frequencies in the parental generation to predict the offspring's genotype frequencies. According to the Hardy–Weinberg law, the genotype frequencies are predicted to fit the $p^2 + 2pq + q^2 = 1$ relationship. If they do not, then one or more of the assumptions are invalid for the population in question. We will use the *CCR5* genotypes of a population in Britain to demonstrate the Hardy–Weinberg law. The population includes 283 individuals, of which 223 have genotype *1/1*, 57 have genotype *1/Δ32* and 3 have genotype *Δ32/Δ32*. These numbers represent genotype frequencies of 223/283 = 0.788, 57/283 = 0.201 and 3/283 = 0.011, respectively. From the genotype frequencies, we compute the *CCR51* allele frequency as 0.89 and the frequency of the *CCR5-Δ32* allele as 0.11. From these allele frequencies, we can use the Hardy–Weinberg law to determine whether this population is in equilibrium. The allele frequencies predict the genotype frequencies as follows:

Expected Frequency of genotype *1/1*
$= p^2 = (0.89)^2 = 0.792$

Expected Frequency of genotype *1/Δ32*
$= 2pq = 2(0.89)(0.11) = 0.196$

Expected Frequency of genotype *Δ32/Δ32*
$= q^2 = (0.11)^2 = 0.012$

These expected frequencies are nearly identical to the observed frequencies. Our test of this population has failed to provide evidence that Hardy–Weinberg assumptions are being violated. The conclusion is confirmed by a χ^2 analysis. (See Chapter 9.) The χ^2 value in this case is tiny: 0.00023. To be statistically significant at even the most generous, accepted level, $p = 0.05$, the χ^2 value would have to be 3.84. (In a test for Hardy–Weinberg equilibrium, the degrees of freedom are given by $k - 1 - m$, where k is the number of genotypes and m is the number of independent allele frequencies estimated from the data. Here, $k = 3$ and $m = 1$, since calculating only one allele frequency allows us to determine the other by subtraction. Thus, we have $3 - 1 - 1 = 1$ degree of freedom.)

On the other hand, if the Hardy–Weinberg test had demonstrated that the population is not in equilibrium, it would indicate that one or more assumptions are not being met. To illustrate this, imagine two hypothetical populations, one living on East Island, the other living on West Island. The East Island population all has genotype *1/1*, so the *CCR51* allele constitutes 100 percent of the population. All of the West

Island population has the genotype *Δ32/Δ32* so the frequency of the *CCR5-Δ32* allele is 100 percent. Both island populations are in Hardy–Weinberg equilibrium.

Now imagine that 500 people from each island move to the previously uninhabited Central Island. The allele and genotype frequencies for the source populations and the new population on Central Island appear in Table 25–3. Both the *1* allele and the *Δ32* allele occur at frequencies of 0.5 in the new population. Given these allele frequencies, the Hardy–Weinberg law predicts that the genotype frequencies will be $0.25 + 0.50 + 0.25 = 1$ These expected frequencies obviously differ from the observed frequencies; there are no heterozygotes at all on Central Island. Therefore, the population is not in Hardy–Weinberg equilibrium (and a χ^2 test would confirm this).

Two assumptions are violated in the new population. First, everyone living on Central Island is a migrant. Second, Central Island's population is not the products of random mating. Everyone on Central Island moved there and has either two parents from East Island or two parents from West Island. However, it would take only one generation of random mating on Central Island to bring the offspring to the expected allele frequencies, as shown in Figure 25–5.

Therese Markow and colleagues documented a real human population that is not in Hardy–Weinberg equilibrium. These researchers studied 122 Havasupai, a population of Native Americans in Arizona. They determined the genotype of each Havasupai individual at two loci in the major histocompatibility complex (MHC). These genes, *HLA-A* and *HLA-B*, encode proteins that are involved in the immune system's discrimination between self and nonself. The immune systems of individuals heterozygous at MHC loci appear to recognize a greater diversity of foreign invaders and thus may be better able to fight disease. Markow and colleagues observed significantly more individuals who are heterozgous at both loci and significantly fewer homozygous individuals than would be expected under the Hardy–Weinberg law. Violation of either, or both, of two Hardy–Weinberg assumptions could explain the excess of heterozygotes among the Havasupai. First, Havasupai fetuses, children, and adults who are heterozygous for *HLA-A* and *HLA-B* may have higher rates of survival than individuals who are homozygous (i.e., selection is occurring). Second, rather than choosing their mates randomly, Havasupai people may somehow prefer mates whose MHC genotypes differ from their own.

Eggs \ Sperm	fr(*1*) = 0.5	fr(*Δ32*) = 0.5
fr(*1*) = 0.5	fr(+/*1*) = 0.25	fr(*1*/*Δ32*) = 0.25
fr(*Δ32*) = 0.5	fr(*Δ32*/*1*) = 0.25	fr(*Δ32*/*Δ32*) = 0.25

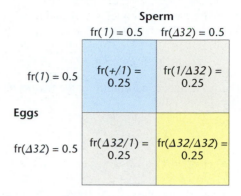

FIGURE 25–5 Hardy–Weinberg equilibrium. If mating is random in a population that also meets the other assumptions of the Hardy–Weinberg law, equilibrium will be reached in one generation. Here the genotype frequencies (fr) after one generation of random mating are 0.25, 0.5, and 0.25; compare these with the genotype frequencies shown in Table 25–3.

25.4 Extensions of the Hardy-Weinberg Law

We commonly find several alleles of a single locus in a population. The ABO blood group in humans (discussed in Chapter 10) is such an example. The locus *I* (isoagglutinin) has three alleles I^A, I^B, *and* I^O, yielding six possible genotypic combinations (I^AI^A, I^BI^B, I^OI^O, I^AI^B, I^AI^O, I^BI^O). Recall that in this case I^A and I^B are codominant alleles, and both of these are dominant to I^O. The result is that homozygous I^AI^A and heterozygous I^AI^O individuals are phenotypically identical, as are I^BI^B and I^BI^O individuals, so we can distinguish only four phenotypic combinations.

By adding another variable to the Hardy–Weinberg equation, we can calculate both the genotype and allele frequencies for the situation involving three alleles. Let *p*, *q*, and *r*

TABLE 25–3 Frequencies of *CCR51* and *CCR5-Δ32* Alleles in a Hypothetical Population Composed of 500 *1/1* Individuals and 500 *Δ32/Δ32* Individuals

	Allele Frequencies		*Genotype Frequencies*		
Population	*CCR51*	*CCR5-Δ32*	*1/1*	*1/Δ32*	*Δ32/Δ32*
Source populations					
East Island	1.0	0	1.0	0	0
West Island	0	1.0	0	0	1.0
New population on Central Island					
Observed	$p = 0.5$	$q = 0.5$	0.5	0	0.5
Expected			$p^2 = 0.25$	$2pq = 0.50$	$q^2 = 0.25$

represent the frequencies of alleles A, B, and O, respectively. Note that because there are three alleles

$$p + q + r = 1$$

Under Hardy–Weinberg assumptions, the frequencies of the genotypes are given by

$$(p + q + r)^2 =$$

$$p^2 + q^2 + r^2 + 2pq + 2pr + 2qr = 1$$

If we know the frequencies of blood types for a population, we can then estimate the frequencies for the three alleles of the ABO system. For example, in one population sampled, the following blood-type frequencies are observed: A = 0.53, B = 0.13, O = 0.26. Because the I^O allele is recessive, the population's frequency of type O blood equals the proportion of the recessive genotype r^2. Thus,

$$r^2 = 0.26$$

$$r = \sqrt{0.26}$$

$$r = 0.51$$

Using r, we can estimate the allele frequencies for the I^A and I^B alleles. The I^A allele is present in two genotypes, I^AI^A *and* I^AI^O. allelles The frequency of the I^AI^A genotype is represented by p^2 and the I^AI^O genotype by $2pr$. Therefore, the combined frequency of type A blood and type O blood is given by

$$p^2 + 2pr = r^2 = 0.53 + 0.26$$

If we factor the left side of the equation and take the sum of the terms on the right, we get

$$(p + r)^2 = 0.79$$

$$p + r = \sqrt{0.79}$$

$$p = 0.89 - r$$

$$p = 0.89 - 0.51 = 0.38$$

Having estimated p and r, the frequencies of allele I^A and allele I^O, we can now estimate the frequency for the I^B allele:

$$p + q + r = 1$$

$$q = 1 - p - r$$

$$= 1 - 0.38 - 0.51$$

$$= 0.11$$

The phenotypic frequencies and genotypic frequencies for this population are summarized in Table 25–4.

25.5 Using the Hardy-Weinberg Law: Calculating Heterozygote Frequency

One application, the Hardy–Weinberg law allows us to estimate the frequency of heterozygotes in a population. The frequency of a recessive trait can usually be determined by counting such individuals in a sample of the population. With this information and the Hardy–Weinberg law, we can then calculate the allele and genotype frequencies.

Cystic fibrosis, an autosomal recessive trait, has an incidence of about 1/2500 = 0.0004 in people of northern European ancestry. Individuals with cystic fibrosis are easily distinguished from the population at large by such symptoms as salty sweat, excess amounts of thick mucus in the lungs, and susceptibility to bacterial infections. Because this is a recessive trait, individuals with cystic fibrosis must be homozygous. Their frequency in a population is represented by q^2, provided that mating has been random in the previous generation. The frequency of the recessive allele therefore is

$$q = \sqrt{q^2} = \sqrt{0.0004} = 0.02$$

Since $p + q = 1$, then the frequency of p is

$$p = 1 - q = 1 - 0.02 = 0.98$$

In the Hardy–Weinberg equation, the frequency of heterozygotes is $2pq$. Thus,

$$2pq = 2(0.98)(0.02)$$

$$= 0.04 \text{ or 4 percent, or } 1/25$$

Thus, heterozygotes for cystic fibrosis are rather common in the population (4 percent), even though the incidence of homozygous recessives is only 1/2500, or 0.04 percent.

In general, the frequencies of all three genotypes can be estimated once the frequency of either allele is known and Hardy–Weinberg assumptions are invoked. The relationship between genotype and allele frequency is shown in Figure 25–6. It is important to note that heterozygotes increase rapidly in a population as the values of p and q move from 0 or 1. This observation confirms our conclusion that when a recessive trait such as cystic fibrosis is rare, the majority of those carrying the allele are heterozygotes. In populations in which the frequencies of p and q are between 0.33 and 0.67, heterozygotes occur at higher frequency than either homozygote.

25.6 Natural Selection

We have noted that the Hardy–Weinberg law establishes an ideal population that allows us to estimate allele and genotype frequencies at a given locus in populations in which the assumptions of random mating, absence of selection, mutation, equal viability and fertility hold. Obviously, it is difficult

TABLE 25–4 Calculating Genotype Frequencies for Multiple Alleles Where the Frequency of Allele I^A = 0.38, Allele I^B = 0.11, and Allele I^O = 0.51

Genotype	*Genotype Frequency*	*Phenotype*	*Phenotype Frequency*
I^AI^A	$p^2 = (0.38)^2 = 0.14$	A	0.53
I^AI^O	$2pr = 2(0.38)(0.51) = 0.39$		
I^BI^B	$q^2 = (0.11)^2 = 0.01$	B	0.12
I^BI^O	$2qr = 2(0.11)(0.51) = 0.11$		
I^AI^B	$2pq = 2(0.38)(0.11) = 0.084$	AB	0.08
I^OI^O	$r^2 = (0.51)^2 = 0.26$	O	0.26

to find natural populations in which all these assumptions hold for all loci. In nature, populations are dynamic, and changes in size and gene pool are common. The Hardy–Weinberg law allows us to investigate populations that vary from the ideal. In this and next sections, we discuss factors that prevent populations from reaching Hardy–Weinberg equilibrium, or that drive populations toward a different equilibrium, and the relative contribution of these factors to evolutionary change.

Natural Selection

The first assumption of the Hardy–Weinberg law is that individuals of all genotypes have equal rates of survival and equal reproductive success. If this assumption does not hold, allele frequencies may change from one generation to the next. To see why, let's imagine a population of 100 individuals in which the frequency of allele *A* is 0.5 and that of allele *a* is 0.5. Assuming the previous generation mated randomly, we find that the genotype frequencies in the present generation are $(0.5)^2 = 0.25$ for *AA*, $2(0.5)(0.5) = 0.5$ for *Aa*, and $(0.5)2 = 0.25$ for *aa*. Because our population contains 100 individuals, we have 25 *AA* individuals, 50 *Aa* individuals, and 25 *aa* individuals. Now suppose that individuals with different genotypes have different rates of survival: All 25 *AA* individuals survive to reproduce, 90 percent or 45 of the *Aa* individuals survive to reproduce, and 80 percent or 20 of the *aa* individuals survive to reproduce. When the survivors reproduce, each contributes two gametes to the new gene pool, giving us $2(25) + 2(45) + 2(20) = 180$ gametes. What are the frequencies of the two alleles in the surviving population? We have 50 *A* gametes from *AA* individuals, plus 45 *A* gametes from *Aa* individuals, so the frequency of allele *A* is $(50 + 45)/180 = 0.53$ We have 45 *a* gametes from *Aa* individuals, plus 40 *a* gametes from *aa* individuals, so the frequency of allele *a* is $(45 + 40)/180 = 0.47$

These differ from the frequencies we started with. Allele *A* has increased, while allele *a* has declined. A difference among individuals in survival or reproduction rate (or both) is called **natural selection**. Natural selection is the principal force that shifts allele frequencies within large populations and is one of the most important factors in evolutionary change.

Fitness and Selection

Selection occurs whenever individuals with a particular genotype enjoy an advantage in survival and reproduction over other genotypes. However, selection may vary from less than 1% to 100% for a lethal gene. In the previous example, selection was strong. Weak selection might involve just a fraction of a percent difference in the survival rates of different genotypes. Advantages in survival and reproduction ultimately translate into increased genetic contribution to future generations. An individual's genetic contribution to future generations is called its **fitness**. Thus, genotypes associated with high rates of reproductive success are said to have high fitness, whereas genotypes associated with low reproductive success are said to have low fitness.

Hardy–Weinberg analysis also allows us to examine fitness. By convention, population geneticists use the letter *w* to represent fitness. Thus, w_{aa} represents the relative fitness of genotype *AA*, w_{Aa} the relative fitness of genotype *Aa*, and w_{aa} the relative fitness of genotype *aa*. Assigning the values $w_{AA} = 1$, $w_{Aa} = 0.9$, and $w_{aa} = 0.8$ could mean, for example, that all *AA* individuals survive, 90 percent of the *Aa* individuals survive, and 80 percent of the *aa* individuals survive, as in the previous example.

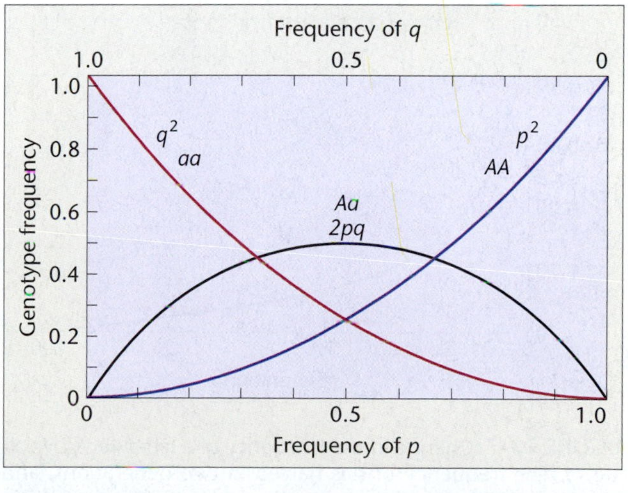

FIGURE 25–6 The relationship between genotype and allele frequencies derived from the Hardy–Weinberg equation.

Let's consider selection against deleterious alleles. Fitness values $w_{AA} = 1$, $w_{Aa} = 1$, and $w_{aa} = 0$ and describe a situation in which allele *a* is a lethal recessive. As homozygous recessive individuals die without leaving offspring, the frequency of allele *a* will decline. The decline in the frequency of allele *a* is described by the equation

$$q_g = \frac{q_0}{1 + gq_0}$$

where q_g is the frequency of allele *a* in generation g, gq_0 is the starting frequency of *a* (i.e., the frequency of *a* in generation zero), and g is the number of generations that have passed.

Figure 25–7 shows what happens to a lethal recessive allele with an initial frequency of 0.5. At first, because of the high percentage of *aa* genotypes, the frequency of allele *a* declines rapidly. The frequency of *a* is halved in only two generations. By the sixth generation, the frequency is halved again. By now, however, the majority of *a* alleles are carried by heterozygotes. Because *a* is recessive, these heterozygotes are not selected against. This means that as time continues to pass, the frequency of allele *a* declines ever more slowly. As long as heterozygotes continue to mate, it is difficult for selection to completely eliminate a recessive allele from a population.

Generation	*p*	*q*	p^2	2*pq*	q^2
0	0.50	0.50	0.25	0.50	0.25
1	0.67	0.33	0.44	0.44	0.12
2	0.75	0.25	0.56	0.38	0.06
3	0.80	0.20	0.64	0.32	0.04
4	0.83	0.17	0.69	0.28	0.03
5	0.86	0.14	0.73	0.25	0.02
6	0.88	0.12	0.77	0.21	0.01
10	0.91	0.09	0.84	0.15	0.01
20	0.95	0.05	0.91	0.09	<0.01
40	0.98	0.02	0.95	0.05	<0.01
70	0.99	0.01	0.98	0.02	<0.01
100	0.99	0.01	0.98	0.02	<0.01

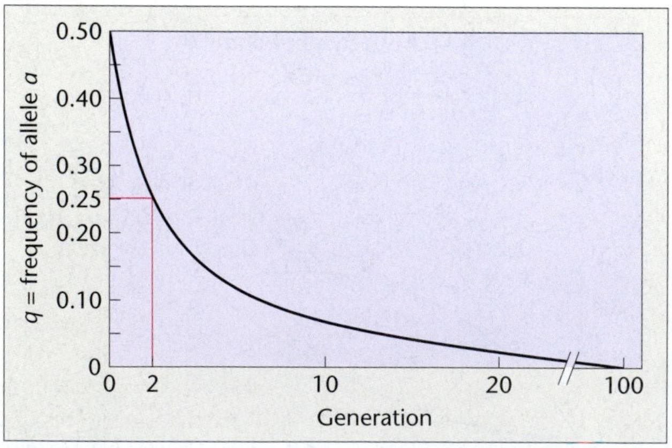

FIGURE 25–7 Change in the frequency of a lethal recessive allele, *a*. The frequency of *a* is halved in two generations, and halved again by the sixth generation. Subsequent reductions occur slowly because the majority of *a* alleles are carried by heterozygotes.

Of course, a deleterious allele need not be recessive; many other scenarios are possible. A deleterious allele may be codominant, so that heterozygotes have intermediate fitness. Or an allele may be deleterious in the homozygous state, but beneficial in the heterozygous state, like the allele for sickle-cell anemia. Figure 25–8 shows examples in which a deleterious allele is codominant, but the intensity of selection varies from strong to weak. In each case, the frequency of the deleterious allele, *a*, starts at 0.99 and declines over time. However, the rate of decline depends heavily on the strength of selection. When only 90 percent of the heterozygotes and 80 percent of the *aa* homozygotes survive (red curve), the frequency of allele *a* drops from 0.99 to less than 0.01 in 85 generations. However, when 99.8 percent of the heterozygotes and 99.6 percent of the *aa* homozygotes survive (blue curve), it takes 1000 generations for the frequency of allele *a* to drop from 0.99 to 0.93. Two important conclusions can be drawn from this graph. First, given thousands of generations, even weak selection can cause substantial changes in allele frequencies; because evolution generally occurs over a large number of generations, selection is a powerful force of evolution. Second, for selection to produce rapid changes in allele frequencies, changes measurable within a human life span, the differences in fitness among genotypes must be large, or generation time must be short.

The manner in which selection affects allele frequencies allows us to make some inferences about the *CCR5-Δ32* allele that we discussed earlier. Because individuals with genotype *Δ32/Δ32* are resistant to most sexually transmitted strains of HIV-1, while individuals with genotypes *1/1*

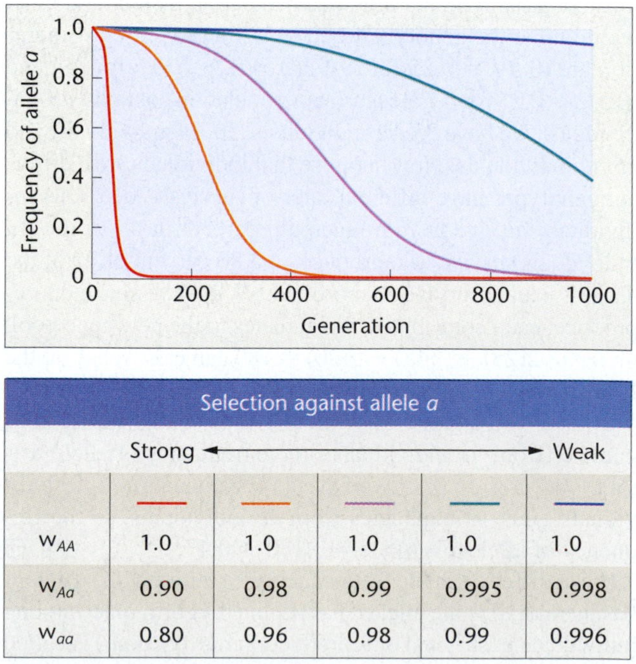

FIGURE 25–8 The effect of selection on allele frequency. The rate at which a deleterious allele is removed from a population depends heavily on the strength of selection.

and *1/Δ32* are susceptible, we might expect AIDS to act as a selective force causing the frequency of the *Δ32* allele to increase over time. Indeed it probably will, but the increase in frequency is likely to be slow in human terms.

We can do a rough calculation as follows: Imagine a population in which the current frequency of the *Δ32* allele is 0.10. Under Hardy–Weinberg assumptions, the genotype frequencies in this population are 0.81 for *1/1*, 0.18 for *1/Δ32* and 0.01 for *Δ32/Δ32* Imagine also that 1 percent of the *1/1* and *1/Δ32* individuals in this population will contract HIV and die of AIDS.

Based on our assumptions, we can assign fitness levels to the genotypes as follows: $w_{1/1} = 0.99$; $w_{1/\Delta 32} = 0.99$; $w_{1/\Delta 32} = 0.99$; $w_{\Delta 32/\Delta 32} = 1.0$. Given the assigned fitness, we can predict that the frequency of the *CCR5-Δ32* allele in the next generation will be 0.100091. In fact, it will take about 100 generations (about 2000 years) for the frequency of the *Δ32* allele to reach just 0.11 (Figure 25–9). In other words, the frequency of the *Δ32* allele will probably not change much over the next few generations in most populations that currently harbor it. A population genetic perspective sheds light on the *CCR5-Δ32* story in other ways as well. Two research groups have analyzed genetic variation at marker loci closely linked to the *CCR5* gene. Both groups concluded that most, if not all present-day copies of the *Δ32* allele are descended from a single ancestral copy that appeared in northeastern Europe at most a few thousand years ago. In fact, one group estimates that the common ancestor of all *Δ32* alleles existed just 700 years ago. How could a new allele rise from a frequency of virtually zero to as high as 20 percent in roughly 30 generations?

It seems that there must have been strong selection in favor of the *Δ32* allele, most likely in the form of an infectious disease. The agent of selection cannot have been HIV-1, because HIV-1 moved from chimpanzees to humans too recently. Because selection occurred about 700 years ago, J. C. Stephens suggests that the agent of selection was bubonic plague. During the Black Death of 1346–1352, between a quarter and a third of all Europeans died from plague. Bubonic plague is caused by the bacterium *Yersinia pestis*. This bacterium manufactures a protein that kills some kinds of white blood cells. Stephens hypothesizes that the process through which the bacterial protein kills white cells involves the *CCR5* gene product. If true, some mechanism makes individuals homozygous for the *Δ32* allele more likely to survive plague epidemics.

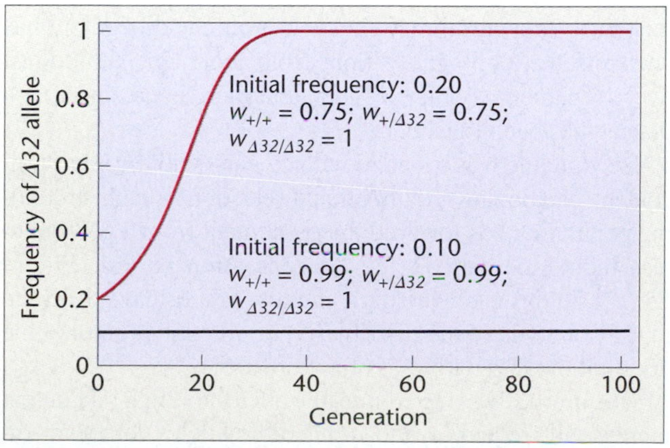

FIGURE 25–9 The rate at which the frequency of the *CCR5-Δ32* allele changes in hypothetical populations with different initial frequencies and different fitnesses.

Based on their study of the *CCR5* locus in Jewish and European populations, William Klitz and his colleagues have suggested that the *Δ32* allele may have arisen as early as the eighth century. They also suggest that smallpox may be the selective agent responsible for the rapid increase in the frequency of this allele. Both HIV and variola, the virus responsible for smallpox, use the CCR5 receptor for infecting cells, and with a fatality rate of 25 percent, waves of smallpox would be a powerful selective agent.

If past epidemics are responsible for the high frequency of the *CCR5-Δ32* allele in European populations, then the virtual absence of the allele in non-European populations is at first somewhat puzzling. Perhaps plague has been more common in Europe than elsewhere. Alternatively, perhaps other populations have different alleles of the *CCR5* gene that also confer protection against the plague. Teams of researchers looking for other alleles of the *CCR5* gene in various populations have found a total of 20 mutant alleles, including *Δ32*. Sixteen of these alleles encode proteins different in structure from that encoded by the *CCR51* allele. Some, like *Δ32* are loss-of-function alleles. Some of the alleles appear confined to Asian populations, others to African populations. Their frequencies are as high as 3–4 percent. Together, these discoveries are consistent with the hypothesis that alteration or loss of the CCR5 protein protects against an as-yet-unidentified infectious disease or diseases.

Selection in Natural Populations

Geneticists have done a great deal of research on the effect of natural selection on allele frequencies in both laboratory and natural populations. Among the most detailed studies of natural populations are those that involve insects exposed to pesticides. Christine Chevillon and colleagues, for example, studied the effect of the insecticide chlorpyrifos on allele frequencies in populations of house mosquitoes (Figure 25–10). Chlorpyrifos kills mosquitoes by interfering with the function of the enzyme acetylcholinesterase (ACE), which under normal circumstances breaks down the neurotransmitter acetylcholine. An allele of ACE called Ace^R encodes a slightly altered version of ACE that is immune to interference by chlorpyrifos.

Chevillon measured the frequency of the Ace^R allele in nine populations. In the first four locations, chlorpyrifos had been used to control mosquitoes for 22 years; in the last five locations, chlorpyrifos had never been used. Chevillon predicted that the frequency of Ace^R would be higher in the exposed populations. The researchers also predicted that the frequencies of alleles for enzymes unrelated to the physiological effects of chlorpyrifos would show no such pattern. Among the control enzymes studied was aspartate amino

FIGURE 25–10 Pesticides used to control insects act as selective agents, changing allele frequencies of resistance genes in the populations exposed to the pesticide. *(Larsh K. Bristol/Visuals Unlimited, Inc.)*

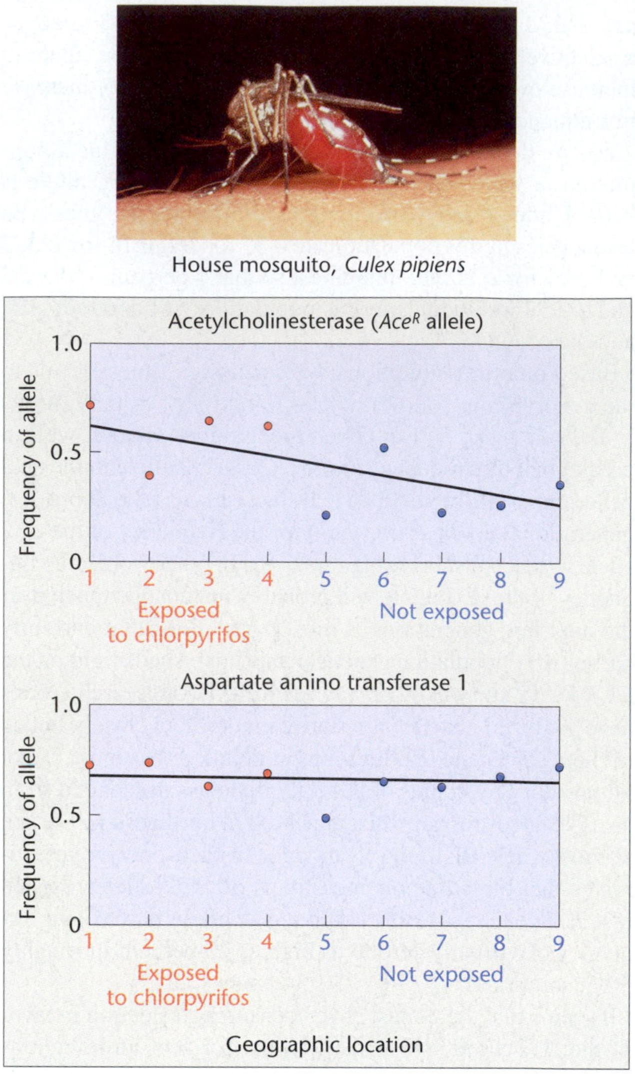

FIGURE 25–11 The effect of selection on allele frequencies in natural populations. (a) The frequency of the Ace^R allele, which confers resistance to the insecticide chlorpyrifos, is higher in house mosquito populations exposed to chlorpyrifos. (b) The frequency of an allele for an enzyme unrelated to chlorpyrifos metabolism (aspartate amino transferase 1) shows no such pattern. *(Photo: Hans Pfletschinger/Peter Arnold, Inc.)*

transferase 1. The results appear in Figure 25–11. As the researchers predicted, the frequency of the Ace^R allele was significantly higher in the exposed populations. Also as predicted, the frequencies of the most common alleles of the control enzyme gene showed no such trends. The explanation is that during the 22 years of exposure to the pesticide, mosquitoes had higher rates of survival if they carried the Ace^R allele. In other words, the Ace^R allele had been favored by natural selection.

Natural Selection and Quantitative Traits

Most phenotypic traits are controlled not by alleles at a single locus, but by the combined influence of the individual's genotype at many different loci and the environment. Because selection is a consequence of the organism's genotypic/phenotypic combination, polygenic or quantitative traits (those controlled by a number of genes that might be susceptible to environmental influences) also respond to selection. Such quantitative traits, including adult body height and weight in humans, often demonstrate a continuously varying distribution resembling a bell-shaped curve. Selection for such traits can be classified as (1) directional, (2) stabilizing, or (3) disruptive.

In **directional selection** (important to plant and animal breeders), desirable traits, often representing phenotypic extremes, are selected. If the trait is polygenic, the most extreme phenotypes that the genotype can express will appear in the population only after prolonged selection. An example of directional selection is the long-running experiment at the State Agricultural Laboratory in Illinois selecting for high and low oil content in corn kernels. At the start of the experiment in 1896, a population of 163 corn ears was surveyed. The 24 ears highest in oil content were used to produce the next generation. In each succeeding generation, the ears with the highest oil content were used to breed. In this line of upward directional selection, the oil content has been raised after 50 generations from about 4 percent to just over 16 percent (Figure 25–12), and there is no sign that a plateau has been reached.

A second line was founded with 12 ears with the lowest oil content, and downward directional selection from this founding population has lowered the oil content from 4 percent to less than 1 percent in the 50-generation period (Figure 25–12). In this case, the high and low graphs have diverged to the point where the distribution of oil content in one line does not overlap with that of the other one.

In the upwardly selected line, the alleles for high oil content increase in frequency and replace the alleles for lower oil content. At some point, all individuals in the line will have the genotype for the highest oil content, and all will express the most extreme phenotype for high oil content. In the

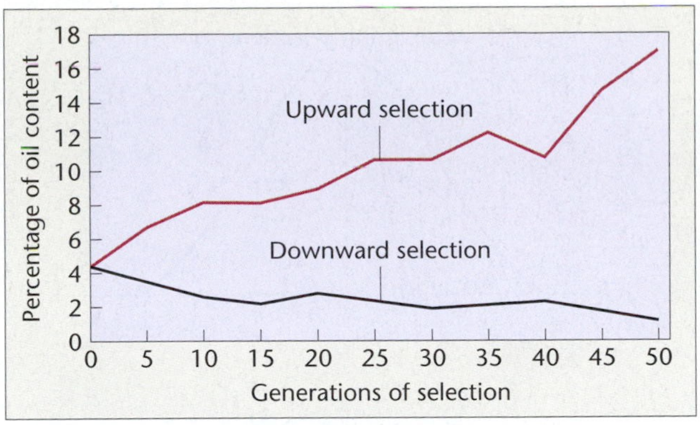

FIGURE 25–12 Long-term selection to alter the oil content of corn kernels.

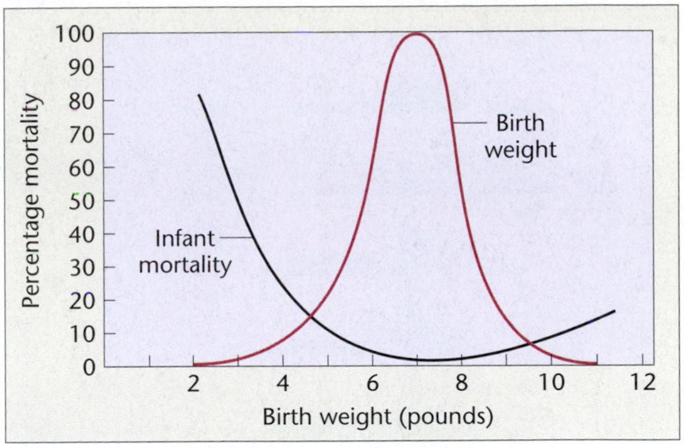

FIGURE 25–13 Relationship between birth weight and mortality in humans.

downwardly selected line, the opposite situation will occur, eventually producing a population with a genotype and phenotype for the lowest oil content. When the lines fail to respond to further selection, no genetic variance for oil content will remain, and each line will have a homozygous genotype for oil content.

In nature, directional selection can occur when one of the phenotypic extremes becomes selected for or against, usually as a result of changes in the environment. A carefully documented example comes from research by Peter and Rosemary Grant and their colleagues, who used the medium ground finches (*Geospiza fortis*) of Daphne Major island in the Galapagos Islands. The beak size of these birds varies enormously. In 1976, for example, some birds in the population had beaks less than 7 mm deep, while others had beaks more than 12 mm deep. Beak size is heritable, which means that large-beaked parents tend to have large-beaked offspring, and small-beaked parents tend to have small-beaked offspring. In 1977, a severe drought on Daphne killed some 80 percent of the finches. Big-beaked birds survived at higher rates than small-beaked birds, because when food became scarce, the big-beaked birds were able to eat a greater variety of seeds. When the drought ended in 1978 and the survivors paired off and bred, the offspring inherited their parents' big beaks. Between 1976 and 1978, the beak depth of the average finch in the Daphne Major population increased by just over 0.5 mm, shifting the average beak size toward one phenotypic extreme.

Stabilizing selection, in contrast, tends to favor intermediate types, with both extreme phenotypes being selected against. One of the clearest demonstrations of stabilizing selection is provided by the data of Mary Karn and Sheldon Penrose on human birth weight and survival for 13,730 children born over an 11-year period. Figure 25–13 shows the distribution of birth weight and the percentage of mortality at 4 weeks of age. Infant mortality increases on either side of the optimal birth weight of 7.5 pounds and quite dramatically so at the low end. At the genetic level, stabilizing selection acts to keep a population well adapted to its environment. In this situation, individuals closer to the average for a given trait will have higher fitness.

Disruptive selection is selection against intermediates and for both phenotypic extremes. It can be viewed as the opposite of stabilizing selection because the intermediate types are selected against. In one set of experiments, John Thoday applied disruptive selection to a population of *Drosophila* on the basis of bristle number. In every generation, he allowed only the flies with high- or low-bristle numbers to breed. After several generations, most of the flies could be easily placed in a low- or high-bristle category (Figure 25–14). In natural populations, such a situation might exist for a population in a heterogeneous environment. The types and effects of selection are summarized in Figure 25–15.

25.7 Mutation

Within a population, the gene pool is reshuffled each generation to produce new genotypes in the offspring. Because the number of possible genotypic combinations is so large, population members alive at any given time represent only a fraction of all possible genotypes. The enormous genetic reserve present in the gene pool allows Mendelian assortment and recombination to produce new genotypic combinations continuously. But assortment and recombination do not produce new alleles. **Mutation** alone acts to create new alleles. It is important to keep in mind that mutational events occur at random—that is, without regard for any possible benefit or disadvantage to the organism. In this section, we consider whether mutation is, by itself, a significant factor in causing allele frequencies to change.

To determine whether mutation is a significant force in changing allele frequencies, we measure the rate at which mutations are produced. As most mutations are recessive, it is difficult to observe mutation rates directly in diploid organisms. Indirect methods that use probability and statistics or large-scale screening programs are employed. For certain dominant mutations, however, a direct method of measurement can be used. To ensure accuracy, several conditions must be met:

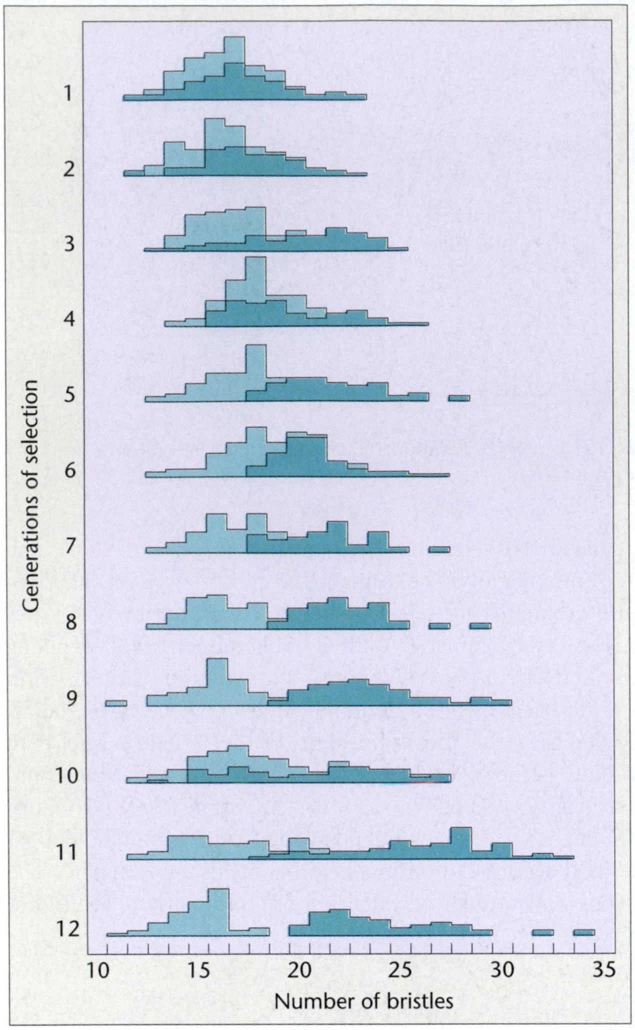

FIGURE 25–14 The effect of disruptive selection on bristle number in *Drosophila*. When individuals with the highest and lowest bristle number were selected, the population showed a nonoverlapping divergence in only 12 generations.

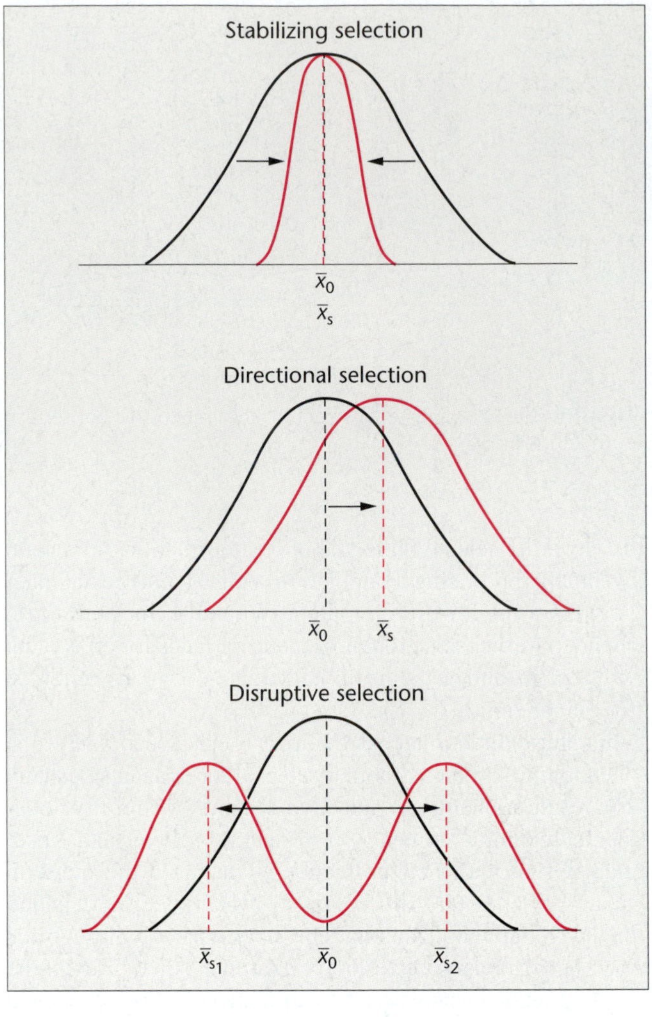

FIGURE 25–15 The impact of stabilizing, directional, and disruptive selection. In each case, the mean of an original population (black) and the mean of the population following selection (red) is shown.

1. The allele must produce a distinctive phenotype that can be distinguished from similar phenotypes produced by recessive alleles.
2. The trait must be fully expressed or completely penetrant so that mutant individuals can be identified.
3. An identical phenotype must never be produced by nongenetic agents such as drugs or chemicals.

Mutation rates can be stated as the number of new mutant alleles per given number of gametes. Suppose that for a given gene that undergoes mutation to a dominant allele, 2 out of 100,000 births exhibit a mutant phenotype. In these two cases, the parents are phenotypically normal. Because the zygotes that produced these births each carry two copies of the gene, we have actually surveyed 200,000 copies of the gene (or 200,000 gametes). If we assume that the affected births are each heterozygous, we have uncovered 2 mutant alleles out of 200,000. Thus, the mutation rate is 2/200,000 or 1/100,000, which in scientific notation is written as 1×10^{-5}.

In humans, a dominant form of dwarfism known as **achondroplasia** fulfills the requirements for measuring mutation rates. Individuals with this skeletal disorder have an enlarged skull, short arms and legs, and can be diagnosed by X-ray examination at birth. In a survey of almost 250,000 births, the mutation rate (μ) for achondroplasia has been calculated as

$$\mu = 1.4 \times 10^{-5} \pm 0.5 \times 10^{-5}$$

Knowing the rate of mutation, we can estimate the extent to which mutation can cause allele frequencies to change from one generation to the next. We represent the normal allele as *d* and the allele for achondroplasia as *D*.

Imagine a population of 500,000 individuals in which everyone has genotype *dd*. The initial frequency of *d* is 1.0, and the initial frequency of *D* is 0. If each individual contributes 2 gametes to the gene pool, the gene pool will contain 1,000,000 gametes, all carrying allele *d*. While the gametes are in the gene pool, 1.4 of every 100,000 *d* alleles mutates into a *D* allele. The frequency of allele *d* is now (1,000,000

− 14)/1,000,000 = 0.999986, and the frequency of D is 14/1,000,000 = 0.000014. From these numbers, it will clearly be a long time before mutation, by itself, causes any appreciable change in the allele frequencies in this population.

More generally, if we have two alleles, A with frequency p and a with frequency q, and if μ represents the rate of mutations converting A into a, then the frequencies of the alleles in the next generation are given by

$$p_{g+1} = p_g - \mu p_g \quad \text{and} \quad q_{g+1} = q_g + \mu p_g$$

where p_g+_1 and q_g+_1 represent the allele frequencies in the next generation, and p_g and q_g represent the allele frequencies in the present generation.

Figure 25–16 shows the replacement rate (change over time) in allele A for a population in which the initial frequency of A is 1.0 and the rate of mutation (μ) converting A into a is 1.0×10^{-5}. At this mutation rate, it will take about 70,000 generations to reduce the frequency of A to 0.5. Even if the rate of mutation increases through exposure to higher levels of radioactivity or chemical mutagens, the impact of mutation on allele frequencies will be extremely weak. The ultimate source of the genetic variability, mutation provides the raw material for evolution, but *by itself* plays a relatively insignificant role in changing allele frequencies. Instead, the fate of alleles created by mutation is more likely to be determined by natural selection (discussed previously) and genetic drift (discussed later).

An evolutionary perspective on mutation can lead to medical discoveries. The autosomal recessive disease cystic fibrosis which we discussed previously in relation to calculating heterozygote frequencies, is caused by a loss-of-function mutation in the gene for a cell-surface protein called the cystic fibrosis transmembrane conductance regulator (CFTR).

The frequency for the mutant alleles causing cystic fibrosis is about 2 percent in European populations. Until recently, most individuals with two mutant alleles died before reproducing, meaning that selection against homozygous recessive individuals was rather strong. This creates a puzzle: In the face of selection against them, what has maintained the mutant alleles at an overall frequency of 2 percent?

One hypothesis, the mutation-selection balance hypothesis, posits that mutation is constantly creating new alleles to replace the ones being eliminated by selection. However, for this scenario to work, the rate of mutations creating new alleles would have to be rather high, on the order of 5×10^{-4} to counteract the effect of selection. Many evolutionary geneticists instead prefer an alternative explanation, the heterozygote superiority hypothesis. According to the heterozygote superiority hypothesis, selection against homozygous mutant individuals is counterbalanced by selection in favor of heterozygotes. The most popular agent of selection in favor of heterozygotes is resistance to an as-yet-unidentified disease.

Recent work suggests that cystic fibrosis heterozygotes may have enhanced resistance to typhoid fever. Typhoid fever is caused by the bacterium *Salmonella typhi*, which infiltrates cells of the intestinal lining. In laboratory studies, mouse intestinal cells that were heterozygous for *CFTR-Δ508,* the analog of the most common cystic fibrosis mutation in humans, acquired 86 percent fewer bacteria than did cells homozygous for the wild-type allele. Whether humans heterozygous for *CFTR-Δ508* also enjoy resistance to typhoid fever remains to be established. If they do, then cystic fibrosis will join sickle-cell anemia as an example of heterozygote superiority.

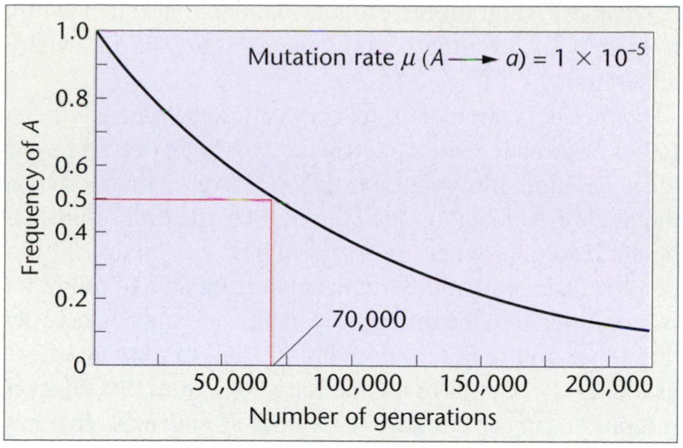

FIGURE 25–16 Replacement rate of an allele by mutation alone, assuming an average mutation rate of 1.0×10^{-5}.

25.8 Migration

Occasionally, a species divides into populations that to some extent are separated geographically. Various evolutionary forces, including selection can establish different allele frequencies in such populations. **Migration** occurs when individuals move between the populations. Imagine a species in which a single locus has two alleles, A and a. There are two populations of this species, one on a mainland and one on an island. The frequency of A on the mainland is represented by p_m, and the frequency of A on the island is p_i. Under the influence of migration from the mainland to the island, the frequency of A in the next generation on the island ($P_{i'}$) is given by

$$P_{i'} = (1 - m)p_i + mp_m$$

where m represents migrants from the mainland to the island.

Under these conditions, the frequency of A in the next generation on the island ($p_{i'}$) will be affected by migration. For example, assume that $p_i = 0.4$ and $p_m = 0.6$ and that 10 percent of the parents of the next generation are migrants from the mainland, so that $m = 0.1$. In the next generation, the frequency of allele A on the island will be

$$p_{i'} = [(1 - 0.1)34]1(0.130.6)$$
$$= 0.36 + 0.06$$
$$= 0.42$$

In this case, migration from the mainland has changed the frequency of A on the island from 0.40 to 0.42 in a single generation.

If either m is large or p_m is very different from p_i, then a rather large change in the frequency of A can occur in a single generation. If migration is the only force acting to change the allele frequency on the island, then an equilibrium will be attained only when $p_i = p_m$. These calculations reveal that the change in allele frequency attributable to migration is proportional to the differences in allele frequency between the donor and recipient populations and to the rate of migration. As m can have a wide range of values, the effect of migration can substantially alter allele frequencies in populations, as shown for the *B* allele of the ABO blood group in Figure 25–17. Although migration can be difficult to quantify, it can often be estimated.

Migration can also be regarded as the flow of genes between populations that were once, but are no longer, geographically isolated. Esteban Parra and colleagues measured allele frequencies for several different DNA sequence polymorphisms in African-American and European-American populations and in African and European populations representative of the ancestral populations from which the two American populations are descended. One locus they studied, a restriction site polymorphism called *FY-NULL*, has two alleles, *FY-NULL*1* and *FY-NULL*2*. Figure 25–18 shows the frequency of *FY-NULL*1* in each population. The frequency of this allele is 0 in the three African populations and 1.0 in the three European populations, but lies between these extremes in all African-American and European-American populations. The simplest explanation for these data is that genes have mixed between American populations with predominantly African ancestry and American populations with predominantly European ancestry. Based on *FY-NULL* and several other loci, researchers estimate that African-American populations derive between 11.6 and 22.5 percent of their ancestry from Europeans and that European-American populations derive between 0.5 and 1.2 percent of their ancestry from Africans.

25.9 Genetic Drift

In laboratory crosses, one condition essential to realizing theoretical genetic ratios (e.g., 3:1, 1:2:1, 9:3:3:1) is a fairly large sample size. A large sample size is also important to the study of population genetics as allele and genotype frequencies are examined or predicted. For example, if a population consists of 1000 randomly mating heterozygotes (*Aa*), the next generation will consist of approximately 25 percent *AA*, 50 percent *Aa*, and 25 percent *aa* genotypes. Provided that the initial and subsequent populations are large, only minor deviations from this mathematical ratio will occur, and the frequencies of *A* and *a* will remain about equal (at 0.50), demonstrating a Hardy-Weinberg relationship.

However, if a population consists of only one set of heterozygous parents and they produce only two offspring, the allele frequency can change drastically. We can also predict genotypes and allele frequencies in the offspring in this case. As Table 25–5 shows, in 10 of 16 times such a cross is made, the allele frequencies would be altered. In 2 of 16 times, either the *A* or *a* allele would be eliminated in a single generation.

Founding a population with only one set of heterozygous parents that produce two offspring is an extreme example, but it illustrates the point that large interbreeding populations

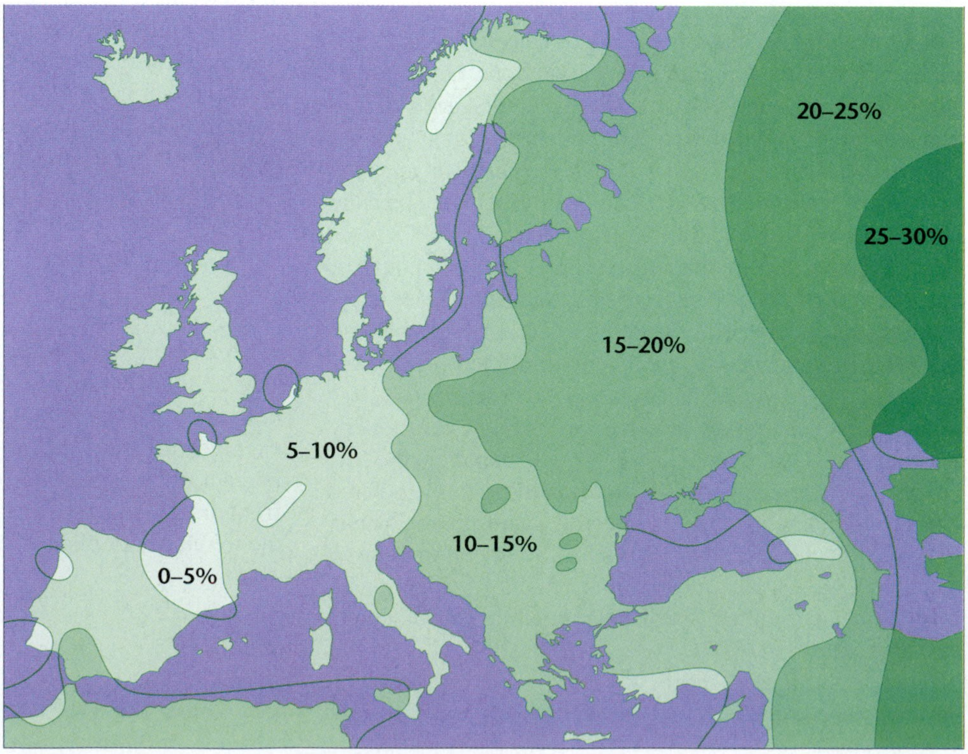

FIGURE 25–17 Migration as a force in evolution. The *B* allele of the *ABO* locus is present in a gradient from east to west. This allele shows the highest frequency in Central Asia and the lowest is in northeastern Spain. The gradient parallels the waves of Mongol migration into Europe following the fall of the Roman Empire and is a genetic relic of human history.

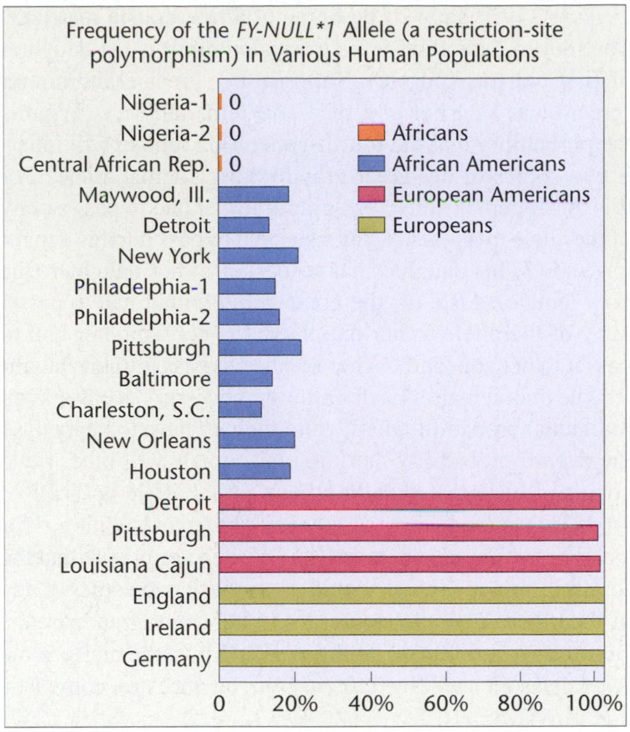

FIGURE 25–18 Frequency histogram of a restriction site polymorphism allele in several populations. These and other data demonstrate that African-American and European-American populations are of mixed ancestry.

are essential to maintain Hardy–Weinberg equilibrium. In small populations, significant random fluctuations in allele frequencies are possible by chance deviation. The degree of fluctuation increases as the population size decreases, a situation known as **genetic drift**. In the extreme case, genetic drift can lead to the chance fixation of one allele to the exclusion of another allele.

To study genetic drift in laboratory populations of *Drosophila melanogaster*, Warwick Kerr and Sewall Wright set up over 100 lines, with four males and four females as the parents for each line. Within each line, the frequency of the sex-linked bristle mutant *forked* (f) and its wild-type allele (f^+) was 0.5. In each generation, four males and four females were chosen at random to parent the next generation. After 16 generations, the complete loss of one allele and the fixation of the other had occurred in 70 lines—29 in which only the *forked* allele was present and 41 in which the wild-type allele had become fixed. The remaining lines were still segregating the two alleles or had gone extinct. If fixation had occurred randomly, then an equal number of lines should have become fixed for each allele. In fact, the experimental results do not differ statistically from the expected ratio of 35:35, demonstrating that alleles can spread through a population and eliminate other alleles by chance alone.

How are small populations created in nature? In one scenario, a large population may be split by some event (like war), creating a small isolated subpopulation. A disaster such as an epidemic might occur, leaving a small number of survivors to breed. Or, a small group might emigrate from the

TABLE 25–5 All Possible Pairs of Offspring Produced by Two Heterozygous Parents (Aa × Aa)

Possible Genotype of Two Offspring	*Probability*	*Allele Frequency* A	*a*
AA and *AA*	(1/4)(1/4) = 1/16	1.00	0.00
aa and *aa*	(1/4)(1/4) = 1/16	0.00	1.00
AA and *aa*	2(1/4)(1/4) = 2/16	0.50	0.50
Aa and *Aa*	(2/4)(2/4) = 4/16	0.50	0.50
AA and *Aa*	2(1/4)(2/4) = 4/16	0.75	0.25
aa and *Aa*	2(1/4)(2/4) = 4/16	0.25	0.75

larger population and become founders in a new environment, such as a volcanically created island.

Allele frequencies in certain isolated human populations demonstrate the role of drift as an evolutionary force in natural populations. The Pingelap atoll in the western Pacific Ocean (lat. 6° N, long. 160° E) has in the past been devastated by typhoons and famine, and around 1780, there were only about nine surviving males. Today there are fewer than 2000 inhabitants, all of whose ancestry can be traced to the typhoon survivors. About 4–10 percent of the current population is blind from infancy. These people are affected by an autosomal recessive disorder, **achromatopsia**, which causes ocular disturbances, a form of color blindness, and cataract formation. The disorder is extremely rare in the human population as a whole. However, the mutant allele is present at a relatively high frequency in the Pingelap population. Genealogical reconstruction shows that one of the original survivors (about 30 individuals) was heterozygous for the condition. If we assume that he was the only carrier in the founding population, the initial gene frequency was 1/60, or 0.016. On average, about 7 percent of the current population is affected (a homozygous recessive genotype), so the frequency of the allele in the present population has increased to 0.26. The story of the inhabitants of Pingelap is told by Oliver Sacks, in his book, *The Island of the Colorblind.*

Another example of genetic drift involves the Dunkers, a small, isolated religious community who emigrated from the German Rhineland to Pennsylvania. Because their religious beliefs do not permit marriage with outsiders, the population has grown only through marriage within the group. When the frequencies of the ABO and MN blood group alleles are compared, significant differences are found among the Dunkers, the German population, and the U.S. population. The frequency of blood group A in the Dunkers is about 60 percent, in contrast to 45 percent in the U.S. and German populations. The I^B allele is nearly absent in the Dunkers. Type M blood is found in about 45 percent of the Dunkers, compared with about 30 percent in the U.S. and German populations. As there is no evidence for a selective advantage of these alleles, it is apparent that the observed frequencies are the result of chance events in a relatively small isolated population.

25.10 Nonrandom Mating

We have explored how violations of the first four assumptions of the Hardy–Weinberg law, in the form of selection, mutation, migration, and genetic drift, can cause allele frequencies to change. The fifth assumption is that the members of a population mate at random. Nonrandom mating itself does not directly alter the frequencies of alleles. It can, however, alter the frequencies of genotypes in a population and thereby indirectly affect the course of evolution.

The most important form of nonrandom mating, and the form we focus on here, is **inbreeding**, mating between relatives. For a given allele, inbreeding increases the proportion of homozygotes in the population. Over time, with complete inbreeding, only homozygotes will remain in the population. To demonstrate this concept, we consider the most extreme form of inbreeding, **self-fertilization**.

Inbreeding

Figure 25–19 shows the results of four generations of self-fertilization, starting with a single individual heterozygous for one pair of alleles. By the fourth generation, only about 6 percent of the individuals are still heterozygous, and 94 percent of the population is homozygous. Note, however, that alleles *A* and *a* still remain at 50 percent.

In humans, inbreeding (called **consanguineous mating**) is related to population size, mobility, and social customs governing marriages among relatives. To describe the amount of inbreeding in a population, Sewall Wright devised the coefficient of inbreeding. Expressed as *F*, the **coefficient of inbreeding** is defined as the probability that the two alleles of a given gene in an individual are identical *because they are descended from the same single copy of the allele in an ancestor*. If $F = 1$, all individuals are homozygous, and both alleles in every individual are derived from the same ancestral copy. If $F = 0$, no individual has two alleles derived from a common ancestral copy.

P_1: Self-fertilization *Aa*

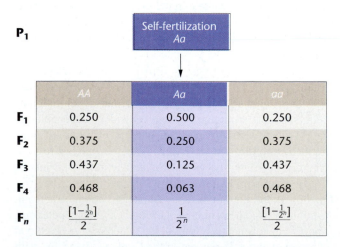

	AA	*Aa*	*aa*
F_1	0.250	0.500	0.250
F_2	0.375	0.250	0.375
F_3	0.437	0.125	0.437
F_4	0.468	0.063	0.468
F_n	$\frac{[1-\frac{1}{2^n}]}{2}$	$\frac{1}{2^n}$	$\frac{[1-\frac{1}{2^n}]}{2}$

FIGURE 25–19 Reduction in heterozygote frequency brought about by self-fertilization. After *n* generations, the frequencies of the genotypes can be calculated according to the formulas in the bottom row.

Figure 25–20 shows a pedigree of a first-cousin marriage. The fourth-generation female (shaded pink) is the daughter of first cousins (purple). Suppose her great-grandmother (green) was a carrier of a recessive lethal allele, *a*. What is the probability that the fourth-generation female will inherit two copies of her great-grandmother's lethal allele? For this to happen, (1) the great-grandmother had to pass a copy of the allele to her son, (2) her son had to pass it to his daughter, and (3) his daughter has to pass it to her daughter (the pink female). Also, (4) the great-grandmother had to pass a copy of the allele to her daughter, (5) her daughter had to pass it to her son, and (6) her son has to pass it to his daughter (the pink female). Each of the six necessary events has an individual probability of 1/2, and they *all* have to happen, so the overall probability that the pink female will inherit two copies of her great-grandmother's lethal allele is $(1/2)^6 = 1/64$. This takes us most of the way toward calculating *F* for a child of a first-cousin marriage. We need only note that the fourth-generation female could also inherit two copies of any of the other three alleles present in her great-grandparents. Because any of four possibilities would give the pink female two alleles identical by descent from an ancestral copy, $F = 4 \times (1/64) = 1/16$.

Genetic Effects of Inbreeding

Inbreeding results in the production of individuals homozygous for recessive alleles that were previously concealed in heterozygotes. Because many recessive alleles are deleterious when homozygous, one consequence of inbreeding is an increased chance that an individual will be homozygous for a recessive deleterious allele. Inbred populations often have a lowered mean fitness. **Inbreeding depression** is a measure of the loss of fitness caused by inbreeding. In domesticated plants and animals, inbreeding and selection have been used for thousands of years, and these organisms already have a high degree of homozygosity at many loci. Further inbreeding will usually produce only a small loss of fitness. However, inbreeding among individuals from large, randomly mating populations can produce high levels of inbreeding depression. This effect can be seen by examining the mortality rates in offspring of inbred animals in zoo populations (Table 25–6). To counteract inbreeding depression, many zoos use DNA fingerprinting to estimate the relatedness of animals used in breeding programs and choose the least related animals as parents.

As natural populations of endangered species decrease, the concern about inbreeding is a factor in designing programs to restore these species. One example is a project in Scandinavian zoos to preserve the Fennoscandic wolf. Conservationists established a wolf population bred in captivity with four founding individuals. Later, two Russian wolves were added to the population to reduce the level of inbreeding, a strategy some conservationists later came to regret because it introduced foreign alleles into the Fennoscandic gene pool. Because the captive-wolf line was founded by such a small number of individuals, it is severely inbred, and individuals show inbreeding depression in the form of smaller body

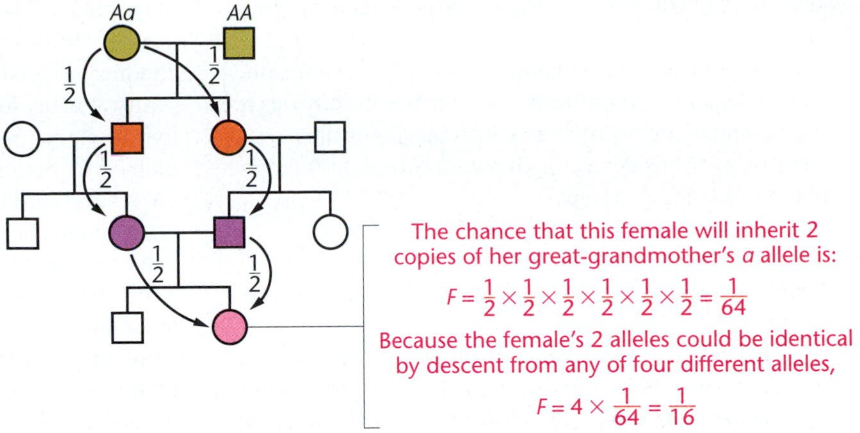
Aa
AA
1/2
The chance that this female will inherit 2 copies of her great-grandmother's a allele is:
F = 1/2 × 1/2 × 1/2 × 1/2 × 1/2 × 1/2 = 1/64
Because the female's 2 alleles could be identical by descent from any of four different alleles,
F = 4 × 1/64 = 1/16

FIGURE 25–20 Calculating the coefficient of inbreeding (F) for the offspring of a first-cousin marriage.

weight, reduced reproductive success, and reduced longevity. Furthermore, a number of wolves in the line are blind.

Researchers concluded that blindness in the captive-wolf line is caused by an autosomal recessive allele at a single locus. The bad news is that all the remaining purebred Fennoscandic wolves (i.e., individuals with no genes from the two Russian wolves) have at least a 6 percent chance of being carriers for blindness. Some purebred individuals have a 67 percent chance of being carriers. The good news is that individuals with a greater than 30 percent chance of being carriers can be removed from the breeding population without reducing the remaining genetic variation in the population by more than 10 percent. Removing the likely carriers would reduce the frequency of the blindness allele from 14 percent to 7 percent, and would improve the long-term prospects of maintaining a viable Fennoscandic wolf population.

In humans, inbreeding increases the risk of spontaneous abortions, neonatal deaths, congenital deformities and recessive genetic disorders. Although less common than in the past, inbreeding occurs in many regions of the world where social customs favor marriage between first cousins. Alan Bittles and James Neel analyzed data from numerous studies on different cultures. They found that the rate of child mortality (i.e., death in the first several years of life) varies dramatically from culture to culture. No matter what the baseline mortality rate for children of unrelated parents, however, children of first cousins virtually always have a higher death rate—typically by about 4.5 percentage points. Over many generations, inbreeding should eventually reduce the frequency of deleterious recessive alleles (if homozygotes die before reproducing). However, some studies indicate that parents who are first cousins tend to have more children to compensate for those lost to genetic disorders. On average, two-thirds of the surviving offspring are heterozygous carriers of the deleterious allele.

It is important to note that inbreeding is not always harmful. Indeed, inbreeding has long been recognized as a useful tool for breeders of domesticated plants and animals. When an inbreeding program is initiated, homozygosity increases, and some breeding stocks become fixed for favorable alleles and others for unfavorable alleles. By selecting the more viable and vigorous plants or animals, the proportion of individuals carrying desirable traits can be increased.

If members of two inbred lines are mated, hybrid offspring are often more vigorous in desirable traits than either of the parental lines. This phenomenon is called **hybrid vigor**. When such an approach was used in breeding programs established for maize, crop yields increased tremendously. Unfortunately, as a consequence of segregation, the hybrid vigor extends only through the first generation. Many hybrid lines are sterile, and those that are fertile show subsequent declines in yield. Consequently, the hybrids must be

TABLE 25–6 Mortality in Offspring of Inbred Zoo Animals

Species	*N*		*Non-inbred*	*Inbred*	*Inbreeding Coefficient*
Zebra	32	Lived:	20	3	0.250
		Died:	7	2	
Eld's deer	24	Lived:	13	0	0.250
		Died:	4	7	
Giraffe	19	Lived:	11	2	0.250
		Died:	3	3	
Oryx	42	Lived:	35	0	0.250
		Died:	2	5	
Dorcas gazelle	92	Lived:	36	17	0.269
		Died:	14	25	

regenerated each time by crossing the original inbred parental lines.

Hybrid vigor has been explained in two ways. The first theory, the **dominance hypothesis**, incorporates the obvious reversal of inbreeding depression, which inevitably must occur in outcrossing. Consider a cross between two strains of maize with the following genotypes:

Strain A		Strain B		F_1
aaBBCCddee	×	*AAbbccDDEE*	→	*AaBbCcDdEe*

The F_1 hybrids are heterozygotes at all loci shown. The deleterious recessive alleles present in the homozygous form in the parents are masked by the more favorable dominant alleles in the hybrids. Such masking is thought to cause hybrid vigor.

The second theory, **overdominance**, holds that in many cases the heterozygote is superior to either homozygote. This may relate to the fact that in the heterozygote two forms of a gene product may be present, providing a form of biochemical diversity. Thus, the cumulative effect of heterozygosity at many loci accounts for the hybrid vigor. Most likely, hybrid vigor results from a combination of phenomena explained by both hypotheses.

We have seen that nonrandom mating can drive the genotype frequencies in a population away from their expected values under the Hardy–Weinberg law. This can indirectly affect the course of evolution. As in stocks purposely inbred by animal and plant breeders, inbreeding in a natural population may increase the frequency of homozygotes for a deleterious recessive allele. With domestic stocks, this increases the efficiency with which selection removes the deleterious allele from the population. Consequently, once deleterious genes are removed, inbreeding no longer causes problems.

Chapter Summary

1. Populations evolve as a result of changes in an allele frequency at a number of loci over a period of time. A key insight in understanding evolution was the recognition that changes in the relative abundance of different phenotypes can be tied to changes in the relative abundance of the alleles that influence the phenotypes. This insight led to the development of population genetics, which studies the factors that cause allele frequencies to change.
2. The Hardy–Weinberg law specifies what will happen to allele frequencies at a given locus in a population under a set of simple assumptions. If there is no selection, no mutation, and no migration, if the population is large, and if individuals mate at random, then allele frequencies will not change from one generation to the next. The population will not evolve.
3. The Hardy–Weinberg formula can be used to investigate whether or not a population is in evolutionary equilibrium at a given locus and to estimate the frequency of heterozygotes in a population from the frequency of homozygous recessives.
4. By specifying the conditions under which allele frequencies will not change, the Hardy–Weinberg law identifies the forces that can cause evolution. Selection, mutation, migration, and genetic drift can cause allele frequencies to change and are thus forces of evolution. Nonrandom mating does not alter the frequencies of alleles, but by altering the frequencies of genotypes, nonrandom mating can indirectly affect the course of evolution.
5. Natural selection is the most powerful of the forces that can alter the frequencies of alleles in a population. The rate of evolution under natural selection depends on the initial frequencies of the alleles and on the relative fitness of different genotypes or the relative strength of selection. In small populations, random genetic drift can also be an important force altering the frequencies of alleles.
6. Mutation and migration introduce new alleles into a population, but by themselves mutation and migration usually have little effect on allele frequencies. Instead, the retention or loss of introduced alleles depends on the fitness they confer and the action of selection.
7. Genetic drift is change in allele frequencies as a result of chance events; it is an important force of evolution in small populations.
8. The most important form of nonrandom mating is inbreeding, or mating between relatives. Inbreeding increases the frequency of homozygotes in a population and decreases the frequency of heterozygotes.

Insights and Solutions

1. Tay–Sachs disease is caused by loss-of-function mutations in a gene on chromosome 15 that encodes a lysosomal enzyme. Tay–Sachs is inherited as an autosomal recessive condition. Among Ashkenazi Jews of central European ancestry, about 1 in 3600 children is born with the disease. What fraction of the individuals in this population are carriers?

Solution: If we let p represent the frequency of the wild-type enzyme allele and q the total frequency of recessive loss-of-function alleles, and if we assume that the population is in Hardy–Weinberg equilibrium, then the frequencies of the genotypes are given by p^2 for homozygous normal, $2pq$ for carriers, and q^2 for individuals with Tay–Sachs. The frequency of Tay–Sachs alleles is thus

$$q = \sqrt{q^2} = \sqrt{\frac{1}{3600}} = 0.017$$

Since $p + q = 1$, we have

$$p = 1 - q = 1 - 0.017 = 0.983$$

Therefore, we can estimate that the frequency of carriers is

$$2pq = 2(0.983)(0.017) = 0.033 \quad \text{or} \quad 1 \text{ in } 30.$$

2. Eugenics is the term employed for the selective breeding of humans to bring about improvements in populations. As a eugenic measure, it has been suggested that individuals suffering from serious genetic disorders should be prevented (sometimes by force of law) from reproducing (by sterilization, if necessary) in order to reduce the frequency of the disorder in future generations. Suppose that such a recessive trait were present in the population at a frequency of 1 in 40,000 and that affected individuals did not reproduce. In 10 generations, or about 250 years, what would be the frequency of the condition? Are the eugenic measures effective in this case?

Solution: Let q represent the frequency of the recessive allele responsible for the disorder. Because the disorder is recessive, we can estimate that

$$q = \sqrt{\frac{1}{40{,}000}} = 0.005$$

If all affected individuals are prevented from reproducing, then in an evolutionary sense, the disorder is lethal: Affected individuals have zero fitness. This means that we can predict the frequency of the recessive allele 10 generations in the future by using the following equation:

$$q_g = q_0(1 + gq_0)$$

Here, $q_0 = 0.005$, and $g = 10$, so we have

$$q_{10} = \frac{(0.005)}{[1 + (10 \times 0.005)]}$$

$$= 0.0048$$

If $q_{10} = 0.0048$, then the frequency of homozygous recessive individuals will be roughly

$$(q_{10})^2 = (0.0048)^2 = 0.000023 = \frac{1}{43{,}500}$$

The frequency of the genetic disorder has been reduced from 1 in 40,000 to 1 in 43,500 in 10 generations, indicating that this eugenic measure has limited effectiveness.

Problems and Discussion Questions

1. The ability to taste the compound PTC is controlled by a dominant allele *T*, while individuals homozygous for the recessive allele *t* are unable to taste PTC. In a genetics class of 125 students, 88 can taste PTC and 37 cannot. Calculate the frequency of the *T* and *t* alleles in this population and the frequency of the genotypes.
2. Calculate the frequencies of the *AA*, *Aa*, and *aa* genotypes after one generation if the initial population consists of 0.2 *AA*, 0.6 *Aa*, and 0.2 *aa* genotypes and meets the requirements of the Hardy-Weinberg relationship. What genotype frequencies will occur after a second generation?
3. Consider rare disorders in a population caused by an autosomal recessive mutation. From the frequencies of the disorder in the population given, calculate the percentage of heterozygous carriers:
   (a) 0.0064 (b) 0.000081 (c) 0.09
   (d) 0.01 (e) 0.10
4. What must be assumed in order to validate the answers in Problem 3?
5. In a population where only the total number of individuals with the dominant phenotype is known, how can you calculate the percentage of carriers and homozygous recessives?
6. Determine whether the following two sets of data represent populations that are in Hardy–Weinberg equilibrium (use χ^2 analysis if necessary):
   (a) *CCR5* genotypes: *1/1*, 60 percent; *1/Δ32*, 35.1 percent; *Δ32 /Δ32*, 4.9 percent
   (b) Sickle-cell hemoglobin: *AA*, 75.6 percent; *AS*, 24.2 percent; *SS*, 0.2 percent
7. If 4 percent of a population in equilibrium expresses a recessive trait, what is the probability that the offspring of two individuals who do not express the trait will express it?
8. Consider a population in which the frequency of allele *A* is $p = 0.7$ and the frequency of allele *a* is $q = 0.3$, and where the alleles are codominant. What will be the allele frequencies after one generation if the following occurs?
   (a) $w_{AA} = 1$, $w_{Aa} = 0.9$, and $w_{aa} = 0.8$
   (b) $w_{AA} = 1$, $w_{Aa} = 0.95$, $w_{aa} = 0.9$

(c) $w_{AA} = 1, w_{Aa} = 0.99, w_{aa} = 0.98$
(d) $w_{AA} = 0.8, w_{Aa} = 1, w_{aa} = 0.8$

9. If the initial allele frequencies are $p = 0.5$ and $q = 0.5$ and allele a is a lethal recessive, what will be the frequencies after 1, 5, 10, 25, 100, and 1000 generations?

10. Determine the frequency of allele A in an island population after one generation of migration from the mainland under the following conditions:
(a) $p_i = 0.6; p_m = 0.1; m = 0.2$
(b) $p_i = 0.2; p_m = 0.7; m = 0.3$
(c) $p_i = 0.1; p_m = 0.2; m = 0.1$

11. Assume that a recessive autosomal disorder occurs in 1 of 10,000 individuals (0.0001) in the general population and that in this population about 2 percent (0.02) of the individuals are carriers for the disorder. Estimate the probability of this disorder occurring in the offspring of a marriage between first cousins. Compare this probability to the population at large.

12. What is the basis of inbreeding depression?

13. Describe how inbreeding can be used in the domestication of plants and animals. Discuss the theories underlying these techniques.

14. Evaluate the following statement: Inbreeding increases the frequency of recessive alleles in a population.

15. In a breeding program to improve crop plants, which of the following mating systems should be employed to produce a homozygous line in the shortest possible time?
(a) self-fertilization
(b) brother–sister matings
(c) first-cousin matings
(d) random matings
Illustrate your choice with pedigree diagrams.

16. If the recessive trait albinism (a) is present in 1/10,000 individuals in a population at equilibrium, calculate the frequency of
(a) the recessive mutant allele
(b) the normal dominant allele
(c) heterozygotes in the population
(d) matings between heterozygotes

17. One of the first Mendelian traits identified in humans was a dominant condition known as *brachydactyly*. This gene causes an abnormal shortening of the fingers or toes (or both). At the time, it was thought by some that the dominant trait would spread until 75 percent of the population would be affected (since the phenotypic ratio of dominant to recessive is 3:1). Show that the reasoning is incorrect.

18. Achondroplasia is a dominant trait that causes a characteristic form of dwarfism. In a survey of 50,000 births, five infants with achondroplasia were identified. Three of the affected infants had affected parents, while two had normal parents. Calculate the mutation rate for achondroplasia and express the rate as the number of mutant genes per given number of gametes.

19. A prospective groom, who is normal, has a sister with cystic fibrosis, an autosomal recessive disease state. Their parents are normal. The brother plans to marry a woman who has no history of cystic fibrosis (CF) in her family. What is the probability that they will produce a CF child? They are both Caucasian and the overall frequency of CF in the Caucasian population is 1/2500—that is, 1 affected child per 2500. (Assume the population meets the Hardy–Weinberg assumptions)

Extra-Spicy Problems

20. A form of dwarfism known as Ellis–van Creveld syndrome was first discovered in the late 1930s, when Richard Ellis and Simon van Creveld shared a train compartment on the way to a pediatrics meeting. In the course of conversation, they discovered that they each had a patient with this syndrome. They published a description of the syndrome in 1940. Affected individuals have a short-limbed form of dwarfism and often have defects of the lips and teeth, and polydactyly (extra fingers). The largest pedigree for the condition was reported in an Old Order Amish population in eastern Pennsylvania by Victor McKusick and his colleagues (1964). In that community, about 5 per 1000 births are affected, and in the population of 8000, the observed frequency is 2 per 1000. All affected individuals have unaffected parents, and all affected cases can trace their ancestry to Samuel King and his wife, who arrived in the area in 1774. It is known that neither King nor his wife were affected with the disorder. There are no cases of the disorder in other Amish communities, such as those in Ohio or Indiana.
(a) From the information provided, derive the most likely mode of inheritance of this disorder. Using the Hardy–Weinberg law, calculate the frequency of the mutant allele in the population and the frequency of heterozygotes, assuming Hardy–Weinberg conditions.
(b) What is the most likely explanation for the high frequency of the disorder in the Pennsylvania Amish community and its absence in other Amish communities?

21. The graph shows the variation in the frequency of a particular allele (A) that occurred over time in two relatively small, independent populations exposed to very similar environmental conditions. A student has analyzed the graph and concluded that the best explanation for these data is that the selection for allele A is occurring in Population 1. Is the student's conclusion correct? Explain.

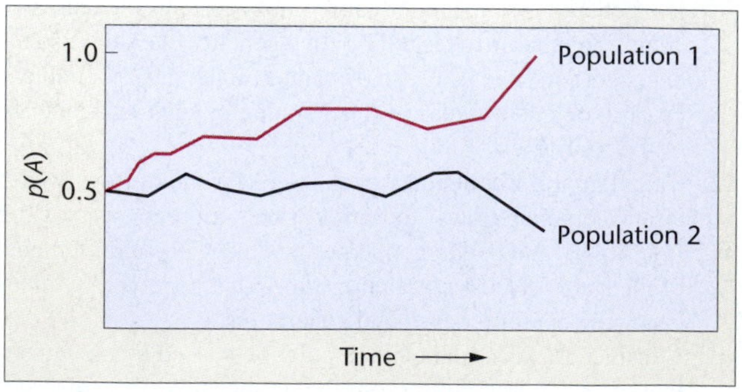

Selected Readings

Ansari-Lari, M.A., et al. 1997. The extent of genetic variation in the *CCR5* gene. *Nature Genetics* 16:221–22.

Ballou, J., and Ralls, K. 1982. Inbreeding and juvenile mortality in small populations of ungulates: A detailed analysis. *Biol. Conserv.* 24:239–72.

Barton, N.H., and Turelli, M. 1989. Evolutionary quantitative genetics: How little do we know? *Annu. Rev. Genet.* 23:337–70.

Bittles, A.H., and Neel, J.V. 1994. The costs of human inbreeding and their implications for variations at the DNA level. *Nature Genet.* 8:117–21.

Carrington, M., and Kissner, T., et al. 1997. Novel alleles of the chemokine-receptor gene *CCR5*. *Am. J. Hum. Genet.* 61:1261–67.

Chevillon, C., et al. 1995. Population structure and dynamics of selected genes in the mosquito *Culex pipiens*. *Evolution* 49: 997–1007.

Crow, J.F. 1986. *Basic concepts in population, quantitative and evolutionary genetics*. New York: W. H. Freeman.

Dudley, J.W. 1977. 76 generations of selection for oil and protein percentage in maize. In *Proceedings of the International Conference on Quantitative Genetics*, ed. E. Pollack, et al., pp. 459–73. Ames, IA: Iowa State University Press.

Fisher, R.A. 1930. *The genetical theory of natural selection*. Oxford, U.K.: Clarendon Press. (Reprinted Dover Press, 1958.)

Freeman, S., and Herron, J.C. 2001. *Evolutionary analysis*. Upper Saddle River, NJ: Prentice Hall.

Freire-Maia, N. 1990. Five landmarks in inbreeding studies. *Am. J. Med. Genet.* 35:118–20.

Gayle, J.S. 1990. *Theoretical population genetics*. Boston, MA: Unwin-Hyman.

Grant, P.R. 1986. *Ecology and evolution of Darwin's finches*. Princeton, NJ: Princeton University Press.

Hartl, D.L. and Clark, A.G. 1997. *Principles of population genetics*. 3d ed. Sunderland, MA: Sinauer Associates.

Jones, J.S. 1981. How different are human races? *Nature* 293:188–90.

Karn, M.N., and Penrose, L.S. 1951. Birth weight and gestation time in relation to maternal age, parity and infant survival. *Ann. Eugen.* 16:147–64.

Kerr, W.E., and Wright, S. 1954. Experimental studies of the distribution of gene frequencies in very small populations of *Drosophila melanogaster*. I. *Forked. Evolution* 8:172–77.

Khoury, M.J., and Flanders, W.D. 1989. On the measurement of susceptibility to genetic factors. *Genet. Epidemiol.* 6:699-701.

Laikre, L., Ryman, N., and Thompson, E.A. 1993. Hereditary blindness in a captive wolf (*Canis lupus*) population: Frequency reduction of a deleterious allele in relation to gene conservation. *Conservation Biol.* 7:592–601.

Leibert, F., et al. 1998. The *DCCR5* mutation conferring protection against HIV-1 in Caucasian populations has a single and recent origin in northeastern Europe. *Hum. Molec. Genet.* 7:399–406.

Liu, R., et al. 1996. Homozygous defect in HIV-1 coreceptor accounts for resistance in some multiply-exposed individuals to HIV-1 infection. *Cell* 86:367–77.

Lucotte, G., and Mercier, G. 1998. Distribution of the *CCR5* gene 32-bp deletion in Europe. *J. Acquired Immune Deficiency Syndrome and Human Retrovirology* 19:174–77.

Markow, T., et al. 1993. HLA polymorphism in the Havasupai: Evidence for balancing selection. *Am. J. Hum. Genet.* 53:943–52.

Martinson, J.J., et al. 1997. Global distribution of the *CCR5* gene 32-bp deletion. *Nature Genet.* 16:100–103.

Mettler, L.E., Gregg, T., and Schaffer, H.E. 1988. *Population genetics and evolution*. 2d ed. Englewood Cliffs, NJ: Prentice-Hall.

Parra, E.J., et al. 1998. Estimating African American admixture proportions by use of population-specific alleles. *Am. J. Hum. Genet.* 63:1839–51.

Pier, G.B., et al. 1998. *Salmonella typhi* uses CFTR to enter intestinal epithelial cells. *Nature* 393:79–82.

Quillent, C., et al. 1998. HIV-1-resistance phenotype conferred by combination of two separate inherited mutations of *CCR5* gene. *The Lancet* 351:14–18.

Renfrew, C. 1994. World linguistic diversity. *Sci. Am.* (Jan.) 270:116–23.

Roberts, D.F. 1988. Migration and genetic change. *Hum. Biol.* 60:521–39.

Samson, M., et al. 1996. Resistance to HIV-1 infection in Caucasian individuals bearing mutant alleles of the *CCR-5* chemokine receptor gene. *Nature* 382:722–25.

Spiess, E.B. 1989. *Genes in populations*. 2d ed. New York: John Wiley.

Stephens, J.C., et al. 1998. Dating the origin of the *CCR5-D32* AIDS-resistance allele by the coalescence of haplotypes. *Am. J. Hum. Genet.* 62:1507–15.

Wallace, B. 1989. One selectionist's perspective. *Quart. Rev. Biol.* 64:127–45.

Woodworth, C.M., Leng, E.R., and Jugenheimer, R.W. 1952. Fifty generations of selection for protein and oil in corn. *Agron. J.* 44:60–66.

Yudin, N.S., et al. 1998. Distribution of *CCR5-D32* gene deletion across the Russian part of Eurasia. *Hum. Genet.* 102:695–98.

GENETICS MediaLab

The resources that follow will help you achieve a better understanding of the concepts presented in this chapter. These resources can be found either on the CD packaged with this textbook or on the Companion Web site found at **http://www.prenhall.com/klug**

CD Resources:

Module 25.1: Population Genetics

Web Problem 1:

Time for completion = 10 minutes

How does a population remain in Hardy–Weinberg Equilibrium (HWE)? Study the page on "The Hardy–Weinberg Equilibrium Model" and Chapter 25 of your text to help you answer the following questions: If a population with gene frequencies of $p = q = 0.5$ actually met all the assumptions of HWE, what would the allele frequencies be after 100 generations? Which of the assumptions of HWE is violated in all natural populations? If you knew only the proportion of nonalbinos in a population, could you determine the frequency of the allele for albinism? To complete this exercise, visit Web Problem 1 in Chapter 25 of your Companion Web site, and select the keyword **HARDY-WEINBERG**.

Web Problem 2:

Time for completion = 10 minutes

How does population size affect gene frequencies over time? Use the "Hardy–Weinberg Simulator" and Chapter 25 of your text to investigate how genetic drift caused by finite population size affects evolution. First, read the introduction in the "Background" section, and then read about the simulation. Select a population in the upper right-hand side of the screen, and run the simulation. After it is completed, take a look at your results before hitting Reset. Did you see a substantial change in the frequency of either allele, *f*(*A*) or *f*(*a*)? Click Reset, and the "Genetic Drift" box. This changes the initial population from 4000 to 400. Run the simulation. Compare this run with the first. How fast did it take for one allele to reach fixation? Is there more genetic variation *within* populations of large size or small size? Is there more genetic variation between populations of large size or small size? To complete this exercise, visit Web Problem 2 in Chapter 25 of your Companion Web site, and select the keyword **GENETIC DRIFT**.

Web Problem 3:

Time for completion = 15 minutes

How does natural selection interact with dominance to affect gene frequencies over time? You will use the "Hardy–Weinberg Simulator" introduced in the last exercise, along with Chapter 25 of your text, to explore these problems. First, consider selection against a deleterious dominant allele. Reset the simulator, click the box for population in the upper right and selection in the lower right, and then click "Run". The simulator will now favor the homozygous recessive (*aa*), which is blue, and selects against the yellow (*AA*) and green (*Aa*). How long do you think it will take before the a allele will be eliminated? Run the simulation and find out. After examining your results, click on "frequency" to determine how many generations it took to eliminate *AA* and *Aa* from the population. Was one genotype eliminated more quickly? How did this scenario compare with the one without selection or with a smaller population or migration? Run the simulation and find out. Do these examples illustrate stabilizing, directional, or disruptive selection? To complete this exercise, visit Web Problem 3 in Chapter 25 of your Companion Web site, and select the keyword **DOMINANCE SELECTION**.

Light and dark forms of the peppered moth *Biston betularia* on light tree bark. (*Breck K. Kent/-Animals Animals*)

26
Genetics and Evolution

KEY CONCEPTS

To begin our discussion of evolution, let's consider three scenarios: (1) In late 1986, a Florida dentist tested positive for HIV. Several months later, he was diagnosed with AIDS. He continued to practice general dentistry for two more years, until one of his patients, a woman with no known risk factors, discovered that she, too, was infected with HIV. When the dentist publicly urged his other patients to have themselves tested, several more were found to be HIV positive. Did this dentist transmit HIV to his patients, or did the patients become infected by some other means? (2) Fossil evidence indicates that Neanderthals (*Homo neanderthalensis*) lived in Europe and western Asia from 300,000 to 30,000 years ago. For some of that time, Neanderthals coexisted with anatomically modern humans (*Homo sapiens*). Since they lived in the same places at the same time, did Neanderthals and modern humans interbreed, so that descendants of the Neanderthals are alive today, or did the Neanderthals die off, so that their lineage is now extinct? (3) The mitochondria that inhabit our cells have their own DNA. The organization and function of the mitochondrial genome bear many similarities to the organization and function of bacterial genomes. Did our mitochondria originate as free-living bacteria that took up residence inside other cells, or is the mitochondrial genome ultimately derived from nuclear chromosomes?

These scenarios pose three apparently unrelated questions. However, these questions in fact have something in common: They all can be addressed by using genetic data to reconstruct evolutionary history. The use of such data in this regard is the focus of the current chapter. We will use the methods of genetic analysis and the reconstruction of evolutionary history to answer those questions at the end of the chapter.

Evolutionary history involves two processes: the transformation and splitting of lineages, which together produce a diversity of populations and species. In Chapter 25, we described the evolution of populations in terms of changes in allele frequencies and outlined the forces that can cause such frequencies to change. Mutation, migration, selection, and drift, individually and collectively, alter allele frequencies and bring about evolutionary divergence and the formation of species. The process depends not only upon changes in allele frequencies, but also upon environmental or ecological diversity. If a population is spread over a geographic range encompassing a number of sub-environments, or niches, the populations occupying these niches will adapt and become genetically differentiated. Differentiated populations are dynamic: They may remain in existence, become extinct, merge with the parental population, or continue to diverge from the parental population until they become reproductively isolated and form new species.

It is difficult to define the exact time when a new species forms. A **species** is a group of interbreeding or potentially interbreeding populations that is reproductively isolated in nature from all other such groups. In sexually reproducing organisms, speciation divides a single gene pool into two or more separate gene pools. Changes in morphology, physiology, and adaptation to an ecological niche may also occur, but are not necessary components of speciation, which can take place gradually or within a few generations.

Because the divergence of populations and species is accompanied by genetic differentiation, we can use patterns of genetic differences in the various groups to reconstruct evolutionary history. After exploring the genetic structure of populations, their divergence across space and time, and the process of speciation, we will discuss how genetic data can be used to answer questions that have an evolutionary context, such as those in the three scenarios at the beginning of the chapter.

26.1 Evolutionary History Includes the Transformation and Divergence of Lineages

Figure 26–1 shows an evolutionary tree, or **phylogeny**, that describes the history of several hypothetical lizard species. The passage of time is plotted horizontally, and changes in the phenotype are plotted vertically. The graph traces the phenotype of the average lizard in each species over time.

The history of these lizards begins with species 1. For some time, that species experiences evolutionary **stasis**; that is, it does not change. Species 1 then undergoes a process of steady transformation, called **phyletic evolution**, or **anagenesis**, and becomes species 2. During this process, there is only one species present at all times. After species 2 forms, it undergoes **cladogenesis**, whereby it gives rise to two distinct and independent daughter species. Further transformation of these daughter species produces the species we see today, species 3 and species 4.

In his 1859 book, *On the Origin of Species*, Charles Darwin amassed evidence that all species derive from a single common ancestor by transformation and speciation:

> ... all living things have much in common, in their chemical composition, their germinal vesicles, their cellular structure, and their laws of growth and reproduction. ... Therefore I should infer ... that probably all the organic beings which have ever lived on this earth have descended from some one primordial form. ...

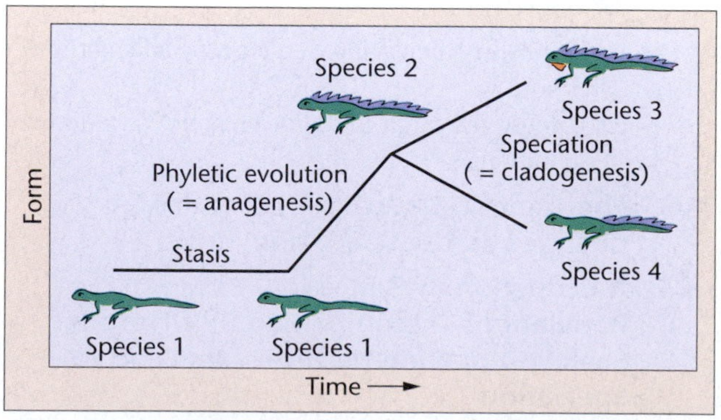

FIGURE 26–1 In phyletic evolution, or anagenesis, one species is transformed over time into another species. At all times, only one species exists. In cladogenesis, one species splits into two or more species.

Everything biologists have since learned supports Darwin's conclusion that only one tree of life exists. To understand evolution, we must understand the mechanisms that transform one species into another and that split one species into two or more.

Chapter 25 covered the mechanisms responsible for the transformation of species. Chief among them is natural selection, discovered independently by Darwin and by Alfred Russel Wallace. The Wallace–Darwin concept of natural selection can be summarized as follows:

1. Individuals of a species exhibit variations in phenotype, including differences in size, agility, coloration, or ability to obtain food.
2. Many of these variations, even small and seemingly insignificant ones, are heritable and passed on to offspring.
3. Organisms tend to reproduce in an exponential fashion. More offspring are produced than can survive. This causes members of a species to engage in a struggle for survival, competing with other members of the species for scarce resources.
4. In the struggle for survival, individuals with particular phenotypes will be more successful than individuals with others, allowing the former to survive and reproduce at higher rates.

As a consequence of natural selection, species change. The phenotypes that confer improved ability to survive and reproduce become more common, and the phenotypes that confer poor prospects for survival and reproduction disappear.

Although Wallace and Darwin proposed that natural selection explains how evolution occurs, they could not explain either the origin of the variations that provided the raw material for evolution or how such variations are passed from parents to offspring. In the 20th century, as biologists applied the principles of Mendelian genetics to populations, the source of variation (mutation) and the mechanism of inheritance (segregation of alleles) were both explained. We now view evolution as due to changes in allele frequencies in populations over time. This union of population genetics with natural selection generated a new view of the evolutionary process, called *neo-Darwinism*.

In this chapter, we consider the mechanisms responsible for speciation. As with natural selection, our understanding of speciation is built on key insights about genetic variation.

26.2 Most Populations and Species Harbor Considerable Genetic Variation

At first glance, it seems that members of a well-adapted population should be highly homozygous because the most favorable allele at each locus has become fixed. Certainly, an examination of most populations of plants and animals reveals many phenotypic similarities among individuals. However, a large body of evidence indicates that most populations contain a high degree of heterozygosity. This built-in genetic diversity is concealed, so to speak, because it is not necessarily apparent in the phenotype; hence, detecting it is not a simple task. Nevertheless, with the use of the techniques discussed next, such investigation has been successful.

Protein Polymorphisms

Gel electrophoresis (see Appendix A) separates protein molecules on the basis of differences in size and electrical charge. If a nucleotide variation in a structural gene results in the substitution of a charged amino acid, such as glutamic acid, for an uncharged amino acid, such as glycine, the net electrical charge on the protein will be altered. This difference in charge can be detected as a change in the rate at which proteins migrate through an electrical field. In the mid-1960s, John Hubby and Richard Lewontin used gel electrophoresis to measure protein variation in natural populations of *Drosophila*. Now researchers routinely use the technique to study genetic variation in a wide range of organisms (Table 26–1).

The electrophoretically distinct forms of a protein produced by different alleles are called *allozymes*. As shown in the table, a surprisingly large percentage of loci examined from diverse species produce distinct allozymes. Of the populations listed, approximately 30 loci per species were examined, and about 30 percent of the loci were polymorphic, with an average of 10 percent allozyme heterozygosity per diploid genome.

These values apply only to genetic variation detectable by altered protein migration in an electric field. Electrophoresis probably detects only about 30 percent of the actual variation due to amino acid substitutions, because many substitutions do not change the net electric charge on the molecule. Richard Lewontin estimated that about two-thirds of all loci in a population are polymorphic. In any individual within the population, about one-third of the loci exhibit genetic variation in the form of heterozygosity.

The significance of genetic variation detected by electrophoresis is controversial. Some argue that allozymes are functionally equivalent and therefore do not play any role in adaptive evolution. We address this argument later in this section.

Variations in Nucleotide Sequence: Analysis of Genes

The most direct way to estimate genetic variation is to compare the nucleotide sequences of genes carried by individuals in a population. With the development of techniques for cloning and sequencing DNA, and with the proliferation of genome projects, nucleotide sequence variations have been catalogued for an increasing number of genes and genomes. In an early study, Alec Jeffreys used restriction enzymes to detect polymorphisms in 60 unrelated individuals. His aim was to estimate the total number of DNA sequence variants in humans. His results showed that within the genes of the β-globin cluster, 1 in 100 base pairs showed polymorphic

TABLE 26–1 Allozyme Heterozygosity at the Protein Level

Species	*Populations Studied*	*Loci Examined*	*Polymorphic Loci per Population (%)*	*Heterozygotes per Locus (%)*
Homo sapiens (humans)	1	71	28	6.7
Mus musculus (mouse)	4	41	29	9.1
Drosophila pseudoobscura (fruit fly)	10	24	43	12.8
Limulus polyphemus (horseshoe crab)	4	25	25	6.1

Note: A polymorphic locus is one for which a population harbors more than one allele.
Source: From Lewontin, (1974, p. 117).

variation. If that region is representative of the genome, this study indicates that at least 3×10^7 nucleotide variants per genome are possible.

In another study, Martin Kreitman examined the *alcohol dehydrogenase* locus (*Adh*) in *Drosophila melanogaster* (Figure 26–2). This locus encodes two allozymic variants: the *Adh-f* and the *Adh-s* alleles. These differ by only a single amino acid (thr versus lys at codon 192). To determine whether the amount of genetic variation detectable at the protein level (one amino acid difference) corresponds to the variation at the nucleotide level, Kreitman cloned and sequenced *Adh* loci from five natural populations of *Drosophila.*

The 11 cloned loci contained 43 nucleotide variations from the consensus *Adh* sequence of 2721 base pairs. These variations are distributed throughout the gene: 14 in the exon coding regions, 18 in the introns, and 11 in the nontranslated and flanking regions. Of the 14 variations in coding regions, only one leads to an amino acid replacement—the one in codon 192, producing the two observed electrophoretic variants. The other 13 nucleotide substitutions do not lead to amino acid replacements. Are the differences in the number of allozyme and nucleotide variants the result of natural selection? If so, of what significance is this fact? We discuss these questions next.

Among the most intensively studied loci to date is the locus encoding the cystic fibrosis transmembrane conductance regulator (CFTR). Recessive loss-of-function mutations in the *CFTR* locus cause cystic fibrosis, a disease with symptoms including salty skin, the production of an excess of thick mucus in the lungs, and susceptibility to bacterial infections. Geneticists have examined the *CFTR* locus in some 30,000 chromosomes from individuals with cystic fibrosis and have found over 500 different mutations that can cause the disease. Among these mutations are missense mutations, amino acid deletions, nonsense mutations, frameshifts, and splice defects.

Figure 26–3 shows a map of the 27 exons in the *CFTR* locus, with most exons identified by function. The histogram above the map shows the locations of some of the disease-causing mutations and the number of copies of each that have been found. A single mutation, a 3-bp deletion in exon 10 called *ΔF508* accounts for 67 percent of all mutant cystic fibrosis alleles, but several other mutations are found in at least 100 of the chromosomes surveyed. In populations of European ancestry, between 1 in 44 and 1 in 20 individuals are heterozygous carriers of mutant alleles. Note that Figure 26–3 includes only the sequence variants that alter the function of the *CFTR* protein. There are undoubtedly many more *CFTR* alleles with silent sequence variants that do not change the structure of the protein and that do not affect its function. The *CFTR* locus shows considerable genetic variation.

Studies of other organisms, including the rat and the mouse, have produced similar estimates of nucleotide diversity. These studies indicate that there is an enormous reservoir of genetic variability within most populations and that, at the level of DNA, most, and perhaps all, genes exhibit diversity from individual to individual.

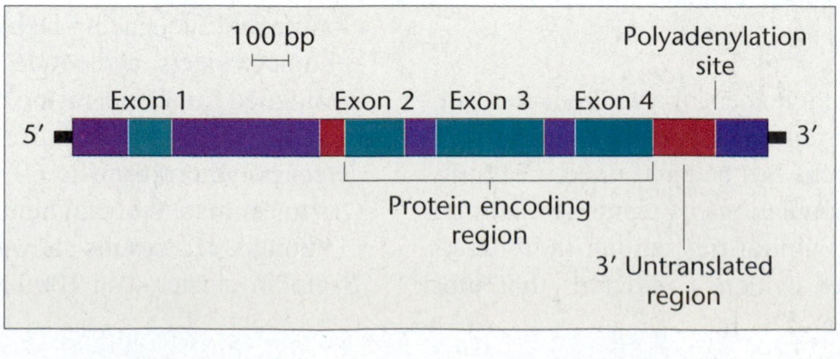

FIGURE 26–2 Organization of the *Adh* locus of *Drosophila melanogaster.*

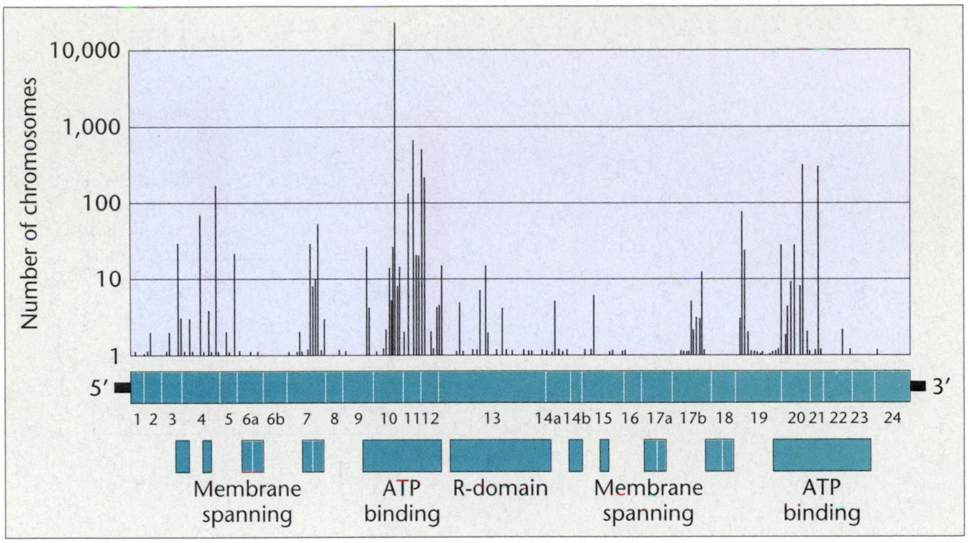

FIGURE 26–3 The locations of disease-causing mutations in the cystic fibrosis gene. The histogram shows the number of copies of each mutation geneticists have found. (The vertical axis is on a logarithmic scale.) The genetic map below the histogram shows the locations and relative sizes of the 27 exons of the *CFTR* locus. The boxes at the bottom indicate the functions of different domains of the CFTR protein. *(Reprinted from* Trends Genet. *8:392–98, L. Tsui, The spectrum of cystic fibrosis mutations, 1992, with permission from Elsevier Science.)*

Variations in Nucleotide Sequence: SNPs and the Genome

Until information from genome projects became available, studies of genetic variation were limited to individual genes, as just described in the previous section. To study genetic variation on a larger scale, nucleotide sequence data from the Human Genome Project is being used to estimate the total genetic variation in human populations. One of the most useful variants being studied is called a **single nucleotide polymorphism** or **SNP** (pronounced snip). What are SNPs? They are sites on a DNA molecule at which there is nucleotide variation from person to person [Figure 26.4(a)]. Most SNPs have only two alleles. To be considered a polymorphism, one allele must have a frequency of at least 1 percent in the population. In the late 1970s and early 1980s, SNPs were detected with restriction enzymes, and were visualized as differences in restriction fragment lengths. Genomic analysis reveals that in comparing two different human DNA sequences, SNPs occur every 1000 to 2000 nucleotides, which translates into 1.6–3.2 million SNPs per genome. A genomic map with 1.42 million SNPs is now available. The map shows that nucleotide diversity varies across the genome. For example, the SNP map of chromosome 21 shows peaks and valleys, presumably representing regions with different rates of mutation, recombination, and selection [(Figure 26.4(b)].

The molecular variations represented by SNPs are being used to trace the evolutionary and geographical origins of our species and the history and migrations of populations. A Human Genome Diversity Project is underway to coordinate the study of human genomic variation and our evolutionary history. The Project coordinates the ethical collection of DNA samples, distributes samples to laboratories, and aids in the analysis of results.

Explaining the High Level of Genetic Variation in Populations

As mentioned earlier, the finding that populations harbor considerable genetic variation at the amino acid and nucleotide levels came as a surprise to many evolutionary biologists. The early consensus was that selection would favor a single optimal allele at each locus and that, as a result, populations would have high levels of homozygosity. This expectation was obviously wrong, and considerable research and argument ensued concerning the forces that maintain genetic variation.

One view about the high degree of genetic variation argues that it reflects primarily the action of mutation and genetic drift. The **neutral theory** of molecular evolution, proposed by Motoo Kimura, argues that mutations leading to amino acid substitutions are rarely favorable. They are sometimes detrimental, but most often are neutral or genetically equivalent to the allele that is replaced. Those few polymorphisms that are favorable or detrimental are preserved or removed from the population, respectively, by natural selection. However, the vast majority of mutations are neutral genetic changes that are not affected by selection. The frequency of these neutral alleles in a population will be determined by mutation rates and random genetic drift. Some neutral mutations will drift to fixation in the population; other neutral mutations will be lost. At any given time, the population may contain several neutral alleles at any particular locus. The diversity of alleles at most polymorphic loci does not, however, reflect the action of natural selection.

Opposed to the proponents of the neutral theory are the **selectionists**. These geneticists point to examples in which enzyme or protein polymorphisms are associated with adaptation to certain environmental conditions. The well-known advantage of sickle-cell anemia heterozygotes in infection by malarial parasites is such an example, and another was discussed in Chapter 25, when we considered evidence that polymorphism at the *CFTR* locus may reflect the superior fitness of heterozygotes in areas where typhoid fever is common.

Selectionists also stress that enzyme polymorphisms may often appear to offer no advantage, but exist in such a frequency that they are impossible to explain as random occurrences. Thus, even though no currently available

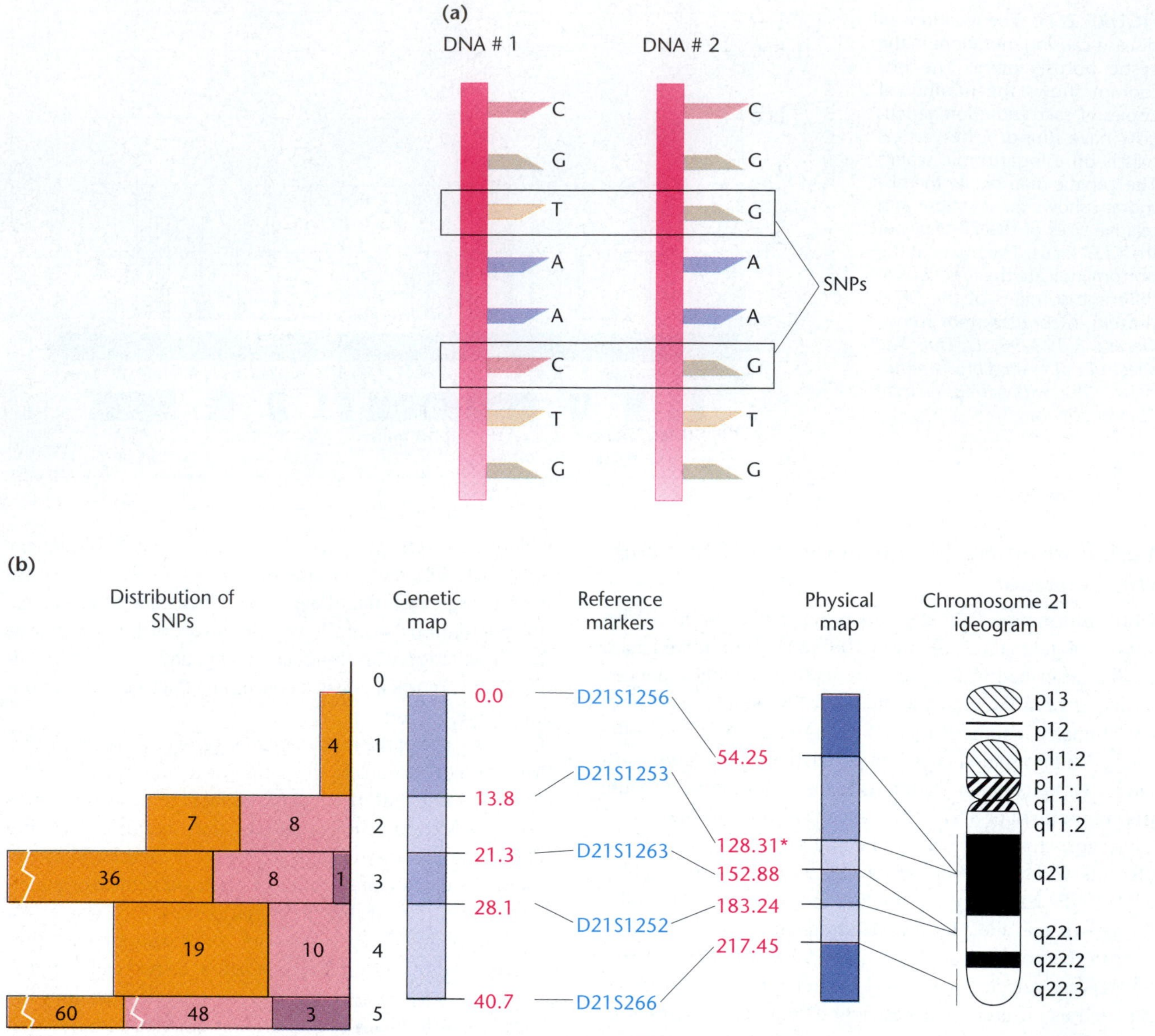

FIGURE 26–4 (a) Single nucleotide polymorphisms (SNPs) are nucleotide differences between two different DNA sequences at a given position. (b) A SNP map of chromosome 21. Each column represents 1 Mb of DNA. The number of SNPs (*y*-axis) is plotted against the approximate cytogenetic location (*x*-axis). Differences in SNP density are thought to reflect differences in mutation rates, recombination, and the effect of selection. Part B is from a public database at: http://ncbi.nlm.nih.gov/SNP/

technique can detect any physiological difference, there may be some slight advantage associated with any given amino acid substitution. In fact, there may be an advantage to having two forms of a given protein: Perhaps the combination allows optimum performance under a wider range of cellular conditions.

The neutralist and selectionist perspectives should not be viewed as mutually incompatible. Instead, they are best seen as the end points of a spectrum of possibilities. The neutralists do not discount natural selection as a guiding force in evolution; rather, they suggest that some features of organisms' genotypes are nonadaptive, fluctuate randomly, and may have been fixed by genetic drift. Still, the selectionists certainly do not deny that genetic drift is an important factor in establishing differences in allele frequencies. It is difficult to argue against the notion that *some* genetic variation must be neutral. The difference between the two theories is in the degree of neutrality that they posit.

Current data are insufficient to determine what fraction of molecular genetic variation is neutral and what fraction is subject to selection. The neutral theory nonetheless serves a crucial function: By pointing out that some genetic variation is expected simply as a result of mutation and drift, the neutral theory provides a working hypothesis for studies of molecular evolution. In other words, biologists must find positive evidence that selection is acting on allele frequencies at a particular locus before they can reject the simpler assumption that only mutation and drift are at work.

26.3 The Genetic Structure of Populations Changes across Space and Time

As geneticists discovered that most populations harbor considerable genetic diversity, they also found that the genetic structure of populations varies across space and time. To illustrate, we consider studies on *Drosophila pseudoobscura* conducted by Theodosius Dobzhansky and his colleagues. This species is found over a wide range of environmental habitats, including the western and southwestern United States. Although the flies throughout this range are morphologically similar, Dobzhansky's team discovered that populations from different locations vary in the arrangement of genes on chromosome 3. They found several different inversions in this chromosome that can be detected by loop formations in larval polytene chromosomes. Each inversion sequence is named after the locale in which it was first discovered (e.g., AR = Arrowhead, British Columbia, and CH = Chiricahua Mountains, Arizona). The inversion sequences were compared with one standard sequence, designated ST.

Figure 26–5 compares the frequencies of three arrangements at different elevations in the Sierra Nevada Mountains of California. The ST arrangement is most common at low elevations; at 8000 feet, AR is the most common and ST least common. In the populations studied, the frequency of the CH arrangement gradually increases with elevation, a phenomenon that is probably the result of natural selection and that parallels the gradual environmental changes occurring at ascending elevations.

Dobzhansky's team also found that, for populations collected at a single site throughout the year, inversion frequencies change as well. That is, cyclic variations in chromosome arrangements occur as the seasons change, as shown in Figure 26–6. Such variation was consistently observed over a period of several years. The frequency of ST always declines during the spring, and that of CH increases in the spring.

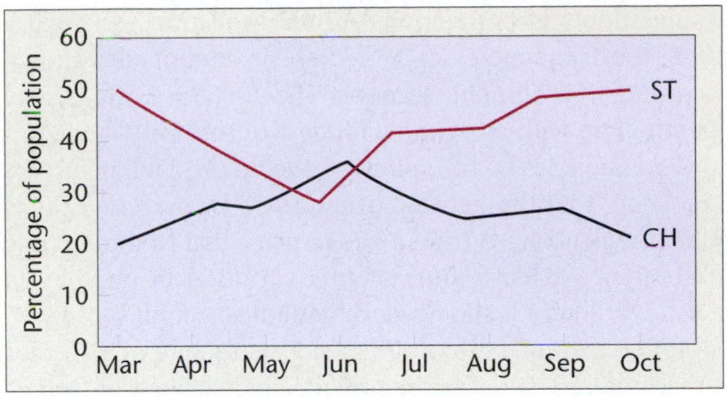

FIGURE 26–6 Changes in the ST and CH arrangements in *D. pseudoobscura* throughout the year.

To test the hypothesis that this cyclic change is a response to natural selection, Dobzhansky and his group devised a laboratory experiment. They constructed a large population cage from which samples of *D. pseudoobscura* could be removed periodically and studied. They began with a population that was 88 percent CH and 12 percent ST. The flies were maintained at 25°C and sampled over a 1-year period. As shown in Figure 26–7, the frequency of ST increased gradually until it was present at a level of 70 percent. At that point, an equilibrium between ST and CH was reached. When the same experiment was performed at 16°C, no change in inversion frequency occurred. The researchers concluded that the equilibrium reached at 25°C was in response to the elevated temperature, the only variable in the experiment.

The results of the study indicate that a balance in the frequency of the two inversions and the gene arrangements they contain is superior to either inversion by itself. The equilibrium reached in the experiment presumably represents the highest mean fitness in the population under controlled laboratory conditions. This interpretation of the experiment suggests that natural selection is the driving force maintaining the diversity in chromosome 3 inversions.

In a more extensive study, Dobzhansky and his colleagues sampled *D. pseudoobscura* populations over a broad geographic range. They found 22 different chromosome

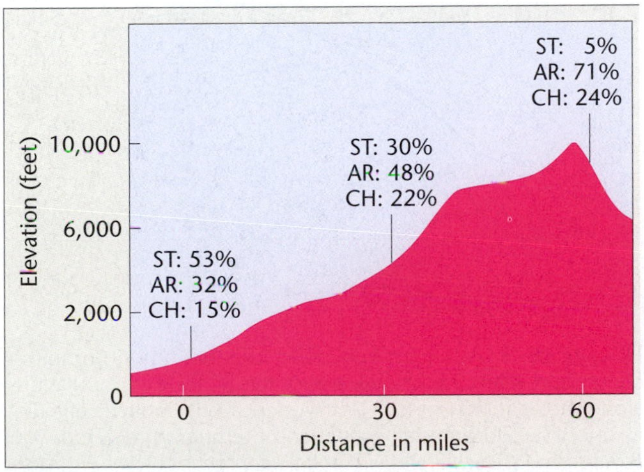

FIGURE 26–5 Inversions in chromosome 3 of *D. pseudoobscura* at different elevations in the Sierra Nevada mountain range near Yosemite National Park.

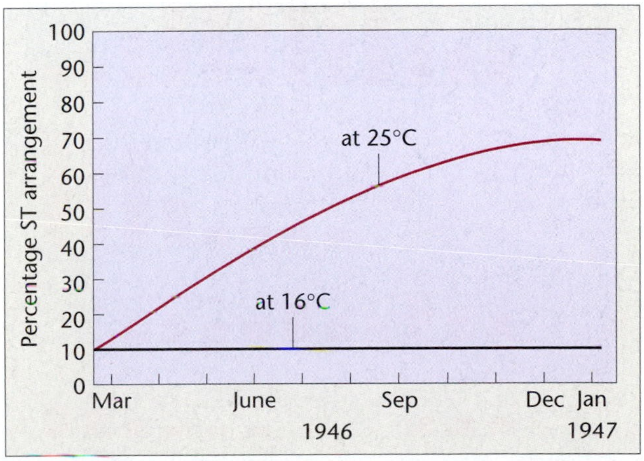

FIGURE 26–7 Increase in the ST arrangement of *D. pseudoobscura* throughout the year

arrangements in populations from 12 locations. In Figure 26–8, the frequencies of 5 of these inversions are shown according to geographic location. The differences are largely quantitative, with most populations differing only in the relative frequencies of inversions. Collectively, Dobzhansky's data show that the genetic structure of *D. pseudoobscura* populations changes from place to place and from one time to another. At least some of this variation in population genetic structure is the result of natural selection.

Another example of how the genetic structure of a species varies among populations is provided by the work of Dennis A. Powers and Patricia Schulte on the mummichog (*Fundulus heteroclitus*), a small fish (5 to 10 cm long) that lives in inlets, bays, and estuaries along the Atlantic coast of North America from Florida to Newfoundland. These workers measured allele frequencies at the locus encoding the enzyme lactate dehydrogenase-B (LDH-B), which is made in the liver, heart, and red skeletal muscle. LDH-B converts lactate to pyruvate and is thus pivotal in both the manufacture of glucose and aerobic metabolism. Two allozymes of LDH-B are seen on gels; they differ at two amino acid positions. The alleles encoding the allozymes are $Ldh\text{-}B^a$ and $Ldh\text{-}B^b$.

Frequencies of the *Ldh-B* alleles vary dramatically among mummichog populations [Figure 26–9(a)]. In northern populations, where the mean water temperature is about 6°C, $Ldh\text{-}B^b$ predominates. In southern populations, where the mean water temperature is about 21°C, $Ldh\text{-}B^a$ predominates. Between the geographic extremes, allele frequencies are intermediate.

To determine whether the geographic variation in *Ldh-B* allele frequencies is due to natural selection, Powers and Schulte studied the biochemical properties of the LDH-B allozymes. They found that the enzyme encoded by $Ldh\text{-}B^b$ has higher catalytic efficiency at low temperatures, whereas the gene product of $Ldh\text{-}B^a$ is more efficient at high temperatures [Figure 26–9(b)]. A mixture of the two forms has intermediate efficiency at all temperatures.

In addition to the functional differences between the LDH-B allozymes, the *Ldh-B* alleles have nucleotide sequence differences in their regulatory regions. One such difference results in the $Ldh\text{-}B^b$ allele's rate of transcription being more

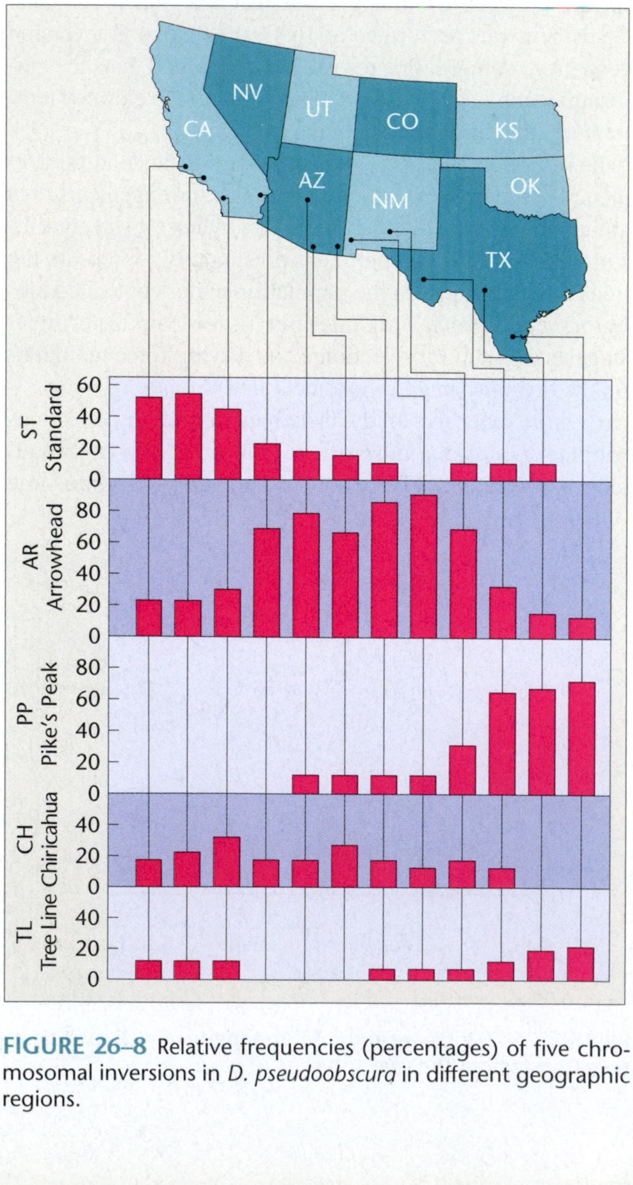

FIGURE 26–8 Relative frequencies (percentages) of five chromosomal inversions in *D. pseudoobscura* in different geographic regions.

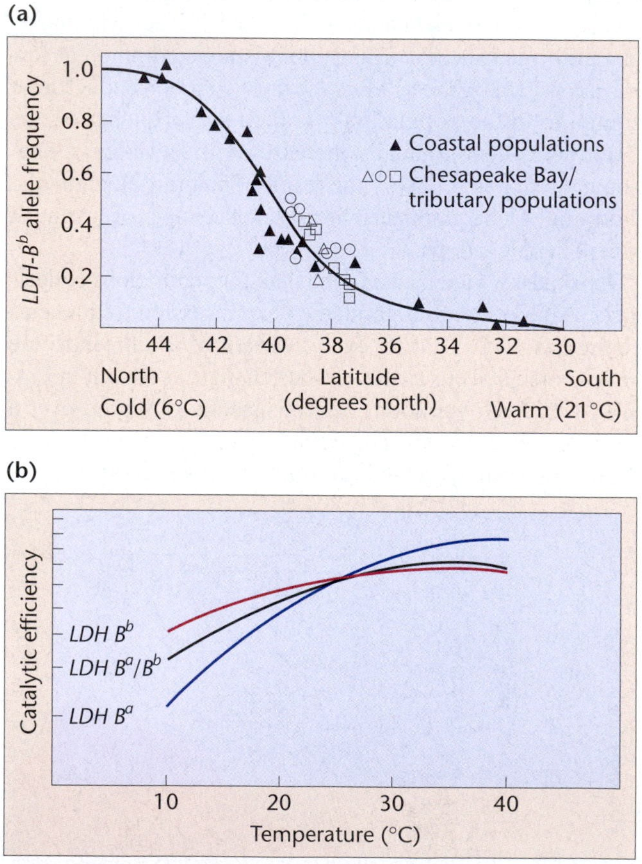

FIGURE 26–9 Variation in genetic structure among mummichog populations. (a) Frequencies of the B^b allele in populations along the Atlantic coast of North America. (b) Catalytic efficiency of LDH-B allozymes as a function of temperature. *(From D.A. Powers and P.M. Schulte. Evolutionary adaptations of gene structure and expression in natural populations in relation to a changing environment,* The Journal of Experimental Zoology, *© 1998. Reprinted by permission of Wiley-Liss, Inc., a subsidiary of John Wiley & Sons, Inc., and with permission from* Annu. Rev. Genet. *25, © 1991 by Annual Reviews.)*

than twice that of the *Ldh-B*[a] allele. As a result, northern fish have higher concentrations of the LDH-B enzyme in their cells.

Differences in the catalytic efficiency and transciption rate of the two *Ldh-B* alleles are consistent with the hypothesis that mummichog populations are adapted to the temperatures at which they live. The fish are ectotherms: Their body temperature is determined by their environment. In general, low body temperatures slow an ectotherm's metabolic rate. The higher transcription rate of the *Ldh-B*[b] allele and the superior low-temperature catalytic efficiency of its gene product appear to help northern fish compensate for the tendency of their cold environment to reduce their metabolic rate. The superior high-temperature catalytic efficiency and lower transcription rate associated with the *Ldh-B*[a] allele appear to allow southern fish to economize on the resources devoted to glucose production and aerobic metabolism. On the basis of this and other evidence, Powers and Schulte suggest that differences among mummichog populations in *Ldh-B* allele frequencies are the result of natural selection.

26.4 A Reduction in Gene Flow between Populations, Accompanied by Divergent Selection or Genetic Drift, Can Lead to Speciation

We have learned that most populations harbor considerable genetic variation and that different populations within a species may have different alleles or allele frequencies at a variety of loci. The genetic divergence of these populations can be caused by natural selection, genetic drift, or both. In Chapter 25, we saw that the migration of individuals between populations, together with the gene flow that accompanies that migration, tends to homogenize allele frequencies among populations. In other words, migration counteracts the tendency of populations to diverge.

When gene flow between populations is reduced or absent, the populations may diverge to the point that members of one population are no longer able to interbreed successfully with members of the other. When populations reach the point where they are reproductively isolated from one another, they have become different species.

The barriers that prevent interbreeding between populations can be physiological, behavioral, or mechanical. The biological and behavioral properties of organisms that prevent or reduce interbreeding are called **reproductive isolating mechanisms** and are classified in Table 26–2.

Prezygotic isolating mechanisms prevent individuals from mating in the first place. Individuals from different populations may not find each other at the right time, may not recognize each other as suitable mates, or may try to mate, but find that they are unable to do so.

Postzygotic isolating mechanisms create reproductive isolation even when the members of two populations are willing and able to mate with each other. For example, genetic divergence may have reached the stage where the viability or fertility of hybrids is reduced. Hybrid zygotes may be formed, but all or most may be inviable. Alternatively, the hybrids may be viable, but be sterile or suffer from reduced fertility. Yet again, the hybrids themselves may be fertile, but their progeny may have lowered viability or fertility. These postzygotic mechanisms act at or beyond the level of the zygote and are generated by genetic divergence.

TABLE 26–2 Reproductive Isolating Mechanisms

Prezygotic Mechanisms (prevent fertilization and zygote formation)
1. Geographic or ecological: The populations live in the same regions but occupy different habitats
2. Seasonal or temporal: The populations live in the same regions but are sexually mature at different times
3. Behavioral (only in animals): The populations are isolated by different and incompatible behavior before mating
4. Mechanical: Cross-fertilization is prevented or restricted by differences in reproductive structures (genitalia in animals, flowers in plants)
5. Physiological: Gametes fail to survive in alien reproductive tracts
Postzygotic Mechanisms (fertilization takes place and hybrid zygotes are formed, but these are nonviable or give rise to weak or sterile hybrids)
1. Hybrid nonviability or weakness
2. Developmental hybrid sterility: Hybrids are sterile because gonads develop abnormally or meiosis breaks down before completion
3. Segregational hybrid sterility: Hybrids are sterile because of abnormal segregation into gametes of whole chromosomes, chromosome segments, or combinations of genes
4. F_2 breakdown: F_1 hybrids are normal, vigorous, and fertile, but the F_2 contains many weak or sterile individuals

Source: From G. Ledyard Stebbins, *Processes of organic evolution*, 3rd ed., copyright 1977, p. 143. Reprinted by permission of Prentice-Hall, Upper Saddle River, NJ.

In some models of speciation, postzygotic isolation evolves first and is followed by prezygotic isolation. Postzygotic isolating mechanisms waste gametes and zygotes and lower the reproductive fitness of hybrid survivors. Selection will therefore favor the spread of alleles that reduce the formation of hybrids, leading to the development of prezygotic isolating mechanisms, which in turn prevent interbreeding and the formation of hybrid zygotes and offspring. In animal evolution, the most effective prezygotic mechanism is behavioral isolation, involving courtship behavior.

Examples of Speciation

In this section, we consider two examples of speciation, one from a laboratory study and the other from a field study. Diane Dodd and her colleagues studied the evolution of digestive physiology in *Drosophila pseudoobscura*. They collected flies from a wild population and established several separate laboratory populations. Some of the laboratory populations were raised on starch-based medium and others on maltose-based medium. Both food sources were stressful for the flies. It was only after several months of evolution by natural selection that the populations became adapted to their artificial diets and began to thrive.

Dodd's team wanted to know whether the starch-adapted populations and the maltose-adapted populations, which diverged under strong selection and in the absence of gene flow, had become different species. Roughly a year after the populations were established, a series of mating trials were performed. For each trial, 48 flies were placed in a bottle: 12 males and 12 females from a starch-adapted population and 12 males and 12 females from a maltose-adapted population. The researchers then noted which flies mated.

Dodd's group predicted that if the populations adapted to different media had speciated, then the flies would prefer to mate with members of their own population. If the populations had not speciated, then the flies would mate at random. The results appear in Table 26–3. Roughly 600 of the 900 matings observed were between males and females from the same population. In other words, the differently adapted fly populations showed partial premating isolation. Dodd and her colleagues concluded that the populations had begun to speciate, but had not yet completed the process.

The Isthmus of Panama, which created a land bridge connecting North and South America and simultaneously separated the Carribbean Sea from the Pacific Ocean, formed roughly 3 million years ago. Nancy Knowlton and colleagues took advantage of a natural experiment that the formation of the Isthmus of Panama had performed on several species of snapping shrimps (Figure 26–10). After identifying seven Caribbean species of snapping shrimp, they matched each one with a Pacific species to form a pair. Members of each pair were more similar to each other in structure and appearance than either was to any other species in its own ocean. Analysis of allozyme allele frequencies and mitochondrial DNA sequences confirmed that the members of each pair were one another's closest genetic relatives.

The Knowlton team's interpretation of these data is that, prior to the formation of the isthmus, the ancestors of each pair were a single species. When the isthmus closed, the seven ancestral species were each divided into two separate populations, one in the Caribbean and the other in the Pacific.

Meeting in a dish in Knowlton's lab for the first time in 3 million years, would Caribbean and Pacific members of a species pair recognize each other as suitable mates? Knowlton placed males and females of a species pair together and noted their behavior toward each other. She then calculated the relative inclination of Caribbean–Pacific couples to mate vs. that of Caribbean–Caribbean or Pacific–Pacific couples. For three of the seven species pairs, transoceanic couples refused to mate altogether. For the other four species pairs, transoceanic couples were 33, 45, 67, and 86 percent as likely to mate with each other as were same-ocean pairs. Of the same-ocean couples that mated, 60 percent produced viable clutches of eggs. Of the transoceanic couples that mated, only 1 percent produced viable clutches. We can conclude from these results that 3 million years of separation has resulted in complete or nearly complete speciation, involving strong pre- and postzygotic isolating mechanisms, for all seven species pairs.

The Minimum Genetic Divergence Required for Speciation

How much genetic separation is required between two populations before they become different species? We shall consider two examples, one from an insect and the other from a plant, which demonstrate that in some cases the answer is "not very much."

Researchers estimate that *Drosophila heteroneura* and *D. silvestris*, found only on the island of Hawaii, diverged from a common ancestral species only about 300,000 years ago.

TABLE 26–3 Incipient Speciation in Laboratory Populations of *Drosophila pseudoobscura* (Number of Matings Involving Each Kind of Male-Female Pair)

		Female	
		Starch-adapted	***Maltose-adapted***
Male	Starch-adapted	290	153
	Maltose-adapted	149	312

Source: Compiled from Dodd, D. M. B. 1989. Reproductive isolation as a consequence of adaptive divergence in *Drosophila pseudoobscura*. *Evolution* 43: 1308–11.

FIGURE 26–10 A snapping shrimp (genus *Alpheus*). *(Carl C. Hansen/Nancy Knowlton/Smithsonian Institution Photo Services.)*

The two species are thought to be descended from *D. planitibia* colonists from the older island of Maui (Figure 26–11). The two species are clearly separated from each other by different and incompatible courtship and mating behaviors (a prezygotic isolating mechanism), by morphology, and by body and wing pigmentation (Figure 26–12).

In spite of their morphological and behavioral divergence, demonstrating significant differences between these species in chromosomal inversion patterns or protein polymorphisms is difficult. DNA hybridization studies carried out on the two species indicate that the sequence diversity between them is only about 0.55 percent (Figure 26–13). Thus, nucleotide sequence diversity may precede the development of protein or chromosomal polymorphisms.

Genetic evidence suggests that the differences between *D. heteroneura* and *D. silvestris* are controlled by a relatively small number of genes. For example, as few as 15 to 19 major loci may be responsible for the morphological differences between the species, demonstrating that the process of speciation likely involves only a small number of genes.

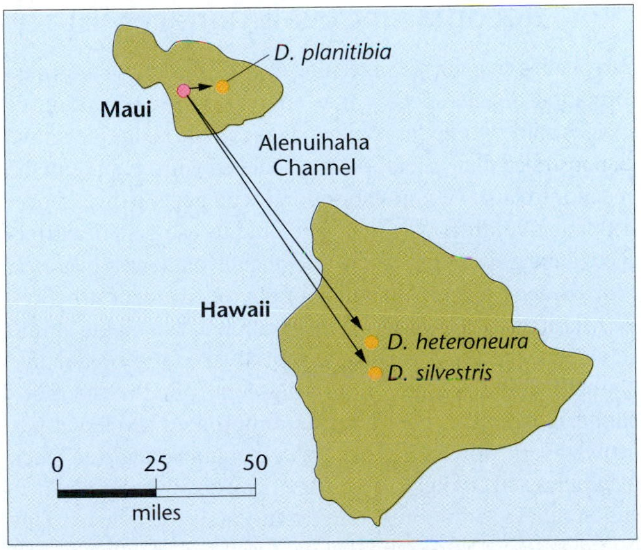

FIGURE 26–11 Proposed pathway for Hawaii's colonization by members of the *D. planitibia* species. The red circle represents a population ancestral to the three present-day species.

Studies using two closely related species of the monkey flower, a plant that grows in the Rocky Mountains and areas west, confirm that species can be separated by only a few genetic differences. One species, *Mimulus cardinalis*, is fertilized by hummingbirds and does not interbreed with *Mimulus lewisii*, which is fertilized by bumblebees. H. D. Bradshaw and his colleagues studied genetic differences related to reproduction in the two species—namely, flower shape, size, and color and nectar production. For each trait, a difference in a single gene provided at least 25 percent of the variation observed among laboratory-created *M. cardinalis* × *lewisii* hybrids. *M. cardinalis* makes 80 times more nectar than does *M. lewisii*, and a single gene is responsible for at least half the difference. A single gene also controls a large part of the differences in flower color between the two species (Figure 26–14). In this case, as in the Hawaiian *Drosophila*, species differences can be traced to a relatively small number of genes.

In at Least Some Instances, Speciation Is Rapid

How much time is required for speciation? In many cases, speciation takes place slowly over a long period. In other cases, however, speciation can be surprisingly rapid.

The Rift Valley lakes of East Africa support hundreds of species of cichlid fish. Lake Victoria (Figure 26–15), for example, has over 400 species. Cichlids have some morphological diversity, but, more dramatically, they are highly specialized for different niches (Figure 26–16). Some eat algae floating on the water's surface, whereas others are bottom feeders, insect feeders, mollusk eaters, and predators on other fish species. Lake Tanganyika and Lake Malawi have a similar array of species. Genetic analyses indicate that the species in a given lake are all more closely related to each other than to species from other lakes. The implication is that most or all the species in, say, Lake Tanganyika are descended from a single common ancestor and that they evolved within their home lake.

Lake Victoria is between 250,000 and 750,000 years old, and there is evidence that the lake may have dried out nearly completely less than 14,000 year ago. Is it possible that the 400 species in the lake today evolved from a common ancestral species in less than 14,000 years?

In a study of cichlid origins in Lake Tanganyika, Norihiro Okada and colleagues examined the insertion of a novel family of SINES (see Chapter 4) into the genomes of cichlid species in Lake Tanganyika. SINES are short, interspersed repetitive DNA sequences inserted into the genome at random. They are a type of retroposon, and integration of a SINE at a locus is an irreversible event. If a SINE is present at the same locus in the genome of all species examined, this is strong evidence that all of those species descend from a common ancestor. Using a SINE called AFC, Okada's team screened 33 species of cichlids belonging to four groups (called species tribes). In each tribe, the SINE was present at all the sites tested, indicating that the species in each tribe are descended from a single ancestral species (Figure 26–17). If further research using SINES and other molecular markers produces similar results with the Lake Victoria species, it

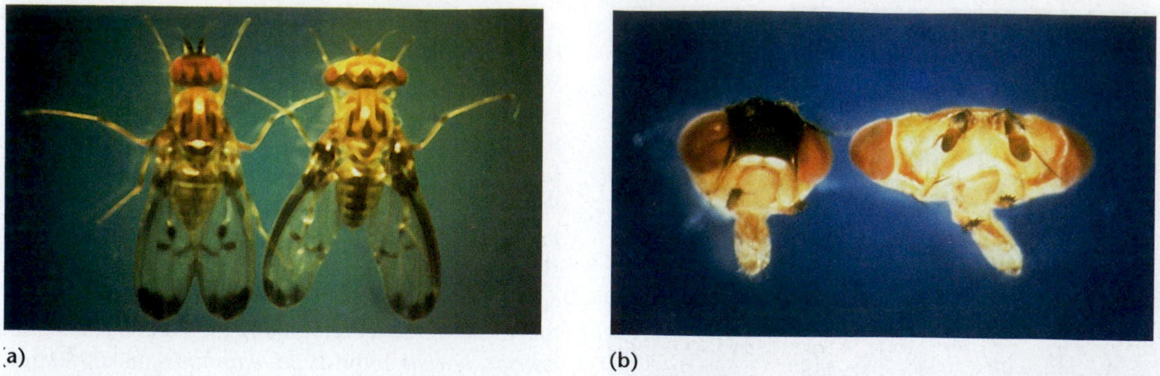

FIGURE 26–12 (a) Differences in pigmentation patterns in *D. sylvestris* (left) and *D. heteroneura* (right). (b) Head morphology in *D. sylvestris* (left) and *D. heteroneura* (right). *(Kenneth Kaneshiro/University of Hawaii/CCRT.)*

means that the 400 cichlid species in this lake evolved in less than 14,000 years. If confirmed, this finding would represent the fastest evolutionary radiation ever documented in vertebrates.

Even faster speciation is possible through the mechanism of polyploidy. The formation of animal species by polyploidy is rare, but is an important factor in plant evolution. It is estimated that one-half of all flowering plants have evolved by this mechanism. One form of polyploidy is allopolyploidy (see Chapter 13), produced by doubling the chromosome number in a hybrid formed by crossing two species.

If two species of related plants have the genetic constitution SS and TT (where S and T represent the haploid set of chromosomes in each species), the F_1 hybrid would have the chromosome constitution ST. Normally, such a plant would be sterile, because there are few or no homologous chromosome pairs and aberrations would arise during meiosis. However, if the hybrid undergoes a spontaneous doubling of its chromosome number, a tetraploid SSTT plant would be produced. This chromosomal aberration might occur during mitosis in somatic tissue, giving rise to a partially tetraploid plant that produces some tetraploid flowers. Alternatively, aberrant meiotic events may produce ST gametes that, when fertilized, would yield SSTT zygotes. The SSTT plants would be fertile because they would possess homologous chromosomes producing viable ST gametes. This new, true-breeding tetraploid would have a combination of characteristics derived from the parental species and would be reproductively isolated from them because F_1 hybrids would be triploids and consequently sterile.

The tobacco plant *Nicotiana tabacum* (2n = 48) is the result of a doubling of the chromosome number in the hybrid between *N. otophora* (2n = 24) and *N. silvestris* (2n = 24) (Figure 26–18). The origin of *N. tabacum* is an example of virtually instantaneous speciation.

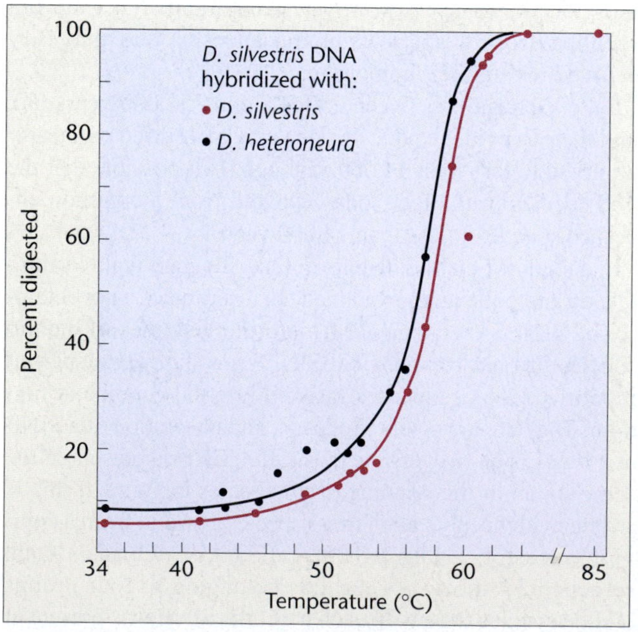

FIGURE 26–13 Nucleotide sequence diversity in the *Drosophila planitibia* species complex. The shift to the left by the heterologous hybrids indicates the degree of nucleotide sequence divergence.

26.5 We Can Use Genetic Differences among Populations or Species to Reconstruct Evolutionary History

Early in this chapter, we noted that evolution involves the transformation of one species into another and the splitting of species into two or more new species. Our examples have demonstrated that speciation is associated with changes in the genetic structure of populations and with genetic divergence. If this is true, then we should be able to use genetic differences among species to reconstruct their evolutionary histories.

In an important early example of phylogeny reconstruction, W. M. Fitch and E. Margoliash assembled data on the amino acid sequence for cytochrome c in a variety of organisms. **Cytochrome c** is a respiratory pigment found in the mitochondria of eukaryotes, and its amino acid sequence has evolved very slowly. For example, the amino acid sequence in humans and chimpanzees is identical, and humans and rhesus monkeys show only one amino acid difference. This is remarkable, considering that the fossil record indicates that the lines leading to humans and monkeys diverged from a common ancestral species approximately 20 million years ago.

(a)

(b)

FIGURE 26–14 Flowers of two closely related species of monkey flowers, (a) *Mimulus cardinalis* and (b) *Mimulus lewisii*. *(Courtesy of Toby Bradshaw and Doug W. Schemske, University of Washington. Photo by Jordan Rehm.)*

Column (a) of Table 26–4 shows the number of amino acid differences between cytochrome c in humans and a variety of other species. The table is broadly consistent with our intuitions about how closely related we are to these other species. For example, we are more closely related to other mammals than we are to insects, and we are more closely related to insects than we are to fungi. Likewise, our cytochrome c differs in 10 amino acids from that of dogs, in 24 amino acids from that of moths, and in 38 amino acids from that of yeast.

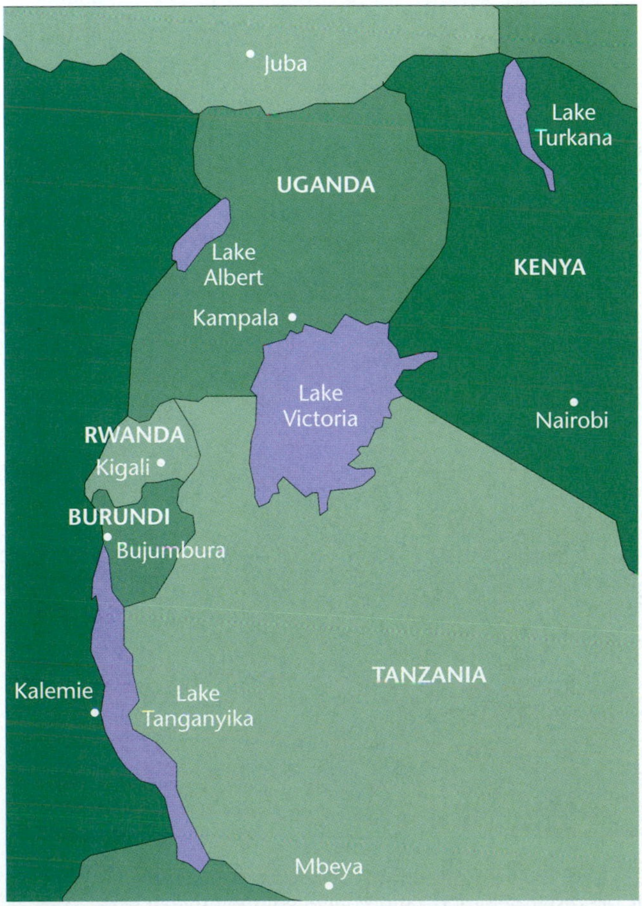

FIGURE 26–15 Lake Victoria in the Rift Valley of East Africa is home to more than 400 species of cichlids.

However, more than one nucleotide change may be required to change a given amino acid. When the nucleotide changes necessary for all amino acid differences observed in a protein are totaled, the **minimal mutational distance** between the genes of any two species is established. Column (b) in Table 26–4 shows such an analysis of the genes encoding cytochrome c. As expected, these values are larger than the corresponding number of amino acids separating humans from the other nine organisms listed.

Fitch used data on the minimal mutational distances between the cytochrome c genes of 19 organisms to reconstruct their evolutionary history. The result is an estimate of the evolutionary tree, or phylogeny, that unites the species (Figure 26–19). The black dots on the tips of the twigs represent the extant species, which are connected to the inferred common ancestors represented by red dots. The ancestral species evolved and diverged to produce the modern species. The common ancestors are connected to still earlier common ancestors, culminating in a single common ancestor for all the species on the tree, represented by the red dot on the extreme left.

A Method for Estimating Evolutionary Trees from Genetic Data

Many methods use data on genetic differences to estimate phylogenies. It is beyond the scope of this chapter to review all of them. Instead, we shall present just one method, called the *u*nweighted *p*air *g*roup *m*ethod using *a*rithmetic averages, or UPGMA. This method is not the most powerful one for estimating phylogenies from genetic data, but, in spite of its name, it is intuitively straightforward. Furthermore, UPGMA works reasonably well under many circumstances.

The starting point for UPGMA is a table of genetic distances among a group of species [Figure 26–20(a)]. To demonstrate the technique, we will use data from DNA hybridization studies by Charles Sibley and Jon Alquist, who computed genetic distances between humans and four species of apes: the common chimpanzee, the gorilla, the siamang, and the common gibbon.

The method uses the following steps:

FIGURE 26–16 Cichlids occupy a diverse array of niches, and each species is specialized for a distinct food source. *(Dr. Paul V. Loiselle.)*

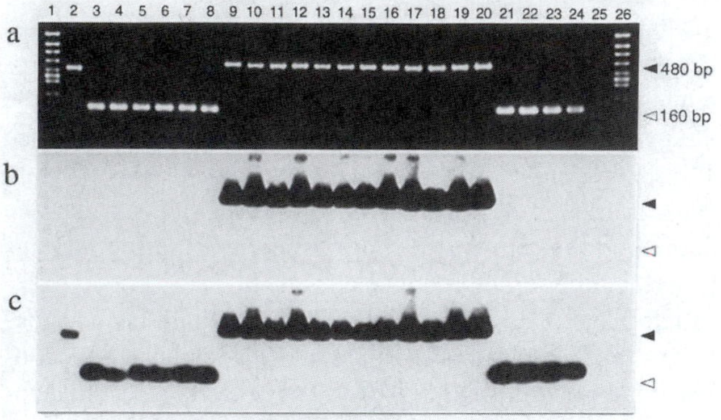

FIGURE 26–17 Tracing specific relationships in cichlids from Lake Tanganyika using repetitive DNA elements. (a) Photograph of an agarose gel showing PCR fragments generated from primers that flank the AFC family of SINES. Large fragments containing this family are present in all samples of genomic DNA from members of the Lamprologini tribe of cichlids (lanes 9–20). DNA from the species in lane 2 has a similar, but shorter fragment, which may represent another repetitive sequence. DNA from other species (lanes 3–8 and 21–24) produce short, non-SINE-containing fragments. (b) A Southern blot of the gel from (a) probed with DNA from the AFC family of SINES. AFC is present in DNA from all species in the Lamprologini tribe (lanes 9–20), but not in DNA from other species (lanes 3–8 and 21–24). (c) A second Southern blot of the gel from (a) probed with the genomic sequence at which the AFC SINE inserts. All species examined (lanes 2–24) contain the insertion sequence. The larger fragments in Lamprologini DNA (lanes 9–20) correspond to those in the previous blot, showing that the SINE is inserted at the same site in all tribal species. In sum, these data show that the AFC SINE is present only in members of this tribe, is present in all member species, and is inserted at the same site in all cases. These results are interpreted as showing a common origin for the species of the tribe in question. *(From Takahashi, K., et al. 1998. A novel family of short interspersed repetitive elements from cichlids: The pattern of insertion of SINES at orthologous loci support the supposed monophyly of four major groups of cichlid fishes in Lake Tanganyika.* Molec. Biol. Evol. *15(4):391–407. Taken from Figure 5, part A, p. 400.)*

1. Search for the smallest genetic distance between any pair of species. In Figure 26–20(a), this distance is 1.628, the distance between human and chimpanzee. Once identified, the corresponding species pair is placed on neighboring branches of an evolutionary tree [Figure 26–20(b)]. The length of each branch is half the genetic distance between the species (1.68/2), so the branches connecting human and chimpanzee to their common ancestor are each about 0.81.

2. Recalculate the genetic distances between this species pair and all other species [Figure 26–20(c)]. The genetic distance between the human–chimpanzee (Hu–Ch)

FIGURE 26–18 The cultivated tobacco plant *Nicotiana tabacum* is the result of hybridization between two other species. *(C.B. & D.W. Frith/Bruce Coleman, Inc.)*

TABLE 26–4 Amino Acid Differences and the Minimal Mutational Distances between Cytochrome c in Humans and Other Organisms

Organism	*(a)* *Amino Acid Differences*	*(b)* *Minimal Mutational Distance*
Human	0	0
Chimpanzee	0	0
Rhesus monkey	1	1
Rabbit	9	12
Pig	10	13
Dog	10	13
Horse	12	17
Penguin	11	18
Moth	24	36
Yeast	38	56

Source: From W. M. Fitch and E. Margoliash, Construction of phylogenetic trees, *Science* 155:279–84, 20 January 1967. Copyright 1967 by the American Association for the Advancement of Science

cluster* and the other species is the average of the distances between each member of the cluster and the other species. For example, the genetic distance between the human–chimp cluster and the gorilla is the average of the human–gorilla distance and the chimp–gorilla distance, or $(2.267 + 2.21)/2 = 2.2385$.

3. Repeat steps 1 and 2 until all the species have been added to the tree.

The smallest distance in the recalculated table is 1.95, the distance between siamang and gibbon [Figure 26–20(c)]. To build the next section of the tree, siamang and gibbon are placed on neighboring branches, with lengths equal to half of 1.95, or 0.98. [Figure 26–20(d)].

Recalculating the table, we now find that there are two clusters plus the gorilla [Figure 26–20(e)]. The genetic distance between the human–chimp cluster and the siamang–gibbon cluster is the average of four distances: human–siamang, human–gibbon, chimp–siamang, and chimp–gibbon.

*We say "cluster" even though what we really have is a pair; the reason is that we can (and will) form larger groups and it is convenient to call them by the same, more inclusive name—ergo cluster.

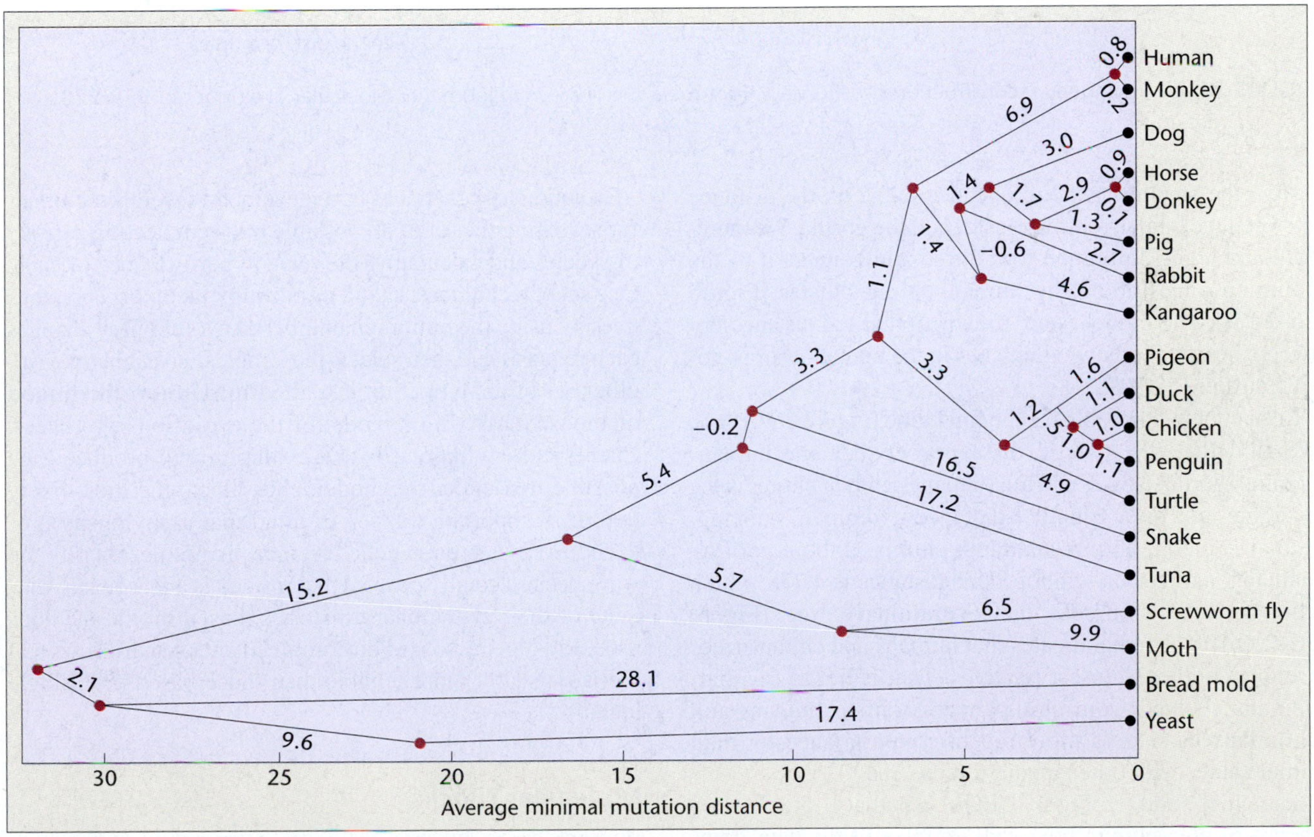

FIGURE 26–19 Phylogeny constructed by comparing homologies in cytochrome c amino acid sequences. *(From W.M. Fitch and E. Margoliash, Construction of phylogenetic trees,* Science *155:279–84, 20 January 1967. Copyright 1967 by the American Association for the Advancement of Science.)*

(a)

	Human	Chimp	Gorilla	Siamang	Gibbon
Human	–				
Chimp	**1.628**	–			
Gorilla	2.267	2.21	–		
Siamang	4.7	5.133	4.543	–	
Gibbon	4.779	4.76	4.753	1.95	–

(c)

	Hu-Ch	Gorilla	Siamang	Gibbon
Hu-Ch	–			
Gorilla	2.2385	–		
Siamang	4.9165	4.543	–	
Gibbon	4.7695	4.753	**1.95**	–

(e)

	Hu-Ch	Gorilla	Si-Gi
Hu-Ch	–		
Gorilla	**2.239**	–	
Si-Gi	4.843	4.648	–

(g)

	Hu-Ch-Go	Si-Gi
Hu-Ch-Go	–	
Si-Gi	**4.778**	–

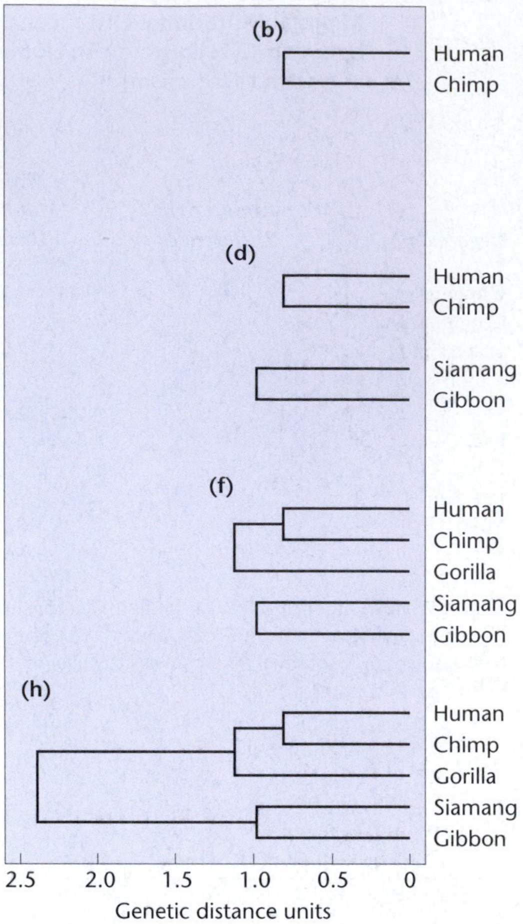

FIGURE 26–20 Phylogeny reconstruction by UPGMA. *(From Table 1, p. 126, from* Journal of Molecular Evolution *26:99–122.)*

Now the smallest genetic distance is 2.239, the distance between the human–chimp cluster and the gorilla. We therefore add a gorilla branch to the tree and connect it to the common ancestor of the human–chimp cluster [Figure 26–20(f)]. The branches are drawn so that the distance between the tips of any two branches in the human–chimp–gorilla cluster is 2.239.

Recalculating the table for the final time [Figure 26–20(g)], we find that the genetic distance between the human–chimp–gorilla cluster and the siamang–gibbon cluster is the average of six genetic distances: human–siamang, human–gibbon, chimp–siamang, chimp–gibbon, gorilla–siamang, and gorilla–gibbon. This distance is 4.778, which allows us to complete our evolutionary tree [Figure 26–20(h)]. The tree indicates that humans and chimpanzees are one another's closest relatives. That is not to say that humans evolved from chimpanzees; rather, humans and chimpanzees share a more recent common ancestor than either shares with other species on the tree.

The chief shortcoming of UPGMA is that it provides no means of determining how well its trees fit the data, compared with other possible trees. For example, how much better does the tree we just created fit our data than a tree that shows chimpanzees and gorillas as closest relatives?

Evolutionary geneticists have developed several techniques for searching the set of all possible trees connecting a group of species and calculating the relative performance of each. One set of techniques, called **parsimony** methods, compares trees by using the minimum number of evolutionary changes each requires and then selects the simplest possible tree. Another set of analytical tools, called **maximum-likelihood** methods, starts with a model of the evolutionary process, calculates how likely it is that evolution will produce each possible tree under the model, and selects the most likely tree. It is important to keep in mind that a phylogeny produced from data on genetic distances is not necessarily the way species actually evolved, but instead is a reasonable estimate of their evolutionary histories, based on the methods used. Phyolgenies based on combined data sets from several loci are usually more reliable than those based on a single locus.

Molecular Clocks

In many cases, we would like to estimate not only which members of a set of species are most closely related, but also when their common ancestors lived. Sometimes we can do so, thanks to **molecular clocks**—amino acid sequences or nucleotide sequences in which evolutionary changes accumulate at a constant rate over time.

Research by Walter M. Fitch and colleagues on the influenza A virus shows how molecular clocks are used. Fitch and his associates sequenced part of the hemagglutinin gene from flu viruses that had been isolated at different times over a 20-year period. They calculated the number of nucleotide differences among the various viruses and constructed an evolutionary tree [Figure 26–21(b)]. Most strains have gone extinct, leaving no descendants among the more recently isolated viruses. Fitch's group then plotted the number of nucleotide substitutions between the first virus and each subsequent virus against the year in which the virus was isolated [Figure 26–21(a)]. The points all fall very close to a straight line, indicating that nucleotide substitutions in this gene have accumulated at a steady rate. The hemagglutinin gene thus serves as a molecular clock. Molecular clocks are used to compare the sequences of new flu viruses as they appear each year and to estimate the time that has passed since each diverged from a common ancestor.

Molecular clocks must be carefully calibrated and used with caution. For example, Fitch's data indicate that strains of influenza A that jump from birds to humans have evolved much more rapidly than strains that have remained in birds. Hence, a molecular clock calibrated from human strains of the virus would be highly misleading if applied to bird strains.

Figure 26–22 shows a phylogeny based on Sibley's complete data set of genetic differences among humans and apes. Using a molecular clock established via the fossil record, Sibley placed this information on a time scale that estimates ages for the common ancestors of each species cluster. Sibley's tree suggests that the most recent common ancestor of humans and chimps lived between 5 and 10 million years ago.

26.6 Reconstructing Evolutionary History Allows Us to Answer a Variety of Questions

Evolutionary analysis addresses a wide range of questions, some of which relate to evolution directly, while others deal with contemporary issues and even criminal activity. To conclude this chapter, we provide answers to the questions posed in the scenarios set forth in the beginning.

Transmission of HIV from a Dentist to His Patients

In late 1986, a Florida dentist tested positive for HIV. Several months later, he was diagnosed with AIDS. He continued to practice general dentistry for two more years, until one of his patients, a young woman with no known risk factors, discovered that she, too, was infected with HIV. When

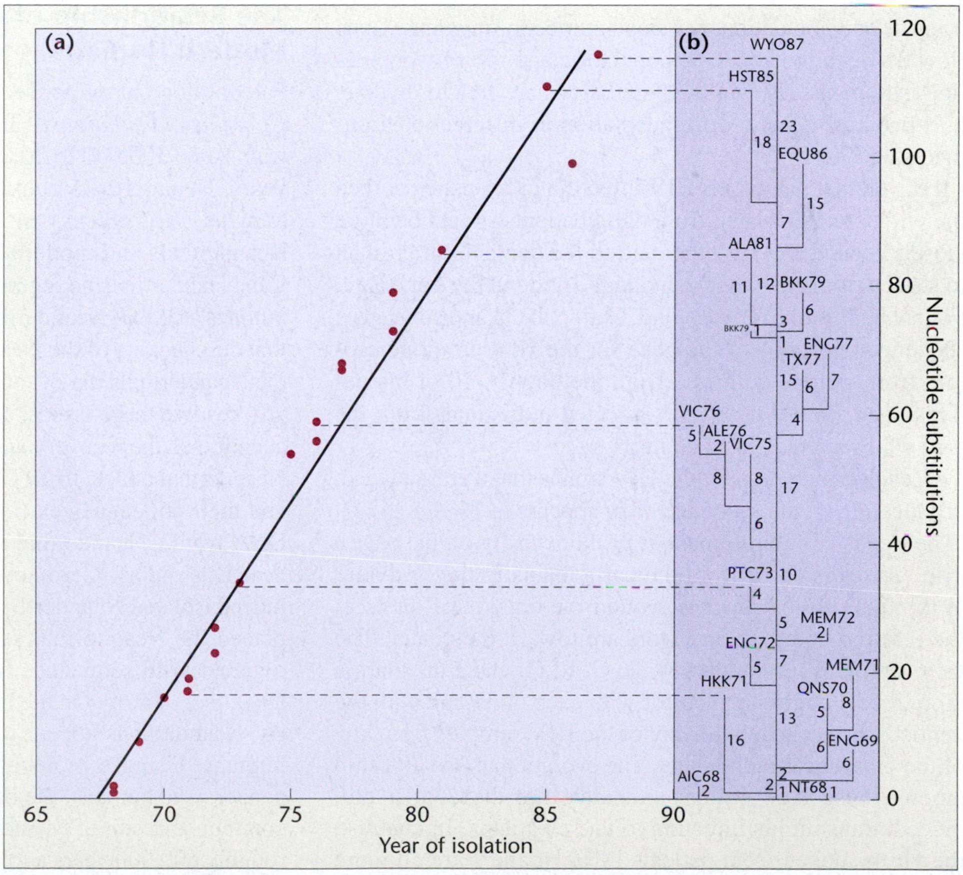

FIGURE 26–21 Molecular clock in the influenza A hemagglutinin gene. (a) Estimate of the phylogeny of the isolates. (b) Number of nucleotide differences between the first isolate and each subsequent isolate as a function of year of isolation. *(From Fitch et al. 1991. Positive Darwinian evolution in human influenza A viruses.* PNAS *88:4270–73.)*

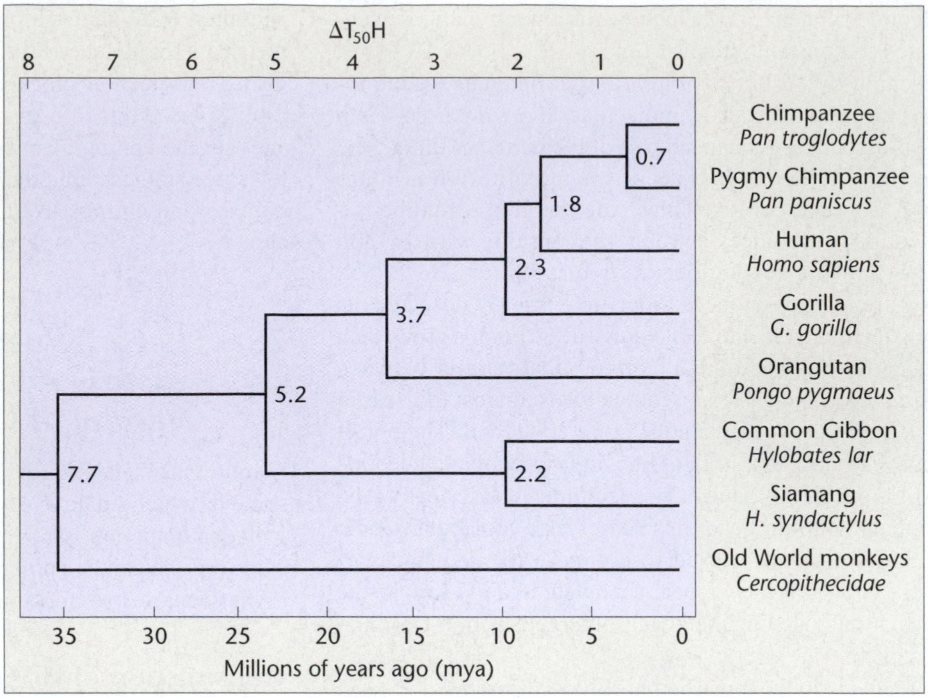

FIGURE 26–22 Phylogenetic sequence of hominoid primates and Old World monkeys, as estimated by DNA hybridization. The numbers at the branch points are $\Delta T_{50}H$ measurements. The evolutionary branch points are dated by means of the fossil record and nucleotide divergence.

the dentist publicly urged his other patients to have themselves tested, several more were found to be HIV positive. Did this dentist transmit HIV to his patients, or did the patients become infected by some other means?

At first glance, this situation appears to be a long way from a discussion of evolution; however, the movement of a virus from one individual to another is similar to the founding of a new island population by a small number of migrants. Gene flow between the ancestral population and the new population is nonexistent, and the populations are free to diverge, as a result of genetic drift, adaptation to different environments, or both.

If the dentist passed his HIV infection to his patients, then the viral strains isolated from these patients should be more closely related to each other and to the dentist's strain than to strains from any other individuals living in the same area. Following this line of reasoning, Chin-Yih Ou and colleagues sequenced portions of the gene for the HIV envelope protein from viruses collected from the dentist, 10 of his patients, and several other HIV-infected individuals from the area who served as local controls.

An evolutionary tree for the HIV strains that were analyzed, produced from the sequence data, appears in Figure 26–23. (The tree is circular to make it fit more easily on the page.) The viruses isolated from the participants in the study are at the tips of the branches around the outside of the tree; the inferred common ancestors are toward the center. The HIV strains from patients A, B, C, E, G, and I all share a more recent common ancestor with each other and with the dentist's strains than with any of the HIV samples from any of the other local individuals. The evolutionary relationship among these HIV strains indicates that the dentist did, indeed, transmit his infection to those patients. In contrast, the viruses taken from patients D, F, H, and J are all more closely related to strains from local controls (LC) than they are to the dentist's strains. These patients got their infection from someone other than the dentist. Patient J, in fact, seems to have acquired his infection from two different sources.

The Relationship of Neanderthals to Modern Humans

Paleontological evidence indicates that the Neanderthals, *Homo neanderthalensis*, lived in Europe and western Asia from some 300,000 to 30,000 years ago. For at least 30,000 years, Neanderthals coexisted with anatomically modern humans (*H. sapiens*) in several areas. Questions about Neanderthals and modern humans remain unanswered: (1) Can Neanderthals be regarded as direct ancestors of modern humans? (2) Did Neanderthals and *H. sapiens* interbreed, so that descendants of the Neanderthals are alive today? Or did the Neanderthals die off and become extinct?

To resolve these issues, several groups have focused their attention on the recovery and analysis of DNA extracted from Neanderthal bones. In 1997, Svante Pääbo, Matthias Krings, and their colleagues extracted fragments of mitochondrial DNA from a Neanderthal skeleton found in Feldhofer cave near Düsseldorf, Germany. After confirming that they had indeed isolated Neanderthal gene fragments, the researchers placed the Neanderthal sequences on a phylogenetic tree together with sequences from over 2000 modern humans [Figure 26–24(a)]. On the basis of Pääbo and Krings's analysis, Neanderthals appear to be a distant relative of modern humans. Using a molecular clock calibrated with chimpanzees and humans, the researchers calculated that the last common ancestor of Neanderthals and modern humans lived roughly 600,000 years ago, four times as long ago as the last

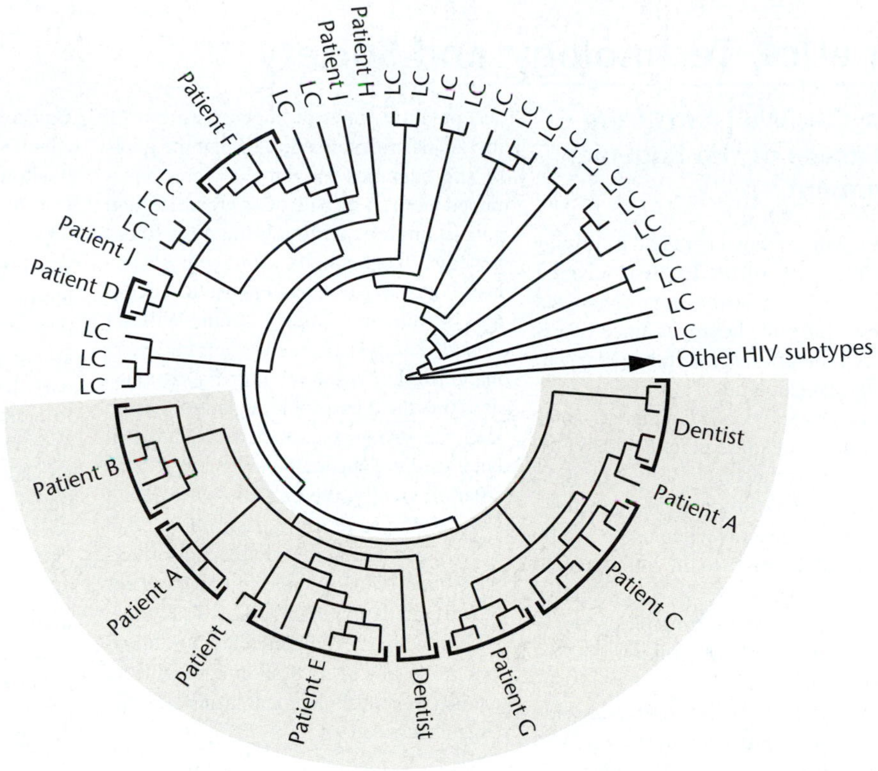

FIGURE 26–23 A phylogeny of HIV strains taken from a dentist, his patients, and several local controls (LC). The group of viral strains in the shaded portion of the phylogeny are all derived from a common ancestral strain. *(From Hillis (1998),* Current Biology *7:R129–R131.)*

common ancestor of all modern humans. However, considerable caution is required when drawing conclusions based on a single Neanderthal specimen.

More recently, Igor Ovchinnikov and his colleagues analyzed mitochondrial DNA recovered from Neanderthal remains discovered in Mezmaiskaya cave in the Caucasus Mountains east of the Black Sea. Although the two Neanderthal sequences (i.e., Pääbo and Krings's, on the one hand, and Ovchinnikov's, on the other) are from individuals from different geographic regions more than 1000 miles apart, they vary only by about 3.5 percent, indicating that they derive from a single gene pool. Further, the amount of variation between the two Neanderthal sequences is comparable to that seen among modern humans. Phylogenetic analysis places the two Neanderthals in a group that is distinct from modern humans [Figure 26–23 (b)]; the conclusion to be drawn from the two studies is that, although Neanderthals and humans have a common ancestor, the Neanderthals were a separate hominid line and did not contribute mitochondrial genes to *H. sapiens*. Taken together, the studies suggest that when the Neanderthals disappeared, their lineage went extinct with them.

The Origin of Mitochondria

Mitochondria are cellular organelles that contain their own genome in the form of a circular DNA molecule. In humans, the mitochondrial genome encodes 13 proteins required for oxidative phosphorylation and ATP production, 22 tRNA molecules, and 2 rDNA genes. The organization and function

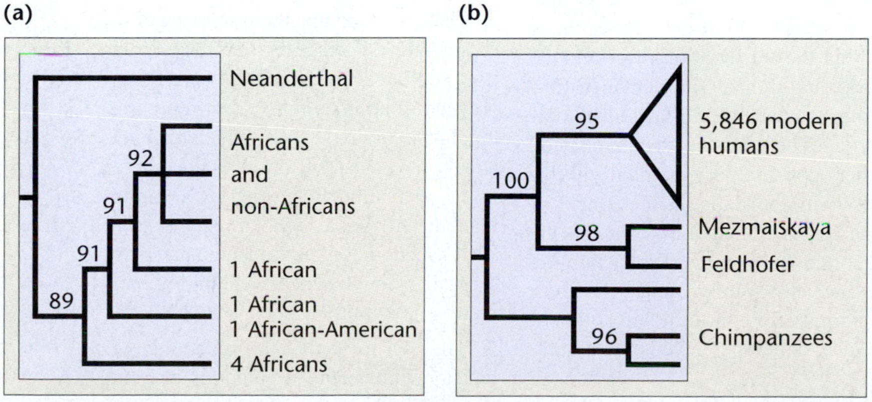

FIGURE 26–24 (a) A phylogeny estimated from mitochondrial DNA sequences of one Neanderthal and over 2000 modern humans. *(From M. Krings, A. Stone, R.W. Schmitz, H. Krainitzki, M. Stoneking, and S. Pääbo. 1997. ©1997 Cell Press. Cell 90:19–30. Fig. 7A, p. 26.)* (b) A phylogeny of the Feldhofer and Mezmaiskaya Neanderthal specimens, compared with over 5000 modern humans. *(From Ovchinnikov, I. V., et al., 2000. Molecular analysis of Neanderthal DNA from the northern Caucasus.* Nature *404:490–493.)*

Genetics, Technology, and Society

What Can We Learn from the Failure of the Eugenics Movement?

The eugenics movement had its origins in the ideas of the English scientist Francis Galton, who became convinced from his study of the appearance of geniuses within families (including his own) that intelligence is inherited. Galton concluded in his 1869 book *Hereditary Genius* that it would be "quite practicable to produce a highly-gifted race of men by judicious marriages during several consecutive generations." The term eugenics, coined by Galton in 1883, refers to the improvement of the human species by such selective mating. Once Mendel's principles were rediscovered in 1900, the eugenics movement flourished.

The eugenicists believed that a wide range of human attributes were inherited as Mendelian traits, including many aspects of behavior, intelligence, and moral character. Their overriding concern was that the presumed genetically "feebleminded" and immoral in the population were reproducing faster than the genetically superior, and that this differential birthrate would result in the progressive deterioration of the intellectual capacity and moral fiber of the human race. Several remedies were proposed. Positive eugenics called for the encouragement of especially "fit" parents to have more children. More central to the goals of the eugenicists, however, was the negative eugenics approach aimed at discouraging the reproduction of the genetically inferior or, better yet, eliminating it altogether.

In the United States, the eugenics movement enjoyed wide popular support for a time and had a significant impact on public policy. Partially at the urging of prominent eugenicists, 30 states passed laws compelling the sterilization of criminals, epileptics, and inmates in mental institutions; most states enacted laws invalidating marriages between the "feebleminded" and others considered eugenically unfit. The crowning legislative achievement of the eugenics movement, however, was the passage of the Immigration Restriction Act of 1924, which severely limited the entry of immigrants from eastern and southern Europe due to their perceived mental inferiority.

Throughout the first two decades of the century, most geneticists passively accepted the views of eugenicists, but by the 1930s critics recognized that the goals of the eugenics movement were determined more by racism, class prejudice, and anti-immigrant sentiment than by sound genetics. Increasingly, prominent geneticists began to speak out against the eugenics movement, among them William Castle, Thomas Hunt Morgan, and Hermann Muller. When the horrific extremes to which the Nazis took eugenics became known, a strong reaction developed that all but ended the eugenics movement.

Paradoxically, the eugenics movement arose at the same time basic Mendelian principles were being developed, principles that eventually undermined the theoretical foundation of eugenics. Today, any student who has completed an introductory course in genetics should be able to identify several fundamental mistakes the eugenicists made:

1. They assumed that complex human traits such as intelligence and personality were strictly inherited, completely disregarding any environmental contribution to the phenotype. Their reasoning was that because certain traits ran in families, they must be genetically determined.

2. They assumed that these complex traits were determined by single genes with dominant and recessive alleles. This belief persisted despite research showing that multiple gene pairs contribute to many phenotypes.

3. They assumed that a single ideal genotype existed in humans. Presumably, such a genotype would be highly homozygous in order to be sustained. This precept runs counter to current evidence suggesting that a high level of homozygosity is often deleterious, supporting the superiority of the heterozygote.

4. They assumed that the frequency of recessively inherited defects in the population could be significantly lowered by preventing homozygotes from reproducing. In fact, for recessive traits that are relatively rare, most of the recessive alleles in the population are carried by asymptomatic heterozygotes who are spared from such selection. Negative eugenic practices, no matter how harsh, are relatively ineffective at eliminating such traits.

5. They assumed that those deemed genetically unfit in the population were outreproducing those thought to be genetically fit. This is the exact reverse of the Darwinian concept of fitness, which equates reproductive success with fitness. (Galton should have understood this, being Darwin's first cousin!)

More than seven decades have passed since the eugenics movement was in full bloom. We now have a much more sophisticated understanding of genetics, as well as a greater awareness of its potential misuses. But the application of current genetic technologies makes possible a "new eugenics" of a scope and power that Francis Galton could not have imagined. In particular, prenatal genetic screening and *in vitro* fertilization enable the selection of children according to their genotype, a power that will dramatically increase as more and more genes are associated with inherited diseases and perhaps even behaviors.

As we move into this new genetic age, we must not forget the mistakes made by the early eugenicists. We must remember that phenotype is a complex interaction between the genotype and the environment, and not lapse into a new hereditarianism that treats a person as only a collection of genes. We must keep in mind that many genes may contribute to a particular phenotype, whether a disease or a behavior, and that the alleles of these genes may interact in unpredictable ways. We must not fall prey to the assumption that there is an ideal genotype. The success of all populations in nature is correlated with genetic diversity. And most of all, we must not use genetic information to advance ideological goals. We may find that there is a fine line between the legitimate uses of genetic technologies, such as having healthy children, and other eugenic practices. It will be up to us to decide exactly where the line falls.

References

Allen, G. E., Jacoby, R., and Glauberman, N. (eds.). 1995. Eugenics and American social history, 1880-1950. Genome 31:885-889.

Hartl, D. L. 1988. A primer of population genetics. 2nd ed. Sunderland, MA: Sinauer.

Kevles, D. J. 1985. In the name of eugenics: Genetics and the uses of human heredity. Berkeley: Univ. of California Press.

of the mitochondrial genome bear many similarities to those of bacterial genomes. Therefore, mitochondria are thought to derive from bacteria; that is, they are descended from free-living prokaryotic organisms that became intracellular symbionts in nucleated host cells. Once incorporated into a nucleated cell, the genome of the mitochondrion became smaller over time, and the reduced genome we now see in mitochondria is thought to be the product of gene loss and gene transfer to the nucleus of the host cell. However, many questions about the origin and evolution of mitochondria remain. Among them is one relating to the bacterial origins of mitochondria: Which group of bacteria are they descended from?

Researchers investigating this question have used information derived from nucleotide sequencing of the mitochondrial genes for small-subunit ribosomal RNAs (SSU rRNA). These sequences are under strong functional constraint and so are highly conserved. Phylogenetic analysis of the sequences identified a group of bacteria known as the α-proteobacteria (purple bacteria) as the closest living relatives of mitochondria. More recent work has divided the α-proteobacteria into two subclasses, one of which is called the rickettsial subdivision.

Based on the SSU phylogeny, the rickettsia have been identified as the bacteria most closely related to mitochondria. Interestingly, rickettsia, like mitochondria, live only inside eukaryotic cells.

Unlike mitochondria, however, rickettsia are disease-causing parasites, responsible for typhus, one of the most serious diseases in the history of our species. Genomic sequencing of one rickettsial species, *Rickettsia prowazekii*, provides additional insight into the relationship between these bacteria and mitochondria. Using the genome sequence information for genes involved in ATP synthesis derived from *R. prowazekii*, other bacteria, and mitochondria from several sources, Siv Andersson and colleagues constructed a phylogenetic tree (Figure 26–25). Their tree indicates a close evolutionary relationship between *R. prowazekii* and mitochondria. Evidence from SSU analysis results in a similar tree. Thus, two lines of evidence, from SSUs and proteins involved in ATP synthesis, provide strong evidence that the mitochondria we carry in our cells derive from an ancient ancestor of the rickettsia.

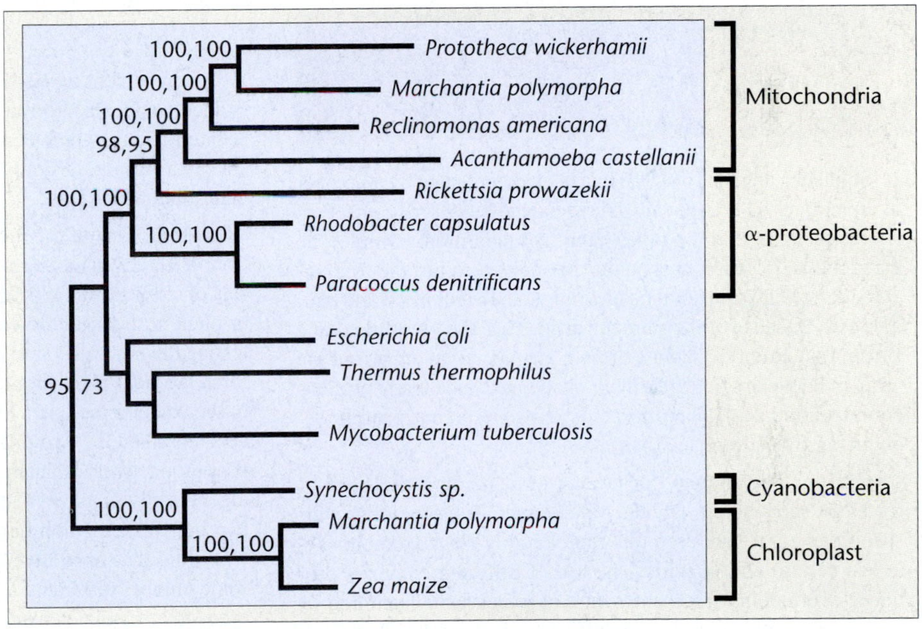

FIGURE 26–25 The evolutionary relationships of mitochondria, α-proteobacteria, cyanobacteria, and chloroplasts. Note that, although *R. prowazekii* is an α-proteobacterium, it is more closely related to mitochondria than to other α-proteobacteria. *(From Anderssen, S.G.E., et al., 1998. The genome sequence of* Rickettsia prowazekii *and the origin of mitochondria.* Nature *396:133–143.)*

Chapter Summary

1. Today's organisms are the products of an evolutionary history that included the transformation, splitting, and divergence of lineages. Alfred Russel Wallace and Charles Darwin formulated the theory of natural selection, which provides a mechanism for the transformation of lineages. The genetic basis of evolution and the role of natural selection in changing allele frequencies were discovered in the 20th century.
2. When population geneticists began studying the genetic structure of populations, they found that most populations harbor considerable genetic diversity, which becomes apparent in electrophoretic studies of proteins and at the level of DNA sequences. Whether the genetic diversity of populations is maintained primarily by mutation plus genetic drift or by natural selection is a matter of some debate.
3. The geographic ranges of most species encompass a degree of environmental diversity. As a result of both adaptation to different environments and genetic drift, different populations within a species may have different alleles or allele frequencies at many loci.
4. Gene flow among populations tends to homogenize their genetic composition. When gene flow is reduced, genetic drift and adaptation to different environments can cause populations to diverge. Eventually, populations may become so different that the individuals in one population either will not or cannot mate with the individuals in the other. At this point, the divergent populations have become different species.
5. Because speciation is associated with genetic divergence, we can use the genetic differences among species to infer their evolutionary history. By comparing amino acid or nucleotide sequences, we can determine the genetic distances among species. We can then use the genetic distances to reconstruct evolutionary trees. The simplest methods for reconstructing phylogenies are based on the assumption that the least divergent species are one another's closest relatives.
6. The reconstruction of evolutionary trees is a key technique in answering a diversity of interesting questions. Examples include tracing the route of transmission of a disease in an epidemic, deciphering recent events in human evolution, and determining the evolutionary relationships among all organisms.

Insights and Solutions

1. Sequence analysis of DNA can be accomplished by a number of techniques. (See Appendix A for a detailed description.) Protein sequencing, on the other hand, is made more complex by the fact that 20 different subunits, as opposed to just the four nucleotides of DNA, need to be unambiguously identified and enumerated. Because of their unique properties, the N-terminal and C-terminal amino acids in a protein are easy to identify, but the array in between offers a difficult challenge, since many proteins contain hundreds of amino acids. How is it that protein sequencing is accomplished?

Solution: The strategy for protein sequencing is the same as for DNA sequencing: Divide and conquer. To accomplish this, specific enzymes are used that reproducibly cleave proteins between certain amino acids. The use of different enzymes produces overlapping fragments, each of which is isolated and its amino acid sequence determined by chemical means. Sequences from overlapping fragments are then assembled to give a sequence for the entire protein. Alternatively, researchers can first sequence the gene that encodes the protein and then use the genetic code to infer the protein's amino acid sequence.

2. A single plant twice the size of others in the same population suddenly appears. Normally, plants of that species reproduce by self-fertilization and by cross-fertilization. Is this new giant plant simply a variant, or could it be a new species? How would you determine which it is?

Solution: One of the most widespread mechanisms of speciation in higher plants is polyploidy, the multiplication of entire sets of chromosomes. The result of polyploidy is usually a larger plant with larger flowers and seeds. There are two ways of testing the new variant to determine whether it is a new species. First, the giant plant should be crossed with a normal-sized plant to see whether the former produces viable, fertile offspring. If it does not, then the two different types of plants would appear to be reproductively isolated. Second, the giant plant should be cytogenetically screened to examine its chromosome complement. If it has twice the number of its normal-sized neighbors, it is a tetraploid that may have arisen spontaneously. If the chromosome number differs by a factor of two and the new plant is reproductively isolated from its normal-sized neighbors, it is a new species.

Problems and Discussion Questions

1. Discuss the rationale behind the statement that inversions in chromosome 3 of *Drosophila pseudoobscura* represent genetic variation.
2. Describe how populations with substantial genetic differences can form. What is the role of natural selection?
3. What types of nucleotide substitutions will not be detected by electrophoretic studies of a gene's protein product?
4. The following data, excerpted from a paper by Heui-Soo Kim and Osamu Takenaka, are short DNA sequences from five species: human, chimpanzee, gorilla, orangutan, and baboon:

```
Human       AGAGGTTTTTCAGTGAATGAAGCTATTTTTAAGGGAGTGTGATTGCTGCC
Chimpanzee                    C
Gorilla                T      C                            C
Orangutan    T        GT   C  C                 C          C
Baboon       C C  G  G T      C        G       GC     C    C G
```

The sequences, each 50 bases long, represent a short piece of a gene for testis-specific protein Y, which is located on the Y chromosome. The complete human sequence is given. The sequences for the other four species are shown only in the spots where they differ from the human sequence. Calculate the genetic difference between each pair of species. For example, chimpanzees and gorillas differ in 2 out of 50 bases, or 4 percent. Thus, the genetic difference between those two species is 0.04. Then use UPGMA to reconstruct the phylogeny for the five species listed. Is your phylogeny consistent with the ones shown in Figure 26–20 and Figure 26–22?

5. Shown in the following diagram are two homologous lengths of the alpha and beta chains of human hemoglobin:

Alpha:	Ala	Val	Ala	His	Val	Asp	Asp	Met	Pro
Beta:	Gly	Leu	Ala	His	Leu	Asp	Asn	Leu	Lys

Consult the genetic code dictionary (Figure 5–7), and determine how many amino acid substitutions may have occurred as a result of a single nucleotide substitution. For any that cannot occur as the result of a single change, determine the minimal mutational distance.

6. Determine the minimal mutational distances between the following amino acid sequences of cytochrome c from various organisms:

Human:	Lys	Glu	Glu	Arg	Ala	Asp
Horse:	Lys	Thr	Glu	Arg	Glu	Asp
Pig:	Lys	Gly	Glu	Arg	Glu	Asp
Dog:	Thr	Gly	Glu	Arg	Glu	Asp
Chicken:	Lys	Ser	Glu	Arg	Val	Asp
Bullfrog:	Lys	Gly	Glu	Arg	Glu	Asp
Fungus:	Ala	Lys	Asp	Arg	Asn	Asp

Compare the distance between humans and each organism.

7. The genetic difference between two species of *Drosophila, D. heteroneura* and *D. sylvestris*, as measured by nucleotide diversity, is about 1.8 percent. The difference between chimpanzees (*Pan troglodytes*) and humans (*Homo sapiens*) is about the same, yet the latter species are classified in different genera. In your opinion, is this valid? If so, why; if not, why not?

8. As an extension of Problem 7, consider the following: In sorting out the complex taxonomic relationships among birds, species with $\Delta T_{50}H$ values of 4.0 are placed in the same genus, even by traditional taxonomy based on morphology. From the data in Figure 26–21, construct a classification that obeys this rule. Use any of the appropriate genus names (*Pongo, Pan, Homo*), or create new ones.

9. The use of nucleotide-sequence data to measure genetic variability is complicated by the fact that the genes of higher eukaryotes are complex in organization and contain 5′ and 3′ flanking regions as well as introns. In a comparison of the nucleotide sequences of two cloned alleles of the γ-globin gene from a single individual, Slightom and colleagues found a variation of 1 percent. Those differences include 13 substitutions of one nucleotide for another and three short DNA segments inserted into one allele or deleted from the other. None of the changes take place in the exons (coding regions) of the gene. Why do you think this is so, and should it change the concept of genetic variation?

10. Discuss the arguments supporting the neutralist hypothesis. What counterarguments are proposed by selectionists?

11. Of what value to our understanding of genetic variation and evolution is the debate concerning the neutralist hypothesis?

Extra-Spicy Problems

12. Native tribes of North America, as well as those of South America, are descendants of one or a few groups of Mongoloid peoples who immigrated from the Asian landmass sometime between 11,000 and 40,000 years ago.

The HLA genes in humans code for the histocompatibility antigens found on the surfaces of most cells. These genes are the most polymorphic genes known in humans. For example, over 40 alleles of the HLA B gene have been identified. In contrast to Old World populations, however, native Americans do not show this phenomenal allelic diversity. Instead, a limited set of alleles of all of the HLA genes are found in varying frequencies in all tribes, regardless of the location of the tribe. What are the various forces that determine allele frequencies? What is the best explanation for these observations?

Another interesting observation about the HLA alleles of native Americans is that South American tribes have alleles of the HLA B gene that are not found in the present-day Asian population. Provide two possible explanations for this observation.

13. Nauru is a remote Pacific atoll occupied by 5000 Micronesians. Colonization by Britain, Australia, and New Zealand, as well as income from phosphate mining, has changed the lifestyle of the Nauruans. Although these people formerly depended on fishing and subsistence farming and had an active lifestyle, food is now imported and high in energy content, and obesity occurs at a high frequency in the population. The multifactorial disease non-insulin-dependent diabetes mellitus (NIDDM) results from a combination of genetic and environmental influences. This disease used to be nonexistent on this island, but after 1950, the prevalence of the disease increased from 1 percent in the 1950s to 21 percent in the mid-1970s and then dropped to 9 percent in 1987. At one point in the 1970s, a severe form striking many young adults reached epidemic proportions in that population. Diabetic women had more stillbirths and less than half as many live births as nondiabetic women. Although the NIDDM epidemic on Nauru has passed its peak, the drop-off cannot be attributed to a decline in environmental risk factors, since the Nauruans lifestyle has not changed since the 1970s.

Consider the foregoing data carefully, and propose an explanation for the increase and then subsequent decrease in the incidence of NIDDM in the young-adult Nauruan population.

Selected Readings

Adcock, G.J., Dennis, E.S., Easteal, S., Huttley, G.A., Jermiin, L.S., Peacock, W.J., and Thorne, A. 2001. Mitochondrial DNA sequences in ancient Australians: Implications for modern human origins. *Proc. Nat. Acad. Sci.* 98: 537–42.

Anderson, W., et al. 1975. Genetics of natural populations: XLII. Three decades of genetic change in *Drosophila pseudoobscura. Evolution* 29:24–36.

Armour, J.A., et al. 1996. Minisatellite diversity supports a recent African origin for modern humans. *Nat. Genet.* 13:154–60.

Avise, J.C. 1990. Flocks of African fishes. *Nature* 347:512–13.

Ayala, F.J., 1984. Molecular polymorphism: How much is there, and why is there so much? *Dev. Genet.* 4:379–91.

Bradshaw, H.D., Jr., et al. 1998. Quantitative trait loci affecting differences in floral morphology between two species of monkeyflower (*Mimulus*). *Genetics* 149:367–82.

Collard, M., and Wood, B. 2000. How reliable are human phylogenetic hypotheses? *Proc. Nat. Acad. Sci.* 97:5003–6.

Cavelli-Sforza, L.I. 1998. The DNA revolution in population genetics. *Trends in Genet.*. 14:60–65.

Diamond, J. 1992. *The third chimpanzee: The evolution and future of the human animal.* New York: HarperCollins.

Dobzhansky, T. 1947. Adaptive changes induced by natural selection in wild populations of *Drosophila. Evolution* 1:1–16.

———. 1948. Genetics of natural populations, XVI. Altitudinal and seasonal changes produced by natural selection in certain populations of *Drosophila pseudoobscura* and *Drosophila persimilis. Genetics* 33:158–76.

———. 1955. *Genetics of the evolutionary process.* New York: Columbia University Press.

Dobzhansky, T., et al. 1966. Genetics of natural populations: XXXVIII. Continuity and change in populations of *Drosophila pseudoobscura* in western United States. *Evolution* 20:418–27.

Fitch, W.M. 1973. Aspects of molecular evolution. *Annu. Rev. Genet.* 7:343–80.

Fitch, W.M., and Margoliash, E. 1967. Construction of phylogenetic trees. *Science* 155:279–84.

———. 1970. The usefulness of amino acid and nucleotide sequences in evolutionary studies. *Evol. Biol.* 4:67–109.

Freeman, S. , and Herron, J.C. 2001. *Evolutionary analysis.* Upper Saddle River, NJ: Prentice Hall.

Fryer, G. 1997. Biological implications of a suggested Late Pleistocene desiccation of Lake Victoria. *Hydrobiologia* 354:177–82.

Gray, I.C., Campbell, D.A., and Spurr, N.K. 2000. Single nucleotide polymorphisms as tool in human genetics. *Hum. Mol. Genet.* 9:2403–08

Gray, M.W., et al. 1999. Mitochondrial evolution. *Science* 283:1476–81.

Gould, S.J. 1982. Darwinism and the expansion of evolutionary theory. *Science* 216:380–87.

Hardison, R. 1999. The evolution of hemoglobin. *Amer. Scient.* 87:126–37.

Hillis, D.M. 1998. Phylogenetic analysis. *Curr. Biol.* 7:R129–31.

Hillis, D.M., Huelsenbeck, J.P., and Cunningham, C.W. 1994. Application and accuracy of molecular phylogenies. *Science* 264:671–77.

Höss, M. 2000. Neanderthal population genetics. *Nature* 404:453–54.

Hunt, J., et al. 1981. Evolution distance in Hawaiian *Drosophila. J. Mol. Evol.* 17:361–67.

International SNP Map Working Group. 2001. A map of human genome sequence variation containing 1.42 million single nucleotide polymorphisms. *Nature* 409:928–33.

Ki, Y., Su, B., Song, X., Lu, D., Chen, L., Li, H., Qi, C., Marzuki, S., et. al. 2001. African origin of modern humans in East Asia: A tale of 12,000 Y chromosomes. *Science* 292: 1151–53.

Kimura, M. 1979a. Model of effectively neutral mutations in which selective constraint is incorporated. *Proc. Natl. Acad. Sci. USA* 76:3440–44.

———. 1979b. The neutral theory of molecular evolution. *Sci. Am.* (Nov.) 241:98–126.

———. 1989. The neutral theory of molecular evolution and the world view of the neutralists. *Genome* 31:24–31.

King, M.C., and Wilson, A.C. 1975. Evolution at two levels: Molecular similarities and biological differences between humans and chimpanzees. *Science* 188:107–16.

Knowlton, N., et al. 1993. Divergence in proteins, mitochondrial DNA, and reproductive compatibility across the Isthmus of Panama. *Science* 260: 1629–32.

Kreitman, M. 1983. Nucleotide polymorphism at the alcohol dehydrogenase locus of *Drosophila melanogaster. Nature* 304:412–17.

Krings, M., et al. 1997. Neandertal DNA sequences and the origin of modern humans. *Cell* 90:19–30.

Lewontin, R.C., and Hubby, J.L. 1966. A molecular approach to the study of genic heterozygosity in natural populations: II. Amount of variation and degree of heterozygosity in natural populations of *Drosophila pseudoobscura. Genetics* 54:595–609.

Meyer, A., Kocher, T.D., Basasibwaki, P., and Wilson, A.C. 1990. Monophyletic origin of Lake Victoria cichlid fishes suggested by mitochondrial DNA sequences. *Nature* 347:550–53.

Müller, M., and Martin, W. 1999. The genome of *Rickettsia prowazekii* and some thoughts on the origin of mitochondria and hydrogenosomes. *Bioessays* 21:377–81.

Ou, C.-Y., et al. 1992. Molecular epidemiology of HIV transmission in a dental practice. *Science* 256:1165–71.

Ovchinnikov, I. V., et al. 2000. Molecular analysis of Neanderthal DNA from the northern Caucasus. *Nature* 404:490–493.

Pääbo, S. 1993. Ancient DNA. *Sci. Am.* (Nov.) 269:86–92.

Powers, D.A., and Schulte, P.M. 1998. Evolutionary adaptations of gene structure and expression in natural populations in relation to a changing environment: A multidisciplinary approach to address the million-year saga of a small fish. *J. Exper. Zool.* 282:71–94.

Relethford, J.H. 2001. Ancient DNA and the origin of modern humans. *Proc. Nat. Acad. Sci.* 98: 390–91.

Schemske, D.W., and Bradshaw, H.D., Jr. 1999. Pollinator preference and the evolution of floral traits in monkeyflowers. *Proc. Nat. Acad. Sci.* 96:11910–15.

Scholz, M., et al. 2000. Genomic differentiation of Neanderthals and anatomically modern man allows a fossil-DNA-based classification of morphologically indistinguishable hominid bones. *Am. J. Hum. Genet.* 66:1927–32.

Sibley, C., and Ahlquist, J. 1984. The phylogeny of the hominoid primates, as indicated by DNA–DNA hybridization. *J. Mol. Evol.* 20:2–15.

GENETICS MediaLab

The resources that follow will help you achieve a better understanding of the concepts presented in this chapter. These resources can be found either on the CD packaged with this textbook or on the Companion Web site found at **http://www.prenhall.com/klug**

CD Rsources:

Self-grading Chapter Problems

Web Resources

Web Destinations in Genetics

Self-grading Chapter Resources

Chapter Search Terms

Genetics Newsgourps

Student Bulletin Board

Web Problem 1:

Time for completion = 10 minutes

What are the genetic changes that take place during speciation? John Werren studies speciation in wasps of the genus Nasonia. After reading about the genetics of speciation among these parasitoids, you should be able to answer the following questions: Which species are allopatric and which are sympatric? What are the isolating mechanisms between *Nasonia vitripennis* and *Nasonia giraulti*? Why does haplodiploidy facilitate the study of differences among these species? Are there interactions between nuclear or cytoplasmic factors that reduce the fitness of hybrids? To complete this exercise, visit Web Problem 1 in Chapter 26 of your Companion Web site, and select the keyword **SPECIATION**.

Web Problem 2:

Time for completion = 10 minutes

Why is speciation so common following the geographic isolation of populations? How important is the role of genetic drift? Use the "EvoTutor" simulation to demonstrate how reducing gene flow can lead to dramatic divergence in a population. Read the introduction, click on Simulation, and then click Run. The simulation begins with an initial population of interbreeding colored dots, with different colors representing genetic variation. At a given point in time, the population is divided and gene flow stops. Eventually, the two populations reconnect and are free to interbreed, unless they have diverged substantially enough to create intrinsic isolating mechanisms. Change the parameters for the initial population size (*N*) and time of geographic separation (*G*), rerun the simulation, and answer the following questions: Which conditions led to the quickest genetic divergence? What was the relative importance of population size versus time of geographic separation? If the initial population size was large enough, did you see periods of divergence followed by gene flow, which homogenized the genetic divergence? To complete this exercise, visit Web Problem 2 in Chapter 26 of your Companion Web site, and select the keyword **EVO TUTOR**.

Web Problem 3:

Time for completion = 15 minutes

How can we determine the pattern of evolutionary relationships among a group of species? The study of phylogenetics has benefited from the increased availability of molecular sequence data. Follow the tutorial on the reconstruction of phylogenies from molecular sequence data, and then answer the following questions. Why do all phylogenetic trees eventually converge on a common ancestor? What is the relationship between mutation and time of divergence? Are all mutations equally useful to estimating phylogenies? What processes might cause different rates of evolution for "essential" vs. "irrelevant" mutations? To complete this exercise, visit Web Problem 3 in Chapter 26 of your Companion Web site, and select the keyword **PHYLOGENY**.

Sibley, C.G., Comstock, J.A., and Ahlquist, J.E. 1990. DNA evidence of hominoid phylogeny: A reanalysis of the data. *J. Mol. Evol.* 30:202–36.

Stiassny, M.L.J., and Meyer, A. 1999. Cichlids of the Rift Lakes. *Sci. Am.* (Feb.) 280:64–69.

Stoneking, M. 1995. Ancient DNA: How do you know when you have it and what can you do with it? *Am. J. Hum. Genet.* 57:1259–62.

Takahashi, K., et al. 1998. A novel family of short interspersed repetitive elements (SINEs) from cichlids: The pattern of insertion of SINEs at orthologous loci support the proposed monophyly of four major groups of cichlid fishes in Lake Tanganyika. *Mol. Biol. Evol.* 15:391–407.

Thorne, A.G., and Wolpoff, M.H. 1992. The multiregional evolution of humans. *Sci. Am.* (Apr.) 266:76–83.

Val, F.C. 1977. Genetic analysis of the morphological differences between two interfertile species of Hawaiian *Drosophila. Evolution* 31:611–29.

Weiss, K.M. 1998. In search of human variation. *Genome Res.*. 8:691–97.

Wilson, A.C., and Cann R.L. 1992. The recent African genesis of humans. *Sci. Am.* (Apr.) 266:68–73.

Humpback whales, the "poster children" of endangered species. (*James Watt/Animals/Animals*)

27
Conservation Genetics

KEY TERMS

As we begin the 21st century, the diversity of life on the earth is under increasing pressure from the direct and indirect effects of explosive human population growth. Approximately 10 million *Homo sapiens* lived on the planet 10,000 years ago. This number grew to 100 million 2000 years ago and to 2.5 billion by 1950. Within the span of a single lifetime, the world's human population more than doubled to 5.5 billion in 1993 and is projected to reach as high as 19 billion by 2100 (Figure 27–1).

The effect of accelerating human population growth on other species has been dramatic. Data from the World Conservation Union (IUCN) show that, globally, 25 percent of all mammal species, 11 percent of birds, 20 percent of reptiles, 25 percent of amphibians, and 34 percent of fish species are vulnerable or endangered. The situation for plants is no better. Based on IUCN surveys, 12 percent of all vascular plant species worldwide—some 34,000 species—are threatened. Not just wild species are at risk. Genetic diversity in domesticated plants and animals is also being lost as many traditional crop varieties and livestock breeds disappear. The Food and Agriculture Organization (FAO) estimates that, since 1900, 75 percent of the genetic diversity in agricultural crops has been lost; also, out of approximately 5000 different breeds of domesticated farm animals worldwide, one-third are at risk of being lost.

Why should we be concerned about losing biodiversity—that is, the biological variation represented by these different plants and animals? As the fossil record shows, unrelated to human influences, many different plants and animals that once inhabited the planet have become extinct over millions of years, and others have taken their places. Biologists are concerned, however, at the accelerated rate of species extinctions we are witnessing today, all of which can be ascribed to direct or indirect human impacts. Deliberate hunting or harvesting of plants and animals by humans, habitat destruction through human development activities, and the indirect effects of global climate change are making it increasingly difficult for many species to survive.

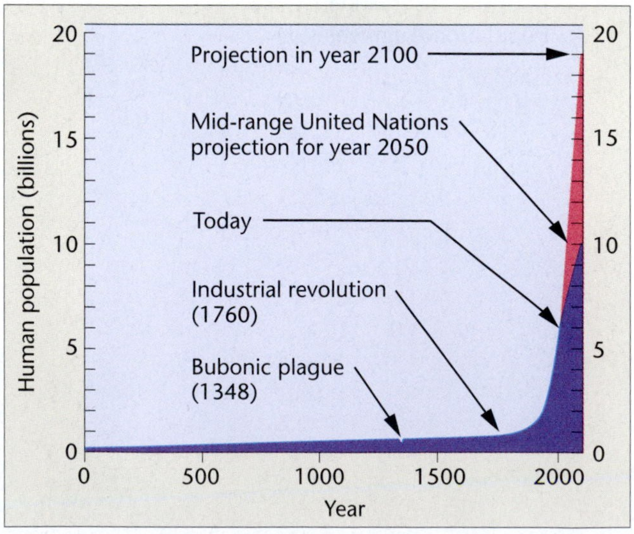

FIGURE 27–1 Growth in human population over the past 2000 years.

Some biologists fear that ecosystems—the complex webs of interdependence of different plants and animals found together in the same environment—may collapse if key sustaining species are lost. This portends possible consequences for our own long-term survival. Other scientists have pointed out that we may be losing the unknown economic potential or other benefits of unexploited plants and animals if we allow them to go extinct. A much-publicized, recent example is the Pacific yew (*Taxus brevifolia*), a rare tree found in coastal forests of the western United States. This slow-growing species was ignored due to its small size and poor timber qualities and was frequently destroyed during logging operations as a "trash tree." In 1991, it was found to be a source of taxol, a compound now widely used as a powerful drug for treating cancer. Quite apart from practical benefits, scientists and nonscientists have made the case that human life will be diminished both spiritually and aesthetically if we do not strive to maintain the fascinating and often beautiful variety of living organisms with which we share the planet.

Conservation biologists work to understand and maintain biodiversity, studying the factors that lead to species decline and the ways in which species can be preserved. The new field of **conservation genetics** has emerged in the last 20 years as scientists have begun to recognize that genetics will be an important tool in maintaining and restoring population viability. The applications of genetics to conservation biology are multifaceted; in this chapter, we explore but a few. Underlying the increasingly important role of genetics in conservation biology, however, is the recognition that biodiversity depends on genetic diversity and that maintaining biodiversity in the long term is unlikely if genetic diversity is lost.

27.1 Genetic Diversity Is at the Heart of Conservation Genetics

While biodiversity encompasses the variation represented by all existing species of plants and animals on this planet at any given time, **genetic diversity** is not as easy to define and study. Genetic diversity can be considered on two levels: **interspecific diversity** and **intraspecific diversity**. Diversity between species (interspecific diversity) is reflected in the number of different plant and animal species present in an ecosystem.

Some ecosystems have a very high level of species diversity, such as a tropical rainforest in which hundreds of different plant and animal species may be found within a few square meters [Figure 27–2(a)]. Other ecosystems, especially where plants and animals must adapt to a harsh environment, may have much lower levels of interspecific diversity [Figure 27–2(b)]. Lists of the different plant and animal species found in a particular environment are used to compile species inventories. Such inventories are used to identify diversity hotspots: geographic areas with especially high levels of interspecific diversity, where conservation efforts can be focused. Conservation biologists working at the ecosystem level are interested in preventing species from being lost and in restoring species that were once part of the system, but

(a) (b)

FIGURE 27-2 (a) Tropical rainforest: an ecosystem with high interspecific diversity. (b) coastal marsh in North Carolina: an ecosystem with low interspecific diversity. *[Photos: (a) David Austen/ Stone; (b) Sarah M. Ward]*

that are no longer present. The reestablishment of the grey wolf (*Canis lupus*) in Yellowstone National Park and the release of captive-bred California condors (*Gymnogyps californianus*) into the mountain ranges they previously occupied in the southwest United States are examples of current attempts by conservation biologists to restore missing species to their ecosystems.

Intraspecific diversity (diversity within a species) is reflected in the level of genetic variation occurring between individuals within a single population of a given species (intrapopulation diversity) or between different populations of the same species (interpopulation diversity). Genetic variation within populations can be measured as the frequency of individuals in the population that are heterozygous at a given locus, or as the number of different alleles at a locus that are present in the population gene pool. When DNA-profiling techniques are used, the percentage of polymorphic loci—those that are represented by varying bands on the DNA profile in different individuals—can be calculated to indicate the extent of genetic diversity in a population. In outbreeding species, most intraspecific genetic diversity is found at the intrapopulation level.

Significant interpopulation diversity can occur if populations are separated geographically and there is no migration or exchange of gametes between them. On the other hand, predominantly inbreeding species (such as self-fertilizing plants) tend to have greater levels of interpopulation than intrapopulation diversity, with a limited number of genotypes dominating an individual population, but greater variation between different populations. Understanding the breeding system and the distribution of genetic variation in an endangered species is important to its conservation, not only to ensure continued production of offspring, but also to determine the best strategy for maintaining intraspecific diversity. For example, would it be more effective to preserve a few large populations or many small, distinct ones? This information can then be used to guide conservation or restoration efforts.

Loss of Genetic Diversity

Loss of genetic diversity in nondomesticated species is usually associated with a reduction in population size. This may be due to excessive hunting or harvesting; for example, biologists blame commercial overfishing for the collapse in the early 1990s of the deep-sea-cod populations off the Newfoundland coast—once one of the most productive fisheries in the world. Habitat loss is also a major cause of population decline. As the global human population increases, more land is developed for housing and transport systems or is put into agricultural production, reducing or eliminating areas that were once home to wild plants and animals. The shrinking available habitat not only reduces populations of wild species, but often also isolates them from each other as individual populations become trapped in pockets of undeveloped land surrounded by areas taken over for agriculture, urban development, or other human uses. This process is known as **population fragmentation**. When populations are no longer in contact with each other, gene flow through migration or gamete exchange between them ceases, and an important mechanism for maintaining genetic variation is lost. We discuss this in more detail later in the chapter.

Loss of genetic diversity in domesticated species is not usually the result of habitat loss or collapsing population numbers; there is little risk that cows or corn as species will become extinct any time soon. Reduction in diversity within domesticated species can be traced to changes in agricultural practice and consumer demand. Modern farming techniques have greatly increased production levels, but they have also led to greater genetic uniformity. As farmers switch to new crop varieties or improved livestock strains on a large scale, they abandon cultivation of many older local types, which may then disappear if efforts are not made to preserve them.

For example, in 1900, more than 100 different kinds of potato were available in the United States. Today, three-quarters of commercial potato production in this country depends on just 9 varieties, with a single type, Russet Burbank, making up 43 percent of the total acreage planted. Modern crop varieties or livestock strains may be better adapted to meet the demands of modern agricultural production, but older types often contain useful genes that can still play a vital role in survival functions, such as resistance to disease, cold, or drought. When the Russian wheat aphid (*Diuraphis noxia*), an insect that causes serious damage to cereal crops, invaded the United States in 1986, plant breeders eventually found genes conferring resistance to this pest—not in modern American wheat lines, but in old traditional varieties from the former Soviet Union, which fortunately had been collected and preserved. The old Russian wheats were crossed with U.S. wheat plants to transfer the resistance genes, creating new commercial wheat varieties that resist Russian wheat aphid attack and saving U.S. wheat growers from crop losses in the millions of dollars.

Identifying Genetic Diversity

For many years, population geneticists based estimates of intraspecific diversity on phenotypic differences between individuals, such as different colors of seeds or flowers, or variation in markings (Figure 27–3). More recently, techniques have been developed for analyzing intraspecific diversity with greater precision at the molecular level. The more traditional method of the techniques used is isozyme analysis. Isozymes are multiple versions of a single enzyme occurring within a single species. Isozymes catalyze the same reaction, but differences in the polypeptide chains forming the bulk of the molecule result in various forms of the enzyme that differ in size or net electrical charge and can be separated by electrophoresis. The presence of isozyme variation in a population can be used as an indicator of genetic variation, as the different versions of the molecule are encoded by different alleles at a locus.

The use of DNA profiles has now become the dominant tool for detecting and assessing intrapopulation and interpopulation genetic variation between individuals drawn from a population. Nuclear, mitochondrial, and chloroplast DNA can all be analyzed to determine levels of genetic variation. Earlier, we described applications in human genetics of DNA-fingerprinting techniques, using either restriction enzymes or polymerase chain reaction (PCR) (See Chapter 18). These techniques have been used with DNA from many different species that are of concern to conservation biologists.

For example, amplified fragment length polymorphisms (AFLPs) are detected by a recently developed DNA-profiling technique, using both restriction enzymes and PCR (Figure 27–4). This method is especially powerful at detecting very small amounts of genetic diversity in a population and is therefore valuable for work with threatened and endangered species. AFLPs are generated by first cutting genomic DNA into fragments by using two restriction enzymes, then attaching short single-stranded pieces of DNA to the cut ends of the fragments. Because the sequence of the attached DNA oligonucleotide is known, a subset of the genomic fragments can then be amplified through PCR by using primers that match the attached piece, but are 1–3 bases longer. Only those fragments that have bases at each end complementary to the oligonucleotide extension will be amplified. AFLPs can thus produce different fingerprint patterns for individuals whose DNA sequence differs by as little as a single base pair.

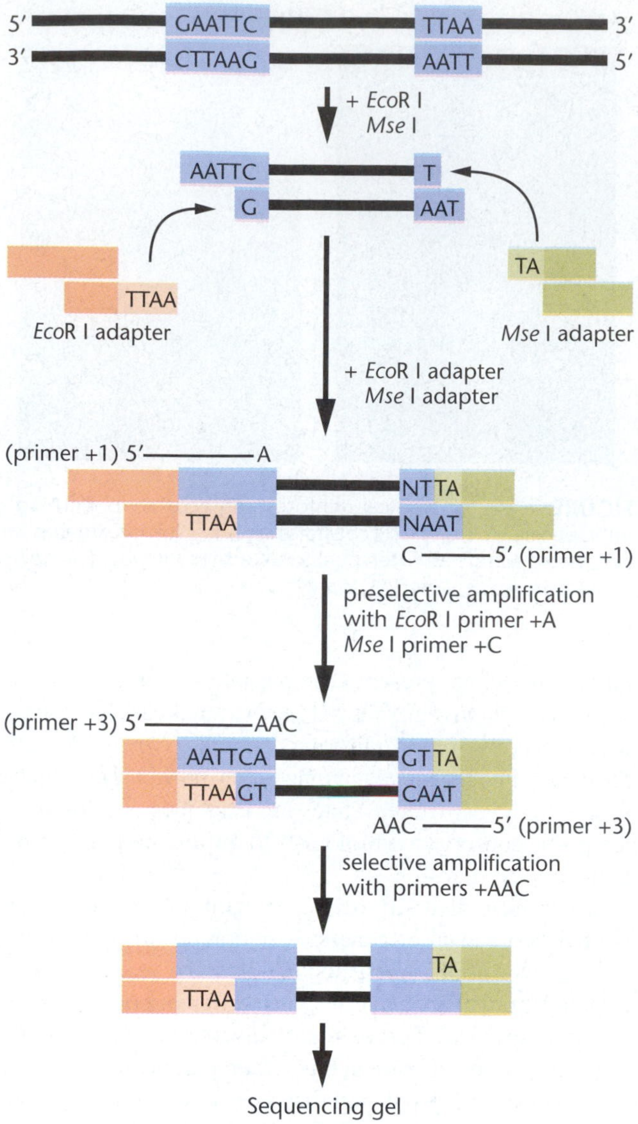

FIGURE 27–4 The procedure used in preparing AFLP-profiling samples.

AFLP profiling (Figure 27–5) has been used extensively to analyze genetic variation in crop species and some livestock and is now being used to examine variation and guide conservation efforts in nondomesticated species. In a recent study, AFLP analysis was performed on the only remaining population of a critically endangered plant species (*Limonium cavanillesii*) growing on the Mediterranean coast of Spain. Researchers found very low levels of diversity, but did detect genetically distinct individual plants within the population from which seed could be collected.

In another recent study, AFLP analysis of the southwestern willow flycatcher, an endangered migratory bird species that nests in the southwestern United States, found significant levels of genetic diversity within populations at different breeding sites, but little interpopulation diversity. This indicated to scientists studying the flycatcher that migration of

FIGURE 27–3 Phenotypic variation in seed color and markings in common bean (*Phaseolus vulgaris*) reveals high levels of intraspecific diversity. *(Sarah M. Ward)*

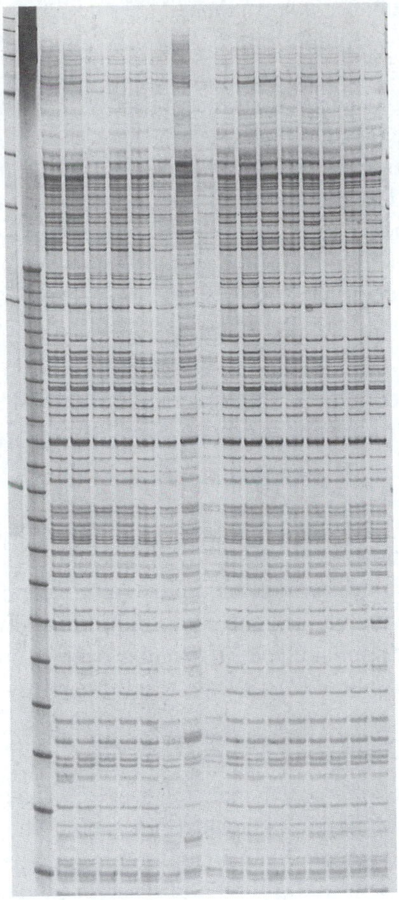

FIGURE 27–5 A representative AFLP gel used in DNA profiling, derived from analysis of jointed goat grass. *(Gel image courtesy of Todd A. Pester and Sarah M. Ward, Colorado State University)*

individual birds between breeding populations is important in maintaining diversity in the species, and conservation efforts should be directed to preserving nesting sites that allow such migration to continue.

We read in the Palo Verde story in Chapter 16 (see the "Genetics, Technology, and Society" essay) about the use of DNA fingerprinting of plants to help solve a murder. Forensic applications of DNA fingerprinting are also used in conservation biology to uncover illegal trade in endangered species that are protected by law. Since 1986, an international moratorium on commercial whaling has been in place, with only limited hunting of a few species permitted. In 1994, scientists based in New Zealand and Hawaii used PCR to amplify mitochondrial DNA sequences extracted from meat on sale in markets in Japan, where whale meat is prized as a delicacy. The DNA fingerprints revealed several meat samples from humpback and fin whales, which are protected species. Two meat samples turned out not to be whale at all, but dolphin! This information assisted Japanese customs officials in tracking down sources of illegal whale-meat imports. Unlawful trade in animal products also occurs in the United States. In a recent study by scientists at the University of Florida, amplification of mitochondrial DNA sequences from turtle meat sold in Louisiana and Florida revealed that one-quarter of the samples were actually alligator, not turtle at all. Most of the rest were from small freshwater turtles, indicating that populations of the larger freshwater-turtle species were declining through overharvesting and in need of protection.

27.2 Population Size Has a Major Impact on Species Survival

Some species have never been numerous, especially those that are adapted to survive in unusual habitats; biologists refer to such species as *naturally rare*. "*Newly rare*" species, on the other hand, are those whose numbers are in decline due to pressures such as habitat loss. Populations of such species may not only be small in number, but also fragmented and isolated from other populations. Both decreased population size and increased population isolation have important genetic consequences, leading to an increased risk of loss of diversity.

How small must a population be before it is considered endangered? It varies somewhat with species, but small populations can quickly become vulnerable to genetic phenomena that increase the risk of extinction. In general, a population of less than 100 individuals is considered extremely sensitive to these problems, including genetic drift, inbreeding, and reduction in gene flow. The effects of such problems on species survival are substantial, and we discuss them in more detail next. Studies of bighorn sheep, for example, have shown that populations of less than 50 are highly likely to go extinct within 50 years. Projections based on computer models show that, for all species, populations of fewer than 10,000 are likely to be limited in adaptive genetic variation, and at least 100,000 individuals must be present if a population is to show long-term sustainability.

Determining the number of individuals a population must contain for it to have long-term sustainability is complicated by the fact that not all members of a population are equally likely to produce offspring: Some will be infertile, too young, or too old. The effective population size (N_e) is defined as the number of individuals in a population having an equal probability of contributing gametes to the next generation. N_e is almost always smaller than the absolute population size (N). The effective population size can be calculated in different ways, depending on the factors that are preventing all individuals in a population from contributing equally to the next generation. In a sexually reproducing population that contains different numbers of males and females, for example, the effective population size is calculated as

$$N_e = \frac{4(N_m N_f)}{N_m + N_f}$$

where N_m is the number of males and N_f the number of females in the population.

A population of 100 males and 100 females would have an effective size of $4(100 \times 100)/(100 + 100) = 200$. In contrast, if there were 180 males and only 20 females, the effective population size would be $4(180 \times 20)/(180 + 20) = 72$.

Effective population size is also affected by fluctuations in absolute population size from one generation to the next. Here, the effective population size is the harmonic mean of the numbers in each generation, so

$$N_e = \frac{1}{\frac{1}{t}\left(\frac{1}{N_1} + \frac{1}{N_2 + \ldots + N_t}\right)}$$

where t = the total number of generations being considered. For example, if a population went through a temporary reduction in size in generation 2, so that $N_1 = 100$, $N_2 = 10$, and $N_3 = 100$, then

$$N_e = \frac{1}{\frac{1}{3}\left(\frac{1}{100} + \frac{1}{10} + \frac{1}{100}\right)} = \frac{1}{0.04} = 25$$

In this case, although the mean actual number of individuals in the population over three generations was 70, the effective population size during that time was only 25. A severe temporary reduction in size such as this is known as a population bottleneck. Bottlenecks occur when a population or species is reduced to a few reproducing individuals whose offspring then increase in numbers over subsequent generations to reestablish the population. Although the number of individuals may be restored to healthier levels, genetic diversity in the newly expanded population is often severely reduced, as gametes from the handful of surviving individuals functioning as parents do not represent all the different allelic frequencies present in the original gene pool.

Captive breeding programs, in which a few surviving individuals from an endangered species are removed from the wild and their offspring raised in a protected environment to rebuild the population, inevitably create population bottlenecks. Bottlenecks also occur naturally when a small number of individuals from one population migrate to establish a new population elsewhere. When a new population derived from a small subset of individuals has significantly less genetic diversity than the original population, it is said to be exhibiting a founder effect. Reduced levels of genetic diversity due to a founder effect can persist for many generations, as shown by studies of two species of Antarctic fur seal (*Arctocephalus gazella* and *Arctocephalus tropicalis*). Seal hunters in the 18th and 19th centuries severely reduced populations of these species, eliminating them from parts of their natural range in the now Southern Ocean.* Although the number of Antarctic fur seals has now rebounded and the two species have recolonized much of their original habitat, mtDNA fingerprinting reveals that a founder effect can still be detected in *A. gazella*, with reduced genetic variation in current populations that are descended from a handful of surviving individuals.

*The International Hydrographic Organization decided to add a fifth world ocean from the southern portions of the Atlantic Ocean, Indian Ocean, and Pacific Ocean and named it the Southern Ocean.

The cheetah (*Acinonyx jubatus*) (Figure 27–6) is a another well-studied example of a species with reduced genetic variation as the result of at least one severe population bottleneck having occurred in its recent history. Isozyme studies of South African cheetah populations have shown levels of genetic variation less than 10 percent of those found in other mammals. The abnormal spermatozoa and poor reproductive rates commonly observed in cheetahs are thought to be linked to the lack of genetic diversity, although when and how the population bottleneck occurred in this species is still unclear.

27.3 Genetic Effects Are More Pronounced in Small, Isolated Populations

Small isolated populations, such as those found in threatened and endangered species, are especially vulnerable to genetic drift, inbreeding, and reduction in gene flow. These phenomena act on the gene pool in different ways, but ultimately have similar effects in that they can all further reduce genetic diversity and long-term species viability.

Genetic Drift

If the number of breeding individuals in a population is small, fewer gametes will form the next generation. The alleles carried by those gametes may not be a representative sample of all those present in the population; purely by chance, some alleles may be underrepresented or not present at all, which will result in changes in allele frequency over time. This phenomenon, known as **genetic drift**, has been previously described in Chapter 25.

A serious result of genetic drift in populations with a small effective population size is the loss of genetic variation. Genetic drift is a random process, so both deleterious and advantageous alleles can become fixed within a small population. This means a useful allele can be lost even if it has the potential to increase fitness or long-term adaptability. If we consider a single locus with two alleles, *A* and *a*, genetic drift

FIGURE 27–6 The cheetah (*Acinonyx jubatus*), a species with reduced genetic variation following population bottlenecks. (*Johnny Johnson/ DRK Photo*)

may result in one of the alleles eventually disappearing, while the other becomes fixed—in other words, it becomes the only version of that gene present in the gene pool of the population. A simple correlation describes the likelihood of the fixation or loss of an allele. The probability that an allele will be fixed through drift is the same as its initial frequency. So, if $p(A) = 0.8$, the probability of A being fixed is 0.8 or 80 percent, and the probability of A being lost through drift is $(1 - 0.8) = 0.2$ or 20 percent. The theoretical effects of genetic drift are examined in Figure 27–7.

Inbreeding

In small populations, the chance of **inbreeding** (matings between closely related individuals) is increased. As described in Chapter 25, inbreeding increases the proportion of homozygotes in a population, thus increasing the possibility that an individual may be homozygous for a deleterious allele. The **inbreeding coefficient** (F) measures the extent of inbreeding occurring in a population; F measures the probability that two alleles of a given gene in an individual are derived from a common ancestral allele. The inbreeding coefficient is inversely related to the frequency of heterozygotes in the population, and can be calculated as

$$F = \frac{2pq - H}{2pq}$$

where $2pq$ is the expected frequency of heterozygotes based on the Hardy–Weinberg principle, and H is the actual frequency of heterozygotes in the population. In a declining population that has become small enough for drift to occur, heterozygosity (H) will decrease with each generation. The smaller the effective population size, the more rapid the decrease in H and the resulting increase in F, as can be seen from the equation

$$H_t = (1 - \tfrac{1}{2}Ne)^t H_0$$

where H_0 = initial frequency of heterozyogtes, H_t = frequency of heterozygotes after t generations, and N_e = effective population size. Figure 27–8 compares rates of increase in F for different effective population sizes.

What effect does inbreeding have on the long-term survival of a population? In some cases, inbreeding may not immediately reduce the amount of genetic variation present in the overall gene pool of a species in which numbers of individuals remain high. Self-pollinating plants, for example, often show high levels of homozygosity and relatively little genetic variation within a single population. But they do tend to have considerable variation between different populations, each of which has adapted to slightly different local environmental conditions. On the other hand, in most outcrossing species, including all mammals, inbreeding is associated with reduced fitness and lower survival rates among offspring. This is referred to as **inbreeding depression**, which can result from increased homozyosity for deleterious alleles. The number of deleterious alleles present in the gene pool of a population is referred to as the **genetic load** (or genetic burden).

In some species, inbreeding accompanied by selection against less fit individuals homozygous for deleterious alleles has resulted in the elimination of these alleles from the gene pool, a process known as *purging the genetic load.* Species that have successfully purged their genetic load do not show continued reduction in fitness, even with many generations of inbreeding. This is true of many domesticated species, especially self-pollinating plants, such as wheat. However, computer simulation experiments have shown that it may take 50 generations or more to complete the purging process, during which time inbreeding depression will still occur.

Alternatively, inbreeding depression can result from heterozygous individuals having a higher level of fitness than

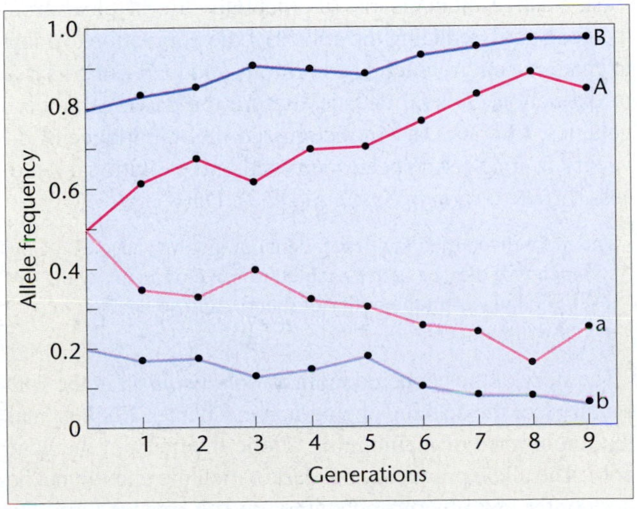

FIGURE 27–7 Change in frequencies over 10 generations for two sets of alleles, A/a and B/b in a theoretical population subject to genetic drift.

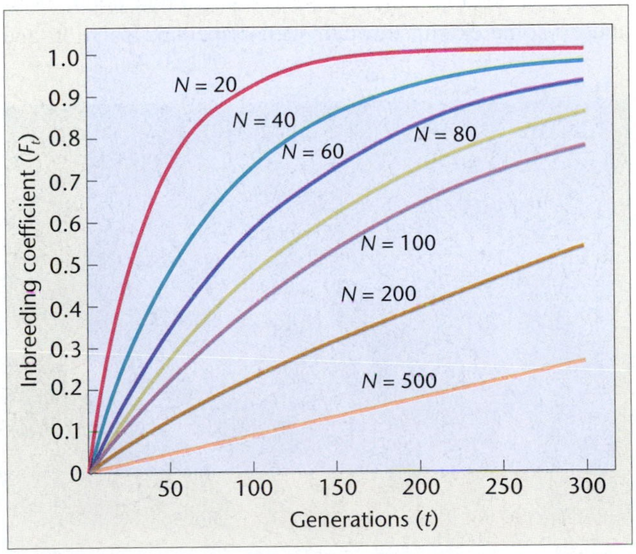

FIGURE 27–8 Increase in inbreeding coefficient (F) in theoretical populations with different effective population size N_e.

either of the corresponding homozygotes. In this case, the long-term survival of the population requires that inbreeding be avoided and that levels of all alleles in the gene pool be maintained—both of which can be difficult in a species that has already suffered a significant reduction in population size.

The effects of inbreeding depression and loss of genetic variation in a small isolated population have been documented in the case of the Isle Royale gray wolves (Figure 27–9). Around 1950, a pair of gray wolves apparently crossed an ice bridge from the Canadian mainland to Isle Royale in Lake Superior. The island had no other wolves and an abundance of moose, which became the wolves' main food source. By 1980, the Isle Royale wolf population had increased to over 50 individuals. Over the next decade, however, wolf numbers declined to fewer than a dozen with no new litters being born, despite plentiful food and no apparent sign of disease. The genetic variation of the remaining wolves was examined by mtDNA analysis and nuclear DNA fingerprinting. It was found that the Isle Royale wolves had levels of homozygosity twice as high as wolves in an adjacent mainland population. Furthermore, the wolves all possessed the same mtDNA genotype, consistent with descent from the same female. Hence, the degree of relatedness between individual Isle Royale wolves was equivalent to that of full siblings, suggesting that the wolves' reproductive failure was due to inbreeding depression, a phenomenon that has also been seen in captive wolf populations.

Reduction in Gene Flow

Gene flow, the gradual exchange of alleles between two populations, is brought about by the dispersal of gametes or the migration of individuals. It is an important mechanism for introducing new alleles into a gene pool and increasing genetic variation. Migration is the main route for gene flow to occur in animals. We have previously examined some of the genetic effects of migration in Chapter 25. In plants, gene flow occurs not through movement of individuals, but as a result of cross-pollination between different populations and, to some extent, through seed dispersal. Isolation and fragmentation of populations in rare and declining species significantly reduces gene flow and the potential for maintaining genetic diversity. As we have already seen, habitat loss is a major threat to species survival. It is not unusual for a threatened or endangered species to be restricted to small separate pockets of the remaining habitat. This isolates and fragments the surviving populations so that movement of individuals can no longer occur between them, preventing gene flow.

FIGURE 27–9 Isle Royale gray wolf (*Canis lupus*). (*Art Wolfe/Stone*).

One species in which the effects of gene-flow reduction have been studied is the North American brown bear (*Ursus arctos*). A team of Canadian and U.S. researchers measured heterozygosity in different brown-bear populations, using amplified microsatellite markers to create DNA profiles for individual bears. The researchers found that levels of heterozygosity in the brown-bear population living in and around Yellowstone Park were only two-thirds as high as those in Canadian and mainland Alaskan brown-bear populations. They concluded that this was due to the isolation of the Yellowstone bears and their reduced migration. The habitat of the Canadian and Alaskan bear populations was much less fragmented, allowing migration of individuals and consequent gene flow between populations. The researchers also examined an island population of brown bears on the Kodiak Archipelago off the Alaskan coast. Here, heterozygosity was even lower—less than one-half that of the mainland populations—providing further evidence for the effect of restricted migration and gene flow on reduced genetic diversity.

27.4 Genetic Erosion Diminishes Genetic Diversity

The loss of previously existing genetic diversity from a population or a species is referred to as **genetic erosion**. Why does it matter if a population loses genetic diversity, especially if the numbers of individuals remain high?

Genetic erosion has two important effects on a population. First, it can result in the loss of potentially useful alleles from the gene pool, reducing the ability of the population to adapt to changing environmental conditions and increasing its risk of extinction. Several decades before the dawn of modern genetics, Charles Darwin recognized the importance of diversity to long-term species survival and evolutionary success. In *The Origin of Species* (1859), Darwin wrote,

> The more diversified the descendants of any one species . . . by so much will they be better enabled to seize on many widely diversified places in the polity of nature, and so enabled to increase in numbers.

The story of the peppered moth *Biston betularia* in the 19th century (see the opening photograph in Chapter 26) illustrates the importance of maintaining allelic diversity in the gene pool. The allele producing the darker melanic phenotype did not confer any obvious advantage to the species, until the moth's environment in parts of Great Britain was significantly altered as a result of the increasing industrialization and the accompanying pollution. The continued presence of the melanic allele in the moth's gene pool, however, enabled it to adapt to the new environmental conditions and survive, just as the persistence of the nonmelanic allele in the population now allows the moth to readapt as its environment changes once again. What might have happened to the peppered moth if the melanic allele had been lost from its gene pool before the Industrial Revolution?

The second important effect of genetic erosion is a reduction in levels of heterozygosity. At the population level, reduced heterozygosity will be seen as an increase in the number of individuals homozygous at a given locus. At the individual level, a decrease in the number of heterozygous loci within the genotype of a particular plant or animal will occur. As we have seen, loss of heterozygosity is a common consequence of reduced population size. Alleles may be lost through genetic drift, or because individuals carrying them die without reproducing. Smaller populations also increase the likelihood of inbreeding, which inevitably increases homozygosity.

Obviously, once an allele is lost from a gene pool, the potential for heterozygosity is greatly reduced, or completely eliminated if there are only two alleles at the locus in question and one is now fixed. The level of homozygosity that can be tolerated varies with species. But studies of populations showing higher than normal levels of homozygosity have documented a range of deleterious effects, including reduced sperm viability and reproductive abnormalities in African lions, increased offspring mortality in elephant seals, and reduced nesting success in spotted woodpeckers.

As yet, no evidence has emerged from field studies conclusively linking the extinction of a wild population to genetic erosion. However, laboratory studies using the fruit fly, *Drosophila melanogaster*, to model evolutionary events show that loss of genetic variation does reduce the ability of a population to adapt to changing environmental conditions. Fruit flies, with their small size and rapid generation time of only 10–14 days, are a useful model organism for studying evolutionary events, especially since large populations can be readily developed and maintained through multiple generations.

In one set of experiments carried out by scientists at Macquarie University in Sydney, fruit-fly populations were reduced to a single pair for up to three generations to create a population bottleneck and then allowed to rebuild in number. The capacity of the bottlenecked populations to tolerate increasing levels of sodium chloride was compared with that of normal outbred populations. Researchers found that the bottlenecked populations went extinct at lower salt concentrations (Figure 27–10).

Experiments comparing inbred and outbred fruit-fly populations exposed to environmental stresses, such as high temperatures and ethanol, have also found that populations with even a low level of inbreeding have a much greater probability of extinction at lower levels of environmental stress than the outbred populations. Experimental results such as these indicate that genetic erosion does indeed reduce the long-term viability of a population by reducing its capacity to adapt to changing environmental conditions.

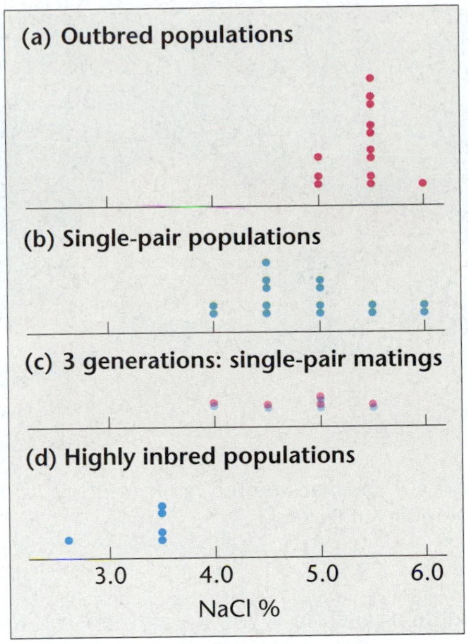

FIGURE 27–10 Effects of bottlenecks in various populations on evolutionary potential in fruit flies, as shown by distributions of NaCl concentrations at extinction.

27.5 Conservation of Genetic Diversity Is Essential to Species Survival

Scientists working to maintain biological diversity face several dilemmas. Should they focus on preserving individual populations, or should they take a broader approach by trying to conserve not just one species, but all the interdependent plants and animals in an ecosystem? How can genetic diversity be maintained in a species whose numbers are declining? Can genetic diversity lost from a population be restored? Early conservation efforts often focused simply on the population size of an endangered species. Biologists now recognize that a complex interplay of different factors must be considered during conservation efforts, including the need to examine the habitat and role of a species within an ecosystem, as well as the importance of genetic variation for long-term survival.

Ex Situ Conservation

Ex situ (literally "off-site") **conservation** involves the removal of plants or animals from their original habitat to an artificially maintained location such as a zoo or botanic garden. Living collections such as these have played a major role in the conservation and restoration of several species close to extinction.

One example is the recovery of the black-footed ferret (*Mustela nigripes*) (Figure 27–11), once widespread throughout the plains of the western United States, but by the 1970s, thought to be extinct after decades of trapping and poisoning of both the ferret itself and its main prey, the prairie dog. A small colony of black-footed ferrets was

FIGURE 27–11 The black-footed ferret (*Mustela nigripes*). *(Jim Brandenburg/Minden Pictures)*

discovered on a ranch in Wyoming in 1981 and transferred to a captive-breeding facility. Starting with the 18 animals originally captured, the breeding program has produced over 3000 ferrets, and this species is now being reintroduced to parts of its original range. With such a small founder group, the population has obviously passed through a severe bottleneck, and the risk of inbreeding and further genetic erosion in the captive-breeding program is very high.

Genetic management strategies, such as using DNA markers to identify the most genetically varied individuals and maintaining careful pedigree records to avoid mating closely related animals, have helped maintain the genetic diversity that remained in the species. Conservation geneticists working on the recovery project estimate that all black-footed ferrets existing today share about 12 percent of their genome, roughly the equivalent of being first cousins. The long-term effects of this degree of genetic similarity in the expanding ferret population remain to be seen. So far, there is no evidence of reduced fitness through inbreeding.

Another form of ex situ conservation is provided by **gene banks**. In contrast to housing entire animals or plants, these collections instead provide long-term storage and preservation for reproductive components, such as sperm, ova, and frozen embryos in the case of animals or seed, pollen, and cultured tissue in the case of plants. Many more individual genotypes can be preserved for longer periods in a gene bank than in a living collection. Cryopreserved gametes or seeds can be used to reconstitute lost or endangered animals or plants after many years in storage. Because they are expensive to construct and maintain, most gene banks are used to conserve these components of domesticated species of economic value.

Gene banks have now been established in many countries to help preserve genetic material of agricultural importance, such as traditional crop varieties that are no longer grown or old livestock breeds that are becoming rare. One of the most important ex situ collections in the United States is the National Seed Storage Laboratory, a U.S. Department of Agriculture facility in Fort Collins, Colorado, which maintains more than 275,000 different accessions of crop varieties and related wild species. Some of the accessions are stored as seeds, and others as cryogenically preserved tissue from which whole plants can be regenerated (Figure 27–12).

Ex situ conservation, while often vital, has a number of disadvantages. A major problem with gene banks is that even large collections cannot contain all the genetic variation present in a species. Conservation geneticists attempt to address this problem by identifying a **core collection** for a species. The core collection is a subset of individual genotypes that represents as much as possible of the genetic variation within a species; preserving the core collection takes priority over collecting and preserving large numbers of genotypes at random. Another disadvantage of ex situ conservation is that the artificial conditions under which a species is preserved in a living collection or gene bank often create their own selection pressures. When seeds of a rare plant species are maintained in cold storage, for example, selection may occur for those genotypes better adapted to withstand the lower temperatures, and genotypes may be lost that would actually have greater fitness in the natural environment. Still, an additional problem posed by ex situ conservation is that, while the greatest biological diversity in both domesticated and nondomesticated species is frequently found in underdeveloped countries, most ex situ collections are in developed countries that have the resources to establish and maintain them. This leads to conflict over who owns and has access to the potentially valuable genetic resources maintained in such collections.

In Situ Conservation

In situ (literally "on-site") **conservation** attempts to preserve the population size and biological diversity of a species while maintaining it in its original habitat. The use of species inventories to identify diversity hotspots, as we learned earlier, is an important tool for determining the best places to establish parks and reserves where plants and animals are protected from hunting or collecting and where their habitat can be maintained.

FIGURE 27–12 Cryogenically preserved seeds of rare crop varieties at the U.S. Department of Agriculture National Seed Storage Laboratory. *(Photo by Scott Bauer/ARS Photo Unit/USDA)*

Genetics, Technology, and Society

Gene Pools and Endangered Species: The Plight of the Florida Panther

In the last 400 years, more than 700 of Earth's animal and plant species have become extinct. In the United States alone, at least 30 species have suffered extinction in the last decade. In addition, hundreds of genetically distinct plant and animal species are now endangered. The rates of species and ecosystem losses are accelerating rapidly and are predicted to continue unabated throughout this century. Such a dramatic loss of biological diversity is a direct consequence of human activity. As we humans increase our numbers and spread over Earth's surface, we harness more of Earth's resources for our use—clearing the forests, polluting the water, and inexorably and permanently altering Earth's natural balance.

Although the destruction continues, we are beginning to understand the implications of our actions and to make efforts to save some of Earth's more endangered life forms. However, despite our best efforts to save threatened species, we often intercede after population numbers have suffered severe declines. This means that much of the genetic diversity that existed in the species is lost, and the population must be rebuilt from a reduced gene pool. The resulting genetic uniformity can reduce fitness, as well as expose genetic diseases. The reduced fitness may further deplete the numbers of the threatened plant or animal, and the population spirals downward to extinction.

The story of the Florida panther provides a poignant example of how an animal can be brought to the verge of extinction, and how the efforts of scientists and the general public might restore this creature to a healthy place in the ecosystem. The Florida panther, *Felis concolor coryi*, is one of 30 subspecies of cougar (puma or mountain lion) and is one of the most endangered mammals in the world. Florida panthers once roamed the southeastern corner of North America, from South Carolina and Arkansas to the southern tip of Florida. As people settled the southern and eastern states, panthers were killed as potential threats to livestock and humans. In the 19th and 20th centuries, panthers were killed by hunting, highway collisions, poisoning, and loss of habitat. Today, only 30–50 Florida panthers remain, isolated in southern Florida in the Big Cypress Swamp and Everglades National Park regions. Population projections indicate that, unless drastic measures are taken, there is an 85 percent likelihood that the Florida panther will be extinct in 25 years.

Together, the geographical isolation and inbreeding of the few remaining Florida panthers have resulted in the loss of genetic variability and declining health. The panther has been separated from other cougar subspecies for 15 to 25 generations. As a result, Florida panthers have the lowest levels of genetic heterozygosity of any subspecies of cougar. The loss of genetic diversity has manifested itself in the appearance of a number of severe genetic defects. Almost 80 percent of panther males born after 1989 in the Big Cypress region show a rare, heritable (autosomal dominant or sex-linked recessive) condition known as cryptorchidism, manifested as the failure of one or both testicles to descend. This defect is associated with lower testosterone levels and reduced sperm count. In general, Florida panther males have the poorest seminal quality of any cat species or any cougar subspecies, with approximately 93 percent abnormal sperm. Life-threatening congenital heart defects are also appearing in the Florida panther population, possibly due to an autosomal dominant gene defect. In addition, some immune deficiencies are appearing, and these may have a genetic component. Reduced immunity could leave the small panther population susceptible to diseases that could wipe out the population. Other less serious genetic features have appeared, such as a kink in the tail and a whorl of fur on the back.

Over the last two decades, a faint glimmer of hope has appeared for the Florida panther. Federal and state agencies, as well as private individuals, are implementing a "Florida Panther Recovery Program." The goal of the plan is to exceed 130 breeding animals (wild and captive) soon after the year 2000, and to approach 500 by the year 2010. If successful, the plan would grant the panther a 95 percent probability of survival, while retaining 90 percent of its genetic diversity. The plan is multifaceted and includes a captive-breeding program, strict protection, increasing and improving the panther habitat, and educating the public and private landowners. Wildlife underpasses have been constructed on highways in panther territory, and these have significantly reduced panther highway fatalities (which account for half of panther deaths).

In 1995, a genetic restoration program began. In order to introduce genetic diversity into the Florida panther population and to retard the detrimental effects of inbreeding, eight female wild Texas panthers (a related subspecies from western Texas) were released into Florida panther territory. Two of the females have been killed—one from an automobile collision and one from gunshot wounds. However, the remaining Texas females have given birth to 12 healthy kittens (as of the summer of 1998), and one of these F_1 offspring has produced three kittens of her own. None of the kittens appear to have the "kinked" tail of the inbred Florida panther.

The survival of the Florida panther is far from certain. Its recovery will require years of monitoring and frequent intervention. In addition, people must be willing to share their land with wild creatures that do not directly further their self-interests. However, as public support for the return of the Florida panther has been strong, there may be hope for this unique, impressive animal.

References

Fergus, C. 1991. The Florida panther verges on extinction. *Science* 251: 1178–80.

Hedrick, P.W. 1995. Gene flow and genetic restoration: The Florida panther as a case study. *Conservation Biology* 9: 996–1007.

O'Brien, S.J. 1944. A role for molecular genetics in biological conservation. *Proc. Natl. Acad. Sci. USA* 91: 5748–55.

Web site

Florida Panther Net
www.panther.state.fl.us/

For domesticated species, there is increasing interest in "on-farm" preservation, by farmers are encouraged through provision of additional resources and financial incentives to maintain traditional crop varieties and livestock breeds. Nondomesticated species with economic potential have also been targeted for in situ preservation. In 1998, the U.S. Department of Agriculture established its first in situ conservation sites for a wild plant, protecting populations of the native rock grape (*Vitis rupestris*) in several eastern states. The rock grape is prized by winegrowers not for its fruit, but for its roots; grape vines grafted onto wild rock-grape rootstock are resistant to phylloxera, a serious pest of wine grapes.

The advantage of in situ conservation for the rock grape, as with other species, is that larger populations with greater genetic diversity can be maintained. Species conserved in situ will also continue to live and reproduce in the environments to which they are adapted, reducing the likelihood that novel selection pressures will produce undesirable changes in allele frequency. However, as the global human population continues to rise, setting aside suitable areas for in situ conservation becomes a greater challenge. As we have already seen in the case of the North American brown bear, even in large preserves like Yellowstone National Park, a species' migration and gene flow may be eliminated with the consequent loss of genetic diversity. This problem is even more acute in smaller and more fragmented areas of protected habitat.

Population Augmentation

What genetic considerations should accompany efforts to restore populations or species that are in decline? As we have already seen, populations that go through bottlenecks continue to suffer from low levels of genetic diversity, even after their numbers have recovered. We have also seen that inbreeding and drift contribute to genetic erosion in small populations, and fragmentation interrupts migration and gene flow, further reducing diversity. Captive-breeding programs to restore a critically endangered species from a few surviving individuals risk genetic erosion in the renewed population from founder effect and inbreeding, as in the case of the black-footed ferret.

An alternative strategy used by conservation biologists is **population augmentation**—boosting the numbers of a declining population by transplanting and releasing individuals of the same species captured or collected from more numerous populations elsewhere. Population augmentation projects in the United States have involved bighorn sheep and grizzly bears in the Rocky Mountains. It has also been employed with the Florida panther (*Felis coryi*), an isolated population of less than 50 animals confined to the area around the Big Cypress Swamp and Everglades National Park in south Florida. As discussed in the essay at the end of this chapter, DNA-profiling patterns show high levels of inbreeding in the Florida panther, with reduced fitness due to severe reproductive abnormalities and increased susceptibility to parasite infections. Seven unrelated animals from a captive population of South and North American panthers were released into the Everglades in the 1960s and have interbred with the Florida population, producing more genetically diverse family groups. Further population augmentation in this species has been controversial, however, as some biologists argue that the unique features that allow the Florida panther to be classified as a separate subspecies will be lost if mating with other panthers takes place.

Despite such controversies, population augmentation would appear to be a valuable restoration tool. This strategy cannot only increase population numbers, but also genetic diversity, if the transplanted individuals are unrelated to those in the population to which they are introduced. A potential problem with population augmentation, however, is the recently identified phenomenon of **outbreeding depression**, in which reduced fitness occurs in the progeny from matings between genetically diverse individuals. Outbreeding depression occurring in the F_1 generation is thought to be due to the offspring being less well adapted to local environmental conditions than the parents.

This phenomenon has been documented in some plant species in which seed of the same species, but from a different location, was used to revegetate a damaged area. Outbreeding depression that occurs in the F_2 and later generations is thought to be due to the disruption of coadapted gene complexes—groups of alleles that have evolved to work together to produce the best level of fitness in an individual. This type of outbreeding depression has been documented in F_2 hybrid offspring from matings between fish from different salmon populations in Alaska. Studies such as this suggest that restoring the most beneficial type and amount of genetic diversity in a population is more complicated than previously thought, reinforcing the argument that the best long-term strategy for species survival is to prevent the loss of diversity in the first place.

Chapter Summary

1. Biodiversity is being lost as an increasing number of plant and animal species are threatened with extinction. Conservation genetics applies principles of population genetics to the preservation and restoration of threatened species. A major concern of conservation geneticists is the maintenance of genetic diversity.
2. Genetic diversity includes interspecific differences (reflected by the number of different species present in an ecosystem) and intraspecific differences (reflected by genetic variation within a population or between different populations of the same species). Genetic diversity can be measured by examining different phenotypes in the population or, at the molecular level, by using isozyme analysis or DNA-profiling techniques.
3. Major declines in a species' population numbers reduce genetic diversity and contribute to their risk of extinction as a result of genetic drift, inbreeding, or loss of gene flow. Populations that suffer severe reductions in effective size and then

recover are said to have passed through a population bottleneck and often show reduced genetic diversity.

4. Loss of genetic diversity reduces the capacity of a population to adapt to changing environmental conditions as useful alleles may disappear from the gene pool. Reduced genetic diversity also results in greater levels of homozygosity in a population, often leading to an accumulation of deleterious alleles and inbreeding depression.
5. Conservation of genetic diversity depends on ex situ methods, such as living collections, captive-breeding programs, and gene banks, and in situ approaches, such as the establishment of parks and preserves.
6. Population augmentation, in which individuals are transplanted into a declining population from a more numerous population of the same species located elsewhere, can be used to increase numbers and genetic diversity. However, the risk of outbreeding depression accompanies this process.

Insights and Solutions

1. Is a rare species found as several fragmented subpopulations more vulnerable to extinction than an equally rare species found as one larger population? What factors should be considered when managing fragmented populations of a rare species?

Solution: A rare species in which the remaining individuals are divided among smaller isolated subpopulations can appear to be less vulnerable. If one subpopulation becomes extinct through local causes, such as disease or habitat loss, then the remaining subpopulations may still survive. However, genetic drift will cause smaller populations to experience more rapidly increasing homozygosity over time compared with larger populations. Even with random mating, the change in heterozygosity from one generation to the next due to drift can be calculated as

$$H_1 = H_0(1 - {}^1\!/_2N)$$

where H_0 is the frequency of heterozygotes in the present generation, H_1 is the frequency of heterozygotes in the next generation, and N is the number of individuals in the population. Thus, in a small population of 50 individuals with an initial heterozygote frequency of 0.5, in just one generation, heterozygosity will decline to $0.5(1 - 1/100) = 0.495$, a loss of 0.5%. In a larger population of 500 individuals and the same initial heterozygote frequency, after one generation, heterozygosity will be $0.5(1 - 1/1000) = 0.4995$, a loss of only 0.05%.

This means that smaller populations are likely to show the effects of homozygosity for deleterious alleles sooner than larger populations, even with random mating. If populations are fragmented so that movement of individuals or gametes between them is prevented, management options could include transplanting individuals from one subpopulation to another to enable gene flow to occur. Establishment of "wildlife corridors" of undisturbed habitat that will connect fragmented populations could be considered. In captive populations, exchange of breeding adults (or their gametes through shipment of preserved semen or pollen) can be undertaken. Management for increased population numbers, however, is vital to prevent further genetic erosion through drift.

Problems and Discussion Questions

1. A wildlife biologist studied four generations of a population of rare Ethiopian jackals. When the study began, there were 47 jackals in the population, and analysis of microsatellite loci from these animals showed a heterozygote frequency of 0.55. In the second generation, an outbreak of distemper occurred in the population, and only 17 animals survived to adulthood. These produced 20 surviving offspring, which in turn gave rise to 35 progeny in the fourth generation.
   (a) What was the effective population size for the four generations of this study?
   (b) Based on this effective population size, what is the heterozygote frequency of the jackal population in generation 4?
   (c) Assuming an inbreeding coefficient of zero at the beginning of the study, no change in microsatellite allele frequencies in the gene pool, and random mating in all generations, what is the inbreeding coefficient in generation 4?
2. Chondrodystrophy, a lethal form of dwarfism, has recently been reported in captive populations of the California condor, and it has killed embryos in 5 out of 169 fertile eggs. Chondrodystrophy in condors appears to be caused by an autosomal recessive allele with an estimated frequency of 0.09 in the gene pool of this species. How do you think California condor populations should be managed in the future to minimize the effect of this lethal allele? What are the advantages and disadvantages of attempting to eliminate it from the gene pool?
3. A geneticist is studying three loci, each with one dominant and one recessive allele, in a small population of rare plants. She estimates the frequencies of the alleles at each of these loci as follows:

$A = 0.75$	$a = 0.25$
$B = 0.80$	$b = 0.20$
$C = 0.95$	$c = 0.05$

   What is the probability that all the recessive alleles will be lost from the population through genetic drift?
4. How are genetic drift and inbreeding similar in their effects on a population? How are they different?

5. You are the manager of a game park in Africa with a native herd of just 16 black rhinos, an endangered species worldwide. Describe how you would manage this herd to establish a viable population of black rhinos in the park. What genetic factors would you take into consideration in your management plan?
6. Compare the causes and effects of inbreeding depression and outbreeding depression.
7. Cloning, using the techniques similar to those pioneered by the Scottish scientists who produced Dolly the sheep, has been proposed as a way to increase the numbers of some highly endangered mammal species. Discuss the advantages and disadvantages of using such an approach to aid long-term species survival.

Extra-Spicy Problems

8. In a population of wild poppies found in a remote region of the mountains of eastern Mexico, almost all members have pale yellow flowers. Breeding experiments show that pale yellow is recessive to deep orange. Using the tools of the conservation geneticist (described in this chapter), how could you determine experimentally whether the prevalence of the recessive phenotype among the eastern Mexican poppy population is due to natural selection or simply due to the effects of genetic drift and/or inbreeding?

Selected Readings

Avise, J.C., and Hamrick, J.L., eds. 1996. *Conservation genetics: Case histories from nature*. New York: Chapman & Hall.

Baker, C.S., Palumbi, S.R. 1994. Which whales are hunted? A molecular genetic approach to monitoring whaling. *Science* 265:1538–1539.

Beatiie, A., and Erlich, P. 2001. *Wild solutions: How biodiversity is money in the bank*. New Haven, CT: Yale University Press.

Bongaarts, J., and Bulatao, R.A. eds. 2000. *Beyond six billion: Forecasting the world's population*. Washington, DC: National Academy Press.

Bonnell, M.L., Selander, R.K. 1974. Elephant seals: genetic variation and near extinction. *Science* 184:908–90.

Frankham, R. 1995. Conservation genetics. *Ann. Rev. Genet.* 29:305–327.

———. 1999. Quantitative genetics in conservation biology. *Genet. Res.* 74:237–44.

Gerber, L.R., DeMaster, D.P., and Roberts, S.P. 2000. Measuring success in conservation. *Amer. Scient.* 88:316–21.

Gharrett, A.J., and Smoker, W.W. 1991. Two generations of hybrids between even-year and odd-year pink salmon (*Oncorhynchus gorbuscha*): A test for outbreeding depression? *Canadian J. Fish. Aquat. Sci.* 48:426–238.

Lacy, R.C. 1997. Importance of genetic variation to the viability of mammalian populations. *J. Mammalogy* 78:320–335.

O'Neill, B.C., MacKellar, F.L., and Lutz, W. 2001. *Population and climate change*. Cambridge, U.K.: Cambridge University Press.

Paetkau, D., Waits, L.P., Clarkson, P.L., Craighead, L., Vyse, E., Ward, R., and Strobeck, C. 1998. Variation in genetic diversity across the range of North American brown bears. *Conserv. Biol.* 12:418–429.

Palacios, C., and Gonzalez-Candelas, F. 1999. AFLP analysis of the critically endangered *Limonium cavanillesii*. *J. Heredity* 90:485–489.

Ralls, K., Ballou, J.D., Rideout B.A., and Frankham, R. 2000. Genetic management of chondrodystrophy in California condors. *Animal Conservation* 3:145–153.

Roman, J., and Bowen, B.W. 2000. The mock turtle syndrome: genetic identification of turtle meat purchased in the southeastern United States of America. *Animal Conservation* 3:61–65.

Storfer, A. 1996. Quantitative genetics: A promising approach for the assessment of genetic variation in endangered species. *Trends Ecol. Evol.* 11:343–47.

Wayne, R.K., et al. 1991. Conservation genetics of the endangered Isle Royale gray wolf. *Conserv. Biol.* 5:41–51.

Wilson, E.O., ed. 1988. *Biodiversity*. Washington, DC: National Academy of Sciences.

Wynen, L.P., et al. 2000. Postsealing genetic variation and population structure of two species of fur seal. *Molecular Ecology* 9:299–314.

GENETICS MediaLab

The resources that follow will help you achieve a better understanding of the concepts presented in this chapter. These resources are on the Companion Web site for this book at **http://www.prenhall.com/klug**

Web Problem 1:

Time for completion = 10 minutes

What can genetics tell us about the behavior of populations in the wild? Traditionally, observational field studies have provided the data for reaching conclusions about relatedness, reproductive success, range, etc. In one ongoing study, a population of grizzly bears was periodically sampled by capture in order to genotype individuals. Read the description of the research at the linked Web site (keyword **GRIZZLIES**) in order to answer the following questions: What conclusions can be drawn about hereditary relationships with the new information? What conclusions from observations are contradicted by the new information? What is responsible for maintaining genetic diversity in this population of bears, and what steps might be taken to promote diversity in the future? To complete this exercise, visit Web Problem 1 in Chapter 27 of your Companion Web site, and select the keyword **GRIZZLIES**.

Web Problem 2:

Time for completion = 10 minutes

How does habitat fragmentation affect maintenance of diversity? Fragmentation of habitat may be more harmful than we think for groups of closely related species in nearly overlapping ranges. Recently, researchers analyzing mitochondrial sequences and satellite markers have established the presence of two distinct species of sage grouse in the Western United States. Read the report and related links at the Web site (keyword **SAGE GROUSE**), and use the material in Chapter 27 to answer the following questions: What is the impact on genetic diversity if two groups are found to be reproductively isolated? How effective is the addition of further fragmented habitat? What can be done to mitigate the effects of isolation? To complete this exercise, visit Web Problem 2 in Chapter 27 of your Companion Web site, and select the keyword **SAGE GROUSE**.

Web Problem 3:

Time for completion = 15 minutes

What are the consequences of supplementing wild populations with captive-bred stock? Salmon hatcheries provide a familiar, though controversial, example. Hatcheries frequently use a brood stock of fish not indigenous to the particular river, and economies of scale favor uniform animals. What impact do these choices have on native runs? Should we let native runs dwindle? Read the article in the linked Web site (keyword **SALMON**), and critique both points of view. How does population augmentation compare with habitat restoration? To complete this exercise, visit Web Problem 3 in Chapter 27 of your Companion Web site, and select the keyword **SALMON**.

Appendix A
Experimental Methods

In addition to the techniques of genetic analysis, physical and chemical techniques for the separation and analysis of macromolecular components of the cell nucleus and cytoplasm have been instrumental in advancing our understanding of genetics at the molecular level. In this appendix, we will describe the background and theoretical basis of some techniques that have been important in molecular genetics.

Isotopes

Isotopes are forms of an element that have the same number of protons and electrons, but differ in the number of neutrons contained in the atomic nucleus. For example, the most common form of carbon has an atomic number of 6 (the number of protons in the nucleus) and an atomic weight of 12 (the sum of the protons and neutrons in the nucleus). In a very small percentage of carbon atoms, a seventh neutron is present, producing an atom with an atomic weight of 13. This is an example of a so-called **heavy isotope**. Since the number of protons and electrons, and hence the net charge, has not changed, the atom has the same chemical properties as **carbon-12** (^{12}**C**) and differs only in mass. **Carbon-13** (^{13}**C**) is thus a stable, heavy isotope of carbon.

Although the addition of neutrons does not alter the chemical properties of an atom, it can produce instabilities in the atomic nucleus. If another neutron is added to a carbon-13 atom, the isotope **carbon-14** (^{14}**C**) results. However, the presence of eight neutrons and six protons is an unstable condition, and the atom undergoes a nuclear reaction in which radiation is emitted during the transition to a more stable condition. Therefore, carbon-14 is a **radioactive isotope** of carbon.

The type of radiation emitted and the rate at which these nuclear events take place are characteristic of the element. Table A–1 lists types of radioactivity. The rate at which a radioactive isotope emits radiation is expressed as its **half-life**, which is the time required for a given amount of a radioactive substance to lose one-half of its radioactivity. Table A–2 lists some of the isotopes available for use in research.

Detection of Isotopes

The choice of which isotope to use as a tracer in biological experiments depends on a combination of its physical and chemical properties, which enable the investigator to quantitate the amount of radioactivity or measure the ratio of heavy to light isotopes. For the detection of heavy isotopes, two methods are commonly employed: **mass spectrometry** and **equilibrium density gradient centrifugation**. (Each of these methods is next discussed in the section.) Although the use of heavy isotopes has been more restricted than that of radioisotopes, they have been instrumental in several basic advances in molecular biology [e.g., demonstrating the semiconservative nature of DNA replication and the existence of messenger RNA (mRNA)].

There are a number of methods to detect radioisotopes, the foremost being **liquid scintillation spectrometry**, which provides quantitative information about the amount of radioactive isotope present in a sample, and **autoradiography**, which is used to demonstrate the cytological distribution and localization of radioactively labeled molecules.

TABLE A–1 Properties of Ionizing Radiation

Type	*Relative Penetration*	*Relative Ionization*	*Range in Biological Tissue*
Alpha particle (2 protons + 2 neutrons)	1	10,000	Microns
Beta particle (electron)	100	100	Microns-mm
Gamma ray	> 1000	< 1	∞

TABLE A–2 Some Isotopes Used in Research

Element	*Isotope*	*Half-life*	*Radiation*
H	^{2}H	—	Stable
	^{3}H	12.3 years	β
C	^{13}C	—	Stable
	^{14}C	5700 years	β
N	^{15}N	—	Stable
O	^{18}O	—	Stable
P	^{32}P	14 days	β
S	^{35}S	87 days	β
K	^{40}K	1.2×10^{9} years	β, gamma
Fe	^{59}Fe	45 days	β, gamma
I	^{125}I	60 days	β, gamma
	^{131}I	8 days	β, gamma

In recording radioactivity by liquid scintillation counting, a small sample of the material to be counted is solubilized and immersed in a solution containing a **phosphor** (an organic compound that emits a flash of light after it absorbs energy released by the decay of the radioactive compound). The counting chamber of the liquid scintillation spectrometer is equipped with very sensitive photomultiplier tubes that record the light flashes emitted by the phosphor. The data are recorded as counts of radioactivity per unit time and are displayed on a printout or can be fed into a computer for storage and analysis.

In autoradiography, a gel, a chromatogram, a plant, or an animal part is placed against a sheet of photographic film. Radioactive decay from the incorporated isotope behaves just as light energy does and reduces silver gains in the emulsion. After exposure, the sheet of film is developed and fixed, revealing a deposit of silver grains corresponding to the location of the radioactive substance (Figure A–1).

Alternatively, to record the subcellular localization of an incorporated labeled isotope, cells or chromosomal preparations that have been incubated with radioactively labeled compounds are affixed to microscope slides and covered with a thin layer of liquid photographic emulsion. Following exposure, the slides are processed to develop and fix the reduced silver grains in the emulsion. After staining, microscopic examination reveals the location and extent of labeled isotope incorporation (Figure A–2).

Centrifugation Techniques

The centrifugation of biological macromolecules is widely employed to provide information about their physical characteristics (e.g., size, shape, density, and molecular weight) and to purify and concentrate cells, organelles, and their molecular components.

Differential centrifugation is commonly used to separate materials, such as cell homogenates, according to size. Initially, the homogenate is distributed uniformly in the centrifuge tube. Following a period of centrifugation, the pellet obtained is enriched for the largest and most dense particles, such as nuclei, in the homogenate. After each step, the supernatant can be recentrifuged at higher speeds to pellet the next heavier component. A typical fractionation scheme for cell homogenates is shown in Figure A–3. Further fractionation using density gradient techniques can be employed to purify any of the fractions obtained by differential centrifugation.

Rate zonal centrifugation is used to separate particles on the basis of differences in their sedimentation rates. Mixtures of macromolecules, such as proteins or nucleic acids, or cellular organelles (e.g., mitochondria) may be separated. In this technique, which employs a medium of increasing density, the rate at which particles sediment depends on the size, shape, density, and frictional resistance of the solvent.

In addition to the preparation and purification of macromolecules and cellular components, rate zonal centrifugation can be used to determine the **sedimentation coefficients** and **molecular weights** of biological macromolecules. If a purified molecule such as a protein is spun in a centrifugal field, the molecule will eventually sediment toward the bottom at a constant velocity. At this point, the molecular weight (***M***) can be calculated as

$$M = f \times v/\omega^{2} r$$

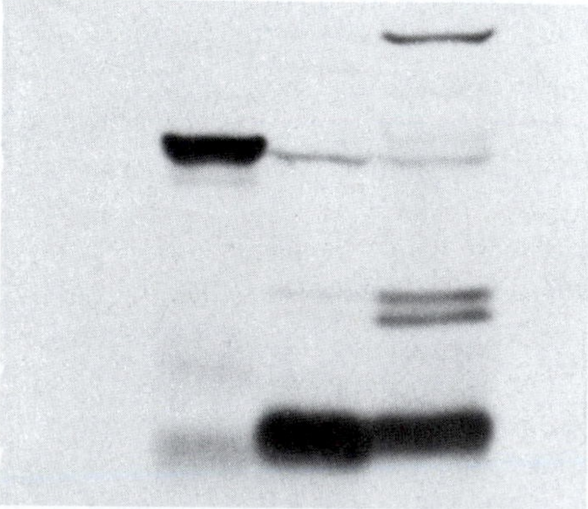

FIGURE A–1 Autoradiogram of radioactive bacterial proteins synthesized in a minicell system.

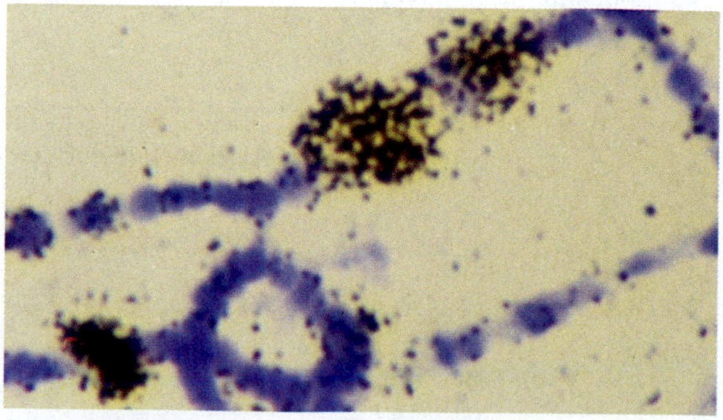

FIGURE A–2 Autoradiogram of RNA synthesis in salivary glands of *Drosophila* larva. Silver grains are deposited over sites of RNA synthesis at chromosome puffs

Step 1
Centrifuge cell homogenate at 700 xg for 10 minutes

Pellet	**Supernatant**
Nuclei	Mitochondria Plasma membrane fragments Lysosomes Soluble cell sap

Step 2
Centrifuge supernatant at 10,000 xg for 10 minutes

Pellet	**Supernatant**
Membranes Mitochondria Lysosomes	Ribosomes Soluble cell sap

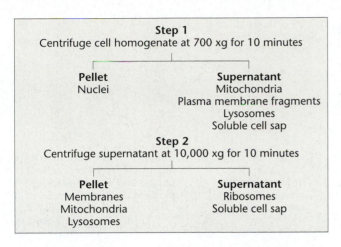

FIGURE A–3 Fractionation scheme for cell homogenates.

where f is the frictional coefficient of the solvent system (which has been calculated from other measurements) and v/ω^2r is the rate of sedimentation per unit applied centrifugal field. The latter value is given the symbol S, or sedimentation coefficient. The S value for most proteins is between 1×10^{-13} sec and 2×10^{-11} sec. A sedimentation coefficient of 1×10^{-13} is defined as one **Svedberg unit (S)**; this unit is named for The Svedberg, a pioneer in the field of centrifugation. Thus, a protein with a sedimentation value of 2×10^{-11} sec would have a value of 200 S. Figure A–4 shows the S values of selected molecules and particles.

Isopycnic centrifugation, or **equilibrium density gradient centrifugation**, is one of the most widely used techniques in genetics and molecular biology. In this technique, the solvent varies in density from one end of the tube to the other. A mixture of molecules layered on top and centrifuged through this gradient will migrate toward the bottom of the tube until each particle reaches its isopycnic point—that is, the place in the gradient at which the density of the solvent equals the buoyant density of the particle.

When each molecular species in the mixture migrates to its own characteristic isopycnic point, it no longer moves and is at equilibrium no matter how much longer the centrifugal field is applied. After separation by this method, components may be recovered by puncturing the bottom of the tube and collecting fractions. Two methods of forming density gradients are commonly employed (Figure A–5). One uses preformed gradients of sucrose or soluble salts of heavy metals such as cesium chloride or cesium sulfate; in the second method, the gradient is formed by the action of the centrifugal field on the salt solution.

Renaturation and Hybridization of Nucleic Acids

The ability of separated complementary strands of nucleic acids to unite and form stable, double-stranded molecules has been used to measure the relatedness of nucleic acids from different parts of the cell, different organs, and even different species. If both strands are DNA, the process is known as **renaturation** or **reassociation**. If one strand is RNA and the other DNA, the technique is called **hybridization**. Renaturation and hybridization involve two steps: (1) a rate-limiting step, in which collision or nucleation between two homologous strands initiates base pairing, and (2) a rapid pairing of complementary bases, or ``zippering'' of the strands, to form a double-stranded molecule. For DNA

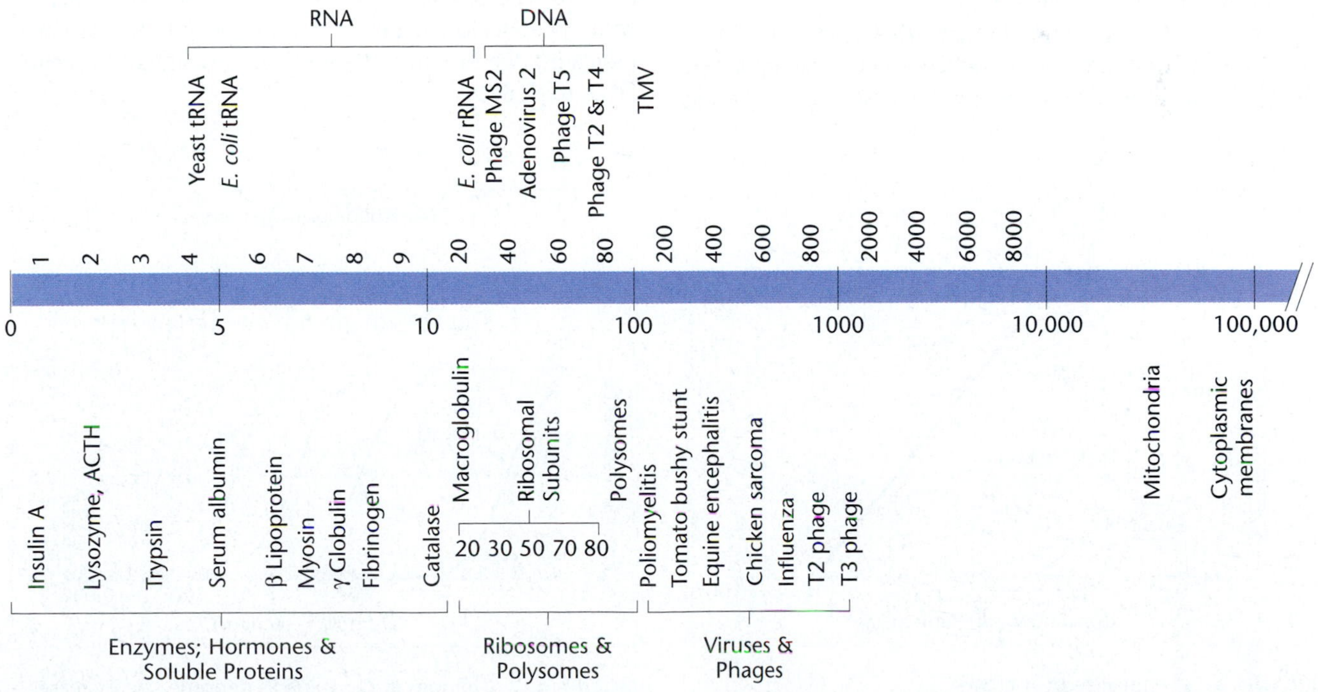

FIGURE A–4 S values of some common biological molecules and particles.

FIGURE A–5 Two methods of forming density gradients.

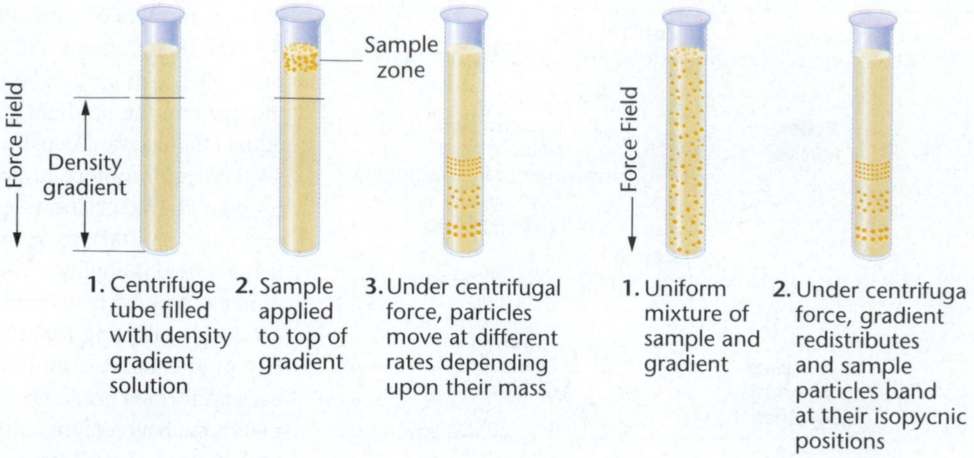

renaturation, the formation of double-stranded molecules can be assayed at any time during an experiment. A sample is passed over a column of **hydroxyapatite**, which selectively binds double-stranded DNA, but allows single-stranded molecules to pass through. The double-stranded molecules can be released from the column by raising the temperature or salt concentration.

The process of renaturation follows second-order kinetics according to the equation

$$\frac{C}{C_0} = \frac{1}{1 + k\,C_0 t}$$

where C is the single-stranded DNA concentration at time t, C_0 is the total DNA concentration, and k is a second-order rate constant. It is usually convenient to express the data from renaturation experiments as the fraction of single-stranded DNA at any time t versus the product of total DNA concentration and time, as shown in Figure A–6.

In the process of renaturation, the DNA is initially single stranded and is renatured to double-stranded structures in the final state. The time necessary to reassociate half of the DNA in a sample at a given concentration should be proportional to the number of different pieces of DNA present. Consequently, the half-reassociation time should be proportional to the DNA content of the genome, with smaller genomes having shorter half-renaturation rates. Figure A–7 confirms this expectation and shows increases in C_0t values as the size of the genome increases. This proportional relationship between C_0t and genome size is valid only for cases in which repetitive DNA sequences are absent from the DNA being studied. DNA from calf thymus (and many other eukaryotic sources) exhibits a complex pattern of reassociation, indicating that bovine DNA contains some sequences (in this case, 40 percent of the total DNA) that reassociate rapidly and others that reassociate more slowly. The rapidly reassociating fraction must therefore contain sequences that are present in many copies. The more slowly renaturing DNA, however, contains sequences present in only one copy per genome. The *E. coli* DNA shown in Figure A–8 renatures with a pattern close to that of an ideal second-order reaction, indicating the absence of significant amounts of repeated DNA sequences.

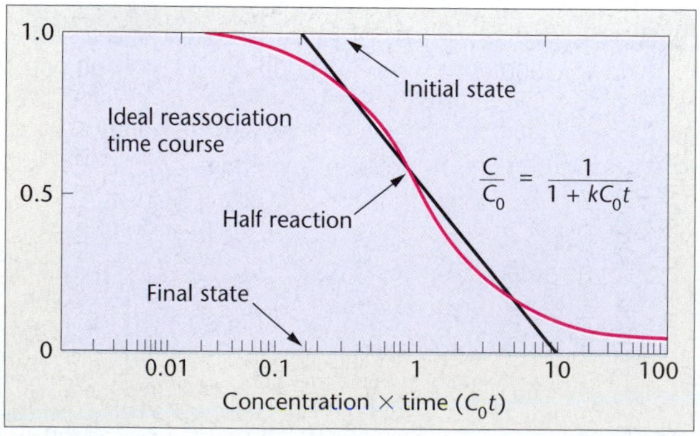

FIGURE A–6 Idealized C_0t curve.

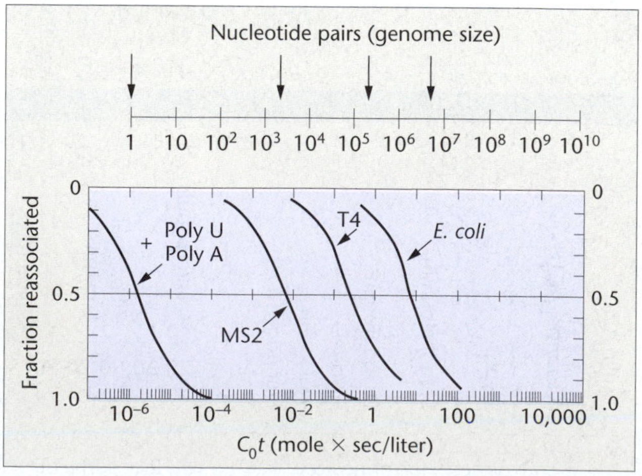

FIGURE A–7 Changes in C_0t value as genome size increases.

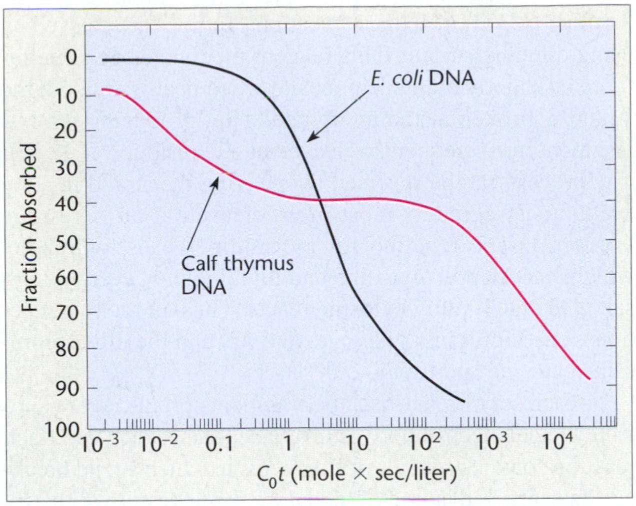

FIGURE A–8 Renaturation curve for *E. coli* DNA, containing no repetitive DNA sequences, and for calf thymus DNA, containing several families of repetitive DNA.

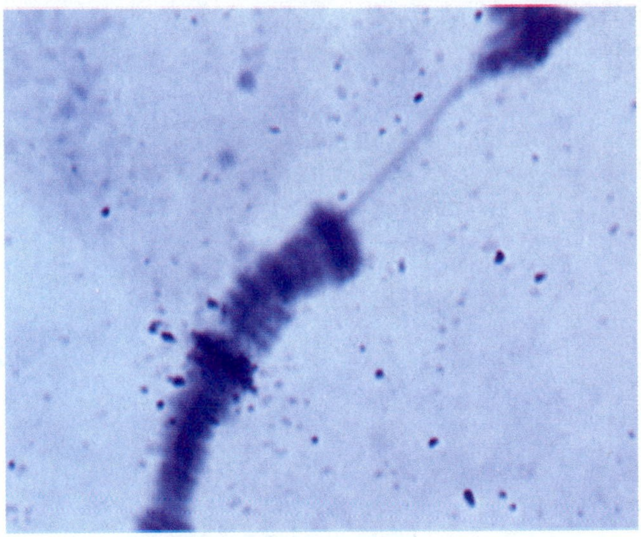

FIGURE A–9 Light micrograph of *in situ* hybridization of radioactive 5*S* RNA to a single band of polytene chromosome in the salivary gland of a *Drosophila* larva.

In the case of hybridization between DNA and RNA, two approaches can be used—either the RNA or the DNA can be in excess. RNA-excess hybridization is usually preferred, because most double-stranded molecules that are formed are RNA:DNA hybrids. Since DNA is present in low concentration, and since single-stranded RNA molecules lack complementary RNA strands in the mixture, the number of RNA:RNA and DNA:DNA hybrids is negligible.

The extent of hybridization can be followed by using hydroxyapatite columns or by using radioactively labeled RNA or DNA. In practice, one of the components of the hybridization reaction is usually immobilized on a substrate such as nitrocellulose paper. For example, DNA may be sheared to a uniform size, denatured to single strands, immobilized on nitrocellulose filters, and hybridized with an excess of labeled RNA. After hybridization, the unbound RNA is removed by washing, and the hybrids are assayed by liquid scintillation counting. DNA:RNA hybridization can also be performed using cytological preparations, a technique called ***in situ* hybridization**. DNA, which is a part of an intact chromosome preparation fixed to a slide, can be denatured and hybridized to radioactive RNA. Hybrid formation is detected using autoradiography (Figure A–9).

Electrophoresis

Electrophoresis is a technique that measures the rate of migration of charged molecules in a liquid-containing medium when an electrical field is applied to the liquid. Negatively charged molecules (anions) will migrate toward the positive electrode (anode), and positively charged molecules (cations) will migrate toward the negative electrode (cathode). Several factors affect the rate of migration, including the strength of the electrical field and the molecular sieving action of the medium (paper, starch, or gel) in which migration takes place. Since proteins and nucleic acids are electrically charged, electrophoresis has been used extensively to provide information about the size, conformation, and net charge of these macromolecules. On a larger scale, electrophoresis provides a method of fractionation that can be used to isolate individual components in mixtures of proteins or nucleic acids.

More recently, the analytical separation of proteins has been enhanced with the development of two-dimensional electrophoresis techniques, in which separation in the first dimension is by net charge and in the second dimension by size and molecular weight.

In practice, electrophoresis employs a buffer system, a medium (paper, cellulose acetate, starch gel, agarose gel, or polyacrylamide gel), and a source of direct current. Samples are applied, and current is passed through the system for an appropriate time. Following migration of the molecules, the gel or paper may be treated with selective stains to reveal the location of the separated components (Figure A–10).

DNA Sequencing

The ability to determine directly the sequence of DNA segments has added immensely to our understanding of gene structure and the mechanisms of gene regulation. Although techniques for determining the base composition of DNA were available in the 1940s, it was not until the 1960s that methods providing direct chemical analysis of nucleotide sequence were developed and used. These first methods were based on those employed in determining the amino acid sequence of proteins, and were time consuming and laborious. For example, in 1965, Robert Holley determined the sequence of a tRNA molecule consisting of 74 nucleotides, a task that required almost a year of concentrated effort.

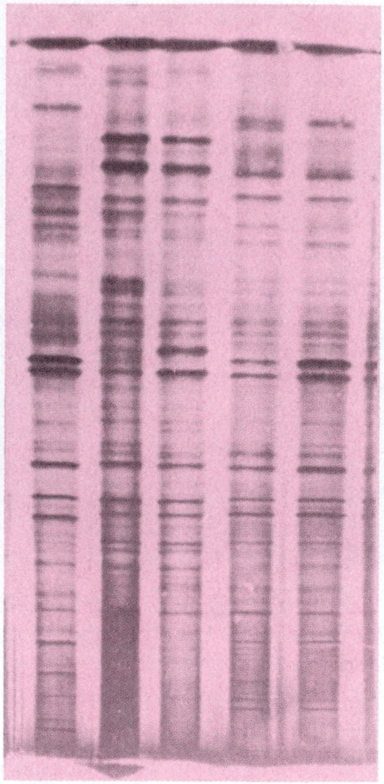

FIGURE A–10 Coomassie blue-stained protein slab gel.

In the 1970s, more efficient and direct methods of nucleotide sequencing were developed. These methods advanced in parallel with recombinant DNA techniques that allow the isolation of large quantities of purified DNA segments from any organism. A chemical method, developed by Allan Maxam and Walter Gilbert, cleaves DNA at specific bases. A second method, developed by Fred Sanger and colleagues, synthesizes a stretch of DNA that terminates at a given base. In both methods, the DNA to be sequenced is subjected to four individual reactions (one for each base). The products of the four reactions are a series of DNA fragments that differ in length by only one nucleotide. These reaction products are electrophoretically separated in four adjacent lanes on a gel. Each band on the gel corresponds to a base, and the sequence of the DNA segment can be read from the bands on the gel.

In the Maxam–Gilbert method, the complementary strands of the DNA fragment to be sequenced are separated and recovered. The strand to be sequenced is labeled at its 5′ end with radioactive ^{32}P, using the enzyme polynucleotide kinase. This provides a means of identifying specific DNA fragments after gel electrophoresis. In the next step, aliquots of the DNA are subjected to each of four chemical treatments that cleave the strand at a specific nucleotide. The reaction is carried out for a limited time so that any given molecule is cleaved at only a small number of the target nucleotides. The result is a collection of fragments, all labeled at the 5′ end, but differing in length, depending on the point of cleavage.

Four different reactions cleave DNA at guanine (G > A), adenine (A > G), cytosine alone (C), or cytosine and thymine (C + T). In these reactions, purines are cleaved by using dimethylsulfate. This reagent methylates guanine far more efficiently than adenine, and when heat is applied, the strand is broken at the methylated site, producing a DNA fragment most frequently cleaved at a G residue (G > A). The process can be reversed by cleaving the strand in acid, producing fragments most often cleaved at an A residue (A > G). The reaction for pyrimidines uses hydrazine, which cleaves both cytosine and thymine; however, in high salt (2M NaCl), only cytosine reacts. Thus, in the two reactions, one represents only cytosine (C) and the other represents cytosine and thymine (C + T).

For each set of reactions, the fragments produced are subjected to gel electrophoresis in adjacent lanes on a gel. Each reaction contains a series of fragments in which strand breakage has occurred at one of the target bases, resulting in a series of fragments that differ in length. Under the influence of the electric field during electrophoresis, the different-sized fragments separate from each other, with the smallest fragments migrating the farthest. Because the DNA fragments contain a radioactive 5′ end, the gel is subjected to autoradiography by placing a sheet of X-ray film over it for an appropriate exposure time. The DNA sequence is then analyzed by reading the bands on the gel from the bottom up, reading across all four lanes. In the example shown in Figure A–11, the first bases on the gel read GCGG.

The Sanger method for DNA sequencing relies on an initial enzymatic treatment. The DNA strand to be sequenced is used as a template for the synthesis of a new DNA strand catalyzed by DNA polymerase I. This method employs a modified dideoxynucleoside triphosphate to generate a series of DNA fragments. The dideoxynucleoside triphosphates lack a 3′-OH group, which allows them to be added to a DNA strand undergoing synthesis. But because they lack a 3′-OH group, no nucleotide can be added to them, causing termination of strand synthesis and producing a DNA

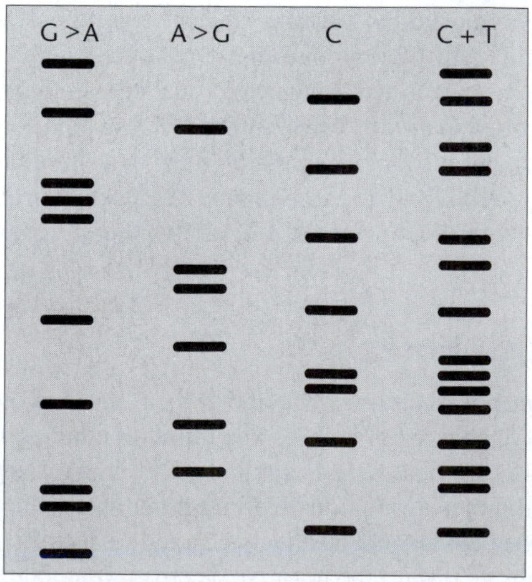

FIGURE A–11 Nucleotide sequence derived by the Maxam–Gilbert method.

fragment. Four separate reaction mixtures are set up, each with a template DNA strand, a primer, all four radioactive nucleoside triphosphates, and a small amount of a single dideoxyribonucleoside triphosphate. Each of the four reactions contains a different dideoxynucleoside triphosphate that acts as a chain terminator. Because only a small amount of the modified nucleoside is used in each reaction, the newly synthesized strands are randomly terminated, producing a collection of fragments.

After synthesis, the radioactive fragments are separated by electrophoresis in four adjacent lanes, one corresponding to each of the reactions. The fragments are visualized by autoradiography, and the gel is analyzed by reading the sequence from the bottom, as in the Maxam–Gilbert reaction. In the example shown in Figure A–12, the first band is in the lane with ddT, so it is a T residue, and the next few bands have the sequence GCAATCG.

DNA sequencing is the heart of genome projects. Using a combination of recombinant DNA techniques and nucleotide sequencing, the genomes of more than 50 species of prokaryotes have been sequenced, with hundreds more projects underway. The Human Genome Project sponsored by the U.S. Department of Energy and the National Institutes of Health and a private project sponsored by the biotechnology company Celera have reported a draft sequence of the more than 3 billion nucleotides in the human genome. Genome projects for a small, but growing, number of other eukaryotic species have also been completed with the use of DNA sequencing technology.

The Polymerase Chain Reaction

The polymerase chain reaction (PCR) technique relies on the fact that the enzyme DNA polymerase requires a double-stranded primer region of DNA to initiate DNA synthesis. In the PCR technique, that primer is supplied in the form of a synthetic oligonucleotide, allowing the replication of a specific region of DNA. In other words, DNA polymerase can be directed to replicate only one specific region from an entire genome. Because these same primer sites are copied onto the newly synthesized strands of DNA, they can be used as primers after each round of replication. By repeating the replication process several to many times, a region of DNA can be amplified millions or billions of times, producing enough copies of the DNA for experimental purposes without the need for cloning (Figure A–13).

Instead of serving as a single technique, the PCR reaction has become a versatile tool used in combination with other methods and has become the basis for a technical revolution in molecular genetics. The PCR method is routinely used to detect mutations, produce *in vitro* mutations, diagnose genetic disorders, prepare DNA for sequencing, identify viruses and bacteria in infectious diseases, amplify DNA from fossils, analyze genetic defects in gametes and single cells from human embryos, and a host of other applications. Two such applications are described next.

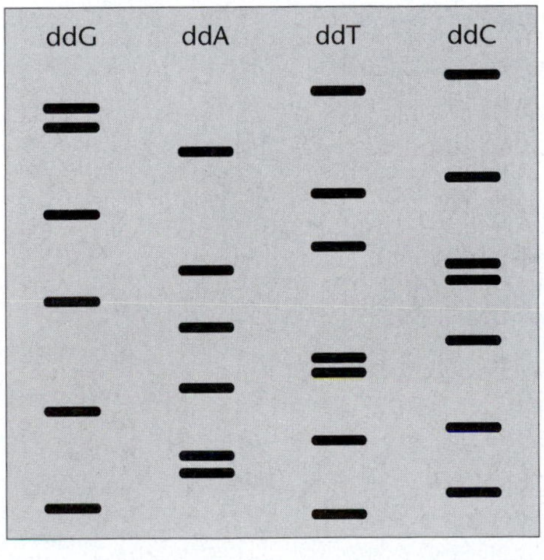

FIGURE A–12 Nucleotide sequence derived by the Sanger method.

Mutation Detection

A PCR-based method called single-strand conformation polymorphism (SSCP) can be used to screen for mutations caused by single base substitutions. Under certain conditions, single-stranded DNA fragments fold into nucleotide-sequence-dependent conformations. These conformations affect the mobility of the fragment during electrophoresis. Single base substitutions change the conformation of the DNA, altering its electrophoretic mobility. The method is fast, is simple, and does not require DNA sequencing to detect the single base changes.

To detect mutations using SSCP, the DNA of interest is first amplified by PCR. After amplification, the DNA is converted to single strands by boiling and is then loaded onto a gel for electrophoresis. DNA from a normal gene is similarly amplified, denatured, and loaded in adjacent lanes. After electrophoresis, any variations in base sequence in the DNA being tested can be detected as a shifted band. The method gives best results when the DNA fragments being tested are around 200 base pairs, so screening an entire gene requires the use of several to many fragments, each about 200 base pairs in length.

In the development of new PCR-based techniques, one technique is often piggybacked onto another. While SSCP is a rapid and simple technique, its accuracy is somewhat limited. DNA sequencing is a more accurate technique for detecting single base changes, but it is relatively slow and laborious. Recently, the two techniques have been combined into a method called *dideoxy fingerprinting (ddF)*. The technique begins by amplifying a DNA segment by PCR. The amplified segment is used as a template for one of the Sanger dideoxy sequencing reactions, and the gel is run under conditions in which the position of the band reflects both size (like DNA sequencing) and conformation (like SSCP). This combined technique is three times faster than sequencing alone and, although more complex than SSCP, has a degree of accuracy approaching 100 percent.

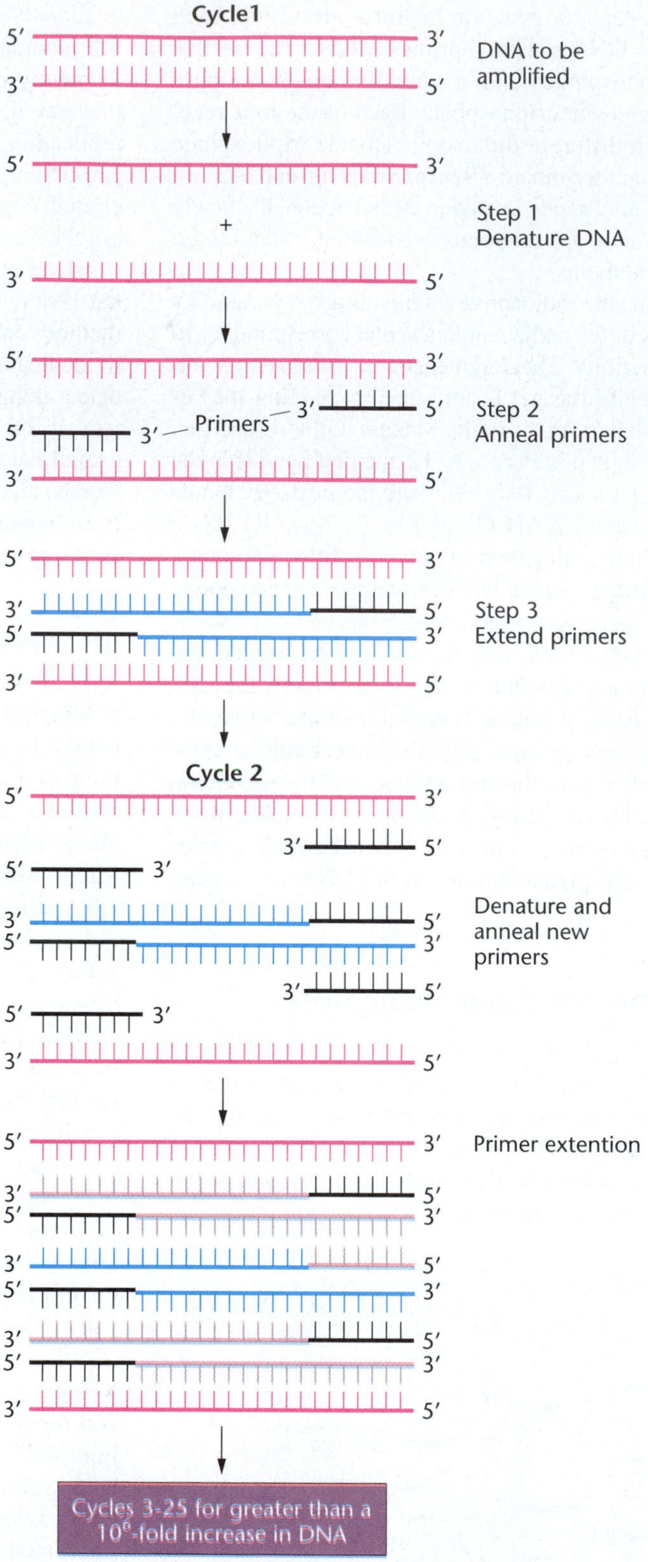

FIGURE A–13 PCR amplification. In the polymerase chain reaction (PCR) method, first the DNA to be amplified is denatured into single strands, and then each strand is annealed to a different primer. The primers are synthetic oligonucleotides complementary to sequences flanking the region to be amplified. DNA polymerase is added along with nucleotides to extend the primers in the 3′ direction, resulting in a double-stranded DNA molecule with the primers incorporated into the newly synthesized strand. In a second PCR cycle, the products of the first cycle are denatured into single strands, primers are added, and they are annealed and extended by DNA polymerase. Repeated cycles can amplify the original DNA sequence more than a million times.

PCR Cycle Sequencing

Conventional DNA sequencing requires subcloning of DNA into vectors, followed by growth of a host colony and extraction and purification of the cloned DNA. This DNA is then used as a template in the four reactions for DNA sequencing. The process is completed by gel electrophoresis. The technique of cycle sequencing incorporates parts of the PCR reaction with parts of the standard reaction for the dideoxy method of DNA sequencing. A double-stranded DNA segment to be sequenced is first heated to form single strands that will serve as templates. Primers are added along with DNA polymerase, which begins the process of DNA replication. Extension of the primer continues until a labeled dideoxynucleotide is incorporated. The labeled strand is separated from the template (by heating) and is then used in another round of replication. Next, the amplified extension products are analyzed by gel electrophoresis. This adaptation of DNA sequencing is faster than conventional sequencing, requires only a very small amount of template, and can be used to directly sequence plasmid or phage clones. To further simplify the process, commercial kits are available that provide all required materials and solutions.

Versatility: The Advantage of PCR

The preceding examples illustrate the versatility of the polymerase chain reaction and some of its applications in molecular genetics. This remarkable technique has so many adaptations and variations, that several journals are devoted to reports on PCR.

Appendix B
Glossary

A-DNA An alternative form of the right-handed double-helical structure of DNA in which the helix is more tightly coiled, with 11 base pairs per full turn of the helix. In this form, the bases in the helix are displaced laterally and tilted in relation to the longitudinal axis. It is not yet clear whether this form has biological significance.

abortive transduction An event in which transducing DNA fails to be incorporated into the recipient chromosome. (See *transduction*.)

acentric chromosome Chromosome or chromosome fragment with no centromere.

acquired immunodeficiency syndrome (AIDS) An infectious disease caused by a retrovirus designated as human immunodeficiency virus (HIV). The disease is characterized by a gradual depletion of T lymphocytes, recurring fever, weight loss, multiple opportunistic infections, and rare forms of pneumonia and cancer associated with collapse of the immune system.

acridine dyes A class of organic compounds that bind to DNA and intercalate into the double-stranded structure, producing local disruptions of base pairing. The disruptions result in additions or deletions in the next round of replication.

acrocentric chromosome Chromosome with the centromere located very close to one end. Human chromosomes 13, 14, 15, 21, and 22 are acrocentric.

active immunity Immunity gained by direct exposure to antigens followed by antibody production.

active site The portion of a protein, usually an enzyme, whose structural integrity is required for function (e.g., the substrate binding site of an enzyme).

adaptation A heritable component of the phenotype that confers an advantage in survival and reproductive success. Also, the process by which organisms adapt to the current environmental conditions.

additive genes See *polygenic inheritance*.

additive variance Genetic variance that is attributed to the substitution of one allele for another at a given locus. This variance can be used to predict the rate of response to phenotypic selection in quantitative traits.

albinism A condition caused by the lack of melanin production in the iris, hair, and skin. In humans, it is most often inherited as an autosomal recessive condition.

aleurone layer In seeds, the outer layer of the endosperm.

alkaptonuria An autosomal recessive condition in humans that is caused by lack of the enzyme homogentisic acid oxidase. Urine of homozygous individuals turns dark upon standing, due to oxidation of excreted homogentisic acid. The cartilage of homozygous adults blackens from deposition of a pigment derived from homogentisic acid. Such individuals often develop arthritic conditions.

allele One of the possible mutational states of a gene, distinguished from other alleles by phenotypic effects.

allele frequency Measurement of the proportion of individuals in a population that carry a particular allele.

allele-specific nucleotide (ASO) Synthetic nucleotides, usually 15 to 20 bp in length, that, under carefully controlled conditions, will hybridize only to a complementary sequence with a perfect match. Under these same conditions, ASOs with a one-nucleotide mismatch will not hybridize.

allelic exclusion In a plasma cell that is heterozygous for an immunoglobulin gene, the selective action of only one allele.

allelism test See *complementation test*.

allolactose A lactose derivative that acts as the inducer for the *lac* operon.

allopatric speciation Process of speciation associated with geographic isolation.

allopolyploid Polyploid condition formed by the union of two or more distinct chromosome sets, with a subsequent doubling of the number of chromosomes.

allosteric effect Conformational change in the active site of a protein, brought about by interaction with an effector molecule.

allotetraploid Diploid for two genomes derived from different species.

allozyme An allelic form of a protein that can be distinguished from other forms by electrophoresis.

alpha fetoprotein (AFP) A 70-kd glycoprotein synthesized in embryonic development by the yolk sac. High levels of this protein in the amniotic fluid are associated with neural-tube defects such as spina bifida. Lower than normal levels may be associated with Down syndrome.

alternative splicing The generation of different protein molecules from the same pre-mRNA by changing the number and order of exons in the mRNA product.

***Alu* sequence** An interspersed DNA sequence of approximately 300 bp found in the genome of primates that is cleaved by the restriction enzyme *Alu* I. *Alu* sequences are composed of a head-to-tail dimer, with the first monomer having approximately 140 bp and the second having approximately 170 bp. In humans, the sequences are dispersed throughout the genome and are present in 300,000 to 600,000 copies, constituting some 3 to 6 percent of the genome. (See *SINEs*.)

amber codon The codon UAG, which does not code for insertion of an amino acid, but for chain termination.

Ames test An assay developed by Bruce Ames to detect mutagenic and carcinogenic compounds, using reversion to histidine independence in the bacterium *Salmonella typhimurium.*

amino acid Any of the subunit building blocks that are covalently linked to form proteins.

aminoacyl tRNA Covalently linked combination of an amino acid and a tRNA molecule.

amniocentesis A procedure used to test for fetal defects, in which fluid and fetal cells are withdrawn from the amniotic layer surrounding the fetus.

amphidiploid See *allotetraploid.*

anabolism The metabolic synthesis of complex molecules from less complex precursors.

analog A chemical compound structurally similar to another chemical compound, but differing by a single functional group (e.g., 5-bromodeoxyuridine is an analog of thymidine).

anaphase Stage of cell division in which chromosomes begin moving to opposite poles of the cell.

anaphase I The stage in the first meiotic division during which members of homologous pairs of chromosomes separate from one another.

aneuploidy A condition in which the number of chromosomes is not an exact multiple of the haploid set.

angstrom Unit of length equal to 10^{-10} meter. Abbreviated Å.

annotation Analysis of genomic nucleotide sequence data in order to identify the protein-coding genes, the non-protein-coding genes, the genes' regulatory sequences, and the genes' function(s).

antibody Protein (immunoglobulin) produced in response to an antigenic stimulus, with the capacity to bind specifically to the antigen.

anticipation A phenomenon first observed in myotonic dystrophy, whereby the severity of the symptoms increases from generation to generation and the age of onset decreases from generation to generation. This phenomenon is caused by the expansion of trinucleotide repeats within or near a gene.

anticodon The nucleotide triplet in a tRNA molecule that is complementary to, and binds to, the codon triplet in an mRNA molecule.

antigen A molecule, often a cell surface protein, that is capable of eliciting the formation of antibodies.

antiparallel Describing molecules in parallel alignment, but running in opposite directions. Most commonly used to describe the opposite orientations of the two strands of a DNA molecule.

apoptosis A genetically controlled program of cell death, activated as part of normal development or as a result of cell damage.

ascospore A meiotic spore produced in certain fungi.

ascus In fungi, the sac enclosing the four or eight ascospores.

asexual reproduction Production of offspring in the absence of any sexual process.

assortative mating Nonrandom mating between males and females of a species. Selection of mates with the same genotype is positive; selection of mates with opposite genotypes is negative.

ATP Adenosine triphosphate.

attached-X chromosome Two conjoined X chromosomes that share a single centromere.

attenuator A nucleotide sequence between the promoter and the structural gene of some operons that can act to regulate the transit of RNA polymerase, reducing transcription of the related structural gene.

autogamy A process of self-fertilization that results in homozygosis.

autoimmune disease The production of antibodies that results from an immune response to one's own molecules, cells, or tissues. Such a response results from the inability of the immune system to distinguish self from nonself. Examples of autoimmune diseases include arthritis, scleroderma, systemic lupus erythematosus, and juvenile-onset diabetes.

autonomously replicating sequences (ARS) Origins of replication, about 100 nucleotides in length, found in yeast chromosomes. ARS elements are also present in organelle DNA.

autopolyploidy Polyploid condition resulting from the replication of one diploid set of chromosomes.

autoradiography Production of a photographic image by radioactive decay. Used to localize radioactively labeled compounds within cells and tissues.

autosomes Chromosomes other than the sex chromosomes. In humans, there are 22 pairs of autosomes.

autotetraploid An autopolyploid condition composed of four similar genomes. In this situation, genes with two alleles (*A* and *a*) can have five genotypic classes: *AAAA* (quadraplex), *AAAa* (triplex), *Aaaa* (duplex), *Aaaa* (simplex), and *aaaa* (nulliplex).

auxotroph A mutant microorganism or cell line which requires a substance for growth that can be synthesized by wild-type strains.

back-cross A cross involving an F_1 heterozygote and one of the P_1 parents (or an organism with a genotype identical to that of one of the parents).

bacteriophage A virus that infects bacteria. (Synonym is *phage.*)

bacteriophage mu A group of phages whose genetic material behaves as an insertion sequence that can cause inactivation of host genes and rearrangement of host chromosomes.

balanced lethals Recessive, nonallelic lethal genes, each carried on different homologous chromosomes. When organisms carrying balanced lethal genes are interbred, only organisms with genotypes identical to those of the parents (heterozygotes) survive.

balanced polymorphism Genetic polymorphism maintained in a population by natural selection.

Barr body Densely staining nuclear mass seen in the somatic nuclei of mammalian females. Discovered by Murray Barr, this body is thought to represent an inactivated X chromosome.

base analog See *analog.*

base substitution A single base change in a DNA molecule that produces a mutation. There are two types of substitutions: *transitions*, in which a purine is substituted for a purine or a pyrimidine is substituted for a pyrimidine; and *transversions*, in which a purine is substituted for a pyrimidine or vice versa.

β-galactosidase A bacterial enzyme encoded by the *lacZ* gene that converts lactose into galactose and glucose.

bidirectional replication A mechanism of DNA replication in which two replication forks move in opposite directions from a common origin of replication.

biodiversity The genetic diversity present in populations and species of plants and animals.

biometry The application of statistics and statistical methods to biological problems.

biotechnology Commercial and/or industrial processes that use biological organisms or products.

bivalents Synapsed homologous chromosomes in the first prophase of meiosis.

Bombay phenotype A rare variant of the ABO system in which affected individuals do not have A or B antigens and thus appear as blood type O, even though their genotype may carry unexpressed alleles for the A or B antigens.

bottleneck Fluctuation in allele frequency that occurs when a population undergoes a temporary reduction in size.

BrdU (5-bromodeoxyuridine) A mutagenically active analog of thymidine in which the methyl group at the 5′ position in thymine is replaced by bromine.

buoyant density A property of particles (and molecules) that depends upon their actual density, as determined by partial specific volume and degree of hydration. Provides the basis for density gradient separation of molecules or particles.

CAAT box A highly conserved DNA sequence found in the untranslated promoter region of eukaryotic genes. This sequence is recognized by transcription factors.

cAMP Cyclic adenosine monophosphate. An important regulatory molecule in both prokaryotic and eukaryotic organisms.

CAP Catabolite activator protein. A protein that binds cAMP and regulates the activation of inducible operons.

carcinogen A physical or chemical agent that causes cancer.

carrier An individual who is heterozygous for a recessive trait.

catabolism A metabolic reaction in which complex molecules are broken down into simpler forms, often accompanied by the release of energy.

catabolite activator protein See *CAP*.

catabolite repression The selective inactivation of an operon by a metabolic product of the enzymes encoded by the operon.

cdc* mutation** A class of ***cell ***d***ivision ***c***ycle mutations in yeast that affect the timing and progression through the cell cycle.

cDNA DNA synthesized from an RNA template by the enzyme reverse transcriptase.

cDNA library A collection of cloned cDNA sequences.

cell cycle Sum of the phases of growth of an individual cell type; divided into G1 (gap 1), S (DNA synthesis), G2 (gap 2), and M (mitosis).

cell-free extract A preparation of the soluble fraction of cells, made by lysing cells and removing the particulate matter, such as nuclei, membranes, and organelles. Often used to carry out the synthesis of proteins by the addition of specific, exogenous mRNA molecules.

CEN In yeast, fragments of chromosomal DNA, about 120 bp in length, that, when inserted into plasmids, confer the ability to segregate during mitosis. These segments contain at least three types of sequence elements associated with centromere function.

centimeter A unit of length equal to 10^{-2} meter. Abbreviated as cm.

centimorgan A unit of distance between genes on chromosomes. One centimorgan represents a value of 1 percent crossing over between two genes. Abbreviated as cM.

central dogma The concept that information flow progresses from DNA to RNA to proteins. Although exceptions are known, this idea is central to an understanding of gene function.

centric fusion See *Robertsonian translocation*.

centriole A cytoplasmic organelle composed of nine groups of microtubules, generally arranged in triplets. Centrioles function in the generation of cilia and flagella and serve as foci for the spindles in cell division.

centromere Specialized region of a chromosome to which sister chromatids remain attached after replication, and the site to which spindle fibers attach during cell division. Location of the centromere determines the shape of the chromosome during the anaphase portion of cell division. Also known as the primary constriction.

centrosome Region of the cytoplasm that contains the centriole.

chaperone A protein that regulates the folding of a polypeptide into a functional conformation.

character An observable phenotypic attribute of an organism.

charon phages A group of genetically modified lambda phages designed to be used as vectors (carriers) for cloning foreign DNA. Named after the ferryman in Greek mythology who carried the souls of the dead across the River Styx.

chemotaxis Negative or positive response to a chemical gradient.

chiasma (pl. **chiasmata**) The crossed strands of nonsister chromatids seen in diplotene of the first meiotic division. Regarded as the cytological evidence for exchange of chromosomal material, or crossing over.

chi-square (χ^2) analysis Statistical test to determine if an observed set of data fits a theoretical expectation.

chloroplast A cytoplasmic self-replicating organelle containing chlorophyll. The site of photosynthesis.

chorionic villus sampling (CVS) A technique of prenatal diagnosis that intravaginally retrieves fetal cells from the chorion and uses them to detect cytogenetic and biochemical defects in the embryo.

chromatid One of the longitudinal subunits of a replicated chromosome, joined to its sister chromatid at the centromere.

chromatin Term used to describe the complex of DNA, RNA, histones, and nonhistone proteins that make up uncoiled chromosomes characteristic of the eukaryotic interphase nucleus.

chromatography Technique for the separation of a mixture of solubilized molecules by their differential migration over a substrate.

chromocenter An aggregation of centromeres and heterochromatic elements of polytene chromosomes.

chromomere A coiled, beadlike region of a chromosome, most easily visible during cell division. The aligned chromomeres of polytene chromosomes are responsible for their distinctive banding pattern.

chromosomal aberration Any change resulting in the duplication, deletion, or rearrangement of chromosomal material.

chromosomal mutation See *chromosomal aberration.*

chromosomal polymorphism Alternative structures or arrangements of a chromosome that are carried by members of a population.

chromosome In prokaryotes, an intact DNA molecule containing the genome; in eukaryotes, a DNA molecule complexed with RNA and proteins to form a threadlike structure containing genetic information arranged in a linear sequence and visible during mitosis and meiosis.

chromosome banding Technique for the differential staining of mitotic or meiotic chromosomes to produce a characteristic banding pattern or for the selective staining of certain chromosomal regions such as centromeres, the nucleolus organizer regions, and GC- or AT-rich regions. Not to be confused with the banding pattern present in polytene chromosomes, which is produced by the alignment of chromomeres.

chromosome map A diagram showing the location of genes on chromosomes.

chromosome puff A localized uncoiling and swelling in a polytene chromosome, usually regarded as a sign of active transcription.

chromosome theory of inheritance The idea put forward by Walter Sutton and Theodore Boveri that chromosomes are the carriers of genes and the basis for the Mendelian mechanisms of segregation and independent assortment.

chromosome walking A method for analyzing long stretches of DNA, in which the end of a cloned segment of DNA is subcloned and used as a probe to identify other clones that overlap the first clone.

***cis* configuration** The arrangement of two genes or two mutant sites within a gene on the same homolog, such as

$$\frac{a^1 \quad a^2}{+ \quad\; +}$$

Contrasts with a *trans* arrangement, where the mutant alleles are located on opposite homologs.

***cis-trans* test** A genetic test to determine whether two mutations are located within the same cistron.

cistron The portion of a DNA molecule that codes for a single polypeptide chain; defined by a genetic test as a region within which two mutations cannot complement each other.

cline A gradient of genotype or phenotype distributed over a geographic range.

clonal selection Theory of the immune system which proposes that antibody diversity precedes exposure to the antigen and that the antigen functions to select the cells containing its specific antibody to undergo proliferation.

clone Identical molecules, cells, or organisms all derived from a single ancestor by asexual or parasexual methods—for example, a DNA segment that has been enzymatically inserted into a plasmid or chromosome of a phage or a bacterium and replicated to form many copies.

cloned library A collection of cloned DNA molecules representing all or part of an individual's genome.

code See *genetic code.*

codominance Condition in which the phenotypic effects of a gene's alleles are fully and simultaneously expressed in the heterozygote.

codon A triplet of nucleotides that specifies or encodes the information for a single amino acid. Sixty-one codons specify the amino acids used in proteins, and three codons signal termination of growth of the polypeptide chain.

coefficient of coincidence A ratio of the observed number of double-crossovers divided by the expected number of such crossovers.

coefficient of inbreeding The probability that two alleles present in a zygote are descended from a common ancestor.

coefficient of selection A measurement of the reproductive disadvantage of a given genotype in a population. If, for genotype *aa*, only 99 of 100 individuals reproduce, then the selection coefficient (s) is 0.1.

colchicine An alkaloid compound that inhibits spindle formation during cell division. Used in the preparation of karyotypes to collect a large population of cells inhibited at the metaphase stage of mitosis.

colinearity The linear relationship between the nucleotide sequence in a gene (or the RNA transcribed from it) and the order of amino acids in the polypeptide chain specified by the gene.

competence In bacteria, the transient state or condition during which the cell can bind and internalize exogenous DNA molecules, making transformation possible.

complementarity Chemical affinity between nitrogenous bases as a result of hydrogen bonding. Responsible for the base pairing between the strands of the DNA double helix.

complementation test A genetic test to determine whether two mutations occur within the same gene. If two mutations are introduced into a cell simultaneously and produce a wild-type phenotype (i.e., they complement each other), they are often nonallelic. If a mutant phenotype is produced, the mutations are noncomplementing and are often allelic.

complete linkage A condition in which two genes are located so close to each other that no recombination occurs between them.

complex locus A gene within which a set of functionally related pseudoalleles can be identified by recombinational analysis (e.g., the *bithorax* locus in *Drosophila*).

complexity The total number of nucleotides or nucleotide pairs in a population of nucleic acid molecules, as determined by reassociation kinetics.

concatemer A chain or linear series of subunits linked together. The process of forming a concatemer is called concatenation (e.g., multiple units of a phage genome produced during replication).

concordance Pairs or groups of individuals identical in their phenotype. In studies of twins, a condition in which both twins exhibit or fail to exhibit a trait under investigation.

conditional mutation A mutation that expresses a wild-type phenotype under certain (permissive) conditions and a mutant phenotype under other (restrictive) conditions.

conjugation Temporary fusion of two single-celled organisms for the sexual transfer of genetic material.

consanguineous Related by a common ancestor within the previous few generations.

consensus sequence A common, although not necessarily identical, sequence of nucleotides in DNA or amino acids in proteins.

conservation genetics The branch of genetics concerned with the preservation and maintenance of wild species of plants and animals in their natural environments.

continuous variation Phenotype variation exhibited by quantitative traits distributed from one phenotypic extreme to another in an overlapping or continuous fashion.

cosmid A vector designed to allow cloning of large segments of foreign DNA. Cosmids are hybrids composed of the *cos* sites of lambda inserted into a plasmid. In cloning, the recombinant DNA molecules are packaged into phage protein coats, and after infection of bacterial cells, the recombinant molecule replicates and can be maintained as a plasmid.

coupling conformation See *cis configuration.*

covalent bond A nonionic chemical bond formed by the sharing of electrons.

cri-du-chat syndrome A clinical syndrome in humans produced by a deletion of a portion of the short arm of chromosome 5. Afflicted infants have a distinctive cry which sounds like that of a cat.

crossing over The exchange of chromosomal material (parts of chromosomal arms) between homologous chromosomes by breakage and reunion. The exchange of material between nonsister chromatids during meiosis is the basis of genetic recombination.

cross-reacting material (CRM) Nonfunctional form of an enzyme, produced by a mutant gene, which is recognized by antibodies made against the normal enzyme.

C-terminal amino acid The terminal amino acid in a peptide chain that carries a free carboxyl group.

C terminus The end of a polypeptide that carries a free carboxyl group of the last amino acid. By convention, the structural formula of polypeptides is written with the C terminus at the right.

***C* value** The haploid amount of DNA present in a genome.

***C*-value paradox** The apparent paradox that there is no relationship between the size of the genome and the evolutionary complexity of species. For example, the *C* value (haploid-genome size) of amphibians varies by a factor of 100.

cyclins A class of proteins found in eukaryotic cells that are synthesized and degraded in synchrony with the cell cycle and regulate passage through stages of the cycle.

cytogenetics A branch of biology in which the techniques of both cytology and genetics are used to study heredity.

cytokinesis The division or separation of the cytoplasm during mitosis or meiosis.

cytological map A diagram showing the location of genes at particular chromosomal sites.

cytoplasmic inheritance Non-Mendelian form of inheritance involving genetic information transmitted by self-replicating cytoplasmic organelles, such as mitochondria and chloroplasts.

cytoskeleton An internal array of microtubules, microfilaments, and intermediate filaments that confers shape and the ability to move on a eukaryotic cell.

dalton (Da) A unit of mass equal to that of the hydrogen atom, which is 1.67×10^{-24} gram. A unit used in designating molecular weights.

Darwinian fitness See *fitness.*

deficiency (deletion) A chromosomal mutation involving the loss or deletion of chromosomal material.

degenerate code Term used to describe the genetic code, in which a given amino acid may be represented by more than one codon. Some amino acids (leucine) have six codons; others (isoleucine) have three; and so on.

deletion See *deficiency.*

deme A local interbreeding population.

denatured DNA DNA molecules that have been separated into single strands.

de novo Newly arising; synthesized from less complex precursors rather than having been produced by modification of an existing molecule.

density gradient centrifugation A method of separating macromolecular mixtures by use of centrifugal force and solutions of varying density. In buoyant density gradient centrifugation, using cesium chloride, the cesium solution establishes a gradient under the influence of the centrifugal field, and the macromolecules, such as DNA, sediment in the gradient until the density of the cesium chloride solution equals that of the macromolecules.

deoxyribonuclease A class of enzymes that breaks down DNA into oligonucleotide fragments by introducing single-stranded breaks into the double helix.

deoxyribonucleic acid (DNA) A macromolecule usually consisting of antiparallel polynucleotide chains held together by hydrogen bonds, in which the sugar residues are deoxyribose. The primary carrier of genetic information.

deoxyribose The five-carbon sugar associated with the deoxyribonucleotides found in DNA.

dermatoglyphics The study of the surface ridges of the skin, especially of the hands and feet.

determination A regulatory event that establishes a specific pattern of future gene activity and developmental fate for a given cell.

diakinesis The final stage of meiotic prophase I in which the chromosomes become tightly coiled and compacted and move toward the periphery of the nucleus.

dicentric chromosome A chromosome having two centromeres.

dideoxynucleotide A nucleotide containing a dexoyribose sugar that lacks a 3′ hydroxyl group. Stops further chain elongation when incorporated into a growing polynucleotide; used in the Sanger method of DNA sequencing.

differentiation The process of complex changes by which cells and tissues attain their adult structure and functional capacity.

dihybrid cross A genetic cross involving two characters, in which the parents possess different forms of each character (e.g., tall, round × short, wrinkled peas).

diploid A condition in which each chromosome exists in pairs; having two of each chromosome.

diplotene A stage of meiotic prophase immediately after pachytene. In diplotene, one pair of sister chromatids begins separating from the other, and chiasmata become visible. These overlaps move laterally toward the ends of the chromatids (terminalization).

directional selection A selective force that changes the frequency of an allele in a given direction, either toward fixation or toward elimination.

discontinuous replication of DNA The synthesis of DNA in discontinuous fragments on the lagging strand of the replication fork. The fragments, known as Okazaki fragments, are joined by DNA ligase to form a continuous strand.

discontinuous variation Phenotypic data that fall into two or more distinct, nonoverlapping classes.

discordance In studies of twins, a situation where one twin expresses a trait, but the other does not.

disjunction The separation of chromosomes at the anaphase stage of cell division.

disruptive selection Simultaneous selection for phenotypic extremes in a population, usually resulting in the production of two phenotypically discontinuous strains.

dizygotic twins Twins produced from separate fertilization events; two ova fertilized independently. Also known as fraternal twins.

DNA See *deoxyribonucleic acid.*

DNA footprinting See *footprinting.*

DNA gyrase One of the DNA topoisomerases that functions during DNA replication to reduce molecular tension caused by supercoiling. DNA gyrase produces and then seals double-stranded breaks.

DNA ligase An enzyme that forms a covalent bond between the 5′ end of one polynucleotide chain and the 3′ end of another polynucleotide chain. Also called polynucleotide-joining enzyme.

DNA polymerase An enzyme that catalyzes the synthesis of DNA from deoxyribonucleotides and a template DNA molecule.

DNase Deoxyribonucleosidase, an enzyme that degrades or breaks down DNA into fragments or constitutive nucleotides.

dominance The expression of a trait in the heterozygous condition.

dominant suppression A form of epistasis in which a dominant allele at one locus suppresses the effect of a dominant allele at another locus, resulting in a 13:3 phenotypic ratio.

dosage compensation A genetic mechanism that regulates the levels of gene products at certain loci on the X chromosome in mammals such that males and females have equal amounts of a gene product. In mammals, this effect is accomplished by random inactivation of one X chromosome.

double-crossover Two separate events of chromosome breakage and exchange occurring within the same tetrad.

double helix The model for DNA structure proposed by James Watson and Francis Crick, involving two antiparallel, hydrogen-bonded polynucleotide chains wound into a right-handed helical configuration, with 10 base pairs per full turn of the double helix. Often called B-DNA.

Duchenne muscular dystrophy An X-linked recessive genetic disorder caused by a mutation in the gene for dystrophin, a protein found in muscle cells.

duplication A chromosomal aberration in which a segment of the chromosome is repeated.

dyad The products of tetrad separation or disjunction at the first meiotic prophase. Consists of two sister chromatids joined at the centromere.

dystrophin See *Duchenne muscular dystrophy.*

effective population size The number of individuals in a population that have an equal probability of contributing gametes to the next generation.

effector molecule Small, biologically active molecule that acts to regulate the activity of a protein by binding to a specific receptor site on the protein.

electrophoresis A technique used to separate a mixture of molecules by their differential migration through a stationary phase (such as a gel) in an electrical field.

endocytosis The uptake by a cell of fluids, macromolecules, or particles by pinocytosis, phagocytosis, or receptor-mediated endocytosis.

endomitosis Chromosomal replication that is not accompanied by either nuclear or cytoplasmic division.

endonuclease An enzyme that hydrolyzes internal phosphodiester bonds in a polynucleotide chain or nucleic acid molecule.

endoplasmic reticulum A membranous organelle system in the cytoplasm of eukaryotic cells. The outer surface of the membranes may be ribosome-studded (rough) or smooth ER.

endopolyploidy The increase in chromosome sets that results from endomitotic replication within somatic nuclei.

endosymbiont theory The proposal that self-replicating cellular organelles such as mitochondria and chloroplasts were originally free-living organisms that entered into a symbiotic relationship with nucleated cells.

enhancer Originally identified as a 72-bp sequence in the genome of SV40, the virus that increases the transcriptional activity of nearby structural genes. Similar sequences that enhance transcription have been identified in the genomes of eukaryotic cells. Enhancers can act over a distance of thousands of base pairs and can be located 5′, 3′ or internal to the gene that they affect and thus are different from promoters.

environment The complex of geographic, climatic, and biotic factors within which an organism lives.

enzyme A protein or complex of proteins that catalyzes a specific biochemical reaction.

epigenesis The idea that an organism develops by the appearance and growth of new structures. Opposed to preformationism, which holds that development is the growth of structures already present in the egg.

episome A circular genetic element in bacterial cells that can replicate independently of the bacterial chromosome or integrate and replicate as part of the chromosome.

epistasis Nonreciprocal interaction between genes such that one gene interferes with or prevents the expression of another gene. For example, in *Drosophila*, the recessive gene *eyeless*, when homozygous, prevents the expression of eye-color genes present in the genome.

epitope The portion of a macromolecule or cell that acts to elicit an antibody response; an antigenic determinant. A complex molecule or cell can contain several such sites.

equational division A division of each chromosome into longitudinal halves that are distributed into two daughter nuclei. Chromosome division in mitosis is an example of equational division.

equatorial plate See *metaphase plate.*

euchromatin Chromatin or chromosomal regions that are lightly staining and are relatively uncoiled during the interphase portion of the cell cycle. Euchromatic regions contain most of the structural genes.

eugenics The improvement of the human species by selective breeding. Positive eugenics refers to the promotion of breeding of those with favorable genes, and negative eugenics refers to the discouragement of breeding among those with undesirable traits.

eukaryotes Organisms that have true nuclei and membranous organelles and whose cells demonstrate mitosis and meiosis.

euphenics Medical or genetic intervention to reduce the impact of defective genotypes.

euploid Polyploid with a number of chromosomes that is an exact multiple of the number of chromosomes in a basic chromosome set.

evolution The origin of plants and animals from preexisting types. Descent with modifications.

excision repair Removal of damaged DNA segments, followed by repair. Excision can include the removal of individual bases (base repair) or a stretch of damaged nucleotides (nucleotide repair). The gap created by excision is filled by polymerase, and the ends are ligated to form an intact molecule.

exon (extron) The one or more DNA segments of a gene that are transcribed and translated into protein.

exonuclease An enzyme that breaks down nucleic acid molecules by breaking the phosphodiester bonds at the 3′ or 5′ terminal nucleotides.

expressed sequence tags (ESTs) All or part of the nucleotide sequence of cDNA clones. Used as markers in construction of genetic maps.

expression vector Plasmids or phages carrying promoter regions designed to cause expression of inserted DNA sequences.

expressivity The degree or range in which a phenotype for a given trait is expressed.

extranuclear inheritance Transmission of traits by genetic information contained in cytoplasmic organelles such as mitochondria and chloroplasts.

F^+ cell A bacterial cell that contains a fertility (F) factor. Acts as a donor in bacterial conjugation.

F^- cell A bacterial cell that does not contain a fertility (F) factor. Acts as a recipient in bacterial conjugation.

F factor An episome in bacterial cells that confers the ability to act as a donor in conjugation.

F′ factor A fertility (F) factor that contains a portion of the bacterial chromosome.

F_1 generation First filial generation; the progeny resulting from the first cross in a series.

F_2 generation Second filial generation; the progeny resulting from a cross of the F_1 generation.

F pilus See *pilus*.

facultative heterochromatin A chromosome or chromosome region that becomes heterochromatic only under certain conditions (e.g., the mammalian X chromosome).

familial trait A trait transmitted through and expressed by members of a family.

fate map A diagram or ``map'' of an embryo, showing the location of cells whose developmental fate is known.

fertility (F) factor See *F factor*.

fiber FISH A form of fluorescence *in situ* hybridization in which a probe is hybridized to the stretched chromatin fibers of a chromosome.

filial generations See *F_1 generation* and *F_2 generation*.

fingerprint The pattern of ridges and whorls on the tip of a finger; the pattern obtained by enzymatically cleaving a protein or nucleic acid and subjecting the digest to two-dimensional chromatography or electrophoresis.

FISH See *fluorescence in situ hybridization*.

fitness A measure of the relative survival and reproductive success of a given individual or genotype.

fixation In population genetics, a condition in which all members of a population are homozygous for a given allele.

fluctuation test A statistical test developed by Salvadore Luria and Max Delbrück to determine whether bacterial mutations arise spontaneously or are produced in response to selective agents.

fluorescence *in situ* hybridization (FISH) A method of *in situ* hybridization that uses probes labeled with a fluorescent tag, causing the site of hybridization to fluoresce when viewed in ultraviolet light under a microscope.

flush-crash cycle A period of rapid population growth followed by a drastic reduction in population size.

fmet See *formylmethionine*.

folded-fiber mode A model of eukaryotic chromosome organization in which each sister chromatid consists of a single fiber, composed of double-stranded DNA and protein, that is wound up like a tightly coiled skein of yarn.

footprinting A technique for identifying a DNA sequence that binds to a particular protein, based on the idea that the phosphodiester bonds in the region covered by the protein are protected from digestion by deoxyribonucleases.

formylmethionine (fmet) A molecule derived from the amino acid methionine by attachment of a formyl group to its terminal amino group. This molecule is the first amino acid inserted in all bacterial polypeptides. Also known as *N*-formyl methionine.

founder effect A form of genetic drift. The establishment of a population by a small number of individuals whose genotypes carry only a fraction of the different kinds of alleles in the parental population.

fragile site A heritable gap or nonstaining region of a chromosome that can be induced to generate chromosome breaks.

fragile X syndrome A genetic disorder caused by the expansion of a CGG trinucleotide repeat and a fragile site at Xq27.3, within the *FMR-1* gene.

frameshift mutation A mutational event leading to the insertion of one or more base pairs in a gene, shifting the codon reading frame in all codons following the mutational site.

fraternal twins See *dizygotic twins*.

G1 checkpoint A point in the G1 phase of the cell cycle when a cell becomes committed to initiate DNA synthesis and continue the cycle or withdraw into the G0 resting stage.

G0 The G-0 stage of the cell cycle is a point in the G1 phase at which cells withdraw from the cell cycle and enter a nondividing, but metabolically active, state.

gain-of-function mutation A mutation that confers a new function on a protein, producing a new phenotype. In humans, the HD mutation responsible for Huntington disease may be a gain-of-function mutation.

gamete A specialized reproductive cell with a haploid number of chromosomes.

gap genes Genes expressed in contiguous domains along the anterior–posterior axis of the *Drosophila* embryo, which regulate the process of segmentation in each domain.

gene The fundamental physical unit of heredity, whose existence can be confirmed by allelic variants and which occupies a specific chromosomal locus. A DNA sequence coding for a single polypeptide.

gene amplification The process by which gene sequences are selected and differentially replicated either extrachromosomally or intrachromosomally.

gene conversion The process of nonreciprocal recombination by which one allele in a heterozygote is converted into the corresponding allele.

gene duplication An event in replication leading to the production of a tandem repeat of a gene sequence.

gene flow The gradual exchange of genes between two populations, brought about by the dispersal of gametes or the migration of individuals.

gene frequency The percentage of alleles of a given type in a population.

gene interaction Production of novel phenotypes by the interaction of alleles of different genes.

gene mutation See *point mutation.*

gene pool The total of all alleles possessed by reproductive members of a population.

generalized transduction The transduction of any gene in the bacterial genome by a phage.

genetic anticipation The phenomenon of a progressively earlier age of onset and increasing severity of symptoms for a genetic disorder in successive generations.

genetic background All genes carried in the genome other than the one being studied.

genetic code The nucleotide triplets that code for the 20 amino acids or for chain initiation or termination.

genetic counseling Analysis of risk for genetic defects in a family and the presentation of options available to avoid or ameliorate possible risks.

genetic drift Random variation in allele frequency from generation to generation. Most often observed in small populations.

genetic engineering The technique of altering the genetic constitution of cells or individuals by the selective removal, insertion, or modification of individual genes or gene sets.

genetic equilibrium Maintenance of allele frequencies at the same value in successive generations. A condition in which allele frequencies are neither increasing nor decreasing.

genetic erosion The loss of genetic diversity from a population or a species.

genetic fine structure Intragenic recombinational analysis that provides mapping information at the level of individual nucleotides.

genetic load Average number of recessive lethal genes carried in the heterozygous condition by an individual in a population.

genetic polymorphism The stable coexistence of two or more discontinuous genotypes in a population. When the frequencies of two alleles are carried to an equilibrium, the condition is called balanced polymorphism.

genetics The branch of biology that deals with heredity and the expression of inherited traits.

genome The array of genes carried by an individual.

genomic imprinting A condition where the expression of a trait depends on whether the trait has been inherited from a male or a female parent.

genomics The study of genomes, including nucleotide sequence, gene content, organization, and gene number.

genotype The specific allelic or genetic constitution of an organism; often, the allelic composition of one or a limited number of genes under investigation.

germplasm Hereditary material transmitted from generation to generation.

Goldberg–Hogness box A short nucleotide sequence 20 to 30 bp 5′ to the initiation site of eukaryotic genes to which RNA polymerase II binds. The consensus sequence is TATAAAA. Also known as TATA box.

graft-versus-host disease (GVHD) In transplants, reaction by immunologically competent cells of the donor against the antigens present on the cells of the host. In human bone marrow transplants, often a fatal condition.

gynandromorph An individual composed of cells with both male and female genotypes.

gyrase One of a class of enzymes known as topoisomerases. Gyrase converts closed circular DNA to a negatively supercoiled form prior to replication, transcription, or recombination.

H substance The carbohydrate group present on the surface of red blood cells. When unmodified, it results in blood type O; when modified by the addition of monosaccharides, it results in types A, B, and AB.

haploid A cell or organism having a single set of unpaired chromosomes. The gametic number of chromosomes.

haplotype The set of alleles from closely linked loci carried by an individual and usually inherited as a unit.

Hardy–Weinberg law The principle that both gene and genotype frequencies will remain in equilibrium in an infinitely large population in the absence of mutation, migration, selection, and nonrandom mating.

heat shock A transient response following exposure of cells or organisms to elevated temperatures. The response involves activation of a small number of loci, inactivation of some previously active loci, and selective translation of heat shock mRNA. Appears to be a nearly universal phenomenon observed in organisms ranging from bacteria to humans.

helicase An enzyme that participates in DNA replication by unwinding the double helix near the replication fork.

helix-turn-helix motif The structure of a region of DNA-binding proteins in which a turn of four amino acids holds two alpha helices at right angles to each other.

hemizygous Conditions where a gene is present in a single dose in an otherwise diploid cell. Usually applied to genes on the X chromosome in heterogametic males.

hemoglobin (Hb) An iron-containing, oxygen-carrying protein occurring chiefly in the red blood cells of vertebrates.

hemophilia An X-linked trait in humans associated with defective blood-clotting mechanisms.

heredity Transmission of traits from one generation to another.

heritability A measure of the degree to which observed phenotypic differences for a trait are genetic.

heterochromatin The heavily staining, late-replicating regions of chromosomes that are condensed in interphase. Thought to be devoid of structural genes.

heteroduplex A double-stranded nucleic acid molecule in which each polynucleotide chain has a different origin. These structures may be produced as intermediates in a recombinational event or by the *in vitro* reannealing of single-stranded, complementary molecules.

heterogametic sex The sex that produces gametes containing unlike sex chromosomes.

heterogeneous nuclear RNA (hnRNA) The collection of RNA transcripts in the nucleus, representing precursors and processing intermediates to rRNA, mRNA, and tRNA. Also represents RNA transcripts that will not be transported to the cytoplasm, such as snRNA (small nuclear RNA).

heterokaryon A somatic cell containing nuclei from two different sources.

heterosis The superiority of a heterozygote over either homozygote for a given trait.

heterozygote An individual with different alleles at one or more loci. Such individuals will produce unlike gametes and therefore will not breed true.

Hfr A strain of bacteria exhibiting a high frequency of recombination. Such strains have a chromosomally integrated F factor that is able to mobilize and transfer part of the chromosome to a recipient F^- cell.

histocompatibility antigens See *HLA*.

histones Proteins complexed with DNA in the nucleus. These proteins are rich in the basic amino acids arginine and lysine and function in the coiling of DNA to form nucleosomes.

HLA Cell surface proteins that are produced by histocompatibility loci and involved in the acceptance or rejection of tissue and organ grafts and transplants.

hnRNA See *heterogeneous nuclear RNA*.

holandric A trait transmitted from males to males. In humans, genes on the Y chromosome are holandric.

Holliday structure An intermediate in bidirectional DNA recombination, seen under the transmission electron microscope as an X-shaped structure showing four single-stranded DNA regions.

homeobox A sequence of about 180 nucleotides that encodes a 60-amino-acid sequence called a *homeodomain*, which is part of a DNA-binding protein that acts as a transcription factor.

homeotic mutation A mutation that causes a tissue normally determined to form a specific organ or body part to alter its differentiation and form another structure. Alternatively spelled as "homoeotic."

homogametic sex The sex which produces gametes that do not differ with respect to sex-chromosome content; in mammals, the female is homogametic.

homogeneously staining regions (hsr) Segments of mammalian chromosomes that stain lightly with Giemsa following exposure of cells to a selective agent. These regions arise in conjunction with gene amplification and are regarded as the structural locus for the amplified gene.

homologous chromosome Chromosomes that synapse or pair during meiosis. Chromosomes that are identical with respect to their genetic loci and centromere placement.

homozygote An individual with identical alleles at one or more loci. Such individuals will produce identical gametes and will therefore breed true.

homunculus The miniature individual imagined by preformationists to be contained within the sperm or egg.

human immunodeficiency virus (HIV) A human retrovirus associated with the onset and progression of acquired immunodeficiency syndrome (AIDS).

hybrid An individual produced by crossing two parents of different genotypes.

hybrid vigor See *heterosis*.

hybridoma A somatic cell hybrid produced by the fusion of an antibody-producing cell and a cancer cell, specifically, a myeloma. The cancer cell contributes the ability to divide indefinitely, and the antibody cell confers the ability to synthesize large amounts of a single antibody.

hydrogen bond An electrostatic attraction between a hydrogen atom bonded to a strongly electronegative atom, such as oxygen or nitrogen, and another atom that is electronegative or contains an unshared electron pair.

hypervariable regions The regions of antibody molecules that attach to antigens. These regions have a high degree of diversity in amino acid content.

identical twins See *monozygotic twins*.

Ig See *immunoglobulin*.

imaginal disc Discrete groups of cells, set aside during embryogenesis in holometabolous insects, that are determined to form the external body parts of the adult.

immunoglobulin The class of serum proteins that have the properties of antibodies.

inborn error of metabolism A biochemical disorder that is genetically controlled—usually, an enzyme defect that produces a clinical syndrome.

inbreeding Mating between closely related organisms.

inbreeding depression A decrease in viability, vigor, or growth in progeny following several rounds of inbreeding.

incomplete dominance Expression of a heterozygous phenotype that is distinct from, and often intermediate to, that of either parent.

incomplete linkage Occasional separation of two genes on the same chromosome by a recombinational event.

independent assortment The independent behavior of each pair of homologous chromosomes during the segregation of the pairs in meiosis I. The random distribution of maternal and paternal homologs into gametes.

inducer An effector molecule that activates transcription.

inducible enzyme system An enzyme system under the control of a regulatory molecule, or inducer, which acts to block a repressor and allow transcription.

initiation codon The triplet of nucleotides (AUG) in an mRNA molecule that codes for the insertion of the amino acid methionine as the first amino acid in a polypeptide chain.

insertion sequence See *IS element*.

***in situ* hybridization** A technique for the cytological localization of DNA sequences complementary to a given nucleic acid or polynucleotide.

intercalary deletion A form of chromosome deletion where material is lost from within the chromosome. Deletions that involve the end of the chromosome are called terminal deletions.

intercalating agent A compound that inserts between bases in a DNA molecule, disrupting the alignments and pairing of bases in the complementary strands (e.g., acridine dyes).

interference A measure of the degree to which one crossover affects the incidence of another crossover in an adjacent region of the same chromatid. Negative interference increases the chances of another crossover; positive interference reduces the probability of a second crossover event.

interphase The portion of the cell cycle between divisions.

intervening sequence See *intron*.

intron A portion of DNA between coding regions in a gene that is transcribed, but does not appear in the mRNA product.

inversion A chromosomal aberration in which the order of a chromosomal segment has been reversed.

inversion loop The chromosomal configuration resulting from the synapsis of homologous chromosomes, one of which carries an inversion.

in vitro Literally, in glass; outside the living organism; occurring in an artificial environment.

in vivo Literally, in the living; occurring within the living body of an organism.

IS element A mobile DNA segment that is transposable to any of a number of sites in the genome.

isoagglutinogen An antigenic factor or substance present on the surface of cells that is capable of inducing the formation of an antibody.

isochromosome An aberrant chromosome with two identical arms and homologous loci.

isolating mechanism Any barrier to the exchange of genes between different populations of a group of organisms. In general, isolation can be classified as spatial, environmental, or reproductive.

isotopes Forms of a chemical element that have the same number of protons and electrons, but differ in terms of the number of neutrons contained in the atomic nucleus.

isozyme Any of two or more distinct forms of an enzyme that have identical or nearly identical chemical properties, but differ in terms of some property such as net electrical charge, pH optima, number and type of subunits, or substrate concentration.

kappa particles DNA-containing cytoplasmic particles found in certain strains of *Paramecium aurelia*. When these self-reproducing particles are transferred into the growth medium, they release a toxin, paramecin, that kills other, sensitive strains. A nuclear gene, *K*, is responsible for maintaining kappa particles in the cytoplasm.

karyokinesis The process of nuclear division.

karyotype The chromosome complement of a cell or an individual. Often used to refer to the arrangement of metaphase chromosomes in a sequence according to length and position of the centromere.

kilobase (kb) A unit of length consisting of 1000 nucleotides. Abbreviated kb.

kinetochore A fibrous structure with a size of about 400 nm, located within the centromere. It appears to be the site of microtubule attachment during division.

Klenow fragment A part of bacterial DNA polymerase that lacks exonuclease activity, but retains polymerase activity. It is produced by enzymatic digestion of the intact enzyme.

Klinefelter syndrome A genetic disorder in human males caused by the presence of an extra X chromosome; Klinefelter males are XXY instead of XY. This syndrome is associated with enlarged breasts, small testes, sterility, and, occasionally, mild mental retardation.

knockout mice In producing knockout mice, a cloned normal gene is inactivated by the insertion of a marker, such as an antibiotic resistance gene. The altered gene is transferred to embryonic stem cells, where the altered gene will replace the normal gene (in some cells). These cells are injected into a blastomere embryo, producing a mouse that is bred to yield mice that are homozygous for the mutated gene.

***lac* repressor protein** A protein that binds to the operator in the *lac* operon and blocks transcription.

lagging strand In DNA replication, the strand synthesized in a discontinuous fashion, 5′ to 3′ away from the replication fork. Each short piece of DNA synthesized in this fashion is called an Okazaki fragment.

lampbrush chromosomes Meiotic chromosomes characterized by extended lateral loops, which reach maximum extension during diplotene. Although most intensively studied in amphibians, these structures occur in meiotic cells of organisms ranging from insects to humans.

lariat structure A structure formed by an intron via a 5′–2′ bond during processing and removal of that intron from an mRNA molecule.

leader sequence The portion of an mRNA molecule from the 5′ end to the beginning codon; may contain regulatory or ribosome binding sites.

leading strand During DNA replication, the strand synthesized continuously 5′ to 3′ from the origin of replication toward the replication fork.

leptotene The initial stage of meiotic prophase I, during which the chromosomes become visible and are often arranged in a bouquet configuration, with one or both ends of the chromosomes gathered at one spot on the inner nuclear membrane.

lethal gene A gene whose expression results in death.

leucine zipper A structural motif in a DNA-binding protein that is characterized by a stretch of leucine residues spaced at every seventh amino acid residue, with adjacent regions of positively charged amino acids. Leucine zippers on two polypeptides may interact to form a dimer that binds to DNA.

LINEs Long interspersed elements; repetitive sequences found in the genomes of higher organisms, such as the 6-kb L1 sequences found in primate genomes.

linkage Condition in which two or more nonallelic genes tend to be inherited together. Linked genes have their loci along the same chromosome and do not assort independently, but can be separated by crossing over.

linkage group A group of genes whose loci are on the same chromosome.

linking number The number of times that two strands of a closed, circular DNA duplex cross over each other.

locus The site or place on a chromosome where a particular gene is located.

lod score A statistical method used to determine whether two loci are linked or unlinked. A lod (log of the odds) score of four indicates that linkage is 10,000 times more likely than nonlinkage. By convention, lod scores of three to four are taken as signs of linkage.

long terminal repeat (LTR) Sequence of several hundred base pairs found at the ends of retroviral DNAs.

loss-of-function mutation A mutation that eliminates the function of the encoded protein, producing a mutant phenotype.

Lutheran blood group One of a number of blood group systems inherited independently of the ABO, MN, and Rh systems. Alleles of this group determine the presence or absence of antigens on the surface of red blood cells. Gene is on human chromosome 19.

Lyon hypothesis The random inactivation of the maternal or paternal X chromosome in somatic cells of mammalian females early in development. All daughter cells will have the same X chromosome inactivated, producing a mosaic pattern of expression of genes on the X chromosome.

lysis The disintegration of a cell brought about by the rupture of its membrane.

lysogenic bacterium A bacterial cell carrying the DNA of a temperate bacteriophage integrated into its chromosome.

lysogeny The process by which the DNA of an infecting phage becomes repressed and integrated into the chromosome of the bacterial cell it infects.

lytic phase The condition in which a temperate bacteriophage loses its integrated status in the host chromosome (i.e., becomes induced), replicates, and lyses the bacterial cell.

major histocompatibility loci See *MHC*.

map unit A measure of the genetic distance between two genes, corresponding to a recombination frequency of 1 percent. See *centimorgan*.

mapping functions Map distance estimates from recombination when the recombination frequency in a region exceeds 15–20 percent and double-crossovers are undetectable.

maternal effect Phenotypic effects on the offspring that are produced by the maternal genome. Factors transmitted through the egg cytoplasm that produce a phenotypic effect in the progeny.

maternal influence See *maternal effect*.

maternal inheritance The transmission of traits via cytoplasmic genetic factors such as mitochondria or chloroplasts.

mean The arithmetic average.

median The value in a group of numbers below and above which there are equal numbers of data points or measurements.

meiosis The process in gametogenesis or sporogenesis during which one replication of the chromosomes is followed by two nuclear divisions, to produce four haploid cells.

melting profile See T_m.

merozygote A partially diploid bacterial cell containing, in addition to its own chromosome, a chromosome fragment introduced into the cell by transformation, transduction, or conjugation.

messenger RNA See *mRNA*.

metabolism The sum of chemical changes in living organisms by which energy is generated and used.

metacentric chromosome A chromosome with a centrally located centromere, producing chromosome arms of equal lengths.

metafemale In *Drosophila*, a poorly developed female of low viability in which the ratio of X chromosomes to sets of autosomes exceeds 1.0. Previously called a superfemale.

metamale In *Drosophila*, a poorly developed male of low viability in which the ratio of X chromosomes to sets of autosomes is less than 0.5. Previously called a supermale.

metaphase The stage of cell division in which the condensed chromosomes lie in a central plane between the two poles of the cell and in which the chromosomes become attached to the spindle fibers.

metaphase plate The arrangement of mitotic or meiotic chromosomes at the equator of the cell during metaphase.

MHC Major histocompatibility loci. In humans, the HLA complex; in mice, the H2 complex.

micrometer A unit of length equal to 1×10^{-6} meter. Previously called a micron. Abbreviated μm.

micron See *micrometer*.

migration coefficient An expression of the proportion of migrant genes entering the population per generation.

millimeter A unit of length equal to 1×10^{-3} meter. Abbreviated nm.

minimal medium A medium containing only those nutrients that will support the growth and reproduction of wild-type strains of an organism.

mismatch repair A process of excision repair during which one or more unpaired bases are excised, followed by the synthesis of a new segment, using the complementary strand as a template.

missense mutation A mutation that alters a codon to that of another amino acid, causing an altered translation product to be made.

mitochondrion Found in the cells of eukaryotes, a cytoplasmic, self-reproducing organelle that is the site of ATP synthesis.

mitogen A substance that stimulates mitosis in nondividing cells (e.g., phytohemagglutinin).

mitosis A form of cell division resulting in the production of two cells, each with the same chromosome and genetic complement as of the parent cell.

mode In a set of data, the value occurring with the greatest frequency.

monohybrid cross A genetic cross between two individuals, involving only one character (e.g., $AA \times aa$).

monosomic An aneuploid condition in which one member of a chromosome pair is missing; having a chromosome number of $2n - 1$.

monozygotic twins Twins produced from a single fertilization event; the first division of the zygote produces two cells, each of which develops into an embryo. Also known as identical twins.

mRNA An RNA molecule transcribed from DNA and translated into the amino acid sequence of a polypeptide.

mtDNA Mitochondrial DNA.

mu phage A phage group in which the genetic material behaves like an insertion sequence, capable of insertion, excision, transposition, inactivation of host genes, and induction of chromosomal rearrangements.

multigene family A gene set descended from a common ancestor by duplication and subsequent divergence. The globin genes represent a multigene family.

multiple alleles Three or more alleles of the same gene.

multiple-factor inheritance See *polygenic inheritance*.

multiple infection Simultaneous infection of a bacterial cell by more than one bacteriophage, often of different genotypes.

mutagen Any agent that causes an increase in the rate of mutation.

mutant A cell or organism carrying an altered or mutant gene

mutation The process that produces an alteration in DNA or chromosome structure; the source of most alleles.

mutation rate The frequency with which mutations take place at a given locus or in a population.

muton The smallest unit of mutation in a gene, corresponding to a single base change.

nanometer A unit of length equal to 1×10^{-9} meter. Abbreviated nm.

natural selection Differential reproduction of some members of a species, resulting from variable fitness conferred by genotypic differences.

neutral mutation A mutation with no immediate adaptive significance or phenotypic effect.

nonautonomous transposon A transposable element that lacks a functional transposase gene.

noncrossover gamete A gamete that contains no chromosomes that have undergone genetic recombination.

nondisjunction An error during cell division in which the homologous chromosomes (in meiosis) or the sister chromatids (in mitosis) fail to separate and migrate to opposite poles; responsible for monosomy and trisomy defects, among others.

nonsense codon The nucleotide triplet in an mRNA molecule that signals the termination of translation. Three such codons are known: UGA, UAG, and UAA.

nonsense mutation A mutation that changes an amino acid codon into a termination codon: UAG (amber codon), UAA (ochre codon), or UGA (opal codon). Leads to premature termination during translation of mRNA.

NOR See *nucleolar organizer region.*

normal distribution A probability function that approximates the distribution of random variables. The normal curve, also known as a Gaussian or bell-shaped curve, is the graphic display of the normal distribution.

***N*-terminal amino acid** The terminal amino acid in a peptide chain that carries a free amino group.

***N* terminus** The end of a polypeptide that carries a free amino group of the first amino acid. By convention, the structural formula of polypeptides is written with the *N* terminus at the left.

nu body See *nucleosome.*

nuclease An enzyme that breaks bonds in nucleic acid molecules.

nucleoid The DNA-containing region within the cytoplasm in prokaryotic cells.

nucleolar organizer region (NOR) A chromosomal region containing the genes for rRNA; most often found in physical association with the nucleolus.

nucleolus A nuclear organelle that is the site of ribosome biosynthesis; usually associated with or formed in association with the NOR.

nucleoside A purine or pyrimidine base covalently linked to a ribose or deoxyribose sugar molecule.

nucleosome A complex of four histone molecules, each present in duplicate, wrapped by two turns of a DNA molecule. One of the basic units of eukaryotic chromosome structure. Also known as a nu body.

nucleotide A nucleoside covalently linked to a phosphate group. Nucleotides are the basic building blocks of nucleic acids. The nucleotides commonly found in DNA are deoxyadenylic acid, deoxycytidylic acid, deoxyguanylic acid, and deoxythymidylic acid. The nucleotides in RNA are adenylic acid, cytidylic acid, guanylic acid, and uridylic acid.

nucleotide pair The pair of nucleotides (A and T, or G and C) in opposite strands of the DNA molecule that are hydrogen bonded to each other.

nucleus The membrane-bounded cytoplasmic organelle of eukaryotic cells that contains the chromosomes and nucleolus.

null allele A nonexpressed allele; a mutation that completely eliminates function of a gene.

null hypothesis Used in statistical tests, it states that there is no difference between the observed and expected data sets. Statistical methods such as chi-square analysis are used to test the probability of this hypothesis.

nullisomic Describes an individual with a chromosomal aberration in which both members of a chromosome pair are missing.

ochre codon The codon UAA, which does not code for insertion of an amino acid, but for chain termination.

Okazaki fragment The small, discontinuous strands of DNA produced during DNA synthesis on the lagging strand.

oligonucleotide A linear sequence of nucleotides (about 10–20) connected by 5′ to 3′ phosphodiester bonds.

oncogene A gene whose activity promotes uncontrolled proliferation in eukaryotic cells.

opal codon The codon UGA, which does not code for insertion of an amino acid, but for chain termination.

open reading frame (ORF) The nucleotides between the start and stop codons that encode amino acids.

operator region A region of a DNA molecule that interacts with a specific repressor protein to control the expression of an adjacent gene or gene set.

operon A genetic unit that consists of one or more structural genes (that code for polypeptides) and an adjacent operator gene that controls the transcriptional activity of the structural gene or genes.

origin of replication (ori) Sites along the length of the chromosome where DNA replication begins.

ortholog A similar DNA sequence or nonallelic chromosome region between species. For example, the *sry* gene in the mouse is an ortholog of the *SRY* gene in humans. See *paralog.*

outbreeding depression Reduction in fitness in the offspring produced by mating genetically diverse parents. It is thought to result from a lowered adaptation to local environmental conditions.

overdominance A phenomenon in which heterozygotes have a phenotype that is more extreme than either homozygous genotype.

overlapping code A genetic code, first proposed by George Gamow, in which any given nucleotide is shared by three adjacent codons.

pachytene The stage in meiotic prophase I when the synapsed homologous chromosomes split longitudinally (except at the centromere), producing a group of four chromatids, called a tetrad.

pair-rule genes Genes expressed as stripes around the blastoderm embryo during development of the *Drosophila* embryo.

palindrome A word, number, verse, or sentence that reads the same backward or forward (e.g., *able was I ere I saw elba*). In nucleic acids, a sequence in which the base pairs read the same on complementary strands in the 5′-to-3′ direction—for example,

5′GAATTC3′

3′CTTAAG5′

Palindromes often occur as sites for restriction endonuclease recognition and cutting.

pangenesis A discarded theory of development that postulated the existence of pangenes, small particles from all parts of the body that concentrated in the gametes, passing traits from generation to generation and blending the traits of the parents in the offspring.

paracentric inversion A chromosomal inversion that does not include the centromere.

paralog A similar DNA sequence or nonallelic chromosome region within a species. The two α-globin genes in humans are paralogs. Their similarity indicates an evolutionary relationship and descent from a common ancestral sequence. See *ortholog*.

parasexual Condition describing recombination of genes from different individuals that does not involve meiosis, gamete formation, or zygote production. The formation of somatic cell hybrids is an example.

parental gamete See *noncrossover gamete*.

parthenogenesis Development of an egg without fertilization.

partial diploids See *merozygote*.

partial dominance See *incomplete dominance*.

patroclinous inheritance A form of genetic transmission in which the offspring have the phenotype of the father.

pedigree In human genetics, a diagram showing the ancestral relationships and transmission of genetic traits over several generations in a family.

P element Transposable DNA elements found in *Drosophila* that are responsible for hybrid dysgenesis.

penetrance The frequency (expressed as a percentage) with which individuals of a given genotype manifest at least some degree of a specific mutant phenotype associated with a trait.

peptide bond The covalent bond between the amino group of one amino acid and the carboxyl group of another amino acid.

pericentric inversion A chromosomal inversion that involves both arms of the chromosome and thus involves the centromere.

phage See *bacteriophage*.

phenocopy An environmentally induced phenotype (nonheritable) that closely resembles the phenotype produced by a known gene.

phenotype The observable properties of an organism that are genetically controlled.

phenylketonuria (PKU) A hereditary condition in humans associated with an inability to metabolize the amino acid phenylalanine. The most common form is caused by lack of the enzyme phenylalanine hydroxylase.

Philadelphia chromosome The product of a reciprocal translocation that contains the short arm of chromosome 9 carrying the *c-abl* oncogene and the long arm of chromosome 22 carrying *bcr*.

phosphodiester bond In nucleic acids, the covalent bond between a phosphate group and adjacent nucleotides, extending from the 5′ carbon of one pentose (ribose or deoxyribose) to the 3′ carbon of the pentose in the neighboring nucleotide. Phosphodiester bonds form the backbone of nucleic acid molecules.

photoreactivation enzyme (PRE) An exonuclease that catalyzes the light-activated excision of ultraviolet-induced thymine dimers from DNA.

photoreactivation repair Light-induced repair of damage caused by exposure to ultraviolet light. Associated with an intracellular enzyme system.

phyletic evolution The gradual transformation of one species into another over time; vertical evolution.

pilus A filamentlike projection from the surface of a bacterial cell. Often associated with cells possessing F factors.

plaque A clear area on an otherwise opaque bacterial lawn, caused by the growth and reproduction of phages.

plasmid An extrachromosomal, circular DNA molecule (often carrying genetic information) that replicates independently of the host chromosome.

pleiotropy Condition in which a single mutation simultaneously affects several characters.

ploidy Term referring to the basic chromosome set or to multiples of that set.

point mutation A mutation that can be mapped to a single locus. At the molecular level, a mutation that results in the substitution of one nucleotide for another.

polar body A cell produced at either the first or second meiotic division in females that, as a result of an unequal cytokinesis, contains almost no cytoplasm.

polycistronic mRNA A messenger RNA molecule that encodes the amino acid sequence of two or more polypeptide chains in adjacent structural genes.

polygenic inheritance The transmission of a phenotypic trait whose expression depends on the additive effect of a number of genes.

polylinker A segment of DNA that has been engineered to contain multiple sites for restriction enzyme digestion. Polylinkers are usually found in engineered vectors such as plasmids.

polymerase chain reaction (PCR) A method for amplifying DNA segments that uses cycles of denaturation, annealing to primers, and DNA polymerase-directed DNA synthesis.

polymerases The enzymes that catalyze the formation of DNA and RNA from deoxynucleotides and ribonucleotides, respectively.

polymorphism The existence of two or more discontinuous, segregating phenotypes in a population.

polynucleotide A linear sequence of more than 20 nucleotides, joined by 5′-to-3′ phosphodiester bonds. See *oligonucleotide*.

polypeptide A molecule made up of amino acids joined by covalent peptide bonds. This term is used to denote the amino acid chain before it assumes its functional three-dimensional configuration.

polyploid A cell or individual having more than two sets of chromosomes.

polyribosome See *polysome*.

polysome A structure composed of two or more ribosomes associated with mRNA, engaged in translation. Formerly called polyribosome.

polytene chromosome A chromosome that has undergone several rounds of DNA replication without separation of the replicated chromosomes, forming a giant, thick chromosome with aligned chromomeres that produce a characteristic banding pattern.

population A local group of individuals that belong to the same species and are actually or potentially interbreeding.

position effect Change in expression of a gene associated with a change in the gene's location within the genome.

postzygotic isolation mechanism Factors that prevent or reduce inbreeding by acting after fertilization to produce nonviable, sterile hybrids or hybrids of lowered fitness.

preadaptive mutation A mutational event that later becomes of adaptive significance.

preformationism The discredited idea that an organism develops by growth of structures already present in the egg or sperm.

prezygotic isolation mechanism Factors that reduce inbreeding by preventing courtship, mating, or fertilization.

Pribnow box A 6-bp sequence 5′ to the beginning of transcription in prokaryotic genes, to which the sigma subunit of RNA polymerase binds. The consensus sequence for this box is TATAAT.

primary protein structure Refers to the sequence of amino acids in a polypeptide chain.

primary sex ratio Ratio of males to females at fertilization.

primer In nucleic acids, a short length of RNA or single-stranded DNA that is necessary for the functioning of polymerases.

prion An infectious pathogenic agent devoid of nucleic acid and composed mainly of a protein, PrP, with a molecular weight of 27,000 to 30,000 daltons. Prions are known to cause scrapie, a degenerative neurological disease in sheep, and bovine spongiform encephalopathy (BSE, or mad cow disease) in cattle and are thought to cause similar diseases in humans, such as Kuru and Creutzfeldt–Jakob disease.

probability Ratio of the frequency of a given event to the frequency of all possible events.

proband An individual in whom a genetically determined trait of interest is first detected. Formerly known as a propositus.

probe A macromolecule such as DNA or RNA that has been labeled and can be detected by an assay such as autoradiography or fluorescence microscopy. Probes are used to identify target molecules, genes, or gene products.

product law The law which holds that the probability of two independent events occurring simultaneously is the product of the events' independent probabilities.

progeny The offspring produced from a mating.

prokaryotes Organisms lacking nuclear membranes, meiosis, and mitosis. Bacteria and blue–green algae are examples of prokaryotic organisms.

promoter A region that has a regulatory function and to which RNA polymerase binds prior to the initiation of transcription.

proofreading A molecular mechanism for correcting errors in replication, transcription, or translation. Also known as editing.

prophage A phage genome integrated into a bacterial chromosome. Bacterial cells carrying prophages are said to be lysogenic.

propositus (female, **proposita**) See *proband.*

protein A molecule composed of one or more polypeptides, each composed of amino acids covalently linked together.

proteomics The study of the expressed proteins in a cell at a given time.

protooncogene A cellular gene that normally functions to initiate or maintain cell division. Protooncogenes can be converted to oncogenes by alterations in structure or expression.

protoplast A bacterial or plant cell with the cell wall removed. Sometimes called a spheroplast.

prototroph A strain (usually microorganisms) that is capable of growth on a defined, minimal medium. Wild-type strains are usually regarded as prototrophs.

pseudoalleles Genes that behave as alleles to one another by complementation, but that can be separated from one another by recombination.

pseudoautosomal inheritance Inheritance of alleles located within the regions of the Y chromosome that are homologous to the X chromosome. Because these alleles are located on both the X and Y chromosome, their pattern of inheritance is indistinguishable from that of autosomal inheritance.

pseudodominance The expression of a recessive allele on one homolog as caused by the deletion of the dominant allele on the other homolog.

pseudogene A nonfunctional gene with sequence homology to a known structural gene present elsewhere in the genome. The genes differ from their functional relatives by insertions or deletions and by the presence of flanking direct repeat sequences of 10 to 20 nucleotides.

punctuated equilibrium A pattern in the fossil record of brief periods of species divergence punctuated with long periods of species stability.

quantitative inheritance See *polygenic inheritance.*

quantitative trait loci (QTL) Two or more genes that act on a single polygenic trait.

quantum speciation Formation of a new species within a single generation or a few generations by a combination of selection and drift.

quaternary protein structure Types and modes of interaction between two or more polypeptide chains within a protein molecule.

R point The point (also known as the restriction point) during the G1 stage of the cell cycle when either a commitment is made to DNA synthesis and another cell cycle or the cell withdraws from the cycle and becomes quiescent.

race A genotypically or geographically distinct subgroup within a species.

rad A unit of absorbed dose of radiation with an energy equal to 100 ergs per gram of irradiated tissue.

radioactive isotope One of the forms of an element, differing from the other forms in atomic weight and possessing an unstable nucleus that emits ionizing radiation during decay.

random amplified polymorphic DNA (RAPD) A PCR method that uses random primers about 10 nucleotides in length to amplify unknown DNA sequences.

random mating Mating between individuals without regard to genotype.

reading frame Linear sequence of codons (groups of three nucleotides) in a nucleic acid.

reannealing Formation of double-stranded DNA molecules from dissociated single strands.

recessive Term describing an allele that is not expressed in the heterozygous condition.

reciprocal cross A paired cross in which the genotype of the female in the first cross is present as the genotype of the male in the second cross, and vice versa.

reciprocal translocation A chromosomal aberration in which nonhomologous chromosomes exchange parts.

recombinant DNA A DNA molecule formed by the joining of two heterologous molecules. Usually applied to DNA molecules produced by in vitro ligation of DNA from two different organisms.

recombinant gamete A gamete containing a new combination of genes produced by crossing over during meiosis.

recombination The process that leads to the formation of new gene combinations on chromosomes.

recon A term used by Seymour Benzer to denote the smallest genetic units between which recombination can occur.

reductional division The chromosome division that halves the number of diploid chromosomes. The first division of meiosis is a reductional division.

redundant genes Gene sequences present in more than one copy per haploid genome (e.g., ribosomal genes).

regulatory site A DNA sequence having to do with the control of expression of other genes, usually involving an interaction with another molecule.

rem "Radiation equivalent in man"; the dosage of radiation that will cause the same biological effect as one roentgen of X rays.

renaturation The process by which a denatured protein or nucleic acid returns to its normal three-dimensional structure.

repetitive DNA sequences DNA sequences present in many copies in the haploid genome.

replicating form (RF) Double-stranded nucleic acid molecules present as an intermediate during the reproduction of certain viruses.

replication The process of DNA synthesis.

replication fork The Y-shaped region of a chromosome, associated with the site of replication.

replicon A chromosomal region or free genetic element containing the DNA sequences necessary for the initiation of DNA replication.

replisome The term used to describe the complex of proteins, including DNA polymerase, that assembles at the bacterial replication fork to synthesize DNA.

repressible enzyme system An enzyme or group of enzymes whose synthesis is regulated by the intracellular concentration of certain metabolites.

repressor A protein that binds to a regulatory sequence adjacent to a gene and blocks transcription of the gene.

reproductive isolation Absence of interbreeding between populations, subspecies, or species. Reproductive isolation can be brought about by extrinsic factors, such as behavior, and intrinsic barriers, such as hybrid inviability.

resistance transfer factor (RTF) A component of R plasmids that confers the ability for cell-to-cell transfer of the R plasmid by conjugation.

resolution In an optical system, the shortest distance between two points or lines at which they can be perceived to be two points or lines.

restriction endonuclease Nuclease that recognizes specific nucleotide sequences in a DNA molecule and cleaves or nicks the DNA at that site. Derived from a variety of microorganisms, the enzymes that cleave both strands of the DNA are used in the construction of recombinant DNA molecules.

restriction fragment length polymorphism (RFLP) Variation in the length of DNA fragments generated by a restriction endonuclease. Such variations are caused by mutations that create or abolish cutting sites for restriction enzymes. RFLPs are inherited in a codominant fashion and can be used as genetic markers.

restrictive transduction See *specialized transduction.*

retrovirus Viruses with RNA as genetic material that use the enzyme reverse transcriptase during their life cycle.

reverse transcriptase A polymerase that uses RNA as a template to transcribe a single-stranded DNA molecule as a product.

reversion A mutation that restores the wild-type phenotype.

R factor (R plasmid) Bacterial plasmids that carry antibiotic resistance genes. Most R plasmids have two components: an r-determinant, which carries the antibiotic resistance genes, and the resistance transfer factor (RTF).

RFLP See *restriction fragment length polymorphism.*

Rh factor An antigenic system first described in the rhesus monkey. Recessive *r*/*r* individuals produce no Rh antigens and are classified as Rh negative, while *R*/*R* and *R*/*r* individuals have Rh antigens on the surface of their red blood cells and are classified as Rh positive.

ribonucleic acid A nucleic acid characterized by the sugar ribose and the pyrimidine uracil, usually a single-stranded polynucleotide. Several forms are recognized, including ribosomal RNA, messenger RNA, transfer RNA, and heterogeneous nuclear RNA.

ribose The five-carbon sugar associated with the ribonucleotides found in RNA.

ribosomal RNA See *rRNA.*

ribosome A ribonucleoprotein organelle consisting of two subunits, each containing RNA and protein. Ribosomes are the site of translation of mRNA codons into the amino acid sequence of a polypeptide chain.

RNA See *ribonucleic acid.*

RNA editing Alteration of the nucleotide sequence of an mRNA molecule after transcription and before translation. There are two main types of editing: *substitution editing*, which changes individual nucleotides, and *insertion–deletion editing*, in which individual nucleotides are added or deleted.

RNA polymerase An enzyme that catalyzes the formation of an RNA polynucleotide strand, using the base sequence of a DNA molecule as a template.

RNase A class of enzymes that hydrolyzes RNA.

Robertsonian translocation A form of chromosomal aberration that involves the fusion of long arms of acrocentric chromosomes at the centromere.

roentgen A unit of measure of the amount of radiation corresponding to the generation of 2.083×10^9 ion pairs in one cubic centimeter of air at 0°C at an atmospheric pressure of 760 mm of mercury. Abbreviated R.

rolling-circle model A model of DNA replication in which the growing point or replication fork rolls around a circular template strand; in each pass around the circle, the newly synthesized strand displaces the strand from the previous replication, producing a series of contiguous copies of the template strand.

rRNA The RNA molecules that are the structural components of the ribosomal subunits. In prokaryotes, they are the 16*S*, 23*S*, and 5*S* molecules; in eukaryotes, they are the 18*S*, 28*S*, and 5*S* molecules.

RTF See *resistance transfer factor.*

S_1 nuclease A deoxyribonuclease that cuts and degrades single-stranded molecules of DNA.

satellite DNA DNA that forms a minor band when genomic DNA is centrifuged in a cesium salt gradient. This DNA usually consists of short sequences repeated many times in the genome.

SCE See *sister chromatid exchange.*

secondary protein structure The alpha-helical or pleated-sheet form of a protein molecule brought about by the formation of hydrogen bonds between amino acids.

secondary sex ratio The ratio of males to females at birth.

secretor An individual who has soluble forms of the blood group antigens A or B present in saliva and other body fluids. This condition is caused by a dominant, autosomal gene unlinked to the *ABO* locus (*I* locus).

sedimentation coefficient See *Svedberg coefficient unit.*

segment polarity genes Genes that regulate the spatial pattern of differentiation within each segment of the developing *Drosophila* embryo.

segregation The separation of homologous chromosomes into different gametes during meiosis.

selection The force that brings about changes in the frequency of alleles and genotypes in populations through differential reproduction.

selection coefficient (*s*) A quantitative measure of the relative fitness of one genotype as compared with another.

selfing In plant genetics, the fertilization of ovules of a plant by pollen produced by the same plant. Reproduction by self-fertilization.

semiconservative replication A model of DNA replication in which a double-stranded molecule replicates in such a way that the daughter molecules are composed of one parental (old) and one newly synthesized strand.

semisterility A condition in which a proportion of all zygotes are inviable.

sex chromatin body See *Barr body.*

sex chromosome A chromosome, such as the X or Y chromosome in humans, that is involved in sex determination.

sexduction Transmission of chromosomal genes from a donor bacterium to a recipient cell by the F factor.

sex-influenced inheritance Phenotypic expression that is conditioned by the sex of the individual. A heterozygote may express one phenotype in one sex and the alternative phenotype in the other sex.

sex-limited inheritance A trait that is expressed in only one sex, even though the trait may not be X linked.

sex ratio See *primary sex ratio* and *secondary sex ratio.*

sexual reproduction Reproduction through the fusion of gametes, which are the haploid products of meiosis.

Shine–Dalgarno sequence The nucleotides AGGAGG present in the leader sequence of prokaryotic genes that serve as a ribosome binding site. The 16*S* RNA of the small ribosomal subunit contains a complementary sequence to which the mRNA binds.

shotgun experiment The cloning of random fragments of genomic DNA into a vehicle such as a plasmid or phage, usually to produce a library from which clones of specific interest can be selected.

sibling species Species that are morphologically almost identical, but reproductively isolated from one another.

sickle-cell anemia A genetic disease in humans that is caused by an autosomal recessive gene and is fatal in the homozygous condition if untreated. It is caused by an alteration in the amino acid sequence of the beta chain of globin.

sickle-cell trait The phenotype exhibited by individuals heterozygous for the sickle-cell gene.

sigma factor A polypeptide subunit of the RNA polymerase that recognizes the binding site for the initiation of transcription.

SINEs Stands for "short interspersed elements"; repetitive sequences found in the genomes of higher organisms, such as the 300-bp *Alu* sequence.

single-stranded binding proteins (SSBs) In DNA replication, proteins that bind to and stabilize single-stranded regions of DNA that result from the action of unwinding proteins.

sister-chromatid exchange (SCE) A crossing-over event that can occur in meiotic and mitotic cells; involves the reciprocal exchange of chromosomal material between sister chromatids (joined by a common centromere). Such exchanges can be detected cytologically after BrdU incorporation into the replicating chromosomes.

site-directed mutagenesis A process that uses a synthetic oligonucleotide containing a mutant base or sequence as a primer for inducing a mutation at a specific site in a cloned gene.

small nuclear RNA (snRNA) Species of RNA molecules ranging in size from 90 to 400 nucleotides. The abundant snRNAs are present in 1×10^4 to 1×10^6 copies per cell. snRNAs are associated with proteins and form RNP particles known as snRNPs, or *snurps*. Six uridine-rich snRNAs known as U1–U6 are located in the nucleoplasm, and the complete nucleotide sequence of these snRNAs is known. snRNAs have been implicated in the processing of pre-mRNA and may have a range of cleavage and ligation functions.

snurps See *small nuclear RNA (snRNA).*

solenoid structure A level of eukaryotic chromosome structure generated by the supercoiling of nucleosomes.

somatic cell genetics The use of cultured somatic cells to investigate genetic phenomena by parasexual techniques involving the fusion of cells from different organisms.

somatic cells All cells other than the germ cells or gametes in an organism.

somatic mutation A mutational event occurring in a somatic cell. In other words, such mutations are not heritable.

somatic pairing The pairing of homologous chromosomes in somatic cells.

SOS response The induction of enzymes to repair damaged DNA in *E. coli*. The response involves activation of an enzyme that cleaves a repressor, activating a series of genes involved in DNA repair.

spacer DNA DNA sequences found between genes, usually repetitive DNA segments.

specialized transduction Genetic transfer of only specific host genes by transducing phages.

speciation The process by which new species of plants and animals arise.

species A group of actually or potentially interbreeding individuals that is reproductively isolated from other such groups.

spheroplast See *protoplast*.

spindle fibers Cytoplasmic fibrils formed during cell division that are involved with the separation of chromatids at anaphase and their movement toward opposite poles in the cell.

spliceosome The nuclear macromolecule complex within which splicing reactions occur to remove introns from pre-mRNAs.

spontaneous mutation A mutation that is not induced by a mutagenic agent.

spore A unicellular body or cell encased in a protective coat that is produced by some bacteria, plants, and invertebrates; is capable of survival in unfavorable environmental conditions; and can give rise to a new individual upon germination. In plants, spores are the haploid products of meiosis.

SRY A gene found near the pseudoautosomal boundary of the Y chromosome (sex-determining region of the Y chromosome). Accumulated evidence indicates that this gene is the testis-determining factor (TDF).

stabilizing selection Preferential reproduction of individuals that have genotypes close to the mean for the population. A selective elimination of genotypes at both extremes.

standard deviation A quantitative measure of the amount of variation in a sample of measurements from a population.

standard error A quantitative measure of the amount of variation in a sample of measurements from a population.

sterility The condition of being unable to reproduce; free from contaminating microorganisms.

strain A group with common ancestry that has physiological or morphological characteristics of interest for genetic study or domestication.

structural gene A gene that encodes the amino acid sequence of a polypeptide chain.

sublethal gene A mutation causing lowered viability, with death before maturity in less than 50 percent of the individuals carrying the gene.

submetacentric chromosome A chromosome with the centromere placed so that one arm of the chromosome is slightly longer than the other.

subspecies A morphologically or geographically distinct interbreeding population of a species.

sum law A law which holds that the probability of one or the other of two mutually exclusive events occurring is the sum of the events' individual probabilities.

supercoiled DNA A form of DNA structure in which the helix is coiled upon itself. Such structures can exist in stable forms only when the ends of the DNA are not free, as in a covalently closed circular DNA molecule.

superfemale See *metafemale*.

supermale See *metamale*.

suppressor mutation A mutation that acts to restore (completely or partially) the function lost by a previous mutation at another site.

Svedberg coefficient unit A unit of measure for the rate at which particles (molecules) sediment in a centrifugal field. This unit is a function of several physicochemical properties, including size and shape. A sedimentation value of 1×10^{-13} sec is defined as one Svedberg coefficient (*S*) unit.

symbiont An organism coexisting in a mutually beneficial relationship with another organism.

sympatric speciation Process of speciation involving populations that inhabit, at least in part, the same geographic range.

synapsis The pairing of homologous chromosomes at meiosis.

synaptonemal complex (SC) An organelle consisting of a tripartite nucleoprotein ribbon that forms between the paired homologous chromosomes in the pachytene stage of the first meiotic division.

syndrome A group of signs or symptoms that occur together and characterize a disease or abnormality.

synkaryon The nucleus of a zygote that results from the fusion of two gametic nuclei. Also used in somatic cell genetics to describe the product of nuclear fusion.

syntenic test In somatic cell genetics, a method for determining whether or not two genes are on the same chromosome.

T_m The temperature at which a population of double-stranded nucleic acid molecules is half-dissociated into single strands. This value is taken to be the melting temperature for that species of nucleic acid.

TATA box See *Goldberg–Hogness box*.

tautomeric shift A reversible isomerization in a molecule, brought about by a shift in the localization of a hydrogen atom. In nucleic acids, tautomeric shifts in the bases of nucleotides can cause changes in other bases at replication and are a source of mutations.

TDF (testis-determining factor) The product of a gene on the Y chromosome that controls the developmental switch point for the development of the indifferent gonad into a testis. See *SRY*.

telocentric chromosome A chromosome in which the centromere is located at the end of the chromosome.

telomerase The enzyme that adds short, tandemly repeated DNA sequences to the ends of eukaryotic chromosomes.

telomere The terminal chromomere of a chromosome.

telophase The stage of cell division in which the daughter chromosomes reach the opposite poles of the cell and re-form nuclei. Telophase ends with the completion of cytokinesis.

telophase I In the first meiotic division, when duplicated chromosomes reach the poles of the dividing cell.

temperate phage A bacteriophage that can become a prophage and confer lysogeny upon the host bacterial cell.

temperature-sensitive mutation A conditional mutation that produces a mutant phenotype at one temperature range and a wild-type phenotype at another temperature range.

template The single-stranded DNA or RNA molecule that specifies the nucleotide sequence of a strand synthesized by a polymerase molecule.

terminalization The movement of chiasmata toward the ends of chromosomes during the diplotene stage of the first meiotic division.

tertiary protein structure The three-dimensional structure of a polypeptide chain, brought about by folding upon itself.

test cross A cross between an individual whose genotype at one or more loci may be unknown and an individual who is homozygous recessive for the genes in question.

tetrad The four chromatids that make up paired homologs in the prophase of the first meiotic division. The four haploid cells produced by a single meiotic division.

tetrad analysis Method for the analysis of gene linkage and recombination, using the four haploid cells produced in a single meiotic division.

tetranucleotide hypothesis An early theory of DNA structure which proposed that the molecule is composed of repeating units, each consisting of the four nucleotides containing adenine, thymine, cytosine, and guanine.

theta structure An intermediate in the bidirectional replication of circular DNA molecules. At about midway through the cycle of replication, the intermediate resembles the Greek letter theta.

thymine dimer A pair of adjacent thymine bases in a single polynucleotide strand that have chemical bonds formed between carbon atoms 5 and 6. This lesion, usually caused by exposure to ultraviolet light, inhibits DNA replication unless repaired by the appropriate enzymes.

topoisomerase A class of enzymes that convert DNA from one topological form to another. During DNA replication, these enzymes facilitate the unwinding of the double-helical structure of DNA.

totipotent The ability of a cell or embryo part to give rise to all adult structures. This capacity is usually progressively restricted during development.

trait Any detectable phenotypic variation of a particular inherited character.

***trans* configuration** The arrangement of two mutant alleles on opposite homologs, such as

$$\frac{a^1 \quad +}{+ \quad a^2}$$

Contrasts with a *cis* arrangement, where the alleles are located on the same homolog.

transcription Transfer of genetic information from DNA by the synthesis of an RNA molecule copied from a DNA template.

transcriptome The set of mRNA molecules present in a cell at any given time.

transdetermination Change in developmental fate of a cell or group of cells.

transduction Virally mediated gene transfer from one bacterium to another, or the transfer of eukaryotic genes mediated by retrovirus.

transfer RNA See *tRNA*.

transformation Heritable change in a cell or an organism as brought about by exogenous DNA.

transgenic organism An organism whose genome has been modified by the introduction of external DNA sequences into the germ line.

transition A mutational event in which one purine is replaced by another, or one pyrimidine is replaced by another.

translation The derivation of the amino acid sequence of a polypeptide from the base sequence of an mRNA molecule in association with a ribosome.

translocation A chromosomal mutation associated with the transfer of a chromosomal segment from one chromosome to another. Also used to denote the movement of mRNA through the ribosome during translation.

transmission genetics The field of genetics concerned with the mechanisms by which genes are transferred from parent to offspring.

transposable element A DNA segment that translocates to other sites in the genome, essentially independent of sequence homology. Usually, such elements are flanked by short, inverted repeats of 20 to 40 base pairs at each end. Insertion into a structural gene can produce a mutant phenotype. Insertion and excision of transposable elements depend on two enzymes, transposase and resolvase. Such elements have been identified in both prokaryotes and eukaryotes.

transversion A mutational event in which a purine is replaced by a pyrimidine, or a pyrimidine is replaced by a purine.

trinucleotide repeat A tandemly repeated cluster of three nucleotides (such as CTG) in or near a gene that undergo an expansion in copy number, resulting in a disease phenotype.

triploidy The condition in which a cell or organism possesses three haploid sets of chromosomes.

trisomy The condition in which a cell or organism possesses two copies of each chromosome, except for one, which is present in three copies. The general form for trisomy is therefore $2n + 1$.

tRNA Transfer RNA; a small ribonucleic acid molecule which contains a three-base segment (anticodon) that recognizes a codon in mRNA, a binding site for a specific amino acid, and recognition sites for interaction with the ribosomes and the enzyme that links it to its specific amino acid.

tumor suppressor gene A gene which encodes a gene product that normally functions to suppress cell division. Mutations in tumor suppressor genes result in the activation of cell division and tumor formation.

Turner syndrome A genetic condition in human females that is caused by a 45,X genotype (XO). Such individuals are phenotypically female, but are sterile, because of undeveloped ovaries.

unequal crossing over A crossover between two improperly aligned homologs, producing one homolog with three copies of a region and the other with one copy of that region.

unique DNA DNA sequences that are present only once per genome. Single-copy DNA.

universal code The assumption that the genetic code is used by all life forms. In general, this assumption is true; some exceptions are found in mitochondria, ciliates, and mycoplasmas.

unwinding proteins Nuclear proteins that act during DNA replication to destabilize and unwind the DNA helix ahead of the replicating fork.

variable-number tandem repeats (VNTRs) Short repeated DNA sequences (2–20 nucleotides) present as tandem repeats between two restriction enzyme sites. Variations in the number of repeats creates DNA fragments of differing lengths following restriction enzyme digestion.

variable region Portion of an immunoglobulin molecule that exhibits many amino acid sequence differences between antibodies of differing specificities.

variance A statistical measure of the variation of values from a central value, calculated as the square of the standard deviation.

variegation Patches of differing phenotypes, such as color, in a piece of tissue.

vector In recombinant DNA, an agent, such as a phage or plasmid, into which a foreign DNA segment will be inserted.

viability The measure of the number of individuals in a given phenotypic class that survive, relative to that for another class (usually a wild-type class).

virulent phage A bacteriophage that infects and lyses the host bacterial cell.

VNTR See *variable-number tandem repeats*.

W, Z chromosomes Sex chromosomes in species for which the female is the heterogametic sex (WZ).

Western blot A technique in which proteins are separated by gel electrophoresis and transferred by capillary action to a nylon membrane or nitrocellulose sheet. A specific protein can be identified through hybridization to a labeled antibody.

wild type The most commonly observed phenotype or genotype, designated as the norm or standard.

wobble hypothesis An idea proposed by Francis Crick stating that the third base in an anticodon can align in several ways in order to allow it to recognize more than one base in the codons of mRNA.

writhing number The number of times that the axis of a DNA duplex crosses itself by supercoiling.

X inactivation In mammalian females, the random cessation of transcriptional activity of one X chromosome. This event, which occurs early in development, is a mechanism of dosage compensation. The molecular basis of inactivation is unknown, but involves a region called the X-chromosome inactivation center (XIC) on the proximal end of the *p* arm. Some loci on the tip of the short arm of the XIC can escape inactivation. See *Barr body*, *Lyon hypothesis*.

XIST A locus in the X-chromosome inactivation center that may control inactivation of the X chromosome in mammalian females.

X linkage The pattern of inheritance resulting from genes located on the X chromosome.

X-ray crystallography A technique used to determine the three-dimensional structure of molecules through diffraction patterns produced by X-ray scattering by crystals of the molecule under study.

YAC A cloning vector in the form of a yeast artificial chromosome, constructed using chromosomal elements including telomeres (from a ciliate), centromeres, origin of replication, and marker genes from yeast. YACs are used to clone long stretches of eukaryotic DNA.

Y chromosome Sex chromosome in species for which the male is heterogametic (XY).

Y linkage Mode of inheritance shown by genes located on the Y chromosome.

Z-DNA An alternative structure of DNA in which the two antiparallel polynucleotide chains form a left-handed double helix. Z-DNA has been shown to be present along with B-DNA in chromosomes and may have a role in regulation of gene expression.

zein Principal storage protein of maize endosperm, consisting of two major proteins, with molecular weights of 19,000 and 21,000 daltons, respectively.

zinc finger A DNA-binding domain of a protein which has a characteristic pattern of cysteine and histidine residues that complex with zinc ions, throwing intermediate amino acid residues into a series of loops or fingers.

zygote The diploid cell produced by the fusion of haploid gametic nuclei.

zygotene A stage of meiotic prophase I in which the homologous chromosomes synapse and pair along their entire length, forming bivalents. The synaptonemal complex forms at this stage.

Appendix C
Answers to Selected Problems

Chapter 1

2. *Epigenesis* refers to the theory that organisms are derived from the assembly and reorganization of substances in the *egg* which eventually lead to the development of the adult. *Preformationism* is a 17th century theory, which states that the sex cells (eggs or sperm) contain miniature adults, called homunculi, which grow in size to become the adult. Each postulates a fundamental difference in the manner in which organisms develop from hereditary determiners.

4. Their theory of natural selection proposed that more offspring are produced than can survive, and that in the competition for survival, those with favorable variations survive. Darwin did not understand the nature of heredity and variation which led him to lean toward older theories of pangenesis and inheritance of acquired characteristics.

6. The "trinity" of molecular genetics refers to the relationships among DNA, RNA, and protein. The processes of *transcription* and *translation* are integral to understanding these relationships.

8. *Basic* research involves the study of the fundamental mechanisms of genetics as described in the answer to question #7. *Applied* research makes use of the applications in agriculture and medicine as described in the text.

10. *Positive eugenics* encouraged parents displaying favorable characteristics to have large families while *negative eugenics* attempted to restrict reproduction for parents displaying unfavorable characteristics. *Euphenics* refers to medical genetic intervention designed to reduce the impact of defective genes on individuals.

12. There is promise that a certain amount of human suffering will be minimized by the application of genetics to crop production (disease resistance, protein content, growth conditions) and medicine. Major medical areas of activity include genetic counseling, gene mapping and identification, disease diagnosis, and genetic engineering.

Chapter 2

2. Prior to 1940 most of the interest in genetics centered on the transmission of similarity and variation from parents to offspring (transmission genetics). While some experiments examined the possible nature of the hereditary material, abundant knowledge of the structural and enzymatic properties of proteins generated a bias which worked to favor proteins as the hereditary substance. In addition, proteins were composed of as many as twenty different subunits (amino acids) thereby providing ample structural and functional variation for the multiple tasks which must be accomplished by the genetic material. The tetranucleotide hypothesis (structure) provided insufficient variability to account for the diverse roles of the genetic material.

4. Specific degradative enzymes, proteases, RNase, and DNase were used to selectively eliminate components of the extract and, if transformation is concomitantly eliminated, then the eliminated fraction is the transforming principle. DNase eliminates DNA and transformation, therefore it must be the transforming principle.

6. Actually phosphorus is found in approximately equal amounts in DNA and RNA. Therefore labeling with ^{32}P would "tag" both RNA and DNA. However, the T2 phage, in its mature state, contains very little if any RNA, therefore DNA would be interpreted as being the genetic material in T2 phage.

8. The early evidence would be considered indirect in that at no time was there an experiment, like transformation in bacteria, in which genetic information in one organism was transferred to another using DNA. Rather, by comparing DNA content in various cell types (sperm and somatic cells) and observing that the *action* and *absorption* spectra of ultraviolet light were correlated, DNA was considered to be the genetic material. This suggestion was supported by the fact that DNA was shown to be the genetic material in bacteria and some phage. Direct evidence for DNA being the genetic material comes from a variety of observations including gene transfer which has been facilitated by recombinant DNA techniques.

10. The structure of deoxyadenylic acid is given below and in the text. Linkages among the three components require the removal of water (H_2O).

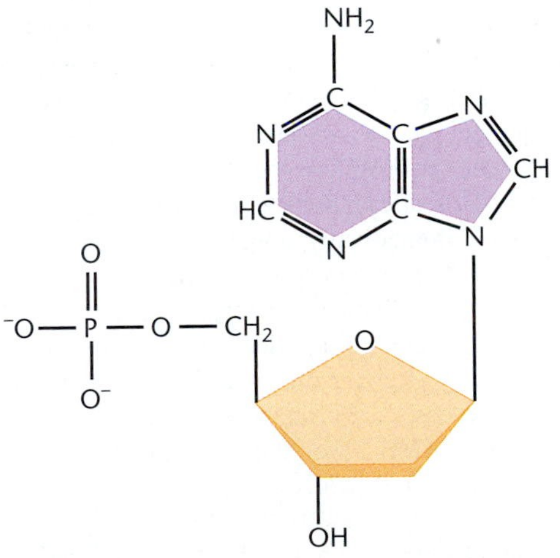

12. Guanine: 2-amino-6-oxypurine
Cytosine: 2-oxy-4-aminopyrimidine
Thymine: 2,4-dioxy-5-methylpyrimidine
Uracil: 2,4-dioxypyrimidine

16. Because in double-stranded DNA, A = T and G = C (within limits of experimental error), the data presented would have indicated a lack of pairing of these bases in favor of a single-stranded structure or some other nonhydrogen-bonded structure.

Alternatively, from the data it would appear that A = G and T = C which would require purines to pair with purines and pyrimidines to pair with pyrimidines. In that case, the DNA would have contradicted the data from Franklin and Watkins which called for a constant diameter for the double-stranded structure.

18. Three main differences between RNA and DNA are the following:
 (1) uracil in RNA replaces thymine in DNA,
 (2) ribose in RNA replaces deoxyribose in DNA, and
 (3) RNA often occurs as both single- and partially double-stranded forms whereas DNA most often occurs in a double-stranded form.

20. The nitrogenous bases of nucleic acids (nucleosides, nucleotides, and single- and double-stranded polynucleotides), absorb UV light maximally at wavelengths 254 to 260 nm. Using this phenomenon, one can often determine the presence and concentration of nucleic acids in a mixture. Since proteins absorb UV light maximally at 280 nm, this is a relatively simple way of dealing with mixtures of biologically important molecules.
UV absorption is greater in single-stranded molecules (hyperchromic shift) as compared to double-stranded structures, therefore one can easily determine, by applying denaturing conditions, whether a nucleic acid is in the single- or double-stranded form. In addition, A-T rich DNA denatures more readily than G-C rich DNA, therefore one can estimate base content by denaturation kinetics.

22. Guanine and cytosine are held together by three hydrogen bonds whereas adenine and thymine are held together by two. Because G-C base pairs are more compact, they are more dense than A-T pairs. The percentage of G-C pairs in DNA is thus proportional to the buoyant density of the molecule.

24. Repetitive sequences renature relatively quickly because the likelihood of complementary strands interacting increases. For curve A in the problem, there is evidence for a rapidly renaturing species (repetitive) and a slowly renaturing species (unique). The fraction which reassociates faster than the *E. coli* DNA is highly repetitive and the last fraction (with the highest $C_ot_{1/2}$ value) contains primarily unique sequences. Fraction B contains mostly unique, relatively complex DNA.

26. Because G-C base pairs are formed with three hydrogen bonds while A-T base pairs by two such bonds, it takes more energy (higher temperature) to separate G-C pairs.

28. In one sentence of Watson and Crick's paper in *Nature*, they state, "It has not escaped our notice that the specific pairing we have postulated immediately suggests a possible copying mechanism for the genetic material."
The model itself indicates that unwinding of the helix and separation of the double-stranded structure into two single strands immediately exposes the specific hydrogen bonds through which new bases are brought into place.

30. The ratios for MS-2 would be as follows:
 $0.5/10^5 = 0.001/X$
 or
 $X/0.001 = 10^5/0.5$
 $X = (0.001)(10^5)/0.5$
 $X = 200$ base pairs
The ratios for *E. coli* would be as follows:
 $0.5/10^5 = 10.0/X$ or
 $X/10.0 = 10^5/0.5$
 $X = (10.0)(10^5)/0.5$
 $X = 2 \text{ X } 10^6$ base pairs

32. Left side (a) = left, right side (b) = right.

34. Under this condition, the hydrolyzed 5-methyl cytosine becomes thymine.

36. **(a)** Heat application would yield a hyprochromic shift if the DNA is double-stranded. One could also get a rough estimation of the GC content from the kinetics of denaturation and the degree of sequence complexity from comparative renaturation studies.
 (b) Determination of base content by hydrolysis and chromatography could be used for comparative purposes and could also provide evidence as to the strandedness of the DNA.
 (c) Antibodies for Z-DNA could be used to determine the degree of left-handed structures, if present.
 (d) Sequencing the DNA from both viruses would indicate sequence homology. In addition, through various electronic searches readily available on the Internet (wed site: *blast @ ncbi . nlm . nih . gov*, for example) one could determine whether similar sequences exist in other viruses or in other organisms.

Chapter 3

2. A comparison of the density of DNA samples at various times in the experiment (initial ^{15}N culture, and subsequent cultures grown in the ^{14}N medium) showed that after one round of replication in the ^{14}N medium, the DNA was half as dense (intermediate) as the DNA from bacteria grown only in the ^{15}N medium. In a sample taken after two rounds of replication in the ^{14}N medium, half of the DNA was of the intermediate density and the other half was as dense as DNA containing only ^{14}N DNA.

4. Refer to the text for an illustration of the labeling of *Vicia* chromosomes under a Taylor, Woods, and Hughes experimental design. Notice that only those cells which pass through the S phase in the presence of the ^3H-thymidine are labeled and that each double helix (per chromatid) is "half-labeled."
 (a) Under a conservative scheme all of the newly labeled DNA will go to one sister chromatid, while the other sister chromatid will remain unlabeled. In contrast to a semiconservative scheme, the first replicative round would produce one sister chromatid which has label on both strands of the double helix.
 (b) Under a dispersive scheme all of the newly labeled DNA will be interspersed with unlabeled DNA. Because these preparations (metaphase chromosomes) are highly coiled and condensed structures derived from the "spread out" form at interphase (which includes the S phase) it is impossible to detect the areas where label is not found. Rather, both sister chromatids would appear as evenly labeled structures.

6. The *in vitro* replication requires a DNA template, a divalent cation (Mg^{++}), and all four of the deoxyribonucleoside triphosphates: dATP, dCTP, dTTP, and dGTP. The lower case "d" refers to the deoxyribose sugar.

8. Two general analytical approaches showed that the products of DNA polymerase I were probably copies of the template DNA. Because *base composition* can be similar without reflecting sequence similarity, the least stringent test was the comparison of base composition. By comparing *nearest neighbor frequencies*, Kornberg determined that there is a very high likelihood that the product was of the same base sequence as the template.

10. The *in vitro* rate of DNA synthesis using DNA polymerase I is slow, being more effective at replicating single-stranded DNA than double-stranded DNA. In addition, it is capable of degrading as well as synthesizing DNA. Such degradation suggested that it functioned as a repair enzyme. In addition, DeLucia and Cairns discovered a strain of *E. coli* (*pol*A1) which still replicated its DNA but was deficient in DNA polymerase I activity.

12. As stated in the text, *biologically active* DNA implies that the DNA is capable of supporting typical metabolic activities of the cell or organism and is capable of faithful reproduction.

14. As shown in the text, DNA polymerase I and DNA ligase are used to synthesize and label an RF duplex. DNase is used to nick one of the two strands. The heavy (^{32}P/BU-containing) DNA is isolated by denaturation and centrifugation and DNA polymerase and DNA ligase are used to make a synthetic complementary strand. When isolated, this synthetic strand is capable of transfecting *E. coli* protoplasts, from which new ϕX174 phages are produced.

16. All three enzymes share several common properties. First, none can *initiate* DNA synthesis on a template but all can *elongate* an existing DNA strand assuming there is a template strand as shown in the figure below. Polymerization of nucleotides occurs in the 5' to 3' direction where each 5' phosphate is added to the 3' end of the growing polynucleotide.

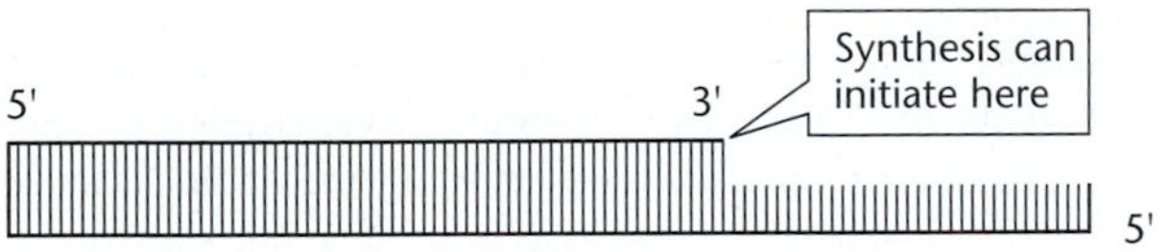

All three enzymes are large complex proteins with a molecular weight in excess of 100,000 daltons and each has 3' to 5' exonuclease activity.

DNA polymerase I:
- exonuclease activity
- present in large amounts
- relatively stable
- removal of RNA primer

DNA polymerase II:
- possibly involved in repair function

DNA polymerase III:
- exonuclease activity
- essential for replication
- complex molecule

18. Given a stretch of double-stranded DNA, one could initiate synthesis at a given point and either replicate strands in one direction only (unidirectional) or in both directions (bidirectional) as shown below. Notice that in the text the synthesis of complementary strands occurs in a *continuous* $5' > 3'$ mode on the leading strand in the direction of the replication fork, and in a *discontinuous* $5' > 3'$ mode on the lagging strand opposite the direction of the replication fork. Such discontinuous replication forms Okazaki fragments.

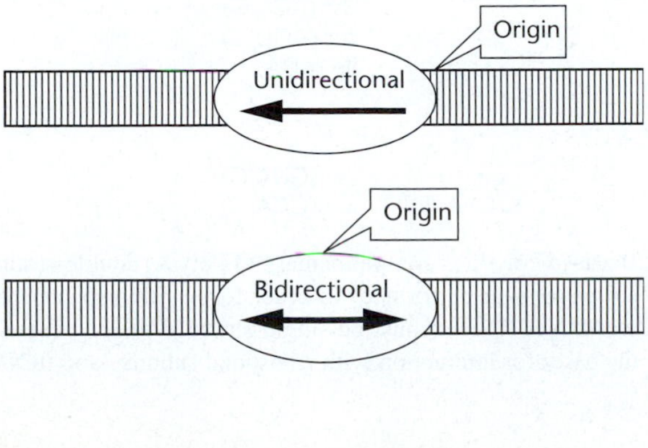

20. *Okazaki fragments* are relatively short (1000 to 2000 bases in prokaryotes) DNA fragments which are synthesized in a discontinuous fashion on the lagging strand during DNA replication. Such fragments appear to be necessary because template DNA is not available for $5' > 3'$ synthesis until some degree of continuous DNA synthesis occurs on the leading strand in the direction of the replication fork. The isolation of such fragments provides support for the scheme of replication shown in the text. DNA *ligase* is required to form phosphodiester linkages in gaps which are generated when DNA polymerase I removes RNA primer and meets newly synthesized DNA ahead of it. Notice in the text the discontinuous DNA strands are ligated together into a single continuous strand. *Primer* RNA is formed by RNA primase to serve as an initiation point for the production of DNA strands on a DNA template. None of the DNA polymerases are capable of initiating synthesis without a free 3′ hydroxyl group. The primer RNA provides that group and thus can be used by DNA polymerase III.

22. Eukaryotic DNA is replicated in a manner which is very similar to that of *E. coli*. Synthesis is bidirectional, continuous on one strand and discontinuous on the other, and the requirements of synthesis (four deoxyribonucleoside triphosphates, divalent cation, template, and primer) are the same. Okazaki fragments of eukaryotes are about one-tenth the size of those in bacteria. Because there is a much greater amount of DNA to be replicated and DNA replication is slower, there are multiple initiation sites for replication in eukaryotes (and increased DNA polymerase per cell) in contrast to the single replication origin in prokaryotes. Replication occurs at different sites during different intervals of the S phase. The proposed functions of four DNA polymerases are described in the text.

24. **(a)** In *E. coli*, 100kb are added to each growing chain per minute. Therefore the chain should be about 4,000,000bp.

(b) Given

$$(4 \times 10^6 \text{ bp}) \times 0.34\text{nm/bp} = 1.36 \times 10^6\text{nm or } 1.3\text{mm}$$

26. *Gene conversion* is likely to be a consequence of genetic recombination in which nonreciprocal recombination yields products in which it appears that one allele is "converted" to another. Gene conversion is now considered a result of heteroduplex formation which is accompanied by mismatched bases. When these mismatches are corrected, the "conversion" occurs.

28. Telomerase activity is present in germ line tissue to maintain telomere length from one generation to the next. In other words, telomeres can not shorten indefinitely without eventually eroding genetic information.

30.

Initial Labeled Base	***Labeled Base after Spleen Phosphodiesterase Digestion***	
	ANTIPARALLEL	**PARALLEL**
G	A,T	C,T
C	G,A,G	G,A
T	C,T, G	C,T,A,G
A	T,C,A,T	T,C,A,G

One can determine which model occurs in nature by comparing the pattern in which the labeled phosphate is shifted following spleen phosphodiesterase digestion. Focus your attention on the antiparallel model and notice that the frequency which "C" (for example) is the 5′ neighbor of "G" is not necessarily the same as the frequency which "G" is the 5′ neighbor of "C." However, in the parallel model (b), the frequency which "C" is the 5′ neighbor of "G" is the same as the frequency which "G" is the 5′ neighbor of "C." By examining such "digestion frequencies" it can be determined that DNA exists in the opposite polarity.

32.

(a) DNA polymerase would catalyze a bond between the 5′ end of the last nucleotide added and the 3′ end of the incoming nucleotide. In this reaction, the energy would be provided by the cleavage of the gamma- and beta- phosphates of the last nucleotide added to the chain rather than of the incoming nucleotide.

(b) If DNA polymerase removed a base, it would not be able to add any more bases to the chain because the penultimate base would have a monophosphate rather than a triphosphate and there would be no source of energy for the polymerization reaction.

Chapter 4

2. By having a circular chromosome, no free ends present the problem of linear chromosomes, namely complete replication of terminal sequences.

4. Puffs represent active genes as evidenced by staining and uptake of labeled RNA precursors as assayed by autoradiography.

6. While greater DNA content per cell is associated with eukaryotes, one can not universally equate genomic size with an increase in organismic complexity. There are numerous examples where DNA content per cell varies considerably among closely related species. Because of the diverse cell types of multicellular eukaryotes, a variety of gene products is required, which may be related to the increase in DNA content per cell. In addition, the advantage of diploidy automatically increases DNA content per cell. However, seeing the question in another way, it is likely that a much higher *percentage* of the genome of a prokaryote is actually involved in phenotype production than in a eukaryote.

Eukaryotes have evolved the capacity to obtain and maintain what appears to be large amounts of "extra" perhaps "junk" DNA. This concept will be examined in subsequent chapters of the text. Prokaryotes on the other hand, with their relatively short life cycle, are extremely efficient in their accumulation and use of their genome.

Given the larger amount of DNA per cell and the requirement that the DNA be partitioned in an orderly fashion to daughter cells during cell division, certain mechanisms and structures (mitosis, nucleosomes, centromeres, *etc.*) have evolved for packaging and distributing the DNA. In addition, the genome is divided into separate entities (chromosomes) to perhaps facilitate the partitioning process in mitosis and meiosis.

8. *Heterochromatin* is chromosomal material which stains deeply and remains condensed when other parts of chromosomes, euchromatin, are otherwise pale and decondensed. Heterochromatic regions replicate late in S phase and are relatively inactive in a genetic sense because there are few genes present or if they are present, they are repressed. Telomeres and the areas adjacent to centromeres are composed of heterochromatin.

10. The first step of this solution is to convert all of the given values to cubic Å remembering that 1 mm = 10,000 Å. Using the formula πr^2 for the area of a circle and $4/3\ \pi r^3$ for the volume of a sphere, the following calculations apply:

Volume of DNA:

$3.14 \times 10\text{Å} \times 10\ \text{Å} \times (50 \times 10^4\ \text{Å}) = 1.57 \times 10^8\ \text{Å}^3$

Volume of cuspid:

$4/3\ (3.14 \times 400\ \text{Å} \times 400\ \text{Å} \times 400\ \text{Å}) = 2.67 \times 10^8\ \text{Å}^3$

Because the capsid head has a greater volume than the volume of DNA, the DNA will fit into the capsid.

12. Volume of the nucleus $= 4/3\ \pi r^3$
$= 4/3 \times 3.14 \times (5 \times 10^3\text{nm})^3$
$= 5.23 \times 10^{11}\text{nm}^3$

Volume of the chromosome $= \pi r^2 \times$ length
$= 3.14 \times 5.5\text{nm} \times 5.5\text{nm} \times (2 \times 10^9\text{nm})$
$= 1.9 \times 10^{11}\text{nm}^3$

Therefore, the percentage of the volume of the nucleus occupied by the chromatin is
$= 1.9 \times 10^{11}\text{nm}^3 / 5.23 \times 10^{11}\text{nm}^3 \times 100$
= about 36.3%

Chapter 5

2. No. The the term "reading frame" refers to the number of bases contained in each codon. The reason that (+++) or (−−−) restored the reading frame is because the code is triplet. By having the (+++) or (−−−), the translation system is "out of phase" until the third "+" or "−" is encountered. If the code contained six nucleotides (a sextuplet code), then the translation system is "out of phase" until the sixth "+" or "−" is encountered.

4. Assume that you have introduced a copolymer (ACACACAC...) to a cell free protein synthesizing system. There are two possibilities for establishing the reading frames: ACA, if one starts at the first base and CAC if one starts at the second base. These would code for two different amino acids (ACA = threonine; CAC = histidine) and would produce repeating polypeptides which would alternate *thr-his-thr-his*. . . or *his-thr-his-thr*. . .

Because of a triplet code, a trinucleotide sequence will, once initiated, remain in the same reading frame and produce the same code all along the sequence regardless of the initiation site. Given the sequence CUACUACUACUA, notice the different reading frames producing three different sequences each containing the same amino acid.

6. From the repeating polymer ACACA. . . one can say that threonine is either CAC or ACA. From the polymer CAACAA. . . with ACACA. . . , ACA is the only codon in common. Therefore, threonine would have the codon ACA.

8. The basis of the technique is that if a trinucleotide contains bases (a codon) which are complementary to the anticodon of a charged tRNA, a relatively large complex is formed which contains the ribosome, the tRNA, and the trinucleotide. This complex is trapped in the filter whereas the components by themselves are not trapped. If the amino acid on a charged, trapped tRNA is radioactive, then the filter becomes radioactive.

10. Apply the most conservative pathway of change.

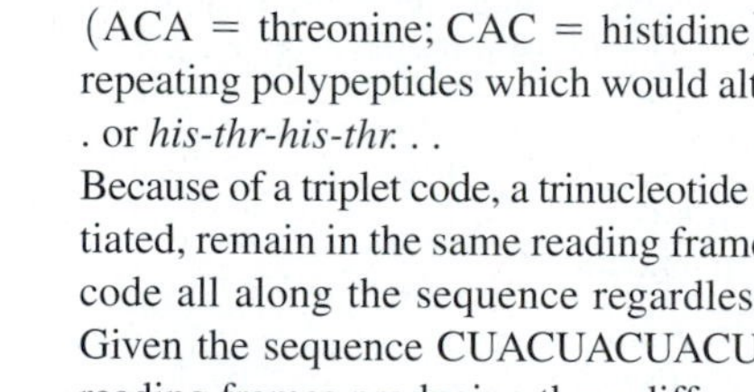

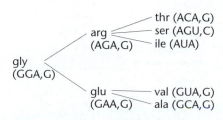

12. Because Poly U is complementary to Poly A, double-stranded structures will be formed. In order for an RNA to serve as a messenger RNA it must be single-stranded thereby exposing the bases for interaction with ribosomal subunits and tRNAs.

14. (a)

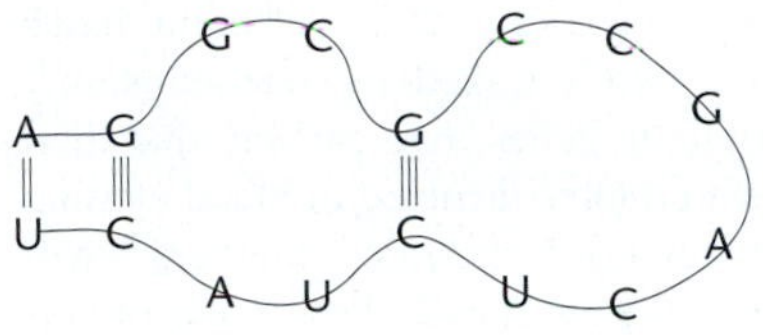

(b) TCCGCGGCTGAGATGA (use complementary bases, substituting T for U)

(c) GCU

(d) Assuming that the AGG. . . is the 5′ end of the mRNA, then the sequence would be *arg-arg-arg-leu-tyr*

16. (a) Starting from the 5′ end and locating the AUG triplets one finds two initiation sites leading to the following two sequences:

met-his-tyr-glu-thr-leu-gly
met-arg-pro-leu-gly

(b) In the shorter of the two reading sequences (the one using the internal AUG triplet), a UGA triplet was introduced at the second codon. While not in the reading frames of the longer polypeptide (using the first AUG codon) the UGA triplet eliminates the product starting at the second initiation codon.

18. The central dogma of molecular genetics and to some extent, all of biology, states that DNA produces, through transcription, RNA, which is "decoded" (during translation) to produce proteins.

20. RNA polymerase from *E. coli* is a complex, large (almost 500,000 daltons) molecule composed of subunits $(\alpha,\beta,\beta',\sigma)$ in the proportion $\alpha 2,\beta,\beta',\sigma$ for the holoenzyme. The β subunit provides catalytic function while the sigma (σ) subunit is involved in recognition of specific promoters. The core enzyme is the protein without the sigma.

22. While some folding (from complementary base pairing) may occur with mRNA molecules, they generally exist as single-stranded structures which are quite labile. Eukaryotic mRNAs are generally processed such that the 5′ end is "capped" and the 3′ end has a considerable string of adenine bases. It is thought that these features protect the mRNAs from degradation. Such stability of eukaryotic mRNAs probably evolved with the differentiation of nuclear and cytoplasmic functions. Because prokaryotic cells exist in a more unstable environment (nutritionally and physically, for example) than many cells of multicellular organisms, rapid genetic response to environmental change is likely to be adaptive. To accomplish such rapid responses, a labile gene product (mRNA) is advantageous.

A pancreatic cell, which is developmentally stable (differentiated) and existing in a relatively stable environment could produce more insulin on stable mRNAs for a given transcriptional rate.

24.

Proline:	C_3, and one of the C_2A triplets
Histidine:	one of the C_2A triplets
Threonine:	one C_2A triplet, and one A_2C triplet
Glutamine:	one of the A_2C triplets
Asparagine:	one of the A_2C triplets
Lysine:	A_3

26. (a,b) Use the code table to determine the number of triplets which code each amino acid, then construct a graph and plot such as this one :

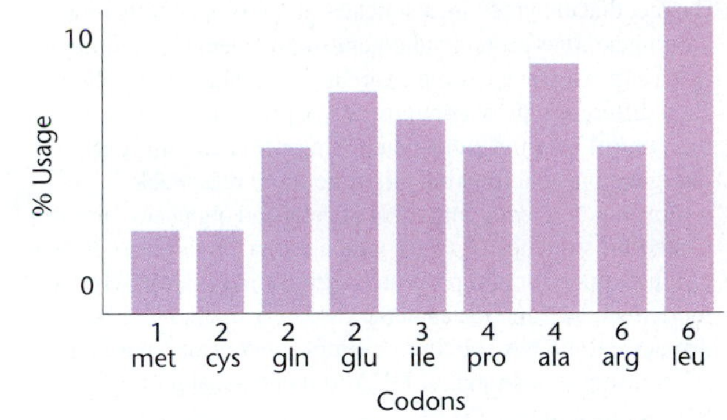

(c) There appears to be a weak correlation between the relative frequency of amino acid usage and the number of triplets for each.

(d) To continue to investigate this issue one might examine additional amino acids in a similar manner. In addition, different phylogenetic groups use code synonyms differently. It may be possible to find situations in which the relationships are more extreme. One might also examine more proteins to determine whether such a weak correlation is stronger with different proteins.

Chapter 6

2. Transfer RNAs are "adaptor" molecules in that they provide a way for amino acids to interact with sequences of bases in nucleic acids. Amino acids are specifically and individually attached to the 3' end of tRNAs which possess a three-base sequence (the anticodon) to base-pair with three bases of mRNA. Messenger RNA, on the other hand, contains a copy of the triplet codes which are stored in DNA. The sequences of bases in mRNA interact, three at a time, with the anticodons of tRNAs.

Enzymes involved in transcription include the following: RNA polymerase (*E. coli*), and RNA polymerase I, II, III (eukaryotes). Those involved in translation include the following: aminoacyl tRNA synthetases, peptidyl transferase, and GTP-dependent release factors.

4. The sequence of base triplets in mRNA constitutes the sequence of codons. A three-base portion of the tRNA constitutes the anticodon.

6. The steps involved in tRNA charging are outlined in the text. An amino acid in the presence of ATP, Mg++, and a specific aminoacyl synthetase produces an amino acid-AMP enzyme complex (+ PPi). This complex interacts with a specific tRNA to produce the aminoacyl tRNA.

8. Phenylalanine is an amino acid which, like other amino acids, is required for protein synthesis. While too much phenylalanine and its derivatives cause PKU in phenylketonurics, too little will restrict protein synthesis.

10. Tyrosine is a precursor to melanin, skin pigment. Individuals with PKU fail to convert phenyalanine to tyrosine and even though tyrosine is obtained from the diet, at the population level, individuals with PKU have a tendency for less skin pigmentation.

12. The pathway therefore would be as follows:

AA → IGP → I → TRY

The metabolic blocks are created at the following locations.

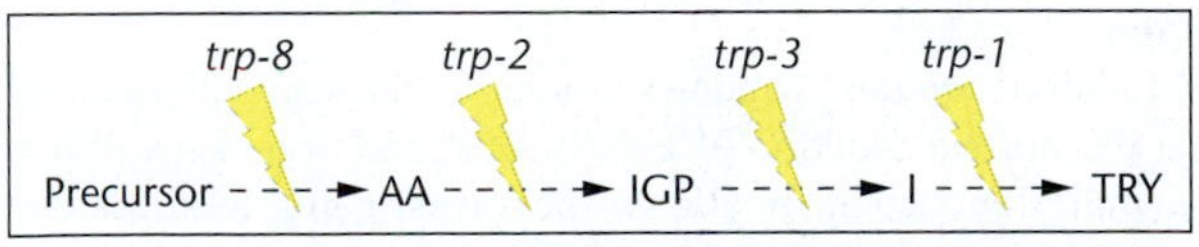

ne fact that enzymes are a subclass of the general term protein, a one-gene:one-protein statement might seem to be more appropriate. However, some proteins are made up of subunits, each different type of subunit (polypeptide chain) being under the control of a different gene. Under this circumstance, the one-gene:one-polypeptide might be more reasonable.

It turns out that many functions of cells and organisms are controlled by stretches of DNA which either produce no protein product (operator and promoter regions, for example) or have more than one function as in the case of overlapping genes and differential mRNA splicing. A simple statement regarding the relationship of a stretch of DNA to its physical product is difficult to formulate.

16. The following types of normal hemoglobin are presented in the text:

Hemoglobin	*Polypeptide chains*
HbA	$2\alpha 2\beta$ (alpha, beta)
HbA2	$2\alpha 2\delta$ (alpha, delta)
HbF	$2\alpha 2\gamma$ (alpha, gamma)
Gower 1	$2\zeta 2\varepsilon$ (zeta, epsilon)

The alpha and beta chains contain 141 and 146 amino acids, respectively. The zeta chain is similar to the alpha chain while the other chains are like the beta chain.

18. In the late 1940's Pauling demonstrated a difference in the electrophoretic mobility of HbA and HbS (sickle-cell hemoglobin) and concluded that the difference had a chemical basis. Ingram determined that the chemical change occurs in the primary structure of the globin portion of the molecule using the fingerprinting technique. He found a change in the 6th amino acid in the β chain.

20. *Colinearity* refers to the sequential arrangement of subunits, amino acids and nitrogenous bases, in proteins and DNA, respectively. Sequencing of genes and products in MS2 phage and studies on mutations in the A subunit of the *tryptophan synthetase* gene indicate a colinear relationship.

24. Enzymes function to regulate catabolic and anabolic activities of cells. They influence (lower) the energy of activation thus allowing chemical reactions to occur under conditions which are compatible with living systems. Enzymes possess active sites and/or other domains which are sensitive to the environment. The active site is considered to be a crevice, or pit, which binds reactants, thus enhancing their interaction. The other domains mentioned above may influence the conformation and therefore function of the active site.

26. All of the substitutions involve one base change.

28. With the codes for valine being GUU, GUC, GUA, and GUG, single base changes from glutamic acid's GAA and GAG can cause the glu>>>val switch. The normal glutamic acid is a negatively charged amino acid whereas valine carries no net charge and lysine is positively charged. Given these significant charge changes one would predict some, if not considerable, influence on protein structure and function. Such changes could stem from internal changes in folding or interactions with other molecules in the RBC, especially other hemoglobin molecules. HbC homozygotes suffer mild hemolytic anemia (a benign hemoglobinopathy). Recent studies indicate that HbC may be protective against severe forms of malaria.

Chapter 7

2. Mutations are the "windows" through which geneticists look at the normal function of genes, cells, and organisms. When a mutation occurs it allows the investigator to formulate questions as to the function of the normal allele of that mutation. For example, hemophilia is an inherited blood-clotting disease. Because there are three different inherited forms of the disease, two X-linked and one autosomal, all determined by non-allelic genes, one can say that there are at least three different proteins involved in blood clotting. At a different level, mutations provide "markers" with which biologists can study the genetics and dynamics of populations.

4. It is true that *most* mutations are thought to be deleterious to an organism. A gene is a product of perhaps a billion or so years of evolution and it is only natural to suspect that random changes will probably yield negative results. However, *all* mutations may not be deleterious. Those few, rare variations which are beneficial will provide a basis for possible differential propagation of the variation. Such changes in gene frequency represent the basis of the evolutionary process.

6. If one unit of output from the normal gene gives the same phenotype as in the normal homozygote, where there are two units of output, the allele is considered "recessive."

	Phenotype, if mutant is:	
Genotypes	***recessive***	***dominant***
wild/wild	wild	wild
wild/mutant	wild	mutant
mutant/mutant	mutant	mutant

8. Mutations constantly occur in both somatic and gametic tissues. When in somatic tissue, the mutation will not be passed to the next generation, however the physiological or structural role of the mutant cell may be compromised, but of little or no impact. If a somatic mutation occurs early in development and if numerous progeny cells are produced from the mutant cell, then a significant tissue mass may have abnormal function. In addition, mutation can alter the normal regulatory aspects of a cell cycle. When such regulation is compromised, then abnormal rates of cell proliferation, cancer, may result. Mutations in the gametic tissue may pose problems for future generations.

10. All three of the agents are mutagenic because they cause base substitutions. Deaminating agents oxidatively deaminate bases such that cytosine is converted to uracil and adenine is converted to hypoxanthine. Uracil pairs with adenine and hypoxanthine pairs with cytosine. Alkylating agents donate an alkyl group to the amino or keto groups of nucleotides, thus altering base-pairing affinities. 6-ethyl guanine acts like adenine, thus pairing with thymine. Base analogues such as 5-bromouracil and 2-amino purine are incorporated as thymine and adenine respectively yet they base pair with guanine and cytosine respectively.

12. X-rays are of higher energy and shorter wavelength than UV light. They have greater penetrating ability and can create more disruption of DNA.

14. *Photoreactivation* can lead to repair of UV-induced damage. An enzyme, photoreactivation enzyme, will absorb a photon of light to cleave thymine dimers. *Excision repair* involves the products of several genes, DNA polymerase I, and DNA ligase to clip out the UV-induced dimer, fill in, and join the phosphodiester backbone in the resulting gap. The excision repair process can be activated by damage which distorts the DNA helix.

Recombinational repair is a system which responds to DNA that has escaped other repair mechanisms at the time of replication. If a gap is created on one of the newly synthesized strands a "rescue operation or SOS response" allows the gap to be filled. Many different gene products are involved in this repair process: *recA*, *lexA*. In SOS repair, the proofreading by DNA polymerase III is suppressed and this therefore is called an "error-prone system."

16. Each involves a ballooning of tribnucleotide repeats. Genetic anticipation is the occurrence of an earlier age of onset of a genetic disease in successive generations.

18. In *site-directed mutagenesis* the goal is usually to change one or more of the nucleotides within a gene. A piece of DNA complementary to the gene being studied is obtained which contains the desired altered base sequence. The DNA with the altered base sequence will hybridize with the original (unaltered) DNA strand and upon replication, two types of duplexes will be formed; one like the original, unaltered sequence, the other containing the altered sequence.

20. Transitions involve pyrimidine → pyrimidine and purine → purine substitutions while transversions involve pyrimidine → purine and purine → pyrimidine substitutions. The four types of transitions are the following.

A → G	G → A
T → C	C → T

The eight types of transversions are the following:

A →	T, C
G →	T, C
T →	G, A
C →	G, A

22. Each organism mentioned in the problem possesses a variety of transposable elements. Bacteria possess insertion sequences (about 800 to 1500 base pairs in length) as well as transposons which are larger. Both are mobile in bacterial, viral, and plasmid DNAs and both have repeated base sequences at their ends. In maize, Barbara McClintock described the genetic behavior of mobile elements (*Ds* and *Ac*). *Ds* can move if *Ac* is present, thus *transposable controlling elements* exist.

An *Ac* element is 4563 base pairs long and similar in structure to some bacterial transposons. Transposons often code for transposase enzymes which are essential for transposition. *Copia* elements in *Drosophila* may be present in numerous copies in the genome and contain direct and inverted terminal repeats. *P* elements, also in *Drosophila* are responsible for a phenomenon called hybrid dysgenesis. Humans possess a variety of transposable elements including the *Alu* family of short interspersed elements (SINES) which are between 200 to 300 base pairs long and may exist in 300,000 copies per genome. Long interspersed elements (LINES) also occur in the human genome and seem to be capable of movement. Such elements share common structural features, are often mobile, and may influence gene activity.

24. It is likely that the reverse transcriptase, in making DNA, provides a DNA segment which is capable of integrating into the yeast chromosome as other types of DNA are known to do.

26. Your study should include examination of the following short-term aspects: immediate assessment of radiation amounts distributed in a matrix of the bomb sites as well as a control area not receiving bomb-induced radiation, radiation exposure as measured by radiation sickness and evidence of radiation poisoning from tissue samples, abortion rates, birthing rates, and chromosomal studies. Long-term assessment should include: sex-ratio distortion (males being more influenced by X-linked recessive lethals than females), chromosomal studies, birth and abortion rates, cancer frequency and type and, genetic disorders. In each case data should be compared to the control site to see if changes are bomb-related. In addition, to attempt to determine cause-effect, it is often helpful to show a dose response.

28. **(a)** For those organisms that generate energy by aerobic respiration, a process occurs which involves the reduction of molecular oxygen. Partially reduced species are produced as intermediates and by-products of such molecular action: O_2^-, H_2O_2, and OH^-. These species are potent electrophilic oxidants that escape mitochondria and attack numerous cellular components. Collectively, these are called reactive oxygen species (ROS).

(b) Hydrogen bonding can occur to any other base, including self pairs. Homopurine (A:A, G:G) and heteropurine (A:G) pairs represent anomalous base pairing possibilities even with non-altered bases. While G:C is undoubtedly the most stable, several mispairs are actually stronger than the A-T pair.

Base pairing is complicated by the fact that the purines possess two H-bonding faces, the Watson-Crick face, involving ring positions 1 and 6 for Adenine and, 1, 2, and 6 for Guanine, and the Hoogsteen face involving ring positions 6 and 7. The typical pairing mode is indicated as *wc* where pairing occurs on the Watson-Crick face in the normal orientation, even for the mispair A:G. Alteration of pairing and favoring of the Hoogsteen face can be favored with the alteration generated by oxoGuanine. Indeed, triple helix configurations commonly involve the Hoogsteen face.

(c) If not repaired, in the first round of replication involves the pairing of oxoG to Adenine, while in the next round of replication, Adenine pairs with its normal Thymine. Therefore, if one starts with a G:C pair, one ends up with an A:T pair.

(d) It turns out that G:G>T:A transversions are quite commonly found in human cancers and are especially prevalent in the tumor suppressor gene *p53*. One component is a triphosphatase that cleanses the nucleotide precursor pool by removing the two outermost phosphates from oxo-dGTP. Another involves a DNA glycosylase that initiates repair of misreplicated oxoG:A by hydrolyzing the glycosidic bond linking the adenine base to the sugar. Another is a DNA gycosylase/lyase system that recognizes oxoG opposite cytosine. Of the three systems, the DNA glycosylases are probably the most effective.

Chapter 8

2. Chromosomes which are homologous share many properties including: *overall length, position of the centromere, banding patterns, type and location of genes, and autoradiographic pattern. Diploidy* is a term often used in conjunction with the symbol 2*n*. It means that both members of a homologous pair of chromosomes are present. *Haploidy* specifically refers to the fact that each haploid cell contains *one chromosome of each homologous pair of chromosomes.*

8. Regulatory checkpoints include those in G1/S, G2/M, and mitosis. Each allows checks on intracellular and environmental conditions. *Autoradiography* is a technique which can be used to determine that there is DNA synthesis during the interphase. If DNA precursors are available early in interphase, they can be incorporated into replicating DNA during the S phase. Autoradiographs made after the S phase will show radioactive material in the DNA. Complexed cdk and cyclin proteins provide the mechanism for passage from one stage of the cell cycle to the next.

10. Compared with mitosis which maintains a chromosomal constancy, meiosis provides for a reduction in chromosome number, and an opportunity for exchange of genetic material between homologous chromosomes. In mitosis there is no change in chromosome number or kind in the two daughter cells whereas in meiosis numerous potentially different haploid (n) cells are produced. During oogenesis, only one of the four meiotic products is functional; however, four of the four meiotic products of spermatogenesis are potentially functional.

12. Sister chromatids are genetically identical, except where mutations may have occurred during DNA replication. Nonsister chromatids are genetically similar if on homologous chromosomes or genetically dissimilar if on nonhomologous chromosomes. If crossing over occurs, then chromatids attached to the same centromere will no longer be identical.

14. (a) If there are 16 chromosomes there should be 8 tetrads.

(b) After meiosis I and in the second meiotic prophase, there are as many dyads as there are pairs of chromosomes. There will be 8 dyads.

(c) Because the monads migrate to opposite poles during meiosis II (from the separation of dyads) there should be 8 monads migrating to *each* pole.

(d) $(1/2)^8$

16. First, through independent assortment of chromosomes at anaphase I of meiosis, daughter cells (secondary spermatocytes and secondary oocytes) may contain different sets of maternally- and paternally-derived chromosomes. Second, crossing over, which happens at a much higher frequency in meiotic cells as compared to mitotic cells, allows maternally- and paternally-derived chromosomes to exchange segments thereby increasing the likelihood that daughter cells (that is, secondary spermatocytes and secondary oocytes) are genetically unique.

18. One half of each tetrad will have a maternal homolog: $(1/2)^{10}$.

20. In angiosperms, meiosis results in the formation of microspores (male) and megaspores (female) which give rise to the haploid male and female gametophyte stage. Micro- and megagametophytes produce the pollen and the ovules respectively. Following fertilization, the sporophyte is formed.

22. The transition is at the end of interphase and the beginning of mitosis (prophase) when the chromosomes are in the condensation process. This eventually leads to the typically shortened and "fattened" metaphase chromosome.

24. Mutations symbolized as *cdc* are those involved in regulating the cell division cycle. Through the use of such mutations, kinases and their involvement with cyclins were also discovered. In addition, *cdc* mutations helped in the discovery of cell cycle checkpoints.

26. They would probably be homologous chromosomes and contain similar (but not identical) genetic information. Their centromeres would most likely be in the same position relative to chromosome arm lengths and any physical characteristics such as secondary constrictions or bands would be similar. They would have a similar sequence of nitrogenous bases. They would most likely replicate synchronously during the S phase of the cell cycle.

Chapter 9

2. (a) The parents are both normal, therefore they could be either *AA* or *Aa*. The fact that they produce an albino child requires that each parent provides an *a* gene to the albino child; thus the parents must both be heterozygous (*Aa*).

(b) The female is *aa* and the male is either *AA* or *Aa*. However, if the male is *Aa,* the likelihood of having six children, all normal is only 1/64.

(c) The female must be *aa*. The fact that half of the children are normal and half are albino indicates a typical "test cross" in which the*Aa* male is mated to the *aa* female.

(d)

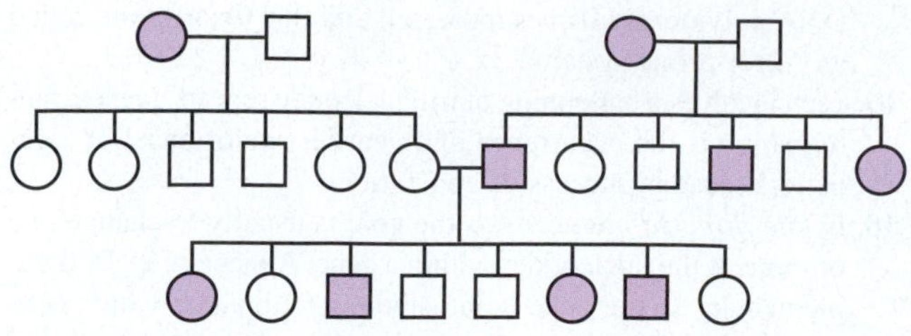

6. Genotypes of all individuals:

P_1 *Cross*	F_1 *Progeny* *Checkered*	*Plain*
(a) *PP* × *PP*	*PP*	
(b) *PP* × *pp*	*Pp*	
(c) *pp* × *pp*		*pp*
(d) *PP* × *pp*	*Pp*	
(e) *Pp* × *pp*	*Pp*	*pp*
(f) *Pp* × *Pp*	*PP*,*Pp*	*pp*
(g) *PP* × *Pp*	*PP*, *Pp*	

8. $WWgg = 1/16$

10. A test cross involves a cross of an organism with an unknown genotype to a fully homozygous recessive organism. In Problem #9, (d) fits this description.

12. Mendel's four postulates are related to the diagram below.

1. Factors occur in pairs. Notice *A* and *a*.
2. Some genes have dominant and recessive alleles. Notice *A* and *a*.
3. Alleles segregate from each other during gamete formation. When homologous chromosomes separate from each other at anaphase I, alleles will go to opposite poles of the meiotic apparatus.
4. One gene pair separates independently from other gene pairs. Different gene pairs on the same homologous pair of chromosomes (if far apart) or on non-homologous chromosomes will separate independently from each other during meiosis.

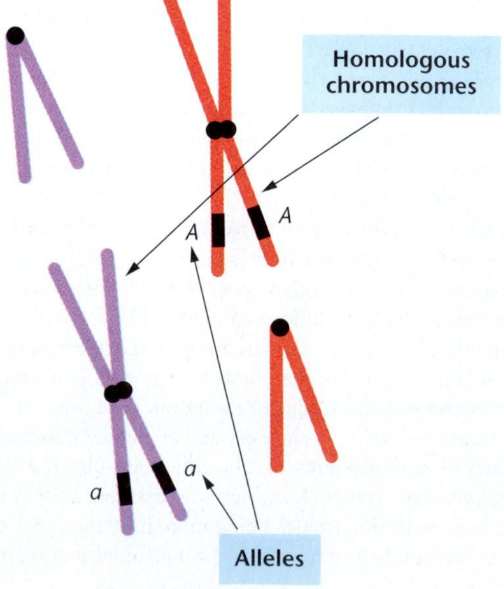

14. Homozygosity refers to a condition where both genes of a pair are the same (*i.e. AA* or *GG* or *hh*) whereas heterozygosity refers to the condition where members of a gene pair are different (*i.e. Aa* or *Gg or Bb*).

16. The general formula for determining the number of kinds of gametes produced by an organism is 2^n where n = number of *heterozygous* gene pairs.

(a) 4: *AB, Ab, aB, ab*
(b) 2: *AB, aB*
(c) 8: *ABC, ABc, AbC, Abc, aBC, aBc, abC, abc*
(d) 2: *ABc, aBc*
(e) 4: *ABc, Abc, aBc, abc*
(f) $2^5 = 32$

18. Set up the symbols as follows: *G* = yellow seeds, *g* = green seeds. We know that the gene for yellow seeds is dominant to that for green seeds because the F_1 is all yellow, not green.

Phenotypes	*Genotypes*
P_1: Yellow × green	*GG* × *gg*
F_1: all yellow	*Gg*
F_2: 6022 yellow	1/4 *GG*; 2/4 *Gg*
2001 green	1/4 *gg*

Of the yellow F_2 *offspring*, notice that 1/3 of them are *GG* and 2/3 are *Gg*. If you selfed the 1/3 *GG* types then all the offspring (the 166) would breed true whereas the others (353 which are *Gg*) should produce offspring in a 3:1 ratio when selfed.

GG × *GG* = all *GG*

Gg × *Gg* = 1/4 *GG*; 2/4 *Gg*; 1/4 *gg*

20. Symbols:

Seed shape	*Seed color*
W = round	*G* = yellow
w = wrinkled	*g* = green

P_1: *WWgg* × *wwGG*
F_1: *WwGg* cross to *wwgg*
(which is a typical testcross)
The offspring will occur in a typical 1:1:1:1 as
1/4 *WwGg* (round, yellow)
1/4 *Wwgg* (round, green)
1/4 *wwGg* (wrinkled, yellow)
1/4 *wwgg* (wrinkled, green)

22. **(a)** $\chi^2 = \Sigma\,(o - e)^2/e = .064$
probability (*p*) value between 0.9 and 0.5.
We would therefore say that there is a "good fit" between the observed and expected values.
(b) $\chi^2 = 0.39$
The *p* value in the table for 1 degree of freedom is still between 0.9 and 0.5, however because the χ^2 value is larger in (b) we should say that the deviations from expectation are greater.

24. For the test of a 3:1 ratio, the χ^2 value is 33.3 with an associated *p* value of less than 0.01 for 1 degree of freedom. For the test of a 1:1 ratio, the χ^2 value is 25.0 again with an associated *p* value of less than 0.01 for 1 degree of freedom. Based on these probability values, both null hypotheses should be rejected.

26.

I-1 (*Aa*), I-2 (*aa*), I-3 (*Aa*), I-4(*Aa*)
II-1 (*aa*), II-2 (*Aa*), II-3 (*aa*), II-4 (*Aa*), II-5 (*Aa*),
II-6 (*aa*), II-7 (*AA* or *Aa*), II-8 (*AA* or *Aa*)
III-1 (*AA* or *Aa*), III-2 (*AA* or *Aa*), III-3 (*AA* or *Aa*),
III-4 (*aa*), III-5 (probably *AA*), III-6 (*aa*)
IV-1 through IV-7 all *Aa*.

28. 1/8

30. 3/4

32. 4/9

34. 1/9

36. $P = [8!\,(3/4)^6(1/4)^2]/6!2!$

38. **(a)**

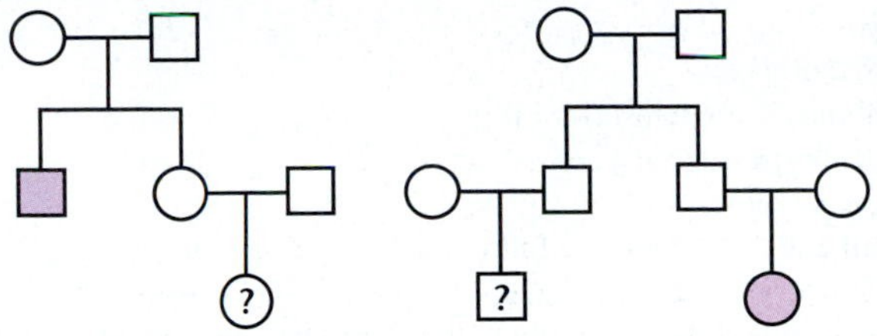

(b) 1/12
(c) 1/2
(d) 5/12

40. Because there are three possibilities within the group, it probably represents a case of incomplete dominance or codominance.

Chapter 10

2. *Incomplete dominance* can be viewed more as a quantitative phenomenon where the heterozygote is intermediate (approximately) between the limits set by the homozygotes. Pink is intermediate between red and white. *Codominance* can be viewed in a more qualitative manner where both of the alleles in the heterozygote are expressed.

4. *S* = short, *s* = long

Cross 1:
Ss × *ss*
1/2 *Ss* (short), 1/2 *ss* (long)
Cross 2:
Ss × *Ss*
1/4 *SS* (lethal), 2/4 *Ss* (short), 1/4 *ss* (long)

6. 1(A):1(B):1(AB):1(O).

8. Symbolism:
Se = secretor = nonsecretor
$I^A\,I^B$ *Sese* × $I^O\,I^O$ *Sese*
(a)

	$I^O Se$	$I^O\ se$
I^A *Se*	A, secretor	A, secretor
I^A *se*	A, secretor	A, nonsecretor
I^B *Se*	B, secretor	B, secretor
I^B *se*	B, secretor	B, nonsecretor

Overall ratio: 3/8 A, secretor
1/8 A, nonsecretor
3/8 B, secretor
1/8 B, nonsecretor
(b) 1/4 of all individuals will have blood type O.

10.

(a)
Phenotypes:
Himalayan × Himalayan ⇒ albino
Genotypes: c^hc^a $c^h\,c^a$ c^ac^a
The Himalayan parents must both be heterozygous to produce an albino offspring.
Phenotypes:
full color × albino ⇒ chinchilla
Genotypes: Cc^{ch} c^ac^a $c^{ch}c^a$

Because of the cc albino parent, the genotype of the chinchilla F_1 must be $c^{ch}c^a$. Also in order to have a chinchilla offspring at all, the full color parent must he heterozygous for chinchilla. Therefore the cross of albino with chinchilla would be as follows:

$c^ac^a \times c^{ch}c^a$

1/2 chinchilla; 1/2 albino

(b)

Phenotypes:

albino × chinchilla ⇒ albino

Genotypes: c^ac^a $c^{ch}c^a$ c^ac^a

Phenotypes:

full color × albino ⇒ full color

Genotypes: $C\_$ c^ac^a Cc^a

It is impossible to determine the complete genotype of the full color parent but the full color offspring must be as indicated, Cc^a. Therefore the cross of the albino with full color would be as follows:

$c^ac^a \times Cc^a \rightarrow$ 1/2 full color; 1/2 albino

(c)

Phenotypes:

chinchilla × albino ⇒ Himalayan

Genotypes: $c^{ch}c^h$ c^ac^a c^hc^a

The chinchilla parent must be heterozygous for Himalayan because of the Himalayan offspring.

Phenotypes:

full color × albino ⇒ Himalayan

Genotypes: Cc^h c^ac^a c^hc^a

Therefore a cross between the two Himalayan types would produce the following offpsring:

$c^hc^a \times c^hc^a$

⇓

3/4 Himalayan; 1/4 albino

12.

Flower color:

RR = red; Rr = pink; rr = white

Flower shape:

P = personate; p = peloric

Plant height:

D = tall; d = dwarf

(a)

$RRPPDD \times rrppdd$ ⌉

$RrPpDd$ (pink, personate, tall)

(b) Use *components* of the forked line method as follows:

2/4 pink × 3/4 personate × 3/4 tall

= 18/64

14. **(a)** This is a case of incomplete dominance in which, as shown in the third cross, the heterozygote (palomino) produces a typical 1:2:1 ratio. Therefore one can set the following symbols:

$C^{ch}C^{ch}$ = chestnut

C^cC^c = cremello

$C^{ch}C^c$ = palomino

(b) The F_1 resulting from matings between cremello and chestnut horses would be expected to be all palomino. The F_2 would be expected to fall in a 1:2:1 ratio as in the third cross in part (a) above.

16.

(a) In a cross of

$AACC \times aacc$,

the offpsring are all $AaCc$ (agouti) because the C allele allows pigment to be deposited in the hair and when it is it will be agouti. F_2 offspring would have the following "simplified" genotypes with the corresponding phenotypes:

$A\_C\_$ = 9/16 (agouti)

A_cc = 3/16

(colorless because cc is epistatic to A)

$aaC\_$ = 3/16 (black)

$aacc$ = 1/16

(colorless because cc is epistatic to aa)

The two colorless classes are phenotyically indistinguishable, therefore the final ratio is 9:3:4.

(b) Results of crosses of female agouti ($A\_C\_$) × $aacc$ (males) are given in three groups:

(1) To produce an even number of agouti and colorless offspring, the female parent must have been $AACc$ so that half of the offspring are able to deposit pigment because of C and when they do, they are all agouti(having received only A from the female parent).

(2) To produce an even number of agouti and black offspring the mother must have been Aa and so that no colorless offspring were produced, the female must have been CC. Her genotype must have been $AaCC$.

(3) Notice that half of the offspring are colorless, therefore the female must have been Cc. Half of the pigmented offspring are black and half are agouti, therefore the female must have been Aa. Overall, the $AaCc$ genotype seems appropriate.

18. **(a)**

$AaBbCc$ ⇒ gray (C allows pigment)

(b)

A_B_Cc ⇒ gray (C allows pigment)

(c) Use the forked line method for this portion

Combining the phenotypes gives (always count the proportions to see that they add up to 1.0):

16/32 albino;

9/32 gray;

3/32 yellow;

3/32 black;

1/32 cream

(d) Use the forked line method for this portion

3/4 A_
- 3/4 B_
  - 1/2 Cc ⟶ 9/32 gray
  - 1/2 cc ⟶ 9/32 albino
- 1/4 bb
  - 1/2 Cc ⟶ 3/32 yellow
  - 1/2 cc ⟶ 3/32 albino

1/4 aa
- 3/4 B_
  - 1/2 Cc ⟶ 3/32 black
  - 1/2 cc ⟶ 3/32 albino
- 1/4 bb
  - 1/2 Cc ⟶ 1/32 cream
  - 1/2 cc ⟶ 1/32 albino

3/4 A_ ⟶ 1/1 BB
- 3/4 C_ ⟶ 9/16 (gray)
- 1/4 cc ⟶ 3/16 (albino)

1/4 aa ⟶ 1/1 BB
- 3/4 C_ ⟶ 3/16 (black)
- 1/4 cc ⟶ 1/16 (albino)

Combining the phenotypes gives (always count the proportions to see that they add up to 1.0):

9/16 (gray);

3/16 (black);

4/16 (albino)

(e) Use the forked line method for this portion

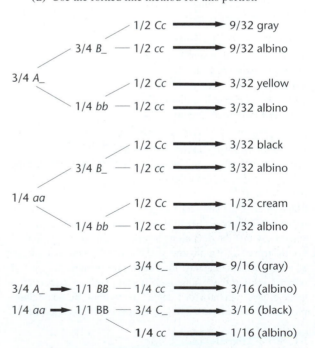

1/1 *AA* → 3/4 *B_* → 1/2 *Cc* → 3/8 (gray)
1/1 *AA* → 3/4 *B_* → 1/2 *cc* → 3/8 (albino)
1/1 *AA* → 1/4 *bb* → 1/2 *Cc* → 1/8 (yellow)
1/1 *AA* → 1/4 *bb* → 1/2 *cc* → 1/8 (albino)

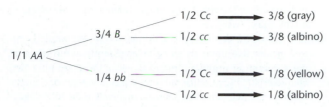

The final ratio would be
3/8 (grey);
1/8 (yellow);
4/8 (albino)

20. **(a)** Assign the phenotypes from cross (C) as indicated:

*A_B_* = 9/16 (green)
A_bb = 3/16 (brown)
*aaB_* = 3/16 (grey)
aabb = 1/16 (blue)

Cross A:
P_1: *AABB* × *aaBB*
F_1: *AaBB*
F_2: 3/4 *A_BB*: 1/4 *aaBB*

Cross B:
P_1: *AABB* × *AAbb*
F_1: *AABb*
F_2: 3/4 *AAB_* : 1/4 *AAbb*

Cross C:
P_1: *aaBB* × *AAbb*
F_1: *AaBb*
F_2: 9/16 *A_B_* : 3/16 *A_bb*:
3/16 *aaB_* : 1/16 *aabb*

(b) This question is exactly as that in Cross C. The genotype of the unknown P_1 individual would be *AAbb* (brown) while the F_1 would be *AaBb* (green).

22. **(a)** Assign the phenotypes as given, then see if patterns emerge.

*A_B_* = 9/16 (yellow)
A_bb = 3/16 (blue)
*aaB_* = 3/16 (red)
aabb = 1/16 (mauve)

From this information, the genotypes for the various phenotypes and the solution to the problem become clear. As stated in the problem all colors *may* be true-breeding. See that each type can exist as a full homozygote. If plants with blue flowers (homozygotes) are crossed to red-flowered homozygotes, the F_1 plants would have yellow flowers. Also as stated in the problem, if yellow-flowered plants are crossed with mauve-flowered plants, the F_1 plants are yellow and the F_2 will occur in a 9:3:3:1 ratio. All of the observations fit the model as proposed.

(b) If one crosses a true-breeding red plant (*aaBB*) with a mauve plant (*aabb*), the F_1 should be red (*aaBb*). The F_2 would be as follows:
aaBb × *aaBb*
3/4 *aaB_* (red): 1/4 *aabb* (mauve)

24. **(a)** 1/4
(b) 1/2
(c) 1/4
(d) zero

26. Symbolism: Normal wing margins = sd^+; scalloped = *sd*
(a) P_1: $X^{sd}X^{sd} \times X^+/Y$
F_1: 1/2 X^+X^{sd} (female, normal)
1/2 X^{sd}/Y (male, scalloped)
F_2: 1/4 X^+X^{sd} (female, normal)
1/4 $X^{sd}X^{sd}$ (female, scalloped)
1/4 X^+/Y (male, normal)
1/4 X^{sd}/Y (male, scalloped)

(b) P_1: $X^+/X^+ \times X^{sd}/Y$
F_1: 1/2 X^+X^{sd} (female, normal)
1/2 X^+/Y (male, normal)
F_2: 1/4 X^+X^+ (female, normal)
1/4 X^+X^{sd} (female, normal)
1/4 X^+/Y (male, normal)
1/4 X^{sd}/Y (male, scalloped)

If the *scalloped* gene were not X-linked, then all of the F_1 offspring would be wild (phenotypically) and a 3:1 ratio of normal to scalloped would occur in the F_2.

28. Set up the symbolism and the cross in the following manner:
P_1: X^+X^+; *su-v*/*su-v* × X^v/Y; $su\text{-}v^+/su\text{-}v^+$
F_1: 1/2 X^+X^v; $su\text{-}v^+/su\text{-}v$ (female, normal)
1/2 X^+/Y; $su\text{-}v^+/su\text{-}v$ (male, normal)

F_2: 2/4 females, $X^+/\_$ → 3/4 $su\text{-}v^+/\_$; 1/4 *su-v*/*su-v*
1/4 males, X^+/Y → 3/4 $su\text{-}v^+/\_$; 1/4 *su-v*/*su-v*
1/4 males, X^v/Y → 3/4 $su\text{-}v^+/\_$; 1/4 *su-v*/*su-v*

8/16 wild type females; (none of the females are homozygous for the *vermilion* gene)
5/16 wild type males; (4/16 because they have no *vermilion* gene and 1/16 because the X-linked, hemizygous *vermilion* gene is suppressed by *su-v*/*su-v*)
3/16 vermilion males; (no suppression of the *vermilion* gene)

30.
(a) $w/w;se^+/se^+ \times w^+/Y;se/se$
F_1:
$w^+/w;se^+/se$ = wild females
$w/Y;se^+/se$ = white-eyed males
F_2:
1/4 w^+/Y
1/4 w/Y
1/4 w^+/w
1/4 w/w
× 3/4 $se^+/\_$; 3/4 *se*/*se*

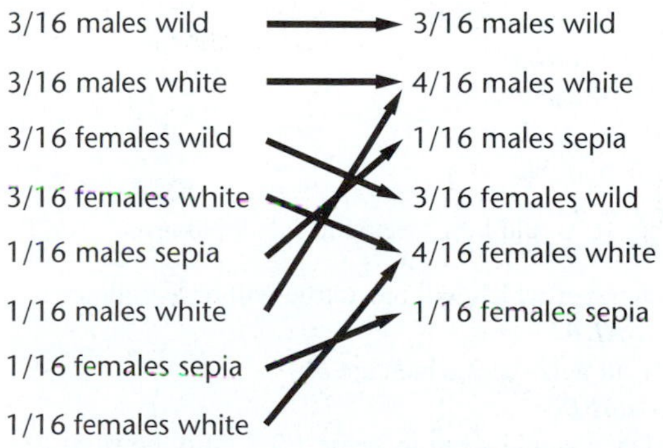

(b)
$w^+/w^+;se/se$ X $w/Y;se^+/se^+$
⇓
F_1: $w^+/w;se^+/se$ = wild females
$w^+/Y;se^+/se$ = wild males
F_2:

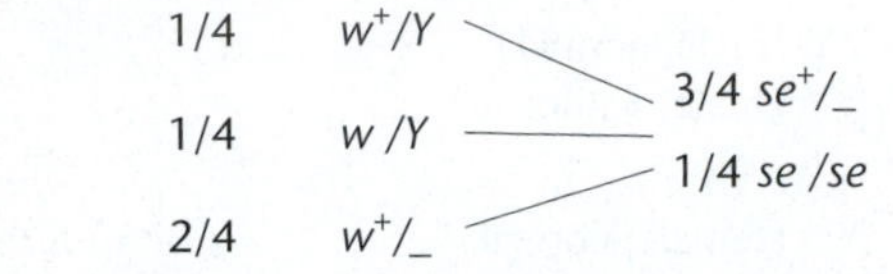

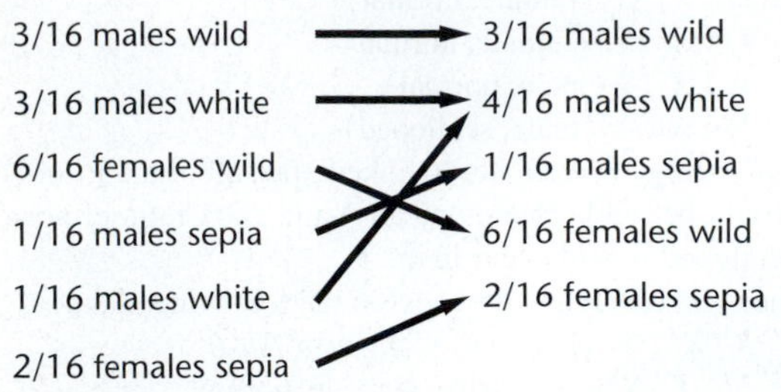

32.

RR = red, *Rr* = red in females,
Rr = mahogany in males,
rr = mahogany.
P_1:
female: *RR* (red) × male: *rr* (mahogany)
F_1:
Rr = females red; males mahogany
1/2 females (red)
1/2 males (mahogany)
F_2:
1/4 *RR*; 2/4 *Rr*; 1/4 *rr*
Because half of the offspring are males and half are females, one could, for clarity, rewrite the F_2 as:

	1/2 females	*1/2 males*
1/4 *RR*	1/8 red	1/8 red
2/4 *Rr*	2/8 red	2/8 mahogany
1/4 *rr*	1/8 mahogany	1/8 mahogany

34. Passage of X-linked genes typcially occurs from carrier mother to effected son. The fact that the father in couple #2 has hemophilia would not predispose his son to hemophilia. The #1 couple has no valid claim.

36. *Penetrance* refers to the percentage of individuals which express the mutant phenotype while *expressivity* refers to the range of expression of a given phenotype.

38. **(a)** *AAB-* × *aaBB* (other configurations possible but each must give all offspring with *A* and *B* dominant alleles)
(b) *AaB-* × *aaBB* (other configurations are possible but no types can be produced)
(c) *AABb* × *aaBb*
(d) *AABB* × *aabb*
(e) *AaBb* × *Aabb*
(f) *AaBb* × *aabb*
(g) *aaBb* × *aaBb*
(h) *AaBb* × *AaBb*

Those genotypes which will breed true will be as follows:
black = *AABB*
golden = all genotypes which are *bb*
brown = *aaBB*

40. A first glance would seem to favor a 9:7 ratio, however, the phenotypes would have to be reversed for such a result to fit. Therefore, one must consider an alternative explanation. A 27:9:9:9:3:3:3:1 ratio fits very well with A_B_C_ being purple and any homozygous recessive combination giving white. Thus a 27(purple):37(white) ratio fits well. To test this hypothesis, one might take the purple F's and cross them to the pure breeding (*aabbcc*) white type. Such a cross should give a 1(purple):7(white).

42. **(a)** Because the denominator in the ratios is 64 one would begin to consider that there are three independently assorting gene pairs operating in this problem. Because there are only two characteristics (eye color and croaking) however, one might hypothesize that two gene pairs are involved in the inheritance of one trait while one gene pair is involved in the other.
(b) Notice that there is a 48:16 (or 3:1) ratio of rib-it to knee-deep and a 36:16:12 (or 9:4:3) ratio of blue to green to purple eye color. Because of these relationships one would conclude that croaking is due to one (dominant/recessive) gene pair while eye color is due to two gene pairs. Because there is a (9:4:3) ratio regarding eye color, some gene interaction (epistasis) is indicated.
(c,d) Symbolism:
Croaking: *R_* = rib-it; *rr* = knee-deep
Eye color:
Since the most frequent phenotype is blue eye, let *A_B_* represent the genotypes. For the purple class, "a 3/16 group" use the *A_bb* genotypes. The "4/16" class (green) would be the *aaB_* and the *aabb* groups.
(e) The cross involving a blue-eyed, knee-deep frog and a purple-eyed, rib-it frog would have the genotypes:

AABBrr × *AAbbRR*

which would produce an F_1 of *AABbRr* which would be blue-eyed and rib-it. The F_2 will follow a pattern of a 9:3:3:1 ratio because of homozygosity for the A locus and heterozygosity for both the B and R loci.

9/16 *AAB_R_* = blue-eyed, rib – it
3/16 *AAB_rr* = blue-eyed, knee-deep
3/16 *AAbbR_* = purple-eyed, rib-it
1/16 *AAbbrr* = purple-eyed, knee-deep

(f) The different results can arise because of the genetic variety possible in producing the green-eyed frogs. Since there is no dependence on the *B* locus, the following genotypes can define the green phenotype:

aaBB, aaBb, aabb

(g) Notice that the ratio of purple-eyed to green-eyed frogs is 3:1, therefore expect the parents to be heterozygous for the *A* locus. Because the ratio of rib-it to knee-deep is also 3:1 expect both parents to be heterozygous at the *R* locus. The *B* locus would have the *bb* genotype because both parents are purple-eyed as given in the problem. Both parents would therefore be *AabbRr*.

44.

(a) P_1: *YYBB* × *yWbb*
F_1: *YyBb* and *YWBb*
Crossing these F_1's gives the observed ratios in the F_2.
(b) Given a blue male with the genotype *yyBb* and a green female with the genotype *YWBb*, the offspring are as given part **b** of the question.

46.

(a)		F1:	*AABbCC* =	speckled
	F2:	3	*AAB_CC* =	speckled
		1	*AAbbCC* =	yellow
(b)		F1:	*AABbCc* =	speckled
	F2:	9	*AAB_C_* =	speckled
		3	*AAB_cc* =	green
		3	*AAbbC_* =	yellow } 4
		1	*AAbbcc* =	yellow
(c)	F1:		*AaBBCc* =	speckled
	F2:	9	*A_BBC_* =	speckled

3	A_BBcc =	green	
3	$aaBBC\_$ =	colorless	} 4
1	$aaBBcc$ =	colorless	

Chapter 11

2. The *Protenor* form of sex determination involves the XX/XO condition while the *Lygaeus* mode involves the XX/XY condition.

4. In *primary* nondisjunction half of the gametes contain two X chromosomes while the complementary gametes contain no X chromosomes. Fertilization, by a Y-bearing sperm cell, of those female gametes with two X chromosomes would produce the XXY Klinefelter syndrome. Fertilization of the "no-X" female gamete with a normal X-bearing sperm will produce the Turner syndrome.

6. Males and females share a common placenta and therefore hormonal factors carried in blood. Hormones and other molecular species (transcription factors perhaps) triggered by the presence of a Y chromosome lead to a cascade of developmental events which both suppress female organ development and enhance mastulinization. Other mammals also exhibit a variety of similar effects depending on the sex of their uterine neighbors during development.

8. Because synapsis of chromosomes in meiotic tissue is often accompanied by crossing over, it would be detrimental to sex-determining mechanisms to have sex-determining loci on the Y chromosome transferred, through crossing over, to the X chromosome.

10.

Klinefelter syndrome (XXY) = 1
Turner syndrome (XO) = 0
47, XYY = 0
47, XXX = 2
48,XXXX = 3

12. Females will display mosaic retinas with patches of defective color perception. Under these conditions, their color vision may be influenced.

14. Many organisms have evolved over millions of years under the fine balance of numerous gene products. Many genes required for normal cellular and organismic function in *both* males and females are located on the X chromosome. These gene products have nothing to do with sex determination or sex differentiation.

16. One could account for the significant departures from a 1:1 ratio of males to females by suggesting that at anaphase I of meiosis, the Y chromosome more often goes to the pole which produces the more viable sperm cells. One could also speculate that the Y-bearing sperm has a higher likelihood of surviving in the female reproductive tract, or that the egg surface is more receptive to Y-bearing sperm. At this time the mechanism is unclear.

18. (See figure at top of next column.

20. **(a)** Something is missing from the male-determining system of sex determination either at the level of the genes, gene products, or receptors, *etc.*.

(b) The *SOX9* gene or its product is probably involved in male development. Perhaps it is activated by *SRY*.

(c) There is probably some evolutionary relationship between the *SOX9* gene and *SRY*. There is considerable evidence that many other genes and pseudogenes are also homologous to *SRY*.

(d) Normal female sexual development does not require the *SOX9* gene or gene product(s).

(e) *SRY* may activate *SOX9*, which is also required for normal skeletal development. A good review of SRY and its

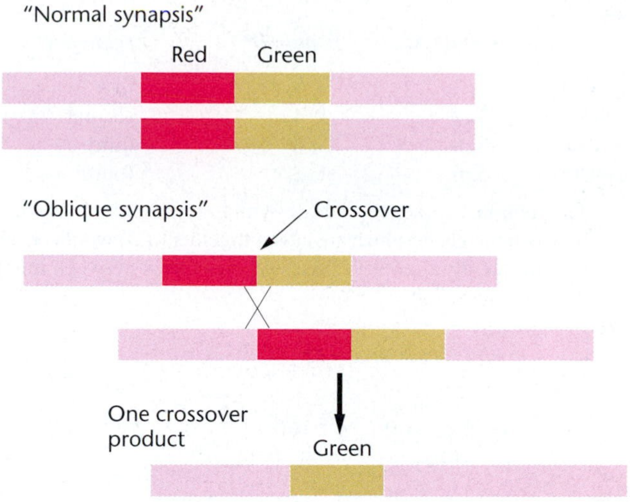

relatives among a variety of vertebrates is found in Jeyasuria and Pace. 1998. *Journal of Experimental Zoology* 281:428-449

22. In snapping turtles, sex determination is strongly influenced by temperature such that males are favored in the 26–34°C range. Lizards, on the other hand, appear to have their sex determined by factors other than temperature in the 20–40°C range.

Chapter 12

4. Because crossing over occurs at the four-strand stage of the cell cycle (that is, after S phase) notice that each single crossover involves only two of the four chromatids.

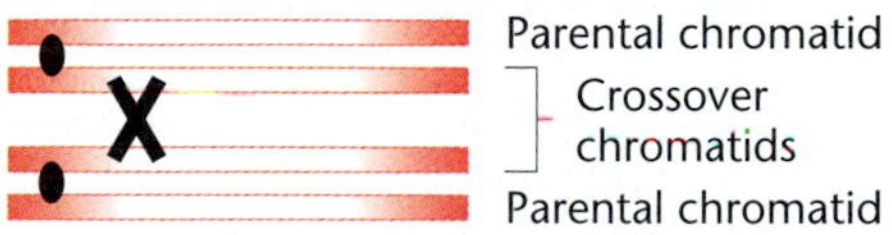

6. Positive interference occurs when a crossover in one region of a chromosome interferes with crossovers in nearby regions. Such interference ranges from zero (no interference) to 1.0 (complete interference). Interference is often explained by a physical rigidity of chromatids such that they are unlikely to make sufficiently sharp bends to allow crossovers to be close together.

8. Since the distance between *dp* and *ap* is greatest, they must be on the "outside" and *cl* must be in the middle. The genetic map would be as follows:

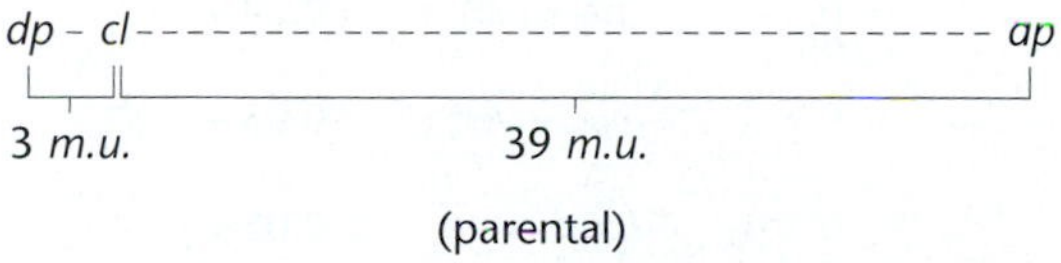

10. *RY*/*ry* × *ry*/*ry*
Seeing that there are 20 crossover progeny among the 200, or 20/200, the map distance would be 10 map units.

12. *PZ*/*pz* × *pz*/*pz*
Adding the crossover percentages together (6.9 + 7.1) gives 14% which would be the map distance between the two genes.

14.

	female A:	*female B:*	*Frequency:*
NCO	3, 4	7, 8	first
SCO	1, 2	3, 4	second
SCO	7, 8	5, 6	third
DCO	5, 6	1, 2	fourth

The single crossover classes which represent crossovers between the genes which are closer together (*d-b*) would occur less frequently than the classes of crossovers between more distant genes (*b-c*).

16.

(a) *y w*+/++ *ct* × *y w* +/Y

(b)

$$y - w = \frac{9 + 6 + 0 + 0}{1000} \times 100$$

$$= 1.5 \text{ map units}$$

$$w - ct = \frac{90 + 95 + 0 + 0}{1000} \times 100$$

$$= 18.5 \text{ map units}$$

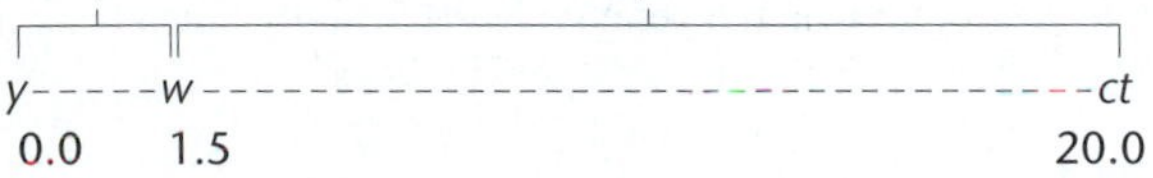

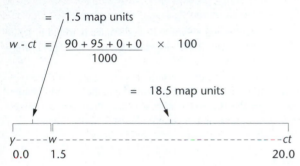

(c) There were
$.185 \times .015 \times 1000 = 2.775$
double crossovers expected.

(d) Because the cross to the F_1 males included the normal (wild type) gene for *cut wings* it would not be possible to unequivocally determine the genotypes from the F_2 phenotypes for all classes.

18.

P_1:
females: + / + ; *p e*/*p e*
×
males:*dp*/*dp*;++/++
F_1:
females:+/*dp*;++/*p e*
×
males:*dp*/*dp*; *p e*/*p e*

Female gametes: use a modification of the forked-line method for determining the types of gametes to be produced. The *dumpy* locus will give .5+ and .5 *dp* to the gametes because of independent assortment (on a different chromosome) and the other two loci will segregate with 20% (map units) being the recombinants, and 80% being the parentals.

0.5 +	0.4 +	+ (parental)	= 0.20 + + +
	0.1 +	*e* (crossover)	= 0.05 + + *e*
	0.1 *p*	+ (crossover)	= 0.05 + *p* +
	0.4 *p*	*e* (parental)	= 0.20 + *p e*
0.5 *dp*	0.4 +	+ (parental)	= 0.20 *dp* + +
	0.1 +	*e* (crossover)	= 0.05 *dp* + *e*
	0.1 *p*	+ (crossover)	= 0.05 *dp p* +
	0.4 *p*	*e* (parental)	= 0.20 *dp p e*

Crossed with *dp p e* from the male gives the following offspring:
0.20 wild type
0.05 ebony
0.05 pink
0.20 pink, ebony
0.20 dumpy
0.05 dumpy, ebony
0.05 dumpy, pink
0.20 dumpy, pink, ebony

For the reciprocal cross:
F_1:
males: + /*dp*;++/*p e*
×
females:*dp*/*dp*; *p e*/*p e*
there would be no crossover classes.

0.5 +	0.5 +	+ (parental)	= 0.25 + + +
	0.5 *p*	*e* (parental)	= 0.25 + *p e*
0.5 *dp*	0.5 +	+ (parental)	= 0.25 *dp* + +
	0.5 *p*	*e* (parental)	= 0.25 *dp p e*

Crossed with *dp p e* from the female gives the following offspring:
.25 wild type
.25 pink, ebony
.25 dumpy
.25 dumpy, pink, ebony

The results would change because of no crossing over in males.

20. (a,b) +*b c*/ *a*++
a − *b* = 7 map units
b − *c* = 2 map units

(c) The progeny phenotypes that are missing are ++*c* and *a b* +, which, of 1000 offspring, 1.4 (.07 × .02 × 1000) would be expected. Perhaps by chance or some other unknown selective factor, they were not observed.

22. Because sister chromatids are genetically identical (with the exception of rare new mutations) crossing over between sisters provides no increase in genetic variability. Individual genetic variability could be generated by somatic crossing over because certain patches on the individual would be genetically different from other regions. This variability would be of only minor consequence in all likelihood. Somatic crossing over would have no influence on the offspring produced.

24. (a) There would be $2^n = 8$ genotypic and phenotypic classes and they would occur in a 1:1:1:1:1:1:1:1 ratio.

(b) There would be two classes and they would occur in a 1:1 ratio.

(c) There are 20 map units between the *A* and *B* loci and locus *C* assorts independently from both *A* and *B* loci.

26. Assign the following symbols for example:

R = Red *r* = yellow
O = Oval *o* = long

Progeny A: *Ro*/*rO* × *rroo* = 10 map units
Progeny B: *RO*/*ro* × *rroo* = 10 map units

28. 10 map units

30.

For Cross 1:

$$\frac{36 + 14}{100} = 50 \text{ map units}$$

Because there are 50 map units between genes *a* and *b*, they are not linked.

For Cross 2:

$$\frac{3 + 9}{100} = 12 \text{ map units}$$

Because genes *a* and *b* are not linked they could be on non-homologous chromosomes or far apart (50 map units or more) on the same chromosome. Because genes *c* and *b* are linked and therefore on the same chromosome, it is also possible that genes *a* and *c* are on different chromosome pairs. Under that condition, the NP and P (parental ditypes) would be equal; however, there is a possibility that the following arrangement occurs and that genes *a* and *c* are linked.

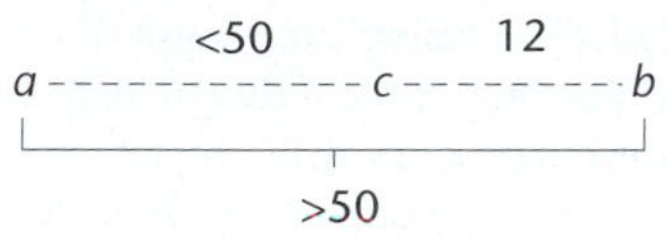

34.

MDH1: chromosome 2
PEPS: chromosome 4
PMG1: chromosome 1

36. It is important to remember that there is no crossing over in males and that if two genes are on the same chromosome, there will be complete linkage of the genes in the male gametes. In females, crossing over will produce parental and crossover gametes. What you will have is the following gametes from the females (left) and males (right):

$bw^+ st^+$	1/4	$bw^+ st^+$	1/2
$bw^+ st$	1/4	$bw\ st$	1/2
$bw\ st^+$	1/4		
$bw\ st$	1/4		

Combining these gametes will give the ratio presented in the table of results.

38.

Mapping the distance between *B* and *m* would be as follows:

$(57 + 64)/(226 + 218 + 57 + 64) \times 100 =$
$121/565 \times 100 = 21.4$ map units.

We would conclude that the *ebony* locus is either far away from *B* and *m* (50 map units or more) or it is on a different chromosome. In fact, *ebony* is on a different chromosome.

Chapter 13

4. The fact that there is a significant maternal age effect associated with Down syndrome indicates that nondisjunction in older females contributes disproportionately to the number of Down syndrome individuals. In addition, certain genetic and cytogenetic marker data indicate the influence of female nondisjunction.

6. Because an allotetraploid has a possibility of producing bivalents at meiosis I, it would be considered the most fertile of the three. Having an even number of chromosomes to match up at the metaphase I plate, autotetraploids would be considered to be more fertile than autotriploids.

8. American cultivated cotton has 26 pairs of chromosomes; 13 large, 13 small. Old world cotton has 13 pairs of large chromosomes and American wild cotton has 13 pairs of small chromosomes. It is likely that an interspecific hybridization occurred followed by chromosome doubling. These events probably produced a fertile amphidiploid (allotetraploid). Experiments have been conducted to reconstruct the origin of American cultivated cotton.

10. While there is the appearance that crossing over is suppressed in inversion "heterozygotes" the phenomenon extends from the fact that the crossover chromatids end up being abnormal in genetic content. As such they fail to produce viable (or competitive) gametes or lead to zygotic or embryonic death.

12.

Females: N^+/N × males: B/Y

14/	N^+/B	females; Bar
1/4	N/B	females; Notch, Bar
1/4	N^+/Y	males; wild
1/4	N/Y	**lethal**

The final phenotypic ratio would be 1:1:1 for the phenotypes shown above.

If a gene exists in a duplicated state and if that gene's product is required for survival, then mutation in either (but not both) the original gene or its duplicate will not ordinarily threaten the survival of the organism. Duplication of a gene provides a buffer to mutation.

14. It is likely that when certain combinations of genes are of selective advantage in a specific and stable environment, it would be beneficial to the organism to protect that gene combination from disruption through crossing over. By having the genes in an inversion, crossover chromatids are not recovered and therefore are not passed on to future generations.

Translocations offer an opportunity for new gene combinations by associations of genes from nonhomologous chromosomes. Under certain conditions such new combinations may be of selective advantage and meiotic conditions have evolved so that segregation of translocated chromosomes yields a relatively uniform set of gametes.

16. The primrose, *Primula kewensis*, with its 36 chromosomes, is likely to have formed from the hybridization and subsequent chromosome doubling of a cross between the two other species, each with 18 chromosomes.

18. The rare double crossovers in the boundaries of a paracentric or pericentric inversion produce only minor departures from the standard chromosomal arrangement as long as the crossovers involve the same two chromatids. With two-strand double crossovers, the second crossover negates the first. However, three-strand and four-strand double crossovers have consequences which lead to anaphase bridges as well as a high degree of genetically unbalanced gametes.

20. In the trisomic, segregation will be "2 X 1" as illustrated below:

P_1:

$b/b/b \times b^+/b^+$

gametes bb b b^+

F_1:

$b^+/b^1/b^2$ × b^+/b

(normal bristles) (normal bristles)

Notice that there are several segregation patterns created by the trivalent at anaphase I.

gametes:

b^+	b^+
b^1b^2	b
b^+b^1	
b^+b^2	
b^2	
b^1	

F_2:

b^+b^+ = normal bristles
$b^+b^1b^2$ = normal bristles
$b^+b^+b^1$ = normal bristles
$b^+b^+b^2$ = normal bristles

b^+b^2 = normal bristles
b^+b^1 = normal bristles
b^+b = normal bristles
bb^1b^2 = bent bristles
b^+bb^1 = normal bristles
b^+bb^2 = normal bristles
bb^2 = bent bristles
bb^1 = bent bristles

22. Given some of the information in the above problem the expression would be as follows:
$(35/36W{:}1/36w)(35/36A{:}1/36a)$
$(35/36)^2$ W—A—
$35/(36)^2$ W—$aaaa$
$35/(36)^2$ $wwwwA$—
$1/(36)^2$ $wwwwaaaa$

24.

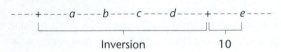

----+----a----b----c----d----+----e----

Inversion 10

26. **(a)** The father must have contributed the abnormal X-linked gene.
(b) Since the son is XXY and heterozygous for anhidrotic dysplasia, he must have received both the defective gene and the Y chromosome from his father. Thus non-disjunction must have occurred during meiosis I.
(c) This son's mosaic phenotype is caused by X-chromosome inactivation, a form of dosage compensation in mammals.

Chapter 14

2. The mt^+ strain (resistant for the nuclear and chloroplast genes) contributes the "cytoplasmic" component of streptomycin resistance which would negate any contribution from the mt^- strain. Therefore, all the offspring will have the streptomycin resistance phenotype. In the reciprocal cross, with the mt^+ strain being streptomycin sensitive, all the offspring will be sensitive.

4.
(a) neutral
(b) segregational (nuclear mutations)
(c) suppressive

6. Examine the text and notice that the inheritance patterns for the two, *segregational* and *neutral*, are quite different. The segregational mode is dependent on nuclear genes while that of the neutral type is dependent on cytoplasmic influences, namely mitochondria. If the two are crossed as stated in the problem, then one would expect, in the diploid zygote, the *segregational* allele to be "covered" by normal alleles from the neutral strain. On the other hand, as the nuclear genes are again "exposed" in the haploid state of the ascospores, one would expect a 1:1 ratio of normals to petites. The petite phenoytpe is caused by the nuclear, segregational gene.

8. **(a)** There are many similarities among mitochondrial, chloroplast, and prokaryotic molecular systems. It is likely that mitochondria and chloroplasts evolved from bacteria in a symbiotic relationship, therefore it is not surprising that certain antibiotics which influence bacteria will also influence all mitochondria and chloroplasts.
(b) Clearly, the mt^+ strain is the donor of the *cp*DNA since the inheritance of resistance or sensitivity is dependent on the status of the mt^+ gene.

10. In a maternal effect, the *genotype* of the mother influences the *phenotype* of her immediate offspring in a non-Mendelian manner. The fact that all of the offspring (F_1) showed a dextral coiling pattern indicates that one of the parents (maternal parent) contains the *D* allele. Taking these offspring and seeing that their progeny (call these F_2) occur in a 1:1 ratio indicates that half of the offspring (F_1) are *dd*. In order to have these results, one of the original parents must have been *Dd* while the other must have been *dd*.
Parents: $Dd \times dd$
Offspring (F_1): 1/2 *Dd*, 1/2 *dd*
(all dextral because of the maternal genotype)
Progeny (F_2):
All those from *Dd* parents will be dextral while all those from *dd* parents will be sinistral.

12. Since there is no evidence for segregation patterns typical of chromosomal genes and Mendelian traits, some form of extranuclear inheritance seems possible. If the *lethargic* gene is dominant then a maternal effect may be involved. In that case, some of the F_2 progeny would be hyperactive because maternal effects are only temporary, affecting only the immediate progeny. If the lethargic condition is caused by some infective agent, then perhaps injection experiments could be used. If caused by a mitochondrial defect, then the condition would persist in all offspring of lethargic mothers, through more than one generation.

14. **(a)** The presence of bcd^-/bcd^- males can be explained by the maternal effect: mothers were bcd^+/bcd^-. **(b)** The cross female $bcd^+/bcd^- \times$ male bcd^-/bcd^-
will produce an F_1 with normal embryogenesis because of the maternal effect. In the F_2, any cross having bcd^+/bcd^- mothers will have phenotypically normal embryos. Any cross involving homozygous bcd^-/bcd^- mothers will have problems with embryogenesis.

16. **(a)** A locus, *Segregation Distortion* (*SD*), is present on the wild type chromosome. Some aspect of *SD* causes a shift in the segregation ratio by allowing sperm to carry the *SD* chromosome at the expense of the homologue.
(b) One could use this *SD* chromosome in a variety of crosses and determine that the abnormal segregation is based on a particular chromosomal element. One could even map the *SD* locus on the second chromosome (as has been done).
(c) Segregation Distortion describes a condition in which typical Mendelian segregation is distorted from the 50:50.

Chapter 15

2. **(a)** The requirement for physical contact between bacterial cells during conjugation was established by placing a filter in a U-tube such that the medium can be exchanged but the bacteria can not come in contact. Under this condition, conjugation does not occur.
(b) By treating cells with streptomycin, an antibiotic, it was shown that recombination would not occur if one of the two bacterial strains was inactivated. However, if the other was similarly treated, recombination would occur. Thus, directionality was suggested, with one strain being a donor strain and the other being the recipient.
(c) An F^+ bacterium contains a circular, double-stranded, structurally independent, DNA molecule which can direct recombination.

4. Mapping the chromosome in an Hfr $\times$ F^- cross takes advantage of the oriented transfer of the bacterial chromosome through the conjugation tube. For each F type, the point of insertion and the direction of transfer are fixed, therefore breaking the conjugation tube at different times produces partial diploids with corresponding portions of the donor chromosome being transferred. The length of the chromosome being transferred is contingent on the duration of conjugation, thus mapping of genes is based on time.

6. As shown in the text, the F^+ element can enter the host bacterial chromosome and upon returning to its independent state, it may pick up a piece of a bacterial chromosome. When combined with a bacterium with a complete chromosome, a partial diploid, or merozygote, is formed.

8. In the first data set, the transformation of each locus, a^+ or b^+, occurs at a frequency of .031 and .012 respectively. To determine if there is linkage one would determine whether the frequency of double transformants a^+b^+ is greater than that expected by a multiplication of the two independent events. Multiplying .031 × .012 gives .00037 or approximately 0.04%. From this information, one would consider no linkage between these two loci. Notice that this frequency is approximately the same as the frequency in the second experiment, where the loci are transformed independently.

10. In their experiment a filter was placed between the two auxotrophic strains which would not allow contact. F-mediated conjugation requires contact and without that contact, such conjugation can not occur. The treatment with DNase showed that the filterable agent was not naked DNA.

12. In *generalized transduction* virtually any genetic element from a host strain may be included in the phage coat and thereby be transduced. In *specialized (restricted) transduction* only those genetic elements of the host which are closely linked to the insertion point of the phage can be transduced. Specialized transduction involves the process of lysogeny.
Because only certain genetic elements are involved in specialized transduction, it is not useful in determining linkage relationships. Cotransduction of genes in generalized transduction allows linkage relationships to be determined.

14. Viral recombination occurs when there is a sufficiently high number of infecting viruses so that there is a high likelihood that more than one type of phage will infect a given bacterium. Under this condition phage chromosomes can recombine by crossing over.

16. **(a)** The concentration of phage is greater than 10^4.
(b) The concentration of phage is around 1.4×10^6.
(c) The concentration of phage is less than 10^6.

18. Because there are only two complementation groups in the *r*II region one would have the following groupings:

Group A: 1,4,5 Group B: 2,3

(a) Therefore the result of testing
2 X 3 = no lysis;
2 X 4 = lysis;
3 X 4 = lysis.

(b) Because mutant 5 failed to complement with mutations in either cistron, it probably represents a major alteration in the gene such that both cistrons are altered. A deletion which overlaps both cistrons could cause such a major alteration.

(c) $2(5 \times 10^1)/(2 \times 10^5) = 5 \times 10^{-4}$

(d) Because mutant 6 complemented mutations 2 and 3, it is likely to be in the cistron with mutants 1,4, and 5. A lack of recombinants with mutant 4 indicates that mutant 6 is a deletion which overlaps mutation 4. Recombinants with 1 and 5 indicates that the deletion does not overlap these mutations.

20. **(a)**

Combination	Complementation
1, 2	–
1, 3	+
2, 4	+
4, 5	–

(b) $2(8 \times 10^2/4 \times 10^7) = 4 \times 10^{-5}$
(c) The dilution would be 10^{-3} and the colony number would be 8×10^3.
(d) Mutant 7 might well be a deletion spanning parts of both A and B cistrons.

22. **(a)** Rifampicin eliminates the donor strain which is rif^s.
(b) *b a* ——— *c* ——— F
(c) To determine the location of the *rif* gene one could use a donor strain which was rif^r but sensitive to another antibiotic (ampicillin for example). The interrupted mating experiment is conducted as usual on an ampicillin-containing medium but the recombinants must be replated on a rifampicin medium to determine which ones are sensitive.

Chapter 16

2. *Reverse transcriptase* is often used to promote the formation of cDNA (complementary DNA) from a mRNA molecule. Eukaryotic mRNAs typically have a 3′ polyA tail. The poly dT segment provides a double-stranded section which serves to prime the production of the complementary strand.

4. 1.38×10^5

6. Given that there is only one site for the action of *Hind*III, then the following will occur. Cuts will be made such that a four base single-stranded set of sticky ends will be produced. For the antibiotic resistance to be present, the ligation will reform the plasmid into its original form. However, two of the plasmids can join to form a dimer as indicated in the diagram below.

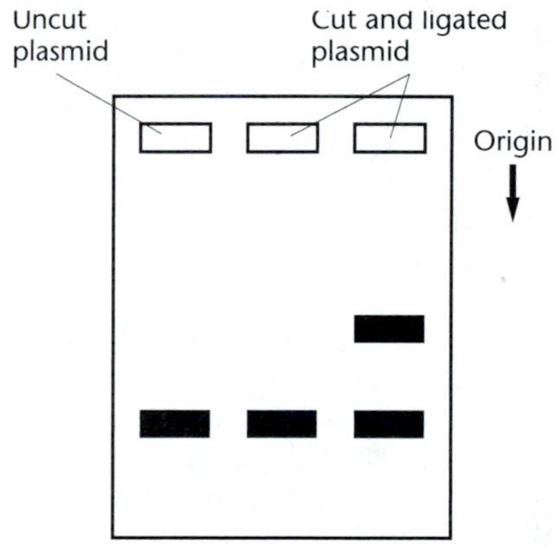

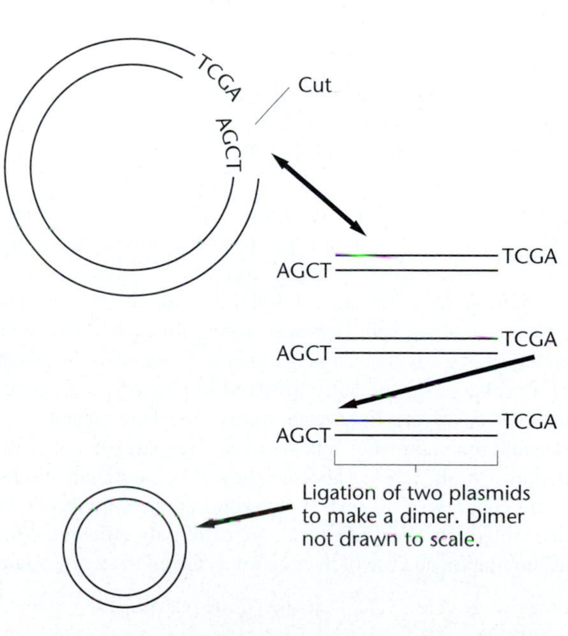

8. All other factors being equal (appropriate cloning sites and selectable markers), it is important to consider the size of the foreign DNA which can be cloned into the vector. Generally for large genomes, it is best to use a vector which will accept relatively large fragments.

10. Assuming a random distribution of all four bases, the four-base sequence would occur (on average) every 256 base pairs (4^4), the six-base sequence every 4096 base pairs (4^6), and the eight-base sequence every 65,536 base pairs (4^8). One might use an eight-base restriction enzyme to produce a relatively few large fragments. If one wanted to construct a eukaryotic genomic library, such large fragments would have to be cloned into special vectors, such as yeast artificial chromosomes.

12.

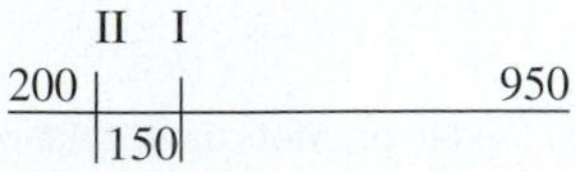

14. There may be several factors contributing to the lack of representation of the 5′ end of the mRNA. One has to deal with the possibility that the reverse transcriptase may not completely synthesize the DNA from the RNA template. The other reason may be that the 3′ end of the copied DNA tends to fold back on itself thus providing a primer for the DNA polymerase. Additional preparation of the cDNA requires some digestion at the folded region. Since this folded region corresponds to the 5′ end of the mRNA, some of the message is often lost.

16. Option (b) fits the expectation because both bands in the offspring are found in both parents. Option (d) would be a possibility, however since the primers were stated as being highly polymorphic in humans, it is likely that they are also highly polymorphic in other primates. Thus, the likelihood of such a match is expected to be low in the general population.

18. An appropriate sequence would be the following: 3, 1, 7, 8, 9, 6. Note that (2) may be used to make restriction maps for determining the direction of the walk and the relationships of overlapping clones.

20. **(a)** The overall size of the fragment is 12kb. From the A + N digest, sites A and N must be 1kb apart. N must be 2kb from an E site. Pattern #5 is the likely choice. Notice that digest A + N breaks up the 6kb E fragment.

(b) By drawing lines though sections that hybridize to the probe, one can see that the only place of consistent overlap to the probe is the 1kb fragment between A and N.

Chapter 17

2. While greater DNA content per cell is associated with eukaryotes, one can not universally equate genomic size with an increase in organismic complexity. There are numerous examples where DNA content per cell varies considerably among closely related species. Because of the diverse cell types of multicellular eukaryotes, a variety of gene products is required, which may be related to the increase in DNA content per cell. In addition, the advantage of diploidy automatically increases DNA content per cell. However, seeing the question in another way, it is likely that a much higher *percentage* of the genome of a prokaryote is actually involved in phenotype production than in a eukaryote. Eukaryotes have evolved the capacity to obtain and maintain what appears to be large amounts of "extra" perhaps "junk" DNA. This concept will be examined in subsequent chapters of the text. Prokaryotes on the other hand, with their relatively short life cycle, are extremely efficient in their accumulation and use of their genome. Given the larger amount of DNA per cell in eukaryotes and the requirement that the DNA be partitioned in an orderly fashion to daughter cells during cell division, certain mechanisms and structures (mitosis, nucleosomes, centromeres, *etc.*) have evolved for *packaging* the DNA. In addition, the genome is divided into separate entities (chromosomes) to perhaps facilitate the partitioning process in mitosis and meiosis.

4. While the β-globin gene family is a relatively large (60kb) sequence and restriction analyses show that it is composed of six genes, one is a pseudogene and therefore does not produce a product. The five functional genes each contain two similarly-sized introns which when included with non-coding flanking regions (5′ and 3′), and spacer DNA between genes, accounts for the 95% mentioned in the question.

6. Many plants have genomes larger than *Arabidopsis* but have about the same number of genes. Much of the extra DNA in large-genome plants is in the form of repetitive DNA, often transposons. Because plants closely related to *Arabidopsis* often have large genomes, it is thought that the small genome of *Arabidopsis* arose through genomic contraction.

Chapter 18

2. Glyphosate (a herbicide) inhibits EPSP, a chloroplast enzyme involved in the synthesis of the amino acids phenylalanine, tyrosine, and tryptophan. To generate glyphosate resistance in crop plants a fusion gene was created which introduced a viral promoter to control the EPSP synthetase gene. The fusion product was placed into the Ti vector and transferred to *A. tumifaciens* which was used to infect crop cells. Calluses were selected on the basis of their resistance to glyphosate. Resistant calluses were later developed into transgenic plants. There is a remote possibility that such an "accident" can occur. However, in retracing the steps to generate the resistant plant in the first place, it seems more likely that the trait will not "escape" from the plant; rather that the engineered *A. tumifaciens* may escape, infect and transfer glyphosate resistance to pest species.

4. **(a,b)** One of the main problems with gene therapy is delivery of the desired virus to the target tissue in an effective manner. Several of the problems involving the use of retroviral vectors are the following. (1) Integration into the host must be cell specific so as not to damage non-target cells. (2) Retroviral integration into host cell genomes only occurs if the host cell is replicating. (3) Insertion of the viral genome might influence non-target but essential genes. (4) Retroviral genomes have a low cloning capacity and can not carry large inserted sequences as are many human genes. (5) There is a possibility that recombination with host viruses will produce an infectious virus which may do harm.

(c) The question posed here plays on the practical versus the ethical. It would certainly be more efficient (although perhaps more difficult technically) to engineer germ tissue, for once it is done in a family, the disease would be eliminated. However, there are considerable ethical problems associated with germ plasm therapy. It recalls previous attempts of the eugenics movements of past decades which involved the use of selective breeding to purify the human stock. Some present-day biologists have said publically that germ line gene therapy will *not* be conducted.

6. Positional cloning is a technique whereby the linkage group of a genetic disorder is determined by association of certain RFLP (as markers). A more precise location is obtained by association (linkage studies in kindreds) with additional RFLP markers. Once the general region of the gene is located, sequencing is employed to identify the genes within the general region. Comparison of the sequences in individuals with and without the affliction allows the identification of the gene responsible for the disease. Difficulties in positional cloning relate to the availability of sufficient RFLP markers in the region of the gene and sufficient kindreds to do the actual genetic association of the gene in question to the RFLP markers. Mutations which would speed the process would be those which are most easily seen, perhaps those which change the banding patterns of chromosomes (deletions, duplication, inversions. translocations) or those which cause sufficient base changes that would alter probe hybridization.

8. Even though you have developed a method for screening seven of the mutations described, it is possible that negative results can occur even though the person carries the gene for CF. In other words, the specific probes (or allele-specific oligonucleotides) that have been developed will not necessarily be useful for screening all mutant genes. In addition, the cost-effectiveness of such a screening proposal would need to be considered.

10. It will hybridize by base complementation to the normal DNA sequence.

12. One method is to use the amino acid sequence of the protein to produce the gene synthetically. Alternatively, since the introns are spliced out of the hnRNA in the production of mRNA, if mRNA can be obtained, it can be used to make DNA (cDNA) through the use of reverse transcriptase.

14. The child in question is a carrier of the deletion in the beta-globin gene, just as the parents are carriers. Its genotype is therefore $\beta^A\beta^o$.

Chapter 19

2. Under *negative* control, the regulatory molecule interferes with transcription while in *positive* control, the regulatory molecule stimulates transcription. Negative control is seen in the *lactose* and *tryptophan* systems as well as a portion of the *arabinose* regulation. Catabolite repression and a portion of the *arabinose* regulatory systems are examples of positive control.

4. **(a)** Due to the deletion of a base early in the *lac* Z gene there will be "frameshift" of all the reading frames downstream from the deletion. It is likely that either premature chain termination of translation will occur (from the introduction of a nonsense triplet in a reading frame) or the normal chain termination will be ignored. Regardless, a mutant condition for the *Z* gene will be likely. If such a cell is placed on a lactose medium, it will be incapable of growth because β-galactosidase is not available. **(b)** If the deletion occurs early in the A gene, one might expect impaired function of the A gene product, but it will not influence the use of lactose as a carbon source.

6.

$I^+O^+Z^+$ = Because of the function of the active repressor from the I^+ gene, and no lactose to influence its function, there will be **No Enzyme Made**.

$I^+O^cZ^+$ = There will be a **Functional Enzyme Made** because of the constitutive operator is in *cis* with a *Z* gene. The lactose in the medium will have no influence because of the constitutive operator. The repressor can not bind to the mutant operator.

$I^-O^+Z^-$ = There will be a **Nonfunctional Enzyme Made** because with $I^-$the system is constitutive but the *Z* gene is mutant. The absence of lactose in the medium will have no influence because of the non-functional repressor. The mutant repressor can not bind to the operator.

$I^-O^+Z^-$ = There will be a **Nonfunctional Enzyme Made** because with *I* the system is constitutive but the *Z* gene is mutant. The lactose in the medium will have no influence because of the non-functional repressor. The mutant repressor can not bind to the operator.

$I^-O^+Z^+/F'I^+$ = There will be **No Enzyme Made** because in the absence of lactose, the repressor product of the I^+ gene will bind to the operator and inhibit transcription.

$I^+O^cZ^+/F'O^+$ = Because there is a constitutive operator in *cis* with a normal *Z* gene, there will be **Functional Enzyme Made**. The lactose in the medium will have no influence because of the mutant operator.

$I^+O^+Z^-/F'I^+O^+Z^+$ = Because there is lactose in the medium, the repressor protein will not bind to the operator and transcription will occur. The presence of a normal *Z* gene allows a **Functional and Non-functional Enzyme to be Made**. The repressor protein is diffusable, working in *trans*.

$I^-O^+Z^-/F'I^+O^+Z^+$ = Because there is no lactose in the medium, the repressor protein (from I^+) will repress the operators and there will be **No Enzyme Made**.

$I^s\,O^+Z^+/F'O^+$ = With the product of I^s there is binding of the repressor to the operator and therefore **No Enzyme Made**. The lack of lactose in the medium is of no consequence because the mutant repressor is insensitive to lactose.

$I^+O^cZ^+/F'O^+Z^+$ = The arrangement of the constitutive operator (O^c) with the *Z* gene will cause a **Functional Enzyme to be Made**.

8. In order to understand this question, it is necessary that you understand the negative regulation of the *lactose* operon by the *lac* repressor as well as the positive control exerted by the CAP protein. Remember, if lactose is present, it inactivates the *lac* repressor. If glucose is present, it inhibits adenyl cyclase thereby reducing, through a lowering of *c*AMP levels, the positive action of CAP on the *lac* operon.

(a) With no lactose and no glucose, the operon is off because the *lac* repressor is bound to the operator and although CAP is bound to its binding site, it will not override the action of the repressor.

(b) With lactose added to the medium, the *lac* repressor is inactivated and the operon is transcribing the structural genes. With no glucose, the CAP is bound to its binding site, thus enhancing transcription.

(c) With no lactose present in the medium, the *lac* repressor is bound to the operator region, and since glucose inhibits adenyl cyclase, the CAP protein will not interact with its binding site. The operon is therefore "off."

(d) With lactose present, the *lac* repressor is inactivated, however since glucose is also present, CAP will not interact with its binding site. Under this condition transcription is severely diminished and the operon can be considered to be "off."

10. First notice that in the first row of data, the presence of tm in the medium causes the production of active enzyme from the wild type arrangement of genes. From this one would conclude that the system is *inducible*. To determine which gene is the structural gene, look for the *IE* function and see that it is related to *C*. Therefore *C* codes for the **structural gene**. Because when *B* is mutant, no enzyme is produced, *B* must be the **promoter**.

Notice that when genes *A* and *D* are mutant, constitutive synthesis occurs, therefore one must be the operator and the other gene codes for the repressor protein. To distinguish these functions, one must remember that the repressor operates as a diffusible substance and can be on the host chromosome or the F factor (functioning in *trans*). However, the operator can only operate in *cis*. In addition, in *cis*, the constitutive operator is dominant to its wild type allele, while the mutant repressor is recessive to its wild type allele.

Notice that the mutant *A* gene is dominant to its wild type allele, whereas the mutant *d* allele is recessive (behaving as wild type in the first row). Therefore, the *A* locus is the **operator** and the *D* locus is the **repressor** gene.

12. The first two sentences in the problem indicate an inducible system where oil stimulates the production of a protein(?) which turns on (positive control) genes to metabolize oil. The different results in strains #2 and #4 suggest a *cis*-acting system. Because the operon by itself (when mutant as in strain #3) gives constitutive synthesis of the structural genes, *cis*-acting system is also supported. The *cis*-acting element is most likely part of the operon.

14. You will need to identify the complementary regions. You will find four regions which "fit." To get started, find the CACUUCC sequence. It pairs, with one mismatch with a second region. Hint: The third region is composed of seven bases and starts with an AG.

Chapter 20

2. ***Chromatin remodeling***: Changes in DNA/chromosome structure can influence overall gene output. DNA methylation also influences transcription efficiency.

Transcription: There are several factors which are known to influence transcription: *promoters*, TATA, CAAT, and GC boxes, as well as other upstream regulatory sequences; *enhancers*, which are *cis*-acting sequences that act at various locations and orientations; *transcription factors*, with various structural motifs (zinc fingers, homeodomains, and leucine zippers) which bind DNA and influence transcription; *receptor-hormone complexes* which influence transcription.

Processing and transport types of regulation involve the efficiency of hnRNA maturation as related to capping, polyA tail addition, intron removal, mRNA stability.

Translation: After mRNAs are produced from the processing of hnRNA, they have the potential of being translated. The stability of the mRNAs appears to be an additional regulatory control point.

4. Transcription factors are proteins which are *necessary* for the initiation of transcription. However, they are not *sufficient* for the initiation of transcription. To be activated, RNA polymerase II requires four or five transcription factors. Transcription factors contain at least two functional domains: one binds to the DNA sequences of promoters and/or enhancers, the other interacts with RNA polymerase or other transcription factors.

6. Both the *lac* and *gal* systems are influenced by catabolite repression, however, the *lac* system is under negative control whereas the *gal* system is under positive control. Both systems are inducible.

8. The work of Cleveland and colleagues allowed selective changes to be made in the *met-arg-glu-lys* sequence. Only the engineered mRNA sequences which caused an amino acid substitution negated the autoregulation, indicating that it is the sequence of the amino acids, not the mRNA which is critical in the process of autoregulation. Notice that code degeneracy allows for changes in mRNA sequence without changes in the amino acid sequence. Therefore, the model which depicts binding of factors to the nascent polypeptide chain is supported. There are a variety of experiments which could be used to substantiate such a model. One might stabilize the proposed MREI-protein complex with "crosslinkers," treat with RNAse to digest mRNA and to break up polysomes, then isolate individual ribosomes. One may use some specific antibody or other method to determine whether tubulin subunits contaminate the ribosome population.

Chapter 21

2. Many of the appendages of the head, including the mouth parts and the antennae are evolutionary derivatives of ancestral leg structures. In *spineless aristapedia* the distal portion of the antenna is replaced by its ancestral counterpart, the distal portion of the leg (tarsal segments). Because the replacement of the arista (end of the antenna) can occur by a mutation in a single gene, one would consider that one "selector" gene distinguishes aristal from tarsal structures. Notice that a "one-step" change is involved in the interchange of leg and antennal structures.

4. There are several somewhat indirect methods for determining transcriptional activity of a given gene in different cell types. First, if protein products of a given gene are present in different cell types, it can be assumed that the responsible gene is being transcribed. Second, if one is able to actually observe, microscopically, gene activity, as is the case in some specialized chromosomes (polytene chromosomes), gene activity can be inferred by the presence of localized chromosomal puffs.

A more direct and common practice to assess transcription of particular genes is to use labeled probes. If a labeled probe can be obtained which contains base sequences that are complementary to the transcribed RNA, then such probes will hybridize to that RNA if present in different tissues. This technique is called *in situ* hybridization and is a powerful tool in the study of gene activity during development.

6. Because in *ftz/ftz* embryos, the engrailed product is absent and in *en/en* embryos *ftz* expression is normal, one can conclude that the *ftz* gene product regulates, either directly or indirectly, *en*. Because the *ftz* gene is expressed normally in *en/en* embryos, the product of the *engrailed* gene does not regulate expression of *ftz*.

8. The fact that nuclei from almost any source remain transcriptionally and translationally active substantiates the fact that the genetic code and the ancillary processes of transcription and translation are compatible throughout the animal and plant kingdoms. Because the egg represents an isolated, "closed" system which can be mechanically, environmentally, and to some extent biochemically manipulated, various conditions may be developed which allow one to study facets of gene regulation. For instance, the influence of transcriptional enhancers and suppressors may be studied along with factors which impact on translational and post-translational processes. Combinations of injected nuclei may reveal nuclear-nuclear interactions which could not normally be studied by other methods.

10. The typical developmental sequence for axis and segment formation in *Drosophila* proceeds from the gap genes to the pair-rule genes to the segment polarity genes. The fact that *fushi-tarazu* (*ftz*) is affected by early (anterior-posterior determining genes) and gap genes indicates that *ftz* functions after those genes. That segment polarity genes are influenced by *ftz* indicates that *ftz* functions earlier, thus placing *ftz* in the pair-rule group of genes.

Chapter 22

2. Review Chapter 8 in the text and note that the following stages of the cell cycle are discussed: G1, G0, S, G2. The G1 stage begins after mitosis and is involved in the synthesis of many cytoplasmic elements. In the S phase DNA synthesis occurs. G2 is a period of growth and preparation for mitosis. Most cell cycle time variation is caused by changes in the duration of G1. G0 is the non-dividing state.

4. Kinases regulate other proteins by adding phosphate groups. Cyclins bind to the kinases, switching them on and off. Several cyclins, including D and E, can move cells from G1 to S. At the G2/mitosis border a CDK1 (cyclin dependent kinase) combines with another cyclin (cyclin B). Phosphorylation occurs bringing about a series of changes in the nuclear membrane, cytoskeleton, and histone 1.

6. A tumor suppressor gene is a gene that normally functions to suppress cell division. Since tumors and cancers represent a significant threat to survival and therefore Darwinian fitness, strong evolutionary forces would favor a variety of co-evolved and perhaps complex conditions in which mutations in these suppressor genes would be recessive. Looking at it in another way, if a tumor suppressor gene makes a product that regulates the cell cycle favorably, cellular conditions have evolved in such a way that sufficient quantities of this gene product are made from just one gene (of the two present in each diploid individual) to provide normal function.

8. Oncogenes are genes that induce or maintain uncontrolled cellular proliferation associated with cancer. They are mutant forms of proto-oncogenes which normally function to regulate cell division. They may be formed through point mutations, gene amplification, translocations, repositioning of regulatory sequences, *etc.*.

10. Unfortunately, it is common to spend enormous amounts of money on dealing with diseases after they occur rather than concentrating on disease prevention. Too often pressure from special interest groups or lack of political stimulus retards advances in education and prevention. Obviously, it is less expensive, both in terms of human suffering and money, to seek preventive measures for as many diseases as possible. However, having gained some understanding of the mechanisms of disease, in this case cancer, it must also be stated that no matter what preventive measures are taken it will be impossible to completely eliminate disease from the human population. It is extremely important, however, that we increase efforts to educate and protect the human population from as many hazardous environmental agents as possible.

12. **(a)** The mRNA triplet for Gln is CAG(A). The mRNA triplet which specifies a stop is one of three: UAA, UAG, or UGA. The strand of DNA which codes for the CAG(A) would be the following: 3′-GTC(T)-5′. Therefore, if the G mutated to an A (transition), then the DNA strand would be 3′-ATC(T)-5′ which would cause a UAG(A) triplet to be produced and this would cause the stop.

(b) Tumor suppression because loss-of-function causes predisposition to cancer.

(c) Some women may carry genes (perhaps mutant) which "spare" for the *BRCA1* gene product. Some women may have immune systems which recognize and destroy precancerous cells or they may have mutations in breast signal transduction genes so that cell division suppression occurs in the absence of *BRCA1*.

Chapter 23

2. Typically four disulfide bridges hold the two heavy and two light chains together in each antibody molecule. The disulfide bridges thereby contribute to the three dimensional structure and are dependent on cysteine residues.

4. V × D × J × C. Thus in this case the answer would be 10 V ×30 D × 50 J × 3 C = 45,000.

6. They produce double-stranded breaks between the RSS and coding DNA. When rejoining occurs, non-complementary base pairing may occur followed by trimming and filling. These processes create additional molecular diversity.

8. The distribution of genes is therefore quite non-uniform, with a small fraction of the genome containing the bulk of the genes.

10. In addition to pairing of homologous chromosome during meiosis I, additional molecular events are unique. In meiotic yeast cells, the Scc1 component of cohesin is replaced by a meiosis-specific subunit called Rec8 which is broken down by separin.

12. It is likely to have little direct effect on G1 because sister chromatids are not present as yet. However, overexpression of separin after S phase would probably cause a destruction of securins. This would result in premature chromatid separation and aneuploidy.

Chapter 24

4. **(a)**

rrSsTtuu × *RrSsTtUu*
(moderate) (moderate)

Offspring from this cross can range from very tall *RrSSTTUu* (12 "tall" units) to very short *rrssttuu* (8 "small" units).

(b) If the individual with a minimum height, *rrssttuu*, is married to an individual of intermediate height *RrSsTtUu*, the offspring can be no taller than the height of the tallest parent. Notice that there is no way of having more than four uppercase alleles in the offspring.

6. **(a)** Because the extreme phenotypes (6cm and 30cm) each represent 1/64 of the total, it is likely that there are three gene pairs in this cross. Also, the fact that there are seven categories of phenotypes, which, because of the relationship $2n + 1 = 7$, would give the number of gene pairs (n) of 3. The genotypes of the parents would be combinations of alleles which would produce a 6cm (*aabbcc*) tail and a 30cm (*AABBCC*) tail while the 18cm offspring would have a genotype of *AaBbCc*.

(b) A mating of an *AaBbCc* (for example) pig with the 6cm *aabbcc* pig would result in the following offspring:

Gametes (18cm tail)	***Gamete (6cm tail)***	***Offspring***
ABC		*AaBbCc* (18cm)
ABc		*AaBbcc* (14cm)
AbC		*AabbCc* (14cm)
Abc	*abc*	*Aabbcc* (10cm)
aBC		*aaBbCc* (14cm)
aBc		*aaBbcc* (10cm)
abC		*aabbCc* (10cm)
abc		*aabbcc* (6cm)

In this example, a 1:3:3:1 ratio is the result. However, had a different 18cm tailed-pig been selected, a different ratio would occur:

AABbcc × *aabbcc*

Gametes (18cm tail)	*Gamete (6cm tail)*	*Offspring*
ABc	*abc*	*AaBbcc* (14cm)
Abc		*Aabbcc* (10cm)

8. *Monozygotic twins* are derived from a single fertilized egg and are thus genetically identical to each other. They provide a method for determining the influence of genetics and environment on certain traits. *Dizygotic twins* arise from two eggs fertilized by two sperm cells. They have the same genetic relationship as siblings. The role of genetics and the role of the environment can be studied by comparing the expression of traits in monozygotic and dizygotic twins. The higher concordance value for monozygotic twins as compared to the value for dizygotic twins indicates a significant genetic component for a given trait.

10. (a) Mean = 140 cm.

(b) Variance = 374.18

(c) The *standard deviation* is the square root of the variance or 19.34.

(d) The *standard error* of the mean is the standard deviation divided by the square root of n, or about 0.70.

The plot approximates a normal distribution. Variation is continuous.

12. The formula for estimating heritability is

$$H^2 = V_G/V_P$$

where V_G and V_P are the genetic and phenotypic components of variation, respectively. V_P is the combination of genetic and environmental variance. Because the two parental strains are inbred, they are assumed to be homozygous and the variance of 4.2 and 3.8 considered to be the result of environmental influences. The average of these two values is 4.0. The F_1 is also genetically homogeneous and gives us an additional estimation of the environmental factors. By averaging with the parents

$$[(4.0 + 5.6)/2 = 4.8]$$

we obtain a relatively good idea of environmental impact on the phenotype. The phenotypic variance in the F_2 is the sum of the genetic (V_G) and environmental (V_E) components. We have estimated the environmental input as 4.8, so 10.3 (V_P) minus 4.8, gives us an estimate of (V_G) which is 5.5. Heritability then becomes 5.5/10.3 or 0.53. This value, when viewed in percentage form indicates that about 53% of the variation in plant height is due to genetic influences.

14. $h^2 = (7.5 - 8.5/6.0 - 8.5) = 0.4$

Selection will have little relative influence on olfactory learning in *Drosophila*.

16. Chromosome 2 seems to confer considerable resistance to the insecticide, somewhat in the heterozygous state and more in the homozygous state. Thus, some partial dominance is occurring.

18. (a)

3″ = 1	4″ = 12	5″ = 66
6″ = 220	7″ = 495	8″ = 792
9″ = 924	10″ = 792	11″ = 495
12″ = 220	13″ = 66	14″ = 12
15″ = 1		

(b)

3″ = 1	4″ = 6	5″ = 15
6″ = 20	7″ = 15	8″ = 6
9″ = 1		

Chapter 25

2.

p = frequency of A
= 0.2 + .3
= 0.5

$q = 1 - p = 0.5$

Frequency of *AA* $= p^2 = (.5)^2 = .25$ or 25%

Frequency of *Aa* $= 2pq = 2(.5)(.5) = .5$ or 50%

Frequency of *aa* $= q^2 = (.5)^2 = .25$ or 25%

The initial population was not in equilibrium, however, after one generation of mating under the Hardy-Weinberg assumptions, the population is in equilibrium and will continue to be so (and not change) until one or more of the Hardy-Weinberg assumptions is not met. Note that *equilibrium* does not necessarily mean p and q equal 0.5.

4. In order for the Hardy-Weinberg equations to apply, the population must be in equilibrium.

6. (a)

The equilibrium values will be as follows:

Frequency of +/+ $= p^2 = (.7755)^2 = .6014$ or 60.14%

Frequency of +/Δ32 $= 2pq = 2(.7755)(.2245) = .3482$ or 34.82%

Frequency of Δ32/Δ32 $= q^2 = (.2245)^2 = .0504$ or 5.04%

Comparing these equilibrium values with the observed values strongly suggests that the observed values are drawn from a population in equilibrium.

(b) The equilibrium values will be as follows:

Frequency of *AA* $= p^2 = (.877)^2 = .7691$ or 76.91%

Frequency of *AS* $= 2pq = 2(.877)(.123) = .2157$ or 21.57%

Frequency of *SS* $= q^2 = (.123)^2 = .0151$ or 1.51%

$\chi^2 = 1.47$

In calculating degrees of freedom in a test of gene frequencies, the "free variables" are reduced by an additional degree of freedom because one estimated a parameter (p or q) used in determining the expected values. Therefore, there is one degree of freedom even though there are three classes. Checking the χ^2 table with 1 degree of freedom gives a value of 3.84 at the 0.05 probability level.

Since the χ^2 value calculated here is smaller, the null hypothesis (the observed values fluctuate from the equilibrium values by chance and chance alone) should not be rejected. Thus the frequencies of *AA*, *AS*, *SS* sampled a population which is in equilibrium.

8. (a)

$$q_{g+1} = [.9(.7)(.3)+.8(.3)^2/[1(.7)^2+.9(2)(.7)(.3)+.8(.3)^2]$$

$q_{g+1} = .278$ $p_{g+1} = .722$

(b) $q_{g+1} = .289$ $p_{g+1} = .711$

(c) $q_{g+1} = .298$ $p_{g+1} = .702$

(d) $q_{g+1} = .319$ $p_{g+1} = .681$

10. **(a)** $p_1 = 0.6 + 0.2(0.1 - 0.6) = 0.5$
(b) $p_1 = 0.2 + 0.3(0.7 - 0.2) = 0.35$
(c) $p_1 = 0.1 + 0.1(0.2 - 0.1) = 0.11$

12. *Inbreeding depression* refers to the reduction in fitness observed in populations which are inbred. With inbreeding comes an increase in the number of homozygous individuals and a decrease in genetic variability. Genetic variability is necessary for a genetic response to environmental change. As deleterious genes become homozygous, more individuals are less fit in the population.

14. While inbreeding increases the frequency of homozygous individuals in a population, it does not change the *gene* frequencies. There will be fewer heterozygotes in the population to compensate for the additional homozygotes.

16.
(a) q is 0.01
(b) $p = 1 - q$ or .99
(c) $2pq = 2(.01)(.99)$
$= 0.0198$ (or about 1/50)
(d) $2pq \times 2pq$
$= 0.0198 \times 0.0198$
$= 0.000392$ or about 1/255

18. There are 50,000 births, therefore 100,000 gametes (genes) involved. The frequency of mutation is therefore given as follows: 2/100,000 or 2×10^{-5}

20. **(a)** For the population, since $q^2 = .022$, then $q = .045$, $p = .955$, and $2(pq) = 0.086$. For the community, since $q^2 = .005$, $q = .07$, $p = .93$, and $2(pq) = 0.132$.
(b) The "founder effect" is probably operating here. Relatively small, local populations which are relatively isolated in a reproductive sense, tend to show differences in gene frequencies when compared to larger populations. In such small populations, homozygosity is increased as a gene has a higher probability of "meeting itself."

Chapter 26

2. During speciation, individuals or groups of potentially interbreeding organisms become genetically distinct from other members of the species. Members of different populations with substantial genetic divergence are, at first, not reproductively isolated from each other although gene flow may be restricted. The distinction between such groups is not absolute in that one group may blend with other groups of the species. Any process which favors changes in gene frequencies has the potential of generating substantial genetic differences.
Factors such as selection, migration, genetic drift, or even mutation, may be important in generating significant genetic change. One would certainly include geographic isolation as a major barrier to gene flow and thus an important process in such formation.
Natural selection occurs when there is non-random elimination of individuals from a population. Since such selection is a strong force in changing gene frequencies, it should also be considered as a significant factor in subspecies formation.

4. See figure in top right column.

6. Approach this problem by writing the possible codons for all the amino acids (except Arg and Asp which show no change) in the human cytochrome c chain. Then determine the minimum number of nucleotide substitutions required for each changed amino acid in the various organisms. Once listed, then count up the numbers for each organism: horse, 3; pig, 2; dog, 3; chicken, 3; bullfrog, 2; fungus, 6.

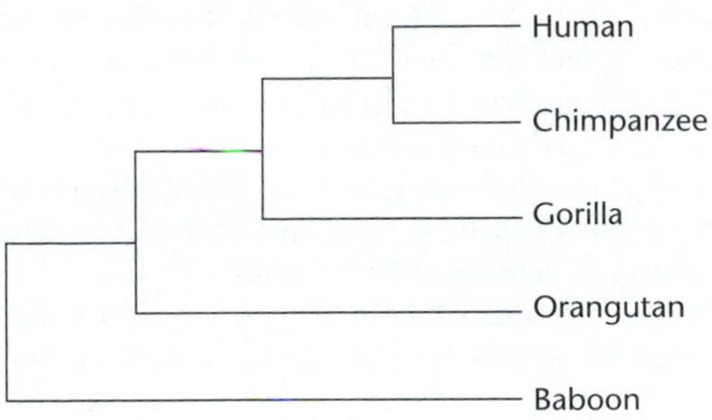

8. In looking at the figure, notice that the $\Delta T_{50}H$ value of 4.0 on the right could be used as a decision point such that any group which diverged above that line would be considered in the same genus while any group below would be in a different genus. Under this rule, one would have the chimpanzee, pygmy chimpanzee, human, gorilla and orangutan in the same genus. If one assumed that 3.7 is close enough to be considered above 4.0, given considerable experimental error, one could provide a scheme where the orangutan is not included with the chimpanzee, pygmy chimpanzee, human, and gorilla.

10. The text lists several cornerstones of the *neutral mutation theory:*
(a) the relatively uniform rate of amino acid substitution in different organisms (under different types of selection);
(b) there is no particular pattern to the substitutions indicating that selection is not eliminating some variations;
(c) the rate of mutation is relatively high and has remained relatively constant for millions of years even though environments have fluctuated greatly over that period of time;
(d) certain regions of molecules and certain functions of those molecules should logically be less likely to have amino acid substitutions influence the phenotype.
(e) the rate of amino acid substitution in some proteins is much too high to have been produced by selection. The *selectionists* suggest that even though amino acid substitutions *appear* to be neutral, it is more likely that their influence has just not been determined. In addition, they point out that many polymorphisms are clearly maintained in the population *by* selection. Thus, the issues listed above don't really challenge present views of genetic variation.

12. Given the small range of *HLA* diversity, one might conclude that the Native Americans descended from a relatively small population either because of a small number who originally arrived or because of significant population crashes over time. Additionally, either the South American groups received an influx of genes from other populations or only small bands of North Americans survived.

Chapter 27

2. The frequency of the lethal gene in the captive population ($q^2 = 5/169$ and $q = .172$) is approximately double that in the gene pool as a whole ($q = 0.09$). Applying the formula

$$q_n = q_o/(1 + nq_o)$$

one can estimate that it would take 10 generations to reduce the lethal gene's frequency to .063 in the captive population with no intervention (random mating assumed). Since condors produce very few eggs per year, a more proactive approach seems justified.
First, if detailed records are kept of the breeding partners of the captive birds, then knowledge of heterozygotes should be available. Breeding programs could be established to restrict matings between those carrying the lethal gene. Such "kinship management" is often used in captive populations. If kinship

records are not available, it is often possible to establish kinship using genetic markers such as DNA microsatellite polymorphisms. Using such markers, one can often identify mating partners and link them to their offspring.

By coupling knowledge of mating partners with the likelihood of producing a lethal genetic combination, selective matings can often be used to minimize the influence of a deleterious gene. In addition, such markers can be used to establish matings which optimize genetic mixing, thus reducing inbreeding depression.

4. Both genetic drift and inbreeding tend to drive populations toward homozygosity. Genetic drift is more common when the effective breeding size of the population is low. When this condition prevails, inbreeding is also much more likely. They are different in that inbreeding can occur when certain population structures or behaviors favor matings between relatives, regardless of the effective size of the population. Inbreeding tends to increase the frequency of both homozygous classes at the expense of the heterozygotes. Genetic drift can lead to fixation of one allele or the other, thus producing a single homozygous class.

6. Inbreeding depression, over time, reduces the level of heterozygosity, usually a selectively advantageous quality of a species. When homozygosity increases (through loss of heterozygosity) deleterious alleles are likely to become more of a load on a population. Outbreeding depression occurs when there is a reduction in fitness of progeny from genetically diverse individuals. It is usually attributed to offspring being less well-adapted to the local environmental conditions of the parents.

8. Often, molecular assays of overall heterozygosity can indicate the degree of inbreeding and/or genetic drift. An allele whose frequency is dictated by inbreeding and/or will not be uniquely influenced. That is, other alleles would be characterized by decreased heterozygosity as well. So, if the genome in general has a relatively high degree of heterozygosity, the gene is probably influenced by selection rather than inbreeding and/or genetic drift.

Credits

Chapter 1: Introduction to Genetics

CO1 Omikron/Science Source/Photo Researchers, Inc. **F1.1** The Metropolitan Museum of Art. Gift of John D. Rockefeller, Jr., 1932. (32.143.3) Photograph © 1983 The Metropolitan Museum of Art **F1.2** Eugene Delacroix, frescos from the spandrels of the main hall of the Assemblée Nationale, Paris, France. © Photograph by Erich Lessing/Art Resource, NY **F1.3** Hartsoeker, N. Essay de dioptrique, Paris, 1694. p. 230/National Library of Medicine **F1.4** ©ARCHIV/Photo Researchers, Inc. **F1.5** M. Wurtz/Biozentrum, University of Basel/Science Photo Library/Photo Researchers, Inc. **F1.6** Biophoto Associates/Science Source/Photo Researchers, Inc. **F1.7** Sovereign/Phototake NYC **F1.12** Dan McCoy/Rainbow **F1.13** Photo courtesy of Roslin Institute **F1.14** Zig Leszczynski/Animals Animals/Earth Scenes **F1.15** Grant Hellman/Grant Hellman Photgraphy, Inc. **F1.16** Renee Lynn/Photo Researchers, Inc. **F1.17** From "Our genes: What the Map Shows" from *TIME*, January 17, 1994. Copyright © 1994 by Time, Inc. Reprinted by permission.

Chapter 2: DNA Structure and Analysis

CO2 Richard Megna/Fundamental Photographs **F2.3(a and b)** Bruce Iverson **F2.5(b)** Oliver Meckes/Max-Planck-Institut-Tubingen/Photo Reserachers, Inc. **F2.8(b)** Runk/Schoenberger/Grant Hellman Photography, Inc. **F2.8** © Biology Media/Photo Researchers, Inc. **F2.13** Photo by M.H.F. Wilkins, Courtesy of Biophysics Department, King's College, London, England **F2.17** Ken Edward/Science Source/Photo Researchers, Inc. **F2.22** Ventana Medical Systems, Inc.

Chapter 3: DNA Replication and Recombination

CO3 Evelin Schrock, Stan du Manoir, and Tom Reid, National Institutes of Health **F3.2(a)** Courtesy of the Greenwood Genetic Center, Greenwood, SC **F3.5(a)** Walter H. Hoge/Peter Arnold, Inc. **F3.5(b)** Figure from "Molecular Genetics," Pt. 1 pp. 74075 , J.H. Taylor (ed). Copyright © 1963 and renewed 1991 by Academic Press, reproduced by permission of the publisher. **F3.5** © Figure from "Molecular Genetics," Pt. 1 pp. 74075 , J.H. Taylor (ed). Copyright © 1963 and renewed 1991 by Academic Press, reproduced by permission of the publisher. **F3.6** Reprinted from CELL, Vol. 25, 1981, pp. 659, Sundin and Varshavsky, (1 figure), with permission from Elsevier Science. Courtesy of A. Varshavsky. **F3.9** Maria Schnos and Ross Inman/Institue for Molecular Virology/University of Wisconsin-Madison. **F3.15** Reproduced by permission from H.J. Kreigstein and D.S. Hogness, Preceedings of the National Academy of Sciences 71:136 (1974), p. 137, Fig. 2 **F3.16** D r. Harold Weintraub, Howard Hughes Medical Institute, Fred Hutchinson Cancer Center/"Essential Molecular Biology" 2e, Freifelder and Malachinski, Jones & Bartlett, Fig. 7-24, pp. 141. **F3.19(b)** David Dressler, Oxford University, England

Chapter 4: Chromosome Structure and DNA Sequence Organization

CO4 Dr. Gopal Murti/Science Photo Library/Photo Researchers, Inc **CO4** Science VU/BMRL/Visuals Unlimited **F4.1(a and b)** Dr. M. Wurtz/Biozentrum, University of Basel/Science Photo Library/Photo Researchers, Inc. **F4.2** Science Source/Photo Researchers, Inc. **F4.3** Dr. Gopal Murti/Science Photo Library/Photo researchers, Inc. **F4.5** Image courtesy of Brian Harmon and John Sedat, University of California, San Francisco **F4.6** Science Source/Photo Researchers, Inc. **F4.7(a and b)** Omikron/Photo Researchers, Inc. **F4.8a** From Olins and Olins, 1978, Fig. 1 and 4 **F4.8b** From Olins and Olins, 1978, Fig. 1 and 4 **F4.11** Chen and Ruddle, 1971, p. 54. **F4.11** Dr. David Adler, Department of Pathology, University of Washington **F4.12** Dept. of Clinical Cytogenetics, Addenbrookes Hospital, Cambridge/Science Photo Library/Photo Researchers, Inc. **F4.12** Dr. David Adler, Department of Pathology, University of Washington **F4.16** From Pardue & Gall, 1972. © Keter.

Chapter 5: The Genetic Code and Transcription

CO5 Prof. Oscar L. Miller/Science Photo Library/Photo Researchers, Inc. **F5.11** Bert W. O'Malley, M.D., Baylor College of Medicine **F5.15c** O.L. Miller, Jr. Barbara A. Hamkalo, C.A. Thomas, Jr. Science 169:392-395, 1970 by the American Association for the Advancement of Science. **F:5.15d** O.L. Miller, Jr. B.R. Beatty, Journal of Cellular Physiology, Vol. 74 (1969). Reprinted by permission of Wiley-Liss, Inc. inc., a subsidiary of John Wiley & Sons, Inc

Chapter 6: Translation and Proteins

CO6 Reprinted from the font cover of Science, Vol. 292, May 4 2001 "Crystal structure of a Thermus thermophilus 70S ribosome containing three bound transfer RNAs (top) and exploded views showing its different molecular components (middle and bottom), Inage provided by Dr. Albion Baucom (baucom@biology.ucsc.edu). Copyright American Association for the Advancement of Science. **F6.9a** "The Structure and Function of Polyribosomes." Alexander Rich, Jonathan R. Warner, and Howard M. Goodman, 1963. Reproduced by permission of the Cold Spring Harbor Laboratory Press. Cold Spring Harbor Symp. Quant. Biol. 28 (1963) fig. 4C (top), p. 273 © 1964. **F6.13b** Francis Leroy/Biocosmos/Science Photo Library/Photo Researchers, Inc.

Chapter 7: Gene Mutation, DNA Repair, and Transposable Elements

CO7 Francis Leroy/Biocosmos/Science Photo Library/Photo Researchers, Inc. **F7.3** Sue Ford /Science Photo Library/Photo Researchers, Inc. **F7.4** Mary Evans Picture Library/Photo Researchers,

Inc. **F7.20(a and b)** W. Clark Lambert M.D., Ph.D./University of Medicine & Dentistry of New Jersey **F7.24** Stanley Cohen/Science Photo Library/Photo Researchers, Inc. **F7.25** Dr. Nina Federoff/Dept. of Embryology, Carnegie Institute of Washington

Chapter 8: Mitosis and Meiosis

C08 Dr. Andrew S. Bajer, University of Oregon **F8.2** Dr. Kevin P. Campbell, University of Iowa. Images acquired using a Leitz Diaplan fluorescence microscope with attached Optronics DEI-750 camera, Central Microscopy Research Facility, University of Iowa, Iowa City, IA **F8.3** CNRI/Science Photo Library/Photo Researchers, Inc. **F8.5(a)** Cytographics/Visuals Unlimited. **F8.5(b)** L. Lisco/D.W. Fawcett/Visuals Unlimited **F8.9** Dr. Andrew S. Bajer, University of Oregon **F8.10** Schil Gerius/Springer-Verlag GmbH & Co. KG/from Mathew Schibler, Protoplasma, 137:29-44 (1987) **F8.18(a)** Biophoto Associates/Photo Researchers, Inc. **F8.18(b)** Biophoto Associates/Science Source/Photo Researchers, Inc. **F8.18** © Biophoto Associates/Science Source/Photo Researchers, Inc. **F8.19(a)** Reproduced with permission, from "Annual Review of Genetics," 6. © 1972 by Annual Reviews, Inc. Photo from D. von Wettstein.

Chapter 9: Mendelian Genetics

CO9 Courtesy of Allan Gotthelf. **PDQ** Joyce Photographics/Photo Researchers, Inc. **PDQ** .J. Erwin/Photo Researchers, Inc.

Chapter 10: Extensions of Mendelian Ratios

CO10 Tom Cerniglio/Oak Ridge National Laboratory **F10.1** John D. Cunningham/Visuals Unlimited **F10.3(a)** Photo courtesy of Stanton K. Short (The Jackson Laboratory, Bar Harbor, ME) **F10.3(b)** Photo courtesy of The Jackson Laboratory, Bar Harbor, Maine **F10.9** Irene Vandermolen/Animals Animals/Earth Scenes **F10.10** Steve Mathews/QT Rabbitry **F10.12** Carolina Biological Supply Co./Phototake NYC **F10.14** Mary Teresa Giancoli **F10.15** Hans Reinhard/Bruce Coleman, Inc. **F10.16** Debra P. Hershkowitz/Bruce Coleman, Inc. **F10.17(a)** Tanya Wolff/University of California at Berkeley **F10.17(b)** Joel C. Eisenberg, Ph.D., Dept. of Biochemistry, St. Louis University Medical Center **F10.17(c)** Tanya Wolff/University of California at Berkeley **F10.18(a)** Tanya Wolff/University of California at Berkeley **F10.18(b)** Dr. Steven Henikoff, Howard Hughes Medical Institute, Fred Hutchinson Cancer Research Center, Seattle, WA **F10.19(a)** Jane Burton/Bruce Coleman, Inc. **F10.19(b)** Steve Klug/Dr. William S. Klug **F10.10a (PDQ)** Chinchilla: Grant Heilman/Grant Heilman Photography, Inc. **F10.10b** Himalayan: American Rabbit Breeders Association, Inc. **F10.38 (PDQ)** William H. Mullins/Photo Researchers, Inc. **F10.41a (PDQ)** Nigel J.H. Smith/Animals Animals/Earth Sciences **F10.41b (PDQ)** Francois Gohier/Photo Researchers, Inc. **F10.44 (PDQ)** Robert Pearcy/Animals Animals/Earth Sciences **F10.1 a and c (IS)** Dr. Ralph Somes **F10.1 b and d (IS)** J. James Bitgood/University of Wisconsin/Animal Sciences Dept.

Chapter 11: Sex Determination and Sex Chromosomes

CO11 James King Holmes/Science Photo Library/Photo Researchers, Inc. **F11.1** Biophoto Assoc./Photo Reserachers, Inc. **F11.3** BillBeatty/Visuals Unlimited **F11.4** Dr. Maria Gallegos, University of California, San Francisco **F 11.6a** Courtesy of the Greenwood Genetic Center, Greenwood, SC. **F 11.6b** Courtesy of the Greenwood Genetic Center, Greenwood, SC **F 11.7a** Catherine G. Palmer, Indiana University **F11.7b** Catherine G. Palmer, Indiana University **F 7.9a** Stuart Kenter Associates **F11.9b** Stuart Kenter Associates **F11.11a** W. Layer/Okapia/Photo Researchers, Inc. **F 11.11b** Reed/Williams/Animals Animals/Earth Scenes

Chapter 12: Linkage, Crossing Over and Mapping in Eukaryotes

CO12 B. John Cabisco/Visuals Unlimited **F 12.7** ©Biological Photo Service **F12.17** Dr. Sheldon Wolff & Jody Bodycote/Laboratory of Radiology and Environmental Health, University of California, San Francisco **F12.18(b)** M.I. Walker/Photo Researchers, Inc. **F 12.18(c)** Dwight Kuhn Photography **F12.19(b)** John D. Cunningham/Visuals Unlimited **F12.19c** Science Source/Photo Researchers, Inc. **F12.20** James W. Richardson/Visuals Unlimited

Chapter 13: Chromosome Mutations: Variation in Chromosome Number and Arrangement

CO13 Evelin Schrock, Stan du Manoir and Tom Reid, National Institutes of Health **F13.2** Courtesy of University of Washington Medical Center Pathology **F13.3(b)** Richard Shiell/Animals Animals/Earth Scenes **F13.5(a)** Courtesy of the Greenwood Genetic Center, Greenwood, SC **F13.5(b)** William McCoy/Rainbow **F 13.7** David D. Weaver, M.D., Indiana University **F13.8(a)** David D. Weaver, M.D., Indiana University **F13.12** Ken Wagner/Phototake NYC **F13.13(b)** Pfizer, Inc./Phototake NYC **F13.18(a, b, and c)** Mary Lilly, Carnegie Institution of Washington **F13.25** Dr. Jorge Yunis, From Yunis and Chandler, 1979 **F 13.26** Science VU/Visuals Unlimited

Chapter 14: Extranuclear Inheritance

CO14 Thomas A. Steitz, Yale University, Department of Molecular Biophysics and Biochemistry **F14.4a** Dr. Ronald A. Butow, Department of Molecular Biology and Oncology, University of Texas Southwestern Medical Center **F14.4b** Dr. Ronald A. Butow, Department of Molecular Biology and Oncology, University of Texas Southwestern Medical Center **F14.6** Dr. Richard G. Kolodnar, Dana-Farber Cancer Institute **F14.7** Don W. Fawcett/Kahri/Dawid/Science Source/Photo Researchers, Inc. **F14.9 (a and b)** Dr. Alan Pestronk, Dept. of Neurology, Washington University School of Medicine, St. Louis **F14.10** John D. Cunningham/Visuals Unlimited **F14.13** Robert & Linda Mitchell Photography

Chapter 15: Genetics of Bacteria and Bacteriophages

CO15 Dr. L. Caro/Science Photo Library/Photo Researchers, Inc. **F15.2** Michael G. Gabridge/Visuals Unlimited **F15.5** Dennis Kunkal/Phototake NYC. **F15.12** K.G. Murti/Visuals Unlimited **F15.13** M. Wurtz/Biozentrum, University of Basel/cience Photo Lbrary/Photo Researchers, Inc. **F15.16** Bruce Iverson.

Chapter 16: Recombinant DNA Technology

CO16 Michael Gabridge/Visuals Unlimited **F16.4** K.G. Murti/Visuals Unlimited **F16.6** Michael Gabridge/Visuals Unlimited **F16.7** M. Wurtz/Biozentrum, University of Basel/Science Photo Library/Photo Researchers, Inc. **F16.9** Jack Griffith, University of North Carolina **F16.15** Jon Gordon/Phototake NYC **F16.18** Bio-Rad Laboratories Diagnostics Group **F16.22** National Institutes of Health/Custom Medical Stock Photo, Inc. **F16.25** Dr. Suzanne McCutcheon **F16.25a and b** Reprinted with permission from *Journal of Food Protection*, Vol. 59, No. 6, 1996, Pages 573. Copyright held by the International Association for Food Protection, Des Moines, Iowa, U.S.A. Courtesy of Dr. Pina M. Fratamico. **F16.26** Dr. Suzanne McCutcheon

Chapter 17: Genomics, Bioinformatics, and Proteomics

CO 17 J. Mylne & J. Botella, Department of Botany, University of Queensland Australia **Table 17.1** Table from "On the Total Number of Genes and Their Length Distribution in Complete Microbial

Genomes" by M. Skovgaard, et al., *Current Trends in Genetics*, 17 (8), 2001, pp. 425–428. Copyright © 2000 by Macmillan Magazines Ltd. Reprinted by permission. **F17.4** From "DNA Sequence of both chromosomes of *vibrio cholerae*" by J.F. Heidelberg et al., *Nature* 406 (2000) , pp. 477-484. Copyright © 2000 by Macmillan Magazines Ltd. Reprinted by permission. **F17.6** Figure 1 from "The Complete Genome of the Hyperthermiaphilic Bacterium, *aquifex aeolicus*" by G. Deckert et al., in *Nature*, 392, 1998. Copyright © 2000 by Macmillan Magazines Ltd. Reprinted by permission. **F17.8c** Figure 8, p.37 from "Complete Nucleotide Sequence of the Rabbit B-Like Globin Gene Cluster. Analysis of Intergenic Sequences and Comparison with the Human B-Like Gene Cluster" by J.B. Margot et al., *Journal of Molecular Biology*, 1989, 205, pp. 15-40. Copyright © 1989 by Academic Press. Reprinted by permission. **F17.8d** Figs. 1 & 3 from "Nested Retrotransposons in the Intergenic Regions of the Maize Genome" by San Miguel et al., *Science* 274, 1996, pp. 765-758. Copyright © 1996 by American Association for the Advancement of Science. Reprinted by permission. **Table 17.6** From Table 4, p. 1458, "The Complete Genome Sequence of *Escherichia coli* K-12" by F. Blattner et al., *Science* 277, 1997, pp. 1453-1462. Copyright © 1997 by American Association for the Advancement of Science. **Table 17.6** From Table 2, p. 400, "The Minimal Gene Complement of *Mycoplasma genitalium* K-12" by C. M. Fraser et al., *Science* 270, 1995.5Copyright © 1997 by American Association for the Advancement of Science. **F17.14** Fig. 1 p. 536, in "Genes isochors and bands in human chromosomes 21 and 22" by S. Saccone in *Chromosome Research* 9, 2001, pp. 533-539. Copyright © by Kluwer Academic Publishers, Inc. Reprinted with permission. **F17.15** Figs. 1&2, p. 30755 from "Crystal Structure of Human Frataxin" by S. Dhe-Paganon, et al. in *Journal of Biological Chemistry*, 2000, 275, pp. 30753–30756. Reprinted by permission of Dr. Steven E. Shoelson, Joslin Diabetes Center, Boston, MA. **F17.16** Fig. 2, p. R382 from "Genomics Leprosy. A degenerative disease of the genome" by D. Young and B. Robertson in *Current Biology*, 11, 2001, pp. R381-R383. Copyright © 2001 with permission from Elsevier Science. **F17.22** Dr. Carl Merril/Laboratory of Biochemical Genetics of the National Institute of Mental Health, NIH. **F17.25** Fig. 3B, p. 394 in "The Nuclear Pore Complex: From Molecular Architecture to Funcational Dynamics" by D. Stoffler in *Current Opinions in Cellular Biology*, 1999, 11, pp. 391–401. Copyright © with permission from Elsevier Science. **F17.26** Richard Kessel/Visuals Unlimited **F17.27** Fig. 3B, p. 394, in "The Nuclear Pore Complex: From Molecular Architecture to Funcational Dynamics" by D. Stoffler in *Current Opinions in Cellular Biology* , 1999, 11, pp. 391-401. Copyright © 2001 with permission from Elsevier Science.

Chapter 18: Applications and Ethics of Genetic Technology

CO18 Affymetrix, Inc. **F18.4** From Barker, D. et al. 1987. "Gene for Recklinghausen Neurofibromatosis is in the Pericentromeric Region of Chromosome 17". *Science* 236:1100-1102, Fig. 1. © 1987 by the American Association for the Advancement of Science. **F18.5a** Dorothy Warburton/Phototake NYC **F18.5b** Courtesy of Werner Schempp; Glaser et al. (1998). From Chromosome Research 6:841-846 (1998), Figure 4, published by Kluwer Academic Publishers: www.kluweronline.nl **F18.11** Image courtesy of Affymetrix, Inc.**F18.12** Cancer Genetics Branch/National Human Genome Research Institute/NIH **F18.14** Baylor College of Medicine/Peter Arnold, Inc. **F18.17** Dr. William S. Klug **F18.20** Hank Morgan/Photo Researchers, Inc. **F18.21** Figure 6, p.513 from "Gene Chips and Functional Genomics" written by H. Hamadeh and C. A. Afshari, illustrated by Barbara Aulicino, in American Scientist, 2000, 88, pp. 506–515. Copyright © 2000. Reprinted by permission of Scientific Reserach Society. **F18.25** Charles J. Arntzen, President & CEO, Boyce Thompson Institute, Ithaca, NY

Chapter 19: Gene Regulation in Prokaryotes

CO19 *Nature* vol 401 Sept. 16, 1999 (Cover). Article "Structure of the trp RNA-binding attenuation protein, TRAP, bond to RNA" by Alfred A. Antson, Eleanor J. Dodson, Guy Dodson, Richard B. Greaves, Xiao-ping Chen & Paul Gollnick. **F19.11a** Science Lewis et al/Johnson Research Foundation **F19.11b** Science Lewis, et al 271, pg. 1247–1254/Johnson Research Foundation. **F19.11c** Science Lewis et al/Johnson Research Foundation. **F19.14a** *Nature* vol 401 Sept. 16, 1999 (Cover). Article "Structure of the trp RNA-binding attenuation protein, TRAP, bond to RNA" by Alfred A. Antson, Eleanor J. Dodson, Guy Dodson, Richard B. Greaves, Xiao-ping Chen & Paul Gollnick.

Chapter 20: Gene Regulation in Eukaryotes

CO20 Yale University and the Howard Hughes Medical Institute, Dr Paul Sigler **F20.2** From Qumsiyeh, Mazin B. 1999 "Structure and function of the nucleus: anatomy and physiology of chromatin" *Cellular Molecular Life Science* Vol. 55, pp. 1129-1140, Fig. 1C, pg. 1132. **F20.2 a and b** From Lamond, A.I. and Earnshaw W.C. 1998. "Structure and Function in the Nucleus" in *Science*, 280: 547–553, Fig. 1, c, d, p. 547. Copyright American Association for the Advancement of Science. Provided by Dr. Amanda G. Fisher/Lymphocyte Development Group, MRC Clinical Sciences Centre, Imperial College School of Medicine, Hammersmith Hoppital, London. **F20.9** D r. Paul B. Sigler

Chapter 21: Developmental Genetics

CO21 Courtesy of Edward B. Lewis, California Institute of Technology Archives **F21.1(a)** F.R. Turner/Visuals Unlimited **F21.1(b)** R. Calentine/Visuals Unlimited **F21.3** Fig. 1, p. 575 from "Crosstalk between cell cycle regulators and the myogenic factor MyoD in skeltal myoblasts" by M. Kitzmann and A. Fenandez in *Cellular and Molecular Life Sciences*, 58, (2001), pp. 571–579. Copyright © 2001. Reprinted by permission of Birkhaeuser Publishers Ltd. **F21.4** Fig. 7a, p. 693 from "EGF Receptor and Notch Signaling Act Upstream of Eyeless/PAX6 to Control Eye Specification" by J.P. Kumar and K. Moses from *Cell*, 104, 2001, pp. 687–697. Copyright © 2001 with permission of Elsevir Science. **Table 21.5** Fig. 7, p. 235 from "Signaling Pathways in Development" by J. Gearhart in *Teratology*, 60, 1999. Copyright © 1999. Reprinted by permission of Wiley–liss, Inc., a subsidiary of John Wiley & Sons, Inc. **F21.6** Dr. William S. Klug, The College of New Jersey **F21.7** Fig. 7a, p. 693 from "EGF Receptor and Notch Signaling Act Upstream of Eyeless/PAX6 to Control Eye Specification" by J.P. Kumar and K. Moses in *Cell*, 2001, 104, pp. 687-697. Copyright © 2001 with permission of Elsevier Science. **F21.8** F. Rudolph Turner, Indiana University **F21.12** Jim Langeland, Stephen Paddock, and Sean Carroll, University of Wisconsin at Madison **F21.14a** Peter A. Lawrence and P. Johnson, Development 105:761-67 (1989). **F21.14b** Peter A. Lawrence "The Making of a Fly", Blackwell Scientific, 1992. **F21.15** Jim Langeland, Stephen Paddock, and Sean Carroll, University of Wisconsin at Madison **F21.16a** Reproduced by permission from From T. Kaufmann, et al. Advanced Genetics 27:309-62, 1990. Image courtesy of F. Rudolph Turner, Indiana University. **F21.16b** Reproduced by permission from From T. Kaufmann, et al. *Advanced Genetics* 27:309-62, 1990. Image courtesy of F. Rudolph Turner, Indiana University. **F21.21** Elliot M. Meyerowitz/California Institute of Technology, Division of Biology **F21.24a** Elliot M. Meyerowitz/California Institute of Technology, Division of Biology **F21.24b** J. Bowman/California Institute of Technology, Division of Biology **F21.25** From "Signaling Pathways in Development" by J. Gearhart in *Teratology*, 60, 1999. Copyright © 1999. Reprinted by permission of Wiley–liss, Inc., a subsidiary of John Wiley & Sons, Inc. **F21.30** Fig. 7, p. 235 from "1998 working lecture: Signaling Pathways in Development" by J. Gearhart in *Teratology*, 60, 1999. Copyright © 1999. Reprinted by permission of Wiley–liss, Inc., a subsidiary of John Wiley & Sons, Inc.

Chapter 22: Genetics of Cancer

CO22 AMC/Albany Medical College/Custom Medical Stock Photo, Inc. **F22.10** Sung-Hou Kim, University of California, Dept. of Chemistry and Lawrence Berkeley National Laboratory, Berkeley, CA **F22.11** Albert Paglialunga/Phototake NYC **F22.14** From Fig. 5A, p. M16 in "BCR-ABL Inhibition as a Modality of CML Therapeutics" by E. Buchdunger et al., in Biochimica et Biophysica Acta, 2001, 1551, pp. M11–M18. Copyright © 2001. By permission of Wiley-Liss, Inc., a subsidiary of John Wiley & Sons, Inc. **F22.15** Fig. from p. 2232 by Lore Leighton in "Why Some Leukemia Cells Resist STI-571" by J. Marx in *Science*, 2001, 292, (5525), pp. 2231–2231. Copyright © 2001 by Lore Leighton. Reprinted by permission of Lore Leighton.

Chapter 23: Genes, Chromosomes, and Chromatin

CO23 Science VU/BMRL/Visuals Unlimited **F23.8** Reproduced by permission from McBlane, F. and Boyes, J. Simulation of V (D)J recombination by histone acetylation. Current Biology Vol. 10:483–486, Fig. 3a, pg. 484. Copyright © Elsevier Science Ltd. Image courtesy of Dr. Fraser McBlane. **F23.7** Fig. 2, p.189 from "How To Make Ends Meet in V(D)J Recombination" H. Grawunder and E. Harfst in *Current Opinions in Immunology*, 13, 2001, pp. 186–194. Copyright © 2001 with permission of Elsevier Science. **F23.9** Reproduced by permission from McBlane, F. and Boyes, J. Simulation of V (D)J recombination by histone acetylation. Current Biology Vol. 10:483–486, Fig. 3a, pg. 484. Copyright © Elsevier Science Ltd. Image courtesy of Dr. Fraser McBlane. **F23.10** Fig. 1, p. 4 from "Isochores and the Evolutionary Genomes of Vertebrates" by G. Bernardi in *Gene*, 241, 2000, pp. 3–17. Copyright © 2001 with permission of Elsevier Science. **F23.11** Fig.5, p. 53 of "Isochore Chromosome Maps of Eulcaryotic Genomes" by J. Oliver et al., in *Gene*, 276, 2001, pp. 47–56. Copyright © 2001 with permission of Elsevier Science. **F23.12** Fig. 3, p. 1856 from "A Compositional Map of Human Chromosome 21" by K. Gardiner et al., in *The Embo Journal*, 9, 1990, pp. 1853–1858. Copyright © 1990. Reprinted by permission of Oxford University Press. **F23.13** Fig. 4 & 5, p. 6 from "Isochores and the Evolutionary Genomes of Vertebrates" by G. Bernardi, in *Gene*, 241, 2000, pp. 3–17. Copyright © 2001 with permission of Elsevier Science. **F23.14** Fig. 1, p. 536 in "Genes, Isochores and Bands in Human Chromosomes 21 and 22" Chromosome Research, 2001, 9, pp. 533–539. Copyright © 2001. Reprinted with kind permission from Kluwer Academic Publishers. **F23.15** Fig.1, p. E13 from "Together Until Separin Do Us Part" by A. Amon, *Nature Cell Biology*, 3, 2001, pp. E12–E14. Copyright © 2001. Reprinted by permission from the publisher. **F23.16** Fig. 7, p. 41 in "Sister Chromatid Separation at Anaphase is Promoted by Cleavage of the Cohesion Subunit Section 1" by F. Uhlmann et al., *Nature*, vol. 400, 1999, pp. 37–42. Copyright © 1999. Reprinted by permission of Nature Publishing Group.

Chapter 24: Quantitative Genetics

CO24 Lysbeth Corsi/Visuals Unlimited. Earl Roberge/ Photo Researchers, Inc.**F 24.1** Bildarchiv Okapia/Photo Researchers, Inc. **F 24.2** Norm Thomas/Photo Researchers, Inc.

Chapter 25: Population Genetics

CO25 Edward S. Ross/California Academy of Sciences **F25.1** Michel Samson, Frederick Libert et al., "Resistance to HIV-1 infection in Caucasian individuals bearing alleles of the CCR-5 chemokine receptor gene." Reprinted with permission from *Nature* 382, Aug. 22, 1996, p. 725, Fig. 3. © 1996 Macmillan Magazines Ltd. 25.10 Larsh K. Bristol/Visuals Unlimited **F25.11** Hans Pfletschinger/Peter Arnold, Inc.

Chapter 26: Genetics and Evolution

CO26 Breck P. Kent/Animals Animals/Earth Scenes **F26.9** Carl C. Hansen/Nancy Knowlton/Smithsonian Institution Photo Services **F26.11a** Kenneth Kaneshiro/University of Hawaii/CCRT **F26.11b** Kenneth Kaneshiro/University of Hawaii/CCRT **F26.14a** Courtesy of Toby Bradshaw and Doug W. Schemske, University of Washington. Photo by Jordan Rehm. **F26.14b** Courtesy of Toby Bradshaw and Doug W. Schemske, University of Washington. Photo by Jordan Rehm. **F 26.15a** Dr. Paul V. Loiselle **F26.15b** Dr. Paul V. Loiselle **F26.17** Fig. 5 in Molecular Biology Evolution 15 (4) pages 391-407, Kazuhiko Takahasi/Norihiro Okada. The Society for Molecular Biology and Evolution. **F26.18** C.B. & D.W. Frith/Bruce Coleman, Inc.

Chapter 27: Conservation of Genetics

CO27 James Watt/Animals Animals/Earth Scenes **F27.2a** David Austen/Getty Images Inc. **F27.2b** Sarah Ward/Department of Soil and Crop Services/Colorado State University **F27.3** Sarah Ward/Department of Soil and Crop Services/Colorado State University **F27.5** Todd A. Pester/Monsanto Company, Ecological Technology Center **F27.6** Johnny Johnson/DRK Photo **F27.9** Art Wolfe/Getty Images Inc. **F27.11** Jim Bradenburg/Minden Pictures **F27.12** Scott Bauer/USDA/ARS/Agricultural

Appendix A: Experimental Methods

FA.1 Ronald E. Law/ Dr. William S. Klug **FA.2** and **FA.9** Ralph Sinbaldi/Dr. William S. Klug **FA.10** Dr. Suzanne McCutcheon

Index

*__Bold__ page number indicates a figure.
Italic page numbers indicate a table.

C

Y

Z

William S. Klug and Michael R. Cummings
Genetics: Student CD-ROM
SITE LICENSE AGREEMENT
Prentice-Hall, Inc.

YOU SHOULD CAREFULLY READ THE TERMS AND CONDITIONS BEFORE USING THE CD-ROM PACKAGE. USING THIS CD-ROM PACKAGE INDICATES YOUR ACCEPTANCE OF THESE TERMS AND CONDITIONS.

Prentice-Hall, Inc. provides this program and licenses its use. You assume responsibility for the selection of the program to achieve your intended results, and for the installation, use, and results obtained from the program. This license extends only to use of the program in the United States or countries in which the program is marketed by authorized distributors.

LICENSE GRANT
You hereby accept a nonexclusive, nontransferable, permanent license to install and use the program ON A SINGLE COMPUTER at any given time. You may copy the program solely for backup or archival purposes in support of your use of the program on the single computer. You may not modify, translate, disassemble, decompile, or reverse engineer the program, in whole or in part.

TERM
The License is effective until terminated. Prentice-Hall, Inc. reserves the right to terminate this License automatically if any provision of the License is violated. You may terminate the License at any time. To terminate this License, you must return the program, including documentation, along with a written warranty stating that all copies in your possession have been returned or destroyed.

LIMITED WARRANTY
THE PROGRAM IS PROVIDED "AS IS" WITHOUT WARRANTY OF ANY KIND, EITHER EXPRESSED OR IMPLIED, INCLUDING, BUT NOT LIMITED TO, THE IMPLIED WARRANTIES OR MERCHANTABILITY AND FITNESS FOR A PARTICULAR PURPOSE. THE ENTIRE RISK AS TO THE QUALITY AND PERFORMANCE OF THE PROGRAM IS WITH YOU. YOU (AND NOT PRENTICE-HALL, INC. OR ANY AUTHORIZED DEALER) ASSUME THE ENTIRE COST OF ALL NECESSARY SERVICING, REPAIR, OR CORRECTION. NO ORAL OR WRITTEN INFORMATION OR ADVICE GIVEN BY PRENTICE-HALL, INC., ITS DEALERS, DISTRIBUTORS, OR AGENTS SHALL CREATE A WARRANTY OR INCREASE THE SCOPE OF THIS WARRANTY.

SOME STATES DO NOT ALLOW THE EXCLUSION OF IMPLIED WARRANTIES, SO THE ABOVE EXCLUSION MAY NOT APPLY TO YOU. THIS WARRANTY GIVES YOU SPECIFIC LEGAL RIGHTS AND YOU MAY ALSO HAVE OTHER LEGAL RIGHTS THAT VARY FROM STATE TO STATE.

Prentice-Hall, Inc. does not warrant that the functions contained in the program will meet your requirements or that the operation of the program will be uninterrupted or error-free.

However, Prentice-Hall, Inc. warrants the CD-ROM(s) on which the program is furnished to be free from defects in material and workmanship under normal use for a period of ninety (90) days from the date of delivery to you as evidenced by a copy of your receipt.

The program should not be relied on as the sole basis to solve a problem whose incorrect solution could result in injury to person or property. If the program is employed in such a manner, it is at the user's own risk and Prentice-Hall, Inc. explicitly disclaims all liability for such misuse.

LIMITATION OF REMEDIES
Prentice-Hall, Inc.'s entire liability and your exclusive remedy shall be:
1. the replacement of any CD-ROM not meeting Prentice-Hall, Inc.'s "LIMITED WARRANTY" and that is returned to Prentice-Hall, or
2. if Prentice-Hall is unable to deliver a replacement CD-ROM that is free of defects in materials or workmanship, you may terminate this agreement by returning the program.

IN NO EVENT WILL PRENTICE-HALL, INC. BE LIABLE TO YOU FOR ANY DAMAGES, INCLUDING ANY LOST PROFITS, LOST SAVINGS, OR OTHER INCIDENTAL OUT OF THE USE OR INABILITY TO USE SUCH PROGRAM EVEN IF PRENTICE-HALL, INC. OR AN AUTHORIZED DISTRIBUTOR HAS BEEN ADVISED OF THE POSSIBILITY OF SUCH DAMAGES, OR FOR ANY CLAIM BY ANY OTHER PARTY.

SOME STATES DO NOT ALLOW FOR THE LIMITATION OR EXCLUSION OF LIABILITY FOR INCIDENTAL OR CONSEQUENTIAL DAMAGES, SO THE ABOVE LIMITATION OR EXCLUSION MAY NOT APPLY TO YOU.

GENERAL
You may not sublicense, assign, or transfer the license of the program. Any attempt to sublicense, assign or transfer any of the rights, duties, or obligations hereunder is void.

This Agreement will be governed by the laws of the State of New York.

Should you have any questions concerning this Agreement, you may contact Prentice-Hall, Inc. by writing to:
ESM Media Development
Higher Education Division
Prentice-Hall, Inc.
1 Lake Street
Upper Saddle River, NJ 07458

Should you have any questions concerning technical support, you may write to:

New Media Production
Higher Education Division
Prentice-Hall, Inc.
1 Lake Street
Upper Saddle River, NJ 07458

YOU ACKNOWLEDGE THAT YOU HAVE READ THIS AGREEMENT, UNDERSTAND IT, AND AGREE TO BE BOUND BY ITS TERMS AND CONDITIONS. YOU FURTHER AGREE THAT IT IS THE COMPLETE AND EXCLUSIVE STATEMENT OF THE AGREEMENT BETWEEN US THAT SUPERSEDES ANY PROPOSAL OR PRIOR AGREEMENT, ORAL OR WRITTEN, AND ANY OTHER COMMUNICATIONS BETWEEN US RELATING TO THE SUBJECT MATTER OF THIS AGREEMENT.

Minimum System Requirements:

PC:
- Windows 98/2000/NT/ME
- Pentium 200 MHz processor
- 64 MB RAM
- 4x CD-ROM drive
- 832 x 624 pixel screen resolution
- Color monitor running "Thousands of Colors" or higher
- Netscape Navigator/Communicator 4.x (Netscape Communicator 4.78 installer is included on CD-ROM) or Internet Explorer 5.0 and higher. **NOTE: Netscape 6 is NOT supported.
- Macromedia Shockwave 8 player or later (installer included on CD-ROM) or Flash 5 player or later

Macintosh:
- Power PC with OS 8.6, 9.2
- 32 Mb RAM
- 832 x 624 pixel screen resolution
- Color monitor running "Thousands of Colors" or higher
- 4x CD-ROM drive
- Netscape Navigator/Communicator 4.x (Netscape Communicator 4.78 installer is included on CD-ROM) or Internet Explorer 5.0 and higher. **NOTE: Netscape 6 is NOT supported.
- Macromedia Shockwave 8 player or later (installer included on CD-ROM) or Flash 5 player or later

To Begin:

Place this CD in your CD-ROM drive then select the "start.html" file. This file will automatically launch your preferred Web browser and display the opening page.

This CD relies heavily on technology-rich content, which requires Macromedia Flash v.5 or Macromedia Shockwave v.8. You may find that you are not able to proceed unless you install one of these programs. The installation is quick and easy. The installer for Shockwave is included on your CD-ROM in the "Shockwave Installer" folder. Once installed, restart the program as before.